Obesity and Diabetes

Joel Faintuch • Salomão Faintuch
Editors

Obesity and Diabetes

Scientific Advances and Best Practice

Second Edition

Springer

Editors
Joel Faintuch
Hospital das Clinicas
Sao Paulo University Medical
School
Sao Paulo, Brazil

Salomão Faintuch
Harvard Medical School
Beth Israel Deaconess Medical Center
Boston, MA, USA

ISBN 978-3-030-53369-4 ISBN 978-3-030-53370-0 (eBook)
https://doi.org/10.1007/978-3-030-53370-0

This Springer imprint is published by the registered company Springer Nature Switzerland AG.
The registered company address is: Gewerbestrasse 11, 6330 Cham, Switzerland

Foreword

Obesity and diabetes mellitus are two interrelated, chronic, debilitating, incurable diseases that have increased in incidence worldwide. In the USA, over 60% of adult Americans are overweight or obese. Most countries around the world are also seeing dramatically increasing rates of obesity. Although not all people with type 2 diabetes are overweight or obese, the majority are. In addition, while most individuals who suffer from obesity are not diabetic, the incidence of T2DM increases as adiposity increases. So intertwined are these seemingly independent metabolic disorders that some refer to them together as "diabesity."

Despite spending billions of dollars on research and the development of a multitude of different treatments, we have failed to control either disease and find ourselves heading off the cliff. While there has been an explosion of new medications, surgical procedures, and the introduction of metabolic/bariatric devices, the incidence of both diseases continues to rise. It has been estimated that within a few short years, the cost to treat the various related medical conditions of these two diseases will be financially unsustainable for most countries.

To better manage this impending crisis, clinicians and researchers need to broaden our understanding of the pathophysiology of both diabetes and obesity. It is far too simplistic to blame patient behavior, societal pressures, or the environment for the epidemic. One must consider other possibilities as well. Recently, researchers have taken a closer look at potential causes or contributors such as bile salt composition, genomics, microbiomes, inflammation, and the deep brain. A better understanding of the pathophysiology is necessary to guide the development of more efficacious and cost-effective treatments. In addition, to be the most beneficial, the research findings need to be properly organized and be available worldwide.

This new book entitled, *Obesity and Diabetes*, edited by Joel Faintuch, MD, a surgeon and a highly regarded nutrition support specialist, and Salomao Faintuch, MD, a Harvard Director of Interventional Radiology, should accomplish these goals. The book is extremely thorough and very well organized. Its nine sections (blocks) cover a wide range of current and cutting-edge information from the epidemiology of diabetes and obesity to bariatric and metabolic surgery to treat both disorders. The blocks and chapters of the book are strategically organized. They begin with basic science and epidemiologic chapters and conclude with clinical treatments and ongoing developments. As previously stated, we as a human race cannot continue as we have, partially understanding and inadequately managing obesity and diabetes mellitus. We cannot sustain the costs nor the effects on health and quality of life. *Obesity and Diabetes* may become a valuable tool for researchers and clinicians alike.

Harvard Medical School,
Center for Metabolic and Bariatric Surgery,
Brigham and Women's Hospital, Boston, MA, USA

Scott A. Shikora

Preface

About 6 years ago, the first edition of this book was being organized. At that time, obesity and diabetes were already counted among the leading non-communicable diseases in the world; however, the global burden had not been elucidated yet. As underscored in the current Introduction chapter, the figures were soon unearthed: USD2 trillion for obesity in 2014 and USD1.3 trillion for diabetes (2015), summing up to USD3.3 trillion. Only major addictions fluctuate in the same range. If one adds up the values for smoking (USD2.1 trillion) and alcoholism (USD1.4 trillion), the total will reach USD3.5 trillion. However, one should consider that whereas alcoholism is a mounting concern, the smokers' curve is receding in many countries, and could turn flat in the coming decades. Metabolic diseases in turn, with obesity and diabetes spearheading the trend, continue to grow unabatedly, so that it is projected that half of the world will be overweight or obese by 2030.

No other disease group comes close. Does that mean that medical schools should sidestep gastroenterology, rheumatology, and gynecology, concentrating instead on obesity/diabetes, and prevention of substance abuse (interspersed with classes on emerging infectious epidemics)? The healthcare domain is not that much utilitarian. Results-based performance management is important; however, it does not supersede the requirements for a global technical and scientific professional background. Moreover, one is not allowed to forget the ethical commitment toward the patient as an integral human being, not as a collection of disease labels. All these factors notwithstanding, practically all specialties will be touched, if not overwhelmed by the new scenario. Orthopedists are already having difficulty to cope with knees and hips damaged by excessive body weight, whereas ophthalmologists, nephrologists, neurologists, and even pediatricians are facing a deluge of diabetics in their daily practice, to mention just a few examples.

The same is true for the allied healthcare professions. Becoming familiar with the intricacies of obesity and diabetes is a must for nurses, psychologists, dietitians, hospital social workers, and specialists in biomedical engineering, given the sheer numbers of this population. Of course, the commonalities as well as the diversities of the illnesses need to be understood, from the vantage point of new achievements in an array of fields, from drug therapy to bariatric and metabolic surgery, and from appetite control to artificial pancreas. Among the cutting-edge science highlights of this edition, omics biomarkers originated from genomic, metabolomic, and immunologic profiling are dissected within the framework of diagnosis, disease monitoring, prognosis, and new therapeutic targets, in obesity and diabetes. Anti-obesity vaccines ? Outlandish and whimsical as they may sound, and past false starts notwithstanding, there are robust experimental investigation lines in this arena. Similarly microbiomic pathways and signatures are becoming indispensable in the assessment and follow-up of metabolic diseases and treatments. There is compelling evidence that part of the surgical response to bariatric and metabolic procedures is microbiome dependent, analogously to an array of other treatments and circumstances.

Both type 1 and 2 diabetes are benefitting from genetic sequencing and immune profiling, with results which can be harnessed for diagnostic and therapeutic purposes. Bariatric and endoscopic interventions are in rapid mutation, including robotic platforms and innovative approaches, as described by respected authorities. Cell therapy and nanoparticle carriers are put into context. Public health initiatives were not neglected either. From government-mandated taxes for unhealthy foods to scrutinized school canteen menus and the contrast between personalized dietetic counseling and general guidance for the population, including Internet-based initiatives, smartphones, and wearable devices, no stone was left unturned in the effort to bring the latest and most practical information.

Sao Paulo, Brazil — Joel Faintuch
Boston, MA, USA — Salomão Faintuch

Contents

Part I Epidemiology, Inheritance, Environment and Pathways

1 **Introduction to Obesity and Diabetes: The Windows of Opportunity** 3
Joel Faintuch and Salomao Faintuch

2 **Nutrition Transition and Obesity Trends in Argentina Within the Latin American Context** 9
Sonia Alejandra Pou, Natalia Tumas, and Laura Rosana Aballay

3 **Challenges and Economic Burden of Diabetes in Africa** 21
Camille Maadjhou Mba and Jean Claude Mbanya

4 **Brown Adipose Tissue in Obesity and Diabetes** 35
Martín Alcalá, Laura Herrero, Dolors Serra, and Marta Viana

5 **White and Brown Adipose Tissue in Obesity and Diabetes** 55
Brooks P. Leitner and Borja Martinez-Tellez

6 **microRNAs in Obesity and Metabolic Diseases** 71
Giuseppe Iacomino, Fabio Lauria, Antonella Venezia, Nunzia Iannaccone, Paola Russo, and Alfonso Siani

7 **From Adipogenic Viruses to Antidiabetic Drug: A Translational Journey** 97
Vijay Hegde and Nikhil V. Dhurandhar

8 **Genetic Profiles in the Obese Population** 107
Ana Carolina Proença da Fonseca, Patrícia Torres Bozza, and Pedro Hernán Cabello

9 **The Upstream Environment for the Obesity Epidemic** 127
Peter Congdon

10 **Inflammation and Its Role in Obesity-Related Complications** 137
Mohit Singh, Jiten R. Sharma, Hina Agraval, and Umesh C. S. Yadav

11 Crosstalk Between Adipose Tissue, Macrophages, and Other Immune Cells: Development of Obesity and Inflammation-induced Metabolic Diseases 151
Rohit Patel, Johnna Francis Varghese, and Umesh Chand Singh Yadav

12 Diabetes Forecasts and Statistics for the Coming Decades 165
Alexandre Assuane Duarte and Olga Golubnitschaja

13 Pancreatic Islets of Langerhans: Adapting Cell and Molecular Biology to Changes of Metabolism 175
Fernanda Ornellas, Iara Karise, Marcia Barbosa Aguila, and Carlos Alberto Mandarim-de-Lacerda

14 Targeting Advanced Glycation End Products (esRAGE and sRAGE) for Obesity, Diabetes, and its Associated Complications 191
Chinedum Eleazu and Mahaneem Mohamed

15 Anti-incretin Effect: A Missing Link between Obesity, Diabetes, and Metabolic Surgery 199
Theocharis Koufakis, Spyridon N. Karras, and Kalliopi Kotsa

Part II Genomics, Metabolomics and Other Omics

16 An Update on Mendelian Forms of Obesity and their Personalized Treatments 207
Selene Chen and David Meyre

17 The Genetic Basis of Diabetic Kidney Disease 221
Christopher A. Simeone, Jose M. Lazaro-Guevara, and Marcus G. Pezzolesi

18 Epigenetics and Chronic Inflammation: Role in Early Detection of Type 2 Diabetes 237
Meenu Ghai

19 Salivary and Urinary Metabolome in Pediatric Obesity and Metabolic Syndrome 249
Jacopo Troisi, Francesca Marciano, Giovanni Scala, Elizabeth Plunk, Luca Pierri, and Angelo Colucci

20 'State-of-the-Art' Metabolomics Investigations of Type 2 Diabetes 265
Benita Percival, Justine Leenders, and Martin Grootveld

21 The Gut Microbiome and Type 2 Diabetes Mellitus 283
Shruti Panwar, Samriddhi Arora, Sapna Sharma, and Prabhanshu Tripathi

22 Immunotherapeutic Approach to the Treatment and Prevention of Obesity 297
Tatsuhiko Azegami and Hiroshi Itoh

Part III Associated Disorders and Obesity Paradoxes

23 Metabolic Syndrome Update 305
Uzma Zafar

24 Non-alcoholic Fatty Liver Disease: A Global Public Health Issue 321
Eda Kaya and Yusuf Yilmaz

25 Nonalcoholic Fatty Pancreatic Disease (NAFPD) 335
Alhareth Al Juboori, Harleen Chela, Sami Samiullah, and Veysel Tahan

26 Normal-weight Obesity: A Hidden Pandemic 347
Shajith Anoop and Nitin Kapoor

27 Obesity in Children/Adolescents and Obesity-Related Comorbidities 361
Taisa Kohut, Jennifer Robbins, Elizabeth Prout, Lorraine Katz, Julie Brothers, Sandra Vazquez-Diaz, Kimberly Genuario, and Jennifer Panganiban

28 Obesity and Cancer: Linked Molecular Mechanisms 385
Erika Di Zazzo, Monica Rienzo, Maria Michela Marino, Donatella Fiore, Chiara Piscopo, Amelia Casamassimi, Bruno Moncharmont, and Ciro Abbondanza

29 Cognitive Impairment in Obesity and Diabetes 399
Cristina Carvalho and Paula I. Moreira

30 Metabolic Pathways Underlying Neuropsychiatric Disorders and Obesity 415
Laís Bhering Martins, Silvia Fernandes Mauricio, Adaliene Versiani Matos Ferreira, and Antônio Lúcio Teixeira

31 Heart Failure and the Obesity Paradox 427
Marijana Tadic and Cesare Cuspidi

32 Inflammatory Mechanisms in Diabetic Kidney Disease 437
Radica Z. Alicic, Emily J. Cox, Joshua J. Neumiller, and Katherine R. Tuttle

33 The Epidemiology of the Diabetes: Depression Comorbidity in Brazil—Inequality and Interaction 457
Finn Diderichsen

34 Contribution of Hyperglycemia and Unhealthy Diet to Cardiovascular Mortality 471
Jian Zhang and Dong Li

35 Gestational Diabetes Mellitus 479
Elpida Vounzoulaki and Samuel Seidu

36 Update on Gestational Diabetes 493
Megan Jones and Allison Akers

37 **Vascular Complications in Type 2 Diabetes** 509
Chih Hao Chen Ku

Part IV General Therapy and Prevention

38 **What We Know and Don't About High-Intensity Sweeteners** 529
Clipper F. Young, Anne Lee, and Joy Dugan

39 **Personalised Molecular Feedback for Weight Loss** 541
Shilpa Tejpal

40 **Diabetes and Obesity in the Child and Adolescent: Guidelines and Challenges** 553
Suzanne Cuda and Marisa Censani

41 **Type 2 Diabetes: An Unresolved Disease** 567
Sarah Cuschieri

42 **Glucose Control in the Intensive Care Unit** 579
Jan Gunst and Greet Van den Berghe

43 **Continuous Glucose Monitors as Wearable Lifestyle Behavior Change Tools in Obesity and Diabetes** 591
Susan M. Schembre, Yue Liao, and Michelle R. Jospe

Part V Innovative Endoscopic, Cell Therapy and Other Interventions

44 **Endoscopic Techniques for Obesity and Diabetes** 607
Vitor Ottoboni Brunaldi, João Almiro Ferreira Filho, and Daniel Martone

45 **Bariatric Embolization: A Possible Non-surgical Option for Weight Reduction** 619
Muhammad A. Latif and Clifford R. Weiss

46 **Progress in Noninvasive Beta-Cell Mass Imaging** 631
Bluma Linkowski Faintuch and Salomao Faintuch

47 **Cancer Staging with 18F-FDG PET/CT in Hyperglycemic Patients** 641
Monica Finessi, Virginia Liberini, and Désirée Deandreis

48 **Pancreatic Islet Transplantation: A Surgical Approach to Type 1 Diabetes Treatment** 655
Samuel Rodriguez, Michael Alexander, and Jonathan R. T. Lakey

49 **Pancreatic Transplantation in Diabetes: Indications, Contraindications and Perspectives** 665
Vinicius Rocha-Santos and Carlos Andres Pantanali

Part VI Pharmacological Therapy and Cost Containment

50 Pharmacotherapy of Type 2 Diabetes 679
Jaskiran Kaur, Paras Famta, Navneet Khurana, Manish Vyas, and Gopal L. Khatik

51 GLP1-Receptor Agonists in Diabetes: Drugs, General Effects, and Cardiovascular Impact 695
Ikaro Breder and Andrei C. Sposito

52 Diabetes and Treatments 705
Milan Obradovic, Emina Sudar-Milovanovic, Zoran Gluvic, Takashi Gojobori, Magbubah Essack, and Esma R. Isenovic

53 Treatment of Diabetes and Heart Failure 719
Bradley Brochu and Michael Chan

54 Cardiovascular Impact of Newer Diabetes Medications 735
Ravi Kant, Mc Anto Antony, Arshpreet Kaur, Nathan A. Gilreath, Rashmi Chandra, and Vipin Verma

55 Pathogenesis and Molecular Targets in Treatment of Diabetic Wounds 747
Satish Patel, Pragati, Shradha Devi Dwivedi, Krishna Yadav, Jagat R. Kanwar, Manju Rawat Singh, and Deependra Singh

56 Efficiency in Public Health Through the Promotion of Diabetes Day Hospitals: A Regional Proposal 759
Ascensión Barroso, Ramón Sanguino, Victoria Barroso, and M. Isabel Sánchez-Hernández

Part VII Bariatric and Metabolic Surgery

57 Surgical Options in Obesity and Diabetes 767
Jaime Ruiz-Tovar and Lorea Zubiaga

58 One Anastomosis Gastric Bypass in the Treatment of Obesity: Effects on Body Weight and the Metabolome 777
Adriana Mika, Tomasz Sledzinski, Monika Proczko-Stepaniak, and Faidon Magkos

59 Robotic Bariatric Surgery 791
Candice Silverman, Anna Isaacs, and Nan Guang Tan

60 The Role of Obesity and Bariatric Surgery in the Management of Knee and Hip Osteoarthritis 813
Katelyn M. Mellion and Shanu N. Kothari

61 Relapse of Diabetes After Metabolic/Bariatric Surgery 827
Lívia Porto Cunha da Silveira, Tarissa Zanata Petry, and Ricardo Vitor Cohen

62 Malnutrition in Obesity 835
Sonmoon Mohapatra and Capecomorin S. Pitchumoni

Part VIII New Apps, Drugs, Artificial Intelligence and Public Health Initiatives

63 Sugar-Sweetened Beverage Taxes: Origins, Mechanisms, and Current Worldwide Status 851
Fabrizio Ferretti

64 Mobile Health Interventions for Weight Management in Overweight and Obese Populations 865
Lynnette Lyzwinski

65 Biomarkers and Machine Learning Applications in Obesity 883
Olga Tsave and Ioannis Kavakiotis

66 Current Approaches in Diabetes Mellitus Prediction: Applications of Machine Learning and Emerging Biomarkers 893
Sergey A. Solodskikh, Vladimir M. Dudenkov, Viktor Yu. Glanz, Anna V. Panevina, Vasily N. Popov, and Alexey S. Velikorodny

67 Lifestyle Interventions for Sarcopenic Obesity in Polycystic Ovary Syndrome 907
Philip D. Chilibeck, Maryam Kazemi, Laura E. McBreairty, and Gordon A. Zello

68 Drug Pipeline for Obesity Therapy 921
Azania T. Panicker, Priyanshi Desai, Sanika Karnik, and Maushmi S. Kumar

69 MC4R as a Target for Pharmacotherapeutic Treatment of Obesity and Type 2 Diabetes 935
Juliana Pereira Lopes Gonçalves, Daniel Palmer, and Morten Meldal

70 Nanotechnology: Can It Be a Crusader in Diabesity? 947
Alexis Marie Speer and Mahua Choudhury

Part IX Obesity Devices and Supplemental Material

71 Old and New Anti-obesity Devices for Medical, Surgical, and Endoscopical Use 959
Joel Faintuch and Salomao Faintuch

72 Electroceutical Approaches for Gastroparesis 967
Niranchan Paskaranandavadivel, Recep Avci, Nipuni Nagahawatte, Aydin Farajidavar, and Leo K. Cheng

73 Useful Online Resources Hosting Information on Obesity and Diabetes 983
Joel Faintuch and Jacob J. Faintuch

Correction to: Obesity and Diabetes C1
Joel Faintuch and Salomão Faintuch

Glossary 989

Part I

Epidemiology, Inheritance, Environment and Pathways

Introduction to Obesity and Diabetes: The Windows of Opportunity

1

Joel Faintuch and Salomao Faintuch

Contents

The Problem 4
The Solution 4
Pediatric and Adult Obesity 4
Obesity-Induced Prediabetes or Diabetes 4
State of the Art 5
The Three (or Five) Trillion Dollar Question 5
Market Stratification 5
Ongoing Initiatives 6
Final Considerations 6
References 7

Abstract

Obesity and diabetes are not just associated with heart attack, stroke, kidney failure, shortened life span, and sudden death. They represent the three trillion question of the twenty-first century. Indeed the combined world financial burden of obesity and diabetes already exceeds USD3 trillion, and could get 50% higher or more around 2030. No other disease category comes close. Healthcare budgets which are already strained could become overwhelmed worldwide. The purposes of this introduction and this book are not stern advices or scary threats. On the contrary, the aims are to present and discuss feasible, sensible, and up-to-date approaches for prevention, diagnosis, and handling of obesity and diabetes.

Keywords

Weight loss · Metabolic diseases · Comorbidities · Financial cost · Obesogenic environment · Pediatric obesity · Recent onset diabetes · Recent onset obesity

J. Faintuch (✉)
Hospital das Clinicas, Sao Paulo University Medical School, Sao Paulo, Brazil
e-mail: j.faintuch@hc.fm.usp.br; blfaintuch@hotmail.com

S. Faintuch
Harvard Medical School, Beth Israel Deaconess Medical Center, Boston, MA, USA

J. Faintuch and S. Faintuch (eds.), *Obesity and Diabetes*, https://doi.org/10.1007/978-3-030-53370-0_1

The Problem

Type 2 diabetes (T2D) is the major cause of blindness, kidney failure, heart attack, and stroke. Obesity is currently the number one predisposing factor to T2D and is independently associated with cardiovascular and kidney disease, thus closing the circle. However, that is not all. Both shorten the life span, predispose to sudden death (Chen et al. 2019; Aune et al. 2018), and their clinical trajectory is rarely free from the mentioned comorbidities, along with myriads of others (Bray et al. 2018; Aga et al. 2019). Diabetics can suffer from as much as 97.5% prevalence of multimorbidity (Aga et al. 2019). Among the obese the metabolically healthy individual, exempt from associated illnesses, is a recognized category, however many suspects this status to be intrinsically unstable and short lived, eventually migrating to multimorbidity as well (Bray et al. 2018).

The Solution

Weight loss ameliorates these diseases and most comorbidities in a dose-related manner—the more weight lost, the better the outcome (Bray et al. 2018). With a reductionist approach, one could affirm that this entire book is redundant and dispensable. Why worry about biomarkers, omics, and complex diagnostic and prognostic algorithms, if health can in many circumstances be restored just by shedding the unnecessary pounds of weight? Who should be interested in specialized and expensive teamwork, pharmacologic breakthroughs, or ingenious surgical and endoscopic interventions, if reconfiguring one's diet will suffice?

The fact is that both obesity and diabetes are "curable," or at least amenable to prolonged, drug-free remission. Intensive lifestyle interventions can do the trick, along with bariatric and metabolic surgery (Ang 2018; Hirukawa et al. 2018; Willmer and Salzmann-Erikson 2018; Chen et al. 2018). What are the odds? And what does prolonged remission really mean?

Pediatric and Adult Obesity

Children and adolescents are among the best candidates for obesity "cure." Exaggerated adiposity is an unfavorable omen concerning adult life, however the prognosis is not necessarily ominous. Spontaneous normalization of body weight is still a possibility at these age brackets, particularly if a durable passion for sports and active life arises. Of course, structured dietary and lifestyle changes would be ideal, and the family should be actively engaged, not leaving the responsibility on the immature shoulders of the youngster (Ferraro and Adamo 2015).

Stress- or lifestyle-induced, recent onset obesity in adults is also a comparatively fertile ground for successful interventions. As long as the precipitating factors can be neutralized, never-obese people who for a short while accumulated adipose tissue will rather easily revert to their previous, eutrophic condition. The expanded fat cell population will shrink, however, it is unlikely it will actually vanish. Adipocyte size, number, and metabolic status respond to lifestyle and pharmacologic interventions with tissue remodeling, sometimes quite remarkable, however the potential for weight regain survives. Consequently, lifelong surveillance and prevention measures will be advised (Moreira-Pais et al. 2020).

Obesity-Induced Prediabetes or Diabetes

There are reasons to believe that recent onset, obesity-related T2D in reasonably young people, much more than in the lean, the aged, and those with a diabetes history >5 years, is the most amenable to satisfactory remission, whether achieved by lifestyle (Ried-Larsen et al. 2019), drug (Hirukawa et al. 2018; Bohula et al. 2018) or surgical means (Hirukawa et al. 2018). Conversion of prediabetes to T2D is inhibited, new cases become more sparse, and even established disease can be fully compensated. How long will the honeymoon last?

The longest available follow-ups are after bariatric surgery. The Swedish Obesity Study, which tracks a large cohort of adults submitted to a variety of operative techniques, has documented 38.1% and 30.4% remission after 10 and 15 years of follow up respectively (Sjostrom et al. 2014). Definition of prolonged remission is at variance with the recommendations of the American Diabetes Association, namely glycated hemoglobin HbA1c < 5.7% and fasting plasma glucose (FPG) < 100 mg/dL) without drugs for at least 5 years (Buse et al. 2009). In the alluded to the protocol, the adopted criterion was more lax, namely FPG < 110 mg/dL in the absence of antidiabetic treatment (Sjostrom et al. 2014). Discrepancies notwithstanding these results are respectable, not only because they robustly exceed those observed in the controls, but also because micro and macrovascular complications were similarly benefitted.

State of the Art

The massive research efforts directed at obesity and diabetes in recent decades, confirmed what many had long suspected, namely that these diseases, in keeping with other metabolic aberrations, are preventable and potentially reversible, at least within certain limits. The requirements are early detection, removal of known drivers, and notably the implementation of a healthy and supervised lifestyle, with emphasis on diet, weight control, elimination of addictions, regular exercise, and combat of sedentarism.

The Three (or Five) Trillion Dollar Question

For how long is that feasible with full compliance of the individuals, and how many would actually be touched by such policies? That is not the million-dollar question. It already involves over three trillion dollars, and in the coming decade could advance beyond five trillion, as outlined below. During the last century, and especially a couple of decades after World War II, humanity has unwillingly cradled an obesogenic and hedonic environment, which started in industrialized countries and is spreading to all latitudes. What could be more gratifying than palatable food and drink on demand, especially because it is available, affordable, breaks no laws, and is not followed by a terrible hangover on the next day? Of course, the bathroom scale will protest, larger clothes will eventually be purchased, and sooner or later an appointment at a primary care facility will have to be scheduled. However, who cares about the future?

Market Stratification

The global diabetes care drugs market reached USD69.7 billion in 2019 and is growing at a compound annual growth rate (CAGR) of 4.6% (www.mordorintelligence.com/industry-reports/diabetes-drugs-market). The blood glucose monitoring systems market is smaller (USD10.1 billion in 2018), however it is expanding more rapidly (CAGR of 6.7%), thus the forecast for 2026 is USD17.1 billion (www.fortunebusinessinsights.com/industry-reports/blood-glucose-monitoring-market-100648).

The obesity prescription market is currently tiny, with few and not exceptionally effective agents. Only in 2026 is it expected to cross the billion-dollar barrier (www.bloomberg.com/press-releases/2019-06-18/anti-obesity-prescription-drugs-market-to-cross-us-1-000-million-by-2026-says-tmr). In turn, the weight management market boasts a CAGR of 8.2% and will be worth a more hefty USD442 billion in 2026 (www.marketwatch.com/press-release/weight-management-market-2019-size-statistics-growth-revenue-analysis-trends-industry-forecast-report%2D%2D2025market-research-engine-2019-12-19).

Wearable medical devices, an important share of which is devoted to obesity, diabetes, and weight management, was worth USD9 billion in 2018 and is growing at the exponential CAGR rate of 39% (www.gminsights.com/industry-analysis/wearable-medical-devices-market). In one considers other lifestyle products and services, along with more specialized treatments such as bariatric operations and devices, the total market cap for these metabolic derangements could advance substantially beyond half-trillion dollars until 2025/2026.

Calculations for the absolute global economic burden are more staggering. For diabetes, they are believed to increase from USD1.3 trillion in 2015, up to $2.5 trillion according to the past trends, by 2030. This translates as a share of global GDP from 1.8% in 2015 to a maximum of 2.2% (Bommer et al. 2018).

Obesity weighs more, in all senses. Its 2014 global economic impact was USD2.0 trillion or 2.8% of the global gross domestic product (GDP). One should not be surprised if's up to 50% more by 2030, given the current epidemiological trajectory. The only current combined burden of smoking (USD2.1 trillion) and alcohol addiction (1.4 trillion) is in the same range (Dobbs et al. 2014).

No other disease category comes close. The industry forecast for inflammatory bowel disease drugs is in the range of USD22 billion for 2026 (www.grandviewresearch.com/press-release/global-inflammatory-bowel-disease-ibd-treatment), hepatitis C treatment is estimated at USD11 billion (2025) (www.worldhepatitisalliance.org/news/sep-2015/cost-comprehensive-global-viral-hepatitis-prevention-and-treatment-effort-might-peak/), and the oncology drug market should not exceed a comparatively negligible USD176.5 million (2025) (www.alliedmarketresearch.com/oncology-cancer-drugs-market). Devices, palliative assistance, and ancillary care, along with indirect costs (absenteeism, premature death) could tip the scales a lot further, however still far away from the figures mentioned above.

Ongoing Initiatives

Of course academic and government healthcare planners and scientists are not staying idle. Major local or nationwide diabetes and obesity initiatives are going on in the USA, the UK, Australia, Canada, New Zealand, Austria, India, China, Israel, and United Arab Emirates, among others. From taxes and legislation inhibiting intake of risky foods and drinks, not overlooking planned cities and buildings that stimulate walking and stair climbing, to targeted anti-obesity and antidiabetic lifestyle and pharmacological interventions, policy makers have their hands full (Galaviz et al. 2018).

What are the prospects for a "magic pill" in the obesity and type 2 diabetes context? Even though the industry is putting out safer and more powerful drugs nearly every year, and invasive as well as intensive lifestyle interventions improve, prolonged remission is still not within the grasp of the masses, only of limited groups (Ang 2018; Hirukawa et al. 2018; Willmer and Salzmann-Erikson 2018; Chen et al. 2018; Ferraro and Adamo 2015). Incidentally, there is much hope also for candidates with type 1 diabetes, as stem cell therapy and immunotherapy are making large strides (https://labiotech.eu/features/immunotherapy-type-1-diabetes. Breakthrough type 1 diabetes).

Final Considerations

In the first chapter of The Descent of Man, and Selection in Relation to Sex, published in 1871, Charles Darwin argued that some anatomical structures were useless, and would eventually disappear. These included toes, wisdom teeth, paranasal sinuses, cervical ribs, and even body hair (Darwin 1981). He was not the only one to engage in phenotypic futures studies. In more recent century, specifically in 1993, Nebel and Wright predicted that man will eventually increase brain capacity, and undergo leg atrophy (Nebel and Wright 1993). Indeed the more gray and white matter, the better to deal with sophisticated computers and to analyze big datasets. On the other hand, who needs to carry around long and heavily muscled legs if elevators, escalators, cars, and other conveyances are ubiquitous?

Both prophecies failed to include obesity and diabetes. Men and women are becoming smarter, communicating instantly, and moving around at increasing speed, however they are paying dearly in the form of metabolic illnesses. If one does not intend to pass on these scourges to the coming generations, one has to act here and now. This book represents an effort by highly regarded specialists and respected scholars from different continents, to present new ways and means not only to identify and manage obesity and diabetes, but also to prevent it before it moves out of control.

References

Aga F, Dunbar SB, Kebede T, Gary RA (2019) The role of concordant and discordant comorbidities on performance of self-care behaviors in adults with type 2 diabetes: a systematic review. Diabetes Metab Syndr Obes 12:333–356

Ang GY (2018) Reversibility of diabetes mellitus: narrative review of the evidence. World J Diabetes 9 (7):127–131

Aune D, Schlesinger S, Norat T, Riboli E (2018) Diabetes mellitus and the risk of sudden cardiac death: a systematic review and meta-analysis of prospective studies. Nutr Metab Cardiovasc Dis 28(6):543–556

Bohula EA, Scirica BM, Inzucchi SE, DK MG, Keech AC, Smith SR, Kanevsky E, Murphy SA, Leiter LA, Dwyer JP, Corbalan R, Hamm C, Kaplan L, Nicolau JC, Ophuis TO, Ray KK, Ruda M, Spinar J, Patel T, Miao W, Perdomo C, Francis B, Dhadda S, Bonaca MP, Ruff CT, Sabatine MS, Wiviott SD, CAMELLIA-TIMI 61 Steering Committee Investigators (2018) Effect of lorcaserin on prevention and remission of type 2 diabetes in overweight and obese patients (CAMELLIA-TIMI 61): a randomised, placebo-controlled trial. Lancet 392(10161):2269–2279

Bommer C, Sagalova V, Heesemann E, Manne-Goehler J, Atun R, Bärnighausen T, Davies J, Vollmer S (2018) Global economic burden of diabetes in adults: projections from 2015 to 2030. Diabetes Care 41 (5):963–970

Bray GA, Heisel WE, Afshin A, Jensen MD, Dietz WH, Long M, Kushner RF, Daniels SR, Wadden TA, Tsai AG, Hu FB, Jakicic JM, Ryan DH, Wolfe BM, Inge TH (2018) The science of obesity management: an endocrine society scientific statement. Endocr Rev 39 (2):79–132

Buse JB, Caprio S, Cefalu WT, Ceriello A, Del Prato S, Inzucchi SE, McLaughlin S, Phillips GL, Robertson RP, Rubino F, Kahn R, Kirkman MS (2009) How do we define cure of diabetes? Diabetes Care 32 (11):2133–2135

Chen JC, Hsu NY, Lee WJ, Chen SC, Ser KH, Lee YC (2018) Prediction of type 2 diabetes remission after metabolic surgery: a comparison of the individualized metabolic surgery score and the ABCD score. Surg Obes Relat Dis 14(5):640–645

Chen H, Deng Y, Li S (2019) Relation of body mass index categories with risk of sudden cardiac death. Int Heart J 60(3):624–630

Darwin C (1981) The descent of man and selection in relation to sex. Princeton Science Library, Princeton, NJ

Dobbs R, Sawers C, Thompson F, Manyika J, Woetzel JR, Child P, McKenna S, Spatharou A (2014) Overcoming obesity: an initial economic analysis. McKinsey Global Institute, Jakarta, ID. www.mckinsey.com/mgi/our-research/discussion-papers-and-briefings. Accessed 24 Mar 2020

Ferraro ZM, Adamo KB (2015) Maternal, paternal, and societal efforts are needed to "cure" childhood obesity. Mayo Clin Proc 90(4):555

Galaviz KI, Venkat Narayan KM, Lobelo F, Weber MB (2018) Lifestyle and the prevention of type 2 diabetes: a status report. Am J Lifestyle Med 12(1):4–20

Hirukawa H, Hashiramoto M, Tanizawa Y, Kaku K (2018) Pioglitazone and sulfonylurea remission from type 2 diabetes mellitus and anti-atherosclerosis in Japan (PREVENT-J) study group. Remission of hyperglycemia after withdrawal of oral antidiabetic drugs in Japanese patients with early-stage type 2 diabetes. J Diabetes Investig 9 (5):1119–1127

https://labiotech.eu/features/immunotherapy-type-1-diabetes/. Breakthrough type 1 diabetes

Moreira-Pais A, Ferreira R, Neves JS, Vitorino R, Moreira-Gonçalves D, Nogueira-Ferreira R (2020) Sex differences on adipose tissue remodeling: from molecular mechanisms to therapeutic interventions. J Mol Med (Berl) 98(4):483–493

Nebel BJ, Wright RT (1993) Environmental science: the way the world works 1993. Available at https://thepopculturecompany.com. Accessed 30 Mar 2020

Ried-Larsen M, Johansen MY, MacDonald CS, Hansen KB, Christensen R, Wedell-Neergaard AS, Pilmark NS, Langberg H, Vaag AA, Pedersen BK, Karstoft K (2019) Type 2 diabetes remission 1 year after an intensive lifestyle intervention: a secondary analysis of a randomized clinical trial. Diabetes Obes Metab 21 (10):2257–2266

Sjostrom L, Peltonen M, Jacobson P, Ahlin S, Andersson-Assarsson J, Anveden A, Bouchard C, Carlsson B, Karason K, Lonroth H, Naslund I, Sjostrom E, Taube M, Wedel H, Svensson PA, Sjoholm K, Carlsson LM (2014) Association of bariatric surgery with long-term remission of type 2 diabetes and with microvascular and macrovascular complications. JAMA 311:2297–2304

Willmer M, Salzmann-Erikson M (2018) The only chance of a normal weight life: a qualitative analysis of online forum discussions about bariatric surgery. PLoS One 13(10):e0206066

www.alliedmarketresearch.com/oncology-cancer-drugs-market. Accessed 30 Mar 2020

www.bloomberg.com/press-releases/2019-06-18/anti-obesity-prescription-drugs-market-to-cross-us-1-000-million-by-2026-says-tmr. Accessed 30 Mar 2020
www.fortunebusinessinsights.com/industry-reports/blood-glucose-monitoring-market-100648. Accessed 30 Mar 2020
www.gminsights.com/industry-analysis/wearable-medical-devices-market. Accessed 30 Mar 2020
www.grandviewresearch.com/press-release/global-inflammatory-bowel-disease-ibd-treatment. Accessed 30 Mar 2020
www.marketwatch.com/press-release/weight-management-market-2019-size-statistics-growth-revenue-analysis-trends-industry-forecast-report%2D%2D2025market-research-engine-2019-12-19. Accessed 30 Mar 2020
www.mordorintelligence.com/industry-reports/diabetes-drugs-market. Accessed 30 Mar 2020
www.worldhepatitisalliance.org/news/sep-2015/cost-comprehensive-global-viral-hepatitis-prevention-and-treatment-effort-might-peak/. Accessed 30 Mar 2020

Nutrition Transition and Obesity Trends in Argentina Within the Latin American Context

2

Sonia Alejandra Pou, Natalia Tumas, and Laura Rosana Aballay

Contents

Understanding the Path to the Obesity Epidemic 10
The Population Context 10
Worldwide Changes in Recent Decades 11
The Burden of Obesity Worldwide 12
General Dietary Patterns 13
Findings in Latin America 13
Focus on Argentina 13
Highly Processed Foods 14
Physical Activity Patterns 14
Disease Burden 14
Public Health Interventions and Future Perspectives 15
Childhood Undernutrition and Adult Obesity 16
References 17

Abstract

The world population has undergone a rapid shift in dietary and physical activity patterns, from traditional diets to a dietary pattern characterized by highly processed foods and sugar-sweetened beverages, coupled with increasingly sedentary lifestyles. These shifts have been concomitant with demographic, macroeconomic, and technological changes, and closely related to the widespread obesity

S. A. Pou (✉)
Faculty of Medical Sciences, Institute for Investigations in Health Sciences (INICSA), National University of Cordoba, National Council for Scientific and Technical Investigations (CONICET), Cordoba, Argentina

Faculty of Medical Sciences, Statistics and Biostatistics, School of Nutrition, National University of Cordoba, Cordoba, Argentina
e-mail: pousonia@conicet.gov.ar

N. Tumas
Center for Investigations and Studies on Culture and Society (CIECS), National Council for Scientific and Technical Investigations (CONICET), National University of Cordoba, Cordoba, Argentina

Faculty of Health Sciences, School of Nutrition, Catholic University of Cordoba, Cordoba, Argentina

L. R. Aballay
Faculty of Medical Sciences, Statistics, Biostatistics and Informatics Applied to Nutrition, School of Nutrition, National University of Cordoba, Cordoba, Argentina

J. Faintuch and S. Faintuch (eds.), *Obesity and Diabetes*, https://doi.org/10.1007/978-3-030-53370-0_2

epidemic. Barry Popkin (University of North Carolina at Chapel Hill, USA) has defined the entire process as Nutrition Transition and has written extensively about its stages, drivers, and consequences. Based on this literature and on our previous work, in this chapter we present the key elements of the nutrition transition process in developing countries, describing its distinctive features in the Latin America region, especially in Argentina. We also describe obesity trends in this context. Finally, we discuss public health interventions in the developing world, and future perspectives to deal with a still unresolved consequence of the nutrition transition, the obesity and noncommunicable diseases epidemic.

Keywords

Obesity burden · Obesity epidemic · Obesity trends · Obesity public health · Highly processed foods · Nutrition transition

Understanding the Path to the Obesity Epidemic

The nutrition transition theory describes the process of shifts in the diet and in the physical activity patterns that result in changes in stature, body composition and body size of populations (Popkin 1994, 2004). Both quantitative and qualitative dimensions are involved in these changes (Popkin 2002a, 2009).

Several societies seem to converge on a shift away from traditional diets toward a more globalized pattern characterized by less legumes, vegetables, and fruits, and more animal-source foods, edible oils, and processed foods—high in refined carbohydrates, added sugars, and sweeteners (Popkin 2002a, 2015, 2017; United Nations System, Standing Committee on Nutrition 2010). Physical activity decreases rapidly, driven by more sedentary jobs and leisure activities, and the increasing use of vehicles (Popkin 2006, 2015). In regard to the nutritional status, a shift from increased prevalence of undernutrition toward a high prevalence of overweight and obesity is observed (Popkin 1994, 2004).

The Population Context

Two historic processes affect and are affected by nutritional transition: the demographic transition and the epidemiologic transition (Popkin 2002a). The demographic transition is defined as the shift from a population pattern of high fertility and high mortality, to another of low fertility and low mortality, with the consequent life expectancy increase (Notestein 1945). Closely related to these demographic changes is the epidemiological transition, which refers to the evolution from a pattern of high prevalence of infectious diseases, to predominance of noncommunicable diseases (NCDs) (Omran 1971).

These transitions have some elements in common: they describe the changes observed in the demographic, health and diet dynamics of the populations (mainly in developed countries) through a succession of stages; they share common pathways; they consider that the stages (with more or less delay) will occur in all regions worldwide; and they were designed with the purpose of making forecasts and planning new socioeconomic, food, and health policies (Popkin et al. 2012; Nicolau-Nos and Pujol-Andreu 2011). The interplay among these epidemiologic, demographic, and social changes is such, that it determines the nature and pace of nutritional shifts (Popkin 2001).

Five broad nutrition patterns were proposed by the lead author of this theory, Barry Popkin (Popkin 1993, 2002a), to describe the nutrition transition. Although the patterns were outlined as historical stages, they are not restricted to the periods in which they first arose. In point of fact, some nutrition transition patterns coexist at the same time, with spatial and socioeconomic variations. We summarize these broad nutrition transition patterns in Table 2.1.

The shifts involved in these patterns occur in different regions at different rates; however, there is a consensus regarding that the most rapid

Table 2.1 Summary of the nutrition transition patterns according to the dimensions involved

Nutrition transition pattern	Diet	Physical activity	Nutritional status	Other features
Collecting food	Varied diet, high in carbohydrates and fiber, and low in saturated fat	High level of physical activity	Robust, lean, low malnutrition prevalence	Typical of hunter–gatherer
Famine	Less varied diet, cereals predominant. Periods of food scarcity	Little modification in physical activity levels	Nutritional deficiencies, stature reduction	Agriculture, livestock farming Deepening of social stratification
Receding famine	Less starchy staples and more fruit, vegetables and animal protein intake. Lower variety of diet and famines	Physical activity levels started to decrease	Several nutritional deficiencies disappear, stature grows	Crop rotation, fertilizer use, industrial revolution. Women join the labor force
Nutrition-related noncommunicable diseases	Higher in fat (specially saturated), cholesterol, refined carbohydrates and processed foods. Lower in fiber and polyunsaturated fatty acids	Increasing sedentary jobs and lifestyles	Obesity prevalence increases, many disabling conditions	Technology revolution. Characteristics of high-income societies and increasing low-income societies
Behavioral change	More carbohydrates, fruits and vegetables; less fat and processed food	High leisure exercises, sedentary jobs	Lower body-fat and obesity; better bone health	Service sector mechanization. Changes aimed to increase the disability-free life expectancy

Data from Popkin (1993), (2001) (2002a), and (2004)

change is observed in developing countries (Popkin 2009; Hawkes et al. 2017). Specifically, the shifts in diet and physical activity patterns seem to be particularly accelerated in the low- and middle-income countries (LMICs) (Popkin et al. 2012; Popkin 2002b).

It is worth mentioning that Popkin has written extensively about the influence of changes in the food system and macroeconomic factors, that underlie many of these worldwide dietary shifts (Popkin 2009, 2017; Popkin et al. 2012), remarking that the major influence on the obesity epidemic must be viewed as environmental rather than personal or genetic (Popkin 2001). In line with this assumption, other authors state that along with income, relative food prices and preferences are a major determinant of dietary patterns (Finaret and Masters 2019), and therefore of the obesity outcome.

Worldwide Changes in Recent Decades

Four interrelated macroeconomic and technological factors have been pointed out. A critical driver is the *technology* that affects economic and domestic works, the production and processing of foods, and the expansion of mass marketing, which leads to changes in dietary patterns and related health conditions. Additionally, transportation facilities (e.g., trains) and leisure sedentary-promoting devices (e.g., television, mobile phones, and computers) had a central impact on the reduction of physical activity levels (Popkin 2006, 2015).

Another widely studied driver is *urbanization*. This demographic process is associated with a greater variety and availability of food, more marketing activities on foods, higher food processing, and increased intake of food away from home (Since jobs are more frequently incompatible with home food preparation). The urbanization process has experienced an enormous acceleration, and nowadays approximately 55% of the world's population is urban (United Nations 2017). The Latin American and Caribbean region are the most urbanized in the world (about 80% rate), and in Argentina, around 91% of the population live in urban areas (Pou et al. 2017).

More robust per capita *income* and *expansion of global trade* are great drivers of the nutrition transition process. While income has increased worldwide, the prices of many foods have diminished, thus leading to lower proportions of income allocated to food. Besides, income increase allows acquiring labor-saving devices and others that foster a sedentary lifestyle (Popkin 2015). The opening of global trade in goods and services provides more opportunities to access modern media, food retail, food services, and technologies that reduce physical activity (Popkin 2006, 2015). A recent analysis about trade openness and the obesity epidemic that included 175 countries during the 1975–2016 period, concluded that trade openness was positively associated with country obesity prevalence, and its influence concentrated among developing nations (An et al. 2019).

Despite the general tendencies, wide inequalities remain in many LMIC countries, where undernutrition (underweight, stunting, micronutrient deficiencies) persists and coexists with increased prevalence of overweight, obesity, and associated NCDs (United Nations System, Standing Committee on Nutrition 2010; Popkin et al. 2012; Ng et al. 2014). This is the so-called "double burden of malnutrition," which can be observed at the country, community, household, and even at the individual levels (United Nations System, Standing Committee on Nutrition 2010; Popkin et al. 2012). This complex nutritional and epidemiological scenario constitutes an enormous challenge for the public health of these countries, and particularly for the Latin American region, as described in detail below.

The Burden of Obesity Worldwide

Most of the world's population lives in countries where overweight and obesity kill more people than underweight, and it is estimated that at least 2.8 million people die each year as a consequence of overweight or obesity (World Health Organization 2017, 2018a), even though these nutritional conditions are preventable.

In the Americas, in 2016 the prevalence of overweight and obesity in adults was 66.2% and 59.3% in men and women, respectively. Among the countries with the highest prevalence are the USA (68%), Mexico (65%), Canada, and Bahamas (64%) (Organización Panamericana de la Salud 2019). In Latin America and the Caribbean, it has been noted that 7.5% of children under 5 years of age live with overweight, whereas worldwide prevalence is 5.9% (FAO, OPS, WFP, UNICEF 2019).

Obesity is not only a disease, but also a metabolic risk factor associated with other NCDs such as cardiovascular problems, diabetes, and cancer (Finucane et al. 2011; Ford et al. 2017), which are the main causes of death and disability worldwide. The World Health Organization (WHO) reported that 71% of the global disease burden is due to NCDs and that these diseases disproportionately affect people in LMICs (World Health Organization 2018b). In Argentina, NCDs were responsible for the largest proportion of deaths in 2016 (almost 80% of total mortality) (World Health Organization 2018c). Based on the last National Survey of Risk Factors, it was estimated that 61.6% of the adult population living in urban areas of Argentina in 2018 have overweight and among them, 25.4% present obesity. This level of prevalence is similar to developed countries and represents a growth of almost double in 13 years (from an obesity prevalence of 14.6% in 2005) (Ministerio de Salud y Desarrollo Social de la Nación 2019).

According to the World School Health Survey (Ministerio de Salud y Desarrollo Social de la Nación 2018), the prevalence of overweight and obesity in students aged between 13 and 17 in Argentina was 30.3% and 7.4%, respectively, with higher values in men than in women. Excess weight in the specific group of adolescents (aged 13–15 years old) has increased progressively throughout the three editions of the World School Health Survey; the overweight prevalence increased from 24.5% (2007) to 28.6% (2012) and 33.1% (2018) and the percentage of students with obesity was about 4.4% in 2007, with values to 5.9% and 7.8% in 2012 and 2018, respectively

(Ministerio de Salud y Desarrollo Social de la Nación 2018).

General Dietary Patterns

The dietary pattern descriptions are based on extensive research carried out in higher-income countries, while the scientific evidence is comparatively scarce in LMICs. Yet Popkin (2015) sheds light on this matter and points out that, nowadays, there is sufficient data available about LMICs to document this generalized dietary trend in all urban areas and, increasingly, in rural ones: (1) a huge increase in the consumption of vegetable oils and the practice of frying food, instead of using traditional healthier cooking methods; (2) very high levels of sugar-sweetened beverages and fruit juice intakes; (3) increased consumption of animal-source foods; (4) a diet shift toward the intake of highly processed food products and a reduction of legumes, fruits, and vegetables; (5) away-from-home meals, frequent snacking and a rise in eating frequency (Popkin 2009, 2011, 2015).

Findings in Latin America

Although these trends are widespread even in Latin American countries, there is much heterogeneity between and within countries (by age, gender, and sociodemographic conditions) (Popkin 2002a; Kovalskys et al. 2019). A multicenter cross-sectional study assessing food consumption in adults (15–65 years old) in an urban sample from 8 Latin American countries (Argentina, Brazil, Chile, Colombia, Costa Rica, Ecuador, Peru, and Venezuela), called ELANS (Latin American Study of Nutrition and Health) (Kovalskys et al. 2019), indicates in general, deficiencies for nutrient-dense food groups; healthy food tends to be more consumed by high socioeconomic persons and older people; vegetables and red meat remain the two most consumed food groups in the region, although their amounts are below and over recommendations, respectively, in most of the countries (Kovalskys et al. 2019).

Focus on Argentina

According to the ELANS (Kovalskys et al. 2019), Argentina was the leading consumer of sugar-sweetened beverages (mean of 1245 g/day) and red meat (prevalence of consumption of 82.3%, average consumption of 129.7 g/day), during 2014–2015. Particularly in Argentina and Chile, processed meat consumption was more common among low socioeconomic groups. Inversely, Argentina is among the countries with less legumes consumption in the region. Regarding fruits and vegetables mean intakes were markedly below current recommendations (at least five servings/day), as in all the countries of the region. This is consistent with the results of the last National Surveys of Risk Factors that reported a mean intake of only two servings/day of these vegetable -source foods in the Argentinian adult population in 2018 (Ministerio de Salud y Desarrollo Social de la Nación 2019). Nevertheless, both fruits and vegetables were consumed in greater amounts among the low socioeconomic groups in most of the Latin American countries, including Argentina (Kovalskys et al. 2019).

In the 1996–2013 National Survey of Household Expenditure (Zapata et al. 2016; Zapata and Rovirosa 2016), the traditional diet of the Argentinian people in the past (mainly unprocessed or minimally processed foods) is now moving to a diet rich in processed foods. Similar to the ELANS results, but from a dynamic perspective, the authors highlight the higher and increased apparent consumption of sugar-sweetened soft drinks or juices and processed meats (This consumption has doubled and tripled, respectively, in the last 20 years). Moreover, total fruit consumption decreased by 41% in the study period. Yet when income increases, consumption and fruit diversity improve as well. The consumption of ready-to-eat foods also grows as household income becomes more elevated (Zapata et al. 2016; Zapata and Rovirosa 2016).

Highly Processed Foods

Processed food products have been defined as "substances extracted and refined from unprocessed or minimally processed foods that are 'ready-to-eat' or 'ready-to-heat,' made from industrially prepared ingredients and additives, usually highly palatable and intensively marketed, and often high in free sugars, trans-fats and low in micronutrients" (Sievert et al. 2019). Although the health costs or benefits of food processing are still discussed, a report of the Pan American Health Organization on this matter discourages their consumption; it also indicates that sales of ultra-processed food products (also called highly processed food products) are associated with weight gain and obesity in Latin America (PAHO-OMS 2015). It was estimated that the volume sales of ultra-processed drinks per capita is about 184.5 kg/capita/year in Argentina, which ranks among the highest values, along with the USA (238.8 kg/capita/year) and Mexico (188.5 kg/capita/year) (Vandevijvere et al. 2019).

Regarding food selection among Argentinian adolescents, a tendency to low consumption of fruits and vegetables was observed from the World School Health Survey 2018 (Ministerio de Salud y Desarrollo Social de la Nación 2018). Only 21% and 10.5% of them consumed fruits and vegetables, respectively, two or more times per day. A public health issue that needs to be considered is that approximately 10% of those adolescents currently eat fast food 3 or more days a week away from home, around 50% more than in 2012 when only 6.8% of the total of number of participants aged between 13–15 years old adopted this dietary practice. Moreover, a third of the adolescents declared a consumption of sugar-sweetened drinks of 1 or more times a day.

Physical Activity Patterns

A sedentary lifestyle or low level of physical activity is an indicator of low energy expenditure and constitutes one of the main modifiable risk factors of most NCDs (World Health Organization 2018b). Some researchers argue that physical inactivity or sedentary lifestyle is the fourth-ranking risk factor for global mortality (6% of deaths recorded worldwide) (Guthold et al. 2018).

The results of the National Surveys of Risk Factors show that people from this country have a low level of physical activity, increasing its prevalence from 54.7% in 2013 to 65% in 2018 (Ministerio de Salud y Desarrollo Social de la Nación 2019).

The World School Health Survey confirmed that only 16.5% of students aged between 13 and 17 were physically active in 2018, and 55.3% of them spent at least 3 h a day sitting, outside school hours. This is a clear indicator of sedentary behavior in this population. Meanwhile, the percentages of physical inactivity among young people remain high, mainly among women (Ministerio de Salud y Desarrollo Social de la Nación 2018).

Disease Burden

Latin America, along with the Middle East and North Africa, is one of the low- and middle-income regions with the highest burden of obesity (Popkin and Reardon 2018). Changes in diet and activity patterns lead to the emergence of chronic disease problems and increased disability (Popkin and Gordon-Larsen 2004; Popkin et al. 1996). In Latin America (Popkin 2002a), many countries entered into the so-called nutrition related-NCDs stage far earlier than others. Haiti and some Central American subpopulations, in 2002 were still in the receding famine stage. In contrast, Mexico experienced an accelerated transition in the 1990s (Popkin 2002a). While obesity prevalence continued to increase among all socioeconomic groups, the highest burden was among disadvantaged women (Pérez-Ferrer et al. 2018).

Argentina, an upper-middle income country, with a current total population that exceeds 40 million people (40,117,096, census 2010) is ranked as the fifth most populous country in this region. The nutrition transition process presents geographical differences within this country. In

Table 2.2 Summary of the nutrition transition profiles in Argentina (2005–2013) according to the dimensions involved

Nutrition transition profiles	Nutritional features	Sociodemographic characteristics	Provinces in cluster	Related patterns of nutrition transition theory
Socionutritional lag	High prevalence of stunting in children; low prevalence of childhood obesity	High proportion of poverty households and population without health insurance; low proportion of population with higher education; relatively high infant mortality rates.	Misiones, Corrientes, Tucumán, Santiago del Estero, Jujuy, Formosa, Río Negro, San Juan, Chaco, Salta	Receding famine
Double burden of malnutrition	High prevalence of childhood and adult obesity; intermediate prevalence of stunting in children	High proportion of urban households	Catamarca, La Rioja, Santa Fe, Buenos Aires, Neuquén, Chubut, Santa Cruz, Tierra del Fuego	Moving from "receding famine" to "degenerative diseases" stage
Incipient socionutritional improvement	Low prevalence of stunting in children and adult obesity	Low proportion of poverty households	Córdoba, San Luis, Mendoza, La Pampa, Autonomous City of Buenos Aires, Entre Ríos	Behavioral change

Data from Tumas et al. (2019)

our recent work (Tumas et al. 2019), we identified three profiles, which were named "socionutritional lag" (characterized by undernutrition and socioeconomically disadvantaged conditions), "double burden of malnutrition" (undernutrition by stunting and overweight in highly urbanized scenarios), and "Incipient socionutritional improvement" (low prevalence of malnutrition and more favorable poverty indicator values). These profiles allowed us to differentiate the Argentinian provinces into three groups according to the nutritional status and sociodemographic characteristics of their populations. The key issues in each of the profiles identified are summarized in Table 2.2.

As discussed by Tumas et al. (2019), the "socionutritional lag" profile is closely related to the so-called "receding famine" pattern proposed in the NT theory, given that child undernutrition problems are distinctive characteristics in both scenarios. The "double burden of malnutrition," in turn, could reflect a transitional situation of these populations between the "receding famine" and the "degenerative diseases" stage stated by Popkin, which is also a distinctive characteristic of many developing regions. On the other hand, the "incipient socionutritional improvement" profile could be linked to the last stage of the nutrition transition, the "behavioral change" pattern, because the low prevalence of malnutrition (both under- and over-nutrition), together with favorable socioeconomic conditions were the most dominant features. Interestingly, this was the only profile that showed no associations with the obesity burden in Argentina.

These results highlight the important role of sociodemographic factors such as urbanization and poverty levels in shaping nutrition transition profiles in Argentina (Tumas et al. 2019). Previous studies focused on the link between specific dietary patterns and obesity or NCDs in urban populations of Argentina already found an association between unhealthy dietary patterns with obesity (Pou et al. 2016), as well as with several diet-related cancers (Niclis et al. 2015; Pou et al. 2012, 2014a, b; Tumas et al. 2014).

Public Health Interventions and Future Perspectives

In response to the growing burden of NCDs, the global community has worked through the World Health Organization and the United Nations, aiming to reduce premature mortality from NCDs by 25% until 2025 (World Health Organization 2018b). However, the results of the global consensus and derived interventions have not been successful.

As Popkin anticipated, achieving the pattern of behavioral change (Popkin 2009), seems to be quite difficult in the current scenario of the Latin American region.

The accelerated speed at which nutritional and epidemiological changes occur in developing countries often exceeds the national capacity of these countries to address the rapid increase of NCDs and to engender a healthy transition (Popkin 2002b; *Lancet* 2017). In addition, the rapidly increasing burden of overweight and obesity coupled with increased waist circumferences and major diet shifts, adversely affect the burden of diabetes and other metabolic disorders (Popkin 2015), which are capital public health problems in the developing world.

In LMICs, it is also noticeable that the double burden of malnutrition underlies a context of food insecurity coupled with energy imbalance (Popkin 2002b); this may reflect persistent social inequalities in health distribution that should be addressed.

Childhood Undernutrition and Adult Obesity

Current hypotheses of the developmental origins of adult disease merit special attention in these regions, where obesogenic environments are expanding while undernutrition persists (Popkin et al. 2012; Wells et al. 2020). According to Wells et al. (2020), early undernutrition in life (during fetal and infant development) followed by later energy imbalance, impose a high metabolic load on a depleted capacity for homeostasis; this may exacerbate the health costs of adult obesity especially among individuals who have previously suffered from undernutrition. Public health policy interventions and programs should simultaneously address malnutrition caused by energy imbalance linked to nutritional deficiencies (i.e., stunting, anemia, and other nutritional deficiencies), along with an excess of body weight (i.e., overweight or obesity). These need to focus on the early stages of life and to adopt a life-course perspective.

Tirado et al. (2016) found that all 18 countries Latin American countries under study had relevant policies to address malnutrition, especially undernutrition, and micronutrient deficiencies, but only some of them had policies to address overweight and obesity (Tirado et al. 2016). Especially in LMICs, it has been highlighted the central role of breastfeeding (Wells et al. 2020). This practice, economically affordable for the population living in poverty, increases the chances of achieving a healthy weight and growth during early childhood.

Ongoing initiatives include taxation of sugar-sweetened beverages, marketing control especially concerning child-oriented, front-of-package labeling profiles with a positive or negative logo, and special regulations related to schools and/or other public facilities (Popkin 2017). Tirado et al. (2016) report that regulatory frameworks to address overweight and obesity have been introduced. However, the authors highlight the scarcity of data on the allocation of human and financial resources to promote balanced nutrition (a crucial element in terms of the efficacy of public policies), and that most of the countries studied had food-based dietary guidelines, but lacked the legislation to increase access to healthy food and/or address the obesogenic environment (Tirado et al. 2016).

We want to emphasize the relevance of timely feedbacks on current policies. Besides, we highlight the need to improve systems and instruments for data collection on nutrition and diet in developing countries, especially from Latin America. In this regard, Walls et al. (2018) reported that the standardized instruments commonly used to assess diets in LMICs are not appropriate for measuring, for example, the consumption of ultra-processed foods (2018). In accordance with Popkin, we believe that in order to tackle obesity and NCDs epidemics, we must focus our major efforts on finding "environmental solutions" (Popkin 1998). Individual-level interventions conceived from a biomedical approach, are too simplistic for addressing these complex and multifactorial diseases. An intersectoral approach is essential for the definition and implementation of public policies; this means involve multiple actors, including in the first place the food and beverage industries as well as government officials.

References

An R, Guan C, Liu J, Chen N, Clarke C (2019) Trade openness and the obesity epidemic: a cross-national study of 175 countries during 1975–2016. Ann Epidemiol 37:31–36. https://doi.org/10.1016/j.annepidem.2019.07.002

FAO, OPS, WFP, UNICEF (2019) Panorama de la seguridad alimentaria y nutricional en América Latina y el Caribe 2019. FAO, Santiago, 135 p

Finaret AB, Masters WA (2019) Beyond calories: The new economics of nutrition. Ann Rev Resour Econ 11:237–259. https://doi.org/10.1146/annurev-resource-100518-094053

Finucane MM, Stevens GA, Cowan MJ et al (2011) National, regional, and global trends in body-mass index since 1980: systematic analysis of health examination surveys and epidemiological studies with 960 country-years and 9•1 million participants. Lancet 377(9765):557–567. https://doi.org/10.1016/S0140-6736(10)62037-5

Ford ND, Patel SA, Narayan KM (2017) Obesity in low- and middle-income countries: burden, drivers, and emerging challenges. Annu Rev Public Health 38:145–164. https://doi.org/10.1146/annurev-publhealth-031816-044604

Guthold R, Stevens GA, Riley LM, Bull FC (2018) Worldwide trends in insufficient physical activity from 2001 to 2016: a pooled analysis of 358 population-based surveys with 1·9 million participants. Lancet Glob Health 6(10):e1077–e1086. https://doi.org/10.1016/S2214-109X(18)30357-7

Hawkes C, Harris J, Gillespie S (2017) Urbanization and the nutrition transition. In: Global food policy report, vol 4. IFPRI, Washington, DC, pp 34–41

Kovalskys I, Rigotti A, Koletzko B et al (2019) Latin American consumption of major food groups: Results from the ELANS study. PLoS One 14(12):e0225101. https://doi.org/10.1371/journal.pone.0225101

Ministerio de Salud y Desarrollo Social de la Nación (2018) Encuesta Mundial de Salud Escolar. Argentina 2018. Resumen ejecutivo total nacional. CABA: Secretaría de Gobierno de Salud, 13 p

Ministerio de Salud y Desarrollo Social de la Nación (2019) 4ta Encuesta Nacional de Factores de Riesgo. Informe definitivo. Buenos Aires: Dirección Nacional de Promoción de la Salud y Control de Enfermedades Crónicas No Transmisibles, 262 p

NCD Risk Factor Collaboration (NCD-RisC) (2017) Worldwide trends in body-mass index, underweight, overweight, and obesity from 1975 to 2016: a pooled analysis of 2416 population-based measurement studies in 128·9 million children, adolescents, and adults. Lancet 390(10113):2627–2642. https://doi.org/10.1016/S0140-6736(17)32129-3

Ng M, Fleming T, Robinson M et al (2014) Global, regional, and national prevalence of overweight and obesity in children and adults during 1980–2013: a systematic analysis for the Global Burden of Disease Study 2013. Lancet 384:766–781. https://doi.org/10.1016/S0140-6736(14)60460-8

Niclis C, Román MD, Osella AR, Eynard AR, Díaz MP (2015) Traditional dietary pattern increases risk of prostate cancer in Argentina: results of a multilevel modeling and bias analysis from a case-control study. J Cancer Epidemiol Rev 2015:179562. https://doi.org/10.1155/2015/179562

Nicolau-Nos R, Pujol-Andreu J (2011) Aspectos políticos y científicos del modelo de la transición nutricional: evaluación crítica y nuevos desarrollos. Documentos del Trabajo de la Sociedad española de Historia Agraria, pp 11–15

Notestein F (1945) Population: the long view. In: Schultz TW (ed) Food for the world. University of Chicago Press, Chicago, IL, pp 36–57

Omran AR (1971) The epidemiologic transition. A theory of the epidemiology of population change. Milbank Mem Fund Q 49(4):509–538

Organización Panamericana de la Salud (2019) Indicadores básicos 2019: Tendencias de la salud en las Américas. OPS, Washington, DC, 21 p

PAHO-OMS (2015) Ultra-processed food and drink products in Latin America: TRENDS, impact on obesity, policy implications. PAHO, Washington, DC, 76 p

Pérez-Ferrer C, McMunn A, Zaninotto P, Brunner EJ (2018) The nutrition transition in Mexico 1988-2016: the role of wealth in the social patterning of obesity by education. Public Health Nutr 21(13):2394–2401. https://doi.org/10.1017/S1368980018001167

Popkin BM (1993) Nutritional patterns and transitions. Popul Dev Rev 19:138–157

Popkin BM (1994) The nutrition transition in low-income countries: an emerging crisis. Nutr Rev 52:285–298. https://doi.org/10.1111/j.1753-4887.1994.tb01460.x

Popkin BM (1998) The nutrition transition and its health implications in lower-income countries. Public Health Nutr 1(1):5–2. https://doi.org/10.1079/phn19980004

Popkin BM (2001) The nutrition transition and its relationship to demographic change. In: Nutrition and health in developing countries. Humana Press, Totowa, NJ, pp 427–445

Popkin BM (2002a) An overview of the nutrition transition and its health implications: the Bellagio meeting. Public Health Nutr 5(1A):93–103. https://doi.org/10.1079/phn2001280

Popkin BM (2002b) The shift in stages of the nutrition transition in the developing world differs from past experiences! Public Health Nutr 5(1A):205–214. https://doi.org/10.1079/PHN2001295
Popkin BM (2004) The nutrition transition: an overview of world patterns of change. Nutr Rev 62:3–40. https://doi.org/10.1111/j.1753-4887.2004.tb00084.x
Popkin BM (2006) Technology, transport, globalization and the nutrition transition. Food Policy 31:554–569
Popkin BM (2009) Global changes in diet and activity patterns as drivers of the nutrition transition. In: Emerging societies-coexistence of childhood malnutrition and obesity, vol 63. Karger, New York, pp 1–14
Popkin BM (2011) Contemporary nutritional transition: determinants of diet and its impact on body composition. Proc Nutr Soc 70(1):82–91. https://doi.org/10.1017/S0029665110003903
Popkin BM (2015) Nutrition transition and the global diabetes epidemic. Curr Diab Rep 15(9):64. https://doi.org/10.1007/s11892-015-0631-4
Popkin BM (2017) Relationship between shifts in food system dynamics and acceleration of the global nutrition transition. Nutr Rev 75(2):73–82. https://doi.org/10.1093/nutrit/nuw064
Popkin BM, Gordon-Larsen P (2004) The nutrition transition: worldwide obesity dynamics and their determinants. Int J Obes Relat Metab Disord 28 (Suppl 3):S2–S9. https://doi.org/10.1038/sj.ijo.0802804
Popkin BM, Reardon T (2018) Obesity and the food system transformation in Latin America. Obes Rev 19 (8):1028–1064. https://doi.org/10.1111/obr.12694
Popkin BM, Richards MK, Montiero CA (1996) Stunting is associated with overweight in children of four nations that are undergoing the nutrition transition. J Nutr 126(12):3009–3016. https://doi.org/10.1093/jn/126.12.3009
Popkin BM, Adair L, Ng S (2012) Global nutrition transition and the pandemic of obesity in developing countries. Nutr Rev 70(1):3–21. https://doi.org/10.1111/j.1753-4887.2011.00456.x
Pou SA, Díaz Mdel P, Osella AR (2012) Applying multilevel model to the relationship of dietary patterns and colorectal cancer: an ongoing case-control study in Córdoba, Argentina. Eur J Nutr 51(6):755–764. https://doi.org/10.1007/s00394-011-0255-7
Pou SA, Niclis C, Eynard AR, Díaz MP (2014a) Dietary patterns and risk of urinary tract tumors: a multilevel analysis of individuals in rural and urban contexts. Eur J Nutr 53(5):1247–1253. https://doi.org/10.1007/s00394-013-0627-2
Pou SA, Niclis C, Aballay LR, Tumas N, Román MD, Muñoz SE, Coquet JB, Díaz MP (2014b) Cancer and its association with dietary patterns in Córdoba (Argentina). Nutr Hosp 29(3):618–628. https://doi.org/10.3305/nh.2014.29.3.7192
Pou SA, Del Pilar DM, De La Quintana AG, Forte CA, Aballay LR (2016) Identification of dietary patterns in urban population of Argentina: study on diet-obesity relation in population-based prevalence study. Nutr Res Pract 10(6):616–622. https://doi.org/10.4162/nrp.2016.10.6.616
Pou SA, Tumas N, Sanchez Soria D, Ortiz P, Díaz MP (2017) Large-scale societal factors and noncommunicable diseases: Urbanization, poverty and mortality spatial patterns in Argentina. Appl Geogr 86:32–40. https://doi.org/10.1016/j.apgeog.2017.06.022
Sievert K, Lawrence M, Naika A, Baker P (2019) Processed foods and nutrition transition in the Pacific: regional trends, patterns and food system drivers. Nutrients 11(6):E1328. https://doi.org/10.3390/nu11061328
Tirado MC, Galicia L, Husby HM et al (2016) Mapping of nutrition and sectoral policies addressing malnutrition in Latin America. Rev Panam Salud Publica 40 (2):114–123
Tumas N, Niclis C, Aballay LR, Osella AR, Díaz MP (2014) Traditional dietary pattern of South America is linked to breast cancer: an ongoing case-control study in Argentina. Eur J Nutr 53(2):557–566. https://doi.org/10.1007/s00394-013-0564-0
Tumas N, Rodríguez Junyent C, Aballay LR, Scruzzi GF, Pou SA (2019) Nutrition transition profiles and obesity burden in Argentina. Public Health Nutr 22 (12):2237–2247. https://doi.org/10.1017/S1368980019000429
United Nations, Population Division of the Department of Economic and Social Affairs (2017) Revision of World Population Prospects, United Nations, New York
United Nations System, Standing Committee on Nutrition (2010) 6th report on the world nutrition situation–progress in nutrition. United Nations, 134 p
Vandevijvere S, Jaacks LM, Monteiro CA et al (2019) Global trends in ultraprocessed food and drink product sales and their association with adult body mass index trajectories. Obes Rev 20(Suppl 2):10–19. https://doi.org/10.1111/obr.12860
Walls HL, Johnston D, Mazalale J, Chirwa EW (2018) Why we are still failing to measure the nutrition transition. BMJ Glob Health 3(1):e000657. https://doi.org/10.1136/bmjgh-2017-000657
Wells JC, Sawaya AL, Wibaek R et al (2020) The double burden of malnutrition: aetiological pathways and consequences for health. Lancet 395(10217):75–88. https://doi.org/10.1016/S0140-6736(19)32472-9
World Health Organization (2017) 10 facts on obesity. Available from: https://www.who.int/features/factfiles/obesity/en/
World Health Organization (2018a). Obesity and overweight. Available from: https://www.who.int/news-room/fact-sheets/detail/obesity-and-overweight

World Health Organization (2018b) Noncommunicable diseases. Available from: https://www.who.int/en/news-room/fact-sheets/detail/noncommunicable-diseases
World Health Organization (2018c) Noncommunicable diseases country profiles 2018. Available from: https://www.who.int/nmh/countries/en/
Zapata ME, Rovirosa A (2016) Carmuega. La mesa Argentina en las últimas dos décadas: cambios en el patrón de consumo de alimentos y nutrientes 1996–2013. Ciudad Autónoma de Buenos Aires: Centro de Estudios sobre Nutrición Infantil—CESNI, 204 p
Zapata ME, Rovirosa A, Carmuega E (2016) [Changes in the food and beverage consumption pattern in Argentina, 1996–2013]. Salud Colect 12(4):473–486. Spanish. https://doi.org/10.18294/sc.2016.936

3 Challenges and Economic Burden of Diabetes in Africa

Camille Maadjhou Mba and Jean Claude Mbanya

Contents

Introduction 22

Social and Financial Burden 22

Challenges to Diabetes Management in Africa 23
Epidemiological transitions: 23

Access to Health Care 23

Knowledge, Beliefs, and Health Seeking Behaviors in Africa 24

Lack of Political Will 24

Under-Allocation of Resources 25

Complications and Mortality Attributable to Diabetes 25

Prevalence and Incidence of Diabetes Complications 25

Disability and Mortality 26

Economic Burden of Diabetes Mellitus 26

Costs Attributable to Diabetes Mellitus in Africa 26

Purchasing Power Parity Calculations 27

Direct Cost 28

Profile of African Countries 28

Indirect Cost 30

C. M. Mba
MRC Epidemiology Unit, University of Cambridge, Cambridge, UK

Department of Public Health, Faculty of Medicine and Biomedical Sciences, University of Yaoundé 1, Yaoundé, Cameroon

J. C. Mbanya (✉)
Department of Internal Medicine and Specialties, Faculty of Medicine and Biomedical Sciences, University of Yaoundé 1, Yaoundé, Cameroon
e-mail: jcmbanya@yahoo.co.uk

J. Faintuch and S. Faintuch (eds.), *Obesity and Diabetes*, https://doi.org/10.1007/978-3-030-53370-0_3

Intangible Cost ... 30
Cost of Diabetes Complications ... 30
Comparative Health Care Expenditures in Africa ... 31
Perspectives ... 32
National Diabetes Programs (NDPs) ... 32
Sustainable Development Goals ... 32
Universal Health Coverage (UHC) ... 32
References ... 32

Abstract

Diabetes imposes a huge financial burden on affected individuals, their families, society, and country. Undesirable changes in lifestyle coupled with inadequate health care systems are some of the major drivers of diabetes burden in sub-Saharan Africa (SSA), where there will be the greatest increase of diabetes in the world of 143%, within the next 25 years. Only 5.5% of the GDP in SSA was spent on health in 2016 compared to the global estimate of 10%. Although SSA has witnessed a recent increase in diabetes incidence, the current diabetes-related expenditure of US$9.5 billion represents only 1% of the overall global expenditure on this disease. These include the direct costs associated with medical care, and the indirect costs resulting from cessation or reduced productivity because of disease-related disability/premature mortality. Indirect cost per patient with diabetes in SSA is higher than direct expenditures, contributing over 60% of the total cost.

Keywords

Diabetes burden · Diabetes cost · Indirect cost · Direct cost · Sub Saharan Africa

Introduction

Nearly 80% of noncommunicable disease (NCDs) related deaths including diabetes, occur in low- and middle- income countries, which constitute the majority of countries in sub-Saharan Africa (SSA) (WHO 2019a). Available estimates show that 7.1% of adults living in Africa or 33 million people suffer from diabetes, although in 2019 the International Diabetes Federation estimated that diabetes affected over 19 million people in the region, while an additional 45 million adults had impaired glucose tolerance (IGT) (IDF Diabetes Federation 2019a). Nigeria is among the top ten countries in the world with the highest number of people with IGT (8.2 million people). If current trends persist, both the number of people with type 2 diabetes (T2D) and IGT in SSA will increase by 143% by 2045; the highest global predicted increase. However, the true number could be much higher, given that as many as two thirds of those with the disease remain undiagnosed (IDF Diabetes Federation 2019a).

This rising burden of diabetes in SSA has been triggered by increasing obesity and changing lifestyles in both rural and urban communities (Mbanya et al. 2014). As in other parts of the world, more than 90% of patients have type 2 diabetes with diagnosis often made late in the course of disease progression, or at the time of complications. Also, approximately 10,300 children and adolescents are diagnosed each year with type 1 diabetes (T1D) which carries a high mortality rate (IDF Diabetes Federation 2019a).

Social and Financial Burden

Delayed diagnosis and poor metabolic control contribute to the high morbidity and premature mortality in the region (Dalal et al. 2011). The mortality rate is increasing and occurring at younger ages in SSA compared with the other regions; three of four deaths due to diabetes occur in the active age group less than 60 years old (IDF Diabetes Federation 2019a; Miranda et al. 2019). Total disability adjusted life-years (DALYs) due to diabetes increased by 126.4% between 1990 and 2017, and the Lancet Diabetes and Endocrinology Commission estimated that in 2015, the overall cost of diabetes in sub-Saharan Africa was US$19.45 billion, or 1.2% of cumulative gross domestic product (GDP) (Gouda et al. 2019; Atun et al. 2017).

Despite progress in health coverage, access to medications especially insulin and quality healthcare facilities remain a challenge, with wide disparities between African countries (Mutyambizi et al. 2018). Poor access to insulin is the major cause of mortality in children with T1D (Atun and Gale 2015; Choukem et al. 2019). This has a huge financial implication for the economy of households and the already overburdened healthcare systems.

Few African countries have a national diabetes plan and there is a lack of access to full coverage for essential health services (IDF Diabetes Federation 2019a). Due to the lack of primary data and the high proportion of undiagnosed diabetes to estimate the real-world cost of diabetes in the SSA, public health interventions have relied on results from modeling methods (Afroz et al. 2018). Africa's health service delivery capacity has improved, yet this is still insufficient to meet current and future needs of the care of diabetes (WHO 2019b).

Challenges to Diabetes Management in Africa

These challenges are of two magnitudes: the increase in diabetes growth and those related to the implementation of effective diabetes care in Africa (Fig. 3.1).

Epidemiological transitions:

In the 1980s, diabetes was still uncommon in Africa, but today it has exploded in many communities. This trend in T2D has been strongly linked to obesity, suboptimal diet, and physical inactivity, resulting from the rapid epidemiologic shifts involving urbanization and changing lifestyles in both the rural and urban communities of SSA (Mbanya et al. 2014). The age-standardized diabetes prevalence increased in parallel with body mass index from 3.4 to 8.5% in men, and 4.1 to 8.9% in women between 1980 and 2014 (NCD Risk Factor Collaboration (NCD-RisC)—Africa Working Group 2017). Global rise in BMI is usually linked to the increase in BMI in rural communities, except in SSA where BMI rose faster among urban women, and at a similar rate among urban and rural men (NCD Risk Factor Collaboration (NCD-RisC) 2019). The expansion of supermarkets with access to highly processed food and sweetened beverages as well as reduced physical activity could be driving the rural-urban differences (Pastakia et al. 2017; Assah et al. 2011).

Access to Health Care

Many African countries are not prepared to effectively manage diabetes and they provide only suboptimal care to patients. Early identification, management, and control are important determinants for the reduction of diabetes complications, thus leading to a substantial reduction of the economic burden (Atun and Gale 2015; Kapwata and Manda 2018).

Timely diagnosis and prevention rely mostly on the availability of health care practitioners, regular screening, and management of comorbidities (Pastakia et al. 2017). The proximity of healthcare facilities has been strongly linked to increasing utilization of these services in developing countries, especially in rural communities (Feikin et al. 2009). Financial

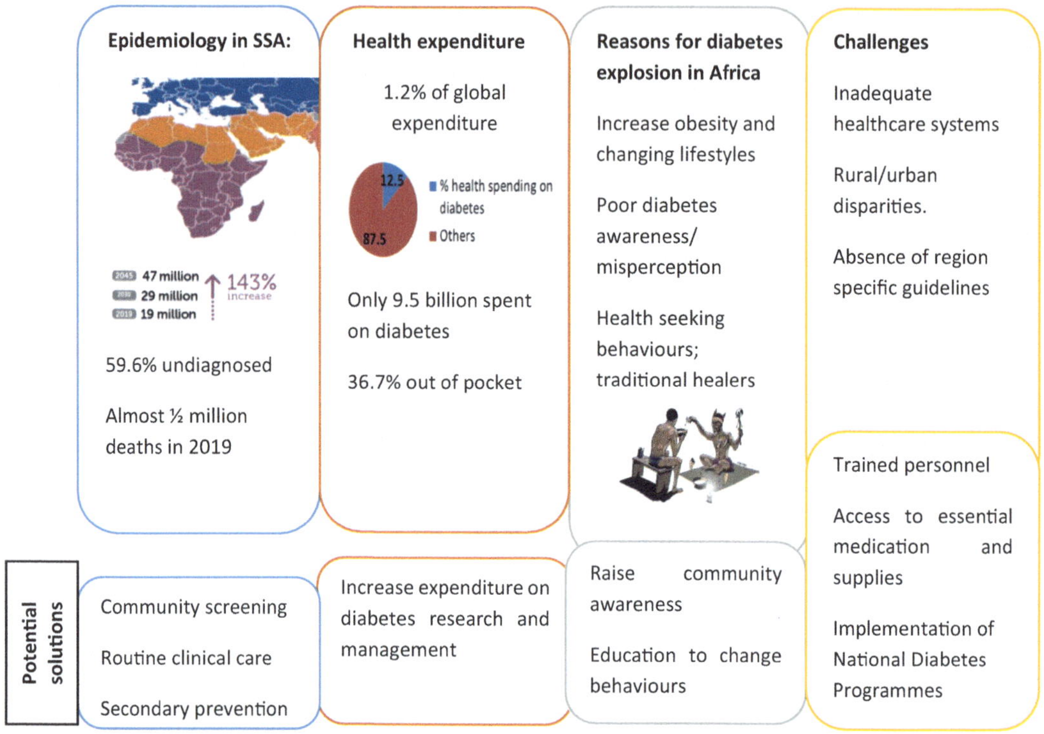

Fig. 3.1 Socioeconomic burden of diabetes and healthcare challenges in Africa (IDF Diabetes Federation 2019a; Choukem et al. 2019)

availability and accessibility of health services, transportation facilities, and cost, as well as the receptiveness of the healthcare professionals to the cultural and social norms of the community are equally relevant (Huerta Munoz and Källestål 2012). Poor access to insulin is the leading cause of mortality in patients with T1D (Gale 2006). Among the 400 million people globally who do not have access to basic healthcare services, the majority live in SSA (Sanogo et al. 2019).

Knowledge, Beliefs, and Health Seeking Behaviors in Africa

Obesity which is a major risk factor for T2D is still perceived by some people as a sign of wealth and good living, while physical activity like regularly walking is a sign of poverty. Although most patients recognize that health care facilities are the ideal places to turn to, it is not unusual for patients to seek solely or as complementary care, the traditional healers (Choukem et al. 2019). Diverse knowledge, beliefs, and perceptions in addition to finances are key factors that influence a patient's decision to seek medical advice during an illness. Moreover, misconceptions associated with hospital facilities contribute to increasing the total cost associated with diabetes, as some patients tend to be managed at home or seek traditional healers, thereby presenting to health facilities mostly at the time of complications (Choukem et al. 2019).

Lack of Political Will

There is still a lack of appropriate action taken by key decision leaders in SSA, in order to halt the progression of diabetes (Atun and Gale 2015). The misconception that diabetes is a disease of the rich and the elderly, hampers commitments to mobilize adequate resources for the management of diabetes and its complications. In fact, less than 30% of African countries have operational integrated policies for diabetes care (Kengne et al. 2013).

Under-Allocation of Resources

Inadequate financing of NCDs like diabetes constitutes a major barrier to the early diagnosis and management of diabetes complications (Atun and Gale 2015). Although infections, maternal and perinatal mortality continue to threaten African healthcare systems, the idea that infectious diseases should be controlled before tackling NCDs is mostly misleading. According to WHO estimates in 2015 infectious diseases, maternal, perinatal, and nutritional conditions accounted for 59.1% of all age total DALYs, while NCDs were responsible for 30.7% (WHO 2019a). In addition, NCDs are responsible for up to 37% of total loss of productivity in SSA, compared to only 27% for infectious diseases (WHO 2019b).

Complications and Mortality Attributable to Diabetes

Noncommunicable diseases like diabetes and cardiovascular account for 41 million deaths each year; the leading cause of mortality globally. Over 85% of these deaths occur in low- and middle-income countries (WHO 2019a), and the acute and chronic complications associated with the disease make it very costly to manage (Kengne et al. 2013). Diabetes is a major cause of non-traumatic lower limb amputation, blindness, kidney failure, heart attacks, and stroke (IDF Diabetes Federation 2019a).

Prevalence and Incidence of Diabetes Complications

Chronic complications of diabetes may be present at the time of diagnosis especially in patients with T2D, but can also appear soon after the diagnosis in patients with T1D (IDF Diabetes Federation 2019a). In early 2000, macrovascular complications of diabetes were considered rare in Africa, with coronary heart disease affecting 5–8% of type 2 diabetic patients (Mbanya and Sobngwi 2003). However, there is increasing evidence suggesting an increase in the burden of cardiovascular disease in SSA (Kengne et al. 2013), as a result of poor glycemic control in addition to other cardiovascular risk factors such as high blood pressure and dyslipidemia (Kengne et al. 2005).

Five percent of patients with diabetes have cerebrovascular accidents at diagnosis, and up to 9.5% and 25% of patients with type 1 and 2 diabetes respectively also have retinopathy at diagnosis. Depending on the glycemic control and duration of diabetes the prevalence of diabetic retinopathy varies between 13 and 55 %, with severe retinopathy representing 15% of all cases (Mbanya and Sobngwi 2003; Kengne et al. 2005). Eight to 49% of patients with diabetes exhibit nephropathy and neuropathy is present in 11–66% depending on the diagnostic method used (Dalal et al. 2011). Long-standing complications in patients with T1D do not differ greatly from those with T2D (Motala et al. 2001).

Lower extremity amputation varies from 1.5 to 7% (Mbanya and Sobngwi 2003). Diabetes complications in Africa are increasing as the prevalence and incidence of diabetes continue to rise. In Ghana, the incidence of diabetes-related lower limb amputations increased from 0.6 per 1000 follow up years in 2010 to 10.9 in 2015 (Sarfo-Kantanka et al. 2019). This is in contrast to industrialized countries where the complication rates are reducing (IDF Diabetes Federation 2019a). The coexistence of multiple chronic complications of diabetes is common, like in Nigeria where 51.7% of patients with diabetes and renal complications also had the peripheral vascular disease (Agaba 2004). Interactions between peripheral damaged organs and infection occur in up to 53% of patients with diabetes (Dalal et al. 2011). Heavy investment in the health care system like dialysis units and coronary care units are necessary to take care of these diabetes-related complications.

Disability and Mortality

In 2017, NCDs accounted for 80% of disabilities globally with diabetes being the fourth leading cause (Mbanya et al. 2014). The age-standardized DALY rates for diabetes are higher amongst men than women (Gouda et al. 2019). Disability and the premature mortality caused by diabetes occur at a younger age in SSA compared to other regions. Globally in 2019, diabetes was responsible for 366,200 deaths with 3 of 4 deaths occurring before 60 years whereas, in 2009, diabetes-related mortality was highest in the age group of 30–39 years (IDF Diabetes Federation 2019a, b). High blood pressure which is common in blacks, was found to be about twice more common in people with diabetes compared to age-matched controls, while dyslipidemia ranged from 16 to 89% and overweight 50–83%. Cardiovascular complications are amongst the leading cause of mortality from chronic complications in patients with T2D (Kengne et al. 2005).

Although there is poor documentation on the causes of deaths in Africa, available estimates suggest that diabetes-related mortality ranges between 8 and 41% of all-cause mortality, and this is primarily due to acute complications and infections (Peer et al. 2014). In Cameroon, acute complications were the leading cause of mortality (22.2%) followed by cardiovascular complications (16.7%) and nephropathy (14.8%) (Foryoung et al. 2018). The life expectancy in children with T1D could be as low as 1% in Africa (Beran and Yudkin 2006). In South Africa, the predominant cause of mortality in patients with T1D is nephropathy, accounting for over 40% of all deaths (Gill et al. 2005). Table 3.1 shows the number of deaths in adults aged 20–79 years for some SSA countries.

Economic Burden of Diabetes Mellitus

Three alternative approaches exist to estimate the economic burden of public health problems like diabetes including the willingness to pay approach, the macroeconomic/production function approach, and the cost of illness approach (COI). The last one is the most commonly employed and takes into account the direct cost, indirect cost, and intangible cost (Fig. 3.2).

Costs Attributable to Diabetes Mellitus in Africa

Since 2001, there has been a steady rise in the gross domestic product (GDP) of many sub-Saharan African countries, reaching its peak in 2014. In 2016, the mean GDP per capita spent by SSA countries was equivalent to US$1511.3 with a wide range from US$282.1 in Burundi to US$15,077.8 in Seychelles (The World Bank 2019). Overall, less than 5.6% of the GDP was spent on healthcare in the low- and middle-income countries of SSA. The World Health Organization (WHO) estimates that up to 11 million Africans fall into poverty each year, due to high out of pocket payments for healthcare (WHO 2019b). Few studies have quantified the economic burden of diabetes at the regional level. A major difficulty in collating data for the cost of illness studies is that authors report on only one type of cost; mostly the direct cost (Afroz et al. 2018). In South Africa, where the diabetes age-adjusted prevalence is 12.7%, mean diabetes-related expenditure was US$1245.0 per patient (Table 3.1), while in Benin, the age-adjusted diabetes prevalence was 1% and mean expenditure was US$163.8 pp.

Diabetes costs have increased from 1.4 billion in 2010 to 9.5 billion in 2019. This is due to both an increase in the prevalence of diabetes by 57%, and an increase in the cost of persons living with diabetes from US$222.6 to US$509 (IDF Diabetes Federation 2019a, b). According to current reports, the cost of care for people with diabetes in Africa accounts for 12.5% of healthcare expenses, with the highest percentage in South Africa (23.0%) (IDF Diabetes Federation 2019a).

Table 3.1 Prevalence estimates and economic indicators of diabetes for some African countries

Country	Age-adjusted diabetes prevalence for the age group 20–79 years (%)[a]	Mean diabetes-related expenditure per person (20–79) with diabetes (PPP)[a]	GDP per capita (current US$)[b]	Current Health Expenditure per capita in US$[c]	Number of diabetes-related deaths (20–79)[a]
Benin	1.0	163.8	829.5	31	692.8
Botswana	5.8	1417.6	7893.2	466	1674.5
Burkina Faso	7.3	177.6	642.4	44	9675.2
Cameroon	6.0	311.3	1421.6	68	13,744.3
Central African Republic[d]	6.0	72.0	449.8	24	3162.9
Chad[d]	6.0	135.3	664.3	30	5706.5
Comoros	12.3	173.6	1320.5	59	357.4
Cote d'Ivoire[d]	2.4	327.0	1557.2	70	5207.1
Equatorial Guinea[d]	6.0	1305.8	9738.4	301	491.5
Ethiopia	4.3	113.3	768.4	25	23,156.8
Gabon[d]	6.0	1015.2	7212.5	204	922.6
Gambia	1.9	119.3	679.8	23	193.3
Ghana	2.5	262.2	2025.9	67	5397.8
Kenya	3.1	324	1568.2	77	8080.5
Mauritania	7.1	213.0	1145.5	49	1776.5
Mozambique	3.3	101.8	461.4	21	9485.0
Namibia	4.5	1871.8	5646.5	447	1095.3
Niger[d]	2.4	87.7	375.9	29	3180.6
Nigeria	3.1	468.6	1968.6	74	63,957.7
Sierra Leone[d]	2.4	587.9	499.4	66	19.3
South Africa	12.7	1245.0	6132.5	499	89,834.4
Tanzania	5.7	170.1	1004.8	34	18,031.7
Uganda	2.5	191	631.5	39	6288.0
Zimbabwe	1.8	540.9	1602.4	110	2621.9

[a]IDF Atlas, 2019 (IDF Diabetes Federation 2019a)
[b]World Bank indicators, 2017 (The World Bank 2019)
[c]WHO Global expenditure database, 2017 (WHO 2019c)
[d]Indicates absence of in-country data sources on diabetes

Purchasing Power Parity Calculations

Nineteen million adults with diabetes recorded in Africa in 2019 resulted in a total economic cost of US$9.5 billion. Using the 2005 purchasing power parity (PPP), Kirigia and colleagues estimated the cost of diabetes per patient at Int$3633, giving a total burden of Int$25.51 billion, with indirect cost contributing 68% of it (Kirigia et al. 2009). Consistent with this, other reviews have reported the indirect costs of diabetes per patient to be higher than the direct costs (Mutyambizi et al. 2018). Nevertheless, Bommer and colleagues found a lower contribution of indirect cost (44.4%) (Bommer et al. 2017).

Cost of medication and hospitalization are the major causes of direct expenses (Afroz et al. 2018). A review reported the cost of medication to contribute as high as 50% of the total cost (Mutyambizi et al. 2018).

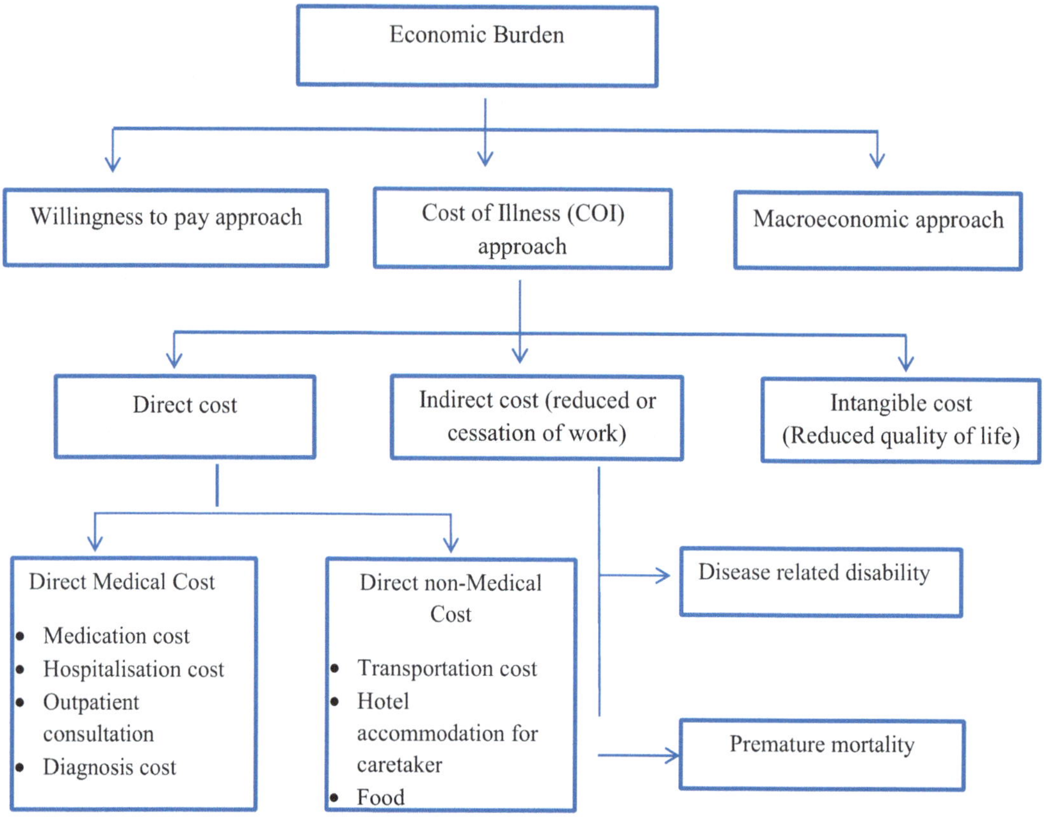

Fig. 3.2 Measurements approaches to the economic burden of diabetes (Adapted from Kirigia et al. (2009))

Direct Cost

The direct cost of a disease refers to resources used for the treatment of the disease regardless of whether they are born by patients themselves, by private or public payers, or by the government (IDF Diabetes Federation 2019a). This is usually divided into the direct medical cost which are expenses, borne as a result of direct medical treatment like medications, outpatient visits, laboratory tests, and hospitalizations. The second is the direct nonmedical cost which are expenses incurred as a result of transportation, food, lodgment. In COI studies, cost of medications, diagnostic cost, cost of the consultation, and inpatient cost is the most commonly reported (Mutyambizi et al. 2018).

Profile of African Countries

The direct cost of diabetes differs greatly between African countries with reported costs estimated from Int\$3.5 billion to Int\$4.5 billion and over US\$10.8 billion per annum (Mutyambizi et al. 2018; Bommer et al. 2017) (Table 3.2). There are inconsistent findings of the major contributors, with some reviews reporting the cost of hospitalization as predominant, and other medication costs (Mutyambizi et al. 2018; Afroz et al. 2018). For outpatients in Nigeria, the cost of medication varied between Int\$420.63 to Int \$1025.00 per patient per annum for type 1 and 2 diabetes respectively, thus contributing as high as 90% of all treatment cost. For hospitalized patients, the cost of medication ranged from Int

Table 3.2 Summary of studies on diabetes costs from Africa (adapted from Mapa et al. (2019))

Year of costing	Setting/Reference	Perspectives	Type of diabetes	Direct cost (US$)	Indirect costs (US$)	Total cost (US $)
2019	Africa (IDF Diabetes Federation 2019a)	Societal	1 and 2			9.5 billion
2018	Africa (Mutyambizi et al. 2018)	All	1 and 2	Int$3.5 billion to Int$4.5 billion per annum		
2015	Africa (Bommer et al. 2017)	All	1 and 2	10.8 billion	8.6 billion	19.4 billion
2015	Guinea (Alouki et al. 2015)	Patients	2	126–1093 pp		
	Mali	Patients	2	137–869 pp		
	Benin	Patients	2	212–828 pp		
	Burkina Faso	Patients	2	224–859 pp		
2013–2015	Benin (Ogle et al. 2016)	Family	1	588 pp		
	Burkina Faso	Family	1	1185 pp		
	Central African Republic	Family	1	737 pp		
	Ivory coast	Family	1	553 pp		
	Malawi	Family	1	999 pp		
	Mauritania	Family	1	630 pp		
2013	Morocco (Boutayeb et al. 2013)	Societal	1 and 2	259–830 pp or 0.47 billion to 1.5 billion	1113 pp or 2 billion	
2011–2012	Nigeria (Suleiman and Festus 2015)	Societal	2	1.64 billion		
2009–2011	Cameroon (Ngassam et al. 2012)	Patients	2	148 pp		
2010	Nigeria (Abdulganiyu and Fola 2014)	Patients and hospital	2	1.5 million		
2009	Mali (Bermudez-Tamayo et al. 2017)	Patient	2	365.48 pp	762.20 pp	1127 pp
2009	Ghana (Quaye et al. 2015)	Health system	1 and 2	423 pp		
2005	Sudan (Elrayah-Eliadarous et al. 2010)	Patients	2	175[a]		
2005	Africa (Kirigia et al. 2009)	Societal	1 and 2	Int$17.41 billion	Int$8.1 billion	Int $25.51 billion
2003–2004	Nigeria (Suleiman et al. 2006)	Patients and hospital	1 and 2	1.07 billion		
1995	Sudan (Elrayah et al. 2005)	Patients	1	283[b]		

Abbreviation: *pp* per person

Unless otherwise stated, costs are annual cost in US$

Common perspectives used in COI studies are the patient perspective, the employer perspective (loss of productivity), health system perspective, government perspective and societal perspective

Int$: International dollar (local cost converted to the purchasing power of one US$)

[a]Annual cost of minimum care diabetes supplies

[b]Median cost

$553 to 1438 per admission in Nigeria and Int$80 to 461 in South Africa. In Cameroon, the cost of medication was estimated at I$3.9 per patient per month (Mutyambizi et al. 2018). The greatest expense incurred for diabetes medications comes from insulin. Indeed, the annual treatment for diabetes with insulin is US$104.4 compared to US$26.9 with metformin (Atun et al. 2017).

Several factors are associated with the direct cost of diabetes include socioeconomic factors like age, level of education and income class, factors related to diabetes like duration of diabetes, number of complications, mode of treatment, hospitalizations, and surgery (Afroz et al. 2018). As shown in Table 3.2, the cost of T1D is consistently higher than for type 2 diabetes across the studies.

Indirect Cost

The indirect cost is the opportunity cost resulting from reduced or loss of productivity at work by patients and their accompanying persons, because of the disease-related disability or premature mortality (Afroz et al. 2018). This includes the productive time lost for medical reasons like traveling for medical consultations and waiting to be admitted, hospital admissions for complications, absence from work for disease-related disability, and years of work lost due to premature deaths. Depending on the approach used to calculate the indirect cost of diabetes, the economic burden might be overstated. Some methods estimate the indirect cost for the unemployed while others use a more conservative approach like the friction-cost approach (Van den Hout 2010).

The COI study in 2015 used four sources of indirect costs: labor-force drop out; mortality; absenteeism; and presenteeism (Bommer et al. 2017). Globally, the first two contribute 48.5% and 45.5% respectively of indirect cost, however, this is different in LMICs where mortality dominates, contributing 63.6% of the indirect cost in middle-income countries and 90.6% in low-income countries. As shown in Table 3.2, the annual indirect cost of diabetes per patient was estimated at US$762.20 in Mali, for an annual total cost per patient of US$365.48, while in Morocco, the indirect cost per patient per month was US$1113 (Bermudez-Tamayo et al. 2017; Boutayeb et al. 2013). Because most studies report only on the direct cost, there is a dearth of information on the indirect cost of diabetes in Africa. Nonetheless, its contribution to total cost could contribute to over 60% of the total cost (Kirigia et al. 2009; Mapa-Tassou et al. 2019).

Intangible Cost

Intangible cost relates to the reduced quality of life as a result of the physical, psychological, and social handicap of disease. The quality of life can be severely affected by chronic diseases and examples include pain, worry, anxiety, loss of independence, and lack of participation in social events (Afroz et al. 2018).

Intangible costs are the most difficult to assess given that they are related to emotions. Because the quality of life is difficult to express in terms of money, many studies do not report on the intangible cost. Intangible cost of diabetes may range between US$30.8 per month if the patient had no complications, to US$143.9 if four or more complications were present. If patients had blood glucose levels within targets, the intangible cost was estimated at 10.3 per month, compared to US$41.1, if poorly controlled, with an annual estimate of US$496.7 (Afroz et al. 2018).

Cost of Diabetes Complications

The major contributor to direct costs for diabetes-related complications is hospitalization, and for indirect cost are absenteeism, disability, and premature mortality. There is a linear relationship between expenditures and the number of complications present. According to the International Diabetes Federation estimates, the treatment of diabetes-related complications accounts for over 50% of the direct cost attributable to diabetes, and this expenditure is highest in the

60–70 years age group (IDF Diabetes Federation 2019a).

Nephropathy, diabetic foot, and stroke are the complications associated with the highest expenses in Africa (Mutyambizi et al. 2018). Patients with diabetes are likely to spend up to four times more if they have diabetes-related complications compared to those with no complication, and this varies greatly by countries. In Mali, up to Int$1346.5 per patient was spent on nephropathy while only Int$33.03 was spent in Guinea. For acute complications like diabetic ketoacidosis, the cost of management differs in those with and without chronic complications. In Guinea, the cost of treatment for diabetic ketoacidosis ranged from Int$22.1 to Int$39.2, and in Burkina Faso from Int$628.27 to Int $1085.4 (Mutyambizi et al. 2018).

Mean annual health expenditures for people with four or more complications are expected to be about 20 times more than in people without complications (IDF Diabetes Federation 2019a). Consistent with this, the presence of one or more complications increased the cost ratio by 1.1 for retinopathy and up to 4.3 for an infected diabetes foot in another study. This is because, in Africa, foot complications in patients with diabetes often progress to infection or gangrene, resulting in longer hospitalizations and higher mortality rates (Mutyambizi et al. 2018).

Comparative Health Care Expenditures in Africa

In 2016, 5.5% of the GDP in Africa was spent on health against 10% at the global level, with SSA having the highest out of pocket expenses (36.7%) (WHO 2019b; The World Bank 2019). In SSA, the GDP percent spent on health is not related to the national expenditure. For example, Sierra Leone with GDP per capita of US$501.4 spent 16.53% of its budget on healthcare, while others like Equatorial Guinea and Nigeria having a GDP of US$9738.4 and 1968.6 spent only 3.38% and 3.35% respectively. Similarly, out of pocket expenses on health care were highest in Equatorial Guinee and Nigeria (72.83% and 75.21% respectively) (WHO 2019c).

Globally, at least US$760 billion is expected to be spent on diabetes annually, and it could reach US$845 billion by 2045 (IDF Diabetes Federation 2019a). The expenditure per person in high-income countries (US$5339) was over 30-folds and almost sevenfold higher than in low- and middle-income countries respectively in 2019. Reduced or missed production opportunities due to diabetes contribute an additional 35% to the annual global health expenditures. Overall, direct expenses on diabetes have increased by 4.5% over the last 2 years, and it is predicted to increase by 11.2% over the next 25 years (IDF Diabetes Federation 2019a). It is possible that the real estimates are more important as actual projections are conservative, and based on the assumption that diabetes prevalence and expenditures are not changing. One can find substantially higher estimates for the global economic burden of diabetes at US$1.31 trillion, equivalent to 1.8% of the world GDP, with indirect cost contributing 34.7% of the overall burden in a COI study of 2015 (Bommer et al. 2017).

In 2015, North America had the highest economic burden at US$499.40 billion (2.6% of its total GDP), while Africa had the lowest diabetes-related expenditure; US$9.45 billion (1.2% of its total GDP). The contribution of the indirect cost was more important in high-income countries compared to LMICs (40.0% versus 33.5%) (Bommer et al. 2017). While the cost of medication is the major contributor to direct cost in Africa, there are conflicting reports between the cost medication or cost of hospitalization at the global level (Ng et al. 2014). Indirect cost contributed as high as 44.6% in SSA and 57.4% in South Asia as opposed to only 21.9% in Latin America and the Caribbean, although there is a report of a higher contribution of the indirect cost of 68% to the total cost in SSA (Bommer et al. 2017). (Kirigia et al. 2009).

The lowest annual expenditure per person spent in SSA was in the Central African Republic, with similar estimates in Bangladesh (US$64), and Nepal (US$80). Overall, slightly higher expenditure was spent in 2019 in women than men across all the regions (US$382.6 billion versus US$377.6 billion) and this difference is predicted to be maintained over the years (IDF Diabetes Federation 2019a).

Perspectives

Countries with the lowest GDP have the highest burden of NCDs in terms of DALYs, therefore if nothing is done, the rapid increase in diabetes will badly affect the African region. The potential strategies to address the economic burden of diabetes are listed below.

National Diabetes Programs (NDPs)

Early management and lifestyle changes can delay or prevent T2D in 80% of individuals (IDF Diabetes Federation 2019a). Therefore, the development and implementation of diabetes programs may reduce the incidence of diabetes and mortality. Unfortunately, less than one-third of SSA countries have a national diabetes plan (IDF Diabetes Federation 2019a). Important areas for the focus of NDPs are conducting community cost-effective interventions, aiming at raising awareness on diabetes and its risk factors, increasing physical activity, losing weight, eating healthily, along with routine screening and clinical care, and access to essential medication and supplies.

Sustainable Development Goals

The third UN summit in New York in September 2018 recognized the challenges posed by diabetes and other NCDs on the socioeconomic development. To achieve the 2030 targets related to sustainable development goal 3, especially item 3.4 which aims at reducing premature mortality from NCDs by at least 33%, there is a need for governments, leaders, and stakeholders in SSA to urgently react to the threat imposed by diabetes and other NCDs. SSA healthcare systems continue to prioritize infectious diseases and currently, less than 30% of ministries in SSA have integrated policies for the management of diabetes and other NCDs (Kengne et al. 2013). Given that over 70% of deaths globally are due to NCDs, there is value in investing in strengthening the health care system and undertake more translational research to reduce the burden of diabetes (WHO 2019a). Furthermore, because diabetes mortality is highest in low-income countries, prioritization of the limited resources using cost-effective strategies, like the WHO Best Buys interventions (e.g., taxation on alcohol and tobacco products) to change behaviors may improve health outcomes (Atun et al. 2017; Kengne et al. 2013).

Universal Health Coverage (UHC)

Most countries in SSA do not have a national health care insurance coverage, and therefore out of pocket payment of care limits access for individuals with low income. Such an objective could be achieved by ensuring an equitable tax payment system, and compulsory health insurance (Sanogo et al. 2019).

References

Abdulganiyu G, Fola T (2014) What is the cost of illness of type II diabetes mellitus in a developing economy? Int J Pharm Pharm Sci 6:5

Afroz A, Alramadan MJ, Hossain MN, Romero L, Alam K, Magliano DJ et al (2018) Cost-of-illness of type 2 diabetes mellitus in low and lower-middle income countries: a systematic review. BMC Health Serv Res 18(1):972

Agaba EI (2004) Characteristics of type 2 diabetics presenting with end stage renal disease at the Jos University Teaching Hospital, Nigeria. West Afr J Med 23(2):142–145

Alouki K, Delisle H, Besançon S, Baldé N, Sidibé-Traoré A, Drabo J et al (2015) Simple calculator to estimate the medical cost of diabetes in sub-Saharan Africa. World J Diabetes 6(16):1312–1322

Assah FK, Ekelund U, Brage S, Mbanya JC, Wareham NJ (2011) Urbanization, physical activity, and metabolic health in sub-Saharan Africa. Diabetes Care 34 (2):491–496

Atun R, Gale EAM (2015) The challenge of diabetes in sub-Saharan Africa. Lancet Diabetes Endocrinol 3 (9):675–677

Atun R, Davies JI, Gale EAM, Bärnighausen T, Beran D, Kengne AP et al (2017) Diabetes in sub-Saharan Africa: from clinical care to health policy. Lancet Diabetes Endocrinol 5(8):622–667
Beran D, Yudkin JS (2006) Diabetes care in sub-Saharan Africa. Lancet 368(9548):1689–1695
Bermudez-Tamayo C, Besançon S, Johri M, Assa S, Brown JB, Ramaiya K (2017) Direct and indirect costs of diabetes mellitus in Mali: a case-control study. PLoS One 12(5):e0176128
Bommer C, Heesemann E, Sagalova V, Manne-Goehler J, Atun R, Bärnighausen T et al (2017) The global economic burden of diabetes in adults aged 20-79 years: a cost-of-illness study. Lancet Diabetes Endocrinol 5 (6):423–430
Boutayeb W, Lamlili MEN, Boutayeb A, Boutayeb S (2013) Estimation of direct and indirect cost of diabetes in Morocco. J Biomed Sci Eng 06(07):732–738
Choukem SP, Dimala CA, Maadjhou C, Mbanya JC (2019) Diabetes management in Africa. In: Rodriguez-Saldana J (ed) The diabetes textbook. Springer, Cham, pp 273–288
Dalal S, Beunza JJ, Volmink J, Adebamowo C, Bajunirwe F, Njelekela M et al (2011) Non-communicable diseases in sub-Saharan Africa: what we know now. Int J Epidemiol 40(4):885–901
Elrayah H, Eltom M, Bedri A, Belal A, Rosling H, Östenson C-G (2005) Economic burden on families of childhood type 1 diabetes in urban Sudan. Diabetes Res Clin Pract 70(2):159–165
Elrayah-Eliadarous H, Yassin K, Eltom M, Abdelrahman S, Wahlström R, Ostenson C-G (2010) Direct costs for care and glycaemic control in patients with type 2 diabetes in Sudan. Exp Clin Endocrinol Diabetes 118(4):220–225
Feikin DR, Nguyen LM, Adazu K, Ombok M, Audi A, Slutsker L et al (2009) The impact of distance of residence from a peripheral health facility on pediatric health utilisation in rural western Kenya. Trop Med Int Health 14(1):54–61
Foryoung JB, Ditah C, Fon PN, Mboue-Djieka Y, Nebongo DN, Mbango ND et al (2018) Long-term mortality in outpatients with type 2 diabetes in a reference hospital in Cameroon: a retrospective cohort study. BMJ Open 8(2):e019086
Gale EAM (2006) Dying of diabetes. Lancet 368 (9548):1626–1628
Gill GV, Huddle KRL, Monkoe G (2005) Long-term (20 years) outcome and mortality of Type 1 diabetic patients in Soweto, South Africa. Diabetes Med 22 (12):1642–1646
Gouda HN, Charlson F, Sorsdahl K, Ahmadzada S, Ferrari AJ, Erskine H et al (2019) Burden of non-communicable diseases in sub-Saharan Africa, 1990–2017: results from the Global Burden of Disease Study 2017. Lancet Glob Health 7(10):e1375–e1387
Huerta Munoz U, Källestål C (2012) Geographical accessibility and spatial coverage modeling of the primary health care network in the Western Province of Rwanda. Int J Health Geogr 11:40
IDF Diabetes Federation (2019a) IDF Diabetes Atlas, 9th ed. International Diabetes Federation, Brussels (cited 2019 Dec 19). http://www.diabetesatlas.org/
IDF Diabetes Federation (2019b) International diabetes atlas, 4th ed. International Diabetes Federation, Brussels, Belgium 2009 (cited 2019 Dec 19). https://www.idf.org/e-library/epidemiology-research/diabetes-atlas/21-atlas-4th-edition.html
Kapwata T, Manda S (2018) Geographic assessment of access to health care in patients with cardiovascular disease in South Africa. BMC Health Serv Res 18 (1):197
Kengne AP, Amoah AGB, Mbanya J-C (2005) Cardiovascular complications of diabetes mellitus in sub-Saharan Africa. Circulation 112(23):3592–3601
Kengne AP, June-Rose Mchiza Z, Amoah AGB, Mbanya J-C (2013) Cardiovascular diseases and diabetes as economic and developmental challenges in Africa. Prog Cardiovasc Dis 56(3):302–313
Kirigia JM, Sambo HB, Sambo LG, Barry SP (2009) Economic burden of diabetes mellitus in the WHO African region. BMC Int Health Hum Rights 9:6
Mapa-Tassou C, Katte J-C, Mba Maadjhou C, Mbanya JC (2019) Economic impact of diabetes in Africa. Curr Diabetes Rep 19(2):5
Mbanya J-C, Sobngwi E (2003) Diabetes in Africa. Diabetes microvascular and macrovascular disease in Africa. J Cardiovasc Risk 10(2):97–102
Mbanya JC, Assah F, Saji J, Atanga EN (2014) Obesity and type 2 diabetes in Sub-Sahara Africa. Curr Diabetes Rep 14:1–8
Miranda JJ, Barrientos-Gutiérrez T, Corvalan C, Hyder AA, Lazo-Porras M, Oni T et al (2019) Understanding the rise of cardiometabolic diseases in low- and middle-income countries. Nat Med 25(11):1667–1679
Motala AA, Pirie FJ, Gouws E, Amod A, Omar MAK (2001) Microvascular complications in South African patients with long-duration diabetes mellitus. S Afr Med J 91(11):987–992
Mutyambizi C, Pavlova M, Chola L, Hongoro C, Groot W (2018) Cost of diabetes mellitus in Africa: a systematic review of existing literature. Glob Health 14(1):3
NCD Risk Factor Collaboration (NCD-RisC) (2019) Rising rural body-mass index is the main driver of the global obesity epidemic in adults. Nature 569 (7755):260–264
NCD Risk Factor Collaboration (NCD-RisC)—Africa Working Group (2017) Trends in obesity and diabetes across Africa from 1980 to 2014: an analysis of pooled population-based studies. Int J Epidemiol 46(5): 1421–1432
Ng CS, Lee JYC, Toh MP, Ko Y (2014) Cost-of-illness studies of diabetes mellitus: a systematic review. Diabetes Res Clin Pract 105(2):151–163
Ngassam E, Nguewa J-L, Ongnessek S, Foutko A, Mendane F, Balla V et al (2012) P318 Cout de la

prise en charge du diabète de type 2 a l'hopital central de yaounde. Diabetes Metab 38:A105

Ogle GD, Kim H, Middlehurst AC, Silink M, Jenkins AJ (2016) Financial costs for families of children with Type 1 diabetes in lower-income countries. Diabetes Med 33(6):820–826

Pastakia SD, Pekny CR, Manyara SM, Fischer L (2017) Diabetes in sub-Saharan Africa—from policy to practice to progress: targeting the existing gaps for future care for diabetes. Diabetes Metab Syndr Obes 10:247–263

Peer N, Kengne A-P, Motala AA, Mbanya JC (2014) Diabetes in the Africa Region: an update. Diabetes Res Clin Pract 103(2):197–205

Quaye EA, Amporful EO, Akweongo P, Aikins MK (2015) Analysis of the financial cost of diabetes mellitus in four cocoa clinics of Ghana. Value Health Reg Issues 7:49–53

Sanogo NA, Fantaye AW, Yaya S (2019) Universal health coverage and facilitation of equitable access to care in Africa. Front Public Health 7:102

Sarfo-Kantanka O, Sarfo FS, Kyei I, Agyemang C, Mbanya JC (2019) Incidence and determinants of diabetes-related lower limb amputations in Ghana, 2010–2015—a retrospective cohort study. BMC Endocr Disord 19(1):27

Suleiman IA, Festus JA (2015) Cost of illness among diabetes mellitus patients in Niger Delta, Nigeria. J Pharm Health Serv Res 6:53–60

Suleiman IA, Fadeke OF, Okubanjo OO (2006) Pharmacoeconomic evaluation of anti-diabetic therapy in a Nigerian Tertiary Health Institution. Ann Afr Med 5:132–137

The World Bank (2019) Data indicators (cited 2019 Dec 19). https://data.worldbank.org/indicator

Van den Hout WB (2010) The value of productivity: human-capital versus friction-cost method. Ann Rheum Dis 69(1):89–91

WHO (2019a) NCD mortality and morbidity (cited 2019 Dec 19). http://www.who.int/gho/ncd/mortality_morbidity/en/

WHO (2019b) A heavy burden: the productivity cost of illness in Africa. WHO, Regional Office for Africa (cited 2019 Dec 19). https://www.afro.who.int/publications/heavy-burden-productivity-cost-illness-africa

WHO (2019c) Global health expenditure database 2017 (cited 2019 Dec 19). Available from: https://apps.who.int/nha/database/Select/Indicators/en

Brown Adipose Tissue in Obesity and Diabetes

4

Martín Alcalá, Laura Herrero, Dolors Serra, and Marta Viana

Contents

Adipose Tissue Classification: Types, Location, and Function 37

Clinical Measurement of Brown Adipose Tissue 38
Positron Emission Tomography (PET) Imaging 38
Magnetic Resonance (MR) 39
Preclinical Imaging Techniques (Thermometry, Ultrasound, and Near-Infrared Spectroscopy) 39

Secretory Activity of Brown Adipose Tissue 40
Brown Adipokines or "Batokines" 40
Brown Adipose Tissue Signaling to Other Tissues 41

Neuroendocrine Regulation of Brown Adipose Tissue 44
Sympathetic BAT Innervation 45
Sensory Innervation of BAT 45

Adipose Tissue Dysfunction in Obesity and Diabetes 46
Sick Fat 46
The WAT Expandability Hypothesis 46
Brown Adipose Tissue in Obesity and Diabetes 47

M. Alcalá · M. Viana (✉)
Department of Chemistry and Biochemistry, Facultad de Farmacia, Universidad San Pablo-CEU, CEU Universities, Madrid, Spain
e-mail: mviana@ceu.es

L. Herrero · D. Serra
Department of Biochemistry and Physiology, School of Pharmacy and Food Sciences, Institut de Biomedicina de la Universitat de Barcelona (IBUB), Universitat de Barcelona, Barcelona, Spain

Centro de Investigación Biomédica en Red de Fisiopatología de la Obesidad y la Nutrición (CIBEROBN), Instituto de Salud Carlos III, Madrid, Spain

J. Faintuch and S. Faintuch (eds.), *Obesity and Diabetes*, https://doi.org/10.1007/978-3-030-53370-0_4

Therapeutic Perspectives for Brown and White Adipose Tissue Manipulation ... 47
Chemical Activation of BAT ... 47
Bat Transplantation ... 48
Direct BAT Grafting ... 49
Promises and Pitfalls of Human BAT Transplantation ... 49

References ... 50

Abstract

The scientific community is leading efforts to find efficient therapies for obesity and diabetes, which led to the rediscovery in 2009 of the presence of functional brown adipose tissue (BAT) depots in adult humans. In this chapter, we describe the main characteristics of this tissue and the image techniques available for its detection. In addition, we discuss the secretory capacity of BAT that is responsible for crosstalk with other organs and confers a regulatory role in several processes beyond the canonical generation of heat. Finally, we explain the mechanisms through which obesity and diabetes interrupt the physiological role and development of BAT and we summarize the latest results of the studies that use BAT transplant as a new therapy in the treatment of metabolic diseases.

Abbreviations

Ad-MSCs	Adipose tissue derived-mesenchymal stem cells
ARC	Arcuate nucleus
BAT	Brown adipose tissue
BMP	Bone morphogenetic protein
BOLD	Blood-oxygen-level-dependent
^{14}C-FBnTP	^{14}C-fluorobenzyl triphenyl phosphonium
CGRP	Calcitonin gene-related peptide
^{11}C-MBR	(S,S)-O-methyreboxetine
CNS	Central nervous system
^{11}C-PBR28	Peripheral benzodiazepine receptor tracker
CT	Computational tomography
CXCL14	Chemokine (C-X-C motif) ligand 14
12, 13-diHOME	12, 13-dyhidroxy-9Z-octadecenoic acid
DIO	Diet-induced obesity
DMH	Dorsomedial hypothalamic nucleus
ECM	Extracellular matrix
eWAT	Epididymal white adipose tissue
FA	Fatty acids
FGF21	Fibroblast growth factor 21
^{18}F-FDG	^{18}F-fluorodeoxyglucose
^{18}F-THA	^{14}(R,S)-fluoro-6-thiaheptadecanoic acid
GDF15	Growth differentiation factor-15
HFD	High-fat diet
IL-6	Interleukin-6
iPSCs	Induced pluripotent stem cells
IRT	Infrared thermography
LepRB	Leptin receptor
MnPO	Median preoptic subnucleus
MR	Magnetic resonance
MSCs	Mesenchymal stem cells
NE	Norepinephrine
NGF	Nerve growth factor
NIRTRS	Near-infrared time-resolved spectroscopy
NOD	Non-obese diabetic
NRG4	Neuregulin 4
NTS	Nucleus of solitary track
PET	Positron emission tomography
POMC	Precursor proopiomelanocortin
PRV	Pseudorabies virus
scWAT	Subcutaneous white adipose tissue
SNS	Sympathetic nervous system
SUV	Standardized uptake value
TH	Tyrosine hydroxylase
TNFα	Tumor necrosis factor α
UCP1	Uncoupling protein 1
VEGF-α	Vascular endothelial growth factor-α
WAT	White adipose tissue

Adipose Tissue Classification: Types, Location, and Function

Traditionally, adipose tissue has been classified by its energy profile. According to this classification, white adipose tissue (WAT) is composed of adipocytes that are mainly comprised of a single, large lipid droplet that constrains the nuclei against the plasma membrane. This tissue mainly acts as an energy store, although it also has a secretory role, exerting autocrine, paracrine, and endocrine functions. In humans, there are major depots in subcutaneous (abdominal, gluteofemoral) and visceral regions (omental, mesenteric, mediastinal, and epicardial) while inclusions of minor depots can be observed in organs such as the liver, muscle, or kidneys.

Brown adipose tissue (BAT) has a completely distinct developmental origin as it comes from $Pax7^+/Myf5^+$ stem cells (in contrast to WAT that comes from stem cells that do not express either of these progenitors). The whole tissue is densely vascularized and innervated, but its main feature is the presence of multilocular adipocytes that contain small lipid droplets and have an elevated mitochondrial content, which is responsible for its thermogenic function. This is due to the characteristic presence of uncoupling protein 1 (UCP1). UCP1 is a channel located in the inner mitochondrial membrane that returns protons pumped by the electron transport chain from the intermembrane space to the mitochondrial matrix, avoiding their use by ATP synthase, so the released energy is dissipated as heat. In humans, it was thought that the presence of this tissue was mainly restricted to interscapular and perirenal regions in babies and infants, that require this thermogenic function to maintain body temperature until they acquire the shivering reflex. However, in 2009, five independent researchers rediscovered the presence of BAT in cervical, supraclavicular, para-vertebral, suprarenal, and periaortic regions in adults and, more importantly, determined that this tissue can still be activated. This discovery rapidly boosted interest in BAT as a potential therapeutic tool against metabolic diseases (Cypess et al. 2009; Zingaretti et al. 2009; van Marken Lichtenbelt et al. 2009; Virtanen et al. 2009; Saito et al. 2009) (Fig. 4.1).

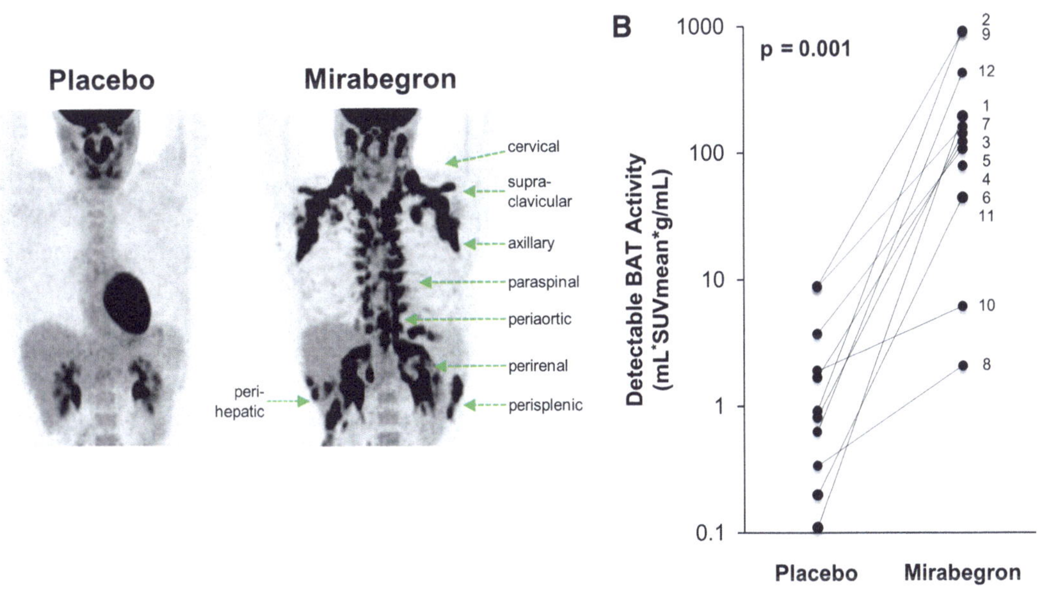

Fig. 4.1 Brown adipose tissue detection and activation by ^{18}F-FDG uptake. (**a**) PET images of a 21-year-old man who was given a placebo (left) or 200 mg of the β3-AR agonist mirabegron (right). Twelve male subjects were given a placebo or 200 mg mirabegron. (**b**) BAT metabolic activity as reflected by ^{18}F-FDG uptake. Reproduced with permission from Cypess et al. (2015)

Finally, it is known that under certain conditions, such as cold or a high-fat diet (HFD), in some WAT depots a population of UCP1 positive adipocytes can emerge, although they come from a $Pax7^{-}$/$Myf5^{-}$ origin. These adipocytes are called brite (brown-in-white) or beige, but also "inducible" since they are responsible for adaptative thermogenesis or "browning." As an example of this adaptation capacity, if BAT is removed from mice, there is a significant "browning" process triggered in WAT as a compensatory mechanism (Piao et al. 2018).

Clinical Measurement of Brown Adipose Tissue

There are currently 91 clinical trials registered in the USA that target BAT in obesity, diabetes, or cardiovascular disease. In this context, it is essential to have imaging-based tools that unequivocally identify the presence and effect of regulators on BAT activity in humans. Below, the most commonly used techniques will be described, together with some promising new approaches. We recommend as additional reading the reviews by Sampath et al. (2016) and Ong et al. (2018).

Positron Emission Tomography (PET) Imaging

^{18}F-fluorodeoxyglucose PET (^{18}F-FDG PET) has been used in oncology to detect tumors, since FDG is internalized by glucose receptors and trapped intracellularly by the action of hexokinase in areas with an elevated consumption of glucose. On several occasions, the signal was detected in areas other than tumors. The combination of this result with morphological information obtained from computational tomography (CT) scans, allowed the co-localization of FDG-positive areas with a high-lipid density. In 2009, the set of independent research that triggered BAT rediscovery used ^{18}F-FDG PET/CT to track BAT mass and activity in humans (Cypess et al. 2009; van Marken Lichtenbelt et al. 2009; Virtanen et al. 2009; Saito et al. 2009). As this technique relies on glucose uptake by active tissue, non-stimulated adult humans can sometimes be wrongly categorized as BAT-negative because of a lack of signal in the PET, when they would be better classified as BAT-inactive. In fact, cold exposure or drug-driven activation in these patients usually leads to an increase in the standardized uptake value (SUV) (Garcia et al. 2006).

Although ^{18}F-FDG PET/CT is probably the most common technique used for BAT visualization, there are several limitations to this approach. First, it requires the use of ionizing radiation and trained personal. Second, the technical protocols to measure BAT activation also differ (from temperature-controlled rooms or cooling vests to adrenergic stimulation by drugs), which leads to wide dispersion of observed effects. In this context, a panel of experts established the BARCIST recommendations to enhance comparability across experiments (Chen et al. 2016). The technique is based on glucose uptake, which is not even the preferential energy substrate in BAT and can be modified by disease (obesity, diabetes), drugs (β-blockers, anxiolytics), outer temperature or individual factors such as age, gender, or ethnicity (Soderlund et al. 2007; Pfannenberg et al. 2010; Pace et al. 2011; Rakheja et al. 2011).

Consequently, the use of other molecular trackers has been proposed, although most of them rely on BAT activity. For instance, (^{14}C)-FBnTP (^{14}C-F-fluorobenzyl triphenyl phosphonium) is used as a tracker of membrane potential, which decreases after BAT activation (Madar et al. 2011). ^{11}C-MRB (^{11}C-(S,S)-O-methylreboxetine) (Hwang et al. 2015) and ^{18}F-THA (14 (R,S)-^{18}fluoro-6-thiaheptadecanoic acid) (Ouellet et al. 2012) are used as trackers of fatty acid uptake and oxidation, and $^{15}O\text{-}O_2$ detects increased mitochondrial oxygen consumption after BAT activation (Mueez et al. 2016). More recently, ^{11}C-PBR28 (^{11}C-peripheral benzodiazepine receptor tracker, an outer membrane protein in the mitochondria) has shown promising results, detecting BAT mass independently of its activation degree (Ran et al. 2018).

Magnetic Resonance (MR)

MR is a noninvasive technique based on the use of magnetic fields and electromagnetic radiation to detect mainly the hydrogen, carbon, or xenon nucleus. In the presence of a magnetic field, protons will align with it in two differential spin states: a stable, low-energy spin and a high-energy spin. With the application of electromagnetic radiation, low-energy nuclei suffer a transition to high-energy protons, absorbing a discrete amount of energy called precession frequency, that can be detected. This value differs in aqueous or lipid environments, allowing the detection of fat depots. The application of this technique to BAT detection has been brilliantly reviewed by Karampinos et al. (2019).

Due to the lack of ionizing radiation, this technique can be safely used in infants or to establish protocols using repeated measurements, such as pre- and post-intervention studies or time-course analysis. This methodology can structurally identify BAT and even distinguish it from WAT (Hamilton et al. 2011), although MR fat fraction measurements have certain limitations as the spatial resolution is restricted to 1 mm^3. For instance, it may not be possible to resolve differences between intracellular water and non-lipid structures, such as vessels or connective tissue. In addition, it is not possible to differentiate white, brown, and beige adipocytes when they coexist in the same depot, as occurs in obese patients or in the supraclavicular regions. Nevertheless, modifications in the protocol have been published that solve this issue, based on the smaller size of BAT lipid droplets (Branca and Warren 2011).

In addition, MR technology can be adapted to determine BAT activity. Water-fat composition studies can be used to determine intracellular lipid extinction, used as a fuel to generate heat during cold-induced BAT activation. The increase in blood flow can be tracked using a technique known as the blood-oxygen-level-dependent (BOLD) effect, that relies on the differential paramagnetic properties of oxyhemoglobin and deoxyhemoglobin (Iris Chen et al. 2013). Other emerging but still experimental approaches involve hyperpolarized MR imaging, using the lipophilic accumulation of ^{129}Xe to measure BAT perfusion or the breakdown of hyperpolarized ^{13}C-pyruvate into bicarbonate and lactate (Branca et al. 2014).

Preclinical Imaging Techniques (Thermometry, Ultrasound, and Near-Infrared Spectroscopy)

The increase in skin temperature associated with the thermogenic activity of the BAT can be directly measured using infrared thermography (IRT). This approach is widely used in animal models in preclinical studies. In the last decade, the technique has also been applied in humans due to its simplicity, reduced cost, and safety, which allows repetitive measurements and use in sensitive populations. In humans, IRT is used to detect changes in skin temperature, although it is limited to the supraclavicular region, where BAT is more superficial. Several authors have demonstrated that cold-induced activation of BAT determined by ^{18}F-FDG PET/CT correlates with an increase in skin temperature detected by IRT (Law et al. 2018).

However, this approach has several limitations. First, it is a method based on tissue activation, which implies that inactive BAT will not be detected unless there is a change in temperature generation. Second, it currently lacks a standardized protocol in terms of patient preparation (acclimatization, cold intensity, and duration) or data analysis (image acquisition and analysis of the region of interest). Finally, but crucially, the determination of supraclavicular BAT-generated temperature can be highly limited by (i) the shielding effect of the surrounding tissues and (ii) the contribution to skin temperature of increased blood flow after BAT activation. For instance, the thickness of the subcutaneous adipose tissue in obese patients hampers the diffusion of the temperature toward the skin, misleading the signal acquisition. To our knowledge, only two reports have evaluated BAT activation using this approach in obese patients, and both of them explicitly acknowledge the limitations of the protocol (El Hadi et al. 2016; Hartwig et al. 2017).

Ultrasonography

Ultrasounds are based on identifying increased blood flow in activated BAT by detecting microbubbles injected into patients. This technique has obtained promising results in mice, in which norepinephrine activated BAT was detected and estimated BAT mass was obtained, that correlates with the observed results after necropsy. However, its application in humans is currently limited to a small pilot project in 13 lean patients, in whom ultrasound-detected BAT mass activation was similar to that detected by ^{18}F-FDG PET/CT (Flynn et al. 2015).

Near-Infrared Spectroscopy (NIRTRS)

NIRTRS is based on the relative transparency of tissues to near-infrared wavelengths, and the differential absorption of hemoglobin and myoglobin in relation to oxygenation state. The technique can be used to detect hemoglobin oxygenation in the microvasculature, which is inversely related to oxidative activation of BAT. However, while the first studies in humans reported comparable results in the detection of supraclavicular BAT to the gold standard ^{18}F-FDG PET/CT (Nirengi et al. 2015), later reports noted that this technique is apparently unable to detect changes after BAT activation (Hartwig et al. 2017; Acosta et al. 2019), which would potentially limit the use of this simple, noninvasive method.

Secretory Activity of Brown Adipose Tissue

The secretory profile of WAT in lean, obese, and diabetic individuals has been thoroughly studied in the past two decades (Lee et al. 2019). However, a focus on the BAT secretome has been more recent. Due to the distinct roles of both adipose depots in energy metabolism, it is not likely that the type, amount, roles in signaling pathways and targets of the molecules secreted by BAT are similar (Villarroya et al. 2019a).

Brown Adipokines or "Batokines"

The term "batokines" (BAT-secreted mediators or brown adipokines) refers to the proteins, peptides, metabolites, regulatory factors, or miRNAs released by brown fat tissue, including BAT-resident cells other than brown adipocytes such as immune cells, preadipocytes, neurons, etc. They can be classified according to their chemical nature, their role in the regulation of the thermogenic process (Villarroya et al. 2017a, b, 2019a), or their mechanism of action in cellular communication (autocrine, paracrine, or endocrine) (Villarroya et al. 2017b). Table 4.1 summarizes the BAT-secreted signaling molecules that have been identified in basal conditions.

Using a proteomics approach, two recent studies (Ali Khan et al. 2018; Villarroya et al. 2019b) reported a set of proteins secreted after cAMP and noradrenaline stimulation of brown adipose murine cells. The studies show 60% coincidence, but interestingly both identified extracellular matrix (ECM) components (distinct types of collagens) or regulators (matricellular proteins: non-structural proteins in the ECM with a regulatory role) as members of the secretome of thermogenically activated brown adipocytes. This highlights the importance of BAT remodeling when the tissue has to be expanded in response to thermogenic activation (Villarroya et al. 2019a). In humans, a pilot study using comparative proteomics between white and brown adipocytes identified 101 proteins that are exclusively secreted by the BAT (Deshmukh et al. 2019).

Batokines and Transplantation

A direct way to identify the endocrine secretion of batokines is through BAT transplantation studies. For example, BAT transplantation in animal models of type I diabetes (Gunawardana and Piston 2012, 2015) has been shown to improve glycemia. In type II diabetes (high-fat diet-induced and ob./ob mice), BAT transplantation increased energy expenditure and improved glucose tolerance (Stanford et al. 2013; Liu et al. 2013, 2015; Zhu et al. 2014). BAT

Table 4.1 Secreted molecules by BAT after cold/adrenergic activation with recognized function

Gene	Protein	Function
Adipoq	Adiponectin	Thermogenesis inactivation.
Angptl8	Angiopoietin-like protein-8	Thermogenesis inactivation.
Agt	Angiotensinogen	Enhances BAT differentiation.
Bmp8b	Bone morphogenetic protein-8b	Enhances BAT differentiation. Increases sympathetic tone in BAT.
Cxcl14	C-X-C motif chemokine 14	Thermogenesis activation by M2 macrophage recruitment.
Edn-1	Endothelin-1	Thermogenesis inactivation.
Fgf2	Fibroblast growth factor-2	Enhances BAT differentiation.
Fgf21	Fibroblast growth factor-21	Thermogenesis activation.
Fst	Follistatin	Thermogenesis activation.
Gdf15	Growth and differentiation factor-15	Acts in immune cells to reduce inflammation.
Igf1	Insulin-like growth factor-1	Glucose homeostasis. BAT differentiation.
Il6	Interleukin-6	Glucose homeostasis. Regulation of Inflammation.
Ptgds	Lipocalin prostaglandin D synthase	BAT activation and WAT browning. Control of substrate choice in BAT (glucose vs lipids)
Lr11	Low-density Lipoprotein receptor relative, soluble form	Thermogenesis activation
Metrnl	Meteorin-like	Thermogenesis activation through immune cells-mediated catecholamine release
Mstn	Myostatin/growth and differentiation factor-8	Thermogenesis inactivation
Ngf	Nerve growth factor	Enhances BAT activity by increasing innervation
Nrg4	Neuregulin-4	Hepatic lipogenesis, fuel oxidation. Promotes BAT activation by increasing angiogenesis.
Pm20d1	Peptidase M20 domain-containing protein 1	Thermogenesis activation UCP1-independent.
S100b	S100-B protein	Enhances BAT activity by increasing innervation.
Slit2	C-terminal fragment of SLIT2 protein	Thermogenesis activation.
Vegfa	Vascular endothelial growth factor A	Enhances BAT activity by increasing vascularization.
Non-peptidic		
	Adenosine	Thermogenesis activation.
	(9Z)-12,13-Dihydroxy-9-octadecenoic acid (12,13-diHOME)	Promotes fatty acid uptake by BAT.
	Endocannabinoids	Thermogenesis inactivation.
	Hydrogen peroxide (H2O2)	vasodilation and second messenger in BAT differentiation.
	miR-99b	Inhibits hepatic FGF21 expression.
	Nitric oxide	Vasodilation in BAT and activation of thermogenesis.
	Prostaglandins	BAT activation and WAT browning. Control of substrate choice in BAT (glucose vs lipids)
	Triiodothyronine	Thermogenesis activation

transplantation can increase sympathetic tone and affect the endogenous BAT, WAT, heart, and muscle of recipient mice (Zhu et al. 2014). These effects cannot be exclusively explained by an increase in thermogenesis, so the consensus conclusion is that the BAT secretome should account for the observed phenotypical improvements.

Brown Adipose Tissue Signaling to Other Tissues

There is still little evidence about the role of the BAT endocrine secretome and its effects on target organs (Fig. 4.2). Here we highlight some of the more recent and relevant findings.

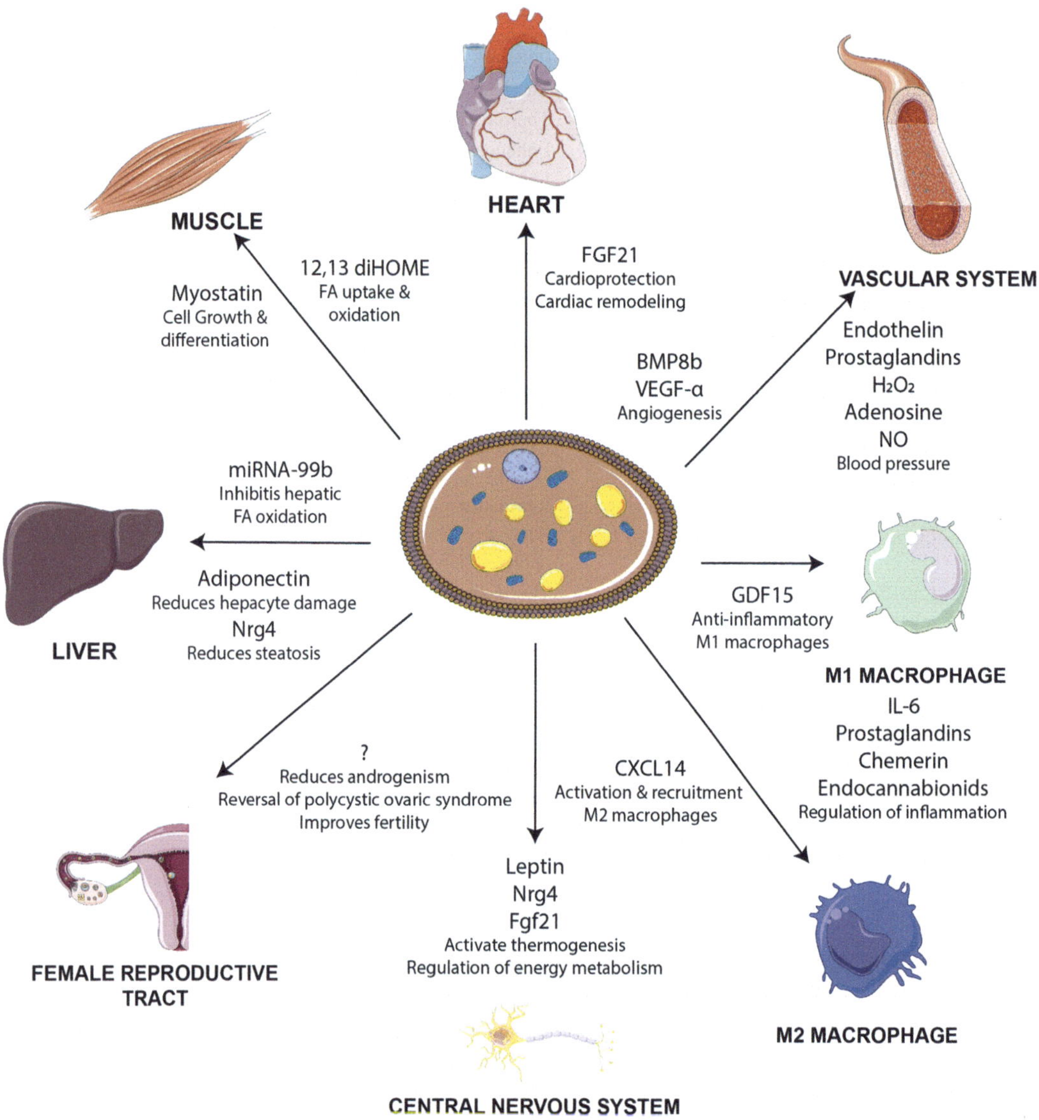

Fig. 4.2 Brown adipose tissue crosstalk. The activation of BAT stimulates the secretion of signaling molecules that target different organs, exerting regulatory roles in fatty acid oxidation and glucose uptake by the muscle, cardioprotective effects in the heart, regulation of blood pressure and arterial stiffness in blood vessels, the promotion of an anti-inflammatory phenotype in immune cells, regulation of lipid metabolism in the liver, regulation of ovulation and participation in the regulatory mechanism of the metabolism by the central nervous system

Muscle Physical exercise is a common non-pharmacological approach to treat metabolic diseases. Besides enhancement of muscular function per se, the study of bidirectional communication between muscle and BAT is beginning to reveal new agents that regulate the beneficial effects of exercise. For instance, exercise increases WAT browning through the secretion of myokines (Boström et al. 2012), and mildly increases BAT thermogenic activity in a lactate-driven mechanism (De Matteis et al. 2013). In the other direction, BAT seems to modify skeletal muscle function through the secretion of myostatin, which inhibits skeletal muscle cell growth and differentiation. Under thermoneutral temperature, BAT inactivation led to increased myostatin secretion, reducing the exercise capacity of skeletal muscle. However, BAT activation lowered myostatin levels and favored exercise performance (Kong et al. 2018). BAT also participates in fatty acid uptake and oxidation by muscle, through the secretion of 12,13-dihydroxy-9Z-octadecenoic acid (12,13-diHOME) after exercise (Stanford et al. 2018).

Cardiovascular System Under thermogenic activation, BAT strongly releases fibroblast growth factor 21 (FGF21) (Hondares et al. 2011). Tissues and organs such as WAT, brain, pancreas, and heart are known to be sensitive to FGF21 action. In the heart, FGF21 seems to have a strong cardioprotective effect (Planavila et al. 2013) by attenuating hypertensive cardiac remodeling (Thoonen et al. 2015; Ruan et al. 2018).

There is also increasing evidence of the role of BAT signaling to blood vessels. For instance, bone marrow protein 8 b (BMP8b, one of the most abundantly expressed genes after BAT activation) and vascular endothelial growth factor (VEGF-α) promote angiogenesis, which is required to increase vascular density and thermogenesis in the BAT (Bagchi et al. 2013; Pellegrinelli et al. 2018). In addition, BAT removal in lean mice caused aortic stiffness, which suggests a key contribution to vascular remodeling (Grunewald et al. 2018).

Finally, several proteins and non-proteins secreted by BAT play a key role in maintaining blood pressure both by causing contraction (endothelin, some prostaglandins) and/or relaxation (adenosine, H_2O_2, nitric oxide). Although this indicates that BAT is an important regulator of vascular function, more research is required to understand its specific contribution.

Immune Cells Polarization of the infiltrated immune cells within BAT or beige adipocytes conditions the thermogenic programming of the tissue (Villarroya et al. 2018a, b). While the recruitment of M2-alternative activated macrophages promotes BAT function, M1-pro-inflammatory macrophages are negatively associated with the thermogenic activity (Villarroya et al. 2019a).

In normal circumstances, stimulated brown adipocytes can decrease the inflammatory response in macrophages in a paracrine fashion (Dowal et al. 2017). The release of chemokine (C-X-C motiv) ligand 14 (CXCL14) batokine leads to the alternative activation and recruitment of M2-macrophages in response to noradrenergic stimulation (Cereijo et al. 2018), while growth and differentiation factor-15 (GDF15) exerts anti-inflammatory effects on M1-macrophages (Verdeguer et al. 2015; Campderrós et al. 2019).

One of the most studied cytokines, interleukin-6 (IL-6) plays a dual pro-inflammatory and anti-inflammatory role. In BAT, it is secreted in response to cold and noradrenergic stimulation and it has been reported as crucial to maintain the thermogenic response. As an example, BAT transplantation from IL-6 KO mice loses the capability of promoting a healthy metabolic response in HFD-fed mice (Stanford et al. 2013).

Other factors secreted by BAT exhibit pro- and anti-inflammatory functions, such as endocannabinoids, chemerin, or prostaglandins, whose endocrine role and implication in BAT crosstalk to other tissues is yet to be defined.

Liver Considering that BAT activation relies mainly on glucose and fatty acid flux rather than on intracellular stored fuels, it seems reasonable that there is crosstalk between both organs, as the liver is the key regulatory organ in glucose and lipid homeostasis. FGF21 has been identified as a key molecule in this communication. In normal conditions, most circulating FGF21 is synthesized in the liver, where it participates in fatty acid oxidation and insulin sensitivity (Markan et al. 2014). To maximize fuel efflux for thermogenic activity under adrenergic or cold stimulation, BAT responds in a dual mode. First, it induces the expression of miR-99b, which, along with other miRNAs, is packed in exosomes and travels to the liver, where it inhibits FGF21 expression (Thomou et al. 2017) to minimize fuel consumption by the liver.

The second effect is the induction of BAT FGF21 expression, which will act in an autocrine fashion enhancing triglyceride-rich lipoprotein uptake and oxidation (Hondares et al. 2011; Schlein et al. 2016).

Another batokine, neuregulin 4 (Nrg4) reduces steatosis by controlling *de novo* lipogenesis in the liver. It may be a useful strategy to treat non-alcoholic fatty liver disease, as Nrg4 levels are downregulated in obesity (Wang et al. 2014). In a similar manner, adiponectin secreted by the BAT may reduce hepatocyte damage and death in alcohol-induced steatosis (Shen et al. 2019).

Female Reproductive Tract In ovariectomized mice, the development of obesity appears to be partially mediated by loss of BAT activity (Bartness and Wade 1984; Pedersen et al. 2001), probably related to the estrogen receptor present in BAT (Wade and Gray 1978). Women with polycystic ovary syndrome report lower BAT activity (Flávia et al. 2019), whereas transplantation of BAT in rats reversed polycystic ovaries and hyperandrogenism, and reduced infertility by reversing anovulation (Yuan et al. 2016). Whether this improvement is related to the thermogenic activity or the secretory role of BAT is still to be determined.

Brain Some of the autocrine/paracrine factors released by BAT act in the brain to activate neural circuits that innervate adipose tissues and cooperatively regulate lipid metabolism and energy homeostasis. One of these well-defined adipose-derived secretory factors is leptin. The expression of this gene is similar in WAT and BAT in lean mice, but WAT expresses significantly greater levels than BAT in obese mice. Leptin acts through leptin receptor (LepRb) to enhance the excitability of neurons of different parts of the brain including the arcuate nucleus (ARC), dorsomedial hypothalamic nucleus (DMH), median preoptic subnucleus (MnPO) and the nucleus of the solitary tract (NTS), implicated in the regulation of BAT (Zhang et al. 2011).

The ARC nucleus in the hypothalamus is in contact with the blood-brain barrier and is considered one of the gates for the brain to sense circulating factors. Leptin action in ARC increased thermogenesis in BAT by activating the efferent sympathetic nervous system (SNS) (Rahmouni and Morgan 2007; Harlan et al. 2011), through the induction of anorexigenic neuropeptide precursor proopiomelanocortin (POMC) expression and increasing the release of POMC products in second-order neurons (Yasuda et al. 2004; Dodd et al. 2015).

Another population of ARC neurons, the GABAergic RIP-Cre (Cre-mediated expression of rat insulin II promoter) are also activated by leptin and are involved in the neural circuits that activate BAT thermogenesis (Kong et al. 2012). Mice fed with a high-fat diet had impaired BAT sympathetic neuronal activation in response to systemic leptin, but the leptin intraparenchymal injection in diet-induced obese (DIO) mice increased BAT temperature (Enriori et al. 2011). Other batokines such as FGF21 (Mutsnaini et al. 2019) and Ngr4 (Rosell et al. 2014) interact with neural circuits to modulate BAT activity and cooperatively regulate whole-body energy metabolism.

Systemic Influences Other hormonal and nutritional signals and stimuli (temperature, immunologic responses, stress, and hypoxia) may have neurochemical influences on the central nervous system (CNS) networks that govern BAT sympathetic nerve activity. They may drive sympathetic BAT activation or inhibition and control BAT thermogenesis and energy expenditure, and contribute to whole-body energy homeostasis (see reviews by Morrison et al. 2014; Contreras et al. 2017).

Neuroendocrine Regulation of Brown Adipose Tissue

The central nervous system (CNS) integrates a wide array of environmental stimuli: diet, fasting, exercise, temperature, and behavioral states, sending outward neural signals via the autonomic nervous system that directly contact multiple tissues to control metabolism and energy balance. Crosstalk between adipose tissues (BAT and different WATs) and CNS has emerged as an important issue for maintaining energy balance and promoting survival in response to metabolic challenges such as obesity. Therefore, intense research is focused on understanding the bidirectional communication between fad pads and brain. This encompasses endocrine signals from adipose tissues to CNS and neural signals through efferent sympathetic nerves and afferent sensory nerves and provides a framework for understanding overall BAT thermogenesis by the CNS (recently reviewed in François et al. 2018; Blaszkiewicz et al. 2019; Zhu et al. 2019).

Sympathetic BAT Innervation

The sympathetic innervation of BAT was described nearly a century ago. The physiological effect of sympathetic stimulation of neurons that innervate BAT is increased thermogenesis. The secretion of norepinephrine at the terminal nerves that surround adipocytes activates β3-adrenergic receptor and the signaling pathways that promote lipolysis, fatty acid (FA) oxidation, UCP1 protein expression and heat dissipation of the energy of FA (Cannon and Nedergaard 2004; Contreras et al. 2017). Several technologies have been used to demonstrate sympathetic innervation of BAT. Bartness and colleagues have provided important histological and functional evidence of brain sympathetic neuronal connectivity with adipose tissues by using pseudorabies virus (PRV), a retrograde trans-synaptic neuronal tracer that only traces sympathetic neurons that are synaptically connected.

They showed the hierarchical neuronal pathways defining pre- and postganglionic sympathetic innervation of BAT (Bamshad et al. 1999; Song et al. 2008) in Siberian hamsters and rats, and the description of a higher order of neurons located in different parts of the brain such as the hypothalamus and raphe pallidus (Bamshad et al. 1999; Oldfield et al. 2002; Cano et al. 2003). In addition, immunohistochemical labeling with tyrosine hydroxylase (TH), a specific marker for sympathetic nerve fibers involved in the biosynthesis of norepinephrine, have shown dense innervation of BAT (Giordano et al. 1996; Murano et al. 2009). Under cold exposure, it has been shown that TH increases at BAT parenchymal nerve fibers with a concomitant increase of UCP1 expression, which indicates enhanced sympathetic and thermogenic activity (Brito et al. 2008; Murano et al. 2009).

Experimental Denervation

Other techniques such as surgical or chemical denervation of BAT decreased UCP1 expression and highlighted the functional relevance of BAT innervation in regulating thermogenesis, energy expenditure, body mass, and browning (Dulloo and Miller 1984; Klingenspor et al. 1994; Nguyen et al. 2017). In addition, BAT sympathetic denervation can activate the sympathetic activity of other adipose tissues. For example, it can promote browning in inguinal WAT and suggests tissue crosstalk with the brain to modulate energy balance (Nguyen et al. 2017). Diet can modulate the sympathetic activity of BAT. Acute HFD treatments increase the norepinephrine (NE) turnover in BAT but this decreases after long treatments with HFD (Levin et al. 1983). More recently, it has been shown that diet-induced obesity increases UCP1 protein in mice and rats (Fromme and Klingenspor 2011; Alcalá et al. 2017) and BAT temperature after 20 weeks of diet (Enriori et al. 2011), which suggests that sympathetic activity in BAT increases to promote energy expenditure in conditions of excess calories.

Sensory Innervation of BAT

The first studies that suggest the presence of sensory neurons in BAT were performed by immunohistochemical detection of sensory nerve associated markers, such as calcitonin gene-related peptide (CGRP) and substance P (Giordano et al. 1996). More recent studies with anterograde transneuronal tracer have confirmed the presence of sensory nerve projections from BAT to dorsal root ganglia, which propagate their signal to the CNS (Bartness et al. 2010; Ryu et al. 2015). In addition, it has been demonstrated that there is considerable crosstalk between adipose tissues and the CNS. BAT and WAT can share some sympathetic outflows from the CNS, and sensory information circuits of BAT and sympathetic circuits of WAT and vice versa have important overlaps in the CNS (Nguyen et al. 2017; Ryu et al. 2017).

In addition, it has been shown that sensory information about activated lipolysis in WAT can influence the efferent CNS circuits to BAT, by activating thermogenesis in this tissue (Garretson et al. 2016). All these data suggest a feedback mechanism that sensory information from each adipose tissue is communicated to the brain by an afferent pathway, which integrates the signal and regulates the efferent sympathetic outputs to each specific adipose tissue. Nevertheless, the nature of the signals that activate sensory neurons in different adipose tissues is unknown. Only a few brown adipokines (nerve growth factor [NGF], NRG4) have been identified as an outgrowth of promoters of sympathetic neurites (Néchad et al. 1994; Murano et al. 2009).

Adipose Tissue Dysfunction in Obesity and Diabetes

Since the mid-1990s, WAT has been seen not just as a storage depot, but also as an active, secretory tissue that establishes crosstalk with other organs. Dysregulation in this communication is a typical feature in obesity and diabetes and is partly responsible for adverse metabolic outcomes in organs such as muscle, liver, and blood vessels.

Sick Fat

WAT malfunction in obesity is characterized by the following anatomical and functional processes:

- Structural changes in the size (hypertrophy) and number (hyperplasia) of adipocytes and stiffness of the extracellular matrix
- Infiltration of immune cells, together with increased inflammatory secretion
- Oxidative and endoplasmic reticulum stress
- Defective lipolysis and increased energy metabolism
- Insulin resistance
- Mitochondrial dysfunction

These processes are not isolated events, but they are interconnected. For instance, oxidative stress increases endoplasmic reticulum stress, promotes inflammation, and participates in insulin resistance.

The WAT Expandability Hypothesis

The WAT expandability hypothesis tries to establish a time-course description of the events that begin with the increased demand for triglyceride storage (Tan and Vidal-Puig 2008). This hypothesis claims that there is an individual limit for fat accumulation within WAT. As long as this limit is not exceeded, the expansion of WAT will be healthy, which explains the presence of approximately 20% of morbidly obese patients who do not exhibit the expected comorbidities (dyslipidemia, type 2 diabetes or hypertension).

However, if the limit is surpassed, fat accumulation begins in ectopic tissues, leading for example to fatty liver or cirrhosis or loss of insulin sensitivity in the muscle. In this unhealthy expansion model, the extracellular matrix is remodeled and becomes fibrotic, impeding the hypertrophy of adipocytes, which get isolated from oxygen and food supply from blood vessels. As a response, the recruitment of immune cells begins through adipokine secretion, with a triple mission: (i) to attract blood monocytes to the tissue, which will become pro-inflammatory macrophages to phagocyte the apoptotic adipocytes; (ii) to generate new blood vessels to maximize oxygen and food supply to the hypertrophic adipocytes, whose size can become bigger than the oxygen diffusion limit (~100 μm) and (iii) to reduce insulin sensitivity, in order to limit the accumulation of triglycerides mediated by the insulin lipogenic effect. However, when obesity becomes chronic, these changes in the biology of the adipocyte are not self-controlled, and the secretory profile of WAT negatively affects other tissues (Sun et al. 2011).

Brown Adipose Tissue in Obesity and Diabetes

In spite of the technical problems when it comes to measuring BAT amount and activity with current imaging methods, particularly in obese or type 2 diabetic patients, it is well-established that both the basal activity and its response to stimuli (cold, adrenergic agonists) is considerably diminished in these patients (Cypess et al. 2009; Eriksson et al. 2019).

BAT Dysfunction During Obesity

Recent studies in animal models have demonstrated that BAT from obese individuals undergoes similar pathological processes during obesity development, such as changes in the extracellular matrix, infiltration of immune cells and pro-inflammatory secretion, oxidative stress and changes in energy metabolism. Nonetheless, in comparison to WAT, the appearance of these processes is delayed since WAT preferentially uptakes the circulating fat (McGregor et al. 2013; Alcalá et al. 2017). However, once fat begins to accumulate within the brown adipocyte, a clear change in the phenotype is observed (commonly known as "whitening"). The change is characterized by an increase in the size of the multilocular lipid droplets, which gives the adipocyte a "white" appearance that is translated into a loss of thermogenic potential.

Reduced adrenergic stimulation, the role of adipokines/batokines, oxidative and endoplasmic reticulum stress, and some miRNAs are potentially responsible for the obesity-related decrease in BAT mass (Alcalá et al. 2019). Infiltration of activated M1 macrophages and pro-inflammatory cytokines promote a decline in UCP1 expression, which alters thermogenic activity (Sakamoto et al. 2016; Villarroya et al. 2018a). Oxidative stress is also responsible for BAT activity decline (Cui et al. 2019) (Fig. 4.3).

Therapeutic Perspectives for Brown and White Adipose Tissue Manipulation

As one of the regulators of body temperature, BAT controls triglyceride levels and uses glucose and lipids to fuel thermogenesis to produce heat. Therefore, the process of thermogenesis is associated with several metabolic pathways such as mitochondrial respiration, lipolysis, lipogenesis, or FA uptake, transport, storage, and oxidation (Calderon-Dominguez et al. 2016). In experimental animals, activated BAT is the major plasma lipid-clearing organ and has the highest FA oxidation rate (Doh et al. 2005; Bartelt et al. 2011).

The fact that aging, obesity, and diabetes have been associated with a decrease in the presence and activity of BAT in humans has highlighted BAT as a potential therapeutic target for metabolic diseases (Hany et al. 2002; Nedergaard et al. 2007; Cypess et al. 2009; Zingaretti et al. 2009; van Marken Lichtenbelt et al. 2009; Virtanen et al. 2009; Saito et al. 2009). These studies indicated that, in addition to its role in cold-induced thermogenesis, BAT also plays a key role in diet-induced thermogenesis. Efforts to enhance either the activity or the mass of thermogenic adipocytes include the increase in brown adipocyte differentiation or *de novo* brown adipogenesis, the transdifferentiation of white into beige and thermogenic adipocytes, boosting BAT activity, and the transplantation of thermogenic adipocytes.

Chemical Activation of BAT

In addition to natural activators of thermogenesis such as cold or exercise, several other activators have been described, such as bone morphogenetic protein 7 (BMP7) (Tseng et al. 2008; Boon et al. 2013) and BMP8b (Whittle et al. 2012), irisin (Boström et al. 2012), natriuretic peptides (Bordicchia et al. 2012), norepinephrine (Villarroya and Vidal-Puig 2013), meteorin-like (Rao et al. 2014), bile acids (Watanabe et al. 2006), adenosine (Gnad et al. 2014), FGF21

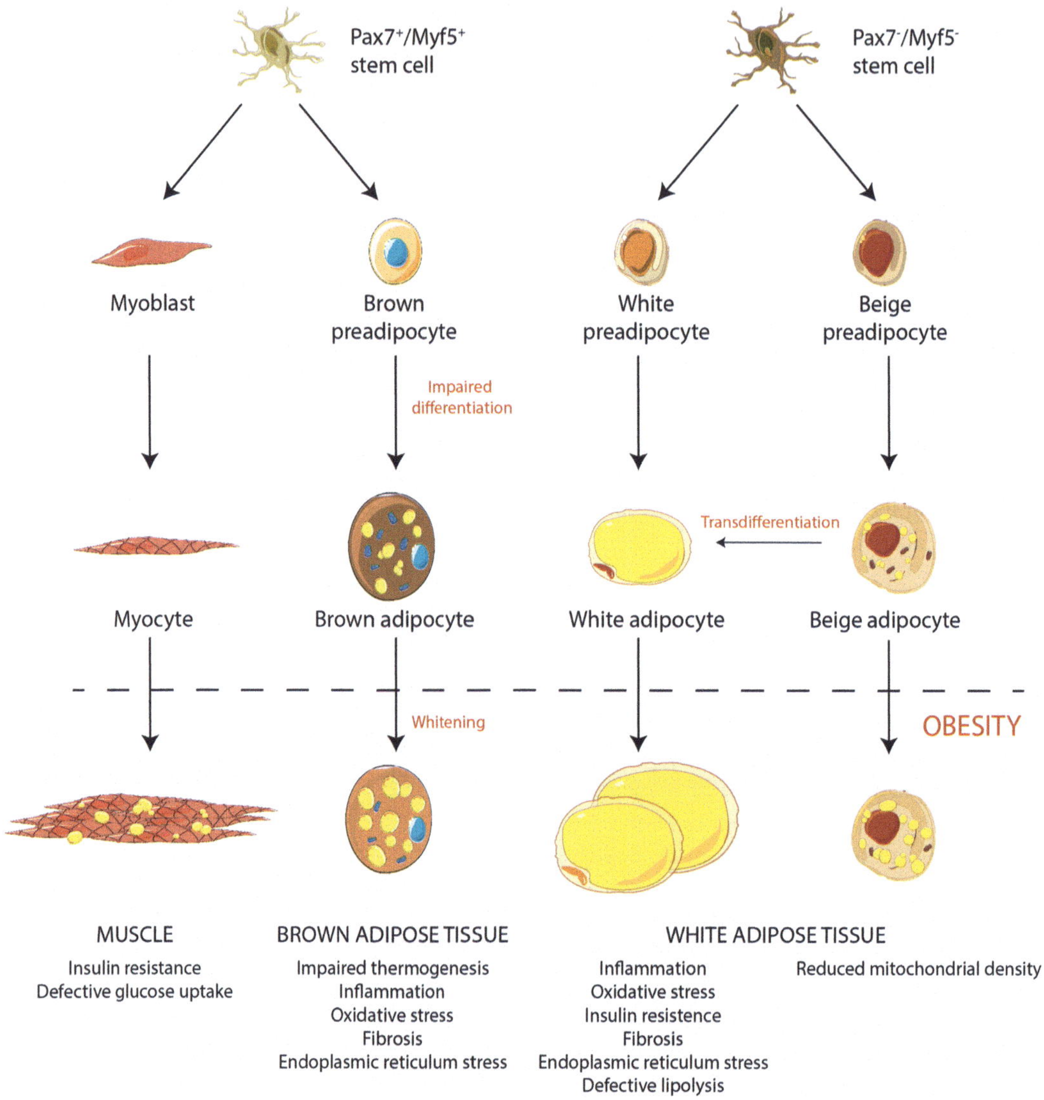

Fig. 4.3 Effects of obesity and diabetes in the development and function of adipose tissue. The pathological mechanisms that characterize obesity and diabetes can disturb the differentiation of brown adipocytes from precursor cells. They can also interfere in transdifferentiation processes, initially by promoting browning of white adipocytes to overcome the excess of nutrients and finally shifting the phenotype of brown adipocytes toward a white adipocyte (whitening). The thermogenic activation of BAT is also interrupted

(Hondares et al. 2010), or the β3-adrenergic receptor agonist mirabegron (Cypess et al. 2015). These factors have been shown to boost the browning of adipose tissue, *i.e.*, increased expression and activity of UCP1 within WAT, and have been extensively reviewed (Bonet et al. 2013; Wu et al. 2013; Calderon-Dominguez et al. 2016).

Bat Transplantation

The potential benefits of transplanting functional BAT are currently under study in animal models (Soler-Vázquez et al. 2018; White et al. 2019). In the case of adipocyte implantation, adipocytes could be obtained from mesenchymal stem cells (MSCs). MSC sources include embryonic stem cells, and stem cells isolated from bone marrow, umbilical cord, peripheral blood, adipose tissue, bone marrow, cartilage, tendon, muscle, and dental pulp. Adipose tissue derived-MSCs (Ad-MSCs) are currently predominant in the study of metabolic diseases since a large number of MSCs can be isolated, and because they have a higher capacity to differentiate into mature adipocytes (Rashnonejad et al. 2018). Lee *et al.* injected human MSC-derived brown adipocytes intraperitoneally into obese mice for 10 weeks (Lee et al. 2017). This treatment successfully decreased bodyweight, triglyceride and cholesterol levels, hepatic steatosis, glucose intolerance, and inflammation.

Silva et al. found a reduction in body weight and glucose levels in non-obese diabetic (NOD) mice fed an HFD, that had been transplanted with differentiated Ad-MSCs isolated from human mediastinal brown fat depots (Silva et al. 2014). Improvements in glucose and lipid homeostasis have also been described in obese and diabetic mice subcutaneously transplanted with brown adipocytes, generated by cellular reprogramming procedures using induced pluripotent stem cells (iPSC) or from mouse embryonic fibroblasts (Kishida et al. 2015). Another example of the successful effects of adipocyte implantation on metabolic phenotype is an elegant study by Min et al. (2016). Beige adipocytes were derived from human capillary network progenitors and activated *in vitro* with adrenergic stimuli. The subcutaneous implantation of these thermogenic adipocytes into NOD mice treated with HFD led to an improvement in systemic glucose tolerance.

Direct BAT Grafting

Several studies have performed direct BAT transplantation in mouse models of diabetes and/or obesity (Gunawardana and Piston 2012, 2015; Stanford et al. 2013, 2015; Liu et al. 2013, 2015; Zhu et al. 2014). Regardless of the BAT donor (embryonic or adult), the location of the transplantation (subcutaneous or the visceral cavity), or the type of transplanted fat (exercise-induced beige fat or BAT), these studies showed a reduction in glucose and insulin levels, adiposity, inflammation, hepatic steatosis and an increase in oxygen consumption. Finally, BAT transplantation has shown metabolic improvements in other derangements such as Parkinson's disease (Wu et al. 2017), cardiovascular disease (Thoonen et al. 2015), polycystic ovary syndrome (Yuan et al. 2016) and atherosclerosis (Kikai et al. 2017).

Promises and Pitfalls of Human BAT Transplantation

First, in humans, there might be differences in the MSCs source, isolation method, duration, and capacity for differentiation to mature adipocytes, and the obtained outcomes. Second, due to its body distribution, adipose tissue in humans might be less accessible than in rodents. Ad-MSCs can be isolated from WAT during elective surgical interventions such as liposuction. Third, the beneficial effects of BAT transplantation might include several mechanisms, such as the endocrine release of batokines that should be further investigated. Fourth, most of the transplantation experiments have been performed with mice housed under thermal stress (22 °C), in order to activate BAT. Cannon and colleagues correctly claim that if animal experiments are to be translated into humans, mice should be housed at thermoneutrality (30 °C for mice; 22 °C for humans) since humans usually live at thermoneutrality throughout the year (Fischer et al. 2019).

As thermogenesis is tightly adjusted to ambient temperature, it is not known whether BAT thermogenesis might reach a threshold. Thus, the effectiveness might be questionable if the individual is at thermoneutrality. Finally, BAT activation could be associated with excessive body temperature, or perhaps enhanced food intake as a compensatory effect to normalize energy balance. Clinical administration of the BAT-stimulating β3-adrenergic receptor agonist Mirabegron to young lean males was associated with increased glucose uptake into BAT, elevated resting metabolic rate, and tachycardia (Cypess et al. 2015), even though the protocol is criticized because the dosage was atypically high. Thus, the search for potential therapies that aim to increase BAT mass or activity at thermoneutrality to fight against obesity and its associated metabolic diseases should continue.

Acknowledgments This study was supported by the Ministry of Spain (MINECO) (SAF2017-83813-C3-1-R to DS and LH, cofounded by the ERDF), the Centro de Investigación Biomédica en Red de Fisiopatología de la Obesidad y la Nutrición (CIBEROBN) (Grant CB06/03/0001 to DS), the Government of Catalonia (2017SGR278 to DS), the Universidad CEU San Pablo-Banco Santander (FUSP-BS-PPC-USP02/2017) to MA and the Fundació La Marató de TV3 (201627-30 to DS).

References

Acosta FM, Berchem J, Martinez-Tellez B et al (2019) Near-infrared spatially resolved spectroscopy as an indirect technique to assess brown adipose tissue in young women. Mol Imaging Biol 21:328–338

Alcalá M, Calderon-Dominguez M, Bustos E et al (2017) Increased inflammation, oxidative stress and mitochondrial respiration in brown adipose tissue from obese mice. Sci Rep 7:16082

Alcalá M, Calderon-Dominguez M, Serra D et al (2019) Mechanisms of impaired brown adipose tissue recruitment in obesity. Front Physiol 10:94

Ali Khan A, Hansson J, Weber P et al (2018) Comparative secretome analyses of primary murine white and brown adipocytes reveal novel adipokines. Mol Cell Proteomics 17:2358–2370

Bagchi M, Kim LA, Boucher J et al (2013) Vascular endothelial growth factor is important for brown adipose tissue development and maintenance. FASEB J 27.3257–3271

Bamshad M, Song CK, Bartness TJ (1999) CNS origins of the sympathetic nervous system outflow to brown adipose tissue. Am J Physiol 276:R1569–R1578

Bartelt A, Bruns OT, Reimer R et al (2011) Brown adipose tissue activity controls triglyceride clearance. Nat Med 17:200–206

Bartness TJ, Wade GN (1984) Effects of interscapular brown adipose tissue denervation on body weight and energy metabolism in ovariectomized and estradiol-treated rats. Behav Neurosci 98:674–685

Bartness TJ, Vaughan CH, Song CK (2010) Sympathetic and sensory innervation of brown adipose tissue. Int J Obes 34:S36–S42

Blaszkiewicz M, Willows JW, Johnson CP, Townsend KL (2019) The importance of peripheral nerves in adipose tissue for the regulation of energy balance. Biology (Basel) 8:10

Bonet ML, Oliver P, Palou A (2013) Pharmacological and nutritional agents promoting browning of white adipose tissue. Biochim Biophys Acta - Mol Cell Biol Lipids 1831:969–985

Boon MR, van den Berg SAA, Wang Y et al (2013) BMP7 activates brown adipose tissue and reduces diet-induced obesity only at subthermoneutrality. PLoS One 8:e74083

Bordicchia M, Liu D, Amri EZ et al (2012) Cardiac natriuretic peptides act via p38 MAPK to induce the brown fat thermogenic program in mouse and human adipocytes. J Clin Invest 122:1022–1036

Boström P, Wu J, Jedrychowski MP et al (2012) A PGC1-alpha-dependent myokine that drives brown-fat-like development of white fat and thermogenesis. Nature 481:463–468

Branca RT, Warren WS (2011) In vivo brown adipose tissue detection and characterization using water-lipid intermolecular zero-quantum coherences. Magn Reson Med 65:313–319

Branca RT, He T, Zhang L et al (2014) Detection of brown adipose tissue and thermogenic activity in mice by hyperpolarized xenon MRI. Proc Natl Acad Sci USA 111:18001–18006

Brito NA, Brito MN, Bartness TJ (2008) Differential sympathetic drive to adipose tissues after food deprivation, cold exposure or glucoprivation. Am J Physiol Integr Comp Physiol 294:R1445–R1452

Calderon-Dominguez M, Mir JF, Fucho R et al (2016) Fatty acid metabolism and the basis of brown adipose tissue function. Adipocyte 5:98–118

Campderrós L, Moure R, Cairó M et al (2019) Brown adipocytes secrete GDF15 in response to thermogenic activation. Obesity 27:1606–1616

Cannon B, Nedergaard J (2004) Brown adipose tissue: function and physiological significance. Physiol Rev 84:277–359

Cano G, Passerin AM, Schiltz JC et al (2003) Anatomical substrates for the central control of sympathetic outflow to interscapular adipose tissue during cold exposure. J Comp Neurol 460:303–326

Cereijo R, Gavaldà-Navarro A, Cairó M et al (2018) CXCL14, a brown adipokine that mediates brown-fat-to-macrophage communication in thermogenic adaptation. Cell Metab 28:750–763.e6

Chen KY, Cypess AM, Laughlin MR et al (2016) Brown adipose reporting criteria in imaging studies (BARCIST 1.0): recommendations for standardized FDG-PET/CT experiments in humans. Cell Metab 24:210–222

Contreras C, Nogueiras R, Diéguez C et al (2017) Traveling from the hypothalamus to the adipose tissue: the thermogenic pathway. Redox Biol 12:854–863

Cui X, Xiao W, You L et al (2019) Age-induced oxidative stress impairs adipogenesis and thermogenesis in brown fat. FEBS J 286:2753–2768

Cypess AM, Lehman S, Williams G et al (2009) Identification and importance of brown adipose tissue in adult humans. N Engl J Med 360:1509–1517

Cypess AM, Weiner LS, Roberts-Toler C et al (2015) Activation of human brown adipose tissue by a β3-adrenergic receptor agonist. Cell Metab 21:33–38

De Matteis R, Lucertini F, Guescini M et al (2013) Exercise as a new physiological stimulus for brown adipose tissue activity. Nutr Metab Cardiovasc Dis 23:582–590

Deshmukh AS, Peijs L, Beaudry JL et al (2019) Proteomics-based comparative mapping of the secretomes of human Brown and White adipocytes reveals EPDR1 as a novel batokine. Cell Metab 30:963–975.e7

Dodd GT, Decherf S, Loh K et al (2015) Leptin and insulin act on POMC neurons to promote the browning of white fat. Cell 160:88–104

Doh K-O, Kim Y-W, Park S-Y et al (2005) Interrelation between long-chain fatty acid oxidation rate and carnitine palmitoyltransferase 1 activity with different isoforms in rat tissues. Life Sci 77:435–443

Dowal L, Parameswaran P, Phat S et al (2017) Intrinsic properties of brown and white adipocytes have differential effects on macrophage inflammatory responses. Mediators Inflamm 2017:1–11

Dulloo AG, Miller DS (1984) Energy balance following sympathetic denervation of brown adipose tissue. Can J Physiol Pharmacol 62:235–240

El Hadi H, Frascati A, Granzotto M et al (2016) Infrared thermography for indirect assessment of activation of brown adipose tissue in lean and obese male subjects. Physiol Meas 37:N118–N128

Enriori PJ, Sinnayah P, Simonds SE et al (2011) Leptin action in the dorsomedial hypothalamus increases sympathetic tone to brown adipose tissue in spite of systemic leptin resistance. J Neurosci 31:12189–12197

Eriksson O, Selvaraju RK, Berglund M, Espes D (2019) Metabolically active brown adipose tissue is found in adult subjects with type 1 diabetes. Int J Mol Sci 20:5827

Fischer AW, Cannon B, Nedergaard J (2019) The answer to the question "What is the best housing temperature to translate mouse experiments to humans?" is: thermoneutrality. Mol Metab 26:1–3

Flávia RO, Mamede M, Bizzi MF et al (2019) Brown adipose tissue activity is reduced in women with polycystic ovary syndrome. Eur J Endocrinol 181:473–480

Flynn A, Li Q, Panagia M et al (2015) Contrast-enhanced ultrasound: a novel noninvasive, nonionizing method for the detection of brown adipose tissue in humans. J Am Soc Echocardiogr 28:1247–1254

François M, Qualls-Creekmore E, Berthoud H-R et al (2018) Genetics-based manipulation of adipose tissue sympathetic innervation. Physiol Behav 190:21–27

Fromme T, Klingenspor M (2011) Uncoupling protein 1 expression and high-fat diets. Am J Physiol Integr Comp Physiol 300:R1–R8

Garcia CA, Van Nostrand D, Atkins F et al (2006) Reduction of brown fat 2-deoxy-2-[F-18]fluoro-D-glucose uptake by controlling environmental temperature prior to positron emission tomography scan. Mol Imaging Biol 8:24–29

Garretson JT, Szymanski LA, Schwartz GJ et al (2016) Lipolysis sensation by white fat afferent nerves triggers brown fat thermogenesis. Mol Metab 5:626–634

Giordano A, Morroni M, Santone G et al (1996) Tyrosine hydroxylase, neuropeptide Y, substance P, calcitonin gene-related peptide and vasoactive intestinal peptide in nerves of rat periovarian adipose tissue: an immunohistochemical and ultrastructural investigation. J Neurocytol 25:125–136

Gnad T, Scheibler S, Von Kugelgen I et al (2014) Adenosine activates brown adipose tissue and recruits beige adipocytes via A2A receptors. Nature 516:395–399

Grunewald ZI, Winn NC, Gastecki ML et al (2018) Removal of interscapular brown adipose tissue increases aortic stiffness despite normal systemic glucose metabolism in mice. Am J Physiol Regul Integr Comp Physiol 314:R584–R597

Gunawardana SC, Piston DW (2012) Reversal of type 1 diabetes in mice by brown adipose tissue transplant. Diabetes 61:674–682

Gunawardana SC, Piston DW (2015) Insulin-independent reversal of type 1 diabetes in nonobese diabetic mice with brown adipose tissue transplant. Am J Physiol Metab 308:E1043–E1055

Hamilton G, Smith DL Jr, Bydder M et al (2011) MR properties of brown and white adipose tissues. J Magn Reson Imaging 34:468–473

Hany TF, Gharehpapagh E, Kamel EM et al (2002) Brown adipose tissue: a factor to consider in symmetrical tracer uptake in the neck and upper chest region. Eur J Nucl Med 29:1393–1398

Harlan SM, Morgan DA, Agassandian K et al (2011) Ablation of the leptin receptor in the hypothalamic arcuate nucleus abrogates leptin-induced sympathetic activation. Circ Res 108:808–812

Hartwig V, Guiducci L, Marinelli M et al (2017) Multimodal imaging for the detection of brown adipose tissue activation in women: a pilot study using NIRS and infrared thermography. J Heal Eng 2017:5986452

Hondares E, Rosell M, Gonzalez FJ et al (2010) Hepatic FGF21 expression is induced at birth via PPARα in response to milk intake and contributes to thermogenic activation of neonatal brown fat. Cell Metab 11:206–212

Hondares E, Iglesias R, Giralt A et al (2011) Thermogenic activation induces FGF21 expression and release in brown adipose tissue. J Biol Chem 286:12983–12990

Hwang JJ, Yeckel CW, Gallezot JD et al (2015) Imaging human brown adipose tissue under room temperature conditions with (11)C-MRB, a selective norepinephrine transporter PET ligand. Metabolism 64:747–755

Iris Chen Y-C, Cypess AM, Chen Y-C et al (2013) Measurement of human brown adipose tissue volume and activity using anatomic MR imaging and functional MR imaging. J Nucl Med 54:1584–1587

Karampinos DC, Weidlich D, Wu M et al (2019) Techniques and applications of magnetic resonance imaging for studying brown adipose tissue morphometry and function. Handb Exp Pharmacol 251:299–324

Kikai M, Yamada H, Wakana N et al (2017) Transplantation of brown adipose tissue inhibits atherosclerosis in apoE-/- mice: contribution of the activated FGF-21-adiponectin axis. Cardiovasc Res. https://doi.org/10.1093/cvr/cvx212

Kishida T, Ejima A, Yamamoto K et al (2015) Reprogrammed functional brown adipocytes ameliorate insulin resistance and dyslipidemia in diet-induced obesity and type 2 diabetes. Stem Cell Rep 5:569–581

Klingenspor M, Meywirth A, Stöhr S, Heldmaier G (1994) Effect of unilateral surgical denervation of brown adipose tissue on uncoupling protein mRNA level and cytochrom-c-oxidase activity in the Djungarian hamster. J Comp Physiol B 163:664–670

Kong D, Tong Q, Ye C et al (2012) GABAergic RIP-Cre neurons in the arcuate nucleus selectively regulate energy expenditure. Cell 151:645–657

Kong X, Yao T, Zhou P et al (2018) Brown adipose tissue controls skeletal muscle function via the secretion of myostatin. Cell Metab 28:631–643.e3

Law J, Morris DE, Izzi-Engbeaya C et al (2018) Thermal imaging is a noninvasive alternative to PET/CT for measurement of brown adipose tissue activity in humans. J Nucl Med 59:516–522

Lee C-W, Hsiao W-T, Lee OK-S (2017) Mesenchymal stromal cell-based therapies reduce obesity and metabolic syndromes induced by a high-fat diet. Transl Res 182:61–74.e8

Lee M-W, Lee M, Oh K-J (2019) Adipose tissue-derived signatures for obesity and type 2 diabetes: adipokines, batokines and microRNAs. J Clin Med 8:854

Levin BE, Triscari J, Sullivan AC (1983) Altered sympathetic activity during development of diet-induced obesity in rat. Am J Physiol Regul Integr Comp Physiol 13:R347–R355

Liu X, Zheng Z, Zhu X et al (2013) Brown adipose tissue transplantation improves whole-body energy metabolism. Cell Res 23:851–854

Liu X, Wang S, You Y et al (2015) Brown adipose tissue transplantation reverses obesity in Ob/Ob mice. Endocrinology 156:2461–2469

Madar I, Isoda T, Finley P et al (2011) 18F-fluorobenzyl triphenyl phosphonium: a noninvasive sensor of brown adipose tissue thermogenesis. J Nucl Med 52:808–814

Markan KR, Naber MC, Ameka MK et al (2014) Circulating FGF21 is liver derived and enhances glucose uptake during refeeding and overfeeding. Diabetes 63:4057–4063

McGregor RA, Kwon E-Y, Shin S-K et al (2013) Time-course microarrays reveal modulation of developmental, lipid metabolism and immune gene networks in intrascapular brown adipose tissue during the development of diet-induced obesity. Int J Obes 37:1524–1531

Min SY, Kady J, Nam M et al (2016) Human "brite/beige" adipocytes develop from capillary networks, and their implantation improves metabolic homeostasis in mice. Nat Med 22:312–318

Morrison SF, Madden CJ, Tupone D (2014) Central neural regulation of brown adipose tissue thermogenesis and energy expenditure. Cell Metab 19:741–756

Mueez UD, Raiko J, Saari T et al (2016) Human brown adipose tissue [(15)O]O2 PET imaging in the presence and absence of cold stimulus. Eur J Nucl Med Mol Imaging 43:1878–1886

Murano I, Barbatelli G, Giordano A, Cinti S (2009) Noradrenergic parenchymal nerve fiber branching after cold acclimatisation correlates with brown adipocyte density in mouse adipose organ. J Anat 214:171–178

Mutsnaini L, Kim C-S, Kim J et al (2019) Fibroblast growth factor 21 deficiency aggravates obesity-induced hypothalamic inflammation and impairs thermogenic response. Inflamm Res 68:351–358

Néchad M, Ruka E, Thibault J (1994) Production of nerve growth factor by brown fat in culture: relation with the in vivo developmental stage of the tissue. Comp Biochem Physiol Comp Physiol 107:381–388

Nedergaard J, Bengtsson T, Cannon B (2007) Unexpected evidence for active brown adipose tissue in adult humans. Am J Physiol Endocrinol Metab 293:E444–E452

Nguyen NLT, Barr CL, Ryu V et al (2017) Separate and shared sympathetic outflow to white and brown fat coordinately regulates thermoregulation and beige adipocyte recruitment. Am J Physiol Regul Integr Comp Physiol 312:R132–R145

Nirengi S, Yoneshiro T, Sugie H et al (2015) Human brown adipose tissue assessed by simple, noninvasive near-infrared time-resolved spectroscopy. Obes (Silver Spring) 23:973–980

Oldfield B, Giles M, Watson A et al (2002) The neurochemical characterisation of hypothalamic pathways projecting polysynaptically to brown adipose tissue in the rat. Neuroscience 110:515–526

Ong FJ, Ahmed BA, Oreskovich SM et al (2018) Recent advances in the detection of brown adipose tissue in adult humans: a review. Clin Sci 132:1039–1054

Ouellet V, Labbé SM, Blondin DP et al (2012) Brown adipose tissue oxidative metabolism contributes to energy expenditure during acute cold exposure in humans. J Clin Invest 122:545–552

Pace L, Nicolai E, D'Amico D et al (2011) Determinants of physiologic 18F-FDG uptake in brown adipose tissue in sequential PET/CT examinations. Mol Imaging Biol 13:1029–1035

Pedersen SB, Bruun JM, Kristensen K, Richelsen B (2001) Regulation of UCP1, UCP2, and UCP3 mRNA expression in brown adipose tissue, white adipose tissue, and skeletal muscle in rats by estrogen. Biochem Biophys Res Commun 288:191–197

Pellegrinelli V, Peirce VJ, Howard L et al (2018) Adipocyte-secreted BMP8b mediates adrenergic-induced remodeling of the neuro-vascular network in adipose tissue. Nat Commun 9:4974

Pfannenberg C, Werner MK, Ripkens S et al (2010) Impact of age on the relationships of brown adipose tissue with sex and adiposity in humans. Diabetes 59:1789–1793

Piao Z, Zhai B, Jiang X et al (2018) Reduced adiposity by compensatory WAT browning upon iBAT removal in mice. Biochem Biophys Res Commun 501:807–813

Planavila A, Redondo I, Hondares E et al (2013) Fibroblast growth factor 21 protects against cardiac hypertrophy in mice. Nat Commun 4:2019
Rahmouni K, Morgan DA (2007) Hypothalamic arcuate nucleus mediates the sympathetic and arterial pressure responses to leptin. Hypertension 49:647–652
Rakheja R, Ciarallo A, Alabed YZ, Hickeson M (2011) Intravenous administration of diazepam significantly reduces brown fat activity on 18F-FDG PET/CT. Am J Nucl Med Mol Imaging 1:29–35
Ran C, Albrecht DS, Bredella MA et al (2018) PET imaging of human brown adipose tissue with the TSPO tracer [(11)C]PBR28. Mol Imaging Biol 20:188–193
Rao RR, Long JZ, White JP et al (2014) Meteorin-like is a hormone that regulates immune-adipose interactions to increase beige fat thermogenesis. Cell 157:1279–1291
Rashnonejad A, Ercan G, Gunduz C et al (2018) Comparative analysis of human UCB and adipose tissue derived mesenchymal stem cells for their differentiation potential into brown and white adipocytes. Mol Biol Rep. https://doi.org/10.1007/s11033-018-4156-1
Rosell M, Kaforou M, Frontini A et al (2014) Brown and white adipose tissues: intrinsic differences in gene expression and response to cold exposure in mice. Am J Physiol Endocrinol Metab 306:E945–E964
Ruan CC, Kong LR, Chen XH et al (2018) A2A receptor activation attenuates hypertensive cardiac remodeling via promoting brown adipose tissue-derived FGF21. Cell Metab 28:476–489.e5
Ryu V, Garretson JT, Liu Y et al (2015) Brown adipose tissue has sympathetic-sensory feedback circuits. J Neurosci 35:2181–2190
Ryu V, Watts AG, Xue B, Bartness TJ (2017) Bidirectional crosstalk between the sensory and sympathetic motor systems innervating brown and white adipose tissue in male Siberian hamsters. Am J Physiol Regul Integr Comp Physiol 312:R324–R337
Saito M, Okamatsu-Ogura Y, Matsushita M et al (2009) High incidence of metabolically active brown adipose tissue in healthy adult humans: effects of cold exposure and adiposity. Diabetes 58:1526–1531
Sakamoto T, Nitta T, Maruno K et al (2016) Macrophage infiltration into obese adipose tissues suppresses the induction of UCP1 level in mice. Am J Physiol Metab 310:E676–E687
Sampath SC, Sampath SC, Bredella MA et al (2016) Imaging of brown adipose tissue: state of the art. Radiology 280:4–19
Schlein C, Talukdar S, Heine M et al (2016) FGF21 lowers plasma triglycerides by accelerating lipoprotein catabolism in white and Brown adipose tissues. Cell Metab 23:441–453
Shen H, Jiang L, Lin JD et al (2019) Brown fat activation mitigates alcohol-induced liver steatosis and injury in mice. J Clin Invest 129:2305–2317
Silva FJ, Holt DJ, Vargas V et al (2014) Metabolically active human brown adipose tissue derived stem cells. Stem Cells 32:572–581
Soderlund V, Larsson SA, Jacobsson H (2007) Reduction of FDG uptake in brown adipose tissue in clinical patients by a single dose of propranolol. Eur J Nucl Med Mol Imaging 34:1018–1022
Soler-Vázquez MC, Mera P, Zagmutt S et al (2018) New approaches targeting brown adipose tissue transplantation as a therapy in obesity. Biochem Pharmacol 155:346–355
Song CK, Vaughan CH, Keen-Rhinehart E et al (2008) Melanocortin-4 receptor mRNA expressed in sympathetic outflow neurons to brown adipose tissue: neuroanatomical and functional evidence. Am J Physiol Regul Integr Comp Physiol 295:R417–R428
Stanford KI, Middelbeek RJW, Townsend KL et al (2013) Brown adipose tissue regulates glucose homeostasis and insulin sensitivity. J Clin Invest 123:215–223
Stanford KI, Middelbeek RJW, Townsend KL et al (2015) A novel role for subcutaneous adipose tissue in exercise-induced improvements in glucose homeostasis. Diabetes 64:2002–2014
Stanford KI, Lynes MD, Takahashi H et al (2018) 12,13-diHOME: an exercise-induced lipokine that increases skeletal muscle fatty acid uptake. Cell Metab 27:1111–1120.e3
Sun K, Kusminski CM, Scherer PE (2011) Adipose tissue remodeling and obesity. J Clin Invest 121:2094–2101
Tan CY, Vidal-Puig A (2008) Adipose tissue expandability: the metabolic problems of obesity may arise from the inability to become more obese. Biochem Soc Trans 36:935–940
Thomou T, Mori MA, Dreyfuss JM et al (2017) Adipose-derived circulating miRNAs regulate gene expression in other tissues. Nature 542:450–455
Thoonen R, Ernande L, Cheng J et al (2015) Functional brown adipose tissue limits cardiomyocyte injury and adverse remodeling in catecholamine-induced cardiomyopathy. J Mol Cell Cardiol 84:202–211
Tseng Y-H, Kokkotou E, Schulz TJ et al (2008) New role of bone morphogenetic protein 7 in brown adipogenesis and energy expenditure. Nature 454:1000–1004
van Marken Lichtenbelt WD, Vanhommerig JW, Smulders NM et al (2009) Cold-activated brown adipose tissue in healthy men. N Engl J Med 360:1500–1508
Verdeguer F, Soustek MS, Hatting M et al (2015) Brown adipose YY1 deficiency activates expression of secreted proteins linked to energy expenditure and prevents diet-induced obesity. Mol Cell Biol 36:184–196
Villarroya F, Vidal-Puig A (2013) Beyond the sympathetic tone: the new brown fat activators. Cell Metab 17:638–643
Villarroya F, Cereijo R, Villarroya J, Giralt M (2017a) Brown adipose tissue as a secretory organ. Nat Rev Endocrinol 13:26–35
Villarroya F, Gavaldà-Navarro A, Peyrou M et al (2017b) The lives and times of brown adipokines. Trends Endocrinol Metab 28:855–867
Villarroya F, Cereijo R, Gavaldà-Navarro A et al (2018a) Inflammation of brown/beige adipose tissues in obesity and metabolic disease. J Intern Med 284:492–504

Villarroya F, Cereijo R, Villarroya J et al (2018b) Toward an understanding of how immune cells control brown and beige adipobiology. Cell Metab 27:954–961
Villarroya J, Cereijo R, Gavaldà-Navarro A et al (2019a) New insights into the secretory functions of brown adipose tissue. J Endocrinol 243:R19–R27
Villarroya J, Cereijo R, Giralt M, Villarroya F (2019b) Secretory proteome of brown adipocytes in response to cAMP-mediated thermogenic activation. Front Physiol 10:67
Virtanen KA, Lidell ME, Orava J et al (2009) Functional brown adipose tissue in healthy adults. N Engl J Med 360:1518–1525
Wade GN, Gray JM (1978) Cytoplasmic 17β-[3H]estradiol binding in rat adipose tissues. Endocrinology 103:1695–1701
Wang G-X, Zhao X-Y, Meng Z-X et al (2014) The brown fat-enriched secreted factor Nrg4 preserves metabolic homeostasis through attenuation of hepatic lipogenesis. Nat Med 20:1436–1443
Watanabe M, Houten SM, Mataki C et al (2006) Bile acids induce energy expenditure by promoting intracellular thyroid hormone activation. Nature 439:484–489
White JD, Dewal RS, Stanford KI (2019) The beneficial effects of brown adipose tissue transplantation. Mol Aspects Med 68:74–81
Whittle AJ, Carobbio S, Martins L et al (2012) BMP8B increases brown adipose tissue thermogenesis through both central and peripheral actions. Cell 149:871–885
Wu J, Cohen P, Spiegelman BM (2013) Adaptive thermogenesis in adipocytes: is beige the new brown? Genes Dev 27:234–250
Wu R, Liu XM, Sun JG et al (2017) DJ-1 maintains energy and glucose homeostasis by regulating the function of brown adipose tissue. Cell Discov 3:1–18
Yasuda T, Masaki T, Kakuma T, Yoshimatsu H (2004) Hypothalamic melanocortin system regulates sympathetic nerve activity in brown adipose tissue. Exp Biol Med 229:235–239
Yuan X, Hu T, Zhao H et al (2016) Brown adipose tissue transplantation ameliorates polycystic ovary syndrome. Proc Natl Acad Sci USA 113:2708–2713
Zhang Y, Kerman IA, Laque A et al (2011) Leptin-receptor-expressing neurons in the dorsomedial hypothalamus and median preoptic area regulate sympathetic brown adipose tissue circuits. J Neurosci 31:1873–1884
Zhu Z, Spicer EG, Gavini CK et al (2014) Enhanced sympathetic activity in mice with brown adipose tissue transplantation (transBATation). Physiol Behav 125:21–29
Zhu Q, Glazier BJ, Hinkel BC et al (2019) Neuroendocrine regulation of energy metabolism involving different types of adipose tissues. Int J Mol Sci 20:2707
Zingaretti MC, Crosta F, Vitali A et al (2009) The presence of UCP1 demonstrates that metabolically active adipose tissue in the neck of adult humans truly represents brown adipose tissue. FASEB J 23:3113–3120

White and Brown Adipose Tissue in Obesity and Diabetes

5

Brooks P. Leitner and Borja Martinez-Tellez

Contents

Distribution and Functions of Human Adipose Tissue 56
White Adipose Tissue 56
Brown Adipose Tissue 57
Anatomical Sites 57
Beige Adipose Tissue 58

White Adipose Tissue and Insulin Resistance 58
Glucose Homeostasis Crosstalk 59
Inflammatory Adiposopathy 59
Ectopic Fat Depots 59

Activation of Human Brown Adipose Tissue 60
Cold Exposure Techniques 60
Exercise Activation 62
Food Intake and Thermoregulation 62

Role of Brown Fat in Human Metabolism in Healthy, Obese, and Diabetic Individuals 63
BAT Hormone-Like Secretions 63
Correlations with Obesity and Diabetes 63

Methods for Detecting Human Brown Adipose Tissue 64
Monitoring of Skin Temperature 64
Radiolabeled Glucose Analog 64
Tracer and Imaging Alternatives 65
Immunohistochemical or Gene-Expression Confirmation 65

Future Investigations into Human Brown Adipose Tissue Physiology 66
Dietary Components 66

B. P. Leitner (✉)
Department of Internal Medicine and Cellular and Molecular Physiology, Yale University, New Haven, CT, USA
e-mail: brooks.leitner@yale.edu

B. Martinez-Tellez
Department of Medicine, Division of Endocrinology, and Einthoven Laboratory for Experimental Vascular Medicine, Leiden University Medical Center, Leiden, The Netherlands

J. Faintuch and S. Faintuch (eds.), *Obesity and Diabetes*, https://doi.org/10.1007/978-3-030-53370-0_5

Pharmaceutical Compounds 66
Gut Microbiota as a New Target to Combat T2D 66

Future Perspectives 67

References 67

Abstract

It has become increasingly evident that the diversity of adipose tissue types may play distinct and potentially important roles in human physiology. White adipose tissue (WAT), the primary energy storage tissue, is important in homeostasis but when found in excess can predispose individuals to severe insulin resistance and diabetes. Human life is compatible with WAT composing between 4 and 60% of total body mass, pointing to its incredible adaptability. In contrast, brown adipose tissue (BAT), typically thought to occur only in hibernating animals and babies, is becoming recognized for its role in human energy expenditure, hormone production, immune regulation, among others. BAT usually comprises around 0–2% of total body mass; however, recent scientific advances in noninvasive imaging and molecular biology have allowed to explore unknown functions of this dynamic tissue. We discuss here the anatomy and physiology of WAT and BAT in the lean and obese human, and potential mechanisms regarding the interplay within obesity and Type 2 diabetes. Later, we review the current methodology for measuring and detecting WAT and BAT, how to experimentally modulate BAT activity, as well as future investigations to yield greater insight into BAT's functional role in human health.

Keywords

Brown adipose tissue · White adipose tissue · Brown fat · White fat · Insulin resistance

Distribution and Functions of Human Adipose Tissue

It is estimated that in 2025, global obesity rates will reach 18% in men and 21% in women (NCD Risk Factor Collaboration 2016), with diabetes prevalence reaching nearly 400 million people globally (Nathan 2015). Obesity, defined clinically by a body mass index (BMI) of greater than 30 kg/m^2, is accompanied by an excess of white adipose tissue (WAT). Obesity is the result of long-term energy intake exceeding energy expenditure, being one of the main comorbidities Type 2 Diabetes (T2D). A diagnosis of T2D is most often made with a glycosylated hemoglobin (HbA1C) of $\geq$6.5% or a fasting plasma glucose of $\geq$126 mg/dl (Association, A. D. 2 2019). T2D is more likely to be obesity-related and can progress to loss of pancreatic beta-cell insulin secretion, often in the background of insulin resistance. Because the risk for T2D is increased substantially by obesity, the mechanisms that we will discuss throughout the chapter will pertain to both obesity and T2D, unless stated otherwise.

White Adipose Tissue

Humans have several types of adipose tissue, although the most studied are white adipose tissue (WAT) and brown adipose tissue (BAT). WAT is normally distributed subcutaneously and in the visceral compartments, the latter known as visceral adipose tissue (VAT). The amount of WAT in humans varies a lot between individuals, with minimal levels sustainable with life around 2% of body mass for men and 6% of body mass for women, and maximal levels reaching over 60% of body mass. In simple terms, the higher the BMI, the higher the WAT accumulation in both compartments. Subcutaneous WAT is found in all areas beneath the skin surface but has a propensity to be stored in central, or proximal, locations. The distribution of subcutaneous WAT differs between men, who typically have an "android" distribution and women, with a "gynoid" distribution (Min and Min 2015).

VAT is also found in central locations, however within the abdominal cavity, surrounding vital organs. VAT contributes a much smaller percentage to total body fat (~5% in men and ~3% in women) (Sasai et al. 2015), yet the accumulation of VAT confers greater risk for T2D and cardiovascular disease than subcutaneous WAT. This phenomenon partly explains the occasional metabolically unhealthy, normal weight phenotype, where individuals that do not have BMIs classified as obese suffer from metabolic syndrome comparable to individuals with much higher body mass. Some data show that physical exercise can selectively reduce VAT, even in the absence of significant weight loss.

As WAT is a tissue for energy storage, it is evolutionarily designed with the ability to expand to a great extent, so that high energy density lipids can be available even in times of food shortage. In the modern era with food shortage being less of a concern, this expansion has occurred without periods of energy deficit. Though WAT exists in excess in patients with obesity, WAT in itself is not causative of insulin resistance (the inability for insulin to perform its usual functions). Lessons from patients with lipodystrophy demonstrate this fact, as they have nearly nonexistent WAT, but often develop severe T2D. The contributions of excess adiposity to diabetes may be a result of adipose-derived cytokines (adipokines) or accumulation of ectopic lipids in skeletal muscle and the liver that disrupt insulin signaling. Much has been learned since the discovery of WAT as an endocrine organ (the discovery of leptin, as an adipokine) (Zhang et al. 1994), yet there is still much to be investigated.

Brown Adipose Tissue

The main function of BAT is to generate heat (thermogenesis), principally through dissipation of the proton gradient in the inner mitochondrial membrane by way of uncoupling protein 1 (UCP1) (Cannon and Nedergaard 2004). Thus, as fuel sources including glucose and fatty acids are utilized by the brown adipocyte, rather than the generation of adenosine triphosphate (ATP), heat is released, resulting in energy-depleting thermogenesis. BAT is extremely important in maintaining euthermia in the hibernating animal during prolonged periods of fasting in cold temperatures.

Though the depots of human BAT were extensively defined nearly 50 years ago by autopsy (Heaton 1972), advances in noninvasive imaging have brought a resurgence in the interest in BAT. Classically thought only to exist in the interscapular adipose tissue region of human newborns, ^{18}F-Fluorodeoxyglucose (^{18}F-FDG) positron emission tomography (PET) combined with X-ray computed tomography (CT) studies have shown that active BAT exists in a large portion of the adult population, and can be reliably activated by cold (Cypess et al. 2009; van Marken Lichtenbelt et al. 2009) and by other means. Indeed, over 450 publications on PubMed have studied adult BAT in humans since just 2008. BAT's functions include substrate metabolism, immune system modulation, endocrine functions, and others. To begin a thorough understanding of BAT function in adults, we have characterized the depots of adult men and women with and without obesity (Leitner et al. 2017; Martinez-Tellez et al. 2019a).

Anatomical Sites

Generally, BAT lines visceral locations surrounding major vasculature in the abdomen, neck, and axilla, vital organs including the kidneys and adrenal glands, and alongside the sympathetic trunk. The highest proportion of BAT exists in the supraclavicular region, making it the most commonly studied BAT depot across various studies, followed by the paraspinal, cervical, abdominal (mostly retroperitoneal), axillary, and mediastinal (Leitner et al. 2017). Previous evidence showed that newborns have huge amounts of subcutaneous BAT in the dorsocervical area, as adequate skeletal muscle has not yet been developed for shivering or activity thermogenesis. It has been thought that this subcutaneous BAT disappeared when the newborn grew older; however, recent evidence suggests that it could be still present in young healthy adults, mostly in women (Martinez-Tellez et al. 2019a; Kim et al. 2019). The discovery of this new depot will boost research.

Beige Adipose Tissue

Beige and/or brite adipose tissue has been discovered as an inducible form of brown adipose tissue, with morphological features between those of white adipocytes and brown adipocytes, however physiological functions (thermogenesis) of BAT (Wu et al. 2012; Jespersen et al. 2013) remain under intense investigation. In humans, very few studies have been able to characterize anatomical differences throughout the human body, with both functional and biochemical confirmation (Cypess et al. 2013). We have also shown that depots outside of these listed likely have very little capacity for browning or the conversion of white to brown adipocytes; however, depots currently harnessing metabolically active adipocytes comprise up to 5% of total body fat, thus making BAT an organ of appreciable size in adult humans.

Though anatomical characterization has come to higher resolution in recent years, it is unclear whether or not there are differential functions harbored by white adipocytes or brown adipocytes that exist in different regions. As most functions seem to operate at an organism-wide level, impactful studies can still be conducted without this knowledge. Yet, as noninvasive imaging and other biochemical techniques become more advanced, perhaps soon we will be able to create a functional atlas of human WAT and BAT. It is interesting that most anatomy textbooks do not have a chapter that maps adipose tissue. Maybe that points to our need to better understand adipocytes in all forms (Zwick et al. 2018).

White Adipose Tissue and Insulin Resistance

WAT can account for as little as 4% of total body mass in patients with congenital generalized lipodystrophy (Lima et al. 2016), a disease characterized by the absence of WAT, and as much as nearly 60% of body mass in patients with morbid obesity (Shah and Braverman 2012). Adipose tissue expansion by hypertrophy (increase in adipocyte size) or hyperplasia (increase in adipocyte number) is frequently followed by unfavorable metabolic consequences, such as insulin resistance and inflammation, and as such, WAT plays a major role in the development of insulin resistance.

Insulin, a major anabolic hormone, mediates an integrated metabolic response in the presence of nutrients, and insulin signaling is highly active in the adipocytes of lean and obese humans. Most importantly, insulin's role in the white adipocyte is to suppress lipolysis and stimulate glucose uptake. In the lean individual, these functions are tightly regulated to ensure no excessive freeing of fatty acids and the disposal of glucose. However, leading up to the development of and in the setting of T2D, insulin resistance ensues, manifesting as both blunted insulin signaling in adipocytes, and reduced insulin receptor content (Petersen and Shulman 2018). Functionally, insulin's main roles in the adipocyte, to stimulate glucose uptake and to suppress lipolysis are less effective. The effects of insulin resistance lead to an excess of plasma insulin, non-esterified fatty acids, and glucose levels, and implicates WAT as a major player in the pathogenesis of insulin resistance.

Interestingly, WAT plays a small role in whole-body glucose disposal (less than 5%), however has a significant impact on the development of insulin resistance. Several mechanisms have been proposed for WAT's role in the pathogenesis of T2D, despite a seemingly minor participation in glucose disposal. Most convincingly, WAT dysfunction includes altered crosstalk with skeletal muscle and liver at maintaining energetic homeostasis and chronic low-grade inflammation.

Glucose Homeostasis Crosstalk

There is substantial coordination between the skeletal muscle, liver, and WAT to maintain normoglycemia in the fed and the fasted states. As humans transition to the fasted state (hours after the previous feeding), leptin and the hypothalamic–pituitary–adrenal (HPA) axis become activated to regulate WAT lipolysis, in order to free fatty acids to the plasma and to the liver for gluconeogenesis (Perry et al. 2018). This integrated response causes a whole-body shift from carbohydrate to fat oxidation. Studies have shown that this coordinated response may be altered in obese men, suggesting that nonesterified fatty acids are not released as readily, resulting in a net storage of fat. In addition, insulin's roles in suppressing WAT lipolysis and endogenous glucose production (from the liver) become less effective, which has been suggested as a link between WAT inflammation and the development of hyperglycemia (Roden and Shulman 2019).

Inflammatory Adiposopathy

Inflammation in WAT has been characterized by inadequate tissue vascularization, fibrosis, and hypoxia, which develop in the setting of increased WAT mass and cell size. Thus, obesity and the accompanying excess of WAT has been characterized as a state of chronic low-grade inflammation, with possible impaired immune function and activation of stress pathways. Much research has been dedicated to elucidate the mechanisms of WAT inflammation, including upregulation of the inflammatory cytokines tumor necrosis factor (TNF alpha), interleukin (IL)-1B, and IL-6 within adipose depots. Evidence for treatability of obesity-induced inflammation came from a clinical trial in obese humans, where Amlexanox, an anti-inflammatory drug typically used in the treatment of asthma, improved glycemic control in a subset of obese, T2D (Oral et al. 2017). Clearly, excess WAT and obesity are accompanied by multiple mechanisms that factor into the development of insulin resistance and diabetes, yet it is important to mention that obesity is not required for the development of these conditions.

Ectopic Fat Depots

WAT is not the only player in the development of T2D. Patients with lipodystrophy lacking adipocytes, and thus all adipose-derived hormones including leptin, often become hyperphagic due to the absence of leptin's hunger-suppressing effect. With the incapacity to store excess food intake in adipocytes, ectopic lipids are stored in skeletal muscle and the liver, sometimes to extreme extents. In the setting of massively increased ectopic lipid storage, patients without treatment often develop severe insulin resistance and poorly controlled diabetes, which highlights the role of ectopic lipids rather than excess WAT in the causal pathway of insulin resistance (Grundy 2015). It is possible that this same mechanism could underlie non-lipodystrophic patients who develop T2D, with contributions from genetics, epigenetics, and other factors.

Brown Adipose Tissue and Insulin Resistance Preclinical models have suggested that BAT is a great tool to combat adiposity and T2D. Some of these studies have demonstrated that the transplantation of interscapular BAT from lean to obese mice, induced a significant weight loss and improved levels of HbA1C, alleviating insulin resistance. This is the main reason why this tissue has aroused a great interest in human physiology. Although human clinical studies have developed interventions, based on cold exposure, to activate and recruit the amount of BAT in obese and T2D individuals, they have failed to replicate the positive effects of preclinical models (Jensen 2015). The main reason could be that human BAT can represent as little as 1% of total body mass, whereas in preclinical models BAT reaches up to 30–40% of the total body mass.

Moreover, it is important to note that BAT consumes glucose and fatty acids, and brown adipocytes have glucose transporter 1 (GLUT1, non-insulin-dependent) and 4 (GLUT4, insulin-dependent). The most used technique to quantify BAT in humans is ^{18}F-FDG-PET/CT scan, which is based on a glucose tracer. This technique is also used to quantify BAT in T2D. As one can imagine, in the scenario of insulin resistance this technique is far less accurate for the quantification of such tissue, which hampers understanding of the role of BAT in insulin resistance in humans.

Activation of Human Brown Adipose Tissue

The interest in BAT lies in its previously mentioned ability to uncouple ATP production from the electron transport chain, providing a means to "waste" energy in the form of heat. Studies that have attempted to examine the maximal energy expending capacity for BAT have been difficult to perform in humans, but have ranged from 5 to 730 kcal/day at maximal activation (Ruiz et al. 2018; Marlatt et al. 2018). These measures range from clinically meaningless, to having a measurable effect on reducing weight.

Because BAT is inactive in thermoneutral resting conditions, maximal thermogenesis must be measured in the active state. Researchers have attempted many approaches to BAT activation, including cold exposure, exercise, and dietary/pharmacological modulation (Ruiz et al. 2018) (see Fig. 5.1). Most often, cold exposure is used to study active BAT. This is a reasonable method to use, as it highlights BAT's most obvious role in nature: non-shivering thermogenesis to defend against the cold. Cold is sensed by free nerve endings in the skin, and the hypothalamus begins a coordinated whole-body response to defend core body temperature: intense vasoconstriction occurs at the distal extremities to shunt warm blood to vital organs, contributing to a slowed heart rate.

Even at temperatures too low to activate detectable shivering, metabolic rate increases (up to 17% above resting in lean men and 6% above resting in obese men) (Brychta et al. 2019). The blunted response in non-shivering thermogenesis in obese men is likely due to a thicker "shell," and perhaps greater insulation from cold exposure. This insulation requires less heat production in response to the same cold stimulus, as the heat is trapped in the core, and loss is prevented. If this is the sole mechanism for reduced BAT activity in obese men, it may be expected that pharmacological agents may yield more success for energy expenditure than what cold exposure can reveal.

Cold Exposure Techniques

Methods for studying cold-induced thermogenesis and BAT in humans include ambient cold air exposure, the wearing of liquid-perfused cooling suits, and placing the feet on ice blocks. It is uncertain whether all of these methods are comparable, yet studies suggest that the degree of BAT activation is likely reproducible when maintaining a standardized protocol (Fraum et al. 2019). All methodologies considered, BAT is quickly activated in response to cold exposure (Leitner et al. 2018), it can contribute to cold-induced energy expenditure in the absence of shivering, and BAT metabolizes a combination of both fatty acids and glucose when active.

In addition, cold-acclimation studies have demonstrated that BAT volume and metabolic activity can be increased over the course of a week in both lean and obese individuals (Lee et al. 2014; Hanssen et al. 2016). Detectable BAT is found in most obese patients, though consistently to a lesser extent than their lean counterparts. One of the only studies that has rigorously measured BAT in type 2 diabetic

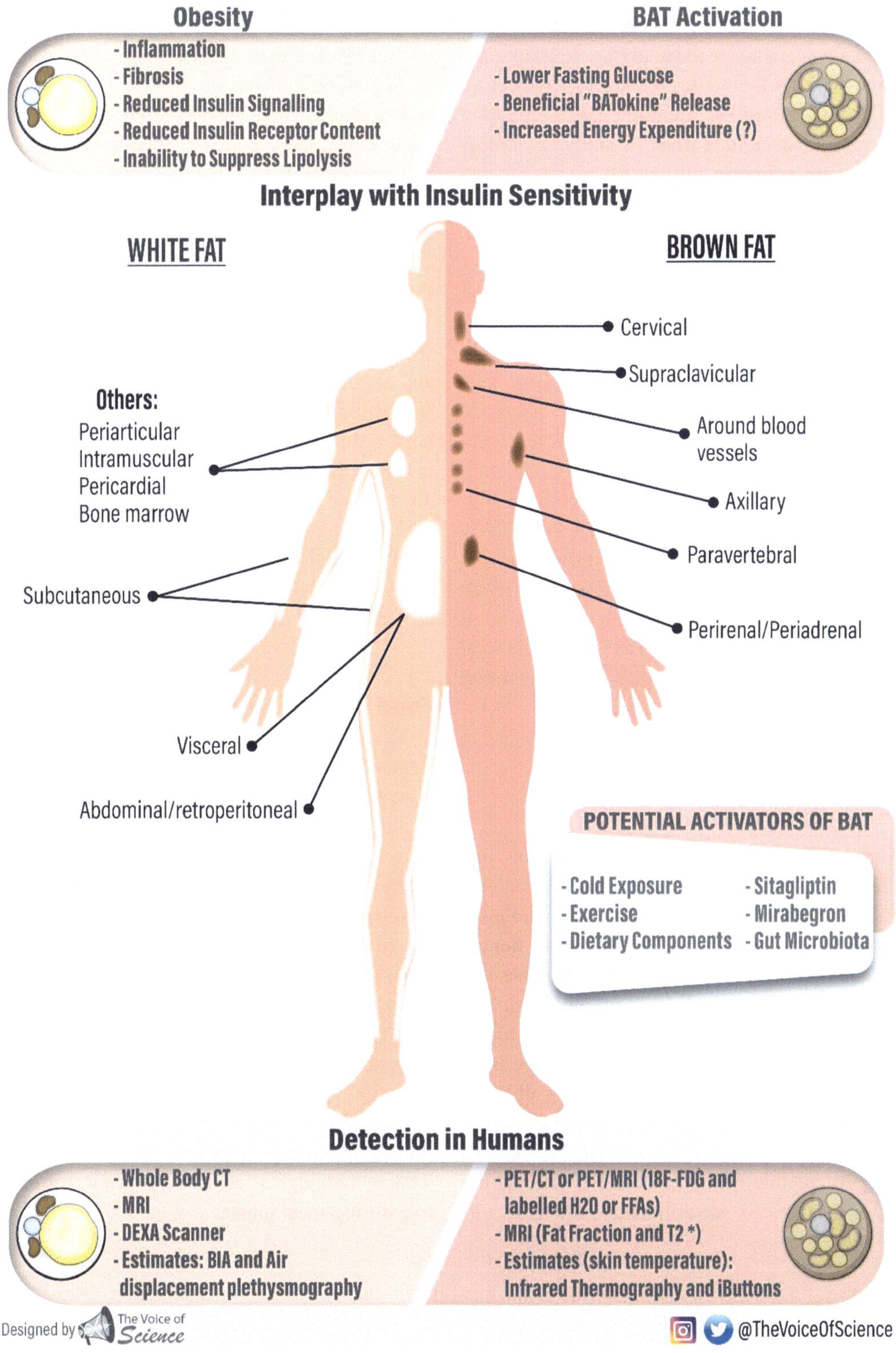

Fig. 5.1 This schematic figure summarizes the main issues discussed in the present chapter. *18F-FDG* 18F-fluorodeoxyglucose, *BAT* brown adipose tissue, *CT* computed tomography, *DEXA* Dual-energy X-ray

patients has shown that an improvement in insulin sensitivity following chronic cold exposure, can be mostly attributed to skeletal muscle adaptation rather than BAT glucose uptake (Hanssen et al. 2015), thus the utility of chronic BAT activation for energy expenditure must certainly be considered carefully. Importantly, nearly all studies that have examined subcutaneous WAT for features of BAT in response to chronic cold exposure, have yielded negative results.

Recently, a new study nicely shows that human BAT is rapidly activated upon cold exposure, achieving its maximum activity after 35 min of cold exposure (Oreskovich et al. 2019). This study was performed by a dynamic magnetic resonance imaging (MRI) scan. After 2 h of cold exposure, a warm-up phase during 30 min followed, and no replenishment of the lipid content of the BAT depots ensued. This study revealed that BAT is only active during the first part of the cooling protocols, questioning much evidence published so far. To date the most used cooling protocols to activate BAT applied a total duration of 2 h; therefore, they have quantified BAT when it was not truly activated.

Exercise Activation

Exercise is a non-pharmacological tool that induces a myriad of physiological adaptations, enhancing human cardiovascular health and insulin sensitivity in almost every single human being. Therefore, it is biologically plausible that exercise will affect BAT function, although it is unknown in which direction. Preclinical studies have suggested that exercise could increase BAT function and induce browning in almost all WAT depots (Stanford et al. 2015). However, it seems that most of these studies were not performed under strictly thermoneutral conditions, rendering it impossible to discriminate whether this browning effect is caused by the exercise, by the temperature, or by the combination of both stimuli (McKie et al. 2019).

Regarding humans, the evidence is even more contradictory. There are only a few cross-sectional studies that showed that trained-individuals had lower levels of BAT ^{18}F-FDG uptake in comparison to sedentary controls. Moreover, we recently demonstrated that objectively measured physical activity by accelerometers and cardiorespiratory fitness protocols, were not associated with BAT ^{18}F-FDG uptake (Acosta et al. 2019). Surprisingly, we observed that handgrip strength was positively and slightly related with BAT ^{18}F-FDG uptake, suggesting that maybe resistance training could have a different role in terms of BAT activation in comparison to endurance training. However, much more studies are needed (Martinez-Tellez et al. 2019b). Lastly, only a few studies performed exercise interventions including BAT ^{18}F-FDG uptake and browning markers in humans, and again the results are quite controversial.

Food Intake and Thermoregulation

Recently, we demonstrated that BAT activity is not related to energy intake assessed by an ad libitum meal and by 24-h dietary recalls (Sanchez-Delgado et al. 2020). Moreover, we did not detect the role of BAT in the regulation of appetite-related sensations either before or after the subjects ate. In a submitted study, we also observed that BAT activity was not related to the adherence to a Mediterranean diet, which is one of the healthiest diets. On the other hand, an independent study showed that the oxygen consumption of BAT depots increased after a single meal intake. It seems that BAT is not playing a role in appetite perception, energy, or nutrition intakes, although it could be playing a minimal role during meal intake.

Fig. 5.1 (continued) absorptiometry, *FFA* free fatty acids, *H20* water, *MRI* magnetic resonance imaging, *PET* positron emission tomography. This image is courtesy by The Voice Of Science

Role of Brown Fat in Human Metabolism in Healthy, Obese, and Diabetic Individuals

One option to prevent and combat obesity and related comorbidities is to increase thermogenesis, which leads to increased energy expenditure. In mice, BAT is responsible for 20% (Blondin et al. 2017) of both resting metabolic rate and adaptive thermogenesis, i.e., cold-induced thermogenesis (CIT) and meal-induced thermogenesis (MIT) (Nathan 2015), and can account for up to 60% of total energy expenditure when fully stimulated. Human BAT becomes metabolically active upon cold exposure in most individuals. However, a recent study nicely demonstrated that the contribution to the increase observed in CIT is almost negligible (Din et al. 2016; Carpentier et al. 2018). The same research group, demonstrated that BAT was active after eating a carbohydrate-rich meal to a similar extent than during mild cold exposure, yet its contribution to the increase in MIT is quite low. These novel studies confirm that BAT is contributing to human thermogenesis, but in a lower proportion than was thought at the beginning. What is the main role of BAT in the human metabolism? After 10 years of thoughtful research in human trials and mice studies we still do not know, although several new roles have been attributed, such as a secretory organ orchestrating the shivering and non-shivering responses or immune system modulation, among others.

BAT Hormone-Like Secretions

Some specialists suggest that the role of BAT in the human metabolism is negligible because it represents only 1% of total body mass. On the other hand, humans have similarly small organs (i.e., thyroid, adrenal, pituitary) that have very important roles in metabolism, so it could be reasonably expected that BAT has unknown and relevant functions. BAT has the ability to secrete several hormones (batokines) that could coordinate/orchestrate the responses of other tissues to different stimuli. For instance, we know that when humans are exposed to cold there is an increase of the energy expenditure, which could be explained by the contribution of two thermogenic tissues (BAT and skeletal muscles) (Villarroya et al. 2017). One of the hypotheses that is open for discussion is that BAT could be releasing a set of batokines that could boost the involvement of skeletal muscle during cold exposure.

BAT has been proposed to interact with bone (via IGFBP2), WAT, brain, pancreas, heart (via FGF-21 or IL6), and liver (IGF1), and therefore, impact tissue plasticity and metabolism. A research group from Copenhagen performed the first human BAT secretome. In this particular study, they discovered that ependymin-related protein 1 (EPDR1) was a novel batokine important for brown fat commitment (Deshmukh et al. 2019). They suggest that this batokine could be involved in the regulation of human metabolism.

Correlations with Obesity and Diabetes

Originally, it has been thought that obesity is negatively related to the amount of BAT; however, we recently discovered that obesity was positively related to BAT volume in 150 young healthy adults. This controversial finding could be explained by several factors: (1) BAT normally is measured with a glucose analog. Since the individuals of our study were quite young (<25 years old), they might not have developed insulin resistance yet; (2) most studies that found a negative relation between the degree of obesity and BAT activity, were performed without cold stimulation; and (3) another possible explanation is that the prevalence of T2D or insulin resistance increases with age, and those studies that were showing the inverse relationship between the degree of obesity and BAT activity did not include age as a possible confounder.

In an alternative study, it was shown that BAT measured with glucose, in T2D patients, was diminished in comparison with a free fatty acid tracer, suggesting that BAT could be playing a role in obese and T2D individuals (Blondin et al.

2015). Moreover, there is a group of obese individuals characterized by a lower risk of obesity-related cardiometabolic complications, the so-called metabolically healthy obese (MHO) (Blüher 2010). MHO individuals have a normal metabolic profile, i.e., do not have dyslipidemia, hyperglycemia, hypertension, or T2D, whereas their counterparts that present any of these conditions are known as metabolically unhealthy obese (MUO).

In an unpublished study, we observed that MHO individuals had higher BAT activity, cold and meal-induced thermogenic responses in comparison to MUO individuals (Sánchez-Delgado 2018). We also observed that MHO perceived the temperature as colder than MUO controls (Martinez-Tellez 2018). Both groups were similar in body mass index, although they were different in terms of the amount of VAT.

Methods for Detecting Human Brown Adipose Tissue

The contribution of BAT to adult human energy expenditure, thermogenesis, and other systemic functions remains difficult to fully uncover. One major reason for this difficulty is the variety of methods used to examine BAT in vivo. In general, methods either employ a noninvasive imaging strategy or a measurement of another physiological parameter as a surrogate of BAT activity.

Monitoring of Skin Temperature

The impetus for using skin temperature as a measure of BAT activity derives from BAT's inherent function as a thermogenic organ, which, by definition generates heat to be dissipated from the body. This heat dissipation is thought to be captured at the surface of the skin, as a reflection of heat transfer from the deeper tissues to dilated blood vessels superficial to BAT. The inherent shortcoming of this technique is that skin vasculature is impacted by a number of other systemic factors, including catecholamines and nitric oxide among others, such that the blood flow beneath a skin surface measurement may be reflecting physiological process unrelated to underlying BAT activity.

A major benefit of using skin temperature is the ability to capture physiological shifts in long-term bouts, free-living conditions, and during multiple environmental settings (such as exercise, and entering cold or warm rooms). Skin temperature is commonly measured with either infrared thermography, where photos are taken of exposed skin, or with surface temperature probes like iButton chips (www.maximintegrated.com). As supraclavicular BAT is detected in the highest proportion of individuals, and because it is the depot least obstructed by other structures such as bone and muscle, infrared photos and surface probes often capture this location on subjects.

It is important to take into account that supraclavicular fossa is a complex anatomic region that is contiguous with the neck above and axilla below (Kellman et al. 1987). There are several structures in this region, such as the scalene and omohyoid muscles, subclavian vessels, and brachial plexus. Other contents of the supraclavicular fossa include small branches of the subclavian vessels, fat, lymph nodes, and the posterior lung apex. The benefits of supraclavicular skin temperature measurements are the ease of use, minimal discomfort to study participants, the ability to be employed in a variety of settings, and relatively low cost. However, observers still deal with the inability to rule out other physiological factors that influence skin blood flow and therefore temperature (Jimenez-Pavon et al. 2019; Martinez-Tellez et al. 2019c, d).

Radiolabeled Glucose Analog

The most commonly used technique for detecting BAT is ^{18}F-FDG PET/CT, which utilizes a radioactive glucose analog coupled with anatomic and radiodensity data from X-ray computed tomography. This technique has likely been studied most due to a number of factors including active BAT's avidity for glucose, the current availability in many clinical settings because of its utility in cancer staging, and the largest breadth of preprocessing and image processing protocols associated with it. In general, subjects undergo a protocol in order to activate BAT (cold exposure, dietary or physical intervention, pharmacological intervention) and are then injected with the ^{18}F-FDG glucose analog tracer. The tracer accumulates in metabolically active tissue by entering the cell and becoming phosphorylated by hexokinase, then trapped as ^{18}F-FDG-6-phosphate.

The concentration of the trapped glucose analog can be quantified as a standardized uptake value (SUV) relative to the subject's body weight or lean mass and represents a reflection of tissue-specific glucose uptake. Static ^{18}F-FDG PET/CTs give no information regarding metabolic flux rates, nor the fate of glucose within the tissue (whether or not the tissue utilizes the glucose for glycolysis, oxidative phosphorylation, or anaplerosis), which constitute some of the method's largest limitations for studying in vivo physiology. Though ^{18}F-FDG still provides rich information about whole-body metabolic activity, it is important that a CT scan accompanies the PET scan.

The CT scan is necessary to determine the density and structure of the underlying tissue.

Attempts to classify metabolically active tissue (obtained from PET) as BAT requires that the CT detects this tissue within the range of densities known to be adipose tissue. On CT alone, brown and white adipose tissue have nearly indistinguishable densities, with BAT being slightly denser due to its higher vascularization, lower lipid content, and higher mitochondrial content. Thus, CT alone cannot reliably distinguish WAT from BAT. Similarly, metabolic activity from the PET scan cannot differentiate any tissue from another, as the only output is tracer accumulation. Only the combination of the density of adipose tissue (from CT) and the metabolic activity above a certain threshold (from PET) are able to quantify active BAT (Fig. 5.1).

Tracer and Imaging Alternatives

The goal of these alternatives, such as fatty acids, is to reduce radioactivity exposure to subjects (i.e., with MRI), or examine metabolic flux rates (i.e., with dynamic PET/CT imaging). Studies from animal models have suggested that the primary fuel for BAT thermogenesis is lipid, thus a glucose tracer misses much of the information. Tracers to measure fatty acid uptake (Blondin et al. 2017) have been used, and labeled water (Muzik et al. 2013) can also be considered to examine adipose tissue perfusion. MRI has been used both on its own or in combination with PET, in order to detect BAT. MRI can be used in many ways in order to examine features of living tissue, and most commonly fat fraction has been employed, in order to distinguish WAT from BAT. Limitations include the possible presence of beige adipose tissue, long scanning duration in which physiological changes in BAT activation may impact the fat fraction and very few validation studies to date (Rasmussen et al. 2013). Dynamic PET imaging holds promise but awaits further validation. In summary, all noninvasive imaging methods have positive and negative aspects, none of which outperforming all others in accuracy.

Immunohistochemical or Gene-Expression Confirmation

Ideally, evidence from all studies should be complemented with immunohistochemical evaluation of BAT, WAT, or beige adipose tissue markers from biopsy (Cypess et al. 2013). Due to the technical complexity and discomfort of obtaining adipose tissue biopsies from deep locations, these studies are quite limited. Additionally, not all studies that have performed adipose tissue biopsies in the abdomen have found expression of thermogenic genes or presence of browning in WAT, even after a cold exposure intervention, which is the best stimulus to activate BAT.

Future Investigations into Human Brown Adipose Tissue Physiology

Since 2009 researchers have been looking for alternative strategies to activate human BAT without exposing participants to cold, which is not very comfortable (Fig. 5.1).

Dietary Components

Capsinoids are substances naturally present in chili peppers and they are particularly abundant in *C.annuum L.* or "CH-19 Sweet" (non-pungent red pepper). Capsinoids include capsiate, dihydrocapsiate, and nordihydrocapsiate. Although capsinoids are structurally similar to capsaicin, they are 1000 times less pungent but are as potent as capsaicin in increasing thermogenesis. The thermogenic activation pathways of capsinoids include transient receptor potential cation channel (TRPV1), which have possible mechanisms of action on BAT.

Tea catechins are polyphenolic components present in green tea. The most abundant and bioactive component is epigallocatechin gallate. The thermogenic effect of tea catechins has repeatedly been shown in humans. The mechanisms of action are similar to capsinoids, via TRP channels. However, there is no solid evidence that capsinoids and tea catechins can activate and recruit BAT in humans. Therefore, studies are needed to elucidate the role of dietary components as possible BAT activators (Osuna-Prieto et al. 2019).

Pharmaceutical Compounds

Mirabegron, an antimuscarinic drug, is a β3-adrenergic receptor agonist normally used to treat overactive bladder in adults. Cypess et al. showed that the acute ingestion of 50 or 200 mg of mirabegron was able to increase the ^{18}F-FDG uptake by BAT in young healthy adults (Baskin et al. 2018). Chronic administration of mirabegron was also able to increase the levels of BAT ^{18}F-FDG uptake. In an alternative study, we have observed that a single dose of 200 mg of Mirabegron was not able to induce a decrease of the fat fraction of BAT measured by magnetic resonance imaging. Moreover, we observed that mirabegron did not have any effect of the lipidomic profile in humans. However, there is still a debate ongoing about whether mirabegron is directly activating human BAT. Recently, Blondin et al. demonstrated that human brown adipocytes lacked β3-receptors, suggesting that mirabegron has specificity for others β-receptors (Blondin et al. 2019).

The antidiabetic agent *Sitagliptin* is a dipeptidyl peptidase-4 (DPP-4) inhibitor, which increases the production of insulin and decreases the production of glucagon by the pancreas. Recently we aimed to study the long-term effect (12 weeks) of 100 mg/day of sitagliptin on the levels of BAT in overweight individuals. Although we observed that insulin resistance of the participants was attenuated, BAT ^{18}F-FDG uptake was not affected by sitagliptin (Nahon et al. 2018). Curiously, sitagliptin induced an increase in subcutaneous WAT in the abdomen, suggesting a possible increase in browning markers. We also demonstrated that sitagliptin induced upregulation of the mitochondrial gene *PGC1B* in skeletal muscle.

Gut Microbiota as a New Target to Combat T2D

Case–control studies have shown differences in gut microbiota composition, diversity and function between healthy people and obese individuals, and have suggested that modifying the gut microbiota composition of obese individuals toward a healthy phenotype might be a useful therapy to combat T2D. In that respect, Plovier et al. showed that fecal microbiota transplantation (FMT) from healthy donor mice to obese recipient mice reduces adiposity and energy intake, while improving lipoprotein metabolism and glucose tolerance (Plovier et al. 2017). The bacteria *Akkermansia municiphila* appeared to be the cause of these beneficial effects, and it was successfully used and tested in a human trial. They also showed that modifying the gut microbiota composition had a strong impact on the parameters of cardiometabolic health in humans.

Gut microbiota regularly produces short-chain fatty acids (SCFA) (e.g., butyrate or acetate), which can play an important role in the regulation of metabolic health, mitigating T2D. Butyrate also acts on the gut–brain neural circuit to improve energy metabolism, by activating brown adipose tissue (BAT) in APOE*3-Leiden. CETP male mice (Li et al. 2018). Further studies are needed to confirm this hypothesis in a human setting.

Future Perspectives

- BAT is present in obese and T2D individuals but its role is unknown.
- Cooling protocols for BAT activation may not be longer than 1 h based on the new evidence, which shows that BAT could be only activated during the first 35 min of cold exposure.
- BAT depots are not easily biopsied. However, several studies suggest that humans could have a subcutaneous BAT in the dorsocervical area, which is an accessible depot. The confirmation of this finding could mean a step forward in the field.
- Nuclear medicine techniques should be further developed in order to better quantify the thermogenic activity of BAT and WAT during dynamic processes. This new technique could be applied to cold exposure, but also to acute responses such as meal-intake, drugs, or even exercise interventions.
- Supraclavicular skin temperature should be interpreted as the result of several thermogenic tissues and blood vessels that are activated in response to the stimulus, not only BAT activation. The fact that supraclavicular skin temperature did not decrease after a cold exposure is disturbing, and further studies are needed that help to understand which are the main tissues that could be explaining this lack of a decrease.
- There is a lack of translational models' studies, promoting the transition of findings from preclinical to human studies.

Acknowledgments This chapter was made possible by support from the NIH Medical Scientist Training Program Training Grant T32GM007205 and a postdoctoral grant from the Fundación Alfonso Martin Escudero.

References

Acosta FM et al (2019) Association of objectively measured physical activity with brown adipose tissue volume and activity in young adults. J Clin Endocrinol Metab 104:223–233

Association, A. D. 2 (2019) Classification and diagnosis of diabetes: standards of medical care in diabetes—2019. Diabetes Care 42:S13–S28

Baskin AS et al (2018) Regulation of human adipose tissue activation, gallbladder size, and bile acid metabolism by a β3-adrenergic receptor agonist. Diabetes db180462. https://doi.org/10.2337/db18-0462

Blondin DP et al (2015) Selective impairment of glucose but not fatty acid or oxidative metabolism in brown adipose tissue of subjects with type 2 diabetes. Diabetes 64:2388–2397

Blondin DP et al (2017) Inhibition of intracellular triglyceride lipolysis suppresses cold-induced brown adipose tissue metabolism and increases shivering in humans. Cell Metab 25:438–447

Blondin DP et al (2019) Human BAT thermogenesis is stimulated by the β $_2$-adrenergic receptor. https://papers.ssrn.com/abstract=3440265

Blüher M (2010) The distinction of metabolically 'healthy' from 'unhealthy' obese individuals. Curr Opin Lipidol 21:38

Brychta RJ et al (2019) Quantification of the capacity for cold-induced thermogenesis in young men with and without obesity. J Clin Endocrinol Metab. https://doi.org/10.1210/jc.2019-00728

Cannon B, Nedergaard J (2004) Brown adipose tissue: function and physiological significance. Physiol Rev 84:277–359

Carpentier AC et al (2018) Brown adipose tissue energy metabolism in humans. Front Endocrinol 9:447

Cypess AM et al (2009) Identification and importance of brown adipose tissue in adult humans. N Engl J Med 360:1509–1517

Cypess AM et al (2013) Anatomical localization, gene expression profiling and functional characterization of adult human neck brown fat. Nat Med 19:635–639

Deshmukh AS et al (2019) Proteomics-based comparative mapping of the secretomes of human brown and white

adipocytes reveals EPDR1 as a novel batokine. Cell Metab 30:963–975.e7
Din UM et al (2016) Human brown adipose tissue [(15)O] O2 PET imaging in the presence and absence of cold stimulus. Eur J Nucl Med Mol Imaging 43:1878–1886
Fraum TJ et al (2019) Repeatability of quantitative brown adipose tissue imaging metrics on positron emission tomography with 18F-fluorodeoxyglucose in humans. Cell Metab 30:212–224.e4
Grundy SM (2015) Adipose tissue and metabolic syndrome: too much, too little or neither. Eur J Clin Invest 45:1209–1217
Hanssen MJW et al (2015) Short-term cold acclimation improves insulin sensitivity in patients with type 2 diabetes mellitus. Nat Med 21:863–865
Hanssen MJW et al (2016) Short-term cold acclimation recruits brown adipose tissue in obese humans. Diabetes 65:1179–1189
Heaton JM (1972) The distribution of brown adipose tissue in the human. J Anat 112:35–39
Jensen MD (2015) Brown adipose tissue – not as hot as we thought. J Physiol 593:489–490
Jespersen NZ et al (2013) A classical brown adipose tissue mRNA signature partly overlaps with brite in the supraclavicular region of adult humans. Cell Metab 17:798–805
Jimenez-Pavon D et al (2019) Infrared thermography for estimating supraclavicular skin temperature and BAT activity in humans: a systematic review. Obes Silver Spring Md 27:1932–1949
Kellman GM et al (1987) MR imaging of the supraclavicular region: normal anatomy. AJR Am J Roentgenol 148:77–82
Kim K et al (2019) Whole body and regional quantification of active human brown adipose tissue using 18F-FDG PET/CT. JoVE J Vis Exp e58469. https://doi.org/10.3791/58469
Lee P et al (2014) Temperature-acclimated brown adipose tissue modulates insulin sensitivity in humans. Diabetes 63:3686–3698
Leitner BP et al (2017) Mapping of human brown adipose tissue in lean and obese young men. Proc Natl Acad Sci 114:8649–8654
Leitner BP et al (2018) Kinetics of human brown adipose tissue activation and deactivation. Int J Obes 1. https://doi.org/10.1038/s41366-018-0104-3
Li Z et al (2018) Butyrate reduces appetite and activates brown adipose tissue via the gut-brain neural circuit. Gut 67:1269–1279
Lima JG et al (2016) Clinical and laboratory data of a large series of patients with congenital generalized lipodystrophy. Diabetol Metab Syndr 8:23
Marlatt KL, Chen KY, Ravussin E (2018) Is activation of human brown adipose tissue a viable target for weight management? Am J Physiol-Regul Integr Comp Physiol 315:R479–R483
Martinez-Tellez B (2018) Role of brown adipose tissue on the thermoregulatory system and physical fitness: the Actibate study. Universidad de Granada
Martinez-Tellez B et al (2019a) Evidence of high 18F-fluorodeoxyglucose uptake in the subcutaneous adipose tissue of the dorsocervical area in young adults. Exp Physiol 104:168–173
Martinez-Tellez B, Sanchez-Delgado G, Amaro-Gahete FJ, Acosta FM, Ruiz JR (2019b) Relationships between cardiorespiratory fitness/muscular strength and 18 F-fluorodeoxyglucose uptake in brown adipose tissue after exposure to cold in young, sedentary adults. Sci Rep 9:1–9
Martinez-Tellez B et al (2019c) Concurrent validity of supraclavicular skin temperature measured with iButtons and infrared thermography as a surrogate marker of brown adipose tissue. J Therm Biol 82: 186–196
Martinez-Tellez B et al (2019d) Supraclavicular skin temperature measured by iButtons and 18F-fluorodeoxyglucose uptake by brown adipose tissue in adults. J Therm Biol 82:178–185
McKie GL et al (2019) Housing temperature affects the acute and chronic metabolic adaptations to exercise in mice. J Physiol 597:4581–4600
Min K-B, Min J-Y (2015) Android and gynoid fat percentages and serum lipid levels in United States adults. Clin Endocrinol (Oxf) 82:377–387
Muzik O et al (2013) 15O PET measurement of blood flow and oxygen consumption in cold-activated human brown fat. J Nucl Med 54:523–531
Nahon KJ et al (2018) Effect of sitagliptin on energy metabolism and brown adipose tissue in overweight individuals with prediabetes: a randomised placebo-controlled trial. Diabetologia. https://doi.org/10.1007/s00125-018-4716-x
Nathan DM (2015) Diabetes: advances in diagnosis and treatment. JAMA 314:1052–1062
NCD Risk Factor Collaboration (2016) Trends in adult body-mass index in 200 countries from 1975 to 2014: a pooled analysis of 1698 population-based measurement studies with 19·2 million participants. The Lancet 387:1377–1396
Oral EA et al (2017) Inhibition of IKKε and TBK1 improves glucose control in a subset of patients with type 2 diabetes. Cell Metab 26:157–170.e7
Oreskovich SM et al (2019) MRI reveals human brown adipose tissue is rapidly activated in response to cold. J Endocr Soc 3:2374–2384
Osuna-Prieto FJ et al (2019) Activation of human brown adipose tissue by capsinoids, catechins, ephedrine, and other dietary components: a systematic review. Adv Nutr Bethesda Md 10:291–302
Perry RJ et al (2018) Leptin mediates a glucose-fatty acid cycle to maintain glucose homeostasis in starvation. Cell 172:234–248.e17
Petersen MC, Shulman GI (2018) Mechanisms of insulin action and insulin resistance. Physiol Rev 98:2133–2223
Plovier H et al (2017) A purified membrane protein from Akkermansia muciniphila or the pasteurized bacterium improves metabolism in obese and diabetic mice. Nat. Med. 23:107–113

Rasmussen JM et al (2013) Brown adipose tissue quantification in human neonates using water-fat separated MRI. PLoS One 8:e77907

Roden M, Shulman GI (2019) The integrative biology of type 2 diabetes. Nature 576:51–60

Ruiz JR et al (2018) Role of human brown fat in obesity, metabolism and cardiovascular disease: strategies to turn up the heat. Prog Cardiovasc Dis. https://doi.org/10.1016/j.pcad.2018.07.002

Sánchez-Delgado G (2018) Brown adipose tissue and exercise: implications on human energy balance and metabolism. The Actibate study. Universidad de Granada

Sanchez-Delgado G et al (2020) Brown adipose tissue volume and 18F-fluorodeoxyglucose uptake are not associated with energy intake in young human adults. Am J Clin Nutr. https://doi.org/10.1093/ajcn/nqz300

Sasai H et al (2015) Does visceral fat estimated by dual-energy X-ray absorptiometry independently predict cardiometabolic risks in adults? J Diabetes Sci Technol 9:917–924

Shah NR, Braverman ER (2012) Measuring adiposity in patients: the utility of body mass index (BMI), percent body fat, and leptin. PLoS One 7:e33308

Stanford KI, Middelbeek RJW, Goodyear LJ (2015) Exercise effects on white adipose tissue: Beiging and metabolic adaptations. Diabetes 64:2361–2368

van Marken Lichtenbelt WD et al (2009) Cold-activated brown adipose tissue in healthy men. N Engl J Med 360:1500–1508

Villarroya F, Cereijo R, Villarroya J, Giralt M (2017) Brown adipose tissue as a secretory organ. Nat Rev Endocrinol 13:26–35

Wu J et al (2012) Beige adipocytes are a distinct type of thermogenic fat cell in mouse and human. Cell 150: 366–376

Zhang Y et al (1994) Positional cloning of the mouse obese gene and its human homologue. Nature 372:425

Zwick RK, Guerrero-Juarez CF, Horsley V, Plikus MV (2018) Anatomical, physiological, and functional diversity of adipose tissue. Cell Metab 27:68–83

microRNAs in Obesity and Metabolic Diseases

6

Giuseppe Iacomino, Fabio Lauria, Antonella Venezia, Nunzia Iannaccone, Paola Russo, and Alfonso Siani

Contents

Obesity: The Worldwide Epidemic 73

White Adipose Tissue: An Overview 73
Adipose Tissue Proliferation 73

miRNAs: Small RNAs with a Big Role in Gene Regulation 74
Epigenetic Control Mechanisms 74
miRNAs in Multicellular Organisms 74

miRNA Biogenesis and Action Mechanisms 75

Detection of miRNAs 77
miRNAs Target Prediction 77

miRNA in Obesity and Obesity-Associated Comorbidities 79

Role of miRNAs in Adipose Tissue 79
Adipogenesis Pathways 81
Let-7 miRNA and Diabetes Prevention 81
Other Tissue-Specific miRNAs and Metabolic Diseases 81
Adipose Tissue Cross Talk and miRNAs 82
Intercellular Message Signaling 83

Role of miRNAs in the Pancreas 83
Additional Pancreatic Activities of miRNAs 84

This chapter is an update from the previously published article by Iacomino G, Siani A. Role of microRNAs in obesity and obesity-related diseases. Genes Nutr 12, 23 (2017). https://doi.org/10.1186/s12263-017-0577-z.

G. Iacomino (✉) · F. Lauria · N. Iannaccone · P. Russo · A. Siani
Institute of Food Sciences, National Research Council, Avellino, Italy
e-mail: piacomino@isa.cnr.it

A. Venezia
LILT – Istituto Nazionale Tumori IRCCS Fondazione G. Pascale, Napoli, Italy

J. Faintuch and S. Faintuch (eds.), *Obesity and Diabetes*, https://doi.org/10.1007/978-3-030-53370-0_6

Role of miRNAs in the Muscle 84
Insulin Resistance 84
Myocardial Metabolism 85

Role of miRNAs in the Liver 85
Additional miRNAs 86

Circulating miRNAs 86
Anti Nuclease Protection 86
Circulating miRNA in Obesity, Diabetes and Associated Conditions 86
Blood Glucose, Blood Pressure and Metabolic Syndrome 87
Bariatric Surgery 87
Pediatric Protocols 87
Translational Considerations 88

References 88

Abstract

microRNAs (miRNAs) have emerged as key regulators of metabolic processes, playing critical roles in maintaining energy balance and metabolic homeostasis. Dysregulation of miRNAs may contribute to metabolic abnormalities since they play pivotal roles in body fat accumulation, obesity, and obesity-related diseases by directly affecting the status and functions of adipose tissue, pancreas, liver, and muscle. The discovery of circulating miRNAs highlighted their potential as either endocrine signaling molecules or disease indicators. This chapter aims to provide recent findings concerning the contribution of miRNAs to obesity and obesity-related morbid conditions.

Keywords

Obesity, metabolic disease · Disease biomarkers · microRNA, miRNA

List of Abbreviations

ABCA1	ATP-binding cassette transporter 1
AGO	Argonaute
Apo	Apolipoprotein
BMI	Body mass index
BMPs	Bone morphogenetic proteins
C/EBPs	CCAAT/enhancer-binding proteins
c-miRNAs	Circulating miRNAs
Creb1	cAMP-Responsive element binding protein 1
ERK	Extracellular signal-regulated kinases
FABP4	Fatty acid-binding protein 4
FASN	Fatty acid synthase
FGF21	Fibroblast growth factor 21
Foxa2	Forkhead box protein A2
FOXO1	Forkhead box O1
HbA1c	Glycated hemoglobin A1c
HNF	Hepatocyte nuclear factor
HOMA-IR score	Homeostatic model assessment for insulin resistance
INSR	Insulin receptor
IRS-1	Insulin receptor substrate 1
LETFs	Liver-enriched transcription factors
LXRα	Liver X receptor α
miRNAs	microRNAs
ncRNAs	Small noncoding RNAs
N-CoR	Nuclear receptor corepressor
Ngn3	Neurogenin3
NGS	Next-generation sequencing
NOD	Nonobese diabetic mice
NPM-1	Nucleophosmin-1
Pdx1	Pancreatic and duodenal homeobox 1
PEPCK	Phosphoenolpyruvate carboxykinase
PI3K	Insulin/phosphoinositide-3 kinase
PPARγ	Proliferator-activated receptor-γ
RB	Retinoblastoma susceptibility protein
RISCs	RNA-induced silencing complexes

RNAseq	RNA sequencing
RXRα	Retinoid X receptor alpha
SFA	Saturated fatty acids
SHIP1	SH2 (Src homology 2)-containing inositol phosphatase-1
SIRT1	Sirtuin 1
SMRT	Silencing mediator for retinoid and thyroid hormone receptors
SREBP1	Sterol regulatory element-binding protein
T2D	Type-2 diabetes
TGF-β	Transforming growth factor beta

Obesity: The Worldwide Epidemic

Up to four million deaths worldwide can be connected to overweight and obesity, and more than two-thirds of these deaths are due to cardiovascular diseases (Hurt et al. 2010; Branca et al. 2007; D'Adamo and Caprio 2011; Ahima and Lazar 2013; Bluher 2013). The obesity epidemic and the resulting consequences threaten both developed and developing countries, although a plateau in prevalence rates has been reported for childhood obesity in several Western countries (WHO Library Cataloguing 2016). Besides the secular changes in energy intake and expenditure invoked as the underlying cause, a number of loci on the human genome have been consistently associated with obesity (Berndt et al. 2013; Kunej et al. 2013; Loos and Yeo 2014; Fall and Ingelsson 2014). Currently, it is well accepted that "common" obesity is the result of a mixed effect of environmental risk factors and multiple gene variants, additively conferring a degree of susceptibility (polygenic obesity) (Stranger et al. 2011). Actually, the individual susceptibility to weight gain may vary significantly due to the interaction between genetic background, lifestyle, and environmental factors.

Most of the genes associated with obesity predisposition are also correlated with food intake and energy balance control (O'Rahilly and Farooqi 2008; Rankinen and Bouchard 2006; Locke et al. 2015). Nowadays, despite extensive study of obesity genetics, most of the genetic variability in BMI remains unexplained and the effect of a single candidate gene is quite limited.

White Adipose Tissue: An Overview

White adipose tissue is the main tissue for energy storage in humans. Besides the storage function, this tissue is metabolically active by releasing hundreds of different factors, including hormones as leptin and adiponectin, growth factors such as IGF-1 and PDGF, and cytokines such as IL-6, IL-8 or TNF-α acting as inflammatory mediators (Booth et al. 2016), possibly involved in the regulation of appetite and insulin sensitivity. Growing evidence supports the notion that chronic low-grade inflammation is a basic feature of obesity that contributes to insulin resistance in target organs, including adipose tissue, liver, muscle, and the vascular system (Jung and Choi 2014).

Nutrient excess, particularly fats and carbohydrates may concurrently trigger inflammatory responses, which further interfere with metabolic function, stress, and inflammation. Therefore, the nutrition-immunity theory suggests that obesity induced by overnutrition triggers low-level inflammatory processes into adipose tissue (Wellen and Hotamisligil 2005; Kammoun et al. 2014). This state is related to the large macrophage recruitment and enhanced immune cell proliferation/activation/infiltration associated with adipocyte hypertrophy and impaired adipogenesis (Kraakman et al. 2014). The latter process is closely controlled by a combination of regulatory signals including extracellular circulating hormones, endocellular transcription factors, and post-transcriptional modulators of gene expression (Arner and Kulyte 2015; Kim et al. 2014; Li et al. 2015; Rottiers and Naar 2012; Xie et al. 2009; Tang and Lane 2012).

Adipose Tissue Proliferation

Adipogenesis is a process in which fibroblast-like pre-adipocytes are differentiated into mature adipocytes, a complex cascade that embraces cell commitment, clonal expansion, and terminal differentiation (Ali et al. 2013). More in-depth, the differentiation from pre-adipocytes to mature adipocytes is a step coordinated by several transcription factors such as the peroxisome proliferator-activated receptor-γ (PPARγ) and CCAAT/enhancer-binding proteins (C/EBPs), which direct the regulation of gene expression required for adipocyte phenotypes. C/EBPβ and C/EBPδ are induced by adipogenic stimuli and are the main adipogenesis regulators. C/EBPβ and C/EBPδ targets are the promoters of genes encoding essential adipogenic factors such as C/EBPα and PPARγ, as well as the sterol regulatory element-binding protein (SREBP1), the primary lipogenic gene regulator. The C/EBPα intronless gene encodes a leucine zipper transcription factor, that can bind as a homodimer to specific promoters and enhancers, or it can form heterodimers with the related C/EBPβ and C/EBPγ. As an example, C/EBPα binds to the promoter and modifies the leptin gene expression, providing a key role in body weight homeostasis.

C/EBPα is sufficient to cause differentiation of pre-adipocytes into mature adipocytes (Linhart et al. 2001). Peculiarly, PPARγ directly triggers endogenous C/EBPα transcription. Besides, through a virtuous circle, C/EBPα activates the PPARγ gene, thus supporting adipogenesis (Tang and Lane 2012). PPARγ and C/EBPα concurrently support the expression of genes implicated in insulin sensitivity, lipogenesis and lipolysis, and finally in the terminal differentiation of mature adipocytes. The dynamic series of events regulating cell commitment and adipocyte differentiation also includes an anti-adipogenic signaling cascade mediated by Wnt, BMPs, TGF-β, and hedgehog (Moseti et al. 2016). As expected, an increase in adipocyte volume (hypertrophy) and number (hyperplasia) results in higher fat mass and energy storage levels in adipose tissue, likely contributing to an increased risk of obesity.

miRNAs: Small RNAs with a Big Role in Gene Regulation

At present, hundreds of post-translational modifications of core histones including acetylation, methylation, phosphorylation, and ubiquitylation have been discovered. A wide number of dynamic regulatory systems are responsible for histone modifications and chromatin remodeling mechanisms, contributing to a complex epigenetic regulation network of gene activity (Iacomino et al. 2001, 2006; Venkatesh and Workman 2015). Essentially, any step in the gene expression flow is wisely controlled, and the discovery of small noncoding RNAs (ncRNAs) has added new players to the broad range of existing mechanisms (Catalanotto et al. 2016).

Epigenetic Control Mechanisms

Epigenetics is usually defined as heritable modifications in a gene that occur in the DNA without sequence modification. This definition encloses the concept that epigenetics is a heritable trait (Ling and Ronn 2019). Nevertheless, the study of epigenetics contribution to obesity/altered metabolic conditions is still in early infancy.

Based on their working mechanism, synthesis, and structure ncRNAs are classified as small interfering RNAs, PIWI-interacting RNAs, endogenous small interfering RNAs, promoter-associated RNAs, small nucleolar RNAs, and microRNAs (miRNAs). In a few years, miRNA study has progressed from a single article reporting the discovery of a noncoding RNA in *C. elegans* (Lee et al. 1993), to thousands of publications describing their critical association to a variety of physiological processes and diseases (Rupaimoole and Slack 2017).

miRNAs in Multicellular Organisms

miRNAs are small ncRNAs with a length of 20–24 nucleotides involved in gene expression regulation (Bartel 2009; Ghildiyal and Zamore 2009; Ha and Kim 2014). Today more than 2000 different miRNAs have been identified, showing them to be one of the most abundant classes of gene regulatory molecules in multicellular organisms, and their number is still increasing in the miRBase database (Kozomara and Griffiths-Jones 2014). Release 22 of the repository contains 38,589 entries representing hairpin precursor miRNAs, expressing 48,885 mature miRNA products, in 271 species. miRNAs are essential elements of the cellular epigenetic machinery and behave as specific gene silencers by base pairing to 3′ untranslated sequence of a target mRNA. However, they have also been evidenced to bind anyplace along the nucleotide sequence. miRNAs exert their actions by either inhibiting translation or by affecting mRNA stability and degradation (Ebert and Sharp 2012).

The bases in position 2–8 of a miRNA are essential for base-pairing with a target mRNA and are referred to as "seed sequence." New advances in a transcriptome-wide method of mapping miRNA binding sites revealed that a large proportion of miRNA–targets during in vivo interactions are mediated not only through the canonical "seed sites," but also via noncanonical seed-like motifs (Seok et al. 2016). miRNAs have been clustered into families built on seed region similarity, which is responsible for the ability to target common clusters of transcripts (Wang et al. 2016). Some miRNAs show a tissue-specific localization, but most miRNAs have a wide-ranging tissue distribution (Li et al. 2013).

Transcriptome Sculpting

Interestingly, a single miRNA can regulate at the same time large cohorts of transcripts, and a distinct mRNA usually includes multiple interaction sites for diverse miRNAs leading to intricate regulatory circuits (Ebert and Sharp 2012; Krek et al. 2005). A single miRNA usually has a weak modulatory effect on a distinct target, but often its action coordinately affects multiple transcripts in a signaling pathway, so exercising significant cumulative effects in complex regulatory networks. In this view, the metazoan miRNAs have been defined as the "sculptors" of the transcriptome (Bartel 2018).

Scientific literature supports the view that endogenous miRNAs influence the expression of up to 60% of mouse and human genes (Friedman et al. 2009; Lewis et al. 2005). Accordingly, miRNAs have been described to be involved in an extraordinarily large array of physiologic processes, both in health and in disease conditions (van Rooij 2011; Bartel 2004; Iacomino et al. 2020). While some miRNAs are expressed ubiquitously in the body, other miRNAs are specifically enriched in certain tissues. A compilation of knockout phenotypes showing crucial biological functions of miRNAs has been recently published (Bartel 2018). Finally, expanding evidence suggests that miRNAs dysregulation acts as an etiologic factor and/or indicator of numerous diseases including cancer (Mendell and Olson 2012; Paul et al. 2018; Bracken et al. 2016; Li et al. 2014; Pedroza-Torres et al. 2019).

miRNA Biogenesis and Action Mechanisms

In recent years, the miRNAs biogenesis has been comprehensively described (Fig. 6.1) (Finnegan and Pasquinelli 2013). miRNA coding genes are evolutionary conserved and are located in the introns or exons of protein-coding genes (about 40% of mammalian miRNAs are found within introns of protein-coding genes), as well as in intergenic regions (Kim et al. 2009a). In the cell nucleus, pri-miRNAs are transcribed using the genomic DNA as a template by RNA polymerase II together with the host gene (intragenic miRNAs), or independently of the host gene with the use of their promoter (intergenic miRNAs), as large primary transcripts long several hundred bases (Gulyaeva and Kushlinskiy 2016). miRNAs transcription, as well as for protein-coding genes, is regulated by transcription factors (Kim et al. 2009a).

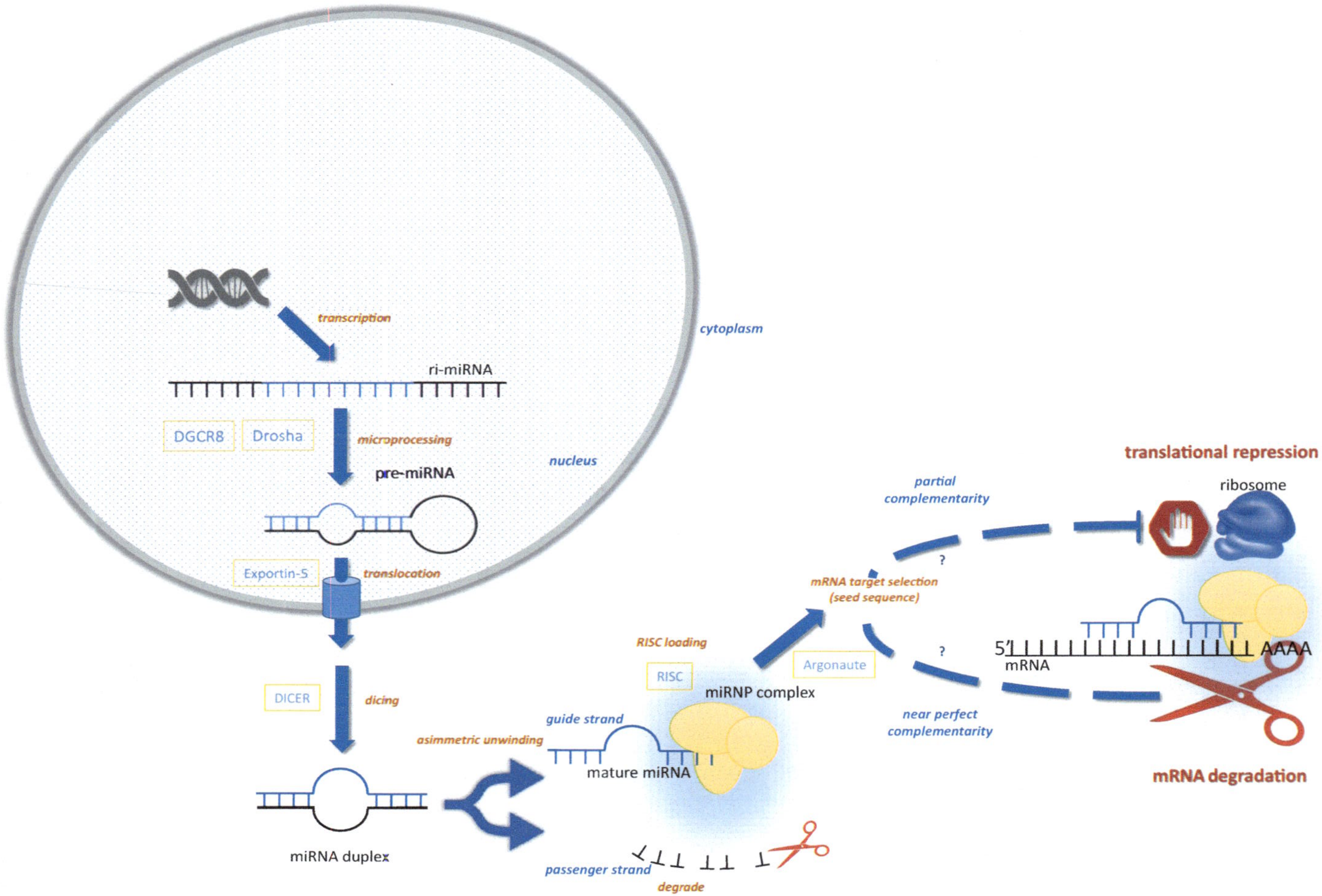

Fig. 6.1 Single miRNAs are transcribed from genomic DNA by RNA polymerase II (pri-miRNA). Drosha-DGCR8 processes the pri-miRNA to pre-miRNA. Exportin-5 moves the pre-miRNA to the cytoplasm where it is recognized and cleaved to create a miRNA duplex by the DICER complex. The duplex unwinds and the mature miRNA assembles into RISC. The miRNA guide strand base-pairs with target mRNA and determines gene silencing via mRNA cleavage or translation repression depending on the degree of homology between the miRNA "seed" to the 3′ UTR target sequence of the mRNA

miRNAs are often encoded by multiple loci, some of which are structured in co-transcribed clusters. The pri-miRNA exhibits an exclusive structure which is represented by a hairpin and three spiral turns, flanked by a single-stranded RNA. This arrangement can be recognized by an enzymatic complex containing the RNA-binding cofactor DiGeorge Syndrome Critical Region 8 (DGCR8, Pasha) and the RNase III Drosha (a double-strand endonuclease). The resulting processing produces a 60–70 nucleotide long precursor miRNA (pre-miRNA) that, interacting with Exportin-5 and Ran GTPase, is exported to the cytoplasm, where it is subsequently processed by the enzyme Dicer RNase III that cuts pre-miRNA into 19–24 nucleotide long mature miRNA duplex (Okada et al. 2009).

The duplex is then unwound and only one strand produces the mature sequence that is integrated into the RISCs' (RNA-induced silencing complexes) ribonuclear complexes containing the Argonaute (AGO) family proteins, that target mRNAs by complementary base pairing, resulting in a posttranscription gene silencing either by translation repression or degradation of the target transcript (Xiong et al. 2015). When the miRNA and the target mRNA are fully complementary Ago2 activity cuts the target RNA between nucleotides complementary to positions 10 and 11 of the guide strand (Pratt and MacRae 2009). Otherwise, as with almost all miRNA–target interactions, RISC promotes a cascade of inhibitory steps that lead the targets to canonical degradation pathways (Filipowicz et al. 2008).

For some exonic miRNAs, such as miR-22, miR-155, and miR-146a, both spliced and unspliced transcripts may act as primary miRNA transcripts. Since these are partly localized in the cytoplasm they are not completely accessible for handling. The integration of splicing and transport to the cytoplasm may denote an extra mechanism for the modulation and functioning of exonic miRNA (Pawlicki and Steitz 2008). Genetic and somatic mutations in miRNA motifs can disrupt the interactions between miRNAs and their targets. However, mutations can also make the miRNA able to interact with a set of new targets, rewiring the miRNA regulatory networks.

Detection of miRNAs

A variety of approaches have been developed to determine miRNAs in biological samples using different technologies including Next-Generation Sequencing (NGS), microarray, and reverse transcription-quantitative PCR (Mestdagh et al. 2014). Characterization of miRNAs is generally more difficult to perform compared to mRNA profiling, because techniques should be able to discriminate miRNAs that differ by as little as one nucleotide, also taking into account variations between mature miRNAs and their precursors. Likewise, the profiling of circulating miRNAs may be complex due to low concentration, undesirable inhibitors that may interfere with downstream quantification procedures (i.e., hemoglobin in blood-derived samples) and, eventually, cellular components which may lead to an additional bias. Consequently, the inconsistencies identified among different studies maybe at least partly explained by different extraction and detection procedures, experimental setup, and finally, data analysis, emphasizing the need for well-standardized procedures (El-Khoury et al. 2016).

miRNAs Target Prediction

Due to multiple target recognition, the identification of a miRNA interaction with a target mRNA remains a challenge (Riffo-Campos et al. 2016; Seitz 2017). This difficulty is usually circumvented through bioinformatics tools, aimed at a subsequent bench validation of predicted interactors. Computational methods identify miRNA targets, typically selecting binding sites that have been retained evolutionarily. A partial list of the in silico prediction tools for miRNA targets identification is reported in Table 6.1. Existing approaches, however, often uncover difficulties in achieving effective

Table 6.1 Relevant "web-based" tools for miRNAs analysis

Tool	URL	Description
Diana Tools	http://diana.imis.athena-innovation.gr/DianaTools/index	Databases and software for interpreting and archiving data in a systematic framework ranging from the analysis of expression regulation from deep sequencing data, the annotation of miRNA regulatory elements and targets to the interpretation of the role of ncRNAs in various diseases and pathways.
MirGeneDB	https://mirgenedb.org/	A catalog of miRNA genes that have been validated and annotated.
miRTar	http://mirtar.mbc.nctu.edu.tw/human/	A complete toolbox for identifying the biological function and regulatory relationships between a group of known or putative miRNA and their targets. It also provides an integrative pathway analysis.
miRandola	http://mirandola.iit.cnr.it/index.php	A comprehensive manually curated classification of different extracellular circulating noncoding RNA types.
miRanda	http://www.microrna.org/microrna/getGeneForm.do	A comprehensive resource of microRNA target predictions and expression profiles.
miRò2	http://microrna.osumc.edu/miro/	A computational knowledge base for the inference of miRNA associations.
miRSystem	http://mirsystem.cgm.ntu.edu.tw/	A database that integrates well-known miRNA target gene prediction programs: DIANA, miRanda, miRBridge, PicTar, PITA, rna22, and TargetScan.
HMDD	http://www.cuilab.cn/hmdd	A human miRNAs disease database which contains miRNA names, disease names, dysfunction evidences, and relative literature.
PROGmiR	http://xvm145.jefferson.edu/progmir/	A prognostic database for cancers based on their miRNA expression signature.
VIRmiRNA	http://crdd.osdd.net/servers/virmirna/	A viral RNA database including functional targets.
miRIAD	https://www.bioinfo.mochsl.org.br/miriad/	A web search tool designed to help users to access integrated information concerning intragenic miRNAs and their host genes.
Exocarta	http://www.exocarta.org/	ExoCarta, an exosome database, provides with the contents that were identified in exosomes in multiple organisms.
microPIR	http://www4a.biotec.or.th/micropir/	An integrated database of miRNA targets within human promoters.
miRNAtools3	http://mirnatools.eu/index.html	A catalog of utilities for miRNA analysis.
miRDeep	http://www.australianprostatecentre.org/research/software/mirdeep-star	An integrated standalone application useful for the prediction of miRNAs from small RNAseq data.
Ingenuity® Pathway	https://www.qiagenbioinformatics.com/products/ingenuity-pathway-analysis/	An integrated web server for identifying miRNA–target interactions. Commercial software.
miR2GO	http://compbio.uthsc.edu/miR2GO/home.php	Comparative functional analysis for microRNAs.
miRTargetLink Human	https://ccb-web.cs.uni-saarland.de/mirtargetlink/	Detailed information on human microRNA–mRNA interactions in the form of interactive interaction networks.
mirDIP	http://ophid.utoronto.ca/mirDIP/searchSummary.jsp	Find miRNAs that target a gene, or genes targeted by a miRNA, in Homo sapiens.
miRmine	http://guanlab.ccmb.med.umich.edu/mirmine/index.html	Human miRNA expression database.
miTALOS2	http://mips.helmholtz-muenchen.de/mitalos/#/search	Localize miRNA targets within cell signaling pathways.

(continued)

Table 6.1 (continued)

Tool	URL	Description
miRGator	mirgator.kobic.re.kr/	miRNA portal encompassing miRNA diversity, expression profiles, target relationships, and various supporting tools.
miRDB	http://mirdb.org/miRDB/	An online database for miRNA target prediction and functional annotations.
PITA	https://genie.weizmann.ac.il/pubs/mir07/mir07_prediction.html	PITA test the UTR for potential miRNA targets.
TargetScan	http://www.targetscan.org/vert_71/	Predicts biological targets of miRNAs by searching for the presence seed region of each miRNA.
PhenomiR	http://mips.helmholtz-muenchen.de/phenomir/	Provides information about differentially regulated miRNA expression in diseases and several biological processes.
GeneTrail2	https://genetrail2.bioinf.uni-sb.de/start.html	Statistical analysis of molecular signatures. Enrichment analysis based on miRNA categories like miRDB or miRTarBase, or map miRNAs to gene targets.

performance given the high false-positive and false-negative rates of discovery (Liu et al. 2014).

miRNA in Obesity and Obesity-Associated Comorbidities

miRNAs are useful tools in clinical applications (Keller and Meese 2016), with the majority of reports focusing on "circulating" miRNAs as cancer theranostic biomarkers (Pratt and MacRae 2009; Filipowicz et al. 2008; Pawlicki and Steitz 2008; Mestdagh et al. 2014; El-Khoury et al. 2016; Riffo-Campos et al. 2016; Seitz 2017; Liu et al. 2014; Keller and Meese 2016; Calin and Croce 2006; Matsuzaki and Ochiya 2017; Torres et al. 2020). Yet, recent studies have demonstrated that changes in miRNA profiles of various tissues (e.g., pancreas, adipose tissue, and liver) are also related to obesity (Fig. 6.2) (Kunej et al. 2013) and several metabolic diseases (Jiang et al. 2009; Dumortier et al. 2013).

miRNAs are affected by nutrition and lifestyle factors (Slattery et al. 2017), and many microRNA families have been associated with dietetic interventions (Palmer et al. 2014). Lastly, dietary miRNAs have been supposed to survive gastrointestinal digestion (Liang et al. 2014). However, the role and ability of food-related miRNAs to interfere with cross-species mRNA remains puzzling (Pastrello et al. 2016).

Role of miRNAs in Adipose Tissue

The first evidence of a role for miRNAs in fat cells originated from a study of miR-14 function in Drosophila, negatively affecting the fat metabolism by targeting p38 and MAPK (Xu et al. 2003). Then, a wide-ranging spectrum of miRNAs activities in regulating glucose and lipid metabolism was recognized, by mediating adipocyte differentiation, regulating cell mass and insulin-signaling pathways in physiological and pathological conditions in humans (Arner and Kulyte 2015; Hartig et al. 2015).

There are many elements involved in regulating the multiple steps of adipogenesis and emerging evidence support miRNAs' contribution to adipose tissue development and pathophysiology (Yi et al. 2020). Nowadays, adequate information regarding miRNAs acting mechanisms remains still limited (Zhang et al. 2013a). As a working example, it has been shown that the miRNAs reported in Fig. 6.3a interfere with adipocytes differentiation, while others promote adipogenesis (Fig. 6.3, panel b) (Price and Fernandez-Hernando 2016).

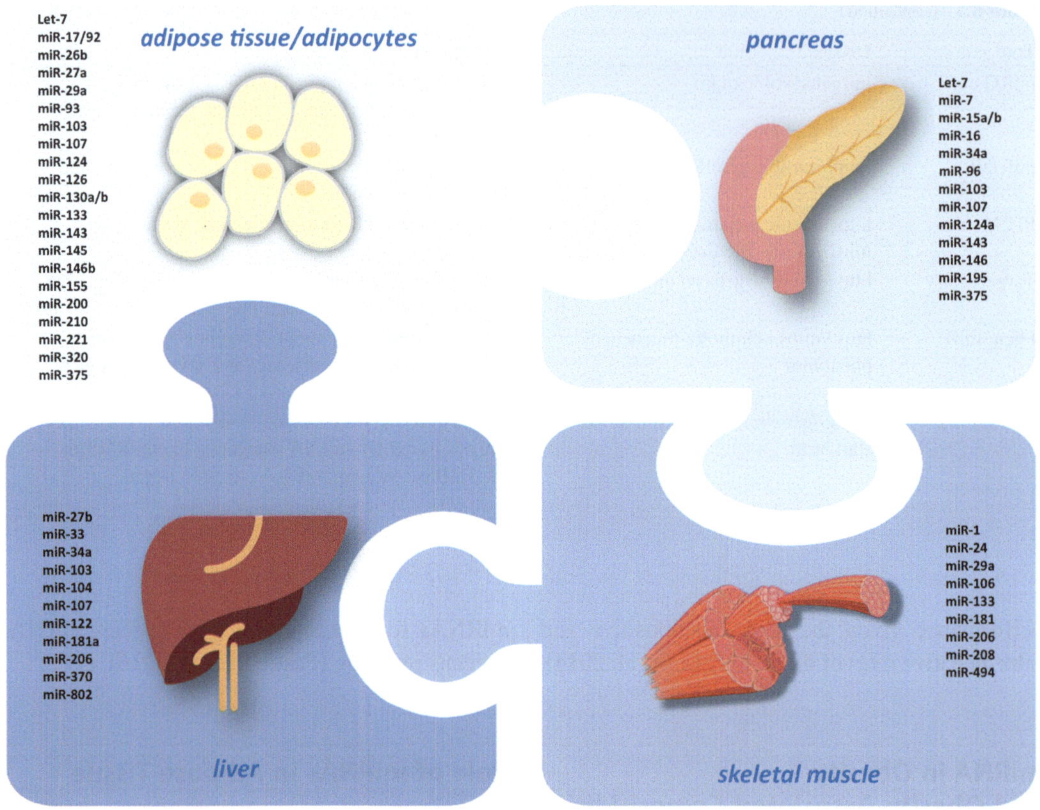

Fig. 6.2 Changes in miRNA profiles in different tissues are linked to obesity and metabolic diseases

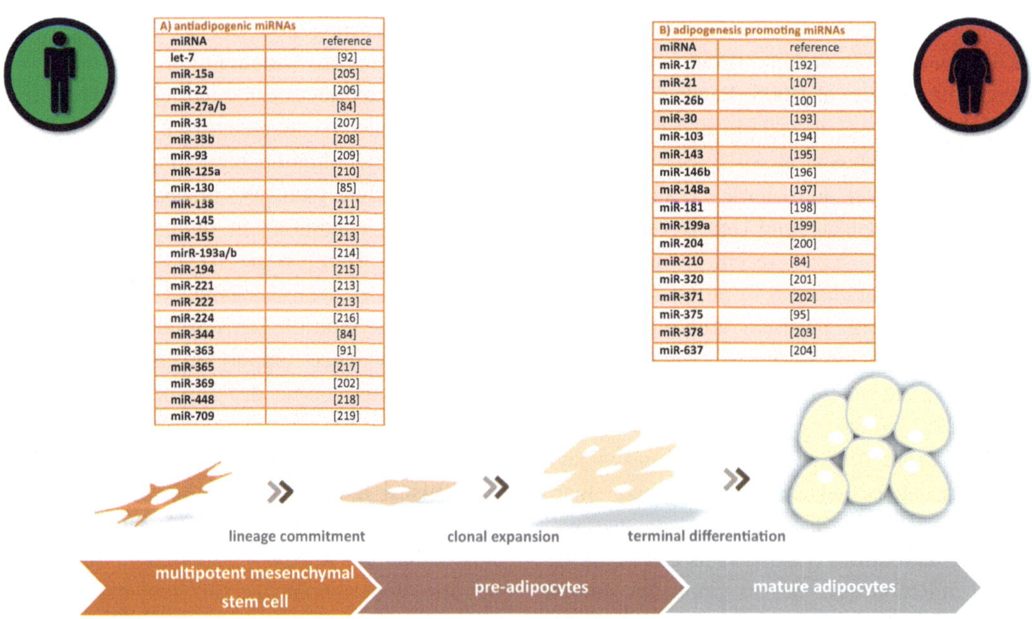

A) antiadipogenic miRNAs

miRNA	reference
let-7	[92]
miR-15a	[205]
miR-22	[206]
miR-27a/b	[84]
miR-31	[207]
miR-33b	[208]
miR-93	[209]
miR-125a	[210]
miR-130	[85]
miR-138	[211]
miR-145	[212]
miR-155	[213]
mirR-193a/b	[214]
miR-194	[215]
miR-221	[213]
miR-222	[213]
miR-224	[216]
miR-344	[84]
miR-363	[91]
miR-365	[217]
miR-369	[202]
miR-448	[218]
miR-709	[219]

B) adipogenesis promoting miRNAs

miRNA	reference
miR-17	[192]
miR-21	[107]
miR-26b	[100]
miR-30	[193]
miR-103	[194]
miR-143	[195]
miR-146b	[196]
miR-148a	[197]
miR-181	[198]
miR-199a	[199]
miR-204	[200]
miR-210	[84]
miR-320	[201]
miR-371	[202]
miR-375	[95]
miR-378	[203]
miR-637	[204]

Fig. 6.3 Emerging evidence supports miRNAs' contribution to adipose tissue development and physiopathology. A restricted example is here reported with some miRNAs interfering with adipocytes differentiation (panel **a**) while others promoting adipogenesis (panel **b**)

Adipogenesis Pathways

miR-130 and miR-143 are the best studied of adipogenesis-related miRNAs. miR-143 and miR-145 are frequently analysed together because they are closely located and can be co-transcribed. miR-143 was recognized as a positive human adipocyte differentiation regulator acting through ERK5 signaling. Expression of miR-143 and miR-145 is also increased in the liver of a mouse model of obesity; miR-143 overexpression adversely affects insulin-stimulated AKT activation and homeostasis of glucose. Besides, mice missing the miR-143–145 cluster do not develop insulin resistance associated with obesity (Jordan et al. 2011).

Another remarkable example is represented by miR-27a and miR-130a, overexpression of which suppresses adipocyte differentiation through PPARγ downregulation (Qin et al. 2010; Lee et al. 2011). Lower levels of miR-130a and -130b have been documented in abdominal subcutaneous adipose tissue of obese women (Ortega et al. 2013). Furthermore, circulating miR-130b is higher in obese children (Prats-Puig et al. 2013). A study from Wang et al. documented miR-130b as a potential biomarker for overweight, hypertriacylglycerolemia, and metabolic syndrome, suggesting a mechanism connecting obesity and obesity-related metabolic diseases, through a circulating miRNAs crosstalk between adipose and muscle tissues (Wang et al. 2013a).

The authors also established that the addition of TGF-β to mature adipocytes (3T3-L1) significantly increased the level of miR-130b in the culture medium and slightly decreased intracellular level, confirming that this miRNA is released from adipocytes during adipogenesis. Other miRNAs affect lineage determination. As an example, miR-124 has a pro-adipogenic effect by targeting Dlx5, a pro-osteogenic transcription factor that governs cell fate in human bone marrow-derived mesenchymal stem cells (Qadir et al. 2013). The best-characterized polycistronic miRNA cluster is miR-17-92 (encoding for miR-17, miR-18a, miR-19a, miR-20a, miR-19b-1, and miR-92a), essential for many developmental and pathogenic processes (Concepcion et al. 2012). This cluster is overexpressed in the clonal expansion of adipocytes and acts by directly targeting the RB family Rb2/p130 so controlling the cell checkpoint mediated by RB-E2F. In the same path, miR-363 inhibits the differentiation of adipocytes by targeting E2F and simultaneously reducing C/EBPα and PPARγ (Chen et al. 2014).

Let-7 miRNA and Diabetes Prevention

Let-7 was the first identified human miRNA. It belongs to a well-conserved family of 11 members connected with numerous critical cell functions such as apoptosis, proliferation, and cell cycle checkpoints. Members of this miRNA family directly regulate oncogenes such as RAS and HMGA2 and play a key role in developmental processes and pathogenesis of metabolic diseases, being important regulators of glucose metabolism and peripheral insulin resistance by addressing IGF1R, insulin receptor (INSR), and insulin receptor substrate-2. Also, let-7 negatively controls adipogenesis by modulating the expression of the high mobility group AT-hook2.

Let-7 is upregulated in the model of 3T3-L1 adipogenesis. The ectopic introduction of let-7 in 3T3-L1 and 3T3-F442A reduced the cell clonal expansion as well as the terminal differentiation (Sun et al. 2009). This miRNA serves as an anti-adipogenic factor regulating the transition from clonal expansion to terminal differentiation. Moreover, let-7 is directly involved in the metabolism of glucose and insulin resistance by acting on the insulin/IGF-1R pathway (Zhu et al. 2011). Knockout mice with reduced expression of let-7 do not develop insulin resistance despite a diet-induced obesity, supporting the potential of let-7 in developing new therapeutic strategies for diabetes (Frost and Olson 2011).

Other Tissue-Specific miRNAs and Metabolic Diseases

miR-375 has been reported to promote adipocyte differentiation through its ability to concurrently induce *C/EBPα*, *PPARγ2*, the adipocyte fatty acid-binding protein (*aP2*), and, ultimately, to promote triglycerides accumulation. Conversely, miR-375 suppresses phosphorylation of ERK1/2 (Ling et al. 2011).

Many papers reported the central role of miR-206 in the growth and development of the skeletal muscle by promoting myogenic differentiation. Besides, it has been correlated to several diseases, including heart failure, chronic obstructive pulmonary disease, Alzheimer's disease, and some cancers (Novak et al. 2014). Interestingly, miR-206 is downregulated in most of these conditions, suggesting a role of this miRNA as a "disease-avoiding" molecule (Novak et al. 2014). This miRNA is highly expressed in brown adipocytes but missing in white adipocytes (Walden et al. 2009). miR-206 suppresses the liver X receptor α (*LXRα*), a key target of PPAR, thus inhibiting lipogenesis and modulating lipid metabolism (Zhong et al. 2013).

miR-26b inhibits human preadipocytes differentiation (Nishimura et al. 2008; Song et al. 2014), while knockdown of miR-26b improves adipocyte differentiation with an establishment of adipocyte-specific transcripts and lipid droplets.

Recent findings indicate that miR-146b is markedly expressed during adipogenesis (Ahn et al. 2013). Sirtuin 1 (SIRT1) is negatively regulated by miR-146b. SIRT1 controls gene activity by deacetylating various transcription factors, including forkhead box O1 (FOXO1). Forkhead box O transcription factor family converts the multiplicity of extracellular stimuli, including growth factors, nutrients, and oxidative stress, into different biological responses through modulation of specific genes. SIRT1, by interacting with PPARγ co-repressors N-CoR and SMRT, inhibits PPARγ and precludes adipogenesis (SIRT1 level is reduced during adipogenesis). Accordingly, overexpression of miR-146b induces differentiation of 3T3-L1 cells, while inhibition of miR-146b reduces adipocyte differentiation (Jing et al. 2007).

The highly conserved miR-8/miR-200 family consists of a single ortholog (miR-8) in *Drosophila melanogaster*, and of five members (miR-200a, miR-200b, miR-200c, miR-141, and miR-429) in vertebrates (Trumbach and Prakash 2015). miR-8/miR-200 have been reported as repressors of the evolutionarily conserved Wnt/Wingless pathway in Drosophila eye and in mouse mesenchymal stem cells, controlling the eye size and the differentiation of the mesenchymal stem cells into adipocytes, correspondingly (Kennell et al. 2008). Drosophila miR-8 and human miR-200 family also prevent the expression of an inhibitor of the insulin/phosphoinositide-3 kinase (PI3K) signaling in adipose and liver cells, thus affecting cell growth and proliferation (Hyun et al. 2009). Overexpression of members of this miRNA family increases the level of the fatty acid-binding protein 4 (FABP4), promotes lipid accumulation and, finally, induces adipogenesis.

In 3T3-L1 cells, the expression of miR-210 increases significantly during adipogenesis (Liang et al. 2013). Transfection of miR-210 miRNA mimics into cells resulted in the expression of adipogenic markers through activation of the PI3K/Akt pathway by addressing SHIP1, a negative regulator of the pathway, and in the terminal differentiation. Moreover, ectopic inhibition of the endogenous miR-210 during adipogenesis halts adipocyte differentiation (Liang et al. 2013). Similarly, miR-21 significantly promotes 3T3-L1 adipocyte differentiation, increasing adiponectin expression, and decreasing AP-1 protein level. Also, miR-21 may enhance differentiation of human adipose-derived stem cells by directly inhibiting TGF-β receptor 2 expression (Kim et al. 2009b).

Adipose Tissue Cross Talk and miRNAs

Recently, Thomou et al. identified new potential implications of adipose tissue in the mechanism of the cell crosstalk (Thomou et al. 2017). Mice with an adipose-tissue-specific knockout of Dicer miRNA-processing enzyme, as well as humans with lipodystrophy, show a significant decrease in circulating miRNA levels. White and brown adipose tissue transplantation can restore levels of many circulating miRNAs, an event associated with increased glucose tolerance and decreased hepatic fibroblast growth factor 21 (FGF21) mRNA and circulating FGF21. This factor plays a major role in stimulating the oxidation of fatty acids in the liver and the absorption of glucose in adipocytes. Interestingly, FGF21 levels are significantly increased in patients with T2D, nonalcoholic fatty liver disease, and are correlated with BMI, so indicating obesity as a possible FGF21-resistant condition (Thomou et al. 2017).

Intercellular Message Signaling

Li et al. found that adipose tissue-secreted extracellular vesicles (containing miRNAs) deliver intercellular signaling. miR-221-3p mediates a mechanism by which perivascular adipose tissue triggers early-stage vascular remodeling in the context of obesity-associated inflammation (Li et al. 2019).

Role of miRNAs in the Pancreas

A decrease in β-cell function/mass has been linked to falling levels of insulin in T2D (Butler and Dhawan 2015). Dedifferentiation and cell identity processes of β-cells can likewise contribute to the decline in insulin production. The first indication of the relevance of miRNAs in hormone secretion came from a cloning approach of small RNAs from the β cell-derived cell line MIN6 (Poy et al. 2004). It has been recognized that miR-124a2 and miR-375 are involved in regulating the differentiation and establishment of pancreatic islets (Kaviani et al. 2016; Sebastiani et al. 2017), and high levels of miR-7 and miRNA-375 are expressed during pancreatic development. A panel of 40 miRNAs primarily expressed in islets has been identified, comparing islet-cell miRNAs profile with those of other human tissues (van de Bunt et al. 2013).

miR-375 is the most abundant miRNA in pancreatic islets and its presence is mandatory for normal tissue differentiation. The wide expression of miR-375 is observed during the development of pancreatic islet cells, while β-cell activity is connected to its reduction (Wei et al. 2013). During pancreas development, miR-375 targets various genes associated with cell growth (Joglekar et al. 2009) in addition to many transcription factors such as PDX1, HNF6, and INSM1, which are involved in the functioning of islets (Avnit-Sagi et al. 2009). Notably, the transcription factor neurogenin3 (Ngn3), which had a prominent role in the development of pancreatic islet (Gu et al. 2002) also interferes with miR-375 expression. miR-15a, miR-15b, miR-16, and miRNA-195 likewise target Ngn3. Of note, miR-375 has been reported to be involved in modulating insulin secretion in stimulated cell line MIN6 (Poy et al. 2004), whereas other miRNAs (let-7, miR-103, and -107) have an impact on rodent insulin sensitivity.

miR-375 decreases glucose-stimulated insulin secretion by downregulating mRNA of myotrophin (involved in insulin-granules cell membrane fusion) and thus inhibits exocytosis. miR-375 simultaneously affects insulin expression by targeting the phosphoinositide-dependent kinase-1 in INS1-E cells (El Ouaamari et al. 2008). Besides, Sedgeman et al. found that pancreatic beta-cells export miR-375-3p to HDL, and that this process is inversely related to insulin secretion (Sedgeman et al. 2019).

Additional miRNAs expressed in pancreatic islets, such as let-7 and miR-143, are connected to glucose homeostasis by targeting key components of the insulin signaling pathway (Zhu et al. 2011). There is evidence for miR-7 involvement in pancreas development (Kredo-Russo et al. 2012), and high levels of miR-7 are detectable in the pancreas both in the developing and adult phases. Overexpression of miR-7 in pancreatic progenitors has been shown to impair the differentiation of α- and β-cells and is connected with the repression of Pax6. On the contrary, knockdown of miR-7 during early embryonic life determines a downregulation of insulin production, a reduction in the number of β-cells, and the post-natal onset of glucose intolerance. Additionally, the in vitro inhibition of miR-7 promotes β-cell death in explanted pancreatic buds. In brief, dysregulation of the miR-7 signaling network in response to metabolic stress or cellular insults contributes to the loss of β-cell identity and T2D progress (Martinez-Sanchez et al. 2016).

Other miRNAs, such as miR-34a and miR-146a, are overexpressed in the pancreas only during the differentiation processes, and partially contribute to cytokine-mediated adjacent cell dysfunction in nonobese diabetic (NOD) mice, during the first stage of type 1 diabetes (Dumortier et al. 2013).

Additional Pancreatic Activities of miRNAs

miR-96 and miR-124a control the expression of proteins involved in insulin exocytosis and affect β-cell secretion (Lovis et al. 2008). miR-124a is upregulated during the late pancreas development. This miRNA targets the cAMP-responsive element-binding protein 1 (Creb1) and the forkhead box protein A2 (Foxa2) mRNAs. The last transcriptional factor modulates the expression of the insulin gene in multiple pathways responsible for the secretion of this hormone, primarily through the regulation of pancreatic and duodenal homeobox 1 (Pdx1) which is critical for glucose balance and pancreas development and, together with Ngn3, is indispensable for β-cell differentiation. Furthermore, miR124a promotes SNAP25, Rab3A, and synapsin-1A levels and decreases those of Rab27A and Noc2, key components involved in the exocytotic release of insulin granules (Lovis et al. 2008).

Noteworthy, Islet-specific miRNAs remain of singular interest for identifying islet stress or damage. It has been shown that during prediabetes, islets undergo stress and release miRNAs with consistent miRNA signatures specific to different stages (Vasu et al. 2019).

Role of miRNAs in the Muscle

Striated muscle tissue is the largest consumer of glucose in the human body since skeletal muscle makes up to 40% of the total body mass, contributing to nearly 75% of insulin-mediated glucose adsorption. Regular physical activity and fitness lead to successful improvements in health status partly due to positive effects on insulin resistance and T2D.

Several miRNAs, known as myomiR family, display enriched expression in the skeletal muscle and operate as modulators of skeletal and cardiac myogenesis, proliferation, metabolism, exercise as well as hypertrophy. MyomiRs include miR-1, miR-133a, miR-133b, miR-206, miR-208a, miR-208b, miR-486, miR-499 and are under the transcriptional control of myogenic regulatory factors (McCarthy 2011).

miR-206 is expressed primarily in the skeletal muscle, while miR-208a is cardio-specific; however, most of these miRNAs are co-expressed in the cardiac and skeletal muscles (Kirby and McCarthy 2013). MyomiRs directly target pathways that control skeletal muscle homeostasis. Their deregulation occurs in several muscular disorders and wasting, but is also as biomarker of muscle status, regeneration, and therapy effect (Polakovicova et al. 2016).

Reduced expression of miR-133 was established in cardiac hypertrophy models of mouse and human models (Feng et al. 2010). Acute exercise leads to an increase in the levels of miR-1, miR-133a, and miR-206 (Gomes et al. 2014), relevant miRNAs that potentially contribute to cell-to-cell communication.

Insulin Resistance

miR-29a induces insulin resistance by addressing PPARδ in skeletal muscle cells in rodents. The overexpression of miR-29a inhibits PPARδ finally affecting the expression of its PGC-1α coactivator. PPARδ/PGC-1α-dependent signaling results in a reduction in glucose transporter levels 4, the main transporter of glucose in the skeletal muscle, along with a reduction in insulin-dependent glucose intake and adenosine triphosphate (ATP) levels (Zhou et al. 2016). Analogously, miR-29a levels are noticeably elevated in diabetic mice (db/db) liver. Overexpression of this miRNA leads to an attenuation of insulin action on phosphoenolpyruvate carboxykinase (PEPCK) expression in the liver, a phenomenon connected to an unrestrained hepatic glucose output. A predominant insulin action in the liver is the inhibition of gluconeogenesis by a cascade of events that terminates with a diminished PEPCK activity, an occurrence that has been shown to be abolished in diabetes, leading to an uncontrolled hepatic glucose output (Pandey et al. 2011). High fat diet induces the expression of miR-29a in myocytes, an event that impairs the insulin signaling pathway and glucose uptake through a pronounced decrease in insulin receptor substrate 1 (IRS-1), suggesting that upregulation of miR-29a by saturated fatty acids (SFA) is causally associated with the development of muscle insulin resistance (Yang et al. 2014).

miR-106b was associated with muscle insulin resistance and T2D. miR-106b is highly expressed in the diabetic subjects' muscle. High level of this miRNA provokes mitochondrial dysfunction and insulin resistance in C2C12 myotubes by addressing mitofusin-2 (a mitochondrial membrane protein participating in mitochondrial fusion in mammalian cells, induced during myogenesis and involved in the maintenance of the mitochondrial network). Remarkably, expression of miR-106b is improved after treatment with TNF-α, suggesting that its increased expression under chronic low-grade inflammation could be a link between mitochondrial alteration and T2D (Zhang et al. 2013b).

Myocardial Metabolism

An interesting field of research is the pleiotropic regulatory functions of miR-208a, a heart-specific miRNA that also affects glucose metabolism and energy homeostasis. Cardiac muscle tissue helps to regulate systemic energy homeostasis through MED13 (Baskin et al. 2014), a subunit of the Mediator complex, which governs transcription by the thyroid hormone and additional nuclear hormone receptors (Baskin et al. 2015). MED13 level is negatively controlled by miR-208a. Oligonucleotides anti-miR-208 provide resistance to diet-induced obesity and improve glucose tolerance in the mouse model (Grueter et al. 2012).

Role of miRNAs in the Liver

miR-122 is a dominant hepatocyte-specific miRNA representing about 75% of total miRNA expression in hepatocytes with about 135,000 copies, making it one of the highly expressed miRNAs in humans. Levels of miR-122 are controlled by liver-enriched transcription factors (LETFs), including hepatocyte nuclear factor (HNF) 6 and 4a. The miR-122 regulatory network has been involved in many functions in the liver, ranging from cholesterol metabolism, stress responses, viral infection, cancer, and circadian hepatic gene regulation (Szabo and Bala 2013; Tsai et al. 2012). This miRNA also plays a role in metabolic syndrome and other liver disorders, such as alcohol-related liver inflammation, autoimmune processes, and the development of liver fibrosis in both humans and in animal models. Pathological suppression of miR-122 has been identified in hepatocellular carcinoma (Burchard et al. 2010), nonalcoholic steatohepatitis (Cheung et al. 2008), and liver cirrhosis (Szabo and Bala 2013). Due to its role in the metabolism of cholesterol, this miRNA is intensively investigated. Inhibition via antisense of miR-122 in healthy mice results in lower rates of serum cholesterol, LDL and serum triglyceride, increased oxidation of hepatic fatty acids, and decreased expression of key genes involved both in the metabolism of fatty acids and biosynthesis of cholesterol, including the rate-limiting enzyme 3-hydroxy-3-methylglutaryl-CoA-reductase (Esau et al. 2006). Antisense inhibition of this miRNA in chimpanzee results in a reduction in plasma cholesterol supporting its central role in maintaining liver metabolism (Elmen et al. 2008). miR-122 can also be detected in blood, and circulating miR-122 has been suggested as a liver injury biomarker in hepatitis B and C, nonalcoholic fatty liver disease, and drug-induced liver disease (Bandiera et al. 2015).

miR-33, whose coding sequence is embedded within an intron of the SREBP gene, has been shown to regulate cholesterol metabolism, fatty acids metabolism, and glucose metabolism/insulin signaling (Davalos et al. 2011). miRNA targets include molecules involved both in cholesterol transport to HDL/reverse cholesterol transport and the transport of cholesterol to bile.

Additional miRNAs

miR-27b controls lipid metabolism and is impaired in dyslipidemia, so disrupting both heart and liver functions (Vickers et al. 2013). miR-27a is considered a critical post-transcriptional hub interconnecting different components of the lipid metabolism, being able to target RXRα, ABCA1, FASN, SREBP1, SREBP2, PPARα, PPARγ, ApoA1, ApoB100, and ApoE3 (Vickers et al. 2013).

miR-34a affects hepatic levels of SIRT1, and miR-34a upregulation with a concomitant decrease in Sirt1 levels has been reported in fatty livers of mice with diet-induced obesity (Lee et al. 2010). In addition, miR-370 targets carnitine palmitoyl transferase, a mitochondrial enzyme involved in the transport of long-chain fatty acids across the membrane, and concomitantly controls lipid metabolism (Iliopoulos et al. 2010).

Circulating miRNAs

While miRNAs were first detected in cells, a growing number of miRNAs are found in high concentrations in blood and other body fluids such as in serum, plasma, urine, saliva, and maternal milk (Turchinovich et al. 2011). Stability in blood and body fluids has been questioned in the past (Cortez and Calin 2009), given the ubiquity of nucleases. Nevertheless, this finding stimulated interest in the possibility that changes in cell-free miRNAs could be used as noninvasive biomarkers (Kim 2015). Because of their availability, the most typical miRNA sources are whole blood, serum, and plasma (Grasedieck et al. 2013).

Anti Nuclease Protection

Circulating miRNAs (c-miRNAs) are not, as predictable, naked molecules and two major mechanisms to shield them from nuclease activity have been identified. The first one consists in the formation of complexes of specific proteins, such as Argonaute 2 (AGO-2) (involved in the RNA silencing complex) (Arroyo et al. 2011), with high-density lipoproteins (Vickers et al. 2011), or with nucleophosmin-1 (NPM-1) (an RNA-binding protein involved in the nuclear export of ribosomes) (Wang et al. 2010). The second proposed mechanism stems from the detection of c-miRNAs inside circulating microvesicles or exosomes (Mittelbrunn et al. 2011), originating either from the cell plasma membrane or from the endosomal compartments (Camussi et al. 2011). While a defined mechanism for the release of miRNAs from donor cells remains largely unknown, growing evidence supports the possibility that extracellular miRNAs, organized in either exosomes or protein complexes, can be transferred to recipient cells where they can regulate the translation of the target/s gene/s (Weber et al. 2010).

Circulating miRNA in Obesity, Diabetes and Associated Conditions

Differential c-miRNA profiles have been identified in subjects with obesity and T2D (Guay and Regazzi 2013) and levels of specific c-miRNAs vary from one setting to another. miR-126 is reduced in T2D (Zampetaki et al. 2010); elevated serum levels of microRNAs 192 and 194 are associated with incident T2D (Jaeger et al. 2018); miR-1, miR-21, miR-133a, and miR-208 are enriched in the plasma after myocardial infarction (Zile et al. 2011); miR-122 is increased in hepatic injury and steatosis (Cermelli et al. 2011), in addition to let-7e in hypertension (Li et al. 2011).

A meta-analysis by Zhu and Leung showed that miR-29a, miR-34a, miR-103, miR-107, miR-132, miR-142-3p, miR-144, and miR-375 represent potential circulatory biomarkers for T2D (Zhu and Leung 2015). In the same analysis, miR-199a-3p and miR-223 were found as potential tissue biomarkers of T2D. An additional meta-analysis reported that levels of circulating miR-27a, miR-29a, miR-142-3p, miR-222, miR-320a, and miR-375 were increased and levels of miR-17, miR-20b, miR-197, and miR-652 decreased in individuals with T2D (Villard et al. 2015).

Diabetic subjects are known to exhibit increased platelet activity. Platelet-derived miRNAs could be exploited as biomarkers in inflammatory diseases, including T2D and CVD, given the abundance of platelets in the blood and their wide contribution to the c-miRNAs pool (Pordzik et al. 2018). In a study by Stratz et al. (2014) miRNA relative levels between the T2D cohort and controls were not significantly different in platelets; nevertheless, functional implications of these molecules can be relevant in the pathogenesis of T2D (Pordzik et al. 2019).

Recently, Ghai et al. identified several miRNAs encapsulated in circulating extracellular vesicles that were affected by metformin treatment in T2D patients, suggesting that these miRNAs may serve as possible biomarkers for assessing drug treatment response (Ghai et al. 2019).

Besides, miR-191 was positively associated with glycemic impairment and progression in the high-risk Asian Indian ethnic group; conversely, miR-122, miR-15a, miR-197, miR-320a, miR-423, miR-486 exhibited an inverse relationship with odds for glycemic progression after 2.5 years compared to those who remained stable (Flowers et al. 2015).

Blood Glucose, Blood Pressure and Metabolic Syndrome

In a large population-based study ($n = 871$), levels of miR-144-5p, miR-122-5p, miR-148a-3p, miR-589-5p, and let-7a-5p were associated with glycemic status (Mononen et al. 2019). In the same study, miR-144-5p and miR-148a-3p were linked to glucose levels, while miR-144-5p, miR-122-5p, miR-184, and miR-339-3p were associated with insulin levels and HOMA-IR score. miR-148a-3p, miR-15b-3p, miR-93-3p, miR-146b-5p, miR-221-3p, miR-18a-3p, miR-642a-5p, and miR-181-2-3p were linked to HbA1c levels.

Besides, high blood pressure was associated with circulating miR-130a and miR-195 (Karolina et al. 2012). Changes in circulating miR 23a, miR 27a, miR 130, miR 195, miR 197, miR 320a, and miR 509-5p were associated with metabolic syndrome (Karolina et al. 2012; Deiuliis 2016). c-miRNA profiles showed a sex-specific association with metabolic syndrome (Wang et al. 2013b). Moreover, it was proposed that let-7b, miR-143, and miR-221 control both atherogenic and adipogenic progressions (Hulsmans et al. 2011).

Bariatric Surgery

Ortega et al. identified a significant increase in circulating miR-140-5p, miR-142-3p, miR-222, and decrease of miR-532-5p, miR-125b, miR-130b, miR-221, miR-15a, miR-423-5p, and miR-520c-3p in morbidly obese patients. Bariatric surgery resulted in a significant reduction in circulating miR-140-5p, miR-122, miR-193a-5p, and miR-16-1, and an increase in miR-221 and miR-199a-3p levels (Ortega et al. 2013). Besides, levels of circulating miR-17-5p and miR-132 were decreased in obesity, reflecting the pattern of expression of miRNAs in omental fat from the same group of obese people (Heneghan et al. 2011).

Pediatric Protocols

Differential c-miRNAs signatures occur in overweight/obese children and adolescents as compared to normal weight (Prats-Puig et al. 2013; Iacomino et al. 2016, 2019; Can et al. 2016; Thompson et al. 2017; Iacomino and Siani 2017). Accordingly, our group has reported that a panel of miRNAs is differentially expressed in overweight/obese European children, miR551a and miR-501-5p resulted upregulated; miR-10b-5p, miR-191-3p, miR-215-5p, and miR-874-3p resulted downregulated (Iacomino et al. 2019). In a further analysis by Oses et al. miR-122 and miR-34a, were overexpressed in obese children and nonalcoholic fatty liver disease (NAFLD) and/or insulin resistance (Oses et al. 2019).

Mounting evidence was also found in newborns, pregestational maternal obesity, and gestational obesity (Carreras-Badosa et al. 2015). Among babies born to obese or lean mothers, the rates of some c-miRNAs, including miR-155, miR-181a, and miR-221, differ significantly. Such miRNAs were also suggested as useful predictors of obesity and increased risk of developing disorders in children born to obese women (Mendez-Mancilla et al. 2018).

Several c-miRNAs were found down- and upregulated in pregestational and gestational obesity; some were associated with pregnancy weight gain, others were closely associated with metabolic parameters during gestation and were independent predictors of pre- and post-natal growing (Carreras-Badosa et al. 2015). Also, levels of some c-miRNAs were associated with gestational diabetes, especially in prepregnancy overweight women (Wander et al. 2017). Recent data showed that the supply of miRNAs in maternal milk was likewise altered in overweight/obese lactating mothers, pointing to a plausible molecular communication between mother and infant through miRNAs (Zamanillo et al. 2019).

Other studies also investigated how diet, exercise, weight loss, and bariatric surgery impact miRNAs profiles in obese patients. A significant modulation of specific c-miRNAs was observed in overweight/obese subjects after low or high glycemic index diet and low-fat diet (Giardina et al. 2019). Moreover, miR-21, miR-126, miR-130b, miR-221, and miR-222 are upregulated in both obese and normal-weight subjects following acute aerobic exercise (Bao et al. 2018). Several studies have shown that gastric bypass deeply impacts c-miRNA levels from the preoperative baseline with time-dependent changes (Alkandari et al. 2018). The combination of gastric bypass surgery with exercise likewise affected the pattern of several c-miRNAs (Nunez Lopez et al. 2017). A summary of several bariatric surgery studies was recently reported in a meta-analysis from Langi et al. (2019).

Translational Considerations

miRNA related drugs have been synthesized and systemically administered for therapeutic purposes, and one of them, the small-interfering RNA (siRNA) agent Patisiran (Onpattro, Alnylam Pharmaceuticals, Cambridge, MA, USA), was FDA approved for a rare polyneuropathy caused by hereditary transthyretin-mediated amyloidosis. Given the ability of miRNA mimics to overexpress the transcript, and of miRNA repressors to silence it, they are well-positioned to modulate peripheral tissues regarding genetic/epigenetic derangements. Indeed, it is estimated that 60% of human protein-coding genes are targeted or regulated by miRNAs. In this sense, several trials are going on to develop additional drugs for benign and malignant diseases, including oncological chemoresistance in certain cancer modalities (Hanna et al. 2019).

Ordinary diabetes and obesity have not been given similar attention, perhaps because there are comparatively simple and well-established treatment options in the market. Nevertheless, applications for metabolic disease screening, prognosis, and early diagnosis of complications are envisaged. As mentioned, selected miRNAs can be detected years before conventional markers become positive. In this sense, miRNA-based predictive algorithms could be adopted in the near future, to improve patient management in the clinical routine (Jiménez-Lucena et al. 2018).

References

Ahima RS, Lazar MA (2013) Physiology. The health risk of obesity--better metrics imperative. Science 341 (6148):856–858. https://doi.org/10.1126/science.1241244

Ahn J, Lee H, Jung CH, Jeon TI, Ha TY (2013) MicroRNA-146b promotes adipogenesis by suppressing the SIRT1-FOXO1 cascade. EMBO Mol

Med 5(10):1602–1612. https://doi.org/10.1002/emmm.201302647
Ali AT, Hochfeld WE, Myburgh R, Pepper MS (2013) Adipocyte and adipogenesis. Eur J Cell Biol 92 (6–7):229–236. https://doi.org/10.1016/j.ejcb.2013.06.001
Alkandari A, Ashrafian H, Sathyapalan T, Sedman P, Darzi A et al (2018) Improved physiology and metabolic flux after Roux-en-Y gastric bypass is associated with temporal changes in the circulating microRNAome: a longitudinal study in humans. BMC Obes 5:20. https://doi.org/10.1186/s40608-018-0199-z
Arner P, Kulyte A (2015) MicroRNA regulatory networks in human adipose tissue and obesity. Nat Rev Endocrinol 11(5):276–288. https://doi.org/10.1038/nrendo.2015.25
Arroyo JD, Chevillet JR, Kroh EM, Ruf IK, Pritchard CC et al (2011) Argonaute2 complexes carry a population of circulating microRNAs independent of vesicles in human plasma. Proc Natl Acad Sci U S A 108 (12):5003–5008. https://doi.org/10.1073/pnas.1019055108
Avnit-Sagi T, Kantorovich L, Kredo-Russo S, Hornstein E, Walker MD (2009) The promoter of the pri-miR-375 gene directs expression selectively to the endocrine pancreas. PLoS One 4(4):e5033. https://doi.org/10.1371/journal.pone.0005033
Bandiera S, Pfeffer S, Baumert TF, Zeisel MB (2015) miR-122--a key factor and therapeutic target in liver disease. J Hepatol 62(2):448–457. https://doi.org/10.1016/j.jhep.2014.10.004
Bao F, Slusher AL, Whitehurst M, Huang CJ (2018) Circulating microRNAs are upregulated following acute aerobic exercise in obese individuals. Physiol Behav 197:15–21. https://doi.org/10.1016/j.physbeh.2018.09.011
Bartel DP (2004) MicroRNAs: genomics, biogenesis, mechanism, and function. Cell 116(2):281–297
Bartel DP (2009) MicroRNAs: target recognition and regulatory functions. Cell 136(2):215–233
Bartel DP (2018) Metazoan microRNAs. Cell 173 (1):20–51. https://doi.org/10.1016/j.cell.2018.03.006
Baskin KK, Grueter CE, Kusminski CM, Holland WL, Bookout AL et al (2014) MED13-dependent signaling from the heart confers leanness by enhancing metabolism in adipose tissue and liver. EMBO Mol Med 6 (12):1610–1621. https://doi.org/10.15252/emmm.201404218
Baskin KK, Winders BR, Olson EN (2015) Muscle as a "mediator" of systemic metabolism. Cell Metab 21 (2):237–248. https://doi.org/10.1016/j.cmet.2014.12.021
Berndt SI, Gustafsson S, Magi R, Ganna A, Wheeler E et al (2013) Genome-wide meta-analysis identifies 11 new loci for anthropometric traits and provides insights into genetic architecture. Nat Genet 45 (5):501–512
Bluher M (2013) Adipose tissue dysfunction contributes to obesity related metabolic diseases. Best Pract Res Cl En 27(2):163–177
Booth A, Magnuson A, Fouts J, Foster MT (2016) Adipose tissue: an endocrine organ playing a role in metabolic regulation. Horm Mol Biol Clin Invest 26 (1):25–42. https://doi.org/10.1515/hmbci-2015-0073
Bracken CP, Scott HS, Goodall GJ (2016) A network-biology perspective of microRNA function and dysfunction in cancer. Nat Rev Genet 17(12):719–732. https://doi.org/10.1038/nrg.2016.134
Branca F, Nikogosian H, Lobstein T, World Health Organization. Regional Office for Europe (2007) The challenge of obesity in the WHO European region and the strategies for response summary
Burchard J, Zhang C, Liu AM, Poon RT, Lee NP et al (2010) microRNA-122 as a regulator of mitochondrial metabolic gene network in hepatocellular carcinoma. Mol Syst Biol 6:402. https://doi.org/10.1038/msb.2010.58
Butler AE, Dhawan S (2015) beta-Cell identity in type 2 diabetes: lost or found? Diabetes 64(8):2698–2700. https://doi.org/10.2337/db15-0550
Calin GA, Croce CM (2006) MicroRNA signatures in human cancers. Nat Rev Cancer 6(11):857–866. https://doi.org/10.1038/nrc1997
Camussi G, Deregibus MC, Bruno S, Grange C, Fonsato V et al (2011) Exosome/microvesicle-mediated epigenetic reprogramming of cells. Am J Cancer Res 1 (1):98–110
Can U, Buyukinan M, Yerlikaya FH (2016) The investigation of circulating microRNAs associated with lipid metabolism in childhood obesity. Pediatr Obes 11 (3):228–234. https://doi.org/10.1111/ijpo.12050
Carreras-Badosa G, Bonmati A, Ortega FJ, Mercader JM, Guindo-Martinez M et al (2015) Altered circulating miRNA expression profile in pregestational and gestational obesity. J Clin Endocrinol Metab 100(11): E1446–E1456. https://doi.org/10.1210/jc.2015-2872
Catalanotto C, Cogoni C, Zardo G (2016) MicroRNA in control of gene expression: an overview of nuclear functions. Int J Mol Sci 17(10):1712. https://doi.org/10.3390/ijms17101712
Cermelli S, Ruggieri A, Marrero JA, Ioannou GN, Beretta L (2011) Circulating microRNAs in patients with chronic hepatitis C and non-alcoholic fatty liver disease. PLoS One 6(8):e23937. https://doi.org/10.1371/journal.pone.0023937
Chen L, Cui J, Hou J, Long J, Li C et al (2014) A novel negative regulator of adipogenesis: microRNA-363. Stem Cells 32(2):510–520. https://doi.org/10.1002/stem.1549
Cheung O, Puri P, Eicken C, Contos MJ, Mirshahi F et al (2008) Nonalcoholic steatohepatitis is associated with altered hepatic MicroRNA expression. Hepatology 48 (6):1810–1820. https://doi.org/10.1002/hep.22569
Concepcion CP, Bonetti C, Ventura A (2012) The microRNA-17-92 family of microRNA clusters in development and disease. Cancer J 18(3):262–267. https://doi.org/10.1097/PPO.0b013e318258b60a
Cortez MA, Calin GA (2009) MicroRNA identification in plasma and serum: a new tool to diagnose and monitor diseases. Expert Opin Biol Ther 9(6):703–711. https://doi.org/10.1517/14712590902932889

D'Adamo E, Caprio S (2011) Type 2 diabetes in youth: epidemiology and pathophysiology. Diabetes Care 34 (Suppl 2):S161–S165. https://doi.org/10.2337/dc11-s212
Davalos A, Goedeke L, Smibert P, Ramirez CM, Warrier NP et al (2011) miR-33a/b contribute to the regulation of fatty acid metabolism and insulin signaling. Proc Natl Acad Sci U S A 108(22):9232–9237. https://doi.org/10.1073/pnas.1102281108
Deiuliis JA (2016) MicroRNAs as regulators of metabolic disease: pathophysiologic significance and emerging role as biomarkers and therapeutics. Int J Obes 40 (1):88–101. https://doi.org/10.1038/ijo.2015.170
Dumortier O, Hinault C, Van Obberghen E (2013) MicroRNAs and metabolism crosstalk in energy homeostasis. Cell Metab 18(3):312–324. https://doi.org/10.1016/j.cmet.2013.06.004
Ebert MS, Sharp PA (2012) Roles for microRNAs in conferring robustness to biological processes. Cell 149(3):515–524
El Ouaamari A, Baroukh N, Martens GA, Lebrun P, Pipeleers D et al (2008) miR-375 targets 3'-phosphoinositide-dependent protein kinase-1 and regulates glucose-induced biological responses in pancreatic beta-cells. Diabetes 57(10):2708–2717. https://doi.org/10.2337/db07-1614
El-Khoury V, Pierson S, Kaoma T, Bernardin F, Berchem G (2016) Assessing cellular and circulating miRNA recovery: the impact of the RNA isolation method and the quantity of input material. Sci Rep 6:19529. https://doi.org/10.1038/srep19529
Elmen J, Lindow M, Schutz S, Lawrence M, Petri A et al (2008) LNA-mediated microRNA silencing in non-human primates. Nature 452(7189):896–899. https://doi.org/10.1038/nature06783
Esau C, Davis S, Murray SF, Yu XX, Pandey SK et al (2006) miR-122 regulation of lipid metabolism revealed by in vivo antisense targeting. Cell Metab 3 (2):87–98. https://doi.org/10.1016/j.cmet.2006.01.005
Fall T, Ingelsson E (2014) Genome-wide association studies of obesity and metabolic syndrome. Mol Cell Endocrinol 382(1):740–757. https://doi.org/10.1016/j.mce.2012.08.018
Feng B, Chen S, George B, Feng Q, Chakrabarti S (2010) miR133a regulates cardiomyocyte hypertrophy in diabetes. Diabetes Metab Res Rev 26(1):40–49. https://doi.org/10.1002/dmrr.1054
Filipowicz W, Bhattacharyya SN, Sonenberg N (2008) Mechanisms of post-transcriptional regulation by microRNAs: are the answers in sight? Nat Rev Genet 9(2):102–114. https://doi.org/10.1038/nrg2290
Finnegan EF, Pasquinelli AE (2013) MicroRNA biogenesis: regulating the regulators. Crit Rev Biochem Mol Biol 48(1):51–68. https://doi.org/10.3109/10409238.2012.738643
Flowers E, Gadgil M, Aouizerat BE, Kanaya AM (2015) Circulating micrornas associated with glycemic impairment and progression in Asian Indians. Biomark Res 3:22. https://doi.org/10.1186/s40364-015-0047-y
Friedman RC, Farh KK, Burge CB, Bartel DP (2009) Most mammalian mRNAs are conserved targets of microRNAs. Genome Res 19(1):92–105. https://doi.org/10.1101/gr.082701.108
Frost RJ, Olson EN (2011) Control of glucose homeostasis and insulin sensitivity by the Let-7 family of microRNAs. Proc Natl Acad Sci U S A 108 (52):21075–21080. https://doi.org/10.1073/pnas.1118922109
Ghai V, Kim TK, Etheridge A, Nielsen T, Hansen T et al (2019) Extracellular vesicle encapsulated micrornas in patients with type 2 diabetes are affected by metformin treatment. J Clin Med 8(5):617. https://doi.org/10.3390/jcm8050617
Ghildiyal M, Zamore PD (2009) Small silencing RNAs: an expanding universe. Nat Rev Genet 10(2):94–108
Giardina S, Hernandez-Alonso P, Diaz-Lopez A, Salas-Huetos A, Salas-Salvado J et al (2019) Changes in circulating miRNAs in healthy overweight and obese subjects: effect of diet composition and weight loss. Clin Nutr 38(1):438–443. https://doi.org/10.1016/j.clnu.2017.11.014
Gomes CP, Oliveira-Jr GP, Madrid B, Almeida JA, Franco OL et al (2014) Circulating miR-1, miR-133a, and miR-206 levels are increased after a half-marathon run. Biomarkers 19(7):585–589. https://doi.org/10.3109/1354750X.2014.952663
Grasedieck S, Sorrentino A, Langer C, Buske C, Dohner H et al (2013) Circulating microRNAs in hematological diseases: principles, challenges, and perspectives. Blood 121(25):4977–4984. https://doi.org/10.1182/blood-2013-01-480079
Grueter CE, van Rooij E, Johnson BA, DeLeon SM, Sutherland LB et al (2012) A cardiac microRNA governs systemic energy homeostasis by regulation of MED13. Cell 149(3):671–683. https://doi.org/10.1016/j.cell.2012.03.029
Gu G, Dubauskaite J, Melton DA (2002) Direct evidence for the pancreatic lineage: NGN3+ cells are islet progenitors and are distinct from duct progenitors. Development 129(10):2447–2457
Guay C, Regazzi R (2013) Circulating microRNAs as novel biomarkers for diabetes mellitus. Nat Rev Endocrinol 9(9):513–521. https://doi.org/10.1038/nrendo.2013.86
Gulyaeva LF, Kushlinskiy NE (2016) Regulatory mechanisms of microRNA expression. J Transl Med 14(1):143. https://doi.org/10.1186/s12967-016-0893-x
Ha M, Kim VN (2014) Regulation of microRNA biogenesis. Nat Rev Mol Cell Biol 15(8):509–524
Hanna J, Hossain GS, Kocerha J (2019) The potential for microRNA therapeutics and clinical research. Front Genet 10:478. https://doi.org/10.3389/fgene.2019.00478. eCollection 2019
Hartig SM, Hamilton MP, Bader DA, McGuire SE (2015) The miRNA interactome in metabolic homeostasis. Trends Endocrinol Metab 26(12):733–745. https://doi.org/10.1016/j.tem.2015.09.006
Heneghan HM, Miller N, McAnena OJ, O'Brien T, Kerin MJ (2011) Differential miRNA expression in omental adipose tissue and in the circulation of obese patients identifies novel metabolic biomarkers. J Clin

Endocrinol Metab 96(5):E846–E850. https://doi.org/10.1210/jc.2010-2701

Hulsmans M, De Keyzer D, Holvoet P (2011) MicroRNAs regulating oxidative stress and inflammation in relation to obesity and atherosclerosis. FASEB J 25(8):2515–2527. https://doi.org/10.1096/fj.11-181149

Hurt RT, Kulisek C, Buchanan LA, McClave SA (2010) The obesity epidemic: challenges, health initiatives, and implications for gastroenterologists. Gastroenterol Hepatol (N Y) 6(12):780–792

Hyun S, Lee JH, Jin H, Nam J, Namkoong B et al (2009) Conserved MicroRNA miR-8/miR-200 and its target USH/FOG2 control growth by regulating PI3K. Cell 139(6):1096–1108. https://doi.org/10.1016/j.cell.2009.11.020

Iacomino G, Siani A (2017) Role of microRNAs in obesity and obesity-related diseases. Genes Nutr 12:23. https://doi.org/10.1186/s12263-017-0577-z

Iacomino G, Tecce MF, Grimaldi C, Tosto M, Russo GL (2001) Transcriptional response of a human colon adenocarcinoma cell line to sodium butyrate. Biochem Biophys Res Commun 285(5):1280–1289. https://doi.org/10.1006/bbrc.2001.5323

Iacomino G, Medici MC, Napoli D, Russo GL (2006) Effects of histone deacetylase inhibitors on p55CDC/Cdc20 expression in HT29 cell line. J Cell Biochem 99(4):1122–1131. https://doi.org/10.1002/jcb.21014

Iacomino G, Russo P, Stillitano I, Lauria F, Marena P et al (2016) Circulating microRNAs are deregulated in overweight/obese children: preliminary results of the I.Family study. Genes Nutr 11:7. https://doi.org/10.1186/s12263-016-0525-3

Iacomino G, Russo P, Marena P, Lauria F, Venezia A et al (2019) Circulating microRNAs are associated with early childhood obesity: results of the I.Family Study. Genes Nutr 14:2. https://doi.org/10.1186/s12263-018-0622-6

Iacomino G, Lauria F, Russo P, Marena P, Venezia A, et al (2020) Circulating miRNAs are associated with sleep duration in children/adolescents: results of the I. Family Study. Exp Physiol 105(2):347–356. https://doi.org/10.1113/EP088015

Iliopoulos D, Drosatos K, Hiyama Y, Goldberg IJ, Zannis VI (2010) MicroRNA-370 controls the expression of microRNA-122 and Cpt1alpha and affects lipid metabolism. J Lipid Res 51(6):1513–1523. https://doi.org/10.1194/jlr.M004812

Jaeger A, Zollinger L, Saely CH, Muendlein A, Evangelakos I et al (2018) Circulating microRNAs -192 and -194 are associated with the presence and incidence of diabetes mellitus. Sci Rep 8(1):14274. https://doi.org/10.1038/s41598-018-32274-9

Jiang Q, Wang Y, Hao Y, Juan L, Teng M et al (2009) miR2Disease: a manually curated database for microRNA deregulation in human disease. Nucleic Acids Res 37(Database issue):D98–D104. https://doi.org/10.1093/nar/gkn714

Jiménez-Lucena R, Camargo A, Alcalá-Diaz JF (2018) A plasma circulating miRNAs profile predicts type 2 diabetes mellitus and prediabetes: from the CORDIOPREV study. Exp Mol Med 50:1–12. https://doi.org/10.1038/s12276-018-0194-y

Jing E, Gesta S, Kahn CR (2007) SIRT2 regulates adipocyte differentiation through FoxO1 acetylation/deacetylation. Cell Metab 6(2):105–114. https://doi.org/10.1016/j.cmet.2007.07.003

Joglekar MV, Joglekar VM, Hardikar AA (2009) Expression of islet-specific microRNAs during human pancreatic development. Gene Expr Patterns 9(2):109–113. https://doi.org/10.1016/j.gep.2008.10.001

Jordan SD, Kruger M, Willmes DM, Redemann N, Wunderlich FT et al (2011) Obesity-induced overexpression of miRNA-143 inhibits insulin-stimulated AKT activation and impairs glucose metabolism. Nat Cell Biol 13(4):434–446. https://doi.org/10.1038/ncb2211

Jung UJ, Choi MS (2014) Obesity and its metabolic complications: the role of adipokines and the relationship between obesity, inflammation, insulin resistance, dyslipidemia and nonalcoholic fatty liver disease. Int J Mol Sci 15(4):6184–6223. https://doi.org/10.3390/ijms15046184

Kammoun HL, Kraakman MJ, Febbraio MA (2014) Adipose tissue inflammation in glucose metabolism. Rev Endocr Metab Disord 15(1):31–44. https://doi.org/10.1007/s11154-013-9274-4

Karolina DS, Tavintharan S, Armugam A, Sepramaniam S, Pek SL et al (2012) Circulating miRNA profiles in patients with metabolic syndrome. J Clin Endocrinol Metab 97(12):E2271–E2276. https://doi.org/10.1210/jc.2012-1996

Kaviani M, Azarpira N, Karimi MH, Al-Abdullah I (2016) The role of microRNAs in islet beta-cell development. Cell Biol Int 40(12):1248–1255. https://doi.org/10.1002/cbin.10691

Keller A, Meese E (2016) Can circulating miRNAs live up to the promise of being minimal invasive biomarkers in clinical settings? Wiley Interdiscip Rev RNA 7(2):148–156. https://doi.org/10.1002/wrna.1320

Kennell JA, Gerin I, MacDougald OA, Cadigan KM (2008) The microRNA miR-8 is a conserved negative regulator of Wnt signaling. Proc Natl Acad Sci U S A 105(40):15417–15422. https://doi.org/10.1073/pnas.0807763105

Kim YK (2015) Extracellular microRNAs as biomarkers in human disease. Chonnam Med J 51(2):51–57. https://doi.org/10.4068/cmj.2015.51.2.51

Kim VN, Han J, Siomi MC (2009a) Biogenesis of small RNAs in animals. Nat Rev Mol Cell Biol 10(2):126–139. https://doi.org/10.1038/nrm2632

Kim YJ, Hwang SJ, Bae YC, Jung JS (2009b) MiR-21 regulates adipogenic differentiation through the modulation of TGF-beta signaling in mesenchymal stem cells derived from human adipose tissue. Stem Cells 27(12):3093–3102. https://doi.org/10.1002/stem.235

Kim H-J, Cho H, Alexander R, Patterson HC, Gu M et al (2014) MicroRNAs are required for the feature

maintenance and differentiation of brown adipocytes. Diabetes 63(12):4045–4056
Kirby TJ, McCarthy JJ (2013) MicroRNAs in skeletal muscle biology and exercise adaptation. Free Radic Biol Med 64:95–105. https://doi.org/10.1016/j.freeradbiomed.2013.07.004
Kozomara A, Griffiths-Jones S (2014) miRBase: annotating high confidence microRNAs using deep sequencing data. Nucleic Acids Res 42(Database issue):D68–D73. https://doi.org/10.1093/nar/gkt1181
Kraakman MJ, Murphy AJ, Jandeleit-Dahm K, Kammoun HL (2014) Macrophage polarization in obesity and type 2 diabetes: weighing down our understanding of macrophage function? Front Immunol 5:470. https://doi.org/10.3389/fimmu.2014.00470
Kredo-Russo S, Mandelbaum AD, Ness A, Alon I, Lennox KA et al (2012) Pancreas-enriched miRNA refines endocrine cell differentiation. Development 139 (16):3021–3031. https://doi.org/10.1242/dev.080127
Krek A, Grun D, Poy MN, Wolf R, Rosenberg L et al (2005) Combinatorial microRNA target predictions. Nat Genet 37(5):495–500
Kunej T, Jevsinek Skok D, Zorc M, Ogrinc A, Michal JJ et al (2013) Obesity gene atlas in mammals. J Genomics 1:45–55
Langi G, Szczerbinski L, Kretowski A (2019) Meta-analysis of differential miRNA expression after bariatric surgery. J Clin Med 8(8):1220. https://doi.org/10.3390/jcm8081220
Lee RC, Feinbaum RL, Ambros V (1993) The C-elegans heterochronic gene lin-4 encodes small rnas with antisense complementarity to lin-14. Cell 75(5):843–854
Lee J, Padhye A, Sharma A, Song G, Miao J et al (2010) A pathway involving farnesoid X receptor and small heterodimer partner positively regulates hepatic sirtuin 1 levels via microRNA-34a inhibition. J Biol Chem 285(17):12604–12611. https://doi.org/10.1074/jbc.M109.094524
Lee EK, Lee MJ, Abdelmohsen K, Kim W, Kim MM et al (2011) miR-130 suppresses adipogenesis by inhibiting peroxisome proliferator-activated receptor gamma expression. Mol Cell Biol 31(4):626–638. https://doi.org/10.1128/MCB.00894-10
Lewis BP, Burge CB, Bartel DP (2005) Conserved seed pairing, often flanked by adenosines, indicates that thousands of human genes are microRNA targets. Cell 120(1):15–20. https://doi.org/10.1016/j.cell.2004.12.035
Li S, Zhu J, Zhang W, Chen Y, Zhang K et al (2011) Signature microRNA expression profile of essential hypertension and its novel link to human cytomegalovirus infection. Circulation 124(2):175–184. https://doi.org/10.1161/CIRCULATIONAHA.110.012237
Li P, Hua X, Zhang Z, Li J, Wang J (2013) Characterization of regulatory features of housekeeping and tissue-specific regulators within tissue regulatory networks. BMC Syst Biol 7:112. https://doi.org/10.1186/1752-0509-7-112
Li XJ, Ren ZJ, Tang JH (2014) MicroRNA-34a: a potential therapeutic target in human cancer. Cell Death Dis 5:e1327. https://doi.org/10.1038/cddis.2014.270
Li J, Zhou C, Li J, Su Z, Sang H et al (2015) Global correlation analysis for microRNA and gene expression profiles in human obesity. Pathol Res Pract 211 (5):361–368
Li X, Ballantyne LL, Yu Y, Funk CD (2019) Perivascular adipose tissue-derived extracellular vesicle miR-221-3p mediates vascular remodeling. FASEB J 33 (11):12704–12722. https://doi.org/10.1096/fj.201901548R
Liang WC, Wang Y, Wan DC, Yeung VS, Waye MM (2013) Characterization of miR-210 in 3T3-L1 adipogenesis. J Cell Biochem 114(12):2699–2707. https://doi.org/10.1002/jcb.24617
Liang G, Zhu Y, Sun B, Shao Y, Jing A et al (2014) Assessing the survival of exogenous plant microRNA in mice. Food Sci Nutr 2(4):380–388. https://doi.org/10.1002/fsn3.113
Ling C, Ronn T (2019) Epigenetics in human obesity and type 2 diabetes. Cell Metab 29(5):1028–1044. https://doi.org/10.1016/j.cmet.2019.03.009
Ling HY, Wen GB, Feng SD, Tuo QH, Ou HS et al (2011) MicroRNA-375 promotes 3T3-L1 adipocyte differentiation through modulation of extracellular signal-regulated kinase signalling. Clin Exp Pharmacol Physiol 38(4):239–246. https://doi.org/10.1111/j.1440-1681.2011.05493.x
Linhart HG, Ishimura-Oka K, DeMayo F, Kibe T, Repka D et al (2001) C/EBPalpha is required for differentiation of white, but not brown, adipose tissue. Proc Natl Acad Sci U S A 98(22):12532–12537
Liu B, Li J, Cairns MJ (2014) Identifying miRNAs, targets and functions. Brief Bioinform 15(1):1–19. https://doi.org/10.1093/bib/bbs075
Locke AE, Kahali B, Berndt SI, Justice AE, Pers TH et al (2015) Genetic studies of body mass index yield new insights for obesity biology. Nature 518(7538):197–U401
Loos RJF, Yeo GSH (2014) The bigger picture of FTO: the first GWAS-identified obesity gene. Nat Rev Endocrinol 10(1):51–61
Lovis P, Gattesco S, Regazzi R (2008) Regulation of the expression of components of the exocytotic machinery of insulin-secreting cells by microRNAs. Biol Chem 389(3):305–312. https://doi.org/10.1515/BC.2008.026
Martinez-Sanchez A, Rutter GA, Latreille M (2016) MiRNAs in beta-cell development, identity, and disease. Front Genet 7:226. https://doi.org/10.3389/fgene.2016.00226
Matsuzaki J, Ochiya T (2017) Circulating microRNAs and extracellular vesicles as potential cancer biomarkers: a systematic review. Int J Clin Oncol 22(3):413–420. https://doi.org/10.1007/s10147-017-1104-3
McCarthy JJ (2011) The MyomiR network in skeletal muscle plasticity. Exerc Sport Sci Rev 39 (3):150–154. https://doi.org/10.1097/JES.0b013e31821c01e1

Mendell JT, Olson EN (2012) MicroRNAs in stress signaling and human disease. Cell 148(6):1172–1187
Mendez-Mancilla A, Lima-Rogel V, Toro-Ortiz JC, Escalante-Padron F, Monsivais-Urenda AE et al (2018) Differential expression profiles of circulating microRNAs in newborns associated to maternal pregestational overweight and obesity. Pediatr Obes 13(3):168–174. https://doi.org/10.1111/ijpo.12247
Mestdagh P, Hartmann N, Baeriswyl L, Andreasen D, Bernard N et al (2014) Evaluation of quantitative miRNA expression platforms in the microRNA quality control (miRQC) study. Nat Methods 11(8):809–815. https://doi.org/10.1038/nmeth.3014
Mittelbrunn M, Gutierrez-Vazquez C, Villarroya-Beltri C, Gonzalez S, Sanchez-Cabo F et al (2011) Unidirectional transfer of microRNA-loaded exosomes from T cells to antigen-presenting cells. Nat Commun 2:282. https://doi.org/10.1038/ncomms1285
Mononen N, Lyytikainen LP, Seppala I, Mishra PP, Juonala M et al (2019) Whole blood microRNA levels associate with glycemic status and correlate with target mRNAs in pathways important to type 2 diabetes. Sci Rep 9(1):8887. https://doi.org/10.1038/s41598-019-43793-4
Moseti D, Regassa A, Kim WK (2016) Molecular regulation of adipogenesis and potential anti-adipogenic bioactive molecules. Int J Mol Sci 17(1):124. https://doi.org/10.3390/ijms17010124
Nishimura R, Hata K, Ikeda F, Ichida F, Shimoyama A et al (2008) Signal transduction and transcriptional regulation during mesenchymal cell differentiation. J Bone Miner Metab 26(3):203–212. https://doi.org/10.1007/s00774-007-0824-2
Novak J, Kruzliak P, Bienertova-Vasku J, Slaby O, Novak M (2014) MicroRNA-206: a promising theranostic marker. Theranostics 4(2):119–133. https://doi.org/10.7150/thno.7552
Nunez Lopez YO, Coen PM, Goodpaster BH, Seyhan AA (2017) Gastric bypass surgery with exercise alters plasma microRNAs that predict improvements in cardiometabolic risk. Int J Obes 41(7):1121–1130. https://doi.org/10.1038/ijo.2017.84
Okada C, Yamashita E, Lee SJ, Shibata S, Katahira J et al (2009) A high-resolution structure of the pre-microRNA nuclear export machinery. Science 326(5957):1275–1279. https://doi.org/10.1126/science.1178705
O'Rahilly S, Farooqi IS (2008) Human obesity: a heritable neurobehavioral disorder that is highly sensitive to environmental conditions. Diabetes 57(11):2905–2910. https://doi.org/10.2337/db08-0210
Ortega FJ, Mercader JM, Catalan V, Moreno-Navarrete JM, Pueyo N et al (2013) Targeting the circulating microRNA signature of obesity. Clin Chem 59(5):781–792. https://doi.org/10.1373/clinchem.2012.195776
Oses M, Margareto Sanchez J, Portillo MP, Aguilera CM, Labayen I (2019) Circulating miRNAs as biomarkers of obesity and obesity-associated comorbidities in children and adolescents: a systematic review. Nutrients 11(12):2890. https://doi.org/10.3390/nu11122890
Palmer JD, Soule BP, Simone BA, Zaorsky NG, Jin L et al (2014) MicroRNA expression altered by diet: can food be medicinal? Ageing Res Rev 17:16–24. https://doi.org/10.1016/j.arr.2014.04.005
Pandey AK, Verma G, Vig S, Srivastava S, Srivastava AK et al (2011) miR-29a levels are elevated in the db/db mice liver and its overexpression leads to attenuation of insulin action on PEPCK gene expression in HepG2 cells. Mol Cell Endocrinol 332(1-2):125–133. https://doi.org/10.1016/j.mce.2010.10.004
Pastrello C, Tsay M, McQuaid R, Abovsky M, Pasini E et al (2016) Circulating plant miRNAs can regulate human gene expression in vitro. Sci Rep 6:32773. https://doi.org/10.1038/srep32773
Paul P, Chakraborty A, Sarkar D, Langthasa M, Rahman M et al (2018) Interplay between miRNAs and human diseases: a review. J Cell Physiol 233(3):2007–2018. https://doi.org/10.1002/jcp.25854
Pawlicki JM, Steitz JA (2008) Primary microRNA transcript retention at sites of transcription leads to enhanced microRNA production. J Cell Biol 182(1):61–76. https://doi.org/10.1083/jcb.200803111
Pedroza-Torres A, Romero-Córdoba SL, Justo-Garrido M, Salido-Guadarrama I, Rodríguez-Bautista R et al (2019) MicroRNAs in tumor cell metabolism: roles and therapeutic opportunities. Front Oncol 9:1404. https://doi.org/10.3389/fonc.2019.01404
Polakovicova M, Musil P, Laczo E, Hamar D, Kyselovic J (2016) Circulating microRNAs as potential biomarkers of exercise response. Int J Mol Sci 17(10):1553. https://doi.org/10.3390/ijms17101553
Pordzik J, Pisarz K, De Rosa S, Jones AD, Eyileten C et al (2018) The potential role of platelet-related microRNAs in the development of cardiovascular events in high-risk populations, including diabetic patients: a review. Front Endocrinol (Lausanne) 9:74. https://doi.org/10.3389/fendo.2018.00074
Pordzik J, Jakubik D, Jarosz-Popek J, Wicik Z, Eyileten C et al (2019) Significance of circulating microRNAs in diabetes mellitus type 2 and platelet reactivity: bioinformatic analysis and review. Cardiovasc Diabetol 18(1):113. https://doi.org/10.1186/s12933-019-0918-x
Poy MN, Eliasson L, Krutzfeldt J, Kuwajima S, Ma X et al (2004) A pancreatic islet-specific microRNA regulates insulin secretion. Nature 432(7014):226–230. https://doi.org/10.1038/nature03076
Prats-Puig A, Ortega FJ, Mercader JM, Moreno-Navarrete JM, Moreno M et al (2013) Changes in circulating microRNAs are associated with childhood obesity. J Clin Endocrinol Metab 98(10):E1655–E1660. https://doi.org/10.1210/jc.2013-1496
Pratt AJ, MacRae IJ (2009) The RNA-induced silencing complex: a versatile gene-silencing machine. J Biol Chem 284(27):17897–17901. https://doi.org/10.1074/jbc.R900012200

Price NL, Fernandez-Hernando C (2016) miRNA regulation of white and brown adipose tissue differentiation and function. Biochim Biophys Acta 1861(12 Pt B):2104–2110. https://doi.org/10.1016/j.bbalip.2016.02.010

Qadir AS, Woo KM, Ryoo HM, Baek JH (2013) Insulin suppresses distal-less homeobox 5 expression through the up-regulation of microRNA-124 in 3T3-L1 cells. Exp Cell Res 319(14):2125–2134. https://doi.org/10.1016/j.yexcr.2013.04.020

Qin L, Chen Y, Niu Y, Chen W, Wang Q et al (2010) A deep investigation into the adipogenesis mechanism: profile of microRNAs regulating adipogenesis by modulating the canonical Wnt/beta-catenin signaling pathway. BMC Genomics 11:320. https://doi.org/10.1186/1471-2164-11-320

Rankinen T, Bouchard C (2006) Genetics of food intake and eating behavior phenotypes in humans. Annu Rev Nutr 26:413–434. https://doi.org/10.1146/annurev.nutr.26.061505.111218

Riffo-Campos AL, Riquelme I, Brebi-Mieville P (2016) Tools for sequence-based miRNA target prediction: what to choose? Int J Mol Sci 17(12):1987. https://doi.org/10.3390/ijms17121987

Rottiers V, Naar AM (2012) MicroRNAs in metabolism and metabolic disorders. Nat Rev Mol Cell Biol 13 (4):239–250

Rupaimoole R, Slack FJ (2017) MicroRNA therapeutics: towards a new era for the management of cancer and other diseases. Nat Rev Drug Discov 16(3):203–222. https://doi.org/10.1038/nrd.2016.246

Sebastiani G, Nigi L, Grieco GE, Mancarella F, Ventriglia G et al (2017) Circulating microRNAs and diabetes mellitus: a novel tool for disease prediction, diagnosis, and staging? J Endocrinol Investig 40(6):591–610. https://doi.org/10.1007/s40618-017-0611-4

Sedgeman LR, Beysen C, Ramirez Solano MA, Michell DL, Sheng Q et al (2019) Beta cell secretion of miR-375 to HDL is inversely associated with insulin secretion. Sci Rep 9(1):3803. https://doi.org/10.1038/s41598-019-40338-7

Seitz H (2017) Issues in current microRNA target identification methods. RNA Biol 14(7):831–834. https://doi.org/10.1080/15476286.2017.1320469

Seok H, Ham J, Jang ES, Chi SW (2016) MicroRNA target recognition: insights from transcriptome-wide non-canonical interactions. Mol Cells 39(5):375–381. https://doi.org/10.14348/molcells.2016.0013

Slattery ML, Herrick JS, Mullany LE, Stevens JR, Wolff RK (2017) Diet and lifestyle factors associated with miRNA expression in colorectal tissue. Pharmgenomics Pers Med 10:1–16. https://doi.org/10.2147/PGPM.S117796

Song G, Xu G, Ji C, Shi C, Shen Y et al (2014) The role of microRNA-26b in human adipocyte differentiation and proliferation. Gene 533(2):481–487. https://doi.org/10.1016/j.gene.2013.10.011

Stranger BE, Stahl EA, Raj T (2011) Progress and promise of genome-wide association studies for human complex trait genetics. Genetics 187(2):367–383. https://doi.org/10.1534/genetics.110.120907

Stratz C, Nuhrenberg T, Fiebich BL, Amann M, Kumar A et al (2014) Controlled type II diabetes mellitus has no major influence on platelet micro-RNA expression. Results from micro-array profiling in a cohort of 60 patients. Thromb Haemost 111(5):902–911. https://doi.org/10.1160/TH13-06-0476

Sun T, Fu M, Bookout AL, Kliewer SA, Mangelsdorf DJ (2009) MicroRNA let-7 regulates 3T3-L1 adipogenesis. Mol Endocrinol 23(6):925–931. https://doi.org/10.1210/me.2008-0298

Szabo G, Bala S (2013) MicroRNAs in liver disease. Nat Rev Gastroenterol Hepatol 10(9):542–552. https://doi.org/10.1038/nrgastro.2013.87

Tang QQ, Lane MD (2012) Adipogenesis: from stem cell to adipocyte. Annu Rev Biochem 81:715–736. https://doi.org/10.1146/annurev-biochem-052110-115718

Thomou T, Mori MA, Dreyfuss JM, Konishi M, Sakaguchi M et al (2017) Adipose-derived circulating miRNAs regulate gene expression in other tissues. Nature 542(7642):450–455. https://doi.org/10.1038/nature21365

Thompson MD, Cismowski MJ, Serpico M, Pusateri A, Brigstock DR (2017) Elevation of circulating microRNA levels in obese children compared to healthy controls. Clin Obes 7(4):216–221. https://doi.org/10.1111/cob.12192

Torres R, Lang UE, Hejna M, Shelton SJ, Joseph NM et al (2020) MicroRNA ratios distinguish melanomas from nevi. J Invest Dermatol 140(1):164–173.e7. https://doi.org/10.1016/j.jid.2019.06.126

Trumbach D, Prakash N (2015) The conserved miR-8/miR-200 microRNA family and their role in invertebrate and vertebrate neurogenesis. Cell Tissue Res 359 (1):161–177. https://doi.org/10.1007/s00441-014-1911-z

Tsai WC, Hsu SD, Hsu CS, Lai TC, Chen SJ et al (2012) MicroRNA-122 plays a critical role in liver homeostasis and hepatocarcinogenesis. J Clin Invest 122 (8):2884–2897. https://doi.org/10.1172/JCI63455

Turchinovich A, Weiz L, Langheinz A, Burwinkel B (2011) Characterization of extracellular circulating microRNA. Nucleic Acids Res 39(16):7223–7233

van de Bunt M, Gaulton KJ, Parts L, Moran I, Johnson PR et al (2013) The miRNA profile of human pancreatic islets and beta-cells and relationship to type 2 diabetes pathogenesis. PLoS One 8(1):e55272. https://doi.org/10.1371/journal.pone.0055272

van Rooij E (2011) The art of microRNA research. Circ Res 108(2):219–234

Vasu S, Kumano K, Darden CM, Rahman I, Lawrence MC et al (2019) MicroRNA signatures as future biomarkers for diagnosis of diabetes states. Cells 8(12):1533. https://doi.org/10.3390/cells8121533

Venkatesh S, Workman JL (2015) Histone exchange, chromatin structure and the regulation of transcription. Nat Rev Mol Cell Biol 16(3):178–189. https://doi.org/10.1038/nrm3941

Vickers KC, Palmisano BT, Shoucri BM, Shamburek RD, Remaley AT (2011) MicroRNAs are transported in plasma and delivered to recipient cells by high-density lipoproteins. Nat Cell Biol 13(4):423–433. https://doi.org/10.1038/ncb2210

Vickers KC, Shoucri BM, Levin MG, Wu H, Pearson DS et al (2013) MicroRNA-27b is a regulatory hub in lipid metabolism and is altered in dyslipidemia. Hepatology 57(2):533–542. https://doi.org/10.1002/hep.25846

Villard A, Marchand L, Thivolet C, Rome S (2015) Diagnostic value of cell-free circulating microRNAs for obesity and type 2 diabetes: a meta-analysis. J Mol Biomark Diagn 6(6):251. https://doi.org/10.4172/2155-9929.1000251

Walden TB, Timmons JA, Keller P, Nedergaard J, Cannon B (2009) Distinct expression of muscle-specific microRNAs (myomirs) in brown adipocytes. J Cell Physiol 218(2):444–449

Wander PL, Boyko EJ, Hevner K, Parikh VJ, Tadesse MG et al (2017) Circulating early- and mid-pregnancy microRNAs and risk of gestational diabetes. Diabetes Res Clin Pract 132:1–9. https://doi.org/10.1016/j.diabres.2017.07.024

Wang K, Zhang S, Weber J, Baxter D, Galas DJ (2010) Export of microRNAs and microRNA-protective protein by mammalian cells. Nucleic Acids Res 38 (20):7248–7259. https://doi.org/10.1093/nar/gkq601

Wang YC, Li Y, Wang XY, Zhang D, Zhang H et al (2013a) Circulating miR-130b mediates metabolic crosstalk between fat and muscle in overweight/obesity. Diabetologia 56(10):2275–2285. https://doi.org/10.1007/s00125-013-2996-8

Wang YT, Tsai PC, Liao YC, Hsu CY, Juo SH (2013b) Circulating microRNAs have a sex-specific association with metabolic syndrome. J Biomed Sci 20:72. https://doi.org/10.1186/1423-0127-20-72

Wang Y, Luo J, Zhang H, Lu J (2016) microRNAs in the same clusters evolve to coordinately regulate functionally related genes. Mol Biol Evol 33(9):2232–2247. https://doi.org/10.1093/molbev/msw089

Weber JA, Baxter DH, Zhang S, Huang DY, Huang KH et al (2010) The microRNA spectrum in 12 body fluids. Clin Chem 56(11):1733–1741. https://doi.org/10.1373/clinchem.2010.147405

Wei R, Yang J, Liu GQ, Gao MJ, Hou WF et al (2013) Dynamic expression of microRNAs during the differentiation of human embryonic stem cells into insulin-producing cells. Gene 518(2):246–255. https://doi.org/10.1016/j.gene.2013.01.038

Wellen KE, Hotamisligil GS (2005) Inflammation, stress, and diabetes. J Clin Invest 115(5):1111–1119. https://doi.org/10.1172/JCI25102

WHO Library Cataloguing (2016) WHO Library Cataloguing in Publication Data Report of the commission on ending childhood obesity. WHO Library Cataloguing, Geneva

Xie H, Lim B, Lodish HF (2009) MicroRNAs induced during adipogenesis that accelerate fat cell development are downregulated in obesity. Diabetes 58 (5):1050–1057

Xiong XP, Vogler G, Kurthkoti K, Samsonova A, Zhou R (2015) SmD1 modulates the miRNA pathway independently of its pre-mRNA splicing function. PLoS Genet 11(8):e1005475. https://doi.org/10.1371/journal.pgen.1005475

Xu P, Vernooy SY, Guo M, Hay BA (2003) The Drosophila microRNA Mir-14 suppresses cell death and is required for normal fat metabolism. Curr Biol 13 (9):790–795

Yang WM, Jeong HJ, Park SY, Lee W (2014) Induction of miR-29a by saturated fatty acids impairs insulin signaling and glucose uptake through translational repression of IRS-1 in myocytes. FEBS Lett 588 (13):2170–2176. https://doi.org/10.1016/j.febslet.2014.05.011

Yi X, Liu J, Wu P, Gong Y, Xu X et al (2020) The key microRNA on lipid droplet formation during adipogenesis from human mesenchymal stem cells. J Cell Physiol 235(1):328–338. https://doi.org/10.1002/jcp.28972

Zamanillo R, Sanchez J, Serra F, Palou A (2019) Breast milk supply of microRNA associated with leptin and adiponectin is affected by maternal overweight/obesity and influences infancy BMI. Nutrients 11(11):2589. https://doi.org/10.3390/nu11112589

Zampetaki A, Kiechl S, Drozdov I, Willeit P, Mayr U et al (2010) Plasma microRNA profiling reveals loss of endothelial miR-126 and other microRNAs in type 2 diabetes. Circ Res 107(6):810–817. https://doi.org/10.1161/CIRCRESAHA.110.226357

Zhang R, Wang D, Xia Z, Chen C, Cheng P et al (2013a) The role of microRNAs in adipocyte differentiation. Front Med 7(2):223–230. https://doi.org/10.1007/s11684-013-0252-8

Zhang Y, Yang L, Gao YF, Fan ZM, Cai XY et al (2013b) MicroRNA-106b induces mitochondrial dysfunction and insulin resistance in C2C12 myotubes by targeting mitofusin-2. Mol Cell Endocrinol 381(1-2):230–240. https://doi.org/10.1016/j.mce.2013.08.004

Zhong D, Huang G, Zhang Y, Zeng Y, Xu Z et al (2013) MicroRNA-1 and microRNA-206 suppress LXRalpha-induced lipogenesis in hepatocytes. Cell Signal 25(6):1429–1437

Zhou Y, Gu P, Shi W, Li J, Hao Q et al (2016) MicroRNA-29a induces insulin resistance by targeting PPARdelta in skeletal muscle cells. Int J Mol Med 37(4):931–938. https://doi.org/10.3892/ijmm.2016.2499

Zhu H, Leung SW (2015) Identification of microRNA biomarkers in type 2 diabetes: a meta-analysis of controlled profiling studies. Diabetologia 58(5):900–911. https://doi.org/10.1007/s00125-015-3510-2

Zhu H, Shyh-Chang N, Segre AV, Shinoda G, Shah SP et al (2011) The Lin28/let-7 axis regulates glucose metabolism. Cell 147(1):81–94. https://doi.org/10.1016/j.cell.2011.08.033

Zile MR, Mehurg SM, Arroyo JE, Stroud RE, DeSantis SM et al (2011) Relationship between the temporal profile of plasma microRNA and left ventricular remodeling in patients after myocardial infarction. Circ Cardiovasc Genet 4(6):614–619. https://doi.org/10.1161/CIRCGENETICS.111.959841

From Adipogenic Viruses to Antidiabetic Drug: A Translational Journey

7

Vijay Hegde and Nikhil V. Dhurandhar

Contents

Introduction 98
Infectobesity 98
SMAM-1 Adenovirus 99
Human Adipogenic Virus: Adenovirus 36 99
Ad36 Improves Glycemic Control 99
Mechanism of Ad36 Action 100
Virus-Induced Obesity: A Causative Role in Humans 101
Exploiting the Antidiabetic Property of Ad36 103
Possibilities for Therapeutic Administration 104
Summary 104
References 105

Abstract

Obesity is a complex disease with multifactorial etiology requiring cause-specific treatment approaches. Among the various causes, obesity due to microbial infections has been reported since 1982, linking several microbes to obesity. Among them, avian adenovirus SMAM-1 and the human adenovirus Ad36 have been extensively studied for the past 25 years. Experimental Ad36 infection causes obesity in animal models, yet paradoxically improves glycemic control. The E4orf1 gene of Ad36 has been shown to be necessary and sufficient for adipogenesis and also responsible for better glycemic control. Animal models show that E4orf1 may be a candidate to treat type 1 or type 2 diabetes or nonalcoholic fatty liver disease. Collectively these studies show a causational and correlational link of Ad36 to animal and human obesity. Overall, developing vaccines to prevent virus-induced obesity and harnessing the antihyperglycemic potential of E4orf1 are the two long-term goals of this line of investigation.

V. Hegde · N. V. Dhurandhar (✉)
Department of Nutritional Sciences, Obesity and Metabolic Health Laboratory, Texas Tech University, Lubbock, TX, USA
e-mail: nikhil.dhurandhar@ttu.edu

J. Faintuch and S. Faintuch (eds.), *Obesity and Diabetes*, https://doi.org/10.1007/978-3-030-53370-0_7

Keywords

Ad36 virus · Adipogenesis · Viral obesity · Obesity vaccine · E4orf1 gene · Obesity control · Diabetes control

Introduction

The prevalence of obesity worldwide has nearly tripled since 1975 and continues to grow at a pandemic rate (Collaboration NCDRF 2016, 2017). Since 2013, obesity has been recognized as a disease by the American Medical Association (AMA) (Pollack 2013). The AMA declaration makes it more effective to address this complex disease and associated comorbidities, as obesity substantially increases the risk of metabolic diseases (for example, type 2 diabetes mellitus (T2D) and fatty liver disease), cardiovascular diseases (hypertension, myocardial infarction, and stroke), musculoskeletal disease (osteoarthritis), Alzheimer disease, depression, and some types of cancer (breast, ovarian, prostate, liver, kidney, and colon).

The fundamental cause of obesity is considered as an energy imbalance between calories consumed and calories expended. Hence, the core treatment recommendations rely almost exclusively on eating less and moving more to create a negative energy balance for inducing weight loss. However, at the individual level, weight-loss interventions aimed at reducing calorie intake and increasing energy expenditure are frequently not successful in the long term. Besides eating and physical activity, there are several putative factors such as infection, epigenetic changes, increased maternal age, endocrine disruptions, and intrauterine effects that might contribute to obesity (McAllister et al. 2009). This highlights that obesity has a multifactorial etiology, yet its current treatment is not cause-specific. For example, sleep disorders or sleep apnea is linked with obesity (Bayon et al. 2014), so an effective and comprehensive weight-loss treatment should address sleep as well. It is known that infections can influence obesity and obesity can influence susceptibility to or severity of infections (Dhurandhar et al. 2015). This chapter describes a novel translational journey from the identification and characterization of an adipogenic adenovirus Ad36, to exploiting its unique antidiabetic property for therapeutic application.

Infectobesity

Studies from several laboratories over the past 30 years have indicated that certain infections might promote the development of obesity, which is termed as "*Infectobesity*," (Dhurandhar 2011; Akheruzzaman et al. 2019) or obesity of infectious origin. Induction of obesity in mice experimentally infected with canine distemper virus was the first report of *Infectobesity* (Lyons et al. 1982). Since then, several animal and human viruses, bacteria, parasites, and scrapie agents have been reported to cause obesity in insects, chickens, rodents, and nonhuman primate models (Pasarica and Dhurandhar 2007). In a longitudinal birth cohort study, it was shown that infection, but not antibiotic use, during infancy is associated with a risk of childhood obesity (Li et al. 2017). Several additional microbes including *Chlamydia pneumoniae* (Dart et al. 2002), *Selenomonas noxia* (Goodson et al. 2009), *Helicobacter pylori* (Arslan et al. 2009), the gut microflora (Ley et al. 2006), and a collective load of infection with herpes simplex virus 1 or 2, enterovirus and *C. pneumoniae* (Fernandez-Real et al. 2007) are all associated with human obesity. The mechanism by which these microbes impose their adipogenic action varies, ranging from an effect on the central nervous system (canine distemper virus, Borna disease virus), to metabolism (gut microbiota), and adipose tissue (adenoviruses) (Pasarica and Dhurandhar 2007). The only human pathogens causatively and correlatively linked with animal and human obesity, respectively, are adenoviruses that directly influence the adipose tissue.

SMAM-1 Adenovirus

The first adipogenic adenovirus identified was an avian adenovirus, SMAM-1. Experimental SMAM-1 infection of chickens increased adiposity, yet reduced serum cholesterol and triglyceride levels compared to uninfected controls (Dhurandhar et al. 1990, 1992). Another set of uninfected chickens sharing cages with SMAM-1 infected chickens acquired infection and increased adiposity. Screening for the presence of SMAM-1 antibodies as an indicator of past SMAM-1 infection in humans, found 20% seropositivity (Dhurandhar et al. 1997). This was a surprising finding as humans are not considered susceptible to avian adenovirus infection. Phenotypically, the SMAM-1 seropositive subjects had significantly greater body mass index (BMI) as an indicator of adiposity, but lower serum cholesterol and triglyceride levels. The similarities found among animal and human studies suggested a causative role of SMAM-1 for increasing adiposity in humans, without conclusively establishing such a role.

Human Adipogenic Virus: Adenovirus 36

Human adenovirus 36 (Ad36) was the first human adipogenic virus reported, which was earlier isolated from the fecal sample of a girl suffering from enteritis (Wigand 1980). The presence of neutralizing antibodies is an indication of current or past infection with the virus. Compared with other human adenoviruses, Ad36 is antigenically distinct, which generates neutralizing antibodies that are specific to Ad36 infection. This greatly helps in specifically identifying past Ad36 infection in humans. A series of different experiments in animal models including chickens, mice, rats, and marmosets showed that Ad36 increases adiposity (Dhurandhar 2002; Dhurandhar et al. 2000, 2001, 2002; Pasarica et al. 2006; Na and Nam 2012). About 60–100% of the animals experimentally infected with Ad36 develop obesity (adiposity >85th percentile) compared with the uninfected control group. Ad36 infected animals have about 35–60% greater adipose tissue mass, a level of fat gain which is very similar to that increased by many genetic models of obesity (Chida et al. 2006; Butler et al. 2000; Lang et al. 2008). In rodents, Ad36 infection showed significantly greater adiposity in just 4 days post inoculation (Pasarica et al. 2008a), which persisted even 6–7 months later (Pasarica et al. 2006; Dhurandhar et al. 2002). In chow-fed mice, the adipogenic effect of Ad36 is evident; however, in mice on 60% fat diet, Ad36 infection did not increase adiposity over that induced by the high fat diet in control mock, uninfected mice (Krishnapuram et al. 2011). Ad36 infection and obesity can be transferred from an infected set of animals to uninfected recipients, thus fulfilling a Koch's postulate (Dhurandhar et al. 2001).

Ad36 Improves Glycemic Control

Rats experimentally infected with Ad36 gained adipose tissue, yet improved systemic glycemic control by lowering fasting serum insulin levels compared to control, and improved insulin sensitivity as determined by lower HOMA-index (Pasarica et al. 2006). In chow-fed and high-fat-fed mice, Ad36 infection improved glycemic control by lowering fasting serum insulin levels (Krishnapuram et al. 2011). Ad36 in fact inhibited the proximal insulin signaling and yet, increased the distal insulin signaling leading to increased cellular glucose uptake (Fig. 7.1). In adipose tissue, skeletal muscle, and liver, Ad36 infection selectively activated the Ras-PI3K pathway (distal insulin signaling) by increasing the abundance of glucose transporters (GLUT 4 and GLUT 1) in skeletal muscle and adipose tissue and their translocation to the cell membrane. In the liver, Ad36 attenuated GLUT2 expression, which suggested reduced hepatic glucose release.

This in vivo finding supported the in vitro study which demonstrated reduced glucose release in Ad36-infected hepatocytes

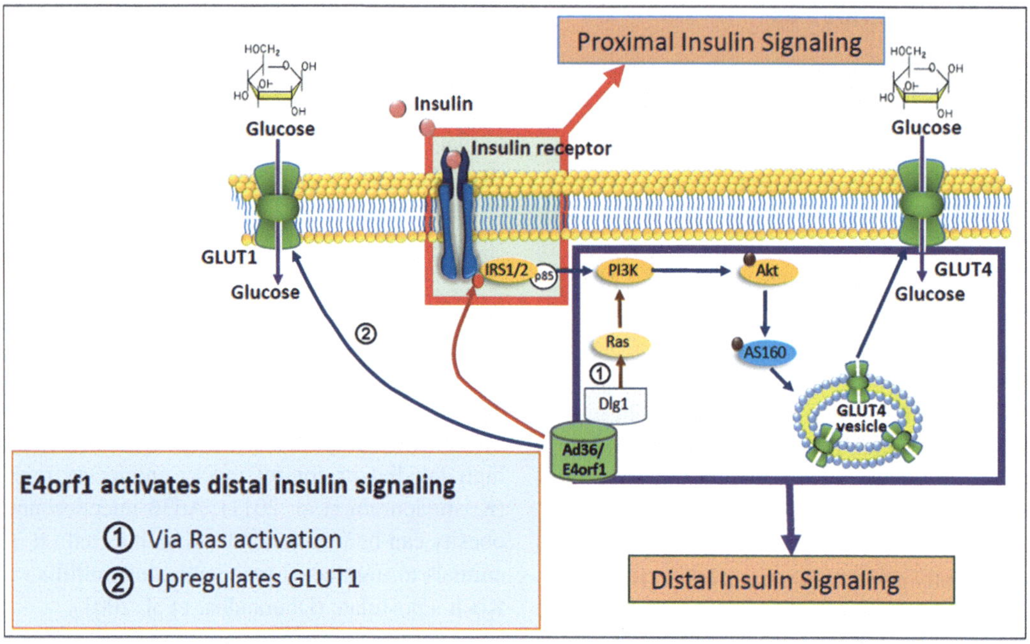

Fig. 7.1 Ad36/E4orf1 enhances cellular glucose uptake independent of insulin signaling. Ad36 and its E4orf1 protein increases cellular glucose uptake by increasing *GLUT1* and *GLUT4* gene expression and protein abundance, which appears to be mediated via *Ras*-activated *PI 3-kinase* pathway and independent of insulin signaling. More specifically, Ad36/E4orf1 expression blocks the proximal insulin signaling by increasing serine phosphorylation of insulin receptor and insulin receptor substrate 1/2 (IRS1/2), while increasing cellular glucose uptake by activating Ras/PI3K mediated phosphorylation of PKB/Akt leading to translocation of Glut4 from the cytoplasm to the cell membrane, collectively known as the distal insulin signaling

(Krishnapuram et al. 2011; Rogers et al. 2008a; Wang et al. 2008). Thus, Ad36 appears to improve systemic glycemic control by increasing cellular glucose uptake in skeletal muscle and adipose tissue, and by decreasing hepatic glucose output, thereby reducing circulating glucose levels. A similar signaling mechanism was observed when human skeletal muscle cells (HSKM) and adipose tissue explants were experimentally infected in vitro (Wang et al. 2008; Rogers et al. 2008b). Interestingly, Ad36-infected high-fat-fed mice significantly lowered hepatic lipid content and increased glycogen compared to control (Krishnapuram et al. 2011) indicating that Ad36 infection protects the liver from hepatic steatosis and inflammation. Ad36 reduced hepatic glucose release, lipogenesis, inflammation (downregulated inflammatory markers IL-6, INFγ, TNF-α), insulin resistance, and fibrosis and upregulated lipid oxidation and export in the livers of mice (Krishnapuram et al. 2011).

Mechanism of Ad36 Action

The adipogenic effect of Ad36 is evident in the visceral adipose tissue of Ad36-infected rats, showing 3–60-fold greater expression of adipogenic genes compared to that in uninfected rats matched for body fat (Pasarica et al. 2006). Studies revealed that Ad36 infects resident stem cells/stromal cells from adipose tissue and induces adipogenic commitment, replication, and differentiation in these cells leading to increased lipid accumulation (Rogers et al. 2008a; Vangipuram et al. 2004; Rathod et al. 2009; Pasarica et al. 2008b) (Fig. 7.2). Thus, Ad36 appears to increase adipose tissue

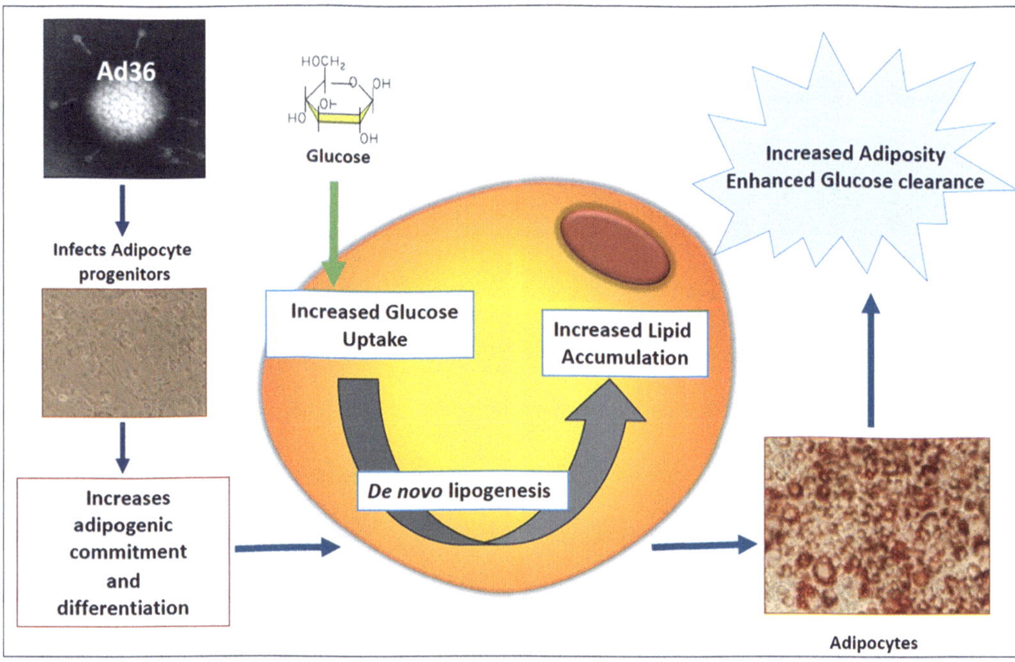

Fig. 7.2 Mechanism of Ad36 action. Ad36 infects resident stem cells/stromal cells (adipogenic progenitors) from adipose tissue and induces adipogenic commitment, replication, and differentiation in these cells, leading to increased glucose uptake, de novo lipogenesis, and lipid accumulation. Collectively this causes increased adiposity accompanied with enhanced glucose clearance

hyperplasia and hypertrophy to increase adiposity—a hypothesis that was indirectly tested in humans. Adipocyte progenitors obtained from human subjects had about eight-fold greater potential to differentiate into adipocytes, when the cells harbored Ad36 DNA in their adipose tissue, compared to their counterparts without the viral DNA (Pasarica et al. 2008b). This indicated an association of Ad36 infection with greater adipogenic potential in humans. Ad36 increases lipid accumulation only in the infected cells as observed in human adipose-derived stem cells (Pasarica et al. 2008b) (Fig. 7.3).

Adenoviral genes are divided into early genes, denoted by the letter E and late genes, denoted by the letter L. The early genes are transcribed earlier in virus replication, and the late genes are for structural proteins. The early gene 4 open reading frame 1 (E4orf1) of Ad36 was identified as to be necessary and sufficient for the Ad36 mediated induction of adipogenesis (Rogers et al. 2008a).

Virus-Induced Obesity: A Causative Role in Humans

Obesity has an insidious onset and progress, which makes it challenging to determine when an individual might have been infected. Also, obesity can have multiple etiological factors, which makes it difficult to identify one single cause. It is possible that for an individual, a combination of infection with other factors may be involved in obesity development. Lastly, while animal experiments convincingly demonstrated that Ad36 increases adiposity, experimental Ad36 infection of humans is not possible due to ethical considerations. For these reasons, it is challenging to establish a causative role for Ad36 in human obesity. Instead, a framework has been proposed to indirectly assess such a role of Ad36 (Dhurandhar 2011).

One key component is to test the hypothesis that humans who are naturally infected with Ad36

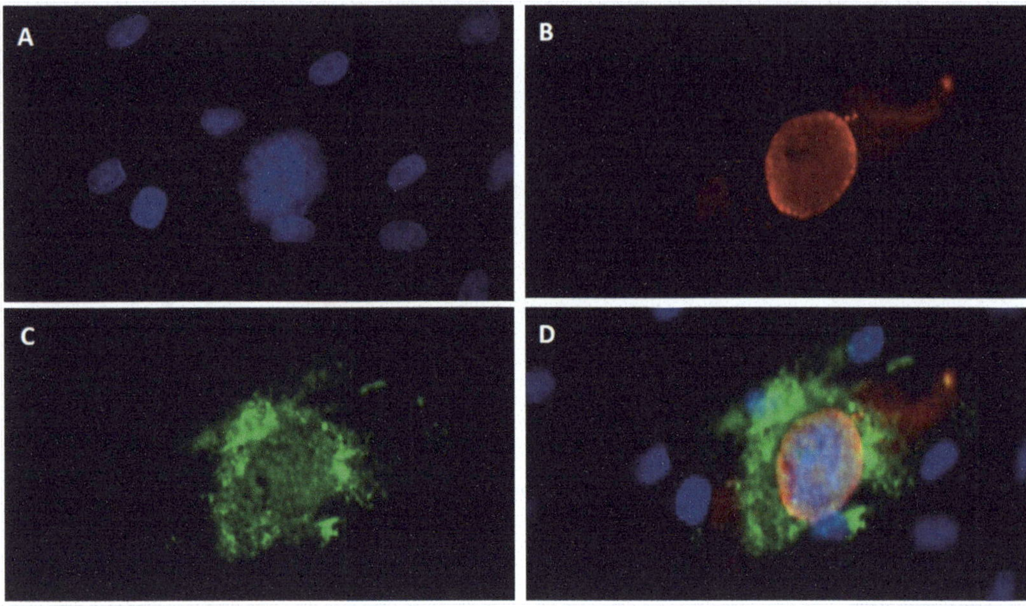

Fig. 7.3 Ad36 infection induces lipid accumulation in human adipose-derived stem/stromal cells (hASC). Lipid accumulation is observed specifically only in Ad36-infected hASC. (**a**) shows DAPI stain for nuclei of multiple cells (blue), and (**b**) Ad36 hexon protein antibodies are seen in one infected cell (red). (**c**) Lipid-specific BODIPY stain shows cellular lipid accumulation (green), (**d**) only in the Ad36-infected cell (merge) 6 days post inoculation

will exhibit a phenotype similar to that observed in animals that are experimentally infected with Ad36. In Ad36 infected animals, the major phenotypic changes include greater adiposity, yet better metabolic profile including lower serum cholesterol and triglycerides, better glycemic control, and less hepatic steatosis (Pasarica and Dhurandhar 2007; Pasarica et al. 2006; Na and Nam 2012; Krishnapuram et al. 2011). Interestingly, natural Ad36 infection appears to mirror associations with these phenotypes in humans. Nearly a dozen studies from Korea (Na et al. 2010, 2012), Italy (Trovato et al. 2009), and the United States of America (Dhurandhar 2011) have reported the association of Ad36 with overweight and/or obesity in adults and children. A study of 1400 Caucasian, Black, or Hispanic men, women, or children from four separate cohorts showed significantly better glycemic control and lower hepatic fat levels in individuals who were naturally infected with Ad36 (Krishnapuram et al. 2011). The association between Ad36 infection and obesity has been conclusively shown in a series of meta-analysis studies. Yamada et al. showed that Ad36 infection was positively associated with significantly higher BMI and obesity (Yamada et al. 2012). Two other meta-analysis studies with 5739 and 10,191 subjects respectively, also showed the association between Ad36 infection and increased risk of obesity (Xu et al. 2015; Shang et al. 2014).

Most twins share very similar body profiles due to their shared genetic makeup. Therefore, twins who are discordant for Ad36 infection could provide evidence about the possible adipogenic role of Ad36 in humans. Twins with natural Ad36 infection had significantly greater BMI and body fat, compared to their co-twins who were uninfected (Atkinson et al. 2005).

Associations do not establish causations. A 10-year follow-up of over 1400 Hispanic men and women, those who were naturally infected with Ad36 at baseline, showed significantly greater percent body fat at baseline and a

Table 7.1 Similarity between animal and human studies of Ad36

Experimental infection	Natural exposure in humans
(a) Adiposity and adipogenesis	(a) Obesity and greater body fat gain
(b) Better glycemic control	(b) Better glycemic control
(c) Less liver fat	(c) Less liver fat

significant gain in adiposity and better glycemic control over 10 years. Natural Ad36 infection was also positively associated with obesity but negatively associated with nonalcoholic fatty liver disease (NAFLD) in Italian populations (Trovato et al. 2009, 2010, 2012) (Table 7.1).

The implications for identifying and characterizing Ad36 induced obesity is that unique prevention and treatment strategies can be identified. Additionally, these strategies could be beneficial for the treatment of other adipogenic pathogens. For example, a long-term goal of this line of investigation is to develop a vaccine to prevent obesity that is specifically caused by Ad36 infection.

Exploiting the Antidiabetic Property of Ad36

In vitro studies identified E4orf1, a 125 amino acid protein of Ad36, as required and essential for the observed beneficial effects with Ad36. In vitro studies show that E4orf1 increases glucose uptake in pre-adipocytes, adipocytes, and myoblasts, while reducing glucose release from hepatocytes (Dhurandhar et al. 2011a). E4orf1 selectively bypasses the proximal insulin signaling and upregulates the distal insulin signaling pathway via increased expression of PI3K, PKB/AKT, and GLUT4 translocation to increase cellular glucose uptake (Na et al. 2016a; Shastri et al. 2018). Inflammatory cytokines generally impair glycemic control. However, E4orf1 improves cellular glucose uptake even in the presence of inflammatory cytokines (Na et al. 2016b). In hepatocytes, E4orf1 reduces glucose output and de-novo lipogenesis, increases complete fatty acid oxidation, and promotes lipid transport, which are considered the key determinants of hepatic lipid storage (Dhurandhar et al. 2012). Collectively, these in vitro studies show that E4orf1 can improve cellular glucose metabolism even in the presence of insulin resistance, impaired insulin signaling, and inflammation.

Cell signaling studies also indicated that E4orf1 reduces de-novo lipogenesis, and promotes fatty acid oxidation and the export of lipid from hepatocytes (Dhurandhar et al. 2012), which may explain its contribution to reduced hepatic steatosis in mice (McMurphy et al. 2017; Kusminski et al. 2015). Functionally, Ad36 or E4orf1 increase glucose uptake in preadipocytes, adipocytes, myocytes, and reduce glucose release from hepatocytes (Dhurandhar et al. 2011a, 2012; Krishnapuram et al. 2013; Na et al. 2013), which likely collectively contributes to better glycemic control induced by E4orf1 in vivo (McMurphy et al. 2017; Kusminski et al. 2015; Hegde et al. 2016).

Ad36 or E4orf1 robustly upregulate the expression of PPAR-gamma and adiponectin in adipocytes, which in turn may help enhance cellular glucose uptake (Rogers et al. 2008b; Dubuisson et al. 2011). However, E4orf1 is capable of enhancing cellular glucose uptake independently of peroxisome proliferator-activated receptor (PPAR) gamma (Dubuisson et al. 2011). Based on what was known about E4orf1 action and our findings, the molecular events that lead to E4orf1-induced glucose uptake can be surmised as follows (Dhurandhar et al. 2011a, b, 2012; Kusminski et al. 2015; Frese et al. 2003; Kong et al. 2014; Kumar et al. 2014): E4orf1 protein is synthesized in a cell, after Ad36 infection, hence it is unlikely to have a cell surface receptor. In its interaction with the insulin signaling pathway, it bypasses proximal insulin signaling, which includes the binding of insulin to its cell receptor and the signal transduction through tyrosine phosphorylation of insulin receptor substrates (IRS). In fact, E4orf1 inhibits the proximal insulin signaling by reducing tyr-phosphorylation of IRS, and by increasing its ser-phosphorylation. Instead, E4orf1 enters insulin signaling at the level of phosphatidyl inositol 3-kinase (PI3K) activation.

E4orf1 has a PDZ domain-binding motif (PBM) at the C-terminal. Through its PBM, the peptide binds to PDZ proteins, which are scaffolding proteins that facilitate protein–protein interactions. E4orf1 requires its PBM to bind to Dlg (Drosophila disc large)-1 protein. After the binding, the complex travels to the cell membrane, where it activates Ras and subsequently, upregulates PI3K. E4orf1 requires intact PBM for Ras induction, and for the activation of PI3K-AKT signaling, leading to greater membrane translocation of Glut4, and glucose uptake (Fig. 7.1). This understanding prompted efforts to therapeutically exploit the properties of E4orf1.

Possibilities for Therapeutic Administration

Being an exogenous, virus-derived protein, E4orf1 does not have any endogenous cellular receptors, and delivering it in vivo is challenging. Hence, the efficacy of E4orf1 was tested in vivo using different approaches: (1) Retrovirus mediated expression, (2) Adeno-associated vector expressing E4orf1, and (3) Transgenic mice with adipose tissue-specific inducible expression of E4orf1. All three approaches conducted in separate laboratories improved glycemic control in mice models (McMurphy et al. 2017; Kusminski et al. 2015; Hegde et al. 2016).

Retrovirus-mediated expression of E4orf1 in mice reproducibly improved glucose clearance following a glucose challenge despite a high-fat diet (Hegde et al. 2016). Further, E4orf1 also improved tissue-specific molecular signaling in mice, mainly greater protein expression of adiponectin, p-AKT, and glucose transporter Glut4. Hepatic expression of AAV-E4orf1 in a genetic model of diabetes (db/db mice), dietary model of insulin resistance (DIO mice), and normoglycemic animals (wild-type mice), completely alleviated hyperglycemia and robustly improved glycemic control without significantly increasing liver fat accumulation or hepatic steatosis (McMurphy et al. 2016). Transgenic expression of E4orf1 in the adipose tissue of mice upon doxycycline induction improved glucose clearance, despite a high-fat diet challenge, by enhancing the Ras-ERK-MAPK signaling in transgenic adipocytes (Kusminski et al. 2015).

Glucose and insulin responses to glucose load were determined at the same time in these mice, which showed near lack of insulin response in the E4orf1 group, indicating that the enhanced glucose disposal by E4orf1 seems to reduce the need of endogenous insulin (Kusminski et al. 2015). This phenomenon is termed as the "*insulin sparing effect*" of E4orf1 (Dhurandhar 2013). Hydrodynamic gene delivery of E4orf1 with an adipocyte-targeting sequence (ATS-E4orf1) in the liver of high-fat-fed and streptozotocin-injected mice (disease models of type 2 and type 1 diabetes, respectively), improved the ability of these mice to eliminate excess glucose from the blood and ameliorate liver function in both disease models.(Yoon et al. 2017; Voss et al. 2015) Therefore, E4orf1 expression, systemic or tissue-specific, in mice improves glycemic control independent of high-fat diet intake and impaired proximal insulin signaling, by reducing insulin levels in vivo.

This highlights that E4orf1 is not an insulin secretagogue or insulin mimetic, rather it displays endogenous insulin sparing action to clear glucose. This insulin-independent ability of E4orf1and its protection of liver against lipid accumulation makes it a novel and effective therapeutic candidate for type 2 diabetes, type 1 diabetes, and NAFLD.

Summary

Either the E4orf1 protein itself or its small molecule analogs may be useful for developing as therapeutic agent(s) to improve hyperglycemia. An appropriate delivery mechanism, such as nanoparticle-mediated delivery of E4orf1 protein or small molecule analogs, is needed to effectively and specifically deliver E4orf1 to targeted tissues, in order to harness its therapeutic effects.

References

Akheruzzaman M, Hegde V, Dhurandhar NV (2019) Twenty-five years of research about adipogenic adenoviruses: a systematic review. Obes Rev 20(4):499–509

Arslan E, Atilgan H, Yavasoglu I (2009) The prevalence of helicobacter pylori in obese subjects. Eur J Intern Med 20(7):695–697

Atkinson RL, Dhurandhar NV, Allison DB et al (2005) Human adenovirus-36 is associated with increased body weight and paradoxical reduction of serum lipids. Int J Obes 29(3):281–286

Bayon V, Leger D, Gomez-Merino D, Vecchierini M-F, Chennaoui M (2014) Sleep debt and obesity. Ann Med 46(5):264–272

Butler AA, Kesterson RA, Khong K et al (2000) A unique metabolic syndrome causes obesity in the melanocortin-3 receptor-deficient mouse. Endocrinology 141(9):3518–3521

Chida D, Osaka T, Hashimoto O, Iwakura Y (2006) Combined interleukin-6 and interleukin-1 deficiency causes obesity in young mice. Diabetes 55(4):971–977

Collaboration NCDRF (2016) Trends in adult body-mass index in 200 countries from 1975 to 2014: a pooled analysis of 1698 population-based measurement studies with 19.3872 million participants. Lancet (10026):1377–1396

Collaboration NCDRF (2017) Worldwide trends in body-mass index, underweight, overweight, and obesity from 1975 to 2016: a pooled analysis of 2416 population-based measurement studies in 128.9 million children, adolescents, and adults. Lancet 390 (10113):2627–2642

Dart AM, Martin JL, Kay S (2002) Association between past infection with chlamydia pneumoniae and body mass index, low-density lipoprotein particle size and fasting insulin. Int J Obes Relat Metab Disord 26 (4):464–468

Dhurandhar NV (2002) Infections and body weight: an emerging relationship? Int J Obes Relat Metab Disord 26(6):745–746

Dhurandhar NV (2011) A framework for identification of infections that contribute to human obesity. Lancet Infect Dis 11:963–969

Dhurandhar NV (2013) Insulin sparing action of adenovirus 36 and its E4orf1 protein. J Diabetes Complicat 27 (2):191–199

Dhurandhar N, Kulkarni P, Ajinkya S, Sherikar A (1990) Avian adenovirus leading to pathognomonic obesity in chicken. J Bom Vet College 2:131–132

Dhurandhar NV, Kulkarni P, Ajinkya SM, Sherikar A (1992) Effect of adenovirus infection on adiposity in chicken. Vet Microbiol 31(2-3):101–107

Dhurandhar NV, Kulkarni PR, Ajinkya SM, Sherikar AA, Atkinson RL (1997) Association of adenovirus infection with human obesity. Obes Res 5(5):464–469

Dhurandhar NV, Israel BA, Kolesar JM, Mayhew GF, Cook ME, Atkinson RL (2000) Increased adiposity in animals due to a human virus. Int J Obes Relat Metab Disord 24(8):989–996

Dhurandhar NV, Israel BA, Kolesar JM, Mayhew G, Cook ME, Atkinson RL (2001) Transmissibility of adenovirus-induced adiposity in a chicken model. Int J Obes Relat Metab Disord 25(7):990–996

Dhurandhar NV, Whigham LD, Abbott DH et al (2002) Human adenovirus Ad-36 promotes weight gain in male rhesus and marmoset monkeys. J Nutr 132 (10):3155–3160

Dhurandhar EJ, Dubuisson O, Mashtalir N, Krishnapuram R, Hegde V, Dhurandhar NV (2011a) E4orf1: a novel ligand that improves glucose disposal in cell culture. PLoS One 6(8):e23394

Dhurandhar EJ, Dong C, Ye J, NV D (2011b) Ad36 E4orf1: a novel ligand to attenuate hepatic steatosis without weight loss? Obesity 19(Suppl 1):S86

Dhurandhar EJ, Krishnapuram R, Hegde V et al (2012) E4orf1 improves lipid and glucose metabolism in hepatocytes: a template to improve steatosis & hyperglycemia. PLoS One 7(10):e47813

Dhurandhar NV, Bailey D, Thomas D (2015) Interaction of obesity and infections. Obes Rev 16(12):1017–1029

Dubuisson O, Dhurandhar EJ, Krishnapuram R et al (2011) PPARgamma-independent increase in glucose uptake and adiponectin abundance in fat cells. Endocrinology 152(10):3648–3660

Fernandez-Real JM, Ferri MJ, Vendrell J, Ricart W (2007) Burden of infection and fat mass in healthy middle-aged men. Obesity (Silver Spring) 15(1):245–252

Frese KK, Lee SS, Thomas DL et al (2003) Selective PDZ protein-dependent stimulation of phosphatidylinositol 3-kinase by the adenovirus E4-ORF1 oncoprotein. Oncogene 22(5):710–721

Goodson JM, Groppo D, Halem S, Carpino E (2009) Is obesity an oral bacterial disease? J Dent Res 88 (6):519–523

Hegde V, Na HN, Dubuisson O et al (2016) An adenovirus-derived protein: a novel candidate for anti-diabetic drug development. Biochimie 121:140–150

Kong K, Kumar M, Taruishi M, Javier RT (2014) The human adenovirus E4-ORF1 protein subverts discs large 1 to mediate membrane recruitment and dysregulation of phosphatidylinositol 3-kinase. PLoS Pathog 10(5):e1004102

Krishnapuram R, Dhurandhar EJ, Dubuisson O et al (2011) Template to improve glycemic control without reducing adiposity or dietary fat. Am J Physiol Endocrinol Metab 300(5):E779–E789

Krishnapuram R, Dhurandhar EJ, Dubuisson O, Hegde V, Dhurandhar NV (2013) Doxycycline-regulated 3T3-L1 preadipocyte cell line with inducible, stable expression of adenoviral E4orf1 gene: a cell model to study insulin-independent glucose disposal. PLoS One 8(3):e60651

Kumar M, Kong K, Javier RT (2014) Hijacking discs large 1 for oncogenic phosphatidylinositol 3-kinase activation in human epithelial cells is a conserved mechanism of human adenovirus E4-ORF1 proteins. J Virol 88(24):14268–14277

Kusminski CM, Gallardo-Montejano VI, Wang ZV et al (2015) E4orf1 induction in adipose tissue promotes insulin-independent signaling in the adipocyte. Mol Metab 4(10):653–664

Lang P, van Harmelen V, Ryden M et al (2008) Monomeric tartrate resistant acid phosphatase induces insulin sensitive obesity. PLoS One 3(3):e1713

Ley RE, Turnbaugh PJ, Klein S, Gordon JI (2006) Microbial ecology: human gut microbes associated with obesity. Nature 444(7122):1022–1023

Li DK, Chen H, Ferber J, Odouli R (2017) Infection and antibiotic use in infancy and risk of childhood obesity: a longitudinal birth cohort study. Lancet Diabetes Endocrinol 5(1):18–25

Lyons MJ, Faust IM, Hemmes RB, Buskirk DR, Hirsch J, Zabriskie JB (1982) A virally induced obesity syndrome in mice. Science 216(4541):82–85

McAllister EJ, Dhurandhar NV, Keith SW et al (2009) Ten putative contributors to the obesity epidemic. Crit Rev Food Sci Nutr 49(10):868–913

McMurphy TB, Huang W, Xiao R, Liu X, Dhurandhar NV, Cao L (2016) Hepatic expression of adenovirus 36 E4ORF1 improves glycemic control and promotes glucose metabolism via AKT activation. Diabetes 66:358–371

McMurphy TB, Huang W, Xiao R, Liu X, Dhurandhar NV, Cao L (2017) Hepatic expression of adenovirus 36 E4ORF1 improves glycemic control and promotes glucose metabolism through AKT activation. Diabetes 66(2):358–371

Na HN, Nam JH (2012) Adenovirus 36 as an obesity agent maintains the obesity state by increasing MCP-1 and inducing inflammation. J Infect Dis 205(6):914–922

Na HN, Hong YM, Kim J, Kim HK, Jo I, Nam JH (2010) Association between human adenovirus-36 and lipid disorders in Korean schoolchildren. Int J Obes 34 (1):89–93

Na HN, Kim J, Lee HS et al (2012) Association of human adenovirus-36 in overweight Korean adults. Int J Obes 36(2):281–285

Na HN, Dubuisson O, Hegde V, Nam JH, NV d (2013) Human adenovirus Ad36 and its E4orf1 gene enhance cellular glucose uptake even in presence of inflammatory cytokines. Obesity Abstract Book S210

Na HN, Hegde V, Dubuisson O, Dhurandhar NV (2016a) E4orf1 enhances glucose uptake independent of proximal insulin signaling. PLoS One 11(8):e0161275

Na HN, Dubuisson O, Hegde V, Nam JH, Dhurandhar NV (2016b) Human adenovirus Ad36 and its E4orf1 gene enhance cellular glucose uptake even in the presence of inflammatory cytokines. Biochimie 124:3–10

Pasarica M, Dhurandhar NV (2007) Infectobesity: obesity of infectious origin. Adv Food Nutr Res 52:61–102

Pasarica M, Shin AC, Yu M et al (2006) Human adenovirus 36 induces adiposity, increases insulin sensitivity, and alters hypothalamic monoamines in rats. Obesity (Silver Spring) 14(11):1905–1913

Pasarica M, Loiler S, Dhurandhar NV (2008a) Acute effect of infection by adipogenic human adenovirus Ad36. Arch Virol 153(11):2097–2102

Pasarica M, Mashtalir N, McAllister EJ et al (2008b) Adipogenic human adenovirus Ad-36 induces commitment, differentiation, and lipid accumulation in human adipose-derived stem cells. Stem Cells 26 (4):969–978

Pollack A (2013) A.M.A. recognizes obesity as a disease. www.nytimes.com/2013/06/19/business/ama-recognizes-obesity-as-a-disease.html (Accessed 5 Sept 2018).

Rathod MA, Rogers PM, Vangipuram SD, McAllister EJ, Dhurandhar NV (2009) Adipogenic cascade can be induced without adipogenic media by a human adenovirus. Obesity (Silver Spring) 17(4):657–664

Rogers PM, Fusinski KA, Rathod MA et al (2008a) Human adenovirus Ad-36 induces adipogenesis via its E4 orf-1 gene. Int J Obes 32(3):397–406

Rogers PM, Mashtalir N, Rathod MA et al (2008b) Metabolically favorable remodeling of human adipose tissue by human adenovirus type 36. Diabetes 57 (9):2321–2331

Shang Q, Wang H, Song Y et al (2014) Serological data analyses show that adenovirus 36 infection is associated with obesity: a meta-analysis involving 5739 subjects. Obesity (Silver Spring) 22(3):895–900

Shastri AA, Hegde V, Peddibhotla S, Feizy Z, Dhurandhar NV (2018) E4orf1: a protein for enhancing glucose uptake despite impaired proximal insulin signaling. PLoS One 13(12):e0208427

Trovato GM, Castro A, Tonzuso A et al (2009) Human obesity relationship with Ad36 adenovirus and insulin resistance. Int J Obes 33(12):1402–1409

Trovato GM, Martines GF, Garozzo A et al (2010) Ad36 adipogenic adenovirus in human non-alcoholic fatty liver disease. Liver Int 30(2):184–190

Trovato GM, Martines GF, Trovato FM et al (2012) Adenovirus-36 seropositivity enhances effects of nutritional intervention on obesity, bright liver, and insulin resistance. Dig Dis Sci 57:535–544

Vangipuram SD, Sheele J, Atkinson RL, Holland TC, Dhurandhar NV (2004) A human adenovirus enhances preadipocyte differentiation. Obes Res 12(5):770–777

Voss JD, Atkinson RL, Dhurandhar NV (2015) Role of adenoviruses in obesity. Rev Med Virol 25 (6):379–387

Wang ZQ, Cefalu WT, Zhang XH et al (2008) Human adenovirus type 36 enhances glucose uptake in diabetic and nondiabetic human skeletal muscle cells independent of insulin signaling. Diabetes 57(7):1805–1813

Wigand R (1980) Age and susceptibility of Swiss mice for mouse adenovirus, strain FL. Arch Virol 64 (4):349–357

Xu MY, Cao B, Wang DF et al (2015) Human adenovirus 36 infection increased the risk of obesity: a meta-analysis update. Medicine (Baltimore) 94(51):e2357

Yamada T, Hara K, Kadowaki T (2012) Association of adenovirus 36 infection with obesity and metabolic markers in humans: a meta-analysis of observational studies. PLoS One 7(7):e42031

Yoon IS, Park S, Kim RH, Ko HL, Nam JH (2017) Insulin-sparing and fungible effects of E4orf1 combined with an adipocyte-targeting sequence in mouse models of type 1 and type 2 diabetes. Int J Obes

Genetic Profiles in the Obese Population 8

Ana Carolina Proença da Fonseca, Patrícia Torres Bozza, and Pedro Hernán Cabello

Contents

Introduction 108

Monogenic Obesity 108
Homeostatic Pathways Involved in the Regulation of Appetite and Satiety 108
Non-syndromic Form of Obesity 109
Leptin Gene Aberrations 110

The Melanocortin Pathway 116
Syndromic Form of Obesity 118

Polygenic or Common Obesity 119
Relevant Candidates 120

Ongoing Studies and Future Perspectives 120

References 121

Abstract

Based on the genetic etiology, obesity could be divided into two main categories: polygenic and monogenic (syndromic and non-syndromic) forms. The polygenic form of obesity is caused by a combination of environmental factors and several genetic variants, with minor effects. The monogenic forms are caused by mutations in single genes with major effects. We focus on the recent advances and future perspectives in the field of genetic obesity, to provide a comprehensive review covering the underlying mechanisms involved in the development of obesity.

Keywords

Obesity · Genetic of obesity · Leptin-melanocortin pathway · Mutation · BMI · Severe early-onset obesity

A. C. P. da Fonseca (✉)
Laboratory of Immunopharmacology, Oswaldo Cruz Institute, Oswaldo Cruz Foundation, Rio de Janeiro, Brazil

Human Genetics Laboratory, Oswaldo Cruz Institute, Oswaldo Cruz Foundation, Rio de Janeiro, Brazil
e-mail: ana.proenca@ioc.fiocruz.br

P. T. Bozza
Laboratory of Immunopharmacology, Oswaldo Cruz Institute, Oswaldo Cruz Foundation, Rio de Janeiro, Brazil

P. H. Cabello
Human Genetics Laboratory, Oswaldo Cruz Institute, Oswaldo Cruz Foundation, Rio de Janeiro, Brazil

Human Genetics Laboratory, Grande Rio University, Rio de Janeiro, Brazil

J. Faintuch and S. Faintuch (eds.), *Obesity and Diabetes*, https://doi.org/10.1007/978-3-030-53370-0_8

Introduction

Obesity can cause adverse health effects due to its association with several diseases, such as type 2 diabetes (T2D), hypertension, metabolic syndrome, cardiovascular diseases, and some cancers (Garvey et al. 2016; WHO 2019). These comorbidities associated with obesity contribute to reducing lifespan, and are responsible for three million deaths per year (Finucane et al. 2011; De Lorenzo et al. 2016).

Globally, 650 million adults and 124 million children and adolescents (aged 5–19 years) were obese in 2016. Currently, the global prevalence is 13% and will reach 20% by 2025, if the post-2000 trends continue (Garvey et al. 2016; WHO 2019). One of the major contributors to population weight gain is energy imbalance, caused by the ingestion of high-calorie foods that exceeds energy expenditure. However, not all subjects with obesogenic lifestyle develop this disorder, suggesting a strong genetic component (Pigeyre and Meyre 2018). The percentage of obesity that can be attributed to genetics ranges from 40 to 70%, according to previous studies using monozygotic and dizygotic twins (Wardle et al. 2008).

The etiology of obesity is extremely complex, being influenced by multiple interactions among environmental, hormonal, lifestyle, metabolic, nutritional, and genetic factors (De Lorenzo et al. 2016). Common forms of obesity are multifactorial and polygenic, caused by an interaction of environmental factors with multiple genetic variants with minor effects. However, rare monogenic forms were identified in humans and mice, which are caused by mutations with major effects in a single gene (Fig. 8.1) (da Fonseca et al. 2017). Several genes have been associated with human obesity, in which they could be related to either polygenic, monogenic, or both forms (Locke et al. 2015; Saeed et al. 2015; Hasnain 2016; Tunç et al. 2017).

Monogenic Obesity

Monogenic forms of obesity are caused by a loss-of-function mutation in a single gene (Albuquerque et al. 2015). These single-gene disorders result from variants of disturbing genes that encode receptors or ligands involved in appetite and satiety regulation as well as metabolism (da Fonseca et al. 2017). They can be divided into two different groups: syndromic obesity is usually associated with cognitive delay, dysmorphic characteristics and, organ-specific developmental abnormalities; whereas monogenic non-syndromic obesity is associated with intense hyperphagia, early-onset severe obesity and in some cases neuroendocrine dysfunctions. These forms can have a Mendelian autosomal (dominant and recessive) or X-linked inheritance (Pigeyre and Meyre 2018). Although monogenic forms of obesity are rare in humans, the identification of affected individuals allows for personalized clinical management, genetic counseling and, in some cases, specific pharmacological therapy.

Homeostatic Pathways Involved in the Regulation of Appetite and Satiety

In the early 1900s, clinical and experimental studies have postulated the involvement of the hypothalamus in the regulation of satiety, appetite, and body weight. Firstly, mice and cats were submitted to chemical and electrical experiments that damaged their hypothalamus, which result in an increase or loss of bodyweight depending on the size and region of the lesion (Hetherington and Ranson 1940; Anand and Brobeck 1951). In humans, patients with brain tumors in the hypothalamic area can become obese (Daousi et al. 2005). Later, the existence of a specific system in the hypothalamus that can regulate energy homeostasis called leptin-melanocortin pathway was discovered (Fig. 8.2) (Cone 2005; Morton et al. 2006; Larder et al. 2014).

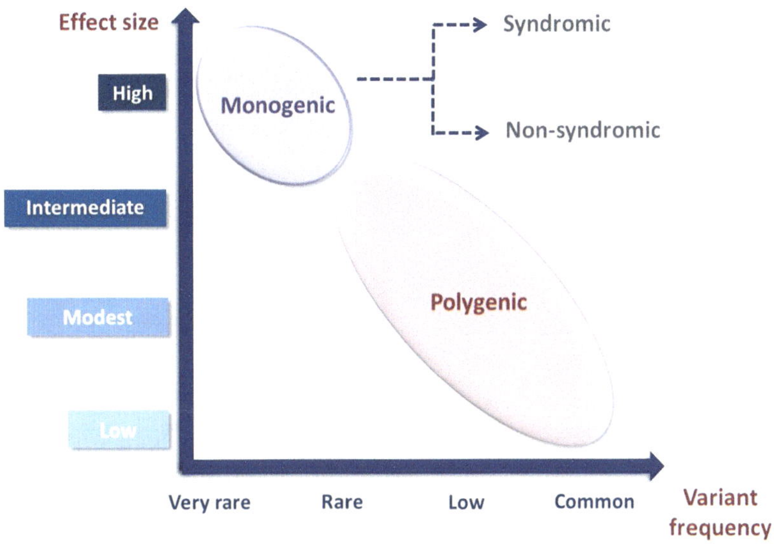

Fig. 8.1 Polygenic obesity occurs due to a combination of multiple genetic variants with common, low, and rare frequency that have to low up to intermediate effect. However, monogenic obesity is caused by rare or very rare mutations with high effects

The brain controls energy balance through obtaining information about the energy stock by peripheral hormones, and then it is used to regulate the neuronal circuit into the hypothalamus (Cone 2005; Larder et al. 2014). During fasting and low energy stock, the stomach secretes ghrelin that binds to its receptor in the orexigenic neurons, stimulating the expression of neuropeptide Y (NPY) and agouti-related protein (AgRP). Orexigenic neurons project from the arcuate nucleus (ARC) to the paraventricular nucleus (PVN) of hypothalamus, stimulating food intake, and suppressing energy expenditure (Cone 2005; Vetter et al. 2010; Sutton et al. 2016). However, leptin is secreted by white adipose tissue when the body accumulates adiposity and/or energy replacement occurs. It crosses the blood–brain barrier and binds to leptin receptor (LEPR), localized in the ARC.

Leptin acts suppressing NPY and AgRP expression as well as inducing the proopiomelanocortin (POMC), cocaine, and amphetamine-related transcript production in anorexigenic neurons (Vetter et al. 2010; Albuquerque et al. 2015; da Fonseca et al. 2017). POMC is cleaved and processed by prohormone convertase 1/3 (PC1/3, encoded by proprotein convertase subtilisin/kexin type 1 [*PCSK1*] gene), which produces α-melanocyte-stimulating hormone (α-MSH) (Freemark 2018). These products bind to the melanocortin-4 receptor (MC4R) in the PVN, resulting in a decreased energy intake and an increased basal energy expenditure (Vetter et al. 2010; Larder et al. 2014; da Fonseca et al. 2017; Freemark 2018). Interestingly, genetic variants involved in leptin-melanocortin pathway have been associated with the development of severe early-onset obesity (Wabitsch et al. 2014; Saeed et al. 2015; da Fonseca et al. 2019b).

Non-syndromic Form of Obesity

Over the past two decades, a large number of different genetic variants in the ligands or receptors, which disrupt the signaling of the leptin-melanocortin pathway were identified, resulting in hyperphagia, severe obesity in early childhood and in some cases, endocrine and immune disorders (Asai et al. 2013; Tunç et al. 2017). All these mutations were found in thirteen genes, including *LEP*, *LEPR*, *POMC*, *PCSK1*, and *MC4R* (Table 8.1). The non-syndromic form of obesity can affect up to 2–5% of the severely obese subjects and is transmitted under Mendelian autosomal recessive or dominant category

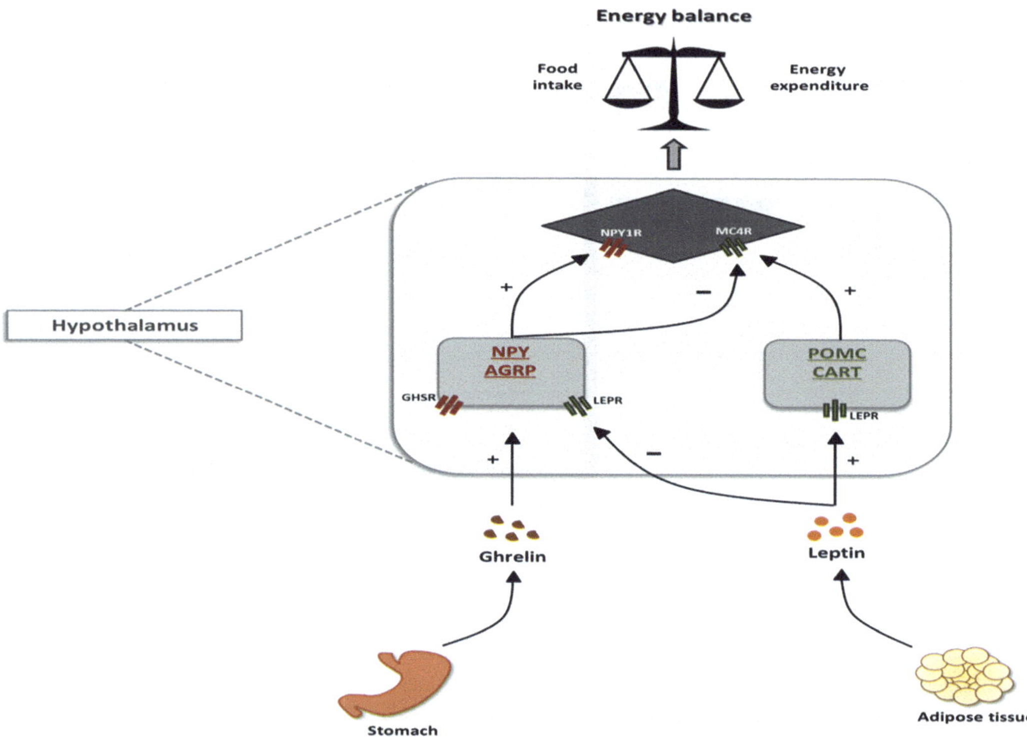

Fig. 8.2 Schematic presentation of hypothalamic leptin-melanocortin pathway. Leptin is secreted by adipose tissue and binds to its receptor located in two sets of neurons in the arcuate nuclei of hypothalamus. Leptin activates the anorexigenic pathway and inhibits orexigenic system, resulting in decreased food intake and increased basal energy expenditure. In contrast, ghrelin is produced by stomach and activates the orexigenic neurons, increasing the caloric intake and suppressing the energy expended. LEPR, receptor of leptin; GHSR, growth hormone secretagogue receptor; NPY, neuropeptide Y; AGRP, agouti-related protein; NPY1R, neuropeptide Y receptor Y1; POMC, proopiomelanocortin; CART, cocaine and amphetamine-related transcript; MC4R, melanocortin-4 receptor

(da Fonseca et al. 2017; Freemark 2018; Saeed et al. 2018).

Leptin Gene Aberrations

Congenital leptin deficiency (OMIM#614962) is a very rare autosomal recessive disorder caused by homozygous mutation in *LEP* gene on chromosome 7q31.3. In 1997, Montague and collaborators described for the first time a human leptin deficiency caused by a frameshift homozygous mutation (c.398delG; p. Gly133_fs*14). This variant was identified in two severely obese children who are cousins, within a highly consanguineous family of Pakistani origin. The same mutation was reported in another two children of Pakistani origin but unrelated to the earlier found patients (Farooqi et al. 2002; Gibson et al. 2004). Eleven other pathogenic mutations have since been reported. These genetic variants were identified in patients from consanguineous families in Colombia, Pakistan, German, Turkey, India, and Egypt and from a non-consanguineous family in Austria. With the exception of two mutations (p. Asn103Lys and p.Arg105Trp), all pathogenic *LEP* variants found are population specific and have not been identified in other ethnic groups (Saeed et al. 2015; Hasnain 2016).

To date, less than 50 subjects have been described to be homozygous for loss-of-function mutations in *LEP* gene. These patients usually exhibit very low or undetectable leptin levels

Table 8.1 Pathogenic mutations identified in *LEP*, *LEPR*, *POMC*, and *PCSK1* genes

Gene/Pathogenic mutations	*n*	Ethnicity	Type of mutation	Sample	Consanguinity	Analysis method	References
LEP							
c.1-44del42	1	Pakistani	Splicing defect	Child	Yes	Next generation sequencing (candidate genes)	Saeed et al. (2015)
c.104_106delTCA (p. Ile35del)	2	Pakistani	Inframe	Child	Yes	Sanger sequencing	Fatima et al. (2011), Saeed et al. (2012)
c.163C>T (p.Gln55Stop)	1	Indian	Nonsense	Child	Yes	Sanger sequencing	Thakur et al. (2014)
c.272T>C (p.Leu72Ser)	1	Austrian	Missense	Child	No	Sanger sequencing	Fischer-Posovszky et al. (2010)
c.298G>T (p.Asp100Try)	1	Turkish	Missense	Child	Yes	Sanger sequencing	Wabitsch et al. (2014)
c.309C>A (p.Asn103Lys)	2	Egyptian	Missense	Child	Yes	Sanger sequencing	Mazen et al. (2009)
	1	Pakistani			NA	PCR–RFLP	Hasnain (2016)
	2	German				Sanger sequencing	Wabitsch et al. (2015)
c.313C>T (p.Arg105Trp)	3	Turkish	Missense	Child/adult	Yes	Sanger sequencing	Strobel et al. (1998)
	2	Pakistani		Child		Next generation sequencing (candidate genes)	Saeed et al. (2015)
c.350G>A (p.Cys117Tyr)	1	Pakistani	Missense	Child	Yes	Next generation sequencing (candidate genes)	Saeed et al. (2015)
c.350G>T (p.Cys117Phe)	2	Colombian	Missense	Adult	Yes	Next generation sequencing (candidate genes)	Yupanqui-Lozno et al. (2019)
c.398delG (p. Gly133_fs*14)	22	Pakistani	Frameshift	Child	Yes	Sanger sequencing and next generation sequencing (candidate genes)	Montague et al. (1997), Farooqi et al. (2002), Gibson et al. (2004), Fatima et al. (2011), Saeed et al. (2012, 2015)
c.481_482delCT (p. Leu161fs*17)	1	Pakistani	Frameshift	Child	Yes	Sanger sequencing	Fatima et al. (2011)
LEPR							
c.2396-1G>T	4	Pakistani	Splicing defect	Child	Yes	Next generation sequencing (candidate genes)	Saeed et al. (2014, 2015)

(continued)

Table 8.1 (continued)

Gene/Pathogenic mutations	*n*	Ethnicity	Type of mutation	Sample	Consanguinity	Analysis method	References
c.2597+1G>A	3	Algerian	Splicing defect	Child and adult	Yes	Single-stranded conformation-dependent polymorphism (SSCP)	Clement et al. (1998)
4-bp deletion in codon 22	3	Bangladeshi	Frameshift	Child	Yes	Sanger sequencing	Farooqi et al. (2007b)
c.93G>A (p.Trp31*)	3	Southern European	Nonsense	Adult	Yes	Sanger sequencing	Farooqi et al. (2007b)
11-bp deletion in codon 70	2	Turkish	Frameshift	Adolescent and adult	Yes	Sanger sequencing	Farooqi et al. (2007b)
66-bp deletion in codon 514	1	Iranian	Frameshift	Child	Yes	Sanger sequencing	Farooqi et al. (2007b)
c.479delA (p. His160Leufs*9)	2	Sudanese	Frameshift	Child	Yes	Whole exome sequencing	Gill et al. (2014)
c.556delT (p. Cys186Alafs*27)	1	Guinean	Frameshift	Child	Yes	Whole exome sequencing	Gill et al. (2014)
c.946C>A (p.Pro316Thr)	2	Egyptian	Missense	Child	Yes	Sanger sequencing	Mazen et al. (2011)
c.1226C>A (p.Ala409Glu)	1	Turkish	Missense	Child	Yes	Sanger sequencing	Farooqi et al. (2007b)
c.1675G>A (p.Trp558*)	2	Pakistani	Nonsense	Child	Yes	Next generation sequencing (candidate genes)	Saeed et al. (2014, 2015)
c.1810T>A (p.Cys604Ser)	2	Pakistani	Missense	Child	Yes	Next generation sequencing (candidate genes)	Saeed et al. (2015)
c.1810T>G (p.Cys604Gly)	1	French	Missense	Adult	Yes	Sanger sequencing	Huvenne et al. (2015)
c.1871dup (p. Asn624Lys*21)	2	French	Splicing defect	Child and adult	No	Sanger sequencing	Le Beyec et al. (2013), Huvenne et al. (2015)
c.1992G>C (p.Trp664Arg)	1	Norwegian	Missense	Child	Yes	Sanger sequencing	Farooqi et al. (2007b)
c.2051A>C (p.His684Pro)	1	United Kingdom	Missense	Child	No	Sanger sequencing	Farooqi et al. (2007b)
c.2357T>C (p.Leu786Pro)	1	Portuguese	Missense	Adult	Yes	Sanger sequencing	Huvenne et al. (2015)
c.2491G>A (p. Asp831Asn)	1	Turkish	Missense	Child	Yes	Sanger sequencing	Huvenne et al. (2015)
Δexon6–8 (p. Pro166Cysfs*7)	5	French (Reunion Island)	Large deletion	Child, adolescent and adult	NA	Sanger sequencing	Huvenne et al. (2015)

Δexon6–8 (p. Pro166Cysfs*7) and c.1604–1G>A	1	French (Reunion Island)	Large deletion and splicing defect	Adolescent	NA	Sanger sequencing	Huvenne et al. (2015)
1-bp deletion in codon 15 and c.1835G>A (p. Arg612His)	1	United Kingdom	Frameshift and missense	Adult	Yes	Sanger sequencing	Farooqi et al. (2007b)
c.1264T>C (p.Tyr422His) and c.2131dup (p. Tyr711Asnfs*18)	2	French	Missense and Frameshift	Adult	No	Sanger sequencing	Huvenne et al. (2015)
c.946C>A (p.Pro316Thr) and c.1938G>C (p. Trp646Cys)	1	Turkish	Missense	Child	Yes	Sanger sequencing	Andiran et al. (2011)
c.1604-8A>G (p. Lys597Serfs*34) and p. Val596Aspfs*3	1	Dutch	Splicing defect	Adolescent	NA	Whole exome sequencing	Hannema et al. (2016)
c.1753–1dupG (p. Met585Aspfs*2) and c.2168C>T (p.Ser723Phe)	1	Dutch	Splicing defect	Child	NA	Whole exome sequencing	Hannema et al. (2016)
1322bp incl intron 3 and 58857bp/incl exon 4–18 small insertions	1	Pakistani	CNV	Child	Yes	Next generation sequencing (candidate genes)	Saeed et al. (2015)
POMC							
c.-11C>A	1	German	Out-of-frame translation initiation codon	Child	No	Sanger sequencing	Krude et al. (1998)
	1	Netherland			NA		Krude et al. (2003)
	1	Scottish-German			Yes		Aslan et al. (2014)
c.64delA	1	Turkish	Frameshift	Child	Yes	Unknown	Özen et al. (2015)
c.133-2A>C	2	Iraqi	Splicing defect	Child	Yes	Unknown	Ozsu and Bahm (2017)
c.202C>T (p.Gln68Ter)	1	Egyptian	Nonsense	Child	No	Sanger sequencing	Cirillo et al. (2012)
c.206delC (p. Pro69Leufs*2)	2	Turkish	Frameshift	Child	Yes/no	Sanger sequencing	Farooqi et al. (2006), Çetinkaya et al. (2018)
c.223dupC	1	North African	Frameshift	Adolescent	Yes	Sanger sequencing	Clément et al. (2008)
c.231C>A (p.Tyr77Ter)	1	Hispanic	Nonsense	Child	No	Sanger sequencing	Mendiratta et al. (2011)
c.256C>T (p.Arg86Ter)	1	Indian	Nonsense	Child	Yes	Sanger sequencing	Hung et al. (2012)

(continued)

Table 8.1 (continued)

Gene/Pathogenic mutations	*n*	Ethnicity	Type of mutation	Sample	Consanguinity	Analysis method	References
c.-11C>A and c.403_404insGG (p. Lys136Alafs*23)	2	Switzerland	Out-of-frame translation initiation codon and Frameshift		NA	Sanger sequencing	Krude et al. (2003)
c.-11C>A and c.251G>A (p.Trp84Ter)	1	Unknown	Out-of-frame translation initiation codon and nonsense	Child	No	Sanger sequencing	Anisimova et al. (2016), Çetinkaya et al. (2018)
c.151A>T and c.299delC	1	Slovenian	Missense and Frameshift		NA	Sanger sequencing	Krude et al. (2003)
c.313G>T and c.433delC	1	Unknown	Frameshift and missense	Child	NA	Sanger sequencing	Krude et al. (1998)
PCSK1							
c.1095+1G>T	1	Turkish	Splicing defect	Adolescent	Yes	Sanger sequencing	Martín et al. (2013)
c.1095+1G>A	1	Turkish	Splicing defect	Child	Yes	Sanger sequencing	Martín et al. (2013)
c.544–2A>G	1	Turkish	Splicing defect	Child	No	Sanger sequencing	Härter et al. (2016)
c.2T>C (p.Met1?)	2	Arab	Start lost	Child and adolescent	Yes	Sanger sequencing	Martín et al. (2013)
c.238C4T (p.Arg80*)	3	French	Nonsense	Child and adult	No	Next generation sequencing (candidate genes)	Philippe et al. (2015)
c.693C>G (p.Tyr231*)	1	Turkish	Nonsense	Child	Yes	Sanger sequencing	Martín et al. (2013)
c.920C>T (p.Ser307Leu)	1	Libyan	Missense	Child	Yes	Sanger sequencing	Farooqi et al. (2007a)
c.927C>G (p.Asn309Lys)	1	Arab	Missense	Child	Yes	Exome sequencing	Wilschanski et al. (2014)
c.1009C>T (p.Gln337*)	1	Turkish	Nonsense	Child	Yes	Sanger sequencing	Martín et al. (2013)
c.1213C>T (p.Arg405*)	2	African	Nonsense	Child	No	Sanger sequencing	Martín et al. (2013)
c.1269C>A (p.Asn423Lys)	1	Canadian	Missense	Adolescent	No	Sanger sequencing	Martín et al. (2013)
c.1350delGGAT (p. Val450fs*1)	1	Arab	Frameshift	Child	Yes	Sanger sequencing	Martín et al. (2013)
c.1643T>C (p.Phr548Ser)	1	Turkish	Missense	Child	Yes	Sanger sequencing	Martín et al. (2013)
c.1777G>A (p.Gly593Arg)	1	Hispanic	Missense	Child	Yes	Sanger sequencing	Martín et al. (2013)
c.620+4A>C and c.1777G>A (p.Gly593Arg)	1	NA	Splicing defect and missense	Adult	NA	Single-stranded conformation-dependent polymorphism (SSCP)	Jackson et al. (1997)

c.748G>T (p.Glu250*) and c.638_640delCAG (p. Ala213del)	1	Caucasian	Nonsense and inframe	Child	No	Sanger sequencing	Jackson et al. (2003)
474 kb deletion including the entire gene and c.1024delT (Trp342Glyfs*92)	1	NA	Whole gene deletion and frameshift	Child	NA	Sanger sequencing	Frank et al. (2013)
c.625G>A (p.Gly209Arg) and c.772C>A (p. Pro258Thr)	1	Indian	Missense	Child	Yes	Sanger sequencing	Martín et al. (2013)

Note: *NA*, not available

into the circulation as a result of impairment in either synthesis or secretion of this adipocytokine (Saeed et al. 2015; Hasnain 2016). However, Wabitsch and collaborators (2014, 2015) described three probands carrying different homozygous *LEP* mutations (p.Asp100Tyr and p.Asp103Lys) with hyperleptinemia. In these cases, the mutant leptin is produced and secreted but is not functional.

Leptin Receptor Disorders

Leptin receptor deficiency (OMIM#614963) is another autosomal recessive disorder, caused by homozygous/compound heterozygous mutations in *LEPR* gene on chromosome 1p31. Clement and collaborators (1998) reported the first cases of *LEPR* deficiency in three siblings with early-onset morbid obesity from a consanguineous family of Kabilian origin (north of Algeria). After that, several *LEPR* pathogenic mutations have been reported in probands from European, Egyptian, Pakistani, and Turkish origin (Gill et al. 2014; Saeed et al. 2014, 2015; Huvenne et al. 2015).

Gill et al. (2014) have performed whole-exome sequencing in five individuals with extreme obesity of four consanguineous families (four probands and one affected sibling). They identified two novel *LEPR* frameshift mutations (p.His160Leufs*9 and p.Cys186Alafs*27), resulting in a truncated receptor that lacks the domain required for leptin signal transduction. Interestingly, one novel large deletion was observed in six unrelated subjects with early-onset morbid obesity from Reunion Island (Indian Ocean, France), suggesting a founder effect. Due to the high frequency of this large deletion, a quantitative real-time PCR was developed to detect heterozygous carriers in this population (Huvenne et al. 2015).

Clinical Profile

Since *LEPR* deficiency also disrupts leptin signaling, clinical hallmarks of affected patients are similar to those with homozygous *LEP* mutations. These patients had typical features including severe early-onset obesity, hyperphagia, hypogonadotropic hypogonadism, impaired immune system, and neuroendocrine/metabolic dysfunction (Le Beyec et al. 2013; Gill et al. 2014).

Selected LEP-deficient probands could be pharmacologically treated with recombinant leptin (metreleptin/Myalept), leading to reduction in food intake, fat mass, and metabolic and endocrine abnormalities (Wabitsch et al. 2014; Roth et al. 2018). However, the drug is not approved for this indication, just for lipodystrophy.

LEPR deficiency patients usually have elevated circulation leptin concentrations, reflecting the loss of sensitivity of the receptor. There is no evidence that metreleptin could be useful; however, Setmelanotide is currently undergoing advanced clinical trials for such indication, reportedly meeting all defined endpoints (Globenewswire.com 2019).

The Melanocortin Pathway

Proopiomelanocortin Gene

Similarly to *LEP* and *LEPR*, rare pathogenic mutations described in *POMC* and *PCSK1* were associated with severe early-onset obesity, transmitted under Mendelian autosomal recessive. *POMC* gene is located on chromosome 2p23.3 and encodes a complex pro-peptide which is processed post-transcriptionally to produce several bioactive peptides (adrenocorticotropin [ACTH], β-endorphin, and α-,β-, γ- MSH). These peptides bind to melanocortin receptors and act on energy balance, skin pigmentation, adrenal steroidogenesis, and thermoregulation. Alpha- and beta-MSH activate MC3R and MC4R as well as antagonize the action of AgRP, involved in the control of satiety and appetite (Cawley et al. 2016). Homozygous or heterozygous pathogenic mutations have been reported in 17 probands from Netherland, Slovenian, German, Switzerland, Indian, Turkish, Iraqi, and Hispanic origins (Anisimova et al. 2016; Ozsu and Bahm 2017; Çetinkaya et al. 2018).

The first cases of severe obesity due to POMC deficiency (OMIM#609734) were observed in two unrelated children from Germany. The first proband carried compound heterozygote variants (c.433delC and c.313G>T), which disrupt the corrected synthesis of ACTH and α-MSH, and the second proband was homozygous for a mutation in the splice site (c.-11C>A) that eliminates POMC translation. Both children presented severe early-onset obesity, adrenal insufficiency, and red hair pigmentation. Later, the same group described three new cases with similar clinical phenotype (Krude et al. 2003). The hypopigmentation is due to inadequate activation of MC1R by POMC peptides in the skin. However, Farooqi et al. (2006) identified a novel *POMC* homozygous frameshift mutation (c.206delC) in a 2-year-old male proband with severe obesity, hypocortisolism, and brown hair with dark red roots. This result suggests that patients with severe early-onset obesity and adrenal insufficiency should be evaluated to *POMC* mutations even in the absence of red hair pigmentation (Mendiratta et al. 2011). Despite POMC deficiency being a rare disorder, several studies have reported that the defective allele in the heterozygous state may contribute to the obesity susceptibility (Farooqi et al. 2001; Challis et al. 2002; Lee et al. 2006; Samuels et al. 2013).

Proprotein Convertase Gene

Congenital proprotein convertase (PC1/3) deficiency (OMIM#600955) is caused by loss-of-function mutations in the proprotein convertase subtilisin /kexin type 1 (*PCSK1*) gene located on chromosome 5q15 (Frank et al. 2013; Martín et al. 2013; Härter et al. 2016). PC1/3 is a member of the serine endoproteases family, which is expressed in endocrine and neuronal cells. In the hypothalamus, PC1/3 is involved in the POMC processing, which results in the production of α-MSH (Smith and Funder 1988; Tao 2005; Vetter et al. 2010). Homozygous and compound heterozygous mutations in the *PCSK1* gene were described in several patients with hyperphagia, severe malabsorptive diarrhea, central diabetes insipidus, growth hormone deficiency, primary hypogonadism, and other endocrines abnormalities (Frank et al. 2013; Martín et al. 2013; Wilschanski et al. 2014; Härter et al. 2016). The functional consequences of *PCSK1* mutations may have an important role in the severity and variety of the clinical phenotype (Martín et al. 2013). Interestingly, Philippe et al. (2015) reported a novel nonsense heterozygous mutation in *PCSK1* that was co-segregate with obesity in a French family, following a dominant mode of inheritance. Therefore, rare pathogenic mutations in *PCSK1* may be associated with an autosomal dominant or recessive form of Mendelian obesity (Martín et al. 2013; Philippe et al. 2015).

Melanocortin Receptor Genes

Melanocortin 4 receptor (MC4R) deficiency (OMIM#618406) is the most common cause of non-syndromic monogenic obesity. This disorder is caused by heterozygous, heterozygous compound, and homozygous deleterious variants in *MC4R* gene, located on chromosome 18q21.32 (Doulla et al. 2014; da Fonseca et al. 2017; Drabkin et al. 2018). Several rare pathogenic variants were identified in the *MC4R* gene, which could lead to partial or complete loss of receptor function (Collet et al. 2017). *MC4R* variations are inherited in an autosomal codominant model, with a high degree of expressivity and penetrance in heterozygous carriers (Farooqi et al. 2003; Farooqi 2015). Previous studies have reported that patients with heterozygous mutations have a less severe phenotype compared to homozygous carriers (Farooqi et al. 2003; Drabkin et al. 2018). *MC4R* is expressed in the PVN and is the receptor for α-MSH, regulating appetite and satiety. Consistent with this role, loss-of-function *MC4R* variants are associated with hyperphagia and severe early-onset obesity. Moreover, some patients can also exhibit severe hyperinsulinemia and increased linear growth (Farooqi et al. 2003; Melchior et al. 2012; Doulla et al. 2014; da Fonseca et al. 2019b).

Recently, our group described for the first time a Brazilian case with *MC4R* start lost mutation (p. Met1?). This variant was identified in a 29-years-old female patient with childhood-onset obesity, moderate binge-eating disorder, and high caloric intake (4900 calories per day) (da Fonseca et al. 2019b). Since pharmacological agents are being developed to treat patients with MC4R deficiency, it is extremely important to assess the sequence of this gene in severe early-onset obesity individuals in order to identify pathogenic variations that would benefit from this treatment (Chen et al. 2015; Collet et al. 2017).

Setmelanotide has entered two phase III clinical trials in 2017, including pro-opiomelanocortin (POMC) deficiency obesity, besides leptin receptor (LEPR) deficiency obesity, with encouraging results (Globenewswire.com 2019).

Other Genes

Besides MC4R protein, BDNF haploinsufficiency has also been associated with severe early-onset obesity. This protein is a member of the nerve growth factor family, involved in the proliferation, survival, and differentiation of neurons (Tapia-Arancibia et al. 2004). A previous study has also indicated that BDNF acts on leptin-melanocortin pathway and may be a downstream target of MC4R signaling (Xu et al. 2003). Gray et al. (2006) reported a child harboring a de novo chromosomal inversion (46, XX, inv (11) (p13p15.3), which disrupts one copy of *BDNF* gene located on chromosome 11p13. Additionally, deletion of entire *BDNF* gene in a mother and a child were associated with severe obesity and developmental delay (Harcourt et al. 2018). However, data regarding point pathogenic mutations in the *BDNF* are scarce in the literature. Recently, Serra-Juhé et al. (2019) have analyzed 463 severe early-onset obese patients and 480 controls from the Spanish population. They identified three rare variants in four patients and none in normal-weight individuals. Two of these mutations were predicted to be pathogenic (p. Ile123Val and p.Cys141Gly) by in silico analyses. These probands presented obesity and mild to severe hyperphagia. Two of these patients also exhibited liver steatosis and insulin resistance with dyslipidemia. None of them showed congenital and behavioral issues.

Syndromic Form of Obesity

Syndromic forms of obesity are used to describe obese subjects with particular phenotypes, such as mental retardation, dysmorphic features, and organ-specific developmental abnormalities. At least four rare syndromes are associated with severe hyperphagia and obesity, caused by chromosomal abnormalities and genetic variations in genes that are involved in appetite and satiety control, suggesting an origin at the level of the central nervous system (Bell et al. 2005). Syndromic obesity includes Prader Willi (OMIM#176270), Bardet-Biedl (OMIM#209900), Alstrom (OMIM#203800), and Cohen (OMIM#216550) syndromes (Albuquerque et al. 2015). The genetic basis of these syndromes is extremely heterogeneous, and may be inherited in either an autosomal dominant or a recessive fashion.

Prader Willi Syndrome

Prader Willi syndrome (PWS) is the most common form (1 in 10,000–15,000 live births), affecting an estimated 350,000–400,000 individuals in the world (Irizarry and Haqq 2018). PWS is an autosomal dominant disorder characterized by peculiar facial features, infantile hypotonia, hypothalamic hypogonadism, short stature, mild mental retardation, and behavioral problems and hyperphagia resulting in severe obesity. Most of the cases are caused by a paternally inherited deletion at the chromosomal region 15q11–q13 (~70%) and less frequently by maternal uniparental disomy 15 (~20–30%). Additionally, rare cases (~5%) are affected by imprinting defects and chromosome 15 translocations or rearrangements (Bell et al. 2005; Butler 2011; Irizarry and Haqq 2018).

The responsible gene(s) for PWS development as well as hyperphagia and obesity phenotype in these patients is still unknown, though several candidate genes have been reported. Interestingly, patients with PWS exhibit high plasma levels of ghrelin at any age, suggesting an interaction among increased appetite, obesity, and leptin-melanocortin pathway (Feigerlová et al. 2008).

However, ghrelin-reducing agents do not diminish hyperphagia or body mass in subjects with PWS (De Waele et al. 2008). Recently, Burnett and colleagues performed an elegant study showing that neuroendocrine PWS-associated phenotypes may be a result of reduced expression of prohormone convertase 1, which impacts on the processing of proinsulin, pro-GH releasing hormone and proghrelin (Burnett et al. 2017). Further studies are necessary to clarify the role of PC1 in PWS cases.

Bardet–Biedl Syndrome

Bardet–Biedl syndrome (BBS) is a rare disorder (1 in 125,000–160,000 in Europe) characterized by early-onset obesity, progressive rod-cone dystrophy, polydactyly, learning disabilities, progressive renal dysfunction, hypogonadism, and genital abnormalities (Irizarry and Haqq 2018). It has a typically a autosomal recessive inheritance, but some families may have a triallelic mode of transmission (i.e., interaction of two or more *loci*) (Beales et al. 2003; Khan et al. 2016). BBS is a genetically heterogeneous disorder caused by mutations in at least 19 *loci* (also called BBS genes), all of which have a key role in ciliary development or intraflagellar transport (Fan et al. 2004; Khan et al. 2016; Irizarry and Haqq 2018).

Previous functional studies in single-cell organisms have shown that certain BBS genes are specific to ciliated cells, which are important for mammalian development, contributing to correctly position the organs in the body. Dysfunction in these ciliated cells may explain the alterations in pigmentary epithelial and structural abnormalities; however, the exact mechanisms leading to the obesity phenotype are still unknown (Irizarry and Haqq 2018). Experimental studies have shown that BBS mice are leptin resistant with decreased expression of POMC, suggesting that BBS genes may have an important role in maintaining leptin sensitivity in POMC neurons localized into the hypothalamus. These results might suggest that BBS genes can impact on the anorexigenic pathway leading to hyperphagia and obesity (Rahmouni et al. 2008).

Alstrom and Cohen Syndrome

Alstrom syndrome (AS) is an extremely rare autosomal recessive disorder caused by mutations in the *ALMS1* gene on chromosome 2p13. As cases exhibit progressive cone-rod dystrophy leading to juvenile blindness, sensorineural hearing loss, and early-onset obesity (Maffei et al. 2017). Screening for *ALMS1* gene in these affected patients showed a strong clustering of pathogenic mutations in exons 8, 10, and 16. Moreover, the majority of disease-causing mutations were nonsense or frameshift, which result in premature protein truncation (Irizarry and Haqq 2018). Similar to BBS, AS is caused by mutations disrupting neuronal cilia function, which may impact on appetite and satiety responses (Davenport et al. 2007).

Finally, Cohen syndrome is also an autosomal recessive disease caused by mutations in *COH1* gene (also called *VPS13B*) located at chromosome 8q22 (Albuquerque et al. 2015). This gene encodes a transmembrane protein that has an important role in the development and function of eye, hematological system, and central nervous system. The mechanism by which abnormalities in this protein lead to the Cohen syndrome phenotype is still unknown; however, the affected patients show mental retardation, obesity, hypotonia and craniofacial, ocular, and limb abnormalities. Pathogenic mutations in *COH1* are rare around the world, but it is overrepresented in some populations, such as Finnish, Japanese, Caucasian, and Jewish (Rodrigues et al. 2018).

Polygenic or Common Obesity

The common form of obesity is multifactorial and polygenic, caused by an association between environmental factors and multiple genetic variants that have a slight effect on body weight. Several approaches have been used to find genetic variants associated with obesity or obesity-related traits, particularly genome-wide association study (GWAS). This approach evaluates a huge number of polymorphisms across the genome in a large sample and through statistical analysis, it allows identifying the loci associated with shifts of BMI or other anthropometric measurements (for example, the waist–hip ratio [WHR]) (Albuquerque et al. 2015; da Fonseca et al. 2017). To date, GWAS discovered more than 100 obesity-associated loci, in which genetic variants explain only a small portion of heritability in obesity (Speliotes et al. 2010; Locke et al. 2015; Wu et al. 2018).

Recently, Wu et al. (2018) performed GWAS for BMI and WHR using 242 monozygotic and 140 dizygotic twin pairs. A total of 291 loci were nominally associated with BMI-WHR, playing an important role in different pathways, such as homeostasis, olfactory transduction, and platelet production. Despite the robust studies and the vast number of obesity-associated loci, most of the genetic variability in BMI remains unexplained.

Relevant Candidates

Among the loci identified by GWAS, *FTO* gene has been consistently associated with obesity in different populations (Frayling et al. 2007; Albuquerque et al. 2013; da Fonseca et al. 2019a). FTO is a nucleic acid demethylase that has an important role in energy homeostasis, suggesting that this enzyme influences obesity at the epigenetic level (Gerken et al. 2007). Interestingly, polymorphisms in the intron 1 of *FTO* may have an effect on appetite/satiety, food choices, expended energy, and body weight (McCaffery et al. 2012; Qi et al. 2014; Magno et al. 2018; Rivas et al. 2018). Additionally, several common genetic variants of *BDNF* gene were associated with obesity or obesity-related traits (Beckers et al. 2008; Thorleifsson et al. 2009; Speliotes et al. 2010). A polymorphism (rs4074134) near *BDNF* gene was associated with the reduction of BMI and waist circumference in the Northern Han Chinese ancestry (Han et al. 2013). This polymorphism was also associated with obesity in the Japanese population; however, this result was not observed in a Brazilian population (Hotta et al. 2009; da Fonseca et al. 2019a). The discrepancy of the results may be explained by differences in the genetic background of the populations.

Besides *FTO* and *BDNF*, GWAS have also identified other obesity-associated loci that are expressed or known to act in the central nervous system, involved in regulation of food intake and energy expenditure. Several of these genes have been previously associated with severe early-onset obesity (*MC4R*, *POMC*, *SH2B1*, and *BBS4*), suggesting an overlap of the genetic background between monogenic and polygenic form of obesity (Speliotes et al. 2010; Locke et al. 2015).

Ongoing Studies and Future Perspectives

The use of genetic knowledge to predict subjects at high risk of developing obesity, especially in an obesogenic environment, may be useful in preventing this disease (Srivastava et al. 2016). Currently, patients with suspicion of monogenic obesity (syndromic and non-syndromic) can be screened based on diagnostic hypothesis using chromosomal microarray analysis, candidate gene sequencing, whole genome/exome sequencing, conventional cytogenetic, and/or fluorescence in situ hybridization. Theses analyses can contribute to identifying the causal variants allowing molecular diagnosis, clinical management, genetic counseling, and in some cases, specific treatment (Pigeyrc ct al. 2016).

Obesity treatment based on genotype (personalized medicine) is being pursued in patients with congenital leptin deficiency, and especially leptin receptor deficiency. caused by pathogenic mutations in *LEPR* gene.

The MC4R agonist (Setmelanotide) has been shown to decrease food intake and body weight in patients POMC deficiency as well. This agonist binds to MC4R in these patients, replacing the missing αMSH activity in the leptin-melanocortin system (Kühnen et al. 2016; Collet et al. 2017). Kühnen et al. (2016) have treated two POMC deficient subjects using subcutaneous injection of setmelanotide. Both patients showed a substantial reduction in hunger and weight loss (patient 1: 51.0 kg after 42 weeks; patient 2: 20.5 kg after 12 weeks), improving their quality of life. In addition, setmelanotide can also reduce body weight in MC4R deficient patients (on average 0.6 kg/week). It is speculated that setmelanotide may be effective in treating other genetic obesity disorders, including PCSK1 deficiency and Prader-Willi syndrome (Kühnen et al. 2016).

References

Albuquerque D, Nóbrega C, Manco L (2013) Association of FTO polymorphisms with obesity and obesity-related outcomes in Portuguese children. PLoS One 8:e54370. https://doi.org/10.1371/journal.pone.0054370

Albuquerque D, Stice E, Rodríguez-López R et al (2015) Current review of genetics of human obesity: from molecular mechanisms to an evolutionary perspective. Mol Gen Genomics 290:1991. https://doi.org/10.1007/s00438-015-1015-9

Anand BK, Brobeck JR (1951) Hypothalamic control of food intake in rats and cats. Yale J Biol Med 24:123–140

Andiran N, Çelik N, Andiran F (2011) Homozygosity for two missense mutations in the leptin receptor gene (P316T; W646C) in a Turkmenian girl with severe early-onset obesity. J Pediatr Endocrinol Metab 24:1043–1045

Anisimova AS, Rubtsov PM, Akulich KA et al (2016) Late diagnosis of POMC deficiency and in vitro evidence of residual translation from allele with c.-11C>a mutation. J Clin Endocrinol Metab 102:359–362

Asai M, Ramachandrappa S, Joachim M et al (2013) Loss of function of the melanocortin 2 receptor accessory protein 2 is associated with mammalian obesity. Science 341:275–278. https://doi.org/10.1126/science.1233000

Aslan IR, Ranadive SA, Valle I et al (2014) The melanocortin system and insulin resistance in humans: insights from a patient with complete POMC deficiency and type 1 diabetes mellitus. Int J Obes 38:148

Beales PL, Badano JL, Ross AJ et al (2003) Genetic interaction of BBS1 mutations with alleles at other BBS loci can result in non-Mendelian Bardet-Biedl syndrome. Am J Hum Genet 72:1187–1199

Beckers S, Peeters A, Zegers D et al (2008) Association of the BDNF Val66Met variation with obesity in women. Mol Genet Metab 95:110–112

Bell CG, Walley AJ, Froguel P (2005) The genetics of human obesity. Nat Rev Genet 6:221–234. https://doi.org/10.1038/nrg1556

Burnett LC, LeDuc CA, Sulsona CR et al (2017) Deficiency in prohormone convertase PC1 impairs prohormone processing in Prader-Willi syndrome. J Clin Invest 127:293–305

Butler MG (2011) Prader-Willi syndrome: obesity due to genomic imprinting. Curr Genomics 12:204–215

Cawley NX, Li Z, Loh YP (2016) 60 YEARS OF POMC: biosynthesis, trafficking and secretion of pro-opiomelanocortin-derived peptides. J Mol Endocrinol JME-15

Çetinkaya S, Güran T, Kurnaz E et al (2018) A patient with proopiomelanocortin deficiency: an increasingly important diagnosis to make. J Clin Res Pediatr Endocrinol 10:68

Challis BG, Pritchard LE, Creemers JWM et al (2002) A missense mutation disrupting a dibasic prohormone processing site in pro-opiomelanocortin (POMC) increases susceptibility to early-onset obesity through a novel molecular mechanism. Hum Mol Genet 11:1997–2004

Chen KY, Muniyappa R, Abel BS et al (2015) RM-493, a melanocortin-4 receptor (MC4R) agonist, increases resting energy expenditure in obese individuals. J Clin Endocrinol Metab 100:1639–1645. https://doi.org/10.1210/jc.2014-4024

Cirillo G, Marini R, Ito S et al (2012) Lack of red hair phenotype in a North-African obese child homozygous for a novel POMC null mutation: nonsense-mediated decay RNA evaluation and hair pigment chemical analysis. Br J Dermatol 167:1393–1395

Clement K, Vaisse C, Lahlou N et al (1998) A mutation in the human leptin receptor gene causes obesity and pituitary dysfunction. Nature 392:398–401. https://doi.org/10.1038/32911

Clément K, Dubern B, Mencarelli M et al (2008) Unexpected endocrine features and normal pigmentation in a young adult patient carrying a novel homozygous mutation in the POMC gene. J Clin Endocrinol Metab 93:4955–4962. https://doi.org/10.1210/jc.2008-1164

Collet T, Dubern B, Mokrosinski J et al (2017) Evaluation of a melanocortin-4 receptor (MC4R) agonist (Setmelanotide) in MC4R de fi ciency. Mol Metab:1–9. https://doi.org/10.1016/j.molmet.2017.06.015

Cone RD (2005) Anatomy and regulation of the central melanocortin system. Nat Neurosci 8:571–578

da Fonseca ACP, Mastronardi C, Johar A et al (2017) Genetics of non-syndromic childhood obesity and the use of high-throughput DNA sequencing technologies. J Diabetes Complicat 31:1549–1561

da Fonseca ACP, Abreu GM, Zembrzuski VM et al (2019a) The association of the fat mass and obesity-associated gene (FTO) rs9939609 polymorphism and the severe obesity in a Brazilian population. Diabetes, Metab Syndr Obes Targets Ther 12:667

da Fonseca ACP, de Medeiros AG, Zembrzuski VM et al (2019b) Identification of the MC4R start lost mutation in a morbidly obese Brazilian patient. Diabetes, Metab Syndr Obes Targets Ther 12:257

Daousi C, Dunn AJ, Foy PM et al (2005) Endocrine and neuroanatomic features associated with weight gain and obesity in adult patients with hypothalamic damage. Am J Med 118:45–50. https://doi.org/10.1016/j.amjmed.2004.06.035

Davenport JR, Watts AJ, Roper VC et al (2007) Disruption of intraflagellar transport in adult mice leads to obesity and slow-onset cystic kidney disease. Curr Biol 17:1586–1594

De Lorenzo A, Soldati L, Sarlo F et al (2016) New obesity classification criteria as a tool for bariatric surgery indication. World J Gastroenterol 22:681

De Waele K, Ishkanian SL, Bogarin R et al (2008) Long-acting octreotide treatment causes a sustained decrease in ghrelin concentrations but does not affect weight, behaviour and appetite in subjects with Prader–Willi syndrome. Eur J Endocrinol 159:381–388

Doulla M, McIntyre AD, Hegele RA, Gallego PH (2014) A novel MC4R mutation associated with childhood-onset obesity: a case report. Paediatr Child Health 19:515–518

Drabkin M, Birk OS, Birk R (2018) Heterozygous versus homozygous phenotype caused by the same MC4R mutation: novel mutation affecting a large consanguineous kindred. BMC Med Genet 19:135

Fan Y, Esmail MA, Ansley SJ et al (2004) Mutations in a member of the Ras superfamily of small GTP-binding proteins causes Bardet-Biedl syndrome. Nat Genet 36:989

Farooqi SI (2015) Genetic, molecular and physiological mechanisms involved in human obesity: society for endocrinology medal lecture 2012. Clin Endocrinol 82:23–28. https://doi.org/10.1111/cen.12588

Farooqi IS, Keogh JM, Kamath S et al (2001) Metabolism: partial leptin deficiency and human adiposity. Nature 414:34

Farooqi IS, Matarese G, Lord GM et al (2002) Beneficial effects of leptin on obesity, T cell hyporesponsiveness, and neuroendocrine/metabolic dysfunction of human congenital leptin deficiency. J Clin Invest 110:1093–1103. https://doi.org/10.1172/JCI15693

Farooqi IS, Keogh JM, Yeo GSH et al (2003) Clinical spectrum of obesity and mutations in the melanocortin 4 receptor gene. N Engl J Med 348:1085–1095. https://doi.org/10.1056/NEJMoa022050

Farooqi IS, Drop S, Clements A et al (2006) Heterozygosity for a POMC-null mutation and increased obesity risk in humans. Diabetes 55:2549–2553. https://doi.org/10.2337/db06-0214

Farooqi IS, Volders K, Stanhope R et al (2007a) Hyperphagia and early-onset obesity due to a novel homozygous missense mutation in prohormone convertase 1/3. J Clin Endocrinol Metab 92:3369–3373. https://doi.org/10.1210/jc.2007-0687

Farooqi SI, Wangensteen T, Collins S et al (2007b) Clinical and molecular genetic Spectrum of congenital deficiency of the leptin receptor. N Engl J Med 356:237–247. https://doi.org/10.1056/NEJMoa063988.Clinical

Fatima W, Shahid A, Imran M et al (2011) Leptin deficiency and leptin gene mutations in obese children from Pakistan. Int J Pediatr Obes 6:419–427

Feigerlová E, Diene G, Conte-Auriol F et al (2008) Hyperghrelinemia precedes obesity in Prader-Willi syndrome. J Clin Endocrinol Metab 93:2800–2805

Finucane MM, Stevens GA, Cowan MJ et al (2011) National, regional, and global trends in body-mass index since 1980: systematic analysis of health examination surveys and epidemiological studies with 960 country-years and 9·1 million participants. Lancet 377:557–567. https://doi.org/10.1016/S0140-6736(10)62037-5

Fischer-Posovszky P, von Schnurbein J, Moepps B et al (2010) A new missense mutation in the leptin gene causes mild obesity and hypogonadism without affecting T cell responsiveness. J Clin Endocrinol Metab 95:2836–2840

Frank GR, Fox J, Candela N et al (2013) Severe obesity and diabetes insipidus in a patient with PCSK1 deficiency. Mol Genet Metab 110:191–194. https://doi.org/10.1016/j.ymgme.2013.04.005

Frayling TM, Timpson NJ, Weedon MN et al (2007) A common variant in the FTO gene is associated with body mass index and predisposes to childhood and adult obesity. Science 316:889–894

Freemark MS (2018) Pediatric obesity: etiology, pathogenesis and treatment. Springer, New York

Garvey WT, Mechanick JI, Brett EM et al (2016) American Association of Clinical Endocrinologists and American College of Endocrinology comprehensive clinical practice guidelines for medical care of patients with obesity. Endocr Pract 22:1–203

Gerken T, Girard CA, Tung YL et al (2007) The obesity-associated FTO gene encodes a 2-oxoglutarate – dependent nucleic acid demethylase. Science 318:1469–1472. https://doi.org/10.1126/science.1151710.The

Gibson WT, Farooqi IS, Moreau M et al (2004) Congenital leptin deficiency due to homozygosity for the delta-133G mutation: report of another case and evaluation of response to four years of leptin therapy. J Clin Endocrinol Metab 89:4821–4826. https://doi.org/10.1210/jc.2004-0376

Gill R, Cheung YH, Shen Y et al (2014) Whole-exome sequencing identifies novel LEPR mutations in individuals with severe early onset obesity. Obesity (Silver Spring) 22:576–584. https://doi.org/10.1002/oby.20492

www.globenewswire.com/news-release/2019/11/04/1940636/0/en/Rhythm-Pharmaceuticals-Announces-Late-breaking-Data-from-Phase-3-Trials-of-Setmelanotide-in-POMC-and-LEPR-Deficiency-Obesities-at-ObesityWeek-2019.html. (2019) Accessed 18 Jan 2020

Gray J, Yeo GSH, Cox JJ et al (2006) Hyperphagia, severe obesity, impaired cognitive function, and hyperactivity associated with functional loss of one copy of the brain-derived neurotrophic factor (BDNF) gene. Diabetes 55:3366–3371. https://doi.org/10.2337/db06-0550

Han X, Luo Y, Zhang X et al (2013) Rs4074134 near BDNF gene is associated with type 2 diabetes mellitus in Chinese Han population independently of body mass index. PLoS One 8:e56898

Hannema SE, Wit JM, Houdijk ME et al (2016) Novel leptin receptor mutations identified in two girls with severe obesity are associated with increased bone mineral density. Horm Res Paediatr 85:412–420

Harcourt BE, Bullen DVR, Kao K et al (2018) Maternal inheritance of BDNF deletion, with phenotype of obesity and developmental delay in mother and child. Am J Med Genet Part A 176:194–200

Härter B, Fuchs I, Müller T et al (2016) Early clinical diagnosis of PC1/3 deficiency in a patient with a novel homozygous PCSK1 splice-site mutation. J Pediatr Gastroenterol Nutr 62:577–580. https://doi.org/10.1097/MPG.0000000000001018

Hasnain S (2016) The p. N103K mutation of leptin (LEP) gene and severe early onset obesity in Pakistan. Biol Res 49:23

Hetherington AW, Ranson SW (1940) Hypothalamic lesions and adiposity in the rat. Anat Rec 78:149–172. https://doi.org/10.1002/ar.1090780203

Hotta K, Nakamura M, Nakamura T et al (2009) Association between obesity and polymorphisms in SEC16B, TMEM18, GNPDA2, BDNF, FAIM2 and MC4R in a Japanese population. J Hum Genet 54:727–731

Hung C-N, Poon W-T, Lee C-Y et al (2012) A case of early-onset obesity, hypocortisolism, and skin pigmentation problem due to a novel homozygous mutation in the proopiomelanocortin (POMC) gene in an Indian boy. J Pediatr Endocrinol Metab 25:175–179

Huvenne H, Le Beyec J, Pépin D et al (2015) Seven novel deleterious LEPR mutations found in early-onset obesity: a ΔExon6–8 shared by subjects from Reunion Island, France, suggests a founder effect. J Clin Endocrinol Metab 100:E757–E766

Irizarry KA, Haqq AM (2018) Syndromic obesity. Pediatric obesity. Springer, New York, pp 153–182

Jackson RS, Creemers JWM, Ohagi S et al (1997) Obesity and impaired prohormone processing associated with mutations in the human prohormone convertase 1 gene. Nat Genet 16:303–306

Jackson RS, Creemers JWM, Farooqi IS et al (2003) Small-intestinal dysfunction accompanies the complex endocrinopathy of human proprotein convertase 1 deficiency. J Clin Invest 112:1550–1560. https://doi.org/10.1172/JCI200318784

Khan SA, Muhammad N, Khan MA et al (2016) Genetics of human Bardet–Biedl syndrome, an updates. Clin Genet 90:3–15

Krude H, Biebermann H, Luck W et al (1998) Severe early-onset obesity, adrenal insufficiency and red hair pigmentation caused by POMC mutations in humans. Nat Genet 19:155–157. https://doi.org/10.1038/509

Krude H, Biebermann H, Schnabel D et al (2003) Obesity due to proopiomelanocortin deficiency: three new cases and treatment trials with thyroid hormone and ACTH4-10. J Clin Endocrinol Metab 88:4633–4640. https://doi.org/10.1210/jc.2003-030502

Kühnen P, Clément K, Wiegand S et al (2016) Proopiomelanocortin deficiency treated with a melanocortin-4 receptor agonist. N Engl J Med 375:240–246

Larder R, Lim CT, Coll AP (2014) Genetic aspects of human obesity. In: Eric Fliers MK, JARBT-H of CN (ed) Clinical neuroendocrinology. Elsevier, London, pp 93–106

Le Beyec J, Cugnet-Anceau C, Pépin D et al (2013) Homozygous Leptin receptor mutation due to uniparental disomy of chromosome 1: response to bariatric surgery. J Clin Endocrinol Metab 98:E397–E402

Lee YS, Challis BG, Thompson DA et al (2006) A POMC variant implicates beta-melanocyte-stimulating hormone in the control of human energy balance. Cell Metab 3:135–140. https://doi.org/10.1016/j.cmet.2006.01.006

Locke A, Kahali B, Berndt S et al (2015) Genetic studies of body mass index yield new insights for obesity biology. Nature 518:197–206. https://doi.org/10.1038/nature14177.Genetic

Maffei P, Favaretto F, Milan G, Marshall JD (2017) Alström syndrome. Diabetes associated with single gene defects and chromosomal abnormalities. Karger, Basel, pp 134–144

Magno FCCM, Guaraná HC, Fonseca ACP et al (2018) Influence of FTO rs9939609 polymorphism on appetite, ghrelin, leptin, IL6, TNFα levels, and food intake of women with morbid obesity. Diabetes Metab Syndr Obes 11:199–207. https://doi.org/10.2147/DMSO.S154978

Martín MG, Solorzano-Vargas RS, Wang J et al (2013) Congenital Proprotein Convertase 1/3 deficiency causes malabsorptive diarrhea and other endocrinopathies in a pediatric cohort. Gastroenterology 145:138–148. https://doi.org/10.1016/j.freeradbiomed.2008.10.025.The

Mazen I, El-Gammal M, Abdel-Hamid M, Amr K (2009) A novel homozygous missense mutation of the leptin gene (N103K) in an obese Egyptian patient. Mol Genet Metab 97:305–308

Mazen I, El-Gammal M, Abdel-Hamid M et al (2011) Homozygosity for a novel missense mutation in the leptin receptor gene (P316T) in two Egyptian cousins with severe early onset obesity. Mol Genet Metab 102:461–464

McCaffery JM, Papandonatos GD, Peter I et al (2012) Obesity susceptibility loci and dietary intake in the look AHEAD trial. Am J Clin Nutr 95:1477–1486

Melchior C, Schulz A, Windholz J et al (2012) Clinical and functional relevance of melanocortin-4 receptor variants in obese German children. Horm Res Paediatr 78:237–246. https://doi.org/10.1159/000343816

Mendiratta MS, Yang Y, Balazs AE et al (2011) Early onset obesity and adrenal insufficiency associated with

a homozygous POMC mutation. Int J Pediatr Endocrinol 2011:5. https://doi.org/10.1186/1687-9856-2011-5
Montague CT, Farooqi IS, Wareham NJ et al (1997) Congenital leptin deficiency is associated with severe early-onset obesity in humans. Nature 387:903–908. https://doi.org/10.1038/43185
Morton GJ, Cummings DE, Baskin DG et al (2006) Central nervous system control of food intake and body weight. Nature 443:289–295
Özen S, Özcan N, Uçar SK et al (2015) Unexpected clinical features in a female patient with proopiomelanocortin (POMC) deficiency. J Pediatr Endocrinol Metab 28:691–694
Ozsu E, Bahm A (2017) Delayed diagnosis of proopiomelanocortin (POMC) deficiency with type 1 diabetes in a 9-year-old girl and her infant sibling. J Pediatr Endocrinol Metab 30:1137–1140
Philippe J, Stijnen P, Meyre D et al (2015) A nonsense loss-of-function mutation in PCSK1 contributes to dominantly inherited human obesity. Int J Obes 39:1–30. https://doi.org/10.1038/ijo.2014.96
Pigeyre M, Meyre D (2018) Monogenic obesity. Pediatric obesity. Springer, pp 135–152
Pigeyre M, Yazdi FT, Kaur Y, Meyre D (2016) Recent progress in genetics, epigenetics and metagenomics unveils the pathophysiology of human obesity. Clin Sci 130:943–986. https://doi.org/10.1042/CS20160136
Qi Q, Kilpeläinen TO, Downer MK et al (2014) FTO genetic variants, dietary intake and body mass index: insights from 177 330 individuals. Hum Mol Genet 23:6961–6972
Rahmouni K, Fath MA, Seo S et al (2008) Leptin resistance contributes to obesity and hypertension in mouse models of Bardet-Biedl syndrome. J Clin Invest 118:1458–1467
Rivas AMO, Santos JL, Valladares MA et al (2018) Association of the FTO fat mass and obesity–associated gene rs9939609 polymorphism with rewarding value of food and eating behavior in Chilean children. Nutrition 54:105–110
Rodrigues JM, Fernandes HD, Caruthers C et al (2018) Cohen syndrome: review of the literature. Cureus 10
Roth CL et al (2018) Changes in satiety hormones in response to leptin treatment in a patient with leptin deficiency. Hormone research in paediatrics 90.6:424–430
Saeed S, Butt TA, Anwer M et al (2012) High prevalence of leptin and melanocortin-4 receptor gene mutations in children with severe obesity from Pakistani consanguineous families. Mol Genet Metab 106:121–126. https://doi.org/10.1016/j.ymgme.2012.03.001
Saeed S, Bonnefond A, Manzoor J et al (2014) Novel LEPR mutations in obese Pakistani children identified by PCR-based enrichment and next generation sequencing. Obesity (Silver Spring) 22:1112–1117. https://doi.org/10.1002/oby.20667
Saeed S, Bonnefond A, Manzoor J et al (2015) Genetic variants in LEP, LEPR, and MC4R explain 30% of severe obesity in children from a consanguineous population. Obesity (Silver Spring) 23:1687–1695. https://doi.org/10.1002/oby.21142
Saeed S, Arslan M, Froguel P (2018) Genetics of obesity in consanguineous populations: toward precision medicine and the discovery of novel obesity genes. Obesity 26:474–484
Samuels ME, Gallo-Payet N, Pinard S et al (2013) Bioinactive ACTH causing glucocorticoid deficiency. J Clin Endocrinol Metab 98:736–742. https://doi.org/10.1210/jc.2012-3199
Serra-Juhé C, Martos-Moreno GÁ, de Pieri FB et al (2019) Heterozygous rare genetic variants in non-syndromic early-onset obesity. Int J Obes 1
Smith AI, Funder JW (1988) Proopiomelanocortin processing in the pituitary, central nervous system, and peripheral tissues. Endocr Rev 9:159–179. https://doi.org/10.1210/edrv-9-1-159
Speliotes EK, Willer CJ, Berndt SI et al (2010) Association analyses of 249,796 individuals reveal 18 new loci associated with body mass index. Nat Genet 42:937–948
Srivastava A, Srivastava N, Mittal B (2016) Genetics of obesity. Indian J Clin Biochem 31:361–371
Strobel A, Issad T, Camoin L et al (1998) A leptin missense mutation associated with hypogonadism and morbid obesity. Nat Genet 18:213
Sutton AK, Myers MG, Olson DP (2016) The role of PVH circuits in leptin action and energy balance. Annu Rev Physiol 78:207–221. https://doi.org/10.1146/annurev-physiol-021115-105347
Tao Y-X (2005) Molecular mechanisms of the neural melanocortin receptor dysfunction in severe early onset obesity. Mol Cell Endocrinol 239:1–14. https://doi.org/10.1016/j.mce.2005.04.012
Tapia-Arancibia L, Rage F, Givalois L, Arancibia S (2004) Physiology of BDNF: focus on hypothalamic function. Front Neuroendocrinol 25:77–107. https://doi.org/10.1016/j.yfrne.2004.04.001
Thakur S, Kumar A, Dubey S et al (2014) A novel mutation of the leptin gene in an Indian patient. Clin Genet 86:391–393
Thorleifsson G, Walters GB, Gudbjartsson DF et al (2009) Genome-wide association yields new sequence variants at seven loci that associate with measures of obesity. Nat Genet 41:18
Tunç S, Demir K, Tükün FA et al (2017) Melanocortin-4 receptor gene mutations in a group of Turkish obese children and adolescents. J Clin Res Pediatr Endocrinol 9:216
Vetter ML, Faulconbridge LF, Webb VL, Wadden TA (2010) Behavioral and pharmacologic therapies for obesity. Nat Rev Endocrinol 6:578–588. https://doi.org/10.1038/nrendo.2010.121
Wabitsch M, Funcke J-B, Lennerz B et al (2014) Biologically inactive Leptin and early-onset extreme obesity. N Engl J Med 372:48–54. https://doi.org/10.1056/NEJMoa1406653

Wabitsch M, Funcke J-B, Lennerz B et al (2015) Biologically inactive Leptin and early-onset extreme obesity. N Engl J Med 372:48–54. https://doi.org/10.1056/NEJMoa1406653

Wardle J, Carnell S, Haworth CMA, Plomin R (2008) Evidence for a strong genetic influence on childhood adiposity despite the force of the obesogenic environment. Am J Clin Nutr 87:398–404

Wilschanski M, Abbasi M, Blanco E et al (2014) A novel familial mutation in the PCSK1 gene that alters the oxyanion hole residue of proprotein convertase 1/3 and impairs its enzymatic activity. PLoS One 9:1–7. https://doi.org/10.1371/journal.pone.0108878

World Health Organization (2019) Obesity and overweight. http://www.who.int/mediacentre/factsheets/fs311/en/. Accessed 11 Nov 2019

Wu Y, Duan H, Tian X et al (2018) Genetics of obesity traits: a bivariate genome-wide association analysis. Front Genet 9:179

Xu B, Goulding EH, Zang K et al (2003) Brain-derived neurotrophic factor regulates energy balance downstream of melanocortin-4 receptor. Nat Neurosci 6:736

Yupanqui-Lozno H, Bastarrachea RA, Yupanqui-Velazco ME et al (2019) Congenital leptin deficiency and leptin gene missense mutation found in two Colombian sisters with severe obesity. Genes (Basel) 10:342

The Upstream Environment for the Obesity Epidemic

9

Peter Congdon

Contents

Introduction: The Obesity Epidemic 128
Comorbidities of Obesity 128
Environment and Obesity 129
Sprawl 129
Food Access 129
Exercise Access and Physical Activity 130
Neighbourhood Social Environment 130
Environment and Diabetes 130
Spatio-Temporal Aspects of Obesity 131
Conclusion: Unresolved Issues and Future Perspectives 132
References 133

Abstract

While individual behaviour and genetics are major risk factors for obesity, for health prevention it is important also to consider upstream influences on obesity, stemming from the urban physical and social environment. Environmental features such as urban sprawl, food deserts, and varying access to exercise and greenspace, affect both activity and diet, and hence bodyweight. Thus many studies show environmental impacts on obesity, and on related outcomes such as diabetes, with these impacts being socially and spatially differentiated: with worse food and exercise access in lower income areas. Analytical frameworks for the obesity epidemic should reflect wide geographic differences in incidence and prevalence between regions and small areas, including spatial clustering. A case study of changing obesity in US counties shows widening inequalities and persistently high obesity clustering in some regions.

P. Congdon (✉)
School of Geography, Queen Mary University of London, London, UK
e-mail: p.congdon@qmul.ac.uk

J. Faintuch and S. Faintuch (eds.), *Obesity and Diabetes*, https://doi.org/10.1007/978-3-030-53370-0_9

Keywords

Obesity epidemic · Diabetes epidemic · Food deserts · Urban sprawl · Exercise access · Spatio-temporal · Spatial clustering

Introduction: The Obesity Epidemic

The obesity epidemic continues across the world, and is now recognised as a major public health issue, though this recognition is only relatively recent (James 2009). In the United States, the prevalence of obesity in 2015–2016 was 40%, affecting about 93 million adults, though the upward trend is irregular (Flegal et al. 2012). The trend extends to the third world, with an estimating doubling in global obesity levels since 1980 (Bhurosy and Jeewon 2014; Fox et al. 2019). Obesity is a major risk factor for cardiovascular disease, diabetes, and hypertension, and hence a major factor for increased health care costs (Biener et al. 2017).

While genetic factors, possibly linked to ethnicity, and adverse health behaviours, influence susceptibility to obesity (Stryjecki et al. 2018; Riveros-McKay et al. 2019), they cannot in themselves account for the rapid increases in obesity. Instead, behavioural changes linked to changes in the food and exercise environments are also relevant to increased obesity (Faith and Kral 2006; Caballero 2007; Cummins and Macintyre 2006; Egger and Swinburn 1997). Physical activity is declining, and sedentary behaviour increasing, both in developed societies, and lower and middle-income countries (Gaskin and Orellana 2018). Dietary changes, such as growing fast food consumption and the "nutrition transition" (Popkin 2015) are also relevant.

Comorbidities of Obesity

Obesity is a risk factor for increased morbidity and mortality, particularly for cardiovascular disease, diabetes, and hypertension, but also for cancer and other chronic diseases, including liver and kidney disease and osteoarthritis. The impact of obesity and overweight on health and mortality is recognised in disease burden studies (Must et al. 1999).

Regarding overall mortality, there is some debate about the impacts of overweight (BMI 25–<30) and grade 1 obesity (BMI 30–<35), with Flegal et al. (2013) reporting excess mortality only at obesity grades 2 and above. However, the study by the Global BMI Mortality Collaboration (2016) reports no mortality benefit from any grade of overweight or obesity, and finds a log-linear increase in mortality as BMI increases compared to a reference BMI of 20–25.

Among the diseases linked to obesity, the recent spread of diabetes has been most evident, and like obesity has been characterised as an epidemic (e.g. Zimmet 2017; Unnikrishnan et al. 2017). The diabetes epidemic affects low and middle-income countries, such as India and China, as well as higher-income developed societies. In the United States, around 9% of adults (or 30 million) have diabetes according to a report by CDC (2017), and similar levels are reported in China (Yang et al. 2010) and India (Health Issues India 2019). In LMIC countries diabetes is generally higher in urban settings, and hence is associated with urbanisation (Gassasse et al. 2017; Ramachandran et al. 1999).

The impact of obesity on the risk of diabetes, particularly type 2 diabetes (T2D), is well-established, though diabetes is multifactorial, and many other risk factors than obesity or BMI per se are relevant. For example, among normal-weight subjects, diabetes incident risk is especially related to waist size (Carnethon et al. 2012). Population ethnic composition is also relevant to variations of diabetes risk. Thus raised diabetes risk among south Asian groups has been attributed to increased insulin resistance, even adjusting for adiposity (McKeigue et al. 1991). There is also increased research on diabetes risk through epigenetic changes that can be transmitted from one generation to another thus reinforcing increases in diabetes.

Despite such qualifications, the majority of T2D subjects are obese, and higher obesity grades elevate the risk of diabetes, as does longer established obesity. Abdullah et al. (2010) report an overall relative risk (RR) of 7.2 for diabetes among obese persons compared to those of normal weight, and for overweight a RR of 3.0. Among different obesity indices, the waist-to-height ratio may offer better predictions of diabetes risk than waist circumference or BMI (Ashwell and Browning 2011). Potential biophysical mechanisms include:

- Abdominal obesity causing fat cells to release pro-inflammatory reducing sensitivity to insulin.
- Obesity triggering metabolic changes that cause adipose tissue to release increased amounts of fatty acids, glycerol, hormones, and pro-inflammatory cytokines involved in developing insulin resistance.

Environment and Obesity

A major research effort has focused on the impacts on obesity of the urban built environment and its obesogenic aspects. Thus Cummins and Macintyre (2006) refer to the "over-emphasis on the role of individual health behaviours, which has tended to ignore the influence of the complex social and physical contexts in which individual behavioural decisions are made. Such critiques have led to a new focus on 'environmental' exposures that encourage excessive food intake and discourage physical activity". Environmental exposures are partly invoked to explain wide contrasts in obesity prevalence between income groups. Prominent themes in this work are the impacts of urban sprawl, of food deserts and access to healthy food outlets, and of inequity in access to exercise opportunities.

Sprawl

Impacts of the built environment on both physical activity and diet have been framed, especially in terms of neighbourhoods with high walkability and connectivity, as opposed to car-dependent environments associated with urban sprawl. Sprawl relates to land use patterns associated with recent suburban car-dependent development, though the concept has many facets and varying definitions. Thus a literature and conceptual review of sprawl by Galster et al. (2001) found no common definition and relatively few attempts at operationalisation. These authors posit eight distinct dimensions: density, continuity, concentration, clustering, centrality, nuclearity, mixed uses, and proximity.

Thus sprawl is generally characterised as low-density suburban development, with low connectivity, car dependence, low walkability, and discouraged physical activity. This land-use pattern is distinct from that in more central city areas, or in compact mixed-use development, with relatively high connectivity and walkability, and high levels of active commuting (Kashef 2011). As examples of land-use effects on obesity (and diabetes), Creatore et al. (2016) find adverse trends in these outcomes in less walkable Ontario neighbourhoods over the period 2001–2012, but no such trends in the most walkable neighbourhoods; while Flint and Cummins (2016) find obesity linked to car-only commuting.

Food Access

The food environment and access to healthy food outlets are another feature of the urban environment that has been linked to growing obesity, and also to sprawl (Mead 2008; Hamidi 2019). Food type is relevant to the concept of food deserts, typified as areas with low access to supermarkets, and hence to fresh fruit and vegetables, and more likely to be deprived or ethnic majority areas (Fleischhacker et al. 2011). For example, Larsen and Gililand (2008) find low-income residents of inner-city neighbourhoods in London (Ontario) had poorer access to supermarkets than high-income residents, and that inequalities in access to supermarkets had increased.

A concomitant concept is that of "food swamps" (Cooksey-Stowers et al. 2017) whereby low-income areas have food access biased to fast food outlets and convenience stores offering especially foods high in fat, salt, or sugar

(HFSS) (Food Foundation 2019). For child obesity, school location in relation to fast food outlets is therefore potentially important (Aviola et al. 2014).

Consumption of healthy food in deprived areas is also affected by relative costs of healthy food as against HFSS foods (Drewnowski and Specter 2004). Thus Jones et al. (2014) report that since 2002, healthier foods have been consistently more expensive than less healthy ones, with a growing gap between them.

Exercise Access and Physical Activity

Access to exercise opportunities and natural space is associated with greater physical activity and improved health; for example, Angraal et al. (2019) show an adverse impact on cardiovascular mortality of diminished access to exercise opportunities. However, access to such opportunities is socially and spatially unequal. Thus the location of urban green space and parks is typically biased to areas containing higher income and white ethnic groups. In particular, in the United States, nonwhite and low-income groups generally live in the urban core or in inner suburbs where green space is limited and/or poorly maintained, while upper-income groups live on the suburban periphery with more abundant green space (Wolch et al. 2014). Regarding exercise opportunities (and hence levels of physical activity vs. sedentary behaviour), Timperio et al. (2010) show that BMI among older children is reduced by a higher density of public open spaces designated for sport or recreation.

Neighbourhood Social Environment

Neighbourhood impacts on obesity and diabetes extend to the social environment, including area deprivation, social capital, crime perceptions, and neighbourhood safety (Lee et al. 2019a). Thus impacts of area deprivation may occur even after controlling for adverse impacts of lower individual socio-economic status (Diez-Roux et al. 1997), sometimes described as "deprivation-amplification" (Cummins and Macintyre 2006). The widespread adoption of multilevel perspectives is based on a premise that characteristics of communities are potentially related to cardiovascular outcomes, and risk factors, independently of individual-level variables. Hence one should seek to elucidate their independent and combined effects.

Regarding crime, there are plausible pathways through which neighbourhood crime levels influence obesity, with impacts on activity and sedentary behaviour acting as mediating or intermediate variables in the association (Richardson et al. 2017; An et al. 2017).

Regarding social capital, Wu et al. (2018) find that adults with higher network diversity and high generalized trust have lower obesity risk, while social capital also impacts obesity-related behaviours, including smoking, diet, and physical activity. Social capital is generally lower in deprived areas, and social capital may mediate the typically negative relationship between neighbourhood deprivation and self-rated health (Verhaeghe and Tampubolon 2012).

Environment and Diabetes

Given the downstream obesity response to environmental factors, similar environmental impacts on diabetes have been reported, though with some specificity also.

Thus Bravo et al. (2019) consider quality of the built environment as a potential influence on T2D, beyond the usual research focus on food and physical activity access. Environmental quality is likely to mediate the impact of area deprivation on obesity and diabetes prevalence.

Den Braver et al. (2018) also mention urban residence per se (as opposed to rural residence) as a risk factor for diabetes. This may be related to adverse impacts of the urban physical environment, especially pollution, on diabetes risk (Rajagopalan and Brook 2012; O'Donovan and Cadena-Gaitán 2018). Migration to city areas may also enhance diabetes risk (Ruiz-Alejos et al. 2018).

Other studies report similar risk factors in environment-diabetes ecological studies as in environment–obesity studies. Thus Dendup et al. (2018) refer to walkability, physical activity resources, and access to green space as protective factors reducing diabetes risk.

Spatio-Temporal Aspects of Obesity

Trends to higher levels of obesity are highly disparate at subnational levels. For example, at the state level in the United States, adult obesity rates in 2018 range nearly twofold, from 39.5% in Mississippi to 23% in Colorado. Variations at small area scales, such as county or zip code level, are much wider.

In addition, many studies show spatial clustering in obesity (and related outcomes) at small area levels (e.g. Huang et al. 2015; Schuurman et al. 2009), and such spatial concentrations tend to be persistent (Joost et al. 2016). Such spatial clustering reflects clustering in risk factors, namely sociodemographic aspects of different areas (compositional impacts in the terminology of multilevel studies), and contextual effects, such as food and exercise access, and sprawl patterns (Schneider et al. 2017).

From a spatio-temporal perspective, one may be interested in the persistence of high obesity risk in certain areas ((Joost et al. 2016), in future spatial predictions (Guo et al. 2019), and the role of collective vs. individual behaviour in the obesity epidemic (e.g. Gallos et al. 2012).

To exemplify spatial clustering and persistent clustering, we consider obesity prevalence data for 3141 US counties from the Behavioral Risk Factor Surveillance System (BRFSS) for four periods, 2005–2007, 2008–2010, 2011–2013, and 2014–2016. Figure 9.1 maps out obesity prevalence in the final period (for the US continental counties), with break points at quintiles.

We first consider spatial clustering in 2014–2016 using the local Moran statistic as a local index of spatial association or LISA (Anselin et al. 2006). In particular, this assesses which areas can be classed as the centre of "high-high" clusters, whereby both the area itself and the areas surrounding it have significantly raised obesity. Figure 9.2 shows the location of these clusters as well as other possible cluster types under the LISA cluster scheme. Figure 9.2 shows that whereas county clusters with high obesity concentrate in southeast United States, clusters with low obesity tend to be in the west and mountain states.

To exemplify spatio-temporal modelling of the obesity epidemic in the United States, we use a Bayesian disease mapping approach (Lawson and Lee 2017) in BUGS (Lunn et al. 2009). Denote

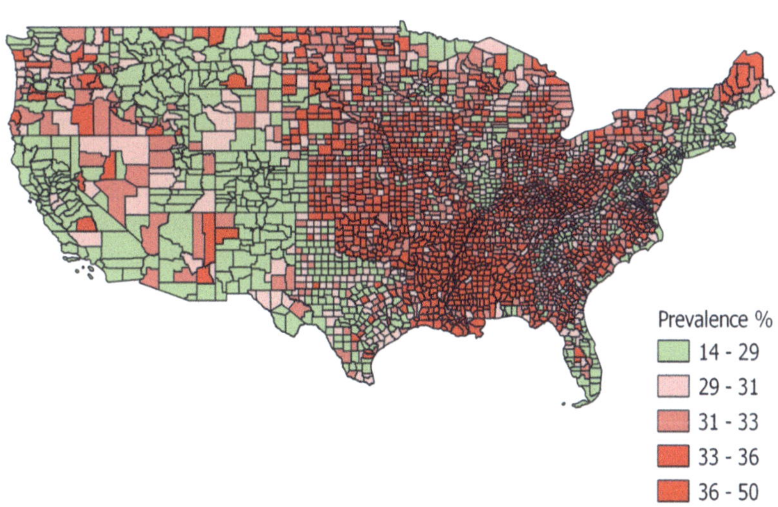

Fig. 9.1 Obesity prevalence, US continental counties (2014–2016)

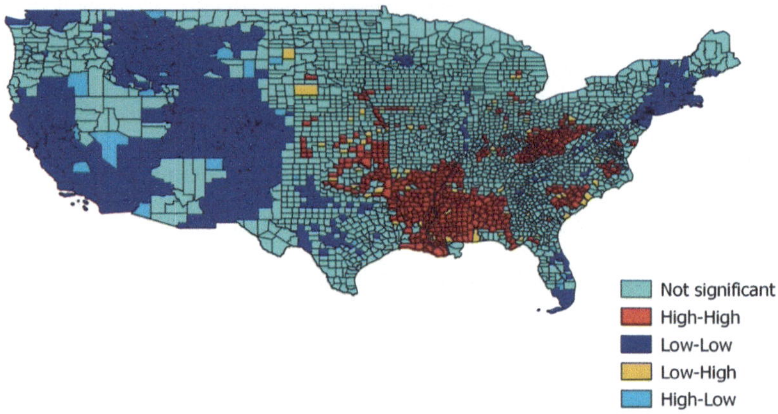

Fig. 9.2 US continental counties, obesity 2014–2016, cluster patterns

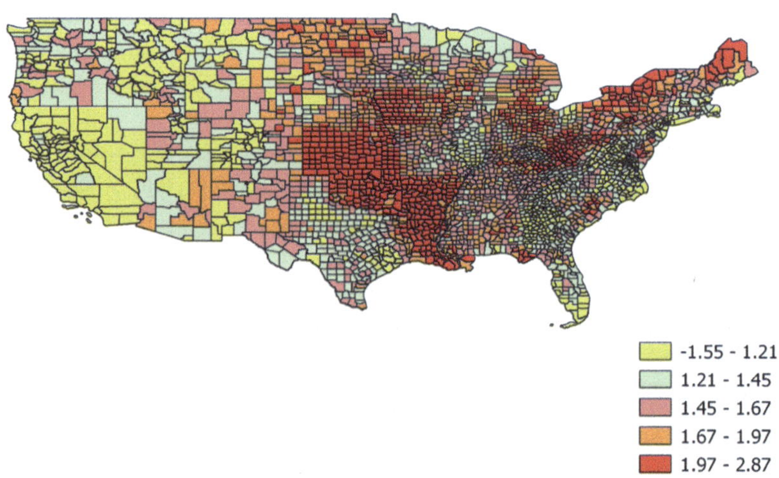

Fig. 9.3 Obesity prevalence growth rates (%), US continental counties, 2005–2007 to 2014–2016

age-standardised obesity rates in county i (=1, …, 3141) and period t (=1, …, 4) as r_{it}. These are modelled as

$$r_{it} = \beta_0 + s_i + d_i t + u_{it},$$

where β_0 is the intercept, and s_i are spatially correlated effects measuring the obesity level in period 1, as in the scheme of Besag et al. (1991). The d_i are spatially correlated linear growth rates, while the u_{it} are iid normal errors. This is an analogue of the varying intercepts, varying slopes model used in analysing longitudinal and multi-level data, but allows for spatial clustering (as evident in Figs. 9.1 and 9.2).

Figure 9.3 maps out the estimated growth rates d_i, and shows that the highest growth in obesity to be in the same counties which have the highest obesity according to Figs. 9.1 and 9.2. This suggests both persistent spatial inequalities in obesity, and a widening of such inequalities.

Conclusion: Unresolved Issues and Future Perspectives

There is by no means a consensus in the literature about the impacts of the urban environment on obesity and related outcomes, and debate continues. For example, on the fundamental question of the relative importance of diet and physical activity, some studies emphasize physical activity (Fisher et al. 2013: Blair et al. 2013), while some stress dietary changes (Malhotra et al. 2015; Luke and Cooper 2013).

Regarding debate on specific environmental impacts, and in particular the impact of sprawl, Eid et al. (2008) consider panel survey data on obesity status and particularly the role of propensity to obesity. They characterise the sprawl effect in terms of levels of mixed use, and find a negative correlation between mixed-use and obesity, after controlling for observable individual characteristics. However, they report that "once we take advantage of the panel dimension of our data to control for unobserved propensity to be obese, the correlation between obesity and mixed-use also vanishes". In related work considering self-selection of residential location, Plantinga and Bernell (2007) find that individuals who move to denser locations lose weight, but that BMI affects the choice of a dense or sprawling location, such that dense locations are unlikely to be selected by high BMI individuals.

Regarding food access and location of healthy as against unhealthy food outlets [a supply side influence], sceptical studies have instead focused on the reliance on observational data to establish the role of the food environment (Hall 2018), and on demand differences between income groups. Thus Allcott et al. (2017) considered supermarket location, and concluded that "the causal impact of unhealthful food supply is small, relative to either the overall obesity rate or the nutrition-income relationship." Other studies have reported equivocal findings on the impact of access to fast food outlets, with Fraser et al. (2010) reporting "conflicting results between obesity/overweight and fast food outlet availability". Snowdon (2018) found a lack of evidence to support official policies (in the United Kingdom) to restrict location of fast food outlets near schools. In a review of 74 studies of the relationship between the density and proximity of fast food outlets and the prevalence of obesity, Snowdon reports that only fifteen (20%) found a positive association between the proximity and/or density of fast food outlets and obesity/body mass, while 44 (60%) found no positive association.

Therefore future research has a role in clarifying such unresolved aspects of the environment–obesity nexus. In terms of quantitative analysis of obesity, especially its spatio-temporal aspects, there is more scope to investigate geographic heterogeneity, for example in terms of spatially varying impacts of ecological factors on obesity prevalence and obesity-related mortality (Lee et al. 2019b; Wen et al. 2010). Spatio-temporal modelling might also be used to analyse the varying pace of the obesity epidemic, and the considerable national variation in obesity levels (Hruby and Hu 2015).

References

Abdullah A, Peeters A, de Courten M, Stoelwinder J (2010) The magnitude of association between overweight and obesity and the risk of diabetes: a meta-analysis of prospective cohort studies. Diabetes Res Clin Pract 89(3):309–319

Allcott H, Diamond R Dubé, J (2017) The geography of poverty and nutrition: food deserts and food choices across the United States. NBER working paper series. National Bureau of Economic Research, Cambridge

An R, Yang Y, Hoschke A, Xue H, Wang Y (2017) Influence of neighbourhood safety on childhood obesity: a systematic review and meta-analysis of longitudinal studies. Obes Rev 18(11):1289–1309

Angraal S, Gupta A, Khera R, Nasir K, Desai N (2019) Association of access to exercise opportunities and cardiovascular mortality. Am Heart J 212:152–156

Anselin L, Syabri I, Kho Y (2006) GeoDa: an introduction to spatial data analysis. Geogr Anal 38:5–22

Ashwell M, Browning L (2011) The increasing importance of waist-to-height ratio to assess cardiometabolic risk: a plea for consistent terminology. Open Obes J 3 (1):70–77

Aviola P, Nayga R, Thomsen M, Danforth D, Smartt J (2014) The effect of fast-food restaurants on childhood obesity: a school level analysis. Econ Hum Biol 12:110–119

Besag J, York J, Mollié A (1991) Bayesian image restoration, with two applications in spatial statistics. Ann Inst Stat Math 43(1):1–20

Bhurosy T, Jeewon R (2014) Overweight and obesity epidemic in developing countries: a problem with diet, physical activity, or socioeconomic status? Sci World J 2014:964236. https://doi.org/10.1155/2014/964236

Biener A, Cawley J, Meyerhoefer C (2017) The high and rising costs of obesity to the US health care system. J Gen Intern Med 32(Suppl 1):6–8

Blair S, Archer E, Hand G (2013) Commentary: Luke and Cooper are wrong: physical activity has a crucial role

in weight management and determinants of obesity. Int J Epidemiol 42:1836–1838
Bravo M, Anthopolos R, Miranda M (2019) Characteristics of the built environment and spatial patterning of type 2 diabetes in the urban core of Durham, North Carolina. J Epidemiol Community Health 73(4):303–310
Caballero B (2007) The global epidemic of obesity: an overview. Epidemiol Rev 29(1):1–5
Carnethon M, De Chavez P, Biggs M, Lewis C, Pankow J, Bertoni A (2012) Association of weight status with mortality in adults with incident diabetes. JAMA 308 (6):581–590
Centre for Disease Control (CDC) (2017) National Diabetes Statistics Report, 2017: estimates of diabetes and its burden in the United States. CDC, Washington
Cooksey-Stowers K, Schwartz M, Brownell K (2017) Food swamps predict obesity rates better than food deserts in the United States. Int J Environ Res Public Health 14(11):1366
Creatore M, Glazier R, Moineddin R, Fazli G, Johns A, Gozdyra P, Booth G (2016) Association of neighborhood walkability with change in overweight, obesity, and diabetes. JAMA 315:2211–2222
Cummins S, Macintyre S (2006) Food environments and obesity—neighbourhood or nation? Int J Epidemiol 35 (1):100–104
Den Braver N, Lakerveld J, Rutters F, Schoonmade L, Brug J, Beulens J (2018) Built environmental characteristics and diabetes: a systematic review and meta-analysis. BMC Med 16(1):12
Dendup T, Feng X, Clingan S, Astell-Burt T (2018) Environmental risk factors for developing type 2 diabetes mellitus: a systematic review. Int J Environ Res Public Health 15(1):78
Diez-Roux AV, Nieto FJ, Muntaner C, Tyroler HA, Comstock GW, Shahar E (1997) Neighborhood environments and coronary heart disease: a multilevel analysis. Am J Epidemiol 146(1):48–63
Drewnowski A, Specter S (2004) Poverty and obesity: the role of energy density and energy costs. Am J Clin Nutr 79:6–16
Egger G, Swinburn B (1997) An "ecological" approach to the obesity pandemic. Br Med J 315(7106):477–480
Eid J, Overman H, Puga D, Turner M (2008) Fat city: questioning the relationship between urban sprawl and obesity. J Urban Econ 63(2):385–404
Faith M, Kral T (2006) Social environmental and genetic influences on obesity and obesity-promoting behaviors: fostering research integration. In: Hernandez L, Blazer D (eds) Genes, behavior, and the social environment. National Academies Press, Washington
Fisher G, Hunter G, Allison D (2013) Commentary: physical activity does influence obesity risk when it actually occurs in sufficient amount. Int J Epidemiol 42: 1845–1848
Flegal K, Carroll M, Kit B, Ogden C (2012) Prevalence of obesity and trends in the distribution of body mass index among US adults, 1999-2010. JAMA 307(5): 491–497
Flegal K, Kit B, Orpana H, Graubard B (2013) Association of all-cause mortality with overweight and obesity using standard body mass index categories: a systematic review and meta-analysis. JAMA 309(1):71–82
Fleischhacker S, Evenson K, Rodriguez D, Ammerman A (2011) A systematic review of fast food access studies. Obes Rev 2011(12):e460–e471
Flint E, Cummins S (2016) Active commuting and obesity in mid-life: cross-sectional, observational evidence from UK Biobank. Lancet Diabetes Endocrinol 4(5): 420–435
Fox A, Feng W, Asal V (2019) What is driving global obesity trends? Globalization or "modernization"? Glob Health 15(1):32
Fraser L, Edwards K, Cade J, Clarke G (2010) The geography of fast food outlets: a review. Int J Environ Res Public Health 7(5):2290–2308
Gallos L, Barttfeld P, Havlin S, Sigman M, Makse H (2012) Collective behavior in the spatial spreading of obesity. Nat Sci Rep 2:454. https://www.nature.com/articles/srep00454
Galster G, Hanson R, Ratcliffe M, Wolman H, Coleman S, Freihage J (2001) Wrestling sprawl to the ground: defining and measuring an elusive concept. Hous Policy Debate 12(4):681–717
Gaskin C, Orellana L (2018) Factors associated with physical activity and sedentary behavior in older adults from six low-and middle-income countries. Int J Environ Res Public Health 15(5):908
Gassasse Z, Smith D, Finer S, Gallo V (2017) Association between urbanisation and type 2 diabetes: an ecological study. BMJ Glob Health 2(4):e000473
Guo C, Wang H, Feng G, Li J, Su C, Zhang J (2019) Spatiotemporal predictions of obesity prevalence in Chinese children and adolescents: based on analyses of obesogenic environmental variability and Bayesian model. Int J Obes 43(7):1380–1390
Hall K (2018) Did the food environment cause the obesity epidemic? Obesity 26(1):11–13
Hamidi S (2019) Urban sprawl and the emergence of food deserts in the USA. Urban Studies. https://doi.org/10.1177/0042098019841540
Health Issues India (2019) Diabetes: an Indian epidemic. https://www.healthissuesindia.com/2019/10/12/diabetes-an-indian-epidemic/
Hruby A, Hu FB (2015) The epidemiology of obesity: a big picture. PharmacoEconomics 33(7):673–689
Huang R, Moudon A, Cook A, Drewnowski A (2015) The spatial clustering of obesity: does the built environment matter? J Hum Nutr Diet 28(6):604–612
James W (2009) WHO recognition of the global obesity epidemic. Int J Obes 32(S7):S120
Jones N, Conklin A, Suhrcke M, Monsivais P (2014) The growing price gap between more and less healthy foods: analysis of a novel longitudinal UK dataset. PLoS One 9(10):e109343

Joost S, Duruz S, Marques-Vidal P, Bochud M, Stringhini S, Paccaud F (2016) Persistent spatial clusters of high body mass index in a Swiss urban population as revealed by the 5-year GeoCoLaus longitudinal study. BMJ Open 6(1):e010145

Kashef M (2011) Walkability and residential suburbs: a multidisciplinary perspective. J Urban 4(1):39–56

Larsen K, Gilliland J (2008) Mapping the evolution of'food deserts' in a Canadian city: supermarket accessibility in London, Ontario, 1961–2005. Int J Health Geogr 7(1):16

Lawson A, Lee D (2017) Bayesian disease mapping for public health. In: Rao ARS, Pyne S, Rao CR (eds) Handbook of statistics, vol 36. Elsevier, London, pp 443–481

Lee A, Cardel M, Donahoo W (2019a) Social and environmental factors influencing obesity. In: Feingold K, Anawalt B, Boyce A et al. (eds) Endotext [Internet]. South Dartmouth

Lee K, Heo J, Jayaraman R, Dawson S (2019b) Proximity to parks and natural areas as an environmental determinant to spatial disparities in obesity prevalence. Appl Geogr 112:102074

Luke A, Cooper R (2013) Physical activity does not influence obesity risk: time to clarify the public health message. Int J Epidemiol 42:1831–1836

Lunn D, Spiegelhalter D, Thomas A, Best N (2009) The BUGS project: evolution, critique and future directions. Stat Med 28:3049–3082

Malhotra A, Noakes T, Phinney S (2015) It is time to bust the myth of physical inactivity and obesity: you cannot outrun a bad diet. Br J Sports Med 49:967–968

McKeigue P, Shah B, Marmot M (1991) Relation of central obesity and insulin resistance with high diabetes prevalence and cardiovascular risk in South Asians. Lancet 337:382–386

Mead M (2008) Urban issues: the sprawl of food deserts. Environ Health Perspect 116(8):A335

Must A, Spadano J, Coakley E, Field A, Colditz G, Dietz W (1999) The disease burden associated with overweight and obesity. JAMA 282(16):1523–1529

O'Donovan G, Cadena-Gaitán C (2018) Air pollution and diabetes: it's time to get active! Lancet Planet Health 2 (7):e287–e288

Plantinga A, Bernell S (2007) The association between urban sprawl and obesity: is it a two-way street? J Reg Sci 47(5):857–879

Popkin B (2015) Nutrition transition and the global diabetes epidemic. Curr Diab Rep 15:64

Rajagopalan S, Brook R (2012) Air pollution and type 2 diabetes: mechanistic insights. Diabetes 61(12): 3037–3045

Ramachandran A, Snehalatha C, Latha E, Manoharan M, Vijay V (1999) Impacts of urbanisation on the lifestyle and on the prevalence of diabetes in native Asian Indian population. Diabetes Res Clin Pract 44(3):207–213

Richardson AS, Troxel WM, Ghosh-Dastidar M, Hunter GP, Beckman R, Colabianchi N (2017) Pathways through which higher neighbourhood crime is longitudinally associated with greater body mass index. Int J Behav Nutr Phys Act 14(1):155. https://doi.org/10.1186/s12966-017-0611-y

Riveros-McKay F, Mistry V, Bounds R, Hendricks A, Keogh JM, Thomas H (2019) Genetic architecture of human thinness compared to severe obesity. PLoS Genet 15(1):e1007603

Ruiz-Alejos A, Carrillo-Larco R, Miranda J, Anderson C, Gilman R, Smeeth L, Bernabé-Ortiz A (2018) Addressing the impact of urban exposure on the incidence of type 2 diabetes mellitus: the PERU MIGRANT study. Sci Rep 8(1):5512

Schneider S, Diehl K, Görig T, Schilling L, De Bock F, Hoffmann K (2017) Contextual influences on physical activity and eating habits-options for action on the community level. BMC Public Health 17(1):760. https://doi.org/10.1186/s12889-017-4790-x

Schuurman N, Peters P, Oliver L (2009) Are obesity and physical activity clustered? A spatial analysis linked to residential density. Obesity 17(12):2202–2209

Snowdon C (2018) Fast food outlets and obesity: what is the evidence? Institute of Economic Affairs, London

Stryjecki C, Alyass A, Meyre D (2018) Ethnic and population differences in the genetic predisposition to human obesity. Obes Rev 19(1):62–80

The Food Foundation (2019) Written evidence submitted by the Food Foundation to the Health and Social Care Committee's Inquiry on Childhood Obesity, Oct 2019. https://foodfoundation.org.uk/

Timperio A, Jeffery R, Crawford D, Roberts R, Giles-Corti B, Ball K (2010) Neighbourhood physical activity environments and adiposity in children and mothers: a three-year longitudinal study. Int J Behav Nutr Phys Act 7(1):18

Unnikrishnan R, Pradeepa R, Joshi S, Mohan V (2017) Type 2 diabetes: demystifying the global epidemic. Diabetes 66(6):1432–1442

Verhaeghe P, Tampubolon G (2012) Individual social capital, neighbourhood deprivation, and self-rated health in England. Soc Sci Med 75(2):349–357

Wen T, Chen D, Tsai M (2010) Identifying geographical variations in poverty-obesity relationships: empirical evidence from Taiwan. Geospat Health:257–265

Wolch J, Byrne J, Newell J (2014) Urban green space, public health, and environmental justice: the challenge of making cities 'just green enough'. Landsc Urban Plan 125:234–244

Wu Y, Moore S, Dube L (2018) Social capital and obesity among adults: longitudinal findings from the Montreal neighbourhood networks and healthy aging panel. Prev Med 111:366–370

Yang WY, Lu JM, Weng JP et al (2010) Prevalence of diabetes among men and women in China. N Engl J Med 362:1090–1101

Zimmet P (2017) Diabetes and its drivers: the largest epidemic in human history? Clin Diabetes Endocrinol 3(1):1

Inflammation and Its Role in Obesity-Related Complications

10

Mohit Singh, Jiten R. Sharma, Hina Agraval, and Umesh C. S. Yadav

Contents

Introduction 138

Inflammation & Innate Immunity 138

Intertwined Relation of Obesity and Inflammation 139
Intestinal Dysbiosis and Obesity 139
High-Calorie Diet and Inflammation 140
Nonbacterial Dietetic Immunogens and Inflammation 140

Factors Responsible for Inflammation in Obesity 140
Genetic and Epigenetic Factors 140
MicroRNAs 140

Cellular Links Between Obesity and Inflammation 141
Adipocytes 141
Treg Cells 142
γδ T Cells 142
B Lymphocytes 143
Myeloid Cells 143

Molecular Mediators of Obesity and Inflammation 143
TNF-α 143
IL-1β 143
Adiponectin and Leptin 144
Secreted Frizzled-Related Protein 5 144

Obesity, Inflammation and Cancer 144
Liver Cancer 144
Pancreatic Cancer 145
Colorectal Cancer 145

M. Singh · J. R. Sharma · H. Agraval
School of Life Sciences, Central University of Gujarat, Gandhinagar, Gujarat, India

U. C. S. Yadav (✉)
Special Centre for Molecular Medicine, Jawaharlal Nehru University, New Delhi, India
e-mail: ucsyadav@mail.jnu.ac.in

J. Faintuch and S. Faintuch (eds.), *Obesity and Diabetes*, https://doi.org/10.1007/978-3-030-53370-0_10

Therapeutic Approaches for Obesity and Inflammation 145
Probiotics and Phytochemicals 145
Cell-based Therapy 146

References 147

Abstract

The pathophysiology of inflammation has emerged as a player in the quest to understand the fundamental functions involved in obesity-enabled diseases. The dysregulated adipose tissue homeostasis and its endocrine function due to the excessive consumption of high-calorie diet, compounded with a modern sedentary lifestyle and predisposing effects of environmental and genetic factors, result in the initiation and the development of low-grade chronic inflammation in obese and overweight patients. The increased inflammatory cytokines and chemokines further complicate the already challenged physiology of the person with obesity and manifest into several associated comorbidities, including T2DM, heart diseases, cancer, and early aging. This chapter discusses the intricate association among obesity, inflammation, and their resultant and associated comorbidities. Additionally, some of the current and conventional therapeutic approaches and potential future interventional strategies targeting newer molecular players are also addressed.

Keywords

Obesity · Chronic low-grade inflammation · High-calorie diet · Comorbidities · Body mass index · Therapeutic targets

Introduction

A sustained high-calorie diet intake leads to the modification of the adipose tissues (AT), where hypertrophy and hyperplasia of the adipocytes are observed (Jo et al. 2009). This modification leads to alteration in the functions of the adipocytes, and contributes to the recruitment of the immune cells, which then secrete inflammatory cytokines and chemokines (Kumari et al. 2019; Kumari and Yadav 2018). Inflammation of adipocytes results in chronic low low-grade local inflammation, which ultimately becomes systemic (Reilly and Saltiel 2017).

Inflammation & Innate Immunity

Inflammation is the defensive reaction of the body in face of toxic effects, in conjunction with a healing process to restore damaged tissue. Several factors like microbial invasion, tissue injury, and many exogenous and endogenous antigens cause inflammation (Takeuchi and Akira 2010). Classically, there are five cardinal signs of inflammation: heat (*calor*), pain (*dolor*), redness (*rubor*), swelling (*tumor*), and loss of tissue function (*functio laesa*). In the fifth century BC, Hippocrates coined terms such as edema (swelling), which are still used to describe inflammation, and also observed inflammation in the early stage of the healing process after tissue damage.

Enhanced vascular endothelium permeability allows serum outflow and immune cell extravasation. Under ideal conditions, the inflammatory response rapidly proceeds and tissue damage is repaired. Nevertheless, if it lasts for an extended period, inflammation could lead to multiple complications and even death.

Inflammation engages immune cells, including macrophages and dendritic cells (DCs) and non-immune cells, including epithelial cells, endothelial cells, and fibroblasts, which play important roles and contribute to innate immunity (Hamada et al. 2019). Proinflammatory cytokines inclusive of tumor necrosis factor-alpha (TNF-α) and interleukin-1 (IL-1), facilitate leukocyte extravasation by growing leukocyte adhesion molecules in endothelial cells. Innate immune cells, including DCs, macrophages, and neutrophils, are activated at the site of infection or injury, remove foreign particles and host debris by phagocytosis, plus secrete cytokines that activate the slower, lymphocyte-mediated adaptive immune response (Newton and Dixit 2012). Germline-encoded receptors of pattern recognition (PRRs) sense the existence of a structurally conserved area of microbial species known as pathogen-associated molecular patterns (PAMPs). Current findings suggest that PRRs also identify endogenous molecules secreted from damaged cells, called damage/danger-associated molecular patterns (DAMPs) (Singh et al. 2019).

There are currently listed four diverse classes of PRR families. These families are toll-like receptors (TLRs) and C-type lectin receptors (CLRs), which are transmembrane proteins, while the other two are cytoplasmic proteins, retinoic acid-inducible gene (RIG)-I-like receptors (RLRs) and NOD receptors (NLRs). These receptors are found in other nonprofessional immune cells apart from macrophages and DCs. PRR senses PAMPs or DAMPs and upregulates gene transcription involving inflammatory responses, except for certain NLRs. Such genes encode proinflammatory cytokines, chemokines, type I interferons (IFNs), and antimicrobial proteins involved in the regulation of many uncharacterized proteins and signals for PRR. The inducible gene expression patterns differ among activated PRRs. Inflammatory cytokines, for instance, TNF-α, IL-1, and IL-6 control inflammatory response. Such pro-inflammatory mediators are pleiotropic proteins that control inflammatory tissue apoptosis, modify endothelial vascular permeability, use inflamed tissue blood cells, and induce the development of acute-phase proteins (Kim et al. 2016).

Intertwined Relation of Obesity and Inflammation

Elevated levels of proinflammatory adipokines and cytokines including resistin, leptin, IL-6, TNF-α in the visceral adipocytes, and associated immune cells contribute to obesity pathophysiology. Inflammation of the white adipose tissue (WAT) correlates with the obese or overweight phenotype (Ellulu et al. 2017). WAT, a physiological storage site of lipids, is recognized as a dynamic contributor in several physiological and pathophysiological processes such as insulin resistance (IR), decreased glucose tolerance, and diabetes. Obese WAT is infiltrated by macrophages, which may be a significant source of locally produced pro-inflammatory cytokines (Lauterbach and Wunderlich 2017). WAT macrophage infiltration is diminished with a reduction of body weight in obese or overweight patients.

Intestinal Dysbiosis and Obesity

Obesity is also associated with chronic low-grade inflammation and endotoxemia originated in the gut. It is thus akin to a disease state with a complex interaction of genetic and environmental factors. A high-calorie diet has a profound effect on the gut microflora. The lipopolysaccharide (LPS), a pathogen-associated molecular pattern (PAMP), and a principal constituent of the Gram-negative bacterial cell wall, in physiological conditions, is found to be hiked in the intestinal lumen and low or nil in the plasma (Gnauck et al. 2016). In case the LPS concentration increases in the plasma, it has a direct effect on the gut barrier function, enhancing tight junction permeability in the enterocyte cells of the intestine by expression and localization of CD14 and TLR-4 in the plasma membrane of those epithelial cells (Guo et al. 2013). Further, LPS-induced tight junction permeability has been reported to act via LPS/TLR-4/FAK/MyD88 signaling axis (Guo et al. 2015). Thus, being an endotoxin, LPS mediates gut barrier dysfunction via incorporation of focal adhesion kinase (FAK) and MyD88 adapter proteins.

TLRs are upregulated with inflammation in the tissues, and in addition to intake of dietary lipids that are influenced by the action of the gut microflora, help in stimulating the innate immune response arm (Schoeler and Caesar 2019). There is evidence that inflammation caused by the gut microbiota can help in the activation of the signaling pathways of IKKβ, JNK, and PI3K, thereby regulating the progression of obesity, inflammation of the AT, and resistance to insulin (Hullar and Fu 2014; Miro-Blanch and Yanes 2019; Ramos-Molina et al. 2019).

High-Calorie Diet and Inflammation

High-calorie Western diets have also been reported to induce residential immune cells of the gut, specifically macrophages, and alter their phenotype from tolerogenic to inflammatory. This phenotype change has further been implicated in the loss of Tregs cell population and function (Bain and Schridde 2018; Wang et al. 2019). The immune cells of the innate immune system are prone to transcriptional and epigenetic modulation in order to show a trained immune response, thus enhancing the immune response. The same is true even in the case of a sterile immunogen derived from a high-calorie diet (Netea et al. 2016).

Nonbacterial Dietetic Immunogens and Inflammation

Plenty of sterile bioactive molecules are now known that activate the innate immune response such as oxidized LDL, urate crystals, and cholesterol crystals. A high-calorie diet is linked to increased levels of oxLDL in circulation (Chatauret et al. 2014). In a study, Bekkering et al. showed that peripheral blood monocytes from healthy individuals, when treated with exogenous oxLDL, had a modulation in their function and phenotype. Furthermore, these cells showed a foamy appearance with enhanced production of cytokines such as TNFα, IL-6, IL-8, and IL-18 (Bekkering et al. 2014). This observation was further confirmed by Varghese et al., who implicated the role of a transcription factor sterol regulatory element-binding protein-1 or SREBP-1 in the enhancement of proinflammatory and foam cell phenotype of the U937 monocytes-derived macrophages (Varghese et al. 2019).

The oxLDL has also been implicated to modulate the innate immune response and generate a proinflammatory phenotype through the activation of the NLRP3 inflammasome (Liu et al. 2014; Sheedy et al. 2013; Singh et al. 2019). Inflammasomes are critical immune sensors that perceive environmental agents in the form of PAMPs and DAMPs and cause a release of proinflammatory cytokines. Uric acid and cholesterol crystals could activate the NLRP3 inflammasome complex and ultimately, the release of the proinflammatory cytokines (Beydoun et al. 2018; Stodle et al. 2018). Christ et al. have recently shown that the high-calorie diet sets off an innate immune response in *Ldlr*$^{-/-}$ mice, via NLRP3 inflammasome complex activation. The infiltration of myeloid cells like monocytes in the obese adipocytes, specifically the WAT, indicates the same fact (Christ et al. 2018). The infiltrating monocytes and their site-specific differentiation into macrophages are yet another cause of an inflammatory phenotype evident in obese people (Bai and Sun 2015) (Fig. 10.1).

Factors Responsible for Inflammation in Obesity

Genetic and Epigenetic Factors

In modern civilizations, where high-calorie food consumption and lack of physical activity, along with controlled climate encourages weight gain, polygenic obesity is the most prevalent modality (Alam et al. 2012; Heo et al. 2018; Raffan et al. 2016). However, in early severe obesity in children, monogenic forms are not uncommon. Even though genetic and epigenetic mechanisms mostly respond to nonspecific inflammatory changes, a more direct contribution is envisaged for epigenetics (Raghuraman et al. 2016), possibly involving gut microbiota (Ramos-Molina et al. 2019).

MicroRNAs

MicroRNAs (miRNA) are endogenous small single-stranded noncoding RNA molecules

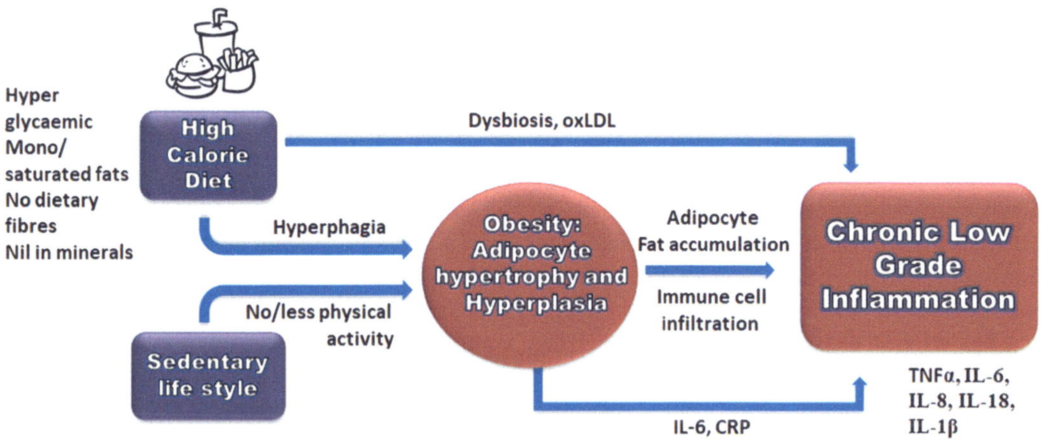

Fig. 10.1 Schematic representation of High-calorie diet-induced obesity and chronic low-grade inflammation. Development of an obese/overweight phenotype occurs upon chronic consumption of a high-calorie diet and leads to development of chronic low-grade inflammation in the body

(containing ~22 nucleotides), and their function is to regulate gene expression post-transcriptionally. They have also been involved in obesity and diabetes, by targeting various pathways of the metabolic function or adipocyte differentiation. miR-21 functions in the human AT and is related to BMI. miRNAs can act as modulators of the inflammatory reaction by selectively suppressing NF-kB pathway components. miR-146b-5p, which is known to regulate inflammatory signaling cascades, is reduced in obese and Type 2 diabetes mellitus (T2DM) individual monocytes. miR-107 inhibition caused by toll-like receptor 4 (TLR4) is dysregulated in obesity, resulting in an increased inflammatory response in macrophages (Olivieri et al. 2013). In peripheral blood mononuclear cells, gene overexpression for miR-103 and miR-143 was shown in a T2DM rat model (Vatandoost et al. 2015).

Both miRNAs were related to dysregulated adipogenesis, affecting (AT) development in the case of abundance of substrates, and miR-193b regulated adipocyte secretion of chemokine CCL2, reflecting a pathway seen in macrophages. LncRNA E33 overexpression was connected with the increased expression of several pro-inflammatory genes in RAW 264.7 cells, while the silencing of lncRNA E33 reversed these effects (Reddy et al. 2014).

Cellular Links Between Obesity and Inflammation

AT includes various types of cells, including pre-adipocytes, adipocytes, fibroblasts, and endothelial cells. The cell–cell cross talk that takes place, both locally with associated immune cells, and also systemically with other organs, is mediated not only by intercellular contact, but also by cytokines, chemokines, and other biomolecules (Liu and Nikolajczyk 2019).

Adipocytes

During obesity, the distance between adipocytes and AT vasculature increases due to adipocyte enlargement, potentially resulting in hypoxia. Hypoxia leads to the activation of hypoxia-inducible factor 1 (HIF-1) expression in adipocytes, to protect the AT from hyperplasia-related programmed cell death. It also causes the secretion of differentiation cues for the resident immune cells. When macrophages infiltrate these hypoxic ATs to clear the apoptotic and necrotic cells, they further amplify the inflammatory response via the production of IL-6 and TNF-α locally. Hypoxic ATs release differential signals acquired by residential or infiltrating macrophages, resulting in an overall change from the M2 anti-inflammatory pattern, to M1 pro-inflammatory macrophage cells.

An elevated M1-like macrophage count and their secretome are found to be positively correlated with IR during obesity. AT inflammation has also been associated with a change in the T lymphocytes (Dam et al. 2016). The T lymphocyte subpopulation change includes an activation state of the $CD8^+$ T cells, a reduction of the Tregs and Th2 cells, an elevation in the number of Th1 and Th17, all being the hallmark of proinflammatory milieu shift. During obesity-induced hypoxic conditions, it has been reported that the Th17/Treg ratio gets modulated (Liu and Nikolajczyk 2019). One of the proposed mechanisms is the proteasomal degradation of Foxp3, that leads to the activation of RORγt protein, an essential molecule for Treg development. Thus, Treg cells' development gets suppressed aiding in the inflammation of the adipocytes. HIF-1 also induces IFNγ secretion from Th1 cells. This is followed by an activation event of transcription factors such as STAT3. STAT3 is further reported to be a major transcription factor for TH17 activity (Raphael and McGeachy 2018).

Activation of $CD8^+$ T cells is also an occurrence in the inflammatory response to obesity that results in activation/infiltration of M1-like macrophage in AT. The population of both effector $CD8^+$ T cells and memory cells is found to be increased in visceral adipose tissue (VAT), in comparison to subcutaneous adipose tissue. Under the hypoxic WAT microenvironment, the influx of $CD8^+$ T cells is related to enhanced expression of HIF-1. This information indicates that hypoxic adipocyte microenvironment regulates an infiltration of $CD8^+$ T cells before M1 phenotype changes in the macrophages.

Treg Cells

Tregs cells are anti-inflammatory T cells that are comparatively common in lean and their numbers decrease with the development of obesity. VAT-resident Tregs express IL-10, which prevents proinflammatory IL-6, RANTES (Regulated on activation, normal T Cell produced and secreted), SAA-3 (Serum amyloid A3) and MMP-3 (Matrix metalloproteinase-3) from being produced via TNF-α. IL-10 also establishes the phosphorylation of insulin receptor substrate (IRS)-1 and thus recovers glucose transporter GLUT4 expression in adipocytes, following TNF-α administration. In VAT, IL-33 receptor regulates IL-33 ligand Tregs-ST2 (also known as IL1RL1 and IL-1 IL1RAP receptor accessory protein). IL-33 also contributes to the stimulation of innate lymphoid cells type 2 (ILC2) and eosinophilic aggregation in murine VAT, indicating that these functions work in tandem with VAT-associated Treg anti-inflammatory impact (Han et al. 2015; Kolodin et al. 2015; Liu and Nikolajczyk 2019; McLaughlin et al. 2014).

Adipocyte lipid metabolites, stressed by growing fatty acids content in obesity, deliver antigens present in CD1 lymphocytes, which stimulate iNKT cells to reduce AT inflammation. Adipose iNKTs can also help Tregs by supplying IL-2, consistently with the results that CD1d knockout mice lowered the production of IL-2 in AT, and it was related to lean AT Tregs during interim high-fat diet (HFD) eating (Park et al. 2018; Satoh and Iwabuchi 2018).

γδ T Cells

γ-T cells express the surface T cell receptor (TCR) genes, and are mostly found among intestinal intraepithelial lymphocytes. In mice fed a HFD to induce obesity, there is a reduction in both ordinary αβ and γδ lymphochytes. Diet-induced weight loss restores both categories, and also improves outcome in experimental colitis. Thus both types of lymphocytes seem to enhance inflammation, when underexpressed in obesity.

The Treg population in AT is negatively regulated by IL-17, which originated from M1 macrophages and lymphocytes. IL-17 and other pro-inflammatory cytokines, including interleukin-1β (IL-1β), tumor necrosis factor α, and IL-6 which are associated with the progression of simple obesity toward metabolic syndrome. Cytolytic functions, involving $CD8^+$ T cells, may also help prevent malignancy in obesity (Le Menn et al. 2018; Singh et al. 2016).

B Lymphocytes

B cells are known to infiltrate the growing AT, where they secrete cytokines and autoreactive antibodies (IgG), that interfere in the inflammatory microenvironment. B cell subtypes comprise B1 and B2 cells, which are subdivided additionally into unknown regulatory B cells (Bregs)/B10 subsets. The development of IL-10 and a variety of surface markers also define them. Each of these subpopulations of B cells has been involved in AT inflammation associated with obesity (Liu and Nikolajczyk 2019). The secreted antibodies from the B cell subsets have different functions: B1 cells are found to secrete anti-inflammatory or natural antibodies (IgM) and B2 cells the proinflammatory autoreactive IgG.

Natural antibodies are in reverse association with circulating levels of MCP-1 (CCL2). MCP-1 is related to Th2 response and controls the chemotaxis of macrophages, thereby regulating chronic inflammation (Engin 2017; Frasca and Blomberg 2017; Lu et al. 2019).

B cells provide antigens to T cells and thereby trigger macrophages as well as DCs. Even though B cells' secretion of obesity-related cytokines is less, their participation in inflammation is enhanced by having unique antigen concentration capability, via epitope-targeted acceptance/processing to control the function of T cells. Also, in obesity/T2DM, B cells regulate anti-inflammatory Treg cells. B cell-knockout mice on HFD show rare $CD4^+$ and $CD8^+$ T cells generating IFNα, indicating that other mechanisms may affect the consequence of B–T cell signaling in obesity (Liu and Nikolajczyk 2019).

Myeloid Cells

Myeloid dendritic cells and macrophages are the most capable antigen-presenting cells (APCs). At first, local AT macrophages network with naive $CD4^+$ T cells through Class II MHC molecules. Subsequently, a defined inflammatory milieu brings $CD11c^+$ macrophages into AT which, in effect, stimulates traditional T cells to multiply and thus maintain adaptive immune response, which leads to AT inflammation. Chronic high-calorie diet intake leads to the modulation of the adipocyte functioning further, interfering with β catenin and PPARγ, important protein responsible for adipocyte homeostasis. It was shown that a reduction in the expression levels of PPARγ and β-catenin in conventional DCs of AT leads to the prevalence of a proinflammatory state (Macdougall et al. 2018).

Molecular Mediators of Obesity and Inflammation

TNF-α

The chronic low-grade inflammation, features release of proinflammatory cytokines such as TNF-α, IL-6, and IL-1β, along with the dysregulated secretion of proinflammatory adipokines such as resistin and lipocalin-2 from the adipocytes. These are risk factors for the development of type 2 diabetes as well (Saltiel and Olefsky 2017).

In obese individuals, specifically in males, the level of TNF-α has been found to be elevated in plasma (El-Haggar and Mostafa 2015). TNF-α is known to enhance the phosphorylation of insulin receptor substrate (IRS)-1 and aggravate IR. Further, in the visceral adipocytes, TNF-α triggers IR via activation of JNK1/2 and its cross talk with the IRS signaling pathway (Fernández-Veledo et al. 2009). Since peripheral IR also characterizes T2DM in skeletal muscle, AT and liver, the prevalence of chronic low-grade inflammation in these organs may lead to IR, metabolic syndrome, and diabetes (Honka et al. 2018).

IL-1β

Obesity induces danger signals such as glucose and free fatty acids, and causes defective cell organelle function such as that of mitochondria, which ultimately result in reactive oxygen species (ROS) generation. The increased level of ROS activates NLRP3 inflammasome complex in the peripheral tissues such as adipocytes, skeletal muscle, islets cells, and also in macrophages (Sharma et al. 2018). IL-1β is released, contributing to IR. Further, IL-1β induced expression of suppressor of cytokine signaling protein further suppresses insulin-dependent β cell survival.

There are several factors associated with obesity-linked inflammation, which include cytokines and adipokines, which propel the development of cardiovascular diseases, diabetes, neurodegenerative diseases, and several types of cancer (Han and Lean 2016).

Adiponectin and Leptin

Adiponectin prevents atherosclerosis owing to its protective action on the function of the endothelial cells. Since it is evident that reduced bioavailability of endothelial NO synthase (eNOS) causes an increased leukocyte adhesion and platelet activation, it ultimately leads to compromised endothelial cell barrier function (Siragusa and Fleming 2016). It was shown that adiponectin induced NO production in the endothelial cells via AMPK mediated expression of the eNOS gene, thus improving the endothelial dysfunction in cardiovascular diseases (Deng et al. 2010).

Leptin induced expression of TNF-α and IL-6 is thought to promote endothelial cell dysfunction and ultimately cardiovascular diseases (Tahergorabi and Khazaei 2015). In a study by Lee et al. of this cascade, IL-6 and TNF-α were implicated in oxidative stress-induced coronary endothelial cell dysfunction in obese rodents (Lee et al. 2017).

Secreted Frizzled-Related Protein 5

Secreted frizzled-related protein 5 (Sfrp5) is a soluble anti-inflammatory adipokine. Similar to adiponectin, the expression of Sfrp5 has also been found to diminish in obesity and cardiovascular diseases (Tong et al. 2019). Sfrp5 is a negative modulator of the secreted Wnt proteins, specifically Wnt5a, which is known to induce macrophages to produce proinflammatory cytokines. However, non-canonical Wnt5a-mediated JNK1 signaling cascade is suppressed, maintaining a Th2 type anti-inflammatory profile. Wnt5a has been found to promote inflammation in the endothelial cells and found present in atherosclerotic lesions, and Sfrp5 could participate in retarding obesity, IR, and atherosclerosis. It is questioned whether Wnt5A noncanonical signalling pathways are the only ones, as the canonical β-catenin pathway can also be activated with receptor availability. Inflammation is a common substrate of obesity, T2DM, and CVD, which makes it possible for Sfrp5 to be at the crossroad between such conditions (Wang et al. 2020).

Obesity, Inflammation and Cancer

Liver Cancer

Elevated levels of TNF-α/IL-6 during obesity and consequent low-grade chronic inflammation promote the IKK/JNK and AMPK/TORC-1 signaling cascades that underlie the obesity-induced inflammatory liver damage. Along with other insults, engaging diet and microbiome aberrations, the development of nonalcoholic fatty liver disease and nonalcoholic steatohepatitis embodies a dysfunctional state of the hepatocytes, which then ultimately die due to the prolonged insult, and trigger the regeneration mechanism of the liver (Kholodenko and Yarygin 2017). Thus, the obese phenotype and accumulation of resultant pro-inflammatory cytokines released by the adipocytes and infiltrated immune cells via a portal vein in the liver, may ultimately lead to the development of hepatocellular carcinoma (HCC). Moreover, the same low-grade inflammation with the constant buildup of adipokines, cytokines, and growth factors can fuel tumorigenesis in the peripheral tissues.

Pancreatic Cancer

Cohort-based studies have directly implicated an increased mortality rate of obese individuals with pancreatic ductal adenocarcinoma (PDAC) (Genkinger et al. 2015). Bariatric surgery could be protective in such patients (Xu et al. 2018). Owing to the chronic low-grade inflammation in obese individuals, elevated levels of IL-6 and TNF-α (Kern et al. 2018) and other cytokines and chemokines are found in the circulation. IL-6 is known to be the activator of JAK-STAT (signal transducers and activator of transcription) pathway. Development of obesity-induced IR via accumulation of visceral fat and inflammation is thought to be via IL-6/JAK/STAT signaling. PDAC is thought to involve KRAS oncogene expression, which further activates STAT3 and NF-κB signaling cascades leading to the development of PDAC (Ling et al. 2012; Pramanik et al. 2018). Furthermore, STAT-3 activation is implicated to be necessary for the development of PDAC (Fukuda et al. 2011; Lesina et al. 2011). Moreover, obesity-induced recruitment of Tregs cells in the pancreas with neoplasia causes the suppression of the immune clearance of the malignant cells and promotes tolerance of neoplastic cells aiding to PDAC complications (Cox and Olive 2012).

Colorectal Cancer

Obesity is associated with colorectal cancer (CRC), most common form of malignancy in men and women (White et al. 2018). The connecting link could be partly related to adipokine and cytokine levels, specifically that of adiponectin, leptin, IL-6, and TNF-α, even though the participation of dysbiosis and gut epithelial barrier dysfunction should not be overlooked. Adiponectin, being anti-inflammatory, is also an anticancer in nature (Monks et al. 2019). Adiponectin-knockout mice were found much more susceptible to inflammation-induced CRC as compared to their wild-type counterparts (Saxena et al. 2012). Mechanistically it is known to suppress mTOR protein via AMPK activation and thus reduces cell growth in human colorectal cancer cells. Leptin, a pro-inflammatory adipokine that is overexpressed in obese individuals, induces cell proliferation and anti-apoptotic properties via p42/44 mitogen-activated protein kinase (MAPK) and PI3K/Akt/mammalian target of rapamycin (mTOR) axis. In colorectal cancer specimens, leptin receptor (LEPR) expression was associated with tumor stage, neoangiogenesis index, regional and distant metastases (Vuletic et al. 2019).

IL-6 is a potential driver of several malignancies, including IL-6 driven PDAC, and also involves CRC. It is overexpressed in both obese and CRC patients (Zeng et al. 2017). TNF-α is overexpressed in animal models of high-fat diet-induced obesity (Wu et al. 2016). The role of TNF-α in IL-6 production and development of CRC has been anticipated (Chung et al. 2017) (Fig. 10.2).

Therapeutic Approaches for Obesity and Inflammation

The therapeutic approaches currently in practice involve pharmacological approaches, probiotic/prebiotics, phytochemicals, gene therapy and stem cell-based therapy. Some of these approaches are discussed below briefly.

Probiotics and Phytochemicals

Some *Lactobacillus sps* (*L. rhamnosus*, and *L. plantarum*) and *Bifidobacterium* (*B. infantis* and *B. longum*) are being used in probiotic formulations for obesity treatment. Some species of probiotic bacteria, including *Lactobacillus* and *Bifidobacterium* genera have the capabilities to

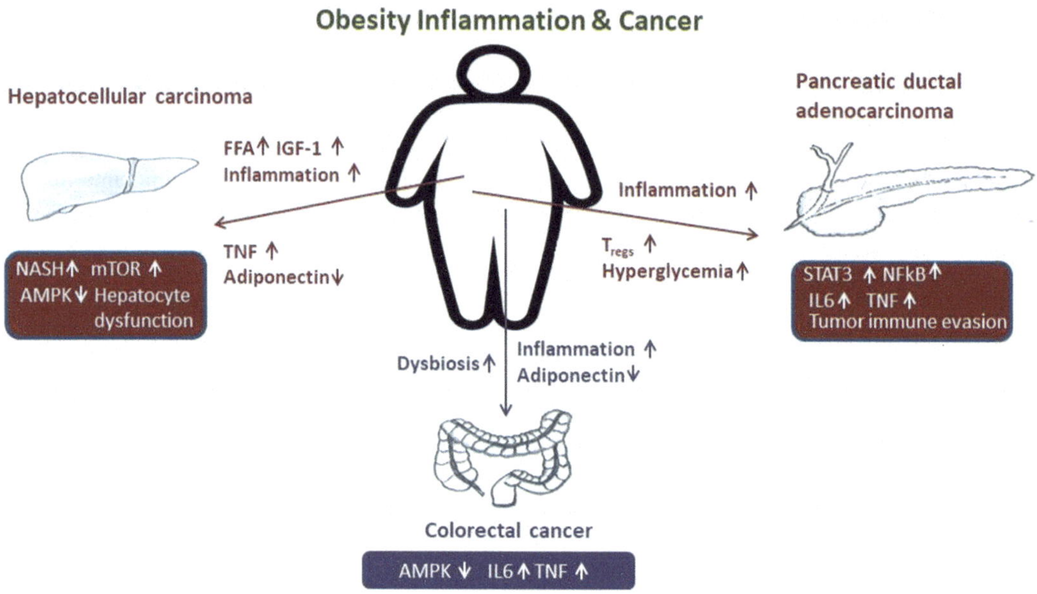

Fig. 10.2 Schematic representing association of obesity, chronic low-grade inflammation, and cancer. Several mechanisms are involved in the development of different forms of cancers owing to obesity and chronic low-grade inflammation (*FAA* free fatty acid, *IGF-1* Insulin-like growth factor-1, *TNF* tumor necrosis factor, *AMPK* adenosine monophosphate-activated protein kinase, *IL-6* Interleukin-6, Signal transducer and activator of transcription 3, T_{regs} T regulatory cells)

decrease body weight while other species might have deleterious effects. Prebiotics and probiotics are endowed with antiobesity and antidiabetic properties as shown in multiple animal models. However, outcomes of human trials are still not sufficiently robust to be endorsed in official guidelines and recommendations (Cerdó et al. 2019).

Plant-derived phytochemicals are also promising as anti-inflammatory molecules in the treatment of obesity. Curcumin experimentally modulates different cellular pathways and could act as an anti-inflammatory agent in obesity and associated comorbidities (Kandhari et al. 2018; Balaji et al. 2016; Firdous 2014. Nevertheless, a meta-analysis failed to confirm benefits for inflammatory markers in various contexts, including metabolic syndrome (White et al. 2019). Resveratrol has been shown to reduce levels of COX-2, IL-6, and TNF-α in adipocytes. Other flavones like apigenin and luteolin can reduce the TNF-α induced expression of NF-κB, and thus can be potential drugs in decreasing inflammation in obesity-related and other metabolic complications (Koh et al. 2018).

Cell-based Therapy

Cell-based approaches including brown adipose tissue transplantation, cell lysates, exosomes, and stem cell administration are currently being investigated in the treatment of obesity and related complications (Payab et al. 2018). Immunomodulatory potential of adipose tissue stem cells (ASCs) derived exosomes were also analyzed in one study on C57BL/6 male obese mice. ASCs derived exosomes polarized macrophages to M2 phenotype and further decreased inflammation in WAT. In one study, mice treated with adipose tissue-derived stem cells (ADSCs) exhibited lower WAT inflammation and obesity progression (Shang et al. 2015; Zhao et al. 2018). Mesenchymal stem cells isolated from human AT were helpful in regulating metabolic dysregulation induced by a high-fat diet in obese mice (Shree et al. 2019).

Acknowledgments MS and JS acknowledge UGC, Govt. of India for Junior Research Fellowship. HA acknowledges the Department of Science and Technology, Govt. of India, for INSPIRE fellowship. Ramanujan Fellowship by SERB to UCSY is duly acknowledged.

References

Alam I, Ng TP, Larbi A (2012) Does inflammation determine whether obesity is metabolically healthy or unhealthy? The aging perspective. Mediat Inflamm 2012:14. https://doi.org/10.1155/2012/456456

Bai Y, Sun Q (2015) Macrophage recruitment in obese adipose tissue. Obes Rev 16(2):127–136. https://doi.org/10.1111/obr.12242

Bain CC, Schridde A (2018) Origin, differentiation, and function of intestinal macrophages. Front Immunol 9:2733–2733. https://doi.org/10.3389/fimmu.2018.02733

Balaji M, Ganjayi MS et al (2016) A review on possible therapeutic targets to contain obesity: the role of phytochemicals. Obes Res Clin Pract 10(4):363–380. https://doi.org/10.1016/j.orcp.2015.12.004

Bekkering S, Quintin J, Joosten LA, van der Meer JW, Netea MG, Riksen NP (2014) Oxidized low-density lipoprotein induces long-term proinflammatory cytokine production and foam cell formation via epigenetic reprogramming of monocytes. Arterioscler Thromb Vasc Biol 34(8):1731–1738. https://doi.org/10.1161/atvbaha.114.303887

Beydoun MA, Fanelli-Kuczmarski MT, Canas JA, Beydoun HA, Evans MK, Zonderman AB (2018) Dietary factors are associated with serum uric acid trajectory differentially by race among urban adults. Br J Nutr 120(8):935–945. https://doi.org/10.1017/s0007114518002118

Cerdó T, García-Santos JA, Bermúdez G, M., & Campoy, C. (2019) The role of probiotics and prebiotics in the prevention and treatment of obesity. Nutrients 11 (3):635. https://doi.org/10.3390/nu11030635

Chatauret N, Favreau F, Giraud S, Thierry A, Rossard L, Le Pape S et al (2014) Diet-induced increase in plasma oxidized LDL promotes early fibrosis in a renal porcine auto-transplantation model. J Transl Med 12:76. https://doi.org/10.1186/1479-5876-12-76

Christ A, Günther P, Lauterbach MAR, Duewell P, Biswas D, Pelka K et al (2018) Western diet triggers NLRP3-dependent innate immune reprogramming. Cell 172(1):162–175.e114. https://doi.org/10.1016/j.cell.2017.12.013

Chung SS, Wu Y, Okobi Q, Adekoya D, Atefi M, Clarke O et al (2017) Proinflammatory cytokines IL-6 and TNF-α increased telomerase activity through NF-κB/STAT1/STAT3 activation, and Withaferin a inhibited the signaling in colorectal cancer cells. Mediat Inflamm 2017:5958429–5958429. https://doi.org/10.1155/2017/5958429

Cox AD, Olive KP (2012) Silencing the killers: paracrine immune suppression in pancreatic cancer. Cancer Cell 21(6):715–716. https://doi.org/10.1016/j.ccr.2012.05.029

Dam V, Sikder T, Santosa S (2016) From neutrophils to macrophages: differences in regional adipose tissue depots. Obes Rev 17(1):1–17. https://doi.org/10.1111/obr.12335

Deng G, Long Y, Yu YR, Li MR (2010) Adiponectin directly improves endothelial dysfunction in obese rats through the AMPK–eNOS pathway. Int J Obes 34(1):165–171. https://doi.org/10.1038/ijo.2009.205

El-Haggar SM, Mostafa TM (2015) Adipokines and biochemical changes in Egyptian obese subjects: possible variation with sex and degree of obesity. Endocrine 48 (3):878–885. https://doi.org/10.1007/s12020-014-0390-z

Ellulu MS, Patimah I, Khaza'ai H, Rahmat A, Abed Y (2017) Obesity and inflammation: the linking mechanism and the complications. Arch Med Sci 13 (4):851–863. https://doi.org/10.5114/aoms.2016.58928

Engin A (2017) The pathogenesis of obesity-associated adipose tissue inflammation. Adv Exp Med Biol 960:221–245. https://doi.org/10.1007/978-3-319-48382-5_9

Fernández-Veledo S, Vila-Bedmar R, Nieto-Vazquez I, Lorenzo M (2009) c-Jun N-terminal kinase 1/2 activation by tumor necrosis factor-α induces insulin resistance in human visceral but not subcutaneous adipocytes: reversal by liver X receptor agonists. J Clin Endocrinol Metabol 94(9):3583–3593. https://doi.org/10.1210/jc.2009-0558

Firdous SM (2014) Phytochemicals for treatment of diabetes. EXCLI J 13:451–453

Frasca D, Blomberg BB (2017) Adipose tissue inflammation induces B cell inflammation and decreases B cell function in aging. Front Immunol 8:1003–1003. https://doi.org/10.3389/fimmu.2017.01003

Fukuda A, Wang SC, Morris JP, Folias AE, Liou A, Kim GE et al (2011) Stat3 and MMP7 contribute to pancreatic ductal adenocarcinoma initiation and progression. Cancer Cell 19(4):441–455. https://doi.org/10.1016/j.ccr.2011.03.002

Genkinger JM, Kitahara CM, Bernstein L, Berrington de Gonzalez A, Brotzman M, Elena JW et al (2015) Central adiposity, obesity during early adulthood, and pancreatic cancer mortality in a pooled analysis of cohort studies. Ann Oncol 26(11):2257–2266. https://doi.org/10.1093/annonc/mdv355

Gnauck A, Lentle RG, Kruger MC (2016) The characteristics and function of bacterial lipopolysaccharides and their endotoxic potential in humans. Int Rev Immunol 35(3):189–218. https://doi.org/10.3109/08830185.2015.1087518

Guo S, Al-Sadi R, Said HM, Ma TY (2013) Lipopolysaccharide causes an increase in intestinal tight junction permeability in vitro and in vivo by inducing enterocyte membrane expression and localization of

TLR-4 and CD14. Am J Pathol 182(2):375–387. https://doi.org/10.1016/j.ajpath.2012.10.014

Guo S, Nighot M, Al-Sadi R, Alhmoud T, Nighot P, Ma TY (2015) Lipopolysaccharide regulation of intestinal tight junction permeability is mediated by TLR4 signal transduction pathway activation of FAK and MyD88. J Immunol (Baltimore) 195(10):4999–5010. https://doi.org/10.4049/jimmunol.1402598

Hamada A, Torre C, Drancourt M, Ghigo E (2019) Trained immunity carried by non-immune cells. Front Microbiol 9:3225–3225. https://doi.org/10.3389/fmicb.2018.03225

Han TS, Lean MEJ (2016) A clinical perspective of obesity, metabolic syndrome and cardiovascular disease. JRSM Cardiovasc Dis 5:2048004016633371

Han JM, Wu D, Denroche HC, Yao Y, Verchere CB, Levings MK (2015) IL-33 reverses an obesity-induced deficit in visceral adipose tissue $ST2^+$ T regulatory cells and ameliorates adipose tissue inflammation and insulin resistance. J Immunol *194*(10):4777. https://doi.org/10.4049/jimmunol.1500020

Heo J-W, Yoo S-Z, No M-H, Park D-H, Kang J-H, Kim T-W et al (2018) Exercise training attenuates obesity-induced skeletal muscle remodeling and mitochondria-mediated apoptosis in the skeletal muscle. Int J Environ Res Public Health 15(10):2301. https://doi.org/10.3390/ijerph15102301

Honka MJ, Latva-Rasku A, Bucci M, Virtanen KA, Hannukainen JC, Kalliokoski KK, Nuutila P (2018) Insulin-stimulated glucose uptake in skeletal muscle, adipose tissue and liver: a positron emission tomography study. Eur J Endocrinol 178(5):523–531. https://doi.org/10.1530/eje-17-0882

Hullar MA, Fu BC (2014) Diet, the gut microbiome, and epigenetics. Cancer J 20(3):170–175. https://doi.org/10.1097/ppo.0000000000000053

Jo J, Gavrilova O, Pack S et al (2009) Hypertrophy and/or hyperplasia: dynamics of adipose tissue growth. PLoS Comput Biol 5(3):e1000324. https://doi.org/10.1371/journal.pcbi.1000324

Kandhari K, Agraval H, Sharma A, Yadav UCS, Singh RP (2018) Flavonoids and cancer stem cells maintenance and growth. In: Rani V, Yadav UCS (eds) Functional food and human health. Springer Singapore, Singapore, pp 587–622

Kern L, Mittenbühler MJ, Vesting AJ, Ostermann AL, Wunderlich CM, Wunderlich FT (2018) Obesity-induced TNFα and IL-6 signaling: the missing link between obesity and inflammation-driven liver and colorectal cancers. Cancers 11(1):24. https://doi.org/10.3390/cancers11010024

Kholodenko IV, Yarygin KN (2017) Cellular mechanisms of liver regeneration and cell-based therapies of liver diseases. Biomed Res Int 2017:17. https://doi.org/10.1155/2017/8910821

Kim B, Park KY, Ji Y, Park S, Holzapfel W, Hyun CK (2016) Protective effects of lactobacillus rhamnosus GG against dyslipidemia in high-fat diet-induced obese mice. Biochem Biophys Res Commun 473 (2):530–536. https://doi.org/10.1016/j.bbrc.2016.03.107

Koh YC, Yang G, Lai CS, Weerawatanakorn M, Pan MH (2018) Chemopreventive effects of phytochemicals and medicines on M1/M2 polarized macrophage role in inflammation-related diseases. Int J Mol Sci 19(8): E2208. https://doi.org/10.3390/ijms19082208

Kolodin D, van Panhuys N, Li C, Magnuson AM, Cipolletta D, Miller CM et al (2015) Antigen- and cytokine-driven accumulation of regulatory T cells in visceral adipose tissue of lean mice. Cell Metab 21 (4):543–557. https://doi.org/10.1016/j.cmet.2015.03.005

Kumari B, Yadav U (2018) Adipokine Visfatin's role in pathogenesis of diabesity and related metabolic derangements. Curr Mol Med 18(2):116–125

Kumari B, Singh M, Rajput PK, Yadav UC (2019) *In silico* characterization of visfatin and its additive effect with phytochemical hesperetin on insulin synthesis. RJLBPCS 2019. https://doi.org/10.26479/2019.0502.27

Lauterbach MAR, Wunderlich FT (2017) Macrophage function in obesity-induced inflammation and insulin resistance. Pflugers Archiv 469(3-4):385–396. https://doi.org/10.1007/s00424-017-1955-5

Le Menn G, Sibille B, Murdaca J, Rousseau A-S, Squillace R, Vergoni B et al (2018) Decrease in αβ/γδ T-cell ratio is accompanied by a reduction in high-fat diet–induced weight gain, insulin resistance, and inflammation. FASEB J 33(2):2553–2562. https://doi.org/10.1096/fj.201800696RR

Lee J, Lee S, Zhang H, Hill MA, Zhang C, Park Y (2017) Interaction of IL-6 and TNF-α contributes to endothelial dysfunction in type 2 diabetic mouse hearts. PLoS One 12(11):e0187189. https://doi.org/10.1371/journal.pone.0187189

Lesina M, Kurkowski MU, Ludes K, Rose-John S, Treiber M, Klöppel G et al (2011) Stat3/Socs3 activation by IL-6 transsignaling promotes progression of pancreatic intraepithelial neoplasia and development of pancreatic cancer. Cancer Cell 19(4):456–469. https://doi.org/10.1016/j.ccr.2011.03.009

Ling J, Kang Y a, Zhao R, Xia Q, Lee D-F, Chang Z et al (2012) KrasG12D-induced IKK2/β/NF-κB activation by IL-1α and p62 feedforward loops is required for development of pancreatic ductal adenocarcinoma. Cancer Cell 21(1):105–120. https://doi.org/10.1016/j.ccr.2011.12.006

Liu R, Nikolajczyk BS (2019) Tissue immune cells fuel obesity-associated inflammation in adipose tissue and beyond. Front Immunol 10:1587–1587. https://doi.org/10.3389/fimmu.2019.01587

Liu W, Yin Y, Zhou Z, He M, Dai Y (2014) OxLDL-induced IL-1beta secretion promoting foam cells formation was mainly via CD36 mediated ROS production leading to NLRP3 inflammasome activation. Inflamm Res 63(1):33–43. https://doi.org/10.1007/s00011-013-0667-3

Lu J, Zhao J, Meng H, Zhang X (2019) Adipose tissue-resident immune cells in obesity and type 2 diabetes. Front Immunol 10:1173–1173. https://doi.org/10.3389/fimmu.2019.01173

Macdougall CE, Wood EG, Loschko J, Scagliotti V, Cassidy FC, Robinson ME et al (2018) Visceral adipose tissue immune homeostasis is regulated by the crosstalk between adipocytes and dendritic cell subsets. Cell Metab 27(3):588–601.e584. https://doi.org/10.1016/j.cmet.2018.02.007

McLaughlin T, Liu LF, Lamendola C, Shen L, Morton J, Rivas H et al (2014) T-cell profile in adipose tissue is associated with insulin resistance and systemic inflammation in humans. Arterioscler Thromb Vasc Biol 34 (12):2637–2643. https://doi.org/10.1161/atvbaha.114.304636

Miro-Blanch J, Yanes O (2019) Epigenetic regulation at the interplay between gut microbiota and host metabolism. Front Genet 10(638). https://doi.org/10.3389/fgene.2019.00638

Monks M, Irakleidis F, Tan PH (2019) Complex interaction of adiponectin-mediated pathways on cancer treatment: a novel therapeutic target. J Cancer Metastasis Treat 5:24

Netea MG, Joosten LAB, Latz E, Mills KHG, Natoli G, Stunnenberg HG et al (2016) Trained immunity: a program of innate immune memory in health and disease. Science 352(6284):aaf1098. https://doi.org/10.1126/science.aaf1098

Newton K, Dixit VM (2012) Signaling in innate immunity and inflammation. Cold Spring Harb Perspect Biol 4 (3). https://doi.org/10.1101/cshperspect.a006049

Olivieri F, Rippo MR, Prattichizzo F, Babini L, Graciotti L, Recchioni R, Procopio AD (2013) Toll like receptor signaling in "inflammaging": microRNA as new players. Immun Ageing 10(1):11–11. https://doi.org/10.1186/1742-4933-10-11

Park YJ, Park J, Huh JY, Hwang I, Choe SS, Kim JB (2018) Regulatory roles of invariant natural killer T cells in adipose tissue inflammation: defenders against obesity-induced metabolic complications. Front Immunol 9:1311–1311. https://doi.org/10.3389/fimmu.2018.01311

Payab M, Goodarzi P, Heravani N, Hadavandkhani M, Zarei Z, Falahzadeh K et al (2018) Stem cell and obesity: current state and future perspective, vol 1089

Pramanik KC, Makena MR, Bhowmick K, Pandey MK (2018) Advancement of NF-κB signaling pathway: a novel target in pancreatic cancer. Int J Mol Sci 19 (12):3890. https://doi.org/10.3390/ijms19123890

Raffan E, Dennis RJ, O'Donovan CJ, Becker JM, Scott RA, Smith SP et al (2016) A deletion in the canine POMC gene is associated with weight and appetite in obesity-prone Labrador retriever dogs. Cell Metab 23 (5):893–900. https://doi.org/10.1016/j.cmet.2016.04.012

Raghuraman S, Donkin I, Versteyhe S, Barrès R, Simar D (2016) The emerging role of epigenetics in inflammation and immunometabolism. Trends Endocrinol Metab 27(11):782–795. https://doi.org/10.1016/j.tem.2016.06.008

Ramos-Molina B, Sánchez-Alcoholado L, Cabrera-Mulero A, Lopez-Dominguez R, Carmona-Saez P, Garcia-Fuentes E et al (2019) Gut microbiota composition is associated with the global DNA methylation pattern in obesity. Front Genet 10:613–613. https://doi.org/10.3389/fgene.2019.00613

Raphael I, McGeachy MJ (2018) STAT3 regulation of effector Th17 cells and its implications for treatment of autoimmunity. J Immunol 200(Suppl 1):121.125

Reddy MA, Chen Z, Park JT, Wang M, Lanting L, Zhang Q et al (2014) Regulation of inflammatory phenotype in macrophages by a diabetes-induced long noncoding RNA. Diabetes 63(12):4249–4261. https://doi.org/10.2337/db14-0298

Reilly SM, Saltiel AR (2017) Adapting to obesity with adipose tissue inflammation. Nat Rev Endocrinol 13:633. https://doi.org/10.1038/nrendo.2017.90

Saltiel AR, Olefsky JM (2017) Inflammatory mechanisms linking obesity and metabolic disease. J Clin Invest 127(1):1–4. https://doi.org/10.1172/jci92035

Satoh M, Iwabuchi K (2018) Role of natural killer T cells in the development of obesity and insulin resistance: insights from recent Progress. Front Immunol 9:1314–1314. https://doi.org/10.3389/fimmu.2018.01314

Saxena A, Chumanevich A, Fletcher E, Larsen B, Lattwein K, Kaur K, Fayad R (2012) Adiponectin deficiency: role in chronic inflammation induced colon cancer. Biochim Biophys Acta (BBA) – Mol Basis Dis 1822(4):527–536. https://doi.org/10.1016/j.bbadis.2011.12.006

Schoeler M, Caesar R (2019) Dietary lipids, gut microbiota and lipid metabolism. Rev Endocr Metabol Disord. https://doi.org/10.1007/s11154-019-09512-0

Shang Q, Bai Y, Wang G, Song Q, Guo C, Zhang L, Wang Q (2015) Delivery of adipose-derived stem cells attenuates adipose tissue inflammation and insulin resistance in obese mice through remodeling macrophage phenotypes. Stem Cells Dev 24(17):2052–2064. https://doi.org/10.1089/scd.2014.0557

Sheedy FJ, Grebe A, Rayner KJ, Kalantari P, Ramkhelawon B, Carpenter SB et al (2013) CD36 coordinates NLRP3 inflammasome activation by facilitating intracellular nucleation of soluble ligands into particulate ligands in sterile inflammation. Nat Immunol 14(8):812–820. https://doi.org/10.1038/ni.2639

Sharma A, Tate M et al (2018) Oxidative stress and NLRP3-inflammasome activity as significant drivers of diabetic cardiovascular complications: therapeutic implications. Front Physiol 9:114–114. https://doi.org/10.3389/fphys.2018.00114

Shree N, Venkategowda S, Venkatranganna MV, Datta I, Bhonde RR (2019) Human adipose tissue mesenchymal stem cells as a novel treatment modality for correcting obesity induced metabolic dysregulation.

Int J Obes 43(10):2107–2118. https://doi.org/10.1038/s41366-019-0438-5

Singh UP, Mehrpooya P, Marpe B, Singh NP, Murphy EA, Mishra MK et al (2016) High fat diet influences T cell homeostasis and macrophage phenotype to maintain chronic inflammation. J Immunol 196(Suppl 1):197.115

Singh M, Kumari B, Yadav UCS (2019) Regulation of oxidized LDL-induced inflammatory process through NLRP3 inflammasome activation by the deubiquitinating enzyme BRCC36. Inflamm Res 68 (12):999–1010. https://doi.org/10.1007/s00011-019-01281-5

Siragusa M, Fleming I (2016) The eNOS signalosome and its link to endothelial dysfunction. Pflugers Arch - Eur J Physiol 468(7):1125–1137. https://doi.org/10.1007/s00424-016-1839-0

Stodle GS, Silva GB, Tangeras LH, Gierman LM, Nervik I, Dahlberg UE et al (2018) Placental inflammation in pre-eclampsia by nod-like receptor protein (NLRP)3 inflammasome activation in trophoblasts. Clin Exp Immunol 193(1):84–94. https://doi.org/10.1111/cei.13130

Tahergorabi Z, Khazaei M (2015) Leptin and its cardiovascular effects: focus on angiogenesis. Adv Biomed Res 4:79–79. https://doi.org/10.4103/2277-9175.156526

Takeuchi O, Akira S (2010) Pattern recognition receptors and inflammation. Cell 140(6):805–820. https://doi.org/10.1016/j.cell.2010.01.022

Tong S, Ji Q, Du Y, Zhu X, Zhu C, Zhou Y (2019) Sfrp5/Wnt pathway: a protective regulatory system in atherosclerotic cardiovascular disease. J Interf Cytokine Res 39(8):472–482. https://doi.org/10.1089/jir.2018.0154

Varghese JF, Patel R, Yadav UCS (2019) Sterol regulatory element binding protein (SREBP) -1 mediates oxidized low-density lipoprotein (oxLDL) induced macrophage foam cell formation through NLRP3 inflammasome activation. Cell Signal 53:316–326. https://doi.org/10.1016/j.cellsig.2018.10.020

Vatandoost N, Amini M, Iraj B, Momenzadeh S, Salehi R (2015) Dysregulated miR-103 and miR-143 expression in peripheral blood mononuclear cells from induced prediabetes and type 2 diabetes rats. *Gene 572* (1):95–100. https://doi.org/10.1016/j.gene.2015.07.015

Vuletic M, Milosevic V, Jancic S, Zujovic J, Krstic M, Vukmirovic F (2019) Clinical significance of leptin receptor (LEPR) and endoglin (CD105) expressions in colorectal adenocarcinoma. J BUON 24 (6):2448–2457

Wang S, Ye Q, Zeng X, Qiao S (2019) Functions of macrophages in the maintenance of intestinal homeostasis. J Immunol Res 2019:8. https://doi.org/10.1155/2019/1512969

Wang D, Zhang Y, Shen C (2020) Research update on the association between SFRP5, an anti-inflammatory adipokine, with obesity, type 2 diabetes mellitus and coronary heart disease. J Cell Mol Med. https://doi.org/10.1111/jcmm.15023

White A, Ironmonger L, Steele RJC, Ormiston-Smith N, Crawford C, Seims A (2018) A review of sex-related differences in colorectal cancer incidence, screening uptake, routes to diagnosis, cancer stage and survival in the UK. BMC Cancer 18(1):906–906. https://doi.org/10.1186/s12885-018-4786-7

White CM, Pasupuleti V, Roman YM, Li Y, Hernandez AV (2019) Oral turmeric/curcumin effects on inflammatory markers in chronic inflammatory diseases: a systematic review and meta-analysis of randomized controlled trials. Pharmacol Res 146:104280. https://doi.org/10.1016/j.phrs.2019.104280

Wu P, Zhang F, Dai Y, Han L, Chen S (2016) Serum TNF-α, GTH and MDA of high-fat diet-induced obesity and obesity resistant rats. Saudi Pharm J 24(3):333–336. https://doi.org/10.1016/j.jsps.2016.04.011

Xu M, Jung X et al (2018) Obesity and pancreatic cancer: overview of epidemiology and potential prevention by weight loss. Pancreas 47(2):158–162. https://doi.org/10.1097/MPA.0000000000000974

Zeng J, Tang Z-H, Liu S, Guo S-S (2017) Clinicopathological significance of overexpression of interleukin-6 in colorectal cancer. World J Gastroenterol 23 (10):1780–1786. https://doi.org/10.3748/wjg.v23.i10.1780

Zhao H, Shang Q, Pan Z, Bai Y, Li Z, Zhang H et al (2018) Exosomes from adipose-derived stem cells attenuate adipose inflammation and obesity through polarizing M2 macrophages and Beiging in white adipose tissue. Diabetes 67(2):235–247. https://doi.org/10.2337/db17-0356

Crosstalk Between Adipose Tissue, Macrophages, and Other Immune Cells: Development of Obesity and Inflammation-induced Metabolic Diseases

11

Rohit Patel, Johnna Francis Varghese, and Umesh Chand Singh Yadav

Contents

Introduction 152

Adipose Tissue as a Source of Inflammation 153
Tumor Necrosis Factor α 153
Interleukin-6 153
Monocyte Chemoattractant Protein-1 (MCP-1) 154

Adipocytokines: A Linking Mediator Between Adipose Tissue, Inflammation, and Immunity 155
Normal vs. Sick Fat 155

Orchestration of Inflammatory Responses 155
Immune Cell Recruitment 155
Macrophages 156
Dendritic Cells (DCs) 158
Lymphocytes 158
Regulatory T (Treg) Cells 158
Unconventional Lymphocytes 159

Metabolic Regulation of Adipose Tissue-residing Immune Cells in Different Disease Conditions 160
Arterial Hypertension 160
Atherosclerosis 160
Insulin Resistance and Diabetes 161

References 161

Abstract

During metabolic disorders, the adipocytes increase in number and size along with the alteration in their functions. They secrete various types of chemical factors which contribute to inflammation and various physiological dysregulations. The adipose tissue acts as a reservoir of the immune cells such as macrophages, T-cells, and B-cells that

R. Patel · J. F. Varghese
School of Life Sciences, Central University of Gujarat, Gandhinagar, Gujarat, India

U. C. S. Yadav (✉)
Special Centre for Molecular Medicine, Jawaharlal Nehru University, New Delhi, India
e-mail: ucsyadav@mail.jnu.ac.in

J. Faintuch and S. Faintuch (eds.), *Obesity and Diabetes*, https://doi.org/10.1007/978-3-030-53370-0_11

maintain the homeostasis of body's metabolic processes. However, the dysfunctional adipocytes and overpopulated immune cells secrete various cytokines and chemokines which promote low-grade chronic inflammation in the body causing the development of several metabolic diseases including diabetes, vascular diseases, and cancer.

Keywords

Adipose tissue · Obesity · Insulin resistance · Inflammation · Macrophage · Adipocytokines

Introduction

The adipose tissue comprises of resident preadipocytes which differentiate into mature adipocytes throughout their lifespan, and thus keep expanding the adipose tissue based on nutritional status and increased storage requirements. Mature adipocytes have a tendency to expand in size to accommodate storage demands and in overnutrition situation, they become hypertrophic. Adipocytes lack uniformity and vary in terms of size and population based on the localization in the body, capacity to secrete adipocytokines, as well as the cellular compositions, number of immune cells, blood vessels, and stromal cells forming it (Coelho et al. 2013).

Adipose tissue is both morphologically and physiologically differentiated, for example, mammalian adipose tissue is present in two forms: white adipose tissue (WAT) and brown adipose tissue (BAT). Mostly, the WAT is represented as a reservoir of energy whereas BAT is widely present in human neonates and is involved in regulating the body temperature via thermogenesis (Villarroya et al. 2019). WAT comprises of preadipocytes (adipocytes that are not yet lipid-loaded), fibroblasts, endothelial cells, and leukocytes, most importantly macrophages. WAT has a range of functions, which include release of hormones such as resistin, visfatin, and adiponectin. Along with this, cytokines such as interleukin (IL)-6 and tumor necrosis factor-α (TNF-α) released from WAT are involved in modifying insulin sensitivity and inflammatory processes, WAT also releases leptin and angiotensin which participate in controlling the nutritional uptake (Coelho et al. 2013).

Apart from these, several micro RNAs (miRs) have been found dysregulated in WAT during obesity, and establish a close association with metabolic disorders associated with obesity. The miRs are small noncoding around 20–22 nucleotide long RNA sequences, which bind to their target mRNA at 3'UTRs and regulate the translation of mRNA. Some miRs are also involved in regulating inflammation in WAT. For example, miR-221 and miR-222 have been reported to have a positive correlation with the expression of TNF-α released by WAT. The exosomal release of the miRs from adipose tissues such as miR-99b also regulates the metabolism of other tissues. MiR-99b induces suppression of hepatic Fgf21 by binding to its 3'UTR. Fgf21 is primarily involved in maintaining the metabolic regulation in obese and diabetic models by increasing insulin sensitization and lowering glucose levels (Tezze et al. 2019).

Brown adipocytes are comparatively smaller than WAT. They are abundant with cytoplasmic content, possess varied cytoplasmic lipid droplets and numerous mitochondria which are involved in release of heat via oxidation of stored fatty acids. BAT is characterized by extensive vascularization, and the blood vessels assist the delivery of components for storage and oxidation, as well as disperse the heat generated to different parts of the body (Coelho et al. 2013). Under obese conditions, release of exosomal miRs from BAT such as miR-325 and miR-743b is presumed to target the uncoupling protein 1 (UCP-1) expression, and miR-98 targets peroxisome proliferator-activated receptor-gamma coactivator alpha (PGC1α), both these proteins are involved in adaptive thermogenesis. They contribute towards the metabolic syndrome by whitening of the BAT (Tezze et al. 2019).

Adipose Tissue as a Source of Inflammation

Obesity influences the quantity, nature, and subtypes of the immune cells that are harboured in adipose tissue and thus the later has emerged as an active immunological organ (Grant and Dixit 2015).

Adipose tissue secretes a variety of adipokines and cytokines which may be both pro- (TNF-α and IL-1β) and anti-inflammatory (IL-10) in nature. Approximately 30% of circulating IL-6, a proinflammatory cytokine, originates from adipose tissue. With increased adiposity, the level of IL-6 increases, and is further stimulated by TNF-α and IL-1β (Grant and Dixit 2015). In obese condition, adipose tissue is poorly oxygenated due to proportional reduction of blood flow. This hypoxic condition activates HIF-1α and NF-κB followed by alteration in the expression of several proinflammatory adipokines (Ge et al. 2014). These molecules have the ability to affect insulin sensitivity of adipose tissue and promote insulin refractoriness, participating in the development of type-2-diabetes mellitus (T2DM), cardiovascular diseases (CVDs) and other metabolic disorders.

Tumor Necrosis Factor α

Adipocytes synthesize a 26 kDa transmembrane protein pro-TNF-α, which upon stimulation by various oxidative and inflammatory stimuli gets cleaved by matrix metalloproteinases (MMPs) into a soluble and active form of TNF-α, a 17 kDa protein, and released into circulation. It is known to increase serum glucose concentrations and induce insulin resistance in *in vivo* experimental models of obesity. TNF-α is also a potential activator of NF-κB-mediated proinflammatory signaling pathway, which is implicated in numerous inflammatory diseases. Increased levels of TNF-α, have been attributed to the presence of adipose tissue-residing macrophages especially the M1 type. As shown in Fig. 11.1, insulin impairment in the presence of TNF-α is mediated via activation of c-Jun-N-terminal kinase (JNK) or inhibitor of NF-κB kinase (IKK) and increased expression of suppressor of cytokine signaling (SOCS3), which then inhibits the signaling via insulin receptor substrate (IRS).

MiR-155 expressed in obese individuals inhibits the transcripts of proteins which act as repressors of TNF- α (Fig. 11.1). Transgenic mouse models with an overexpression of miR-155 exhibit enhanced production of TNF-α. Similarly, under hyperglycemic conditions, an increased expression of miR-155 and miR-146a is responsible for activation of TNF-α/TGF-β1/NF-κB proinflammatory signaling pathway (Marques-Rocha et al. 2015). Another regulatory pathway for TNF- α production involves miR-145-mediated activation of p65, an integral part of NF-κB complex. Along with this miR-145 downregulates the expression of protease ADAM17, leading to an increased membrane bound fraction of biologically active form of TNF-α (Marques-Rocha et al. 2015).

Interleukin-6

IL-6 along with the other proinflammatory cytokines is secreted by the NK cells residing in the adipose tissue (Lu et al. 2016). Elevated

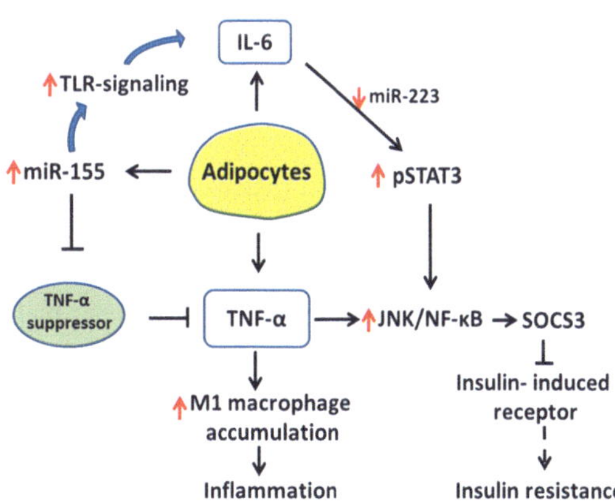

Fig. 11.1 Schematic diagram showing cytokines production from adipocyte and its effect on inflammation and insulin resistance. Adipocytes during obesity secrete proinflammatory cytokines (TNF-α and IL-6) which induce inflammation via accumulation of macrophages and insulin resistance via JNK/NF-κB-mediated SOCS3 activation. Adipocytes also express miR-155 that regulates the release of both TNF-α and IL-6. Further, decreased levels of miR-223 during obesity results in increased IL-6-mediated STAT signaling leading to activation of NF-κB

circulating level of IL-6 in obese subjects is strongly correlated with insulin resistance (Varghese et al. 2018). IL-6 levels also positively correlate to the body mass and plasma free fatty acids in the patients with T2DM. IL-6 participates in glucose homeostasis and regulation and activation of the immune system (Kurauti et al. 2017). IL-6 directly inhibits the signal transduction of insulin receptor thereby inhibiting insulin activity, leading to insulin resistance (Zand et al. 2017). IL-6 acts on the insulin signaling by upregulating SOCS3 expression, which impairs the insulin-induced receptor and phosphorylates the IRS-1 in hepatocytes and adipocytes (Fig. 11.1). It is also actively involved in regulation of insulin-degrading enzyme (IDE), an important molecule involved in insulin clearance in the body. Low level of IL-6 leads to decreased expression and activity of IDE which could result in hyperinsulinemia (Kurauti et al. 2017).

Blocking the IL-6 signaling leads to depreciation of larger adipocytes replaced by smaller adipocytes. In addition to this, IL-6-mediated signaling promotes the chemotaxis and infiltration of macrophages into the adipose tissue as well as their polarization into the M1 type (Kraakman et al. 2015). As observed in TNF-α regulation, miR-155 also plays an important role in mediating increased expression of IL-6 by potentiating TLR signaling, via negatively regulating SH2 domain-containing inositol-5-phosphatase 1 (SHIP1) and suppressor of cytokine signaling 1 (SOCS1) (Marques-Rocha et al. 2015). IL-6/STAT3 (signal transducers and activators of transcription 3) is responsible for complex inflammatory pathways, where IL-6 promotes phosphorylation and activation of STAT3. IL-6 promotes STAT3 by inhibiting miR-223, a negative regulator of STAT3 (Marques-Rocha et al. 2015).

Monocyte Chemoattractant Protein-1 (MCP-1)

A member of CC chemokine superfamily, this cytokine recruits and activates circulating monocytes during angiogenesis and acute inflammation. MCP-1 levels are augmented in obese individuals, and gene expression of MCP-1 is elevated in both visceral and subcutaneous adipose tissues. Elevated MCP-1 attracts more leukocytes to the inflammatory site, which further aggravates inflammation by secreting excessive levels of TNF-α and IL-6, contributing to insulin refractoriness and diabetes. Infiltration of monocytes into the adipose tissue is also contributed by the adipocytes, i.e., by secreting MCP-1 along with macrophage inflammatory protein-1α (MIP-1α), macrophage colony-stimulating factor (M-CSF), macrophage migration inhibition factor (MIF-1), and chemokine CCL5 (RANTES), thereby facilitating the diversification and maturation of monocytes into macrophages (Fig. 11.2) (Coelho et al. 2013). Thus, inclining the polarization of macrophages into M1 type, increased plasma levels of MCP-1 via CCR2 stimulate the phosphorylation of ERK1/2. Activated ERK is then responsible for decreased tyrosine phosphorylation of IR-β and increased serine-based phosphorylation of IRS-1, leading to insulin resistance in hepatocytes and skeletal muscles (Chen et al. 2015).

When elevated, miR-467b is found to target and inhibit MCP-1 and TNF-α via downregulating lipoprotein lipase. However, downregulated expression of miR-467b in macrophages under obese conditions leads to an enhanced expression of MCP-1 (Tian et al. 2012). Another miR identified to target MCP-1 secretion in macrophages is miR-125-5b-5p, which is

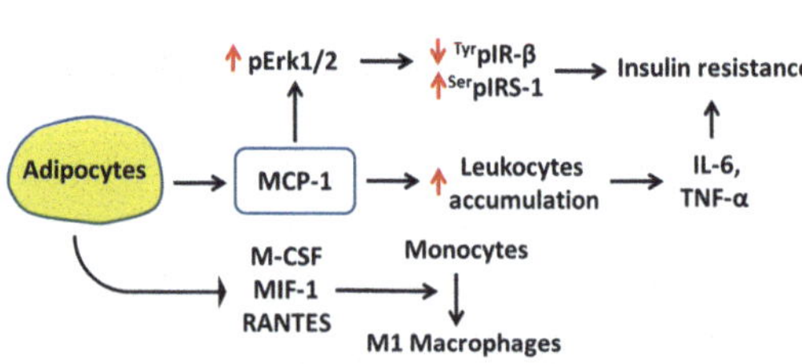

Fig. 11.2 Schematic representation of MCP-1-mediated pathway in adipocyte. MCP-1 released from adipocyte during obesity causes leukocyte accumulation, which further leads to IL-6 and TNF-α production. MCP-1 also activates Erk1/2 signaling leading to insulin resistance. Further, adipocytes also cause M1 macrophage accumulation via M-CSF, MIF-1, and RANTES production

also found to be downregulated under obese conditions. This miR targets beta lactamase protein (LACTB), an active-site serine protein, involved in maintaining metabolic circuit thereby regulating the levels of MCP-1. However, the downregulation of miR-125-5b-5p leads to upregulation of LACTB, followed by increased MCP-1 secretion in macrophages (Lu et al. 2016).

Adipocytokines: A Linking Mediator Between Adipose Tissue, Inflammation, and Immunity

Adipocytokines encompass molecules like leptin, resistin, visfatin, and adiponectin. Other cytokines released by the adipose tissue (IL-1, IL-6, MCP-1, and TNF-α) are not considered as adipocytokines, even though taking part in the immune modulation. Granulocyte-macrophage colony-stimulating factor (GM-CSF), IL-4, and IL-13, which induce proliferation of macrophages, should also be added to the list (Braune et al. 2017), as certain chemokines including CCR5, C-X-C motif chemokine 12 (CXCL12), and semaphorin 3E promote macrophage infiltration and recruitment.

Blocking IL-6-induced signaling cascade does not completely abrogate insulin resistance or obesity-induced weight gain (Kraakman et al. 2015). This indicates the role of other cytokines and adipokines in the obesity-related pathogenesis. Further, IL-6 signaling also assists the retention of macrophages by macrophage-adipocyte adhesion, and this inhibits conversion of WAT to beige adipose tissue (Chung et al. 2017), thus continues to promote adipose tissue-mediated inflammation.

Normal vs. Sick Fat

Adipocytes from lean individuals, akin to normal fat, preferably secrete adipokines that are anti-inflammatory in nature, which include *transforming growth factor beta* (TGF-β), IL-10, IL-4, IL-13, and IL-1 receptor antagonist. In contrast to this, adipose tissue from obese individuals, sick fat, secretes proinflammatory cytokines such as TNF-α, visfatin, resistin, plasminogen activator inhibitor 1, leptin, and IL-6, which promote inflammatory condition and pathogenesis (Ouchi et al. 2011). The proinflammatory adipokines modulate insulin sensitivity directly or indirectly via stimulation of inflammatory pathways which affects insulin-mediated signaling. For example, phosphorylation of serine residue of the IRS by the adipokines through the JNK and I-kappa B kinase β (IKKβ)/NFκB pathway, disrupts the insulin signaling leading to insulin resistance.

Orchestration of Inflammatory Responses

Inflammation in the adipose tissue induces upregulation of proinflammatory adipocytokines. The connection between adipocytokines and miRs appears to be reciprocal where cytokine signaling has an impact on the miRs production. Likewise, the miRs also target mRNA of cytokines and cytokine signaling. For example IL-1β and TNF-α are critical stimulators of miR-155 and miR-146a while IL-6-mediated suppression of miR-200c directs the indispensable activation of inflammatory signaling cascade (Marques-Rocha et al. 2015). In general miR-185, miR-103, miR107, and miR143 get downregulated in obese individuals whereas they are upregulated in lean subjects (Marques-Rocha et al. 2015).

Immune Cell Recruitment

Neutrophils, eosinophils, basophils, mast cells, B and T cells infiltrate and reside in the adipose tissue. The immune cell population and their associated response may vary based on various factors such as diet, cold exposure, body weight, feeding and fasting (Grant and Dixit 2015). The *in vivo* experiment in mice model showed that in high-fat diet (HFD) fed animals, adipose tissues were infiltrated by neutrophils within 3 days, followed by macrophages in 2 weeks, and recruitment of B and T cells by the end of 4 weeks (Winer et al. 2011; Talukdar et al. 2012).

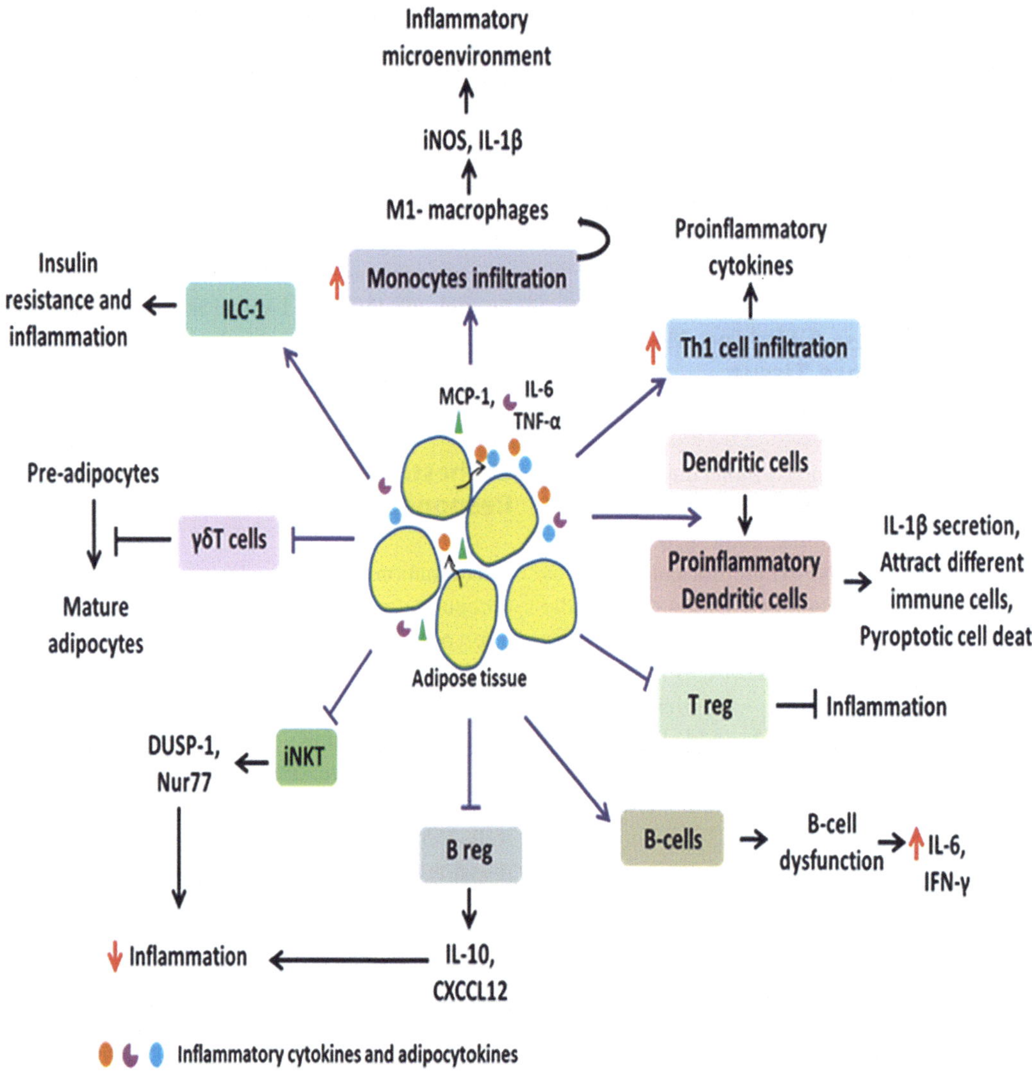

Fig. 11.3 Schematic diagram showing interaction of macrophages and other immune cells with adipose tissue. Various adipocytokines (IL-6, MCP-1, IL-1β, TNF-α) released from adipocytes and adipose-residing immune cells during obesity increases inflammation-mediated metabolic dysregulation. These inflammatory mediators attract and activate various immune cells which further secrete various chemokines and cytokines causing excessive inflammation and metabolic complications

The cells tend to change their number and characteristics based on different conditions and microenvironment in the tissue. Insulin resistance and obesity are certain conditions that have the ability to skew the adipocyte-residing immune cells, from anti-inflammatory subtype to the proinflammatory type. This also includes the macrophage polarization where M2 type of cells transform into M1 type, which is also accompanied by the loss of regulatory T cells (Fig. 11.3).

Macrophages

Approximately 50% of the immune cells residing in the obese adipose tissue are macrophages, while in lean individuals it accounts for about 10%. The tissue macrophage population is directly correlated to the total adiposity and adipose cell size and are involved in secreting the majority of inflammatory cytokines. Chemokines secreted by the adipocytes such as MCP-1,

MIF-1, chemokine CCL5 (RANTES), and M-CSF, attract and facilitate the infiltration of monocytes into the adipose tissue where they eventually transform into macrophages (Coelho et al. 2013). As observed in the obese mice models, macrophages when infiltrated in the adipose tissue, undergo transformation of phenotype from M2 type with cluster of differentiation (CD)-206, arginase (Arg)-1, and CD301 markers, to proinflammatory M1 phenotype with expression of nitric oxide synthase (NOS)2, TNF-α, and integrin alpha X chain protein (Itgax) (Appari et al. 2018; Rondini and Granneman 2020).

The M1 type prominently expresses inducible nitric oxide synthase (iNOS), IL-1β, and CD11c on the cell surface whereas the M2 macrophages express IL-10, Arg, and chitinase-like protein Ym-1. This switching of macrophages from M2 to M1 type is at the very core of adipose tissue. In obese conditions, the presence of M1 type of adipose tissue macrophages reflects the futile clearance of dead adipocytes. Generally, efficient efferocytosis is accompanied with the presence of M2 type of macrophages, establishing an anti-inflammatory microenvironment (Boutens and Stienstra 2016). Under obese conditions, it is the M1 macrophages that populate the adipogenic clusters, facilitating adipogenesis and angiogenesis as well as activation of the proinflammatory signaling needed for efficient adipogenesis (Boutens and Stienstra 2016).

The adipose tissue inflammation and resultant resistance to insulin, lipid oxidation, and mitochondrial biogenesis are regulated by PPARγ and PPAR coactivator-1β (PGC1β). Knockdown of PPARγ specifically exacerbates the inflammation in macrophages and promotes HFD-induced obesity (Odegaard et al. 2007). Adipose tissue-residing macrophages derived from obese mice have the tendency to induce systemic glucose tolerance and insulin resistance by downregulating PPARγ (Ying et al. 2017). Obesity induces lysosome biogenesis in macrophages residing in the adipose tissue and inhibition of the lysosome function impairs the lipid metabolism in the cells thus, reducing the whole adipose tissue lipolysis (Xu et al. 2013).

Increased content of lipids in the cells and alteration in lipid metabolism are associated with a low-grade inflammation during obesity (Xu et al. 2013). One of the critical signaling pathways that controls lipid metabolism in the macrophages is AMPK. It is an antagonist of the biosynthetic pathways, i.e., it suppresses fatty acid synthesis as well as increases their oxidation via phosphorylation of acetyl-CoA carboxylase (ACC) (Pearce and Everts 2015). The AMPKβ1-deficient macrophages lead to increase in the levels of diacylglycerol and inflammatory markers, and suppress the expression of mitochondrial enzymes and phosphorylation of ACC. This conducts to increased lipid accumulation (Galic et al. 2011).

In AMPKβ1$^{-/-}$ mice, administration of HFD enhanced inflammation of adipose tissue residing macrophages, with subsequent insulin resistance (Galic et al. 2011). Presence of saturated fatty acids such as palmitate instigates the signal transduction pathways pertaining to inflammation and lipid metabolism. These saturated fatty acids bind to cell surface receptors such as TLRs followed by proinflammatory signaling cascade activation, excessive inflammatory cytokine production, and internalization of palmitate. This is subsequently followed by activation of p62 and PPARγ proteins, which promote the metabolic phenotype of macrophages (Kratz et al. 2014). Studies state that deficiency of fatty acid synthase (FAS) prevents the recruitment of macrophages in the adipose tissue. It is also associated with the retention of cholesterol in the plasma membrane and disruption of Rho GTPase trafficking, which modulates cell adhesion, activation, and migration (Wei et al. 2016).

Macrophage transcriptome studies indicate that adipose tissue-residing macrophages in the obese adopt a peculiar metabolic program, which includes glycolysis and oxidative phosphorylation (Boutens et al. 2018). Inflammation-induced excessive glycolysis in macrophages contributes to increased production of inflammatory cytokines. Inhibition of glycolysis with 2-deoxyglucose (2-DG) reduced lactate production and glucose oxidation by macrophages (Boutens et al. 2018).

Dendritic Cells (DCs)

In obese mice models depletion of adipose-residing DCs resulted in recovering the insulin sensitivity and decrease in the levels of proinflammatory cytokines (Zand et al. 2017). High-fat diet skews the DCs towards the proinflammatory phenotype with an increase in the secretion of IL-1β and expression of IL-1R, toll-like receptor (TLR) 4 and caspase-1. The proinflammatory phenotype of DCs and their secretions attract more inflammatory immune cells into the adipose tissue. Further, increased levels of caspase-1 and IL-1β promote pyroptotic cell death, releasing more danger molecules in the milieu which add significantly to ongoing onslaught of inflammatory bursts. Unlike the conventional type, the plasmacytoid DCs (pDCs) possess a low capacity of processing soluble antigen. pDCs recruitment in the adipose tissue induces obesity and insulin resistance, whereas its depletion is followed by reduced metabolic complications (Ghosh et al. 2016; Hannibal et al. 2017).

The cDCs residing in the visceral adipose tissue (VAT) have the ability to acquire a tolerogenic phenotype, i.e., antigen-specific activation of T-cell tolerance. This phenomenon in cDCs is attributed to the activation of *peroxisome proliferator-activated receptor gamma* (PPARγ) and β-catenin pathways (Macdougall et al. 2018). Inhibition of the β-catenin and PPARγ pathways does not affect the phenotypical features of the animal such as adipose tissue content, food intake, or weight gain, however it aggravates the local inflammatory responses and obesity-induced insulin resistance (Macdougall et al. 2018). Tolerogenic properties of adipose tissue-residing cDCs serve as inflammation check points. cDCs are likely to undergo phenotypical and functional changes based on the level of obesity, providing a foundation to authenticate the role of cDCs in obesity and associated metabolic disorders.

Lymphocytes

Genes of lymphocytes involved in recombination and diversity for antibody production are metabolically active. Recombination-activating gene (RAG)-1-deficient mice tend to develop insulin resistance (Winer et al. 2009). Approximately 5–10% of the total immune cell population is composed of T cells which mainly reside in the subcutaneous and visceral adipose tissue, and act as a major hub for memory T cells inducing proliferation and effectiveness (Han et al. 2017). In obese mice, the population of T cells increases specifically when the subset $CD8^+$ population is greater than the $CD4^+$ in the adipose tissue. Insulin resistance is attributed to the presence of $CD8^+$ cells. Infiltration of these cells into adipose tissue is followed by the recruitment and accumulation of macrophages. Genetic depletion of $CD8^+$ T cells depreciates inflammation of adipose tissue and ameliorates insulin resistance (Nishimura et al. 2009).

Regulatory T (Treg) Cells

Treg population keeps a check on the inflammation in the adipose tissue. These cells are generally found in fat depots in normal states and their numbers tend to strikingly reduce under the obese conditions. Apart from the role of T cells, B cells also play a significant role in adipose tissue. Depleting B cells in HFD-fed mice improved tolerance towards glucose as well as reduced adipose tissue inflammation (Winer et al. 2011). B cells tend to secrete more proinflammatory cytokines such as interferon gamma (IFN-γ), IL-6, and IL-8, than anti-inflammatory cytokines which include IL-10 and IL-5. B cells exert detrimental effects by producing pathogenic IgG type of antibodies.

Nutrient rich microenvironment in the adipose tissue instigates Treg cells to take up lipids. This property is generally not seen in the conventional or lymphoid-residing Treg cells. The PPAR-γ-mediated fatty acid metabolism is a critical factor that regulates the phenotype of adipose tissue-residing Treg cells, depicting a PPARγ-dependent expression of forkhead box P3 (Foxp3) and GATA3 (Cipolletta et al. 2012). Lipid metabolism gets augmented in Treg cells in the presence of pioglitazone, a synthetic agonist of PPARγ. This leads to an increase in the expression of fatty acid transporters CD36, enzymes involved in fatty acid synthesis (Slc27a2, Stearoyl-CoA desaturase (Scd) 1 and Lipe), triglyceride synthesis (Dgat1), fatty acid oxidation (Cpt1a), and lipid droplet-associated protein (Plin2) (Cipolletta et al. 2012). Thus, it is possible that Treg cells could store lipids by regulating the lipolysis to generate free fatty acids for both anabolic and catabolic purposes. These findings go hand in hand with the fact that Tregs preferably utilize lipid oxidation for energy in comparison to effector T-cells that utilize the energy generated from glycolysis (Michalek et al. 2011).

Treg cells are attuned to the local microenvironment while residing in the adipose tissue, and this affects the metabolic profile, function, and phenotype. However, some features of adipose Treg cells are similar to Treg cells residing in other microenvironments. For example, signaling pathway that controls the Treg metabolism is mediated via the lipid phosphatase (PTEN) (Newton et al. 2016). Native $CD4^+$ T-cells undergo transformation into Treg cells when PTEN is overexpressed, leading to the formation of $Foxp3^+$ Treg cells in adipose tissue (Kalin et al. 2017). Studies state that deletion of PTEN leads to a hyper-glycolytic state, which disrupts the Treg lineage affecting the CD25 and Foxp3 expression.

A subtype of B cells, i.e., adipose tissue regulatory B (Breg) cells, are observed to restrain the inflammation of adipose tissue and positively affect the insulin sensitivity. These cells are involved in the survival and homeostasis of the adipose tissue by releasing IL-10, CXCL12, and free fatty acids (Nishimura et al. 2013).

Unconventional Lymphocytes

Unconventional lymphocytes include the invariant natural killer cells (iNKT), γδ T cells, mucosal-associated invariant T (MAIT) cells, and innate lymphoid cells.

iNKT Cells

A significant number of invariant natural killer cells (iNKT) are present in the adipose tissue. These are innate T cell types and represent 15–20% of the entire T cells. Mice that lack iNKT cells show metabolic derangements including weight gain, insulin resistance, larger adipocytes, and fatty livers. Increasing the number of iNKT cells improves the condition of weight loss and metabolism. iNKT cell-mediated amelioration of insulin sensitivity is attributed to its association with Th2 type cytokine production (Lynch et al. 2012). Adipose iNKT cells have a unique transcriptional signaling with an overexpression of mitogen-activated protein (MAP) kinase phosphatase Dusp1 and Nur77, a nuclear transcription factor, which then regulates inflammation, cell growth, glucose metabolism, and insulin resistance (Lynch et al. 2015).

Unlike iNKT cells that are found in other tissues and express poxvirus finger transcription factor, the adipose tissue iNKT cells express the basic leucine zipper transcription factor E4 promoter-binding protein 4 (E4BP4) which controls the production of IL-10. Enrichment of regulatory iNKT cells in adipose tissue controls inflammation and maintains it in quiescent state, and regulates the homeostasis of other anti-inflammatory immune cells such as Treg and M2 macrophages (Lynch et al. 2015).

γδ T Cells

These are unconventional type of T cells and represent 1–10% of the entire circulating T cells, in mice as well as humans. They are reduced in obese patients and have a negative correlation with the body mass index (Costanzo et al. 2015). They inhibit the differentiation of preadipocytes 3T3-L1 to mature adipocytes, and impair glucose uptake in mature adipocytes (Zuniga et al. 2010). These cells are preferably accumulated in the adipose deposits of the inguinal, colon, and small intestine.

Innate Lymphoid Cells (ILCs)

ILCs have a classic lymphoid morphology, however they are identified by the expression of cytokines and functional molecules released by them (ILC1 and ILC2) (Artis and Spits 2015). During obesity and T2DM, the ILC2 in VAT secretes IL-5 and IL-13, which regulate the levels of eosinophils and alternatively activated macrophages, thereby controlling insulin resistance and HFD-induced obesity (Hams et al. 2013; Molofsky et al. 2013). Adipose tissue in obese subjects exhibited a reduced number of ILC2s (Brestoff et al. 2015). Anti-inflammatory cytokine IL-25 elicited ILC2 response regulates metabolic homeostasis, by controlling lower weight gain when subjected to HFD (Hams et al. 2013).

In contrast, IFN-γ induces the ILC to polarize towards ILC1 type, where they contribute towards insulin resistance in HFD-induced obesity (O'Sullivan et al. 2016). Decreasing their number results in reduced cytotoxicity via alteration in the conversion of proinflammatory macrophages to anti-inflammatory population. Metabolic disorders are exacerbated in cases where there is an adoptive transfer of ILC1 population of cells (Boulenouar et al. 2017).

In mouse models, HFD appreciably increases the number of neutrophils and macrophages. Expression of neutrophil-derived elastase and myeloperoxidase increases when a HFD is administered along with neutrophil infiltration.

Metabolic Regulation of Adipose Tissue-residing Immune Cells in Different Disease Conditions

In adipose tissue of obese organisms, macrophages with a much more complex immunophenotype are observed which does not conform to the traditional M1 and M2 polarization type (Kratz et al. 2014). This morphology is induced by external stimuli such as glucose, insulin, and palmitate that get elevated during metabolic disorders. Metabolic markers follow, such as Perilipin 2(Plin2) and ATP-binding cassette transporter (Abca) 1, along with release of proinflammatory cytokines TNF-α and IL-1β (Kratz et al. 2014). Dead adipocytes capable of inducing the expression of IL-1β and IL-6, are cleared by the macrophages, with inhibition of inflammation in obesity (Coats et al. 2017).

Since macrophages are smaller than adipocytes, they clear the adipocytes via lysosomal exocytosis, driven via NADPH oxidase 2 (Coats et al. 2017). Lysosomes biogenesis occurs during obesity, and inhibition of the macrophage lysosome function may increase the lipid content in macrophages, reducing adipocyte lipolysis (Xu et al. 2013). Macrophages contribute towards dysregulated lipid metabolism independent of their phenotype.

Arterial Hypertension

Many hypertensive patients are not obese, however they present the features of metabolic dysregulation. Visceral adipose tissue (VAT) inflammation can be identified, partaking in the pathogenesis of vascular dysfunction (Mikolajczyk et al. 2016). Loss of protective function of the VAT triggers a loss of endothelium-dependent vasodilation and leads to stiffness of blood vessels (Nosalski and Guzik 2017). Morphological alterations induce a more proinflammatory VAT, which dedifferentiates and becomes metabolically active. Increased levels of resistin and visfatin and decreased levels of adiponectin and leptin can be observed. Additionally, increased production of chemokines such as RANTES or CXCL10, a key molecule for recruitment of activated $CD8^+$ T cells, monocytes, and macrophages, are also implicated (Nosalski and Guzik 2017).

Atherosclerosis

Augmented levels of adipokines such as vaspin, leptin, visfatin, and chemerin due to VAT dysfunction are associated with the development of atherosclerosis. At the early stages, immune cells such as macrophages, dendritic cells, and T cells infiltrate the adipose tissue that surrounds the vasculature (Skiba et al. 2017). This is followed by endothelial dysfunction and induction of oxidative stress. Endothelial dysfunction is a critical event in the atherosclerotic plaque formation (Patel et al. 2019). These conditions, however, can be reversed by targeting metabolic functions of proteins such as Angiotensin (Ang) 1–7 (Skiba et al. 2017). Adipose tissue inflammation is continuously observed over various stages of the disease. Aged $ApoE^{-/-}$ mice showed high number of leukocytes in the aorta, which is indicative of advanced atherosclerosis. Along with this, T cell population could be correlated with the lesion size, as their depletion has been shown to prevent atherosclerosis (Meng et al. 2016). Epicardial adipose tissue is a depot of paracrine-acting adipokines and can contribute towards vascular atherogenesis.

Insulin Resistance and Diabetes

Elevated numbers of M1 type macrophages affect insulin signaling, and are attributed to the increased $CD8^+/CD4^+$ ratio and a decrease in Treg cells. This population tilt is mediated via the regulation of signal transducer and activator of transcription3 (STAT3) signaling (Priceman et al. 2013). Neutrophils, though present in small numbers, can impair insulin signaling as well. Neutrophil elastase induces glucose intolerance and insulin resistance via increased degradation of insulin receptor substrate-1. Inhibition or genetic knockdown of this enzyme reverses the adversity. Its inhibition also ameliorates the inflammation in adipose tissue by decreasing TLR4-dependent expression of proinflammatory markers (Talukdar et al. 2012).

During obesity and diabetes, mast cells secrete IL-6 and IFN-γ, which also promote and contribute to proinflammatory effect. The pathogenic response receptors such as TLR4, and the NLR family inflammasome such as cryporin/NLRP3, induce the secretion of active IL-1β and IL-18 in macrophages that recognize the ligands upregulated during obesogenic conditions (Varghese et al. 2019). The downstream effect of this activation includes the regulation of inflammatory intracellular lipids such as ceramides and sphingolipids, which are then responsible for inducing insulin resistance and diabetes (Summers 2010).

Acknowledgments Senior Research Fellowship to Dr. Patel from SERB, DST, Govt. of India and to Dr. Varghese from GSBTM, Govt. of Gujarat, India are thankfully acknowledged. The grant support to Dr. Yadav from GSBTM (sanction no.GSBTM/MD/PROJECTS/SSA/5050/2016-17) and SERB (sanction no.ECR/2015/000266) is highly acknowledged.

References

Appari M, Channon KM, McNeill E (2018) Metabolic regulation of adipose tissue macrophage function in obesity and diabetes. Antioxid Redox Signal 29 (3):297–312. https://doi.org/10.1089/ars.2017.7060

Artis D, Spits H (2015) The biology of innate lymphoid cells. Nature 517(7534):293–301

Boulenouar S, Michelet X et al (2017) Adipose type one innate lymphoid cells regulate macrophage homeostasis through targeted cytotoxicity. Immunity 46(2): 273–286

Boutens L, Stienstra R (2016) Adipose tissue macrophages: going off track during obesity. Diabetologia 59(5):879–894

Boutens L, Hooiveld GJ et al (2018) Unique metabolic activation of adipose tissue macrophages in obesity promotes inflammatory responses. Diabetologia 61 (4):942–953

Braune J, Weyer U et al (2017) IL-6 regulates M2 polarization and local proliferation of adipose tissue macrophages in obesity. J Immunol 198(7):2927–2934

Brestoff JR, Kim BS et al (2015) Group 2 innate lymphoid cells promote beiging of white adipose tissue and limit obesity. Nature 519(7542):242–246

Chen L, Chen R et al (2015) Mechanisms linking inflammation to insulin resistance. Int J Endocrinol 2015: 508409

Chung KJ, Chatzigeorgiou A et al (2017) A self-sustained loop of inflammation-driven inhibition of beige adipogenesis in obesity. Nat Immunol 18(6):654–664

Cipolletta D, Feuerer M et al (2012) PPAR-gamma is a major driver of the accumulation and phenotype of adipose tissue Treg cells. Nature 486(7404):549–553

Coats BR, Schoenfelt KQ et al (2017) Metabolically activated adipose tissue macrophages perform detrimental and beneficial functions during diet-induced obesity. Cell Rep 20(13):3149–3161

Coelho M, Oliveira T et al (2013) Biochemistry of adipose tissue: an endocrine organ. Arch Med Sci 9(2): 191–200

Costanzo AE, Taylor KR et al (2015) Obesity impairs gammadelta T cell homeostasis and antiviral function in humans. PLoS One 10(3):e0120918

Galic S, Fullerton MD et al (2011) Hematopoietic AMPK beta1 reduces mouse adipose tissue macrophage inflammation and insulin resistance in obesity. J Clin Invest 121(12):4903–4915

Ge Q, Brichard S et al (2014) microRNAs as a new mechanism regulating adipose tissue inflammation in obesity and as a novel therapeutic strategy in the metabolic syndrome. J Immunol Res 2014:987285

Ghosh AR, Bhattacharya R et al (2016) Adipose recruitment and activation of plasmacytoid dendritic cells fuel metaflammation. Diabetes 65(11):3440–3452

Grant RW, Dixit VD (2015) Adipose tissue as an immunological organ. Obesity (Silver Spring) 23(3): 512–518

Hams E, Locksley RM et al (2013) Cutting edge: IL-25 elicits innate lymphoid type 2 and type II NKT cells that regulate obesity in mice. J Immunol 191(11): 5349–5353

Han SJ, Glatman Zaretsky A et al (2017) White adipose tissue is a reservoir for memory T cells and promotes protective memory responses to infection. Immunity 47(6):1154–1168

Hannibal TD, Schmidt-Christensen A et al (2017) Deficiency in plasmacytoid dendritic cells and type I interferon signalling prevents diet-induced obesity and insulin resistance in mice. Diabetologia 60(10): 2033–2041

Kalin S, Becker M et al (2017) A Stat6/Pten axis links regulatory T cells with adipose tissue function. Cell Metab 26(3):475–492

Kraakman MJ, Kammoun HL et al (2015) Blocking IL-6 trans-signaling prevents high-fat diet-induced adipose tissue macrophage recruitment but does not improve insulin resistance. Cell Metab 21(3):403–416

Kratz M, Coats BR et al (2014) Metabolic dysfunction drives a mechanistically distinct proinflammatory phenotype in adipose tissue macrophages. Cell Metab 20 (4):614–625

Kurauti MA, Costa-Júnior JM et al (2017) Interleukin-6 increases the expression and activity of insulin-degrading enzyme. Sci Rep 7:46750

Lu JB, Yao XX et al (2016) MicroRNA-125b-5p attenuates lipopolysaccharide-induced monocyte chemoattractant protein-1 production by targeting inhibiting LACTB in THP-1 macrophages. Arch Biochem Biophys 590:64–71

Lynch L, Nowak M et al (2012) Adipose tissue invariant NKT cells protect against diet-induced obesity and metabolic disorder through regulatory cytokine production. Immunity 37(3):574–587

Lynch L, Michelet X et al (2015) Regulatory iNKT cells lack expression of the transcription factor PLZF and control the homeostasis of T(reg) cells and macrophages in adipose tissue. Nat Immunol 16(1):85–95

Macdougall CE, Wood EG et al (2018) Visceral adipose tissue immune homeostasis is regulated by the crosstalk between adipocytes and dendritic cell subsets. Cell Metab 27(3):588–601

Marques-Rocha JL, Samblas M et al (2015) Noncoding RNAs, cytokines, and inflammation-related diseases. FASEB J 29(9):3595–3611

Meng X, Yang J, Dong M et al (2016) Regulatory T cells in cardiovascular diseases. Nat Rev Cardiol 13 (3):167–179

Michalek RD, Gerriets VA et al (2011) Cutting edge: distinct glycolytic and lipid oxidative metabolic programs are essential for effector and regulatory CD4+ T cell subsets. J Immunol 186(6):3299–3303

Mikolajczyk TP, Nosalski R et al (2016) Role of chemokine RANTES in the regulation of perivascular inflammation, T-cell accumulation, and vascular dysfunction in hypertension. FASEB J 30(5):1987–1999

Molofsky AB, Nussbaum JC et al (2013) Innate lymphoid type 2 cells sustain visceral adipose tissue eosinophils and alternatively activated macrophages. J Exp Med 210(3):535–549

Newton R, Priyadharshini B et al (2016) Immunometabolism of regulatory T cells. Nat Immunol 17(6): 618–625

Nishimura S, Manabe I et al (2009) CD8+ effector T cells contribute to macrophage recruitment and adipose tissue inflammation in obesity. Nat Med 15(8):914–920

Nishimura S, Manabe I et al (2013) Adipose natural regulatory B cells negatively control adipose tissue inflammation. Cell Metab 18(5):759–766

Nosalski R, Guzik TJ (2017) Perivascular adipose tissue inflammation in vascular disease. Br J Pharmacol 174 (20):3496–3513

Odegaard JI, Ricardo-Gonzalez RR et al (2007) Macrophage-specific PPARgamma controls alternative activation and improves insulin resistance. Nature 447 (7148):1116–1120

O'Sullivan TE, Rapp M et al (2016) Adipose-resident group 1 innate lymphoid cells promote obesity-associated insulin resistance. Immunity 45(2):428–441

Ouchi N, Parker JL et al (2011) Adipokines in inflammation and metabolic disease. Nat Rev Immunol 11(2):85–97

Patel R, Varghese JF et al (2019) Induction of endothelial dysfunction by oxidized low-density lipoproteins via downregulation of Erk-5/Mef2c/KLF2 signaling: amelioration by Fisetin. Biochimie 163:152–162

Pearce EJ, Everts B (2015) Dendritic cell metabolism. Nat Rev Immunol 15(1):18–29

Priceman SJ, Kujawski M et al (2013) Regulation of adipose tissue T cell subsets by Stat3 is crucial for diet-induced obesity and insulin resistance. Proc Natl Acad Sci U S A 110(32):13079–13084

Rondini EA, Granneman JG (2020) Single cell approaches to address adipose tissue stromal cell heterogeneity. Biochem J 477(3):583–600. https://doi.org/10.1042/BCJ20190467

Skiba DS, Nosalski R et al (2017) Anti-atherosclerotic effect of the angiotensin 1-7 mimetic AVE0991 is mediated by inhibition of perivascular and plaque inflammation in early atherosclerosis. Br J Pharmacol 174(22):4055–4069

Summers SA (2010) Sphingolipids and insulin resistance: the five Ws. Curr Opin Lipidol 21(2):128–135

Talukdar S, Oh DY et al (2012) Neutrophils mediate insulin resistance in mice fed a high-fat diet through secreted elastase. Nat Med 18(9):1407–1412

Tezze C, Romanello V et al (2019) FGF21 as modulator of metabolism in health and disease. Front Physiol 10:419

Tian GP, Chen WJ et al (2012) MicroRNA-467b targets LPL gene in RAW 264.7 macrophages and attenuates lipid accumulation and proinflammatory cytokine secretion. Biochimie 94(12):2749–2755

Varghese JF, Patel R et al (2018) Novel insights in the metabolic syndrome-induced oxidative stress and

inflammation-mediated atherosclerosis. Curr Cardiol Rev 14(1):4–14
Varghese JF, Patel R et al (2019) Sterol regulatory element binding protein (SREBP) -1 mediates oxidized low-density lipoprotein (oxLDL) induced macrophage foam cell formation through NLRP3 inflammasome activation. Cell Signal 53:316–326
Villarroya J, Cereijo R et al (2019) New insights into the secretory functions of brown adipose tissue. J Endocrinol 243:R19–R27
Wei X, Song H et al (2016) Fatty acid synthesis configures the plasma membrane for inflammation in diabetes. Nature 539(7628):294–298
Winer S, Chan Y et al (2009) Normalization of obesity-associated insulin resistance through immunotherapy. Nat Med 15(8):921–929
Winer DA, Winer S et al (2011) B cells promote insulin resistance through modulation of T cells and production of pathogenic IgG antibodies. Nat Med 17 (5):610–617
Xu X, Grijalva A et al (2013) Obesity activates a program of lysosomal-dependent lipid metabolism in adipose tissue macrophages independently of classic activation. Cell Metab 18(6):816–830
Ying W, Riopel M et al (2017) Adipose tissue macrophage-derived exosomal miRNAs can modulate in vivo and in vitro insulin sensitivity. Cell 171(2): 372–384
Zand H, Morshedzadeh N et al (2017) Signaling pathways linking inflammation to insulin resistance. Diabetes Metab Syndr 11(Suppl 1):S307–S309
Zuniga LA, Shen WJ et al (2010) IL-17 regulates adipogenesis, glucose homeostasis, and obesity. J Immunol 185(11):6947–6959

Diabetes Forecasts and Statistics for the Coming Decades

12

Alexandre Assuane Duarte and Olga Golubnitschaja

Contents

Introduction 166
Global Profile of the Diabetes Epidemic 166
Decreased Diabetes Incidence in the USA 167
The Elderly Population 167
Gender and Social/economic Profiles 168
Demographic Patterns 168
Worldwide Estimations for 2040 169
Public Health Challenges 169
Diabetes Complications 170
Direct and Indirect Costs 170
Outlook for the Next 30 Years 171
Improved Screening, Diagnosis, and Prognosis 172
References 173

Abstract

Diabetes has become one of the foremost diseases worldwide, even though growth and prevalence are geographically heterogeneous, according to such variables as ethnic profile, diet, and selected lifestyle variables. Traditionally a concern of the affluent Western society, it is rapidly spreading to developing societies, with potentially disastrous public-health consequences. In recent decades, the regions with the steepest elevation of prevalence were the Eastern Mediterranean, Africa, and South

A. Assuane Duarte (✉)
University Center IMEPAC Araguari, Araguari, Brazil
e-mail: alexandreassuane@hotmail.com

O. Golubnitschaja
Radiological Clinic, Excellence University of Bonn, Bonn, Germany

Breast Cancer Research Centre, Excellence University of Bonn, Bonn, Germany

Centre for Integrated Oncology, Cologne-Bonn, Excellence University of Bonn, Bonn, Germany

J. Faintuch and S. Faintuch (eds.), *Obesity and Diabetes*, https://doi.org/10.1007/978-3-030-53370-0_12

East Asia. In contrast, Europe and particularly the USA, perhaps on account of long experience and more abundant healthcare resources, have achieved more success in curbing the epidemic. Both modifiable and nonmodifiable risk factors are operative. Obesity and unhealthy lifestyle are prominent in the first modality, whereas aging deserves attention among nonmodifiable precipitants, as the world population is becoming older. Given the possibility of 600 million diabetics in 2045, from current 422 million, the need for implementation of preventive policies cannot be overemphasized.

Keywords

Diabetes mellitus · Epidemic · Comorbidities · Prevalence · Medical care · Costs · Economy · Health policy · Predictive preventive personalised medicine

Introduction

Diabetes mellitus (DM) affects 422 million people in all continents (WHO 2019). The outlook is becoming worse, as regions and age brackets that were relatively free from the disease, especially children, are progressively becoming involved, in tandem with lifestyle changes and longer life expectations (Fig. 12.1). A relevant example is children and teenagers, whose glucose homeostasis deteriorates in parallel with overweight and obesity (Golubnitschaja 2013). Such young patients will eventually pay a high price in the form of cardiovascular disease, cancer, neurological disorders, and premature death (Golubnitschaja 2013).

Global Profile of the Diabetes Epidemic

A worldwide evaluation encompassing 34 years (1980–2014) (World Health Organization 2016) revealed that few regions were able to curb disease growth (Fig. 12.2). Growth accelerates most in less-affluent areas like the Eastern Mediterranean, with a 2.3 times jump in prevalence between 1980 and 2014, and Africa, with a similar proportion of elevation. (Duarte et al. 2018)

South-East Asia region does not come far behind, with a 2.1 times stronger prevalence in 2014. In contrast, the illness is nearly under control in Europe, with a comparatively modest 1.4

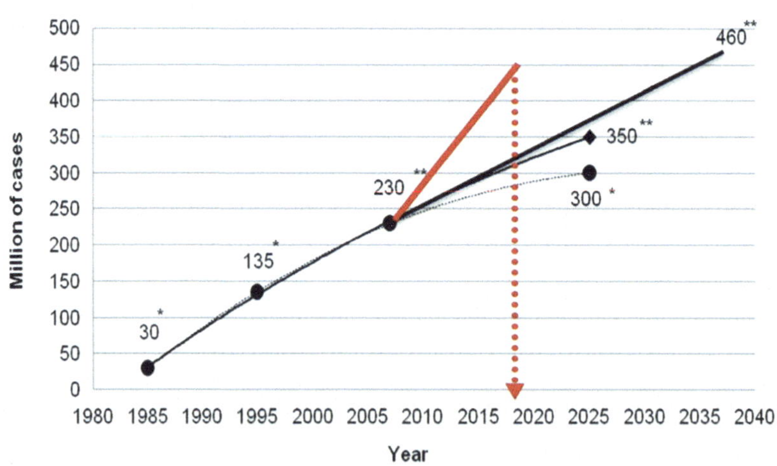

Fig. 12.1 Single asterisks are applied to prognoses from the 1980s, and double ones for those of the early 2000s. Actual prevalence figures, collected in recent years (in red) are even more worrisome (Golubnitschaja 2013). Reprinted with Permission from Duarte AA, Mohsin S, Golubnitschaja O. Diabetes care in figures: current pitfalls and future scenario. EPMA J 2018; 9(2): 126–131

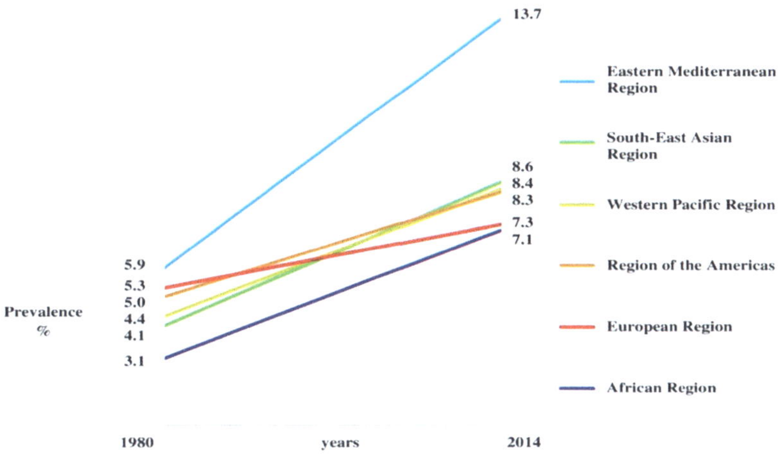

Fig. 12.2 Prevalence of DM on a global scale: retrospective comparative analysis in the period of time from the year 1980 to 2014 (NCD Risk Factor Collaboration (NCD-RisC) 2016). Reprinted with Permission from Duarte AA, Mohsin S, Golubnitschaja O. Diabetes care in figures: current pitfalls and future scenario. EPMA J 2018; 9(2): 126–131

times elevation in the observed period. Differently from Latin America, the Caribbean, and Canada, where the increase is robust, thus influencing the global profile of the Americas, the USA featured a reduction in the incidence of diabetes in the last decade. According to certain authors, this could be due to immigration of younger populations (Benoit et al. 2019).

Decreased Diabetes Incidence in the USA

The announcement comes from the Centers for Disease Control and Prevention (CDC). After peaking in 2009, with 1.7 million new cases a year, the incidence of diabetes in the USA shrunk by 35%, to 1.3 million in 2017. Prevalence remained stable during this period, a finding attributed to longer survival for such populations (Benoit et al. 2019; Media Relations 2019).

Does that mean that at least in the context of richer countries, with countrywide diabetes prevention programs, the tide of diabetes can be stemmed? There are both favorable and ominous evidence. All-cause and cardiovascular mortality in Americans with diabetes are following the same positive trend (International Diabetes Federation 2017). In turn, obesity spread is unrelenting,

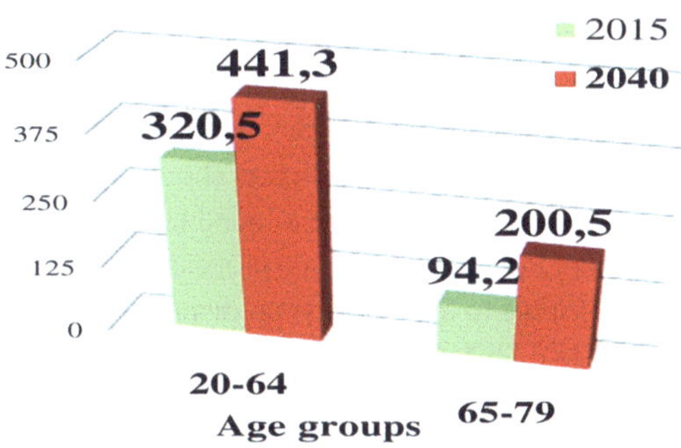

Fig. 12.3 The entire DM patient's pool (in million of patients) stratified by the age as registered in the year 2015; prognostic estimations for the corresponding age-groups are made for the year 2040 (International Diabetes Federation 2015). Reprinted with Permission from Duarte AA, Mohsin S, Golubnitschaja O. Diabetes care in figures: current pitfalls and future scenario. EPMA J 2018; 9(2): 126–131

among nearly all social classes and age groups, which strongly impacts the future of type 2 diabetes (International Diabetes Federation 2017).

The Elderly Population

Life expectancy is improving in most countries, a finding consistent with more diabetic cases (World Health Organization 2016). The 2.1 times anticipated growth of the condition in the old-age population (65–79 years) till 2040, actually exceeds by 50% of the prognosis for those in the 20–64 age bracket (Fig. 12.3) (International Diabetes Federation 2017). Younger people are currently the clear majority, however, the opposite is predicted for 2040, with more than two-thirds of that population being represented by the elderly people (68.8%) (International Diabetes Federation 2017).

Gender and Social/economic Profiles

Type 2 diabetes (DM2) is a multifactorial disorder, driven by both genetic and environmental factors. The impact of obesity is the most highlighted, with its accompanying array of sedentarism, lack of physical exercise, and high-fat (Western) diet, yet other predisposing factors point toward diabetes. Besides aging, as alluded to, smoking and alcohol addiction should be mentioned (Luo et al. 2018). The alluded to epigenetic factors could act in the aging process and therefore contribute to the diabetes predisposition (Luo et al. 2018). Differences in prevalence regarding gender and financial income, partially reflect the summed-up effects of the mentioned circumstances.

Demographic Patterns

Demographic patterns of diabetes are still conflicting in the literature, depending on geography, ethnic background, economic status, and other characteristics, yet a global forecast for 2040 addressing the living environment has already been outlined (Fig. 12.4) (International Diabetes Federation 2015).

Such a progressive concentration of the disease in the city milieu is consistent with the global urbanization trend, even though rural obesity is not a novelty anymore. Indeed, pediatric obesity rates in certain peripheral locations can be higher than in urban areas (Hosseini and Yilmaz 2019).

Significantly, gender shifts in the diabetes trend for the next two decades are unlikely (International Diabetes Federation 2017). Even though

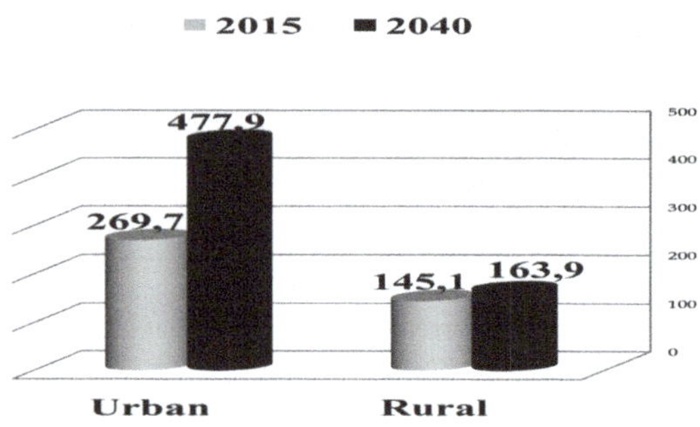

Fig. 12.4 DM patient distribution profiles (in million) between the urban and rural areas as documented for the year 2015 and prognosed for the year 2040 (World Health Organization 2016; International Diabetes Federation 2017, 2015). Reprinted with Permission from Duarte AA, Mohsin S, Golubnitschaja O. Diabetes care in figures: current pitfalls and future scenario. EPMA J 2018; 9(2): 126–131

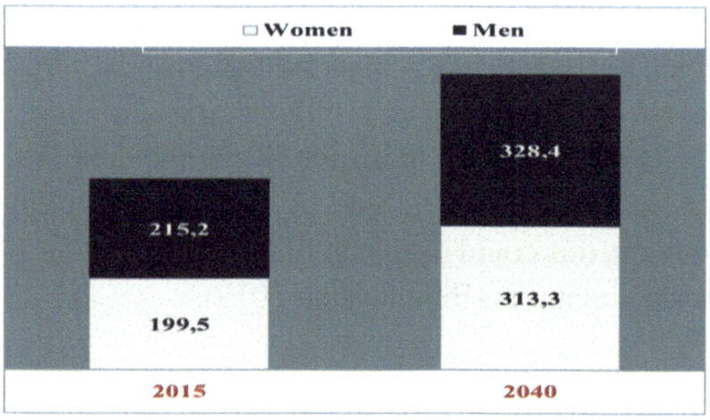

Fig. 12.5 Gender-related patient distribution profiles (in million): actually (year 2015) registered patients versus prognoses for the year 2040 (International Diabetes Federation 2017). Reprinted with Permission from Duarte AA, Mohsin S, Golubnitschaja O. Diabetes care in figures: current pitfalls and future scenario. EPMA J 2018; 9(2): 126–131

data on diabetes registered a higher prevalence among men (Fig. 12.5), women were more obese or overweight than men. Only in the age group of 70–79 women have a higher prevalence of diabetes than men (International Diabetes Federation 2017). The higher the income of the country, the higher were the rates of overweight and obesity, which are strong well-known risk factors for diabetes (World Health Organization 2016).

Current (2017) statistics reveal a higher and increasing prevalence of the disorder in populations aged 20–64 years. The prognosis for the prevalence of diabetes in the range of 65–79 years of age will be more than double in 2045, with a prevalence shift toward the elderly population (International Diabetes Federation 2015).

Worldwide Estimations for 2040

In most of the Western Asia-Pacific and South-East Asia regions, until just a couple of decades ago, diabetes was a secondary concern (Duarte et al. 2018). Now it is forecasted that they will carry the largest burden on the planet, at least in absolute numbers, responding for 56% of the diabetes population (World Health Organization 2016).

Projections are less ominous for Europe, and even less for the USA, as already signaled (World Health Organization 2016). Yet the region of North America and the Caribbean will continue to deteriorate, predominantly on account of the contribution of Mexico, but also of Canada and some Caribbean islands (World Health Organization 2016; Benoit et al. 2019).

Concerning expansion compared to current burden, no area will suffer more than Sub-Saharan Africa, followed by North Africa, and Middle East, with 3.0 and 2.4 elevation, respectively, until 2040 (Duarte et al. 2018) (Fig. 12.6).

Public Health Challenges

As emphasized in Table 12.1, the USA and Western European countries still have to deal with large numbers of diabetics, requiring major financial and labor investments (Statista 2019). Yet the speedometer is moving faster in the South and East of the planet, with non-Western countries being predominantly overwhelmed by the epidemic (International Diabetes Federation 2015). This could partly be due to the susceptibility of South Asians, and possibly Asians in general, to lesser degrees of obesity, acquiring precocious comorbidities, particularly type 2 diabetes (Pomeroy et al. 2019). Atavic low lean mass is an ethnic feature of some of these populations,

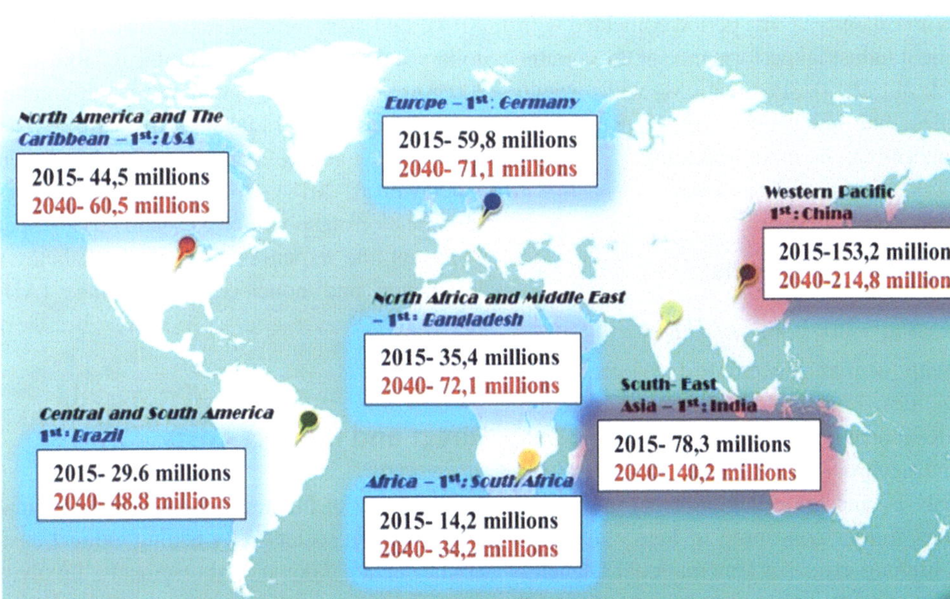

Fig. 12.6 Prognosis of diabetes prevalence in 2040 (International Diabetes Federation 2017, 2015). Reprinted with Permission from Duarte AA, Mohsin S, Golubnitschaja O. Diabetes care in figures: current pitfalls and future scenario. EPMA J 2018; 9(2): 126–131

Table 12.1 Top ten countries regarding diabetes in 2017

Country	Millions of adult diabetic patients
China	114.4
India	72.9
United States of America	30.2
Brazil	12.5
Mexico	12.0
Indonesia	10.3
Russian Federation	8.5
Egypt	8.2
Germany	7.5
Pakistan	7.5

Statista (2019)

and could underlie early deterioration of glucose homeostasis, when exposed to the Western diet and lifestyle (Pomeroy et al. 2019).

The unrelenting demands for massive diabetes-related budgets in those regions, potentially threatening the whole healthcare economy, are illustrated by Chinese figures. In 1993, only 0.07% of the gross domestic product (GDP) was channeled to diabetes. Figures 15 years later underwent a 300% jump (0.21% of GDP in 2008) (Mao et al. 2019). Specifically for China, this was no calamity, as the period coincided with an excellent industrial performance of the country (Mao et al. 2019). For nations with fragile economies and less impressive GDP growth, consequences can be more unsettling.

Diabetes Complications

Micro- and macrovascular troubles are notorious, along with neurodegenerative and oncological comorbidities (International Diabetes Federation 2017). Retinopathy potentially leading to blindness, and nephropathy with attending needs for hemodialysis and transplantation in advanced cases, can further contribute to heavy acute as well as lifelong expenses (International Diabetes Federation 2015). Chronic leg ulcers and impaired wound healing, besides demanding occasional surgery and amputation, further interfere with health budgets and quality of life (WHO 2019).

Ischemic heart disease is traditionally listed as one the most frequent, and remains among the most costly complications, both in Western and Eastern countries (Gerdtham et al. 2009; Ward et al. 2014; ADA 2018). Fatal myocardial infarction incurs a USD 4067 bill in Taiwan (Cheng et al. 2018). Less common, however, expensive disorders include ischemic stroke, with an annual event cost of 42,119 USD in the USA (Ward et al. 2014), end-stage renal disease, reaching 28,551 dollars annual costs in Australia (Clarke et al. 2008), and leg amputation, with annual expenditures of 14,949 euros in Sweden (Gerdtham et al. 2009) and USD 7677 in Taiwan (Cheng et al. 2018). It has been estimated that complications respond for 25% and 45%, respectively, of the expenses of treating diabetes in emergency and nonemergency settings (ADA 2013).

Direct and Indirect Costs

Another way of financially analyzing the impact of diabetes on society is by adding indirect costs, to the better known disbursements involving screening, diagnosis, therapy, and long-term management of the illness and its direct complications (Seuring et al. 2015). Such less prominent, however, no less tangible items

Table 12.2 Per capita diabetes-related direct and indirect costs

Country	Direct	Indirect
USA	4221–9346	518–4050
Spain	907–4690	556–3379
Italy	4588	208
Netherlands	2780	195
Sweden	1507–2855	2174–2606
Norway	2061	650
Paraguay	2587	245
Jamaica	2439	549
Nicaragua	2145	1105
Iran	2142	2199
Serbia	1610	187
India	1557	305
Thailand	1082	649
Latin America/Caribbean[a]	1011–1677	4628–6560
Mexico	263–1072	64–7135
Pakistan	620	65

In US dollars (Seuring et al. 2015)
[a]Estimated range for the entire region

encompass absenteeism, reduced productivity for the labor sector, as well as for those unemployed, however, active in household and other informal activities, disease-related disability, and premature deaths triggering lost productivity (Seuring et al. 2015).

As displayed in Table 12.2, estimates significantly differ among countries, as a consequence of disparate cost- of- living, healthcare practices, and computed variables, such as just drug consumption, or also hospitalization, diagnostic work-up, and rehabilitation, among direct costs for instance (Seuring et al. 2015). Nevertheless, the numbers are almost always substantial, notably for poor countries. The most striking discrepancies concern indirect costs, as there is less consensus about how to define and quantify those values. Usually lower than direct costs, in some studies, they come out higher (Seuring et al. 2015).

In the USA, diabetes-related costs including medical expenses and diminished productivity were roughly 300 billion USD in 2010. A 71% increase until 2025 (514 billion USD) is anticipated, more conspicuous than expected inflation or GDP shifts in the period (ADA 2013). The same growth curve, projected for 2040, would render a USA bill of 772 billion, and a worldwide expense of 8.5 trillion, assuming 650 million diabetics at that time (Duarte et al. 2018). Such values are unfeasible, particularly in the developing world, forecasting a crisis in the management of the diabetes population.

According to the American Diabetes Association (ADA), not less than USD 237 billion direct and 90 billion indirect costs, were incurred by the USA in 2017 with the disease (ADA 2018). This amounts to one-fourth of the healthcare budget for total expenditures, or half as much for specific diabetes costs (ADA 2018)

In other words, each diagnosed patient requires a total disbursement around USD 16,750 per year, and 57% corresponds to specific diabetes costs (USD 9600 per capita) (ADA 2018). Such outcomes exceed the average results obtained by Seuring et al. (2015), and displayed for that country in Table 12.2.

Outlook for the Next 30 Years

The growth of obesity continues uncurbed in the entire world, including the USA. This means that despite the respite in the American diabetes epidemic (Benoit et al. 2019; Media Relations 2019), the future for much of the planet still looks bleak. The increase in lifespan is another negative variable concerning diabetes, as old age is detrimental to pancreas physiology, consistent with an elevation of new-onset cases. At the same time, as existing diabetics live longer, prevalence will be impacted, even if incidence remains the same. As a consequence, the number of diabetics in the USA could double or triple till 2050, affecting as many as one-third of the population, from 10% of current population (CDC 2019).

In synthesis, there is no way of avoiding the fact that type 2 diabetes is a catastrophic enemy, one that has eluded eradication, even though responding to an arsenal of glucose-lowering drug and lifestyle therapies. One death in seven is attributed to the disease, which is also behind the majority of instances of blindness (<75 years of age), renal failure, and nontraumatic lower limb amputations (CDC 2019).

The same way that in politics, the price of liberty is eternal vigilance, diabetes explosive growth will only be curtailed employing sensible, incremental steps, aiming to act on preventive and predictive strategies, avoiding complications and promoting quality of life, possibly benefiting from advances in artificial intelligence, machine learning, databanks, microbiomics, and all other omics (multi-omics), integrated into the innovative concept of predictive, preventive, and personalized medicine (3PM or PPPM) (Golubnitschaja 2013).

Improved Screening, Diagnosis, and Prognosis

In 2019, Issa et al. investigated a possible predictive marker for the risks and complications of DM2, namely angiopoietin-like protein-8 - (ANGPTL-8) (Issa et al. 2019). The role that this liver-secreted amino acid plays has been subject of investigation, controversy, and lack of consensus in the past decades (Issa et al. 2019). This study confirmed that high ANGPTL-8 levels are a risk factor for having DM2 (four times risk [OR 4.03 unadjusted; 6.26 adjusted]). Another interesting finding was a strong correlation between ANGPTL-8 levels and impaired kidney function, possibly predicting diabetic nephropathy. Larger follow-up studies are needed to validate this association (Issa et al. 2019).

Wang et al. (2019) developed in Central China an affordable and personalized method to predict the diabetes risk of patients and to analyze the relationship between the studied risk factors and DM2. The 3-year cohort was composed of 5557 nondiabetic patients, with a validation cohort of 1870 (Wang et al. 2019). The so-called nomogram was based on age, blood pressure, blood sugar, and cholesterol levels. Simple instruments like these can generate large scale identification of individuals at-risk in a suboptimal health status (SHS) (Wang et al. 2019; Ge et al. 2019). Therefore, health policies are better implemented and resources can be wisely used in a targeted manner for a specific population. The above-mentioned method can be easily implemented in multiple countries creating immense databanks that can be further useful globally (Wang et al. 2019).

China is the country with the highest absolute number of individuals with DM2, and urban disease increase in the last two decades (Ge et al. 2019). Ge et al. (2019) used a screening questionnaire assessing the suboptimal health status condition which is characterized by fatigue and lack of energy for 3 months or more, through such variables as fatigue, organ systems, and mental status (suboptimal health status questionnaire-25, SHSQ-25), indicating that SHS could be a strong predictor of DM2 (Ge et al. 2019). Unhealthy lifestyle habits and physical inactivity were definitive factors contributing to a worse suboptimal health state.

The most prominent approach to reverse current trends in the diabetes epidemic is the innovative concepts of predictive, preventive, and personalized medicine (PPPM or 3PM), promoted by EPMA and EPMA Journal (Golubnitschaja 2013). This translates into predictive diagnostics, targeted prevention, and person- tailored treatments in the context of so-called suboptimal health status, which is the state before the onset of the disease, thus preventing complications of NCDs (noncommunicable diseases) and particularly diabetes (Golubnitschaja 2013; Duarte et al. 2018).

Acknowledgment This chapter was based on the original article "Diabetes care in figures: current pitfalls and future scenario" published by Duarte et al. (2018).

References

ADA (2013) Economic costs of diabetes in the U.S. in 2012. Diabetes Care 36:103

ADA (2018) Economic Costs of Diabetes in the U.S. in 2017. Diabetes Care 41(5):917–928

Benoit SR, Hora I, Albright AL, Gregg EW (2019) New directions in incidence and prevalence of diagnosed diabetes in the USA. BMJ Open Diabetes Res Care 7 (1):e000657

CDC (2019) Number of Americans with diabetes projected to double or triple by 2050. www.cdc.gov/media/pressrel/2010/r101022.html. Accessed 5 Sept 2019

Cheng SW, Wang CY, Chen JH, Ko Y (2018) Healthcare costs and utilization of diabetes-related complications in Taiwan-A claims database analysis. Medicine 97 (31):e11602

Clarke P, Leal J, Kelman C et al (2008) Estimating the cost of complications of diabetes in Australia using administrative health-care data. Value Health 11:199–206

Duarte AA, Mohsin S, Golubnitschaja O (2018) Diabetes care in figures: current pitfalls and future scenario. EPMA J 9(2):126–131

Ge S, Xu X, Zhang J et al (2019) Suboptimal health status as an independent risk factor for type 2 diabetes mellitus in a community-based cohort: the China suboptimal health cohort study. EPMA Journal 10:65–72. https://doi.org/10.1007/s13167-019-0159-9

Gerdtham UG, Clarke P, Hayes A et al (2009) Estimating the cost of diabetes mellitus-related events from inpatient admissions in Sweden using administrative hospitalization data. Pharmacoeconomics 27:81–90

Golubnitschaja O (2013) Three levels of prediction, prevention and individualised treatment algorithms to advance diabetes care: integrative approach. In: Mozaffari M (ed) New strategies to advance pre/diabetes care: integrative approach by PPPM. Springer, Dordrecht, pp 11–28

Hosseini H, Yilmaz A (2019) Using Telehealth to address pediatric obesity in rural Pennsylvania. Hosp Top 97 (3):107–118

International Diabetes Federation (2015) IDF Diabetes Atlas, 7th edn. International Diabetes Federation, Brussels, Belgium. www.diabetesatlas.org; Accessed 20 Aug 2019

International Diabetes Federation (2017) IDF Diabetes Atlas, 8th ed. www.diabetesatlas.org/resources/2017-atlas.html. Accessed 30 Aug 2019

Issa YA, Abd ElHafeez SS, Amin NG (2019) The potential role of angiopoietin-like protein-8 in type 2 diabetes mellitus: a possibility for predictive diagnosis and targeted preventive measures? EPMA Journal 10:239–248. https://doi.org/10.1007/s13167-019-00180-3

Luo B, Zhang J, Hu Z, Gao F, Zhou Q, Song S, Qin L, Xu H (2018) Diabetes-related behaviours among elderly people with pre-diabetes in rural communities of Hunan, China: a cross-sectional study. BMJ Open 8 (1):e015747

Mao W, Yip CW, Chen W (2019) Complications of diabetes in China: health system and economic implications. BMC Public Health 19(1):269

Media Relations (2019). www.cdc.gov/media/releases/2019/p0529-diabetes-cases-decline.html. Accessed 29 August 2019

NCD Risk Factor Collaboration (NCD-RisC) (2016) Worldwide trends in diabetes since 1980: a pooled analysis of 751 population-based studies with 4.4 million participants. Lancet 387(10027):1513–1530

Pomeroy E, Mushrif-Tripathy V, Cole TJ, Wells JCK, Stock JT (2019) Ancient origins of low lean mass among South Asians and implications for modern type 2 diabetes susceptibility. Sci Rep 9(1):10515

Seuring T, Archangelidi O, Suhrcke M (2015) The economic costs of type 2 diabetes: a global systematic review. Pharmaco Economics 33(8):811–831

Statista (2019). www.statista.com/statistics/281082/countries-with-highest-number-of-diabetics. Accessed 5 Sept 2019

Wang K, Gong M, Xie S et al (2019) Nomogram prediction for the 3-year risk of type 2 diabetes in healthy mainland China residents. EPMA J 10:227–237. https://doi.org/10.1007/s13167-019-00181-2

Ward A, Alvarez P, Vo L et al (2014) Direct medical costs of complications of diabetes in the United States: estimates for event-year and annual state costs (USD 2012). J Med Econ 17:176–178

WHO (2019). www.who.int/health-topics-diabetes. Accessed 29 Aug 2019

World Health Organization (2016) Global report on diabetes. https://apps.who.int/iris/bitstream/handle/10665/204871/9789241565257_eng.pdf. Accessed 29 Aug 2019

Pancreatic Islets of Langerhans: Adapting Cell and Molecular Biology to Changes of Metabolism

13

Fernanda Ornellas, Iara Karise, Marcia Barbosa Aguila, and Carlos Alberto Mandarim-de-Lacerda

Contents

The Pancreatic Islet 176

Cytoarchitecture, Vascularization, Innervation, and Paracrine Interactions 176

Postnatal Pancreas 178
Role of Nonbeta Islet Cells 179

Beta-cell Proliferation 179

Insulin Synthesis 181
Insulin Output Modulation 181

Signaling Pathways Related to Insulin 181
Glucose Transporters 183

Islet Transplantation and Stem-cell Differentiation 184
Stem Cell Sources 184

Alpha-cell Plasticity 184
Regenerative Medicine 185

The Beta-cells Mass 185
Beta Cell Mass Increase 185

Pancreatic Islets in Obesity and Intermittent Fasting 185
Fasting and Glucotoxicity 186
Cellular Effects 186

Fasting and Refeeding Sequence 187

References 187

F. Ornellas · I. Karise · M. B. Aguila · C. A. Mandarim-de-Lacerda (✉)
Laboratório de Morfometria, Metabolismo & Doença Cardiovascular, Centro Biomédico, Instituto de Biologia, Universidade do Estado do Rio de Janeiro, Rio de Janeiro, Brazil
e-mail: mandarim@uerj.br

J. Faintuch and S. Faintuch (eds.), *Obesity and Diabetes*, https://doi.org/10.1007/978-3-030-53370-0_13

Abstract

Langerhans published in 1869, the thesis entitled "Contributions to the microscopic anatomy of the pancreas" (translated from German). In a healthy adult man, the islet mass is usually 1–2% of the pancreas mass, and can reach 500,000–1 million islets with 50–250 μm of diameter. Nowadays, at least five types of cells were identified in the pancreatic islets, which are responsible for the secretion of hormones: alpha, beta, delta, PP, and epsilon cells. Although pancreatic progenitor cells can be differentiated from stem cells, progenitor cells treated with some combinations of signaling factors can generate different cell types. Beta-cells are hard to produce, and people should be informed on how important it is to have a healthy lifestyle, in order to avoid exposing beta cells to the toxicity of sustained hyperglycemia. In the future, beta-cell neogenesis could repopulate the injured islets of the diabetic individual.

Keywords

Alpha-cell · Beta-cell · Diabetes · Obesity · Insulin signaling

The Pancreatic Islet

The year of 2019 corresponds to the 150th anniversary of the description of the pancreatic islet by Paul Langerhans (1847–1888). In February 1869, Langerhans presented his thesis entitled "Beiträge zur mikroskopischen Anatomie der Bauchspeicheldrüse" (Contributions to the microscopic anatomy of the pancreas) (Langerhans 1869), in which he described the presence of "islets" of clear cells throughout the gland, with staining properties different from the surrounding tissues. However, Langerhans did not suggest any function, except for the erroneous assumption that the islets might be like lymph nodes (Barach 1952). After 150 years, the pancreatic islets are better known, and cell composition and secretions in physiological and pathological situations were determined (Mandarim-de-Lacerda 2019).

It was Gustave-Edouard Laguesse (1861–1927) in Lille who named in 1893, the "islets of Langerhans," and postulated that they could act in digestion. Laguesse also hypothesized that the model of diabetes found in the pancreatectomized dog could be explained by the absence of these islets, which suggests their intervention in an internal secretion. Laguesse created the term "endocrine" (*endo* = inside, and *krino* = I secrete), to designate this type of cells pouring in the circulating blood to their product of synthesis (Wémeau et al. 2018).

A thin layer of connective tissue surrounds the islets, and separates them from the exocrine pancreas. In mature mammals, the islet mass is usually 1–2% of the pancreas mass and, in a healthy adult man, that can mean 500,000 to 1 million islets, with 50 to 250 μm in diameter (Hellman 1959a). The tail of the pancreas concentrates a higher density of islets than the head and body of the pancreas (Hellman 1959b).

Cytoarchitecture, Vascularization, Innervation, and Paracrine Interactions

The human and rodent islets differ substantially by several characteristics, including architecture, cell composition, gene expression, and some aspects of insulin secretion. Pancreatic islets of mice are highly vascularized, with endothelial and endocrine islet cell interactions essential for islet cell differentiation and function. Human islets have five times fewer vessels per island than islets of mice (Brissova et al. 2015).

Five types of cells could be identified in the pancreatic islets, responsible for the secretion of hormones. Considering rodent islets, we can find the following subsets of endocrine cells: (Kim et al. 2009)

1. Alpha-cells (releasing glucagons), correspond to 20%;
2. Beta-cells (insulin and amylin / islet amyloid polypeptide), cosecreting these molecules with an amylin-insulin ratio of 1: 100, (Adeghate and Kalasz 2011) correspond to 70%;

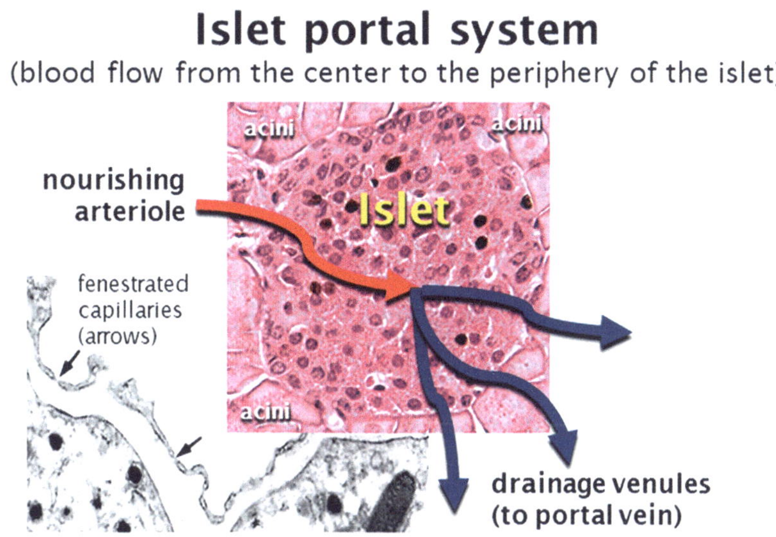

Fig. 13.1 The vascularization of the pancreatic islet. Arterial branches ultimately derived from the celiac artery reach the pancreas, and one or more arterioles nourish the islet from its central part, subsequently undergoing capillarization (the capillaries of this region are fenestrated). The blood is collected by small venules that eventually reach the portal vein (modified from Leung 2010)

3. Delta-cells (producing somatostatin), represent less than 10%;
4. Epsilon-cells (generating ghrelin) amount to less than 1%;
5. PP cells (also known as gamma-cells or F cells, providing pancreatic polypeptide), represent less than 5%;
6. Additionally, serotonin-producing enterochromaffin cells are present in pancreatic islets of mammals, at least at a given stage of development (Wierup et al. 2014).

The human islets have insulin-immunoreactive beta-cells, glucagon-immunoreactive alpha-cells, and somatostatin-containing delta cells dispersed in the islet (Brissova et al. 2005; Folli et al. 2018). Rodent islets exhibit a predominance of insulin-producing beta-cells located in the islet core, while the periphery has some numbers of alpha-, delta-, and PP cells (Abdulreda et al. 2016). The human islets contain proportionately fewer beta-cells and more alpha-cells. In human islets, most beta-, alpha-, and delta-cells are aligned along the blood vessels, without any order or arrangement, indicating that islet microcirculation probably does not determine the order of paracrine interactions (Fig. 13.1) (Cabrera et al. 2006; Leung 2010).

In smaller human islets (<60 μm diameter), beta-cells are in the islet core, while alpha-cells are peripherally located, parallel to the vessels. In larger human islets, alpha-cells are also found along the vessels that enter and branch within the islets. A three-dimensional analysis showed that the islet cells are distributed in an organization of trilaminar epithelial plaques, surrounded by vessels on both sides. In the islet core, the ratio between beta- and alpha-cells is higher than in the periphery, and decreases with increasing islet diameter (Bosco et al. 2010). In epithelial plaques, beta-cells are frequently located in a central position, however, generally display cytoplasmic extensions between peripheral nonbeta-cells (Baetens et al. 1979).

The vascularization and innervation of the islet are highly interconnected, and are essential for intercellular communication. Both participate in control of blood flow, a vital role for endocrine secretion (Penicaud 2017). Sympathetic innervation is necessary for the formation of pancreatic islets, as well as their functional maturation. Ablation of sympathetic innervation or beta-adrenergic signaling during development, results in a change in islet architecture with reduced insulin secretion, and impaired glucose tolerance in mice (Borden et al. 2013).

The beta-cells in the islet show excellent communication with each other via gap junctions, and with other cell types via diffusible chemical

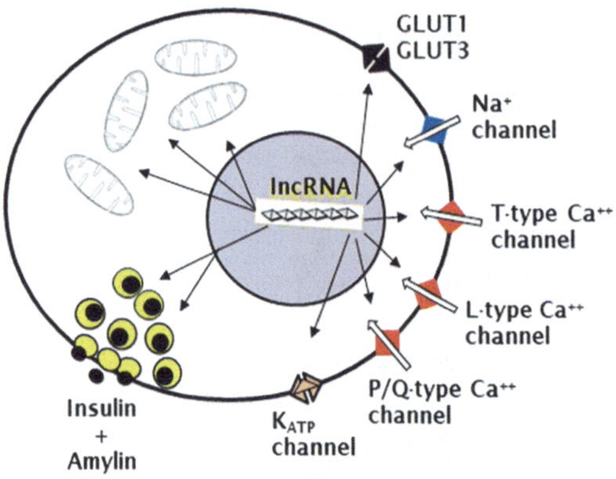

Fig. 13.2 The beta-cell. Schematic representation of the beta-cell, highlighting the channels and membrane receptors that act on cell function, and the production of insulin and amylin. We see that three types of Ca^{++} channels are present (T-type or low-voltage, L-type or voltage-dependent, and P/Q-type or high-voltage). All are controlled by long noncoding RNAs (lncRNAs) that play epigenetic regulatory roles in key molecular processes (gene expression, genetic imprinting, histone modification, chromatin dynamics), interacting with all kinds of molecules (modified from Esguerra and Eliasson 2014)

messengers (Benninger and Hodson 2018). The islet endocrine cells use a sophisticated system of endocrine, paracrine, and autocrine signals to synchronize their activities, which allows rapid and accurate control not only of hormone release, but also of cell differentiation and survival. Moreover, the different categories of paracrine/autocrine signals are complemented by neurotransmitters and neuropeptides, which are controlled by long noncoding RNAs (lncRNA transcripts with lengths exceeding 200 nucleotides, not translated into protein) (Esguerra and Eliasson 2014) (Fig. 13.2).

Like neurons, endocrine cells synthesize, accumulate, and release neurotransmitters in the islets, and have receivers capable of decoding these signals (Di Cairano et al. 2016). Several neurotransmitters have been proposed to function as islet paracrine signals, but most of them have not yet met the criteria to be considered authentic paracrine signals, especially in human islets (the results of other species cannot be directly translated into the human context) (Caicedo 2013).

Postnatal Pancreas

During fetal and early postnatal life, rapid beta-cell proliferation results in daily doubling of their population, associated with vascular growth (Nikolova et al. 2006). In the human fetus, beta-cells already secrete insulin in response to secretagogues (Adam et al. 1969), however in rodents, this occurs only after birth (Kervran et al. 1979). At birth, the islet architecture appears similar to adult animals, while their total population continues to increase. Beta- and alpha-cells continue rapidly growing between days 4 to 10 after birth, until day 28, due to demand for insulin. Most islet cells develop in the tail and the dorsal pancreas (Dhawan et al. 2007).

Apoptosis, as well as proliferation, modulates the development of the endocrine pancreas. Apoptosis of beta-cells is rare during embryogenesis, however not after birth, with islet remodeling and changes in beta-cell maturation. The frequency of apoptotic beta-cells in adult rats is approximately 0.5%, with reduction until 6 months after birth (Scaglia et al. 1997).

Mitochondria play a role in islet protection, preventing apoptosis against oxidative stress. Also, insulin-like growth factor 2 (IGF2), a survival factor that potentiates beta-cell growth, maturation, and function, is another growth factor that prevents apoptosis of beta-cells, expressed in beta-cells at the beginning of rodent life (Reusens and Remacle 2006). Besides, neuropeptide Y (NPY) receptors play a role in the control of beta-cell mass. Activation of the NPY receptor functionally protects the islets, by restoring glucose reactivity after chemical-induced damage. Thus, NPY receptor activation might attenuate beta-cell apoptosis, preserving functional beta-cell mass, and lessening hyperglycemia phenotype in a low-dose streptozotocin diabetes model (Franklin et al. 2018).

There is a balance between the growth of new beta-cells and the loss of old beta-cells, which can be evaluated through the assessment of the total mass (Marinho et al. 2019a). There are various sources for the endogenous production of new beta-cells from existing cells. Beta-cells, long

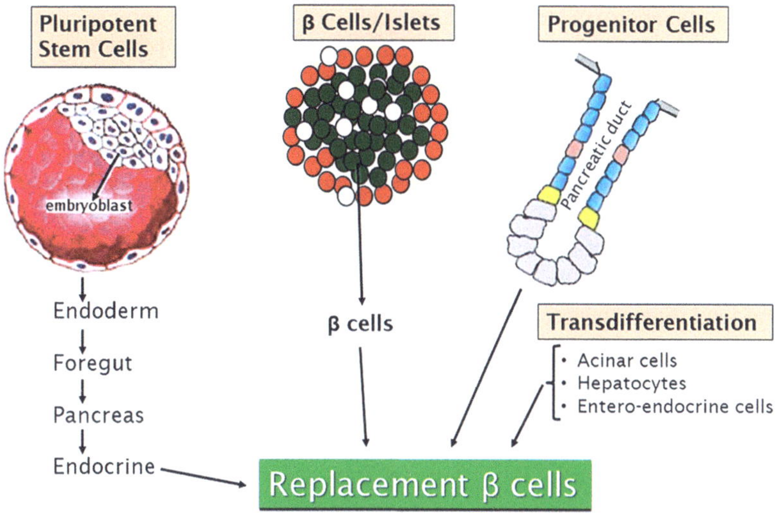

Fig. 13.3 Expansion of the beta-cell mass in adults: possible mechanisms (modified from Nichols et al. 2014)

considered postmitotic, have demonstrated the potential for regenerative capacity. Likewise, the presence of facultative endocrine progenitor cells of the pancreas has been established. Also, the malleability of cell identity has enabled the generation of beta-cells from other differentiated cell types (Nichols et al. 2014).

The expansion of the beta-cell mass in adults continues, as their ability to regenerate decreases with age, and involves three possible mechanisms: (Soggia et al. 2011) (a) the proliferation of preexisting beta-cells; (b) neogenesis from progenitor cells or undefined adult stem cells; (c) transdifferentiation from end-stage cells (especially in association with conditions such as obesity and pregnancy) (Fig. 13.3).

Role of Nonbeta Islet Cells

Alpha-cells usually produce glucagon, however when beta-cells are damaged, alpha-cells also provide glucagon-like peptide type 1 (GLP-1), a beta-cell growth and survival factor (Habener and Stanojevic 2013). In the mouse, some glucagon-producing alpha-cells and somatostatin-producing delta-cells become insulin-expressing cells, after the ablation of insulin-secreting beta-cells, thus promoting the recovery from diabetes.

Therefore, islet nonbeta-cells (alpha-cells and polypeptide-producing gamma-cells) can be reprogrammed by transcription factors, to produce and secrete insulin in response to glucose. Once transplanted into diabetic mice, converted human alpha-cells reverse diabetes and continue to produce insulin, even after 6 months (Furuyama et al. 2019). Also, transformed alpha-cells acquire the electrophysiological characteristics of beta-cells, and conduct insulin secretion stimulated by glucose. Pathways regulated by aristaless-related homeobox (Arx) protein, and by DNA methyltransferase-1 (Dnmt1) enzyme, are enough to achieve the targeted generation of beta-cells from adult pancreatic alpha-cells (Chakravarthy et al. 2017).

Beta-cell Proliferation

About 1–2 g is the beta-cell mass in the adult human pancreas, essential for glucose homeostasis (Saisho et al. 2013). In early life, beta-cells are primarily derived from precursor cells, after which the proliferation of beta-cells speeds up to generate functional beta-cell mass (Jiang et al. 2018). Increment of beta-cell mass is noted to peak within the first 2 years of life, and then to rapidly decrease in childhood (Mezza et al. 2014).

After birth, approximately 3% of beta-cells are in the active cell cycle, whereas in adults, human beta-cell proliferation is low (Mezza et al. 2014).

Transforming growth factor-beta (TGFbeta) isoforms are potent regulators of growth and differentiation. Moreover, TGFbeta receptor signaling pathway is involved in pancreatic development (Daneshmandi et al. 2017), promoting endocrine cell differentiation and maturation, and inhibiting acinar cell growth in embryogenesis. Beta-cell proliferation might occur after pancreatic duct ligation, depending on infiltrating macrophages releasing TGFbeta1 to upregulate SMAD 7 protein, an antagonist of the TGFbeta receptor signaling pathway. The inflammation-induced beta-cell proliferation is linked to the enhanced levels of Cyclin D1 and Cyclin D2 proteins, as well as nuclear exclusion of p27 protein (regulated by SMAD 7) (Xiao et al. 2014). TGFbeta1, SMAD 2, and SMAD 3 are expressed in adult islets, while SMAD 7 is not observed in insulin, glucagon, and somatostatin cells in healthy individuals; however, it is overexpressed after pancreatic injury (Xiao et al. 2016).

Figure 13.4 illustrates the canonical TGFbeta/SMAD signaling pathway. On the cell membrane, TGFbeta binds to type II receptors, recruiting type I receptors, then it phosphorylates SMAD 2 and SMAD 3. Notably, translocation from the cytoplasm to the nucleus regulates downstream gene expression, after forming a complex with SMAD 4 (Jiang et al. 2018). Besides, the TGFbeta receptor signaling pathway interferes on noncanonical pathways, involving mitogen-activated protein kinase (MAPK) pathways, and

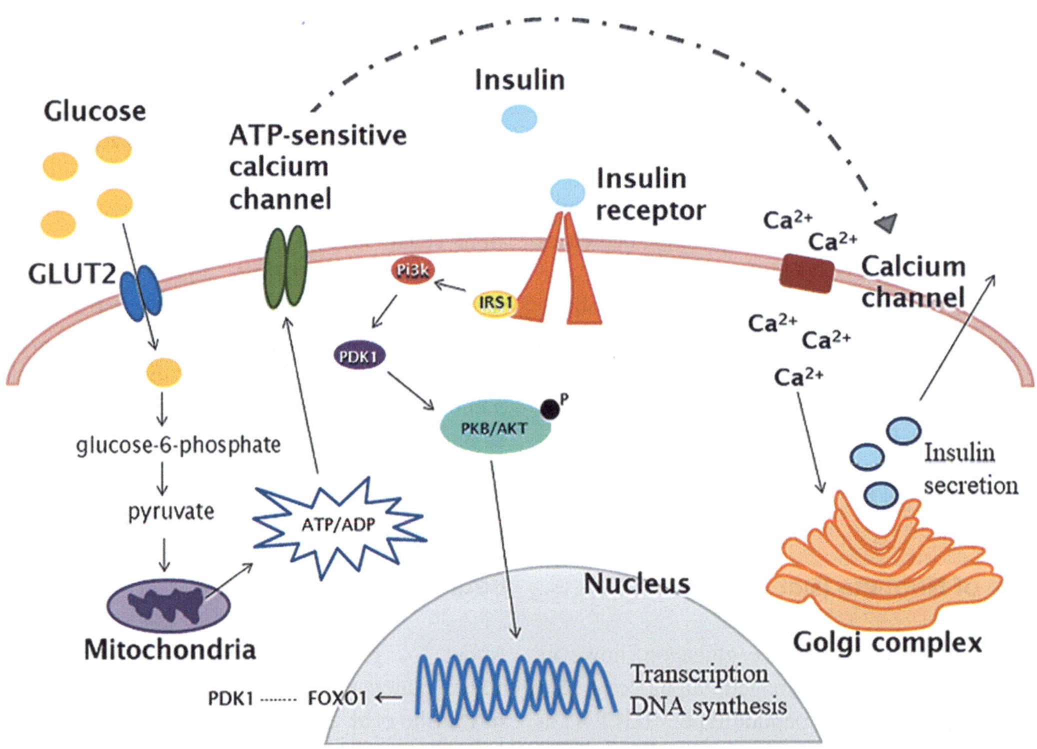

Fig. 13.4 PI3K / AKT insulin signaling pathway in pancreatic beta cells. Physiologically, in the postprandial period, insulin binds to its receptor on the alpha subunit, leading to the phosphorylation of its membrane receptor beta subunit. The phosphorylation of tyrosine residues activates IRS1, leading to AKT phosphorylation and FOXO2 transcription, which in turn triggers PDX1 transcription. In fasting periods, AKT-p promotes FOXO1 transcription, which inhibits PDX1 transcription. Abbreviations: AKT, protein kinase B; FOXO2, human Forkhead box protein O2; IRS1, insulin receptor substrate 1; PDX1, pancreatic and duodenal homeobox 1

the phosphatidylinositol 3′-kinase (PI3K)/protein kinase B (AKT) pathway (Hamidi et al. 2017).

The WNTs are part of a family of secreted signaling proteins, that perform their functions through autocrine and paracrine mechanisms. These proteins interact with receptors and trigger a cascade reaction, resulting in the accumulation of beta-catenin in the cytoplasm. The postnatal activation of beta-catenin in beta-cells, results in increased proliferation of this cell type, whereas increased expression of Axin 2 (WNT/beta-catenin pathway antagonist) has an opposite effect (Rulifson et al. 2007).

Insulin Synthesis

Preproinsulin is a high molecular weight molecule, consisting of four different domains: C-peptide, A and B chains (insulin), and a signal peptide, which is responsible for its rapid penetration into the endoplasmic reticulum (Suckale and Solimena 2010). In the rough endoplasmic reticulum, the preproinsulin undergoes cleavage of its signal peptide, giving rise to proinsulin. The proinsulin molecule consists of two chains, the alpha-carboxyterminal and the beta-aminoterminal, joined by peptide C. The alignment of the disulfide bonds that bind the two chains is the primary function of peptide C, allowing adequate molecular folding and cleavage. Microvesicles transport proinsulin to the Golgi complex, an ATP-dependent process (Vakilian et al. 2019).

Within the Golgi complex, up to the formation of secretory granules, proinsulin is converted into insulin by cleavage of the C-peptide (predominantly at the junction with the beta chain), by specific endopeptidases (proconvertases 2 and 3), and an exopeptidase (carboxypeptidase H) (Liu et al. 2014). Once separated, insulin and C peptide are packed into secretory granules. Insulin molecules, in the presence of zinc and acidic pH, aggregate and form hexamers, initiating the crystallization process. Under normal conditions, 95% of the secreted hormone is insulin, and 5% proinsulin (Fig. 13.5) (Liu et al. 2018).

Insulin Output Modulation

The beta-cell can modulate the synthesis and secretion of insulin according to metabolic demand. Increases in the extracellular concentration of glucose, neurotransmitters, nutrients, and hormones stimulate the synthesis of proinsulin; however, glucose concentration does not affect its conversion into insulin. The glucose threshold to enhance insulin synthesis is between 2 and 4 mmol/L and, to boost its secretion, 4–6 mmol/L. Among the nutrients that can stimulate insulin synthesis, ribose, and some amino acids like leucine are involved (Suckale and Solimena 2010).

Growth hormone, cortisol, glucagon, GLP-1, gastric inhibitory polypeptide (GIP), secretin, cholecystokinin, gastrin, situations such as pregnancy and obesity stimulate insulin synthesis. Somatostatin and epinephrine inhibit insulin secretion, but somatostatin does not affect proinsulin synthesis. In the case of excessive production of insulin by the beta-cell, the granules are degraded by lysosomes, through proteolytic action, a process that is known as crinophagia (Liu et al. 2014).

Signaling Pathways Related to Insulin

There are several pathways of insulin signaling: the phosphatidylinositol 3-kinase (Pi3k) pathway, with formation of the CAP/Cble complex, and the RAS pathway—Mitogenic Activated Protein Kinase (RAS-MAPK). The Pi3k path is the most important, leading to metabolic actions such as glucose transport, glycolysis, glycogen synthesis, lipid metabolism, and protein synthesis, besides cell growth and inhibition of apoptosis. Inhibition of Pi3k blocks almost all metabolic responses stimulated by insulin (Saad et al. 1992).

Pi3k is an enzyme composed of two subunits, one regulatory (p85) and one catalytic (p110). The catalytic subunit is unstable and only detected in association with the regulatory subunit. It plays an essential role in cell differentiation and mitogenesis. The regulatory subunit

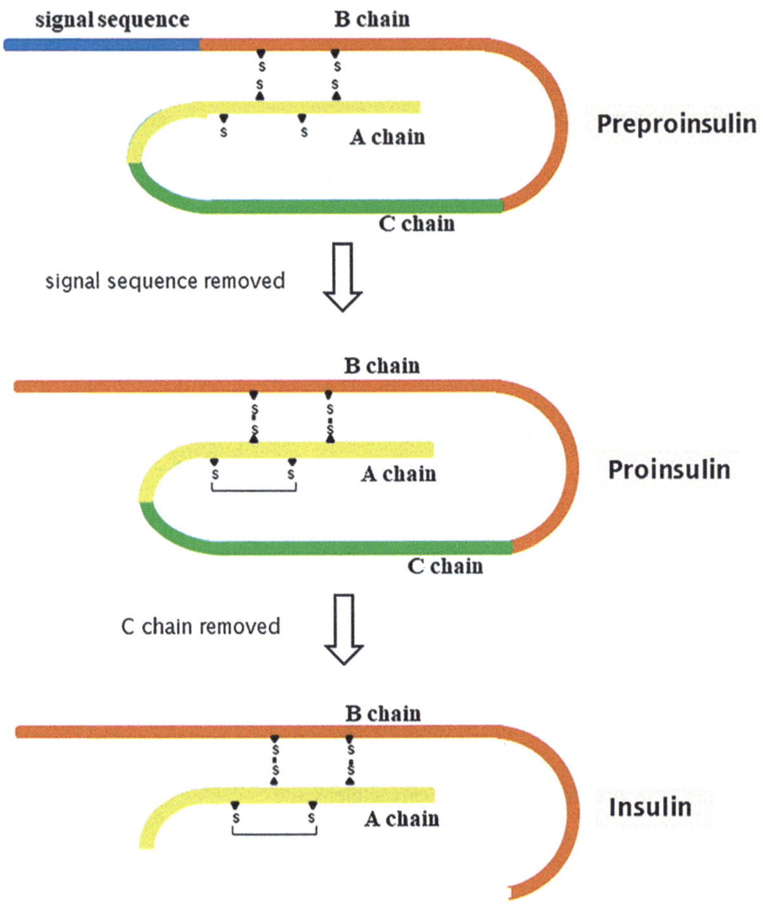

Fig. 13.5 Insulin synthesis and processing. The mature insulin hormone is produced from the processing of its preprohormone, called preproinsulin. The initial cleavage of a 23 amino acid amino-terminal amino acid, called the signal sequence, and the formation of hydrogen bridges, culminating in the production of proinsulin. The removal of C peptide (C chain) by proteolysis produces insulin with its A and B chains

contains two SH2 domains, which allow its binding to the YMXM and YXXM sites (Y = tyrosine, M = methionine, and X = any amino acid), phosphorylated by the insulin receptor substrate (IRS) proteins, activating the associated catalytic domain (Lietzke et al. 2000). The enzyme catalyzes the phosphorylation of phosphoinositides at the 3-position of the inositol ring, producing phosphatidylinositol-3-phosphate, phosphatidylinositol-3,4-diphosphate, and phosphatidylinositol-3,4,5-triphosphate (PIP3).

PIP3 recruits other serine/threonine kinases for the plasma membrane, such as phosphoinositol-dependent kinase 1 (a protein homologous to the murine thymoma v-AKT viral oncogene) (Kim and Park 2018). In pancreatic beta-cells, insulin and insulin growth factor 1 (IGF-1) activate AKT in a manner dependent on the activation of Pi3k. In the Pi3k/AKT insulin pathway, the insulin receptor substrate type 1 (IRS1), is the primary substrate of the insulin receptor in the beta-cells. When IRS1 expression decreases in insulin resistance, this substrate can be partially offset by the increase of IRS2 (Kim et al. 2007).

The binding of insulin to the alpha subunits of its membrane receptor activates the beta subunits, promoting autophosphorylation, which in turn enables the intrinsic tyrosine kinase function that catalyzes the phosphorylation of IRS1, which interacts with Pi3k, stimulating the recruitment of the significant molecule of the cascade, AKT (Brunet et al. 1999). The AKT protein, once phosphorylated, positively regulates Forkhead transcription factors (FOX), whereas FOXO2 inhibits FOXO1 (Katoh and Katoh 2004). With the decrease of AKT expression, FOXO2 drops and FOXO1 is more expressed, with consequent reduction of the pancreatic and duodenal homeobox 1 transcription factor (PDX1), controlled by the earliest identified gene in the embryonic development of the pancreas, characterizing a state of cellular dysfunction (Kitamura et al. 2002). PDX1 is a crucial transcription factor, that is expressed at several stages of pancreatic development and in the process of differentiation of beta-cells. The expression of PDX1 is linked to the preservation of pancreatic beta-cell function, and clinical studies have shown a decrease in PDX1 expression during insulin resistance (Yang et al. 2012).

Glucose Transporters

Glucose transporters (GLUT) are also critical factors in the response of insulin secretion. In physiological conditions, GLUT2, present in the cytoplasm of pancreatic beta-cells, is stimulated to translocate into the plasma membrane by the increased concentration of extracellular glucose, allowing the entry of glucose into the cell, with the consequent participation of this molecule in energy metabolism (Beamish et al. 2017). Glucose serves as a substrate in the glycolysis pathway, producing pyruvate molecules that participate in the tricarboxylic acid cycle, and the electron transport chain in the mitochondria, generating adenosine triphosphate (ATP—), the chief energy coin that leads to the closure of voltage-dependent potassium channels.

These in turn act by causing depolarization of the plasma membrane over the calcium channels, leading to the influx of this ion into the cytoplasm, which promotes the conversion of proinsulin into insulin, regulating the secretion of this hormone (Dai et al. 2012). The CAP/Cbl pathway is also required for insulin to stimulate glucose transport (Cbl is a proto-oncogene, which is associated with CAP adapter protein). The CAP-Cbl complex is phosphorylated, migrates to the cell membrane, and promotes a second signal for the translocation of GLUT4, parallel to the Pi3k pathway (Fig. 13.6) (Chiang et al. 2001).

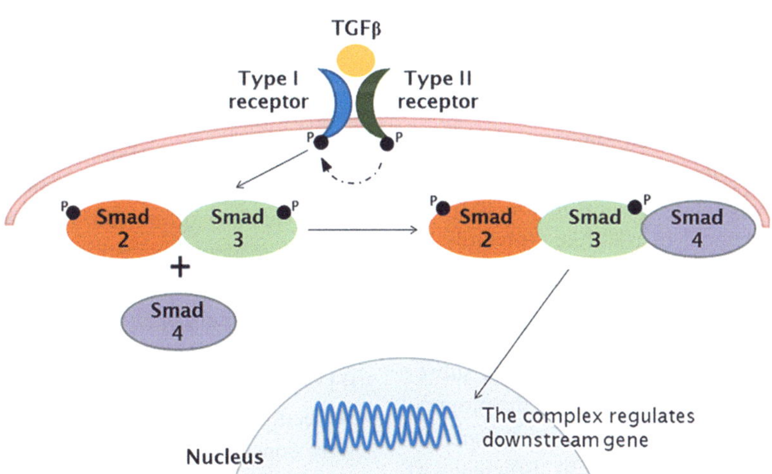

Fig. 13.6 Canonical TGFβ/SMAD signaling pathway. On the cell membrane, TGFβ binds to type II receptors, recruiting type I receptors, then phosphorylates SMAD2 and SMAD3 to form a complex with SMAD4. The complex translocates to the nucleus, regulating downstream gene expression. Abbreviations: SMAD, refer to the homologies to the *C. elegans* SMA ("small" worm phenotype) and *Drosophila* MAD ("Mothers Against Decapentaplegic") family of genes; TGF, transforming growth factor

Islet Transplantation and Stem-cell Differentiation

Insulin-secreting beta-cells and glucagon-secreting alpha-cells cooperate to regulate blood glucose levels. Destruction or dysfunction of beta-cells is known to lead to diabetes, and hence problems with insulin production. Unfortunately, no treatment yet can permanently stop the progression of diabetes and its various complications, even though relatively prolonged remission is possible with intensive lifestyle interventions, as well as bariatric/ metabolic surgery in a proportion of cases (Taylor et al. 2019).

Islet transplantation can often normalize blood glucose levels for several years, and thus prevent complications from diabetes, in insulin-dependent patients (Wojtusciszyn et al. 2019). The procedure is minimally invasive or mini-surgical, involving purified islets which are infused through the portal vein of the liver. The procedure often needs to be repeated, so that the individual no longer needs long-term insulin.

Stem Cell Sources

Pancreatic islets for transplantation are scarce. Stem cell-derived cells could be an alternative. Differentiation of stem cells into mature alpha-cells and beta-cells is one of the principal challenges, besides subsequent isolation, purification, and arrangement in islet-like structures for transplantation (Anazawa et al. 2019).

Stem cell unicellular RNA sequencing (scRNA-seq) at each stage of differentiation, helps in the identification of the types of subsequently generated cells, thus guiding the progression of pancreatic progenitor cells into specific cell lines (Schiebinger et al. 2019). Efficiency of current approaches to achieve the desired aims is not ideal because of cell heterogeneity, and insufficient knowledge of all the necessary molecular signaling factors (Theis and Lickert 2019).

Pancreatic progenitor cells can be differentiated from stem cells, and subsequently to hormone- and nonhormone-expressive cell types (Veres et al. 2019), confirming that parental cell specification is the key to produce the targeted cell types. The desired endocrine cell subtypes should be isolated, purified, and then reassembled into pseudo-islets, to provide mature stem cell-derived islet cells. A simple dissociation and reassembly procedure, can remove most nonendocrine proliferative cells from cell culture (Balboa et al. 2019). Stem cell-derived pancreatic progenitor cells have already been transplanted to patients with type 1 diabetes. However, they need to further differentiate into beta-cells and insulin, upon glucose stimulation (Clinicaltrials 2017). Alpha-cells should be transplanted as well, as they control hormone secretion, and contribute to glucose regulation (Veres et al. 2019).

Alpha-cell Plasticity

Insulin production is stimulated in nonsmoothed alpha-cells, when the combination of beta-cell loss or inhibition of insulin signaling, and gamma cell inactivation or mild inhibition of alpha-cells occurs (Chera and Herrera 2016). Animal studies have shown that after ablation of almost all beta-cells in the pancreas of adult mice, 1–2% of alpha-cells expressing glucagon, and gamma cells expressing somatostatin, spontaneously secrete insulin, leading beta-cell mass to significant regeneration and normoglycemia (Chera et al. 2014). The mechanism that fully controls insulin expression is unknown; however, studies have identified insulin signaling pathways in alpha-cells and surprisingly also in gamma cells, as regulators of alpha-cell identity and conversion into insulin-producing cells (Thorel et al. 2011).

It seems that maintaining cellular identity is an active, constant process mediated by repressive signals, that are released by neighboring cells, and prevent cells from differentiating after adulthood (Thowfeequ 2007). Half a century of research into the determination and maintenance of cell identity, has revealed that adult cells are not terminally differentiated but retain some potential for plasticity, even in higher organisms (Chera et al. 2014). The spontaneous conversion of adult cells is considered a sporadic event, which is highly regulated and activated exclusively after injury, and whose efficiency correlates with mechanisms that preserve a specific cellular identity. In contrast, knowledge of the regulation of "brake signals" responsible for maintaining the integrity of adult cells is still limited (Thowfeequ 2007).

Regenerative Medicine

The fate of a cell and its functionality are regulated by genetic and epigenetic factors, that become "blocked" due to self-regulatory feedback, or through the action of regulatory signals from the microenvironment in which the cells reside, namely neighboring cells and extracellular matrix (Chera and Herrera 2016; Chera et al. 2014). Thus, it is these factors that determine the plasticity of a cell. Changes in the identity of adult cells, mainly if triggered by stress responses, are a basis for regenerative medicine (Jessen et al. 2015). Studies with stem cells have helped to clarify what happens inside and outside the cell, for this plasticity to occur (Chakravarthy et al. 2017).

Massive beta-cell damage may lead to the release of local signals, that act as modulators of change or maintenance of alpha-cells, causing a small amount of these cells to assume the function of lost cells (Pagliuca et al. 2014). Thus, the stability of cellular identity is context-dependent, and has several levels of control over changes in phenotype (Thorel et al. 2011). Maintaining cellular status is an active process of repressive signals released from neighboring cells, blocking an intrinsic tendency of cells to modify their phenotype and functional characteristics (Holmberg and Perlmann 2012).

The Beta-cells Mass

In insulin-resistant states, pancreatic islets often respond by increasing insulin secretion to maintain normoglycemia, a process called beta-cell compensation (Zimmet et al. 2001), and it is evident from rodent studies that both the expansion of beta-cell mass, and the increased function of these cells are essential (Korc 2019). In experimental studies, it is possible to quantify beta-cell mass and monitor pancreatic islet injury or recovery (Mandarim-de-Lacerda 2019; Marinho et al. 2019a). Stimulating factors related to beta-cell mass development include increased nutrient supply (mainly glucose and fatty acids), insulin, and other growth factor signals, and increased levels of sensitivity to incretin hormones such as GLP-1 (Chen et al. 2016). Increasing the enteric nutrients, particularly in the form of fat, may also result in the expansion of beta-cell mass, by increasing GLP-1 produced from L cells in the gut (Nauck and Meier 2018).

Beta Cell Mass Increase

As mentioned (Fig. 13.3), a variety of sources can originate the endogenous production of new beta-cells from existing cells: (a) beta-cells (long thought to be postmitotic) are now considered to have potential for regeneration, (b) pancreatic facultative endocrine progenitor cells, and (c) generating beta-cells from other differentiated cell types (Nichols et al. 2014). In mice, alpha and delta cells can be lineage-traced and reprogrammed by the transcription factors PDX1 and MAFA, to produce and secrete insulin in response to glucose (Furuyama et al. 2019).

Pancreatic Islets in Obesity and Intermittent Fasting

The prevention of obesity-induced type 2 diabetes mellitus (T2D) by dietary modifications, including daily calorie restriction and intermittent fasting, has been described in human clinical trials (Tuomilehto et al. 2001). Caloric restriction regimens are still the most common dietary strategies, reducing energy intake by 20–50% of the needs (Omodei and Fontana 2011). Figure 13.7 illustrates the islet modification due to obesity and intermittent fasting.

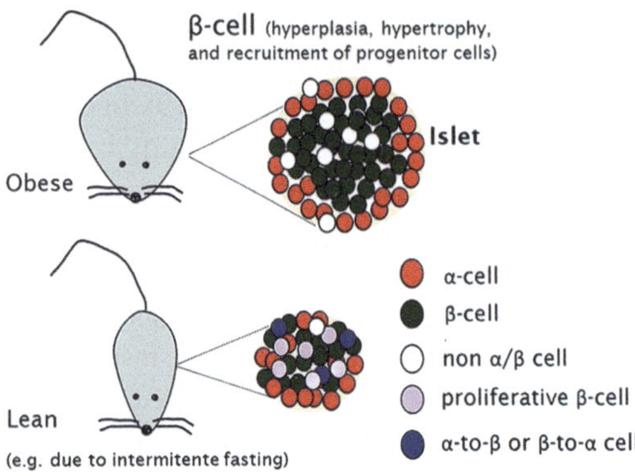

Fig. 13.7 Adaptation of the pancreatic islet and cells. Obesity represents a stress factor for the pancreatic islet, which increases in size, and the beta-cell is stimulated to produce more insulin to maintain blood glucose control. The beta-cell undergoes hyperplasia, hypertrophy, and progenitor cells are recruited. With time and established insulin resistance, there is depletion of beta-cells. Weight loss improves pancreatic islet stress, which tends to regain cell arrangement and function

For many, intermittent fasting seems to require less effort than ordinary calorie restriction diets, is cost-effective, and is a useful tool in promoting weight loss and maintenance (Das et al. 2007). Intermittent fasting has shown promise in achieving weight loss goals, and can prevent and cure rodent models of T2D (Mattson et al. 2017). Intermittent fasting differs from caloric restriction, as it only requires an individual to restrict energy 1–3 days to a week, and allows for free food consumption on the nonrestriction days (Varady 2011). When C57BL/6 mice are fed a high-fat diet *ad libitum*, they develop hyperinsulinemia, obesity, and systemic inflammation (Fraulob et al. 2010), all of which are prevented by intermittent fasting (Marinho et al. 2019b). Interestingly, the antidiabetic effect is not due to caloric restriction, because mice of the intermittent fasting group consume the same amount of food as control mice fed *ad libitum* (Mattson et al. 2017).

Fasting and Glucotoxicity

Beta-cells are highly sensitive to nutrient availability. They are endowed with a low rate of replication, and neogenesis rarely occurs in the adult pancreas. Thus, beta-cell depletion and the consequent loss of insulin secretion, represent a severe stage of diabetes (Meier et al. 2008). In islets of mice with T2D, irreversible beta-cell damage can be elicited by chronic exposure to supraphysiological glucose concentrations (glucotoxicity) (Robertson et al. 2003).

The cellular and molecular mechanisms by which intermittent fasting treats diabetes involve increased sensitivity of insulin receptor signaling. Thus, insulin more readily stimulates glucose uptake by muscle, hepatocytes, and other cell types, including neurons (Sequea et al. 2012). Different signaling pathways might be affected by intermittent fasting: (a) reductions of mTOR signaling and up-regulation of cAMP-responsive element-binding protein (Hatori et al. 2012); (b) improved mitochondrial function (Descamps et al. 2005); (c) stimulation of mitochondrial biogenesis (Cheng et al. 2016); (d) brain-derived neurotrophic factor (Yuen and Sander 2014); (e) autophagy pathways (Longo and Mattson 2014).

Cellular Effects

Another feature of diabetic beta-cell failure is the dedifferentiation of beta-cells, which results in increased nonhormone-producing cells within pancreatic islets (i.e., cells producing neither insulin nor glucagon, known as nonalpha/beta) (Talchai et al. 2012). Intermittent fasting can improve the markers of transitional alpha-to-beta or beta-to-alpha-cells that coexpress both alpha- (i.e., glucagon) and beta biomolecules (i.e., PDX-1 and insulin). Also, differentiated or committed cells can be diminished, followed by the induction of transitional cells, and significant increases in the proliferation and quantity of insulin-generating beta-cells (Cheng et al. 2017).

Neurogenin 3 (NGN3—expressed by 2–10% of acinar and duct cells in the healthy adult human pancreas), is a marker for progenitors giving rise to beta-cells, and intermittent fasting preserves or restores beta-cell function, once NGN3-driven beta-cell regeneration occurs in diabetic mice

(Wei et al. 2018). Also, autophagic flux causing cell death is enhanced, stimulating progenitor, or other cells to regenerate beta-cells in order to reverse the beta-cell loss, despite continued high-fat intake in an experimental model, thus improving glucose tolerance by enhancing glucose-stimulated insulin secretion and NGN3 expression (Liu et al. 2017).

Fasting and Refeeding Sequence

Intermittent fasting promotes FOXO-1 and its transcriptional targets in pancreatic islets, transitionally inducing the expression of progenitor cell marker NGN3+ upon refeeding, which leads to beta-cell regeneration, and changes in a wide range of cytokines associated with beta-cell recovery (Chakravarthy et al. 2017; Cigliola et al. 2018). Thus, intermittent fasting and refeeding generate the complex and highly coordinated conditions, required for the generation of stable insulin-producing beta-cells, and reversion of severe beta-cell depletion (Cheng et al. 2017). It not only suppresses the circulating cytokines associated with beta-cell damage (e.g., TNF-alpha and IL-12), but also increases those associated with regeneration (e.g., IL-2 and IL-10) (Dirice et al. 2014). Inflammation decreases, and a cytokine profile beneficial for the restoration of insulin secretion and the reversal of hyperglycemia is achieved (Cheng et al. 2017).

Conflict of interest There is no conflict of interest in this manuscript.

Funding Conselho Nacional de Desenvolvimento Científico e Tecnológico (Brazil) (CNPq, Grant No 302.920/2016-1 to CAML, and 305.865/2017-0 to MBA), Fundação Carlos Chagas Filho de Amparo à Pesquisa do Estado do Rio de Janeiro (Faperj, Grant No E-26/202.935/2017 to CAML, and E-26/202.795/2017 to MBA). These foundations had no interference in the accomplishment and submission of the manuscript.

References

Abdulreda MH, Rodriguez-Diaz R, Cabrera O, Caicedo A, Berggren PO (2016) The different faces of the pancreatic islet. Adv Exp Med Biol 938:11–24

Adam PA, Teramo K, Raiha N, Gitlin D, Schwartz R (1969) Human fetal insulin metabolism early in gestation. Response to acute elevation of the fetal glucose concentration and placental tranfer of human insulin-I-131. Diabetes 18:409–416

Adeghate E, Kalasz H (2011) Amylin analogues in the treatment of diabetes mellitus: medicinal chemistry and structural basis of its function. Open Med Chem J 5:78–81

Anazawa T, Okajima H, Masui T, Uemoto S (2019) Current state and future evolution of pancreatic islet transplantation. Ann Gastroenterol Surg 3:34–42

Baetens D, Malaisse-Lagae F, Perrelet A, Orci L (1979) Endocrine pancreas: three-dimensional reconstruction shows two types of islets of langerhans. Science 206:1323–1325

Balboa D, Saarimaki-Vire J, Otonkoski T (2019) Concise review: human pluripotent stem cells for the modeling of pancreatic beta-cell pathology. Stem Cells 37:33–41

Barach JH (1952) Paul Langerhans, 1847–1888. Diabetes 1:411–413

Beamish CA, Zhang L, Szlapinski SK, Strutt BJ, Hill DJ (2017) An increase in immature beta-cells lacking Glut2 precedes the expansion of beta-cell mass in the pregnant mouse. PLoS One 12:e0182256

Benninger RKP, Hodson DJ (2018) New understanding of beta-cell heterogeneity and in situ islet function. Diabetes 67:537–547

Borden P, Houtz J, LeachSD KR (2013) Sympathetic innervation during development is necessary for pancreatic islet architecture and functional maturation. Cell Rep 4:287–301

Bosco D, Armanet M, Morel P et al (2010) Unique arrangement of alpha- and beta-cells in human islets of Langerhans. Diabetes 59:1202–1210

Brissova M, Fowler MJ, Nicholson WE et al (2005) Assessment of human pancreatic islet architecture and composition by laser scanning confocal microscopy. J Histochem Cytochem 53:1087–1097

Brissova M, Shostak A, Fligner CL et al (2015) Human islets have fewer blood vessels than mouse islets and the density of islet vascular structures is increased in type 2 diabetes. J Histochem Cytochem 63:637–645

Brunet A, Bonni A, Zigmond MJ et al (1999) Akt promotes cell survival by phosphorylating and inhibiting a Forkhead transcription factor. Cell 96:857–868

Cabrera O, Berman DM, KenyonNS RC, Berggren PO, Caicedo A (2006) The unique cytoarchitecture of human pancreatic islets has implications for islet cell function. Proc Natl Acad Sci U S A 103:2334–2339

Caicedo A (2013) Paracrine and autocrine interactions in the human islet: more than meets the eye. Semin Cell Dev Biol 24:11–21

Chakravarthy H, Gu X, Enge M et al (2017) Converting adult pancreatic islet alpha cells into beta cells by targeting both Dnmt1 and Arx. Cell Metab 25:622–634

Chen C, Chmelova H, Cohrs CM et al (2016) Alterations in beta-cell calcium dynamics and efficacy outweigh islet mass adaptation in compensation of insulin resistance and prediabetes onset. Diabetes 65:2676–2685

Cheng A, Yang Y, Zhou Y et al (2016) Mitochondrial SIRT3 mediates adaptive responses of neurons to exercise and metabolic and excitatory challenges. Cell Metab 23:128–142
Cheng CW, Villani V, Buono R et al (2017) Fasting-mimicking diet promotes Ngn3-driven beta-cell regeneration to reverse diabetes. Cell 168:775–788. e12
Chera S, Herrera PL (2016) Regeneration of pancreatic insulin-producing cells by in situ adaptive cell conversion. Curr Opin Genet Dev 40:1–10
Chera S, Baronnier D, Ghila L et al (2014) Diabetes recovery by age-dependent conversion of pancreatic delta-cells into insulin producers. Nature 514:503–507
Chiang SH, BaumannCA KM et al (2001) Insulin-stimulated GLUT4 translocation requires the CAP-dependent activation of TC10. Nature 410:944–948
Cigliola V, Ghila L, Thorel F et al (2018) Pancreatic islet-autonomous insulin and smoothened-mediated signalling modulate identity changes of glucagon(+) alpha-cells. Nat Cell Biol 20:1267–1277
Clinicaltrials (2017). https://clinicaltrials.gov/ct2/show/NCT03163511. Accessed 5 Sept 2019.
Dai XQ, Manning Fox JE, Chikvashvili D et al (2012) The voltage-dependent potassium channel subunit Kv2.1 regulates insulin secretion from rodent and human islets independently of its electrical function. Diabetologia 55:1709–1720
Daneshmandi S, Karimi MH, Pourfathollah AA (2017) TGF-beta engineered mesenchymal stem cells (TGF-beta/MSCs) for treatment of Type 1 diabetes (T1D) mice model. Int Immunopharmacol 44:191–196
Das SK, Gilhooly CH, Golden JK et al (2007) Long-term effects of 2 energy-restricted diets differing in glycemic load on dietary adherence, body composition, and metabolism in CALERIE: a 1-y randomized controlled trial. Am J Clin Nutr 85:1023–1030
Descamps O, Riondel J, Ducros V, Roussel AM (2005) Mitochondrial production of reactive oxygen species and incidence of age-associated lymphoma in OF1 mice: effect of alternate-day fasting. Mech Ageing Dev 126:1185–1191
Dhawan S, Georgia S, Bhushan A (2007) Formation and regeneration of the endocrine pancreas. Curr Opin Cell Biol 19:634–645
Di Cairano ES, Moretti S, Marciani P et al (2016) Neurotransmitters and neuropeptides: new players in the control of islet of langerhans' cell mass and function. J Cell Physiol 231:756–767
Dirice E, Kahraman S, Jiang W et al (2014) Soluble factors secreted by T cells promote beta-cell proliferation. Diabetes 63:188–202
Esguerra JL, Eliasson L (2014) Functional implications of long non-coding RNAs in the pancreatic islets of Langerhans. Front Genet 5:209
Folli F, La Rosa S, Finzi G et al (2018) Pancreatic islet of Langerhans' cytoarchitecture and ultrastructure in normal glucose tolerance and in type 2 diabetes mellitus. Diabetes Obes Metab 20(Suppl 2):137–144
Franklin ZJ, Tsakmaki A, Fonseca Pedro P et al (2018) Islet neuropeptide Y receptors are functionally conserved and novel targets for the preservation of beta-cell mass. Diabetes Obes Metab 20:599–609
Fraulob JC, Ogg-Diamantino R, Santos CF, Aguila MB, Mandarim-de-Lacerda CA (2010) A mouse model of metabolic syndrome: insulin resistance, fatty liver and non-alcoholic fatty pancreas disease (NAFPD) in C57BL/6 mice fed a high fat diet. J Clin Biochem Nutr 46:212–223
Furuyama K, Chera S, van Gurp L et al (2019) Diabetes relief in mice by glucose-sensing insulin-secreting human alpha-cells. Nature 567:43–48
Habener JF, Stanojevic V (2013) Alpha cells come of age. Trends Endocrinol Metab 24:153–163
Hamidi A, Song J, Thakur N et al (2017) TGF-beta promotes PI3K-AKT signaling and prostate cancer cell migration through the TRAF6-mediated ubiquitylation of p85alpha. Sci Signal 10:eaal4186
Hatori M, Vollmers C, Zarrinpar A et al (2012) Time-restricted feeding without reducing caloric intake prevents metabolic diseases in mice fed a high-fat diet. Cell Metab 15:848–860
Hellman B (1959a) Actual distribution of the number and volume of the islets of Langerhans in different size classes in non-diabetic humans of varying ages. Nature 184(Suppl 19):1498–1499
Hellman B (1959b) The frequency distribution of the number and volume of the islets Langerhans in man. I. Studies on non-diabetic adults. Acta Soc Med Ups 64:432–460
Holmberg J, Perlmann T (2012) Maintaining differentiated cellular identity. Nat Rev Genet 13:429–439
Jessen KR, Mirsky R, Arthur-Farraj P (2015) The role of cell plasticity in tissue repair: adaptive cellular reprogramming. Dev Cell 34:613–620
Jiang Y, Fischbach S, Xiao X (2018) The role of the TGFbeta receptor signaling pathway in adult beta cell proliferation. Int J Mol Sci 19:3136
Katoh M, Katoh M (2004) Human FOX gene family (review). Int J Oncol 25:1495–1500
Kervran A, Randon J, Girard JR (1979) Dynamics of glucose-isnduced plasma insulin increase in the rat fetus at different stages of gestation. Effects of maternal hypothermia and fetal decapitation. Biol Neonate 35:242–248
Kim C, Park S (2018) IGF-1 protects SH-SY5Y cells against MPP(+)-induced apoptosis via PI3K/PDK-1/Akt pathway. Endocr Connect 7:443–455
Kim JJ, Kido Y, SchererPE WMF, Accili D (2007) Analysis of compensatory beta-cell response in mice with combined mutations of Insr and Irs2. Am J Physiol Endocrinol Metab 292:E1694–E1701
Kim A, Miller K, Jo J, Kilimnik G, Wojcik P, Hara M (2009) Islet architecture: a comparative study. Islets 1:129–136

Kitamura T, Nakae J, Kitamura Y et al (2002) The forkhead transcription factor Foxo1 links insulin signaling to Pdx1 regulation of pancreatic beta cell growth. J Clin Invest 110:1839–1847

Korc M (2019) Pathogenesis of pancreatic cancer-related diabetes mellitus: Quo Vadis? Pancreas 48:594–597

Langerhans P (1869) Beiträge zur mikroskopischen Anatomie der Bauchspeicheldrüse. Inaugural-dissertation. Medicinischen Facultät der Friedrich Wilhelm Universität, Berlin, Gustav Lange

Leung PS (2010) Physiology of the pancreas. Adv Exp Med Biol 690:13–27

Lietzke SE, Bose S, Cronin T et al (2000) Structural basis of 3-phosphoinositide recognition by pleckstrin homology domains. Mol Cell 6:385–394

Liu M, Wright J, Guo H, Xiong Y, Arvan P (2014) Proinsulin entry and transit through the endoplasmic reticulum in pancreatic beta cells. Vitam Horm 95:35–62

Liu H, Javaheri A, Godar RJ et al (2017) Intermittent fasting preserves beta-cell mass in obesity-induced diabetes via the autophagy-lysosome pathway. Autophagy 13:1952–1968

Liu M, Weiss MA, Arunagiri A et al (2018) Biosynthesis, structure, and folding of the insulin precursor protein. Diabetes Obes Metab 20(Suppl 2):28–50

Longo VD, Mattson MP (2014) Fasting: molecular mechanisms and clinical applications. Cell Metab 19:181–192

Mandarim-de-Lacerda CA (2019) Pancreatic islet (of Langerhans) revisited. Histol Histopathol 34:985–993

Marinho TS, Aguila MB, Mandarim-de-Lacerda CA (2019a) Pancreatic islet stereology: estimation of beta cells mass. Int J Morphol 37:1131–1134

Marinho TS, Ornellas F, Barbosa-da-Silva S, Mandarim-de-Lacerda CA, Aguila MB (2019b) Beneficial effects of intermittent fasting on steatosis and inflammation of the liver in mice fed a high-fat or a high-fructose diet. Nutrition 65:103–112

Mattson MP, Longo VD, Harvie M (2017) Impact of intermittent fasting on health and disease processes. Ageing Res Rev 39:46–58

Meier JJ, Butler AE, Saisho Y et al (2008) Beta-cell replication is the primary mechanism subserving the postnatal expansion of beta-cell mass in humans. Diabetes 57:1584–1594

Mezza T, Muscogiuri G, Sorice GP et al (2014) Insulin resistance alters islet morphology in nondiabetic humans. Diabetes 63:994–1007

Nauck MA, Meier JJ (2018) Incretin hormones: their role in health and disease. Diabetes Obes Metab 20(Suppl 1):5–21

Nichols RJ, New C, Annes JP (2014) Adult tissue sources for new beta cells. Transl Res 163:418–431

Nikolova G, Jabs N, Konstantinova I et al (2006) The vascular basement membrane: a niche for insulin gene expression and Beta cell proliferation. Dev Cell 10:397–405

Omodei D, Fontana L (2011) Calorie restriction and prevention of age-associated chronic disease. FEBS Lett 585:1537–1542

Pagliuca FW, Millman JR, Gurtler M et al (2014) Generation of functional human pancreatic beta cells in vitro. Cell 159:428–439

Penicaud L (2017) Autonomic nervous system and pancreatic islet blood flow. Biochimie 143:29–32

Reusens B, Remacle C (2006) Programming of the endocrine pancreas by the early nutritional environment. Int J Biochem Cell Biol 38:913–922

Robertson RP, Harmon J, Tran PO, Tanaka Y, Takahashi H (2003) Glucose toxicity in beta-cells: type 2 diabetes, good radicals gone bad, and the glutathione connection. Diabetes 52:581–587

Rulifson IC, KarnikSK HPW et al (2007) Wnt signaling regulates pancreatic beta cell proliferation. Proc Natl Acad Sci U S A 104:6247–6252

Saad MJ, Araki E, Miralpeix M, Rothenberg PL, White MF, Kahn CR (1992) Regulation of insulin receptor substrate-1 in liver and muscle of animal models of insulin resistance. J Clin Invest 90:1839–1849

Saisho Y, Butler AE, Manesso E, Elashoff D, Rizza RA, Butler PC (2013) Beta-cell mass and turnover in humans: effects of obesity and aging. Diabetes Care 36:111–117

Scaglia L, Cahill CJ, Finegood DT, Bonner-Weir S (1997) Apoptosis participates in the remodeling of the endocrine pancreas in the neonatal rat. Endocrinology 138:1736–1741

Schiebinger G, Shu J, Tabaka M et al (2019) Optimal-transport analysis of single-cell gene expression identifies developmental trajectories in reprogramming. Cell 176:1517

Sequea DA, Sharma N, Arias EB, Cartee GD (2012) Calorie restriction enhances insulin-stimulated glucose uptake and Akt phosphorylation in both fast-twitch and slow-twitch skeletal muscle of 24-month-old rats. J Gerontol A Biol Sci Med Sci 67:1279–1285

Soggia A, Hoarau E, Bechetoille C, Simon MT, Heinis M, Duvillie B (2011) Cell-based therapy of diabetes: what are the new sources of beta cells? Diabetes Metab 37:371–375

Suckale J, Solimena M (2010) The insulin secretory granule as a signaling hub. Trends Endocrinol Metab 21:599–609

Talchai C, Xuan S, Lin HV, Sussel L, Accili D (2012) Pancreatic beta cell dedifferentiation as a mechanism of diabetic beta cell failure. Cell 150:1223–1234

Taylor R, Al-Mrabeh A, Sattar N (2019) Understanding the mechanisms of reversal of type 2 diabetes. Lanc Diab Endocrinol 7:726–736

Theis FJ, Lickert H (2019) A map of beta-cell differentiation pathways supports cell therapies for diabetes. Nature 569:342–343

Thorel F, Damond N, Chera S et al (2011) Normal glucagon signaling and beta-cell function after near-total alpha-cell ablation in adult mice. Diabetes 60:2872–2882

Thowfeequ S, Myatt EJ, Tosh D (2007) Transdifferentiation in developmental biology, disease, and in therapy. Develop Dynam Off Pub Amer Associ Anatom 236:3208–3217

Tuomilehto J, Lindstrom J, Eriksson JG et al (2001) Prevention of type 2 diabetes mellitus by changes in lifestyle among subjects with impaired glucose tolerance. N Engl J Med 344:1343–1350

Vakilian M, Tahamtani Y, Ghaedi K (2019) A review on insulin trafficking and exocytosis. Gene 706:52–61

Varady KA (2011) Intermittent versus daily calorie restriction: which diet regimen is more effective for weight loss? Obes Rev 12:e593–e601

Veres A, Faust AL, Bushnell HL et al (2019) Charting cellular identity during human in vitro beta-cell differentiation. Nature 569:368–373

Wei S, Han R, Zhao J et al (2018) Intermittent administration of a fasting-mimicking diet intervenes in diabetes progression, restores beta cells and reconstructs gut microbiota in mice. Nutr Metab (Lond) 15:80

Wémeau J-L, Crépin G, Dubois G, Triboulet J-P, Wattel F (2018) Les grands Académiciens lillois. Bull Acad Natle Méd 202:1229–1258

Wierup N, Sundler F, Heller RS (2014) The islet ghrelin cell. J Mol Endocrinol 52:R35–R49

Wojtusciszyn A, Branchereau J, Esposito L et al (2019) Indications for islet or pancreatic transplantation: statement of the TREPID working group on behalf of the Societe francophone du diabete (SFD), Societe francaise d'endocrinologie (SFE), Societe francophone de transplantation (SFT) and Societe francaise de nephrologie—dialyse—transplantation (SFNDT). Diabetes Metab 45:224–237

Xiao X, Gaffar I, Guo P et al (2014) M2 macrophages promote beta-cell proliferation by up-regulation of SMAD7. Proc Nat Sci U S A 111:E1211–E1220

Xiao X, Fischbach S, Song Z et al (2016) Transient suppression of TGFbeta receptor signaling facilitates human islet transplantation. Endocrinology 157:1348–1356

Yang BT, Dayeh TA, Volkov PA et al (2012) Increased DNA methylation and decreased expression of PDX-1 in pancreatic islets from patients with type 2 diabetes. Mol Endocrinol 26:1203–1212

Yuen AW, Sander JW (2014) Rationale for using intermittent calorie restriction as a dietary treatment for drug resistant epilepsy. Epilepsy Behav 33:110–114

Zimmet P, Alberti KG, Shaw J (2001) Global and societal implications of the diabetes epidemic. Nature 414:782–787

14 Targeting Advanced Glycation End Products (esRAGE and sRAGE) for Obesity, Diabetes, and its Associated Complications

Chinedum Eleazu and Mahaneem Mohamed

Contents

Introduction 192

Concept of Obesity 192
Type 2 Diabetes Mellitus 192
Hypertension, Dyslipidemia, and Cardiovascular Disease 192
Atherosclerotic Cardiac Disease 193
Cancer 193

Other Comorbidities 193
Lifestyle, Pharmacotherapy, and Surgery 193

The Advanced Glycation End Products (AGEs)—Receptor for Advanced Glycation End Products (RAGE) System 193
Oxidative Damage 193
Dietary Imbalance 193
The RAGE Family 194

esRAGE and sRAGE Signaling 194
Correlation with Comorbidities 194
Obesity and Diabetes 195

Therapeutic Interventions 196

References 196

C. Eleazu
Department of Physiology, School of Medical Sciences, Universiti Sains Malaysia, Penang, Malaysia

Department of Chemistry, Biochemistry and Molecular Biology, Alex Ekwueme Federal University, Ndufu-Alike, Ikwo, Ebonyi State, Nigeria

M. Mohamed (✉)
Department of Physiology, School of Medical Sciences, Universiti Sains Malaysia, Penang, Malaysia
e-mail: mahaneem@usm.my

J. Faintuch and S. Faintuch (eds.), *Obesity and Diabetes*, https://doi.org/10.1007/978-3-030-53370-0_14

Abstract

Options currently being used for the treatment of obesity abound, however, few have successfully and sustainably reversed the disease and its comorbidities. This has led to the search for alternative approaches. Targeting of the signaling of the soluble and endogenous secretory receptors for advanced glycation end products (sRAGE and esRAGE) could be the cornerstone, in the management of obesity and its associated complications. This chapter critically reviews available evidence and interventions in this field.

Keywords

Obesity · Metabolic dysfunction · Receptor for advanced glycation end products · Advanced glycation end products · Nutrition

Abbreviations

AGEs	Advanced glycation end products
$A\beta_{1-42}$	Amyloid β1-42
esRAGE	Endogenous secretory receptor for advanced glycation end products
IKKβ	Inhibitory kappa B (IκB) kinase β
MAPKs	Mitogen-activated protein kinases
MCP-1	Monocyte chemoattractant peptide 1
NFκB	Nuclear factor kappa-light-chain-enhancer of activated B cells
NADPH-oxidase	Nicotinamide adenine dinucleotide phosphate oxidase
RAGE	Receptor for Advanced Glycation End products
Ros	Reactive oxygen species
sRAGE	Soluble receptor for advanced glycation end products
TNF- α	Tumor necrosis factor alpha
T2DM	Type 2 diabetes mellitus
4-HNE	4-hydroxy-nonenal

Introduction

Few therapies have been able to reverse obesity and its complications (Eleazu et al. 2019). Given the increasing rates of this condition (Chowdhury et al. 2016), different avenues have been pursued. Targeting of the signaling of the soluble and endogenous secretory receptors for advanced glycation end products (sRAGE and esRAGE) could be promising (Eleazu et al. 2019).

Concept of Obesity

Obesity results in hyperplasia and hypertrophy of the adipocytes (Eleazu et al. 2019), engaging endogenous and environmental mechanisms (Emami et al. 2016; Eleazu et al. 2019).

Type 2 Diabetes Mellitus

One of the closest associations for excess body fat is the risk of type 2 diabetes mellitus (T2DM), also known as "diabesity" (Leitner et al. 2017). Obesity induces increased release of free fatty acids (FFAs) and secretion of proinflammatory cytokines, such as interleukin 6 and TNF-α from visceral fat, with a high lipolytic rate, and possibly abdominal adipose depots as well, which trigger insulin resistance, desensitizing cells to insulin response following hepatic glucose production, and leading to increased rate of glucose production. On the muscle, the increased free fatty acids also inhibit pancreatic insulin secretion and derange glucose uptake (Boden, 2008). A BMI as low as 22 kg/m^2 can induce diabetes (Someya et al. 2019).

Hypertension, Dyslipidemia, and Cardiovascular Disease

A BMI above 26 kg/m^2 can be followed by 2–3 times more risk for hypertension. An elevation in stroke volume, along with hypertrophy of the left ventricle, can aggravate cardiovascular morbidity and mortality. With systemic vascular resistance unchanged or elevated, blood pressure rises. Insulin resistance can also enhance sympathetic tone, andrenal sodium, and water reabsorption. Genetic risk is also worth mentioning (Dlamini et al. 2019).

Atherosclerotic Cardiac Disease

Visceral adipose tissue is linked with increased plasma triacylglycerol (TAG), decreased HDL cholesterol, increased production of VLDL cholesterol, impaired clearance of TAG rich lipoproteins, and higher lipolysis in the adipose tissue. Lipolysis can also precipitate fatty liver (Sarwar et al. 2018).

Insulin resistance favors expression of proinflammatory adipokines including interleukin-6, IL-1β, and tumor necrosis factor-α, further worsening insulin resistance, promoting increased flux of FFAs into the liver, mitochondrial dysfunction, reactive oxygen species production, hepatic inflammation, and fibrogenesis (Lambertucci et al. 2018; Divella et al. 2019).

Cancer

At least one cancer in five is more frequent among the obese (Wolin et al. 2010; Roberts et al. 2010), perhaps as a consequence of insulin resistance and systemic inflammation, along with fatty liver disease (Divella et al. 2019). Disturbed sex steroid signaling, along with insulin/insulin-like growth factors are also related to cell proliferation (Roberts et al. 2010).

Other Comorbidities

Kidney disease (Navarro et al. 2016) and premature death, primarily from cardiovascular diseases are currently well recognized in obesity. Overweight carries a milder danger (Kyrouu et al. 2018). BMI is obviously a major determinant of metabolic syndrome (He et al. 2014; Miranda et al. 2017; Miranda et al. 2018), whereas visceral fat seems more relevant for cardiovascular risk than BMI (Ferrier 2014; Eleazu et al. 2019).

Lifestyle, Pharmacotherapy, and Surgery

Lifestyle and pharmacotherapy are standard approaches for milder cases (Ferrier 2014; Eleazu et al. 2019, Apovian et al. 2015), contrasting with bariatric surgery, reserved for class II or III obesity. (Ferrier 2014; Eleazu et al. 2019).

The Advanced Glycation End Products (AGEs)—Receptor for Advanced Glycation End Products (RAGE) System

AGEs are toxic compounds that are formed through nonenzymatic glycoxidation reactions, induced by the nucleophilic addition of free amino groups from proteins, lipids or nucleic acids to the carbonyl groups of monosaccharides (Yamagishi et al. 2015; Nowotny et al. 2015; Eleazu et al. 2019). The Maillard reaction leads to the generation of reversible Schiff base adducts that spontaneously rearrange to form more stable, covalently bound Amadori products, which progress to AGEs (Nowotny et al. 2015; Kehm et al. 2019; Eleazu et al. 2019). AGEs can be derived from sugars, lipids, and amino acids (Uribarri et al. 2015) (Fig. 14.1).

Cross-linked products between AGEs in basement membrane of extracellular matrix (ECM) and AGE-RAGE binding can damage collagen, vitronectin, and laminin, potentially inducing vascular stiffening and myocardial dysfunction (Cheng et al. 2017).

Oxidative Damage

Nuclear factor kappa-beta (NFκB) and nicotinamide adenine dinucleotide phosphate oxidase (NADPH-oxidase) can be part of the negative response, with elevated oxidative stress (Luevano-Contreras and Chapman-Novakofski 2010; Vlassara and Uribarri 2014). Further inflammation, apoptosis and metabolic dysfunction can be the result (Cheng et al. 2017; Yang et al. 2018) (Fig. 14.2).

Dietary Imbalance

Diet supplies not only substrates, but also AGEs themselves (carbohydrates, lipids, and proteins). About 10% of exogenously derived AGEs are

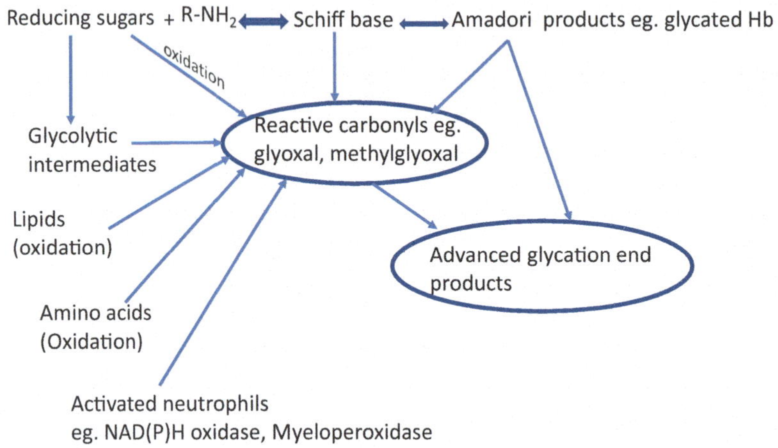

Fig. 14.1 Formation of advanced glycation end products (AGEs) from different pathways. AGEs are formed through the initial reaction between reducing sugars and the free amino group of a protein going through the stages of Schiff base and Amadori product formation. Other pathways through lipid oxidation or activated neutrophils may also lead to the formation of AGEs, even in the absence of glucose. Sources: Mostafa et al. (2007), Uribarri et al. (2015), Eleazu et al. (2019)

absorbed, and two-thirds are retained (Yamagishi et al. 2015). Lipoxidation and glycation are features of excessive body weight, including malondialdehyde (MDA), acrolein, 4-hydroxy-nonenal (4-HNE), and glycation products (glyoxal and methyl glyoxal) (Aldini et al. 2007). Insulin resistance and diabetes are negatively impacted (Eleazu et al. 2019; Vasudevan et al. 2011).

The RAGE Family

RAGE is a cell-surface immunoglobulin, including soluble molecules, full spelling RAGE (F-RAGE), (Chiavaroli et al. 2012) and soluble RAGE (sRAGE) (D'Adamo et al. 2011). RAGE, is considered as a pattern-recognition receptor (PRR), linking with AGEs, amyloid β-peptide, DNA-binding protein high mobility group box-1 (HMGBl)/amphoterin), and S100/calgranulins (Eleazu et al. 2019; Chuah et al. 2013).

RAGE occurs in kidney, neurons, skin, and lungs (Eleazu et al. 2019; Chuah et al. 2013), promoting transduction of kinases, and inducing oxidative stress via NFκB and NADPH-oxidase (Kehm et al. 2019). Increased ROS also increases AGEs formation, which further activates NFkB, and cytokines, which could form a vicious cycle for AGEs (Nowotny et al. 2015; Chiavaroli et al. 2012; Giannini et al. 2012; Pierine et al. 2014; Miranda et al. 2018; Eleazu et al. 2019) (Fig. 14.2).

esRAGE and sRAGE Signaling

Soluble receptors for AGEs attract extracellular RAGE ligands (Yang et al. 2018), inhibiting RAGE formation via "decoy receptors". This phenomenon along with RAGE antagonists or antibodies could antagonize deleterious clinical developments (Eleazu et al. 2019; Miranda et al. 2018). Less AGE binding to full length receptor (F-RAGE) (Lue et al. 2009) counteracts RAGE formation (Younessi and Yoonessi 2011; Eleazu et al. 2019).

Correlation with Comorbidities

In the experience of Falcone et al. (2005) reduced sRAGE could signal cardiovascular disease. According to D'Adamo et al. (2011), low esRAGE and sRAGE were markers of fatty liver in prepubertal obesity. Low esRAGE coexists

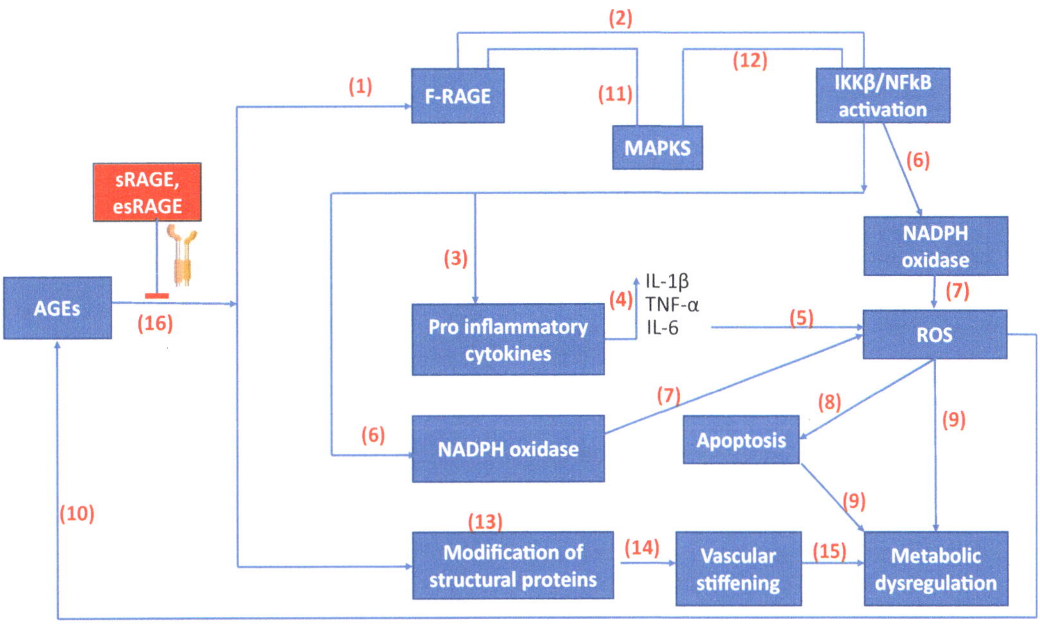

Fig. 14.2 Mechanism of action of advanced glycation end products (AGEs) and interaction between AGEs and soluble receptor AGE (sRAGE). F-RAGE- Full Length RAGE; sRAGE-soluble receptor for advanced glycation end products, esRAGE-endogenous secretory receptor for advanced glycation end products; MAPKS-Mitogen-activated protein kinases. (1) Binding of AGEs to full spelling (F-RAGE), (2) triggers the activation of IKKβ/NFkB pathway. (3) Activation of NFkB further transcribes and activates other (4) inflammatory cytokines including interleukin-1 beta, interleukin-6, tumor necrosis factor, (5) leading to the formation of ROS or (6) it activates NADPH oxidase, (7) generating reactive oxygen species in the process. (8) Increased ROS formation could affect the integrity of the membrane, triggering apoptosis, (9) leading to metabolic dysregulation. (10) Increased ROS also increases AGEs formation through a feed forward mechanism, which further activates NFkB. This activation increases the expression of proinflammatory cytokines such as interleukin-6, tumor necrosis factor alpha (TNF- α), and in addition to RAGE itself, thereby intensifying the inflammatory response and leading to metabolic dysregulation. (11) Activation of F-RAGE can also promote the activation of phosphatidyl inosito-3-kinase (PI3K)/protein kinase B (Akt) and the mitogen-activated protein kinase (MAPKs), (12) generating inflammatory mediators, ROS, apoptotis, and metabolic dysregulation. (13) AGEs formation could also promote the modification of structural proteins like collagen, vitronectin, and laminin, weakening the integrity of vascular structure and leading to vascular stiffening (14), myocardial dysfunction, and (15) metabolic dysregulation. (16) Endogenous secretory receptor for AGEs (esRAGE) and soluble receptor for AGEs (sRAGE) bind to AGEs, removing them from circulation and making them unavailable for binding to full length RAGE to trigger AGE-RAGE proinflammatory signaling and pathological effects. By this, they downregulate the expression of RAGE and attenuate AGEs/RAGE-mediated metabolic dysfunction (Nowotny et al. 2015)

with depressed adiponectin and higher thromboxane-dependent platelet activation, which can lead to elevated cardiovascular and metabolic risk (Vazzana et al. 2012).

sRAGE concentrations point toward metabolic syndrome components in male, however, not in female subjects (He et al. 2014; Eleazu et al. 2019). Also Guclu et al. (2016) found a negative relationship between sRAGE level and metabolic syndrome.

Obesity and Diabetes

Diminished sRAGE and T2DM are a recognized pattern (Miranda et al. 2017), analogously to excessive body weight (Dozio et al. 2017; Miranda et al. 2017; Miranda et al. 2018). Total sRAGE is endowed with similar properties (Dozio et al. 2017; Miranda et al. 2017; Miranda et al. 2018). During weight loss esRAGE values recover (Miranda et al. 2018).

Therapeutic Interventions

RAGE and sRAGE agonists have been used in Alzheimer's disease combined with diabetes/insulin resistance. Azeliragon (vTv Therapeutics, High Point, NC, USA) crosses the blood-brain barrier (Bongarzone et al. 2017), antagonizing the linking of sRAGE to amyloid β1-42 ($A\beta_{1-42}$). An ongoing trial addresses Alzheimer's disease and high HbA1c.

Many other potential applications have been envisaged in cell and animal models encompassing diabetic vascular dysfunction, cardiovascular disease, some modalities of cancer, rheumatoid arthritis, inflammatory bowel disease, endotoxemia, and shock (Hudson and Lippman 2018); however, clinical studies are lacking so far.

References

Aldini G, Dalle-Donne I, Facino RM, Milzani A, Carini M (2007) Intervention strategies to inhibit protein carbonylation by lipoxidation-derived reactive carbonyls. Med Res Rev 27:817–868

Apovian CM, Aronne LJ, Bessesen DH, McDonnell ME, Murad MH, Pagotto U, Ryan DH, Still CD (2015) Pharmacological management of obesity: an endocrine society clinical practice guideline. J Clin Endocrinol Metab 100:342–362

Boden G (2008) Obesity and free fatty acids (FFA). Endocrinol Metab Clin North Am 37(3):635–ix. https://doi.org/10.1016/j.ecl.2008.06.007

Bongarzone S, Savickas V, Luzi F, Gee AD (2017) Targeting the receptor for advanced glycation endproducts (RAGE): a medicinal chemistry perspective. J Med Chem 60:7213–7232

Cheng HS, Ton SH, Kadir KA (2017) Therapeutic agents targeting at AGE-RAGE axis for the treatment of diabetes and cardiovascular disease: a review of clinical evidence. Clin Diab Res 1:16–34

Chiavaroli V, D'adamo E, Giannini C, De-Giorgis T, De Marco S, Chiarelli F, Mohn A (2012) Serum levels of receptors for advanced glycation end products in normal-weight and obese children born small and large for gestational age. Diabetes Care 35:1361–1363

Chowdhury SS, Lecomte V, Erlich JH, Maloney CA, Morris MJ (2016) Paternal high- fat diet in rats leads to renal accumulation of lipid and tubular changes in adult offspring. Nutrients 8:521

Chuah YK, Basir R, Talib H, Tie TH, Nordin N (2013) Receptor for advanced glycation end products and its involvement in inflammatory diseases. Int J Inflamm 2013:403460

D'Adamo E, Giannini C, Chiavaroli V, De-Giorgis T, Verrotti A, Chiarelli F, Mohn A (2011) What is the significance of soluble and endogenous secretory receptor for advanced glycation end products in liver steatosis in obese pre-pubertal children? Antioxid Redox Signal 14:1167–1172

Divella R, Mazzocca A, Daniele A, Sabba C, Paradiso A (2019) Obesity, nonalcoholic fatty liver disease and adipocytokines network in promotion of cancer. Int J Biol Sci 15:610–616

Dlamini Z, Hull R, Makhafola TJ, Mbele M (2019) Regulation of alternative splicing in obesity-induced hypertension. Diabetes Metab Syndr Obes 12:1597–1615

Dozio E, Briganti S, Delnevo A, Vianello E, Ermetici F, Secchi F, Sardanelli F, Morricone L, Malavazos AE, Corsi Romanelli MM (2017) Relationship between soluble receptor for advanced glycation end products (sRAGE), body composition and fat distribution in healthy women. Eur J Nutr 56:2557e64

Eleazu C, Omar N, Lim OZ, Yeoh BS, Hussain NHN, Mohamed M (2019) Obesity and comorbidity: could simultaneous targeting of esRAGE and sRAGE be the panacea? Front Physiol 10:787

Emami SR, Jafari M, Haghshenas R, Ravasi A (2016) Impact of eight weeks endurance training on biochemical parameters and obesity-induced oxidative stress in high fat diet-fed rats. J Exerc Nutr Biochem 20(1):029–035

Falcone C, Emanuele E, D'Angelo A, Buzzi MP, Belvito C, Cuccia M, Geroldi D (2005) Plasma levels of soluble receptor for advanced glycation end products and coronary artery disease in nondiabetic men. Arterioscler Thromb Vasc Biol 25:1032–1037

Ferrier DR (2014) Lippincott's illustrated reviews: biochemistry, 6th edn. Lippincott Williams and Wilkins, Philadelphia, PA

Giannini C, D'Adamo E, de Giorgis T, Chiavaroli V, Verrotti A, Chiarelli F, Mohn A (2012) The possible role of esRAGE and sRAGE in the natural history of diabetic nephropathy in childhood. Pediatr Nephrol 27:269–275

Guclu M, Ali A, Eroglu DU, Büyükuysal SO, Cander S, Ocak N (2016) Serum levels of sRAGE are associated with body measurements, but not glycemic parameters in patients with prediabetes. Metab Syndr Relat Disord 14:33–39

He CT, Lee CH, Hsieh CH, Hsiao FC, Kuo P, Chu NF, Hung YJ (2014) Soluble form of receptor for advanced glycation end products is associated with obesity and metabolic syndrome in adolescents. Int J Endocrinol 2014:7

Hudson BI, Lippman ME (2018) Targeting RAGE signaling in inflammatory disease. Annu Rev Med 69:349–364

Kehm R, Rückriemen J, Weber D, Deubel S, Grune T, Hohn A (2019) Endogenous advanced glycation end products in pancreatic islets after short-term carbohydrate intervention in obese, diabetes-prone mice. Nutr Diab 9:9

Kyrouu I, Randeva HS, Tsigos C, Kaltsas G, Weickert MO (2018) Clinical problems caused by obesity. Endotext—NCBI Bookshelf. MDText.com, Inc, South Dartmouth, MA, p 16

Lambertucci F, Arboatti A, Sedlmeier MG, Motiño O, Alvarez MDL, Ceballos MP, Villar SR, Roggero E, Monti JA, Pisani G, Quiroga AD, Martín-Sanz P, Carnovale CE, Francés DE, Ronco MT (2018) Disruption of tumor necrosis factor alpha receptor 1 signaling accelerates NAFLD progression in mice upon a high-fat diet. J Nutr Biochem 58:17–27. https://doi.org/10.1016/j.jnutbio.2018.04.013

Leitner DR, Frühbeck G, Yumuk V, Schindler K, Micic D, Woodward E, Toplak H (2017) Obesity and type 2 diabetes: two diseases with a need for combined treatment strategies—EASO can lead the way. Obes Facts 10:483–492

Lue LF, Walker DG, Jacobson S, Sabbagh M (2009) Receptor for advanced glycation end products: its role in Alzheimer's disease and other neurological diseases. Future Neurol 4:167–177

Luevano-Contreras C, Chapman-Novakofski K (2010) Dietary advanced glycation end products and aging. Nutrients 2:1247–1265

Miranda ER, Somal VS, Mey JT, Blackburn BK, Wang E, Farabi S, Karstoft K, Fealy CE, Kashyap S, Kirwan JP, Quinn L, Solomon TPJ, Haus JM (2017) Circulating soluble RAGE isoforms are attenuated in obese, impaired glucose tolerant individuals and are associated with the development of type 2 diabetes. Am J Physiol Endocrinol Metab 313:E631e40

Miranda ER, Fuller KNZ, Perkins RK, Kroeger CM, Trepanowski JF, Varady KA, Haus JM (2018) Endogenous secretory RAGE increases with improvements in body composition and is associated with markers of adipocyte health. Nutr Metab Cardiovasc Dis 28:1155–1165

Mostafa AA, Randell EW, Vasdev SC, Gill VD, Han Y, Gadag V, Raouf AA, El Said H (2007) Plasma protein advanced glycation end products, carboxymethyl cysteine, and carboxyethyl cysteine, are elevated and related to nephropathy in patients with diabetes. Mol Cell Biochem 302(1–2):35-42. https://doi.org/10.1007/s11010-007-9422-9

Navarro ME, Santos KC, Nascimento AF, Francisqueti FV, Minatel IO, Pierine DT, Luvizotto RA, Ferreira AL, Campos DH, Corrêa CR (2016) Renal inflammatory and oxidative and metabolic changes after 6 weeks of cafeteria diet in rats. J Bras Nefrol 38:9–14

Nowotny K, Jung T, Höhn A, Weber D, Grune T (2015) Advanced glycation end products and oxidative stress in type 2 diabetes mellitus. Biomol Ther 5 (1):194–222

Pierine DT, Navarro MEL, Minatel IO, Luvizotto RAM, Nascimento AF, Ferreira ALA, Yeum KJ, Correa CR (2014) Lycopene supplementation reduces TNF-α via RAGE in the kidney of obese rats. Nutr and Diab 4: e142

Roberts DL, Dive C, Renehan AG (2010) Biological mechanisms linking obesity and cancer risk: new perspectives. Annu Rev Med 61:301–316

Sarwar R, Pierce N, Koppe S (2018) Obesity and nonalcoholic fatty liver disease: current perspectives. Diabetes Metab Syndr Obes 11:533–542

Someya Y, Tamura Y, Kohmura Y, Aoki K, Kawai S, Daida H, Naito H (2019) A body mass index over 22 kg/m^2 at college age is a risk factor for future diabetes in Japanese men. PLoS One 14(1):e0211067

Uribarri J, del Castillo MD, de la Maza MP, Filip R, Gugliucci A, Luevano-Contreras C, Macías-Cervantes MH, Markowicz Bastos DH, Medrano A, Menini T, Portero-Otin M, Rojas A, Sampaio GR, Wrobel K, Wrobel K, Garay-Sevilla ME (2015) Dietary advanced glycation end products and their role in health and disease. Adv Nutr 6:461–473

Vasudevan DM, Sreekumari S, Vaidyanathan K (2011) Textbook of biochemistry for medical students, 7th edn. Jaypee Brothers Medical Publishers Pvt. Ltd, New Delhi, p 123

Vazzana N, Guagnano MT, Cuccurullo C, Ferrante E, Lattanzio S, Liani R, Romano M, Davi G (2012) Endogenous secretory RAGE in obese women: association with platelet activation and oxidative stress. J Clin Endocrinol Metab 97:E1726–E1730

Vlassara H, Uribarri J (2014) Advanced glycation end products (AGE) and diabetes: cause, effect, or both? Curr Diab Rep 14:453

Wolin KY, Carson K, Colditz GA (2010) Obesity and cancer. Oncologist 15:556–565

Yamagishi S, Fukami K, Matsui T (2015) Crosstalk between advanced glycation end products (AGEs)-receptor RAGE axis and dipeptidyl peptidase-4-incretin system in diabetic vascular complications. Cardiovasc Diabetol 14:1–12

Yang G, Huang Y, Wu X, Lin X, Xu J, Chen X, Bai X, Li Q (2018) Endogenous secretory receptor for advanced glycation end products protects endothelial cells from AGEs induced apoptosis. Biomed Res Int 2018:8

Younessi P, Yoonessi A (2011) Advanced glycation end products and their receptor mediated roles: inflammation and oxidative stress. Iran J Med Sci 36:154–166

Anti-incretin Effect: A Missing Link between Obesity, Diabetes, and Metabolic Surgery

15

Theocharis Koufakis, Spyridon N. Karras, and Kalliopi Kotsa

Contents

Introduction 200

The Anti-incretin Theory 200
Early Response to Dysglycemia after Bariatric Surgery 200
Post-bariatric Hypoglycemia 201

Potential Pathways and Molecules 201
Dopamine 201
Ghrelin 202
Other Molecules 202

Correlation with Diabetes and Obesity Remission Following Metabolic Surgery .. 202
Enhancement of Pancreatic Beta-cell Mass 202

Future Perspectives 203

References 203

Abstract

The exact changes in physiology of individuals undergoing metabolic surgery are still under investigation. According to the foregut hypothesis, surgical bypass of certain parts of the gastrointestinal (GI) tract leads to the downregulation of biological factors that are secreted by the upper GI system, and compete against incretin actions (anti-incretins). Those molecules have a negative effect on insulin secretion, action, and sensitivity. Various biological factors have been hypothesized to exert anti-incretin properties, including dopamine, ghrelin, enterostatin, and oxyntomodulin. Disruption of the balance between incretins and anti-incretins could be considered as a pathway leading to the development of obesity and insulin resistance.

Keywords

Incretins · Anti-incretins · Type 2 diabetes · Obesity · Insulin resistance · Metabolic surgery

T. Koufakis · S. N. Karras · K. Kotsa (✉)
Division of Endocrinology and Metabolism and Diabetes Center, First Department of Internal Medicine, Medical School, Aristotle University of Thessaloniki, AHEPA University Hospital, Thessaloniki, Greece
e-mail: kalmanthou@yahoo.gr

J. Faintuch and S. Faintuch (eds.), *Obesity and Diabetes*, https://doi.org/10.1007/978-3-030-53370-0_15

Introduction

Incretins are gut hormones which play a key role in the preservation of glucose homeostasis in humans. Gastric inhibitory polypeptide (GIP) and glucagon-like peptide-1 (GLP-1) amplify the glucose-dependent secretion of insulin by the pancreatic beta-cells, presenting at the same time extrapancreatic actions. GLP-1 is known to inhibit postprandial glucagon secretion and delay gastric emptying, duodenal, and small intestine motility, thus resulting in delayed sugar absorption (Seino et al. 2010; Karras et al. 2012). Incretins mediate the well-described incretin effect, according to which orally digested glucose triggers a stronger insulin response than that resulting from the intravenous administration of the same amount of glucose (Rehfeld and Stadil 1973). The incretin effect has been recognized to be reduced in people with type 2 diabetes (T2D) compared with healthy controls, although it is debated whether this is a cause or a consequence of the diabetic state (Knop et al. 2007).

Changes in incretin biodynamics have been recognized as part of the mechanisms explaining the positive effects of metabolic surgery on obesity and T2D. Several studies have established increased concentrations of both GLP-1 and GIP after bariatric surgery procedures (Bose et al. 2010; Honka et al. 2018), related to improved insulin secretion and lower insulin resistance (IR). However, the exact changes in the physiology of individuals undergoing metabolic surgery are still under investigation and intensive research is currently being conducted to elucidate the full spectrum of the implicated mechanisms (Karras et al. 2019).

According to the foregut hypothesis (Rubino and Gagner 2002), surgical bypass of certain parts of the gastrointestinal (GI) tract leads to the downregulation of biological factors that are secreted by the upper GI system, and compete with incretins actions (anti-incretins). Those molecules have a negative effect on insulin secretion, action, and sensitivity (Rubino and Amiel 2014).

The Anti-incretin Theory

The anti-incretin theory suggests that the presence of food in the GI tract, apart from activating the cascade of incretin-related biological actions, also stimulates a negative feedback adaptation, which is the release of anti-incretins (Kamvissi et al. 2010). Anti-incretins aim to balance the glucose-lowering effects of incretins, in order to prevent excessive insulin release, potentially leading to hypoglycemia (Rubino and Gagner 2002; Kamvissi et al. 2010; Rubino and Amiel 2014).

Incretins also stimulate beta-cell proliferation (Lavine and Attie 2010; Lindqvist et al. 2014). A similar phenomenon occurs with modern GLP-1 analogues such as liraglutide, which inhibits beta-cell apoptosis through stimulation of PI3-kinase-dependent AKT

phosphorylation (Kapodistria et al. 2018). Albeit protective (Lavine and Attie 2010), these responses might be potentially harmful in case of unlimited proliferation, due to the absence of appropriate regulatory mechanisms. Animal studies have showed that Roux-en-Y gastric bypass (RYGB) results in increased beta-cell mass postoperatively (Lindqvist et al. 2014), suggesting that RYGB abolishes gut-derived signals that prevent excessive growth of beta-cells, in face of elevated incretins (Rubino and Gagner 2002; Holst 2007; Rubino and Amiel 2014).

Early Response to Dysglycemia after Bariatric Surgery

Up to 89% of bariatric patients experience early T2D remission (Wickremesekera et al. 2005). As this appears during the first postoperative days, weight loss cannot play a significant role. Surgical exclusion of specific parts of the GI tract from nutrient transit, could abolish anti-incretin signals, thus resulting in improved insulin action and sensitivity (Rubino and Gagner 2002; Kamvissi et al. 2010; Rubino and Amiel 2014).

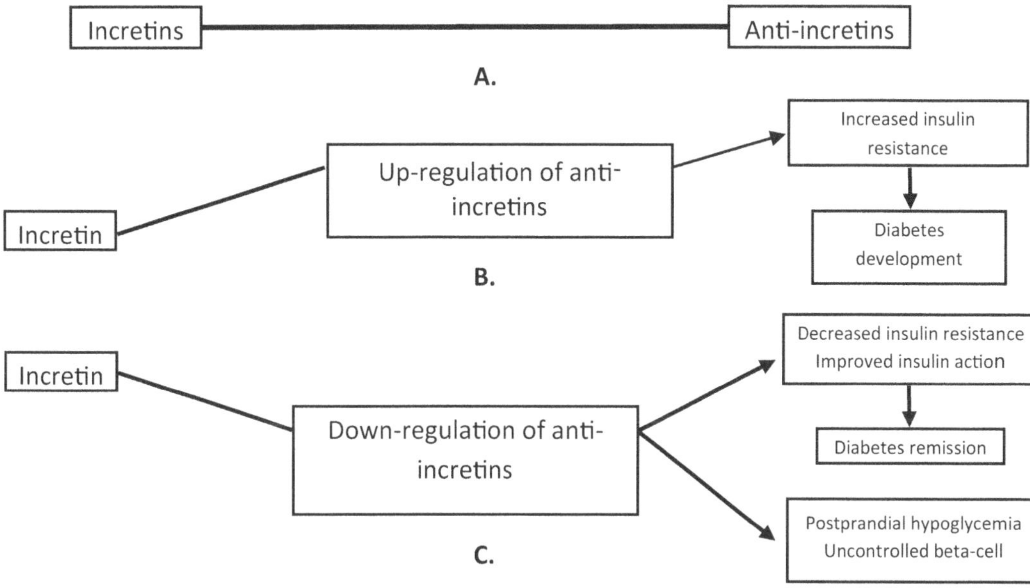

Fig. 15.1 A schematic overview of the balance between incretins and anti-incretins. (**A**) The normal state, in which the balance between the two mechanisms results in the preservation of normal glucose homeostasis, (**B**) Type 2 Diabetes, where an exaggeration of the anti-incretin effect could contribute to the development of insulin resistance and hyperglycemia, (**C**) Following metabolic surgery, where down-regulation of anti-incretins could lead to diabetes remission, and rarely to postprandial hypoglycemia and unrestricted beta-cell growth

Post-bariatric Hypoglycemia

On the other hand, postprandial hypoglycemia possibly linked to unrestrained beta-cell proliferation, occasionally occurs after RYGB (Cummings 2005). It is also postulated that Western-pattern diets or food additives might promote the overexpression of anti-incretins and through this pathway, to deteriorate beta-cell function and sensitivity (Rubino and Gagner 2002; Kamvissi et al. 2010). Figure 15.1 summarizes current knowledge about these shifts.

Potential Pathways and Molecules

Any anti-incretin should gather a number of features, including: (a) downregulation of insulin production and release from the pancreatic beta-cells, (b) production by an enteroendocrine source as a result of exposure to nutrients, and c. response to fasting or carbohydrate restraint (Alfa et al. 2015).

Dopamine

Dopamine exerts regulatory effects on glucose metabolism, given that it has the ability to downregulate incretin-mediated enhanced glucose-dependent phosphoprotein signaling procedures, as well as to inhibit *in vitro* GLP-1-promoted islet and beta-cell proliferation (Maffei et al. 2015). Tyrosine hydroxylase, the rate-limiting enzyme of dopamine synthesis, is highly expressed in parietal and ileal epithelial cells, as well as Lieberkuhn crypts (Kozicz and Arimura 2002). Of note, D2-like receptors which are believed to be the key mediators of dopamine-related suppression of insulin secretion, are expressed in beta-cells (Rubi et al. 2005). As shown in a study conducted with a rodent beta-cell line (INS-1 832/12), insulin secretion was amplified following the knock-out of the aforementioned receptors (Wu et al. 2008), further supporting the implication of dopamine in an autocrine circuit of neurotransmitter regulation, which acts as a limiting factor of glucose-stimulated insulin secretion (Chaudhry et al. 2016). There is an experimental evidence that diet-derived tyrosine, a precursor of L-DOPA and dopamine, participates in postprandial glucose homeostasis. This circuit could be involved in protection against hypoglycemia via inhibition of glucose-stimulated β-cell insulin secretion, a response consistent with the anti-incretin hypothesis (Korner et al. 2019).

Ghrelin

Ghrelin has been also postulated to play an anti-incretin role, albeit relevant data are more limited here. Ghrelin has been demonstrated to prevent GLP-1-mediated stimulation of cyclic adenosine monophosphate signaling and insulin secretion in beta-cells (Damdindorj et al. 2012), and suppress the production of adiponectin (Cummings 2009), whereas synthetic ghrelin administration in humans can restrain insulin secretion (Tong et al. 2010). However, the effect of metabolic surgery on ghrelin is controversial, with both increase and stability being observed (Cummings 2009).

Other Molecules

The amino acid sensor G protein receptor 142 (GP142) integrates anti-incretin properties, since it stimulates pancreatic glucagon secretion, but also exerts adjusting effects on insulin-mediated glucose uptake in fat and muscle, and hepatic glucagon-mediated glucose production (Lin et al. 2018).

Oxyntomodulin has been also reported to upregulate glucagon secretion both *in vitro* and *in vivo* (Baldissera et al. 1988), having at the same time the ability to decrease glucagon secretion through the activation of the GLP-1 receptor (Holst et al. 2018a, b). Those GP142 and oxyntomodulin actions could be considered as part of the physiological response antagonizing incretin-induced hypoglycemia and preserving normal glucose levels, through the modulation of glucagon hepatic production, peripheral glucose utilization, and perhaps other mechanisms (Karras et al. 2019).

Correlation with Diabetes and Obesity Remission Following Metabolic Surgery

Salinari et al. (2017) investigated the impact of metabolic surgery (biliopancreatic diversion) on glucose homeostasis variables. Participants with obesity exhibited lower insulin sensitivity, whereas in normal individuals, insulin sensitivity was lower during oral compared to intravenous glucose administration. The more robust difference between oral and intravenous insulin sensitivity in obese individuals was attenuated following biliopancreatic diversion. It was predicted that hypoglycemia would occur during an OGTT, in case insulin sensitivity remained in the range of intravenous glucose administration, particularly in case of obesity and IR. According to the anti-incretin model, postprandial amplification of IR protects against hypoglycemia, following the elevation of insulin levels as a result of food intake.

In an experimental study (Salinari et al. 2014), nonobese, diabetic Goto-Kakizaki rats were subjected to resection or bypass of various intestinal segments. Duodenal-jejunal bypass (DJB) as well as jejunectomy, although improving insulin sensitivity, did not have an impact on GIP and GLP-1 concentrations, indicating no role for elevated incretin levels. Downregulation of specific duodenum- and jejunum-derived molecules that negatively affect insulin sensitivity might be responsible instead. It could also be hypothesized that an exaggerated anti-incretin effect is a component of T2D pathophysiology, even in the absence of obesity.

Enhancement of Pancreatic Beta-cell Mass

Lindqvist et al. (2014) achieved improvements in glycemic homeostasis following RYGB in a porcine model, driven by an increase in beta-cell mass, islet number, and number of extra islet beta-cells, suggesting that the operation resulted in the suppression of signals that inhibit beta-cell proliferation. This is in agreement with the anti-incretin theory, albeit different studies have failed to demonstrate an increased beta-cell mass after bariatric procedures (Meier et al. 2006). Inappropriately, improved beta-cell function, rather than changes in beta-cell mass, might explain the post-operative hypoglycemia occasionally seen in people subjected to RYGB.

A different study (Salinari et al. 2013), assessed the effects of protein extracts, derived from the duodenum-jejunum conditioned-medium (CM) of diabetic rodents on insulin sensitivity, both *in vitro* and *in vivo*. Jejunal proteins negatively affected muscle insulin signaling, through promotion of Akt ^{473}Ser phosphorylation in L6 cells, possibly via mTOR Complex 2 (mTORC2) or TSC activation, and Akt recruitment to plasma membrane. The authors indicated the proximal small bowel of animals with diabetes as a source of diabetogenic factors that have the potential to impair peripheral insulin sensitivity; however, these interesting findings need to be replicated by studies in humans.

Future Perspectives

The idea that T2D and obesity could be effectively managed or even put into remission following metabolic surgery is gaining popularity, due to increasing evidence supporting the effectiveness of surgical approaches, not only in reducing weight loss and improving glucose outcomes (Batterham and Cummings 2016), but also in preventing obesity-related macrovascular complications (Yan et al. 2019). The exact mechanisms mediating the effects of bariatric surgery are still hotly debated, as these are complex and involve interactions between numerous physiological and pathogenetic pathways (Karras et al. 2019).

The anti-incretin theory provides an alternative theoretical framework to explain the mechanisms behind the effects of metabolic surgery on T2D, and highlights the significance of the GI tract in the homeostatic regulation of food intake and energy balance in humans. Disruption of this two-way feedback loop, could be considered as an additional pathway leading to the development of obesity and IR.

Obviously, the anti-incretin theory should be studied in the context of other mechanisms that explain the impact of bariatric surgery on metabolic outcomes. For example, changes in incretin bioavailability (Holst et al. 2018a, b), differentiation of the vagus nerve anatomy (de Lartigue et al. 2014), altered expression of bile acids (Penney et al. 2015), and modified gut microbiome (Liou et al. 2013), have been all shown to contribute to the metabolic benefits of bariatric surgery.

Anti-incretin molecules and pathways, as well as genetic, molecular or biochemical markers able to monitor such phenomenon in obesity and diabetes, are worthwhile topics for future research (Karras et al. 2019).

References

Alfa RW, Park S, Skelly KR et al (2015) Suppression of insulin production and secretion by a decretin hormone. Cell Metab 21:323–334

Baldissera FG, Holst JJ, Knuhtsen S et al (1988) Oxyntomodulin (glicentin-(33-69)): pharmacokinetics, binding to liver cell membranes, effects on isolated perfused pig pancreas, and secretion from isolated perfused lower small intestine of pigs. Regul Pept 21:151–166

Batterham RL, Cummings DE (2016) Mechanisms of diabetes improvement following bariatric/metabolic surgery. Diabetes Care 39:893–901

Bose M, Teixeira J, Olivan B et al (2010) Weight loss and incretin responsiveness improve glucose control independently after gastric bypass surgery. J Diabetes 2:47–55

Chaudhry S, Bernardes M, Harris PE et al (2016) Gastrointestinal dopamine as an anti-incretin and its possible role in bypass surgery as therapy for type 2 diabetes with associated obesity. Minerva Endocrinol 41:43–56

Cummings DE (2005) Gastric bypass and nesidioblastosis—too much of a good thing for islets? N Engl J Med 353:300–302

Cummings DE (2009) Endocrine mechanisms mediating remission of diabetes after gastric bypass surgery. Int J Obes (Lond) 33:S33–S40

Damdindorj B, Dezaki K, Kurashina T et al (2012) Exogenous and endogenous ghrelin counteracts GLP-1 action to stimulate cAMP signaling and insulin secretion in islet beta-cells. FEBS Lett 586:2555–2562

Holst JJ (2007) The physiology of glucagon-like peptide 1. Physiol Rev 87:1409–1439

Holst JJ, Albrechtsen NJW, Gabe MBN et al (2018a) Oxyntomodulin: Actions and role in diabetes. Peptides 100:48–53

Holst JJ, Madsbad S, Bojsen-Møller KN et al (2018b) Mechanisms in bariatric surgery: gut hormones, diabetes resolution, and weight loss. Surg Obes Relat Dis 14:708–714

Honka H, Koffert J, Kauhanen S et al (2018) Liver blood dynamics after bariatric surgery: the effects of mixed-meal test and incretin infusions. Endocr Connect 7:888–896

Kamvissi V, Salerno A, Bornstein SR et al (2010) Incretins or anti-incretins? A new model for the "entero-pancreatic axis". Horm Metab Res 47:84–87

Kapodistria K, Tsilibary EP, Kotsopoulou E et al (2018) Liraglutide, a human glucagon-like peptide-1 analogue, stimulates AKT-dependent survival signalling and inhibits pancreatic β-cell apoptosis. J Cell Mol Med 22:2970–2980

Karras S, Goulis DG, Mintziori G et al (2012) The effects of incretins on energy homeostasis: physiology and implications for the treatment of type 2 diabetes mellitus and obesity. Curr Vasc Pharmacol 10:781–791

Karras SN, Koufakis T, Mustafa OG et al (2019) Anti-incretin effect: the other face of Janus in human glucose homeostasis. Obes Rev 20:1597–1607

Knop FK, Vilsbøll T, Højberg PV et al (2007) Reduced incretin effect in type 2 diabetes: cause or consequence of the diabetic state? Diabetes 56:1951–1959

Korner J, Cline GW, Slifstein M, Barba P, Rayat GR, Febres G, Leibel RL, Maffei A, Harris PE (2019) A role for foregut tyrosine metabolism in glucose tolerance. Mol Metab 23:37–50

Kozicz T, Arimura A (2002) Distribution of urocortin in the rat's gastrointestinal tract and its colocalization with tyrosine hydroxylase. Peptides 23:515–521

de Lartigue G, Ronveaux CC, Raybould HE (2014) Deletion of leptin signaling in vagal afferent neurons results in hyperphagia and obesity. Mol Metab 3:595–607

Lavine JA, Attie AD (2010) Gastrointestinal hormones and the regulation of β-cell mass. Ann N Y Acad Sci 1212:41–58

Lin HV, Wang J, Wang J et al (2018) GPR142 prompts glucagon-like Peptide-1 release from islets to improve β-cell function. Mol Metab 11:205–211

Lindqvist A, Spégel P, Ekelund M et al (2014) Gastric bypass improves b-cell function and increases b-cell mass in a porcine model. Diabetes 63:1665–1671

Liou AP, Paziuk M, Luevano JM Jr et al (2013) Conserved shifts in the gut microbiota due to gastric bypass reduce host weight and adiposity. Sci Transl Med 5:178ra41

Maffei A, Segal AM, Alvarez-Perez JC et al (2015) Anti-incretin, anti-proliferative action of dopamine on beta-cells. Mol Endocrinol 29:542–557

Meier JJ, Butler AE, Galasso R et al (2006) Hyperinsulinemic hypoglycemia after gastric bypass surgery is not accompanied by islet hyperplasia or increased beta-cell turnover. Diabetes Care 29:1554–1559

Penney NC, Kinross J, Newton RC et al (2015) The role of bile acids in reducing the metabolic complications of obesity after bariatric surgery: a systematic review. Int J Obes 39:1565–1574

Rehfeld JF, Stadil F (1973) The effect of gastrin on basal- and glucose-stimulated insulin secretion in man. J Clin Invest 52:1415–1426

Rubi B, Ljubicic S, Pournourmohammadi S et al (2005) Dopamine D2-like receptors are expressed in pancreatic beta-cells and mediate inhibition of insulin secretion. J Biol Chem 280:36824–36832

Rubino F, Amiel SA (2014) Is the gut the "sweet spot" for the treatment of diabetes? Diabetes 63:2225–2228

Rubino F, Gagner M (2002) Potential of surgery for curing type 2 diabetes mellitus. Ann Surg 236:554–559

Salinari S, Debard C, Bertuzzi A et al (2013) Jejunal proteins secreted by db/db mice or insulin-resistant humans impair the insulin signaling and determine insulin resistance. PloS One 8:e56258

Salinari S, le RouxCW BA et al (2014) Duodenal-jejunal bypass and jejunectomy improve insulin sensitivity in Goto-Kakizaki diabetic rats without changes in incretins or insulin secretion. Diabetes 63:1069–1078

Salinari S, Mingrone G, Bertuzzi A et al (2017) Down regulation of insulin sensitivity after oral glucose administration: evidence for the anti-incretin effect. Diabetes 66:2756–2763

Seino Y, Fukushima M, Yabe D (2010) GIP and GLP-1, the two incretin hormones: similarities and differences. J Diabetes Investig 1:8–23

Tong J, Prigeon RL, Davis HW et al (2010) Ghrelin suppresses glucose-stimulated insulin secretion and deteriorates glucose tolerance in healthy humans. Diabetes 59:2145–2151

Wickremesekera K, Miller G, Naotunne TD et al (2005) Loss of insulin resistance after Roux-en-Y gastric bypass surgery: a time course study. Obes Surg 15:474–481

Wu W, Shang J, Feng Y et al (2008) Identification of glucose-dependant insulin secretion targets in pancreatic beta-cells by combining defined-mechanism compound library screening and siRNA gene silencing. J Biomol Screen 13:128–134

Yan G, Wang J, Zhang J et al (2019) Long-term outcomes of macrovascular diseases and metabolic indicators of bariatric surgery for severe obesity type 2 diabetes patients with a meta-analysis. PLoS One 14:e0224828

Part II

Genomics, Metabolomics and Other Omics

An Update on Mendelian Forms of Obesity and their Personalized Treatments

16

Selene Chen and David Meyre

Contents

Introduction 208
Leptin Deficiency 209
Leptin Receptor Deficiency 209
Proopiomelanocortin Deficiency (POMC) 209
Proprotein Convertase 1 Deficiency (PC1) 209
Melanocortin 3 Receptor Deficiency (MC3R) 210
Melanocortin 4 Receptor Deficiency (MC4R) 210
Kinase Suppressor of Ras 2 Deficiency (KSR2) 210
Melanocortin 2 Receptor Accessory Protein 2 (MRAP2) Deficiency 210
Adenylate Cyclase 3 (ADCY3) Deficiency 210
Steroid Receptor Coactivator-1 (SRC-1) Deficiency 211
Semaphorin 3 Signaling Deficiency 211
An Overview of Current Treatments of Obesity in Child and Adult Populations .. 211
Lifestyle Modifications 211
Behavioral Modifications 212

S. Chen
Department of Health Research Methods, Evidence, and Impact, McMaster University, Hamilton, ON, Canada

D. Meyre (✉)
Department of Health Research Methods, Evidence, and Impact, McMaster University, Hamilton, ON, Canada

Michael DeGroote Centre for Learning & Discovery, McMaster University, Hamilton, ON, Canada

Department of Pathology and Molecular Medicine, McMaster University, Hamilton, ON, Canada
e-mail: meyred@mcmaster.ca

J. Faintuch and S. Faintuch (eds.), *Obesity and Diabetes*, https://doi.org/10.1007/978-3-030-53370-0_16

Pharmaceuticals 212
Bariatric Surgery 212

Diagnosis and Treatments of Mendelian Forms of Obesity 213
Diagnosis of Nonsyndromic Monogenic Obesity 213
Personalized Management of Nonsyndromic Monogenic Forms of Obesity 213
Bariatric Surgery in Monogenic Obesity 214

References 214

Abstract

Obesity has a strong inherited component (40–75% of the risk is driven by genetic factors), including monogenic syndromic or not, oligogenic, and polygenic modalities. Monogenic nonsyndromic/oligogenic genes in the leptin-melanocortin pathway are analyzed, as well as their role in energy balance. Early genetic testing along with traditional and ongoing attempts to clinically handle several relevant categories is similarly addressed.

Keywords

Obesity · Monogenic · Leptin/melanocortin pathway · Genetic testing · Mereleptin · Setmelanotide

Introduction

Obesity is a serious pandemic and an often neglected concern (NCD Risk Factor Collaboration (NCD-RisC) 2017). It has nearly tripled in a few decades to reach 650 million adults and 124 children in 2016 (NCD Risk Factor Collaboration (NCD-RisC) 2017). Practically, all ages and socioeconomic groups are affected. The disease is associated with the development of multiple comorbidities: depression, sleep apnea, osteoarthritis, diabetes, hypertension, cardiovascular disease, and some cancer (Apovian 2016). Obesity and its many comorbidities result in poorer quality of life and reduced disability adjusted life years and life expectancy (Ng et al. 2014). Current treatments are useful for selected groups; however, they did not effectively change the obesity epidemic (Bray et al. 2016). This modest success in treating obesity may be attributed to a "one size fits all" management and may not be in the best interest of patients who come from diverse ethnic, socioeconomic, and health backgrounds (Bruce et al. 2007).

In 2010, Auffray et al. proposed the concept of P4 (predictive, preventive, personalized, and participatory), also known as precision medicine (Auffray et al. 2010). Precision medicine is a move away from standardized treatment, to the one that adapts therapies to achieve the best outcomes for individual patients (Auffray et al. 2010). Precision medicine proposes to make use of an array of biological and 'multi omics' information, such as patient genome, epigenome, transcriptome, proteome, and microbiome, to cite a few, for more precise prevention and care (Auffray et al. 2010). This concept of treatment may apply successfully to the management of complex disorders such as obesity (Blakemore and Froguel 2010).

Obesity is multifactorial in essence (Reddon et al. 2016). Environmental risk exposures include poor diet, lack of exercise, altered sleep patterns, psychosocial stress, endocrine disruptors, pharmaceutical iatrogenesis, and controlled ambient temperatures (McAllister et al. 2009), while biological factors such as *in utero* programing, demographic variables, and comorbidities such as depression, gut microbiome, epigenetics, and genes also influence susceptibility to obesity (Pigeyre et al. 2016). Up to 40–75% of genetic propensity toward fat accumulation emerges from twin, family, and population studies (Stryjecki et al. 2018; Yang et al. 2015a), and an array of monogenic (i.e., Mendelian) syndromic, nonsyndromic, oligogenic, and polygenic transmissions are identified (Pigeyre et al. 2016; Tam et al. 2018; Kaur et al. 2017).

At the individual level, nature and nurture contribute to obesity in varying degrees, from predominantly environmental (e.g., Sumo wrestlers) to purely genetic (e.g., monogenic condition) origins (Kanehisa et al. 1998; Lunsky and Meyre 2019). Nonsyndromic monogenic adiposity could represent excellent targets for the application of precision medicine (Blakemore and Froguel 2010). Not only do monogenic defects cause striking body weight gain, but also their simple biological origin (i.e., partial/complete inactivation of one gene) allows for actionable and effective treatments for a subset of patients (Blakemore and Froguel 2010; Lunsky and Meyre 2019).

Leptin Deficiency

Leptin controls appetite and calorie balance. Leptin deficiency is a rare autosomal recessive disorder, registered in different parts of the world (Pigeyre and Meyre 2018; Yupanqui-Lozno et al. 2019). Homozygous or heterozygous compound loss-of-function mutations (*LEP*) are observed in human populations. Truncated transcription and virtual absence of leptin can be the result (Montague et al. 1997; Haglund et al. 2018). Nonfunctioning protein without altering circulating levels is another possibility (Haglund et al. 2018; Wabitsch et al. 2015a). Hyperphagia and severe early obesity are typical of complete leptin deficiency (Montague et al. 1997; Kohlsdorf et al. 2018). T-cell compromise and immune deficiency are also featured (Farooqi et al. 2002). Late puberty and aberrant hormonal cycles can be observed (Strobel et al. 1998). Heterozygous carriers of *LEP* mutations can also display increased body fat (Farooqi et al. 2001).

Leptin Receptor Deficiency

Loss-of-function coding mutations in the leptin receptor (*LEPR*) gene have been documented in human populations as leptin deficiency (Pigeyre and Meyre 2018). LEPRb is the longest of the six leptin isoforms coded by *LEPR* and most commonly contains the mutations triggering obesity (Wasim et al. 2016). Disorders include severe and precocious weight gain, along with immune deficiency, hormonal changes, stunted growth, and diabetes (Clement et al. 1998; Farooqi et al. 2007a; Hannema et al. 2016; Armagan et al. 2019; Niazi et al. 2018). Mutations in *LEPR* can be relatively common among families with a high inbreeding rate (Saeed et al. 2015). Heterozygous mutations in *LEPR* are followed by comparable BMI; however, body fat mass is more abundant than that in control population (Farooqi et al. 2007a).

Proopiomelanocortin Deficiency (POMC)

In certain circumstances, *POMC* is not translated, or derived hormones (ACTH, α-MSH, and β-MSH) are incorrectly formed (Krude et al. 1998; Clement et al. 2008; Farooqi et al. 2006). Both homozygous and compound heterozygous cases are described in literature (Pigeyre and Meyre 2018; Hilado and Randhawa 2018). Excessive weight gain is often detected shortly after birth, along with diabetes and multiple hormonal dysfunctions (Krude et al. 1998; Clement et al. 2008; Farooqi et al. 2006; Ozsu and Bahm 2017). Red hair, noticed in Caucasian individuals with light skin (Clement et al. 2008; Cirillo et al. 2012), is putatively connected with the α-MSH pathway. Excessive adiposity in heterozygotes has also been reported (Farooqi et al. 2006; Challis et al. 2002; Lee et al. 2006; Dubern et al. 2008).

Proprotein Convertase 1 Deficiency (PC1)

Failure of PC1 can be linked to major weight gain after birth, deranged glucose homeostasis, intestinal function and circulating hormones including POMC (O'Rahilly et al. 1995). An initial affected case was confirmed as heterozygous compound for the *PCSK1* gene encoding the PC1 protein (Jackson et al. 1997), followed by additional subjects (Pigeyre and Meyre 2018). *PCSK1* is therefore admitted as a monogenic recessive mechanism of uncontrolled weight gain, within the PC1 pathway, triggering protean hormonal and nonhormonal derangements (Jackson et al. 1997, 2003; Farooqi et al. 2007b; Martin et al. 2013; Frank et al. 2013). PC1 mutations impair important hormone cascades; consequently, heterozygous mutations in *PCSK1* precipitate both oligogenic and monogenic dominant cases of elevated body mass index (BMI) (Creemers et al. 2012; Philippe et al. 2015).

Melanocortin 3 Receptor Deficiency (MC3R)

From Singapore (Lee et al. 2007, b) to Poland (Demidowich et al. 1863), however not in the USA (Ehtesham et al. 2019), partial or complete mutations related to MC3R have been associated with obesity, with a threefold elevation of the risk on aveage (Ehtesham et al. 2019), consistent with experimental results (Chen et al. 2000; Butler et al. 2000). MC3R functions differently from the melanocortin 4 receptor (MC4R), depending on kinases including ERK1 and ERK 2 (Yang et al. 2015b).

Melanocortin 4 Receptor Deficiency (MC4R)

Mutations in the *MC4R* gene are the most often recognized (0.2–5.8%) among cases with excessive body weight (Lunsky and Meyre 2019). Food intake is controlled via α-MSH and agouti-related peptide (Farooqi and O'Rahilly 2008), involving signals from POMC neurons (Farooqi and O'Rahilly 2008). β-arrestin and MAPK are equally emphasized in this context (Lotta et al. 2019). Homozygous and heterozygous compound cases suffer from hyperphagia, preference for high-fat instead of high-sucrose food, and major shifts in body composition and arterial pressure (Lunsky and Meyre 2019). Heterozygotes suffer from less obvious repercussions, depending on clinical and environmental circumstances (Stutzmann et al. 2008; Drabkin et al. 2018; Saeed et al. 2015).

Kinase Suppressor of Ras 2 Deficiency (KSR2)

This biomolecule modulates a number of kinases, (Frodyma et al. 2017) encompassing AMPK, which is relevant for energy homeostasis (Costanzo-Garvey et al. 2009; Guo et al. 2017). Reduced energy expenditure despite hyperactivity, insulin resistance, and glucose intolerance is observed in fully deficient laboratory animals (Costanzo-Garvey et al. 2009; Brommage et al. 2008; Revelli et al. 2011), with less robust responses upon milder deficiency (Revelli et al. 2011). Brain knockout conducts to similar however milder consequences than in systemic knockout mice (Guo et al. 2017). Usual features of *KSR2* human mutation carriers encompass weight gain shortly after birth, glucose homeostasis changes, and diminished consumption of energy substrates (Pearce et al. 2013).

Melanocortin 2 Receptor Accessory Protein 2 (MRAP2) Deficiency

This predominantly brain protein is relevant for MC4R and other receptors (Chan et al. 2009). In mice, a codominant form of obesity is elicited by the deficiency (Asai et al. 2013). Several *MRAP2* mutations including E24X variant lead to advanced obesity in humans (Asai et al. 2013). Heterozygous carriers can be at risk as well (Schonnop et al. 2016). The Q174R mutation diminishes MC4R signaling *in vitro* (Schonnop et al. 2016). Large-scale sequencing of *MRAP2* exons along with functional assessment of 23 rare *MRAP2* variants confirmed that loss-of-function heterozygous mutations precipitate monogenic hyperphagic childhood and adult obesity, hyperglycemia, and hypertension (Baron et al. 2019).

Adenylate Cyclase 3 (ADCY3) Deficiency

Common variants in the *ADCY3* gene have been associated with body mass index variations in diverse human populations (Speliotes et al.

2010; Wen et al. 2012; Monda et al. 2013). A missense mutation in *Adcy3* is associated with body weight variation in rats (Keele et al. 2018). Whereas Adcy3 deficiency in rodents favors excessive adiposity, gain-of-function mutation has the opposite effect (Pitman et al. 2014; Tong et al. 2016). Homozygous loss-of-function mutations in *ADCY3*, hyperphagia, and severe obesity were confirmed in humans (Saeed et al. 2018). A founder essential splice mutation in *ADCY3* also resulted in higher BMI (Grarup et al. 2018). Homozygous mutations in *ADCY3* are not very rare in certain obese populations (Saeed et al. 2018; Grarup et al. 2018; Bjerregaard and Jorgensen 2013). ADCY3 colocalizes with MC4R, and impairment is recognized as a factor for weight gain (Siljee et al. 2018).

Steroid Receptor Coactivator-1 (SRC-1) Deficiency

SRC-1 is secreted in the hypothalamus and enhances *POMC* transcription via the STAT3 pathway (Yang et al. 2019). Complete SRC-1 deficiency in mice leads to reduced energy expenditure and significantly higher weight gain upon high-fat diet (Picard et al. 2002). Concordantly, deletion of SRC-1 leads to decreased *POMC* expression, increased food intake, and high-fat diet induced obesity (Yang et al. 2019; Wang et al. 2006). Fifteen heterozygous missense mutations in *SRC-1* were found in cases with obesity at young age. Dysfunction of the STAT3 and POMC reporter pathways was detected (Yang et al. 2019). In contrast, four missense mutations found in nonobese controls did not impact SRC-1 function (Yang et al. 2019). In an experimental model of SRC-1 mutation (p.L1376P), exaggerated adiposity coincided with defective POMC response (Yang et al. 2019).

Semaphorin 3 Signaling Deficiency

Class 3 Semaphorins (SEMA3A-G) and their receptors (PLXNA1-4; Neuropilin (NRP1-2)) are linked to GnRH neurons in the central nervous system (Alto and Terman 2017). SEMA3 signaling contributes to melanocortin circuits and energy balance (van der Klaauw et al. 2019). In mouse models and also *in vitro*, it was found that SEMA3s acting via NRP2 direct the development of POMC projections in deep brain nuclei (van der Klaauw et al. 2019). Deletion of NRP2 receptors in POMC neurons of mice disrupted POMC projections, causing reduced energy expenditure and weight gain (van der Klaauw et al. 2019).

Experimental evidence confirms the role of energy homeostasis of genes in the SEMA 3 signaling pathway (van der Klaauw et al. 2019). Among a cohort of European ancestry, 40 rare variants of 13 genes related to SEMA3 signaling were found in 573 cases with severe BMI elevation at young age (van der Klaauw et al. 2019; Hendricks et al. 2017). Different molecular mechanisms were identified (van der Klaauw et al. 2019). Very rare (minor allele frequency $<0.025\%$) variants were prominently featured by affected cases in this large series (van der Klaauw et al. 2019).

An Overview of Current Treatments of Obesity in Child and Adult Populations

Lifestyle Modifications

Research has found that dietary changes along with exercise are associated with less adiposity in children and adults with high BMI (Dombrowski et al. 2014; Peirson et al. 2014, 2015a). Lifestyle modifications also have a beneficial impact on cardio-metabolic outcome, lipid profile, and glucose homeostasis (Ho et al. 2013; Baillot et al. 2015). A significant problem with these methods is poor adherence, which ultimately hinders treatment effectiveness

(Burgess et al. 2017). If weight loss is achieved following lifestyle modifications, another challenge is to maintain this benefit in the long run (Peirson et al. 2015b).

Behavioral Modifications

Behavioral interventions involve patients learning to self-monitor nutrition and exercise, identifying barriers to weight loss and supporting one another on their journey (Jin 2018). These interventions often last for 1–2 years and have shown moderate treatment effects in terms of lowering BMI (Peirson et al. 2015a; Jin 2018). Furthermore, pairing these approaches with lifestyle modification has shown to increase adherence to the program and modestly improve weight loss outcomes in children and adults (Dombrowski et al. 2014; O'Connor et al. 2017). A benefit of these treatments is that they involve minimal side effects and as such can be easily applied into care settings (Peirson et al. 2015a; O'Connor et al. 2017). Another advantage is the possibility to use media devices to promote behavioral changes, even though the efficacy of eHealth interventions has yet to be demonstrated (Hutchesson et al. 2015). A notable concern is that behavioral intervention effects may not been maintained in the long run (Peirson et al. 2015a).

Pharmaceuticals

Orlistat prevents the uptake and digestion of triglycerides. It is a widely used agent with a good safety profile and has been shown to provide clinically significant and ongoing decreases in BMI when used in conjunction with a slightly hypocaloric diet and exercise (Khera et al. 2016). Common side effects of orlistat are steatorrhea (i.e., excess of fat in feces) and diarrhea, caused by the body's limited absorption of dietary fats in response to the drug (Bansal and Al Khalili 2019). Other side effects include abdominal pain and fecal spotting (Smith et al. 2013). Phentermine-topiramate is a drug combination prescribed to severely obese adults. Phentermine acts as an appetite suppressant, stimulating norepinephrine and epinephrine release, while topiramate increases the effect on gamma-aminobutyric acid-A receptors, altering sensations of satiety (Allison et al. 2012). Common adverse outcomes include dry mouth, constipation, dysgeusia (i.e., distortion of the sense of taste), and insomnia (Allison et al. 2012). Overall, the pair demonstrates beneficial dose-dependent effects on weight outcomes in obese adults after 1 year (Allison et al. 2012). Liraglutide is an efficacious drug used to treat both obesity and type 2 diabetes (T2D). It is a glucagon-like peptide 1 receptor agonist, suppressing appetite and energy intake in addition to delaying gastric emptying (Astrup et al. 2009). The drug was found to be more effective than orlistat in reducing weight outcomes and blood pressure at all doses in addition to decreasing the prevalence of prediabetes in patients (Astrup et al. 2009). Adverse events aside from nausea and vomiting were often short-lived with users of liraglutide, highlighting its efficacy over others as a treatment for obesity (Astrup et al. 2009). Lorcaserin is another drug that acts on serotonin 5-HT2C receptors in the central nervous system to inhibit appetite in obese individuals (Fidler et al. 2011). Dizziness, headaches, and nausea are common adverse effects of the medication. Lorcaserin administered alongside a lifestyle modification program was found to provide the best weight outcomes in adults (Fidler et al. 2011).

Bariatric Surgery

While these interventions are invasive and have potential for mortality, this risk is relatively low (0–1.5%) (Arterburn and Courcoulas 2014). Information from randomized clinical trials has shown bariatric procedures to be more effective than other nonsurgical interventions (Noria and Grantcharov 2013). Some studies also cite improvements to quality of life measures in those who were surgically treated (Moulla et al. 2018). Another benefit of bariatric surgery is its ability to improve various obesity-associated comorbidities (Picot et al. 2009). Many patients with T2D have reported normal levels of plasma glucose following their surgeries and lowered cardiovascular risk (Noria and Grantcharov 2013; Moulla et al. 2018; Picot et al. 2009). Even though increasingly popular, this modality is still not available to many candidates (Arterburn and Courcoulas 2014; Noria and Grantcharov 2013; Moulla et al. 2018; Picot et al. 2009).

Diagnosis and Treatments of Mendelian Forms of Obesity

Diagnosis of Nonsyndromic Monogenic Obesity

Both dysmorphic, intellectually disabled and normal cases can be seen (Pigeyre and Meyre 2018). *LEP* and *LEPR* mutations are associated with more severe and precocious diseases, contrasting with much more benign heterozygous *MC4R* mutations (Kohlsdorf et al. 2018; Vazquez-Moreno et al. 2019). In contrast, adults with heterozygous *MC4R* mutations present higher BMI than nonmonogenic obese adults (Stutzmann et al. 2008). Targeting children and adults with moderate to extreme obesity is a priority. Clinical and biochemical features along with family assessment may guide the genetic diagnosis (Pigeyre and Meyre 2018). Hypopigmentation of hair and skin, intestinal dysfunction, and a Mendelian recessive pattern of inheritance for obesity may signal the presence of homozygous/heterozygous compound mutations in the *POMC* and *PCSK1* genes (Choquet and Meyre 2010). Leptin, ACTH, and cortisol should be monitored. Leptin is low in congenital deficiency and can substantially increase in mutations in *LEPR*. (Montague et al. 1997; Clement et al. 1998; Farooqi et al. 2007a). Inactive "normal" leptin levels are possible in some circumstances (Wabitsch et al. 2015b, c). ACTH deficiency including clinical symptoms is a possibility in *POMC* and *PCSK1* mutations. Tailored metabolic and hormonal panels are being designed for some of these cases, with wider coverage and reduced costs (Shabana 2016).

Genetic sequencing, preferably addressing complete exome panels, is the advised approach in possibly monogenic cases. Major targets are *LEP, LEPR, POMC, PCSK1, MC4R, MC3R, ADCY3, MRAP2*, and *KSR2*, and explanatory algorithms are available at www.acmg.net and other sites. Variant prediction and pathogenicity tools are also available (Creemers et al. 2012; Vazquez-Moreno et al. 2019; Tarnanas et al. 2015). *In vitro* and *in vivo* functional experiments, while labor- and resource-intensive, remain the gold standards to assess the biological consequences of mutations (Creemers et al. 2012; Richards et al. 2015). Importantly, they often provide divergent conclusions from *in silico* prediction tools used alone (Heikkinen et al. 2009). For monogenic obesity genes contributing to an important burden of disease (e.g., *MC4R*), it may be relevant to assess *in vitro* the functional consequences of all possible missense variants once and for all, as has been recently performed for the *PPARG* gene (Meyre et al. 2019). Testing of other family members to establish cosegregation with obesity is recommended, especially when a new pathogenic variant is discovered.

Personalized Management of Nonsyndromic Monogenic Forms of Obesity

Lifestyle attention is crucial in monogenic obesity mutation carriers. *MC4R* and *POMC* heterozygous loss-of-function mutations are initially responsive, even though a high risk of relapse should be accounted for (Majithia et al. 2016; Reinehr et al. 2009; Santoro et al. 2006). The environment can contribute to penetrance (Stutzmann et al. 2008). Better socio-economic status is protective in some countries (Morell-Azanza et al. 2019). However, this association is

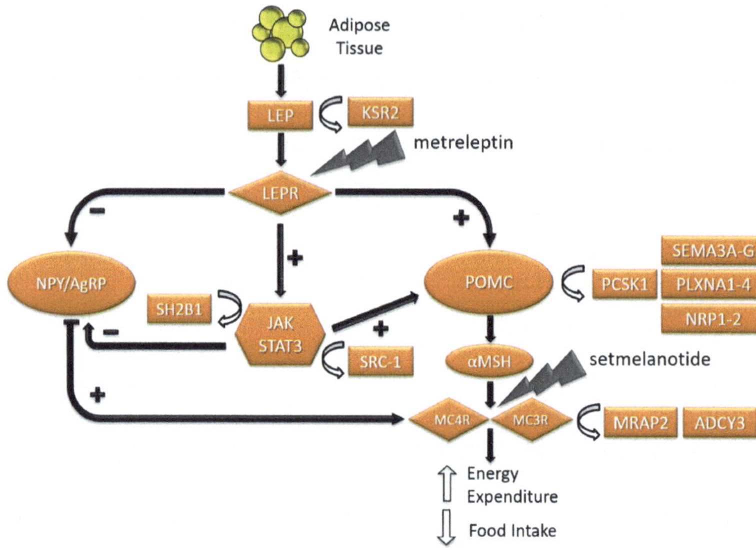

Fig. 16.1 Genes located in the leptin-melanocortin pathway, which have been associated with monogenic/oligogenic obesity

not observed in *MC4R* heterozygous mutation carriers, suggesting that higher education is not always beneficial for lifestyle (Stutzmann et al. 2008). Biologically driven hunger can be overpowering (Newton et al. 2017).

Classic anti-obesity drugs such as sibutramine, liraglutide, or serotonin and noradrenaline reuptake inhibitor induce weight reduction and improve cardiometabolic health risks in *MC4R* mutation carriers (Kleinendorst et al. 2017; Hainerova and Lebl 2013). Other targeted agents are being evaluated as well (Fig. 16.1). Complete POMC deficiency can respond to setmelanotide (Iepsen et al. 2018), similar to *LEPR* homozygous and *MC4R* heterozygous mutations (Kuhnen et al. 2016; Clement et al. 2018; Collet et al. 2017).

Bariatric Surgery in Monogenic Obesity

Homozygous *LEPR* and compound heterozygous *MC4R* mutations (Le Beyec et al. 2013; Huvenne et al. 2015; Aslan et al. 2011a; Jelin et al. 2016) could be valid surgical candidates, as well as heterozygous *MC4R* functional mutations, even though with less favorable outcomes (Meyre et al. 2014; Moore et al. 2014; Censani et al. 2014; Bonnefond et al. 2016; Elkhenini et al. 2014; Hatoum et al. 2012; Aslan et al. 2011b).

Funding Sources DM holds a Canada Research Chair in Genetics of Obesity.

Declarations of Interest The authors declare that they have no competing interests.

References

Allison DB, Gadde KM, Garvey WT et al (2012) Controlled-release phentermine/topiramate in severely obese adults: a randomized controlled trial (EQUIP). Obesity (Silver Spring) 20:330–342

Alto LT, Terman JR (2017) Semaphorins and their signaling mechanisms. Methods Mol Biol 1493:1–25

Apovian CM (2016) Obesity: definition, comorbidities, causes, and burden. Am J Manag Care 22:s176–s185

Armagan C, Yilmaz C, Koc A et al (2019) A toddler with a novel LEPR mutation. Hormones (Athens) 18:237–240

Arterburn DE, Courcoulas AP (2014) Bariatric surgery for obesity and metabolic conditions in adults. BMJ 349: g3961

Asai M, Ramachandrappa S, Joachim M et al (2013) Loss of function of the melanocortin 2 receptor accessory protein 2 is associated with mammalian obesity. Science 341:275–278

Aslan IR, Ranadive SA, Ersoy BA, Rogers SJ, Lustig RH, Vaisse C (2011a) Bariatric surgery in a patient with complete MC4R deficiency. Int J Obes 35:457–461

Aslan IR, Campos GM, Calton MA, Evans DS, Merriman RB, Vaisse C (2011b) Weight loss after Roux-en-Y gastric bypass in obese patients heterozygous for MC4R mutations. Obes Surg 21:930–934

Astrup A, Rossner S, Van Gaal L et al (2009) Effects of liraglutide in the treatment of obesity: a randomised, double-blind, placebo-controlled study. Lancet 374:1606–1616

Auffray C, Charron D, Hood L (2010) Predictive, preventive, personalized and participatory medicine: back to the future. Genome Med 2:57

Baillot A, Romain AJ, Boisvert-Vigneault K et al (2015) Effects of lifestyle interventions that include a physical activity component in class II and III obese individuals: a systematic review and meta-analysis. PLoS One 10: e0119017

Bansal AB, Al Khalili Y (2019) Orlistat. StatPearls, Treasure Island, FL

Baron M, Maillet J, Huyvaert M et al (2019) Loss-of-function mutations in MRAP2 are pathogenic in hyperphagic obesity with hyperglycemia and hypertension. Nat Med 25:1733–1738

Bjerregaard P, Jorgensen ME (2013) Prevalence of obesity among Inuit in Greenland and temporal trend by social position. Am J Hum Biol 25:335–340

Blakemore AI, Froguel P (2010) Investigation of Mendelian forms of obesity holds out the prospect of personalized medicine. Ann N Y Acad Sci 1214:180–189

Bonnefond A, Keller R, Meyre D et al (2016) Eating behavior, low-frequency functional mutations in the melanocortin-4 receptor (MC4R) gene, and outcomes of bariatric operations: a 6-year prospective study. Diabetes Care 39:1384–1392

Bray MS, Loos RJ, McCaffery JM et al (2016) NIH working group report-using genomic information to guide weight management: From universal to precision treatment. Obesity (Silver Spring) 24:14–22

Brommage R, Desai U, Revelli JP et al (2008) High-throughput screening of mouse knockout lines identifies true lean and obese phenotypes. Obesity (Silver Spring) 16:2362–2367

Bruce MA, Sims M, Miller S, Elliott V, Ladipo M (2007) One size fits all? Race, gender and body mass index among U.S. adults. J Natl Med Assoc 99:1152–1158

Burgess E, Hassmen P, Welvaert M, Pumpa KL (2017) Behavioural treatment strategies improve adherence to lifestyle intervention programmes in adults with obesity: a systematic review and meta-analysis. Clin Obes 7:105–114

Butler AA, Kesterson RA, Khong K et al (2000) A unique metabolic syndrome causes obesity in the melanocortin-3 receptor-deficient mouse. Endocrinology 141:3518–3521

Censani M, Conroy R, Deng L et al (2014) Weight loss after bariatric surgery in morbidly obese adolescents with MC4R mutations. Obesity (Silver Spring) 22:225–231

Challis BG, Pritchard LE, Creemers JW et al (2002) A missense mutation disrupting a dibasic prohormone processing site in pro-opiomelanocortin (POMC) increases susceptibility to early-onset obesity through a novel molecular mechanism. Hum Mol Genet 11:1997–2004

Chan LF, Webb TR, Chung TT et al (2009) MRAP and MRAP2 are bidirectional regulators of the melanocortin receptor family. Proc Natl Acad Sci U S A 106:6146–6151

Chen AS, Marsh DJ, Trumbauer ME et al (2000) Inactivation of the mouse melanocortin-3 receptor results in increased fat mass and reduced lean body mass. Nat Genet 26:97–102

Choquet H, Meyre D (2010) Genomic insights into early-onset obesity. Geno Med 2:36

Cirillo G, Marini R, Ito S et al (2012) Lack of red hair phenotype in a North-African obese child homozygous for a novel POMC null mutation: nonsense-mediated decay RNA evaluation and hair pigment chemical analysis. Br J Dermatol 167:1393–1395

Clement K, Vaisse C, Lahlou N et al (1998) A mutation in the human leptin receptor gene causes obesity and pituitary dysfunction. Nature 392:398–401

Clement K, Dubern B, Mencarelli M et al (2008) Unexpected endocrine features and normal pigmentation in a young adult patient carrying a novel homozygous mutation in the POMC gene. J Clin Endocrinol Metab 93:4955–4962

Clement K, Biebermann H, Farooqi IS et al (2018) MC4R agonism promotes durable weight loss in patients with leptin receptor deficiency. Nat Med 24:551–555

Collet TH, Dubern B, Mokrosinski J et al (2017) Evaluation of a melanocortin-4 receptor (MC4R) agonist (Setmelanotide) in MC4R deficiency. Molecul Metabol 6:1321–1329

Costanzo-Garvey DL, Pfluger PT, Dougherty MK et al (2009) KSR2 is an essential regulator of AMP kinase, energy expenditure, and insulin sensitivity. Cell Metab 10:366–378

Creemers JW, Choquet H, Stijnen P et al (2012) Heterozygous mutations causing partial prohormone convertase 1 deficiency contribute to human obesity. Diabetes 61:383–390

Demidowich AP, Jun JY, Yanovski JA (1863) Polymorphisms and mutations in the melanocortin-3 receptor and their relation to human obesity. Biochim Et Biophys Acta Molecul Basis Dis 2017:2468–2476

Dombrowski SU, Knittle K, Avenell A, Araujo-Soares V, Sniehotta FF (2014) Long term maintenance of weight loss with non-surgical interventions in obese adults: systematic review and meta-analyses of randomised controlled trials. BMJ 348:g2646

Drabkin M, Birk OS, Birk R (2018) Heterozygous versus homozygous phenotype caused by the same MC4R mutation: novel mutation affecting a large consanguineous kindred. BMC Med Genet 19:135

Dubern B, Lubrano-Berthelier C, Mencarelli M et al (2008) Mutational analysis of the pro-opiomelanocortin gene in French obese children led to the identification of a novel deleterious heterozygous mutation located in the alpha-melanocyte stimulating hormone domain. Pediatr Res 63:211–216

Ehtesham S, Qasim A, Meyre D (2019) Loss-of-function mutations in the melanocortin-3 receptor gene confer risk for human obesity: a systematic review and meta-analysis. Obes Rev Off J Internat Assoc Study Obes 20:1085–1092

Elkhenini HF, New JP, Syed AA (2014) Five-year outcome of bariatric surgery in a patient with melanocortin-4 receptor mutation. Clin Obes 4:121–124

Farooqi IS, O'Rahilly S (2008) Mutations in ligands and receptors of the leptin-melanocortin pathway that lead to obesity. Nat Clin Pract Endocrinol Metab 4:569–577

Farooqi IS, Keogh JM, Kamath S et al (2001) Partial leptin deficiency and human adiposity. Nature 414:34–35

Farooqi IS, Matarese G, Lord GM et al (2002) Beneficial effects of leptin on obesity, T cell hyporesponsiveness, and neuroendocrine/metabolic dysfunction of human congenital leptin deficiency. J Clin Invest 110:1093–1103

Farooqi IS, Drop S, Clements A et al (2006) Heterozygosity for a POMC-null mutation and increased obesity risk in humans. Diabetes 55:2549–2553

Farooqi IS, Wangensteen T, Collins S et al (2007a) Clinical and molecular genetic spectrum of congenital deficiency of the leptin receptor. N Engl J Med 356:237–247

Farooqi IS, Volders K, Stanhope R et al (2007b) Hyperphagia and early-onset obesity due to a novel homozygous missense mutation in prohormone convertase 1/3. J Clin Endocrinol Metab 92:3369–3373

Fidler MC, Sanchez M, Raether B et al (2011) A one-year randomized trial of lorcaserin for weight loss in obese and overweight adults: the BLOSSOM trial. J Clin Endocrinol Metab 96:3067–3077

Frank GR, Fox J, Candela N et al (2013) Severe obesity and diabetes insipidus in a patient with PCSK1 deficiency. Mol Genet Metab 110:191–194

Frodyma D, Neilsen B, Costanzo-Garvey D, Fisher K, Lewis R (2017) Coordinating ERK signaling via the molecular scaffold Kinase Suppressor of Ras. F1000Res 6:1621

Grarup N, Moltke I, Andersen MK et al (2018) Loss-of-function variants in ADCY3 increase risk of obesity and type 2 diabetes. Nat Genet 50:172–174

Guo L, Costanzo-Garvey DL, Smith DR, Neilsen BK, MacDonald RG, Lewis RE (2017) Kinase suppressor of ras 2 (KSR2) expression in the brain regulates energy balance and glucose homeostasis. Molecul Metabol 6:194–205

Haglund E, Nguyen L, Schafer NP, Lammert H, Jennings PA, Onuchic JN (2018) Uncovering the molecular mechanisms behind disease-associated leptin variants. J Biol Chem 293:12919–12933

Hainerova IA, Lebl J (2013) Treatment options for children with monogenic forms of obesity. World Rev Nutr Diet 106:105–112

Hannema SE, Wit JM, Houdijk ME et al (2016) Novel leptin receptor mutations identified in two girls with severe obesity are associated with increased bone mineral density. Hormon Resear Paediat 85:412–420

Hatoum IJ, Stylopoulos N, Vanhoose AM et al (2012) Melanocortin-4 receptor signaling is required for weight loss after gastric bypass surgery. J Clin Endocrinol Metab 97:E1023–E1031

Heikkinen S, Argmann C, Feige JN et al (2009) The Pro12Ala PPARgamma2 variant determines metabolism at the gene-environment interface. Cell Metab 9:88–98

Hendricks AE, Bochukova EG, Marenne G et al (2017) Rare variant analysis of human and rodent obesity genes in individuals with severe childhood obesity. Sci Rep 7:4394

Hilado MA, Randhawa RS (2018) A novel mutation in the proopiomelanocortin (POMC) gene of a Hispanic child: metformin treatment shows a beneficial impact on the body mass index. J Pediat Endocrinol Metabol: JPEM 31:815–819

Ho M, Garnett SP, Baur LA et al (2013) Impact of dietary and exercise interventions on weight change and metabolic outcomes in obese children and adolescents: a systematic review and meta-analysis of randomized trials. JAMA Pediat 167:759–768

Hutchesson MJ, Rollo ME, Krukowski R et al (2015) eHealth interventions for the prevention and treatment of overweight and obesity in adults: a systematic review with meta-analysis. Obes Rev 16:376–392

Huvenne H, Le Beyec J, Pépin D et al (2015) Seven novel deleterious LEPR mutations found in early-onset obesity: a ΔExon6–8 shared by subjects from reunion Island, France, suggests a founder effect. J Clin Endocrinol Metab 100:E757–EE66

Iepsen EW, Zhang J, Thomsen HS et al (2018) Patients with obesity caused by melanocortin-4 receptor mutations can be treated with a glucagon-like peptide-1 receptor agonist. Cell Metab 28:23–32

Jackson RS, Creemers JW, Ohagi S et al (1997) Obesity and impaired prohormone processing associated with mutations in the human prohormone convertase 1 gene. Nat Genet 16:303–306

Jackson RS, Creemers JW, Farooqi IS et al (2003) Small-intestinal dysfunction accompanies the complex endocrinopathy of human proprotein convertase 1 deficiency. J Clin Invest 112:1550–1560

Jelin EB, Daggag H, Speer AL et al (2016) Melanocortin-4 receptor signaling is not required for short-term weight loss after sleeve gastrectomy in pediatric patients. Int J Obes 40:550–553

Jin J (2018) Behavioral interventions for weight loss. JAMA 320:1210

Kanehisa H, Kondo M, Ikegawa S, Fukunaga T (1998) Body composition and isokinetic strength of

professional Sumo wrestlers. Eur J Appl Physiol Occup Physiol 77:352–359
Kaur Y, de Souza RJ, Gibson WT, Meyre D (2017) A systematic review of genetic syndromes with obesity. Obes Rev Off J Internat Assoc Study Obes 18:603–634
Keele GR, Prokop JW, He H et al (2018) Genetic fine-mapping and identification of candidate genes and variants for adiposity traits in outbred rats. Obesity (Silver Spring) 26:213–222
Khera R, Murad MH, Chandar AK et al (2016) Association of pharmacological treatments for obesity with weight loss and adverse events: a systematic review and meta-analysis. JAMA 315:2424–2434
van der Klaauw AA, Croizier S, Mendes de Oliveira E et al (2019) Human semaphorin 3 variants link melanocortin circuit development and energy balance. Cell 176:729–742
Kleinendorst L, van Haelst MM, van den Akker ELT (2017) Young girl with severe early-onset obesity and hyperphagia. BMJ Case Rep 2017:bcr2017221067
Kohlsdorf K, Nunziata A, Funcke JB et al (2018) Early childhood BMI trajectories in monogenic obesity due to leptin, leptin receptor, and melanocortin 4 receptor deficiency. Int J Obes 42:1602–1609
Krude H, Biebermann H, Luck W, Horn R, Brabant G, Gruters A (1998) Severe early-onset obesity, adrenal insufficiency and red hair pigmentation caused by POMC mutations in humans. Nat Genet 19:155–157
Kuhnen P, Clement K, Wiegand S et al (2016) Proopiomelanocortin deficiency treated with a melanocortin-4 receptor agonist. New England J Med 375:240–246
Le Beyec J, Cugnet-Anceau C, Pepin D et al (2013) Homozygous leptin receptor mutation due to uniparental disomy of chromosome 1: response to bariatric surgery. J Clin Endocrinol Metab 98:E397–E402
Lee YS, Challis BG, Thompson DA et al (2006) A POMC variant implicates beta-melanocyte-stimulating hormone in the control of human energy balance. Cell Metab 3:135–140
Lee YS, Poh LKS, Kek BLK, Loke KY (2007) The role of melanocortin 3 receptor gene in childhood obesity. Diabetes 56:2622–2630
Lotta LA, Mokrosinski J, Mendes de Oliveira E et al (2019) Human gain-of-function MC4R variants show signaling bias and protect against obesity. Cell 177:597–607
Lunsky IO, Meyre D (2019) Decoding Mendelian obesity. Curr Opin Endocr Metab Res 4:21–28
Majithia AR, Tsuda B, Agostini M et al (2016) Prospective functional classification of all possible missense variants in PPARG. Nat Genet 48:1570–1575
Martin MG, Lindberg I, Solorzano-Vargas RS et al (2013) Congenital proprotein convertase 1/3 deficiency causes malabsorptive diarrhea and other endocrinopathies in a pediatric cohort. Gastroenterology 145:138–148
McAllister EJ, Dhurandhar NV, Keith SW et al (2009) Ten putative contributors to the obesity epidemic. Crit Rev Food Sci Nutr 49:868–913
Meyre D, Froguel P, Horber FF, Kral JG (2014) Comment on: Valette et al. Melanocortin-4 receptor mutations and polymorphisms do not affect weight loss after bariatric surgery. PLOS ONE 2012; 7(11):E48221. PLoS One 9:e93324
Meyre D, Andress EJ, Sharma T et al (2019) Contribution of rare coding mutations in CD36 to type 2 diabetes and cardio-metabolic complications. Sci Rep 9:17123
Monda KL, Chen GK, Taylor KC et al (2013) A meta-analysis identifies new loci associated with body mass index in individuals of African ancestry. Nat Genet 45:690–696
Montague CT, Farooqi IS, Whitehead JP et al (1997) Congenital leptin deficiency is associated with severe early-onset obesity in humans. Nature 387:903–908
Moore BS, Mirshahi UL, Yost EA et al (2014) Long-term weight-loss in gastric bypass patients carrying melanocortin 4 receptor variants. PLoS One 9:e93629
Morell-Azanza L, Ojeda-Rodriguez A, Giuranna J et al (2019) Melanocortin-4 receptor and lipocalin 2 gene variants in Spanish children with abdominal obesity: effects on BMI-SDS after a lifestyle intervention. Nutrients 11:960
Moulla Y, Lyros O, Bluher M, Simon P, Dietrich A (2018) Feasibility and safety of bariatric surgery in high-risk patients: a single-center experience. J Obes 2018:7498258
NCD Risk Factor Collaboration (NCD-RisC) (2017) Worldwide trends in body-mass index, underweight, overweight, and obesity from 1975 to 2016: a pooled analysis of 2416 population-based measurement studies in 128.9 million children, adolescents, and adults. Lancet 390:2627–2642
Newton S, Braithwaite D, Akinyemiju TF (2017) Socio-economic status over the life course and obesity: systematic review and meta-analysis. PLoS One 12: e0177151
Ng M, Fleming T, Robinson M et al (2014) Global, regional, and national prevalence of overweight and obesity in children and adults during 1980–2013: a systematic analysis for the Global Burden of Disease Study 2013. Lancet 384:766–781
Niazi RK, Gjesing AP, Hollensted M et al (2018) Identification of novel LEPR mutations in Pakistani families with morbid childhood obesity. BMC Med Genet 19:199
Noria SF, Grantcharov T (2013) Biological effects of bariatric surgery on obesity-related comorbidities. Can J Surg 56:47–57
O'Connor EA, Evans CV, Burda BU, Walsh ES, Eder M, Lozano P (2017) Screening for obesity and intervention for weight management in children and adolescents: evidence report and systematic review for the us preventive services task force. JAMA 317:2427–2444
O'Rahilly S, Gray H, Humphreys PJ et al (1995) Brief report: impaired processing of prohormones associated with abnormalities of glucose homeostasis and adrenal function. N Engl J Med 333:1386–1390

Ozsu E, Bahm A (2017) Delayed diagnosis of proopiomelanocortin (POMC) deficiency with type 1 diabetes in a 9-year-old girl and her infant sibling. J Pediat Endocrinol Metabol: JPEM 30:1137–1140

Pearce LR, Atanassova N, Banton MC et al (2013) KSR2 mutations are associated with obesity, insulin resistance, and impaired cellular fuel oxidation. Cell 155:765–777

Peirson L, Douketis J, Ciliska D, Fitzpatrick-Lewis D, Ali MU, Raina P (2014) Treatment for overweight and obesity in adult populations: a systematic review and meta-analysis. CMAJ Open 2:E306–E317

Peirson L, Fitzpatrick-Lewis D, Morrison K, Warren R, Usman Ali M, Raina P (2015a) Treatment of overweight and obesity in children and youth: a systematic review and meta-analysis. CMAJ Open 3: E35–E46

Peirson L, Fitzpatrick-Lewis D, Ciliska D, Usman Ali M, Raina P, Sherifali D (2015b) Strategies for weight maintenance in adult populations treated for overweight and obesity: a systematic review and meta-analysis. CMAJ Open 3:E47–E54

Philippe J, Stijnen P, Meyre D et al (2015) A nonsense loss-of-function mutation in PCSK1 contributes to dominantly inherited human obesity. Int J Obes 39:295–302

Picard F, Gehin M, Annicotte J et al (2002) SRC-1 and TIF2 control energy balance between white and brown adipose tissues. Cell 111:931–941

Picot J, Jones J, Colquitt JL et al (2009) The clinical effectiveness and cost-effectiveness of bariatric (weight loss) surgery for obesity: a systematic review and economic evaluation. Health Technol Assess 13:1–190, 215–357, iii–iv

Pigeyre M, Meyre D (2018) Monogenic obesity. In: Freemark M (ed) Pediatric obesity: etiology, pathogenesis and treatment, 2nd edn. Humana Press, Cham, pp 135–152

Pigeyre M, Yazdi FT, Kaur Y, Meyre D (2016) Recent progress in genetics, epigenetics and metagenomics unveils the pathophysiology of human obesity. Clin Sci (Lond) 130:943–986

Pitman JL, Wheeler MC, Lloyd DJ, Walker JR, Glynne RJ, Gekakis N (2014) A gain-of-function mutation in adenylate cyclase 3 protects mice from diet-induced obesity. PLoS One 9:e110226

Reddon H, Gueant JL, Meyre D (2016) The importance of gene-environment interactions in human obesity. Clin Sci (Lond) 130:1571–1597

Reinehr T, Hebebrand J, Friedel S et al (2009) Lifestyle intervention in obese children with variations in the melanocortin 4 receptor gene. Obesity (Silver Spring) 17:382–389

Revelli JP, Smith D, Allen J et al (2011) Profound obesity secondary to hyperphagia in mice lacking kinase suppressor of ras 2. Obesity (Silver Spring) 19:1010–1018

Richards S, Aziz N, Bale S et al (2015) Standards and guidelines for the interpretation of sequence variants: a joint consensus recommendation of the American College of Medical Genetics and Genomics and the Association for Molecular Pathology. Gene Med 17:405–424

Saeed S, Bonnefond A, Manzoor J et al (2015) Genetic variants in LEP, LEPR, and MC4R explain 30% of severe obesity in children from a consanguineous population. Obesity (Silver Spring) 23:1687–1695

Saeed S, Bonnefond A, Tamanini F et al (2018) Loss-of-function mutations in ADCY3 cause monogenic severe obesity. Nat Genet 50:175–179

Santoro N, Perrone L, Cirillo G et al (2006) Weight loss in obese children carrying the proopiomelanocortin R236G variant. J Endocrinol Investig 29:226–230

Schonnop L, Kleinau G, Herrfurth N et al (2016) Decreased melanocortin-4 receptor function conferred by an infrequent variant at the human melanocortin receptor accessory protein 2 gene. Obesity (Silver Spring) 24:1976–1982

Shabana HS (2016) The p. N103K mutation of leptin (LEP) gene and severe early onset obesity in Pakistan. Biol Res 49:23

Siljee JE, Wang Y, Bernard AA et al (2018) Subcellular localization of MC4R with ADCY3 at neuronal primary cilia underlies a common pathway for genetic predisposition to obesity. Nat Genet 50:180–185

Smith SM, Meyer M, Trinkley KE (2013) Phentermine/topiramate for the treatment of obesity. Ann Pharmacother 47:340–349

Speliotes EK, Willer CJ, Berndt SI et al (2010) Association analyses of 249,796 individuals reveal 18 new loci associated with body mass index. Nat Genet 42:937–948

Strobel A, Issad T, Camoin L, Ozata M, Strosberg AD (1998) A leptin missense mutation associated with hypogonadism and morbid obesity. Nat Genet 18:213–215

Stryjecki C, Alyass A, Meyre D (2018) Ethnic and population differences in the genetic predisposition to human obesity. Obes Rev Off J Internat Assoc Study Obes 19:62–80

Stutzmann F, Tan K, Vatin V et al (2008) Prevalence of melanocortin-4 receptor deficiency in Europeans and their age-dependent penetrance in multigenerational pedigrees. Diabetes 57:2511–2518

Tam V, Turcotte M, Meyre D (2018) Established and emerging strategies to crack the genetic code of obesity. Obes Rev Off J Internat Assoc Study Obes 20:212–240

Tarnanas I, Tsolaki A, Wiederhold M, Wiederhold B, Tsolaki M (2015) Five-year biomarker progression variability for Alzheimer's disease dementia prediction: can a complex instrumental activities of daily living marker fill in the gaps? Alzheimer Dement 1:521–532

Tong T, Shen Y, Lee HW, Yu R, Park T (2016) Adenylyl cyclase 3 haploinsufficiency confers susceptibility to diet-induced obesity and insulin resistance in mice. Sci Rep 6:34179

Vazquez-Moreno M, Zeng H, Locia-Morales D et al (2019) The melanocortin 4 receptor p.Ile269Asn mutation is associated with childhood and adult obesity in Mexicans. J Clin Endocrinol Metab 105: dgz276

Wabitsch M, Funcke JB, Lennerz B et al (2015a) Biologically inactive leptin and early-onset extreme obesity. N Engl J Med 372:48–54

Wabitsch M, Funcke J-B, Lennerz B et al (2015b) Biologically inactive leptin and early-onset extreme obesity. New England J Med 372:48–54

Wabitsch M, Funcke JB, von Schnurbein J et al (2015c) Severe early-onset obesity due to bioinactive leptin caused by a p.N103K mutation in the leptin gene. J Clin Endocrinol Metab 100:3227–3230

Wang Z, Qi C, Krones A et al (2006) Critical roles of the p160 transcriptional coactivators p/CIP and SRC-1 in energy balance. Cell Metab 3:111–122

Wasim M, Awan FR, Najam SS, Khan AR, Khan HN (2016) Role of leptin deficiency, inefficiency, and leptin receptors in obesity. Biochem Genet 54:565–572

Wen W, Cho YS, Zheng W et al (2012) Meta-analysis identifies common variants associated with body mass index in east Asians. Nat Genet 44:307–311

Yang J, Bakshi A, Zhu Z et al (2015a) Genetic variance estimation with imputed variants finds negligible missing heritability for human height and body mass index. Nat Genet 47:1114–1120

Yang F, Huang H, Tao YX (2015b) Biased signaling in naturally occurring mutations in human melanocortin-3 receptor gene. Int J Biol Sci 11:423–433

Yang Y, van der Klaauw AA, Zhu L et al (2019) Steroid receptor coactivator-1 modulates the function of Pomc neurons and energy homeostasis. Nat Commun 10:1718

Yupanqui-Lozno H, Bastarrachea RA, Yupanqui-Velazco ME et al (2019) Congenital leptin deficiency and leptin gene missense mutation found in two Colombian sisters with severe obesity. Gene 10:342

The Genetic Basis of Diabetic Kidney Disease

17

Christopher A. Simeone, Jose M. Lazaro-Guevara, and Marcus G. Pezzolesi

Contents

Introduction 222

Pathophysiology of DKD in T1D and T2D 223

Nondiabetic Kidney Deterioration 223

Disease Trajectory 223

The Genetic Architecture of DKD through GWASs 225

Combined Databases 229

Coding vs Noncoding Variants 230

Ethnic Patterns 230

Contributions of Low Frequency and Rare Variations in DKD Patients 231

Statistical Challenges 231

From Genetic Causes of DKD to Personalized Therapies 232

C. A. Simeone · J. M. Lazaro-Guevara
Division of Nephrology and Hypertension, Department of Internal Medicine, University of Utah, Salt Lake City, UT, USA

Department of Human Genetics, University of Utah, Salt Lake City, UT, USA

M. G. Pezzolesi (✉)
Division of Nephrology and Hypertension, Department of Internal Medicine, University of Utah, Salt Lake City, UT, USA

Department of Human Genetics, University of Utah, Salt Lake City, UT, USA

Diabetes and Metabolism Research Center, University of Utah, Salt Lake City, UT, USA
e-mail: Marcus.Pezzolesi@hsc.utah.edu

J. Faintuch and S. Faintuch (eds.), *Obesity and Diabetes*, https://doi.org/10.1007/978-3-030-53370-0_17

Angiotensin Modulators 232
Future Directions 232
Epigenomics 233
References 233

Abstract

Diabetic kidney disease (DKD) is a microvascular complication of progressive renal decline in patients with diabetes, which is the leading cause of end-stage renal disease (ESRD) globally. Early epidemiological and genetic studies demonstrated that DKD is a heritable condition, indicative of an underlying genetic component for its increased susceptibility. Since then, significant work has been conducted to unravel causal genes implicated in DKD development and progression. With the advancement of genomic technologies, genome-wide association studies (GWASs) and next-generation sequencing (NGS) approaches continue to expand our knowledge of the genetic architecture of DKD and uncover novel biological pathways implicated in disease pathogenesis. Additionally, the establishment of large international collaborations has led to significantly increased cohort sizes to improve overall statistical power to detect novel associations.

Keywords

Diabetic kidney disease (DKD) · End-stage renal disease (ESRD) · Diabetes · Genetics · Genome-wide association studies (GWAS) · Rare variants

Introduction

The rise in incidence of diabetes and its micro- and macrovascular complications, including diabetic kidney disease (DKD), have become a financial and healthcare burden globally. DKD is a progressive, microvascular complication that affects 30–40% of patients with type 1 (T1D) or type 2 (T2D) diabetes and is the leading cause of end-stage renal disease (ESRD) (Gu, 2019, Alicic et al., 2017, Li and Pezzolesi, 2018, Reutens, 2013). DKD initiation and progression arise through metabolic dysregulation associated with diabetes, e.g., hyperglycemia and hyperlipidemia, and activation of the renin-angiotensin system (Reidy et al., 2014, Alicic et al., 2017). Prescribed medications, such as angiotensin-converting enzyme inhibitors (ACEi) and angiotensin II receptor blockers (ARBs), in conjunction with intense glycemic control interventions, can improve overall clinical outcomes (Brenner et al., 2001). Despite the near universal implementation of renoprotective therapies, there has been little reduction in the rate of DKD and ESRD, and DKD continues to be associated with excess morbidity and premature mortality in patients with diabetes.

Early epidemiological studies demonstrated that DKD is a heritable condition shown to aggregate in families, which thereby provided a foundation for studies aimed at examining the genetic factors underlying its susceptibility (Borch-Johnsen et al., 1992, Quinn et al., 1996, Seaquist et al., 1989, Fioretto et al., 1999, Faronato et al., 1997). Current estimates of the contribution of genetics to the overall risk of DKD (i.e., heritability or h^2) range from 0.3 to 0.49 (Freedman et al., 2007, Sandholm et al., 2017), suggesting that genetic variation may contribute up to 49% of the risk of DKD. Completion of efforts to sequence the human genome, through the Human Genome Project (Lander et al., 2001), and characterization of a haplotype map of the genome, through the International HapMap Project (Altshuler et al., 2005), enabled investigators to begin examining sources of genetic variation, such as single nucleotide polymorphisms (SNPs), at a genome-wide scale using genome-wide association studies (GWASs).

Early GWASs of DKD focused on the role of common variation (i.e., SNPs with a minor allele frequency (MAF) > 5%) and identified several SNPs strongly associated with DKD. More recently, large-scale collaborative efforts have led to improved GWASs of DKD, which include larger sample sizes, and facilitated studies that are better powered to detect robust and more reproducible genetic signals. In addition to common variation, which likely explains only a fraction of the associated risk of DKD, next-generation sequencing (NGS) studies involving whole genome sequencing (WGS) and whole exome sequencing (WES) are emerging to help explain some of the 'missing heritability' of DKD (Manolio et al. 2009).

Pathophysiology of DKD in T1D and T2D

DKD patients present with numerous histological and clinical phenotypes that vary with the type and duration of diabetes. With regard to renal histology in T1D DKD, glomerulopathy is apparent with an expansion of the mesangium, basement membrane thickening, and podocyte loss, along with tubular atrophy and tubulointerstitial fibrosis (Alicic et al., 2017; Reidy et al., 2014; Fioretto et al. 2008). Furthermore, increasing proteinuria and decreasing estimated glomerular filtration rate (eGFR) occur in T1D DKD patients with advanced glomerulopathy. Contrarily, renal histology in T2D DKD patients is varied and occasionally does not manifest the T1D DKD phenotype. A subset of patients presents similar histological patterns observed in T1D DKD patients; others may, however, present with just some or absent glomerulosclerosis.

Nondiabetic Kidney Deterioration

Early reports of nondiabetic forms of kidney disease (i.e., NDKD), including immunoglobulin A (IgA) nephropathy, focal segmental glomerulosclerosis, and other glomerulopathies, have been reported in < 10% of T2D proteinuric patients (Fioretto et al. 2008) yet, recent estimates from multiple centers report the incidence of NDKD in T2D patients from 33% to 73% (Anders et al., 2018). Unfortunately, the only way to distinguish DKD from NDKD in patients with diabetes is through kidney biopsy and histological analysis. This process is the "gold standard" for diagnosing DKD, albeit physicians rarely perform kidney biopsies on diabetic patients.

Instead, clinical biomarkers are used to assess renal damage and functional decline in patients with diabetes. By measuring urinary albumin excretion and serum or urinary creatinine, physicians can monitor the severity of renal injury and functional impairment to renal insufficiency, respectively. Measurements of progressive renal damage include calculations of urinary albumin-creatinine ratios (ACRs) and renal decline with eGFR, where persistent macroalbuminuria/proteinuria is defined as ACR > 300 mg/g and severe kidney damage is defined by an eGFR < 30 mL/min/1.73m^2 (Reutens, 2013).

Disease Trajectory

The natural history of DKD is thought to proceed first with normal glomerular function or hyperfiltration and then with urinary microalbuminuria progressing into macroalbuminuria, which is accompanied by decreasing eGFR and, ultimately, results in ESRD (Alicic et al., 2017, Macisaac et al., 2014, Reutens, 2013) (Fig. 17.1). However, it is difficult to determine the degree of renal decline by the presence of albuminuria alone. Progressive renal impairment can occur without albuminuria in T1D DKD patients who have advanced glomerular lesions (Caramori et al., 2003). In T2D DKD patients, the absence of the transition from microalbuminuria to macroalbuminuria and albuminuria without renal decline can occur (Retnakaran et al., 2006, Gaede et al., 2004). Additionally, individuals can revert to normoalbuminuric status and still have renal insufficiency (Perkins et al., 2010).

For extreme phenotypes, overt macroalbuminuria indicates sustained renal damage and a likely progression to ESRD. Because of

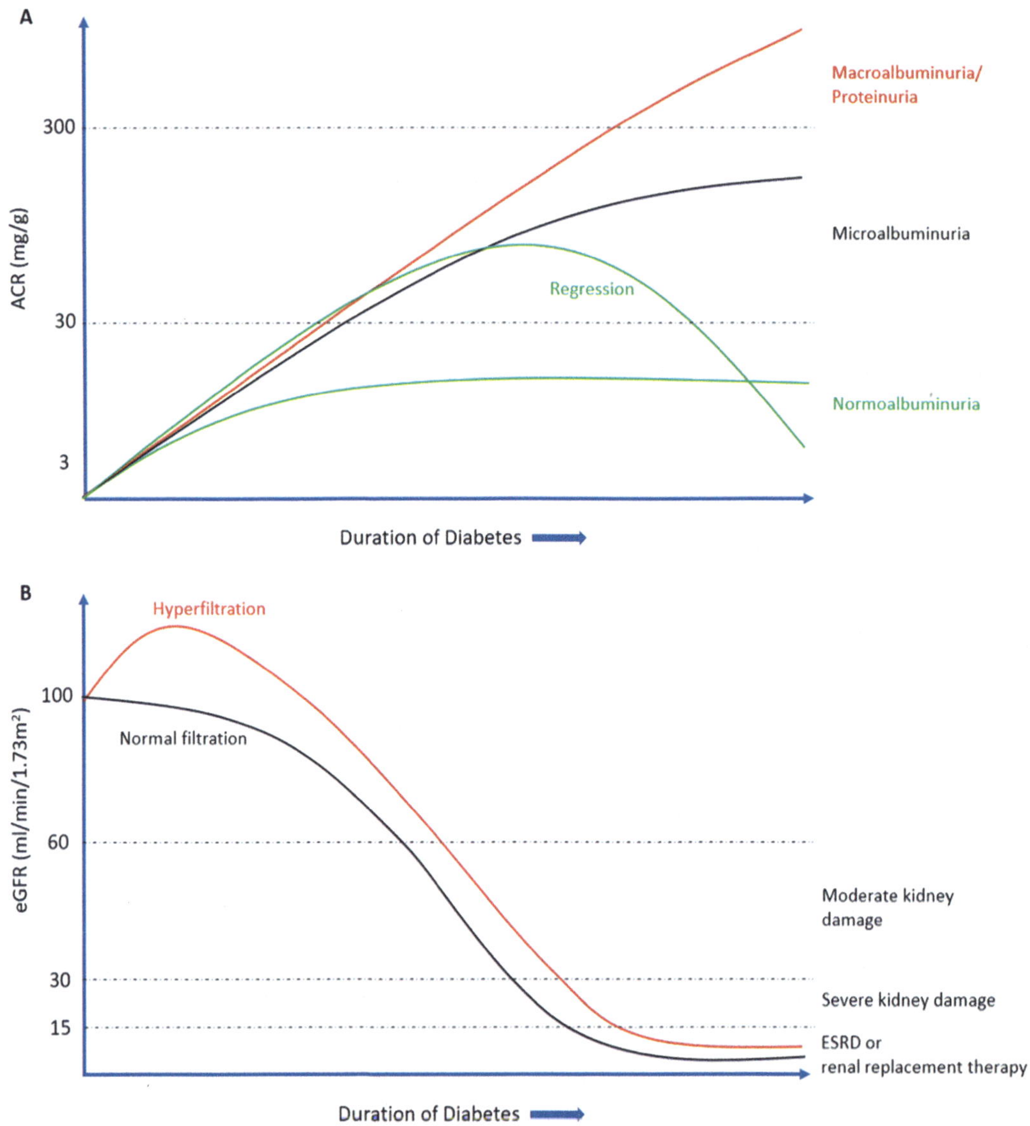

Fig. 17.1 Natural History of DKD. Graphical representations demonstrating the progression of DKD using clinical markers of (**A**) renal injury (albumin-creatinine ratio, ACR) and (**B**) renal function decline (estimated glomerular filtration rate, eGFR) in patients with diabetes. Macroalbuminuria/proteinuria is defined as ACR > 300 mg/g and severe kidney damage eGFR < 30 mL/min/1.73m^2. Regression from microalbuminuria to normoalbuminuria illustrates an example of the phenotypic heterogeneity that can occur with DKD

the dynamic presentation of albuminuria, eGFR and its longitudinal trajectory (i.e., rate of renal function decline) are considered better predictors of progression of DKD (Skupien et al., 2012). Importantly, patients with poor glycemic control do not always progress or ever develop DKD, suggesting that other influences, such as individual genetic risk, may play a role in both susceptibility and progression (Reidy et al., 2014).

When designing genetic studies of DKD, researchers primarily rely on ACR and eGFR measurements to define cohorts and examine the association of various albuminuria- or eGFR-

A

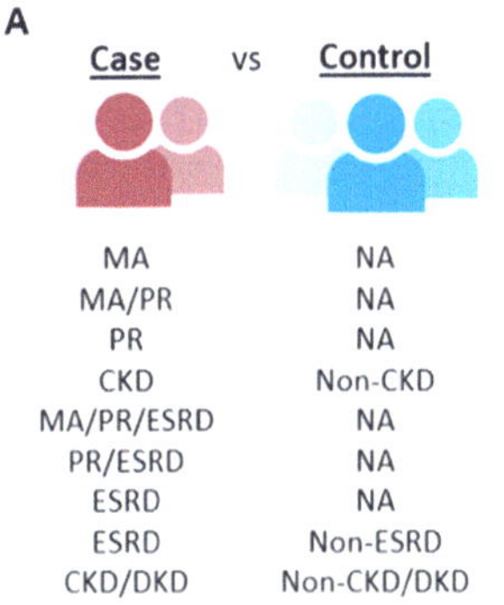

B

Clinical Measurement	Phenotype	Clinical Parameters
Albuminuria-based phenotype (uACR; mg/g)	NA	< 30
	MA	30 – 300
	PR	> 300
eGFR-based phenotype (ml/min/1.73m^2)	Non CKD	≥ 60
	CKD	< 60
Combined phenotype	Non CKD-DKD	eGFR ≥ 60 and uACR < 30
	CKD-DKD	eGFR < 45 and uACR ≥ 30
	ESRD	Renal replacement therapy or transplant

C

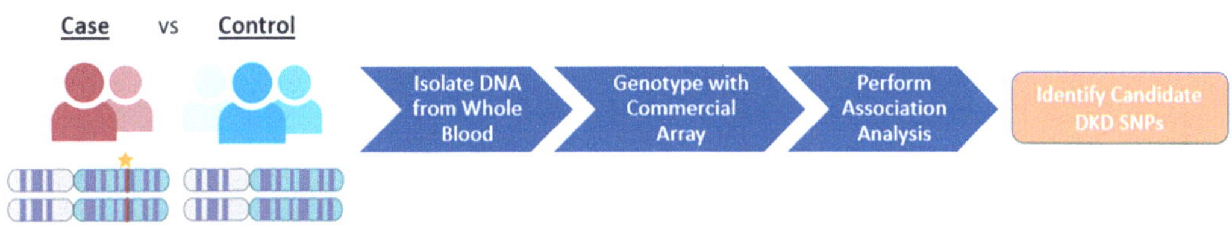

Fig. 17.2 GWASs of DKD. (**A**) Dichotomous phenotypes examined in case-control-based approaches with presented clinical parameters for each ACR-based, eGFR-based, or combined phenotype in (**B**). (**C**) General approaches for conducting a GWAS to identify candidate DKD SNPs. *NA*, normoalbuminuria; *MA*, microalbuminuria; *PR*, proteinuria; *CKD*, chronic kidney disease; *ESRD*, end-stage renal disease; *eGFR*, estimated glomerular filtration rate; *DKD*, diabetic kidney disease

based phenotypes with genetic variants (i.e., SNPs) (Fig. 17.2). Genetic studies typically dichotomize patients based on these phenotypes and perform case-control association analyses. For example, DKD 'case' subjects could be defined as those who have proteinuria or ESRD, and the frequency of SNPs observed in these patients could be compared with frequencies observed in diabetic 'control' patients with normoalbuminuria. Studies may also examine associations between SNPs and quantitative traits for ACR or eGFR, instead of dichotomizing cases and controls for association analyses. Unfortunately, no genetic analysis of renal function decline has been conducted to date (Li and Pezzolesi, 2018).

The Genetic Architecture of DKD through GWASs

While early genetic studies employed positional cloning and linkage analysis to identify potential causal regions for DKD susceptibility, current genetic approaches utilize GWASs to identify susceptibility markers. GWASs detect associations between the frequency of different alleles of SNPs and a disease trait within a given population and have been very powerful in complex, common disease genetics. Generally, GWASs analyze common SNP variation within the genomes of unrelated individuals. Here, we present the major findings from recent GWASs of T1D and T2D DKD patients (Table 17.1).

Early GWASs of DKD were primarily performed on T1D DKD cohorts, as there are less compounding risk factors compared to T2D DKD, such as cardiovascular health and age,

Table 17.1 Current DKD loci identified through GWAS approaches in various populations and renal phenotypes examined

Chr	Reported / Nearby Loci	SNP	*P*-value	Odds Ratio	Phenotype Examined	Diabetes	Population	References
2	*AFF3*	rs7583877	1.2×10^{-8}	1.29	ESRD vs. nonESRD	T1D	Caucasian	Sandholm et al. (2012)
2	*HS6ST1*	rs13427836	6.3×10^{-7}	0.19	uACR	T1D / T2D	Caucasian	Teumer et al. (2016)
2	*ERBB4*	rs7588550	2.1×10^{-7}	0.66	Prot./ESRD vs normo.	T1D	Caucasian	Sandholm et al. (2012)
2	*COL4A3*	rs55703767	5.34×10^{-12}	0.79	DN vs. normo.	T1D	Caucasian	Salem et al. (2019)
2	*COL4A3*	rs55703767	3.38×10^{-10}	0.83	All albuminuria vs. normo.	T1D	Caucasian	Salem et al. (2019)
2	*COL4A3*	rs55703767	5.30×10^{-9}	0.77	CKD + DN vs. eGFR $\geq$ 60, normo.	T1D	Caucasian	Salem et al. (2019)
2	*COL4A3*	rs55703767	9.28×10^{-9}	0.78	Macro. vs. normo.	T1D	Caucasian	Salem et al. (2019)
2	*COLEC11*	rs12615970	9.43×10^{-9}	0.76	CKD vs. no CKD	T1D	Caucasian	Salem et al. (2019)
2	*SSB*	rs1974990	1.4×10^{-6}	3.17	eGFR	T1D / T2D	Caucasian / Asian	van Zuydam et al. (2018)
2	*RND3/RBM43*	rs72858591	1.32×10^{-8}	1.55	Prot./ESKD vs. nondiabetic, nonESKD	T2D	African American	Guan et al. (2019)
3	*TAMM41*	rs142823282	8.32×10^{-10}	6.73	Micro. vs. normo.	T1D	Caucasian	Salem et al. (2019)
3	*LINC01266*	rs115061173	4.07×10^{-8}	9.40	ESRD vs. normo.	T1D	Caucasian	Salem et al. (2019)
3	*STAC*	rs116216059	1.37×10^{-8}	8.73	ESRD vs. nonESRD	T1D	Caucasian	Salem et al. (2019)
3	*SLITRK3*	rs58627064	4.38×10^{-11}	1.84	Prot./ESKD vs. nondiabetic, nonESKD	T2D	African American	Guan et al. (2019)
4	*PTPN13*	rs61277444	6.0×10^{-6}	1.42	ESRD vs. no DKD	T1D	Caucasian	Sandholm et al. (2017)
4	*HAND2-AS1*	rs145681168	2.06×10^{-7}	5.53	Micro. vs. normo.	T1D	Caucasian	Salem et al. (2019)
4	*MUC7*	rs191449639	1.32×10^{-8}	32.42	DN vs. normo.	T1D	Caucasian	Salem et al. (2019)
4	*GLRA3*	rs10011025	1.5×10^{-9}	0.21	AER	T1D	Caucasian	Sandholm et al. (2014)
5	*SNCAIP*	rs149641852	1.37×10^{-8}	9.01	CKD extreme vs. eGFR $\geq$ 60	T1D	Caucasian	Salem et al. (2019)
6	*SCAF8/ CNKSR3*	rs955333	1.3×10^{-8}	0.73	Prot./ESRD vs normo.	T2D	Trans-ethnic	Iyengar et al. (2015)
6	*SCAF8/ CNKSR3*	rs12523822	5.7×10^{-9}	0.57	Prot./ESRD vs normo.	T2D	Trans-ethnic	Iyengar et al. (2015)

6	*DDR1*	rs118124843	4.42×10^{-8}	3.79	Micro. vs normo.	T1D	Caucasian	Salem et al. (2019)
6	*GABRR1*	rs9942471	4.5×10^{-8}	1.25	Micro. vs. normo.	T1D / T2D	Caucasian / Asian	van Zuydam et al. (2018)
7	*CHN2/ CPVL*	rs39059	5.0×10^{-6}	1.39	Prot./ESRD vs normo.	T1D	Caucasian	Pezzolesi et al. (2009b)
7	*ELMO1*	rs741301	8.0×10^{-6}	2.67	Prot./ESRD vs normo.	T2D	Japanese	Shimazaki et al. (2005)
7	*ELMO1*	rs11769038, rs1882080	1.7×10^{-3}, 3.2×10^{-3}	1.24, 1.23	Prot./ESRD vs normo.	T1D	Caucasian	Pezzolesi et al. (2009a)
7	*CNTNAP2*	rs1989248	6.0×10^{-7}	1.26	CKD + DKD vs. eGFR ≥ 60, normo.	T1D	Caucasian	Sandholm et al. (2017)
7	*CNTNAP2*	rs1989248	1.8×10^{-6}	1.29	ESRD vs. no DKD	T1D	Caucasian	Sandholm et al. (2017)
7	*CNTNAP2*	rs731565	5.82×10^{-2}	1.22	Prot/ESRD vs normo.	T2D	Trans-ethnic	Iyengar et al. (2015)
7	*MBLAC1*	rs77273076	1.04×10^{-8}	9.16	Micro. vs normo.	T1D	Caucasian	Salem et al. (2019)
7	*PRKAG2*	rs10224002	2.7×10^{-8}	2.01	eGFR	T1D / T2D	Caucasian / Asian	van Zuydam et al. (2018)
8	*PRNCR1*	rs551191707	4.39×10^{-8}	1.70	ESRD vs. macro.	T1D	Caucasian	Salem et al. (2019)
9	*FRMD3*	rs10868025	5.0×10^{-7}	1.45	Prot./ESRD vs normo.	T1D	Caucasian	Pezzolesi et al. (2009b)
10	*NRG3*	rs72809865	7.4×10^{-6}	1.17	Combined DKD vs normo.	T1D	Caucasian	Sandholm et al. (2017)
10	*SORBS1*	rs1326934	5.69×10^{-7}	0.84	Prot./ESRD vs normo.	T1D	Caucasian	Germain et al. (2015)
11	*CARS*	rs451041	3.1×10^{-6}	1.36	Prot./ESRD vs normo.	T1D	Caucasian	Pezzolesi et al. (2009b)
11	*RAB38*	rs649529	5.8×10^{-7}	−0.14	ACR	T1D / T2D	Caucasian	Teumer et al. (2016)
11	*PLEKHA7*	rs183937294	1.65×10^{-8}	17.22	Micro. vs normo.	T1D	Caucasian	Salem et al. (2019)
13	*MYO16/ IRS2*	rs1411766/ rs17412858	1.8×10^{-6}	1.41	Prot./ESRD vs normo.	T1D	Caucasian	Pezzolesi et al. (2009b)
14	*STXBP6*	rs61983410	9.84×10^{-8}	0.79	Micro. vs normo.	T1D	Caucasian	Salem et al. (2019)
14	*PAPLN*	rs113554206	5.39×10^{-7}	4.60	Macro. vs. normo.	T1D	Caucasian	Salem et al. (2019)
15	*RGMA/ MCTP2*	rs12437854	2.0×10^{-9}	1.80	Prot./ESRD vs. nonESRD	T1D	Caucasian	Sandholm et al. (2012)
16	*UMOD*	rs11864909	2.1×10^{-12}	2.22	eGFR	T1D / T2D	Caucasian / Asian	van Zuydam et al. (2018)

(continued)

Table 17.1 (continued)

Chr	Reported / Nearby Loci	SNP	*P*-value	Odds Ratio	Phenotype Examined	Diabetes	Population	References
16	*FTO*	rs56094641	7.74×10^{-10}	1.23	Prot./ESRD vs normo.	T2D	Japanese	Taira et al. (2018)
17	*PRCD*	rs895157	7.70×10^{-8}	1.28	Prot./ESRD vs normo.	T2D	Japanese	Taira et al. (2018)
17	*ENPP7*	rs142563193	3.27×10^{-8}	0.69	Prot./ESRD vs. nondiabetic, nonESKD	T2D	African American	Guan et al. (2019)
17	*ENPP7*	rs142671759	5.53×10^{-9}	2.26	Prot./ESRD vs. nondiabetic, nonESKD	T2D	African American	Guan et al. (2019)
18	*18p11*	rs185299109	1.28×10^{-8}	20.75	CKD vs no CKD	T1D	Caucasian	Salem et al. (2019)
19	*GNG7*	rs4807299	3.21×10^{-8}	1.67	Prot./ESRD vs. nondiabetic, nonESKD	T2D	African American	Guan et al. (2019)
20	*BMP7*	rs144434404	2.67×10^{-9}	6.78	Micro. vs normo.	T1D	Caucasian	Salem et al. (2019)
20	*PLCB4*	rs2206136	2.1×10^{-8}	1.20	CKD vs no CKD	T1D / T2D	Caucasian/ Asian	van Zuydam et al. (2018)
22	*MHY9*	rs5750250	7.7×10^{-8}	1.27	Prot./ESRD vs normo.	T2D	Trans-ethnic	Iyengar et al. (2015)
22	*APOL1*	rs136161	5.23×10^{-7}	1.36	Prot./ESRD vs normo.	T2D	Trans-ethnic	Iyengar et al. (2015)
22	*APOL1*	rs9622363	1.42×10^{-10}	0.77	Prot./ESRD vs. nondiabetic, nonESKD	T2D	African American	Guan et al. (2019)

Chr, chromosome; *SNP*, single nucleotide polymorphism; *Normo,* normoalbuminuria; *Micro*, microalbuminuria; *Macro/Prot*, macroalbuminuria/proteinuria; *DKD/DN*, diabetic kidney disease; *CKD*, chronic kidney disease; *ESRD/ESKD*, end-stage renal/kidney disease; *eGFR*, estimated glomerular filtration rate; *ACR*, albumin-creatinine ratio; *AER*, albumin excretion rate which can increase phenotypic heterogeneity. One of the first T1D DKD GWASs identified *FRMD3* and *CARS* in 820 diabetic cases with either persistent macroalbuminuria or ESRD (Pezzolesi et al., 2009b). The original discovery cohort signals did not reach genome-wide significance (defined as a *P*-value $< 5 \times 10^{-8}$), most likely due to its modest sample size. Associations at these loci were reproducible in a cohort composed of patients from the Diabetes Control and Complications Trial/Epidemiology of Diabetes Interventions and Complications Study (DCCT/EDIC) (Pezzolesi et al., 2009b).

Another early GWAS analysis and a subsequent replication study identified another reproducible candidate DKD gene, *ELMO1*, in both T2D and T1D study cohorts (Shimazaki et al., 2005, Pezzolesi et al., 2009a). As genetic associations at both *FRMD3* and *ELMO1* were independently replicated, the functional roles of both genes in DKD susceptibility have been examined further. In a study by Martini et al., one SNP located in the promoter region of *FRMD3* was computationally and functionally demonstrated to show increased binding of transcription factors, which might downregulate FRMD3 expression and lead to increased DKD susceptibility (Martini et al., 2013). Interestingly, for *ELMO1*, most functional studies have demonstrated that increased renal expression of ELMO1 occurs in DKD mice, overexpression of ELMO1 can promote increased expression of extracellular matrix proteins *in vitro,* and increased expression can cause pathological features of DKD such as glomerular changes and urinary excretion of albumin (Shimazaki et al., 2005, Hathaway et al., 2016). Contrarily, another study challenged the role of ELMO1 in DKD, demonstrating that introducing additional ELMO1 mRNA in PDX1 morphant diabetic zebrafish is renoprotective, by restoring structural integrity in the pronephros (Sharma et al., 2016).

Since these early genetic studies, realizing that common variants have modest impact on disease risk (e.g., odd ratio (OR) typically 1.10–1.40) and that larger sample sizes can improve detection of these signals, collaborative efforts from investigators across the world have empowered larger GWASs. In 2012, the Genetics of Nephropathy—an International Effort (GENIE) consortium—identified two genome-wide significant loci for associations with ESRD (an intronic SNP rs7583877 in *AFF3* and an intergenic SNP rs12437854 located between *RGMA* and *MCTP2*), in a meta-analysis of >12,000 individuals (Sandholm et al., 2012). Additionally, although not genome-wide significant, a strong association with diabetic nephropathy (defined as persistent macroalbuminuria or ESRD) was also seen (intronic SNP rs7588550 in the *ERBB4* gene).

In addition to increasing sample sizes, to overcome limitations of these early GWASs due to imprecise and variable phenotypic heterogeneity, Sandholm et al. analyzed a range of ACR- and eGFR-based subphenotypes, which span different stages and severities of DKD in a meta-analysis of 4 large cohorts (Sandholm et al., 2017). However, despite being well-powered to detect variants with a MAF $\geq$ 10% and an OR $\geq$ 1.55, no SNPs reached genome-wide significance in this study. Suggestive associations for ESRD-based phenotypes ("ESRD vs. no DKD" and "CKD + DKD vs. eGFR >60, normoalbuminuric") were detected in *CNTNAP2* (rs1989248), *PTPN13* (rs61277444), and *AFF3* (rs7562121). As many of the samples included in this study overlap with those in the previous GWAS from the GENIE consortium, this suggestive association at *AFF3* does not provide independent replication of the previously reported association. Interestingly, independent replication was observed in this study for signals reported at *ELMO1* (Shimazaki et al., 2005), the *MYO16-IRS2* locus on chromosome 13q (Pezzolesi et al., 2009b), and *SIK1* (Sambo et al., 2014).

Combined Databases

Following a study design similar to that of Sandholm et al. (Sandholm et al., 2017), the SUrrogate markers for Micro- and Macrovascular hard endpoints for Innovative diabetes Tools (SUMMIT) consortium performed the largest T2D DKD GWAS composed of > 20,000 T2D patients using eight dichotomous and quantitative DKD subphenotypes (van Zuydam et al., 2018). To further increase sample size, a combined T1D and T2D meta-analysis that included samples of European and Asian ancestry was performed. In total, the sample size of this analysis exceeded 40,000 subjects. Despite this, the yield for discovery of novel loci was modest. One SNP, rs2206136 in *PLCB4*, reached significance with the "CKD" phenotype (P-value $= 2.1 \times 10^{-8}$) in discovery GWAS but not in replication analyses. A novel signal at rs9942471, a position 7 kb upstream of *GABRR1*, was associated with increased microalbuminuria risk in European subjects. Additionally, the combined meta-analysis for eGFR identified a new genome-wide association mapping near *SSB*. Interestingly, as the authors note, while this study aimed to overcome some of the limitations of earlier studies, its significantly larger sample size came with increased phenotypic and, likely genetic, heterogeneity that may have impacted its ability to detect additional genetic associations.

Coding vs Noncoding Variants

SNPs included in these studies mostly map to nonprotein coding regions, which limits the interpretation of the candidate SNP's role in disease susceptibility. Similarly, identified SNPs may be in linkage disequilibrium with causal variants some distance from the identified association, e.g., a causal variant may localize to a gene several hundred kilo-basepair away from the detected association, thereby complicating investigations into the functionality of an associated SNP. Importantly, completion of the National Heart, Lungs, and Blood Institute's Exome Sequencing Project led to the development of a new generation of commercially available genotyping arrays that, in addition to common noncoding variation, now includes coding or 'exome' variants with MAFs as low as 0.5% across the genome (Auer et al., 2016).

Using this technology, the Juvenile Diabetes Research Foundation funded Diabetic Nephropathy Collaborative Research Initiative (JDRF-DNCRI) recently performed the largest T1D DKD GWAS, with 19,406 T1D individuals with and without kidney disease, and identified 16 novel genome-wide significant associations with various dichotomous phenotypes (Salem et al., 2019). The strongest association (rs55703767, a common missense variant in *COL4A3*) was a protective variant against DN, linked to lower glomerular basement membrane thickness. In contrast to the SUMMIT study, which included T1D and T2D individuals of European and Asian ancestry from 16 different studies genotyped on various platforms, the JDRF-DNCRI study benefited from uniform genotyping, quality control procedures, and standardized phenotype definitions. Adoption of this study design, coupled with an increased sample size, helped foster the discovery of several novel associations with DKD in this study.

Ethnic Patterns

Although the majority of genetic studies of DKD have focused on subjects of European ancestry, several studies have included individuals of different ethnicities and demonstrated potential population-specific signals. The majority of these efforts have, however, been in patients with T2D. In the only trans-ethnic meta-analysis of T2D DKD patients, conducted as part of the Family Investigation of Nephropathy and Diabetes (FIND) study, Iyengar et al. identified SNP rs955333, located between *SCAF8* and *CNKSR3*, among patients of American Indian, Europeans, and Mexican ancestries (Iyengar et al., 2015). No SNP reached genome-wide significance in African American T2D patients, although suggestive associations were seen in *APOL1* and *MYH9* genes. As *APOL1* is a well-known ESRD susceptibility locus in African Americans, this signal may be due to nonDKD influences.

In addition to early GWASs that first reported associations at *ELMO1*, a more recent GWAS in T2D DKD Japanese patients identified a genome-wide significant association with rs56094641 in *FTO* as a susceptibility marker for DKD (Taira et al., 2018). The *FTO* locus has repeatedly been reported to be associated with obesity; however, its association with DKD was not affected by adjustment for BMI in this study. Interestingly, this association was not replicated in European T2D DKD patients. Many studies of T2D attributed ESRD in African Americans have reported additional significant loci, including *SLITRK3*, *ENPP7*, *GNG7*, *SFI1*, and *LIMK2* (Guan et al., 2019, Palmer et al., 2014). With regard to T1D DKD, there have not been any studies to date to address the genetics of T1D DKD in African Americans; intense effort is needed to expand this area of research.

Contributions of Low Frequency and Rare Variations in DKD Patients

Similar to other common, complex diseases, most of the variants identified through GWASs of DKD only explain a small proportion of its overall risk. It has been proposed that some of this 'missing heritability,' which is not accounted for by common variation, may be attributed to additional risk conferred by low frequency ($0.5\% \leq$ MAF $< 5\%$) and rare (MAF $< 0.5\%$) variation (Lee et al., 2014; Manolio et al. 2009). Although next-generation GWASs are beginning to explore their role in DKD susceptibility, challenges for investigating low frequency and rare variation persist in all complex disease genetics. Among these, the sample size requirement needed to detect a robust association with low frequency and rare variation remains a major hurdle.

Statistical Challenges

For single variant association tests, the sample sizes to achieve 80% power (at a significance level of 5×10^{-8} with an OR $= 1.4$) would need to be 6400, 54,000, and 540,000 for variants with MAFs $= 0.1$, 0.01, and 0.001, respectively (Lee et al., 2014). Unfortunately, current DKD cohorts do not exceed 50,000 patients. Additionally, while newer commercial genotyping arrays do include as many as 200,000–250,000 low frequency and rare variants, these arrays are limited to the variants included in the array's design, such that novel variants and those not selected for inclusion cannot be assessed. Finally, while NGS-based approaches (e.g., WES and WGS) are ideal for discovering and interrogating low frequency and rare variation, these approaches still remain costly at scale, despite their continuous declining cost and, thereby, have had limited utility in the field.

Despite these challenges, WES and GWASs including low frequency and rare variants have been utilized with some success to explore the contribution of these classes of variants in DKD. In 2017, Sandholm et al. performed WES in 997 T1D DKD subjects (Sandholm et al., 2017). While no variants reached exome-wide significance (*P*-value $< 2.5 \times 10^{-7}$) for single variant and gene aggregation tests, nominal (*P*-value $< 9 \times 10^{-5}$) associations with rare missense alleles were identified; the strongest association was observed at an intronic SNP within *NVL* (rs188427269, MAF $= 0.2\%$, and *P*-value $= 3.3 \times 10^{-7}$). Importantly, the lack of exome-wide significance in this study is likely due to its modest sample size, as 997 patients may not be enough to detect significant associations with rare variants. In the T1D DKD GWAS from the JDRF-DNCRI, Salem et al. reported several associations with low frequency variants included on a commercial genotyping array associated with ESRD, microalbuminuria, and DN (Salem et al., 2019). The lowest *P*-value was observed at low frequency intronic variants in *PAPLN* (rs113554206, MAF $= 1.2\%$, and *P*-value $= 5.39 \times 10^{-7}$).

Given the excess rate of ESRD in African Americans relative to European Americans, Guan et al. evaluated the impact of low-frequency variants in 47 candidate genes involved in kidney structure (e.g., podocyte, glomerular basement membrane, and renal tubular cell genes) with ESRD in approximately 5000 African American subjects (Guan et al., 2016). After excluding carriers of *APOL1* risk alleles, statistically significant associations were observed at two missense variants, located in *CLDN8* and *COL4A3*. A subsequent exome-wide association study by the same group identified suggestive associations at *OTUD7B*, *IFITN3*, and *DLGAP5* (Guan et al., 2018).

While these studies are the first to begin investigating the role of low frequency and rare variants in DKD, these associations should be interpreted with caution due to an overall lack of genome-wide significance due to their limited sample sizes. These studies may suggest that low frequency or rare variants do not contribute significantly to the underlying genetic component of DKD, as would be seen in monogenic conditions, but may still modestly impact its susceptibility.

From Genetic Causes of DKD to Personalized Therapies

Although progress in personalized therapeutics in DKD has been limited, genetic studies in DKD have shown promise toward translating robust findings to improved patient care. Pharmacogenomics examines how genetic variation can impact beneficial or adverse reactions to therapies and, using genetic data to predict therapeutic response, is beginning to emerge in clinical practice (Roden et al., 2019). This approach has been applied to evaluate reactions to therapies for DKD patients.

One of the earliest studied pharmacogenomic loci in DKD is the insertion-deletion polymorphism of the *ACE* gene (rs4646994), where the insertion (*I*) or deletion (*D*) of a 287 basepair Alu repeat in intron 16 of *ACE* affects patient responses to angiotensin converting enzyme (ACE) inhibitors, a first-line treatment for DKD. In T1D DKD patients, early reports indicated that individual homozygous for the deletion allele (*DD*) was more likely to progress to ESRD and lacked a response to renoprotective therapies. However, a study from the Steno Diabetes Center demonstrated that there is no difference in the renoprotective effects in either *II* or *DD* individuals (Andersen et al., 2003).

Angiotensin Modulators

These conflicting results in T1D DKD patients could result from the modest sample sizes of these studies or the efficacies of the drugs examined. More recently, Wang et al. investigated the renoprotective efficacy of valsartan, an angiotensin receptor blocker (ARB), in T2D patients carrying the *ACE I/D* polymorphism and reported an increased risk of developing DKD among carriers of the *D* allele (Wang et al., 2016). They also observed a beneficial decrease in albuminuria levels for patients with the *ID* and *DD* genotypes when treated with valsartan. This observation is likely due to decreased plasma ACE activity in patients carrying this polymorphism.

While no genetic studies of DKD have reported a significant association with *CYP2C9*, two coding polymorphisms, *2 (Arg144Cys, rs1799853) and *3 (Ile359Leu, rs1057910), can influence CYP2C9 metabolism of ARBs. As ARBs can improve renal function in DKD patients, some studies have examined pharmacological responses with these variants. Lajer et al. described that patients without the CYP2C9*3 polymorphism had a significant change in systolic blood pressure after 4 months compared to *3 carriers. However, both *3 and non*3 carriers had no significant difference in urinary albumin excretion during the study period (Lajer et al., 2007).

Future Directions

Overcoming DKD's phenotypic heterogeneity has proven to be among the most challenging. As a progressive disease, the lack of consistent findings across studies can, in part, be attributed to the array of clinical definitions established by different studies. Since most genetic approaches are inherently cross-sectional, investigators typically examine multiple phenotypes that span all stages and presentations of renal disease. This approach has drawbacks, as a patient's renal status could change over time (i.e., a microalbuminuric patient reverts to normoalbuminuria or, alternatively, progresses to proteinuria or ESRD), leading to misclassification. Misclassification bias then could result in a significant number of false-negative associations. Indeed, longitudinal investigations of DKD have shed new light on the natural history of this disease and its heterogeneous nature (Perkins et al., 2010, Skupien et al., 2012, Krolewski and Bonventre, 2012, Krolewski, 2015). Additionally, the rate of renal function decline varies widely among DKD patients; perhaps, genetic factors influence whether this decline is slow, where ESRD is reached after decades of diabetes, or decline is rapid and progression to ESRD occurs in as little as a few years after the onset of diabetes (Skupien et al., 2012, Krolewski and Bonventre, 2012, Krolewski, 2015). Because of the heterogeneity of the subphenotypes of DKD, considering such disparate patients simultaneously likely poses challenges to uncovering the genetic basis of DKD.

Additionally, it is possible that a subset of DKD patients have nondiabetic kidney disease (NDKD) coincident with diabetes or concurrent DKD and NDKD. Renal biopsies from patients with diabetes and kidney disease suggest that as many as 30–80% of patients diagnosed with DKD actually had kidney disease attributed to nondiabetic causes (Haider et al., 2011, Sharma et al., 2013, Zhuo et al., 2013). Importantly, this has major implications in terms of patient care, monitoring, and family planning, as well as for research studies aimed at understanding the factors that contribute to DKD. To begin investigating this, we have recently shown that a subset of 'DKD' patients carry rare pathogenic variants indicative of NDKD, e.g., variants in *COL4A3* (associated with focal segmental glomerulosclerosis, thin basement membrane nephropathy, and Alport Syndrome) and *REN* (associated with autosomal dominant tubulointerstitial kidney disease) (Lazaro-Guevara et al., 2020).

Large research consortiums are now bringing together investigators across the globe to combine resources and enhance the cohort sizes of current genetic studies, such as the Juvenile diabetes research foundation-Diabetic nephropathy collaborative research initiative (JDRF-DNCRI) most recent T1D DKD GWAS. Even though establishing large cohorts increases statistical power, how many samples are sufficient to accurately identify all of the contributing genetic variation underlying DKD? Is there a sample size threshold that diminishes any further efforts of larger GWASs?

Epigenomics

In addition to variation at the DNA sequence level, increasing research shows that epigenetic and epigenomic mechanisms, i.e., heritable changes in gene expression, including DNA methylation and histone modifications, which do not alter the sequence of DNA, are involved in DKD pathogenesis (Kato and Natarajan, 2019). Nucleic acid and chromatin histone complex modifications elicit changes in gene expression by affecting the binding of potential transcriptional regulators. It has been demonstrated that "metabolic memory" of previous metabolic dysregulation exists, where patients with longstanding diabetes can still experience complications even after intense glucose control, highlighting prolonged modification of gene expression patterns in DKD risk (Reidy et al., 2014). Therefore, identifying epigenetic signatures could lead to new treatments and diagnostic biomarkers for DKD initiation and progression.

References

Alicic RZ, Rooney MT, Tuttle KR (2017) Diabetic kidney disease: challenges, progress, and possibilities. Clin J Am Soc Nephrol 12:2032–2045

Altshuler D, Brooks LD, Chakravarti A, Collins FS, Daly MJ, Donnelly P (2005) A haplotype map of the human genome. Nature 437:1299–1320

Anders HJ, Huber TB, Isermann B, Schiffer M (2018) CKD in diabetes: diabetic kidney disease versus non-diabetic kidney disease. Nat Rev Nephrol 14:361–377
Andersen S, Tarnow L, Cambien F, Rossing P, Juhl TR, Deinum J, Parving HH (2003) Long-term renoprotective effects of losartan in diabetic nephropathy: interaction with ACE insertion/deletion genotype? Diabetes Care 26:1501–1506
Auer PL, Reiner AP, Wang G, Kang HM, Abecasis GR et al (2016) Guidelines for large-scale sequence-based complex trait association studies: lessons learned from the NHLBI exome sequencing project. Am J Hum Genet 99:791–801
Borch-Johnsen K, Norgaard K, Hommel E, Mathiesen ER, Jensen JS, Deckert T, Parving HH (1992) Is diabetic nephropathy an inherited complication? Kidney Int 41:719–722
Brenner BM, Cooper ME, De Zeeuw D, Keane WF, Mitch WE et al (2001) Effects of losartan on renal and cardiovascular outcomes in patients with type 2 diabetes and nephropathy. N Engl J Med 345:861–869
Caramori ML, Fioretto P, Mauer M (2003) Low glomerular filtration rate in normoalbuminuric type 1 diabetic patients: an indicator of more advanced glomerular lesions. Diabetes 52:1036–1040
Faronato PP, Maioli M, Tonolo G, Brocco E, Noventa F et al (1997) Clustering of albumin excretion rate abnormalities in Caucasian patients with NIDDM. The Italian NIDDM Nephropathy Study Group. Diabetologia 40:816–823
Fioretto P, Steffes MW, Barbosa J, Rich SS, Miller ME, Mauer M (1999) Is diabetic nephropathy inherited? Studies of glomerular structure in type 1 diabetic sibling pairs. Diabetes 48:865–869
Fioretto P, Caramori ML, Mauer M (2008) The kidney in diabetes: dynamic pathways of injury and repair. The Camillo Golgi Lecture 2007. Diabetologia 51:1347
Freedman BI, Bostrom M, Daeihagh P, Bowden DW (2007) Genetic factors in diabetic nephropathy. Clin J Am Soc Nephrol 2:1306–1316
Gaede P, Tarnow L, Vedel P, Parving HH, Pedersen O (2004) Remission to normoalbuminuria during multifactorial treatment preserves kidney function in patients with type 2 diabetes and microalbuminuria. Nephrol Dial Transplant 19:2784–2788
Germain M, Pezzolesi MG, Sandholm N, Mcknight AJ, Susztak K et al (2015) SORBS1 gene, a new candidate for diabetic nephropathy: results from a multi-stage genome-wide association study in patients with type 1 diabetes. Diabetologia 58:543–548
Gu HF (2019) Genetic and epigenetic studies in diabetic kidney disease. Front Genet 10:507
Guan M, Ma J, Keaton JM, Dimitrov L, Mudgal P et al (2016) Association of kidney structure-related gene variants with type 2 diabetes-attributed end-stage kidney disease in African Americans. Hum Genet 135:1251–1262
Guan M, Keaton JM, Dimitrov L, Hicks PJ, Xu J et al (2018) An exome-wide association study for type 2 diabetes-attributed end-stage kidney disease in African Americans. Kidney Int Rep 3:867–878
Guan M, Keaton JM, Dimitrov L, Hicks PJ, Xu J et al (2019) Genome-wide association study identifies novel loci for type 2 diabetes-attributed end-stage kidney disease in African Americans. Hum Genomics 13:21
Haider DG, Peric S, Friedl A, Fuhrmann V, Wolzt M, Horl WH, Soleiman A (2011) Kidney biopsy in patients with diabetes mellitus. Clin Nephrol 76:180–185
Hathaway CK, Chang AS, Grant R, Kim HS, Madden VJ, Bagnell CR Jr, Jennette JC, Smithies O, Kakoki M (2016) High Elmo1 expression aggravates and low Elmo1 expression prevents diabetic nephropathy. Proc Natl Acad Sci U S A 113:2218–2222
Iyengar SK, Sedor JR, Freedman BI, Kao WH, Kretzler M et al (2015) Genome-wide association and trans-ethnic meta-analysis for advanced diabetic kidney disease: family investigation of nephropathy and diabetes (FIND). PLoS Genet 11:e1005352
Kato M, Natarajan R (2019) Epigenetics and epigenomics in diabetic kidney disease and metabolic memory. Nat Rev Nephrol 15:327–345
Krolewski AS (2015) Progressive renal decline: the new paradigm of diabetic nephropathy in type 1 diabetes. Diabetes Care 38:954–962
Krolewski AS, Bonventre JV (2012) High risk of ESRD in type 1 diabetes: new strategies are needed to retard progressive renal function decline. Semin Nephrol 32:407–414
Lajer M, Tarnow L, Andersen S, Parving HH (2007) CYP2C9 variant modifies blood pressure-lowering response to losartan in Type 1 diabetic patients with nephropathy. Diabet Med 24:323–325
Lander ES, Linton LM, Birren B, Nusbaum C, Zody MC et al (2001) Initial sequencing and analysis of the human genome. Nature 409:860–921
Lazaro-Guevara, J., Morales, J. F., Wright, A. H., Gunville, R., Frodsham, S. G., et al. 2020. Targeted next-generation identifies pathogenic variants in kidney-disease related gene in phatients with diabetic kidney disease. https://doi.org/10.1101/2020.05.18.102822.
Lee S, Abecasis GR, Boehnke M, Lin X (2014) Rare-variant association analysis: study designs and statistical tests. Am J Hum Genet 95:5–23
Li M, Pezzolesi MG (2018) Advances in understanding the genetic basis of diabetic kidney disease. Acta Diabetol 55:1093–1104
Macisaac RJ, Ekinci EI, Jerums G (2014) Markers of and risk factors for the development and progression of diabetic kidney disease. Am J Kidney Dis 63:S39–S62
Manolio TA, Collins FS, Cox NJ, Goldstein DB, Hindorff LA et al (2009) Finding the missing heritability of complex diseases. Nature 461:747–753
Martini S, Nair V, Patel SR, Eichinger F, Nelson RG et al (2013) From single nucleotide polymorphism to transcriptional mechanism: a model for FRMD3 in diabetic nephropathy. Diabetes 62:2605–2612

Palmer ND, Ng MC, Hicks PJ, Mudgal P, Langefeld CD, Freedman BI, Bowden DW (2014) Evaluation of candidate nephropathy susceptibility genes in a genome-wide association study of African American diabetic kidney disease. PLoS One 9:e88273

Perkins BA, Ficociello LH, Roshan B, Warram JH, Krolewski AS (2010) In patients with type 1 diabetes and new-onset microalbuminuria the development of advanced chronic kidney disease may not require progression to proteinuria. Kidney Int 77:57–64

Pezzolesi MG, Katavetin P, Kure M, Poznik GD, Skupien J, Mychaleckyj JC, Rich SS, Warram JH, Krolewski AS (2009a) Confirmation of genetic associations at ELMO1 in the GoKinD collection supports its role as a susceptibility gene in diabetic nephropathy. Diabetes 58:2698–2702

Pezzolesi MG, Poznik GD, Mychaleckyj JC, Paterson AD, Barati MT et al (2009b) Genome-wide association scan for diabetic nephropathy susceptibility genes in type 1 diabetes. Diabetes 58:1403–1410

Quinn M, Angelico MC, Warram JH, Krolewski AS (1996) Familial factors determine the development of diabetic nephropathy in patients with IDDM. Diabetologia 39:940–945

Reidy K, Kang HM, Hostetter T, Susztak K (2014) Molecular mechanisms of diabetic kidney disease. J Clin Invest 124:2333–2340

Retnakaran R, Cull CA, Thorne KI, Adler AI, Holman RR (2006) Risk factors for renal dysfunction in type 2 diabetes: U.K. Prospective Diabetes Study 74. Diabetes 55:1832–1839

Reutens AT (2013) Epidemiology of diabetic kidney disease. Med Clin North Am 97:1–18

Roden DM, Mcleod HL, Relling MV, Williams MS, Mensah GA, Peterson JF, Van Driest SL (2019) Pharmacogenomics. Lancet 394:521–532

Salem RM, Todd JN, Sandholm N, Cole JB, Chen WM et al (2019) Genome-wide association study of diabetic kidney disease highlights biology involved in glomerular basement membrane collagen. J Am Soc Nephrol 30:2000–2016

Sambo F, Malovini A, Sandholm N, Stavarachi M, Forsblom C et al (2014) Novel genetic susceptibility loci for diabetic end-stage renal disease identified through robust naive Bayes classification. Diabetologia 57:1611–1622

Sandholm N, Salem RM, Mcknight AJ, Brennan EP, Forsblom C et al (2012) New susceptibility loci associated with kidney disease in type 1 diabetes. PLoS Genet 8:e1002921

Sandholm N, Forsblom C, Makinen VP, Mcknight AJ, Osterholm AM et al (2014) Genome-wide association study of urinary albumin excretion rate in patients with type 1 diabetes. Diabetologia 57:1143–1153

Sandholm N, Van Zuydam N, Ahlqvist E, Juliusdottir T, Deshmukh HA et al (2017) The genetic landscape of renal complications in type 1 diabetes. J Am Soc Nephrol 28:557–574

Seaquist ER, Goetz FC, Rich S, Barbosa J (1989) Familial clustering of diabetic kidney disease. Evidence for genetic susceptibility to diabetic nephropathy. N Engl J Med 320:1161–1165

Sharma SG, Bomback AS, Radhakrishnan J, Herlitz LC, Stokes MB, Markowitz GS, D'agati VD (2013) The modern spectrum of renal biopsy findings in patients with diabetes. Clin J Am Soc Nephrol 8:1718–1724

Sharma KR, Heckler K, Stoll SJ, Hillebrands JL, Kynast K et al (2016) ELMO1 protects renal structure and ultrafiltration in kidney development and under diabetic conditions. Sci Rep 6:37172

Shimazaki A, Kawamura Y, Kanazawa A, Sekine A, Saito S et al (2005) Genetic variations in the gene encoding ELMO1 are associated with susceptibility to diabetic nephropathy. Diabetes 54:1171–1178

Skupien J, Warram JH, Smiles AM, Niewczas MA, Gohda T, Pezzolesi MG, Cantarovich D, Stanton R, Krolewski AS (2012) The early decline in renal function in patients with type 1 diabetes and proteinuria predicts the risk of end-stage renal disease. Kidney Int 82:589–597

Taira M, Imamura M, Takahashi A, Kamatani Y, Yamauchi T et al (2018) A variant within the FTO confers susceptibility to diabetic nephropathy in Japanese patients with type 2 diabetes. PLoS One 13:e0208654

Teumer A, Tin A, Sorice R, Gorski M, Yeo NC et al (2016) Genome-wide association studies identify genetic loci associated with albuminuria in diabetes. Diabetes 65:803–817

Van Zuydam NR, Ahlqvist E, Sandholm N, Deshmukh H, Rayner NW et al (2018) A genome-wide association study of diabetic kidney disease in subjects with type 2 diabetes. Diabetes 67:1414–1427

Wang Y, Peng W, Zhang X, Qiao H, Wang L, Xu Z, Wu C (2016) The association of ACE gene polymorphism with diabetic kidney disease and renoprotective efficacy of valsartan. J Renin-Angiotensin-Aldosterone Syst 17:1470320316666749

Zhuo L, Ren W, Li W, Zou G, Lu J (2013) Evaluation of renal biopsies in type 2 diabetic patients with kidney disease: a clinicopathological study of 216 cases. Int Urol Nephrol 45:173–179

Epigenetics and Chronic Inflammation: Role in Early Detection of Type 2 Diabetes

18

Meenu Ghai

Contents

Introduction 238
Metabolic Memory 238

Adipose Tissue Changes and Cytokine Expression 238

Chronic Inflammation and Insulin Resistance 239

Epigenetics of T2D 239
Hyperglycemia and Epigenetics 239
Investigations in Twins 239
Epigenetic Wide Association Studies 240

Noncoding RNA 240

MicroRNAs in Type 2 Diabetes and Inflammation 241

Long Noncoding RNA in Type 2 Diabetes 241

Noncoding RNAs as Therapeutics Targets 241
High Fat Diet and Polyphenols 242
Beta Cell Dysfunction 242

Early Detection/Relative Risk Tests 242

Ongoing Clinical Trials 244

References 244

Abstract

Low-grade inflammation initiates even before type 2 diabetes (T2D) is diagnosed. As a response to change in inflammatory profile of cells and tissues, epigenetics modifications such as DNA methylation (DNAm) and non-coding RNAs provide early cues about forth-coming discord in the bodily system. Thus, targeting epigenetic alterations that occur in the initial phases leading to T2D can provide therapeutics, which will prevent or halt the cascade of T2D and associated long-term complications. New biomarkers need to be identified for prediction and monitoring the progression of T2D. Early detection of

M. Ghai (✉)
Department of Genetics, School of Life Sciences, Westville Campus, University of Kwa-Zulu Natal, Durban, South Africa
e-mail: ghai@ukzn.ac.za

J. Faintuch and S. Faintuch (eds.), *Obesity and Diabetes*, https://doi.org/10.1007/978-3-030-53370-0_18

diabetes or prediabetes will also protect individuals from undesirable effects of chronic inflammation and metabolic memory.

Keywords

Type 2 diabetes · Chronic inflammation · DNAm · Noncoding RNAs · Metabolic memory

Introduction

The risk factors of type 2 diabetes (T2D) include genetic predisposition, age, and ethnicity; however, environmental factors should similarly be taken into account. T2D is characterized by hyperglycemia and combination of insulin resistance and/or insufficient insulin production by beta cells, which leads to metabolic derangements and widespread complications (Rawshani et al. 2017; Sladek 2018).

Chronic inflammation plays a role in T2D, sometimes even before disease onset, even though it is not clinically prominent in most cases (Reddy et al. 2019; Tsalamandris et al. 2019). It is characterized by increased pro-inflammatory cytokines, histological changes, and immune shifts (Gregor and Hotamisligil 2011; Hotamisligil 2017). "Self-"antigens can emerge and promote autoimmune activation. Insulin secretion and hyperglycemia can also be further impaired.

Metabolic Memory

Prolonged low-grade inflammation could mediate metabolic memory (Ceriello 2009; Testa et al. 2017) (Fig. 18.1). Intensive glycemic control is often recommended for the management of glucose homeostasis and microvascular complications (Holman et al. 2008; Gaede et al. 2008).

Demonstrated genetic variants are not responsible for more than 10% of the heritable risk for T2D, while rare variants, gene–environment interactions, gene–gene interactions, and epigenetic factors (Kirchner et al. 2013) could be underestimated. Epigenetic mechanisms such as DNA methylation (DNAm) and noncoding RNAs regulate inflammatory gene expression and should deserve much attention (Naidoo et al. 2018).

Adipose Tissue Changes and Cytokine Expression

Obesity leads to changes in tissue cell distribution and inflammatory activity, including pro-inflammatory cytokines and chemokines (Esser et al. 2014) (Fig. 18.1).

Both subcutaneous adipose tissue (SAT) and visceral adipose tissue (VAT) are associated with metabolic risk profile; however, the role of VAT is far more relevant (Kwon et al. 2017).

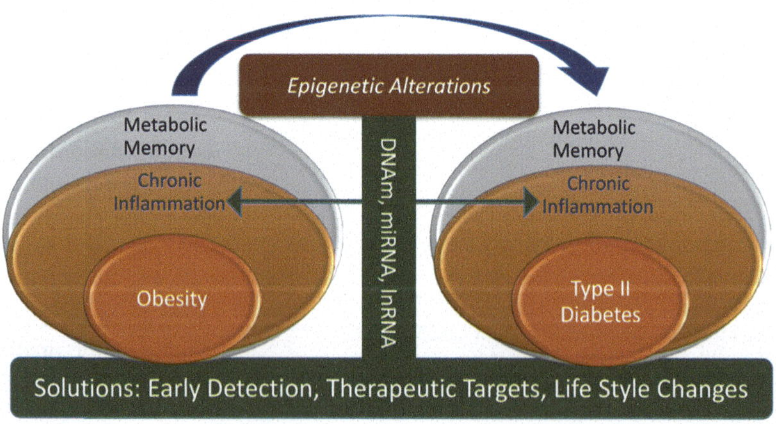

Fig. 18.1 Simplified depiction of development of chronic inflammation and "undesirable" metabolic memory in obesity and T2D. Epigenetic alterations provide early cues to metabolic imbalance and hence can be an ideal target for early detection of chronic inflammation in obesity and T2D

Macrophages, mast cells, neutrophils, and T and B lymphocytes in VAT seem to be key players (Mraz and Haluzik 2014). Adipose tissue (AT) macrophages exhibit intense pro-inflammatory (M1) characteristics (Verboven et al. 2018).

Macrophage activation in AT could precipitate insulin resistance (IR), along with M1 macrophage- or B-cell-mediated inflammation (Russo and Lumeng 2018), with little interference by subcutaneous fat (Verboven et al. 2018).

Chronic Inflammation and Insulin Resistance

Chronic inflammation is associated with insulin resistance in the primary insulin target organs, mainly adipose tissue, muscle, and liver (Gregor and Hotamisligil 2011; Reilly and Saltiel 2017). Inflammatory cytokines compromise insulin signaling in adipocytes by inducing inflammatory pathways (Lee 2013). Dysregulation of preadipocyte/adipocyte functions promotes ectopic fat deposition and insulin resistance (Samuel and Shulman 2012; Wu and Ballantyne 2017). Communication between M1 and M2 macrophage polarizing is detected in this context, with predominance of the alternatively activated macrophage (M2) phenotype and activation of transcription molecules nuclear factor-kappa beta (NF-κB) and activator protein 1 (AP1). Insulin resistance-mediated epigenetic changes are recognized in monocytes and macrophages (Reddy et al. 2014; Hanson and Godfrey 2015).

Epigenetics of T2D

At least 144 genetic variants (not completely independent) at 129 loci, identified by means of genome wide association studies (GWAS), are associated with T2D (Morris et al. 2012; Flannick and Florez 2016). Novel 39 common variants and 4 rare variants should be added (Xue et al. 2018). Gene expression of three genes, which play a role in T2D pathogenesis, namely, *CAMK1D*, *TP53INP1*, and *ATP5G*, was regulated by DNAm associated with genetic variants (Xue et al. 2018).

Age, obesity, diet, physical activity, and fetal and early postnatal development are among the environmental drivers that should not be overlooked. Epigenetics modifications comprise DNAm, histone modification, and noncoding RNA, which regulate gene expression in a tissue specific manner. Environmental influences and DNA sequence variations or single nucleotide polymorphisms (SNPs), such as CpG-SNP, structural variations, and gene–gene interactions, also deserve to be mentioned.

Hyperglycemia and Epigenetics

Hyperglycemia can induce a variety of epigenetic changes that can even persist after the normalization of glucose levels, mainly through the involvement of inflammatory genes (Reddy et al. 2015; Al-Haddad et al. 2016). Hyperglycemia can alter the activity of DNA methyltransferases (DNAmet), with irreversible changes over time. Epigenetic modifications may explain the long-term harmful effects of metabolic memory (Reddy et al. 2015).

DNAm is a well-characterized epigenetic modification. Unlike genotypic variation, DNAm intensity patterns are liable to change over time, with age or following disease or other exposure. Naidoo et al. (2018) elaborately reviewed cell and tissue specific methylation changes associated with inflammation and T2D.

Investigations in Twins

Monozygotic twins present an ideal model to study epigenetic contributions to the disease, as they are genetically the same and genetic influences on epigenetic variations can be ruled out. Because of tissue specificity of epigenetic regulation, it is essential to investigate a homogenous tissue or cell type from a target organ, which is influenced by alterations due to diabetes. Ribel-Madsen et al. (2012) investigated global DNAm differences in skeletal muscle and subcutaneous adipose tissues from 12 monozygotic twins, discordant for type 2 diabetes. A total of 26,850 cytosine-guanine dinucleotide (CpG) sites in the promoters of 14,279 genes were analyzed. Small intratwin pair methylation differences between diabetic and nondiabetics occurred; however, promoters of genes *PPARGC1A* in muscle and *HNF4A* in adipose tissue showed increased methylation in type 2 diabetes.

Increased expression of genes involved in inflammation and glycan degradation was observed in diabetes discordant monozygotic twins, with most differentially expressed genes being *ELOVL6*, *GYS2*, *FADS1*, *SPP1* (OPN), *CCL18*, and *IL1RN* (Nilsson et al. 2014). Significant DNAm differences were modest; however, 1410 sites showed differential DNAm. Also 15,627 sites, representing 7046 genes including *PPARG*, *KCNQ1*, *TCF7L2*, and *IRS1*, showed differential DNAm in adipose tissue.

Genome-wide methylated DNA immunoprecipitation sequencing (MeDIP-seq) in monozygotic twins reported hypermethylated DMRs in diabetic groups, with high hypermethylation in promoter of the *MALT1* gene, involved in insulin and glycemic pathways and related to taurocholate levels in blood (Yuan et al. 2014). Liu et al. (2015) conducted bioinformatics analysis of DNAm differences in muscle tissues of monozygotic twins, observing 38 differentially methylated genes, including Sirtuin 1, N-acetyltransferase 6, phospholipase A2 group XIIB, and nuclear factor of activated T cell calcineurin-dependent 1. Hwang et al. (2018), reported *ELOVL5* as an epigenetic biomarker for diabetes risk.

Epigenetic Wide Association Studies

Epigenetic modifications can precede disease manifestation or occur as a result of disease diagnosis, and hence, prospective studies are better suited for studying casual changes. In the large epigenetic wide association study (EWAS) by Demerath et al. (2015), 37 variable methylation sites in leucocyte DNA were associated with body mass index (BMI) and one with waist circumference (WC), including sites in genes *CPT1A*, *ABCG1*, and *SREBF1*. Cross-tissue association study using adipose tissue DNA revealed wide agreement in BMI– and WC–methylation associations. Replication using blood DNA was similarly achieved for 37 BMI probes and one WC probe. Sixteen of these sites also replicated in adipose tissue, including 15 novel methylation findings near genes involved in lipid metabolism, immune response/cytokine signaling, and other diverse pathways, including LGALS3BP, KDM2B, PBX1, and BBS2, among others. Additional EWAS showing link between DNAm and T2D has also been reported (Yuan et al. 2014; Nilsson et al. 2014; Dayeh et al. 2014)

In a recent study, Cardona et al. (2019) identified 15 novel methylation variable positions (MVPs) with significant associations with incident and prevalent T2D and also confirmed three previously identified MVPs. The study used whole blood DNA collected at baseline, up to 11 years before T2D onset, to investigate early epigenetic changes and their role in onset of T2D. The research was an incident T2D case-cohort study from the European Prospective Investigation into Cancer and Nutrition (EPIC)-Norfolk cohort. Integration of genome –wide genetic data with methylation data revealed one MVP, cg00574958 at *CPT1A*, which could possibly play direct casual role in T2D. Interestingly, none of the genes with identified MVPs had been identified by genetic association studies, thus indicating use of novel mechanisms/pathways for epigenetic alterations.

Noncoding RNA

Noncoding RNAs do not translate into proteins but bring about epigenetic changes and function to regulate gene expression at the transcriptional and posttranscriptional level. Several microRNA (miRNA) and long noncoding RNA (lncRNA) have been implicated in beta cell development, insulin sensitivity/resistance, insulin production/secretion, and insulin signaling. Noncoding RNAs are actively involved in early inflammation of the visceral adipose tissue and regulate inflammatory profiles of TNF-α, IL-1, IL-6, IL-18, intercellular adhesion molecule 1, VCAM-1, and plasminogen activator inhibitor 1 (Hamar 2012; Marques-Rocha et al. 2015).

MicroRNAs in Type 2 Diabetes and Inflammation

miRNAs are linked with obesity and inflammation, either directly or indirectly (through regulatory elements such as transcription factors) (Arner and Kulyté 2015). Microarray analysis of human WAT tissues revealed several potentially relevant miRNAs (Heneghan et al. 2011; Keller et al. 2011; Arner and Kulyté 2015).

To state few examples, miR-103 and miR-107 have been implicated in glucose homeostasis and insulin sensitivity (Trajkovski et al. 2011); miR-107 may also play a role in proinflammatory events (Foley and O'Neill 2012); miR-221 promotes white adipose tissue inflammation and decreases insulin sensitivity (Peng et al. 2018). miR 122 is implicated in liver steatosis and elevation in plasma cholesterol levels, circulating miR-122 is strongly associated with metabolic syndrome and T2D (Willeit et al. 2017), miR-146a expression is negatively correlated with fasting blood glucose (FBG), HbA1c, insulin resistance, proinflammatory signals such as TRAF6 and NFκβ mRNA levels, and also proinflammatory cytokines TNFα and IL-6 (Balasubramanyam et al. 2011), and finally, decreased serum level of miR-146a is a sign of chronic inflammation in T2D patients (Baldeón et al. 2014).

A detailed literature survey by Yaribeygi et al. (2018) reports different noncoding microRNAs related to glucose metabolism and pathophysiology of T2D. The authors propose four mechanisms via which miRNAs may alter glucose metabolism: beta cell development, insulin sensitivity/resistance, insulin production/secretion, and insulin signaling. Gottmann et al. (2018) combined QTL, miRNA prediction, and transcriptomics to design a computational framework (miR-QTL-Scan) to identify miRNAs that are regulated in obesity and T2D. The study discovered 170 miRNA regulative elements in human adipose tissue and blood cells.

Long Noncoding RNA in Type 2 Diabetes

Human β-cell transcriptome analysis has shown dynamic regulation and abnormal expression of lncRNAs in T2D (Saeidi et al. 2018). A recent study addressed expression level of lncRNAs in peripheral blood mononuclear cells (PBMCs). Significantly increased levels of lncRNAs HOTAIR, MEG3, LET, MALAT1, MIAT, CDKN2BAS1/ANRIL, XIST, PANDA, GAS5, Linc-p21, ENST00000550337.1, PLUTO, and NBR2 were observed in T2D patients. lncRNA expression patterns of THRIL and SALRNA1 diminished in T2D. The majority of altered lncRNAs were positively correlated with poor glycemic control, insulin resistance, and transcriptional markers of senescence (Sathishkumar et al. 2018).

Circulating lncRNA GAS5 less than 10 ng/μl was found to predict twelve times higher odds of having diabetes (Carter et al. 2015). Wang et al. (2017) identified 55 lncRNAs and 202 mRNAs to be differentially expressed in the blood samples from T2D group. miRNAs were mainly associated with immune regulation, inflammation, and insulin resistance. Newly emerged research interestingly indicates the use of exosomes miRNA and lncRNAs as potential biomarkers for detection or treatments of diabetes and diabetic complications (Chang and Wang 2019). Noncoding RNAs present an attractive target for T2D detection at early stages and may serve as clinical biomarkers.

Noncoding RNAs as Therapeutics Targets

Several studies have reported the inhibition of miRNAs for therapeutic purposes (Nunez Lopez et al. 2016; Nigi et al. 2018; Jiménez-Lucena et al. 2018). Inhibition of miRNAs by antisense oligonucleotides in mice models has proved to improve glucose homeostasis and insulin sensitivity.

Inhibition of miR-103 and miR-107 by antisense oligonucleotides in liver and in adipose tissue of obese mice improved glucose homeostasis and insulin sensitivity. In contrast, injections in liver of an adenoviral vector expressing miR-107 resulted in elevated fasting blood glucose and insulin levels, impaired glucose tolerance, decreased insulin sensitivity, and increased hepatic glucose production (Trajkovski et al. 2011). Antisense oligonucleotides also demonstrated the potential therapeutic effects of liver-specific miR-122 inhibition and downregulation of miR-181a (Esau et al. 2006; Zhou et al. 2012). Lucia La Sala et al. (2019) demonstrated microRNA-21 as a biomarker for early detection of prediabetes and glucose imbalances. There was upregulation in impaired glucose tolerance (IGT) and in T2D subjects. Inhibition of miR-200c and miR-21 has also been suggested against T2D (Geach 2015; Seeger et al. 2014).

Tissue-specific miR-199a-3p and miR-223 are potential tissue biomarkers for pancreas and liver, respectively, in the context of T2D (Zhu and Leung 2015).

High Fat Diet and Polyphenols

Dietary modifications have also been shown to indirectly modulate miRNA expression (Parra et al. 2010). Experimental high fat diet increased the expression of miR-103 and miR-107, with reversal by plant-derived polyphenols. Dietary polyphenols fed with high fat diet also decreased the expression of miR-122, reduced weight gain, liver steatosis, and insulin resistance (Joven et al. 2012).

Beta Cell Dysfunction

In the relationship between Insulin sensitivity and Cardiovascular disease (RISC) cohort, biomarkers of IGT/reduced β-cell glucose sensitivity were mainly adiponectin, alpha-1-antitrypsin (known to regulate adiponectin levels), endocan, miR-181a, miR-342, and miR-323. The proteins identified were adiponectin, endocan, and sialoadhesin (Belongie et al. 2017).

Despite success in animal models, techniques for *in vivo* specific delivery of "therapeutic" miRNAs will need to be tested and optimized, in order to restore physiological levels of a miRNA deregulated in a specific tissue. miRNA combination therapy could be necessary to address multiple dysregulated miRNAs associated with T2D.

Early Detection/Relative Risk Tests

The International Diabetes Federation (IDF) has estimated that among 642 million cases by 2040, half of them will not be diagnosed (Ogurtsova et al. 2017). Early detection tests/screening of the general population are very important to facilitate risk identification and to curb the damage at early stages. Selective or targeted screening of T2D is performed in high risk cohorts selected by age, body weight, or ethnic origin. Opportunistic screening is conducted by health care professionals, for any other reason, during routine appointments. The most useful glucose homeostasis test is still debated (Vatandoost et al. 2015).

According to Barry et al. (2017), glycated hemoglobin A1c (HbA1c) was neither sensitive nor specific for prediabetes; fasting glucose was specific however not sensitive. Divergence between the criteria of the American Diabetes Association and WHO leads to identification of different populations with limited overlap. The study also suggested that "screen and treat" policies alone are unlikely to halt the epidemic of type 2 diabetes. In the experience of Laiteerapong et al. (2019), newly diagnosed diabetes and an unstable initial 10-year HbA1c predisposed to microvascular complications, even after adjusting for HbA1c control in year 1.

Repeated measurements of fasting plasma glucose seem to better predict diabetes than a single

Table 18.1 List of ongoing clinical trials on T2D and epigenetics

Title of the study	Type of study	Stage/ No. of participants	Major outcomes(epigenetics)	Location (Country)
Epigenetic Contribution to the Pathogenesis of Diabetic Nephropathy in Qatari Population	Observational/ prospective	Recruiting N =240	Assessment of DNA methylation in monocytes, Assessment of urinary micro-RNAs	Qatar
Network-based Epigenome-Wide association Study in Obesity precision Medicine: NEWTON Clinical Trial	observational / Case control/ prospective	Not yet Recruiting N =100	(a) A disease module containing the crucial differentially methylated genes in obese patients and obese patients with T2D compared to controls. (b) Identification of differentially expressed micro-RNA and mRNA	Italy
Sex-specific Relationship of Epigenetic-Based Modifications in the Saliva and Blood With the Occurrence of Type 2 Diabetes	Observational/ Prospective	Recruiting N =224	Correlation of miRNAs in the saliva with the amount of intrahepatic liver fat, ectopic lipids in the heart, and skeletal muscle.	Austria
Unravelling the Role of Mitochondrial DNA Methylation in Type 2 Diabetes	Observational/ Prospective	Recruiting N=36	Mitochondrial DNA methylation and D-loop changes in insulin resistant states such as obesity and type 2 diabetes	United States
Impact of Metabolic Health on Sperm Epigenetic Marks in Humans.	Interventional/ Nonrandomized	Recruiting N =40	Changes in epigenetic markers (DNA methylation and noncoding RNAs) in overweight men with type 1 or type 2 diabetes with a 3 month lifestyle intervention program.	United States
Perturbation of Interactome Through MicroRNA And Methylome Analyses In Diabetes Endophenotypes: the PIRAMIDE Study Design	Observational/ case-control/ Prospective	Recruiting N =35	(a) Identification of differentially methylated genes in both CD4+ and CD8+ cells from T2D patients and controls. (b) Identification of differentially expressed microRNA and mRNA target in both CD4+ and CD8+ cells isolated from T2D patients and controls. (c) Putative epigenetics interactions in the presence or absence of T2D-related cardiovascular dysfunction.	Italy
Investigating the Underlying Mechanisms of Exercise Resistance in Individuals With Type 2 Diabetes	Interventional/ nonrandomized.	Active N=33	Relationship between the basal promoter methylation status of key genes involved in fuel metabolism and activated by exercise in skeletal muscle tissue and cells.	United States
The Interaction Between Protein Intake, Gut Microbiota, and Type 2 Diabetes in Subjects With Different Ethnic Backgrounds (MICRODIET)	Interventional/ Randomized	Recruiting N =120	Epigenetic modifications in monocytes of patients at baseline and end of dietary intervention	France

(continued)

Table 18.1 (continued)

Title of the study	Type of study	Stage/ No. of participants	Major outcomes(epigenetics)	Location (Country)
Who Will Benefit From Bariatric Surgery for Diabetes? Using Fat Distribution Measurement, Gut Hormone Profiles, and Genetic Data to Predict Diabetes Remission	Observational/ prospective	Recruiting N =210	Epigenetic changes in visceral adipose tissue of T2D patients on mixed meal remission, who underwent Roux-en-Y gastric bypass surgery	United Kingdom
Consequences of Hypoglycemia on Cardiovascular and Inflammatory Responses (HCIR)	Interventional, nonrandomized, clinical trial	Recruiting N =112	Epigenetic modifications due to hypoglycemia in the promoter regions of the pro-inflammatory cytokines in monocytes	Netherlands
Association Between Urinary and Serum Levels of miRNA 192 and miRNA 25 and Glomerular Filtration and Albuminuria in Patients With and Without Type 2 Diabetes.	Observational/ retrospective	Recruiting N =300	Urine and serum expression of miRNA192 and miRNA 25	Italy
Prognostic Predictors of Response to Hypoglycemic Therapy	Interventional, randomized	Recruiting N =800	Serum level of microRNA-126, microRNA-21, microRNA-27, miRNA-125, and miRoRNA-155 from baseline to 6 and 12 months after intervention	Russia

test (Inoue et al. 2012). In the opinion of Jackson et al. (2017), a 50-g oral glucose followed by 1-h glucose measurement is superior to other screening tests. If abnormal, an oral glucose tolerance test could follow. Such strategy could be cost effective in high risk populations.

Feldman et al. (2017) were able to diagnose diabetes 4.6 years earlier, by screening asymptomatic individuals, and these cases had a more favorable clinical course than clinically detected diabetes after diagnosis. Screening for both T2D and prediabetes should begin at age 45–50, with follow-up every 5 years, and targeted screening appears to be more cost-effective compared to routine evaluation (Einarson et al. 2017).

Ongoing Clinical Trials

Table 18.1 depicts a few ongoing clinical trials on T2D and epigenetics. The recent research will accelerate the interventions into clinical practice and improve early detection of T2D and arrest of associated complications.

References

Al-Haddad R, Karnib N, Assaad RA et al (2016) Epigenetic changes in diabetes. Neurosci Lett 625:64–69. https://doi.org/10.1016/j.neulet.2016.04.046

Arner P, Kulyté A (2015) MicroRNA regulatory networks in human adipose tissue and obesity. Nat Rev Endocrinol 11(5):276–288. https://doi.org/10.1038/nrendo.2015.25

Balasubramanyam M, Aravind S, Gokulakrishnan K et al (2011) Impaired miR-146a expression links subclinical inflammation and insulin resistance in Type 2 diabetes. Mol Cell Biochem 351(1–2):197–205. https://doi.org/10.1007/s11010-011-0727-3

Baldeón RL, Weigelt K, de Wit H et al (2014) Decreased serum level of miR-146a as sign of chronic inflammation in type 2 diabetic patients. PLoS One 9(12): e115209. https://doi.org/10.1371/journal.pone.0115209

Barry E, Roberts S, Oke J, Vijayaraghavan S, Normansell R, Greenhalgh T (2017) Efficacy and effectiveness of screen and treat policies in prevention of type 2 diabetes: systematic review and meta-analysis of screening tests and interventions. BMJ 356:i6538. https://doi.org/10.1136/bmj.i6538

Belongie KJ, Ferrannini E, Johnson K, Andrade-Gordon P, Hansen MK, Petrie JR, Wagner B (2017) Identification of novel biomarkers to monitor β-cell function and enable early detection of type 2 diabetes risk. PLoS One 12(8):e0182932

Cardona A, Day FR, Perry JRB et al (2019) Epigenome-wide association study of incident type 2 diabetes in a British population: EPIC-Norfolk Study. Diabetes 68 (12):2315–2326. https://doi.org/10.2337/db18-0290

Carter G, Miladinovic B, Patel AA, Deland L, Mastorides S, Patel NA (2015) Circulating long noncoding RNA GAS5 levels are correlated to prevalence of type 2 diabetes mellitus. BBA Clin 4:102–107. https://doi.org/10.1016/j.bbacli.2015.09.001

Chang W, Wang J (2019) Exosomes and their noncoding RNA cargo are emerging as new modulators for diabetes mellitus. Cells 8(8):853

Ceriello A (2009) Hypothesis: the "metabolic memory", the new challenge of diabetes. Diabetes Res Clin Pract 86(Suppl 1):S2–S6. https://doi.org/10.1016/S0168-8227(09)70002-6

Dayeh T, Volkov P, Salö S et al (2014) Genome-wide DNAm analysis of human pancreatic islets from type 2 diabetic and non-diabetic donors identifies candidate genes that influence insulin secretion. PLoS Genet 10 (3):e1004160. https://doi.org/10.1371/journal.pgen.1004160

Demerath EW, Guan W, Grove ML et al (2015) Epigenome-wide association study (EWAS) of BMI, BMI change and waist circumference in African American adults identifies multiple replicated loci. Hum Mol Genet 24(15):4464–4479. https://doi.org/10.1093/hmg/ddv161

Einarson TR, Bereza BG, Acs A, Jensen R (2017) Systematic literature review of the health economic implications of early detection by screening populations at risk for type 2 diabetes. Curr Med Res Opin 33(2):331–358. https://doi.org/10.1080/03007995.2016.1257977

Esau C, Davis S, Murray SF et al (2006) miR-122 regulation of lipid metabolism revealed by in vivo antisense targeting. Cell Metab 3(2):87–98. https://doi.org/10.1016/j.cmet.2006.01.005

Esser N, Legrand-Poels S, Piette J, Scheen AJ, Paquot N (2014) Inflammation as a link between obesity, metabolic syndrome and type 2 diabetes. Diabetes Res Clin Pract 105(2):141–150. https://doi.org/10.1016/j.diabres.2014.04.006

Feldman AL, Griffin SJ, Fhärm E et al (2017) Screening for type 2 diabetes: do screen-detected cases fare better? Diabetologia 60(11):2200–2209. https://doi.org/10.1007/s00125-017-4402-4

Flannick J, Florez JC (2016) Type 2 diabetes: genetic data sharing to advance complex disease research. Nat Rev Genet 17(9):535–549. https://doi.org/10.1038/nrg.2016.56

Foley NH, O'Neill LA (2012) miR-107: a toll-like receptor-regulated miRNA dysregulated in obesity and type II diabetes. J Leukoc Biol 92(3):521–527. https://doi.org/10.1189/jlb.0312160

Gaede P, Lund-Andersen H, Parving HH, Pedersen O (2008) Effect of a multifactorial intervention on mortality in type 2 diabetes. N Engl J Med 358 (6):580–591. https://doi.org/10.1056/NEJMoa0706245

Geach T (2015) Preventing β-cell apoptosis and T2D with microRNAs—a role for miR-200? Nat Rev Endocrinol 11(8):444. https://doi.org/10.1038/nrendo.2015.92

Gottmann P, Ouni M, Saussenthaler S et al (2018) A computational biology approach of a genome-wide screen connected miRNAs to obesity and type 2 diabetes. Mol Metab 11:145–159. https://doi.org/10.1016/j.molmet.2018.03.005

Gregor MF, Hotamisligil GS (2011) Inflammatory mechanisms in obesity. Annu Rev Immunol 29:415–445. https://doi.org/10.1146/annurev-immunol-031210-101322

Hamar P (2012) Role of regulatory micro RNAs in type 2 diabetes mellitus-related inflammation. Nucleic Acid Ther 22(5):289–294. https://doi.org/10.1089/nat.2012.0381

Hanson MA, Godfrey KM (2015) Genetics: epigenetic mechanisms underlying type 2 diabetes mellitus. Nat Rev Endocrinol 11(5):261–262. https://doi.org/10.1038/nrendo.2015.31

Heneghan HM, Miller N, McAnena OJ, O'Brien T, Kerin MJ (2011) Differential miRNA expression in omental adipose tissue and in the circulation of obese patients identifies novel metabolic biomarkers. J Clin Endocrinol Metab 96(5):E846–E850. https://doi.org/10.1210/jc.2010-2701

Holman RR, Paul SK, Bethel MA, Neil HA, Matthews DR (2008) Long-term follow-up after tight control of blood pressure in type 2 diabetes. N Engl J Med 359 (15):1565–1576. https://doi.org/10.1056/NEJMoa0806359

Hotamisligil GS (2017) Inflammation, metaflammation and immunometabolic disorders. Nature 542 (7640):177–185. https://doi.org/10.1038/nature21363

Hwang JY, Lee HJ, Go MJ et al (2018) Genome-wide methylation analysis identifies ELOVL5 as an epigenetic biomarker for the risk of type 2 diabetes mellitus. Sci Rep 8(1):14862. https://doi.org/10.1038/s41598-018-33238-9

Inoue K, Inoue M, Matsumoto M, Akimoto K (2012) Persistent fasting hyperglycaemia is more predictive of type 2 diabetes than transient fasting hyperglycaemia. Diabet Med 29(7):e75–e81. https://doi.org/10.1111/j.1464-5491.2011.03536.x

Jackson SL, Safo SE, Staimez LR et al (2017) Glucose challenge test screening for prediabetes and early diabetes. Diabet Med 34(5):716–724. https://doi.org/10.1111/dme.13270

Jiménez-Lucena R, Camargo A, Alcalá-Diaz JF et al (2018) A plasma circulating miRNAs profile predicts type 2 diabetes mellitus and prediabetes: from the CORDIOPREV study. Exp Mol Med 50(12):168. https://doi.org/10.1038/s12276-018-0194-y

Joven J, Espinel E, Rull A et al (2012) Plant-derived polyphenols regulate expression of miRNA paralogs miR-103/107 and miR-122 and prevent diet-induced fatty liver disease in hyperlipidemic mice. Biochim Biophys Acta 1820(7):894–899. https://doi.org/10.1016/j.bbagen.2012.03.020

Keller P, Gburcik V, Petrovic N et al (2011) Gene-chip studies of adipogenesis-regulated microRNAs in mouse primary adipocytes and human obesity. BMC Endocr Disord 11:7. https://doi.org/10.1186/1472-6823-11-7

Kirchner H, Osler ME, Krook A, Zierath JR (2013) Epigenetic flexibility in metabolic regulation: disease cause and prevention? Trends Cell Biol 23(5):203–209. https://doi.org/10.1016/j.tcb.2012.11.008

Kwon H, Kim D, Kim JS (2017) Body fat distribution and the risk of incident metabolic syndrome: a Longitudinal Cohort Study. Sci Rep 7(1):10955. https://doi.org/10.1038/s41598-017-09723-y

La Sala L, Mrakic-Sposta S, Tagliabue E et al (2019) Circulating microRNA-21 is an early predictor of ROS-mediated damage in subjects with high risk of developing diabetes and in drug-naïve T2D. Cardiovasc Diabetol 18(1):18. https://doi.org/10.1186/s12933-019-0824-2

Laiteerapong N, Ham SA, Gao Y et al (2019) The legacy effect in type 2 diabetes: impact of early glycemic control on future complications (the diabetes & aging study). Diabetes Care 42(3):416–426. https://doi.org/10.2337/dc17-1144

Liu F, Sun Q, Wang L, Nie S, Li J (2015) Bioinformatics analysis of abnormal DNA methylation in muscle samples from monozygotic twins discordant for type 2 diabetes. Mol Med Rep 12(1):351–356. https://doi.org/10.3892/mmr.2015.3452

Lee J (2013) Adipose tissue macrophages in the development of obesity-induced inflammation, insulin resistance and type 2 diabetes. Arch Pharm Res 36 (2):208–222. https://doi.org/10.1007/s12272-013-0023-8

Marques-Rocha JL, Samblas M, Milagro FI, Bressan J, Martínez JA, Marti A (2015) Noncoding RNAs, cytokines, and inflammation-related diseases. FASEB J 29(9):3595–3611. https://doi.org/10.1096/fj.14-260323

Morris AP, Voight BF, Teslovich TM et al (2012) Large-scale association analysis provides insights into the genetic architecture and pathophysiology of type 2 diabetes. Nat Genet 44(9):981–990. https://doi.org/10.1038/ng.2383

Mraz M, Haluzik M (2014) The role of adipose tissue immune cells in obesity and low-grade inflammation. J Endocrinol 222(3):R113–R127. https://doi.org/10.1530/JOE-14-0283

Naidoo V, Naidoo M, Ghai M (2018) Cell- and tissue-specific epigenetic changes associated with chronic inflammation in insulin resistance and type 2 diabetes mellitus. Scand J Immunol 88(6):e12723. https://doi.org/10.1111/sji.12723

Nigi L, Grieco GE, Ventriglia G et al (2018) MicroRNAs as regulators of insulin signaling: research updates and potential therapeutic perspectives in type 2 diabetes. Int J Mol Sci 19(12):3705. https://doi.org/10.3390/ijms19123705

Nilsson E, Jansson PA, Perfilyev A et al (2014) Altered DNAm and differential expression of genes influencing metabolism and inflammation in adipose tissue from subjects with type 2 diabetes. Diabetes 63 (9):2962–2976. https://doi.org/10.2337/db13-1459

Nunez Lopez YO, Garufi G, Seyhan AA (2016) Altered levels of circulating cytokines and microRNAs in lean and obese individuals with prediabetes and type 2 diabetes. Mol BioSyst 13(1):106–121. https://doi.org/10.1039/c6mb00596a

Ogurtsova K, da Rocha Fernandes JD, Huang Y et al (2017) IDF diabetes atlas: global estimates for the prevalence of diabetes for 2015 and 2040. Diabetes Res Clin Pract 128(6):40–50. https://doi.org/10.1016/j.diabres.2017.03.024

Parra P, Serra F, Palou A (2010) Expression of adipose microRNAs is sensitive to dietary conjugated linoleic acid treatment in mice. PLoS One 5(9):e13005. https://doi.org/10.1371/journal.pone.0013005

Peng J, Zhou Y, Deng Z et al (2018) miR-221 negatively regulates inflammation and insulin sensitivity in white adipose tissue by repression of sirtuin-1 (SIRT1). J Cell Biochem 119(8):6418–6428. https://doi.org/10.1002/jcb.26589

Rawshani A, Rawshani A, Franzén S et al (2017) Mortality and cardiovascular disease in type 1 and type 2 diabetes. N Engl J Med 376(15):1407–1418. https://doi.org/10.1056/NEJMoa1608664

Reddy MA, Chen Z, Park JT et al (2014) Regulation of inflammatory phenotype in macrophages by a diabetes-induced long noncoding RNA. Diabetes 63 (12):4249–4261. https://doi.org/10.2337/db14-0298

Reddy MA, Zhang E, Natarajan R (2015) Epigenetic mechanisms in diabetic complications and metabolic memory. Diabetologia 58(3):443–455. https://doi.org/10.1007/s00125-014-3462-y

Reddy P, Lent-Schochet D, Ramakrishnan N, McLaughlin M, Jialal I (2019) Metabolic syndrome is an inflammatory disorder: A conspiracy between adipose tissue and phagocytes. Clin Chim Acta 496:35–44. https://doi.org/10.1016/j.cca.2019.06.019

Reilly SM, Saltiel AR (2017) Adapting to obesity with adipose tissue inflammation. Nat Rev Endocrinol 13 (11):633–643. https://doi.org/10.1038/nrendo.2017.90

Ribel-Madsen R, Fraga MF, Jacobsen S et al (2012) Genome-wide analysis of DNAm differences in muscle and fat from monozygotic twins discordant for type 2 diabetes. PLoS One 7(12):e51302. https://doi.org/10.1371/journal.pone.0051302

Russo L, Lumeng CN (2018) Properties and functions of adipose tissue macrophages in obesity. Immunology 155(4):407–417. https://doi.org/10.1111/imm.13002

Saeidi L, Ghaedi H, Sadatamini M et al (2018) Long non-coding RNA LY86-AS1 and HCG27_201

expression in type 2 diabetes mellitus. Mol Biol Rep 45 (6):2601–2608. https://doi.org/10.1007/s11033-018-4429-8
Samuel VT, Shulman GI (2012) Mechanisms for insulin resistance: common threads and missing links. Cell 148(5):852–871. https://doi.org/10.1016/j.cell.2012.02.017
Sathishkumar C, Prabu P, Mohan V, Balasubramanyam M (2018) Linking a role of lncRNAs (long non-coding RNAs) with insulin resistance, accelerated senescence, and inflammation in patients with type 2 diabetes. Hum Genomics 12(1):41. https://doi.org/10.1186/s40246-018-0173-3
Seeger T, Fischer A, Muhly-Reinholz M, Zeiher AM, Dimmeler S (2014) Long-term inhibition of miR-21 leads to reduction of obesity in db/db mice. Obesity (Silver Spring) 22(11):2352–2360. https://doi.org/10.1002/oby.20852
Sladek R (2018) The many faces of diabetes: addressing heterogeneity of a complex disease. Lancet Diabetes Endocrinol 6(5):348–349. https://doi.org/10.1016/S2213-8587(18)30070-6
Testa R, Bonfigli AR, Prattichizzo F, La Sala L, De Nigris V, Ceriello A (2017) The "metabolic memory" theory and the early treatment of hyperglycemia in prevention of diabetic complications. Nutrients 9 (5):437. https://doi.org/10.3390/nu9050437
Trajkovski M, Hausser J, Soutschek J et al (2011) MicroRNAs 103 and 107 regulate insulin sensitivity. Nature 474(7353):649–653. https://doi.org/10.1038/nature10112
Tsalamandris S, Antonopoulos AS, Oikonomou E et al (2019) The role of inflammation in diabetes: current concepts and future perspectives. Eur Cardiol 14 (1):50–59. https://doi.org/10.15420/ecr.2018.33.1
Vatandoost N, Amini M, Iraj B, Momenzadeh S, Salehi R (2015) Dysregulated miR-103 and miR-143 expression in peripheral blood mononuclear cells from induced prediabetes and type 2 diabetes rats. Gene 572 (1):95–100. https://doi.org/10.1016/j.gene.2015.07.015
Verboven K, Wouters K, Gaens K et al (2018) Abdominal subcutaneous and visceral adipocyte size, lipolysis and inflammation relate to insulin resistance in male obese humans. Sci Rep 8(1):4677. https://doi.org/10.1038/s41598-018-22962-x
Wang X, Chang X, Zhang P, Fan L, Zhou T, Sun K (2017) Aberrant expression of long non-coding RNAs in newly diagnosed type 2 diabetes indicates potential roles in chronic inflammation and insulin resistance. Cell Physiol Biochem 43(6):2367–2378. https://doi.org/10.1159/000484388
Willeit P, Skroblin P, Moschen AR et al (2017) Circulating microRNA-122 Is associated with the risk of new-onset metabolic syndrome and type 2 diabetes. Diabetes 66(2):347–357. https://doi.org/10.2337/db16-0731
Wu H, Ballantyne CM (2017) Skeletal muscle inflammation and insulin resistance in obesity. J Clin Invest 127 (1):43–54. https://doi.org/10.1172/JCI88880
Xue A, Wu Y, Zhu Z et al (2018) Genome-wide association analyses identify 143 risk variants and putative regulatory mechanisms for type 2 diabetes. Nat Commun 9(1):2941. https://doi.org/10.1038/s41467-018-04951-w
Yaribeygi H, Katsiki N, Behnam B, Iranpanah H, Sahebkar A (2018) MicroRNAs and type 2 diabetes mellitus: Molecular mechanisms and the effect of antidiabetic drug treatment. Metabolism 87:48–55. https://doi.org/10.1016/j.metabol.2018.07.001
Yuan W, Xia Y, Bell CG et al (2014) An integrated epigenomic analysis for type 2 diabetes susceptibility loci in monozygotic twins. Nat Commun 5:5719. https://doi.org/10.1038/ncomms6719
Zhou B, Li C, Qi W et al (2012) Downregulation of miR-181a upregulates sirtuin-1 (SIRT1) and improves hepatic insulin sensitivity. Diabetologia 55 (7):2032–2043. https://doi.org/10.1007/s00125-012-2539-8
Zhu H, Leung SW (2015) Identification of microRNA biomarkers in type 2 diabetes: a meta-analysis of controlled profiling studies. Diabetologia 58(5):900–911. https://doi.org/10.1007/s00125-015-3510-2B62

Salivary and Urinary Metabolome in Pediatric Obesity and Metabolic Syndrome

19

Jacopo Troisi, Francesca Marciano, Giovanni Scala, Elizabeth Plunk, Luca Pierri, and Angelo Colucci

Contents

Introduction 250

Metabolic Features of Pediatric Obesity and MetS 251

Urine and Salivary Pathways and Biomolecules 253
Saliva Production 253
Urine Production 253

Salivary and Urinary Metabolome 254
Salivary Metabolome 254
Urinary Metabolome 255
Small Intestinal Bacterial Overgrowth 257

Potential Targets for New Drug Development 258
Phytotherapeutic Agents 258
Gut Microbiota Interventions 258

Ongoing Studies 258

References 260

J. Troisi (✉)
Department of Medicine and Surgery and Dentistry, "Scuola Medica Salernitana", University of Salerno, Salerno, Italy

Theoreo srl—Spin-Off Company of the University of Salerno, Salerno, Italy

European Biomedical Research Institute of Salerno (EBRIS), Salerno, Italy
e-mail: troisi@theoreosrl.com

F. Marciano
European Biomedical Research Institute of Salerno (EBRIS), Salerno, Italy

G. Scala
Theoreo srl—Spin-Off Company of the University of Salerno, Salerno, Italy

Hosmotic srl, Vico Equense, Italy

E. Plunk
Department of Biological and Environmental Sciences, University of Tennessee-Chattanooga, Chattanooga, TN, USA

L. Pierri
Department of Translational Medical Sciences, Section of Pediatrics, University of Naples Federico II, Naples, Italy

A. Colucci
Department of Medicine and Surgery and Dentistry, "Scuola Medica Salernitana", University of Salerno, Salerno, Italy

J. Faintuch and S. Faintuch (eds.), *Obesity and Diabetes*, https://doi.org/10.1007/978-3-030-53370-0_19

Abstract

Childhood obesity is reaching alarming rates in many countries, posing an urgent and serious challenge. Adipose tissue is a key endocrine organ that releases several pro- or anti-inflammatory adipokines that are involved in the pathogenesis of obesity complications. Moreover, it is suggested that gut microbiota is also involved in the development of these complications, including obesity-related liver disease and metabolic syndrome.

Metabolomics has recently started to pave the way to a better pathomechanistic understanding of hepatometabolic complications, leading to a more efficient diagnosis and better therapeutic approaches. Several metabolites and metabolic pathways contribute to a complex metabolic fingerprint of obesity, metabolic syndrome, and obesity-related non-alcoholic fatty liver disease (NAFLD). Indeed, these conditions seem to have different salivary and urinary metabolomic signatures. The aim of this chapter is to provide an overview of these signatures, metabolomic-driven findings, and perspectives for obesity and obesity-related condition diagnoses and treatments.

Keywords

Pediatric obesity · metabolic syndrome · non-alcoholic fatty liver disease · metabolomics

Introduction

Obese children and adolescents are at high risk of developing several metabolic disorders that were previously considered to be "adult" diseases, including insulin resistance (IR), metabolic syndrome (MetS), type-2 diabetes mellitus (T2DB), hypertension, dyslipidemia, obstructive sleep apnea (OSA), and non-alcoholic fatty liver disease (NAFLD) (Kumar and Kelly 2017; WHO 2016; Morales Camacho et al. 2019; Dietz and Robinson 2005; Lakshman et al. 2012). This is the first time in human history that we are meeting these phenomena in pediatric populations, and the effects on subsequent adulthood are not easily predictable (Rosenbaum 2018).

Adipose tissue is a key endocrine organ that releases several pro- or anti-inflammatory adipokines, which are involved in the pathogenesis of obesity complications. Moreover, it is suggested that gut microbiota is also involved in the development of these complications, including obesity-related liver disease and obesity itself (Vajro et al. 2013a).

NAFLD is a clinical–pathological condition characterized by abnormal lipid deposition in hepatocytes in the absence of excess alcohol intake. It includes a disease spectrum ranging from isolated hepatic steatosis to more severe forms (Non-alcoholic Steatohepatitis, NASH) with inflammation, liver fibrosis, cirrhosis, and eventually hepatocellular carcinoma (HCC) (Tiniakos et al. 2010).

The majority of patients with NAFLD are obese: the obesity-related chronic mild systemic inflammation blocks insulin signaling pathways and leads to insulin resistance, which plays a likely role in the liver disease pathomechanism (Sanyal et al. 2001; Fabbrini et al. 2010).

NAFLD is now considered as the hepatic manifestation of MetS,(Abenavoli et al. 2017; Buzzetti et al. 2016) that is essentially characterized by the clustering of metabolic abnormalities such as glucose intolerance, obesity, dyslipidemia, and hypertension (Hosseinpour-Niazi et al. 2019; Bussler et al. 2017).

While the pathogenesis of MetS and the role of each of its components is complex and not completely elucidated (Fig. 19.1), abdominal obesity and insulin resistance are considered important causative factors (Alberti et al. 2006).

The International Diabetes Federation (IDF) provided the Worldwide Consensus definition of MetS, identifying abdominal obesity (defined as waist circumference) as a prerequisite of MetS (Alberti et al. 2006; Zimmet et al. 2007). In addition, diagnosis is defined by the presence of at least two of the following features: high blood triglycerides (TG), reduced HDL cholesterol

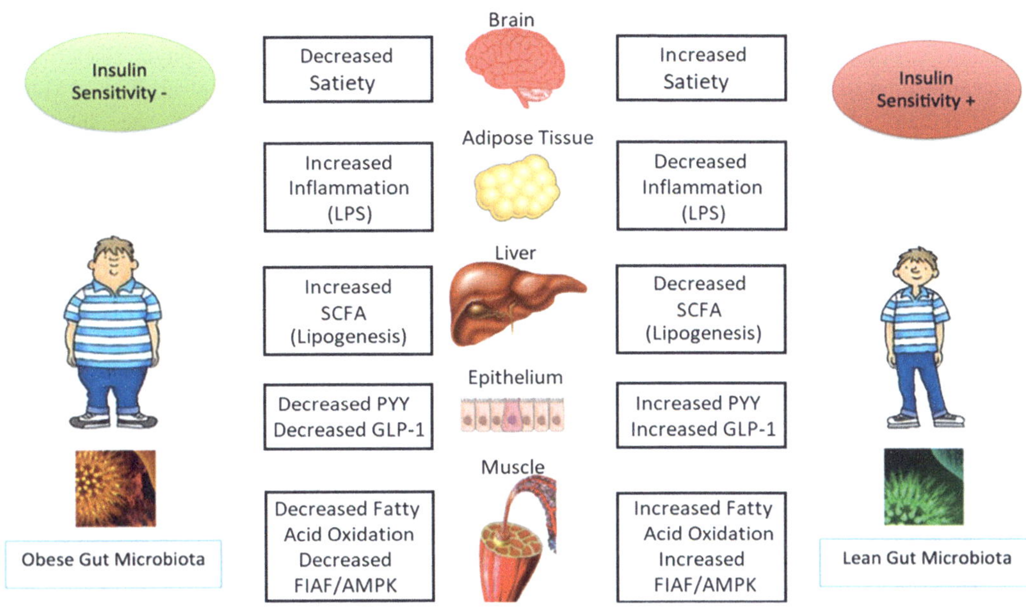

Fig. 19.1 Pathophysiological changes related to insulin resistance in various organs

(HDL-C), increased blood pressure, or increased fasting blood glucose (FBG) (Zafar et al. 2018).

Although there is no agreement on a MetS definition for children and adolescents (Bussler et al. 2017), prevalence is approximately 30%. A current MetS definition in pediatric population is based on the presence of 3 or more of following criteria: (1) TG $\geq$ 110 mg/dL or (WHO 2016); HDL-C <40 mg/dL (Morales Camacho et al. 2019) ;FPG $\geq$ 100 mg/dL (Genuth et al. 2003); (Dietz and Robinson 2005) Systolic blood pressure or diastolic blood pressure $\geq$ 90th percentile for sex, age, and height; and (Lakshman et al. 2012) WC $\geq$ 90th percentile for age and sex (Hosseinpour-Niazi et al. 2019; Kelishadi et al. 2007).

MetS severity in childhood has been associated with type-2 diabetes and cardiovascular disease (CVD) in adulthood; therefore, it is necessary to implement, during childhood, interventions for a healthy lifestyle (Lakshman et al. 2012; Eckel et al. 2005).

Metabolic profile analysis, also known as metabolic phenotyping, can be useful to characterize metabolic patterns of cells, tissues, organisms, and biological fluids. It allows the identification of specific features (or biomarkers) of obesity and MetS, in order to assess and track metabolic risk. A good understanding of the biochemical mechanisms underlying obesity and NAFLD is required, to develop suitable screening strategies (Spruijt-Metz 2011; Troisi et al. 2017; Bervoets et al. 2018).

Metabolic Features of Pediatric Obesity and MetS

Significantly different metabolome signatures were identified in normal-weight and obese adolescents' biofluids. Using combined untargeted and targeted metabolomics, it has been demonstrated that inflammation-driven insulin resistance, ammonia toxicity, and oxidative stress could represent crucial metabolomic signatures in obese adolescents (Troisi et al. 2019; Cho et al. 2017a).

A complex metabolic fingerprint characterizes obesity, especially visceral obesity and obesity-related NAFLD (Troisi et al. 2017). Pathogenesis of MetS possibly depends on the interaction between obesity, insulin resistance, and inflammation (Wittcopp and Conroy 2016).

Accumulation of free fatty acids (FFA) in the liver of pediatric obese subjects is responsible for impaired insulin signaling and consequently insulin resistance, reducing its effect on suppression of glucose production (D'Adamo et al. 2009). The consequence is hyperinsulinemia that induces the liver to produce fat, therefore causing hypertriglyceridemia. In addition adipocytes, because of insulin resistance, induce an increase in lipolysis, raising non-esterified fatty acids (NEFA) levels in blood (Das et al. 2014).

Differences in metabolites observed in obese children with and without NAFLD (Miccheli et al. 2015), could also be related to the microbiota and the gut–liver axis (GLA).

Different metabolites and their pathways derived from mammalian and gut microbial co-metabolism have been associated with phenotypes of MetS, including diabetes, hypertension, and obesity (Dumas et al. 2014). It is widely believed that gut microbiota plays an important role in obesity, MetS, and fatty liver by increasing energy harvesting, when a condition of intestinal dysbiosis, and more rarely when some instances of small intestinal bacterial overgrowth (SIBO) occur (Rutigliano et al. 2017; Tresaco et al. 2005). Changes in the abundance or composition of bacterial species of the gastrointestinal tract are reported in obesity and other metabolic conditions (Vajro et al. 2013a). A damaged intestinal barrier due to local or systemic inflammation could establish a condition known as "leaky gut," in which the GLA can amplify the common interactions between intestinal bacteria/bacterial products and hepatic receptors (Nejatinamini et al. 2015), promoting the portal flow of a large number of metabolites (Goodson et al. 2014).

Gut microbiota deeply changes among individuals, and its composition varies with different conditions. In particular, obesity is associated with significant variations in the structure of intestinal microbiota (Fig. 19.2), in experimental circumstances changing the ratio between the major bacterial phyla Bacteroides and Firmicutes (Turnbaugh et al. 2006). Moreover, the gene richness of the intestinal microbiome and MetS are correlated (Le Chatelier et al. 2013) and can be modulated by diet, reinforcing the role of intestinal ecology quality.

A class of metabolites, known as methylamines, are produced by gut microbiota via bacterial degradation of choline into trimethylamine (TMA), later metabolized into trimethylamine-N-oxide (TMAO) by flavin-containing monooxygenase 3 (FMO3) and other isoforms. This class of metabolites includes TMA, TMAO, and dimethylamine (DMA), identified as markers of NAFLD in murine models (Dumas et al. 2006; Cobbold et al. 2009). In addition, elevated values of homeostatic model assessment-insulin resistance (HOMA-IR) (Tresaco et al. 2005) and of serum uric acid (UA) (Nejatinamini et al. 2015) are associated with MetS and/or hepatic steatosis.

Elevated urinary/blood levels of aromatic amino acids (AAAs) and branched-chain amino acids (BCAAs) occurred in samples from obese patients, and appear to be correlated with IR and the risk of developing obesity-related MetS (Miccheli et al. 2015; Wiklund et al. 2014; Wurtz et al. 2012; Newgard et al. 2009).

Also pertinent to NAFLD, gamma-glutamyl dipeptides indicate increased glutathione (GSH) synthesis as part of an oxidative stress response (Soga et al. 2011).

Lipid metabolism, tyrosine (Jin et al. 2016), alanine, and the urea cycle (Martos-Moreno et al. 2014), acylcarnitine catabolism with and without changes in nucleotides, lysolipids, and inflammation markers are also reported in obesity and related disorders (Butte et al. 2015).

MetS and obesity share a pro-inflammatory state (Wisse 2004; Dandona et al. 2005). Insulin features anti-inflammatory properties, and insulin resistance could contribute to elevated mediators such as TNF-alpha, C-reactive protein, and IL-6 in MetS (Wahba and Mak 2007).

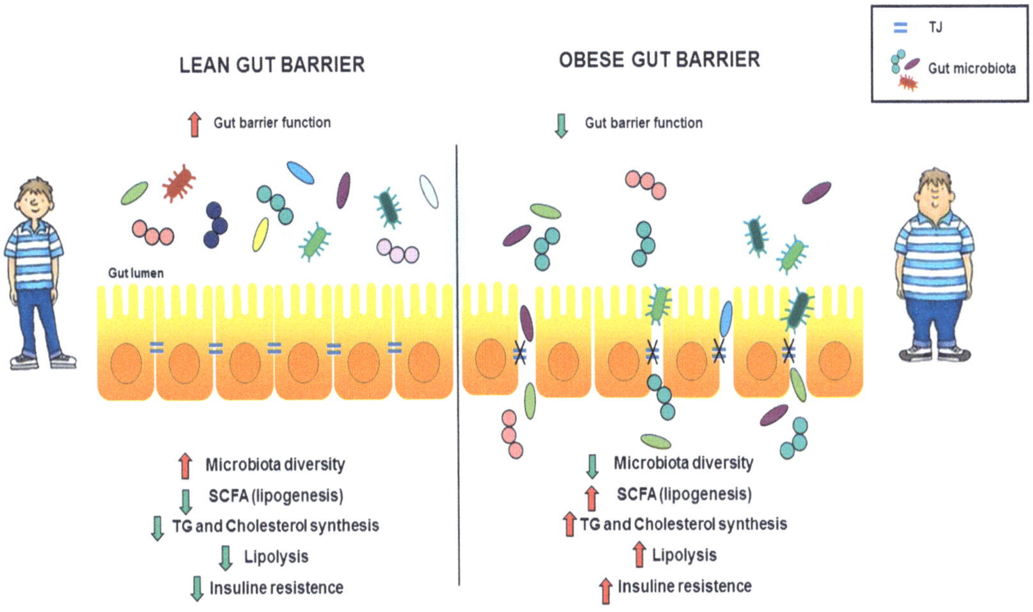

Fig. 19.2 Pictured are the differences in the gut barrier of lean and obese individuals

Urine and Salivary Pathways and Biomolecules

Saliva Production

Saliva is an exocrine solution in which water represents almost 99%, while the remaining 1% consists of a variety of electrolytes and proteins (de Almeida et al. 2008). It is also made up of a complex mixture of secretory products (organic and inorganic molecules) that originated from the salivary glands. Saliva compounds come from the oropharynx, upper airways, gastrointestinal reflux, gingival sulcus fluid, food deposits, and blood-derived compounds (Dodds et al. 2005). The inorganic part includes weak and strong ions (Na^+, K^+, Cl^-, Ca^{2+}, HCO^3, Mg^{2+}, and NH_3); the organic part contains components such as body secretion molecules (urea, uric acid, and creatinine), lipids like cholesterol and fatty acids and more than 400 types of protein. The most important proteins have a glandular origin (alpha-amylase, lysozyme, mucins, and proline-rich proteins) or are plasma-derived (albumin, secretory immunoglobulin A, and transferrin) (Hofman 2001).

Saliva is a mirror of the normal internal characteristics and disease state of an individual (Pfaffe et al. 2011; Nunes et al. 2015; Desai and Mathews 2014; Garrett et al. 1998; Chiappin et al. 2007) and can be defined as a natural filtrate of blood containing potentially diagnostic and prognostic “omic” features such as small molecules, metals, proteins, and DNA.

It can be conveniently and noninvasively collected, it is easy to store, and its collection, compared to that of other biological fluids, such as blood, are not expensive.

Urine Production

Urine production involves two processes: renal–glomerular filtration and tubular reabsorption, commanded by systemic hydration, and electrolyte balance. The glomerulus represents the site of formation of the primary urine while the tubule is involved in the transformation of pre-urine in urine (Ogobuiro and Tuma 2019).

Salivary and Urinary Metabolome

Salivary Metabolome

Pediatric obesity, its related liver disease, and MetS seem to have different salivary metabolomic signatures. The differences can be primarily found in metabolites involved in energy, amino acids, and organic acids metabolism, along with intestinal bacteria metabolism, all of which could be influenced by diet, microbiota, and intestinal mucins (Troisi et al. 2019).

Salivary metabolic distinguish children with reduced numbers of MetS criteria, from those with many indexes, who are at higher risk of hepato-metabolic complications (Troisi et al. 2019).

A partial least squares discriminant analysis (PLS-DA) model can be devised to classify the patient classes based on their salivary metabolomic signature, paving the bases for their use for both early disease diagnosis and monitoring in those with MetS (Troisi et al. 2019). Moreover, it is shown that salivary testing of uric acid (UA), glucose, insulin, and HOMA together with selected anthropometric parameters, may help to noninvasively screen obese children with hepatic steatosis and/or having MetS components (Troisi et al. 2018). Similarly to adults (Goodson et al. 2014; Mussavira et al. 2015; Fekete et al. 1993) and adolescents (Hartman et al. 2016), glucose and insulin are higher in the saliva of obese subjects with or without MetS than in healthy controls. Insulin resistance (IR) is associated with the components of MetS among obese children and adolescents (Nejatinamini et al. 2015; Cho et al. 2017b), and seems superior to glucose for MetS characterization (Costa et al. 2012).

In addition, data of salivary HOMA index (calculated by substituting the glucose and insulin serum values with the corresponding salivary values), indicated that values for obese subjects were higher than those for controls, and the trend is amplified when obesity was complicated by a fatty liver rather than by the number of MetS components.

Even though salivary UA levels were higher in obese patients, no differences were observed with or without a fatty liver. A close relationship seems to exist between UA levels and MetS; in fact, despite hyperuricemia not representing one of the components of MetS, it is demonstrated that UA levels tend to increase in parallel with the number of MetS components. Saliva of obese children had higher levels of palmitic acid, myristic acid, urea, N-acetyl galactosamine, maltose, gluconic acid and isoleucine, and lower levels of hydroxybutyric acid and malic acid, which appear prevalent in patients without steatosis. On the other hand, lauric acid, maltose, and methyl maleic acid were featured by those with steatosis (Troisi et al. 2019).

The levels of valine, mannose, acetopyruvic acid, palmitic acid, triethylene glycol, gluconic acid, citric acid, scyllo-inositol, deoxyglucose, psicopyranose, myoinositol, and cycloserine were higher in obese patients. On the contrary, normal-weight patients showed higher levels of 1,2,3-butanetriol, 2-oxovaleric acid, 2-palmitoylglycerol, di-n-octyl phthalate, itaconic acid, methyl galactoside, stearic acid, 2-piperidinone, maltose, 2-deoxy-D-ribose, pentane dioic acid, glycerol, pentitol, glyceric acid, methyl maleic acid, 2-deoxypentofuranose, β-hydroxy pyruvic acid, 2-hydroxy-methylcyclopentanol, and L-serine. In obese patients with steatosis elevated D-glucuronic acid γ-lactone, 2-deoxy-ribolactone, 2-hydroxyisocaproic acid, pyroglutamic acid, and propanoic acid occurred. In the absence of steatosis butanoic acid, maltose, thiamine, glucopyranose, 2-hydroxybutyric acid, and mannose predominated (Troisi et al. 2018).

The network of salivary molecules that are differently produced in lean and obese patients, with or without MetS/NAFLD comorbidities, was also characterized by higher levels of palmitic acid and myristic acid, which are two saturated fatty acids, generally elevated in patients with steatosis (Troisi et al. 2019). Myristic acid seems to be associated with obesity but not with MetS (Takato et al. 2017).

An increase in activity of C16 Δ9-desaturase and C18 Δ9-desaturase with a decreased Δ5-desaturase activity can be responsible for altered fatty acid metabolism (Kang et al. 2017).

The correlation among lipid profile, glucose, and insulin levels suggests a conserved ability in adapting to a caloric challenge, compared with metabolically unhealthy subjects (Badoud et al. 2015; Aristizabal et al. 2016). Important advances notwithstanding, the predictive performance of these salivary markers remains unsatisfactory (Troisi et al. 2018).

In conclusion, different salivary metabolites and metabolic pathways are involved in a complex metabolic fingerprint of obesity, obesity-related NAFLD, and obesity-related MetS (Magge et al. 2017). Table 19.1 reports a summary of the principal papers about metabolomics application in pediatric obesity.

Urinary Metabolome

Urinary metabolome analysis similarly detects typical obesity-related dietary behavior as well as GLA alteration (Troisi et al. 2017). Intestinal permeability, small intestine bacterial overgrowth (SIBO), and diet preferences can be relevant in this context.

Lower urinary xylitol and phenylacetic acid (PAA) was noticed in the obese population.

Xylitol is not endogenously produced by humans, but it can be found in several types of fruits and vegetables. Its intake can be considered important in preventing the development of obesity and MetS. Amo et al. (2011) demonstrated that long-term intake of xylitol is able to suppress the accumulation of visceral fat, and contextually increase insulin and lipids levels in the plasma of rats fed a high-fat diet.

Xylitol is more elevated in normal-weight patients, supporting the idea that it could protect against obesity-related liver damage (Troisi et al. 2017).

Also, PPA was lower in obese children (Troisi et al. 2017). PPA is synthesized from the amino acid phenylalanine (PA) via phenylpyruvate; phenolic acids are generally associated with beneficial effects for human health (Pandey and Rizvi 2009), whereas low urinary levels of PPA can be related to an unhealthy diet due to a low intake of plant fibers.

Furthermore, Troisi et al. (2017) reported higher urinary glucose levels in obese children than in the normal-weight controls. This evidence is in accordance with findings in both obese children (Urakami et al. 2007) and adults (Elliott et al. 2015).

Methyl histidine concentration in urine was elevated, especially in patients with NAFLD, contrasting with low levels of xylitol (Troisi et al. 2017). Moreover, this metabolite has previously been associated with BMI in obese individuals (Elliott et al. 2015). The same association was found with 4-cresyl sulfate (PCS) that can be considered directly linked to gut microbial activity (Elliott et al. 2015). Consistently, urinary PCS was elevated in obese children without NAFLD and showed a negative correlation with the presence of SIBO.

There is an inverse relationship between BMI and urinary 3-hydroxymandelate, a metabolite of the alpha-adrenergic agonist p-synephrine (Stohs et al. 2011) that is strictly related to tyrosine metabolism.

Indeed, pseudouridine (PSI) level in urine was higher in obese individuals, especially in those with steatosis. It represents a measure of protein turnover, a metabolic process that may be targeted by oxidative DNA stress that is common in pediatric obesity, obesity-related NAFLD, and in other conditions of high dietary energy intake (Topp et al. 2008).

GLA dysfunction, which plays a role in obesity and obesity-related hepatic complications, is responsible for typical signals of metabolome signature, especially in obese children with NAFLD (Troisi et al. 2017). In particular, increased intestinal permeability (IP), gut microbiota dysbiosis, and SIBO have been extensively associated with severity of obesity-related liver damage (Guercio Nuzio et al. 2017; Boursier et al. 2016).

Table 19.1 Principal studies regarding the metabolomics application in pediatric obesity

Author	Population	Sample	Platform	Statistical analysis	Outcome
Topp et al. (2008)	44 normal-weight children (3–18 years old)	Urine	HPLC (targeted approach)	Wilcoxon Mann Whitney	Close relationships between whole-body RNA and protein degradation and metabolic rate. The relationship between 8-oxodG excretion and metabolic rate, however, is less strong suggesting that factors other than metabolic rate considerably affect the oxidative stress to DNA.
Pastore et al. (2014)	63 NAFLD patients	Plasma	HPLC (targeted approach)	Kolmogorov-Smirnov tests	Data demonstrated a defective hepatic sulfur metabolism in children with NAFLD, and that high levels of Hcy and Cys probably correlate with a pattern of more severe histological liver damage, due to mechanisms that require further studies.
Butte et al. (2015)	353 normal weight (NW) Hispanic children and 450 obese children	Plasma	GC-MS (untargeted approach)	PCA and ANOVA	Global metabolomic profiling in nonobese and obese children showed increased BCAA and acylcarnitine catabolism and changes in nucleotides, lysolipids, and inflammation markers. Strong signature of reduced fatty acid catabolism and increased steroid derivatives.
Jin et al. (2016)	39 Hispanic-American, obese adolescents aged 11–17 years (30 had hepatic steatosis $\geq$ 5% and 9 were matched controls with hepatic steatosis < 5%)	Plasma	UPLC-Q/Orbitrap (untargeted approach)	Manhattan plots and t-test	Metabolic pathways of several amino acids are significantly disturbed in adolescents with elevated hepatic steatosis. This is a novel finding and suggests that these pathways may be integral in the mechanisms of NAFLD.
Cho et al. (2017a, b)	91 nonobese and 93 obese adolescent subjects	Urine	UPLC-Q/ToF (untargeted approach) DI-Q/ToF (targeted approach)	PCA, Wilcoxon signed ranks test, Pearson correlation, and linear regression	Increase of inflammation-related metabolites is a crucial metabolomic signature of obesity compared with normal weight in adolescents. These metabolites may be useful for predicting cardiovascular risk. Furthermore, the metabolomic differences observed in this study suggested that adolescent obesity induces impending insulin resistance, ammonia toxicity, and oxidative stress.

(continued)

Table 19.1 (continued)

Author	Population	Sample	Platform	Statistical analysis	Outcome
Gawlik et al. (2016)	87 obese children with and without insulin resistance (IR)	Urine	GC-MS (targeted approach)	K-means clustering and ANOVA	The steroidal metabolomic signature of IR in obese children is characterized by enhanced secretion of steroids from all three adrenal pathways.
Bervoets et al. (2018)	65 obese and 37 normal-weight children	Plasma	NMR spectroscopy (untargeted approach)	PLSDA	Increased levels of lipids, N-acetyl glycoproteins, and lactate. • Decreased levels of several amino acids, α-ketoglutarate, glucose, citrate, and chlorinated phospholipids as compared with normal-weight children.
Troisi et al. (2017)	22 obese children and 14 NW	Urine	GC-MS (untargeted approach)	PLSDA	Obese children had: • Higher levels of glucose/1-methylhistidine • Lower levels of xylitol, phenyl acetic acid, and hydroquinone The metabolic pathways of BCAA and/or their metabolites correlated with an excess of visceral fat centimeters.
Goffredo et al. (2017)	78 obese adolescents with and without insulin resistance (IR)	Plasma	LC-Q/Trap (targeted approach)	Random Forest analysis	Dysregulation of BCAA metabolism characterizes obese adolescents with NAFLD independently of obesity and insulin resistance.
Leal-Witt et al. (2018)	35 prepubertal children with obesity enrolled 6-month-long lifestyle intervention program	Plasma	UPLC-Q/ToF (untargeted approach)	PCA	A sphingolipid metabolism-related signature was identified as the major contributor.
Troisi et al. (2019)	23 obese and 18 NW children	Saliva	GC-MS (untargeted approach)	PLSDA	Differences in metabolites involved in energy, amino and organic acid metabolism.

Small Intestinal Bacterial Overgrowth

A direct relationship was also observed between BMI and urinary excretion of leucine, valine, and isoleucine; branched-chain amino acids (BCAAs) and/or their metabolites showed a correlation with visceral fat, increased IP and SIBO (Troisi et al. 2017). SIBO-positive children had elevated levels of glycolic acid and mannose while in controls valine, PCS, butyrate, and adipic acid were more abundant (Troisi et al. 2017). In addition, urinary cysteine levels were found to be elevated, in accordance with the data obtained from plasma of obese children (Pastore et al. 2014) and adults (Kalhan et al. 2011) with NAFLD.

Finally, the increase in the concentration of both N-methyl nicotinate and hydroquinone (HQ) in urine appears to be protective against hepatic steatosis, probably due to their cytoprotective role (Jin et al. 2016; Dinkova-Kostova and Wang 2011). Glucose and xylitol metabolism, tyrosine metabolism, nicotinate and nicotinamide (vitamin B3) metabolism, and the pentose-6-phosphate pathway are among the most prominent in pediatric obesity (Troisi et al. 2017).

Potential Targets for New Drug Development

It is suggested that nutritional behaviors characterized by elevated fruit and vegetable intake and high monounsaturated fats intake are inversely associated with MetS, and they could improve this condition, especially regarding blood lipids, IR, and liver performance (Kushner 2014; Patti et al. 2015; Pitsavos et al. 2007; Esposito et al. 2007; Camacho and Camacho 2017).

Phytotherapeutic Agents

In many disease models polyphenols exert beneficial effects (Tenore et al. 2014). Annurca apple flesh polyphenols have shown significant hypolipidemic potential due to their ability to reducing cholesterol and low-density lipoproteins (LDL), and increasing apolipoprotein A1 (ApoA1) expression, the major protein component of high-density lipoproteins (HDL) in plasma (Tenore et al. 2014).

The molecules of polyphenolic extract that exert a hypolipidemic effect are procyanidrins. In vitro and in vivo experiments have provided evidence of their potential role in reducing postprandial hypertriglyceridemia by pancreatic lipase inhibition (Tenore et al. 2014). This mechanism induces the suppression of triglyceride absorption (Tenore et al. 2011).

Chlorogenic acid is a phenolic compound from the hydroxycinnamic acid family (an apple-derived polyphenol) that seems to protect against liver injury (Santana-Gálvez et al. 2017). It has been observed to have an impact on many transcription factors and enzymes which are involved in lipid metabolism, being associated with positive effects on obesity and dyslipidemia (Santana-Gálvez et al. 2017). It significantly reduced body weight, visceral fat mass, plasma leptin, and insulin levels, triglycerides in liver and cholesterol in adipose tissue (Cho et al. 2010). In addition, chlorogenic acid appears to be involved in the prevention of weight gain and in liver lipid accumulation reduction, delaying liver steatosis onset and blocking high-fat diet-induced IR (Ma et al. 2015) (Fig. 19.3).

Patti et al. (2015) evaluated the effect of a mixed supplement that contains chlorogenic acid, *Curcuma longa*, silymarin, guggul, and inulin on MetS in humans, reporting significant reductions in body weight, body mass index, waist circumference, fasting glucose, and total cholesterol. Within the same context, Annurca flesh polyphenols could play a beneficial role in MetS prevention and treatment (Tenore et al. 2014).

Gut Microbiota Interventions

Another possible strategy could be based on the administration of probiotics or prebiotics, or fecal transplantation (Saint-Georges-Chaumet and Edeas 2015). Probiotics and related dietary supplements which are employed to enhance and fortify gut microbiota do not have a simple nutritive value (Ewaschuk and Dieleman 2006). Indeed, they are able to change the resident gut microbiota composition and the gut lumen, favoring an anti-inflammatory environment, and improving gut barrier integrity. Similarly, the concept of "prebiotics" is referred to some fermentable carbohydrates that, through their metabolism by gut microorganisms, selectively modulate the composition and/or activity of the gut microbiota (Bluemel et al. 2016; Vajro et al. 2013b).

Ongoing Studies

Urine and saliva have a high diagnostic potential in obesity and MetS, especially in pediatric population, and could make blood sampling unnecessary in a number of circumstances (Troisi et al. 2017, 2018; Belmonte et al. 2017; Pierri et al. 2018). On this basis, an interesting clinical study has been proposed by the University of Salerno: "N.a.p.o.l.i. Study" (Non-invasive Approach for Pediatric Obesity Liver and metabolic Involvement) that aims to achieve a solution for obesity

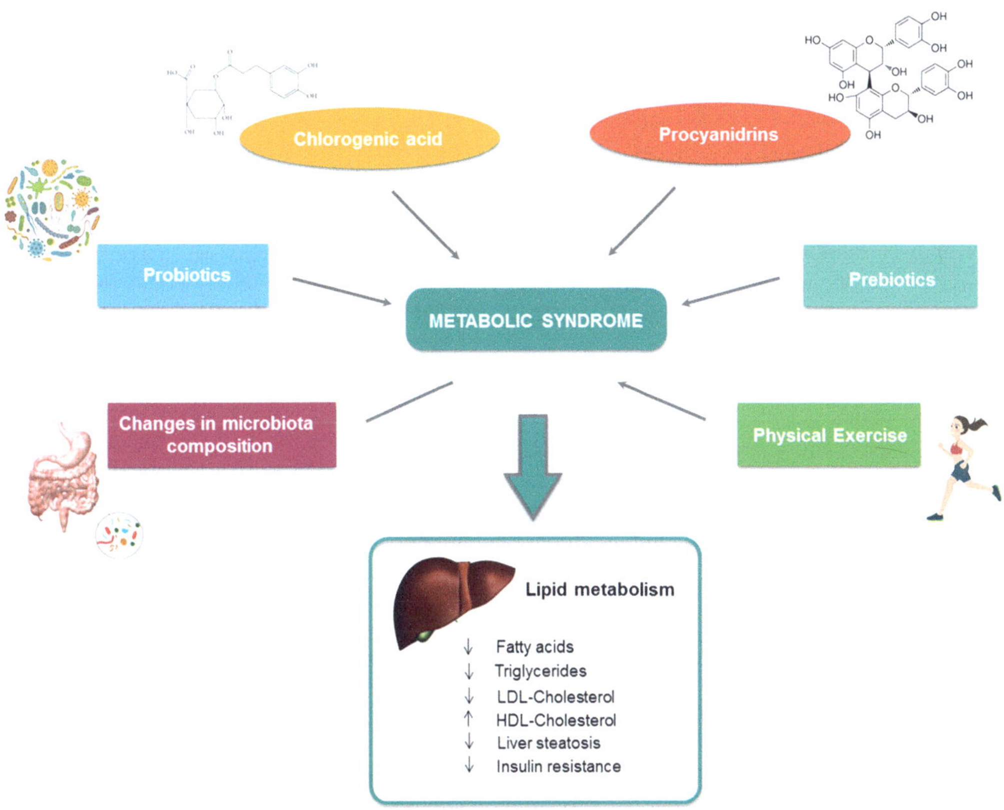

Fig. 19.3 Factors influencing the development of metabolic syndrome and the implications of metabolic syndrome on lipid metabolism

management and monitoring in a pediatric population (Napoli Study 2019). The goal is to develop an integrated system that is able to minimize obesity incidence during the pediatric age, and specifically to monitor comorbidities related to obesity and overweight (hepatic steatosis and MetS). The objectives are (1) Designing and validating a noninvasive system able to identify basal metabolic conditions in pediatric subjects, and to characterize a metabolic profile that is generally related to hepatometabolic conditions; (2) Designing and prototyping a decision support system able to manage the enrollment, the profiling, and the personalization of a diet-therapeutic plan, based on annurca apple supplementation in pediatric subjects.

The study expects to enroll 1000 pediatric patients (500 overweight or obese and 500 normal weight) from different countries in Europe. The inclusion criteria are age between 5 and 16 years; normal weight (BMI percentile between 5th and 85th), overweight (BMI > 85th) or obesity (BMI > 95th) conditions; adherence to the study and signature of informed consent. The exclusion criteria are history of chronic or acute not obesity-related diseases, and/or pathologies that do not allow saliva collection.

Other ongoing studies include Science and Technology in childhood Obesity Policy (STOP) (2019) the EPHORT Childhood Obesity Project (2016), and Childhood Obesity Research Demonstration (CORD) 3.0. (2019) STOP is working to identify policy changes that will address childhood obesity across Europe, until May 2022. It will assess from before birth through childhood, the key components of becoming obese, while putting emphasis on vulnerable and socially disadvantaged children. The goal is to have effective and sustainable solutions for childhood obesity to be implemented in the EU.

The EPHORT Childhood Obesity Project is taking place from 2014 to 2020. Eight areas of action have been identified: (1) Support a healthy start in life, (2) Promote healthier environments, especially in schools and pre-schools, (3) Make the healthy option the easier option, (4) Restrict marketing and advertising to children, (5) Inform and empower families, (6) Encourage physical activity, (7) Monitor and evaluate, and (8) Increase research. The first task for this project was to identify the prevalence of obesity per country in the EU. Then referring to the Areas of Actions, policies, and activities were made. After this, each country was evaluated to see if the actions had been taken, and the prevalence of obesity was measured. Task three included an overview of the engagement of the relevant states, the Commission, and the international organizations. Once this was complete the team is proceeding to assess the Action Plan and make recommendations for the second half-period.

CORD 3.0 will be funded by Massachusetts General Hospital, The Miriam Hospital, Stanford University, University of Nebraska, and Washington University in St. Louis from 2019 to 2024. The goal of the project is for these five research teams to focus on adapting, testing, and packaging effective programs to reduce childhood obesity among children in lower-income families.

References

Abenavoli L, Greco M, Milic N et al (2017) Effect of Mediterranean diet and antioxidant formulation in non-alcoholic fatty liver disease: a randomized study. Nutrients 9(8):870

Alberti KGMM, Zimmet P, Shaw J (2006) Metabolic syndrome—a new world-wide definition. A consensus statement from the international diabetes federation. Diabet Med 23(5):469–480

de Almeida PDV, Gregio AM, Machado MA, De Lima AA, Azevedo LR (2008) Saliva composition and functions: a comprehensive review. J Contemp Dent Pract 9(3):72–80

Amo K, Arai H, Uebanso T et al (2011) Effects of xylitol on metabolic parameters and visceral fat accumulation. J Clin Biochem Nutr 49:1106150086

Aristizabal JC, Barona J, Gonzalez-Zapata LI, Deossa GC, Estrada A (2016) Fatty acid content of plasma triglycerides may contribute to the heterogeneity in the relationship between abdominal obesity and the metabolic syndrome. Metab Syndr Relat Disord 14 (6):311–317

Badoud F, Lam KP, Perreault M, Zulyniak MA, Britz-McKibbin P, Mutch DM (2015) Metabolomics reveals metabolically healthy and unhealthy obese individuals differ in their response to a caloric challenge. PLoS One 10(8):e0134613

Belmonte F, Bisogno A, Troisi J et al (2017) Salivary levels of uric acid, insulin and HOMA: a promising field of study to non-invasively identify obese children at risk of metabolic syndrome and fatty liver. Dig Liver Dis 49:e247. https://doi.org/10.1016/j.dld.2017.09.017

Bervoets L, Massa G, Guedens W, Reekmans G, Noben J-P, Adriaensens P (2018) Identification of metabolic phenotypes in childhood obesity by (1)H NMR metabolomics of blood plasma. Fut Sci OA 4(6): FSO310. https://doi.org/10.4155/fsoa-2017-0146

Bluemel S, Williams B, Knight R, Schnabl B (2016) Precision medicine in alcoholic and nonalcoholic fatty liver disease via modulating the gut microbiota. Am J Phys Gastrointest Liver Phys 311(6):G1018–G1036

Boursier J, Mueller O, Barret M et al (2016) The severity of nonalcoholic fatty liver disease is associated with gut dysbiosis and shift in the metabolic function of the gut microbiota. Hepatology 63(3):764–775

Bussler S, Penke M, Flemming G et al (2017) Novel insights in the metabolic syndrome in childhood and adolescence. Horm Res Paediatr 88(3–4):181–193. https://doi.org/10.1159/000479510

Butte NF, Liu Y, Zakeri IF et al (2015) Global metabolomic profiling targeting childhood obesity in the Hispanic population. Am J Clin Nutr 102 (2):256–267. https://doi.org/10.3945/ajcn.115.111872

Buzzetti E, Pinzani M, Tsochatzis EA (2016) The multiple-hit pathogenesis of non-alcoholic fatty liver disease (NAFLD). Metabolism 65(8):1038–1048

Camacho JDH, Camacho MH (2017) Clinical update on metabolic syndrome. Rev Esp Nutri Hum Diet 21 (4):384–392

Centers for Disease Control and Prevention (CDC) (2019) CDC's Childhood Obesity Research Demonstration (CORD) Project 3.0. https://www.cdc.gov/obesity/strategies/healthcare/cord3.html. Accessed 26 Nov 2019

Chiappin S, Antonelli G, Gatti R, Elio F (2007) Saliva specimen: a new laboratory tool for diagnostic and basic investigation. Clin Chim Acta 383(1-2):30–40

Cho A-S, Jeon S-M, Kim M-J et al (2010) Chlorogenic acid exhibits anti-obesity property and improves lipid metabolism in high-fat diet-induced-obese mice. Food Chem Toxicol 48(3):937–943

Cho K, Moon JS, Kang J-H et al (2017a) Combined untargeted and targeted metabolomic profiling reveals urinary biomarkers for discriminating obese from normal-weight adolescents. Pediatr Obes 12 (2):93–101. https://doi.org/10.1111/ijpo.12114

Cho J, Hong H, Park S, Kim S, Kang H (2017b) Insulin resistance and its association with metabolic syndrome in Korean children. Biomed Res Int 2017:7

Cobbold JF, Anstee QM, Goldin RD et al (2009) Phenotyping murine models of non-alcoholic fatty liver disease through metabolic profiling of intact liver tissue. Clin Sci 116(5):403–413
Costa RF, Santos NS, Goldraich NP, Barski TF, de Andrade KS, Kruel LFM (2012) Metabolic syndrome in obese adolescents: a comparison of three different diagnostic criteria. J Pediatr 88(4):303–309. https://doi.org/10.2223/JPED.2200
D'Adamo E, Santoro N, Caprio S (2009) Metabolic syndrome in pediatrics: old concepts revised, new concepts discussed. Endocrinol Metab Clin 38 (3):549–563
Dandona P, Aljada A, Chaudhuri A, Mohanty P, Garg R (2005) Metabolic syndrome: a comprehensive perspective based on interactions between obesity, diabetes, and inflammation. Circulation 111(11):1448–1454
Das VL, Purohit P, Sharma P, Das A (2014) Metabolic syndrome in pediatric age-a group requiring intensive review. J Obes Metabol Res 1(4):238
Desai GS, Mathews ST (2014) Saliva as a non-invasive diagnostic tool for inflammation and insulin-resistance. World J Diabetes 5(6):730–738. https://doi.org/10.4239/wjd.v5.i6.730
Dietz WH, Robinson TN (2005) Overweight children and adolescents. N Engl J Med 352(20):2100–2109
Dinkova-Kostova AT, Wang XJ (2011) Induction of the Keap1/Nrf2/ARE pathway by oxidizable diphenols. Chem Biol Interact 192(1-2):101–106
Dodds MW, Johnson DA, Yeh C-K (2005) Health benefits of saliva: a review. J Dent 33(3):223–233
Dumas M-E, Barton RH, Toye A et al (2006) Metabolic profiling reveals a contribution of gut microbiota to fatty liver phenotype in insulin-resistant mice. Proc Natl Acad Sci 103(33):12511–12516
Dumas M-E, Kinross J, Nicholson JK (2014) Metabolic phenotyping and systems biology approaches to understanding metabolic syndrome and fatty liver disease. Gastroenterology 146(1):46–62
Eckel RH, Grundy SM, Zimmet PZ (2005) The metabolic syndrome. Lancet 365(9468):1415–1428
Elliott P, Posma JM, Chan Q et al (2015) Urinary metabolic signatures of human adiposity. Sci Transl Med 7 (285):285ra62
EPHORT Childhood Obesity Project (2016) RIVM DNI for PH and the E. https://ec.europa.eu/health/sites/health/files/nutrition_physical_activity/docs/ev_20161123_co05_en.pdf. Accessed 26 Nov 2019
Esposito K, Ciotola M, Giugliano D (2007) Mediterranean diet and the metabolic syndrome. Mol Nutr Food Res 51(10):1268–1274
Ewaschuk JB, Dieleman LA (2006) Probiotics and prebiotics in chronic inflammatory bowel diseases. World J Gastroenterol 12(37):5941
Fabbrini E, Sullivan S, Klein S (2010) Obesity and nonalcoholic fatty liver disease: biochemical, metabolic, and clinical implications. Hepatology 51(2):679–689
Fekete Z, Korec R, Feketeova E et al (1993) Salivary and plasma insulin levels in man. Biochem Mol Biol Int 30 (4):623–629
Garrett JR, Ekström J, Anderson LC (1998) Glandular mechanisms of salivary secretion, vol 10. Karger Medical and Scientific Publishers, Basel
Gawlik A, Shmoish M, Hartmann MF, Malecka-Tendera E, Wudy SA, Hochberg ZE (2016) Steroid metabolomic disease signature of nonsyndromic childhood obesity. J Clin Endocrinol Metab 101(11):4329–4337. https://doi.org/10.1210/jc.2016-1754
Genuth S, Alberti KG, Bennett P, Buse J, Defronzo R, Kahn R, Kitzmiller J, Knowler WC, Lebovitz H, Lernmark A, Nathan D, Palmer J, Rizza R, Saudek C, Shaw J, Steffes M, Stern M, Tuomilehto J, Zimmet P (2003) Follow-up report on the diagnosis of diabetes mellitus diabetes. Diabetes Care 26:3160–3167
Goffredo M, Santoro N, Tricò D, Giannini C, D'Adamo E, Zhao H, Peng G, Yu X, Lam TT, Pierpont B, Caprio S, Herzog RI (2017) A branched-chain amino acid-related metabolic signature characterizes obese adolescents with non-alcoholic fatty liver disease. Nutrients 9 (7):642. https://doi.org/10.3390/nu9070642. PMID: 28640216; PMCID: PMC5537762
Goodson JM, Kantarci A, Hartman M-L et al (2014) Metabolic disease risk in children by salivary biomarker analysis. PLoS One 9(6):e98799. https://doi.org/10.1371/journal.pone.0098799
Guercio Nuzio S, Di Stasi M, Pierri L et al (2017) Multiple gut–liver axis abnormalities in children with obesity with and without hepatic involvement. Pediat Obes 12 (6):446–452
Hartman M-L, Goodson JM, Barake R et al (2016) Salivary biomarkers in pediatric metabolic disease research. Pediatr Endocrinol Rev 13(3):602–611
Hofman LF (2001) Human saliva as a diagnostic specimen. J Nutr 131(5):1621S–1625S
Hosseinpour-Niazi S, Bakhshi B, Betru E, Mirmiran P, Darand M, Azizi F (2019) Prospective study of total and various types of vegetables and the risk of metabolic syndrome among children and adolescents. World J Diabetes 10(6):362
Jin R, Banton S, Tran VT et al (2016) Amino acid metabolism is altered in adolescents with nonalcoholic fatty liver disease—an untargeted, high resolution metabolomics study. J Pediatr 172:14–19
Kalhan SC, Guo L, Edmison J et al (2011) Plasma metabolomic profile in nonalcoholic fatty liver disease. Metabolism 60(3):404–413. https://doi.org/10.1016/j.metabol.2010.03.006
Kang M, Lee A, Yoo HJ et al (2017) Association between increased visceral fat area and alterations in plasma fatty acid profile in overweight subjects: a cross-sectional study. Lipids Health Dis 16(1):248
Kelishadi R, Gouya MM, Ardalan G et al (2007) First reference curves of waist and hip circumferences in an Asian population of youths: CASPIAN study. J Trop Pediatr 53(3):158–164
Kumar S, Kelly AS (2017) Review of childhood obesity: from epidemiology, etiology, and comorbidities to clinical assessment and treatment. Mayo Clin Proc 92:251–265
Kushner RF (2014) Weight loss strategies for treatment of obesity. Prog Cardiovasc Dis 56(4):465–472

Lakshman R, Elks CE, Ong KK (2012) Childhood obesity. Circulation 126(14):1770–1779
Le Chatelier E, Nielsen T, Qin J et al (2013) Richness of human gut microbiome correlates with metabolic markers. Nature 500(7464):541
Leal-Witt MJ, Llobet M, Samino S, Castellano P, Cuadras D, Jimenez-Chillaron JC, Yanes O, Ramon-Krauel M, Lerin C (2018) Lifestyle intervention decreases urine trimethylamine N-Oxide levels in prepubertal children with obesity. Obesity 26:1603–1610. https://doi.org/10.1002/oby.22271
Ma Y, Gao M, Liu D (2015) Chlorogenic acid improves high fat diet-induced hepatic steatosis and insulin resistance in mice. Pharm Res 32(4):1200–1209
Magge SN, Goodman E, Armstrong SC (2017) The metabolic syndrome in children and adolescents: shifting the focus to cardiometabolic risk factor clustering. Pediatrics 140:e20171603. https://doi.org/10.1542/peds.2017-1603
Martos-Moreno GA, Sackmann-Sala L, Barrios V et al (2014) Proteomic analysis allows for early detection of potential markers of metabolic impairment in very young obese children. Int J Pediatr Endocrinol 2014 (1):9. https://doi.org/10.1186/1687-9856-2014-9
Miccheli A, Capuani G, Marini F et al (2015) Urinary (1) H-NMR-based metabolic profiling of children with NAFLD undergoing VSL#3 treatment. Int J Obes 39 (7):1118–1125. https://doi.org/10.1038/ijo.2015.40
Morales Camacho WJ, Molina Díaz JM, Plata Ortiz S, Plata Ortiz JE, Morales Camacho MA, Calderón BP (2019) Childhood obesity: aetiology, comorbidities, and treatment. Diabetes Metab Res Rev 35:e3203
Mussavira S, Dharmalingam M, Omana SB (2015) Salivary glucose and antioxidant defense markers in type II diabetes mellitus. Turk J Med Sci 45(1):141–147
Napoli Study (2019) University of Salerno. November. https://napolistudy.org.
Nejatinamini S, Ataie-Jafari A, Qorbani M et al (2015) Association between serum uric acid level and metabolic syndrome components. J Diab Metab Disord 14:70. https://doi.org/10.1186/s40200-015-0200-z
Newgard CB, An J, Bain JR et al (2009) A branched-chain amino acid-related metabolic signature that differentiates obese and lean humans and contributes to insulin resistance. Cell Metab 9(4):311–326
Nunes S, Alessandro L, Mussavira S, Sukumaran BO (2015) Clinical and diagnostic utility of saliva as a non-invasive diagnostic fluid: a systematic review. Biochem Med Biochem Med 25(2):177–192
Ogobuiro I, Tuma F (2019) Physiology, renal. In: StatPearls [Internet]. StatPearls Publishing.
Pandey KB, Rizvi SI (2009) Plant polyphenols as dietary antioxidants in human health and disease. Oxidative Med Cell Longev 2(5):270–278
Pastore A, Alisi A, di Giovamberardino G et al (2014) Plasma levels of homocysteine and cysteine increased in pediatric NAFLD and strongly correlated with severity of liver damage. Int J Mol Sci 15 (11):21202–21214
Patti AM, Al-Rasadi K, Katsiki N et al (2015) Effect of a natural supplement containing curcuma longa, guggul, and chlorogenic acid in patients with metabolic syndrome. Angiology 66(9):856–861
Pfaffe T, Cooper-White J, Beyerlein P, Kostner K, Punyadeera C (2011) Diagnostic potential of saliva: current state and future applications. Clin Chem 57 (5):675–687
Pierri L, Troisi J, Bisogno A et al (2018) A metabolomic salivary signature of pediatric obesity related liver disease. J Pediatr Gastroenterol Nutr 66:642–643
Pitsavos C, Panagiotakos DB, Tzima N et al (2007) Diet, exercise, and C-reactive protein levels in people with abdominal obesity: the ATTICA epidemiological study. Angiology 58(2):225–233
Rosenbaum M (2018) Special considerations relevant to pediatric obesity. In: Endotext [Internet]. MDText.com, Inc.
Rutigliano I, Vinci R, De Filippo G et al (2017) Metabolic syndrome, hepatic steatosis, and cardiovascular risk in children. Nutrition 36:1–7. https://doi.org/10.1016/j.nut.2016.10.017
Saint-Georges-Chaumet Y, Edeas M (2015) Microbiota–mitochondria inter-talk: consequence for microbiota–host interaction. FEMS Pathog Dis 74(1):ftv096
Santana-Gálvez J, Cisneros-Zevallos L, Jacobo-Velázquez DA (2017) Chlorogenic acid: recent advances on its dual role as a food additive and a nutraceutical against metabolic syndrome. Molecules 22(3):358
Sanyal AJ, Campbell-Sargent C, Mirshahi F et al (2001) Nonalcoholic steatohepatitis: association of insulin resistance and mitochondrial abnormalities. Gastroenterology 120(5):1183–1192
Soga T, Sugimoto M, Honma M et al (2011) Serum metabolomics reveals γ-glutamyl dipeptides as biomarkers for discrimination among different forms of liver disease. J Hepatol 55(4):896–905. https://doi.org/10.1016/j.jhep.2011.01.031
Spruijt-Metz D (2011) Etiology, treatment, and prevention of obesity in childhood and adolescence: a decade in review. J Resear Adoles 21(1):129–152
Stohs SJ, Preuss HG, Shara M (2011) A review of the receptor-binding properties of p-synephrine as related to its pharmacological effects. Oxidative Med Cell Longev 2011:482973
STOP (2019) Science and Technology in childhood Obesity Policy. http://www.stopchildobesity.eu/. Accessed 26 Nov 2019
Takato T, Iwata K, Murakami C, Wada Y, Sakane F (2017) Chronic administration of myristic acid improves hyperglycaemia in the Nagoya–Shibata–Yasuda mouse model of congenital type 2 diabetes. Diabetologia 60(10):2076–2083
Tenore GC, Troisi J, Di Fiore R, Manfra M, Novellino E (2011) Nutraceutical value and toxicological profile of selected red wines from Morocco. Food Chem 129 (3):792–798. https://doi.org/10.1016/j.foodchem.2011.05.022

Tenore GC, Giorgio C, Stiuso P, Ritieni A, Giannetti D, Novellino E (2014) Effects of annurca apple polyphenols on lipid metabolism in HepG2 cell lines: a source of nutraceuticals potentially indicated for the metabolic syndrome. Food Res Int 63:252–257

Tiniakos DG, Vos MB, Brunt EM (2010) Nonalcoholic fatty liver disease: pathology and pathogenesis. An Rev Pathol Mech Dis 5:145–171

Topp H, Fusch G, Schöch G, Fusch C (2008) Noninvasive markers of oxidative DNA stress, RNA degradation and protein degradation are differentially correlated with resting metabolic rate and energy intake in children and adolescents. Pediatr Res 64(3):246

Tresaco B, Bueno G, Pineda I, Moreno LA, Garagorri JM, Bueno M (2005) Homeostatic model assessment (HOMA) index cut-off values to identify the metabolic syndrome in children. J Physiol Biochem 61 (2):381–388

Troisi J, Pierri L, Landolfi A et al (2017) Urinary metabolomics in pediatric obesity and NAFLD identifies metabolic pathways/metabolites related to dietary habits and gut-liver axis perturbations. Nutrients 9(5):E485. https://doi.org/10.3390/nu9050485

Troisi J, Belmonte F, Bisogno A et al (2018) Salivary markers of hepato-metabolic comorbidities in pediatric obesity. Dig Liver Dis 51:P516–P523. https://doi.org/10.1016/j.dld.2018.11.009

Troisi J, Belmonte F, Bisogno A et al (2019) Metabolomic salivary signature of pediatric obesity related liver disease and metabolic syndrome. Nutrients 11(2):274. https://doi.org/10.3390/nu11020274

Turnbaugh PJ, Ley RE, Mahowald MA, Magrini V, Mardis ER, Gordon JI (2006) An obesity-associated gut microbiome with increased capacity for energy harvest. Nature 444(7122):1027–1031. https://doi.org/10.1038/nature05414

Urakami T, Morimoto S, Nitadori Y, Harada K, Owada M, Kitagawa T (2007) Urine glucose screening program at schools in Japan to detect children with diabetes and its outcome-incidence and clinical characteristics of childhood type 2 diabetes in Japan. Pediat Resear 61(2):141

Vajro P, Paolella G, Fasano A (2013a) Microbiota and gut-liver axis: their influences on obesity and obesity-related liver disease. J Pediatr Gastroenterol Nutr 56 (5):461–468. https://doi.org/10.1097/MPG.0b013e318284abb5

Vajro P, Paolella G, Fasano A (2013b) Microbiota and gut-liver axis: a mini-review on their influences on obesity and obesity related liver disease. J Pediatr Gastroenterol Nutr 56(5):461

Wahba IM, Mak RH (2007) Obesity and obesity-initiated metabolic syndrome: mechanistic links to chronic kidney disease. Clin J Am Soc Nephrol 2(3):550–562

WHO (2016) Ending childhood obesity: report of the Commission on Ending Childhood Obesity (ECHO). http://apps.who.int/iris/bitstream/10665/204176/1/9789241510066_eng.pdf?ua=1.

Wiklund PK, Pekkala S, Autio R et al (2014) Serum metabolic profiles in overweight and obese women with and without metabolic syndrome. Diabetol Metab Syndr 6(1):40. https://doi.org/10.1186/1758-5996-6-40

Wisse BE (2004) The inflammatory syndrome: the role of adipose tissue cytokines in metabolic disorders linked to obesity. J Am Soc Nephrol 15(11):2792–2800

Wittcopp C, Conroy R (2016) Metabolic syndrome in children and adolescents. Pediatr Rev 37(5):193

Wurtz P, Makinen V-P, Soininen P et al (2012) Metabolic signatures of insulin resistance in 7,098 young adults. Diabetes 61(6):1372–1380. https://doi.org/10.2337/db11-1355

Zafar U, Khaliq S, Ahmad HU, Manzoor S, Lone KP (2018) Metabolic syndrome: an update on diagnostic criteria, pathogenesis, and genetic links. Hormones 17 (3):299–313

Zimmet P, Alberti KGM, Kaufman F et al (2007) The metabolic syndrome in children and adolescents—an IDF consensus report. Pediatr Diabetes 8(5):299–306. https://doi.org/10.1111/j.1399-5448.2007.00271.x

'State-of-the-Art' Metabolomics Investigations of Type 2 Diabetes

20

Benita Percival, Justine Leenders, and Martin Grootveld

Contents

The Role of Metabolomics 266
The Normal and Pathological Metabolome 266
Targeted vs. Untargeted Approach 266

Biomarkers of Type 2 Diabetes 267
Serum and Plasma Biomarkers 267
Saliva 272
Urine 273

Disturbed Metabolic Pathways Involved in T2D Pathogenicity 275
Personal Experience: Retrospective Network Pathway Analysis of Published Data . . 275

Prognostic Fingerprints 276
Short vs. Long-term Glycemic Homeostasis 278
Kidney Dysfunction and Ketosis 278

Potential Bedside/'Point-of-Care' Metabolomics 278

Future Developments 279
Bedside Devices 279
Metabolomic Panels 280
Portable Oral Prosthesis and Breath Volatolomic Analysis 280
Artificial Pancreas and Wearable Technology 280

References 280

Abstract

Metabolomics represents the global investigation of complex biomolecular patterns in living systems, i.e. in organs, cells, tissues or biofluids, and the understanding of how clinically significant deviations from such patterns may facilitate our knowledge of the pathogenesis of human diseases. This now widely expanded research area has been extensively applied to the prognostic monitoring of type 2 diabetes (T2D), particularly in human biofluids. The achievements of T2D metabolomics investigations include the identification of key biomarkers in human biofluid samples, which may also be employed for determining individuals at risk of this condition. T2D biomarkers arise from disturbed

B. Percival · J. Leenders · M. Grootveld (✉)
Leicester School of Pharmacy, De Montfort University, Leicester, UK
e-mail: mgrootveld@dmu.ac.uk

J. Faintuch and S. Faintuch (eds.), *Obesity and Diabetes*, https://doi.org/10.1007/978-3-030-53370-0_20

metabolic pathways, which include those distinct from glycolysis and lipid metabolism. Additionally, prognostic fingerprints and biomarkers of insulin resistance have been explored using metabolomics analysis, and these strategies have been critically evaluated for their possible future 'point-of-care' uses in clinical settings. The potential advancement for such 'bedside' metabolomics tests and the future direction of metabolomics in T2D is discussed in detail.

Keywords

Diabetes biomarkers · Plasma metabolome · Saliva metabolome · Urine metabolome · Diabetes metabolic pathways · Diabetes prognostic fingerprints · Bedside metabolome

The Role of Metabolomics

Although the metabolome was first explored centuries ago, through organoleptic tasting analysis and smell of urine as a means of diagnosing diseases, applications of bioanalytical techniques to simultaneously analyse more than one class of biomolecules in urine were first performed by Pauling et al. (1971), and the human metabolome was first described by Oliver et al. (1998). Over the last 20 years, the metabolome itself and means for the utilisation of multianalyte bioanalytical techniques that can be applied to it, have been further defined by others such as Nicholson and Lindon (2008), and many developments have been explored in the literature, with a progressive history of techniques employed for these studies, along with their establishment as key investigatory tools (documented by Kell and Oliver (2016)).

The Normal and Pathological Metabolome

The study of the metabolome allows for the monitoring of metabolic processes within a system by assessment of low-molecular-mass metabolites within biofluids, tissues, cells or other biosamples. Indeed, the metabolome can be affected by many biological processes. These can either be through external stimuli such as medical intervention, i.e. medication, diet or an exercise regimen, or through internal stimuli. An internal stimulus can be introduced by the modification of gene expression using techniques such as cell transfection, both *in vivo* and *in vitro*. Moreover, metabolomics techniques assess these changes by providing a 'snapshot' of the status of biological processes occurring at a specific point in time. This responsive information can provide a high level of detail regarding cellular metabolic processes, may facilitate phenotypic evaluations, and therefore yields an overall 'picture' or 'fingerprint' of a disease's chemopathological status.

Even more valuably, metabolomics is able to probe changing disease status, e.g. the effects of metformin therapy or the regression of inflammatory conditions, in type 2 diabetes (T2D), as can the influence of dietary or physical exercise interventions. Hence, these strategies may be successfully employed to monitor disease severity and progression, information which further increases our understanding of the aetiology, manifestation and progression of particular conditions (German et al. 2005). A combination of metabolomic strategies has been previously applied to pre-diabetes (Zeng et al. 2017), type 1 diabetes (Haukka et al. (2018)), T2D and gestational diabetes (Bentley-Lewis et al. 2015).

Targeted vs. Untargeted Approach

Leading metabolomics techniques are largely untargeted ones and aim to simultaneously detect all possible metabolites present in biological samples. This multicomponent quantitative bioanalytical approach differs from targeted metabolomics investigations, which focus on the analysis of only limited numbers or classes of metabolites that are linked via specific biochemical pathways. Typical techniques include nuclear magnetic resonance (NMR) and liquid chromatography-mass spectrometric (LC-MS) analyses, which are readily applicable to a wide range of biosamples (Zhang et al. 2012). Briefly, high-resolution NMR analysis is able to detect and quantify metabolites in a high-throughput, simultaneous, non-destructive manner requiring minimal sample preparation. LC-MS strategies, however, can target specific metabolites more sensitively than NMR, whilst lacking the specificity of the latter for absolute metabolite determinations (Sandlers 2017). Therefore, a combination of both LC-MS and NMR methodologies is important for a global understanding of metabolic effects involved in disease processes, and can be essential for thorough determinations of the metabolic status of biofluids and tissue biopsies. Alternative analytical techniques include Fourier transform infrared (FTIR) and Raman spectroscopies (Lindon et al. 2011).

Biomarkers of Type 2 Diabetes

One of the first diagnoses of diabetes was performed in fifth/sixth century BC using a urine sample, and this arose from a physician, Sushruta, acknowledging frequent urination in subjects with this condition, and ants being attracted to the urine (Sharma and Chandola 2011). This knowledge is still applied today in rural areas in different parts of the world, for example in Vietnam (Mull, Nguyen and Mull 2001). Diabetes mellitus became the designated terminology for these patients, which is Latin for abundant flux (of urine) and 'honeyed' or 'sweet,' respectively.

Currently, common biomarkers that are detected in clinical settings include glucose and ketone bodies in urine and plasma, as well as glycosylated haemoglobin A1c (HbA1c) in whole blood. HbA1c reflects the success of therapeutic control by T2D patients self-administering injectable insulin or oral hypoglycaemic agents such as metformin, throughout a 6–8 week period.

However, there are many more markers that can provide further information regarding the status of T2D, e.g. ketone bodies and branched-chain amino acids (BCAAs). These alternative biomarkers can provide valuable information regarding the early detection of T2D, i.e. prior to its full onset, a development which can give rise to improved outcomes for patients.

To date, a combination of biofluids and tissues has been investigated in animals and humans, including blood serum (Drogan et al. 2015; Floegel et al. 2013), blood plasma, urine and saliva. Table 20.1 provides insight into biomarkers that have been suggested to be clinically significant, regarding T2D diagnosis in serum or plasma, urine and saliva.

Serum and Plasma Biomarkers

One of the largest metabolomics studies exploring diabetes was performed with samples collected during the European Prospective Investigation into Cancer and Nutrition (EPIC) (Floegel et al. 2013; Drogan et al. 2015). Floegel et al. (2013) used flow injection analysis mass spectrometry/mass spectrometry (FIA-MS/MS) to determine serum metabolites from control ($n = 1482$) and T2D ($n = 800$) participants. The global metabolome of this cohort was monitored over a mean 7-year period to observe the increased or decreased risk of T2D form the status of biomolecular analytes. Upregulations in hexose, phenylalanine and diacyl-phosphatidylcholines (C32:1, C36:1, C38:3 and C40:5), along with downregulations in glycine, sphingomyelin (C16:1), acyl-alkyl-phosphatidylcholines (C34:3, C40:6, C42:5, C44:4 and C44:5) and lysophosphatidylcholine (C18:2) were observed in T2D patients. Drogan et al. (2015) observed 300 individuals who provided serum samples at the beginning of a trial, and also after a median of 6 years, in an untargeted LC-MS investigation. Significant metabolite markers included lysophosphatidylcholine (demethyl 16:0) and phosphatidylcholine (oleoyl-20:0/oleoyl-20:0), which were all downregulated in T2D. As expected, hexose sugars such as glucose were elevated in T2D plasma. In another protocol, glyoxylate was observed to be upregulated

Table 20.1 A summary of metabolite markers that are disturbed in Type 2 Diabetes (T2D) whole blood, serum and plasma, urine and saliva in humans

Potential Marker	Biofluid	Increased/ decreased	Technique	References
Lipids, fatty acids				
Diacyl-phosphatidylcholines C32:1, C36:1, C38:3 and C40:5	Serum	↓	MS	Floegel et al. (2013)
Lysophosphatidylcholine C18:2	Serum	↑	MS	Floegel et al. (2013)
Lysophosphatidylcholine (dm16:0) and phosphatidylcholine (O-20:0/O-20:0)	Serum	↓	MS	Drogan et al. (2015)
Stearate	Serum	↑	MS	Xu et al. (2013)
Palmitate	Serum	↑	MS	Xu et al. (2013)
Glycerophospholipids	Serum	↓	MS	Xu et al. (2013)
Sphingomyelins	Serum	↓	MS	Xu et al. (2013)
Myristic Acid	Serum	↑	MS	Xu et al. (2013)
15-Methyl palmitate	Plasma	↓	MS	Menni et al. (2013)
10-Heptadecanoate	Plasma	↓	MS	Menni et al. (2013)
Adrenate (22:4 (n-6))	Plasma	↑	MS	Menni et al. (2013)
Arachidonate (20:4 (n-6))	Plasma	↑	MS	Menni et al. (2013)
Myristate (14:0)	Plasma	↓	MS	Menni et al. (2013)
Myristoleate (14:1 (n-5))	Plasma	↓	MS	Menni et al. (2013)
Palmitoleate (16:1 (n-5))	Plasma	↓	MS	Menni et al. (2013)
Pentanedionate (15:0)	Plasma	↓	MS	Menni et al. (2013)
5-Dodecanoate (12:1 (n=7))	Plasma	↓	MS	Menni et al. (2013)
Heptanoate (7:0)	Plasma	↓	MS	Menni et al. (2013)
Pelargonate (9:0)	Plasma	↓	MS	Menni et al. (2013)
Palmitoyl Sphingomyelin	Plasma	↓	MS	Menni et al. (2013)
Cholesterol	Plasma	↓	MS	Menni et al. (2013)
Low-molecular-mass Molecules (e.g. BCAAs and TCA cycle)				
1-Methylhistidine	Plasma Urine	↓ ↓	MS MS	Yousri et al. (2015)
2-Deoxyglucose	Plasma	↑	MS	Suhre et al. (2010)
2-Hydroxybutyrate	Plasma Serum Urine Saliva	↑ ↑ ↑ ↑ ↓	MS MS MS MS MS	Yousri et al. (2015) Menni et al. (2013) Xu et al. (2013) Yousri et al. (2015) Barnes et al. (2014)
3-D-Hydroxybutyrate	Plasma Urine	↑ ↑	MS ^{1}H NMR MS	Yousri et al. (2015) Suhre et al. (2010) Salek et al. (2007) Yousri et al. (2015)
3-Hydroxyproline	Urine	↑	MS	Yousri et al. (2015)
3-Methoxytyrosine	Plasma Urine	↓ ↓	MS MS	Yousri et al. (2015) Yousri et al. (2015)
3-Methyl-2-oxobutyrate	Plasma	↑	MS	Menni et al. (2013)
3-Methyl-3-oxovalerate	Plasma	↑	MS	Menni et al. (2013)
4-Hydroxyphenylpyruvate	Plasma Urine	↑ ↑	MS MS	Yousri et al. (2015) Yousri et al. (2015)
4-Methyl-2-oxopentanoate	Plasma	↑	MS	Menni et al. (2013)
5-Oxoproline	Urine	↓	MS	Yousri et al. (2015)
Acetate	Urine	↑	^{1}H NMR	Salek et al. (2007)
Acetoacetate	Urine	↑	^{1}H NMR MS	Salek et al. (2007) Yousri et al. (2015)

(continued)

Table 20.1 (continued)

Potential Marker	Biofluid	Increased/ decreased	Technique	References
Adipate	Urine	↓	MS	Yousri et al. (2015)
Alanine	Serum Urine	↓ ↑ ↑	^{1}H NMR MS MS ^{1}H NMR	Coco et al. (2019) Xu et al. (2013) Yousri et al. (2015) Urpi-Sarda et al. (2019)
Allantoin	Urine	↓	^{1}H NMR	Salek et al. (2007)
Aminohippurate	Urine	↓	^{1}H NMR	Salek et al. (2007)
Betaine	Urine	↑	MS	Svingen et al. (2016)
Carnitine	Serum	↓	^{1}H NMR	Coco et al. (2019)
Citrate	Serum Urine	↓ ↑	^{1}H NMR ^{1}H NMR	Coco et al. (2019) Urpi-Sarda et al. (2019)
Citrulline	Plasma	↓	MS	Yousri et al. (2015) Menni et al. (2013)
Creatine	Serum	↓	^{1}H NMR	Coco et al. (2019)
Creatinine	Serum Urine	↓ ↓	^{1}H NMR ^{1}H NMR/ MS	Coco et al. (2019) Salek et al. (2007) Yousri et al. (2015)
Cysteine	Urine	↑	MS	Yousri et al. (2015)
Dimethylamine	Urine	↑	^{1}H NMR	Salek et al. (2007)
Dimethylglycine	Urine	↑	^{1}H NMR MS	Salek et al. (2007) Urpi-Sarda et al. (2019) Urpi-Sarda et al. (2019)
Ethanolamine	Urine	↓	MS	Yousri et al. (2015)
Fumarate	Urine	↓	^{1}H NMR	Salek et al. (2007)
Glutamate	Serum Urine	↓ ↓	^{1}H NMR MS ^{1}H-NMR	Coco et al. (2019) Yousri et al. (2015) Urpi-Sarda et al. (2019)
Glycerol	Serum	↑	^{1}H-NMR	Aloha-Ollie et al. (2019)
Glycine	Serum	↓	MS	Floegel et al. (2013)
Glutamine	Serum Serum	↓ ↑	^{1}H NMR MS	Coco et al. (2019) Xu et al. (2013)
Glyoxylate		↑	MS	Nikiforova et al. (2014)
Heptanoate	Plasma	↓	MS	Yousri et al. (2015)
Hippurate	Urine	↓	^{1}H NMR	Salek et al. (2007) Urpi-Sarda et al. (2019)
Homocitrulline	Urine	↓	MS	Yousri et al. (2015)
Homocysteine	Urine	↓	MS	Yousri et al. (2015)
Homovanillate	Urine	↓	MS	Yousri et al. (2015)
Isobutyrylcarnitine	Urine	↓	MS	Yousri et al. (2015)
Isoleucine	Serum Plasma Urine	↓ ↑ ↑ ↑ ↑	^{1}H NMR MS ^{1}H NMR MS MS	Coco et al. (2019) Xu et al. (2013) Aloha-Ollie et al. (2019) Menni et al. (2013) Yousri et al. (2015)
Kynurenate	Plasma Urine	↓ ↓	MS MS	Yousri et al. (2015) Yousri et al. (2015)

(continued)

Table 20.1 (continued)

Potential Marker	Biofluid	Increased/ decreased	Technique	References
Leucine	Serum Plasma Urine	↓ ↑ ↑ ↑ ↑	^{1}H NMR MS ^{1}H NMR MS MS	Coco et al. (2019) Xu et al. (2013) Aloha-Ollie et al. (2019) Menni et al. (2013) Yousri et al. (2015)
Lysine	Serum	↓	^{1}H NMR	Coco et al. (2019)
L-Octanoylcarnitine	Plasma	↓	MS	Menni et al. (2013)
Malate	Plasma	↑	MS	Yousri et al. (2015) Menni et al. (2013)
Methionine	Serum	↓	^{1}H NMR	Coco et al. (2019)
Myo-inositol	Urine	↓	MS	Yousri et al. (2015)
N-acetylaspartate	Urine	↑	^{1}H NMR	Salek et al. (2007)
N-acetyl groups (including glycoproteins)	Urine	↓	^{1}H NMR	Salek et al. (2007)
N-acetylglycine	Plasma	↓	MS	Menni et al. (2013)
N-acetyl-β-alanine	Urine	↑	MS	Yousri et al. (2015)
N-butyrate	Urine	↑	^{1}H NMR	Salek et al. (2007)
N-methylnicotinate	Urine	↓	^{1}H NMR	Salek et al. (2007)
N-methylnicotinamide	Urine	↑	^{1}H NMR	Salek et al. (2007)
Ornithine	Plasma Urine	↓ ↑	MS MS	Yousri et al. (2015) Yousri et al. (2015)
Phenylalanine	Serum Plasma Urine	↓ ↑ ↑ ↑ ↑	^{1}H NMR MS MS ^{1}H NMR MS MS ^{1}H NMR	Coco et al. (2019) Floegel et al. (2013) Xu et al. (2013) Aloha-Ollie et al. (2019) Suhre et al. (2010) Yousri et al. (2015) Urpi-Sarda et al. (2019)
Pipecolate	Urine	↑	MS	Yousri et al. (2015)
Proline	Serum Plasma Urine	↑ ↑ ↑	MS MS MS	Xu et al. (2013) Menni et al. (2013) Yousri et al. (2015)
Pyroglutamine	Serum Plasma Urine Saliva	↓ ↓ ↓ ↓	MS MS MS MS	Xu et al. (2013) Yousri et al. (2015) Yousri et al. (2015) Barnes et al. (2014)
Sarcosine	Urine	↑	MS	Svingen et al. (2016)
Succinate	Urine	↓	^{1}H NMR	Salek et al. (2007)
Trans-urocanate	Urine	↓	MS	Yousri et al. (2015)
Tyrosine	Serum	↓	^{1}H NMR	Coco et al. (2019)
Urea	Plasma	↑	MS	Menni et al. (2013)
Valine	Serum Plasma	↓ ↑ ↑	^{1}H NMR MS MS	Coco et al. (2019) Xu et al. (2013) Menni et al. (2013)
Vanillylmandelate	Urine	↓	MS	Yousri et al. (2015)
α-Ketobutyrate	Plasma Urine	↑ ↑	MS MS	Yousri et al. (2015) Yousri et al. (2015)
β-Hydroxypyruvate	Plasma Urine	↑ ↑	MS MS	Yousri et al. (2015) Yousri et al. (2015)

(continued)

Table 20.1 (continued)

Potential Marker	Biofluid	Increased/ decreased	Technique	References
Carbohydrates and Derivatives				
1,3-Dihydroxyacetone	Plasma	↑	MS	Yousri et al. (2015)
1,5-Anhydroglucitol	Plasma Saliva	↓ ↓	MS MS	Yousri et al. (2015) Suhre et al. (2010) Barnes et al. (2014) Menni et al. (2013) Yousri et al. (2015)
2-Methylcitrate	Urine	↓	MS	Yousri et al. (2015)
3-Hydroxyisobutyrate	Plasma	↓	MS	Yousri et al. (2015)
Arabinose	Plasma	↑	MS	Menni et al. (2013)
Arabitol	Urine	↓	MS	Yousri et al. (2015)
Fructose	Plasma	↑	MS	Yousri et al. (2015) Menni et al. (2013)
Gluconate	Plasma	↑	MS	Yousri et al. (2015)
Glucose	Plasma Serum Urine Saliva	↑ ↑ ↑ ↑	MS ^{1}H NMR MS MS	Yousri et al. (2015) Barnes et al. (2014) Suhre et al. (2010) Menni et al. (2013) Coco et al. (2019) Yousri et al. (2015) Barnes et al. (2014)
Lactate	Plasma Serum Urine	↑ ↓ ↑	MS ^{1}H NMR MS	Yousri et al. (2015) Menni et al. (2013) Coco et al. (2019) Yousri et al. (2015)
Mannose	Plasma Urine	↑ ↑	MS MS	Yousri et al. (2015) Suhre et al. (2010) Menni et al. (2013) Yousri et al. (2015)
Pyruvate	Plasma	↑	MS	Yousri et al. (2015)
Ribose	Urine	↓	MS	Yousri et al. (2015)
Threonate	Urine	↓	MS	Yousri et al. (2015)
Xylonate	Urine	↓	MS	Yousri et al. (2015)

The arrows represent upregulations or downregulations in T2D relative to control samples
Abbreviations: NMR, Nuclear Magnetic Resonance; MS, Mass Spectrometry

using GC-MS, and recommended it as early marker for T2D (Nikiforova et al. 2014).

Amino Acids and Phospholipids

^{1}H NMR analysis has also been applied to serum metabolites in T2D patients (n = 26) and controls (n = 7) (Coco et al. 2019). Serum alanine, glutamine, glutamate, leucine, lysine, methionine, tyrosine and phenylalanine were all downregulated in T2D participants (Coco et al. 2019). Interestingly, these biomarker findings contrast with those established by other researchers (Table 20.1).

Similar findings to Drogan et al. (2015) were observed in a LC-MS and GC-MS metabolomics investigation, and significantly elevated fructose, α-hydroxybutyrate, alanine, proline, phenylalanine, glutamine, leucine, isoleucine, valine, mysterate, palmitate and stearate levels were observed, together with diminished pyroglutamate, glycerophospholipids and sphingomyelins, in the serum of groups of patients with either impaired fasting glucose levels or T2D (Xu et al. 2013).

Menni et al. (2013) observed 447 metabolites from fasted plasma samples from females with T2D (n = 115), individuals with impaired fasting glucose (n = 192) and controls (n = 1897) using ultraperformance liquid chromatography–mass spectrometry (UPLC-MS). Of 447 metabolites analysed, 42 were observed to be significantly different on comparison of these two groups (Menni et al. 2013), as noted in Table 20.1.

Prognostic Metabolites

Aloha-Ollie et al. (2019) quantified over 229 metabolites, including amino acids, fatty acids, glycolysis-related metabolites and lipoproteins using ^{1}H NMR analysis in 11,896 individuals, including T2D (n = 293) and controls (n = 11,603) from four Finnish cohorts. The biomarkers that were identified as the strongest indication for this condition included elevated aromatic amino acids, vLDL particles and triacylglycerols (TAGs). Biomarkers shown to be significant for individuals at risk of T2D included elevated glycerol and mono-unsaturated fatty acids and 'acute-phase' glycoproteins. Proposed biomarkers associated with a lower risk of developing T2D included glutamine, linoleate and HDL-triacylglycerols.

Liu et al. (2017) used a combination of liquid chromatography-mass spectrometry (LC-MS), and NMR combined with a least absolute shrinkage and selection operator regression approach, to identify biomarkers that were present in T2D blood samples (n = 349), compared to age-matched controls (n = 3998). In the 14 year follow-up study, similar findings to those of Aloha-Ollie et al. (2019) were found. There were clear disturbances in amino acid metabolism, including pathways featuring isoleucine and tyrosine. Further metabolic differences included those in triacylglycerols (TAGs), lactate, glycerol, pyruvate and vLDL-TAGs.

Some of the markers for insulin resistance include glucose, mannose, α-hydroxybutyrate (α-HB) and α-tocopherol, in addition to selected amino acids and 1-linoleoyl glycerophosphocholine (Peddinti et al. 2017). Elevated α-hydroxybutyrate (HB) is also a marker of glucose intolerance (Peddinti et al. 2017). Branched-chain amino acids (BCAAs) are particularly important since they have been demonstrated to predict T2D 12 years prior to its onset (Wang et al. 2011).

Saliva

Saliva is often proposed as an ideal biofluid for non-invasive analysis of diseases in view of its easy sample collection and its ability to reflect the contents of blood (Yousri et al. 2015; Barnes et al. 2014). Glucose is one of the metabolites that is elevated in the saliva of T2D patients (Gupta et al. 2014; Kadashetti et al. 2015). To determine salivary metabolites, Gupta et al. (2014) collected postprandial, unstimulated whole saliva. Participants were requested to rinse with water thoroughly and not to swallow for a period of 5 min before spitting into a sterile container. Kadashetti et al. (2015) collected fasting salivary samples, also requesting participants to wash their mouths with water. Yousri et al. (2015) collected salivary samples ~2 hr. post-prandially, and critically evaluated the limitations of non-fasted samples in their investigation.

Acetone and Lactate

Acetone has been determined as significantly increased in patients with diabetes in ketoacidosis (Fuji et al. 2014). Other metabolic changes that have been observed include the ratio of D- to L-lactate levels in dried salivary spots from controls (n = 68), pre-diabetic (n = 30) and T2D (n = 5) patients, in which T2D participants had a significantly higher ratio (Numako et al. 2016). The levels of D- and L-lactate were targeted in view of an upregulation of its methylglyoxal precursor, which arises from diabetic complications (Numako et al. 2016). The D- to L-lactate ratio showed good agreement with Hba1c values of blood in patients with diabetes;

however, the correlation of Hba1c values with D- to L-lactate ratio was poor in pre-diabetic samples (Numako et al. 2016). As in the study by Yousri et al. (2015), salivary samples were collected ~2 h. post-prandially (Numako et al. 2016).

Other Biomolecules

Additional disturbances have been observed in glycated proteins, electrolytes, albumin and enzymes (e.g. amylase (Abd-Elraheem et al. 2017), lysozyme and peroxidase) (Negrato and Tarzia 2010), however, and not using NMR and MS metabolomics-based approaches described here. Using an MS strategy, Yousri et al. (2015) found three significant metabolites in diabetic saliva including downregulations in pyroglutamine and 1,5-anhydroglucitol (1,5-AG). Some disturbances in unidentified metabolites in saliva were also documented as part of this investigation (Yousri et al. 2015). Significant salivary metabolites might present as a viable option for point-of-care sampling in a clinical setting.

Oral Diseases and Other Confounding Factors

Dental hygiene status can, of course, markedly influence salivary results, and periodontal diseases show some significant differences in salivary profiles when compared to healthy controls (Aimetti, et al. 2012). Epidemiological studies have shown that periodontitis was observed to be 3 times higher in diabetes patients, specifically poorly controlled individuals (Mealey and Ocampo 2000). Nevertheless, an investigation showed that controls with/without periodontitis and diabetic patients with/without periodontitis all have unique metabolomic profiles when assessed using GC-MS/LC-MS (Barnes et al. 2014). Moreover, Barnes et al. (2014) detected significantly different levels of 1,5-AG, glucose and α-hydroxybutyrate in the salivary profiles of T2D participants from those of healthy controls. These levels had larger fold-changes than those observed in the plasma profiles. Significant alterations were observed in saliva samples with diabetes and gingivitis as a co-morbidity when compared to diabetes patients without gingivitis, and predominantly increased salivary levels of fatty acids and sphingomyelins, along with changes in purine degradation and antioxidant status (Barnes et al. 2014).

Sample Contamination

Food debris, nasal mucus and traces of blood severely impact metabolomics results, if samples are not carefully collected. The overall outlook for salivary metabolomics does, however, show much potential, as patients are much more comfortable providing this sample type of sample than others involving more invasive collection regimens.

Urine

Urine is also very easy to collect and provides high sample volumes. Moreover, this biofluid contains many hundreds of metabolites that are readily detectable.

LC-MS investigations have been performed on urinary samples of T2D (Svingen et al. 2016), with downregulation of betaine, dimethylglycine and sarcosine characteristic of T2D samples. ^{1}H NMR investigations also found T2D downregulations of betaine and dimethylglycine, as well as of acetoacetate, acetate, n-butyrate, 3-hydroxybutyrate, dimethylamine, N-methylnicotinate (NMN) amide and N-acetyl aspartate (NAA), along with downregulated creatinine, N-acetyl functions, nicotinamide mononucleotide acid (NMN), aminohippurate, phenylacetylglycine (PAG), hippurate, allantoin, fumarate and succinate. Twelve control participants provided 7 samples each (total n = 84 samples) and 30 T2D participants provided 1–3 samples (total n = 50 samples). Figures 20.1 and 20.2 demonstrate how the metabolite information can be simultaneously obtained, in a typical diabetic urine sample, using NMR analysis with both 1D and 2D profiling approaches. In a different investigation, urinary betaine concentrations have also been demonstrated to be lower in participants with T2D; this metabolite was determined by LC-MS/MS analysis (Svingen et al. 2016), and this observation is consistent

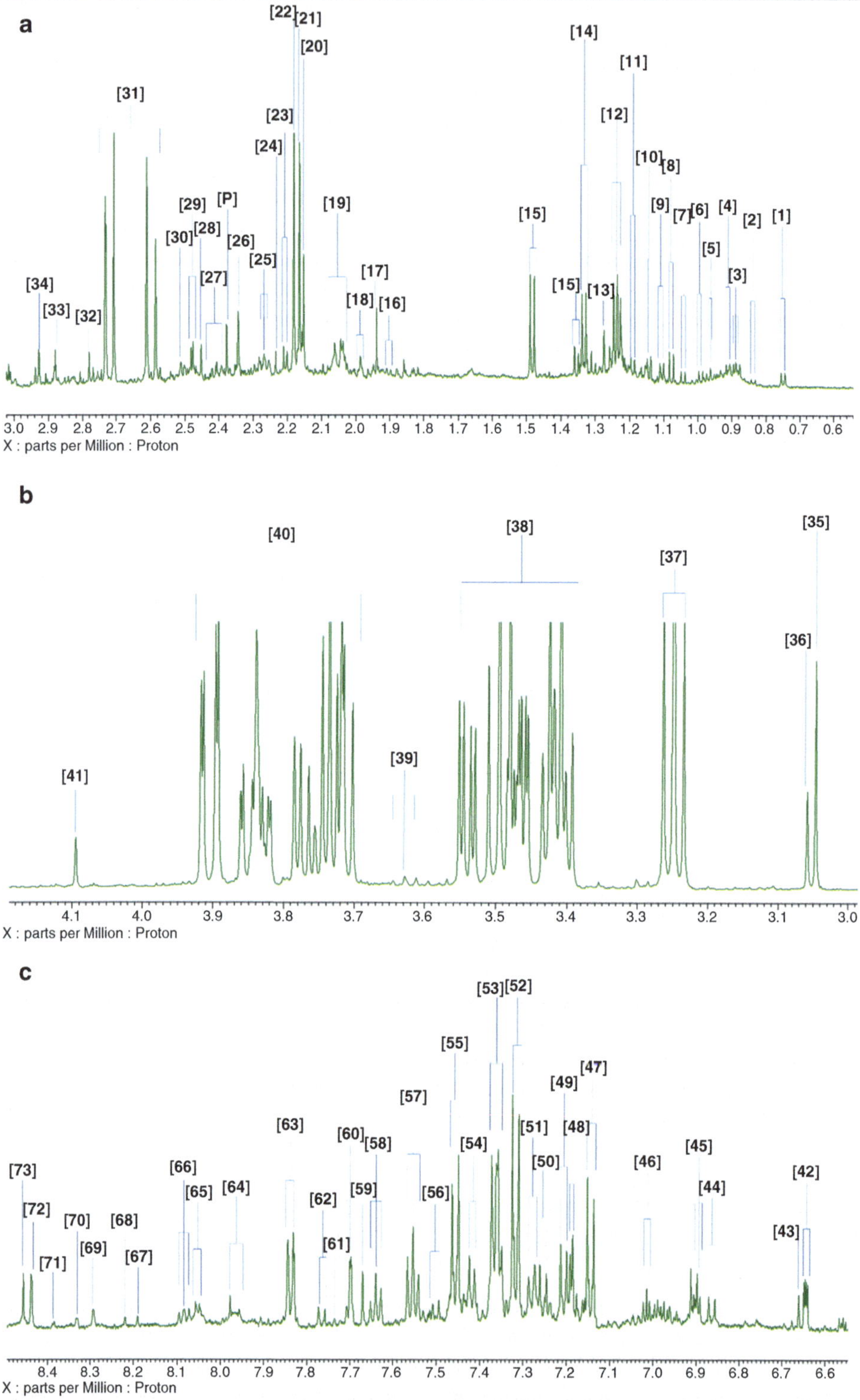

Fig. 20.1 (**a**), (**b**) and (**c**), Expanded 0.60–3.00, 3.00–4.15 and 6.50–8.50 ppm regions, respectively, of the 600 MHz ^{1}H NMR profile of a urine sample collected from a T2D patient receiving metformin treatment. Spectra acquired revealed that this patient was also taking paracetamol as an analgesic. Signals attributable to urea and water removed.

with those of other investigations. A full listing of significant ^{1}H NMR-detectable metabolites identified is also observed in the urinary profiles (Table 20.1).

Disturbed Metabolic Pathways Involved in T2D Pathogenicity

Energy metabolism, i.e. the tricarboxylic acid cycle or Krebs cycle (TCA), is unsurprisingly altered and this can be observed by metabolic changes in lactate, for example. Other disturbed pathways in saliva, urine and plasma include glycolysis, for example, pyruvate levels. However, these markers are not specific to T2D and are featured markers of many other diseases.

Lipid metabolism and amino acid metabolism are also observed to be modified in T2D chemopathogenesis. High levels of ketones are observed in T2D, which arise from the use of lipids as a source of body fuel rather than glucose, and also from the deamination of amino acids. Diabetic ketoacidosis arises from such elevated blood plasma ketone bodies. BCAAs, e.g. isoleucine and leucine, are also elevated in diabetes and are ascribable to a lack of uptake of energy sources, a further impairment of metabolic control (Suhre et al. 2010).

Personal Experience: Retrospective Network Pathway Analysis of Published Data

Using *MetExplore* in a qualitative context, we aimed to identify the possible metabolic sub-networks involved in the pathophysiology of T2D (Cottret et al. 2018). Based on this analysis, the predominant pathways involved in T2D are those involved in the biosynthesis and/or metabolism of amino acids, specifically BCAA metabolism; phenylalanine and tryptophan metabolism; tyrosine biosynthesis; alanine, aspartate and glutamate metabolism; glycine, serine and threonine metabolism; and arginine biosynthesis.

Fig. 20.1 (continued) Assignments: [1] Unassigned Isopropyl-$CH(C\underline{H}_3)_2$ function; [2] 2-Hydroxyisovalerate-δ-$C\underline{H}_3$; [3] 2-Hydroxybutyrate-CH_3; [4] Isoleucine-CH_3; [5] Leucine-$C\underline{H}_3$; [6] and [7] Valine-$C\underline{H}_3$s; [8] Methylsuccinate-$C\underline{H}_3$; [9] 2-Oxoisovalerate-$C\underline{H}_3$s; [10] 3-Aminoisobutyrate-CH_3; [11] isobutyrate-CH_3s; [12] 3-D-Hydroxybutyrate-CH_3; [13] 2,2-Dimethyl succinate-CH_3s; [14] Lactate-$C\underline{H}_3$; [15] Acetoin-$C\underline{H}_3$; [X] Alanine-CH_3; [16] 2-Hydroxyglutarate-1/2-$C\underline{H}_2$; [17] Acetate-$C\underline{H}_3$; [18] 2-Hydroxyglutarate-1/2-$C\underline{H}_2$; [19] N-Acetyl-$C\underline{H}_3$ functions of N-acetyl sugars, N-acetylamino acids (sharp resonances) and carbohydrate side-chain residues of trace level urinary N-acetylated glycoproteins (broader resonances); [20] -$NHCOC\underline{H}_3$ protons of paracetamol and its L-cysteinyl metabolite; [21] -$NHCOC\underline{H}_3$ protons of paracetamol glucuronide metabolite; [22] -$NHCOC\underline{H}_3$ protons of paracetamol sulphate metabolite; [23] 2-Hydroxyglutarate-1/2-$C\underline{H}_2$; [24] Acetone-$C\underline{H}_3$; [25] 2-Hydroxyglutarate-1/2-$C\underline{H}_2$; [26] Oxaloacetate-$C\underline{H}_3$; [P] Pyruvate-$C\underline{H}_3$; [27] Glutamine-γ-$C\underline{H}_2$; [28] Succinate-$C\underline{H}_2$s; [29] 3-Hydroxymethylglutarate-1/2-$C\underline{H}_2$; [30] Methylamine-N-$C\underline{H}_3$; [31] Citrate-$C\underline{H}_{2A/B}$; [32] Dimethylamine-N$(C\underline{H}_3)_2$; [33] Trimethylamine-$C\underline{H}_3$; [34] Dimethylglycine-N$(C\underline{H}_3)_2$; [35] Metformin-N-$C\underline{H}_3$; [36] Creatinine/Creatine-N-$C\underline{H}_3$; [37], [38] and [40] Glucose-C2-C6H ring protons; [39] Dimethylglycine-CH_2; [41] Creatinine-$C\underline{H}_2$; [42] Unassigned; [43] Unassigned; [44] Tyrosine aromatic ring (Ar)-$\underline{H}$; [45] Paracetamol-Ar-$\underline{H}$; [46] Xanthurenate-Ar-$\underline{H}$ [47] and [53] Paracetamol glucuronide metabolite Ar-H ([53] also Indoxyl sulphate-Ar-$\underline{H}$); [48] and [49] Indoxyl sulphate-Ar-$\underline{H}$; [50] Tyrosine-Ar-$\underline{H}$; [51] Indoxyl sulphate-Ar-H; [52] and [55] Paracetamol sulphate metabolite-Ar-$\underline{H}$; [54] Phenylglycine-Ar-CH and Benzoate-Ar-$C\underline{H}$; [56] Indoxyl sulphate-Ar-$C\underline{H}$; [57] Hippurate-Ar-$C\underline{H}$; [58] Hippurate-Ar-$C\underline{H}$; [59] 1-Methylhistidine imidazole ring (IR)-$C\underline{H}$; [60] Indoxyl sulphate-Ar-$C\underline{H}$; [61] Xanthine-Ar-$C\underline{H}$; [62] 4-Hydroxybenzoate-Ar-$C\underline{H}$; [63] Hippurate-$C\underline{H}$; [64] Quinolinate-Ar-$C\underline{H}$; [H] Histidine AR-$C\underline{H}$; [65] 6-Hydroxynicotinate-Ar-CH; [66] Trigonelline-Ar-$C\underline{H}$; [67] Imidazole-IR-$C\underline{H}$; [68] Hypoxanthine-C2/C7-$C\underline{H}$; [69] and [72] 1-Methyladenine-Ar-$C\underline{H}$; [70] Nicotinate-Ar-$C\underline{H}$; [71] Unassigned; [73] Formate-$C\underline{H}$

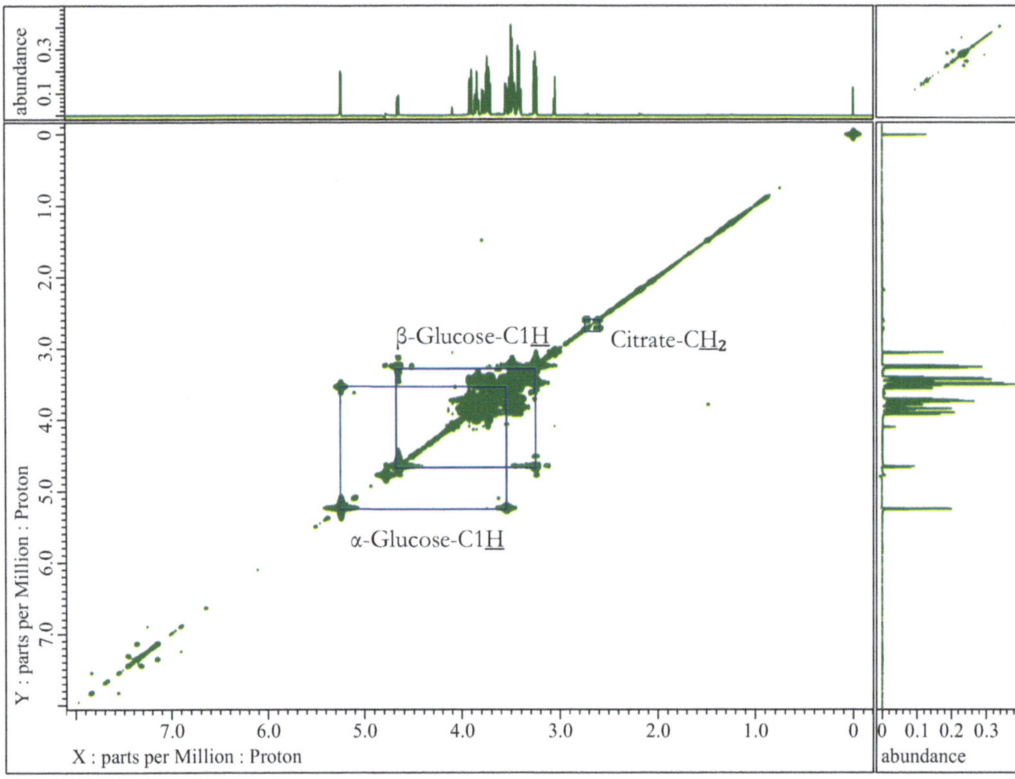

Fig. 20.2 2D ^{1}H-^{1}H NMR correlation spectroscopy (COSY) analysis of a T2D urine sample acquired using a 600 MHz facility (De Montfort University, Leicester, UK). Samples were prepared using 450 μL of urine, 50 μL of D_2O with TSP (0.5% (w/v)), 50 μL of sodium azide solution (0.04% (w/v)) and 50 μL of phosphate buffer solution (1.00 mol./L). Two clear connections are observed in the 2D NMR spectral profile ascribable to α- and β-Glucose-C1H resonances at $\delta = 5.21$ and 4.65 ppm, respectively. 1D projections are observable on the x- and y-axis

Further pathways featured are the biosynthesis of carbohydrates, glyoxylate and dicarboxylate metabolism and 2-oxocarboxylic acid metabolism. Figure 20.3 highlights the relationship between these pathways by displaying a sub-network using the lightest path algorithm, which forms the lightest paths between each pair of blood serum metabolites available in Table 20.1, and allows visualisation of metabolic paths between pairs of elements of interest.

The metabolic pathways highlighted in Fig. 20.3 are consistent with previous studies performed. For example, disturbances in BCAA metabolism have been demonstrated to arise from impaired insulin resistance (Zhao et al. 2016). Sun et al. (2019) also found alanine, aspartate and glutamate metabolism, glycine, serine and threonine metabolism and phenylalanine, tyrosine and tryptophan biosynthesis to be significant ($p<0.05$) pathways in their investigation, which assessed metabolomic signatures in T2D, using *MetaboAnalyst* pathway analysis. Interestingly, arginine biosynthesis and 2-oxocarboxylic acid metabolism pathways have also been demonstrated to be significantly enriched in downregulated T2D-associated genes (Wu et al. 2017).

Prognostic Fingerprints

There is considerable interest in the prognosis of T2D, which includes:

Fig. 20.3 Network pathway analysis of blood serum metabolome alterations in T2D patients, based on the biomarker listings in Table 20.1. Results shown are the sub-network, obtained by using the lightest path algorithm. This network highlights relationships between the TCA cycle, lipid metabolism and amino acid biosynthesis/metabolism in T2D. Enzyme abbreviations: R00149, Carbon-dioxide:ammonia ligase; R00200, ATP:pyruvate 2-O-phosphotransferase; R00220, L-serine ammonia-lyase; R00243, L-glutamate:NAD+ oxidoreductase (deamination); R00245, L-glutamate gamma-semialdehyde:NAD+ oxidoreductase; R00248, L-glutamate:NADP+ oxidoreductase (deaminating); R00253, L-Glutamate:ammonia ligase (ADP-forming); R00256, L-glutamine amidohydrolase; R00258, L-Alanine:2-oxoglutarate aminotransferase; R00342, (S)-malate:NAD+ oxidoreductase; R00351, acetyl-CoA: oxaloacetate C-acetyltransferase (thioester-hydrolysing); R00369, L-Alanine:glyoxylate aminotransferase; R00372, Glycine:2-oxoglutarate aminotransferase; R00551, L-Arginine amidinohydrolase; R00557, L-arginine,NADPH:oxygen oxidoreductase (nitric-oxide-forming); R00565, L-Arginine:glycine amidinotransferase; R00575, HCO3-:L-glutamine amido-ligase (ADP-forming, carbamate-phosphorylating); R00577, acyl-CoA:L-glutamine N-acyltransferase; R00585, L-Serine:pyruvate aminotransferase; R00658, 2-phospho-D-glycerate hydro-lyase (phosphoenolpyruvate-forming); R00667, L-Ornithine:2-oxo-acid aminotransferase; R00707, (S)-1-pyrroline-5-carboxylate:NAD + oxidoreductase; R00708, (S)-1-pyrroline-5-carboxylate:NADP+ oxidoreductase; R00709, isocitrate:NAD+ oxidoreductase (decarboxylating); R00945, 5,10-Methylenetetrahydrofolate:glycine hydroxymethyltransferase; R01221, glycine synthase; glycine cleavage system; R01251, L-Proline:NADP+ 5-oxidoreductase; R01252, L-Proline,2-oxoglutarate:oxygen oxidoreductase (4-hydroxylating); R01324, citrate hydroxymutase; R01392, D-Glycerate:NADP+ 2-oxidoreductase; R01398, Carbamoyl-phosphate:L-ornithine carbamoyltransferase; R01883, S-Adenosyl-L-methionine: guanidinoacetate N-methyltransferase; R03295, trans-4-hydroxy-L-proline:quinone oxidoreductase; R04444, L-1-pyrroline-3-hydroxy-5-carboxylate:NAD+ oxidoreductase; R04445, L-1-pyrroline-3-hydroxy-5-carboxylate:NADP+ oxidoreductase; R05052, L-erythro-4-hydroxyglutamate:2-oxoglutarate aminotransferase; R08572, ATP:(R)-glycerate 2-phosphotransferase

- Prediction and early assessment of pre-diabetes, acute T2D diabetes and chronic diabetes;
- Full assessment (deep phenotyping) of the diabetic condition itself, i.e. diabetes sufferers can present a range of other symptoms and disorders, including diabetic kidney damage, ocular problems, cardiovascular and other metabolic syndrome diseases, and, of course, diabetic ketosis/ketoacidosis.

Metabolomics can successfully be employed to evaluate the onset and diagnosis of T2D. As noted above, Yousri et al. (2015) performed an extensive study observing global changes in the T2D metabolome; however, more importantly, this investigation defined metabolites associated with acute, short- and long-term dysregulation of glycemic control (Yousri et al. 2015). For this investigation, 3-hydroxybutyrate (an intermediate in valine metabolism) and amino acid (e.g., leucine) levels in urine were the best indicators of acute glycemic dysregulation.

Short vs. Long-term Glycemic Homeostasis

Short-term glycemic dysregulation could be further determined by monitoring the salivary, plasma and urinary levels of a combination of metabolites, including those of urinary glycolate, plasma 3-hydroxyisobutyrate and salivary 1,5-Anhydroglucitol (1,5-AG) (Yousri et al. 2015). Blood serum 1,5-AG was also proposed as a short-term glycemic marker in an earlier investigation (Suhre et al. 2010). However, carbohydrate levels in plasma were the best indicators of long-term dysregulation (e.g. those of glucose anomers) (Yousri et al. 2015). Another longitudinal investigation using ultra-performance liquid chromatography-quadrupole time-of-flight mass spectrometry demonstrated plasma levels of iso-valeraldehyde, linoleate, lyso-phosphatidylcholine (18:1), 2-pyrroloylglycine and dityrosine as reliable indications for the development of T2D (Zeng et al. 2017).

Markers for insulin resistance have also been explored as part of metabolomics investigations, and BCAAs have shown promise as biomarkers for screening, enabling early intervention in the development of T2D by dietary or exercise interventions (Roberts, Koulman and Griffin 2014).

Kidney Dysfunction and Ketosis

Further validation is required in order to assess renal function in diabetic patients (Pena et al. 2015). Over 1 in 4 diabetic patients suffer from diabetic kidney disease (Pena et al. 2015), and hence the stratification of biomarkers for this purpose is currently an important consideration. Candidate biomarkers include urinary glutamine and tyrosine, and plasma butenoylcarnitine and histidine (Pena et al. 2014). However, the response of these biomarkers to approved treatments, a process required for their full validation, is yet to be explored.

Ketoacidosis is more prominent in type 1 diabetes; however, it can occur in T2D. Indeed, mild ketosis has been observed in metabolomics investigations with stable elderly T2D, which was indicated by increased 3-D-hydroxybutyrate blood plasma concentrations (Suhre et al. 2010).

Potential Bedside/'Point-of-Care' Metabolomics

Dipstick testing for glucose and ketone body levels, along with electronic meters, which can simultaneously detect and provide a reference range for plasma and urinary samples, is well established. This can be performed in a home or clinical setting. Ketones can be alternatively monitored in the blood using handheld devices; however, currently these are mostly used in veterinary medicine (Wang et al. 2015).

Needle-free flash glucose monitors can also be employed as a non-invasive option (Distiller, Cranston and Mazze 2016). To test for T2D, the usual recommendations encompass whole blood HbA1c and fasting plasma glucose levels, along with a 2 h plasma glucose after a 75g oral glucose load (oral glucose tolerance test/ OGTT) (American Diabetes Association 2018). In a paediatric setting, blood spots from the heel or finger pricks are often utilised.

Future Developments

The benefits of adopting metabolomics into personalised care (precision medicine) include pre-diagnosis, predictive outcomes and indicated treatment regimens.

Metabolic disturbances routinely monitored in diabetes are essentially limited to blood glucose, HbA1c, ketone body and lipid profiles. This neglects the TCA cycle, metabolites and those involving selected amino acids (Table 20.1). The unfortunate lack of translation of metabolomic technologies into clinical practice is largely ascribable to the size, required resources and expense of instruments used to obtain simultaneous metabolic fingerprinting information on biofluid samples. Additionally, dataset sizes acquired from metabolomics investigations are large, probing the status of often hundreds of metabolites, and hence it may be unfeasible for their large-scale applications within clinical practices (Blümuch 2019). Finally, the restraints on metabolomics investigations include ensuring that a strict experimental design is employed, which requires the selected participants adhering to particular criteria, i.e. fasting status, age range, sex, exercise regimen, dietary regimen, etc. (Ghini et al. 2019).

Bedside Devices

As technology advances, there is some promise for low-field (LF), near-portable benchtop NMR devices to become more commonplace, rather than the high-field facilities that are restricted to large industrial and research centres (Blümuch 2019). Moreover, smaller patterns of metabolites might be more feasible to monitor and interpret within a clinical setting. For the first time, Percival et al. (2019) presented the potential 'point-of-care' applications of urinary testing of T2D patients using a benchtop ^{1}H NMR-linked metabolomics strategy. There are clear benefits of this approach over the use of high-field facilities, including the lower operating, maintenance and initial costs of the facility, the ability to detect and quantify glucose and other metabolites simultaneously, and perform metabolomics investigations on biofluid samples, which would benefit clinical management. The disadvantages of this study are evident: the 60 MHz operating frequency of the benchtop NMR facility is not comparable to high-field facilities that have been utilised in previous studies, since the levels of metabolites detectable are significantly reduced (approximately 4–5 times).

The investigation would also benefit from a larger sample size in control and T2D cohorts. However, this shows the scope and potential applications, since technological advances are continually being accomplished in NMR, with higher operating frequencies now achievable with benchtop facilities (i.e. up to 100 or more MHz), and automation is made possible for these facilities with robotic autosamplers. Further developments in the miniaturisation of NMR spectrometers in order to increase sensitivity and translate into clinical settings have been explored (Wishart 2019; Blümuch 2019). This gives potential access to healthcare centres, GP surgeries, dental practices, community pharmacies and hospitals, and such NMR facilities are substantially more practical and cost-effective.

One solution to overcome the experimental design challenges of diurnal variation in metabolic profiles includes the collection of pooled urine or plasma samples, in order to provide an average profile for healthy or diseased individuals.

Sensible options for the translation of metabolomics strategies into clinical settings comprise selection of the most significant metabolites for such clinical applications, and the development of reliable linear combinations of biomarkers or algorithms in order to provide a cheaper, efficient solution for early detection, diagnosis and prognosis.

Metabolomic Panels

The idea of a blood based-panel has shown some success by Varvel et al. (2014) who monitored 19 markers and demonstrated that this was an improved approach over the routine, established monitoring of HbA1c and fasting glucose levels for T2D patients. It would be ideal to use markers of pre-diabetes as demonstrated by Varvel et al. (2014), in order to control or avoid the onset of T2D, and therefore provide improved treatment outcomes and healthcare cost savings.

Portable Oral Prosthesis and Breath Volatolomic Analysis

Other devices that have potential use for the diagnosis and monitoring of T2D include salivary mouthguard biosensors with telemetry that are able to detect glucose concentrations from 5–1000 μmol/L (Arakawa et al. 2016). Other biosensors have been developed for the detection and determination of glucose in saliva include optical biosensors (Soni and Jha 2015). However, potential contaminations are possible from blood and food debris. This is a technology that could be further developed with wider metabolomics applications. Breath testing for ketones has also been demonstrated using gas-chromatography/mass-spectrometry, and the methodology has shown a high sensitivity and specificity (Qiao et al. 2014).

Another technology, which comes full circle from the history of the diagnosis of diabetes, is the development of an 'electronic nose' that is able to detect urinary levels of ammonia, ethyl methyl ketone and acetone. Indeed, the statistical testing provided a successful classification of diabetic and control patients using acetone alone (Seesard, Sriphrapradang, Kitiyakara and Kerdcaroen 2016).

Metabolomics has also been shown to support genotypic investigations (Kim et al. 2016), and could easily complement those involving the genetic profiling of T2D, in an integrative omics approach.

Artificial Pancreas and Wearable Technology

In the artificial pancreas context, continuous glucose monitoring has been conducted, and these systems release insulin if levels are above specified limits (Olczuk and Priefer 2017). Naturally, these devices are geared at advanced disease, primarily type 1 diabetes, and are not serving to inform a potentially preventable pre-condition, e.g. pre-diabetes. Moreover, they determine a single, albeit the most important metabolite, glucose, bypassing up- or downregulated markers in T2D as described here. Other further developments are likely to be observed in the smartphone, smartwatches and the smart application industries in general, with huge potential for monetary savings in future healthcare centres and systems (Wang and Kricker 2018).

References

Abd-Elraheem SE et al (2017) Salivary changes in type 2 diabetic patients. Diabetes Metab Syndr Clin Res Rev 11(2):S637–S641

Aimetti M et al (2012) Metabonomic analysis of saliva reveals generalized chronic periodontitis signature. Metabolomics 8(3):465–474

Aloha-Ollie AV et al (2019) Circulating metabolites and the risk of type 2 diabetes: a prospective study of 11,896 young adults from four Finnish cohorts. Diabetolgia 62:2298–2309

American Diabetes Association (2018) 2. Classification and diagnosis of diabetes: standards of medical care in diabetes—2018. Diabetes Care 41(1):S13–S27

Arakawa T et al (2016) Mouthguard biosensor with telemetry system for monitoring of saliva glucose: a novel cavitas sensor. Biosens Bioelectron 84(15):106–111

Barnes VN et al (2014) Global metabolomic analysis of human saliva and plasma from healthy and diabetic subjects, with and without periodontal disease. PLoS One 9(11):e114091

Bentley-Lewis R et al (2015) Metabolomic profiling in the prediction of gestational diabetes mellitus. Diabetologia 58(6):1329–1332

Blümuch B (2019) Low-field and benchtop NMR. J Magn Reson 306:27–35

Coco LD et al (2019) NMR-based metabolomic approach tracks potential serum biomarkers of disease progression in patients with type 2 diabetes mellitus. J Clin Med 8:1–20

Cottret L et al (2018) MetExplore: collaborative edition and exploration of metabolic networks. Nucleic Acids Res 46:W495–W502

Distiller LA, Cranston I, Mazze R (2016) First clinical experience with retrospective flash glucose monitoring (FGM) analysis in South Africa: characterizing glycemic control with ambulatory glucose profile. J Diabetes Sci Technol 10:1294–1302

Drogan D et al (2015) Untargeted metabolic profiling identifies altered serum metabolites of type 2 diabetes mellitus in a prospective, Nested Case Control Study. Clin Chem 61(3):487–497

Floegel A et al (2013) Identification of serum metabolites associated with risk of type 2 diabetes using a targeted metabolomic approach. Diabetes 62(2):639–648

Fuji S et al (2014) Determination of acetone in saliva by reversed-phase liquid chromatography with fluorescence detection and the monitoring of diabetes mellitus patients with ketoacidosis. Clin Chim Acta 430:140–144

German JB et al (2005) Metabolomics: building on a century of biochemistry to guide human health. Metabolomics 1(1):3–9

Ghini V et al (2019) New for sample quality assessment in metabolomics. New Biotechnol 52:25–34

Gupta S et al (2014) Comparison of salivary and serum glucose levels in diabetic patients. J Diabetes Sci Technol 9(1):91–96

Haukka JK et al (2018) Metabolomic profile predicts development of microalbuminuria in individuals with type 1 diabetes. Sci Rep 8:1–10

Kadashetti V et al (2015) Glucose level estimation in diabetes mellitus by saliva: a bloodless revolution. Rom J Intern Med 53(3):248–252

Kell DB, Oliver SG (2016) The metabolome 18 years on: a concept comes of age. Metabolomics 12:1–8

Kim Y-J et al (2016) Association of metabolites with obesity and type 2 diabetes based on FTO genotype. PLoS One 11(6):1–11

Lindon JC, Nicolson J, Holmes E (2011) The handbook of metabonomics and metabolomics. Elsevier, Amsterdam, pp 11–17

Liu J et al (2017) Metabolomics based markers predict type 2 diabetes in a 14-year follow-up study. Metabolomics 13:1–12

Mealey BL, Ocampo GL (2000) Diabetes mellitus and periodontal disease. Periodontol 44:127–153

Menni C et al (2013) Biomarkers for type 2 diabetes and impaired fasting glucose using a nontargeted metabolomics approach. Diabetes 62(12):4270–4276

Mull SD, Nguyen N, Mull DJ (2001) Vietnamese diabetic patients and their physicians what ethnography can teach us. West J Med 175(5):307–311

Negrato CA, Tarzia O (2010) Buccal alterations in diabetes mellitus. Diabetol Metab Syndr 2(3):1–11

Nicholson JK, Lindon JC (2008) Systems Biology: Metabonomics. Nature 455:1045–1056

Nikiforova V-J et al (2014) Glyoxylate, a new marker metabolite of type 2 diabetes. J Diabetes Res 2014:1–9

Numako M et al (2016) Dried saliva spot (DSS) as a convenient and reliable sampling for bioanalysis: an application for the diagnosis of diabetes mellitus. Anal Chem 88:635–639

Olczuk D, Priefer R (2017) A history of continuous glucose monitors (CGMs) in self-monitoring of diabetes mellitus. Diabetes Metab Syndr 12:181–187

Oliver SG et al (1998) Systematic functional analysis of the yeast genome. Trends Biotechnol 16:373–378

Pauling I et al (1971) Quantitative analysis of urine vapor and breath by gas liquid partition chromatography. Proc Natl Acad Sci 68:2374–2376

Peddinti G et al (2017) Early metabolic markers identify potential targets for the prevention of type 2 diabetes. Diabetologia 60:1740–1750

Pena M et al (2014) Urine and plasma metabolites predict the development of diabetic nephropathy in individuals with type 2 diabetes mellitus. Diabet Med 31:1138–1147

Pena M et al (2015) Prognostic clinical and molecular biomarkers of renal disease in type 2 diabetes. Nephrol Dial Transplant 30(4):iv86–iv95

Percival B et al (2019) Low-field, benchtop NMR spectroscopy as a potential tool for point-of-care diagnostics of metabolic conditions: validation, protocols and computational models. High Throughput 8:2

Qiao Y et al (2014) Breath ketone testing: a new biomarker for diagnosis and therapeutic monitoring of diabetic ketosis. Biomed Res Int 869186:1–5

Roberts DL, Koulman A, Griffin JL (2014) Towards metabolic biomarkers of insulin resistance and type 2 diabetes: progress from the metabolome. Lancet Diabet Endocrinol 2(1):65–75

Salek RM et al (2007) A metabolomic comparison of urinary changes in type 2 diabetes in mouse, rat and human. Physiol Genomics 29:99–108

Sandlers Y (2017) The future perspective: metabolomics in laboratory medicine for inborn errors of metabolism. Transl Res. 189:65–75

Seesard T, Sriphrapradang C, Kitiyakara T, Kerdcaroen T (2016) Self-screening for diabetes by sniffing urine samples based on a hand-held electronic nose 2016. 9th Biomedical Engineering International Conference (BMEiCON) https://doi.org/10.1109/BMEiCON.2016.7859586

Sharma H, Chandola HM (2011) Prameha in ayurveda: correlation with obesity, metabolic syndrome, and diabetes mellitus. Part 1—etiology, classification, and pathogenesis. J Altern Complement Med 17 (6):491–496

Soni A, Jha SK (2015) A paper strip based non-invasive glucose biosensor for salivary analysis. Biosens Bioelectron 67:763–768

Suhre K et al (2010) Metabolic footprint of diabetes: a multiplatform metabolomics study in an epidemiological setting. PLoS One 5(11):e13953

Sun Y et al (2019) Metabolomics signatures in type 2 diabetes: a systematic review and integrative analysis. J Clin Endocrinol Metab 105:1000–1008

Svingen G-F et al (2016) Prospective associations of systemic and urinary choline metabolites with incident type 2 diabetes. Clin Chem 62(5):755–765

Urpi-Sarda M et al (2019) Non-targeted metabolomic biomarkers and metabotypes of type 2 diabetes: a cross-sectional study of PREDIMED trial participants. Diabetes Metab 45(2):167–174

Varvel SA et al (2014) Comprehensive biomarker testing of glycemia, insulin resistance, and beta cell function has greater sensitivity to detect diabetes risk than fasting glucose and HbA1c and is associated with improved glycemic control in clinical practice. J Cardiovasc Transl Res 7(6):597–606

Wang P, Kricker LJ (2018) Current and emerging trends in point-of-care technology and strategies for clinical validation and implementation. Clin Chem 64 (10):1439–1452

Wang TJ et al (2011) Metabolite profiles and the risk of developing diabetes. Nat Med 17:448–453

Wang X, Zhao W, Neethirajan S, Duffield T (2015) Microfluidic biosensor for β-Hydroxybutyrate (βHBA) determination of subclinical ketosis diagnosis. J Nanobiotechnol 13:13

Wishart DS (2019) NMR metabolomics: A look ahead. J Magn Reson 306:155–161

Wu C et al (2017) Transcriptional profiles of type 2 diabetes in human skeletal muscle reveal insulin resistance, metabolic defects, apoptosis, and molecular signatures of immune activation in response to infections. Biochem Biophys Res Commun 482:282–288

Xu F et al (2013) Metabolic signature shift in type 2 diabetes mellitus. J Clin Endocrinol Metab 98:E1060–E1065

Yousri NA et al (2015) A systems view of type 2 diabetes-associated metabolic perturbations in saliva, blood and urine at different timescales of glycaemic control. Diabetologia 58:1855–1867

Zeng H et al (2017) Metabolic Biomarkers for Prognostic Prediction of Pre-diabetes: results from a longitudinal cohort study. Sci Rep 7:1–12

Zhang A et al (2012) Modern analytical techniques in metabolomics analysis. Analyst 2(137):293–300

Zhao X et al (2016) The relationship between branched-chain amino acid related metabolomic signature and insulin resistance: a systematic review. J Diabetes Res 2016:2794591

The Gut Microbiome and Type 2 Diabetes Mellitus

21

Shruti Panwar, Samriddhi Arora, Sapna Sharma, and Prabhanshu Tripathi

Contents

Introduction 284
The Disruptive Role of Antibiotics 285
Lipopolysaccharide (LPS)-associated Endotoxemia 285

Metabolic Syndrome and Obesity 286

Inflammation 286

Composition of Gut Microbiota 286

Factors Involved in T2D Pathogenesis 286
Incretins 286
Metabolic Functions 287
Epithelial Barrier Impairment and Pro-Inflammatory Cytokines 289
Bile Acid Transformation 289
Gut Microbiome in Pathogenesis of T2D 289

Current Therapeutics 290
Bariatric Surgery 290
Prebiotics, Probiotics, and Synbiotics 290

S. Panwar
Translational Health Science and Technology Institute, Faridabad, India

S. Arora
Laboratory of Calciomics and Systemic Pathophysiology, Regional Centre for Biotechnology, Faridabad, India

S. Sharma
Gene Regulation Laboratory, School of Biotechnology, Jawaharlal Nehru, University, New Delhi, India

P. Tripathi (✉)
Translational Health Science and Technology Institute, Faridabad, India

Food Drug and Chemical Toxicology, CSIR-Indian Institute of Toxicology Research, Lucknow, Uttar Pradesh, India
e-mail: prabhanshu@thsti.res.in; p.tripathi1@iitr.res.in

J. Faintuch and S. Faintuch (eds.), *Obesity and Diabetes*, https://doi.org/10.1007/978-3-030-53370-0_21

Supplementation Meta-analyses 291
Other Perspectives 291

Plant Extract-based Therapy 292

References 293

Abstract

Type 2 diabetes (T2D) is a highly prevalent metabolic disorder associated with many other diseases like obesity, insulin resistance, atherosclerosis, hypertension, arthritis, and various types of cancers. Recent findings suggest that gut microbiome plays an important physiological role in many aspects like host defense, digestion, absorption and fermentation of complex sugars, short chain fatty acid (SCFA) production, metabolic signaling pathways, and reshaping the intestinal barrier. SCFA, a ligand to G protein couple receptors Gpr41 and Gpr43, can secrete peptide tyrosine (PYY), which increases gut motility and nutrient absorption. Gut microbiome could play a role in hepatic and muscle insulin sensitivity via the AMP kinase and other pathways.

Keywords

Microbiota · Short chain fatty acids · Intestinal barrier · Energy metabolism · Carbohydrate metabolism · PYY · Gut hormones · Insulin sensitivity

Introduction

Asia is reported to top the charts to have most rapidly emerging cases of type 2 diabetes (T2D) especially in countries like China and India (Kassi et al. 2011; Reaven 1988; Zheng et al. 2017). The major factors held responsible for the global T2D epidemic include obesity, sedentary lifestyle, and increased consumption of processed food, refined grains, and high sugar containing beverages. Among T2D cases, cardiovascular disease (CVD) is the major cause of mortality and morbidity, and additionally, kidney failure is also prevalent in such patients. Metabolic disorders are common in developing countries, attributed not just to increase in energy intake and decrease in physical activity in recent decades but also due to alterations in gut microbiota.

The gut harbors trillions of bacteria that are involved in regulating digestion, metabolism, extraction of nutrients from ingested food, synthesis of vitamins, prevention against pathogens, and immunomodulation (Brestoff and Artis 2013). In the past few years, different findings have confirmed the role of disbalanced gut microbiome in various metabolic disorders. It has also been observed that changes in gut microbiome could be a risk factor in diseases like T2D and CVD. Additionally, low grade inflammation, as well as modification in the secretion of incretin and butyrate formation, has been proposed to be a potential mechanism linking gut microbiota and T2D (Moreno-Indias et al. 2014).

The gut of healthy adults remains stable over time and is transiently altered under the effect of external influences such as diet, disease, and environment. Diet provides nutrients for both host and the residing microflora. Alterations in dietary intake account for 57% of total structural variation in gut microbiome as compared to genetics, which only contribute 12% (Zhang et al. 2010). Microbiota Accessible Carbohydrates (MACs) have been shown to play a role in shaping the microbial ecosystem and are significantly reduced with consumption of a 'western' diet (high in fat and carbohydrates and low in fiber) as compared to traditional diet (low in fat and carbohydrates and high in fiber).

Western diet/traditional diet also impacts the ratio of two important bacterial phyla, Firmicutes and Bacteroidetes , possibly resulting in impairment/establishment of various gut-associated physiological functions (Tidjani Alou et al. 2016; Turnbaugh et al. 2007).

European children have high levels of Bacteroidetes as well as of *Enterobacteriaceae* as compared to rural African children, which may be attributed to high fiber intake by Africans (Jacquemont et al. 2011). Studies have reported that a healthy individual has substantial concentrations of bacteria from Clostridiales order (*Roseburia* species and *Faecalibacterium prausnitzii*), which are efficient producers of the small chain fatty acid (SCFA) butyrate through the fermentation process (Parks et al. 2013; Karlsson et al. 2013). It is instrumental in fulfilling energy requirements of gut epithelium, in regulating inflammatory responses in the gut by serving as a ligand to GPCRs on immune cells, and in maintaining gut epithelial barrier (Parada Venegas et al. 2019).

The Disruptive Role of Antibiotics

Antibiotics exert selective pressure on important mediators of gut biology, leading to long term reduction in their diversity (Jernberg et al. 2010). Even veterinarian antibiotics such as avoparcin (a glycopeptide structurally similar to vancomycin) could play a role in weight increment in patients treated with vancomycin, due to colonization by *Lactobacillus spp*, which once exposed to that agent can become resistant to glycopeptides and are found in high concentrations in feces of obese individuals (Thuny et al. 2010; Brugman et al. 2006). One advance in the molecular investigation of metabolic disorders was the introduction of genome wide association studies (GWASs), which have helped identify candidate genes involved in increasing/decreasing insulin resistance in diabetes and variable metabolic phenotypes (Jacquemont et al. 2011).

A common feature that connects T2D and obesity is low grade inflammation in tissues involved in regulating metabolism. This inflammation is characterized by moderate rise in cytokines such as IL-6, IL-1, and TNF-alpha, which hampers insulin signaling and leads to insulin resistance in both conditions (Kahn et al. 2006; Hotamisligil 2006).

Lipopolysaccharide (LPS)-associated Endotoxemia

LPS is an endotoxin released by Gram negative bacteria. Changes in the proportion of Gram negative bacteria or disruption of gut epithelial barrier are blamed for release of LPS in systemic circulation (Musso et al. 2010). LPS-mediated endotoxemia can direct the release of pro-inflammatory cytokines by immune cells. LPS binds to CD14/TLR4 receptor present on macrophages to mount an inflammatory response and also promotes the expression of transcription factor NFKB and protein kinase MAPK in adipocytes. An opposite, protective mechanism attributed to gut microbes involves increase in production of incretins such as glucagon like peptide-1 (GLP-1). It is evident from several studies that rise in *Bifidobacterium spp* is accompanied by increase in levels of GLP-1 and PYY secreted by intestine (Cani et al. 2007a).

These two molecules have an important effect on decreasing insulin resistance and increasing functionality of beta cells. GLP-2, a nonincretin however produced by the same intestinal cells, plays a role in increasing expression of zona occludens-1 (ZO) protein, essential for maintaining the gut barrier and preventing LPS-associated endotoxemia (Cani et al. 2009). Another mechanism related to T2D pathophysiology, as alluded to, could be shifts in SCFA (butyrate) producing bacteria. Butyrate is an important source of energy for colonic cells and is produced with the help of *C. coccoides* and *Eubacterium rectale* groups, along with other butyrate fermenting bacteria in gut. Reduction of such bacteria has been reported to initiate increase in opportunistic pathogens deranging insulin signaling and predisposing to insulin resistance (Caricilli and Saad 2013).

Metformin and Gut Ecology

Metformin is popularly used to treat diabetes and is reported to increase the levels of butyrate and propionate producing bacteria. High fiber diet increases the abundance of *Prevotella*, while bariatric surgery has been shown to affect bacterial composition in the gut, which may contribute to BMI independent amelioration of glucose metabolism (Brunkwall and Orho-Melander 2017).

Metabolic Syndrome and Obesity

The subclinical stages leading to metabolic syndrome are marked by alteration in gut microbiome, which was shown through an interesting finding demonstrating the relevance of increased abundance of *Akkermansia muciniphila* in maintaining adequate levels of free fatty acids (FFA) in humans, while subjects with subclinical metabolic alterations had decreased population of *A. muciniphila* and surprisingly increased levels of *Lactobacillus* , resulting in elevated factors such as FFA, pro-inflammatory cytokines, and serum IL-6 (Reaven et al. 1988; Randle et al. 1963; Rodriguez-Carrio et al. 2017).

Inflammation

Some probiotics have shown to be effective in modulating the immune system against altered gut microbiota associated with chronic low level inflammation, which can contribute to T2D (Zhang and Zhang 2013). One such probiotic, *Lactobacillus casei* (Matsuzaki et al. 1997), exhibited antidiabetic properties on oral administration to a noninsulin-dependent diabetes mellitus (NIDDM) model using KK-Ay mice. Other such examples include *Bifidobacterium longum* (Chen et al. 2011).

Composition of Gut Microbiota

The resident gut microflora is shaped by important factors during the course of our lifetime, which include early colonization depending on type of birth (vaginal vs c-section), age, lifestyle, host genetics, health/disease status, medications, and, most obviously, the dietary pattern. The main bacterial phyla residing in our gut are Firmicutes (Gram-positive), Bacteroidetes (Gram-negative), and Actinobacteria (Gram-positive). Firmicutes encompass more than 200 genera, the most predominant of which are *Bacillus and Clostridium.*

The infant microflora is modulated on consuming solid foods around the age of 2–3 years, progressively resembling stable adult microbiota with predominance of Firmicutes and Bacteroidetes (Odamaki et al. 2016) (Fig. 21.1).

Factors Involved in T2D Pathogenesis

Incretins

Incretins like glucagon like peptide-1 (GLP-1) and gastric inhibitory polypeptide (GIP) that are released by small bowel endocrine L cells and K cells, respectively, stimulate the pancreatic beta cells and increase insulin production and secretion. GLP-1 influences various functions like satiety, reduced dietary intake, and delayed gastric emptying, which are relevant therapeutic targets for patients suffering from both obesity and T2D (Nauck and Meier 2018). Often, patients suffering from T2D experience gut microbial dysbiosis, which leads to GLP-1 resistance by reduced expression of GLP-1R and neuronal nitric oxide synthase. There is a break in the gut brain axis for the control of insulin secretion and gastric emptying, controlled by a pattern recognition receptor (PRR)-dependent mechanism expressed by the enteric nervous system (Yamane and Inagaki 2018).

Short Chain Fatty Acids (SCFAs)

Major products derived from microbial fermentation are short chain fatty acids (SCFAs) such as acetate, propionate, and butyrate. These are taken up by host organs and have different metabolic fates. They similarly act as signaling molecules and are known ligands for at least two G protein coupled receptors, i.e., Gpr41 and Gpr43, (Greiner and Bäckhed 2016) which are dominantly expressed on colon, adipocytes, and distal small intestine. On binding GPCRs, they induce the secretion of peptide YY (PYY) (Karra et al. 2009) and leptin. PYY is a peptide hormone released by enteroendocrine cells of intestine and is

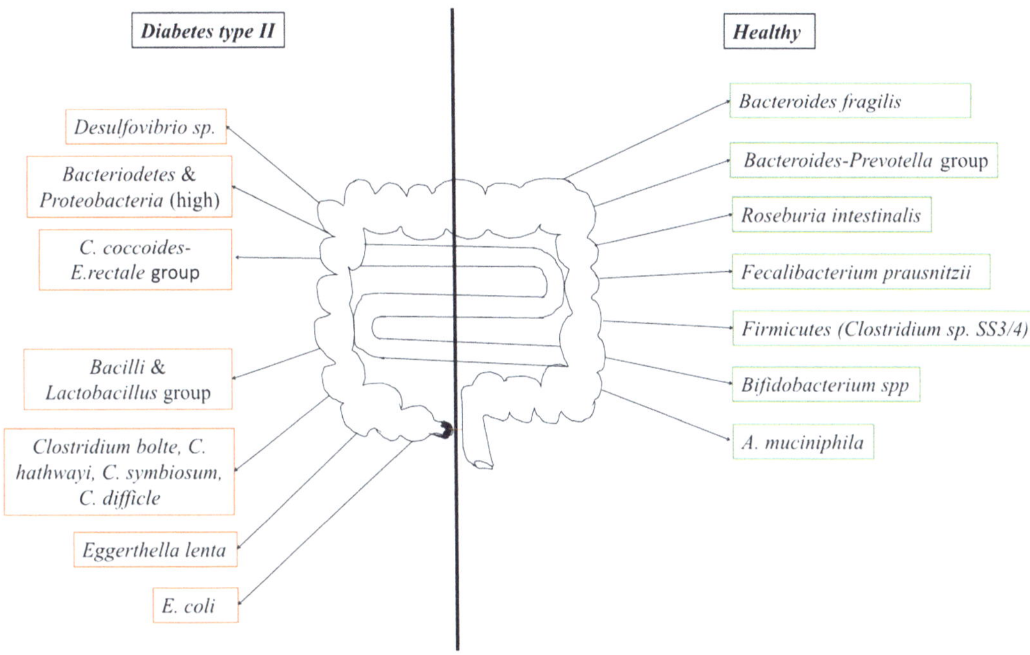

Fig. 21.1 Status of gut microbiome in type 2 diabetes (left side) and healthy individual (right side). *L. acidophilus, L. gasseri, L. salivarius* are increased inT2D

functionally important in gut motility and increased expression of liver fatty acid binding protein (L-FABP), a peptide responsible for intracellular free fatty acid (FFA) distribution, eventually promoting intestinal nutrient absorption. Various other roles of SCFAs are depicted in Fig. 21.2.

AMP-activated Protein Kinase (AMPK)

AMPK enzyme is expressed in liver and skeletal muscle and maintains the cellular energy homeostasis. Increased expression of AMPK increases beta oxidation in liver and muscles and favors weight loss. AMPK activation targets carnitine palmitoyltransferase-1 through Acyl-coA carboxylase activity, which disfavors anabolic pathways such as glycogen storage and improves insulin sensitivity (Angin et al. 2016). Gut dysbiosis affects the host health by inhibiting AMPK, which negatively affects the beta oxidation, promotes synthesis of cholesterol and triglycerides, and enhances the rate of lipogenesis. There are experimental evidences that suggest that these factors favor the host toward weight gain and obesity, which may be followed by T2D (Boulangé et al. 2016).

Metabolic Functions

Experimental evidences suggest that the gut microbes perform important metabolic functions by regulating the expression of host genes. For example, the gut commensal *Bacteroides thetaiotaomicron* found in the intestine of mice and humans maintains the expression of genes involved in several functions like nutrient absorption, mucosal barrier fortification, xenobiotic metabolism, angiogenesis (Stappenbeck et al. 2002), postnatal maturation of intestine, and increased capillaries in small intestine, which aid in absorption (Hooper et al. 2001). This might help indirectly in the digestion and absorption in an healthy individual as compared to a diabetic patient.

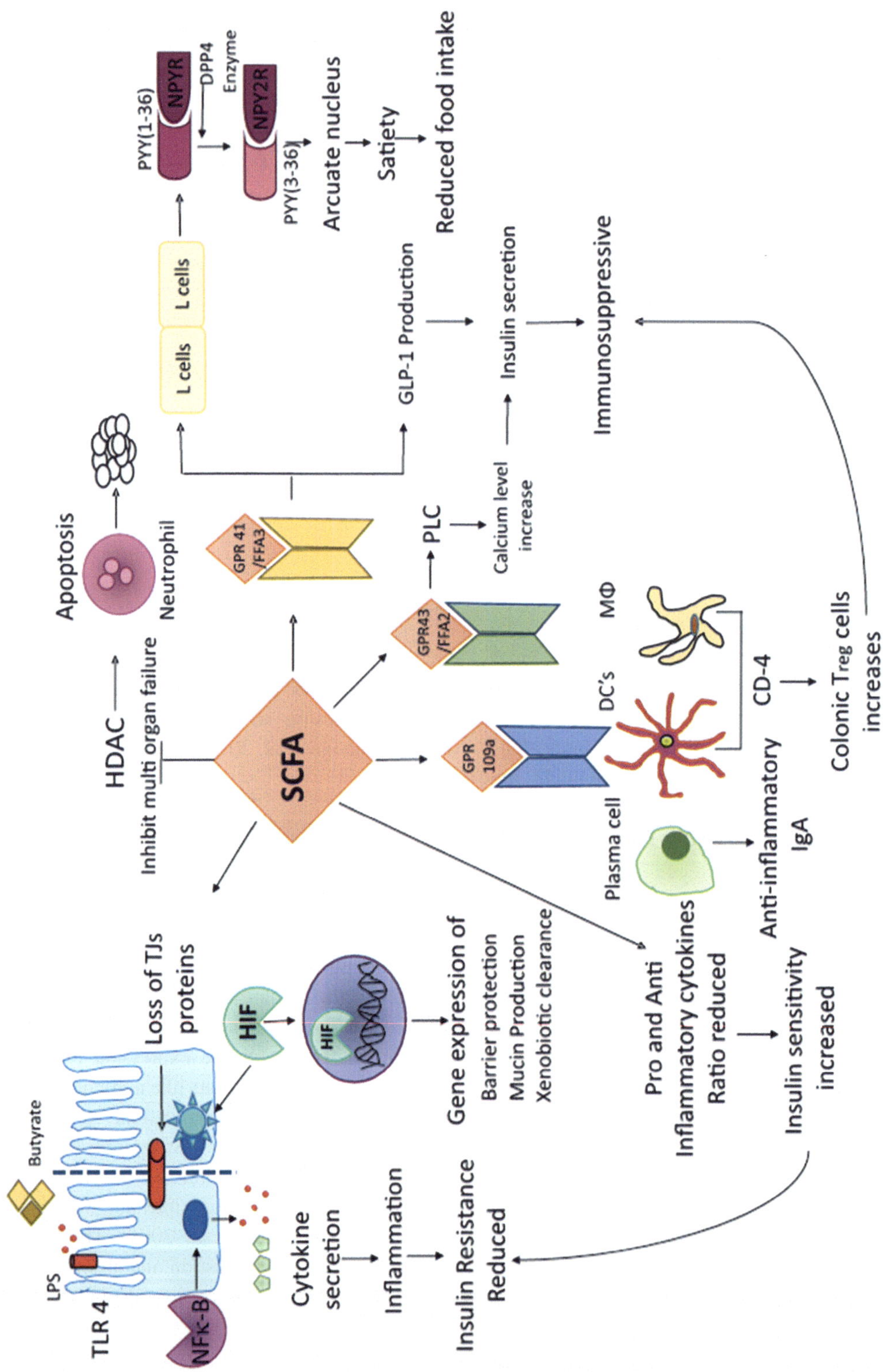

Fig. 21.2 Diverse roles of bacterially fermented short chain fatty acids (SCFAs) in various signaling pathways

Epithelial Barrier Impairment and Pro-Inflammatory Cytokines

Microbial dysbiosis, which is a main feature in correlating gut microbiome with T2D pathogenesis, is related to gut epithelial barrier impairment causing increased circulatory levels of tissue MAMPs such as LPS and PG (Salguero et al. 2019). This can cause low grade inflammation by releasing pro-inflammatory cytokines and can be a possible cause for obesity and T2D.

Bile Acid Transformation

Primary bile acids, cholic acid (CA) and chenodeoxycholic acid (CDSA), can function as ligand for nuclear transcription factor farnesoid X receptor (FXR), suppress hepatic gluconeogenesis and lipogenesis, increase fatty acid oxidation, and also play a role in improvement of extrahepatic insulin sensitivity. These are synthesized from cholesterol in liver and conjugated to taurine or glycine and on entering the GI tract is further metabolized by gut microbiota through cycle of oxidation, esterification, desulfation, and epimerization to form secondary bile acids deoxycholic acid (DCA) and lithocholic acid (LCA) in humans. Secondary bile acids are product of bacterial metabolism like which involves predominantly *Bacteroides fragilis, Bacteroides intestinalis*, and *E. coli.* (Ferrell and Chiang 2019). The bile acid metabolism is in loop with gut microbiota as bile acids control gut bacteria overgrowth and protect against inflammation, while the gut microbiota plays a role in biotransformation of bile acids (Ikegami and Honda 2018), bile acid composition, (Chiang and Ferrell 2018) and enterohepatic circulation of bile.

Gut Microbiome in Pathogenesis of T2D

T2D is accompanied by insulin resistance and beta cell dysfunction. Insulin resistance is defined as failure of uptake of glucose by peripheral tissues of the body, leading to imbalanced glucose homeostasis. During conditions of IR, insulin is produced in excessive amount to compensate for the loss of insulin sensitivity, termed as hyperinsulinemia. When this compensatory feedback loop is impaired or becomes nonfunctional, this can potentially result in T2D. The reason for this failure could be either change in beta cell function and mass or failure to respond adequately to secretory molecules (which stimulates beta cell to release insulin). Beta cell must be able to compensate for the decrease in insulin sensitivity for diabetes to occur.

The first evidence linking gut microbiota with glucose metabolism was shown in studies done on germ free mice in 2004 (Backhed et al. 2004). Subsequent studies done by several groups using metagenomics platform and 16S rRNA sequencing analysis have reported altered ratio of two important phyla: *Firmicutes* (Gram positive) and *Bacteroidetes* (Gram negative). Gut microflora interacts with host metabolism, leading to development of insulin resistance and T2D. One of the mechanisms linking insulin resistance and T2D is the low grade inflammation, which is a hallmark characteristic commonly observed in T2D and obesity, resulting due to prevalence of Gram-negative bacteria. Gram-negative bacteria have LPS on their outer membrane, which is a strong activator of Toll-like receptors that further activate increased production of inflammatory responses and pro-inflammatory cytokine in the bloodstream.

Several studies in the past have reported disrupted gut barrier function, leading to translocation of LPS into the circulation and resulting in microbiota-derived inflammation or endotoxemia contributing to pathogenesis of T2DM and obesity (Amar et al. 2011; Brun et al. 2007; Burcelin et al. 2012). Increased gut permeability is due to reduced expression of Zonula occludens-1 (ZO-1), claudin, and occludin, proteins that are required to restrict bacteria and substances from crossing intestinal lumen and enter the circulation. Reduction in SCFAs such as acetate, butyrate, and propionate also affects expression of tight junction proteins contributing to disturbed gut barrier function.

Breakdown of gut barrier leads to a release of LPS in blood, resulting in inflammation and its related metabolic disorders. LPS binds to TLR4 (TLR4 is a group of toll-like receptors expressed on most of the cells and macrophages and recognizes Pathogen-associated molecular patterns) to activate extensive downstream signaling, resulting in increased state of pro-inflammation as compared to TLR4 deficiency where these responses are attenuated (experiments performed on TLR4 and CD14 knockout mice). TLR4 inactivation in different tissues such as bone marrow derived cells and liver-specific TLR4 knockout mice showed improved glucose tolerance, and the absence of it in muscle prevents from diet-induced insulin resistance, which is an indication of protection against insulin resistance (Velloso et al. 2015).

One of the important roles of gut microflora is the production of nongaseous SCFAs in metabolizing dietary fiber. The significant commensal gut bacteria involved in SCFA generation are *Akkermansia muciniphila, Prevotella spp., Ruminococcus spp., Coprococcus spp., Faecalibacterium prausnitzii, Eubacterium rectale, and Roseburia spp* (Rinninella et al. 2019).It performs plethora of functions such as regulation of inflammatory response, production of intestinal hormone and lipogenesis, and most importantly maintenance of gut health. Butyrate is a source of energy for colonic epithelial cells and is also important for cell differentiation and growth (Ardawi and Newsholme 1985). On the other hand, acetate is a precursor of cholesterol and fatty acids and propionate may act as a precursor for gluconeogenesis (Sa'ad et al. 2010; Delzenne and Williams 2002).

Butyrate also increases the mucus production and impacts expression of tight junction proteins which leads to decrease in gut barrier permeability (Bordin et al. 2004; Peng et al. 2007). It has been observed that upon administration of acetate in mice fed with high fat diet, there is activation of AMPK resulting in increased fat oxidation and energy expenditure which in turn improve insulin resistance and glucose tolerance (Kahn et al. 2005). It has been shown that reduction of SCFA production can lead to increase in pro-inflammatory cytokine production, resulting in development of T2DM. SCFAs are ligand for G-protein-coupled receptors such as GPR41 and GPR43, affecting crucial functions like inflammation, expression of tight junction proteins, and enteroendocrine regulation (Kimura et al. 2014). Upon GPCR activation, there is induction of peptide YY (PYY) and leptin. Leptin is secreted by adipocyte with diverse effects on appetite and energy metabolism.

Current Therapeutics

Bariatric Surgery

Bariatric surgery has proven to achieve an effective weight loss, along with improvement of metabolic and inflammatory diseases . It has also helped to partially clarify the controversy about excess of Firmicutes in the obese population. Even though the Firmicutes/Bacteroidetes imbalance typical of obese animals is still debated regarding humans, a systematic review confirmed reduction of Firmicutes in parallel to marked weight loss during the first postoperative year. This is compatible with a role of this phylum in obesity pathogenesis. An increase in Proteobacteria as well as *Bacteroides* also occurred, with less obvious implications Luijten et al. (2019).

Prebiotics, Probiotics, and Synbiotics

Nondigestible polysaccharides including inulin, oligosaccharides, fructooligosaccharides, and galactooligosaccharides (prebiotics) promote enrichment of healthy bacteria such as *Bifidobacteria* and *Lactobacilli* over unhealthy bacteria and thereby increase the production of short-chain fatty acids (SCFAs). This results in reduction of inflammation, maintaining integrity of intestinal membrane and improvement in absorption of nutrients (Colantonio et al. 2019). Mice fed with high fat diet have reduced levels of *Bifidobacterium* as compared to control diet, which results in increased endotoxemia and could be a potential reason for altered gut microbiota (Cani et al. 2007b). Enriching the feed of HF diet mice with prebiotic oligofructose restores the level of *Bifidobacterium* and is positively correlated with improved glucose tolerance, glucose-induced insulin secretion, and maintaining normalized inflammatory tone, which favor the prevention of high fat diet-induced metabolic disorders such as obesity and T2D. In another study, oligofructose has been shown to exert antidiabetic effect and increase the secretion of glucagon like peptide 1 (GLP-1) (Cani et al. 2006).

The presence of oligofructose improves glucose tolerance, fasting blood glucose, glucose-stimulated insulin secretion, and insulin sensitive hepatic glucose production, with reduced body weight gain (Cani et al. 2006). Treating type 2 diabetes (T2D) patients with transglucosidase (which generates prebiotic fibers, including oligosaccharides) was followed by reduction in levels of hyperglycemia and weight gain. This is believed to be the result of production of oligosaccharides and manipulation in gut microbiome (increased Bacteroidetes-to-Firmicutes ratio) (Cani et al. 2006).

Most of the probiotic products commercially available have *Bifidobacterium* and *Lactobacillus* as their chief constituents, eventually including other lactic acid bacteria. Synbiotics in turn convey mixtures of both probiotics (live microorganisms) and prebiotics (fermentable fibers that induce elevated proliferation of probiotics and other favorable species) (He and Shi 2017).

Supplementation Meta-analyses

Recent systematic reviews of randomized controlled trials have attempted to shed light on this highly discussed therapeutic area. Prebiotics and synbiotics for T2D triggered improvements for fasting blood glucose, HbA1c, and HDL cholesterol, whereas in a subgroup analysis, advantages were unearthed for additional lipid fractions (triglycerides and total cholesterol) (Mahboobi et al. 2018).

Probiotics prescribed to diabetics elicited similar results concerning fasting blood glucose, glycated hemoglobin, insulin, and homeostatic model of insulin resistance (Koutnikova et al. 2019; Tao et al. 2020). For overweight, however, not for obese participants, body weight and additional anthropometric measurements diminished (Koutnikova et al. 2019). Inflammatory and oxidative stress biomarkers are similarly listed among the targets of probiotics and synbiotics in the diabetic population, as high sensitivity C-reactive protein and malonyldialdehyde decreased, whereas total antioxidant capacity, nitric oxide, and glutathione were enhanced (Zheng et al. 2019).

With such encouraging responses notwithstanding, studies continue to be heterogeneous on account of different doses, bacterial strains, and treatment regimens. Outcomes are still associated with specific probiotic and synbiotic preparations and with certain countries (Koutnikova et al. 2019), an important reason why few societies include them in official guidelines.

Other Perspectives

Fecal Microbiota Transplant (FMT)

The procedure dates back to fourth century AD, when the Chinese doctor Ge Hong first documented fecal suspension ingestion in his handbook. It was used to treat patients with food poisoning or severe diarrhea. A Chinese doctor, Li Shizhen administered a cocktail of dry feces, fermented, fresh fecal suspension, and infant feces for treatment of severe diarrhea, fever, pain, vomiting, and constipation. In 1958 Ben Eisemen, an American surgeon, successfully treated with fecal enemas four patients who had developed severe pseudomembranous enterocolitis after antibiotic use.

The description of the multiple metabolic functions of the gut microbiota has raised the hypothesis of employing FMT for metabolic disorders. A few encouraging results were reported; however, further exploration is needed for testing its efficacy (Bakker and Nieuwdorp 2017).

Table 21.1 Impact of dietary patterns on gut microbiome

Diet	Impact on gut microbiome
Fat Diet	
High-saturated fat and low fiber diet (mice)	Decrease in Bacteroidetes Increase in Firmicutes and Proteobacteria
High fat diet (mice)	Increase in *Lactococcus* and *Allobaculum* and decrease in *Akkermansia*
High fat diet in healthy adults	Increase in *Alistipes and Bacteroides* and decrease in *Faecalibacterium*
Protein Diet	
Long term protein rich diet	*Bacteroides* enterotype
Short-term protein rich diet	Increase in bile tolerant bacterial species (*Alistipes* and *Bilophile)* Decrease in saccharolytic microbes (*Roseburia, Eubacteria rectale, and Ruminococcus bromii*)
Plant-based protein diet	Increase in lactobacilli and bifidobacteria (enhancing SCFA production)
Carbohydrate Diet	
Complex carbohydrate (long term)	Increase in *Prevotella*
Nondigestible carbohydrate (overweight individuals)	Increase in phylum Firmicutes (including *Ruminococcus* species)

Personalized Nutrient/Diet

Reshaping host microbiome by modulation of the dietary exposure opens a new horizon for "personalized diet". It is based on manipulation of host–microbiota interaction through dietary intervention in order to induce change in host physiology, including development and progression of diseases. Diet can bring changes in composition of gut microbiome and function in an individual specific manner. Big data platforms and machine learning tools are being applied to datasets combining microbiome and clinical features , in order to relate specific foods to physiology, including external influences from drugs and antibiotics and specific outcomes such as cardiovascular risk (Kolodziejczyk et al. 2019; Wan et al. 2019; Dudek-Wicher et al. 2018) (Table 21.1).

Plant Extract-based Therapy

Dietary polyphenols are one of the major plant fiber families, identified in tea, coffee, vegetables, and fruits. They are grouped into flavonoids and nonflavonoids. Similar to other potentially bioactive/phytotherapeutic agents, a small proportion of polyphenols are absorbed in the small intestine, while most of them collect in the large intestinal lumen, where they are acted upon by gut microbiota for further enzymatic processing.

Conversion of polyphenols and other fibers into bioactive compounds in gut can influence the population of gut microbiota and affect host health, including cardiovascular risk, obesity, and other metabolic diseases (Brochot et al. 2019). Most available microbiome investigations were conducted in mice; however, a few reviews and meta-analyses in humans, not necessarily addressing metabolic diseases, are starting to appear.

One of them focused on all types of fiber. Concentrations of *Bifidobacterium spp.* and *Lactobacillus spp.* were elevated , along with fecal butyrate. Fructans and galacto-oligosaccharides were endowed with the most robust effects of the microbiome (So et al. 2018). The favorable impact of prebiotic fibers concerning *Bifidobacterium* was confirmed in another meta-analysis (Wilson et al. 2019).

As regards metabolic syndrome, cardiovascular and neurodegenerative disorders, reviews with polyphenols and resveratrol suggest possible benefit, however, insisting that data are preliminary and sometimes inconsistent (Potì et al. 2019; Chaplin et al. 2018).

References

Amar J, Chabo C, Waget A et al (2011) Intestinal mucosal adherence and translocation of commensal bacteria at the early onset of type 2 diabetes: molecular mechanisms and probiotic treatment. EMBO Molecul Med 3:559–572

Angin Y, Beauloye C, Horman S, Bertrand L (2016) Regulation of carbohydrate metabolism, lipid metabolism, and protein metabolism by AMPK. Exp Suppl 107:23–43

Ardawi M, Newsholme E (1985) Fuel utilization in colonocytes of the rat. Biochem J 231:713–719

Backhed F, Ding H, Wang T et al (2004) The gut microbiota as an environmental factor that regulates fat storage. PNAS 101:15718–15723

Bakker GJ, Nieuwdorp M (2017) Fecal microbiota transplantation: therapeutic potential for a multitude of diseases beyond clostridium difficile. Microbiol Spectr 5:28840809

Bordin M, D'Atri F, Guillemot L, Citi S (2004) Histone deacetylase inhibitors up-regulate the expression of tight junction Proteins1 1Swiss Cancer League, Swiss National Science Foundation, Ministry for Italian University and Research, ERASMUS Program (M. Bordin), and Roche Research Foundation fellowship (L. Guillemot). Mol Cancer Res 2:692–701

Boulangé CL, Neves AL, Chilloux J, Nicholson JK, Dumas M-E (2016) Impact of the gut microbiota on inflammation, obesity, and metabolic disease. Genome Medicine 8:42

Brestoff JR, Artis D (2013) Commensal bacteria at the interface of host metabolism and the immune system. Nat Immunol 14:676–684

Brochot A, Azalbert V, Landrier JF, Tourniaire F, Serino M (2019) A two-week treatment with plant extracts changes gut microbiota, caecum metabolome, and markers of lipid metabolism in ob/ob mice. Mol Nutr Food Res 63:e1900403

Brugman S, Klatter FA, Visser JT et al (2006) Antibiotic treatment partially protects against type 1 diabetes in the bio-breeding diabetes-prone rat. Is the gut flora involved in the development of type 1 diabetes? Diabetologia 49:2105–2108

Brun P, Castagliuolo I, Leo VD et al (2007) Increased intestinal permeability in obese mice: new evidence in the pathogenesis of nonalcoholic steatohepatitis. Am J Phys-Gastroint Liver Physiol 292:G518–GG25

Brunkwall L, Orho-Melander M (2017) The gut microbiome as a target for prevention and treatment of hyperglycaemia in type 2 diabetes: from current human evidence to future possibilities. Diabetologia 60:943–951

Burcelin R, Garidou L, Pomié C (2012) Immunomicrobiota cross and talk: the new paradigm of metabolic diseases. Semin Immunol 24:67–74

Cani PD, Knauf C, Iglesias MA, Drucker DJ, Delzenne NM, Burcelin R (2006) Improvement of glucose tolerance and hepatic insulin sensitivity by oligofructose requires a functional glucagon-like peptide 1 receptor. Diabetes 55:1484–1490

Cani PD, Amar J, Iglesias MA et al (2007a) Metabolic endotoxemia initiates obesity and insulin resistance. Diabetes 56:1761

Cani PD, Neyrinck AM, Fava F et al (2007b) Selective increases of bifidobacteria in gut microflora improve high-fat-diet-induced diabetes in mice through a mechanism associated with endotoxaemia. Diabetologia 50:2374–2383

Cani PD, Possemiers S, Van de Wiele T et al (2009) Changes in gut microbiota control inflammation in obese mice through a mechanism involving GLP-2-driven improvement of gut permeability. Gut 58:1091

Caricilli MA, Saad JAM (2013) The role of gut microbiota on insulin resistance. Nutrients 5:829–851

Chaplin A, Carpéné C, Mercader J (2018) Resveratrol, metabolic syndrome, and gut microbiota. Nutrients 10 (11):1651

Chen JJ, Wang R, X-f L, Wang R-l (2011) Bifidobacterium longum supplementation improved high-fat-fed-induced metabolic syndrome and promoted intestinal Reg I gene expression. Exp Biol Med 236:823–831

Chiang JYL, Ferrell JM (2018) Bile acid metabolism in liver pathobiology. Gene Expr 18:71–87

Colantonio AG, Werner SL, Brown M (2019) The effects of prebiotics and substances with prebiotic properties on metabolic and inflammatory biomarkers in individuals with type 2 diabetes mellitus: a systematic review. J Acad Nutr Diet 120:587–607

Delzenne NM, Williams CM (2002) Prebiotics and lipid metabolism. Curr Opin Lipidol 13:61–67

Dudek-Wicher RK, Junka A, Bartoszewicz M (2018) The influence of antibiotics and dietary components on gut microbiota. Prz Gastroenterol 13:85–92

Ferrell JM, Chiang JYL (2019) Understanding bile acid signaling in diabetes: from pathophysiology to therapeutic targets. Diabetes Metab J 43:257–272

Greiner TU, Bäckhed F (2016) Microbial regulation of GLP-1 and L-cell biology. Molecul Metabol 5:753–758

He M, Shi B (2017) Gut microbiota as a potential target of metabolic syndrome: the role of probiotics and prebiotics. Cell Biosci 7:54

Hooper LV, Wong MH, Thelin A, Hansson L, Falk PG, Gordon JI (2001) Molecular analysis of commensal host-microbial relationships in the intestine. Science 291:881

Hotamisligil GS (2006) Inflammation and metabolic disorders. Nature 444:860–867

Ikegami T, Honda A (2018) Reciprocal interactions between bile acids and gut microbiota in human liver diseases. Hepatol Res 48:15–27

Jacquemont S, Reymond A, Zufferey F et al (2011) Mirror extreme BMI phenotypes associated with gene dosage at the chromosome 16p11.2 locus. Nature 478:97–102

Jernberg C, Lofmark S, Edlund C, Jansson JK (2010) Long-term impacts of antibiotic exposure on the human intestinal microbiota. Microbiology 156:3216–3223

Kahn BB, Alquier T, Carling D, Hardie DG (2005) AMP-activated protein kinase: ancient energy gauge provides clues to modern understanding of metabolism. Cell Metab 1:15–25

Kahn SE, Hull RL, Utzschneider KM (2006) Mechanisms linking obesity to insulin resistance and type 2 diabetes. Nature 444:840–846

Karlsson FH, Tremaroli V, Nookaew I et al (2013) Gut metagenome in European women with normal, impaired and diabetic glucose control. Nature 498:99

Karra E, Chandarana K, Batterham RL (2009) The role of peptide YY in appetite regulation and obesity. J Physiol 587:19–25

Kassi E, Pervanidou P, Kaltsas G, Chrousos G (2011) Metabolic syndrome: definitions and controversies. BMC Med 9:48

Kimura I, Inoue D, Hirano K, Tsujimoto G (2014) The SCFA receptor GPR43 and energy metabolism. Front Endocrinol 5:85

Kolodziejczyk AA, Zheng D, Elinav E (2019) Diet–microbiota interactions and personalized nutrition. Nat Rev Microbiol 17:742–753

Koutnikova H, Genser B, Monteiro-Sepulveda M, Faurie JM, Rizkalla S, Schrezenmeir J, Clément K (2019) Impact of bacterial probiotics on obesity, diabetes and non-alcoholic fatty liver disease related variables: a systematic review and meta-analysis of randomised controlled trials. BMJ Open 9(3):e017995

Luijten JCHBM, Vugts G, Nieuwenhuijzen GAP, Luyer MDP (2019) The importance of the microbiome in bariatric surgery: a systematic review. Obes Surg 29:2338–2349

Mahboobi S, Rahimi F, Jafarnejad S (2018) Effects of prebiotic and synbiotic supplementation on glycaemia and lipid profile in type 2 diabetes: a meta-analysis of randomized controlled trials. Adv Pharm Bull 8 (4):565–574

Matsuzaki T, Yamazaki R, Hashimoto S, Yokokura T (1997) Antidiabetic effects of an oral administration of Lactobacillus casei in a non-insulin-dependent diabetes mellitus (NIDDM) model using KK-Ay mice. Endocr J 44:357–365

Moreno-Indias I, Cardona F, Tinahones FJ, Queipo-Ortuño MI (2014) Impact of the gut microbiota on the development of obesity and type 2 diabetes mellitus. Front Microbiol 5:190

Musso G, Gambino R, Cassader M (2010) Obesity, diabetes, and gut microbiota. Diabetes Care 33:2277

Nauck MA, Meier JJ (2018) Incretin hormones: their role in health and disease. Diabetes Obes Metab 20(Suppl 1):5–21

Odamaki T, Kato K, Sugahara H et al (2016) Age-related changes in gut microbiota composition from newborn to centenarian: a cross-sectional study. BMC Microbiol 16:90

Parada Venegas D, De la Fuente MK, Landskron G et al (2019) Short chain fatty acids (SCFAs)-mediated gut epithelial and immune regulation and its relevance for inflammatory bowel diseases. Front Immunol 10:277

Parks BW, Nam E, Org E et al (2013) Genetic control of obesity and gut microbiota composition in response to high-fat, high-sucrose diet in mice. Cell Metab 17:141–152

Peng L, He Z, Chen W, Holzman IR, Lin J (2007) Effects of butyrate on intestinal barrier function in a Caco-2 cell monolayer model of intestinal barrier. Pediatr Res 61:37

Potì F, Santi D, Spaggiari G, Zimetti F, Zanotti I (2019) Polyphenol health effects on cardiovascular and neurodegenerative disorders: a review and meta-analysis. Int J Mol Sci 20(2):E351

Randle PJ, Garland PB, Hales CN, Newsholme EA (1963) The glucose fatty-acid cycle. Its role in insulin sensitivity and the metabolic disturbances of diabetes mellitus. Lancet 1:785–789

Reaven GM (1988) Banting lecture 1988. Role of insulin resistance in human disease. Diabetes 37:1595–1607

Reaven GM, Hollenbeck C, Jeng CY, Wu MS, Chen YD (1988) Measurement of plasma glucose, free fatty acid, lactate, and insulin for 24 h in patients with NIDDM. Diabetes 37:1020–1024

Rinninella E, Raoul P, Cintoni M et al (2019) What is the healthy gut microbiota composition? A changing ecosystem across age, environment, diet, and diseases. Microorganisms 7:14

Rodriguez-Carrio J, Salazar N, Margolles A et al (2017) Free fatty acids profiles are related to gut microbiota signatures and short-chain fatty acids. Front Immunol 8:823

Sa'ad H, Peppelenbosch MP, Roelofsen H, Vonk RJ, Venema K (2010) Biological effects of propionic acid in humans; metabolism, potential applications and underlying mechanisms. Biochim Biophys Acta Mol Cell Biol Lipids 1801:1175–1183

Salguero MV, Al-Obaide MAI, Singh R, Siepmann T, Vasylyeva TL (2019) Dysbiosis of gram-negative gut microbiota and the associated serum lipopolysaccharide exacerbates inflammation in type 2 diabetic patients with chronic kidney disease. Exper Therapeut Med 18:3461–3469

So D, Whelan K, Rossi M, Morrison M, Holtmann G, Kelly JT, Shanahan ER, Staudacher HM, Campbell KL (2018) Dietary fiber intervention on gut microbiota composition in healthy adults: a systematic review and meta-analysis. Am J Clin Nutr 107(6):965–983

Stappenbeck TS, Hooper LV, Gordon JI (2002) Developmental regulation of intestinal angiogenesis by indigenous microbes via Paneth cells. Proc Natl Acad Sci 99:15451

Tao YW, Gu YL, Mao XQ, Zhang L, Pei YF (2020) Effects of probiotics on type II diabetes mellitus: a meta-analysis. J Transl Med 18(1):30

Thuny F, Richet H, Casalta J-P, Angelakis E, Habib G, Raoult D (2010) Vancomycin treatment of infective endocarditis is linked with recently acquired obesity. PLoS One 5:e9074

Tidjani Alou M, Lagier J-C, Raoult D (2016) Diet influence on the gut microbiota and dysbiosis related to nutritional disorders. Human Microbio J 1:3–11

Turnbaugh PJ, Ley RE, Hamady M, Fraser-Liggett CM, Knight R, Gordon JI (2007) The human microbiome project. Nature 449:804–810

Velloso LA, Folli F, Saad MJ (2015) TLR4 at the crossroads of nutrients, gut microbiota, and metabolic inflammation. Endocr Rev 36:245–271

Wan Y, Wang F, Yuan J et al (2019) Effects of dietary fat on gut microbiota and faecal metabolites, and their relationship with cardiometabolic risk factors: a 6-month randomised controlled-feeding trial. Gut 68:1417–1429

Wilson B, Rossi M, Dimidi E, Whelan K (2019) Prebiotics in irritable bowel syndrome and other functional bowel disorders in adults: a systematic review and meta-analysis of randomized controlled trials. Am J Clin Nutr 109(4):1098–1111

Yamane S, Inagaki N (2018) Regulation of glucagon-like peptide-1 sensitivity by gut microbiota dysbiosis. J Diabet Invest 9:262–264

Zhang Y, Zhang H (2013) Microbiota associated with type 2 diabetes and its related complications. Food Sci Human Wellness 2:167–172

Zhang C, Zhang M, Wang S et al (2010) Interactions between gut microbiota, host genetics and diet relevant to development of metabolic syndromes in mice. ISME J 4:232–241

Zheng Y, Ley SH, Hu FB (2017) Global aetiology and epidemiology of type 2 diabetes mellitus and its complications. Nat Rev Endocrinol 14:88

Zheng HJ, Guo J, Jia Q, Huang YS, Huang WJ, Zhang W, Zhang F, Liu WJ, Wang Y (2019) The effect of probiotic and synbiotic supplementation on biomarkers of inflammation and oxidative stress in diabetic patients: a systematic review and meta-analysis of randomized controlled trials. Pharmacol Res 142:303–313

22 Immunotherapeutic Approach to the Treatment and Prevention of Obesity

Tatsuhiko Azegami and Hiroshi Itoh

Contents

Introduction 298
Lifestyle Interventions 298
Pharmacological Management 298
The Role of Vaccines 298

Features of Therapeutic Vaccines 298

Ghrelin Vaccines 299
Vaccine Against Synthetic Ghrelin Peptides 300
Virus Like Particles (VLPs) 300

Nanogel-based vaccine 300

Glucose-dependent Insulinotropic Polypeptide (GIP) Vaccines 300

Adipocyte Vaccines 300
Somatostatin Vaccine 301
Adenovirus 36 (Ad36) Vaccine 301

Future Possibilities 301

References 301

T. Azegami (✉)
Keio University Health Center, Yokohama-shi, Japan

Department of Internal Medicine, Keio University School of Medicine, Tokyo, Japan
e-mail: t.azegami-1114@z2.keio.jp

H. Itoh
Department of Internal Medicine, Keio University School of Medicine, Tokyo, Japan

J. Faintuch and S. Faintuch (eds.), *Obesity and Diabetes*, https://doi.org/10.1007/978-3-030-53370-0_22

Abstract

Clinical availability of anti-obesity drugs is limited, and their inadequate effectiveness and safety concerns sometimes discourage widespread use. A therapeutic vaccine has the potential to be an attractive tool for preventing and treating obesity, because of the possibility of prolonged therapeutic effect and low frequency of administration. Experimental investigations have shown that vaccines targeting endogenous molecules that promote obesity could be a viable alternative. Recent novelties in drug-delivery systems and biotechnology will support further progress in vaccine development. This chapter provides an overview of recent advances in the area.

Keywords

Adenovirus 36 · Adipocyte · Glucose-dependent Insulinotropic Polypeptide · Ghrelin · Immunotherapy · Somatostatin · Vaccine

Introduction

The worldwide prevalence of obesity has markedly increased (nearly tripled in 40 years) (Flegal et al. 2013; Wilson et al. 2002; WHO 2020), and it is predicted that the condition will cost the U.S. healthcare system \$48–\$66 billion a year by 2030 (Wang et al. 2011). To combat the cardiometabolic comorbidities of obesity and their economic burden, global strategies for obesity prevention and treatment are required.

Lifestyle Interventions

Lifestyle initiatives targeting energy intake and physical activity are generally the first approach to weight reduction (Bray et al. 2016). They can trigger a 5–8% reduction in body weight (Heymsfield and Wadden 2017). However, obese persons often lack the motivation to continue with lifestyle management for weight reduction, because obesity is usually asymptomatic. Therefore, long-term weight management in obese persons remains a difficult task, and weight regain is a common problem following weight loss intervention (Heymsfield and Wadden 2017).

Pharmacological Management

The following prescription agents are approved by the U.S. Food and Drug Administration: phentermine (sympathomimetic amine), orlistat (lipase inhibitor), phentermine–topiramate (sympathomimetic amine and an antiepileptic drug), lorcaserin (5-HT2c receptor agonist), naltrexone–bupropion (opioid antagonist and aminoketone antidepressant), and liraglutide (glucagon-like peptide-1 [GLP1] receptor agonist). Yet relatively few patients receive sustained pharmacologic therapy for reasons that encompass insufficient weight loss, side effects, and prompt recurrence of obesity after treatment be interrupted (Heymsfield and Wadden 2017).

The Role of Vaccines

A therapeutic vaccine may be a potential candidate for improving treatment adherence. In general, a vaccine has prolonged therapeutic effects and low frequency of administration, when it succeeds in inducing neutralizing antibodies against a target molecule (Table 22.1).

Features of Therapeutic Vaccines

Therapeutic vaccines for the treatment of chronic diseases have the potential to increase treatment adherence, reduce healthcare costs, and offer enhanced specificity for target molecules. When administration of a therapeutic vaccine successfully induces antibodies that bind to the target self-molecule and inhibit its

Table 22.1 Potential vaccines targeting obesity

Target	Antigen	Species/References	Effect
Ghrelin	Ghr1 (1–10 a.a.)—KLH Ghr2 (13–28 a.a.)—KLH Ghr3 (1–28 a.a.) KLH	Male Wistar rat (Zorrilla et al. 2006)	No change in food intake 20% reduction in weight gain (Ghr1, Ghr3)
	Ghrelin (1–10 a.a.)—BSA	Male/female piglet (Vizcarra et al. 2007)	15% reduction in food intake 10% reduction in weight gain
	Ghrelin–NS1 (VLP-based)	DIO male C57BL6/J mouse (Andrade et al. 2013)	Decrease in food intake Increase in energy expenditure No change in weight gain
Adipocyte	Pig adipose tissue	Male/female human adult (Bourinbaiar and Jirathitikal 2010)	No change in body weight 7.6% reduction in waist size 25.9% increase in HDL-C
	Mouse adipocytes	Male/female Sprague Dawley rat (Lai et al. 2010)	About 50% reduction in weight gain
Somatostatin	Somatostatin-CAT	DIO male C57BL6/J Mouse (Haffer 2012)	12–13% reduction in body weight No change in food intake
Adenovirus 36	Inactivated Ad36	Ad36-infected C57BL6/J mouse (Na and Nam 2014)	No change in food intake 17% reduction in weight gain 20% reduction in epididymal fat

BSA, bovine serum albumin; DIO, diet-induced obesity; KLH, keyhole limpet hemocyanin; NS1, non-structural protein 1; VLP, virus-like particle

function, the vaccine should have a long-term therapeutic effect. In the case of hypertension vaccines, three doses of a peptide vaccine targeting angiotensin II type 1 receptor reduced blood pressures for 24 weeks after final immunization in hypertensive rats (Azegami et al. 2012), and a DNA vaccine targeting angiotensin II decreased blood pressure for at least 6 months in rats (Koriyama et al. 2015). The prolonged therapeutic effects of such vaccines will mean a low frequency of administration and may result in increased treatment adherence and savings in medication costs.

In addition to their prolonged effect, therapeutic vaccines have the potential to have greater specificity to target molecules compared with conventional low-molecular-weight drugs (Hansel et al. 2010). In general, high specificity to the target tends to result in few off-target effects and low rates of drug–drug interaction, and it may reduce the incidence of side effects (Hansel et al. 2010). Therapeutic vaccines also have some advantages over monoclonal antibody therapy, including lower production costs, no possibility of inducing anti-drug antibodies, and less frequent dosing (Hansel et al. 2010).

Ghrelin Vaccines

The especial feature of the ghrelin peptide is the *O*-acylation at the Ser^3 residue. It uniquely stimulates food ingestion with diminished energy expenditure (Kojima et al. 1999; Nakazato et al. 2001; Tschop et al. 2000; Wortley et al. 2004). Peripheral ghrelin, which is produced mainly in the gastric X/A-like cells, modulates the nucleus tractus solitarius via the vagus nerve. This results in an increase in noradrenaline in the arcuate nucleus of the hypothalamus and consequent appetite stimulation (Date et al. 2006). *O*-acylation at Ser^3 with octanoate, which is mediated by ghrelin *O*-acyltransferase (GOAT), provides the orexigenic action of ghrelin (Yang et al. 2008); unacylated ghrelin negatively impacts appetite and body weight (Asakawa et al. 2005).

Laboratory experiments demonstrate that antagonization of ghrelin ameliorates obesity. Genetic deletion of ghrelin increases energy expenditure and locomotor activity in mice, which are less prone to diet-induced obesity (Wortley et al. 2005). Genetic deletion of ghrelin receptor (growth-hormone secretagogue receptor: GHSR), GHSR antagonists, and GOAT inhibitors attenuate diet-induced obesity in mice (Zigman et al. 2005; Maletinska et al. 2011; Barnett et al. 2010). However, as of 2019, no anti-obesity drug that targets ghrelin function—such as a ghrelin inhibitor, GHSR antagonist, or GOAT inhibitor—has been clinically available.

Vaccine Against Synthetic Ghrelin Peptides

Zorrilla et al. (2006) synthesized three ghrelin peptides (Ghr1, Ghr2, and Ghr3) for vaccine development in 2006. Keyhole limpet hemocyanin (KLH) and a couple of adjuvants were attached, and antigen-specific antibodies were elicited. Only Ghr1-KLH and Ghr3-KLH induced weight reduction (by 20%) (Zorrilla et al. 2006).

Also in a porcine model, vaccine comprising the N-terminal residues (1–10) of porcine ghrelin, combined with additional molecules, diminished appetite and weight accrual in piglets (Vizcarra et al. 2007).

Virus Like Particles (VLPs)

VLPs are viral proteins that reliably self-assemble and exhibit antigenic epitopes (Kushnir et al. 2012). Clinical trials in other areas have confirmed the concept (Maurer et al. 2005; Ambuhl et al. 2007). Also a ghrelin VLP vaccine was arranged and, in rodents, diminished appetite however not body weight (Andrade et al. 2013). More extensive studies are necessary in the area.

Nanogel-based vaccine

Advances in nanomaterial development have also been applied to the development of innovative anti-obesity vaccines. To avoid the risk of localized skin adverse events and psychological and physiological stress, when vaccines are administered by using injectable delivery methods, nanoparticle drug delivery to mucosal surfaces is an option (Lamichhane et al. 2014). Nanometer-sized polymer hydrogels (nanogels) can incorporate various proteins through hydrophobic interactions and subsequently allow their release in their native form (Azegami et al. 2018). The cationic type of cholesteryl-group-bearing pullulan (cCHP) nanogel is adequate for local application and interaction with nasal dendritic cells (Nochi et al. 2010). A recombinant fusion protein based on mouse ghrelin and PspA (pneumococcal surface protein A), paved the way for an intranasal vaccine using cCHP nanogel. In mice, serum IgG antibodies were elicited and weight gain was reduced (by 7%). Mitochondrial uncoupling protein 1 in brown adipose tissue was upregulated (Azegami et al. 2017).

Glucose-dependent Insulinotropic Polypeptide (GIP) Vaccines

GIP, as one of incretins, promotes pancreatic insulin secretion in a glucose-dependent manner (Sadry and Drucker 2013), and increases adipose tissue (Miyawaki et al. 2002). In animal experiments, a GIP receptor antagonist attenuated diet-induced obesity and subsequently improved glucose metabolism (McClean et al. 2007). GIP combined with bacteriophage Qβ VLPs allowed the development of a vaccine. In a mice model, weight gain was inhibited (by 35%), without hyperglycemia. The anti-obesity response was attributed to elevation of energy expenditure (Fulurija et al. 2008).

Adipocyte Vaccines

A small clinical trial with oral tablets of pig adipose tissue was conducted for 3 months (Bourinbaiar and Jirathitikal 2010). Waist circumference decreased (by 7.6%), however not body weight. In turn, an intraperitoneally injected xenogeneic adipocyte vaccine in rats was followed by adipocyte apoptosis and 50% amelioration of weight increase (Lai et al. 2010).

Somatostatin Vaccine

Pharmacological supplementation with growth hormone (GH) decreases body fat in obese adults (Kim et al. 1999); however, the clinical applications of GH are limited by its very short half-life. Vaccine inhibition of somatostatin, which blocks GH secretion, was attempted by means of intraperitoneal immunization with somatostatin and carrier protein chloramphenicol acetyltransferase. Weight reduction (by 12–13%) and IgG production were achieved (Haffer 2012).

Adenovirus 36 (Ad36) Vaccine

In some large series, many obese persons tested positive for antibodies against Ad36 (Atkinson et al. 2005), and animals respond to Ad36 challenge with increased adiposity (Dhurandhar et al. 2002). An Ad36-based vaccine in mice was successful in preventing body fat elevation (Na and Nam 2014).

Future Possibilities

As therapeutic vaccines elicit the production of antigen-specific antibodies to neutralize biomolecules that promote weight gain, not only ghrelin, GIP, adipocytes, and somatostatin, but also other endogenous molecules can be targeted. The therapeutic vaccine approach is feasible for pathways with which no small-molecule drug has yet been able to interact. Brain and intracellular molecules that cannot be reached by neutralizing antibodies would not be suitable targets. In addition, enhancement of the immune reaction is a trade-off between preferable humoral immune responses and undesired autoimmune reactions.

Better selection of epitope sequences from self-antigens and adjuvants designed to elicit Th2 activity will help to induce Th2-dominant humoral immune responses. Enhanced drug-delivery systems and biotechnology will also help to develop effective and safe vaccines.

References

Ambuhl PM, Tissot AC, Fulurija A, Maurer P, Nussberger J, Sabat R et al (2007) A vaccine for hypertension based on virus-like particles: preclinical efficacy and phase I safety and immunogenicity. J Hypertens 25(1):63–72

Andrade S, Pinho F, Ribeiro AM, Carreira M, Casanueva FF, Roy P et al (2013) Immunization against active ghrelin using virus-like particles for obesity treatment. Curr Pharm Des 19(36):6551–6558

Asakawa A, Inui A, Fujimiya M, Sakamaki R, Shinfuku N, Ueta Y et al (2005) Stomach regulates energy balance via acylated ghrelin and desacyl ghrelin. Gut 54(1):18–24

Atkinson RL, Dhurandhar NV, Allison DB, Bowen RL, Israel BA, Albu JB et al (2005) Human adenovirus-36 is associated with increased body weight and paradoxical reduction of serum lipids. Int J Obes 29(3):281–286

Azegami T, Sasamura H, Hayashi K, Itoh H (2012) Vaccination against the angiotensin type 1 receptor for the prevention of L-NAME-induced nephropathy. Hypertens Res 35(5):492–499

Azegami T, Yuki Y, Sawada S, Mejima M, Ishige K, Akiyoshi K et al (2017) Nanogel-based nasal ghrelin vaccine prevents obesity. Mucosal Immunol 10 (5):1351–1360

Azegami T, Yuki Y, Nakahashi R, Itoh H, Kiyono H (2018) Nanogel-based nasal vaccines for infectious and lifestyle-related diseases. Mol Immunol 98:19–24

Barnett BP, Hwang Y, Taylor MS, Kirchner H, Pfluger PT, Bernard V et al (2010) Glucose and weight control in mice with a designed ghrelin O-acyltransferase inhibitor. Science 330(6011):1689–1692

Bourinbaiar AS, Jirathitikal V (2010) Effect of oral immunization with pooled antigens derived from adipose tissue on atherosclerosis and obesity indices. Vaccine 28(15):2763–2768

Bray GA, Fruhbeck G, Ryan DH, Wilding JP (2016) Management of obesity. Lancet 387(10031):1947–1956

Date Y, Shimbara T, Koda S, Toshinai K, Ida T, Murakami N et al (2006) Peripheral ghrelin transmits orexigenic signals through the noradrenergic pathway from the hindbrain to the hypothalamus. Cell Metab 4 (4):323–331

Dhurandhar NV, Whigham LD, Abbott DH, Schultz-Darken NJ, Israel BA, Bradley SM et al (2002) Human adenovirus Ad-36 promotes weight gain in male rhesus and marmoset monkeys. J Nutr 132 (10):3155–3160

Flegal KM, Kit BK, Orpana H, Graubard BI (2013) Association of all-cause mortality with overweight and obesity using standard body mass index categories: a systematic review and meta-analysis. JAMA 309 (1):71–82

Fulurija A, Lutz TA, Sladko K, Osto M, Wielinga PY, Bachmann MF et al (2008) Vaccination against GIP for the treatment of obesity. PLoS One 3(9):e3163

Haffer KN (2012) Effects of novel vaccines on weight loss in diet-induced-obese (DIO) mice. J Anim Sci Biotechnol 3(1):21

Hansel TT, Kropshofer H, Singer T, Mitchell JA, George AJ (2010) The safety and side effects of monoclonal antibodies. Nat Rev Drug Discov 9(4):325–338

Heymsfield SB, Wadden TA (2017) Mechanisms, pathophysiology, and management of obesity. N Engl J Med 376(3):254–266

Kim KR, Nam SY, Song YD, Lim SK, Lee HC, Huh KB (1999) Low-dose growth hormone treatment with diet restriction accelerates body fat loss, exerts anabolic effect and improves growth hormone secretory dysfunction in obese adults. Horm Res 51(2):78–84

Kojima M, Hosoda H, Date Y, Nakazato M, Matsuo H, Kangawa K (1999) Ghrelin is a growth-hormone-releasing acylated peptide from stomach. Nature 402 (6762):656–660

Koriyama H, Nakagami H, Nakagami F, Osako MK, Kyutoku M, Shimamura M et al (2015) Long-term reduction of high blood pressure by angiotensin II DNA vaccine in spontaneously hypertensive rats. Hypertension 66(1):167–174

Kushnir N, Streatfield SJ, Yusibov V (2012) Virus-like particles as a highly efficient vaccine platform: diversity of targets and production systems and advances in clinical development. Vaccine 31(1):58–83

Lai QG, Jiang BQ, Zhou XH, Xu X, Gao ZN, Yuan KF (2010) The effects and mechanism of xenogeneic adipocyte vaccine for the prevention of obesity in rats. J Int Med Res 38(5):1700–1707

Lamichhane A, Azegami T, Kiyono H (2014) The mucosal immune system for vaccine development. Vaccine 32 (49):6711–6723

Maletinska L, Matyskova R, Maixnerova J, Sykora D, Pychova M, Spolcova A et al (2011) The Peptidic GHS-R antagonist [D-Lys(3)]GHRP-6 markedly improves adiposity and related metabolic abnormalities in a mouse model of postmenopausal obesity. Mol Cell Endocrinol 343(1-2):55–62

Maurer P, Jennings GT, Willers J, Rohner F, Lindman Y, Roubicek K et al (2005) A therapeutic vaccine for nicotine dependence: preclinical efficacy, and Phase I safety and immunogenicity. Eur J Immunol 35 (7):2031–2040

McClean PL, Irwin N, Cassidy RS, Holst JJ, Gault VA, Flatt PR (2007) GIP receptor antagonism reverses obesity, insulin resistance, and associated metabolic disturbances induced in mice by prolonged consumption of high-fat diet. Am J Physiol Endocrinol Metab 293(6):E1746–E1755

Miyawaki K, Yamada Y, Ban N, Ihara Y, Tsukiyama K, Zhou H et al (2002) Inhibition of gastric inhibitory polypeptide signaling prevents obesity. Nat Med 8 (7):738–742

Na HN, Nam JH (2014) Proof-of-concept for a virus-induced obesity vaccine; vaccination against the obesity agent adenovirus 36. Int J Obes 38(11):1470–1474

Nakazato M, Murakami N, Date Y, Kojima M, Matsuo H, Kangawa K et al (2001) A role for ghrelin in the central regulation of feeding. Nature 409(6817):194–198

Nochi T, Yuki Y, Takahashi H, Sawada S, Mejima M, Kohda T et al (2010) Nanogel antigenic protein-delivery system for adjuvant-free intranasal vaccines. Nat Mater 9(7):572–578

Sadry SA, Drucker DJ (2013) Emerging combinatorial hormone therapies for the treatment of obesity and T2DM. Nat Rev Endocrinol 9(7):425–433

Tschop M, Smiley DL, Heiman ML (2000) Ghrelin induces adiposity in rodents. Nature 407 (6806):908–913

Vizcarra JA, Kirby JD, Kim SK, Galyean ML (2007) Active immunization against ghrelin decreases weight gain and alters plasma concentrations of growth hormone in growing pigs. Domest Anim Endocrinol 33 (2):176–189

Wang YC, McPherson K, Marsh T, Gortmaker SL, Brown M (2011) Health and economic burden of the projected obesity trends in the USA and the UK. Lancet 378 (9793):815–825

WHO (2020). https://www.who.int/news-room/fact-sheets/detail/obesity-and-overweight.

Wilson PW, D'Agostino RB, Sullivan L, Parise H, Kannel WB (2002) Overweight and obesity as determinants of cardiovascular risk: the Framingham experience. Arch Intern Med 162(16):1867–1872

Wortley KE, Anderson KD, Garcia K, Murray JD, Malinova L, Liu R et al (2004) Genetic deletion of ghrelin does not decrease food intake but influences metabolic fuel preference. Proc Natl Acad Sci U S A 101(21):8227–8232

Wortley KE, del Rincon JP, Murray JD, Garcia K, Iida K, Thorner MO et al (2005) Absence of ghrelin protects against early-onset obesity. J Clin Invest 115 (12):3573–3578

Yang J, Brown MS, Liang G, Grishin NV, Goldstein JL (2008) Identification of the acyltransferase that octanoylates ghrelin, an appetite-stimulating peptide hormone. Cell 132(3):387–396

Zigman JM, Nakano Y, Coppari R, Balthasar N, Marcus JN, Lee CE et al (2005) Mice lacking ghrelin receptors resist the development of diet-induced obesity. J Clin Invest 115(12):3564–3572

Zorrilla EP, Iwasaki S, Moss JA, Chang J, Otsuji J, Inoue K et al (2006) Vaccination against weight gain. Proc Natl Acad Sci U S A 103(35):13226–13231

Part III

Associated Disorders and Obesity Paradoxes

Metabolic Syndrome Update 23

Uzma Zafar

Contents

Introduction 306

Diagnosis: Conflicts and Regional Criteria 306

Worldwide Growth of Metabolic Syndrome 306

Involved Organs and Pathways 308
Mechanism of Insulin Action 308
Insulin Resistance and Molecular Events 309
Insulin Resistance and Cardiovascular Disease 309
Skeletal Muscle IR 310
Hepatic IR 311
Adipose Tissue IR 312
Lipotoxicity and Glucotoxicity 312
Role of Inflammation in Met S 313
Proinflammatory Cytokines 314
Thrombotic Mediators and Prooxidants Associated with Metabolic Syndrome 315

Is the Gut Microbiome Involved? 315
Microbe and Host Cell Cross Talk 315

Genetic Links Determine Susceptibility to IR and Met S 317
Nutritional Genomics and Epigenetics in Met S 317

Possibilities of Prevention and Therapy 317

Ongoing Studies and Future Perspectives 318

References 318

Abstract

Metabolic syndrome (Met S) has emerged as one of the major public health concerns of the modern times. Identification of this condition is based upon existence of certain features of clinical and metabolic disarray such as central obesity, hypertension, elevated plasma glucose, and dyslipidemias. Pathogenesis of Met S is an interplay of genomic and acquired factors that alter pathways of insulin action, along with inflammation, pro-oxidative, and

U. Zafar (✉)
Lahore Medical and Dental College, Lahore, Pakistan

Department of Physiology & Cell Biology, University of Health Sciences, Lahore, Pakistan
e-mail: Uzma.zargham@gmail.com

J. Faintuch and S. Faintuch (eds.), *Obesity and Diabetes*, https://doi.org/10.1007/978-3-030-53370-0_23

thrombotic cascades. Although genetic predisposition determines the susceptibility to various environmental modulators, lifestyle and dietary habits are usually central in the onset and progression of type 2 diabetes mellitus, hypertension, and dyslipidemias. Nutritional genetics might help in understanding the effects of different food constituents on gene regulation resulting in diet-related metabolic disorders. A healthy gut microbiome is also important in the prevention of type 2 diabetes mellitus, Met S, and cardiovascular diseases.

Keywords

Dyslipidemia · Diabetes mellitus · Cardiovascular disease · Gut Microbiome · Lipotoxicity · Glucotoxicity · Inflammation

Introduction

Metabolic syndrome (Met S) refers to the aggregation of clinical and biochemical derangements that considerably increase the risk of coronary artery disease (CAD), diabetes mellitus (DM), and hepatic steatosis. These include high blood pressure, central obesity, dyslipidemias, and impaired blood glucose levels (Roberts et al. 2013). Met S was described for the first time by the Swedish physician Eskil Kylin in 1923 as an association of gout with impaired glycemic indices, obesity, and hypertension (Nilsson 2001). The hypothesis of insulin resistance (IR) as a core defect in Met S was forwarded by Gerald Reaven in 1988 when he delivered Banting lecture at the American Diabetes Association meeting and named this entity as Syndrome X. In the following years, Met S was described as "the deadly quartet" and "a secret killer" and besides the concept of IR, central or upper body obesity independent of whole body obesity was regarded to be a leading risk factor for Met S (Kaplan 1989; Foster 1989). Later on, various other studies also confirmed the association of visceral obesity with Met S and the related traits (Katzmarzyk et al. 2006).

Diagnosis: Conflicts and Regional Criteria

In 2005, the International Diabetes Federation (IDF) emphasized upon measurement of waist circumference as a simple screening method for assessment of IR and formulated its guidelines for the identification of Met S. These included abdominal obesity as a mandatory component, along with the presence of any two of the following features: impaired fasting glucose (IFG) or impaired glucose tolerance (IGT), hypertension, low HDL-c, and high triglycerides. It was observed that some ethnic groups showed features of Met S and IR at lower waist circumference levels (Roberts et al. 2013). In 2005, American Heart Association and National Heart Lung and Blood Institute (AHA/NHLBI) reconciled the different diagnostic criteria of Met S. Their recommendations showed difference from IDF guidelines regarding the consideration of waist circumference as a mandatory component of Met S but agreed upon the other four points of IDF definition. Finally, IDF and AHA/NHLBI conceded upon a common criterion. It required three out of the following five factors to be present: ethnic-specific criterion of waist circumference, IFG or IGT, hypertension, low HDL-c, and high triglycerides. These and other criteria for Met S are summarized in Table 23.1, and guidelines for waist circumference can be found in Table 23.2 (Alberti et al. 2009; Zafar et al. 2018; Aguilar-Salinas and Viveros-Ruiz 2019).

Worldwide Growth of Metabolic Syndrome

About 25% of the adult population in most of the countries is suffering from this metabolic derangement (Ranasinghe et al. 2017). Prevalence of obesity and T2DM is in parallel, although not synonymous with Met S (O'neill and O'driscoll 2015). In accordance with the revised guidelines of IDF, prevalence of Met S

Table 23.1 Diagnostic guidelines for metabolic syndrome

WHO, 1998	EGIR, 1999	NCEP: ATPIII, 2001	IDF, 2006	Common statement (IDF-NHBLI/ AHA 2009)
Evidence for IR (High blood insulin levels) along with any two of the successive features:	Above 75th percentile or top 25% of the fasting insulin values among non-diabetics with any two of the following	Any three or more of the following	[a]Central obesity (a mandatory component); established by ethnic/racial, specific WC plus any two of the following	Any three of the following five features
Abdominal/central obesity: W:H more or equal to; 0.9 in men 0.85 in women or BMI more than or equal to 30 kg/m^2	Abdominal/central obesity WC: more or equal to; 94 cm in men 80 cm in women	Abdominal/central obesity WC more or equal to; 102 cm in men 88 cm in women	Abdominal/central obesity; established by [b]ethnic/racial, specific WC	Abdominal/ central obesity; established by [b]ethnic/racial, specific WC
Triglycerides; more or equal to 150 mg/dl or on lipid-lowering agent	Triglycerides; more or equal to 150 mg/dl or on lipid-lowering agent	Triglycerides; more or equal to 150 mg/dl or on lipid-lowering agent	Triglycerides; more or equal to 150 mg/dl or on lipid-lowering agent	Triglycerides; more or equal to 150 mg/dl or on lipid-lowering agent
HDL-C less or equal to 35 mg/dl in males and 39 mg/dl in females or on lipid-lowering agent	HDL-C less or equal to 39 mg/dl or on lipid-lowering agent	HDL-C less or equal to 40 mg/dl in males and 50 mg/dl in females or on lipid-lowering agent	HDL-C; less than 40 mg/dl for men less than 50 mg/dl for women or on treatment for dyslipidemias	HDL-C; less than 40 mg/dl for men less than 50 mg/dl for women or on treatment for dyslipidemias
BP more or equal to 140/90 mm Hg or on anti-hypertensive	BP more or equal to 140/90 mm Hg or on anti-hypertensive	BP more or equal to 130/85 or on anti-hypertensive	BP more or equal to 130/85 mm Hg or on anti-hypertensive	BP more or equal to 130/85 mm Hg or on anti-hypertensive
Impaired FPG and GT	FPG; more or equal to 110 mg/dl	FPG; more or equal to 110 mg/dl	FPG; more or equal to 100 mg/dl or on treatment for DM	FPG; more or equal to 100 mg/dl or on treatment for DM
Micro-albuminuria urine albumin more or equal to 20 µg per min	None	None	None	None

[a]If BMI more than 30 kg/m^2, central obesity can be assumed to be present irrespective of the WC measurement. [b]Ethnic-specific WC measurements given in Table 23.2 (Alberti et al. 2009; Zafar et al. 2018; Aguilar-Salinas and Viveros-Ruiz 2019)

List of abbreviations; BMI, body mass index; BP, blood pressure; DM, diabetes Mellitus; EGSIR, European Group for the Study of Insulin Resistance; FPG, fasting plasma glucose; HDL-C, high-density lipoprotein cholesterol; GT, glucose tolerance; IDF, International Diabetes Federation; ATPIII, Adult Treatment Panel; NHLBI/AHA, National Heart Lung and Blood Institute/ American Heart association; WC, waist circumference; WHO, World Health Organization; W:H, waist-to-hip ratio

Table 23.2 Ethnic-based values for waist circumference

Ethnicity/country	Males	Females
Europids	More than or equal to 94 cm	More than or equal to 80 cm
South Asians	More than or equal to 90 cm	More than or equal to 80 cm
Chinese/Japanese	More than or equal to 90 cm	More than or equal to 80 cm

Alberti et al. (2009)

in the United States ranges between 35 and 39%, whereas in Europe, it is reported to be 18–30% (Aguilar et al. 2015). In a meta-analysis in the Asian Pacific region, the lowest burden was found in the Philippines (11.9%) and highest was recorded from Aga Khan institutions (49.0%), Pakistan (Ranasinghe et al. 2017). Prevalence of Met S increases with advancing age in a sex-specific manner: below the age of 40 years about 20% of males and 16% of females, between 40 and 60 years 42% of males and 36% of females, and above 60 years 60% of males and 62% females. Aging is generally accompanied by decrease in lean muscle mass and deposition of abdominal fat; both of these increase IR. Moreover, with the aging process, specific defects in fatty acid oxidation in muscles also occur which further decrease insulin sensitivity (Ervin 2009).

Involved Organs and Pathways

Phenotypic expression of Met S is the result of acquired and genetic factors: (1) insulin resistance, (2) visceral obesity and adipose dysfunction, and (3) oxidative stress, proinflammatory, and prothrombotic state. One of the core defects in Met S is resistance to insulin actions (McCracken et al. 2018). It is associated with accelerated atherosclerotic cardiovascular disease, hyperuricemia/gout, chronic kidney disease, and obstructive sleep apnea.

Mechanism of Insulin Action

Insulin on reaching the target area rapidly interacts with its transmembrane tetrameric tyrosine kinase receptor, having two alpha and two beta domains. Alpha domains are extracellular and exhibit insulin-binding site. Beta domains are extracellular, transmembrane, and intracellular. Binding of insulin with the alpha domain results in auto-phosphorylation of tyrosine molecules in the beta domains at three sites: 1158, 1162, and 1163. This activated insulin receptor-kinase results in further phosphorylation of tyrosine molecules of the insulin receptor substrate (IRS) and acts as a docking site for IRS and other downstream effector proteins such as growth factor receptor-binding-protein 2 (Grb2), shc, and many others.

Insulin receptor phosphorylation and excitation leads to the activation of the following two downstream post-receptor signal transduction pathways: (1) Phosphoinositide kinase 3 (PI3-K)/protein-kinase-B (Akt)/molecular target for rapamycin (mTOR) and (2) ras-raf-mitogen-activated protein kinase (MAPK). PI3-k on phosphorylation by IRS-1 further activates phosphoinositide diphosphate (PIP2) to phosphoinositide triphosphate (PIP3), followed by the Akt activation and its recruitment to the cell membrane (Boucher et al. 2014).This activated Akt is responsible for various metabolic actions of insulin such as translocation of insulin-sensitive glucose transporter 4 (GLUT4) expressed by adipocytes and muscles, and inactivation of the enzyme glycogen synthase kinase (GSK), making it incapable of inhibiting the enzyme glycogen synthase, thus promoting conversion of glucose to glycogen. GSK is also a potent inhibitor of eukaryotic initiation factor (eIF) 2B required for the synthesis of cellular proteins. This process is reversed by the insulin-mediated activation of Akt, thereby increasing the production of proteins (Fig. 23.1). PI3-K/Akt pathway is responsible for various cellular

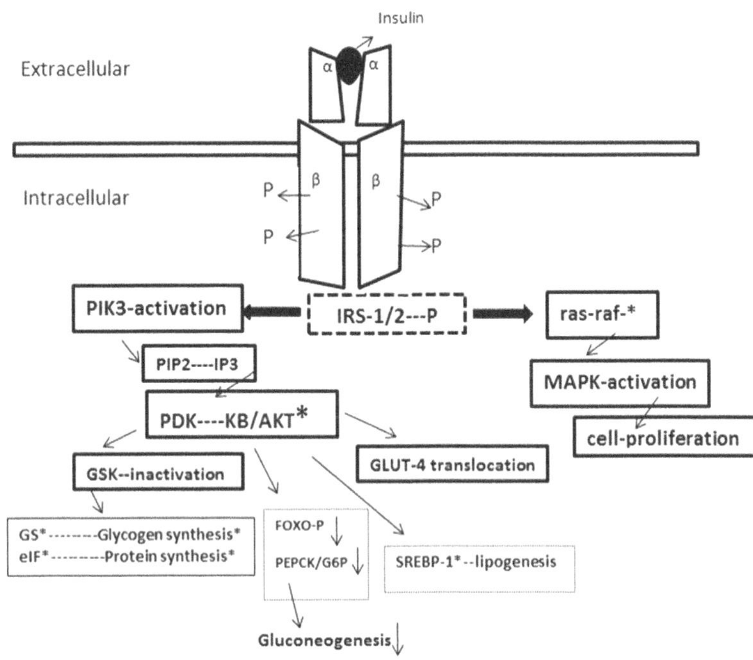

Fig. 23.1 Interaction of insulin with its receptor and post receptor events. *GLUT 4* glucose transporter 4, *PIK3* Phosphoinositide kinase 3, *PIP2–IP3* Phosphoinositide diphosphate—inositidetriphosphate, *MAPk* mitogen activated protein kinase, *PDK—PKB/AKT** phosphoinositide dependent kinase B or AkT activation, *GSK* glycogen synthase kinase, *GS** glycogen synthase activation, *eIF** eukaryotic initiation factor activation, *FOXO-P* forkhead box protein O phosphorylation, *PEPCK/G6P* phosphoenolpyruvate-carboxykinase/glucose6phosphatase, *SREBP* sterol regulatory element binding protein, *IRS-1/2* insulin receptor substrate 1/2

metabolic responses to insulin, while ras-raf-MAPK is responsible for gene expression and mitogenesis (Draznin 2010).

Insulin Resistance and Molecular Events

IR is described as impairment of insulin-mediated actions at its target areas such as skeletal muscle, liver, and adipocytes. In addition to these classical insulin responsive areas, IR also occurs in brain, gastrointestinal tract, pancreatic beta cells, and vascular endothelial cells (Nigi et al. 2018). There is decrease in tissue response to insulin-stimulated glucose uptake and its utilization, impaired glycogen synthesis, and inability to suppress lipid hydrolysis (Ormazabal et al. 2018). In T2DM and Met S, IR is mainly attributed to post-receptor signal transduction events. Instead of tyrosine, there is serine phosphorylation of insulin receptor substrate-1 (IRS-1) that leads to the decreased association of p-85KDA subunit of PI3-K with IRS-1, resulting in diminished activity of the enzyme PI3-K and impaired activation of the downstream signaling cascade.

This altered association of PI3-K with IRS-1 is highly correlated with decreased insulin-stimulated disposal of glucose by the skeletal muscles and also reduced activity of the glycogen synthase. There is also impaired PI3-K gene regulation in the skeletal muscles and similar derangement in adipose tissue of type2 diabetics. In 80–90% of type 2 diabetics, impaired phosphorylation of IRS-1 and PI3-K has been observed, despite of normal insulin receptor tyrosine phosphorylation (Wilcox 2005).

Insulin Resistance and Cardiovascular Disease

Insulin stimulates endothelial nitric oxide synthase (NOS) through PI3-K/Akt pathway, which in turn leads to the increased production

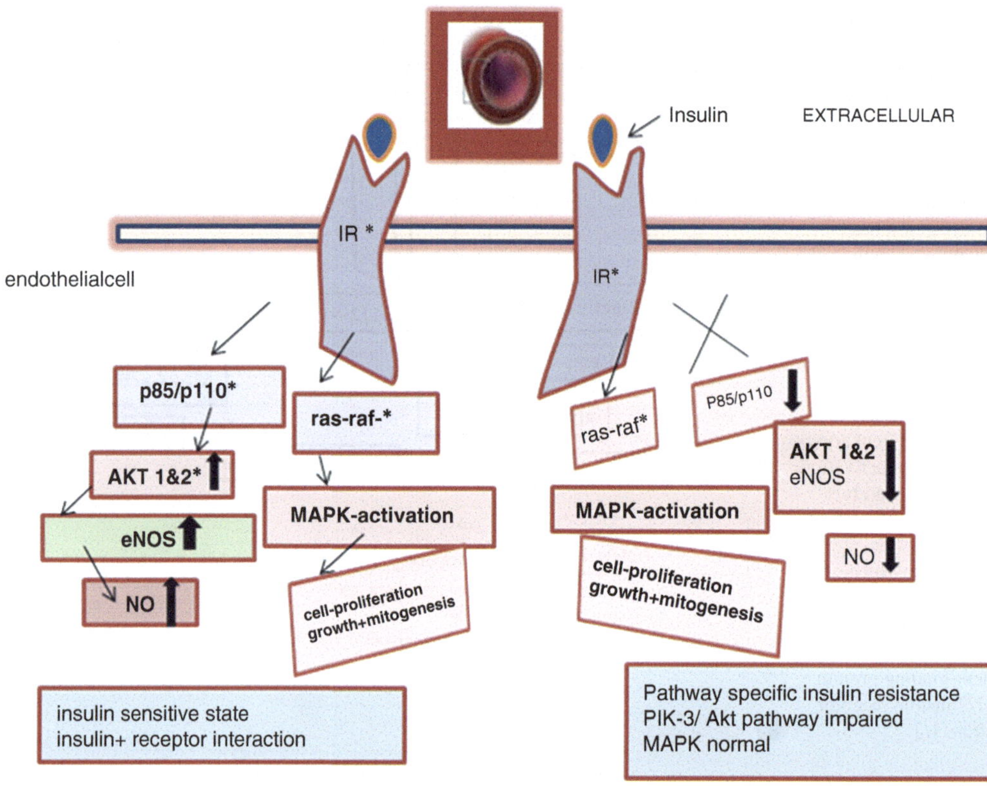

Fig. 23.2 Pathway dependent insulin resistance in endothelium. *eNOS* endothelial Nitric oxide synthase, *IR* insulin receptor, *MAPk* mitogen activated protein kinase, *NO* nitric oxide, *PIK3* Phosphoinositide kinase

of NO in the blood vessels. Insulin also activates MAPK pathway, thus promoting vascular smooth muscle proliferation. In states of IR, there is impairment of PI3-K activation while MAPK pathway remains intact (Fig. 23.2). This results in the attenuation of NO-mediated inflammation buffering actions, while increased MAPK-stimulated cellular proliferation sets the stage for endothelial dysfunction and vascular disease (Fulton 2009).

Skeletal Muscle IR

Skeletal tissue is the major area for the removal of excessive blood glucose. In IR and T2DM, insulin-stimulated uptake of glucose is reduced through GLUT4 as there is uncoupling of IRS-1 from insulin receptor, due to the phosphorylation of threonine/serine residues of IRS-1 (Fig. 23.3). Intracellular glucose utilization and synthesis of glycogen are also markedly compromised (DeFronzo and Tripathy 2009; Ruiz-Alcaraz et al. 2013). The exact cause of the skeletal tissue IR is not known.

Previous studies have linked this insulin-resistant state to the accumulation of intra-myocellular lipid metabolites such as diacylglycerol (DAG) and ceramides. Increased deposition of fat within the skeletal muscles could be a consequence of the excess supply of free fatty acids, or reduced rate of lipid oxidation within the muscles. Various studies have revealed intrinsic defects in mitochondrial capacity to oxidize fatty acids independent of the plasma free fatty acid levels, in T2DM and obesity. Insulin-resistant normo-glycemic offspring of type 2 diabetics are unable to increase the

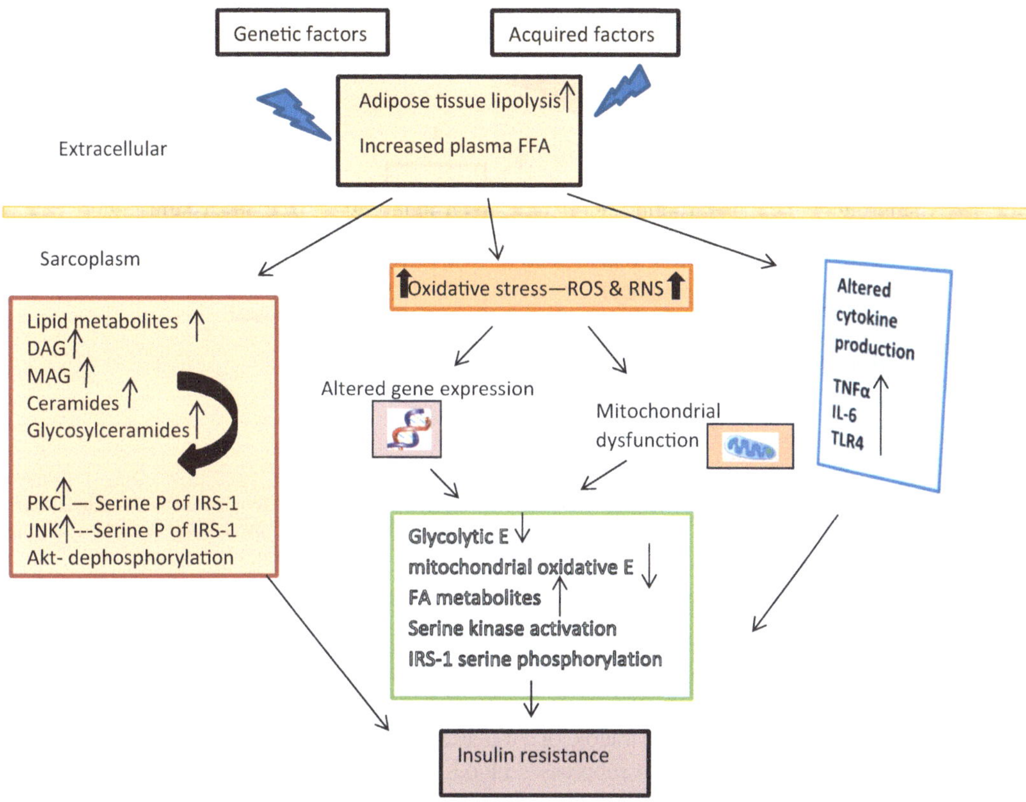

Fig. 23.3 Mechanisms linking lipotoxicity with skeletal insulin resistance. DAG (diacylglycerol); E (enzyme); FFA (free fatty acids); IL-6 (interleukin 6); IRS-1(insulin receptor substrate 1); JNK (janus kinase); MAG (monoacylglycerol); PKC (protein kinase C); RNS (reactive nitrogen species); ROS (reactive oxygen species); TLR4 (toll like receptor 4); TNFα (tumor necrosis factor alpha)

mitochondrial oxidative ability. Spectroscopic and biochemical studies provide evidence for the impaired activity of electron transport chain, decreased activity of mitochondrial enzymes and rate of synthesis of ATP (Abdul-Ghani and DeFronzo 2010).

Hepatic IR

Central obesity, adipose dysfunction, and muscle IR result in the diversion of metabolites toward the liver and accumulation of DAG in the hepatocytes. Lipolysis in the adipose organs is an important source of circulating free fatty acids that promote hepatic IR and steatosis. Non-alcoholic fatty liver disease (NAFLD) occurs when lipid transport to the liver exceeds its capacity to oxidize or export the lipids. Hepatic inflammation is the outcome; however, it can also be one of the causes of IR, especially in the presence of gut dysbiosis and lipopolysaccharide translocation. Due to accumulation of DAG in the hepatocytes, there is activation and translocation of protein kinase C to the cell membrane that results in altered phosphorylation of IRS-2, and diminished activation of PI3-K and Akt. Transcriptional activity of the enzymes involved in gluconeogenesis, i.e., glucose-6-phosphatase and PEPCK gene promoters is regulated by the forkhead box protein O (FOXO) 1 and 3 class of transcription factors (Fig. 23.4).

These factors are inactivated and phosphorylated by the insulin responsive Akt. In IR, due to impaired Akt activation, there is decreased FOXO-1 factor phosphorylation. As a

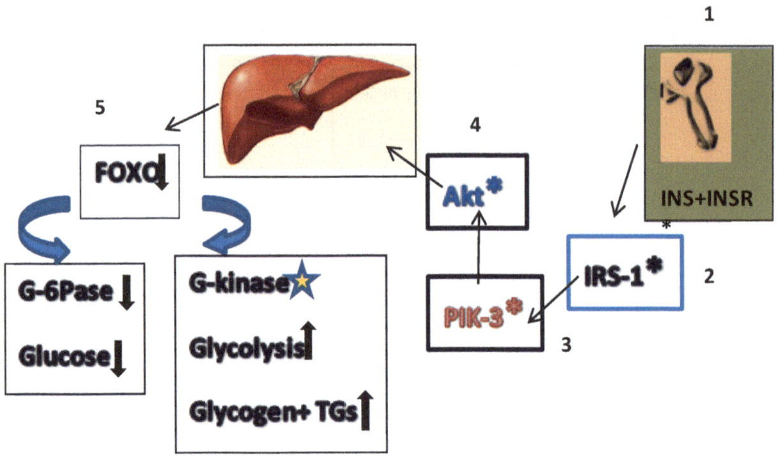

Fig. 23.4 Regulation of FOXO transcription factors by insulin. *AKT** protein kinase B activation, *FOXO* forkhead-box-protein-O inhibited, *G6Pase* glucose 6 phosphatase, *GKinase* Glucokinase activated, *INS+INSR** insulin + insulin receptor substrate activation, *PIK-3** phosphoinositide 3 kinase activation, *TGs* triglycerides

result, this transcription factor enters the nucleus and switches on the activity of the abovementioned rate determining enzymes, resulting in overt hyperglycemia. A paradox of selective IR is the inability of insulin to suppress hepatic glucose production, whereas its capability to activate lipogenesis and synthesis of triglycerides remains intact (Perry et al. 2014; Sparks et al. 2012). In IR and type 2 DM, there is hypersecretion of very low-density lipoproteins (VLDL) by the liver due to accelerated synthesis of Apo B, and increased hepatic influx of free fatty acids released from the adipose tissue. Persistent nuclear localization of FOXO-1 transcription factor also increases the expression of microsomal transfer protein (MTP) and apoC-111 by the hepatocytes (Samuel and Shulman 2016). There is increased hepatic production of inflammatory mediators such as plasminogen activator-inhibitor (PA-I) and C-reactive protein (CRP). These metabolic abnormalities contribute to the process of atherosclerosis (Rehman and Akash 2016).

Adipose Tissue IR

Adipose tissue influences lipid and glucose metabolism by the release of FFA, adipokines, and other inflammation provoking factors. Insulin acts on adipocytes by (1) stimulating uptake of glucose and synthesis of triglycerides, (2) suppressing breakdown of triglycerides and release of free fatty acids and glycerol into the blood. In adipose tissue IR, there is impaired suppression of triglyceride hydrolysis with increased release of FFA into the blood that results in altered skeletal muscle insulin signaling, glucose intolerance, uncontrolled hepatic gluconeogenesis and increased triglyceride synthesis (Gastaldelli et al. 2017). In states of central obesity or insulin resistance infiltration of white adipose tissue (WAT) with macrophages promotes lipolysis, resulting in increased release and delivery of FFA to the liver. This further enhances synthesis of triglycerides by the liver and aggravates hyperlipidemia. Moreover, increased delivery of FFA to the liver also accelerates hepatic gluconeogenesis that further worsens IR (Samuel and Shulman 2016).

Lipotoxicity and Glucotoxicity

In IR and obesity, insulin-mediated inhibition of lipolysis in the adipocytes is impaired which results in the increased release of free fatty acid in the blood. Persistent rise in plasma free fatty acids leads to their deposition in non-adipose

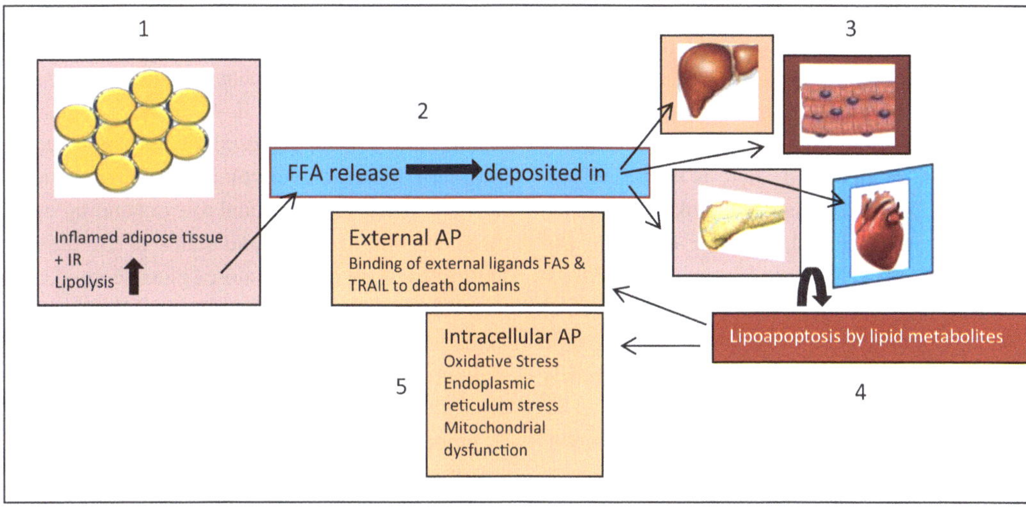

Fig. 23.5 Lipoapoptosis mediated by lipid metabolites. *AP* apoptotic pathway, *FFA* free fatty acids, *IR* insulin resistance

tissues (ectopic fat). The detrimental effect of lipid metabolites in skeletal muscles, liver, kidney, pancreas, and heart is one of the forms of "lipotoxicity" (Del Prato 2009) whereas "glucotoxicity" implies the injurious effects of glucose on the cells (Mota et al. 2016). In chronic hyperglycemia, due to increased glucose accumulation in the cells equilibrium between the defensive anti-oxidative mechanisms and synthesis of reactive oxygen species (ROS) is deranged. Lipotoxicity and glucotoxicity induce metabolic stress through diverse pathways such as ROS-mediated caspase activation followed by apoptosis, endoplasmic reticulum stress, insult of the mitochondrial oxidative pathways, alteration of insulin signaling by modification of protein kinase C, and dysregulation of renin-angiotensin cycle in the tissues. ROS-mediated alteration of cellular proteins (Fig. 23.5) results in various vascular complications of diabetes mellitus (Berdja et al. 2016).

Role of Inflammation in Met S

In obesity, excessive fat deposition in peri-visceral and omental areas is closely associated with chronic inflammation and immune response; however obesity-related inflammation occurs in other organs as well, i.e., liver, skeletal muscles, pancreas and brain (Saltiel and Olefsky 2017). Adipose tissue hypoxia in hypertrophied adipose organs can initiate inflammation by the activation of hypoxia inducible factor-1 gene and upregulation of various other pro-inflammatory genes (Lee et al. 2014). Another proposed mechanism is that in obesity intestinal permeability is increased that results in higher blood levels of lipopolysaccharide (LPS) from intestinal bacteria. LPS can trigger inflammation by activation of toll-like receptor 4 (TLR) and other pattern recognition receptors located in the adipocytes and the liver (Amar et al. 2011). Moreover, increased circulating levels of various lipid-derived factors such as free fatty acids can initiate inflammation by binding with TLR 4 and 2, that trigger activation of nuclear factor kappa B and janus kinase 1.

These pathways on activation result in the increased production of chemokines such as MCP-1 (monocyte chemoattractant protein-1) by the adipocytes and the liver, which lead to the infiltration of these organs with the inflammatory macrophages.

In the lean adipose tissue, 10% of the total cell mass is constituted by macrophages while in obese fat the count is increased up to 40%.

These resident macrophages are categorized as M1 (classically activated) and M2 (alternatively activated) macrophages. M2 secrete insulin-sensitizing adipokines and anti-inflammatory factor as IL-10 and arginase (Saltiel and Olefsky 2017). On the other hand, there is increased production of pro-inflammatory cytokines such as tumor necrosis factor alpha (TNF-α), IL-6, and IL-12 by the M1 (Navarro-Gonzalez and Mora-Fernández 2008).

In the obese adipose tissue, the M1 to M2 ratio is considerably increased and M1 is intimately linked with inflammation and IR (Castoldi et al. 2016). In conclusion, inflammation follows obesity and it plays a significant role in the onset of IR, defective insulin secretion, and disruption of metabolic homeostasis (Saltiel and Olefsky 2017).

Adipose tissue a multifunctional metabolically active endocrine entity is involved in the production of a wide range of bioactive molecules such as cytokines, prostaglandins, hormones, and growth factors (Schmidt et al. 2015). In central obesity due to infiltration of visceral AT with M1 as compared to the M2 macrophages, secretion of proinflammatory cytokines such as TNF-α and IL-6 is increased whereas production of anti-inflammatory cytokines is suppressed.

Proinflammatory Cytokines

TNF-α was the first cytokine found to be positively associated with IR and obesity (Makki et al. 2013). TNF-α directly interferes with insulin signal transduction pathways by serine phosphorylation of IRS-1. It induces IR indirectly by alteration of adipocyte maturation and lipid homeostasis. TNF-α also aggravates the immune response by increasing the production of IL-6, a pro-inflammatory cytokine, and suppressing the expression of adiponectin, an adipose-derived insulin-sensitizing hormone (Makki et al. 2013).

IL-6 a multifarious peptide plays a central role in the regulation of immune mechanisms, inflammation, and host defense response. About one-third of the IL-6 circulating in the blood is derived from skeletal muscles, a small proportion is released from adipocytes, and the rest being originated from macrophages and other cells. Transient increase in IL-6 levels assists in maintaining normal glucose homeostasis whereas long-term increase in circulating IL-6 may lead to IR. IL-6 seems to play a dual role depending upon the metabolic demands and states of the tissues. During exercise IL-6 promotes glucose uptake by the skeletal muscles and increases AMP-mediated oxidation of fatty acids, leading to myogenesis and muscle hypertrophy. However, IL-6 exerts pro-inflammatory effects on adipocytes and liver by impairing insulin receptor and IRS-1 signaling, due to increased activation of SOCs3 (suppressor of cytokine signaling 3) (Senn et al. 2003). It worsens IR by inhibiting adipogenesis and differentiation of the stem cells to their mature state, dysregulating fatty acid metabolism, and downregulating the expression and production of adiponectin by the adipocytes (Pricola et al. 2009).

Leptin is an adipose-derived adipokine which under physiological states suppresses appetite, increases sympathetic activity and energy expenditure, and modulates neuroendocrine activity, by acting on the leptin receptor (ObRb) in the hypothalamus. Leptin exerts its insulin-sensitizing effects by promoting glucose uptake and fatty acid oxidation in the skeletal muscles (Mantzoros et al. 2011). However, elevated leptin levels are associated with abdominal obesity, increased risk of cardiovascular complications, and insulin resistance, being a biomarker for Met S (Li et al. 2011). In obese subjects, despite the high leptin levels, its effect as anorexic hormone is reduced due to leptin resistance that results from defects in molecular mechanisms associated with leptin receptor signaling, or decreased permeability of leptin through the blood-brain barrier (Gruzdeva et al. 2019).

Adiponectin, another adipose-derived hormone, acts as an intrinsic insulin sensitizer. It inhibits gluconeogenesis in the liver and promotes fatty acid oxidation and glucose uptake by the skeletal muscles. Adiponectin also reduces plasma concentration of low-density lipoprotein and apolipoprotein B, and low blood adiponectin levels are found to be associated with hypertension. Adiponectin expression is suppressed by the pro-inflammatory cytokines, implying that inflammation is an important factor leading to hypo-adiponectinemia in obese and insulin-resistant states (Zhang et al. 2015). Most of the adipose-derived factors with the exception of adiponectin and adipsin show a positive association with total adipose mass and insulin resistance. The altered adipokine profile results in profound changes in insulin sensitivity, metabolic health, and energy homeostasis (Kumari et al. 2019).

Thrombotic Mediators and Prooxidants Associated with Metabolic Syndrome

Plasma thrombotic factors apo111 and CRP, uric acid, and ferritin levels also deserve attention (Thaman and Arora 2013). Plasminogen activator inhibitor 1 (PAI-1) is released from adipocytes, endothelial cells, and vascular smooth muscles. In insulin-resistant states, its secretion is increased due to various prooxidant and inflammatory factors. Increased circulating PAI-1 levels are intimately associated with deep venous thrombosis and myocardial infarction. Uric acid, an end-product of purine metabolism, is predominantly produced by the liver (McGillicuddy et al. 2009). In conditions of inflammation and IR, its levels in the blood represent oxidative tissue damage. Various studies have demonstrated that serum uric acid levels significantly correlate with Met S-related traits and associated future comorbidities, such as stroke and coronary artery disease (CAD). In the extracellular environment, such as blood vessels, uric acid acts as a powerful antioxidant by scavenging free oxygen species and chelating metals, whereas inside the cells, it exhibits prooxidant properties. It accelerates lipid peroxidation that decreases cholesterol transport, leading to inflammation. It induces vascular damage by increasing TNF-α secretion and decreasing NO-mediated vasodilation (McGillicuddy et al. 2009).

Is the Gut Microbiome Involved?

Various studies have reported the significant contribution of gut microbiome as an environmental factor toward the pathogenesis of IR and adiposity (Kho and Lal 2018). There is colonizing of the mucosal lining of human body by distinct microbial ecosystems (Arora and Bäckhed 2016). In addition to bacteria, these include viruses, phages, fungi, and yeasts. Dietary factors can modify the gene expression of gut microbes, that in turn influence host metabolism. Different macronutrients present in the food serve as substrates for the gut bacteria and results in the production of short-chain fatty acids (SCFAs) and bile acid metabolites. The SCFAs contributed by the gut microbial population include butyrate, propionate, acetate, and hexanoate. SCFAs increase the gut glucose absorption by enhancing the expression of sodium-glucose transporters in the enterocytes (Fig. 23.6).

These bacterial products also induce the release of incretins, leptin, and peptide YY by the enteroendocrine cells that affect satiety and modulate host energy homeostasis (Festi et al. 2014). SCFAs influence host lipid metabolism by promoting lipogenesis and inhibiting free fatty acid oxidation. Besides the synthesis of SCFAs, gut microbiota also synthesize micronutrients such as vitamin K1, K2, B5, and B12. Gut microbes also play an essential role in the interaction of bile acid metabolites with the host (Kho and Lal 2018).

Microbe and Host Cell Cross Talk

In the gastrointestinal tract, there is constant interaction between the lining epithelial cells and the gut microbiota. The host cells transmit key information from the gut bacteria to the immune cells located in the *lamina propria.* In normal conditions, the gut barrier (tight epithelial junctions, protective thick mucosal lining with various antimicrobial factors, inter-epithelial immune cells, and production of immunoglobulin

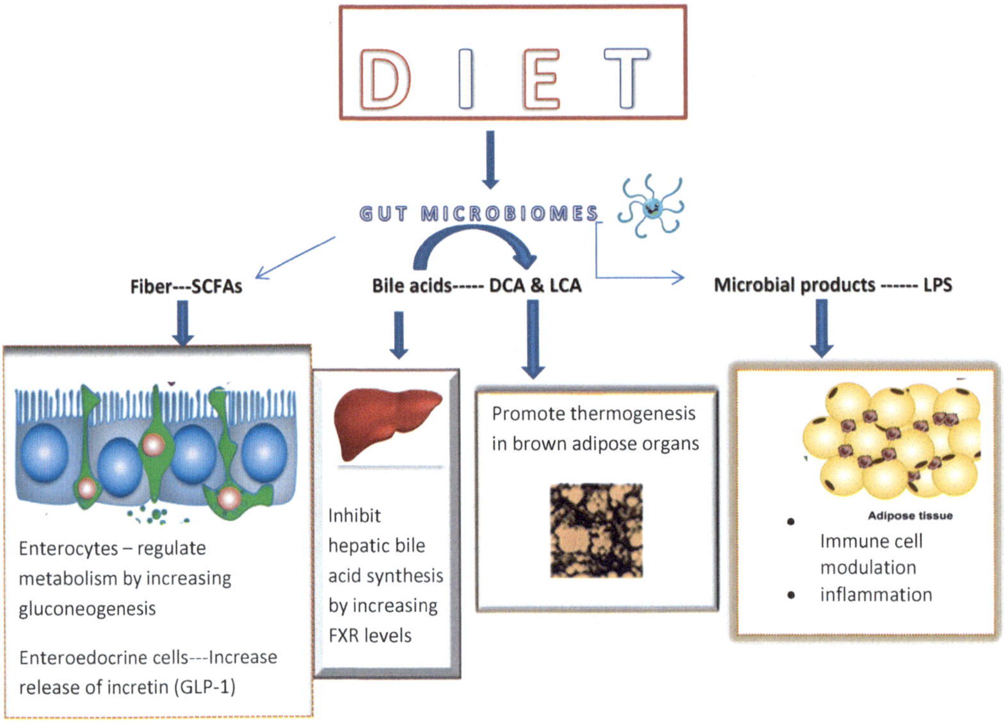

Fig. 23.6 Interaction between diet and gut microbes. *DCA* deoxycholic acid, *FXR* farnesoid × receptor, *GLP-1* glucagon like peptide-1, *LCA* lithocolic acid, *LPS* lipopolysaccharides, *SCFAs* short chain fatty acids

A) prevents translocation of microbes and deleterious metabolites. With bowel inflammation or unbalanced diet, certain bacterial components such as lipopolysaccharides (LPS) gain access through the protective mucosal barrier, and via interaction with the gut innate immune system can initiate low-grade inflammation and insulin resistance. The condition of increased gut microbe-derived lipopolysaccharides in the blood is termed as metabolic endotoxemia.

Immune Activation in Dysbiosis

The cells of innate immune system have various microbe monitoring sites such as toll-like (TOLs) receptors and nucleotide-binding oligomerization domain (NODs) receptors. TOLs are transmembrane proteins in the lysosomal compartment, whereas NODs are intracytosolic. These receptors can detect pathogen linked or danger linked molecular patterns from the microbes such as lipopolysaccharides, peptidoglycans, and lipoteichoic acid (Vidal 2019). In T2DM, there is reduction in butyrate-producing microbiome, whereas *Lactobacillus sp.* levels are increased (Arora and Bäckhed 2016). The microbe-derived product trimethylamine (TMA) plays a potential role in the development of atherosclerosis and cardiovascular diseases (CVD). TMA is readily converted to TMAO (trimethylamine-N-oxide) by the liver, and it is a biomarker for atherosclerotic lesions and CVD. Metagenomic studies have revealed that certain bacterial taxa observed in atherosclerotic plaques have also been found in the gut and oral cavity of the same subjects. This highlights the possibility that the gut microbiota may be the source of bacteria in the atherosclerotic vascular areas and in turn contribute to the pathogenesis of CVD.

Genetic Links Determine Susceptibility to IR and Met S

Met S and its heterogenous features have polymorphic genetic etiology (Ziki and Mani 2016).

The role of genetics has been estimated at between 44 and 65% for T2DM and dyslipidemias, and it is also very high for the other features of Met S, i.e., obesity and hypertension (Brunetti et al. 2014). Many of the genetic loci identified by candidate gene analysis and GWAS are linked with beta cell function, insulin secretion, insulin resistance, and lipid or energy homeostasis (Brown and Walker 2016). There is significant role of genetic pool and familial predisposition in the age of onset of Met S and its associated components, progression to the complications and response to the pharmacotherapy (Roomi et al. 2014).

Nutritional Genomics and Epigenetics in Met S

Nutritional genetics is a combination of nutrigenomics and nutrigenetics. Nutrigenomics deal with the effects of different food constituents on gene regulation and expression, while nutrigenetics is the study of individual variations regarding the effect of food constituents on gene expression. Nutritional genetics consider food as a major factor that determines the genetic response through various epigenetic mechanisms such as DNA methylation, histone remodeling, and micro RNA regulation. Stress, lack of exercise, and concomitant endocrine mechanisms also modify gene expression through molecular epigenetic pathways (Kuneš et al. 2015).

Possibilities of Prevention and Therapy

Dietary care in accordance with ATP111 (Alberti et al. 2009; Zafar et al. 2018) recommends to avoid or reduce the consumption of cholesterol, trans, and saturated fat, reduce the intake of refined carbohydrates, and increase the consumption of fruits, vegetables, and whole grains. There is an emergent concept of personalized nutrition which emphasizes that individuals vary in their response to macronutrients depending upon the genotype, microbiome, immune system, and phenotypic profile (Ludwig et al. 2018). Weight reduction (7–10% body weight over 6–12 months) is also widely adopted (Ley et al. 2006).

In order to maintain the desired weight, regular physical activity (30 min of moderate activity daily) cannot be overemphasized (Paley and Johnson 2018). Common suggestions include (1) introduction of shorts bouts of physical activity such as 10–15 min of brisk walking, jogging, cycling, or using treadmill and (2) reducing screen time and other sedentary leisure modalities (Naugle et al. 2014).

Use of statins in combination with fibrates especially fenofibrate is attractive to achieve the ATP111 target for LDL as well as HDL-cholesterol. Fenofibrate unlike gemfibrozil does not interfere with the statin metabolism in the liver so chance of myopathy is considerably less.

Another goal is to keep blood pressure (BP) < 120/80. It is preferable to use angiotensin I blockers or angiotensin receptor blockers for the control of BP, as some clinical trials recommend that they have advantages over the other drugs in DM. However, the majority of the studies suggest that risk reduction is with lowering the BP alone irrespective of the drug advised (Abraham et al. 2015; January et al. 2019).

There is an emerging interest in the possibility that lowering IR with metformin also delays the onset of DM and associated comorbidities such as cardiovascular complications in insulin-resistant states. Other considerations include sodium-glucose co-transporter inhibitors and incretin receptor agonists. Good glycemic control by keeping the HbA1c < 7% can delay the onset of cardiovascular complications. Choice of hypoglycemic agent if needed, beyond the lifestyle therapy depends upon the clinical assessment (Chaudhury et al. 2017).

Ongoing Studies and Future Perspectives

- Better recognition of metabolic and genetic factors contributing to the evolvement of IR and associated morbidities. Studies are going on to determine the potential therapeutic role of gene editing tool CRISPR/Cas9 in inflammatory and various immune-mediated diseases (Rodríguez-Rodríguez et al. 2019).
- Bariatric surgery and modern incisionless endoscopic alternatives are endowed with a conspicuously beneficial impact on Met S and its individual components. Despite relatively limited access in some countries, and eventual risks, they should be considered in patients refractory to lifestyle and pharmacologic interventions.

References

Abdul-Ghani MA, DeFronzo RA (2010) Pathogenesis of insulin resistance in skeletal muscle. Biomed Res Int 2010:476279

Abraham HMA, White CM, White WB (2015) The comparative efficacy and safety of the angiotensin receptor blockers in the management of hypertension and other cardiovascular diseases. Drug Saf 38:33–54

Aguilar-Salinas CA, Viveros-Ruiz T (2019) Recent advances in managing/understanding the metabolic syndrome. F1000Res 8:F1000

Aguilar M, Bhuket T, Torres S et al (2015) Prevalence of the metabolic syndrome in the United States, 2003–2012. JAMA 313:1973–1974

Alberti KG, Eckel RH, Grundy SM et al (2009) Harmonizing the metabolic syndrome: a joint interim statement of the international diabetes federation task force on epidemiology and prevention; national heart, lung, and blood institute; American heart association; world heart federation; international atherosclerosis society; and international association for the study of obesity. Circulation 120:1640–1645

Amar J, Chabo C, Waget A et al (2011) Intestinal mucosal adherence and translocation of commensal bacteria at the early onset of type 2 diabetes: molecular mechanisms and probiotic treatment. EMBO Molecul Med 3:559–572

Arora T, Bäckhed F (2016) The gut microbiota and metabolic disease: current understanding and future perspectives. J Intern Med 280:339–349

Berdja S, Smail L, Saka B et al (2016) Glucotoxicity induced oxidative stress and inflammation in vivo and in vitro in Psammomys obesus: involvement of aqueous extract of Brassica rapa rapifera. Evid Based Complement Alternat Med 2016:3689208

Boucher J, Kleinridders A, Kahn CR (2014) Insulin receptor signaling in normal and insulin-resistant states. Cold Spring Harb Perspect Biol 6:a009191

Brown AE, Walker M (2016) Genetics of insulin resistance and the metabolic syndrome. Curr Cardiol Rep 18:75

Brunetti A, Chiefari E, Foti D (2014) Recent advances in the molecular genetics of type 2 diabetes mellitus. World J Diabetes 5:128

Castoldi A, Naffah de Souza C, Câmara NOS et al (2016) The macrophage switch in obesity development. Front Immunol 6:637

Chaudhury A, Duvoor C, Dendi R et al (2017) Clinical review of antidiabetic drugs: implications for type 2 diabetes mellitus management. Front Endocrinol 8:6

Coltell O, Asensio EM, Sorlí JV, Barragán R, Fernández-Carrión R, Portolés O, Ortega-Azorín C, Martínez-LaCruz R, González JI, Zanón-Moreno V et al (2019) Genome-wide association study (GWAS) on bilirubin concentrations in subjects with metabolic syndrome: sex-specific GWAS analysis and gene-diet interactions in a mediterranean population. Nutrients 11(1):90

DeFronzo RA, Tripathy D (2009) Skeletal muscle insulin resistance is the primary defect in type 2 diabetes. Diabetes Care 32:S157–S163

Del Prato S (2009) Role of glucotoxicity and lipotoxicity in the pathophysiology of Type 2 diabetes mellitus and emerging treatment strategies. Diabet Med 26:1185–1192

Draznin B (2010) Mitogenic action of insulin: friend, foe or 'frenemy'? Diabetologia 53:229

Ervin RB (2009) Prevalence of metabolic syndrome among adults 20 years of age and over, by sex, age, race and ethnicity, and body mass index; United States, 2003–2006. Natl Health Stat Rep 5:131–138

Festi D, Schiumerini R, Eusebi LH et al (2014) Gut microbiota and metabolic syndrome. World J Gastroenterol: WJG 20:16079

Foster DW (1989) Insulin resistance—A secret killer? N Engl J Med 320:733–734

Fulton DJ (2009) Mechanisms of vascular insulin resistance: a substitute Akt? Circ Res 109:1035–1037

Gastaldelli A, Gaggini M, DeFronzo RA (2017) Role of adipose tissue insulin resistance in the natural history of T2DM: results from the San Antonio Metabolism Study. Diabetes 66:815–822

Gruzdeva O, Borodkina D, Uchasova E et al (2019) Leptin resistance: underlying mechanisms and diagnosis. Diab Meta Synd Obes 12:191

January CT, Wann LS, Calkins H et al (2019) 2019 AHA/ACC/HRS focused update of the 2014 AHA/ACC/HRS guideline for the management of patients with atrial fibrillation: a report of the American College of Cardiology/American Heart Association

Task Force on Clinical Practice Guidelines and the Heart Rhythm Society. J Am Coll Cardiol 74:104–132
Kaplan NM (1989) The deadly quartet: upper-body obesity, glucose intolerance, hypertriglyceridemia, and hypertension. Arch Intern Med 149:1514–1520
Katzmarzyk PT, Janssen I, Ross R et al (2006) The importance of waist circumference in the definition of metabolic syndrome: prospective analyses of mortality in men. Diabetes Care 29:404–409
Kho ZY, Lal SK (2018) The human gut microbiome—a potential controller of wellness and disease. Front Microbiol 9:1835
Kuneš J, Vaněčková I, Mikulášková B et al (2015) Epigenetics and a new look on metabolic syndrome. Physiol Res 64:611–620
Kumari R, Kumar S, Kant R (2019) An update on metabolic syndrome: metabolic risk markers and adipokines in the development of metabolic syndrome. Diabetol Metab Syndr 13(4):2409–2417
Lee YS, J-w K, Osborne O et al (2014) Increased adipocyte O_2 consumption triggers HIF-1α, causing inflammation and insulin resistance in obesity. Cell 157:1339–1352
Ley RE, Turnbaugh PJ, Klein S et al (2006) Microbial ecology: human gut microbes associated with obesity. Nature 444:1022
Li W-C, Hsiao K-Y, Chen I-C et al (2011) Serum leptin is associated with cardiometabolic risk and predicts metabolic syndrome in Taiwanese adults. Cardiovasc Diabetol 10:36
Ludwig DS, Willett WC, Volek JS et al (2018) Dietary fat: From foe to friend? Science 362:764–770
Makki K, Froguel P, Wolowczuk I (2013) Adipose tissue in obesity-related inflammation and insulin resistance: cells, cytokines, and chemokines. ISRN Inflam 2013:139239
Mantzoros CS, Magkos F, Brinkoetter M et al (2011) Leptin in human physiology and pathophysiology. Am J Physiol Endocrinol Metab 301:E567–E584
McCracken E, Monaghan M, Sreenivasan S (2018) Pathophysiology of the metabolic syndrome. Clin Dermatol 36(1):14–20
McGillicuddy FC, de la Llera MM, Hinkle CC et al (2009) Inflammation impairs reverse cholesterol transport in vivo. Circulation 119:1135
Mota M, Banini BA, Cazanave SC et al (2016) Molecular mechanisms of lipotoxicity and glucotoxicity in nonalcoholic fatty liver disease. Metabolism 65:1049–1061
Naugle KE, Naugle KM, Wikstrom EA (2014) Cardiovascular and affective outcomes of active gaming: using the nintendo wii as a cardiovascular training tool. J Stren Cond Res Nat Stren Cond Associ 28:443
Navarro-Gonzalez JF, Mora-Fernández C (2008) The role of inflammatory cytokines in diabetic nephropathy. J Am Soc Nephrol 19:433–442
Nigi L, Grieco GE, Ventriglia G et al (2018) MicroRNAs as regulators of insulin signaling: research updates and potential therapeutic perspectives in Type 2 diabetes. Int J Mol Sci 19:3705
Nilsson S (2001) Research contributions of Eskil Kylin. Sven Med Tidskr 5(1):15–28
O'neill S, O'driscoll L (2015) Metabolic syndrome: a closer look at the growing epidemic and its associated pathologies. Obes Rev 16:1–12
Ormazabal V, Nair S, Elfeky O et al (2018) Association between insulin resistance and the development of cardiovascular disease. Cardiovasc Diabetol 17:1–14
Paley CA, Johnson MI (2018) Abdominal obesity and metabolic syndrome: exercise as medicine? BMC Spor Sci Med Rehabil 10:7
Perry RJ, Samuel VT, Petersen KF et al (2014) The role of hepatic lipids in hepatic insulin resistance and type 2 diabetes. Nature 510:84
Pricola KL, Kuhn NZ, Haleem-Smith H et al (2009) Interleukin-6 maintains bone marrow-derived mesenchymal stem cell stemness by an ERK1/2-dependent mechanism. J Cell Biochem 108:577–588
Ranasinghe P, Mathangasinghe Y, Jayawardena R et al (2017) Prevalence and trends of metabolic syndrome among adults in the asia-pacific region: a systematic review. BMC Public Health 17:101
Rehman K, Akash MSH (2016) Mechanisms of inflammatory responses and development of insulin resistance: how are they interlinked? J Biomed Sci 23:87
Roberts CK, Hevener AL, Barnard RJ (2013) Metabolic syndrome and insulin resistance: underlying causes and modification by exercise training. Compr Physiol 3:1–58
Rodríguez-Rodríguez DR, Ramírez-Solís R, Garza-Elizondo MA et al (2019) Genome editing: a perspective on the application of CRISPR/Cas9 to study human diseases. Int J Mol M 43:1559–1574
Roomi MA, Lone KP, Madassar A (2014) Vitamin D and cardiometabolic risk factors in adult non-diabetic offspring of type 2 diabetic parents. J Pak Med Assoc 64:1229–1234
Ruiz-Alcaraz AJ, Lipina C, Petrie JR et al (2013) Obesity-induced insulin resistance in human skeletal muscle is characterised by defective activation of p42/p44 MAP kinase. PLoS One 8:e56928
Saltiel AR, Olefsky JM (2017) Inflammatory mechanisms linking obesity and metabolic disease. J Clin Invest 127:1–4
Samuel VT, Shulman GI (2016) The pathogenesis of insulin resistance: integrating signaling pathways and substrate flux. J Clin Invest 126:12–22
Schmidt FM, Weschenfelder J, Sander C et al (2015) Inflammatory cytokines in general and central obesity and modulating effects of physical activity. PLoS One 10:e0121971
Senn JJ, Klover PJ, Nowak IA et al (2003) Suppressor of cytokine signaling-3 (SOCS-3), a potential mediator of interleukin-6-dependent insulin resistance in hepatocytes. J Biol Chem 278:13740–13746

Sparks JD, Sparks CE, Adeli K (2012) Selective hepatic insulin resistance, VLDL overproduction, and hypertriglyceridemia. Arterioscler Thromb Vasc Biol 32:2104–2112

Thaman R, Arora G (2013) Metabolic syndrome: definition and pathophysiology—the discussion goes on. J Physiol Pharmacol Advan 3:48–56

Vidal H (2019) Human gut microbiome: hopes, threats and promises. 21st European Congress of Endocrinology. BioScientifica.

Wilcox G (2005) Insulin and insulin resistance. Clin Biochem Rev 26:19

Zafar U, Khaliq S, Ahmad HU et al (2018) Metabolic syndrome: an update on diagnostic criteria, pathogenesis, and genetic links. Hormones 17:299–313

Zhang J, Kho P, Kuliesh O (2015) Adiponectin, resistin and leptin: possible markers of metabolic syndrome. Endocrinol Metab Syndr 4:1–4

Ziki MDA, Mani A (2016) Metabolic syndrome: genetic insights into disease pathogenesis. Curr Opin Lipidol 27:162

Non-alcoholic Fatty Liver Disease: A Global Public Health Issue

24

Eda Kaya and Yusuf Yilmaz

Contents

Worldwide Growth of Non-alcoholic Fatty Liver Disease 322

Correlations with Obesity, Diabetes, and Metabolic Syndrome 323

Obese and Nonobese NAFLD 323

NAFLD and Diabetes 323

The Natural Course of NAFLD 324
Coffee and Alcohol 325

Noninvasive Diagnostic Methods vs. Liver Biopsy 325
Single vs. Multiple Criteria 326
Imaging Tools 326
First-line Liver Fibrosis Assessment 326

Pathophysiology 327
Subsequent Steps 327
The Liver Damage Cascade 327
The Microbiota and the Gut Barrier Function 327
Disease-specific Variants 328

Lifestyle and Pharmacological Interventions 328
Bariatric Interventions 329
Drug Therapy 329

References 330

E. Kaya
Helios Hospital Schleswig, Academical Educational Hospital of Kiel and Lübeck Universities, Schleswig, Germany

Y. Yilmaz (✉)
Department of Gastroenterology, School of Medicine, Marmara University, Istanbul, Turkey

Liver Resarch Unite, Institute of Gastroenterology, Marmara University, Istanbul, Turkey
e-mail: dryusufyilmaz@gmail.com

J. Faintuch and S. Faintuch (eds.), *Obesity and Diabetes*, https://doi.org/10.1007/978-3-030-53370-0_24

Abstract

Non-alcoholic fatty liver disease (NAFLD) is the most common liver disease, with a worldwide prevalence of 25%. NAFLD represents a spectrum of liver disease severity from simple hepatic steatosis to non-alcoholic steatohepatitis, which is a potentially progressive liver condition with ongoing liver injury. It is strongly associated with obesity, insulin resistance, metabolic syndrome, genetic factors, and lifestyle factors. However, the exact pathophysiology of NAFLD remains to be established. Moreover, the diagnosis of NASH also represents a major challenge. The reference standard in the diagnosis of NASH is considered liver biopsy. Due to its invasive nature, there is a major effort to optimize biochemical and imaging methods in terms of replacing liver biopsy. To date, a combination of noninvasive tests based on blood tests with imaging methods is recommended. Although there is no consensus about the pharmacological therapy of NASH, the international guidelines recommend achieving significant weight loss and lifestyle modification. In conclusion, the combination of high prevalence and its high burden warrant further research in that area.

Keywords

Non-alcoholic fatty liver disease · Non-alcoholic steatohepatitis · Fibrosis · Liver failure · Liver transplantation

Worldwide Growth of Non-alcoholic Fatty Liver Disease

Non-alcoholic fatty liver disease (NAFLD) represents an umbrella term of clinical and pathologic entities, which encompasses simple hepatic steatosis, non-alcoholic fatty liver (NAFL), to non-alcoholic steatohepatitis (NASH), which according to compelling evidence is associated with fibrosis, cirrhosis, and, in some cases, hepatocellular carcinoma (HCC), in the absence of significant alcohol ingestion or other secondary causes of hepatic fat accumulation (Chalasani et al. 2018; European Association for the Study of the Liver (EASL); European Association for the Study of Diabetes (EASD); European Association for the Study of Obesity (EASO) 2016; Kaya and Yılmaz 2019a; Yilmaz 2012a).

There are different NAFLD prevalence reported around the world, mainly due to different study populations, but also due to the heterogeneity of the methods used to diagnose NAFLD. There are numerous noninvasive methods such as conventional ultrasonography, computerized tomography, magnetic resonance imaging (MRI), controlled attenuation parameter (CAP) by transient elastography (TE), and MRI-derived proton density fat fraction (PDFF-MRI). Depending on the accuracy, they can lead to over- or under-diagnosis (Kaya and Yılmaz 2019b). In a comprehensive meta-analysis, NAFLD is recognized as the most common chronic liver disease, with a prevalence of 25% worldwide. The highest prevalence was reported from the Middle East and South America, reaching 32% and 30%, respectively, whereas the lowest corresponds to Africa with 13% (Younossi et al. 2016).

From those 25% of the adult world population, approximately 20% (one-fifth) were classified as NASH corresponding to 16.5 million people (Estes et al. 2018a). Considering the future progress, by 2030 NAFLD prevalence was projected at 33.5%. As a reflection of the ageing population and disease progression, NASH was expected to increase by 27%. In this scenario, NASH is predicted to increase the number of liver-related deaths until 2030 by 178%. Therefore, NAFLD represents a significant public health issue worldwide (Sayiner et al. 2019).

NASH, the more progressive subtype of NAFLD, has an increased risk of proceeding to decompensated liver cirrhosis, hepatocellular carcinoma (HCC), and finally liver failure. Moreover, NAFLD has been recognized as a major contributor to the global burden of HCC. In the coming future, considering the trends of increase in NAFLD prevalence, NAFLD is expected to become the leading cause of HCC (Younossi et al. 2019a). In addition to this, NASH is the most rapidly growing aetiology for liver transplantation (Estes et al. 2018b).

Correlations with Obesity, Diabetes, and Metabolic Syndrome

The prevalence of NAFLD has increased in many countries both in the paediatric and adult populations as a result of the spread of sedentary lifestyle and westernized diet (Golabi et al. 2018). NAFLD is mostly associated with increased visceral obesity and metabolic abnormalities such as insulin resistance, diabetes mellitus, and dyslipidaemia. It became increasingly clear that NASH is the hepatic manifestation of the metabolic syndrome. Additionally, the prevalence of NASH is higher among diabetic and obese individuals. The risk of developing NASH was also reported to be in line with an increased number of components of metabolic syndrome (Chalasani et al. 2018; Younossi et al. 2019a). Moreover, being female, age over 50 years, obesity, metabolic syndrome, hypertension, dyslipidaemia, and hypertriglyceridaemia were reported as risk factors for developing NASH (Povsic et al. 2019).

According to a recent meta-analysis, more than 80% of NASH patients were overweight or obese (Chalasani et al. 2018). Obesity is known to increase the risk of developing NAFLD (Younossi et al. 2018; Singh et al. 2015). In fact, the prevalence of NAFLD is directly proportional to the increase in body mass index (BMI) (Younossi et al. 2019b). Given that NAFLD is associated with visceral obesity, the mutual evaluation of BMI and waist circumference can be more accurate in the classification of obesity in terms of NAFLD (Mongraw-Chaffin et al. 2015). NAFLD appears to increase at the same rate as obesity (World Health Organization (WHO) 2020; Fan et al. 2017). Although the prevalence of NAFLD was estimated to be around 25%, it could reach 90% in morbidly obese individuals, making possible even further prevalence elevations of NAFLD and related comorbidities (Younossi et al. 2019b).

Obese and Nonobese NAFLD

The majority of the patients undergoing bariatric surgery due to severe obesity have NAFLD, with a prevalence of approximately 95% (Sasaki et al. 2015; Subichin et al. 2015). On the other hand, a subgroup of patients, so-called lean NAFLD, were also reported to have NAFLD, although the BMI is normal (Akyuz et al. 2015; Yılmaz et al. 2019). As known, overweight is defined as >23 kg/m^2 for Asian and >25 kg/m^2 for Caucasian ethnicities (Consultation WHOE; WHO Expert Consultation 2004). Epidemiological data show a prevalence of 5–26% in Asian and 7–20% in Western populations (Consultation WHOE; WHO Expert Consultation 2004). Recent studies showed that NAFLD without the presence of obesity implies added severity, with increased rates of diabetes mellitus, metabolic syndrome and cardiovascular disorders (Sung et al. 2009).

NAFLD and Diabetes

Type 2 diabetes mellitus appears to be a major driver in the development of NAFLD. Moreover, type 2 diabetes mellitus accelerates the progression of liver disease (Stefan et al. 2019). According to a recent meta-analysis, the global prevalence of NAFLD among patients with type 2 diabetes mellitus was estimated to be over 55%, which corresponds to 4.7% of the global population. The prevalence of NASH among the diabetic population is estimated to be around 37%. In the same study, among the biopsied diabetic patients with NAFLD, the prevalence of advanced fibrosis was reported as 17% (Younossi et al. 2019c). Therefore, noninvasive screening of diabetic patients for both NASH and liver fibrosis is recommended, with the use of available noninvasive tools (Arrese et al. 2019).

Additionally, those patients had higher rates of other metabolic comorbidities. The vast majority of diabetic patients with NAFLD met the criteria of metabolic syndrome. More than half of them had hyperlipidaemia and hypertension; approximately one-fourth had cardiovascular disease

(Kalra et al. 2013; Targher and Byrne 2015; Firneisz 2014). In fact, NAFLD itself is significantly associated with an increased risk of cardiovascular events and cardiovascular-related deaths (Targher et al. 2016).

NAFLD is considered the hepatic manifestation of metabolic syndrome due to its frequent coexistence with metabolic syndrome. However, the presence of metabolic syndrome is not sufficient to explain the coexistence of NAFLD. Because neither all the patients with NAFLD fulfil the criteria of metabolic syndrome nor all metabolic syndrome patients have NAFLD (Yilmaz 2012b). In fact, 20–80% of the patients with NAFLD have sufficient criteria for metabolic syndrome (Green 2003). In a recent community-based study, NAFLD alone was not associated with mortality, but those NAFLD patients with metabolically abnormal profiles had an increased risk of mortality (Niriella et al. 2019).

The Natural Course of NAFLD

NAFLD is associated with increased all-cause mortality. However, isolated hepatic steatosis without hepatic inflammation behaves differently than NASH, exhibiting no increase in liver-related mortality and minimal risk for disease progression. Approximately 80% of the NAFLD patients remain as non-alcoholic fatty liver for life. The remaining 20% develop NASH, and from those, approximately 7% progress to HCC over 6.5 years (Torres DM1, Williams CD, Harrison SA. 2012). The annual development of HCC among NAFLD-associated cirrhosis patients was approximately 3% (Ascha et al. 2010). In another 8 years, 30% of those cirrhotic subjects are estimated to progress to hepatic decompensation (Torres DM1, Williams CD, Harrison SA. 2012). Patients with NASH showed progression to significant fibrosis at a rate of 13% in 17 years, to cirrhosis 2% in 20 years and without NASH to significant fibrosis at a rate of 10% in 19 years, to cirrhosis 3.1% in 7.6 years (Hagström et al. 2017a; Calzadilla Bertot and Adams 2016). The probability of developing cirrhosis is estimated to be annually 6% for F2 fibrosis and 8% for F3 fibrosis (Singh et al. 2015; Argo et al. 2009; Singh et al. 2015). The natural course of NAFLD is schematized in Fig. 24.1.

Based on a meta-analysis with 11 cohort studies including 411 biopsy-proven NAFLD patients, and over 2145.5 person-years of follow-up evaluation, 33.6% of the subjects showed progression in fibrosis, 43.1% remained stable, and 22.3% had improvement regarding fibrosis stage. Progression of fibrosis at stage 1 was determined as 14.3 years for patients with NAFL and 7.1 years with NASH. Among NASH patients, 20% had rapid progression rates in the fibrosis stage (Singh et al. 2015). In a meta-analysis of Dulai et al., fibrosis stage was found to be directly associated with all-cause mortality and liver-related mortality. This increase is more

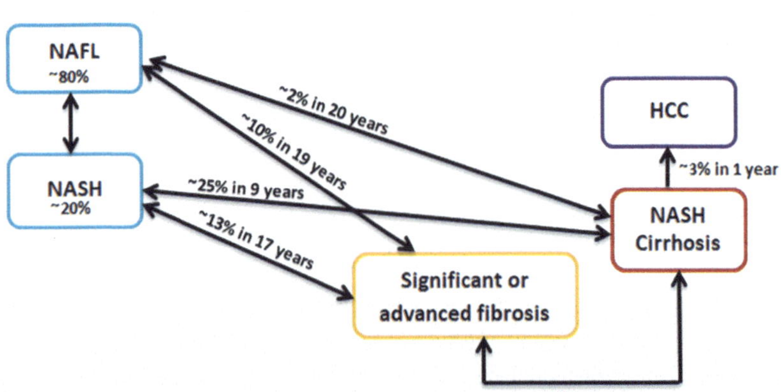

Fig. 24.1 Natural course of NAFLD

pronounced in patients with fibrosis stages of 3–4 (Dulai et al. 2017).

Liver-related mortality is not the leading cause of death among patients with NASH. In decreasing order, the most common aetiologies for mortality are cardiovascular disease, malignancies and then liver-related causes (Francque et al. 2011; Dam-Larsen et al. 2009; Söderberg et al. 2010). Diabetes mellitus, severe insulin resistance, increased BMI, significant weight gain (Adams et al. 2005), and cigarette smoking (Enc et al. 2019) are associated with disease severity. On the other hand, the effect of increased serum transaminases at disease severity remains controversial. Studies have shown that although increased serum aminotransferases indicate the probability of NAFLD, the absence of increased aminotransferases does not exclude NAFLD (Mofrad et al. 2003). NAFLD patients with normal liver enzymes are characterized by a severe metabolic profile, however, similar rates of advanced fibrosis compared to subjects with elevated aminotransferases (Ulasoglu et al. 2019).

Coffee and Alcohol

Coffee consumption is considered a protective factor for NAFLD (Yesil and Yilmaz 2013). A meta-analysis showed a significantly decreased risk of NAFLD among coffee drinkers, including less risk of liver fibrosis among patients with NAFLD, who consumed coffee regularly, which can make coffee a preventive factor for NAFLD and NAFLD-related complications (Wijarnpreecha et al. 2017). There are also studies showing the benefits of low amounts of alcohol consumption. Up to 13 units per week were shown to be associated with a lower stage of fibrosis (Hagström et al. 2017b).

One unit (10 ml ethanol) roughly corresponds to 25 ml of distilled drinks, 80 ml of wine, and 170 ml of beer. However, there is insufficient evidence about the benefits of alcohol consumption. Heavy episodic drinking may accelerate fibrosis progression and increase the risk of hepatocellular carcinoma. Therefore, considering the overlap between the pathophysiological mechanisms of NAFLD and alcoholic fatty liver disease, alcohol consumption is discouraged (Ajmera et al. 2017). In parallel, recent data showed there is no safe level or beneficial level of alcohol consumption. The safe level must be no consumption at all (Burton and Sheron 2018).

Noninvasive Diagnostic Methods vs. Liver Biopsy

Liver biopsy is considered the gold standard in approach for identifying the presence of NASH and a histological classification of the disease in terms of fibrosis. However, due to major limitations such as high cost, sampling errors, patients' discomfort, and the presence of procedure-related morbidity and even mortality, its use in the clinical practice remains questionable. Therefore, there has been significant interest in developing noninvasive methodologies to predict the severity of the disease (Musso et al. 2011). Risk factors for NAFLD are summarized in Table 24.1.

The fibrosis stage is specifically associated with the prognosis of the disease. Stages 3–4

Table 24.1 Risk factors for non-alcoholic fatty liver disease

Risk factors for NAFLD
Female gender
>50 years of age
Obesity
Metabolic syndrome
Diabetes mellitus type 2
Dyslipidaemia
Genetic predisposition
Dysbiosis

increased the risk of liver-related mortality by 50–80%. In this context, noninvasive detection of the fibrosis stage has been a major focus. The most common noninvasive scores are NAFLD Fibrosis Score (NFS), the Fibrosis-4 Score (FIB-4), the AST to Platelet Ratio Index (APRI), and the BARD Score, which includes several clinical and biochemical parameters obtained by routine clinical examinations (Drescher et al. 2019). The NAFLD guideline of the American Association for the Study of Liver Diseases (AASLD) recommends FIB-4 and NFS as clinically useful tools for the prediction of advanced fibrosis (F≥3) (Chalasani et al. 2018), Kaya et al. 2020. However, those scores have a high negative predictive value rather than positive predictive value, which means they can be confidently used for exclusion of advanced fibrosis rather than detection of it as also recommended in the guideline of European Association for the Study of the Liver (EASL) (European Association for the Study of the Liver (EASL); European Association for the Study of Diabetes (EASD); European Association for the Study of Obesity (EASO) 2016; Kaya et al. 2019).

Single vs. Multiple Criteria

Moreover, those scores include a grey zone, in which NAFLD patients are classified as having an indeterminate risk for advanced fibrosis. Another clinically important question concerns the management of those patients. Therefore, noninvasive scores are useful in the first-line stratification of the disease. Owing to the high-negative predictive values of those tests, patients classified as low risk for advanced fibrosis are recommended to be managed in primary care, whereas those at high risk should be directly referred to secondary care. For the patients in the indeterminate zone, Fibroscan examination is recommended for detecting the future management strategy. A noninvasive fibrosis marker or score, a diagnostic test, or an algorithm incorporating a panel of biomarkers is not capable of making a comprehensive statement of the disease outcome. Therefore, the combined use of noninvasive scores with imaging methods has been advised to increase diagnostic accuracy (Jafarov et al. 2019; Zhou et al. 2019).

Imaging Tools

Conventional ultrasonography is recommended for the diagnosis of moderate-to-severe steatosis, with a sensitivity of 85% and a specificity of 94% (Hernaez et al. 2011). However, it is incapable of detecting steatosis lower than 20% or in morbidly obese patients (Chalasani et al. 2018). Computerized tomography is more effective for evaluating hepatic steatosis; however, it remains also limited by insufficient accuracy for mild-to-moderate hepatic steatosis, and it involves radiation exposure (Schwenzer et al. 2009). Alternatively, controlled attenuation parameter (CAP) by Fibroscan, which measures the hepatic fat quantity by attenuation of the shear waves, and magnetic resonance-derived proton density fat fraction (MRI-PDFF), are more accurate and up-to-date methods for detection of hepatic steatosis. MRI-PDFF is a robust noninvasive method to monitor the treatment effect by means of hepatic fat quantification (Caussy et al. 2018). It is considered more accurate, reproducible, and reliable than liver histology. However, its use remains limited due to high costs, need for expertise, and long examination duration (Dulai et al. 2016). In terms of quick accessibility and cost-effectiveness, CAP (Fibroscan) is preferred over MRI-PDFF in clinical routine (Leung et al. 2013).

First-line Liver Fibrosis Assessment

Vibration-controlled transient elastography (VCTE) is the first Food and Drug Administration-approved modality by FibroScan employing ultrasound-based technology, which measures the velocity of a shear wave that is emitted by a probe in the intercostal space into the liver (Leung et al. 2013). Magnetic Resonance Elastography (MRE) is an excellent method as well for assessment of fibrosis stage in NAFLD. According to recent information, MRE performs better than VCTE for the identification of significant fibrosis; on the other hand, it performed equally well for the quantification of advanced fibrosis (Loomba et al. 2014).

Pathophysiology

The development of NAFLD classically relies on the "multiple hit and organ theory," in agreement with a large spectrum of metabolic dysfunctions caused by the interaction of genetic and environmental factors (Buzzetti et al. 2016). Liver fat accumulation, caused by obesity and insulin resistance, seems to represent the "first hit" (Fang et al. 2018). Hepatic fat accumulation in the liver is mainly comprised of triacylglycerol (TAG) derived from the esterification of glycerol and free fatty acids (FFA) (Buzzetti et al. 2016). TAG in hepatic fat is hydrolysed and secreted into the blood circulation as very-low-density lipoprotein particles. Disruption of those pathways can result in hepatic steatosis (Musso et al. 2013). Peroxisome proliferator-activated receptor-α (PPAR-α) plays a significant role in the regulation of β-oxidation in hepatocytes. The downregulation of PPAR-α was significantly associated with NAFLD and NASH (Tanaka et al. 2017). De novo lipogenesis in hepatocytes is mainly regulated by activation of transcriptional factor sterol regulatory element-binding protein 1c (SREBP-1c), which is enhanced by hyperglycaemia, and explains the close association between NAFLD and insulin resistance (Tanaka et al. 2019).

Subsequent Steps

Hepatic fat accumulation does not represent alone strong toxicity for the liver. There is no association between the degree of steatosis and NAFLD severity. However, TAG-derived molecules and its precursors such as palmitate, diacylglycerol (DAG), and ceramide are likely to be a detriment to hepatocytes. Palmitate increases oxidative stress leading to lipoapoptosis. DAG activates protein kinase C disrupting the insulin signalling pathway, and ceramide promotes the production of palmitate (Akazawa and Nakao 2018; Jiang et al. 2015; Gan et al. 2014). Furthermore, overloaded TAG storage creates metabolic distress and following this lipotoxicity, which increases together oxidative stress.

Normally, oxidation of FFAs ensues through α-, β-, and ω-oxidation. Mitochondrial β-oxidation and peroxisomal α-, β-oxidation are normal metabolic processes. However, if these metabolic pathways are impaired, ω-oxidation occurs in endoplasmic reticulum leading to reactive oxygen species production. It was observed that ω-oxidation and oxidation via NADPH oxidase in Kupffer cells are increased in NAFLD patients, causing inhibition of mitochondrial β-oxidation. All these metabolic processes lead to DNA damage in cell nucleus and mitochondria, and release of cytokines which promotes hepatocellular injury (Yao et al. 2019).

The Liver Damage Cascade

Hepatocellular injury causes the release of several pro-inflammatory mediators recruiting immune cells and activating Kupffer cells, which result in the release of bioactive molecules further damaging hepatocytes (Zhang et al. 2015). Chronic hepatocyte death or impaired hepatocyte regeneration leads to alternative replacement by fibres and extracellular matrix, resulting in significant scar tissue formation and remodelling of the normal structure of hepatic lobules. Fibrogenesis mainly in the perisinusoidal space is relatively specific to steatohepatitis (Lee et al. 2015).

The Microbiota and the Gut Barrier Function

There is increasing research interest on gut-liver axis dysfunction including intestinal dysbiosis, bacterial overgrowth, and alteration of gastrointestinal mucosa permeability. In the development of NAFLD, gastrointestinal microbiota play a significant role in maintaining barrier integrity and intestinal permeability. There are several mechanisms associated with alteration in gastrointestinal microbiota and development of NAFLD: Deterioration in microbiota can damage the intestinal epithelium and tight junction proteins in the gut, which allow harmful substances such as bacteria, ethanol, and endotoxins entering into the portal circulation. Furthermore, microbiota digest and ferment the excessive dietary energy into short-chain fatty acids and produces ethanol, affecting the liver similarly to chronic alcoholism (Doulberis et al. 2017).

According to the study of Zhu et al., there is a significantly increased population of alcohol-producing bacteria in patients with NASH compared to obese and healthy subjects. Furthermore, a significantly increased serum ethanol concentration was observed in NASH patients, although there was no difference between obese and healthy subjects (Zhu et al. 2013). In a more recent study, high alcohol-producing Klebsiella pneumonia was found to be associated with up to 60% of individuals of the study cohort. Moreover, clinical isolates of high alcohol-producing Klebsiella pneumonia transferred via oral gavage or faecal transplant also induced NAFLD in mice, supporting the strong association with NAFLD development (Yuan et al. 2019).

Disease-specific Variants

NAFLD is also associated with a genetic predisposition. Although only a minority of the genetic modifiers of NAFLD has been validated, there are several genetic associations to mention. The initial genome was identified as PNPLA3, which was validated in different ethnic groups as a modifier of NAFLD severity. In obese children and adolescents, the PNPLA3 rs738409 variant was also suspected to affect the histological severity in NAFLD (Valenti et al. 2010). Additionally, PNPLA3 has been accepted globally as a major determinant of not only steatosis but also the severity of NASH, fibrosis stage, and probability of HCC development (Anstee and Day 2013). Recently, the TM6SF2 gene has been reported as another disease modifier, which might be clinically useful in the future for estimating the disease progression severity (Dongiovanni et al. 2015).

TM6SF2 rs58542926 T-allele mediates hepatic accumulation of triglycerides and cholesterol creating a predisposition to NAFLD-related fibrosis, whereas C-allele carriage protects liver excreting VLDL from the liver, at a price of increased atherosclerosis and cardiovascular disease (Kahali et al. 2015). More recently, HSD17B13 was found to play role in the progression of liver disease from steatosis to later stages of non-alcoholic steatohepatitis, fibrosis, and cirrhosis, since the reduced activity of HSD17B13 was associated with a lower risk of progression from steatosis to steatohepatitis (Abul-Husn et al. 2018). MBOAT gene was also associated with disease severity in both NAFLD and alcoholic fatty liver disease (Caussy et al. 2019).

Lifestyle and Pharmacological Interventions

NAFLD has been associated with a diet of high caloric amount, which contains excessive saturated fats, refined carbohydrates, and high fructose (Barrera and George 2014), along with sedentary behaviour (Gerber et al. 2012). Therefore, lifestyle modification is the cornerstone treatment of NAFLD and has an important impact on the natural course of the disease (Kugelmas et al. 2003). Weight loss between 5 and 7% can diminish fat accumulation, whereas a 7–10% weight loss is significantly associated with improvement in NASH and fibrosis, with a chance of 64% and 50%, respectively. Weight loss of >10% was associated with up to 90% chance of NASH resolution and up to 81% of regression in fibrosis (Vilar-Gomez et al. 2015).

According to the recent studies and expert opinions, following a Mediterranean diet can reduce liver fat even without weight loss, which is characterized by reduced carbohydrate and increased monounsaturated and ω-3 fatty acid intake (Romero-Gomez et al. 2017). Therefore, it has been also recommended as the preferred diet type in the major guidelines (Chalasani et al. 2018; European Association for the Study of the Liver (EASL); European Association for the Study of Diabetes (EASD); European Association for the Study of Obesity (EASO) 2016). In addition to diet modifications, exercise should also be recommended. Aerobic exercise training of 150 min per week contributed to weight reduction and had an impact on intrahepatic fat (Hashida et al. 2017).

Zelber-Sagi et al. demonstrated that weight gain was a predictor of NAFLD development even among nonobese individuals. Loss of 5% of the initial body weight resulted in remission of NAFLD in 75% of those individuals (Zelber-Sagi et al. 2012). Both lean and obese individuals seem to similarly benefit from weight loss in terms of NAFLD remission (Varol Hamurcu et al. 2020).

Bariatric Interventions

In line with weight reduction, bariatric surgery is an excellent method to recommend for morbidly obese patients. After a one-year follow-up, in more than 85% of the patients, resolution of NASH has been observed. Moreover, in those patients, pathological features were also ameliorated (Laursen et al. 2019). In a meta-analysis, steatosis, inflammation, and ballooning mostly improved or completely resolved. An extensive review emphasizes that in most patients, liver fibrosis diminishes, whereas refractory cases exist in which both fibrosis and inflammation persist or progress (Laursen et al. 2019).

Drug Therapy

There is no specifically licensed pharmacological treatment for NAFLD. Yet, given the close association between NAFLD and type 2 diabetes mellitus, pioglitazone has been utilized in NASH targeting both adipose tissue metabolism and inflammation through PPAR-γ. Thus, it reduces hepatic steatosis, inflammation, and ballooning increasing uptake of fatty acid by adipocytes. However, because of its side effects such as weight gain and the conflict of its effect on the improvement of hepatic fibrosis, its use recommended only in selected diabetic patients (Cusi et al. 2016).

Although there is no convincing evidence about vitamin E and improvement of liver fibrosis, given the antioxidative and anti-inflammatory effect of vitamin E, it has been investigated as a therapeutic option in NASH. However, due to lack of further data, vitamin E is not recommended in NASH patients with diabetes, NAFLD without liver biopsy, NASH cirrhosis, or cryptogenic cirrhosis. The risks and benefits must be discussed for each patient individually (Chalasani et al. 2018).

There are several pharmaceutical agents with completed or ongoing phase III controlled trials. Obeticholic acid improves hepatic insulin sensitivity and decreases lipogenesis, inflammation, and fibrosis. The interim analysis of the phase 3 REGENERATE Study showed that obeticholic acid use resulted in fibrosis improvement without worsening of NASH (Younossi et al. 2019d). However, due to its side effects such as LDL increase and pruritus, its use in clinical routine remains limited considering that 13% discontinued therapy and 17% exhibited increased LDL levels (Younossi et al. 2019d). Another agent selonsertib inhibits apoptosis signal-regulating kinase 1, which reduces hepatocyte apoptosis and fibrosis. Still selonsertib failed to meet endpoint in STELLAR-3 and STELLAR-4 phase III clinical trials (Trial Site 2019; Clinical Trials 2019). Also Emricasan, an oral pan-caspase inhibitor, suppresses apoptosis failed to meet endpoint, also worsening liver histology (Shiffman et al. 2019). The ongoing phase III trials are with elafibranor, a PPARα and PPARδ agonist, which also improves insulin sensitivity and hepatic inflammation (Yuan et al. 2019), and cenicriviroc, a dual C-C chemokine receptor type 2 and 5 (CCR2/5) inhibitor (Yuan et al. 2019). The results of these trials are expected to be released through the midyear of 2020. In conclusion, diet and exercise remain still the cornerstone of NAFLD therapy.

References

Abul-Husn NS, Cheng X, Li AH, Xin Y, Schurmann C, Stevis P et al (2018) A protein-truncating HSD17B13 variant and protection from chronic liver disease. N Engl J Med 378(12):1096–1106

Adams LA, Sanderson S, Lindor KD, Angulo P (2005) The histological course of nonalcoholic fatty liver disease: a longitudinal study of 103 patients with sequential liver biopsies. Hepatology 42:132–138

Ajmera VH, Terrault NA, Harrison SA (2017) Is moderate alcohol use in nonalcoholic fatty liver disease good or bad? A critical review. Hepatology 65(6):2090–2099

Akazawa Y, Nakao K (2018) To die or not to die: death signaling in nonalcoholic fatty liver disease. J Gastroenterol 53:893–906

Akyuz U, Yesil A, Yilmaz Y (2015) Characterization of lean patients with nonalcoholic fatty liver disease: potential role of high hemoglobin levels. Scand J Gastroenterol 50(3):341–346

Anstee QM, Day CP (2013) The genetics of NAFLD. Nat Rev Gastroenterol Hepatol 10(11):645–655

Argo CK, Northup PG, Al-Osaimi AM, Caldwell SH (2009) Systematic review of risk factors for fibrosis progression in non-alcoholic steatohepatitis. J Hepatol 51(2):371–379

Arrese M, Barrera F, Triantafilo N, Arab JP (2019) Concurrent nonalcoholic fatty liver disease and type 2 diabetes: diagnostic and therapeutic considerations. Expert Rev Gastroenterol Hepatol 13(9):849–866

Ascha MS, Hanouneh IA, Lopez R, Tamimi TA, Feldstein AF, Zein NN (2010) The incidence and risk factors of hepatocellular carcinoma in patients with nonalcoholic steatohepatitis. Hepatology 51(6):1972–1978

Barrera F, George J (2014) The role of diet and nutritional intervention for the management of patients with NAFLD. Clin Liver Dis 18:91–112

Burton R, Sheron N (2018) No level of alcohol consumption improves health. Lancet 392(10152):987–988

Buzzetti E, Pinzani M, Tsochatzis EA (2016) The multiple-hit pathogenesis of non-alcoholic fatty liver disease (NAFLD). Metabolism 65:1038–1048

Calzadilla Bertot L, Adams LA (2016) The natural course of non-alcoholic fatty liver disease. Int J Mol Sci 20: E774

Caussy C, Reeder SB, Sirlin CB, Loomba R (2018) Non-invasive, quantitative assessment of liver fat by MRIPDFF as an endpoint in NASH Trials. Hepatology 68:763–772

Caussy C, Bhargava M, Villesen IF, Gudmann NS, Leeming DJ, Karsdal MA et al (2019) Collagen formation assessed by N-terminal propeptide of type 3 procollagen is a heritable trait and is associated with liver fibrosis assessed by magnetic resonance elastography. Hepatology 70(1):127–141

Chalasani N, Younossi Z, Lavine JE, Charlton M, Cusi K, Rinella M et al (2018) The diagnosis and management of nonalcoholic fatty liver disease: practice guidance from the American Association for the Study of Liver Diseases. Hepatology 67:328–357

Clinical Trials (2019). https://www.clinicaltrialsarena.com/news/gileads-selonsertib-fails/. Accessed 23 Nov 2019

Consultation WHOE; WHO Expert Consultation (2004) Appropriate body-mass index for Asian populations and its implications for policy and intervention strategies. Lancet 363(9403):157–163

Cusi K, Orsak B, Bril F, Lomonaco R, Hecht J, Ortiz-Lopez C et al (2016) Long-term pioglitazone treatment for patients with nonalcoholic steatohepatitis and prediabetes or type 2 diabetes mellitus. Ann Intern Med 165:305–315

Dam-Larsen S, Becker U, Franzmann MG, Larsen K, Christoffersen P, Bendtsen F (2009) Final results of a long-term, clinical follow-up in fatty liver patients. Scand J Gastroenterol 44:1236–1243

Dongiovanni P, Petta S, Maglio C, Fracanzani AL, Pipitone R, Mozzi E et al (2015) Transmembrane 6 superfamily member 2 gene variant disentangles nonalcoholic steatohepatitis from cardiovascular disease. Hepatology 61:506–514

Doulberis M, Kotronis G, Gialamprinou D, Kountouras J, Katsinelos P (2017) Non-alcoholic fatty liver disease: An update with special focus on the role of gut microbiota. Metabolism 71:182–197

Drescher HK, Weiskirchen S, Weiskirchen R (2019) Current status in testing for nonalcoholic fatty liver disease (NAFLD) and nonalcoholic steatohepatitis (NASH). Cell 8(8):E845. https://doi.org/10.3390/cells8080845

Dulai PS, Sirlin CB, Loomba R (2016) MRI and MRE for non-invasive quantitative assessment of hepatic steatosis and fibrosis in NAFLD and NASH: clinical trials to clinical practice. J Hepatol 65(5):1006–1016

Dulai PS, Singh S, Patel J, Soni M, Prokop LJ, Younossi Z et al (2017) Increased risk of mortality by fibrosis stage in nonalcoholic fatty liver disease: systematic review and meta-analysis. Hepatology 65(5):1557–1565

Enc FY, Ulasoglu C, Bakir A, Yilmaz Y (2019) The interaction between current smoking and hemoglobin on the risk of advanced fibrosis in patients with biopsy-proven nonalcoholic fatty liver disease. Eur J Gastroenterol Hepatol 32:597–600. https://doi.org/10.1097/MEG.0000000000001536

Estes C, Razavi H, Loomba R, Younossi Z, Sanyal AJ (2018a) Modeling the epidemic of nonalcoholic fatty liver disease demonstrates an exponential increase in burden of disease. Hepatology 67:123–133

Estes C, Anstee QM, Arias-Loste MT, Bantel H, Bellentani S, Caballeria J et al (2018b) NAFLD disease burden in China, France, Germany, Italy, Japan, Spain, United Kingdom, and United States for the period 2016–2030. J Hepatol 69:896–904

European Association for the Study of the Liver (EASL); European Association for the Study of Diabetes (EASD); European Association for the Study of Obesity (EASO) (2016) EASL-EASD-EASO clinical practice guidelines for the management of non-alcoholic fatty liver disease. J Hepatol 64 (6):1388–1402

Fan JG, Kim SU, Wong VW (2017) New trends on obesity and NAFLD in Asia. J Hepatol 67(4):862–873

Fang YL, Chen H, Wang CL, Liang L (2018) Pathogenesis of non-alcoholic fatty liver disease in children and adolescence: from "two hit theory" to "multiple hit model". World J Gastroenterol 24(27):2974–2983

Firneisz G (2014) Non-alcoholic fatty liver disease and type 2 diabetes mellitus: the liver disease of our age? World J Gastroenterol 20:9072–9089

Francque S, De Maeght S, Adler M, Deltenre P, de Galocsy C, Orlent H et al (2011) High prevalence of advanced fibrosis in association with the metabolic syndrome in a Belgian prospective cohort of NAFLD patients with elevated ALT: results of the Belgian NAFLD registry. Acta Gastroenterol Belg 74:9–16

Gan LT, Van Rooyen DM, Koina ME, McCuskey RS, Teoh NC, Farrell GC (2014) Hepatocyte free cholesterol lipotoxicity results from JNK1-mediated mitochondrial injury and is HMGB1 and TLR4-dependent. J Hepatol 61:1376–1384

Gerber L, Otgonsuren M, Mishra A, Escheik C, Birerdinc A, Stepanova M et al (2012) Non-alcoholic fatty liver disease (NAFLD) is associated with low level of physical activity: a population-based study. Aliment Pharmacol Ther 36:772–781

Golabi P, Otgonsuren M, de Avila L, Sayiner M, Rafiq N, Younossi ZM (2018) Components of metabolic syndrome increase the risk of mortality in nonalcoholic fatty liver disease (NAFLD). Medicine (Baltimore) 97 (13):e0214

Green RM (2003) NASH—hepatic metabolism and not simply the metabolic syndrome. Hepatology 38:14–17

Hagström H, Nasr P, Ekstedt M, Hammar U, Stål P, Hultcrantz R, Kechagias S (2017a) Fibrosis stage but not NASH predicts mortality and time to development of severe liver disease in biopsy-proven NAFLD. J Hepatol 67(6):1265–1273

Hagström H, Nasr P, Ekstedt M, Kechagias S, Önnerhag K, Nilsson E et al (2017b) Low to moderate lifetime alcohol consumption is associated with less advanced stages of fibrosis in non-alcoholic fatty liver disease. Scand J Gastroenterol 52(2):159–165

Hashida R, Kawaguchi T, Bekki M, Omoto M, Matsuse H, Nago T et al (2017) Aerobic vs. resistance exercise in non-alcoholic fatty liver disease: a systematic review. J Hepatol 66(1):142–152

Hernaez R, Lazo M, Bonekamp S, Kamel I, Brancati FL, Guallar E, Clark JM (2011) Diagnostic accuracy and reliability of ultrasonography for the detection of fatty liver: a meta-analysis. Hepatology 54:1082–1090

Jafarov F, Kaya E, Bakir A, Eren F, Yilmaz Y (2019) The diagnostic utility of fibrosis-4 or non-alcoholic fatty liver disease fibrosis score combined with liver stiffness measurement by Fibroscan in assessment of advanced liver fibrosis: a biopsy-proven nonalcoholic fatty liver disease study. Euro J Gastroent Hepatol 32:642–649. https://doi.org/10.1097/MEG.0000000000001573

Jiang C, Xie C, Li F, Zhang L, Nichols RG, Krausz KW, Cai J, Qi Y, Fang ZZ, Takahashi S, Tanaka N, Desai D, Amin SG, Albert I, Patterson AD, Gonzalez FJ (2015) Intestinal farnesoid X receptor signaling promotes nonalcoholic fatty liver disease. J Clin Invest 125:386–402

Kahali B, Liu YL, Daly AK, Day CP, Anstee QM, Speliotes EK (2015) TM6SF2: catch-22 in the fight against nonalcoholic fatty liver disease and cardiovascular disease? Gastroenterology 148(4):679–684

Kalra S, Vithalani M, Gulati G, Kulkarni CM, Kadam Y, Pallivathukkal J et al (2013) Study of prevalence of nonalcoholic fatty liver disease (NAFLD) in type 2 diabetes patients in India (SPRINT). J Assoc Physicians India 61:448–453

Kaya E, Yılmaz Y (2019a) Non-alcoholic fatty liver disease: a growing public health problem in Turkey. Turk J Gastroenterol 10:865–871

Kaya E, Yılmaz Y (2019b) Türkiye'de ve dünyada nonalkolik yağlı karaciğer hastalığı epidemiyolojisi. In: Sonsuz A (ed) Nonalkolik Yağlı Karaciğer Hastalığı, 1st edn. Türkiye Klinikleri, Ankara, pp 1–7

Kaya E, Bakir A, Kani HT, Demirtas CO, Keklikkiran C, Yilmaz Y (2019) Simple noninvasive scores are clinically useful to exclude, not predict, advanced fibrosis: a study in Turkish patients with biopsy-proven nonalcoholic fatty liver disease. Gut Liver 14:486–491. https://doi.org/10.5009/gnl19173

Kaya E, Bakir E, Eren F, Yilmaz Y (2020). The utility of noninvasive scores in non-alcoholic fatty liver disease patients with normal and elevated serum transaminases. Hepatol Forum 1(1):8–13

Kugelmas M, Hill DB, Vivian B, Marsano L, McClain CJ (2003) Cytokines and NASH: a pilot study of the effects of lifestyle modification and vitamin E. Hepatology 38:413–419

Laursen TL, Hagemann CA, Wei C, Kazankov K, Thomsen KL, Knop FK, Grønbæk H (2019) Bariatric surgery in patients with non-alcoholic fatty liver disease—from pathophysiology to clinical effects. World J Hepatol 11(2):138–149

Lee YA, Wallace MC, Friedman SL (2015) Pathobiology of liver fibrosis: a translational success story. Gut 64:830–841

Leung VY, Shen J, Wong VW, Abrigo J, Wong GL, Chim AM, Chu SH, Chan AW, Choi PC, Ahuja AT, Chan HL, Chu WC (2013) Quantitative elastography of liver fibrosis and spleen stiffness in chronic hepatitis B carriers: comparison of shear-wave elastography and transient elastography with liver biopsy correlation. Radiology 269:910–918

Loomba R, Wolfson T, Ang B, Hooker J, Behling C, Peterson M et al (2014) Magnetic resonance elastography predicts advanced fibrosis in patients with nonalcoholic fatty liver disease: a prospective study. Hepatology 60:1920–1928

Mofrad P, Contos MJ, Haque M, Sargeant C, Fisher RA, Luketic VA et al (2003) Clinical and histologic spectrum of nonalcoholic fatty liver disease associated with normal ALT values. Hepatology 37:1286–1292

Mongraw-Chaffin M, Golden SH, Allison MA, Ding J, Ouyang P, Schreiner PJ et al (2015) The sex and race specific relationship between anthropometry and body fat composition determined from computed tomography: evidence from the multi-ethnic study of atherosclerosis. PLoS One 10:e0139559

Musso G, Gambino R, Cassader M, Pagano G (2011) Meta-analysis: natural history of non-alcoholic fatty liver disease (NAFLD) and diagnostic accuracy of non-invasive tests for liver disease severity. Ann Med 43:617–649

Musso G, Gambino R, Cassader M (2013) Cholesterol metabolism and the pathogenesis of non-alcoholic steatohepatitis. Prog Lipid Res 52:175–191

Niriella MA, Kasturiratne A, Beddage TU, Withanage SA, Goonatilleke DC, Abeysinghe CP et al (2019) Metabolic syndrome, but not non-alcoholic fatty liver disease, increases 10-year mortality: a prospective, community-cohort study. Liver Int 40:101–106. https://doi.org/10.1111/liv.14237

Povsic M, Wong OY, Perry R, Bottomley J (2019) A structured literature review of the epidemiology and disease burden of non-alcoholic steatohepatitis (NASH). Adv Ther 36:1574–1594

Romero-Gomez M, Zelber-Sagi S, Trenell M (2017) Treatment of NAFLD with diet, physical activity and exercise. J Hepatol 67(4):829–846

Sasaki A, Nitta H, Otsuka K, Umemura A, Baba S, Obuchi T et al (2015) Bariatric surgery and non-alcoholic fatty liver disease: current and potential future treatments. Front Endocrinol 5:164

Sayiner M, Golabi P, Younossi ZM (2019) Disease burden of hepatocellular carcinoma: a global perspective. Dig Dis Sci 64(4):910–917

Schwenzer NF, Springer F, Schraml C, Stefan N, Machann J, Schick F (2009) Non-invasive assessment and quantification of liver steatosis by ultrasound, computed tomography and magnetic resonance. J Hepatol 51:433–445

Shiffman M, Freilich B, Vuppalanchi R, Watt K, Chan JL, Spada A et al (2019) Randomised clinical trial: emricasan versus placebo significantly decreases ALT and caspase 3/7 activation in subjects with non-alcoholic fatty liver disease. Aliment Pharmacol Ther 49(1):64–73

Singh S, Allen AM, Wang Z, Prokop LJ, Murad MH, Loomba R (2015) Fibrosis progression in nonalcoholic fatty liver vs nonalcoholic steatohepatitis: a systematic review and meta-analysis of paired-biopsy studies. Clin Gastroenterol Hepatol 13(4):643–654

Söderberg C, Stål P, Askling J, Lindberg G, Marmur J, Hultcrantz R (2010) Decreased survival of subjects with elevated liver function tests during a 28-year follow-up period. Hepatology 51:595–602

Stefan N, Haring HU, Cusi K (2019) Non-alcoholic fatty liver disease: causes, diagnosis, cardiometabolic consequences, and treatment strategies. Lancet Diabetes Endocrinol 7(4):313–324

Subichin M, Clanton J, Makuszewski M, Bohon A, Zografakis JG, Dan A (2015) Liver disease in the morbidly obese: a review of 1000 consecutive patients undergoing weight loss surgery. Surg Obes Relat Dis 11:137–141

Sung KC, Ryan MC, Wilson AM (2009) The severity of nonalcoholic fatty liver disease is associated with increased cardiovascular risk in a large cohort of non-obese Asian subjects. Atherosclerosis 203:581–586

Tanaka N, Aoyama T, Kimura S, Gonzalez FJ (2017) Targeting nuclear receptors for the treatment of fatty liver disease. Pharmacol Ther 179:142–157

Tanaka N, Kimura T, Fujimori N, Nagaya T, Komatsu M, Tanaka E (2019) Current status, problems, and perspectives of non-alcoholic fatty liver disease research. World J Gastroenterol 25(2):163–177

Targher G, Byrne CD (2015) A perspective on metabolic syndrome and nonalcoholic fatty liver disease. Metab Syndr Relat Disord 13:235–238

Targher G, Byrne CD, Lonardo A, Zoppini G, Barbui C (2016) Non-alcoholic fatty liver disease and risk of incident cardiovascular disease: a meta-analysis. J Hepatol 65(3):589–600

Torres DM, Williams CD, Harrison SA (2012) Features, diagnosis, and treatment of nonalcoholic fatty liver disease. Clin Gastroenterol Hepatol 10(8):837–858

Trial Site (2019). https://www.trialsitenews.com/gilead-reports-phase-3-stellar-3-of-selonsertib-for-bridging-fibrosis-f3-due-to-nonalcoholic-steatohepatitis-fails-to-meet-endpoints/. Accessed 23 Nov 2019

Ulasoglu C, Enc Yilmaz F, Kaya E, Yilmaz Y (2019) Characterization of patients with biopsy-proven non--alcoholic fatty liver disease and normal aminotransferase levels. J Gastrointest Liver Dis 28(4):427–431

Valenti L, Alisi A, Galmozzi E, Bartuli A, Del Menico B, Alterio A et al (2010) I148M patatin-like phospholipase domain-containing 3 gene variant and severity of pediatric nonalcoholic fatty liver disease. Hepatology 52:1274–1280

Varol Hamurcu P, Kaya E, Alphan E, Yilmaz Y (2020) Role of intensive dietary and lifestyle interventions in the treatment of lean nonalcoholic fatty liver disease patients. Eur J Gastroenterol Hepatol 32:1352–1357

Vilar-Gomez E, Martinez-Perez Y, Calzadilla-Bertot L, Torres-Gonzalez A, Gra-Oramas B, Gonzalez-Fabian L et al (2015) Weight loss through lifestyle modification significantly reduces features of nonalcoholic steatohepatitis. Gastroenterology 149(2):367–378

Wijarnpreecha K, Thongprayoon C, Ungprasert P (2017) Coffee consumption and risk of nonalcoholic fatty liver disease: a systematic review and meta-analysis. Eur J Gastroenterol Hepatol 29(2):e8–e12

World Health Organization (WHO) (2020) Obesity and overweight. Obtained from the world wide web. http://www.who.int/news-room/fact-sheets/detail/obesity-and-overweight. Accessed 23 Nov 2019

Yao Y, Cai J, She Z, Li H (2019) Insights into the epidemiology, pathogenesis, and therapeutics of nonalcoholic fatty liver diseases. Adv Sci (Weinh) 6 (4):1801585

Yesil A, Yilmaz Y (2013) Review article: coffee consumption, the metabolic syndrome and non-alcoholic fatty liver disease. Aliment Pharmacol Ther 38(9):1038–1044

Yilmaz Y (2012a) Review article: is non-alcoholic fatty liver disease a spectrum, or are steatosis and non-alcoholic steatohepatitis distinct conditions? Aliment Pharmacol Ther 36(9):815–823
Yilmaz Y (2012b) Is nonalcoholic fatty liver disease the hepatic expression of the metabolic syndrome? World J Hepatol 4:332–334
Yılmaz Y, Kanı HT, Demirtaş CÖ, Kaya E, Sapmaz AF, Qutranji L et al (2019) Growing burden of non-alcoholic fatty liver disease in Turkey: a single-center experience. Turk J Gastroenterol 30 (10):892–898
Younossi ZM, Koenig AB, Abdelatif D, Fazel Y, Henry L, Wymer M (2016) Global epidemiology of nonalcoholic fatty liver disease-Meta-analytic assessment of prevalence, incidence, and outcomes. Hepatology 64:73–84
Younossi Z, Anstee QM, Marietti M, Hardy T, Henry L, Eslam M, George J, Bugianesi E (2018) Global burden of NAFLD and NASH: trends, predictions, risk factors and prevention. Nat Rev Gastroenterol Hepatol 15 (1):11–20
Younossi Z, Stepanova M, Ong JP, Jacobson IM, Bugianesi E, Duseja A et al (2019a) Nonalcoholic steatohepatitis is the fastest growing cause of hepatocellular carcinoma in liver transplant candidates. Clin Gastroenterol Hepatol 17:748–755
Younossi Z, Tacke F, Arrese M, Sharma BC, Mostafa I, Bugianesi E, Wong VW, Yilmaz Y, George J, Fan J, Vos MB (2019b) Global perspectives on non-alcoholic fatty liver disease and non-alcoholic steatohepatitis. Hepatology 69(6):2672–2682
Younossi ZM, Golabi P, de Avila L, Paik JM, Srishord M, Fukui N, Qiu Y, Burns L, Afendy A, Nader F (2019c) The global epidemiology of NAFLD and NASH in patients with type 2 diabetes: a systematic review and meta-analysis. J Hepatol 71(4):793–801
Younossi Z, Ratziu V, Loomba R et al (2019d) Positive results from REGENERATE: a phase 3 international, randomized, placebo-controlled study evaluating obeticholic acid treatment for NASH. J Hepatol 70: e5. https://doi.org/10.1016/S0618-8278(19)30006-4
Yuan J, Chen C, Cui J, Lu J, Yan C, Wei X et al (2019) Fatty liver disease caused by high-alcohol-producing klebsiella pneumoniae. Cell Metab 30(4):675–688
Zelber-Sagi S, Lotan R, Shlomai A, Webb M, Harrari G, Buch A et al (2012) Predictors for incidence and remission of NAFLD in the general population during a seven-year prospective follow-up. J Hepatol 56:1145–1151
Zhang L, Nichols RG, Correll J, Murray IA, Tanaka N, Smith PB, Hubbard TD, Sebastian A, Albert I, Hatzakis E, Gonzalez FJ, Perdew GH, Patterson AD (2015) Persistent organic pollutants modify gut microbiota-host metabolic homeostasis in mice through aryl hydrocarbon receptor activation. Environ Health Perspect 123:679–688
Zhou JH, Cai JJ, She ZG, Li HL (2019) Noninvasive evaluation of nonalcoholic fatty liver disease: current evidence and practice. World J Gastroenterol 25 (11):1307–1326
Zhu L, Baker SS, Gill C, Liu W, Alkhouri R, Baker RD, Gill SR (2013) Characterization of gut microbiomes in nonalcoholic steatohepatitis (NASH) patients: A connection between endogenous alcohol and NASH. Hepatology 57(2):601–609

25 Nonalcoholic Fatty Pancreatic Disease (NAFPD)

Alhareth Al Juboori, Harleen Chela, Sami Samiullah, and Veysel Tahan

Contents

Introduction 336
The Spectrum of Metabolic Syndrome 336
Historical Notes 336
Nomenclature 337

Epidemiology 337

Pathogenesis of NAFPD—Obesity 337
Age 338
Metabolic Syndrome 338
β-Cell Function 338
Other Possible Etiologies 339

Diagnosis 339
Confounding Variables 339
Transabdominal Ultrasound 339
Endoscopic Ultrasound (EUS) 340
CT Scan 340
MRI and MRS 340
Histology 340

Possible Complications—Metabolic Syndrome 341
Acute and Chronic Pancreatitis 341
Pancreatic Insufficiency 341
Pancreatic Fibrosis 341
Pancreatic Cancer 342
Surgical Risk 342

Treatment 342
Weight Loss 342
Oral Hypoglycemic Agents 342

References 342

A. Al Juboori · H. Chela · S. Samiullah · V. Tahan (✉)
Division of Gastroenterology and Hepatology,
Department of Internal Medicine, University of Missouri
School of Medicine, Columbia, MO, USA
e-mail: tahanv@health.missouri.edu

J. Faintuch and S. Faintuch (eds.), *Obesity and Diabetes*, https://doi.org/10.1007/978-3-030-53370-0_25

Abstract

NAFPD is a novel disease entity that is gaining popularity similar to NAFLD. It has an association with features of the metabolic syndrome, especially obesity. Given increased awareness, it is being diagnosed more commonly and at times is discovered as an incidental finding on imaging. Research thus far has shown that it carries clinical significance and can lead to pancreatitis as well as surgical risk, and even potentially pancreatic cancer. Though treatment strategies are not well established, given the links to metabolic syndrome one of the common approaches is weight loss. Medications such as oral hypoglycemic agents have been studied in animals but there have yet to be studied in humans. As NAFPD still remains a mystery in many ways, it requires further research into the pathophysiology as well as the long-term consequences and management.

Keywords

Non alcoholic pancreatic steatosis · Alcoholic pancreatic steatosis · Metabolic syndrome · Chronic pancreatitis · Pancreatic cancer · Ectopic fat

Abbreviations

AT1R	Angiotensin type-1 receptor
CLE	carboxyl-esterase lipase mutation
CT	Computed Tomography
EUS	Endoscopic ultrasound
MRI	magnetic resonance imaging
MRS	magnetic resonance spectroscopy
NAFLD	nonalcoholic fatty liver disease
NAFPD	Nonalcoholic Fatty Pancreatic Disease
PFF	pancreatic fat fraction
WHO	World health organization

Introduction

There has been a tenfold increase in obesity among children and adolescents in the last 4 decades, and the trend is expected to continue by 2022 (WHO 2020). Comorbidities associated with obesity such as coronary artery disease, cerebrovascular accidents, type 2 diabetes mellitus, and cancers are primary etiologies for many preventable deaths (Pi-Sunyer 2009). There is a medical spending difference in obesity, and it can be as high as $1429 per patient/year. In the United States, the estimated medical spending in the obesity population was close to $147 billion dollars in 2008 (Finkelstein et al. 2009). Adult females hold higher prevalence in obesity in Asians, Hispanics, and African American populations (CDC 2020).

The Spectrum of Metabolic Syndrome

Metabolic syndrome comprises five risk factors according to the National Institutes of Health namely obesity, hypertriglyceridemia, low HDL, hypertension, and fasting hyperglycemia or diabetes mellitus (Younossi 2019). The link between metabolic syndrome and nonalcoholic fatty liver disease (NAFLD) is very well-established, as it can lead to nonalcoholic steatohepatitis and eventually cirrhosis, obviously without any significant alcohol intake (Fazel et al. 2016). Unlike nonalcoholic fatty liver disease and metabolic syndrome, pancreatic steatosis or fat deposition in the pancreas has just earned attention in the last few years.

Historical Notes

John Schaefer in 1926 was probably the first to identify the normal weight of the pancreas in the adult population (Schaefer 1926). He was followed by Ogilvie in 1933, who noticed in a case-control study of 19 cadavers of obese and

Table 25.1 Nomenclature for pancreatic steatosis

Nomenclature	Definition
Pancreatic lipomatosis Pancreatic steatosis Fatty pancreas	Nonspecific terminologies
Lipomatous pseudo-hypertrophy	Extreme pancreatic fat accumulation
Fatty replacement	Adipocytes associated with acinar extinction
Fatty infiltration	Obesity-induced adipocyte accumulation
Nonalcoholic fatty pancreas disease	Metabolic syndrome-associated fat accumulation
Nonalcoholic fatty steatopancreatitis	Pancreatitis due to fat deposition

Adapted from Smits and Van Greenen (2011)

lean patients, and increase in the mean pancreatic fat deposition (17.1%, range 0–48.5%), compared to the lean group (9.3%, range 2.5–23.6%) (Ogilvie 1933). Olson in 1978, analyzing 394 autopsies, identified strong correlation between older age/overweight status and the degree of lipomatosis of the pancreas (Olsen 1978). Association was found between other elements of metabolic syndrome, including diabetes mellitus and dyslipidemia, with pancreatic fatty deposition.

Nomenclature

Smits and van Greenen suggested to standardize nomenclature for better definition of pancreatic steatosis as listed below (Smits and Geenen 2011) (Table 25.1).

Epidemiology

Wang et al reported in 2014 prevalence of pancreatic steatosis in 16% of a cohort which included 8097 patients; it was significantly associated with diabetes (12.6% versus 5.2%) (Wang et al. 2014). Pham et al performed a retrospective chart review of 232 pediatric patients (age 2–18) in Texas who obtained computed tomography imaging in an emergency room visit. The calculated pancreatic steatosis prevalence was 10% (Pham et al. 2016).

Singh et al in 2017 reported the pooled prevalence of NAFPD from 11 studies (12,675 patients), reaching 33%. There was association between NAFPD and metabolic syndrome, diabetes mellitus, and hypertension (Singh et al. 2017). Hispanics seem to exhibit higher pancreatic fat fraction (PFF) than African Americans, and increasingly so with age. Also males have higher PFF than females. PFF was related to visceral adipose tissue, circulating free fatty acids, and hepatic fat fraction (Le et al. 2011). In another study, pancreatic triglyceride levels in the Hispanic and Caucasian population were higher when compared to African Americans (Szczepaniak et al. 2012).

Pathogenesis of NAFPD—Obesity

Obesity leads to develop nonalcoholic fatty pancreatic disease (Sepe et al. 2011). Visceral and ectopic fat have been recognized, as adipokines which promote oxidative stress initiate local inflammatory cascades and eventually organ damage (Tilg and Hotamisligil 2006), analogously to nonalcoholic fatty liver disease and steatohepatitis (Schaefer 1926; Mathur et al. 2007a).

One of the main accepted mechanisms that can lead to pancreatic steatosis is acinar cell death, which is subsequently replaced by adipocytes and fat deposition (intracellular triglycerides), mainly in type 2 diabetes mellitus and obesity (Yu and Wang 2017). The pancreas is thought to have higher sensitivity to fatty infiltration than the liver (Yu and Wang 2017). Pancreatic steatosis majorly happens via adipocyte infiltration to the pancreatic parenchyma, yet intracellular triglyceride accumulation has been noticed in both acinar and islet cells (Smits and Geenen 2011). The high-fat diet responsible for high insulin levels, oxidative stress, and cellular influx of free fatty acids with decreased beta-oxidation, could result in triglyceride deposition (Fraulob et al. 2010). In a prospective study, it was shown that the risk for pancreatic steatosis will increase by 5% for every 1 kg/m^2 increment in BMI (Sepe et al. 2011).

Age

Increasing fatty infiltration of the pancreas with aging is well established (Rössle 1921; Ogiu et al. 1997). The ratio between fat to acinar cells increases with age, and pancreatic atrophy could simultaneously happen (Saisho et al. 2007). Sonographic studies confirm increased pancreatic steatosis with aging (Glaser and Stienecker 2000). Moreover, fat-pancreatic ratio increases steadily in both genders (Saisho et al. 2007). Olson et al had reported that 15% of the elderly between age of 60 and 69 years had severe pancreatic fatty infiltration; meanwhile, no cases of severe pancreatic fatty infiltration were observed below age of 40 (Olsen 1978). Age independently from body weight is considered as a significant element for pancreatic steatosis development (Schmitz-Moormann et al. 1981).

Metabolic Syndrome

Metabolic syndrome is strongly linked to increasing risk of cardiovascular disease, diabetes mellitus, cerebrovascular accidents, and mortality by coronary disease (Huang 2009; Tenenbaum and Fisman 2011). NAFLD is a relevant hepatic manifestation for metabolic syndrome, as visceral fat is more pathologically and inflammatorily significant than subcutaneous fat. Nonalcoholic steatohepatitis, hepatocellular carcinoma, and cirrhosis are accepted complications in NAFLD patients (Perumpail et al. 2015). Wu et al found in a fatty pancreatic group higher hemoglobin A1c, fasting blood glucose, total cholesterol, triglyceride, and LDL-C ($p<0.03$). Higher systolic blood pressure rather than diastolic blood pressure was registered as well (Wu and Wang 2013). In an EUS-based evaluation of 284 subjects, 110 were found to have pancreatic steatosis. Associations included age older than 60 years, BMI more than 25 kg/m2, diabetes mellitus, hypertension, dyslipidemia, and fatty liver. Visceral adipose tissue was correlated with pancreatic steatosis independently from BMI ($p<0.01$) (Choi 2010).

β-Cell Function

It is debated whether pancreatic fatty infiltration contributes to pathogenesis of type 2 diabetes mellitus (Zhao et al. 2015). Animal studies suggested that pancreatic steatosis could impact β-cell function and accelerate their apoptosis. Zhao et al reported that a high-fat, high-sucrose diet in Bama minipigs leads to enlarged islet cells and β-cell injury, through increased lipid peroxidation and decreased pancreatic antioxidant enzyme function (Zhao et al. 2015).

In eleven nondiabetic children with heterozygous carboxyl-esterase lipase mutation (CLE), i.e., a rare mutation involved with diabetes, pancreatic exocrine dysfunction, and adult pancreatic steatosis, pancreatic steatosis was found before the occurrence of diabetes. Decreased first-phase insulin secretion also occurred (Raeder et al. 2007).

Pancreatic fatty content is higher among adult men with type 2 diabetes, as measured by MRS. Pancreatic lipomatosis was negatively linked with β cell function in nondiabetic men, however not in those with type 2 diabetes (Tushuizen et al. 2007). Heni et al reported an inverse relationship between pancreatic fatty content and β cell function among patients with impaired oral glucose tolerance/impaired fasting glucose. Pancreatic fat was associated with lower levels of insulin secretion (Heni et al. 2010). In Hong Kong, 110/685 healthy subjects exhibited NAFPD. Those subjects displayed higher insulin resistance and when combined with NAFLD, the correlation with insulin resistance was even higher, independently from BMI (Wong et al. 2014).

Other Possible Etiologies

Alcohol is an important factor. Consumption that exceeds 14 grams a week can lead to pancreatic steatosis (Tariq et al. 2016). By causing pancreatitis, alcohol can upregulate transcription factors responsible for cholesterol and triglyceride expression (Al-Haddad et al. 2009). Rosiglitazone, corticosteroids, gemcitabine, and octreotide are linked as well to pancreatic steatosis /fatty pancreas development (Ye and Liu 2017).

Iron overload is also a potential mechanism. One of the most common causes of iron overload is hemochromatosis, a hereditary condition in which iron accumulates in the reticuloendothelial system, liver, heart, skin, pancreas, and exocrine glands. In the pancreas, iron toxicity through oxidative stress leads to acinar and islet cell apoptosis with ultimate replacement with adipocytes. This was also well-recognized in patients with Cooley's anemia and myelodysplastic syndrome since they receive frequent blood transfusions (Midiri et al. 1999; Lin et al. 2007). Viruses like HIV, hepatitis B, and reovirus have also been linked to pancreatic steatosis. Malnutrition could predispose to pancreatic steatosis especially in conditions like alcoholism, AIDS, and kwashiorkor, by altering pancreatic structure followed by replacement with fat (Chehter et al. 2000). Congenital disorders like cystic fibrosis, heterozygous carboxyl-ester-lipase mutations, Johnson blizzard syndrome, and Shwachman-Diamond syndrome are associated with pancreatic steatosis (Raeder et al. 2007; Daentl et al. 1979). In cystic fibrosis, mucous plugs inside pancreatic ductules injure pancreatic cells, which will eventually be replaced by fat (Walters 1966).

Diagnosis

Imaging findings of a fatty pancreas alone are not sufficient to make the diagnosis. NAFPD entails a fatty infiltration of the pancreas along with the presence of components of the metabolic syndrome (as per the International Diabetes Federation or the National Cholesterol Education Program Adult Treatment Panel III). Independent diseases that lead to fatty replacement of the pancreas should be excluded (Pinte et al. 2019; Grundy 2019). Though histological examination can provide information on the degree of fatty infiltration of the pancreas, it is too invasive for routine use. A histological classification system known as the pancreatic lipomatosis score, established by Olsen in 1978, could confirm the severity of pancreatic steatosis, in selected cases (Olsen 1978). Hence, the primary modalities used for diagnosis are imaging studies.

Confounding Variables

Due to the variable morphology of the pancreas, the margins may not be clearly delineated and fat infiltration may not be homogenous, interfering with assessment (Matsumoto et al. 1995). Moreover, there is normal fatty replacement of the pancreas as an age-related physiological process (Kühn et al. 2015; Chantarojanasiri et al. 2014). Body habitus can be a challenge, as transabdominal ultrasound may not provide good visualization if there is substantial abdominal fat. Study-related factors like cost and accessibility also need to be considered.

Transabdominal Ultrasound

Transabdominal ultrasound is a readily available, cost-effective, fast, and noninvasive. An increased echogenicity is seen in the pancreatic tissue when there is fat infiltration, and the kidney or the liver is used for comparison (Yu and Wang 2017; Al-Haddad et al. 2009; Romana et al. 2018). It does have inherent limitations such as impaired visualization of the pancreatic tissue if the patient is obese, or if there are overlying shadows due to bowel gas. It is also an operator-dependent technique, which can impact the accuracy of the test and the quality of the images. Fibrosis of the pancreatic tissue is also seen as increased echogenicity, and this could be difficult to differentiate from steatosis (Ustundag 2011). Some attempt at using quantitative techniques to evaluate the echogenicity pancreas like the pancreato-perihepatic fat index (linked to metabolic syndrome) (Jeong et al. 2014).

Endoscopic Ultrasound (EUS)

As the probe is of a higher frequency and is able to get in closer approximation to the pancreatic tissue, it allows for an enhanced quality of the images obtained (Pinte et al. 2019). Another benefit is that it allows for a more complete visualization of the pancreas, and for simultaneous comparison of the echogenicity to adjacent organs (Al-Haddad et al. 2009). However, it is an invasive test with an inherent risk of complications (Polkowski et al. 2017). It is similarly operator dependent, with some difficulty in differentiation from fibrosis[50.] Real-time elastography during EUS has been tried as a measure of pancreatic steatosis; however, it did not show an association between elasticity and steatosis like it did for fibrosis (Kuwahara et al. 2017; Barreto et al. 2018). NAFPD can be divided into four grades of severity on EUS when comparison is made of the echogenicity of the pancreas and the main pancreatic duct, against that of the spleen. Normal pancreatic tissue is associated with grades I and II and fatty infiltration seen with grades II and IV (Sepe et al. 2011).

CT Scan

CT scan employs Hounsfield units, and areas of fatty infiltration will be hypodense in comparison to the spleen or the liver (Pinte et al. 2019). When CT with contrast is performed, the normal pancreatic tissue will appear enhanced as compared to the regions of fatty infiltration (Kim et al. 2014a). CT scan is endowed with correlation to histology as well (Pinte et al. 2019). Measurements such as the pancreas-to-spleen attenuation ratio and fat/ parenchyma ratio can be derived from the CT scan and used for quantification of pancreatic fat (Saisho et al. 2007; Kim et al. 2014a). Among the downsides, radiation exposure and potential contrast exposure require attention.

MRI and MRS

Magnetic resonance imaging (MRI) and magnetic resonance spectroscopy (MRS) are among the preferred imaging modalities. Advantages include lack of radiation exposure and of invasiveness (Pinte et al. 2019). There is a good correlation with the amount of fatty infiltration detected on MRS and the amount of triglycerides present in the islet cells, and hence, MRS can be used as a surrogate for quantitative assessment of fat deposition within the islet cells (Lee et al. 2009; Lingvay et al. 2009). Single-voxel MR spectroscopy has been shown to be most comparable to histology, being considered the most accurate imaging modality (Kim et al. 2014a; Lee et al. 2009; Lingvay et al. 2009; Catanzaro et al. 2016; Ma et al. 2014). A novel MRI modality known as three-dimensional iterative decomposition with echo asymmetry and least-squares estimation (IDEAL) is stated to be superior to MRS for detection of pancreatic lipomatosis (Hu et al. 2009). Though it has several advantages including greater accuracy, there are drawbacks such as being more expensive as compared to the other techniques and more time consuming to perform, with limited availability.

Histology

In nonalcoholic fatty liver disease, there is deposition of triglyceride within the cells however in NAFPD the main mechanism of steatosis is the infiltration by adipocytes into the pancreatic tissue (Catanzaro et al. 2016). Intracellular accumulation of triglycerides in the islet and acinar cells can still occur (Catanzaro et al. 2016)· Both forms of fat accumulation can be damaging to the pancreatic tissue. The adipocytes can exert a paracrine effect on the acinar and islet cells and induce injury, whereas the intracellular accumulation can lead to direct damage (Catanzaro et al. 2016). Scoring systems such as the pancreatic lipomatosis score formulated in 1978 by Olsen can be used to grade the severity of steatosis (Olsen 1978). A modification of this score by van Geenen et al has also been developed (Pinte et al. 2019).

Possible Complications—Metabolic Syndrome

Shared disease entities encompass fatty liver disease, diabetes as well as the potential for cardiovascular disease (Pinte et al. 2019; Bi et al. 2019). Increased fatty accumulation has been seen in the pancreas of those with impaired glucose tolerance, and this could eventually progress to the development of overt type II diabetes (Li et al. 2017; Ou et al. 2013). Cardiovascular risk is also seen to be increased in patients with pancreatic steatosis, in the form of atherosclerosis. NAFPD has been associated with increased epicardial adipose tissue and increased intima-media thickness of the aorta, and NAFPD could be a surrogate for the presence of subclinical atherosclerosis (Kul et al. 2019). In non-obese type II diabetic patients, there was increased prevalence of carotid artery plaque and vascular stiffness (Kim et al. 2014b). Patients with NAFLD also tend to have NAFPD but the reverse association is not as clear (Lesmana et al. 2015; Li et al. 2017).

Acute and Chronic Pancreatitis

Obesity and other aspects of the metabolic syndrome are considered to be predisposing factors to acute pancreatitis (Pinte et al. 2019). The degree of pancreatic steatosis has been associated with the severity of the inflammation in acute pancreatitis (Smits and Geenen 2011). By promoting the release of cytokines such as IL-1β and TNF-α, pancreatic steatosis exacerbates the inflammatory response in acute pancreatitis and hence the severity of the injury (Papachristou et al. 2006). Acinar cells respond to cytotoxicity by the release on enzymes including lipases, with causes lipolysis yielding free fatty acids (Catanzaro et al. 2016; Acharya et al. 2014). Unsaturated fatty acids further exert direct damage to the acinar cells resulting in necrosis, and areas surrounding necrotizing fat endure more severe injury, known as peri-fat acinar necrosis (Acharya et al. 2013; Navina et al. 2011).

The association between pancreatic steatosis and chronic pancreatitis is not as clear, and there may be a negative association between the two (Mathur et al. 2007b). In cases of chronic pancreatitis with recurrent acute episodes and features of metabolic syndrome, the mechanism is not due to fatty infiltration but instead to fibrosis. Decrease in normal pancreatic parenchyma and fatty replacement could be just a consequence (Pinte et al. 2019).

Pancreatic Insufficiency

Fatty infiltration of the pancreas potentially causes exocrine pancreatic insufficiency by inducing lipotoxicity to the acinar cells, when there is intracellular accumulation of fat (Catanzaro et al. 2016). The adipocytes can also exert an adverse paracrine influence, and the death of acinar cells can occur (Catanzaro et al. 2016). However, there is limited evidence, and this is only hypothetical.

Pancreatic Fibrosis

Long-standing inflammation in NASH eventually progresses to liver fibrosis, and it is speculated that a similar progression would occur in nonalcoholic steatopancreatitis (NASP) (Catanzaro et al. 2016). Diabetic rats given a high-fat diet suffered intracellular fat deposition in the acinar cells and eventually fibrosis (Matsuda et al. 2014). Another study conducted on swine did not show a link between pancreatic fat accumulation and fibrotic changes (Fullenkamp et al. 2011). In over 900 human autopsies, there was no correlation between pancreatic steatosis and fibrosis (van Geenen et al. 2011). A study performed by Mathur et al revealed that there was a negative association between pancreatic steatosis and fibrosis (Mathur et al. 2007b).

Pancreatic Cancer

Obesity is a known predisposing factor for pancreatic cancer, and NAFPD has been shown to be an independent risk factor for its development as well (Catanzaro et al. 2016). An association between pancreatic steatosis (intra- and extralobular deposition) and pancreatic intraepithelial neoplasia has been seen (Rebours et al. 2015). A proposed mechanism for carcinogenesis in NAFPD may be similar to that in the liver, in that the increase in adipocytes can cause inflammation and NASP, which can predispose to malignant changes over time (Smits and Geenen 2011). Pancreatic cancers that develop as a consequence of NAFPD may have a worse prognosis compared to those without a fatty pancreas (Catanzaro et al. 2016). NAFPD can promote the spread of the tumor and increase the associated mortality, by causing changes in the microenvironment of the tumor giving it more aggressive properties (Mathur et al. 2009).

Surgical Risk

With pancreatic steatosis, there is a higher likelihood of blood loss during surgery and formation of a pancreatic fistula postoperatively (Catanzaro et al. 2016; Mathur et al. 2007b). With regard to pancreatic transplant, obesity and older age (elderly patients that are obese likely have steatosis) of the transplant donor and recipient are associated with pancreatic fistula, graft pancreatitis, and thrombosis as well as infections (Catanzaro et al. 2016). Interestingly, however, obesity does not impact allograft failure or rejection (Hanish et al. 2005). Hence, preoperative investigations aimed to detect any pancreatic steatosis are helpful for decreasing perioperative risks and salvaging organs for ideal candidates (Pinte et al. 2019; Catanzaro et al. 2016).

Treatment

As it is linked to metabolic syndrome and type II diabetes mellitus, similar approaches to treatment with weight loss and dietary modifications have been anticipated (Pinte et al. 2019). Oral hypoglycemic agents used in type II diabetes patients such as metformin, DPP-4 inhibitors, and thiazolidinediones have been preliminarily assessed and show some promise (Pinte et al. 2019).

Weight Loss

Weight loss is associated with a reduction in the amount of pancreatic fat and increased insulin sensitivity (Honka et al. 2015; Lim et al. 2011). Postbariatric surgery, there is a significant reduction in pancreatic fat and fatty acid uptake in pancreatic tissue with preservation of pancreatic blood flow, and no alteration of the pancreatic fat-free volume(Honka et al. 2015). In another bariatric protocol, eight weeks after the surgery, there was an improvement in the first-phase insulin response and pancreatic fat content only in those with diabetes, however not in the other group. Hence, the decrease in the pancreatic fat was linked to the underlying presence of diabetes and not simply a consequence of reduction in total body fat (Steven et al. 2015).

Oral Hypoglycemic Agents

Metformin reduces the oxidative stress on islet cells and may even impact insulin secretion favorably (Piro et al. 2012). It may decrease visceral fat and insulin resistance (Tajima et al. 2017). Incretin-based therapies like DPP-4 inhibitors such as Sitagliptin can be used alone or in addition to metformin for decreasing pancreatic fat (Reimer et al. 2014).

References

Acharya C, Cline R, Jaligama D et al (2013) Fibrosis reduces severity of acute-on-chronic pancreatitis in humans. Gastroenterology 145(2):466–475

Acharya C, Navina S, Singh V (2014) Role of pancreatic fat in the outcomes of pancreatitis. Pancreatology 14 (5):403–408

Al-Haddad M, Khashab M, Zyromski N, Pungpapong S, Wallace MB, Scolapio J, Woodward T, Noh K, Raimondo M (2009) Risk factors for hyperechogenic pancreas on endoscopic ultrasound: a case-control study. Pancreas 38:672–675

Barreto S, Dirkzwager I, Windsor J, Pandanaboyana S (2018) Predicting post-operative pancreatic fistulae using preoperative pancreatic imaging: a systematic review. ANZ J Surg 89(6):659–665

Bi Y, Wang J, Li M, Zhou J, Sun X (2019) The association between pancreas steatosis and metabolic syndrome: a systematic review and meta-analysis. Diabetes Metab Res Rev 35(5):e3142

Catanzaro R, Cuffari B, Italia A, Marotta F (2016) Exploring the metabolic syndrome: nonalcoholic fatty pancreas disease. World J Gastroenterol 22 (34):7660–7675

CDC (2020) Products—Data Briefs—Number 288—October 2017 [Internet]. Cdc.gov. 2020 [cited 2020 Jan 24]. https://www.cdc.gov/nchs/products/databriefs/db288.htm

Chantarojanasiri T, Hirooka Y, Ratanachu-ek T, Kawashima H, Ohno E, Goto H (2014) Evolution of pancreas in aging: degenerative variation or early changes of disease? J Med Ultrason 42(2):177–183

Chehter EZ, Longo MA, Laudanna AA, Duarte MIS (2000) Involvement of the pancreas in AIDS: a prospective study of 109 post-mortems. AIDS 14 (13):1879–1886

Choi C (2010) Associated factors for a hyperechogenic pancreas on endoscopic ultrasound. World J Gastroenterol 16(34):4329

Daentl DL, Frías JL, Gilbert EF, Opitz JM, Kaye C (1979) The Johanson-Blizzard syndrome: case report and autopsy findings. Am J Med Genet 3(2):129–135

Fazel Y, Koenig AB, Sayiner M, Goodman ZD, Younossi ZM (2016) Epidemiology and natural history of non-alcoholic fatty liver disease. Metabolism 65 (8):1017–1025

Finkelstein EA, Trogdon JG, Cohen JW, Dietz W (2009) Annual medical spending attributable to obesity: payer-and service-specific estimates. Health Aff 28: w822–w831

Fraulob JC, Ogg-Diamantino R, Fernandes-Santos C, Aguila MB, Mandarim-De-Lacerda CA (2010) A mouse model of metabolic syndrome: insulin resistance, fatty liver and non-alcoholic fatty pancreas disease (NAFPD) in C57BL/6 mice fed a high fat diet. J Clin Biochem Nutr 46(3):212–223

Fullenkamp A, Bell L, Robbins R et al (2011) Effect of different obesogenic diets on pancreatic histology in Ossabaw miniature swine. Pancreas 40(3):438–443

van Geenen E, Smits M, Schreuder T, van der Peet D, Bloemena E, Mulder C (2011) Smoking is related to pancreatic fibrosis in humans. Am J Gastroenterol 106 (6):1161–1166

Glaser J, Stienecker K (2000) Pancreas and aging: a study using ultrasonography. Gerontology 46(2):93–96

Grundy S (2019) Then and now: ATP III vs. IV—American College of Cardiology [Internet]. American College of Cardiology. 2013. https://www.acc.org/latest-in-cardiology/articles/2014/07/18/16/03/then-and-now-atp-iii-vs-iv

Hanish S, Petersen R, Collins B et al (2005) Obesity predicts increased overall complications following pancreas transplantation. Transplant Proc 37 (8):3564–3566

Heni M, Machann J, Staiger H et al (2010) Pancreatic fat is negatively associated with insulin secretion in individuals with impaired fasting glucose and/or impaired glucose tolerance: a nuclear magnetic resonance study. Diabetes Metab Res Rev 26(3):200–205

Honka H, Koffert J, Hannukainen J et al (2015) The effects of bariatric surgery on pancreatic lipid metabolism and blood flow. J Clin Endocrinol Metabol 100 (5):2015–2023

Hu H, Kim H, Nayak K, Goran M (2009) Comparison of fat–water MRI and single-voxel MRS in the assessment of hepatic and pancreatic fat fractions in humans. Obesity 18(4):841–847

Huang PL (2009) A comprehensive definition for metabolic syndrome. Dis Model Mech 2(5-6):231–237

Jeong H, Lee M, Kim M (2014) Quantitative analysis of pancreatic echogenicity on transabdominal sonography: correlations with metabolic syndrome. J Clin Ultrasound 43(2):98–108

Kim S, Kim H, Cho J et al (2014a) Quantitative assessment of pancreatic fat by using unenhanced CT: pathologic correlation and clinical implications. Radiology 271(1):104–112

Kim M, Chun H, Park J et al (2014b) The association between ectopic fat in the pancreas and subclinical atherosclerosis in type 2 diabetes. Diabetes Res Clin Pract 106(3):590–596

Kühn J, Berthold F, Mayerle J et al (2015) Pancreatic steatosis demonstrated at MR imaging in the general population: clinical relevance. Radiology 276 (1):129–136

Kul S, Karadeniz A, Dursun I, Şahin S, Faruk Çırakoğlu Ö, Raşit Sayın M, Turan T, Hakan AA (2019) Non-alcoholic fatty pancreas disease is associated with increased epicardial adipose tissue and aortic intima-media thickness. Acta Cardiol Sin 35 (2):118–125

Kuwahara T, Hirooka Y, Kawashima H et al (2017) Usefulness of endoscopic ultrasonography-elastography as a predictive tool for the occurrence of pancreatic fistula after pancreatoduodenectomy. J Hepato-Biliary-Panc Sci 24(12):649–656

Le K-A, Ventura EE, Fisher JQ et al (2011) Ethnic differences in pancreatic fat accumulation and its relationship with other fat depots and inflammatory markers. Diabetes Care 34(2):485–490

Lee Y, Lingvay I, Szczepaniak L, Ravazzola M, Orci L, Unger R (2009) Pancreatic steatosis: harbinger of type 2 diabetes in obese rodents. Int J Obes 34(2):396–400

Lesmana C, Pakasi L, Inggriani S, Aidawati M, Lesmana L (2015) Prevalence of non-alcoholic fatty pancreas disease (NAFPD) and its risk factors among adult medical check-up patients in a private hospital: a large cross sectional study. BMC Gastroenterol 15(1):174

Li S, Su L, Lv G, Zhao W, Chen J (2017) Transabdominal ultrasonography of the pancreas is superior to that of the liver for detection of ectopic fat deposits resulting from metabolic syndrome. Medicine 96(37):e8060

Lim E, Hollingsworth K, Aribisala B, Chen M, Mathers J, Taylor R (2011) Reversal of type 2 diabetes: normalisation of beta cell function in association with decreased pancreas and liver triacylglycerol. Diabetologia 54(10):2506–2514

Lin W-C, Chen J-H, Lin C-H, Shen W-C (2007) Rapidly progressive pancreatic lipomatosis in a young adult patient with transfusion-dependent myelodysplastic syndrome. J Formos Med Assoc 106(8):676–679

Lingvay I, Esser V, Legendre J et al (2009) Noninvasive quantification of pancreatic fat in humans. J Clin Endocrinol Metabol 94(10):4070–4076

Ma J, Song Z, Yan F (2014) Detection of hepatic and pancreatic fat infiltration in type II diabetes mellitus patients with IDEAL-Quant using 3.0T MR: comparison with single-voxel proton spectroscopy. Chin Med J 127(20):3548–3552

Mathur A, Marine M, Lu D et al (2007a) Nonalcoholic fatty pancreas disease. HPB 9(4):312–318

Mathur A, Pitt HA, Marine M, Sazena R, Schmidt CM, Howard TJ et al (2007b) Fatty pancreas: a factor in postoperative pancreatic fistula. Ann Surg 246 (6):1058–1064

Mathur A, Zyromski N, Pitt H et al (2009) Pancreatic steatosis promotes dissemination and lethality of pancreatic cancer. J Am Coll Surg 208(5):989–994

Matsuda A, Makino N, Tozawa T et al (2014) Pancreatic fat accumulation, fibrosis, and acinar cell injury in the Zucker diabetic fatty rat fed a chronic high-fat diet. Pancreas 43(5):735–743

Matsumoto S, Mori H, Miyake H et al (1995) Uneven fatty replacement of the pancreas: evaluation with CT. Radiology 194(2):453–458

Midiri M, Casto AL, Sparacia G et al (1999) MR imaging of pancreatic changes in patients with transfusion-dependent beta-thalassemia major. Am J Roentgenol 173(1):187–192

Navina S, Acharya C, DeLany J et al (2011) Lipotoxicity causes multisystem organ failure and exacerbates acute pancreatitis in obesity. Sci Transl Med 3 (107):107ra110

Ogilvie RF (1933) The islands of langerhans in 19 cases of obesity. J Pathol Bacteriol 37(3):473–481

Ogiu N, Nakamura Y, Ijiri I, Hiraiwa K, Ogiu T (1997) A statistical analysis of the internal organ weights of normal Japanese people. Health Phys 72(3):368–383

Olsen TS (1978) Lipomatosis of the pancreas in autopsy material and its relation to age and overweight. Acta Pathol Microbiol Scand A Pathol 86A(1-6):367–373

Ou H, Wang C, Yang Y, Chen M, Chang C (2013) The association between nonalcoholic fatty pancreas disease and diabetes. PLoS One 8(5):e62561

Papachristou G, Papachristou D, Avula H, Slivka A, Whitcomb D (2006) Obesity increases the severity of acute pancreatitis: performance of APACHE-O Score and correlation with the inflammatory response. Pancreatology 6(4):279–285

Perumpail R, Wong R, Ahmed A, Harrison S (2015) Hepatocellular carcinoma in the setting of non-cirrhotic nonalcoholic fatty liver disease and the metabolic syndrome: US experience. Dig Dis Sci 60 (10):3142–3148

Pham YH, Bingham BA, Bell CS et al (2016) Prevalence of pancreatic steatosis at a pediatric tertiary care center. South Med J 109(3):196–198

Pinte L, Balaban D, Băicuş C, Jinga M (2019) Non-alcoholic fatty pancreas disease—practices for clinicians. Rom J Intern Med 57(3):209–219

Piro S, Rabuazzo AM, Renis M, Purrelo F (2012) Effects of metformin on oxidative stress, adenine nucleotides balance, and glucose-induced insulin release impaired by chronic free fatty acids exposure in rat pancreatic islets. J Endocrinol Investig 35(5):504–510

Pi-Sunyer X (2009) The medical risks of obesity. Postgrad Med 121(6):21–33

Polkowski M, Jenssen C, Kaye P et al (2017) Technical aspects of endoscopic ultrasound (EUS)-guided sampling in gastroenterology: European Society of Gastrointestinal Endoscopy (ESGE) Technical Guideline—March 2017. Endoscopy 49(10):989–1006

Raeder H, Haldorsen I, Ersland L et al (2007) Pancreatic lipomatosis is a structural marker in nondiabetic children with mutations in carboxyl-ester lipase. Diabetes 56(2):444–449

Rebours V, Gaujoux S, d'Assignies G et al (2015) Obesity and fatty pancreatic infiltration are risk factors for pancreatic precancerous lesions (PanIN). Clin Cancer Res 21(15):3522–3528

Reimer R, Grover G, Koetzner L, Gahler R, Lyon M, Wood S (2014) Combining sitagliptin/metformin with a functional fiber delays diabetes progression in Zucker rats. J Endocrinol 220(3):361–373

Romana B, Chela H, Dailey F, Nassir F, Tahan V (2018) Non-alcoholic fatty pancreas disease (NAFPD): a silent spectator or the fifth component of metabolic syndrome? A literature review. Endocr Metab Immune Disord Drug Targets 18(6):547–554

Rössle R (1921) Beiträge zur Kenntnis der gesunden und der kranken Bauchspeicheldrüse. Beitr Path Anat 69:163

Saisho Y, Butler A, Meier J et al (2007) Pancreas volumes in humans from birth to age one hundred taking into account sex, obesity, and presence of type-2 diabetes. Clin Anat 20(8):933–942

Schaefer JH (1926) The normal weight of the pancreas in the adult human being: a biometric study. Anat Rec 32 (2):119–132

Schmitz-Moormann P, Pittner P, Heinze W (1981) Lipomatosis of the pancreas. Pathol—Res Pract 173 (1-2):45–53

Sepe PS, Ohri A, Sanaka S et al (2011) A prospective evaluation of fatty pancreas by using EUS. Gastrointest Endosc 73(5):987–993

Singh RG, Yoon HD, Wu LM, Lu J, Plank LD, Petrov MS (2017) Ectopic fat accumulation in the pancreas and its clinical relevance: a systematic review, meta-analysis, and meta-regression. Metabolism 69:1–13

Smits MM, Geenen EJMV (2011) The clinical significance of pancreatic steatosis. Nat Rev Gastroenterol Hepatol 8(3):169–177

Steven S, Hollingsworth K, Small P et al (2015) Weight loss decreases excess pancreatic triacylglycerol specifically in type 2 diabetes. Diabetes Care 39(1):158–165

Szczepaniak LS, Victor RG, Mathur R et al (2012) Pancreatic steatosis and its relationship to—cell dysfunction in humans: racial and ethnic variations. Diabetes Care 35(11):2377–2383

Tajima K, Shirakawa J, Okuyama T et al (2017) Effects of metformin on compensatory pancreatic β-cell hyperplasia in mice fed a high-fat diet. Am J Phy Endocrinol Metabol 313(3):E367–E380

Tariq H, Nayudu S, Akella S, Glandt M, Chilimuri S (2016) Non-alcoholic fatty pancreatic disease: a review of literature. Gastroenterol Res 9(6):87–91

Tenenbaum A, Fisman E (2011) "The metabolic syndrome… is dead": These reports are an exaggeration. Cardiovasc Diabetol 10(1):11

Tilg H, Hotamisligil GS (2006) Nonalcoholic fatty liver disease: cytokine-adipokine interplay and regulation of insulin resistance. Gastroenterology 131(3):934–945

Tushuizen M, Bunck M, Pouwels P et al (2007) Pancreatic fat content and—cell function in men with and without type 2 diabetes. Diabetes Care 30(11):2916–2921

Ustundag Y (2011) Pancreatic hyperechogenicity on endoscopic ultrasound examination. World J Gastroenterol 17(15):2061

Walters MN-I (1966) Adipose atrophy of the exocrine pancreas. J Pathol Bacteriol 92(2):547–557

Wang CY, Ou HY, Chen MF, Chang TC, Chang CJ (2014) Enigmatic ectopic fat: prevalence of nonalcoholic fatty pancreas disease and its associated factors in a Chinese population. J Am Heart Assoc 3(1):e000297

WHO (2020) Obesity and overweight [Internet]. Who.int. 2020 [cited 2020 Jan 24]. https://www.who.int/news-room/fact-sheets/detail/obesity-and-overweight

Wong V, Wong G, Yeung D et al (2014) Fatty pancreas, insulin resistance, and β-cell function: a population study using fat-water magnetic resonance imaging. Am J Gastroenterol 109(4):589–597

Wu WC, Wang CY (2013) Association between non-alcoholic fatty pancreatic disease (NAFPD) and the metabolic syndrome: case-control retrospective study. Cardiovasc Diabetol 12:77

Ye T, Liu R (2017) Effect of octreotide on pancreatic fibrosis of high-fat diet induced obesity in rats. Gastroenterology 152(5):4784–4794

Younossi ZM (2019) Non-alcoholic fatty liver disease—a global public health perspective. J Hepatol 70 (3):531–544

Yu T-Y, Wang C-Y (2017) Impact of non-alcoholic fatty pancreas disease on glucose metabolism. J Diab Invest 8(6):735–747

Zhao Z, Xin L, Xia J, Yang S, Chen Y, Li K (2015) Long-term high-fat high-sucrose diet promotes enlarged islets and β-cell damage by oxidative stress in Bama Minipigs. Pancreas 44(6):888–895

Normal-weight Obesity: A Hidden Pandemic

26

Shajith Anoop and Nitin Kapoor

Contents

Introduction 348

Healthy Versus Unhealthy Lean and Obese Subjects 348

Prevalence of Normal-weight Obesity 349

Diagnostic Criteria for Normal weight obesity 349
Possible Contribution of Free Fatty Acids (FFAs) 350

Pathogenesis and Cardiometabolic Risk of Normal-weight Obesity 350

Metabolic Consequences of Normal-weight Obesity 351
Liver Fat Accumulation 351
Fatty Pancreas 352

Normal-weight Central Obesity (NWCO) - A Similar Entity as Normal Weight Obesity 352
Abdominal Obesity 352
Epicardial Adipose Tissue (EAT) 353
Involved Pathways 353

Diagnostic Tools: Body Composition Analysis 353
Other Asian Populations 354
Metabolomic Research in Normal-weight Obesity 354
Genomic Sequencing 355

S. Anoop
Department of Endocrinology, Diabetes and Metabolism, Christian Medical College, Vellore, India

N. Kapoor (✉)
Department of Endocrinology, Diabetes and Metabolism, Christian Medical College, Vellore, India

Non Communicable Disease Unit, Melbourne School of Population and Global Health, Faculty of Medicine, Dentistry and Health Sciences, University of Melbourne, Parkville, VIC, Australia
e-mail: nitin.kapoor@cmcvellore.ac.in

J. Faintuch and S. Faintuch (eds.), *Obesity and Diabetes*, https://doi.org/10.1007/978-3-030-53370-0_26

Therapeutics and Prevention of Normal-weight Obesity 355
Conclusion 355
References 355

Abstract

There is a large proportion of individuals affected with type 2 diabetes and metabolic disorders in the south Asian population. However, as per the recent global burden of disease study, there are very few individuals with obesity located in this region. This paradox has reinforced the concept of normal-weight obesity, actually originated in 1981 as metabolically obese normal-weight individual (MONW). It is defined as presence of normal weight despite presence of an increased body fat percentage. The several deterrents in the diagnosis and management of normal-weight obesity, as well as of metabolically normal obese subjects, are addressed in this chapter.

Keywords

Metabolically healthy obesity · Normal-weight obesity · Body mass index cutoff points · Obesity/South Asians · Diabetes/ South Asians

Introduction

Due to a unique phenotype of higher body fat and the occurrence of type 2 diabetes, dyslipidemia, and other cardiovascular morbidities at a lower range of body mass index (BMI) (Conus et al. 2007; Ashwell et al. 2014) and waist circumference (WC) in South Asians, lower cutoff values have been defined :BMI: 23–24.9 kg/m^2 for overweight and $\geq$ 25 kg/m^2 for obesity (Gray et al. 2011). A consensus guideline for these revised measures has been developed for Asian Indians (Misra et al. 2009).

In Ethiopian adults, the optimal cutoff for obesity using body mass index was 22.2 kg/m^2 for males and 24.5 kg/m^2 for females with waist circumference values of 83.7 cm and 78.0 cm, respectively. For females, BMI cutoffs for metabolic syndrome markers ranged from 24.8 kg/m^2 to 26.8 kg/m^2, and for WC, 82.1–96.0 cm. As regards, waist/ height ratio (WHt-R) 0.47–0.56 was proposed for the healthy population, and for markers of metabolic syndrome, 0.78–0.89. For males, the optimal BMI cutoffs for metabolic syndrome markers ranged from 21.0 kg/m^2 to 23.5 kg/m^2 (Sinaga et al. 2018). More recent evidence from the Indian subcontinent shows that upto one-third of the south Indian population could have normal weight obesity (Kapoor et al. 2020a).

BMI does not account for age and gender and does not discriminate between individuals with higher lean body mass and low body fat (Pasco et al. 2014). WC, waist-to-hip ratio (WHR), Wht-R, and abdominal and supra-iliac skinfold thickness could be helpful. However, in individuals with metabolic obesity, anthropometric parameters such as WC, WHR, and Wht-R would not be accurate (Madeira et al. 2013).

Healthy Versus Unhealthy Lean and Obese Subjects

As alluded to, normal-weight individuals with metabolic derangements were first postulated in 1981 (Ruderman et al. 1981). In contrast, certain obese individuals remain insulin sensitive and metabolically healthy (Goossens 2017). Metabolically obese, normal-weight individuals with BMI less than 25 kg/m^2 present with hyperinsulinemia or insulin resistance, along with

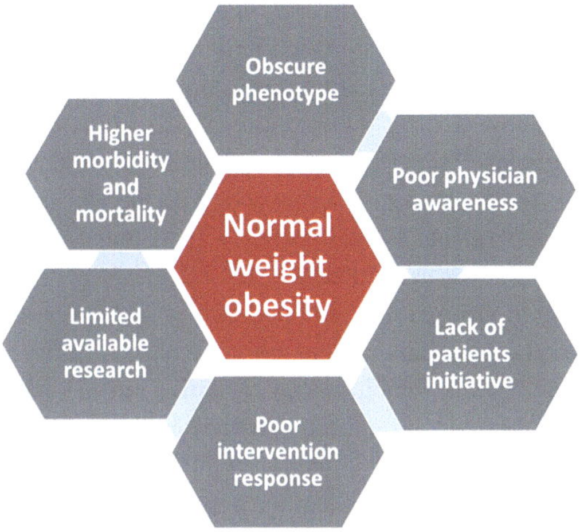

Fig. 26.1 Deterrents of normal-weight obesity in the society

hypertriglyceridemia and hypertension which predispose such individuals to a higher risk of cardiovascular diseases. On the other hand, obese individuals with either extreme insulin sensitivity or insulin resistance are a usual paradox. The term "metabolically healthy obesity"(MHO) describes the absence of any metabolic disorder including type two diabetes, dyslipidemia, and hypertension in an obese individual (Conus et al. 2007; Phillips 2017). However, there are no clear cutoffs demarcating insulin-sensitive and insulin-resistant groups in metabolically healthy obese persons. The several deterrents in the diagnosis and management of normal-weight obesity are summarized in Fig. 26.1.

The major difference between obese, metabolically unhealthy, and healthy subjects is high insulin sensitivity, high HDL levels, low ectopic fat, low triglyceride levels, and low degree of subclinical inflammation. Further, lower visceral, liver, and muscle fat contents are present in MHOB subjects than insulin-resistant obese individuals.

Prevalence of Normal-weight Obesity

The prevalence ranges between 5% and 45%, due to differences in phenotype, sample size effects, social and demographic factors, and lack of consensus on diagnostic criteria (Ding et al. 2016). The first study to define metabolic obesity targeted high body fat percentage (>30%) with no metabolic derangement (De Lorenzo et al. 2006). Meta-analysis studies have shown about 30% worldwide prevalence of metabolic obesity in normal-weight subjects (Wang et al. 2015), or 10–37% (Badoud et al. 2015). In American Hispanics, 43.1% were classified as metabolically obese (Benziger et al. 2015), and in Chinese, 9.5% for men and 6.1% for women were observed at the BMI range < 24.0 kg/m^2 (Jia et al. 2018).

In Koreans with BMI < 23 kg/m^2, Kim et al. detected 36% in men, with body fat % > 20.6% and 29% in women, with body fat % > 33.4% (Kim et al. 2014), while only 9.1% was reported in young Latin Americans (20–55 years of age) (Madeira et al. 2013). Just 2.2% was noticed in Spain by Goday et al. (Ding et al. 2016). In South India, the prevalence was 15.1% and 14.8% in obese subjects (Geetha et al. 2011). Among 6854 multi-ethnic women, the prevalence was 5.4%, especially in Indian women (Moy and Loh 2015)

Diagnostic Criteria for Normal weight obesity

Classically, a major criterion used is insulin sensitivity using the upper quartile of an insulin sensitivity index and Homeostasis model of assessment of insulin resistance (HOMA-IR). A lower level of insulin sensitivity was noted in overweight and obese groups; however, the metabolically normal, obese group with normal insulin sensitivity was similar to the normal-weight group for HOMA-IR value. Further, the metabolically normal, obese group had significantly lower mean total cholesterol and fasting glucose levels suggestive of an important role of hepatic fat, rather than visceral fat in insulin resistance (Zhao et al. 2016; Stefan et al. 2008). A hyperinsulinemic-euglycemic clamp (HEC) study in nonobese (BMI < 25 kg/m^2) Japanese subjects to evaluate muscle and hepatic insulin sensitivity has suggested impaired insulin sensitivity in peripheral tissue and low hepatic glucose production. This is probably due to decreased endogenous insulin sensitivity, in the metabolically obese group, which had significantly higher subcutaneous fat area and total fat than the metabolically nonobese group. Furthermore, oral glucose tolerance test (OGTT) studies showed lower values of Matsuda index but high values of fasting insulin and HOMA-IR in the metabolically obese group (Kaga et al. 2017). MHO subjects feature low amounts of visceral adipose tissue with higher levels of high-density lipoprotein (HDL) cholesterol and lower levels of fasting triglycerides, fasting glucose, fasting insulin, as well as reduced glucose and insulin area under the curve during an oral glucose tolerance test (Karelis et al. 2005).

The HOMA-IR has been combined with other metabolic criteria in order to identify MHO subjects. This implies the presence of fewer than two of the cardiometabolic abnormalities namely systolic/diastolic blood pressure above 130/80 mm Hg, triglycerides above 1.7 mmol/L, fasting glucose levels above or equal to 5.6 mmol/L, C reactive protein (CRP) above 0.1 mg/L, HDL levels less than 1.3 mmol/L (Wildman et al. 2008). In contrast, fulfilling four out of the five metabolic parameters namely HOMA index less than or equal to 2.7 mmol/L, triglycerides less than or equal to 1.7 mmol/L, and low-density lipoprotein cholesterol (LDL) less than or equal to 2.6 mmol/L, and CRP levels less than or equal to 3.0 mg/L were deemed sufficient to identify normal-weight metabolic obesity (MONW) with risk for cardiovascular diseases (Karelis and Rabasa-Lhoret 2008).

Possible Contribution of Free Fatty Acids (FFAs)

FFAs released by lipolysis from the adipocytes are elevated in subjects with metabolic obesity, as in obese subjects. FFAs combine with glycerol to form triglycerides (TG) and are then consolidated by the liver in water-soluble form as very low-density lipoprotein (VLDL) particles. These particles are large and mainly contain TG, along with some cholesterol. Triglyceride, being an energy substrate, is hydrolysed in various tissues. Particles become smaller and dense gradually, forming intermediate-density lipoprotein (IDL) and low-density lipoprotein (LDL) cholesterol particles (Sniderman et al. 2001). High levels of FFAs are strongly associated with hepatic and peripheral insulin resistance, T2DM and cardiovascular diseases (Rachek 2014), due to generation of deleterious lipid metabolites (Boden et al. 2001), and pro-inflammatory cytokines (Boden et al. 2005), which induce oxidative stress in the endoplasmic reticulum (Manna and Jain 2015) and endothelial dysfunction (Ghosh et al. 2017). Thus, it is deemed ideal to include estimation of fasting FFAs as criterion to define metabolic obesity in subjects with normal weight or low BMI.

Pathogenesis and Cardiometabolic Risk of Normal-weight Obesity

Intra-abdominal adipose tissue (IAAT), or visceral adipose tissue (VAT), is a determinant of insulin resistance and cardiovascular disease risk, differently from gluteo-femoral adiposity (Kn 2001; Ibrahim 2010). The IAAT is located primarily in the mesentery and omentum and drains directly through the portal circulation to the liver. The IAAT is more cellular, vascular, innervated, and contains a larger number of inflammatory and immune cells, lesser pre-adipocyte differentiating capacity and a greater percentage of large adipocytes, glucocorticoid, and androgen receptors, when compared to subcutaneous adipose tissue (SCAT). The adipocytes in IAAT are metabolically active, more sensitive to lipolysis, and more insulin resistant than SCAT adipocytes (Fig. 26.2). IAAT has a greater capacity to generate free fatty acids and to uptake glucose than SCAT and is more sensitive to adrenergic stimulation (Goossens 2008), while SCAT is more avid in absorption of circulating free fatty acids and triglycerides.

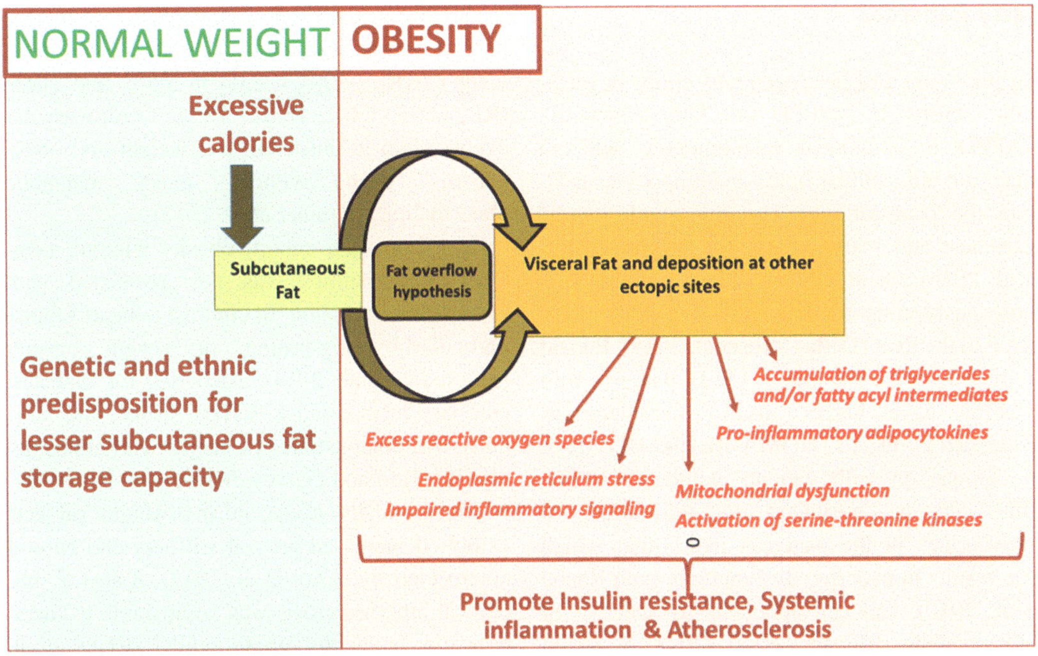

Fig. 26.2 Pathogenesis of normal-weight obesity

Metabolic Consequences of Normal-weight Obesity

Insulin resistance due to deranged lipid parameters, deficient adipose tissue FFA turnover, increased CRP levels, and oxidative stress along with possible influence of age, dietary, and physical activity patterns lead to increased risk for type 2 diabetes (Jean et al. 2014) and cardiovascular diseases (Romero-Corral et al. 2010a). In normal-weight subjects, ectopic fat in liver, skeletal muscles, and the pancreas are additional risk factors (Shulman 2014).

Liver Fat Accumulation

Non-alcoholic fatty liver disease (NAFLD) affects nearly 25% of the global population (Zhu et al. 2015), with insulin resistance being a signature mark, irrespective of body weight. Specifically, peripheral insulin resistance is a robust correlate of liver cirrhosis in NAFLD, more than visceral adiposity (Rosso et al. 2016). In patients with NAFLD, glucagon secretion, and signaling is dysregulated resulting in fasting and postprandial hyperglucagonemia (Foghsgaard et al. 2017), due to impaired paracrine functions of pancreatic beta cells (Unger and Cherrington 2012) and increased hepatic glucose production (Girard 2017). This metabolic state undergoes a vicious feedback cycle leading to homeostatic stress and impaired glycemic status, which has significant implications for renal and cardiovascular diseases (Petta et al. 2019). In addition, circulating FFAs (the major energy substrate for all tissues except brain) are markedly high during fasting state, due to adipose tissue lipolysis and lipoprotein spillover coupled with reduced plasma clearance. FFAs are re-esterified to triglycerides and accumulate in the muscles (intra-myocellular) and liver (intra-hepatocellular) and result in insulin resistance. Increased FFA levels result in elevated production of pro-atherogenic proteins through activation of the IKK/I B–/NF B and JNK pathways.

Fatty Pancreas

An important and heterogeneous entity in metabolic obesity is NAFPD or "fatty pancreas." NAFPD is a clinical phenomenon, wherein increased intracellular lipids and adipocytes infiltrate the pancreatic tissue, along with insulin resistance and beta-cell secretory defect (Tariq et al. 2016). Of note, the pancreas seems to be more susceptible to fat deposition when compared to the liver (Della Corte et al. 2015). Recent studies have shown that NAFLD patients with NAFPD have higher insulin resistance as evidenced by OGTT, HOMA-IR, and HOMA-b, in comparison with patients without NAFPD. This can be attributed to infiltration of triglycerides in the pancreas over time, which also results in exocrine dysfunctions (van Raalte et al. 2010), and cell apoptosis irrespective of body weight (Romana et al. 2018).A recent meta-analysis of 13 studies has shown that non-alcoholic fatty pancreas disease (NAFPD) was associated with a significant increased risk of metabolic syndrome (RR = 2.25; 95% CI, 2.00–2.53; $P < 0.0001$), hypertension, diabetes mellitus, and central obesity (RR = 1.91; 95% CI, 1.67–2.19; $P < 0.0001$) (Bi et al. 2019).

Normal-weight Central Obesity (NWCO) - A Similar Entity as Normal Weight Obesity

NWCO corresponds to a BMI of 18.5–23.9 kg/m^2 and a waist circumference (WC) >85 cm in males or >80 cm in females, a waist-to-height ratio (WHt-R) of ≥0.5, and a waist-to-hip ratio (WHR) of ≥0.9 in males or ≥0.85 in females. Among Chinese adults, it became much more prevalent from 1993 to 2011. Cardiometabolic risks encompass hypertension, diabetes, insulin resistance, decreased insulin sensitivity, and elevated low high-density lipoprotein and triglyceride levels (Song et al. 2019). In a large meta-analysis, an increased risk for cardiovascular events (RR, 1.24; 95% CI, 1.02 to 1.55) was perceived with ≥10 years of follow-up. All metabolically unhealthy groups were considered: normal weight (RR, 3.14; CI, 2.36 to 3.93), overweight (RR, 2.70; CI, 2.08 to 3.30), and obese (RR, 2.65; CI, 2.18 to 3.12). Cardiovascular events seem to cluster in overweight and obese persons even without overt metabolic abnormalities (Kramer et al. 2013).

In a European cohort, obesity without metabolic abnormalities was not associated with myocardial infarction, in contrast to heart failure, particularly long-lasting or severe obesity (Mørkedal et al. 2014). Also risk for diabetes, coronary heart disease (CHD), stroke, and mortality was demonstrated in people with unhealthy obesity (Guo and Garvey 2016).

Metabolically obese, normal-weight subjects exhibited increased arterial stiffness and carotid atherosclerosis (Choi et al. 2013). Risk for subclinical atherosclerosis was comparable to metabolically unhealthy obese adults (Rhee et al. 2014). In the Women´s Health Initiative, following 161,808 postmenopausal women for over a decade, a cumulative effect of excessive body weight and aberrant metabolic profile occurred. Diabetes risk approximately doubled among those with metabolically unhealthy normal weight, as well as the metabolically healthy obese, whereas diabetes affected nearly four times more those both obese and metabolically unhealthy (Cordola Hsu et al. 2020).

Abdominal Obesity

Abdominal obesity leads to cardiomyocyte hypertrophy, myocardial fibrosis and activation of inflammatory pathways relating to macrophage infiltration and cytokine gene expression (Murase et al. 2012). Specifically, low muscle mass and abdominal obesity in normal-weight individuals pose an adverse cardiometabolic risk profile, when compared to those without abdominal obesity. Metabolically, such individuals had higher values for glucose, insulin, HOMA-IR, branched-chain amino acids, and a higher prevalence of metabolic syndrome (Beijers et al. 2017). Hypertriglyceridemia, low and dysfunctional HDL, and increased small

LDL levels, and characteristics of atherogenic dyslipidemia are also relatively common (Chapman et al. 2011). Much of the cholesterol might be embedded in VLDL, IDL, and other remnant lipoproteins, including chylomicron remnants in the non-fasting state. These particles are equally potent as LDL in predicting cardiovascular events and all-cause mortality (Varbo et al. 2015).

Epicardial Adipose Tissue (EAT)

The EAT is located between the myocardium and visceral pericardium, mainly on the right ventricle surface and the anterior wall of the left ventricle of the heart. EAT volume and thickness are considered as surrogate markers of insulin resistance. EAT thickness of 9.5 mm or more in males and 7.5 mm or more in females has a high sensitivity and specificity for prediction of metabolic syndrome (Iacobellis et al. 2008).

A meta-analysis of 38 studies has shown significant correlations of EAT with components of metabolic syndrome specifically fasting blood glucose (FBG), systolic blood pressure (SBP), TGs and HDL-C, independent of obesity. A strong correlation of EAT with VAT was noted (Rabkin 2014). Magnetic resonance imaging in nondiabetic individuals with normal ventricular function has shown positive association of peripheral insulin resistance and glucose tolerance status with cardiac fat accumulation (Sironi et al. 2004). Similarly, EAT thickness measured by ultrasound was increased inversely to glucose disposal, as observed in euglycemic-hyperinsulinemic clamp study in obese subjects (Iacobellis and Leonetti 2005).

Involved Pathways

The EAT serves as a local reservoir of excess of FFAs to maintain myocardial energy supply and prevent toxic effects. During metabolic stress, EAT releases FFAs at high rate and regulates homeostasis of FFAs through the coronary circulation, including expression of transporters fatty acid-binding protein-4 [FABP-4] (Furuhashi et al. 2016). It also controls the vascular tension and the effect of insulin on microcirculatory networks (Meijer et al. 2015).

The EAT decreases vascular tension through the release of adipocytokines such as adiponectin, adrenomedullin, omentin, and other adipocyte-derived relaxing factors to retard vascular remodeling (Gruzdeva et al. 2017). Cytokines released from dysfunctional myocardial endothelium impair insulin signaling (Salgado-Somoza et al. 2012). Shifts in glucose uptake and lipogenesis along with more abundant FFA occur in both CAD and heart failure (HF), even in nondiabetic subjects (Burgeiro et al. 2016). Hyperglycemia and hyperinsulinemia favor accumulation of myocardial lipids, potentially leading to cardiac steatosis (Winhofer et al. 2012).

Diagnostic Tools: Body Composition Analysis

Fat mass, fat-free mass, lean mass, and body mineral density are precisely estimated on dual-energy X-ray absorptiometry (DEXA), while on bioimpedance, it is approximated on the basis of resistance of body to a small electrical impulse to estimate body water content and composition (Peppa et al. 2017; Teixeira et al. 2015) . Many years ago, the WHO has defined the cutoff values for body fat % in male and female obesity as $\geq$25% and $\geq$35%, respectively (WHO 1995), although the topic is controversial and different values can be found in the literature. Moreover, technological discrepancies exist regarding body fat percentage, among bioimpedance analysis and more direct methods such as DEXA or magnetic resonance imaging (MRI).

Phenotypically, Asians have a smaller body frame, based on the wrist circumference or elbow width, and are shorter in height when compared with Caucasians (Zhang et al. 2005). Importantly, South Asians feature higher body fat, lesser lean mass, higher ectopic fat with higher odds of metabolic obesity when compared to Caucasians, Hispanics, and other ethnic groups (Patel et al. 2016). In this purview, cutoff values for fat mass and lean mass have been defined by Misra et al., (Joseph et al. 2011) for Asian Indians, using the DEXA technique, exclusively for males and females. Accordingly, the appropriate cutoff values for total body fat (%), fat mass (kg), total lean mass (%), lean mass (kg), and fat-free mass (kg) were 25.5, 15.1, 73.7, 46.3, and 48.8 for males and 38.0, 20.3, 59.0, 32.6, and 34.8 for females. Corresponding values for fat arm (%), fat mass arm (kg), fat leg (%), fat mass leg (kg), fat trunk (%), and fat mass trunk (kg) were 19.8, 1.4, 21.9, 4.1, 30.1, and 8.8 for males and 43.3, 3.1, 38.9, 6.9, 38.5, and 9.5 for females (Joseph et al. 2011).

Based on MRI, appropriate cutoff values for abdominal adipose tissue compartments have also been defined for Asian Indians. Accordingly, the gender-specific cutoff limits for total abdominal adipose tissue, intra-abdominal adipose tissue, and subcutaneous abdominal adipose tissue were 245.6 cm^2 for males and 203.46 cm^2 in females, 135.3 cm^2 (male) and 75.73 cm^2 (female), and 110.74 cm^2 (male) and 134.02 cm^2 (female), respectively. For detection of CV risk factors, distinct gender differences were seen in ORs for SCAT [3.54 (95% confidence interval [CI], 1.10–11.46) and 6.6 (95% CI, 1.75–24.85) in males and females, respectively] but not for intra-abdominal adipose tissue, a result that conflicts with other experiences (Misra et al. 2010).

Other Asian Populations

A DEXA-based study on 2798 Japanese individuals aged between 20 and 79 years determined cutoff values for fat mass %, truncal fat mass, and truncal fat mass-leg fat mass ratio. Accordingly, the cutoff points were 24% for fat mass %, 8 kg for truncal fat mass and 1.6 for truncal fat mass-leg-fat mass ratio in males, and 35% (fat mass %) 9 kg (truncal fat mass) and 1.4 (truncal-fat mass ratio) females. WHR and truncal fat mass/leg fat mass ratio most accurately detected cardiovascular risk factors (Ito et al. 2003). A bioimpedance-based study on 1687 Sri Lankan individuals aged above 20.6 years, determined total body fat % and fat mass index thresholds of 25.6% and 7.0 kg/m^2 in men, and 39.0% and 11.9 kg/m^2 in women, as cutoff values to detect risk of metabolic syndrome (Ramírez-Vélez et al. 2017). Subsequently, another study on Sri Lankans aged between 18.5 and 25.1 kg/m^2 validated the International Diabetes Federation cutoff values for muscle-to-fat ratio to screen for metabolic syndrome. The cutoff value of fat-to-muscle ratio for metabolic syndrome was 0.225 kg and 0.495 kg for men and women, respectively (Ramírez-Vélez et al. 2018).

A cross-sectional study on 383 Chinese subjects aged between 18 and 24 years and BMI between 18.5 and 23.9 kg/m^2, defined metabolic obesity as a body fat percentage of > 20% or > 30% in male and female students, respectively. Individuals with higher body fat had low physical fitness due to low muscle mass (Zhang et al. 2018). Overall, it is evident that there is potential health risk for individuals with metabolic obesity. While there are significant differential effects due to inter-ethnic differences in comparative studies, due to methodological variations, confounding effects of age, gender, and sample sizes, there is a thrusting need for prospective studies to refine the diagnostic criteria for different populations (Franco et al. 2016).

Metabolomic Research in Normal-weight Obesity

Linoleic acid, β-alanine, histidine, carnitine, and aspartate/asparagine metabolism pathways are upregulated in normal-weight obese and overweight/obesity adults. Metabolites of lysine, glycine, serine, and the urea cycle were elevated only in case of overweight/obesity (Bellissimo et al. 2019). A higher profile of branched-chain amino acids, tyrosine, glutamic acid, and diacyl-phosphatidylcholines C32:1 and C38:3 was characteristic of normal-weight metabolically obese subjects, whereas a reduced pattern of acylcarnitine C18:2 and acyl-lysophosphatidylcholine C18:1 occurred in the metabolically healthy obese subjects (Bagheri et al. 2018).

Genomic Sequencing

Genome Wide Association Studies (GWAS) raise the possibility that *FTO* gene and other loci regulate body fat (Huang et al. 2018). Associations have also been detected for the *IRS1*, *FTO*, and *SPRY2* loci, along with nine novel loci (Morris et al. 2012). One allele at the *IRS1* locus was involved with lower HDL-C, higher TG levels, and elevated insulin resistance. The effect size for each allele was −0.20% and −0.06% of body fat for men and women, respectively. A higher VAT/SAT ratio resulted in men only. In addition, adiponectin diminished in men, without changes of leptin in either gender (Miki et al. 2001). Four other loci (in or near *COBLL1/GRB14, IGF2BP1, PLA2G6, CRTC1*) have been linked to body fat %. Lower expression of *GRB14* in subcutaneous and omental fat corresponded to one of the fat-reducing loci. The loci near *IRS1* and *PLA2G6* had more impact on men, contrasting with those close to *TMEM18* and CRTC1 which were more relevant for women (Lu et al. 2016).

Therapeutics and Prevention of Normal-weight Obesity

A longitudinal follow-up of normal-weight obese school children revealed increased cardiometabolic risk in early adulthood, seven years after diagnosis. In previously published results from our center, in relatively older individuals, a significant reduction in some but not all cardiovascular risk factors occurred (Wiklund et al. 2017). In another Swedish cohort of 3010 adults, it was found that normal-weight obesity was associated with higher levels of serum triglycerides, low-density lipoprotein cholesterol, C-reactive protein, apolipoprotein B, and blood pressure, along with lower high-density lipoprotein cholesterol. In our data, at least two of these parameters, namely HDL cholesterol and systolic blood pressure, significantly improved with lifestyle intervention (Berg et al. 2015). These findings are pertinent as normal-weight obesity increases the risk of developing atherosclerosis and macrovascular disease (Kim et al. 2015). At this point, it is unclear if individuals with normal-weight obesity would respond after a longer duration of lifestyle change, or they are inherently resistant. The potential impact of drugs like metformin and pioglitazone (insulin sensitizers) on improving the cardiometabolic profile of this unique phenotype also deserves attention.

NWO women are to metabolic syndrome (Marques-Vidal et al. 2010). Higher blood pressure, cholesterol, and fasting blood glucose were unveiled, when compared to low body fat counterparts. In the same study elevated leptin, interleukin-1a, interleukin-1b, and interleukin-8-levels occurred. In the NHANES III database, women with normal-weight obesity showed an independent association with cardiovascular mortality (Romero-Corral et al. 2010b). The association was maintained after adjustment for risk factors such as hypertension, dyslipidemia, and waist circumference. A more recent study revealed that with a peer-led lifestyle intervention only a minimal favourable change in systolic blood pressure and LDL was noted after 2 years (Kapoor et al. 2020b).

Conclusion

Normal-weight obesity is an obscure yet a common disorder prevalent not only in the south Asian population, but also in other parts of the world. Little is known about its therapeutic and prevention aspects. Moreover, the genetic predisposition of this phenotype needs to be more extensively explored. Improving the awareness of this clinical entity among physicians would promote earlier diagnosis and prevention of its associated complications.

References

Ashwell M, Mayhew L, Richardson J, Rickayzen B (2014) Waist-to-height ratio is more predictive of years of life lost than body mass index. PLoS One 9(9):e103483

Badoud F, Perreault M, Zulyniak MA, Mutch DM (2015) Molecular insights into the role of white adipose tissue in metabolically unhealthy normal weight and metabolically healthy obese individuals. FASEB J Off Publ Fed Am Soc Exp Biol 29(3):748–758
Bagheri M, Farzadfar F, Qi L et al (2018) Obesity-related metabolomic profiles and discrimination of metabolically unhealthy obesity. J Proteome Res 17 (4):1452–1462
Beijers RJHCG, van de Bool C, van den Borst B, Franssen FME, Wouters EFM, Schols AMWJ (2017) Normal weight but low muscle mass and abdominally obese: implications for the cardiometabolic risk profile in chronic obstructive pulmonary disease. J Am Med Dir Assoc 18(6):533–538
Bellissimo MP, Cai Q, Ziegler TR et al (2019) Plasma high-resolution metabolomics differentiates adults with normal weight obesity from lean individuals. Obes Silver Spring Md 27(11):1729–1737
Benziger CP, Bernabé-Ortiz A, Gilman RH et al (2015) Metabolic abnormalities are common among South American Hispanics subjects with normal weight or excess body weight: the CRONICAS Cohort Study. PLoS One 10(11):e0138968
Berg C, Strandhagen E, Mehlig K, Subramoney S, Lissner L, Bjorck L (2015) Normal weight adiposity in a Swedish population: how well is cardiovascular risk associated with excess body fat captured by BMI? Obes Sci Pract 1(1):50–58
Bi Y, Wang J-L, Li M-L, Zhou J, Sun X-L (2019) The association between pancreas steatosis and metabolic syndrome: a systematic review and meta-analysis. Diabetes Metab Res Rev 35(5):e3142
Boden G, Lebed B, Schatz M, Homko C, Lemieux S (2001) Effects of acute changes of plasma free fatty acids on intramyocellular fat content and insulin resistance in healthy subjects. Diabetes 50(7):1612–1617
Boden G, She P, Mozzoli M et al (2005) Free fatty acids produce insulin resistance and activate the proinflammatory nuclear factor-kappaB pathway in rat liver. Diabetes 54(12):3458–3465
Burgeiro A, Fuhrmann A, Cherian S et al (2016) Glucose uptake and lipid metabolism are impaired in epicardial adipose tissue from heart failure patients with or without diabetes. Am J Physiol Endocrinol Metab 310(7): E550–E564
Chapman MJ, Ginsberg HN, Amarenco P et al (2011) Triglyceride-rich lipoproteins and high-density lipoprotein cholesterol in patients at high risk of cardiovascular disease: evidence and guidance for management. Eur Heart J 32(11):1345–1361
Choi KM, Cho HJ, Choi HY et al (2013) Higher mortality in metabolically obese normal-weight people than in metabolically healthy obese subjects in elderly Koreans. Clin Endocrinol 79(3):364–370
Conus F, Rabasa-Lhoret R, Peronne F (2007) Characteristics of metabolically obese normal-weight (MONW) subjects. Appl Physiol Nutr Metab 32 (1):4–12
Cordola Hsu AR, Ames SL, Xie B, Peterson DV, Garcia L, Going SB, Phillips LS, Manson JE, Anton-Culver H, Wong ND (2020) Incidence of diabetes according to metabolically healthy or unhealthy normal weight or overweight/obesity in postmenopausal women: the Women's Health Initiative. Menopause 24:640–647
De Lorenzo A, Martinoli R, Vaia F, Di Renzo L (2006) Normal weight obese (NWO) women: an evaluation of a candidate new syndrome. Nutr Metab Cardiovasc Dis NMCD 16(8):513–523
Della Corte C, Mosca A, Majo F et al (2015) Nonalcoholic fatty pancreas disease and Nonalcoholic fatty liver disease: more than ectopic fat. Clin Endocrinol 83 (5):656–662
Ding C, Chan Z, Magkos F (2016) Lean, but not healthy: the "metabolically obese, normal-weight" phenotype. Curr Opin Clin Nutr Metab Care 19(6):408–417
Foghsgaard S, Andreasen C, Vedtofte L et al (2017) Nonalcoholic fatty liver disease is prevalent in women with prior gestational diabetes mellitus and independently associated with insulin resistance and waist circumference. Diabetes Care 40(1):109–116
Franco LP, Morais CC, Cominetti C (2016) Normal-weight obesity syndrome: diagnosis, prevalence, and clinical implications. Nutr Rev 74(9):558–570
Furuhashi M, Fuseya T, Murata M et al (2016) Local production of fatty acid-binding protein 4 in epicardial/perivascular fat and macrophages is linked to coronary atherosclerosis. Arterioscler Thromb Vasc Biol 36(5):825–834
Geetha L, Deepa M, Anjana RM, Mohan V (2011) Prevalence and clinical profile of metabolic obesity and phenotypic obesity in Asian Indians. J Diabetes Sci Technol 5(2):439–446
Ghosh A, Gao L, Thakur A, Siu PM, Lai CWK (2017) Role of free fatty acids in endothelial dysfunction. J Biomed Sci 24(1):50
Girard J (2017) Glucagon, a key factor in the pathophysiology of type 2 diabetes. Biochimie 143:33–36
Goossens GH (2008) The role of adipose tissue dysfunction in the pathogenesis of obesity-related insulin resistance. Physiol Behav 94(2):206–218
Goossens GH (2017) The metabolic phenotype in obesity: fat mass, body fat distribution, and adipose tissue function. Obes Facts 10(3):207–215
Gray LJ, Yates T, Davies MJ et al (2011) Defining obesity cut-off points for migrant South Asians. PLoS One 6 (10):e26464
Gruzdeva OV, Akbasheva OE, Dyleva YA et al (2017) Adipokine and cytokine profiles of epicardial and subcutaneous adipose tissue in patients with coronary heart disease. Bull Exp Biol Med 163(5):608–611
Guo F, Garvey WT (2016) Cardiometabolic disease risk in metabolically healthy and unhealthy obesity: stability of metabolic health status in adults. Obes Silver Spring Md 24(2):516–525
Huang LO, Loos RJF, Kilpeläinen TO (2018) Evidence of genetic predisposition for metabolically healthy

obesity and metabolically obese normal weight. Physiol Genomics 50(3):169–178

Iacobellis G, Leonetti F (2005) Epicardial adipose tissue and insulin resistance in obese subjects. J Clin Endocrinol Metab 90(11):6300–6302

Iacobellis G, Willens HJ, Barbaro G, Sharma AM (2008) Threshold values of high-risk echocardiographic epicardial fat thickness. Obes Silver Spring Md 16 (4):887–892

Ibrahim MM (2010) Subcutaneous and visceral adipose tissue: structural and functional differences. Obes Rev Off J Int Assoc Study Obes 11(1):11–18

Ito H, Nakasuga K, Ohshima A et al (2003) Detection of cardiovascular risk factors by indices of obesity obtained from anthropometry and dual-energy X-ray absorptiometry in Japanese individuals. Int J Obes 27 (2):232–237

Jean N, Somers VK, Sochor O, Medina-Inojosa J, Llano EM, Lopez-Jimenez F (2014) Normal-weight obesity: implications for cardiovascular health. Curr Atheroscler Rep 16(12):464

Jia A, Xu S, Xing Y et al (2018) Prevalence and cardiometabolic risks of normal weight obesity in Chinese population: a nationwide study. Nutr Metab Cardiovasc Dis NMCD 28(10):1045–1053

Joseph L, Wasir JS, Misra A et al (2011) Appropriate values of adiposity and lean body mass indices to detect cardiovascular risk factors in Asian Indians. Diabetes Technol Ther 13(9):899–906

Kaga H, Tamura Y, Takeno K et al (2017) Correlates of insulin clearance in apparently healthy non-obese Japanese men. Sci Rep 7(1):1462

Kapoor N et al (2020a) Prevalence of normal weight obesity and its associated cardio-metabolic risk factors – Results from the baseline data of the Kerala Diabetes Prevention Program (KDPP). PLoS One. 15(8): e0237974. https://doi.org/10.1371/journal.pone.0237974. PMID: 32841271; PMCID: PMC7446975

Kapoor N et al (2020b) Effect of a peer-led lifestyle intervention on individuals with normal weight obesity: insights from the Kerala Diabetes Prevention Program. Clin Ther 42(8):1618–1624. https://doi.org/10.1016/j.clinthera.2020.06.007. Epub 2020 Jul 24. PMID: 32718783

Karelis AD, Rabasa-Lhoret R (2008) Inclusion of C-reactive protein in the identification of metabolically healthy but obese (MHO) individuals. Diabetes Metab 34(2):183–184

Karelis AD, Faraj M, Bastard J-P et al (2005) The metabolically healthy but obese individual presents a favorable inflammation profile. J Clin Endocrinol Metab 90 (7):4145–4150

Kim MK, Han K, Kwon H-S et al (2014) Normal weight obesity in Korean adults. Clin Endocrinol 80 (2):214–220

Kim S, Kyung C, Park JS, Lee SP, Kim HK, Ahn CW et al (2015) Normal-weight obesity is associated with increased risk of subclinical atherosclerosis. Cardiovasc Diabetol 14:58

Kn F (2001) Adipose tissue and the insulin resistance syndrome [Internet]. Proc Nutr Soc [cited 2019 Dec 3]. https://pubmed.ncbi.nlm.nih.gov/11681812/

Kramer CK, Zinman B, Retnakaran R (2013) Are metabolically healthy overweight and obesity benign conditions?: A systematic review and meta-analysis. Ann Intern Med 159(11):758–769

Lu Y, Day FR, Gustafsson S et al (2016) New loci for body fat percentage reveal link between adiposity and cardiometabolic disease risk. Nat Commun 7:10495

Madeira FB, Silva AA, Veloso HF et al (2013) Normal weight obesity is associated with metabolic syndrome and insulin resistance in young adults from a middle-income country. PLoS One 8(3):e60673

Manna P, Jain SK (2015) Obesity, oxidative stress, adipose tissue dysfunction, and the associated health risks: causes and therapeutic strategies. Metab Syndr Relat Disord 13(10):423–444

Marques-Vidal P, Pecoud A, Hayoz D, Paccaud F, Mooser V, Waeber G et al (2010) Normal weight obesity: relationship with lipids, glycaemic status, liver enzymes and inflammation. Nut Metabol Cardiovascul Dis 20(9):669–675

Meijer RI, Serné EH, Korkmaz HI et al (2015) Insulin-induced changes in skeletal muscle microvascular perfusion are dependent upon perivascular adipose tissue in women. Diabetologia 58(8):1907–1915

Miki H, Yamauchi T, Suzuki R et al (2001) Essential role of insulin receptor substrate 1 (IRS-1) and IRS-2 in adipocyte differentiation. Mol Cell Biol 21 (7):2521–2532

Misra A, Chowbey P, Makkar BM et al (2009) Consensus statement for diagnosis of obesity, abdominal obesity and the metabolic syndrome for Asian Indians and recommendations for physical activity, medical and surgical management. J Assoc Physicians India 57:163–170

Misra A, Wasir JS, Vikram NK, Pandey RM, Kumar P (2010) Cutoffs of abdominal adipose tissue compartments as measured by magnetic resonance imaging for detection of cardiovascular risk factors in apparently healthy adult Asian Indians in North India. Metab Syndr Relat Disord 8(3):243–247

Mørkedal B, Vatten LJ, Romundstad PR, Laugsand LE, Janszky I (2014) Risk of myocardial infarction and heart failure among metabolically healthy but obese individuals: HUNT (Nord-Trøndelag Health Study). Norway J Am Coll Cardiol 63(11):1071–1078

Morris AP, Voight BF, Teslovich TM et al (2012) Large-scale association analysis provides insights into the genetic architecture and pathophysiology of type 2 diabetes. Nat Genet 44(9):981–990

Moy FM, Loh DA (2015) Cardiometabolic risks profile of normal weight obese and multi-ethnic women in a developing country. Maturitas 81(3):389–393

Murase T, Hattori T, Ohtake M et al (2012) Cardiac remodeling and diastolic dysfunction in DahlS.Z-Lepr (fa)/Lepr(fa) rats: a new animal model of metabolic syndrome. Hypertens Res Off J Jpn Soc Hypertens 35 (2):186–193

Pasco JA, Holloway KL, Dobbins AG, Kotowicz MA, Williams LJ, Brennan SL (2014) Body mass index and measures of body fat for defining obesity and underweight: a cross-sectional, population-based study. BMC Obes 1:9

Patel SA, Shivashankar R, Ali MK et al (2016) Is the "South Asian Phenotype" Unique to South Asians?: Comparing cardiometabolic risk factors in the CARRS and NHANES studies. Glob Heart 11(1):89–96

Peppa M, Stefanaki C, Papaefstathiou A, Boschiero D, Dimitriadis G, Chrousos GP (2017) Bioimpedance analysis vs. DEXA as a screening tool for osteosarcopenia in lean, overweight and obese Caucasian postmenopausal females. Horm Athens Greece 16 (2):181–193

Petta S, Gastaldelli A, Rebelos E et al (2019) Pathophysiology of non alcoholic fatty liver disease. Int J Mol Sci 17(12):2082. https://www.ncbi.nlm.nih.gov/pmc/articles/PMC5187882/

Phillips CM (2017) Metabolically healthy obesity across the life course: epidemiology, determinants, and implications. Ann N Y Acad Sci 1391(1):85–100

van Raalte DH, van der Zijl NJ, Diamant M (2010) Pancreatic steatosis in humans: cause or marker of lipotoxicity? Curr Opin Clin Nutr Metab Care 13 (4):478–485

Rabkin SW (2014) The relationship between epicardial fat and indices of obesity and the metabolic syndrome: a systematic review and meta-analysis. Metab Syndr Relat Disord 12(1):31–42

Rachek LI (2014) Free fatty acids and skeletal muscle insulin resistance. Prog Mol Biol Transl Sci 121:267–292

Ramírez-Vélez R, Correa-Bautista JE, Sanders-Tordecilla A et al (2017) Percentage of body fat and fat mass index as a screening tool for metabolic syndrome prediction in Colombian University students. Nutrients 9 (9):1009

Ramírez-Vélez R, Carrillo HA, Correa-Bautista JE et al (2018) Fat-to-muscle ratio: a new anthropometric indicator as a screening tool for metabolic syndrome in Young Colombian People. Nutrients 10(8):1027

Rhee E-J, Lee MK, Kim JD et al (2014) Metabolic health is a more important determinant for diabetes development than simple obesity: a 4-year retrospective longitudinal study. PLoS One 9(5):e98369

Romana BS, Chela H, Dailey FE, Nassir F, Tahan V (2018) Non-alcoholic fatty pancreas disease (NAFPD): a silent spectator or the fifth component of metabolic syndrome? A literature review. Endocr Metab Immune Disord Drug Targets 18(6):547–554

Romero-Corral A, Somers VK, Sierra-Johnson J et al (2010a) Normal weight obesity: a risk factor for cardiometabolic dysregulation and cardiovascular mortality. Eur Heart J 31(6):737–746

Romero-Corral A, Somers VK, Sierra-Johnson J, Korenfeld Y, Boarin S, Korinek J et al (2010b) Normal weight obesity: a risk factor for cardiometabolic dysregulation and cardiovascular mortality. Eur Heart J 31(6):737–746

Rosso C, Mezzabotta L, Gaggini M et al (2016) Peripheral insulin resistance predicts liver damage in nondiabetic subjects with nonalcoholic fatty liver disease. Hepatol Baltim Md 63(1):107–116

Ruderman NB, Schneider SH, Berchtold P (1981) The "metabolically-obese," normal-weight individual. Am J Clin Nutr 34(8):1617–1621

Salgado-Somoza A, Teijeira-Fernández E, Rubio J, Couso E, González-Juanatey JR, Eiras S (2012) Coronary artery disease is associated with higher epicardial retinol-binding protein 4 (RBP4) and lower glucose transporter (GLUT) 4 levels in epicardial and subcutaneous adipose tissue. Clin Endocrinol 76(1):51–58

Shulman GI (2014) Ectopic fat in insulin resistance, dyslipidemia, and cardiometabolic disease. N Engl J Med 371(12):1131–1141

Sinaga M, Worku M, Yemane T et al (2018) Optimal cut-off for obesity and markers of metabolic syndrome for Ethiopian adults. Nutr J 17(1):109

Sironi AM, Gastaldelli A, Mari A et al (2004) Visceral fat in hypertension: influence on insulin resistance and beta-cell function. Hypertens Dallas Tex 44 (2):127–133

Sniderman AD, Bergeron J, Frohlich J (2001) Apolipoprotein B versus lipoprotein lipids: vital lessons from the AFCAPS/TexCAPS trial. Can Med Assoc J Assoc Medicale Can 164(1):44–47

Song P, Li X, Bu Y et al (2019) Temporal trends in normal weight central obesity and its associations with cardiometabolic risk among Chinese adults. Sci Rep 9(1):5411

Stefan N, Kantartzis K, Machann J et al (2008) Identification and characterization of metabolically benign obesity in humans. Arch Intern Med 168(15):1609–1616

Tariq H, Nayudu S, Akella S, Glandt M, Chilimuri S (2016) Non-alcoholic fatty pancreatic disease: a review of literature. Gastroenterol Res 9(6):87–91

Teixeira TFS, Alves RDM, Moreira APB, Peluzio M (2015) do CG. Main characteristics of metabolically obese normal weight and metabolically healthy obese phenotypes. Nutr Rev 73(3):175–190

Unger RH, Cherrington AD (2012) Glucagonocentric restructuring of diabetes: a pathophysiologic and therapeutic makeover. J Clin Invest 122(1):4–12

Varbo A, Freiberg JJ, Nordestgaard BG (2015) Extreme nonfasting remnant cholesterol vs extreme LDL cholesterol as contributors to cardiovascular disease and all-cause mortality in 90000 individuals from the general population. Clin Chem 61(3):533–543

Wang B, Zhuang R, Luo X et al (2015) Prevalence of metabolically healthy obese and metabolically obese but normal weight in adults worldwide: a meta-analysis. Horm Metab Res Horm Stoffwechselforschung Horm Metab 47(11):839–845

WHO (1995) Physical status: the use and interpretation of anthropometry. Report of a WHO Expert Committee. World Health Organ Tech Rep Ser 854:1–452

Wiklund P, Tormakangas T, Shi Y, Wu N, Vainionpaa A, Alen M et al (2017) Normal-weight obesity and

cardiometabolic risk: a 7-year longitudinal study in girls from prepuberty to early adulthood. Obesity (Silver Spring, Md) 25(6):1077–1082
Wildman RP, Muntner P, Reynolds K et al (2008) The obese without cardiometabolic risk factor clustering and the normal weight with cardiometabolic risk factor clustering: prevalence and correlates of 2 phenotypes among the US population (NHANES 1999–2004). Arch Intern Med 168(15):1617–1624
Winhofer Y, Krssák M, Jankovic D et al (2012) Short-term hyperinsulinemia and hyperglycemia increase myocardial lipid content in normal subjects. Diabetes 61(5):1210–1216
Zhang Y-Y, Liu P-Y, Lu Y et al (2005) Race and sex differences and contribution of height: a study on bone size in healthy Caucasians and Chinese. Am J Hum Biol Off J Hum Biol Counc 17(5):568–575
Zhang M, Schumann M, Huang T, Törmäkangas T, Cheng S (2018) Normal weight obesity and physical fitness in Chinese university students: an overlooked association. BMC Public Health 18(1):1334
Zhao L, Ni Y, Ma X, et al. (2016) A panel of free fatty acid ratios to predict the development of metabolic abnormalities in healthy obese individuals. Sci Rep [Internet] [cited 2019 Dec 2];6. https://www.ncbi.nlm.nih.gov/pmc/articles/PMC4921829/
Zhu J-Z, Dai Y-N, Wang Y-M, Zhou Q-Y, Yu C-H, Li Y-M (2015) Prevalence of nonalcoholic fatty liver disease and economy. Dig Dis Sci 60(11):3194–3202

Obesity in Children/Adolescents and Obesity-Related Comorbidities

27

Taisa Kohut, Jennifer Robbins, Elizabeth Prout, Lorraine Katz, Julie Brothers, Sandra Vazquez-Diaz, Kimberly Genuario, and Jennifer Panganiban

Contents

Pediatric Obesity: A Worldwide Concern 362
Definitions of Obesity 363
Epidemiologic Trends 363
Risk Factors 363
Primary vs. Secondary Causes of Obesity 364
Drug-induced Secondary Obesity 365
Genetic Risk Scores and Epigenetics 365
Clinical Evaluation 365
Lifestyle Interventions 366
Pharmacotherapy 366
Metabolic and Bariatric Surgery 369

The original version of this chapter was revised. A correction to this chapter can be found at https://doi.org/10.1007/978-3-030-53370-0_74

T. Kohut · J. Robbins · E. Prout · J. Panganiban (✉)
Division of Gastroenterology, Hepatology, and Nutrition, The Children's Hospital of Philadelphia, Philadelphia, PA, USA
e-mail: panganibaj@email.chop.edu

L. Katz · S. Vazquez-Diaz
Division of Endocrinology, The Children's Hospital of Philadelphia, Philadelphia, PA, USA

J. Brothers
Division of Cardiology, The Children's Hospital of Philadelphia, Philadelphia, PA, USA

K. Genuario
Division of Behavioral Health, The Children's Hospital of Philadelphia, Philadelphia, PA, USA

J. Faintuch and S. Faintuch (eds.), *Obesity and Diabetes*, https://doi.org/10.1007/978-3-030-53370-0_27

Comorbidities and Long-term Follow-up ... 371
Cardiovascular Risk ... 371
Natural History of Atherosclerosis ... 371
Dyslipidemia ... 371
Management of Dyslipidemia ... 372
Hypertension ... 372
Salty Foods ... 372
Semi-vegetarian Diets ... 372
Insulin Resistance/Type 2 Diabetes ... 372
Severity of Adolescent T2D ... 373
Pediatric Prediabetes ... 374
Metabolic Syndrome ... 374
Polycystic Ovarian Syndrome ... 375
Nonalcoholic Fatty Liver Disease (NAFLD) ... 376
NAFLD Management ... 376
Non-cardiometabolic Comorbidities ... 376
Obstructive Sleep Apnea (OSA) ... 376
Ongoing Studies and Future Perspective ... 377
References ... 378

Abstract

Childhood obesity is a growing concern, which may result in major comorbidities, previously felt to be adult-specific diseases. Genetics, environment, and behavior count among the pillars of this condition. Lifestyle modification is paramount, whereas indications for metabolic and bariatric surgery are becoming popular as well for severely affected adolescents. The scarcity of approved drugs progressively leads to off-label adjunct prescriptions in those refractories to diet and exercise. Comorbidities can persist into later life- impacting morbidity and mortality. Besides prevention and treatment of ongoing problems, long-term sequelae should be avoided to allow for a long and healthy adult life.

Keywords

Childhood/adolescent obesity · Lifestyle modification · Pharmacotherapy · Bariatric surgery · Obesity-related comorbidities

Pediatric Obesity: A Worldwide Concern

Obesity is one of the leading chronic diseases affecting children across the globe and contributes to increased morbidity across the lifespan. This complex disease, characterized primarily by excess body weight for height, requires thorough assessment for etiologies both common and rare. There are many risk factors for excess weight gain including genetic profile, prenatal environments, and postnatal exposures. Though most treatment options for pediatric obesity have focused on behavioral modifications to combat our increasingly obesogenic environments, treatment options for severe obesity have increased over the last decade, including pharmacological and surgical options. It is important that obesity and associated comorbidities are treated and addressed in the clinical setting to decrease the persistence into adulthood; however, it is likely that public health and policy measures will be most effective in meaningful decreases in pediatric obesity prevalence at the national scale.

Table 27.1 Weight classification by BMI percentile for children aged 2–20

Weight Status		CDC BMI Percentile
Underweight		<5th Percentile
Normal Weight		5th percentile < 85th percentile
Overweight		85th percentile—< 95th percentile
Obesity	Class I Obesity	95th percentile—< 120% of the 95th percentile
Severe Obesity	Class II Obesity	120% of the 95th percentile—< 140% of the 95th percentile, or a BMI of 35 (whichever is lower)
	Class III Obesity	140% of the 95th percentile and above, or a BMI of 40 (whichever is lower)

Definitions of Obesity

For children greater than 2 years of age, Body Mass Index (BMI) percentiles are used to define pediatric weight status. The raw BMI is an indirect measure of body fat and is calculated as [weight (kg)/height (m)^2]. The CDC percentiles can be used as outlined in Table 27.1 to determine weight classification for children 2 years and older (Flegal et al. 2009).

Pediatric obesity has been redefined, as rates of severe obesity have increased, where "severe obesity" is typically defined as either Class II or Class III obesity. It is helpful to use newly designed CDC growth charts for children and adolescents with obesity, to track BMI trajectories over time (Gulati et al. 2012). Obesity classifications allow for more precise characterization of a patient's degree of excess weight and also easily convert to adult classifications of obesity (Racette et al. 2017) (Table 27.2).

Table 27.2 Adult weight classification by BMI

Adult weight classification	BMI
Overweight	25.0–29.9
Class I Obesity	30.0–34.9
Class II Obesity	35.0–39.9
Class III Obesity	>40

BMI: kg/ m^2

Epidemiologic Trends

From 1975 to 2016, age-standardized prevalence of obesity increased in girls from 0.7% to 5.6% and in boys from 0.9% to 7.8%. Obesity prevalence has increased in every region of the world and has disproportionately affected low-income regions. Though mean childhood BMIs have seemed to plateau in high-income countries, it is apparent that BMIs are continuing to increase across most of Asia and other lower-income regions (NCD Risk Factor Collaboration (NCD-RisC) 2017).

In the United States, overweight and obesity affect over one in three children aged 2–17 years, with current obesity prevalence of 17.4%. Rates of severe obesity have also increased, with 10% of adolescents having Class II or III obesity using the NHANES 2013–2014 data (Skinner et al. 2018).

Disparities also exist, such that in highly developed countries, low income is generally associated with higher rates of obesity (NCD Risk Factor Collaboration (NCD-RisC) 2017; Shrewsbury and Wardle 2008). There are also racial disparities; African American and Hispanic youth have higher rates of obesity than white and Asian Americans (Weaver et al. 2019).

Risk Factors

During the prenatal period, both maternal obesity and excessive gestational weight gain increase risk of large for gestational age and neonatal adiposity (Starling et al. 2015; Goldstein et al. 2017). Maternal smoking during pregnancy can also increase risk of obesity in childhood (Suzuki et al. 2009; Dubois and Girard 2006).

In the postnatal period, excessive weight gain early in life as well as early adiposity rebound has been shown to increase risk of adolescent obesity, persistence of obesity into adulthood, and worse cardiometabolic outcomes (Geserick et al. 2018; Aris et al. 2019). Breastfeeding has been shown to decrease risk of obesity, with suggestion that there is a dose-response relationship in protection (Horta et al. 2015; Gibson et al. 2017).

There are also many risk factors which include broadly: dietary intake high in energy-dense foods, fast food consumption, physical inactivity, excessive screen time, poor sleep, and parenting styles (Mihrshahi and Baur 2018; Melis Yavuz and Selcuk 2018; Vos and Welsh 2010).

Primary vs. Secondary Causes of Obesity

Primary obesity accounts for most pediatric cases and generally occurs when multifactorial influences result in excess weight gain. It is often related to dietary quality, physical inactivity, psychological factors, and environmental factors in a manner that affects the ratio of energy intake to energy expenditure.

Secondary causes of obesity include a wide variety of underlying disorders that impact a person's pathophysiology, leading to excessive weight gain. Specific medical causes of obesity are generally related to endocrinopathies, specific genetic mutations, or syndromic causes, with the most commonly considered etiologies listed in Table 27.3. These should be suspected and further investigated when related findings and clinical history are present, when weight gain appears to be severe, or when the presentation does not support primary obesity.

Note that for many of the genetic and endocrine causes of obesity, presence of short stature and/or developmental delay can be useful distinguishing features (Table 27.4) (Stipančić 2018). It is notable that neither Melanocortin 4 receptor (MC4R) deficiency nor POMC deficiency is related to decreased height or developmental delay. Defects in MC4R may account for up to 6% of severe pediatric obesity and are

Table 27.3 Endocrine, monogenic, and syndromic causes of obesity

Endocrine	Monogenic	Syndromic
• Hypothyroidism • Cushing disease • Polycystic ovarian syndrome • Growth hormone deficiency • Hypothalamic obesity • Hypogonadism • Pseudohypoparathyroidism • Insulinoma	• POMC deficiency • Melanocortin 4 receptor deficiency • Leptin or Leptin receptor deficiency	• Prader-Willi syndrome • Bardet-Biedl syndrome • Beckwith-Wiedemann syndrome • Alstrom-Hallgren syndrome • Carpenter syndrome • Cohen syndrome

Table 27.4 Key Findings in Select Secondary Causes of Obesity

	Developmental Delay?	Short stature?	Key Findings
Prader-Willi Syndrome	X	X	Hypotonia, hypogonadism, small hands, and feet
Bardet-Biedl Syndrome	X	X	Polydactyly, syndactyly, retinitis pigmentosa, hypogonadism, kidney disorders
Carpenter Syndrome	X	X	Polydactyly, syndactyly, hypogonadism in boys
Cohen Syndrome	X	X	Microcephaly, marked central incisors, long thin fingers
Hypothyroidism		X	Fatigue, constipation, dry skin, and hair
Growth Hormone deficiency		X	Centripetal obesity, young appearance
Cushing's Syndrome		X	Round face, acne, striae, hirsutism, hypertension
Hypothalamic Obesity		X	Signs of increased intracranial pressure, hypopituitarism, history of neuropathology
Alstrom Syndrome		X	Blindness, deafness, type 2 diabetes, acanthosis nigricans

commonly found with accelerated linear growth and hyperinsulinemia (Farooqi et al. 2003). POMC deficiency is notable for presence of adrenal insufficiency and hypopigmentation (Stipančić 2018).

Drug-induced Secondary Obesity

Many anti-diabetic medications such as insulin or sulfonylureas are known to cause weight gain. Centrally acting antipsychotics, antidepressants, and antiepileptic medications are also commonly used in pediatric populations and may especially lead to weight gain when used near the time of puberty. Children receiving high-dose steroids are also prone to excessive weight gain. It is important for the clinician to be aware of these interactions and to consider weight-neutral options when appropriate (Leslie et al. 2007).

Genetic Risk Scores and Epigenetics

Though specific monogenic and syndromic forms of obesity are detailed in Table 27.3, a person with primary obesity may still have an increased risk of obesity due to polygenic factors. Genome-wide studies have shown that specific genetic variations are associated with higher BMI, such that those with the highest risk scores were 25 times as likely to have severe obesity as compared to those with the lowest risk scores. Though BMI differences based on risk score appear even by age 3, research is still growing in this and it is not commonly used in clinical practice (Khera et al. 2019).

There is also increasing evidence that epigenetics, such as DNA methylation, modifications to microRNA, and histone tails, are important in the development of obesity. This is especially convincing as more evidence suggests that perinatal factors, exposure to environmental pollutants or obesogens, and specific gut microbiota affect both epigenetic factors and obesity risk (Lopomo et al. 2016).

Table 27.5 Clinical evaluation of an obese child and adolescent

History	Physical Exam
• Dietary history • Physical activity assessment • Review of medications • Developmental history • Sleep history • Review of systems • Family history of obesity and related comorbidities • Psychosocial screening	• Measurement of height and weight for BMI, BMI percentile, and BMI z-score calculations • Blood pressure measurement with appropriate sized cuff • Assessment for dysmorphic and for Cushingoid features suggestive of a primary cause
Laboratory screening	
Overweight (BMI 85th–95th %) All: • Fasting lipid profile If ≥10 years AND ≥1 risk factor(s)[a]: • Fasting blood glucose or hemoglobin A1c • Aminotransferases (AST, ALT) • Repeat fasting lipid profile every 2 years	Obese (BMI ≥95th %) All: • Fasting lipid profile If ≥10 years regardless of risk factors: • Fasting blood glucose or hemoglobin A1c • Aminotransferases (AST, ALT) • Repeat fasting lipid profile every 2 years

[a]Risk factors: elevated blood pressure, elevated lipid levels, current tobacco use, or family history of obesity-related diseases

Clinical Evaluation

The clinical evaluation of an obese child includes a complete history and physical exam directed at identifying the cause and related comorbidities (Table 27.5) (Kumar and Kelly 2017; Armstrong et al. 2016). The 2017 U.S. Preventive Services Task Force recommends BMI measurement to screen for obesity in children ≥6 years; however, the American Academy of Pediatrics (AAP) recommends annual plotting of BMI on a growth chart for all patients 2 years and older (Quattrin and Wilfley 2017). While no consensus exits about laboratory screening tests and when to screen in children with obesity, general recommendations are summarized in Table 27.5 (Kumar and Kelly 2017; Krebs et al. 2007). Universal screening with a fasting lipid panel has been recommended by the US National Heart, Lung, and Blood Institute 2011 expert panel

Table 27.6 Stepwise approach to weight management in children and adolescents

Stage 1 Prevention Plus: Lifestyle intervention provided by a primary care provider
Stage 2 Structured Weight Management: Monthly visits with a primary care physician and support from a registered dietician
Stage 3 Comprehensive Multidisciplinary[a] *Intervention:* Intensive weight loss program composed of weekly visits for a minimum of 8–12 weeks at a pediatric weight management center
Stage 4 Tertiary Care Intervention: Use of medical diets, medications, and surgery in addition to Stage 3 interventions

Adapted from Table 8 in Barlow, 2007 (see ref. (Barlow 2007))
[a]Multidisciplinary team approach includes a medical provider, registered dietician, exercise specialist, mental health professional, nurse, and social worker.

between 9 and 11 years of age and again between 17 and 21 years of age, although these guidelines remain controversial (Gillman and Daniels 2012). Children with clinical evaluation suggestive of a genetic or endocrine cause or comorbidity may need additional specific testing.

Lifestyle Interventions

Behavioral changes (engaging the family), dietary regimen, and a higher activity level are the official guidelines (Table 27.6) (Kumar and Kelly 2017; Barlow 2007; Spear et al. 2007). Results with low glycemic index and low-carbohydrate diets are similar to those achieved with portion controlling, yet dropout rate is high (Ebbeling et al. 2003; Sondike et al. 2003). The Mediterranean-style diet (MSD) is highly regarded for grown-ups, inhibiting the advent and progression of metabolic syndrome (MetSyn) (Grosso et al. 2014; Babio et al. 2009), and could be an option for the pediatric age bracket (Velázquez-López et al. 2014). The low-fructose diet could attenuate elevated LDL-C and nonalcoholic fatty liver disease (NAFLD); however, weight loss did not materialize (Mager et al. 2015). Family involvement in the utilization of less high calorie density snacks and drinks, and more nutrient-dense components, could be a winning strategy (Epstein et al. 1990; Epstein et al. 1994).

According to the American Academy of Pediatrics (AAP), daily pediatric needs include at least 60 minutes of substantial activities, with no screen exposure under age 2, and at most 2 hours for older children (Hassink et al. 2015).

For cases with advanced adiposity lifestyle, corrections are often insufficient (Ryder et al. 2018; Greenway 2015; MacLean et al. 2011); nevertheless, they alleviate comorbidities (Savoye et al. 2014; Ryder et al. 2013). The government, society, and other stakeholders should join in a common effort to combat the obesogenic environment and its consequences.

Pharmacotherapy

Pharmacotherapy, in conjunction with lifestyle modification, may be the next logical step in the treatment of pediatric obesity refractory to lifestyle modification alone (Ryder et al. 2018; Kelly and Fox 2017; Srivastava et al. 2019). Suggested BMI criteria for pharmacotherapy initiation in adolescents include BMI $\geq$95th% (or BMI $\geq$30 kg/m^2, whichever is lower) plus at least one obesity-related comorbidity; or BMI $\geq$120% of the 95th% (or BMI $\geq$35 kg/m^2, whichever is lower) irrespective of comorbidity. There is no upper BMI threshold for initiation of pharmacotherapy. Obesity pharmacotherapy can also be considered in patients meeting criteria for bariatric surgery; wherein surgery may not be appropriate or possible at that time, or when medications are recommended as adjunctive therapy (Srivastava et al. 2019). While there has been a surge in FDA-approved medications for adult obesity treatment, pharmacologic options for obese youth remain limited. Table 27.7 provides a summary of FDA-approved adult obesity medications and off-label drug use of medications in the pediatric obesity population (Srivastava et al. 2019; Shettar et al. 2017).

Table 27.7 Summary of FDA-approved adult obesity medications and off-label drug use of medications in the pediatric obesity population

Drug name	MOA	FDA indication	Off-label drug use	Side effects	Contraindications & Warnings
FDA approved					
Orlistat	Pancreatic and gastric lipase inhibitor	Obesity ≥12yo	Not indicated	Flatulence, oily stools, diarrhea, vitamin/mineral deficiency	Chronic malabsorption syndrome, cholestasis
Phentermine	Sympatho-mimetic amine	Obesity >16yo "short term"; combination phentermine/ topiramate ER approved for long-term treatment of obesity in adults	<16yo or long-term; beneficial in obesity with low energy states, sleep apnea, hunger, decreased satiety	Increases HR, BP, dry mouth, insomnia, constipation, anxiety, irritability	Cardiovascular disease, hyperthyroidism, active drug use, glaucoma, agitated states
Non-FDA approved					
Metformin	Activation of protein kinase pathway	≥10yo for T2DM	PCOS, insulin resistance, prediabetes, MetSyn, antipsychotic-medication-induced weight gain, stress eating/ emotional eating	Bloating, diarrhea, flatulence	Hold 48hrs prior to contrast, lactic acidosis
Topiramate	Modulation of various NTs	Treatment of epilepsy >2yo and migraines >12yo; combination phentermine/ Topiramate ER approved for long-term treatment of obesity in adults	Weight loss in adult and pediatric patients; useful adjunct in binge eating disorders and weight regain postbariatric surgery	Cognitive dysfunction, kidney stones, metabolic acidosis, teratogenic—adolescents pregnancy counseling (decreased efficacy of OCPs)	Inborn errors of metabolism with hyperammonemia and encephalopathy, acute myopia and secondary angle-closure glaucoma, rapid withdrawal can precipitate seizures, neuropsychiatric dysfunction, metabolic acidosis
Exenatide	GLP-1 agonist	T2DM in adults	<18yo for polygenic obesity (presence of diabetes, hypothalamic, syndromic)	Bloating, nausea & vomiting, abdominal pain, elevation pancreatic amylase & lipase	Post-marketing reports: pancreatitis, renal impairment, severe GI disease
Liraglutide	GLP-1 agonist	Saxenda (3mg dose) approved for obesity in adults, Victoza approved for T2DM in adults	<18yo	Abdominal pain, nausea & vomiting, diarrhea, potential hypoglycemia	Post-marketing reports: pancreatitis, renal impairment, severe GI disease

(continued)

Table 27.7 (continued)

Drug name	MOA	FDA indication	Off-label drug use	Side effects	Contraindications & Warnings
Lisdexamfetamine	Central nervous system stimulant	≥6yo for ADHD, short-term use of binge eating disorder in adults	Beneficial for younger children with ADHD and obesity or binge eating disorder	Anorexia, decreased appetite, decreased weight, diarrhea, dizziness, dry mouth, irritability, anxiety, insomnia, nausea & vomiting, abdominal pain	Serious cardiovascular reactions such as sudden death, BP and HR increases, psychiatric adverse reactions, suppression of growth, peripheral vasculopathy (Raynaud's), serotonin syndrome with serotonergic agents
No pediatric data					
Lorcaserin	5-hydroxy-triptamine-Receptor (R) 2C agonist	Long-term treatment of obesity in adults	<18yo with obesity	Headache, dizziness, fatigue, dry mouth, constipation, headache, back pain, cough in diabetes patients	Serotonin syndrome or neuroleptic malignant-like syndrome when co-administered with other serotonergic or anti-dopaminergic agents, discontinue with signs of valvular heart disease
Naltrexone/ bupropion SR	Blockade of opioid-R-mediated POMC auto-inhibition/ selective inhibition of reuptake of dopamine and nor-adrenaline	Long-term treatment of obesity in adults	Children and adolescents: warning for increased suicidal ideation	Nausea, constipation, headache, dizziness, insomnia, dry mouth, diarrhea	Uncontrolled HTN, seizures, anorexia nervosa, bulimia, active alcohol or chronic opioid use, angle-closure glaucoma, increase suicidal thought, and ideation
Pending new FDA approval					
Setmelanotide	Melano-cortin-4 receptor agonist	Phase 3 trials for monogenic obesity, FDA approval pending for monogenic obesity in adults and children	None; will be approved for pediatric patients as well	Dry mild induration at injection site, darkening of skin nevi	Caution use in structural heart disease and arrhythmias because of potential to increase HR and BP

Adapted from Table 3 in Srivastava et al., 2019 (see ref. (Srivastava et al. 2019))
MOA: mechanism of action; HR: heart rate; BP blood pressure; T2DM: type 2 diabetes mellitus; NTs: neurotransmitters; GLP: glucagon-like 1 receptor; POMC: proopiomelanocortin; OCPs: Oral contraceptives;

Table 27.8 Comorbidity indications for metabolic bariatric surgery and surgical outcomes

Comorbidities	% Remission at 5 years
Type 2 diabetes, prediabetes, insulin resistance	86 (Inge et al. 2019)
Hypertension	68 (Inge et al. 2019)
Low HDL	78 (Inge et al. 2019)
High triglyceride	81 (Inge et al. 2019)
Obstructive Sleep Apnea with apnea/hypopnea index (AHI) ≥ 5	60 (Amin et al. 2017)
Idiopathic intracranial hypertension (pseudotumor cerebri)	
Nonalcoholic fatty liver disease and nonalcoholic steatohepatitis	85 (Manco et al. 2017)
Gastroesophageal reflux disease	
Slipped capital femoral epiphysis (SCFE) or Blount's disease	
Impaired quality of life	

Pratt et al. (2018), Inge et al. (2019), Amin et al. (2017), Manco et al. (2017)

Remission informed whenever available; Blount's disease: tibia bowing, bowed legs

Metabolic and Bariatric Surgery

Metabolic and bariatric surgery (MBS) is currently the most effective sustainable method for weight loss and comorbidity management in children/adolescents with severe obesity. According to the most recent guidelines by the American Society for Metabolic and Bariatric Surgery (ASMBS) published in 2018, and the AAP guidelines published in 2019, bariatric surgery should be considered for adolescents ages 10–18 years (ASMBS) and ≥ 13 years (AAP) Class III obesity (BMI 140% > 95th percentile) without any comorbidities, or Class II obesity (BMI 120% > 95th percentile) with severe comorbidities (Table 27.8) (Pratt et al. 2018; Armstrong et al. 2019; Inge et al. 2019; Amin et al. 2017; Manco et al. 2017). The current procedures utilized in adolescents include the Roux-en-Y gastric bypass (RYGB) and the vertical sleeve gastrectomy (VSG).

Initially, the mechanical and digestive effects of the surgeries (restriction and malabsorption) were the primary considerations, but it is now known that hormonal regulation and metabolism are altered as well (Akkary 2012; Peterli et al. 2012). MBS disrupts hormonal signals between the gastrointestinal tract, the brain stem, and the hypothalamus (Akkary 2012; Peterli et al. 2012). Additionally, MBS impacts the physiological regulation of weight through the alteration of food intake, food absorption, appetite and satiety, energy balance, and behavioral response to food (Table 27.9) (Akkary 2012; Shikora et al. 2007).

RYGB was first performed as an open procedure in 1967 and until 2014, was the most common MBS operation in adults and adolescents (Abraham et al. 2016). The mechanism of the RYGB for weight loss includes restriction, malabsorption, and hormonal changes. A small pouch (20–30 mL) is created (restriction) bypassing >90% of the stomach, and the proximal intestine including the duodenum and the common bile duct. A Roux limb of intestine is created connecting the pouch to the jejunum

Table 27.9 Influence of metabolic bariatric surgery on gut hormones that influence appetite and satiety

Hormones	Pathway location	Hormone effect on appetite	RYGB effect on hormones	VSG effect on hormones	AGB effect on hormones
Ghrelin	V, BS, HT	↑	↓	↓	↑↑
GLP-1	V, BS, HT	↓	↑	↑	No Δ
PYY	V, BS, HT	↓	↑	↑	No Δ
PPP	V, BS	↓	↑	↑	No Δ
CCK	V, BS, HT	↓	No Δ	↑	No Δ

Adapted from Table 2 in Akkary, 2012 (see ref. (Akkary 2012))

RYGB: Roux-en-y gastric bypass; VSG: vertical sleeve gastrectomy; AGB: adjustable gastric band; GLP-1: glucagon-like peptide 1; PYY: peptide YY, PPP: pancreatic polypeptide; CCK: cholecystokinin; V: vagus nerve; BS: brain stem; HT: hypothalamus;

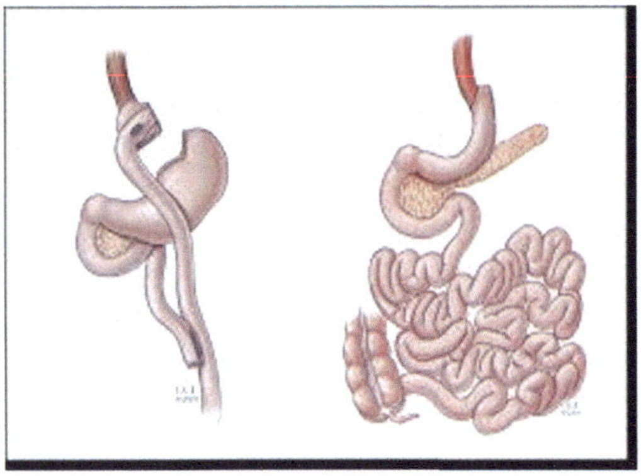

Fig. 27.1 Roux-en-Y gastric bypass (left) and vertical sleeve gastrectomy (right)

(Fig. 27.1) (Göthberg et al. 2014). Bypassing the majority of the stomach and proximal intestine causes malabsorption, including carbohydrates and fat. Additionally, multiple hormones are affected to increase satiety (Table 27.9) (Akkary 2012). RYGB results in approximately 30% total weight loss or 70% excess weight loss. Excess weight loss is the pre-surgery weight—the weight loss divided by the ideal body weight (or BMI=25), depending on the source.

A relatively common side effect of RYGB is dumping syndrome, sometimes precipitated by high carbohydrate and fat consumption, and featuring diarrhea, hypotension, presyncope, and nausea (Göthberg et al. 2014). These symptoms decrease over time, but diarrhea to some degree can persist. Other risks include small bowel obstruction, hernia, overstretching the pouch, and occasionally the need for revisional surgery, especially if weight regain refractory to prescription drugs is detected after some time.

VSG removes 75–80% of the stomach and is currently the most common procedure (80% in

Table 27.10. Nutritional deficiencies, as a complication of metabolic bariatric surgery

Procedure Type	RYGB, VSG	RYGB	RYGB	
Site of nutrient absorption in GI Tract	Stomach	Duodenum	Jejunum	Ileum
Vitamins & Minerals	Water	[a]Calcium	[a]Calcium	Vitamin C
	Copper	[a]Iron	[a]Iron	Folate
	Iodide	Phosphorus	[a]Zinc	[b]Vitamin B12
	Fluoride	Magnesium	Phosphorus	Vitamin D
	Molybdenum	Copper	Magnesium	Vitamin K
	(Intrinsic factor)	Selenium	Thiamine	Magnesium
		Thiamine	Riboflavin	(Bile salts)
		Riboflavin	Niacin	
		Niacin	Biotin	
		Biotin	Folate	
		Folate	Vitamins A, D, E, K	
		Vitamins A, D, E, K	Vitamin C	
			Vitamin B6	
			Manganese	
			Chromium	
			Molybdenum	
			Amino acids	

Adapted from Table 2 and Figure 4 in Shikora et al., 2007 (see ref. (Shikora et al. 2007))

Vitamin and mineral supplementation for life is advised after all bariatric and metabolic procedures. Dosage should be higher for malabsorptive than restrictive modalities.

VSG: vertical sleeve gastrectomy—vitamin B12 deficiency due to the lack of the production of intrinsic factor and fluoride deficiency. Additional vitamin supplementation is recommended due to limited intake of food and possible rapid transit of food during digestion.

[a]Requires hydrogen chloride (HCL) for absorption

[b]Affected secondary due to decreased intrinsic factor production

some countries) performed in adults and adolescents. VSG results in approximately 30% weight loss (Fig. 27.1) (Parrott et al. 2017). VSG restricts the amount of food that can be consumed, decreases ghrelin secretion significantly, and results in alterations of other hormonal pathways (Table 27.9) (Akkary 2012; Peterli et al. 2012). VSG is associated with fewer nutritional complications compared with the RYGB, but still requires monitoring for vitamin deficiencies long-term (Table 27.10) (Shikora et al. 2007). Specific side effects include leak and stricture. Side effects for all bariatric procedures include deep vein thrombosis and pulmonary emboli. Weight loss and improvements in obesity-related comorbidities are similar with both mentioned interventions (Table 27.8) (Pratt et al. 2018; Armstrong et al. 2019; Inge et al. 2019; Amin et al. 2017; Manco et al. 2017). Weight regain can happen in 25% of patients for either procedure, which generally responds to weight loss medication. MBS should be considered in adolescents with severe obesity.

Comorbidities and Long-term Follow-up

A child who is overweight or obese is at increased risk for several comorbidities in the short and long term. The severity of these comorbidities, particularly cardiometabolic, directly correlates with obesity severity (Kumar and Kelly 2017; Skinner et al. 2015). Key cardiometabolic and non-cardiometabolic risks of childhood obesity are described. These patients require long-term follow-up to screen for and, in some cases, to ultimately manage these comorbid conditions.

Cardiovascular Risk

More humans die from cardiovascular disease (CVD) than from any other cause (Mozaffarian et al. 2015), and premonitory endothelial lesions are documented in early childhood. Excessive adiposity and disturbed lipid profile precipitate precocious CVD. Recognized risks occur in most adolescents with excessive body weight, and one in ten accumulates three or more factors (Daniels 2009; Freedman et al. 1999; Steinberger et al. 2009).

Natural History of Atherosclerosis

Fatty streaks in the intima of the aorta can be identified even in infants and although many disappear, progression to fibrous plaques and clinical complications cannot be ruled out (Magnussen et al. 2009), as revealed by the classic Bogalusa Heart Study (Berenson et al. 1998) and Muscatine Study, both including long-term follow-up (Mahoney et al. 2001). Carotid intima-media abnormalities and deranged endothelial response have similarly been demonstrated in later years (Freedman et al. 2004; Urbina et al. 2009; Urbina et al. 2010).

Higher cardiac output and blood volume can follow as well, along with sleep apnea and hypoventilation, which do not bode well for heart and lung physiology (Speiser et al. 2005). Left ventricular mass also increases in obese youth, irrespective of systemic hypertension (Urbina et al. 2009; Yoshinaga et al. 1995; Daniels et al. 1998), potentially aggravating CVD morbidity and mortality (Flynn and Alderman 2005).

Dyslipidemia

Excessive body weight in adolescence mostly continues in later life (Freedman et al. 2007; Juonala et al. 2008), and a deranged lipid profile is common (Kit et al. 2015), usually defined by combined dyslipidemia of obesity (CDO). CDO aberrations include hypertriglyceridemia with less prominent deviation of LDL cholesterol than seen in older ages, yet with small, dense particles. One-third of sixth graders with excessive adiposity exhibited TG/HDL-C ratio > 3.0, and 11.2% had a non-HDL-C >145 mg/dL (Mietus-Snyder et al. 2013).

Future propensity toward myocardial infarction and/or death from CVD has been predicted on the basis of a CDO (Robins et al. 2011) or similar pattern (Morrison et al. 2012).

Management of Dyslipidemia

Diet and physical exercise are mainstays of lipid profile correction, especially when counseling for the whole family is provided by a dietitian (National Heart L and BI 2012). When excessive body weight is simultaneously diminished, all age brackets are benefited (Nemet et al. 2005; Kirk et al. 2012; Ebbeling et al. 2007; Ebbeling et al. 2012; Meyer et al. 2006). Simple sugars and easily hydrolyzed starches are closely linked to triglyceride concentration (Sondike et al. 2003; Ebbeling et al. 2007; Pieke et al. 2000; Pereira et al. 2004), and reduction could also lead to better weight control (Dornas et al. 2015). Fiber and low-mercury fish should be encouraged instead (2X week). Pharmacological agents should basically be considered for refractory cases (De Ferranti et al. 2019).

Hypertension

Blood pressure < 90th percentile for age, sex, and height should always be sought (National Heart L and BI 2012). Values up to the 95th percentile for systolic or diastolic measurements (> 120/80 for adolescents) point toward prehypertension, whereas higher readings confirm the clinical condition. As with adults, three similar findings are required to establish diagnosis.

Excessive body weight predisposes to the condition, and weight loss ameliorates it (Haynes 1986; Rocchini et al. 1988). Sodium, caffeine, and fat foods are deemed deleterious, contrasting with potassium, calcium, and magnesium (He and MacGregor 2006; Dwyer et al. 1998; Sinaiko et al. 1993; Knuiman et al. 1988; Falkner et al. 2000; Simons-Morton et al. 1997).

Salty Foods

Commercial and industrial food carry the largest share of sodium intake, not domestic seasoning, and not more than 1.2 g/day (ages 4–8) or 1.5 g/day (> 8 years old) is allowed (Appel et al. 2006).

The Dietary Approaches to Stop Hypertension (DASH) diet, notably with sodium restriction, is a highly regarded option for grown-ups (Sacks et al. 2001). Young children with reduced consumption of fruits, vegetables, and low-fat dairy foods are prone to hypertension (Moore et al. 2005). Among teenagers, the DASH protocol has also conducted to less deranged systolic blood pressure (Couch et al. 2008).

Semi-vegetarian Diets

More recently, the adoption of a plant-based, no meat, and no added fat diet was shown to significantly decrease systolic blood pressure (along with other cardiovascular risk factors), in a small cohort of children compared to baseline (Macknin et al. 2015). Indeed, dietary changes can significantly impact cardiovascular risk factors in youth.

Insulin Resistance/Type 2 Diabetes

Glucose homeostasis is determined by beta-cell insulin secretion and the response to insulin in peripheral tissues, including the liver, skeletal muscle, and adipose tissue (Vidal and Kahn 2001). Insulin resistance (IR) refers to the state of reduced tissue response to insulin-mediated glucose metabolism (Levy-Marchal et al. 2010). In the setting of reduced tissue sensitivity, insulin secretion must increase to preserve euglycemia (Bergman et al. 1985). Abnormalities of glucose homeostasis may occur when insulin secretion is not adequate to compensate for the degree of IR (Arslanian et al. 2018a). IR is one of the key components for the development of type

2 diabetes (T2D), and it is strongly associated with obesity (Abrams et al. 2013).

Measurement of insulin sensitivity in children can be done by several methods. The hyperinsulinemic-euglycemic clamp is considered the gold standard (DeFronzo et al. 1979). It consists of a predetermined insulin infusion delivered simultaneously with a glucose infusion, with the goal of maintaining the glucose concentration at a "clamped" fasting level (Ighbariya and Weiss 2017; Brar et al. 2013). Clamps are expensive, time-consuming, and invasive, and are used mainly for research purposes (Levy-Marchal et al. 2010; Ighbariya and Weiss 2017; Brar et al. 2013). The frequently sampled intravenous glucose tolerance test (FSIGT) is an alternative method validated in children, that is based on serial insulin and glucose levels after an intravenous glucose administration (Brar et al. 2013). Simpler surrogate measures have been developed using oral glucose tolerance tests (Matsuda Index), and fasting samples such as the Homeostatic model for assessment of IR (HOMA-IR) (Ighbariya and Weiss 2017). HOMA-IR has been validated in obese children and adolescents with prediabetes and T2D (Brar et al. 2013; George et al. 2011).

Adolescents experience a characteristic phase of reduced insulin sensitivity during Tanner stage II-IV, followed by recovery by Tanner stage V (Moran et al. 1999). The mechanism of this phenomenon is not completely understood and may involve changes in body composition and variations in growth hormone secretion (Amiel et al. 1986; Goran and Gower 2001; Moran et al. 2002). The reduced insulin sensitivity seen in adolescents worsens in the presence of excess adipose tissue (Abrams et al. 2013). A dramatic increase in the prevalence of pediatric obesity has contributed to the escalating incidence of T2D in this population (Collaboration et al. 2017; Arslanian et al. 2018b; Mayer-Davis et al. 2017). Mayer-Davis et al (2017) reported a 7.1% annual increase in the incidence of T2D among youths 10–19 years of age during 2011–2012, compared to 2002–2003 (Mayer-Davis et al. 2017).

Table 27.11 Diagnostic criteria for prediabetes and diabetes (Arslanian et al. 2018a)

Prediabetes
• HbA1C 5.7% to 6.5% (39 to 48 mmol/mol)[a] • Fasting glucose ≥100 but <126 mg/dL (≥5.6 but <7.0 mmol/L). • OGTT[b]: 2-hour plasma glucose ≥140 but <200 mg/dL (≥7.8 but <11.1 mmol/L)
Diabetes
• HbA1C >6.5% (>48 mmol/mol)[a] OR • Fasting plasma glucose ≥126 mg/dL (7.0 mmol/L) OR • OGTT[b]: 2-hour plasma glucose ≥200 mg/dL (11.1 mmol/L) OR • Symptoms of hyperglycemia (polyuria, polydipsia, polyphagia) plus a random plasma glucose ≥200 mg/dL (11.1 mmol/L)

[a]The test should be performed in a laboratory using a method that is NGSP certified and standardized to the DCCT assay
[b]The test should be performed as described by the World Health Organization, using a glucose load containing the equivalent of 1.75 mg/kg (maximum 75 g) anhydrous glucose dissolved in water
Fasting is defined as no caloric intake for at least 8 hours
HbA1C: hemoglobin A1C; OGTT: oral glucose tolerance test; NGSP: national glycohemoglobin standardization program; DCCT: diabetes control and complications trials;

Severity of Adolescent T2D

There is evidence that youth-onset T2D has a more rapid progression and increased prevalence of complications in comparison with adult-onset T2D (Arslanian et al. 2018b; Hannon and Arslanian 2015; Levitt Katz et al. 2011; McKay et al. 2012). For those reasons, screening for glucose abnormalities is important to identify children at risk and subsequently to implement early interventions that can modify progression. The American Diabetes Association (ADA) recommends to screen for prediabetes and/or T2D in asymptomatic children and adolescents (after onset of puberty or after age 10), who are overweight or obese (BMI ≥85th % for age and sex) and have 1 or more additional risk factors including: (1) first- or second-degree relative with T2D, (2) minority race or ethnicity, (3) signs of IR (acanthosis nigricans) or comorbidities such as hypertension, dyslipidemia, polycystic ovarian syndrome, or (4) mother with history of diabetes or gestational diabetes during child's pregnancy (ADA 2019).

Pediatric Prediabetes

Prediabetes refers to a condition of blood glucose above normal, however not at the level considered diagnostic for diabetes (Table 27.11) (Arslanian et al. 2018a; Khokhar et al. 2017). Whereas in adults, the progression from prediabetes to overt T2D occurs gradually in about 5–10 years; in obese youth, this transition may only take 2.5 years (Weiss et al. 2005). Lifestyle and drug-based interventions can restore glucose to normal levels and decrease the risk for diabetes in adults; however, large-scale data in youth are limited (Khokhar et al. 2017; Knowler et al. 2002). The Diabetes Prevention Program (DPP) was a large randomized clinical trial conducted on adults with prediabetes that demonstrated a significant reduction in the risk for T2D with intensive lifestyle changes and treatment with Metformin (Knowler et al. 2002). The HEALTHY study trial of nutrition, physical education, and behavior change in middle school children showed significant reduction in obesity and mean insulin level (Foster et al. 2010).

Metabolic Syndrome

The concept of a group of metabolic disorders associated with increased risk for cardiovascular disease was postulated since the 1920s (Eckel et al. 2005). Gerald Raeven in 1988 introduced the term "syndrome X" and recognized IR as the common feature (Reaven 1988). Central obesity was not included as part of the syndrome until the late 1940s, and the term "metabolic syndrome" was preferred as the unifying definition (Eckel et al. 2005). The International Diabetes Federation defines metabolic syndrome (MetSyn) in adults as central obesity (ethnicity specific and measured by waist circumference unless BMI >30 mg/m2) plus any two of the following: 1)

Table 27.12 Diagnostic criteria for metabolic syndrome in children and adolescents

Authors	3 or more criteria
Cook et al. (2003)	• Waist circumference ≥90th percentile • Systolic or diastolic blood pressure ≥90th percentile • Triglycerides ≥1.24 mmol/L or HDL-C ≤1.03 mmol/L • Fasting glucose ≥6.11 mmol/L
Cruz and Goran (2004)	• Waist circumference ≥90th percentile • Blood pressure ≥90th percentile • Triglycerides ≥90th percentile or HDL ≤ 10th percentile • Glucose intolerance (ADA criteria)
Weiss et al. (2004)	• BMI z-score ≥2.0 • Blood pressure >95th percentile • HDL <5th percentile • Triglycerides >95th percentile • Glucose intolerance (ADA criteria)
De Ferranti et al. (2004)	• Waist circumference >75th percentile • Blood pressure >90th percentile • Triglycerides ≥1.1 mmol/L or HDL <1.2 mmol/L (girls), <1.3 mmol/L (boys • Fasting glucose ≥6.1 mmol/L
Viner et al. (2005)	• BMI ≥95th percentile • Systolic blood pressure ≥95th percentile • Triglycerides ≥11.69 mmol/L or HDL ≤0.91 mmol/L or total cholesterol ≥95th percentile • Insulin ≥104.2 pmol/L or fasting glucose ≥5.55 mmol/L
Zimmet et al. (2007)	• Waist circumference ≥90th percentile • Systolic blood pressure ≥130 mm Hg or diastolic blood pressure ≥85 mm Hg • Triglycerides ≥1.7 mmol/L or HDL ≤1.03 mmol/L • Fasting glucose ≥5.6 mmol/L
Ahrens et al. (2014)	Monitoring level (action level) • Waist circumference ≥90th (95th) percentile • Systolic/Diastolic blood pressure ≥90th (95th) percentile • Triglycerides >90th (95th) percentile or HDL ≤10th (5th) percentile • HOMA-IR[a] ≥90th (95th) percentile or fasting glucose ≥90th (95th) percentile

[a]Homeostatic model assessment of insulin resistance

Raised Triglycerides $\geq$150 mg/dL (1.7 mmol/L), 2) Reduced HDL cholesterol < 40 mg/dL (1.03 mmol/L) in males < 50 mg/dL (1.29 mmol/L) in females, 3) Raised blood pressure systolic BP $\geq$130 or diastolic BP $\geq$85 mm Hg, and 4) Raised fasting plasma glucose (FPG) $\geq$ 100 mg/dL (5.6 mmol/L) (Alberti et al. 2006).

Several expert groups have established diagnostic criteria for MetSyn in adults; however, there is lack of consensus regarding a definition in children and adolescents. The scientific interest to adapt a definition for children grew with the recognition of the remarkable increase in pediatric obesity worldwide (Collaboration et al. 2017). Despite conflicting definitions (Table 27.12), most authors agree that the key components are abnormalities in glucose metabolism, central obesity, hypertension, and dyslipidemia (Cook et al. 2003; Cruz and Goran 2004; Weiss et al. 2004; De Ferranti et al. 2004; Viner et al. 2005; Zimmet et al. 2007; Ahrens et al. 2014).

Visceral and ectopic fat deposition are hallmarks in the pathogenesis of the MetSyn. Lipid accumulation in the liver stimulates secretion of adipocytokines and consequently activation of inflammatory pathways (Magge et al. 2017; Bussler et al. 2017). It has been postulated that this low-grade inflammatory state may explain the IR and endothelial dysfunction seen in the MetSyn (Yudkin 2007).

Recently, the focus on MetSyn has shifted toward differentiating it as a distinct pathophysiology or a cluster of cardiometabolic risk factors (Kahn et al. 2006; Bayturan et al. 2010). Some studies showed that the CVD progression and mortality in patients with MetSyn are driven by its individual components rather than by the presence of the syndrome itself (Bayturan et al. 2010; Wang et al. 2007).

This new approach may be particularly important in the pediatric population for which there is less evidence on the definition and utility of the syndrome. In pediatrics, intensive interventions should target children with multiple component risks who are in great need of risk reduction (Magge et al. 2017).

Table 27.13 Diagnostic Criteria for PCOS in Adult Women and Adolescents

Adult women
National Institutes of Health (NIH) criteria (1990) Requires both criteria: • Oligo-ovulation/anovulation • Clinical and/or biochemical signs of hyperandrogenism
Rotterdam Criteria (2003)[a] Requires 2 criteria: • Oligo-ovulation/anovulation • Clinical and/or biochemical signs of hyperandrogenism • Polycystic ovaries (by ultrasound)
Androgen Excess Society (2006) Requires both criteria: • Clinical and/or biochemical signs of hyperandrogenism • Oligo-ovulation/anovulation and/or polycystic ovaries
Adolescents
Amsterdam Criteria (2012) Requires all three criteria: • Oligo-ovulation/anovulation • Clinical and/or biochemical signs of hyperandrogenism • Polycystic ovaries (by ultrasound)
Endocrine Society Criteria (2013) Requires both criteria: • Oligomenorrhea • Clinical and/or biochemical signs of hyperandrogenism
Pediatric Endocrine Society Criteria (2015) • Abnormal uterine bleeding pattern – Abnormal for age or gynecological age – Persistent for 1–2 years • Evidence of hyperandrogenism – Persistent elevated testosterone levels – Moderate-severe hirsutism – Moderate-severe inflammatory acne vulgaris

Polycystic Ovarian Syndrome

Polycystic Ovarian Syndrome (PCOS) is a heterogenous endocrine disorder that affects a significant proportion of reproductive-aged women, including adolescents. It is defined by a combination of clinical features of androgen excess (hirsutism, moderate-severe acne), ovarian dysfunction (oligo-ovulation), and/or polycystic ovarian morphology that are not otherwise explained (Escobar-Morreale 2018). Additionally, PCOS is associated with metabolic disorders including T2D, obesity, dyslipidemia, hypertension, and MetSyn (Tehrani and Amiri 2019).

PCOS is caused by a complex interaction between genetic and environmental factors (Rosenfield 2015; Witchel et al. 2019). IR plays an important role in the pathogenesis of PCOS and is seen in both lean and obese women (Tehrani and Amiri 2019; Witchel et al. 2019).

There is consensus for diagnostic criteria of PCOS in adult women (Table 27.13). Diagnosis in adolescents becomes problematic as some of the features (menstrual irregularities, acne, mild hyperandrogenism, or multifollicular ovarian morphology) can be seen in normal pubertal development (Witchel et al. 2019). In addition, symptoms of PCOS are often evolving and may not be clearly evident in adolescents (Table 27.13). A comprehensive assessment of the history, physical exam, and laboratory data is important to identify the adolescents at risk for PCOS, and if indicated, to initiate treatment that can significantly impact their quality of life.

Nonalcoholic Fatty Liver Disease (NAFLD)

Within a few decades, this has become the number one chronic liver condition for all age brackets (Wong et al. 2015), although the impact on children with excessive adiposity is still not well established (1.7–85%). One pediatric case in four displays nonalcoholic steatohepatitis (NASH), a more inflammatory modality which can advance to cirrhosis (7–10%) and even require transplantation (Molleston et al. 2002; Rubinstein et al. 2008). Alanine aminotransferase increase (boys: ALT $\geq$50 mg/dL, girls: ALT $\geq$44 mg/dL) at age $\geq$10 years has been employed as a screening tool in populations with increased adiposity, only severe cases requiring biopsy (Vos et al. 2017). There is more experience with imaging tools such as Fibroscan in adults than in children (Xanthakos et al. 2017).

NAFLD Management

Healthier food intake and weight loss are advised (Nobili et al. 2006; Nobili et al. 2008), even though no particular diet has been adopted (Ramon-Krauel et al. 2013). Elimination of high sugar drinks looks promising (Ebbeling et al. 2012; De Ruyter et al. 2012), along with aerobic and resistance exercise (Lee et al. 2012). Prescription of obeticholic acid for adults was beneficial (Younossi et al. 2019), and an FDA new drug application for alleviation of fibrosis in adults (NASH) has been accepted. Within the 8–17 age bracket, Metformin was disappointing whereas Vitamin E was beneficial for histology, however not for liver enzymes (Lavine et al. 2011). Other pharmacologic protocols are going on (Konerman et al. 2018), and metabolic/bariatric surgery could also be an alternative for adolescents with excessive adiposity and NASH.

Non-cardiometabolic Comorbidities

Obstructive Sleep Apnea (OSA)

Childhood obesity is associated with a much higher prevalence of OSA (Spilsbury et al. 2015). Anderson et al. recently showed OSA prevalence of 44.6% in overweight/obese children compared with 9.1% in those normal weight (Andersen et al. 2019). The rate of weight gain is an important predictor of severe OSA. A retrospective cohort study showed that yearly change in weight among obese adolescents with severe OSA was significantly higher than those without severe OSA (Johnson et al. 2019). While adenotonsillectomy is effective at reversing failure to thrive in OSA, it is not as effective at treating obesity-associated OSA, for which new and different treatment options need to be considered (Keefe et al. 2018).

Musculoskeletal (MSK) issues

Obese children tend to reduce their level of physical activity and have an increased risk of MSK injuries and pain. Contributing factors include gait/postural impairments, increased load/impact to the lower extremities and back, lower extremity malalignment, and reduced bone mineral density (Chan and Chen 2009; Steinberg et al. 2018). This is illustrated by a study showing obese children walk with increased patellofemoral loads, enhancing risk of developing chronic knee pain (Kim et al. 2019). Obese children should be screened for these biomechanical factors prior to any exercise program to reduce risk for future MSK issues (Steinberg et al. 2018).

Vitamin D (VitD) deficiency

VitD deficiency has been reported in 34% and 49% of obese and severely obese children, respectively. Low 25-hydroxy-Vitamin D (25-OHD) in obese may be explained by inadequate dietary consumption, decreased sun exposure, and trapping of 25-OHD in adipose tissue (Nowicki et al. 2019; Corica et al. 2019). The bone/calcemic effects of VitD are well known, and more recently, the non-bone/non-calcemic effects have been better elucidated. VitD plays an essential role in the regulation of glucose homeostasis, insulin secretion mechanisms, and inflammation associated with obesity (Zakharova et al. 2019). Low 25-OHD has been shown to negatively influence beta-cell function and insulin sensitivity in obese nondiabetic children (Corica et al. 2019). Higher cholecalciferol doses are likely needed to achieve serum 25-OHD targets in obese children (Zakharova et al. 2019).

Psychological Impact and Considerations

Youth with obesity are more likely than their average-weight peers to experience weight-related stigma, bullying, depression, anxiety, low self-esteem, and difficulty with self-regulation including attention deficit hyperactivity disorder (ADHD) (Zeller et al. 2004). Family and parental characteristics play an important role in overweight and obesity in children. Poor family communication and high family conflict are linked to overeating and obesity in children. Maternal depression, overly permissive or overly restrictive parenting practices, and high parental distress about their child's weight are associated with obesity in children. On the other hand, authoritative parenting styles, characterized by high expectations, warmth, and responsiveness are associated with healthier eating habits and average weight in children (Halliday et al. 2014). Parental modeling of healthy lifestyle behaviors is an important indicator of child and adolescent health behaviors.

Mental health diagnoses, such as autism spectrum disorder, ADHD, depression, and anxiety, are more common in obese children. Such concordant diagnoses can significantly interfere with patient/family adherence to weight management interventions. Appropriate behavioral health referrals for mental health diagnoses are imperative to the success of any weight management intervention (Small and Aplasca 2016). Even if a child does not meet criteria for a specific mental health diagnosis, it is important to take the child and family psychosocial functioning into account, as the most effective interventions for weight management center around behavior modification. Common barriers to lifestyle modification adherence include child disruptive behavior, emotional eating, conflict around food, and food sneaking behaviors. Therefore, the most effective multidisciplinary weight management programs are those who have the support of behavioral health professionals, who focus on the child's self-esteem and quality of life without negatively impacting family functioning (Ward et al. 2012).

Ongoing Studies and Future Perspective

As rates of pediatric obesity remain high globally, we are beginning to see consequences affect the health of whole populations. Obesity that persists into adulthood is associated with decreased quality of life, higher morbidity, and shorter life expectancy (Grover et al. 2015). For this reason, public health interventions that focus on obesity prevention should be prioritized, including policy measures that improve food and beverage advertising practices to children, sugar-sweetened beverage taxation, and strengthening school food and physical activity policies, all of which have shown promise in various settings (Kelly et al.

2010; Yoshida and Simoes 2018; Wharton et al. 2008).

Addressing pediatric obesity in the clinical setting continues to evolve with increased use of medication and surgical treatment as outlined above for severe cases, and a growing body of research supporting optimal use. There is also significant ongoing research on the use of health technologies in treatment, including mobile-health tools, activity, and nutrition tracking applications (Turner et al. 2015; Quelly et al. 2016).

There is also much interest in time-restricted feeding and intermittent fasting in treatment of adult obesity and comorbidity management; however, research in adolescents is limited (Stockman et al. 2018). It is also recommended that all obesity interventions directed at children and adolescents be conducted with the utmost sensitivity, awareness of weight bias and stigma, and in a way that supports psychological well-being, so as to avoid potential unintended negative consequences of weight management in children (Pont et al. 2017).

References

Abraham A, Ikramuddin S, Jahansouz C, et al (2016) Trends in bariatric surgery: procedure selection, revisional surgeries, and readmissions. Obes Surg. https://doi.org/10.1007/s11695-015-1974-2

Abrams P, Levitt Katz LE, Moore RH et al (2013) Threshold for improvement in insulin sensitivity with adolescent weight loss. J Pediatr. https://doi.org/10.1016/j.jpeds.2013.04.003

ADA (2019) 13. Children and adolescents: standards of medical care in diabetesd 2019. Diabetes Care. https://doi.org/10.2337/dc19-S013

Ahrens W, Moreno L, Mårild S et al (2014) Metabolic syndrome in young children: definitions and results of the IDEFICS study. Int J Obes. https://doi.org/10.1038/ijo.2014.130

Akkary E (2012) Bariatric surgery evolution from the malabsorptive to the hormonal era. Obes Surg 22:827–831

Alberti KGMM, Zimmet P, Shaw J (2006) Metabolic syndrome—a new world-wide definition. A consensus statement from the International Diabetes Federation. Diabet Med 23:469–480

Amiel SA, Sherwin RS, Simonson DC, et al (1986) Impaired insulin action in puberty. A contributing factor to poor glycemic control in adolescents with diabetes. N Engl J Med. https://doi.org/10.1056/NEJM198607243150402

Amin R, Simakajornboon N, Szczesniak R, Inge T (2017) Early improvement in obstructive sleep apnea and increase in orexin levels after bariatric surgery in adolescents and young adults. Surg Obes Relat Dis. https://doi.org/10.1016/j.soard.2016.05.023

Andersen IG, Holm JC, Homøe P (2019) Obstructive sleep apnea in children and adolescents with and without obesity. Eur Arch Otohino-Laryngol. https://doi.org/10.1007/s00405-019-05290-2

Appel LJ, Brands MW, Daniels SR et al (2006) Dietary approaches to prevent and treat hypertension: A scientific statement from the American Heart Association. Hypertension

Aris IM, Rifas-Shiman SL, Li LJ et al (2019) Patterns of body mass index milestones in early life and cardiometabolic risk in early adolescence. Int J Epidemiol. https://doi.org/10.1093/ije/dyy286

Armstrong S, Lazorick S, Hampl S et al (2016) Physical examination findings among children and adolescents with obesity: an evidence-based review. Pediatrics 137:e20151766

Armstrong SC, Bolling CF, Michalsky MP, et al (2019) Pediatric metabolic and bariatric surgery: evidence, barriers, and best practices. Pediatrics. https://doi.org/10.1542/peds.2019-3223

Arslanian S, Bacha F, Grey M et al (2018a) Evaluation and management of youth-onset type 2 diabetes: A position statement by the American diabetes association. Diabetes Care

Arslanian S, Kim JY, Nasr A, et al (2018b) Insulin sensitivity across the lifespan from obese adolescents to obese adults with impaired glucose tolerance: who is worse off? Pediatr Diabetes. https://doi.org/10.1111/pedi.12562

Babio N, Bulló M, Salas-Salvadó J (2009) Mediterranean diet and metabolic syndrome: the evidence. Public Health Nutr. https://doi.org/10.1017/S1368980009990449

Barlow SE (2007) Expert committee recommendations regarding the prevention, assessment, and treatment of child and adolescent overweight and obesity: summary report. Pediatrics. https://doi.org/10.1542/peds.2007-2329C

Bayturan O, Tuzcu EM, Lavoie A et al (2010) The metabolic syndrome, its component risk factors, and progression of coronary atherosclerosis. Arch Intern Med. https://doi.org/10.1001/archinternmed.2009.551

Berenson GS, Srinivasan SR, Bao W et al (1998) Association between multiple cardiovascular risk factors and atherosclerosis in children and young adults. N Engl J Med. https://doi.org/10.1056/NEJM199806043382302

Bergman RN, Finegood DT, Ader M (1985) Assessment of insulin sensitivity in vivo. Endocr Rev. https://doi.org/10.1210/edrv-6-1-45

Brar PC, Koren D, Gallagher PR et al (2013) Comparison of oral and intravenous glucose tolerance test derived sensitivity and secretory indices in obese adolescents. Clin Pediatr (Phila). https://doi.org/10.1177/0009922812472251

Bussler S, Penke M, Flemming G et al (2017) Novel insights in the metabolic syndrome in childhood and adolescence. Horm Res Paediatr 88:181–193

Chan G, Chen CT (2009) Musculoskeletal effects of obesity. Curr Opin Pediatr 21:65–70

Collaboration NRF, Lewington S, Clarke R et al (2017) Worldwide trends in body-mass index, underweight, overweight, and obesity from 1975 to 2016: A pooled analysis of 2416 population-based measurement studies in 128.9 million children, adolescents, and adults. Lancet. https://doi.org/10.1016/S0140-6736(16)31919-5

Cook S, Weitzman M, Auinger P et al (2003) Prevalence of a metabolic syndrome phenotype in adolescents: Findings from the Third National Health and Nutrition Examination Survey, 1988–1994. Arch Pediatr Adolesc Med. https://doi.org/10.1001/archpedi.157.8.821

Corica D, Zusi C, Olivieri F et al (2019) Vitamin D affects insulin sensitivity and β-cell function in obese non-diabetic youths. Eur J Endocrinol. https://doi.org/10.1530/eje-19-0369

Couch SC, Saelens BE, Levin L et al (2008) The Efficacy of a Clinic-Based Behavioral Nutrition Intervention Emphasizing a DASH-Type Diet for Adolescents with Elevated Blood Pressure. J Pediatr. https://doi.org/10.1016/j.jpeds.2007.09.022

Cruz ML, Goran MI (2004) The metabolic syndrome in children and adolescents. Curr Diab Rep 369:2059–2061

Daniels SR (2009) Complications of obesity in children and adolescents. Int J Obes. https://doi.org/10.1038/ijo.2009.20

Daniels SR, Loggie JMH, Khoury P, Kimball TR (1998) Left ventricular geometry and severe left ventricular hypertrophy in children and adolescents with essential hypertension. Circulation. https://doi.org/10.1161/01.CIR.97.19.1907

De Ferranti SD, Gauvreau K, Ludwig DS et al (2004) Prevalence of the metabolic syndrome in American adolescents: findings from the Third National Health and Nutrition Examination Survey. Circulation. https://doi.org/10.1161/01.CIR.0000145117.40114.C7

De Ferranti SD, Steinberger J, Ameduri R, et al (2019) Cardiovascular Risk Reduction in High-Risk Pediatric Patients: A Scientific Statement From the American Heart Association. Circulation. https://doi.org/10.1161/CIR.0000000000000618

De Ruyter JC, Olthof MR, Seidell JC, Katan MB (2012) A trial of sugar-free or sugar-sweetened beverages and body weight in children. N Engl J Med. https://doi.org/10.1056/NEJMoa1203034

DeFronzo RA, Tobin JD, Andres R (1979) Glucose clamp technique: A method for quantifying insulin secretion and resistance. Am J Physiol Endocrinol Metab Gastrointest Physiol. https://doi.org/10.1152/ajpendo.1979.237.3.e214

Dornas WC, de Lima WG, Pedrosa ML, Silva ME (2015) Health Implications of High-Fructose Intake and Current Research. Adv Nutr. https://doi.org/10.3945/an.114.008144

Dubois L, Girard M (2006) Early determinants of overweight at 4.5 years in a population-based longitudinal study. Int J Obes. https://doi.org/10.1038/sj.ijo.0803141

Dwyer JH, Dwyer KM, Scribner RA et al (1998) Dietary calcium, calcium supplementation, and blood pressure in African American adolescents. Am J Clin Nutr. https://doi.org/10.1093/ajcn/68.3.648

Ebbeling CB, Leidig MM, Sinclair KB et al (2003) A reduced-glycemic load diet in the treatment of adolescent obesity. Arch Pediatr Adolesc Med. https://doi.org/10.1001/archpedi.157.8.773

Ebbeling CB, Leidig MM, Feldman HA et al (2007) Effects of a low-glycemic load vs low-fat diet in obese young adults: A randomized trial. J Am Med Assoc. https://doi.org/10.1001/jama.297.19.2092

Ebbeling CB, Feldman HA, Chomitz VR et al (2012) A randomized trial of sugar-sweetened beverages and adolescent body weight. N Engl J Med. https://doi.org/10.1056/NEJMoa1203388

Eckel RH, Grundy SM, Zimmet PZ (2005) The metabolic syndrome. Lancet 365:1415–1428

Epstein LH, Valoski A, Wing RR, Mccurley J (1990) Ten-year follow-up of behavioral, family-based treatment for obese children. J Am Med Assoc. https://doi.org/10.1001/jama.1990.03450190051027

Epstein LH, Valoski A, Wing RR, McCurley J (1994) Ten-year outcomes of behavioral family-based treatment for childhood obesity. Health Psychol. https://doi.org/10.1037/0278-6133.13.5.373

Escobar-Morreale HF (2018) Polycystic ovary syndrome: definition, aetiology, diagnosis and treatment. Nat Rev Endocrinol 14:270–284

Falkner B, Sherif K, Michel S, Kushner H (2000) Dietary nutrients and blood pressure in urban minority adolescents at risk for hypertension. Arch Pediatr Adolesc Med. https://doi.org/10.1001/archpedi.154.9.918

Farooqi IS, Keogh JM, Yeo GSH et al (2003) Clinical spectrum of obesity and mutations in the melanocortin 4 receptor gene. N Engl J Med. https://doi.org/10.1056/NEJMoa022050

Flegal KM, Wei R, Ogden CL et al (2009) Characterizing extreme values of body mass index-for-age by using the 2000 Centers for Disease Control and Prevention growth charts. Am J Clin Nutr. https://doi.org/10.3945/ajcn.2009.28335

Flynn JT, Alderman MH (2005) Characteristics of children with primary hypertension seen at a referral center. Pediatr Nephrol. https://doi.org/10.1007/s00467-005-1855-3

Foster GD, Linder B, Baranowski T et al (2010) A School-based intervention for diabetes risk reduction. N Engl J Med. https://doi.org/10.1056/NEJMoa1001933

Freedman DS, Dietz WH, Srinivasan SR, Berenson GS (1999) The relation of overweight to cardiovascular risk factors among children and adolescents: The Bogalusa Heart Study. Pediatrics. https://doi.org/10.1542/peds.103.6.1175

Freedman DS, Dietz WH, Tang R et al (2004) The relation of obesity throughout life to carotid intima-media thickness in adulthood: the Bogalusa Heart Study. Int J Obes. https://doi.org/10.1038/sj.ijo.0802515

Freedman DS, Mei Z, Srinivasan SR, et al (2007) Cardiovascular Risk Factors and Excess Adiposity Among Overweight Children and Adolescents: The Bogalusa Heart Study. J Pediatr. https://doi.org/10.1016/j.jpeds.2006.08.042

George L, Bacha F, Lee SJ et al (2011) Surrogate estimates of insulin sensitivity in obese youth along the spectrum of glucose tolerance from normal to prediabetes to diabetes. J Clin Endocrinol Metab. https://doi.org/10.1210/jc.2010-2813

Geserick M, Vogel M, Gausche R et al (2018) Acceleration of BMI in early childhood and risk of sustained obesity. N Engl J Med. https://doi.org/10.1056/NEJMoa1803527

Gibson LA, Alava MH, Kelly MP, Campbell MJ (2017) The effects of breastfeeding on childhood BMI: A propensity score matching approach. J Public Heal UK. https://doi.org/10.1093/pubmed/fdw093

Gillman MW, Daniels SR (2012) Is universal pediatric lipid screening justified? J Am Med Assoc 307:259–260

Goldstein RF, Abell SK, Ranasinha S et al (2017) Association of gestational weight gain with maternal and infant outcomes: a systematic review and meta-analysis. J Am Med Assoc. https://doi.org/10.1001/jama.2017.3635

Goran MI, Gower BA (2001) Longitudinal study on pubertal insulin resistance. Diabetes. https://doi.org/10.2337/diabetes.50.11.2444

Göthberg G, Gronowitz E, Flodmark CE et al (2014) Laparoscopic Roux-en-Y gastric bypass in adolescents with morbid obesity-Surgical aspects and clinical outcome. Semin Pediatr Surg. https://doi.org/10.1053/j.sempedsurg.2013.10.015

Greenway FL (2015) Physiological adaptations to weight loss and factors favouring weight regain. Int J Obes 39:1188–1196

Grosso G, Mistretta A, Marventano S et al (2014) Beneficial effects of the Mediterranean diet on metabolic syndrome. Curr Pharm Des. https://doi.org/10.2174/13816128196661312061121 44

Grover SA, Kaouache M, Rempel P et al (2015) Years of life lost and healthy life-years lost from diabetes and cardiovascular disease in overweight and obese people: a modelling study. Lancet Diabetes Endocrinol. https://doi.org/10.1016/S2213-8587(14)70229-3

Gulati AK, Kaplan DW, Daniels SR (2012) Clinical tracking of severely obese children: a new growth chart. Pediatrics. https://doi.org/10.1542/peds.2012-0596

Halliday JA, Palma CL, Mellor D et al (2014) The relationship between family functioning and child and adolescent overweight and obesity: a systematic review. Int J Obes 38:480–493

Hannon TS, Arslanian SA (2015) The changing face of diabetes in youth: Lessons learned from studies of type 2 diabetes. Ann N Y Acad Sci. https://doi.org/10.1111/nyas.12939

Hassink SG, Daniels SR, Abrams SA et al (2015) The role of the pediatrician in primary prevention of obesity. Pediatrics. https://doi.org/10.1542/peds.2015-1558

Haynes RB (1986) Is weight loss an effective treatment for hypertension? The evidence against. Can J Physiol Pharmacol. https://doi.org/10.1139/y86-142

He FJ, MacGregor GA (2006) Importance of salt in determining blood pressure in children: meta-analysis of controlled trials. Hypertension. https://doi.org/10.1161/01.HYP.0000245672.27270.4a

Horta BL, Loret De Mola C, Victora CG (2015) Long-term consequences of breastfeeding on cholesterol, obesity, systolic blood pressure and type 2 diabetes: a systematic review and meta-analysis. Acta Paediatr Int J Paediatr 04:30–37

Ighbariya A, Weiss R (2017) Insulin resistance, prediabetes, metabolic syndrome: what should every pediatrician know? J Clin Res Pediatr Endocrinol 9:49–57

Inge TH, Courcoulas AP, Jenkins TM et al (2019) Five-year outcomes of gastric bypass in adolescents as compared with adults. N Engl J Med. https://doi.org/10.1056/NEJMoa1813909

Johnson RF, Hansen A, Narayanan A et al (2019) Weight gain velocity as a predictor of severe obstructive sleep apnea among obese adolescents. Laryngoscope. https://doi.org/10.1002/lary.28296

Juonala M, V̂iikari JSA, Rönnemaa T et al (2008) Associations of dyslipidemias from childhood to adulthood with carotid intima-media thickness, elasticity, and brachial flow-mediated dilatation in adulthood: the Cardiovascular Risk in Young Finns Study. Arterioscler Thromb Vasc Biol. https://doi.org/10.1161/ATVBAHA.108.163329

Kahn R, Buse J, Ferrannini E, Stern M (2006) The metabolic syndrome: time for a critical appraisal. Arterial'naya Gipertenz ("Arterial Hypertens"). https://doi.org/10.18705/1607-419x-2006-12-2-99-116

Keefe KR, Patel PN, Levi JR (2018) The shifting relationship between weight and pediatric obstructive sleep apnea: a historical review. Laryngoscope 129:2414–2419

Kelly AS, Fox CK (2017) Pharmacotherapy in the management of pediatric obesity. Curr Diab Rep 17:55

Kelly B, Halford JCG, Boyland EJ et al (2010) Television food advertising to children: a global perspective. Am J Public Health 100:1730–1736

Khera AV, Chaffin M, Wade KH et al (2019) Polygenic prediction of weight and obesity trajectories from birth to adulthood. Cell. https://doi.org/10.1016/j.cell.2019.03.028

Khokhar A, Umpaichitra V, Chin VL, Perez-Colon S (2017) Metformin use in children and adolescents with prediabetes. Pediatr Clin North Am.

Kim N, Browning RC, Lerner ZF (2019) The effects of pediatric obesity on patellofemoral joint contact force during walking. Gait Posture. https://doi.org/10.1016/j.gaitpost.2019.07.307

Kirk S, Brehm B, Saelens BE et al (2012) Role of carbohydrate modification in weight management among obese children: A randomized clinical trial. J Pediatr. https://doi.org/10.1016/j.jpeds.2012.01.041

Kit BK, Kuklina E, Carroll MD et al (2015) Prevalence of and trends in dyslipidemia and blood pressure among us children and adolescents, 1999–2012. JAMA Pediatr. https://doi.org/10.1001/jamapediatrics.2014.3216

Knowler WC, Barrett-Connor E, Fowler SE et al (2002) Reduction in the incidence of type 2 diabetes with lifestyle intervention or metformin. N Engl J Med. https://doi.org/10.1056/NEJMoa012512

Knuiman JT, Hautvast JGAJ, Zwiauer KFM et al (1988) Blood pressure and excretion of sodium, potassium, calcium and magnesium in 8- and 9-year old boys from 19 European centres. Eur J Clin Nutr

Konerman MA, Jones JC, Harrison SA (2018) Pharmacotherapy for NASH: Current and emerging. J Hepatol 68:362–375

Krebs NF, Himes JH, Jacobson D et al (2007) Assessment of child and adolescent overweight and obesity. Pediatrics 120:S193–S228

Kumar S, Kelly AS (2017) Review of childhood obesity: from epidemiology, etiology, and comorbidities to clinical assessment and treatment. Mayo Clin Proc 92:251–265

Lavine JE, Schwimmer JB, Van Natta ML et al (2011) Effect of Vitamin e or metformin for treatment of nonalcoholic fatty liver disease in children and adolescents the tonic randomized controlled trial. J Am Med Assoc. https://doi.org/10.1001/jama.2011.520

Lee SJ, Bacha F, Hannon T et al (2012) Effects of aerobic versus resistance exercise without caloric restriction on abdominal fat, intrahepatic lipid, and insulin sensitivity in obese adolescent boys a randomized, controlled trial. Diabetes. https://doi.org/10.2337/db12-0214

Leslie WS, Hankey CR, Lean MEJ (2007) Weight gain as an adverse effect of some commonly prescribed drugs: a systematic review. QJM 100:395–404

Levitt Katz LE, Magge SN, Hernandez ML, et al (2011) Glycemic control in youth with type 2 diabetes declines as early as two years after diagnosis. J Pediatr. https://doi.org/10.1016/j.jpeds.2010.07.011

Levy-Marchal C, Arslanian S, Cutfield W et al (2010) Insulin resistance in children: Consensus, perspective, and future directions. J Clin Endocrinol Metab. https://doi.org/10.1210/jc.2010-1047

Lopomo A, Burgio E, Migliore L (2016) Epigenetics of obesity. Prog Mol Biol Transl Sci 140:151–184. https://doi.org/10.1016/bs.pmbts.2016.02.002

Macknin M, Kong T, Weier A et al (2015) Plant-based, no-added-fat or American heart association diets: Impact on cardiovascular risk in obese children with hypercholesterolemia and their parents. J Pediatr. https://doi.org/10.1016/j.jpeds.2014.12.058

MacLean PS, Bergouignan A, Cornier MA, Jackman MR (2011) Biology's response to dieting: the impetus for weight regain. Am J Phys 301:R581–R600

Mager DR, Iñiguez IR, Gilmour SYJ (2015) The effect of a low fructose and low glycemic index/load (FRAGILE) dietary intervention on indices of liver function, cardiometabolic risk factors, and body composition in children and adolescents with nonalcoholic fatty liver disease (NAFLD). J Parenter Enter Nutr 39:73–84

Magge SN, Goodman E, Armstrong SC et al (2017) The metabolic syndrome in children and adolescents: shifting the focus to cardiometabolic risk factor clustering. Pediatrics. https://doi.org/10.1542/peds.2017-1603

Magnussen CG, Venn A, Thomson R, et al (2009) The Association of Pediatric low- and high-density lipoprotein cholesterol dyslipidemia classifications and change in dyslipidemia status with carotid intima-media thickness in adulthood. evidence from the cardiovascular risk in young Finns study, the Bog. J Am Coll Cardiol. https://doi.org/10.1016/j.jacc.2008.09.061

Mahoney LT, Burns TL, Stanford W et al (2001) Usefulness of the Framingham risk score and body mass index to predict early coronary artery calcium in young adults (Muscatine Study). Am J Cardiol. https://doi.org/10.1016/S0002-9149(01)01728-3

Manco M, Mosca A, De Peppo F et al (2017) The benefit of sleeve gastrectomy in obese adolescents on nonalcoholic steatohepatitis and hepatic fibrosis. J Pediatr. https://doi.org/10.1016/j.jpeds.2016.08.101

Mayer-Davis EJ, Lawrence JM, Dabelea D, et al (2017) Incidence trends of type 1 and type 2 diabetes among youths, 2002–2012. N Engl J Med. https://doi.org/10.1056/NEJMoa1610187

McKay S, Anderson B, Bush C, et al (2012) A clinical trial to maintain glycemic control in youth with type 2 diabetes. N Engl J Med. https://doi.org/10.1056/NEJMoa1109333

Melis Yavuz H, Selcuk B (2018) Predictors of obesity and overweight in preschoolers: the role of parenting styles and feeding practices. Appetite. https://doi.org/10.1016/j.appet.2017.10.001

Meyer AA, Kundt G, Lenschow U et al (2006) Improvement of Early Vascular Changes and Cardiovascular Risk Factors in Obese Children After a Six-Month Exercise Program. J Am Coll Cardiol. https://doi.org/10.1016/j.jacc.2006.07.035

Mietus-Snyder M, Drews KL, Otvos JD et al (2013) Low-density lipoprotein cholesterol versus particle number in middle school children. J Pediatr. https://doi.org/10.1016/j.jpeds.2013.01.012

Mihrshahi S, Baur LA (2018) What exposures in early life are risk factors for childhood obesity? J. J Paediatr Child Health 13:161–173

Molleston JP, White F, Teckman J, Fitzgerald JF (2002) Obese children with steatohepatitis can develop

cirrhosis in childhood. Am J Gastroenterol. https://doi.org/10.1016/S0002-9270(02)04360-5
Moore LL, Singer MR, Bradlee ML et al (2005) Intake of fruits, vegetables, and dairy products in early childhood and subsequent blood pressure change. Epidemiology. https://doi.org/10.1097/01.ede.0000147106.32027.3e
Moran A, Jacobs DR, Steinberger J et al (1999) Insulin resistance during puberty: Results from clamp studies in 357 children. Diabetes. https://doi.org/10.2337/diabetes.48.10.2039
Moran A, Jacobs DR, Jr., Steinberger J et al. (2002) Association between the insulin resistance of puberty and the insulin-like growth factor-I/growth hormone axis. J Clin Endocrinol Metab 87:4817–4820
Morrison JA, Glueck CJ, Wang P (2012) Childhood risk factors predict cardiovascular disease, impaired fasting glucose plus type 2 diabetes mellitus, and high blood pressure 26 years later at a mean age of 38 years: The Princeton-lipid research clinics follow-up study. Metabolism. https://doi.org/10.1016/j.metabol.2011.08.010
Mozaffarian D, Benjamin EJ, Go AS et al (2015) Heart disease and stroke statistics-2015 update: A report from the American Heart Association. Circulation. https://doi.org/10.1161/CIR.0000000000000152
National Heart L and BI (2012) Expert panel on integrated guidelines for cardiovascular health and risk reduction in children and adolescents. Pediatrics 128:S213–S256
NCD Risk Factor Collaboration (NCD-RisC) (2017) Worldwide trends in body-mass index, underweight, overweight, and obesity from 1975 to 2016: a pooled analysis of 2416 population-based measurement studies in 128·9 million children, adolescents, and adults. Lancet 390:2627–2642. https://doi.org/10.1016/S0140-6736(17)32129-3
Nemet D, Barkan S, Epstein Y et al (2005) Short- and long-term beneficial effects of a combined dietary-behavioral- physical activity intervention for the treatment of childhood obesity. Pediatrics. https://doi.org/10.1542/peds.2004-2172
Nobili V, Marcellini M, Devito R et al (2006) NAFLD in children: a prospective clinical-pathological study and effect of lifestyle advice. Hepatology. https://doi.org/10.1002/hep.21262
Nobili V, Manco M, Devito R et al (2008) Lifestyle intervention and antioxidant therapy in children with nonalcoholic fatty liver disease: a randomized, controlled trial. Hepatology. https://doi.org/10.1002/hep.22336
Nowicki P, Kemppainen J, Maskill L, Cassidy J (2019) The role of obesity in pediatric orthopedics. JAAOS Glob Res Rev. https://doi.org/10.5435/jaaosglobal-d-19-00036
Parrott J, Frank L, Rabena R et al (2017) American Society for Metabolic and Bariatric Surgery Integrated Health Nutritional guidelines for the surgical weight loss patient 2016 update: micronutrients. Surg Obes Relat Dis. https://doi.org/10.1016/j.soard.2016.12.018
Pereira MA, Swain J, Goldfine AB et al (2004) Effects of a low-glycemic load diet on resting energy expenditure and heart disease risk factors during weight loss. J Am Med Assoc. https://doi.org/10.1001/jama.292.20.2482
Peterli R, Steinert RE, Woelnerhanssen B et al (2012) Metabolic and hormonal changes after laparoscopic Roux-en-Y gastric bypass and sleeve gastrectomy: a randomized, prospective trial. Obes Surg. https://doi.org/10.1007/s11695-012-0622-3
Pieke B, Von Eckardstein A, Gülbahçe E et al (2000) Treatment of hypertriglyceridemia by two diets rich either in unsaturated fatty acids or in carbohydrates: Effects on lipoprotein subclasses, lipolytic enzymes, lipid transfer proteins, insulin and leptin. Int J Obes. https://doi.org/10.1038/sj.ijo.0801440
Pont SJ, Puhl R, Cook SRSW (2017) Stigma experienced by children and adolescents with obesity. Pediatrics 140: e20173034. https://doi.org/10.1542/peds.2017-3034
Pratt JSA, Browne A, Browne NT et al (2018) ASMBS pediatric metabolic and bariatric surgery guidelines, 2018. Surg Obes Relat Dis 14:882–901
Quattrin T, Wilfley DE (2017) The promise and opportunities for screening and treating childhood obesity: USPSTF recommendation statement. JAMA Pediatr 171:733–735
Quelly SB, Norris AE, Dipietro JL (2016) Impact of mobile apps to combat obesity in children and adolescents: a systematic literature review. J Spec Pediatr Nurs 21:5–17
Racette SB, Yu L, DuPont NC, Clark BR (2017) BMI-for-age graphs with severe obesity percentile curves: tools for plotting cross-sectional and longitudinal youth BMI data. BMC Pediatr. https://doi.org/10.1186/s12887-017-0885-x
Ramon-Krauel M, Salsberg SL, Ebbeling CB et al (2013) A low-glycemic-load versus low-fat diet in the treatment of fatty liver in obese children. Child Obes. https://doi.org/10.1089/chi.2013.0022
Reaven GM (1988) Role of insulin resistance in human disease. Diabetes. https://doi.org/10.2337/diab.37.12.1595
Robins SJ, Lyass A, Zachariah JP et al (2011) Insulin resistance and the relationship of a dyslipidemia to coronary heart disease: The framingham heart study. Arterioscler Thromb Vasc Biol. https://doi.org/10.1161/ATVBAHA.110.219055
Rocchini AP, Katch V, Anderson J et al (1988) Blood pressure in obese adolescents: effect of weight loss. Pediatrics
Rosenfield RL (2015) The diagnosis of polycystic ovary syndrome in adolescents. Pediatrics 136:1154–1165
Rubinstein E, Lavine JE, Schwimmer JB (2008) Hepatic, cardiovascular, and endocrine outcomes of the histological subphenotypes of nonalcoholic fatty liver disease. Semin Liver Dis 28:380–385
Ryder JR, Vega-López S, Ortega R et al (2013) Lifestyle intervention improves lipoprotein particle size and distribution without weight loss in obese Latino

adolescents. Pediatr Obes. https://doi.org/10.1111/j.I2047T-6310.201Y3.00162.x

Ryder JR, Fox CK, Kelly AS (2018) Treatment options for severe obesity in the pediatric population: current limitations and future opportunities. Obesity 26:951–960

Sacks FM, Svetkey LP, Vollmer WM et al (2001) Effects on blood pressure of reduced dietary sodium and the dietary approaches to stop hypertension (dash) diet. N Engl J Med. https://doi.org/10.1056/NEJM200101043440101

Savoye M, Caprio S, Dziura J et al (2014) Reversal of early abnormalities in glucose metabolism in obese youth: results of an intensive lifestyle randomized controlled trial. Diabetes Care. https://doi.org/10.2337/dc13-1571

Shettar V, Patel S, Kidambi S (2017) Epidemiology of obesity and pharmacologic treatment options. Nutr Clin Pract 32:441–462

Shikora SA, Kim JJ, Tarnoff ME (2007) Nutrition and gastrointestinal complications of bariatric surgery. Nutr Clin Pract 22:29–40

Shrewsbury V, Wardle J (2008) Socioeconomic status and adiposity in childhood: a systematic review of cross-sectional studies 1990–2005. Obesity 16:275–284

Simons-Morton DG, Hunsberger SA, Van Horn L et al (1997) Nutrient intake and blood pressure in the dietary intervention study in children. Hypertension. https://doi.org/10.1161/01.HYP.29.4.930

Sinaiko AR, Gomez-Marin O, Prineas RJ (1993) Effect of low sodium diet or potassium supplementation on adolescent blood pressure. Hypertension. https://doi.org/10.1161/01.HYP.21.6.989

Skinner AC, Perrin EM, Moss LA, Skelton JA (2015) Cardiometabolic risks and severity of obesity in children and young adults. N Engl J Med. https://doi.org/10.1056/NEJMoa1502821

Skinner AC, Ravanbakht SN, Skelton JA et al (2018) Prevalence of obesity and severe obesity in US children, 1999–2016. Pediatrics 141:e20173459. https://doi.org/10.1542/peds.2017-3459

Small L, Aplasca A (2016) Child obesity and mental health. A complex interaction. Child Adolesc Psychiatr Clin N Am 25:269–282

Sondike SB, Copperman N, Jacobson MS (2003) Effects of a low-carbohydrate diet on weight loss and cardiovascular risk factors in overweight adolescents. J Pediatr. https://doi.org/10.1067/mpd.2003.4

Spear BA, Barlow SE, Ervin C et al (2007) Recommendations for treatment of child and adolescent overweight and obesity. Pediatrics 120:S254–S288

Speiser PW, Rudolf MCJ, Anhalt H et al (2005) Consensus statement: childhood obesity. J Clin Endocrinol Metab 90:1871–1887

Spilsbury JC, Storfer-Isser A, Rosen CL, Redline S (2015) Remission and incidence of obstructive sleep apnea from middle childhood to late adolescence. Sleep. https://doi.org/10.5665/sleep.4318

Srivastava G, Fox CK, Kelly AS et al (2019) Clinical considerations regarding the use of obesity pharmacotherapy in adolescents with obesity. Obesity 27:190–204

Starling AP, Brinton JT, Glueck DH et al (2015) Associations of maternal BMI and gestational weight gain with neonatal adiposity in the Healthy Start study. Am J Clin Nutr. https://doi.org/10.3945/ajcn.114.094946

Steinberg N, Nemet D, Pantanowitz M, Eliakim A (2018) Gait pattern, impact to the skeleton and postural balance in overweight and obese children: a review. Sports. https://doi.org/10.3390/sports6030075

Steinberger J, Daniels SR, Eckel RH, et al (2009) Progress and challenges in metabolic syndrome in children and adolescents. A Scientific Statement from the American Heart Association Atherosclerosis, Hypertension, and Obesity in the Young Committee of the Council on Cardiovascular Disease in the Young; Circulation

Stipančić G (2018) Secondary causes of obesity in children and adolescents. Cent Eur J Paediatr. https://doi.org/10.5457/p2005-114.193

Stockman M-C, Thomas D, Burke J, Apovian CM (2018) Intermittent fasting: is the wait worth the weight? Curr Obes Rep 7:172–185. https://doi.org/10.1007/s13679-018-0308-9

Suzuki K, Ando D, Sato M et al (2009) The association between maternal smoking during pregnancy and childhood obesity persists to the age of 9–10 years. J Epidemiol. https://doi.org/10.2188/jea.JE20081012

Tehrani FR, Amiri M (2019) Polycystic ovary syndrome in adolescents: Challenges in diagnosis and treatment. Int J Endocrinol Metab 17:e91554

Turner T, Spruijt-Metz D, Wen CKF, Hingle MD (2015) Prevention and treatment of pediatric obesity using mobile and wireless technologies: a systematic review. Pediatr Obes 10:403–409

Urbina EM, Kimball TR, McCoy CE et al (2009) Youth with obesity and obesity-related type 2 diabetes mellitus demonstrate abnormalities in carotid structure and function. Circulation. https://doi.org/10.1161/CIRCULATIONAHA.108.830380

Urbina EM, Kimball TR, Khoury PR et al (2010) Increased arterial stiffness is found in adolescents with obesity or obesity-related type 2 diabetes mellitus. J Hypertens. https://doi.org/10.1097/HJH.0b013e32833a6132

Velázquez-López L, Santiago-Díaz G, Nava-Hernández J et al (2014) Mediterranean-style diet reduces metabolic syndrome components in obese children and adolescents with obesity. BMC Pediatr. https://doi.org/10.1186/1471-2431-14-175

Vidal J, Kahn SE (2001) Regulation of Insulin Secretion in Vivo

Viner RM, Segal TY, Lichtarowicz-Krynska E, Hindmarsh P (2005) Prevalence of the insulin resistance syndrome in obesity. Arch Dis Child. https://doi.org/10.1136/adc.2003.036467

Vos MB, Welsh J (2010) Childhood obesity: update on predisposing factors and prevention strategies. Curr Gastroenterol Rep 12:280–287

Vos MB, Abrams SH, Barlow SE et al (2017) NASPGHAN Clinical Practice Guideline for the Diagnosis and Treatment of Nonalcoholic Fatty Liver Disease in Children: Recommendations from the Expert Committee on NAFLD (ECON) and the North American Society of Pediatric Gastroenterology. Hepatol Nu J Pediatr Gastroenterol Nutr. https://doi.org/10.1097/MPG.0000000000001482

Wang J, Ruotsalainen S, Moilanen L et al (2007) The metabolic syndrome predicts cardiovascular mortality: a 13-year follow-up study in elderly non-diabetic Finns. Eur Heart J. https://doi.org/10.1093/eurheartj/ehl524

Ward W, Baughcum A, Beck A, Cadieux A, Dreyer M, Getzoff E, Gray J, Isreal E, Meagher SM, Nyamora C, Passeri C, Sallinen B, Santos M, Shaffer L (2012) Children's Hospital Association. Expert guidelines: role of psychologists in assessment and treatment of obese youth. Child Hosp Assoc 11:100

Weaver RG, Brazendale K, Hunt E et al (2019) Disparities in childhood overweight and obesity by income in the United States: an epidemiological examination using three nationally representative datasets. Int J Obes. https://doi.org/10.1038/s41366-019-0331-2

Weiss R, Dziura J, Burgert TS et al (2004) Obesity and the metabolic syndrome in children and adolescents. N Engl J Med. https://doi.org/10.1056/NEJMoa031049

Weiss R, Taksali SE, Tamborlane W V., et al (2005) Predictors of changes in glucose tolerance status in obese youth. Diabetes Care. https://doi.org/10.2337/diacare.28.4.902

Wharton CM, Long M, Schwartz MB (2008) Changing nutrition standards in schools: The emerging impact on school revenue. J Sch Health. https://doi.org/10.1111/j.1746-1561.2008.00296.x

Witchel SF, Oberfield SE, Peña AS (2019) Polycystic ovary syndrome: pathophysiology, presentation, and treatment with emphasis on adolescent girls. J Endocr Soc. https://doi.org/10.1210/js.2019-00078

Wong RJ, Aguilar M, Cheung R et al (2015) Nonalcoholic steatohepatitis is the second leading etiology of liver disease among adults awaiting liver transplantation in the United States. Gastroenterology. https://doi.org/10.1053/j.gastro.2014.11.039

Xanthakos SA, Trout AT, Dillman JR (2017) Magnetic resonance elastography assessment of fibrosis in children with NAFLD: Promising but not perfect. Hepatology 66:1373–1376

Yoshida Y, Simoes EJ (2018) Sugar-sweetened beverage, obesity, and type 2 diabetes in children and adolescents: policies, taxation, and programs. Curr Diab Rep 18:–31

Yoshinaga M, Yuasa Y, Hatano H et al (1995) Effect of total adipose weight and systemic hypertension on left ventricular mass in children. Am J Cardiol. https://doi.org/10.1016/S0002-9149(99)80227-6

Younossi ZM, Ratziu V, Loomba R, Rinella M, Anstee QM, Goodman Z, Bedossa P, Geier A, Beckebaum S, Newsome PN, Sheridan D, Sheikh MY, Trotter J, Knapple W, Lawitz E, Abdelmalek MF, Kowdley KV, Montano-Loza AJ, Boursier J, Mathurin P, Bugianesi E, Mazzella SARSI (2019) Obeticholic acid for the treatment of non-alcoholic steatohepatitis: interim analysis from a multicentre, randomised, placebo-controlled phase 3 trial. Lancet 394 (10215):2184–2196. https://doi.org/10.1016/S0140-6736(19)33041-7

Yudkin JS (2007) Insulin resistance and the metabolic syndrome—or the pitfalls of epidemiology. Diabetologia 50:1576–1586

Zakharova I, Klimov L, Kuryaninova V et al (2019) Vitamin D insufficiency in overweight and obese children and adolescents. Front Endocrinol (Lausanne) 10:10

Zeller MH, Saelens BE, Roehrig H et al (2004) Psychological adjustment of obese youth presenting for weight management treatment. Obes Res. https://doi.org/10.1038/oby.2004.197

Zimmet P, Alberti GKMM, Kaufman F et al (2007) The metabolic syndrome in children and adolescents—An IDF consensus report. Pediatr Diabetes 8:299–306

Obesity and Cancer: Linked Molecular Mechanisms

28

Erika Di Zazzo, Monica Rienzo, Maria Michela Marino, Donatella Fiore, Chiara Piscopo, Amelia Casamassimi, Bruno Moncharmont, and Ciro Abbondanza

Contents

Introduction 386

Comorbidities 386
Biological Mechanisms Linking Obesity to Cancer 386
Hyperinsulinemia/Insulin Resistance 388

Sex Hormones 388
Adipokines and Cancer 388
Adiponectin and Cancer 389

Hypoadiponectinemia 389
Breast Cancer 389
Gastrointestinal Malignancies 389
Colorectal Cancer 390
Ovarian Cancer 390
Endometrial Cancer 390
Prostate Cancer 391

Low Chronic Inflammation 391

The Obesity Paradox in Cancer 391

Obesity and Therapy 392
Drug Dosage 392

Specific Cancer Therapies for Obese and Diabetic Patients? 392
Metformin 392

E. Di Zazzo (✉) · B. Moncharmont
Department of Medicine and Health Sciences "V. Tiberio", University of Molise, Campobasso, Italy
e-mail: erika.dizazzo@unimol.it

M. Rienzo · M. M. Marino · A. Casamassimi · C. Abbondanza
Department of Precision Medicine, University of Campania "Luigi Vanvitelli", Naples, Italy

D. Fiore · C. Piscopo
Department of Pharmacy, University of Salerno, Fisciano, (SA), Italy

J. Faintuch and S. Faintuch (eds.), *Obesity and Diabetes*, https://doi.org/10.1007/978-3-030-53370-0_28

Thiazolidinediones (TZDs) and Other Agents 393

Therapies Targeting APN 393
Oral Adiponectin Receptor Agonist 393
Statins and Assorted Drugs 393

References 394

Abstract

Obesity is related to metabolic defects that may promote not only cancer initiation, but also its progression. The molecular basis for the association between obesity and cancer is not fully understood; however, many pathways are being investigated including hyperinsulinemia/insulin resistance (IR) and abnormalities of the insulin-like growth factor-1 (IGF-1) signaling, sex hormones biosynthesis and pathway, alterations in adipokines pathophysiology, and subclinical chronic low-grade inflammation. In this chapter, we analyze the current knowledge on the proposed biological mechanisms, especially focusing on the role of adiponectin (APN).

Keywords

Obesity · Cancer · Insulin-like growth factor-1 signaling · Sex hormones · adipokines · adiponectin · Low-grade inflammation

Introduction

White adipose tissue (WAT) was formerly studied in connection with caloric reserves, mechanic and thermal insulation, and sexual attraction. However, it is unquestionably a complex endocrine organ (Coelho et al. 2013). Adipocytes constitute 90% of the cells, whereas the stromal-vascular components include endothelial cells (10–20% of cells), pericytes (3–5%), fibroblasts, and others (15–30%). Stem and progenitor cells (0.1%) should not be neglected, along with T- and B-lymphocytes, macrophages, dendritic cells, and others (Cozzo et al. 2017). Sex steroid hormones, insulin resistance, growth factors, cytokines, and adipokines are all influenced by this cellular environment (Osborn and Olefsky 2012). Many cancer types are impacted by the same molecules and pathways (Sung et al. 2019).

Excessive body weight promotes elevated free fatty acids (FFA), triglycerides, glucose, insulin resistance, and insulin production, some of which could also stimulate cancer progression (Osborn and Olefsky 2012).

Comorbidities

Obesity predisposes to metabolic, cardiovascular diseases, as well as of several malignancies (Sung et al. 2019; Global BMI Mortality Collaboration et al. 2016; Goodwin and Stambolic 2015; GBD 2015 Obesity Collaborators et al. 2017), a phenomenon called as "adiponcosis" (Bifulco and Ciaglia 2017).

Cancers of digestive organs, like colon cancer, and tissues with endocrine links, such as breast, ovarian, endometrial, and prostate cancers, receive most attention (Sung et al. 2019; Kyrgiou et al. 2017; Tumminia et al. 2019). Cancer is the number two most fatal human condition, after cardiovascular disease, whereas obesity and related aberrations could be aggravating oncological mechanisms (Sung et al. 2019; Goodwin and Stambolic 2015).

Biological Mechanisms Linking Obesity to Cancer

Possible biological mechanisms comprise hyperinsulinemia/insulin resistance (IR) and abnormalities of the insulin-like growth factor-1 (IGF-1) signaling, sex hormones biosynthesis and

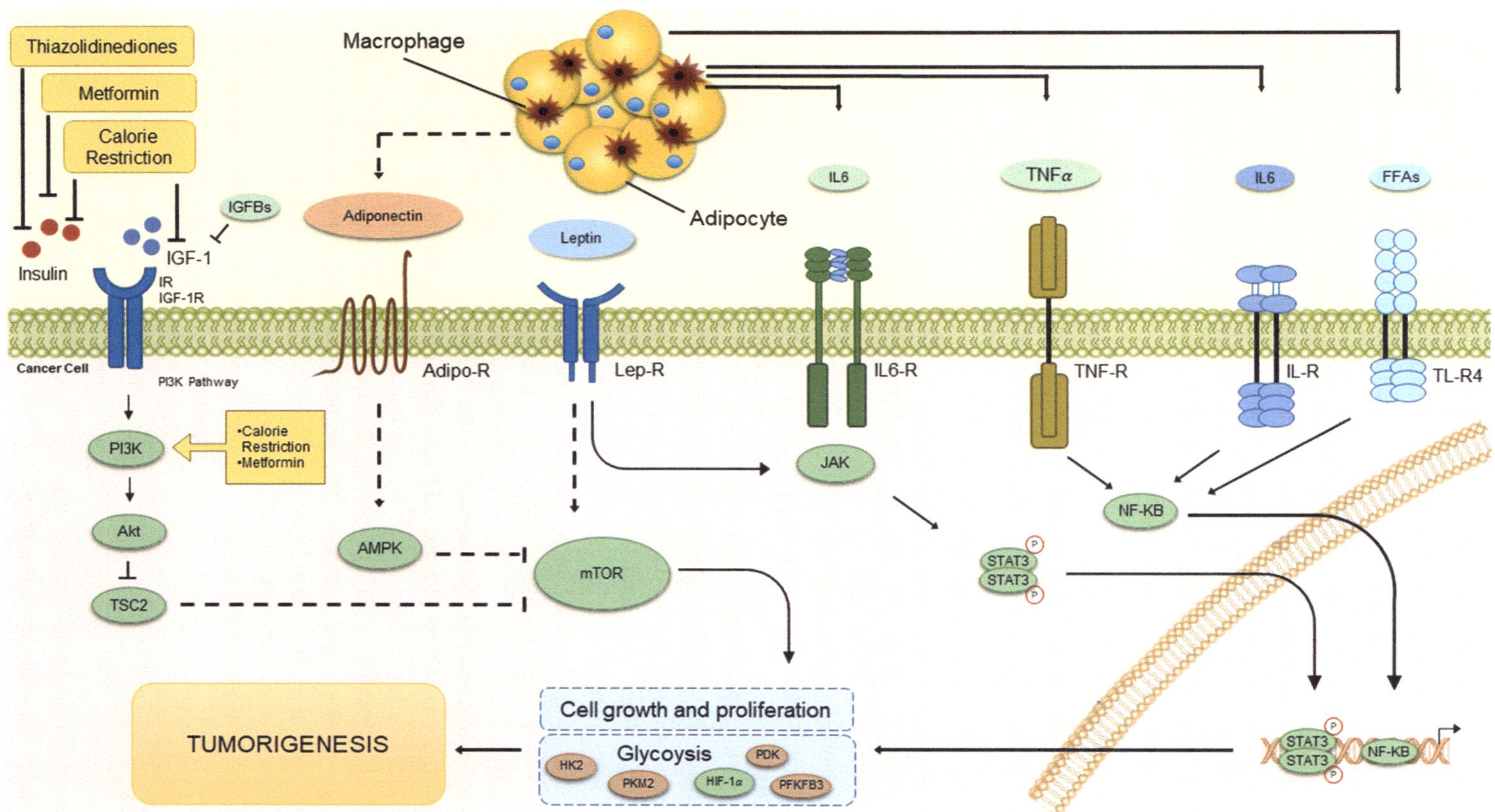

Fig. 28.1 Molecular pathways linking obesity and cancer. Insulin and insulin-like growth factor 1 (IGF-1) trigger off the phosphatidylinositol 3-kinase (PI3K) pathway, which in turn upregulates glycolysis and other metabolic pathways to generate energy needs for proliferation. Adipose tissue in the obese condition is a site of profound inflammatory activity using immune cells that secrete an abundance of cytokines that can influence on neighboring cancer cells to promote tumorigenesis through different metabolic levels.

pathway, alterations in adipokines pathophysiology, and subclinical chronic low-grade inflammation (Fig. 28.1).

Hyperinsulinemia/Insulin Resistance

Insulin resistance is a common feature of obese patients resulting in a compensatory increase in systemic insulin and IGF-1 levels (Ackerman et al. 2017). The excess of body weight is not the only parameter determining hyperinsulinemia: the distribution of the extra weight and of visceral adipose tissue is also a key variable. Hyperinsulinemia induces the production of IGF-1 by hepatocytes and downregulates the secretion of IGF-1 binding proteins (IGFBPs), resulting in an increase in bioavailable IGF-1. Several cancer cell types respond to insulin and IGF-1 receptor binding, by activating PI3K and MAPK pathways or other downstream networks, promoting cell survival, growth, and motility (Di Zazzo et al. 2014; Ackerman et al. 2017).

Insulin resistance, hyperinsulinemia, and high IGF-1 levels increase the risk of endometrial, breast, colorectal, and prostate cancer (Goodwin and Stambolic 2015). When colorectal cancer-susceptible mice were treated with an IGF-1-receptor inhibitor disrupting downstream signaling, tumor burden was significantly reduced (Ackerman et al. 2017).

In adipose tissue, insulin controls lipid storage and inhibition of lipolysis. Once glucose and insulin levels are elevated for prolonged periods as a consequence of overeating or lack of exercise, insulin can induce lipid storage in non-adipose tissue such as skeletal muscle. This inappropriate accumulation of intramuscular fatty acids contributes to altered insulin signaling.

Sex Hormones

Peripheral adipose tissue performs steroids aromatization: androgens and androgenic precursors are converted to estradiol by aromatase. In obesity, the excess of adipose tissue and the increased aromatase activity lead to a higher conversion rate, resulting in higher levels of circulating estrogens (Gérard and Brown 2018). In men and in postmenopausal women, the conversion of androgens to estrogens is dependent upon adipose tissue mass. Estrogen signaling has several consequences in tumor growth promotion. Indeed, estradiol is able to stimulate cell proliferation of breast epithelial and endometrial cells, to inhibit apoptosis, and to induce angiogenesis (Andò et al. 2019). Additionally, obesity is associated with lower plasma levels of sex hormone-binding globulin (SSBG), thus increasing bioavailable concentrations of estradiol and testosterone, resulting in a greater cancer risk (Di Zazzo et al. 2018).

Adipokines and Cancer

Adipokines are bioactive molecules released by WAT that acts locally in an autocrine and paracrine manner, but they also have a systemic, endocrine effect through blood circulation (Diedrich et al. 2015). Adipokines include adiponectin (APN), leptin, resistin, omentin, and chemerin; under physiological conditions, they regulate several processes such as energy balance, lipid metabolism, insulin sensitivity, inflammation, innate and adaptive immunity, angiogenesis, hematopoiesis, and cell proliferation (Schrover et al. 2016).

Abnormal accumulation of adipose tissue induces adipocyte dysfunctions, resulting into the alteration of its endocrine functions and consequently affecting the secretion of different adipokines, usually leading to increased leptin and decreased APN blood concentration. Abnormal adipokine secretion often represents a factor leading to cancer (Fig. 28.1) (Di Zazzo et al. 2019). Upon analysis of a panel of 14 obesity-related hormones, cytokines, coagulation factors, and other biomarkers, C-peptide, IL-6, and TNF-α pointed toward clear-cell renal carcinoma risk (Wang et al. 2019).

Adiponectin and Cancer

APN behaves as insulin-sensitizing, anti-apoptotic, and immune regulatory (Ackerman et al. 2017). Various homo-oligomers are recognized, and as adiposity accumulates, APN diminishes in the same proportion (Di Zazzo et al. 2019). APN usual receptors, AdipoR1 and AdipoR2, are physiologically highly relevant. Hexameric and multimeric APN binds to a different receptor, from the cadherin superfamily (Di Zazzo et al. 2019).

Signaling pathways for cellular responses include AMPK, mTOR, PI3K/AKT, MAPK, STAT3, and NF-kB, (Di Zazzo et al. 2019). AdipoR1 and AdipoR2 also impact energy balance, including insulin sensitization (Di Zazzo et al. 2019). Ceramidase stimulatory activity, which depletes intracellular ceramide, inhibits apoptosis (Holland et al. 2011).

Hypoadiponectinemia

Poor response to APN during excessive adiposity could be explained by low receptor values (Yamauchi et al. 2014). The same tends to occur during inflammation linked to diabetes and atherosclerosis (Di Zazzo et al. 2019). APN is usually classified as an anti-cancer molecule, with anti-inflammatory, anti-proliferative, and pro-apoptotic impact (Porcile et al. 2014; Di Zazzo et al. 2019).

Breast Cancer

Breast Cancer (BC) risk is heightened during advanced adiposity: (i) aromatization of androgens to estrogens stimulates growth of mammary cells; (ii) interference with APN physiology prevents its protective response (Avgerinos et al. 2019). Low APN levels are certainly deleterious in this context (Ye et al. 2014), including cancer growth and virulence. In the ERα-negative human BC cell line MDA-MB-231, Genes involved in cell cycle progression and apoptosis are one of the possible mechanisms (Mauro et al. 2014).

In an ERα-positive cell culture, APN leads to expression of ERα and cell growth (Mauro et al. 2014). Cross talk between APN/AdipoR1, IGF-IR, and ERα seems relevant (Mauro et al. 2015). Insulin and estrogen elevation, which favor cancer progression, contrast with depressed APN, including postmenopausal and ER-positive breast tumors (Di Zazzo et al. 2014; Mauro et al. 2014). APN modulates cell migration and invasion (Jia et al. 2014), AdipoR2 signals lymphatic and vascular penetration, and all of them are markers of metastatic disease (Jeong et al. 2011).

Gastrointestinal Malignancies

Strong correlation between obesity and the development of gastrointestinal (GI) cancers is reported (Murphy et al. 2018). The largest adipose intra-abdominal depot is the omentum, the so-called fatty apron connecting the stomach to the colon, enriched with immune cells and responsible for local inflammation. Obesity disrupts the homeostatic profiles of innate and adaptive immune cell populations within the omentum (O'Sullivan et al. 2018). APN plasma levels were found decreased in patients with GI cancers (Nagaraju et al. 2015). APN is suggested to be involved in esophageal mucosa remodeling, and it might have a protective role against cancer transformation, contributing to the link between obesity and lower esophageal carcinoma (EC) (Almers et al. 2015). Patients with metabolic syndrome, associated with increased leptin and decreased APN serum levels, are prone to Barrett's esophagus, a metaplastic change occurring in response to gastroesophageal reflux disease that can potentially lead to the EC. Moreover, high levels of LMW-APN are associated with a decreased risk of Barrett's esophagus (Tilg and Moschen 2014).

Beales and colleagues demonstrated that in OE33, an esophageal carcinoma cell line, leptin was able to induce proliferation, invasion, and migration and inhibit apoptosis in a STAT3-dependent manner and, by contrast, APN inhibited leptin-stimulated proliferation *via* AdipoR1 (Beales et al. 2014). In gastric cancer, APN inhibited proliferation in *vitro* with most prominent effects in AZ521 and HGC27 gastric cell lines expressing high levels of AdipoR1/R2 mRNA. Moreover, higher concentrations of APN significantly reduce tumor volume and peritoneal metastases in *vivo* (Ishikawa et al. 2007), consistent with previous findings suggesting an anti-angiogenic and tumor suppressive role for this adipokine (Bråkenhielm et al. 2004). In the same models, APN elicits its biological effects through AdipoR1/R2 activation, whose expression was significantly associated with histological type and overall survival (Ishikawa et al. 2007). Negative AdipoR1 immunostaining was found in patients with lymphatic metastasis and peritoneal dissemination, while positive AdipoR1 expression corresponded to a longer survival rate (Tsukada et al. 2011).

Circulating APN levels have been also associated with increased risk of pancreatic cancer (Pothuraju et al. 2018). APN exerted its inhibitory effect through modulation of the β-catenin signaling pathway. In BxPC-3 and CFPAC-1, pancreatic cell lines both expressing the AdipoRs, APN reduced serum-induced phosphorylation of GSK-3β, decreased the nuclear accumulation of β-catenin, and downregulated the expression of cyclin D1. Knockdown of AdipoRs abolished the growth-inhibiting effect induced by APN in *vitro* and in xenograft models (Jiang et al. 2019).

Colorectal Cancer

Low APN is a marker of colorectal cancer (CRC) (Di Zazzo et al. 2019). Both AdipoR1 and AdipoR2 in turn could signal lymphatic metastasis (Ayyildiz et al. 2014), as well as (for AdipoR1) the ability to survive the illness (Choe et al. 2018). APN negatively controls CRC growth, by inhibiting the mechanistic target of rapamycin (mTOR) *via* AMPK phosphorylation, and decreasing PI3K and Akt phosphorylation (Parida et al. 2019).

In CRC models, APN knockdown resulted in increased multiplicity of aggressive colorectal polyps. In an APN-deficient mice model, APN treatment inhibited cancer progression and angiogenesis (Moon et al. 2013). APN deficiency also aggravated azoxymethane-induced colon cancer in C57BL/6J mice (Mutoh et al. 2011).

APN conferred protection against inflammation-induced colon cancers by preventing apoptosis in the goblet cells and promoting conversion of epithelial to goblet cells (Saxena et al. 2012). In HCT116, HT29, and LoVo CRC cell lines, APN induced G1/S cell cycle arrest with concomitant overexpression of p21 and p27 *via* AMPK phosphorylation; inhibition of AdipoRs released cells from APN-induced growth blockade (Kim et al. 2010). APN anti-cancer effect is glucose-dependent, possibly explaining why CRC survival is enhanced in a low glucose medium; however, the opposite occurs with APN in high glucose conditions (Habeeb et al. 2011).

Ovarian Cancer

Ovarian cancer patients have a lower blood concentration of APN than healthy women (Jin et al. 2016); low APN concentration was associated with longer progression-free survival times and a better tumor responsiveness to chemotherapy (Diaz et al. 2013; Slomian et al. 2019). In addition, AdipoR1 expression level in ovarian cancer tissues could represent a marker of prognosis, being positively associated with overall patient's survival (Li et al. 2017). Low APN plasma levels may favor ovarian cancer growth, induced by persistent activation of PI3K/Akt/mTOR signaling. Moreover, APN is able to repress human ovarian cancer cell growth and reverse the stimulatory effects of 17β-estradiol and IGF-1 on cell proliferation through the downregulation of their receptors (Hoffmann et al. 2018).

Endometrial Cancer

Low APN blood levels were associated with an increased risk and a worse prognosis of endometrial cancer (EMC). Low expression of AdipoR1 in endometrial cancer cells is associated with advanced tumor stage (Tumminia et al. 2019). Several hypotheses have been formulated the role of APN implying: (i) activation of AMPK (resulting in cell growth suppression and apoptosis); (ii) extracellular signal-regulated protein kinase (ERK) and Akt pathway inhibition; (iii) reduction of Cyclin D1 expression (Moon et al. 2011). Association seems more evident for type II EMC, and APN modulators are being explored for potential therapeutic avenues (Garikapati et al. 2019).

Prostate Cancer

Conflicting reports about APN exist (Hu et al. 2016). APN concentration in prostate cancer patients was low and connected to disease advent (Goktas et al. 2005). A lower AdipoR1 and AdipoR2 expression was observed in prostate neoplastic tissues compared with healthy prostate tissue (Michalakis et al. 2007). Growing evidence indicates that APN exerts an anti-proliferative action in prostate cancer cells, inhibiting dihydrotestosterone-activated cell proliferation (Bub et al. 2006). The ectopic overexpression of APN in prostate cancer cell lines inhibited mTOR-mediated neoplastic cell proliferation (Gao et al. 2015).

No links are also advocated (Baillargeon et al. 2006), or a significant positive correlation between APN concentrations and incidence of low- or intermediate-risk prostate cancer (Ikeda et al. 2015). Higher APN plasma levels were detected in subjects with cancer stage T3 (advanced) than in subjects with T2 (confined within the prostate). AdipoR2 findings could be a signal of cancer progression and metastasis (Rider et al. 2015).

Low Chronic Inflammation

Large adipocytes become distant from the blood supply, which can trigger chronic hypoxia and inflammation (Boutari and Mantzoros 2018).

Neutrophils and mast cells enhance inflammation whereas the opposite is expected from eosinophils and myeloid-derived suppressor cells. B- and T-lymphocytes and natural-killer cells are also engaged in the process, as well as M1 pro-inflammatory macrophages (ATMs) (Ouchi et al. 2011).

Macrophages constituting crown like structures elicit nuclear factor-kappa B (NF-κB) aggravating chronic inflammation (Ouchi et al. 2011). The cross talk between adipocytes and cancer engages IL-1, IL-6, and TNF-α, ROS generation, adipokines, and other molecules (Fig. 28.1) (Avgerinos et al. 2019). Cancer cells can also induce phenotype alteration of adjacent adipocytes, encompassing reduction of their lipid content and release of adipokine and matrix metalloproteinases (Dirat et al. 2011).

Inflammasome is an innate immune pathway activating proinflammatory cytokines, including IL-1 and IL-18 (Lamkanfi and Dixit 2014). Inflammasome-related genes are encountered in adipocytes (Yin et al. 2014), potentially stimulating cancer growth (Guo et al. 2016). Obesity treatment tends to ameliorate this negative profile (Hagman et al. 2017).

The Obesity Paradox in Cancer

Obesity negatively influences cancer recurrence, prognosis, and survival (Lennon et al. 2016). However, opposite evidence in certain circumstances suggests that obesity reduces cancer incidence and improves survival. Obesity has been ruled out as a risk for cancer mortality (Kuk et al. 2018). Body fatness reduced the nerve sheath tumor risk (Wiedmann et al. 2017); overweight or obesity attenuated mortality of bladder (Pavone et al. 2018); and lung cancer (Zhang et al. 2017) after surgery or chemotherapy. A meta-analysis also showed that obese patients with esophageal cancer had better long-term survival (Kayani et al. 2012).

The use of BMI as a measure of adiposity could partly explain the discrepancies, as it does not fully characterize the intricate biology and physiology of excess body fat. BMI cannot differentiate between lean mass and adipose tissue and depends upon gender, age, ethnicity, and race. Additionally, BMI does not estimate the visceral adipose tissue (VAT), which seems to be metabolically more relevant. Computed tomography and MRI are better alternatives for quantification of VAT; however, they are not always feasible (Allott and Hursting 2015). Anthropometric measures, such as waist circumference (WC) and waist-to-hip ratio (WHR), are viable techniques (Sung et al. 2019), however, not perfect ones, as both VAT and subcutaneous abdominal tissue (SAT) are lumped together in the results (Avgerinos et al. 2019).

Obesity and Therapy

Obesity and comorbidities like dysglycemia, hypertension, and dyslipidemia could reduce the response to chemotherapy. Elevated BMI was associated with poor prognosis in patients affected by colon cancer who received surgical resection of primary tumor and adjuvant chemotherapy with capecitabine and oxaliplatin (Lashinger et al. 2014). In bevacizumab-treated metastatic CRC patients, high visceral fat and BMI were significantly associated with absence of a response and increased progression. BMI was negatively associated with response to standard first-line chemotherapy with platinum and taxanes in ovarian cancer patients (Califano et al. 2014). These results could be related to the expression by adipose tissue of angiogenic factors (in particular VEGF) (Ottaiano et al. 2018). Hyperinsulinemia, a known growth factor, could trigger chemoresistance to 5-fluorouracil, anthracyclines, taxanes, and other drugs upregulating P-glycoprotein (Wei et al. 2015).

Drug Dosage

Pharmacokinetic studies rarely address the obese. The most common strategy is dose-capping or dose-fixed regimens. The consequence of this "depotentiation" attitude may be the use of subtherapeutic strategies, conducting to disease recurrence and mortality. According to the American Society of Clinical Oncology, full weight should be adopted for dosage calculation.

Specific Cancer Therapies for Obese and Diabetic Patients?

A balanced and healthy diet may control factors that sustain obesity-related disease (i.e., IGF-1, insulin, leptin) (Avgerinos et al. 2019). In addition, vigorous aerobic exercise leads to a peak of circulating APN levels (Saunders et al. 2012). Moreover, there is an increasing interest in testing diabetes and cholesterol-lowering drugs for cancer therapy.

Metformin

Metformin could decrease incidence and mortality of cancer by inducing hepatic gluconeogenesis and reducing IR of peripheral tissues, resulting in lower insulin and IGF-1 levels (Gallagher and LeRoith 2015). Metformin blocks tumor growth and induces apoptosis through insulin-independent mechanisms (Safe et al. 2018). Furthermore, metformin decreases cancer recurrence by directly inducing cancer stem cell death (Gallagher and LeRoith 2015).

Glycemic control with metformin can restore adipokine concentrations, increasing APN, and decreasing pro-inflammatory adipokine levels in both humans and mice (Avgerinos et al. 2019). For prostate and breast malignancies, large meta-analyses failed to demonstrate benefits for metformin regarding cancer risk and mortality (Feng et al. 2019; Au Yeung and Schooling 2019; Wang et al. 2020). Yet in other contexts, such as rectal cancer treated by neoadjuvant chemoradiotherapy, better cancer response and lower risk of recurrence were elicited (Kim et al. 2020). Clinical trials are going on. Twelve of them address the following cancers: breast (five), along with head and neck, thyroid, endometrial, multiple myeloma, lymphocytic leukemia, and various gynecologic/ solid tumors (one of each) (www.cancer.gov, 2020).

Thiazolidinediones (TZDs) and Other Agents

Activation of PPAR-γ by TZDs could restrict cell proliferation by decreasing insulin concentration and also influencing key pathways of the Insulin/IGF-1 axis, such as MAPK, PI3K/mTOR, and Glycogen synthase kinase (GSK)3-β/Wnt/β-catenin cascades, which modulate cancer cell survival and differentiation. Additionally, the PPARγ agonists TZDs, rosiglitazone, and pioglitazone augment the circulating level of APN directly, enhancing its gene and protein expression in a dose-dependent manner (Parida et al. 2019). Yet, meta-analysis results are conflicting.

For bladder cancer, an increased risk with pioglitazone was announced, possibly dose and time dependent (Tang et al. 2018). As regards colorectal cancer, no advantages were detected with pioglitazone, whereas other PZDs seemed moderately protective (Liu et al. 2018). A lack of association with breast cancer risk is reported (Du et al. 2018). This last outcome is consistent with a meta-analysis addressing digestive cancers, in which the risk did not differ for incretin mimetics, insulin, metformin, sodium-glucose co-transporter 2, sulfonylureas, TZDs, alpha-glucosidase inhibitors, or placebo (Chai et al. 2019).

Therapies Targeting APN

Increasing plasma APN levels or mimicking some of its cancer-protective properties could mitigate the deleterious effects of metabolic dysfunctions on tumor development and progression (Vansaun 2013; Tumminia et al. 2019). Therefore, pharmacological increase in serum APN levels, up-regulation of AdipoRs expression, or synthesis of AdipoRs agonists could represent promising therapeutic strategies.

Using a high-throughput assay, several natural compounds showing AdipoRs agonist activity were identified. These compounds, acting preferably on AdipoR1 (e.g., matairesinol, arctiin, arctigenin, gramine) or AdipoR2 (e.g., syringin, parthenolide, taxifolion, deoxyschizandrin), shared important anti-cancer properties, including anti-proliferative and anti-inflammatory effects (Sun et al. 2013). ADP355, a peptide-based APN receptor agonist, prevented the proliferation of AdipoRs-positive cancer cell lines. ADP 355 showed high affinity with AdipoR1, and through the regulation of the canonical APN-regulated pathways (i.e., AMPK, Akt, STAT3, and ERK1/2), reduced breast tumor growth both *in vitro* and *in vivo* (Otvos Jr et al. 2015). Additionally, three peptides BHD1028, BHD43, and BHD44 have been designed to mimic APN actions. BHD1028 showed the highest affinity with AdipoR1 and the main activation of AMPK already at low concentration, more than ADP 355 (Kim et al. 2018).

Oral Adiponectin Receptor Agonist

AdipoRon (AdipoR) is the first oral AdipoRs agonist able to bind and activate AdipoR1 and AdipoR2 that successfully re-established APN functions, mainly activating AMPK and PPARγ pathways in obesity-related type 2 diabetes (Okada-Iwabu et al. 2013). Initial reports have also investigated the possible anti-cancer role of AdipoR in preclinical models, especially in pancreatic and ovarian cancer (Akimoto et al. 2018; Ramzan et al. 2019). However, modifying AdipoRs interactions could also result in infertility, cardiac damage, and reduced bone density (Holland and Scherer 2013).

Statins and Assorted Drugs

Statins have been reported to be effective in increasing circulating APN levels. Statins function by releasing cellular oxidative stress, resulting in increased APN multimerization and secretion. Angiotensin converting enzyme inhibitors as well as angiotensin receptor antagonists Ramipril, Quinapril, Losartan, Telmisartan, Irbesartan, and Candesartan have similarly shown promising results in clinical trials. They function by enhancing APN secretion *via* PPARγ, though some of them are also known to induce transcription. Other potential drugs include non-statin anti-hyperlipidemic drugs like Fenofibrate and Zetia, non-TZD anti-diabetic drugs, such as Acarbose and the sulfonylurea Glimepiride and Sulfonylureas (Parida et al. 2019).

References

Ackerman SE, Blackburn OA, Marchildon F, Cohen P (2017) Insights into the Link Between Obesity and Cancer. Curr Obes Rep 6(2):195–203. https://doi.org/10.1007/s13679-017-0263-x

Akimoto M, Maruyama R, Kawabata Y, Tajima Y, Takenaga K (2018) Antidiabetic adiponectin receptor agonist AdipoRon suppresses tumour growth of pancreatic cancer by inducing RIPK1/ERK-dependent necroptosis. Cell Death Dis 9(8):804

Allott EH, Hursting SD (2015) Obesity and cancer: mechanistic insights from transdisciplinary studies. Endocr Relat Cancer 22:R365–R386. https://doi.org/10.1530/ERC-15-0400

Almers LM, Graham JE, Havel PJ, Corley DA (2015) Adiponectin may modify the risk of Barrett's Esophagus in patients with gastroesophageal reflux disease. Clin Gastroenterol Hepatol 13:2256–2264. https://doi.org/10.1016/j.cgh.2015.01.009

Andò S, Gelsomino L, Panza S, Giordano C, Bonofiglio D, Barone I, Catalano S (2019) Obesity, leptin and breast cancer: epidemiological evidence and proposed mechanisms. Cancers (Basel) 11(1):E62. https://doi.org/10.3390/cancers11010062

Au Yeung SL, Schooling CM (2019) Impact of glycemic traits, type 2 diabetes and metformin use on breast and prostate cancer risk: a Mendelian randomization study. BMJ Open Diabetes Res Care 7(1):e000872. https://doi.org/10.1136/bmjdrc-2019-000872

Avgerinos KI, Spyrou N, Mantzoros CS, Dalamaga M (2019) Obesity and cancer risk: emerging biological mechanisms and perspectives. Metabolism 92:121–135. https://doi.org/10.1016/j.metabol.2018.11.001

Ayyildiz T, Dolar E, Ugras N, Adim SB, Yerci O (2014) Association of adiponectin receptor (Adipo-R1/-R2) expression and colorectal cancer. Asian Pac J Cancer Prev 15:9385–9390

Baillargeon J, Platz EA, Rose DP, Pollock BH, Ankerst DP, Haffner S, Higgins B, Lokshin A, Troyer D, Hernandez J et al (2006) Obesity, adipokines, and prostate cancer in a prospective population-based study. Cancer Epidemiol Biomark Prev 15:1331–1335

Beales ILP, Garcia-Morales C, Ogunwobi OO, Mutungi G (2014) Adiponectin inhibits leptin-induced oncogenic signalling in oesophageal cancer cells by activation of PTP1B. Mol Cell Endocrinol 382:150–158. https://doi.org/10.1016/j.mce.2013.08.013

Bifulco M, Ciaglia E (2017) Updates on "adiponcosis": More new incoming evidence strengthening the obesity-cancer link. Eur J Intern Med 41:e19–e20

Boutari C, Mantzoros CS (2018) Inflammation: a key player linking obesity with malignancies. Metabolism 81:A3–A6. https://doi.org/10.1016/j.metabol.2017.12.015

Bråkenhielm E, Veitonmäki N, Cao R, Kihara S, Matsuzawa Y, Zhivotovsky B, Funahashi T, Cao Y (2004) Adiponectin-induced antiangiogenesis and antitumor activity involve caspase-mediated endothelial cell apoptosis. Proc Natl Acad Sci U S A 101:2476–2481

Bub JD, Miyazaki T, Iwamoto Y (2006) Adiponectin as a growth inhibitor in prostate cancer cells. Biochem Biophys Res Commun 340:1158–1166

Califano D, Pignata S, Losito NS, Ottaiano A, Greggi S, De Simone V, Cecere S, Aiello C, Esposito F, Fusco A, Chiappetta G (2014) High HMGA2 expression and high body mass index negatively affect the prognosis of patients with ovarian cancer. J Cell Physiol 229:53–59

Chai S, Yu S, Yang Z, Wu S, Gao L, Wang H, Zhang Y, Zhan S, Ji L, Sun F (2019) Effect of incretin-based therapies on cancers of digestive system among 101 595 patients with type 2 diabetes mellitus: a systematic review and network meta-analysis combining 84 trials with a median duration of 30 weeks. BMJ Open Diabetes Res Care 7(1):e000728. https://doi.org/10.1136/bmjdrc-2019-000728

Choe EK, Yi JW, Chai YJ, Park KJ (2018) Upregulation of the adipokine genes ADIPOR1 and SPP1 is related to poor survival outcomes in colorectal cancer. J Surg Oncol 117:1833–1840. https://doi.org/10.1002/jso.25078

Clinical Trials Using Metformin Hydrochloride (2020). www.cancer.gov/about-cancer/treatment/clinical-trials/intervention/metformin-hydrochloride. Accessed 17 March 2020

Coelho M, Oliveira T, Fernandes R (2013) Biochemistry of adipose tissue: an endocrine organ. Arch Med Sci 9:191–200. https://doi.org/10.5114/aoms.2013.33181

Cozzo AJ, Fuller AM, Makowski L (2017) Contribution of adipose tissue to development of cancer. Compr Physiol 8:237–282. https://doi.org/10.1002/cphy.c170008

Di Zazzo E, Feola A, Zuchegna C, Romano A, Donini CF, Bartollino S, Costagliola C, Frunzio R, Laccetti P, Di Domenico M et al (2014) The p85 regulatory subunit of PI3K mediates cAMP-PKA and insulin biological effects on MCF-7 cell growth and motility. Sci World J 2014:565839

Di Zazzo E, Galasso G, Giovannelli P, Di Donato M, Castoria G (2018) Estrogens and their receptors in prostate cancer: therapeutic implications. Front Oncol 8:2. https://doi.org/10.3389/fonc.2018.00002

Di Zazzo E, Polito R, Bartollino S, Nigro E, Porcile C, Bianco A, Daniele A, Moncharmont B (2019) Adiponectin as link factor between adipose tissue and cancer. Int J Mol Sci 20:839

Diaz ES, Karlan BY, Li AJ (2013) Obesity-associated adipokines correlate with survival in epithelial ovarian cancer. Gynecol Oncol 129:353–357

Diedrich J, Gusky HC, Podgorski I (2015) Adipose tissue dysfunction and its effects on tumor metabolism. Horm Mol Biol Clin Invest 21:17–41. https://doi.org/10.1515/hmbci-2014-0045

Dirat B, Bochet L, Dabek M, Daviaud D, Dauvillier S, Majed B et al (2011) Cancer associated adipocytes exhibit an activated phenotype and contribute to breast cancer invasion. Cancer Res 71:2455–2465

Du R, Lin L, Cheng D, Xu Y, Xu M, Chen Y, Wang W, Bi Y, Li D, Lu J (2018) Thiazolidinedione therapy and breast cancer risk in diabetic women: a systematic review and meta-analysis. Diabetes Metab Res Rev 34(2). https://doi.org/10.1002/dmrr.2961

Feng Z, Zhou X, Liu N, Wang J, Chen X, Xu X (2019) Metformin use and prostate cancer risk: a meta-analysis of cohort studies. Medicine (Baltimore) 98 (12):e14955. https://doi.org/10.1097/MD.0000000000014955

Gallagher EJ, LeRoith D (2015) Obesity and diabetes: the increased risk of cancer and cancer-related mortality. Physiol Rev 95:727–748

Gao Q, Zheng J, Yao X, Peng B (2015) Adiponectin inhibits VEGF-A in prostate cancer cells. Tumour Biol 36:4287–4292

Garikapati KK, Ammu VVVRK, Krishnamurthy PT, Chintamaneni PK, Pindiprolu SKSS (2019) Type-II endometrial cancer: role of adipokines. Arch Gynecol Obstet 300(2):239–249. https://doi.org/10.1007/s00404-019-05181-1

GBD 2015 Obesity Collaborators, Afshin A, Forouzanfar MH, Reitsma MB, Sur P, Estep K, Lee A, Marczak L, Mokdad AH, Moradi-Lakeh M et al (2017) Health effects of overweight and obesity in 195 countries over 25 years. N Engl J Med 377:13–27

Gérard C, Brown KA (2018) Obesity and breast cancer—role of estrogens and the molecular underpinnings of aromatase regulation in breast adipose tissue. Mol Cell Endocrinol 466:15–30. https://doi.org/10.1016/j.mce.2017.09.014

Global BMI Mortality Collaboration, Di Angelantonio E, Bhupathiraju SN, Wormser D, Gao P, Kaptoge S, Berrington de Gonzalez A, Cairns BJ, Huxley R, Jackson CL et al (2016) Body-mass index and all-cause mortality: Individual-participant-data meta-analysis of 239 prospective studies in four continents. Lancet 388:776–786

Goktas S, Yilmaz MI, Caglar K, Sonmez A, Kilic S, Bedir S (2005) Prostate cancer and adiponectin. Urology 65:1168–1172

Goodwin PJ, Stambolic V (2015) Impact of the obesity epidemic on cancer. Annu Rev Med 66:281–296

Guo B, Fu S, Zhang J, Liu B, Li Z (2016) Targeting inflammasome/IL-1 pathways for cancer immunotherapy. Sci Rep 6:36107

Habeeb BS, Kitayama J, Nagawa H (2011) Adiponectin supports cell survival in glucose deprivation through enhancement of autophagic response in colorectal cancer cells. Cancer Sci 102:999–1006

Hagman DK, Larson I, Kuzma JN, Cromer G, Makar K, Rubinow KB, Foster-Schubert KE, van Yserloo B, Billing PS, Landerholm RW, Crouthamel M, Flum DR, Cummings DE, Kratz M (2017) The short-term and long-term effects of bariatric/metabolic surgery on subcutaneous adipose tissue inflammation in humans. Metabolism 70:12–22. https://doi.org/10.1016/j.metabol.2017.01.030

Hoffmann M, Gogola J, Ptak A (2018) Adiponectin reverses the proliferative effects of estradiol and IGF-1 in human epithelial ovarian cancer cells by downregulating the expression of their receptors. Hormones Cancer 9:166–174

Holland WL, Scherer PE (2013) Cell biology. Ronning after the adiponectin receptors. Science 342:1460–1461. https://doi.org/10.1126/science.1249077

Holland WL, Miller RA, Wang ZV, Sun K, Barth BM, Bui HH, Davis KE, Bikman BT, Halberg N, Rutkowski JM, Wade MR, Tenorio VM, Kuo MS, Brozinick JT, Zhang BB, Birnbaum MJ, Summers SA, Scherer PE (2011) Receptor-mediated activation of ceramidase activity initiates the pleiotropic actions of adiponectin. Nat Med 17:55–63

Hu MB, Xu H, Hu JM, Zhu WH, Yang T, Jiang HW, Ding Q (2016) Genetic polymorphisms in leptin, adiponectin and their receptors affect risk and aggressiveness of prostate cancer: evidence from a meta-analysis and pooled-review. Oncotarget 7:81049–81061

Ikeda A, Nakagawa T, Kawai K, Onozawa M, Hayashi T, Matsushita Y, Tsutsumi M, Kojima T, Miyazaki J, Nishiyama H (2015) Serum adiponectin concentration in 2,939 Japanese men undergoing screening for prostate cancer. Prostate Inte 3:87–92

Ishikawa M, Kitayama J, Yamauchi T, Kadowaki T, Maki T, Miyato H, Yamashita H, Nagawa H (2007) Adiponectin inhibits the growth and peritoneal metastasis of gastric cancer through its specific membrane receptors AdipoR1 and AdipoR2. Cancer Sci 98:1120–1127. https://doi.org/10.1111/j.1349-7006.2007.00486.x

Jeong YJ, Bong JG, Park SH, Choi JH, Oh HK (2011) Expression of leptin, leptin receptor, adiponectin and adiponectin receptor in ductal carcinoma in situ and invasive breast cancer. J Breast Cancer 14:96–103

Jia Z, Liu Y, Cui S (2014) Adiponectin induces breast cancer cell migration and growth factor expression. Cell Biochem Biophys 70:1239–1245

Jiang J, Fan Y, Zhang W, Shen Y, Liu T, Yao M, Gu J, Tu H, Gan Y (2019) Adiponectin suppresses human pancreatic cancer growth through attenuating the β-catenin signaling pathway. Int J Biol Sci 15:253–264. https://doi.org/10.7150/ijbs.27420

Jin JH, Kim HJ, Kim CY, Kim YH, Ju W, Kim SC (2016) Association of plasma adiponectin and leptin levels with the development and progression of ovarian cancer. Obstetrics Gynecol Sci 59:279–285

Kayani B, Okabayashi K, Ashrafian H, Harling L, Rao C, Darzi A, Kitagawa Y, Athanasiou T, Zacharakis E (2012) Does obesity affect outcomes in patients undergoing esophagectomy for cancer? A meta-analysis. World J Surg 36:1785–1795. https://doi.org/10.1007/s00268-012-1582-4

Kim AY, Lee YS, Kim KH, Lee JH, Lee HK, Jang SH, Kim SE, Lee GY, Lee JW, Jung SA et al (2010) Adiponectin represses colon cancer cell proliferation via AdipoR1-and-R2-mediated AMPK activation. Mol Endocrinol 24:1441–1452

Kim S, Lee Y, Kim JW, Son YJ, Ma MJ, Um JH, Kim ND, Min SH, Kim DI, Kim BB (2018) Discovery of a novel potent peptide agonist to adiponectin receptor 1. PLoS One 13:e0199256. https://doi.org/10.1371/journal.pone.0199256

Kim JM, Park JW, Lee JH, Park YH, Park SJ, Cheon JH, Kim WH, Kim TI (2020) Survival benefit for metformin through better tumor response by neoadjuvant CCRT in rectal cancer. Dis Colon Rectum. https://doi.org/10.1097/DCR.0000000000001624

Kuk JL, Rotondi M, Sui X, Blair SN, Ardern CI (2018) Individuals with obesity but no other metabolic risk factors are not at significantly elevated all-cause mortality risk in men and women. Clin Obes 8:305–312. https://doi.org/10.1111/cob.12263

Kyrgiou M, Kalliala I, Markozannes G, Gunter MJ, Paraskevaidis E, Gabra H et al (2017) Adiposity and cancer at major anatomical sites: umbrella review of the literature. BMJ 356:j477

Lamkanfi M, Dixit VM (2014) Mechanisms and functions of inflammasomes. Cell 157:1013–1022

Lashinger LM, Rossi EL, Hursting SD (2014) Obesity and resistance to cancer chemotherapy: interacting roles of inflammation and metabolic dysregulation. Clin Pharmacol Ther 96:458–463

Lennon H, Sperrin M, Badrick E, Renehan AG (2016) The obesity paradox in cancer: a review. Curr Oncol Rep 18:56. https://doi.org/10.1007/s11912-016-0539-4

Li X, Yu Z, Fang L, Liu F, Jiang K (2017) Expression of adiponectin receptor-1 and prognosis of epithelial ovarian cancer patients. Med Sci Monit 23:1514–1521

Liu Y, Jin PP, Sun XC, Hu TT (2018) Thiazolidinediones and risk of colorectal cancer in patients with diabetes mellitus. Meta-Analysis 24(2):75–81. https://doi.org/10.4103/sjg.SJG_295_17

Mauro L, Pellegrino M, De Amicis F, Ricchio E, Giordano F, Rizza P, Catalano S, Bonofiglio D, Sisci D, Panno ML et al (2014) Evidences that estrogen receptor α interferes with adiponectin effects on breast cancer cell growth. Cell Cycle 13:553–564

Mauro L, Naimo GD, Ricchio E, Panno ML, Andò S (2015) Cross-talk between adiponectin and IGF-IR in breast cancer. Front Oncol 5:157

Michalakis K, Williams CJ, Mitsiades N, Blakeman J, Balafouta-Tselenis S, Giannopoulos A, Mantzoros CS (2007) Serum adiponectin concentrations and tissue expression of adiponectin receptors are reduced in patients with prostate cancer: a case control study. Cancer Epidemiol Biomark Prev 16:308–313

Moon HS, Chamberland JP, Aronis K, Tseleni-Balafouta-S, Mantzoros CS (2011) Direct role of adiponectin and adiponectin receptors in endometrial cancer: in vitro and ex vivo studies in humans. Mol Cancer Ther 10:2234–2243

Moon HS, Liu X, Nagel JM, Chamberland JP, Diakopoulos KN, Brinkoetter MT, Hatziapostolou M, Wu Y, Robson SC, Iliopoulos D et al (2013) Salutary effects of adiponectin on colon cancer: in vivo and in vitro studies in mice. Gut 62:561–570

Murphy N, Jenab M, Gunter MJ (2018) Adiposity and gastrointestinal cancers: epidemiology, mechanisms and future directions. Nat Rev Gastroenterol Hepatol 15:659–670. https://doi.org/10.1038/s41575-018-0038-1

Mutoh M, Teraoka N, Takasu S, Takahashi M, Onuma K, Yamamoto M, Kubota N, Iseki T, Kadowaki T, Sugimura T et al (2011) Loss of adiponectin promotes intestinal carcinogenesis in min and wild-type mice. Gastroenterology 140:2000–2008

Nagaraju GP, Aliya S, Alese OB (2015) Role of adiponectin in obesity related gastrointestinal carcinogenesis. Cytokine Growth Factor Rev 26:83–93. https://doi.org/10.1016/j.cytogfr.2014.06.007

O'Sullivan J, Lysaght J, Donohoe CL, Reynolds JV (2018) Obesity and gastrointestinal cancer: the interrelationship of adipose and tumour microenvironments. Nat Rev Gastroenterol Hepatol 15:699–714. https://doi.org/10.1038/s41575-018-0069-7

Okada-Iwabu M, Yamauchi T, Iwabu M, Honma T, Hamagami K, Matsuda K, Yamaguchi M, Tanabe H, Kimura-Someya T, Shirouzu M, Ogata H, Tokuyama K, Ueki K, Nagano T, Tanaka A, Yokoyama S, Kadowaki T (2013) A small-molecule AdipoR agonist for type 2 diabetes and short life in obesity. Nature 503:493–499. https://doi.org/10.1038/nature12656

Osborn O, Olefsky JM (2012) The cellular and signaling networks linking the immune system and metabolism in disease. Nat Med 18:363–374. https://doi.org/10.1038/nm.2627

Ottaiano A, De Divitiis C, Capozzi M, Avallone A, Pisano C, Pignata S, Tafuto S (2018) Obesity and cancer: biological links and treatment implications.

Curr Cancer Drug Targets 18:231–238. https://doi.org/10.2174/1568009617666170330125619

Otvos L Jr, Kovalszky I, Olah J, Coroniti R, Knappe D, Nollmann FI, Hoffmann R, Wade JD, Lovas S, Surmacz E (2015) Optimization of adiponectin-derived peptides for inhibition of cancer cell growth and signaling. Pept Sci 104:156–166

Ouchi N, Parker JL, Lugus JJ, Walsh K (2011) Adipokines in inflammation and metabolic disease. Nat Rev Immunol 11:85–97. https://doi.org/10.1038/nri2921

Parida S, Siddharth S, Sharma D (2019) Adiponectin, obesity, and cancer: clash of the bigwigs in health and disease. Int J Mol Sci 20:E2519. https://doi.org/10.3390/ijms20102519

Pavone C, Candela L, Fontana D, Simonato A (2018) Postoperative complications and 90-day mortality in radical cystectomy in high-risk patients: a monocentric retrospective observational study. Urologia 85:111–117. https://doi.org/10.1177/0391560317751600

Porcile C, Di Zazzo E, Monaco ML, D'Angelo G, Passarella D, Russo C, Di Costanzo A, Pattarozzi A, Gatti M, Bajetto A, Zona G, Barbieri F, Oriani G, Moncharmont B, Florio T, Daniele A (2014) Adiponectin as novel regulator of cell proliferation in human glioblastoma. J Cell Physiol 229:1444–1454. https://doi.org/10.1002/jcp.24582

Pothuraju R, Rachagani S, Junker WM, Chaudhary S, Saraswathi V, Kaur S, Batra SK (2018) Pancreatic cancer associated with obesity and diabetes: an alternative approach for its targeting. J Exp Clin Cancer Res 37:319. https://doi.org/10.1186/s13046-018-0963-4

Ramzan AA, Bitler BG, Hicdingks D et al (2019) Adiponectin receptor agonist AdipoRon induces apoptotic cell death and suppresses proliferation in human ovarian cancer cells. Mol Cell Biochem 461:37–46

Rider JR, Fiorentino M, Kelly R, Gerke T, Jordahl K, Sinnott JA, Giovannucci EL, Loda M, Mucci LA, Finn S et al (2015) Tumor expression of adiponectin receptor 2 and lethal prostate cancer. Carcinogenesis 36:639–647

Safe S, Nair V, Karki K (2018) Metformin-induced anticancer activities: recent insights. Biol Chem 399:321–335

Saunders TJ, Palombella A, McGuire KA, Janiszewski PM, Després JP, Ross R (2012) Acute exercise increases adiponectin levels in abdominally obese men. J Nutr Metab 2012:148729. https://doi.org/10.1155/2012/148729

Saxena A, Chumanevich A, Fletcher E, Larsen B, Lattwein K, Kaur K, Fayad R (2012) Adiponectin deficiency: role in chronic inflammation induced colon cancer. Biochim Biophys Acta Mol basis Dis 1822:527–536

Schrover IM, Spiering W, Leiner T, Visseren FL (2016) Adipose tissue dysfunction: clinical relevance and diagnostic possibilities. Horm Metab Res 48:213–225. https://doi.org/10.1055/s-0042-103243

Slomian GJ, Nowak D, Buczkowska M, Glogowska-Gruszka A, Slomian SP, Roczniak W, Janyga S, Nowak P (2019) The role of adiponectin and leptin in the treatment of ovarian cancer patients. Endokr Polska 70:57–63

Sun Y, Zang Z, Zhong L, Wu M, Su Q, Gao X, Zan W, Lin D, Zhao Y, Zhang Z (2013) Identification of adiponectin receptor agonist utilizing a fluorescence polarization based high throughput assay. PLoS One 8:e63354. https://doi.org/10.1371/journal.pone.0063354

Sung H, Siegel RL, Torre LA, Pearson-Stuttard J, Islami F, Fedewa SA, Goding Sauer A, Shuval K, Gapstur SM, Jacobs EJ, Giovannucci EL, Jemal A (2019) Global patterns in excess body weight and the associated cancer burden. CA Cancer J Clin 69:88–112. https://doi.org/10.3322/caac.21499

Tang H, Shi W, Fu S, Wang T, Zhai S, Song Y, Han J (2018) Pioglitazone and bladder cancer risk: a systematic review and meta-analysis. Cancer Med 7(4):1070–1080. https://doi.org/10.1002/cam4.1354

Tilg H, Moschen AR (2014) Mechanisms behind the link between obesity and gastrointestinal cancers. Best Pract Res Clin Gastroenterol 28:599–610. https://doi.org/10.1016/j.bpg.2014.07.006

Tsukada T, Fushida S, Harada S, Terai S, Yagi Y, Kinoshita J, Oyama K, Tajima H, Fujita H, Ninomiya I, Fujimura T, Ohta T (2011) Adiponectin receptor-1 expression is associated with good prognosis in gastric cancer. J Exp Clin Cancer Res 30:107. https://doi.org/10.1186/1756-9966-30-107

Tumminia A, Vinciguerra F, Parisi M, Graziano M, Sciacca L, Baratta R, Frittitta L (2019) Adipose tissue, obesity and adiponectin: role in endocrine cancer risk. Int J Mol Sci 20:E2863. https://doi.org/10.3390/ijms20122863

Vansaun MN (2013) Molecular pathways: adiponectin and leptin signaling in cancer. Clin Cancer Res 19:1926–1932

Wang Q, Tu H, Zhu M, Liang D, Ye Y, Chang DW, Long Y, Wu X (2019) Circulating obesity-driven biomarkers are associated with risk of clear cell renal cell carcinoma: a two-stage, case-control study. Carcinogenesis 40:1191–1197

Wang Y, Liu X, Yan P, Tang J, Chen T, Sun Y, Zhou W, Bi Y, Zhang ZJ (2020) Effect of metformin on the risk of prostate cancer in patients with type 2 diabetes by considering different confounding factors: a meta-analysis of observational studies. Eur J Cancer Prev 29(1):42–52. https://doi.org/10.1097/CEJ.0000000000000514

Wei Z, Liang L, Junsong L, Rui C, Shuai C, Guanglin Q, Shicai H, Zexing W, Jin W, Xiangming C, Shufeng W (2015) The impact of insulin on chemotherapeutic sensitivity to 5-fluorouracil in gastric cancer cell lines SGC7901, MKN45 and MKN28. J Exp Clin Cancer Res 34:64

Wiedmann MKH, Brunborg C, Di Ieva A, Lindemann K, Johannesen TB, Vatten L, Helseth E, Zwart JA (2017) Overweight, obesity and height as risk factors for meningioma, glioma, pituitary adenoma and nerve sheath tumor: a large population-based prospective cohort study. Acta Oncol 56:1302–1309. https://doi.org/10.1080/0284186X.2017.1330554

Yamauchi T, Iwabu M, Okada-Iwabu M, Kadowaki T (2014) Adiponectin receptors: a review of their structure, function and how they work. Best Pract Res Clin Endocrinol Metab 28:15–23

Ye J, Jia J, Dong S, Zhang C, Yu S, Li L, Mao C, Wang D, Chen J, Yuan G (2014) Circulating adiponectin levels and the risk of breast cancer: a meta-analysis. Eur J Cancer Prev 23:158–165

Yin Z, Deng T, Peterson LE, Yu R, Lin J, Hamilton DJ, Reardon PR, Sherman V, Winnier GE, Zhan M et al (2014) Transcriptome analysis of human adipocytes implicates the NOD-like receptor pathway in obesity-induced adipose inflammation. Mol Cell Endocrinol 394:80–87

Zhang X, Liu Y, Shao H, Zheng X (2017) Obesity paradox in lung cancer prognosis: evolving biological insights and clinical implications. J Thorac Oncol 12:1478–1488. https://doi.org/10.1016/j.jtho.2017.07.022

Cognitive Impairment in Obesity and Diabetes

29

Cristina Carvalho and Paula I. Moreira

Contents

Introduction 400

Obesity, Cognitive Decline and Dementia 400

Type 2 Diabetes, Cognitive Decline and Dementia 401

Insulin Alterations, Inflammation and Oxidative Stress in Obesity/T2D 401
Additional Pathways 402

The Intestinal Microbiota and the Gut-brain Axis 403
Metabolic Endotoxemia 403
Neurotransmitters 404

Vascular Alterations in Obesity and/or T2D: More than an Oxygen Issue 404

Obesity and/or T2D are Associated with AD 405

Can Anti-obesity and Anti-T2D Therapy be Repurposed for AD? 406
Non Pharmacological Approaches 406
Pharmacological Approaches 407

References 408

Abstract

Alzheimer's disease (AD), the most common form of dementia, affects 55 million people worldwide, and this number is expected to increase to 88 million by 2050. Understanding the molecular mechanisms underpinning pathological cognitive deficits and AD onset and progression is urgently needed, to identify biomarkers and therapeutic targets. Besides

C. Carvalho
CNC—Center for Neuroscience and Cell Biology, University of Coimbra, Coimbra, Portugal

Institute of Interdisciplinary Research, University of Coimbra, Coimbra, Portugal

CIBB—Center for Innovative Biomedicine and Biotechnology, University of Coimbra, Coimbra, Portugal

P. I. Moreira (✉)
CNC—Center for Neuroscience and Cell Biology, University of Coimbra, Coimbra, Portugal

CIBB—Center for Innovative Biomedicine and Biotechnology, University of Coimbra, Coimbra, Portugal

Laboratory of Physiology, Faculty of Medicine, University of Coimbra, Coimbra, Portugal
e-mail: venta@ci.uc.pt; pimoreira@fmed.uc.pt

J. Faintuch and S. Faintuch (eds.), *Obesity and Diabetes*, https://doi.org/10.1007/978-3-030-53370-0_29

ageing, type 2 diabetes (T2D) and obesity are considered major risk factors for cognitive decline and AD. Here, we discuss the mechanistic relationship between AD and metabolic disorders, particularly T2D and obesity. The repurposing of anti-diabetes and anti-obesity approaches in the treatment of AD will be also debated.

Keywords

Type 2 diabetes · Obesity · Cognitive impairment · Alzheimer's disease · Anti-diabetes and anti-obesity approaches

Introduction

By the age of 60, about 12–18% of the population suffers from mild cognitive impairment (MCI) (Petersen 2016), and 14% of cases present a clinical panel of dementia (Morley et al. 2015).

The increase in the number of individuals with cognitive defects represents an enormous economic burden and a significant challenge for patients, families, caregivers and clinicians (Morley et al. 2015). Over the last decades, type 2 diabetes (T2D) and obesity emerged as major risk factors for cognitive decline and dementia (Pugazhenthi et al. 2017). Indeed, a convergence in risk, comorbidities and pathophysiology seems to occur between obesity, T2D and cognitive impairment, this convergence being exacerbated in diabesity (co-occurrence of diabetes and obesity) (Newcombe et al. 2018; Cardoso and Moreira 2019) (Fig. 29.1).

T2D has a 1.2- to 1.7-fold increased risk for development of neurodegenerative conditions, particularly Alzheimer's disease (AD), the most common form of dementia in the elderly. Obesity can be responsible not only for a significant decline in cognition later in life, but can also lead to a decrease in memory performance in the short-term, particularly during middle adulthood (Dye et al. 2017).

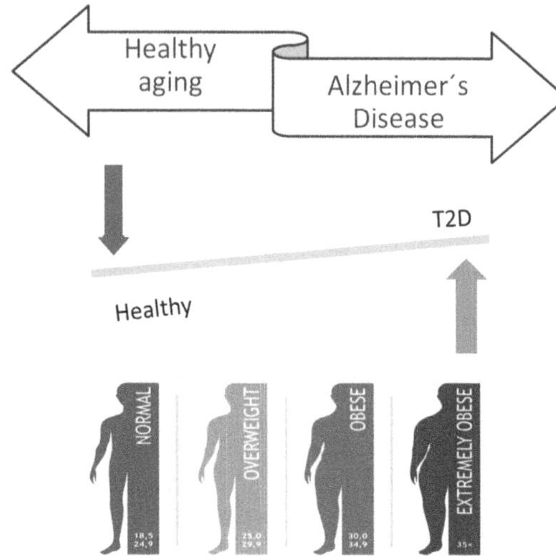

Fig. 29.1 Correlation between obesity and/or T2D and the risk for AD development. Over the last decades, type 2 diabetes (T2D) and obesity emerged as two major risk factors for cognitive decline and dementia, including Alzheimer's disease (AD), the risk being exacerbated when both metabolic diseases co-occurrence

Obesity, Cognitive Decline and Dementia

An estimated number of 41 million of children under the age of 5 are overweight or obese, according to the World Health Organization (WHO 2020), numbers that are expected to rise to approximately 541 million by 2030 (Kelly et al. 2008). Obesity "kills" more people than undernutrition (Fruhbeck et al. 2013). The increased risk for dementia in obese individuals was further confirmed by several studies (Pedditzi et al. 2016; Kivimaki et al. 2018). It has been reported that obesity can be responsible for impaired episodic and working memory, verbal learning and stimulus reward learning (Cournot et al. 2006; Cheke et al. 2017; Coppin et al. 2014).

Nonetheless, a careful analysis should be performed when describing obesity as a risk factor for dementia, since there seems to exist an "obesity paradox" (Pegueroles et al. 2018). During middle life, the risk for dementia is elevated;

while in advanced ages, obesity can have opposite effects, conferring some protection against cognitive decline (Anjum et al. 2018). Evidence for the risk is robust < 65 years however inconclusive between the ages of 65 and 74 (Albanese et al. 2017; Gustafson et al. 2003; Tolppanen et al. 2014; Bowman et al. 2019). Obese adults between 30 and 39 years seem to have a 3.5-fold increase in the relative risk for dementia, and after this age range, the risk seems to decrease in a stepwise fashion till 70 years, becoming significantly reduced from 80 years onward (Dye et al. 2017). An abrupt weight loss precedes dementia diagnosis by approximately 6–10 years (LeBlanc et al. 2017).

One of the major limitations of the studies aimed at exploring the interrelation between obesity and dementia is the gap in gender-inherent differences. It has been postulated that gender-associated susceptibility to cognitive decline and dementia does exist (Candeias et al. 2017; Fisher et al. 2018; Pontifex et al. 2018; Nyarko et al. 2018) and future studies should fill this gap.

Type 2 Diabetes, Cognitive Decline and Dementia

T2D is a complex metabolic disorder characterized by hyperglycaemia, hyperinsulinaemia, dyslipidaemia, as well as lipotoxicity, culminating in a progressive decline of insulin secretion and insulin action (Yazici and Sezer 2017).

Alterations in brain cortical thickness and white matter integrity alongside with decreased psychomotor speed performance are more pronounced in overweight/obese T2D individuals than in those with normal weight (Cardoso and Moreira 2019; Yoon et al. 2017). Moreover, T2D has been associated with reduced performance in verbal memory, processing speed, attention, spatial working memory, verbal fluency, executive function and alterations that are correlated with global brain atrophy (Wisse et al. 2014), sustaining the hypothesis of T2D as an accelerator of cognitive ageing (Dye et al. 2017; Karvani et al. 2019).

T2D can exist independently of obesity driven by genetics, fatty liver disease, inflammation, autoimmunity or stress (Benedict and Zhang 2017; de Candia et al. 2019). In Goto-Kakizaki rats, a non-obese T2D animal model, it was shown that spatial memory impairment, evaluated by the Y-maze test, and hippocampal synaptic dysfunction, evaluated by long-term potentiation (LTP) assays, was correlated with a reduction in synaptosomal nerve-associated protein 25 (SNAP25) and synaptophysin levels, suggesting synaptic dysregulation in this T2D animal model (Duarte et al. 2018; Spinelli et al. 2019).

Insulin Alterations, Inflammation and Oxidative Stress in Obesity/T2D

Insulin has a pivotal role in the central nervous system (CNS). It can also be produced by neuronal and glial cells (Blazquez et al. 2014; Gray et al. 2014). There is clear evidence that insulin resistance (IR) contributes to neurodegenerative events impacting brain function such as memory and learning (Holscher 2019; Folch et al. 2019). In the CNS, insulin modulates the phosphatidylinositol-3-kinase (PI3K)/protein kinase B (Akt) pathway and mitogen-activated protein kinases/Ras (MAPK/Ras) pathway. These pathways are highly involved in the regulation of neuronal survival, neuroplasticity and glucose uptake (Grillo et al. 2009). Furthermore, a role for insulin in the modulation of N-methyl-D-aspartate (NMDA) expression in the cell membrane, induction of long term potentiation (LTP) (Lee et al. 2011) and modulation of acetylcholine and norepinephrine levels (Arnold et al. 2018; Blazquez et al. 2014; Gray et al. 2014; Maciejczyk et al. 2019) has also been reported.

Male Wistar rats exposed to a 50% fat diet gained body weight, which contributed to peripheral IR and cerebral insulin deficiency, resulting in cognitive decline (Nameni et al. 2017). The decrease in cognitive performance may result from the deficient binding of insulin to its receptors that seem to be mainly situated in synapses of brain areas intimately related with memory formation (Spinelli et al. 2019). Brain cortical IR in obese patients may contribute to decreased performance in episodic memory tasks (Cheke et al. 2017). High-fat diet combined to sugar intake induced obesity in male C57BL/6NHsd mice, which contributed to a significant decrease in tyrosine phosphorylation of the insulin receptor and an increase in serine phosphorylation of the insulin receptor substrate (IRS) 1, resulting in neuronal IR (Kothari et al. 2017). Insulin alterations were deeply associated with alterations in inflammatory [Nuclear factor-κB (NFκB), c-Jun N-terminal kinase (JNK)] and stress (p38 MAPK and C/EBP Homologous Protein) responses. Obesity is able to activate inflammatory-related negative regulators of IRS proteins, with emphasis on the suppressor of the 23 cytokine signalling (Socs) protein, which has the capacity to bind phosphorylated insulin receptors blocking the activation of IRS proteins (Johnson and Olefsky 2013).

Using magnetoencephalography, Tschritter and co-workers (2006) evaluated spontaneous and insulin-stimulated cerebrocortical activity in lean and obese humans during a two-step hyperinsulinaemic-euglycaemic clamp. Obese individuals did not respond to insulin infusion (Tschritter et al. 2006). Nevertheless, weight loss and improved IR fail to change grey matter volume or cortical thickness (Drummen et al. 2019).

Obesity can lead to an overproduction of ceramides by the liver, which can cross the blood-brain barrier (BBB) and interfere with insulin receptors causing brain IR (Arboleda et al. 2007). Kabadi (2017) reported that alterations in insulin secretory capacity are more pronounced in lean T2D subjects than IR, although these alterations also affect glucose tolerance. Both obesity and T2D-related IR can affect at least two different mechanisms: (1) peripheral insulin transport across the BBB (described as a receptor-mediated saturable process) and (2) intraneuronal defects in the insulin signalling cascade of either a primary (i.e., genetic) or secondary (i.e., due to an obesity-related factor) nature (Ono 2019).

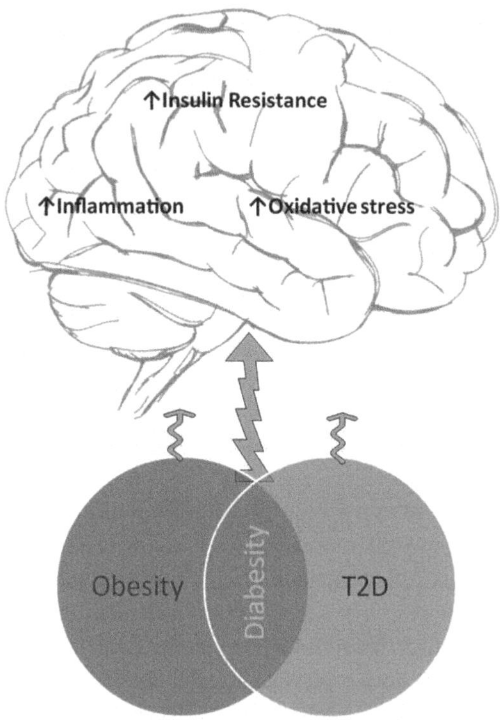

Fig. 29.2 The impact of obesity and/or T2D in the brain. Obesity and type 2 diabetes (T2D) increase brain insulin resistance, inflammation and oxidative stress, these alterations being potentiated when both pathologies co-occur, the so called "diabesity." Together, these alterations increase the risk of cognitive decline and dementia

Additional Pathways

Brain IR induced by obesity or T2D (Fig. 29.2) also causes the loss of the inhibitory function mediated by FoxO transcription factors, which normally modulate cellular metabolism and autophagy, increase amyloid-beta (Aβ) formation and deposition and phosphorylation of tau protein and neurofibrillary tangle formation and decrease synaptic plasticity, predisposing to cognitive impairment (Kothari et al. 2017), among other alterations, which affect mood and memory (Kullmann et al. 2016). The impairment of memory performance caused by IR in humans has been associated with a clear reduction in grey matter volume (Willette et al. 2013) suggesting that insulin signalling disturbances in the brain can underlie cognitive impairment (Talbot et al. 2012). It has been reported that brain IR can be overcome by the intranasal insulin administration, a strategy that has been reported to no interfere with peripheral glucose homoeostasis (Cheke et al. 2017). Furthermore, structural alterations (e.g., hippocampal atrophy) have also been shown in individuals with impaired glucose tolerance and IR (Ursache et al. 2012).

Obesity induces an increase in circulating free fatty acids along with systemic inflammation due to the secretion of pro-inflammatory adipokines by the adipose tissue, and a decrease in adiponectin levels, known for its anti-inflammatory role (Dye et al. 2017; Nigro et al. 2014). In obese and T2D individuals, peripheral macrophages can reach the brain, mainly due to the increased permeability of BBB (Van Dyken and Lacoste 2018), causing low-grade inflammation (Buckman et al. 2014). Kumar and co-workers (2014) reported an increase in mRNA and protein levels of key chemokines in the brains of diabetic db/db mice, compared to wild-type counterparts. Once in the brain, macrophages activate microglia causing hormonal dysregulation (leptin and insulin resistance), increased immune sensitivity and cognitive impairment (Van Dyken and Lacoste 2018).

Mitochondrial dysfunction and oxidative stress are also important players in T2D and obesity, being closely associated with IR and inflammation (Fig. 29.2). Inherited defects leading to decreased mitochondrial capacity to oxidize fatty acids cause the overproduction of mitochondrial reactive oxygen species (ROS) and oxidative stress that, in turn, predispose to inflammation and IR (Dayre et al. 2016). Associated with increased oxidative stress is the decrease in glutathione metabolism, the principal brain antioxidant and regenerator of other free radical scavengers (e.g., vitamins C and E) (Maciejczyk et al. 2019). Increase in ROS production and oxidative stress leads to increased cell membrane permeability, ATP depletion and accumulation of protein aggregates. Importantly, increased ROS production seems to activate pro-inflammatory enzymes contributing to brain inflammation (Pugazhenthi et al. 2017; Sripetchwandee et al. 2018).

The Intestinal Microbiota and the Gut-brain Axis

Hippocrates, the father of medicine said that "all diseases begin in the gut." In fact, evidence suggests that obesity and/or T2D are influenced by dysfunction in the gut-brain axis (Fig. 29.3). The axis consists of a bidirectional communication (Heiss and Olofsson 2018) that integrates neural, hormonal and immunological signalling between the gut and the brain (Collins et al. 2012). The brain communicates with the gut microbiota directly via neurotransmitters (e.g., catecholamines, 5-hydroxytryptamine (5-HT), and γ-aminobutyric acid (GABA)) that can be sensed by the microbes, or indirectly via vagus nerves that modulate intestinal secretion and motricity, and thus its internal milieu (Collins et al. 2012). Gut microbiota can also send signals to the brain via blood-borne substances or afferent spinal and vagal nerves, regulating the release and activity of hormones involved in energy balance such as peptide YY (PYY) and glucagon-like peptide-1 (GLP-1) (Covasa et al. 2019).

Gut microbiota has emerged as an environmental factor that modulates the energy balance of the host (Heiss and Olofsson 2018), eventually causing low-grade inflammation, excess lipid accumulation and loss of insulin sensitivity, which potentiate the risk of developing metabolic diseases and comorbidities (Boulange et al. 2016).

Metabolic Endotoxemia

As previously mentioned, both T2D and obesity are characterized by a low-grade inflammatory state. This inflammatory response can be mediated by the altered gut microbiota (called dysbiosis), through an increase in lipopolysaccharide (LPS) production and its release to

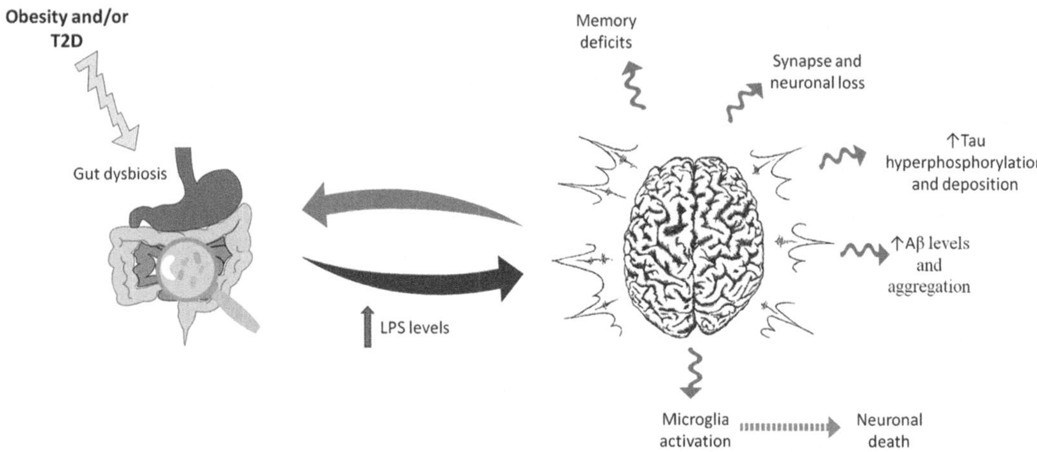

Fig. 29.3 Obesity and/or T2D affects gut-brain axis and predispose to dementia. Throughout life, changes in the intestinal ecosystem can occur leading to development of several pathologies. Also, gut microbiota changes (called dysbiosis) resulting from the consumption of high calories diets cause an increase in lipopolysaccharide (LPS) production and its release to circulating plasma. LPS can easily access the brain activating microglia that can cause neuronal death through increased nitrosative/oxidative stress, inflammation, formation and aggregation of amyloid β (Aβ) and tau proteins and phagocytosis of synapses and neurons. All these alterations may contribute to cognitive decline and dementia

plasma due to an increased permeability of the intestinal barrier (Cani et al. 2008). LPS can then access the brain due to an altered BBB permeability (Cardoso et al. 2012) leading to the activation of microglia (Fig. 29.3). Upon activation, microglia can cause neuronal death through oxidant and cytokine production or by phagocytosis of synapses and neurons (Brown 2019). Moreover, it has already been shown that LPS can trigger Aβ and tau proteins aggregation (Asti and Gioglio 2014; Gardner et al. 2016), causing the loss of brain synapses and neurons and memory deficits (Fig. 29.3), and predisposing to neurodegenerative events (Brown 2019).

Neurotransmitters

Studies involving germ-free animals support a pivotal role for gut microbiota in brain development and function including its influence in exploratory and learning processes (Neufeld et al. 2011; Diaz Heijtz et al. 2011). Germ-free animals are holders of significant alterations in anxiety-related neurotransmitters such as noradrenaline, dopamine and 5-HT (Clarke et al. 2013; Neufeld et al. 2011), as well as altered levels of proteins involved in normal development and function of neuronal synapses, as, for example, the levels of synaptophysin and post-synaptic density protein 95 (PSD95) (Diaz Heijtz et al. 2011). After transfection of gut microbiota from a patient with Alzheimer´s disease to germ-free C57BL/6N mice, Object Location Test and Object Recognition Test were significantly reduced, along with faecal excretion of metabolites related to the nervous system, including γ-aminobutyrate, taurine and valine (Fujii et al. 2019).

Vascular Alterations in Obesity and/or T2D: More than an Oxygen Issue

The quote "the way you breathe affects your memory" suggests the importance of oxygen to brain health. In this line, any alterations compromising the proper delivery of oxygen to the brain impacts its homoeostasis. Body weight is related to alterations in brain structure, white matter and BBB, compromising the homoeostatic regulation of central energy metabolism (Anjum et al. 2018). Increased body fat content compromises vascular integrity, not only through the disruption of BBB integrity, but also via an accentuated decrease in oxygen and glucose delivery to brain cells (Jain et al. 2010). Obesity is involved with endothelial dysfunction, which impacts vascular tone, microvascular insulin resistance, secretion of nitric oxide and endothelin, influencing vascular structure and perivascular inflammation. Microvascular dysfunction in obesity could predispose to chronic diseases, including microvascular dementia (Sorop et al. 2017). Also in diabetic patients undergoing cardiac surgery, a widening of the physiological saturation gap between regional cerebral oxygen saturation ($rScO_2$) and central venous oxygen saturation ($ScvO_2$) was observed during general anaesthesia. $rScO_2$ was lower before induction, and this difference was maintained throughout the operation (Sudy et al. 2019). Decreased cerebral perfusion can lead to vascular damages, flow alteration, abnormal expression of amyloid β and tau proteins, as well as behavioural and cognitive deficits (Dong et al. 2018).

Obesity is responsible for a decrease in endothelial nitric oxide (NO) synthesis and action causing an increased nitrosative/oxidative stress that affects endothelial function (Toda et al. 2014). *In vitro* and animal studies from our group showed that T2D compromises BBB integrity and aorta function through mitochondrial dysfunction and increased oxidative/nitrosative stress potentiating the development of AD-like brain changes (Carvalho et al. 2013, 2014; Sena et al. 2015). Other studies show that metabolic syndrome and hypercholesterolemia also cause alterations in vascular function (Kivipelto et al. 2002; Campos-Pena et al. 2017), which may also contribute to brain alterations.

The two-hit vascular hypothesis postulates that cerebrovascular damage is sufficient to cause neuronal injury and AD-like neurodegeneration (Nelson et al. 2016). Thus, events that compromise vascular health could initiate a cascade of deleterious events that culminate in AD development. It must be emphasized that other neurodegenerative diseases (e.g., Parkinson and Huntington diseases) are also characterized by vascular alterations (Al-Bachari et al. 2017; Lee and Pienaar 2014; Lin et al. 2013; St-Amour et al. 2015).

Obesity and/or T2D are Associated with AD

AD is the fifth cause of death among those with 65 years or older (Alzheimer's Association 2020). The co-occurrence of obesity and T2D increases the risk of AD by 65 % (relative risk in T2D is 1.46) (Caruana et al. 2015). Several cross-sectional studies also show that T2D patients have lower scores in cognitive tests, especially those regarding verbal memory and complex information processing (Zilliox et al. 2016). In fact, AD is also considered a metabolic disorder characterized by deregulated glucose control, described in 80% of the cases, together with impairments in brain insulin responsiveness, glucose utilization and energy metabolism, which lead to increased oxidative stress, and inflammation that worsens IR, and is even designated as type 3 diabetes (Dye et al. 2017).

The existence of common mechanisms between T2D and AD has been supported by clinical studies showing that T2D patients exhibit regional brain atrophy, evaluated by fluorodeoxyglucose-positron emission tomography (FDG-PET), and cognitive impairment, even if they are not diagnosed as having AD (McNay and Recknagel 2011). Interestingly, Boersma and co-workers (2018) reported significantly higher brain glucose uptake rate in T2D and prediabetes subjects, confirming FDG-PET imaging in T2D rats (Ryu et al. 2019; Barriere et al. 2018).

Obesity/T2D-associated insulin signalling impairment and AD have been emphasized (Arnold et al. 2018). mRNA/protein expression of insulin receptors, insulin-like growth factors (IGF1 and IGF2), IRS1 and PI3K/Akt decrease in an inverse proportion to AD neuropathological markers, suggesting a key role for IR in the progression of AD (Maciejczyk et al. 2019). Obesity is associated with thinner brain cortex, along with reduced total cerebral volume (Caunca et al. 2019). A similar picture of diminished cortical thickness is true for diabetes, with an indirect pathway linking T2D and cognitive decline (Moran et al. 2019). Formation of amyloid plaques (APs) and neurofibrillary tangles (NFTs) is related to the insulin signalling pathway in the brain, including GSK3β, JNK, CamKII, CDK5, CK1, MARK4, PLK2, Syk, DYRK1A, PPP, and P70S6K pathways. NFTs and APs lead to the impairment of synaptogenesis, neurotrophy and apoptosis, which are regulated by insulin, cholesterol and glucose metabolism (Rad et al. 2018). Studies from our laboratory show that T2D mice and the triple transgenic mouse model of AD (3xTg-AD) present similar behavioural and cognitive alterations (Carvalho et al. 2013), associated with alterations in the levels of pre- and post-synaptic protein markers as well as increased levels of Aβ and phosphorylated tau proteins (Carvalho et al. 2012, 2015).

There is also evidence that obesity per se or associated with T2D also impact the brain. Besides IR and inflammation, it was recently shown that excessive body fat affects the fornix white matter compromising its role in connecting the hippocampus to other brain regions, which contribute to AD development (Metzler-Baddeley et al. 2019).

Leptin, a hormone produced by adipocytes, has been shown to be involved in obesity/AD relationship in the absence of diabetes. Obese subjects also present increased levels of circulating leptin, usually related to leptin resistance. Leptin influences the functioning of the hippocampus, a region that degenerates in AD. Leptin has cognitive enhancing properties that could facilitate the cellular events that underlie hippocampal-dependent learning and memory. Reductions in leptin ability to regulate hippocampal synaptic function occur with age and also entail increased risk of AD. Leptin administration was associated with beneficial effects in animal models of AD (McGregor and Harvey 2018).

Nevertheless, further studies are required to establish the connections between obesity and/or T2D and neurodegenerative diseases, particularly AD, as inconsistencies and controversial results are recognized (Arnold et al. 2018; McGregor and Harvey 2018).

Can Anti-obesity and Anti-T2D Therapy be Repurposed for AD?

Non Pharmacological Approaches

Some longitudinal studies show that weight loss in midlife seems to reduce the incidence of AD, yet the knowledge about this topic is scarce (Sun et al. 2018). In fact, even small reductions in body weight associated with alterations in diet, bariatric surgery and/or exercise can restore or prevent cognitive dysfunction (Dye et al. 2017). Caloric restriction (CR) or changes in diet composition have been shown to restore blood glucose levels, increase life expectancy and decrease diabetes-related complications (Fontana and Partridge 2015). Several studies also show that the Mediterranean diet (enriched in fresh fruits, vegetables, nuts, fish and olive oil) ameliorates cognition and reduces the risk of developing dementia and AD (Wengreen et al. 2013; Tsivgoulis et al. 2013; Singh et al. 2014; Psaltopoulou et al. 2013). However, other studies show no association between Mediterranean diet and protection against cognitive decline (Dye et al. 2017).

Physical exercise plays also a major role in preventing or reducing obesity and/or T2D and associated complications (Teixeira-Lemos et al. 2011; Yang et al. 2019; Bernardo et al. 2016). Interestingly, CR and physical exercise reduce Aβ deposition associated with selective promotion of anti-amylogenic α-secretase activity and activation of astrocytes in AD animal models (Mouton et al. 2009; Wang et al. 2005). Moreover, both CR and physical exercise seem to reduce the deleterious effects of oxidative stress and IR, promoting an improvement of glucose metabolism as well as neurotrophic functions, thus facilitating neurogenesis and synaptogenesis and improving memory and cognitive functions in animal models of AD (Chen et al. 2016; Paillard et al. 2015). Likewise, also in humans CR and exercise seem to improve neurocognitive functions of healthy or overweight patients (Smith et al. 2010; Il'yasova et al. 2018). Indeed, verbal memory scores increase about 20% after CR and seem to be correlated with decreases in fasting plasma levels of insulin and C-reactive protein, improved insulin sensitivity and reduced inflammatory activity, which contribute to enhanced synaptic plasticity and stimulation of neurofacilitatory pathways in the brain (Witte et al. 2009). Additionally, aerobic exercise and dietary modifications in individuals with high blood pressure who were sedentary and overweight or obese lead to improvements in executive function, memory, learning and psychomotor speed (Smith et al. 2010).

Bariatric surgery can improve white and grey matter integrity as well as memory and executive function, a benefit that persists for several years (Spitznagel et al. 2015; Nota et al. 2020). The positive outcome could be the result of adiposity reduction and increased GLP-1 and peptide YY levels, improved inflammatory status and insulin sensitivity or increased diversity in gut microbiota composition which results in altered composition of short-chain fatty acids that influence host metabolism, including gut hormone secretion and insulin sensitivity (Nota et al. 2020). Much less is known about the response of older adults, as well as those with genetical risk for AD. A longitudinal protocol addressing brain function and structure after bariatric surgery is now going on (Vreeken et al. 2019). It is worth mentioning that all the above non-pharmacological approaches may originate distinct outcomes depending on age, sex and disease stage. Therefore, their prescription should involve personalized health recommendations to achieve desirable results without detrimental side effects (Flannery and Trushina 2019).

Pharmacological Approaches

Intranasal administration of insulin was shown to boost mental activity, improve attention, cognitive functioning and verbal memory in patients with early AD (Claxton et al. 2015), improvements that remained after 4 months of intranasal insulin in MCI and AD patients (Craft et al. 2017). Also in obese and/or T2D patients, the administration of intranasal insulin caused significant improvements in memory and cognitive function (Novak et al. 2014; Claxton et al. 2015). In older T2D adults, intranasal insulin also rescued the network communication between the hippocampus and multiple default mode network regions (a large-scale brain network of interacting brain regions known to have activity highly correlated with each other and distinct from other networks) (Zhang et al. 2015). In fact, intranasal insulin administration may modify the functional connectivity among brain regions regulating memory and complex cognitive behaviours (Zhang et al. 2015; Schilling et al. 2014). Nevertheless, the "Study of Nasal Insulin in the Fight Against Forgetfulness" clinical trial, communicated in the "Clinical Trials on Alzheimer's Disease Conference" held in 2018, shows no effect of intranasal insulin on cognition, although the leader investigator claims that the lack of positive results may result from a change in the delivery device during the trial (Trials of Diabetes-Related Therapies: Mainly a Bust 2018).

Other metabolic hormone-based therapies have been explored in the context of T2D-related neurodegenerative disorders, particularly AD. Pramlintide, a synthetic analogue of the pancreatic hormone amylin, was recently shown to recover Aβ-related decreases in LTP in hippocampal slices from AD mice (Grizzanti et al. 2018).

Pioglitazone has been shown to significantly reduce the incidence of dementia in people with diabetes (Heneka et al. 2015). Likewise, pioglitazone showed significant benefits in early stages of AD and mild-to-moderate AD, causing a significant reduction in Aβ and tau pathology (Cheng et al. 2016). The poor blood-brain barrier (BBB) permeability and important side effects limited its success. Intranasal nano lipid carriers of pioglitazone are being developed to circumvent such shortcomings (Jojo et al. 2019).

Metformin has been shown to modulate the ageing process by reducing oxidative stress and improving memory performance (Campbell et al. 2018). A meta-analysis performed by Zhang and co-workers (2020) showed that metformin therapy significantly improves cognitive function in patients with T2D. However, at least one study reported that long-term use of metformin can actually increase the risk of developing AD (Imfeld et al. 2012). It has been suggested that the effect of metformin may depend on the different profiles of comorbidities of the patient receiving the drug (Wang et al. 2017).

GLP-1 analogues confer protection against mitochondrial deficits, including improvements in ATP synthesis, and oxidative stress leading to the up-regulation of hippocampal expression of neurotrophic tyrosine kinase receptor type 2 (NTRK2) and mammalian target of rapamycin (mTOR) genes, involved in the modulation of synaptic plasticity and long-term potentiation in T2D and/or AD (Bomba et al. 2013; Lennox et al. 2014). A recent study from our laboratory also shows that exendin-4, a GLP-1 analogue, is able to protect the brains of T2D GK rats through the regulation of autophagy (Candeias et al. 2018). Bomba and co-workers (2013) reported that exenatide restored cognitive function of presenilin-1 knock-in mice, a mice model of AD. GLP-1 analogue liraglutide reduces Aβ accumulation by partially reverting inflammation and oxidative stress (Duarte et al. 2020). Also, liraglutide administration promoted antiapoptotic effects and ameliorated insulin synthesis in db/db T2D mice (Shao et al. 2014). Liraglutide was able to attenuate T2D-related loss of brain plasticity, disrupted energy homoeostasis, including reductions in mitochondrial transcription factor A (TFAM), sirtuin 1 and 5' AMP-activated protein kinase (AMPK) phosphorylation (Agrawal et al. 2014).

References

Agrawal R, Zhuang Y, Cummings BP, Stanhope KL, Graham JL, Havel PJ, Gomez-Pinilla F (2014) Deterioration of plasticity and metabolic homeostasis in the brain of the UCD-T2DM rat model of naturally occurring type-2 diabetes. Biochim Biophys Acta 1842 (9):1313–1323. https://doi.org/10.1016/j.bbadis.2014.05.007

Al-Bachari S, Vidyasagar R, Emsley HC, Parkes LM (2017) Structural and physiological neurovascular changes in idiopathic Parkinson's disease and its clinical phenotypes. J Cereb Blood Flow Metab 37 (10):3409–3421. https://doi.org/10.1177/0271678X16688919

Albanese E, Launer LJ, Egger M, Prince MJ, Giannakopoulos P, Wolters FJ, Egan K (2017) Body mass index in midlife and dementia: Systematic review and meta-regression analysis of 589,649 men and women followed in longitudinal studies. Alzheimers Dement (Amst) 8:165–178. https://doi.org/10.1016/j.dadm.2017.05.007

Alzheimer's Association (2020) 2020 Alzheimer's Disease Facts and Figures. www.alz.org/media/Documents/alzheimers-facts-and-figures.pdf

Anjum I, Fayyaz M, Wajid A, Sohail W, Ali A (2018) Does obesity increase the risk of dementia: a literature review. Cureus 10(5):e2660. https://doi.org/10.7759/cureus.2660

Arboleda G, Huang TJ, Waters C, Verkhratsky A, Fernyhough P, Gibson RM (2007) Insulin-like growth factor-1-dependent maintenance of neuronal metabolism through the phosphatidylinositol 3-kinase-Akt pathway is inhibited by C2-ceramide in CAD cells. Eur J Neurosci 25(10):3030–3038. https://doi.org/10.1111/j.1460-9568.2007.05557.x

Arnold SE, Arvanitakis Z, Macauley-Rambach SL, Koenig AM, Wang HY, Ahima RS, Craft S, Gandy S, Buettner C, Stoeckel LE, Holtzman DM, Nathan DM (2018) Brain insulin resistance in type 2 diabetes and Alzheimer disease: concepts and conundrums. Nat Rev Neurol 14(3):168–181. https://doi.org/10.1038/nrneurol.2017.185

Asti A, Gioglio L (2014) Can a bacterial endotoxin be a key factor in the kinetics of amyloid fibril formation? J Alzheimers Dis 39(1):169–179. https://doi.org/10.3233/JAD-131394

Barriere DA, Noll C, Roussy G, Lizotte F, Kessai A, Kirby K, Belleville K, Beaudet N, Longpre JM, Carpentier AC, Geraldes P, Sarret P (2018) Combination of high-fat/high-fructose diet and low-dose streptozotocin to model long-term type-2 diabetes complications. Sci Rep 8(1):424. https://doi.org/10.1038/s41598-017-18896-5

Benedict M, Zhang X (2017) Non-alcoholic fatty liver disease: an expanded review. World J Hepatol 9 (16):715–732. https://doi.org/10.4254/wjh.v9.i16.715

Bernardo TC, Marques-Aleixo I, Beleza J, Oliveira PJ, Ascensao A, Magalhaes J (2016) Physical exercise and brain mitochondrial fitness: the possible role against Alzheimer's disease. Brain Pathol 26 (5):648–663. https://doi.org/10.1111/bpa.12403

Blazquez E, Velazquez E, Hurtado-Carneiro V, Ruiz-Albusac JM (2014) Insulin in the brain: its pathophysiological implications for States related with central insulin resistance, type 2 diabetes and Alzheimer's disease. Front Endocrinol (Lausanne) 5:161. https://doi.org/10.3389/fendo.2014.00161

Boersma GJ, Johansson E, Pereira MJ, Heurling K, Skrtic S, Lau J, Katsogiannos P, Panagiotou G,

Lubberink M, Kullberg J, Ahlstrom H, Eriksson JW (2018) Altered glucose uptake in muscle, visceral adipose tissue, and brain predict whole-body insulin resistance and may contribute to the development of type 2 diabetes: a combined PET/MR study. Horm Metab Res 50(8):e10. https://doi.org/10.1055/a-0831-3620

Bomba M, Ciavardelli D, Silvestri E, Canzoniero LM, Lattanzio R, Chiappini P, Piantelli M, Di Ilio C, Consoli A, Sensi SL (2013) Exenatide promotes cognitive enhancement and positive brain metabolic changes in PS1-KI mice but has no effects in 3xTg-AD animals. Cell Death Dis 4:e612. https://doi.org/10.1038/cddis.2013.139

Boulange CL, Neves AL, Chilloux J, Nicholson JK, Dumas ME (2016) Impact of the gut microbiota on inflammation, obesity, and metabolic disease. Genome Med 8(1):42. https://doi.org/10.1186/s13073-016-0303-2

Bowman K, Thambisetty M, Kuchel GA, Ferrucci L, Melzer D (2019) Obesity and longer term risks of dementia in 65–74 year olds. Age Ageing 48 (3):367–373. https://doi.org/10.1093/ageing/afz002

Brown GC (2019) The endotoxin hypothesis of neurodegeneration. J Neuroinflammation 16(1):180. https://doi.org/10.1186/s12974-019-1564-7

Buckman LB, Hasty AH, Flaherty DK, Buckman CT, Thompson MM, Matlock BK, Weller K, Ellacott KL (2014) Obesity induced by a high-fat diet is associated with increased immune cell entry into the central nervous system. Brain Behav Immun 35:33–42. https://doi.org/10.1016/j.bbi.2013.06.007

Campbell JM, Stephenson MD, de Courten B, Chapman I, Bellman SM, Aromataris E (2018) Metformin use associated with reduced risk of dementia in patients with diabetes: a systematic review and meta-analysis. J Alzheimers Dis 65(4):1225–1236. https://doi.org/10.3233/JAD-180263

Campos-Pena V, Toral-Rios D, Becerril-Perez F, Sanchez-Torres C, Delgado-Namorado Y, Torres-Ossorio E, Franco-Bocanegra D, Carvajal K (2017) Metabolic syndrome as a risk factor for Alzheimer's disease: is abeta a crucial factor in both pathologies? Antioxid Redox Signal 26(10):542–560. https://doi.org/10.1089/ars.2016.6768

Candeias E, Duarte AI, Sebastiao I, Fernandes MA, Placido AI, Carvalho C, Correia S, Santos RX, Seica R, Santos MS, Oliveira CR, Moreira PI (2017) Middle-aged diabetic females and males present distinct susceptibility to Alzheimer disease-like pathology. Mol Neurobiol 54(8):6471–6489. https://doi.org/10.1007/s12035-016-0155-1

Candeias E, Sebastiao I, Cardoso S, Carvalho C, Santos MS, Oliveira CR, Moreira PI, Duarte AI (2018) Brain GLP-1/IGF-1 signaling and autophagy mediate exendin-4 protection against apoptosis in type 2 diabetic rats. Mol Neurobiol 55(5):4030–4050. https://doi.org/10.1007/s12035-017-0622-3

de Candia P, Prattichizzo F, Garavelli S, De Rosa V, Galgani M, Di Rella F, Spagnuolo MI, Colamatteo A, Fusco C, Micillo T, Bruzzaniti S, Ceriello A, Puca AA, Matarese G (2019) Type 2 diabetes: how much of an autoimmune disease? Front Endocrinol (Lausanne) 10:451. https://doi.org/10.3389/fendo.2019.00451

Cani PD, Bibiloni R, Knauf C, Waget A, Neyrinck AM, Delzenne NM, Burcelin R (2008) Changes in gut microbiota control metabolic endotoxemia-induced inflammation in high-fat diet-induced obesity and diabetes in mice. Diabetes 57(6):1470–1481. https://doi.org/10.2337/db07-1403

Cardoso S, Moreira PI (2019) Diabesity and brain disturbances: a metabolic perspective. Mol Asp Med 66:71–79. https://doi.org/10.1016/j.mam.2018.10.002

Cardoso FL, Kittel A, Veszelka S, Palmela I, Toth A, Brites D, Deli MA, Brito MA (2012) Exposure to lipopolysaccharide and/or unconjugated bilirubin impair the integrity and function of brain microvascular endothelial cells. PLoS One 7(5):e35919. https://doi.org/10.1371/journal.pone.0035919

Caruana EJ, Roman M, Hernandez-Sanchez J, Solli P (2015) Longitudinal studies. J Thor Dis 7(11):E537–E540. https://doi.org/10.3978/j.issn.2072-1439.2015.10.63

Carvalho C, Cardoso S, Correia SC, Santos RX, Santos MS, Baldeiras I, Oliveira CR, Moreira PI (2012) Metabolic alterations induced by sucrose intake and Alzheimer's disease promote similar brain mitochondrial abnormalities. Diabetes 61(5):1234–1242. https://doi.org/10.2337/db11-1186

Carvalho C, Machado N, Mota P, Correia S, Cardoso S, Santos R, Santos M, Oliveira C, Moreira P (2013) Type 2 diabetic and Alzheimer's disease mice present similar behavioral, cognitive and vascular anomalies. J Alzheimer Dis 35(3):623–635

Carvalho C, Katz PS, Dutta S, Katakam PV, Moreira PI, Busija DW (2014) Increased susceptibility to amyloid-beta toxicity in rat brain microvascular endothelial cells under hyperglycemic conditions. J Alzheimers Dis 38(1):75–83. https://doi.org/10.3233/JAD-130464

Carvalho C, Santos MS, Oliveira CR, Moreira PI (2015) Alzheimer's disease and type 2 diabetes-related alterations in brain mitochondria, autophagy and synaptic markers. Biochim Biophys Acta 1852 (8):1665–1675. https://doi.org/10.1016/j.bbadis.2015.05.001

Caunca MR, Gardener H, Simonetto M, Cheung YK, Alperin N, Yoshita M, DeCarli C, Elkind MSV, Sacco RL, Wright CB, Rundek T (2019) Measures of obesity are associated with MRI markers of brain aging: The Northern Manhattan Study. Neurology 93 (8):e791–e803. https://doi.org/10.1212/WNL.0000000000007966

Cheke LG, Bonnici HM, Clayton NS, Simons JS (2017) Obesity and insulin resistance are associated with reduced activity in core memory regions of the brain. Neuropsychologia 96:137–149. https://doi.org/10.1016/j.neuropsychologia.2017.01.013

Chen WW, Zhang X, Huang WJ (2016) Role of physical exercise in Alzheimer's disease. Biomedical Report 4 (4):403–407. https://doi.org/10.3892/br.2016.607

Cheng H, Shang Y, Jiang L, Shi TL, Wang L (2016) The peroxisome proliferators activated receptor-gamma agonists as therapeutics for the treatment of Alzheimer's disease and mild-to-moderate Alzheimer's disease: a meta-analysis. Int J Neurosci 126(4):299–307. https://doi.org/10.3109/00207454.2015.1015722

Clarke G, Grenham S, Scully P, Fitzgerald P, Moloney RD, Shanahan F, Dinan TG, Cryan JF (2013) The microbiome-gut-brain axis during early life regulates the hippocampal serotonergic system in a sex-dependent manner. Mol Psychiatry 18 (6):666–673. https://doi.org/10.1038/mp.2012.77

Claxton A, Baker LD, Hanson A, Trittschuh EH, Cholerton B, Morgan A, Callaghan M, Arbuckle M, Behl C, Craft S (2015) Long-acting intranasal insulin detemir improves cognition for adults with mild cognitive impairment or early-stage Alzheimer's disease dementia. J Alzheimers Dis 44(3):897–906. https://doi.org/10.3233/JAD-141791

Collins SM, Surette M, Bercik P (2012) The interplay between the intestinal microbiota and the brain. Nat Rev Microbiol 10(11):735–742. https://doi.org/10.1038/nrmicro2876

Coppin G, Nolan-Poupart S, Jones-Gotman M, Small DM (2014) Working memory and reward association learning impairments in obesity. Neuropsychologia 65:146–155. https://doi.org/10.1016/j.neuropsychologia.2014.10.004

Cournot M, Marquie JC, Ansiau D, Martinaud C, Fonds H, Ferrieres J, Ruidavets JB (2006) Relation between body mass index and cognitive function in healthy middle-aged men and women. Neurology 67 (7):1208–1214. https://doi.org/10.1212/01.wnl.0000238082.13860.50

Covasa M, Stephens RW, Toderean R, Cobuz C (2019) Intestinal sensing by gut microbiota: targeting gut peptides. Front Endocrinol (Lausanne) 10:82. https://doi.org/10.3389/fendo.2019.00082

Craft S, Claxton A, Baker LD, Hanson AJ, Cholerton B, Trittschuh EH, Dahl D, Caulder E, Neth B, Montine TJ, Jung Y, Maldjian J, Whitlow C, Friedman S (2017) Effects of regular and long-acting insulin on cognition and Alzheimer's disease biomarkers: a pilot clinical trial. J Alzheimers Dis 57(4):1325–1334. https://doi.org/10.3233/JAD-161256

Dayre A, Pouvreau C, Butkowski E, de Jong B, Jelinek H (2016) Diabesity Increases Inflammation and Oxidative Stress. Int J Pharmaceut Sci Develop Res 2:012–018. https://doi.org/10.17352/ijpsdr.000006

Diaz Heijtz R, Wang S, Anuar F, Qian Y, Bjorkholm B, Samuelsson A, Hibberd ML, Forssberg H, Pettersson S (2011) Normal gut microbiota modulates brain development and behavior. Proc Natl Acad Sci U S A 108 (7):3047–3052. https://doi.org/10.1073/pnas.1010529108

Dong S, Maniar S, Manole DM, Sun D (2018) Cerebral hypoperfusion and other shared brain pathologies in Ischemic Stroke and Alzheimer's disease. Transl Stroke Res 9(3):238–250. https://doi.org/10.1007/s12975-017-0570-2

Drummen M, Heinecke A, Dorenbos E, Vreugdenhil A, Raben A, Westerterp-Plantenga MS, Adam TC (2019) Reductions in body weight and insulin resistance are not associated with changes in grey matter volume or cortical thickness during the PREVIEW study. J Neurol Sci 403:106–111. https://doi.org/10.1016/j.jns.2019.06.017

Duarte JMN, Skoug C, Silva HB, Carvalho RA, Gruetter R, Cunha RA (2018) Impact of caffeine consumption on type 2 diabetes-induced spatial memory impairment and neurochemical alterations in the hippocampus. Front Neurosci 12:1015. https://doi.org/10.3389/fnins.2018.01015

Duarte AI, Candeias E, Alves IN, Mena D, Silva DF, Machado NJ, Campos EJ, Santos MS, Oliveira CR, Moreira PI (2020) Liraglutide protects against brain amyloid-beta1-42 accumulation in female mice with early Alzheimer's Disease-like pathology by partially rescuing oxidative/nitrosative stress and inflammation. Int J Mol Sci 21(5):1746. https://doi.org/10.3390/ijms21051746

Dye L, Boyle NB, Champ C, Lawton C (2017) The relationship between obesity and cognitive health and decline. Proc Nutr Soc 76(4):443–454. https://doi.org/10.1017/S0029665117002014

Fisher DW, Bennett DA, Dong H (2018) Sexual dimorphism in predisposition to Alzheimer's disease. Neurobiol Aging 70:308–324. https://doi.org/10.1016/j.neurobiolaging.2018.04.004

Flannery PJ, Trushina E (2019) Mitochondrial dynamics and transport in Alzheimer's disease. Mol Cell Neurosci 98:109–120. https://doi.org/10.1016/j.mcn.2019.06.009

Folch J, Olloquequi J, Ettcheto M, Busquets O, Sanchez-Lopez E, Cano A, Espinosa-Jimenez T, Garcia ML, Beas-Zarate C, Casadesus G, Bullo M, Auladell C, Camins A (2019) The involvement of peripheral and brain insulin resistance in late onset Alzheimer's dementia. Front Aging Neurosci 11:236. https://doi.org/10.3389/fnagi.2019.00236

Fontana L, Partridge L (2015) Promoting health and longevity through diet: from model organisms to humans. Cell 161(1):106–118. https://doi.org/10.1016/j.cell.2015.02.020

Fruhbeck G, Toplak H, Woodward E, Yumuk V, Maislos M, Oppert JM, Executive Committee of the European Association for the Study of O (2013) Obesity: the gateway to ill health—an EASO position statement on a rising public health, clinical and scientific challenge in Europe. Obes Facts 6(2):117–120. https://doi.org/10.1159/000350627

Fujii Y, Nguyen TTT, Fujimura Y, Kameya N, Nakamura S, Arakawa K, Morita H (2019) Fecal metabolite of a gnotobiotic mouse transplanted with gut microbiota from a patient with Alzheimer's disease. Biosci Biotechnol Biochem 83(11):2144–2152. https://doi.org/10.1080/09168451.2019.1644149

Gardner LE, White JD, Eimerbrink MJ, Boehm GW, Chumley MJ (2016) Imatinib methanesulfonate reduces hyperphosphorylation of tau following repeated peripheral exposure to lipopolysaccharide. Neuroscience 331:72–77. https://doi.org/10.1016/j.neuroscience.2016.06.007

Gray SM, Meijer RI, Barrett EJ (2014) Insulin regulates brain function, but how does it get there? Diabetes 63 (12):3992–3997. https://doi.org/10.2337/db14-0340

Grillo CA, Piroli GG, Hendry RM, Reagan LP (2009) Insulin-stimulated translocation of GLUT4 to the plasma membrane in rat hippocampus is PI3-kinase dependent. Brain Res 1296:35–45. https://doi.org/10.1016/j.brainres.2009.08.005

Grizzanti J, Corrigan R, Casadesus G (2018) Neuroprotective effects of amylin analogues on Alzheimer's disease pathogenesis and cognition. J Alzheimers Dis 66(1):11–23. https://doi.org/10.3233/JAD-180433

Gustafson D, Rothenberg E, Blennow K, Steen B, Skoog I (2003) An 18-year follow-up of overweight and risk of Alzheimer disease. Arch Intern Med 163 (13):1524–1528. https://doi.org/10.1001/archinte.163.13.1524

Heiss CN, Olofsson LE (2018) Gut microbiota-dependent modulation of energy metabolism. J Inn Immun 10 (3):163–171. https://doi.org/10.1159/000481519

Heneka MT, Fink A, Doblhammer G (2015) Effect of pioglitazone medication on the incidence of dementia. Ann Neurol 78(2):284–294. https://doi.org/10.1002/ana.24439

Holscher C (2019) Insulin signaling impairment in the brain as a risk factor in Alzheimer's disease. Front Aging Neurosci 11:88. https://doi.org/10.3389/fnagi.2019.00088

Il'yasova D, Fontana L, Bhapkar M, Pieper CF, Spasojevic I, Redman LM, Das SK, Huffman KM, Kraus WE, Investigators CS (2018) Effects of 2 years of caloric restriction on oxidative status assessed by urinary F2-isoprostanes: The CALERIE 2 randomized clinical trial. Aging Cell 17(2):e12719. https://doi.org/10.1111/acel.12719

Imfeld P, Bodmer M, Jick SS, Meier CR (2012) Metformin, other antidiabetic drugs, and risk of Alzheimer's disease: a population-based case-control study. J Am Geriatr Soc 60(5):916–921. https://doi.org/10.1111/j.1532-5415.2012.03916.x

Jain V, Langham MC, Wehrli FW (2010) MRI estimation of global brain oxygen consumption rate. J Cereb Blood Flow Metab 30(9):1598–1607. https://doi.org/10.1038/jcbfm.2010.49

Johnson AM, Olefsky JM (2013) The origins and drivers of insulin resistance. Cell 152(4):673–684. https://doi.org/10.1016/j.cell.2013.01.041

Jojo GM, Kuppusamy G, De A, Karri VVSNR (2019) Formulation and optimization of intranasal nanolipid carriers of pioglitazone for the repurposing in Alzheimer's disease using Box-Behnken design. Drug Dev Ind Pharm 45(7):1061–1072. https://doi.org/10.1080/03639045.2019.1593439

Kabadi UM (2017) Major pathophysiology in prediabetes and type 2 diabetes: decreased insulin in lean and insulin resistance in obese. J Endo Soc 1(6):742–750. https://doi.org/10.1210/js.2016-1116

Karvani M, Simos P, Stavrakaki S, Kapoukranidou D (2019) Neurocognitive impairment in type 2 diabetes mellitus. Hormones (Athens) 18(4):523–534. https://doi.org/10.1007/s42000-019-00128-2

Kelly T, Yang W, Chen CS, Reynolds K, He J (2008) Global burden of obesity in 2005 and projections to 2030. Int J Obes 32(9):1431–1437. https://doi.org/10.1038/ijo.2008.102

Kivimaki M, Luukkonen R, Batty GD, Ferrie JE, Pentti J, Nyberg ST, Shipley MJ, Alfredsson L, Fransson EI, Goldberg M, Knutsson A, Koskenvuo M, Kuosma E, Nordin M, Suominen SB, Theorell T, Vuoksimaa E, Westerholm P, Westerlund H, Zins M, Kivipelto M, Vahtera J, Kaprio J, Singh-Manoux A, Jokela M (2018) Body mass index and risk of dementia: analysis of individual-level data from 1.3 million individuals. Alzheimers Dement 14(5):601–609. https://doi.org/10.1016/j.jalz.2017.09.016

Kivipelto M, Laakso MP, Tuomilehto J, Nissinen A, Soininen H (2002) Hypertension and hypercholesterolaemia as risk factors for Alzheimer's disease: potential for pharmacological intervention. CNS Drugs 16(7):435–444

Kothari V, Luo Y, Tornabene T, O'Neill AM, Greene MW, Geetha T, Babu JR (2017) High fat diet induces brain insulin resistance and cognitive impairment in mice. Biochim Biophys Acta Mol basis Dis 1863 (2):499–508. https://doi.org/10.1016/j.bbadis.2016.10.006

Kullmann S, Heni M, Hallschmid M, Fritsche A, Preissl H, Haring HU (2016) Brain insulin resistance at the crossroads of metabolic and cognitive disorders in humans. Physiol Rev 96(4):1169–1209. https://doi.org/10.1152/physrev.00032.2015

Kumar M, Roe K, Nerurkar PV, Orillo B, Thompson KS, Verma S, Nerurkar VR (2014) Reduced immune cell infiltration and increased pro-inflammatory mediators in the brain of Type 2 diabetic mouse model infected with West Nile virus. J Neuroinflammation 11:80. https://doi.org/10.1186/1742-2094-11-80

LeBlanc ES, Rizzo JH, Pedula KL, Yaffe K, Ensrud KE, Cauley J, Cawthon PM, Cummings S, Hillier TA, Study of Osteoporotic Fractures Research Group (2017) Weight trajectory over 20 years and likelihood of mild cognitive impairment or dementia among older women. J Am Geriatr Soc 65(3):511–519 https://doi.org/10.1111/jgs.14552

Lee H, Pienaar IS (2014) Disruption of the blood-brain barrier in Parkinson's disease: curse or route to a cure? Front Biosci (Landmark Ed) 19:272–280

Lee CC, Huang CC, Hsu KS (2011) Insulin promotes dendritic spine and synapse formation by the PI3K/Akt/mTOR and Rac1 signaling pathways. Neuropharmacology 61(4):867–879. https://doi.org/10.1016/j.neuropharm.2011.06.003

Lennox R, Flatt PR, Gault VA (2014) Lixisenatide improves recognition memory and exerts neuroprotective actions in high-fat fed mice. Peptides 61:38–47. https://doi.org/10.1016/j.peptides.2014.08.014

Lin CY, Hsu YH, Lin MH, Yang TH, Chen HM, Chen YC, Hsiao HY, Chen CC, Chern Y, Chang C (2013) Neurovascular abnormalities in humans and mice with Huntington's disease. Exp Neurol 250:20–30. https://doi.org/10.1016/j.expneurol.2013.08.019

Maciejczyk M, Zebrowska E, Chabowski A (2019) Insulin resistance and oxidative stress in the brain: what's new? Int J Mol Sci 20(4):874. https://doi.org/10.3390/ijms20040874

McGregor G, Harvey J (2018) Food for thought: Leptin regulation of hippocampal function and its role in Alzheimer's disease. Neuropharmacology 136:298–306. https://doi.org/10.1016/j.neuropharm.2017.09.038

McNay EC, Recknagel AK (2011) Reprint of: 'Brain insulin signaling: A key component of cognitive processes and a potential basis for cognitive impairment in type 2 diabetes'. Neurobiol Learn Mem 96(4):517–528. https://doi.org/10.1016/j.nlm.2011.11.001

Metzler-Baddeley C, Mole JP, Sims R, Fasano F, Evans J, Jones DK, Aggleton JP, Baddeley RJ (2019) Author correction: Fornix white matter glia damage causes hippocampal gray matter damage during age-dependent limbic decline. Sci Rep 9(1):15164. https://doi.org/10.1038/s41598-019-51737-1

Moran C, Beare R, Wang W, Callisaya M (2019) Srikanth V; Alzheimer's disease neuroimaging initiative (ADNI). Type 2 diabetes mellitus, brain atrophy, and cognitive decline. Neurology 92(8):e823–e830. https://doi.org/10.1212/WNL.0000000000006955

Morley JE, Morris JC, Berg-Weger M, Borson S, Carpenter BD, Del Campo N, Dubois B, Fargo K, Fitten LJ, Flaherty JH, Ganguli M, Grossberg GT, Malmstrom TK, Petersen RD, Rodriguez C, Saykin AJ, Scheltens P, Tangalos EG, Verghese J, Wilcock G, Winblad B, Woo J, Vellas B (2015) Brain health: the importance of recognizing cognitive impairment: an IAGG consensus conference. J Am Med Dir Assoc 16(9):731–739. https://doi.org/10.1016/j.jamda.2015.06.017

Mouton PR, Chachich ME, Quigley C, Spangler E, Ingram DK (2009) Caloric restriction attenuates amyloid deposition in middle-aged dtg APP/PS1 mice. Neurosci Lett 464(3):184–187. https://doi.org/10.1016/j.neulet.2009.08.038

Nameni G, Farhangi MA, Hajiluian G, Shahabi P, Abbasi MM (2017) Insulin deficiency: a possible link between obesity and cognitive function. Int J Dev Neurosci 59:15–20. https://doi.org/10.1016/j.ijdevneu.2017.02.008

Nelson AR, Sweeney MD, Sagare AP, Zlokovic BV (2016) Neurovascular dysfunction and neurodegeneration in dementia and Alzheimer's disease. Biochim Biophys Acta 1862(5):887–900. https://doi.org/10.1016/j.bbadis.2015.12.016

Neufeld KM, Kang N, Bienenstock J, Foster JA (2011) Reduced anxiety-like behavior and central neurochemical change in germ-free mice. Neurogastroenterol Motil Off J Eur Gastroint Motil Soc 23(3):255–264. https://doi.org/10.1111/j.1365-2982.2010.01620.x

Newcombe EA, Camats-Perna J, Silva ML, Valmas N, Huat TJ, Medeiros R (2018) Inflammation: the link between comorbidities, genetics, and Alzheimer's disease. J Neuroinflammation 15(1):276. https://doi.org/10.1186/s12974-018-1313-3

Nigro E, Scudiero O, Monaco ML, Palmieri A, Mazzarella G, Costagliola C, Bianco A, Daniele A (2014) New insight into adiponectin role in obesity and obesity-related diseases. Biomed Res Int 2014:658913. https://doi.org/10.1155/2014/658913

Nota MHC, Vreeken D, Wiesmann M, Aarts EO, Hazebroek EJ, Kiliaan AJ (2020) Obesity affects brain structure and function- rescue by bariatric surgery? Neurosci Biobehav Rev 108:646–657. https://doi.org/10.1016/j.neubiorev.2019.11.025

Novak V, Milberg W, Hao Y, Munshi M, Novak P, Galica A, Manor B, Roberson P, Craft S, Abduljalil A (2014) Enhancement of vasoreactivity and cognition by intranasal insulin in type 2 diabetes. Diabetes Care 37(3):751–759. https://doi.org/10.2337/dc13-1672

Nyarko JNK, Quartey MO, Pennington PR, Heistad RM, Dea D, Poirier J, Baker GB, Mousseau DD (2018) Profiles of beta-amyloid peptides and key secretases in brain autopsy samples differ with sex and apoe epsilon4 status: impact for risk and progression of Alzheimer disease. Neuroscience 373:20–36. https://doi.org/10.1016/j.neuroscience.2018.01.005

Ono H (2019) Molecular mechanisms of hypothalamic insulin resistance. J Mol Sci 20(6):E1317. https://doi.org/10.3390/ijms20061317

Paillard T, Rolland Y, de Souto BP (2015) Protective effects of physical exercise in Alzheimer's disease and parkinson's disease: a narrative review. J Clin Neurol 11(3):212–219. https://doi.org/10.3988/jcn.2015.11.3.212

Pedditzi E, Peters R, Beckett N (2016) The risk of overweight/obesity in mid-life and late life for the development of dementia: a systematic review and meta-analysis of longitudinal studies. Age Ageing 45(1):14–21. https://doi.org/10.1093/ageing/afv151

Pegueroles J, Jimenez A, Vilaplana E, Montal V, Carmona-Iragui M, Pane A, Alcolea D, Videla L, Casajoana A, Clarimon J, Ortega E, Vidal J, Blesa R, Lleo A, Fortea J, Alzheimer's Disease Neuroimaging I (2018) Obesity and Alzheimer's disease, does the obesity paradox really exist? A magnetic resonance imaging study. Oncotarget 9(78):34691–34698. https://doi.org/10.18632/oncotarget.26162

Petersen RC (2016) Mild cognitive impairment. Continuum 22:404–418. https://doi.org/10.1212/CON.0000000000000313

Pontifex M, Vauzour D, Minihane AM (2018) The effect of APOE genotype on Alzheimer's disease risk is influenced by sex and docosahexaenoic acid status. Neurobiol Aging 69:209–220. https://doi.org/10.1016/j.neurobiolaging.2018.05.017

Psaltopoulou T, Sergentanis TN, Panagiotakos DB, Sergentanis IN, Kosti R, Scarmeas N (2013) Mediterranean diet, stroke, cognitive impairment, and depression: a meta-analysis. Ann Neurol 74(4):580–591. https://doi.org/10.1002/ana.23944

Pugazhenthi S, Qin L, Reddy PH (2017) Common neurodegenerative pathways in obesity, diabetes, and Alzheimer's disease. Biochim Biophys Acta Mol basis Dis 1863(5):1037–1045

Rad SK, Arya A, Karimian H, Madhavan P, Rizwan F, Koshy S, Prabhu G (2018) Mechanism involved in insulin resistance via accumulation of β-amyloid and neurofibrillary tangles: link between type 2 diabetes and Alzheimer's disease. Drug Des Devel Ther 12:3999–4021. https://doi.org/10.2147/DDDT.S173970

Ryu JC, Zimmer ER, Rosa-Neto P, Yoon SO (2019) Consequences of metabolic disruption in Alzheimer's disease pathology. Neurotherapeut J Am Soc Exper NeuroTherapeut. https://doi.org/10.1007/s13311-019-00755-y

Schilling TM, Ferreira de Sa DS, Westerhausen R, Strelzyk F, Larra MF, Hallschmid M, Savaskan E, Oitzl MS, Busch HP, Naumann E, Schachinger H (2014) Intranasal insulin increases regional cerebral blood flow in the insular cortex in men independently of cortisol manipulation. Hum Brain Mapp 35 (5):1944–1956. https://doi.org/10.1002/hbm.22304

Sena CM, Pereira AM, Carvalho C, Fernandes R, Seica RM, Oliveira CR, Moreira PI (2015) Type 2 diabetes aggravates Alzheimer's disease-associated vascular alterations of the aorta in mice. J Alzheimers Dis 45 (1):127–138. https://doi.org/10.3233/JAD-141008

Shao Y, Yuan G, Feng Y, Zhang J, Guo X (2014) Early liraglutide treatment is better in glucose control, beta-cell function improvement and mass preservation in db/db mice. Peptides 52:134–142. https://doi.org/10.1016/j.peptides.2013.11.011

Singh B, Parsaik AK, Mielke MM, Erwin PJ, Knopman DS, Petersen RC, Roberts RO (2014) Association of mediterranean diet with mild cognitive impairment and Alzheimer's disease: a systematic review and meta-analysis. J Alzheimers Dis 39(2):271–282. https://doi.org/10.3233/JAD-130830

Smith PJ, Blumenthal JA, Babyak MA, Craighead L, Welsh-Bohmer KA, Browndyke JN, Strauman TA, Sherwood A (2010) Effects of the dietary approaches to stop hypertension diet, exercise, and caloric restriction on neurocognition in overweight adults with high blood pressure. Hypertension 55(6):1331–1338. https://doi.org/10.1161/HYPERTENSIONAHA.109.146795

Sorop O, Olver TD, van de Wouw J, Heinonen I, van Duin RW, Duncker DJ, Merkus D (2017) The microcirculation: a key player in obesity-associated cardiovascular disease. Cardiovasc Res 113(9):1035–1045. https://doi.org/10.1093/cvr/cvx093

Spinelli M, Fusco S, Grassi C (2019) Brain insulin resistance and hippocampal plasticity: mechanisms and biomarkers of cognitive decline. Front Neurosci 13:788. https://doi.org/10.3389/fnins.2019.00788

Spitznagel MB, Hawkins M, Alosco M, Galioto R, Garcia S, Miller L, Gunstad J (2015) Neurocognitive effects of obesity and Bariatric surgery. Eur Eat Disord Rev J Eat Disord Ass 23(6):488–495. https://doi.org/10.1002/erv.2393

Sripetchwandee J, Chattipakorn N, Chattipakorn SC (2018) Links between obesity-induced brain insulin resistance, brain mitochondrial dysfunction, and dementia. Front Endocrinol (Lausanne) 9:496. https://doi.org/10.3389/fendo.2018.00496

St-Amour I, Aube B, Rieux M, Cicchetti F (2015) Targeting cerebrovascular impairments in Huntington's disease: a novel treatment perspective. Neurodegener Dis Management 5(5):389–393. https://doi.org/10.2217/nmt.15.41

Sudy R, Petak F, Schranc A, Agocs S, Blaskovics I, Lengyel C, Babik B (2019) Differences between central venous and cerebral tissue oxygen saturation in anaesthetised patients with diabetes mellitus. Sci Rep 9(1):1974. https://doi.org/10.1038/s41598-019-56221-4

Sun L, Ma L, Ma Y, Zhang F, Zhao C, Nie Y (2018) Insights into the role of gut microbiota in obesity: pathogenesis, mechanisms, and therapeutic perspectives. Protein Cell 9(5):397–403. https://doi.org/10.1007/s13238-018-0546-3

Talbot K, Wang HY, Kazi H, Han Li Y, Bakshi KP, Stucky A, Fuino RL, Kawaguchi KR, Samoyedny AJ, Wilson RS, Arvanitakis Z, Schneider JA, Wolf BA, Bennett DA, Trojanowski JQ, Arnold SE (2012) Demonstrated brain insulin resistance in Alzheimer's disease patients is associated with IGF-1 resistance, IRS-1 dysregulation, and cognitive decline. J Clin Invest 122(4):1316–1338. https://doi.org/10.1172/JCI59903

Teixeira-Lemos E, Nunes S, Teixeira F, Reis F (2011) Regular physical exercise training assists in preventing type 2 diabetes development: focus on its antioxidant and anti-inflammatory properties. Cardiovasc Diabetol 10:12. https://doi.org/10.1186/1475-2840-10-12

Toda N, Ayajiki K, Okamura T (2014) Obesity-induced cerebral hypoperfusion derived from endothelial dysfunction: one of the risk factors for Alzheimer's disease. Curr Alzheimer Res 11(8):733–744

Tolppanen AM, Ngandu T, Kareholt I, Laatikainen T, Rusanen M, Soininen H, Kivipelto M (2014) Midlife and late-life body mass index and late-life dementia: results from a prospective population-based cohort. J Alzheimers Dis 38(1):201–209. https://doi.org/10.3233/JAD-130698

Trials of Diabetes-Related Therapies: Mainly a Bust (2018). www.alzforum.org/news/conference-coverage/trials-diabetes-related-therapies-mainly-bust. Accessed 24 March 2020

Tschritter O, Preissl H, Hennige AM, Stumvoll M, Porubska K, Frost R, Marx H, Klosel B, Lutzenberger W, Birbaumer N, Haring HU, Fritsche A (2006) The cerebrocortical response to hyperinsulinemia is reduced in overweight humans: a magnetoencephalographic study. Proc Natl Acad Sci U S A 103(32):12103–12108. https://doi.org/10.1073/pnas.0604404103

Tsivgoulis G, Judd S, Letter AJ, Alexandrov AV, Howard G, Nahab F, Unverzagt FW, Moy C, Howard VJ, Kissela B, Wadley VG (2013) Adherence to a Mediterranean diet and risk of incident cognitive impairment. Neurology 80(18):1684–1692. https://doi.org/10.1212/WNL.0b013e3182904f69

Ursache A, Wedin W, Tirsi A, Convit A (2012) Preliminary evidence for obesity and elevations in fasting insulin mediating associations between cortisol awakening response and hippocampal volumes and frontal atrophy. Psychoneuroendocrinology 37 (8):1270–1276. https://doi.org/10.1016/j.psyneuen.2011.12.020

Van Dyken P, Lacoste B (2018) Impact of Metabolic Syndrome on Neuroinflammation and the Blood-Brain Barrier. Front Neurosci 12:930. https://doi.org/10.3389/fnins.2018.00930

Vreeken D, Wiesmann M, Deden LN, Arnoldussen IAC, Aarts E, Kessels RPC, Kleemann R, Hazebroek EJ, Aarts EO, Kiliaan AJ (2019) Study rationale and protocol of the BARICO study: a longitudinal, prospective, observational study to evaluate the effects of weight loss on brain function and structure after bariatric surgery. BMJ Open 9(1):e025464. https://doi.org/10.1136/bmjopen-2018-025464

Wang J, Ho L, Qin W, Rocher AB, Seror I, Humala N, Maniar K, Dolios G, Wang R, Hof PR, Pasinetti GM (2005) Caloric restriction attenuates beta-amyloid neuropathology in a mouse model of Alzheimer's disease. FASEB J 19(6):659–661. https://doi.org/10.1096/fj.04-3182fje

Wang CP, Lorenzo C, Habib SL, Jo B, Espinoza SE (2017) Differential effects of metformin on age related comorbidities in older men with type 2 diabetes. J Diabetes Complicat 31(4):679–686. https://doi.org/10.1016/j.jdiacomp.2017.01.013

Wengreen H, Munger RG, Cutler A, Quach A, Bowles A, Corcoran C, Tschanz JT, Norton MC, Welsh-Bohmer KA (2013) Prospective study of dietary approaches to stop hypertension- and Mediterranean-style dietary patterns and age-related cognitive change: the Cache County Study on Memory, Health and Aging. Am J Clin Nutr 98(5):1263–1271. https://doi.org/10.3945/ajcn.112.051276

Willette AA, Xu G, Johnson SC, Birdsill AC, Jonaitis EM, Sager MA, Hermann BP, La Rue A, Asthana S, Bendlin BB (2013) Insulin resistance, brain atrophy, and cognitive performance in late middle-aged adults. Diabetes Care 36(2):443–449. https://doi.org/10.2337/dc12-0922

Wisse LE, de Bresser J, Geerlings MI, Reijmer YD, Portegies ML, Brundel M, Kappelle LJ, van der Graaf Y, Biessels GJ, Utrecht Diabetic Encephalopathy Study G, Group S-MS (2014) Global brain atrophy but not hippocampal atrophy is related to type 2 diabetes. J Neurol Sci 344(1-2):32–36. https://doi.org/10.1016/j.jns.2014.06.008

Witte AV, Fobker M, Gellner R, Knecht S, Floel A (2009) Caloric restriction improves memory in elderly humans. Proc Natl Acad Sci U S A 106 (4):1255–1260. https://doi.org/10.1073/pnas.0808587106

World Health Organization (2020). www.who.int/dietphysicalactivity/childhood/en/. Accessed 24 March 2020

Yang D, Yang Y, Li Y, Han R (2019) Physical exercise as therapy for type 2 diabetes mellitus: from mechanism to orientation. Ann Nutr Metab 74(4):313–321. https://doi.org/10.1159/000500110

Yazici D, Sezer H (2017) Insulin resistance, obesity and lipotoxicity. Adv Exp Med Biol 960:277–304. https://doi.org/10.1007/978-3-319-48382-5_12

Yoon S, Cho H, Kim J, Lee DW, Kim GH, Hong YS, Moon S, Park S, Lee S, Lee S, Bae S, Simonson DC, Lyoo IK (2017) Brain changes in overweight/obese and normal-weight adults with type 2 diabetes mellitus. Diabetologia 60(7):1207–1217. https://doi.org/10.1007/s00125-017-4266-7

Zhang H, Hao Y, Manor B, Novak P, Milberg W, Zhang J, Fang J, Novak V (2015) Intranasal insulin enhanced resting-state functional connectivity of hippocampal regions in type 2 diabetes. Diabetes 64 (3):1025–1034. https://doi.org/10.2337/db14-1000

Zhang QQ, Li WS, Liu Z, Zhang HL, Ba YG, Zhang RX (2020) Metformin therapy and cognitive dysfunction in patients with type 2 diabetes: a meta-analysis and systematic review. Medicine 99(10):e19378. https://doi.org/10.1097/MD.0000000000019378

Zilliox LA, Chadrasekaran K, Kwan JY, Russell JW (2016) Diabetes and cognitive impairment. Curr Diab Rep 16 (9):87. https://doi.org/10.1007/s11892-016-0775-x

Metabolic Pathways Underlying Neuropsychiatric Disorders and Obesity

30

Laís Bhering Martins, Silvia Fernandes Mauricio, Adaliene Versiani Matos Ferreira, and Antônio Lúcio Teixeira

Contents

Introduction 416
Neuropsychiatric Disease and Obesity 417
Neuropsychiatric Disease and Obesity 417

Biological Mechanisms 418
Inflammation 418

Obesity-associated Insulin Resistance 419
Depression and Type 2 Diabetes 419
Alzheimer's Disease, Insulin Resistance and Type 2 Diabetes 420
Hypercortisolaemia 420
Leptin / Leptin Receptor 420
Leptin and Dementia 421
Intestinal Microbiota 421
Microbiota Fingerprints in Metabolic and Psychiatric Disorders 421
Probiotics in Psychiatric Diseases 422

References 423

L. B. Martins
Department of Psychiatry and Behavioral Sciences, Baylor College of Medicine, Houston, TX, USA

S. F. Mauricio · A. V. M. Ferreira
Department of Nutrition, Nursing School, Universidade Federal de Minas Gerais, Belo Horizonte, Brazil

A. L. Teixeira (✉)
Department of Psychiatry and Behavioral Sciences, Baylor College of Medicine, Houston, TX, USA

Neuropsychiatry Program, Department of Psychiatry and Behavioral Sciences, McGovern Medical School, The University of Texas Health Science Center at Houston UT Health, Houston, TX, USA
e-mail: antonio.l.teixeira@uth.tmc.edu

J. Faintuch and S. Faintuch (eds.), *Obesity and Diabetes*, https://doi.org/10.1007/978-3-030-53370-0_30

Abstract

The prevalence of both obesity and neuropsychiatric disorders is growing worldwide. While sociocultural changes have been implicated in this phenomenon, biological factors might also play a role. This chapter will address the intricate relationship between obesity and neuropsychiatric disorders, focusing on mood disorders, schizophrenia and major neurocognitive disorder or dementia. Obese people have a higher risk of developing depression than non-obese subjects. Conversely, depression contributes to unhealthy lifestyle (e.g. sedentarism) and unhealthy eating habits, which favour weight gain. The prevalence of obesity and overweight is elevated among patients with schizophrenia. Unhealthy behaviours and, notably, the use of antipsychotics can contribute to weight gain in this population. High BMI and central obesity in midlife may increase the risk of dementia later in life. After midlife, however, lower BMI has been associated with faster progression of dementia. Therefore, specific mechanisms can be involved in the relationship between obesity and each neuropsychiatric disorder, including inflammation, alterations in neurotransmitter biosynthesis, endocrine pathways (e.g. insulin, leptin and cortisol production and/or sensitization) and changes in gut microbiota. The treatment of obesity must consider the potential comorbidity with neuropsychiatric disorders.

Keywords

Obesity · Depression · Schizophrenia · Dementia · Inflammation · Gut microbiota

Abbreviations

BDNF	brain-derived neurotrophic factor
BH2	dihydrobiopterin
BH4	tetrahydrobiopterin
BMI	body mass index
DM2	diabetes mellitus type 2
GLP-1	glucagon-like peptide-1
GTP	guanosine-triphosphate
GTP-CH1	GTP-cyclohylase 1
HPA	hypothalamic-pituitary-adrenal
IDO	indoleamine 2,3-dioxygenase
IGF-I and IGF-II	growth factor type I and II
IRS	insulin receptor substrate
IFN	interferon
LPS	lipopolysaccharide
PAMPs	Pathogen-associated molecular patterns
PI-3K	phosphatidylinositol-3 kinase
TNF	Tumour necrosis factor alpha

Introduction

At least 30% of the population experiences a mental disorder, such as major depression and generalized anxiety disorder, at some point of their life, which can affect the quality of life and cause disability (Vigo et al. 2016; Steel et al. 2014). Five different psychiatric disorders or mental illnesses appear in the top 20 causes of the Global Burden of Disease study, 2013: major depression (2nd), anxiety disorders (7th), schizophrenia (11th), dysthymia (16th) and bipolar disorder (17th) (Vigo et al. 2016).

Technological development and increase in life expectancy have been hypothesized as contributing factors to growing prevalence of neuropsychiatric disorders in the last decades. Ageing is a major factor underlying the increased prevalence of Alzheimer's disease and other forms of dementia (Haines 2018).

According to Twenge et al. (2017), social media and electronic devices have a major impact on adolescent mental health. The relationship between time spent on electronics and mental health problems seems to be straightforward: the more time a teenager spends on electronic devices, the greater the propensity for mental health problems. On the other hand, adolescents who have high social interaction through sports and/or religious services are less likely to develop psychiatric disorders (Twenge et al. 2017). Interestingly, screen media exposure has also been associated with increased risk of developing obesity (Robinson et al. 2017). Previous studies have shown the strong relationship between time spent in front of television and prevalence of obesity (Gortmaker et al. 1996). Obesity prevalence has strongly increased in the last decades in all age groups, causing a huge health and economic burden worldwide (Engin 2017).

Neuropsychiatric Disease and Obesity

Obese people have 60% more chance to develop depression than non-obese subjects (Strine et al. 2008). Adiposity also seems to have an effect on cognition (Boeka and Lokken 2008; Fergenbaum et al. 2009; Gunstad et al. 2006). When comparing obese and non-obese adolescents, the former group had worse cognitive performance in tests evaluating attention and mental flexibility (Yau et al. 2012). Moreover, longitudinal studies have suggested an association between obesity and dementia later in life (Pedditzi et al. 2016). On the other hand, weight loss promoted by bariatric surgery or diet restriction has been associated with cognitive improvement in adults (Siervo et al. 2011).

Neuropsychiatric Disease and Obesity

Mood Disorders and Obesity

Mood disorders involve changes in emotions and motivation (Rakofsky and Rapaport 2018). Major depressive disorder and bipolar disorder are the most relevant. Major depressive disorder includes recurrent and/or persistent depressive symptoms, whereas with bipolar disorder mania (pathological euphoria and/or irritability) (bipolar I) or hypomania (bipolar II) can alternate with depressive episodes (Martins et al. 2019).

Obese people are prone to depression (Macedo et al. 2015). It further contributes to sedentarism and unhealthy eating habits, eventually leading to additional obesity (Strine et al. 2008; Luppino et al. 2010). Carbohydrate-craving or consumption of more palatable and caloric foods is not uncommon, two further detrimental habits for excessive adiposity (Ventura et al. 2014; Corsica and Spring 2008).

Depressive symptoms correlate with unhealthy food consumption, such as sweets, cookies, snacks and fast food (El Ansari et al. 2014). Conversely, in a recent cross-sectional study involving more than 13,000 participants, Jackson et al. (2019) observed that the consumption of dark chocolate decreases depressive symptoms (Jackson et al. 2019). Chocolate components, such as analogues of anandamide (cannabinoid agonist), phenylethylamine (a neuromodulator) and flavonoids (anti-inflammatory effect), may contribute to the improvement of depressive symptoms. The pleasurable experience of eating tasty food could activate hedonistic neural pathways by the release of dopamine, serotonin and endorphins (Jackson et al. 2019). Also carbohydrates, via the tryptophan/serotonin pathway, could favour a hedonic response (Ventura et al. 2014; Corsica and Spring 2008). Also the motivational system through endogenous opioids could participate (Ventura et al. 2014; Corsica and Spring 2008). As both obesity and mood interfere with each other, therapy should consider both possibilities.

Schizophrenia and Obesity

Psychotic symptoms (hallucinations and delusions), negative symptoms (avolition, flat affect, social withdrawal) and cognitive dysfunction are the hallmarks of schizophrenia (Kahn et al. 2015). Obesity and overweight are common, possibly related to sedentary lifestyle, unhealthy eating behaviours and, notably, the use of antipsychotics (Citrome and Vreeland 2008).

Antipsychotic treatment often triggers weight gain (Tarricone et al. 2010), notably second-generation drugs, like olanzapine and clozapine (Manu et al. 2015). Appetite increases and eating preferences shift, conducting to excessive caloric intake. Olanzapine may reduce glucagon-like peptide-1 (GLP-1) levels (Smith et al. 2011), which is involved in satiation signalling (Smith et al. 2011; Shin et al. 2008; Shah and Vella

2014). Orexigenic protein expression (neuropeptide Y) can increase, along with reduction in the anorexigenic protein expression (POMC) in the arcuate nucleus and upregulation of ghrelin signalling, all favouring hyperphagia (Manu et al. 2015; Weston-Green et al. 2012; Hsu et al. 2015; Dipasquale et al. 2013) (Fig. 30.1)

Excess of saturated fat and diminished fruit and dietary fibre are frequent in schizophrenia, paving the way for metabolic syndrome, diabetes mellitus type 2 (DM2), and reduced life expectancy (Dipasquale et al. 2013). Brain insulin resistance may be involved in memory impairment, suggesting both psychiatric and metabolic approaches (Wijtenburg et al. 2019).

Dementia and Obesity

Alzheimer's disease is the main cause of dementia, followed by vascular dementia and other neurodegenerative conditions. Increase in β-amyloid plaques and neurofibrillary tangles in cortical regions are typically featured (Kivipelto et al. 2018).

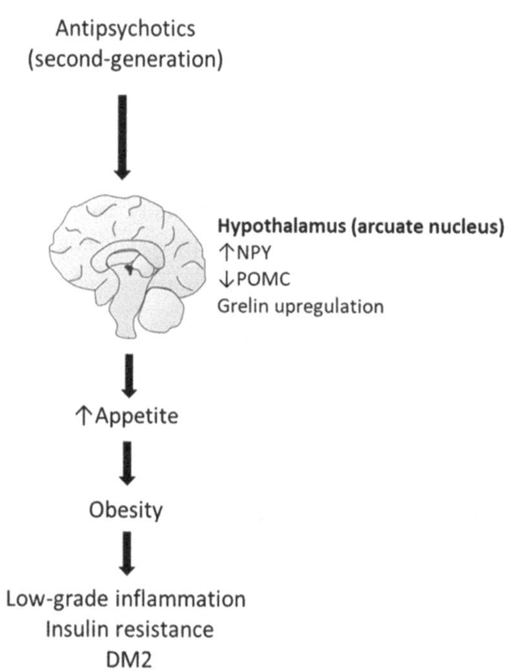

Fig. 30.1 Central effects of antipsychotics in the appetite and metabolic consequences. Second-generation antipsychotics may increase orexigenic protein (NPY, neuropeptide Y) and decrease the anorexigenic protein (POMC) expression in the arcuate nucleus. Second-generation antipsychotics also upregulate ghrelin signalling, leading to hyperphagia which contribute to obesity and metabolic complications development

Prevalence reaches 35% at age 85. By 2050, 131.5 million people could be affected (ADI 2015). This is also the sixth principal cause of death (Haines 2018).

In midlife, elevated BMI is associated with dementia later on, as observed in a large cohort of women (Gustafson et al. 2003), with about 1.4-fold risk increase (Pedditzi et al. 2016). Mechanisms could include inflammatory mediators and adipokines; vascular dysfunction and the consequences on cerebral perfusion; gut microbiota and inflammation, disturbing the brain-gut axis and insulin resistance (Anjum et al. 2018; Solas et al. 2017).

After midlife, dementia is more frequent in those with lower BMI (Kiliaan et al. 2014), consistent with a reversed trend (Gustafson 2012; Whitmer et al. 2008; Whitmer et al. 2007; Gustafson et al. 2009). In a series of 1,349,857 adults, higher BMI 20 years earlier favoured dementia, whereas 10 years earlier the risk diminished (Kivimaki et al. 2018). Frailty could be involved as an independent predictor of dementia (Rogers et al. 2017). The frail elderly suffer more falls, hospitalizations, disability, institutionalization and death (Rogers et al. 2017).

Biological Mechanisms

Inflammation

Adipocyte hypertrophy, recruitment of immune cells and derangements of cytokines and adipocytokines released by adipose tissue can lead to low-grade inflammation during obesity (Kusminski et al. 2016). Circulating cytokines can reach the central nervous system, triggering neuroinflammation, especially in the hypothalamus and hippocampus, related to appetite and memory/learning (Castanon et al. 2015; Dinel et al. 2011). Upregulation of genes responsible for microglia activation and hippocampal inflammation is impar for cognitive function in experimental models (Valcarcel-Ares et al. 2018).

Impairment of learning, memory and executive functioning is associated with obesity (Miller and Spencer 2014). High saturated fat diet has been linked to hypothalamic inflammation (Maric et al. 2014), potentially precipitating apoptosis of anorexigenic neurons in the arcuate nucleus, involved with satiety (Velloso 2009). Glial activation and inflammatory markers in hypothalamus have been noticed in experimental models (Maric et al. 2014). A diet with elevated inflammatory score could trigger depressive symptoms, anxiety and lower likelihood of well-being (Phillips et al. 2018).

Inflammation can alter neurotransmitter signalling involved in emotion and cognitive function, as cytokines are able to interfere with indoleamine 2,3-dioxygenase (IDO) and Guanosine-triphosphate-cyclohydrolase-1 (GTP-CH1), related to serotonin and dopamine production (Castanon et al. 2015). The enzyme IDO catabolizes tryptophan into kynurenine reducing serotonin production. Cancer chemotherapy with interferon-alpha (IFN-α) has been linked to depressive symptoms via the kynurenine pathway (Raison et al. 2010). IDO activity becomes elevated in adipose tissue in the course of obesity, along with decreased plasma tryptophan, in parallel with immune activation (Brandacher et al. 2007; Favennec et al. 2015; Wolowczuk et al. 2012). Besides reducing serotonin levels, the increase of IDO activity could induce depressive symptoms (Qin et al. 2018)

Immune activation enhances GTP-CH1 activity, catalysing the conversion of guanosine-triphosphate (GTP) into dihydrobiopterin (BH2). BH2 is then converted into neopterin due to tetrahydrobiopterin (BH4) production, which is part of dopamine and serotonin synthesis pathways (Fig. 30.2). In major depression, besides diminished serotonin, increase in neopterin correlates with the number of depressive episodes (Celik et al. 2010). Systemic inflammation and neurologic inflammation have been linked to mood disorders, schizophrenia and dementia (Colpo et al. 2018; Noto et al. 2015; Modabbernia et al. 2013; Kohler et al. 2017; Bauer and Teixeira 2018; Keller et al. 2017; Lara et al. 2013).

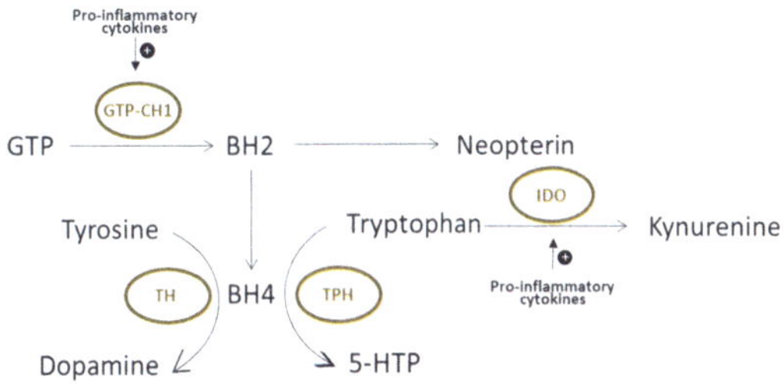

Fig. 30.2 Inflammatory mediators such as pro-inflammatory cytokines increase the enzyme GTP-cyclohydrolase-1 activity which catalyses the conversion of guanosine-triphosphate (GTP) into dihydrobiopterin (BH2) which is then converted into neopterin at the expense of tetrahydrobiopterin (BH4) (a cofactor of the tryptophan hydroxylase (TPH) and tyrosine hydroxylase (TH)). Inflammatory cytokines also activate the enzyme indoleamine 2,3-dioxygenase (IDO) with catabolizes tryptophan into kynurenine, diverting it away from 5-HTP (serotonin) production

Obesity-associated Insulin Resistance

Poor insulin signalling could be a result of an increase in pro-inflammatory cytokines release, such as tumour necrosis factor alpha (TNFα), which leads to serine kinases activation and insulin receptor substrate (IRS) 1 phosphorylation, diminishing insulin-induced activation of phosphatidylinositol-3 kinase (PI-3K) pathways, involved in the glucose uptake (Guo 2014). Deranged glucose homeostasis and specifically diabetes mellitus type 2 (DM2) are associated with major depression and Alzheimer's disease (Lyra e Silva et al. 2019). Insulin acts in hypothalamus and hippocampus, along with other areas (Lyra e Silva et al. 2019).

Depression and Type 2 Diabetes

DM2 can favour depression (20% increased risk) while depression also precedes (60% additional risk) the onset of DM2 (Nouwen et al. 2009; Mezuk et al. 2008; Moulton et al. 2015). The immune-inflammatory response, HPA axis

(Keller et al. 2017), neuroendocrine regulation and insulin resistance (Moulton et al. 2015; Grillo et al. 2019) could underlie such intertwining. Activation of innate immunity and inflammatory response is mentioned in DM2 pathophysiology, possibly contributing to bridge the gap between cognition and mood, and glucose homeostasis (Moulton et al. 2015).

Alzheimer's Disease, Insulin Resistance and Type 2 Diabetes

Previous studies that examined post-mortem brain tissue of Alzheimer's disease patients observed abnormalities in insulin and insulin-like growth factor type I and II (IGF-I and IGF-II) signalling mechanisms (Steen et al. 2005; Rivera et al. 2005). These findings led some authors to propose that Alzheimer's disease might be considered a "type 3 diabetes" (Steen et al. 2005; Rivera et al. 2005). These data also suggest that insulin resistance and DM2 observed in obese patients may contribute to the development of neurodegenerative diseases.

Hypercortisolaemia

Increase in inflammatory cytokine release can also activate the hypophysis pituitary adrenal (HPA) axis (Moulton et al. 2015). HPA axis is similarly activated in both obesity and neuropsychiatric disorders. BMI and waist-to-hip ratio have been correlated with the concentration of hair cortisol, the main stress hormone produced by the adrenal glands (Stalder et al. 2017). Hypercortisolaemia is also observed in early stage of Alzheimer's disease, suggesting activation of the HPA axis (Notarianni 2017). Cortisol plays an important role in glycaemic control. Cortisol suppresses glucose uptake by muscle cells and adipocytes through the inhibition of glucose transporter 4 (GLUT 4) translocation to the cell surface, causing insulin resistance (Kamba et al. 2016). Hypercortisolaemia is implicated in impaired glycose tolerance and pre-diabetes. Therefore, it has been suggested that hypercortisolaemia in Alzheimer's disease has a diabetogenic effect (Notarianni 2017).

Leptin / Leptin Receptor

Leptin is an adipocytokine (cytokine produced by adipocytes) whose primary role is to provide information on energy stores (adipose tissue) to the central nervous system, reducing appetite and controlling energy expenditure (Enriori et al. 2006). Leptin production occurs proportionally to the adipose tissue size; therefore, obese people usually have high levels of serum leptin. But obese people are "leptin resistant" since leptin enters in the brain through a saturable transport mechanism. In other words, hyperleptinaemia in obese people does not promote the expected central effects of reducing appetite (Farr et al. 2015). Actually, previous studies have shown that cerebrospinal fluid level of leptin is low in obese people, despite the high serum levels (Farr et al. 2015; Caro et al. 1996; Schwartz et al. 1996).

Leptin receptor is expressed in various regions of the brain, including cortex, hypothalamus, olfactory bulb, the dorsal raphe nucleus, hippocampus and the nucleus of the solitary tract (Farr et al. 2015; Zou et al. 2019). Leptin plays a role in brain development, organization and maturation of the nervous system (Farr et al. 2015). Leptin replacement therapy given to three genetically leptin-deficient adults resulted in increase of grey matter volume in cerebellum, inferior parietal and anterior cingulate cortices (Matochik et al. 2005).

Leptin also influences the release of brain-derived neurotrophic factor (BDNF), a factor that acts in mood regulation and memory/learning (Martins et al. 2017). BDNF acts in synaptic plasticity and neuronal survival in the hippocampus and other brain regions (Farr et al. 2015). Changes in the cerebral expression of BNDF, i.e. decreased levels and mutations in the gene, have been associated with obesity in human and animal models (Sandrini et al. 2018). Experimental studies have shown that nutritional strategies that prevent weight gain have been associated with "normalizing" BDNF levels in hippocampus and improving cognitive changes (Moy and McNay 2013).

Alterations in serum levels of leptin in neuropsychiatric disorders, such as mood disorders and schizophrenia, have been observed (Atmaca et al. 2002; Jow et al. 2006; Kraus et al. 2001). In an experimental model, leptin receptor ablation in hippocampus caused depressive behaviour (Guo et al. 2013). In humans, an inverse correlation between leptin and depressive symptoms has been reported regardless of body fat and BMI (Jow et al. 2006; Miller et al. 2003; Lawson et al. 2012). Recently, Xu et al. (2018) observed that patients with schizophrenia have higher plasma leptin levels than healthy controls, and there was a negative correlation between leptin levels and depressive symptoms (Xu et al. 2018).

In this specific context, it is important to acknowledge that antipsychotics used to treat schizophrenia increase leptin levels and contribute to weight gain (Ragguett et al. 2017). In a longitudinal study, Venkatasubramanian et al. (2010) observed that antipsychotic-naïve schizophrenia patients had significantly lower leptin levels than controls. After treatment, leptin levels and BMI increased significantly (Venkatasubramanian et al. 2010). Opposite findings have been reported in female antipsychotic-naïve schizophrenic patients (Wang et al. 2007). Therefore, the role of leptin in schizophrenia remains to be clearly defined.

Leptin and Dementia

Besides playing an important role in the normal development of the hypothalamus, leptin participates in learning and memory processes through its action on hippocampal neurons (Anjum et al. 2018; Kiliaan et al. 2014; Carro 2009). Experimental studies have shown reduction in β-amyloid burden after peripheral leptin administration (Kiliaan et al. 2014; Carro 2009). In humans, a negative correlation was observed between the risk for Alzheimer's disease and blood leptin levels in older adults (Fewlass et al. 2004; Lieb et al. 2009). These latter results suggest the participation of leptin in prevention of dementia.

Intestinal Microbiota

Inflammatory mechanisms and HPA axis seem to link gut microbiota and brain function (Jiang et al. 2015). Intestinal microbiota responds to diet, medications, diseases and other environmental influences (Baothman et al. 2016). Dysbiosis, characterized by an imbalance in abundance and composition of microbial populations, may activate immune cells in the intestinal mucosa with repercussions on the gut-brain axis, comprising brain, gut, enteric nervous system and vagus nerve (Jiang et al. 2015; Daulatzai 2014; Round and Mazmanian 2009; Ley et al. 2005; Ley et al. 2006; Rodrigues-Amorim et al. 2018; Jiang et al. 2017).

Microbiota Fingerprints in Metabolic and Psychiatric Disorders

Aberrant gut permeability to macromolecules is an often addressed phenomenon (Fasano 2017). Abnormalities of gut handling of microbial-associated molecular patterns (MAMPs), such as lipopolysaccharide (LPS), a peptidoglycan of gramnegative bacteria cells, can cause activation of immune cells and release of inflammatory markers (Zhang et al. 2009; Rieder et al. 2017). Obese people have higher LPS plasma levels than non-obese people, and after bariatric surgery, LPS levels significantly reduce (Trøseid et al. 2013).

Besides the fact that obesity *per se* contributes to development of proinflammatory state, alterations in gut microbiota in obese individuals may increase LPS and immune cells activation, exacerbating the release of inflammatory mediators. As previously discussed, these mediators can lead to neuroinflammation and consequently to changes in neurotransmitter production and HPA axis functioning (Daulatzai 2014; Rea et al. 2016). The role of microbiota in behaviour has been investigated in germ-free (GF) animals, which have shown enhanced depressive-like behaviours. Interestingly enough, when GF mice were colonized with microbiota from depressive patients, depressive-like behaviours increased (Zheng et al. 2016), along with excessive adiposity (Walker and Parkhill 2013).

Probiotics in Psychiatric Diseases

In a clinical trial conducted with healthy subjects, there was reduction in depressive symptoms following the use of *Lactobacillus helveticus* R0052 and *Bifidobacterium longum* R0175 (3×10^9 colony-forming units/stick) for 30 days compared to placebo (Messaoudi et al. 2011). Findings from other clinical trials support the benefits of probiotic use on the improvement of depressive symptoms (Kuo and Chung 2019) (Huang et al. 2016). In addition to depression, several studies have linked the use of probiotics with cognitive improvement in Alzheimer's disease (Akbari et al. 2016) and schizophrenia (Nemani et al. 2015; Tomasik et al. 2015; Dickerson et al. 2014). Figure 30.3 summarizes the putative mechanisms involved in the

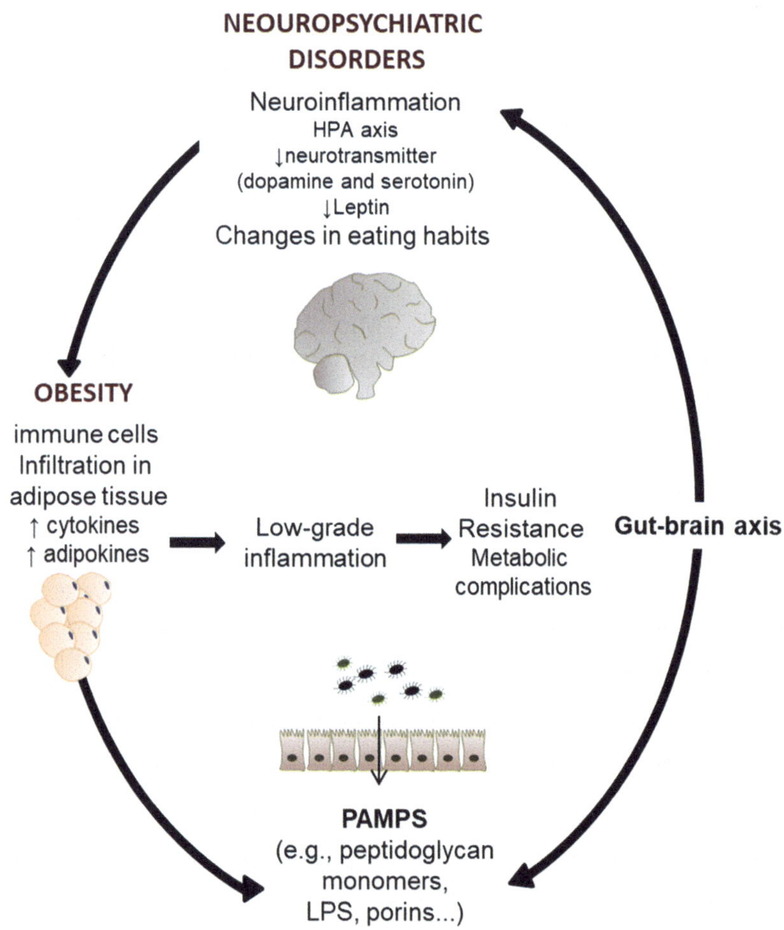

Fig. 30.3 Putative mechanisms involved in the relationship between obesity and neuropsychiatric disorders. Adipose tissue hypertrophy is associated with immune cells infiltration and chronic low-grade inflammation characterized by increased cytokines and adipokines secretion. Inflammatory cytokines impair insulin sensitivity while the excessive production of leptin (an adipokine involving in satiety signalling) causes a central leptin resistance. Obesity is also associated with changes in gut microbiota composition, increased intestinal permeability and activation of immune response through microbial- or pathogen-associated molecular patterns (PAMPs), such as lipopolysaccharide (LPS). The inflammatory molecules can reach the circulation and the central nervous system, where it may cause neuroinflammation and alterations in hypothalamus-pituitary-adrenal (HPA) axis and neurotransmitter production, including serotonin and dopamine, which influences cognition and behaviour

relationship between obesity and neuropsychiatric disorders.

References

ADI (2015) World Alzheimer Report 2015: the global impact of dementia

Akbari E et al (2016) Effect of probiotic supplementation on cognitive function and metabolic status in Alzheimer's disease: a randomized, double-blind and controlled trial. Front Aging Neurosci 8:256

Anjum I et al (2018) Does obesity increase the risk of dementia: a literature review. Cureus 10(5):1–6

Atmaca M et al (2002) Serum leptin and cholesterol levels in patients with bipolar disorder. Neuropsychobiology 46(4):176–179

Baothman OA et al (2016) The role of Gut Microbiota in the development of obesity and Diabetes. Lipids Health Dis 15:108

Bauer ME, Teixeira AL (2018) Inflammation in psychiatric disorders: what comes first? Ann N Y Acad Sci 1437:57–67

Boeka AG, Lokken KL (2008) Neuropsychological performance of a clinical sample of extremely obese individuals. Arch Clin Neuropsychol 23(4):467–474

Brandacher G et al (2007) Chronic immune activation underlies morbid obesity: is IDO a key player? Curr Drug Metab 8(3):289–295

Caro JF et al (1996) Decreased cerebrospinal-fluid/serum leptin ratio in obesity: a possible mechanism for leptin resistance. Lancet 348(9021):159–161

Carro EM (2009) Therapeutic approaches of leptin in Alzheimer's disease. Recent Pat CNS Drug Discov 4 (3):200–208

Castanon N, Luheshi G, Laye S (2015) Role of neuroinflammation in the emotional and cognitive alterations displayed by animal models of obesity. Front Neurosci 9:229

Celik C et al (2010) The association between serum levels of neopterin and number of depressive episodes of major depression. Prog Neuro-Psychopharmacol Biol Psychiatry 34(2):372–375

Citrome L, Vreeland B (2008) Schizophrenia, obesity, and antipsychotic medications: what can we do? Postgrad Med 120(2):18–33

Colpo GD et al (2018) Immune-based strategies for mood disorders: facts and challenges. Expert Rev Neurother 18(2):139–152

Corsica JA, Spring BJ (2008) Carbohydrate craving: a double-blind, placebo-controlled test of the self-medication hypothesis. Eat Behav 9(4):447–454

Daulatzai J (2014) Obesity and gut's dysbiosis promote neuroinflammation, cognitive impairment, and vulnerability to Alzheimer's disease: new directions and therapeutic implications. J Mol Genet Med S1:5

Dickerson FB et al (2014) Effect of probiotic supplementation on schizophrenia symptoms and association with gastrointestinal functioning: a randomized, placebo-controlled trial. Prim Care Companion CNS Disord 16(1):13m01579

Dinel AL et al (2011) Cognitive and emotional alterations are related to hippocampal inflammation in a mouse model of metabolic syndrome. PLoS One 6(9):e24325

Dipasquale S et al (2013) The dietary pattern of patients with schizophrenia: a systematic review. J Psychiatr Res 47(2):197–207

El Ansari W, Adetunji H, Oskrochi R (2014) Food and mental health: relationship between food and perceived stress and depressive symptoms among university students in the United Kingdom. Cent Eur J Public Health 22(2):90–97

Engin A (2017) The definition and prevalence of obesity and metabolic syndrome. Adv Exp Med Biol 960:1–17

Enriori PJ et al (2006) Leptin resistance and obesity. Obesity (Silver Spring) 14(Suppl 5):254S–258S

Farr OM, Tsoukas MA, Mantzoros CS (2015) Leptin and the brain: influences on brain development, cognitive functioning and psychiatric disorders. Metabolism 64 (1):114–130

Fasano A (2017) Gut permeability, obesity, and metabolic disorders: who is the chicken and who is the egg? Am J Clin Nutr 105(1):3–4

Favennec M et al (2015) The kynurenine pathway is activated in human obesity and shifted toward kynurenine monooxygenase activation. Obesity (Silver Spring) 23(10):2066–2074

Fergenbaum JH et al (2009) Obesity and lowered cognitive performance in a Canadian First Nations population. Obesity (Silver Spring) 17(10):1957–1963

Fewlass DC et al (2004) Obesity-related leptin regulates Alzheimer's Abeta. FASEB J 18(15):1870–1878

Gortmaker SL et al (1996) Television viewing as a cause of increasing obesity among children in the United States, 1986–1990. Arch Pediatr Adolesc Med 150 (4):356–362

Grillo CA et al (2019) Insulin resistance and hippocampal dysfunction: Disentangling peripheral and brain causes from consequences. Exp Neurol 318:71–77

Gunstad J et al (2006) Obesity is associated with memory deficits in young and middle-aged adults. Eat Weight Disord 11(1):e15–e19

Guo S (2014) Insulin signaling, resistance, and the metabolic syndrome: insights from mouse models into disease mechanisms. J Endocrinol 220(2):T1–T23

Guo M et al (2013) Selective deletion of leptin receptors in adult hippocampus induces depression-related behaviours. Int J Neuropsychopharmacol 16(4): 857–867

Gustafson DR (2012) Adiposity and cognitive decline: underlying mechanisms. J Alzheimers Dis 30:S97–S112

Gustafson D et al (2003) An 18-year follow-up of overweight and risk of Alzheimer disease. Arch Intern Med 163(13):1524–1528

Gustafson DR et al (2009) Adiposity indicators and dementia over 32 years in Sweden. Neurology 73 (19):1559–1566

Haines JL (2018) Alzheimer disease: perspectives from epidemiology and genetics. J Law Med Ethics 46(3): 694–698

Hsu TM et al (2015) Hippocampus ghrelin signaling mediates appetite through lateral hypothalamic orexin pathways. eLife 4:e11190

Huang R, Wang K, Hu J (2016) Effect of probiotics on depression: a systematic review and meta-analysis of randomized controlled trials. Nutrients 8(8):483

Jackson SE et al (2019) Is there a relationship between chocolate consumption and symptoms of depression? A cross-sectional survey of 13,626 US adults. Depress Anxiety 36(10):987–995

Jiang H et al (2015) Altered fecal microbiota composition in patients with major depressive disorder. Brain Behav Immun 48:186–194

Jiang C et al (2017) The gut microbiota and Alzheimer's disease. J Alzheimers Dis 58(1):1–15

Jow GM, Yang TT, Chen CL (2006) Leptin and cholesterol levels are low in major depressive disorder, but high in schizophrenia. J Affect Disord 90(1):21–27

Kahn RS et al (2015) Schizophrenia. Nat Rev Dis Primers 1:15067

Kamba A et al (2016) Association between higher serum cortisol levels and decreased insulin secretion in a general population. PLoS One 11(11):e0166077

Keller J et al (2017) HPA axis in major depression: cortisol, clinical symptomatology and genetic variation predict cognition. Mol Psychiatry 22(4):527–536

Kiliaan AJ, Arnoldussen IA, Gustafson DR (2014) Adipokines: a link between obesity and dementia? Lancet Neurol 13(9):913–923

Kivimaki M et al (2018) Body mass index and risk of dementia: Analysis of individual-level data from 1.3 million individuals. Alzheimers Dement 14(5): 601–609

Kivipelto M, Mangialasche F, Ngandu T (2018) Lifestyle interventions to prevent cognitive impairment, dementia and Alzheimer disease. Nat Rev Neurol 14:653–666

Kohler CA et al (2017) Peripheral cytokine and chemokine alterations in depression: a meta-analysis of 82 studies. Acta Psychiatr Scand 135(5):373–387

Kraus T et al (2001) Low leptin levels but normal body mass indices in patients with depression or schizophrenia. Neuroendocrinology 73(4):243–247

Kuo PH, Chung YE (2019) Moody microbiome: Challenges and chances. J Formos Med Assoc 118 (Suppl 1):S42–S54

Kusminski CM, Bickel PE, Scherer PE (2016) Targeting adipose tissue in the treatment of obesity-associated diabetes. Nat Rev Drug Discov 15(9):639–660

Lara VP et al (2013) High cortisol levels are associated with cognitive impairment no-dementia (CIND) and dementia. Clin Chim Acta 423:18–22

Lawson EA et al (2012) Leptin levels are associated with decreased depressive symptoms in women across the weight spectrum, independent of body fat. Clin Endocrinol 76(4):520–525

Ley RE et al (2005) Obesity alters gut microbial ecology. Proc Natl Acad Sci U S A 102(31):11070–11075

Ley RE et al (2006) Microbial ecology: human gut microbes associated with obesity. Nature 444 (7122):1022–1023

Lieb W et al (2009) Association of plasma leptin levels with incident Alzheimer disease and MRI measures of brain aging. JAMA 302(23):2565–2572

Luppino FS et al (2010) Overweight, obesity, and depression: a systematic review and meta-analysis of longitudinal studies. Arch Gen Psychiatry 67(3):220–229

Lyra e Silva NM et al (2019) Insulin resistance as a shared pathogenic mechanism between depression and type 2 diabetes. Front Psychiatry 10:57

Macedo TTS et al (2015) Obese people's perception of their own bodies. Esc Anna Nery 19(3):505–510

Manu P et al (2015) Weight gain and obesity in schizophrenia: epidemiology, pathobiology, and management. Acta Psychiatr Scand 132(2):97–108

Maric T, Woodside B, Luheshi GN (2014) The effects of dietary saturated fat on basal hypothalamic neuroinflammation in rats. Brain Behav Immun 36: 35–45

Martins LB, Teixeira AL, Domingues RB (2017) Neurotrophins and migraine. Vitam Horm 104: 459–473

Martins LB et al (2019) Pathways linking obesity to neuropsychiatric disorders. Nutrition 66:16–21

Matochik JA et al (2005) Effect of leptin replacement on brain structure in genetically leptin-deficient adults. J Clin Endocrinol Metab 90(5):2851–2854

Messaoudi M et al (2011) Assessment of psychotropic-like properties of a probiotic formulation (Lactobacillus helveticus R0052 and Bifidobacterium longum R0175) in rats and human subjects. Br J Nutr 105 (5):755–764

Mezuk B et al (2008) Depression and type 2 diabetes over the lifespan: a meta-analysis. Diabetes Care 31(12): 2383–2390

Miller AA, Spencer SJ (2014) Obesity and neuroinflammation: a pathway to cognitive impairment. Brain Behav Immun 42:10–21

Miller GE et al (2003) Pathways linking depression, adiposity, and inflammatory markers in healthy young adults. Brain Behav Immun 17(4):276–285

Modabbernia A et al (2013) Cytokine alterations in bipolar disorder: a meta-analysis of 30 studies. Biol Psychiatry 74(1):15–25

Moulton C, Pickup J, Ismail K (2015) The link between depression and diabetes: the search for shared mechanisms. Lancet Diabetes Endocrinol 3:461–471

Moy GA, McNay EC (2013) Caffeine prevents weight gain and cognitive impairment caused by a high-fat diet while elevating hippocampal BDNF. Physiol Behav 109:69–74

Nemani K et al (2015) Schizophrenia and the gut-brain axis. Prog Neuro-Psychopharmacol Biol Psychiatry 56:155–160

Notarianni E (2017) Cortisol: Mediator of association between Alzheimer's disease and diabetes mellitus? Psychoneuroendocrinology 81:129–137
Noto C et al (2015) High predictive value of immune-inflammatory biomarkers for schizophrenia diagnosis and association with treatment resistance. World J Biol Psychiatry 16(6):422–429
Nouwen A, Lloyd CE, Pouwer F (2009) Depression and type 2 diabetes over the lifespan: a meta-analysis. Response to Mezuk et al. Diabetes Care 32(5):e56
Pedditzi E, Peters R, Beckett N (2016) The risk of overweight/obesity in mid-life and late life for the development of dementia: a systematic review and meta-analysis of longitudinal studies. Age Ageing 45(1):14–21
Phillips CM et al (2018) Dietary inflammatory index and mental health: a cross-sectional analysis of the relationship with depressive symptoms, anxiety and well-being in adults. Clin Nutr 37(5):1485–1491
Qin Y et al (2018) IDO and TDO as a potential therapeutic target in different types of depression. Metab Brain Dis 33(6):1787–1800
Ragguett RM et al (2017) Association between antipsychotic treatment and leptin levels across multiple psychiatric populations: an updated meta-analysis. Hum Psychopharmacol 32(6)
Raison CL et al (2010) CSF concentrations of brain tryptophan and kynurenines during immune stimulation with IFN-alpha: relationship to CNS immune responses and depression. Mol Psychiatry 15(4):393–403
Rakofsky J, Rapaport M (2018) Mood disorders. Behav Neurol Psych 24:804–827
Rea K, Dinan TG, Cryan JF (2016) The microbiome: A key regulator of stress and neuroinflammation. Neurobiol Stress 4:23–33
Rieder R et al (2017) Microbes and mental health: a review. Brain Behav Immun 66:9–17
Rivera EJ et al (2005) Insulin and insulin-like growth factor expression and function deteriorate with progression of Alzheimer's disease: link to brain reductions in acetylcholine. J Alzheimers Dis 8(3):247–268
Robinson TN et al (2017) Screen media exposure and obesity in children and adolescents. Pediatrics 140 (Suppl 2):S97–S101
Rodrigues-Amorim D et al (2018) The role of the gut microbiota in schizophrenia: current and future perspectives. World J Biol Psychiatry 19:1–15
Rogers NT, Steptoe A, Cadar D (2017) Frailty is an independent predictor of incident dementia: evidence from the English Longitudinal Study of Ageing. Sci Rep 7(1):15746
Round JL, Mazmanian SK (2009) The gut microbiota shapes intestinal immune responses during health and disease. Nat Rev Immunol 9(5):313–323
Sandrini L et al (2018) Association between obesity and circulating brain-derived neurotrophic factor (bdnf) levels: systematic review of literature and meta-analysis. Int J Mol Sci 19(8):2281
Schwartz MW et al (1996) Cerebrospinal fluid leptin levels: relationship to plasma levels and to adiposity in humans. Nat Med 2(5):589–593
Shah M, Vella A (2014) Effects of GLP-1 on appetite and weight. Rev Endocr Metab Disord 15(3):181–187
Shin YK et al (2008) Modulation of taste sensitivity by GLP-1 signaling. J Neurochem 106(1):455–463
Siervo M et al (2011) Intentional weight loss in overweight and obese individuals and cognitive function: a systematic review and meta-analysis. Obes Rev 12(11): 968–983
Smith GC, Vickers MH, Shepherd PR (2011) Olanzapine effects on body composition, food preference, glucose metabolism and insulin sensitivity in the rat. Arch Physiol Biochem 117(4):241–249
Solas M et al (2017) Inflammation and gut-brain axis link obesity to cognitive dysfunction: plausible pharmacological interventions. Curr Opin Pharmacol 37:87–92
Stalder T et al (2017) Stress-related and basic determinants of hair cortisol in humans: A meta-analysis. Psychoneuroendocrinology 77:261–274
Steel Z et al (2014) The global prevalence of common mental disorders: a systematic review and meta-analysis 1980-2013. Int J Epidemiol 43(2):476–493
Steen E et al (2005) Impaired insulin and insulin-like growth factor expression and signaling mechanisms in Alzheimer's disease--is this type 3 diabetes? J Alzheimers Dis 7(1):63–80
Strine TW et al (2008) Depression and anxiety in the United States: findings from the 2006 Behavioral Risk Factor Surveillance System. Psychiatr Serv 59 (12):1383–1390
Tarricone I et al (2010) Weight gain in antipsychotic-naive patients: a review and meta-analysis. Psychol Med 40 (2):187–200
Tomasik J et al (2015) Immunomodulatory effects of probiotic supplementation in Schizophrenia patients: a randomized, placebo-controlled trial. Biomark Insights 10:47–54
Trøseid M et al (2013) Plasma lipopolysaccharide is closely associated with glycemic control and abdominal obesity: evidence from bariatric surgery. Diabetes Care 36(11):3627–3632
Twenge JM et al (2017) Increases in depressive symptoms, suicide-related outcomes, and suicide rates among U.S. adolescents after 2010 and links to increased new media screen time. Clin Psychol Sci 6(1):3–17
Valcarcel-Ares MN et al (2018) Obesity in aging exacerbates neuroinflammation, dysregulating synaptic function-related genes and altering eicosanoid synthesis in the mouse hippocampus: potential role in impaired synaptic plasticity and cognitive decline. J Gerontol A Biol Sci Med Sci 74:290–298
Velloso LA (2009) The brain is the conductor: diet-induced inflammation overlapping physiological control of body mass and metabolism. Arq Bras Endocrinol Metabol 53(2):151–158
Venkatasubramanian G et al (2010) A longitudinal study on the impact of antipsychotic treatment on serum leptin in schizophrenia. Clin Neuropharmacol 33(6): 288–292
Ventura T et al (2014) Neurobiologic basis of craving for carbohydrates. Nutrition 30(3):252–256

Vigo D, Thornicroft G, Atun R (2016) Estimating the true global burden of mental illness. Lancet Psychiatry 3 (2):171–178

Walker AW, Parkhill J (2013) Microbiology. Fighting obesity with bacteria. Science 341(6150):1069–1070

Wang HC et al (2007) Increased plasma leptin in antipsychotic-naive females with schizophrenia, but not in males. Neuropsychobiology 56(4):213–215

Weston-Green K, Huang XF, Deng C (2012) Alterations to melanocortinergic, GABAergic and cannabinoid neurotransmission associated with olanzapine-induced weight gain. PLoS One 7(3):e33548

Whitmer RA et al (2007) Body mass index in midlife and risk of Alzheimer disease and vascular dementia. Curr Alzheimer Res 4(2):103–109

Whitmer RA et al (2008) Central obesity and increased risk of dementia more than three decades later. Neurology 71(14):1057–1064

Wijtenburg SA et al (2019) Brain insulin resistance and altered brain glucose are related to memory impairments in schizophrenia. Schizophr Res 208:324–330

Wolowczuk I et al (2012) Tryptophan metabolism activation by indoleamine 2,3-dioxygenase in adipose tissue of obese women: an attempt to maintain immune homeostasis and vascular tone. Am J Phys Regul Integr Comp Phys 303(2):R135–R143

Xu J et al (2018) Increased plasma leptin as a novel predictor for psychopathological depressive symptoms in chronic schizophrenia. Gen Psychiatr 31(3):e100018

Yau PL et al (2012) Obesity and metabolic syndrome and functional and structural brain impairments in adolescence. Pediatrics 130(4):e856–e864

Zhang H et al (2009) Human gut microbiota in obesity and after gastric bypass. Proc Natl Acad Sci U S A 106(7): 2365–2370

Zheng P et al (2016) Gut microbiome remodeling induces depressive-like behaviors through a pathway mediated by the host's metabolism. Mol Psychiatry 21(6): 786–796

Zou X et al (2019) Role of leptin in mood disorder and neurodegenerative disease. Front Neurosci 13:378

Heart Failure and the Obesity Paradox 31

Marijana Tadic and Cesare Cuspidi

Contents

Current Status of Chronic Heart Failure 428

Heart Failure and Obesity 428

Prognosis in Decompensated Patients 428

Outcomes in Obese Patients 429

Possible Involved Pathways and Biomolecules 429
Anthropometric Measurements 430
Other Heart Chambers 430
Biohumoral and Neurohormonal Changes 431

Obesity Paradox 431
Cardiovascular Mortality 431

Risks of Bias in the Interpretation of the Paradox 432

Recommended Lines for Further Investigation 432

References 433

Abstract

Heart failure (HF) represents a complex cardiovascular disease with a wide range of risk factors, different pathophysiology, and various manifestations. Obesity is one of the most important comorbidities and one of the risk factors in HF patients. It seems that obesity has a more important role in development of HF patients with preserved left ventricular ejection fraction (HFpEF) than in HF patients with reduced ejection fraction (HFrEF). The role of obesity on HF with mid-range ejection fraction (HFmrEF) is a matter of debate. The impact of obesity on outcome in HF patients is still unclear. The relationship varies from the linear to the U-shaped. The "obesity paradox," which suggests the reduced risk in mildly overweight subjects in comparison with normal- and underweight individuals, drew significant attention. The mechanisms that relate obesity and HF vary from obesity-induced hemodynamic changes to biohumoral systems

M. Tadic (✉)
University Hospital "Dr. Dragisa Misovic - Dedinje"
Department of Cardiology, Belgrade, Serbia
e-mail: marijana_tadic@hotmail.com

C. Cuspidi
Department of Cardiology, Belgrade, Serbia

J. Faintuch and S. Faintuch (eds.), *Obesity and Diabetes*, https://doi.org/10.1007/978-3-030-53370-0_31

such as adipocytokines, renin-angiotensin-aldosterone and sympathetic nervous systems, natriuretic peptide, and oxidative stress. In the absence of a satisfactory pharmacological approach, which would improve the outcome of this large group of patients, alternative methods such as weight loss and physical activity seem to provide encouraging results.

Keywords

Obesity · Heart failure · Pathophysiology · Obesity paradox

Current Status of Chronic Heart Failure

Heart failure (HF) represents a complex cardiovascular disease. HF definitions and diagnostic criteria have largely changed over the years, and therefore its incidence and prevalence significantly varied in the last few decades. For a long time, HF was defined primarily as the reduction of left ventricular (LV) systolic function defined by ejection fraction (HFrEF), and typical HF symptoms (lung congestion, fluid retention, dizziness, weakness, fatigue) (Tanai and Frantz 2015). Considering the fact that these symptoms are unspecific, a new biochemical marker was recently introduced—brain natriuretic peptide (BNP) and its precursor pro-BNP. This biomarker is currently used in emergency rooms for differential diagnosis of patients with shortness of breath (Senthong et al. 2017).

About two decades ago, a new entity called HF with preserved LV ejection fraction (LVEF), or HFpEF was introduced, with conserved LV pump function and symptoms comparable to HFrEF (Tanai and Frantz 2015). HF with midrange LV ejection fraction (HFmrEF) was subsequently included, with LVEF of 40–50% with symptoms and increased pro-BNP (Ponikowski et al. 2016). The authors are still in doubt if HFmrEF represents a new entity or only an overlap group between HFpEF and HFrEF; however, follow-up studies indicate differences in prevalence and survival.

The American Heart Association found among 110,621 HF patients, 50% with HFrEF, 14% with HFmEF, and 36% with HFpEF (Benjamin et al. 2018). Others observed 25.6% with HFpEF, 21% with HFmrEF and 53.5% with HFrEF (Webb et al. 2018). The OPTIMIZE-HF trial involved 54% with HFrEF, 20% with HFmrEF, and 27% with HFpEF (Fonarow et al. 2007). In contrast, the GWTG-HF protocol announced 47% displaying HFpEF, 39% HFrEF, and 14% HFmrEF (Cheng et al. 2014).

Heart Failure and Obesity

Overweight (1.9 billion) and obesity (600 million) are growing in the world (World Health Organization 2013). Obesity represents one the most important and modifiable causes of cardiovascular mortality (World Health Organization 2013; Kenchaiah et al. 2002; Baena-Díez et al. 2010). Obesity is related with decreased cardiorespiratory fitness, which is an important characteristic of HF. Nevertheless, the prognosis of HF is less ominous in this population, a phenomenon is known as an obesity paradox (Carbone et al. 2017).

Obesity underpins an increment of fat mass; however, obese patients can also have increased amount of lean mass. The latter could potentially explain the obesity paradox, as it is related with improved cardiorespiratory fitness, an important prognostic parameter. However, fat mass could provoke high systemic inflammation that could negate the protective effects.

Prognosis in Decompensated Patients

Mortality rates seem to be higher in HFrEF patients, however, similar among HFmrEF and HFpEF patients (Fonarow et al. 2007; Cheng et al. 2014; Sweitzer et al. 2008; Bhatia et al. 2006). The OPTIMIZE-HF study reported 3.9% mortality in HFrEF, 3.0% in HFmrEF, and 2.9% in HFpEF (Fonarow et al. 2007). In the ADHERE trial, rates were 4.7% with LVEF <25%, 3.4% with LVEF 25–40%, 3.2% when LVEF was

41–54%, and 3.0% with LVEF >55% (Sweitzer et al. 2008). During the GWTG-HF investigation, 30-day and 1-year mortality rates in HFrEF of 9.5% and 37.5% were encountered. In cases with HFmrEF they were 8.2% and 35.1% and in HFpEF, 8.5% and 35.6%, respectively (Cheng et al. 2014). In the Canadian experience, mortality at the same intervals in HFrEF was 7.1% and 25.5%, whereas in HFmrEF and in HFpEF they were 5.1% and 21.3%, and 5.3% and 22.2%, respectively (Bhatia et al. 2006). In both previous studies, there was no difference between HFmrEF and HFpEF (Cheng et al. 2014; Bhatia et al. 2006).

According to a large meta-analysis, adjusted mortality risks deteriorate with every 5–10% reduction in LVEF below 40%, however not with LVEF >40% (Meta-analysis Global Group in Chronic Heart Failure (MAGGIC) 2012). The CHARM trial confirms an adjusted hazard ratio of 1.31 for all-cause mortality per 10% reduction in LVEF <45% (Solomon et al. 2005), however, not when LVEF <40% or >40% (Pocock et al. 2013).

In the GWTGHF series, unadjusted in-hospital mortality for HFpEF diminished between 2005 and 2010, however, not for HFmrEF or HFrEF (Steinberg et al. 2012). In those admitted for acute decompensated HF, almost 49% had HFpEF (EF $\geq$ 50%), 37% had HFrEF (EF $\leq$ 40%), and 14% had HFmrEF (EF 41–49%) (Coles et al. 2014). Survival rates were better at 1, 2, and 5 years for HFpEF (Coles et al. 2014). Prognosis at 1 year improved for all HF groups; however, later on mortality for HFrEF and HFpEF remained elevated (Coles et al. 2014).

Outcomes in Obese Patients

Abdominal obesity has a stronger link to HFpEF than HFrEF. However, the impact of obesity on outcome in HFpEF, HFrEF, and HFmrEF is still a matter of debate. In the series of Savji et al. with a 12-year follow up, BMI correlated more with risk of HFpEF than of HFrEF, notably for women (Savji et al. 2018). Insulin resistance was related only to HFpEF development, not to HFrEF. There is some consensus that the BMI curve for all-cause mortality (including HF) is U-shaped, however not about the nadir, which can change from 32–33 kg/m^2 (Zhang et al. 2019) to 26.5–30.9 kg/m^2 (Haass et al. 2011). Values <23.5 kg/m^2 and >35 kg/m^2 are reported as risky (Haass et al. 2011). Again abdominal obesity is highlighted when all-cause, cardiovascular and noncardiovascular mortality are considered in the HF context (Tsujimoto and Kajio 2017). Yet BMI is not entirely overlooked, as findings >25 kg/m^2 signal risk, which seems higher for HFpEF than HFrEF (Pandey et al. 2017).

Among postmenopausal women, excessive adiposity was related with HFpEF only, not HFrEF, particularly for African American women. Hypertension (40.9%) and obesity (25.8%), were the paramount prognostic markers (Eaton et al. 2016).

In contrast, higher BMI indicated lower risk of symptomatic HF in another series, different from waist-to-height ratio (WHR), which was a positive prognostic marker for the composite outcome (Chandramouli et al. 2019). Low BMI paired with high WHR had a more ominous course than elevated BMI and normal WHR (Chandramouli et al. 2019).

Possible Involved Pathways and Biomolecules

Obesity is strongly related with systemic arterial hypertension, diabetes, insulin resistance, metabolic syndrome, and obstructive sleep apnea, all of which contribute to the development of HF. On the other hand, reduced cardiorespiratory fitness in HF patients could be the cause of obesity and therefore the "circulus vitiosus" is difficult to interrupt. Obesity induces hemodynamic, structural, neurohormonal changes, and systemic inflammation that induce functional and structural cardiac remodeling (Fig. 31.1), including HFpEF, and its impact on outcome in HFrEF patients is significant. However, the influence of obesity on HFmrEF is less clear.

The hemodynamic changes in obesity consider elevated central blood volume, reduced systemic

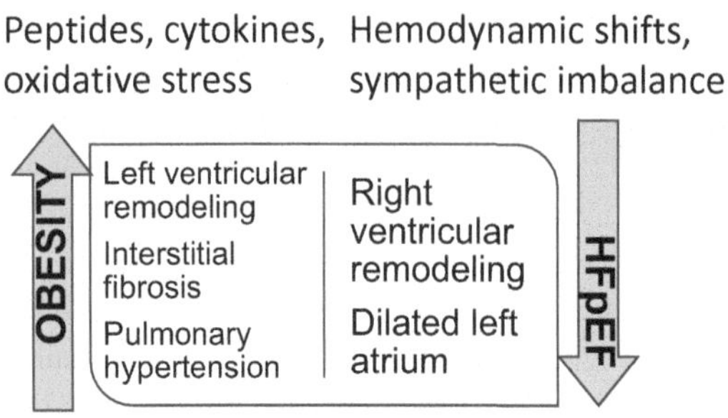

Fig. 31.1 Local and systemic changes associated with obesity

vascular resistance and stroke volume, and ultimately elevated cardiac output proportional to the degree of obesity but not to body surface area (Carbone et al. 2017; Tadic and Cuspidi 2019) (Fig. 31.1). Lean (fat-free) mass in obese patients is one of the most important parameters responsible for the development of a hyperdynamic state. Heart rate is only slightly increased; thus, cardiac output follows elevations of left ventricular (LV) stroke volume.

Cardiac effort, and particularly LV effort, in obese patients exceeds predicted values due to elevated LV stroke work. LV end-diastolic pressure and pulmonary capillary wedge pressure are commonly elevated in moderate-to-severe body adiposity.

The persistent increase in cardiac output due to elevated preload induces an initial LV dilatation, according to the Frank-Starling law, followed by a compensatory hypertrophic response confirmed at the cellular and organ level. LV hypertrophy represents a compensatory mechanism to decrease wall stress and oxygen demand according to LaPlace's law.

Pulmonary artery pressure is often increased due to LV failure, and retrograde transmission of increased LV filling pressure over pulmonary veins to pulmonary circulation, resulting in the increase of pulmonary vascular resistance particularly in those with sleep apnea or pulmonary hypoventilation (Alpert et al. 2018). RV filling pressure is frequently increased, which further induces elevated venous returns to the pulmonary circulation. Dyspnea becomes common in such context.

At the beginning, increased LV filling pressures and impairment in LV diastolic function were considered as the result of LV hypertrophy. However, studies found that obese patients may have LV diastolic dysfunction without LV hypertrophy (Dote et al. 2012; Russo et al. 2011), which indicates that some other mechanisms of LV remodeling besides LV hypertrophy may be involved (Fig. 31.1). LV hypertrophy could be even a protective response in face of increased preload by reduction of wall stress. Obese patients with HFpEF had increased relative wall thickness and smaller cardiac volumes even in the absence of increased LV mass, which demonstrates LV concentric remodeling (Yagmur et al. 2017; Zile et al. 2011).

Anthropometric Measurements

The visceral adipose tissue points toward decreased cardiac output and increased systemic vascular resistance, with opposite correlations for lower body fat. Abdominal subcutaneous fat seems to play a secondary role regarding LV concentric remodeling, only visceral adipose tissue being relevant (Shah et al. 2014). LV hypertrophy and LV mass have strong links with excessive body weight, eccentric hypertrophy being more common than the concentric phenotype (Neeland et al. 2013).

Other Heart Chambers

Right ventricular (RV) remodeling assessment by echocardiography is challenging in obese patients due to the position of the RV in the chest and increased thickness of chest wall, yet reduced function without overt heart disease has been perceived (Cuspidi et al. 2014), including larger volume and mass (Wong et al. 2006). Left atrial

(LA) dilatation, connected with HFpEF, has also been linked to elevated adiposity (Chahal et al. 2012). LA volume seems to follow visceral, yet not subcutaneous abdominal fat (Chahal et al. 2012). LA function and mechanics can deteriorate as well in HFpEF (Oliver et al. 2017).

Biohumoral and Neurohormonal Changes

Activation of the renin-angiotensin-aldosterone and the sympathetic nervous system, natriuretic peptide, systemic inflammation, and oxidative stress are reported. Cytokines such as interleukin (IL) 1 beta, tumor necrosis factor alpha, and IL-1 alpha, experimentally induce diastolic dysfunction (Van Tassell et al. 2013; O'Brien et al. 2014), and anti-inflammatory could ameliorate HFpEF and HFrEF. An IL-1 receptor antagonist improved cardiorespiratory fitness in patients with HFrEF and HFpEF, with increase of peak oxygen consumption (Van Tassell et al. 2012, 2014).

All these mechanisms might induce cardiac interstitial fibrosis that further decreases cardiac elasticity and compliance, and subsequently increases energy consumption during LV diastole—an extremely energy-demanding process. Moreover, these biohumoral and neurohormonal mechanisms promote enhanced atherosclerosis, which also may induce HF.

Obesity Paradox

In addition to an excess of fat mass, obese patients usually also have more lean mass. Patients with heart failure with increased amount of lean mass showed better cardiopulmonary fitness, a major determinant of clinical outcomes in the general population, and therefore lower mortality risk, which could be related to the "obesity paradox" phenomenon. Elevated lean mass represents a stronger predictor in heart failure in comparison with fat mass, whereas elevated fat mass in patients with coronary heart disease could have protective effects when fat mass is not related with increased systemic inflammation (Carbone et al. 2019). Obese patients with decreased amount of lean mass, despite increased BMI and fat mass, namely, sarcopenic obesity, suffer increased mortality. This suggests that lean mass plays a key role, and therefore exercise training and dietary interventions could represent important strategies.

Cardiovascular Mortality

The "obesity paradox" does not reduce risk in morbidly obese patients, only in mildly overweight cases. Nevertheless, there is still no agreement regarding the U-shape relationship between BMI and mortality or risk for HF development (Table 31.1). Mildly overweight patients had the lowest rates of death or cardiovascular hospitalization (Haass et al. 2011), and absence of obesity was an independent predictor of mortality in both HFpEF and HFrEF (Ather et al. 2012). On the other hand, Mohammed et al. reported that obesity was related with better outcome after adjustment for age, sex, and other comorbidities, and this association did not show U-shape (Mohammed et al. 2012).

BMI < 30 kg/m^2 pointed toward less 30-day mortality for HFpEF in one experience (Powell-Wiley et al. 2018). A modest association was found in HFrEF subjects, with no risk increase when BMI > 30 kg/m^2. The "obesity paradox"was not present in HFpEF (Table 31.1).

In another series, lack of obesity was related with mortality in HFpEF and HFrEF (Iorio et al. 2018). Further confirmation that mortality risk had a U-shaped form in both HFpEF and HFrEF is available, however, with the nadir at 30.0–34.9 kg/m^2 (Padwal et al. 2014).

One potential mechanism that could explain the "obesity paradox" is adiponectin, an adipocyte-specific cytokine, inversely related

Table 31.1 The association between obesity paradox and HFpEF

Refs.	Sample size	HFpEF/ HFrEF	Study type	Main findings
I-PRESERVE (2011)	4109 patients	HFpEF	Follow-up (4 years)	Overweight was related with lower mortality Or cardiovascular hospitalization. Underweight And severely overweight patients had higher Mortality rates than overweight
Ather et al. (2012)	942 patients	HFpEF/ HFrEF	2-year follow-up	Nonobese patients had higher mortality in HFpEF and HFrEF patients
Powell-Wiley et al. (2018)	39,647 patients	HFpEF/ HFrEF	30-day and 1-year follow-up	The obesity paradox for 30-day mortality Existed at all BMI levels only in HFrEF patients
Iorio et al. (2018)	2314 patients	HFpEF/ HFrEF	31-month follow-up	Nonobese patients had higher overall mortality In HFpEF and HfrEF patients and mortality rate Between these groups was similar
Padwal et al. (2014)	23,967 subjects	HFpEF/ HFrEF	Meta-analysis (3-year follow-up)	U-shaped form of relationship between mortality And BMI was found in HFpEF and HFrEF patients With highly positioned nadir at 30.0–34.9 kg/m^2

BMI body mass index, *HFpEF* heart failure with preserved ejection fraction, *HFrEF* heart failure with reduced ejection fraction

with BMI in HF patients. Lower adiponectin levels are associated with increased mortality, which means that patients with increased BMI have higher adiponectin and lower mortality. The other mechanisms of "obesity paradox" involve anti-inflammatory effects of elevated lipoproteins, and decreased response of the renin-angiotensin-aldosterone system.

Risks of Bias in the Interpretation of the Paradox

HF is characterized by decreased and/or inadequate cardiac output and elevated systemic vascular resistance, which is why the obese patients with preserved skeletal muscle mass have a better prognosis than HF patients with reduced lean mass, independent of stroke volume and cardiac output (Yancy et al. 2013). Furthermore, obese patients usually have lower systemic vascular resistance, particularly if they do not exhibit arterial hypertension (Alpert et al. 2014; Lavie et al. 2009). Afterload reduction results in improved forward flow and better cardiac output.

Recommended Lines for Further Investigation

Large randomized studies that evaluate intentional weight loss are still missing. Investigation of the impact of fat mass and lean mass on HF occurrence and outcome should be encouraged. It might be reasonable to hypothesize that increase of lean mass would be associated with greater muscular strength and better cardiorespiratory fitness, which altogether improve outcome in HF patients.

References

Alpert MA, Lavie CJ, Agrawal H, Aggarwal KB, Kumar SA (2014) Obesity and heart failure: epidemiology, pathophysiology, clinical manifestations, and management. Transl Res 164(4):345–356

Alpert MA, Karthikeyan K, Abdullah O, Ghadban R (2018) Obesity and cardiac remodleing in adults: mechanisms and clinical implications. Prog Cardiovasc Dis 61(2):114–123

Ather S, Chan W, Bozkurt B, Aguilar D, Ramasubbu K, Zachariah AA, Wehrens XH, Deswal A (2012) Impact of non-cardiac comorbidities on morbidity and mortality in a predominantly male population with heart failure and preserved versus reduced ejection fraction. J Am Coll Cardiol 59(11):998–1005

Baena-Díez JM, Byram AO, Grau M et al (2010) Obesity is an independent risk factor for heart failure: zona Franca cohort study. Clin Cardiol 33(12):760–764

Benjamin EJ, Virani SS, Callaway CW, Chamberlain AM, Chang AR, Cheng S, Chiuve SE, Cushman M, Delling FN, Deo R, de Ferranti SD, Ferguson JF, Fornage M, Gillespie C, Isasi CR, Jiménez MC, Jordan LC, Judd SE, Lackland D, Lichtman JH, Lisabeth L, Liu S, Longenecker CT, Lutsey PL, Mackey JS, Matchar DB, Matsushita K, Mussolino ME, Nasir K, O'Flaherty M, Palaniappan LP, Pandey A, Pandey DK, Reeves MJ, Ritchey MD, Rodriguez CJ, Roth GA, Rosamond WD, UKA S, Satou GM, Shah SH, Spartano NL, Tirschwell DL, Tsao CW, Voeks JH, Willey JZ, Wilkins JT, Wu JH, Alger HM, Wong SS, Muntner P, American Heart Association Council on Epidemiology and Prevention Statistics Committee and Stroke Statistics Subcommittee (2018) Heart disease and stroke statistics-2018 update: a report from the American Heart Association. Circulation 137(12): e67–e492

Bhatia RS, Tu JV, Lee DS, Austin PC, Fang J, Haouzi A, Gong Y, Liu PP (2006) Outcome of heart failure with preserved ejection fraction in a population-based study. N Engl J Med 355:260–269

Carbone S, Lavie CJ, Arena R (2017) Obesity and heart failure: focus on the obesity paradox. Mayo Clin Proc 92(2):266–279

Carbone S, Canada JM, Billingsley HE, Siddiqui MS, Elagizi A, Lavie CJ (2019) Obesity paradox in cardiovascular disease: where do we stand? Vasc Health Risk Manag 15:89–100

Chahal H, McCleland RL, Tandai H, Jain A, Turkbey EB, Hundley WG, Barr RG, Kizer J, Lima JAC, Bluemke DA, Kawut SM (2012) Obesity and right ventricular structure and function: the MESA-right ventricle study. Chest 141:388–395

Chandramouli C, Tay WT, Bamadhaj NS, Tromp J, Teng TK, Yap JJL, MacDonald MR, Hung CL, Streng K, Naik A, Wander GS, Sawhney J, Ling LH, Richards AM, Anand I, Voors AA, Lam CSP, ASIAN-HF Investigators (2019) Association of obesity with heart failure outcomes in 11 Asian regions: a cohort study. PLoS Med 16(9):e1002916

Cheng RK, Cox M, Neely ML, Heidenreich PA, Bhatt DL, Eapen ZJ, Hernandez AF, Butler J, Yancy CW, Fonarow GC (2014) Outcomes in patients with heart failure with preserved, borderline, and reduced ejection fraction in the Medicare population. Am Heart J 168:721–730

Coles AH, Fisher K, Darling C, Yarzebski J, McManus DD, Gore JM, Lessard D, Goldberg RJ (2014) Long-term survival for patients with acute decompensated heart failure according to ejection fraction findings. Am J Cardiol 114:862–868

Cuspidi C, Rescaldani M, Sala C, Grassi G (2014) Left-ventricular hypertrophy and obesity: a systematic review and meta-analysis of echocardiographic studies. J Hypertens 32(1):16–25

Dote K, Miyasaka Y, Tsujimoto S et al (2012) Obesity as an independent risk for left ventricular diastolic dysfunction in 692 Japanese patients. Obes Res Clin Pract 6(3):e175–e262

Eaton CB, Pettinger M, Rossouw J, Martin LW, Foraker R, Quddus A, Liu S, Wampler NS, Hank Wu WC, Manson JE, Margolis K, Johnson KC, Allison M, Corbie-Smith G, Rosamond W, Breathett K, Klein L (2016) Risk factors for incident hospitalized heart failure with preserved versus reduced ejection fraction in multiracial cohort of postmenopausal women. Circ Heart Fail 9(10):e002883

Fonarow GC, Stough WG, Abraham WT, Albert NM, Gheorghiade M, Greenberg BH, O'Connor CM, Sun JL, Yancy CW, Young JB, OPTIMIZE-HF Investigators and Hospitals (2007) Characteristics, treatments, and outcomes of patients with preserved systolic function hospitalized for heart failure: a report from the OPTIMIZE-HF registry. J Am Coll Cardiol 50:768–777

Haass M, Kitzman DW, Anand IS, Miller A, Zile MR, Massie BM et al (2011) Body mass index and adverse cardiovascular outcomes in heart failure patients with preserved ejection fraction: results from the Irbesartan in heart failure with preserved ejection fraction (I-PRESERVE) trial. Circ Heart Fail 4:324–331

Iorio A, Senni M, Barbati G, Greene SJ, Poli S, Zambon E, Di Nora C, Cioffi G, Tarantini L, Gavazzi A, Sinagra G, Di Lenarda A (2018) Prevalence and prognostic impact of non-cardiac co-morbidities in heart failure outpatients with preserved and reduced ejection fraction: a community-based study. Eur J Heart Fail 20 (9):1257–1266

Kenchaiah S, Evans JC, Levy D et al (2002) Obesity and the risk of heart failure. N Engl J Med 347(5):305–313

Lavie CJ, Milani RV, Ventura HO (2009) Obesity and cardiovascular disease: risk factor, paradox, and impact of weight loss. J Am Coll Cardiol 53(21):1925–1932

Meta-analysis Global Group in Chronic Heart Failure (MAGGIC) (2012) The survival of patients with heart failure with preserved or reduced left ventricular

ejection fraction: an individual patient data meta-analysis. Eur Heart J 33:1750–1757
Mohammed SF, Borlaug BA, Roger VL, Mirzoyev SA, Rodeheffer RJ, Chirinos JA, Redfield MM (2012) Comorbidity and ventricular and vascular structure and function in heart failure with preserved ejection fraction: a community-based study. Circ Heart Fail 5 (6):710–719
Neeland IJ, Gupta S, Ayers CR, Turer AT, Rame JE, Das SR, Berry JD, Khera A, McGuire DK, Vega GL, Grundy SM, de Lemos JA, Drazner MH (2013) Relation of regional fat distribution to left ventricular structure and function. Circ Cardiovasc Imaging 6:800–807
O'Brien LC, Mezzaroma E, Van Tassell BW et al (2014) Interleukin-18 as a therapeutic target in acute myocardial infarction and heart failure. Mol Med 20:221–229
Oliver W, Matthews G, Ayers CR, Garg S, Gupta S, Neeland IJ, Drazner MH, Berry JD, Matulevicius S, de Lemos JA (2017) Factors associated with left atrial remodeling in the general population. Circ Cardiovasc Imaging 10(2):e005047
Padwal R, McAlister FA, McMurray JJ, Cowie MR, Rich M, Pocock S, Swedberg K, Maggioni A, Gamble G, Ariti C, Earle N, Whalley G, Poppe KK, Doughty RN, Bayes-Genis A, Meta-analysis Global Group in Chronic Heart Failure (MAGGIC) (2014) The obesity paradox in heart failure with patients with preserved versus reduced ejection fraction: a meta-analysis of individual patient data. Int J Obes 38 (8):1110–1114
Pandey A, LaMonte M, Klein L, Ayers C, Psaty BM, Eaton CB, Allen NB, de Lemos JA, Carnethon M, Greenland P, Berry JD (2017) Relationship between physical activity, body mass index, and risk of heart failure. J Am Coll Cardiol 69(9):1129–1142
Pocock SJ, Ariti CA, McMurray JJ et al (2013) Predicting survival in heart failure: a risk score based on 39 372 patients from 30 studies. Eur Heart J 34:1404–1413
Ponikowski P, Voors AA, Anker SD, Bueno H, Cleland JGF, Coats AJS, Falk V, González-Juanatey JR, Harjola VP, Jankowska EA, Jessup M, Linde C, Nihoyannopoulos P, Parissis JT, Pieske B, Riley JP, Rosano GMC, Ruilope LM, Ruschitzka F, Rutten FH, van der Meer P, ESC Scientific Document Group (2016) 2016 ESC guidelines for the diagnosis and treatment of acute and chronic heart failure: the task force for the diagnosis and treatment of acute and chronic heart failure of the European Society of Cardiology (ESC)developed with the special contribution of the heart failure association (HFA) of the ESC. Eur Heart J 37(27):2129–2200
Powell-Wiley TM, Ngwa J, Kebede S, Lu D, Schulte PJ, Bhatt DL, Yancy C, Fonarow GC, Albert MA (2018) Impact of body mass index on heart failure by race/ethnicity from the get with the guidelines-heart failure (GWTG-HF) registry. JACC Heart Fail 6(3):233–242
Russo C, Jin Z, Homma S et al (2011) Effect of obesity and overweight on left ventricular diastolic function: a community based study in an elderly cohort. J Am Coll Cardiol 57(12):1368–1374
Savji N, Meijers WC, Bartz TM, Bhambhani V, Cushman M, Nayor M, Kizer JR, Sarma A, Blaha MJ, Gansevoort RT, Gardin JM, Hillege HL, Ji F, Kop WJ, Lau ES, Lee DS, Sadreyev R, van Gilst WH, Wang TJ, Zanni MV, Vasan RS, Allen NB, Psaty BM, van der Harst P, Levy D, Larson M, Shah SJ, de Boer RA, Gottdiener JS, Ho JE (2018) The association of obesity and cardiometabolic traits with incident HFpEF and HFrEF. JACC Heart Fail 6 (8):701–709
Senthong V, Kirsop JL, Tang WH (2017) Clinical phenotyping of heart failure with biomarkers: current and future perspectives. Curr Heart Fail Rep 14 (2):106–116
Shah AM, Shah SJ, Anand IS, et al; TOPCAT investigators (2014) Cardiac structure and function in heart failure with preserved ejection fraction: baseline findings from the echocardiographic study of the treatment of preserved cardiac function heart failure with an aldosterone antagonist trial. Circ Heart Fail 7 (1):104–115
Solomon SD, Anavekar N, Skali H et al (2005) Influence of ejection fraction on cardiovascular outcomes in a broad spectrum of heart failure patients. Circulation 112:3738–3744
Steinberg BA, Zhao X, Heidenreich PA, Peterson ED, Bhatt DL, Cannon CP, Hernandez AF, Fonarow GC, Get With the Guidelines Scientific Advisory Committee and Investigators (2012) Trends in patients hospitalized with heart failure and preserved left ventricular ejection fraction: prevalence, therapies, and outcomes. Circulation 126:65–75
Sweitzer NK, Lopatin M, Yancy CW, Mills RM, Stevenson LW (2008) Comparison of clinical features and outcomes of patients hospitalized with heart failure and normal ejection fraction (> or.55%) versus those with mildly reduced (40% to 55%) and moderately to severely reduced (<40%) fractions. Am J Cardiol 101:1151–1156
Tadic M, Cuspidi C (2019) Obesity and heart failure with preserved ejection fraction: a paradox or something else? Heart Fail Rev 24(3):379–385
Tanai E, Frantz S (2015) Pathophysiology of heart failure. Compr Physiol 6(1):187–214
Tsujimoto T, Kajio H (2017) Abdominal obesity is associated with an increased risk of all-cause mortality in patients with HFpEF. J Am Coll Cardiol 70 (22):2739–2749
Van Tassell BW, Arena RA, Toldo S et al (2012) Enhanced interleukin-1 activity contributes to exercise intolerance in patients with systolic heart failure. PLoS One 7(3):e33438
Van Tassell BW, Toldo S, Mezzaroma E, Abbate A (2013) Targeting interleukin-1 in heart disease. Circulation 128(17):1910–1923
Van Tassell BW, Arena R, Biondi-Zoccai G et al (2014) Effects of interleukin-1 blockade with anakinra on

aerobic exercise capacity in patients with heart failure and preserved ejection fraction (from the D-HART pilot study). Am J Cardiol 113(2):321–327

Webb J, Draper J, Fovargue L, Sieniewicz B, Gould J, Claridge S, Barton C, Smith S, Tondel K, Rajani R, Kapetanakis S, Rinaldi CA, McDonagh TA, Razavi R, Carr-White G (2018) Is heart failure with mid range ejection fraction (HFmrEF) a distinct clinical entity or an overlapgroup? Int J Cardiol Heart Vasc 21:1–6

Wong CY, O'Moore-Sullivan T, Leano R, Hakins C, Jenkins C, Marwick TH (2006) Association of subclinical right ventricular dysfunction with obesity. J Am Coll Cardiol 47:611–616

World Health Organization (2013) Obesity and overweight fact sheet. World Health Organization website. http://www.who.int/mediacentre/factsheets/fs311/en/index.html

Yagmur J, Cansel M, Kurtoglu E, Hidayet S, Acıkgoz N, Ermis N, Ozyalin F (2017) Assessment of left atrial volume and function by real time three-dimensional echocardiography in obese patients. Echocardiography 34(2):210–216

Yancy CW, Jessup M, Bozkurt B et al (2013) 2013 ACCF/AHA guideline for the management of heart failure: a report of the American College of Cardiology Foundation/American Heart Association task force on practice guidelines. J Am Coll Cardiol 62(16):e147–e239

Zhang J, Begley A, Jackson R, Harrison M, Pellicori P, Clark AL, Cleland JGF (2019) Body mass index and all-cause mortality in heart failure patients with normal and reduced ventricular ejection fraction: a dose-response meta-analysis. Clin Res Cardiol 108 (2):119–132

Zile MR, Gottdiener JS, Hetzel SJ, et al; I-PRESERVE investigators (2011) Prevalence and significance of alterations in cardiac structure and function in patients with heart failure and a preserved ejection fraction. Circulation 124(23):2491–2501

Inflammatory Mechanisms in Diabetic Kidney Disease

32

Radica Z. Alicic, Emily J. Cox, Joshua J. Neumiller, and Katherine R. Tuttle

Contents

Introduction 438

The Inflammatory Pathogenesis of Diabetic Kidney Disease 438

Conventional Therapeutic Interventions for the Treatment of Diabetic Kidney Disease 441

Arterial Hypertension 443

Healthy Lifestyle 445

New Therapeutic Approaches: Glucagon-Like Peptide-1 Receptor Agonists 445

Glucagon-Like Peptide-1 Receptor Agonist Trials 446

Dulaglutide 446

Liraglutide 446

Semaglutide 448

Lixisenatide 448

Exenatide 448

Glucagon-Like Peptide-1 Receptor Agonist Trials: Ongoing Kidney-Outcome Studies 448

Putative Mechanism of Observed Kidney Outcomes 449

Future Perspectives 450

References 450

R. Z. Alicic (✉) · K. R. Tuttle
Providence Medical Research Center, Providence Health Care, Spokane, WA, USA

University of Washington School of Medicine, Seattle, WA, USA
e-mail: radica.alicic@providence.org

E. J. Cox
Providence Medical Research Center, Providence Health Care, Spokane, WA, USA

J. J. Neumiller
Washington State University College of Pharmacy and Pharmaceutical Sciences, Spokane, WA, USA

Abstract

Up to half of all patients with diabetes will develop diabetic kidney disease (DKD), the leading cause of end-stage kidney disease worldwide. The excess morbidity and mortality experienced by individuals with diabetes are primarily driven by the development of this microvascular complication. Recent research findings implicate inflammation as a major mechanism underlying kidney damage, including the pathological changes associated

J. Faintuch and S. Faintuch (eds.), *Obesity and Diabetes*, https://doi.org/10.1007/978-3-030-53370-0_32

with DKD. Historically, there has been a paucity of therapies that are effective in arresting the development and progression of DKD. Recently, however, glucose-lowering agents in the glucagon-like peptide-1 receptor agonist (GLP-1 RA) class of medications have been shown to preserve declining kidney function and reduce albuminuria in patients with diabetes. In addition to controlling modifiable risk factors such as hyperglycemia, blood pressure, and weight, it appears that GLP-1 signaling imparts direct anti-inflammatory, antifibrotic, and antioxidant effects. Thus, a number of clinical trials are underway to examine the impact of GLP-1 RAs on DKD development and progression.

Keywords

Diabetic kidney disease · Inflammation · GLP-1 receptor agonists · Kidney outcome trials

Introduction

The end of the twentieth century and the first two decades of the twenty-first century are marked by a global epidemic of diabetes, with over 500 million individuals projected to have diabetes worldwide by the year 2030 (Whiting et al. 2011). This pandemic is largely driven by the staggering burden of obesity. Globally, the number of people with obesity increased from 104 million in 1975 to 641 million in 2014 (Collaboration NCDRF 2016), with the adjusted prevalence of obesity in the United States (U.S.) climbing to 35% and 40% among men and women, respectively (Flegal et al. 2016).

A microvascular complication of diabetes, diabetic kidney disease (DKD), occurs in up to half of patients with diabetes and is now the leading cause of end-stage kidney disease (ESKD) (Reutens 2013; Alicic et al. 2017). Clinically, DKD is identified by persistently elevated urinary albumin excretion (urinary albumin-to-creatinine ratio [UACR] >30 mg/g), or a reduction of kidney function (estimated glomerular filtration rate [eGFR] <60 mL/min/1.73 m^2) (National Kidney Foundation 2012). A diagnosis of obesity often precedes the diagnosis of diabetes and DKD, with obesity acting as a continuous risk factor (Macisaac et al. 2014).

The development of DKD amplifies the already-increased cardiovascular risk of individuals with diabetes, and also puts these individuals at greater risk for infections and cancer (Cheung et al. 2017; Rao Kondapally Seshasai et al. 2011). In fact, most of the excess risk for both all-cause and cardiovascular-related mortality in patients with diabetes is related to the presence of DKD (Afkarian et al. 2013). Even early stages of DKD are associated with a three-fold increase in all-cause mortality, and a 16-year loss in life expectancy (Wen et al. 2017). Indeed, only a minority of DKD patients live long enough to reach ESKD and require kidney replacement therapy. Unfortunately for those who do survive long enough to require kidney replacement therapy, dialysis and kidney transplant are cost prohibitive in many parts of the world, thus equating a diagnosis of ESKD to a death sentence (Couser et al. 2011).

The number of deaths attributed to DKD almost doubled between 1990 and 2012 (Lozano et al. 2012). These data underscore the compelling need for both an improved understanding of the pathomechanisms behind the genesis and perpetuation of the disease, and the development of more effective therapies to prevent and treat DKD.

The Inflammatory Pathogenesis of Diabetic Kidney Disease

Activation of the innate immune response and associated inflammatory pathways is increasingly recognized as the origin of the functional and structural changes observed in DKD (Alicic et al. 2017, 2018). Inflammatory mediators are detectable in the urine of patients diagnosed with diabetes long before and after development of DKD, thus indicating a role of inflammation in both instigation and progression of the disease (Alicic et al. 2018; Van et al. 2017; Goldberg 2009). While the specific mechanisms responsible for obesity-induced kidney damage remain to be fully elucidated, it appears that endothelial

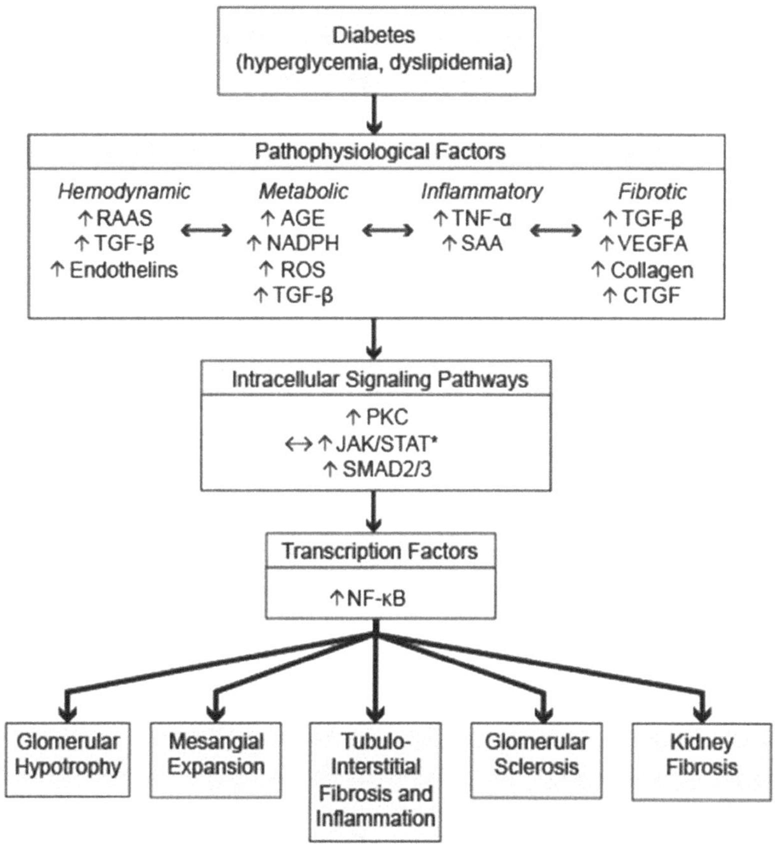

Fig. 32.1 Inflammatory pathways in diabetic kidney disease. *AGE* advanced glycation end-product, *JAK-STAT* Janus kinase/signal transducer and activator of transcription, *KMNPS* kidney mononuclear phagocyte system, *NADPH* nicotinamide adenine dinucleotide phosphate-oxidase, *NF-κB* nuclear factor kappa B transcription factor, *PKC* protein kinase C, *RAS* renin-angiotensin system, *RAGE* receptor for advanced glycation end-products, *ROS* reactive oxygen species, *TGF-β* transforming growth factor beta. ↑ upregulated; ↔ unchanged. *JAK/STAT signaling can be unchanged or upregulated in early and later stages of diabetes, respectively. (Reprinted with permission from (Alicic et al. 2018). This figure was published in *Advances in Chronic Kidney Disease,* volume 25, Alicic RZ, Johnson EJ, Tuttle KR. Inflammatory Mechanisms as New Biomarkers and Therapeutic Targets for Diabetic Kidney Disease, pages 181–191, Copyright by the National Kidney Foundation, Inc. (2017))

injury and activation of inflammatory pathways are important contributors (Alicic et al. 2017, 2018; Van et al. 2017; Goldberg 2009; D'Agati et al. 2016).

The phenotype of diabetes—hyperglycemia, oxidative stress, and elevated production of advanced glycation end-products—collectively activates the interstitial population of macrophages and dendritic cells (the *mononuclear phagocyte system*) in the kidney (Fig. 32.1) (Pichler et al. 2017; Navarro-Gonzalez et al. 2011). Once activated, these innate immune cells are engaged in the kidney and recruit additional cells, resulting in excessive immune cell invasion of the kidney and the destruction of important structures. This self-perpetuating cycle of activation and recruitment is mediated by the release of proinflammatory cytokines from activated immune cells, which recruits other circulating populations of immune cells, including monocytes, macrophages, and mast cells (Fig. 32.2) (Alicic et al. 2018).

Recruitment of mast cells in particular perpetuates this cycle through release of inflammatory cytokines, as well as release of chymase, which converts angiotensin I to II, thus directly perturbing glomerular hemodynamics (Maeda et al. 2010; Reilly et al. 1982). The degree of

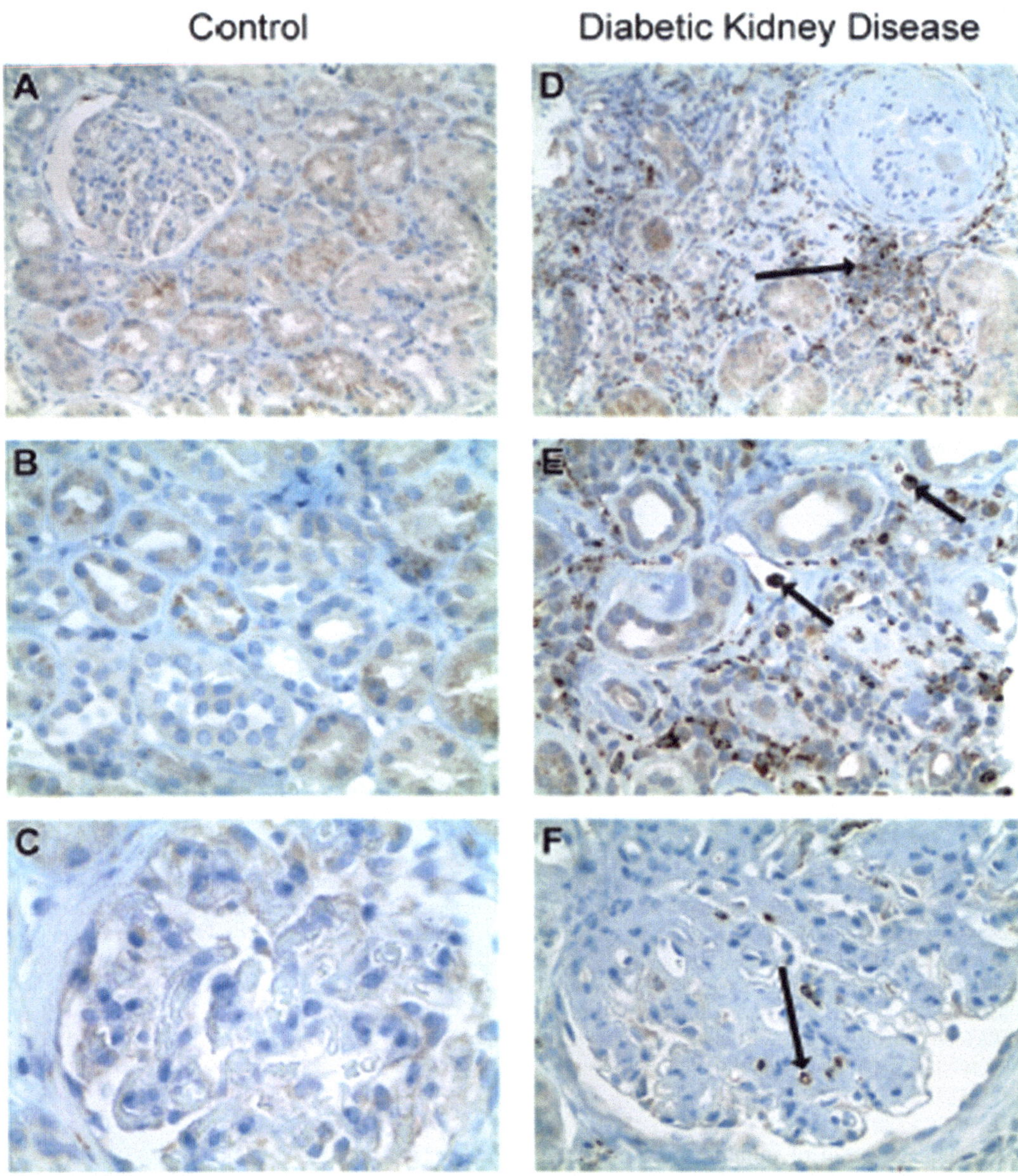

Fig. 32.2 Anti-CD68 (KP-1) immunohistochemistry of kidney biopsies from diabetic patients without and with histological features of diabetic kidney disease. CD68 antibodies highlight influx of macrophage lineage cells in different kidney structures. Immunohistochemistry with anti-CD68 (KP-1) antibody (Roche Diagnostics, Indianapolis, IN) used at a concentration of 0.4 μg/mL. (**a–c**) Human kidney from diabetic patient without histological features of diabetic kidney disease (control); magnification: 200× (**a**), 400× (**b**), 600× (**c**); (**d**) Interstitial macrophage lineage cells infiltrates in diabetic kidney disease (magnification: 200×). (**e**) Macrophage lineage cells in peritubular capillaries (magnification: 400×). (**f**) Macrophage lineage cells in glomerular capillary (magnification: 600×). (Reprinted with permission from (Alicic et al. 2018). This figure was published in *Advances in Chronic Kidney Disease,* volume 25, Alicic RZ, Johnson EJ, Tuttle KR. Inflammatory Mechanisms as New Biomarkers and Therapeutic Targets for Diabetic Kidney Disease, pages 181–191, Copyright by the National Kidney Foundation, Inc. (2017))

tubulointerstitial inflammation, glomerulosclerosis, and rate of loss of eGFR are closely related with the degree of macrophage infiltration and mast cell degranulation (Hiromura et al. 1998; Okon and Stachura 2007; Ninichuk et al. 2007; Furuta et al. 1993).

The proinflammatory cytokines are a group of polypeptide signaling molecules that promote autocrine, paracrine, and juxtacrine signaling to coordinate the innate immune response. Their release by activated immune cells is associated with numerous abnormalities observed in DKD. Interleukin (IL)-1 contributes to perturbed hemodynamics via increased production of prostaglandin E (Pfeilschifter and Muhl 1990), and increased endothelial permeability in the kidney vasculature (Royall et al. 1989). IL-6 contributes to basement membrane thickening and albuminuria (Nosadini et al. 2000; Dalla Vestra et al. 2005; Nagayama et al. 2014) and worsens interstitial invasion by neutrophils (Kaplanski et al. 2003; Jones 2005). Through mechanisms that are not fully understood, IL-18 contributes to microvascular damage in the diabetic kidney, and its levels correlate with early signs of kidney impairment and macroalbuminuria (UACR >300 mg/g) (Fujita et al. 2012; Araki et al. 2007).

The inflammatory cytokine tumor necrosis factor α (TNFα) is released by activated immune cells, and is elevated in the urine and serum of patients with DKD compared to that of patients without diabetes or with uncomplicated diabetes. TNF-α is cytotoxic to kidney cells, alters glomerular hemodynamics, and increases oxidative stress (Navarro et al. 2005; Navarro and Mora-Fernandez 2006).

Monocyte-chemotactic protein-1 (MCP-1)/chemokine C-C motif-ligand 2 (CCL2), C-X3-C motif chemokine (CX3CL1), and C-C motif chemokine 5 constitute a subgroup of cytokines known as *chemoattractant* molecules, which recruit inflammatory cells to the kidney, furthering tissue damage (Navarro-Gonzalez et al. 2011; Furuta et al. 1993). Janus kinase/signal transducer and activator of transcription (JAK/STAT), an intracellular cytokine-associated signaling pathway, is upregulated in the glomerular cells of patients with DKD and correlates inversely with eGFR (Table 32.1) (Berthier et al. 2009; Brosius et al. 2016).

In sum, inflammatory markers have a clear association with the development, progression, and severity of DKD. The histological damage and functional impairment associated with DKD are primarily driven by the excessive and misdirected immune cell invasion of the kidney, which damages important kidney structures, and results in a self-exacerbating cycle of inflammation (Fig. 32.3).

Conventional Therapeutic Interventions for the Treatment of Diabetic Kidney Disease

Optimized blood glucose and blood pressure control has historically been the cornerstone of therapy, as poor glycemic control is clearly associated with the development and progression of DKD. Early in the course of diabetes, intensive glycemic control in patients with type 1 diabetes (HbA1c 6–6.9) reduces the risk of developing DKD (Nathan and Group DER 2014; Group DER et al. 2011; UK Prospective Diabetes Study (UKPDS) 1991). For instance, an absolute difference in HbA1c of 0.9% is associated with a 20% relative risk reduction in the development of macroalbuminuria (UACR >300 mg/g), ESKD, kidney-related death, or eGFR <30 mL/min/1.73 m^2 (Zoungas et al. 2017; Retnakaran et al. 2006).

It is hypothesized that the prolonged beneficial effects of early intensive glycemic control, often referred to as the "legacy effect" or "metabolic memory," can be explained by prevention of metabolism-driven epigenetic modifications (Tonna et al. 2010; Thomas 2014). Unfortunately, once diabetes complications develop, intensive glycemic control (targeting HbA1c levels <7%) places patients at risk for hypoglycemic episodes, and does not appear to decrease risk of cardiovascular complications (Group AC et al. 2008; Duckworth et al. 2009; Gerstein et al. 2008). Lower HbA1c may be even more detrimental to people with DKD, who showed 30% and 40% higher risk for all-cause mortality and

Table 32.1 Recognized actions of inflammatory molecules and pathways in diabetic kidney disease (DKD)

Innate immune cells	
Cell type	*Actions/biological effect*
Kidney Mononuclear Phagocytic System (coexpressing markers of macrophages and dendritic cells)	• Release cytokines and paracrine signaling molecules (Pichler et al. 2017; Navarro-Gonzalez et al. 2011) • Recruit circulating monocytes and macrophages (Nelson et al. 2012; Chow et al. 2005) • Magnitude of infiltration correlates with glomerulosclerosis, loss of eGFR, and tubulointerstitial inflammation (Ninichuk et al. 2007; Furuta et al. 1993; Nguyen et al. 2006; Lin et al. 2012)
Mast cells	• Conversion of angiotensin I to angiotensin II (Reilly et al. 1982) • Release many inflammatory mediators and proteolytic enzyme (Maeda et al. 2010)
Interleukin and tumor necrosis family	
Adhesion molecules	*Actions/biological effect*
Interleukin -1 (IL-1)	• Augment the expression of chemokines and adhesion molecules (Elsherbiny and Al-Gayyar 2016) • Stimulate prostaglandin E2 formation leading to alteration in glomerular hemodynamics (Pfeilschifter and Muhl 1990) • Increase vascular endothelial cells permeability and mesangial cell proliferation (Vesey et al. 2002) • Upsurge glomerular cellularity by elevating hyaluronan production (Jones et al. 2001)
Interleukin -6 (IL-6)	• Instigate podocyte hypertrophy, thickening of GBM (Dalla Vestra et al. 2005; Nagayama et al. 2014) • Cause alterations in plasma flow rate (Roscioni et al. 2014)
Interleukin -18 (IL-18)	• Promote production of cytokines by mesangial cells (Schrijvers et al. 2004) • Induce tubulointerstitial lesions (Turner et al. 2014) • Associated with albuminuria in early stages of DKD (Kim et al. 2012)
Tumor necrosis factor alpha (TNF-α)	• Regulate immune cells and cytokine release (Navarro and Mora-Fernandez 2006) • Alter glomerular hemodynamics (Baud et al. 1998) • Directly induce production of ROS (Koike et al. 2007) • Related to kidney hypertrophy and sodium retention (DiPetrillo et al. 2003) • Induce cell necrosis and apoptosis (O'Brien et al. 1998)
Chemokines	
Adhesion molecules	*Actions/biological effect*
C-C motif chemokine 2 (CCL2) or monocyte chemoattractant protein 1 (MCP-1)	• Involved in release of monocytes from bone marrow (Moser and Loetscher 2001) • Increased levels correlate with kidney macrophage accumulation (Wang et al. 2000) • Angiotensin II probably induces MCP/CCL2 expression (Mizuno et al. 2006; Amann et al. 2003) • In vitro, induce podocyte proliferation, migration, and affects podocyte motility (Burt et al. 2007) • Urine levels correlate with albuminuria (Lee et al. 2009)
CX3-C motif chemokine 1 (CX3CL1)	• Chemoattractant for monocytes, T-cells, and natural killer cells (Umehara et al. 2004) • Induce cell adhesion (Umehara et al. 2004)

(continued)

Table 32.1 (continued)

C-C motif chemokine 5 (CCL5) RANTES	• Chemoattractant for immune cells (Appay and Rowland-Jones 2001) • Expressed by fibroblasts mesangial and tubular epithelial cells (Appay and Rowland-Jones 2001) • Expression in tubular cells directly correlated with the degree of albuminuria (Herder et al. 2006)
Adhesion molecules	
Adhesion Molecules	*Actions/biological effect*
Intercellular adhesion molecule 1 (ICAM-1)	• Involved in leukocyte endothelial transmigration (Gu et al. 2012) • Associated with cardiovascular and all-cause mortality (Astrup et al. 2008) • Levels correlated with severity of albuminuria (Rubio-Guerra et al. 2009)
Vascular cell adhesion protein 1 (VCAM-1)	• Promote adhesion of immune cells (Ina et al. 1999) • Expressed by endothelial and smooth muscle cells (Ina et al. 1999) • In patients with HTN and DM 2 circulating levels correlate with albuminuria (Rubio-Guerra et al. 2009) • High plasma concentration correlates with increased risk of death, independent of other risk factors (Stehouwer et al. 2002)
Endothelial cell-selective adhesion molecule (ESAM)	• Selectively expressed by glomerular endothelial cells (Nasdala et al. 2002) • Regulate vascular permeability and adhesion (Nasdala et al. 2002) • Suggested that its downregulation may promote albuminuria (Nasdala et al. 2002)
E-selectin (CD62E)	• Selectively expressed on endothelial cells following exposure to cytokines (Navarro-Gonzalez et al. 2011) • Soluble levels correlate with the albuminuria and presence of cardiovascular disease in DM 1 patients (Soedamah-Muthu et al. 2006) • Strongly correlated with presence of DKD (Lopes-Virella et al. 2008)
A-Actinin 4	• Exclusively expressed in podocytes C (Kimura et al. 2008) • Expression inversely correlated with the extent of mesangial expansion (Kimura et al. 2008)

Reprinted with permission from (Alicic et al. 2018). This article was published in *Advances in Chronic Kidney Disease*, volume 25, Alicic RZ, Johnson EJ, Tuttle KR. Inflammatory Mechanisms as New Biomarkers and Therapeutic Targets for Diabetic Kidney Disease, pages 181–191, Copyright by the National Kidney Foundation, Inc. (2017)

cardiovascular-related mortality, respectively, with intensive glycemic control compared with standard therapy (Papademetriou et al. 2015). Control of hyperglycemia is therefore recommended to be stratified based on clinical and demographic factors, and patients with diabetes and kidney disease should generally not be treated to an HbA1c target of <7.0% (National Kidney Foundation 2012; ADA 2020).

Arterial Hypertension

Hemodynamic insults to the kidney accelerate the progression of DKD. For instance, after 15 years of follow up, individuals newly diagnosed with type 2 diabetes and treated to blood pressure <150/85 mmHg had significant reduction in overall risk of microvascular complications (37%, $p = 0.009$) compared to those treated to a target blood pressure <180/105 mmHg (Adler et al. 2000). Each 10-mmHg increase in mean

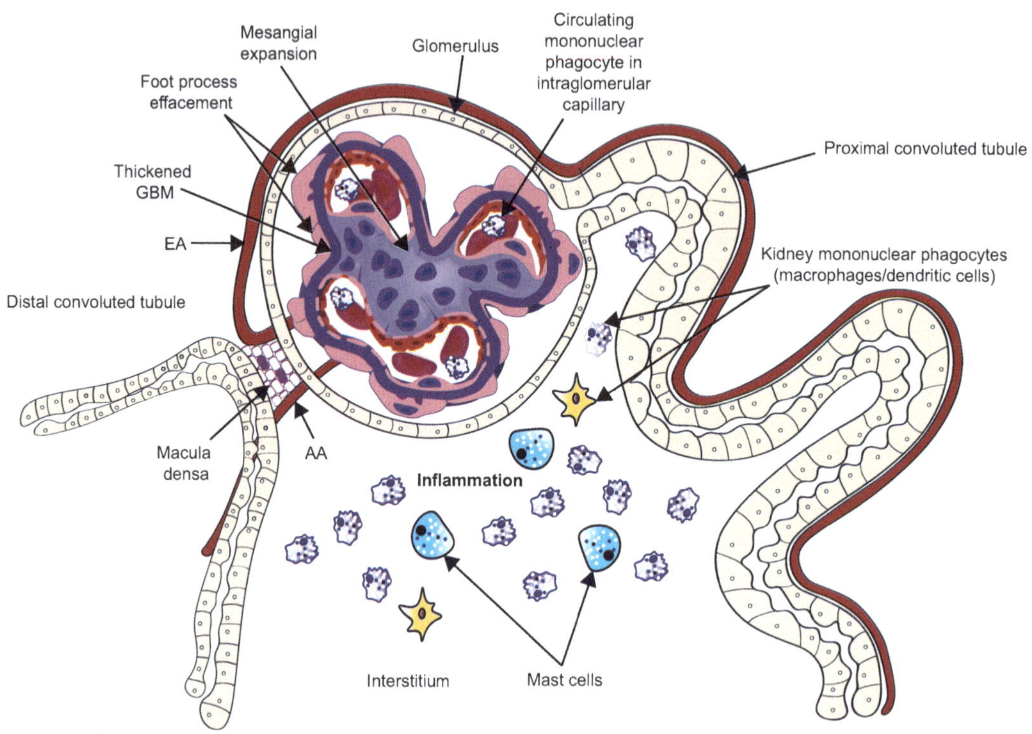

Fig. 32.3 Conceptual model of immunity and inflammation in diabetic kidney disease. Metabolic and hemodynamic abnormalities induced by diabetes instigate cytokine production and activate resident macrophages, dendritic cells, and mast cells. This leads to recruitment of the additional immune cells and further cytokine release. Biological effects of the inflammatory process include podocyte foot process effacement, thickening of the glomerular basement membrane, mesangial expansion, extracellular matrix remodeling, and phenotypic transformation of a variety of kidney cells to myofibroblasts. (Reprinted with permission from (Alicic et al. 2018). This figure was published in *Advances in Chronic Kidney Disease,* volume 25, Alicic RZ, Johnson EJ, Tuttle KR. Inflammatory Mechanisms as New Biomarkers and Therapeutic Targets for Diabetic Kidney Disease, pages 181–191, Copyright by the National Kidney Foundation, Inc. (2017))

systolic blood pressure was associated with a 13% ($p < 0.0001$) increase in hazard ratio for development of both micro- (UACR $\geq$30 to $\leq$300 mg/g) and macroalbuminuria and impaired kidney function, defined as eGFR <60 mL/min/1.73 m^2, or doubling of the blood creatinine level (Adler et al. 2000).

There is a clear evidence that blood pressure control with renin-angiotensin system (RAS) blockade, with either an angiotensin-converting enzyme (ACE) inhibitor or an angiotensin receptor blocker (ARB), reduces the progression of established DKD in patients with macroalbuminuria (Adler et al. 2000; Bakris et al. 2003; Pohl et al. 2005; Brenner et al. 2001). Current treatment guidelines recommend that the majority of patients with diabetes should be treated to a systolic blood pressure (SBP) goal of <140 mmHg and a diastolic blood pressure (DBP) <90 mmHg. Lower SBP and DBP goals, such as <130/80 mmHg, may be appropriate for individuals at high risk of cardiovascular disease, with use of a maximally tolerated dose of either an ACE inhibitor or ARB in patients with diabetes and UACR $\geq$30 mg/g creatinine, and particularly in those with UACR $\geq$300 mg/g creatinine. ACE inhibitors and ARBs can essentially be considered interchangeable, but should not be given together due to higher risks of hyperkalemia and acute kidney injury, without additional benefits for DKD (National Kidney Foundation 2012; ADA 2020; de Boer et al. 2017; Fried et al. 2013).

Healthy Lifestyle

The Verona Diabetes Study reported that obese participants with type 2 diabetes and preserved kidney function (baseline eGFR $\geq$60 mL/min/1.73 m^2) exhibited significantly faster age-adjusted annual eGFR decline during 10 years of follow-up (-1.2 ± 0.1 mL/min/1.73 m^2/year) compared with normal weight individuals (Zoppini et al. 2012). Lifestyle modifications such as regular physical activity, reduction of dietary fat consumption, and smoking cessation are also known to reduce incidence of DKD, and attenuate loss of kidney function (Pongrac Barlovic et al. 2019; Gaede et al. 2003, 2008).

New Therapeutic Approaches: Glucagon-Like Peptide-1 Receptor Agonists

GLP-1 RAs are analogs of the pleiotropic hormone, GLP-1. GLP-1 is produced by enteroendocrine (EEC) intestinal L- cells and is released post-prandially. In vivo, the hormone is quickly degraded by DPP-4, a proteolytic enzyme. Intestinal L-cells are regulated by nutrients, cytokines, bacterial metabolites, and endotoxins (Drucker 2016). Endogenous GLP-1 stimulates glucose-dependent insulin secretion from pancreatic β-cells, suppresses inappropriately elevated glucagon secretion from pancreatic α-cells, delays gastric emptying, and induces satiety via a direct action in the central nervous system (Neumiller 2012). The GLP-1 receptor is most abundantly expressed in the pancreas and duodenum, followed by the stomach and kidney, with very low expression in the heart, lung, and distal intestine, and undetectable expression in the liver and thyroid (Mima et al. 2012; Park et al. 2007; Fan et al. 2019). Currently available GLP-1 RAs are resistant to the DPP-4 enzyme, and can be divided into two main structural categories: shorter-acting exendin-4-based (exenatide and lixisenatide), and longer-acting human GLP-1-based analogs (liraglutide, dulaglutide, and semaglutide) (Table 32.2) (Neumiller 2012; Ahren 2011; Trujillo and Nuffer 2014).

Stemming from the 2008 FDA Guidance for Industry (FDA 2008), all newly marketed glucose-lowering therapies must undergo assessment for cardiovascular safety in a designated cardiovascular outcome trial (CVOT). Implementation of this guidance has provided the medical community reassurance that newly marketed glucose-lowering therapies are safe from a cardiovascular perspective. In CVOTs for the GLP-1 receptor agonists, a multitude of unexpected beneficial biological and clinical effects beyond

Table 32.2 Summary of Currently Available GLP-1 RAs

Entity	Route of administration	Administration frequency	Half-life	Recommended renal dose adjustment
Exenatide	SC Injection	Twice daily	~2.4 h	• Not recommended with CrCl <30 mL/min • Caution recommended with initiating or escalating the dose with CrCl 30–50 mL/min
Lixisenatide	SC Injection	Once daily	~3 h	• Not recommended with CrCl <15 mL/min
Liraglutide	SC Injection	Once daily	~13 h	• No dosage adjustments recommended
Exenatide XR	SC Injection	Once weekly	~1 w	• Not recommended with eGFR <45 mL/min/1.73 m^2 or ESKD
Dulaglutide	SC Injection	Once weekly	~5 d	• No dosage adjustments recommended
Semaglutide	SC Injection	Once weekly	~1 w	• No dosage adjustments recommended
	Oral	Once daily		

Abbreviations: *CrCl* creatinine clearance, *d* days, *eGFR* estimated glomerular filtration rate, *ESKD* end-stage kidney disease, *GLP-1* glucagon-like peptide-1, *h* Hours, *SC* subcutaneous, *w* weeks, *XR* extended release. Information from (Exenatide (Byetta®) Injection 2018; Lixisenatide (Adlyxin®) Injection 2019; Liraglutide (Victoza®) Injection 2019; Exenatide Extended-Release (Bydureon®) Injectable Suspension 2019; Dulaglutide (Trulicity®) Injection 2019; Semaglutide (Ozempic®) Injection 2019; Semaglutide (Rybelsus®) Tablets 2019) glycemic control were observed, including reduction of macroalbuminuria and attenuation of kidney function loss in patients with type 2 diabetes, with or without chronic kidney disease (Greco et al. 2019; Pfeffer et al. 2015; Mann et al. 2017; Marso et al. 2016a, b; Tuttle et al. 2018a).

Glucagon-Like Peptide-1 Receptor Agonist Trials

Dulaglutide

The AWARD-7 trial enrolled 577 individuals with moderate-to-severe DKD (mean eGFR: 38 mL/min/1.73 m^2) who were randomized to receive (1:1:1) dulaglutide 1.5 mg once weekly, dulaglutide 0.75 mg once weekly, or insulin glargine. All participants received insulin lispro for prandial glucose control. Following 52 weeks of treatment, both dulaglutide treatment groups experienced less eGFR decline when compared to those participants in the insulin glargine group: average eGFR decline was −3.3 mL/min/1.73 m^2 in the insulin-treated group and −0.7 mL/min/1.73 m^2 in both higher (1.5 mg weekly) and lower (0.75 mg weekly) dose dulaglutide-treated groups (Tuttle et al. 2018a). Among AWARD-7 patients with macroalbuminuria at high risk for progression, attenuation of mean eGFR decline was maintained (−5.5 mL/min/1.73 m^2 in insulin glargine group compared with −0.7 mL/min/1.73 m^2 and 0.5 mL/min/1.73 m^2 in dulaglutide 0.75 mg and 1.5 mg groups, respectively) (Tuttle et al. 2018a).

Notably, fewer patients in the higher dose dulaglutide group reached the composite endpoint of end-stage kidney disease or >40% eGFR decline, compared to the insulin glargine group (5.2% versus 10.8%, $p = 0.038$) (Fig. 32.4) (Tuttle et al. 2018b). The lesser decline was not influenced by body weight loss (Tuttle et al. 2019a). An additional post hoc analysis found that a composite endpoint of ≥40% eGFR decline or ESKD was lower by approximately half in participants receiving dulaglutide 1.5 mg once-weekly versus those receiving insulin glargine (Tuttle et al. 2019b).

More recently, the Researching Cardiovascular Events with a Weekly Incretin in Diabetes (REWIND) trial findings were reported (Gerstein et al. 2019). Approximately 20% and 35% of the participants had eGFR <60 mL/min/1.73 m^2 and microalbuminuria, respectively. REWIND included a predefined secondary kidney composite outcome inclusive of new-onset macroalbuminuria, a sustained decline in eGFR of ≥30% from baseline, or chronic kidney replacement therapy. Significantly less participants receiving dulaglutide experienced a kidney-related event when compared with placebo (17% vs. 20%, respectively), thus supporting the kidney protective effects observed in AWARD-7 (Fig. 32.4) (Gerstein et al. 2019).

Liraglutide

The Liraglutide Effect and Action in Diabetes: Evaluation of Cardiovascular Outcome Results (LEADER) trial enrolled participants with type 2 diabetes and high cardiovascular risk (Marso et al. 2016a). Approximately 23% of participants had moderate-to-severe DKD. Prespecified secondary kidney outcome was a composite of new-onset persistent macroalbuminuria, persistent doubling of the serum creatinine level, end-stage kidney disease, or death due to disease (Mann et al. 2017). Liraglutide treatment was associated with a lower rate of kidney events compared to those receiving placebo (161 vs. 215 patients; hazard ratio (HR): 0.74; 95% CI: 0.60 to 0.91; $p = 0.004$). The progressive decline in eGFR at 36 months was 7.44 and 7.82 mL/min/1.73 m^2 in the liraglutide and placebo groups, respectively, corresponding to a 2% less decrease observed with liraglutide (Fig. 32.4) (Mann et al. 2017). This finding is supported by several other studies in patients with type 2 diabetes reporting a reduction in albuminuria with liraglutide (Zavattaro et al. 2015; von Scholten et al. 2015). However, results from the Liraglutide Versus Placebo as Add-on to Glucose-Lowering Therapy in Patients With Type 2 Diabetes and Moderate Renal Impairment (LIRA-RENAL) trial did not report a beneficial effect of liraglutide treatment

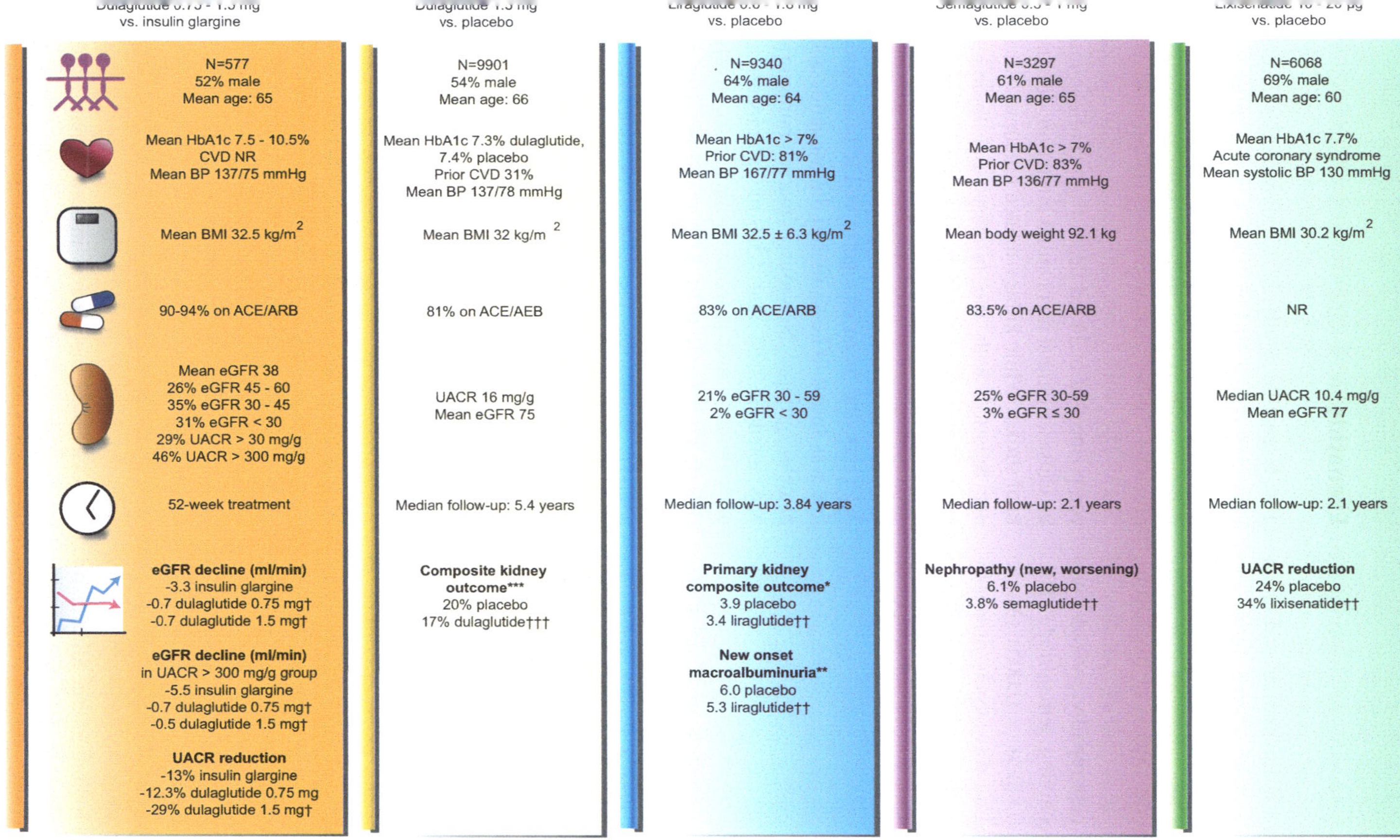

Fig. 32.4 Kidney outcomes in major clinical trials for glucagon-like peptide-1 receptor agonists. Outcomes from clinical trials evaluating members of the glucagon-like peptide-1 receptor agonist class of medications in patients with type 2 diabetes, as well as cardiovascular and renal disease. *Primary composite outcome, new-onset persistent macroalbuminuria, persistent doubling of serum/creatinine and eGFR ≤45 mL/min/1.73 m^2, requirement for renal replacement therapy, or renal death in number of patients (rate per 1000 patient-years of observation). **Rate per 1000 patient-years of observation. ***Composite of new-onset macroalbuminuria sustained eGFR decline ≥30 mL/min/1.73 m^2, or chronic kidney replacement therapy. †Significant from insulin glargine ($p < 0.05$). †† Significant from placebo ($p < 0.05$). ††† Significant from placebo ($p = 0.0004$). Abbreviations: *ACE/ARB* angiotensin-converting enzyme inhibitors or angiotensin receptor blockers, *BMI* body mass index, *BP* blood pressure, *CVD* cardiovascular disease, *DKD* diabetic kidney disease, *eGFR* estimated glomerular filtration rate in units of mL/min/1.73 m^2, *GLP-1* glucagon-like peptide-1, *NR* not reported, *UACR* urine albumin-to-creatinine ratio. Adapted with permission from Alicic RZ, Cox EJ, Tuttle KR. Emergence of GLP-1 Receptor Agonists as a Therapy for Diabetic Kidney Disease. ASN Kidney News. 2019 (Alicic et al. 2019)

on eGFR decline over a shorter period of time of 26 weeks (Davies et al. 2016). There were no differences in eGFR decline between treatment groups (Muskiet et al. 2018).

Semaglutide

The Trial to Evaluate Cardiovascular and Other Long-term Outcomes with Semaglutide in Subjects with Type 2 Diabetes (SUSTAIN-6) was designed to assess the cardiovascular safety of injectable semaglutide (Marso et al. 2016b). SUSTAIN-6 included a secondary outcome of new or worsening nephropathy, which was lower in the group receiving semaglutide (HR: 0.64; 95% CI: 0.46–0.88; $p = 0.005$) (Marso et al. 2016b). This finding was largely driven by an observed reduction in new-onset macroalbuminuria in the semaglutide (2.5%) versus placebo (4.9%) groups (Fig. 32.4) (Marso et al. 2016b).

Lixisenatide

The Evaluation of Lixisenatide in Acute Coronary Syndrome (ELIXA) trial enrolled patients with type 2 diabetes and a recent history of myocardial infarction or hospitalization for unstable angina (Fig. 32.4) (Pfeffer et al. 2015). An exploratory analysis of ELIXA, which included 4441 participants with normoalbuminuria (UACR <30 mg/g), 1148 participants with microalbuminuria, and 389 participants with macroalbuminuria, investigated the percentage change in albuminuria and eGFR according to baseline albuminuria status, and examined the time to new-onset macroalbuminuria and doubling of serum creatinine. After adjustment for traditional risk factors such as HbA1c, lixisenatide was shown to reduce the progression of albuminuria in macroalbuminuric participants (−39.18%; 95% CI: −68.53 to −9.84; $p = 0.0070$). It was also associated with a lower risk for new-onset macroalbuminuria adjusted for baseline, and on-trial HbA1C (HR: 0.808; 95% CI: 0.660–0.991; $p = 0.0404$ and HR: 0.815; 95% CI: 0.665–0.999; $p = 0.0491$, respectively).

Exenatide

The Exenatide Study of Cardiovascular Event Lowering (EXSCEL) trial was a large cardiovascular outcome study, enrolling a total of 14,752 patients with type 2 diabetes with or without previous cardiovascular disease (Holman et al. 2017). The primary outcome was a cardiovascular composite that included death from cardiovascular causes, nonfatal myocardial infarction, or nonfatal stroke. While kidney-related outcomes were not reported with the primary trial findings, results for additional prespecified microvascular analyses were recently published (Bethel et al. 2020). After median follow-up period of 3.2 years, the mean change in eGFR from baseline was found to be similar in the exenatide and placebo groups. New-onset macroalbuminuria was noted in 2.2% and 2.5% of participants in the exenatide and placebo groups, respectively (HR: 0.87; 95% CI: 0.70–1.07; $p = 0.19$) (Bethel et al. 2020). Another recently reported *post-hoc* analysis comparing the effects of twice-daily exenatide versus titrated insulin glargine on kidney function and albuminuria in patients with type 2 diabetes without overt nephropathy, likewise was not shown to affect kidney function decline or progression of albuminuria (Muskiet et al. 2019).

Glucagon-Like Peptide-1 Receptor Agonist Trials: Ongoing Kidney-Outcome Studies

In consideration of the benefits observed in the GLP-1 RA trials summarized, several key studies are ongoing that will shed additional light on the potential benefits of GLP-1 receptor agonism on kidney-related outcomes, in individuals with moderate-to-severe kidney disease. The Effect of Semaglutide Versus Placebo on the Progression of Renal Impairment in Subjects with Type 2 Diabetes and Chronic Kidney Disease (FLOW) is an ongoing randomized clinical trial expected to complete in August 2024 (Clinicaltrials.gov 2019). An estimated 3160 participants will be enrolled and randomized to treatment with weekly injectable semaglutide or placebo. The overall objective of FLOW is to determine the effect of semaglutide on the progression of CKD in patients with type 2 diabetes. Enrolled participants will have renal impairment defined by:

1. An eGFR $\geq$50 and $\leq$75 mL/min/1.73 m^2 and UACR >300 and <5000 mg/g; or
2. An eGFR $\geq$25 and <50 mL/min/1.73 m^2 and UACR >100 and <5000 mg/g.

Participants will additionally be treated with a maximally titrated or tolerated ACE inhibitor or ARB. FLOW will examine time to first occurrence of a composite of eGFR decline of $\geq$50% from baseline, ESRD, or death from kidney or cardiovascular disease, as well as prespecified secondary outcome measures (annual rate of change in eGFR, time to occurrence of all-cause death, time to occurrence of each individual component of the primary composite outcome and relative change in UACR).

Putative Mechanism of Observed Kidney Outcomes

The exact mechanisms responsible for the observed kidney protective effects are subjects of ongoing inquiry. Benefits are likely exerted through influences on modifiable risk factors such as glucose reduction, blood pressure control, and weight loss, as well as direct biological effects on the kidney. The distribution of GLP-1 receptor in kidney is still under investigation, but receptors have been detected in arterial vasculature, glomerular capillaries, endothelial cells, macrophages, juxtaglomerular cells, and cells of the proximal tubules (Fujita et al. 2014; Kodera et al. 2011; Schlatter et al. 2007; Korner et al. 2007; Pyke et al. 2014).

Addition of a GLP-1 RA to background antihyperglycemic therapy leads to additional reductions in HbA1C, ranging from −0.3% to −1.9%, compared to controls in the clinical trials evaluating cardiovascular safety (Pfeffer et al. 2015; Marso et al. 2016a, b; Madsbad 2016). In patients with moderate-to-severe chronic kidney disease (CKD) at stages 3 and 4, active treatment with insulin glargine and dulaglutide resulted in similar reductions in HbA1c (Tuttle et al. 2018a).

Treatment with liraglutide or semaglutide has been associated with SBP reductions of approximately 2–5 mmHg compared to placebo or active treatment with insulin or a sulfonylurea (Sun et al. 2015a). Unfortunately, this effect is blunted or absent in people with moderate-to-severe kidney disease (Tuttle et al. 2018a; Bode 2012).

GLP-1 RAs additionally promote reductions in waist circumference and body weight, with weight reductions of approximately 3 kg on average achieved with GLP-1 RA treatment when compared to other glucose-lowering treatments (Vilsboll et al. 2012; Potts et al. 2015; Sun et al. 2015b). Moreover, the weight loss effect with GLP-1 RA treatment is preserved in patients with CKD (Tuttle et al. 2018a; Davies et al. 2016).

Some of the proposed direct effects on the kidney include alteration in fluid and electrolyte metabolism, through reduction of Na^+/H^+ exchanger 3 (NHE3)-dependent proximal tubule sodium reabsorption, and decrease in sodium bicarbonate and water reabsorption. This is the probable mechanism behind natriuresis and diuresis observed in rats, mice, and humans following GLP-1 receptor activation (Farah et al. 2016; Skov et al. 2013, 2016).

GLP-1 has general anti-inflammatory effects, which may protect the kidney from inflammatory events that induce histological changes and perturb glomerular hemodynamics (Lee and Jun 2016; Deb et al. 2017). Treatment with dulaglutide lowers serum levels of C-reactive protein (CRP) by approximately 1 mg/L, and with exenatide by a mean of 0.5 mg/L (Bunck et al. 2010; Ferdinand et al. 2014). A meta-analysis of randomized controlled trials with GLP-1 RA treatment in patients with type 2 diabetes found an approximately 2 mg/L reduction in serum CRP (Mazidi et al. 2017). In experimental

DKD models, GLP-1 RA treatment exhibits anti-inflammatory and antioxidative effects, resulting in protection from endothelial injury and reduction of proteinuria (Park et al. 2007; Kodera et al. 2011; Hendarto et al. 2012). Plausible candidates for the anti-inflammatory and anti-oxidative mechanisms are increased production of cyclic adenosine monophosphate (cAMP) and protein kinase A (PKA). Even the downstream target of cAMP and PKA activation is unclear. The probable candidate is NAD(P)H oxidase, a major source of oxidative stress in DKD (Jha et al. 2014; Dieter et al. 2018).

In rats, liraglutide has additionally been shown to attenuate oxidative stress, expression of transforming growth factor–beta and fibronectin, and reduce albuminuria (Hendarto et al. 2012). Furthermore, exenedin and liraglutide infusions in rats and humans prevent macrophage infiltration, and inhibit release of proinflammatory cytokines from macrophages, decrease protein levels of intercellular adhesion molecule-1 (ICAM-1) and type IV collagen, decrease TGF-β1 and TGF-β1R expression with inhibit activation of downstream Smad3 and ERK1/2 signaling, activating nuclear factor-κB in kidney tissue (Kodera et al. 2011; Hogan et al. 2014).

Future Perspectives

Major organizations such as the American Diabetes Association (ADA) now recommend GLP-1 RAs with proven cardiovascular benefits be considered for use in patients with indicators of high risk or established atherosclerotic cardiovascular disease, CKD or heart failure independent of baseline HbA1C or individualized glycemic targets (American Diabetes Association 2020). The multitude of biological effects exerted include antihyperglycemic effects, weight loss, and blood pressure reduction.

In addition, a mounting body of knowledge indicates that the GLP-1 RAs address the principal pathomechanism of DKD through their anti-inflammatory, antioxidant, and antifibrotic effects. AWARD-7 suggested that dulaglutide attenuates eGFR decline and progression of albuminuria in patients with type 2 diabetes and moderate-to-severe CKD, providing the first evidence to suggest a benefit of GLP-1 receptor agonism on kidney disease progression in this population (Tuttle et al. 2018a). Additional evidence is needed, however, to determine GLP-1 RAs impact on kidney outcomes in at-risk patients, and effects on cardiovascular risk in this group of patients. Findings from ongoing clinical trials, such as FLOW, will continue to inform our understanding of the role of GLP-1 RAs in patients with established kidney disease. Hopefully, ongoing trials will confirm GLP-1 RAs as one of the most promising therapeutic interventions, to mitigate the risk of development and progression of DKD.

References

ADA (2020) 10. Cardiovascular disease and risk management: standards of medical care in diabetes—2020. Diabetes Care 43(Supplement 1):S111–S134

Adler AI, Stratton IM, Neil HA et al (2000) Association of systolic blood pressure with macrovascular and microvascular complications of type 2 diabetes (UKPDS 36): prospective observational study. BMJ 321 (7258):412–419

Afkarian M, Sachs MC, Kestenbaum B et al (2013) Kidney disease and increased mortality risk in type 2 diabetes. J Am Soc Nephrol 24(2):302–308

Ahren B (2011) GLP-1 for type 2 diabetes. Exp Cell Res 317(9):1239–1245

Alicic RZ, Rooney MT, Tuttle KR (2017) Diabetic kidney disease: challenges, progress, and possibilities. Clin J Am Soc Nephrol 12(12):2032–2045

Alicic RZ, Johnson EJ, Tuttle KR (2018) Inflammatory mechanisms as new biomarkers and therapeutic targets for diabetic kidney disease. Adv Chronic Kidney Dis 25(2):181–191

Alicic RZ, Cox EJ, Tuttle KR (2019) Emergence of GLP-1 receptor agonists as a therapy for diabetic kidney disease. ASN Kidney News, Washington, DC

Amann B, Tinzmann R, Angelkort B (2003) ACE inhibitors improve diabetic nephropathy through suppression of renal MCP-1. Diabetes Care 26 (8):2421–2425

American Diabetes Association (2020) 9. Pharmacologic approaches to glycemic treatment: standards of medical Care in Diabetes-2020. Diabetes Care 43(Suppl 1): S98–S110

Appay V, Rowland-Jones SL (2001) RANTES: a versatile and controversial chemokine. Trends Immunol 22 (2):83–87

Araki S, Haneda M, Koya D et al (2007) Predictive impact of elevated serum level of IL-18 for early renal dysfunction in type 2 diabetes: an observational follow-up study. Diabetologia 50(4):867–873
Astrup AS, Tarnow L, Pietraszek L et al (2008) Markers of endothelial dysfunction and inflammation in type 1 diabetic patients with or without diabetic nephropathy followed for 10 years: association with mortality and decline of glomerular filtration rate. Diabetes Care 31 (6):1170–1176
Bakris GL, Weir MR, Shanifar S et al (2003) Effects of blood pressure level on progression of diabetic nephropathy: results from the RENAAL study. Arch Intern Med 163(13):1555–1565
Baud L, Fouqueray B, Suberville S, Doublier S (1998) Interleukin 10: a logical candidate for suppressing glomerular inflammation? Exp Nephrol 6(1):22–27
Berthier CC, Zhang H, Schin M et al (2009) Enhanced expression of Janus kinase-signal transducer and activator of transcription pathway members in human diabetic nephropathy. Diabetes 58(2):469–477
Bethel MA, Mentz RJ, Merrill P et al (2020) Microvascular and cardiovascular outcomes according to renal function in patients treated with once-weekly Exenatide: insights from the EXSCEL trial. Diabetes Care 43(2):446–452
Bode B (2012) An overview of the pharmacokinetics, efficacy and safety of liraglutide. Diabetes Res Clin Pract 97(1):27–42
Brenner BM, Cooper ME, de Zeeuw D et al (2001) Effects of losartan on renal and cardiovascular outcomes in patients with type 2 diabetes and nephropathy. N Engl J Med 345(12):861–869
Brosius FC, Tuttle KR, Kretzler M (2016) JAK inhibition in the treatment of diabetic kidney disease. Diabetologia 59(8):1624–1627
Bunck MC, Diamant M, Eliasson B et al (2010) Exenatide affects circulating cardiovascular risk biomarkers independently of changes in body composition. Diabetes Care 33(8):1734–1737
Burt D, Salvidio G, Tarabra E et al (2007) The monocyte chemoattractant protein-1/cognate CC chemokine receptor 2 system affects cell motility in cultured human podocytes. Am J Pathol 171(6):1789–1799
Cheung CY, Ma MKM, Chak WL, Tang SCW (2017) Cancer risk in patients with diabetic nephropathy: a retrospective cohort study in Hong Kong. Medicine (Baltimore) 96(38):e8077
Chow FY, Nikolic-Paterson DJ, Ozols E, Atkins RC, Tesch GH (2005) Intercellular adhesion molecule-1 deficiency is protective against nephropathy in type 2 diabetic db/db mice. J Am Soc Nephrol 16 (6):1711–1722
Clinicaltrials.gov (2019) A research study to see how semaglutide works compared to placebo in people with type 2 diabetes and chronic kidney disease (FLOW). https://clinicaltrials.gov/ct2/show/NCT03819153?term=semaglutide+FLOW&draw=2&rank=1. Accessed 18 Dec 2019
Collaboration NCDRF (2016) Trends in adult body-mass index in 200 countries from 1975 to 2014: a pooled analysis of 1698 population-based measurement studies with 19.2 million participants. Lancet 387 (10026):1377–1396
Couser WG, Remuzzi G, Mendis S, Tonelli M (2011) The contribution of chronic kidney disease to the global burden of major noncommunicable diseases. Kidney Int 80(12):1258–1270
D'Agati VD, Chagnac A, de Vries AP et al (2016) Obesity-related glomerulopathy: clinical and pathologic characteristics and pathogenesis. Nat Rev Nephrol 12(8):453–471
Dalla Vestra M, Mussap M, Gallina P et al (2005) Acute-phase markers of inflammation and glomerular structure in patients with type 2 diabetes. J Am Soc Nephrol 16(Suppl 1):S78–S82
Davies MJ, Bain SC, Atkin SL et al (2016) Efficacy and safety of Liraglutide versus placebo as add-on to glucose-lowering therapy in patients with type 2 diabetes and moderate renal impairment (LIRA-RENAL): a randomized clinical trial. Diabetes Care 39(2):222–230
de Boer IH, Bangalore S, Benetos A et al (2017) Diabetes and hypertension: a position statement by the American Diabetes Association. Diabetes Care 40 (9):1273–1284
Deb DK, Bao R, Li YC (2017) Critical role of the cAMP-PKA pathway in hyperglycemia-induced epigenetic activation of fibrogenic program in the kidney. FASEB J 31(5):2065–2075
Dieter BP, Alicic RZ, Tuttle KR (2018) GLP-1 receptor agonists in diabetic kidney disease: from the patient-side to the bench-side. Am J Physiol Ren Physiol 315 (6):F1519–F1525
DiPetrillo K, Coutermarsh B, Gesek FA (2003) Urinary tumor necrosis factor contributes to sodium retention and renal hypertrophy during diabetes. Am J Physiol Ren Physiol 284(1):F113–F121
Drucker DJ (2016) The cardiovascular biology of glucagon-like Peptide-1. Cell Metab 24(1):15–30
Duckworth W, Abraira C, Moritz T et al (2009) Glucose control and vascular complications in veterans with type 2 diabetes. N Engl J Med 360(2):129–139
Dulaglutide (Trulicity®) Injection (2019) Prescribing information. Eli Lilly and Company, Indianapolis, IN
Elsherbiny NM, Al-Gayyar MM (2016) The role of IL-18 in type 1 diabetic nephropathy: the problem and future treatment. Cytokine 81:15–22
Exenatide (Byetta®) Injection (2018) Prescribing information. AstraZeneca Pharmaceuticals LP, Wilmington, DE
Exenatide Extended-Release (Bydureon®) Injectable Suspension (2019) Prescribing information. AstraZeneca Pharmaceuticals LP, Wilmington, DE
Fan Y, Yi Z, D'Agati VD et al (2019) Comparison of kidney transcriptomic profiles of early and advanced diabetic nephropathy reveals potential new mechanisms for disease progression. Diabetes 68 (12):2301–2314

Farah LX, Valentini V, Pessoa TD, Malnic G, McDonough AA, Girardi AC (2016) The physiological role of glucagon-like peptide-1 in the regulation of renal function. Am J Physiol Ren Physiol 310(2): F123–F127

FDA (2008) Guidance for industry: diabetes mellitus—evaluating cardiovascular risk in new antidiabetic therapies to treat type 2 diabetes. Food and Drug Administration, USA

Ferdinand KC, White WB, Calhoun DA et al (2014) Effects of the once-weekly glucagon-like peptide-1 receptor agonist dulaglutide on ambulatory blood pressure and heart rate in patients with type 2 diabetes mellitus. Hypertension 64(4):731–737

Flegal KM, Kruszon-Moran D, Carroll MD, Fryar CD, Ogden CL (2016) Trends in obesity among adults in the United States, 2005 to 2014. JAMA 315 (21):2284–2291

Fried LF, Emanuele N, Zhang JH et al (2013) Combined angiotensin inhibition for the treatment of diabetic nephropathy. N Engl J Med 369(20):1892–1903

Fujita T, Ogihara N, Kamura Y et al (2012) Interleukin-18 contributes more closely to the progression of diabetic nephropathy than other diabetic complications. Acta Diabetol 49(2):111–117

Fujita H, Morii T, Fujishima H et al (2014) The protective roles of GLP-1R signaling in diabetic nephropathy: possible mechanism and therapeutic potential. Kidney Int 85(3):579–589

Furuta T, Saito T, Ootaka T et al (1993) The role of macrophages in diabetic glomerulosclerosis. Am J Kidney Dis 21(5):480–485

Gaede P, Vedel P, Larsen N, Jensen GV, Parving HH, Pedersen O (2003) Multifactorial intervention and cardiovascular disease in patients with type 2 diabetes. N Engl J Med 348(5):383–393

Gaede P, Lund-Andersen H, Parving HH, Pedersen O (2008) Effect of a multifactorial intervention on mortality in type 2 diabetes. N Engl J Med 358(6):580–591

Gerstein HC, Miller ME, et al Action to Control Cardiovascular Risk in Diabetes Study Group (2008) Effects of intensive glucose lowering in type 2 diabetes. N Engl J Med 358(24):2545–2559

Gerstein HC, Colhoun HM, Dagenais GR et al (2019) Dulaglutide and cardiovascular outcomes in type 2 diabetes (REWIND): a double-blind, randomised placebo-controlled trial. Lancet 394(10193):121–130

Goldberg RB (2009) Cytokine and cytokine-like inflammation markers, endothelial dysfunction, and imbalanced coagulation in development of diabetes and its complications. J Clin Endocrinol Metab 94 (9):3171–3182

Greco EV, Russo G, Giandalia A, Viazzi F, Pontremoli R, De Cosmo S (2019) GLP-1 receptor agonists and kidney protection. Medicina (Kaunas) 55(6):233

Group AC, Patel A, MacMahon S et al (2008) Intensive blood glucose control and vascular outcomes in patients with type 2 diabetes. N Engl J Med 358 (24):2560–2572

Group DER, de Boer IH, Sun W et al (2011) Intensive diabetes therapy and glomerular filtration rate in type 1 diabetes. N Engl J Med 365(25):2366–2376

Gu HF, Ma J, Gu KT, Brismar K (2012) Association of intercellular adhesion molecule 1 (ICAM1) with diabetes and diabetic nephropathy. Front Endocrinol (Lausanne) 3:179

Hendarto H, Inoguchi T, Maeda Y et al (2012) GLP-1 analog liraglutide protects against oxidative stress and albuminuria in streptozotocin-induced diabetic rats via protein kinase A-mediated inhibition of renal NAD (P)H oxidases. Metabolism 61(10):1422–1434

Herder C, Peltonen M, Koenig W et al (2006) Systemic immune mediators and lifestyle changes in the prevention of type 2 diabetes: results from the Finnish diabetes prevention study. Diabetes 55(8):2340–2346

Hiromura K, Kurosawa M, Yano S, Naruse T (1998) Tubulointerstitial mast cell infiltration in glomerulonephritis. Am J Kidney Dis 32(4):593–599

Hogan AE, Gaoatswe G, Lynch L et al (2014) Glucagon-like peptide 1 analogue therapy directly modulates innate immune-mediated inflammation in individuals with type 2 diabetes mellitus. Diabetologia 57 (4):781–784

Holman RR, Bethel MA, Mentz RJ et al (2017) Effects of once-weekly Exenatide on cardiovascular outcomes in type 2 diabetes. N Engl J Med 377(13):1228–1239

Ina K, Kitamura H, Okeda T et al (1999) Vascular cell adhesion molecule-1 expression in the renal interstitium of diabetic KKAy mice. Diabetes Res Clin Pract 44(1):1–8

Jha JC, Gray SP, Barit D et al (2014) Genetic targeting or pharmacologic inhibition of NADPH oxidase nox4 provides renoprotection in long-term diabetic nephropathy. J Am Soc Nephrol 25(6):1237–1254

Jones SA (2005) Directing transition from innate to acquired immunity: defining a role for IL-6. J Immunol 175(6):3463–3468

Jones S, Jones S, Phillips AO (2001) Regulation of renal proximal tubular epithelial cell hyaluronan generation: implications for diabetic nephropathy. Kidney Int 59 (5):1739–1749

Kaplanski G, Marin V, Montero-Julian F, Mantovani A, Farnarier C (2003) IL-6: a regulator of the transition from neutrophil to monocyte recruitment during inflammation. Trends Immunol 24(1):25–29

Kim SS, Song SH, Kim IJ et al (2012) Clinical implication of urinary tubular markers in the early stage of nephropathy with type 2 diabetic patients. Diabetes Res Clin Pract 97(2):251–257

Kimura M, Toyoda M, Kato M et al (2008) Expression of alpha-actinin-4 in human diabetic nephropathy. Intern Med 47(12):1099–1106

Kodera R, Shikata K, Kataoka HU et al (2011) Glucagon-like peptide-1 receptor agonist ameliorates renal injury through its anti-inflammatory action without lowering blood glucose level in a rat model of type 1 diabetes. Diabetologia 54(4):965–978

Koike N, Takamura T, Kaneko S (2007) Induction of reactive oxygen species from isolated rat glomeruli by protein kinase C activation and TNF-alpha stimulation, and effects of a phosphodiesterase inhibitor. Life Sci 80(18):1721–1728

Korner M, Stockli M, Waser B, Reubi JC (2007) GLP-1 receptor expression in human tumors and human normal tissues: potential for in vivo targeting. J Nucl Med 48(5):736–743

Lee YS, Jun HS (2016) Anti-inflammatory effects of GLP-1-based therapies beyond glucose control. Mediat Inflamm 2016:3094642

Lee EY, Chung CH, Khoury CC et al (2009) The monocyte chemoattractant protein-1/CCR2 loop, inducible by TGF-beta, increases podocyte motility and albumin permeability. Am J Physiol Ren Physiol 297(1):F85–F94

Lin M, Yiu WH, Wu HJ et al (2012) Toll-like receptor 4 promotes tubular inflammation in diabetic nephropathy. J Am Soc Nephrol 23(1):86–102

Liraglutide (Victoza®) Injection (2019) Prescribing information. Novo Nordisk, Inc, Plainsboro, NJ

Lixisenatide (Adlyxin®) Injection (2019) Prescribing information. Sanofi-aventis U.S., LLC, Bridgewater, NJ

Lopes-Virella MF, Carter RE, Gilbert GE et al (2008) Risk factors related to inflammation and endothelial dysfunction in the DCCT/EDIC cohort and their relationship with nephropathy and macrovascular complications. Diabetes Care 31(10):2006–2012

Lozano R, Naghavi M, Foreman K et al (2012) Global and regional mortality from 235 causes of death for 20 age groups in 1990 and 2010: a systematic analysis for the global burden of disease study 2010. Lancet 380 (9859):2095–2128

Macisaac RJ, Ekinci EI, Jerums G (2014) Markers of and risk factors for the development and progression of diabetic kidney disease. Am J Kidney Dis 63(2 Suppl 2):S39–S62

Madsbad S (2016) Review of head-to-head comparisons of glucagon-like peptide-1 receptor agonists. Diabetes Obes Metab 18(4):317–332

Maeda Y, Inoguchi T, Takei R et al (2010) Inhibition of chymase protects against diabetes-induced oxidative stress and renal dysfunction in hamsters. Am J Physiol Ren Physiol 299(6):F1328–F1338

Mann JFE, Orsted DD, Brown-Frandsen K et al (2017) Liraglutide and renal outcomes in type 2 diabetes. N Engl J Med 377(9):839–848

Marso SP, Daniels GH, Brown-Frandsen K et al (2016a) Liraglutide and cardiovascular outcomes in type 2 diabetes. N Engl J Med 375(4):311–322

Marso SP, Bain SC, Consoli A et al (2016b) Semaglutide and cardiovascular outcomes in patients with type 2 diabetes. N Engl J Med 375(19):1834–1844

Mazidi M, Karimi E, Rezaie P, Ferns GA (2017) Treatment with GLP1 receptor agonists reduce serum CRP concentrations in patients with type 2 diabetes mellitus: a systematic review and meta-analysis of randomized controlled trials. J Diabetes Complicat 31 (7):1237–1242

Mima A, Hiraoka-Yamomoto J, Li Q et al (2012) Protective effects of GLP-1 on glomerular endothelium and its inhibition by PKCbeta activation in diabetes. Diabetes 61(11):2967–2979

Mizuno M, Sada T, Kato M, Fukushima Y, Terashima H, Koike H (2006) The effect of angiotensin II receptor blockade on an end-stage renal failure model of type 2 diabetes. J Cardiovasc Pharmacol 48(4):135–142

Moser B, Loetscher P (2001) Lymphocyte traffic control by chemokines. Nat Immunol 2(2):123–128

Muskiet MHA, Tonneijck L, Huang Y et al (2018) Lixisenatide and renal outcomes in patients with type 2 diabetes and acute coronary syndrome: an exploratory analysis of the ELIXA randomised, placebo-controlled trial. Lancet Diabetes Endocrinol 6 (11):859–869

Muskiet MHA, Bunck MC, Heine RJ et al (2019) Exenatide twice-daily does not affect renal function or albuminuria compared to titrated insulin glargine in patients with type 2 diabetes mellitus: a post-hoc analysis of a 52-week randomised trial. Diabetes Res Clin Pract 153:14–22

Nagayama Y, Braun GS, Jakobs CM et al (2014) Gp130-dependent signaling in the podocyte. Am J Physiol Ren Physiol 307(3):F346–F355

Nasdala I, Wolburg-Buchholz K, Wolburg H et al (2002) A transmembrane tight junction protein selectively expressed on endothelial cells and platelets. J Biol Chem 277(18):16294–16303

Nathan DM, Group DER (2014) The diabetes control and complications trial/epidemiology of diabetes interventions and complications study at 30 years: overview. Diabetes Care 37(1):9–16

National Kidney Foundation (2012) KDOQI clinical practice guideline for diabetes and CKD: 2012 update. Am J Kidney Dis 60(5):850–886

Navarro JF, Mora-Fernandez C (2006) The role of TNF-alpha in diabetic nephropathy: pathogenic and therapeutic implications. Cytokine Growth Factor Rev 17(6):441–450

Navarro JF, Milena FJ, Mora C et al (2005) Tumor necrosis factor-alpha gene expression in diabetic nephropathy: relationship with urinary albumin excretion and effect of angiotensin-converting enzyme inhibition. Kidney Int Suppl 99:S98–S102

Navarro-Gonzalez JF, Mora-Fernandez C, Muros de Fuentes M, Garcia-Perez J (2011) Inflammatory molecules and pathways in the pathogenesis of diabetic nephropathy. Nat Rev Nephrol 7(6):327–340

Nelson PJ, Rees AJ, Griffin MD, Hughes J, Kurts C, Duffield J (2012) The renal mononuclear phagocytic system. J Am Soc Nephrol 23(2):194–203

Neumiller JJ (2012) Incretin pharmacology: a review of the incretin effect and current incretin-based therapies. Cardiovasc Hematol Agents Med Chem 10 (4):276–288

Nguyen D, Ping F, Mu W, Hill P, Atkins RC, Chadban SJ (2006) Macrophage accumulation in human progressive diabetic nephropathy. Nephrology (Carlton) 11 (3):226–231

Ninichuk V, Khandoga AG, Segerer S et al (2007) The role of interstitial macrophages in nephropathy of type 2 diabetic db/db mice. Am J Pathol 170(4):1267–1276

Nosadini R, Velussi M, Brocco E et al (2000) Course of renal function in type 2 diabetic patients with abnormalities of albumin excretion rate. Diabetes 49 (3):476–484

O'Brien JB, Piddington DL, Voelkel-Johnson C, Richards DJ, Hadley LA, Laster SM (1998) Sustained phosphorylation of cytosolic phospholipase A2 accompanies cycloheximide- and adenovirus-induced susceptibility to TNF. J Immunol 161(3):1525–1532

Okon K, Stachura J (2007) Increased mast cell density in renal interstitium is correlated with relative interstitial volume, serum creatinine and urea especially in diabetic nephropathy but also in primary glomerulonephritis. Pol J Pathol 58(3):193–197

Papademetriou V, Lovato L, Doumas M et al (2015) Chronic kidney disease and intensive glycemic control increase cardiovascular risk in patients with type 2 diabetes. Kidney Int 87(3):649–659

Park CW, Kim HW, Ko SH et al (2007) Long-term treatment of glucagon-like peptide-1 analog exendin-4 ameliorates diabetic nephropathy through improving metabolic anomalies in db/db mice. J Am Soc Nephrol 18(4):1227–1238

Pfeffer MA, Claggett B, Diaz R et al (2015) Lixisenatide in patients with type 2 diabetes and acute coronary syndrome. N Engl J Med 373(23):2247–2257

Pfeilschifter J, Muhl H (1990) Interleukin 1 and tumor necrosis factor potentiate angiotensin II- and calcium ionophore-stimulated prostaglandin E2 synthesis in rat renal mesangial cells. Biochem Biophys Res Commun 169(2):585–595

Pichler R, Afkarian M, Dieter BP, Tuttle KR (2017) Immunity and inflammation in diabetic kidney disease: translating mechanisms to biomarkers and treatment targets. Am J Physiol Ren Physiol 312(4):F716–F731

Pohl MA, Blumenthal S, Cordonnier DJ et al (2005) Independent and additive impact of blood pressure control and angiotensin II receptor blockade on renal outcomes in the irbesartan diabetic nephropathy trial: clinical implications and limitations. J Am Soc Nephrol 16(10):3027–3037

Pongrac Barlovic D, Tikkanen-Dolenc H, Groop PH (2019) Physical activity in the prevention of development and progression of kidney disease in type 1 diabetes. Curr Diab Rep 19(7):41

Potts JE, Gray LJ, Brady EM, Khunti K, Davies MJ, Bodicoat DH (2015) The effect of glucagon-like peptide 1 receptor agonists on weight loss in type 2 diabetes: a systematic review and mixed treatment comparison meta-analysis. PLoS One 10(6):e0126769

Pyke C, Heller RS, Kirk RK et al (2014) GLP-1 receptor localization in monkey and human tissue: novel distribution revealed with extensively validated monoclonal antibody. Endocrinology 155(4):1280–1290

Rao Kondapally Seshasai S, Kaptoge S, Thompson A et al (2011) Diabetes mellitus, fasting glucose, and risk of cause-specific death. N Engl J Med 364(9):829–841

Reilly CF, Tewksbury DA, Schechter NM, Travis J (1982) Rapid conversion of angiotensin I to angiotensin II by neutrophil and mast cell proteinases. J Biol Chem 257 (15):8619–8622

Retnakaran R, Cull CA, Thorne KI, Adler AI, Holman RR, UKPDS Study Group (2006) Risk factors for renal dysfunction in type 2 diabetes:: U.K. Prospective Diabetes Study 74. Diabetes 55(6):1832–1839

Reutens AT (2013) Epidemiology of diabetic kidney disease. Med Clin North Am 97(1):1–18

Roscioni SS, Heerspink HJ, de Zeeuw D (2014) The effect of RAAS blockade on the progression of diabetic nephropathy. Nat Rev Nephrol 10(2):77–87

Royall JA, Berkow RL, Beckman JS, Cunningham MK, Matalon S, Freeman BA (1989) Tumor necrosis factor and interleukin 1 alpha increase vascular endothelial permeability. Am J Phys 257(6 Pt 1):L399–L410

Rubio-Guerra AF, Vargas-Robles H, Lozano Nuevo JJ, Escalante-Acosta BA (2009) Correlation between circulating adhesion molecule levels and albuminuria in type-2 diabetic hypertensive patients. Kidney Blood Press Res 32(2):106–109

Schlatter P, Beglinger C, Drewe J, Gutmann H (2007) Glucagon-like peptide 1 receptor expression in primary porcine proximal tubular cells. Regul Pept 141 (1–3):120–128

Schrijvers BF, De Vriese AS, Flyvbjerg A (2004) From hyperglycemia to diabetic kidney disease: the role of metabolic, hemodynamic, intracellular factors and growth factors/cytokines. Endocr Rev 25(6):971–1010

Semaglutide (Ozempic®) Injection (2019) Prescribing information. Novo Nordisk, Inc, Plainsboro, NJ

Semaglutide (Rybelsus®) Tablets (2019) Prescribing information. Novo Nordisk, Inc, Plainsboro, NJ

Skov J, Dejgaard A, Frokiaer J et al (2013) Glucagon-like peptide-1 (GLP-1): effect on kidney hemodynamics and renin-angiotensin-aldosterone system in healthy men. J Clin Endocrinol Metab 98(4):E664–E671

Skov J, Pedersen M, Holst JJ et al (2016) Short-term effects of liraglutide on kidney function and vasoactive hormones in type 2 diabetes: a randomized clinical trial. Diabetes Obes Metab 18(6):581–589

Soedamah-Muthu SS, Chaturvedi N, Schalkwijk CG et al (2006) Soluble vascular cell adhesion molecule-1 and soluble E-selectin are associated with micro- and macrovascular complications in type 1 diabetic patients. J Diabetes Complicat 20(3):188–195

Stehouwer CD, Gall MA, Twisk JW, Knudsen E, Emeis JJ, Parving HH (2002) Increased urinary albumin excretion, endothelial dysfunction, and chronic low-grade inflammation in type 2 diabetes: progressive, interrelated, and independently associated with risk of death. Diabetes 51(4):1157–1165

Sun F, Wu S, Guo S et al (2015a) Impact of GLP-1 receptor agonists on blood pressure, heart rate and hypertension among patients with type 2 diabetes: a systematic review and network meta-analysis. Diabetes Res Clin Pract 110(1):26–37

Sun F, Wu S, Guo S et al (2015b) Effect of GLP-1 receptor agonists on waist circumference among type 2 diabetes patients: a systematic review and network meta-analysis. Endocrine 48(3):794–803

Thomas MC (2014) Glycemic exposure, glycemic control, and metabolic karma in diabetic complications. Adv Chronic Kidney Dis 21(3):311–317

Tonna S, El-Osta A, Cooper ME, Tikellis C (2010) Metabolic memory and diabetic nephropathy: potential role for epigenetic mechanisms. Nat Rev Nephrol 6 (6):332–341

Trujillo JM, Nuffer W (2014) GLP-1 receptor agonists for type 2 diabetes mellitus: recent developments and emerging agents. Pharmacotherapy 34(11):1174–1186

Turner CM, Arulkumaran N, Singer M, Unwin RJ, Tam FW (2014) Is the inflammasome a potential therapeutic target in renal disease? BMC Nephrol 15:21

Tuttle KR, Lakshmanan MC, Rayner B et al (2018a) Dulaglutide versus insulin glargine in patients with type 2 diabetes and moderate-to-severe chronic kidney disease (AWARD-7): a multicentre, open-label, randomised trial. Lancet Diabetes Endocrinol 6 (8):605–617

Tuttle KR, Lakshmanan M, Rayner B, Zimmermann AG, Woodward B, Botros FT (2018b) CKD outcomes in type 2 diabetes and moderate-to-severe CKD treated with dulaglutide versus insulin glargine: AWARD-7 [Oral presentation]. Poster TH-OR033. Paper presented at: American Society of Nephrology Kidney Week, San Diego, CA

Tuttle KR, Lakshmanan MC, Rayner B, Zimmermann AG, Woodward B, Botros FT (2019a) Body weight and eGFR during dulaglutide treatment in type 2 diabetes and moderate-to-severe chronic kidney disease (AWARD-7). Diabetes Obes Metab 21(6):1493–1497

Tuttle KR, Rayner B, Lakshmanan M et al (2019b) Clinical events in type 2 diabetes and moderate-to-severe CKD by albuminuria status: dulaglutide vs. insulin glargine. Abstract SA-OR081. ASN Kidney Week 2019, Washington, DC

UK Prospective Diabetes Study (UKPDS) (1991) VIII. Study design, progress and performance. Diabetologia 34(12):877–890

Umehara H, Bloom ET, Okazaki T, Nagano Y, Yoshie O, Imai T (2004) Fractalkine in vascular biology: from basic research to clinical disease. Arterioscler Thromb Vasc Biol 24(1):34–40

Van JA, Scholey JW, Konvalinka A (2017) Insights into diabetic kidney disease using urinary proteomics and bioinformatics. J Am Soc Nephrol 28(4):1050–1061

Vesey DA, Cheung C, Cuttle L, Endre Z, Gobe G, Johnson DW (2002) Interleukin-1beta stimulates human renal fibroblast proliferation and matrix protein production by means of a transforming growth factor-beta-dependent mechanism. J Lab Clin Med 140(5):342–350

Vilsboll T, Christensen M, Junker AE, Knop FK, Gluud LL (2012) Effects of glucagon-like peptide-1 receptor agonists on weight loss: systematic review and meta-analyses of randomised controlled trials. BMJ 344: d7771

von Scholten BJ, Hansen TW, Goetze JP, Persson F, Rossing P (2015) Glucagon-like peptide 1 receptor agonist (GLP-1 RA): long-term effect on kidney function in patients with type 2 diabetes. J Diabetes Complicat 29(5):670–674

Wang SN, LaPage J, Hirschberg R (2000) Role of glomerular ultrafiltration of growth factors in progressive interstitial fibrosis in diabetic nephropathy. Kidney Int 57(3):1002–1014

Wen CP, Chang CH, Tsai MK et al (2017) Diabetes with early kidney involvement may shorten life expectancy by 16 years. Kidney Int 92(2):388–396

Whiting DR, Guariguata L, Weil C, Shaw J (2011) IDF diabetes atlas: global estimates of the prevalence of diabetes for 2011 and 2030. Diabetes Res Clin Pract 94(3):311–321

Zavattaro M, Caputo M, Sama MT et al (2015) One-year treatment with liraglutide improved renal function in patients with type 2 diabetes: a pilot prospective study. Endocrine 50(3):620–626

Zoppini G, Targher G, Chonchol M et al (2012) Predictors of estimated GFR decline in patients with type 2 diabetes and preserved kidney function. Clin J Am Soc Nephrol 7(3):401–408

Zoungas S, Arima H, Gerstein HC et al (2017) Effects of intensive glucose control on microvascular outcomes in patients with type 2 diabetes: a meta-analysis of individual participant data from randomised controlled trials. Lancet Diabetes Endocrinol 5(6):431–437

33 The Epidemiology of the Diabetes: Depression Comorbidity in Brazil—Inequality and Interaction

Finn Diderichsen

Contents

Introduction: Why Diabetes and Depression? 458
Diabetes: Increasingly a Disorder of Poverty 459
The Social Patterning of Diabetes, Obesity and Depression in Brazil 460
Mechanisms of Health Inequalities 461
Differential Consequences 461
Interacting Comorbidity 462
Clustering 462
Differential Exposure 463
Obesity Patterns in Brazil 464
Gene Expression and Miscellaneous Mechanisms 464
Differential Vulnerability 464
The Impact of Obesity 465
Summarizing 466
References 467

Abstract

Type 2 diabetes has worldwide increasingly become a disease clustering among the low educated. The steep negative gradient of disability across levels of education among both men and women, contrasts with a positive gradient for both diabetes and obesity among men. It is found that both diabetes and depression clusters among the low educated, and the two disorders also cluster to each other. The effect of diabetes on disability is much stronger among people also suffering from depression, and that interaction is particularly remarkable among men. It is also stronger among low educated, particularly among low-educated women.

The unequal prevalence of diabetes is partly due to a similar patterning of obesity, but it is

F. Diderichsen (✉)
University of Copenhagen, Joao Pessoa, PB, Brazil
e-mail: fidi@sund.ku.dk

J. Faintuch and S. Faintuch (eds.), *Obesity and Diabetes*, https://doi.org/10.1007/978-3-030-53370-0_33

found that there exists a differential susceptibility, in which the association between obesity and diabetes is stronger among the low educated. That seems partly due to higher susceptibility to obesity among those suffering from comorbid depression.

There is thus evidence not only of a vicious circle between diabetes—depression—obesity—diabetes, but the actual impact on people's daily life and capability of these comorbidities, is also aggravated by an interaction between the two disorders and with low socioeconomic status. Diabetes and depression thus show a pattern of syndemics in the context of large social inequalities of Brazil.

Keywords

Poverty · Disability · Depression · Diabetes · Social inequality · Disease burden

Introduction: Why Diabetes and Depression?

Some comorbidities are more interesting and important than others—for populations and for patients. The combination of diabetes and depression is one of them. For at least five reasons. Both disorders occur with high and—and for diabetes at least—increasing prevalence. They have both substantial effect on disability and disease burden. Thirdly, both occur with considerable class and gender disparities, in patterns that however differs across populations and social contexts. The two disorders tend further not only to co-occur but also to cluster—i.e., suffering from one of the disorders increases the risk of having the other. Finally, there is evidence that they might interact, i.e., one disease influence the course and consequences of the other.

The research interest in this comorbidity has been extensive. The first published paper on the association between diabetes and depression came in 1949, and since 2010 the annual number of publications is 100–200 (Sweileh 2018). Most of them from high-income countries. In this chapter we shall, with data from the latest Brazilian Health Survey (PNS 2013—see Box 33.1), analyse the social epidemiology of diabetes, and examine how depression is associated with the occurrence, social disparities, and consequences of diabetes.

Box 33.1 Analysis of the Brazilian National Health Survey (PNS 2013)

We use the latest Brazilian National Health Survey PNS 2013 (Szwarcwald et al. 2014). It is a cross-sectional household survey designed with a three-stage random sampling procedure. The census tracts are the primary units, the households are the second-stage units, and one resident per household aged 18 years or older is the third-stage unit. The sample includes 81,357 households. A total of 8.1% did not want to participate or could not be contacted. Interviews on health issues were carried out with 60,202 individuals. The survey was approved by the *Comissão Nacional de Ética em Pesquisa* (CONEP—National Commission of Ethics in Research—no. 328,159). PNS 2013 is thus a large survey with sufficient size for the power-demanding interaction analyses; however, it is cross-sectional, which limits any attempt to draw causal inferences.

Disability is here measured as self-reported consequence of long-term diseases. Disability is based on questions raised for each of 13 different chronic disorders (including one "other diseases" category): "to what degree the disease or its complications limit daily activities including work. Those who answer "moderate, serious, or very serious" to at least one of these questions are classified as having *disability*, i.e., limiting activities due to illness. The basic individual exposure is the highest level of *education* achieved.

Prevalent *diabetes* is based on an affirmative answer to the question: 'Has any

(continued)

Box 33.1 (continued)

doctor told you have diabetes?' Compared to laboratory measurements with Hba1c levels >6.5% or medication self-report, it seems to underestimate the prevalence (Malta et al. 2019). Symptoms of depression are based on the patient health questionnaire—9 (PHQ-9 scale) with a sum-score ranging from 0 to 27. Having moderate or severe depressive symptoms (*Depression*) is defined as scoring ten or higher. For the regression analysis, Generalized Linear models with binomial distribution were used in IBM SPSS v25. Associations are thus measured in absolute terms as risk-difference (RD) in percentage units of prevalence. We have chosen this measure of associations since it is more policy relevant and adequate for estimating interaction. The size of interactions between two exposures is estimated as the difference in disease prevalence between those exposed to both compared to those exposed to none, minus the sum of the effects of single exposures. Some confounders are included in the regression model as indicated under the tables, but several other potential confounders are not included since they are not collected in the survey. We have applied weights to adjust for sampling stratification and differential non-response in the survey (Szwarcwald et al. 2014).

Diabetes: Increasingly a Disorder of Poverty

Diabetes is often described as a disease of modernization, including industrialization and urbanization. In upper middle-income countries like Brazil where these processes develop faster than elsewhere, the diabetes prevalence is increasing approximately 2% annually (NCD RisC 2016a). The industrialization of food production and global shift in diets, with growing marketing and consumption of foods and beverages with added sugars and salt, refined carbohydrates, grain-based desserts and savoury snacks, has hit low- and middle-income countries harder than in the industrialized North, and Latin America earlier than many other regions (Popkin et al. 2012; Popkin and Reardon 2018). Time and energy use on physical activity linked to work, home production, travel and leisure time activities have also changed, and contributed to the rise in obesity and type-2 diabetes (Ng and Popkin 2012).

These changes are universal, but the effects on obesity and diabetes have differed, not only between countries at different economic levels, but also within countries between men and women, and between different socioeconomic groups. Diabetes shows, in both European (Espelt et al. 2012; Imkampe and Gulliford 2011) and some Asian countries (Kim and Nam 2017), increasing disparities and occurs now with steep social gradients, even in welfare states with relatively small socioeconomic inequalities (Agardh et al. 2011). In Brazil, increasing educational inequalities in diabetes have also been observed in the period 1998–2013 (Beltran-Sánchez and Andrade 2016), a period when income inequality in Brazil was declining (see Box 33.2). In that sense, diabetes has become more a disease of poverty than a disease of modernization (Mendenhall 2019).

Box 33.2 The Brazilian Context

Brazil is a middle-income country with a gross domestic product per capita that is 7 times lower than in the US and 4.5 times higher than in India. It experienced in the years 1990–2014 a strong economic growth, where income inequality and poverty declined sharply. Since 2015 these indicators have however pointed in the opposite direction with growing inequalities. Brazil implemented since 1990 a fast-expanding public health service programme (SUS) that generated access to care nationwide, but in particular for poor populations in the North of Brazil. Specific

(continued)

Box 33.2 (continued)

programmes for child health and for the control of non-communicable diseases (NCD) like diabetes and hypertension have been implemented. The human development index (HDI) is a composite index of human development, including measures of education, income and longevity. The average HDI of Brazil increased from 0.611 in 1990 to 0.759 in 2017. There are however still large inequalities between Brazilian states. In 2010, HDI varied from 0.631 in Alagoas in the North East region (similar to Honduras), to 0.824 in the Federal District where Brasilia is located (similar to Argentina). These differences in HDI thus correspond to more than 25 years' development in Brazil. States with high HDI tend also to be more industrialized, urbanized and with higher average income. They have at same time smaller income inequalities. Epidemiologically, Brazil has experienced a very fast decline in communicable disorders and, as elsewhere, a slowly increasing prevalence of many NCDs, including diabetes.

The social epidemiology of depression is much less clarified, particularly in low and middle-income countries. There is no strong evidence that the prevalence of depression is actually increasing (Ferrari et al. 2013), but several studies from Brazil and other middle-income countries have shown higher prevalence among women and among underprivileged groups (Silva et al. 2014; Munhoz et al. 2016; Brinda et al. 2016).

The epidemiological development of diabetes in the Brazilian population has made it a major priority in national health policies, with detailed vertical programs to be implemented in primary care (Almeida-Pititto et al. 2015). When, like in Brazil, 70% of diabetes cases have only primary school or less, policies aimed at tackling the diabetes epidemic must address the growing social inequality in diabetes.

The Social Patterning of Diabetes, Obesity and Depression in Brazil

The empirical analysis in this chapter is based on the latest Brazilian National Health Survey (PNS 2013). See text Box 33.1 for details.

Figure 33.1 shows the age-adjusted prevalence of diabetes, obesity, depression and disability across levels of education for men and women.

In Fig. 33.1, it can be observed that there is a steep educational gradient in prevalence of disability for both women and men. Disability is here measured as moderate-to-severe limitation of daily activities. We have calculated for disability among women the Slope Index of Inequality—SII (Mackenbach and Kunst 1997) as −12.5 percentage points (p.p.), with 95% confidence interval (CI: −14.2; −10.8), and for men as −13.8 (CI: −15.5; −12.3). They confirm significant inequality for these samples. In the Brazilian population, both diabetes and depression (Yokota et al. 2016) are important causes of disability. As many as 36% of diabetics report disability, 48% of those with moderate-to-severe depressive symptoms, and 75% suffer from both.

The steep negative gradient in disability might therefore be influenced by social disparities in the prevalence of diabetes and depression, but many other diseases might play a role too. In Fig. 33.1, it can be seen that there indeed is a negative gradient for diabetes (SII = −7.4; CI: −8.5; −6.2), but only among women. For men the association between education and diabetes is weakly positive (0.6 CI: −0.5; 1.7) When compared to the disability gradient it raises the question, whether the disability consequences of diabetes are stronger among men than women, and in particular for low educated men?

The inequality in diabetes is stronger in the rich (and economically more equal) states in the south of Brazil. Slope index of inequality for diabetes in women is −10.8 (CI: −12.7; −8.9) in these states, compared to −4.4 (CI: −6.4; −2.5) in the poorer (and more unequal) less developed North Eastern states. Even within Brazil the economic development seems to transform diabetes from a disease of modernization increasingly to a disease also of poverty.

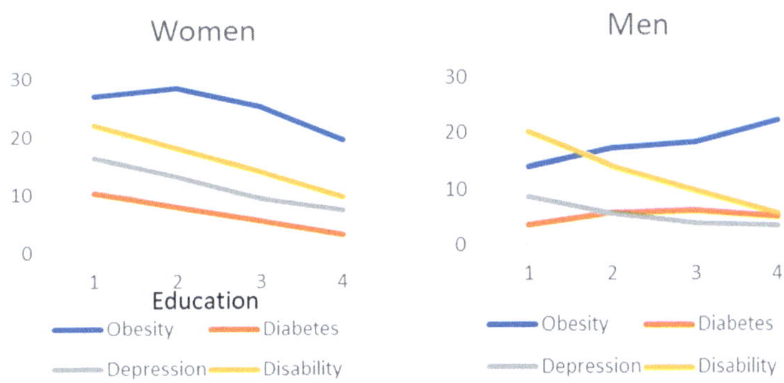

Fig. 33.1 Prevalence (%) of obesity (BMI ≥30), self-reported diabetes, moderate-to-severe symptoms of depression (PHQ-9 ≥ 10) and disability (self-reported moderate to severe limitation of daily activities including work, due to long-term illness). Brazil PNS 2013. Education (1 = less than primary school, 2 = primary school, 3 = secondary school, 4 = tertiary school or higher). Prevalence in percent adjusted for age and state

The curves in Fig. 33.1 raise the question why the association between education and diabetes is so different for men and women. What makes women more susceptible to the protective effect of education on diabetes? It seems in Fig. 33.1 that, at least partly, it could be due to a sex difference in the social patterning of obesity. The educational slope of obesity is negative for women (SII = −10.5 CI: −12.5; -8.5), but positive for men (SII = 7.7 CI: 5.9; 9.5). It would however still not explain that obesity and diabetes are more strongly associated for low-educated women than for others.

Figure 33.1 reveals negative educational gradients for depression with SII = −4.6 CI: -3.5; -5.6) for men, and for women an even steeper gradient with a SII = −9.7 (CI: 8.2; 11.1). As was the case for diabetes, the gradient is stronger in the rich more industrialized states in the South of Brazil, with a SII of −10.6 (CI: -12.9; -8.3) compared to −5.6 (CI: -8.1; -3.1) in the Northeast. Since depression is both a cause and an effect of both obesity and diabetes (Jantaratnotai et al. 2017) the comorbidity with depression might, as we shall see, contribute to some of the questions mentioned here.

To sum up: the simple description of the social patterning of disability, diabetes, depression and obesity in Fig. 33.1 raises at least two questions where the diabetes-depression comorbidity might play a role: (1) Does diabetes have differential consequences in terms of disability across sexes and educational levels? (2) Does obesity have differential effects on diabetes across sexes and levels of education?

Mechanisms of Health Inequalities

These two questions point to two of the generic mechanisms driving disparities in morbidity and disability (Diderichsen et al. 2001). From an ethical point of view, it is relevant to start with the inequality of the disabling consequences of disease. Amartya Sen has argued that if an important quality of human development is greater freedom for people to live the lives they have reason to value, then inequalities in the origins of that un-freedom becomes central. Bad health due to escapable disabilities and premature mortality is a critical source of such un-freedom (Sen 2002).

Differential Consequences

If we go backwards in the causal chain, it is evident that disparities in disability and participation are generated by disparities in underlying morbidities (Schuring et al. 2019). Diabetes and depression are two obvious sources in the Brazilian population, but several other chronic diseases contribute as well, including cardiovascular and musculoskeletal disorders (Malta et al. 2016). Epidemiological studies have since long

Table 33.1 Differential consequences. The interaction between short education (≤ prim educ.) and disease (diabetes or depression) in relation to disability. Percentage points prevalence difference with 95% CI. Adjusted for age and state. Brazil PNS 2013

	Diabetes		Depression	
	Men	Women	Men	Women
≤Prim. educ. + disease	18.3 (16.2;20.5)	24.2 (22.3;26.3)	45.5 (43.3;47.6)	42.7 (41.1;44.4)
≤Prim. educ. 0 disease	6.2 (5.4;7.0)	4.7 (3.8;5.6)	4.9 (4.2;5.7)	2.5 (1.6;3.4)
>Prim. educ. + disease	14.1 (11.4;16.7)	14.4 (11.3;17.5)	31.0 (28.3;33.8)	25.9 (24.0;27.8)
>Prim. educ. 0 disease	0 (ref.)	0 (ref.)	0 (ref.)	0 (ref.)
Interaction educ.*disease	−2.0 (−5.3;1.4)	5.2 (1.6;8.8)	9.5 (6.1;13.0)	14.4 (11.8;16.9)

Prim. educ. Primary education; *ref.* Reference category for the table

indicated that with a given morbidity the rate and degree of disability and participation might differ across levels of education, i.e., *differential consequences* (Burström et al. 2000). That might be a result of not only unequal access to and quality of health care, but also unequal levels of comorbidity and demands at the labour market not compatible with ill-health and short education. Only few studies have analysed the inequalities in disability specifically for diabetes, but they have found clear educational and sex disparities in disability (Ervasti et al. 2016) (Table 33.1).

Table 33.1 unsurprisingly indicates that depression has a very strong association with disability and is stronger than diabetes. The association between disease and disability is more prominent among low educated, particularly for women. That difference is estimated by the interaction between education and disease in the last row of Table 33.1. There are thus indications of *differential consequences* across levels of education—in particular for depression and for women. This may have many explanations. Education might play a role for perceived disability, because the labour market often demands more physical ability among people with low education, and the consequences in terms of reduced workability might therefore be stronger among those suffering from short education (Schuring et al. 2019).

Interacting Comorbidity

Another reason there might be differential consequences of diabetes is the unequal prevalence of comorbid depression, in case depression interacts with diabetes in the effect on disability. Several clinical studies have shown that depression interferes with the care, control, and course of diabetes (Ciechanowski et al. 2000; Pirraglia and Gupta 2007; Petrak et al. 2015). Few studies have however looked at whether there is an interaction between diabetes and depression, in their association with disability (Egede 2004; Kalyani et al. 2017; Diderichsen and Andersen 2019). Table 33.2 demonstrates the interaction.

In Table 33.2, it can be noticed a rather strong interaction between diabetes and depression, particularly among men. We might here have a mechanism contributing to the paradox found in Fig. 33.1, that low-educated men show high levels of disability, however, relatively low prevalence of diabetes and depression. That might thus be due to depression, interfering with the course and consequences of their diabetes.

Clustering

The impact on population health of interaction between disorders will depend on the degree of co-occurrence and clustering. The more clustering there is of two or more disorders to the same individuals, the more opportunities will there be for them to interact. In Brazil, there is indeed a clustering of diabetes and depression. In the alluded-to survey (PNS 2013), the prevalence of diabetes is 6.2% and for depression 7.7%. This means that the prevalence of co-occurrence by chance would be 0.48%. In reality it is 0.92%,

Table 33.2 The interaction between diabetes and depression in relation to disability. Percentage points prevalence difference with 95% CI. Adjusted for age, education and state. Brazil PNS 2013

	Men	Women
+ Diabetes; + depression	65.0 (60.3;69.8)	57.7 (54.2;61.3)
+ Diabetes; 0 depression	9.1 (7.4;10.8)	14.1 (12.4;15.8)
0 Diabetes; + depression	33.8 (32.0;35.6)	32.2 (30.8;33.5)
0 Diabetes; 0 depression	0 (ref.)	0 (ref.)
Interaction diabetes*depression	22.1 (16.8;27.4)	11.5 (7.4;15.5)

ref. Reference category for the table

i.e., 1.9 times higher than expected. This level of clustering has been found in other countries too (Nouwen et al. 2019).

Clustering can be due either to shared causation or to bidirectional causal relationships between the two disorders. As recent reviews have shown, there have been numerous longitudinal studies documenting the bidirectional causal relationship between depression and diabetes (Nouwen et al. 2019; Chireh et al. 2019; Zhao et al. 2019). There is also evidence indicating several shared causes (Chireh and D'Arcy 2019). Shared causation by obesity and short education has been mentioned here, but there are other candidates. A Scandinavian twin study has demonstrated that type 2 diabetes and depression may share a genetic component, albeit with different genetic factors being at play in males and females (Kan et al. 2016).

The dysregulation of the hypothalamic-pituitary-adrenal (HPA) axis is a much-discussed biological link between stress, depression, obesity and type 2 diabetes, even if many mechanistic details are badly understood (Joseph and Golden 2017; Milaneschi et al. 2019). Origins of stress reactions and HPA-dysregulation might therefore be shared, causes, including childhood poverty and maltreatment, generating increased inflammatory vulnerability to the effects of later exposures (Milaniak and Jaffee 2019). Violence, unemployment and living in communities with large income inequalities are examples of particular relevance in Brazil, and might increase stress and thereby both depression and diabetes (Kelly and Ismail 2015; Hammen 2015).

Diseases that cluster and interact with each other and with the social context have been called *syndemics* (Singer et al. 2017). The diabetes/depression comorbidity is an example of a syndemic between non-communicable disorders (Mendenhall et al. 2017; Diderichsen et al. 2019). Mendenhall argues, based on her anthropological studies, that we should consider diabetes invariably in a syndemic context. Poverty and violence are very prevalent in Brazilian slums *('favelas')*, and generate a chronic level of stress that could drive the occurrence of both depression, diabetes and their complications (Mendenhall 2019).

Differential Exposure

If we go a step further back in the causal chains, the inequalities in the incidence of a disease will be driven by unequal distribution of major causes of the disease—i.e., *differential exposure*.

Obesity is the most important cause of type 2 diabetes and might explain a considerable part of the social disparities in diabetes (Smith et al. 2013; Steele et al. 2017). Obesity prevalence has been rising for both men and women in low- and middle-income countries (LMIC) (NCD RisC 2016b); however, the rise has been much more pronounced among low-educated and underprivileged groups, leading to a reversal of the education/obesity relationship from positive in low-income countries to negative in middle- and high-income countries (Jaacks et al. 2019). That trend has been found to be particularly strong among women (Ford et al. 2017). The association between education and body mass index (BMI) seems mainly to be mediated by environmental and psychosocial factors (Blüher 2019; Claassen et al. 2019).

Obesity Patterns in Brazil

As shown in Fig. 33.1, for women in Brazil there is a graded negative association between education and obesity, while it is positive for men. That has not always been the case. In 1975, the association between BMI and income was positive for both men and women in Brazil (Monteiro et al. 2007; Gomes et al. 2019). Since then, the prevalence of obesity has increased faster among low-income than high-income groups for both genders. In the 1990s, the prevalence of obesity actually declined among high-income women. The changing social patterning of obesity clearly plays a major role in the changing patterns of diabetes, but does not fully explain it (Espelt et al. 2012; Kim and Nam 2017). The reason why the profiles in Brazil have become so different for men and women might be due to labour market segregation, as low-educated women to a high-degree work with domestic service, and low-educated men work in agriculture, construction and industry.

Gene Expression and Miscellaneous Mechanisms

Genotype is another underlying cause of type 2 diabetes. Studies so far have not indicated that the genetic risk is associated with education, but they have suggested that education might modify the epigenetic expression, and effect on obesity and diabetes (Liu et al. 2015; van Zon et al. 2017; Frank et al. 2019).

Low birthweight is associated with type 2 diabetes (Knop et al. 2018). Wells has formulated the risk for diabetes as an interaction between low metabolic capacity (influenced by impaired foetal and infant growth), and metabolic load generated by adiposity, sedentary life style and psychosocial stress (Wells 2019). The combination of impaired foetal growth and adult metabolic load is particularly relevant in middle-income countries like Brazil, where changes have happened fast within a lifetime. The health effects of undernutrition early in life and later exposure to obesogenic environments thus represent a dual burden of malnutrition. Fortunately, prevalence of stunting (as a sign of early undernutrition) has declined rapidly in Brazil since the 1970s, and social disparities have been reduced (Monteiro et al. 2010).

Social stress has in several studies been shown to impact on the risk of diabetes, including early adversities (Milaniak and Jaffee 2019) and adult social stress from work (Xu et al. 2017). Poor housing and violence have also been identified as risk factors and potential modifiers of the obesity-diabetes relationship—partly through the metabolic effects of stress (Kelly and Ismail 2015; Volaco et al. 2018; Heraclides et al. 2011).

Stress mobilizes biological responses, including the release of glucose and lipids into the circulation, inflammatory cytokine expression and increased blood pressure. Repeated stress might in the long run generate allostatic load, with dysregulation of glucose metabolism and chronic low-grade inflammation. Stress might in addition influence behaviours, including dietary habits and physical activity (Hackett and Steptoe 2017). Stress can also generate depression, and both stress and depression might work through sleep disturbances in their impact on diabetes (Anothaisintawee et al. 2016).

Differential Vulnerability

As the aforementioned causes might work in the same causal pathway as obesity, they might interact and modify the effect of obesity. Since social stress and childhood adversities are strongly related to education and income, they might generate a *differential vulnerability* to obesity (Diderichsen et al. 2019).

Vulnerability is a concept used in many disciplines with different meanings, and often including many dimensions. One is the lack of capability to change, cope with—or adapt to—determinants of ill health. Since health is a critical constituent of capabilities (Sen 2002), this approach provides a feed-back pathway where ill health and its consequences, through changed capabilities, influence causes and consequences

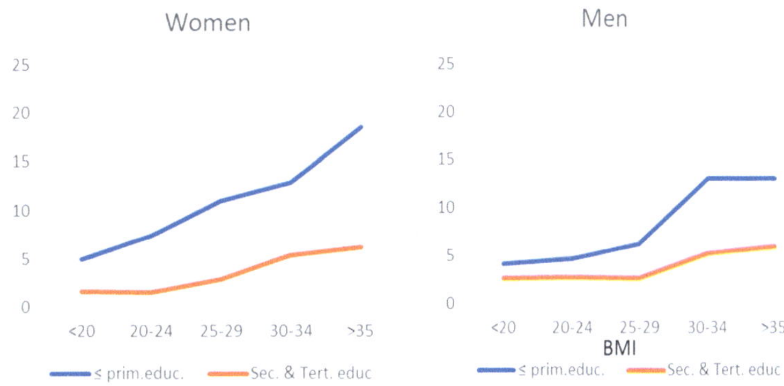

Fig. 33.2 Prevalence (%) of self-reported diabetes (adjusted for age and state) according to BMI stratified for education. PNS Brazil 2013

of disease. Since many different interacting causes in the same pathway might mediate the health effect of social position, the effect of a single cause might differ across social positions, if it interacts with some other cause related to social position—*differential susceptibility*.

It is well known that the association between BMI and type 2 diabetes varies across ethnic groups, with stronger effects among Asian populations (Chiu et al. 2011). Some studies indicate that the obesity-diabetes association is stronger in countries with higher-income levels, compared to low-income countries (Wang et al. 2014). Body fat distribution modifies the BMI-diabetes association and waist circumference has, for the same BMI levels, increased in many countries (NCD RisC 2016b). Stress is associated with increased abdominal adiposity (Surkan et al. 2018) and might contribute to this development.

From an equity point of view, the interesting question is whether the BMI-diabetes association differs across level of education. In Fig. 33.2, data on this are shown for Brazil.

The Impact of Obesity

It can clearly be seen that the association is steeper for the low educated. While ten-point increase in BMI-average is associated with 6 percentage points increased prevalence of diabetes among the low educated, it is only 2–3 percentage points for the better educated. There is in other words an interaction between obesity and short education that can be estimated to 2.6 percentage points (CI: 1.3; 4.0) for women and 4.8 (CI: 3.4; 6.2) for men.

That differential association of BMI and diabetes can be due to interaction between obesity and some other risk factor for diabetes associated with education. The survey we have access to here has measured smoking, dietary habits and physical inactivity and none of them show any interaction with obesity. As the data are cross-sectional, the behavioural factors might act as both causes and effects of diabetes, which makes the interpretation complicated. There is no measure of perceived stress in the PNS, but depression might be an indicator strongly linked with exposure to stress.

Obesity has been found to be associated with depression in a bi-directional causal relationship (Wu and Berry 2018; Stunkard et al. 2003). The relevant question here is to what extent depression modifies the association between obesity and diabetes. Figure 33.3 illustrates that for both men and women there is an interaction between obesity and depression: 3.5 percentage points (CI: 1.4; 5.5) for women and 6.8 (CI: 3.8; 9.8) for men (Fig. 33.3). Similar findings have recently been announced in China (Ning et al. 2019). Since depression is more prevalent among low educated, this might contribute to the differential susceptibility in Fig. 33.2.

Fig. 33.3 Prevalence (%) of self-reported diabetes (adjusted for age and state) according to BMI and stratified for depression. PNS Brazil 2013

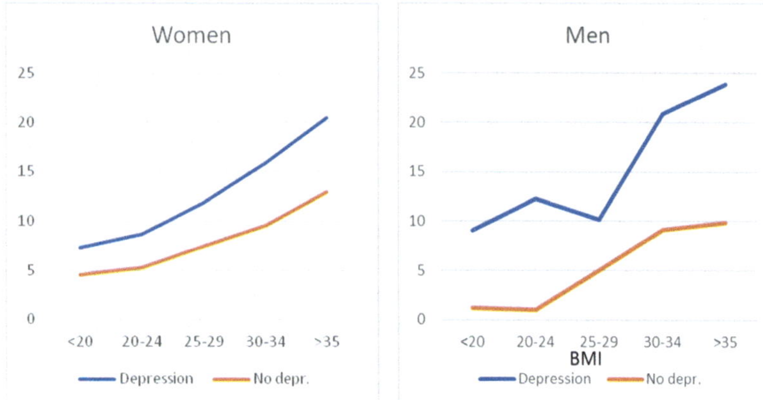

Summarizing

It is well known that the causal sequence of diabetes-depression-obesity-diabetes can create a vicious circle. What this analysis adds is that the disabling effect on people's daily life and activities of the combined diabetes-depression comorbidity is stronger than the sum of each of the two disorders. That interaction between the two disorders is, for underprivileged groups, aggravated by the fact that low education in combination with either diabetes or depression has a strong impact on disability.

We also find that the clustering of two disorders is driven partly by a shared social determination. Being low educated in the highly unequal society of Brazil represents a shared cause of both disorders, particularly for women. Short education also enforces the association of obesity with diabetes, indicating a differential susceptibility to the effect of obesity. We have found the concept of *syndemics* useful when dealing with clustering and interaction, between diseases that mediate the effect of social position on disability. If one disease is associated with low social position, the consequences of another interacting disease might differ across social positions, which means it has differential consequences.

Tackling the vicious circle mentioned here might use different entry points. One is to prevent obesity, which will demand powerful policy initiatives regulating the food industry to produce affordable healthy foods, and planning the physical environment to promote physical activity. Physical activity improves depressive symptoms, insulin sensitivity and glycaemic control. Other entry points are treatment of depression and/or diabetes. A review suggests that treatment of comorbid depression and diabetes is more effective with an integrative approach addressing both conditions together (Darwish et al. 2018).

Most selective serotonin reuptake inhibitor antidepressants have however been found to increase BMI; however, they could on the other hand have a limiting effect on diabetes and complications via reduced depressive symptoms. Several studies on the effect of glycaemic control in diabetes patients with depression or distress have yielded heterogeneous results (Bystritsky et al. 2014). A recent systematic review concluded that psychological interventions with diabetes and elevated diabetes distress, with symptoms of depression and/or anxiety, can improve glycaemic control (Schmidt et al. 2018).

Treatment of obesity might similarly lead to an improvement of depressive symptoms (Sevilla-Gonzalez et al. 2017). Better treatment of depression associated with diabetes is therefore a high priority, and a better understanding of the psychological and physiological mechanisms involved in this comorbidity would be helpful. Diabetes care has improved substantially in recent years in Brazil, but the large inequalities in quality and effect of care still exist (Neves et al. 2018). The epidemiological patterns shown here for Brazil, and the anthropological research on the diabetes-depression syndemic make it clear that the social and economic context is critical.

Funding
This research did not receive any specific grant from funding agencies in the public, commercial, or not-for-profit sectors.
Declaration of Interest
None.

References

Agardh A, Allebeck P, Hallqvist J et al (2011) Type 2 diabetes incidence and socio-economic position: a systematic review and meta-analysis. Int J Epidemiol 40:804–818. https://doi.org/10.1093/ije/dyr029

Almeida-Pititto B, Dias ML, Moraes ACF et al (2015) Type 2 diabetes in Brazil: epidemiology and management. Diab Metab Syndr Obes: Target Therapy 8:17–28. https://doi.org/10.2147/DMSO.S72542

Anothaisintawee T, Reutrakul S, van Cauter E et al (2016) Sleep disturbances compared to traditional risk factors for diabetes development: systematic review and meta-analysis. Sleep Med Rev 30:11e24. https://doi.org/10.1016/j.smrv.2015.10.0

Beltran-Sánchez H, Andrade FCD (2016) Time trends in adult chronic disease inequalities by education in Brazil. Int J Equity Health 15:139. https://doi.org/10.1186/s12939-016-0426-5

Blüher M (2019) Obesity: global epidemiology and pathogenesis. Nat Rev Endocrinol 15(5):289–298. https://doi.org/10.1038/s41574-019-0176-8

Brinda EM, Rajkumar AP, Attermann J et al (2016) Health, social, and economic variables associated with depression among older people in low and middle income countries: world health organization study on global AGEing and adult health. Am J Geriatr Psychiatry 24(12):1196–1208. https://doi.org/10.1016/j.jagp.2016.07.016

Burström B, Diderichsen F, Whitehead M et al (2000) Inequality in the social consequences of illness. Int J Health Serv 30(3):435–451. https://doi.org/10.2190/6PP1-TDEQ-H44D-4LJQ

Bystritsky A, Daniel J, Kronemeyer D (2014) Interactions between diabetes and anxiety and depression. Implications for treatment. Endocrinol Metab Clin N Am 43:269–283. https://doi.org/10.1016/j.ecl.2013.10.001

Chireh B, D'Arcy C (2019) Shared and unique risk factors for depression and diabetes mellitus in a longitudinal study, implications for prevention: an analysis of a longitudinal population sample aged ⩾45 years. Ther Adv Endocrinol Metab 10:1–15. https://doi.org/10.1177/2042018819865828

Chireh B, Li M, D'Arcy C (2019) Diabetes increases the risk of depression: a systematic review, meta-analysis and estimates of population attributable fractions based on prospective studies. Prev Med Rep 14:100822. https://doi.org/10.1016/j.pmedr.2019.100822

Chiu M, Austin PC, Manuel DG et al (2011) Deriving ethnic-specific BMI cutoff points for assessing diabetes risk. Diabetes Care 34(8):1741–1748. https://doi.org/10.2337/dc10-2300

Ciechanowski PS, Katon WJ, Russo JE (2000) Depression and diabetes. Impact of depressive symptoms on adherence, function and costs. Arch Intern Med 160:3278–3285

Claassen MA, Klein O, Bratanova G et al (2019) A systematic review of psychosocial explanations for the relationship between socioeconomic status and body mass index. Appetite 132:208–221. https://doi.org/10.1016/j.appet.2018.07.017

Darwish L, Beroncal E, Sison V (2018) Depression in people with type 2 diabetes: current perspectives. Diabet Metabol Syndrom Obes Target Therapy 11:333–343. https://doi.org/10.2147/DMSO.S106797

Diderichsen F, Andersen I (2019) The syndemics of diabetes and depression in Brazil—an epidemiological analysis. SSM Popul Health 7:002. https://doi.org/10.1016/j.ssmph.2018.11.002

Diderichsen F, Evans T, Whitehead M (2001) The social basis of disparities in health. In: Evans T et al (eds) Challenging inequities in health- from ethics to action. Oxford University Press, New York. https://doi.org/10.1093/acprof:oso/9780195137408.003.0002

Diderichsen F, Hallqvist J, Whitehead M (2019) Differential vulnerability and susceptibility: how to make use of recent development in our understanding of mediation and interaction to tackle health inequalities. Int J Epidemiol 48(1):268–274. https://doi.org/10.1093/ije/dyy167

Egede LE (2004) Diabetes, major depression, and functional disability among U.S. adults. Diabetes Care 27 (2):421–428. https://doi.org/10.2337/diacare.27.2.421

Ervasti J, Kivimäki M, Dray-Spira R et al (2016) Socio-economic gradient in work disability in diabetes: evidence from three occupational cohorts. J Epidemiol Community Health 70:125–131. https://doi.org/10.1136/jech-2015-205943

Espelt A, Kunst AE, Palencia L et al (2012) Twenty years of socio-economic inequalities in type 2 diabetes mellitus prevalence in Spain, 1987–2006. Eur J Pub Health 22(6):765–771. https://doi.org/10.1093/eurpub/ckr158

Ferrari AJ, Somerville AJ, Baxter AJ et al (2013) Global variation in the prevalence and incidence of major depressive disorder: a systematic review of the epidemiological literature. Psychol Med 43:471–481. https://doi.org/10.1017/S0033291712001511

Ford ND, Patel SA, Narayan KMV (2017) Obesity in low- and middle-income countries: burden, drivers, and emerging challenges. Ann Rev Public Health

38:145–164. https://doi.org/10.1146/annurev-publhealth-031816-044604
Frank M, Dragano N, Arendt M et al (2019) A genetic sum score of risk alleles associated with body mass index interacts with socioeconomic position in the Heinz Nixdorf recall study. PLoS One 14(8):e0221252. https://doi.org/10.1371/journal.pone.0221252
Gomes DCK, Sichieri R, Junio EV et al (2019) Trends in obesity prevalence among Brazilian adults from 2002 to 2013 by educational level. BMC Public Health 19:965. https://doi.org/10.1186/s12889-019-7289-9
Hackett RA, Steptoe A (2017) Type 2 diabetes mellitus and psychological stress—a modifiable risk factor. Nat Rev Endocrinol 13(9):547–560. https://doi.org/10.1038/nrendo.2017.64
Hammen CL (2015) Stress and depression: old questions, new approaches. Curr Opin Psychol 4:80–85. https://doi.org/10.1016/j.copsyc.2014.12.024
Heraclides AM, Chandola T, Witte DR et al (2011) Work stress, obesity and the risk of type 2 diabetes: gender-specific bidirectional effect in the Whitehall II study. Obesity 20:428–433. https://doi.org/10.1038/oby.2011.95
Imkampe AK, Gulliford MC (2011) Increasing socio-economic inequality in type 2 diabetes prevalence–repeated cross-sectional surveys in England 1994–2006. Eur J Pub Health 21:484–490. https://doi.org/10.1093/eurpub/ckq106
Jaacks LM, Vandervijveree S, McGowan APCJ et al (2019) The obesity transition: stages of the global epidemic. Lancet Diabetes Endocrinol 7:231–240. https://doi.org/10.1016/S2213-8587(19)30026-9
Jantaratnotai N, Mosikanon K, Lee Y et al (2017) The interface of depression and obesity. Obes Res Clin Pract 11(1):1–10. https://doi.org/10.1016/j.orcp.2016.07.003
Joseph JJ, Golden SH (2017) Cortisol dysregulation: the bidirectional link between stress, depression, and type 2 diabetes mellitus. Ann N Y Acad Sci 1391:20–34. https://doi.org/10.1111/nyas.13217
Kalyani RR, Nan J, Carnethon M et al (2017) Diabetes, Depressive symptoms, and functional disability in African Americans: the Jackson heart study. Arch Med Res 48:582–591. https://doi.org/10.1016/j.arcmed.2018.02.004
Kan C, Pedersen NL, Christensen K et al (2016) Genetic overlap between type 2 diabetes and depression in Swedish and Danish twin registries. Mol Psychiatry 21:903–909. https://doi.org/10.1038/mp.2016.28
Kelly SJ, Ismail M (2015) Stress and type 2 diabetes: a review of how stress contributes to the development of type 2 diabetes. Annu Rev Public Health 36:441–462. https://doi.org/10.1146/annurev-publhealth-031914-12292
Kim GR, Nam CM (2017) Temporal trends in educational inequalities in non-communicable diseases in Korea, 2007-2015. PLoS One 12(12):e0190143. https://doi.org/10.1371/journal.pone.0190143
Knop MR, Geng TT, Gorny MW et al (2018) Birth weight and the risk of type-2 diabetes, cardiovascular disease, and hypeetension in adults. A meta-analysis 7,646,267 pariciapnts from 135 studies. JAHA 7(23): https://doi.org/10.1161/JAHA.118.008870
Liu SY, Walter S, Marden J et al (2015) Genetic vulnerability to diabetes and obesity: does education offset the risk? Soc Sci Med 127:150–158. https://doi.org/10.1016/j.socscimed.2014.09.009
Mackenbach JP, Kunst AE (1997) Measuring the magnitude of socio-economic inequalities in health: an overview of available measures illustrated with two examples from Europe. Soc Sci Med 44:757–771. https://doi.org/10.1016/S0277-9536(96)00073-1
Malta DC, Bernal RTI, Souza MFM et al (2016) Social inequalities in the prevalence of self-reported chronic non-communicable diseases in Brazil: national health survey 2013. Int J Equity Health 15:153. https://doi.org/10.1186/s12939-016-0427-4
Malta DC, Duncan BB, Schmidt MI et al (2019) Prevalence of diabetes mellitus as determined by glycated hemoglobin in the Brazilian adult population, National Health Survey. Rev Bras Epidemiol 22(Suppl 2): E190006.SUPL.2. https://doi.org/10.1590/1980-549720190006.supl.2
Mendenhall E (2019) Rethinking diabetes. Cornell University Press, London
Mendenhall E, Kohrt BA, Norris SA et al (2017) Non-communicable disease syndemics: poverty, depression, and diabetes among low-income populations. Lancet 389:953–963. https://doi.org/10.1016/S0140-6736(17)30402-6
Milaneschi Y, Simmons K, Van Rossum EFC et al (2019) Depression and obesity: evidence of shared biological mechanisms. Molecular Psych 24:18–33. https://doi.org/10.1038/s41380-018-0017-5
Milaniak I, Jaffee SR (2019) Childhood socioeconomic status and inflammation: a systematic review and meta-analysis. Brain Behav Immun 78:161–179. https://doi.org/10.1016/j.bbi.2019.01.018
Monteiro CA, Conde WL, Popkin BM (2007) Income-specific trends in obesity in Brazil: 1975–2003. Am J Public Health 97:1808–1812. https://doi.org/10.2105/AJPH.2006.099630
Monteiro CA, Benicio MHD, Conde WL et al (2010) Narrowing socioeconomic inequality in child stunting: the Brazilian experience, 1974–2007. Bull World Health Organ 88:305–311. https://doi.org/10.2471/BLT.09.069195
Munhoz TN, Nunes BP, Wehermesiter FC et al (2016) A nationwide population-based study of depression in Brazil. J Affective Dis 192:226–233. https://doi.org/10.1016/j.jad.2015.12.038
NCD RisC (2016a) Trends in adult body-mass index in 200 countries from 1975 to 2014: a pooled analysis of 1698 population-based measurement studies with 19.2 million participants. Lancet 387:1377–1396. https://doi.org/10.1016/S0140-6736(16)30054-X

NCD RisC (2016b) Worldwide trends in diabetes since 1980: a pooled analysis of 751 population-based studies with 4.4 million participants. Lancet 387:1513–1530. https://doi.org/10.1016/S0140-6736(16)00618-8

Neves RG, Duro SMS, Flores TR et al (2018) Social inequalities in care for the elderly with diabetes in Brazil. Rev Panam Salud Publica 42:e88. https://doi.org/10.26633/RPSP.2018.88

Ng SW, Popkin BM (2012) Time use and physical activity: a shift away from movement across the globe. Obes Rev 13(8):659–680. https://doi.org/10.1111/j.1467-789X.2011.00982.x

Ning F, Zhang D, Xue B et al (2019) Synergistic effects of depression and obesity on type 2 diabetes incidence in Chinese adults. J Diabetes 12:142. https://doi.org/10.1111/1753-0407.12968

Nouwen A, Adriaanse MC, van Dam K et al (2019) Longitudinal associations between depression and diabetes complications: a systematic review and meta-analysis. Diabet Med 36:1562. https://doi.org/10.1111/dme.14054

Petrak F, Baumeister H, Skinner TC et al (2015) Depression and diabetes: treatment and health-care delivery. Lancet Diabetes Endocrinol 3(6):472–485. https://doi.org/10.1016/S2213-8587(15)00045-5

Pirraglia PA, Gupta S (2007) The interaction of depression and diabetes: a review. Curr Diabetes Rev 3 (4):249–251. https://doi.org/10.2174/157339981070304 0249

Popkin BM, Reardon T (2018) Obesity and the food system transformation in Latin America. Obes Rev 19 (8):1028–1064. https://doi.org/10.1111/obr.12694

Popkin BM, Adair LS, Ng SW (2012) Now and then: the global nutrition transition: the pandemic of obesity in developing countries. Nutr Rev 70(1):3–21. https://doi.org/10.1111/j.1753-4887.2011.00456.x

Schmidt CB, Potter van Loon BJP, Vergouwen ACM et al (2018) Systematic review and meta-analysis of pscholoigcal interventions in people with diabetes and elevated diabetes distress. Diabet Med 35 (9):1157. https://doi.org/10.1111/dme.13709

Schuring M, Schram JLD, Robroek SJW (2019) The contribution of health to educational inequalities in exit from paid employment in five European regions. Scand J Work Environ Health 45(4):346–355. https://doi.org/10.5271/sjweh.3796

Sen A (2002) Why health equity? Health Econ 11:659–666. https://doi.org/10.1002/hec.762

Sevilla-Gonzalez MR, Quntana-Mendoza BM, Aquilar-Salinas CA (2017) Interaction between depression, obesity, and type 2 diabetes: a complex picture. Arch Med Res 48:582e591. https://doi.org/10.1016/j.arcmed.2018.02.004

Silva MT, Galvao TF, Martins SS, Periera MG (2014) Prevalence of depression morbidity among Brazilian adults: a systematic review and meta-analysis. Rev Bras Psiquiatr 36:262–270. https://doi.org/10.1590/1516-4446-2013-1294

Singer M, Bulled N, Ostrach B, Mendenhall E (2017) Syndemics and the biosocial conception of health. Lancet 389(10072):941–950. https://doi.org/10.1016/S0140-6736(17)30003-X

Smith PM, Smith BT, Mustard CA, Lu H, Glazier RH (2013) Estimating the direct and indirect pathways between education and diabetes incidence among Canadian men and women: a mediation analysis. Ann Epidemiol 23(3):143–149. https://doi.org/10.1016/j.annepidem.2012.12.012

Steele CJ, Schöttker B, Marshall AH et al (2017) Education achievement and type 2 diabetes—what mediates the relationship in older adults? Data from the ESTHER study: a population-based cohort study. BMJ Open 7:e013569. https://doi.org/10.1136/bmjopen-2016-013569

Stunkard AJ, Faith MS, Allison KC (2003) Depression and obesity. Biol Psychiatry 54:330–337. https://doi.org/10.1016/S0006-3223(03)00608-5

Surkan P, Sakyi K, Hu A, Olinto M, Gonçalves H, Horta B, Gigante D (2018) Impact of stressful life events on central adiposity in the Pelotas birth cohort. Rev De Saúde Púb 52:61. https://doi.org/10.11606/S1518-8787.2018052000161

Sweileh WM (2018) Analysis of global research output on diabetes depression and suicide. Ann General Psychiatry 17:44. https://doi.org/10.1186/s12991-018-0214-2

Szwarcwald CL, Malta DC, Pereira CA, Vieira MLFP, Conde WL et al (2014) Pesquisa Nacional de Saúde no Brasil: concepção e metodologia de aplicação. Cien Saude Coletiva 19(2):333–342. https://doi.org/10.1590/1413-81232014192.14072012

van Zon SKR, Snieder H, Bültmann U et al (2017) The interaction of socioeconomic position and type 2 diabetes mellitus family history: a cross-sectional analysis of the lifelines cohort and biobank study. BMJ Open 7: e015275. https://doi.org/10.1136/bmjopen-2016-015275

Volaco A, Cavalcanti AM, Filho RP et al (2018) Socioeconomic status: the missing link between obesity and diabetes mellitus? Curr Diabetes Rev 14(4):321–326. https://doi.org/10.2174/1573399813666170621123227

Wang A, Stronks K, Arah OA (2014) Global educational disparities in the associations between body mass index and diabetes mellitus in 49 low-income and middle-income countries. J Epidemiol Community Health 68(8):705–711. https://doi.org/10.1136/jech-2013-203200

Wells JCK (2019) The diabesity epidemic in the light of evolution: insights from the capacity–load model. Diabetologia 62(10):1740–1750. https://doi.org/10.1007/s00125-019-4944-8

Wu YK, Berry DC (2018) Impact of weight stigma on physiological and psychological health outcomes for overweight and obese adults: a systematic review. J Adv Nurs 74(5):1030–1042. https://doi.org/10.1111/jan.13511

Xu T, Magnusson Hanson LL, Lange T et al (2017) Workplace bullying and violence as risk factors for type 2 diabetes: a multicohort study and meta-analysis. Diabetologia 61(1):75–83. https://doi.org/10.1007/s00125-017-4480-3

Yokota RTC, Moura L, Andrade SSCA et al (2016) Contribution of chronic conditions to gender disparities in disability in the older population in Brazil, 2013. Int J Public Health 61:1003–1012. https://doi.org/10.1007/s00038-016-0843-7

Zhao F, Panb A, Yang X et al (2019) Bidirectional association between depressive symptoms and type 2 diabetes mellitus: the China health and retirement longitudinal study. J Diabet Complicat 33:107387. https://doi.org/10.1016/j.jdiacomp.2019.05.018-1056-8727

Contribution of Hyperglycemia and Unhealthy Diet to Cardiovascular Mortality

34

Jian Zhang and Dong Li

Contents

Obesity, a Risk Factor for Cardiovascular Disease 472
Chronic Inflammation 472
Dietary Pattern and Its Impact 472
The Mediterranean Paradigm 472
Metabolic Disease and Dysglycemia 473
Predisposing Factors tor Type 2 Diabetes 473
The Impact of Dyslipidemia 474
Endothelial Damage 474
General Biomarkers and Novel Technologies 474
New Biomarkers 475
Big Data and the Gut Microbiome 475
The Precision Medicine Approach 475
References 475

Abstract

Approximately 463 million adults are living with diabetes worldwide, and another 374 million people are at increased risk of developing type 2 diabetes. Hyperglycemia, one of the significant risk factors for type 2 diabetes mellitus, is also involved in the initiation and progression of major cardiac complications, such as atherosclerotic coronary heart disease. Food as a source of nutrition is closely related to the occurrence of many chronic diseases. A bundle of risk factors has been identified, including a high intake of saturated fat, short on fruits, vegetables, and whole grains, as well as heavy alcohol use. This chapter deals with diabetic care and cardiac mortality due to

J. Zhang
West China Biomedical Big Data Center, West China Hospital, Chengdu, China

D. Li (✉)
Emory Global Diabetes Research Center (EGDRC), Rollins School of Public Health, Emory University, Atlanta, GA, USA

Division of Hospital Medicine, Emory School of Medicine, Atlanta, GA, USA
e-mail: dong.li2@emory.edu

J. Faintuch and S. Faintuch (eds.), *Obesity and Diabetes*, https://doi.org/10.1007/978-3-030-53370-0_34

hyperglycemia and unhealthy diet, along with strategies and practical solutions on how to improve cardiovascular disease prevention and management.

Keywords

Hyperglycemia · Diabetes · Obesity · Cardiovascular risk · Unhealthy diet · Dysglycemia · Dyslipidemia

Obesity, a Risk Factor for Cardiovascular Disease

Epidemiological and clinical evidence demonstrates that obesity will produce a variety of structural and functional adaptations of the cardiovascular system, such as reducing cardiac output and left ventricular systolic function, increases peripheral resistance, left ventricular mass, left ventricular wall thickness, and internal size (Saeedi et al. 2019; Newell et al. 2010; Alpert et al. 2018; Piche et al. 2018). These changes are often associated with high blood pressure and deranged blood lipids, which, in turn, increase the risk of cardiovascular disease (CVD). An unhealthy dietary pattern contributes to obesity (healthy dietary style is shown in Fig. 34.1). Indirect effects are mediated by coexisting cardiovascular risk factors, such as insulin resistance, hyperglycemia, hypertension, and dyslipidemia. As a result, obese people are more likely to develop CVD and manifestations of CVD, especially coronary heart disease (CHD), angina, myocardial infarction (MI), atrial fibrillation, heart failure (HF), and sudden cardiac death (SCD).

Chronic Inflammation

One possible link between diabetes and obesity and subsequent CVD is low-grade inflammation (Duncan et al. 2003). Diabetes and insulin resistance are associated with overexpression of many cytokines in adipose tissue (Shimobayashi et al. 2018). Elevated C—reactive protein (CRP) levels in diabetic patients may be related to endothelial dysfunction. Diabetic patients also suffer from decreased adiponectin secretion, leading to decreased endothelial function (Shimobayashi et al. 2018). Insulin resistance is associated with elevated plasma free fatty acids, which lead to increased triglyceride stores in muscles, increased glucose production in the liver, and increased insulin production in people with type 2 diabetes (Leahy 2005). Insulin resistance in diabetic patients is also associated with hypertrophy of cardiomyocytes and systolic dysfunction (Leahy 2005; Belke et al. 2002).

Dietary Pattern and Its Impact

An unhealthy dietary pattern contributes to obesity (Healthy dietary style is shown in Fig. 34.1). Saturated fats and trans fats will raise cholesterol and low-density lipoprotein (LDL) levels (de Souza et al. 2015; Hooper et al. 2011). Sugar-sweetened beverages are incriminated in overweight, obesity, and may lead to ischemic heart disease (Huang et al. 2014). Higher fruit, vegetable, and fish (omega-3 and omega-6 fatty acids but not fish oil) consumption has been emphasized.

The Mediterranean Paradigm

The Mediterranean diet (higher consumption of fish, legumes, and supplementary nuts or extra virgin olive oil) has been considered one of the healthiest dietary patterns, which proved to reduce CVD mortality by 30%. In this regard eating a diet rich in animal protein, mainly red and processed meat, refined grains, sweets, desserts, and low in fiber, has been linked to an increased risk of cardiovascular death (Chen et al. 2013; Song et al. 2016). Diabetes, high blood pressure, and hypercholesterolemia may be considered intermediate factors in the relationship between dietary patterns and CVD mortality.

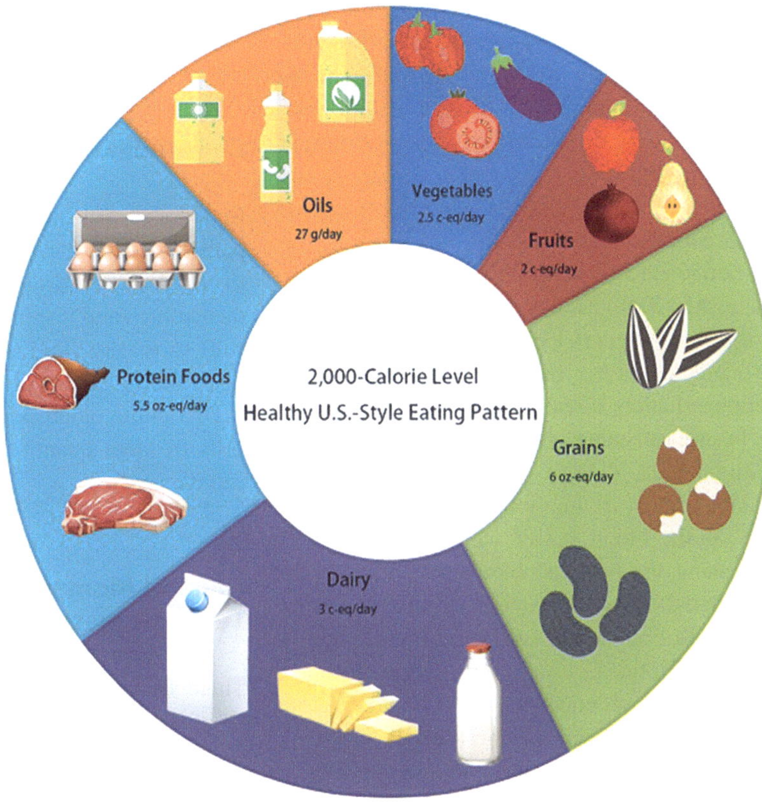

Fig. 34.1 The 2000-Calorie Level of Healthy U. S.-Style Eating Pattern Daily (*2015–2020 Dietary Guidelines for Americans*. 8th Edition). NOTE: Food group amounts shown in cup-(c) or ounce-(oz) equivalents (eq). Oils are shown in grams (g)

Metabolic Disease and Dysglycemia

CVD is a leading cause of death and morbidity in the diabetic population (Matheus et al. 2013). Metabolic syndrome is analogously more prone to CVD events (Ford 2005; Malik et al. 2004; Morrish et al. 2001; Wong 2007). In 2015, the GBD project (Mortality 2015) estimated that diabetes mellitus had become the sixth leading cause of disability worldwide. Approximately 17.7 million people died of cardiovascular disease, accounting for 31% of global deaths, according to WHO. Compared with nondiabetic patients, diabetic patients have a 2–4 times higher risk of cardiovascular disease (Poznyak et al. 2020; Chichareon et al. 2020; Preis et al. 2009). Cardiovascular deaths accounted for 44% of deaths in people with type 1 diabetes and 52% of deaths in people with type 2 diabetes (Morrish et al. 2001). The leading causes of death in diabetic patients include ischemic heart disease, malignant tumors, cerebrovascular disease, and pneumonia/influenza. The most common manifestations of CVD in diabetes are peripheral artery disease (16.2% or more) and heart failure (14.7%), followed by angina and nonfatal MI.

In the SAN Antonio heart study, which followed 4875 patients for 7–8 years, diabetes was significantly associated with increased CVD mortality. The risk of heart failure in diabetes mellitus rose to 40%, compared with an age-adjusted ratio of 2.8 for nondiabetics, representing a two- to threefold elevated risk (Go et al. 2014).

Predisposing Factors tor Type 2 Diabetes

The risk of type 2 diabetes is determined by genetic and metabolic factors, including immutable circumstances such as race, family history, gestational diabetes history, and age, along with modifiable ones such as obesity, unhealthy diet, insufficient physical activity levels, and smoking. Excess fat is often associated with insulin resistance and is a significant risk factor for type 2 diabetes and CVD (Hardy et al. 2012). Sedentary lifestyle, dyslipidemia, and hypertension affect endothelial dysfunction, which can further accelerate atherosclerosis (Rodriguez-Araujo and Nakagami 2018). Inflammatory markers such as C-reactive protein (CRP) in diabetic patients may also predict further risk for cardiovascular and peripheral artery disease.

Insulin resistance is one of the direct causes of cardiovascular disease. Due to the increase of insulin content in the blood, the dilatation function of the blood vessel wall cannot proceed, increasing the patients' blood pressure. On the other hand, the occurrence of insulin resistance can cause endothelial cell function damage in patients. Besides, insulin resistance also can cause injury of endothelial cells and lead to myocardial ischemia, which is the direct pathogenic factor of several cardiovascular diseases.

The Impact of Dyslipidemia

Dyslipidemia is common in diabetic patients and is associated with an increased risk of CVD (Matheus et al. 2008; Incalza et al. 2018), especially for myocardial infarction, atherosclerotic stenosis, and aortic valve stenosis. A large case-control study involving 52 countries demonstrated that for first myocardial infarction (MI), dyslipidemia was the most critical risk factor (Yusuf et al. 2004). Primary prevention randomized controlled trials (RCTs) evidenced that reducing low-density lipoprotein (LDL) lowers the risk of CVD, which aligned with the range of pretreatment LDL levels (Collins et al. 2016). In a well-designed meta-analysis, the degree of reduction in CVD is proportionate to the degree of lowering of LDL cholesterol, till levels as low as 21 mg/dL (0.5 mmol/L) (Sabatine et al. 2018).

Endothelial Damage

Endothelial cell injury is one of the leading pathways toward cardiovascular disease related to metabolic syndrome. Patients with low lipid metabolism and insulin resistance can prompt fat cells to release free fatty acids, which enter the veins and cause an increase in low-density protein (LDL) in the blood. LDL attaches to endothelial cells in large quantities and causes endothelial cell damage.

General Biomarkers and Novel Technologies

There are several categories of CVD biomarkers (Biomarkers Definitions Working G 2001; Bossuyt et al. 2012; Moons 2010; Braunwald 2008) addressing heart failure (brain natriuretic peptide/BNP, N-terminal BNP/NT-BNP, atrial natriuretic peptide/ANP, suppression of tumorigenicity/ST-2), or atherosclerotic coronary artery disease (troponin T or I, creatinine phosphokinase-MB). Additionally, they can target inflammation (C-reactive protein, interleukin/IL-6, fibrinogen, monocyte chemotactic protein-1, tumor necrosis factor/TNF-a), oxidative stress (isoprostanes), and metabolic conditions (low- and high-density lipoprotein, apolipoprotein/ApoB 100, lipoprotein-associated phospholipase A2, homocysteine, vitamin D, fibroblast growth factor 23, adiponectin, glycated hemoglobin, haptoglobin).

ALT, AST, and cardiac-specific troponin I or troponin T levels are clinically used as surrogate predictive markers in CVD progression and prognosis. For myocardial infarction, the ratio of creatine kinase CK-MB1 to CK-MB2 isoforms is specific and a sensitive predictor. The issue remains for the clinical application of these cardiac markers (CK-MB, troponin I, and troponin T), and the profiles have an inevitable delay.

New Biomarkers

Novel biomarkers, including cardiac troponins (cTn), cardiac myosin light-chain kinase (cMLCK), and heart-type fatty acid-binding protein (H-FABP), have emerged (Parsanathan and Jain 2020); however, their clinical value has not yet been fully elucidated. Conventional noninvasive anatomical and electrophysiological biomarkers, namely, carotid intima-media thickness (CIMT) and coronary artery calcification (CAC), continue to be investigated. Carotid intima-media thickness can be used in younger populations to predict CVD events. Coronary CT angiography is often used to measure intimal and medial calcification associated with arterial stiffness. CAC is an independent risk factor to predict subclinical atherosclerosis. Additionally, several microRNAs, including miR-126, miR-223, and miR-197, and a panel of 27 single-nucleotide polymorphisms (SNPs) from genome-wide association studies, have been anticipated as biomarkers to predict CVD. However, the gains are arguably modest, and further research is needed to identify new biomarkers and to determine whether multimarker strategies are a viable approach to better risk stratification.

Big Data and the Gut Microbiome

Big data approaches may deliver clinically relevant phenotypes, scalable insights from real-world evidence (biomedical and omics-data) driving disease prediction, and personalized medicine through advanced analytics. The intestinal microorganisms should not be excluded from this target. Previous studies had shown that *Akkermansia muciniphila* has health benefits, such as reducing obesity, glucose intolerance, insulin resistance, as well as fat accumulation in the liver in animal models. In a clinical protocol with overweight and obese insulin-resistant patients, 3- month administration of the pasteurized bacteria (10^{10}/day) was well tolerated and did not interfere with the general microbiome profile. Effectively insulin sensitivity improved, whereas insulin and cholesterol levels, body weight, and additional anthropometric measurements, as well as inflammatory markers all diminished (Depommier et al. 2019).

The Precision Medicine Approach

Large epidemiologic databases, including CVD information are now available, such as the Global Burden of Disease (GBD) Study (UNC Carolina Population Center 2020; Institute for Health Metrics and Evaluation (IHME) 2020; World Health Organization 2018). Along with large-scale observational and clinical trials, these represent a significant step toward the precision medicine concept; however, more is needed in the realm of multi-omics cardiovascular screening. The genotype, endophenotype, and clinical phenotype in commonly seen cardiovascular diseases has been demonstrated to be more heterogeneous than previously believed, thus representing a challenge to the short-term application of the multi-omics concept (Leopold et al. 2020). Nevertheless, advanced machine-learning methodologies applied to available conventional datasets, such as the classification and regression tree (CART) algorithm, are already permitting the highest accuracy and reliability in the diagnosis of coronary artery disease. In a protocol starting with a 55-independent feature diagnostic panel, the 18 best and also just the 10 highest-rated variables, structured within the CART algorithm, already outperformed previously published models (Ghiasi et al. 2020).

References

Alpert MA, Karthikeyan K, Abdullah O, Ghadban R (2018) Obesity and cardiac remodeling in adults: mechanisms and clinical implications. Prog Cardiovasc Dis 61(2):114–123

Belke DD, Betuing S, Tuttle MJ, Graveleau C, Young ME, Pham M, Zhang D, Cooksey RC, McClain DA, Litwin SE et al (2002) Insulin signaling coordinately regulates

cardiac size, metabolism, and contractile protein isoform expression. J Clin Invest 109(5):629–639
Biomarkers Definitions Working G (2001) Biomarkers and surrogate endpoints: preferred definitions and conceptual framework. Clin Pharmacol Ther 69(3):89–95
Bossuyt PM, Reitsma JB, Linnet K, Moons KG (2012) Beyond diagnostic accuracy: the clinical utility of diagnostic tests. Clin Chem 58(12):1636–1643
Braunwald E (2008) Biomarkers in heart failure. N Engl J Med 358(20):2148–2159
Chen Y, McClintock TR, Segers S, Parvez F, Islam T, Ahmed A, Rakibuz-Zaman M, Hasan R, Sarwar G, Ahsan H (2013) Prospective investigation of major dietary patterns and risk of cardiovascular mortality in Bangladesh. Int J Cardiol 167(4):1495–1501
Chichareon P, Modolo R, Kogame N, Takahashi K, Chang CC, Tomaniak M, Botelho R, Eeckhout E, Hofma S, Trendafilova-Lazarova D, Kőszegi Z, Iñiguez A, Wykrzykowska JJ, Piek JJ, Garg S, Hamm C, Steg PG, Jüni P, Vranckx P, Valgimigli M, Windecker S, Onuma Y, Serruys PW (2020) Association of diabetes with outcomes in patients undergoing contemporary percutaneous coronary intervention: pre-specified subgroup analysis from the randomized GLOBAL LEADERS study. Atherosclerosis 295:45–53
Collins R, Reith C, Emberson J, Armitage J, Baigent C, Blackwell L, Blumenthal R, Danesh J, Smith GD, DeMets D et al (2016) Interpretation of the evidence for the efficacy and safety of statin therapy. Lancet 388 (10059):2532–2561
de Souza RJ, Mente A, Maroleanu A, Cozma AI, Ha V, Kishibe T, Uleryk E, Budylowski P, Schunemann H, Beyene J et al (2015) Intake of saturated and trans unsaturated fatty acids and risk of all cause mortality, cardiovascular disease, and type 2 diabetes: systematic review and meta-analysis of observational studies. BMJ 351:h3978
Depommier C, Everard A, Druart C, Plovier H, Van Hul M, Vieira-Silva S, Falony G, Raes J, Maiter D, Delzenne NM et al (2019) Supplementation with Akkermansia muciniphila in overweight and obese human volunteers: a proof-of-concept exploratory study. Nat Med 25(7):1096–1103
Duncan BB, Schmidt MI, Pankow JS, Ballantyne CM, Couper D, Vigo A, Hoogeveen R, Folsom AR, Heiss G (2003) Atherosclerosis risk in communities S: low-grade systemic inflammation and the development of type 2 diabetes: the atherosclerosis risk in communities study. Diabetes 52(7):1799–1805
Ford ES (2005) Risks for all-cause mortality, cardiovascular disease, and diabetes associated with the metabolic syndrome: a summary of the evidence. Diabetes Care 28(7):1769–1778
Ghiasi MM, Zendehboudi S, Mohsenipour AA (2020) Decision tree-based diagnosis of coronary artery disease: CART model. Comput Methods Prog Biomed 192:105400
Go AS, Mozaffarian D, Roger VL, Benjamin EJ, Berry JD, Blaha MJ, Dai S, Ford ES, Fox CS, Franco S et al (2014) Heart disease and stroke statistics–2014 update: a report from the American Heart Association. Circulation 129(3):e28–e292
Hardy OT, Czech MP, Corvera S (2012) What causes the insulin resistance underlying obesity? Curr Opin Endocrinol Diabetes Obes 19(2):81–87
Hooper L, Summerbell CD, Thompson R, Sills D, Roberts FG, Moore H, Davey Smith G (2011) Reduced or modified dietary fat for preventing cardiovascular disease. Cochrane Database Syst Rev 7:CD002137
Huang C, Huang J, Tian Y, Yang X, Gu D (2014) Sugar sweetened beverages consumption and risk of coronary heart disease: a meta-analysis of prospective studies. Atherosclerosis 234(1):11–16
Incalza MA, D'Oria R, Natalicchio A, Perrini S, Laviola L, Giorgino F (2018) Oxidative stress and reactive oxygen species in endothelial dysfunction associated with cardiovascular and metabolic diseases. Vasc Pharmacol 100:1–19
Institute for Health Metrics and Evaluation (IHME) (2020) GBD results tool. http://ghdx.healthdata.org/gbd-results-tool. Accessed 17 Mar 2020
Leahy JL (2005) Pathogenesis of type 2 diabetes mellitus. Arch Med Res 36(3):197–209
Leopold JA, Maron BA, Loscalzo J (2020) The application of big data to cardiovascular disease: paths to precision medicine. J Clin Invest 130(1):29–38
Malik S, Wong ND, Franklin SS, Kamath TV, L'Italien GJ, Pio JR, Williams GR (2004) Impact of the metabolic syndrome on mortality from coronary heart disease, cardiovascular disease, and all causes in United States adults. Circulation 110(10):1245–1250
Matheus AS, Cobas RA, Gomes MB (2008) Dyslipidemias in type 1 diabetes: a current approach. Arq Bras Endocrinol Metabol 52(2):334–339
Matheus AS, Tannus LR, Cobas RA, Palma CC, Negrato CA, Gomes MB (2013) Impact of diabetes on cardiovascular disease: an update. Int J Hypertens 2013:653789
Moons KG (2010) Criteria for scientific evaluation of novel markers: a perspective. Clin Chem 56 (4):537–541
Morrish NJ, Wang SL, Stevens LK, Fuller JH, Keen H (2001) Mortality and causes of death in the WHO multinational study of vascular disease in diabetes. Diabetologia 44(Suppl 2):S14–S21
Mortality GBD (2015) Causes of Death C: Global, regional, and national age-sex specific all-cause and cause-specific mortality for 240 causes of death, 1990–2013: a systematic analysis for the Global Burden of Disease Study 2013. Lancet 385 (9963):117–171
Newell DG, Koopmans M, Verhoef L, Duizer E, Aidara-Kane A, Sprong H, Opsteegh M, Langelaar M, Threfall J, Scheutz F (2010) Food-borne diseases—the

challenges of 20 years ago still persist while new ones continue to emerge. Int J Food Microbiol 139:S3–S15

Parsanathan R, Jain SK (2020) Novel invasive and noninvasive cardiac-specific biomarkers in obesity and cardiovascular diseases. Metab Syndr Relat Disord 18 (1):10–30

Piche ME, Poirier P, Lemieux I, Despres JP (2018) Overview of epidemiology and contribution of obesity and body fat distribution to cardiovascular disease: an update. Prog Cardiovasc Dis 61(2):103–113

Poznyak A, Grechko AV, Poggio P, Myasoedova VA, Alfieri V, Orekhov AN (2020) The diabetes mellitus-atherosclerosis connection: the role of lipid and glucose metabolism and chronic inflammation. Int J Mol Sci 21(5):1835

Preis SR, Hwang SJ, Coady S, Pencina MJ, D'Agostino RB Sr, Savage PJ, Levy D, Fox CS (2009) Trends in all-cause and cardiovascular disease mortality among women and men with and without diabetes mellitus in the Framingham Heart Study, 1950 to 2005. Circulation 119(13):1728–1735

Rodriguez-Araujo G, Nakagami H (2018) Pathophysiology of cardiovascular disease in diabetes mellitus. Cardiovasc Endocrinol Metab 7(1):4–9

Sabatine MS, Wiviott SD, Im K, Murphy SA, Giugliano RP (2018) Efficacy and safety of further lowering of low-density lipoprotein cholesterol in patients starting with very low levels: a meta-analysis. JAMA Cardiol 3 (9):823–828

Saeedi P, Petersohn I, Salpea P, Malanda B, Karuranga S, Unwin N, Colagiuri S, Guariguata L, Motala AA, Ogurtsova K et al (2019) Global and regional diabetes prevalence estimates for 2019 and projections for 2030 and 2045: results from the International Diabetes Federation Diabetes Atlas, 9(th) edition. Diabetes Res Clin Pract 107843:157

Shimobayashi M, Albert V, Woelnerhanssen B, Frei IC, Weissenberger D, Meyer-Gerspach AC, Clement N, Moes S, Colombi M, Meier JA et al (2018) Insulin resistance causes inflammation in adipose tissue. J Clin Invest 128(4):1538–1550

Song M, Fung TT, Hu FB, Willett WC, Longo VD, Chan AT, Giovannucci EL (2016) Association of animal and plant protein intake with all-cause and cause-specific mortality. JAMA Intern Med 176 (10):1453–1463

UNC Carolina Population Center (2020) Data Sets. www.cpc.unc.edu/projects/china/data/datasets/index.html. Accessed 17 Mar 2020

WHO: World Health Organization (2018) Global Health Observatory (GHO) data. www.who.int/gho/publications/world_health_statistics/2018/en/. Accessed 17 Mar 2020

Wong ND (2007) Metabolic syndrome: cardiovascular risk assessment and management. Am J Cardiovasc Drugs 7(4):259–272

Yusuf S, Hawken S, Ounpuu S, Dans T, Avezum A, Lanas F, McQueen M, Budaj A, Pais P, Varigos J et al (2004) Effect of potentially modifiable risk factors associated with myocardial infarction in 52 countries (the INTERHEART study): case-control study. Lancet 364(9438):937–952

Gestational Diabetes Mellitus

35

Elpida Vounzoulaki and Samuel Seidu

Contents

Gestational Diabetes: A Growing Problem 480
Other Metabolic Features of Pregnancy 480

Screening and Diagnosis 481

Managing Gestational Diabetes 481
Glucose Management 481
Dietary Interventions 481
Exercise Programmes 483
Contraindications to Exercise 483
Glucose-Lowering Medication and Supplements 483
Oral Therapy 484
Vitamins and Other Supplements 484

Future Prognosis for Mother and Child 484
Short-Term Outcomes 484
Long-Term Outcomes 484
Cardiovascular, Metabolic and Other Complications 485
Complications of the Newborn 485

Post-partum Screening and Care 485
Diet, Exercise and Metformin 485
Other Oral Drugs 486
The Multiple Advantages of Breastfeeding 486

Progression to Type 2 Diabetes 487
The Challenges of Effective Screening 487
E-Health and Postpartum Follow-Up 487

Summary 487

References 488

E. Vounzoulaki (✉) · S. Seidu
Diabetes Research Centre, Leicester General Hospital,
Leicester, UK
e-mail: ev63@le.ac.uk

J. Faintuch and S. Faintuch (eds.), *Obesity and Diabetes*, https://doi.org/10.1007/978-3-030-53370-0_35

Abstract

Gestational diabetes mellitus, characterised as glucose intolerance first recognised during the second or third trimester of pregnancy, which is clearly not type 1 or type 2 diabetes, imposes a significant burden on maternal and neonatal health. Women with gestational diabetes are at an increased risk of subsequent progression to type 2 diabetes, a chronic metabolic disorder associated with multiple comorbidities and reduced life expectancy. This chapter will underline the pathophysiological mechanisms involved in gestational diabetes, review the current guidelines and recommendations for screening and disease management, discuss future prognosis for both mother and child, as well as look into post-partum care strategies to delay or prevent type 2 diabetes onset.

Keywords

Gestational diabetes mellitus · Diabetes mellitus · Hyperglycaemia · Pregnancy · Pathophysiology · Screening · Diagnosis · Management

Gestational Diabetes: A Growing Problem

Gestational diabetes mellitus is glucose intolerance with onset or first diagnosis during the second or third trimester of pregnancy that is neither pre-existing type 1 nor type 2 diabetes (American Diabetes Association (ADA) 2019a). Widely recognised risk factors for gestational diabetes include non-white ethnicity, advanced maternal age, obesity, excessive weight gain during pregnancy, family history of diabetes, gestational diabetes in prior pregnancy, foetal macrosomia and unexplained stillbirth (Martis et al. 2018). In 2019, the International Diabetes Federation estimated that, on a global scale, 20.4 million or 15.8% of live births were affected by hyperglycaemia in pregnancy, with 83.6% of those cases being due to gestational diabetes (International Diabetes Federation 2019). This prevalence is expected to rise due to trends in physical inactivity and increased caloric consumption, promoting the spread of the global obesity epidemic (Hunt and Schuller 2007).

Pregnancy is by nature a diabetogenic condition as it is associated with a decrease of insulin sensitivity, which is accompanied by a compensatory increase in insulin secretion. The purpose of these metabolic changes is to ensure that adequate levels of glucose and other nutrients are available for the foetus (Di Cianni et al. 2003). However, obesity and inheritance often interfere with these metabolic shifts, leading to inadequate insulin secretion due to the inability of pancreatic β-cells to adapt, and hence women develop gestational diabetes (Di Cianni et al. 2003).

Other Metabolic Features of Pregnancy

In addition to glucose metabolism, pregnancy induces changes in both lipid and protein metabolism. As there is a need for increased glucose and amino acids supply to the foetus, energy production shifts from an anabolic to a catabolic state, promoting the use of lipids instead of carbohydrates as source (Butte 2000). These alterations lead to fat accumulation, as total cholesterol, low-density lipoprotein (LDL) cholesterol, very low-density lipoprotein (VLDL) and triglycerides are all elevated both during second and third trimesters, while high-density lipoprotein (HDL) cholesterol increases in early pregnancy and remains elevated during its whole duration (Butte 2000; Pusukuru et al. 2016). In pregnancies complicated by gestational diabetes, hyperlipidaemia is exaggerated, with an emphasis on the significantly increased triglyceride levels throughout the whole gestational period (Ryckman et al. 2015). With regard to protein metabolism, plasma levels of branched-chain amino acids (isoleucine, leucine and valine) are elevated in women with gestational diabetes, and these alterations are also positively associated with a higher incident risk of type 2 diabetes (Metzger et al. 1980; Tobias et al. 2018).

To date, the complex pathophysiological mechanisms involved in the development of

Table 35.1 ADA guidelines on screening and diagnosis of gestational diabetes

One-step strategy			
The diagnosis of gestational diabetes is made when any of the following plasma glucose values are met or exceeded:			
Fasting: 92 mg/dL (5.1 mmol/L)			
1 h: 180 mg/dL (10.0 mmol/L)			
2 h: 153 mg/dL (8.5 mmol/L)			
Two-step strategy			
First step: A 50-g non-fasting GLT is performed. If plasma glucose levels at 1 h following the load are ≥130 mg/dL, 135 mg/dL, or 140 mg/dL (7.2 mmol/L,7.5 mmol/L, or 7.8 mmol/L, respectively), perform a 100-g OGTT.			
Second step: A 100-g OGTT is performed at fasting. The patient receives a diagnosis of gestational diabetes if at least two of the following four plasma glucose levels (at fasting state and 1 h, 2 h, 3 h during OGTT) are met or exceeded:			
	Carpenter-Coustan[a]	Or	NDDG
Fasting	95 mg/dL (5.3 mmol/L)		105 mg/dL (5.8 mmol/L)
1 h	180 mg/dL (10.0 mmol/L)		190 mg/dL (10.6 mmol/L)
2 h	155 mg/dL (8.6 mmol/L)		165 mg/dL (9.2 mmol/L)
3 h	140 mg/dL (7.8 mmol/L)		145 mg/dL (8.0 mmol/L)

GLT glucose load test, *NDDG* National Diabetes Data Group, *OGTT* oral glucose tolerance test

[a] The 2020 ADA guidelines recommend that if a two-step approach is used, the Carpenter Coustan criteria should be preferred due to lower diagnostic thresholds

gestational diabetes have not yet been fully understood.

Screening and Diagnosis

According to the American Diabetes Association (ADA), screening for gestational diabetes is performed at 24–28 weeks of gestation in women who have no history of pre-existing diabetes, using either the "One-step" 75-g oral glucose tolerance test (OGTT) or the "Two-step" approach using a 50-g (non-fasting) screen, followed by a 100-g OGTT for those who receive a positive result (American Diabetes Association (ADA) 2019a). The 75-g OGTT is performed in the morning, following an overnight fast of at least 8 h, at the fasting state, 1 and 2 h (American Diabetes Association (ADA) 2019a). In the two-step strategy, the patient receives a 50-g non-fasting glucose load test (GLT), glucose measurement is performed at 1 h, and based on the result, a 100-g OGTT is performed (American Diabetes Association (ADA) 2019a) (Table 35.1).

However, there is a lack of universal agreement on both the screening method as well as the cut-off values, used for the diagnosis of gestational diabetes. This often leads to different estimates of disease prevalence across regions, while different populations are being targeted (Behboudi-Gandevani et al. 2019). Table 35.2 highlights the differences between universal screening guidelines for gestational diabetes. The majority of guidelines support the use of a one-step strategy, while some guidelines suggest both a one-step and a two-step approach.

Managing Gestational Diabetes

Glucose Management

Tight glycaemic control is a key step in the management of gestational diabetes (Fig. 35.1). The Fifth International Workshop-Conference on Gestational Diabetes Mellitus (Metzger et al. 2007) recommends the following target blood glucose levels for women with gestational diabetes during pregnancy:

- Fasting <95 mg/dL (5.3 mmol/L) and either
- 1-h postprandial <140 mg/dL (7.8 mmol/L) or
- 2-h postprandial <120 mg/dL (6.7 mmol/L).

Dietary Interventions

Dietary interventions are the primary strategy for the management of gestational diabetes (Fig. 35.1). Following diagnosis, a nutrition plan

Table 35.2 Universal diagnostic thresholds for gestational diabetes

		Fasting		1-h		2-h		3-h	
Criteria	Method	mg/dL	mmol/L	mg/dL	mmol/L	mg/dL	mmol/L	mg/dL	mmol/L
ADA (2019a)[a]	1-Step	92	5.1	180	10.0	153	8.5	–	–
ADA (2019a)[a]	2-Step	–	–	–	–	–	–	140	–
ACOG (2018)[b]	2-Step	–	–	–	–	–	–	–	7.8
ADA (2019a)[a], ACOG (2018)[b]	2-Step	95	5.3	180	10.0	155	8.6	–	–
ADIPS (2014)[c]	1-Step	92	5.1	180	10.0	153	8.5	–	–
CDA (2018)[d,i]	1-Step 2-Step	–	5.3	–	10.6	–	9.0	–	–
EASD (1996)[e]	1-Step	–	6.0	–	–	–	9.0	–	–
IADPSG (2010)[f]	1-Step	92	5.1	180	10.0	153	8.5	–	–
NICE (2015)[g]	1-Step	–	5.6	–	–	–	7.8	–	–
WHO (2013)[h]	1-Step	92	5.1	180	10.0	153	8.5	–	–

ACOG American College of Obstetricians and Gynecologists, *ADA* American Diabetes Association, *ADIPS* Australasian Diabetes in Pregnancy Society, *CDA* Canadian Diabetes Association, *EASD* European Association for the Study of Diabetes, *IADPSG* International Association of the Diabetes and Pregnancy Study Groups, *NICE* National Institute for Health and Care Excellence, *WHO* World Health Organisation

[a]American Diabetes Association (ADA) (2019a)

[b]The American College of Obstetricians, and Gynecologists (ACOG) (2018)

[c]Nankervis et al. (2014)

[d]Feig et al. (2018)

[e]Brown et al. (1996)

[f]International Association of Diabetes and Pregnancy Study Groups Consensus Panel et al. (2010)

[g]The National Institute for Health and Care Excellence (NICE) (2015)

[h]World Health Organization (2013)

iCut-off values only for two-step strategy using a 50-g glucose challenge test, followed, if abnormal, by a 75-g OGTT (preferred approach), diagnosis of gestational diabetes with one-step strategy with 75-g OGTT: fasting $\geq$5.1 mmol/L, 1-h $\geq$10.0 mmol/L, 2-h $\geq$8.5 mmol/L)

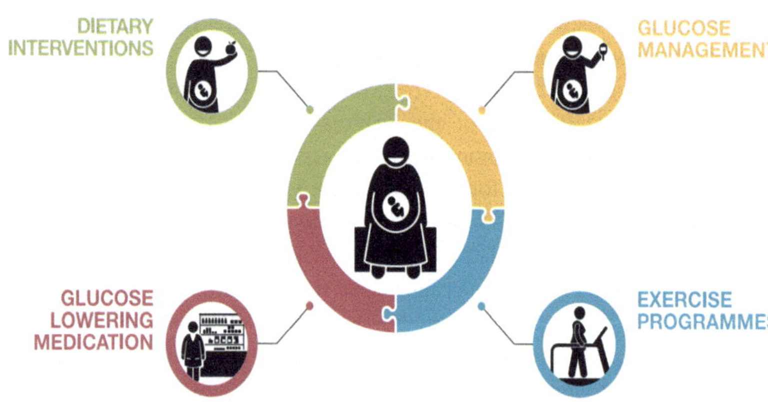

Fig. 35.1 Management of gestational diabetes in pregnancy (Figure created by Michael Bonar, Diabetes Research Centre, Leicester, UK)

is provided to the patient by a registered dietician specialised in gestational diabetes, aiming to achieve sufficient calorie intake for foetal growth and improve maternal health outcomes (American Diabetes Association (ADA) 2019b). Dietary interventions should by no means promote weight loss or excessive gestational weight gain, while they should focus on achieving optimal glycaemic control (Brown et al. 2017). There is no optimal calorie intake for women with gestational diabetes supported by research so far; hence, it is suggested that these women should adhere to the Dietary Reference Intakes for all pregnant women, suggesting a minimum of 175 g of carbohydrates, 71 g of protein and 28 g of fibre (American Diabetes Association (ADA) 2019b).

Overall, medical nutrition guidelines for gestational diabetes recommend adequate protein consumption and selection of low glycaemic-index foods (Tsirou et al. 2019). There are, however, several inconsistencies across the different available medical nutrition therapy guidelines, as recognised by a recent review and critical appraisal, as they often fail to involve patients and dieticians in their development, and be realistic and flexible, while their applicability is low (Tsirou et al. 2019). Future research needs to focus on developing high-quality, evidence-based guidelines that focus on patients, and can therefore be applicable in a realistic clinical setting.

Exercise Programmes

While women with gestational diabetes are first offered dietary interventions, guidelines often recommend the combination of diet and exercise as a more effective approach for disease management (The National Institute for Health and Care Excellence (NICE) 2015; American College of Obstetricians and Gynecologists (ACOG) 2015) (Fig. 35.1). Frequent and sustained exercise can improve insulin sensitivity in skeletal muscle and achieve better glucose control (Martis et al. 2018). Current pregnancy guidelines focus primarily on the prescription of aerobic exercise, and secondarily on resistance strength training (Padayachee 2015). It is generally recommended that the patient should be offered an individualised exercise programme, including moderate-intensity exercise with a duration of at least 20–30 min a day that should be performed daily or on most days of the week, unless there is some contraindication (American College of Obstetricians and Gynecologists (ACOG) 2015). Safe types of physical activity during pregnancy include walking, stationary cycling, swimming and low-impact aerobics (American College of Obstetricians and Gynecologists (ACOG) 2015).

Contraindications to Exercise

There are considerable beneficial outcomes associated with exercise for both the mother and the foetus, while contraindications must be also considered. Performing physical activity in pregnancy can often be hindered due to pregnancy-related pain, as pelvic girdle pain or low back pain prevalence is high and often induces disability (Wu et al. 2004). Contraindications for physical activity during pregnancy as recognised by international guidelines include multiple gestation, pregnancy-induced hypertension or pre-eclampsia, cardiovascular disease, thyroid disease, anaemia, persistent bleeding, premature membrane rupture, premature contractions or labour, and cerclage or incompetent cervix, while a few guidelines recognise morbid obesity, poorly controlled diabetes mellitus, eating disorders and placenta previa as contraindications as well (Evenson et al. 2014).

Glucose-Lowering Medication and Supplements

Women with gestational diabetes who cannot achieve favourable glucose control with lifestyle modifications such as dietary and exercise intervention are administered pharmacological agents (Fig. 35.1). In the United States, ADA recommends insulin as the first-line pharmacological agent (American Diabetes Association (ADA) 2019b). Insulin is administered by either multiple daily injections or continuous subcutaneous infusion. The Food and Drug Administration (FDA) has approved human regular insulin, rapid-acting insulin analogues (lispro and aspart), Neutral Protamine Hagedorn (NPH) long-acting insulin, as well as the long-acting insulin analogue Detemir, classifying them as pregnancy category B, and has established their safety of use during pregnancy (Kintiraki and Goulis 2018).

Oral Therapy

The use of oral medication, including both metformin and glyburide (sulfonylureas), is supported by evidence; however, they are still regarded as a second-line treatment by ADA, as there is a concern about them crossing the placenta, while long-term offspring safety is also questionable (American Diabetes Association (ADA) 2019b). Glyburide has also been linked to both foetal macrosomia and neonatal hypoglycaemia (Malek and Davis 2016; Song et al. 2017). With regard to metformin, it has been found to induce less weight gain compared to insulin; while it has been associated with a lower risk of neonatal hypoglycaemia, it is associated with a higher risk of preterm birth (Balsells et al. 2015; Gui et al. 2013).

Consistently with ADA, the American College of Obstetricians and Gynaecologists (ACOG) guidelines focus on insulin, and they also recommend metformin as well as glyburide (sulfonylureas) as an alternative, while the Fifth International Workshop-Conference in Gestational Diabetes Mellitus focuses on insulin, suggests glyburide as an alternative treatment, and expresses uncertainty about metformin (Metzger et al. 2007; The American College of Obstetricians and Gynecologists (ACOG) 2018). Contrary to that, guidelines by the National Institute for Health and Care Excellence (NICE) recommend metformin as a first-line pharmacological treatment, while insulin is used when there is a contraindication for metformin, and glyburide is also considered an alternative treatment option (The National Institute for Health and Care Excellence (NICE) 2015).

A few advantages of oral medication are the ease of use, wide acceptability as well as lower cost (Ryu et al. 2014). In real-world clinical practice, healthcare professionals need to weigh both favourable outcomes and adverse effects in order to decide upon a treatment option based on the condition of the individual, while more research is required to evaluate long-term treatment-associated outcomes on offspring health.

Vitamins and Other Supplements

The evidence surrounding the efficacy of dietary supplements for the management of gestational diabetes is quite contradicting. A double-blinded randomised controlled trial by Li et al. found that vitamin D supplementation during pregnancy affected by gestational diabetes can improve both fasting plasma glucose and lipid levels (Li and Xing 2016). Contrary to that, results from a recently published systematic review support that there is no adequate evidence on the efficacy of vitamin D supplementation, in improving glucose metabolism or maternal and neonatal outcomes (Rodrigues et al. 2019). Myo-inositol supplementation has drawn research attention over the last few years, with several studies demonstrating its potential in improving insulin resistance in gestational diabetes and reducing the need for insulin treatment (Corrado et al. 2011; Costabile and Unfer 2017; Lubin et al. 2016). The efficacy of both vitamin D and myo-inositol in pregnancy complicated by gestational diabetes needs to be further investigated.

Future Prognosis for Mother and Child

Short-Term Outcomes

A substantial number of maternal complications have been associated, including gestational hypertension and preeclampsia, preterm birth, perineal trauma, caesarean delivery and postpartum bleeding (Kim 2010). With regard to foetal/neonatal health, a diagnosis of gestational diabetes increases the risk of macrosomia, being born large for gestational age, shoulder dystocia, bone fractures, nerve palsy, stillbirth, neonatal hypoglycaemia, hyperbilirubinaemia and respiratory distress syndrome (Martis et al. 2018; Kim 2010).

Long-Term Outcomes

Women affected by gestational diabetes have a nearly 10-fold higher risk of progression to type

2 diabetes than those with a normoglycaemic pregnancy (Vounzoulaki et al. 2020). Incidence of type 2 diabetes has also been reported to markedly increase in the first 5 post-partum years, while it seemed to reach a plateau afterwards (Kim et al. 2002). Among the most common factors associated with an increased risk of progression are the degree of glucose intolerance during pregnancy, maternal age at diagnosis of gestational diabetes, use of insulin during pregnancy, and longer post-partum follow-up periods (Kim et al. 2002).

Cardiovascular, Metabolic and Other Complications

Women with a history of gestational diabetes are at an increased risk of future cardiovascular disease, while they are also more likely to be diagnosed with non-alcoholic fatty liver disease (Goueslard et al. 2016; Foghsgaard et al. 2017), with both these conditions being associated with reduced average life expectancy. A previous diagnosis of gestational diabetes identifies a patient group with a twofold higher risk for future cardiovascular events, and this risk is independent of subsequent progression to type 2 diabetes (Kramer et al. 2019). Additionally, not only glucose, but also mean lipid levels such as total cholesterol, triglycerides, LDL cholesterol, as well as mean systolic blood pressure, are significantly increased in women with previous gestational diabetes (Meyers-Seifer and Vohr 1996).

During pregnancy, women affected by gestational diabetes exhibit impaired microvascular function and skeletal muscle oxygenation as well as blunted cerebral oxygenation, while these impairments can persist in the post-partum period (Dipla et al. 2017; Kintiraki et al. 2018; Vounzoulaki et al. 2019). The impact of gestational diabetes is not limited to physical health but extends to mental health and wellbeing, as women affected by the disease were found to have an over fourfold risk of post-partum depression (Hinkle et al. 2016), with caesarean delivery and gestational weight gain identified as possible risk factors (Nicklas et al. 2013).

Complications of the Newborn

Children are more prone to become overweight or develop metabolic syndrome (Boerschmann et al. 2010; Vääräsmäki et al. 2009), while a recent longitudinal cohort study also demonstrated that exposure to gestational diabetes during pregnancy, diagnosed by 26 weeks of gestation, induces a higher risk of autism spectrum disorder in offspring (Xiang et al. 2015).

Post-partum Screening and Care

Women affected by gestational diabetes during pregnancy require close monitoring in the post-partum period. Post-partum screening is recommended at 4–12 weeks post-partum, using an OGTT (American Diabetes Association (ADA) 2019b). The use of glycated haemoglobin (HbA1c) for post-partum screening is not recommended by ADA, as it can be affected by pregnancy red blood cell turnover or blood loss during labour, while the 75-g OGTT is more sensitive in detecting both prediabetes and type 2 diabetes (American Diabetes Association (ADA) 2019b). If test results are normal, further testing is required every 1–3 years with any recommended glucose test (HbA1c, fasting plasma glucose or OGTT), and frequency is dependent on risk factors such as family history, body mass index (BMI) before pregnancy, and use of insulin or oral medication in pregnancy (American Diabetes Association (ADA) 2019b). Post-partum care in this population mainly focuses on lifestyle modification, pharmacological treatment and breastfeeding (Fig. 35.2).

Diet, Exercise and Metformin

Both lifestyle intervention and metformin have been proven effective in attenuating subsequent progression to type 2 diabetes. Results from the Diabetes Prevention Program Outcomes study have indicated that the cumulative incidence of type 2 diabetes was the lowest in those who received lifestyle intervention, with the onset of

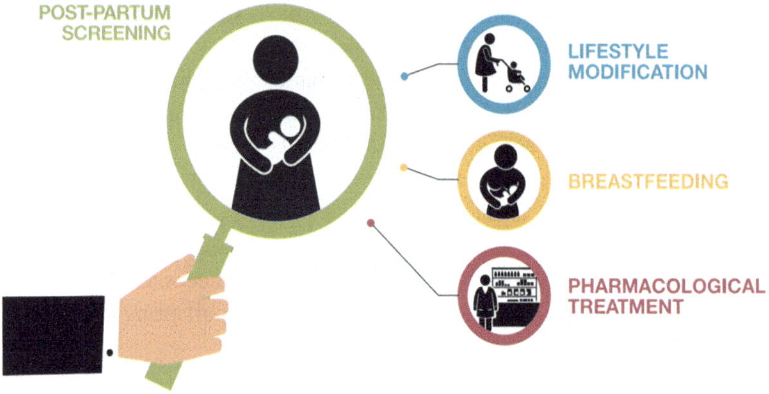

Fig. 35.2 Post-partum management of women with gestational diabetes (Figure created by Michael Bonar, Diabetes Research Centre, Leicester, UK)

diabetes being delayed by about 4 years when lifestyle intervention was implemented and 2 years when metformin was administered compared with placebo (Diabetes Prevention Program Research Group 2009). Similar results were reported from long-term studies conducted in Finland and China (Lindström et al. 2013; Li et al. 2008). In particular, in women with previous gestational diabetes in the Diabetes Prevention Program Outcomes study, both lifestyle intervention and metformin achieved lower rates of progression to type 2 diabetes over a 10-year follow-up period (Aroda et al. 2015).

Another randomised controlled trial implementing lifestyle intervention with a Mediterranean diet and monitored exercise in women with prior gestational diabetes demonstrated that these interventions were associated with lower incidence of impaired glucose regulation and type 2 diabetes (Pérez-Ferre et al. 2015). The efficacy of healthful dietary patterns was also supported by results from the Nurses' Health Study II, where once again, the risk of type 2 diabetes was lower in women who adopted these patterns (Tobias et al. 2012). Lifestyle interventions for the prevention of type 2 diabetes in this high-risk group have been found to be both cost effective and cost saving (Lohse et al. 2011).

Other Oral Drugs

Apart from metformin, troglitazone has been investigated as a potential pharmacological agent and was found to be effective in the prevention of type 2 diabetes in a cohort of Hispanic women with previous gestational diabetes (Buchanan et al. 2002). However, troglitazone was withdrawn from the market in 2000 as it was found to cause hepatotoxicity. In a follow-up of the same patient group where pioglitazone was administered this time, progression rates to type 2 diabetes were once again observed to be lower (Xiang et al. 2006). Previous research has also evaluated the efficacy of acarbose treatment in patients with impaired glucose tolerance and found that it was capable of delaying progression to type 2 diabetes (Chiasson et al. 2002). However, there is a lack of evidence for the efficacy of acarbose for both the management of gestational diabetes during pregnancy and the prevention of type 2 diabetes post-partum.

The Multiple Advantages of Breastfeeding

Breastfeeding should be encouraged among women with previous gestational diabetes, in line with current guidelines (American Diabetes Association (ADA) 2019b). However, breastfeeding is often hindered by the delayed onset of lactogenesis, a cause of both diabetes and obesity that leads to low rates and short duration of breastfeeding (Gunderson 2007). Studies have demonstrated that breastfeeding is associated with reduced post-partum fasting plasma glucose levels in women with previous gestational diabetes (Shub et al. 2019), while it can also contribute to post-partum weight reduction in this population (López-Olmedo et al. 2016). Overall, women with previous gestational diabetes who breastfeed demonstrate improved glucose metabolism and insulin sensitivity that have the potential of reducing the risk of future type 2 diabetes (Gunderson et al. 2012).

Progression to Type 2 Diabetes

Women affected by gestational diabetes during pregnancy have an established higher risk of developing type 2 diabetes in the future. Common risk factors associated with type 2 diabetes incidence in this population include non-European ethnicity, family history of type 2 diabetes, use of insulin during pregnancy and increased BMI (O'Reilly et al. 2011). The association between non-European ethnicity and higher incidence of type 2 diabetes after gestational diabetes has been extensively investigated and is well recognised (Ignell et al. 2013; Kousta et al. 2006). However, research performed in Caucasian women with previous gestational diabetes, who are considered to be at lower risk, demonstrated that incidence of post-partum dysglycaemia was over 10% in the early post-partum period (6 weeks post-partum), highlighting the importance of close monitoring in all women regardless of their ethnicity (Ogonowski and Miazgowski 2009). Time to development of diabetes is an important factor to consider, as development of type 2 diabetes in the early post-partum period is an indicator of marked defects in β-cell function, potentially attributed to genetic predisposition (Kwak et al. 2013).

The Challenges of Effective Screening

Progression to type 2 diabetes can be effectively delayed or prevented when women with prior gestational diabetes attend post-partum screening, as the latter facilitates risk classification for type 2 diabetes and introduction of preventative interventions. However, studies across the world have demonstrated that post-partum screening attendance is suboptimal in this patient group (Goueslard et al. 2017; Kim et al. 2006; Kwong et al. 2009; Blatt et al. 2011). There are several reasons that can justify the low uptake of post-partum screening. First of all, patient surveys have highlighted that there is limited awareness of the need for screening and low-risk perception about future progression to type 2 diabetes (Minsart et al. 2014; Sterne et al. 2011). In addition to that, time restrictions and child responsibilities have also been identified as potential barriers, while the length and nature of the OGTT can also hinder screening attendance (Minsart et al. 2014; Sterne et al. 2011; Van Ryswyk et al. 2016). Healthcare professionals are often unaware of the importance of recommending post-partum screening (Almario et al. 2008).

E-Health and Postpartum Follow-Up

Results from a randomised controlled trial using postal reminders showed that women who received a reminder were almost nine times more likely to get screened (Clark et al. 2009). An e-mail reminder system is a cost-effective and time-efficient means to improve screening attendance, with a 10% improvement in screening uptake for type 2 diabetes (Halperin et al. 2015). In Belgium, reminder letters and emails demonstrated that this strategy improved screening rates (Benhalima et al. 2017). Again, clinics that utilised reminders were two to three times more likely to successful screening (Peticca et al. 2014). All these studies, however, emphasised the fact that even though screening attendance was improved with the use of reminder systems, it still remained low and suboptimal. Improving screening uptake on a global scale will improve health outcomes in this population.

Summary

As gestational diabetes is a multifactorial disease, similar to type 2 diabetes, early detection should be a priority to prevent unfavourable health outcomes. In addition to that, these women should be closely monitored in the post-partum period and receive appropriate care designed to attenuate subsequent progression to type 2 diabetes. A multidisciplinary approach should be applied both during pregnancy and post-partum, being flexible and focusing on individual needs. There are several inconsistencies with regard to the identification of women with gestational diabetes due to different screening methods and diagnostic cut-offs used around the world. This introduces barriers in patient care, as different patients are targeted during pregnancy by the different guidelines and consequently followed up in the post-partum period. Future research needs to address both benefits and harms of different screening methods and diagnostic thresholds on maternal and neonatal outcomes, their effect on maternal wellbeing as well as the associated health costs. Finally, future research should also aim to identify post-partum screening barriers and facilitators that will have the potential to improve screening uptake and reduce progression to type 2 diabetes.

Acknowledgments The authors gratefully acknowledge Mr. Michael Bonar for creating the illustrations featured in this chapter as well as Professor Dimitrios G. Goulis for providing advice on this chapter.

References

Almario CV, Ecker T, Moroz LA, Bucovetsky L, Berghella V, Baxter JK (2008) Obstetricians seldom provide postpartum diabetes screening for women with gestational diabetes. Am J Obstet Gynecol 198:528.e1–528.e5. https://doi.org/10.1016/j.ajog.2007.11.001

American College of Obstetricians and Gynecologists (ACOG) (2015) Physical activity and exercise during pregnancy and the postpartum period. Committee opinion no. 650. Obstet Gynecol 126:135–142. https://doi.org/10.1016/S0140-6736(16)31898-0

American Diabetes Association (ADA) (2019a) Classification and diagnosis of diabetes: standards of medical care in diabetes 2019. Diabetes Care 42:S13–S28. https://doi.org/10.2337/dc19-S002

American Diabetes Association (ADA) (2019b) Management of diabetes in pregnancy: standards of medical care in diabetes 2019. Diabetes Care 42:S165–S172. https://doi.org/10.2337/dc19-S014

Aroda VR, Christophi CA, Edelstein SL, Zhang P, Herman WH, Barrett-Connor E, Delahanty LM, Montez MG, Ackermann RT, Zhuo X, Knowler WC, Ratner RE (2015) The effect of lifestyle intervention and metformin on preventing or delaying diabetes among women with and without gestational diabetes: the diabetes prevention program outcomes study 10-year follow-up. J Clin Endocrinol Metab 100:1646–1653. https://doi.org/10.1210/jc.2014-3761

Balsells M, García-Patterson A, Solà I, Roqué M, Gich I, Corcoy R (2015) Glibenclamide, metformin, and insulin for the treatment of gestational diabetes: a systematic review and meta-analysis. BMJ 350:1–12. https://doi.org/10.1136/bmj.h102

Behboudi-Gandevani S, Amiri M, Bidhendi Yarandi R, Ramezani Tehrani F (2019) The impact of diagnostic criteria for gestational diabetes on its prevalence: a systematic review and meta-analysis. Diabetol Metab Syndr 11:1–18. https://doi.org/10.1186/s13098-019-0406-1

Bellamy L, Casas J, Hingorani AD, Williams D (2009) Type 2 diabetes mellitus after gestational diabetes: a systematic review and meta-analysis. Lancet 373:1773–1779. https://doi.org/10.1016/S0140-6736(09)60731-5

Benhalima K, Verstraete S, Muylle F, Decochez K, Devlieger R, Crombrugge P (2017) Van, Verhaegen a, wens J, Mathieu C (2017) implementing a reminder system in the northern part of Belgium to stimulate postpartum screening for glucose intolerance in women with gestational diabetes: the "sweet pregnancy" project. Int J Endocrinol 2017:1. https://doi.org/10.1155/2017/3971914

Blatt AJ, Nakamoto JM, Kaufman HW (2011) Gaps in diabetes screening during pregnancy and postpartum. Obstet Gynecol 117:61–68. https://doi.org/10.1097/AOG.0b013e3181fe424b

Boerschmann H, Pflüger M, Henneberger L, Ziegler AG, Hummel S (2010) Prevalence and predictors of overweight and insulin resistance in offspring of mothers with gestational diabetes mellitus. Diabetes Care 33:1845–1849. https://doi.org/10.2337/dc10-0139

Brown CJ, Dawson A, Dodds R, Gamsu H, Gillmer M, Hall M, Hounsome B, Knopfler A, Ostler J, Peacock I, Rothman D, Steel J (1996) Report of the pregnancy and neonatal care group. Diabet Med 13:S43–S53

Brown J, Alwan NA, West J, Brown S, Mckinlay CJD, Farrar D, Crowther CA (2017) Lifestyle interventions for the treatment of women with gestational diabetes. Cochrane Database Syst Rev 2017:CD011970. https://doi.org/10.1002/14651858.CD011970.pub2

Buchanan TA, Xiang AH, Peters RK, Kjos SL, Marroquin A, Goico J, Ochoa C, Tan S, Berkowitz K, Hodis HN, Azen SP (2002) Preservation of pancreatic β-cell function and prevention of type 2 diabetes by pharmacological treatment of insulin resistance in high-risk Hispanic women. Diabetes 51:2796–2803. https://doi.org/10.2337/diabetes.51.9.2796

Butte NF (2000) Carbohydrate and lipid metabolism in pregnancy: Normal compared with gestational diabetes mellitus. Am J Clin Nutr 71:1256–1261. https://doi.org/10.1093/ajcn/71.5.1256s

Chiasson JL, Josse RG, Gomis R, Hanefeld M, Karasik A, Laakso M (2002) Acarbose for prevention of type 2 diabetes mellitus: the STOP-NIDDM randomised trial. Lancet 359:2072–2077. https://doi.org/10.1016/S0140-6736(02)08905-5

Clark HD, Graham ID, Karovitch A, Keely EJ (2009) Do postal reminders increase postpartum screening of diabetes mellitus in women with gestational diabetes mellitus? A randomized controlled trial. Am J Obstet Gynecol 200:634.e1–634.e7. https://doi.org/10.1016/j.ajog.2009.01.003

Corrado F, D'Anna R, di Vieste G, Giordano D, Pintaudi B, Santamaria A, di Benedetto A (2011) The effect of myoinositol supplementation on insulin resistance in patients with gestational diabetes. Diabet Med 28:972–975. https://doi.org/10.1111/j.1464-5491.2011.03284.x

Costabile L, Unfer V (2017) Treatment of gestational diabetes mellitus with myo-inositol: analyzing the cutting edge starting from a peculiar case. Eur Rev Med Pharmacol Sci 21:73–76

Di Cianni G, Miccoli R, Volpe L, Lencioni C, Del Prato S (2003) Intermediate metabolism in normal pregnancy and in gestational diabetes. Diabetes Metab Res Rev 19:259–270. https://doi.org/10.1002/dmrr.390

Diabetes Prevention Program Research Group (2009) 10-year follow-up of diabetes incidence and weight loss in the diabetes prevention program outcomes study. Lancet 374:1677–1686. https://doi.org/10.1016/S0140-6736(09)61457-4

Dipla K, Triantafyllou A, Grigoriadou I, Kintiraki E, Triantafyllou GA, Poulios P, Vrabas IS, Zafeiridis A, Douma S, Goulis DG (2017) Impairments in microvascular function and skeletal muscle oxygenation in women with gestational diabetes mellitus: links to cardiovascular disease risk factors. Diabetologia 60:192–201. https://doi.org/10.1007/s00125-016-4129-7

Evenson KR, Barakat R, Brown WJ, Dargent-Molina P, Haruna M, Mikkelsen EM, Mottola MF, Owe KM, Rousham EK, Yeo S (2014) Guidelines for physical activity during pregnancy: comparisons from around the world. Am J Lifestyle Med 8:102–121. https://doi.org/10.1177/1559827613498204

Feig DS, Berger H, Donovan L (2018) 2018 clinical practice guidelines diabetes and pregnancy diabetes Canada clinical practice guidelines expert committee. Can J Diabetes 42:255–282. https://doi.org/10.1016/j.jcjd.2017.10.038

Foghsgaard S, Andreasen C, Vedtofte L, Andersen ES, Bahne E, Strandberg C, Buhl T, Holst JJ, Svare JA, Clausen TD, Mathiesen ER, Damm P, Gluud LL, Knop FK, Vilsbøll T (2017) Nonalcoholic fatty liver disease is prevalent in women with prior gestational diabetes mellitus and independently associated with insulin resistance and waist circumference. Diabetes Care 40:109–116. https://doi.org/10.2337/dc16-1017

Goueslard K, Cottenet J, Mariet AS, Giroud M, Cottin Y, Petit JM, Quantin C (2016) Early cardiovascular events in women with a history of gestational diabetes mellitus. Cardiovasc Diabetol 15:1–7. https://doi.org/10.1186/s12933-016-0338-0

Goueslard K, Cottenet J, Mariet AS, Sagot P, Petit JM, Quantin C (2017) Early screening for type 2 diabetes following gestational diabetes mellitus in France: hardly any impact of the 2010 guidelines. Acta Diabetol 54:645–651. https://doi.org/10.1007/s00592-017-0986-x

Gui J, Liu Q, Feng L (2013) Metformin vs insulin in the Management of Gestational Diabetes: a meta-analysis. PLoS One 8:e64585. https://doi.org/10.1371/journal.pone.0064585

Gunderson EP (2007) Breastfeeding after gestational diabetes pregnancy: subsequent obesity and type 2 diabetes in women and their offspring. Diabetes Care 30:S161. https://doi.org/10.2337/dc07-s210

Gunderson EP, Hedderson MM, Chiang V, Crites Y, Walton D, Azevedo RA, Fox G, Elmasian C, Young S, Salvador N, Lum M, Quesenberry CP, Lo JC, Sternfeld B, Ferrara A, Selby JV (2012) Lactation intensity and postpartum maternal glucose tolerance and insulin resistance in women with recent GDM: the SWIFT cohort. Diabetes Care 35:50–56. https://doi.org/10.2337/dc11-1409

Halperin IJ, Sehgal P, Lowe J, Hladunewich M, Wong BM (2015) Increasing timely postpartum Oral glucose tolerance test completion in women with gestational diabetes: a quality-improvement initiative. Can J Diabetes 39:451–456. https://doi.org/10.1016/j.jcjd.2015.06.004

Hinkle SN, Buck Louis GM, Rawal S, Zhu Y, Albert PS, Zhang C (2016) A longitudinal study of depression and gestational diabetes in pregnancy and the postpartum period. Diabetologia 59:2594–2602. https://doi.org/10.1007/s00125-016-4086-1

Hunt KJ, Schuller KL (2007) The increasing prevalence of diabetes in pregnancy. Obstet Gynecol Clin N Am 34:173–199. https://doi.org/10.1016/j.ogc.2007.03.002

Ignell C, Shaat N, Ekelund M, Berntorp K (2013) The impact of ethnicity on glucose homeostasis after gestational diabetes mellitus. Acta Diabetol 50:927–934. https://doi.org/10.1007/s00592-013-0484-8

International Association of Diabetes and Pregnancy Study Groups Consensus Panel, Metzger BE, Gabbe SG, Persson B, Buchanan TA, Catalano PA, Damm P, Dyer AR, Leiva A, Hod M, Kitzmiler JL, Lowe LP, HD MI, Oats JJ, Omori Y, Schmidt M (2010) International Association of Diabetes and Pregnancy Study Groups recommendations on the diagnosis and classification of hyperglycemia in pregnancy. Diabetes Care 33:676–682. https://doi.org/10.2337/dc09-1848

International Diabetes Federation (2019) IDF Diabetes Atlas—2019

Kim C (2010) Gestational diabetes: risks, management, and treatment options. Int J Women's Health 2:339–351. https://doi.org/10.2147/IJWH.S13333
Kim C, Newton KM, Knopp RH (2002) Gestational diabetes and the incidence of type 2 diabetes. Diabetes Care 25:1862–1868. https://doi.org/10.2337/dc09-1679
Kim C, Bahman P, Burke R, Mcewen LN, Lash RW, Johnson SL, Schwartz KL, Bernstein SJ, Herman WH (2006) Missed opportunities for type 2 diabetes mellitus screening among women with a history of gestational diabetes mellitus. Am J Public Health 96:1643–1648. https://doi.org/10.2105/AJPH.2005.065722
Kintiraki E, Goulis DG (2018) Gestational diabetes mellitus: multi-disciplinary treatment approaches. Metabolism 86:91–101. https://doi.org/10.1016/j.metabol.2018.03.025
Kintiraki E, Dipla K, Triantafyllou A, Koletsos N, Grigoriadou I, Poulakos P, Sachpekidis V, Vrabas IS, Zafeiridis A, Bili E, Douma S, Goulis DG (2018) Blunted cerebral oxygenation during exercise in women with gestational diabetes mellitus: associations with macrovascular function and cardiovascular risk factors. Metabolism 83:25–30. https://doi.org/10.1016/j.metabol.2018.01.009
Kousta E, Efstathiadou Z, Lawrence NJ, Jeffs JAR, Godsland IF, Barrett SC, Doré CJ, Penny A, Anyaoku V, Millauer BA, Cela E, Robinson S, McCarthy MI, Johnston DG (2006) The impact of ethnicity on glucose regulation and the metabolic syndrome following gestational diabetes. Diabetologia 49:36–40. https://doi.org/10.1007/s00125-005-0058-6
Kramer CK, Campbell S, Retnakaran R (2019) Gestational diabetes and the risk of cardiovascular disease in women: a systematic review and meta-analysis. Diabetologia 62:905–914. https://doi.org/10.1007/s00125-019-4840-2
Kwak SH, Choi SH, Jung HS, Cho YM, Lim S, Cho NH, Kim SY, Park KS, Jang HC (2013) Clinical and genetic risk factors for type 2 diabetes at early or late post partum after gestational diabetes mellitus. J Clin Endocrinol Metab 98:744–752. https://doi.org/10.1210/jc.2012-3324
Kwong S, Mitchell RS, Senior PA, Chik CL (2009) Postpartum diabetes screening: adherence rate and the performance of fasting plasma glucose versus oral glucose tolerance test. Diabetes Care 32:2242–2244. https://doi.org/10.2337/dc09-0900
Li Q, Xing B (2016) Vitamin D3-supplemented yogurt drink improves insulin resistance and lipid profiles in women with gestational diabetes mellitus: a randomized double blinded clinical trial. Ann Nutr Metab 68:285–290. https://doi.org/10.1159/000447433
Li G, Zhang P, Wang J, Gregg EW, Yang W, Gong Q, Li H, Li H, Jiang Y, An Y, Shuai Y, Zhang B, Zhang J, Thompson TJ, Gerzoff RB, Roglic G, Hu Y, Bennett PH (2008) The long-term effect of lifestyle interventions to prevent diabetes in the China Da Qing diabetes prevention study: a 20-year follow-up study. Lancet 371:1783–1789. https://doi.org/10.1016/S0140-6736(08)60766-7
Lindström J, Peltonen M, Eriksson JG, Ilanne-Parikka P, Aunola S, Keinänen-Kiukaanniemi S, Uusitupa M, Tuomilehto J (2013) Improved lifestyle and decreased diabetes risk over 13 years: long-term follow-up of the randomised Finnish diabetes prevention study (DPS). Diabetologia 56:284–293. https://doi.org/10.1007/s00125-012-2752-5
Lohse N, Marseille E, Kahn JG (2011) Development of a model to assess the cost-effectiveness of gestational diabetes mellitus screening and lifestyle change for the prevention of type 2 diabetes mellitus. Int J Gynaecol Obstet 115(Suppl):S20–S25. https://doi.org/10.1016/S0020-7292(11)60007-6
López-Olmedo N, Hernández-Cordero S, Neufeld LM, García-Guerra A, Mejía-Rodríguez F, Méndez Gómez-Humarán I (2016) The associations of maternal weight change with breastfeeding, diet and physical activity during the postpartum period. Matern Child Health J 20:270–280. https://doi.org/10.1007/s10995-015-1826-7
Lubin V, Shojai R, Darmon P, Cosson E (2016) A pilot study of gestational diabetes mellitus not controlled by diet alone: first-line medical treatment with myoinositol may limit the need for insulin. Diabetes Metab 42:192–195. https://doi.org/10.1016/j.diabet.2016.01.005
Malek R, Davis SN (2016) Pharmacokinetics, efficacy and safety of glyburide for treatment of gestational diabetes mellitus. Expert Opin Drug Metab Toxicol 12:691–699. https://doi.org/10.1080/17425255.2016.1187131
Martis R, Crowther CA, Shepherd E, Alsweiler J, Downie MR, Brown J (2018) Treatments for women with gestational diabetes mellitus: an overview of Cochrane systematic reviews. Cochrane Database Syst Rev 2018:CD012327. https://doi.org/10.1002/14651858.CD012327.pub2
Metzger BE, Phelps RL, Freinkel N, Navickas IA (1980) Effects of gestational diabetes on diurnal profiles of plasma glucose, lipids, and individual amino acids. Diabetes Care 3:402–409
Metzger BE, Buchanan TA, Coustan DR, De Leiva A, Dunger DB, Hadden DR, Hod M, Kitzmiller JL, Kjos SL, Oats JN, Pettitt DJ, Sacks DA, Zoupas C (2007) Summary and recommendations of the fifth international workshop-conference on gestational diabetes mellitus. Diabetes Care 30(Suppl 2):S251–S260. https://doi.org/10.2337/dc07-s225
Meyers-Seifer CH, Vohr BR (1996) Lipid levels in former gestational diabetic mothers. Diabetes Care 19:1351–1356. https://doi.org/10.2337/diacare.19.12.1351
Minsart AF, Vander Maelen A, Fontaine V, Kirkpatrick C (2014) Social, medical and self-perceived factors influencing postpartum screening of diabetes after

gestational diabetes. J Obstet Gynaecol (Lahore) 34:8–12. https://doi.org/10.3109/01443615.2013.826639
Nankervis A, McIntyre HD, Moses R, Ross GP, Callaway L, Porter C, Jeffries W, Boorman C, De Vries B MA for the AD in PS (2014) ADIPS Consensus guidelines for the testing and diagnosis of hyperglycaemia in pregnancy in Australia and New Zealand. 1–8
Nicklas JM, Miller LJ, Zera CA, Davis RB, Levkoff SE, Seely EW (2013) Factors associated with depressive symptoms in the early postpartum period among women with recent gestational diabetes mellitus. Matern Child Health J 17:1665–1672. https://doi.org/10.1007/s10995-012-1180-y
O'Reilly MW, Avalos G, Dennedy MC, O'Sullivan EP, Dunne F (2011) Atlantic DIP: high prevalence of abnormal glucose tolerance post partum is reduced by breast-feeding in women with prior gestational diabetes mellitus. Eur J Endocrinol 165:953–959. https://doi.org/10.1530/EJE-11-0663
Ogonowski J, Miazgowski T (2009) The prevalence of 6 weeks postpartum abnormal glucose tolerance in Caucasian women with gestational diabetes. Diabetes Res Clin Pract 84:239–244. https://doi.org/10.1016/j.diabres.2009.04.003
Padayachee C (2015) Exercise guidelines for gestational diabetes mellitus. World J Diabetes 6:1033. https://doi.org/10.4239/wjd.v6.i8.1033
Pérez-Ferre N, Del Valle L, Torrejón MJ, Barca I, Calvo MI, Matía P, Rubio MA, Calle-Pascual AL (2015) Diabetes mellitus and abnormal glucose tolerance development after gestational diabetes: a three-year, prospective, randomized, clinical-based, Mediterranean lifestyle interventional study with parallel groups. Clin Nutr 34:579–585. https://doi.org/10.1016/j.clnu.2014.09.005
Peticca P, Shah BR, Shea A, Clark HD, Malcolm JC, Walker M, Karovitch A, Brazeau-Gravelle P, Keely EJ (2014) Clinical predictors for diabetes screening in the first year postpartum after gestational diabetes. Obstet Med 7:116–120. https://doi.org/10.1177/1753495X14528487
Pusukuru R, Shenoi AS, Kyada PK, Ghodke B, Mehta V, Bhuta K, Bhatia A (2016) Evaluation of lipid profile in second and third trimester of pregnancy. J Clin Diagn Res 10:QC12–QC16. https://doi.org/10.7860/JCDR/2016/17598.7436
Rodrigues MRK, Lima SAM, Mazeto GMFDS, Calderon IMP, Magalhães CG, Ferraz GAR, Molina AC, Costa RAA, Nogueira VDSN, Rudge MVC (2019) Efficacy of vitamin D supplementation in gestational diabetes mellitus: systematic review and meta-analysis of randomized trials. PLoS One 14:1–16. https://doi.org/10.1371/journal.pone.0213006
Ryckman KK, Spracklen CN, Smith CJ, Robinson JG, Saftlas AF (2015) Maternal lipid levels during pregnancy and gestational diabetes: a systematic review and meta-analysis. BJOG An Int J Obstet Gynaecol 122:643–651. https://doi.org/10.1111/1471-0528.13261
Ryu RJ, Hays KE, Hebert MF (2014) Gestational diabetes mellitus management with oral hypoglycemic agents. Semin Perinatol 38:508–515. https://doi.org/10.1053/j.semperi.2014.08.012
Shub A, Miranda M, Georgiou HM, McCarthy EA, Lappas M (2019) The effect of breastfeeding on postpartum glucose tolerance and lipid profiles in women with gestational diabetes mellitus. Int Breastfeed J 14:1–5. https://doi.org/10.1186/s13006-019-0238-5
Song R, Chen L, Chen Y, Si X, Liu Y, Liu Y, Irwin DM, Feng W (2017) Comparison of glyburide and insulin in the management of gestational diabetes: a meta-analysis. PLoS One 12:1–18. https://doi.org/10.1371/journal.pone.0182488
Sterne V, Logan T, Palmer MA (2011) Factors affecting attendance at postpartum diabetes screening in women with gestational diabetes mellitus. Pract Diabetes Int 28:64–69. https://doi.org/10.1002/pdi.1559
The American College of Obstetricians, and Gynecologists (ACOG) (2018) Clinical Management Guidelines for Obstetrician–Gynecologists. Gestational diabetes mellitus. ACOG Practice Bulletin No. 190. Obstet Gynecol 131:e49–e64
The National Institute for Health and Care Excellence (NICE) (2015) Diabetes in pregnancy: management from preconception to the postnatal period. 1–67
Tobias DK, Hu FB, Chavarro J, Rosner B, Mozaffarian D, Zhang C (2012) Healthful dietary patterns and type 2 diabetes mellitus risk among women with a history of gestational diabetes mellitus. Arch Intern Med 172:1566–1572. https://doi.org/10.1001/archinternmed.2012.3747
Tobias DK, Clish C, Mora S, Li J, Liang L, Hu FB, Manson JE, Zhang C (2018) Dietary intakes and circulating concentrations of branched-chain amino acids in relation to incident type 2 diabetes risk among high-risk women with a history of gestational diabetes mellitus. Clin Chem 64:1203–1210. https://doi.org/10.1373/clinchem.2017.285841
Tsirou E, Grammatikopoulou MG, Theodoridis X, Gkiouras K, Petalidou A, Taousani E, Savvaki D, Tsapas A, Goulis DG (2019) Guidelines for medical nutrition therapy in gestational diabetes mellitus: systematic review and critical appraisal. J Acad Nutr Diet 119:1320–1339. https://doi.org/10.1016/j.jand.2019.04.002
Vääräsmäki M, Pouta A, Elliot P, Tapanainen P, Sovio U, Ruokonen A, Hartikainen A-L, McCarthy M, Järvelin M-R (2009) Adolescent manifestations of metabolic syndrome among children born to women with gestational diabetes in a general-population birth cohort. Am J Epidemiol 169:1209–1215. https://doi.org/10.1093/aje/kwp020
Van Ryswyk EM, Middleton PF, Hague WM, Crowther CA (2016) Women's views on postpartum testing for type 2 diabetes after gestational diabetes: six month follow-up to the DIAMIND randomised controlled

trial. Prim Care Diabetes 10:91–102. https://doi.org/10.1016/j.pcd.2015.07.003
Vounzoulaki E, Dipla K, Kintiraki E, Triantafyllou A, Grigoriadou I, Koletsos N, Zafeiridis A, Goulis DG, Douma S (2019) Pregnancy and post-partum muscle and cerebral oxygenation during intermittent exercise in gestational diabetes: a pilot study. Eur J Obstet Gynecol Reprod Biol 232:54–59. https://doi.org/10.1016/j.ejogrb.2018.11.012
Vounzoulaki E, Khunti K, Abner SC, Tan BK, Davies MJ, Gillies CL (2020) Progression to type 2 diabetes in women with a known history of gestational diabetes: systematic review and meta-analysis. BMJ m1361. https://doi.org/10.1136/bmj.m1361
World Health Organization (2013) Diagnostic criteria and classification of hyperglycaemia first detected in pregnancy
Wu WH, Meijer OG, Uegaki K, Mens JMA, Van Dieën JH, Wuisman PIJM, Östgaard HC (2004) Pregnancy-related pelvic girdle pain (PPP), I: terminology, clinical presentation, and prevalence. Eur Spine J 13:575–589. https://doi.org/10.1007/s00586-003-0615-y
Xiang AH, Peters RK, Kjos SL, Marroquin A, Goico J, Ochoa C, Kawakubo M, Buchanan TA (2006) Effect of pioglitazone on pancreatic beta-cell function and diabetes risk in Hispanic women with prior gestational diabetes. Diabetes 55:517–522
Xiang AH, Wang X, Martinez MP, Walthall JC, Curry ES, Page K, Buchanan TA, Coleman KJ, Getahun D (2015) Association of maternal diabetes with autism in offspring. JAMA-J Am Med Assoc 313:1425–1434. https://doi.org/10.1001/jama.2015.2707

Update on Gestational Diabetes 36

Megan Jones and Allison Akers

Contents

The Growing Prevalence of Gestational Diabetes 494
Additional Influences 494

Screening and Risk Evaluation 495

Classification of Gestational Diabetes 497

Obstetric and Pharmacologic Management 499
Dietary Guidance 499
Pharmacotherapy 499
Oral Glucose-Lowering Prescriptions: Metformin 500
Glyburide 501
Antenatal Fetal Testing 501
Indications for Ultrasonographic Monitoring 501
Optimal Delivery Time 502
Intrapartum Glucose Homeostasis 502

Complications and Follow Up 503
Additional Neonatal Aberrations 504
Long-Term Complications 504
Maternal Complications 504
Diagnostic Work Up and Long-Term Care 504

Ongoing Studies and Future Directions 505

References 505

M. Jones (✉)
Maternal Fetal Medicine, Dignity Health Perinatal Center,
Ventura, CA, USA
e-mail: mjobgyn@gmail.com

A. Akers
The Ohio State University College of Medicine,
Columbus, OH, USA

J. Faintuch and S. Faintuch (eds.), *Obesity and Diabetes*, https://doi.org/10.1007/978-3-030-53370-0_36

Abstract

Gestational diabetes mellitus (GDM) has become increasingly prevalent in the United States, and its impact on adverse maternal and neonatal outcomes necessitates proper screening, management, and follow up. Although there is a lack of consensus on screening methods for GDM, the American College of Obstetrics and Gynecology (ACOG) recommends the two-step approach between 24 and 28 weeks. In contrast, the International Association of Diabetes and Pregnancy Study Group (IADPSG) recommends a one-step approach. Once the diagnosis of GDM has been made, management may consist of nutritional therapy, alone, or in combination with pharmacologic interventions such as insulin, metformin, or less often, glyburide. Close management is warranted in order to avoid adverse maternal and neonatal outcomes, both during delivery and later in life to both mother and infant.

Keywords

Pregnancy · Diabetes · Screening · Prenatal · Insulin · Delivery · Macrosomia

The Growing Prevalence of Gestational Diabetes

Gestational diabetes mellitus (GDM) is well known to increase the risk of adverse outcomes in mothers and infants. Its additional capacity to contribute to potential maternal and pediatric morbidity later in life prompts the importance of investigating the issue from a public health perspective (Ferrara 2007). In the United States, the prevalence of gestational diabetes appears to be increasing. Between 1989 and 2004, rates increased from 1.9% to 4.2%, with the largest increase found in black women younger than 25 (Getahun et al. 2008). More recent data from 2016 have estimated this prevalence to be even larger at 6.0%, which highlights its pervasiveness over the last 25 years (Deputy et al. 2018). Interestingly, the prevalence of GDM has increased concomitantly with the rise of obesity, as well as with the trend toward older maternal age, decreasing physical activity, and more widespread type 2 diabetes (Dabelea et al. 2005; Ferrara 2007).

In the United States, there is increased prevalence of GDM in minority women as compared to non-Hispanic whites. In a study looking at eligible pregnancies from 1994 to 2002, the prevalence of GDM was highest in Asian and Hispanic women, intermediate in African Americans, and lowest in non-Hispanic whites (Dabelea et al. 2005). According to data from 2016, it is highest in non-Hispanic Asian women at 11.1% (Deputy et al. 2018). Studies in both Europe and Australia have shown analogous disparities with higher GDM prevalence in Asian women (Ferrara 2007). An increased incidence in older mothers (Getahun et al. 2008; Deputy et al. 2018) also extends globally, with an Australian study showing that mothers greater than 40 were more than six times as likely to develop GDM, than their 20–24 year old counterparts (Anna et al. 2008).

Additional Influences

In a study from the early 1990s, it was found that advanced maternal age, family history of diabetes mellitus, nonwhite ethnicity, higher BMI, weight gain in early adulthood, and cigarette smoking predicted an increased risk of GDM (Solomon 1997). Another study from the late 1980s had already observed similar conclusions, however, also saw positive correlations between GDM and parity, history of infertility, preterm delivery, and stillbirth (Berkowitz et al. 1992). Table 36.1 is adapted from the CDC and summarizes the variation in GDM prevalence according to a range of selected maternal characteristics (Deputy et al. 2018).

Despite these previous studies, several epidemiological limitations make it difficult to appropriately study trends in GDM, including dispute with regards to diagnostic criteria and screening

Table 36.1 Pre-existing diabetes and gestational diabetes among women with a live birth (USA, 2016)

Features	Previous diabetes mellitus (%)	Gestational diabetes (%)
Age < 20	0.4	1.9
20–29	0.5–0.7	3.3–5.1
30–39	1.4	7.0–9.6
40+	2.1	12.8
Nulliparous	0.8	5.2
Primiparous	0.8	5.9
Multiparous	1.0	7.1
BMI <18.5	0.3	2.9
BMI 18.5–24.9	0.4	3.6
BMI 25.0–29.9	0.8	6.1
BMI 30.0–34.9	1.3	8.8
BMI 35.0–39.9	2.0	11.2
BMI 40+	3.2	13.9

Modified from Deputy NP, Kim SY, Conrey EJ, Bullard KM. Prevalence and Changes in Preexisting Diabetes and Gestational Diabetes Among Women Who Had a Live Birth—United States, 2012–2016. MMWR Morb Mortal Wkly Rep 2018;67:1201–1207. DOI: https://doi.org/10.15585/mmwr.mm6743a2

Table 36.2 Screening for pregestational or early gestational diabetes

Tests are indicated for BMI > 25 (BMI > 23 for ethnic Asians) with overweight/obesity plus any of these:
Insufficient physical activities
Close relatives with diabetes
High-risk ethnicity (Non-European origin)
History of gestational diabetes or macrosomic newborn
Polycystic ovary syndrome
Arterial hypertension, low HDL-cholesterol (<35 mg/dL), high triglycerides (>250 mg/dL)
Impaired fasting glucose or glucose tolerance test, HbA1c > 5.7%, acanthosis nigricans
Cardiovascular disease

Modified from: American Diabetes Association. Classification and Diagnosis of Diabetes. Diabetes Care 2017;40 (Suppl. 1):S11–S24

techniques (Ferrara 2007). For example, while there is clearly a positive association between severe obesity and GDM, the magnitude of the variation is different among studies. This is likely due to the variations in classifications of obesity as well as diagnostic criteria for GDM (Chu et al. 2007).

Screening and Risk Evaluation

The purpose of screening for GDM is to identify healthy individuals who have a high probability of having or developing the disease. However, identifying these individuals can be confounded by women with probable pre-existing diabetes that is diagnosed early in pregnancy. The American Diabetes Association (ADA) defines GDM as "diabetes diagnosed in the second or third trimester of pregnancy that was not clearly overt diabetes prior to gestation" in order to separate pregnant women with undiagnosed type 2 diabetes (American Diabetes Association 2018a). Screening techniques have also adapted to differentiate GDM from pre-existing diabetes. The American College of Obstetrics and Gynecology (ACOG) recommends that overweight or obese women with diabetic risk factors be screened for undiagnosed type 2 diabetes at the initiation of prenatal care. Table 36.2 is adapted from the American College of Obstetricians and Gynecologists Gestational Diabetes Mellitus Practice Bulletin, and it highlights these characteristics (Gestational Diabetes Mellitus 2018). In women where undiagnosed type

Table 36.3 Diagnostic cut-off points for gestational diabetes (after oral glucose tolerance test)[a]

Fasting	95 mg/dL
1 h	180 mg/dL
2 h	155 mg/dL
3 h	140 mg/dL

[a]Two or more thresholds equaled or exceeded. Modified from American Diabetes Association. Classification and Diagnosis of Diabetes. Diabetes Care 2017; 40(Suppl 1): S11–S24

2 diabetes is suspected, nonpregnant diabetic diagnostic criteria can be used for detection (American Diabetes Association 2018a).

There is a lack of uniformity among national and international standards for screening and diagnosing GDM (HAPO Study Cooperative Research Group 2008). Currently, two strategies, the one-step and two-step approaches, can be used to screen and diagnosis GDM, but controversy still exists between which strategy is best (Saccone et al. 2020). In a meta-analysis investigating differences in perinatal outcomes between both methods, researchers found that the one-step approach was associated with a lower risk of large for gestational age infant, NICU admission, and neonatal hypoglycemia (Saccone et al. 2020). However, additional studies have argued that the one-step approach will not only increase incidence of GDM, but will also increase direct and indirect healthcare costs (Vandorsten et al. 2013). As such, ACOG supports the two-step approach, but many practitioners will choose the one-step approach if felt more appropriate for their patient populations (Gestational Diabetes Mellitus 2018).

The two-step approach includes an initial screening test followed by diagnostic testing (American Diabetes Association 2018a). In women who have not previously been diagnosed with type 1 or 2 diabetes mellitus, current recommendations suggest that all pregnant women are screened for GDM between 24 and 28 weeks' gestation. This can be done via administration of a 50-g oral glucose solution. This screening tool does not require fasting, and it is followed by a 1-h venous glucose determination (Gestational Diabetes Mellitus 2018). The gestational timing of this screening recommendation results from a United States Preventive Services Task Force systematic review that found insufficient evidence to suggest screening earlier than 24 weeks (Moyer and U.S. Preventive Services Task Force 2014). Cutoffs for positive screening values differ among institutions, but thresholds vary between 130 and 140 mg/dL. Some studies argue that the higher threshold of 140 mg/dL may lead to lower rates of false-positive screening tests and thus prevent unnecessary maternal stress, but no randomized controlled trials have determined one cutoff to be more effective than others (Rumbold and Crowther 2002; Gestational Diabetes Mellitus 2018). Therefore, ACOG recommends providers select a single consistent threshold for the 1-h glucose screening test that best fits their patient population (Gestational Diabetes Mellitus 2018).

Abnormal results on the initial 50-g screening test are followed by a 100-g, 3-h diagnostic oral glucose tolerance test (OGTT). Table 36.3 is adapted from the ACOG Gestational Diabetes Mellitus Practice Bulletin and demonstrates diagnostic threshold serum glucose values set by the National Diabetes Data Group and Carpenter and Coustan. Due to the lack of competitive trials between cutoff values, there is no clear consensus on which values to definitively recommend (Gestational Diabetes Mellitus 2018). Historically, two abnormal values from the 3-h OGTT were needed to diagnose GDM according to criteria established in a 1964 study (O'Sullivan and Mahan 1964). More recent data have come to suggest that even a single abnormal value may also be associated with adverse maternal and neonatal outcomes. A systematic review showed women with one abnormal glucose value still had increased risk for poor obstetric outcomes, similar to those who were diagnosed with GDM (Roeckner et al. 2016). Further research is needed to determine if women with one abnormal value would benefit from evaluation and treatment, but currently ACOG guidelines recommend two abnormal values on the 3-h OGTT to establish a

Table 36.4 International Association of Diabetes in Pregnancy Study Group criteria for diagnosis of gestational or overt diabetes in pregnancy (after oral glucose tolerance test)[a]

Fasting	≥92 mg/dL
1 h	≥180 mg/dL
2 h	≥153 mg/dL

[a]Two or more thresholds equaled or exceeded. Modified from International Association of Diabetes and Pregnancy Study Groups Consensus Panel. Diabetes Care 2010; 33(3): 676–682

diagnosis of GDM (Gestational Diabetes Mellitus 2018).

In the one-step approach, a 75-g 2-h OGTT is used to identify GDM. When one plasma glucose measurement exceeds the diagnostic threshold, values set by the International Association of Diabetes and Pregnancy Study Group (IADPSG), a diagnosis of GDM can be made (Table 36.4 from IADPSG) (Metzger 2010). In one study using the IADPSG one-step diagnostic criteria, women diagnosed with GDM had increased adverse pregnancy outcomes such as gestational hypertension, pre-eclampsia, and large for gestational age infants (Caissutti et al. 2017). Long-term follow-up of these women showed that a significantly larger group of patients diagnosed with GDM via the one-step strategy developed prediabetes or type 2 diabetes compared to those women diagnosed with GDM via the two-step method (Lowe et al. 2018). Those who oppose the one-step method argue that, by definition, it increases the incidence and healthcare costs associated with GDM and increases the potential for harm with more obstetric and neonatal intervention (American Diabetes Association 2018a; Lowe et al. 2018; Cheung and Moses 2018). However, this increasing incidence in GDM diagnoses via the one-step method could potentially identify a larger pool of women who would benefit from increased screening for disorders of glucose metabolism later in life (American Diabetes Association 2018a). Ultimately, there is no international consensus on which diagnostic method is best, and rising evidence suggests that no one diagnostic criteria exists but, rather, must be adapted locally with population data (Mcintyre et al. 2018).

Of note, screening for GDM using A1C at 24–28 weeks' gestation is not as effective as the glucose-loading test (American Diabetes Association 2018a). Although glycated hemoglobin levels may be beneficial in helping to screen for diabetes that existed before pregnancy, there is no A1C threshold with both good sensitivity and specificity that make it an appropriate screening test for GDM (Moyer and U.S. Preventive Services Task Force 2014).

Classification of Gestational Diabetes

The White Classification system, first established in 1949, attempted to use alphabetical groups to categorize diabetes in pregnancy in a way that allowed clinicians to stratify patients with risk factors for fetal and neonatal mortality. These characteristics included age at onset of diabetes, duration of disease, and the presence or absence of hypertensive disorders, vascular, or renal complications (Sacks and Metzger 2013). Table 36.5 demonstrates the different classes of the 1949 White Classification system (Bennett et al. 2015).

Ambiguity on how to classify cases of GDM leads to confusion among practitioners (Hare and White 1980). The majority of the population of women in which the White Classification was originally based had diabetes antedating pregnancy, primarily type 1 (Hare and White 1980; Bennett et al. 2015). Therefore, the White Classification was never originally intended to be used with GDM. Upon considering diverse obstetric demographics and the complex nature of GDM, the White Classification revision in 1980 created a separate class for GDM (Hare and White 1980). ACOG further expands the GDM class into A1GDM or A2GDM, with the latter requiring pharmacologic therapy to achieve euglycemia (Gestational Diabetes Mellitus 2018). In class A1GDM, nutritional therapy alone is sufficient (Gestational Diabetes Mellitus 2018). Table 36.6 demonstrates an updated classification system that includes the subcategories of GDM (Gilmartin et al. 2008).

Table 36.5 White classification of diabetes in pregnancy: (A) Initial (1949) version and (B) Final (1980) version

(A) Initial 1949 Version	
Class A	Diagnosis of diabetes made on a glucose tolerance test, which deviates but slightly from the normal
Class B	Duration less than 10 y and Onset age 20 y or older and No vascular disease
Class C	Duration 10–19 y or Onset 10–19 y of age or Minimal vascular disease (e.g., retinal arteriosclerosis or calcified leg vessels)
Class D	Duration 20 y or longer or Onset younger than 10 y of age or More evidence of vascular disease, e.g., retinitis, transitory albuminuria, or transitory hypertension
Class E	Calcified pelvic arteries on X-ray
Class F	Nephritis
(B) Final 1980 Version	
Class A	Diet alone, any duration or onset age
Class B	Onset age 20 years or older and duration less than 10 years
Class C	Onset age 10–19 years or duration 10–19 years
Class D	Onset age younger than 10 years, duration over 20 years, background retinopathy, or hypertension (not pre-eclampsia)
Class R	Proliferative retinopathy or vitreous hemorrhage
Class F	Nephropathy with over 500 mg/d proteinuria
Class RF	Criteria for both Classes R and F coexist
Class H	Arteriosclerotic heart disease clinically evident
Class T	Prior renal transplantation

Table 36.6 Updated White classification

A: Abnormal glucose tolerance test at any age or of any duration treated only by diet therapy
B: Onset at age 20 years or older and duration of less than 10 years
C: Onset at age 10–19 years or duration of 10–19 years
D: Onset before 10 years of age, duration over 20 years, benign retinopathy, or hypertension (not preeclampsia)
– D1: Onset before age 10 years
– D2: Duration over 20 years
– D3: Calcification of vessels of the leg (macrovascular disease)
– D4: Benign retinopathy (microvascular disease)
– D4: Hypertension (not preeclampsia)
R: Proliferative retinopathy or vitreous hemorrhage
F: Renal nephropathy with over 500 mg/d proteinuria
RF: Criteria for both classes R and F
G: Many pregnancy failures
H: Evidence of arteriosclerotic heart disease
T: Prior renal transplant
Gestational diabetes
– A1: Controlled by diet and exercise
– A2: Requires insulin

Obstetric and Pharmacologic Management

Ideal glycemic levels in GDM are another topic of controversy, with no controlled trials coming to a consensus on optimal targets (Gestational Diabetes Mellitus 2018). The ADA and ACOG recommend that fasting glucose levels remain below 95 mg/dL, 1-h postprandial levels remain below 140 mg/dL, and 2-h postprandial levels remain below 120 mg/dL (Gestational Diabetes Mellitus 2018; American Diabetes Association 2018b). There are no clinical guidelines establishing accepted frequencies or timing of glucose monitoring. The general recommendations from ACOG suggest glucose monitoring four times a day, including fasting and after each meal (Gestational Diabetes Mellitus 2018). Additionally, postprandial levels are preferred to preprandial glucose values. One randomized study showed that using postprandial levels rather than preprandial to guide therapy allowed for reduced risk of neonatal hypoglycemia, macrosomia, and cesarean delivery (Veciana et al. 1996). Glucose monitoring may be decreased based on clinical discretion, if the patient has achieved appropriate glycemic control with diet (Gestational Diabetes Mellitus 2018; Turok et al. 2003).

Dietary Guidance

As previously mentioned, GDM can be treated using nutritional therapy alone (class A1GDM) or with additional pharmacologic interventions (class A2GDM). Medical nutrition therapy continues to be a central component of GDM management aimed at establishing normoglycemia, maintaining appropriate weight gain in pregnancy, promoting adequate fetal well-being, and preventing ketonemia (American Diabetes Association 2008). Specific nutritional recommendations vary between patients depending on their pre-pregnancy weights and ideal gestational weight gain, however, typically involve a carbohydrate-controlled meal plan (American Diabetes Association 2008). In general, three moderately sized meals and two to four snacks are recommended, with one snack in the evening to prevent ketosis overnight (Gestational Diabetes Mellitus 2018; American Diabetes Association 2008).

Carbohydrates should be dispersed throughout the day and compose 33–40% of total calories, with a minimum consumption of 175 g (American Diabetes Association 2008; Gestational Diabetes Mellitus 2018; Turok et al. 2003). The remaining calories can be distributed between protein (20%) and fats (40%) (Gestational Diabetes Mellitus 2018). The ADA recommends nutritional counseling at the time of diagnosis with a registered dietitian when possible, in order to provide individualization of therapy (Turok et al. 2003). For example, obese women may benefit from moderate caloric restriction to improve glycemic control (American Diabetes Association 2008). Specifics on medical nutritional therapy should be tailored to the patient needs, with collaboration between physicians and registered dieticians in order to achieve optimal outcomes. Exercise also has a part in management, but there are limited randomized controlled trials regarding its role in patients with GDM. ACOG suggestions reflect general diabetes care, aiming for 30 min of moderate aerobic exercise 5 days a week or 150 min total (Gestational Diabetes Mellitus 2018).

Pharmacotherapy

When nutrition and exercise are not sufficient to establish normoglycemia, pharmacologic management is indicated, as is the case in Class A2GDM. A systematic review found no ideal threshold at which to initiate drug therapy (Nicholson et al. 2008). Table 36.7 specifies the glycemic thresholds used by ACOG and ADA (Gestational Diabetes Mellitus 2018; Gilmartin et al. 2008). Both ACOG and ADA prefer the use of insulin to establish normoglycemia in pregnancy, as it does not cross the placenta and is effective in achieving metabolic control (Gestational Diabetes Mellitus 2018; American

Table 36.7 Glycemic thresholds to begin pharmacologic therapy for GDM

ACOG (mg/dL)	ADA (mg/dL)
Fasting ≥95	Fasting ≥105
1 h post-prandial ≥140	1 h post-prandial ≥155
2 h post-prandial ≥120	2 h post-prandial ≥130

Table 36.8 Pharmacotherapy features of special insulins

Regular insulin	30 min–8 h
NPH insulin	1–18 h
Glargine insulin	1–24 h
Detemir insulin	1–26 h
Lispro, Aspart	1 min–5 h

Modified from Gabbe SG, Graves CR. Management of diabetes mellitus complicating pregnancy. Obstet Gynecol 2003; 102(4): 857–868

Diabetes Association 2018b). The initial daily recommended dosage is 0.7–1.0 units/kg and can be divided into multiple injections, to optimally correct trends of hyperglycemia. NPH insulin, insulin glargine, and insulin detemir are acceptable long-acting options, while insulin lispro and aspart have been utilized for short-acting drugs. Table 36.8 is adopted from ACOG and demonstrates the timing of insulin options in pregnancy (Gestational Diabetes Mellitus 2018). Insulin regimens should be individualized to the patient and simplified for optimal results, and physicians should adjust treatment as pregnancy progresses and insulin resistances increases (Turok et al. 2003).

Oral Glucose-Lowering Prescriptions: Metformin

The Society for Maternal Fetal Medicine (SMFM) states that metformin is also a reasonable first-line alternative to insulin, although many women will still need insulin for glycemic control (SMFM Publications Committee 2018). In one meta-analysis comparing the use of metformin versus insulin, metformin use in GDM was associated with less maternal weight gain, less gestational hypertension, and less neonatal hypoglycemia, suggesting that perhaps there is a benefit to using metformin first line rather than insulin (Gui et al. 2013). However, unlike insulin, metformin crosses the placenta, and little is known about long-term outcomes associated with fetal exposure (SMFM Publications Committee 2018). Current studies looking at neurodevelopmental outcomes in 2-year-olds exposed to insulin and metformin in utero, respectively, show comparable results (Wouldes et al. 2017). Another study determined that metformin during fetal life for maternal GDM does not seem to have an impact on growth and body composition of 8-year-old children (Rø et al. 2012).

Other studies present conflicting results. Metformin has been shown to have cord arterial concentrations at almost twice the level of maternal venous concentrations, and research indicates that the fetus is exposed to concentrations comparable with therapeutic levels in adult populations (Vanky et al. 2005). A New Zealand follow-up of children at age 9 exposed to metformin or insulin in utero, showed that metformin-exposed children were found to have higher weights, arm and waist circumferences, and BMI (Rowan et al. 2018). Maternal tolerance to therapy must also be considered, as metformin can be associated with extensive gastrointestinal side effects (SMFM Publications Committee 2018).

Glyburide

Another option is the sulfonylurea glyburide. However, current data suggest that it is inferior to both insulin and metformin (Balsells et al. 2015). Like metformin, glyburide crosses the placenta and concentrates in cord serum, but there is no evidence to date about long-term effects on fetal exposure in utero (SMFM Publications Committee 2018). Also analogous to metformin, many women still may require insulin to achieve glycemic control (Brown et al. 2017). In a 2017 systematic review, glyburide was associated with increased composite neonatal death and serious morbidity as composed to metformin (Brown et al. 2017). Glyburide has also been correlated with higher birth weight and increased frequency of macrosomia and neonatal hypoglycemia as compared to insulin and metformin (SMFM Publications Committee 2018). Consequently, ACOG currently recommends that glyburide not be used first line, as it has not proven equivalent outcomes as compared to insulin and metformin (Gestational Diabetes Mellitus 2018). Despite data that show glyburide is a less effective choice in treating GDM, studies of commercially insured women showed an increase in glyburide use from 7.4% to 64.5% between 2000 and 2011 (Castillo et al. 2014). The increase in use could be attributed to the fact that oral hypoglycemic agents, including glyburide and metformin, have a lower cost and higher patient acceptance compared to insulin, possibly leading to increased patient satisfaction and compliance (SMFM Publications Committee 2018). Ultimately, clinicians are encouraged to counsel patients on the limited long-term safety data when prescribing glyburide (Gestational Diabetes Mellitus 2018).

Antenatal Fetal Testing

Achieving euglycemia is foundational to the management of GDM, but additional obstetric surveillance is also needed to optimize fetal outcomes. Maternal blood glucose levels are a known risk factor for stillbirth in patients with GDM (Mackin et al. 2019). As such, antenatal fetal testing is initiated at 32 weeks for women with poorly controlled GDM or who require pharmacologic management to achieve euglycemia (Gestational Diabetes Mellitus 2018). In women with A1GDM, risks for stillbirth have not been demonstrated, and consequently, ACOG has found inconclusive evidence regarding the need for antenatal fetal testing. Specific practices often vary by institutional guidelines, but in patients who achieve euglycemia via nutritional therapy alone, antenatal testing can generally be implemented later than in women with poorly controlled GDM (Gestational Diabetes Mellitus 2018). Studies have shown that twice weekly nonstress tests were effective in preventing stillbirth, but currently no antepartum surveillance method or frequency has been proven most successful (Kjos et al. 1995; Turok et al. 2003). Other acceptable fetal testing methods include the modified biophysical profile or full biophysical profile (Turok et al. 2003).

Indications for Ultrasonographic Monitoring

Because of the association of macrosomia and GDM, many clinicians use ultrasonography to assess fetal growth in the third trimester, although specific practices vary (Gestational Diabetes Mellitus 2018). The ADA recommends ultrasound measurements of fetal abdominal circumference every 2–4 weeks starting in the early third trimester and encourages clinicians to use this information to guide management (Metzger et al. 2007). In contrast, the International Federation of Gynecology and Obstetrics vaguely suggests "periodic clinical and sonographic growth assessments from diagnosis until term" (Hod et al. 2015). Ultrasonography can also be used to estimate fetal weight, but has been shown to significantly overestimate the prevalence of large for gestational age (LGA) fetuses in women with GDM. In a study comparing estimated fetal weight with actual birth weight, only 22% of women with a large for gestational age ultrasound diagnosis delivered an LGA neonate (Scifres et al. 2015). Estimated fetal weight via ultrasound has also been found to increase the risk of cesarean delivery (Little et al. 2012). While ultrasound can assist traditional clinical examinations to predict macrosomia, its inherent limitations in the ability to correctly estimate fetal weight account for the variance in obstetrical practices for women with GDM (Johnstone et al. 1996).

Optimal Delivery Time

Timing of delivery is also a challenging topic in GDM, as there is limited quality evidence to support universal guidelines. Delivery practices vary by institution and type of GDM (Gestational Diabetes Mellitus 2018). In Class A1GDM, the timing of delivery should not occur before 39 weeks unless other obstetric complications arise, and expectant management can be utilized through 40 6/7 weeks (Gestational Diabetes Mellitus 2018; Hod et al. 2015). However, some argue that earlier delivery may be indicated when fetal weight is estimated to be greater than 4000 g due to increased risk of shoulder dystocia and subsequent neonatal morbidity (Hod et al. 2015). As previously mentioned, ultrasound's inherent limitations in estimating fetal weight account for discrepancies in its utilization for delivery, and the advantages of this delivery strategy are still debated (Rouse et al. 1997). Clinicians are encouraged to discuss the benefits and risks of elective scheduled deliveries, including cesarean delivery, when there is estimated fetal macrosomia (American College of Obstetricians and Gynecologists' Committee on Practice Bulletins—Obstetrics 2016). For late-term or post-term Class A1GDM pregnancies, induction can be considered to reduce neonatal mortality and maternal morbidity associated with increased cesarean rates (Middleton et al. 2018).

In Class A2GDM that is well controlled, delivery can be considered between 39 0/7 and 39 6/7 weeks (Gestational Diabetes Mellitus 2018). Labor before 40 weeks in women requiring pharmacologic management for GDM may be associated with lower incidence of shoulder dystocia (Lurie et al. 1996). However, management is less clear for women with poorly controlled Class A2GDM, and practices may vary between clinicians and patients (Valent and Caughey 2016). Delivery between 37 0/7 and 38 6/7 weeks may be warranted in these patients (Gestational Diabetes Mellitus 2018). Earlier deliveries in the late preterm should be limited to patients in which aggressive attempts of achieving euglycemia, such as in-hospital therapy, have failed (Gestational Diabetes Mellitus 2018).

Intrapartum Glucose Homeostasis

Intrapartum glycemic monitoring is also necessary in women with GDM in order to avoid neonatal hypoglycemia and associated morbidities that may be caused by maternal hyperglycemia during labor (Ryan and Al-Agha 2013). Class A1GDM women can often be monitored during parturition without the need for additional pharmacologic management for glycemic control (Ryan and Al-Agha 2013). However, women with Class A2GDM often require intrapartum monitoring and pharmacologic intervention, but there is no consensus on specific practices. A recent randomized control trial found that tight maternal glucose control did not result in better outcomes when compared to more liberalized maternal management that allowed for higher maternal glucose levels and less frequent glucose measurements (Hamel et al. 2019). Recommendations from the International Federation of Gynecology and Obstetrics for these women suggest maternal glucose levels of 72–126 mg/dL during labor and delivery to avoid adverse outcomes such as neonatal hypoglycemia, birth asphyxia, and nonreassuring heart rate tracings (Hod et al. 2015).

Complications and Follow Up

Because GDM arises later in pregnancy after organogenesis, it is not associated with the same risks of congenital malformations as pre-existing diabetes (Nold and Georgieff 2004). However, GDM is associated with several short-term and long-term consequences in the fetus and subsequent neonate. The most established ramifications of GDM include large for gestational age, defined as fetal or neonatal weight at or above the 90th percentile for gestational age, and macrosomia, often defined as birth weight greater than or equal to 4500 g (Practice Bulletin No. 173 Summary 2016; Mitanchez 2010). From a physiologic standpoint, this is thought to be due to an increased transplacental transfer of glucose, leading to increased fetal substrate uptake and fetal hyperinsulinemia (Hod et al. 2015). Figure 36.1 highlights the pathophysiology of maternal hypoglycemia on adverse fetal outcomes, including LGA and macrosomia (Hod et al. 2015). In the international, multicenter HAPO study, researchers found that with increasing maternal glucose levels, the frequency of large for gestational age infants increased as well (HAPO Study Cooperative Research Group 2008).

Furthermore, multiple studies have found that increasing maternal hyperglycemia is associated with macrosomia, LGA, and the secondary outcomes associated with macrosomic fetuses such as operative deliveries, postpartum hemorrhage, shoulder dystocia, brachial plexus injury, birth canal tears, and maternal and fetal morbidity (Hod et al. 2015; Kwik et al. 2007; Lazer et al. 1986). However, LGA and macrosomia are multifactorial and are influenced by other factors, such as maternal weight gain. In one study looking at the effects of maternal glucose levels and weight gain on fetal macrosomia risk, 29.3% of women with GDM who gained more than

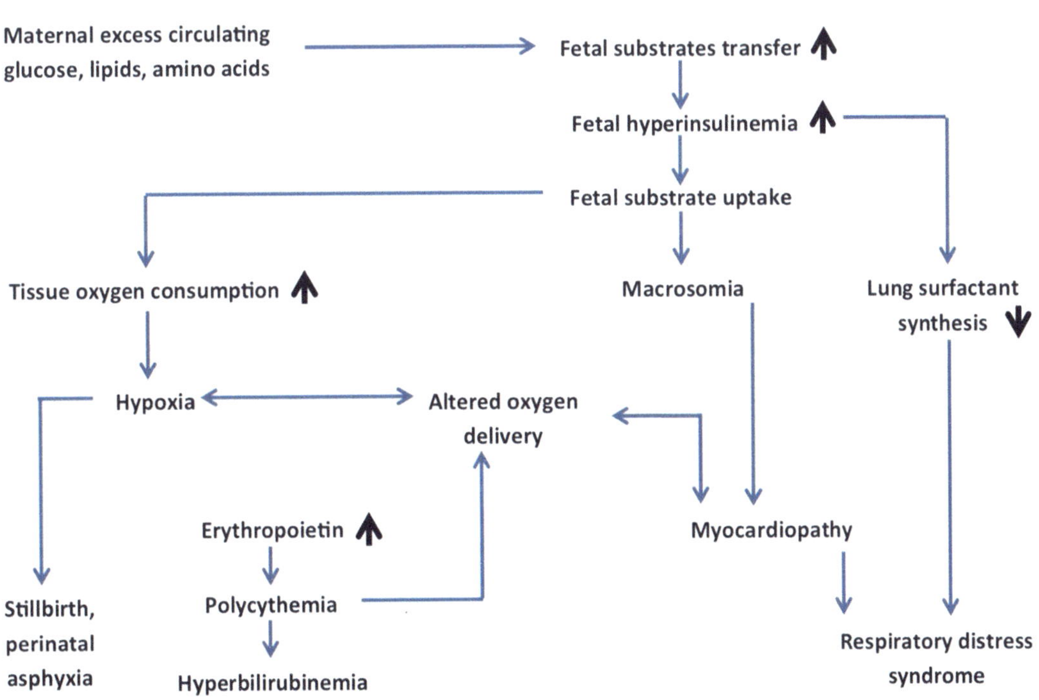

Fig. 36.1 Fetal and neonatal complications arising from intrauterine exposure to maternal hyperglycemia (Modified from: Hod M, Kapur A, Sacks DA, Hadar E, Agarwal M, Di Renzo GC, Cabero Roura L, McIntyre HD, Morris JL, Divakar H. The International Federation of Gynecology and Obstetrics (FIGO) Initiative on gestational diabetes mellitus: A pragmatic guide for diagnosis, management, and care. Int J Gynaecol Obstet. 2015;131 (Suppl 3):S173–211)

40 pounds had a macrosomic newborn compared to only 13.5% of women with GDM who gained appropriate pregnancy weight (Hillier et al. 2009). While confounding factors may influence the association between GDM and fetal size, maternal hyperglycemia remains an important risk factor for LGA and macrosomia.

Additional Neonatal Aberrations

Morbidities associated with GDM include fetal hyperinsulinemia and beta cell hyperplasia, neonatal hyperbilirubinemia, hypoglycemia, polycythemia, and increased intensive neonatal care (HAPO Study Cooperative Research Group 2008; Turok et al. 2003; Hod et al. 2015). Respiratory distress in neonates has also been reported as a complication to GDM in some studies, but the ALPS trial argues that newborns of GDM mothers were not more likely to have severe respiratory complications (Hod et al. 2015; Werner et al. 2018). Similarly, some studies argue that there is an increased risk of stillbirth in GDM mothers, but the association with maternal hyperglycemia is debated (Gestational Diabetes Mellitus 2018).

Long-Term Complications

These include higher risks of obesity, impaired glucose metabolism, and diabetes later in life. Some postulate that this occurs through epigenetic changes in developmental programming. Fetal exposure to maternal hyperglycemia in utero can lead to modification of phenotypic expression in newly formed cells, and in the third trimester, this can lead to aberrant proliferation of fetal adipocytes and muscle cells along with hyperplasia of pancreatic beta cells and neuroendocrine cells. Some argue this physiologic basis to the long-term effects is seen in offspring born to GDM mothers (Hod et al. 2015).

Maternal Complications

Women with GDM are at increased risk of developing pre-eclampsia with increasing risks at higher levels of maternal hyperglycemia (HAPO Study Cooperative Research Group 2008; Joffe et al. 1998). There are risks for developing hypertensive disorders during pregnancy, even after adjusting for potential confounding factors (Joffe et al. 1998). Cesarean delivery is another potential maternal morbidity with 25% of GDM mothers who require pharmacologic therapy undergoing cesarean delivery compared to 9.5% of controls (Gestational Diabetes Mellitus 2018). However, potentially the most concerning maternal consequences of GDM are the long-term associations with cardiometabolic disorders. In women with a previous diagnosis of GDM, 32.9% went on to develop metabolic syndrome within 5–10 years (Varner et al. 2017). Additionally, a history of GDM has been established as a risk factor tor types 1 and 2 diabetes. One study found that younger ages, the need for pharmacologic management, and positivity of autoantibodies in women with GDM predicted a higher risk of subsequent progression to type 1 diabetes (Jarvela et al. 2006).

Type 2 diabetes is a well-known consequence of GDM, with an incidence of almost 20% after 9 years and an estimated 70% within 22–28 years (Feig et al. 2008; Gestational Diabetes Mellitus 2018). Obesity intensifies this risk, as it does in the general population (Baptiste-Roberts et al. 2009). GDM has also been found to be an independent risk factor for long-term cardiovascular morbidity (Kessous et al. 2013).

Diagnostic Work Up and Long-Term Care

Following delivery, screening for diabetes, impaired fasting glucose levels, and impaired glucose tolerance is recommended in all women with GDM between 4–12 weeks postpartum (Gestational Diabetes Mellitus 2018). This is generally done through a fasting plasma glucose and 75-g, 2-h oral glucose tolerance test (Gestational Diabetes Mellitus 2018). Follow up is critical in identifying these women, as many randomized controlled trials have shown the benefit of early interventions such as diet and exercise or pharmacologic therapy, in delaying or preventing type 2 diabetes in women with abnormal glucose tolerance (Kitzmiller et al. 2007). Additionally, women with GDM who have persistent undiagnosed hyperglycemia have high risks in additional pregnancies, including major congenital malformations with maternal hyperglycemia during the first trimester. These risks can be mitigated by early preconception treatment (Kitzmiller et al. 2007). Long-term follow up is essential, and testing is recommended every 1–3 years if initial oral glucose tolerance testing is normal (American Diabetes Association 2018b). However, while routine follow up is critical in recognizing women who have progressed to diabetes, other therapies are available to intervene before diabetes develops.

In a randomized, controlled clinical trial, both metformin and intensive lifestyle therapies were found to be efficacious in reducing the incidence of diabetes by approximately 50%, in women reporting a history of GDM (Ratner et al. 2008). Lifestyle changes may also be effective in reducing future cardiovascular morbidity as well (Kitzmiller et al. 2007). Finally, women with a history of GDM should be counseled on the importance of breastfeeding and its protective role in preventing childhood obesity and maternal diabetes (Hod et al. 2015). Ultimately, a multidisciplinary approach between obstetricians, family physicians, internists, pediatricians, and other health care providers is needed to optimize long-term health outcomes in these women.

Ongoing Studies and Future Directions

Research is still needed in stratification of risk factors to adequately determine high-risk patients who need early pregnancy screening. Additionally, the need exists to develop portable and low-cost point of care testing strategies, to make universal screening for GDM a realistic possibility in areas with limited resources, and the US Preventive Services Task Force maintains that there is a need to look for screening methods that are simpler to use than the current recommendations, such as glycosylated hemoglobin measurement (Final Recommendation Statement [Internet] 2019). Intervention studies regarding optimal glucose control for best outcomes, and pharmacologic studies identifying the efficacy and long-term effects of glucose medications are other areas of future research. Long-term cohort studies following both women with GDM and their children are advised, to determine longitudinal repercussions of diabetes in pregnancy.

Current research being published from the ADA seeks to address many of these issues, such as the effectiveness of metformin in GDM to the economic burden of GDM. Meanwhile, other organizations, such as FIGO, highlight many of the aforementioned research needs, alongside appeals for greater international research collaboration to address these knowledge gaps (Hod et al. 2015).

References

American College of Obstetricians and Gynecologists' Committee on Practice Bulletins—Obstetrics (2016) Practice Bulletin No. 173: fetal macrosomia. Obstet Gynecol 128(5):e195–e209

American Diabetes Association (2008) Nutrition recommendations and interventions for diabetes: a position statement of the American Diabetes Association. Diabetes Care 31(Supplement 1):S61–S78

American Diabetes Association (2018a) Classification and diagnosis of diabetes: standards of medical care in diabetes—2019. Diabetes Care 42(Supplement 1):S13–S28

American Diabetes Association (2018b) 14. Management of diabetes in pregnancy: standards of medical care in diabetes—2019. Diabetes Care 42(Supplement 1):S165–S172

Anna V, Ploeg HPVD, Cheung NW, Huxley RR, Bauman AE (2008) Sociodemographic correlates of the increasing trend in prevalence of gestational diabetes mellitus in a large population of women between 1995 and 2005. Diabetes Care 31(12):2288–2293

Balsells M, Garcia-Patterson A, Sola I, Roque M, Gich I, Corcoy R (2015) Glibenclamide, metformin, and insulin for the treatment of gestational diabetes: a systematic review and meta-analysis. BMJ 350:h102

Baptiste-Roberts K, Barone BB, Gary TL et al (2009) Risk factors for type 2 diabetes among women with gestational diabetes: a systematic review. Am J Med 122 (3):207-214.e4

Bennett SN, Tita A, Owen J, Biggio JR, Harper LM (2015) Assessing White's classification of pregestational diabetes in a contemporary diabetic population. Obstet Gynecol 125(5):1217–1223

Berkowitz GS, Lapinski RH, Wein R, Lee D (1992) Race/ethnicity and other risk factors for gestational diabetes. Am J Epidemiol 135(9):965–973

Brown J, Martis R, Hughes B, Rowan J, Crowther CA (2017) Oral anti-diabetic pharmacological therapies for the treatment of women with gestational diabetes. Cochrane Database Syst Rev 1:CD011967

Caissutti C, Khalifeh A, Saccone G, Berghella V (2017) Are women positive for the one step but negative for the Two Step screening tests for gestational diabetes at higher risk for adverse outcomes? Acta Obstet Gynecol Scand 97(2):122–134

Castillo WC, Boggess K, Stürmer T, Brookhart MA, Benjamin DK, Funk MJ (2014) Trends in glyburide compared with insulin use for gestational diabetes treatment in the United States, 2000–2011. Obstet Gynecol 123(6):1177–1184

Cheung NW, Moses RG (2018) Gestational diabetes mellitus: is it time to reconsider the diagnostic criteria? Diabetes Care 41(7):1337–1338

Chu SY, Callaghan WM, Kim SY et al (2007) Maternal obesity and risk of gestational diabetes mellitus. Diabetes Care 30(8):2070–2076

Dabelea D, Snell-Bergeon JK, Hartsfield CL, Bischoff KJ, Hamman RF, Mcduffie RS (2005) Increasing prevalence of gestational diabetes mellitus (GDM) over time and by birth cohort: Kaiser permanente of colorado GDM screening program. Diabetes Care 28 (3):579–584

Deputy NP, Kim SY, Conrey EJ, Bullard KM (2018) Prevalence and changes in preexisting diabetes and gestational diabetes among women who had a live birth—United States, 2012–2016. MMWR Morb Mortal Wkly Rep 67(43):1201–1207

Feig DS, Zinman B, Wang X, Hux JE (2008) Risk of development of diabetes mellitus after diagnosis of gestational diabetes. Obstet Gynecol Surv 63 (12):759–761

Ferrara A (2007) Increasing prevalence of gestational diabetes mellitus: a public health perspective. Diabetes Care 30(Supplement 2):S141–S146

Final Recommendation Statement [Internet] (2019) Gestational Diabetes Mellitus, Screening—US Preventive Services Task Force. https://www.uspreventiveservicestaskforce.org/Page/Document/RecommendationStatementFinal/gestational-diabetes-mellitus-screening. Accessed 18 Nov 2019

Gestational Diabetes Mellitus (2018) ACOG Practice Bulletin No. 190. American College of Obstetricians and Gynecologists. Obstet Gynecol 131(2):e49–e64

Getahun D, Nath C, Ananth CV, Chavez MR, Smulian JC (2008) Gestational diabetes in the United States: temporal trends 1989 through 2004. Am J Obstet Gynecol 198(5):525.e1-5

Gilmartin A, Ural SH, Repke JT (2008) Gestational diabetes mellitus. Rev Obstet Gynecol 1(3):129–134

Gui J, Liu Q, Feng L (2013) Metformin vs insulin in the management of gestational diabetes: a meta-analysis. PLoS One 8(5):e64585

Hamel MS, Kanno LM, Has P, Beninati MJ, Rouse DJ, Werner EF (2019) Intrapartum glucose management in women with gestational diabetes mellitus. Obstet Gynecol 133(6):1171–1177

HAPO Study Cooperative Research Group (2008) Hyperglycemia and adverse pregnancy outcomes. N Engl J Med 358(19):1991–2002

Hare JW, White P (1980) Gestational diabetes and the White classification. Diabetes Care 3:394

Hillier T, Pedula K, Vesco K et al (2009) Excess gestational weight gain: modifying fetal macrosomia risk associated with maternal glucose. Obstet Anesth Dig 29(4):189

Hod M, Kapur A, Sacks DA et al (2015) The International Federation of Gynecology and Obstetrics (FIGO) Initiative on gestational diabetes mellitus: a pragmatic guide for diagnosis, management, and care. Int J Gynecol Obstet 131:S173–S211

Jarvela IY, Juutinen J, Koskela P et al (2006) Gestational diabetes identifies women at risk for permanent type 1 and type 2 diabetes in fertile age: predictive role of autoantibodies. Diabetes Care 29(3):607–612

Joffe GM, Esterlitz JR, Levine RJ et al (1998) The relationship between abnormal glucose tolerance and hypertensive disorders of pregnancy in healthy nulliparous women. Am J Obstet Gynecol 179 (4):1032–1037

Johnstone FD, Prescott RJ, Steel JM, Mao JH, Chambers S, Muir N (1996) Clinical and ultrasound prediction of macrosomia in diabetic pregnancy. BJOG: Int J Obstet Gynaecol 103(8):747–754

Kessous R, Shoham-Vardi I, Pariente G, Sheiner E (2013) 292: The association between gestational diabetes mellitus and long term maternal cardiovascular morbidity in a follow-up period of more than a decade. Am J Obstet Gynecol 208(1):S132

Kitzmiller JL, Dang-Kilduff L, Taslimi MM (2007) Gestational diabetes after delivery: short-term management and long-term risks. Diabetes Care 30(Supplement 2): S225–S235

Kjos SL, Leung A, Henry OA, Victor MR, Paul RH, Medearis AL (1995) Antepartum surveillance in diabetic pregnancies: predictors of fetal distress in labor. Am J Obstet Gynecol 173(5):1532–1539

Kwik M, Seeho S, Smith C, Mcelduff A, Morris J (2007) Outcomes of pregnancies affected by impaired glucose tolerance. Diabetes Res Clin Pract 77(2):263–268

Lazer S, Biale Y, Mazor M, Lewenthal H, Insler V (1986) Complications associated with the macrosomic fetus. J Reprod Med 31(6):501–505

Little SE, Edlow AG, Thomas AM, Smith NA (2012) Estimated fetal weight by ultrasound: a modifiable risk factor for cesarean delivery? Am J Obstet Gynecol 207(4):309.e1-6

Lowe WL, Scholtens DM, Lowe LP et al (2018) Association of gestational diabetes with maternal disorders of glucose metabolism and childhood adiposity. JAMA 320(10):1005

Lurie S, Insler V, Hagay Z (1996) Induction of labor at 38 to 39 weeks of gestation reduces the incidence of shoulder dystocia in gestational diabetic patients class A2. Am J Perinatol 13(05):293–296

Mackin ST, Nelson SM, Wild SH, Colhoun HM, Wood R, Lindsay RS (2019) Factors associated with stillbirth in women with diabetes. Diabetologia 62(10):1938–1947

Mcintyre HD, Jensen DM, Jensen RC et al (2018) Gestational diabetes mellitus: does one size fit all? a challenge to uniform worldwide diagnostic thresholds. Diabetes Care 41(7):1339–1342

Metzger BE (2010) International association of diabetes and pregnancy study groups recommendations on the diagnosis and classification of hyperglycemia in pregnancy. Diabetes Care 33(3):676–682

Metzger BE, Buchanan TA, Coustan DR et al (2007) Summary and recommendations of the fifth international workshop-conference on gestational diabetes mellitus. Diabetes Care 30(Supplement 2):S251–S260

Middleton P, Shepherd E, Crowther CA (2018) Induction of labour for improving birth outcomes for women at or beyond term. Cochrane Database Syst Rev 5: CD004945

Mitanchez D (2010) Foetal and neonatal complications in gestational diabetes: perinatal mortality, congenital malformations, macrosomia, shoulder dystocia, birth injuries, neonatal complications. Diabetes Metab 36 (6):617–627

Moyer VA, U.S. Preventive Services Task Force (2014) Screening for Gestational Diabetes Mellitus: U.S. Preventive Services Task Force Recommendation Statement. Ann Intern Med 160(6):414–420

Nicholson WK, Wilson LM, Witkop CT et al (2008) Therapeutic management, delivery, and postpartum risk assessment and screening in gestational diabetes. Evid Rep Technol Assessment 162:1–96

Nold JL, Georgieff MK (2004) Infants of diabetic mothers. Pediatr Clin N Am 51(3):619–637

O'Sullivan JB, Mahan CM (1964) Criteria for the oral glucose tolerance test in pregnancy. Diabetes 13:278

Practice Bulletin No. 173 Summary (2016) Fetal Macrosomia. Obstet Gynecol 128(5):1191–1192

Ratner RE, Christophi CA, Metzger BE et al (2008) Prevention of diabetes in women with a history of gestational diabetes: effects of metformin and lifestyle interventions. J Clin Endocrinol Metabol 93 (12):4774–4779

Rø TB, Ludvigsen HV, Carlsen SM, Vanky E (2012) Growth, body composition and metabolic profile of 8-year-old children exposed to metforminin utero. Scand J Clin Lab Invest 72(7):570–575

Roeckner JT, Sanchez-Ramos L, Jijon-Knupp R, Kaunitz AM (2016) Single abnormal value on 3-hour oral glucose tolerance test during pregnancy is associated with adverse maternal and neonatal outcomes: a systematic review and metaanalysis. Am J Obstet Gynecol 215 (3):287–297

Rouse DJ, Owen J, Goldenberg RL, Cliver SP (1997) The effectiveness and costs of elective cesarean delivery for fetal macrosomia diagnosed by ultrasound. Obstet Gynecol Surv 52(6):337–339

Rowan JA, Rush EC, Plank LD et al (2018) Metformin in gestational diabetes: the offspring follow-up (MiG TOFU): body composition and metabolic outcomes at 7–9 years of age. BMJ Open Diabetes Res Care 6(1): e000456

Rumbold AR, Crowther CA (2002) Womens experiences of being screened for gestational diabetes mellitus. Aust N Z J Obstet Gynaecol 42(2):131–137

Ryan EA, Al-Agha R (2013) Glucose control during labor and delivery. Curr Diab Rep 14(1):450

Saccone G, Khalifeh A, Al-Kouatly HB, Sendek K, Berghella V (2020) Screening for gestational diabetes mellitus: one step versus two step approach. A meta-analysis of randomized trials. J Matern Fetal Neonatal Med 33(9):1616–1624

Sacks DA, Metzger BE (2013) Classification of diabetes in pregnancy. Obstet Gynecol 121:345–348

Scifres CM, Feghali M, Dumont T et al (2015) Large-for-gestational-age ultrasound diagnosis and risk for cesarean delivery in women with gestational diabetes mellitus. Obstet Gynecol 126(5):978–986

SMFM Publications Committee (2018) SMFM statement: pharmacological treatment of gestational diabetes. Am J Obstet Gynecol 218(5):1

Solomon CG (1997) A prospective study of pregravid determinants of gestational diabetes mellitus. JAMA 278(13):1078–1083

Turok DK, Ratcliffe SD, Baxley EG (2003) Management of gestational diabetes mellitus. Am Fam Physician 68 (9):1767–1773

Valent A, Caughey A (2016) When to deliver women with diabetes in pregnancy? Am J Perinatol 33 (13):1250–1254

Vandorsten JP, Dodson WC, Espeland MA, Grobman WA, Guise JM, Mercer BM et al (2013) NIH consensus development conference: diagnosing gestational diabetes mellitus. NIH Consens State Sci Statements 29:1–31. (Level III)

Vanky E, Zahlsen K, Spigset O, Carlsen S (2005) Placental passage of metformin in women with polycystic ovary syndrome. Fertil Steril 83(5):1575–1578

Varner MW, Rice MM, Landon MB et al (2017) Pregnancies after the diagnosis of mild gestational

diabetes mellitus and risk of cardiometabolic disorders. Obstet Gynecol 129(2):273–280
Veciana MD, Major CA, Morgan MA et al (1996) Postprandial versus preprandial blood glucose monitoring in women with gestational diabetes mellitus requiring insulin therapy. Obstet Gynecol Surv 51(4):207–208
Werner E, Romano M, Rouse D (2018) 992: The association between gestational diabetes mellitus and neonatal respiratory morbidity. Am J Obstet Gynecol 218(1): S586
Wouldes TA, Battin M, Coat S, Rush EC, Hague WM, Rowan JA (2017) Neurodevelopmental outcome at 2 years in offspring of women randomised to metformin or insulin treatment for gestational diabetes. Obstet Gynecol Surv 72(3):149–151

Vascular Complications in Type 2 Diabetes

37

Chih Hao Chen Ku

Contents

Introduction 510
High-Risk Groups 510
Hypertension 510
Dyslipidemia 511
Diabetic Cardiomyopathy 512
Platelet Dysfunction 512

Clinical Features of Vascular Disease in Patients with Diabetes Mellitus 513
Atherosclerotic Disease 513
Coronary Artery Disease 513
Cerebrovascular Disease 513
Peripheral Artery Disease 514

Treatment 514
Lifestyle Modification 514
Hypertension 515
Dyslipidemia 516
Antiplatelet Therapies 517

Glucose-Lowering Therapies and Cardiovascular Disease 518
Glucose Lowering and Cardiovascular Disease 518
Glucose-Lowering Agents and Cardiovascular Disease 519

References 521

Abstract

It has long been recognized that diabetes is associated with a high cardiovascular risk, although this is not the same in all patients. Diabetes frequently coexists with other risk factors that will enhance this cardiovascular risk, including hypertension, dyslipidemia, and platelet dysfunction. Treatment of all these risk factors is essential to lower the risk in both primary and secondary cardiovascular prevention settings. The choice of antidiabetic agent may also impact this risk, as some classes have been shown to decrease cardiovascular mortality and others will increase the risk of

C. H. C. Ku (✉)
Endocrinology Department, Hospital San Juan de Dios, Caja Costarricense de Seguro Social, San José, Costa Rica

Clinical Pharmacology and Toxicology Department, University of Costa Rica, San José, Costa Rica
e-mail: chenku2409@gmail.com

J. Faintuch and S. Faintuch (eds.), *Obesity and Diabetes*, https://doi.org/10.1007/978-3-030-53370-0_37

hypoglycemia, and may have an impact on cardiovascular events.

Keywords

Cardiovascular complications · Peripheral ischemia · Type 2 diabetes · Cardiomyopathy · Arterial hypertension · Dyslipidemia · Pharmacotherapy

Introduction

It has always been recognized that patients with diabetes have a higher risk of vascular disease, especially cardiovascular complications. These complications are the main cause of death (Dal Canto et al. 2019). Adults with diabetes have 2–4 times increased cardiovascular risk compared to adults without diabetes, despite falling mortality rates in both groups (Rawshani et al. 2018). This trend in reduction in mortality rates is present especially in high-income countries, but not as clear in the middle-to-low-income countries (Dal Canto et al. 2019). On average at the age of 60, patients with diabetes and no cardiovascular disease will have a life expectancy of 6 years less, and those with diabetes and cardiovascular disease will have a 12-year shorter life expectancy (The Emerging Risk Factor Collaboration 2015). This increased risk will be present in patients either with type 1 or type 2 diabetes, and some reports indicate that the relative risk is even higher for type 1 diabetes (Lee et al. 2019).

Since the report by Haffner et al. (1998), diabetes has been considered a high cardiovascular risk group similar to a nondiabetic secondary prevention group. However, other studies have challenged this concept (Afkarian et al. 2016; Schneider et al. 2016). Although patients with diabetes have a higher cardiovascular risk compared to patients without diabetes, this risk can vary with each patient.

High-Risk Groups

Several characteristics will identify subgroups that have a higher risk. These include the presence of microvascular complications such as retinopathy (Sabanayagam et al. 2019), microalbuminuria (Ninomiya et al. 2009), chronic kidney disease (glomerular filtration rate of less than 60 cm^3/min) (Afkarian et al. 2016; Sabanayagam et al. 2019), smoking (Rawshani et al. 2018), and suffering from diabetes for more than 15 years (Hu et al. 2001). Most of these risk groups will also have microvascular alterations that will further increase the risk of macrovascular complications. Recent guidelines support this, recommending risk stratification within patients with diabetes that will tailor treatment goals (Consentino et al. 2019).

There are several reasons as to why patients with diabetes have a higher cardiovascular risk. First, insulin resistance may lay common pathways that will lead to hypertension, diabetes, and cardiovascular disease. Therefore, there is a very close link between hypertension and diabetes. Long-term hyperglycemia may lead not only to microvascular disease but also to changes in the vascular bed, inflammation, lipid oxidation, and an accelerated atherosclerotic process. Additionally, patients with diabetes may develop diabetic cardiomyopathy that will further increase the risk of cardiovascular disease and heart failure. Furthermore, platelet dysfunction is common in patients with diabetes mellitus, and this increases the risk of atherothrombosis and cardiovascular events.

Hypertension

Around 50–80% of patients with diabetes also have hypertension (Chen-Ku et al. 2019; Whelton et al. 2018). Hypertension is a strong factor that increases cardiovascular risk.

Wang et al. (2017) demonstrated that patients with higher insulin levels (as a marker of insulin resistance) have a higher prevalence of hypertension. Insulin resistance will lead to arterial stiffness due to the deposition of extracellular matrix in the artery walls. Smooth muscle cells will generate more reactive oxygen species that will lead to endothelial dysfunction, and there is also an activation of the renin angiotensin system in adipocytes (D'Elia and Strazzullo 2018).

Several proposed mechanisms may link hypertension and diabetes. Insulin resistance leads to an increase in the sympathetic tone, which causes an increase in heart rate, cardiac output, and vascular resistance. Furthermore, there is some evidence that renal tubular cells will have a higher expression of SGLT-2 cotransporters that will lead to a higher sodium retention. This increase of sympathetic tone may lead to an activation of the renin angiotensin system, with higher levels of angiotensin II that will promote vasoconstriction, myocardial fibrosis, salt and water retention.

Additionally, chronic stress may also lead to increased sympathetic activity. The stimuli of alpha 1 receptors generate water and salt retention and insulin resistance. Furthermore, stimuli of alpha 2 receptors in the pancreas will inhibit insulin secretion, which may lead to hyperglycemia. An increased sympathetic tone will enhance ventricular ejection force increasing blood pressure, and if this is sustained for more than 48 h, it can lead to an adaptation of baroreceptors and therefore chronic hypertension (Saxena et al. 2018).

Some studies have shown that in renal tubular cells, intracellular hyperglycemia will increase SGLT-2 expression. A higher-salt diet will also increase tubular SGLT-2 expression (Elliott et al. 2016). Renal tubular cells will increase SGLT2 expression when exposed to norepinephrine, at least in in vitro models. An increase in SGLT2 activity will lead to greater tubular sodium and glucose reabsorption, an additional mechanism that links hypertension and diabetes.

Patients with diabetes are more salt sensitive than healthy adults. Renal response to insulin is maintained in insulin-resistance states, leading to greater renal sodium reabsorption (Uzu 2017).

All of these mechanisms may explain the close correlation between diabetes and hypertension. Hypertension will accelerate the development and progression of microvascular disease (UKPDS 1998). Also, hypertension is a risk factor for vascular complications in diabetes, as patients diagnosed with diabetes and hypertension will have a higher risk of cardiovascular and cerebrovascular complications, along with heart failure (Rawshani et al. 2018). Peripheral artery disease is another vascular complication associated with the presence of hypertension (Mohammedi et al. 2016).

Dyslipidemia

The usual lipid profile that presents in patients with insulin resistance and diabetes is characterized by high triglycerides, low HDL, and an increase in small dense LDL particles. This is also called atherogenic dyslipidemia. In insulin-resistance states, there is a higher efflux of free fatty acids from adipocytes and, under these stimuli, hepatic VLDL production is increased. Since insulin is a coactivator of lipoprotein lipase, when there is insulin resistance there is a decrease in the activity of this enzyme. The increased number of triglyceride-rich lipoproteins competes for a saturable process of lipoproteinlipase-mediated clearance, leading to a greater number of remnant lipoproteins particles (Stahel et al. 2018). Meanwhile, hepatic lipase activity is not affected and it will hydrolyze triglycerides and phospholipids of these remnants, leading to smaller, denser particles of VLDL, VLDL remnants, and LDL, which are more atherogenic (Miksztowicz et al. 2012). On the other hand, there is an increased cholesterol ester transfer protein (CETP)-mediated lipid exchange between HDL and VLDL particles, leading to triglyceride enrichment of VLDL particles, which is a preferred substrate by hepatic lipase, yielding HDL remnants that are cleared by renal glomerular filtration leading to low HDL levels (Stahel et al. 2018).

Diabetes is also characterized as a state of increased production of reactive oxygen species that will oxidize LDL particles, making them more atherogenic. Oxidized LDL particles are easily captured by macrophages, leading to the development and progression of the atherosclerotic plaque. Higher levels of angiotensin II and sympathetic activity will also contribute to the generation of reactive oxygen species (Low Wang et al. 2016).

The role of dyslipidemia in microvascular complications is controversial. The data that support that dyslipidemia treatment will decrease the progression of microvascular complications is not very clear yet. On the other hand, the evidence in favor of statins in cardiovascular and cerebrovascular complications is very clear and strong.

Diabetic Cardiomyopathy

Diabetic cardiomyopathy is defined as the presence of a myocardial structural or functional abnormality in the absence of hypertension, atherosclerotic, or valvular disease. A more recent definition by the American Heart Association and the European Society of Cardiology states that it is a clinical condition, where there is ventricular dysfunction without coronary atherosclerosis and hypertension (Jia et al. 2018). This makes the clinical diagnosis of this entity difficult since the absence of atherosclerosis must be demonstrated, and this often requires invasive tests. Therefore, the true prevalence of diabetic cardiomyopathy is unknown.

Several pathophysiologic mechanisms may explain this condition. Two of the most accepted mechanisms are related to insulin resistance and hyperglycemia. Insulin resistance, as stated previously, will lead to higher levels of free fatty acids (FFA). These FFA will lead to the activation of different pathways, including ceramides that will generate cardiac hypertrophy. Hyperglycemia, through different mechanisms, including the polyol pathways, may promote myocyte proliferation and hypertrophy.

The classic paradigm states that diabetic cardiomyopathy will present initially as diastolic dysfunction (heart failure with preserved ejection fraction), and later will progress to dilation of the left ventricle and a fall in cardiac output (heart failure with reduced ejection fraction). It has long been recognized that patients with diabetes will have a higher risk of heart failure, and this risk is related to both disease duration and degree of hyperglycemia (Erqou et al. 2013). When we look at the data from the cardiovascular outcome trials, it has been shown that heart failure hospitalization rates were at least the same or even more frequent as the rates of ischemic strokes, cardiovascular deaths, or noncardiovascular deaths (Wiviott et al. 2019). Recent trials have also shown that SGLT-2 inhibitors have a great impact on heart failure, and this has raised the awareness and discussion of the importance of heart failure in patients with diabetes.

There are several mechanisms by which hyperglycemia may lead to contractile dysfunction (Zamora and Villena 2019):

- Hexosamine biosynthetic pathway leading to prolonged calcium transients.
- Activation of the protein kinase pathway that will promote cardiac hypertrophy.
- Advanced glycosylation end products pathway with an increase in RAGE expression that promotes oxidative stress, the formation of irreversible crosslinks, fibrosis, collagen deposition, production of inflammatory cytokines and growth factors.
- Polyol flux pathway that will end up in cell apoptosis.

In insulin-resistance states, there is higher free fatty acid efflux from the adipose tissue, leading to a higher fatty acid oxidation in the myocardium. An increase in mitochondrial activity will increase reactive oxygen species formation, and thus, increasing angiotensin I receptor expression (Zamora and Villena 2019).

Platelet Dysfunction

There are several mechanisms by which platelets are dysfunctional in patients with diabetes mellitus, and these mechanisms may also contribute to a higher rate of atherothrombotic events compared to patients without diabetes (Ferreiro and Angiolillo 2011):

- Hyperglycemia may lead to P-selectin expression, activation of PKC and decreased membrane fluidity, by glycation of surface proteins.
- Deficient insulin action by insulin resistance may lead to impaired response to nitric oxide and PGI2 and increased intracellular calcium.
- Associated conditions such as obesity, dyslipidemia, and inflammation may lead to platelet activation.
- There is also upregulation of $P2Y_{12}$ signaling and oxidative stress.

Clinical Features of Vascular Disease in Patients with Diabetes Mellitus

Atherosclerotic Disease

Hypertension and atherogenic dyslipidemia will lead to an accelerated atherosclerotic process and a higher risk of myocardial infarction, stroke, and peripheral artery disease. The burden of disease is not only greater, but it will also be present at a younger age. Other traditional risk factors such as smoking will further enhance this risk. Clinical presentation of atherosclerosis may differ from that of patients without diabetes.

It has also been demonstrated that patients with any profile of vascular disease will have a higher risk of developing vascular diseases in other beds. There is a high frequency of coexistence of different vascular diseases, such as coronary and cerebrovascular disease (Steg et al. 2007).

Coronary Artery Disease

Coronary artery disease may present classically as stable angina, unstable angina, or myocardial infarction. Silent ischemia is more frequent in patients with diabetes compared to the general population. Also, plaque burden tends to be more diffuse. When revascularized, the rate of restenosis is greater and faster in patients with diabetes (Makrilakis and Liatis 2017).

As mentioned previously, some patients with diabetes are at higher risk of coronary artery disease. These circumstances are called risk enhancers and they are outlined in Table 37.1 (Grundy et al. 2019a).

Guidelines do not support routine screening of coronary artery disease either with a stress test, coronary angiogram, or CT angiography if the patient is asymptomatic. In asymptomatic patients, the way to prevent cardiovascular disease is to have a very strict risk factor control, which will be discussed later on.

Clinical suspicion for coronary disease should be very high. Since patients with diabetes can present with silent ischemia, the presence of dyspnea or heart failure obliges to rule out coronary disease. Also, the frequency of atypical chest pain is greater. Therefore, all chest pain in patients with diabetes makes it mandatory to rule out the presence of coronary artery disease.

Cerebrovascular Disease

Strokes are one of the leading causes of morbidity and mortality in patients with diabetes mellitus (Rawshani et al. 2018). Patients with diabetes

Table 37.1 Risk enhancers for coronary artery disease in patients with diabetes mellitus

Long duration of diabetes ($\geq$10 years for patients with type 2 diabetes mellitus or $\geq$20 years for type 1 diabetes mellitus)
Albuminuria $\geq$30 μg of albumin/mg of creatinine
Estimated glomerular filtration rate <60 mL/min/1.73 m^2
Retinopathy
Neuropathy
Ankle brachial index (ABI) <0.9

mellitus have a higher risk of stroke and this is considered in risk scores such as CHA_2DS_2-VASc (Lip et al. 2010). This risk is further enhanced especially by the presence of hypertension, and to a lesser extent dyslipidemia.

Stroke treatment does not differ. Whenever possible, thrombolysis should be performed in ischemic stroke.

Regarding stroke prevention, if the patient has atrial fibrillation, diabetes mellitus, and an additional risk factor, anticoagulation is recommended. Otherwise, stroke prevention should follow guidelines as in the general population, where blood pressure and lipid control are essential.

Peripheral Artery Disease

The risk of peripheral artery disease is inherently higher in patients with diabetes mellitus. The presence of this condition increases the risk of complications associated with diabetic foot disease, lower-extremity amputation, cardiovascular disease, and mortality (Nativel et al. 2018). When defined by abnormal ankle brachial index (ABI), its prevalence may be up to 20% (Selvin and Erlinger 2004), becoming greater with advanced age and diabetes duration.

A small percentage of patients with peripheral artery disease presents with intermittent claudication. In some patients, these symptoms may be confounded with those of diabetic neuropathy and other conditions such as statin-associated myopathy. The symptoms of intermittent claudication may not always be as classically described, and sometimes it may be only lower-extremity myalgia. Any deterioration of either walking quality or speed, as well as fatigue, pain, cramps, discomfort or burns in buttocks, thighs, calves, or feet should raise the suspicion of peripheral artery disease, especially when triggered by exercise and quickly relieved with rest (Nativel et al. 2018).

Different guidelines recommend peripheral artery disease screening, initially with an exhaustive interview, and palpation of pedal pulses (American Diabetes Association 2020). The first line of noninvasive tests when signs or symptoms of lower-extremity artery disease become present, should be the brachial-ankle index. This is a very simple, low-cost test that may be performed in the office and should be conducted at least once a year. If ABI is less than 0.90, it is highly suspicious of peripheral artery disease. The lower the index, the more critical the obstruction. ABI of over 1.4 is also considered abnormal indicating calcified and stiffened arteries. If a low ABI is present, hypertension and lipid treatment goals will change, as this will deem the patient as having a higher risk of other atherosclerotic vascular events (Grundy et al. 2019a; American Diabetes Association 2020).

Treatment

To prevent cardiovascular risk in primary prevention, or lower the risk of a new vascular event in secondary prevention, the main objectives are to control risk factors, including smoking, hypertension, dyslipidemia, and glucose homeostasis.

Lifestyle Modification

Lifestyle modification is the initial approach in all patients with diabetes mellitus, regardless of cardiovascular risk, presence of microvascular complications, or duration of diabetes. However, there have been few trials that have evaluated the impact of lifestyle intervention in cardiovascular endpoints. UKPDS compared intensive pharmacologic intervention with conventional lifestyle intervention, achieving a difference in glucose control favoring pharmacologic intervention. Yet the achieved reduction in cardiovascular events in favor of the pharmacologic intervention group is not evidence that lifestyle interventions are ineffective (UK Prospective Diabetes Study (UKPDS) Group 1998a).

LOOKAHEAD (2007) showed that patients treated with intensive lifestyle modifications achieved an initial reduction in weight and waist circumference, which was paired with an improvement in glucose and blood pressure control. However, this was not sustained in the long term and therefore it was not associated with a decrease in cardiovascular disease. However, the group that was treated with intensive lifestyle modifications required fewer medications (statins, antihypertensive drugs) to achieve different targets (Look AHEAD Research Group 2013).

Hypertension

Early studies have shown that lowering blood pressure will decrease not only cardiovascular disease but also any diabetes-related endpoint (Emdin et al. 2015). It seems that from a cardiovascular perspective, angiotensin-converting enzyme (ACE) inhibitors, angiotensin II receptor blockers (ARB), diuretics, calcium channel blockers, and beta blockers have similar efficacy in reducing cardiovascular events. However, there may be small differences as shown in some meta-analyses, where thiazides are more effective in decreasing heart failure and stroke risk, calcium channel blockers may increase heart failure risk (compared to ACE inhibitors or ARB), and ACE inhibitors and ARB are more effective than beta blockers in stroke and cardiovascular events (Chen et al. 2018).

What has long been debated is the blood pressure treatment goal. Guidelines recommend different targets, especially of systolic blood pressure. Table 37.2 summarizes the blood pressure targets recommended by different professional groups.

Few trials have directly compared blood pressure targets specifically in patients with diabetes. Most data are derived from subgroup analysis. In the UKPDS (1998), lower blood pressure was associated with a lower risk of any diabetes-related endpoints, including cardiovascular disease. However, this lower blood pressure group achieved on average 144/82 mmHg, compared to the less intensive group (154/87 mmHg). In the ACCORD-BP trial, a systolic blood pressure target of 140 mmHg is mentioned, as compared to a target of 120 mmHg. Overall, no difference was observed in cardiovascular events except a slightly lower rate of stroke in the lower blood pressure target group, however, with a higher risk of adverse events, including renal failure and postural hypotension (ACCORD Study Group 2010a).

A meta-analysis (Emdin et al. 2015) has focused on the benefits of starting antihypertensive drugs if initial systolic blood pressure was greater than 140 mmHg compared to less than 140 mmHg. In general, if antihypertensive drugs are started when systolic blood pressure is less than 140 mmHg, only the risk of stroke and development of microalbuminuria were decreased. No difference was observed in myocardial infarction, heart failure, or other cardiovascular events.

Table 37.2 Blood pressure treatment goals recommended by different professional groups

American Diabetes Association (2020)	
High cardiovascular risk (existing atherosclerotic cardiovascular disease or 10-year atherosclerotic disease risk ≥15%)	<130/80 mmHg
Lower cardiovascular disease (10-year atherosclerotic cardiovascular risk <15%	<140/90 mmHg
American Association of Clinical Endocrinologists (AACE) (2019)	
Patients with diabetes mellitus	<130/80 mmHg
European Society of Cardiology (2019)	
Patients with diabetes mellitus	120–130 mgHg Systolic blood pressure, diastolic 70–80 mmHg
Older people (>65 years)	Systolic 130–139 mmHg
Patients at high risk of a cerebrovascular event	Systolic <130 mmHg
American College of Cardiology/American Heart Association (2018)	
Patients with diabetes mellitus	<130/80 mmHg

State of Art

If we want to summarize current treatment recommendations, a blood pressure of less than 140/90 mmHg is recommended for most patients with diabetes mellitus. For those at higher risk such as secondary prevention or having multiple cardiovascular risk factors, a target of less than 130/80 mmHg should be adopted. If overt proteinuria is present, even lower blood pressure targets may be recommended to decrease the magnitude of proteinuria. Therefore, there is not one single blood pressure cutoff point, and this should be individualized.

There is very scarce evidence regarding blood pressure goals in patients over 80 years of age. The only trial performed specifically in this group, the HYVET trial, reached a blood pressure goal of 150/90, and this was associated with a reduction in mortality and cardiovascular events (Beckett et al. 2008). Current guidelines do not make different recommendations based on patient age.

Dyslipidemia

Statins are the gold standard in lipid-lowering therapies. They have been demonstrated to be efficacious in primary and secondary cardiovascular disease prevention (Costa et al. 2006), and the benefit does not differ in patients with diabetes compared to the general population (Cholesterol Treatment Trialists' (CTT) Collaborators 2008).

In primary prevention, CARDS (2004) and HPS-diabetes (2003) included patients with diabetes over 40 years of age with an additional risk factor such as retinopathy, microalbuminuria, hypertension, or smoking. In both of these trials, intervention with either atorvastatin in a dose of 10 mg daily or simvastatin in a dose of 40 mg daily, respectively, decreased the rate of major adverse cardiovascular events (MACE).

In secondary prevention, subgroup analysis of trials (2010) that compared high-intensity versus low-intensity statin showed that patients with diabetes had benefits similar to patients without diabetes.

In the IMPROVE-IT trial (2018), high-risk patients treated with ezetimibe combined with simvastatin compared to simvastatin monotherapy demonstrated a reduction in MACE. Subgroup analysis showed some heterogeneity, where patients with diabetes showed a greater reduction in MACE compared to patients without diabetes. So, it seems that patients with diabetes may derive greater benefits from this combination. However, we must interpret these data with caution since it is a subgroup analysis.

Fibrate Combination

Much has been debated about the role of fibrates either as monotherapy or combined with statins in patients with diabetes. It seems very straightforward that if typical dyslipidemia associated with diabetes is characterized by high triglycerides and low HDL, fibrates should confer beneficial cardiovascular effects. In the FIELD trial (2005), fenofibrate compared to placebo, had the same rate of cardiovascular events. However, there was greater use of statins in the placebo group compared to the fenofibrate group, and this may have attenuated the differences and final cardiovascular results. In the ACCORD-lipid trial (2010b), fenofibrate combined with simvastatin was compared to simvastatin monotherapy and did not show a significant difference in cardiovascular events. Once again, subgroup analysis showed heterogeneity and some benefit, in those patients whose initial triglycerides levels were over 204 mg/dL and HDL of less than 34 mg/dL. On an individual basis, each one of these trials did not show a statistically significant reduction in cardiovascular disease. However, a Cochrane systematic review concluded that fibrates were associated with a slight decrease in cardiovascular events, although not in total mortality (Jakob et al. 2016). Currently, several trials are evaluating the benefit of fenofibrate combined with statin specifically in patients with diabetes, high triglyceride, and low HDL levels.

Omega 3 Fatty Acids

The REDUCE-IT (2019a) trial evaluated the use of a dose of 2 g twice a day of icosanpent ethyl compared to placebo in patients at high cardiovascular risk already taking statins, with high triglycerides (135–500 mg/dL), low HDL (less than 40 mg/dL), and LDL below 100 mg/dL, and showed a decrease in cardiovascular events.

Subgroup analysis of the cardiovascular trials with PCSK9 inhibitors alirocumab and evolocumab (Sabatine et al. 2017) in secondary prevention showed the same benefit in patients with diabetes as in the general population.

State of the Art

Guidelines recommend different LDL targets based on different risks in patients with diabetes (Table 37.3). In those patients without microvascular disease and long-standing diabetes (less than 10 years in type 2 diabetes and 20 years in type 1 diabetes), LDL of less than 100 mg/dL is recommended. In those with microvascular disease and/or long disease duration LDL, the target of less than 70 mg/dL should be sought after. If the patient has an established cardiovascular disease or multiple uncontrolled risk factors or chronic kidney disease, LDL less than 55 mg/dL is recommended. Initial therapy should be statins, and if on the maximally tolerated dose, ezetimibe can be added and then a PCSK9 inhibitor.

Antiplatelet Therapies

Patients with diabetes mellitus have higher platelet reactivity and are also prone to thrombotic events that contribute to their higher atherosclerotic cardiovascular risk. Despite this

Table 37.3 Lipid management recommendations by different professional groups

American Diabetes Association (2020)	
Patients with a 10-year atherosclerotic cardiovascular risk >20%	High-intensity statin (to achieve an LDL reduction of >50%)
40–75 years and those over 75 years without atherosclerotic cardiovascular disease	Moderate-intensity statin (to achieve an LDL reduction of 30–50%)
Patients with atherosclerotic cardiovascular disease on a statin with controlled LDL but with elevated triglycerides (135–499 mg/dL)	Add eicosapent ethyl to statin therapy
American Association of Clinical Endocrinologists (2019)	
Extreme risk (diabetes and established clinical cardiovascular disease)	LDL less than 55 mg/dL, apoB <70 mg/dL
Very high risk (diabetes plus major atherosclerotic risk factor such as hypertension, family history, low HDL, smoking or CKD 3/4)	LDL less than 70 mg/dL, apoB <80 mg/dL
High risk (diabetes without other major risk and/or age less than 40)	LDL <100 mg/dL, apoB <90 mg/dL
European Society of Cardiology (2019)	
Very high risk (atherosclerotic cardiovascular disease either clinical or by imaging, diabetes, and target organ damage, >2 risk factors, early onset of type 1 diabetes of long duration, over 20 years)	LDL less than 55 mg/dL
High risk (diabetes without target organ damage, with diabetes of more than 10 years or other additional risk factors, CKD stage 3)	LDL less than 70 mg/dL
Moderate (young patients with diabetes, T1DM less than 35 years, T2DM less than 50 years with duration of diabetes fewer than 10 years without other risk factors)	LDL less than 100 mg/dL
American College of Cardiology/American Heart Association (2019b)	
In patients with atherosclerotic cardiovascular disease	High-intensity statin to lower LDL >50%
Very high-risk patients (history of multiple atherosclerotic cardiovascular diseases or 1 major atherosclerotic disease plus multiple high-risk conditions such as diabetes)	LDL <70 mg/dL
Patients with diabetes and 40–75 years of age and LDL over 70 mg/dL	Start moderate-intensity statin
Patients with diabetes and 40–75 years of age and LDL over 70 mg/dL at higher risk, especially those with multiple risk factors or those 50–75 years of age	Start high-intensity statin therapy

pathophysiologic mechanism, various recent meta-analyses and trials have shown that aspirin is ineffective in decreasing total mortality and cardiovascular deaths in primary prevention, although there is a slight decrease in myocardial infarction, transient ischemic attack, and stroke. However, there is a higher risk of major bleeding, intracranial bleeding, and gastrointestinal bleeding (Abdelaziz et al. 2019). Therefore, current guidelines do not recommend the routine use of antiplatelet therapy in diabetes, except in patients with multiple risk factors at high risk of cardiovascular events (American Diabetes Association 2020).

In patients with established cardiovascular disease, antiplatelet therapy is effective in decreasing the risk of a new vascular event. A meta-analysis by the antithrombotic Trialist's Collaboration (2009) showed that aspirin was associated with a 20% reduction of major coronary events.

An unanswered question is who should receive dual antiplatelet therapy beyond the first 12 months after cardiac revascularization. The THEMIS-PCI trial (Bhatt et al. 2019b) showed that patients with diabetes and previous percutaneous coronary intervention (PCI), had fewer cardiovascular events when treated long term with aspirin plus ticagrelor; however, patients without PCI did not benefit from this dual therapy. There was no difference in total or cardiovascular mortality. Patients that were treated with dual antiplatelet therapy had a higher risk of major bleeding, and this must be considered when deciding who should receive dual antiplatelet therapy in the long term.

Glucose-Lowering Therapies and Cardiovascular Disease

Glucose Lowering and Cardiovascular Disease

Decreasing Hba1c will lead to a reduction in cardiovascular disease in the long term, however not in the short term. In patients recently diagnosed with either type 1 or type 2 diabetes, UKPDS (2008) and EDIC (2005) showed a reduction in cardiovascular events in the long term. Therefore, glucose control still is important not only to decrease microvascular events but also cardiovascular disease.

In patients with long-term diabetes or established cardiovascular disease, VADT (2009), ACCORD (2008), and ADVANCE (2008) showed that a lower Hba1c target did not decrease cardiovascular events in the short term. When all these trials were pooled in a meta-analysis, it showed that intensive glucose control was associated with a decrease in total coronary heart disease and nonfatal myocardial infarction, however not cardiovascular mortality (Ray et al. 2009). In the ACCORD trial, the intensive treatment group had a higher mortality rate, especially in those patients that started with a higher Hba1c, and who despite multiple treatment optimization strategies were not able to lower Hba1c (Riddle et al. 2010).

Hypoglycemia

One of the caveats of lowering glucose is the risk of hypoglycemia. There have been different studies (Kosiborod et al. 2008) that have shown that hypoglycemia, especially severe hypoglycemia, is associated with a higher rate of cardiovascular mortality. Several mechanisms may explain this link. First, hypoglycemia may lead to electrocardiographic changes, including QT prolongation, and this will make the patient prone to ventricular arrhythmias (Nuryani et al. 2012). Second, during hypoglycemia there is a release of catecholamines, and this may lead to plaque instability and acute coronary events. Third, severe hypoglycemia leads to seizures or loss of consciousness; this may trigger hypoxia, thus precipitating acute coronary events. There is evidence that hypoglycemia in the coronary care unit is associated with chest pain and ECG changes (Desouza et al. 2003). Strategies that lead to lower hypoglycemia risk should be sought, including drug treatment choice, patient education, and dose titration (including insulin and sulfonylureas).

Glucose-Lowering Agents and Cardiovascular Disease

Two classes of drugs have proven to reduce major adverse cardiovascular events and mortality: SGLT2 inhibitors and GLP-1 receptor agonists.

Three SGLT2 inhibitors have been evaluated in cardiovascular outcome trials, empagliflozin, canagliflozin, and dapagliflozin. Inclusion criteria and baseline characteristics differ in these trials and may explain why there are some differences in results. A meta-analysis (Zelniker et al. 2019) showed that overall, in the secondary prevention group, these drugs decreased MACE, especially cardiovascular mortality and heart failure hospitalization. It seems that in patients with multiple risk factors but without established cardiovascular disease, there was no benefit in MACE reduction although there were benefits in heart failure. Also, there were benefits in renal endpoints.

A more recent trial (Perkovic et al. 2019) performed in patients with diabetes and proteinuria showed that canagliflozin 100 mg daily, not only decreased progression of renal disease but also lowered cardiovascular disease, including the primary prevention cohort. In this trial, cardiovascular endpoints were secondary. Dapagliflozin has been recently evaluated in patients with heart failure and an ejection fraction of less than 40%, with and without diabetes, showing a significant reduction in heart failure hospitalization and total mortality. This benefit was observed in patients both with and without diabetes (McMurray et al. 2019).

GLP-1 receptor agonists are a very heterogeneous group. These agents can be classified according to their structure, either based on human GLP-1 (such as liraglutide, semaglutide, albiglutide, or dulaglutide), or based on exendin-4 (exenatide and lixisenatide). Cardiovascular outcome trials have been published with almost all of these agents. Trials with agents based on exendin-4 structure have been neutral in MACE. On the other hand, agents based on human GLP-1 have shown a reduction in MACE and in some of them, all-cause mortality (Marso et al. 2016a, b). Most of these trials have shown benefits in atherosclerotic cardiovascular events, such as myocardial infarction (Hernandez et al. 2018), revascularization (Marso et al. 2016b), and stroke (Marso et al. 2016b; Gerstein et al. 2019) especially in patients at high risk with established cardiovascular disease. The only trial that has shown a benefit in patients in the primary prevention is REWIND with dulaglutide (Gerstein et al. 2019). This is the only trial with enough power to show a benefit in primary prevention settings, and also has had the longest follow up. All GLP-1 receptor agonists have proven to be neutral in heart failure.

Other Antidiabetic Molecules

Metformin is one of the oldest antidiabetic agents. It has been backed up by clinical experience, safety, and low cost. However, there are not that many randomized controlled trials with enough power to evaluate its cardiovascular benefit. UKPDS (1998b) selected patients that were overweight to be randomized to metformin, and not the entire cohort. In this setting, a reduction in myocardial infarction was shown, and this benefit was maintained on the long-term follow-up.

There has been some concern regarding the cardiovascular safety of sulfonylureas since the results of the UGP. Recently, some trials have shown that the newer-generation sulfonylureas seem to be safer. Although not its primary endpoint, in the ADOPT trial (Kahn et al. 2006) in recently diagnosed patients with type 2 diabetes, glyburide had a lower rate of cardiovascular events compared to both metformin and rosiglitazone, although this was mainly due to a lower rate of heart failure. ADVANCE (2008) evaluated a lower Hba1c control based on gliclazide compared to conventional therapy, and the results on cardiovascular endpoints were neutral. TOSCA.IT (2017) randomized patients to pioglitazone or sulfonylureas (mostly glimepiride and gliclazide) and showed comparable rates of cardiovascular events. CAROLINA (2019) was a randomized controlled trial comparing glimepiride and linagliptin, and showed no difference in cardiovascular events. So, it seems that at least gliclazide and glimepiride are neutral in cardiovascular events and comparable to pioglitazone or a DPP4 inhibitor. However, all of these trials also showed a higher rate of hypoglycemia and weight gain with sulfonylureas.

Thiazolidinediones increase the risk of heart failure. Rosiglitazone, albeit initial concerns about its cardiovascular safety, seems to be neutral. Several studies with pioglitazone have shown some benefit in atherosclerosis compared to glimepiride. The primary endpoint of PROACTIVE (2005), the cardiovascular outcome trial of pioglitazone, did not show any clear benefits. The classic MACE was a secondary endpoint in this trial and showed a statistically significant reduction.

DPP-4 inhibitors have been thoroughly evaluated in cardiovascular outcomes. All of them, either in primary or secondary prevention, have been neutral in cardiovascular events. Saxagliptin in the SAVOR trial (Scirica et al. 2014) showed an increase in heart failure hospitalization and alogliptin in the EXAMINE trial (Zannad et al. 2015) showed the same trend, although it did not reach statistical significance. There is no clear explanation of the reason why these two agents may differ from the other ones in the group and confer a higher risk of heart failure. Subgroup analysis in both of these studies showed that this increase in hospitalization for heart failure was present, especially in those patients with higher baseline NT-proBNP levels and without a known history of heart failure.

Insulin Effects

Despite pathophysiologic mechanisms that may link insulin to an increased risk of cardiovascular events, this has not been supported by trial results. Some preclinical data showed insulin accelerating atherosclerosis, water, and salt retention. No trials have evaluated the cardiovascular safety of human insulins. In the UKPDS (1998a), the treatment with human insulin in order to achieve a lower blood glucose control was associated with a decrease in any diabetes-related endpoints, including cardiovascular disease. However, there was a difference in Hba1c in both groups, so it is impossible to discern if the benefit was due to insulin per se or to glucose control. Insulin glargine was compared to placebo in the ORIGIN trial (2012) and showed to be neutral in cardiovascular events. Insulin degludec was compared to insulin glargine in the DEVOTE trial (Marso et al. 2017) and showed similar cardiovascular results. Although DIGAMI (Malmberg 1997) showed initial benefits in acute coronary syndrome with intravenous insulin treatment, this was a small trial. When a larger cohort was investigated, the results were not replicated (Malmberg et al. 2005). Therefore, it seems that insulin is neutral from a cardiovascular standpoint.

In summary, blood glucose control is important to reduce cardiovascular events in the long term. SGLT-2 inhibitors decrease MACE, cardiovascular mortality, renal endpoints, and heart failure. GLP-1 receptor agonists based on human GLP-1 structure decrease atherosclerotic endpoints and mortality. There is some evidence of cardiovascular benefit with metformin. Pioglitazone, saxagliptin, and it seems that alogliptin as well, may increase heart failure hospitalization. Newer-generation sulfonylureas seem to be neutral in cardiovascular events, and the same applies to DPP4 inhibitors. Upon all this evidence, a recent guideline (Consentino et al. 2019) has moved SGLT-2 inhibitors and GLP-1 receptor agonists as first-line treatment in those patients at high cardiovascular risk or with established cardiovascular disease, instead of metformin. All other guidelines still recommend metformin as first-line therapy, and then SGLT-2 inhibitors or GLP-1 receptor agonists especially in patients with established cardiovascular disease, favoring SGLT-2 inhibitors if there is heart failure or chronic kidney disease.

Multifactorial Intervention

Few trials have evaluated the benefits of multifactorial intervention and cardiovascular outcomes. STENO-2 trial randomized patients with type 2 diabetes and microalbuminuria to two groups, one with a strict multifactorial control (total cholesterol less than 190 mg/dL, triglycerides less than 150 mg/dL, blood pressure less than 140/85 mmHg, Hba1c less than 6.5% and aspirin) compared to a less strict control (total cholesterol less than 250 mg/dL, triglycerides less than 195 mg/dL, blood pressure less than 160/95 mmHg, Hba1c less than 7.5%, no routine aspirin use). In the initial report, they showed a great reduction in all diabetes-related endpoints, including microvascular and macrovascular disease (Gaede et al. 2003). On follow up, they pointed out a delay of 7.9 years in the median of mortality and 8.1 years in the appearance of cardiovascular events in the intensive control group, with a risk reduction of 62% regarding cardiovascular events (Gaede et al. 2016).

Bariatric and Metabolic Surgery

There is evidence that bariatric surgery is better compared to optimal medical management, not only in diabetes remission but also in cardiovascular risk factors. However, in the STAMPEDE trial, there was no reduction in cardiovascular events in the bariatric surgery group, probably due to the small number of participants (Schauer et al. 2017). Yet major international associations endorse continued attention to bariatric and metabolic interventions, as an important option for all patients with class III obesity (BMI $\geq$40 kg/m^2), with or without overt diabetes, for those with class II obesity and diabetes (BMI 35.0–39.9 kg/m^2) who do not respond to conventional glucose-lowering therapy, and even in case of just class I obesity patients (BMI 30.0–34.9 kg/m^2), for those who are refractory to optimal antidiabetic treatment (BMI cutoff 2.5 kg/m^2 lower for Asian patients, in all categories) (Rubino et al. 2017).

References

Abdelaziz HK, Saad M, Pothineni NVK, Megaly M, Potluri R, Saleh M et al (2019) Aspirin for primary prevention of cardiovascular events. J Am Coll Cardiol 73(23):2915–2929

ACCORD Study Group, Cushman WC, Evans GW, Byington RP, Goff DC Jr, Grimm RH Jr et al (2010a) Effects of intensive blood-pressure control in type 2 diabetes mellitus. N Engl J Med 362(17):1575–1585

ACCORD Study Group, Ginsberg HN, Elam MB, Lovato LC, Crouse JR 3rd, Leiter LA et al (2010b) Effect of combination lipid therapy in type 2 diabetes mellitus. N Engl J Med 362(17):1563–1574

Action to Control Cardiovascular Risk in Diabetes Study Group, Gerstein HC, Miller ME, Byington RP, Goff DC Jr, Bigger JT et al (2008) Effects of intensive glucose lowering in type 2 diabetes. N Engl J Med 258(24):2545–2559

ADVANCE Study Group, Patel A, MacMahon S, Chalmers J, Neal B, Billot L et al (2008) Intensive blood glucose control and vascular outcomes in patients with type 2 diabetes. N Engl J Med 358 (24):2560–2572

Afkarian M, Katz R, Bansal N, Correa A, Kestenbaum B, Himmelfarb J et al (2016) Diabetes, kidney disease, and cardiovascular outcomes in the Jackson Heart Study. Clin J Am Soc Nephrol 11(8):1384–1391

American Diabetes Association (2020) Standards of medical care in diabetes—2020. Diabetes Care 43(Suppl 1):S1–S204

Antithromobitc Trialists' (ATT) Collaboration, Baigent C, Blackwell L, Collins R, Emberson J, Godwin J et al (2009) Aspirin in the primary and secondary prevention of vascular disease: collaborative meta-analysis of individual participant data from randomised trials. Lancet 373(9678):1849–1860

Beckett NS, Peters R, Fletcher AE, Staessen JA, Jiu L, Dumitrascu D et al (2008) Treatment of hypertension in patients 80 years of age or older. N Engl J Med 358 (18):1887–1898

Bhatt DL, Steg PG, Miller M, Brinton EA, Jacobson TA, Ketchum SB et al (2019a) Effects of eicosapent ethyl on total ischemic events: from REDUCE-IT. J Am Coll Cardiol 73(22):2791–2802

Bhatt DL, Steg PG, Mehta SR, Leiter LA, Simon T, Fox K et al (2019b) Ticagrelor in patients with diabetes and stable coronary artery disease with a history of previous percutaneous coronary intervention (THEMIS-PCI): a phase 3, placebo-controlled, randomised trial. Lancet 394(10204):1169–1180

Chen YJ, Li LJ, Tang WL, Song JY, Qiu R, Li Q et al (2018) First-line drugs inhibiting the renin angiotensin system versus other first-line antihypertensive drug classes for hypertension. Cochrane Database Syst Rev 11:CD008170

Chen-Ku CH, Gonzalez-Galvez G, Vasquez M, Fuente G, Nakazone MA, Silva Giordano AI et al (2019) Vascular complications in patients with type 2 diabetes: prevalence and comorbidities in 6 countries of Latin America (a cohort of the DISCOVER study program). Endocr Pract 25(10):994–1002

Cholesterol Treatment Trialists' (CTT) Collaborators, Kearney PM, Blackwell L, Collins R, Keech A, Simes J et al (2008) Efficacy of cholesterol-lowering therapy in 18686 people with diabetes in 14 randomised trials of statins: a meta-analysis. Lancet 371(9607):117–125

Colhoun HM, Betteridge DJ, Durrington PN, Hitman GA, Neil HA, Livingstone SJ et al (2004) Primary prevention of cardiovascular disease with atorvastatin in type

2 diabetes in the Collaborative Atorvastatin Diabetes Study (CARDS): multicenter randomised placebo-controlled trial. Lancet 364(9435):686–696
Collins R, Armitage J, Parish S, Sleigh P, Peto R (2003) Heart Protection Study Collaborative Group. MRC/BHF Heart Protection Study of cholesterol-lowering with simvastatin in 5963 people with diabetes: a randomised placebo-controlled trial. Lancet 361 (9374):2005–2016
Consentino F, Grant PJ, Aboyans V, Bailey CJ, Ceriello A, Delgado V et al (2019) ESC guidelines on diabetes, pre-diabetes and cardiovascular disease developed in collaboration with the EASD. Eur Heart J 34:1–69
Costa J, Borges M, David C, Vaz Carneiro A (2006) Efficacy of lipid lowering drug treatment for diabetic and non-diabetic patients: meta-analysis of randomised controlled trials. BMJ 332(7550):1115–1124
D'Elia L, Strazzullo P (2018) Excess body weight, insulin resistance and isolated systolic hypertension: potential pathophysiological links. High Blood Press Cardiovasc Prev 25(1):17–23
Dal Canto E, Ceriello E, Rydén L, Ferrini M, Hansen TB, Schnell O et al (2019) Diabetes as a cardiovascular risk factor: an overview of global trends of macro and micro vascular complications. Eur J Prev Cardiol 26 (2_suppl):25–32
Desouza C, Salazar H, Cheong B, Murgo J, Fonseca V (2003) Association of hypoglycemia and cardiac ischemia: a study based on continuous monitoring. Diabetes Care 26(5):1485–1489
Dormandy JA, Charbonnel B, Eckland DJ, Erdmann E, Massi-Benedetti M, Moules IK et al (2005) Secondary prevention of macrovascular events in patients with type 2 diabetes in the PROactive Study (PROspective pioglitAzone Clinical Trial in macroVascular Events): a randomised controlled trial. Lancet 366 (9493):1279–1289
Duckworth W, Abraira C, Moritz T, Reda D, Emanuele N, Reaven PD et al (2009) Glucose control and vascular complications in veterans with type 2 diabetes. N Engl J Med 360(2):129–139
Elliott RH, Mathews VB, Rudnicka C, Schlaich MP (2016) Is it time to think about the sodium glucose co-transporter 2 sympathetically? Nephrology 21 (4):286–294
Emdin CA, Rahimi K, Neal B, Callender T, Perkovic V, Patel A (2015) Blood pressure lowering in type 2 diabetes: a systematic review and meta-analysis. JAMA 313(6):603–615
Erqou S, Lee CT, Suffoletto M, Echuoffo-Tcheugui JB, de Boer RA, van Melle JP et al (2013) Association between glycated hemoglobin and the risk of congestive heart failure in diabetes mellitus: systematic review and meta-analysis. Eur J Heart Fail 15 (2):185–193
Ferreiro JL, Angiolillo DJ (2011) Diabetes and antiplatelet therapy in acute coronary syndrome. Circulation 123 (7):798–813
Gaede P, Vedel P, Larsen N, Jensen GV, Parving HH, Pedersen O (2003) Multifactorial intervention and cardiovascular disease in patients with type 2 diabetes. N Engl J Med 348(5):383–393
Gaede P, Oellgaard J, Carstensen B, Rossing P, Lund-Andersen H, Parving HH et al (2016) Years of life gained by multifactorial intervention in patients with type 2 diabetes mellitus and microalbuminuria: 21 years follow-up on the Steno-2 randomised trial. Diabetologia 59(11):2298–2307
Garber AJ, Abrahamson MJ, Barzilay JI, Blonde L, Bloomgarden ZT, Bush MA et al (2019) Consensus statement by the American Association of Clinical Endocrinologists and American College of Endocrinology on the comprehensive type 2 diabetes management algorithm—2019 executive summary. Endocr Pract 25(1):69–100
Gerstein HC, Colhoun HM, Dagenais GR, Diaz R, Lakshmanan M, Pais P et al (2019) Dulaglutide and cardiovascular outcomes in type 2 diabetes (REWIND): a double-blind, randomised placebo-controlled trial. Lancet 394(10193):121–130
Giugliano RP, Cannon CP, Blazing MA, Nicolau JC, Corbalán R, Spinar J et al (2018) Benefit of adding ezetimibe to statin therapy on cardiovascular outcomes and safety in patients with versus without diabetes mellitus: results from IMPROVE-IT (Improved Reduction of Outcomes: Vytorin Efficacy International trial). Circulation 137(15):1571–1582
Grundy SM, Stone NJ, Baiely AL, Beam C, Birtcher KK, Blumenthal RS et al (2019a) 2018 AHA/ACC/AACVPR/AAPA/ABC/ACPM/ADA/AGS/APhA/ASPC/NLA/PCNA Guideline on the management of blood cholesterol: a report of the American College of Cardiology/American Heart Association Task Force on clinial practice guidelines. Circulation 139(25):e1082–e1143
Grundy SM, Stone NJ, Bailey AL, Beam C, Birtcher KK, Blumenthal RS et al (2019b) 2018 AHA/ACC/AACVPR/AAPA/ABC/ACPM/ADA/AGS/APhA/ASPC/NLA/PCNA Guideline on the management of blood cholesterol: a report of the American College of Cardiology/American Heart Association Task Force on clinical practice guidelines. Circulation 139(25): e1082–e1143
Haffner SM, Lehto S, Rönnemaa T, Pyörälä K, Laakso M (1998) Mortality from coronary heart disease in subjects with type 2 diabetes and nondiabetic subjects with and without prior myocardial infarction. N Engl J Med 339(4):229–234
Hernandez AF, Green JB, Janmohamed S, D'Agostino RB Sr, Granger CB, Jones CP et al (2018) Albiglutide and cardiovascular outcomes in patients with type 2 diabetes and cardiovascular disease (Harmony Outcomes): a double-blind, randomised placebo-controlled trial. Lancet 392(10157):1519–1529
Holman RR, Paul SK, Bethel MA, Mathews DR, Neil HA (2008) 10-year follow-up of intensive glucose control in type 2 diabetes. N Engl J Med 395(15):1577–1589

Hu F, Stamper MJ, Solomon CG, Liu S, Willett WC, Speizer FE et al (2001) The impact of diabetes mellitus on mortality from all causes and coronary heart disease in women: 20 years of follow up. Arch Intern Med 161 (14):1717–1723

Jakob T, Nordmann AJ, Schandelmaier S, Ferreira-González I, Briel M (2016) Fibrates for primary prevention of cardiovascular disease events. Cochrane Database Syst Rev 11:CD009753

Jia G, Hill MA, Sowers JR (2018) Diabetic cardiomyopathy: an update of mechanisms contributing to this clinical entity. Circa Res 122(4):624–638

Kahn SE, Haffner SM, Heise MA, Herman WH, Holman RR, Jones NP et al (2006) Glycemic durability of rosiglitazone, metformin, or glyburide monotherapy. N Engl J Med 355(23):2427–2443

Keech A, Simes RJ, Barter P, Best J, Scott R, Taskinen MR et al (2005) Effects of long-term fenofibrate therapy on cardiovascular events in 9795 people with type 2 diabetes mellitus (the FIELD study): randomised controlled trial. Lancet 366(9500):1849–1861

Kosiborod M, Inzucchi SE, Krumholz HM, Xiao L, Jones PG, Fiske S et al (2008) Glucometrics in patients hospitalized with acute myocardial infarction: defining the optimal outcomes-based measures of risk. Circulation 117(8):1018–1027

LaRosa JC, Deedwania PC, Shepherd J, Wenger NK, Greten H, DeMicco DA et al (2010) Comparison of 80 versus 10 mg of atorvastatin on occurrence of cardiovascular events after the first event (from the Treating to New Targets (TNT) trial). Am J Cardiol 105(3):283–287

Lee YB, Han K, Kim B, Lee SE, Jun JE, Ahn J et al (2019) Risk of early mortality and cardiovascular disease in type 1 diabetes: a comparison with type 2 diabetes, a nationwide study. Cardiovasc Diabetol 18(1):157

Lip GY, Nieuwlaat R, Pisters R, Lane DA, Crijins HJ (2010) Redifining clinical risk stratificaction for predicting stroke and atherothromoboembolism in atrial fibrillation using a novel risk factor-based approach: the euro heart survey on atrial fibrillation. Chest 137 (2):263–272

Look AHEAD Research Group, Pi-Sunyer X, Blackburn G, Brancati FL, Bray GA, Bright R et al (2007) Reduction in weight and cardiovascular disease risk factors in individuals with type 2 diabetes: one-year results of the LOOK AHEAD trial. Diabetes Care 30(6):1374–1383

Look AHEAD Research Group, Wing RR, Bolin P, Brancati FL, Bray GA et al (2013) Cardiovascular effects of intensive lifestyle intervention in type 2 diabetes. N Engl J Med 369(2):145–154

Low Wang CC, Hess CN, Hiatt WR, Goldfine AB (2016) Atherosclerotic cardiovascular disease and heart failure in type 2 diabetes—mechanisms, management, and clinical consideration. Circulation 133(24):2459–2502

Makrilakis K, Liatis S (2017) Cardiovascular screening for the asymptomatic patient with diabetes: more cons than pros. J Diabetes Res 2017:8927473

Malmberg K (1997) Prospective randomised study of intensive insulin treatment on long term survival after acute myocardial infarction in patients with diabetes mellitus. DIGAMI (Diabetes Mellitus, Insulin Glucose Infusion in Acute Myocardial Infarction) Study Group. BMJ 314(7093):1512–1515

Malmberg K, Rydén L, Wedel H, Birkeland K, Bootsma A, Dickstein K et al (2005) Intense metabolic control by means of insulin in patients with diabetes mellitus and acute myocardial infarction (DIGAMI 2): effects on mortality and morbidity. Eur Heart J 26 (7):650–661

Marso SP, Daniels GH, Brown-Frandsen K, Kristensen P, Mann JF, Nauck MA et al (2016a) Liraglutide and cardiovascular outcomes in type 2 diabetes. N Engl J Med 375(4):311–322

Marso SP, Bain SC, Consoli A, Eliaschewitz FG, Jódar E, Leiter LA et al (2016b) Semaglutide and cardiovascular outcomes in patients with type 2 diabetes. N Engl J Med 375(19):1834–1844

Marso SP, McGuire DK, Zinman B, Poulter NR, Emerson SS, Pieber TR et al (2017) Efficacy and safety of degludec versus glargine in type 2 diabetes. N Engl J Med 377(8):723–732

McMurray JJV, Solomon SD, Inzucchi SE, Kober L, Kosiborod MN, Martinez FA et al (2019) Dapagliflozin in patients with heart failure and reduced ejection fraction. N Engl J Med 381(21):1995–2008

Miksztowicz V, Lucero D, Zago V, Cacciagiú L, Lopez G, Gonzalez-Ballerga E et al (2012) Hepatic lipase activity is increased in non-alcoholic fatty liver disease beyond insulin resistance. Diabetes Metab Res Rev 28(6):535–541

Mohammedi K, Woodward M, Hirakawa Y, Zoungas S, Williams B, Lisheng L et al (2016) Microvascular and macrovascular disease and risk for major peripheral artery disease in patients with type 2 diabetes mellitus. Diabetes Care 39(101):1796–1803

Nathan DM, Clearly PA, Backlund JY, Genuth SM, Lachin JM, Orchard TJ et al (2005) Intensive diabetes treatment and cardiovascular disease in patients with type 1 diabetes. N Engl J Med 353 (25):2643–2653

Nativel M, Potier L, Alexandre L, Baillet- Blanco L, Ducasse E, Velho G et al (2018) Lower extremity arterial disease in patients with diabetes: a contemporary narrative review. Cardiovasc Diabetol 17(1):138

Ninomiya T, Perkovic V, de Galan BE, Zoungas S, Pillai A, Jardine M et al (2009) Albuminuria and kidney function independently predict cardiovascular and renal outcomes in diabetes. J Am Soc Nephrol 20 (8):1813–1821

Nuryani N, Ling SS, Nguyen HT (2012) Electrocardiographic signals and swarm-based support vector machine for hypoglycemia detection. Ann Biomed Eng 40(4):934–945

ORIGIN Trial Investigators, Gerstein HC, Bosch J, Dagenais GR, Diaz R, Jung H et al (2012) Basal insulin

and cardiovascular and other outcomes in dysglycemia. N Engl J Med 367(4):319–328
Perkovic V, Jardine MJ, Neal B, Bompoint S, Heerspink HJL, Charytan DM et al (2019) Canagliflozin and renal outcomes in type 2 diabetes and nephropathy. N Engl J Med 380(24):2295–2306
Rawshani A, Rawshani A, Franzén S, Sattar N, Eliasson B, Svensson AM et al (2018) Risk factors, mortality and cardiovascular outcomes in patients with type 2 diabetes. N Engl J Med 379(7):633–644
Ray KK, Seshasai SR, Wijesuriya S, Sivakumaran R, Nethercott S, Preiss D et al (2009) Effect of intensive control of glucose on cardiovascular outcomes and death in patients with diabetes mellitus: a meta-analysis of randomised controlled trials. Lancet 373 (9677):1765–1772
Riddle MC, Ambrosius WT, Brillon DJ, Buse JB, Byington RP, Cohen RM et al (2010) Epidemiologic relationships between A1c and all-cause mortality during a median of 3.4-year follow-up of glycemic treatment in the ACCORD trial. Diabetes Care 33 (5):983–990
Rosenstock J, Kahn SE, Johansen OE, Zinman B, Espeland MA, Woerle HJ et al (2019) Effect of linagliptin vs glimepiride on major adverse cardiovascular outcomes in patients with type 2 diabetes: the CAROLINA randomized controlled trial. JAMA 322 (12):1155–1166
Rubino F, Nathan DM, Eckel RH, Schauer PR, Alberti KG, Zimmet PZ, Del Prato S, Ji L, Sadikot SM, Herman WH, Amiel SA, Kaplan LM, Taroncher-Oldenburg G, Cummings DE, Delegates of the 2nd Diabetes Surgery Summit (2017) Metabolic surgery in the treatment algorithm for type 2 diabetes: a joint statement by international diabetes organizations. Obes Surg 27(1):2–21
Sabanayagam C, Chee ML, Banu R, Cheng CY, Lim SC, Tai ES et al (2019) Association of diabetic retinopathy and diabetic kidney disease with all cause and cardiovascular mortality in multiethnic Asian population. JAMA Netw Open 2(3):e191540
Sabatine MS, Leiter LA, Wiviott SD, Giugliano RP, Deedwania P, De Ferrari GM et al (2017) Cardiovascular safety and efficacy of the PCSK9 inhibitor evolocumab in patients with and without diabetes and the effect of evolocumabon glycaemia and risk of new-onset diabetes: a prespecified analysis of the FOURIER randomized controlled trial. Lancet Diabetes Endocrinol 5(12):941–950
Saxena T, Ali AO, Saxena M (2018) Pathophysiology of essential hypertension: an update. Exp Rev Cardiovasc Ther 16(12):879–887
Schauer PR, Bhatt DL, Kirwan JP, Wolski K, Aminian A, Brethauer SA et al (2017) Bariatric surgery versus intensive medical therapy for diabetes—5 year outcomes. N Engl J Med 376(7):541–551
Schneider C, Coll B, Jick SS, Meier CR (2016) Doubling of serum creatinine and the risk of cardiovascular outcomes in patients with chronic kidney disease and type 2 diabetes mellitus: a cohort study. Clin Epidemiol 8:177–184
Scirica BM, Braunwald E, Raz I, Cavender MA, Morrow DA, Jarolim P et al (2014) Heart failure, saxagliptin, and diabetes mellitus: observations from the SAVOR-TIMI 53 randomized trial. Circulation 130 (18):1579–1588
Selvin E, Erlinger TP (2004) Prevalence of and risk factors for peripheral arterial disease in the United States: results from the National Health and Nutrition Examination Survey, 1999-2000. Circulation 110 (61):738–743
Stahel P, Xiao C, Hegele RA, Lewis GF (2018) The atherogenic dyslipidemia complex and novel approaches to cardiovascular disease prevention in diabetes. Can J Cardiol 34(5):595–604
Steg G, Bhatt DL, Wilson P, D'Agostino R, Ohman EM, Röther J et al (2007) One-year cardiovascular event rates in outpatients with atherothrombosis. JAMA 297 (11):1197–1206
The Emerging Risk Factor Collaboration, Di Angelantonio E, Kaptoge S, Wormser D, Willeit P, Butterworth AS et al (2015) Association of cardiometabolic multimorbidity with mortality. JAMA 314(1):52–60
UK Prospective Diabetes Study (UKPDS) Group (1998a) Intensive blood glucose control with sulphonylureas or insulin compared to conventional treatment and risk of complications in patients with type 2 diabetes (UKPDS 33). Lancet 352(9131):837–853
UK Prospective Diabetes Study (UKPDS) Group (1998b) Effect of intensive blood-glucose control with metformin on complications in overweight patients with type 2 diabetes (UKPDS 34). Lancet 352:854–865
UKPDS (1998) Efficacy of atenolol and captopril in reducing risk of macrovascular and microvascular complications in type 2 diabetes: UKPDS 39. UK Prospective Diabetes Study Group. BMJ 317 (7160):713–720
Uzu T (2017) Salt and hypertension in diabetes. Diabetol Int 8(2):154–159
Vaccaro O, Masulli M, Nicolucci A, Bonora E, Del Prato S, Maggioni AP et al (2017) Effects on the incidence of cardiovascular events of the addition of pioglitazone versus sulfonylureas in patients with type 2 diabetes inadequately controlled with metformin (TOSCA.IT): a randomised, multicentre trial. Lancet Diabetes Endocrinol 5(11):887–897
Wang F, Han L, Hu D (2017) Fasting insulin, insulin resistance and risk of hypertension in the general population: a meta-analysis. Clin Chim Acta 464:57–63
Whelton PK, Carey RM, Aronow WS, Casey DE Jr, Collins KJ, Dennison Himmelfarb C et al (2018) 2017 ACC/AHA/AAPA/ABC/ACPM/AGS/APhA/ASH/NMA/PCNA Guideline for the prevention, detection evaluation and management of high blood pressure in adults: a report of the American College of Cardiology/American Heart Association Task Force on clinical practice guidelines. Hypertension 71(19): e127–e248
Wiviott SD, Raz I, Bonaca MP, Mosenzon O, Kato ET, Cahn A et al (2019) Dapagliflozin and cardiovascular

outcomes in type 2 diabetes. N Engl J Med 380 (4):347–357

Zamora M, Villena JA (2019) Contribution of impaired insulin signaling to the pathogenesis of diabetic cardiomyopathy. Int J Mol Sci 20(11):E2833

Zannad F, Cannon CP, Cushman WC, Bakris GL, Menon V, Perez AT et al (2015) Heart failure and mortality outcomes in patients with type 2 diabetes taking alogliptin versus placebo in EXAMINE: a multicentre, randomised, double-blind trial. Lancet 385(9982):2067–2076

Zelniker TA, Wiviott SD, Raz I, Im K, Goodrich EL, Bonaca MP et al (2019) SGLT2 inhibitors for primary and secondary prevention of cardiovascular and renal outcomes in type 2 diabetes: a systematic review and meta-analysis of cardiovascular outcome trials. Lancet 393(10166):31–39

Part IV

General Therapy and Prevention

What We Know and Don't About High-Intensity Sweeteners

38

Clipper F. Young, Anne Lee, and Joy Dugan

Contents

Introduction 530

High-Intensity Sweeteners 530
HIS: Beyond Calorie Restriction 531
Interactions with Sweet Taste Digestive Receptors and Insulin Secretion 531
HIS on Sweet Taste Receptors in the GI Tract 532
HIS and Insulin Response 532
Discrepancies in Obese Subjects 533
Impact on Appetite 533
Glucose Homeostasis 533
Energy Intake 534
Impact on Body Weight Management 534
Impact on Blood Glucose Management 535
Other Diseases or Abnormalities 535

Ongoing Studies and Future Perspectives 536

References 537

Abstract

The joint position statement by the American Diabetes Association (ADA) and the American Heart Association (AHA) supports the use of high-intensity sweeteners (HIS) or commonly referred to as non-nutritive sweeteners (NNS), replacing sugar-sweetened beverages (SSB) among individuals with prediabetes and diabetes. There is much heterogeneity in results regarding the use of HIS in the literature. Nonhuman models have previously demonstrated obesogenic and carcinogenic potential with high utilization of HIS. However, to date, there is inconsistent evidence regarding HISs and their impacts on obesity and diabetes. In a few well-controlled studies, there appear to be inconsistent effects of HIS on metabolic syndrome. This chapter will describe the different available HISs in the United States. The effects on appetite, taste receptors, and glucose homeostasis will be explored. Furthermore, a review of the potential health impacts from consuming HIS,

C. F. Young (✉) · A. Lee
College of Osteopathic Medicine, Touro University California, Vallejo, CA, USA
e-mail: clipper.young@tu.edu

J. Dugan
College of Education and Health Sciences, Touro University California, Vallejo, CA, USA

J. Faintuch and S. Faintuch (eds.), *Obesity and Diabetes*, https://doi.org/10.1007/978-3-030-53370-0_38

especially on weight and blood glucose, will be discussed.

Keywords

Appetite · Intestinal receptors · Food intake · Insulin response · Weight control · Carcinogenesis · Gut microbiota · Cardiovascular disease

Introduction

One main contributing factor to the obesity pandemic has been an increased intake of energy-dense foods leading to an imbalance between energy consumption and energy expenditure, contributing to the tripling of obesity prevalence worldwide between 1975 and 2016 (World Health Organization 2019). The increasing prevalence of obesity and overweight has also extended into children and adolescents between the ages of 5 and 19, rising from 4% in 1975 to over 18% in 2016 (World Health Organization 2019). These 18% of obese and overweight children and adolescents accounted for over 340 million individuals (World Health Organization 2019).

High-intensity sweeteners (HIS) have been around since the end of the nineteenth century after saccharin was serendipitously discovered (Parker 1978). Human epidemiological studies and animal feeding studies, suggesting that HISs may dysregulate energy balance and cause metabolic syndrome, challenge the traditional belief that these substitutes only deliver a pleasant sweet taste without bringing any negative consequences.

High-Intensity Sweeteners

The United States Food and Drug Administration (FDA) has reviewed scientific data and approved six products to be considered as safe high-intensity sweeteners (HIS) to be used in food products: acesulfame potassium (Ace-K), aspartame, neotame, advantame, saccharin, and sucralose. In addition to these six, there are two substances on which FDA has received Generally Recognized as Safe (GRAS) notices; they are *Siraitia grosvenorii Swingle* (Luo Han Guo/monk fruit) fruit extracts (SGFE) and certain high purity steviol glycosides purified from the leaves of *Stevia rebaudiana Bertoni.* The FDA did not question the notifier's GRAS notices under the intentional condition of use (Food and Drug Administration 2019). Table 38.1 summarizes the acceptable daily intake determined by the FDA and the intensity of each sweetener.

Under the FDA's definition, all eight of these sweeteners are considered as non-nutritive sweeteners, except aspartame (Food and Drug

Table 38.1 Names of high-intensity sweeteners (HIS), regulatory status by the FDA, acceptable daily intake, and intensity (2019)

Names of high-intensity sweetener (HIS)	Regulatory status by the FDA	Acceptable daily intake (kg of body weight, per day) (mg/kg)	Intensity (compared to sucrose)
Saccharin	Approved	15	200–700×
Aspartame	Approved	50	200×
Acesulfame-K	Approved	15	200×
Sucralose	Approved	23	600×
Neotame	Approved	23	7000–13,000×
Advantame	Approved	32.8	20,000×
Steviol Glycosides	GRAS notified	4[a]	200–400×
Siraitia grosvenorii (Luo Han Guo) swingle fruit extracts (SGFE)	GRAS notified	Not specified	100–250×

[a]Determined by the Joint FAO/WHO Expert Committee on Food Additives (JECFA); *FAO* Food and Agriculture Organization of the United Nations

Table 38.2 Professional organizations and their views on sugar substitutes

Association/committee	Snapshot of position statements/recommendations
American Diabetes Association (ADA) (2020) Evert (2019)	→ Replace sugar-sweetened beverages (SSBs) with water as often as possible → When sugar substitutes are used to reduce overall calorie and carbohydrate intake, people should be counseled to avoid compensating with the consumption of additional calories from other food sources → Using sugar substitutes does not make an unhealthy choice healthy; rather, it makes such a choice less unhealthy → ADA supports the American Heart Association science advisory: There is not enough evidence to determine whether sugar substitute use definitively leads to a long-term reduction in body weight or cardiometabolic risk factors, including glycemia
Dietary Guidelines for Americans (2015–2020) USDA (2015)	→ Replacement of sweetened beverages with beverages containing sugar substitutes will reduce caloric intake in the short term → The effectiveness of long-term weight management is uncertain → Sugar substitutes are considered safe for the general population → FDA determined that even for high consumers of these sugar substitutes, their consumption is less than the acceptable daily intake level
Position of the Academy of Nutrition & Dietetics (2012) Fitch and Keim (2012)	→ Nutritive and non-nutritive sweeteners are considered safe as long as consumers do not exceed the federal recommendations on daily consumption → The overall diet plan should be considered along with sugar-substitute-containing products

Administration 2019). Aspartame is a nutritive sweetener since it contains more than 2% of caloric content in the same amount of sugar. These high-intensity sweeteners have been commonly referred to as sugar substitutes, non-nutritive sweeteners, low-calorie sweeteners, or artificial sweeteners. People include HIS in their foods for various reasons. Some may choose to consume HIS in foods and drinks to achieve improved glucose control or their goal weight, while others may enjoy the taste and are not concerned with the presence of HIS.

A joint position statement between the ADA and AHA on HISs, published originally in 2012, states that HIS can be utilized to reduce caloric and carbohydrate consumption for overall diabetes control and to obtain a healthy body weight (Gardner et al. 2012). Table 38.2 has a summary of professional organizations and their positions on HIS.

HIS: Beyond Calorie Restriction

Glucose homeostasis is mainly regulated by hormones and food composition (Björck et al. 1994). However, scientists suspect many other possible factors that can affect the blood glucose control process, including stress level, hot water immersion, environmental factors, body fat mass, and timing of food consumption (Patel and Abate 2013; Leicht et al. 2019; Singh et al. 2019; Yang et al. 2020). As the use of HIS is becoming more widespread in recent years, scientists question the possible adverse effects of these substances on the human body. High-intensity sweeteners are used to replace sugar in drinks or foods to lower the caloric and carbohydrate content.

Interactions with Sweet Taste Digestive Receptors and Insulin Secretion

The preference for sweetness may be innate; however, some studies have demonstrated that this preference may have been conditioned and shaped by the individual's experiences (Ventura and Mennella 2011). It was suspected by scientists that the sweet taste receptor, known as T1R2/T1R3, can be reduced in responsiveness and sensitivity to sweet taste when conditioned to long-term HIS exposure. This adaptation, in turn, would promote the dissociation between sweet taste and caloric content of drinks/foods, causing weakened cephalic responses (Schiffman et al. 1981). As shown in rats, the weakened cephalic response could cause a higher caloric intake. This chain of reaction had led to the hypothesis of HIS indirectly causing weight gain (Davidson et al. 2011).

Swithers and colleagues found that rats taking glucose tolerance tests after ingestion of saccharin-sweetened supplements had a higher glucose level than those who took glucose-sweetened supplements (Swithers et al. 2012). The authors suggested that HIS interfered with the typical relationship between sweet taste and calories suppressing glucagon-like peptide-1 (GLP-1) release, which could potentially alter glucose homeostasis (Swithers et al. 2012). This study supported the theory by Deutsch that when rodents consumed HIS long-term, the cephalic response triggered by sweet taste was weakened (Deutsch 1974). A few human studies on sweet taste receptors in the oral cavity and cephalic-phase insulin release (CPIR) demonstrated mixed results (Teff et al. 1995; Just et al. 2008; Ford et al. 2011). There were no changes in plasma insulin and blood glucose levels among groups that tasted water, sucrose, saccharin, and aspartame, when comparing pre- and post-stimulus area under curves (Teff et al. 1995). In another study, tasting saccharin increased insulin level ($p = 0.031$); however, there was no significant change in blood glucose levels ($p > 0.12$) (Just et al. 2008).

HIS on Sweet Taste Receptors in the GI Tract

Jang et al. (2007) showed that human duodenal L-cells expressed sweet taste receptors, the taste G-protein gustducin, and several other taste transduction elements (Jang et al. 2007). The gut L-cells are responsible for releasing GLP-1 when appropriate. In the gut of mice, sweet taste receptors behave very similar to the oral taste receptors, and the L-cells also express α-gustducin (Jang et al. 2007). When an antagonist to the sweet taste receptors and α-gustducin are present in mice, incretin response is impaired. In vitro, HIS can induce secretion of incretins through the T1R3-dependent mechanism (Jang et al. 2007). It is proposed that in mice, the sweet taste receptors in the gut respond to both nutritive foods (glucose and sucrose) and HIS (sucralose) to affect the release of incretins, increasing SGLT-1 expression and thus potentially altering glucose homeostasis (Margolskee et al. 2007).

HIS and Insulin Response

Previous human studies showed a lack of consensus on the clinical implications of HIS on glucose homeostasis (Sylvetsky and Rother 2016). In healthy adult subjects ($n = 22$), examining the effects of 240 mL of diet soda (sucralose and acesulfame-K) vs. carbonated water, followed by a 75-g glucose tolerance test, diet soda with glucose consumption showed a significantly higher GLP-1 secretion compared to carbonated water with glucose consumption ($p = 0.003$), but no differences in glucose ($p = 0.64$) and insulin levels (at 20 min, $p = 0.20$; at 25 min, $p = 0.28$) (Brown et al. 2012). Interestingly, the same authors did a similar study 3 years later in youths with T1DM ($n = 9$), T2DM ($n = 10$), and healthy controls ($n = 25$), consuming diet soda with glucose compared with drinking carbonated water with glucose. They found that GLP-1 levels increased significantly in T1DM ($p = 0.02$) and healthy control subjects ($p = 0.029$) groups, but without change in the group with T2DM ($p = 0.92$). Glucose and C-peptide levels showed no significant differences between the consumption of the two beverages in any group (Brown et al. 2012). These results indicate that people with T2DM, compared to healthy subjects or people with T1DM, may express a different incretin effect when exposed to HIS or other ingredients in the diet soda used in this study.

Another study compared sucralose, aspartame, and water with glucose consumption between adults with newly diagnosed drug-naïve T2DM (Temizkan et al. 2015). The healthy control group that consumed sucralose had a significantly lower ($p = 0.002$) glucose level compared to the group that drank water before a glucose tolerance test, with no difference between consuming aspartame and water ($p = 0.53$) (Temizkan et al. 2015). GLP-1 was significantly higher ($p = 0.04$) with a sucralose preload compared with a water preload in the healthy control group (Temizkan et al. 2015). Results in subjects with T2DM did not show the same level of difference. In both healthy subjects and subjects with T2DM, the total area under the curve (AUC) for insulin (healthy $p = 0.29$, T2 $p = 0.07$) and c-peptide (healthy $p = 0.87$, T2 $p = 0.46$) were not statistically different between the three consumption settings (Temizkan et al. 2015). Other studies on healthy human subjects have shown a similar result of no significant difference in glycemia and insulin response (Ford et al. 2011; Ma et al. 2010; Wu et al. 2012, 2013). When healthy subjects were studied using HIS (aspartame and sucralose groups) daily for 2 weeks, the total AUC for glucose, insulin, active GLP-1, and leptin were similar to the baseline values (Ahmad et al. 2020).

Discrepancies in Obese Subjects

Pepino and colleagues found that with a preceding sucralose consumption, the glucose tolerance test showed significantly higher plasma glucose concentrations ($p = 0.03$), a greater incremental increase in insulin area under the curve ($p < 0.03$), a greater peak insulin secretion rate ($p < 0.02$), and a decrease in insulin clearance ($p = 0.04$) with no significant difference in plasma active GLP-1 concentrations ($p = 0.72$) (Pepino et al. 2013). A more recent study comparing healthy normal weight and obese individuals ingesting sucralose prior to oral glucose tolerance test (OGTT), resulted in increased glucose AUC in both weight groups; while insulin response to sucralose decreased in healthy weight participants (within 20–40 min) and raised in obese participants (within 90–120 min) (Nichol et al. 2019).

Impact on Appetite

To achieve the "calorie-saving" goal through the consumption of HIS-sweetened items, one must refrain from compensatory energy intake throughout the day. The potential mechanisms relating to appetite and food intake affected by HIS include cephalic phase stimulation, nutritive and osmotic effects, gut peptide response, palatability, informed use leading to overconsumption, loss of signal fidelity, activation of reward systems, and training the palate/learning to like the familiar (Mattes and Popkin 2009). It has been concluded that these potential mechanisms of HIS causing an increase in appetite or food intake were disproved based on the available evidence, or were inconclusive due to insufficient evidence. For example, although adding HIS to products without calories may enhance appetite, the same effect was not observed when HIS was consumed with an energy source (Mattes and Popkin 2009).

The methodologies used in controlled feeding experiments and free-living trials have added complexities, to demystifying the extent of other food item intake due to the consumption of HIS (Gardner et al. 2012). Since controlled feeding experiments cannot simulate real-world situations, such investigations may or may not exhibit an effect on future calorie intake caused by HIS. Education is needed to disseminate that HIS-containing products can assist in weight management only if utilized to replace full-calorie food items, and also if the energy deficiency created for which is not compensated. Many factors exist in the real world, contributing to food choice and calorie intake, so the amount of food consumed after HIS consumption might be confounded by reverse causality (e.g., future intake of higher-calorie food items as a compensation for the previous HIS-containing items) (Gardner et al. 2012).

Glucose Homeostasis

Studies do not support the speculation that HIS stimulates incretin responses. Sucralose, similar to glucose, administered via an intragastric infusion under a fasted state in healthy humans did not stimulate the release of insulin, GLP-1, or gastric inhibitory polypeptide (GIP), thus not compromising gastric emptying (Ma et al. 2009). Sucralose delivered intraduodenaly also in fasted healthy humans, did not affect concentrations of postprandial blood glucose, GLP-1 release, or amount of glucose absorption in the small intestine (Ma et al. 2010). Lastly, GLP-1, insulin, or glucose concentration was also not impacted in healthy human subjects with a fasted state after sucralose being administered orally (Ford et al. 2011), or aspartame, acesulfame K, or sucralose being intragastrically infused (Steinert et al. 2011). In disagreement with such findings, a study done in young, healthy adults revealed that the consumption of HIS in the form of a diet soda before a glucose load, led to a significant increase in GLP-1 secretion but not in insulin or glucose levels (Brown et al. 2009).

Energy Intake

Energy intake has been included as an outcome (either primary or secondary) in experiments comparing HIS vs. caloric sweeteners used in beverages and foods. For example, de la Hunty and colleagues conducted a meta-analysis (de la Hunty et al. 2006), concluding that aspartame was able to create an energy deficit, with 68% of which uncompensated for 24 h, and the energy deficit lasted until the next period of consumption. On the other hand, with the four studies in this meta-analysis examining the effects via beverages alone, the energy deficit created was 85% uncompensated in the subsequent 24 h post-HIS consumption (de la Hunty et al. 2006). These results suggest that substituting sugar-sweetened beverages with HIS-containing beverages might result in a higher calorie net reduction than doing the same in foods.

In a crossover design, when subjects consumed sucrose-sweetened (full calorie) beverages, a lack of energy intake compensation was observed. When HIS-containing beverages were consumed, a net reduction in energy intake was achieved through consuming HIS (Van Wymelbeke et al. 2004). In contrast, a 1-month study generated a similar total energy intake between sugar-containing soft drinks and aspartame-containing soft drinks among normally weighted adults (Reid et al. 2007).

Impact on Body Weight Management

An increased BMI, according to a meta-analysis of 30 cohort studies (n = 405,907), may be associated with routine consumption of HIS (Azad et al. 2017). However, the same meta-analysis demonstrated an inconsistent effect of HIS on BMI in seven randomized control trials (RCTs) (n = 1003) (Azad et al. 2017). A smaller recent RCT (n = 100) demonstrated that two doses of aspartame ingested per day did not affect body weight among healthy lean adults (Higgins et al. 2018). A common myth is that those who consume higher amounts of HIS also consume higher total calories. However, no caloric difference was demonstrated in a 24-h dietary recall study of (n = 14,098) American adults that consumed HIS versus no HIS in the National Health and Nutrition Examination Survey (NHANES) (Malek et al. 2018).

In a smaller study of older adults (n = 749), a linear dose-response increase between HIS-soda consumption and waist circumference occurred (Fowler et al. 2015). A longitudinal cohort study of Hispanics between 12 and 18 years old (n = 98) determined that chronic HIS consumers compared to controls had higher total body fat at baseline and 1-year follow-up (Davis et al. 2018). The children with higher levels of HIS consumption also had higher total energy and carbohydrate consumption. In the UK Millennium Cohort Study, children between 7 and 11 years of age were longitudinally studied to examine the association between HIS consumption and adiposity over 4 years (Laverty et al. 2015). Daily consumption of HIS was associated with higher body fat (+1.18 kg/m^2, 95% confidence intervals: 0.81, 1.54).

Impact on Blood Glucose Management

When BMI is adjusted in both the Nurses' Health Study II ($n = 116{,}000$ women) (Malik et al. 2019) and the Health Professionals Follow-Up Trial ($n = 40{,}000$ men) (de Koning et al. 2012), there is no significant increase in type 2 diabetes mellitus (T2DM) development with HIS intake. Before adjusting for BMI, the Nurses' Health Study II had a 35% higher risk of developing T2DM and 40% higher in the Health Professionals Follow-Up Trial (Malik et al. 2019; de Koning et al. 2012). Higher consumption of HIS in the Health Professionals Follow-Up Trial is hypothesized to be in reaction to weight gain, which is associated with increasing T2DM risk (Malik et al. 2019).

In a double-blind, randomized crossover trial of 60 adults with healthy BMI and without diabetes (Bonnet et al. 2018), all participants underwent a 4-week washout period and then consumed two cans of aspartame and acesulfame K per day or unsweetened, no-calorie beverage for 12 weeks. There was no significant effect on insulin sensitivity or secretion. A recent RCT ($n = 100$) confirmed that two doses of aspartame ingested per day did not affect insulin levels, resting leptin, glucagon-like peptide-1, or gastric inhibitory peptide after 12 weeks among healthy lean adults compared to baseline levels (Higgins et al. 2018).

Other Diseases or Abnormalities

Carcinogenesis

The National Cancer Institute issued a statement affirming the use of HIS is not associated with cancer in humans (National Cancer Institute Website 2019). In many animal models, however, HISs have been shown as carcinogenic agents. Soffritti and colleagues found that after aspartame was administered to Sprague-Dawley rats ($n = 900$) at 0, 80, 400, 2000, 10,000, 50,000, and 100,000 ppm from 8 weeks old to natural death, both male and female mice had increased incidences of cancer (e.g., leukemia and lymphoma), and increased carcinomas of the renal pelvis and ureter were found in female mice (Soffritti et al. 2005). Another study showed a similar result, where male ($n = 457$) and female mice ($n = 396$) were divided into five groups and given different amounts of sucralose (0, 500, 2000, 8000, and 16,000 ppm) from 12 days post-gestation to entire lifespan (Soffritti et al. 2005). The dose-related incidence of malignant tumors/neoplasms was found significantly higher in male mice given doses at 2000 and 16,000 ppm.

However, in a controlled protocol with mice (Mann et al. 2000), three groups of male ($n = 52$) and female ($n = 52$) animals were given a sucralose-containing diet at concentrations of 3000, 10,000, and 30,000 ppm for 104 weeks. The incidence of any tumors was not found to be increased in the sucralose group.

Disruption of Gut Microbiota

An increased ratio between Firmicutes and Bacteroidetes as well as an increase in *Lactobacilli* spp. have been identified following the consumption of HIS, and this composition mirrors that of obese organisms (Suez et al. 2014; Nettleton et al. 2016). As known, aberrations in the gut microbiota have been associated with the development of obesity, diabetes, metabolic syndrome, and cardiovascular disorders (Kho and Lal 2018).

For some, the phenomenon of metabolic endotoxemia could underlie such shifts. This is a low-grade inflammatory response caused by gut dysbiosis (Cani et al. 2007). Lipopolysaccharides (LPS) released from dead bacteria in the gut are taken into the circulatory system, binding to cluster of differentiation (CD)-14 proteins, nucleotide oligomerization domains (NODs), and Toll-like receptors (TLRs) on macrophages and dendritic cells (Tanti et al. 2013). CD14 proteins are modulators in animals for insulin sensitivity, directly linked to hyperglycemia, hyperinsulinemia, and weight gain (Liauchonak et al. 2019). Overproduction of inflammatory cytokines after the activation of the immune cells could trigger subsequent signaling pathways in metabolic cells to operate, ultimately leading to insulin desensitization, downregulating of proteins for glucose transport, and other metabolic effects (Cani et al. 2007).

Cardiovascular and Kidney Disease

The Nurses' Health Study revealed that an increased risk for both coronary heart disease (CHD) and chronic kidney disease (CKD) was associated with the consumption of more than two diet sodas per day, when compared to consuming fewer than one diet soda per month (Fung et al. 2009). After controlling for weight, this association disappeared (Fung et al. 2009; Lin and Curhan 2011). Risks for specific medical conditions only existed when HIS was consumed in beverages (did not include foods with HIS).

Total and Cardiovascular Mortality

A study published in early 2019, including 37,716 men and 80,647 women, examined the relationship between the long-term use of SSBs and HIS, and the risk of mortality in the United States (Malik et al. 2019). A higher risk for total and cardiovascular disease mortality was unearthed (Malik et al. 2019). Besides, a significant association between mortality and HIS consumption was identified in females via cohort-specific analyses, urging further research (Malik et al. 2019).

Reverse causation may explain some of the inconsistencies within the data (Fitch and Keim 2012; Dugan et al. 2019). For example, overweight or obese individuals are more likely to consume HIS-containing foods and beverages. The same population is also more likely to develop type 2 diabetes mellitus (Fitch and Keim 2012; Dugan et al. 2019). Based on the ADA/AHA joint position statement, HIS can safely be used to replace sugar-containing beverages and products. However, due to the inconsistent results in some studies, water may be the best first choice beverage for those attempting to lose weight. More research is needed to fully understand the consequences of long-term HIS consumption.

Ongoing Studies and Future Perspectives

Sylvetskym et al. (2016) suggest eight methodological areas to consider in future studies, which are:

1. Population selection
2. Control group
3. Habitual consumption of HIS
4. Variation among HIS
5. Consumption route
6. Secondary study outcomes other than weight
7. Eucaloric versus lower calorie consumption
8. Translational research

Studies that include a broad range of age, gender, body weight, and race/ethnicity subjects would be helpful for interindividual variations (Sylvetsky et al. 2016). Few studies are evaluating HIS in the perinatal period, children, and elderly (Sylvetsky et al. 2016). A control group in HIS research is often lacking. Many lack controls using water, unflavored seltzer water, or foods without sweetener (Sylvetsky et al. 2016). HIS is mostly examined in the setting of weight loss/management. More reviews on HIS in a population not restricting calories would be helpful.

Evaluating the effects of HISs in specific patient groups with metabolic diseases and underlying inflammatory states would clarify the microbial alterations associated with the disease states (Sylvetsky et al. 2016). Studies that compare the effects of different types of HIS are needed, since, for example, aspartame, sucralose, and stevia may have different impacts on postprandial glycemia, obesity, and other secondary outcomes. Also, most individuals that consume HISs would obtain them from a variety of products (e.g., tea, coffee, sodas, and foods). Further studies should include multiple

HIS-containing products, which will translate into a more real-world representation and application (Toews et al. 2019).

References

Ahmad SY, Friel JK, MacKay DS (2020) The effect of the artificial sweeteners on glucose metabolism in healthy adults: a randomized double-blinded crossover clinical trial. Appl Physiol Nutr Metab 45(6):606–612. https://doi.org/10.1139/apnm-2019-0359

Azad MB, Abou-Setta AM, Chauhan BF et al (2017) Nonnutritive sweeteners and cardiometabolic health: a systematic review and metaanalysis of randomized controlled trials and prospective cohort studies. CMAJ 189:E929–E939

Björck Y, Granfeldt H, Liljeberg J, Tovar NG (1994) Asp. Food properties affecting the digestion and absorption of carbohydrates. Am J Clin Nutr 59:699S–705S

Bonnet F, Tavenard A, Esvan M et al (2018) Consumption of a carbonated beverage with high-intensity sweeteners has no effect on insulin sensitivity and secretion in nondiabetic adults. J Nutr 148:1293–1299

Brown RJ, Walter M, Rother KI (2009) Ingestion of diet soda before a glucose load augments glucagon-like peptide-1 secretion. Diabetes Care 32:2184–2186

Brown RJ, Walter M, Rother KI (2012) Effects of diet soda on gut hormones in youths with diabetes. Diabetes Care 35(5):959–964

Cani PD, Amar J, Iglesias MA, Poggi M, Knauf C, Bastelica D, Neyrinck AM, Fava F, Tuohy KM, Chabo C et al (2007) Metabolic endotoxemia initiates obesity and insulin resistance. Diabetes 56:1761–1772

Davidson TL, Martin AA, Clark K, Swithers SE (2011) Intake of high-intensity sweeteners alters the ability of sweet taste to signal caloric consequences: implications for the learned control of energy and body weight regulation. Q J Exp Psychol 64(7):1430–1441

Davis JN, Asigbee FM, Markowitz AK, Landry MJ, Vandyouse S, Khazaee E, Ghaddar R, Goran MI (2018) Consumption of artificial sweetened beverages associated with adiposity and increasing HbA1c in Hispanic youth. Clin Obes 8:236–243

de Koning L, Malik VS, Kellogg MD, Rimm EB, Hu FB (2012) Sweetened beverage consumption, incident coronary heart disease, and biomarkers of risk of men. Circulation 125:1735–1741

de la Hunty A, Gibson S, Ashwell M (2006) A review of the effectiveness of aspartame in helping with weight control. Nutr Bull 31:115–128

Deutsch R (1974) Conditioned hypoglycemia: a mechanism for saccharin-induced sensitivity to insulin in the rat. J Comp Physiol Psychol 86(2):350–358

Dugan JA, Chiu JW, Shubrook JH, Young CF (2019) What we know—and don't—about non-nutritive sweeteners. J Fam Pract 68(6):310–315

Evert AB, Dennison M, Gardner C, Garvey WT, Lau KHK, MacLeod J, Mitri J, Pereira R, Rawlings K, Robinson S, Saslow L, Uelmen S, Urbanski P, Yancy W (2019) Nutrition therapy for adults with diabetes or prediabetes: a consensus report. Diabetes Care 42:731–754

Fitch C, Keim KS (2012) Position of the academy of nutrition and dietetics: use of nutritive and nonnutritive sweeteners. J Acad Nutr Diet 112(5):739–758

Food and Drug Administration (2019) Additional information about high intensity sweeteners permitted use in food. https://www.fda.gov/food/food-additives-petitions/additional-information-about-high-intensity-sweeteners-permitted-use-food-united-states. Accessed 1 Dec 2019

Ford HE, Peters V, Martin NM et al (2011) Effects of oral ingestion of sucralose on gut hormone response and appetite in healthy normal-weight subjects. Eur J Clin Nutr 65(4):508–513

Fowler SP, Williams K, Hazuda HP (2015) Diet soda intake is associated with long-term increases in waist circumference in a biethnic cohort of older adults: the San Antonio Longitudinal Study of Aging. J Am Geriatr Soc 63(4):708–715

Fung TT, Malik V, Rexrode KM et al (2009) Sweetened beverage consumption and risk of coronary heart disease in women. Am J Clin Nutr 89:1037–1042

Gardner C, Wylie-Rosett J, Gidding SS, Steffen LM, Johnson RK, Reader D, Lichtenstein AH (2012) Nonnutritive sweeteners: current use and health perspective—a scientific statement from the American Heart Association and the American Diabetes Association. Circulation 126:506–519

Higgins KA, Considine RV, Mattes RD (2018) Aspartame consumption for 12 weeks does not affect glycemia, appetite, or bodyweight of healthy, lean adults in a randomized controlled trial. J Nutr 148:650–657

Jang HJ, Kokrashvili Z, Theodorakis MJ, Carlson OD, Kim BJ, Zhou J, Kim HH, Xu X, Chan SL, Juhaszova M, Bernier M, Mosinger B, Margolskee RF, Egan JM (2007) Gut-expressed gustducin and taste receptors regulate secretion of glucagon-like peptide-1. Proc Natl Acad Sci U S A 104 (38):15069–15074

Just T, Pau W, Engel U, Hummel T (2008) Cephalic phase insulin release in healthy humans after taste stimulation? Appetite 51(3):622–627

Kho ZY, Lal SK (2018) The human gut microbiome—a potential controller of wellness and disease. Front Microbiol 9:1835

Laverty AA, Magee L, Monteiro CA, Saxena S, Millett C (2015) Sugar and artificially sweetened beverage consumption and adiposity changes: national longitudinal study. Int J Behav Nutr Phys Act 12:137

Leicht CA, James LJ, Briscoe JHB, Hoekstra SP (2019) Hot water immersion acutely increases postprandial glucose concentrations. Phys Rep 7(20):e14223

Liauchonak I, Qorri B, Dawoud F, Riat Y, Szewczuk MR (2019) Non-nutritive sweeteners and their implications on the development of metabolic syndrome. Nutrients 11:644–663

Lin J, Curhan GC (2011) Associations of sugar and artificially sweetened soda with albuminuria and kidney function decline in women. Clin J Am Soc Nephrol 6:160–166

Ma J, Bellon M, Wishart JM et al (2009) Effect of the artificial sweetener, sucralose, on gastric emptying and incretin hormone release in healthy subjects. Am J Physiol Gastrointest Liver Physiol 296:G735–G739

Ma J, Chang J, Checklin HL et al (2010) Effect of the artificial sweetener, sucralose, on small intestinal glucose absorption in healthy human subjects. Br J Nutr 104:803–806

Malek AM, Hunt KJ, DellaValle DM et al (2018) Reported consumption of low-calorie sweetener in foods, beverages, and food and beverage additions by US adults: NHANES 2007–2012. Curr Dev Nutr 2: nzy054

Malik VS, Li Y, Pan A et al (2019) Long-term consumption of sugar-sweetened and artificially sweetened beverages and risk of mortality in US adults. Circulation 139:2113–2125

Mann S, Yuschak M, Amyes S, Aughton P, Finn J (2000) A carcinogenicity study of sucralose in CD-1 mouse. Food Chem Toxicol 38(2):S91–S98

Margolskee RF, Dyer J, Kokrashvili Z, Salmon KS, Il-egems E, Daly K, Maillet EL et al (2007) T1R3 and gustducin in gut sense sugars to regulate expression of Na+−glucose cotransporter 1. Proc Natl Acad Sci U S A 104(38):15075–15080

Mattes RD, Popkin BM (2009) Nonnutritive sweetener consumption in humans: effects on appetite and food intake and their putative mechanisms. Am J Clin Nutr 89:1–14

National Cancer Institute Website (2019) Artificial sweeteners and cancer. https://www.cancer.gov/about-cancer/causes-prevention/risk/diet/artificial-sweeteners-fact-sheet. Accessed 28 Dec 2019

Nettleton JE, Reimer RA, Shearer J (2016) Reshaping the gut microbiota: impact of low calorie sweeteners and the link to insulin resistance. Physiol Behav 164 (Pt B):488–493

Nichol AD, Salame C, Rother KI, Pepino MY (2019) Effects of sucralose ingestion versus sucralose taste on metabolic responses to an oral glucose tolerance test in participants with normal weight and obesity: a randomized crossover trial. Nutrients 2(1):pii:E29. https://doi.org/10.3390/nu12010029

Parker KJ (1978) Alternative to sugar. The search for an ideal non-nutritive sweetener is almost a century old. Nature 217:493–495

Patel P, Abate N (2013) Body fat distribution and insulin resistance. Nutrients 5(6):2019–2027. https://doi.org/10.3390/nu5062019

Pepino MY, Tiemann CD, Patterson BW, Wice BM, Klein S (2013) Sucralose affects glycemic and hormonal responses to an oral glucose load. Diabetes Care 36 (9):2530–2535

Reid M, Hammersley R, Hill AJ, Skidmore P (2007) Long-term dietary compensation for added sugar: effects of supplementary sucrose drinks over a 4-week period. Br J Nutr 97:193–203

Schiffman SS, Cahn H, Lindley MG (1981) Multiple receptor sites mediate sweetness: evidence from cross adaptation. Pharmacol Biochem Behav 15 (3):377–388

Singh RB, Cornelissen G, Mojto V, Fatima G et al (2019) Effects of circadian restricted feeding on parameters of metabolic syndrome among healthy subjects. Chronobiol Int 17:1–8. https://doi.org/10.1080/07420528.2019.1701817

Soffritti M, Belpoggi F, Esposti D, Lambertini L (2005) Aspartame induces lymphomas and leukaemias in rats. Eur J Oncol 10(2):107–116

Steinert RE, Frey F, Topfer A et al (2011) Effects of carbohydrate sugars and artificial sweeteners on appetite and the secretion of gastrointestinal satiety peptides. Br J Nutr 105:1320–1328

Suez J, Korem T, Zeevi D et al (2014) Artificial sweeteners induce glucose intolerance by altering the gut microbiota. Nature 514:181–186

Swithers SE, Laboy AF, Clark K, Cooper S, Davidson TL (2012) Experience with the high-intensity sweetener saccharin impairs glucose homeostasis and GLP-1 release in rats. Behav Brain Res 233(1):1–14. https://doi.org/10.1016/j.bbr.2012.04.024

Sylvetsky A, Rother K (2016) Trends in the consumption of low-calorie sweeteners. Physiol Behav 164:446–450

Sylvetsky AC, Blau JE, Rother KI (2016) Understanding the metabolic and health effects of low-calorie sweeteners: methodological considerations and implications for future research. Rev Endocr Metab Disord 17(2):187–194

Tanti J-F, Ceppo F, Jager J, Berthou F (2013) Implication of inflammatory signaling pathways in obesity-inducedinsulin resistance. Front Endocrinol 3:181.

Teff KL, Devine J, Engelman K (1995) Sweet taste: effect on cephalic phase insulin release in men. Physiol Behav 57:1089–1095

Temizkan S, Deyneli O, Yasar M, Arpa M, Gunes M, Ya-zici D, Sirikci O, Haklar G, Imeryuz N, Yavuz DG (2015) Sucralose enhances GLP-1 release and lowers blood glucose in the presence of carbohydrate in healthy subjects but not in patients with type 2 diabetes. Eur J Clin Nutr 69(2):162–166

Toews I, Lohner S, de Gaudry DK, Sommer H, Meerphol J (2019) Association between intake of non-sugar sweeteners and health outcomes: systematic review and meta-analyses of randomised and non-randomised controlled trials and observational studies. BMJ 364:k4718

United States Department of Agriculture (2015) 2015–2020 Dietary Guidelines for Americans. https://health.gov/dietaryguidelines/2015/guidelines. Accessed 11 Nov 2019

Van Wymelbeke V, Beridot-Therond ME, de la Gueronniere V, Fantino M (2004) Influence of repeated consumption of beverages containing sucrose or intense sweeteners on food intake. Eur J Clin Nutr 58:154–161

Ventura AK, Mennella JA (2011) Innate and learned preferences for sweet taste during childhood. Curr Opin Clin Nutr Metab Care 14(4):379–384

World Health Organization (2019) Obesity and overweight. https://www.who.int/news-room/fact-sheets/detail/obesity-and-overweight. Accessed 1 Dec 2019

Wu T, Zhao BR, Bound MJ, Checklin HL, Bellon M, Little TJ, Young RL et al (2012) Effects of different sweet preloads on incretin hormone secretion, gastric emptying, and postprandial glycemia in healthy humans. Am J Clin Nutr 95(1):78–83

Wu T, Zhao BR, Bound MJ, Checklin HL, Bellon M, Little TJ, Young RL et al (2013) Artificial sweeteners have no effect on gastric emptying, glucagon-like peptide-1, or glycemia after oral glucose in healthy humans. Diabetes Care 36(12):202–203

Yang BY, Fan S, Thiering E, Seissler J, Nowak D, Dong GH, Heinrich J (2020) Ambient air pollution and diabetes: a systematic review and meta-analysis. Environ Res 180:108817. https://doi.org/10.1016/j.envres.2019.108817

Personalised Molecular Feedback for Weight Loss

39

Shilpa Tejpal

Contents

Importance of Quantitative Biomarkers of Weight Loss 541

Advantages and Limitations of Internet and Mobile Resources 542
Dietary Tracking 542

Biomarkers for BMI and Adiposity 543
Metabolomic Markers 544

Biomarkers for Foods and Drinks 544
Phytotherapeutic Interventions 545
Appetite Regulation 545

Molecular Biomarkers in the Real World 546
Angiotensin-Converting Enzyme 546
Ongoing Investigations and Future Perspectives 547

References 548

Abstract

Exercise, drugs and all sorts of dietary plans are typical weight loss options that overweight or obese individuals are offered, besides more radical actions such as bariatric surgery. However, weight loss is sometimes incremental, especially at the very start of a regimen and this often leads to the people dropping-out of a dietary programme. Hence, there is a need to understand and identify biomarkers that are affected over a short interval, and use them to provide quantitative biofeedback on the efficiency of a diet. It would also allow personalised optimisation of dieting parameters with relevance to short term.

Keywords

Obesity · Biomarker · Molecular feedback · Personalised information

This chapter has been adapted from Shilpa Tejpal, Metabolic Profiling and Identification of Biomarkers for Weight Loss, University of Warwick, Warwick Medical School, April 2019.

S. Tejpal (✉)
Division of Biomedical Sciences, Warwick Medical School, University of Warwick, Coventry, UK
e-mail: tejpalshilpa@gmail.com

J. Faintuch and S. Faintuch (eds.), *Obesity and Diabetes*, https://doi.org/10.1007/978-3-030-53370-0_39

Importance of Quantitative Biomarkers of Weight Loss

The worldwide prevalence of obesity has nearly tripled between 1975 and 2016 (González-Muniesa et al. 2017; Data and Statistics 2015). Obesity is associated with several comorbidities such as type 2 diabetes, cardiovascular diseases (CVDs), sleep apnoea, metabolic syndrome and certain types of cancer (Abdelaal et al. 2017). Obesity is an incredibly complex disease that is associated with metabolic, genetic and behavioural deregulations. They should be studied in combination with environmental factors, socio-economic status, behaviour, education and genotype. All these factors affect caloric intake, thermogenesis, lipid utilisation, nutrient turnover and differential fat storage in tissues (Hruby and Hu 2015).

Several blood and physical markers are used to measure weight loss, such as hip/waist ratio, BMI, body fat and/or weight for physical markers, along with high-density lipoprotein (HDL), low-density lipoprotein (LDL) and fasting glucose in serum. In most diets, the weight lost is rapidly regained after the end of the diet (the "yo-yo" effect); therefore, investigating possibilities to use molecular information to predict whether the weight is kept off on a long-term is an interesting question. There is a range of approaches available to lose weight from surgical or drug-based interventions to different diet programmes.

Weight loss plans involve following a certain diet regime and/or monitoring net caloric intake. To date, no dietary intervention combines molecular measurements with digital technology, tracking life-style parameters such as food intake or exercise, despite the availability of several food tracker apps and websites. All weight loss programmes give a broad description of the diet plan, which is often arbitrary. In addition, every individual has a different metabolism and responds differently to a diet plan (Tejpal et al. 2019a). Thus, personalised optimisation of the diet plan needs to be understood to determine how diet parameters can be adapted to an individual's metabolism.

Advantages and Limitations of Internet and Mobile Resources

Technological support available to help with weight loss includes diet trackers and activity monitors. These resources provide information on nutrition profile and/or caloric content of different food items. They also allow to maintain records of different meals eaten in a day. Other types of devices such as activity monitors record heart rate, number of steps taken in a day, physical activity levels, sleep patterns and blood pressure. There are also several support networks available like Weight Watchers and Slimming World. A support network is a platform where people trying to lose weight come together sharing their stories, challenges, victories, feelings, and providing support to others to help achieve their goals. People also share their food recipes, diet plans and physical activity routines on such networks. Such platforms help people to remain enthusiastic, motivated, set goals, track progress and stay engaged in the weight loss plan being followed by them (Soeliman and Azadbakht 2014). Many studies have shown that dieting is more successful when supported by social interactions, such as exercising with a friend or dieting as a group (Higgs and Thomas 2016; Lemstra et al. 2016).

Dietary Tracking

Tracking and recording eating patterns is a well-known method for effective weight management (Liu et al. 2011, 2012). Traditionally, people used paper forms of the dietary questionnaire, which were considered tiring (Moshfegh et al. 2008). Nowadays, there are several internet (Inc A 2016; Abel et al. 2001) and mobile versions for personal tracking of food eaten (Vereecken et al. 2005; 24th European Congress on Obesity (ECO2017) 2017). Mis-reporting of food intake is a well-known problem (Moshfegh et al. 2008; Conway et al. 2003) in any platform because of its dependency upon honesty, motivation and self-awareness (Liu et al. 2012). There can be under-reporting of the food intake due to unconscious bias of an individual (Scisco et al. 2011). Energy expenditure on the other hand can potentially be tracked without bias using activity monitors, but they do not provide a direct link to weight loss. They may show the number of steps walked or run, calories consumed and/or heart-beat/min (Shilpa Tejpal et al. 2019). Even if a device that accurately measures caloric intake and expenditure is widely available, the information may not be sufficient to motivate users to make changes in their behaviour that would result in weight loss. Also, these devices are not 100% tailored to an individual. They only give an estimate of the calories burnt based on weight and height, sometimes including gender and age, but they do not include any information on fat distribution or hormonal patterns. This information is therefore generic and thus there is need for identification of biomarkers to provide tailored feedback. This may help us understand as to why some individuals fail and/or find it difficult to lose weight, despite of keeping records of food eaten in a day (Teixeira et al. 2012).

Biomarkers for BMI and Adiposity

BMI is often used to identify overweight (25–29.9 kg/m^2), obesity (≥30 kg/m^2) and severe obesity (≥40 kg/m^2); however, it may not be the most efficient method as it is unable to provide information about body fat distribution (Obesity W. Obesity Classification 2019). For example, elderly people lose muscle with age making BMI a less accurate indicator of body composition (Adab et al. 2018). BMI can be complemented with waist circumference (WC) or biometric impedance measurements, which gives a better representation of adiposity distribution to differential fat under the skin (subcutaneous) and intra-abdominal (visceral) (Seven et al. 2016). According to the National Institute for Health and Care Excellence (NICE) guidelines, WC >94 cm (male) and WC ≥85 cm (females) indicates a higher deposition of visceral fat (Obesity W. Obesity Classification 2019). Biometric impedance involves estimation of body composition, particularly body fat. The health risk associated with obesity is identified by combining these methods, as shown in Table 39.1.

Table 39.1 Identification of risk level for obesity related co-morbidities by combining BMI and Waist Circumference

BMI (kg/m^2)	Waist circumference (cm)	Risk of co-morbidities
<18.5	<94 (men) <85 (women)	Low
	≥94 (men) ≥85 (women)	Average
18.5–22.9	<94 (men) <85 (women)	Average
	≥94 (men) ≥85 (women)	Increased
23–24.9	<94 (men) <85 (women)	Increased
	≥94 (men) ≥85 (women)	Moderate
25–29.9	<94 (men) <85 (women)	Moderate
	≥94 (men) ≥85 (women)	Severe
≥30	<94 (men) <85 (women)	Severe
	≥94 (men) ≥85 (women)	Very severe

Each BMI category in association with waist circumference is used to identify the risk for developing obesity related co-morbidities (Obesity W. Obesity Classification 2019)

Normal adipose tissue expresses anti- and pro-inflammatory regulators to allow functioning of adipose tissue (Deng et al. 2016). Interleukin (IL) 4 is a major anti-inflammatory regulator that controls the expression of T helper (TH) type 2 cells, regulatory T cells (Tregs) and macrophages. Macrophage phenotype is divided into groups, namely classically activated macrophages (M1) and alternatively activated macrophages (M2). Expression of M2 macrophages by IL4 promotes systemic insulin sensitivity in lean mice (Odegaard et al. 2008; Ricardo-Gonzalez et al. 2010). Many pro-inflammatory marker levels increase with higher adiposity. A cross-sectional meta-study has found association between C-reactive protein (CRP), a marker for systemic inflammation, and free fat mass (Cox et al. 2015; Choi et al. 2013; Eagan et al. 2010). Increased levels of CRP and triglyceride (TG) are found in overweight women in comparison to normal weight women (Klisic et al. 2014). With nutrition overload, lipid activation and increased energy storage, there is a switch from M2 to M1 (Deng et al. 2016).

Adipocytes secrete adipokines such as leptin and adiponectin, which have opposing effects on immune cell functions. Leptin has pro-inflammatory effects and increases with elevated nutrition uptake and stimulates production of IL 1, 6 and 12, and tumour necrosis factor α (TNFα), while adiponectin has anti-inflammatory effects and decreases with excess adiposity (Deng et al. 2016). The overexpressed pro-inflammatory cytokines are found in obese individuals, and one-third of total circulating concentrations of IL-6 originate from adipose tissue (Ellulu et al. 2017). Levels of IL 6 decrease after 12 months of bariatric surgery in morbidly obese patients (Illán-Gómez et al. 2012). The abundant energy substrates in obesity lead to increased reactive oxygen species (ROS) signalling and mitochondrial dysfunction (McMurray et al. 2016). This is also associated with decreased insulin sensitivity (McMurray et al. 2016). c-Jun N-terminal kinases (JNK) and nuclear factor kappa-light-chain-enhancer of activated B cells (NF-κB) are regulated by ROS and are associated with obesity-induced insulin resistance (Finkel 2011).

Metabolomic Markers

Several metabolomics studies involving untargeted proton (^{1}H) nuclear magnetic resonance spectroscopy (NMR) and ion exchange chromatography (IEC), on obese human and mice urine samples, have identified metabolites associated with BMI and adiposity (Elliott et al. 2015; Du et al. 2013), summarised in Fig. 39.1.

Biomarkers for Foods and Drinks

Double-labelled water (water with isotopes of deuterium and oxygen 18) is used to measure energy expenditure (EE) in an individual (Potischman and Freudenheim 2003). Urine samples are collected to determine the rate of disappearance of each isotope from the body via mass spectrometry. This disappearance rate is further used to indirectly calculate carbon dioxide production to estimate total EE. EE is lower in obese and overweight population in comparison to healthy individuals. To be in an energy balance, energy intake (EI) should be equal to EE. Increased EE is associated with under-reporting of food intake. Urine nitrogen is used as a biomarker for protein intake through dietary means (Bingham 2003). Constant dietary intake over longer periods is also associated with daily nitrogen turn-over and excretion (Bingham 2003). Hydrocarbons in breath are used as a means to measure lipid peroxidation. Peroxidation of *n*-6 and *n*-3 fatty acid releases pentane and ethane, respectively, into breath (Mayne 2003). Supplementation with β carotene reduces breath pentane levels significantly (Gottlieb et al. 1993). Markers for α and β carotene, lycopene and β-cryptoxanthin are correlated with serum cholesterol levels (Fraser et al. 2016). Biomarkers for different foods include *N*-methylnicotinic acid (NMNA), a niacin-related (vitamin B3) metabolite marker for coffee drinking (Lang et al. 2011), proline betaine for citrus fruit consumption (Heinzmann et al. 2010), and *O*-acetyl carnitine for red meat intake (Mozaffarian et al. 2011).

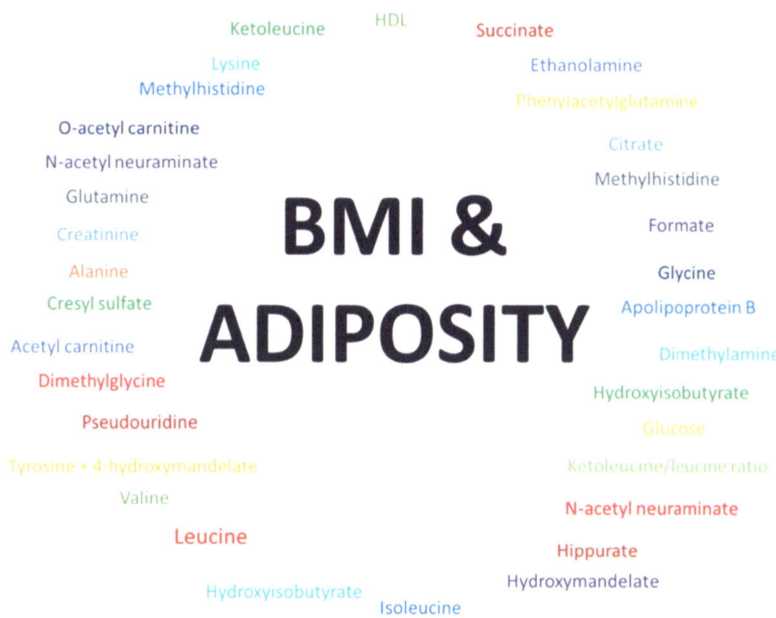

Fig. 39.1 Urinary metabolic signatures of BMI and Adiposity in urine. The information for the metabolic signatures was obtained from (Bouatra et al. 2013)

Phytotherapeutic Interventions

Food components can modulate hunger, satiety and EE. In particular, benefits of plant bioactive compounds on metabolism have become a focus of multidisciplinary studies; and several plant extracts are studied in the prevention of obesity (Lu et al. 2013). Based on their nutritional value, such extracts have physiological benefits and reduce the risk of chronic diseases (Estaquio et al. 2008; Liu et al. 2004). Phenols, anthocyanins and tannins, found in tea, berries and peas, have been demonstrated to decrease lipids, through the inhibition of lipase activity (Sergent et al. 2012). Starch with high glycaemic index is associated with weight gain (Scribner et al. 2007, 2008) by causing increased insulin secretion (Aller et al. 2011).

Diets rich in protein from vegetables (e.g., legumes) induce satiety and suppress intake and appetite by increasing the hormone peptide tyrosine (PYY) and decreasing ghrelin (Nilsson et al. 2013; Lonnie et al. 2018). Raspberries decrease serum glucose, insulin levels, leptin and body weight in mice fed high-fat diet (Prior et al. 2010). Citrus fruit species are associated with decreased inflammation and reduced oxidative stress markers in humans (Coelho et al. 2013). Other studies have found bergamot, grapefruit and orange juice to be associated with decreased total cholesterol, low density lipoprotein (LDL) and glucose concentration in humans (Mollace et al. 2011; Dow et al. 2012; Aptekmann and Cesar 2013). Similar results have also been found in rats after lemon juice administration (Oboh et al. 2015). To summarise, the different plant-based compounds that are present in foods are shown in Table 39.2.

Appetite Regulation

Appetite can be expressed in terms of physical behaviour, peripheral physiology and central nervous system functions (Stafleu et al. 2004). People stop eating due to absence of hunger or because they feel full (Mook and Votaw 1992). Diet-induced thermogenesis (DIT) is the amount of energy used above the basal rate during the breakdown of food in the body. DIT and satiety

Table 39.2 Bioactive compounds currently prescribed or investigated to treat obesity (Solas et al. 2016)

Molecule	Mechanism
Appetite	
Saponins (gingenoside)	Satiety
Fibre and polysaccharides (diverse)	
Terpenes (geniposide)	
Steroidal glycosides (P57A53)	
Polyphenols (proantho-cyanidins)	
Proteins (diverse)	
Energy expenditure	
Alkaloids (capsaicin)	Thermogenesis
Flavonoids (diverse)	
Polyphenols (catechin)	
Fatty acids (MUFA, PUFA, ...)	Beta-oxidation lipolysis/anti-lipogenesis
Curcumin	Induction of brown fat-like phenotype
Metabolism	
Saponins (astragaloside IV, gingenoside)	Lipase inhibition, fat depletion
Fibre	Lipid uptake reduction; reduction of energy dietary value; secretion of anorectic peptides
Pseudo-tetrasaccharide, acarbose	Amylase inhibitor
Polysaccharides (diverse)	Fat depletion
Alkaloids (berberine betaine, piperine, capsaicin)	Browning fat depletion
Polyphenols (resveratrol, proanthocyanidins, epicatechin, diverse from tea)	Adipogenesis inhibition; fat depletion and absorption lipase/amylase inhibition

The table describes the different compounds, their source and mechanism of action

are positively correlated and DIT decreases hunger (Westerterp-Plantenga et al. 1997, 1999; Crovetti et al. 1998). Ghrelin is also associated with hunger. Infusion of ghrelin in healthy subjects leads to enhanced appetite and energy intake levels (Wren et al. 2001). Plasma concentration for ghrelin decreases after administration of glucose (Mizuta et al. 2002) and carbohydrates, (Matthys et al. 2003) but there is no suppression after high-fat administrations (Matthys et al. 2003; Klok et al. 2007). Insulin and glucose are therefore indirectly related to satiety. Neuropeptide Y and Agouti-related protein (AgRP) are located in the hypothalamus, their expression is activated in conditions of fasting and negative energy balance, and they increase the hunger level (Schwartz et al. 2017). Pro-opiomelanocortin (POMC) neurons release the similarly named anorexic neuropeptide that reduces food intake (Schwartz et al. 1997).

Molecular Biomarkers in the Real World

Angiotensin-Converting Enzyme

To find a marker for long-term weight loss, a panel of biomarkers was tested before and after an 8-week diet, and weight loss maintenance was determined after 6 months (Wang et al. 2011). In this study, angiotensin-converting enzyme (ACE) was shown (amongst the extensive blood profiling for diverse protein and steroid hormones) to be the only potential predictor for sustained weight loss (Wang et al. 2011). At the end of the 8-week intervention, individuals with weight loss displayed decreased ACE concentration (~12%) (Wang et al. 2011), which supports the previously reported decreased ACE activity in overweight/obese adults after dieting (Engeli et al. 2005; Harp et al. 2002). ACE is a zinc metallopeptidase enzyme involved in the conversion of angiotensin (Ang) I to angiotensin II (Wang et al. 2011; Brewster and Perazella 2004; Tejpal et al. 2019b). Ang I is obtained by cleavage of angiotensinogen (AGT) with the help of renin. Ang II is well known for its role in increased blood pressure and retention of salt and water (Brewster and Perazella 2004).

Lactate and Insulin

A molecular feedback approach to assist dieting efforts and behavioural responses of people using a web- and mobile-based application was investigated. Skipping a meal in a day regardless of which one resulted in consistent weight loss for that day, in comparison to control days in which any number of meals was allowed. Insulin, ACE and lactate levels in urine showed correlations to BMI, caloric patterns and weight difference (Tejpal et al. 2018, 2019a).

Lactate is the by-product of glucose utilisation by the organs and tissues during hypoxia or glycolysis (Arriarán et al. 2015). It helps to modulate oxygen release in hypoxic tissues and acts as a substrate for gluconeogenesis (Frayn et al. 1990; Kerckhoffs et al. 1998). It also acts as a substrate for lipogenesis in liver and other tissues (Kevin O'Hea and Leveille 1969; Granata et al. 1976; Lopaschuk et al. 1992). Release of lactate in adipose tissue is often attributed to hypoxia, acidosis and stress (Lee Dong et al. 2015; Sestoft et al. 1982). Lactate production is known to increase in vitro in large adipocytes from obese animals, and can reach up to 50–70% of glucose metabolised (Crawford et al. 2010). Indeed, baseline lactate concentrations are higher in obese subjects as compared to lean subjects (Newby et al. 1990).

Insulin is a small-peptide hormone, whose secretion is stimulated by increased levels of glucose in blood (Saltiel and Kahn 2001). Insulin regulates blood glucose homeostasis by increasing glucose uptake in muscles and fat cells and inhibiting its production in liver (Saltiel and Kahn 2001). It varies with food intake in terms of meal composition, timing and quantity (Van Cauter et al. 1992; Stevenson et al. 2005; Scheer et al. 2009). Moreover, its concentration and action are under circadian regulation, giving rise to "afternoon diabetes" (Van Cauter et al. 1992; Scheer et al. 2009; Jarrett and Keen 1969). This finding has been used to suggest that diet plans should incorporate meal timing (Hampton and Johnston 2014) with the breakfast meal being the largest meal, as opposed to the most common behaviour of highest food consumption in the evening (Hampton and Johnston 2014; Gill and Panda 2015). Indeed, it is shown that a high-carbohydrate breakfast promotes greater weight loss than a diet low in carbohydrates by reducing food cravings and increased satiety (Jakubowicz et al. 2012, 2013; Almoosawi et al. 2013).

Ongoing Investigations and Future Perspectives

A large-scale 2-week PREDICT study involving 1100 UK and US adults (60% twins) studied the effects of markers such as blood glucose, fat and insulin to genetics, environment, meal timing, exercise, sleep and gut microbiome. Even for the same meal, the blood markers hugely varied in different individuals. This variation is partially explained by genetic factors corresponding to 50%, 30% and 20% for glucose, insulin and triglyceride, respectively. The identical twins have only 37% of shared gut microbiome, which is only 2% more than the percentage shared by two unrelated individuals (When it Comes to Food, One Size Doesn't Fit All! 2019). Less than 40% of variations between individuals eating the same caloric load are explained by the nutrients content on food labels (Cutler 2019). Time of the meal also affects the individuals' responses.

The data are being used by ZOE company together with universities (ZOE 2019) to develop a machine learning algorithm that will predict a person's response to the food eaten. Another large-scale study has been launched at the Swiss Federal Institute of Technology named Food & You. The aim of the study is to confirm the theory that individuals respond differently to the same meal eaten, as based on the blood glucose measurements (Stéphanie Milliquet 2019). It is a digital study conducted entirely online, where participants record their daily food intake for 2 weeks on MyFoodRepo. They also use a smartwatch or an app to track their physical activity and sleep. Blood sugar levels are also monitored, and a stool sample is collected once to analyse gut microbiome. The data collected will then be used to generate an algorithm that can predict blood sugar levels after a meal.

Personalised Nutrition

An individual having a potato chip triggering a six times higher triglyceride peak, can be steered towards a low-fat snack option based on the personalised information. The current guidelines are based on the questionnaire filled by individuals. The information provided can be mis- or under-reported creating a bias in the data generated. A detailed view of metabolic differences, wearable sensors and machine learning can help structure the personalised guidance options for people.

References

24th European Congress on Obesity (ECO2017) (2017) Porto, Portugal, May 17–20, 2017: Abstracts. Obes Facts 10(suppl 1):1–274

Abdelaal M, le Roux CW, Docherty NG (2017) Morbidity and mortality associated with obesity. Ann Transl Med 5(7):161–161

Abel ED, Peroni O, Kim JK et al (2001) Adipose-selective targeting of the GLUT4 gene impairs insulin action in muscle and liver. Nature 409(6821):729–733

Adab P, Pallan M, Whincup PH (2018) Is BMI the best measure of obesity? BMJ 360:k1274

Aller EEJG, Abete I, Astrup A, Martinez JA, van Baak MA (2011) Starches, sugars and obesity. Nutrients 3 (3):341–369

Almoosawi S, Prynne CJ, Hardy R, Stephen AM (2013) Time-of-day and nutrient composition of eating occasions: prospective association with the metabolic syndrome in the 1946 British birth cohort. Int J Obes 37(5):725–731

Aptekmann NP, Cesar TB (2013) Long-term orange juice consumption is associated with low LDL-cholesterol and apolipoprotein B in normal and moderately hypercholesterolemic subjects. Lipids Health Dis 12:119–119

Arriarán S, Agnelli S, Sabater D, Remesar X, Fernández-López JA, Alemany M (2015) Evidences of basal lactate production in the main white adipose tissue sites of rats. Effects of sex and a cafeteria diet. PLoS One 10(3):e0119572

Bingham SA (2003) Urine nitrogen as a biomarker for the validation of dietary protein intake. J Nutr 133 (3):921S–924S

Bouatra S, Aziat F, Mandal R et al (2013) The human urine metabolome. PLoS One 8(9):e73076

Brewster UC, Perazella MA (2004) The renin-angiotensin-aldosterone system and the kidney: effects on kidney disease. Am J Med 116(4):263–272

Choi J, Joseph L, Pilote L (2013) Obesity and C-reactive protein in various populations: a systematic review and meta-analysis. Obes Rev 14(3):232–244

Coelho RCLA, Hermsdorff HHM, Bressan J (2013) Anti-inflammatory properties of Orange juice: possible favorable molecular and metabolic effects. Plant Foods Hum Nutr 68(1):1–10

Conway JM, Ingwersen LA, Vinyard BT, Moshfegh AJ (2003) Effectiveness of the US Department of Agriculture 5-step multiple-pass method in assessing food intake in obese and nonobese women. Am J Clin Nutr 77(5):1171–1178

Cox AJ, West NP, Cripps AW (2015) Obesity, inflammation, and the gut microbiota. Lancet Diabetes Endocrinol 3(3):207–215

Crawford SO, Hoogeveen RC, Brancati FL et al (2010) Association of blood lactate with type 2 diabetes: the atherosclerosis risk in communities carotid MRI study. Int J Epidemiol 39(6):1647–1655

Crovetti R, Porrini M, Santangelo A, Testolin G (1998) The influence of thermic effect of food on satiety. Eur J Clin Nutr 52:482

Cutler N (2019) Stuy offers more evidence for personalised diets even with identical twins. https://www.nutraingredients.com/Article/2019/06/17/Study-offers-more-evidence-for-personalised-diets-even-with-identical-twins. Accessed 2 Dec 2019

Data and Statistics (2015). http://www.euro.who.int/en/health-topics/noncommunicable-diseases/obesity/data-and-statistics. Accessed 24 Apr 2018

Deng T, Lyon CJ, Bergin S, Caligiuri MA, Hsueh WA (2016) Obesity, inflammation, and cancer. Ann Rev Pathol Mechan Dis 11(1):421–449

Dow CA, Going SB, Chow H-HS, Patil BS, Thomson CA (2012) The effects of daily consumption of grapefruit on body weight, lipids, and blood pressure in healthy, overweight adults. Metab Clin Exp 61(7):1026–1035

Du F, Virtue A, Wang H, Yang X-F (2013) Metabolomic analyses for atherosclerosis, diabetes, and obesity. Biomark Res 1(1):17

Eagan TML, Aukrust P, Ueland T et al (2010) Body composition and plasma levels of inflammatory biomarkers in COPD. Eur Respir J 36(5):1027

Elliott P, Posma JM, Chan Q et al (2015) Urinary metabolic signatures of human adiposity. Sci Transl Med 7 (285):285ra62

Ellulu MS, Patimah I, Khaza'ai H, Rahmat A, Abed Y (2017) Obesity and inflammation: the linking

mechanism and the complications. Arch Med Sci 13 (4):851–863
Engeli S, Böhnke J, Gorzelniak K et al (2005) Weight loss and the renin-angiotensin-aldosterone system. Hypertension 45(3):356–362
Estaquio C, Castetbon K, Kesse-Guyot E et al (2008) The French National Nutrition and health program score is associated with nutritional status and risk of major chronic diseases. J Nutr 138(5):946–953
Finkel T (2011) Signal transduction by reactive oxygen species. J Cell Biol 194(1):7–15
Fraser GE, Jaceldo-Siegl K, Henning SM et al (2016) Biomarkers of dietary intake are correlated with corresponding measures from repeated dietary recalls and food-frequency questionnaires in the Adventist health Study-2. J Nutr 146(3):586–594
Frayn KN, Coppack SW, Smith U (1990) Evidence for lactate production by human adipose tissue in vivo. Diabetologia 33(12):740–741
Gill S, Panda S (2015) A smartphone app reveals erratic diurnal eating patterns in humans that can be modulated for health benefits. Cell Metab 22(5):789–798
González-Muniesa P, Mártinez-González M-A, Hu FB et al (2017) Obesity. Nat Rev Dis Primers 3:17034
Gottlieb K, Zarling EJ, Mobarhan S, Bowen P, Sugerman S (1993) β-Carotene decreases markers of lipid peroxidation in healthy volunteers. Nutr Cancer 19 (2):207–212
Granata AL, Midrio M, Corsi A (1976) Lactate oxidation by skeletal muscle during sustained contraction in vivo. Pflugers Arch 366(2):247–250
Hampton SM, Johnston JD (2014) Probing the diurnal regulation of glycemic control. J Diabetes Complicat 28(6):751–752
Harp JB, Henry SA, DiGirolamo M (2002) Dietary weight loss decreases serum angiotensin-converting enzyme activity in obese adults. Obes Res 10(10):985–990
Heinzmann SS, Brown IJ, Chan Q et al (2010) Metabolic profiling strategy for discovery of nutritional biomarkers: proline betaine as a marker of citrus consumption. Am J Clin Nutr 92(2):436–443
Higgs S, Thomas J (2016) Social influences on eating. Curr Opin Behav Sci 9:1–6
Hruby A, Hu FB (2015) The epidemiology of Obesity: a big picture. PharmacoEconomics 33(7):673–689
Illán-Gómez F, Gonzálvez-Ortega M, Orea-Soler I et al (2012) Obesity and inflammation: change in adiponectin, C-reactive protein, tumour necrosis factor-alpha and Interleukin-6 after bariatric surgery. Obes Surg 22(6):950–955
Inc A (2016) Health & Fitness—app store downloads on iTunes. https://itunes.apple.com/us/genre/ios-health-fitness/id6013?mt=8
Jakubowicz D, Froy O, Wainstein J, Boaz M (2012) Meal timing and composition influence ghrelin levels, appetite scores and weight loss maintenance in overweight and obese adults. Steroids 77(4):323–331
Jakubowicz D, Barnea M, Wainstein J, Froy O (2013) High caloric intake at breakfast vs. dinner differentially influences weight loss of overweight and obese women. Obesity 21(12):2504–2512
Jarrett RJ, Keen H (1969) Diurnal variation of Oral glucose tolerance: a possible pointer to the evolution of diabetes mellitus. Br Med J 2(5653):341–344
Kerckhoffs DAJM, Arner P, Bolinder J (1998) Lipolysis and lactate production in human skeletal muscle and adipose tissue following glucose ingestion. Clin Sci 94 (1):71
Kevin O'Hea E, Leveille G (1969) Significance of adipose tissue and liver as sites of FA synthesis in the pig and the efficiency of utilization of various substrates for lipogenesis. J Nutr 99(3):338–344
Klisic A, Vasiljevic N, Simic T, Djukic T, Maksimovic M, Matic M (2014) Association between C-reactive protein, anthropometric and lipid parameters among healthy normal weight and overweight postmenopausal women in montenegro. Lab Med 45(1):12
Klok MD, Jakobsdottir S, Drent ML (2007) The role of leptin and ghrelin in the regulation of food intake and body weight in humans: a review. Obes Rev 8 (1):21–34
Lang R, Wahl A, Stark T, Hofmann T (2011) Urinary N-methylpyridinium and trigonelline as candidate dietary biomarkers of coffee consumption. Mol Nutr Food Res 55(11):1613–1623
Lee Dong C, Sohn Hyun A, Park Z-Y et al (2015) A lactate-induced response to hypoxia. Cell 161 (3):595–609
Lemstra M, Bird Y, Nwankwo C, Rogers M, Moraros J (2016) Weight loss intervention adherence and factors promoting adherence: a meta-analysis. Patient Prefer Adherence 10:1547–1559
Liu S, Serdula M, Janket S-J et al (2004) A prospective study of fruit and vegetable intake and the risk of type 2 diabetes in women. Diabetes Care 27(12):2993
Liu B, Young H, Crowe FL et al (2011) Development and evaluation of the Oxford WebQ, a low-cost, web-based method for assessment of previous 24 h dietary intakes in large-scale prospective studies. Public Health Nutr 14(11):1998–2005
Liu J, Johns E, Atallah L et al (2012) An intelligent food-intake monitoring system using wearable sensors. Paper presented at: 2012 Ninth International Conference on Wearable and Implantable Body Sensor Networks; 9–12 May 2012
Lonnie M, Hooker E, Brunstrom JM et al (2018) Protein for life: review of optimal protein intake, sustainable dietary sources and the effect on appetite in ageing adults. Nutrients 10(3):360
Lopaschuk G, Collins-Nakai R, Itoi T (1992) Developmental changes in energy substrate use by the heart. Cardiovasc Res 26(12):1172–1180
Lu Y, Xi W, Ding X et al (2013) Citrange fruit extracts alleviate obesity-associated metabolic disorder in high-fat diet-induced obese C57BL/6 mouse. Int J Mol Sci 14(12):23736–23750
Matthys CC, Cummings DE, Callahan HS et al (2003) Roles of leptin and ghrelin in the loss of body weight

caused by a low fat, high carbohydrate diet. J Clin Endocrinol Metab 88(4):1577–1586
Mayne ST (2003) Antioxidant nutrients and chronic disease: use of biomarkers of exposure and oxidative stress status in epidemiologic research. J Nutr 133 (3):933S–940S
McMurray F, Patten DA, Harper M-E (2016) Reactive oxygen species and oxidative stress in Obesity—recent findings and empirical approaches. Obesity 24 (11):2301–2310
Mizuta M, Mondal MS, Matsukura S et al (2002) Plasma ghrelin levels in lean and obese humans and the effect of glucose on ghrelin secretion. J Clin Endocrinol Metab 87(1):240–244
Mollace V, Sacco I, Janda E et al (2011) Hypolipemic and hypoglycaemic activity of bergamot polyphenols: from animal models to human studies. Fitoterapia 82 (3):309–316
Mook DG, Votaw MC (1992) How important is hedonism? Reasons given by college students for ending a meal. Appetite 18(1):69–75
Moshfegh AJ, Rhodes DG, Baer DJ et al (2008) The US Department of Agriculture Automated Multiple-Pass Method reduces bias in the collection of energy intakes. Am J Clin Nutr 88(2):324–332
Mozaffarian D, Hao T, Rimm EB, Willett WC, Hu FB (2011) Changes in diet and lifestyle and long-term weight gain in women and men. N Engl J Med 364 (25):2392–2404
Newby FD, Wilson LK, Thacker SV, DiGirolamo M (1990) Adipocyte lactate production remains elevated during refeeding after fasting. Am J Physiol Endocrinol Metab 259(6):E865–E871
Nilsson A, Johansson E, Ekström L, Björck I (2013) Effects of a brown beans evening meal on metabolic risk markers and appetite regulating hormones at a subsequent standardized breakfast: a randomized cross-over study. PLoS One 8(4):e59985–e59985
Obesity W. Obesity Classification (2019). https://www.worldobesity.org/about/about-obesity/obesity-classification. Accessed 28 Feb 2019
Oboh G, Bello FO, Ademosun AO, Akinyemi AJ, Adewuni TM (2015) Antioxidant, hypolipidemic, and anti-angiotensin-1-converting enzyme properties of lemon (Citrus limon) and lime (Citrus aurantifolia) juices. Comp Clin Pathol 24(6):1395–1406
Odegaard JI, Ricardo-Gonzalez RR, Red Eagle A et al (2008) Alternative M2 activation of Kupffer cells by PPARdelta ameliorates obesity-induced insulin resistance. Cell Metab 7(6):496–507
Potischman N, Freudenheim JL (2003) Biomarkers of nutritional exposure and nutritional status: an overview. J Nutr 133(3):873S–874S
Prior RL, Wilkes S, Rogers T, Khanal RC, Wu X, Howard LR (2010) Purified blueberry anthocyanins and blueberry juice alter development of Obesity in mice fed an obesogenic high-fat diet. J Agric Food Chem 58 (7):3970–3976
Ricardo-Gonzalez RR, Red Eagle A, Odegaard JI et al (2010) IL-4/STAT6 immune axis regulates peripheral nutrient metabolism and insulin sensitivity. Proc Natl Acad Sci U S A 107(52):22617–22622
Saltiel AR, Kahn CR (2001) Insulin signalling and the regulation of glucose and lipid metabolism. Nature 414:799
Scheer FAJL, Hilton MF, Mantzoros CS, Shea SA (2009) Adverse metabolic and cardiovascular consequences of circadian misalignment. Proc Natl Acad Sci 106 (11):4453
Schwartz MW, Seeley RJ, Woods SC et al (1997) Leptin increases hypothalamic pro-opiomelanocortin mRNA expression in the rostral arcuate nucleus. Diabetes 46 (12):2119
Schwartz MW, Seeley RJ, Zeltser LM et al (2017) Obesity pathogenesis: an Endocrine Society scientific statement. Endocr Rev 38(4):267–296
Scisco JL, Muth ER, Dong Y, Hoover AW (2011) Slowing bite-rate reduces energy intake: an application of the bite counter device. J Am Diet Assoc 111 (8):1231–1235
Scribner KB, Pawlak DB, Ludwig DS (2007) Hepatic steatosis and increased adiposity in mice consuming rapidly vs. slowly absorbed carbohydrate. Obesity 15 (9):2190–2199
Scribner KB, Pawlak DB, Aubin CM, Majzoub JA, Ludwig DS (2008) Long-term effects of dietary glycemic index on adiposity, energy metabolism, and physical activity in mice. Am J Physiol Endocrinol Metab 295(5):E1126–E1131
Sergent T, Vanderstraeten J, Winand J, Beguin P, Schneider Y-J (2012) Phenolic compounds and plant extracts as potential natural anti-obesity substances. Food Chem 135(1):68–73
Sestoft L, Bartels PD, Folke M (1982) Pathophysiology of metabolic acidosis: effect of low pH on the hepatic uptake of lactate, pyruvate and alanine. Clin Physiol 2(1):51–58
Seven E, Thuesen Betina H, Linneberg A, Jeppesen JL (2016) Abdominal adiposity distribution quantified by ultrasound imaging and incident hypertension in a general population. Hypertension 68(5):1115–1122
Shilpa Tejpal NS, Manoharan V, Planas-Iglesias J, Myler K, Klein-Seetharaman J (2019) Towards personalised molecular feedback for weight loss. BMC Obesity 6:20
Soeliman FA, Azadbakht L (2014) Weight loss maintenance: a review on dietary related strategies. J Res Med Sci 19(3):268–275
Solas M, Milagro FI, Martínez-Urbistondo D, Ramirez MJ, Martínez JA (2016) Precision Obesity treatments including Pharmacogenetic and Nutrigenetic approaches. Trends Pharmacol Sci 37(7):575–593
Stafleu A, Hendriks HFJ, Smeets PAM, Blom WAM, de Graaf C (2004) Biomarkers of satiation and satiety. Am J Clin Nutr 79(6):946–961
Stéphanie Milliquet JH (2019) EPFL promotes personalized nutrition with Food & You. https://actu.

epfl.ch/news/epfl-promotes-personalized-nutrition-with-food-y-3/. Accessed 2 Dec 2019
Stevenson E, Williams C, Nute M (2005) The influence of the glycaemic index of breakfast and lunch on substrate utilisation during the postprandial periods and subsequent exercise. Br J Nutr 93(6):885–893
Teixeira PJ, Silva MN, Mata J, Palmeira AL, Markland D (2012) Motivation, self-determination, and long-term weight control. Int J Behav Nutr Phys Act 9:22–29
Tejpal S, Bastie C, Seetharaman JK (2018) Lemon juice: a potential source of angiotensin converting enzyme antagonism for weight loss and insulin resistance. Proc Nutr Soc 77(OCE4):E213
Tejpal S, Sanghera N, Manoharan V, Planas-Iglesias J, Myler K, Klein-Seetharaman J (2019a) Towards personalised molecular feedback for weight loss. BMC Obesity 6(1):20
Tejpal S, Sanghera N, Manoharan V, Planas-Iglesias J, Klein-Seetharaman J (2019b) A marker for personalised feedback on dieting: angiotensin converting enzyme. Proc Nutr Soc 78(OCE2):E44
Van Cauter E, Shapiro ET, Tillil H, Polonsky KS (1992) Circadian modulation of glucose and insulin responses to meals: relationship to cortisol rhythm. Am J Physiol Endocrinol Metab 262(4):E467–E475
Vereecken CA, Covents M, Matthys C, Maes L (2005) Young adolescents' nutrition assessment on computer (YANA-C). Eur J Clin Nutr 59(5):658–667
Wang P, Holst C, Andersen MR et al (2011) Blood profile of proteins and steroid hormones predicts weight change after weight loss with interactions of dietary protein level and glycemic index. PLoS One 6(2):e16773
Westerterp-Plantenga MS, Wijckmans-Duijsens NEG, WPHG V-VDV, De Graaf K, Weststrate JA, Van Het Hof KH (1997) Diet-induced thermogenesis and satiety in humans after full-fat and reduced-fat meals. Physiol Behav 61(2):343–349
Westerterp-Plantenga MS, Rolland V, Wilson SAJ, Westerterp KR (1999) Satiety related to 24 h diet-induced thermogenesis during high protein/carbohydrate vs high fat diets measured in a respiration chamber. Eur J Clin Nutr 53:495
When it Comes to Food, One Size Doesn't Fit All! (2019). https://nutrition2me.com/when-it-comes-to-food-one-size-doesnt-fit-all/. Accessed 2 Dec 2019
Wren AM, Murphy KG, Seal LJ et al (2001) Ghrelin enhances appetite and increases food intake in humans. J Clin Endocrinol Metab 86(12):5992–5992
ZOE (2019). https://joinzoe.com/studies/predict-2. Accessed 2 Dec 2019

Diabetes and Obesity in the Child and Adolescent: Guidelines and Challenges

40

Suzanne Cuda and Marisa Censani

Contents

Background ... 554

Overview of Existing Guidelines and Recommendations ... 555

The Maternal and Child Health Bureau, Health Resources and Services Administration: 1998 Obesity Evaluation and Treatment—Expert Committee Recommendations ... 557

AAP Guidance: 2007 Expert Panel and 2015 Algorithm for the Assessment and Management of Childhood Obesity in Patients 2 Years and Older ... 557

AAP Clinical Practice Guideline on the Management of Newly Diagnosed Type 2 Diabetes Mellitus (T2DM) in Children and Adolescents: 2013 ... 558

The Children's Hospital Association Consensus Statement: 2014 ... 558

Endocrine Society Clinical Practice Guideline: 2017 Pediatric Obesity—Assessment, Treatment, and Prevention: An Endocrine Society Clinical Practice Guideline ... 559

The Obesity Medical Association: 2018 Pediatric Obesity Algorithm ... 559

American Diabetes Association (ADA): Evaluation and Management of Youth-Onset Type 2 Diabetes—A Position Statement: 2018 ... 559

AAP Policy Statement: 2019 Pediatric Metabolic and Bariatric Surgery—Evidence, Barriers, and Best Practices ... 560

Issues with the Guidelines/Recommendations ... 560

S. Cuda (✉)
Division of Adolescent Medicine, Pediatric Weight Management, The Children's Hospital of San Antonio, Baylor College of Medicine, San Antonio, TX, USA
e-mail: suzanne.cuda@bcm.edu

M. Censani
Division of Pediatric Endocrinology, Weill Cornell Medicine/New York Presbyterian Hospital, New York, NY, USA

J. Faintuch and S. Faintuch (eds.), *Obesity and Diabetes*, https://doi.org/10.1007/978-3-030-53370-0_40

Media Use ... 560

Failure to Address the Dramatic Increase in Youth Onset Type 2 Diabetes ... 561

The Lack of Pharmacotherapeutic Options for Children and Adolescents ... 562

Obstacles to Bariatric Surgery in the Adolescent with Obesity ... 562

Few Programs Providing Multidisciplinary Treatment for Youth with Obesity . 563

Recognition of Rare Variants of Obesity ... 564

Failure to Recognize the Burden of Risk Factors as Children with Obesity Become Young Adults ... 564

Conclusions ... 565

References ... 565

Abstract

Starting in the 1990s, multiple organizations have published guidelines to address the management of children and adolescents with obesity and diabetes. As the rates of obesity continued to escalate, the guidelines have been constantly evolving to keep up. Significant obstacles include the paucity of controlled trials in children, the bias against managing obesity as a disease in general and especially in children, and the dichotomy in treatment from family based behavioral treatment with modest outcomes to much more aggressive treatment such as surgery. To date, medical options for children with obesity are limited. In this chapter, we chronicle the evolution of the guideline-based management of children with obesity over the past 30 years. We will also review policy on the management of type 2 diabetes in children and adolescents, the incidence of which has increased along with the incidence of severe obesity in children. We will then discuss emerging issues.

Keywords

Obesity · Type 2 diabetes · Guidelines · Children · Adolescent · Algorithm

Background

Childhood obesity is a growing global health problem. One in every three children is either overweight or obese. Management of this chronic debilitating condition has evolved over the past 30 years. Initially, prevention and diagnosis was the focus of management of this developing health crisis. Discussion of the treatment of children and adolescents suffering with obesity emerged as it became apparent that prevention and diagnosis was not going to address the child or adolescent living with obesity. Currently, almost every pediatric health problem is impacted by the coexistence of obesity, even those problems that historically have been associated with low body mass indices (BMI), i.e., type 1 diabetes mellitus.

Management of the pediatric child with obesity has been directed by various publications providing recommendations, guidance, and policy. It has been 22 years since the publication of the first guidance on the management of this complex childhood disease. The field continues to progress with new research and experience, yet suffers from a lack of randomized controlled clinical trials, a continual struggle to find clinical resources to manage this population, and a growing complacency with the existence of obesity in children among both the health care community and the general population.

Overview of Existing Guidelines and Recommendations

Several medical organizations that are focused on the health of children have published guidance on the assessment and management of children with obesity, including the Department of Health and Human Services, the American Academy of Pediatrics (AAP) Institute of Healthy Childhood Weight, the Children's Hospital Association, the Endocrine Society, Pediatric Endocrine Society, and the Obesity Medical Association. Guidelines have also been developed to guide the management of type 2 diabetes (T2DM). In 2013, the AAP published a clinical practice guideline addressing this topic and in 2018 the American Diabetes Association published a position statement.

The timeline for the recommendations and policies on the management of children with obesity and type 2 diabetes published to date is described now and represented in Fig. 40.1.

The first recommendation on the management of children with obesity was published in 1998, when the Department of Health and Human Services convened a committee of pediatric obesity experts (Barlow and Dietz 1998). The AAP updated the 1998 recommendation in 2007 (Barlow et al. 2007), this time using an Expert Committee comprising representatives from 15 professional organizations, appointed experienced scientists, and clinicians in three writing groups. In 2015, the AAP published an update to the 2007 recommendation and called it the "Algorithm for the Assessment and Management of Childhood Obesity in Patients 2 Years and Older" (Institute for Healthy Childhood Weight 2015) (Table 40.1).

In 2014, the Children's Hospital Association published a "consensus statement reviewing management of conditions associated with obesity" (Estrada et al. 2014). In 2017, the Endocrine Society published their "Clinical Practice Guideline: Pediatric Obesity—Assessment, Treatment, and Prevention: An Endocrine Society Clinical Practice Guideline" (Styne et al. 2017), which was an update to their previous publication in

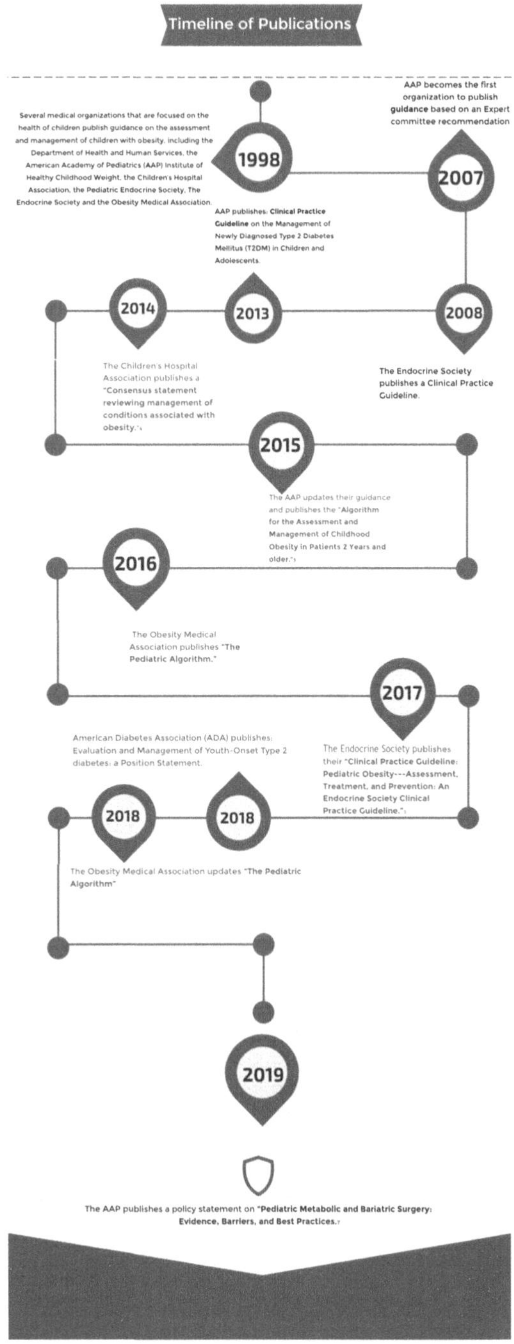

Fig. 40.1 Timeline of publications for recommendations on management of children and adolescents with obesity

2008 (August et al. 2008). In 2016, the Obesity Medical Association published "The Pediatric Algorithm," which was updated in 2018 (Cuda

Table 40.1 Characteristics of recommendations on management of children and adolescents with obesity

	Format	Target audiences	Data source	Discussion of comorbidities	Discussion of management	Discussion of surgery
Maternal and Child Health Bureau 1998	Committee Recommendations	PCPs	Clinical and research experience of the committee members	Minimal	Minimal	None
AAP 2007/2015 Recommendations/ Algorithm	Evidence-based review 2007/ Staged approach	PCPs	Extensive reference	Minimal	Minimal	Minimal
CHA 2014 Consensus Statement	Expert opinion	PCPs	Extensive references	Limited to dyslipidemia, elevated ALT, hypertension, and PCOS	Behavior/family based intervention	None
Endocrine Society 2008/ 2017 CPG	Evidence-Based Review 2017	PCPs and Subspecialists	Extensive references plus 2 systematic reviews	Extensive	Mostly behavior/ Family-based intervention	Minimal
OMA 2016/2018 Algorithm	Evidence-based review 2106/2018	PCPs, pediatric obesity centers and specialists	Extensive references	Extensive	Behavior/ Pharmacotherapy/ Surgery	Some
AAP policy statement on metabolic and bariatric surgery	4 Lead authors from the AAP section on obesity	PCPs, pediatric obesity centers, and subspecialists	Limited references	Limited	Multidisciplinary care/surgery	Extensive

et al. 2016). In 2019, the AAP published a policy statement on "Pediatric Metabolic and Bariatric Surgery: Evidence, Barriers, and Best Practices" (Armstrong et al. 2019).

Publications providing guidance on type 2 diabetes in children and adolescents were developed in 2013 (AAP) (Copeland et al. 2013) and in 2018 (American Diabetes Association) (Arslanian et al. 2018). The 2013 AAP Clinical Practice Guideline was developed in response to the "rapid emergence" of type 2 diabetes, frequently a complication of obesity, which, up until this time, had rarely been seen in children. The 2018 ADA position statement added to the body of knowledge based on new results from two large studies done in adolescents with type 2 diabetes.

The Maternal and Child Health Bureau, Health Resources and Services Administration: 1998 Obesity Evaluation and Treatment—Expert Committee Recommendations

This first recommendation addressing children with obesity reviewed the assessment, medical complications, indications for referral to an obesity specialist, evaluation, and treatment of children with BMI $\geq$ the 85th percentile. At the time of the compilation of this document, data from the National Center for Health Statistics reported a rate of one in five children as being overweight. BMI was characterized by overweight, at risk of overweight, and not at risk for overweight. Triceps skin folds measurement was recommended for the assessment of the degree of overweight. Much of the document focused on the child whose BMI is between the 85th–94th percentile, as this was likely the group presenting most frequently for evaluation at the time. Dietary approach centered on easy ways to reduce caloric intake and the Committee noted that the Food Guide Pyramid was the most helpful guide to healthy eating. Discussion of complications was brief and focused on identifying orthopedic conditions, pseudotumor cerebri, sleep disorders, polycystic ovary syndrome, acanthosis nigricans, hypertension, and dyslipidemia. There was no discussion as to the management of these complications and only a cursory mention of referring children with suspected diabetes to a pediatric endocrinologist. Treatment was centered on family-based behavioral therapy, with provider realization of weight stigma and the need for chronic care.

AAP Guidance: 2007 Expert Panel and 2015 Algorithm for the Assessment and Management of Childhood Obesity in Patients 2 Years and Older

The 2007 recommendation was notable for an evolution from simple identification of obesity to universal screening and assessment, consistent preventive health messages, and early intervention. The hallmark of the 2007 Expert Panel Recommendation was the concept of staged management, with the first stage in primary care and progressing to multidisciplinary care and then to tertiary care at suggested intervals if no clinical improvement. In the 9–10 years since the publication from the Maternal and Child Health Bureau, the prevalence of obesity in children had increased from one in five to closer to one in four, leading to the recommendation for universally assessing children for obesity in primary care settings. The Expert Panel also suggested a change in terminology as compared to the prior recommendation: the term "obesity" should be applied to those children with a BMI $\geq$ the 95th percentile or $\geq$30 kg/m^2, whichever is lower, and "overweight" applied to the 85th–94th percentile, replacing the category of "at risk for overweight." A discussion of the increasing prevalence of severe obesity was included, but no specifics in relation to level of BMI given.

The 2007 Expert panel indicated that management should start with the least intensive stage and progress through the stages based on response to treatment, age, BMI, health risks, and motivation (Barlow et al. 2007). This recommendation goes into more depth on the identification and management of complications and includes problems not previously discussed to include nonalcoholic fatty liver disease, type 2 diabetes, hypothyroidism, Cushing Syndrome, psychiatric disorders, and a very brief mention of genetic syndromes.

The AAP 2015 Algorithm was based on the 2007 Expert Committee Recommendations plus "new evidence and promising practices" (Institute for Healthy Childhood Weight 2015). This two-page document provided a succinct reference for office use.

The four stages discussed are: Prevention Plus, Structured Weight Management, Comprehensive Multi-Disciplinary Intervention, and Tertiary Care Intervention. Progression through the stages is generally at 3–6-month intervals if the child experiences no improvement in BMI/weight status. Treatment is primarily behavior change involving the family. Treatment is targeted to the primary care clinic during the first two stages. At Stage 3, the management moves to a Pediatric Weight Management Clinic/Multi-disciplinary team and the intensity of the care increases. Stage 4 continues in the Pediatric Weight Management Clinic but now involves consideration of the use of medications and bariatric surgery in addition to more frequent and intensive multidisciplinary care.

The 2007 and 2015 guidance included recommendations for laboratory screening: fasting glucose and fasting lipid profile, ALT and AST. According to these publications, the clinical utility of measuring vitamin D and fasting insulin was yet to be determined. These publications also suggest that although there are no guidelines, the age at which laboratory screening should be considered was 2 years of age.

AAP Clinical Practice Guideline on the Management of Newly Diagnosed Type 2 Diabetes Mellitus (T2DM) in Children and Adolescents: 2013

This guideline was developed by a partnership between the AAP, Pediatric Endocrine Society (PES), the American Academy of Family Physicians (AAFP), American Diabetes Association, and the Academy of Nutrition and Dietetics. The document specifically addressed children of ages 10–18 who met the criteria for type 2 diabetes. This publication recommends the use of insulin for those children with HbA1c >9% or blood glucose concentrations greater than 250 mg/dL or in ketoacidosis. For other children with type 2 diabetes, metformin in combination with lifestyle modification was recommended although the partnership recognized that insulin or insulin with metformin can be used in those with more modest levels of hyperglycemia. At the time of this publication, the results of the TODAY (Treatment Options for type 2 Diabetes in Adolescents and Youth) trial were becoming available. Evidence from this trial suggested that metformin in combination with lifestyle management was superior to metformin alone and also suggested that metformin in combination with rosiglitazone was superior to management with metformin alone. However, all groups showed poor glycemic control over time.

The Children's Hospital Association Consensus Statement: 2014

The Consensus statement was developed over a 3–4 year period by an expert committee work group. It was designed to be a management tool for primary care providers. The consensus statement included input from experts working in 25 pediatric obesity centers, a review of literature and existing guidelines, and consultation with pediatric subspecialists.

The consensus statement reviewed screening tests and endorsed the tests recommended by the AAP expert panel and commented that 50–80% of pediatric obesity centers also measure HbA1c, vitamin D 25 OH levels, TSH with reflex-free T4, and fasting insulin. This statement also noted that >80% of pediatric obesity centers refer children for sleep studies who have suggestive symptoms and performed evaluation of PCOS in females with irregular menses and/or signs of hyperandrogenism.

The consensus statement includes treatment guidelines for abnormal LDL and/or triglyceride levels as well as elevated ALT levels, and a recommendation for the management of elevated blood pressure according to existing guidelines at the time of this work. Of note, in 2017, the AAP published an updated clinical practice guideline on the management of children with hypertension (Flynn et al. 2017). Finally, the consensus statement discusses the management of polycystic ovarian syndrome in the context of normal blood glucose levels, prediabetes, and type 2 diabetes.

Endocrine Society Clinical Practice Guideline: 2017 Pediatric Obesity—Assessment, Treatment, and Prevention: An Endocrine Society Clinical Practice Guideline

The Endocrine Society CPG (Styne et al. 2017) was developed using the grading of recommendations, assessment, development, and evaluation approach. The 2017 publication was an update to the publication of the original guidelines in 2008 (August et al. 2008). In addition, two systematic reviews were done to support the guideline: (1) Treatments of Pediatric Obesity: An Umbrella Systematic Review, and (2) The Association of Weight Loss and Cardiometabolic Outcomes in Obese Children: Systematic Review and MetaRegression. The first review looked at data from 133 randomized controlled trials with 30,445 patients. The second review focused on changes in lipids, liver function tests, blood pressure, HbA1c, and fasting blood glucose in children with obesity.

The CPG discusses the diagnosis, prevalence, laboratory evaluation, comorbidities, genetic obesity syndromes, activity recommendations, sleep hygiene, screen time, family involvement, school-based and comprehensive behavioral change interventions, breast feeding of infants, dietary guidance that follows the guidelines of the American Academy of Pediatrics and the US Department of Agriculture, psychosocial issues, pharmacotherapy, and metabolic and bariatric surgery. Each section starts with the recommendation followed by a discussion of the values, evidence and finally, remarks.

The Obesity Medical Association: 2018 Pediatric Obesity Algorithm

A cooperative group from Obesity Medical Association analyzed the evidence concerning diagnosis, evaluation, and management, aiming to provide comprehensive practical help to the healthcare professions. The initial version of the Pediatric Obesity Algorithm was launched in 2016 with an expanded and updated version published in 2018.

American Diabetes Association (ADA): Evaluation and Management of Youth-Onset Type 2 Diabetes—A Position Statement: 2018

This ADA Position statement (Arslanian et al. 2018) starts with a discussion on the differences between onset of type 2 diabetes in youth vs. onset in adulthood. It notes that youth demonstrate greater insulin resistance for any degree of adiposity and greater insulin secretion for any degree of insulin resistance. In addition, this position statement recommends testing for pancreatic autoantibodies to exclude the possibility of autoimmune type 1 diabetes as well as genetic evaluation for monogenic diabetes if suggested by clinical presentation.

Due to the recognition of the low rate of hypoglycemia in youth with type 2 diabetes, including those on insulin, the ADA recommends that HbA1c be $<7\%$ for those on oral agents alone. Increasing evidence for rapidly progressive disease due to loss of beta cell function suggests that even lower HbA1c targets may be necessary.

Recognizing that lower-income and minority youth are the most affected, the ADA recommends culturally sensitive educational materials, assessment of the patient in his or her social context, and preconception counseling.

However, the ADA also recommends a weight loss strategy designed to achieve a 7–10% decrease in excess weight. This should be done through intensive weight management despite the finding in the TODAY (Treatment Options for Type 2 Diabetes in Adolescents and Youth) trial (Flynn et al. 2017) of no long-term difference between those who achieved ≥7% weight loss with lifestyle management and those who were only on metformin monotherapy.

In regards to pharmacotherapy, the only difference between the ADA position statement and the AAP CPG in 2013 was that insulin was recommended if the HbA1c was ≥8.5% instead of 9%. More recently, liraglutide has been recommended for use in adolescents with type 2 diabetes and has shown positive effects on HbA1c, weight status, and fasting plasma glucose (Tamborlan et al. 2019). This was not however mentioned in the ADA position statement as the data were not yet published. Metformin use was recommended for all youth with type 2 diabetes and HbA1c < 8.5%; however, a median time to treatment failure was noted to be a short 11.8 months. The ADA notes that not enough data exist to recommend early use of insulin in an attempt to preserve beta cell function.

The ADA discusses the use of bariatric surgery if the adolescent has uncontrolled hyperglycemia and a BMI ≥35 kg/m^2. However, they also note that 13% of adolescents required a second procedure and an additional 13% required an endoscopic procedure (Inge 2017).

The ADA Position Statement discusses early management of hypertension, albuminuria, screening for neuropathy, retinopathy, evaluation for NAFLD, screening for obstructive sleep apnea, treatment for polycystic ovarian syndrome, and testing and management of dyslipidemia. The discussion of these complications includes concern about the more rapid progression than seen in adults as well as more rapid progression than seen in youth with type 1 diabetes.

AAP Policy Statement: 2019 Pediatric Metabolic and Bariatric Surgery—Evidence, Barriers, and Best Practices

This AAP Policy Statement (Armstrong et al. 2019) reviews the outcomes from several studies on relatively small cohorts of adolescents with severe obesity who have had surgical procedures. The majority of the patients had Roux-en-Y gastric bypass. Outcomes were impressive with average reductions in BMI between 27–29%, and significant resolution of comorbidities at 3 years. Comparison groups of adolescents who were managed with lifestyle all experienced increases in BMI. Surgical complications were reported as infrequent (15%) in the early postoperative period. Eight percent had major perioperative complications. The Policy recommends that surgery be considered at a BMI of 35 kg/m^2 with complications or 40 kg/m^2 with or without complications. Barriers to care are discussed. A disparity of care between groups with lower socioeconomic status and minority race or ethnicity and those with average or high socioeconomic status and white race is noted. Other barriers mentioned include providers who are reluctant to refer their patients for surgery and lack of insurance coverage. The Policy notes that care should occur in a multidisciplinary setting that can provide high-quality pediatric metabolic and bariatric surgery.

Issues with the Guidelines/Recommendations

We have reviewed the existing guidelines and recommendations. The following discussion considers emerging issues of importance in the field of pediatric obesity medicine.

Media Use

AAP Council on Communications and Media policy statement (American Academy of Pediatrics Council on Communications and Media

2016) reviewed publications on television, videos, and mobile or interactive technologies. Below age 2 years adult participation is essential, and excessive media use could be damaging. Below 18 months nothing besides video-chatting should be allowed. Within the bracket of 18--24 months only selected programs and apps should be considered, always in the presence and with assistance of grown ups. Beyond age 2, 1 h or less per day of appropriate movies or videos could be permitted. Bedtime curfew starting 1 h before sleep, with removal of all electronic devices, is another precaution.

Public policy has been historically oriented toward prevention of childhood obesity. Virtually no public policy addresses the treatment and management of the 30% of children and adolescents suffering with obesity. Even attempts to limit marketing to children are difficult to enforce, although in September 2019 the Center for Digital Democracy was able to pass the Children's Online Privacy Protection Act (COPPA) directed at You Tube, which is the first policy limiting media targeted at children. Unfortunately only children aged 12 and younger receive pop ups directing them to You Tube Kids, which is supposed to be a "safer platform for children" (The Children's Online Privacy Protection Act 2019).

Failure to Address the Dramatic Increase in Youth Onset Type 2 Diabetes

The prevalence of youth onset Type 2 diabetes has dramatically increased over the past 20 years, largely as a result of the dramatic increase in youth with obesity and in particular, in youth with severe obesity. According to a recent consensus report (Arslanian et al. 2018) on youth-onset type 2 diabetes, evidence suggests that type 2 diabetes in children is different not only from type 1 diabetes but also from type 2 diabetes in adults. Although type 2 diabetes in youth more closely resembles the pathophysiology in adults with insulin resistance and nonautoimmune beta cell failure, youth-onset type 2 diabetes displays unique features, including rapidly progressive beta cell decline as well as accelerated development of diabetes complications. Several large studies have contributed to most of what we know about youth-onset Type 2 diabetes: The SEARCH for Diabetes in Youth Study (Hamman et al. 2014), the TODAY (Zeitler et al. 2007; TODAY Study Group 2013) (Treatment Options for type 2 Diabetes), and the RISE (Zeitler 2019; Buse et al. 2018) (Restoring Insulin Secretion Study). Unfortunately, these trials have shown that youth who develop type 2 diabetes suffer a more aggressive, rapidly progressive decline in beta cell function as compared to older adults with new-onset type 2 diabetes. In RISE, youth with prediabetes or recently diagnosed type 2 diabetes had 50% lower levels of insulin sensitivity as compared to adults with similar age-adjusted BMI. Not surprisingly, insulin secretion was two- to threefold greater in adolescents as compared to adults. Importantly, even with treatment with either metformin or insulin (glargine), beta cell function significantly declined in 12–15 months from baseline. Despite this, treatment options for type 2 diabetes in adolescents have been extremely limited to two FDA-approved drugs: insulin and metformin, with the recent addition of GLP-1 agonist liraglutide, and the promotion of healthy lifestyles.

The trials mentioned here strongly point to the need to correct dysregulation of glucose and fat metabolism before children and adolescents develop prediabetes and then progress to diabetes. Since progression occurs in spite of pharmacotherapy, weight management becomes the front line of defense against the progression of glycemic dysregulation. The cornerstones of pediatric weight management are diet and lifestyle modification. However, this is usually approached with a "balanced diet," which while typically includes a decrease in carbohydrate consumption from baseline, may not be enough to effect an increase in insulin sensitivity. To date, there is no universal guidance for how to change carbohydrate consumption to effect insulin sensitivity, and each child has a unique level of insulin resistance. Parents and children are mostly focused on a decrease in %BMI or actual weight instead of an increase in insulin sensitivity when the most critical part of their child's treatment could be the prevention of glycemic dysregulation. Historically, we have poorly defined management of this increasingly common complication and we are witnessing the consequences.

The Lack of Pharmacotherapeutic Options for Children and Adolescents

Pharmacotherapy for adults with obesity is a mainstay of treatment. Six medications are currently FDA approved for use in adults with BMI $\geq$ 30 kg/m^2 or those who are overweight (BMI $\geq$ 27 kg/m^2) with obesity-related complications. These medications are covered by most insurance plans for adults, defined as those 18 years or older. However, the only weight loss medication approved by the FDA for children or adolescents is orlistat ($\geq$12 years), and its use is curtailed by significant gastrointestinal side effects. Phentermine is approved for use in older adolescents >16 years of age. No other medications are approved by the FDA for weight loss under the age of 18. There are very limited data on the safety and efficacy of the use of these medications in children and adolescents. Over the past 5 years, pediatric obesity medicine experts have increasingly used weight loss medications "off label." As a group, pediatricians have long used medications "off label," so much so that the American Academy of Pediatrics published a Policy Statement in 2014 on "Off-label Use of Drugs in Children" (AAP Policy Statement 2014). The cardinal principle for the use of off-label medication is to ensure that the benefit of using the medication outweighs the risk.

Evolution of the use of medication (other than orlistat or phentermine) to treat obesity in children and adolescents began in 2012 and 2013 when two studies looked at the use of metformin and lifestyle intervention vs. just lifestyle intervention. Although the results were reported as statistically significant, the weight loss reported was relatively small: 1.3 kg/m^2 and 1.07 kg/m^2 (Mauras et al. 2012; Kendall et al. 2013). Likewise, a trial using a GLP 1 receptor agonist in 2013 only showed a 1.13 kg/m^2 placebo subtracted difference (Kelly et al. 2013).

Pediatric obesity medicine specialists face additional barriers to the use of weight loss medications other than lack of FDA approval. A significant barrier is the lack of insurance coverage for medications leading to out-of-pocket expense for the family. Because of this problem, pediatric obesity medicine specialists who have determined that the use of weight loss medication outweighs the risk default to using phentermine, which is relatively inexpensive to purchase, and with topiramate, which can be prescribed and covered by insurance since it is a medication approved by the FDA for other use in the pediatric population. Similarly, the combination of buproprion and naltrexone or lisexamfetamine (approved in 2015 for use in adults or use for binge-eating disorder) are used. Recently (August 2019), the use of liraglutide was approved for children and adolescents 10 years of age or older, but for the indication of T2DM, not for weight loss.

Metformin is commonly used in the population of children and adolescents, although approved by the FDA only in pediatric patients 10 years of age or older with T2DM. However, many obesity medicine specialists prescribe metformin for PCOS or severe insulin resistance with or without impaired glucose tolerance. Multiple trials (McDonagh et al. 2014; Khokhar et al. 2017; Kaplowitz 2017) have shown that a small amount of weight loss is associated with the use of metformin, especially in the first few months after starting the medication.

Obstacles to Bariatric Surgery in the Adolescent with Obesity

Although the rate of weight loss surgery in adolescents with severe obesity has been increasing, it is still very low. Bariatric surgery in adolescents with severe obesity has been found to be effective in decreasing excess weight and improving comorbidities. However, data on the long-term efficacy and safety of adolescent bariatric surgery are limited. Real barriers to this treatment modality exist. The leap from behavioral management to surgery is too drastic for many families. In addition, although the recent Policy statement from the AAP endorses bariatric surgery as a safe and effective strategy for youths with severe obesity, there are few studies on optimal timing, best preoperative and postoperative care, and adjunctive (including behavioral and medication) therapy. This compounds the lack of insurance coverage, reported as less than half the coverage in adults of similar BMI and comorbid status, the lack of multidisciplinary weight management programs with a surgical tract or referring relationship to a bariatric surgical program, and the low socioeconomic status of many of the adolescents with severe obesity who meet criteria for surgery (Armstrong et al. 2019).

Forty three percent of the adolescents who start the process of evaluation for bariatric surgery made it through the process in one retrospective chart review of 145 adolescent patients (Brode et al. 2018). This statistic is similar to the surgical completion rate for adults. Previous adult studies have found an association between attrition and psychological factors. Interestingly, the study in adolescents by Brode et al. found no difference in assessed psychological factors between those adolescents who completed the program and had surgery and those who did not complete the program. Specifically, there were no differences in current psychotropic medication use, utilization of outpatient therapy, current substance use, or previous inpatient hospitalization for a psychiatric condition. In addition, a recommendation by the clinical team for initiation of psychological treatment did not accurately predict completers vs. noncompleters. Male sex was a significant predictor of noncompletion of surgery, but the reason for this was not clear.

Another significant issue is the management of the change in diet required after surgery in the setting of low-income and minority youth with severe obesity who get a significant amount of their daily nourishment from publically funded meals. Modifying Federally funded School Breakfast and Lunch meals even with a signed letter from a medical provider for a "special diet" is extremely challenging. The new AAP policy advises the consideration of weight loss surgery in adolescents with BMI down to 35 kg/m^2. This will include many adolescents who are fully involved in normal activities, such as sports teams, school clubs, and social events. Many of the adolescents are expected to make "healthy choices." Following a restrictive diet in these situations will be challenging at best.

To complicate matters further, there is a lack of even a basic understanding on the management of adolescents with severe obesity, with or without weight loss surgery. There are no studies detailing dosage of behavioral modification or pharmacotherapy before or after surgery. In addition, studies involving the relative benefit of bariatric surgery performed during adolescence compared to young adulthood (18–25 years old) are needed (Ryder et al. 2018). Historically, pediatricians have been very reticent to use weight loss medications in adolescents with severe obesity. It is difficult to imagine that they will feel more comfortable with managing adolescents with obesity as they progress through the rest of their adolescence post surgery.

Few Programs Providing Multidisciplinary Treatment for Youth with Obesity

In a prospective study collecting data from 31 pediatric weight management programs across the United States, early BMI reduction in the first month of treatment was significantly associated with greater long-term BMI reduction at 6 and 12 months ($\geq$5% BMI reduction from baseline) in adolescents with obesity-seeking treatment (Gross et al. 2019). However, the availability of large multidisciplinary weight management programs with Stage 3 Comprehensive Multi-Disciplinary Intervention and Stage 4 Tertiary Care Intervention is limited with large sections of certain states devoid of Obesity Medicine specialists, particularly those with pediatric obesity experience. In a review of 16 studies from 1995 to 2017 (Zolotarjova et al. 2018) analyzing the effects of multidisciplinary interventions on weight loss and health outcomes in children and adolescents with morbid obesity, positive weight loss results were observed in numerous studies with younger children having a greater reduction in BMI z-score in comparison to adolescents. Authors noted the importance of early incorporation of multidisciplinary interventions in obesity treatment in addition to the lack of standardization of obtaining parameters. Of note, cardiovascular risk parameters were only investigated in three of the studies with improvements noted in this limited subset, and only some of the studies analyzed the effects of intervention on psychosocial well-being. In addition, few reported long-term follow-up outcome measures, limiting the utility of the findings in regards to sustainability and effectiveness of measure studies with a more standardized approach warranted.

Recognition of Rare Variants of Obesity

Obesity is a heterogeneous multifactorial disease, which results from the interrelationship between behavioral, environmental, and genetic factors. Although the genetic causes underlying obesity remain largely unknown, genetic factors are estimated to account for 40–70% of the obesity predisposition of an individual (Zhao and Grant 2011; Ranadive and Vaisse 2008; Serra-Juhe et al. 2019). There are syndromic causes for obesity; however, obesity is often not accompanied by specific phenotypes. There are rare genetic variants affecting several genes such as LEP, LEPR, MC4R, PCSK1, and POMC found in nonsyndromic patients (Serra-Juhe et al. 2019), and it has been estimated that 7% of patients with severe pediatric obesity may have rare abnormalities of chromosomes and/or highly penetrant genetic mutations (Styne et al. 2017; Farooqi and O'Rahilly 2008). MC4R deficiency is the most common monogenic form of obesity accounting for 2–5% of severe obesity in adult and pediatric populations.

Genome-wide association studies (GWAS) have found strongly associated genomic variants with a number of loci reported from GWAS analyses of adult BMI and obesity also having a role in obesity in pediatric populations. Genome-wide approach has revealed genes that underlie the pathogenesis of childhood obesity (Zhao and Grant 2011). Understanding the genetics of childhood obesity is vital in the prevention and treatment of pediatric cases in order to provide families with individualized follow-up and genetic counselling as well as therapeutic approaches for genetic causes of obesity (Chesi and Grant 2015). Although the GWAS approach provides a more comprehensive strategy to locate causal genes related to obesity, the challenge arises from lack of availability of such resources for most patients undergoing evaluation, and limited GWAS-related reports specific for childhood obesity, with most studies uncovering loci in the adult context (Zhao and Grant 2011).

Failure to Recognize the Burden of Risk Factors as Children with Obesity Become Young Adults

Many studies have addressed the presence of risk factors in children and adolescents with obesity. These studies document increased incidence of metabolic syndrome, hypertension, and glycemic dysregulation with increasing severity of obesity. Likewise, in the past two decades, the incidence of cardiovascular disease in young adults (18–50 years) has been steady or has increased, while the incidence of cardiovascular disease in adults over 50 has decreased. A higher BMI in young adulthood has a long-term negative effect on blood pressure and heart structure. These observations suggest that we may see an increase in cardiovascular disease as this younger population ages. Obesity in youth is a major contributing factor to this problem. Not only in developed countries, but also in undeveloped countries, the prevalence of obesity in childhood has increased. In the US, the proportion of children (aged 12–19 years) with a healthy BMI declined between 1999 and 2013, and is now appearing to level off between 65–70%. As children with a body mass index percentile at the >95th percentile have a greater chance of maintaining obesity into adulthood, appropriate counseling and lifestyle modifications to families is needed to prevent the known complications of excess weight gain. The goals of treatment involve the prevention and reduction of the risk of obesity-related sequelae with improvement of health outcomes. A weight loss of 5–10% has been associated with improvement in comorbidities (Kumar et al. 2019).

Other risk factors associated with obesity that directly contribute to cardiovascular disease are physical inactivity, hypertension, diabetes, dyslipidemia, and dietary intake.

In the US, physical activity levels continue to decline in children, and especially in adolescents. Hypertension rates in children with obesity increase as adiposity increases. Prevalence varies between 3.8–24.8% in studies of children with obesity, as compared to the prevalence of 3.5% in the general population. Furthermore, lack of a nocturnal dip in blood pressure is characteristic of children with obesity (Flynn et al. 2017). The increase in incidence of Type 2 diabetes was discussed earlier in this chapter. In the last 40 years, lower triglyceride levels and higher HDL levels have been seen in the general population, but not so in children and adolescents with obesity. In addition, older adults have demonstrated greater improvement in lipid levels than younger adults.

A decrease in the dietary intake of trans fats, total fat, and saturated fat has occurred in the US over the past 30 years. However, these changes were not as profound in children aged 12–19 years as they were in those older than 19 years. NHANES noted that those with a low socioeconomic status, African Americans or Mexican Americans did not have an improvement in dietary fat. In particular, excessive consumption of sugar sweetened beverages and processed meat was strongly associated with cardiometabolic disease–related mortality.

Conclusions

Rates of obesity in children and adolescents have dramatically increased over the last 40 years. In particular, the prevalence of severe obesity continues to climb. Lower socioeconomic and minorities are at highest risk. Management of this global health crisis has not kept up and now younger adults are suffering unprecedented rates of type 2 diabetes and early cardiovascular disease. Family-based behavioral therapy for those children at highest risk requires access to multidisciplinary care in pediatric obesity centers. Such centers are not available in large areas of the country, receive poor institutional support, and often have long waiting times to be seen in the center. An extraordinary amount of time commitment from the family is expected to meet the >25 h of patient contact hours over a 6-month time frame recommended by the USPSTF (US Preventive Services Task Force 2010). The leap from medical treatment to surgical is extreme for many families and insurance coverage is not universal. Randomized controlled trials are necessary to establish optimal treatment.

References

AAP Policy Statement (2014) Off-label use of drugs in children. Pediatrics 133(3):563–567

American Academy of Pediatrics Council on Communications and Media (2016) Media and young minds. Pediatrics 138(5):e20162591. https://doi.org/10.1542/peds.2016-2591

Armstrong SC et al (2019) Pediatric metabolic and bariatric surgery: evidence, barriers, and best practices. Pedatrics 144(6):320193223

Arslanian S et al (2018) Evaluation and management of youth -onset type 2 diabetes; a position statement by the American Diabetes Association. Diabetes Care 41:2648–2668. https://doi.org/10.2337/dci18-0052

August GP et al (2008) Prevention and treatment of pediatric obesity: an Endocrine Society clinical practice guideline based on expert opinion. J Clin Endocrinol Metab 93:4576–4599. https://doi.org/10.1212/jc.2007-2458

Barlow SE, Dietz WH (1998) Obesity evaluation and treatment: expert committee recommendations. Pediatrics 102(3):E29. http://www.pediatrics.org/cgi/content/full/102/3/e29

Barlow SE et al (2007) Expert committee recommendations regarding the prevention, assessment, and treatment of child and adolescent overweight and obesity: summary report. Pediatr Dent 120:S164. https://doi.org/10.1542/peds.2007-2329C

Brode C, Ratcliff M, Reiter-Purtill J, Hunsaker S, Helmrath M, Zeller M (2018) Predictors of preoperative program non-completion in adolescents referred for bariatric surgery. Obes Surg 28(9):2853–2859. https://doi.org/10.1007/s11695-018-3261-5

Buse JB, D'Alessio DA, Riddle MC (2018) Can we RISE to the challenge of youth-onset type 2 diabetes? Diabetes Care 41:1560–1562. https://doi.org/10.2337/dci18-0025

Chesi A, Grant SFA (2015) The genetics of pediatric obesity. Trends Endocrinol Metab 26(12):711–721

Copeland KC et al (2013) Management of newly diagnosied type 2 diabetes mellitus (T2DM) in children and adolescents. Pediatrics 131(2):364–382. https://doi.org/10.1542/peds.2012-3494

Cuda SE et al (2016) Pediatric obesity algorithm, presented by the obesity medical association. www.obesitymedicine.org/childhood-obesity

Estrada E et al (2014) Children's hospital association consensus statements for comorbidities of childhood obesity. Child Obes 4(10):304–317. https://doi.org/10.1089/chi.2013.0120

Farooqi IS, O'Rahilly S (2008) Mutations in ligands and receptors of the leptin-melanocortin pathway that lead to obesity. Nat Clin Pract Endocrinol Metab 4:569–577

Flynn JT et al (2017) Clinical practice guideline for screening and Management of High Blood Pressure in children and adolescents. Pediatrics 140(3): e20171904

Gross AC, Kaizer AM, Kelly AS, Rudser KD, Ryder JF, Borzutzky CR, Santos M, Tucker JM, Yee JK, Fox CK (2019) Long and short of it: early response predicts longer-term outcomes in pediatric weight management. Obesity 27(2):272–279

Hamman RF et al (2014) The SEARCH for diabetes in youth study: rationale, findings, and future directions. Diabetes Care 37:3336–3341

Inge TH (2017) Jenkins ™, Xanthakos SA et al. Lancet Diabetes Endocrinol 5:165–173

Institute for Healthy Childhood Weight (2015) Algorithm for the assessment and management of childhood in patients 2 years and older

Kaplowitz P (2017) Is there a role for metformin in the treatment of childhood obesity? Pediatrics 140: e20171205

Kelly AS, Rudser KD, Nathan BM, Fox CK, Metzig AM, Coombes BJ et al (2013) The effect of glucagon like peptide-1 receptor agonist therapy on body mass index in adolescents with severe obesity: a randomized, placebo controlled, clinical trial. JAMA Pediatr 167 (4):355–360

Kendall D, Vail A, Amin R, Barrett T, Dimitri P, Ivison F et al (2013) Metformin in obese children and adolescents: the MOCA trial. J Clin Endocrinol Metab 98(1):322–329

Khokhar A, Umpaichitra V, Chin VL, Perez-Colon S (2017) Metformin use in children and adolescents with prediabetes. Pediatr Clin N Am 64:1341–1353

Kumar S, King EC, Christison AL, Kelly AS, Ariza AJ, Borzutzky C, Cuda S, Kirk S (2019) Health outcomes of youth in clinical pediatric weight management programs in POWER. J Pediatr 208:57–65

Mauras N, DelGiorno C, Hossain J, Bird K, Killen K, Merinbaum D et al (2012) Metformin use in children with obesity and normal glucose tolerance—effects on cardiovascular markers and intrahepatic fat. J Pediatr Endocrinol Metab 25(1–2):33–40

McDonagh MS, Selph S, Ozpinar A, Foley C (2014) Systematic review of the benefits and risks of metformin in treating obesity in children aged 18 years and older. JAMA Pediatr 168:178–184

Ranadive SA, Vaisse C (2008) Lessons from extreme human obesity: monogenic disorders. Endocrinol Metab Clin N Am 37:733–751

Ryder JR, Fox CK, Kelly AS (2018) Treatment options for severe obesity in the pediatric population: current limitations and future opportunities. Obesity 26:951–960. https://doi.org/10.1002/oby.22196

Serra-Juhe C, Martos-Moreno GA, Bou de Pieri F, Flores R, Chowen JA, Perez-Jurado LA, Argente J (2019) Heterozygous rare genetic variants in non-syndromic early-onset obesity. Int J Obes 44:830. https://doi.org/10.1038/s41366-019-0357-5

Styne DM et al (2017) Pediatric obesity—assessment, treatment and prevention: an endocrine society clinical practice guideline. J Clin Endocrinol Metab 102 (3):709–757

Tamborlan WV et al (2019) Liraglutide in children and adolescents with type 2 diabetes. N Engl J Med 381:637–646. https://doi.org/10.1056/NEJMoa1903822

The Children's Online Privacy Protection Act (2019) Center for digital democracy

TODAY Study Group (2013) Effects of metformin, metformin plus rosiglitazone, and metformin plus lifestyle on insulin sensitivity and beta cell function in TODAY. Diabetes Care 36:1749–1757

US Preventive Services Task Force (2010) Screening for obesity in children and adolescents: US Preventive Services Task Force recommendation statement. Pediatrics 125(2):361–367. https://doi.org/10.1542/PEDS.2009-2037

Zeitler P (2019) Progress in understanding youth-onset diabetes in the United States: recent lessons from clinical trials. World J Pediatr 15:315–321

Zeitler P et al (2007) TODAY study group:treatment options for type 2 diabetes in adolescents and youth: a study of the comparative efficacy of metformin alone or in combination with rosiglitazone or lifestyle intervention in adolescents with type 2 diabetes. Pediatr Diabetes 8:74–87

Zhao J, Grant SFA (2011) Genetics of childhood obesity. J Obes 2011:845148. https://doi.org/10.1155/2011/845148

Zolotarjova J, Ten Velde G, Vreugdenhil ACE (2018) Effects of multidisciplinary interventions on weight loss and health outcomes in children and adolescents with morbid obesity. Obes Rev 19(7):931–946. https://doi.org/10.1111/obr.12680

Type 2 Diabetes: An Unresolved Disease 41

Sarah Cuschieri

Contents

Diabetes: A Historical Disease 568

Diabetes: A Disease of Many Faces 568
Type 1 Diabetes Mellitus 569
Neonatal Diabetes 569
Latent Autoimmune Diabetes of Adults (LADA) 569
Maturity-Onset Diabetes of the Youth 570
Gestational Diabetes Mellitus 570
Type 2 Diabetes Mellitus 570

Type 2 Diabetes Mellitus: A Spectrum of Dysglycaemic Profiles 570
Impaired Glucose Regulation 570
Type 2 Diabetes Mellitus Profile 571
Early Life Genetics and Epigenetics 571
Genetic Element 572
Diagnostic and Screening Testing 573
Obesity 573

Diabetes: From Historical Curiosity to a Serious Epidemic 573
Preventive Screening 576

References 576

Abstract

The existence of diabetes has been reported for centuries dating back to the Egyptian period. As years passed, scientists have identified the pathophysiology of this disease and its management. Diabetes is a multifactorial disease and different types of diabetes have been reported. Even though this is a disease of antiquity, its prevalence has been on the rise, and it is a currently a global epidemic as well as public health emergency. Different stakeholders (ranging from the medical professional, to policy makers and governmental bodies) need to work together to prevent health inequalities and control this growing epidemic.

S. Cuschieri (✉)
Department of Anatomy, Faculty of Medicine & Surgery, University of Malta, Msida, Malta
e-mail: sarah.cuschieri@um.edu.mt

J. Faintuch and S. Faintuch (eds.), *Obesity and Diabetes*, https://doi.org/10.1007/978-3-030-53370-0_41

Keywords

Hyperglycaemia · Type 2 diabetes mellitus · Epidemics · Population at risk · Body weight · Mass screening

Diabetes: A Historical Disease

Diabetes mellitus was recognized around 1550 BC by the Egyptians, between 400 and 500 BC by Hindu pioneers Charaka and Sushruta, as well as by Aretaeus of Cappadocia, in the first century AD. In Greek diabetes means syphon, suggesting the constant elimination of urinary fluid by severely affected patients (International Diabetes Federation 2017; Tattersall 2010).

The pioneer to describe serum hyperglycaemia and the association with sugar in urine was Matthew Dobson from Liverpool, way back in 1775 (Dobson 1776). In 1797 the surgeon John Rollo applied the adjective 'mellitus', meaning 'honey-like' in Latin, to the already established 'diabetes'. Furthermore, he concluded that the serum sugar was originating from vegetables being digested in stomach, and hence proposed a diet high in animal produce as a treatment option for this condition (Rollo 1797). It was in 1815 that glucose was identified as being the leading sugar in the diabetic urine, paving the way for chemical testing, instead of less elegant dipping a finger and tasting it. However, it was only a century later, between 1913 and 1915, that the physician Ivar Christian Bang established a practical micro-method to measure glucose, contributing to the introduction of the glucose tolerance test (Tattersall 2010).

During the same era, the pathophysiology of diabetes was being explored. It was in 1869 that Paul Langerhans made the great discovery that the pancreas parenchyma included 'islands' of different cells (Langerhans 1869). Later on, these 'islets of Langerhans' were reported to be responsible for pancreatic secretions (Laguesse 1893). In the 1900s, the pancreatic secretion was named 'insulin' and was utilized as an injectable treatment for diabetes (De Meyer 1904; Banting et al. 1922). It was in January 1922 that Banting and Best conducted the first human experiment on an adolescent, using injectable insulin. The biological abnormalities and the clinical symptoms related to diabetes were reverted back to normal, making this experiment successful (Banting et al. 1922). However, in the 1930s, it was reported that not every diabetic patient had a positive outcome following injectable insulin. There were those 'sensitive' and 'insensitive' to insulin therapy (Himsworth 1939). The various diabetes classifications and types will be discussed in later sections.

Diabetes management was refined over the years. Designer insulins *lispo*, *aspart*, *glargine* and *detemir* allow custom-tailored prescriptions. The 'insulin pen' made its way in 1981 by John Ireland, which made the management of serum glucose by insulin more acceptable to patients (Paton et al. 1981). Oral hypoglycaemic agents came around in the 1940s, when sulphonylureas were found to be insulin secretagogues and were manufactured (De Franco et al. 2015). Later in 1960 Metformin was introduced on the European market (Tyberghein and Williams 1957). Since then, different classes of oral and injectable hypoglycaemic agents appeared, such as glitazones, glucose-like peptide 1 (GLP-1) agonists, and dipeptidylpeptidase-4 (DPP-4) inhibitors.

Diabetes: A Disease of Many Faces

Himsworth in 1936 pioneered in suggesting the presence of different types of diabetes (Himsworth 1939). Over the years, various expert committees attempted to classify the different forms of diabetes. In 1965, age of onset was considered as the determining factor for the type of diabetes a person develops. The World Health Organization (WHO) divided diabetes into 'Juvenile-onset' and 'Maturity-onset' disease. This classification changed in 1976 to 'Insulin-dependent diabetes mellitus' (IDDM) and 'non-insulin-dependent diabetes mellitus' (NIDDM) (National Diabetes Data Group 1979). However, a revised classification was proposed by the American

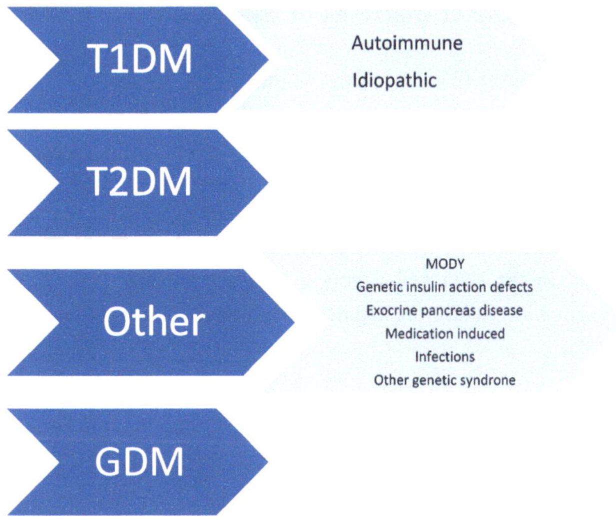

Fig. 41.1 The consensus between ADA (1997) and WHO (1999) aetiological classification of glycaemia disorders (The Expert Committee on the Diagnosis and Classification of Diabetes Mellitus 1997; World Health Organisation Consultation Group 1999) *T1DM* Type 1 diabetes mellitus, *T2DM* Type 2 diabetes mellitus, *MODY* Mature onset disease of the young, *GDM* Gestational diabetes

Diabetes Association (ADA) in 1997 and seconded by the WHO later on in 1999 (The Expert Committee on the Diagnosis and Classification of Diabetes Mellitus 1997; World Health Organisation Consultation Group 1999), as seen in Fig. 41.1.

Type 1 Diabetes Mellitus

Type 1 diabetes mellitus (T1DM) affects individuals usually before 35 years, and is characterized by severely reduced or absent pancreatic beta cells. Hypoinsulinaemia and episodes of acute diabetes ketoacidosis are the usual features. These individuals require insulin therapy for life. Two sub-sets of T1DM have been identified: (1) Autoimmune and (2) Idiopathic. Autoimmune attack of the pancreas is more common in Europids. Autoantibodies, including anti-glutamine acid decarboxylase (anti-GAD), anti-insulin and/or islet cell antibodies are usually detected on presentation (Alberti 2010; Libman et al. 2011). On the other hand, the idiopathic T1DM is more common in non-Europid populations, with the loss of beta cell function but absence of autoantibodies. However, these individuals still have episodes of ketoacidosis and require exogenous insulin for survival (Alberti 2010; Libman et al. 2011).

Neonatal Diabetes

Neonatal diabetes is conventionally diagnosed around 3–6 months of age, in the absence of any autoantibodies (Shield et al. 1997). Two types are present: transient and permanent. The transient type is a rare condition, and is associated with intrauterine growth retardation and the inheritance of two chromosome homologs from the father alone (Sperling 2006a; von Mühlendah and Herkenhoff 1995). Diabetes during adulthood is a risk in this population (Metz et al. 2002). The latter type of neonatal diabetes has been associated with activation of KCNJ11 potassium channel subunit, by Kir 6.2 gene mutation, along with ABCC8 (sulfonylurea receptor 1) and insulin promoter factor (IPF-1) mutations (Sperling 2006b). The treatment of choice for neonatal diabetes is sulphonylureas.

Latent Autoimmune Diabetes of Adults (LADA)

LADA is mostly detected at age 35 or older, along with the presence of GAD autoantibodies. The characteristics of LADA consist of an overlap between type 1 and type 2 phenotypes, with an ultimate exogenous insulin dependence (Tuomi et al. 1993).

Maturity-Onset Diabetes of the Youth

Maturity-onset diabetes of the youth (MODY) is composed of a collection of autosomal dominated inherited disorders that usually present before the age of 25 years. A history of Hyperglycaemia in the family is very common. MODY is non-insulin dependent since it results from a single genetic defect in the beta cell function, rather than the presence of insulin resistance (Tattersall 1974, 1998).

Gestational Diabetes Mellitus

Gestational diabetes mellitus (GDM) is a hyperglycaemic state that presents during pregnancy. At-risk mothers undergo diabetes screening through an oral glucose tolerance test around 28 weeks. Risk factors for GDM include previous GDM, increase in maternal age, type 2 diabetes mellitus in the close family, previous macrosomic babies, and obesity (Metzger et al. 2007; Chan et al. 2002) The plasma glucose level normally goes down to normal levels postpartum, yet the possibility of subsequent type 2 diabetes should not be neglected.

Type 2 Diabetes Mellitus

Type 2 diabetes mellitus (T2DM) is the dominant modality. It is a heterogeneous disease suffering from the impact of environmental and genetic influences. The genetic inheritance is mainly linked to the beta cell function, differently from acquired factors connected to insulin resistance (Tuomilehto et al. 2001).

Insulin sensitivity of the beta cells determines glucose homeostasis. Exhaustion of beta cell output along with insulin resistance results in persistent hyperglycaemia. Increased free fatty acid levels, especially small dense LDL-C and triglycerides, are similarly observed during insulin resistance and hyperglycaemia, making T2DM individuals at a higher cardiovascular risk (Cobb et al. 2013).

Type 2 Diabetes Mellitus: A Spectrum of Dysglycaemic Profiles

The hyperglycaemia state can be described as a spectrum of phenotypes with the end stage being full-blown type 2 diabetes mellitus (T2DM). However, in most individuals prior to reaching this state, a mild asymptomatic hyperglycaemic profile would be present and only picked up on screening.

Impaired Glucose Regulation

A raised plasma glucose level mildly above the normal baseline (>7 mmol/L) is considered as impaired glucose regulation or pre-diabetes. This hyperglycaemic state is still considered as a metabolic condition originating from a combination of pancreatic beta cell dysfunction (primary or secondary) and insulin resistance. Subsequent type 2 diabetes, cardiovascular and dyslipidemic comorbidities cannot be ruled out (Alberti 1996).

Impaired glucose regulation is divided into two different hyperglycaemic states: impaired fasting blood glucose (IFG) and impaired glucose tolerance (IGT). The underlying pathophysiology for both hyperglycaemic states is insulin resistance; however, the site where this occurs varies. Hence, the method of diagnosis also varies. Impaired glucose tolerance (IGT) is investigated by an oral glucose tolerance test (OGTT), with 7.8–11.0 mmol/L of glucose after 2 h. It is uncommon in women >60 years.

IGT signals peripheral insulin resistance of the skeletal muscle, with minimal effect on the liver. In contrast, impaired fasting blood glucose (IFG) points towards hepatic insulin resistance with a normal peripheral pattern. In fact, during an OGTT, there is severe impairment of the first phase of insulin response, but insulin secretion

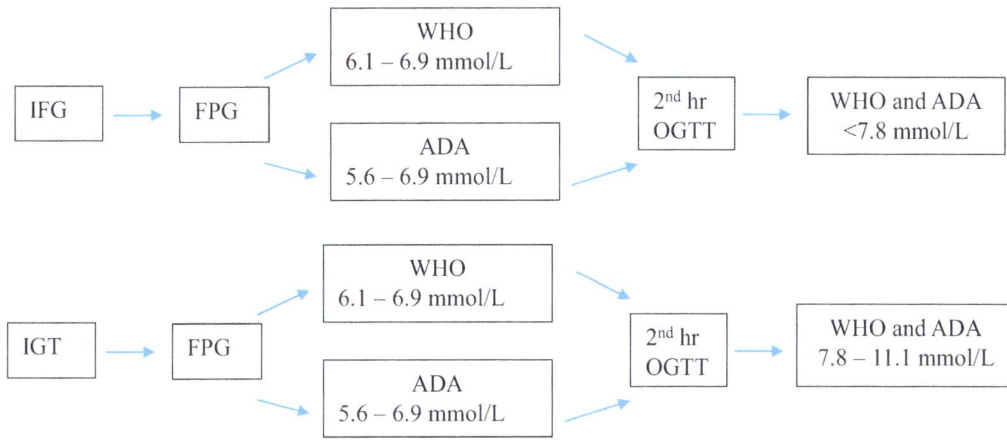

IFG – Impaired fasting glucose
IGT – Impaired glucose tolerance
FPG – Fasting plasma glucose
WHO – World Health Organization
ADA – American Diabetes Association
OGTT – Oral glucose tolerance test

Fig. 41.2 Different diagnostic criteria for pre-diabetes (American Diabetes Association 2018; World Health Organization 2006) *IFG* Impaired fasting glucose, *IGT* Impaired glucose tolerance, *FPG* Fasting plasma glucose, *WHO* World Health Organization, *ADA* American Diabetes Association, *OGTT* Oral glucose tolerance test

improves during the second phase. This condition is more common in men across all age groups; however, prevalence tends to plateau in the middle age (Kanat et al. 2012). The diagnostic criteria for IFG include fasting glucose and OGTT. Disagreements are present between international organizations with regard to the diagnostic cut-off criteria, as seen in Fig. 41.2.

Such disagreements are also present for the cut-off point for the glycated haemoglobin (HbA1C). The World Health Organization (WHO) and International Diabetes Federation (IDF) recommend 6.0–6.5% for high-risk individuals (Kilpatrick et al. 2009). The American Diabetes Association (ADA) considers this state with an HbA1C between 5.7–6.4% (American Diabetes Association 2018). In order to establish a clinical diagnosis of both IFG and IGT, it is important that a repeat OGTT is performed within 3 months of the initial diagnosis. This is essential since a substantial proportion of these individuals would have already developed type 2 diabetes.

During impaired hyperglycaemic states, lifestyle interventions or else initiation of Metformin could reduce the risk of developing type 2 diabetes by about 58% (Tuomilehto et al. 2001; Hamman et al. 2006; Eriksson and Lindgärde 1991; Diabetes Prevention Program Research Group et al. 2009; Lindström et al. 2003; Pan et al. 1997; Ramachandran et al. 2006).

Type 2 Diabetes Mellitus Profile

A multilevel approach is depicted in Fig. 41.3 (Susser 1996).

Early Life Genetics and Epigenetics

The risk to develop type 2 diabetes mellitus can initiate from the intra-uterine period. It has been reported that a hyperglycaemic foetal environment predisposes to insulin resistance, obesity and T2DM in adulthood (Jiang et al. 2013). This

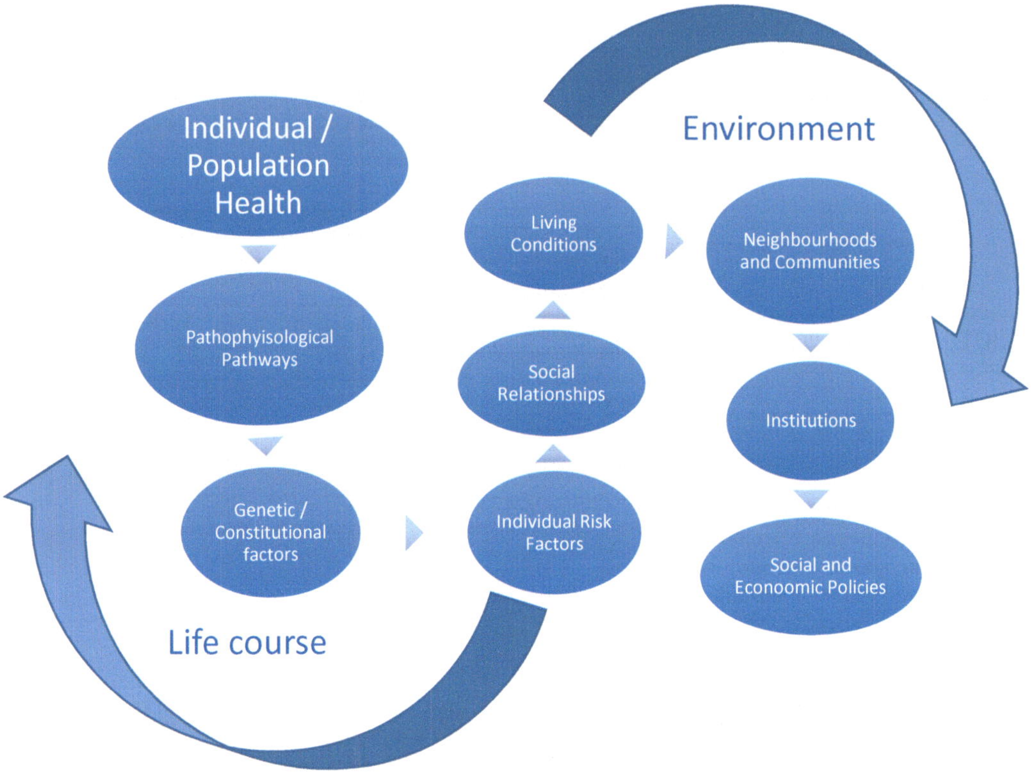

Fig. 41.3 Diabetes mellitus ecologic paradigm over the life course (Kaplan et al. 2000)

is followed by early-life growth trajectories amalgamated with interactions between non-modifiable (genetics) and modifiable (environment) factors that have an effect on insulin resistance, beta cell function and obesity (Kahn et al. 2006; Sladek et al. 2007; Alberti et al. 2006). The birth weight of the child is also a determinant for future T2DM. The *thrifty phenotype hypothesis* suggests that low birth weight following inadequate nutrition in utero leads to biological programming later on in life with a susceptibility to developing T2DM (Randle et al. 1963). On the other hand, the *theory of an association* reported by McCarthy suggests that an association between diabetes and low birth weight is due to pleotropic genes, related to both foetal growth and diabetes onset (McCarthy 1998).

Genetic Element

Twin studies have reported that there is a genetic linkage for the development of T2DM. Monozygotic twins were reported to have a concordance rate of 90%, with a 37% concordance rate for dizygotic twins (Newman et al. 1987; Beck-Nielsen et al. 2003). A person born to one parent afflicted by T2DM has a 40% lifetime risk of progressing to the same condition (Köbberling and Tillil 1982). A number of single-nucleotide polymorphisms (SNPs) have been identified that contribute to the basis for the genetic susceptibility for T2DM (Cuschieri 2019a). Inappropriate lifestyle can also induce diabetes-susceptible genotypes.

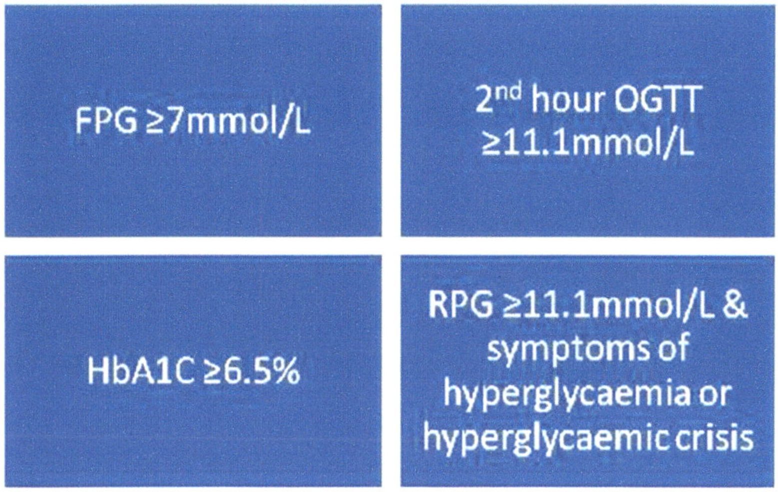

Fig. 41.4 Diagnostic criteria for type 2 diabetes mellitus (American Diabetes Association 2018) *FPG* Fasting plasma glucose, *RPG* Random plasma glucose, *OGTT* 75 g Oral glucose tolerance test, *HbA1C* Haemoglobin A1C

Diagnostic and Screening Testing

The T2DM diagnostic tests are the same as the screening tests for this disease. In 2006 a consensus was reached between WHO, IDF, and ADA, endorsing the same diagnostic cut-off points, as seen in Fig. 41.4.

Obesity

Excess adiposity (obesity) is one of the leading contributing factors to the diabetic epidemic. Both diseases have underlying insulin resistance and inflammatory pathways. The high adiposity status arises from body fat deposition following unhealthy dietary habits, sedentary lifestyles as well as genetic predispositions. The social determinates of health should not be overlooked in the context of obesity, with low socioeconomic status being associated with high adiposity (Cuschieri 2019b). Furthermore, for most of the countries, the typical population living environment is obesogenic in nature, predisposing individuals to unhealthy habits.

Diabetes: From Historical Curiosity to a Serious Epidemic

The (T2DM) burden across different sectors includes disability; morbidity; diminished life expectancy; reduced quality of life; individual and national financial losses; loss of human and social capital; and ultimately increase in mortality (Ali et al. 2010). On evaluating European diabetes-related deaths in 2017, significant mortality rates can be observed across the different countries, as seen in Table 41.1.

A large proportion of this burden is originating from the global ageing of the population. In turn, this has an effect on both the individual and the country level, with an increase in economic burden across healthcare systems. As a matter of fact, this can be observed in Table 41.2, where the mean diabetes expenditure per diabetic person across the European countries was compared for the years 2015 and 2017 (International Diabetes Federation 2015, 2017). Actions targeting the ageing population with appropriate preventive

Table 41.1 Diabetes related deaths in 2017 across Europe (International Diabetes Federation 2015, 2017)

	Diabetes related deaths (20–79 years)
Country	2017
Austria	4963
Belarus	11,564
Czech Republic	9542
Denmark	3165
Finland	3066
Germany	65,178
Greece	5245
Ireland	1092
Israel	2930
Italy	28,695
Malta	280
Portugal	9003
Russian Federation	163,384
Sweden	3730
Switzerland	3100
Turkey	62,473
Ukraine	52,076
United Kingdom	20,846

Table 41.2 Comparison of the mean diabetes expenditure per diabetic person across Europe (International Diabetes Federation 2015, 2017)

	Mean diabetes-related expenditure per person with diabetes (USD)	
Country	2015	2017
Austria	6081	7068
Azerbaijan	602	731
Belarus	583	641
Czech Republic	1552	1763
Denmark	7272	8262
Finland	5043	5876
Germany	5315	6235
Greece	2425	2211
Ireland	5732	6587
Israel	3547	4708
Italy	3450	4006
Malta	2174	2966
Portugal	2101	2420
Russian Federation	1146	1244
Sweden	6776	8872
Switzerland	10,862	12,490
Turkey	846	863
Ukraine	356	259
United Kingdom	4373	5277

interventions and strategies are required. Furthermore, it has been reported that this growing T2DM epidemic is occurring hand in hand with the also exponentiating obesity prevalence, as seen clearly in Table 41.3. Hence targeting both conditions is imperative.

Table 41.3 Prevalence rates of obesity and T2DM across 14 years in Europe

	2003		2006		2010		2013		2015		2017
Country	Obesity (%)	T2DM (%)	Obesity (%)	T2DM (%)	Obesity (%)	T2DM (%)	Obesity (%)	T2DM (%)	Obesity (%)	T2DM (%)	T2DM (%)
Austria	16.2	9.6	17.5	11.1	19.2	11.2	20.5	9.27	21.5	9.5	9.93
Belarus	20.9	6.9	22.1	9.2	23.7	9.1	25.1	6.26	26.1	6.5	7.14
Czech Republic	23.2	9.5	24.3	9.7	25.8	8.7	27.1	9.23	28.0	9.9	9.63
Denmark	16.3	6.9	17.5	7.5	19.0	7.7	20.1	8.58	20.9	9.9	9.86
Finland	19.5	7.2	20.8	8.4	22.4	8.3	23.6	8.85	24.5	9.0	9.56
Germany	19.7	10.2	21.0	11.8	22.8	12.0	24.2	11.95	25.2	10.6	13.4
Greece	20.9	6.1	22.4	8.6	24.4	8.8	25.9	7.01	26.9	7.5	7.71
Ireland	17.9	3.4	19.7	5.6	22.4	5.7	24.7	6.47	26.2	5.3	4.65
Israel	21.9	7.1	23.0	7.8	24.5	7.1	25.6	6.65	26.3	8.5	8.63
Italy	17.9	6.6	19.1	8.7	20.6	8.8	21.7	7.95	22.5	7.9	8.45
Malta	25.6	9.2	27.0	9.7	28.7	9.8	29.9	10.14	30.6	13.9	13.81
Portugal	16.1	7.8	17.7	8.2	19.9	9.9	21.5	12.96	22.7	13.6	14.9
Romania	18.0	9.3	19.1	9.4	21.0	8.4	22.7	5.14	23.9	10.6	12.48
Russian Federation	21.1	9.2	21.9	9.0	23.2	9.0	24.4	10.03	25.2	11.1	8.12
Sweden	16.9	7.3	18.1	7.2	19.6	7.3	20.8	6.36	21.6	6.3	7.2
Switzerland	16.1	9.5	17.2	11.2	18.8	11.3	19.9	7.45	20.8	7.7	7.89
Turkey	22.4	7.0	24.5	7.1	27.4	7.4	29.8	14.58	31.4	12.5	12.54
Ukraine	21.1	9.7	22.0	9.8	23.3	9.6	24.6	2.99	25.6	8.0	8.23
United Kingdom	21.5	3.9	23.2	4.0	25.7	4.9	27.5	6.57	28.9	6.2	5.95

Adapted from WHO Global Observatory data and International Diabetes Federation Atlases (International Diabetes Federation 2017; Cuschieri 2019b; World Health Organization n.d.)

Preventive Screening

The backbone structure for screening criteria was established in 1968 by Wilson and Jungner, to which T2DM fits perfectly. This is because T2DM is a serious disease, with frequent comorbidities and shortened expectancy of life. Diabetes screening tests are relatively cheap with no adverse effects (Evans et al. 2011; Wilson and Jungner 1968). Alas, the setting up of population-based preventive screening has been debated over the years even though diabetes has been a growing disease, affecting both the paediatric and adult population. In 2003, the World Health Organization suggested that if the prevalence of undiagnosed T2DM was very high in a population with an associated cardiovascular disease risk, then population level screening would be beneficial and recommended (International Diabetes Federation 2017). Furthermore, a pan-European population-based screening study (ADDITON—Europe) investigated whether a T2DM screening programme was feasible using a multi-stage screening protocol. The researchers concluded that although this may be challenging, a screening protocol is feasible to implement at a population level (van den Donk et al. 2011). One needs to keep in mind that when screening for T2DM, individuals suffering from impaired fasting blood glucose (IFG) and impaired glucose tolerance (IGT) would be picked up (depending on the screening protocol implemented).

Given the possibility of future T2DM, more adherent follow-up is advised when compared to normoglycaemic individuals. They may be initiated on lifestyle interventions or medication in order to prevent the conversion of their glycaemic status to full-blown diabetes. Hence, screening for T2DM may be visualized as sustaining an added increase in economic burden on the healthcare system. Over time, preventive screening would have a positive outcome on healthcare systems, mortality and morbidity of at-risk individuals, with an increase in productivity and sustainability of the country's economic growth.

References

Alberti KGMM (1996) The clinical implications of impaired glucose tolerance. Diabet Med 13 (11):927–937

Alberti K (2010) The classification and diagnosis of diabetes mellitus. In: Textbook of diabetes. Blackwell Publishing, Hoboken, NJ, pp 24–30

Alberti KGMM, Zimmet P, Shaw J (2006) Metabolic syndrome-a new world-wide definition. A consensus statement from the international diabetes federation. Diabet Med 23(5):469–480

Ali MK, Weber MB, Narayan KM (2010) The global burden of diabetes. In: Textbook of diabetes. Blackwell Publishing, Hoboken, NJ, pp 69–84

American Diabetes Association (2018) Classification and diagnosis of diabetes: standards of medical care in diabetes—2018. Diabetes Care 41(Suppl. 1):S13–S27

Banting FG, Best CH, Collip JB, Campbell WR, Fletcher AA (1922) Pancreatic extracts in the treatment of diabetes mellitus: preliminary report. CMAJ: Canadian Med Assoc J 12:141–146

Beck-Nielsen H, Vaag A, Poulsen P, Gaster M (2003) Metabolic and genetic influence on glucose metabolism in type 2 diabetic subjects—experiences from relatives and twin studies. Best practice & research. Clin Endocrinol Metab 17(3):445–467

Chan LY, Wong SF, Ho LC (2002) Diabetic family history is an isolated risk factor for gestational diabetes after 30 years of age. Acta Obstet Gynecol Scand 81 (2):115–117

Cobb J, Gall W, Adam K et al (2013) A novel fasting blood test for insulin resistance and prediabetes. J Diabetes Sci Technol 7(1):100–110

Cuschieri S (2019a) The genetic side of type 2 diabetes—a review. Diabetes Metab Syndr 13(4):2503–2506. https://doi.org/10.1016/j.dsx.2019.07.012

Cuschieri S (2019b) Type 2 diabetes—an unresolved disease across centuries contributing to a public health emergency. Diabet Metabol Syndr Clin Res Rev 13 (1):450–453

De Franco E, Flanagan SE, Houghton JAL et al (2015) The effect of early, comprehensive genomic testing on clinical care in neonatal diabetes: an international cohort study. Lancet (London, England) 386 (9997):957–963

De Meyer J (1904) Sur la signification physiologique de la secretion interne du pancreas (About the physiological meaning of the internal secretion of the pancreas). Zentralbl Physiologique 18:S826

Diabetes Prevention Program Research Group, Knowler WC, Fowler SE et al (2009) 10-year follow-up of diabetes incidence and weight loss in the diabetes prevention program outcomes study. Lancet (London, England) 374(9702):1677–1686

Dobson M (1776) Experiments and observations on the urine in diabetes. Med Observ Inquir 5:298

Eriksson KF, Lindgärde F (1991) Prevention of type 2 (non-insulin-dependent) diabetes mellitus by diet and physical exercise the 6-year malmö feasibility study. Diabetologia 34(12):891–898

Evans P, Wright C, Pereira D, Langley P (2011) Type 2 diabetes mellitus in family practice: prevention and screening. In: Topics in the prevention, treatment and complications of type 2 diabetes. InTech, New York. http://www.intechopen.com/books/topics-in-the-prevention-treatment-and-complications-of-type-2-diabetes/type-2-diabetes-mellitus-in-family-practice-prevention-and-screening. 10.5772/22451

Hamman RF, Wing RR, Edelstein SL et al (2006) Effect of weight loss with lifestyle intervention on risk of diabetes. Diabetes Care 29(9):2102–2107

Himsworth HK (1939) Insulin-sensitivity and insulin-insensitive types of diabetes mellitus. Clin Sci 4:119–152

International Diabetes Federation (2015) IDF Diabates Atlas—7th edition. International Diabetes Federation, Brussels

International Diabetes Federation (2017) IDF diabetes atlas, 8th edn. International Diabetes Federation, Brussels

Jiang X, Ma H, Wang Y, Liu Y (2013) Early life factors and type 2 diabetes mellitus. J Diabetes Res 2013:485082

Kahn SE, Hull RL, Utzschneider KM (2006) Mechanisms linking obesity to insulin resistance and type 2 diabetes. Nature 444(7121):840–846

Kanat M, Mari A, Norton L et al (2012) Distinct-cell defects in impaired fasting glucose and impaired glucose tolerance. Diabetes 61(2):447–453

Kaplan GA, Everson SA, Lynch JW (2000) The contribution of social and behavioral research to an understanding of the distribution of disease: a multilevel approach. In: Promoting health: intervention strategies from social and behavioral research. National Academies Press, New York, USA, p 42. https://www.ncbi.nlm.nih.gov/books/NBK222836/

Kilpatrick ES, Bloomgarden ZT, Zimmet PZ (2009) International expert committee report on the role of the A1C assay in the diagnosis of diabetes: response to the international expert committee. Diabetes Care 32(12): e159; author reply e16–e159. https://www.ncbi.nlm.nih.gov/pubmed/19940222. https://doi.org/10.2337/dc09-1231

Köbberling J, Tillil H (1982) Empirical risk figures for first degree relatives of non-insulin dependent diabetics. In: Köbberling J, Tattersall R (eds) The genetics of diabetes mellitus. Academic Press, London, pp 201–209

Laguesse G (1893) Sur la formation des ilots de Langerhans dans le pancreas. (About the formation of the Langerhans islets in the pancreas). C R Seances Mem Soc Biol 45:819–820

Langerhans P (1869) Beitrage zur mikroskopischen anatomie der bauchspeicheldruse (Contribution to the microscopic anatomy of the pancreas).

Libman IM, LaPorte RE, Libman AM, Arias P (2011) Non-type 2 diabetes: prevalence, incidence and risk factors. In: Diabetes public health from data to policy. Oxford University Press, Oxford, pp 65–92

Lindström J, Louheranta A, Mannelin M et al (2003) The finnish diabetes prevention study (DPS): lifestyle intervention and 3-year results on diet and physical activity. Diabetes Care 26(12):3230–3236. https://www.ncbi.nlm.nih.gov/pubmed/14633807. https://doi.org/10.2337/diacare.26.12.3230

McCarthy M (1998) Weighing in on diabetes risk. Nat Genet 19(3):209–210

Metz C, Cavé H, Bertrand AM et al (2002) Neonatal diabetes mellitus: chromosomal analysis in transient and permanent cases. J Pediatr 141(4):483–489

Metzger BE, Buchanan TA, Coustan DR et al (2007) Summary and recommendations of the fifth international workshop-conference on gestational diabetes mellitus. Diabetes Care 30(Supplement 2):S251–S260

National Diabetes Data Group (1979) Classification and diagnosis of diabetes mellitus and other categories of glucose intolerance. National diabetes data group. Diabetes 28(12):1039–1057

Newman B, Selby JV, King MC, Slemenda C, Fabsitz R, Friedman GD (1987) Concordance for type 2 (non-insulin-dependent) diabetes mellitus in male twins. Diabetologia 30(10):763–768

Pan XR, Li GW, Hu YH et al (1997) Effects of diet and exercise in preventing NIDDM in people with impaired glucose tolerance the da qing IGT and diabetes study. Diabetes Care 20(4):537–544

Paton JS, Wilson M, Ireland JT, Reith SB (1981) Convenient pocket insulin syringe. Lancet (London, England) 1(8213):189–190

Ramachandran A, Snehalatha C, Mary S et al (2006) The Indian diabetes prevention programme shows that lifestyle modification and metformin prevent type 2 diabetes in Asian Indian subjects with impaired glucose tolerance (IDPP-1). Diabetologia 49 (2):289–297

Randle PJ, Garland PB, Hales CN, Newsholme EA (1963) The glucose fatty-acid cycle its role in insulin sensitivity and the metabolic disturbances of diabetes mellitus. Lancet (London, England) 1(7285):785–789

Rollo J (1797) An account of two cases of the diabetes mellitus, with remarks as they arose during the progress of the cure. C Dilly, London, p 27

Shield JP, Gardner RJ, Wadsworth EJ et al (1997) Aetiopathology and genetic basis of neonatal diabetes. Arch Dis Child Fetal Neonatal Ed 76(1):39

Sladek R, Rocheleau G, Rung J et al (2007) A genome-wide association study identifies novel risk loci for type 2 diabetes. Nature 445(7130):881–885

Sperling MA (2006a) ATP-sensitive potassium channels — neonatal diabetes mellitus and beyond. N Engl J Med 355(5):507–510

Sperling MA (2006b) The genetic basis of neonatal diabetes mellitus. Pediatr Endocrinol Rev 4(Suppl 1):71–75

Susser MSE (1996) Choosing a future for epidemiology: II. From black box to Chinese boxes and eco-epidemiology. Am J Public Health 86:674–677

Tattersall RB (1974) Mild familial diabetes with dominant inheritance. Q J Med 43(170):339–357

Tattersall R (1998) Maturity-onset diabetes of the young: a clinical history. Diabet Med 15(1):11–14

Tattersall RB (2010) The history of diabetes mellitus. In: Textbook of diabetes. Blackwell Publishing, Hoboken, NJ, pp 3–20

The Expert Committee on the Diagnosis and Classification of Diabetes Mellitus (1997) Report of the expert committee on the diagnosis and classification of diabetes mellitus. Diabetes Care 20(7):1183–1197

Tuomi T, Groop LC, Zimmet PZ, Rowley MJ, Knowles W, Mackay IR (1993) Antibodies to glutamic acid decarboxylase reveal latent autoimmune diabetes mellitus in adults with a non-insulin-dependent onset of disease. Diabetes 42(2):359–362

Tuomilehto J, Lindström J, Eriksson JG et al (2001) Prevention of type 2 diabetes mellitus by changes in lifestyle among subjects with impaired glucose tolerance. N Engl J Med 344(18):1343–1350

Tyberghein JM, Williams RH (1957) Metabolic effects of phenethyldiguanide, a new hypoglycemic compound. Exp Biol Med 96(1):29–32

van den Donk M, Sandbaek A, Borch-Johnsen K et al (2011) Screening for type 2 diabetes lessons from the ADDITION-Europe study. Diabet Med 28 (11):1416–1424

von Mühlendah KE, Herkenhoff H (1995) Long-term course of neonatal diabetes. N Engl J Med 333 (11):704–708

Wilson JMG, Jungner G (1968) Principles and practice of screening for disease, vol 34. World Health Organization, Geneva. http://apps.who.int/iris/bitstream/10665/37650/17/WHO_PHP_34.pdf

World Health Organisation Consultation Group (1999) Definition, diagnosis and classification of diabetes mellitus and its complications. part 1: diagnosis and classification of diabetes: report of a WHO consolation. WHO, Geneva

World Health Organization (2006) Definition and diagnosis of diabetes mellitus and intermediate hyperglycemia report of a WHO/IDF consultation. WHO, Geneva

World Health Organization (n.d.) GHO by category overweight/obesity. WHO, Geneva

Glucose Control in the Intensive Care Unit

42

Jan Gunst and Greet Van den Berghe

Contents

Introduction 580

Hyperglycemia and Outcome: A Causal Relationship or Not? 580

Harm by Iatrogenic Hypoglycemia? 581

How to Explain the Discrepant RCT Results? 582
Differences in Blood Glucose Targets 582
Different Protocols to Monitor Glucose and Administer Insulin 583
Different Feeding Protocols 584

Toward Precision Medicine? 585
Do Diabetes Patients Benefit from More Loose Blood Glucose Control? 585
An Individualized Target? 585

Strategies to Optimize Glucose Control 586
Decision-Support Software 586
Continuous Glucose Monitoring 586
Closed-Loop Glucose Control: The Artificial Pancreas 586

Conclusion 587

References 587

Abstract

Most critically ill patients develop hyperglycemia, the degree of which is associated with mortality. However, randomized controlled trials investigating tight glucose control have found a divergent impact on morbidity and mortality, possibly explained by differences in glucose targets, monitoring and titration tools, and nutritional strategies. In patients receiving parenteral nutrition early in the course of critical illness, tight glucose control has shown to be effective and safe when performed with a validated protocol, including accurate glucose measurements. The optimal blood glucose target remains unclear for patients not receiving early parenteral nutrition, and may differ according to the premorbid diabetes status. Efficacy and safety of tight glucose control can be improved by

J. Gunst · G. Van den Berghe (✉)
Department of Cellular and Molecular Medicine, Clinical Division and Laboratory of Intensive Care Medicine, KU Leuven, Leuven, Belgium
e-mail: greet.vandenberghe@kuleuven.be

J. Faintuch and S. Faintuch (eds.), *Obesity and Diabetes*, https://doi.org/10.1007/978-3-030-53370-0_42

using validated decision-support software, continuous glucose monitoring and closed-loop insulin delivery, although widespread implementation of these technologies is precluded by the ongoing controversy on the topic, the added cost, and, for some devices, lack of measurement accuracy.

Keywords

Glucose targets · Intensive insulin therapy · Intensive glucose monitoring · Tight glucose control · Closed-loop glucose control · Continuous glucose monitoring · Hypoglycemia risk · Hypoglycemia prevention

Introduction

The stress response evoked by severe medical illnesses, major trauma, and surgery induces profound metabolic and endocrine alterations, including insulin resistance and resultant hyperglycemia. Numerous observational studies in different subgroups of critically ill patients have shown a clear relationship between spontaneous (i.e., untreated) blood glucose concentrations and subsequent mortality risk, whereby the risk of mortality is lowest in patients with blood glucose in the normal, age-adjusted fasting range, and progressively increases with more severe hyper- and hypoglycemia (Fig. 42.1) (Falciglia et al. 2009; Kosiborod et al. 2005; Egi et al. 2008). In patients with pre-existing diabetes, a similar U-shaped relationship between blood glucose and outcome has been observed, although the curve appears flattened and somewhat shifted to the right. However, whether such associations reflect a causal relationship between blood glucose concentrations and outcome cannot be deducted from observational studies. Indeed, the higher mortality risk in patients with more severe hyperglycemia and hypoglycemia could equally be explained by increased illness severity in these patients, with inherently an increased mortality risk.

Hyperglycemia and Outcome: A Causal Relationship or Not?

Assessment of causality can only be performed by a randomized controlled trial in which patients are randomized to treating stress hyperglycemia

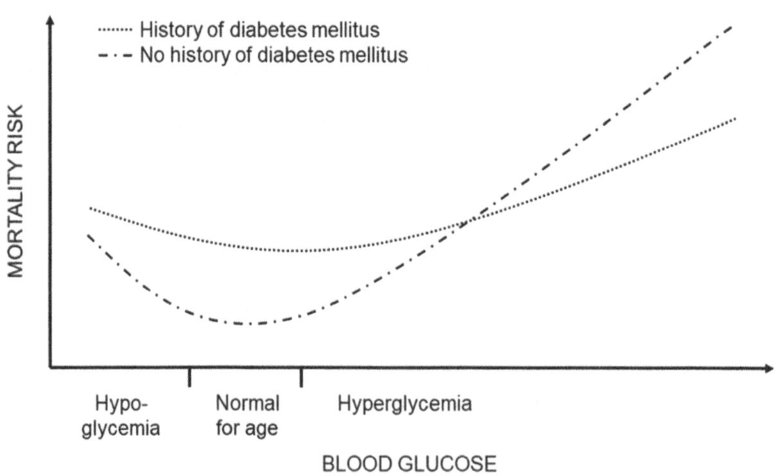

Fig. 42.1 Association of blood glucose concentrations with outcome. In critically ill patients, numerous observational studies have revealed a U-shaped relationship between upon-admission and mean blood glucose concentrations and the risk of mortality, with the lowest mortality risk associated with normal-for-age fasting blood glucose concentrations in patient without pre-existing diabetes. In patients with pre-existing diabetes, the curve is flattened and the nadir somewhat shifted to the right

versus tolerating this condition up to a certain threshold. Randomized controlled trials (RCTs) on this topic have yielded conflicting results, however (Van den Berghe et al. 2001, 2006a; Vlasselaers et al. 2009; Finfer et al. 2009; Brunkhorst et al. 2008; Preiser et al. 2009; Annane et al. 2010; Kalfon et al. 2014; Macrae et al. 2014; Agus et al. 2012, 2017a; Bilotta et al. 2007, 2009; Jeschke et al. 2010).

In three pioneer RCTs performed in Leuven, Belgium, which were performed in the context of early use of parenteral nutrition, strictly targeting the normal, age-adjusted fasting range of blood glucose (80–110 mg/dL [4.4–6.1 mmol/L] for adults, 50–80 mg/dL [2.8–4.4 mmol/L] for infants, and 70–100 mg/dL [3.9–5.6 mmol/L] for children older than 1 year) with insulin therapy, an intervention labeled as tight glucose control, was compared with tolerating hyperglycemia up to the renal threshold (215 mg/dL [11.9 mmol/L]) (Van den Berghe et al. 2001, 2006a; Vlasselaers et al. 2009). The latter threshold was standard care at that time, since more severe hyperglycemia may induce obvious complications, including osmotic diuresis and associated fluid shifts. In the Leuven RCTs, altogether having included 3448 patients, tight glucose control significantly improved morbidity and mortality of critically ill adults and children.

Indeed, among others, the intervention reduced the incidence of acute kidney injury and critical illness polyneuropathy, facilitated weaning from mechanical ventilation, and reduced the incidence of infectious complications, hereby reducing dependency on intensive care, risk of mortality, and costs for healthcare (Van den Berghe et al. 2001, 2006a, b, c; Vlasselaers et al. 2009; Hermans et al. 2007). Importantly, the mortality benefit was maintained on the long-term and in critically ill children, patients had slightly improved neurocognitive outcome 4 years after randomization, despite the increased incidence of severe hypoglycemia (Ingels et al. 2006; Mesotten et al. 2012). Mechanistic studies performed by the Leuven research group attributed the benefit to prevention of cellular glucose overload and associated toxicity, rather than to glycemia-independent effects of insulin (Ellger et al. 2006; Vanhorebeek et al. 2009a, b; Gunst and Van den Berghe 2016). Inspired by the clear benefit of a seemingly "simple" intervention, glucose control strategies were soon implemented in many centers, albeit with varying targets. Subsequently, several implementation studies and small single-center RCTs confirmed clinical benefit by implementing tight glucose control (Bilotta et al. 2007, 2009; Jeschke et al. 2010; Krinsley 2004; Lecomte et al. 2008). However, clinical benefit was not confirmed in multicenter RCTs, and the largest multicenter RCT until now, the NICE-SUGAR study ($n = 6104$), found excess mortality by tight glucose control (Finfer et al. 2009; Brunkhorst et al. 2008; Preiser et al. 2009; Annane et al. 2010; Kalfon et al. 2014; Macrae et al. 2014; Agus et al. 2012, 2017a). As a result, tight glucose control remains debated, and clinical practices vary widely (Gunst and Van den Berghe 2016; Marik 2016; Niven et al. 2015; Krinsley and Preiser 2019).

Harm by Iatrogenic Hypoglycemia?

All RCTs on tight glucose control have shown a significantly increased risk of hypoglycemia (Van den Berghe et al. 2001, 2006a; Vlasselaers et al. 2009; Finfer et al. 2009; Brunkhorst et al. 2008; Preiser et al. 2009; Annane et al. 2010; Kalfon et al. 2014; Macrae et al. 2014; Agus et al. 2012, 2017a; Bilotta et al. 2007, 2009; Jeschke et al. 2010). This has been a concern for clinicians, since prolonged, severe hypoglycemia may be life threatening. Moreover, in a secondary analysis of the NICE-SUGAR study, the investigators attributed the excess mortality by the intervention to an increased risk of hypoglycemia (Finfer et al. 2012). However, increased incidence of hypoglycemia has also been found in RCTs showing benefit of tight glucose control (Van den Berghe et al. 2001, 2006a; Vlasselaers et al. 2009; Bilotta et al. 2007, 2009; Jeschke et al. 2010). Moreover, in a nested-case control study of patients included in the pediatric Leuven RCT, iatrogenic severe hypoglycemia did neither associate with increased biomarkers of neuronal damage nor with impaired neurocognitive outcome (Mesotten et al. 2012; Vanhorebeek et al. 2010). The prevention of hyperglycemia during glucose reperfusion in the Leuven RCT may have played a role, since hyperglycemia during glucose reperfusion is also potentially deleterious (Vanhorebeek et al. 2010; Suh et al. 2007).

In this regard, an animal model has shown that neuronal death was triggered by glucose reperfusion rather than by hypoglycemia per se (Suh et al. 2007). Hence, although prolonged, severe hypoglycemia is harmful and should be avoided, it remains unclear whether short-term iatrogenic hypoglycemia by itself impairs outcome. Nevertheless, confirmation of this hypothesis would require an RCT, which is ethically not justifiable. Hence, one cannot exclude the possibility that iatrogenic hypoglycemia may have offset some of the benefit of tight glucose control in the Leuven RCTs. Therefore, hypoglycemia should be prevented and treated early, while carefully avoiding rebound hyperglycemia.

How to Explain the Discrepant RCT Results?

Most likely, the outcome differences across subsequent RCTs relate to methodological differences among the trials. Indeed, on several key points, the protocols of the Leuven RCTs and subsequent multicenter RCTs substantially differed (Table 42.1). Three key differences include different blood glucose targets in the randomization groups, different protocols to monitor glucose and administer insulin, and different feeding protocols (Van den Berghe et al. 2009).

Differences in Blood Glucose Targets

The pioneer Leuven studies had a worldwide impact on blood glucose control practices (Niven et al. 2015). The shift in standard care drove researchers of subsequent RCTs to compare tight glucose control no longer with tolerating hyperglycemia up to the renal

Table 42.1 Key differences between the pioneer Leuven RCTs and subsequent multicenter RCTs

	Leuven RCTs	Multicenter RCTs
Glucose target		
Control group	<215 mg/dL	In general <180 mg/dL
Intervention group	80–110 mg/dL (adults); 70–100 mg/dL (children older than 1 year); 50–80 mg/dL (infants)	80–110 mg/dL (regardless of age)
Glucose measurement		
Measurement site	Predominantly arterial	Variable; venous and capillary measurements in some RCTs
Measurement device	Predominantly blood gas analyzer	Variable; potentially inaccurate glucometers in some RCTs
Insulin administration		
Infusion practice	Continuous intravenous infusion	Variable; boluses allowed in some RCTs
Dose adjustment	Well-trained nurses, allowing intuitive decision making	Variable; strict, unvalidated protocols in some RCTs
Nutritional management	Full feeding in acute phase of critical illness (including early PN)	Variable

To convert mg/dL to mmol/L, divide by 18
PN parenteral nutrition, *RCT* randomized controlled trial

threshold, but to a lower, intermediate target [in general <180 mg/dL (<10 mmol/L)]. Hence, in most subsequent studies, including NICE-SUGAR, the comparison was different (Finfer et al. 2009). As a result, the difference in achieved blood glucose concentrations was generally much smaller than in the Leuven RCTs, whereby some RCTs may have been underpowered to detect a benefit of tight glucose control. However, this does not explain the observed harm in NICE-SUGAR.

Aggregating the RCT evidence, one could argue that targeting an intermediate range of blood glucose would be optimal. However, no large RCT has directly compared such intermediate target with tolerating hyperglycemia up to the renal threshold. Conversely, a secondary analysis of the adult Leuven RCTs has suggested a dose-dependent effect of blood glucose lowering on outcome. Indeed, regardless of randomization, patients with blood glucose in the tight range had better outcome than patients with blood glucose in the intermediate range, which suggests that tight glucose control may be preferred (Van den Berghe et al. 2006b). Moreover, apart from a different glucose target in the control group, other methodological differences may equally explain the increased mortality risk in NICE-SUGAR (Van den Berghe et al. 2009).

Not only the glucose target in the control group differed between the RCTs. In the pediatric multicenter RCTs on tight glucose control, the target in the intervention group differed as compared to the Leuven RCT (Vlasselaers et al. 2009; Macrae et al. 2014; Agus et al. 2012, 2017a). Indeed, in Leuven, the target range in the intervention group was age-adjusted and lower than in adults, to account for the lower fasting blood glucose concentrations in healthy infants and children (Vlasselaers et al. 2009). In contrast, in the pediatric multicenter RCTs, the higher, adult target was used in the intervention group (Macrae et al. 2014; Agus et al. 2012, 2017a). Consequently, the achieved difference in blood glucose in pediatric multicenter RCTs was considerably lower than in Leuven. Indeed, whereas the mean achieved difference in blood glucose between both randomization groups was 34 mg/dL (1.9 mmol/L) and 45 mg/dL (2.5 mmol/L) for infants, respectively, older children included in the Leuven RCT, the mean difference was 7 mg/dL (0.4 mmol/L), 9 mg/dL (0.5 mmol/L), and 14 mg/dL (0.8 mmol/L) in subsequent multicenter RCTs (Agus et al. 2017b). Hence, the hypothesized effect size, which was extrapolated from the Leuven RCT, was probably not realistic. Consequently, these RCTs were likely underpowered (Gunst and Van den Berghe 2017). Nevertheless, one pediatric multicenter RCT suggested a potential benefit of tight glucose control in the high-risk subgroup of noncardiac surgery patients (Macrae et al. 2014).

Different Protocols to Monitor Glucose and Administer Insulin

Second, the protocol whereby tight glucose control was realized, substantially differed between the consecutive RCTs. In contrast to the Leuven studies, in which the glucose control protocol was well standardized, protocols of some subsequent RCTs, including NICE-SUGAR, lacked standardization (Van den Berghe et al. 2009). Standardization, including frequent, accurate glucose measurements and a validated insulin protocol, is likely important to prevent and early treat hypoglycemia. In Leuven, glucose measurements were predominantly performed on arterial blood using a blood gas analyzer, which results in a fast and accurate measurement of blood glucose (Van den Berghe et al. 2001, 2006a; Vlasselaers et al. 2009). Insulin was administered by continuous intravenous infusion through a central venous line, without administering boluses. The insulin dose was titrated by experienced nurses using a guideline that allowed intuitive decision making. Overall, the Leuven protocol led to a high time-in-target range, and a lower incidence of hypoglycemia than in NICE-SUGAR (Van den Berghe et al. 2009).

In NICE-SUGAR, the lack of standardization with regard to glucose measurements and insulin titration may have increased the risk of the intervention (Gunst and Van den Berghe 2016). Indeed, in contrast to Leuven, a variety of point-of-care glucometers were allowed, as were capillary and venous glucose measurements. This may have resulted in a substantial number of inaccurate glucose measurements. In this regard, a center participating in NICE-SUGAR has reported problematic inaccuracies with the glucometer used at that time, with measurement errors up to 100 mg/dL (5.6 mmol/L) (Cembrowski et al. 2010; Scott et al. 2009). Likewise, capillary and venous glucose measurements are potentially inaccurate, especially in patients with shock and in patients receiving intravenous glucose, respectively (Van den Berghe et al. 2009). Consequently, inaccurate glucose measurements in NICE-SUGAR may have led to undetected episodes of prolonged and severe hypoglycemia, which could explain the excess mortality.

Moreover, inaccurate glucose measurements could have increased glucose variability, which is also associated with impaired outcome (Krinsley et al. 2013; Bagshaw et al. 2009). The risk of hypoglycemia in NICE-SUGAR, both detected and undetected, and of large glucose fluctuations, was potentially aggravated by the insulin administration protocol. Indeed, in contrast to Leuven, insulin boluses were allowed on top of continuous infusion (Finfer et al. 2009). Moreover, insulin was titrated using a strict and unvalidated if-then algorithm. Altogether, these methodological differences between trials indicate that safe and effective implementation of tight glucose control requires a well-standardized and accurate protocol that includes frequent, accurate glucose measurements and a validated insulin protocol.

Different Feeding Protocols

A third important difference between the subsequent glucose control RCTs is the difference in feeding protocols, which is explained by the long-lasting lack of solid evidence on critical care nutrition (Casaer and Van den Berghe 2014). Indeed, for a long time, artificial feeding practices have been based on observational studies and expert opinion. In the Leuven RCTs on tight glucose control, all patients received early parenteral nutrition when enteral nutrition was insufficient to meet the caloric target, which was in line with the European feeding guidelines at that time (Singer et al. 2009). In subsequent multicenter RCTs, feeding protocols differed. In NICE-SUGAR, although not protocolized, on average much less parenteral glucose was administered in the acute phase of critical illness (Finfer et al. 2009). This feeding strategy by itself reduces the degree of hyperglycemia. Recently, two large multicenter RCTs have demonstrated that early parenteral nutrition prolonged intensive care dependency in both critically adults and children, also when feeding-induced hyperglycemia is treated (Casaer et al. 2011; Fivez et al. 2016). Hence, contemporary feeding guidelines now no longer recommend early initiation of parenteral nutrition (Singer et al. 2019).

Apart from attenuating the degree of hyperglycemia, the current feeding practice of withholding early parenteral nutrition also increases the risk of hypoglycemia when tight glucose control is aimed for (Casaer et al. 2011; Fivez et al. 2016). Currently, no large RCT has investigated whether tight glucose control is still beneficial, when performed with a validated protocol, including accurate glucose measurements, in the absence of early parenteral nutrition. A meta-analysis of existing RCTs suggested that the outcome benefit of tight glucose control was only present in RCTs having administered larger amounts of intravenous glucose in the acute phase of critical illness (Marik and Preiser 2010). However, this meta-analysis did not correct for other methodological differences between the included trials. In contrast, a secondary analysis of the adult Leuven RCTs found benefit by tight glucose control also in patients in the lowest tertile of parenteral glucose administration (Van den Berghe et al. 2006b). Hence, in patients not receiving early parenteral nutrition, it currently remains unknown whether stress hyperglycemia should be tolerated up to a certain threshold or treated using a validated protocol. This is being investigated in the multicenter Tight versus Liberal Blood Glucose Control in Adult Critically Ill Patients (TGC-fast) study, which is currently enrolling patients (clinicaltrials.gov, NCT03665207).

Toward Precision Medicine?

Apart from the methodological differences between the RCTs, one could also question whether the divergent results could be explained by inclusion of different patient subgroups. However, the NICE-SUGAR study, including mixed intensive care unit (ICU) patients, did not find a subgroup in whom there was a differential impact of the intervention (Finfer et al. 2009). Likewise, secondary analyses from the adult Leuven RCTs found benefit in all subgroups, with the possible exception of diabetes patients (Van den Berghe et al. 2006b).

Do Diabetes Patients Benefit from More Loose Blood Glucose Control?

Observational studies have shown that the relationship between spontaneous blood glucose concentrations and outcome differs between patient subgroups (Falciglia et al. 2009; Kosiborod et al. 2005; Egi et al. 2008). Indeed, in particular, in patients with pre-existing diabetes mellitus, the U-shaped curve is somewhat flattened and the nadir shifted to the right (Fig. 42.1). This may suggest that patients with pre-existing diabetes mellitus may benefit from less tight glucose control during ICU stay, and that acutely lowering chronic hyperglycemia may be harmful. Although there is some physiological rationale to support this hypothesis, there is no clear RCT evidence to support it. A pooled analysis from the adult Leuven studies found that the mortality benefit of tight glucose control was only present in the subgroup of nondiabetes patients (Van den Berghe et al. 2006b). However, the incidence of acute kidney injury and critical illness polyneuropathy was numerically lower in the group receiving tight glucose control, also in the subgroup of diabetes patients. Likewise, several multicenter RCTs, including NICE-SUGAR, found no different effect of the intervention in diabetes patients versus nondiabetics (Finfer et al. 2009; Brunkhorst et al. 2008; Kalfon et al. 2014).

An Individualized Target?

Observational studies in diabetes patients have indicated that the relationship between blood glucose and outcome differed according to the estimated antecedent blood glucose control (Egi et al. 2011; Plummer et al. 2014). Indeed, the lowest mortality risk was found to associate with blood glucose concentrations close to the average preadmission level, as estimated from upon-admission hemoglobin A1c (Egi et al. 2011; Preiser et al. 2018). Likewise, a study in hospitalized patients has suggested that relative hyperglycemia upon admission, as reflected by an increased ratio of the upon-admission blood glucose to the estimated average blood glucose concentration before hospital admission, may be a better predictor of outcome than absolute hyperglycemia (Roberts et al. 2015). This suggests that the ideal blood glucose target in critical illness may depend on the average preadmission blood glucose concentration.

Nevertheless, confirmation of this hypothesis requires confirmation by an RCT. Published RCTs have not separately reported on outcome in diabetes patients with and without good antecedent glucose control. The multicenter CONTROLING trial was designed to address this knowledge gap. The investigators randomized critically ill patients to individualized blood glucose treatment versus standard care (clinicaltrials.gov, NCT02244073). In the intervention group, the individualized target was calculated based on the upon-admission hemoglobin A1c level. The study was terminated prematurely by the data safety monitoring board. At the time of submission of this manuscript, no results have been published yet.

Strategies to Optimize Glucose Control

The conflicting results between the consecutive RCTs illustrate that tight glucose control is not a "simple" intervention that is easy to implement, but an intervention that requires a validated and accurate glucose control protocol and a skilled nursing team, with prevention of hypoglycemia and large glucose fluctuations.

Decision-Support Software

To further improve the quality and safety of tight glucose control, several computerized algorithms have been developed, with varying design and performance (Salinas and Mendez 2019; Chase et al. 2018). The Leuven research group has developed a computerized algorithm—the LOGIC-Insulin software—based on the experiences of the nurses in Leuven (Van Herpe et al. 2013). The algorithm takes into account demographic data, as well as data on steroids, nutrition, and trends in blood glucose, and consequently advises on the insulin dose and glucose bolus (in case of hypoglycemia), and on the time of the next glucose measurement. In the single-center randomized LOGIC-1 study, the software was found to be superior to nurse-guided tight glucose control, with a higher time in blood glucose target range and virtually no episodes of hypoglycemia (Van Herpe et al. 2013).

Subsequently, the software was validated in a multicenter RCT, which demonstrated a similar high performance in the different centers, with a higher time in target range than in the nurse-controlled group, and without increasing the risk of hypoglycemia (Dubois et al. 2017). Apart from the LOGIC-Insulin algorithm, also other algorithms have shown to improve the time in blood glucose target range, without increasing the risk of hypoglycemia (Salinas and Mendez 2019; Chase et al. 2018). Hence, implementing such algorithms could greatly improve the efficacy and safety of tight glucose control, especially in centers not experienced with tight glucose control.

Continuous Glucose Monitoring

Likewise, (near-)continuous glucose monitoring (CGM) could theoretically improve the quality of tight glucose control. However, the use of subcutaneous sensors measuring interstitial glucose, which are often used in ambulatory patients with diabetes, is complicated in critically ill patients because of potential inaccuracies due to edema and poor capillary perfusion, as well as to the non-negligible time lag between systemic and interstitial glucose concentrations (van Steen et al. 2017). Therefore, intravascular sensors have been developed, and several devices currently have regulatory approval (van Steen et al. 2017). Despite this, most devices do not meet the very strict requirements of point accuracy, as set by a group of critical care experts (Finfer et al. 2013). Theoretically however, this drawback could be overcome by trend accuracy (Bochicchio et al. 2019; Righy Shinotsuka et al. 2016). Nevertheless, the limited durability and added cost of a CGM device, as well as the current controversy about the benefit-risk ratio of tight glucose control, preclude widespread implementation of CGM (Krinsley et al. 2017).

Closed-Loop Glucose Control: The Artificial Pancreas

Theoretically, the ultimate solution for high-quality glucose control is the implementation of closed-loop insulin delivery systems, which combine CGM with automated, software-controlled insulin delivery. Recent RCTs have demonstrated improved glucose control as compared to standard care, by closed-loop insulin delivery in ambulatory and hospitalized diabetes patients (Bally et al. 2018; Brown et al. 2019; Boughton et al. 2019). However in these patients, interstitial glucose sensors were used and insulin was administered subcutaneously, which is potentially hazardous in critically ill patients. To the best of our knowledge, only one company has an available closed-loop system that combines intravascular blood glucose measurement, venous in this case, with venous administration of insulin

(Salinas and Mendez 2019; Okabayashi et al. 2014). Interestingly, the system has been tested in a single-center RCT, in which intermediate glucose control was compared to tight glucose control guided by the closed-loop device, in patients receiving early parenteral nutrition. As compared to the control group, closed-loop glucose control improved clinical outcome (Okabayashi et al. 2014). However, apart from not being commercially available in most countries, the limited durability, extra cost, and controversy about tight glucose control may preclude clinicians to use such device (Salinas and Mendez 2019; Krinsley et al. 2017).

Conclusion

The optimal blood glucose target for critically ill patients remains unclear. When performed in the context of parenteral nutrition administered early in the course of critical illness, and with use of a validated protocol, including accurate glucose measurements, tight glucose control has shown to be effective and safe. The impact of tight glucose control, when performed with a validated protocol in the context of not administering parenteral nutrition in the acute phase of critical illness, remains unclear. The use of clinically validated decision-support software may improve the quality of glucose control and decrease the incidence of hypoglycemia.

There is insufficient evidence to use different blood glucose targets in subgroups of critically ill patients, with the potential exception of poorly controlled diabetes patients, who might benefit from less strict control. In the absence of new evidence, it seems prudent to avoid severe hyperglycemia, hypoglycemia, and large glucose fluctuations for all critically ill patients. A large RCT is currently ongoing to test the impact of tight glucose control in the context of omitting early parenteral feeding, and with use of accurate measurements and a validated insulin-titration computer-guided algorithm (clinicaltrials.gov, NCT03665207).

Acknowledgments JG and GVdB receive a research grant from the Research Foundation—Flanders (FWO; T003617N). JG receives a research grant from the University of Leuven (C24/17/070) and a postdoctoral research fellowship supported by the Clinical Research and Education Council of the University Hospitals Leuven. GVdB receives structural research financing via the Methusalem program funded by the Flemish government through the university of Leuven (METH14/06) and a European Research Council Advanced Grant (ERC-2017-ADG-785809).

References

Agus MS, Steil GM, Wypij D et al (2012) Tight glycemic control versus standard care after pediatric cardiac surgery. N Engl J Med 367:1208–1219

Agus MS, Wypij D, Hirshberg EL et al (2017a) Tight glycemic control in critically ill children. N Engl J Med 376:729–741

Agus MSD, Wypij D, Nadkarni VM (2017b) Tight glycemic control in critically ill children. N Engl J Med 376:e48

Annane D, Cariou A, Maxime V et al (2010) Corticosteroid treatment and intensive insulin therapy for septic shock in adults: a randomized controlled trial. JAMA 303:341–348

Bagshaw SM, Bellomo R, Jacka MJ et al (2009) The impact of early hypoglycemia and blood glucose variability on outcome in critical illness. Crit Care 13: R91

Bally L, Thabit H, Hartnell S et al (2018) Closed-loop insulin delivery for glycemic control in noncritical care. N Engl J Med 379:547–556

Bilotta F, Spinelli A, Giovannini F, Doronzio A, Delfini R, Rosa G (2007) The effect of intensive insulin therapy on infection rate, vasospasm, neurologic outcome, and mortality in neurointensive care unit after intracranial aneurysm clipping in patients with acute subarachnoid hemorrhage: a randomized prospective pilot trial. J Neurosurg Anesthesiol 19:156–160

Bilotta F, Caramia R, Paoloni FP, Delfini R, Rosa G (2009) Safety and efficacy of intensive insulin therapy in critical neurosurgical patients. Anesthesiology 110:611–619

Bochicchio GV, Nasraway SA, Moore LJ et al (2019) Fifteen-minute frequency of glucose measurements and the use of threshold alarms: impact on mitigating dysglycemia in critically ill patients. J Diabetes Sci Technol 2019:1932296819886917

Boughton CK, Bally L, Martignoni F et al (2019) Fully closed-loop insulin delivery in inpatients receiving nutritional support: a two-Centre, open-label, randomised controlled trial. Lancet Diabetes Endocrinol 7:368–377

Brown SA, Kovatchev BP, Raghinaru D et al (2019) Six-month randomized, multicenter trial of closed-

loop control in type 1 diabetes. N Engl J Med 381:1707–1717
Brunkhorst FM, Engel C, Bloos F et al (2008) Intensive insulin therapy and pentastarch resuscitation in severe sepsis. N Engl J Med 358:125–139
Casaer MP, Van den Berghe G (2014) Nutrition in the acute phase of critical illness. N Engl J Med 370:1227–1236
Casaer MP, Mesotten D, Hermans G et al (2011) Early versus late parenteral nutrition in critically ill adults. N Engl J Med 365:506–517
Cembrowski GS, Tran DV, Slater-Maclean L, Chin D, Gibney RT, Jacka M (2010) Could susceptibility to low hematocrit interference have compromised the results of the NICE-SUGAR trial? Clin Chem 56:1193–1195
Chase JG, Desaive T, Bohe J et al (2018) Improving glycemic control in critically ill patients: personalized care to mimic the endocrine pancreas. Crit Care 22:182
Dubois J, Van Herpe T, van Hooijdonk RT et al (2017) Software-guided versus nurse-directed blood glucose control in critically ill patients: the LOGIC-2-multicenter randomized controlled clinical trial. Crit Care 21:212
Egi M, Bellomo R, Stachowski E et al (2008) Blood glucose concentration and outcome of critical illness: the impact of diabetes. Crit Care Med 36:2249–2255
Egi M, Bellomo R, Stachowski E et al (2011) The interaction of chronic and acute glycemia with mortality in critically ill patients with diabetes. Crit Care Med 39:105–111
Ellger B, Debaveye Y, Vanhorebeek I et al (2006) Survival benefits of intensive insulin therapy in critical illness—impact of maintaining normoglycemia versus glycemia-independent actions of insulin. Diabetes 55:1096–1105
Falciglia M, Freyberg RW, Almenoff PL, D'Alessio DA, Render ML (2009) Hyperglycemia-related mortality in critically ill patients varies with admission diagnosis. Crit Care Med 37:3001–3009
Finfer S, Blair D, Bellomo R et al (2009) Intensive versus conventional glucose control in critically ill patients. N Engl J Med 360:1283–1297
Finfer S, Liu B, Chittock DR et al (2012) Hypoglycemia and risk of death in critically ill patients. N Engl J Med 367:1108–1118
Finfer S, Wernerman J, Preiser J-C et al (2013) Clinical review: consensus recommendations on measurement of blood glucose and reporting glycemic control in critically ill adults. Crit Care 17:229
Fivez T, Kerklaan D, Mesotten D et al (2016) Early versus late parenteral nutrition in critically ill children. N Engl J Med 374:1111–1122
Gunst J, Van den Berghe G (2016) Blood glucose control in the ICU: don't throw out the baby with the bathwater! Intensive Care Med 42:1478–1481
Gunst J, Van den Berghe G (2017) Paediatric endocrinology: critical illness—another trial, but are we any wiser? Nat Rev Endocrinol 13:254–256
Hermans G, Wilmer A, Meersseman W et al (2007) Impact of intensive insulin therapy on neuromuscular complications and ventilator dependency in the medical intensive care unit. Am J Respir Crit Care Med 175:480–489
Ingels C, Debaveye Y, Milants I et al (2006) Strict blood glucose control with insulin during intensive care after cardiac surgery: impact on 4-years survival, dependency on medical care, and quality-of-life. Eur Heart J 27:2716–2724
Jeschke MG, Kulp GA, Kraft R et al (2010) Intensive insulin therapy in severely burned pediatric patients: a prospective randomized trial. Am J Respir Crit Care Med 182:351–359
Kalfon P, Giraudeau B, Ichai C et al (2014) Tight computerized versus conventional glucose control in the ICU: a randomized controlled trial. Intensive Care Med 40:171–181
Kosiborod M, Rathore SS, Inzucchi SE et al (2005) Admission glucose and mortality in elderly patients hospitalized with acute myocardial infarction: implications for patients with and without recognized diabetes. Circulation 111:3078–3086
Krinsley JS (2004) Effect of an intensive glucose management protocol on the mortality of critically ill adult patients. Mayo Clin Proc 79:992–1000
Krinsley JS, Preiser JC (2019) Is it time to abandon glucose control in critically ill adult patients? Curr Opin Crit Care 25:299–306
Krinsley JS, Egi M, Kiss A et al (2013) Diabetic status and the relation of the three domains of glycemic control to mortality in critically ill patients: an international multicenter cohort study. Crit Care 17:R37
Krinsley JS, Chase JG, Gunst J et al (2017) Continuous glucose monitoring in the ICU: clinical considerations and consensus. Crit Care 21:197
Lecomte P, Van Vlem B, Coddens J et al (2008) Tight perioperative glucose control is associated with a reduction in renal impairment and renal failure in non-diabetic cardiac surgical patients. Crit Care 12: R154
Macrae D, Grieve R, Allen E et al (2014) A randomized trial of hyperglycemic control in pediatric intensive care. N Engl J Med 370:107–118
Marik PE (2016) Tight glycemic control in acutely ill patients: low evidence of benefit, high evidence of harm! Intensive Care Med 42:1475–1477
Marik PE, Preiser JC (2010) Toward understanding tight glycemic control in the ICU: a systematic review and metaanalysis. Chest 137:544–551
Mesotten D, Gielen M, Sterken C et al (2012) Neurocognitive development of children 4 years after critical illness and treatment with tight glucose control: a randomized controlled trial. JAMA 308:1641–1650
Niven DJ, Rubenfeld GD, Kramer AA, Stelfox HT (2015) Effect of published scientific evidence on glycemic control in adult intensive care units. JAMA Intern Med 175:801–809
Okabayashi T, Shima Y, Sumiyoshi T et al (2014) Intensive versus intermediate glucose control in surgical intensive care unit patients. Diabetes Care 37:1516–1524

Plummer MP, Bellomo R, Cousins CE et al (2014) Dysglycaemia in the critically ill and the interaction of chronic and acute glycaemia with mortality. Intensive Care Med 40:973–980

Preiser JC, Devos P, Ruiz-Santana S et al (2009) A prospective randomised multi-Centre controlled trial on tight glucose control by intensive insulin therapy in adult intensive care units: the Glucontrol study. Intensive Care Med 35:1738–1748

Preiser JC, Lheureux O, Prevedello D (2018) A step toward personalized glycemic control. Crit Care Med 46:1019–1020

Righy Shinotsuka C, Brasseur A, Fagnoul D, So T, Vincent JL, Preiser JC (2016) Manual versus automated moNitoring accuracy of GlucosE II (MANAGE II). Crit Care 20:380

Roberts GW, Quinn SJ, Valentine N et al (2015) Relative hyperglycemia, a marker of critical illness: introducing the stress hyperglycemia ratio. J Clin Endocrinol Metab 100:4490–4497

Salinas PD, Mendez CE (2019) Glucose management technologies for the critically ill. J Diabetes Sci Technol 13:682–690

Scott MG, Bruns DE, Boyd JC, Sacks DB (2009) Tight glucose control in the intensive care unit: are glucose meters up to the task? Clin Chem 55:18–20

Singer P, Berger MM, Van den Berghe G et al (2009) ESPEN guidelines on parenteral nutrition: intensive care. Clin Nutr 28:387–400

Singer P, Blaser AR, Berger MM et al (2019) ESPEN guideline on clinical nutrition in the intensive care unit. Clin Nutr 38:48–79

Suh SW, Gum ET, Hamby AM, Chan PH, Swanson RA (2007) Hypoglycemic neuronal death is triggered by glucose reperfusion and activation of neuronal NADPH oxidase. J Clin Invest 117:910–918

Van den Berghe G, Wouters P, Weekers F et al (2001) Intensive insulin therapy in critically ill patients. N Engl J Med 345:1359–1367

Van den Berghe G, Wilmer A, Hermans G et al (2006a) Intensive insulin therapy in the medical ICU. N Engl J Med 354:449–461

Van den Berghe G, Wilmer A, Milants I et al (2006b) Intensive insulin therapy in mixed medical/surgical intensive care units—benefit versus harm. Diabetes 55:3151–3159

Van den Berghe G, Wouters PJ, Kesteloot K, Hilleman DE (2006c) Analysis of healthcare resource utilization with intensive insulin therapy in critically ill patients. Crit Care Med 34:612–616

Van den Berghe G, Schetz M, Vlasselaers D et al (2009) Intensive insulin therapy in critically ill patients: NICE-SUGAR or Leuven blood glucose target? J Clin Endocrinol Metab 94:3163–3170

Van Herpe T, Mesotten D, Wouters PJ et al (2013) LOGIC-insulin algorithm-guided versus nurse-directed blood glucose control during critical illness: the LOGIC-1 single-center, randomized, controlled clinical trial. Diabetes Care 36:188–194

van Steen SC, Rijkenberg S, Limpens J, van der Voort PH, Hermanides J, DeVries JH (2017) The clinical benefits and accuracy of continuous glucose monitoring systems in critically ill patients-a systematic scoping review. Sensors (Basel) 17(1):146

Vanhorebeek I, Gunst J, Ellger B et al (2009a) Hyperglycemic kidney damage in an animal model of prolonged critical illness. Kidney Int 76:512–520

Vanhorebeek I, Ellger B, De Vos R et al (2009b) Tissue-specific glucose toxicity induces mitochondrial damage in a burn injury model of critical illness. Crit Care Med 37:1355–1364

Vanhorebeek I, Gielen M, Boussemaere M et al (2010) Glucose dysregulation and neurological injury biomarkers in critically ill children. J Clin Endocrinol Metab 95:4669–4679

Vlasselaers D, Milants I, Desmet L et al (2009) Intensive insulin therapy for patients in paediatric intensive care: a prospective, randomised controlled study. Lancet 373:547–556

Continuous Glucose Monitors as Wearable Lifestyle Behavior Change Tools in Obesity and Diabetes

43

Susan M. Schembre, Yue Liao, and Michelle R. Jospe

Contents

Barriers to HbA1c Reductions 592

Continuous Glucose Monitoring 592
CGM Use in Type 1 Diabetes 592
Continuous Glucose Monitoring in Type 2 Diabetes 593
Continuous Glucose Monitoring in Nondiabetic Populations 593
Variability of Glucose Concentrations in Nondiabetic Populations 593
Acceptability of CGM in Nondiabetic Populations 594

Continuous Glucose Monitors as Lifestyle Behavior Change Tools 594
Dietary Modification 594
Physical Activity Behavior 596
Weight Control 597
Hunger Training 597

Theory-Based Insights on Using CGM as Biological Feedback to Motivate Lifestyle Behavior Change 598

Future Opportunities for Continuous Glucose Monitorings 599

References 599

The original version of this chapter was revised. A correction to this chapter can be found at https://doi.org/10.1007/978-3-030-53370-0_74

S. M. Schembre (✉)
Department of Family and Community Medicine, University of Arizona College of Medicine-Tucson, Tucson, AZ, USA
e-mail: sschembre@email.arizona.edu

Y. Liao
Department of Kinesiology, College of Nursing and Health Innovation, The University of Texas at Arlington, Arlington, TX, USA
e-mail: yue.liao@uta.edu

M. R. Jospe
Department of Medicine, The University of Otago, Dunedin, New Zealand
e-mail: michelle.jospe@otago.ac.nz

Abstract

Recent advancements in continuous glucose monitoring (CGM) represent a novel and untapped resource to optimize behavior change interventions for the prevention and treatment of type 2 diabetes and obesity. In this chapter, we provide a brief history about CGM and evidence supporting its use, including nontraditional indications (people with type 2 diabetes and nondiabetic populations). We then discuss current applications for CGM as a tool for dietary modification, physical activity behavior change, and weight control

J. Faintuch and S. Faintuch (eds.), *Obesity and Diabetes*, https://doi.org/10.1007/978-3-030-53370-0_43

as well as insights on the theoretical basis for using CGM as biological feedback to motivate lifestyle behavior change. The chapter concludes with a discussion on the future opportunities for CGM as a wearable lifestyle behavior change tool for the treatment of obesity and diabetes.

Keywords

Wearable electronic device · Smartphone diabetes apps · Glucometer · Insulin pump · Glucose-monitoring device · Type 2 diabetes · Type 1 diabetes · Obesity · Technology-supported lifestyle intervention

At least 38% of U.S. adults are obese (Flegal et al. 2016) and nearly 30 million U.S. adults have diabetes—more than 90% of these patients are diagnosed with type 2 diabetes mellitus (T2DM) (Centers for Disease Control and Prevention 2017). In 2015, 1.5 million new cases of diabetes were diagnosed and another 84 million individuals were estimated to have prediabetes (Centers for Disease Control and Prevention 2017). Projections suggest that by 2050 one in three adults will be affected by this life-limiting chronic condition (Boyle et al. 2010). In 2017, the total estimated cost of diabetes was $327 billion, including $237 billion in direct medical costs (American Diabetes A 2018). These costs are expected to rise to more than $622 billion by 2030 (Rowley et al. 2017).

Among patients with T2DM, glycated hemoglobin (HbA1c) levels are a key indicator of glucose control. Elevated HbA1c levels are associated with all-cause mortality and the risk of cardiovascular disease and cancers of the colon and rectum, stomach, pancreas, breast, and liver (de Beer and Liebenberg 2014). Lowering HbA1c levels to 6.5–7% significantly benefits health outcomes. Efforts to improve diet and exercise remain the foundation of glycemic management (Davies et al. 2018). Meta-analyses demonstrate that intensive lifestyle interventions can reduce HbA1c by 0.29–0.54% (Martos-Cabrera et al. 2020); however, less encouraging results have also been observed (Usman et al. 2018).

Barriers to HbA1c Reductions

The gold standard lifestyle intervention protocol for T2DM management, which was implemented in the Look AHEAD trial, showed favorable results in overweight and obese adults with T2DM; however, 45% of participants were unable to achieve group weight loss goals (>7% reduction) and nearly 50% were unable to reduce their HbA1c to <7% (Pi-Sunyer et al. 2007). Furthermore, it is widely accepted that intensive lifestyle interventions like Look AHEAD require tremendous patient commitment and are resource intensive. In fact, an early evaluation of the National Diabetes Prevention Program, which was the programmatic foundation for Look AHEAD, show that less than 20% of delivery costs ($139 of $800 per beneficiary) are reimbursed (Ritchie and Gritz 2018). More effective and less resource- and time-intensive interventions that significantly improve glycemic control in patients with T2DM need to be considered.

Continuous Glucose Monitoring

CGM is a method of tracking glucose levels 24 h a day, and is often used in place of point-of-care glucometers for self-monitoring of blood glucose (SMBG). Users of CGM technology wear a small, transdermal (or implantable) sensor that measures glucose concentrations in interstitial fluid for an extended wear time (typically 3–14 days). Glucose readings are transmitted wirelessly to a receiver or compatible smart device, where they are displayed continuously or when the user scans the sensor as in flash glucose monitoring. The benefit of CGM versus SMBG is that the displayed data include current glucose levels as well as historical trends. Having information about the rate and direction of change in glucose levels helps inform treatment decisions.

CGM Use in Type 1 Diabetes

It has been more than two decades since the first CGM system was approved by the FDA in 1999 for the management of (type 1) diabetes. Advancements in glucose-monitoring technology made over the past 20 years (i.e., increased sensor duration, improved accuracy, and reduced mean absolute relative difference/MARD to <11%) have continued to revolutionize diabetes management, such that CGM is now widely accepted for the management of T1DM (Didyuk et al. 2020). Several diabetes organizations have reviewed the literature on the appropriate use of CGM in T1DM management and concluded it is a useful educational and disease management tool (Carlson et al. 2017). CGM aids patients and clinicians to visualize the important role that diet, exercise, stress management, and diabetes medications can have in managing diabetes, and facilitates effective shared decision making as part of standard clinical care (Carlson et al. 2017). The most recent CGM innovations take diabetes treatment to an even more exciting level—insulin pump-CGM combinations and closed-loop systems. These two systems provide patients access to insulin pump and CGM data in one device and automatically adjust insulin levels, respectively.

Continuous Glucose Monitoring in Type 2 Diabetes

Indications for using CGM have only begun to be addressed outside of T1DM (Carlson et al. 2017). The relatively high cost-to-benefit ratio, lack of standardized format for displaying results, and uncertainty on how best to use CGM data to make therapeutic decisions have kept CGM from widespread use in T2DM (Kompala and Neinstein 2019; Petrie et al. 2017). Currently, CGM has only been approved by the FDA for use by patients with T2DM who are on an intensive insulin regimen. One reason for the limited approval of CGM for T2DM is there have been few controlled studies of CGM use in T2DM. A recently published meta-analysis found only seven, qualified randomized controlled trials that tested the utility of CGM in patients with T2DM (Ida et al. 2019). Results indicated that HbA1c levels and time spent with hypoglycemia were significantly lower in the CGM group as compared to the SMBG group. Despite the limited number of trials, results demonstrate the feasibility and efficacy of using CGM for T2DM management. As a testament to this potential, the American Association of Clinical Endocrinologists consensus is that more frequent use of CGM in T2DM is anticipated to increasingly replace SMBG (Garber et al. 2019).

Continuous Glucose Monitoring in Nondiabetic Populations

It is anticipated that future generations of commercially available body worn sensors, including CGM, will have the capability to monitor various health-related biological measures. The earliest examples of this are wearable heartrate monitors that, years after their introduction to the market, became a standard feature on most smartwatches and activity trackers targeted toward the health-conscious consumer. More recently, some activity trackers have features to track sleep by actigraphy, heart rate, pulse oximetry, electrocardiogram, and respiration. As CGM becomes smaller and more affordable, health-conscious consumers who are interested in the "quantified self" will have the power to push the existing wearable sensor market to make glucose monitoring available to a wider consumer base. If this is the case, however, there are key questions about the utility and feasibility of glucose monitoring in nondiabetic populations that will become important to companies that distribute CGM.

Variability of Glucose Concentrations in Nondiabetic Populations

In nondiabetic populations, glucose levels are believed to remain within the normal range of 70–100 mg/dL. If glucose levels stray out of this range, the amounts of insulin and glucagon produced by the pancreas are adjusted to bring them back into this range. Despite this, research shows that glucose concentrations have substantial variability among nondiabetic populations despite lacking sizable changes in magnitude typically seen in people with diabetes. CGM data from 57 healthy individuals without a prior diagnosis of diabetes, followed for 2–4 weeks (Hall et al. 2018), showed highly variable intra- and interpersonal patterns of fluctuation. Participants were categorized into low variability, moderate variability, and severe variability based on their CGM data. These glucose patterns were found to be correlated with clinical and metabolic parameters, including HbA1c and triglyceride concentrations.

In another pilot analysis with 13 healthy adults (92% overweight/obese) who wore a CGM, an average of three hyperglycemic events (defined as glucose level >140 mg/dL) occurred across 5 days, and the average mean amplitude of glycemic excursion (MAGE) was 2.2 mmol/L or 40 mg/dL (Liao et al. 2018). This is similar to other studies where the average MAGE ranged from 1.96 to 2.2 mmol/L, or 35–40 mg/dL in healthy individuals (Monnier et al. 2006; Zhou et al. 2011) vs. typically >4 mmol/L or >70 mg/dL in diabetics (Zhu et al. 2019). Analyses of blood samples also show a positive association between HbA1c and 24-h glucose average estimated from CGM data, as well as the frequency of acute hyperglycemic events over the monitoring period. In summary, variability in glucose pattern can be detected by CGM devices in the nondiabetic populations and has clinically relevant implications.

Acceptability of CGM in Nondiabetic Populations

How acceptable is CGM for those with T2DM as a replacement for SMBG and for those without diabetes to wear such a device, particularly given the invasiveness of the current models? Whelan and colleagues (Whelan et al. 2019) showed that adults (n = 45) aged ≥40 years who were at moderate-to-high risk of T2DM (determined by the Leicester Risk Assessment) were willing to wear and engage with a CGM (Freestyle Libre/www.freestylelibre.com) for as long as 6 weeks. A majority of the participants in the study found the CGM comfortable to wear. Participants also expressed an interest in the information about glucose fluctuations provided by the CGM. Similarly, among 30 healthy adults (73% women, 67% overweight/obese, aged 24–64 years) who wore a CGM (Dexcom G4/www.dexcom.com) and a waist accelerometer, and recorded all eating events using a smartphone app for 7 days, >90% participants agreed that the CGM device was easy to use and provided information of interest (Liao and Schembre 2018).

Another pilot intervention study recruited insufficiently active adults with overweight or obesity (n = 19, 84% women, aged 26–55 years, 84% racial minorities). Seventy-four percent of the participants had elevated fasting blood glucose (between 100–125 mg/dL) at their baseline visit. This pilot intervention involved a 10-day self-monitoring period with a CGM (Freestyle Libre) (Liao et al. 2020). All participants agreed that the CGM was easy to use, and more than 90% agreed that the CGM provided information that was of interest to them. All participants also expressed an interest in receiving messages that provide an interpretive summary of their glucose patterns, and how they are associated with their eating and activity behaviors.

Continuous Glucose Monitors as Lifestyle Behavior Change Tools

Through January 2018, only six studies using real-time CGM to promote health-related behaviors, including dietary intake and physical activity, for weight control and HbA1c reductions were published (Ehrhardt and Al Zaghal 2019).

Dietary Modification

The standard-of-care practice for preventing and treating T2DM includes nutrition therapy. According to the American Diabetes Association (Marathe et al. 2017), the primary goals of nutrition therapy for adults with T2DM are to promote and support healthful eating patterns that achieve and maintain a healthy body weight, and to attain individualized glycemic control. However, determining when and what to eat is the most challenging part of treatment for patients (Marathe et al.

2017). Nutrition therapy requires education that is traditionally provided by a registered dietitian who has specialized training in diabetes-specific medical nutrition therapy. Dietitian-delivered nutrition therapy has been associated with improvements in glycemic control as evidenced by 0.5–2% reductions in HbA1c (Marathe et al. 2017). However, many private clinical practices and community clinics are unable to support a full-time dietitian. Further, patients from larger healthcare institutions, where access to certified diabetes educators is more prevalent, are often not referred to nutrition therapy by their primary care provider or they are unwilling to follow through with the referral. As such, the lack of education or continued education is a key barrier to disease management.

Food Diaries and Glucose Monitoring Feedback

As an adjunct to medical nutrition therapy, self-monitoring of glucose levels and food journaling is the standard recommendation for diabetes self-management. Food journaling most often requires a person to record all food and beverages consumed for multiple days. The strategy of documenting the consumption of food throughout the day and correlating it to its caloric or carbohydrate value can encourage an individual to incorporate healthier food choices, through accountability and the ability to identify and modify problem areas. As an additional benefit, journaling is believed to help those who regularly monitor glucose levels to identify how specific foods might cause their levels to fluctuate (Neithercott 2011); however, manually or digitally recording monitored glucose levels before and after each and every meal and snack, to determine what foods cause glucose spikes, is burdensome and cognitively taxing. Dietary self-monitoring, even using digital technology, is highly burdensome and does not lead to significant or sustained changes in behavior (Parkin et al. 2015).

Smartphone Apps

Popular mobile nutrition apps rely solely on tracking calories eaten—a highly unreliable and tedious task, and they most often do not provide feedback relevant to diabetes management. A vast majority of these diet trackers (e.g., MyFitnessPal, FitBit) lack the ability to personalize behavior change guidance, and have subsequently had lackluster effects on improving dietary intake (and physical activity) among users (Husain and Spence 2015). Furthermore, digital journaling tools have not yet been optimized to function with more advanced glucose monitoring technology. Applications that have been designed specifically for diabetes management (e.g., Glooko, mySugr) are similarly suboptimal. A majority, if not all of the existing diabetes self-management mobile apps, are limited in their function as a glucose and diet tracker for carbohydrate counting, and often require manual data entry (Darby et al. 2016). No existing diabetes management tools provide, to patients or providers, automated and personalized guidance on lifestyle behaviors (e.g., diet, exercise, weight control) that are based specifically on person-specific glucose response data (Darby et al. 2016).

The Contribution of CGM Devices

CGM offers users the ability to visualize the effect food has on glycemic responses (Freeman and Lyons 2008). It is hypothesized that comparing postprandial glucose excursions to foods consumed at usual meal times that have similar carbohydrate composition can provide important insights to patients on how to modify dietary behaviors in an effort to manage HbA1c levels. Using CGM data to motivate changes in dietary intake can be done in real time or retrospectively, but it relies greatly on one's ability to collect and interpret the data. The use of real-time CGM to improve glycemic control among people with T2DM has been of recent interest.

A recent systematic review and meta-analysis included five randomized controlled trials with th aim of supporting the use of CGM among people with T2DM (Janapala et al. 2019). Each study looked at the impact of CGM vs. SMBG on HbA1c. A significant pooled mean difference in HbA1c of –0.25% (–0.45, –0.06) favored CGM. However, only one study provided real-time dietary guidance. In a sample of patients ($n = 65$) with poorly controlled T2DM

(8% < HbA1c < 10%), Yoo and colleagues (Yoo et al. 2008) used real-time CGM once a month for 3 days over a period of 12 weeks. Real-time dietary instructions to patients were to "take in little amounts of food" in response to hyperglycemic alarms (>300 mg/dL). Dietary habits were assessed using food diaries covering 3 days of meals at baseline and 3 months later. While the real-time CGM group showed significant reductions in total energy intake over time, the difference was not significantly greater than the SMBG group. Each of the other studies relied on the patients to interpret the CGM data and respond accordingly (e.g., modify future meals that previously resulted in elevated postprandial glucose excursions).

Periodical Guided Analysis of CGM Records

An alternative method of providing behavioral guidance to CGM users is to examine CGM data and behavioral data retrospectively, with a trained health care provider. Retrospective CGM involves the use of a CGM for a predetermined monitoring period. During the monitoring period, which usually spans 3–10 days, patients wear a CGM and record foods consumed and meal times as well as other behavioral data (e.g., exercise and medication). The CGM data are later accessed by the patient's physician and compared with the behavioral data. Physicians skilled in the interpretation of the data can offer behavioral instructions to the patient with the goal of improving clinically relevant outcomes (e.g., time spent in range, HbA1c).

Research examining the impact of retrospective CGM on clinical outcomes is limited (Alfadhli et al. 2016; Lee et al. 2019; Mohan et al. 2016). Rarely, if at all, do these studies describe recommendations made to patients regarding diet or other lifestyle behaviors. None of the studies assessed changes in dietary intake. Only one study (Lee et al. 2019), which conducted 90 min pattern management-based diabetes education sessions in the group randomized to retrospective CGM ($n = 30$) vs. the control group ($n = 30$), assessed perceived changes in self-care behaviors, including diet, as well as self-efficacy (to eat a healthy diet). Scores on both diet measures were significantly improved in the CGM group vs. the control group over the 6-month study period, despite there being no significant differences in changes in HbA1c over time between the groups.

Physical Activity Behavior

CGM-based physical activity counseling in intervention settings, which aims to draw the connection between physical activity and improved glucose control, has been shown to increase physical activity. (Allen et al. 2008, 2011). In healthy individuals, associations were found between sedentary time, physical activity, and glucose variability (e.g., standard deviation/SD and MAGE) in free-living settings (Kingsnorth et al. 2018). Leveraging this type of information, Liao and colleagues tested a pilot intervention that featured the use of CGM (Liao et al. 2020). In this study, sedentary overweight and obese adults ($n = 19$) completed an in-person physical activity education session, followed by a 10-day self-monitoring period with a CGM and a Fitbit (www.fitbit.com). During the physical activity education session, participants received materials about the benefits of physical activity and an overview of different types of physical activity. Participants were also presented with a web-based glucose simulator that demonstrated how physical activity (i.e., walking) an impact glucose levels during the day. After the education session, participants were given a Freestyle Libre CGM to wear for the next 10 days.

Participants were instructed to obtain a glucose reading each morning upon waking, each night before going to sleep, and at least two other times throughout the day by scanning the CGM sensor with the reader. Upon scanning, the reader displays the current glucose level and a graph showing 8 h of glucose history, with an arrow indicating the directional trend. Participants were encouraged to observe how their daily glucose patterns were influenced by their behaviors (e.g., eating and exercising). The study tested whether participants increased their motivation to change their physical activity behavior after completing this intervention by examining stages of change. As described in the transtheoretical model, individuals will go through six stages of change when adopting a behavior: precontemplation (not intending to make changes), contemplation (thinking about making changes), preparation (developing plans to make changes), action (actively engaging in the new behavior), and maintenance (keep engaging in the new behavior for an extended period of time).

The study used the Exercise Stages of Change—Continuous Measure URICA-E2 psychometric questionnaire, to assess the different stages of change with the assumption that physical activity behavior is typically a dynamic process that moves between different stages (Lerdal et al. 2009). Compared to preintervention, participants scored lower in the contemplation stages and higher in the action stage at postintervention ($ps < 0.05$; see Fig. 43.1) suggesting improvements in motivation to be physically active.

Weight Control

Eating in response to hunger facilitates energy homeostasis as an intermediary step in weight regulation (Feig et al. 2018). Yet, in today's permissive food environment, many eating events are unrelated to energy depletion. Rather, they are predicted by nonphysiological determinants of food intake, such as the hedonic properties of palatable foods (Blundell and Finlayson 2004; Lowe and Butryn 2007); individual differences in hedonic eating behavior (e.g., emotional eating) (Schembre et al. 2009; Schembre 2011); sensitivity to food rewards (Cornier 2009, 2011; Versace et al. 2016; Versace and Schembre 2015); and impulsivity and self-control (Batterink et al. 2010; French et al. 2012; Meule et al. 2014; Nederkoorn et al. 2006). Eating behaviors not regulated by physiological signals of hunger and satiety have been empirically linked to weight gain (Bryant et al. 2008; McCrory et al. 2002; Provencher et al. 2003).

Research has shown that people with obesity are hypersensitive to interoceptive signals of hunger and may eat when glucose levels, an indicator of available short-term energy, remain elevated (Schembre and Huh 2014; Schembre et al. 2020), and thus may be less likely to distinguish physiological hunger from their hedonic desire to eat without help (Ciampolini et al. 2013; Ciampolini and Sifone 2011; Jospe et al. 2015; Taylor et al. 2015). Teaching people to self-regulate their energy intake by differentiating between physiological hunger and their hedonic desire to eat without hunger is an empirically supported weight-control strategy (Bacon et al. 2002, 2005; Daubenmier et al. 2016; Gletsu-Miller and McCrory 2014; Kristeller 2005; Kristeller et al. 2006). These data provide a strong evidence base for developing weight control interventions that use glucose monitoring to self-regulate energy intake for the prevention and management of obesity and diabetes.

Hunger Training

Hunger Training teaches people to eat only when their real-time glucose level is at or below a pre-established threshold that reflects a fasted state (Jospe et al. 2015, 2017). The scientific premise of hunger training does not require glucose to be a valid proxy for hunger. Rather, in order to promote energy balance, energy intake should not occur when glucose levels remain elevated. Hunger Training has more recently been conceptualized as a form of intermittent fasting called inter-meal fasting (Schembre et al. 2020).

Historically implemented using glucometers, Hunger Training has been validated in normoglycemic men and women as an effective weight loss method (Ciampolini et al. 2010b; Jospe et al. 2015, 2017). In overweight individuals, as little as 2–4 weeks of hunger training—without additional diet or physical activity guidance—results in weight loss of 7% within 5 months (Ciampolini et al. 2010b) and as much as 1.7% in only 2 weeks (Jospe et al. 2015). When Hunger Training was combined with a single

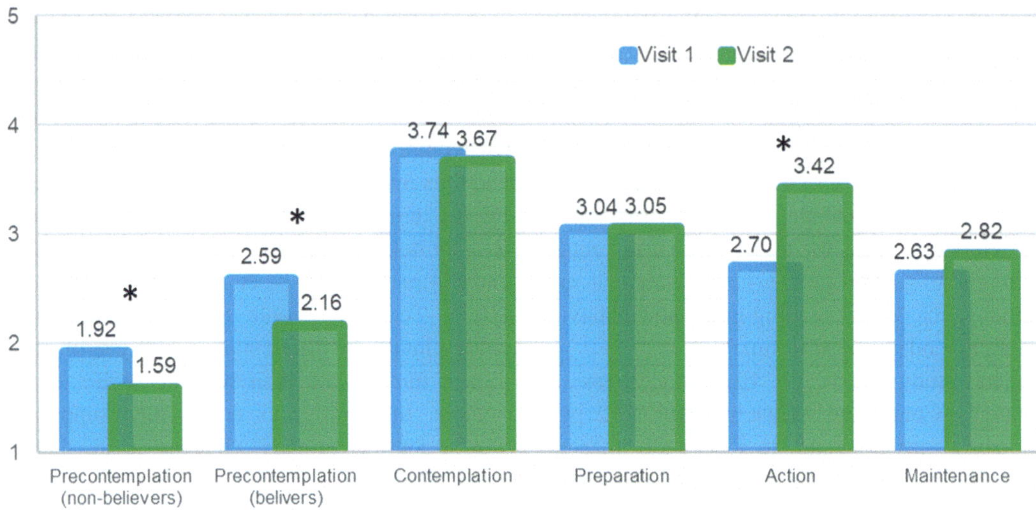

Fig. 43.1 Mean score for each stage of change, captured by the Exercise Stages of Change—Continuous Measure (URICA-E2), before and after a physical activity intervention ($n = 19$) that features a 10-day CGM monitoring. Asterisks indicate a significant difference in mean scores by a two-tailed paired-sample t-test

session of diet and physical activity advice and used consistently, it resulted in weight reductions of 13 kg in males and 5 kg in females at 6 months (Jospe et al. 2017). Participants expressed that hunger training increased their awareness of hungry versus nonhungry eating, improved their ability to recognize feelings of hunger and satiety, and reduced their number of eating occasions (de Bruin et al. 2019).

Hunger Training is as successful for weight loss with CGM devices as it has been with traditional glucometers (Jospe et al. 2020). However, CGM resulted in better adherence to the Hunger Training protocol, with patients generally preferring the ease and convenience of CGM, as well as the detailed feedback that it enables (de Bruin et al. 2019). Future behavior change interventions should consider using CGM instead of glucometers to encourage better adherence.

Restored Glucose Homeostasis

Recommendations to eat at lower glucose levels are generally consistent with preprandial glycemic targets (80–130 mg/dL) recommended by the ADA to manage HbA1c levels among people with diabetes (American Diabetes A 2017). Relatedly, interventions like Hunger Training that use blood glucose as a feedback tool have been shown to have favorable effects on glycemic control, insulin sensitivity, and beta-cell function among normoglycemic individuals (HbA1c $< 5.5\%$) (Allen et al. 2011; Ciampolini et al. 2010a; Kempf et al. 2013; Virdi et al. 2013; Yoo et al. 2008) that are comparable to outcomes of traditional lifestyle interventions for adults with T2DM (Huang et al. 2015; Pillay et al. 2015; Terranova et al. 2015).

Theory-Based Insights on Using CGM as Biological Feedback to Motivate Lifestyle Behavior Change

The use of CGM in health behavior interventions facilitates biological feedback (biofeedback), a designated behavior change technique (Abraham and Michie 2008; Michie et al. 2018). By definition, biofeedback reflects the collection of physiological or biochemical data using a monitoring device that is interpreted by an outside agent who provides feedback to prompt or motivate a desired behavior or behavior change. Traditionally, biofeedback referred to a therapeutic approach that used wearable sensors to monitor and provided real-time feedback on various physiological processes once thought to be involuntary (e.g., heart rate, brain waves, body temperature, muscle contraction). The goal of biofeedback therapy was to teach patients to alter their physiological activity through changes in thinking, emotions, and behavior for the purposes of improving health. Here we are conceptualizing the use of "biofeedback" differently, while maintaining many of the operational characteristics (e.g., providing real-time feedback to users of wearable sensor technology) and end goals (improving health). Specifically, we propose that CGM can be used to provide real-time feedback on the body's acute biological response to health behaviors (i.e., dietary intake and physical activity) and mood states (i.e., psychological stress) that impact acute changes in glucose concentrations, and that this feedback can be used to motivate behavior changes that can have short- and long-term impacts on related health outcomes, if maintained (Schembre et al. 2016).

Consistent with health behavior change theories, it is hypothesized that biofeedback acts as a cue to action (Health Belief Model) (Saghafi-Asl et al. 2020) and increases a person's intrinsic motivation (self-determination theory (Deci and Ryan 2012)) to adopt or maintain a desired behavior change. It is further hypothesized that biofeedback increases someone's perceived susceptibility to acquiring a disease. More specifically, biofeedback about indicators of disease progression (e.g., glucose values and diabetes) could arouse feelings of vulnerability that increase one's intrinsic motivation to take preventive actions (Self-Determination Theory). The experience of Liao and colleagues (Liao et al. 2020, described earlier) showed that among sedentary overweight and obese adults, a brief educational session demonstrating the acute impact of physical activity to reduce glucose levels with simulated CGM data, increased motivation to be physically active.

Future Opportunities for Continuous Glucose Monitorings

The use of CGM for the management and prevention of T2DM is currently hindered by the relatively high cost-to-benefit ratio and uncertainty on how best to use CGM data to make therapeutic decisions (Kompala and Neinstein 2019; Petrie et al. 2017). Manufacturers are hard at work with a goal of reducing the cost of CGM in an effort to expand their market.

CGM can be used to aid dietary choices and nutrition education, for instance, to show how various food choices and portion sizes influences glucose, to increase physical activity and, possibly, to improve sleep and stress management.

There are also research opportunities to optimize understanding of CGM results, which can be overwhelming for some patients (Hilliard et al. 2019; Kruger et al. 2019; Ritholz et al. 2010). Strategies are needed to assist patients with the meaningful interpretation of CGM data in an effort to make health behavior changes that impact on glycemic control.

References

Abraham C, Michie S (2008) A taxonomy of behavior change techniques used in interventions. Health Psychol 27:379–387. https://doi.org/10.1037/0278-6133.27.3.379

Alfadhli E, Osman E, Basri T (2016) Use of a real time continuous glucose monitoring system as an educational tool for patients with gestational diabetes. Diabetol Metab Syndr 8:48. https://doi.org/10.1186/s13098-016-0161-5

Allen NA, Fain JA, Braun B, Chipkin SR (2008) Continuous glucose monitoring counseling improves physical activity behaviors of individuals with type 2 diabetes: a randomized clinical trial. Diabetes Res Clin Pract 80:371–379. https://doi.org/10.1016/j.diabres.2008.01.006

Allen N, Whittemore R, Melkus G (2011) A continuous glucose monitoring and problem-solving intervention to change physical activity behavior in women with type 2 diabetes: a pilot study. Diabetes Technol Ther 13:1091–1099. https://doi.org/10.1089/dia.2011.0088

American Diabetes A (2017) Standards of medical care in diabetes-2017 abridged for primary care providers. Clin Diabetes 35:5–26. https://doi.org/10.2337/cd16-0067

American Diabetes A (2018) Economic costs of diabetes in the U.S. in 2017. Diabetes Care 41:917–928. https://doi.org/10.2337/dci18-0007

Bacon L, Keim NL, Van Loan MD, Derricote M, Gale B, Kazaks A, Stern JS (2002) Evaluating a 'non-diet' wellness intervention for improvement of metabolic fitness, psychological well-being and eating and activity behaviors. Int J Obes Relat Metab Disord 26:854–865

Bacon L, Stern JS, Van Loan MD, Keim NL (2005) Size acceptance and intuitive eating improve health for obese, female chronic dieters. J Am Diet Assoc 105:929–936

Batterink L, Yokum S, Stice E (2010) Body mass correlates inversely with inhibitory control in response to food among adolescent girls: an fMRI study. NeuroImage 52:1696–1703. https://doi.org/10.1016/j.neuroimage.2010.05.059

Blundell JE, Finlayson G (2004) Is susceptibility to weight gain characterized by homeostatic or hedonic risk factors for overconsumption? Physiol Behav 82:21–25

Boyle JP, Thompson TJ, Gregg EW, Barker LE, Williamson DF (2010) Projection of the year 2050 burden of diabetes in the US adult population: dynamic modeling of incidence, mortality, and prediabetes prevalence. Popul Health Metrics 8:1–12. https://doi.org/10.1186/1478-7954-8-29

Bryant EJ, King NA, Blundell JE (2008) Disinhibition: its effects on appetite and weight regulation. Obes Rev 9:409–419

Carlson AL, Mullen DM, Bergenstal RM (2017) Clinical use of continuous glucose monitoring in adults with type 2 diabetes. Diabetes Technol The 19:S4–S11. https://doi.org/10.1089/dia.2017.0024

Centers for Disease Control and Prevention (2017) National Diabetes Statistics Report, 2017. Centers for Disease Control and Prevention, Atlanta, GA

Ciampolini M, Sifone M (2011) Differences in maintenance of mean blood glucose (BG) and their association with response to "recognizing hunger". Int J Gen Med 4:403–412. https://doi.org/10.2147/IJGM.S19035

Ciampolini M et al (2010a) Sustained self-regulation of energy intake: initial hunger improves insulin sensitivity. J Nutr Metab 2010:286952. https://doi.org/10.1155/2010/286952

Ciampolini M, Lovell-Smith D, Sifone M (2010b) Sustained self-regulation of energy intake. Loss of weight in overweight subjects. Maintenance of weight in normal-weight subjects. Nutr Metab (Lond) 7:4. https://doi.org/10.1186/1743-7075-7-4

Ciampolini M, Lovell-Smith HD, Kenealy T, Bianchi R (2013) Hunger can be taught: Hunger Recognition regulates eating and improves energy balance. Int J Gen Med 6:465–478. https://doi.org/10.2147/IJGM.S40655

Cornier MA (2009) The effects of overfeeding and propensity to weight gain on the neuronal responses to visual food cues. Physiol Behav 97:525–530. https://doi.org/10.1016/j.physbeh.2009.03.019

Cornier MA (2011) Is your brain to blame for weight regain? Physiol Behav 104:608–612. https://doi.org/10.1016/j.physbeh.2011.04.003

Darby A, Strum MW, Holmes E, Gatwood J (2016) A review of nutritional tracking mobile applications for diabetes patient use. Diabetes Technol The 18:200–212. https://doi.org/10.1089/dia.2015.0299

Daubenmier J et al (2016) Effects of a mindfulness-based weight loss intervention in adults with obesity: a randomized clinical trial. Obesity (Silver Spring, Md) 24:794–804. https://doi.org/10.1002/oby.21396

Davies MJ et al (2018) Management of hyperglycemia in type 2 diabetes, 2018. A Consensus Report by the American Diabetes Association (ADA) and the European Association for the Study of Diabetes (EASD). Diabetes Care 41(12):2669–2701. https://doi.org/10.2337/dci18-0033

de Beer JC, Liebenberg L (2014) Does cancer risk increase with HbA1c, independent of diabetes? Br J Cancer 110:2361–2368. https://doi.org/10.1038/bjc.2014.150

de Bruin WE, Ward AL, Taylor RW, Jospe MR (2019) Am I really hungry? 'A qualitative exploration of patients' experience, adherence and behaviour change during hunger training: a pilot study. BMJ Open 9: e032248. https://doi.org/10.1136/bmjopen-2019-032248

Deci EL, Ryan RM (2012) Self-determination theory. In: Van Lange PAM, Kruglanski AW, Higgins ET (eds) Handbook of theories of social psychology. Sage Publications Ltd., New York, pp 416-436

Didyuk O, Econom N, Guardia A, Livingston K, Klueh U (2020) Continuous glucose monitoring devices: past, present, and future focus on the history and evolution of technological innovation. J Diabetes Sci Technol 2020:1932296819899394. https://doi.org/10.1177/1932296819899394

Ehrhardt N, Al Zaghal E (2019) Behavior modification in prediabetes and diabetes: potential use of real-time continuous glucose monitoring. J Diabetes Sci Technol 13:271–275. https://doi.org/10.1177/1932296818790994

Feig EH, Piers AD, Kral TVE, Lowe MR (2018) Eating in the absence of hunger is related to loss-of-control eating, hedonichunger, and short-term weight gain in normal-weight women. Appetite 123:317–324. https://doi.org/10.1016/j.appet.2018.01.013

Flegal KM, Kruszon-Moran D, Carroll MD, Fryar CD, Ogden CL (2016) Trends in obesity among adults in the United States, 2005 to 2014. JAMA 315:2284–2291. https://doi.org/10.1001/jama.2016.6458

Freeman J, Lyons L (2008) The use of continuous glucose monitoring to evaluate the glycemic response to food. Diabetes Spect 21:134–137. https://doi.org/10.2337/diaspect.21.2.134

French SA, Epstein LH, Jeffery RW, Blundell JE, Wardle J (2012) Eating behavior dimensions. Associations with energy intake and body weight. A review. Appetite 59:541–549. https://doi.org/10.1016/j.appet.2012.07.001

Garber AJ et al (2019) Consensus statement by the American Association of Clinical Endocrinologists and American College of Endocrinology on the Comprehensive Type 2 Diabetes Management Algorithm—

2019 Executive Summary. Endocr Pract 25:69–100. https://doi.org/10.4158/CS-2018-0535

Gletsu-Miller N, McCrory MA (2014) Modifying eating behavior: novel approaches for reducing body weight, preventing weight regain, and reducing chronic disease risk. Adv Nutr 5:789–791. https://doi.org/10.3945/an.114.006601

Hall H, Perelman D, Breschi A, Limcaoco P, Kellogg R, McLaughlin T, Snyder M (2018) Glucotypes reveal new patterns of glucose dysregulation. PLoS Biol 16: e2005143. https://doi.org/10.1371/journal.pbio.2005143

Hilliard ME et al (2019) Benefits and barriers of continuous glucose monitoring in young children with type 1 diabetes. Diabetes Technol Ther 21:493–498. https://doi.org/10.1089/dia.2019.0142

Huang J, Ding H, McBride S, Ireland D, Karunanithi M (2015) Use of smartphones to estimate carbohydrates in foods for diabetes management. Stud Health Technol Inform 214:121–127. https://doi.org/10.3233/978-1-61499-558-6-121

Husain I, Spence D (2015) Can healthy people benefit from health apps? Brit Med J 350:h1887. https://doi.org/10.1136/bmj.h1887

Ida S, Kaneko R, Murata K (2019) Utility of real-time and retrospective continuous glucose monitoring in patients with type 2 diabetes mellitus: a meta-analysis of randomized controlled trials. J Diabetes Res 2019:4684815. https://doi.org/10.1155/2019/4684815

Janapala RN et al (2019) Continuous glucose monitoring versus self-monitoring of blood glucose in type 2 diabetes mellitus: a systematic review with meta-analysis. Cureus 11:e5634. https://doi.org/10.7759/cureus.5634

Jospe MR, Brown RC, Roy M, Taylor RW (2015) Adherence to hunger training using blood glucose monitoring: a feasibility study. Nutr Metab (Lond) 12:22. https://doi.org/10.1186/s12986-015-0017-2

Jospe MR et al (2017) The effect of different types of monitoring strategies on weight loss: a randomized controlled trial. Obesity (Silver Spring, Md) 25:1490–1498. https://doi.org/10.1002/oby.21898

Jospe MR et al (2020) Teaching people to eat according to appetite – does the method of glucose measurement matter? Appetite 151:104691. https://doi.org/10.1016/j.appet.2020.104691

Kempf K, Tankova T, Martin S (2013) ROSSO-in-praxi-international: long-term effects of self-monitoring of blood glucose on glucometabolic control in patients with type 2 diabetes mellitus not treated with insulin. Diabetes Technol Ther 15:89–96. https://doi.org/10.1089/dia.2012.0213

Kingsnorth AP, Whelan ME, Sanders JP, Sherar LB, Esliger DW (2018) Using digital health technologies to understand the association between movement behaviors and interstitial glucose: exploratory analysis. JMIR Mhealth Uhealth 6:e114. https://doi.org/10.2196/mhealth.9471

Kompala T, Neinstein A (2019) A new era: increasing continuous glucose monitoring use in type 2 diabetes. Am J Manag Care 25:SP123–SP126

Kristeller J (2005) Know your hunger. Spirituality & Health Media, LLC Traverse City, MI

Kristeller JL, Baer RA, Quillian-Wolever R (2006) Mindfulness-based approaches to eating disorders. In: Baer RA (ed) Mindfulness-based treatment approaches: clinician's guide to evidence base and applications. Elsevier, Academic Press, Amsterdam, Boston, p 75

Kruger DF, Edelman SV, Hinnen DA, Parkin CG (2019) Reference guide for integrating continuous glucose monitoring into clinical practice. Diabetes Educ 45:3S–20S. https://doi.org/10.1177/0145721718818066

Lee SK, Shin DH, Kim YH, Lee KS (2019) Effect of diabetes education through pattern management on self-care and self-efficacy in patients with type 2 diabetes. Int J Environ Res Public Health 16:3323. https://doi.org/10.3390/ijerph16183323

Lerdal A et al (2009) Stages of change—continuous measure (URICA-E2): psychometrics of a Norwegian version. J Adv Nurs 65:193–202. https://doi.org/10.1111/j.1365-2648.2008.04842.x

Liao Y, Schembre S (2018) Acceptability of continuous glucose monitoring in free-living healthy individuals: implications for the use of wearable biosensors in diet and physical activity research. JMIR Mhealth Uhealth 6:e11181. https://doi.org/10.2196/11181

Liao Y, Desai N, Li L, Shen Q, Daniel CR, Kopetz S, Schembre SM (2018) Exploring the associations between daily glucose dynamics and cancer risk-related biomarkers in overweight and obese adults. Paper presented at the American Association for Cancer Research Mechanisms Underlying Etiology and Outcomes, Austin, TX

Liao Y, Basen-Engquist K, Urbauer DL, Bevers TB, Hawke E (2020) Design of a novel physical activity intervention in overweight and obese adults using continuous glucose monitoring: a pilot study. Cancer Epidemiol Biomarkers Prev. 4:761-768. doi: 10.1158/1055-9965.EPI-19-0906

Lowe MR, Butryn ML (2007) Hedonic hunger: a new dimension of appetite? Physiol Behav 91:432–439. https://doi.org/10.1016/j.physbeh.2007.04.006

Marathe PH, Gao HX, Close KL (2017) American Diabetes Association standards of medical care in diabetes 2017. J Diabetes 9:320–324. https://doi.org/10.1111/1753-0407.12524

Martos-Cabrera MB, Velando-Soriano A, Pradas-Hernández L, Suleiman-Martos N, Cañadas-De la Fuente GA, Albendín-García L, Gómez-Urquiza JL (2020) Smartphones and apps to control glycosylated hemoglobin (HbA1c) level in diabetes: a systematic review and meta-analysis. J Clin Med 9(3):pii: E693. https://doi.org/10.3390/jcm9030693

McCrory MA, Suen VM, Roberts SB (2002) Biobehavioral influences on energy intake and adult weight gain. J Nutr 132:3830S–3834S
Meule A, Lutz AP, Vogele C, Kubler A (2014) Impulsive reactions to food-cues predict subsequent food craving. Eat Behav 15:99–105. https://doi.org/10.1016/j.eatbeh.2013.10.023
Michie S, West R, Sheals K, Godinho CA (2018) Evaluating the effectiveness of behavior change techniques in health-related behavior: a scoping review of methods used. Transl Behav Med 8:212–224. https://doi.org/10.1093/tbm/ibx019
Mohan V et al (2016) Use of retrospective continuous glucose monitoring for optimizing management of type 2 diabetes in India. J Assoc Physicians India 64:16–21
Monnier L, Mas E, Ginet C, Michel F, Villon L, Cristol JP, Colette C (2006) Activation of oxidative stress by acute glucose fluctuations compared with sustained chronic Hyperglycemia in patients with type 2 diabetes. JAMA J Am Med Assoc 295:1681–1687. https://doi.org/10.1001/jama.295.14.1681
Nederkoorn C, Smulders FT, Havermans RC, Roefs A, Jansen A (2006) Impulsivity in obese women. Appetite 47:253–256
Neithercott T (2011) Dear food diary. Recording what you eat can help with weight and blood glucose control. Diabetes Forecast 64(29):32–34
Parkin CG, Homberg A, Hinzmann R (2015) 8th Annual symposium on self-monitoring of blood glucose (SMBG): April 16–18, 2015. Republic Malta Diabetes Technol Ther 17:832–850. https://doi.org/10.1089/dia.2015.0325
Petrie JR, Peters AL, Bergenstal RM, Holl RW, Fleming GA, Heinemann L (2017) Improving the clinical value and utility of CGM systems: issues and recommendations: a joint statement of the European Association for the Study of Diabetes and the American Diabetes Association Diabetes Technology Working Group. Diabetes Care 40:1614–1621. https://doi.org/10.2337/dci17-0043
Pi-Sunyer X et al (2007) Reduction in weight and cardiovascular disease risk factors in individuals with type 2 diabetes: one-year results of the look AHEAD trial. Diabetes Care 30:1374–1383. https://doi.org/10.2337/dc07-0048
Pillay J, Armstrong MJ, Butalia S, Donovan LE, Sigal RJ, Vandermeer B, Chordiya P, Dhakal S, Hartling L, Nuspl M, Featherstone R, Dryden DM (2015) Share behavioral programs for type 2 diabetes mellitus: a systematic review and network meta-analysis. Ann Intern Med 163(11):848–860. https://doi.org/10.7326/M15-1400
Provencher V, Drapeau V, Tremblay A, Despres JP, Lemieux S (2003) Eating behaviors and indexes of body composition in men and women from the Quebec family study. Obes Res 11:783–792
Ritchie ND, Gritz RM (2018) New medicare diabetes prevention coverage may limit beneficiary access and widen health disparities. Med Care 56:908–911. https://doi.org/10.1097/MLR.0000000000000981
Ritholz MD, Atakov-Castillo A, Beste M, Beverly EA, Leighton A, Weinger K, Wolpert H (2010) Psychosocial factors associated with use of continuous glucose monitoring. Diabet Med 27:1060–1065. https://doi.org/10.1111/j.1464-5491.2010.03061.x
Rowley WR, Bezold C, Arikan Y, Byrne E, Krohe S (2017) Diabetes 2030: insights from yesterday, today, and future. Trends Popul Health Manag 20:6–12. https://doi.org/10.1089/pop.2015.0181
Saghafi-Asl M, Aliasgharzadeh S, Asghari-Jafarabadi M (2020) Factors influencing weight management behavior among college students: an application of the Health Belief Model. PLoS One 15(2):e0228058. https://doi.org/10.1371/journal.pone.0228058
Schembre SM (2011) Weight-related eating behavior questionnaires: applying theory to measurement. In: Preedy VR, Watson RR, Martin CR (eds) Handbook of behavior, food and nutrition. Springer, New York, pp 3487–3506
Schembre SM, Huh J (2014) Using pre-prandial blood glucose data as a marker of maladaptive eating events. Paper presented at the Annals of Behavioral Medicine, Oxford
Schembre S, Greene G, Melanson K (2009) Development and validation of a weight-related eating questionnaire. Eat Behav 10:119–124. https://doi.org/10.1016/j.eatbeh.2009.03.006
Schembre SM, Liao Y, Busaidy N, Li L, Basen-Engquist-K, Huh J (2016) Mobile glucose monitoring as biological feedback on ingestive behavior: implications for dietary intake and weight management. Paper presented at the Society for the Study of Ingestive Behaviors, Porto, Portugal
Schembre SM, Liao Y, Huh J, Keller S (2020) Using pre-prandial blood glucose to assess eating in the absence of hunger in free-living individuals. Eat Behav. 238:101411. doi: 10.1016/j.eatbeh.2020.101411.
Terranova CO, Brakenridge CL, Lawler SP, Eakin EG, Reeves MM (2015) Share Effectiveness of lifestyle-based weight loss interventions for adults with type 2 diabetes: a systematic review and meta-analysis. Diabetes Obes Metab. 17(4):371-8. doi: 10.1111/dom.12430
Taylor RW, Roy M, Jospe MR, Osborne HR, Meredith-Jones KJ, Williams SM, Brown RC (2015) Determining how best to support overweight adults to adhere to lifestyle change: protocol for the SWIFT study. BMC Public Health 15:861. https://doi.org/10.1186/s12889-015-2205-4
Usman M, Gillies CL, Khunti K, Davies MJ (2018) Effects of intensive interventions compared to standard care in people with type 2 diabetes and microalbuminuria on risk factors control and cardiovascular outcomes: a systematic review and meta-analysis of randomised controlled trials. Diabetes Res Clin Pract 146:76–84. https://doi.org/10.1016/j.diabres.2018.10.002

Versace F, Schembre SM (2015) 'Obesogenic' oversimplification. Obes Rev 16:702–703. https://doi.org/10.1111/obr.12301

Versace F, Kypriotakis G, Basen-Engquist K, Schembre SM (2016) Heterogeneity in brain reactivity to pleasant and food cues: evidence of sign-tracking in humans. Soc Cogn Affect Neurosci 11(4):604–611. https://doi.org/10.1093/scan/nsv143

Virdi NS et al (2013) Association of self-monitoring of blood glucose use on glycated hemoglobin and weight in newly diagnosed, insulin-naive adult patients with type 2 diabetes. J Diabetes Sci Technol 7:1229–1242

Whelan ME, Orme MW, Kingsnorth AP, Sherar LB, Denton FL, Esliger DW (2019) Examining the use of glucose and physical activity self-monitoring technologies in individuals at moderate to high risk of developing type 2 diabetes: randomized trial. JMIR Mhealth Uhealth 7:e14195. https://doi.org/10.2196/14195

Yoo HJ et al (2008) Use of a real time continuous glucose monitoring system as a motivational device for poorly controlled type 2 diabetes. Diabetes Res Clin Pract 82:73–79. https://doi.org/10.1016/j.diabres.2008.06.015

Zhou J et al (2011) Establishment of normal reference ranges for glycemic variability in Chinese subjects using continuous glucose monitoring. Med Sci Monit 17:CR9-13. https://doi.org/10.12659/msm.881318

Zhu J, Yuan L, Ni WJ, Luo Y, Ma JH (2019) Association of higher circulating insulin antibody with increased mean amplitude glycemic excursion in patients with type 2 diabetes mellitus: a cross-sectional, retrospective case-control study. J Diabetes Res 2019:7304140. https://doi.org/10.1155/2019/7304140

Part V

Innovative Endoscopic, Cell Therapy and Other Interventions

Endoscopic Techniques for Obesity and Diabetes

44

Vitor Ottoboni Brunaldi, João Almiro Ferreira Filho, and Daniel Martone

Contents

Introduction 608
Gastric Aspiration: Aspire Assist® 608
Space-Occupying Devices: Intragastric Balloon (IGB) 609
Type 2 Diabetes 610
Transpyloric Shuttle® (TPS) 610
Gastric Reduction: Endoscopic Sleeve Gastroplasty (ESG) 611
POSE (Primary Obesity Surgery Endoluminal) 612
Endomina Method 613
Metabolic Procedures: Duodenal Mucosal Resurfacing (Fractyl®) 614
Duodenal Jejunal Bypass Liner/DJBL (EndoBarrier®) 614
Endoscopic Magnetic Partial Jejunal Diversion/EMPJD 615
References 616

V. O. Brunaldi (✉)
Gastrointestinal Endoscopy Unit, Gastroenterology Department, University of São Paulo Medical School, São Paulo, Brazil

Center for Gastrointestinal Endoscopy, Surgery and Anatomy Department, Ribeirão Preto Medical School, University of São Paulo, Ribeirão Preto, São Paulo, Brazil
e-mail: vbrunaldi@gmail.com; vitor.brunaldi@usp.br; vobrunaldi@hcrp.usp.br

J. A. Ferreira Filho · D. Martone
Center for Gastrointestinal Endoscopy, Surgery and Anatomy Department, Ribeirão Preto Medical School, University of São Paulo, Ribeirão Preto, São Paulo, Brazil

J. Faintuch and S. Faintuch (eds.), *Obesity and Diabetes*, https://doi.org/10.1007/978-3-030-53370-0_44

Abstract

Along with the classic intragastric balloon, several endoscopic bariatric therapies (EBTs) are currently available in clinical practice: aspiration therapy, transpyloric shuttle, gastric reduction techniques, metabolic procedures focused on the small bowel, among others. The aim of this chapter is to summarize the current evidence on EBTs and to provide a glimpse into the future of the endoluminal treatment of obesity and metabolic disorders.

Keywords

Obesity · Overweight · Endoscopy · Aspiration therapy · Intragastric balloon · Bariatrics · Type 2 diabetes

Introduction

The interventional gold standard for severe obesity has long been bariatric surgery as it promotes sustainable weight loss and improvement of obesity-related comorbidities (Colquitt et al. 2014; O'Brien et al. 2019).

As obesity escalates, the number of bariatric surgeries also rises. Recent data showed a total of 685,000 bariatric procedures worldwide in 2016 (Angrisani et al. 2018). However, this number is tiny in the face of the many hundreds of millions of patients currently suffering from the disease.

In this context, novel alternatives to bariatric surgery are certainly needed. As such, endoscopic bariatric therapies (EBTs) have recently become a field of interest in most interventional endoscopy groups. Initially, EBTs were mainly composed of space-occupying devices, mimicking the intragastric balloon. However, creative procedures have been proposed over the last 20 years and EBTs currently entail gastric aspiration, gastric reduction methods, endoscopic bypass liners, duodenal mucosal resurfacing, and magnetic partial jejunal diversion.

Gastric Aspiration: Aspire Assist®

The AspireAssist™ is a modified percutaneous gastrostomy device composed of a discreet external port (Fig. 44.1) and a large fenestrated 15-cm-long intragastric tube. The external port connects to a portable aspiration device that allows removal of around 30% of the recently ingested food (Sullivan et al. 2013).

Initially, two noncontrolled series assessed weight loss outcomes at 1 year and both reported excess weight loss (EWL) around 50% (Sullivan et al. 2013; Noren and Forssell 2016). In 2016, the U.S. Food and Drug Administration (FDA) approved the AspireAssist™ after the publication of an American randomized multicenter controlled trial. The study enrolled 171 patients for either aspiration therapy (AT) plus lifestyle intervention versus lifestyle intervention alone in a 2:1 ratio. The technical success rate was 97% (111 of 114 attempts). At hour, the mean EWL and total weight loss (TWL) were 31.5 ± 26.7% and 12.1 ± 9.6% for the AT group versus 9.8 ± 15.5% and 3.5 ± 6.0% for the control group, respectively ($p < 0.001$). The authors also reported amelioration of HbA1c, triglycerides, and low-density lipoprotein levels,

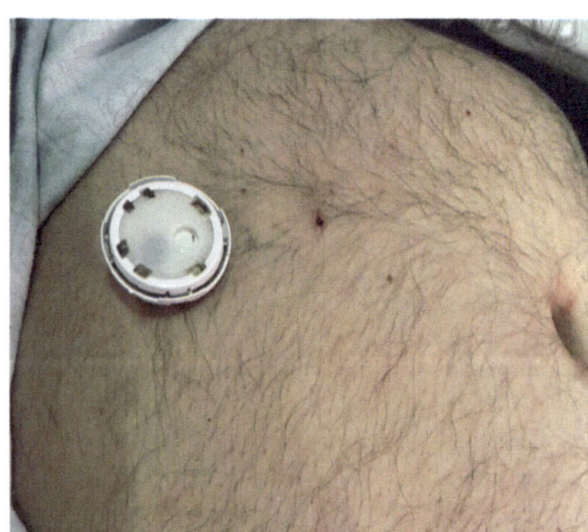

Fig. 44.1 The external port of the AspireAssist™ device. It should be connected to a portable aspirator to partially remove the recently ingested food

reduction in systolic and diastolic blood pressure, and an increase in high-density cholesterol.

The serious adverse events (SAEs) rate was 3.6%, namely, one case of severe abdominal pain, one patient with peritonitis, one gastric ulcer, and one case of port malfunction requiring tube replacement (Thompson et al. 2017). More recently, the same group published the 4-year follow-up results from this cohort. Among the 58 patients who continued in the study, 43 withdrew the AT before completing 4 years. The reasons for withdrawal were either meeting weight loss goals (25 cases) or insufficient weight loss (18 cases). For the 15 completers, the mean %TWL was 18.7 ± 11.7 and the %EWL was 50.8 ± 31.9%. The most common adverse events after the first year were peristomal inflammation and persistent fistula. The latter was ten times more frequent from years 2 to 4 than during year 1 (Thompson et al. 2019).

In another series, Nyström et al also reported 4-year data from a post-market European registry study (Nystrom et al. 2018). Five centers gathered data from 201 individuals undergoing AT. All patients presented a BMI of at least 35 kg/m^2 and had failed conservative weight loss therapies. The mean %TWL at 1, 2, 3, and 4 years were 18.2 ± 9.4% (n = 155/173), 19.8 ± 11.3% (n = 82/114), 21.3 ± 9.6% (n = 24/43), and 19.2 ± 13.1% (n = 12/30), respectively. There were 42 postprocedural AEs, namely, 12 gastric leaks, 10 stomal irritation/granulation tissue, 9 infections or possible infections, 8 buried bumpers, and 3 tube dislodgments.

Current data suggest AspireAssist™ therapy is effective and safe in addressing moderate-to-severe obesity. Nonetheless, it does not seem a perennial therapy and there are still little data on weight maintenance after removal of the device.

Space-Occupying Devices: Intragastric Balloon (IGB)

The intragastric balloon (IGB) is the oldest and the most commonly employed endoscopic bariatric therapy. It is an effective but mainly transitory method to control obesity. Its most attractive advantage is the minimal risk for major complications (Sallet et al. 2004). Besides primary treatment, it may also serve as a bridging therapy to bariatric surgery for patients at high surgical risk (e.g., superobese individuals) (Dumonceau 2008).

Since the early 1980s, when the pioneer air-filled Garren-Edwards Gastric Bubble was firstly released, several modifications have been implemented on the IGBs. Such refinements have been mainly tailored by the experts' suggestions from a conference held in Tampor Springs, in 1987.

Currently, different models of IGBs are available for clinical use. Most are silicone-based spheres, either water or gas-filled (Fig. 44.2), adjustable or nonadjustable, including single, double, or triple balloons. Some IGBs require endoscopic placement, whereas there are IGBs that circumvent such procedure. All devices demand endoscopic removal after 6- or 12-month indwelling.

IGB therapy provides EWL of around 30% and AWL varying from 4 to 28 kg. Patients with higher BMIs tend to present more remarkable results (Imaz et al. 2008). Sham-controlled studies have demonstrated statistical superiority of IGB versus control regarding absolute weight loss, %TWL, and BMI reduction (Moura et al. 2016). A recent meta-analysis has also proven that water-filled balloons provide better weight

Fig. 44.2 The nonadjustable liquid-filled 12-month indwelling intragastric balloon

loss compared to the air-filled ones in a non-head-to-head comparison (Saber et al. 2017). Of note, however, not all patients undergoing IGB implantation lose weight satisfactorily. Up to 40% fail to achieve adequate outcome (defined as %TWL < 10% or %EWL < 25%). Early removal, early adaptation to restrictive symptoms, de novo dietary disorders (e.g., binge eating) could explain such high clinical failure rates (Dumonceau 2008).

Type 2 Diabetes

Chan et al. investigated the impact of this therapy on glycemic levels in non-insulin-dependent overweight patients. The authors demonstrated a significant reduction in HbA1c levels (8.6–7.3%), and the proportion of patients presenting glycemic levels over 100 mg/dL dropped from 50% before IGB to 12% after treatment. Moreover, 58% of the patients had hypertriglyceridemia before IGB versus only 19% after the treatment. Serum transaminases also sharply dropped with the treatment (Chan et al. 2008).

Regarding bridging therapy to bariatric surgery, strong data support IGB as an effective and safe modality (Busetto et al. 2004). Some studies have also demonstrated the bridging sleeve gastrectomy (SG) as an alternative to the IGB in this scenario. However, even less invasive surgeries such as the laparoscopic SG have higher complication rates than most endoscopic techniques, which renders the IGB an attractive option (Genco et al. 2005).

A recent meta-analysis found that nausea and vomits are the most frequent adverse events. A subgroup analysis demonstrated that water-filled balloons elicit symptoms more frequently than the air-filled ones (Trang et al. 2018); however, they seem to be more effective in terms of weight loss (Saber et al. 2017). A large series with more than 2000 procedures showed an overall AEs rate of 2.8%, including rare cases of perforation (0.19%) and gastric outlet obstruction (0.76%) (Genco et al. 2005).

Transpyloric Shuttle® (TPS)

The Food and Drug Administration (FDA) has recently approved the TPS (BARONova Inc, San Carlos, CA) for marketing. Its mechanism of action is intermittent gastric outlet obstruction rather than simple space restriction. The device is composed of two different-sized silicone-based bulbs connected by a flexible tether. The large bulb anchors in the pylorus to prevent migration, while the smaller one stays in the duodenum. Consequently, the device remains in a transpyloric fashion, and the antral pump intermittently dislocates the large bulb to allow transient food passage (Fig. 44.3).

Patients (10 each) were compared in a 3-month vs. 6-month TPS treatment. Three-month patients presented mean EWL and TWL of 25.1 ± 14% and 8.9 ± 5%, respectively. Six-month patients had mean %EWL and AWL of 41 ± 21.1% and 14.5 ± 5.8%. However, 10 patients (50%) presented gastric ulcers, two requiring early removal (Marinos et al. 2014). Such an elevated rate fostered refinements in the device focused on safety improvement.

Recently, a multicenter double-blind trial (NCT02518685) enrolled 270 patients in a 2:1 ratio to receive TPS treatment or sham. The

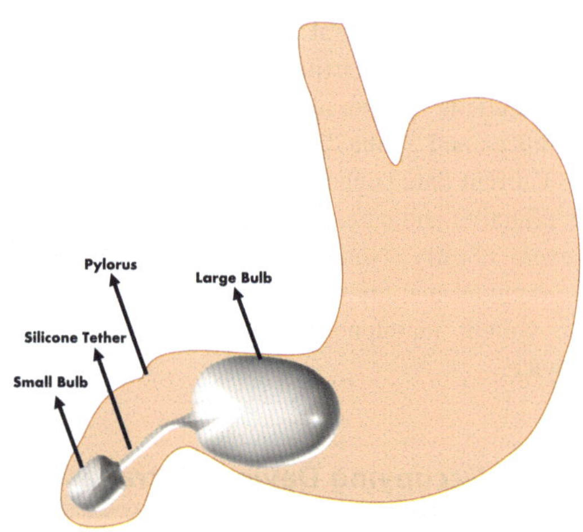

Fig. 44.3 Schematics of the Transpyloric Shuttle device in place

upgraded TPS provided greater TWL than sham at 1 year (9.5% [8.2, 10.8] vs. 2.8% [1.1, 4.5], $p < 0.0001$). Sixty-seven percent of the TPS cohort achieved the predetermined goal of ≥5% TWL. Furthermore, there was a mean BMI reduction of 3.5 kg/m^2 for treated patients versus 1.01 kg/m^2 in the sham group ($p < 0.0001$). Ten percent of patients required early removal (before 12 months). There were no bleeding or perforation cases despite a 10% gastric ulcer rate. Only 2.5% were considered serious adverse events (Rothstein et al. 2018).

In spite of the relatively good performance of the TPS at promoting weight loss, the ulceration and early removal rates are still elevated.

Gastric Reduction: Endoscopic Sleeve Gastroplasty (ESG)

To perform an ESG, an endoscopic-suturing device—the Apollo Overstitch™—is attached to a therapeutic double-channel gastroscope and delivers full-thickness stitches (Fig. 44.4). The suturing pattern varies among centers; however, most perform running sutures from the anterior to the posterior wall passing through the greater curvature, and then back to anterior where the suture is finally cinched (Brunaldi and Galvao Neto 2019). Recently, some authors have also employed reinforcing interrupted stitches in-between the running sutures. The procedure is performed in an antrum-fundus direction, preserving the fundus. That creates a fundal pouch that possibly delays gastric emptying and enhances weight loss by improving satiety (Abu Dayyeh et al. 2017). The final aspect is sleeve-like anatomy with a prominent fundal pouch (Fig. 44.5).

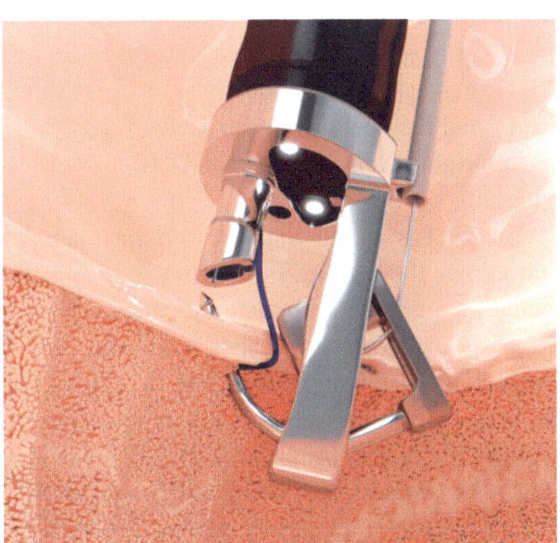

Fig. 44.4 The Apollo Overstitch full-thickness endoscopic suturing device

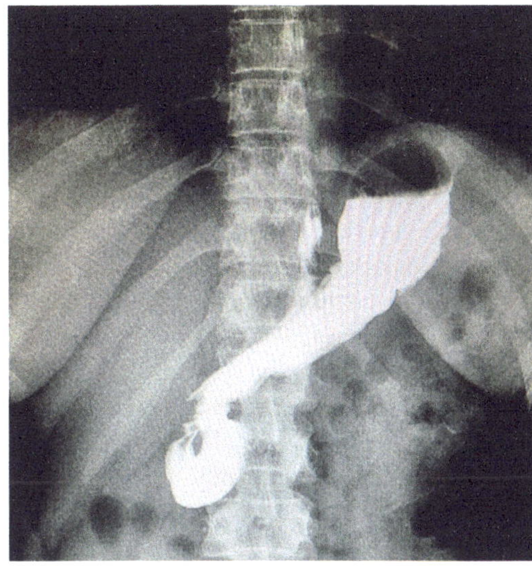

Fig. 44.5 An upper gastrointestinal series showing the endoscopic sleeve gastroplasty anatomy at 3 months postoperatively

Abu Dayehh et al. first described the ESG procedure in 2013 (Abu Dayyeh et al. 2013). Initially, simple interrupted stitches were employed, but as the experience grew worldwide, the technique was refined to the current running suture. Kumar et al. recently summarized such evolution in a comprehensive multicenter series (Kumar et al. 2018).

The ESG is probably the most promising transoral gastric-reduction procedure addressing obesity. Since its first description in 2013, several studies with more than 1500 cases have been reported in the world, including Brazil (Galvao-Neto et al. 2016), USA (Sharaiha et al. 2015), India (Bhandari et al. 2019), Saudi Arabia (Alqahtani et al. 2018), and Spain (Lopez-Nava et al. 2017).

To date, the largest series was published by Alqahtani et al. in 2018 (1000 ESG patients from a single center). Baseline BMI and age were 33.3 ± 4.5 kg/m^2 and 34.4 ± 9.5 years, respectively. Procedure time was 61 ± 16 min, and patients were routinely discharged on the same day. Most complained of mild postoperative abdominal pain and nausea (92.4%), but all symptoms were controlled with oral medications. Twenty-four patients (2.4%) were readmitted due to severe abdominal pain (8), postprocedural bleeding (7), perigastric collection with pleural effusion (4), or postprocedural self-limited fever (5). There was neither mortality nor emergency interventions. Patients presented TWL of $13.7 \pm 6.8\%$ ($n = 369/423$), $15.0 \pm 7.7\%$ ($n = 216/232$), and $14.8 \pm 8.5\%$ ($n = 54/63$) at 6, 12, and 18 months, respectively. All 28 cases of hypertension, 13 of the 17 cases of diabetes, and 18 of the 32 cases of dyslipidemia had complete remission by the third month. During follow-up, eight patients were revised to sleeve gastrectomy due to weight loss failure, and five underwent redo ESG due to weight regain (Alqahtani et al. 2018).

Lopez-Nava et al. have published the longest follow-up to date. In an international multicenter series of 248 cases, the authors reported TWL of 18.6% [95% CI: 15.7–21.5], which was similar for all the three participating centers ($p = 0.7$). In a linear regression analysis, they found that weight loss at 6 months highly predicted weight maintenance and weight loss at 24 months, thus pointing to an important early-on treatment predictor of poor long-term outcome (Lopez-Nava et al. 2017). A congress abstract reported even longer-term data, with AWL of 18.7 kg (10–27.3) and TWL of 14.5% (8.2–20.9) at 5 years. Patients achieved the nadir weight at 24 months and presented a mean weight regain of 2.4 kg by the final follow-up (Hajifathalian et al. 2019).

In a meta-analysis of 8 studies comprising 1772 procedures, TWL was 15.1% [14.29, 16], 16.4% [15.16, 17.82], and 17.1% [14.64, 19.66] at 6, 12, and 18–24 months, respectively. The AEs rate was as low as 2% (Hedjoudje et al. 2020). Even though there are no controlled data yet, retrospective studies have shown ESG to be superior to IGB (Fayad et al. 2019), and similar to laparoscopic sleeve gastrectomy (LSG) in mildly obese individuals (Novikov et al. 2018).

Currently, a Chinese study is comparing ESG to the LSG (NCT03124485). A multicenter open-label trial comparing ESG to lifestyle intervention is also going on. At the end of a 1-year follow-up, patients from the control group will be allowed to cross over to undergo ESG. Eight centers in the USA are participating, and the enrollment goal is 200 not morbidly obese individuals (BMI ≥ 30 and ≤ 40 kg/m^2) (MERIT Trial NCT03406975).

POSE (Primary Obesity Surgery Endoluminal)

The primary obesity surgery endoluminal (POSE) method employs a platform initially developed for natural orifices surgery/NOTES (IOP®, USGI Medical, San Clemente, CA), in order to plicate the fundus and distal body.

Before addressing primary obesity procedures, the IOP system had been tested and approved for gastrointestinal tissue apposition. In this sense, endoluminal cholecystectomy (Swanstrom et al. 2008), gastrotomy closure (Sclabas et al. 2006), repair of gastro-gastric fistulas (Raman et al. 2011), transgastric appendectomy (Horgan et al. 2011), and remodeling of gastrojejunal anastomosis in weight regain (Borao et al. 2010) have been reported.

The IPO is a four-channel working platform that allows the passing of different instruments, including a gripper with a suture cutter, a tissue-driven propeller, a suture anchor catheter, and a channel for a thinner endoscope. It allows the delivery of full-thickness plications under endoscopic guidance. The original POSE procedure (Fig. 44.6) consisted of two parallel lines of 4–5 plications each in the gastric fundus, followed by an additional three or four on the greater curvature of the distal gastric body (Kumar 2017). Therefore, this procedure aims at impairing

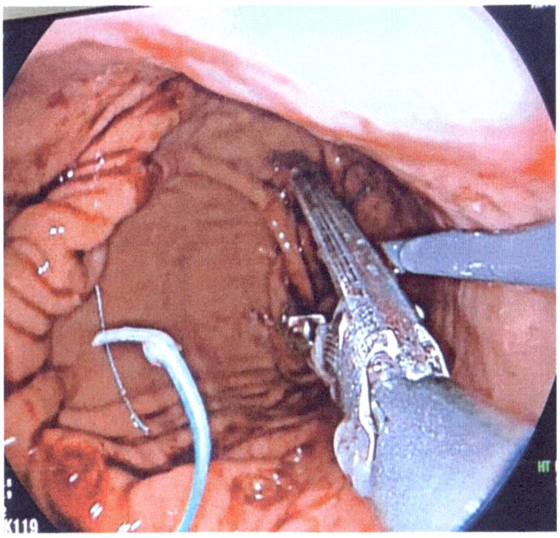

Fig. 44.6 The original POSE procedure. On the left side of the image, note the last fundal plication line. On the right side, the plication line on the greater curvature of the distal body

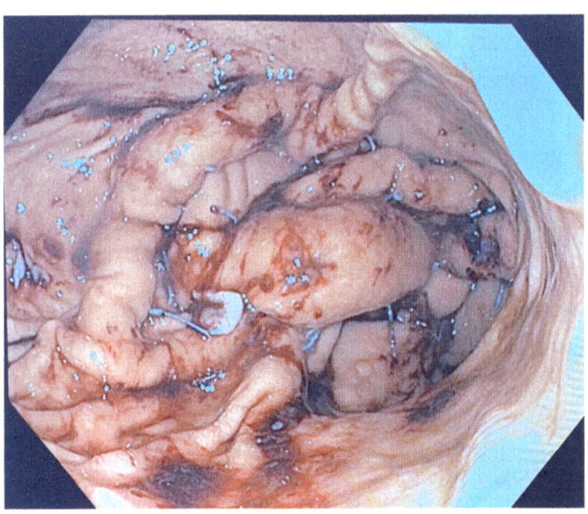

Fig. 44.7 The endoscopic final appearance of the POSE 2.0 procedure

gastric accommodation by reducing the gastric fundus (Kumar 2017).

After exciting results from noncontrolled studies (Espinos et al. 2013; Lopez-Nava et al. 2015), a large multicenter sham-controlled trial failed to demonstrate significant weight loss with the original POSE procedure (Sullivan et al. 2017). In "the ESSENTIAL trial" with 332 moderately obese patients, at 20 months, TWL was only 4.95 $\pm$ 7% in the procedure group versus 1.30 $\pm$ 5.5% in the sham group ($p < 0.0001$). Only 41% of patients presented TWL $\geq$5% at the final follow-up. Despite the statistical superiority, the intervention group did not achieve predetermined thresholds of 5% TWL and responders' rate of 50%.

A newly developed procedure—POSE 2.0—aims at delaying gastric emptying rather than impairing accommodation (Fig. 44.7). There are a few ongoing studies that might become available soon.

Endomina Method

The European regulatory agency—the Conformité Européen—has already approved the triangulation platform named Endomina® (Endotools S.A., Gosselies, Belgium) for clinical use. This is an incisionless single-use device that allows full-thickness suturing to promote gastric volume reduction.

Probably, the most attractive feature of the Endomina is that it attaches to any standard gastroscope, exempting the need for specific endoscopes. However, it requires two operators to deliver stitches: the first controls the handles of the device, while the other manipulates the scope. The procedure begins with a regular upper endoscopy followed by the placement of two guidewires in the duodenum. Then, the endoscope is removed and the platform is gently introduced over the guidewires. The endoscope is finally inserted and fixed to the system, while the guidewires are retrieved (Cauche et al. 2013; Huberty et al. 2019).

Using a 5Fr preloaded needle, the system creates an apposition of opposing walls along the greater curvature from the antrum to the fundus. Bipolar coagulation at the suturing sites has also been employed to enhance the durability and strength of the plication. The total duration of the procedure varied from 126 to 147 min in the first published series (Huberty et al. 2017; Wallstabe et al. 2018).

To date, three human studies assessing the efficacy and safety of the Endomina method have been published. All studies involved a Belgian center that helped develop both device and procedure. Huberty et al. published the first-in-man study in 2016 enrolling 12 obese patients. At 6 months AWL, TWL, and BMI reduction were 10.9 ± 7.3 kg, 11 ± 8%, and 3.9 ± 2.3 kg/m^2, respectively. No serious adverse events were reported (Huberty et al. 2017). The following series demonstrated similar results, but slightly worse outcomes at a longer-term (Wallstabe et al. 2018; Huberty et al. 2018). Controlled data are still lacking, and the real therapeutic potential of the Endomina method is yet to be determined.

Metabolic Procedures: Duodenal Mucosal Resurfacing (Fractyl®)

The duodenal mucosal resurfacing (DMR) is based on the hypothetical central role of the duodenum in regulating glycemic homeostasis. Gut hormones produced by the normal duodenum possibly stimulate the surrounding endocrine pancreatic tissue and systemic insulin receptors, thus helping control serum glucose levels. This is called the incretin effect. Conversely, in patients with type 2 diabetes (T2D), alterations in the proximal bowel epithelium could impair such paracrine and endocrine network, leading to systemic insulin resistance and glucose intolerance (Moran-Atkin et al. 2013). Animal and human studies suggest that bypassing the duodenum improves T2D, which is unrelated to weight loss (Kindel et al. 2009; de Jonge et al. 2013). This observation corroborated the so-called foregut hypothesis and fostered the development of endoscopic alternatives to surgical bypass—such as the DMR.

The DMR procedure promotes a catheter-based hydrothermal ablation of the superficial duodenal mucosa (Fig. 44.8) (de Moura et al. 2019). After re-epithelization, new cells are expected to be healthier and more efficient at producing and releasing incretin hormones, thus improving diabetes control.

The first-in-man trial enrolled 44 individuals to undergo the DMR procedure in a 6-month follow-up study. Patients experienced a reduction of 1.2% in HbA1c (Rajagopalan et al. 2016).

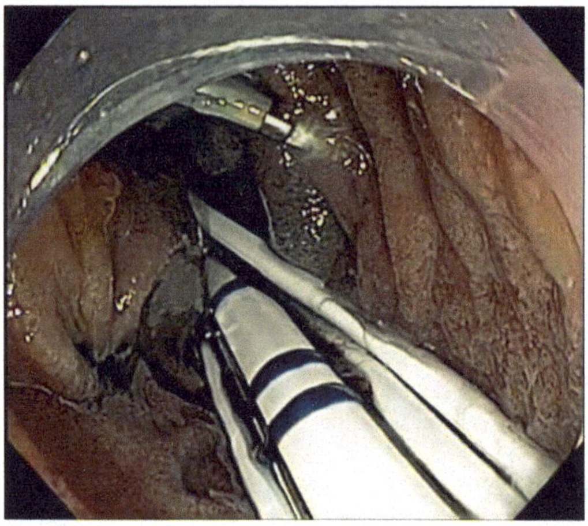

Fig. 44.8 The Duodenal Mucosal Resurfacing procedure. The catheter has two blue external marks 1 cm apart that allow the endoscopic guidance for insertion. On the top of the figure, there is an endoclip demarcating the level of the major papilla. Distally, the duodenal mucosa has already been injected with saline plus methylene blue and superficially ablated (whitish mucosa)

One-year data from a multicenter study with 46 patients showed not only a reduction in HbA1c, but also improvement in fasting plasma glucose and HOMA-IR score. However, 52% of the patients experienced at least one AE related to the procedure, of which one was considered a serious AE (van Baar et al. 2020). Currently, there is an ongoing large multicenter sham-controlled trial with a crossover design (NCT02879383).

Duodenal Jejunal Bypass Liner/DJBL (EndoBarrier®)

This is an endoluminal device composed of three main parts: the anchor, the liner, and the delivery system. The anchor has a large nitinol-based opening with bidirectional barbs for fixation. The liner is a 62-cm-long impermeable fluoropolymer sleeve that avoids contact of the food with the duodenal mucosa (Fig. 44.9). The delivery system is a single-use device preloaded with the implant.

The rationale of the DJBL is based not only on the foregut hypothesis, but also on the hindgut

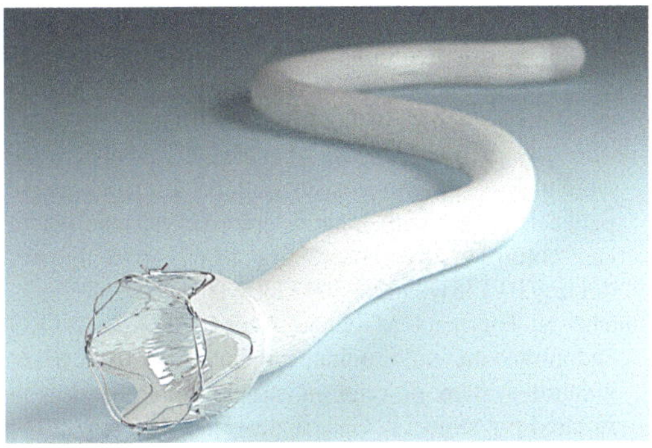

Fig. 44.9 The duodenal jejunal bypass liner device

hypothesis. The latter advocates that the presence of intact bile acids and undigested food in the distal gut triggers the release of incretin hormones (Verdam et al. 2011; Penney et al. 2015). Therefore, bypassing the duodenum with an impermeable liner precludes contact of the food bolus with a possibly impaired duodenal mucosa (foregut hypothesis), and delivers chyme into the distal gut (hindgut hypothesis).

Tarnoff et al. published the first animal study on DJBL in 2008. Six pigs underwent endoscopic deployment of the device and presented significant lower rates of weight gain compared to control animals (0.42 kg/day vs. 0.23 kg/day, $p = 0.01$) (Tarnoff et al. 2008).

Since the first human experience in 2009 (Rodriguez-Grunert et al. 2008), five randomized studies have already been published. Most report a significant impact on HbA1C levels and some also describe amelioration of metabolic parameters. A recent systematic review comprising all RCTs showed a mean reduction by 1.3% in HbA1c levels at the explant. Such a result outweighed by 0.9% the control group. Accordingly, HOMA-IR decreased by 4.6 [2.9, 6.3] (Jirapinyo et al. 2018).

In time, however, the high incidence of SAEs undermined the routine adoption of the DJBL. Some series reported up to 68% of severe complications, including liver abscesses and cholangitis (Forner et al. 2017; Quezada et al. 2018). Recently, Betzel et al. published a systematic review on AEs related to the DJBL and found 891 reports from 1056 procedures. Thirty-three were considered SAEs, including 11 liver abscesses (Betzel et al. 2018). Probably, the sharp anchoring barbs were responsible for most of the related complications, as they ulcerate the duodenal wall, thus creating a solution of continuity that allows bacterial translocation and injures surrounding vessels.

Such an AEs rate deterred further studies on the DBJL, culminating in withdrawal from the market. New prototypes with different anchoring strategies as well as devices deployed in the stomach are under development and might be available soon.

Endoscopic Magnetic Partial Jejunal Diversion/EMPJD

This is also based on the hindgut hypothesis, which is observed after surgical ileal interposition or biliopancreatic diversion. The technique involves the deployment of two self-assembling magnets (one anterograde and the other retrograde), that should be driven to couple across the wall of two intestinal segments. The magnetic compression deters the blood flow leading to round necrosis that works as a wide anastomosis a week after coupling. Usually, the magnets are uneventfully expelled during defecation (Ryou et al. 2011).

The proof-of-concept study was first published in 2016. Five pigs underwent EMPJD however due to the particular anatomy of the porcine intestine, the only possible anastomosis was jejunocolonostomy. The average duration of the procedure was 14.7 min. At 3 months, the authors typically found a 3.5 cm-wide epithelized anastomosis. The animals undergoing EMPJD presented significant weight stabilization compared to litter-paired control pigs (Ryou et al. 2016).

In the Czech Republic, 11 obese individuals who failed conservative weight loss methods were selected to undergo EMPJD. After 12 attempts, 10 cases were technically successful. Per protocol, all procedures were laparoscopy-guided. The laparoscopic forceps was used to assist coupling if sole endoscopic manipulation was not successful. Eight out of ten cases required laparoscopic assistance for coupling magnets. Upper endoscopy confirmed broad anastomoses at 2, 6, and 12 months (Machytka et al. 2017).

Mean TWL was 8.2%, 10.6%, and 14.6% at 3, 6, and 12 months, respectively. Accordingly, EWL was 21.7%, 28.3%, and 40.2%. Diabetic patients experienced a distinct reduction in HbA1c at 12 months compared to baseline (7.8 vs. 5.9). There were significant reductions in postprandial insulin at 2 and 6 months, and a significant increase in PYY at 2 months (Machytka et al. 2017).

Currently, there is an ongoing open-label trial settled in Argentina. As the sole endoscopic coupling of magnets incurs substantial technical difficulties, this method shall be converted to a combined surgical-endoscopic procedure shortly.

References

Abu Dayyeh BK, Rajan E, Gostout CJ (2013) Endoscopic sleeve gastroplasty: a potential endoscopic alternative to surgical sleeve gastrectomy for treatment of obesity. Gastrointest Endosc 78:530–535. https://doi.org/10.1016/j.gie.2013.04.197

Abu Dayyeh BK, Acosta A, Camilleri M et al (2017) Endoscopic sleeve gastroplasty alters gastric physiology and induces loss of body weight in obese individuals. Clin Gastroenterol Hepatol 15:37-43.e1. https://doi.org/10.1016/j.cgh.2015.12.030

Alqahtani A, Al-Darwish A, Mahmoud AE et al (2018) Short-term outcomes of endoscopic sleeve gastroplasty in 1000 consecutive patients. Gastrointest Endosc 89 (6):1132–1138. https://doi.org/10.1016/j.gie.2018.12.012

Angrisani L, Santonicola A, Iovino P et al (2018) IFSO worldwide survey 2016: primary, endoluminal, and revisional procedures. Obes Surg 28(12):3783–3794. https://doi.org/10.1007/s11695-018-3450-2

Betzel B, Drenth JPH, Siersema PD (2018) Adverse events of the duodenal-jejunal bypass liner: a systematic review. Obes Surg 28:3669–3677. https://doi.org/10.1007/s11695-018-3441-3

Bhandari M, Jain S, Mathur W et al (2019) Endoscopic sleeve gastroplasty is an effective and safe minimally invasive approach for treatment of obesity: first Indian experience. Dig Endosc 32(4):541–546. https://doi.org/10.1111/den.13508

Borao F, Gorcey S, Capuano A (2010) Prospective single-site case series utilizing an endolumenal tissue anchoring system for revision of post-RYGB stomal and pouch dilatation. Surg Endosc 24:2308–2313. https://doi.org/10.1007/s00464-010-0919-z

Brunaldi VO, Galvao Neto M (2019) Endoscopic techniques for weight loss and treating metabolic syndrome. Curr Opin Gastroenterol 35:424–431. https://doi.org/10.1097/MOG.0000000000000561

Busetto L, Segato G, De Luca M et al (2004) Preoperative weight loss by intragastric balloon in super-obese patients treated with laparoscopic gastric banding: a case-control study. Obes Surg 14:671–676. https://doi.org/10.1381/096089204323093471

Cauche N, Hiernaux M, Chau A et al (2013) Sa1435 endomina: the endoluminal universal robotized triangulation system: description and preliminary results in isolated pig stomach. Gastrointest Endosc 77:AB204–AB205. https://doi.org/10.1016/j.gie.2013.03.486

Chan AOO, Chow WS, Lam KF et al (2008) The effect of intragastric balloon placement on weight loss and type 2 diabetes control. Aliment Pharmacol Therap 28:162–164

Colquitt JL, Pickett K, Loveman E, Frampton GK (2014) Surgery for weight loss in adults. Cochrane Database Syst Rev 2014:CD003641. https://doi.org/10.1002/14651858.CD003641.pub4

de Jonge C, Rensen SS, Verdam FJ et al (2013) Endoscopic duodenal-jejunal bypass liner rapidly improves type 2 diabetes. Obes Surg 23:1354–1360. https://doi.org/10.1007/s11695-013-0921-3

de Moura EGH, Ponte-Neto AM, Tsakmaki A et al (2019) Histologic assessment of the intestinal wall following duodenal mucosal resurfacing (DMR): a new procedure for the treatment of insulin-resistant metabolic disease. Endosc Int Open 7:E685–E690. https://doi.org/10.1055/a-0862-0263

Dumonceau JM (2008) Evidence-based review of the bioenterics intragastric balloon for weight loss. Obes Surg 18:1611–1617

Espinos JC, Turro R, Mata A et al (2013) Early experience with the Incisionless Operating Platform (IOP) for the treatment of obesity: the Primary Obesity Surgery Endolumenal (POSE) procedure. Obes Surg 23:1375–1383. https://doi.org/10.1007/s11695-013-0937-8

Fayad L, Cheskin LJ, Adam A et al (2019) Endoscopic sleeve gastroplasty versus intragastric balloon insertion: efficacy, durability, and safety. Endoscopy 51 (6):532–539. https://doi.org/10.1055/a-0852-3441

Forner PM, Ramacciotti T, Farey JE, Lord RV (2017) Safety and effectiveness of an endoscopically placed duodenal-jejunal bypass device (EndoBarrier(R)): outcomes in 114 patients. Obes Surg 27:3306–3313. https://doi.org/10.1007/s11695-017-2939-4

Galvao-Neto MDP, Grecco E, de Souza TF et al (2016) Endoscopic sleeve gastroplasty—minimally invasive therapy for primary obesity treatment. Arquivos Brasil Cirurg Digest 29(Suppl 1):95–97. https://doi.org/10.1590/0102-6720201600S10023

Genco A, Bruni T, Doldi SB et al (2005) Bioenterics intragastric balloon: the italian experience with 2,515 patients. Obes Surg 15:1161–1164. https://doi.org/10.1381/0960892055002202

Hajifathalian K, Ang B, Dawod QM et al (2019) 175 Long-term follow up and outcomes after endoscopic sleeve gastroplasty for treatment of obesity (5 year data). Gastrointest Endosc 89:AB58. https://doi.org/10.1016/j.gie.2019.04.019

Hedjoudje A, Dayyeh BA, Cheskin LJ et al (2020) Efficacy and safety of endoscopic sleeve gastroplasty: a systematic review and meta-analysis. Clin Gastroenterol Hepatol 18(5):1043-1053.e4. https://doi.org/10.1016/j.cgh.2019.08.022

Horgan S, Thompson K, Talamini M et al (2011) Clinical experience with a multifunctional, flexible surgery system for endolumenal, single-port, and NOTES procedures. Surg Endosc 25:586–592. https://doi.org/10.1007/s00464-010-1225-5

Huberty V, Ibrahim M, Hiernaux M et al (2017) Safety and feasibility of an endoluminal-suturing device for endoscopic gastric reduction (with video). Gastrointest Endosc 85:833–837. https://doi.org/10.1016/j.gie.2016.08.007

Huberty V, Machytka E, Boskoski I et al (2018) Endoscopic gastric reduction with an endoluminal suturing device: a multicenter prospective trial with 1-year follow-up. Endoscopy 50:1156–1162

Huberty V, Verset L, Deviere J (2019) Endoscopic full-thickness resection of a gastric GI stromal tumor. VideoGIE 4:120–122. https://doi.org/10.1016/j.vgie.2018.11.003

Imaz I, Martínez-Cervell C, García-Alvarez EE et al (2008) Safety and effectiveness of the intragastric balloon for obesity. A meta-analysis. Obes Surg 18:841–846. https://doi.org/10.1007/s11695-007-9331-8

Jirapinyo P, Haas AV, Thompson CC (2018) Effect of the duodenal-jejunal bypass liner on glycemic control in patients with type 2 diabetes with obesity: a meta-analysis with secondary analysis on weight loss and hormonal changes. Diabetes Care 41:1106–1115. https://doi.org/10.2337/dc17-1985

Kindel TL, Yoder SM, Seeley RJ et al (2009) Duodenal-jejunal exclusion improves glucose tolerance in the diabetic, Goto-Kakizaki rat by a GLP-1 receptor-mediated mechanism Ligament of Treitz. J Gastrointest Surg 13(10):1762–1772. https://doi.org/10.1007/s11605-009-0912-9

Kumar N (2017) Gastric Plication. Gastrointest Endosc Clin N Am 27:257–265. https://doi.org/10.1016/j.giec.2016.12.003

Kumar N, Abu Dayyeh BK, Lopez-Nava Breviere G et al (2018) Endoscopic sutured gastroplasty: procedure evolution from first-in-man cases through current technique. Surg Endosc 32:2159–2164. https://doi.org/10.1007/s00464-017-5869-2

Lopez-Nava G, Bautista-Castano I, Jimenez A et al (2015) The Primary Obesity Surgery Endolumenal (POSE) procedure: one-year patient weight loss and safety outcomes. Surg Obes Relat Dis 11:861–865. https://doi.org/10.1016/j.soard.2014.09.026

Lopez-Nava G, Sharaiha RZ, Vargas EJ et al (2017) Endoscopic sleeve gastroplasty for obesity: a multicenter study of 248 patients with 24 months follow-up. Obes Surg 27:2649–2655. https://doi.org/10.1007/s11695-017-2693-7

Machytka E, Buzga M, Zonca P et al (2017) Partial jejunal diversion using an incisionless magnetic anastomosis system: 1-year interim results in patients with obesity and diabetes. Gastrointest Endosc 86:904–912. https://doi.org/10.1016/j.gie.2017.07.009

Marinos G, Eliades C, Raman Muthusamy V, Greenway F (2014) Weight loss and improved quality of life with a nonsurgical endoscopic treatment for obesity: clinical results from a 3- and 6-month study. Surg Obes Relat Dis 10:929–934. https://doi.org/10.1016/j.soard.2014.03.005

Moran-Atkin E, Brody F, Fu SW, Rojkind M (2013) Changes in GIP gene expression following bariatric surgery. Surg Endosc 27:2492–2497. https://doi.org/10.1007/s00464-012-2764-8

Moura D, Oliveira J, De Moura EGH et al (2016) Effectiveness of intragastric balloon for obesity: a systematic review and meta-analysis based on randomized control trials. Surg Obes Relat Dis 12:420–429. https://doi.org/10.1016/j.soard.2015.10.077

Noren E, Forssell H (2016) Aspiration therapy for obesity; a safe and effective treatment. BMC Obes 3:56. https://doi.org/10.1186/s40608-016-0134-0

Novikov AA, Afaneh C, Saumoy M et al (2018) Endoscopic sleeve gastroplasty, laparoscopic sleeve gastrectomy, and laparoscopic band for weight loss: how do they compare? J Gastrointest Surg 22:267–273. https://doi.org/10.1007/s11605-017-3615-7

Nystrom M, Machytka E, Noren E et al (2018) Aspiration therapy as a tool to treat obesity: 1- to 4-year results in a 201-patient multi-center post-market european registry study. Obes Surg 28:1860–1868. https://doi.org/10.1007/s11695-017-3096-5

O'Brien PE, Hindle A, Brennan L et al (2019) Long-term outcomes after bariatric surgery: a systematic review and meta-analysis of weight loss at 10 or more years for all bariatric procedures and a single-centre review of 20-year outcomes after adjustable gastric banding. Obes Surg 29:3–14. https://doi.org/10.1007/s11695-018-3525-0

Penney NC, Kinross J, Newton RC, Purkayastha S (2015) The role of bile acids in reducing the metabolic complications of obesity after bariatric surgery: a systematic review. Int J Obes 39:1565–1574. https://doi.org/10.1038/ijo.2015.115

Quezada N, Munoz R, Morelli C et al (2018) Safety and efficacy of the endoscopic duodenal-jejunal bypass liner prototype in severe or morbidly obese subjects implanted for up to 3 years. Surg Endosc 32:260–267. https://doi.org/10.1007/s00464-017-5672-0

Rajagopalan H, Cherrington AD, Thompson CC et al (2016) Endoscopic duodenal mucosal resurfacing for the treatment of type 2 diabetes: 6-month interim analysis from the first-in-human proof-of-concept study. Diabetes Care 39:2254–2261. https://doi.org/10.2337/dc16-0383

Raman SR, Holover S, Garber S (2011) Endolumenal revision obesity surgery results in weight loss and closure of gastric-gastric fistula. Surg Obes Relat Dis 7:304–308. https://doi.org/10.1016/j.soard.2011.01.045

Rodriguez-Grunert L, Galvao Neto MP, Alamo M et al (2008) First human experience with endoscopically delivered and retrieved duodenal-jejunal bypass sleeve. Surg Obes Relat Dis 4:55–59. https://doi.org/10.1016/j.soard.2007.07.012

Rothstein R, Woodman G, Swain J et al (2018) Weight reduction in patients with obesity using the transpyloric shuttle®: endobesity® II study

Ryou M, Cantillon-Murphy P, Azagury D et al (2011) Smart Self-Assembling MagnetS for ENdoscopy (SAMSEN) for transoral endoscopic creation of immediate gastrojejunostomy (with video). Gastrointest Endosc 73:353–359. https://doi.org/10.1016/j.gie.2010.10.024

Ryou M, Agoston AT, Thompson CC (2016) Endoscopic intestinal bypass creation by using self-assembling magnets in a porcine model. Gastrointest Endosc 83:821–825. https://doi.org/10.1016/j.gie.2015.10.023

Saber AA, Shoar S, Almadani MW et al (2017) Efficacy of first-time intragastric balloon in weight loss: a systematic review and meta-analysis of randomized controlled trials. Obes Surg 27:277–287. https://doi.org/10.1007/s11695-016-2296-8

Sallet JA, Marchesini JB, Paiva DS et al (2004) Brazilian multicenter study of the intragastric balloon. Obes Surg 14:991–998. https://doi.org/10.1381/0960892041719671

Sclabas GM, Swain P, Swanstrom LL (2006) Endoluminal methods for gastrotomy closure in natural orifice transenteric surgery (NOTES). Surg Innov 13:23–30. https://doi.org/10.1177/155335060601300105

Sharaiha RZ, Kedia P, Kumta N et al (2015) Initial experience with endoscopic sleeve gastroplasty: technical success and reproducibility in the bariatric population. Endoscopy 47:164–166. https://doi.org/10.1055/s-0034-1390773

Sullivan S, Stein R, Jonnalagadda S et al (2013) Aspiration therapy leads to weight loss in obese subjects: a pilot study. Gastroenterology 145:1245. https://doi.org/10.1053/j.gastro.2013.08.056

Sullivan S, Swain JM, Woodman G et al (2017) Randomized sham-controlled trial evaluating efficacy and safety of endoscopic gastric plication for primary obesity: the ESSENTIAL trial. Obesity (Silver Spring, Md) 25:294–301. https://doi.org/10.1002/oby.21702

Swanstrom L, Dunst C, Whiteford M et al (2008) Initial experience with NOTES transgastric cholecystectomy

Tarnoff M, Shikora S, Lembo A, Gersin K (2008) Chronic in-vivo experience with an endoscopically delivered and retrieved duodenal-jejunal bypass sleeve in a porcine model. Surg Endosc 22:1023–1028. https://doi.org/10.1007/s00464-007-9652-7

Thompson CC, Abu Dayyeh BK, Kushner R et al (2017) Percutaneous gastrostomy device for the treatment of class ii and class iii obesity: results of a randomized controlled trial. Am J Gastroenterol 112:447–457. https://doi.org/10.1038/ajg.2016.500

Thompson CC, Abu Dayyeh BK, Kushnir V et al (2019) Aspiration therapy for the treatment of obesity: 4-year results of a multicenter randomized controlled trial. Surg Obes Relat Dis 15:1348–1354. https://doi.org/10.1016/j.soard.2019.04.026

Trang J, Lee SS, Miller A et al (2018) Incidence of nausea and vomiting after intragastric balloon placement in bariatric patients—a systematic review and meta-analysis. Int J Surg (London, England) 57:22–29. https://doi.org/10.1016/j.ijsu.2018.06.038

van Baar ACG, Holleman F, Crenier L et al (2020) Endoscopic duodenal mucosal resurfacing for the treatment of type 2 diabetes mellitus: one year results from the first international, open-label, prospective, multicentre study. Gut 69(2):295–303. https://doi.org/10.1136/gutjnl-2019-318349

Verdam FJ, Greve JWM, Roosta S et al (2011) Small intestinal alterations in severely obese hyperglycemic subjects. J Clin Endocrinol Metab 96:379–383. https://doi.org/10.1210/jc.2010-1333

Wallstabe I, Oberaender N, Weimann A, Schiefke I (2018) Endoscopic sleeve gastroplasty using the novel Endomina device for morbidly obese patients. Endoscopy 50:E327–E328. https://doi.org/10.1055/a-0667-7696

Bariatric Embolization: A Possible Non-surgical Option for Weight Reduction

45

Muhammad A. Latif and Clifford R. Weiss

Contents

Introduction 620

Current Techniques 620

Role of Interventional Radiology 620

Basis of Bariatric Embolization for Weight Loss 622
Appetite and Ghrelin Production 622
Stomach as an Endocrine Organ 622
Anatomy of the Gastric Fundus 622
Ghrelin Suppression After Gastric Bypass Surgery 623

Bariatric Embolization 623
Patient Selection 623
Vascular Access 623
Embolic Agents 623
Radiation Dose 626
Post-procedure Immediate Care 626

Long-Term Metabolic Effects 626

Combination Therapy with Clinical and Surgical Maneuvers 626

Ongoing Human Trials and Other Lines of Investigation 626

References 626

Abstract

Gastric arterial embolization has historically been used to treat gastrointestinal hemorrhage. After the discovery of the appetite-inducing hormone, ghrelin, gastric fundal embolization or bariatric embolization has gained importance for weight reduction. Although still in initial experimental stage, the only effective nonsurgical intervention for reduction of ghrelin is bariatric gastric embolization. Preliminary findings from early single-arm clinical trials have demonstrated that bariatric embolization results in significant decrease of plasma ghrelin levels and leads to weight loss.

M. A. Latif · C. R. Weiss (✉)
Russell H. Morgan Department of Radiology and Radiological Science, Division of Interventional Radiology, Johns Hopkins University School of Medicine, Baltimore, MD, USA
e-mail: cweiss@jhmi.edu

J. Faintuch and S. Faintuch (eds.), *Obesity and Diabetes*, https://doi.org/10.1007/978-3-030-53370-0_45

It is a safe, novel, promising technique that has the potential to be an alternative procedure to surgery for patients with body mass index between 30 and 40. Common side effects include transient epigastric pain, nausea, vomiting, and stomach ulcers.

Keywords

Bariatric embolization · Left gastric artery embolization · Ghrelin · Weight loss · Appetite-stimulating hormone

Introduction

The World Health Organization recognizes obesity as a global epidemic (Afshin et al. 2017; WHO 2000; Caballero 2007), and in 2013, the American Medical Association (AMA) categorized obesity as a separate disease entity with multiple pathological aspects, requiring a variety of surgical and nonsurgical interventions (A.M.A. Recognizes Obesity as a Disease 2019; Kosti and Panagiotakos 2006). In a recent report of countries of the Organization of Economic Cooperation and Development (OECD), including the United States, one child out of five was found to be overweight or obese (OBESITY Update 2014). Obesity is a risk factor for type 2 diabetes mellitus (T2DM) (Al-Goblan et al. 2014), cardiovascular diseases (CVD) such as hypertension (DeMarco et al. 2014), stroke, and coronary heart disease (Jin 2013), gall bladder disease (Everhart 1993), and cancer (endometrial, breast, prostate, colon) (Stone et al. 2018), resulting in increased risk of disability or premature death (Smith and Smith 2016; Abdelaal et al. 2017). Obesity is also linked with nonfatal diseases, including gout, respiratory conditions (chronic obstructive pulmonary disease, asthma, obstructive sleep apnea) (Poulain et al. 2006), gastro-esophageal reflux disease, osteoarthritis, and infertility. There are also serious psychosocial health implications of obesity due to societal prejudice against obesity (Simon et al. 2006).

Current Techniques

The mechanisms of weight loss for bariatric surgeries such as Roux-en-Y gastric bypass, sleeve gastrectomy, and gastric banding include restricting meal size, inhibiting nutrient absorption, and altering gastrointestinal hormones involved in metabolic homeostasis (Elder and Wolfe 2007; Beckman et al. 2010; Cummings et al. 2002; Karra et al. 2010; O'Rourke et al. 2006; Paxton et al. 2013; Sumithran et al. 2011). Surgical and endoscopic bariatric interventions can be associated with short- and long-term complications (Neff et al. 2013) (Table 45.1). Leaks, stenoses, bleeding, and venous thromboembolic events (VTE) are listed in the early postoperative period (Lim et al. 2018). Late surgical complications include band slippage, band erosion, megaesophagus or pseudoachalasia, gallstone disease, perforation, bleeding, small bowel obstruction, and internal hernia (Lim et al. 2018).

Role of Interventional Radiology

Recently, there have been advancements in interventional radiology to treat obesity as a minimally invasive alternative (Neff et al. 2013). Current radiological interventions include percutaneous computed tomography (CT)-guided cryovagotomy (Prologo et al. 2019) and image-guided gastric fundal embolization (bariatric embolization) (Arepally et al. 2007). Percutaneous CT-guided cryovagotomy involves the use of cryoablation to freeze the posterior vagal trunk located at the base of the esophagus, which normally transmits the hunger signal from the stomach to the brain (Prologo et al. 2019). Bariatric embolization is a minimally invasive, image-guided procedure that targets the appetite-stimulating endocrine functions of the gastric fundus (Gunn and Oklu 2014; Madoff 2013).

Table 45.1 Current surgical, endoscopic, and interventional radiology methods for weight reduction

Specific category	Procedure detail	Proposed mechanism of action for obesity	Benefits	Risks	% Of 1 year weight loss
Roux-en-Y gastric bypass	Small, upper gastric pouch connected to small bowel; larger lower "remnant" pouch;	Reduction of gastric size with resulting decrease in appetite	Rapid weight loss; No device inside the body; Quick improvement in comorbidities (Diabetes and hypertension)	Long healing; Nutrient absorption Dietary restrictions; Irreversible, possible surgical complications	31.2
Sleeve gastrectomy	Resection of the stomach to decrease its size	Reduction of gastric size with resulting decrease in appetite	Rapid weight loss; No device inside the body; Short recovery time	Irreversible procedure Complications, including leak, bleeding, etc.	25.2
Adjustable gastric band	Inflatable silicone device laparoscopically placed around the top portion of the stomach to decrease food consumption	Reduction of gastric size with resulting decrease in appetite	Lower risk of complications No cutting of the stomach Short hospital stay	Infection , erosion, complications Several adjustments Gradual weight loss	13.7
Biliopancreatic diversion with duodenal switch (BPD/DS) (Baltasar et al. 2001)	Duodenal switch with a vertical subtotal gastrectomy and pylorus preservation	Reduction of gastric size; decrease in appetite;	Drastic, significant weight loss	Dumping, osteoporosis Poor nutrition bleeding, infection, hernia	70.1 (% EWL) [a]
Endoscopic sleeve gastroplasty (accordion procedure) (Sharaiha 2017)	Endoscopic suturing device to reduce stomach size	Reduction of gastric size with resulting decrease in appetite	No incisions, Low cost, Outpatient procedure	Bleeding, leak, venous thrombosis, pulmonary embolism	20
Transpyloric shuttle (TPS)	Silicone rubber balloon endoscopically placed in the stomach for 12 months	Decreased food absorption and appetite	Nonsurgical, Longer effect, Significant weight loss	Bleeding, infection, tearing of the esophagus or stomach	9.5
Endoluminal magnetic partial jejunal diversion	Endoscopically placed magnet for intestine diversion	Nutrients, digestive fluids circumvent part of the small bowel	Significant weight loss	Vitamin deficiencies Abdominal distension Diarrhea Constipation Vomiting	14.6
Bariatric Embolization (Hafezi-Nejad et al. 2019)	Embolization of the LGA	Decrease in ghrelin and appetite	Safe with no major complications	Gastric ulcer/ ischemia/ perforation, splenic infarction, acute pancreatitis	8.7

(continued)

Table 45.1 (continued)

Specific category	Procedure detail	Proposed mechanism of action for obesity	Benefits	Risks	% Of 1 year weight loss
Percutaneous CT-guided cryovagotomy	Argon gas used to freeze the posterior vagal trunk located at the base of the esophagus	Disruption of the hunger signal from the stomach to the brain	No data available	No data available	No data available

[a]The percent of excess weight loss or %EWL (i.e., [initial weight–actual weight/actual weight–ideal weight] × 100) is accepted as a good measure of end results, and %EWL > 50 is considered a success

Basis of Bariatric Embolization for Weight Loss

Appetite and Ghrelin Production

Ghrelin is an orexigenic hormone secreted predominantly by the stomach (Druce et al. 2004; Strader and Woods 2005; Inui et al. 2004; Inui 2001). Ghrelin-producing cells are located in the fundus of the stomach, and directly stimulate appetite resulting in weight gain (Hayashida et al. 2001; Takiguchi et al. 2012). It has been shown by Maksud et al. that ghrelin-forming cells have higher density and expression in the stomach mucosa of obese patients (Maksud et al. 2011). Because of the unique nature of this hormone and its effect on appetite, multiple approaches to modulate ghrelin production have been attempted; however, no direct clinical technique has been shown to be successful (Cummings and Shannon 2003; Hu et al. 2005; Murray et al. 2003; Loftus et al. 2000; Wren et al. 2001). In 2007, Arepally et al. (2007, 2008) described an endovascular procedure that directly targeted the gastric fundus and its function. In this investigation, gastric artery chemical embolization (GACE) was performed by injecting a liquid sclerosing agent, sodium morrhuate, into the main left gastric artery (LGA) with fluoroscopic guidance, resulting in a decrease in the plasma level of ghrelin.

Stomach as an Endocrine Organ

Ghrelin is the only hormone that stimulates food intake. Discovered by Kojima et al. (1999), it activates the growth hormone secretagogue receptor in the hypothalamus or anterior pituitary. Ghrelin inhibition on the central nervous system is promptly followed by anorexia and loss of weight (Hu et al. 2005; Loftus et al. 2000). In humans ghrelin injection elicits hunger and enhanced caloric intake (Cummings et al. 2001).

The tendency to recover lost weight after dieting could be partially explained by rebound ghrelin elevation (Strader and Woods 2005). After meals ghrelin decreases in the normal however not in most obese individuals, potentially triggering hyperphagia (English et al. 2002).

Anatomy of the Gastric Fundus

Seven parts are admitted in the stomach: cardia, fundus, antrum, pylorus, lesser curvature, greater curvature, and incisura angularis. The neuroregulatory pathway involved in satiety and appetite stimulation is fundamentally located at the fundus (Cummings et al. 2002, 2005; Frühbeck et al. 2004; Tritos et al. 2003), which contains 10–20 times more ghrelin than the duodenum, which is second (Druce et al. 2004; Inui et al. 2004). The left gastric artery perfuses the

Fig. 45.1 Normal arterial supply of the stomach

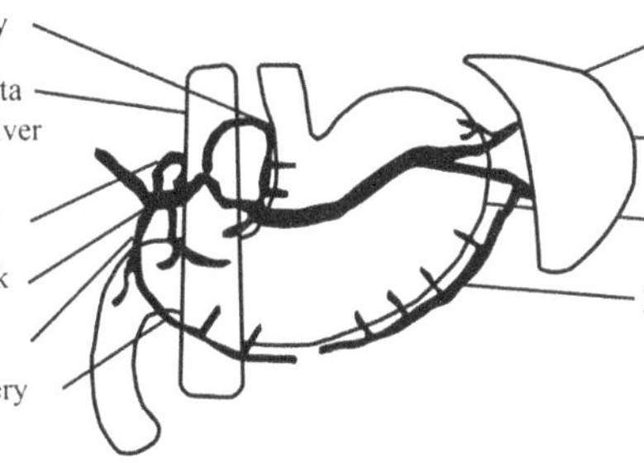

fundus (Fig. 45.1), and it is amenable to percutaneous endovascular catheterization.

The short gastric arteries, and the gastroepiploic artery, which have different origins, have a more modest participation. Gastric arterial embolization is effective in cases of bleeding and is also safe, as existing vascular arcades prevent unplanned ischemia.

Ghrelin Suppression After Gastric Bypass Surgery

Since gastric bypass surgery isolates the gastric fundus from ingested nutrients, ghrelin profiles are shown to be lower by 77% compared to those in controls (Murray et al. 2003; Cummings et al. 2005; Nakazato et al. 2001; Okumura et al. 2002). Furthermore, the normal diurnal pattern of ghrelin is interrupted, and the meal-initiated fluctuations are blunted (Murray et al. 2003; Cummings et al. 2005).

Bariatric Embolization

Gastric embolization, previously used for hemorrhage only, could target the appetite-stimulating endocrine functions of the gastric fundus (Madoff 2013; Lu et al. 2014; Loffroy et al. 2010, 2015; Gunn and Oklu 2014). Spherical embolics (300–500 μm) are delivered via a transcatheter approach into the LGA (Madoff 2013). Fundal mucosal ischemia depresses the production of ghrelin (Paxton et al. 2013; Sumithran et al. 2011; Arepally et al. 2007, 2008). Weight and ghrelin response were documented in both animals (Arepally et al. 2007, 2008; Bawudun et al. 2012) and humans (Bin et al. 2018; Kim et al. 2018; Kipshidze et al. 2015; Syed et al. 2016; Weiss et al. 2017).

Patient Selection

Patients with BMI between 30 kg/m^2 and up are the typical candidates. In case of GI disease (stomach or duodenum ulcer, cancer) or local radiation therapy, exclusion applies. It is not known whether bariatric surgery is safe and feasible after bariatric embolization. Future bariatric surgery should be preoperatively discussed with the patients.

Vascular Access

Transfemoral or transradial arterial access are available. The transradial approach looks more feasible and safe (Pirlet et al. 2019).

Embolic Agents

Historically sodium morrhuate was employed; however, better options are now in the market (Arepally et al. 2007). Bead block microspheres (300–500 μm) (www.bostonscientific.com), polyvinyl alcohol (PVA) particles (500–700 μm) (www.cookmedical.com), and embosphere microspheres (300–500 μm) (www.merit.com) are preferred (Vaidya et al. 2008). Spherical embolics of 300–500 μm size are conventionally injected (Table 45.2). Microspheres are biocompatible, hydrophilic, nonresorbable particles produced from an acrylic polymer and impregnated with porcine gelatin (No Title 2019; Kshirsagar and Saudagar 2016).

Table 45.2 Clinical trials for bariatric embolization

Title	Acronym, status, results	Eligibility	Interventions	Embolic agent	Phase	Cases	Locations
Bariatric Embolization Trial for the Obese Nonsurgical	BET-ON, Not yet recruiting, No results available	Obesity Weight Loss Body Weight Morbid Obesity (BMI >40, or BMI >35 with medical comorbidities)	LGA embolization	450 μm Embozene microparticles	Phase 1	5	University of Calgary, CA
Bariatric Embolization of Arteries for the Treatment of Obesity	BEAT Obesity, Completed, Has results	Morbid Obesity (BMI between 40 and 60)	LGA/LGEA embolization	Artificial Embolization Device Embosphere Microspheres	Not applicable	20	Johns Hopkins Hospital, Baltimore, MD, USA
Bariatric Embolization of Arteries for the Treatment of Nonalcoholic Steatohepatitis	None available, Not yet recruiting, No results available	Obesity Weight Loss Body Weight Nonalcoholic Steatohepatitis Nonalcoholic Fatty Liver Disease (NAFLD) (BMI >35)	LGA/LGEA embolization	Embosphere microspheres	Not applicable	8	Saint Louis University, Saint Louis, Missouri, United States
Bariatric Embolization of Arteries in Obese Patients with HCC to Allow Salvage Liver Transplantation	None available, Suspended, No results available	Obesity Weight Loss Hepatocellular Carcinoma Hepatitis Cirrhosis (BMI >35)	LGA embolization	Embosphere Microspheres	Not applicable	8	Saint Louis University, Saint Louis, Missouri, United States
Transradial Selective Catheterization of the Celiac Artery in Obese Patients	None available, recruiting, No results available	Obesity (BMI >30)	Transradial celiac artery angiography	Feasibility , safety of transradial angiography of the celiac artery	Not Applicable	54	Quebec Heart & Lung Institute (IUCPQ-UL), Quebec City, Quebec, CA
Gastric Artery Embolization Trial for Lessening Appetite Nonsurgical	GETLEAN, Unknown, Ni results available	Obesity (BMI >40)	LGA embolization	300–500 μm Bead Block	Not applicable	5	Dayton Interventional Radiology, Dayton, Ohio, United States
Bariatric Embolization for Morbid Obesity	BAEMO, Unknown, Initial results available	Morbid Obesity (BMI >30)	LGA embolization Healthy diet and exercise	500–710 μm PVA (COOK Inc, Bloomington IN, USA)	Phase 3	50	Zhongda Hospital, Southeast University, Nanjing, Jiangsu, China
Gastric Arterial Embolization for Weight Loss	None available, Unknown, No results	Obesity (BMI >40, or BMI >35 with medical comorbidities)	LGA embolization Embolization of gastroepiploic artery	150–250 μm PVA particles	Not applicable	10	Martin Simons, University Health Network, Toronto, CA
Bariatric Embolization of Arteries with Imaging Visible Embolics	(BEATLES), Recruiting, No results available	Obesity Morbid Obesity Weight Loss (BMI >35)	LGA embolization	Bariatric embolization, imaging-visible embolics	Not applicable	59	Johns Hopkins Hospital Baltimore, MD, USA

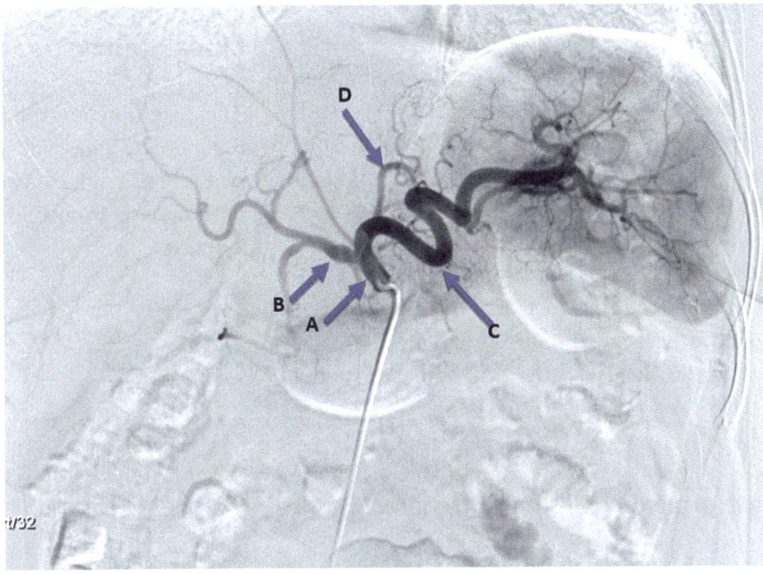

Fig. 45.2 Left gastric artery embolization procedure in a 54-year-old female with morbid obesity. Celiac artery angiogram showing Celiac artery (arrow *A*), Common Hepatic artery (arrow *B*), Splenic artery (arrow *C*), and Left Gastric Artery (arrow *D*)

The exact choice will depend on the arterial anatomy, angiographic findings, types of embolization (localized, proximal, or segmental), duration of occlusion desired, the need for tissue viability, and the patient comorbidities (Lubarsky et al. 2009, 2010; Shin 2012). Bead block microspheres, used in other contexts (Liaw et al. 2012), are a valid option (Kipshidze et al. 2015; Syed et al. 2016). Polyvinyl alcohol (PVA) agents are also a possibility (Bin et al. 2018; Pirlet et al. 2019) (Fig. 45.2).

In the BEAT Obesity trial, 300–500 μm-calibrated embospheres were technically feasible, well tolerated, and followed by short- or intermediate-term weight loss (Weiss et al. 2019). Experimentally smaller microspheres show greater weight gain suppression and fundal ghrelin expression, but with more gastric ulceration (Fu et al. 2018). Radio-opaque embolic beads (Mikhail et al. 2018) have advantages regarding intraprocedural monitoring of target and nontarget embolization, and postprocedural evaluation of embolization (Duran et al. 2016; Thompson et al. 2018). Other embolic agents are listed in Table 45.3 (Gunn and Oklu 2014; Brown et al. 2016; Anton and Rahman 2015).

Table 45.3 Different types of embolic agents

Embolic agents types			
	Permanent		
Temporary	Nonabsorbable microparticles	Mechanical embolic agents	Liquid embolic agents
Gelatin sponge	Polyvinyl alcohol particles (spherical and non-spherical)	Pushable coils	N-butyl-2-cyanoacrylate
Oxidized cellulose and microfibrillar collagen	Tris-acryl gelatin microspheres	Liquid coils	Ethylene vinyl alcohol copolymer
Thrombin	Embozene microspheres	Detachable coils	
		Vascular plugs (covered and non-covered)	

Radiation Dose

Fluroscopic dose follows the recommendations of the Society of Interventional Radiology (SIR), American College of Radiology (ACR), American Association of Physicists in Medicine (AAPM), and National Council on Radiation Protection and Measurements (NCRP) (Stecker et al. 2009; Miller et al. 2004; Radiology AC of 2015; Mahesh 2012; Balter et al. 2012). There is no evidence of skin burns from bariatric embolization in the Beat Obesity trial, and interventions usually take less than 1 h (Weiss et al. 2017).

Post-procedure Immediate Care

Gastric ulcer is a risk, and proton pump inhibitors are recommended before and after the procedure. A subsequent overnight admission can be needed in circumstances of nausea, vomiting, or epigastric pain.

Long-Term Metabolic Effects

Nejad et al. have shown 8.1% weight loss in 1 year (Hafezi-Nejad et al. 2019), less than the 13.7–31.2% outcome after surgical intervention (Arterburn et al. 2018). However, bariatric embolization is more safe and effective especially for patients who have contraindications (Gunn and Oklu 2014; Bin et al. 2018; Hafezi-Nejad et al. 2019; Stahl and Malhotra 2019), or are not interested in bariatric surgery.

Combination Therapy with Clinical and Surgical Maneuvers

The inclusion of diet control and exercise could contribute to weight control on a longer term. Embolization is not proposed as a replacement for bariatric surgery, but as a complementary method to facilitate weight loss with lifestyle modification.

Ongoing Human Trials and Other Lines of Investigation

There are only few clinical trial results for weight reduction, along with others in the recruitment phase (Table 45.2). Most of the safety data have been derived from studies to treat gastrointestinal hemorrhage.

References

A.M.A. Recognizes Obesity as a Disease (2019) The New York Times [Internet]. https://www.nytimes.com/2013/06/19/business/ama-recognizes-obesity-as-a-disease.html?auth=link-dismiss-google1tap&smid=pl-share

Abdelaal M, le Roux CW, Docherty NG (2017) Morbidity and mortality associated with obesity. Ann Transl Med 5(7):161

Afshin A, Forouzanfar MH, Reitsma MB, Sur P, Estep K, Lee A et al (2017) Health effects of overweight and obesity in 195 countries over 25 years. N Engl J Med 377(1):13–27

Al-Goblan AS, Al-Alfi MA, Khan MZ (2014) Mechanism linking diabetes mellitus and obesity. Diabetes Metab Syndr Obes Targets Ther 7:587–591

Anton K, Rahman T (2015) AB-JOW, 2015 undefined. Weight loss following left gastric artery embolization in a human population without malignancy: a retrospective review. pdfs.semanticscholar.org [Internet]. https://pdfs.semanticscholar.org/2e86/ddf2aa3621944f40f460bf254f04224d02ed.pdf. Accessed 18 Mar 2020

Arepally A, Barnett BP, Montgomery E, Patel T (2007) Catheter-directed gastric artery chemical embolization for modulation of systemic ghrelin levels in a porcine model: initial experience. Radiology 244(1):138–143

Arepally A, Barnett BP, Patel TT, Howland V, Boston RC, Kraitchman DL et al (2008) Catheter-directed gastric artery chemical embolization suppresses systemic ghrelin levels in porcine model. Radiology 249:127–133

Arterburn D, Wellman R, Emiliano A, Smith SR, Odegaard AO, Murali S et al (2018) Comparative effectiveness and safety of bariatric procedures for weight loss a pcornet cohort study. Ann Intern Med 169(11):741–750

Baltasar A, Bou R, Bengochea M, Arlandis F, Escrivá C, Miró J et al (2001) Duodenal switch: an effective therapy for morbid obesity-Intermediate results. Obes Surg 11:54–58

Balter S, Miller DL, Schueler BA (2012) Radiation dose measurements and monitoring for fluoroscopically guided interventional procedures. J Am Coll Radiol 9 (8):595–597

Bawudun D, Xing Y, Liu WY, Huang YJ, Ren WX, Ma M et al (2012) Ghrelin suppression and fat loss after left gastric artery embolization in canine model. Cardiovasc Intervent Radiol 35(6):1460–1466

Beckman LM, Beckman TR, Earthman CP (2010) Changes in gastrointestinal hormones and leptin after roux-en-Y gastric bypass procedure: a review. J Am Diet Assoc 110(4):571–584

Bin BZ, Qin YL, Deng G, Zhao GF, Zhong BY, Teng GJ (2018) Bariatric embolization of the left gastric arteries for the treatment of obesity: 9-month data in 5 patients. Obes Surg 28(4):907–915

Brown KT, Do RK, Gonen M, Covey AM, Getrajdman GI, Sofocleous CT et al (2016) Randomized trial of hepatic artery embolization for hepatocellular carcinoma using doxorubicin-eluting microspheres compared with embolization with microspheres alone. J Clin Oncol 34(17):2046–2053

Caballero B (2007) The global epidemic of obesity: an overview. Epidemiol Rev 29:1–5. http://www.ncbi.nlm.nih.gov/pubmed/17569676. Accessed 23 Jul 2019

Cummings DE, Shannon MH (2003) Ghrelin and gastric bypass: is there a hormonal contribution to surgical weight loss? J Clin Endocrinol Metab 88 (7):2999–3002

Cummings DE, Purnell JQ, Frayo RS, Schmidova K, Wisse BE, Weigle DS (2001) A preprandial rise in plasma ghrelin levels suggests a role in meal initiation in humans. Diabetes 50(8):1714–1719

Cummings DE, Weigle DS, Scott Frayo R, Breen PA, Ma MK, Patchen Dellinger E et al (2002) Plasma ghrelin levels after diet-induced weight loss or gastric bypass surgery. N Engl J Med 346(21):1623–1630

Cummings DE, Overduin J, Foster-Schubert KE (2005) Roles for ghrelin in the regulation of appetite and body weight. Curr Opin Endocrinol Diabetes 12(1):72–79

DeMarco VG, Aroor AR, Sowers JR (2014) The pathophysiology of hypertension in patients with obesity. Nat Rev Endocrinol 10:364–376

Druce MR, Small CJ, Bloom SR (2004) Minireview: gut peptides regulating satiety. Endocrinology [Internet] 145(6):2660–2665. https://academic.oup.com/endo/article-lookup/doi/10.1210/en.2004-0089

Duran R, Sharma K, Dreher MR, Ashrafi K, Mirpour S, De Lin M et al (2016) A novel inherently radiopaque bead for transarterial embolization to treat liver cancer—a pre-clinical study. Theranostics 6(1):28–39

Elder KA, Wolfe BM (2007) Bariatric surgery: a review of procedures and outcomes. Gastroenterology 132 (6):2253–2271

English PJ, Ghatei MA, Malik IA, Bloom SR, Wilding JPH (2002) Food fails to suppress ghrelin levels in obese humans. J Clin Endocrinol Metab 87(6):2984

Everhart JE (1993) Contributions of obesity and weight loss to Gallstone disease. Ann Intern Med 119 (10):1029–1035

Frühbeck G, Diez Caballero A, Gil MJ (2004) Fundus functionality and ghrelin concentrations after bariatric surgery. N Engl J Med 350:308–309

Fu Y, Weiss CR, Paudel K, Shin EJ, Kedziorek D, Arepally A et al (2018) Bariatric arterial embolization: effect of microsphere size on the suppression of fundal ghrelin expression and weight change in a swine model. Radiology 289(1):83–89

Gunn AJ, Oklu R (2014) A preliminary observation of weight loss following left gastric artery embolization in humans. J Obes [Internet] 2014:185349. http://www.hindawi.com/journals/jobe/2014/185349/

Hafezi-Nejad N, Bailey CR, Gunn AJ, Weiss CR (2019) Weight loss after left gastric artery embolization: a systematic review and meta-analysis. J Vasc Interv Radiol 30:1593–1603.e3

Hayashida T, Murakami K, Mogi K, Nishihara M, Nakazato M, Mondal MS et al (2001) Ghrelin in domestic animals: distribution in stomach and its possible role. Domest Anim Endocrinol 21(1):17–24

Hu Z, Seung HC, Van Haasteren G, Wang J, Lane MD (2005) Effect of centrally administered C75, a fatty acid synthase inhibitor, on ghrelin secretion and its downstream effects. Proc Natl Acad Sci U S A 102:3972

Inui A (2001) Ghrelin: an orexigenic and somatotrophic signal from the stomach. Nat Rev Neurosci 2:551–560

Inui A, Asakawa A, Bowers CY, Mantovani G, Laviano A, Meguid MM et al (2004) Ghrelin, appetite, and gastric motility: the emerging role of the stomach as an endocrine organ. FASEB J 18(3):439–456

Jin J (2013) Obesity and the heart. JAMA J Am Med Assoc 310(19):2113

Karra E, Yousseif A, Batterham RL (2010) Mechanisms facilitating weight loss and resolution of type 2 diabetes following bariatric surgery. Trends Endocrinol Metab 21(6):337–344

Kim DJ, Raman HS, Salter A, Ramaswamy R, Gunn AJ, Weiss CR et al (2018) Analysis of weight changes after left gastric artery embolization in a cancer-naive population. Diagn Interv Radiol 24(2):94–97

Kipshidze N, Archvadze A, Bertog S, Leon MB, Sievert H (2015) Endovascular bariatrics: first in humans study of gastric artery embolization for weight loss. JACC: Cardiovasc Interv 8(12):1641–1644

Kojima M, Hosoda H, Date Y, Nakazato M, Matsuo H, Kangawa K (1999) Ghrelin is a growth-hormone-releasing acylated peptide from stomach. Nature 402 (6762):656–660

Kosti RI, Panagiotakos DB (2006) The epidemic of obesity in children and adolescents in the world. Cent Eur J Public Health 14(4):151–159

Kshirsagar DS, Saudagar RB (2016) Microsphere: a review. Res J Top Cosmet Sci 7(1):27–37

Liaw JVP, Yun CH, Walker TG, Kalva SP, D'Othée BJ (2012) Comparison of clinical and MR imaging outcomes after uterine fibroid embolization with Bead Block and Embosphere. Eur J Radiol 81 (6):1371–1375. http://www.ncbi.nlm.nih.gov/pubmed/21439744

Lim R, Beekley A, Johnson DC, Davis KA (2018) Early and late complications of bariatric operation. Trauma

Surg Acute Care Open 3:e000219. http://www.ncbi.nlm.nih.gov/pubmed/30402562
Loffroy R, Rao P, Ota S, De Lin M, Kwak BK, Geschwind JF (2010) Embolization of acute nonvariceal upper gastrointestinal hemorrhage resistant to endoscopic treatment: results and predictors of recurrent bleeding. Cardiovasc Intervent Radiol 33(6):1088–1100
Loffroy R, Favelier S, Pottecher P, Estivalet L, Genson PY, Gehin S et al (2015) Transcatheter arterial embolization for acute nonvariceal upper gastrointestinal bleeding: indications, techniques and outcomes. Diagn Interv Imaging 96(7–8):731–744
Loftus TM, Jaworsky DE, Frehywot CL, Townsend CA, Ronnett GV, Daniel Lane M et al (2000) Reduced food intake and body weight in mice treated with fatty acid synthase inhibitors. Science 288:2379–2381
Lu Y, Loffroy R, JYW L, Barkun A (2014) Multidisciplinary management strategies for acute non-variceal upper gastrointestinal bleeding. Br J Surg 101(1):e34–e50
Lubarsky M, Ray CE, Funaki B (2009) Embolization agents-which one should be used when? Part 1: large-vessel embolization. Semin Interv Radiol 26 (4):352–357
Lubarsky M, Ray C, Funaki B (2010) Embolization agents-which one should be used when? Part 2: small-vessel embolization. Semin Interv Radiol 27 (1):99–104. http://www.ncbi.nlm.nih.gov/pubmed/21359018
Madoff DC (2013) Science to practice: can transarterial embolotherapy be used as a viable alternative to treat obesity? Radiology 266(2):369–371
Mahesh M (2012) Radiation dose management for fluoroscopically guided interventional medical procedures. Med Phys 39:5789–5790
Maksud FAN, Alves JS, Diniz MTC, Barbosa AJA (2011) Density of ghrelin-producing cells is higher in the gastric mucosa of morbidly obese patients. Eur J Endocrinol 165:57–62
Mikhail AS, Pritchard WF, Negussie AH, Krishnasamy VP, Amchin DB, Thompson JG et al (2018) Mapping drug dose distribution on CT images following transarterial chemoembolization with radiopaque drug-eluting beads in a rabbit tumor model. Radiology 289(2):396–404. http://www.ncbi.nlm.nih.gov/pubmed/30106347
Miller DL, Balter S, Wagner LK, Cardella J, Clark TWI, Neithamer CD et al (2004) Quality improvement guidelines for recording patient radiation dose in the medical record. J Vasc Interv Radiol 23:P11–P18
Murray CDR, Kamm MA, Bloom SR, Emmanuel AV (2003) Ghrelin for the gastroenterologist: history and potential. Gastroenterology 125(5):1492–1502
Nakazato M, Murakami N, Date Y, Kojima M, Matsuo H, Kangawa K et al (2001) A role for ghrelin in the central regulation of feeding. Nature 409:194–198
Neff KJ, Olbers T, le Roux CW (2013) Bariatric surgery: the challenges with candidate selection, individualizing treatment and clinical outcomes. BMC Med 11(1):8
No Title (2019) [Internet]. https://www.merit.com/wp-content/uploads/2018/11/404690001EN_001-Embosphere-in-a-Syringe_Canada-IFU.pdf. Accessed 10 Dec 2019
O'Rourke RW, Andrus J, Diggs BS, Scholz M, McConnell DB, Deveney CW (2006) Perioperative morbidity associated with bariatric surgery: an academic center experience. Arch Surg 141(3):262–268
OBESITY Update (2014) [cited 2019 Dec 8]; https://doi.org/10.1787/health
Okumura H, Nagaya N, Enomoto M, Nakagawa E, Oya H, Kangawa K (2002) Vasodilatory effect of ghrelin, an endogenous peptide from the stomach. J Cardiovasc Pharmacol 39(6):779–783
Paxton BE, Kim CY, Alley CL, Crow JH, Balmadrid B, Keith CG et al (2013) Bariatric embolization for suppression of the hunger hormone ghrelin in a porcine model. Radiology 266(2):471–479
Pirlet C, Ruzsa Z, Costerousse O, Nemes B, Merkely B, Poirier P et al (2019) Transradial left gastric artery embolization to treat severe obesity: a pilot study. Catheter Cardiovasc Interv 93(3):365–370
Poulain M, Doucet M, Major GC, Drapeau V, Sériès F, Boulet LP et al (2006) The effect of obesity on chronic respiratory diseases: pathophysiology and therapeutic strategies. CMAJ 174(9):1293–1299
Prologo JD, Lin E, Horesh Bergquist S, Knight J, Matta H, Brummer M et al (2019) Percutaneous CT-guided cryovagotomy in patients with class I or class II obesity: a pilot trial. Obesity 27(8):1255–1265
Radiology AC of (2015) ACR–AAPM technical standard for management of the use of radiation in fluoroscopic procedures. 2013
Sharaiha R (2017) Managing obesity with endoscopic sleeve gastroplasty. Gastroenterol Hepatol 13:547–549
Shin JH (2012) Recent update of embolization of upper gastrointestinal tract bleeding. Korean J Radiol 13 (Suppl 1):S31
Simon GE, Von Korff M, Saunders K, Miglioretti DL, Crane PK, Van Belle G et al (2006) Association between obesity and psychiatric disorders in the US adult population. Arch Gen Psychiatry 63(7):824–830
Smith KB, Smith MS (2016) Obesity statistics. Prim Care 43(1):121–135
Stahl JM, Malhotra S (2019) Obesity surgery indications and contraindications [Internet]. StatPearls Publishing. http://www.ncbi.nlm.nih.gov/pubmed/30020657. Accessed 10 Mar 2020
Stecker MS, Balter S, Towbin RB, Miller DL, Vañó E, Bartal G et al (2009) Guidelines for patient radiation dose management. J Vasc Interv Radiol 20(7 Suppl): S263–S273
Stone TW, McPherson M, Gail Darlington L (2018) Obesity and cancer: existing and new hypotheses for a causal connection. EBioMedicine 30:14–28

Strader AD, Woods SC (2005) Gastrointestinal hormones and food intake. Gastroenterology 128(1):175–191

Sumithran P, Prendergast LA, Delbridge E, Purcell K, Shulkes A, Kriketos A et al (2011) Long-term persistence of hormonal adaptations to weight loss. N Engl J Med 365(17):1597–1604

Syed MI, Morar K, Shaikh A, Craig P, Khan O, Patel S et al (2016) Gastric artery embolization trial for the lessening of appetite nonsurgically (GET LEAN): six-month preliminary data. J Vasc Interv Radiol 27 (10):1502–1508

Takiguchi S, Miyata H, Mori M, Adachi S, Nakajima K, Doki Y et al (2012) Mapping analysis of ghrelin producing cells in the human stomach associated with chronic gastritis and early cancers. Dig Dis Sci 57 (5):1238–1246

Thompson JG, van der Sterren W, Bakhutashvili I, van der Bom IM, Radaelli AG, Karanian JW et al (2018) Distribution and detection of radiopaque beads after hepatic transarterial embolization in swine: cone-beam CT versus MicroCT. J Vasc Interv Radiol 29 (4):568–574

Tritos NA, Mun E, Bertkau A, Grayson R, Maratos-Flier E, Goldfine A (2003) Serum ghrelin levels in response to glucose load in obese subjects post-gastric bypass surgery. Obes Res 11(8):919–924

Vaidya S, Tozer KR, Chen J (2008) An overview of embolic agents. Semin Interv Radiol 25:204–215

Weiss CR, Akinwande O, Paudel K, Cheskin LJ, Holly B, Hong K et al (2017) Clinical safety of bariatric arterial embolization: preliminary results of the BEAT obesity trial. Radiology 283(2):598–608

Weiss CR, Abiola GO, Fischman AM, Cheskin LJ, Vairavamurthy J, Holly BP et al (2019) Bariatric embolization of arteries for the treatment of obesity (BEAT obesity) trial: results at 1 year. Radiology 291 (3):792–800

WHO (2000) Obesity: preventing and managing the global epidemic. Report of a WHO consultation. World Heal Organ-Tech Rep Ser 894:252

Wren AM, Small CJ, Abbott CR, Dhillo WS, Seal LJ, Cohen MA et al (2001) Ghrelin causes hyperphagia and obesity in rats. Diabetes 50(11):2540–2547

Progress in Noninvasive Beta-Cell Mass Imaging

46

Bluma Linkowski Faintuch and Salomao Faintuch

Contents

Introduction 632

Main Chronic Disorders Related to Beta-Cells 632
Glucotoxicity 632
Lipotoxicity 632
β-Cell Mass in Human Obesity 633
Beta Cell Mass in T2D 633
Rodent Versus Human β Cells 633
Cell Types 633

Beta-Cell Imaging 634
Histological Estimations 634
Noninvasive Imaging of Pancreatic β Cells 635
Single Diagnostic Tools 635
Combined Diagnostic Tools 636
Clinical Trials 637

References 638

Abstract

Beta-cells play a decisive role in regulating glucose homeostasis, via insulin synthesis and release. Beta cell maintenance (mass, architecture, survival) and function (insulin release) are central to type 2 diabetes (T2D), in which the pancreas is technically normal to begin with. Yet progressive loss of beta cells tends to occur with disease progress, due to glucolipotoxicity and additional mechanisms. Many candidate probes for beta cell imaging have been designed based on the screening of genomic, SNP, or proteomic β cell databanks, but no consensus on the ideal probes has yet been reached. Monitoring the beta-cell mass requires a reliable method for noninvasive utilization. Despite the non-negligible challenges, many advances have been reached in preclinical studies, and human trials are going on.

Keywords

Glucotoxicity · Lipotoxicity · Beta cell mass · Noninvasive imaging · Islet cell histology ·

B. L. Faintuch (✉)
Nuclear and Energy Research Institute, Butantã, Sao Paulo-SP, Brazil
e-mail: blfaintuch@hotmail.com

S. Faintuch
Beth Israel Deaconess Medical Center, Division of Interventional Radiology, Boston, MA, USA

J. Faintuch and S. Faintuch (eds.), *Obesity and Diabetes*, https://doi.org/10.1007/978-3-030-53370-0_46

Optical imaging · Magnetic resonance imaging · Radiopharmaceutical imaging · SPECT/CT imaging · PET/CT imaging

Introduction

In physiological conditions, concentrations of blood glucose are stable within a relatively narrow range, despite large fluctuations in intake (intermittent meals, night fasting) as well as consumption (mild office or household activities, hard-working jobs or sports, sedentary periods, sleeping).

Insulin secretion, dependent on pancreatic beta cell population and function, is often more critical than insulin resistance in T2D, despite the strong association between insulin resistance and obesity, the most frequent precipitant of diabetes onset (Fig. 46.1) (Ashcroft and Rorsman 2012).

Beta cell insults include cytokine-induced inflammation, obesity and insulin resistance, and saturated fat and free fatty acids (FFA), along with hyperglycemia itself. A progressive decline of beta cell function leads to beta cell exhaustion. Even incretin-mimicking drugs like glucagon-like peptide 1 receptor agonists, well-regarded worldwide for their excellent clinical response in T2D, can induce exhaustion in some contexts (Van Raalte and Verchere 2016).

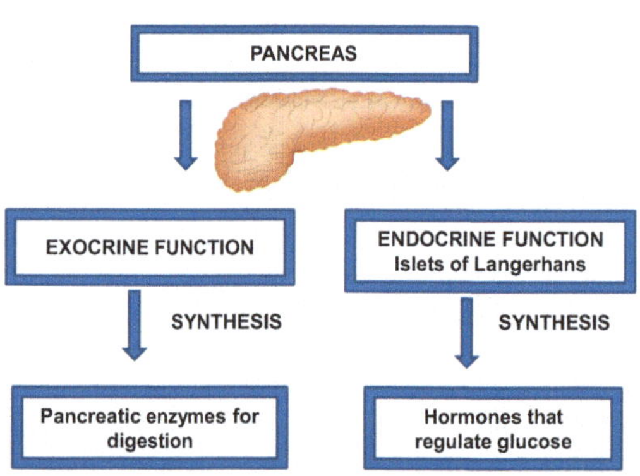

Fig. 46.1 Pancreas and its functions

Main Chronic Disorders Related to Beta-Cells

The fundamental nononcological diseases precipitated by beta-cell aberrations are type 1 and type 2 diabetes. Obesity is intimately related although it should not be listed among the consequences as it is obviously one of the causes of dysglycemia, along with genetics, epigenetics, aging, glucolipotoxicity, and other mechanisms (Li et al. 2020a). Incidentally the interplay with obesity is complex, with different clusters of obese patients exhibiting patterns of insulin responsivity, secretion, and resistance that range from the perfectly healthy to the highly deranged (Li et al. 2020b).

In the course of obesity, both insulin sensitivity and β-cell function tend to decrease (Li et al. 2020a, 2020b). In a healthy organism, there is a continuous feedback relationship between the β-cells and the insulin-sensitive tissues. If adipose tissue, liver, and muscles exhibit a larger glucose turnover, this will lead to a parallel and proportional increase in insulin supply.

Glucotoxicity

Should insulin resistance ensue, β-cells will be overworked in order to maintain the same normal levels of glucose. At some moment pancreatic β-cell demise could occur, partly as a consequence of glucotoxic effects of persistently elevated blood glucose. Further decompensation of glucose homeostasis, potentially leading to prediabetes of full-blown type 2 diabetes, should be anticipated. Reduced proliferation and enhanced beta cell apoptosis have been recognized for some time in T2D (Li et al. 2017).

Lipotoxicity

Plasma nonesterified fatty acids (NEFA) levels play a major role in insulin release, yet the continuous exposure to them interferes with glucose-stimulated insulin secretion pathways, and can result in reduced insulin biosynthesis. Insulin resistance and failure of the compensatory mechanisms of β-cells further contribute to increased amounts of NEFAs. Indeed adipose tissue release of NEFAs into plasma is inhibited by the antilipolytic action of insulin.

During metabolic syndrome and obesity, it is possible to measure both glucose-specific and NEFA-specific insulin resistance, which are clearly different from normal. The combined effect of lipotoxic increases in plasma NEFA levels and the rise of glucose levels might produce the typical combined outcome glucolipotoxicity (Walker et al. 2020).

β-Cell Mass in Human Obesity

In nondiabetic subjects, obesity is associated with a modest expansion of β-cell mass, in parallel with the α-cell mass, so that the proportional elevation of both glucagon and insulin is maintained in this bihormonal, multifaceted disease (Ellenbroek et al. 2017).

Adiposity and particularly ectopic fat in the pancreas, liver, muscle, and other glucose recipient organs will exacerbate insulin resistance, which subsequently impairs beta cell function. Obesity also increases insulin demand; therefore, hyperfunction of beta cells may ultimately exhaust them, as a result of beta cell dysfunction (Cerf 2013).

In those individuals who develop diabetes, beta cell physiological compensation declines due to hyperglycemia resulting in a persistent progressively increasing metabolic load that could be further influenced by aging and an eventually adverse genetic profile.

Current mathematical models try to integrate such diverse factors as fasting plasma glucose, insulin, leptin, glucagon, nonesterified fatty acids (NEFA), and very-low-density lipoprotein triglyceride (VLDLTG), as well as muscle, hepatic, and pancreatic lipids, fat mass, and mass of β-cells. Combined with dietary pattern, the model is likely to predict the trajectory to both fat and lean T2D (Hassell Sweatman 2020).

Beta Cell Mass in T2D

As alluded to, beta cell mass shrinks in T2D; however, differences between lean and overweight/obese diabetics are not remarkable (41% and 38% decrease, respectively). Duration of disease is a robust and widely recognized driver in such circumstances, with 24% reduction in subjects with less than 5 years after diagnosis, versus 54% in those with over 15 years of overt diabetes (Rahier et al. 2008; Butler et al. 2016).

Rodent Versus Human β Cells

The vast majority of experimental investigations concerning beta cell imaging are conducted in rodents. This has been a hindrance for clinical translation, as major anatomical and functional differences separate these nonhuman species from patients (Kang et al. 2020). It is worth emphasizing this point, as nominally successful animal techniques continuously appear, without sufficiently highlighting such critical detail.

Cell Types

Pancreatic islet cells are not dissimilar between man and rodents: β, α, δ, ε, and pancreatic polypeptide cells (Fig. 46.2).

Yet microvascularization, cytoarchitechture, and composition of cell clusters, which are merely histological arrangements yet with significant functional impact, pursue conflicting models.

In rodents, 80% of the islets consist of insulin-expressing β cells, forming a core surrounded by other cell types. In turn, only 50–60% of the islet cells are of the β variety in humans, and their distribution is scattered. About 30% are glucagon-producing α-cells, along with small proportions of δ-cells (somatostatin), γ—or PP cells (pancreatic polypeptide), and ε-cells (ghrelin) (da Silva Xavier 2018). Additional differences in the cells concern calcium ion

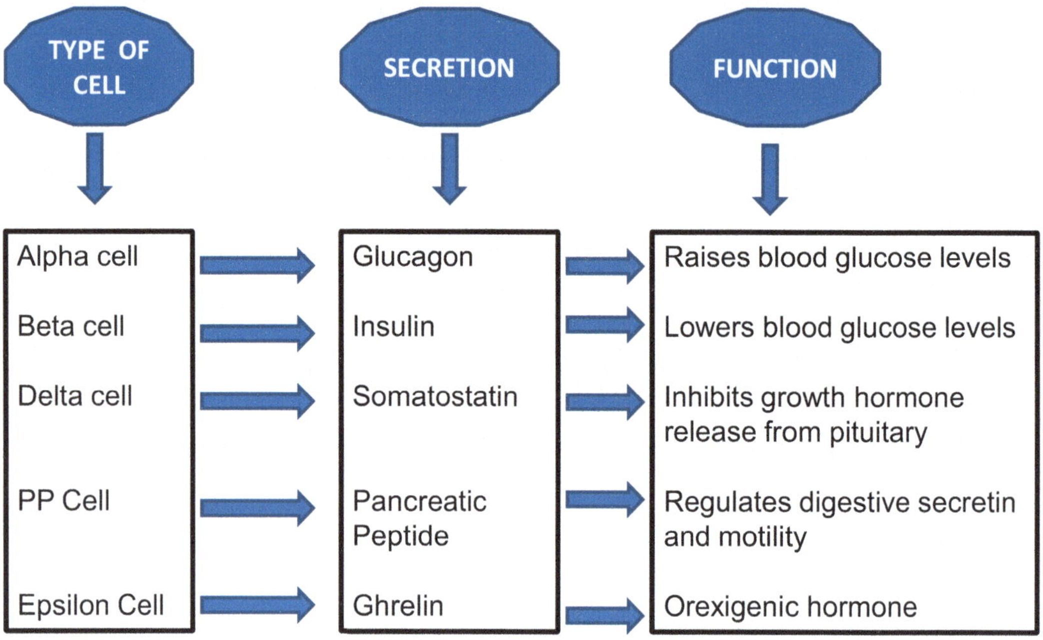

Fig. 46.2 Composition of pancreatic islets

oscillation, and the main transporter for glucose uptake, which is GLUT2 in rodents and GLUT1 in humans. In addition to functional differences (Johnston et al. 2016), such architecture interferes with imaging profiles (Murakami et al. 2019) (Fig. 46.3).

As a result, species-specific expression level and islet distribution of beta cells are inevitable during imaging procedures with positron emission tomography (PET) and single-photon emission computed tomography (SPECT), complicating the interpretation of such diagnostic resources

Beta-Cell Imaging

In front view, the adult pancreas measures about 3.0 cm at the head, 2.5 cm at neck and body, and 2.0 cm at the tail (these values are not applicable to old persons, given the age-related atrophy of the gland). The pancreatic islets constitute 1–2% of this mass and are widely scattered throughout the pancreatic parenchyma. Given an estimated total number of 3.2–14.8 million islets, and an individual diameter of 30 μm to >400 μm, total islet volume could reach 0.5–2.0 cm^3. This is not a sufficiently large size for single islet identification, and a challenge even for global tracing given their widespread distribution in a much larger host organ, yet with specific cell probes, collective beta cell imaging can be envisaged (Huang et al. 2018).

Histological Estimations

Histometric techniques for beta cell mass calculation are well established. Most commonly, islet beta-cell area is defined on pancreatic tissue sections and extrapolated for total pancreatic volume. Yet, no routine clinical application will emerge as multiple biopsies would be required, entailing the risk of pancreatitis, bleeding, pseudocysts, and other serious complications. Such approach is useful as a gold standard, for experimental validation of imaging procedures (Linnemann et al. 2014).

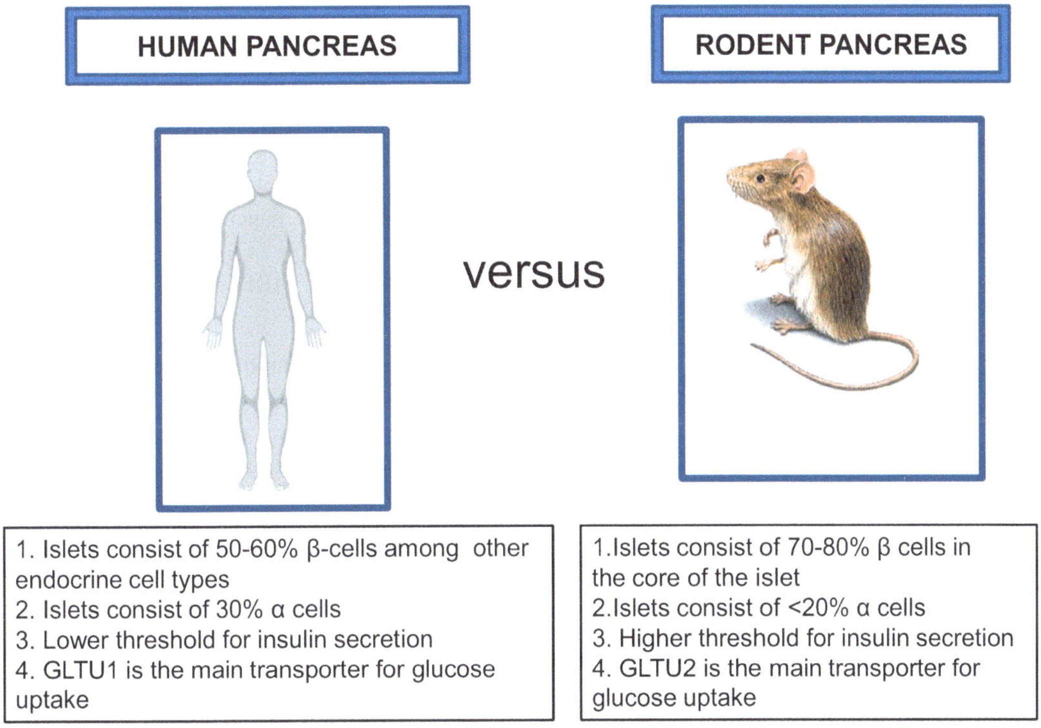

Fig. 46.3 Differences between human and rodent islets

Noninvasive Imaging of Pancreatic β Cells

Diagnosis of beta-cell loss or dysfunction in different phases of diabetes would be conspicuously helped by imaging of the beta cell mass (BCM). Interventions aiming at preservation or regeneration of beta-cells, based on incretin analogs or cell therapy techniques, could be directly monitored. These include cell-based immunomodulatory approaches for type 1 diabetes (Black and Zorina 2020)

As alluded to, when diabetes is diagnosed, a significant proportion of insulin-secreting pancreatic β cells are already dysfunctional or destroyed (Rahier et al. 2008; Butler et al. 2016). Early evaluation of BCM could entail better preservation procedures, in order to prevent such often irreversible loss.

Single Diagnostic Tools

Optical Imaging

Bioluminescence tomography (BLT), fluorescence molecular tomography (FMT), and optical projection tomography (OPT) are not employed yet in the clinical setting, however, have been successful in experimental protocols.

Optical methods rely on the detection and image acquisition of natural or notably induced light emission from a known biological source, with a particular wavelength. Fluorescent proteins have been used for a long time to track cells and proteins, both in vitro and in vivo.

Also, the enzyme luciferase has been transfected to isolated human beta cells, acting as a source of bioluminescence (Kracht et al. 2018). However, no translational applications are foreseen for this complicated intervention in the near future.

Fluorescent probes, after excitation at a specific wavelength, emit an optical signal. Fluorophores range from 510 nm to 900 nm, with better penetration at longer wavelengths. Despite the fact that fluorescence alone is not feasible yet for pancreatic β cell tracking in vivo, encouraging results were announced with a hybrid technique, combining a fluorescence probe with a radioactive marker (18 F). In animals, this enabled the successful identification of pancreatic beta cells with the help of a PET scan (Kang et al. 2019).

Magnetic Resonance Imaging (MRI)

This classic tool uses strong magnetic fields to excite hydrogen atoms. These emit radio frequency signals that can be detected by a receiver and transformed into an image.

MRI contrast agents suffer from limited sensitivity of the systems, and the MRI signal is difficult to quantitate in absolute numbers, in order to calculate the β cell mass. As classic computed tomography and other traditional methods, it lacks specificity and resolution power for the challenge represented by beta cells (Kang et al. 2019).

Radiopharmaceutical Imaging

The physical basis for nuclear medicine imaging is the detection of radiation emitted from a radioactive isotope (biological marker). Many tracers have been tested within the context of pancreatic β cells, particularly monoclonal antibodies and peptides, targeting β cell specific antigens, receptors, or its major secretion molecule, namely, insulin. Table 46.1 lists imaging agents studied in experimental conditions for PET or SPECT imaging of pancreatic beta cells, and radiolabeled with a variety of radioactive isotopes, along with MRI protocols.

Single Photon Emission Computed Tomography (SPECT)

Based on γ-ray-emitting radiotracers, this is one of the most valued nuclear medicine technologies. It offers a high diagnostic sensitivity in the picomolar range, something desirable when dealing with pancreatic islets and related molecules. Combination with computed tomography (SPECT/CT), it could be endowed with the right assessment qualities for BCM, as discussed now.

Combined Diagnostic Tools

Fusion Imaging linking functional and anatomic imaging data via PET/MRI, PET/CT, SPECT/CT, or PET/fluorescence probe, could bypass the weaknesses of single methods by means of methodological synergies, thus paving the way for clinically relevant information (Singh et al. 2017).

Table 46.1 Summary of the main targets and imaging agents

Target	Imaging agent	Type of imaging
VMAT2 Vesicular monoamine transporter 2 neutrotransmission	^{11}C-DTBZ ^{18}F-FP-DTBZ	PET
β-Cell surface antigen	^{111}In-IC2 (mAb)	SPECT
Gangliosides on the plasma membranes of pancreatic β cells	^{125}I-R2D6 (mAb)	SPECT
Dopamine receptors (D2 and D3)	^{11}C-PHNO ^{11}C-5-HDP	PET
GLP1R Glucagon-like-peptide 1 receptor	^{18}F-(Nle14,Lys40)-exendin-4 ^{111}In-DTPA-Lys40-exendin-3 ^{177}Lu-Do3A-VS-Cys40-exendin-4	PET SPECT SPECT
TMEM 27 Transmembrane protein 27	^{89}Zr-8/9 mAb	PET
ZnT8 protein Zinc transporter 8	^{125}I-Ab31 ^{125}I-exendin 4	SPECT
GPR44—Pancreatic protein GPR44 antagonist	11C-AZ12204657	PET
FXY-D2ya	Gd-DOTA-P88	MRI
Zn +@	Gd-DOTA-diBPEN	MRI
Ca+2 Channel	MnCl2	MRI

PET/MRI

PET-MRI is one of these promising approaches. The contrast for MRI was manganese (Mn), an ion gated by voltage-dependent calcium channels, involved in insulin secretion by pancreatic beta cells, along with radiolabeled exendin-4 for PET, a glucagon like peptide 1 receptor agonist, which stimulates beta cell metabolism. In vivo quantification of β-cell function and mass was thus achieved (Michelotti et al. 2020).

SPECT/CT

Creative diagnostic interventions were recently developed in mice, with potential for future use in humans. In one model, pancreatic alpha and beta cells were traced by a dipeptidyl peptidase 6 (DPP6) antibody (nanobody "4hD29"). Subsequent SPECT/CT investigation allowed identification of implanted human pancreatic islet cells in mice. The possibility of employing PET/CT for the same purpose is also envisaged (Demine et al. 2020).

Glucagon-like peptide 1 analog exendin-4, radiolabeled with Indium 111, has been proposed for specific quantification of BCM, by means of a SPECT/CT protocol. Preliminary results in diabetic mice were very positive (Murakami et al. 2019) (Fig. 46.4).

Clinical Trials

A trial was conducted a couple of years ago, using an 18F-FP DTBZ (fluoropropyl-dihydrotetrabenazine) marker, with the aim of evaluating an islet imaging technique and measuring BCM, in both prediabetes and diabetes, compared to healthy obese volunteers (Cline et al. 2018). This marker binds to vesicular monoamine transporter type 2 (VMAT2), which is

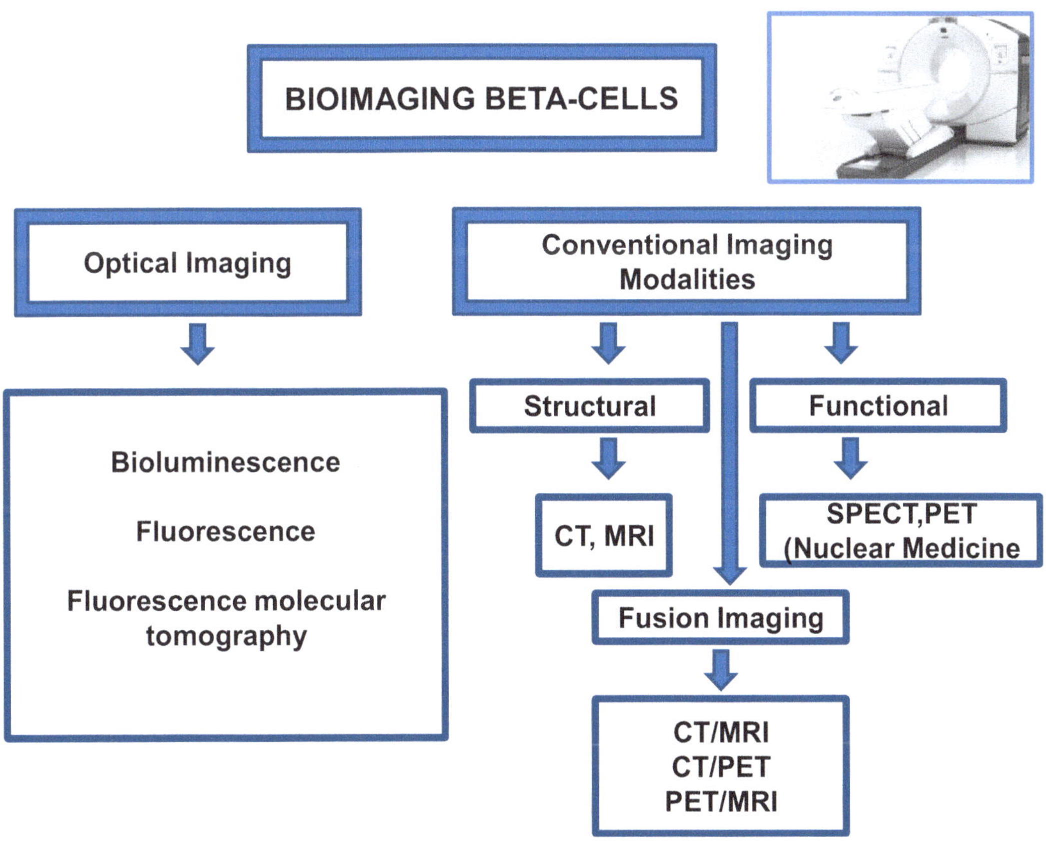

Fig. 46.4 Tools to bioimage beta-cells

expressed by insulin-secreting beta cells. Results were consistent with the hypothesis, and suggested that measured loss of BCM contributed to deficient insulin secretion in humans with type 2 diabetes. Binding potential had previously been tested in type 1 diabetes and repeatable values were observed, anticipating its interest in this population as well (Freeby et al. 2016).

This line of investigation didn't receive unblemished acclamation. In a relatively widely cited criticism by investigators from the University of Pennsylvania, it was recommended that PET scanning be reserved for more substantial cell masses, or for highly concentrated clusters, such as occurring after beta cell transplantation, or hyperinsulinism/insulinoma. The limit for effective PET detection is stated as 8–10 cm^3, something incompatible with islet cell size and histological distribution (Alavi and Werner 2018). Nevertheless the authors stood their ground, in face of their results, and of continuous advances in PET imaging technology (Gotthardt et al. 2018).

References

Alavi A, Werner TJ (2018) Futility of attempts to detect and quantify beta cells by pet imaging in the pancreas: why it is time to abandon the approach. Diabetologia 61(12):2512–2515

Ashcroft FM, Rorsman P (2012) Diabetes mellitus and the β-cell: the last ten years. Cell. 148(6):1160–1171. https://doi.org/10.1016/j.cell.2012.02.010

Black L, Zorina T (2020) Cell-based immunomodulatory therapy approaches for type 1 diabetes mellitus. Drug Discov Today 25(2):380–391

Butler AE, Dhawan S, Hoang J, Cory M, Zeng K, Fritsch H, Meier JJ, Rizza RA, Butler PC (2016) β-cell deficit in obese type 2 diabetes, a minor role of β-cell dedifferentiation and degranulation. J Clin Endocrinol Metab 101(2):523–532

Cerf ME (2013, March 27) Beta cell dysfunction and insulin resistance. Front Endocrinol 4:37. https://doi.org/10.3389/fendo.2013.00037

Cline GW, Naganawa M, Chen L, Chidsey K, Gonzalez SC, Pawlak S, Rossulek M, Zhang Y, Bini J, McCarthy TJ, Carson RE, Calle RA (2018) Decreased VMAT2 in the pancreas of Humans with type 2 diabetes mellitus measured in vivo by PET imaging. Diabetologia 61(12):2598–2607

da Silva Xavier G (2018) The cells of the islets of langerhans. J Clin Med 7:54. https://doi.org/10.3390/jcm7030054. (Review)

Demine S, Garcia Ribeiro R, Thevenet J, Marselli L, Marchetti P, Pattou F, Kerr-Conte J, Devoogdt N, Eizirik DL (2020) A nanobody-based nuclear imaging tracer targeting dipeptidyl peptidase to determine the mass of human beta cell grafts in mice. Diabetologia 63(4):825–836

Ellenbroek JH, Töns HAM, Hanegraaf MAJ, Rabelink TJ, Engelse MA, Carlotti F, de Koning EJP (2017 Dec) Pancreatic α-cell mass in obesity. Diabetes Obes Metab 19(12):1810–1813

Freeby MJ, Kringas P, Goland RS, Leibel RL, Maffei A, Divgi C, Ichise M, Harris PE (2016) Cross-sectional and test-retest characterization of PET with [(18)F]FP-(+)-DTBZ for β cell mass estimates in diabetes. Mol Imaging Biol 18(2):292–301

Gotthardt M, Eizirik DL, Aanstoot HJ, Korsgren O, Mul D, Martin F, Boss B, Jansen TJP, van Lith SAM, Buitinga M, Eriksson O, Cnop M, Brom M (2018) Detection and quantification of beta cells by PET imaging: why clinical implementation has never been closer. Diabetologia 61(12):2516–2519

Hassell Sweatman CZW (2020) Mathematical model of diabetes and lipid metabolism linked to diet, leptin sensitivity, insulin sensitivity and VLDLTG clearance predicts paths to health and type II diabetes. J Theor Biol 486:110037

Huang HH, Harrington S, Stehno-Bittel L (2018) The flaws and future of islet volume measurements. Cell Transplant 27(7):1017–1026

Johnston NR, Mitchell RK, Haythorne E, Pessoa MP, Semplici F, Ferrer J, Piemonti L, Marchetti P, Bugliani M, Bosco D, Berishvili E, Duncanson P, Watkinson M, Broichhagen J, Trauner D, Rutter GA, Hodson DJ (2016) Beta cell hubs dictate pancreatic islet responses to glucose. Cell Metab 24(3):389–401

Kang NY, Soetedjo AAP, Amirruddin NS, Young-Tae Chang YT, Eriksson O, Teo AKK (2019) Tools for Bioimaging pancreatic β cells in diabetes. Trends Mol Med 25(8):708–722. https://doi.org/10.1016/j.molmed.2019.05.004

Kang NY, Lee JY, Lee SH, Song IH, Hwang YH, Kim MJ, Phue WH, Agrawalla BK, Wan SYD, Lalic J, Park SJ, Kim JJ, Kwon HY, Im SH, Bae MA, Ahn JH, Lim CS, Teo AKK, Park S, Kim SE, Lee BC, Lee DY, Chang YT (2020) Multimodal imaging probe development for pancreatic β cells: from fluorescence to PET. J Am Chem Soc 142(7):3430–3439

Kracht MJL, de Koning EJP, Hoeben RC, Roep BO, Zaldumbide A (2018) Bioluminescent reporter assay for monitoring ER stress in human beta cells. Sci Rep 8(1):17738

Li N, Karaca M, Maechler P (2017) Upregulation of UCP2 in beta-cells confers partial protection against both oxidative stress and glucotoxicity. Redox Biol 13:712–727

Li X, Chan LWC, Li X, Liu C, Yang G, Gao J, Dai M, Wang Y, Xie Z, Liu J, Zhou F, Zheng T, Feng D, Guo S, Li H, Sun K, Yang S (2020a) Obesity-induced regulator of calcineurin 1 overexpression leads to β-cell failure through mitophagy pathway inhibition. Antioxid Redox Signal 32(7):413–428

Li L, Song Q, Yang X (2020b) Categorization of β-cell capacity in patients with obesity via OGTT using K-means clustering. Endocr Connect 9(2):135–143

Linnemann AK, Baan M, Davis DB (2014) Pancreatic b-cell proliferation in obesity. Adv Nutr 5:278–288. https://doi.org/10.3945/an.113.005488

Michelotti FC, Bowden G, Küppers A, Joosten L, Maczewsky J, Nischwitz V, Drews G, Maurer A, Gotthardt M, Schmid AM, Pichler BJ (2020) PET/MRI enables simultaneous in vivo quantification of β-cell mass and function. Theranostics. 10(1):398–410

Murakami T, Fujimoto H, Fujita N, Hamamatsu K, Matsumoto K, Inagaki N (2019 Dec 1) Noninvasive evaluation of GPR119 agonist effects on β-cell mass in diabetic male mice using 111In-Exendin-4 SPECT/CT. Endocrinology. 160(12):2959–2968

Rahier J, Guiot Y, Goebbels RM, Sempoux C, Henquin JC (2008) Pancreatic β-cell mass in European subjects with type 2 diabetes. Diabetes Obes Metab 10(Suppl 4):32–42

Singh PP, Gahlot K, Mourya R, Gahlot V, Sagar, MP, Sagar IN (2017) Understanding "the fusion imaging"—Methods of fusion with their advantages and disadvantages and study various fusion combination with their applications. www.myESR.org European Society of Radiology 2017 ECR Congress https://doi.org/10.1594/ecr2017/C1848

Van Raalte DH, Verchere CB (2016) Glucagon-like peptide-1 receptor agonists: beta-cell protection or exhaustion? Trends Endocrinol Metab 27(7):442–445

Walker RE, Ford JL, Boston RC, Savinova OV, Harris WS, Green MH, Shearer GC (2020, Jan 7) Trafficking of non-esterified fatty acids in insulin resistance and relationship to dysglycemia. Am J Physiol Endocrinol Metab 318(3):E392–E404. www.physiology.org, https://doi.org/10.1152/ajpendo.00331.2019

Cancer Staging with 18F-FDG PET/CT in Hyperglycemic Patients

47

Monica Finessi, Virginia Liberini, and Désirée Deandreis

Contents

Clinical Importance of 18F FDG PET/CT in Cancer Staging 642
Indications in Lymphomas 642
Carcinoma and Adenocarcinoma 642
Tumor Uptake and Hyperglycemia 643

Mechanism of [18F]FDG Uptake 643

Intracellular Pathways of Glucose and Labelled Glucose 643

Cancer Cell Pathways 644

Effects of Hyperglycemia on [18F]FDG Distribution and PET/CT Scan Interpretation 646

Diagnostic Strategies in Normal and Disglycemic Patients 647

Conclusions 652

References 652

Abstract

In oncology positron emission tomography/computed tomography (PET/CT) is a standard imaging procedure for cancer staging, restaging, treatment monitoring, and radiation therapy planning. Despite the availability of many radiotracers, 18F-fluoro-2-deoxy-2-D-glucose ([18F]FDG) is still the most used for most cancers staging.

Hyperglycemic conditions, antidiabetic oral medications, or insulin can have an impact on [18F]FDG PET/CT scan accuracy. A correct knowledge of how these conditions influence [18F]FDG distribution is fundamental for patients management before [18F]FDG PET/CT. International PET/CT guidelines propose protocols for patients' adequate preparation and therapy management before [18F]FDG administration, in particular with regard to diabetic patients.

Adapted reprint with permission from: Finessi M, Bisi G, Deandreis D. Hyperglycemia and 18F-FDG PET/CT, issues and problem solving: a literature review. Acta Diabetol 2020; 57(3):253–262 https://doi.org/10.1007/s00592-019-01385-8

M. Finessi (✉) · V. Liberini · D. Deandreis
Division of Nuclear Medicine, AOU Città della Salute e della Scienza, Department of Medical Sciences, University of Turin, Turin, Italy
e-mail: monica.finessi@unito.it

J. Faintuch and S. Faintuch (eds.), *Obesity and Diabetes*, https://doi.org/10.1007/978-3-030-53370-0_47

Keywords

[18F]FDG PET/CT · Cancer staging · Hyperglycemia · Oral medication · Insulin

Clinical Importance of 18F FDG PET/CT in Cancer Staging

In oncology positron emission tomography/computed tomography (PET/CT) is a standard procedure, commonly used in several cancers for staging, restaging, treatment monitoring, and radiation therapy planning. Despite many radiotracers being available, according to different applications, 18F-fluoro-2-deoxy-2-D-glucose ([18F]FDG) is the most widespread. Focusing on cancer TNM staging, local tumor (T) extension is frequently a prerogative of morphological imaging such as CT scan or MRI to evaluate local extension, but [18F]FDG PET can help in a better tumoral mass characterization, especially identifying necrotic or fibrotic areas inside the tumoral mass to guide biopsy or locoregional treatment. Furthermore [18F]FDG is a very important tool for lymph-node (N) and distant metastases (M) detection: baseline [18F]FDG PET/CT is fundamental to assess tumor burden, which not only is a prognostic factor but also is fundamental to plan the optimal therapeutic strategy (Gallamini et al. 2014).

Indications in Lymphomas

[18F]FDG PET/CT is recommended in disease staging in several tumors, in particular both in Hodgkin lymphoma (HL) and in [18F]FDG-avid non-Hodgkin lymphoma (NHL) subtypes. [18F]FDG PET/CT improves the accuracy of staging both for nodal and extranodal sites, allowing a staging change (sometimes upstaging) in 10–30% of patients, in particular for a better definition of splenic involvement and for its high sensitivity in the case of bone marrow involvement: in the case of bone marrow [18F]FDG PET/CT uptake in HL, bone marrow biopsy is in fact no longer indicated (Cheson et al. 2014). In the case of LH or [18F]FDG-avid NHL, [18F]FDG PET/CT is fundamental also in response to therapy assessment to define complete or partial response, stable or progressive disease using a 5-point scale (Deauville score), based on [18F]FDG lesion uptake compared to background (Cheson et al. 2014).

Carcinoma and Adenocarcinoma

[18F]FDG PET/CT has demonstrated a pivotal role also in several [18F]FDG-avid solid tumors. Despite T extension being often provided by conventional imaging (CT, MRI), PET/CT scan could provide additional information: for example solitary pulmonary solid nodules greater or equal to 8 mm can safely be considered benign if the PET-CT scan is negative (Madsen et al. 2016). On the other hand, in the case of advanced NSCLC the main advantage of [18F]FDG PET/CT is the assessment of tumor spread to the pleura (Gallamini et al. 2014). In the case of breast cancer, despite [18F]FDG PET/CT not being recommended for staging localized disease, evidences support its use in locally advanced breast cancer based on improved regional and distant staging, and also for its prognostic role, as baseline tumor glycolytic activity is associated with biological behavior and prognosis (Caresia Aroztegui et al. 2017).

The fundamental role of [18F]FDG PET/CT in solid tumors staging has been widely demonstrated in N and M assessment. For example in the case of NSCLC, [18F]FDG PET/CT is superior to contrast-enhanced CT (CeCT), by adding metabolic information able to disclose morphologically undetectable nodal dissemination: despite this superiority the sensitivity of PET-CT is in general insufficient to rule out mediastinal lymph node metastasis (Madsen et al. 2016) and surgical staging remains the standard (Gallamini et al. 2014). In the case of advanced breast cancer, the imaging-guided sentinel lymph node biopsy with ^{99m}Tc-colloid remains the gold standard, but the high positive predictive value of [18F]FDG PET/CT ($\geq$90%) could guide an axillary lymph node dissection in the case of axillary node uptake (Gallamini et al. 2014).

In the case of melanoma staging, [18F]FDG PET/CT could add clinical information for patients with palpable or macroscopic locoregional nodal metastasis: approximately 60% of patients with locoregional nodal metastasis will develop distant metastases (Perng et al. 2015). In M detection, there's no doubt on [18F] FDG PET/CT leading role, and it is widely supported by international guidelines. For example, in NSCLC no curative-intent treatment should be planned until a PET-CT scan has excluded occult distant metastases (Madsen et al. 2016) or, in the case of melanoma, numerous studies have shown that FDG PET/CT can detect M parameter with high sensitivity (86%) and specificity (91%) (Perng et al. 2015).

Tumor Uptake and Hyperglycemia

Several conditions can impact [18F]FDG PET/CT accuracy, first of all peculiar tumor [18F]FDG avidity related to tumor aggressiveness, histology, and metabolism. Hyperglycemia is the second most important condition that can impact [18F]FDG PET/CT scan results, and an accurate knowledge of [18F]FDG uptake mechanism and management of patient glycemic status before performing the exam are pivotal to obtain the maximum quality of the procedure.

Mechanism of [18F]FDG Uptake

The chemical structure of [18F]FDG is similar to natural glucose, differing in a carbon-2 atom labelled with [18F] (Ido et al. 1978), a radioisotope with a half-life of 109.8 min and a positron emission decay. Cancer tissue presents accelerated glucose metabolism even in the presence of oxygen, with an associated increase in lactate production (Potter et al. 2016). This phenomenon was firstly described by Warburg in 1920 and named "Warburg effect" (Warburg and Negelein 1924). On the basis of the increased lactate production even in the presence of oxygen, he hypothesized a mitochondrial dysfunction that was not supported by further research. Aerobic glycolysis, in fact, is a favorable metabolic pathway for tumoral cell because, despite generally leading to a lower production of ATP molecules compared to mitochondrial oxidative phosphorylation, on the other hand it is much faster and in the same amount of time, produces a higher number of ATP molecules (Chen et al. 2017).

Intracellular Pathways of Glucose and Labelled Glucose

As normal glucose, [18F]FDG is transported by membrane specific glucose transporters (GLUT) into the cell cytosol, and phosphorylated by hexokinase to [18F]FDG 6-phosphate that, differently to glucose 6-phosphate, is not a good substrate for further enzyme action through the glycolytic chain, and it is trapped into the cell and not metabolized to carbon dioxide and water (Larson 2006).

Five different glucose transporters (GLUT) are responsible for [18F]FDG transport across cell membrane: GLUT-1 is ubiquitously expressed in the cell membrane of various tissues and it is up-regulated by several growth factors (Pauwels et al. 2000); GLUT-2 expression is regulated by glucose concentration and it is mainly expressed by intestine, kidney, liver, pancreatic islets, and brain (Theorens et al. 1988); GLUT-3 is expressed in neurons and ensures a glucose supply in the brain even in hypoglycemic conditions (Maher 1995); GLUT-4 is stimulated by insulin and it is expressed in skeletal and cardiac muscle, and in brown and white adipose tissue (Rea and James 1997); GLUT-5 is expressed in the small intestine and is responsible for fructose transport (Kayano et al. 1990).

In euglycemic status, several organs present a physiologic biodistribution of [18F]FDG that reflects GLUTs expression (Fig. 47.1). It is very intense in brain, moderate in liver, and weak in skeletal muscle; furthermore it is variable in the cardiac muscle because myocardial cells primarily use the beta oxidation pathway of fatty acids,

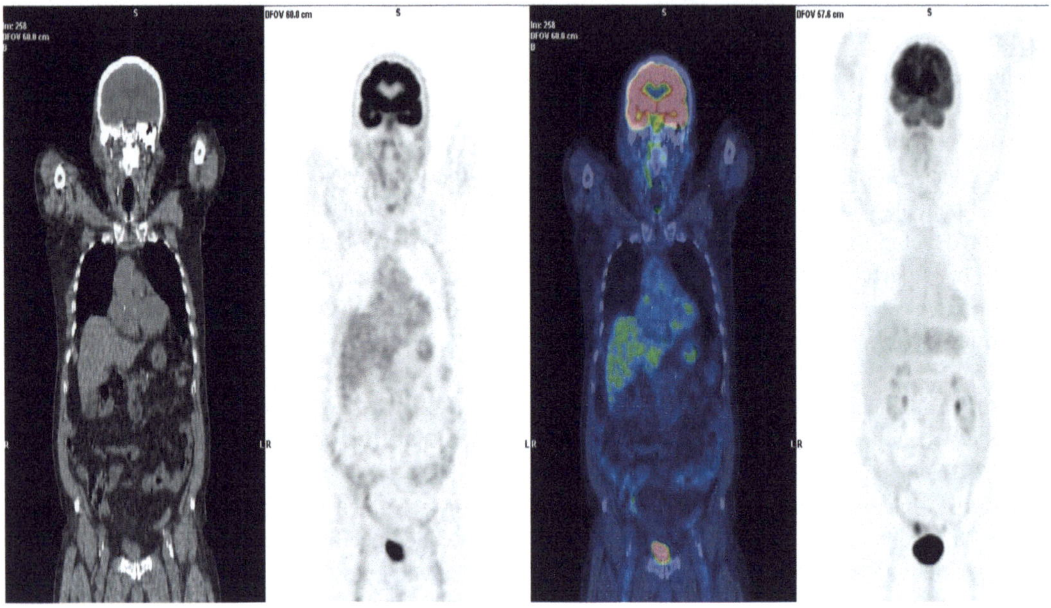

Fig. 47.1 Physiologic biodistribution of [18F]FDG. In fasting and euglycemic conditions [18F]FDG uptake reflects GLUTs expression: [18F]FDG uptake is intense in brain, because of GLUT-3 expression that ensures correct glucose levels even in hypoglycemic conditions, weak in skeletal muscle for GLUT-4 expression stimulated by insulin, variable in cardiac muscle that primarily uses free fatty acids, but may present a metabolic shift to glucose metabolism after glucose load; intense in the urinary system because of [18F]FDG physiological urinary excretion

but following a glucose load may present a metabolic shift toward glucose metabolism; moreover, myocardial cells protect themselves from hypoxic state by means of the "glucose-fatty acid cycle" (Randle et al. 1963).

Finally, urinary system presents high [18F]FDG concentration due to it physiological urinary tracer excretion. Despite the absence of glycosuria in physiological and euglycemic status, renal tubular cells, in addition to GLUTs, also present a sodium-dependent glucose transporter (SGLT) that is responsible for re-uptake of glucose from the filtrate in the proximal tubules, promoting glucose transport against its concentration gradient (Szabo et al. 2006). SGLT presents a lower affinity for [18F]-FDG compared to glucose, due to the replacement of a hydroxy group in D-glucose with a [18F] atom (Moran et al. 1999), and for this reason [18F]-FDG cannot be reabsorbed in the proximal tubules of the kidney and it is accumulated in the urine (Qiao et al. 2007).

Cancer Cell Pathways

Cancer cells [18F]FDG uptake can vary substantially depending on glycolysis levels at one site and it is associated with both increased GLUT and intracellular hexokinase expression (Brown and Wahl 1993); moreover, in cancer cells, glucose 6-phosphatase is markedly downregulated, so its levels are insufficient to break down [18F]FDG 6-phosphate (Larson 2006).

Generally, aggressive and proliferating tumors present a high expression of GLUT and hexokinases, so in these case [18F]FDG uptake is generally very high, but FDG avidity can differ among different histological subtype: Hodgkin Lymphoma, Diffuse Large B Cell Non-Hodgkin Lymphoma (DLBCL) or Follicular Lymphoma

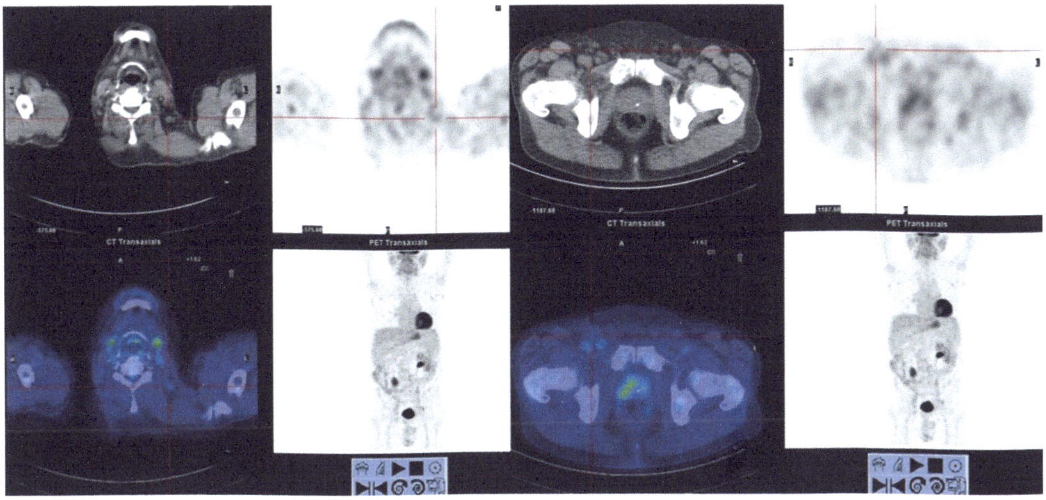

Fig. 47.2 [18F]FDG PET/CT in a patient with marginal zone lymphoma: laterocervical and inguinal lymphnodes that present very low [18F]FDG avidity

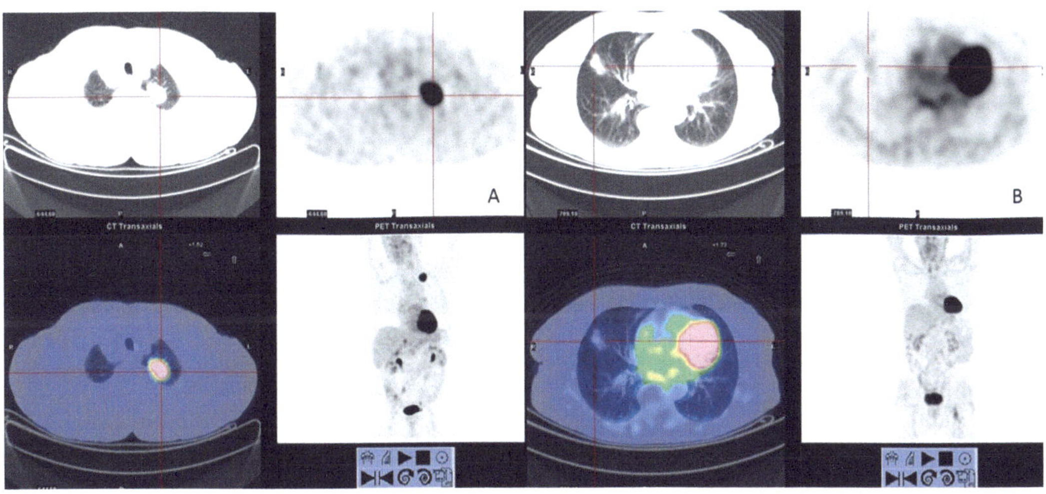

Fig. 47.3 [18F]FDG PET/CT scan in two patients with recent diagnosis of pulmonary nodule prior to histological diagnosis: left apical node of patient A presents very high [18F]FDG uptake compared to medium lobe node of patient B. Subsequent histological diagnosis revealed in case A a NSCLC instead in case B a typical lung carcinoid

presents high [18F]FDG uptake, while other hematologic malignancies, such as Marginal Zone Lymphoma (ML), for example, demonstrate low [18F]FDG uptake (Fig. 47.2). Also solid tumors may have different [18F]FDG avidity. In the case of lung cancer, in fact, Non-Small Cell Lung Carcinoma (NSCLC) presents high [18F]FDG uptake, while tumors with neuroendocrine differentiation (typical carcinoid) present low/absent [18F]FDG uptake (Fig. 47.3). Other well-differentiated tumors that use different metabolic pathway, for example, membrane phospholipids turnover in cell membrane instead of glycolysis, such as Prostatic Carcinoma or Hepatocellular Carcinoma (HCC), present a very low [18F]FDG avidity and should be studied specifically with other PET tracers.

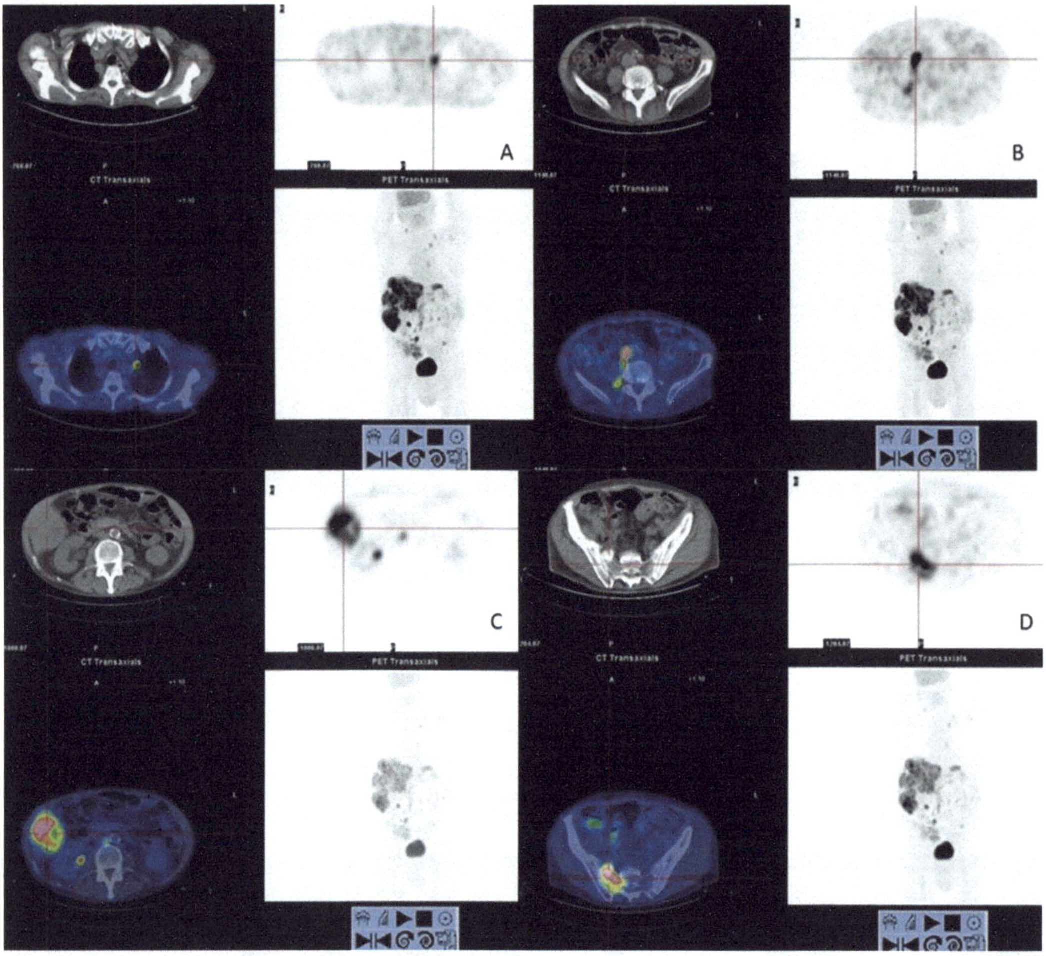

Fig. 47.4 Different [18F]FDG uptake in a patient with DLBCL. (**a**) prevascular lymphnode; (**b**) iliac lymphnode; (**c**) heterogeneous hepatic localization of lymphoma with central necrotic component that reduce [18F]FDG uptake; (**d**) bone localization of lymphoma

[18F]FDG uptake may be influenced not only by tumoral histological characteristics, but also by intra-tumoral heterogeneity in the case of intra-tumoral necrosis, that is responsible for low and heterogeneous uptake (Fig. 47.4) and cell dedifferentiation (undifferentiated neuroendocrine tumors, for example, may present high [18F]FDG uptake). Also different therapies effects may modify [18F]FDG uptake: cytotoxic therapies such as chemotherapy or external radiotherapy may reduce [18F]FDG uptake, but may also cause flogistic reaction with increased [18F] FDG uptake at that site. For these reason an interval of at least three weeks from the last cycle of chemotherapy and three months from the end of radiotherapy is often required, especially in hematologic malignancies, before performing [18F]FDG PET/CT scan to avoid false positive findings.

Effects of Hyperglycemia on [18F]FDG Distribution and PET/CT Scan Interpretation

Several clinical conditions may affect [18F]FDG uptake. Several published studies described the effects of hyperglycemia and hyperinsulinemia on [18F]FDG biodistribution. Hyperglycemia is the leading cause of altered biodistribution because, both in normal and in cancer cells, it leads to a direct competition between plasmatic glucose and [18F]FDG uptake (Di et al. 2018), reducing the binding site of [18F]FDG (Wahl et al. 1992). Reactive hyperinsulinemia to hyperglycemia results in a higher skeletal and myocardial muscle [18F]FDG uptake (Diederichs et al. 1998) for GLUT-4 up-regulation with lower uptake in cancer cells compared to physiological tissue leading to the risk of false negative findings. Also the standardized uptake value (SUV), a semiquantitative parameter expressing FDG concentration in tissues (Juweid and Cheson 2006), may be affected by blood glucose levels. SUV is the a-dimensional semiquantitative expression of the tracer uptake in a region of interest (ROI), for example, tumoral lesion, normalized by total amount of activity administered and body weight (Body weighted SUV—SUV_{bw}) or body surface area (SUV_{bsa}); it is used as SUVmax or mean or peak to compare [18F]FDG uptake on pre and post-therapy scan and to define response to therapy (Wahl et al. 2009). Furthermore, SUVs of healthy tissues, such as liver and mediastinal blood pool, are used as references in international criteria to define disease uptake and to assess response to therapy in lymphoproliferative disease (Barrington and Kluge 2017).

In 2013 Büsing et al. (2013) enrolled 90 patients with Blood Glucose Levels (BGL) ranging from 50 to 372 mg/dl to assess the impact of chronically elevated BGL on [18F]FDG tumor uptake and biodistribution in healthy organs. The authors found a significant association between BGL increase and both cerebral uptake reduction ($p < 0.001$) and muscle uptake increase ($p < 0.001$) and weak associations between BGL and liver uptake ($p = 0.06$), tumoral ($p = 0.133$), fat, lung, and spleen uptake expressed as SUVmax ($p = 0.136$–0.157).

Sprinz et al. in 2018 investigated the effects of glycemia on [18F]FDG uptake in healthy liver, brain, and lungs in 5623 patients that underwent PET/CT, stratified into four groups by serum glucose levels. All organs showed significant differences in mean SUVmax according to different groups ($p < 0.001$) in univariate analysis, while multivariate analysis adjusted for sex, age, and BMI, confirmed significant differences only for brain and liver and not for lung uptake.

The effect of BGL on SUVmax and SUV mean of [18F]FDG uptake was assessed in 8380 patients collected in a recent meta-analysis published in 2019 (Eskian et al. 2019). Patients were divided into 5 groups by BGL, and SUVmax and SUVmean values of tumor, brain, muscle, liver, and blood pool were recorded. Significant inverse correlations ($p < 0.001$) were found between BGL and SUVmax and SUVmean both in brain and in muscle, while positive correlations were found between BGL and SUVmax and SUVmean in liver ($p = 0.001$, $p = 0004$) and blood pool ($p = 0.008$, $p < 0.001$). No significant correlation was found between BGL and SUVmax or SUVmean in tumors.

All hyperglycemic groups compared to the euglycemic group presented significantly lower brain and muscle SUVs ($p < 0.001$ for both), while higher SUVmax and SUVmean in liver ($p = 0.001$, $p = 0004$) and blood pool ($p = 0.008$, $p < 0.001$) were reported. On the contrary, tumoral tissue presented significantly lower SUVmax only in the case of BGL >200 mg/dl. The explanation could be that in tumoral tissue GLUT could be not saturated even in the case of high BGL because tumoral cells overexpress GLUT in order to respond to the hypoxic condition subsequent to induced angiogenesis (Eskian et al. 2019; Yang et al. 2017) demonstrated in various cancer types (Macheda et al. 2005; Carvalho et al. 2011).

Diagnostic Strategies in Normal and Disglycemic Patients

In order to avoid hyperglycemic status guidelines to manage correct patient preparation for [18F] FDG PET/CT scan have been proposed by international societies such as the European Association of Nuclear Medicine (EANM) (Boellaard et al. 2014), the Society of Nuclear Medicine and Molecular Imaging (SNMMI) (Delbeke et al. 2006), the American College of Radiology (ACR) (American College of Radiology 2007), and the National Cancer Institute (NCI) (Shankar et al. 2006).

Non-diabetic patients should fast for at least 4–6 h and parenteral nutrition and intravenous fluids containing glucose should be discontinued for the same time before [18F]FDG injection to prevent high insulinemia. BGL must be measured by a glucometer prior [18F]FDG administration (Boellaard et al. 2014; Delbeke et al. 2006; American College of Radiology 2007; Shankar et al. 2006); for clinical routine the accepted upper BGL threshold is 200 mg/dL, while for research trials ranges between 126 and 150 mg/dL are recommended (Boellaard et al. 2014). In the case of hyperglycemia, it is possible to decrease BGL in patients by hydration until an acceptable level is achieved (Boellaard et al. 2014). The management of diabetic patient is more challenging providing specific recommendations in this setting (Boellaard et al. 2014) for diabetic patients treated by oral medication or insulin.

For diabetic patients treated with oral medication such as metformin, fasting for at least 4–6 h before [18F]-FDG injection in association with adequate hydration is recommended, without oral medication withdrawal to allow for a controlled blood sugar level. PET/CT scan should preferably be performed in the late morning. Metformin decreases BGL by lowering gluconeogenesis, increasing insulin sensitivity and enhancing glucose consumption by enterocytes (Rena et al. 2017). Its known action on the bowel assumes a critical role in [18F]FDG physiological distribution: metformin significantly increases [18F]FDG accumulation in the bowel, in particular in the colon (Martin and Saleem 2014) (Fig. 47.5a and b).

Massolo et al. (2013) in 2013 verified this phenomenon in fifty-three mice that performed dynamic acquisitions for [18F]FDG kinetic evaluation under fasting conditions over a 4-month study period. Mice were divided into 4 groups: untreated mice (group 1), mice exposed to metformin treatment for 48 h before each PET scan (group 2), mice treated for the whole study period (group 3), and mice in which treatment was interrupted 48 h before PET scan (group 4). They found that prolonged drug administration significantly increased bowel [18F]FDG uptake after a relatively long period of treatment and persisted after drug washout.

This increased [18F]FDG uptake has been also demonstrated in clinical practice.

A prospective study published by Gontier at al. (2008) demonstrated an intense, diffuse, and continuous pattern distribution of [18F]FDG along the bowel, strongly predominant in the colon, in patients treated with metformin. They enrolled fifty-five patients under oral medication for diabetes mellitus divided in two groups on the basis of anti-diabetic treatment (group 1a treated with metformin and group 1b in anti-diabetic treatment excluding metformin) and compared to control group (group 2, patients without diabetes mellitus). Patients treated with oral medication presented significantly increased [18F]FDG bowel uptake compared to controls ($p < 0.001$); [18F]FDG bowel uptake was significantly higher in group 1a compared to group 1b ($p < 0.01$).

The effect of metformin on [18F]FDG bowel uptake must kept in mind in the case of PET/CT scan performed for abdominal evaluation, in particular in the case of suspected peritoneal carcinomatosis, colo-rectal or gynecological neoplasm in which increased [18F]FDG bowel uptake could hide pathological uptake and induce false negative results (Gontier et al. 2008). No guideline consensus is available on the oral anti-diabetic treatment management in the case of abdominal evaluation, but several published studies on the timing of metformin discontinuation are available.

A retrospective study published in 2016 (Lee et al. 2016) aimed to assess the impact of metformin discontinuation up to 72 h in [18F]FDG bowel uptake. Two hundred and forty diabetic patients were divided into four groups on the

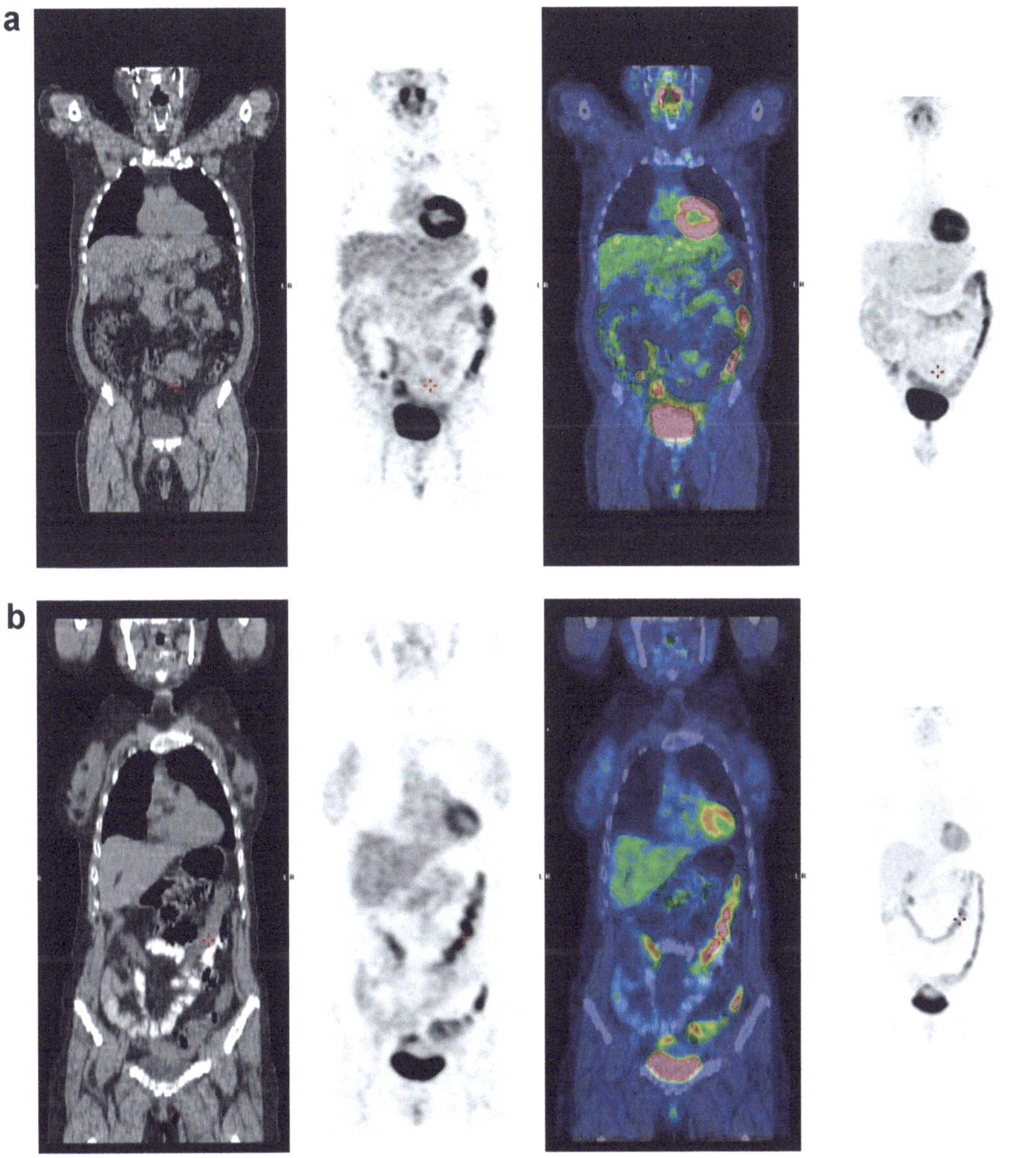

Fig. 47.5 (**a**) [18F]FDG PET/CT scan performed for characterization of a hepatic lesion in a 41 year old diabetic patient treated with metformin. High intestinal uptake is visible in particular in descending colon. (**b**) [18F]FDG PET/CT scan performed for suspected occult lesion in a 51 year old diabetic patient treated with metformin and hepatic lesions suspected for metastases. High intestinal uptake is visible in particular in transverse and sigmoid colon

basis of metformin discontinuation: <24 h (group A), 24–48 h (group B), 48–72 h (group C), and no metformin at all (control group). Compared with the control group, [18F]FDG uptake increased significantly from the ileum to the rectosigmoid colon in group A ($p < 0.001$), from the transverse to the rectosigmoid colon in group B ($p < 0.001$) and from the descending colon to the rectosigmoid colon in group C ($p < 0.001$), highlighting a suboptimal metformin discontinuation <72 h for images interpretation, in particular for the distal colon segments

Conversely, a prospective study published in 2010 (Oh et al. 2010) concluded that discontinuation of metformin for 2 days is feasible to reduce high [18F]FDG bowel uptake. One hundred thirty-eight diabetic patients were divided into two groups: group A treated with metformin and group B in which the regimen did not include metformin and they were compared to patients without diabetes mellitus (control group). Group A was divided into two subgroups on the basis of metformin discontinuation (group A1 continued metformin; group A2 stopped metformin treatment 2 days before PET/CT scan). Ten diabetic patients underwent two consecutive PET/CT scans before and after the discontinuation of metformin. Group A1 compared to group A2 and group B presented a significantly higher ($p < 0.001$) [18F]FDG bowel uptake. In 10 patients who underwent serial PET/CT scans, [18F]FDG bowel uptake decreased by 64% and hidden colorectal malignancies were revealed in two patients after the discontinuation of the drug.

Despite no consensus being available on the timing metformin discontinuation, in the case of [18F]FDG PET/CT scan for abdominal malignancies, a careful evaluation of patient's drug treatment must be conducted, to ensure the optimal patient preparation in order to avoid false negative result preventing the rise of BGL.

In the case of treatment with insulin, European guidelines (Boellaard et al. 2014) suggest different options for scheduling [18F]FDG PET/CT scan on the basis of treatment protocol. Commonly, insulin-dependent patients can be scheduled for PET/CT scan in late morning or midday, [18F]FDG should be injected no sooner than 4 h or 6 h after subcutaneous injection of rapid-acting and short-acting insulin respectively after breakfast in the early morning and subsequent fasting. For patients treated with intermediate-acting and/or long-acting insulin, [18F]FDG injection is not recommended on the same day of insulin administration and PET/CT scan should be scheduled in the early morning after insulin injection the evening before and after night fasting. Particular attention needs long-acting insulin management that could interfere with FDG uptake; thus, intermediate-acting replacement is mostly recommended. In the case of continuous insulin infusion, patients should be scheduled in the early morning and the insulin pump stopped for at least 4 h prior [18F]FDG injection.

The basis of this careful attention on insulin administration is the insulin's affinity for GLUT-4, expressed in the skeletal and cardiac muscle, and in brown and yellow adipose cells (Rea and James 1997): insulin causes the shift of GLUT-4 from intracellular location to the plasma membrane (Huang and Czech 2007; Bryant et al. 2002), promoting both normal glucose and [18F]FDG intracellular uptake and resulting in altered radiotracer biodistribution and suboptimal image quality (Martin and Saleem 2014; Surasi et al. 2014). The aforementioned study by Büsing et al. (2013) also found out that diabetic and insulin treated patients compared to nondiabetics and noninsulin patients presented significant lower mean cerebral SUVmax ($p < 0.001$) and higher mean muscular SUVmax ($p < 0.001$).

An increase up to 50% was also observed in average fat tissue SUVmax and myocardial uptake in diabetic patients and insulin patients, respectively.

Despite this known effect of both endogenous and exogenous insulin on [18F]FDG biodistribution, several studies investigated the impact on image quality of insulin administration before [18F]FDG injection to correct hyperglycemia (Fig. 47.6).

In 2006 Turcotte et al. (2006) assessed the impact of intravenous insulin 60 min before [18F]FDG injection on muscular, liver, or lung [18F]FDG uptake. They compared 53 diabetic patients with BGL >7.0 mmol/l vs 53 euglycemic nondiabetic patients and found no significant difference for the SUV calculated on the lung, liver, heart, and skeletal muscles.

A study published in 2009 (Roy et al. 2009) aimed to assess the clinical impact of intravenous administration of short-acting insulin in 63 patients with glycemia greater than 10 mmol/L according to standardized protocol: 2 units for glycemia of 10.0–12.0 mmol/L, 3 units for glycemia of 12.1–14.0 mmol/L, and 4–6 units for glycemia of 14.1 mmol/L and above to reach a glycemia lower than 10.0 mmol/L. In the case of

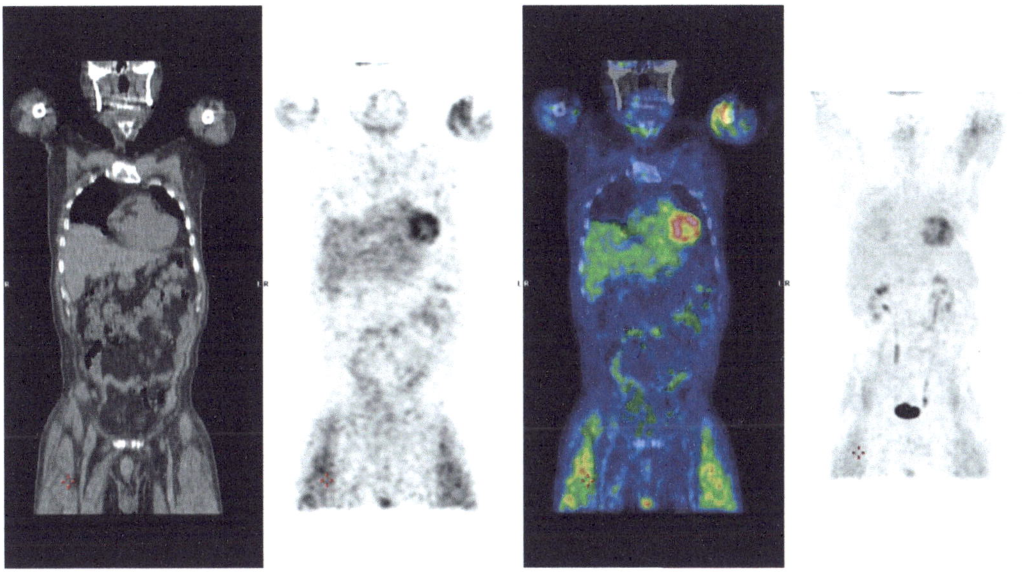

Fig. 47.6 [18F]FDG PET/CT scan performed for restaging of Hodgkin lymphoma after 6 cycles of chemotherapy in a 65 years old diabetic patient treated with short acting insulin. High muscular [18F]FDG uptake is visible in particular in quadriceps muscle

glycemia above 10.0 mmol/L after 30 min, a second insulin dose was given. [18F]FDG was administered at least 60 min after the last insulin administration. After PET/CT scan patients were divided in two groups on the basis of [18F]FDG visual distribution: group A with adequate biodistribution (normal biodistribution, mild muscular uptake or muscular uptake involving more than one muscle group) and group B with altered biodistribution (diffuse muscular uptake of moderate intensity or diffuse, intense muscular uptake resulting in a nondiagnostic examination). [18F]FDG distribution was also semi-quantitatively assessed by SUV *mean* of liver, gluteal muscles, and myocardium. Group A compared to group B presented a significantly longer delay between insulin and [18F]FDG injections ($p < 0.01$), higher glycemia reduction after insulin injection ($p < 0.01$), higher hepatic mean SUV ($p < 0.01$) and lower gluteal muscular mean SUV ($p < 0.01$). This study concluded that an interval of at least 90 min between insulin and [18F]FDG administration should be considered and that hepatic and muscular SUVs could be useful tools to define adequate biodistribution of [18F]FDG.

In 2013 Caobelli et al. (2013) evaluated the usefulness and impact on muscular [18F]FDG uptake of a protocol of intravenous insulin administration before [18F]FDG PET/CT scan in 130 diabetic patients. In 20 patients with BGL >180 mg/dl intravenous insulin was administered 30 min before 18F-FDG injection (group 1); in ten patients with BGL >160 mg/dl and <200 mg/dl, no insulin was injected (group 2); 100 euglycemic patients were used as control group. Biodistribution was adequate in group 2, control group, and in 95% of patients in group 1. No significant differences in gluteal muscle SUVmax were found between groups ($p = 0.20$) and no false negative result was recorded at 6-month follow-up evaluation. In 2013 Song et al. (2013) assessed the impact of intravenous ultra-short insulin administration 60 min before [18F]FDG injection in 105 diabetic patients: 52 patients with BGL >190 mg/dl received 3–5 IU of insulin were compared to the remaining 53 with BGL <190 mg/dl who did not receive insulin and no significant differences in image quality ($p = 0.47$), hepatic SUVmean (p = 0.13), gluteal muscle, and brain uptake (p = 0.71 and p = 0.16, respectively) were found.

A study conducted by Garcia et al. in 2014 concluded that the quality of [18F]FDG PET/CT scan is not affected by subcutaneous administration of rapid-acting insulin if radiotracer is injected at least 4 h later. They enrolled 120 patients divided in 4 groups on the basis of insulin administration and delay to [18F]FDG injection: 30 diabetic patients with BGL<160 mg/dl without further insulin administration (group 1), 30 diabetic patients with BGL ranging from 168 to 260 mg/dl in which subcutaneous rapid-acting insulin was administered and [18F]FDG injection after a delay of 30–115 min with BGL below 160 mg/dl (group 2), 30 diabetic patients with BGL ranging from 192 to 324 mg/dl in which 18F-FDG was injected 4 h after subcutaneous rapid-acting insulin (group 3), and 30 nondiabetic patients with normal BGL (72–104 mg/dl) (control group). For each patient SUVmax of rectus femoris muscle was calculated: in group 2 SUVmax deviated without relation between BGL and [18F]FDG muscle uptake and the quality of PET-CT scan was suboptimal in 60% of patients in group 2, in 13% of patients in group 1, while it was optimal in all patients of group 3.

Despite several studies investigating protocols of intravenous administration of insulin before [18F]FDG administration in order to reduce BGLs, none has yet been validated. EANM and SNMMI guidelines (Boellaard et al. 2014; Delbeke et al. 2006) recommend, in the case of necessity of insulin to correct hyperglycemia, an appropriate delay between [18F]FDG and insulin administration depending on the type and insulin way of administration. Furthermore EANM guidelines (Boellaard et al. 2014) suggest avoiding insulin administration unless this interval is less than 4 h and preferring rapid-acting insulin subcutaneous injection (effective life 2–4 h), while short-acting, intermediate-acting, or long-acting insulin is not recommended for their longer effective life (3–6 h, 12–18 h and 24 h, respectively).

Conclusions

The management of hyperglycemia in both diabetic and non-diabetic patients represents an issue in PET/CT practice, extensively described by the studies mentioned above. Both hyperglycemia and oral and insulin medications showed advantages and disadvantages on PET/CT scan accuracy: the respect of simple rules guarantees a good quality [18F]FDG PET/CT in staging cancer, and it assumes a pivotal role in patient clinical management.

Acknowledgements Reprinted/adapted by permission from "Springer Nature": Springer Milan, *Acta Diabetologica,* "Hyperglycemia and 18F-FDG PET/CT, issues and problem solving: a literature review"—Monica Finessi, Gianni Bisi, Désirée Deandreis, DOI 10.1007/s00592-019-01385-8, ISBN 0123456789, ISSN 14325233.

Conflict of Interest

The authors declare that they have no conflict of interest.

References

American College of Radiology (2007) ACR–SPR practice guideline for performing FDG-PET/CT in oncology. Available from: http://www.acr.org/;/media/71B746780F934F6D8A1BA5CCA5167EDB.pdf. Published

Barrington SF, Kluge R (2017) FDG PET for therapy monitoring in Hodgkin and non-Hodgkin lymphomas. Eur J Nucl Med Mol Imaging. 44(Suppl 1):S97–S110

Boellaard R, Delgado-Bolton R, Oyen WJG, Giammarile F, Tatsch K, Eschner W et al (2014) FDG PET/CT: EANM procedure guidelines for tumour imaging: version 2.0. Eur J Nucl Med Mol Imaging. 42(2):328–354

Brown R, Wahl R (1993) Over expression of glut-1 glucose transporter in human breast cancer: an immunohistochemical study. Cancer. 72:2979–2985

Bryant NJ, Govers R, James DE (2002) Regulated transport of the glucose transporter GLUT4. Nat Rev Mol Cell Biol. 3(4):267–277

Büsing KA, Schönberg SO, Brade J, Wasser K (2013) Impact of blood glucose, diabetes, insulin, and obesity on standardized uptake values in tumors and healthy organs on 18F-FDG PET/CT. Nucl Med Biol [Internet]. 40(2):206–213. Available from: https://doi.org/10.1016/j.nucmedbio.2012.10.014

Caobelli F, Pizzocaro C, Paghera B, Guerra UP (2013) Proposal for an optimized protocol for intravenous administration of insulin in diabetic patients undergoing 18F-FDG PET/CT. Nucl Med Commun. 34(3):271–275

Caresia Aroztegui AP, García Vicente AM, Alvarez Ruiz S, Delgado Bolton RC, Orcajo Rincon J, Garcia Garzon JR et al (2017) 18F-FDG PET/CT in breast cancer: evidence-based recommendations in initial staging. Tumor Biol. 39(10):1–23

Carvalho K, Cunha I, Rocha R, Ayala F, Cajaiba M, Begnami M et al (2011) GLUT1 expression in malignant tumors and its use as an immunodiagnostic marker. Clin (Sao Paulo). 66:965–972

Chen Z, Liu M, Li L, Chen L (May 2017) Involvement of the Warburg effect in non-tumor diseases processes. J Cell Physiol. 2017:2839–2849

Cheson BD, Fisher RI, Barrington SF, Cavalli F, Schwartz LH, Zucca E et al (2014) Recommendations for initial evaluation, staging, and response assessment of hodgkin and non-hodgkin lymphoma: the lugano classification. J Clin Oncol. 32(27):3059–3067

Delbeke D, Coleman RE, Guiberteau MJ, Brown ML, Royal HD, Siegel BA et al (2006) Procedure guideline for tumor imaging with 18F-FDG PET/CT 1.0. J Nucl Med [Internet]. 47(5):885–895. Available from: http://jnm.snmjournals.org/content/47/5/885.short

Di R, Cheng SC, Hudson M, Sakellis C, Van Den Abbeele AD, Kim CK et al (2018) Effects of hyperglycemia on fluorine-18-fluorodeoxyglucose biodistribution in a large oncology clinical practice. Nucl Med Commun 39(5):417–422

Diederichs C, Staib L, Glatting G, Beger H, Reske S (1998) FDG PET: elevated plasma glucose reduces both uptake and detection rate of pancreatic malignancies. J Nucl Med. 39:1030–1033

Eskian M, Alavi A, Khorasanizadeh M, Viglianti BL, Jacobsson H (2019) Effect of blood glucose level on standardized uptake value (SUV) in 18 F- FDG PET-scan: a systematic review and meta-analysis of 20,807 individual SUV measurements. Eur J Nucl Med Mol Imaging. 46:224–237

Gallamini A, Zwarthoed C, Borra A (2014) Positron emission tomography in oncology. Cancers (Basel). 6:1821–1889

Garcia JR, Sanchis A, Juan J, Tomas J, Domenech A, Soler M et al (2014) Influence of subcutaneous administration of rapid-acting insulin in the quality of 18F-FDG PET/CT studies. Nucl Med Commun. 35 (5):459–465

Gontier E, Fourme E, Wartski M, Blondet C, Bonardel G, Le Stanc E et al (2008) High and typical 18F-FDG bowel uptake in patients treated with metformin. Eur J Nucl Med Mol Imaging. 35(1):95–99

Huang S, Czech MP (2007) The GLUT4 glucose transporter. Cell Metab. 5(4):237–252

Ido T, Wan C, Casella J et al (1978) Labeled 2-deoxy-D-glucose analogs: 18F labeled 2-deoxy-2-fluoro-D-glucose, 2-deoxy-2-fluoro-D-mannose and 14C-2-deoxy-2-fluoro-D-glucose. J Label Compd Radiopharm. 14:175–183

Juweid ME, Cheson BD (2006) Positron-emission tomography and assessment of cancer therapy. N Engl J Med. 354:496–507

Kayano T, Burant C, Fukumoto H, Gould G, Fan Y, Eddy R et al (1990) Human facilitative glucose transporters. Isolation, functional characterization, and gene localization of cDNAs encoding an isoform (GLUT5) expressed in small, and adipose tissue and an unusual glucose transporter pseudogene-like sequence (GLUT6). J Biol Chem. 265:13276–13282

Larson SM (2006) 18 F-FDG imaging: molecular or functional? J Nucl Med. 47:31N–32N

Lee SH, Jin S, Lee HS, Ryu JS, Lee JJ (2016) Metformin discontinuation less than 72 h is suboptimal for F-18 FDG PET/CT interpretation of the bowel. Ann Nucl Med. 30(9):629–636

Macheda ML, Rogers S, Best JD (2005) Molecular and cellular regulation of glucose transporter (GLUT) proteins in cancer. J Cell Physiol. 202 (3):654–662

Madsen PH, Holdgaard PC, Christensen JB, Høilund-Carlsen PF (2016) Clinical utility of F-18 FDG PET-CT in the initial evaluation of lung cancer. Eur J Nucl Med Mol Imaging [Internet]. 43(11):2084–2097. Available from: https://doi.org/10.1007/s00259-016-3407-4

Maher F (1995) Immunolocalization of GLUT1 and GLUT3 glucose transporters inprimary cultured neurons and glia. J Neurosci Res. 42:459–469

Martin J, Saleem N (2014) 18F-FDG PET-CT scanning and diabetic patients: what to do? Nucl Med Commun. 35(12):1197–1203

Massollo M, Marini C, Brignone M, Emionite L, Salani B, Riondato M et al (2013) Metformin temporal and localized effects on gut glucose metabolism assessed using 18F-FDG PET in mice. J Nucl Med [Internet]. 54 (2):259–266. Available from: http://jnm.snmjournals.org/cgi/doi/10.2967/jnumed.112.106666

Moran JK, Lee HB, Blaufox MD (1999) Optimization of urinary FDG excretion during PET imaging, J Nucl Med [Internet]. 40(8):1352–1357. Available from: http://www.ncbi.nlm.nih.gov/pubmed/10450688

Oh JR, Song HC, Chong A, Ha JM, Jeong SY, Min JJ et al (2010) Impact of medication discontinuation on increased intestinal FDG accumulation in diabetic patients treated with metformin. Am J Roentgenol. 195(6):1404–1410

Pauwels E, Sturm E, Bombardieri E, Cleton F, Stokkel M (2000) Positron-emission tomographywith [18F] fluoro- deoxy glucose, part 1. Biochemical uptake mechanism and its implication for clinical studies. J Cancer Res Clin Oncol. 126:549–559

Perng P, Marcus C, Subramaniam RM (2015) 18F-FDG PET/CT and melanoma: staging, immune modulation and mutation-targeted therapy assessment, and prognosis. Am J Roentgenol. 205(2):259–270

Potter M, Newport E, Morten KJ (2016) The Warburg effect: 80 years on. Biochem Soc Trans. 44:1499–1505

Qiao H, Bai J, Chen Y, Tian J (2007) Kidney Modelling for FDG Excretion with PET. Int J Biomed Imaging. 2007:1–4

Randle P, Garland P, Hales C, Newsholme E (1963) The glucose fatty-acid cycle. Its role in insulin sensitivity and the metabolic disturbances of diabetes mellitus. Lancet. 1(7285):785–789

Rea S, James D (1997) Moving GLUT4: the biogenesis and trafficking of GLUT4 storage vesicles. Diabetes. 46:1667–1677

Rena G, Hardie DG, Pearson ER (2017) The mechanisms of action of metformin. Diabetologia. 60 (9):1577–1585

Roy F-N, Beaulieu S, Boucher L, Bourdeau I, Cohade C (2009) Impact of intravenous insulin on 18F-FDG PET in diabetic cancer patients. J Nucl Med [Internet]. 50 (2):178–183. Available from: http://jnm.snmjournals.org/cgi/doi/10.2967/jnumed.108.056283

Shankar L, Hoffman J, Bacharach S et al (2006) Consensus recommendations for the use of 18F-FDG PET as an indicator of therapeutic response in patients in National Cancer Institute Trials. J Nucl Med. 47:1059–1066

Song HS, Yoon JK, Lee SJ, Yoon SH, Jo KS, An YS (2013) Ultrashort-acting insulin may improve on18F-FDG PET/CT image quality in patients with uncontrolled diabetic mellitus. Nucl Med Commun. 34 (6):527–532

Sprinz C, Zanon M, Altmayer S, Watte G, Irion K, Marchiori E et al (2018) Effects of blood glucose level on 18F fluorodeoxyglucose (18F-FDG) uptake for PET/CT in normal organs: an analysis on 5623 patients. Sci Rep. 8 (1):6–11

Surasi DS, Bhambhvani P, Baldwin JA, Almodovar SE, O'Malley JP (2014) 18F-FDG PET and PET/CT patient preparation: a review of the literature. J Nucl Med Technol [Internet]. 42(1):5–13. Available from: http://tech.snmjournals.org/cgi/doi/10.2967/jnmt.113.132621

Szabo Z, Xia J, Mathews W, Brown P (2006) Future direction of renal PET. Semin Nucl Med 36(1):36–50

Theorens B, Sarkar H, Kaback H, Lodish H (1988) Cloning and functional expression in bacteria of a novel glucose transporter present in liver, intestine, kidney, and beta-pancreatic islet cells. Cell. 55:281–290

Turcotte E, Leblanc M, Carpentier A, Bénard F (2006) Optimization of whole-body positron emission tomography imaging by using delayed 2-deoxy-2-[F-18]fluoro-d-glucose injection following I.V. insulin in diabetic patients. Mol Imaging Biol. 8(6):348–354

Wahl R, Henry C, Ethier S (1992) Serum glucose: effects on tumor and normal tissue accumulation of 2-[F-18]-fluoro-2-deoxy-D-glucose in rodents with mammary carcinoma. Radiology. 183:643–647

Wahl RL, Jacene H, Kasamon Y, Lodge MA (2009) From RECIST to PERCIST: evolving considerations for PET response criteria in solid tumors. J Nucl Med 50 (Suppl_1):122S–150S

Warburg OP, Negelein E (1924) Uber den stoffwechsel der carcinomzelle. Biochem Zeitschrift. 152:309–335

Yang J, Wen J, Tian T, Lu Z, Wang Y, Wang Z et al (2017) GLUT-1 overexpression as an unfavorable prognostic biomarker in patients with colorectal cancer. Oncotarget [Internet]. 8(7):11788–11796. Available from: http://www.oncotarget.com/fulltext/14352

Pancreatic Islet Transplantation: A Surgical Approach to Type 1 Diabetes Treatment

48

Samuel Rodriguez, Michael Alexander, and Jonathan R. T. Lakey

Contents

Cost and Prevalence of Diabetes 656

Diabetes Pathophysiology 656
Exogenous Insulin Treatment 656
Islet Transplantation 657
Intrahepatic Duct Site 657
Clinical Autologous Islet Engraftment 657
Clinical Islet Allogenic Engraftment 657
Islet Graft Assessment 658
Clinical Islet Xenogeneic Engraftment 658
Long-term Challenges 658
Regulatory Issues 658
Acceptance by Hospital Personnel 659
Advances in Immunosuppression Regimes 659
Advances in Immunomodulatory Drugs 660

Enhancing Immunoisolation Technology 660
Microencapsulation of Islet Graft 660
Macroencapsulation 660

References 661

S. Rodriguez · M. Alexander
Department of Surgery, University of California Irvine, Orange, CA, USA

J. R. T. Lakey (✉)
Department of Surgery, University of California Irvine, Orange, CA, USA

Department of Biomedical Engineering, University of California Irvine, Irvine, CA, USA
e-mail: jlakey@hs.uci.edu

J. Faintuch and S. Faintuch (eds.), *Obesity and Diabetes*, https://doi.org/10.1007/978-3-030-53370-0_48

Abstract

The medical and social burden of T1DM places large constraints on both patients and their families. Transplantation of pancreatic beta cells can be a viable therapeutic alternative. Current challenges are the shortage of donors and the toxicity of immunosuppression. Xenotransplantation and biomaterial protection are currently the most promising technology. Immunoisolation is necessary to protect the recipient from foreign antigens. In turn, shortage of donors ceases to be a barrier. Improvements in certain areas of immunoprotection and graft site neovascularization are the focus of current clinical efforts.

Keywords

Islet engraftment · Autologous pancreatic islet · Xenogeneic pancreatic islet · Immunosuppression pancreatic transplantation · Pancreatic islet protection · Islet site neovascularization

Cost and Prevalence of Diabetes

In 2015 diabetes affected over 9.4% of the US population (5–10% type 1 diabetes), with 1.5 million new cases every year, and many more cases (33.9% of the population) suffered from prediabetes. Worldwide as many as 642 million diabetics will be seen by 2040 (Control 2017; Cho et al. 2018). The world financial burden reaches US$1.31 trillion (Bommer et al. 2017), with the highest expenses corresponding to the United States. They reached $327 billion (1.69% of the 2017 US GDP (Group TWB 2019)), or 26% more than 5 years earlier (2012) (Association AD 2018).

Diabetes Pathophysiology

Pancreatic *β*-cells sense glucose by means of GLUT2 (Thorens 2015) and other pathways, monitoring shifts such as elevations in the post absorptive state (Leibiger et al. 2008). They allow glucose entry in multiple cells enabling glycolysis, glycogenesis, and lipogenesis (Leibiger et al. 2008). Impairment of glucose-stimulated insulin secretion, or GSIS, is one of the central derangements of diabetes (Boland et al. 2017). Insulin resistance and β-cell dysfunction predominate in type 2 diabetes (T2DM) (Petersen and Shulman 2018), differently from type 1 or insulin-dependent diabetes mellitus (T1DM, IDDM), in which autoimmune destruction of pancreatic β-cells is the hallmark.

A Type IV hypersensitivity reaction is described in T1DM (McLaughlin et al. 2016), leading to β-cell destruction (Lampasona et al. 2010; Rewers and Ludvigsson 2016; Pociot and Lernmark 2016). When 90% of β-cell mass is damaged, insulin dependence becomes inevitable (Atkinson et al. 2014). Highly dangerous diabetic ketoacidosis and/or hyperosmolar hyperglycemia are some of the possible consequences of ensuing glucose elevation (Umpierrez and Korytkowski 2016).

Chronically macroangiopathy and particularly microangiopathy can be followed by microvascular and neural troubles, including retinopathy and neuropathy (Madonna et al. 2017). Cardiovascular events and peripheral arterial obstruction are other possibilities, with risk of appendage/limb amputation (Chillarón et al. 2014).

Exogenous Insulin Treatment

Since 1922 (Best 1956), insulin, along with self-monitoring of blood glucose (SMBG), has been introduced in clinical routine (Association AD 2019). In the Diabetes Control and Complication Trial (DCCT), intensive insulin therapy diminished the advent of major complications (Control and Group CTR 1993). Among cases of non-insulin-dependent diabetes mellitus (NIDDM) and T2DM patients in the UK Prospective Diabetes Study (UKPDS), restoration of glucose homeostasis was equally beneficial (Group UPDS 1998; Ohkubo et al. 1995).

Even careful exogenous insulin confers less protection against long-term organ damage than physiological release. Moreover, intensive therapy entails a higher risk of hypoglycemia (McCall 2012). Two hypoglycemic episodes (BG < 50 mg/dL = 2.8 mmol/L) per week have been documented during intensive insulin therapy, with at least one of these potentially disabling (Arabi et al. 2009; MacLeod et al. 1993).

Islet Transplantation

In 1893, Minkowski and von Mering, in Strasbourg, France, performed a subcutaneous partial auto-transplantation of the pancreas in a dog, which marginally improved physiology after subsequent pancreatectomy (Minkowski 1893). Williams subsequently attempted sheep pancreas transplantation in an IDDM adolescent (Williams 1894). The endocrine tissue eventually emerged as the relevant fraction (1–2% of total mass), with acinar tissue potentially precipitating inflammation, necrosis, and graft failure (Fichera 1928; Heuser et al. 2000; Gray 1989).

The Edmonton Protocol clinical trials (NCT00014911) introduced sirolimus (0.2 mg/kg; 0.1 mg/kg/day), low-dose tacrolimus (3–6 ng/ml), and daclizumab (5 × 1 mg/14 days) for immune suppression. (Shapiro et al. 2000; Ryan et al. 2001) A trial adopting the Edmonton protocol achieved 31% insulin independence for over 2 years (Shapiro et al. 2006). Over 50% insulin independence for 5 years was subsequently accomplished (Tekin et al. 2016).

Intrahepatic Duct Site

The intrahepatic duct was selected for some time; however, instant blood-mediated inflammatory reaction (IBMIR), hypoxic apoptosis, and supraphysiological levels of drugs and nutrients resulted in frequent failures (Johansson et al. 2005; Shapiro et al. 2005). Portal hypertension and thrombosis were additional risks, besides the inconveniences of the invasive operation and prolonged immunosuppression (Bottino et al. 2018; Ryan et al. 2005).

Clinical Autologous Islet Engraftment

Over 827 infusions in 819 recipients are available at the Clinical Islet Transplant Registry (CITR) since 1999 (Registry 2017a). Indications mainly address severe and refractory pancreatitis requiring pancreatectomy (Bellin et al. 2017). Total pancreatectomy with islet transplantation (TPIAT) is similar to islet allotransplantation; however, it is free from rejection.

Approximately 90% of TPIAT patients have islet graft function; however, only one-third become insulin independent. Islet autograft attrition may also occur especially in cases >35 years, resulting in low insulin retention after 5 years. In turn when doses of ≥275,000 islet equivalents (IEQs) are administered to those in a lower age bracket (18–35 years), over 70% insulin independence is demonstrated (Registry 2017a). C-peptide responds in all age groups, however not HbA1c (baseline < 7.0%), with favorable outcome only for age groups 12–18 and 18–35 years old. Differently from exogenous insulin prescriptions, hypoglycemia practically never occurs (Registry 2017a).

Clinical Islet Allogenic Engraftment

Allotransplantation uses cadaveric islets, collected from digested pancreatic tissue. As occurs with all allotransplantations, a long waiting list for suitable donors and strict matching criteria mean very substantial delays till treatment. Moreover immunosuppression is necessary and entails an array of potential complications (Shapiro et al. 2017; Registry 2017b).

Post-procurement islet culturing before allotransplantation is currently a routine, from just 35% from 1999 to 2002 to 100% from 2015 to 2018. Culturing time is now longer, consistent with the benefits of stabilizing the donor islets. Total islet graft volume is diminishing, without affecting the IEQs and IEQ/kg ratio, thus rendering the transplant less invasive. Total β-cells and β-cells/kg recipient ratio are higher, with less endotoxin content.

Islet Graft Assessment

Approximately 10,000 IE/kg of body weight is recommended. Multiple donors may be required to this aim. In 2015 the CITR Registry included data on 1086 allogeneic islet transplant recipients (877 islet transplant alone "ITA," 183 islet after kidney "IAK," 24 simultaneous islet kidney "SIK," and 2 kidney after islet "KAI") (Registry 2017b). Insulin independence was 70% and 55% respectively, after one and 5 years, with C-peptide of 90% and 70% after identical follow-up periods.

Blood glucose of 60–140 mg/dL for 5 years was achieved in the majority (80%), without severe hypoglycemic episodes. Re-infusion was often necessary (73%); nevertheless, the procedures were less invasive and with considerably milder post-operative complications than whole pancreas transplantation.

Clinical Islet Xenogeneic Engraftment

Massive demand and limited cadaver donors are a significant barrier for allogenic transplants (Peloso et al. 2015; Ris et al. 2004). Furthermore whole pancreas transplantations are preferred, because of a rather well-established experience. Islet xenotransplantation is a very promising route (Markmann et al. 2016; Cozzi et al. 2009; Van Der Windt et al. 2012). particularly porcine pig islets (Lanza et al. 1991; Bottino et al. 2014).

Human and porcine insulin differ by only one amino acid (Richter and Neises 2003; Han and Tuch 2001). Pure transgenic, pathogen-free donors, available in unlimited numbers and devoid of ethical constraints, are other important points. For ideal islet size, yield, and functionality juvenile porcine islet sources (JPIs) seem the best (Kim et al. 2009; Bottino et al. 2007; Smith et al. 2018; Nagaraju et al. 2015). They are also more resistant to hypoxia (Nagaraju et al. 2015).

While adult porcine islet (API) donors yield large amounts of mature islets (Fig. 48.1), the isolation and digestive process is more extensive and costly than with JPI, even though these are not fully mature and require culturing in nutrient-rich media (Lau et al. 2019; Lau et al. 2018). Porcine islets can maintain viability in vivo, (O'Shea and Sun 1986) with insulin secretion in vivo for up to 174 days (Omer et al. 2003). Also in T1DM patients, encapsulated porcine islets remained active for 9.5 years (O'Shea and Sun 1986).

Long-term Challenges

In keeping with endogenous defense against non-self-antigens in general, hemagglutinating and lympho-cytotoxic antibodies against pig tissue occur in humans, (McKenzie et al. 1968) including α-(1,3)-galactose, incriminated in acute host immune rejection (Omori et al. 2006; Sandrin and Mckenzie 1994), potentially requiring immune suppression. Porcine endogenous retroviruses (PERVs) and other microorganisms could be another source of concern. (Sandrin and Mckenzie 1994) Biocompatible polymers to encapsulate xenogeneic tissues could avoid most of these obstacles. Yet xenografts in the United States are strictly regulated; therefore, most studies proceed in animal models (Cooper et al. 2017).

Regulatory Issues

Some protocols prefer other countries, in order to avoid rigorous FDA scrutiny. In Mexico, two cohorts underwent xenograft implantation of a collagen-covered pre-vascularized scaffold (250,000 islets, 30,000–100,000 Sertoli cells). HbA1c, insulin requirements, and C-peptide responded for up to 4 years; no pig-related microorganisms were detected in these populations (Valdés-González et al. 2005; Valdes-Gonzalez et al. 2010).

DIABECELL microencapsulated islet scaffold (Living Cell Technologies/LCT, Melbourne, Australia) has been intraperitoneally injected for unstable type 1 diabetes. This microencapsulated graft is delivered via laparoscopy into the

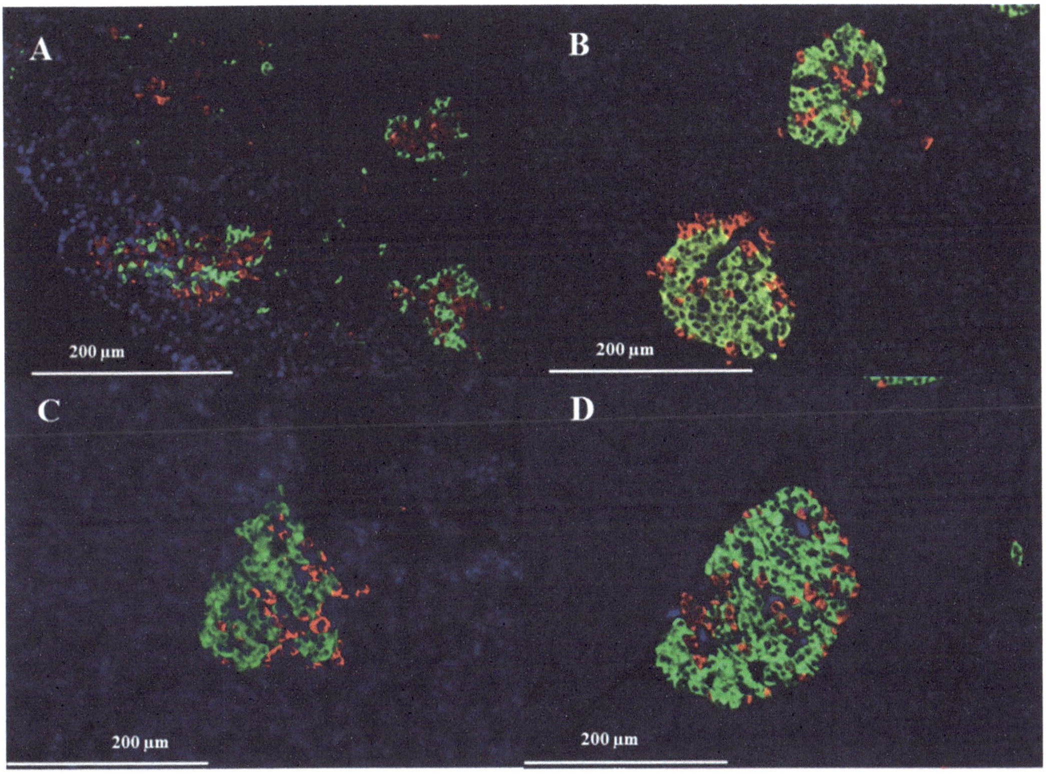

Fig. 48.1 Characterization of Porcine Islet Endocrine Composition. Characterization of porcine pancreatic islet cellular composition using fluorescence immunohistochemistry. Neonatal porcine islets (**a**) demonstrate a lower incidence of insulin positive cells (green) compared to juvenile (**b, c**) and adult pigs (**d**) and are significantly smaller in size ($p = 0.001$). Juvenile porcine islets (weaned (**b**) and pre weaned (**c**)) are significantly smaller ($p = 0.03$, pre weaned; $p = 0.04$, weaned) but do not demonstrate significant differences in insulin content ($p = 0.4$, pre weaned; $p = 0.1$, weaned). Glucagon positive cells stain red. A blue counterstain (DAPI) is used to stain the nuclei

intraperitoneal (IP) cavity, and has been approved in Russia, New Zealand, and Argentina. In the Russian pilot protocol all eight patients partially responded, and two became insulin independent for 4 weeks post-operation (Schuurman 2011). Somewhat analogously in New Zealand, 14 IDDM patients (5000–20,000 IE/kg, wild-type pre-weaned JPI) were followed with no immunosuppression for 52 weeks, with partial response in 57% (8/14) (Matsumoto et al. 2016). As many as 10 bacteria, 15 viruses including porcine endogenous retroviruses, and one protozoan were systematically screened with no transmission observed (Wynyard et al. 2014; Morozov et al. 2017).

Acceptance by Hospital Personnel

Past graft failures have generated negative attitudes against xenotransplantations; however, this could be overcome with better results and more transparent information (Abalovich et al. 2017).

Advances in Immunosuppression Regimes

Robust immune suppressors, such as those adopted in the Edmonton protocol, are compatible with prolonged function, of up to 10–12 years

(Ryan et al. 2005; Brennan et al. 2016). Less positive outcomes are also noticed (Shapiro et al. 2006), with insulin independence after one and 2 years of 44% and 31% respectively. Tacrolimus has recently been the most adopted drug, with a substantial reduction of sirolimus and other mammalian targets of rapamycin (mTORs) (Berney et al. 2018).

Nevertheless calcineurin inhibitors (CI), encompassing tacrolimus, can have diabetogenic effects and β-cell toxicity (Rangel 2014). CI-free immunosuppressant belatacept and the anti-leukocyte antigen-1 antibody efalizumab could be an option (Posselt et al. 2010). Janus kinase inhibitor tofacitinib is also being investigated (Kim et al. 2018).

Advances in Immunomodulatory Drugs

Various modalities of rejection can be inhibited through modulation of immune cells by cytokines, interleukins, and chemokines (Tian et al. 1996; Kovarik 2013). One such cocktail utilized in conjunction with islet allograft adopted T-cell mediator thymoglobulin, an interleukin-1β (Il-1β) blocker called anakinra, and tumor necrosis factor-α (TNF-α) blocker etanercept, with extended survival of the graft (Naziruddin et al. 2018; Matsumoto et al. 2011).

Enhancing Immunoisolation Technology

The use of fibrin has extended islet survival and functionality as well as diminished the marginal islet mass up to 90% (Pappalardo 2019; Riopel et al. 2015).

Microencapsulation of Islet Graft

Alginate and other biocompatible materials have been used to create a semi-permeable capsule, for immunoisolation without impairing glucose sensitization and insulin release (de Vos and Marchetti 2002; Pawar and Edgar 2012; Food and Drug Administration 2019; Sondermeijer et al. 2016; De Vos et al. 1997a; Paques et al. 2014; Goosen et al. 1987). Viscous alginate can be polymerized into a rigid hydrogel capsule, (Paques et al. 2014; Goosen et al. 1987) and further polycations (e.g., poly-l-lysine) (Thu et al. 1996) can be added to increase both the integrity and immunoprotection (De Vos et al. 1997b; Fritschy et al. 1991) (Fig. 48.2).

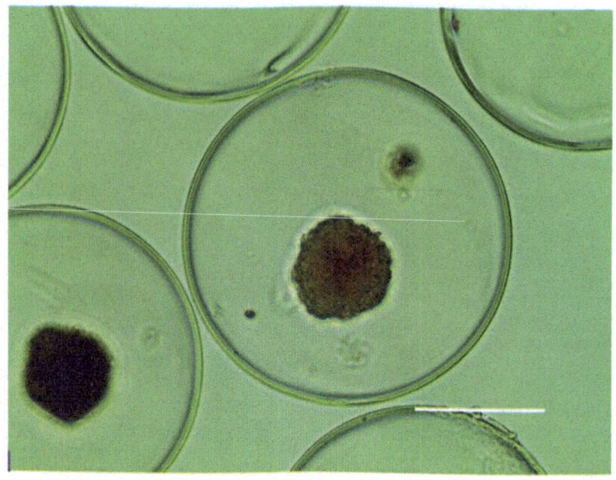

Fig. 48.2 Alginate Encapsulated Porcine Islet with poly-l-lysine (PLL) coating. Image taken by research associates of the Lakey lab at UC Irvine. Porcine Islets stained with dithizone for 15 min. Image taken with inverted-light microscopy (EVOS Microscope system, Thermo Fisher) at 4/10PH lens at 20× magnification. Scale bar = 200 μm

All these advances notwithstanding results in non-human primates (Bottino et al. 2014; Safley et al. 2018) and humans (Matsumoto et al. 2016) are not as favorable as in other species. Lack of metabolite supply to the intraperitoneal transplantation site, spherical aggregation, and pericapsular fibrotic growth (PFO) are significant problems (Krishnan et al. 2017). Capsules often clump together in the Douglas space (recto-uterine pouch). Incidentally the whole peritoneal cavity has been described as a low-oxygen environment, which prevents adequate functioning of the encapsulated graft.

Macroencapsulation

Retrievable macroencapsulation devices have room for large amounts of islets within a biocompatible shell, potentially made with alginate (de Vos et al. 2002; Dufrane et al. 2010; Evron et al. 2018), collagen (Harrington et al. 2017), polycaprolactone (Smink et al. 2017, 2018; Marchioli et al. 2016), polyurethane (Zondervan et al. 1992), and polytetrafluoroethylene (Boettler et al. 2016). The subcutaneous space (SC) has been the suggested site (Bottino et al. 2018). In order to overcome the scarcity of oxygen, one polyurethane shell with inner chamber islet-containing alginate matrix was designed to receive external oxygen infusion (β-Air, Beta-O_2 Technologies. Rosh-Haayin, Israel). Preliminary results indicated no immunization or rejection for 6 months, even though viability was insufficient (Rotem and Avni 2017; Carlsson et al. 2018).

Pre-vascularization of engraftment site, with the help of such agents as vascular endothelial growth factor (VEGF), was attempted to solve hypoxia-induced graft loss with the macroencapsulation technique (Sakata et al. 2014). Good vascularization and islet structure preservation were noticed with the two-step Sernova Cell Pouch procedure (Sernova Corp, London, Ontario, Canada); however, functional end points were not successful (Gala-Lopez et al. 2016).

References

Abalovich A, Matsumoto S, Wechsler CJ et al (2017) Level of acceptance of islet cell and kidney xenotransplants by personnel of hospitals with and without experience in clinical xenotransplantation. Xenotransplantation 24(4):e12315

Arabi YM, Tamim HM, Rishu AH (2009) Hypoglycemia with intensive insulin therapy in critically ill patients: predisposing factors and association with mortality. Crit Care Med 37(9):2536–2544

Association AD (2018) Economic costs of diabetes in the U.S. in 2017. Diabetes Care 41(5):917–928

Association AD (2019) 6. Glycemic targets: standards of medical care in diabetes-2019. Diabetes Care 42(Suppl 1):S61

Atkinson MA, Eisenbarth GS, Michels AW (2014) Type 1 diabetes. Lancet 383(9911):69–82

Bellin MD, Forlenza G, Majumder K et al (2017) Total pancreatectomy with islet autotransplantation resolves pain in young children with severe chronic pancreatitis. J Pediatr Gastroenterol Nutr 64(3):440

Berney T, Andres A, Toso C, Majno P, Squifflet J-P (2018) mTOR inhibition and clinical transplantation: pancreas and islet. Transplantation 102(2S):S30–S31

Best CH (1956) The first clinical use of insulin. Diabetes 5 (1):65–67

Boettler T, Schneider D, Cheng Y et al (2016) Pancreatic tissue transplanted in TheraCyte™ encapsulation devices is protected and prevents hyperglycemia in a mouse model of immune-mediated diabetes. Cell Transplant 25(3):609–614

Boland BB, Rhodes CJ, Grimsby JS (2017) The dynamic plasticity of insulin production in β-cells. Mol Metab 6 (9):958–973

Bommer C, Heesemann E, Sagalova V et al (2017) The global economic burden of diabetes in adults aged 20-79 years: a cost-of-illness study. Lancet Diabetes Endocrinol 5(6):423–430

Bottino R, Balamurugan A, Smetanka C et al (2007) Isolation outcome and functional characteristics of young and adult pig pancreatic islets for transplantation studies. Xenotransplantation 14(1):74–82

Bottino R, Wijkstrom M, Van Der Windt D et al (2014) Pig-to-monkey islet xenotransplantation using multi-transgenic pigs. Am J Transplant 14(10):2275–2287

Bottino R, Knoll MF, Knoll CA, Bertera S, Trucco MM (2018) The future of islet transplantation is now. Front Med 5:202

Brennan DC, Kopetskie H, Sayre P et al (2016) Long-term follow-up of the Edmonton protocol of islet transplantation in the United States. Am J Transplant 16 (2):509–517

Carlsson PO, Espes D, Sedigh A et al (2018) Transplantation of macroencapsulated human islets within the bioartificial pancreas βAir to patients with type 1 diabetes mellitus. Am J Transplant 18(7):1735–1744

Chillarón JJ, Le-Roux JAF, Benaiges D, Pedro-Botet J (2014) Type 1 diabetes, metabolic syndrome and cardiovascular risk. Metabolism 63(2):181–187

Cho NH, Shaw JE, Karuranga S et al (2018) IDF diabetes atlas: global estimates of diabetes prevalence for 2017 and projections for 2045. Diabetes Res Clin Pract 138:271–281

Control CFD (2017) Prevention. National diabetes statistics report, 2017. Atlanta, GA: Centers for Disease Control and Prevention, US Department of Health and Human Services, p 20

Control D, Group CTR (1993) The effect of intensive treatment of diabetes on the development and progression of long-term complications in insulin-dependent diabetes mellitus. N Engl J Med 329(14):977–986

Cooper DK, Pierson RN III, Hering BJ et al (2017) Regulation of clinical xenotransplantation—time for a reappraisal. Transplantation 101(8):1766

Cozzi E, Tallacchini M, Flanagan EB, Sykes M, Vanderpool H (2009) The international xenotransplantation association consensus statement on conditions for undertaking clinical trials of porcine islet products in type 1 diabetes--chapter 1: key ethical requirements and progress toward the definition of an international regulatory framework. Xenotransplantation 16(4):203–214

de Vos P, Marchetti P (2002) Encapsulation of pancreatic islets for transplantation in diabetes: the untouchable islets. Trends Mol Med 8(8):363–366

De Vos P, De Haan B, Wolters G, Strubbe J, Van Schilfgaarde R (1997a) Improved biocompatibility but limited graft survival after purification of alginate for microencapsulation of pancreatic islets. Diabetologia 40(3):262–270

De Vos P, De Haan B, Van Schilfgaarde R (1997b) Effect of the alginate composition on the biocompatibility of alginate-polylysine microcapsules. Biomaterials 18 (3):273–278

de Vos P, Hamel AF, Tatarkiewicz K (2002) Considerations for successful transplantation of encapsulated pancreatic islets. Diabetologia 45 (2):159–173

Dufrane D, Goebbels R-M, Gianello P (2010) Alginate macroencapsulation of pig islets allows correction of streptozotocin-induced diabetes in primates up to 6 months without immunosuppression. Transplantation 90(10):1054–1062

Evron Y, Colton CK, Ludwig B et al (2018) Long-term viability and function of transplanted islets macroencapsulated at high density are achieved by enhanced oxygen supply. Sci Rep 8(1):6508

Fichera G (1928) Implanti omoplastici feto-umani nei cancro e nel diabete. Tumori 14:434

Food and Drug Administration (2019) Food and Drug: Sec. 184.1724 Sodium alginate. In: Food and Drug Administration, ed.

Fritschy WM, Wolters GH, Van Schilfgaarde R (1991) Effect of alginate-polylysine-alginate microencapsulation on in vitro insulin release from rat pancreatic islets. Diabetes 40(1):37–43

Gala-Lopez B, Pepper A, Dinyari P et al (2016) Subcutaneous clinical islet transplantation in a prevascularized subcutaneous pouch—preliminary experience. CellR4 4(5):e2132

Goosen MF, O'Shea GM, Sun AM (1987) Microencapsulation of living tissue and cells. In: Google patents. Mountain View, CA, Google

Gray DW (1989) The role of exocrine tissue in pancreatic islet transplantation. Transpl Int 2(1):41–45

Group TWB (2019). https://databank.worldbank.org/reports.aspx?source=2&series=NY.GDP.MKTP.CD&country=. Accessed 1 May 2020

Group UPDS (1998) Intensive blood-glucose control with sulphonylureas or insulin compared with conventional treatment and risk of complications in patients with type 2 diabetes (UKPDS 33). Lancet 352 (9131):837–853

Han X-G, Tuch BE (2001) Cloning and characterization of porcine insulin gene. Comp Biochem Physiol B: Biochem Mol Biol 129(1):87–95

Harrington S, Williams J, Rawal S, Ramachandran K, Stehno-Bittel L (2017) Hyaluronic acid/collagen hydrogel as an alternative to alginate for long-term immunoprotected islet transplantation. Tissue Eng Part A 23(19–20):1088–1099

Heuser M, Wolf B, Vollmar B, Menger MD (2000) Exocrine contamination of isolated islets of LANGERHANS deteriorates the process of revascularization after free TRANSPLANTATION1. Transplantation 69(5):756–761

Johansson H, Lukinius A, Moberg L et al (2005) Tissue factor produced by the endocrine cells of the islets of Langerhans is associated with a negative outcome of clinical islet transplantation. Diabetes 54(6):1755

Kim HI, Lee SY, Jin SM et al (2009) Parameters for successful pig islet isolation as determined using 68 specific-pathogen-free miniature pigs. Xenotransplantation 16(1):11–18

Kim J-M, Shin J-S, Min B-H et al (2018) Pre-clinical results of islet Allo-transplantation using JAK inhibitor as replacement for Tacrolimus widely used immunosuppressive drug in islet transplantation in Cynomolgus monkeys. Transplantation 102:S746

Kovarik J (2013) From immunosuppression to immunomodulation: current principles and future strategies. Pathobiology 80(6):275–281

Krishnan R, Ko D, Foster CE et al (2017) Immunological challenges facing translation of alginate encapsulated porcine islet xenotransplantation to human clinical trials. In: Cell microencapsulation. Springer, New York, pp 305–333

Lampasona V, Petrone A, Tiberti C et al (2010) Zinc transporter 8 antibodies complement GAD and IA-2 antibodies in the identification and characterization of adult-onset autoimmune diabetes: non insulin requiring autoimmune diabetes (NIRAD) 4. Diabetes Care 33(1):104–108

Lanza RP, Butler DH, Borland KM et al (1991) Xenotransplantation of canine, bovine, and porcine islets in diabetic rats without immunosuppression. Proc Natl Acad Sci 88(24):11100–11104

Lau H, Corrales N, Lee S et al (2018) Exendin-4 improves yield and function of isolated pre-weaned porcine islets. J Endocrinol Diabetes 5(5):1–7

Lau H, Corrales N, Alexander M et al (2019) Necrostatin-1 supplementation enhances young porcine islet maturation and in vitro function. Xenotransplantation 27(1): e12555

Leibiger IB, Leibiger B, Berggren P-O (2008) Insulin signaling in the pancreatic β-cell. Annu Rev Nutr 28:233–251

MacLeod K, Hepburn D, Frier BM (1993) Frequency and morbidity of severe hypoglycaemia in insulin-treated diabetic patients. Diabet Med 10(3):238–245

Madonna R, Balistreri CR, Geng Y-J, De Caterina R (2017) Diabetic microangiopathy: pathogenetic insights and novel therapeutic approaches. Vasc Pharmacol 90:1–7

Marchioli G, Luca AD, de Koning E et al (2016) Hybrid polycaprolactone/alginate scaffolds functionalized with VEGF to promote de novo vessel formation for the transplantation of islets of Langerhans. Adv Healthc Mater 5(13):1606–1616

Markmann JF, Bartlett ST, Johnson P et al (2016) Executive summary of IPITA-TTS opinion leaders report on the future of β-cell replacement. Transplantation 100 (7):e25–e31

Matsumoto S, Takita M, Chaussabel D et al (2011) Improving efficacy of clinical islet transplantation with iodixanol-based islet purification, thymoglobulin induction, and blockage of IL-1β and TNF-α. Cell Transplant 20(10):1641–1647

Matsumoto S, Abalovich A, Wechsler C, Wynyard S, Elliott RB (2016) Clinical benefit of islet xenotransplantation for the treatment of type 1 diabetes. EBioMedicine 12:255–262

McCall AL (2012) Insulin therapy and hypoglycemia. Endocrinol Metab Clin 41(1):57–87

McKenzie I, Stocker J, Ting A, Morris P (1968) Human lymphocytotoxic and haemagglutinating activity against sheep and pig cells. Lancet 292(7564):386–387

McLaughlin KA, Richardson CC, Ravishankar A et al (2016) Identification of tetraspanin-7 as a target of autoantibodies in type 1 diabetes. Diabetes 65 (6):1690–1698

Minkowski O (1893) Untersuchungen über den Diabetes mellitus nach Exstirpation des Pankreas. Archiv für experimentelle Pathologie und Pharmakologie 31 (2–3):85–189

Morozov VA, Wynyard S, Matsumoto S, Abalovich A, Denner J, Elliott R (2017) No PERV transmission during a clinical trial of pig islet cell transplantation. Virus Res 227:34–40

Nagaraju S, Bottino R, Wijkstrom M, Trucco M, Cooper DK (2015) Islet xenotransplantation: what is the optimal age of the islet-source pig? Xenotransplantation 22 (1):7–19

Naziruddin B, Kanak M, Chang C et al (2018) Improved outcomes of islet autotransplant after total pancreatectomy by combined blockade of IL-1β and TNFα. Am J Transplant 18(9):2322–2329

O'Shea GM, Sun A (1986) Encapsulation of rat islets of Langerhans prolongs xenograft survival in diabetic mice. Diabetes 35(8):943–946

Ohkubo Y, Kishikawa H, Araki E et al (1995) Intensive insulin therapy prevents the progression of diabetic microvascular complications in Japanese patients with non-insulin-dependent diabetes mellitus: a randomized prospective 6-year study. Diabetes Res Clin Pract 28 (2):103–117

Omer A, Duvivier-Kali VF, Trivedi N, Wilmot K, Bonner-Weir S, Weir GC (2003) Survival and maturation of microencapsulated porcine neonatal pancreatic cell clusters transplanted into immunocompetent diabetic mice. Diabetes 52(1):69–75

Omori T, Nishida T, Komoda H et al (2006) A study of the xenoantigenicity of neonatal porcine islet-like cell clusters (NPCC) and the efficiency of adenovirus-mediated DAF (CD55) expression. Xenotransplantation 13(5):455–464

Pappalardo D, Mathisen TR, Finne-Wistrand A (2019) Biocompatibility of resorbable polymers: a historical perspective and framework for the future. Biomacromolecules 20(4):1465–1477

Paques JP, van der Linden E, van Rijn CJ, Sagis LM (2014) Preparation methods of alginate nanoparticles. Adv Colloid Interf Sci 209:163–171

Pawar SN, Edgar KJ (2012) Alginate derivatization: a review of chemistry, properties and applications. Biomaterials 33(11):3279–3305

Peloso A, Dhal A, Zambon JP et al (2015) Current achievements and future perspectives in whole-organ bioengineering. Stem Cell Res Ther 6(1):107

Petersen MC, Shulman GI (2018) Mechanisms of insulin action and insulin resistance. Physiol Rev 98 (4):2133–2223

Pociot F, Lernmark Å (2016) Genetic risk factors for type 1 diabetes. Lancet 387(10035):2331–2339

Posselt AM, Szot GL, Frassetto LA et al (2010) Islet transplantation in type 1 diabetic patients using calcineurin inhibitor-free immunosuppressive protocols based on T-cell adhesion or costimulation blockade. Transplantation 90(12):1595

Rangel EB (2014) Tacrolimus in pancreas transplant: a focus on toxicity, diabetogenic effect and drug–drug interactions. Expert Opin Drug Metab Toxicol 10 (11):1585–1605

Registry CIT (2017a) Inaugural report on autologous islet transplantation, Rockville, MD

Registry CIT (2017b) Collaborative islet transplant registry tenth annual report. The Emmes Corporation Rockville, MD: CITR Coordinating Center; January 6, 2017

Rewers M, Ludvigsson J (2016) Environmental risk factors for type 1 diabetes. Lancet 387 (10035):2340–2348

Richter B, Neises G (2003) 'Human'insulin versus animal insulin in people with diabetes mellitus. Cochrane Database Syst Rev 3:CD003816

Riopel M, Trinder M, Wang R (2015) Fibrin, a scaffold material for islet transplantation and pancreatic endocrine tissue engineering. Tissue Eng Part B Rev 21 (1):34–44

Ris F, Toso C, Veith FU, Majno P, Morel P, Oberholzer J (2004) Are criteria for islet and pancreas donors sufficiently different to minimize competition? Am J Transplant 4(5):763–766

Rotem A, Avni Y (2017) Systems and methods for providing oxygen to transplanted cells. In: Google patents. Google, Mountain View, CA

Ryan EA, Lakey JR, Rajotte RV et al (2001) Clinical outcomes and insulin secretion after islet transplantation with the Edmonton protocol. Diabetes 50(4):710–719

Ryan EA, Paty BW, Senior PA et al (2005) Five-year follow-up after clinical islet transplantation. Diabetes 54(7):2060

Safley SA, Kenyon NS, Berman DM et al (2018) Microencapsulated adult porcine islets transplanted intraperitoneally in streptozotocin-diabetic non-human primates. Xenotransplantation 25(6):e12450

Sakata N, Aoki T, Yoshimatsu G et al (2014) Strategy for clinical setting in intramuscular and subcutaneous islet transplantation. Diabetes Metab Res Rev 30 (1):1–10

Sandrin M, Mckenzie IF (1994) Galα (1, 3) gal, the major xenoantigen (s) recognised in pigs by human natural antibodies. Immunol Rev 141(1):169–190

Schuurman HJ (2011) Xenotransplantation: from the lab to the clinic: sunrise symposium at the XXIII international congress of the transplantation society, Vancouver, Canada, August 2010. Clin Transpl 25(4):E415–E421

Shapiro AJ, Lakey JR, Ryan EA et al (2000) Islet transplantation in seven patients with type 1 diabetes mellitus using a glucocorticoid-free immunosuppressive regimen. N Engl J Med 343(4):230–238

Shapiro AJ, Gallant HL, Hao EG et al (2005) The portal immunosuppressive storm: relevance to islet transplantation? Ther Drug Monit 27(1):35–37

Shapiro AJ, Ricordi C, Hering BJ et al (2006) International trial of the Edmonton protocol for islet transplantation. N Engl J Med 355(13):1318–1330

Shapiro AMJ, Pokrywczynska M, Ricordi C (2017) Clinical pancreatic islet transplantation. Nat Rev Endocrinol 13(5):268–277

Smink AM, Li S, Hertsig DT et al (2017) The efficacy of a prevascularized, retrievable poly (d, l,-lactide-co-ε-caprolactone) subcutaneous scaffold as transplantation site for pancreatic islets. Transplantation 101(4): e112–e119

Smink AM, de Haan BJ, Lakey JR, de Vos P (2018) Polymer scaffolds for pancreatic islet transplantation—Progress and challenges. Am J Transplant 18 (9):2113–2119

Smith KE, Purvis WG, Davis MA et al (2018) In vitro characterization of neonatal, juvenile, and adult porcine islet oxygen demand, β-cell function, and transcriptomes. Xenotransplantation 25(6):e12432

Sondermeijer HP, Witkowski P, Woodland D et al (2016) Optimization of alginate purification using polyvinylidene difluoride membrane filtration: effects on immunogenicity and biocompatibility of three-dimensional alginate scaffolds. J Biomater Appl 31 (4):510–520

Tekin Z, Garfinkel MR, Chon WJ et al (2016) Outcomes of pancreatic islet Allotransplantation using the Edmonton protocol at the University of Chicago. Transplant Direct 2(10):e105–e105

Thorens B (2015) GLUT2, glucose sensing and glucose homeostasis. Diabetologia 58(2):221–232

Thu B, Bruheim P, Espevik T, Smidsrød O, Soon-Shiong-P, Skjåk-Bræk G (1996) Alginate polycation microcapsules: I. interaction between alginate and polycation. Biomaterials 17(10):1031–1040

Tian J, Clere-Salzler M, Herschenfeld A et al (1996) Modulating autoimmune responses to GAD inhibits disease progression and prolongs islet graft survival in diabetes–prone mice. Nat Med 2(12):1348–1353

Umpierrez G, Korytkowski M (2016) Diabetic emergencies—ketoacidosis, hyperglycaemic hyperosmolar state and hypoglycaemia. Nat Rev Endocrinol 12(4):222

Valdés-González RA, Dorantes LM, Garibay GN et al (2005) Xenotransplantation of porcine neonatal islets of Langerhans and Sertoli cells: a 4-year study. Eur J Endocrinol 153(3):419–427

Valdes-Gonzalez R, Rodriguez-Ventura A, White D et al (2010) Long-term follow-up of patients with type 1 diabetes transplanted with neonatal pig islets. Clin Exp Immunol 162(3):537–542

Van Der Windt DJ, Bottino R, Kumar G et al (2012) Clinical islet xenotransplantation: how close are we? Diabetes 61(12):3046–3055

Williams PW (1894) Notes on diabetes treated with extract and by grafts of sheep's pancreas. Br Med J 2:1303–1304

Wynyard S, Nathu D, Garkavenko O, Denner J, Elliott R (2014) Microbiological safety of the first clinical pig islet xenotransplantation trial in New Zealand. Xenotransplantation 21(4):309–323

Zondervan G, Hoppen H, Pennings A, Fritschy W, Wolters G, van Schilfgaard R (1992) Design of a polyurethane membrane for the encapsulation of islets of Langerhans. Biomaterials 13(3):136–144

Pancreatic Transplantation in Diabetes: Indications, Contraindications and Perspectives

49

Vinicius Rocha-Santos and Carlos Andres Pantanali

Contents

Introduction 666

Indication 666
Type 1 Diabetes 666
Type 2 Diabetes (T2DM) 666

Contraindications 668
Cardiac and Pelvic Vascular Assessment 668
Surgical Complications 668
Cohort Study 669
Age 670
Obesity 671
Donors 671

Perspectives 672
Robotic Pancreas Transplantation 672
β Cell Replacement 672
Pancreas Islet Transplantation 672

References 673

Abstract

Pancreas transplantation (PT) is the only currently available therapeutic modality capable of definitely treating type I diabetes mellitus (T1DM). The patient can recover endogenous pancreatic function avoiding continuous use of exogenous insulin. Furthermore, stable blood glucose allows avoiding severe complications associated with diabetes. Simultaneous pancreas-kidney transplantation presents better results than isolated pancreas or kidney transplants, being followed by lower incidence of perioperative morbidity such as graft thrombosis and lower immune rejection. These and other modalities, including robotic surgery, stem (beta) cell implants and bioprinting/organoid techniques, are discussed in the chapter.

Keywords

Pancreas transplantation · Pancreas kidney transplantation · Pancreas liver

V. Rocha-Santos (✉) · C. A. Pantanali
Liver and Gastrointestinal Transplant Division, Department of Gastroenterology, University of Sao Paulo Medical School, São Paulo, Brazil
e-mail: vinicius.rocha@hc.fm.usp.br; dr_vinicius@uol.com.br

J. Faintuch and S. Faintuch (eds.), *Obesity and Diabetes*, https://doi.org/10.1007/978-3-030-53370-0_49

transplantation · Pancreatic thrombosis · Type 1 diabetes · Type 2 diabetes

Introduction

The first pancreatic transplantation (PT) (Redfield et al. 2015a) was performed by William Kelly and Richard Lilley at the Minnesota University in 1966 (Kelly et al. 1967). The patient was a young female with T1DM in association with end-stage renal failure. She died a couple of days following transplantation. The 1970s were called "the partial graft period" in the United States and Europe due to the implantation of a partial body and tail pancreas graft in the iliac vessels of the recipient (Gliedman et al. 1973; Dubernard et al. 1978; Sollinger et al. 1983). Only in the middle of the 1980s with the introduction of cyclosporine and the creation of the International Pancreas Transplantation Association (IPTA) did PT treatment for T1DM increase patient and graft survival, becoming an efficient treatment (Starzl et al. 1981; Wadstrom et al. 1995; Brayman et al. 1994; Cantarovich et al. 1997; Stratta et al. 1994, 1996; Knechtle et al. 1991).

T1DM causes renal failure in around 9% of the patients, who should undergo replacement renal therapy such as dialysis or kidney transplant (Mogensen et al. 1983a; de Boer et al. 2007). T1DM patients who are in hemodialysis and have no medical contraindication should be submitted to kidney and pancreas transplant, for better survival. It can be done with a living kidney donor first, followed by cadaver PT alone. Or the simultaneous transplantation of both organs can be performed from the same deceased donor (Gruessner and Gruessner 2018).

In early times PT was associated with a high number of postoperative complications that caused pancreas graft loss in 8–30% of the recipients. In the last two decades advances in surgical techniques, improvement in the immunosuppressor field and better management of postoperative complications contributed to pancreas graft survival comparable to kidney and liver transplants (Redfield et al. 2015b; Lindahl et al. 2013, 2014; Sollinger et al. 2009).

Indication

There are advantages for PT when compared with other therapies in patients with T1DM and end-stage renal disease (Dean et al. 2017; Dunn 2014). It is well recognized that PT is the gold standard treatment to achieve exogenous insulin independence. PT recipients present adequate glucose control, avoiding secondary diabetic complications encompassing retinopathy, nephropathy, and neuropathy (Redfield et al. 2015a; Boggi et al. 2013; Lombardo et al. 2017).

Type 1 Diabetes

Around 425 million of adults present different types of diabetes worldwide. One million children and adolescents have T1DM with a higher incidence in children under 15 years. Geographical variation occurs, with 38.4 cases per 100,000 individuals in Finland, 7.6 in Brazil, and just 0.5 in Korea (International Diabetes Federation 2018).

Microvascular complications such as retinopathy, neuropathy, and nephropathy occur in patients with worse glycemic controls suggesting that glycated hemoglobin has a direct relationship with morbidity (Groop et al. 2009). According to Morgensen et al., in a large number of patients (30–50%) renal function deteriorates (Mogensen et al. 1983b). The author classified kidney injury in five phases: the first one presents proteinuria only whereas the last phase, occurring in 9% of the patients, is considered end-stage renal failure. At this time, only replacement therapy by means of dialysis or kidney transplant can be a lifesaving option.

Type 2 Diabetes (T2DM)

Type 2 diabetes accounts for around 90% of the cases of diabetes, being the cause of renal failure in 30% of the kidney transplant waiting lists (Stumvoll et al. 2005). Pancreas transplant has been performed in nearly 10%, representing 90% of the total of PT (Gruessner and Gruessner 2016). The most import issue in T2DM with end-stage renal disease is a refined selection of these patients. Currently, even in insulin-dependent patients with measurable C-peptide, criteria similar to T1DM cases selected for PT have been applied. Comorbidities, body mass index (BMI), and age are some of the benchmark variables utilized for these selections (Light and Tucker 2013; Light and Barhyte 2005; Stratta et al. 2015).

In the United States, the United Network for Organ Sharing (UNOS) helped the pancreas transplantation society to regulate T2DM patients that should be selected for PT. The criteria are insulin-dependence with positive C-peptide (at least 2 ng/ml), and maximum BMI of 30 kg/m^2, depending of the percentage of the T2DM patients in waiting list. BMI and C-peptide limits in T2DM candidates are still controversial, and debates go on.

The International Pancreas Transplant Registry (IPTR) evaluated the results in T2DM recipients (Gruessner et al. 2017). The data was divided over three eras: 1995–2001, 2002–2008, and 2009–2015. PT in T2DM was performed in 1514 patients. The vast majority of the recipients (97%) were submitted to pancreas transplantation in association with kidney transplant, either simultaneously or pancreas transplant after kidney transplant. The pancreas graft survival improved significantly after 1 and 3 years from 80.2% and 70.5% in era 1–89.0% and 83.3% in era 2 (Al-Qaoud et al. 2018).

Simultaneous Pancreas Kidney (SPK)

The International Registry of Pancreas Transplantation reported more than 50,000 pancreas transplants described by the end of 2016. According to Gruessner et al., from 2011 to 2016, 84% of total PTs were SPKs (Gruessner and Gruessner 2018). In the study from Indiana University, SPKs made up 63% of PT modalities, for those aged 30–39 (Shah et al. 2013). In fact, SPK is well accepted as the main therapy in type 1 diabetes patients with end-stage renal disease, on dialysis or pre-dialysis with glomerular filtration rate (GFR) <20 mL/min/1.73 m^2 (Gruessner and Gruessner 2018). These patients should undergo kidney transplant simultaneously from the same deceased donor. The kidney graft presents better survival when performed with simultaneous PT due to diabetes control, in comparison with cadaveric kidney alone (Gruessner and Sutherland 2005). SPK is particularly recommended in T1DM candidates who lack an available living donor for isolated kidney transplantation first.

Perioperative morbidity such as graft thrombosis and immune rejection during a 3-year period are lower (Gruessner and Sutherland 2005). We can speculate that grafts (pancreas and kidney) from the same donor can reduce exposure to different antigens decreasing the risk of immunological rejection. The anticoagulant effects of uremia could also explain the lower risk of vascular thrombosis in this scenario.

At the Liver and Gastrointestinal Transplant Division, Department of Gastroenterology, University of Sao Paulo Medical School, 89.7% of the pancreas recipients were simultaneous pancreas kidney transplants. In a cohort study, there were no differences between the three transplant modalities concerning graft loss due to thrombosis. However, we notice that global graft failure seems to be higher in the pancreas after kidney modality (Rocha-Santos 2018).

Pancreas After Kidney (PAK)

PAK is the second main modality of PT. It is considered for T1DM patients who have already undergone kidney transplant. These kidney recipients are already under immunosuppression, and can be evaluated for PAK in order to obtain the potential benefits of sustained euglycemia. The selection criteria of patients for PAK include normal renal graft function and it can be performed 4–6 months after the kidney transplant.

In Brazil, the waiting list for a pancreas graft follows chronological criteria including blood type. At Sao Paulo city, the waiting time for SPK transplantation can be greater than 2 years, extending dialysis therapy and increasing risk of death. In this context when there are living kidney donors, kidney transplantation should be conducted first, and then a cadaveric PT. PAK presents slightly worse results when compared with SPK, concerning graft thrombosis and immunological rejection (Gruessner and Gruessner 2018).

Pancreas Transplantation Alone (PTA)

PTA is the minor modality of PT. It is considered for patients with labile T1DM usually characterized by an endocrinologist, including acute and severe metabolic complications such as hypoglycemia, hyperglycemia, or ketoacidosis requiring medical attention; clinical and emotional issues due to exogenous insulin therapy, that are so severe as to be incapacitating; and consistent failure of insulin-based management to prevent acute complications. GFR>60 mL/min/1.73 m^2 is required, since nephrotoxicity caused by calcineurin inhibitors can be associated with progressive nephropathy, accelerating the advent of end-stage renal failure. In such circumstances, subsequent need for a kidney transplant can be complicated by the patient's hypersensitization due to previous PTA. Thus a solitary transplant should be considered early on, before the patient develops end-stage renal disease (Niederhaus 2015; Gruessner and Gruessner 2013).

The number of patients submitted to PTA remains relatively stable, and survival was the highest compared with other categories. Such outcome notwithstanding, there is still hesitancy to consider PTA without severe diabetic secondary complications, given the postoperative surgical and immunosuppression risks.

Contraindications

As pancreas transplant presents a high level of perioperative complications, even in young patients, preoperative evaluation is very important to select the best recipient (Stockland et al. 2009; Fridell et al. 2011; Gilabert et al. 2002). Patients able to tolerate medical or surgical complications should be preferred. Reoperations can be needed, and some candidates suffer from hemodynamic lability due to long periods of diabetes, needing high doses of vasopressors. On the other hand, the waiting list mortality for PT candidates submitted to dialysis has been described as 6.6% at one year and 54% at 5 years (Gruessner et al. 2004). A multidisciplinary transplantation team is essential to critically compare mortality risk in the waiting list versus morbidity and mortality in the postoperative period.

Malignancies, chronic and acute infections, and severe cardiovascular morbidity are contraindications for PT. Older recipients above 50–55 years of age should be avoided due to the long period of diabetes. On the other hand when PT is pondered, biological age could be more relevant than chronological age (Siskind et al. 2014a). Obese patients also present worse post-transplant results.

Cardiac and Pelvic Vascular Assessment

Preoperative cardiovascular evaluation should include cardiac stress testing and cardiac catheterization. T1DM patients, due to the long period of diabetes, often suffer from severe atherosclerosis, and cardiovascular complications are the main cause of death after pancreas transplantation (Samoylova et al. 2019). In the preoperative evaluation, we routinely perform Doppler ultrasonography of the iliac vessels. Upon any alteration of the flow or caliber of the vessel, a noncontrast computed tomography scanning of the pelvis is requested to evaluate the extent of calcification of the iliac vessels. This strategy permits the choice of the best preoperative strategy for PT implantation or even, in case of severe stenosis, to rule out PT.

Psychosocial status should also be evaluated, as these candidates will require permanent treatment after hospital discharge.

Surgical Complications

Pancreas graft failure is higher during the first year of PT. Among the complications of the first 60 days pancreatitis occurs in 3–12%, wound

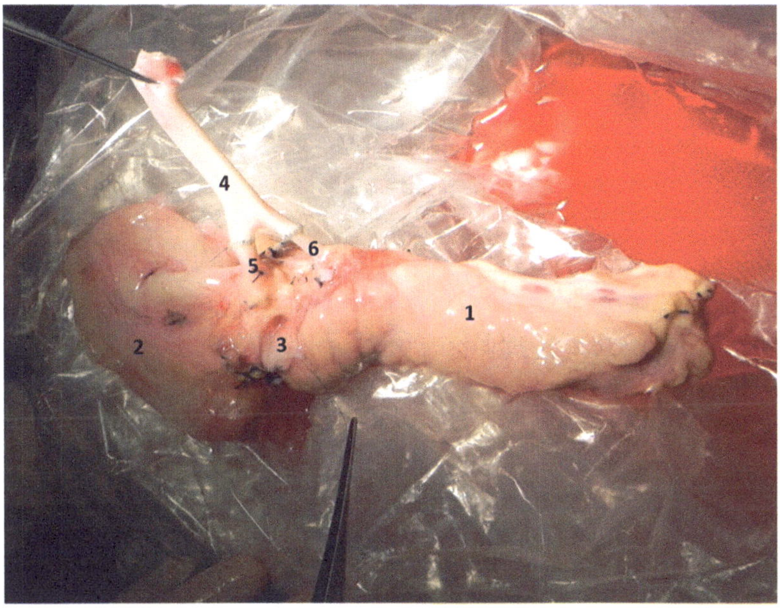

Fig. 49.1 Pancreas graft: (1) body of the pancreas, (2) duodenum, (3) portal vein, (4) y-graft, (5) superior mesenteric artery and (Starzl et al. 1981) splenic artery

infection or abscess in 1–5%, and focal or diffuse pancreas graft necrosis in 12% of the recipients (Grochowiecki et al. 2014).

Major complications, including graft thrombosis, have been demonstrated in 6–17% of PT, duodenal fistula in 0.5–2%, and intra-abdominal hemorrhage in 0.5% of the patients (Grochowiecki et al. 2014; Sansalone et al. 2005; Michalak et al. 2005; Gruessner et al. 2010; Malaise et al. 2005). Reoperations, abdominal wall dehiscence, infection, malnutrition, and graft failure with renal dysfunction and death are also listed. Thrombosis is the most important cause of pancreas graft loss.

Cohort Study

In our Service, the standard surgical technique for pancreas transplantation was adopted (Fig. 49.1), and risk factors that lead to early pancreas graft loss due to thrombosis were the aim (Fig. 49.2). We evaluated the records of 137 individuals in a retrospective design, from March 2000 to May 2017 (Table 49.1).

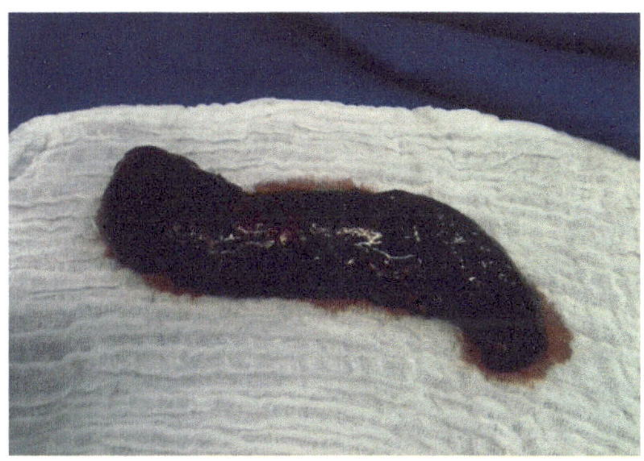

Fig. 49.2 Pancreatic thrombosis

Thirty-five transplants (25.5%) had cold ischemia time over 14 h; in 44 recipients (32.1%) it was less than 10 h, and in the other 58 patients (42.3%) the range was 10–12 h. Fourteen patients (10.2%) underwent isolated transplants, with 13 pancreas transplants after kidney and one PTA (Rocha-Santos 2018). In all univariate analyses, the experience of the harvest team ($p = 0.04$) was the only feature that achieved significance. With multiple logistic regression in addition to the team ($p = 0.021$), age of the recipient less than 30 years, donor intensive care

Table 49.1 Clinical characteristics of the 137 recipients who underwent pancreas transplantation

Gender	*N*	%	CI 95%
Male	83	60.6	52.3–68.5
Female	54	39.4	31.5–47.7
BMI			
Underweight	14	10.2	6.0–16.1
Normal	95	69.3	61.3–76.6
Overweight or obese	28	20.4	14.3–27.8
Dialysis			
No	11	8.0	4.3–13.5
Yes	117	85.4	78.8–90.6
Without information	9	6.6	3.3–11.6
Type of dialysis			
Fistula	90	71.4	63.1–78.8
Catheter	15	11.9	7.1–18.4
Peritoneal	6	4.8	2.0–9.6
Without information	15	11.9	7.1–18.4

N Number of patients, *CI 95%* 95% confidence interval

Table 49.2 Multiple logistic regression analysis of factors influencing graft loss by thrombosis

Variable	*B*	Standard error	*p* value	OR/CI 95%
Age				
≤29 years				1.00
30–40 years	−2.58	1.038	0.013	0.08 (0.01–0582)
≥40 years	−2.16	0.901	0.017	0.12 (0.02–0.676)
Donor ICU stay				
≤10 days				1.00
>10 days	0.21	0.113	0.05	1.24 (0.99–1.542)
Experienced team				
No				1.00
Yes	−1.75	0.881	0.046	0.17 (0.031–0.973)
Donor Glucose	0.01	0.004	0.023	1.01 (1.001–1.019)

B Beta coefficient, *OR (CI 95%)* odds ratio and 95% confidence interval

unit time ($p = 0.033$) and donor blood glucose ($p = 0.023$) were significant (Table 49.2). Pancreas graft survival was 86.7%, 85.7%, and 85.7% after 1, 5, and 10 years respectively. Patient survival at 1, 5, and 10 years was 88.0%, 82.5%, and 78.5%, respectively.

In the late postoperative period, patients can stay euglycemic for a period superior to 10 years. Even when late failure occurs due to so-called "immunological graft loss," there is little invasive surgery risk. Only rarely will the transplanted patient need to undergo a new laparotomy (Sansalone et al. 2005; Michalak et al. 2005; Gruessner et al. 2010).

Age

Most centers worldwide limit PT for patients younger than 50 years (Shah et al. 2013; Sener et al. 2010). According to IPTR, only 2% were performed in those older than 60 years (Fourtounas 2014). Some studies recommend that pancreas transplantation should be performed in patients younger than 45 years, in order to decrease postoperative morbidity and mortality (Freise et al. 1998). On the other hand, the chronological age cannot express the real health condition of the patients. In fact, the upper age cutoff point has been questioned (Heldal et al. 2010; Schmitt et al. 2009).

At the University of Indiana, recipients older than 60 years with simultaneous pancreas kidney transplantation presented no statistical difference concerning postoperative morbidity and cardiovascular events (Shah et al. 2013). In contrast, in the experience of Strata et al. older recipients presented higher graft loss. They seem to die more often due to cardiac events in contrast to younger recipients, who in turn suffer higher immunological graft loss (Tullius et al. 2010).

The higher risk of death due to preoperative morbidity, compared to the risk of immunological pancreas graft loss in young patients, should be balanced. If more aged patients survive the postoperative period with no cardiac events, graft function can increase patient survival for a long period of time.

Outcomes by age using the UNOS database indicated that out of 280 patients over 60 years old who underwent pancreas transplantation, 154 (55%) were submitted to SPK. Survival was shorter in older patients, however, with minimal differences between the various age groups. No difference in one-year graft survival was observed for any age group. Those with 40–49 years achieved 67.8% 5-year graft survival, among those with 50–59 years survival was 59.9%, and for the 18- to 29-year bracket it was not better than 56.8% (Siskind et al. 2014b).

In our series age was mostly 25–40 years and BMI 18–25 kg/m^2. This is a safe range for both age and BMI. Only 5 individuals (3.6%) were older than 50 years. Surprisingly, in the multiple logistic regression model, those under age of 30 had a greater possibility of graft loss. We speculate that although the arteries for the implant are healthier in younger patients, coagulation cascade could be more active. The intrinsic thrombophilia of the diabetic could be the main risk factor, without technical interference (Rocha-Santos 2018).

Obesity

The ONUS database evaluated simultaneous pancreas kidney transplants as related to obesity (Zalewska and Ploeg 2014). Among 5725 patients BMI <25 corresponded to 56%, 33% were overweight (BMI 25–29.9), and 11% obese (BMI 30–40). Postoperative transplant complications, including delayed kidney graft function, pancreas graft thrombosis, and mortality, were higher in the obese group.

The growing prevalence of obesity in the general population accelerates insulin resistance in the T1DM population. In most studies, obesity of the recipient is a risk factor for graft loss (Bumgardner et al. 1995). A few authors have published good results in low-grade obesity (Laurence et al. 2015). In our experience with 23 patients (16.8%) exhibiting BMI above 25, no increased risk for graft thrombosis emerged. In contrast, 16 patients (11.6%) had a BMI below 18. The effects of malnutrition on graft thrombosis are poorly studied.

Donors

There is a consensus that donor advanced age, obesity, certain causes of brain death, and presence of shock are risk factors for thrombosis of the pancreas graft (Fridell et al. 2011). In a multivariate analysis, Troppman et al. concluded that donor age and cardiovascular cause of brain death increase the risk of vascular thrombosis of the pancreas (Troppmann et al. 1996).

These variables are so important that scales assessing the risk of graft loss according to donor features were published. Axelrod et al. analyzed data from 9401 transplants and calculated a donor risk index for graft one-year survival. Variables were age, BMI, height, serum creatinine, gender, ethnicity (African-American, Asian, Caucasian), stroke, donation after cardiac death, transplantation modality. and cold ischemia time. The reference donor was male, 28 years old, white, with a BMI of 24, height of 173 cm, absence of stroke, serum creatinine less than 2.5 mg/dl, pancreas not donated after cardiac death or pancreatitis, with up to 12 h of cold ischemia (Axelrod et al. 2010).

The Eurotransplant Group pre-acquisition pancreas adequacy score (P-PASS considers donor age, body mass index, admission to the intensive care unit (ICU), cardiac arrest, serum sodium, serum amylase, serum lipase, and use of catecholamines. The study concludes that high-score P-PASS donor grafts reduced the risk of organ loss by up to 3 times. However, several studies have shown that these risk scores are not necessarily appropriate for other populations (Schenker et al. 2010; Blok et al. 2016; Woeste et al. 2010).

Perspectives

Robotic Pancreas Transplantation

The first robotic pancreas transplant was performed in 2012 by Boggi et al. (2012). Although they reported only two patients, they proved that it was feasible despite technical challenges, with lower surgical morbidity. On the other hand, ischemia time was higher. Large series are now available (Spaggiari et al. 2013). Robotic transplant surgery is standard at a small number of pioneering centers, (Stiegler and Schemmer 2018) not being routinely adopted.

β Cell Replacement

PT for T1DM includes the exocrine pancreas, which can precipitate pancreatitis and duodenal and pancreatic fistula. New perspectives aim to bypass the exocrine portion of the pancreas.

Pancreas Islet Transplantation

From 1990 to 1999, the International Islet Transplant Registry informed a one-year islet survival of 41%, with 11% insulin free (Brendel et al. 2001). The Edmonton immunosuppression protocol in 2000 improved the outlook. It included corticosteroid-free immunosuppression and islet infusion from different donors (Shapiro et al. 2000). Seven patients became insulin free (1 year), which to some extent continued thereafter (5 years) (Shapiro et al. 2000; Ryan et al. 2005). The technique is accepted as a valid option for beta cell replenishment (Rheinheimer et al. 2015).

A transhepatic catheter inoculates islets into the portal vein. Surgical catheterization of the mesenteric vein is also possible (Mallett and Korbutt 2009). A receptive vascular bed with little rejection and long-term graft function would be ideal; (Rajab 2010) however, few locations fulfill such requirements.

The stomach submucosa, chosen for our experimental study, has already shown a reduced inflammatory response (Echeverri et al. 2009). It is a well-vascularized region with easy approach for the graft and subsequent analysis, which in larger organisms could be done by endoscopy (Echeverri et al. 2009; Wszola et al. 2009). Adequate glycemic control and reduced immunological response are also expected (Xinnong et al. 2014). In our experience, five rats were inoculated in the gastric body submucosa, and four in the fundus submucosa. No immunosuppression was used. The gastric body location was associated with better glycemic control (de Mesquita et al. 2018).

The omentum also enjoys convenient vascularization, and blood flow straight to the portal vein. An in situ adherent, resorbable plasma-thrombin biological scaffold has already been tested in rats. The subcutaneous area could not be overlooked as a potential inoculation region either, as it is convenient, safe, and compatible with easy graft removal, if necessary. Unfortunately blood supply is insufficient (Berman et al. 2016).

Stem Cell-Derived Beta Cells

The B cell replacement from whole organ transplantation continues to be the gold standard therapy for T1DM with end-stage kidney failure. Although this invasive therapy can provide insulin-free independence, it is well recognized the great number of postoperative complication leading a high number of morbidity and mortality. On the other hand, the B cell replacement from islet transplantation presents a higher benign postoperative course, although the insulin-free independence is poor and patients present the morbidity associated with the immunosuppression.

The results displayed above open the doors for others options for B cell replacement. The International Pancreas and Islet Transplant Association, in collaboration with the Harvard Stem Cell Institute, the Juvenile Diabetes Research Foundation (JDRF), and the Helmsley Foundation, held a 2-day Key Opinion Leaders Meeting in Boston in 2016 (Odorico et al. 2018).

Advances concerning the successful transplantation of stem cell–derived beta cells for diabetic patients have been remarkable. The steady generation of pure populations of fully functional beta cells remains obscure and according to this meeting, the question persists as to whether pure or mix of beta cells with other islet endocrine cells can provide superior presentation. Maybe it will be necessary to create an immunoisolation strategy that offers a hospitable environment while preventing immune-mediated injury and to define an ideal local for its location. Despite these challenge, development has been done since several teams working together in early phase trials. Moreover, the growth in the field of genome can help this issue in the future by creating better disease models and producing safer and more effective health cell mass. The meeting concluded that "It seems inevitable that stem cell–derived beta cells will play an important role in the care of diabetic patients within the next decade." (Odorico et al. 2018).

3D Bioprinting

The 3D bioprinting method employs degradable biopolymer scaffolds mimicking the islet architecture (Wang and Rosenberg 1999). Hydrogel capsules provide a semi-permeable membrane with favorable oxygen and nutrient supply (Gibly et al. 2011). Despite some scaffold loss, extracellular matrix proteins are deposited, enhancing islet survival (Berman et al. 2009).

Loss of extracellular matrix interactions is deleterious for beta cell survival. In another protocol in our laboratory, extracellular protein laminin (LN) was combined with a biodritin (alginate) microencapsulation procedure. Satisfactory stability and biocompatibility were achieved in a mouse transplantation model, with long-term islet survival (Campanha-Rodrigues et al. 2015).

References

Al-Qaoud TM, Odorico JS, Redfield RR III (2018) Pancreas transplantation in type 2 diabetes: expanding the criteriaTalal M. Curr Opin Organ Transplant 23:454–460

Axelrod DA, Sung RS, Meyer KH et al (2010) Systematic evaluation of pancreas allograft quality, outcomes and geographic variation in utilization. Am J Transplant 10:837–845

Berman DM, O'Neil JJ, Coffey LCK, Chaffanjon PCJ, Kenyon NM, Ruiz P et al (2009) Long-term survival of nonhuman primate islets implanted in an omental pouch on a biodegradable scaffold. Am J Transplant. 9:91–104

Berman DM, Molano RD, Fotino C et al (2016) Bioengineering the endocrine pancreas: intraomental islet transplantation within a biologic resorbable scaffold. Diabetes. 65(5):1350–1361

Blok JJ, Kopp WH, Verhagen MJ et al (2016) The Value of PDRI and P-PASS as predictors of outcome after pancreas transplantation in a Large European Pancreas Transplantation Center. Pancreas 45:331–336

Boggi U, Signori S, Vistoli F et al (2012) Laparoscopic robot-assisted pancreas transplantation: first world experience. Transplantation 93:201–206

Boggi U, Rosati CM, Marchetti P (2013) Follow-up of secondary diabetic complications after pancreas transplantation. Curr Opin Organ Transplant 18(1):102–110

Brayman KL, Egidi MF, Naji AL et al (1994) Is induction therapy necessary for successful simultaneous pancreas and kidney transplantation in the cyclosporine era? Transplant Proc 26:2525–2527

Brendel MD HB, Schultz AO, Bretzel RG (2001) International islet transplant registry. Newsletter # 9. University Hospital Giessen. http://www.med.uni-giessen.de/itr/newsletter/no_9/news_9.pdf. Accessed June 3 2018

Bumgardner GL, Henry ML, Elkhammas E et al (1995) Obesity as a risk factor after combined pancreas/kidney transplantation. Transplantation. 60:1426–1430

Campanha-Rodrigues AL, Grazioli G, Oliveira TC, Campos-Lisbôa AC, Mares-Guia TR, Sogayar MC (2015) Therapeutic potential of laminin-biodritin microcapsules for type 1 diabetes mellitus. Cell Transplant 24(2):247–261

Cantarovich D, Karam G, Hourmant M et al Randomized comparison of triple therapy venus anti-thymocyte globulin induction following combined pancreas-renal transplantation. The International Congress on Immunosuppression, Orlando, Florida, 11--13 December 1997

de Boer IH et al (2007) Central obesity, incident microalbuminuria, and change in creatinine clearance in the epidemiology of diabetes interventions and complications study. J am Soc nephrol 18:235–243

de Mesquita GHA, Jardim YJ, Iuamoto LR, Suguita FY, Essu FF, Oliveira LT, Meyer A, Crescenzi A, Rocha-Santos V, Galvão FHF, Andraus W, Chaib E, D'Albuquerque LAC (2018, Feb 9) Gastric fundus submucosa as a site for islets transplantation: an experimental study. Ann Med Surg (Lond) 28:1–5

Dean PG, Kukla A, Stegall MD et al (2017) Pancreas transplantation. BMJ 357:j1321

Dubernard JM et al (1978) A new method of preparation of segmental pancreatic grafts for transplantation: trials in dogs and in man. Surgery 84:633–640

Dunn TB (2014) Life after pancreas transplantation: reversal of diabetic lesions. Curr Opin Organ Transplant 19 (1):73–79

Echeverri GJ, McGrath K, Bottino R, Hara H, Dons EM, Van Der Windt DJ, Ekser B, Casu A, Houser S, Ezzelarab M, Wagner R, Trucco M, Lakkis FG, Cooper DKC (2009) Endoscopic gastric submucosal transplantation of islets (ENDO-STI): technique and initial results in diabetic pigs. Am. J. Transplant. 9:2485–2496

Fourtounas C (2014) Transplant options for patients with type 2 diabetes and chronic kidney disease. World J Transplant 4:102–110

Freise CE, Stock PG, Melzer JS (1998) Increased morbidity and mortality of simultaneous pancreas-renal transplantation in patients over 49 years of age. Transplant Proc 30:292

Fridell JA, Mangus RS, Mull AB et al (2011) Early reexploration for suspected thrombosis after pancreas transplantation. Transplantation 91:902–908

Gibly RF, Zhang X, GrahamML HBJ, Kaufman DB, Lowe WL et al (2011) Extrahepatic islet transplantation with microporous polymer scaffolds in syngeneic mouse and allogeneic porcine models. Biomaterials. 32:9677–9684

Gilabert R, Fernandez-Cruz L, Real MI, Ricart MI, Astudillo E, Montana X (2002) Treatment and outcome of pancreatic venous graft thrombosis after kidney–pâncreas transplantation. Br J Surg 89:355–360

Gliedman ML et al (1973) Clinical segmental pancreatic transplantation with ureter-pancreatic duct anastomosis for exocrine drainage. Surgery 74:171–180

Grochowiecki T, Gałązka Z, Madej K et al (2014) Surgical complications related to transplanted pancreas after simultaneous pancreas and kidney transplantation. Transplant Proc 46:2818–2821

Groop PHL et al (2009) The presence and severity of chronic kidney disease predicts al causemortality in type 1 diabetes. Diabetes 58:1651–1658

Gruessner RW, Gruessner AC (2013) Pancreas transplant alone: a procedure coming of age. Diabetes Care 36 (8):2440–2447

Gruessner AC, Gruessner RW (2016) Pancreas transplantation of US and non-US Cases from 2005 to 2014 as reported to the United Network for Organ Sharing (UNOS) and the International Pancreas Transplant Registry (IPTR). Rev Diabet Stud 13:35–58

Gruessner AC, Gruessner RWG (2018) Pancreas transplantation for patients with type 1 and type 2 diabetes mellitus in the United States: a registry report. Gastroenterol Clin N Am 47:417–441

Gruessner AC, Sutherland DE (2005) Pancreas transplant outcomes for United States (US) and non-US cases as reported to the United Network for Organ Sharing (UNOS) and the International Pancreas Transplant Registry (IPTR) as of June 2004. Clin Transplant 19:433–438

Gruessner RWG, Sutherland DER, Gruessner AC (2004) Mortality assessment for pancreas transplants. Am J Transplant 4(12):2018–2026

Gruessner AC, Sutherland DE, Gruessner RW (2010) Pancreas transplantation in the United States: a review. Curr Opin Organ Transplant 15:93–101

Gruessner AC, Laftavi MR, Pankewycz O, Gruessner RWG (2017) Simultaneous pancreas and kidney transplantation: is it a treatment option for patients with type 2 diabetes mellitus? An analysis of the international pancreas transplant registry. Curr Diab Rep 17:44

Heldal K, Hartmann A, Grootendorst DC et al (2010) Benefit of kidney transplantation beyond 70 years of age. Nephrol Dial Transplant 25:1680–1687

International Diabetes Federation (2018) Diabetes Atlas, 8th edn. International Diabetes Federation, Brussels. www.diabetesatlas.org Accessed April 21 2018.

Kelly WD, Lillehei RC et al (1967) Allotransplantation of the pancreas and duodenum along with the kidney in diabetic nephropathy. Surgery 61:827–835

Knechtle SJ, Pirsch JD, Groshek M et al (1991) OKT3 versus ALG induction therapy in combined pancreas-kidney transplantation. Transplant Proc 23:1581–1582

Laurence JM, Marquez MA, Bazerbachi F, Seal JB, Selzner M, Norgate A, McGilvray ID, Schiff J, Cattral MS (2015, Jun) Optimizing pancreas transplantation outcomes in obese recipients. Transplantation. 99 (6):1282–1287

Light JA, Barhyte DY (2005) Simultaneous pancreas-kidney transplants in type I and type II diabetic patients with end-stage renal disease: similar 10-year outcomes. Transplant Proc 37:1283–1284

Light J, Tucker M (2013) Simultaneous pancreas kidney transplants in diabetic patients with end-stage renal disease: the 20-yr experience. Clin Transplant 27: E256–E263

Lindahl JP, Hartmann A, Horneland R et al (2013) Improved patient survival with simultaneous pancreas and kidney transplantation in recipients with diabetic end-stage renal disease. Diabetologia 56:1364–1371

Lindahl JP, Jenssen T, Hartmann A (2014) Long-term outcomes after organ transplantation in diabetic end-stage renal disease. Diabetes Res Clin Pract 105:14–21

Lombardo C, Perrone VG, Amorese G et al (2017) Update on pancreatic transplantation on the management of diabetes. Minerva Med 108(5):405–418

Malaise J, Steurer W, Koenigsrainer A et al (2005) Simultaneous pancreas-kidney transplantation in a large multicenter study: surgical complications. Transplant Proc 37:2859–2860
Mallett AG, Korbutt GS (2009) Alginate modification improves longtermsurvival and function of transplanted encapsulated islets. Tissue Eng Part A 15:1301–1309
Michalak G, Kwiatkowski A, Czerwinski J et al (2005) Surgical complications of simultaneous pancreas-kidney transplantation: a 16-year-experience at one center. Transplant Proc 37:3555–3557
Mogensen CE et al (1983a) The stages in diabetic renal disease. With emphasis on the stage of incipient diabetic nephropathy. Diabetes 32(Suppl 2):64–78
Mogensen CE et al (1983b) The stages in diabetic renal disease. With emphasis on the stage of incipient diabetic nephropathy. Diabetes 32(Suppl 2):64–78
Niederhaus SV (2015) Pancreas transplant alone. Curr Opin Organ Transpl 20(1):115–120
Odorico J, Markmann J, Melton D, Greenstein J, Hwa A, Nostro C et al (2018 August) Report of the key opinion leaders meeting on stem cell-derived beta cells. Transplantation. 102(8):1223–1229
Rajab A (2010) Islet transplantation: alternative sites. Curr. Diabetes Rep. 10:332–337
Redfield RR, Scalea JR, Odorico JS (2015a) Simultaneous pancreas and kidney transplantation: current trends and future directions. Curr Opin Organ Transplant 20:94–102
Redfield RR, Scalea JR, Odorico JS (2015b) Simultaneous pancreas and kidney transplantation: current trends and future directions. Curr Opin Organ Transplant 20:94–102
Rheinheimer J, Bauer AC, Silveiro SP, Estivalet AAF, Bouças AP, Rosa AR et al (2015) Human pancreatic islet transplantation: an update and description of the establishment of a pancreatic islet isolation laboratory. Arch Endocrinol Metab. 59:161–170
Rocha-Santos V (2018) Doctoral Thesis, Liver and Gastrointestinal Transplant Division, Department of Gastroenterology, University of Sao Paulo Medical School
Ryan EA, Paty BW, Senior PA, Bigam D, Alfadhli E, Kneteman NM et al (2005) Five-year follow-up after clinical islet transplantation. Diabetes. 54:2060–2069
Samoylova ML, Borle D, Ravindra KV (2019) Pancreas transplantation indications, techniques, and outcomes. Surg Clin N Am 99:87–101
Sansalone CV, Maione G, Aseni P et al (2005) Surgical complications are the main cause of pancreatic allograft loss in pancreas-kidney transplant recipients. Transplant Proc 37:2651–2653
Schenker P, Vonend O, Ertas N et al (2010) Preprocurement pancreas allocation suitability score does not correlate with long-term pancreas graft survival. Transplant Proc 42:178–180
Schmitt TM, Kumer SC, Pruett TL et al (2009) Advanced recipient age (>60 years) alone should not be a contraindication to liver retransplantation. Transplant Int 22:601–605
Sener A, Cooper M, Bartlett ST (2010) Is there a role for pancreas transplantation in type 2 diabetes mellitus? Transplantation 90:121–123
Shah AP, Mangus RS, Powelson JA et al (2013) Impact of recipient age on whole organ pancreas transplantation. Clin Transplant 27:E49–E55
Shapiro AMJ, Lakey JRT, Ryan EA, Korbutt GS, Toth E, Warnock GL et al (2000) Islet transplantation in seven patients with type 1 diabetes mellitus using a glucocoricoid-free immunosuppressive regimen. N Engl J Med. 343:230–238
Siskind E, Maloney C, Akerman M et al (2014a) An analysis of pancreas transplantation outcomes based on age groupings–an update of the UNOS database. ClinTransplant 28(9):990–994
Siskind E, Maloney C, Akerman M et al (2014b) An analysis of pancreas transplantation outcomes based on age groupings: an update of the UNOS database. Clin Transplant 28:990–994
Sollinger HW et al (1983) Segmental pancreaticallotransplantation with pancreatico-cystostomy and high-dose cyclosporine and low-dose prednisone. Transplant Proc 15:2997
Sollinger HW, Odorico JS, Becker YT, D'Alessandro AM, Pirsch JD (2009) One thousand simultaneous pancreas-kidney transplants at a single center with 22-year follow-up. Ann Surg 250:618–630
Spaggiari M, Tulia KA, Di Bella C et al (2013) Promising outcomes from 5 robotic pancreas transplants in obese patients. Transplantation 96:S1–S155
Starzl TE et al (1981) Liver transplantation with use of cyclosporin a and prednisone. N Engl J Med. 305 (5):266–269
Stiegler P, Schemmer P (2018) Robot-assisted transplant surgery: vision or reality? A comprehensive review. Visc Med 34:24–30
Stockland AH, Willingham DL, Paz-Fumagalli R et al (2009) Pancreas transplant venous thrombosis: role of endovascular interventions for graft salvage. Cardiovasc Intervent Radiol 32:279–282
Stratta RJ, Taylor RJ, Lowell JA et al (1994) OKT3-induction in 100 consecutive pancreas transplants. Mnsplant Proc 26:546–547
Stratta RJ, RJ T, Weide LG et al (1996) A prospective randomized trial of OKT3 versus ATGAM-induction therapy in pancreas transplant recipients. Transplant Proc 28:917–918
Stratta RJ, Rogers J, Farney AC et al (2015) Pancreas transplantation in C-peptide positive patients: does 'type' of diabetes really matter? J Am Coll Surg 220:716–727
Stumvoll M, Goldstein BJ, van Haeften TW (2005) Type 2 diabetes: principles of pathogenesis and therapy. Lancet 365:1333–1346
Troppmann C, Gruessner AC, Benedetti C et al (1996) Vascular graft thrombosis after pancreatic transplantation: univariate and multivariate operative and nonoperative risk factor analysis. J Am Coll Surg 182:285–316

Tullius SG, Tran H, Guleria I et al (2010) The combination of donor and recipient age is critical in determining host immunoresponsiveness and renal transplant outcome. Ann Surg 252:662–674

Wadstrom J, Brekke B, Wramner L et al (1995) Triple versus quadruple induction immunosuppression in pancreas transplantation. Transplant Proc 27:1317–1318

Wang RN, Rosenberg L (1999) Maintenance of beta-cell function and survival following islet isolation requires re-establishment of the islet-matrix relationship. J Endocrinol. 163:181–190

Woeste G, Moench C, Hauser IA et al (2010) Can the preprocurement pancreas suitability score predict ischemia-reperfusion injury and graft survival after pancreas transplantation? Transplant Proc 42:4202–4205

Wszola M, Berman A, Fabisiak M, Domagala P, Zmudzka M, Kieszek R, Perkowska-Ptasinska A, Sabat M, Pawelec K, Kownacki L (2009) Trans Endoscopic Gastric SubMucosa Islet Transplantation (eGSM-ITx) in pigs with streptozotocine induced diabetes—technical aspects of the procedure—preliminary report. Ann. Transplant. 14:45–50

Xinnong L, Jinhang Z, Yang L, Hu S, Guangyonng Z, Wang L (2014) Gastric subserous space islet transplantation: techniques and initial results in diabetic inbred Lewis rats. Ann. Transplant. 19:331–336

Zalewska K, Ploeg R (2014) National standards for organ retrieval from deceased donors. NORS Retrieval Standards, Bristol

Part VI

Pharmacological Therapy and Cost Containment

Pharmacotherapy of Type 2 Diabetes 50

Jaskiran Kaur, Paras Famta, Navneet Khurana, Manish Vyas, and Gopal L. Khatik

Contents

Introduction 680

Major Antidiabetic Drug Families 680
Sulfonylureas 680
Glinides (Meglitinides/D-Phenylalanine Analogs) 681
Biguanides 682
Thiazolidinediones (TZDs) 683
GLP-1 Receptor Agonists (Injectable Drugs) 684
DPP-4 Inhibitors 685
Alpha-Glucosidase Inhibitors 686
Sodium-Glucose Co-transporter-2 Inhibitors/Gliflozins (SGLT-2 Inhibitors) 686
Oral/Nasal Insulin 687

References 692

Abstract

Type 2 diabetes mellitus (T2DM) is the most common form of diabetes, distinguished by insulin resistance and beta-cell dysfunction. Increased body mass index and central adiposity are in a strong relationship with insulin resistance, and are believed to be among the key factors involved in the incidence of T2DM. The prevalence of T2DM is exponentially expanding, and has paralleled that of obesity. The objective of this review is to encapsulate the pharmacology of major antidiabetic drug families along with efficacy, effect on body weight, dose, contraindications, and associated adverse effects. It also focuses on recent pharmacological approaches and drugs under clinical trial, for the successful therapy of T2DM and diabesity in the future.

J. Kaur · N. Khurana
Department of Pharmacology, School of Pharmaceutical Sciences, Lovely Professional University, Phagwara, Punjab, India

P. Famta
Institute of Nano-Science and Technology, Mohali, Punjab, India

M. Vyas
Department of Ayurveda Pharmacy, School of Pharmaceutical Sciences, Lovely Professional University, Phagwara, Punjab, India

G. L. Khatik (✉)
Department of Pharmaceutical Chemistry, School of Pharmaceutical Sciences, Lovely Professional University, Phagwara, Punjab, India
e-mail: gopal_niper@rediffmail.com

J. Faintuch and S. Faintuch (eds.), *Obesity and Diabetes*, https://doi.org/10.1007/978-3-030-53370-0_50

Keywords

Type 2 diabetes · Obesity · Antidiabetic drugs · Pharmacotherapy

Introduction

Type 2 diabetes mellitus (T2DM) can be characterized by insulin resistance and variable insulin secretion that is associated with increased blood glucose levels and glucotoxicity. The number of diabetics globally is expected to reach 642 million by 2040 (Upadhyay et al. 2018). Increased BMI or obesity is found to be in strong association with many pathophysiological conditions, notably T2DM, as both share insulin resistance. Obese individuals can be up to 80 times more likely to develop T2DM in comparison to individuals with BMI < 22 (Al-Goblan et al. 2014; Diabetes.co.uk 2019a). Therapeutic targets include tight control of glycosylated hemoglobin (HbA1c $\leq$ 7.0%) levels, to prevent long-term complications and mortality (Upadhyay et al. 2018). In the case of obesity (BMI $\geq$ 27 for Asian countries, or BMI $\geq$ 30 in the Western world), medications are recommended, and it is a prerequisite to review the weight gain or weight loss effect of antidiabetic medications before prescribing.

A minimum 5% weight loss in obese T2DM patients is desired, and if the goal is not attained, alternative medication with good efficacy and reduced risk of hypoglycemia should be considered (American Association of Diabetes Educators 2016). Weight reduction in overweight and obese individuals could help in improving insulin sensitivity, as well as attenuating the risk of metabolic and cardiovascular complications (Diabetes.co.uk 2019a).

Major Antidiabetic Drug Families

Sulfonylureas

Sulfonylureas (SUs) were the first oral antidiabetic agents approved by FDA, starting in 1958. Their origin is traced to the observation in 1942 that some sulfonamide antibacterial products triggered critical hypoglycemia in typhoid patients (Upadhyay et al. 2018). These are produced from substitution at a para position on the benzene ring, and at one nitrogen residue of urea moiety (Brunton et al. 2013). First-generation SUs are rarely used, as they carry a great risk of hypoglycemia in contrast to modern SUs.

Mechanism of Action

SUs act as insulin secretagogues by modulating ATP sensitive potassium ion channels (K_{ATP} channel), via sulfonylurea receptor present on the pancreatic beta-cell membrane. SUs bind with K_{ATP} channel and cause its inhibition, which leads to an increase in the concentration of K^+ ions in the cytoplasm of beta-cells, and depolarizes calcium ion channels. This leads to the increased influx of calcium ions into the beta-cell that provokes exocytosis of insulin granules and stimulates insulin release. To some extent, SUs also bind and activate Epac2 (guanine nucleotide exchange factor), which further works together with Rap1 protein to increase the approachability of insulin vesicles, to meld with the plasma membrane of the beta cell (Thulé and Umpierrez 2014).

Extra-Pancreatic Actions

Increased expression of glucose transporter GLUT-4 in the periphery and reduced peripheral insulin resistance are recognized. Suppression of sarcolemma and mitochondrial ATP channels on cardiac cells, and K_{ATP} channels on smooth muscle cells, by binding with sulfonylurea receptor SUR2A which causes cardiovascular complications, is also mentioned. Gliclazide causes reduced hepatic glucose output, with anti-oxidant and antiplatelet effect. Glyburide, in turn, aggravates blood pressure due to sympathetic activation (Kalra et al. 2015).

Efficacy

All SUs are equally efficacious at equipotent doses. Therapy is initiated from a lower dose range, on the basis of glycemic control. Longer-acting SUs are administered as a single dose daily and rest is available in extended-release dosage forms (glipizide) or micronized tablets (glyburide), to prolong the duration of action and the dose is kept lower. SUs reduce glycated hemoglobin by 1.5%–2%, with a fasting plasma glucose (FPG) decrease of up to 60–70 mg/dl.

Usually, the primary failure of SU monotherapy is seen in patients with low C-peptide levels and high FPG levels (>250 mg/dl). In such patients, SUs cause less than 30 mg/dl decrease in FPG levels. Almost 75% of the patients come under secondary failure on SU monotherapy, initially responding well (>30 mg/dl reduction in FPG), however failing to reach acceptable glucose control (Brunton et al. 2013; Dipiro et al. 2005).

Effect on Body Weight

Older SU drugs (glimepiride, glyburide, glipizide, gliclazide) are reported to induce weight gain of 1.8–2.6 kg in previous studies (Pappachan et al. 2019). Modern SUs (glimepiride, gliclazide) are widely preferred over SUs due to less risk of hypoglycemia, weight gain, and cardiovascular complications. The Scottish Intercollegiate Guidelines Network (SIGN), International Diabetes Federation (IDF), and National Institute for Health and Care Excellence (NICE) recommend SUs use in patients who are not obese. Weight neutrality is reported with glimepiride ER and gliclazide MR during the period of 1 year. Regular consumption of glimepiride produces weight neutrality for about 1.5 years (Kalra et al. 2015).

Contraindications

SUs are not used in patients with cardiovascular diseases, liver, and renal dysfunction, as well as those exhibiting hypersensitivity to the drug.

Adverse Effects

Due to the narrow therapeutic index, SUs even in small doses can induce hypoglycemia in pediatrics and geriatrics. High doses can cause chronic hypoglycemia (less than 60 mg/dl), rarely leading to death due to neuroglycopenia and cardiotoxicity (Klein-Schwartz et al. 2016).

Glinides (Meglitinides/D-Phenylalanine Analogs)

This class of anti-diabetic drugs includes repaglinide and nateglinide, which act as ATP-sensitive potassium channel blockers. They are also insulin secretagogues, and are sometimes named prandial glucose regulators. They have a similar mechanism of action (MOA) as SUs but are structurally unrelated. The first agent of this class, repaglinide, was approved by FDA in 1997, and nateglinide in 2000. They are rapidly absorbed and have a short duration of action, in contrast to second-generation Sus; thus, flexible dosage can be designed (White 2010).

Mechanism of Action

Both meglitinides, analogously to SUs, block ATP-dependent potassium ion channels, causing insulin release from pancreatic beta-cells. Repaglinide is derived from benzoic acid, while nateglinide is derived from d-phenylalanine (Brunton et al. 2013). Meglitinides bind to sulfonylurea receptor SUR1, and are also reported to bind at a separate site on beta-cells. Nateglinide shows larger selectivity for SUR1 over SUR2, in contrast to repaglinide and SUs. Nateglinide remains bonded to the receptor for a very limited period of about 2 s, contrasting with repaglinide which stays for 3 min, whereas the dissociation from the receptor is 90 times greater than repaglinide. Therefore nateglinide has a very limited on-off impact on insulin discharge (Guardado-Mendoza et al. 2013).

Efficacy

These agents are reported to treat long-term hyperglycemic stages in patients with T2DM. In a 3.6 months clinical trial, repaglinide (0.5–4 mg/meal) reduced glycated hemoglobin level by 1.57%, while nateglinide (60–120 mg/meal) achieved 1.04% decrease. Repaglinide reduced pre-prandial glucose values by 57 mg/dl, while nateglinide by 18 mg/dl, in a period of 16 weeks (Rosenstock et al. 2004).

Effect on Body Weight

Patients on repaglinide suffer a weight gain of 1.8 kg and with nateglinide 0.7 kg, after 3.6 months of therapy (Rosenstock et al. 2004). Meglitinides are indeed associated with minor to moderate weight gain, as per a report by the American Association of Diabetes Educators 2016 (American Association of Diabetes Educators 2016).

Contraindications

They are contraindicated in diabetic ketoacidosis, hypersensitivity, and renal and liver diseases. Repaglinide is contraindicated in patients on lipid-lowering gemfibrozil, as it can increase repaglinide blood levels by 28.6 fold (Grant and Graven 2016).

Adverse Effects

Meglitinides are associated with hypoglycemia (16–31%) and weight gain, however less remarkably than Sus (Hinnen 2015; Hossain and Pervin 2018).

Biguanides

Guanidine is the active ingredient in *Galega officinalis*, a small tree found in Europe and Western Asia, which was prescribed in primitive times for diabetic patients. However, guanidine and derivatives showed toxicity. First-generation biguanides (phenformin and buformin) were introduced for diabetes pharmacotherapy in 1950. In 1970 phenformin was withdrawn due to cardiac mortality and lactic acidosis. Metformin appeared as a safer alternative and was approved in the United States in 1995. Now it is one of the most widely prescribed antidiabetic agents (Hundal and Inzucchi 2003).

Mechanism of Action

Metformin is an AMPK (AMP-dependent protein kinase) activator. When cellular energy stores are reduced, AMPK is stimulated by phosphorylation. Stimulated AMPK initiates the process of fatty acid oxidation and glucose uptake, reducing lipid production and gluconeogenesis. The net outcome of such actions is decreased hepatic glucose output, enhanced insulin sensitivity in fat cells and muscles, enhanced storage of glycogen in skeletal muscles, and reduced blood glucose levels. The exact mechanism by which metformin stimulates AMPK is unknown. Metformin has shown reduced cellular respiration by selectively affecting mitochondrial complex I. It does not disturb the discharge of any islet hormones, and hardly causes hypoglycemia (Brunton et al. 2013).

Efficacy

Metformin decreases glycated Hb level by 1.5%–2%, with fasting glucose reduction by 60–80 mg/dl. It is acceptable even in cases of extremely high postprandial glucose (PPG) (>300 mg/dl), and is also able to reduce triglycerides and LDL-C by 8%–15%, with a slight increase in HDL-C by 2%. It stimulates PAI-1 (plasminogen activator inhibitor-1), and induces a decrease in weight by 2–3 kg (Dipiro et al. 2005).

Effect on Body Weight

Metformin is recommended for diabesity patients on account of suppression of hepatic gluconeogenesis, increased insulin sensitivity in peripheral tissue, and some degree of anorexia. Approximately 31% reduction in episodes of T2DM in highly obese patients has been achieved by metformin as stated by Diabetes Prevention Program (DPP) clinical trial (The Diabetes Prevention Program Research Group 2002). A combination of metformin with lifestyle modification has shown improvement in gestational diabetes (Pappachan et al. 2019).

Contraindications

Caution is advised for patients liable to display metabolic acidosis including renal insufficiency, hepatic failure, alcoholism, congestive heart failure (CHF), sepsis, and surgery. Certain studies reported the safety of metformin with stable CHF, resulting in reduced mortality rates (Upadhyay et al. 2018).

Adverse Effects

Gastrointestinal symptoms (10–25%) can include dyspepsia, diarrhea, nausea, and intestinal muscle spasm. Dosage tapering or an intake with a meal can be helpful. Reduced levels of cobalamin (20–30%) in malabsorption is a possibility. Overdose, or regular use of metformin for years, can rarely precipitate metabolic acidosis (3–6/100,000 patients) (Brunton et al. 2013).

Thiazolidinediones (TZDs)

TZDs (insulin sensitizers) were approved for clinical use by FDA in 1997, though controversies surrounded the first agent of this class, troglitazone. Troglitazone was withdrawn from the market due to liver toxicity (Yki-Jarvinen 2004). Pioglitazone is the only widely used TZD, as rosiglitazone was also withdrawn in some countries in terms of cardiovascular safety (Pappachan et al. 2019).

Mechanism of Action

TZDs or glitazones are the selective synthetic ligands of nuclear peroxisome-proliferator–activated receptor gamma (PPARγ) (Yki-Jarvinen 2004). Such types of receptors are a subfamily of 48-member nuclear–receptors, and regulate transcription of genes linked to glucose and lipid metabolism (Brunton et al. 2013; Yki-Jarvinen 2004). PPARγ is mainly expressed in adipocytes, with minor findings on skeletal, cardiac, smooth muscle cells, pancreatic β-cells, macrophages, and vascular endothelial cells (Brunton et al. 2013). Hence adipose tissue is the main target of PPARγ agonists. However, there is no evidence whether TZDs reduce insulin resistance by directly acting on to the target tissues, or indirectly through the release of secreted products by adipocytes such as adiponectin, that increases insulin sensitivity. TZDs binding with PPARγ receptor induce a conformational change, and form a heterodimer with another nuclear receptor (retinoid X receptor), which then further binds with peroxisome proliferator response element. Such activation causes adipocyte differentiation, increases subcutaneous fat cells, FFA oxidation, FFA uptake into adipocytes, and storage from extra-adipose tissues to adipose tissues. Such effects enhance insulin sensitivity in fat cells, muscles, and liver (Brunton et al. 2013; Dipiro et al. 2005; Yki-Jarvinen 2004).

Efficacy

Pioglitazone or rosiglitazone therapy for 6 months decreases glycated hemoglobin by 1.5%, and PPG level by 60–70 mg/dl at a higher dose. TZDs slowly reduce blood glucose levels for about 3–4 months, after which the highest glucose response is seen. Patients must be counseled regarding such point, so that they continue therapy without any drop-out (Brunton et al. 2013; Dipiro et al. 2005).

Effect on Body Weight

A weight gain of 3.6 kg after 35 months has been registered in a large trial (Dormandy et al. 2005). TZDs lead to weight gain of up to 3–4 kg in 6 months and 5 kg in 3–5 years as per UKPDS (Ko et al. 2017). Weight gain varies on the basis of dose regimen used (Wilding 2006). These agents also affect lipid parameters such as TGs, LDL-C, and HDL-C. Pioglitazone reduces triglyceride levels by 10–20% while rosiglitazone has a neutral effect. Rosiglitazone increases LDL-C up to 5–15%, whereas pioglitazone is devoid of such effect. Both are able to alter minor, dense LDL-C particles to bulky soft particles, which are less dense and less atherogenic, and also increase HDL-C up to 3–9 mg/dl (Dipiro et al. 2005).

Contraindications

In patients with New York Heart Association (NYHA) class III and IV cardiac failure, and even in class I and II patients, caution is advised (Dipiro et al. 2005).

Adverse Effects

TZDs as single oral anti-diabetic agents do not produce hypoglycemia, whereas in combination with SUs can slightly increase the risk (Elte and Blicklé 2007). Other adverse effects include edema (2–4%) due to sodium/water retention and enhanced vascular permeability, cardiac failure, and weight gain. The severity of edema (approx. 15%) increases in combination with insulin. Edema is dose-dependent, so by dropping the dose and using diuretics, alleviation could follow, especially if edema is not severe (Dipiro et al. 2005).

GLP-1 Receptor Agonists (Injectable Drugs)

These are also called incretin mimetics. Incretins are hormones produced by epithelial intestinal L-cells, in response to food ingested: GIP (glucose-dependent insulinotropic polypeptide) and GLP-1 (glucagon-like peptide 1). GLP-1 administration reduces blood glucose levels by mediating insulin release, and also inhibits it when the glucose level is normal, while GIP is not clinically effective in T2DM patients (Brunton et al. 2013; Hinnen 2017; Yammine et al. 2019).

Pancreatic alpha-cells produce and split proglucagon into glucagon and a large peptide (C-terminal) comprising GLPs (GLP-1 and GLP-2), whereas intestinal L-cells transform proglucagon into a large peptide (N-terminal) comprising glucagon or GLPs. GLP-2 is of little importance for diabetes management. It mainly influences epithelial cell proliferation in the digestive tract; hence, it has been approved as an orphan drug (teduglutide; GLP-2 agonist) for the treatment of short bowel syndrome, by both the FDA and the European Medicine Agency (Brunton et al. 2013).

Mechanism of Action

There are two GLP-1 receptor agonists (exenatide and liraglutide) prescribed intravenously for type 2 diabetic subjects. These agents increase insulin secretion which is glucose-dependent, inhibit glucagon, slow gastric emptying, reduce appetite and body weight, and normalize pre-prandial and postprandial glucose levels, by stimulating GLP-1 receptor (Brunton et al. 2013; Meloni et al. 2013). These receptors are G-protein coupled receptors (GPCR) expressed in pancreatic alpha-cells, beta-cells, specific neurons in hindbrain, intestinal cells, and other structures.

Beta-cell functioning is required for incretin-based therapy; else, it could lead to failure (Tripathi 2013). GLP-1 as such has limited therapeutic potential, due to rapid degradation by dipeptidyl peptidase 4 enzyme (DPP-4); consequently, half-life is just 2–3 min. The enzyme is mainly expressed in luminal capillary endothelial cells, kidney, gastrointestinal mucosa, and immune cells. To increase the half-life, DPP-4-resistant GLP-1 agonists have been produced. The use of DPP-4 inhibitors potentiates the pharmacological action of incretin-based therapy (Brunton et al. 2013; Tripathi 2013).

Exenatide: It is a synthetic exendin-4, potent, DPP-4-resistant GLP-1 agonist which is short-acting. It is injected subcutaneously twice daily, with a plasma half-life of 3 hours, and acts for about 6–10 h. It can be used as monotherapy or in combination with metformin, SUs, or TZDs if required. It reduces glycated hemoglobin by 1% in patients with T2DM, after 6.9–18.8 months (Brunton et al. 2013; Tripathi 2013; Soni 2016).

Liraglutide: It is rigidly bound to a plasma protein that accounts for increased plasma half-life of >12 h, and thus acts for >24 h. Due to its prolonged plasma half-life, it is given subcutaneously once daily. It reduces glycated hemoglobin by 30%, which is greater than exenatide. It can be used as monotherapy, or in combination with metformin, SUs, or TZDs (Brunton et al. 2013; Tripathi 2013; Soni 2016).

Effect on Weight

Patients on exenatide show a weight reduction of 2.5–4 kg within 30–82 weeks of therapy, whereas liraglutide can be followed by a weight reduction of 5.6 kg, in obese non-diabetic or diabetic patients. Therefore it can also be used as an anti-obesity drug (Brunton et al. 2013; Tripathi 2013; Soni 2016).

Contraindications

Patients with a family history of medullary carcinoma of the thyroid, and patients with multiple endocrine neoplasias, should avoid such therapy. Caution is advised due to the presence of gastroesophageal reflux (Hinnen 2017).

Adverse Effects

Nausea, vomiting, and diarrhea are common (40–50%) and dose-related. Hypoglycemia is rarely reported with GLP analogs, except when used in combination with Sus (Brunton et al. 2013; Tripathi 2013).

DPP-4 Inhibitors

DPP-4 belongs to the family of prolyl oligopeptidases. It is a serine protease highly expressed on intestinal epithelial cells, brush-border renal cells, pancreatic cells, hepatic cells, endothelial cells, glandular epithelium, and T-cells. It is also called as T-cell differentiation antigen (CD26). DPP-4 protein consists of an extracellular domain, with the catalytic site in the C-terminal attached to the transmembrane protein, and an intracellular tail at the N-terminal. When it is cleaved it discharges the extracellular domain in the blood, in the form of soluble DPP-4, and also circulates in body fluids (seminal and cerebrospinal fluids).

The extracellular domain of the protein also contains glycosylation and cysteine-rich sites that contribute to non-enzymatic actions of DPP-4. Such sites are believed to interact with adenosine deaminase, streptokinase, and plasminogen, and are also a binding site for chemokines. The antidiabetic effect of DPP-4 inhibitors was evaluated after the investigations about inactivation of incretin hormones by DPP-4 (Deacon 2019). DPP-4 inhibitors are identified to enhance plasma half-life of incretin-based therapy. The first DPP-4 inhibitor sitagliptin was approved by FDA in 2006. After this saxagliptin was approved in the United States and vildagliptin in Europe and America, as monotherapy or combination therapy with other anti-diabetics. DPP-4 inhibitors linagliptin and alogliptin are also available (Dicker 2011).

Mechanism of Action

Due to the essential role of DPP-4 enzyme in inactivation of incretin hormones, orally active inhibitor drugs of this enzyme have been produced. These are indirectly acting insulin-secreting agents and are mostly used in combination therapy (Brunton et al. 2013; Tripathi 2013). DPP-4 inhibitors such as sitagliptin, vildagliptin, and saxagliptin are selective competitive inhibitors of DPP-4, with high affinity for the enzyme. The concentration of physiologically produced GLP-1 and GIP after food intake was doubled with DPP-4 inhibitors. Such selective inhibition of enzyme provides an extended safety profile and allows prolonged treatment with DPP-4 inhibitors.

It indirectly increases insulin secretion and inhibition of glucagon in patients with T2DM. DPP-4 inhibitors are also reported to counter-regulate glucagon release by pancreatic alpha-cells under hypoglycemia. Thus, these agents do not cause alpha-cell dysfunction, and improve alpha-cell sensitivity to glucose without affecting appetite. Experimental beta-cell neogenesis was demonstrated, and genetically DPP-4-deficient rats showed resistance to streptozotocin (STZ)-induced beta-cell destruction. These drugs also improve TGs and FFA levels (Brunton et al. 2013; Thornberry and Gallwitz 2009).

Efficacy

Sitagliptin has been shown to reduce glycated hemoglobin by 0.65–1.0% within 2–7 months, and 0.67% upon 12 months of therapy. Saxagliptin reduces glycated hemoglobin by 0.43–1.17%, while vildagliptin reduces by 1.4% in 24 weeks of therapy. In terms of anti-hyperglycemic effect, DPP-4 inhibitors were found to be slightly less effective than SUs, and equivalent to metformin and TZDs (Dicker 2011; Amori et al. 2007).

Effect on Weight

Sitagliptin during 12 months triggered weight reduction of 1.5 kg, up to weight gain of 1.8 kg.

Vildagliptin and saxagliptin were not much different: weight shifts of −1.8 kg to +1.3 kg, and −1.8 kg to +0.7 kg, respectively, after 5 months. In a meta-analysis of 13 studies they were considered to be weight-neutral (Dicker 2011).

Contraindications

Hypersensitive persons and those with chronic pancreatitis and renal diseases are not safe candidates for such therapy (Pathak and Bridgeman 2010).

Adverse Effects

A slight effect on in-vitro immune cell functions has been seen with DPP-4 inhibitors, and allergic reactions such as anaphylaxis, angioedema, Stevens-Johnson syndrome, flu-like symptoms, and skin reactions are reported. Hypoglycemia is observed during combination therapy of DPP-4 inhibitors with SUs and insulin (Brunton et al. 2013; Dicker 2011; Diabetes.co.uk 2019b).

Alpha-Glucosidase Inhibitors

The alpha-glucosidase enzyme is released by the chorionic epithelium of enterocytes, with the function of digestion of carbohydrates and glucose production. Due to its important role in glucose production, various alpha-glucosidase inhibitors were produced as a new class of oral hypoglycemic agents in the 1980s. They are obtained from natural sources, including plants and micro-organisms. The alpha-glucosidase enzymes from rodent intestine and yeast are employed for pharmacological purposes. They include acarbose, miglitol and voglibose (Yin et al. 2014).

Mechanism of Action

Alpha-glucosidase inhibitors retard intestinal absorption of carbohydrates by preventing alpha-glucosidase action in enterocytes of the brush border of the mucosa. Mouth to cecum transit time is reduced and leads to larger evacuation of carbohydrates via stools. Acarbose, produced by actinobacteria *Actinoplanes utahensis,* as well as voglibose, is a reversible inhibitor. These agents also cause the release of GLP-1 in systemic circulation which adds on to the therapeutic action, while miglitol is a powerful inhibitor of the enzyme sucrase (Brunton et al. 2013; Tripathi 2013). Acarbose also inhibits glucoamylase and alpha-amylase enzymes, thus impairing carbohydrate digestion and absorption (Derosa and Maffioli 2012).

Efficacy of Alpha-Glucosidase Inhibitor

Post prandial glucose concentration diminishes by 1–3 millimole/liter and pre-prandial glucose concentration by 1 millimole/l. Also glycated hemoglobin can be lowered by 0.5–0.8%, and even can exceed 1% at higher doses; however, proper maintenance of diet is a must. Body weight and lipid parameters remain unchanged. Long-term administration of these agents in prediabetics can prevent the onset of T2DM and heart disease. Both monotherapy and combination protocols are available (Brunton et al. 2013; Krentz 2018).

Contraindications

Diabetic ketoacidosis, inflammatory bowel disease, other gastrointestinal conditions, and hypersensitivity are contraindications (Access.fda.gov. PRECOSE (acarbose tablets) 2019).

Adverse Effects

Flatulence (78%) and loose stools (14%) are dose-related. In certain cases hepatitis is reported, which reverts when therapy is stopped (Derosa and Maffioli 2012).

Sodium-Glucose Co-transporter-2 Inhibitors/Gliflozins (SGLT-2 Inhibitors)

This is the FDA-approved latest class of anti-diabetics. Primarily introduced in 2013, SGLT-2 inhibitors consist of canagliflozin, dapagliflozin, ertugliflozin, and empagliflozin, which can be used as monotherapy, or in combination with metformin and linagliptin. These agents are reported to reduce the incidence of mortality due to cardiac diseases in patients with T2DM (Fda.gov 2019a, b; Hsia et al. 2017).

Mechanism of Action

The glomerulus of the kidney filters all the circulating glucose, which is reabsorbed into the blood via SGLT-2 of the proximal convoluted tubule. These transporters cause 90% reabsorption of filtered glucose, so these are an interesting target for T2DM. Inhibition of this co-transporter promotes glucouresis, and lowers blood glucose levels along with weight reduction (Tripathi 2013; Hsia et al. 2017). Empagliflozin has maximum specificity for SGLT-2, while canagliflozin is the least specific. These drugs can also be used in obese individuals, due to their weight reduction effect (Hsia et al. 2017).

Efficacy of SGLT-2 Inhibitors

They are useful in all the stages of T2DM because the mechanism of action is independent of beta-cell function. Hypoglycemia can occur if these agents are used in combination with an insulin secretagogue (Zurek et al. 2017). SGLT-2-inhibitors showed a 9.1% reduction of glycated hemoglobin at baseline, 0.9% in 1 month, and 0.8% in 6 months of therapy (Schork et al. 2019).

Effect on Body Weight

Weight loss of 2.6 kg in 6 months of therapy is documented, due to glucouresis and loss of adipose mass. Apart from body fat these agents also reduce epicardial fat. Extracellular fluid diminishes after 28 days, and returns to baseline after 3–6 months of therapy (Schork et al. 2019). Therefore SGLT-2 therapy is associated with conspicuous weight loss.

Contraindications

These are contraindicated in patients with kidney diseases or nephropathy (GFR < 45 mL/min/1.73 m (Al-Goblan et al. 2014; Hsia et al. 2017).

Adverse Effects

They can increase the risk of serious genital anaerobic infections (Fournier's gangrene), and urinary tract infections in diabetic patients (Fda.gov 2019b; Hsia et al. 2017).

Oral/Nasal Insulin

Insulin therapy can be used in all modalities of diabetes, including T1DM, T2DM, and gestational diabetes. It can be taken subcutaneously, intravenously, intramuscularly, and more recently also orally. The subcutaneous route is classic for long-term management of glycemic control, notably in patients with T1DM (Brunton et al. 2013). Oral insulin is a long-acting, basal insulin analog. When target with combination therapy of oral hypoglycemic agents is not achieved, oral insulin therapy is instigated to control glycosylated hemoglobin levels (A1C) > 7% in type 2 diabetic patients (Box 50.1).

Box 50.1 Pharmacotherapical Recommendations for Type 2 Diabetes (Modified from American Diabetes Association) (American Diabetes Association 2019)

(A) Initial therapy in all cases: Metformin (Goal HbA1c $\leq$ 7.0%.

(B) Patients without atherosclerosis or heart/renal failure.

1. With obesity: GLP-1 receptor agonists (liraglutide, semaglutide, exenatide) or SGLT-2 inhibitors (empaglifozin, canaglifozin); if necessary, the two can be combined, or DPP-4 inhibitors (in the absence of GLP-1 receptor agonists), SUs, TZDs, and basal insulin can be included.
2. Weight loss not required: DPP-4 inhibitors, GLP-1 receptor agonists, SUs and TZDs are the first options; in case of insufficient response, two of these can be combined, eventually including SGLT-2 inhibitors; as a third line of therapy, SUs or basal insulin can be added.

(C) Cardiovascular atherosclerosis.

1. GLP-1 receptor agonists or SGLT-2 inhibitors.

(continued)

Box 50.1 (continued)

2. If necessary, the following adjuvants can be introduced: Basal insulin, TZDs, SUs, and also DPP-4 inhibitors (however these are incompatible with GLP receptor agonists).

(D) Heart or renal failure.

1. SGLT-2 inhibitors are the first option, however GLP receptor agonists can be added.
2. Should compensation not be adequate, GLP-1 receptor agonists), SGLT-2 inhibitors, DPP-4 inhibitors (however these are incompatible with GLP receptor agonists), and basal insulin can be used.

Oral insulin at high dose, in insulin-naive patients with T2DM, seems to be effective in controlling FPG (7.1 mmol/L) levels after 8 weeks, comparable to subcutaneous insulin glargine (6.8 mmol/L), with the absence of any serious adverse effects or hypoglycemia (Halberg et al. 2019). Only nasal insulin (Afrezza), for both type 1 and type 2 diabetic patients, is already FDA approved. Afrezza is rapid-acting inhaled insulin and is not a substitute for long-acting insulin. It is taken before a meal or within 20 minutes of starting a meal and can be used along with long-acting insulin in patients with T1DM (Medscape.com 2014). The FDA has also approved the first glucagon therapy (Baqsimi nasal powder) for the emergency management of severe hypoglycemia (Fda.gov 2019c).

Pharmacological Approaches of the American Diabetes Association (ADA)

Metformin is recommended as the first-line agent for the treatment of T2DM. Metformin therapy should be continued till it is tolerated and not contraindicated and in the long run causes vitamin B12 deficiency especially in patients with anemia or peripheral neuropathy. Therefore, vitamin B12 levels should be monitored during metformin therapy. Initially, insulin can be added with anti-diabetic drugs in case of weight loss, hypertriglyceridemia, ketosis, and hyperglycemia ($\geq$300 mg/dl) or glycated hemoglobin ($>$10%). ADA recommends a patient-oriented approach for selecting an appropriate pharmacotherapy to control blood glucose levels, including the efficacy of the drug and patient-related factors as important considerations. Along with active pharmacological agents, lifestyle modification should be considered to improve health. For T2DM patients having atherosclerotic cardiovascular disease, SGLT-2 inhibitors and GLP-1 receptor agonists are recommended as part of an anti-diabetic regimen with its well-known cardiovascular profits. For patients having atherosclerotic cardiovascular disease at risk of cardiac failure or if cardiac failure co-exists SGLT-2 inhibitors are ideal. T2DM patients with chronic renal diseases should be recommended SGLT-2 inhibitors and GLP-1 receptor agonists to prevent the progression of kidney diseases or cardiovascular diseases or both. Among the parenteral drugs to achieve extreme blood glucose reduction, GLP-1 receptor agonists are considered over insulin. After every 3–6 months, anti-diabetic drug regimen should be assessed and adjusted based on the patient's factor. The combination of new anti-diabetic agents with initial therapy should be recommended to reduce glycated hemoglobin level by 0.7–1%. If glycated hemoglobin levels are not controlled in 3 months without atherosclerotic cardiovascular disease or chronic kidney disease, SUs, TZDs, SGLT-2 inhibitors, GLP-1 agonists, DPP-4 inhibitors, and basal insulin can be combined with metformin with special emphases on patient factor and treatment selective effects (American Diabetes Association 2019). For the treatment of T2DM in obese patients, selection of anti-diabetic drug should be based on their effect on body weight. Initially, diet modification should be recommended and can be further combined with exercise and anti-obesity agents (BMI $\geq$ 27 kg/m^2) in T2DM patients. For patients receiving anti-obesity agents with weight loss $<$5% in 3 months or any safety and tolerability issues during the

Table 50.1 Properties of anti-diabetic agents used in the treatment of T2DM

Class	Drugs	Dosage range	HbA1c reduction (%)	Weight change	Hypoglycemia	Duration of action
Oral						
Biguanide	Metformin	500–3000 mg (Waring 2016)	1–2% (Brunton et al. 2013)	Neutral (American Diabetes Association 2019)	No (American Diabetes Association 2019)	24 h (Dipiro et al. 2005)
SGLT-2 inhibitors	Dapagliflozin Canagliflozin Empagliflozin Ertugliflozin	5–10 mg (Drugs.com 2019a) 100–300 mg (Drugs.com 2019b) 10–25 mg (Drugs.com 2019c) 5–15 mg (Drugs.com 2019d)	0.6–1.2% (Bashier et al. 2017)	Loss (American Diabetes Association 2019)	No (American Diabetes Association 2019)	$t_{1/2}$ ~ 12.9 h (Drugs.com 2019e) 24 h (Drugs.com 2019f) $t_{1/2}$ ~ 12.4 h (Drugs.com 2019g) $t_{1/2}$ ~ 16.6 h (Drugs.com 2019h)
Meglitinides	Repaglinide Nateglinide	0.5–16 mg (Brunton et al. 2013) 180–360 mg (Brunton et al. 2013)	1–2% (Brunton et al. 2013)	Gain (Diabetes.co.uk 2019c)	Yes (Diabetes.co.uk 2019c)	2–6 h (Brunton et al. 2013) 2–4 h (Brunton et al. 2013)
DPP-4 inhibitors	Sitagliptin Vildagliptin Saxagliptin Alogliptin Linagliptin	100 mg (Brunton et al. 2013) 50–100 mg (Brunton et al. 2013) 2.5–5 mg (Brunton et al. 2013) 25 mg (Drugs.com 2019j) 5 m (Medscape.com 2019)	0.5–1%(Brunton et al. 2013)	Neutral (American Diabetes Association 2019)	No (American Diabetes Association 2019); Drugs.com (2019i)	24 h (Tripathi 2013) 12–24 h (Tripathi 2013) 24 h (Tripathi 2013) 24 h (Cada et al. 2013) >24 h (Gallwitz 2012)
TZDs	Pioglitazone Rosiglitazone	15–45 mg (Waring 2016)	0.5–1.4% (Brunton et al. 2013)	Gain (American Diabetes Association 2019)	No (American Diabetes Association 2019)	24 h (Dipiro et al. 2005)

(continued)

Table 50.1 (continued)

Class	Drugs	Dosage range	HbA1c reduction (%)	Weight change	Hypoglycemia	Duration of action
		4–8 mg (Waring 2016)				24 h (Dipiro et al. 2005)
Sulfonyl ureas (second generation)	Glyburide Glipizide Gliclazide Glimepiride	1.25–20 mg (Brunton et al. 2013) 5-40 mg (Brunton et al. 2013) 40–320 mg (Medicine.org.uk 2019) 1–8 mg (Brunton et al. 2013)	1–2% (Brunton et al. 2013)	Gain (American Diabetes Association 2019)	Yes (American Diabetes Association 2019); Drugs.com (2019k)	12–24 h (Brunton et al. 2013) 12–18 h (Brunton et al. 2013) 12–24 h (Tripathi 2013) 24 h (Brunton et al. 2013)
Alpha-glucosidase inhibitors	Acarbose Miglitol	25–50/100 mg (Dipiro et al. 2005) 25–50/100 mg (Dipiro et al. 2005)	0.5–0.8% (Brunton et al. 2013)	Neutral/minor loss (American Association of Diabetes Educators 2016)	No (Diabetes.co.uk 2019d)	1–3 h (Dipiro et al. 2005) 1–3 h (Dipiro et al. 2005)
Parenteral						
GLP-1 RAs	Exenatide Liraglutide	2 mg s.c (Drugs.com 2019l) 1.2–1.8 mg s.c (Drugs.com 2019m)	0.9–1.6% (Babenko et al. 2019)	Loss (American Diabetes Association 2019)	No (American Diabetes Association 2019); Diabetes.co.uk (2019e)	6–10 h (Tripathi 2013) 24 h (Drugs.com 2019m)

course of therapy the drugs should be discontinued and alternative pharmacological approaches should be taken into account. Anti-diabetic agents having an influence on body weight include metformin, alpha-glucosidase inhibitors, SGLT-2 inhibitors, GLP-1 receptor agonists, and amylin mimetics. Dipeptidyl peptidase 4 inhibitors are weight-neutral while insulin

Table 50.2 Drug therapy under investigation for the treatment of diabesity and T2DM

Drug	Sponsor	Phase	Status	Study title/ objective	Completion date	References
Trodusquemine (MSI-1436)	Genaera corporation	I	Completed	Single dose, tolerance, and pharmacokinetic study in obese T2DM patients	April 2009	Clinicaltrial.gov (2019a)
SAR425899	Sanofi	I	Completed	Safety and tolerability in overweight to obese and T2DM patients	Oct 2018	Clinicaltrial.gov (2019b)
KRP-104 (novel DPP-4 inhibitor)	ActivX biosciences, Inc.	II	Completed	Randomized, double-blind, placebo-controlled trial to assess safety and effectiveness in T2DM patients	Aug 2008	Clinicaltrial.gov (2019c)
Oral ORMD-0801 (oral insulin)	Oramed, ltd.	II	Completed	Randomized, double-blind, placebo-controlled, parallel-group trial to evaluate safety and efficacy.	April 2016	Clinicaltrial.gov (2019d)
LX4211	Lexicon pharmaceuticals	II	Completed	Assessment of safety, efficacy, and tolerability in T2DM patients	Dec 2009	Zambrowicz et al. (2012); Clinicaltrial.gov (2019e)
HOE901-U300 (new formulation of insulin glargine)	Sanofi	III	Completed	Consideration of efficacy between HOE901-U300 and lantus in T2DM patients	March 2014	Bolli et al. (2015); Clinicaltrial.gov (2019f)
Rimonabant (SR141716)	Sanofi	III	Completed	Assessment of weight reduction and persistence in 12 months with a diabetic diet in obese T2DM patients	May 2004	Clinicaltrial.gov (2019g)
Sitagliptin and metformin therapy	The University of Texas Health Science Center at san Antonio	IV	Completed	Assessment of sitagliptin and metformin effects alone and in combination contrast to placebo on liver gluconeogenesis.	Oct 2012	Solis-Herrera et al. (2013); Clinicaltrial.gov (2019h)

secretagogues, thiazolidinediones, and insulin induce weight gain. For short-term and long-term management of obesity (BMI $\geq$ 27 kg/m^2), all FDA approved agents for weight loss are found to be associated with glycemic control in T2DM patients or patients at risk of T2DM. Five weight-reducing drugs (combination drugs)—orlistat, lorcaserin, phentermine/topiramate (ER), naltrexone/bupropion ER, and liraglutide—have been approved by FDA for long-term use in patients with obesity-associated T2DM, dyslipidemia, and high BP who want to lose weight (Care and Suppl 2019).

General Features of Available and Pipeline Drugs

Table 50.1 summarizes the prescription characteristics of common antidiabetic drugs, whereas Table 50.2 highlights coming products or combinations, for diabetes with or without obesity.

References

Access.fda.gov. PRECOSE (acarbose tablets) (2019). https://www.accessdata.fda.gov/drugsatfda_docs/label/2008/020482s023lbl.pdf. Accessed 27 Aug 2019

Al-Goblan AS, Al-Alfi MA, Khan MZ (2014) Mechanism linking diabetes mellitus and obesity. Diabetes Metab Syndr Obes 7:587–591

American Association of Diabetes Educators (2016) Weight loss medications: AADE Practice Advisory

American Diabetes Association (2019) Pharmacologic approaches to glycemic treatment: standards of medical care in diabetes-2019. Diabetes Care 42:S90–S102

Amori RE, Lau J, Pittas AG (2007) Efficacy and safety of incretin therapy in type 2 diabetes. JAMA 298:194–206

Babenko AY, Savitskaya DA, Kononova YA et al (2019) Predictors of effectiveness of glucagon-like peptide-1 receptor agonist therapy in patients with type 2 diabetes and obesity. J Diabetes Res 2019:1–12

Bashier A, Abdulaziz A, Rashid F, Ibrahim E (2017) Efficacy and safety of SGLT2 inhibitors in reducing glycated hemoglobin and weight in emirati patients with type 2 diabetes. J Clin Med Res 9:499–507

Bolli GB, Riddle MC, Bergenstal RM et al (2015) New insulin glargine 300 U/ml compared with glargine 100 U/ml in insulin-naïve people with type 2 diabetes on oral glucose-lowering drugs: a randomized controlled trial (EDITION 3). Diabetes Obes Metab 17:386–394

Brunton L, Chabner B, Knollman B (2013) Goodman & Gilman's: the pharmacological basis of therapeutics. McGraw-Hill, New York

Cada DJ, Levien TL, Baker DE (2013) Alogliptin. Hosp Pharm 48:580–592

Care D, Suppl SS (2019) 8. Obesity management for the treatment of type 2 diabetes: standards of medical care in diabetes 2019. *Diabetes Care* 42:S81–S89

Clinicaltrial.gov (2019a) A single dose, tolerance and pharmacokinetic study in obese or overweight type 2 diabetic volunteers. https://clinicaltrials.gov/ct2/show/NCT00606112?draw=2. Accessed 8 Aug 2019

Clinicaltrial.gov (2019b) A study on safety and tolerability of SAR425899 in overweight to obese subjects and type 2 diabetes mellitus patients not requiring anti-diabetic pharmacotherapy with an optional 6-month safety extension period. https://clinicaltrials.gov/ct2/show/NCT03414736?term=DRUGS+in+phase+1&cond=Obesity%3B+Endocrine&cntry=US&draw=3. Accessed 8 Aug 2019

Clinicaltrial.gov (2019c) A randomized, double-blind, placebo-controlled study to assess the safety and efficacy of KRP-104 in patients with type 2 diabetes inadequately controlled on metformin alone. https://clinicaltrials.gov/ct2/show/NCT00525330?term=drugs+in+phase+2&cond=Type+2+Diabetes+Mellitus&cntry=US&draw=2&rank=18. Accessed 8 Aug 2019

Clinicaltrial.gov (2019d) Phase II safety and efficacy study of oral ORMD-0801 in patients with type 2 diabetes mellitus. https://clinicaltrials.gov/ct2/show/NCT02496000. Accessed 30 Aug 2019

Clinicaltrial.gov (2019e) Study of LX4211 in subjects with type 2 diabetes mellitus. https://clinicaltrials.gov/ct2/show/NCT00962065?term=drugs+in+phase+2&cond=Type+2+Diabetes+Mellitus&cntry=US&rank=3. Accessed 8 Aug 2019

Clinicaltrial.gov (2019f) Comparison of new formulation of insulin glargine with lantus in patients with type 2 diabetes on non-insulin anti-diabetic therapy. https://clinicaltrials.gov/ct2/show/NCT01676220?term=novel+formulation&cond=Diabetes+Mellitus%2C+Type+2&cntry=US&rank=1. Accessed 8 Aug 2019

Clinicaltrial.gov (2019g) Obese patients with type 2 diabetes (RIO-diabetes). https://clinicaltrials.gov/ct2/show/NCT00029848?term=drugs+in+phase+3&cond=Obesity%3B+Endocrine&cntry=US&draw=4&rank=23. Accessed 8 Aug 2019

Clinicaltrial.gov (2019h) Mechanism of gluocse lowering effects of sitagliptin and metformin alone and in combination in patients with T2DM. https://clinicaltrials.gov/ct2/show/NCT00820573?term=phase+4&cond=T2DM+%28Type+2+Diabetes+Mellitus%29&cntry=US&draw=2&rank=9. Accessed 8 Aug 2019

Deacon CF (2019) Corrigendum: physiology and pharmacology of DPP-4 in glucose homeostasis and the treatment of type 2 diabetes. Front Endocrinol (Lausanne) 10:1–14

Derosa G, Maffioli P (2012) α-Glucosidase inhibitors and their use in clinical practice. Arch Med Sci 8:899–906
Diabetes.co.uk (2019a) Diabetes and obesity. 2019. https://www.diabetes.co.uk/diabetes-and-obesity.html. Accessed 27 Aug 27 2019
Diabetes.co.uk (2019b) DPP-4 inhibitors (Gliptins) - drugs, suitability, benefits & side effects. https://www.diabetes.co.uk/diabetes-medication/dpp-4-inhibitors.html. Accessed 1 Aug 2019
Diabetes.co.uk (2019c) Diabetes medication SIDE EFFECTS. https://www.diabetes.co.uk/features/diabetes-medication-side-effects.html. Accessed 9 Aug 2019
Diabetes.co.uk (2019d) Alpha glucosidase inhibitors - suitability, benefits & side effects. https://www.diabetes.co.uk/diabetes-medication/alpha-glucosidase-inhibitor.html. Accessed 8 Aug 2019
Diabetes.co.uk (2019e) Incretin mimetics (GLP-1 Agonists) - suitability, benefits & side effects. https://www.diabetes.co.uk/diabetes-medication/incretin-mimetics.html. Accessed 8 Aug 2019
Dicker D (2011) DPP-4 inhibitors: impact on glycemic control and cardiovascular risk factors. Diabetes Care 34:276–278
Dipiro JT, Talbert RL et al (2005) Pharmacotherapy: a pathophysiological approach. McGraw-Hill Medical, New York
Dormandy JA, Charbonnel B, Eckland DJ et al (2005) Secondary prevention of macrovascular events in patients with type 2 diabetes in the PROactive Study (PROspective pioglitAzone Clinical Trial In macroVascular Events): a randomised controlled trial. Lancet 366:1279–1289
Drugs.com (2019a) Dapagliflozin dosage guide with precautions. https://www.drugs.com/dosage/dapagliflozin.html. Accessed 23 Aug 2019
Drugs.com (2019b) Invokana dosage guide. https://www.drugs.com/dosage/invokana.html. Accessed 23 Aug 2019
Drugs.com (2019c) Empagliflozin dosage guide with precautions. https://www.drugs.com/dosage/empagliflozin.html. Accessed 23 Aug 2019
Drugs.com (2019d) Ertugliflozin dosage guide with precautions. https://www.drugs.com/dosage/ertugliflozin.html. Accessed 23 Aug 2019
Drugs.com (2019e) Dapagliflozin (professional patient advice). https://www.drugs.com/ppa/dapagliflozin.html. Accessed 2 Sep 2019
Drugs.com (2019f) Canagliflozin (Professional Patient Advice). https://www.drugs.com/ppa/canagliflozin.html. Accessed 2 Sep 2019
Drugs.com (2019g) Empagliflozin (Professional Patient Advice). https://www.drugs.com/ppa/empagliflozin.html Accessed 2 Sep 2019
Drugs.com (2019h) Ertugliflozin (Professional Patient Advice). https://www.drugs.com/ppa/ertugliflozin.html. Accessed 2 Sep 2019
Drugs.com (2019i) Nesina (alogliptin) FDA Approval History. https://www.drugs.com/history/nesina.html. Accessed 8 Aug 2019
Drugs.com (2019j) Alogliptin - FDA prescribing information, side effects and uses. https://www.drugs.com/pro/alogliptin.html. Accessed 22 Aug 2019
Drugs.com (2019k) Gliclazide. https://www.drugs.com/international/gliclazide.html. Accessed 8 Aug 2019
Drugs.com (2019l) Exenatide Dosage guide with precautions. https://www.drugs.com/dosage/exenatide.html. Accessed 22 Aug 2019
Drugs.com (2019m) Victoza dosage guide. https://www.drugs.com/dosage/victoza.html. Accessed 22 Aug 2019
Elte JWF, Blicklé JF (2007) Thiazolidinediones for the treatment of type 2 diabetes. Eur J Intern Med 18:18–25
Fda.gov (2019a) Sodium-glucose Cotransporter-2 (SGLT2) Inhibitors. https://www.fda.gov/drugs/postmarket-drug-safety-information-patients-and-providers/sodium-glucose-cotransporter-2-sglt2-inhibitors. Accessed 1 Aug 2019
Fda.gov (2019b) FDA warns about rare occurrences of a serious infection of the genital area with SGLT2 inhibitors for diabetes. https://www.fda.gov/drugs/drug-safety-and-availability/fda-warns-about-rare-occurrences-serious-infection-genital-area-sglt2-inhibitors-diabetes. Accessed 1 Aug 2019
Fda.gov (2019c) FDA approves first treatment for severe hypoglycemia that can be administered without an injection. https://www.fda.gov/news-events/press-announcements/fda-approves-first-treatment-severe-hypoglycemia-can-be-administered-without-injection. Accessed 30 Aug 2019
Gallwitz B (2012) Linagliptin—a novel dipeptidyl peptidase inhibitor for type 2 diabetes therapy. Clin Med Insights Endocrinol Diabetes 5:1
Grant JS, Graven LJ (2016) Progressing from metformin to sulfonylureas or meglitinides. Workplace Health Saf 64:433–439
Guardado-Mendoza R, Prioletta A, Jiménez-Ceja LM, Sosale A, Folli F (2013) The role of nateglinide and repaglinide, derivatives of meglitinide, in the treatment of type 2 diabetes mellitus. Arch Med Sci 9:936–943
Halberg IB, Lyby K, Wassermann K, Heise T, Zijlstra E, Plum-Mörschel L (2019) Efficacy and safety of oral basal insulin versus subcutaneous insulin glargine in type 2 diabetes: a randomised, double-blind, phase 2 trial. Lancet Diabetes Endocrinol 7:179–188
Hinnen DA (2015) Therapeutic options for the management of postprandial glucose in patients with type 2 diabetes on basal insulin. Clin Diabetes 33:175–180
Hinnen D (2017) Glucagon-like peptide 1 receptor agonists for type 2 diabetes. Diabetes Spectr 30:202–210
Hossain MA, Pervin R (2018) Current antidiabetic drugs: review of their efficacy and safety. Seoul, Elsevier Inc, pp 455–473

Hsia DS, Grove O, Cefalu WT (2017) An update on SGLT2 inhibitors for the treatment of diabetes mellitus. Curr Opin Endocrinol Diabetes Obes 24:73–79

Hundal RS, Inzucchi SE (2003) Metformin: new understandings, new uses. Drugs 63:1879–1894

Kalra S, Aamir A, Das A et al (2015) Place of sulfonylureas in the management of type 2 diabetes mellitus in South Asia: a consensus statement. Indian J Endocrinol Metab 19:577–596

Klein-Schwartz W, Stassinos GL, Isbister GK (2016) Treatment of sulfonylurea and insulin overdose. Br J Clin Pharmacol 81:496–504

Ko KD, Kim KK, Lee KR (2017) Does weight gain associated with thiazolidinedione use negatively affect cardiometabolic health? J Obes Metab Syndr 26:102–106

Krentz AJ (2018) Evolution of glucose-lowering drugs for type 2 diabetes: a new era of cardioprotection. New York, Elsevier, pp 431–454

Medicine.org.uk (2019) Gliclazide 80 mg Tablets BP - Summary of Product Characteristics. https://www.medicines.org.uk/emc/product/5883/smpc. Accessed 9 Aug 2019

Medscape.com (2014) FDA approves inhaled insulin afrezza for diabetes. https://www.medscape.com/viewarticle/827539. Accessed 30 Aug 2019

Medscape.com (2019) Tradjenta (linagliptin) dosing, indications, interactions, adverse effects, and more. https://reference.medscape.com/drug/tradjenta-linagliptin-999652. Accessed 22 Aug 2019

Meloni AR, Deyoung MB, Lowe C, Parkes DG (2013) GLP-1 receptor activated insulin secretion from pancreatic β-cells: mechanism and glucose dependence. Diabetes Obes Metab 15:15–27

Pappachan JM, Fernandez CJ, Chacko EC (2019) Diabesity and antidiabetic drugs. Mol Aspects Med 66:3–12

Pathak R, Bridgeman MB (2010) Dipeptidyl peptidase-4 (DPP-4) inhibitors in the management of diabetes. P T 35:509–513

Rosenstock J, Hassman DR et al (2004) Repaglinide versus nateglinide monotherapy. Diabetes Care 27:1265–1270

Schork A, Saynisch J, Vosseler A et al (2019) Effect of SGLT2 inhibitors on body composition, fluid status and renin-angiotensin-aldosterone system in type 2 diabetes: a prospective study using bioimpedance spectroscopy. Cardiovasc Diabetol 18:1–12

Solis-Herrera C, Triplitt C, de Garduno-Garcia J, Adams J, RA DF, Cersosimo E (2013) Mechanisms of glucose lowering of dipeptidyl peptidase-4 inhibitor sitagliptin when used alone or with metformin in type 2 diabetes. Diabetes Care 36:2756–2762

Soni H (2016) Peptide-based GLP-1/glucagon co-agonists: a double-edged sword to combat diabesity. Med Hypotheses 95:5–9

The Diabetes Prevention Program Research Group (2002) Reduction in the incidence of type 2 diabetes with lifestyle modification or metformin. N Engl J Med 346:393–403

Thornberry NA, Gallwitz B (2009) Mechanism of action of inhibitors of dipeptidyl-peptidase-4 (DPP-4). Best Pract Res Clin Endocrinol Metab 23:479–486

Thulé PM, Umpierrez G (2014) Sulfonylureas: a new look at old therapy topical collection on pharmacologic treatment of type 2 diabetes. Curr Diab Rep 14:1–8

Tripathi KD (2013) Essentials of medical pharmacology. Jaypee Brothers Medical Publishers, New Delhi

Upadhyay J, Polyzos SA, Perakakis N et al (2018) Pharmacotherapy of type 2 diabetes: an update. Metabolism 78:13–42

Waring WS (2016) Antidiabetic drugs. Med (United Kingdom) 44:138–140

White JR (2010) A brief history of the development of diabetes medications. Diabetes Spectr 27:82–86

Wilding J (2006) Thiazolidinediones, insulin resistance and obesity: finding a balance. Int J Clin Pract 60:1272–1280

Yammine L, Kosten TR, Pimenova M, Schmitz JM (2019) Cigarette smoking, type 2 diabetes mellitus, and glucagon-like peptide-1 receptor agonists as a potential treatment for smokers with diabetes: an integrative review. Diabetes Res Clin Pract 149:78–88

Yin Z, Zhang W, Feng F, Zhang Y, Kang W (2014) α-Glucosidase inhibitors isolated from medicinal plants. Food Sci Hum Wellness 3:136–174

Yki-Jarvinen H (2004) Drug therapy: thiazolidinediones. N Engl J Med 351:1106–1118

Zambrowicz B, Freiman J, Brown PM et al (2012) LX4211, a dual SGLT1/SGLT2 inhibitor, improved glycemic control in patients with type 2 diabetes in a randomized, placebo-controlled trial. Clin Pharmacol Ther 92:158–169

Zurek AM, Yendapally R, Urteaga EM (2017) A review of the efficacy and safety of sodium–glucose cotransporter 2 inhibitors: a focus on diabetic ketoacidosis. Diabetes Spectr 30:137–142

GLP1-Receptor Agonists in Diabetes: Drugs, General Effects, and Cardiovascular Impact

51

Ikaro Breder and Andrei C. Sposito

Contents

Cardiovascular Risk of Obesity and Diabetes ... 696
GLP-1RA Drugs ... 696
General Effects ... 696
Neurodegenerative Diseases ... 698
Myocardium ... 698
Blood Pressure and Heart Rate ... 699

Cardiovascular Impact of GLP-1RA Drugs ... 699
Thiazolidinediones ... 699
Sodium Glucose Co-transporter 2 ... 699
GLP-1RA ... 700
International Endorsements ... 701

Future Perspective ... 701

References ... 701

Abstract

Type 2 diabetes mellitus (T2DM) increases the risk of cardiovascular disease (CVD) and frequently presents a phenotype related to other comorbidities, such as obesity, nephropathy, hepatic steatosis, and increased risk of type 3 diabetes, i.e., neurodegenerative diseases associated with metabolic dysfunction. Although previous large clinical trials reported inconsistent reduction in all-cause mortality and CVD after glycemic control, the new pharmacologic classes of sodium-glucose co-transporter 2 inhibitors (SGLT2-i) and glucagon-like peptide-1 receptor agonists (GLP-1RAs) were associated with a reduction of major cardiovascular events (MACE), changing the paradigm of T2DM treatment. These classes of drugs are able to act concomitantly in distinct pathways of the metabolic syndrome. This chapter aims to review: (i) T2DM and obesity as CVD risk factors; (ii) previous clinical trials with neutral, deleterious, or beneficial results; (iii) why GLP-1RAs are an interesting class of drugs in the context of the metabolic syndrome; (iv) up-to-date information about GLP-1RAs in clinical trials; and (v) future perspectives.

I. Breder · A. C. Sposito (✉)
Atherosclerosis and Vascular Biology Laboratory, State University of Campinas (Unicamp), Campinas, Brazil
e-mail: andreisposito@gmail.com; sposito@unicamp.br

J. Faintuch and S. Faintuch (eds.), *Obesity and Diabetes*, https://doi.org/10.1007/978-3-030-53370-0_51

Keywords

GLP-1-RA · Cardiovascular risk · Diabetes · Obesity · Metabolic syndrome

Cardiovascular Risk of Obesity and Diabetes

A cohort study including 190,672 participants free of clinical cardiovascular disease (CVD) found that higher BMI increased the CVD risk for men and women by 21% and 32% in overweight, 67% and 85% in obese, and 314% and 253% in morbidly obese participants (Khan et al. 2018). In a meta-analysis of Mendelian randomized studies that grouped 881,692 participants, each standard deviation (SD) increase in BMI elevated the odds ratio of type 2 diabetes mellitus (T2DM) by 67%, and of coronary artery disease (CAD) by 20% (Riaz et al. 2018).

Visceral obesity is related to epicardial and pericardial fat accumulation, worsening the coronary lesions (Ding et al. 2009) by a pro-inflammatory environment mediated through endoplasmic reticulum (ER) stress, cell death, and production of pro-inflammatory cytokines (Rocha and Libby 2009). Moreover, obesity increases insulin resistance, which is a well-established risk factor for CAD, heart failure (HF), and sudden death. As a result, T2DM patients present a similar or higher risk of CVD events when compared to patients with previous myocardial infarction (Juutilainen et al. 2005).

Contrary to expectations, for many years consecutive clinical trials demonstrated that glycemic control was not a robust means of reducing cardiovascular events (Turner 1998; Patel et al. 2008; Hayward et al. 2015). Therefore, new anti-diabetic treatments that concomitantly reduce mortality or cardiovascular events are highly desired, even more so if they are able to reduce metabolic syndrome compounds, such as obesity, insulin resistance, and elevated blood pressure, while offering a good security profile (Sposito et al. 2018).

GLP-1RA Drugs

The incretin effect was first described in the 1960s (Elrick et al. 1964). It was characterized as a greater insulin release after oral meal intake, compared with intravenous glucose. Later, gastric inhibitory polypeptide (GIP) (1970s) (Brown et al. 1975) and GLP-1 (1980s) (Mojsov et al. 1987) were described as incretin mimetic hormones. In native conditions GLP-1 is released by L-cells presents in the distal ileum and colon, which are stimulated mainly by ingested fat and carbohydrates. The action of GLP-1 happens through its receptor, a G-protein present in the pancreas, gastrointestinal tract, lungs, kidney, brain, arteries, and heart. In natural conditions, the hormone is degraded in 1–2 min by the enzymes dipeptidyl peptidase-4 (DPP-4) and neutral endopeptidase (Mojsov et al. 1987). For this reason, molecular modifications were made in an effort to reduce this quick degradation, allowing GLP-1 to be used medically.

After molecular changes, two distinct groups were produced: (i) Exendin-4 derivatives, synthesized from the saliva of the lizard *Heloderma suspectum*, which have approximately 50% homology with native GLP-1; and (ii) human GLP-1 analogues, with greater than 90% of homology (Sposito et al. 2018; Røder 2018). Subsequently, these groups underwent differentiations in molecular characteristics, which account for the difference in half-life between the first generation (hours) and the second generation (days) (Table 51.1).

General Effects

Although the apparent major effect of GLP-1 is to improve insulin release in a glycemic-dependent manner, many other pleiotropic effects in organs/systems have been reported

Table 51.1 General characteristics of GLP-1RAs

Drugs	Structure	Molecular Weight	Half-life
Exenatide	Exendin-4	4.2 KDa	2–4 h
Lixisenatide	Exendin-4	4.9 KDa	3.0 h
Exenatide ER	Exendin-4	4.2 KDa	7–14 days
Liraglutide	GLP-1 analogue	3.8 KDa	12 h
Dulaglutide	GLP-1 analogue	59.7 KDa	~4 days
Albiglutide	GLP-1 analogue	73 KDa	~5 days
Semaglutide	GLP-1 analogue	4.1 KDa	~7 days

Adapted from Sposito et al. (Sposito et al. 2018)

Pancreas

Activation of GLP-1 receptor leads to AMPc stimulation, which culminates in insulin release (Baggio and Drucker 2007) in a glycemic-dependent manner. Moreover, GLP-1RA reduces endoplasmic reticulum (ER) stress, a process present in diabetes that leads to reduced activity and apoptosis of β-cells (Rowlands et al. 2018). Many experiments have already demonstrated that GLP-1RAs are able to reduce ER stress and therefore protect against β-cell death. In Wolfram syndrome, a rare disease considered a pure model of ER stress, GLP-1 improved β-cell responsiveness inhibiting ER stress (Kondo et al. 2018).

Glucagon secretion in α-cells is also modulated, through both a direct and an indirect mechanism. The indirect action may happen due to insulin release, which exerts a negative feedback on glucagon secretion. The direct mechanism is related to an ability to inhibit glucagon secretion during hyperglycemia and increase glucagon secretion during hypoglycemia. This direct action may explain, at least in part, why patients with type 1 diabetes and no residual β-cells responded to GLP-1RA reducing glucagonemia and fasting glycaemia (Creutzfeldt et al. 1996), and why hypoglycemia is an uncommon adverse event in patients with T2DM using GLP-1RA therapies (Zhang et al. 2019).

Gastrointestinal Tract

GLP-1RAs inhibit gastric emptying through vagal stimulation (Abbott et al. 2005). The slowed gastric emptying explains the indirect reduced post-prandial glycemia, insulinemia, and triglyceridemia. Even in patients with type 1 diabetes, in whom GLP-1 is not able to induce insulin secretion, a reduced glycemic excursion secondary to the lower gastric emptying has been described (Baggio and Drucker 2007; Dupre et al. 1995).

Renal

The LEADER (liraglutide) and SUSTAIN-6 (semaglutide) clinical trials reported approximately 36% reduction of nephropathy events after GLP-1RA treatment (Verma et al. 2019). Recently, a post hoc analysis of REWIND also demonstrated renal protection by dulaglutide, like liraglutide and semaglutide, mainly by reduced albuminuria (Gerstein et al. 2019). Renal protection could happen through both indirect and direct mechanisms. The indirect are: (i) increased urinary sodium excretion, (ii) decreased systemic tumor necrosis factor (TNF)-α, interleukin (IL)-6, and IL-1, (iii) decreased oxidative stress, (iv) decreased protein intake, and (v) reduced blood pressure (DeFronzo 2017). The direct protection of the kidneys by GLP-1RAs was demonstrated by an experiment where podocytes were co-cultured with palmitic acid and GLP-1RA, presenting lower rates of apoptosis and improving the levels of autophagy inhibitor markers, such as nephrin and podocin (Guo et al. 2017).

Liver

GLP-1RAs may improve non-alcoholic fatty liver disease (NAFLD) by direct and indirect mechanisms. The direct are related to AMP-activated protein kinase (AMPK), an enzyme that senses energy expenditure fluctuations, switching the metabolism from lipogenesis to lipid oxidation, targeting ATP formation. When there is triglyceride accumulation the phosphorylation of AMPK is inhibited, worsening insulin resistance, local inflammation, and oxidative stress (Madiraju et al. 2016). Liraglutide, a GLP-1RA, improves AMPK phosphorylation, restoring lipid and glucose metabolism (Yu et al. 2019). Indirect effects are linked to weight loss, triglycerides reduction, and glycemic improvement. Clinical trials have already demonstrated that liraglutide reduces hepatic fat content (Petit et al. 2017; Feng et al. 2017; Armstrong et al. 2013).

Adipose Tissue and Energetic Control

GLP-1 receptors are expressed in the nuclei related to brown adipose tissue (BAT) control, in the brain. Therefore, it seems reasonable that GLP-1RAs could interfere with body temperature. In preclinical studies GLP-1 activated BAT with browning of white adipose tissue (WAT) after intra-cerebral injection. This may happen through AMPK activation and/or by sympathetic stimulation. However when the medicine is injected peripherally results are not so consistent, leading to the hypothesis that peripheral vagal stimulation could interfere with this pathway. In clinical studies the results are also not conclusive, varying from neutral to stimulating effects of BAT and browning of WAT (Beiroa et al. 2014; López et al. 2015).

Neurodegenerative Diseases

At least two clinical trials demonstrated some protective effects of GLP-1RAs on the central nervous system. Exenatide extended-release was tested on patients with Parkinson's disease, resulting in a significant improvement on the Movement Disorder Society-Unified Parkinson's Disease Rating Scale (MDS-UPDRS) part III (motor) scale (Athauda et al. 2017). In another trial with Alzheimer's disease, liraglutide prevented the decline of glucose consumed by the brain, although no effect on cognition was reported (Gejl et al. 2016). Preclinical studies have demonstrated that GLP-1RAs inhibit inflammation, promote mitochondrial biogenesis, stimulate neurogenesis, and restore neuronal insulin signaling inhibiting ER neuronal stress and apoptosis (Panagaki et al. 2017). In a rat model of Wolfram syndrome, 6 months of liraglutide treatment reduced neuroinflammation, ER stress, and the death of retinal ganglion cells (Seppa et al. 2019).

Cardiovascular: Arteries

GLP-1 receptors are present in the microvascular endothelium and smooth muscle cells (Bangshaab et al. 2019), showing direct effects on the artery wall, as demonstrated in cultured cells, animal models, and clinical trials. GLP-1-RAs activate protein kinase A (PKA)/ERK pathways, stimulating nitric oxide synthesis and thereby improving endothelial vasomotor function (Wei et al. 2016). GLP-1RAs also reduce vascular inflammation and oxidative stress, inhibiting the pro-atherogenic environment in vessels, thus stabilizing the plaques against rupture (Balestrieri et al. 2015). In clinical trials, GLP-1RAs consistently reduced carotid pulse wave velocity (PWV), what also suggests improvement in arterial stiffness.

On the other hand, a meta-analysis grouping five studies demonstrated that GLP-1RAs had a neutral effect on vasodilation, as evaluated by Flow-Mediated Dilatation (FMD). It is important to note that these studies had high heterogeneity and small numbers of enrolled patients, and did not perform FMD using a modern method, including a continuous evaluation and the use of stereotaxic probe stabilizer (Batzias et al. 2018; Thijssen et al. 2019).

Myocardium

GLP-1-receptors were found in the myocardium, explaining in part the pleiotropic effects on the cardiovascular system. (i) In isolated hearts of mice, GLP-1RAs led to enhancement of glucose update and coronary flow (Ban et al. 2008). (ii) Intravenous GLP-1RAs infusion improved regional and global left ventricular function in patients with severe systolic dysfunction after coronary angioplasty (Nikolaidis et al. 2004). (iii) Exenatide administered over a 6-month period to patients diagnosed with ST-segment elevation after myocardial infarction resulted in lower creatine kinase myocardial band (CK-MB) and troponin I levels, and improved the global longitudinal strain (Woo et al. 2013). (iv) Treatment-naïve patients diagnosed with T2DM receiving GLP-1-RA showed improved left ventricular strain, left ventricular twisting, and reduced N-terminal-pro hormone brain natriuretic peptide (NT-proBNP) (Lambadiari et al. 2018).

However, in a placebo-controlled crossover protocol of 41 patients diagnosed with T2DM and stable coronary disease, liraglutide did not improved left ventricular ejection, global longitudinal strain, or strain rate (Kumarathurai et al. 2016). Also, the FIGHT trial enrolled a total of 300 recently hospitalized patients diagnosed with established HF and reduced left ventricular failure, and no reduction in deaths, re-hospitalization for HF, or any echocardiographic parameters emerged, when comparing liraglutide to placebo (Margulies et al. 2016). Taking these data together, there is no consensus about GLP-1RA improving HF.

Blood Pressure and Heart Rate

GLP-1RAs have an anti-hypertensive effect that seems to be independent of the improvement of HbA1c or weight reduction, and is additive to anti-hypertensive drugs. A large meta-analysis involving 33 trials demonstrated a reduction of approximately 2.2 mmHg after long-term use of GLP-1RA (Katout et al. 2014). Another meta-analysis grouping 31 trials demonstrated a reduction in systolic blood pressure of approximately 2.4 mmHg (Robinson et al. 2013).

Higher heart rate (HR) is associated with all-cause mortality (Jensen et al. 2013). GLP-1 receptors are also found on the sinoatrial node, stimulating HR. In fact, both short- and long-term administration of GLP-1 leads to higher HR and reduced HR variability (HRV), even after metabolic improvement and weight reduction (Smits et al. 2017; Kumarathurai et al. 2017). However, the exact mechanisms in humans are still unclear and were not associated with a worse cardiovascular outcome (Fig. 51.1).

Cardiovascular Impact of GLP-1RA Drugs

Early clinical trials regarding CVD in diabetes, such as UKPDS (United Kingdom Prospective Diabetes Study) (Holman et al. 2008), ACCORD (Action to Control Cardiovascular Risk in Diabetes) (Gerstein et al. 2008), ADVANCE (Action in Diabetes and Vascular Disease: Preterax and Diamicron-MR Controlled Evaluation) (Patel et al. 2008), and VADT (Veterans Affairs Diabetes Trial) (Hayward et al. 2015), did not find a consistent reduced risk of mortality or cardiovascular events after intensive glycemic control (Eckel et al. 2019).

Thiazolidinediones

In 1999, a meta-analysis grouping 42 clinical trials with rosiglitazone, a thiazolidinedione, found a significantly higher risk of myocardial infarction and a borderline risk of death from cardiovascular causes (Nissen and Wolski 2007). Studies on dipeptidyl peptidase-4 inhibitors (DPP4-i) mainly reported a neutral effect on CVD. A meta-analysis involving 40,781 participants included in DPP4-i trials reinforced that this class of drugs was related to higher or lower all-cause mortality and cardiovascular mortality (Zheng et al. 2018). Saxagliptin, a member of this class, was the only one associated with significantly higher incidence of hospitalization for HF (Scirica et al. 2014).

Sodium Glucose Co-transporter 2

EMPA-REG (empagliflozin) (Steiner 2016), CANVAS (canagliflozin) (Guthrie 2018), and

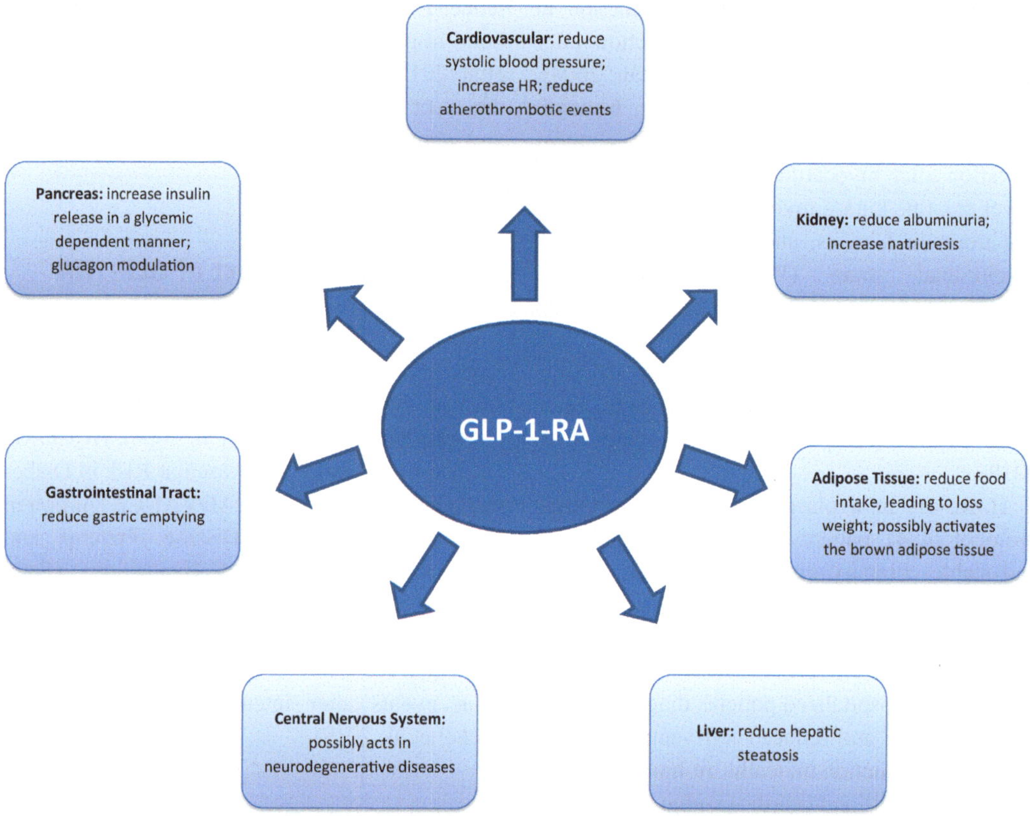

Fig. 51.1 Metabolic effects of GLP-1RAs

DECLARE (dapagliflozin) (Wiviott et al. 2018) studies reported a benefit in CV outcomes, reducing the incidence of HF hospitalization and CV deaths. A meta-analysis including these three randomized controlled trials also found a relative risk reduction for 3-point major adverse cardiovascular events (MACE) in individuals on secondary prevention (Zelniker et al. 2019), suggesting a smaller but significant effect of SGLT2-i on atherothrombotic events.

GLP-1RA

The EXSCEL trial studied exendin-4 derivatives, represented by exenatide, in 14,752 patients, while the ELIXA trial studied lixisenatide in 6068 patients. These drugs reached statistical significance regarding non-inferiority ($p < 0.001$), but not for superiority to placebo ($p = 0.06$ and 0.81, respectively). Among GLP-1 analogues, liraglutide (LEADER trial), subcutaneous semaglutide (SUSTAIN-6 trial), and dulaglutide (REWIND trial) showed significance for both inferiority and superiority, in contrast to oral semaglutide (PIONEER 6). In a meta-analysis of oral semaglutide grouping a total of 9890 patients, no superiority was found in cardiovascular protection compared with active control by liraglutide, empagliflozin, or sitagliptin (Avgerinos et al. 2019).

Another recent meta-analysis combined a total of 56,004 participants and included ELIXA (lixisenatide), EXSCEL (exenatide), LEADER (liraglutide), SUSTAIN-6 (subcutaneous semaglutide), Harmony Outcomes (albiglutide), REWIND (dulaglutide), and PIONEER 6 (oral semaglutide) (Kristensen et al. 2019), concluding that GLP-1RAs improved all the endpoints: MACE, by 12%; death from cardiovascular causes, by 12%; deaths by fatal and non-fatal stroke, by 9%; and all-cause mortality, by 12%. Interestingly, for the first time GLP-1-RAs were demonstrated to reduce in 9% the risk of admission by HF, hypothetically secondary to myocardial infarction reduction.

In addition, our group previously published a meta-analysis (Sposito et al. 2018) comparing the efficacy related to pharmacological composition between the clinical trials of exendin-4-based therapy, ELIXA (lixenatide) and EXSCEL (exenatide), versus GLP-1 analogues LEADER (liraglutide), SUSTAIN-6 (semaglutide), and Harmony (Albiglutide). When evaluating cardiovascular death, non-fatal myocardial infarction, non-fatal stroke, and hospitalization for unstable angina and HF, GLP-1-analogs were superior to placebo, unlike the exendin-4 derivatives (risk ratio 0.96 vs 0.85; 1.00 vs 0.82; 0.92 vs 0.86).

International Endorsements

The European Society of Cardiology indicated that GLP-1RAs or SGLT2-i should be the first line therapy for high cardiovascular risk patients (Kristensen et al. 2019), while the European Association for the Study of Diabetes (EASD)-American Diabetes Association (ADA) consensus suggested the prescription of GLP-1RA in high-risk patients with T2DM (Cosentino et al. 2019; Davies et al. 2018). Featuring a good safety profile, acting on diverse components of the metabolic syndrome puzzle, and improving cardiovascular outcomes, GLP-1RAs seem to be a very promising class of drugs to be offered in clinical practice.

Future Perspective

SUSTAIN-9 (Zinman et al. 2019), AWARD-10 (Ludvik et al. 2018), and DURATION-8 (Frías et al. 2016) reported additive metabolic and glycemic control when GLP-1RAs were prescribed together with SGLT2-i therapies. Although this is provocative, there have been no studies of the concomitant use of both regarding hard cardiovascular outcomes.

Type 1 diabetes seems to be another possible target of GLP-1 use. When liraglutide was added to insulin pump therapy, patients presented a significant reduction in both HbA1c and weight, and improvement in the time in range of glycemia target, without increased hypoglycemic events (Dejgaard et al. 2019).

Finally, phase II trials of two co-agonists GLP-1/GIP and GLP-1/glucagon and three co-agonists (GLP/GIP/glucagon) showed good results regarding glycemic control, glycemic excursions, and weight loss, with promising implications for clinical practice (Petersen et al. 2019).

References

Abbott CR, Monteiro M, Small CJ, Sajedi A, Smith KL, Parkinson JRC et al (2005) The inhibitory effects of peripheral administration of peptide YY 3-36 and glucagon-like peptide-1 on food intake are attenuated by ablation of the vagal-brainstem-hypothalamic pathway. Brain Res. https://doi.org/10.1016/j.brainres.2005.03.011

Armstrong MJ, Houlihan DD, Rowe IA, Clausen WHO, Elbrønd B, Gough SCL et al (2013) Safety and efficacy of liraglutide in patients with type 2 diabetes and elevated liver enzymes: individual patient data meta-analysis of the LEAD program. Aliment Pharmacol Ther. https://doi.org/10.1111/apt.12149

Athauda D, Maclagan K, Skene SS, Bajwa-Joseph M, Letchford D, Chowdhury K et al (2017) Exenatide once weekly versus placebo in Parkinson's disease: a randomised, double-blind, placebo-controlled trial. Lancet. https://doi.org/10.1016/S0140-6736(17)31585-4

Avgerinos I, Michailidis T, Liakos A, Karagiannis T, Matthews DR, Tsapas A et al (2019) Oral semaglutide for type 2 diabetes: a systematic review and meta-analysis. Diabetes Obes Metab. https://doi.org/10.1111/dom.13899

Baggio LL, Drucker DJ (2007) Biology of incretins: GLP-1 and GIP. Gastroenterology. https://doi.org/10.1053/j.gastro.2007.03.054

Balestrieri ML, Rizzo MR, Barbieri M, Paolisso P, D'Onofrio N, Giovane A et al (2015) Sirtuin 6 expression and inflammatory activity in diabetic atherosclerotic plaques: effects of incretin treatment. Diabetes. https://doi.org/10.2337/db14-1149

Ban K, Noyan-Ashraf MH, Hoefer J, Bolz SS, Drucker DJ, Husain M (2008) Cardioprotective and vasodilatory actions of glucagon-like peptide 1 receptor are mediated through both glucagon-like peptide 1 receptor-dependent and -independent pathways. Circulation. https://doi.org/10.1042/CS20171299

Bangshaab M, Gutierrez A, Huynh KD, Knudsen JS, Arcanjo DDR, Petersen AG et al (2019) Different mechanisms involved in liraglutide and glucagon-like peptide-1 vasodilatation in rat mesenteric small arteries. Br J Pharmacol. https://doi.org/10.1111/bph.14534

Batzias K, Antonopoulos AS, Oikonomou E, Siasos G, Bletsa E, Stampouloglou PK et al (2018) Effects of newer Antidiabetic drugs on endothelial function and arterial stiffness: a systematic review and meta-analysis. J Diabetes Res. https://doi.org/10.1155/2018/1232583

Beiroa D, Imbernon M, Gallego R, Senra A, Herranz D, Villarroya F et al (2014) GLP-1 agonism stimulates brown adipose tissue thermogenesis and browning through hypothalamic AMPK. Diabetes. https://doi.org/10.2337/db14-030

Brown JC, Dryburgh JR, Ross SA, Dupre J (1975) Identification and actions of gastric inhibitory polypeptide. Recent Prog Horm Res. https://doi.org/10.1016/b978-0-12-571131-9.50017-7

Cosentino F, Grant PJ, Aboyans V, Bailey CJ, Ceriello A, Delgado V et al (2019) 2019 ESC guidelines on diabetes, pre-diabetes, and cardiovascular diseases developed in collaboration with the EASD. Eur Heart J. https://doi.org/10.1093/eurheartj/ehz486

Creutzfeldt WOC, Kleine N, Willms B, Ørskov C, Holst JJ, Nauck MA (1996) Glucagonostatic actions reduction of fasting hyperglycemia by exogenous glucagon-like peptide I(7-36) amide in type I diabetic patients. Diabetes Care. https://doi.org/10.2337/diacare.19.6.580

Davies MJ, D'Alessio DA, Fradkin J, Kernan WN, Mathieu C, Mingrone G et al (2018) Management of hyperglycemia in type 2 diabetes, 2018. A consensus report by the American Diabetes Association (ADA) and the European Association for the Study of Diabetes (EASD). Diabetes Care. https://doi.org/10.2337/dci18-0033

DeFronzo RA (2017) Combination therapy with GLP-1 receptor agonist and SGLT2 inhibitor. Diabetes Obes Metab. https://doi.org/10.1111/dom.12982

Dejgaard TF, Schmidt S, Frandsen CS, Vistisen D, Madsbad S, Andersen HU et al (2019) Liraglutide reduces hyperglycaemia and body weight in overweight, dysregulated insulin-pump-treated patients with type 1 diabetes: the lira pump trial - a randomized, double-blinded, placebo-controlled trial. Diabetes Obes Metab. https://doi.org/10.1111/dom.13911

Ding J, Hsu FC, Harris TB, Liu Y, Kritchevsky SB, Szklo M et al (2009) The association of pericardial fat with incident coronary heart disease: the multi-ethnic study of atherosclerosis (MESA). Am J Clin Nutr. https://doi.org/10.3945/ajcn.2008.27358

Dupre J, Behme MT, Hramiak IM, McFarlane P, Paul Williamson M, Zabel P et al (1995) Glucagon-like peptide I reduces postprandial glycemic excursions in IDDM. Diabetes. https://doi.org/10.2337/diab.44.6.626

Eckel RH, Farooki A, Henry RR, Koch GG, Leiter LA (2019) Cardiovascular outcome trials in type 2 diabetes: what do they mean for clinical practice? Clin Diabetes. https://doi.org/10.2337/cd19-0001

Elrick H, Stilmmler L, Hlad CJ, Arai Y (1964) Plasma insulin response to oral and intravenous glucose administration. J Clin Endocrinol Metab. https://doi.org/10.1210/jcem-24-10-1076

Feng W, Gao C, Bi Y, Wu M, Li P, Shen S et al (2017) Randomized trial comparing the effects of gliclazide, liraglutide, and metformin on diabetes with non-alcoholic fatty liver disease. J Diabetes 9 (8):800–809

Frías JP, Guja C, Hardy E, Ahmed A, Dong F, Öhman P et al (2016) Exenatide once weekly plus dapagliflozin once daily versus exenatide or dapagliflozin alone in patients with type 2 diabetes inadequately controlled with metformin monotherapy (DURATION-8): a 28 week, multicentre, double-blind, phase 3, randomised control. Lancet Diabetes Endocrinol. https://doi.org/10.1016/S2213-8587(16)30267-4

Gejl M, Gjedde A, Egefjord L, Møller A, Hansen SB, Vang K et al (2016) In Alzheimer's disease, 6-month treatment with GLP-1 analog prevents decline of brain glucose metabolism: randomized, placebo-controlled, double-blind clinical trial. Front Aging Neurosci. https://doi.org/10.3389/fnagi.2016.00108

Gerstein HC, Miller ME, Byington RP, Goff DC, Bigger JT, Buse JB et al (2008) Effects of intensive glucose lowering in type 2 diabetes. N Engl J Med 358 (24):2545–2559

Gerstein HC, Colhoun HM, Dagenais GR, Diaz R, Lakshmanan M, Pais P et al (2019) Dulaglutide and renal outcomes in type 2 diabetes: an exploratory analysis of the REWIND randomised, placebo-controlled trial. Lancet. https://doi.org/10.1016/S0140-6736(19)31150-X

Guo H, Wang B, Li H, Ling L, Niu J, Gu Y (2017) Glucagon-like peptide-1 analog prevents obesity-related glomerulopathy by inhibiting excessive autophagy in podocytes. Am J Physiol Physiol. https://doi.org/10.1152/ajprenal.00302.2017

Guthrie R (2018) Canagliflozin and cardiovascular and renal events in type 2 diabetes. Postgrad Med. https://doi.org/10.1080/00325481.2018.1423852

Hayward RA, Reaven PD, Wiitala WL, Bahn GD, Reda DJ, Ge L et al (2015) Follow-up of glycemic control and cardiovascular outcomes in type 2 diabetes. N Engl J Med. https://doi.org/10.1056/NEJMoa1414266

Holman RR, Paul SK, Bethel MA, Neil HAW, Matthews DR (2008) Long-term follow-up after tight control of blood pressure in type 2 diabetes. N Engl J Med. https://doi.org/10.1056/NEJMoa0806359

Jensen MT, Suadicani P, Hein HO, Gyntelberg F (2013) Elevated resting heart rate, physical fitness and all-cause mortality: a 16-year follow-up in the

Copenhagen male study. Heart. https://doi.org/10.1136/heartjnl-2012-303375

Juutilainen A, Lehto S, Rönnemaa T, Pyörälä K, Laakso M (2005) Type 2 diabetes as a "coronary heart disease equivalent": an 18-year prospective population-based study in Finnish subjects. Diabetes Care. https://doi.org/10.2337/diacare.28.12.2901

Katout M, Zhu H, Rutsky J, Shah P, Brook RD, Zhong J et al (2014) Effect of GLP-1 mimetics on blood pressure and relationship to weight loss and glycemia lowering: results of a systematic meta-analysis and meta-regression. Am J Hypertens. https://doi.org/10.1093/ajh/hpt196

Khan SS, Ning H, Wilkins JT, Allen N, Carnethon M, Berry JD et al (2018) Association of body mass index with lifetime risk of cardiovascular disease and compression of morbidity. JAMA Cardiol. https://doi.org/10.1001/jamacardio.2018.0022

Kondo M, Tanabe K, Amo-Shiinoki K, Hatanaka M, Morii T, Takahashi H et al (2018) Activation of GLP-1 receptor signalling alleviates cellular stresses and improves beta cell function in a mouse model of Wolfram syndrome. Diabetologia 61:2189–2201

Kristensen SL, Rørth R, Jhund PS, Docherty KF, Sattar N, Preiss D et al (2019) Cardiovascular, mortality, and kidney outcomes with GLP-1 receptor agonists in patients with type 2 diabetes: a systematic review and meta-analysis of cardiovascular outcome trials. Lancet Diabetes Endocrinol. https://doi.org/10.1016/S2213-8587(19)30249-9

Kumarathurai P, Anholm C, Nielsen OW, Kristiansen OP, Mølvig J, Madsbad S et al (2016) Effects of the glucagon-like peptide-1 receptor agonist liraglutide on systolic function in patients with coronary artery disease and type 2 diabetes: a randomized double-blind placebo-controlled crossover study. Cardiovasc Diabetol. https://doi.org/10.1186/s12933-016-0425-2

Kumarathurai P, Anholm C, Larsen BS, Olsen RH, Madsbad S, Kristiansen O et al (2017) Effects of liraglutide on heart rate and heart rate variability: a randomized, double-blind, placebo-controlled crossover study. Diabetes Care. https://doi.org/10.2337/dc16-1580

Lambadiari V, Pavlidis G, Kousathana F, Varoudi M, Vlastos D, Maratou E et al (2018) Effects of 6-month treatment with the glucagon like peptide-1 analogue liraglutide on arterial stiffness, left ventricular myocardial deformation and oxidative stress in subjects with newly diagnosed type 2 diabetes. Cardiovasc Diabetol. https://doi.org/10.1186/s12933-017-0646-z

López M, Diéguez C, Nogueiras R (2015) Hypothalamic GLP-1: the control of BAT thermogenesis and browning of white fat. Adipocytes. https://doi.org/10.4161/21623945.2014.983752

Ludvik B, Frías JP, Tinahones FJ, Wainstein J, Jiang H, Robertson KE et al (2018) Dulaglutide as add-on therapy to SGLT2 inhibitors in patients with inadequately controlled type 2 diabetes (AWARD-10): a 24-week, randomised, double-blind, placebo-controlled trial. Lancet Diabetes Endocrinol

Madiraju AK, Alves T, Zhao X, Cline GW, Zhang D, Bhanot S et al (2016) Argininosuccinate synthetase regulates hepatic AMPK linking protein catabolism and ureagenesis to hepatic lipid metabolism. Proc Natl Acad Sci USA. https://doi.org/10.1073/pnas.1606022113

Margulies KB, Hernandez AF, Redfield MM, Givertz MM, Oliveira GH, Cole R et al (2016) Effects of liraglutide on clinical stability among patients with advanced heart failure and reduced ejection fraction: a randomized clinical trial. JAMA. https://doi.org/10.1001/jama.2016.10260

Mojsov S, Weir GC, Habener JF (1987) Insulinotropin: glucagon-like peptide I (7-37) co-encoded in the glucagon gene is a potent stimulator of insulin release in the perfused rat pancreas. J Clin Invest. https://doi.org/10.1172/JCI112855

Nikolaidis LA, Mankad S, Sokos GG, Miske G, Shah A, Elahi D et al (2004) Effects of glucagon-like Peptide-1 in patients with acute myocardial infarction and left ventricular dysfunction after successful reperfusion. Circulation. https://doi.org/10.1161/01.CIR.0000120505.91348.58

Nissen SE, Wolski K (2007) Effect of rosiglitazone on the risk of myocardial infarction and death from cardiovascular causes. N Engl J Med 356:2457–2471

Panagaki T, Michael M, Hölscher C (2017) Liraglutide restores chronic ER stress, autophagy impairments and apoptotic signalling in SH-SY5Y cells. Sci Rep. https://doi.org/10.1038/s41598-017-16488-x

Patel A, MacMahon S, Chalmers J, Neal B, Billot L, Woodward M et al (2008) Intensive blood glucose control and vascular outcomes in patients with type 2 diabetes. N Engl J Med. https://doi.org/10.1056/NEJMoa0802987

Petersen J, Strømgaard K, Frølund B, Clemmensen C (2019) Designing poly-agonists for treatment of metabolic diseases: challenges and opportunities. Drugs 79:1187–1197

Petit JM, Cercueil JP, Loffroy R, Denimal D, Bouillet B, Fourmont C et al (2017) Effect of liraglutide therapy on liver fat content in patients with inadequately controlled type 2 diabetes: the lira-NAFLD study. J Clin Endocrinol Metab. https://doi.org/10.1210/jc.2016-2775

Riaz H, Khan MS, Siddiqi TJ, Usman MS, Shah N, Goyal A et al (2018) Association between obesity and cardiovascular outcomes: a systematic review and meta-analysis of Mendelian randomization studies. JAMA Netw Open. https://doi.org/10.1001/jamanetworkopen.2018.3788

Robinson LE, Holt TA, Rees K, Randeva HS, O'Hare JP (2013) Effects of exenatide and liraglutide on heart rate, blood pressure and body weight: systematic review and meta-analysis. BMJ Open. https://doi.org/10.1136/bmjopen-2012-001986

Rocha VZ, Libby P (2009) Obesity, inflammation, and atherosclerosis. Nat Rev Cardiol. https://doi.org/10.1038/nrcardio.2009.55

Røder ME (2018) Major adverse cardiovascular event reduction with GLP-1 and SGLT2 agents: evidence and clinical potential. Ther Adv Chronic Dis. https://doi.org/10.1177/2040622317735283

Rowlands J, Heng J, Newsholme P, Carlessi R (2018) Pleiotropic effects of GLP-1 and analogs on cell signaling, metabolism, and function. Front Endocrinol (Lausanne). https://doi.org/10.3389/fendo.2018.00672

Scirica BM, Braunwald E, Raz I, Cavender MA, Morrow DA, Jarolim P et al (2014) Heart failure, saxagliptin, and diabetes mellitus: observations from the SAVOR-TIMI 53 randomized trial. Circulation. https://doi.org/10.1161/CIRCULATIONAHA.114.010389

Seppa K, Toots M, Reimets R, Jagomäe T, Koppel T, Pallase M et al (2019) GLP-1 receptor agonist liraglutide has a neuroprotective effect on an aged rat model of Wolfram syndrome. Sci Rep. https://doi.org/10.1038/s41598-019-52295-2

Smits MM, Tonneijck L, Muskiet MHA, Hoekstra T, Kramer MHH, Diamant M et al (2017) Heart rate acceleration with GLP-1 receptor agonists in type 2 diabetes patients: an acute and 12-week randomised, double-blind, placebo-controlled trial. Eur J Endocrinol. https://doi.org/10.1530/EJE-16-0507

Sposito AC, Berwanger O, De Carvalho LSF, Saraiva JFK (2018) GLP-1RAs in type 2 diabetes: mechanisms that underlie cardiovascular effects and overview of cardiovascular outcome data. Cardiovasc Diabetol. https://doi.org/10.1186/s12933-019-0825-1

Steiner S (2016) Empagliflozin, cardiovascular outcomes, and mortality in type 2 diabetes. Zeitschrift fur Gefassmedizin 13:1

Thijssen DHJ, Bruno RM, van Mil ACCM, Holder SM, Faita F, Greyling A et al (2019) Expert consensus and evidence-based recommendations for the assessment of flow-mediated dilation in humans. Eur Heart J 40 (30):2534–2547

Turner R (1998) Intensive blood-glucose control with sulphonylureas or insulin compared with conventional treatment and risk of complications in patients with type 2 diabetes (UKPDS 33). Lancet 352 (9131):837–853

Verma S, Bain SC, Monk Fries T, Mazer CD, Nauck MA, Pratley RE et al (2019) Duration of diabetes and cardiorenal efficacy of liraglutide and semaglutide: a post hoc analysis of the LEADER and SUSTAIN 6 clinical trials. Diabetes Obes Metab. https://doi.org/10.1111/dom.13698

Wei R, Ma S, Wang C, Ke J, Yang J, Li W et al (2016) Exenatide exerts direct protective effects on endothelial cells through the AMPK/Akt/eNOS pathway in a GLP-1 receptor-dependent manner. Am J Physiol Metab. https://doi.org/10.1152/ajpendo.00400.2015

Wiviott SD, Raz I, Bonaca MP, Mosenzon O, Kato ET, Cahn A et al (2018) Dapagliflozin and cardiovascular outcomes in type 2 diabetes. N Engl J Med. https://doi.org/10.1056/NEJMoa1812389

Woo JS, Kim W, Ha SJ, Kim JB, Kim SJ, Kim WS et al (2013) Cardioprotective effects of exenatide in patients with ST-segment-elevation myocardial infarction undergoing primary percutaneous coronary intervention: results of exenatide myocardial protection in revascularization study. Arterioscler Thromb Vasc Biol. https://doi.org/10.1161/ATVBAHA.113.301586

Yu P, Xu X, Zhang J, Xia X, Xu F, Weng J et al (2019) Liraglutide attenuates nonalcoholic fatty liver disease through adjusting lipid metabolism via SHP1/AMPK signaling pathway. Int J Endocrinol. https://doi.org/10.1155/2019/1567095

Zelniker TA, Wiviott SD, Raz I, Im K, Goodrich EL, Bonaca MP et al (2019) SGLT2 inhibitors for primary and secondary prevention of cardiovascular and renal outcomes in type 2 diabetes: a systematic review and meta-analysis of cardiovascular outcome trials. Lancet. https://doi.org/10.1016/S0140-6736(18)32590-X

Zhang Y, Parajuli KR, Fava GE, Gupta R, Xu W, Nguyen LU et al (2019) GLP-1 receptor in pancreatic A-cells regulates glucagon secretion in a glucose-dependent bidirectional manner. Diabetes. https://doi.org/10.2337/db18-0317

Zheng SL, Roddick AJ, Aghar-Jaffar R, Shun-Shin MJ, Francis D, Oliver N et al (2018) Association between use of sodium-glucose cotransporter 2 inhibitors, glucagon-like peptide 1 agonists, and dipeptidyl peptidase 4 inhibitors with all-cause mortality in patients with type 2 diabetes a systematic review and meta-analysis. JAMA. https://doi.org/10.1056/NEJMoa072761

Zinman B, Bhosekar V, Busch R, Holst I, Ludvik B, Thielke D et al (2019) Semaglutide once weekly as add-on to SGLT-2 inhibitor therapy in type 2 diabetes (SUSTAIN 9): a randomised, placebo-controlled trial. Lancet Diabetes Endocrinol. https://doi.org/10.1016/S2213-8587(19)30066-X

Diabetes and Treatments 52

Milan Obradovic, Emina Sudar-Milovanovic, Zoran Gluvic, Takashi Gojobori, Magbubah Essack, and Esma R. Isenovic

Contents

Introduction 706
History 706
Etiology and Pathophysiology of DMT1 and DMT2 707
Type 1 Diabetes 707
Type 2 Diabetes 707
Delivery of Insulin, Islet Transplantation, and Stem Cells 707
Noninvasive Insulin Administration 708
The Artificial Pancreas 708
Drugs 709
Pharmacotherapy of DMT1 709
Non-insulin Pharmacological Agents 710
Pharmacotherapy of DMT2 710
Therapeutic Aims 710
Monotherapy of DMT2 710
Combination Therapy for DMT2 711
Insulin Introduction 711
Exenatide, Liraglutide, Lixisenatide and Semaglutide 711
Sitagliptin, Vildagliptin, Saxagliptin, Linagliptin, and Alogliptin 712

M. Obradovic · E. Sudar-Milovanovic · E. R. Isenovic (✉)
Department of Radiobiology and Molecular Genetics, VINČA Institute of Nuclear Sciences - National Institute of the Republic of Serbia, Belgrade, Serbia
e-mail: isenovic@yahoo.com

Z. Gluvic
Clinic for Internal Medicine, Department of Endocrinology and Diabetes, Zemun Clinical Hospital, School of Medicine, University of Belgrade, Belgrade, Serbia

T. Gojobori · M. Essack
Computer, Electrical and Mathematical Sciences and Engineering Division (CEMSE), Computational Bioscience Research Center, Computer (CBRC), King Abdullah University of Science and Technology (KAUST), Thuwal, Kingdom of Saudi Arabia

Biological and Environmental Sciences and Engineering Division (BESE), King Abdullah University of Science and Technology (KAUST), Thuwal, Kingdom of Saudi Arabia

J. Faintuch and S. Faintuch (eds.), *Obesity and Diabetes*, https://doi.org/10.1007/978-3-030-53370-0_52

Acarbose, Miglitol, and Voglibose 712
Pioglitazone 712
Sulphonylureas and Glinides (Gliclazide, Glipizide, Glimepiride, Repaglinide) . 712
Dapagliflozin, Canagliflozin, and Empagliflozin 712
References 713

Abstract

The diabetes pandemic demands solutions for proper glycemic control and prevention of future chronic complications that could result in organ failure or comorbidities. In this regard, we now know that patients diagnosed with diabetes require individual management plans. Thus, new treatment management strategies have been designed to allow clinicians to tailor the most appropriate therapy for diabetes patients individually. These treatment management plans extend beyond defining the appropriate medications for patients; they provide a directive toward some acute and chronic complications that should be screened for, as they are historically known to occur with diabetes. Observing any of the complications or comorbidities requires the patient medication regimen to be adapted accordingly. This chapter describes such modern treatment plans for the two primary forms of diabetes, type 1 and type 2, based on both basic and clinical studies, later incorporated in various diabetes management guidelines and outlines expected future trends.

Keywords

Diabetes · Delivery of insulin · Islet transplantation · Stem cells therapy · Hyperbaric oxygen therapy · Pharmacotherapy of diabetes

Introduction

Diabetes mellitus (DM) is an endocrine disease, characterized by hyperglycemia and multiple metabolic disorders that cause serious local and systemic effects (Nair 2007; Forbes and Cooper 2013; Katsarou et al. 2017). Nowadays, DM is one of the biggest health problems and has reached pandemic proportions (Forouhi and Wareham 2014). According to the International Diabetes Federation, 425 million people worldwide have DM, with a tendency to be 629 million in 2045 (Cho et al. 2018). There are at least five types of DM, while the two primary forms are DM type 1 (DMT1) and DM type 2 (DMT2) (ADA 2010; Katsarou et al. 2017; Cho et al. 2018). Autoimmunity is a significant factor in the development of DMT1, while genetic predisposition and obesity are leading risk factors for the development of DMT2 (Al-Goblan et al. 2014; Nair 2007).

Long-term anti-diabetic pharmacotherapy and lifestyle adaptations are necessary to achieve glycemic control and decline multisystem disorders in DM patients (Rai et al. 2016; Katsarou et al. 2017).

History

Nearly one century ago, exogenous insulin became available (Banting and Best 1990; Maclean 1926). In the meantime, plenty of oral hypoglycemics were designed, and they achieve desirable results in patients with DMT2 (Butterfield et al. 1957; Krall et al. 1958). However, exogenous insulin is irreplaceable in the treatment of people with DMT1, being required in abundant cases of DMT2 as well (Handelsman et al. 2015). Despite constant progress in terms of the new therapeutic approaches, a comprehensive and efficacious cure for DM is still not accomplished. Current therapies for patients with DM have several shortfalls, including efficacy, timings, and glycemic control (Shah et al. 2016; Castle et al. 2017; Evans et al. 2011; Pathak et al.

2019). Also, these therapies produce numerous side effects like gastric irritation, injection phobia, diarrhea, and loss of appetite (Pathak et al. 2019; Rai et al. 2016; Zaric et al. 2019).

Etiology and Pathophysiology of DMT1 and DMT2

Diabetes mellitus (DM) is a metabolic disorder characterized by alterations and impairment in insulin and glucagon secretion and/or action that lead to hyperglycemia (ADA 2009; Girard 2017). Besides the increased blood glucose level, DM is also characterized by other biochemical disorders arising as a consequence of inadequate regulation of insulin synthesis/actions, in association with long-term injury, dysfunction, and failure of different organs, especially blood vessels and heart (Jovanovic et al. 2017; Obradovic et al. 2017; Sudar-Milovanovic et al. 2015, 2017; Obradovic et al. 2015; Soskic et al. 2011), however also nerves, kidneys, and eyes (ADA 2009).

The overall prevalence of DM among adults over 18 years of age is steadily growing, and it has been increasing more rapidly in countries with low- and middle-income economies. The growth in DM prevalence reflects the rise in overweight and obesity, which are a consequence of physical inactivity and unhealthy diets (Roglic 2016). The World Health Organization (WHO) shows DM prevalence grew from 4.7% (1980) to 8.5% (2014), being the seventh leading cause of death worldwide in 2016 (WHO 2018). Furthermore, it is one of the leading causes of stroke, heart and kidney failure, blindness, as well as lower limb amputation.

Type 1 Diabetes

DMT1 is a multifactorial autoimmune disease that develops under the influence of environmental or/and genetic factors (Atkinson et al. 2014; Ikegami et al. 2011). DMT1 mostly develops in patients at a young age, before the age of 30. The main characteristic of DMT1 is lack of insulin production, and DMT1 patients are dependent on exogenous insulin application.

DMT1 is an immune-mediated type of DM, and typically an autoimmune demolition of the insulin-secreting beta cells is based on DMT1 development. Factors involved in its pathogenesis trigger lymphocyte infiltration in pancreatic beta cells, and the consequent production of different proinflammatory cytokines responsible for the pancreatic beta cells destruction (Fatima et al. 2016).

Increased risk for DMT1 is generally recognized in patients by serological confirmation of an autoimmune process occurring in pancreatic islets/beta cells, considering that this is one of the first pathological alterations, and additionally by genetic marker determination (ADA 2009). Some patients with DMT1 can exhibit detectable insulin secretion, indicating some surviving beta cells, or an ongoing cycle of destruction and regeneration of such cells (Meier et al. 2005).

Type 2 Diabetes

It represents the vast majority (85–90%) of DM cases. DMT2 is a heterogeneous, progressive metabolic and endocrine illness, and it occurs as an interplay of various genetic as well as environmental factors. The underlying origins are insulin resistance in combination with deficient compensatory beta-cell reaction and adequate insulin secretion (ADA 2009). The impairment of insulin-secreting pancreatic beta cells shows progression over time.

When DMT2 patients with normal fasting plasma glucose, however with postprandial hyperglycemia, exhibit impaired insulin action, usually it is caused by a reduction of total insulin receptor numbers (Belfiore et al. 2009; Obradovic et al. 2019).

Delivery of Insulin, Islet Transplantation, and Stem Cells

The management of DM aims to recover glycemic control and reduced micro- and macro-vascular complications, by administrating pharmacological therapy and modifications of lifestyle. Insulin replacement represents the first-line option for insulin-dependent DM patients (Pathak et al. 2019). One of the significant challenges in the treatment of patients with DM is the efficacy of exogenous insulin in achieving long-term normal ranging glycemia (Yeh et al. 2012). One alternative is the use of continuous insulin infusion pumps (Yeh et al. 2012; Heller et al. 2017). Also, the creation of insulin analogs with different times of action, from rapid and short-acting to ultra-long acting, contributed to improvements in therapy (Shah et al. 2016; Pathak et al. 2019). Implementation of these technological advances does not prevent long-term insulin dependence, and adverse effects like invasiveness (Shah et al. 2016).

Noninvasive Insulin Administration

To avoid such complications, researchers proposed new routes for insulin administration beyond the standard subcutaneous route (Rys et al. 2018; Atkinson et al. 2014), such as nasal (Kullmann et al. 2018; Schmid et al. 2018), oral (Fonte et al. 2013), pulmonary (Mastrandrea 2010; Ledet et al. 2015), and transdermal delivery systems (Zaric et al. 2019).

Drug delivery carrier systems protect antidiabetic drugs from enzymatic degradation at the absorption site, and ensure delivery at the optimal and effective concentration for a more extended period at the target site (Rai et al. 2016). The encapsulation of insulin in particles increases its potential and allows its appropriate transport to the specific site to better control of DM (Rai et al. 2016; Zaric et al. 2019). These particles are usually microsized or nanosized. The microparticle system adjusts the pattern of drug release and improves the hypoglycemic effect of oral delivery of insulin (Wong et al. 2018). The main limiting factors of microparticle systems are the size of particles and their hydrophilic/hydrophobic nature that makes difficult their transport through biological membranes (Wong et al. 2018).

The use of nanocarriers reduces these obstacles (Bahman et al. 2019). The small size and structural diversity of nanoparticles increase the potential of insulin through better absorption and distribution, site specificity, and protection from enzymatic degradation (Bahman et al. 2019). Also, this form of application extends the release pattern of insulin and decreases the frequency of dosage, facilitating normoglycemia for a more extended period, up to 22 days (Peng et al. 2012). Other approaches considered for delivery of insulin and antidiabetics include encapsulation in liposomes, vesicular systems, and other nanoparticles (Souto et al. 2019).

The Artificial Pancreas

The use of continuous insulin infusion pumps and glucose monitoring has enabled a constant movement toward artificial pancreas development (Boughton and Hovorka 2019). Pancreas or islet transplantation seems to be the best choice to prevent dependence on insulin; however, donor shortages limit this option. Although islet transplantation was substantially enhanced in the past 2 decades, there are still limitations like instant blood-mediated inflammatory reaction, ischemia-induced loss of islet, harmful effects of immunosuppressive drugs, and apoptosis of transplanted cells (Bottino et al. 2018).

Stem cells generating new beta cells represent a promising approach for long-term treatment of DM (Stanekzai et al. 2012; Cierpka-Kmiec et al. 2019; Aguayo-Mazzucato and Bonner-Weir 2010). Many studies show that human embryonic stem cells can be used for beta cell generation and transplantation in patients with DMT1 (Cierpka-Kmiec et al. 2019; Kalra et al. 2018). Challenges are generation of genetically stable cells, survival rate of the cell, potential of transplanted cells, and ethical issues considering utilization of embryo-derived stem cells (Kalra et al. 2018). Some of these issues are overcome using induced pluripotent stem cells generated from somatic cells of patients with DMT1 to produced functional beta cells (Kalra et al. 2018; Kunisada et al. 2012).

Transplantation of insulin-producing human embryonic stem cells into streptozotocin-induced diabetic mice resulted in long-term normalization of blood glucose levels (Vegas et al. 2016). Yet loss of novel beta cells immediately after portal vein transplantation is in the range of 5–47% (Potter et al. 2014; Naziruddin et al. 2014). A current suggestion is co-transplantation of islets with different cell types, such as mesenchymal stem cells (MSC), bone marrow-derived MSC, endothelial colony-forming cells, and others (Kerby et al. 2013; de Souza et al. 2017; Borg et al. 2014; Corradi-Perini et al. 2017; Jung et al. 2014), to reduce losses of beta cells and increase their potential. This approach was shown to increase islet survival and function, and promote angiogenesis and better glycemic control due to the anti-apoptotic and pro-angiogenic effects of MSC (de Souza et al. 2017; Pathak et al. 2019).

Drugs

Non-insulin agents or those who sensitize insulin action enable the opening of new chapters in diabetes management (Grant and Kirkman 2015; Rowley et al. 2017; Leon and Maddox 2015). Some of the innovations also concern insulin treatment, namely insulin analogs with improved control of glycemia and lowered hypoglycemia risk, as well as the development of non-parenteral route/s of insulin administration (Biester et al. 2017; Fink et al. 2018). Such innovations are placed into clinical practice through precisely defined recommendations, such as those of the American Diabetes Association (ADA) (ADA 2018), American Association of Clinical Endocrinologists (AACE) (Garber et al. 2019), European Association for the Study of Diabetes (EASD) (Davies et al. 2018), and the National Institute for Health and Care Excellence (NICE) (McGuire et al. 2016). Even though these recommendations have subtle differences, they are oriented toward a patient-centered approach, and direct clinicians toward better and more uniform management of the patients (ADA 2018; Garber et al. 2019; Davies et al. 2018; McGuire et al. 2016). Well-controlled diabetes apparently slows down the atherosclerotic process, which is the base of associated micro- and macrovascular diabetes complications and major cardiovascular events, including ones with fatal outcomes (Zoungas et al. 2014).

Pharmacotherapy of DMT1

Current recommendations are focused on individually tailored management of DMT1, consisting of appropriate dietary habits and physical activities, along with administration of insulin. It is strongly advised to adapt insulin treatment to carbohydrate intake, pre-prandial glycaemia, and anticipated physical activity (ADA 2018; Handelsman et al. 2015; National Clinical Guideline Centre 2015).

The mainstay of DMT1 management is insulin, administered by pen device or continuous subcutaneous insulin infusion (CSII). The amount and number of divided insulin doses in children or adults immediately after DMT1 diagnosis are often small (1–2 divided doses), however often increases in time (Biester et al. 2017; Fink et al. 2018; Pickup 2019). In the last decades, there was a tendency to opt for insulin analogs with improved pharmacokinetics, pharmacodynamics, and safety compared to human insulin (Biester et al. 2017; Fink et al. 2018; Heinemann et al. 2017).

Rapid insulin analogs are active immediately after administration, with short-lived effects. Basal analogs have more prolonged effects than classic neutral protamine Hagedorn (NPH) human insulin regarding glycemia control, with lower hypoglycemic risk. The usual dose is 0.5 IU/kg of body weight. During puberty and in exceptional circumstances, insulin requirements are higher. Overall, CSII is widely recommended, and it can be used in patients older than 65 years (ADA 2018; Handelsman et al. 2015; National Clinical Guideline Centre 2015; REPOSE Study Group 2017; Beck et al. 2017).

In addition, pre-prandial inhaled insulin has been shown to be efficient in the form of rapid insulin analogs (B28 aspart-insulin); however, its use in DMT1 patients is not widespread as it was expected (ADA 2018; Heinemann and Parkin 2018).

Non-insulin Pharmacological Agents

Metformin, dipeptidyl peptidase 4 inhibitors (DPP-4i), glucagon-like peptide 1 receptor agonists (GLP-1 RA), and sodium-glucose cotransporter-2 inhibitors (SGLT-2i) are not Food and Drug Administration (FDA)-approved, despite showing beneficial effects in obese DMT1 patients (ADA 2018; Handelsman et al. 2015; Livingstone et al. 2017).

Pramlintide, an analog of the hormone amylin also secreted by pancreatic beta cells, is approved for use of DMT1 adults; however, it exhibits higher risk for hypoglycemia and obliged reduction of prandial insulin (Hieronymus and Griffin 2015). For patients with ineffective glycemic regulation or diabetics referred to renal transplantation, pancreas and islet transplantation could be the better treatment option (Gruessner 2011; Nakamura et al. 2019).

Pharmacotherapy of DMT2

Non-pharmacological measures are usually the preferred initial treatment for DMT2 patients. Apart from lifestyle changes (nutrition and physical activity) and eventual pharmacotherapy, an integral part of DMT2 management is the screening for cardiovascular disease (CVD) risk factors as well as the detection of chronic micro- and macro-vascular complications (ADA 2018; Garber et al. 2019; Davies et al. 2018; McGuire et al. 2016).

DMT2 management aims to achieve clinical and biochemical goals (actual profile and retrograde glycemic regulation), avoiding hypoglycemic episodes and gain in body weight in obese patients, as well as controlling the risk of atherosclerotic CVD (ASCVD) (ADA 2018).

Therapeutic Aims

HbA1c, morning glycemia, and 5-point daily glycemic profile should be adjusted to age, duration of DM, risk of hypoglycemia, and the presence of comorbidities and chronic vascular complications. The level of HbA1c is an essential marker for the assessment of retrograde glycemic control. In DM patients with no comorbidities and low hypoglycemia risk, HbA1c level $\leq 7\%$ (ADA 2018) indicates reasonable retrograde DM control for the previous 90–120 days. On the other hand, in DM patients with severe comorbidities and considerable risk of hypoglycemia, the acceptable level of HbA1c is $<8\%$ (ADA 2018; Garber et al. 2019; Davies et al. 2018; McGuire et al. 2016). A more stringent HbA1c goal of $<6.5\%$ is set in some special populations of individuals suffering from DM (ADA 2018; Garber et al. 2019; Davies et al. 2018; McGuire et al. 2016).

Pharmacotherapy of DMT2 is conducted as mono- or combination-therapy. The therapy chosen depends on biochemical and clinical factors, existing comorbidities, as well as side effects of administered drugs.

Monotherapy of DMT2

For patients with newly diagnosed DMT2 and HbA1c $<9\%$ (6) or $>6.5\%$ (9), metformin is the initial therapy, with a daily dose of 1.5–2.0 g. Metformin contributes to decrease in body weight, and reduces the risk of hypoglycemia. Gastrointestinal side effects are often transitional and dose-dependent. It is contraindicated if glomerular filtration rate (eGFR) is <30 ml/min. (Livingstone et al. 2017; Sanchez-Rangel and Inzucchi 2017).

In metformin-intolerant patients, acceptable alternatives are GLP1-RA, DPP-4i, alpha glucosidase inhibitors (AGi), and SGLT-2i. With the administration of such agents, the risk of hypoglycemia is lower, and there is no weight gain. Other alternatives are thiazolidinediones (TZD) and sulphonylureas (SFU) or glinides, however with more risk of hypoglycemia and weight gain (ADA 2018; Garber et al. 2019; Davies et al. 2018; McGuire et al. 2016).

Combination Therapy for DMT2

In DMT2 patients with HbA1c >9% (ADA 2018) or ≥7.5% (Garber et al. 2019; McGuire et al. 2016), as well as in those for whom metformin is not enough to achieve adequate glycemic regulation, another agent is added to the treatment regime (management intensification). Metformin-intolerant patients are administered two or more agents, with a complementary mechanism of action. There are fixed combinations that include metformin + DPP-4i/TZD/SFU on the market. Additionally, metformin with modified-release could be a suitable alternative for some of the metformin-intolerant. The acceptable level of HBA1c after management intensification should be <7%.

Insulin Introduction

If despite the use of non-pharmacological measures and dual pharmacotherapy, HbA1c is ≥7.5%, basal insulinization should be the solution. Insulin treatment starts with a plan of its administration, and continued use of metformin is recommended in suitable and tolerant patients (ADA 2018; Garber et al. 2019; Davies et al. 2018; McGuire et al. 2016).

Dose tapering or need for administration of other non-insulin glucose-lowering agents are relevant concerns (McGuire et al. 2016). If HbA1c >9%, indicating poor control, the patient should be conducted to dual oral treatment. Furthermore, if the patient initially presented with HbA1c ≥10% or with glycemia >16.7 mmol/L or is clinically symptomatic, combined treatment of insulin and oral antihyperglycemic therapy, or even multiple insulin injections, should be considered (ADA 2018). If an insulinization process of basal insulin introduction along with metformin and/or other non-insulin glucose-lowering agents fails to control DMT2, the addition of one or more doses of rapid-acting insulin or GLP-1RA is a possible alternative (ADA 2018; Garber et al. 2019; Davies et al. 2018; McGuire et al. 2016). Also, if a DMT2 patient is suffering from ASCVD, the addition of a CVD-beneficial agent such as empagliflozin, canagliflozin, dapagliflozin, or liraglutide is worth considering (Arnett et al. 2019; Zelniker et al. 2019; Wiviott et al. 2019; Furtado et al. 2019).

Exenatide, Liraglutide, Lixisenatide and Semaglutide

These are GLP-1RA peptides, structurally homologous to the natural incretin glucagon-like peptide-1 (GLP-1), and currently available for subcutaneous administration. Semaglutide has been approved by FDA in 2019 for oral use. GLP-1RA stimulate glucose-dependent pancreatic insulin secretion as well as reduce glucagon secretion, slowing down gastric emptying. They also significantly reduce body weight and lower HbA1c levels. They are devoid of negative effects on bone metabolism, the appearance of diabetic ketoacidosis (DKA), and congestive heart failure (CHF) deterioration. Some gastrointestinal side effects could be encountered (nausea, vomiting, bloating, gastroesophageal reflux disease, and gastroparesis); however, they often improve with time. The risk of hypoglycemia is low.

In studies on rodents, exenatide intake lead to hyperplasia of C cells and medullary thyroid cancer, while all GLP-1RA were associated with pancreatitis. Exenatide is not indicated in patients with eGFR < 30 ml/min. (ADA 2018; Garber et al. 2019; Davies et al. 2018; McGuire et al. 2016; Deacon 2019).

Sitagliptin, Vildagliptin, Saxagliptin, Linagliptin, and Alogliptin

These represent the family of DPP-4i, blocking the DPP4 enzyme involved in the degradation of incretins such as GLP-1 and gastric inhibitory peptide (GIP). As a result, more elevated incretin levels act on a simultaneous increase in glucose-dependent insulin secretion, and decrease in glucagon secretion. There is no convincing evidence regarding the higher risk of pancreatitis or pancreatic cancer with the use of DPP-4i. DPP-4i exhibit neutral effects regarding hypoglycemia risk and body weight change, and gastrointestinal side effects are not frequent. DPP-4i effectively reduce albuminuria and do not have negative effects on the bones and the appearance of DKA. Saxagliptin is not recommended for CHF patients. Also, renal dose adjustment is required when using all DPP-4i except for linagliptin (ADA 2018; Garber et al. 2019; Davies et al. 2018; McGuire et al. 2016; Deacon 2019; McGuire et al. 2019; Scirica et al. 2013).

Acarbose, Miglitol, and Voglibose

These AGi block the alpha-glucosidase enzyme involved in carbohydrate reabsorption in the gastrointestinal system. Digestive side effects can be frequent, including bloating, diarrhea, abdominal cramps, and mild elevation of liver enzymes. They are neutral regarding hypoglycemia risk, body weight change, bone disease, CHF deterioration, and the appearance of DKA. Renal dose adjustment is not required (ADA 2018; Garber et al. 2019; Davies et al. 2018; McGuire et al. 2016; Liu and Ma 2017).

Pioglitazone

This TZD activates the PPARγ receptors and, through determined signal pathways, decreases insulin resistance in various tissues, predominantly in skeletal muscles (Yki-Jarvinen 2004). The clinicians are under pressure to precisely select the patients for their use because of some associated side effects, such as body weight gain, fluid retention, higher risk of bone fractures, and bladder cancer (Wang et al. 2017; Mehtala et al. 2019). Hypoglycemia risk and body weight change are not common with moderate doses, and there are no associated gastrointestinal side effects. They occasionally exhibit mild negative effects on the bones, CHF deterioration, and the appearance of DKA. Renal dose adjustment is not required; however, TZDs are generally not recommended in any stage of renal failure due to fluid retention (ADA 2018; Garber et al. 2019; Davies et al. 2018; McGuire et al. 2016; Yki-Jarvinen 2004; Wang et al. 2017; Mehtala et al. 2019).

Sulphonylureas and Glinides (Gliclazide, Glipizide, Glimepiride, Repaglinide)

Such drugs mediate insulin secretion after binding to SUR Ki6.2 receptor, the sodium channel (Kalra and Gupta 2015). Significant effects are to be expected regarding hypoglycemia and weight gain. Mild negative effects on CHF deterioration are likely, however not on bones and DKA. Renal dose adjustment is required (ADA 2018; Garber et al. 2019; Davies et al. 2018; McGuire et al. 2016; Wang et al. 2018; Harsch et al. 2018).

Dapagliflozin, Canagliflozin, and Empagliflozin

SGLT-2i reduce proximal tubule glucose reabsorption by binding to the SGLT-2 receptors. Ascending urinary infections, chronic and treatment-resistant urinary and vaginal candidiasis, elevation of LDL-cholesterol, dehydration, and hypotension, as well as DKA have been registered (ADA 2018; Garber et al. 2019; Davies et al. 2018; McGuire et al. 2016; Wanner and Marx 2018; Lupsa and Inzucchi 2018). They are usually safe regarding hypoglycemia risk and contribute to bodyweight reduction. There are no associated gastrointestinal side effects.

SGLT-2i use is not indicated in patients with eGFR < 45 (60) ml/min. Empagliflozin use requires special attention regarding potential negative effects on the bones, as well as the use of all SGLT-2i regarding DKA appearance (ADA 2018; Garber et al. 2019; Davies et al. 2018; McGuire et al. 2016; Singh and Kumar 2018). Some studies point out to positive effects of empagliflozin and dapagliflozin in cases with the occurrence of major CVD, ASCVD, and CHF deterioration (Arnett et al. 2019; Zelniker et al. 2019; Wiviott et al. 2019; Furtado et al. 2019).

In exceptional circumstances, when it is impossible to control DMT2 with the usual mono/dual/combined therapy, colesevelam (Ooi and Loke 2014) and quick-release bromocriptine (Lopez Vicchi et al. 2016) could be used. Their mechanism in blood sugar lowering is not known; however, the risk of hypoglycemia is low.

Acknowledgments This work is part of the collaboration between the Laboratory of Radiobiology and Molecular Genetics, Vinca Institute of Nuclear Sciences, University of Belgrade, Belgrade, Serbia, Department of Endocrinology and Diabetes, Zemun Clinical Hospital, School of Medicine, University of Belgrade, Belgrade, Serbia, and, Computational Bioscience Research Center (CBRC) at King Abdullah University of Science and Technology (KAUST). This work was funded by the Ministry of Education, Science and Technological Development of the Republic of Serbia, and has been supported by the KAUST grant OSR#4129 (awarded to E.R.I. and V.B. B.), which also supported E.R.I., M.O., and E.S.M. M.E. has been supported by the KAUST Office of Sponsored Research (OSR) Award no. FCC/1/1976-17-01, and TG by the King Abdullah University of Science and Technology (KAUST) Base Research Fund (BAS/1/1059-01-01).

References

ADA (2009) Diagnosis and classification of diabetes mellitus. Diabetes Care 32(Suppl 1):S62–S67. https://doi.org/10.2337/dc09-S062

ADA (2010) Diagnosis and classification of diabetes mellitus. Diabetes Care 33(Suppl 1):S62–SS9. https://doi.org/10.2337/dc10-S062

ADA (2018) Standards of medical care in diabetes—2018 abridged for primary care providers. Clin Diabetes 36(1):14–37. https://doi.org/10.2337/cd17-0119

Aguayo-Mazzucato C, Bonner-Weir S (2010) Stem cell therapy for type 1 diabetes mellitus. Nat Rev Endocrinol 6(3):139–148. https://doi.org/10.1038/nrendo.2009.274

Al-Goblan AS, Al-Alfi MA, Khan MZ (2014) Mechanism linking diabetes mellitus and obesity. Diabetes Metab Syndr Obes 7:587–591. https://doi.org/10.2147/DMSO.S67400

Arnett DK, Blumenthal RS, Albert MA, Buroker AB, Goldberger ZD, Hahn EJ et al (2019) 2019 ACC/AHA guideline on the primary prevention of cardiovascular disease: a report of the American College of Cardiology/American Heart Association task force on clinical practice guidelines. J Am Coll Cardiol 74(10):e177–e232. https://doi.org/10.1016/j.jacc.2019.03.010

Atkinson MA, Eisenbarth GS, Michels AW (2014) Type 1 diabetes. Lancet 383(9911):69–82. https://doi.org/10.1016/s0140-6736(13)60591-7

Bahman F, Greish K, Taurin S (2019) Nanotechnology in insulin delivery for management of diabetes. Pharm Nanotechnol 7(2):113–128. https://doi.org/10.2174/2211738507666190321110721

Banting FG, Best CH (1990) Pancreatic extracts. 1922. J Lab Clin Med 115(2):254–272

Beck RW, Riddlesworth TD, Ruedy KJ, Kollman C, Ahmann AJ, Bergenstal RM et al (2017) Effect of initiating use of an insulin pump in adults with type 1 diabetes using multiple daily insulin injections and continuous glucose monitoring (DIAMOND): a multicentre, randomised controlled trial. Lancet Diabetes Endocrinol 5(9):700–708. https://doi.org/10.1016/s2213-8587(17)30217-6

Belfiore A, Frasca F, Pandini G, Sciacca L, Vigneri R (2009) Insulin receptor isoforms and insulin receptor/insulin-like growth factor receptor hybrids in physiology and disease. Endocr Rev 30(6):586–623. https://doi.org/10.1210/er.2008-0047

Biester T, Kordonouri O, Danne T (2017) Pharmacological properties of faster-acting insulin aspart. Curr Diab Rep 17(11):101. https://doi.org/10.1007/s11892-017-0931-y

Borg DJ, Weigelt M, Wilhelm C, Gerlach M, Bickle M, Speier S et al (2014) Mesenchymal stromal cells improve transplanted islet survival and islet function in a syngeneic mouse model. Diabetologia 57(3):522–531. https://doi.org/10.1007/s00125-013-3109-4

Bottino R, Knoll MF, Knoll CA, Bertera S, Trucco MM (2018) The future of islet transplantation is now. Front Med 5:202. https://doi.org/10.3389/fmed.2018.00202

Boughton CK, Hovorka R (2019) Is an artificial pancreas (closed-loop system) for type 1 diabetes effective? Diabet Med 36(3):279–286. https://doi.org/10.1111/dme.13816

Butterfield WJ, Camp JL, Hardwick C, Holling HE (1957) Clinical studies on the hypoglycaemic action of the sulphonylureas. Lancet 272(6972):753–756

Castle JR, DeVries JH, Kovatchev B (2017) Future of automated insulin delivery systems. Diabetes Technol Ther 19(Suppl 3):S-67–S-72. https://doi.org/10.1089/dia.2017.0012

Cho NH, Shaw JE, Karuranga S, Huang Y, da Rocha Fernandes JD, Ohlrogge AW et al (2018) IDF diabetes atlas: global estimates of diabetes prevalence for 2017 and projections for 2045. Diabetes Res Clin Pract

138:271–281. https://doi.org/10.1016/j.diabres.2018.02.023

Cierpka-Kmiec K, Wronska A, Kmiec Z (2019) In vitro generation of pancreatic beta-cells for diabetes treatment. I. Beta-like cells derived from human pluripotent stem cells. Folia Histochem Cytobiol 57 (1):1–14. https://doi.org/10.5603/FHC.a2019.0001

Corradi-Perini C, Santos TM, Camara NOS, Riella MC, Aita CAM (2017) Co-transplantation of xenogeneic bone marrow–derived Mesenchymal stem cells alleviates rejection of pancreatic islets in non-obese diabetic mice. Transplant Proc 49(4):902–905. https://doi.org/10.1016/j.transproceed.2017.01.064

Davies MJ, D'Alessio DA, Fradkin J, Kernan WN, Mathieu C, Mingrone G et al (2018) Management of hyperglycaemia in type 2 diabetes, 2018. A consensus report by the American Diabetes Association (ADA) and the European Association for the Study of Diabetes (EASD). Diabetologia 61(12):2461–2498. https://doi.org/10.1007/s00125-018-4729-5

de Souza BM, Boucas AP, Oliveira FD, Reis KP, Ziegelmann P, Bauer AC et al (2017) Effect of co-culture of mesenchymal stem/stromal cells with pancreatic islets on viability and function outcomes: a systematic review and meta-analysis. Islets 9 (2):30–42. https://doi.org/10.1080/19382014.2017.1286434

Deacon CF (2019) Physiology and pharmacology of DPP-4 in glucose homeostasis and the treatment of type 2 diabetes. Front Endocrinol 10:80. https://doi.org/10.3389/fendo.2019.00080

Evans M, Schumm-Draeger PM, Vora J, King AB (2011) A review of modern insulin analogue pharmacokinetic and pharmacodynamic profiles in type 2 diabetes: improvements and limitations. Diabetes Obes Metab 13(8):677–684. https://doi.org/10.1111/j.1463-1326.2011.01395.x

Fatima N, Faisal SM, Zubair S, Ajmal M, Siddiqui SS, Moin S et al (2016) Role of pro-inflammatory cytokines and biochemical markers in the pathogenesis of type 1 diabetes: correlation with age and glycemic condition in diabetic human subjects. PLoS One 11(8): e0161548. https://doi.org/10.1371/journal.pone.0161548

Fink H, Herbert C, Gilor C (2018) Pharmacodynamics and pharmacokinetics of insulin detemir and insulin glargine 300 U/mL in healthy dogs. Domest Anim Endocrinol 64:17–30. https://doi.org/10.1016/j.domaniend.2018.03.007

Fonte P, Araujo F, Reis S, Sarmento B (2013) Oral insulin delivery: how far are we? J Diabetes Sci Technol 7 (2):520–531. https://doi.org/10.1177/193229681300700228

Forbes JM, Cooper ME (2013) Mechanisms of diabetic complications. Physiol Rev 93(1):137–188. https://doi.org/10.1152/physrev.00045.2011

Forouhi NG, Wareham NJ (2014) Epidemiology of diabetes. Medicine (Abingdon) 42(12):698–702. https://doi.org/10.1016/j.mpmed.2014.09.007

Furtado RHM, Bonaca MP, Raz I, Zelniker TA, Mosenzon O, Cahn A et al (2019) Dapagliflozin and cardiovascular outcomes in patients with type 2 diabetes mellitus and previous myocardial infarction. Circulation 139(22):2516–2527. https://doi.org/10.1161/circulationaha.119.039996

Garber AJ, Abrahamson MJ, Barzilay JI, Blonde L, Bloomgarden ZT, Bush MA et al (2019) Consensus statement by the American association of clinical endocrinologists and American college of endocrinology on the comprehensive type 2 diabetes management algorithm—2019 executive summary. Endocr Pract 25 (1):69–100. https://doi.org/10.4158/cs-2018-0535

Girard J (2017) Glucagon, a key factor in the pathophysiology of type 2 diabetes. Biochimie 143(12):33–36. https://doi.org/10.1016/j.biochi.2017.10.004

Grant RW, Kirkman MS (2015) Trends in the evidence level for the American Diabetes Association's "standards of medical care in diabetes" from 2005 to 2014. Diabetes Care 38(1):6–8. https://doi.org/10.2337/dc14-2142

Gruessner AC (2011) 2011 Update on pancreas transplantation: comprehensive trend analysis of 25,000 cases followed up over the course of twenty-four years at the international pancreas transplant registry (IPTR). Rev Diabet Stud 8(1):6–16. https://doi.org/10.1900/rds.2011.8.6

Handelsman Y, Bloomgarden ZT, Grunberger G, Umpierrez G, Zimmerman RS, Bailey TS et al (2015) American association of clinical endocrinologists and American College of Endocrinology—clinical practice guidelines for developing a diabetes mellitus comprehensive care plan—2015. Endocr Pract 21(Suppl 1):1–87. https://doi.org/10.4158/ep15672.gl

Harsch IA, Kaestner RH, Konturek PC (2018) Hypoglycemic side effects of sulfonylureas and repaglinide in ageing patients - knowledge and self-management. J Physiol Pharmacol 69(4). https://doi.org/10.26402/jpp.2018.4.15

Heinemann L, Parkin CG (2018) Rethinking the viability and utility of inhaled insulin in clinical practice. J Diabetes Res 2018:4568903. https://doi.org/10.1155/2018/4568903

Heinemann L, Fritz I, Khatami H, Edelman SV (2017) Administration of biosimilar insulin analogs: role of devices. Diabetes Technol Ther 19(2):79–84. https://doi.org/10.1089/dia.2016.0263

Heller S, White D, Lee E, Lawton J, Pollard D, Waugh N et al (2017) A cluster randomised trial, cost-effectiveness analysis and psychosocial evaluation of insulin pump therapy compared with multiple injections during flexible intensive insulin therapy for type 1 diabetes: the REPOSE trial. Health Technol Assess 21(20):1–278. https://doi.org/10.3310/hta21200

Hieronymus L, Griffin S (2015) Role of amylin in type 1 and type 2 diabetes. Diabetes Educ 41(1 Suppl):47s–56s. https://doi.org/10.1177/0145721715607642

Ikegami H, Noso S, Babaya N, Kawabata Y (2011) Genetics and pathogenesis of type 1 diabetes: prospects for prevention and intervention. J Diabetes Investig 2 (6):415–420. https://doi.org/10.1111/j.2040-1124.2011.00176.x

Jovanovic A, Sudar-Milovanovic E, Obradovic M, Pitt SJ, Stewart AJ, Zafirovic S et al (2017) Influence of a high-fat diet on cardiac iNOS in female rats. Curr Vasc Pharmacol 15(5):491–500

Jung HS, Kim MJ, Hong SH, Lee YJ, Kang S, Lee H et al (2014) The potential of endothelial colony-forming cells to improve early graft loss after intraportal islet transplantation. Cell Transplant 23(3):273–283. https://doi.org/10.3727/096368912x661364

Kalra S, Gupta Y (2015) Sulfonylureas. JPMA 65 (1):101–104

Kalra K, Chandrabose ST, Ramasamy TS, Kasim N (2018) Advances in the generation of functional beta-cells from induced pluripotent stem cells as a cure for diabetes mellitus. Curr Drug Targets 19 (13):1463–1477. https://doi.org/10.2174/1389450119666180605112917

Katsarou A, Gudbjornsdottir S, Rawshani A, Dabelea D, Bonifacio E, Anderson BJ et al (2017) Type 1 diabetes mellitus. Nat Rev Dis Primers 3:17016. https://doi.org/10.1038/nrdp.2017.16

Kerby A, Jones ES, Jones PM, King AJ (2013) Co-transplantation of islets with mesenchymal stem cells in microcapsules demonstrates graft outcome can be improved in an isolated-graft model of islet transplantation in mice. Cytotherapy 15(2):192–200. https://doi.org/10.1016/j.jcyt.2012.10.018

Krall LP, White P, Bradley RF (1958) Clinical use of the biguanides and their role in stabilizing juvenile-type diabetes. Diabetes 7(6):468–477

Kullmann S, Veit R, Peter A, Pohmann R, Scheffler K, Haring HU et al (2018) Dose-dependent effects of intranasal insulin on resting-state brain activity. J Clin Endocrinol Metab 103(1):253–262. https://doi.org/10.1210/jc.2017-01976

Kunisada Y, Tsubooka-Yamazoe N, Shoji M, Hosoya M (2012) Small molecules induce efficient differentiation into insulin-producing cells from human induced pluripotent stem cells. Stem Cell Res 8(2):274–284. https://doi.org/10.1016/j.scr.2011.10.002

Ledet G, Graves RA, Bostanian LA, Mandal TK (2015) A second-generation inhaled insulin for diabetes mellitus. Am J Health Syst Pharm 72(14):1181–1187. https://doi.org/10.2146/ajhp140540

Leon BM, Maddox TM (2015) Diabetes and cardiovascular disease: epidemiology, biological mechanisms, treatment recommendations and future research. World J Diabetes 6(13):1246–1258. https://doi.org/10.4239/wjd.v6.i13.1246

Liu Z, Ma S (2017) Recent advances in synthetic alpha-Glucosidase inhibitors. Chem Med Chem 12 (11):819–829. https://doi.org/10.1002/cmdc.201700216

Livingstone R, Boyle JG, Petrie JR (2017) A new perspective on metformin therapy in type 1 diabetes. Diabetologia 60(9):1594–1600. https://doi.org/10.1007/s00125-017-4364-6

Lopez Vicchi F, Luque GM, Brie B, Nogueira JP, Garcia Tornadu I, Becu-Villalobos D (2016) Dopaminergic drugs in type 2 diabetes and glucose homeostasis. Pharmacol Res 109:74–80. https://doi.org/10.1016/j.phrs.2015.12.029

Lupsa BC, Inzucchi SE (2018) Use of SGLT2 inhibitors in type 2 diabetes: weighing the risks and benefits. Diabetologia 61(10):2118–2125. https://doi.org/10.1007/s00125-018-4663-6

Maclean H (1926) Some observations on diabetes and insulin in general practice. Postgrad Med J 1(6):73–77

Mastrandrea LD (2010) Inhaled insulin: overview of a novel route of insulin administration. Vasc Health Risk Manag 6:47–58

McGuire H, Longson D, Adler A, Farmer A, Lewin I (2016) Management of type 2 diabetes in adults: summary of updated NICE guidance. BMJ 353:i1575. https://doi.org/10.1136/bmj.i1575

McGuire DK, Alexander JH, Johansen OE, Perkovic V, Rosenstock J, Cooper ME et al (2019) Linagliptin effects on heart failure and related outcomes in individuals with type 2 diabetes mellitus at high cardiovascular and renal risk in CARMELINA. Circulation 139(3):351–361. https://doi.org/10.1161/circulationaha.118.038352

Mehtala J, Khanfir H, Bennett D, Ye Y, Korhonen P, Hoti F (2019) Pioglitazone use and risk of bladder cancer: a systematic literature review and meta-analysis of observational studies. Diabetol Int 10(1):24–36. https://doi.org/10.1007/s13340-018-0360-4

Meier JJ, Bhushan A, Butler AE, Rizza RA, Butler PC (2005) Sustained beta cell apoptosis in patients with long-standing type 1 diabetes: indirect evidence for islet regeneration ? Diabetologia 48(11):2221–2228

Nair M (2007) Diabetes mellitus, part 1: physiology and complications. Br J Nurs 16(3):184–188. https://doi.org/10.12968/bjon.2007.16.3.22974

Nakamura T, Fujikura J, Anazawa T, Ito R, Ogura M, Okajima H et al (2019) Long-term outcome of islet transplantation on insulin-dependent diabetes mellitus: an observational cohort study. J Diabetes Investig. https://doi.org/10.1111/jdi.13128

National Clinical Guideline Centre (2015) Type 1 diabetes in adults: diagnosis and management. London: National Institute for Health and Care Excellence (UK)

Naziruddin B, Iwahashi S, Kanak MA, Takita M, Itoh T, Levy MF (2014) Evidence for instant blood-mediated inflammatory reaction in clinical autologous islet transplantation. Am J Transplant 14(2):428–437. https://doi.org/10.1111/ajt.12558

Obradovic M, Sudar E, Zafirovic S, Stanimirovic J, Labudovic-Borovic M, Isenovic ER (2015) Estradiol in vivo induces changes in cardiomyocytes size in obese rats. Angiology 66(1):25–35. https://doi.org/10.1177/0003319713514477

Obradovic M, Stanimirovic J, Panic A, Bogdanovic N, Sudar-Milovanovic E, Cenic-Milosevic D et al (2017) Regulation of Na+/K+-ATPase by estradiol and IGF-1 in cardio-metabolic diseases. Curr Pharm Des 23 (10):1551–1561

Obradovic M, Zafirovic S, Soskic S, Stanimirovic J, Trpkovic A, Jevremovic D et al (2019) Effects of IGF-1 on the cardiovascular system. Curr Pharm Des. https://doi.org/10.2174/1381612825666191106091507

Ooi CP, Loke SC (2014) Colesevelam for type 2 diabetes mellitus: an abridged Cochrane review. Diabet Med 31 (1):2–14. https://doi.org/10.1111/dme.12295

Pathak V, Pathak NM, O'Neill CL, Guduric-Fuchs J, Medina RJ (2019) Therapies for type 1 diabetes: current scenario and future perspectives. Clin Med Insights Endocrinol Diabetes 12:1179551419844521. https://doi.org/10.1177/1179551419844521

Peng Q, Zhang ZR, Gong T, Chen GQ, Sun X (2012) A rapid-acting, long-acting insulin formulation based on a phospholipid complex loaded PHBHHx nanoparticles. Biomaterials 33(5):1583–1588. https://doi.org/10.1016/j.biomaterials.2011.10.072

Pickup JC (2019) Insulin Pumps. Diabet Technol Therap 21(S1):S32–s41. https://doi.org/10.1089/dia.2019.2503

Potter KJ, Westwell-Roper CY, Klimek-Abercrombie AM, Warnock GL, Verchere CB (2014) Death and dysfunction of transplanted beta-cells: lessons learned from type 2 diabetes? Diabetes 63(1):12–19. https://doi.org/10.2337/db12-0364

Rai VK, Mishra N, Agrawal AK, Jain S, Yadav NP (2016) Novel drug delivery system: an immense hope for diabetics. Drug Deliv 23(7):2371–2390

REPOSE Study Group (2017) Relative effectiveness of insulin pump treatment over multiple daily injections and structured education during flexible intensive insulin treatment for type 1 diabetes: cluster randomised trial (REPOSE). BMJ 356. https://doi.org/10.1136/bmj.j1285

Roglic G (2016) WHO global report on diabetes: a summary. Int J Noncommun Dis 1(1):3–8

Rowley WR, Bezold C, Arikan Y, Byrne E, Krohe S (2017) Diabetes 2030: insights from yesterday, today, and future trends. Popul Health Manag 20(1):6–12. https://doi.org/10.1089/pop.2015.0181

Rys PM, Ludwig-Slomczynska AH, Cyganek K, Malecki MT (2018) Continuous subcutaneous insulin infusion vs multiple daily injections in pregnant women with type 1 diabetes mellitus: a systematic review and meta-analysis of randomised controlled trials and observational studies. Eur J Endocrinol 178(5):545–563. https://doi.org/10.1530/eje-17-0804

Sanchez-Rangel E, Inzucchi SE (2017) Metformin: clinical use in type 2 diabetes. Diabetologia 60(9):1586–1593. https://doi.org/10.1007/s00125-017-4336-x

Schmid V, Kullmann S, Gfrorer W, Hund V, Hallschmid M, Lipp HP et al (2018) Safety of intranasal human insulin: a review. Diabetes Obes Metab 20 (7):1563–1577. https://doi.org/10.1111/dom.13279

Scirica BM, Bhatt DL, Braunwald E, Steg PG, Davidson J, Hirshberg B et al (2013) Saxagliptin and cardiovascular outcomes in patients with type 2 diabetes mellitus. N Engl J Med 369(14):1317–1326. https://doi.org/10.1056/NEJMoa1307684

Shah RB, Patel M, Maahs DM, Shah VN (2016) Insulin delivery methods: past, present and future. Int J Pharm Investig 6(1):1–9. https://doi.org/10.4103/2230-973x.176456

Singh M, Kumar A (2018) Risks associated with SGLT2 inhibitors: an overview. Curr Drug Saf 13(2):84–91. https://doi.org/10.2174/1574886313666180226103408

Soskic SS, Dobutovic BD, Sudar EM, Obradovic MM, Nikolic DM, Djordjevic JD et al (2011) Regulation of inducible nitric oxide synthase (iNOS) and its potential role in insulin resistance, diabetes and heart failure. Open Cardiovasc Med J 5:153–163. https://doi.org/10.2174/1874192401105010153

Souto EB, Souto SB, Campos JR, Severino P, Pashirova TN, Zakharova LY, Silva AM, Durazzo A, Lucarini M, Izzo AA, Santini A (2019) Nanoparticle delivery systems in the treatment of diabetes complications. Molecules 24(23). https://doi.org/10.3390/molecules24234209

Stanekzai J, Isenovic ER, Mousa SA (2012) Treatment options for diabetes: potential role of stem cells. Diabetes Res Clin Pract 98(3):361–368. https://doi.org/10.1016/j.diabres.2012.09.010

Sudar-Milovanovic E, Jovanovic A, Misirkic-Marjanovic-M, Vucicevic L, Janjetovic K, Isenovic ER (2015) Effects of intracerebroventricularly (ICV) injected ghrelin on cardiac inducible nitric oxide synthase activity/expression in obese rats. Exp Clin Endocrinol Diabetes 123(10):581–588. https://doi.org/10.1055/s-0035-1559758

Sudar-Milovanovic E, Zafirovic S, Jovanovic A, Trebaljevac J, Obradovic M, Cenic-Milosevic D et al (2017) Hormonal regulation of nitric oxide (NO) in cardio-metabolic diseases. Curr Pharm Des 23 (10):1427–1434

Vegas AJ, Veiseh O, Gurtler M, Millman JR, Pagliuca FW, Bader AR et al. Long-term glycemic control using polymer-encapsulated human stem cell-derived beta cells in immune-competent mice. Nat Med. 2016;22(3):306–11. doi:10.1038/nm.4030.

Wang W, Zhou X, Kwong JSW, Li L, Li Y, Sun X (2017) Efficacy and safety of thiazolidinediones in diabetes patients with renal impairment: a systematic review and meta-analysis. Sci Rep 7(1):1717. https://doi.org/10.1038/s41598-017-01965-0

Wang LC, Fang FS, Gong YP, Yang G, Li CL (2018) Characteristics of repaglinide and its mechanism of action on insulin secretion in patients with newly diagnosed type-2 diabetes mellitus. Medicine 97(38): e12476. https://doi.org/10.1097/md.0000000000012476

Wanner C, Marx N (2018) SGLT2 inhibitors: the future for treatment of type 2 diabetes mellitus and other chronic diseases. Diabetologia 61(10):2134–2139. https://doi.org/10.1007/s00125-018-4678-z

WHO (2018). http://www.who.int/news-room/fact-sheets/detail/diabetes. Accessed 16 Dec 2019

Wiviott SD, Raz I, Sabatine MS (2019) Dapagliflozin and cardiovascular outcomes in type 2 diabetes. Reply. N Engl J Med 380(19):1881–1882. https://doi.org/10.1056/NEJMc1902837

Wong CY, Al-Salami H, Dass CR (2018) Microparticles, microcapsules and microspheres: a review of recent developments and prospects for oral delivery of insulin. Int J Pharm 537(1–2):223–244. https://doi.org/10.1016/j.ijpharm.2017.12.036

Yeh HC, Brown TT, Maruthur N, Ranasinghe P, Berger Z, Suh YD et al (2012) Comparative effectiveness and safety of methods of insulin delivery and glucose monitoring for diabetes mellitus: a systematic review and meta-analysis. Ann Intern Med 157(5):336–347. https://doi.org/10.7326/0003-4819-157-5-201209040-00508

Yki-Jarvinen H (2004) Thiazolidinediones. N Engl J Med 351(11):1106–1118. https://doi.org/10.1056/NEJMra041001

Zaric BL, Obradovic M, Sudar-Milovanovic E, Nedeljkovic J, Lazic V, Isenovic ER (2019) Drug delivery systems for diabetes treatment. Curr Pharm Des 25(2):166–173. https://doi.org/10.2174/1381612825666190306153838

Zelniker TA, Wiviott SD, Raz I, Im K, Goodrich EL, Bonaca MP et al (2019) SGLT2 inhibitors for primary and secondary prevention of cardiovascular and renal outcomes in type 2 diabetes: a systematic review and meta-analysis of cardiovascular outcome trials. Lancet 393(10166):31–39. https://doi.org/10.1016/s0140-6736(18)32590-x

Zoungas S, Woodward M, Li Q, Cooper ME, Hamet P, Harrap S et al (2014) Impact of age, age at diagnosis and duration of diabetes on the risk of macrovascular and microvascular complications and death in type 2 diabetes. Diabetologia 57(12):2465–2474. https://doi.org/10.1007/s00125-014-3369-7

53 Treatment of Diabetes and Heart Failure

Bradley Brochu and Michael Chan

Contents

Introduction 720
Cardiovascular Disease Profiles 721

Guidelines for Primary and Secondary Prevention of Cardiovascular Disease in Diabetes 721
Physical Activity 721
Weight Loss 722
Smoking Cessation 722
Blood Pressure 722
Antiplatelet Therapy 722
Lipid-Lowering Therapy 722

Glucose-Lowering Drugs and Cardiovascular Disease 723
Insulin 723
Metformin 724
Sulfonylureas 724
Thiazolidinediones 725
Dipeptidyl Peptidase-4 Inhibitors 726
Sodium Glucose Cotransporter 2 (SGLT2) Inhibitors 726
GLP-1 Agonists 728

Can Renal Complications Be Mitigated? 729

Do Guidelines Exist for Pre-Diabetes? 729
Pharmacotherapy 730
Statins and Pre-diabetes 730

Ongoing Studies and Future Perspectives 730
Antidiabetic Agents in Normoglycemic Patients 730

References 731

B. Brochu
Mazankowski Alberta Heart Institute, University of Alberta, Edmonton, AB, Canada

M. Chan (✉)
CK Hui Heart Centre (Robbins Pavilion), Royal Alexandra Hospital, University of Alberta, Edmonton, AB, Canada
e-mail: chanmcy@gmail.com

J. Faintuch and S. Faintuch (eds.), *Obesity and Diabetes*, https://doi.org/10.1007/978-3-030-53370-0_53

Abstract

Accelerated atherosclerosis is a well-known complication of diabetes, and recently the importance of heart failure independent of coronary artery disease has become a focus of research and innovation. A number of recent cardiovascular outcome trials have demonstrated cardiovascular and renal benefits with two novel classes of glucose-lowering drugs: SGLT2 inhibitors and GLP-1 agonists. Work is ongoing to define the underlying mechanisms of these benefits. Interest in demonstrating the cardiovascular safety of older glucose-lowering drugs has also been rekindled. In addition, in the very near future, it is likely that SGLT2 inhibitors will be administered to new patient populations, including those with type 1 diabetes or without diabetes altogether. Efforts in lifestyle management and preventive health continue to be well spent, due to their potential effectiveness and the growing global burden of disease.

Keywords

Heart failure · Diabetes · Cardiovascular outcome trial · SGLT2 inhibitor · GLP-1 agonist

Introduction

Type 1 and 2 diabetes mellitus, referred to hereafter as simply "diabetes," is a major worldwide health issue (Williams 2019). Cardiovascular disease is a well-documented complication and a leading cause of death in these patients (Williams 2019; Grant and Cosentino 2019; Brochu and Chan 2019). Historically, efforts to reduce adverse cardiovascular endpoints have focused on atherosclerotic outcomes, such as myocardial infarction, stroke, and peripheral vascular disease. It is important to recognize that heart failure represents a large burden of disease in patients with diabetes, and is frequently a presenting diagnosis (Shah et al. 2015). The clinical courses of diabetes and heart failure are intertwined: diabetes is known to cause heart failure through a number of mechanisms; however, heart failure can also be a risk factor for the development of type 2 diabetes (Grant and Cosentino 2019).

Diabetes is recognized as a heterogenous disease, and approximately 90% of the patients have type 2 diabetes (Grant and Cosentino 2019). Patients with type 1 diabetes have an absolute deficiency of insulin, and at least early in their disease experience are less subject to the complications of insulin resistance. These patients tend to be younger and are often more motivated: they readily adopt new technology such as insulin pumps and continuous blood glucose monitors (American Diabetes Association 2019a). Glycemic control is of the utmost importance for the prevention of microvascular complications, as patients with type 1 diabetes are faced with a disease experience that is often two to three decades longer than their counterparts with type 2 diabetes.

Adverse cardiovascular outcomes are related to glycemic control and length of diabetes diagnosis (Grant and Cosentino 2019). Patients with type 2 diabetes have a disease course which is dominated by insulin resistance and its subsequent complications. There are frequently comorbid conditions that must be considered when developing appropriate prevention and treatment strategies. Regardless, all patients with diabetes mellitus should benefit from dedicated preventative efforts to reduce the risk of developing or worsening cardiovascular disease.

Over the last 30 years, the target glycated hemoglobin (HbA1c) and the glucose-lowering drugs (GLDs) available for clinical use have changed considerably. The curve of cardiovascular events and glycemic control has been defined as U-shaped by several studies that demonstrated worse outcomes with very aggressive HbA1c targets (The Action to Control Cardiovascular Risk in Diabetes Study Group 2008; Duckworth et al. 2009). Several agents have also been independently associated with adverse cardiovascular events, leading regulatory bodies to impose safety restrictions on the approval of new GLDs (Gerstein et al. 2006; Home et al. 2009; Nissen and Wolski 2007). This requirement has led to the publication of a number of cardiovascular outcome trials (CVOTs) in the last 5 years, including several landmark publications that unexpectedly showed significant benefit with novel GLDs (Zinman et al. 2015; Marso et al. 2016a; Neal et al. 2017; Wiviott et al. 2019). For this reason, the landscape of diabetes and heart failure is rapidly changing.

Cardiovascular Disease Profiles

Heart failure is also a heterogenous disease with at least two well-recognized phenotypes. Heart failure with reduced ejection fraction (HFrEF) is the most intuitive, and is defined by left ventricular systolic dysfunction with a left ventricular ejection fraction (LVEF) of less than 40% (Grant and Cosentino 2019). Ischemic heart disease is felt to be the commonest etiology of HFrEF in patients with diabetes due to accelerated, diffuse and often silent coronary artery disease (Grant and Cosentino 2019). Heart failure with preserved ejection fraction (HFpEF) is defined by symptoms of heart failure with LVEF greater than 50% (Grant and Cosentino 2019). Some guidelines advocate for additional parameters, such as left atrial enlargement or evidence of diastolic dysfunction on echocardiography (Grant and Cosentino 2019). Though there remains debate regarding the precise definition, diabetic cardiomyopathy is an entity postulated to be responsible for these changes, through fibrosis and elevated left ventricular mass in the absence of hypertension or other conditions of increased afterload (Connelly et al. 2018; Dunlay et al. 2019; Verma and McMurray 2018). Some authors advocate for the use of the term "heart failure" with mid-range ejection fraction (HFmrEF) to define patients with symptoms of heart failure and a LVEF between 40 and 49%, though this has not yet been widely adopted (Grant and Cosentino 2019).

Guidelines for Primary and Secondary Prevention of Cardiovascular Disease in Diabetes

Accelerated atherosclerosis is a well-known complication of diabetes and recently the importance of heart failure independent of coronary artery disease has become a focus of research and innovation. Many complications seen with diabetes are irreversible owing to their involvement of micro- or macrovasculature; this highlights the importance of prevention in the management of complications of diabetes. Owing to the evolving landscape of diabetes management, many national and international organizations have published updated clinical practice guidelines (Grant and Cosentino 2019; Connelly et al. 2018; American Diabetes Association 2019b), not overlooking lifestyle measures (Grant and Cosentino 2019; American Diabetes Association 2019c; Sigal et al. 2018; Wharton et al. 2018).

Physical Activity

Physical activity is important in both primary and secondary prevention of cardiovascular disease in patients with diabetes. It has been shown to improve glycemic control, reduce insulin resistance, lower blood pressure, and contribute to weight loss (Grant and Cosentino 2019; American Diabetes Association 2019c; Sigal et al. 2018). Adults with diabetes should participate in a minimum of 150 min per week of moderate intensity activity (Grant and Cosentino 2019; American Diabetes Association 2019c; Sigal et al. 2018). Aerobic activity should be divided into sessions of at least 10 min duration, with the goal of amassing 30 min or more of activity on most days of the week (Grant and Cosentino 2019; American Diabetes Association 2019c; Sigal et al. 2018). Improvements in insulin resistance are best maintained if there are not more than 2 days between scheduled physical activity (American Diabetes Association 2019c). It has been suggested that prolonged periods of sedentary activity be briefly interrupted by standing or performing a physical task approximately every 30 min; wearable fitness accessories may provide helpful reminders (American Diabetes Association 2019c). Patients with diabetes also benefit from resistance and flexibility training, and are encouraged to incorporate this into their planned activity two or three times per week (American Diabetes Association 2019c; Sigal et al. 2018). It is reasonable to suggest that patients considering resistance training obtain expert instruction to avoid injury (American Diabetes Association 2019c).

Weight Loss

Loss of 5% or more of total body weight will lead to a significant reduction in insulin resistance and an improvement in blood pressure control (Wharton et al. 2018).

Smoking Cessation

Smoking is an independent risk factor for cardiovascular disease. Smoking cessation should be encouraged, and assistance should be provided to all patients who use tobacco (Grant and Cosentino 2019; Stone et al. 2018).

Blood Pressure

Blood pressure screening and control are important for both primary and secondary prevention of cardiovascular disease in patients with diabetes (Grant and Cosentino 2019; American Diabetes Association 2019b; Tobe et al. 2018). With the heterogeneity of data available, there is some variation in the suggested blood pressure targets in various published guidelines (Grant and Cosentino 2019; American Diabetes Association 2019b; Tobe et al. 2018). A blood pressure target of <130/80 mmHg is generally recommended in patients with diabetes, provided it can be achieved safely and with few adverse effects (Grant and Cosentino 2019; Tobe et al. 2018). In patients over the age of 65, more conservative targets may be appropriate (Grant and Cosentino 2019). Some groups advocate for a blood pressure target of 140/90 mmHg in most patients with diabetes, reserving the blood pressure target of 130/80 mmHg for higher-risk individuals, including secondary prevention (American Diabetes Association 2019b). It is recommended that antihypertensive therapy be targeted to home or ambulatory blood pressure readings, as these are felt to be more representative of a patients' true blood pressure than office-based measurements (Grant and Cosentino 2019; American Diabetes Association 2019b; Tobe et al. 2018). For patients in whom lifestyle modifications are not sufficient to achieve the desired blood pressure, angiotensin receptor blockers (ARB), angiotensin converting enzyme inhibitors (ACEi), and non-dihydropyridine calcium channel blockers have evidence for reducing cardiovascular endpoints in patients with diabetes (Grant and Cosentino 2019; Tobe et al. 2018).

Antiplatelet Therapy

At the present time, there is insufficient evidence to recommend antiplatelet therapy for the primary prevention of cardiovascular disease in all patients with diabetes (Stone et al. 2018). Daily low-dose acetylsalicylic acid has not been shown to reduce the rates of key cardiovascular endpoints in diabetics who are deemed low risk, and has been demonstrated to significantly increase the rate of gastrointestinal bleeding (Stone et al. 2018). In patients with and without diabetes, guidelines continue to advocate for antiplatelet therapy for the secondary prevention of cardiovascular disease; however, this is largely on the basis of treatment of coronary artery disease (Stone et al. 2018).

Lipid-Lowering Therapy

Lipid-lowering therapy plays a key role in the prevention and treatment of cardiovascular disease in patients with diabetes (Grant and Cosentino 2019; American Diabetes Association

2019b; Stone et al. 2018). Of the agents available for clinical use, HMG-CoA reductase inhibitors (statins) have the largest body of evidence (Grant and Cosentino 2019; Stone et al. 2018). High-intensity statin therapy is generally recommended for primary prevention in patients with diabetes over the age of 40 (American Diabetes Association 2019b; Stone et al. 2018). In patients under 40, guidelines generally endorse the use of moderate- to high-intensity statin therapy in the presence of multiple additional risk factors, a long duration of diabetes, or signs of other end-organ damage attributable to diabetes (Stone et al. 2018). Statin therapy is strongly recommended for secondary prevention of atherosclerotic cardiovascular disease, in all patients with diabetes (American Diabetes Association 2019b; Stone et al. 2018).

In patients who do not achieve a target low-density lipoprotein cholesterol (LDL-C) of less than 1.8 mmol/L, addition of ezetimibe 10 mg daily is recommended (American Diabetes Association 2019b; Stone et al. 2018; Cannon et al. 2015). Proprotein convertase subtilisin/kexin type 9 (PCSK9) inhibitors are the newest and most potent lipid-lowering therapy available (evolocumab, alirocumab), though their role specifically in patients with diabetes remains to be defined (Grant and Cosentino 2019). PCSK9 inhibitors are currently indicated in patients who are unable to tolerate statins or achieve target LDL-C on combination therapy with a statin and ezetimibe (Stone et al. 2018). There is less evidence of a direct benefit of lipid-lowering on heart failure; however, prevention of epicardial and microvascular coronary disease is key in preventing development and progression of left ventricular dysfunction, and therefore felt by the authors to be important in the management of patients with diabetes.

Glucose-Lowering Drugs and Cardiovascular Disease

Insulin

Insulin has prothrombotic and inflammatory effects mediated by mitogen-activated protein kinase (MAPK), and antithrombotic effects mediated by phosphatidylinositol 3-kinase (PI3K) (Dongerkery et al. 2017). It has been demonstrated that PI3K-mediated effects are blunted in states of insulin resistance, theoretically causing excess insulin to induce a prothrombotic state (Dongerkery et al. 2017). In addition, it has been observed that the inflammatory vascular changes in diabetes and the metabolic syndrome, termed endothelial dysfunction, cannot be corrected by achieving tight glycemic control alone (Dongerkery et al. 2017). Therefore, in type 2 diabetes insulin use can be used as a surrogate marker for more significant insulin resistance and endothelial dysfunction; however, it has not been mechanistically linked to adverse outcomes from the metabolic syndrome (Dongerkery et al. 2017). This suggests that it is the underlying disease process, rather than exogenous insulin use, that is responsible for cardiovascular outcomes.

A longer duration of diabetes must be recognized as a confounder, as it has been associated with an increased likelihood of requiring insulin therapy and experiencing adverse cardiovascular outcomes (Dongerkery et al. 2017). To mitigate this effect, as well as other confounders associated with the metabolic syndrome, it is useful to consider patients with type 1 diabetes. In the landmark Diabetes Control and Complications Trial (DCCT), there were few cardiovascular events due to the young age of the patients studied and the relative paucity of comorbid medical conditions; however, a 10-year follow-up study was able to demonstrate a 42% reduction in a composite cardiovascular endpoint in those treated with intensive insulin therapy, compared with conventional twice-daily insulin (The Diabetes Control and Complications Trial Research Group 1993; The Diabetes Control and Complications Trial/Epidemiology of Diabetes Interventions and Complications (DCCT/EDIC) Study Research Group 2005). The HbA1c was 0.5% higher in patients who experienced a cardiovascular endpoint, and this difference met statistical significance (The Diabetes Control and Complications Trial/Epidemiology of Diabetes Interventions and Complications (DCCT/EDIC) Study Research Group 2005).

More recent data from the Outcome Reduction with the Initial Glargine Intervention (ORIGIN) trial suggests that the addition of low-dose insulin glargine does not increase the risk of composite major adverse cardiovascular endpoints (MACE), when compared with standard care. The very large ORIGIN randomized clinical trial reported that cardiovascular outcomes among those who received basal insulin glargine were no better than among patients who received standard care, including those with impaired fasting glucose, impaired glucose tolerance, or early type 2 diabetes (Gerstein et al. 2012). Further to this, ultra-long-acting insulin degludec was non-inferior to insulin glargine for MACE in an RCT of patients with established type 2 diabetes (Marso et al. 2017).

Prevention of Hypoglycemia

Patients who have concomitant medical comorbidities are at higher risk of serious hypoglycemia. Historically, insulin therapy has been felt to be the agent with the highest potential risk of hypoglycemia, and this effect is potentiated by more intensive regimens (The Diabetes Control and Complications Trial/Epidemiology of Diabetes Interventions and Complications (DCCT/EDIC) Study Research Group 2005). Basal-only insulin is felt to confer the lowest risk of hypoglycemia, and recent studies of ultra-long acting insulin degludec demonstrate fewer symptomatic and blood glucose–confirmed hypoglycemic events than insulin glargine (Marso et al. 2017; Heller et al. 2019). In addition to the challenge of clinical management, major hypoglycemic events may be associated with as much as a two-fold increase in adverse cardiovascular outcomes, and should be avoided (Goto et al. 2013).

The authors prefer to use glucose-lowering drugs with proven cardiovascular benefit as first- and second-line choices, reserving insulin for those with difficult to manage hyperglycemia and with type 1 diabetes. In patients with type 2 diabetes who require insulin, we suggest long- or ultra-long acting basal insulin, and the involvement of a multidisciplinary team for comprehensive diabetes management.

Metformin

Metformin is a biguanide that has been used in the management of type 2 diabetes for decades (Rena et al. 2017). Despite its years of clinical use, researchers and clinicians have recently become more aware of the complex and incompletely understood mechanisms of action, including decreased hepatic gluconeogenesis, modulation of intestinal absorption, and reduced insulin resistance (Rena et al. 2017). Metformin is a preferred glucose-lowering drug due to its efficacy, weight neutrality, and low likelihood of inducing serious hypoglycemic events. Historically, there has been concern regarding the risk of lactic acidosis in patients receiving metformin which has limited its use in patients with multiple comorbidities, especially chronic kidney disease (Grant and Cosentino 2019). In a large, observational trial of patients hospitalized with heart failure, metformin has been shown to reduce the risk of all-cause mortality when compared to sulfonylurea or insulin monotherapy (Andersson et al. 2010).

In a recent retrospective cohort study of patients with type 2 diabetes and reduced renal function, metformin monotherapy was found to result in fewer adverse cardiovascular endpoints than sulfonylurea monotherapy (Roumie et al. 2019). Overall, metformin is a recommended glucose-lowering drug in patients with type 2 diabetes and an estimated glomerular filtration rate of greater than 30 mL/min. In patients with a new diagnosis of type 2 diabetes and at high cardiovascular risk, recent international guidelines have advocated that glucose-lowering drugs with demonstrated cardiovascular benefit, such as SGLT2i, be started preferentially over metformin, reserving metformin for patients who do not achieve target HbA1c on monotherapy (Grant and Cosentino 2019).

Sulfonylureas

Along with metformin and insulin, sulfonylureas are the oldest effective glucose-lowering drugs that remain on the market in North America.

These agents act by inhibiting K_{ATP} channels on the cell membrane of pancreatic beta cells, thereby stimulating insulin release (Wexler 2019). Owing to this mechanism of action, sulfonylureas are susceptible to causing hypoglycemic events and this is observed with relative frequency in clinical practice (Wexler 2019). An added consideration is that many sulfonylureas are renally excreted, increasing their potential to cause hypoglycemia in patients with concomitant renal dysfunction.

Early trials with the sulfonylurea tolbutamide showed a 2.5-fold increase in cardiovascular mortality and for this reason, sulfonylureas continue to display a special warning on increased risk of cardiovascular mortality mandated by regulatory bodies (Pfizer Inc 2009). Overall, historical data have been equivocal regarding the cardiovascular safety of other sulfonylureas. Recently, the long-acting sulfonylurea glimepiride was compared to linagliptin, a DPP-4 inhibitor which will be discussed later in this chapter, in an RCT (Rosenstock et al. 2019). Glimepiride was found to be non-inferior to linagliptin for MACE as well as the prespecified secondary outcome hospitalization for heart failure, though it was significantly less effective at meeting and maintaining HbA1c targets (Rosenstock et al. 2019).

It has been proposed that this data can be taken in concert with the findings of a second RCT which compared linagliptin to placebo, to suggest that glimepiride has effective non-inferiority to placebo for adverse cardiovascular outcomes (Wexler 2019; Rosenstock et al. 2019; Martin et al. 2019). While these data do imply the cardiovascular safety of the sulfonylurea glimepiride, the authors of this chapter are reluctant to accept that it is non-inferior to linagliptin based on the comparison of two separate study populations. Additionally, recent registry data suggest that adverse cardiovascular outcomes are more prevalent with sulfonylurea monotherapy as compared with metformin monotherapy (Roumie et al. 2019). With the advent of new glucose-lowering drugs with enhanced safety profiles and demonstrable cardiovascular benefit, sulfonylureas have fallen to third- and fourth-line agents (Grant and Cosentino 2019; Connelly et al. 2018). In the opinion of the authors, sulfonylureas should not be used without a compelling justification, such as extreme needle phobia or irremediable financial constraint.

Thiazolidinediones

Thiazolidinediones (TZDs) are potent glucose-lowering drugs that act by activating the gamma isoform of the peroxisome proliferator-activated receptor (PPARg) and modulating the transcription of several genes involved in glucose and lipid handling (Hauner 2002). They were quickly adopted into clinical practice due to their efficacy and modest patient-reported side effect profile. As the use of TZDs increased, a signal toward increased major adverse cardiovascular outcomes was noted (Nissen and Wolski 2007). An RCT of patients with impaired fasting glucose and/or impaired glucose tolerance demonstrated that while rosiglitazone did prevent progression to a diagnosis of diabetes, it was associated with a sevenfold increase in the risk of heart failure (Gerstein et al. 2006). A meta-analysis published in 2007 found a 43% increase in the risk of myocardial infarction in patients receiving rosiglitazone (Nissen and Wolski 2007).

The following year, an RCT of intensive glycemic control with a target HbA1c of <6.0% compared with standard care demonstrated an increase in the risk of death from any cause, death from a cardiovascular cause and non-fatal myocardial infarction; in this trial, 91.2% of participants in the intensive-therapy arm were prescribed rosiglitazone, compared with 57.5% in the standard therapy arm (The Action to Control Cardiovascular Risk in Diabetes Study Group 2008). Notably, there was also a numerical increase in fatal and non-fatal cases of decompensated heart failure in the intensive control group, though this did not reach statistical significance (The Action to Control Cardiovascular Risk in Diabetes Study Group 2008). Finally, in a randomized, open-label study of patients treated with metformin and a sulfonylurea, the addition of rosiglitazone was found to increase the risk of hospitalization or death from heart failure more than twofold despite no increase in all-cause mortality (Home et al. 2009). The findings of these studies led regulatory bodies to place warnings and restrictions on the use of rosiglitazone. The authors of this chapter discourage the clinical use of TZDs.

Dipeptidyl Peptidase-4 Inhibitors

Dipeptidyl peptidase-4 (DPP-4) is an enzyme responsible for degradation of the endogenous incretins glucagon-like peptide-1 (GLP-1) and glucose-dependent insulinotropic peptide (GIP) (Nauck et al. 2017). DPP-4 inhibitors selectively inhibit this enzyme, potentiating the effect of endogenous incretins and thus the effect of insulin secretion (Nauck et al. 2017). This class of medications had its cardiovascular safety examined by RCTs years after the first agent, sitagliptin, was approved for clinical use. The first of these trials in 2013 assessed saxagliptin in patients with type 2 diabetes, with or at high proximate risk for cardiovascular disease (Scirica et al. 2013). Saxagliptin was found to be non-inferior to placebo for MACE; however, there was a 1.27-fold increase in the rate of hospitalization for heart failure (HHF) (Scirica et al. 2013). Alogliptin was assessed in an RCT of patients post-myocardial infarction and found to be non-inferior for MACE; however, death or hospitalization from heart failure was not reported (White et al. 2013). For these reasons the authors do not recommended saxagliptin or alogliptin in patients with or at risk for heart failure.

Sitagliptin was found to be non-inferior to placebo for MACE in an RCT of patients with type 2 diabetes and established cardiovascular disease (Green et al. 2015). Importantly, there was no increase in the rate of HHF with sitagliptin therapy (Green et al. 2015). In a 2018 RCT, linagliptin was shown to be non-inferior to placebo for MACE and HHF in patients with type 2 diabetes and high cardiovascular risk (Martin et al. 2019). Sitagliptin and linagliptin appear to be neutral in their effect on MACE and HHF, and as such the authors recommend these agents when clinically indicated, as an adjunct to agents with proven cardiovascular benefit.

Sodium Glucose Cotransporter 2 (SGLT2) Inhibitors

Sodium glucose cotransporter 2 (SGLT2) is a member of a family of glucose-sodium cotransporters that relies on a sodium gradient established by the Na/K ATPase within the cell membrane (Kalra 2014). SGLT2 is concentrated in the proximal convoluted tubule of the nephron and is responsible for the majority of the reabsorption of renally-filtered glucose, approximately 350 mg/min (Kalra 2014). SGLT2 inhibitors act by selectively blocking this cotransporter, therefore promoting glycosuria and lowering plasma levels of glucose (Kalra 2014). Though modest in reduction of HbA1c, SGLT2 inhibitors are attractive due to a low risk of major hypoglycemia, and effective weight neutrality (Zinman et al. 2015; Neal et al. 2017; Wiviott et al. 2019).

The first Cardiovascular Outcome Trial (CVOT) to demonstrate superiority of a GLD over placebo was empagliflozin, cardiovascular outcomes and mortality in type 2 diabetes (EMPA-REG OUTCOME), published in 2015 (Zinman et al. 2015). Patients with established cardiovascular disease were randomized to receive empagliflozin or placebo (Zinman et al. 2015). Empagliflozin was found to be superior to placebo for MACE and key prespecified secondary endpoints, including HHF and all-cause mortality (Zinman et al. 2015). There was no increase in major adverse outcomes, although an increase in genital infections in both male and female patients was noted (Zinman et al. 2015). In the Canagliflozin and Cardiovascular and Renal Events (CANVAS) trial, an RCT of patients with type 2 diabetes with or at high risk for cardiovascular disease, canagliflozin was superior to placebo for MACE and HHF; however, it was non-inferior to placebo for cardiovascular death and all-cause mortality (Neal et al. 2017).

Lower Limb Amputation

In addition to increasing the risk of genital infections, treatment with canagliflozin was shown to increase the risk of amputation, and this has since been adopted as a safety parameter for trials of new glucose-lowering drugs (Neal et al. 2017). In the Dapagliflozin and Cardiovascular Outcomes (DECLARE–TIMI 58) trial, an RCT of patients with type 2 diabetes with or at high risk for cardiovascular disease, dapagliflozin was non-inferior to placebo for MACE and superior in the composite endpoint of prevention of cardiovascular death or HHF (Wiviott et al. 2019). An RCT assessing the cardiovascular outcomes with the use of ertugliflozin is underway at the time of writing. It has been postulated that the difference in cardiovascular outcomes observed in the three published CVOTs relates to the patient populations studied, with empagliflozin being assessed only in patients with established cardiovascular disease.

A recent meta-analysis of these trials demonstrated that the superiority for MACE was present only in patients with prior cardiovascular disease. However it confirmed that SGLT2 inhibitors reduced the risk of cardiovascular death or hospitalization for HHF, irrespective of previously diagnosed atherosclerotic cardiovascular disease (Zelniker et al. 2019a). While there was a numerical difference in the composite of HHF or cardiovascular death, it did not reach statistical significance in patients without established cardiovascular disease (Zelniker et al. 2019a).

Involved Pathways

There have been a number of postulated mechanisms for the improvement in cardiovascular outcomes observed with SGLT2 inhibitors. The most widely accepted and intuitive mechanism is an improvement in ventricular loading conditions, caused by the diuretic effects of these medications (Verma and McMurray 2018). Unlike classical diuretics, SGLT2 inhibitors cause glycosuria and osmotic diuresis in addition to natriuresis, assisting in the removal of extravascular fluid with less intravascular volume depletion and compensatory tachycardia; this may also optimize the abnormal pressure-volume relationship in the diabetic heart (Verma and McMurray 2018; Bertero et al. 2018). A second well-supported mechanism is based on an improvement in the bioenergetics of the diabetic heart. In patients with diabetes or heart failure, the heart is less able to utilize various energy sources, and comes to rely on the metabolism of free fatty acids (Verma and McMurray 2018; Bertero et al. 2018). In addition to being a less efficient source of ATP production, the accumulation of free fatty acid intermediates may contribute to diastolic dysfunction (Verma and McMurray 2018; Lopaschuk et al. 2010).

SGLT2 inhibitors induce a mild hyperketonemic state, similar to a period of fasting, increasing the production and myocardial uptake of the ketone b-hydroxybutyrate (Bertero et al. 2018). This ketone has similar energy efficiency to glucose and is not known to produce toxic intermediates or metabolites (Verma and McMurray 2018; Bertero et al. 2018). A third developing hypothesis is that SGLT2 inhibitors may slow the process of myocardial fibrosis through direct effects on cardiac fibroblast phenotype and function (Verma and McMurray 2018). In a recent RCT, empagliflozin was associated with a small reduction of left ventricular mass when compared with placebo, in patients with type 2 diabetes and coronary artery disease (Verma et al. 2019). Encouraging data exist in animal and in vitro models with human cardiac fibroblasts, and may help explain the cardiovascular benefits of SGLT2 inhibitors in patients without diabetes or hyperglycemia (Verma and McMurray 2018).

There are also a number of proposed mechanisms of cardiovascular benefit that are less well defined. SGLT2 inhibitors may modulate cytosolic and mitochondrial calcium concentrations in cardiac myocytes through inhibition of a Na^+/H^+ exchanger; however, the mechanism remains to be defined, as SGLT2 is not expressed in the human heart (Verma and McMurray 2018). SGLT2 inhibitors may also reduce inflammation and endothelial dysfunction through modulation of adipokines (Verma and McMurray 2018). Research is ongoing to determine if these mechanisms may be utilized in the

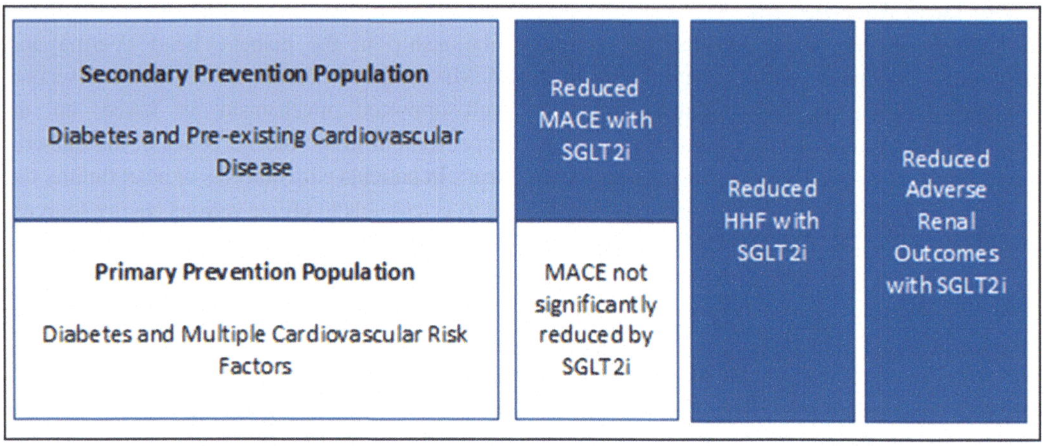

Fig. 53.1 The effect of SGLT2 inhibitors (SGLT2i) on major adverse cardiovascular outcomes (MACE), hospitalization for heart failure (HHF), and adverse renal outcomes, stratified by primary versus secondary prevention

development of novel pharmaceutical agents (Fig. 53.1).

The authors recommend the use of SGLT2 inhibitors in most patients with type 2 diabetes in the absence of contraindications.

GLP-1 Agonists

GLP-1 is an incretin which lowers plasma glucose by enhancing the secretion of endogenous insulin in a glucose-dependent manner (Boyle et al. 2018). This action results in a low risk of major hypoglycemia when mimicked exogenously (Boyle et al. 2018).

In Liraglutide and Cardiovascular Outcomes (LEADER) trial, the first CVOT of a GLP-1 agonist, liraglutide was compared to placebo in patients with type 2 diabetes, with or at high risk for cardiovascular disease (Marso et al. 2016a). Liraglutide was superior to placebo for MACE and all-cause mortality; there was a signal for a decrease in HHF, but this did not reach statistical significance (Marso et al. 2016a). Parenteral semaglutide was superior to placebo for MACE in an RCT of patients with established cardiovascular disease, chronic kidney disease, or both (Marso et al. 2016b). Semaglutide was non-inferior to placebo for HHF and all-cause mortality (Marso et al. 2016b).

The largest CVOT of a GLP-1 agonist compared once-weekly exenatide to placebo patients with type 2 diabetes, with or at high risk of cardiovascular disease (Holman et al. 2017). Exenatide was found to be non-inferior to placebo for MACE and HHF, but did demonstrate a statistically significant reduction in all-cause mortality (Holman et al. 2017). While most trial data exist for parenteral formulations of GLP-1 agonists, a CVOT has recently been completed for an oral formulation of semaglutide (Husain et al. 2019). This agent was shown to be non-inferior to placebo for MACE and HHF, and superior for all-cause mortality (Husain et al. 2019). As a principal limitation of prior agents has been the parenteral route of administration, this formulation of semaglutide has the potential to significantly expand the role of GLP-1 agonists, and is likely to encourage further drug development.

Multiple Cardiovascular Effects

GLP-1 agonists affect multiple organ systems in addition to the pancreas. These agents have been shown to decrease blood pressure, increase heart rate, and enhance myocardial contractility through molecular mechanisms that have not yet been defined (Boyle et al. 2018; Kalra et al. 2016). They may also play a role in reducing vascular inflammation (Boyle et al. 2018). The interplay of these mechanisms in high-risk individuals is a postulated mechanism for the improvements seen in the CVOTs of GLP-1 agonists. These agents also slow gastric emptying, modify glucagon secretion, and enhance renal natriuresis (Boyle et al. 2018; Kalra et al. 2016).

Can Renal Complications Be Mitigated?

Diabetes mellitus is the worldwide leading cause of end-stage renal disease (ESRD) (Perkovic et al. 2019). While many patients with diabetes succumb to cardiovascular disease before developing ESRD, it has been recognized for some time that there are important overlaps in the therapies for these conditions. Blockade of the renin-angiotensin-aldosterone system (RAAS) has long been recognized as a cornerstone of therapy in patients with HRrEF; randomized trial data, such as the Reduction of endpoints in NIDDM with the Angiotensin II Antagonist Losartan (RENAAL) study, has also demonstrated that RAAS blockade reduces progression to ESRD among patients with type 2 diabetes (Brenner et al. 2001). Similarly, lipid-lowering therapy with statins has shown benefit in managing both conditions (Grant and Cosentino 2019).

Several of the recent CVOTs have demonstrated an improvement in renal endpoints, and this finding has led to the development of several renal outcome–focused trials of SGLT2 inhibitors (Zinman et al. 2015; Neal et al. 2017; Wiviott et al. 2019). The recent Canagliflozin and Renal Events in Diabetes with Established Nephropathy Clinical Evaluation (CREDENCE), focusing on cases with albuminuria and significant renal disease (GFR less than 90 ml per/min/ per 1.73 $m^2$2) randomized to canagliflozin or placebo, was stopped early due to benefit in the primary composite endpoint of progression to ESRD, doubling of serum creatine, or renal or cardiovascular death (Perkovic et al. 2019). This is arguably the first large positive RCT using hard renal clinical endpoints in diabetics with kidney disease.

Other renal outcome–focused trials of empagliflozin and dapagliflozin are ongoing. A systematic review and meta-analysis of SGLT2 inhibitors found a 33% reduction in progression to dialysis, renal transplantation, or death due to kidney disease in over 38,000 patients in RCTs, driven largely by the canagliflozin study that was stopped early (Neuen et al. 2019). When the three major completed SGLT2 inhibitor CVOTs were analyzed, a 45% reduction in the same composite renal endpoint occurred (Zelniker et al. 2019a). Importantly, this benefit was seen in both patients with cardiovascular risk factors and those with established atherosclerotic cardiovascular disease (Zelniker et al. 2019a).

Renal outcomes were also assessed in the major CVOTs of GLP-1 agonists (Zelniker et al. 2019b). In meta-analysis of these trials, GLP-1 agonists were found to reduce a composite renal endpoint by 18%; however, this effect was driven by a reduction in progression to macroalbuminuria (Zelniker et al. 2019b). There was no significant difference in doubling of serum creatinine, progression to ESRD, or death from renal causes with GLP-1 agonists (Zelniker et al. 2019b).

In addition to other good general medical management principles, including blood pressure and lipid control, the authors recommend SGLT2 inhibitor therapy in patients with type 2 diabetes who have established atherosclerotic cardiovascular disease or multiple cardiovascular risk factors, to prevent or limit progression of significant renal disease.

Do Guidelines Exist for Pre-Diabetes?

Hyperglycemia and insulin resistance are recognized as a spectrum of disease; while there is a designated threshold for fasting plasma glucose, oral glucose tolerance or HbA1c to make a diagnosis of diabetes, those with abnormal values below these thresholds remain at an increased risk of adverse outcomes that is more difficult to define (Grant and Cosentino 2019). For this reason, patients with pre-diabetes should be managed with the same lifestyle recommendations as patients with diabetes. Similarly, these patients should undergo screening for hypertension, dyslipidemia, albuminuria, and electrocardiographic abnormalities, and any abnormalities should be further investigated and managed per guidelines (Grant and Cosentino 2019; American Diabetes Association 2019b; Stone et al. 2018; Tobe et al. 2018). In the management of hypertension specifically, combination therapy with a beta blocker and diuretic should be avoided, as this has been found to increase progression to a diagnosis of diabetes (Grant and Cosentino 2019).

Pharmacotherapy

A number of glucose-lowering drugs have been used to prevent or delay the progression from pre-diabetes to diabetes. Metformin has been shown to reduce the progression to diabetes in patients with impaired fasting glucose or glucose tolerance when compared to placebo, though it was not as effective as intensive lifestyle intervention (Knowler et al. 2002). Rosiglitazone was found to significantly reduce the progression to diabetes, but its use was associated with an unacceptable increase in adverse cardiovascular outcomes (Gerstein et al. 2006). Insulin glargine has been shown to reduce progression to a diagnosis of diabetes in patients with pre-diabetes by 28%, when compared to standard of care, although this is at the cost of a nearly threefold increase in the risk of major hypoglycemia (Gerstein et al. 2012). Administering a glucose-lowering drug is expected to improve indices of glycemic control and, if effective, should prevent a formal diagnosis of diabetes; the clinical relevance of this remains to be defined.

Statins and Pre-diabetes

Arguably, the diagnosis of diabetes is less important than the outcomes of the metabolic abnormalities or therapeutic interventions. For this reason, statin therapy is recommended in patients with elevated cholesterol and high proximate risk of cardiovascular disease, despite increasing the progression to diabetes, though a causal mechanism has not yet been defined (Grant and Cosentino 2019).

Ongoing Studies and Future Perspectives

Antidiabetic Agents in Normoglycemic Patients

With positive findings in the CVOTs of SGLT2 inhibitors and GLP-1 agonists in the last five years, and a number of postulated mechanisms that are independent of hyperglycemia and insulin resistance, there has been increasing interest in utilizing these agents in patients without diabetes. The Dapagliflozin and Prevention of Adverse Outcomes in Heart Failure (DAPA-HF) trial randomized patients with New York Heart Association class II to IV heart failure and a LVEF less than 40%, to dapagliflozin or placebo in addition to standard medical therapy (McMurray et al. 2019). Approximately 42% of patients in each arm had a diagnosis of diabetes mellitus (McMurray et al. 2019). Dapagliflozin was superior to placebo in a composite primary heart failure endpoint, as well as for cardiovascular mortality and all-cause mortality, regardless of whether the patients had diabetes or not. Interestingly, the hazard ratio for the composite primary endpoint was numerically better in patients without diabetes (McMurray et al. 2019). Studies are ongoing with empagliflozin in patients with heart failure with both preserved and reduced ejection fraction. If empagliflozin is shown superior to placebo in patients with heart failure with preserved ejection fraction, it would be one of the first agents to demonstrate benefit in this population.

Non-insulin Glucose-Lowering Agents in Type 1 Diabetes

While insulin is necessary in this patient population, additional agents that can enhance glycemic control, lower body weight, and perhaps confer cardiovascular benefit are of interest, particularly as cardiovascular disease remains the leading cause of death in patients with type 1 diabetes. While several SGLT2 inhibitors are under review by regulatory bodies for use in type 1 diabetes, none have yet received this indication (Danne et al. 2019). A recent international consensus paper has suggested characteristics of patients with type 1 diabetes who would be ideal candidates for initiation of SGLT2 inhibitor therapy based on a lower overall risk of diabetic ketoacidosis, the most concerning adverse effect of these medications (Danne et al. 2019). From the perspective of the cardiovascular specialist, SGLT2 inhibitor use in this population is attractive, if the potential risks of therapy can be mitigated.

References

American Diabetes Association (2019a) Diabetes technology: standards of medical care in diabetes-2019. Diabetes Care 42:S71–S80

American Diabetes Association (2019b) Cardiovascular disease and risk management: standards of medical care in diabetes-2019. Diabetes Care 42:S103–S123

American Diabetes Association (2019c) Lifestyle management: standards of medical care in diabetes-2019. Diabetes Care 42:S46–S60

Andersson C, Olesen J, Hansen P et al (2010) Metformin treatment is associated with a low risk of mortality in diabetic patients with heart failure: a retrospective nationwide cohort study. Diabetologia 53:2546–2553

Bertero E, Prates Roma L, Ameri P, Maack C (2018) Cardiac effects of SGLT2 inhibitors: the sodium hypothesis. Cardiovasc Res 114:12–18

Boyle JG, Livingstone R, Petrie JR (2018) Cardiovascular benefits of GLP-1 agonists in type 2 diabetes: a comparative review. Clin Sci 132:1699–1709

Brenner BM, Cooper ME, de Zeeuw D et al (2001) Effects of losartan on renal and cardiovascular outcomes in patients with type 2 diabetes and nephropathy. N Engl J Med 345:861–869

Brochu B, Chan M (2019) Treatment of diabetes and heart failure: a paradigm shift for cardiologists? Curr Opin Cardiol 34:207–212

Cannon CP, Blazing MA, Giugliano RP et al (2015) Ezetimibe added to statin therapy after acute coronary syndromes. N Engl J Med 372:2387–2397

Connelly KA, Liu P, Gilbert RE (2018) Treatment of diabetes in people with heart failure: diabetes Canada clinical practice guidelines expert committee. Can J Diabetes 42:S196–S200

Danne T, Garg S, Peters AL et al (2019) International consensus on risk Management of Diabetic Ketoacidosis in patients with type 1 diabetes treated with sodium-glucose Cotransporter (SGLT) inhibitors. Diabetes Care 42:1147–1154

Dongerkery S, Schroeder P, Shomali M (2017) Insulin and its cardiovascular effects: what is the current evidence? Curr Diab Rep 17:1–8

Duckworth W, Abraira C, Moritz T (2009) Glucose control and vascular complications in veterans with type 2 diabetes. J Vasc Surg 49:1084

Dunlay S, Givertz M, Aguilar D et al (2019) Type 2 diabetes mellitus and heart failure: a scientific statement from the American Heart Association and the Heart Failure Society of America. Circulation 140:e294–e324

Gerstein HC, Yusuf S, Bosch J et al (2006) Effect of rosiglitazone on the frequency of diabetes in patients with impaired glucose tolerance or impaired fasting glucose: a randomised controlled trial. Lancet 368:1096–1105

Gerstein HC, Bosch J, Dagenais GR et al (2012) Basal insulin and cardiovascular and other outcomes in Dysglycemia. N Engl J Med 367:319–328

Goto A, Arah OA, Goto M, Terauchi Y, Noda M (2013) Severe hypoglycaemia and cardiovascular disease: systematic review and meta-analysis with bias analysis. BMJ 347:11

Grant PJ, Cosentino F (2019) The 2019 ESC guidelines on diabetes, pre-diabetes, and cardiovascular diseases developed in collaboration with the EASD. Eur Heart J 40:3215–3217

Green JB, Bethel MA, Armstrong PW et al (2015) Effect of sitagliptin on cardiovascular outcomes in type 2 diabetes. N Engl J Med 373:232–242

Hauner H (2002) The mode of action of thiazolidinediones. Diabetes Metab Res Rev 18:S10–S15

Heller SR, DeVries JH, Wysham C, Hansen CT, Hansen MV, Frier BM (2019) Lower rates of hypoglycaemia in older individuals with type 2 diabetes using insulin degludec versus insulin glargine U100: results from SWITCH 2. Diabetes Obes Metab 21:1634–1641

Holman RR, Bethel MA, Mentz RJ et al (2017) Effects of once-weekly exenatide on cardiovascular outcomes in type 2 diabetes. N Engl J Med 377:1228–1239

Home PD, Pocock SJ, Beck-Nielsen H et al (2009) Rosiglitazone evaluated for cardiovascular outcomes in oral agent combination therapy for type 2 diabetes (RECORD): a multicentre, randomised, open-label trial. Lancet 373:2125–2135

Husain M, Birkenfeld AL, Donsmark M et al (2019) Oral semaglutide and cardiovascular outcomes in patients with type 2 diabetes. N Engl J Med 381:841–851

Kalra S (2014) Sodium glucose co-transporter-2 (SGLT2) inhibitors: a review of their basic and clinical pharmacology. Diabetes Ther 5:355–366

Kalra S, Baruah MP, Sahay RK, Unnikrishnan AG, Uppal S, Adetunji O (2016) Glucagon-like peptide-1 receptor agonists in the treatment of type 2 diabetes: past, present, and future. Indian J Endocrinol Metabol 20:254–267
Knowler WC, Barrett-Connor E, Fowler SE et al (2002) Reduction in the incidence of type 2 diabetes with lifestyle intervention or metformin. N Engl J Med 346:393–403
Lopaschuk GD, Ussher JR, Folmes CDL, Jaswal JS, Stanley WC (2010) Myocardial fatty acid metabolism in health and disease. Physiol Rev 90:207–258
Marso SP, Daniels GH, Brown-Frandsen K et al (2016a) Liraglutide and cardiovascular outcomes in type 2 diabetes. N Engl J Med 375:311–322
Marso SP, Bain SC, Consoli A et al (2016b) Semaglutide and cardiovascular outcomes in patients with type 2 diabetes. N Engl J Med 375:1834–1844
Marso SP, McGuire DK, Zinman B et al (2017) Efficacy and safety of degludec versus glargine in type 2 diabetes. N Engl J Med 377:723–732
Martin S, Rode R, Muller C et al (2019) Effect of Linagliptin vs placebo on major cardiovascular events in adults with type 2 diabetes and high cardiovascular and renal risk. JAMA 321:69–79
McMurray JJV, Solomon SD, Inzucchi SE et al (2019) Dapagliflozin in patients with heart failure and reduced ejection fraction. N Engl J Med 381:1995–2008
Nauck M, Meier J, Cavender M, Abd El Aziz M, Drucker D (2017) Cardiovascular actions and clinical outcomes with glucagon-like peptide-1 receptor agonists and dipeptidyl peptidase-4 inhibitors. Circulation 136:849–870
Neal B, Perkovic V, Matthews DR (2017) Canagliflozin and cardiovascular and renal events in type 2 diabetes. N Engl J Med 377:644–657
Neuen BL, Young T, Heerspink HJL et al (2019) SGLT2 inhibitors for the prevention of kidney failure in patients with type 2 diabetes: a systematic review and meta-analysis. Lancet Diabetes Endocrinol 7:845–854
Nissen SE, Wolski K (2007) Effect of rosiglitazone on the risk of myocardial infarction and death from cardiovascular causes. N Engl J Med 356:2457–2471
Perkovic V, Jardine MJ, Neal B et al (2019) Canagliflozin and renal outcomes in type 2 diabetes and nephropathy. N Engl J Med 380:2295–2306
Pfizer Inc (2009) Micronase: glyburide tablets, USP. Pharmacia & Upjohn Company, New York, NY
Rena G, Hardie D, Pearson E (2017) The mechanisms of action of metformin. Diabetologia 60:1577–1585
Rosenstock J, Kahn SE, Johansen OE et al (2019) Effect of linagliptin vs glimepiride on major adverse cardiovascular outcomes in patients with type 2 diabetes. JAMA 322:1155–1166
Roumie CL, Chipman J, Min JY et al (2019) Association of treatment with metformin vs sulfonylurea with major adverse cardiovascular events among patients with diabetes and reduced kidney function. JAMA 322:1167–1111
Scirica BM, Bhatt DL, Braunwald E et al (2013) Saxagliptin and cardiovascular outcomes in patients with type 2 diabetes mellitus. N Engl J Med 369:1317–1326
Shah AD, Langenberg C, Rapsomaniki E et al (2015) Type 2 diabetes and incidence of cardiovascular diseases: a cohort study in 1.9 million people. Lancet Diabetes Endocrinol 3:105–113
Sigal RJ, Armstrong MJ, Bacon SL et al (2018) Physical activity and diabetes: diabetes Canada clinical practice guidelines expert committee. Can J Diabetes 42:S54–S63
Stone JA, Houlden RL, Lin P, Udell JA, Verma S (2018) Cardiovascular protection in people with diabetes: diabetes Canada clinical practice guidelines expert committee. Can J Diabetes 42:S126–S129
The Action to Control Cardiovascular Risk in Diabetes Study Group (2008) Effects of intensive glucose lowering in type 2 diabetes. N Engl J Med 358:2545–2559
The Diabetes Control and Complications Trial Research Group (1993) The effect of intensive treatment of diabetes on the development and progression of long-term complications in insulin-dependent diabetes mellitus. N Engl J Med 329:977–986
The Diabetes Control and Complications Trial/Epidemiology of Diabetes Interventions and Complications (DCCT/EDIC) Study Research Group (2005) Intensive diabetes treatment and cardiovascular disease in patients with type 1 diabetes. N Engl J Med 353:2643–2653
Tobe SW, Gilbert RE, Jones C, Leiter LA, Prebtani AP, Woo V (2018) Treatment of hypertension: diabetes Canada clinical practice guidelines expert committee. Can J Diabetes 42:S186–S189
Verma S, McMurray J (2018) SGLT2 inhibitors and mechanisms of cardiovascular benefit: a state-of-the-art review. Diabetologia 61:2108–2117
Verma S, Mazer C, Yan A et al (2019) Effect of Empagliflozin on left ventricular mass in patients with type 2 diabetes mellitus and coronary artery disease: the EMPA-HEART CardioLink-6 randomized clinical trial. Circulation 140:1693–1702
Wexler DJ (2019) Sulfonylureas and cardiovascular safety. JAMA 322:1147
Wharton S, Pedersen SD, Lau DC, Sharma AM (2018) Weight management in diabetes: diabetes Canada clinical practice guidelines expert committee. Can J Diabetes 42:S124–S129
White WB, Cannon CP, Heller SR et al (2013) Alogliptin after acute coronary syndrome in patients with type 2 diabetes. N Engl J Med 369:1327–1335
Williams R, Colagiuri S et al (2019) DF Atlas 9th Edition. International Diabetes Foundation. Accessed from www.diabetesatlas.org
Wiviott SD, Raz I, Bonaca MP et al (2019) Dapagliflozin and cardiovascular outcomes in type 2 diabetes. N Engl J Med 380:347–357
Zelniker TA, Wiviott SD, Raz I et al (2019a) SGLT2 inhibitors for primary and secondary prevention of

cardiovascular and renal outcomes in type 2 diabetes: a systematic review and meta-analysis of cardiovascular outcome trials. Lancet 393:31–39

Zelniker TA, Wiviott SD, Raz I et al (2019b) Comparison of the effects of glucagon-like peptide receptor agonists and sodium-glucose Cotransporter 2 inhibitors for prevention of major adverse cardiovascular and renal outcomes in type 2 diabetes mellitus. Circulation 139:2022–2031

Zinman B, Wanner C, Lachin JM et al (2015) Empagliflozin, cardiovascular outcomes, and mortality in type 2 diabetes. N Engl J Med 373:2117–2128

Cardiovascular Impact of Newer Diabetes Medications

54

Ravi Kant, McAnto Antony, Arshpreet Kaur, Nathan A. Gilreath, Rashmi Chandra, and Vipin Verma

Contents

Introduction 736
Elevated Morbidity and Mortality 736
Heavy Financial and Social Burden 736
Better Therapeutic Screening 737

Glucagon-like Peptide-1 Receptor Agonists (GLP-1 RAs) and Sodium-Glucose Cotransporter-2 (SGLT-2) Inhibitors 737

Cardiovascular Impact of Glucagon-like Peptide-1 Receptor Agonists (GLP1-RA) 738
Oral GLP1-RA 740

Sodium-Glucose Cotransporter-2 (SGLT-2) Inhibitors 741
Ethnic and Clinical Considerations 742
Renal Protection 742
Official Guidelines 742

Therapeutic Recommendations for DM Patients with High CV Risk 743
Primary Approach 743
Obesity and Renal Comorbidities 743
Contraindications 743
Diabetes End Points and Pharmacologic Alternatives 743

References 744

RaviKant (✉) · M. A. Antony · N. A. Gilreath · R. Chandra · V. Verma
Medical University of South Carolina, Anderson, SC, USA
e-mail: kant.ravi.md@gmail.com

A. Kaur
Division of Endocrinology, Louisiana State University, Shreveport, LA, USA

J. Faintuch and S. Faintuch (eds.), *Obesity and Diabetes*, https://doi.org/10.1007/978-3-030-53370-0_54

Abstract

Cardiovascular disease (CVD) in the diabetic patient is both widespread and fatal in nature. In the United States, roughly two-thirds of deaths in diabetic patients are due to CVD. In addition to the increased morbidity and mortality of CVD in a patient with diabetes, the impact of this comorbidity on health care expenditure is staggering. Unfortunately, researches on the effects of strict glycemic control on cardiovascular (CV) events have been unable to demonstrate a conclusive benefit, despite improvement for microvascular complications. Newer glycemic agents, specifically the glucagon-like peptide-1 receptor agonists (GLP-1 RA) and sodium-glucose cotransporter-2 (SGLT-2) inhibitors, have shown much more promising cardioprotective properties, and are valuable as an add-on therapy for patients who do not reach their hemoglobin A1c goal on metformin alone. Most of the drugs in GLP-1 RA and SGLT-2 inhibitor class have shown strong cardioprotective effects in well-designed cardiovascular outcome trials and all of them have demonstrated CV safety. Multiple SGLT-2 inhibitors have shown renal protective effects as well as prevent hospitalization for heart failure and hence, should be preferred over GLP-1 RAs in patients with concurrent T2DM and heart failure or CKD.

Keywords

Diabetes mellitus · Cardiovascular disease · Cardiovascular outcomes trials · GLP-1 analogue · SGLT2 inhibitors · GLP-1 receptor agonist · T2DM

Introduction

No single diabetes comorbidity is more prevalent than cardiovascular disease (CVD). In a 2019 study by the International Diabetes Federation (IDF), it is estimated that 463 million adults between the ages of 20 and 79 years have diabetes mellitus, accounting for roughly 9.3% of the adult population (International Diabetes Federation 2019). Of those that have type 2 diabetes (T2DM), roughly 32.2%, or more than 134 million people, are also affected by CVD (Einarson et al. 2018a). The relationship between CVD and T2DM is not simply associative, but also predictive in nature. A meta-analysis of 102 prospective studies conducted by the Emerging Risk Factors Collaboration found that when compared to nondiabetics, those with diabetes have an adjusted hazard ratio (HR) of almost 2 for coronary heart disease, ischemic stroke, unclassified stroke, and deaths from vascular diseases (Collaboration TERF 2010). Furthermore, the age of occurrence of CVD in the diabetic population is unique, occurring 14.6 years earlier than in a patient with CVD alone (Booth et al. 2006).

Elevated Morbidity and Mortality

Cardiovascular disease in the diabetic patient is both widespread and fatal in nature. In the United States, roughly two-thirds of deaths in diabetic patients are due to CVD (Wang et al. 2016). Of these deaths, approximately 40% are due to ischemic heart disease and 10% to stroke (Wang et al. 2016). In diabetic patients with CVD hospitalization is also very common, with CVD accounting for 20% of hospital inpatient days spent by all diabetic patients regardless of CVD status in 2012 (American Diabetes Association 2013). When compared to all the chronic complications of diabetes, CVD accounts for roughly 46% of inpatient hospital days, 43.7% of physician office visit, and 38.2% of emergency department visits (American Diabetes Association 2013).

Heavy Financial and Social Burden

CVD is estimated to account for 27% of the total cost of treating diabetes in the United States, which is projected to be 294.6 billion in 2019 (International Diabetes Federation 2019; American Diabetes Association 2013). For an individual with T2DM, CVD greatly increases their individual health care expenditure. A systemic review conducted by Einarson et al. revealed that a type 2 diabetic with CVD has an average health care cost of 112%, or almost $7000, more than a type 2 diabetic without CVD complications (Einarson et al. 2018b). Due to the significant morbidity, mortality, and health care expenditure related to CVD in the diabetic patient, much research has been conducted on the effect of glycemic control on CVD. Unfortunately, researches on the effects of strict glycemic control on cardiovascular (CV) events have been unable to demonstrate a conclusive impact, despite clear benefit to microvascular complications (Kant et al. 2019).

Better Therapeutic Screening

The uncertainty of these results led many researchers to begin looking at the impact of specific glucose-lowering pharmacotherapies on CVD. This research strategy was rewarded when Nissen et al. demonstrated a 43% increased risk of myocardial infarction in patients treated with rosiglitazone (Nissen and Wolski 2007). The results of this experiment not only prompted the United States Food and Drug Administration (US FDA) in 2008 to request CV outcome data for approval of all new drugs used to treat diabetes, but also sparked the interest of researchers to collect CV outcome data on glucose-lowering therapies that were currently being used (https://www.fda.gov/regulatory-information/search-fda-guidance-documents/diabetes-mellitus-developing-drugs-and-therapeutic-biologics-treatment-and-prevention). Metformin was found to be associated with cardioprotective properties; however, many glycemic agents were found to not have the positive effect on CV outcomes that was previously expected (Aldossari 2018). For example, pioglitazone, a thiazolidinedione, has been proven to reduce major adverse CV events, but increases the risk of heart failure (Sesti et al. 2018). Furthermore, two of the dipeptidyl peptidase 4 (DPP-4) inhibitors, saxagliptin and alogliptin, were also found to have the potential to increase the risk of heart failure, and sulfonylureas have been linked to an increased risk of cardiovascular events (Aldossari 2018; Sesti et al. 2018).

Glucagon-like peptide-1 receptor agonists (GLP-1 RA) and sodium-glucose cotransporter-2 (SGLT-2) inhibitors provide much more promising cardioprotective properties, and are valuable as an add-on therapy for patients who do not reach their hemoglobin A1c (A1c) goal on metformin alone (Kant et al. 2019). Therefore, it is imperative for clinicians to be aware of GLP-1 RA and SGLT-2 inhibitor's impact on adverse CV outcomes in patients with concurrent T2DM and CVD.

Glucagon-like Peptide-1 Receptor Agonists (GLP-1 RAs) and Sodium-Glucose Cotransporter-2 (SGLT-2) Inhibitors

GLP1 receptor agonists, which include drugs such as include lixisenatide, liraglutide, dulaglutide, semaglutide, and exenatide, mimic the intestinal secreted incretin hormone responsible for nutrient ingestion. By doing so, GLP-1 acts in a myriad of ways, including stimulating glucose-dependent insulin secretion and decreasing gastric emptying. GLP-1RAs are advisable if weight loss is the goal, but are associated with increased risk of pancreatitis, nausea, and diarrhea. This makes this class of glycemic agents ill-advised for a patient dealing with malnutrition or cachexia (Sesti et al. 2018). SGLT-2 inhibitors, which include drugs such as empagliflozin, canagliflozin, ertugliflozin, and dapagliflozin, act in a much different manner by decreasing the glucose reabsorption in the proximal convoluted tubule of the kidney. SGLT-2 inhibitors are limited by significant reduction in estimated glomerular filtration rate (eGFRs) and have been linked to increased risk of urinary and genital infections (Sesti et al. 2018). Both GLP-1 RAs and SGLT-2 inhibitors are valuable in that they do not increase the risk of hypoglycemia. In regard to their cardioprotective properties, the effect of these agents on the rate of major adverse cardiac events (MACE) has been studied (Tables 54.1

Table 54.1 Cardiovascular outcome trials of GLP-1 receptor agonists

CVOT trial/medication	Notable features and results
ELIXA trial/LIXISENATIDE	– Enrolled 6068 patients with T2DM and recent coronary event within 180 days – No significant difference in MACE-4 – A1c reduced by 0.6%, weight by 0.6 kg, systolic BP by 0.8 mmHg, LDL by 1.5 mg/dl and Tg by 1.8 mg/dl
LEADER trial/LIRAGLUTIDE	– Enrolled 9340 patients with T2DM and high CV risks – 13% reduction in MACE – 22% reduction in death from CV causes – 15% reduction in all-cause mortality – A1c reduced by 0.4%, weight by 2.3 kg, systolic BP by 1.2 mmHg, LDL by 1.5 mg/dl, and Tg by 1.8 mg/dl
EXSCEL trial/EXENATIDE	– Enrolled 14,752 patients with T2DM at a wide range of CV risk (27% of patients without known CV disease) – 9% reduction in MACE[a] – 14% reduction in all-cause mortality[b] – A1c reduced by 0.7%, weight by 1.27 kg, systolic BP by 1.57 mmHg, LDL by 1.5 mg/dl and Tg by 1.8 mg/dl
SUSTAIN-6 trial/INJECTABLE SEMAGLUTIDE	– Enrolled 3297 patients with T2DM and established CV disease or with high CV risks – 26% reduction in MACE – 39% reduction in nonfatal stroke – 26% reduction in nonfatal myocardial infarction[c] – 0.5 mg weekly semaglutide group: A1c reduced by 0.7%, body weight by 2.9 kg and systolic BP by 1.3 mmHg – 1 mg weekly semaglutide group: A1c reduced by 1%, body weight by 4.3 kg and systolic BP by 2.6 mmHg
REWIND trial/DULAGLUTIDE	– Enrolled 9901 patients with T2DM with previous CV event or cardiovascular risk – 12% reduction in MACE – 24% reduction in non-fatal stroke – A1c reduced by 0.61%, body weight by 1.46 kg, BMI by 0.53 kg/m2, systolic BP by 1.7 mmHg, total cholesterol by 0.07 mmol/L and LDL by 0.05 mmol/L
PIONEER-6 trial/ORAL SEMAGLUTIDE	– Enrolled 3183 patients with T2DM and high risk of CV events – No significant reduction in MACE – 51% reduction in death from CV cause – 49% reduction in all-cause mortality – A1c reduced by 1.0% and weight by 4.2 kg

[a]Nonsignificant reduction (Hazard Ratio, 0.91; 95% Confidence Interval, 0.83–1.00; $P < 0.001$ for noninferiority and $P = 0.06$ for superiority)
[b]This difference was not considered to be statistically significant on the basis of the hierarchical testing plan
[c]Nonsignificant reduction (Hazard Ratio, 0.74; 95% Confidence Interval, 0.51–1.08; $P = 0.12$)
CVOT cardiovascular outcome trial; *T2DM* type 2 diabetes mellitus; *CV* cardiovascular; *A1c* hemoglobin A1c; *BP* blood pressure; *Tg* triglycerides; *LDL* low density lipoproteins; *MACE* major adverse cardiovascular events, a composite of death from cardiovascular causes, non-fatal myocardial infarction, or nonfatal stroke; *MACE-4* MACE endpoint as above and hospitalization for unstable angina

and 54.2). MACE encompasses composite of death from cardiovascular causes, nonfatal myocardial infarctions, and nonfatal strokes. Results have shown that many GLP-1RAs and SGLT-2 inhibitors are not only CV safe but are also cardioprotective (Kant et al. 2019).

Cardiovascular Impact of Glucagon-like Peptide-1 Receptor Agonists (GLP1-RA)

In the EXSCEL trial 14,752 diabetic patients with or without CV disease were managed with extended-release exenatide (Holman et al. 2017).

Table 54.2 Cardiovascular outcome trials of SGLT-2 inhibitors

CVOT trial/medication	Notable features and results
EMPA-REG OUTCOME trial/ EMPAGLIFLOZIN	– Enrolled 7028 patients with T2DM and established CV disease – 14% reduction in MACE in pooled empagliflozin group – 38% reduction in death from CV causes – 32% reduction in all-cause mortality – 35% reduction in hospitalization for heart failure – 39% reduction in renal outcomes – A1c reduced by 0.24–0.36%
CANVAS trial/ANAGLIFLOZIN	– Enrolled 9734 patients with T2DM and either established CV disease or high risk of CV disease – 14% reduction in MACE – 33% reduction in hospitalization for heart failure – 40% reduction in renal outcomes – A1c reduced by 0.58%, weight by 1.6 kg and systolic BP by 3.93 mmHg
DECLARE-TIMI 58 trial/ DAPAGLIFLOZIN	– Enrolled 17,160 patients with T2DM and with variable CV risks (40.5% with established CV disease) – No significant difference in MACE – 27% reduction in hospitalization for heart failure – 24% reduction in renal outcomes – A1c reduced by 0.42%, weight by 1.8 kg and systolic BP by 2.7 mmHg

CVOT cardiovascular outcome trial; *T2DM* type 2 diabetes mellitus; *CV* cardiovascular; *A1c* hemoglobin A1c; *BP* blood pressure; *MACE* major adverse cardiovascular events, a composite of death from cardiovascular causes, non-fatal myocardial infarction, or nonfatal stroke

Non-inferiority, however not superiority, regarding placebo was demonstrated, as MACE did not significantly change. In this cardiovascular outcome trial (CVOT) the usual care setting was not changed, and subjects with a wide range of CV risks were included. However many subjects in the exenatide group (43%) prematurely discontinued the trial regimen. These and other difficulties notwithstanding, significance was nearly reached for the primary endpoint MACE. Metabolic and clinical variables also positively responded, with the exception of diastolic blood pressure and heart rate, which mildly increased.

Another major trial (LEADER, with 9340 patients) studied 1.8 mg daily injection of liraglutide (Marso et al. 2016a). Among diabetics with elevated cardiovascular risk, MACE effectively diminished, including less CV death by 22%, and less all-cause mortality by 15%. These benefits were documented within the first one and half years of treatment. When compared to the placebo group, the liraglutide group had fewer patients on insulin and oral antidiabetic medication which led to fewer hypoglycemic episodes, fewer number of obese patients (12.6 vs. 15.2%), and fewer number of patients with a baseline A1c >8.3% (13.7 vs. 16.1%). These differences could possibly explain liraglutide's dominance over placebo. With regard to the metabolic outcomes, from baseline to week 36, the LEADER trial showed a mean reduction in A1c by 0.4%, body weight by 2.3 kg, systolic BP by 1.2 mmHg, LDL by 1.5 mg/dl, and Tg by 1.8 mg/dl; however, similar to EXSCEL trial there was a mean increase in diastolic BP by 0.6 mmHg and heart rate by 3 beats/min.

The ELIXA trial [Lixisenatide in patients with type 2 diabetes and acute coronary syndrome] was the first CVOT among the GLP-1 RA's. It was a multicenter, randomized, double-blind, placebo-controlled trial in which 6068 patients with type 2 diabetes who had an acute coronary syndrome within 6 months before screening were randomly assigned to receive either a daily injection lixisenatide or placebo and were followed for a median of 2.1 years (Pfeffer et al. 2015). Lixisenatide achieved non-inferiority in comparison to placebo for the occurrence of MACE-4; primary composite end point of cardiovascular death, myocardial infarction, stroke, or hospitalization for unstable angina; however, it did not achieve superiority. Occurrence of MACE was similar in lixisenatide and placebo groups. The possible explanation could have been due to enrollment of high-risk patients with recent coronary artery disease and short duration of follow-up. With regard to the metabolic outcomes, from baseline to week 12, the ELIXA trial showed a mean reduction in A1c by 0.6%, body weight by 0.6 kg, systolic BP by 0.8 mmHg, LDL by 1.5 mg/dl, Tg by 1.8 mg/dl, and a mean increase in diastolic BP by 0.6 mmHg and heart rate by 0.4 beats/min.

The SUSTAIN-6 trial [Semaglutide and Cardiovascular Outcomes in Patients with Type 2 Diabetes] was a multicenter, randomized, double-blind, placebo-controlled parallel group trial in which 3297 patients with type 2 diabetes with established CV disease were randomly assigned to receive either the weekly injection of Semaglutide or placebo with a median follow-up of 2.1 years (Marso et al. 2016b). Injectable semaglutide was found to be superior to placebo with 26% relative risk reduction in MACE. Unlike liraglutide, the reduction in MACE with semaglutide was driven by a significant reduction of 39% in nonfatal stroke and a non-significant reduction in nonfatal MI by 26%. Rates of CV death were similar in Semaglutide and control group. Diabetic retinopathy occurred at significantly higher rate in semaglutide-treated patients, with hazard ratio of 1.76. From baseline to week 104, the SUSTAIN-6 trial showed a mean reduction in A1c by 0.7% in the group that received 0.5 mg and 1% with 1 mg weekly semaglutide, a reduction in body weight by 2.9 vs. 4.3 kg according to semaglutide dose, and a mean reduction in systolic BP by 1.3 vs. 2.6 mmHg. Similarly to other GLP-1 RA's CVOT, there was a mean increase in HR by 2 and 2.5 beats/min, depending on semaglutide dose.

The REWIND trial [dulaglutide and cardiovascular outcomes in type 2 diabetes] was a multicenter, randomized, double-blind, placebo-controlled trial in which 9901 patients with type 2 diabetes who had either a previous cardiovascular event or cardiovascular risk factors were randomly assigned to receive either weekly injection of 1.5 mg dulaglutide or placebo, with a median follow-up for 5.4 years (Gerstein et al. 2019). Dulaglutide achieved superiority over placebo in the occurrence of the composite MACE with 12% reduced risk, and this was similar between participants with and without previous CV events. Like the injectable semaglutide, the reduction in MACE was primarily driven by a significant reduction in non-fatal stroke by 24%. There was no significant reduction in CV death and non-fatal MI. It is noteworthy that dulaglutide showed its CV benefits within a year of use. Unlike the other GLP-1 RAs trials, this one had a longer median duration of follow-up (5.4 vs 1.5–3.8 years) which shows that the CV benefits and improvement in metabolic profile by GLP-1 RAs persist for a longer duration than previously reported. Reduction in A1c by 0.61%, body weight by 1.46 kg, BMI by 0.53 kg/m2, systolic BP by 1.7 mmHg, total cholesterol by 0.07 mmol/L, LDL cholesterol by 0.05 mmol/L, and waist-to-hip ratio in men and women was documented. However, there was a mean increase in heart rate by 1.87 beats/min.

Oral GLP1-RA

Oral semaglutide was approved by the US FDA on September 202,019 and is indicated in treatment of adult patients with type 2 diabetes, along with diet and exercise. The tablet contains SNAC [sodium *N*-(8-[2-hydroxylbonzoyl] amino) caprylate] which acts as an absorption enhancer, by alkalinizing the stomach pH leading to increased drug solubility, and it also prevents drug degradation. In the PIONEER (Peptide Innovation for Early Diabetes treatment) trial, 8845 people with T2DM were enrolled across 10 clinical trials. Oral semaglutide was either compared with placebo and/or with one of the FDA-approved glucose-lowering agents. The CVOT is the PIONEER 6 trial in which 3183 patients with T2DM and high risk of CV events were randomly assigned to receive either oral semaglutide or placebo with a median follow-up of 15.9 months (Husain et al. 2019). Like the injectable counterpart, oral semaglutide also showed noninferiority for the occurrence of MACE but failed to show significant reduction in MACE when compared with placebo. However, there was a significant reduction in death from CV cause by 51% and all-cause mortality by 49%. Nonfatal myocardial infarction and nonfatal stroke both occurred at similar rates in the treatment and placebo groups.

Sodium-Glucose Cotransporter-2 (SGLT-2) Inhibitors

Their unique non-insulin-dependent mechanism of action, along with weight loss benefit and a low risk of hypoglycemia, makes them a great option for second-line therapy for T2 DM after metformin. The Empagliflozin Cardiovascular Outcome Event Trial in Type 2 Diabetes Mellitus Patients-Removing Excess Glucose (EMPA-REG OUTCOME), the Canagliflozin Cardiovascular Assessment Study (CANVAS), and the Dapagliflozin Effect on Cardiovascular Events–Thrombolysis in Myocardial Infarction 58 (DECLARE–TIMI 58) were randomized placebo-controlled trials, evaluating the effect of empagliflozin, canagliflozin, and dapagliflozin respectively, on the primary outcome of major adverse cardiovascular (CV) events (MACE) which included death from cardiovascular causes, nonfatal myocardial infarction (MI), or nonfatal stroke (Zinman et al. 2015; Neal et al. 2017; Wiviott et al. 2019).

EMPA-REG OUTCOME trial included 7028 patients with type 2 diabetes, aged 18 years or above, with established CV disease, followed for a median of 3.1 years (Zinman et al. 2015). Empagliflozin was found to be superior to placebo with 14% relative risk reduction in MACE, 38% in CV deaths, 35% in congestive heart failure (CHF) hospitalization, and 32% for all cause-mortality. No difference was noted in the rates of nonfatal myocardial infarctions or stroke. Patients on empagliflozin had significantly more genital infections in both sexes (6.4 vs. 1.8%).

The CANVAS study had two identical trials (CANVAS and CANVAS-Renal) evaluating a total of 9734 patients with type 2 diabetes (A1c 7–10.5%) with either established CV disease (aged 30 years or older) or high risk of CV disease (aged 50 years or older with 2 or more risk factors for atherosclerotic CV disease) (Neal et al. 2017). Median follow-up was 188.2 weeks. Results showed significant decrease of 14% in the primary outcome of MACE, however not for individual components of MACE, nonfatal MI, or nonfatal stroke. There was a significantly higher risk of amputation of toes, feet, or legs in the canagliflozin group (6.3 vs. 3.4 participants with amputation per 1000 patient-years); with highest absolute risk (but not relative risk) in the patients with a previous history of amputation or peripheral vascular disease.

DECLARE–TIMI 58 was the phase 3 trial to study the CV effects of dapagliflozin in patients with type 2 diabetes, and established CV disease or 2 risk factors for atherosclerotic CV disease (men aged 55 years or older or women aged 60 years or older, and one additional risk factor) (Wiviott et al. 2019). This was the largest of the 3 major CV trials for SGLT-2 inhibitors with a total of 17,160 participants, out of which 13,198 patients completed the trial, followed for a median of 4.6 years. Dapagliflozin resulted in a significant reduction of 27% in the composite outcome of CV death and hospitalization for heart failure compared to placebo, which was due to lower rate of hospitalization for heart failure. A significant decrease of 24% was also noted in the renal composite outcome. There was no difference between the groups in the rate of MACE, cardiovascular mortality, or all-cause mortality. Dapagliflozin group had significantly higher rates of diabetic ketoacidosis (0.3% vs. 0.1%) and genital infections leading to discontinuation of trial medication (0.9 vs. 0.1%), but the number of total events was low.

Dapagliflozin and Prevention of Adverse Outcomes in Heart Failure (DAPA-HF) trial was another phase 3, placebo-controlled trial, in 4744 patients, 18 years or older, with ejection fraction (EF) of <40% and NYHA class II, III, or IV heart failure; followed for a median of 18.2 months. Only 42% of the participants had diabetes at the time of screening, and an additional 3% of the patients in each group received a new diagnosis of diabetes (McMurray et al. 2019). Patients in dapagliflozin group (10 mg daily dose in addition to recommended therapy) had 26% lower rates of primary outcome of worsening heart failure (hospitalization or urgent visit resulting in intravenous therapy for heart failure) or cardiovascular death, but the most significant difference was in the number of hospitalizations. Dapagliflozin was equally effective in patients with or without type 2 diabetes.

Ethnic and Clinical Considerations

All of these studies had approximately 70% White population, 20–25% Asians, and only about 5% of Black population. Cardio-renal effects of these medications are also unclear in patients with lower risk of atherosclerotic CV disease since all the trials had patients either with established CV disease, or with high risk of CV disease. The lack of effect on MACE in DECLARE-TIMI 58 trial may likely be due to less patients with established CV disease compared to the other trials. A meta-analysis of the above 3 trials included a total of 34,322 patients with mean age of 63.5 years and 35.1% women, 60.2% with known CV disease and 39.8% with multiple risk factors (0% in EMPA-REG, 34% in CANVAS and 59% in DECLARE-TIMI 58 trials) (Zelniker et al. 2019). SGLT-2 inhibitors reduced the risk of major cardiac event by 11%, CV death and hospitalization for heart failure by 23%, and overall death by 15%. The risk of myocardial infarction was decreased by 11% and CV death by 16%, however with a high level of heterogeneity ($I2 = 79.7\%$). SGLT-2 inhibitors also decreased the composite renal outcome of worsening of renal function, end-stage renal disease (ESRD), or renal death by 45%. All the above effects were mostly restricted to the patients with established CV disease, and no effect was found in patients with multiple risk factors except for a reduction in hospitalization for heart failure by 30%.

Renal Protection

Both EMPA-REG OUTCOME and CANVAS trials showed a significant reduction in secondary renal outcomes (worsening nephropathy or initiation of renal-replacement therapy). The Canagliflozin and Renal Events in Diabetes with Established Nephropathy Clinical Evaluation (CREDENCE), a double-blind, placebo-controlled, randomized trial, was designed to evaluate the renal effects of Canagliflozin (100 mg daily dose) in 4401 patients with type 2 diabetes and albuminuric chronic kidney disease (CKD) (Perkovic et al. 2019). Median follow-up was 2.62 years. Primary composite outcome (ESRD, doubling of serum creatinine level from baseline sustained for 30 days, death from renal or CV disease) rate was significantly lower in the canagliflozin group with a 30% lower relative risk. The effects were consistent across all renal components.

Official Guidelines

Based on the data from above-mentioned trials, the US FDA approved the following new indications for these medications:

- Empagliflozin (December, 2016) to reduce the risk of CV death in adult patients with type 2 diabetes and established CV disease;
- Canagliflozin (October, 2018) to reduce the risk of MACE in adults with type 2 diabetes mellitus and established CV disease;
- Dapagliflozin (October 2019) to reduce the risk of hospitalization for heart failure in adults with type 2 diabetes mellitus and established CV disease or multiple CV risk factors; and
- Canagliflozin (September 2019) to reduce the risk of ESRD, doubling of serum creatinine, cardiovascular death, and hospitalization for heart failure in adults with type 2 diabetes mellitus and diabetic nephropathy with albuminuria.

The exact mechanisms of cardiovascular and renal protective effects of SGLT-2 inhibitors are still unclear but are thought to be multidimensional. Improvement in glycemic control would translate into decrease in microvascular complications. Other non-glucose-dependent mechanisms, including weight reduction, blood pressure reduction, natriuresis, and reduction in intraglomerular pressure, likely contribute toward their overall cardio-renal effects.

Therapeutic Recommendations for DM Patients with High CV Risk

Intensive glycemic control, with target A1c of <6%, in older patients with long-standing T2DM and established CAD or high risk for CVD, has shown to increase mortality with no significant reduction of MACE (Action to Control Cardiovascular Risk in Diabetes Study Group et al. 2008). Patients with diabetes who have experienced hypoglycemic events have higher risk for acute CV events (Johnston et al. 2011). Therefore, avoiding hypoglycemia is one of the vital goals of treating diabetes in patients with or at high risk for CVD.

Primary Approach

Guidelines recommend lifestyle changes and metformin as first-line therapies for treatment of patients with T2DM. Metformin, GLP-1 RAs, and SGLT-2 inhibitors have no to minimal risk of hypoglycemia. The consensus statement from the American Diabetes Association (ADA) and the European Association for the Study of Diabetes (EASD) recommends GLP-1 RAs and SGLT-2 inhibitors after metformin, for medical management of patients with T2DM and clinical CVD (Davies et al. 2018).

Obesity and Renal Comorbidities

GLP-1 RAs are either daily or weekly injectable except recently approved oral semaglutide. GLP-1 RA agonists should be considered as the first injectable agent particularly in obese patients with high CV risk (Davies et al. 2018). In patients with heart failure and CKD, SGLT-2 inhibitors have stronger data demonstrating consistent benefit (Kant et al. 2019). In fact, US FDA has approved dapagliflozin to reduce the risk of hospitalization for heart failure in adults with T2DM and established CV disease or multiple CV risk factors, and canagliflozin to reduce the risk of end-stage kidney disease and CV events in patients with T2DM and CKD. Empagliflozin and canagliflozin are more suitable for patients with CKD if GFR is acceptable as these agents have shown reduction in CKD progression. However, patients with advanced stage III and IV CKD will benefit from GLP-1 receptor agonist as SGLT-2 inhibitors are less effective and not approved to be used in patients with significantly reduced renal function.

Contraindications

Differences in adverse effects and US FDA warnings should be considered while choosing a glycemic agent. GLP-1 RAs are associated with increased gastrointestinal side effects and should be avoided in patients with pancreatitis, gastroparesis, alcohol abuse, and/or high risk for medullary thyroid cancer (Davies et al. 2018). Injectable semaglutide was associated with increased risk of diabetes retinopathy (Marso et al. 2016b). On the other hand, SGLT-2-inhibitors are not advisable in patients with predisposition to fragility fractures, and/or genital or urinary infections. Canagliflozin and ertugliflozin have been associated with increased risk of leg and foot amputations (Neal et al. 2017). Patients should be counseled regarding the preventive foot care, and these medications should be cautiously prescribed in patients with high risk for amputation or known peripheral vascular disease.

Diabetes End Points and Pharmacologic Alternatives

If patients are unable to achieve their target HbA1c or tolerate GLP-1 RA and/or SGLT-

2 inhibitors, other glycemic agents should be considered for add-on therapy. Dipeptidyl peptidase 4 (DPP-4) inhibitors can be used if the patient is not on GLP-1 RA. DPP-4 inhibitors have low risk of hypoglycemia, but saxagliptin and alogliptin have been linked to increased risk of heart failure (Scirica et al. 2013; White et al. 2013; Verma et al. 2017). Similarly, thiazolidinediones (TZDs) should be avoided in patients with heart failure (Lago et al. 2007). Sulfonylureas are among the oldest and cheapest oral hypoglycemic agents; however, they are associated with increased risk of hypoglycemia, adverse CV events, and mortality (Aldossari 2018; Sesti et al. 2018).

References

Action to Control Cardiovascular Risk in Diabetes Study Group, Gerstein HC, Miller ME, Byington RP, Goff DC Jr, Bigger JT, Buse JB, Cushman WC, Genuth S, Ismail-Beigi F, Grimm RH Jr, Probstfield JL, Simons-Morton DG, Friedewald WT (2008) Effects of intensive glucose lowering in type 2 diabetes. N Engl J Med 358(24):2545–2559. https://doi.org/10.1056/NEJMoa0802743

Aldossari KK (2018) Cardiovascular outcomes and safety with antidiabetic drugs. Int J Health Sci (Qassim) 12(5):70–83

American Diabetes Association (2013) Economic costs of diabetes in the U.S. in 2012. Diabetes Care 36(4):1033–1046. https://doi.org/10.2337/dc12-2625

Booth GL, Kapral MK, Fung K, Tu JV (2006) Relation between age and cardiovascular disease in men and women with diabetes compared with non-diabetic people: a population-based retrospective cohort study. Lancet 368(9529):29–36. https://doi.org/10.1016/s0140-6736(06)68967-8

Collaboration TERF (2010) Diabetes mellitus, fasting blood glucose concentration, and risk of vascular disease: a collaborative meta-analysis of 102 prospective studies. Lancet 375(9733):2215–2222. https://doi.org/10.1016/s0140-6736(10)60484-9

Davies MJ, D'Alessio DA, Fradkin J, Kernan WN, Mathieu C, Mingrone G, Rossing P, Tsapas A, Wexler DJ, Buse JB (2018) Management of hyperglycemia in type 2 diabetes, 2018. A consensus report by the American Diabetes Association (ADA) and the European Association for the Study of Diabetes (EASD). Diabetes Care 41:2669–2701

Einarson TR, Acs A, Ludwig C, Panton UH (2018a) Prevalence of cardiovascular disease in type 2 diabetes: a systematic literature review of scientific evidence from across the world in 2007–2017. Cardiovasc Diabetol 17(1). https://doi.org/10.1186/s12933-018-0728-6

Einarson TR, Acs A, Ludwig C, Panton UH (2018b) Economic burden of cardiovascular disease in type 2 diabetes: a systematic review. Value Health 21(7):881–890. https://doi.org/10.1016/j.jval.2017.12.019

Gerstein HC, Colhoun HM, Dagenais GR, Diaz R, Lakshmanan M, Pais P, Probstfield J, Riesmeyer JS, Riddle MC, Rydén L, Xavier D, Atisso CM, Dyal L, Hall S, Rao-Melacini P, Wong G, Avezum A, Basile J, Chung N, Conget I, Cushman WC, Franek E, Hancu N, Hanefeld M, Holt S, Jansky P, Keltai M, Lanas F, Leiter LA, Lopez-Jaramillo P, Cardona Munoz EG, Pirags V, Pogosova N, Raubenheimer PJ, Shaw JE, Sheu WH, Temelkova-Kurktschiev T, REWIND Investigators (2019) Dulaglutide and cardiovascular outcomes in type 2 diabetes (REWIND): a double-blind, randomised placebo-controlled trial. Lancet 394(10193):121–130

Holman RR, Bethel MA, Mentz RJ, Thompson VP, Lokhnygina Y, Buse JB, Chan JC, Choi J, Gustavson SM, Iqbal N, Maggioni AP, Marso SP, Öhman P, Pagidipati NJ, Poulter N, Ramachandran A, Zinman B, Hernandez AF, EXSCEL Study Group (2017) Effects of once-weekly exenatide on cardiovascular outcomes in type 2 diabetes. N Engl J Med 377(13):1228–1239. https://doi.org/10.1056/NEJMoa1612917

Husain M, Birkenfeld AL, Donsmark M, Dungan K, Eliaschewitz FG, Franco DR, Jeppesen OK, Lingvay I, Mosenzon O, Pedersen SD, Tack CJ, Thomsen M, Vilsbøll T, Warren ML, Bain SC, PIONEER 6 Investigators (2019 Aug 29) Oral semaglutide and cardiovascular outcomes in patients with type 2 diabetes. N Engl J Med 381(9):841–851

International Diabetes Federation (2019) IDF diabetes atlas, 9th edn. International Diabetes Federation, Brussels

Johnston SS, Conner C, Aagren M, Smith DM, Bouchard J, Brett J (2011) Evidence linking hypoglycemic events to an increased risk of acute cardiovascular events in patients with type 2 diabetes. Diabetes Care 34:1164e70

Kant R, Munir KM, Kaur A, Verma V (2019) Prevention of macrovascular complications in patients with type 2 diabetes mellitus: review of cardiovascular safety and efficacy of newer diabetes medications. World J Diabetes 10(6):324–332. https://doi.org/10.4239/wjd.v10.i6.324

Lago RM, Singh PP, Nesto RW (2007) Congestive heart failure and cardiovascular death in patients with prediabetes and type 2 diabetes given thiazolidinediones: a meta-analysis of randomised clinical trials. Lancet

Marso SP, Daniels GH, Brown-Frandsen K, Kristensen P, Mann JF, Nauck MA, Nissen SE, Pocock S, Poulter NR, Ravn LS, Steinberg WM, Stockner M, Zinman B, Bergenstal RM, Buse JB, LEADER Steering Committee; LEADER Trial Investigators (2016a) Liraglutide and cardiovascular outcomes in type 2 diabetes. N Engl J Med 375(4):311–322. https://doi.org/10.1056/NEJMoa1603827

Marso SP, Bain SC, Consoli A, Eliaschewitz FG, Jódar E, Leiter LA, Lingvay I, Rosenstock J, Seufert J, Warren ML, Woo V, Hansen O, Holst AG, Pettersson J, Vilsbøll T, SUSTAIN-6 Investigators (2016b) Semaglutide and cardiovascular outcomes in patients with type 2 diabetes. N Engl J Med 375 (19):1834–1844. https://doi.org/10.1056/NEJMoa1607141

McMurray JJV, Scott D, Solomon SD, Inzucchi SE SE (2019) Dapagliflozin in patients with heart failure and reduced ejection fraction. N Engl J Med 381:1995–2008. https://doi.org/10.1056/NEJMoa1911303

Neal B, Perkovic V, Mahaffey KW et al (2017) Canagliflozin and cardiovascular and renal events in type 2 diabetes. N Engl J Med 377:644–657

Nissen SE, Wolski K (2007) Effect of rosiglitazone on the risk of myocardial infarction and death from cardiovascular causes. N Engl J Med 356(24):2457–2471. https://doi.org/10.1056/NEJMoa072761

Perkovic V, Jardine MJ, Neal B et al (2019) Canagliflozin and renal outcomes in type 2 diabetes and nephropathy. N Engl J Med 380:2295–2306

Pfeffer MA, Claggett B, Diaz R, Dickstein K, Gerstein HC, Køber LV, Lawson FC, Ping L, Wei X, Lewis EF, Maggioni AP, McMurray JJ, Probstfield JL, Riddle MC, Solomon SD, Tardif JC, ELIXA Investigators (2015) Lixisenatide in patients with type 2 diabetes and acute coronary syndrome. N Engl J Med 373 (23):2247–2257. https://doi.org/10.1056/NEJMoa1509225

Scirica BM, Bhatt DL, Braunwald E, Steg PG, Davidson J, Hirshberg B, Ohman P, Frederich R, Wiviott SD, Hoffman EB, Cavender MA, Udell JA, Desai NR, Mosenzon O, McGuire DK, Ray KK, Leiter LA, Raz I, SAVOR-TIMI 53 Steering Committee and Investigators (2013) Saxagliptin and cardiovascular outcomes in patients with type 2 diabetes mellitus. N Engl J Med 369(14):1317–1326

Sesti G et al (2018) Management of diabetes in older adults. Nutr Metab Cardiovasc Dis 28(3):206–218. https://doi.org/10.1016/j.numecd.2017.11.007

Verma S, Goldenberg RM, Bhatt DL, Farkouh ME, Quan A, Teoh H, Connelly KA, Leiter LA, Friedrich JO (2017 Feb 24) Dipeptidyl peptidase-4 inhibitors and the risk of heart failure: a systematic review and meta-analysis. CMAJ Open 5(1):E152–E177

Wang CCL, Hess CN, Hiatt WR, Goldfine AB (2016) Clinical update: cardiovascular disease in diabetes mellitus. Circulation 133(24):2459–2502. https://doi.org/10.1161/circulationaha.116.022194

White WB, Cannon CP, Heller SR, EXAMINE Investigators et al (2013) Alogliptin after acute coronary syndrome in patients with type 2 diabetes. N Engl J Med 369:1327–1335

Wiviott SD, Raz I, Bonaca MP et al (2019) Dapagliflozin and cardiovascular outcomes in type 2 diabetes. N Engl J Med 380:347–357. https://doi.org/10.1056/NEJMoa1812389

Zelniker TA, Wiviott SD, Raz I et al (2019 Jan 5) SGLT2 inhibitors for primary and secondary prevention of cardiovascular and renal outcomes in type 2 diabetes: a systematic review and meta-analysis of cardiovascular outcome trials. Lancet 393(10166):31–39. https://doi.org/10.1016/S0140-6736(18)32590-X

Zinman B, Wanner C, Lachin JM et al (2015 Nov 26) Empagliflozin, cardiovascular outcomes, and mortality in type 2 diabetes. N Engl J Med 373(22):2117–2128

Pathogenesis and Molecular Targets in Treatment of Diabetic Wounds

55

Satish Patel, Pragati, Shradha Devi Dwivedi, Krishna Yadav, Jagat R. Kanwar, Manju Rawat Singh, and Deependra Singh

Contents

Introduction 748
Normal Wound Healing 748
Diabetic Foot Ulcer 749

Major Factors and Key Molecular Pathways of Diabetic Wounds 750
Mitochondrial Overproduction of Reactive Oxygen Species 752
Alteration of Growth Factors 752
Matrix Metalloproteinase 9 753
Defective Cytokine Production 753
Abnormal Cellular Activity 754
Neuropathy 754
Nitric Oxide Interventions 754
Vascular Disease 754

Preventive Care 755

Therapeutic Approaches: Standard Care Versus Cells, Biomaterials, and Growth Factors 756

References 756

S. Patel
Drug Standardisation Unit (H), CCRH, Hyderabad, Telangana, India

Pragati · S. D. Dwivedi · K. Yadav · M. R. Singh · D. Singh (✉)
University Institute of Pharmacy, Pt. Ravishankar Shukla University, Raipur, Chhattisgarh, India
e-mail: deependraiop@gmail.com

J. R. Kanwar
Nanomedicine-Laboratory of Immunology and Molecular Biomedical Research (NLIMBR), School of Medicine (SoM), Molecular and Medical Research (MMR) Strategic Research Centre, Faculty of Health, Deakin University, Waurn Ponds, VIC, Australia

J. Faintuch and S. Faintuch (eds.), *Obesity and Diabetes*, https://doi.org/10.1007/978-3-030-53370-0_55

Abstract

Wound healing in diabetes is remarkably delayed due to various underlying pathological processes. Diabetes alters all the stages of wound healing such as remodeling, proliferation, hemostasis, and inflammatory phases. In diabetic patients, minor skin lesions may lead to unhealed chronic ulcers, and ultimately result in infection, gangrene, even amputation. Physiological factors responsible for the delay of wound healing include impaired growth factor and cytokine production, angiogenic response, macrophage and neutrophil function, collagen accumulation, and variation in the ratio of collagen types leading to weakened healing response. Key molecular targets for the local/pharmacologic treatment of wound healing include growth factors and other molecules, absorbable biomaterials, and cell regeneration therapy.

Keywords

Pathogenesis · Diabetic wound · Molecular targets · Healing · Diabetic foot ulcer

Introduction

Diabetes is a major cause for slower healing in every population. Between 2019 and 2045, the global expenditures for diabetes treatment is expected to grow from 760 billion U.S. dollars to 845 billion U.S. dollars (Elflein 2019).

Diabetes is a chronic metabolic disease that affects more than 463 million persons, and it is estimated that 20% of them develop complicated diabetic wounds or foot ulcer (IDF Diabetes Atlas 9th edition 2019; Nunan et al. 2014; Patel et al. 2019). Most acute wounds heal without issue; however, as age increases, impaired blood circulation and other conditions like smoking, obesity, and chronic diseases, such as diabetes, lead to slower healing. Diabetic complications resulting from diabetes include neuropathy, arterial damage, and ischemia, which may complicate diabetic wounds (Nunan et al. 2014). Unhealed wounds are prone to infection and lower-limb amputation. Diabetes is one of the principal causes of nontraumatic lower-extremity amputation.

In diabetes, each stage of healing, i.e., hemostasis, inflammatory, proliferation, and remodeling phase, is altered. Diabetic wounds show signs of impaired healing due to an uncoordinated healing process. Elongated inflammatory phase with hindrance in the mature granulation tissue formation and reduced wound tensile strength is observed in diabetic wounds (Patel et al. 2019).

Normal Wound Healing

Multiple sequential cellular and biochemical phenomena are necessary to restore damaged tissue. Hemostasis and clot formation is conventionally the starting point, triggering the inflammatory phase, commanded by neutrophils and macrophages, during which debris and bacteria are eliminated and growth factors are secreted. True repair only begins with the proliferative phase, including revascularization (angiogenesis) and build-up of the cellular and noncellular matrix. The maturation phase is responsible for tissue strength, encompassing remodeling of the local architecture and vascular abundance (Fadini et al. 2014; Singh et al. 2011, 2013). Differently from acute injuries, chronic wounds are more indolent and don't follow the same phases. As a consequence, healing can be delayed for 12 weeks or more (Anisha et al. 2013; Mohandas et al. 2015).

Diabetic Wounds

In diabetic patients, a minor skin wound often leads to chronic, nonhealing ulcers and ultimately results in infection, gangrene, even amputation. Damage of numerous layers of dermal tissue involving epidermis, dermis, and sometimes the subcutaneous tissue is not uncommon. The prevalence of foot ulcers ranges from 4% to 10% with a lifetime incidence as high as 25% (Dadpay et al. 2012). It is the most common complication of diabetes, greater than retinopathy, nephropathy, heart attack, and stroke combined.

In the Asian continent, the diabetic foot represents a significant health problem, provoked by the high frequency of infection and the ever-rising incidence of diabetes. Insufficiency of diabetic foot care centers, deprived foot care information, approach and practice among diabetic patients, deferred recommendation or reporting to the podiatry centers, and limited income and educational status of the patients contribute significantly to the increased frequency of diabetic foot complications (Viswanathan et al. 2005).

Etiology and Pathogenesis

There are scores of reported physiologic disorders reportedly responsible for wound healing deficiencies in diabetes. Some representative ones are listed below:

- Impaired growth factor production (Qi et al. 2018)
- Impaired cytokine production (Zubair and Ahmad 2019)
- Impaired angiogenic response (Galeano et al. 2011)
- Weakened immune response (Peleg et al. 2007)
- Decreased neuropeptide expression (Theocharidis and Veves 2020)
- Impaired macrophage and neutrophil function (Maruyama et al. 2007)
- Increased serum matrix metalloproteinase-9 (Li et al. 2013)
- Impaired collagen accumulation and variation in the ratio of collagen types (Stolarczyk et al. 2018)
- Dysregulation of procalcitonin, fibrinogen, and IL-6 (Korkmaz et al. 2018)
- Aberrant macrophage polarization and function in wound healing responses (Ganesh and Ramkumar 2020)
- Imbalance between extracellular matrix (ECM) components and remodeling by matrix metalloproteinases (Gooyit et al. 2014)
- Deficiency of thrombin-activable fibrinolysis inhibitor (Verkleij et al. 2010)
- Advanced glycation end products (AGEP) modification of platelet-derived growth factor (PDGF) (Nass et al. 2010)
- Decreased levels of chemokine receptor CXCR3 and its ligand 10, CXCL10 (Bodnar et al. 2009)

Diabetic Foot Ulcer

Peripheral vascular and neuropathy disorders are believed to be crucial for the development of diabetic foot ulcers, compromising the survival and well-being of diabetic patients (Jeffcoate 2011). Predisposing conditions include previous deformations of the foot, reduction in regional oxygenation and perfusion, poor eyesight, and obesity. Poorly compensated diabetes as well as bacterial colonization are aggravating circumstances, along with presence of resistant bacteria, which further worsen the prognosis and make the therapy expensive (Hariono et al. 2018).

Classification of Diabetic Foot Ulcer

- Wagner–Meggit
- Brodsky Depth—Ischemic
- University of Texas
- International Working Group (2019 Guidelines)
- SAD
- PEDIS
- Other classifications

Wagner–Meggit

This is one of the oldest classification systems, created for the dysvascular foot. It includes six grade systems (grade 0 to grade 5), which emphasize ulcer depth, concentration of tissue necrosis, and occurrences of gangrene (Mehraj 2018).

Depth Ischemic

This classification system is the modernized form of the Wagner–Meggit classification, aiming to show a clear difference between lesion and foot vascularity (Mehraj 2018). It consists of three grades which depend upon the presence or absence of ischemia with total or partial gangrene.

University of Texas

The University of Texas San Antonio classification system (UTSA) measures diabetic foot wound based on the depth of wound, infection, and ischemia in the lower limb. The grading system depends upon the wound depth, while stages of the classification depend upon the ischemia occurrence, bioburden of lesions, or merger of both by eliminating neuropathy. Superior grades or stages of a wound are less prone to healing. As compared to the Wagner classification, this system looks more promising and accurate. Nevertheless, optimal use of this and other classifications is still debated (Bravo-Molina et al. 2018).

International Working Group (2019 Guidelines)

The International Working Group has updated its guidelines related to diabetic foot disease on prevention, offloading, peripheral artery disease, infection, wound healing interventions, and classification of diabetic foot ulcers. These six protocols are not included in Table 55.1; however, they can be searched in the literature.

The following variables are endorsed as relevant for classification: patient-related (end-stage renal failure), limb-related (peripheral artery disease and loss of protective sensation), and ulcer-related (area, depth, site, single, or multiple and infection). Thorough wound assessment, including severity of infection and arterial perfusion (need for revascularization), is underscored. Among existing classifications, SINBAD (Site, Ischemia, Neuropathy, Bacterial Infection, and Depth), WIfI (Wound, Ischemia, and Foot Infection), and the guidelines of the Infectious Diseases Society of America are recommended for certain purposes (Lipsky et al. 2012; IWGDF Guidelines Org 2019; Bus et al. 2020a, b; Monteiro-Soares et al. 2020).

SAD

This classification system (Size, Arteriopathy, Denervation) considers size, denervation, sepsis, and arteriopathy. It is designed for hectic clinical practice and doesn't require any specialist techniques. Even though quite old, it has been reasonably well-validated (Monteiro-Soares et al. 2014).

PEDIS

As the acronym indicates, it addresses Perfusion, Extent, Depth/Tissue Loss, Infection, and Sensation. The numerous classification grades make it complex for clinical practice use (Jain and Joshi 2013).

Other Classifications

There is no dearth of diabetic foot ulcer classifications. However, most methods were developed with relatively small series and were not widely adopted, consequently suffering from limited validation. Many are not particularly user-friendly (Monteiro-Soares et al. 2014). Current representative tools are depicted in Table 55.1.

Diagnosis and staging are fundamental; however, instrumental monitoring is not less of a priority. To this aim, neuropathy and autonomic dysfunction, peripheral vascular disease, and eventual osteomyelitis need to be investigated. 3D and hyperspectral wound imaging are equally emphasized as reliable tools for lesion measurement and ulcer (Fernández-Torres et al. 2020).

Major Factors and Key Molecular Pathways of Diabetic Wounds

Molecular factors/targets for the management of diabetic wounds include cytokines, growth factors, clotting factors, prostaglandins, free radicals, nitric oxide, insulin-like growth factor (IGF-1) signaling axis including gangliosides, neuropeptides, mi-RNAs, lactoferrin, stromal cell-derived factor (SDF-1α), Hypoxia inducible factors (HIFs), thymosin beta 4, substance P, endopeptidase cathepsin D, and RANKL (Martí-Carvajal et al. 2015; Dam and Paller 2018; Zubair and Ahmad 2019; Liu et al. 2020). These agents directly or indirectly modulate vascularization, innervation, matrix reconstruction, and

Table 55.1 Examples of classifications of diabetic foot wound

Wagner-Meggit classification system

Grades	Foot wound
0	No open wound or cellulitis
1	Superficial ulcer
2	Deep ulcer upto tendons and joint tissue
3	Deep ulcer with abscess, osteomyelitis, and joints sepsis
4	Local gangrene forefoot or heel
5	Gangrene of entire foot

Depth ischemic classification

Depth grade	Definition	Ischemia grade	Definition
0	At risk, foot with the previous ulcer that may cause a new ulcer	A	No ischemia
1	Superficial noninfected ulcer	B	
2	Deep ulcer with tendon or joint exposed (±infection)	C	Partial forefoot gangrene
3	Extensive ulcer with bone exposed or deep abscess	D	Total foot gangrene

University of Texas Classification

Stages	Grades			
	0	**1**	**2**	**3**
A	Healed pre- or post-ulcerative lesion completely epithelialized	Superficial wound not involving bone, tendon, or capsule	Wound penetrating tendon or capsule	Wound penetrating to bone or joint
B	With infection	With infection	With infection	With infection
C	With ischemia	With ischemia	With ischemia	With ischemia
D	With infection and with ischemia	With infection and with ischemia	With infection and with ischemia	With infection and with ischemia

(AD) SAD system

Grades	Area	Deep	Sepsis	Arteriopathy	Denervation
0	Skin intact	Intact skin		Pedal pulse	Intact
1	Lesion <1 cm^2	Superficial (skin and subcutaneous tissues)	No infected lesions	Pedal pulse reduce or miss	Reduced
2	The lesion from 1 to 3 cm^2	Lesion penetrating to tendon, periosteum, and joint capsule	Cellulitis-associated lesions	Absence of both pedal pulses	Absent
3	Lesion >3 cm^2	A lesion in bone or joint space	Osteomyelitis-associated lesions	Gangrene	Charcot joint

PEDIS classification

Risk factor group	Characteristics
0	No neuropathy, no PVD
1	Neuropathy, no deformity PVD
2	Neuropathy and deformity, and/or PVD
3	History pathology

reepithelialization, including keratinocyte migration and proliferation on an extracellular matrix. All of these can be defective in diabetic wounds (Martí-Carvajal et al. 2015; Dam and Paller 2018).

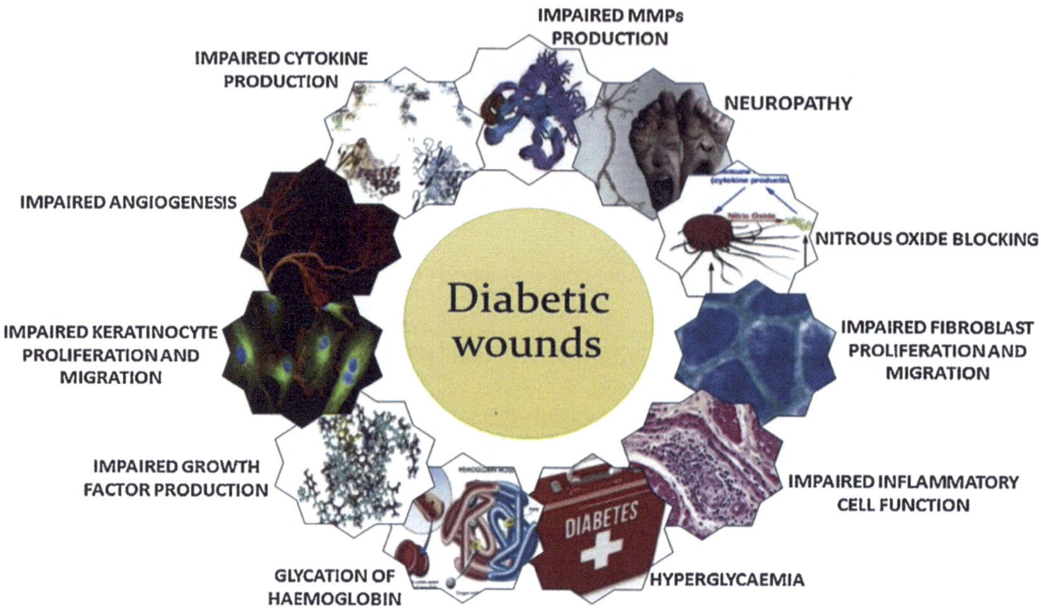

Fig. 55.1 Factors responsible for diabetic wounds [Adopted and reproduced from Patel et al. 2019]

It is important to emphasize that despite encouraging experimental results, few of these biomolecules have been investigated in large randomized trials. Indeed, a Cochrane Systematic Review a few years ago, addressing 11 growth factors for diabetic wounds, concluded that all increased the likelihood of healing of foot ulcers in diabetic patients. Such optimism notwithstanding, the disclaimer was that protocols suffered from high risk of bias, and side effects were possibly underreported (Martí-Carvajal et al. 2015). More clear-cut positive outcomes were detected for recombinant human epidermal growth factor (rhEGF), both intralesionally and topically applied, in a meta-analysis covering 6 trials and 530 patients (Bui et al. 2019). There are several factors that responsible for diabetic wounds are mentioned in Fig. 55.1.

Mitochondrial Overproduction of Reactive Oxygen Species

Advanced glycation end-products (AGEs) are a consequence of deranged glucose homeostasis. Transcription factors involved in inflammation and protein kinase C can be activated in such circumstances, and nerve protein glycation can occur, in conjunction with tissue oxidative stress and ischemia. Wound healing can be impaired because of these aberrations, including diminished local sensation which potentially results in additional injury (Patel et al. 2019; Shaikh-Kader et al. 2019).

Alteration of Growth Factors

Growth factors are pharmacologically active polypeptides. In all phases of wound healing, they promote relevant biological and molecular events. In the granulation phase of tissues, growth factors contribute to the early inflammation stage (Patel et al. 2019). Compromised wounds often demonstrate a defect in the kind and quantity of growth factor, due to alteration in the occurrence, enhancement in the degradation, reduction in the trapping, release, and production. Extracellular matrix (ECM) synthesis is categorized by a balance between matrix formation and matrix degradation, for optimal healing. Some of the factors influencing the formulation of ECM are VEGF,

IGF-I, IGF-II, TGF-β23, KGF24, PDGF25, EGF26, FGF27, TNF-α, and IL-6, which can be diminished in diabetic person, including suppression of receptors along with quick degradation of growth factors (Fui et al. 2019; Su et al. 2019; Patel et al. 2019; Zubair and Ahmad 2019).

Platelet-Derived Growth Factor

At the early stage of wound healing, platelets synthesize the platelet-derived growth factor (PDGF). PGDF is an essential mitogen that promotes the proliferation of fibroblast, matrix production, along with the maturation of connective tissue. In all stages of wound healing, PDGF continuously activates various cellular responses. PGDF binds with a receptor of tyrosine kinase and triggers various signaling pathways, leading to the enhancement of migration and proliferation of the cell. For the inflammatory cell and fibroblast, PDGF acts as a chemoattractant and encourages the production of collagen, glucosamine, and proteoglycan. In diabetic wound patients, there is a reduction in the expression of PDGF and PDGF receptors (Ishihara et al. 2019).

Vascular Endothelial Growth Factor

The wound healing process is affected by the concentration of VEGF as it supports the rate-determining steps in angiogenesis and vasculogenesis. With the help of the protease, it causes the degeneration of a three-dimensional network of an extracellular macromolecule of active vessels. In the case of diabetic wounds, it could enhance the density of capillary and develop the perfusion rate of blood along with the metabolism in wounded tissue. In a small series of diabetic foot ulcers, managed by hyperbaric oxygen therapy (HBOT), VEGF became elevated. The author defends that HBOT aids lesion epithelialization, both directly and indirectly, through VEGF upsurge and TNF-α downturn (Semadi 2019).

Transforming Growth Factor Beta

It has been reported that in diabetic patients there is a decrease in the amount of TGF β in wounded tissue that leads to retardation of the wound healing process. At the promoter site, MMP-encoded genes show the TGF-β1-dependent inhibitory element with a decrease in gene expression. The decrease in expression of TGF-β and upregulation of MMPs lead to destruction of growth factor transcription factors like Smad-2, Smad-3, and Smad-4, which also activate and repress TGF-β target genes. TGF-β1 activates Smad-2 and Smad-3 for the production of collagen (Hozzein et al. 2015). The decrease in the level of TGF-β1 causes increased recruitment of activated inflammatory cells, predisposing to a delayed inflammatory phase till the proliferation phase of the healing process in DWs (Heublein et al. 2015). Decreased levels and expression of those growth factors could contribute to poor and prolonged wound healing processes in diabetes (Patel et al. 2019).

Matrix Metalloproteinase 9

The central role of the extracellular matrix in the tissue remodeling processes involved in wound healing has already been alluded to. Endopeptidase enzymes degrade such matrices, and inappropriate conduction of this phenomenon can seriously affect the healing sequence. Matrix metalloproteinase 9 (MMP9) is actually a cluster of different crystal structures highly expressed in diabetic foot ulcer healing. Inhibitors have been identified and their binding mode was elucidated. In this sense, they could play a pharmacologic role in the handling of such complication (Hariono et al. 2018).

Defective Cytokine Production

Elevated interleukin-6 (IL-6) in diabetic foot ulcers has been demonstrated, and these levels decrease as the ulcers heal (Korkmaz et al. 2018). In experimental animals, similar lesions treated with IL-22 heal more rapidly, due to better vascularization, reepithelialization, granulation tissue formation, and VEGF release, with less keratinocyte differentiation. Pharmacologically, interleukin-22 seems to be superior to PDGF and VEGF, because of gene induction related to reepithelialization, innate host defense mechanism, and tissue remodeling (Zubair and Ahmad 2019).

Abnormal Cellular Activity

At the start of the healing process, neutrophils appear, followed by monocytes which differentiate into macrophages. Endothelial cells, fibroblasts and keratinocytes, are analogously involved in the restoration of damaged tissue (Patel et al. 2019; Krzyszczyk et al. 2018). Macrophages and neutrophils are often increased in diabetic wounds. Macrophages in diabetic patients have reduced clearance activity; reduced capability to phagocyte the dead cells. Decreased T cells, increased B cells, dysregulation of the proliferation of macrophages, fibroblasts, endothelial cells, and keratinocytes are all reported. Infiltration of macrophages and neutrophils is prolonged in diabetes, and macrophages produce a reduced level of cytokines.

Mesenchimal stem cells are a promise to overcome such overlapping deficiencies, given their potential for multilineage differentiation. In-vivo and in-vitro protocols have availed themselves not only of direct cell therapy, but also of indirect intervention with the help of micro RNA (miRNA) and long noncoding RNA (lncRNA) (Li et al. 2020).

Neuropathy

Peripheral neuropathy mainly alters the sensory, motor, and autonomic function. An insensate foot can lead to injury, including skin irritation and pressure sores. Alteration of autonomic function predisposes to a delayed healing process due to arteriovenous shunting, impaired circulation, and edema (Theocharidis and Veves 2020). Due to the lack of protective sensation, many wounds remain unnoticed and progressively become worse.

The polyol pathway, along with many others, has been implied in the pathogenesis of diabetic neuropathy. In this sense, they could serve as potential targets for pharmacotherapy (Dewanjee et al. 2018).

Nitric Oxide Interventions

In diabetes, hyperglycemia can decrease the production of nitric oxide by inhibiting endothelial nitric oxide synthase (NOS) activation, which can favor the accumulation of reactive oxygen species, mainly superoxide. In the presence of metals ion like ferrous or cuprous ions, reactive oxygen species are converted to the highly reactive and damaging hydroxyl radical. In addition, they are also involved in the oxidization of sulfhydryl groups in proteins, lipid peroxidation, and the generation of reactive aldehydes and nitrogen oxide. These radicals disrupt the endothelium, which affects vascular function like vasoconstriction response, platelet aggregation, abnormal growth, and inflammation. Nitric oxide is a potent vasodilator, and local administration could positively influence indolent wound healing.

The hypothesis that a deficiency of NOS in diabetic patients leads to poor vascularization, peripheral neuropathy, and foot ulcers has been entertained (Walton et al. 2019). There is evidence that the genotype eNOS distribution is not different between diabetics with and without foot ulcers. Incidentally, in the same experience VEGF gene polymorphism wasn't different either (Erdogan et al. 2018). Nevertheless, transdermal nitric oxide (NO) treatment, in the form of NO donors, iNOS induction, and other pathways, has received attention in the recent literature (Erdogan et al. 2018; Walton et al. 2019).

Vascular Disease

The association of vascular disease and nonhealing of foot ulcer is well-established. Such circumstances notwithstanding, indications

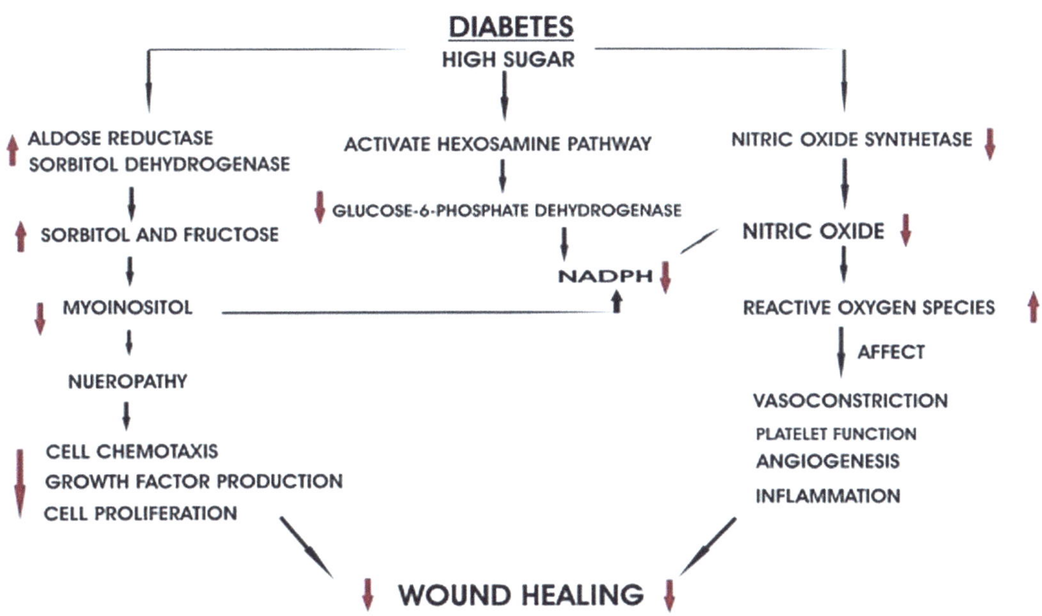

Fig. 55.2 Major pathways responsible for decreased wound healing in diabetes [Adopted and reproduced from Patel et al. 2019]

and outcomes of revascularization surgery are still controversial. One of the largest reviews, conducted by the International Working Group of the Diabetic Foot, covered over 13,000 patients. Results concerning part of this investigation point out towards fairly good response with both open and endovascular operations. Approximately 60% of all wounds were healed after 1 year, and amputation rate was about 10% after the same follow-up period. Revascularization interventions are, therefore, recommended when there is clear evidence of peripheral artery disease and ulcer; however, the best technique is still open to debate (Forsythe et al. 2020). The major mechanisms responsible for the decline in the wound healing process in case of diabetes are mentioned in Fig. 55.2.

Preventive Care

Traditional handling of diabetic foot ulcers is associated with inadequate efficacy, prolonged morbidity resulting in high direct and indirect cost, insufficiently documented side effects, and relapse rate as high as 50%. Hence, prevention of diabetic foot ulcers is the most important challenge.

According to the IWGDF guidelines (Bus et al. 2020b), screening of asymptomatic cases for peripheral neuropathy and arteriopathy is the first concern. If low risk is estimated, education and self-care by the patient should be highlighted, and pre-ulcerative signs should be treated. Additionally for those with moderate to high risk, footwear should be carefully selected, and monitoring of foot skin temperature is advised.

Indeed, footwear that relieves plantar pressure is useful for secondary prevention as well, reducing recurrence rates. Refractory cases should be surgically managed, and access to a multidisciplinary, integrated center should be a priority, whenever feasible (Bus et al. 2020b).

Therapeutic Approaches: Standard Care Versus Cells, Biomaterials, and Growth Factors

Many innovative local approaches are being assayed for patients with complex, recurrent, or refractory lesions, including stem cell therapy, photobiomodulation, and nanotherapy, in order to restore function and prevent amputation. Wound dressings, as well as scaffolds made of absorbable biomaterials, often impregnated with growth factors, nitric oxide and other pharmacologic agents, have been designed (Shu et al. 2018; Zarei et al. 2018; Erdogan et al. 2018; Zubair and Ahmad 2019;Walton et al. 2019).

Secondary infection is not uncommon, sometimes by resistant bacteria, and antibiotics are an integral part of the therapeutic arsenal in such context, within the recommendations of the Infectious Diseases Society of America (Lipsky et al. 2012). Revascularization surgery, as already mentioned, is indicated for those with demonstrated vascular impairment. Debridement and amputation, along with eventual flap rotation and skin grafting, once comparatively common among diabetics, should be reserved for carefully selected acute/chronic cases or for certain acute/urgent foot ulcers (Monteiro-Soares et al. 2014; Forsythe et al. 2020).

Acknowledgments The authors are thankful to the Director, University Institute of Pharmacy, Pt. Ravishankar Shukla, University Raipur, Chhattisgarh, for providing necessary infrastructure facilities and DHR-ICMR project no. V.25011/286-HRD/2016-HR for financial support.

References

Anisha BS, Biswas R, Chennazhi KP, Jayakumar R (2013) Chitosan-hyaluronic acid/nano silver composite sponges for drug resistant bacteria infected diabetic wounds. Int J Biol Macromol 62:310–320. https://doi.org/10.1016/j.ijbiomac.2013.09.011

Bodnar RJ, Yates CC, Rodgers ME, Du X, Wells A (2009) IP-10 induces dissociation of newly formed blood vessels. J Cell Sci 122(12):2064–2077. https://doi.org/10.1242/jcs.048793

Bravo-Molina A, Linares-Palomino JP, Vera-Arroyo B, Salmerón-Febres LM, Ros-Díe E (2018) Inter-observer agreement of the Wagner, University of Texas and PEDIS classification systems for the diabetic foot syndrome. Foot Ankle Surg 24(1):60–64. https://doi.org/10.1016/j.fas.2016.10.009

Bui TQ, Bui QVP, Németh D, Hegyi P, Szakács Z, Rumbus Z, Tóth B, Emri G, Párniczky A, Sarlós P, Varga O (2019) Epidermal growth factor is effective in the treatment of diabetic foot ulcers: meta-analysis and systematic review. Int J Environ Res Public Health 16 (14). https://doi.org/10.3390/ijerph16142584

Bus SA, Van Netten JJ, Hinchliffe RJ, Apelqvist J, Lipsky BA, Schaper NC, IWGDF Editorial Board (2020a) Standards for the development and methodology of the 2019 International Working Group on the Diabetic Foot guidelines. Diabetes Metab Res Rev. https://doi.org/10.1002/dmrr.3267

Bus SA, Lavery LA, Monteiro-Soares M, Rasmussen A, Raspovic A, Sacco ICN, van Netten JJ, International Working Group on the Diabetic Foot (2020b) Guidelines on the prevention of foot ulcers in persons with diabetes (IWGDF 2019 update). Diabetes Metab Res Rev 36:e3269. https://doi.org/10.1002/dmrr.3269

Dadpay M, Sharifian Z, Bayat M, Bayat M, Dabbagh A (2012) Effects of pulsed infra-red low level-laser irradiation on open skin wound healing of healthy and streptozotocin-induced diabetic rats by biomechanical evaluation. J Photochem Photobiol B Biol 111:1–8. https://doi.org/10.1016/j.jphotobiol.2012.03.001

Dam DHM, Paller AS (2018) Gangliosides in diabetic wound healing. Prog Mol Biol Transl Sci 156:229–239. https://doi.org/10.1016/bs.pmbts.2017.12.006

Dewanjee S, Das S, Das AK, Bhattacharjee N, Dihingia A, Dua TK, Kalita J, Manna P (2018) Molecular mechanism of diabetic neuropathy and its pharmacotherapeutic targets. Eur J Pharmacol 833:472–523. https://doi.org/10.1016/j.ejphar.2018.06.034

Elflein J (2019) Global healthcare expenditure due to diabetes in 2019 and 2045. https://www.statista.com/statistics/241820/estimated-global-healthcare-expenditures-to-treat-diabetes/. Accessed 10 Dec 2019

Erdogan M, Kulaksizoglu M, Tetik A, Solmaz S, Kucukaslan AS, Eroglu Z (2018) The relationship of the endothelial nitric oxide synthase (eNOS) and vascular endothelial growth factor (VEGF) gene polymorphism in Turkish type 2 diabeticpatients with and without diabetic foot ulcers. Foot (Edinburgh) 37:5–10. https://doi.org/10.1016/j.foot.2018.06.006

Fadini GP, Albiero M, Millioni R, Poncina N, Rigato M, Scotton R, Boscari F, Brocco E, Arrigoni G, Villano G, Turato C, Biasiolo A, Pontisso P, Avogaro A (2014) The molecular signature of impaired diabetic wound healing identifies serpinB3 as a healing biomarker. Diabetologia 57(9):1947–1956. https://doi.org/10.1007/s00125-014-3300-2

Fernández-Torres R, Ruiz-Muñoz M, Pérez-Panero AJ, García-Romero J, Gónzalez-Sánchez M (2020) Instruments of choice for assessment and monitoring diabetic foot: a systematic review. J Clin Med 9(2). https://doi.org/10.3390/jcm9020602

Forsythe RO, Apelqvist J, Boyko EJ, Fitridge R, Hong JP, Katsanos K, Mills JL, Nikol S, Reekers J, Venermo M, Zierler RE, Hinchliffe RJ, Schaper NC (2020) Effectiveness of revascularisation of the ulcerated foot in patients with diabetes and peripheral artery disease: a systematic review. Diabetes Metab Res Rev 36:e3279. https://doi.org/10.1002/dmrr.327

Fui LW, Lok MPW, Govindasamy V, Yong TK, Lek TK, Das AK (2019) Understanding the multifaceted mechanisms of diabetic wound healing and therapeutic application of stem cells conditioned medium in the healing process. J Tissue Eng Regen Med 13 (12):2218–2233. https://doi.org/10.1002/term.2966

Galeano M, Polito F, Bitto A, Irrera N, Campo GM, Avenoso A, Calò M, Cascio PL, Minutoli L, Barone M, Squadrito F, Altavilla D (2011) Systemic administration of high-molecular weight hyaluronan stimulates wound healing in genetically diabetic mice. Biochim Biophys Acta Mol Basis Dis 1812 (7):752–759. https://doi.org/10.1016/j.bbadis.2011.03.012

Ganesh GV, Ramkumar KM (2020 Mar 7) Macrophage mediation in normal and diabetic wound healing responses. Inflamm Res. https://doi.org/10.1007/s00011-020-01328-y

Gooyit M, Peng Z, Wolter WR, Pi H, Ding D, Hesek D, Lee M, Boggess B, Champion MM, Suckow MA, Mobashery S, Chang M (2014) A chemical biological strategy to facilitate diabetic wound healing. ACS Chem Biol 9(1):105–110. https://doi.org/10.1021/cb4005468

Hariono M, Yuliani SH, Istyastono EP, Riswanto FDO, Adhipandito CF (2018) Matrix metalloproteinase 9 (MMP9) in wound healing of diabetic foot ulcer: molecular target and structure-based drug design. Wound Med 22:1–13. https://doi.org/10.1016/j.wndm.2018.05.003

Heublein H, Bader A, Giri S (2015) Preclinical and clinical evidence for stem cell therapies as treatment for diabetic wounds. Drug Discov Today 20(6):703–717. https://doi.org/10.1016/j.drudis.2015.01.005

Hozzein WN, Badr G, Al Ghamdi AA, Sayed A, Al-Waili NS, Garraud O (2015) Topical application of propolis enhances cutaneous wound healing by promoting TGF-beta/smad-mediated collagen production in a streptozotocin-induced type I diabetic mouse model. Cell Physiol Biochem 37(3):940–954. https://doi.org/10.1159/000430221

IDF Diabetes Atlas 9th edition 2019. https://www.diabetesatlas.org/en/. Accessed 22 February 2020

Ishihara J, Ishihara A, Starke RD, Peghaire CR, Smith KE, McKinnon TAJ, Tabata Y, Sasaki K, White MJV, Fukunaga K, Laffan MA, Lutolf MP, Randi AM, Hubell JA (2019) The heparin binding domain of von Willebrand factor binds to growth factors and promotes angiogenesis in wound healing. Blood 133 (24):2559–2569. https://doi.org/10.1182/blood.2019000510

IWGDF Guidelines Org (2019). https://iwgdfguidelines.org/wp-content/uploads/2019/05/01-IWGDF-practical-guidelines-2019.pdf. Accessed 22 February 2020

Jain A, Joshi S (2013) Diabetic foot classifications: review of literature. Med Sci 2(3):715. https://doi.org/10.5455/medscience.2013.02.8069

Jeffcoate WJ (2011) Stratification of foot risk predicts the incidence of new foot disease, but do we yet know that the adoption of routine screening reduces it? Diabetologia 54(5):991–993. https://doi.org/10.1007/s00125-011-2075-y

Korkmaz P, Koçak H, Onbasi K, Biçici P, Ozmen A, Uyar C, Ozatag DM (2018) The role of serum procalcitonin, interleukin-6, and fibrinogen levels in differential diagnosis of diabetic foot ulcer infection. J Diabetes Res. https://doi.org/10.1155/2018/7104352

Krzyszczyk P, Schloss R, Palmer A, Berthiaume F (2018) The role of macrophages in acute and chronic wound healing and interventions to promote pro-wound healing phenotypes. Front Physiol 9:419. https://doi.org/10.3389/fphys.2018.00419

Li Z, Guo S, Yao F, Zhang Y, Li T (2013) Increased ratio of serum matrix metalloproteinase-9 against TIMP-1 predicts poor wound healing in diabetic foot ulcers. J Diabetes Complicat 27(4):380–382. https://doi.org/10.1016/j.jdiacomp.2012.12.007

Li B, Luan S, Chen J, Zhou Y, Wang T, Li Z, Fu Y, Zhai A, Bi C (2020) The MSC-derived Exosomal lncRNA H19 promotes wound healing in diabetic foot ulcers by upregulating PTEN via MicroRNA-152-3p. Mol Ther Nucleic Acids 19:814–826. https://doi.org/10.1016/j.omtn.2019.11.034

Lipsky BA, Berendt AR, Cornia PB, Pile JC, Peters EJ, Armstrong DG, Deery HG, Embil JM, Joseph WS, Karchmer AW, Pinzur MS, Senneville E, Infectious Diseases Society of America (2012) Infectious Diseases Society of America clinical practice guideline for the diagnosis and treatment of diabetic foot infections. Clin Infect Dis 54(12):e132–e173

Liu D, Liu L, Yao L, Peng X, Li Y, Jiang T, Kuang H (2020) Synthesis of ZnO nanoparticles using radish root extract for effective wound dressing agents for diabetic foot ulcers in nursing care. J Drug Deliv Sci Technol 55:101364. https://doi.org/10.1016/j.jddst.2019.101364

Martí-Carvajal AJ, Gluud C, Nicola S, Simancas-Racines-D, Reveiz L, Oliva P, Cedeño-Taborda J (2015) Growth factors for treating diabetic foot ulcers. Cochrane Database Syst Rev 10:CD008548. https://doi.org/10.1002/14651858.CD008548.pub2

Maruyama K, Asai J, Ii M, Thorne T, Losordo DW, D'Amore PA (2007) Decreased macrophage number

and activation lead to reduced lymphatic vessel formation and contribute to impaired diabetic wound healing. Am J Pathol 170(4):1178–1191. https://doi.org/10.2353/ajpath.2007.060018
Mehraj DM (2018) A review of Wagner classification and current concepts in management of diabetic foot. Int J Orthop Sci 4(1):933–935. https://doi.org/10.22271/ortho.2018.v4.i1n.133
Mohandas A, Anisha BS, Chennazhi KP, Jayakumar R (2015) Chitosan-hyaluronic acid/VEGF loaded fibrin nanoparticles composite sponges for enhancing angiogenesis in wounds. Colloids Surf B Biointerfaces 127:105–113. https://doi.org/10.1016/j.colsurfb.2015.01.024
Monteiro-Soares M, Martins-Mendes D, Vaz-Carneiro A, Sampaio S, Dinis-Ribeiro M (2014) Classification systems for lower extremity amputation prediction in subjects with active diabetic foot ulcer: a systematic review and meta-analysis. Diabetes Metab Res Rev 30 (7):610–622. https://doi.org/10.1002/dmrr.2535
Monteiro-Soares M, Russell D, Boyko EJ, Jeffcoate W, Mills JL, Morbach S, Game F, International Working Group on the Diabetic Foot (IWGDF) (2020) Guidelines on the classification of diabetic foot ulcers (IWGDF 2019). Diabetes Metab Res Rev 36:e3273. https://doi.org/10.1002/dmrr.3273
Nass N, Vogel K, Hofmann B, Presek P, Silber RE, Simm A (2010) Glycation of PDGF results in decreased biological activity. Int J Biochem Cell Biol 42 (5):749–754. https://doi.org/10.1016/j.biocel.2010.01.012
Nunan R, Harding KG, Martin P (2014) Clinical challenges of chronic wounds: searching for an optimal animal model to recapitulate their complexity. DMM Dis Models Mech 7(11):1205–1213. https://doi.org/10.1242/dmm.016782
Patel S, Srivastava S, Singh MR, Singh D (2019) Mechanistic insight into diabetic wounds: pathogenesis, molecular targets and treatment strategies to pace wound healing. Biomed Pharmacother 112(January):108615. https://doi.org/10.1016/j.biopha.2019.108615
Peleg AY, Weerarathna T, McCarthy JS, Davis TME (2007) Common infections in diabetes: pathogenesis, management and relationship to glycaemic control. Diabetes Metab Res Rev 23(1):3–13. https://doi.org/10.1002/dmrr.682
Qi M, Zhou Q, Zeng W, Wu L, Zhao S, Chen W, Luo C, Shen M, Zhang J, Tang CE (2018) Growth factors in the pathogenesis of diabetic foot ulcers. Front Biosci (Landmark Ed) 23:310–317
Semadi NI (2019) The role of VEGF and TNF-alpha on epithelialization of diabetic foot ulcers after hyperbaric oxygen therapy. Open Access Maced J Med Sci 7 (19):3177–3183. https://doi.org/10.3889/oamjms.2019.297
Shaikh-Kader A, Houreld NN, Rajendran NK, Abrahamse H (2019) The link between advanced glycation end products and apoptosis in delayed wound healing. Cell Biochem Funct 37(6):432–442. https://doi.org/10.1002/cbf.3424
Shu X, Shu S, Tang S, Yang L, Liu D, Li K, Dong Z, Ma Z, Zhu Z, Din J (2018) Efficiency of stem cell based therapy in the treatment of diabetic foot ulcer: a meta-analysis. Endocr J 65(4):403–413. https://doi.org/10.1507/endocrj.EJ17-0424
Singh D, Singh MR, Saraf S, Saraf S (2011) Development of delivery cargoes for debriding enzymes effective in wound healing. Trends Appl Sci Res 6(8):863–876. https://doi.org/10.3923/tasr.2011.863.876
Singh MR, Saraf S, Vyas A, Jain V, Singh D (2013) Innovative approaches in wound healing: trajectory and advances. Artificial Cells Nanomed Biotechnol 41(3):202–212. https://doi.org/10.3109/21691401.2012.716065
Stolarczyk A, Sarzyńska S, Gondek A, Cudnoch-Jędrzejewska A (2018) Influence of diabetes on tissue healing in orthopaedic injuries. Clin Exp Pharmacol Physiol 45(7):619–627. https://doi.org/10.1111/1440-1681.12939
Su L, Zheng J, Wang Y, Zhang W, Hu D (2019) Emerging progress on the mechanism and technology in wound repair. Biomed Pharmacother 117:109191. https://doi.org/10.1016/j.biopha.2019.109191
Theocharidis G, Veves A (2020) Autonomic nerve dysfunction and impaired diabetic wound healing: the role of neuropeptides. Auton Neurosci 223. https://doi.org/10.1016/j.autneu.2019.102610
Verkleij CJN, Roelofs JJTH, Havik SR, Meijers JCM, Marx PF (2010) The role of thrombin-activatable fibrinolysis inhibitor in diabetic wound healing. Thromb Res 126(5):442–446. https://doi.org/10.1016/j.thromres.2010.08.008
Viswanathan V, Madhavan S, Rajasekar S, Chamukuttan S, Ambady R (2005) Amputation prevention initiative in South India: positive impact of foot care education. Diabetes Care 28(5):1019–1021. https://doi.org/10.2337/diacare.28.5.1019
Walton DM, Minton SD, Cook AD (2019) The potential of transdermal nitric oxide treatment for diabetic peripheral neuropathy and diabetic foot ulcers. Diabetes Metab Syndr 13(5):3053–3056. https://doi.org/10.1016/j.dsx.2018.07.003
Zarei F, Negahdari B, Eatemadi A (2018) Diabetic ulcer regeneration: stem cells, biomaterials, growth factors. Artif Cells Nanomed Biotechnol 46(1):26–32. https://doi.org/10.1080/21691401.2017.1304407
Zubair M, Ahmad J (2019) Role of growth factors and cytokines in diabetic foot ulcer healing: a detailed review. Rev Endocr Metab Disord 20(2):207–217. https://doi.org/10.1007/s11154-019-09492-1

Efficiency in Public Health Through the Promotion of Diabetes Day Hospitals: A Regional Proposal

56

Ascensión Barroso, Ramón Sanguino, Victoria Barroso, and M. Isabel Sánchez-Hernández

Contents

Efficiency in Public Health 759
Day Hospital Concept 760
Social Impact of Diabetes 761
Diabetes Day Hospital 761
Final Remarks 762
References 763

Abstract

Efficiency in public health implies cost reduction and improvements in citizens' quality of life. Diabetes negatively affects the quality of life of the people suffering from this disease. It is considered one of the main public health problems worldwide because it has a high prevalence and is a direct cause of death, disability, and high expenditures. Focusing the attention on some regional experiences in Spain, we analyze the advantages of the creation of a Diabetes Day Hospital (DDH) as a way to rationalize costs and, at the same time, to promote improvements in the quality of life of people with diabetes. Within the context of general healthcare services, technical efficiency was calculated with reference to the relationship between the resources used and the expected savings per patient as a health output. The strength of this approach is the possibility to ensure a better coordination of the different agents involved in the treatment of the disease and the possibility to reduce costs in the regional healthcare delivery system.

Keywords

Diabetes · Day Hospital · Efficiency · Public Health System

A. Barroso · R. Sanguino (✉) · M. I. Sánchez-Hernández
Faculty of Economics and Business Science, University of Extremadura, Badajoz, Spain
e-mail: sanguino@unex.es

V. Barroso
Faculty of Medicine, University of Extremadura, Badajoz, Spain

Efficiency in Public Health

Everyone deserves equal rights and opportunities and this includes the right to good health. However, inequities in health remain both in

J. Faintuch and S. Faintuch (eds.), *Obesity and Diabetes*, https://doi.org/10.1007/978-3-030-53370-0_56

developing and developed countries that are most of the times avoidable and unnecessary and always unjust.

Public health spending in Spain during 2016 was 71,477 million euros, or 9% of GDP. Public participation was 71.2%, the remainder being in charge of private parties (IDIS Report 2019). This is slightly above the OECD average of 8.9% (Organization for Economic Cooperation and development/OECD Health Statistics 2019). These data are very similar worldwide, around 10% of GDP, according to The Global Health Expenditure Database (2020).

The main stakeholders in the healthcare sector are urging the governments to promote new initiatives to ensure the viability of the national health system, seriously questioned because of its high deficit. Since the 1980s, Spain drags an annual deficit with healthcare around 6000 million Euros, which represents up to 11% difference between the budgeted and the really spent money (El Economista 2018).

Many national governments worldwide must assess the efficiency of their health sectors to ensure that public money is used to best effect. The technical efficiency approach (Cylus et al. 2017) is based on Farrell's concept that a hospital that produces the maximum amount of output from a given input is technically efficient (Farrell 1957). Hospital efficiency is crucial as they are primary consumers of health resources.

Day Hospital Concept

One of the bets to improve efficiency in the public health sector is through the day hospitalization. Day hospital (DH) is a new healthcare modality, both in public and private sphere, which has experienced a great development in recent years, notably increasing efficiency in the attention to patients who were previously admitted to conventional hospitals (Ministry of Health and Social Policy 2009). In Spain, within the network of public hospitals, there are 16,820 day hospital places (36.2 places per 100,000 inhabitants), a resource that is progressively increasing in line with the increase in ambulatory care (Ministry of Health, Consumption, and Social Welfare 2017).

Day Hospital (DH) is defined as the assembly of healthcare procedures which are performed in a hospital for a few hours, either for diagnosis, clinical investigation, or treatment that cannot be performed in a consulting room, yet not requiring hospitalization (Royal Decree 1277/2003, Ministry of Health and Social Policy 2009).

The dissemination and consolidation of the day hospital, the diversity of organizational, structural, and functional configurations, as well as its repercussion in terms of quality and patient safety, prioritize its inclusion in the actions included in the quality plan of the national health system together with the elaboration of quality and safety criteria (Ministry of Health and Social Policy 2009).

There are some documents containing standards and recommendations of quality to set up DH in Spain (Ministry of Health and Consumer Affairs 2008). They serve to key aspects of the DH such as the security and rights of the patient, organization and management, physical structure, and resources. Other countries like United Kingdom, United States, or Canada have similar standards and recommendations .

In Spain, the main applicable guidelines are Law 14/1986 and Law 16/2003. In addition, according to the Royal Decree 1277/2003 (Government of Spain 2003), for the authorization of a DH it would be necessary to make a modification of the plan of the general hospital, which must be previously authorized. In connection with them, two different regulations are observed: (1) on authorization and registration, which assesses healthcare centers before they come into operation, and (2) on accreditation, for evaluation of operational centers.

Only four regions have legislation and official programs for accreditation based on voluntary external assessment. Others have specific accreditation programs (organ transplants, assisted reproduction, hemotherapy treatments, continued training). So far, there is no specific accreditation system for DHs.

Four modalities of DH are currently defined: medical (including oncological), psychiatric, geriatric, and surgical (Ministry of Health, Social Services and Equality 2016, 2017). Recognition of diabetes day hospital (DDHs) is still expected.

Social Impact of Diabetes

Diabetes is considered one of the main public health problems worldwide, mainly due to its high prevalence, the high economic cost to the public health system, and the number of premature deaths (Giusti et al. 2020). Diabetes Mellitus (DM) is a metabolic disorder of multiple etiology characterized by chronic hyperglycemia with alterations in the metabolism of carbohydrates, fats, and proteins, resulting from defects in the secretion of insulin, in the action of insulin, or both (American Diabetes Association 2014). Diabetes requires continuous medical care with multifactorial risk-reduction strategies beyond glycemic control. Ongoing patient self-management education and support are critical to preventing acute complications and reducing the risk of long-term complications (American Diabetes Association 2019).

An improvement in the care of DM would help to increase the life expectancy of the people who suffer it, given that it would reduce the likelihood and frequency of diabetes-related complications. Derived costs would also diminish. Numerous investigations tried to identify the risk factors for diabetes such as lifestyle, environment, or genetic traits (Zhang et al. 2020). The identification of these factors could allow the development of measures of primary prevention, which could have long-term consequences for public health policies (Harder et al. 2009) as they could reduce health spending and increase efficiency. The importance of an efficient use of public health resources for regional economic growth and stability and for the well-being of patients has been studied in recent years (Zhang et al. 2015). According to Crespo et al. (2013), 14.8% of all hospital discharges in Spain include a diagnosis of diabetes, with a cost of 1934 million Euros (33% of the total cost of the disease).

Diabetes Day Hospital

As alluded to, diabetes is a complex, chronic illness requiring continuous medical care with multifactorial risk-reduction strategies beyond glycemic control (American Diabetes Association 2019). Moreover, the identification of the risk factors for diabetes could allow the development of measures of primary prevention, which could have long-term consequences for public health policies (Harder et al. 2009) as they could reduce health spending and increase efficiency. Therefore, DDHs pave the way for an improvement in the quality of life of these cases (Zhang et al. 2015).

The DDH is specialized, as it aims to offer patient care 24 h a day. It requires synchronization with primary care units and the other services in the hospital (Morales 2014). It is normally located close to the endocrinology outpatient clinics because it functionally depends on the endocrinology service. Principal facilities are a waiting room, a medical consultation office, a treatment room, a group education room, a retinography room, a warehouse, and toilets. Figure 56.1 shows an example of how the DDH could be distributed:

The DDHs assist to patients: with recent onset of type 1 diabetes mellitus, with type 2 diabetes mellitus to schedule the insulin treatment, with decompensation in diabetes mellitus, woman with gestational diabetes, to implement and monitor insulin infusion pumps and continuously monitor glucose, to detect and diagnose chronic complications of diabetes mellitus, and to offer therapeutic diabetes education (Government of Extremadura 2007; Government of Andalusia 2013; Health Parc of Barcelona 2015).

The staff of the DDH is an endocrine physician and nurses who are also experts in diabetes nutrition and education. Moreover, the Public Sanitary System also includes psychologists, who can help to maintain therapeutic adhesion if necessary. The DDHs operate uninterruptedly at least 12 h a day. If the patient remains in the DDH during meal times, the nurses prepare the diets and request the corresponding food from the hospital kitchen.

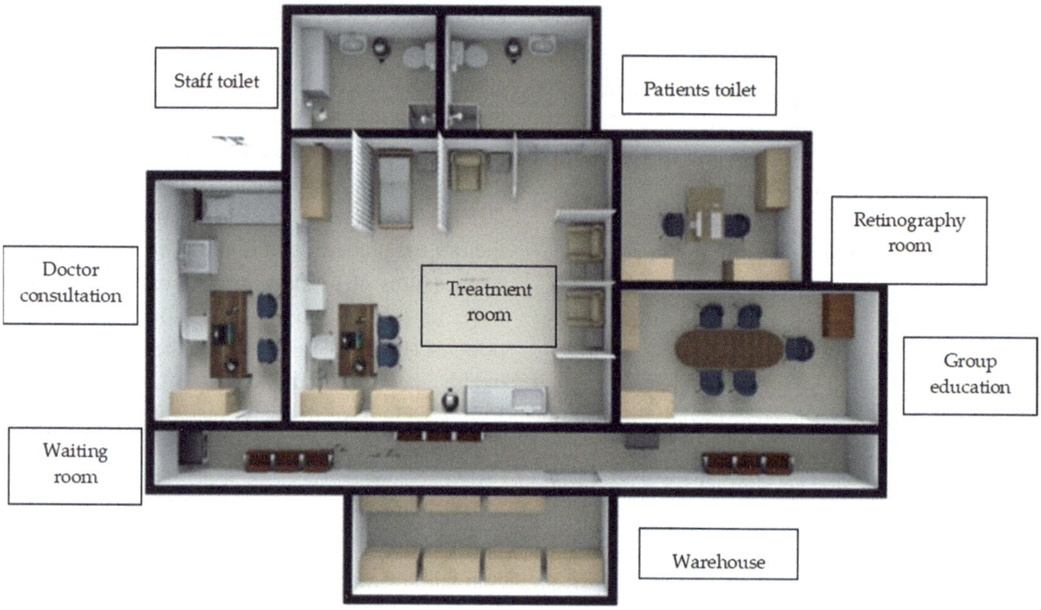

Fig. 56.1 Diabetes day hospital distribution

Barroso et al. (2019) calculated the costs of creating a DDH, considering the total investment imputed annually (material resources 23.655 euros and remodelling 15.000 euros). The cost of human resources was about 155.950 euros and meals (breakfast and lunch) 66.000. The total estimated cost was 860,520 Euros. The cost per patient in this case, following Morales (2014) and encompassing five new consultations as well as 18 returns per day, was 194,82 euros. The expected savings were 645,16 euros per patient per day.

Final Remarks

In this protocol, we point out how the creation of DDHs contributes to the efficiency of the Regional Public Health Services, detailing the structure, functioning, and related costs. A reduction of costs has been demonstrated through the work of Barroso et al. (2019), by evaluating the economical efficiency of the DDH to meet the needs and expectations of healthcare managers and policymakers.

In addition, the DDH offers an excellent opportunity to improve the care of people with diabetes and thus increasing their quality of life, by offering a better coordination between all the parties involved in the monitoring and treatment of the process of the disease.

We want also to emphasize the role of nurses as a key element in the care, education, and nutrition of patients. In fact, the DDH will contribute to the specialization of the nursing staff, creating a group of qualified professionals.

The efficiency in the specific regional healthcare could be improved and must be improved in line with the social responsible culture of the territory. The effective implementation of the DDH should contribute to an improvement in the optimization of resources since unnecessary expenses can be avoided and chronic complications should be effectively treated, thereby reducing the costs derived from them.

According to the previously mentioned percentage of declared diabetes, DDHs seem to be relevant for social justice purposes. For the near future, it is realistic to expect that the next regional public budgets will consider executing day hospitals. Otherwise, failure to pursue the efficiency of the public health system will jeopardize the sustainability of the regional and national health system.

References

American Diabetes Association (2014) Diagnosis and classification of diabetes mellitus. Diabetes Care 37: S81–S90

American Diabetes Association (2019) 2. Classification and diagnosis of diabetes: standards of medical care in diabetes—2019. Diabetes Care 42:S13–S28

Barroso V, Barroso A, Sanguino R, Sánchez-Hernández MI (2019) In favour of regional diabetes day hospitals. Int J Environ Res Public Health 16(13):2293

Crespo C, Brosa M, Soria-Juan A, Lopez-Alba A, López-Martínez N, Soria B (2013) Costes directos de la diabetes mellitus y de sus complicaciones en España (Estudio SECCAID: Spain estimated cost Ciberdem-Cabimer in Diabetes). Adv Diabetol 29(6):182–189. (In Spanish)

Cylus J, Papanicolas I, Smith PC (2017) How to make sense of health system efficiency comparisons? World Health Organization, Regional Office for Europe

El Economista (2018) La reforma del Gobierno amenaza con subir el déficit sanitario a 7500 millones. Available online at www.eleconomista.es (In Spanish)

Farrell M (1957) The measurement of productive efficiency. J R Stat Soc 120:253–281

Giusti A, Maggini M, Colaceci S (2020) The burden of chronic diseases across Europe: what policies and programs to address diabetes? A SWOT analysis. Health Res Policy Syst 18(1):12

Government of Andalusia (2013) Guía Formativa del Residente de Endocrinología y Nutrición. 2013. Available online at www.juntadeandalucia.es (In Spanish)

Government of Extremadura (2007) Plan Integral de Diabetes 2007–2012. In: Consejería de Sanidad y Dependencia (ed) Government of Extremadura: Mérida, Spain (In Spanish)

Government of Spain (2003) Royal Decree 1277/2003, de 10 de Octubre, Por el Que se Establecen las Bases Generales Sobre Autorización de Centros, Servicios y Establecimientos Sanitarios, BOE 254: Madrid, 254, 37893–37902 (In Spanish)

Harder T, Roepke K, Diller N, Stechling Y, Dudenhausen JW, Plagemann A (2009) Birth weight, early weight gain, and subsequent risk of type 1 diabetes: systematic review and meta-analysis. Am J Epidemiol 169 (12):1428–1436

Health Parc of Barcelona (2015) Guía de Acogida del Residente de Endocrinología. 2015. Available online at www.parcdesalutmar.cat (In Spanish)

IDIS Report (2019) Análisis de situación de la sanidad privada 2019. Instituto para el Desarrollo e Integración de la sanidad, Report n° 9. Available online at www.fundacionidis.com (In Spanish)

Ministry of Health and Consumer Affairs (2008) Day hospital unit standards and recommendations. Reports, surveys and research 2008

Ministry of Health and Social Policy (2009) Hospital de Día: Estándares y Recomendaciones. In: Ministerio de Sanidad y Política Social (ed) Ministry of Health and Social Policy, Madrid, Spain (In Spanish)

Ministry of Health, Consumption and Social Welfare (2017) Hospitales, camas en funcionamiento y puestos de Hospital de día. Ministerio de Sanidad, Consumo y Bienestar Social. Available online at www.mscbs.gob.es (In Spanish)

Ministry of Health, Social Services and Equality (2016) Indicadores Hospitalarios. Evolución 2002–2013. Ministerio de Sanidad, Servicios Sociales e Igualdad. Available online at www.msssi.gob.es (In Spanish)

Ministry of Health, Social Services and Equality (2017) Indicadores hospitalarios. Evolución 2010–2015. Ministerio de Sanidad, Servicios Sociales e Igualdad, 2017. Available online at www.msssi.gob.es (In Spanish)

Morales C (2014) Hospital de Día de Diabetes: Experiencia Macarena. Available on line at https://www.saedyn.es (In Spanish)

OECD (2019) Health statistics 2019. Available online at www.oecd.org

The Global Health Expenditure Database (2020) World Health Organization. Global Health Expenditure database. Available on line at apps.who.int/nha/database

Zhang X, Zhao L, Cui Z, Wang Y (2015) Study on equity and efficiency of health resources and services based on key indicators in China. PLoS One 10(12):1–15

Zhang Y, Pan XF, Chen J, Xia L, Cao A, Zhang Y, Wang J, Li H, Yang K, Guo K, He M, Pan A (2020) Combined lifestyle factors and risk of incident type 2 diabetes and prognosis among individuals with type 2 diabetes: a systematic review and meta-analysis of prospective cohort studies. Diabetologia 63(1):21–33

Part VII

Bariatric and Metabolic Surgery

Surgical Options in Obesity and Diabetes 57

Jaime Ruiz-Tovar and Lorea Zubiaga

Contents

Introduction 768
Classification of Bariatric Operations 768
Metabolic Operations 769

Laparoscopic Adjustable Gastric Band (LAGB) 769
Clinical Results 770

Laparoscopic Sleeve Gastrectomy (LSG) 770
Clinical Results 771

Roux-en-Y Gastric Bypass (RYGB) 771
Clinical Results 772

One Anastomosis Gastric Bypass 772
Clinical Results 773

The Biliopancreatic Diversion (BPD) 773
Clinical Results 774

Conclusion 774

References 775

Abstract

Obesity is a chronic, progressive, and multifactorial disease. Hypocaloric diet associated with lifestyle modification and physical exercise represents the first step of treatment for obesity, but weight loss and remission of comorbidities are difficult to maintain in the long term. Bariatric surgery has demonstrated to be the most effective therapy in the long term for severely obese patients and for the treatment of type 2 diabetes mellitus (T2D) in obese subjects. There are different bariatric techniques, with important differences between them, related to the weight-loss effect, complication rates, and postoperative evolution of the different comorbidities and nutritional sequelae. *In this chapter, we will review the actual evidence about the mechanism of action of the different bariatric techniques and their clinical results in*

J. Ruiz-Tovar (✉)
Department of Surgery, Universidad Rey Juan Carlos, Madrid, Spain
e-mail: jruiztovar@gmail.com

L. Zubiaga
Institut National de la Santé et de la Recherche Médicale (INSERM-1190), Lille, France

J. Faintuch and S. Faintuch (eds.), *Obesity and Diabetes*, https://doi.org/10.1007/978-3-030-53370-0_57

terms of weight loss and remission of comorbidities, mostly T2D.

Keywords

Obesity · Type 2 diabetes mellitus · Bariatric surgery · Restrictive procedures · Malabsorptive procedures · Sleeve gastrectomy · Adjustable gastric band · Roux-en-Y gastric bypass · One Anastomosis gastric bypass · Biliopancreatic diversion

Introduction

Obesity is a chronic, progressive, and multifactorial disease involving genetic, metabolic, psychological, and endocrinology-related factors. It is a risk factor for type 2 diabetes mellitus (T2D), high blood pressure, dyslipidemia, cardiovascular and respiratory diseases, psychosocial disorders, and several types of cancer (Salas-Salvado et al. 2007).

Proper hypocaloric diet associated with lifestyle modification and physical exercise represents the first step of treatment for obesity. However, several studies have shown that weight loss and glycemic control are difficult to maintain in the long term, even with intensive lifestyle intervention (Look et al. 2013). Bariatric surgery has demonstrated to be the most effective therapy in the long term, for severely obese patients (BMI $\geq$ 35 kg/m^2) with associated metabolic diseases and for morbidly obese patients (BMI $\geq$ 40 kg/m^2). In addition, there is increasing literature that supports the inclusion of bariatric surgery for the treatment of T2D in obese subjects (Salas-Salvado et al. 2007).

There are different bariatric techniques, which are universally considered as procedures that are efficient regarding obesity-associated comorbidities and weight loss (Daigle et al. 2018).

Classification of Bariatric Operations

Due to differences in nutritional and metabolic outcome, an individualized treatment taking into account the characteristics of each patient is necessary (Magouliotis et al. 2018; Tice et al. 2008; Colquitt et al. 2014; Rachlin and Galvani 2018). Globally, bariatric procedures can be divided into three groups (Colquitt et al. 2014) (Fig. 57.1):

- Restrictive procedures (sleeve gastrectomy, adjustable gastric banding): Imply a functional or anatomic reduction of the gastric volume, and consequently, reduce the amount of food intake.
- Malabsorptive procedures (biliopancreatic diversion, duodenal switch, one anastomosis gastric bypass, single anastomosis duodenum-ileal bypass): All of them coincide in a major change in the intestinal anatomy, bypassing a substantial length of the gut and consequently limiting the absorption of the nutrients.

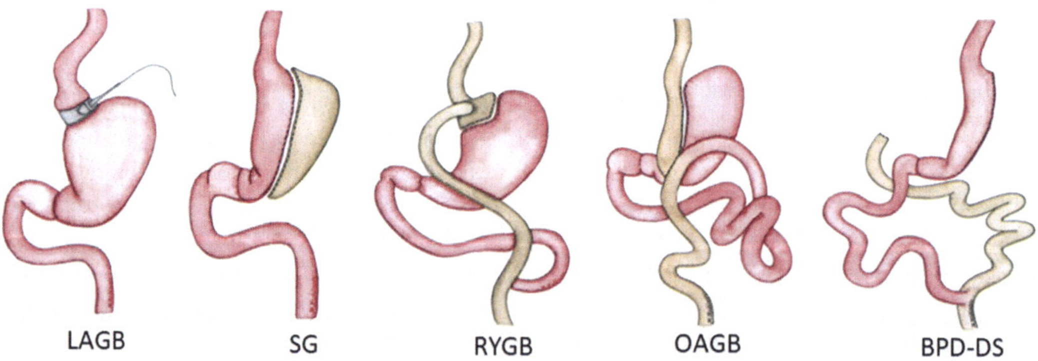

Fig. 57.1 Graphic representation of the main techniques in bariatric/metabolic surgery

- Mixed procedures (Roux-en-Y gastric bypass—classic and modified modalities): Imply a rather balanced combination, with partial reduction of the gastric volume, and some limitation of the absorption of nutrients due to intestinal bypass.

Over the last decades, the Roux-en-Y gastric bypass (RYGB) has been considered the gold standard, as it is associated with low morbidity and mortality rates and obtains excellent short- and mid-term weight loss with remission of comorbidities. However, long-term follow-up series have demonstrated a weight regain at 10-years follow-up, with over 30% of the patients presenting a BMI > 35 kg/m^2, and with less than 20% of the subjects nutritionally healthy (Higa et al. 2011). Although there is extensive literature that claims beneficial health effects with RYGB, these procedures are not exempted from long-term complications related to nutritional deficiencies. In general, all procedures involving malabsorption at any degree will have a risk of micronutrient and macronutrient deficiencies. Consequently, RYGB patients are at risk as well of developing nutrient deficiency-related disorders, such as anemia and certain types of neuropathies or osteoporosis (Kim et al. 2015; Rodriguez-Carmona et al. 2014).

The fear of nutritional consequences has led to the exponential growth of sleeve gastrectomy, a mostly restrictive procedure, which is currently the most performed bariatric intervention worldwide. With this technique, nutritional deficiencies are less frequent; however, it suffers from disadvantages, such as the appearance of de novo acid gastro-esophageal reflux disease and mid- and long-term weight regain (Felsenreich et al. 2019).

Some bariatric surgeons look for durability of the results, even if assuming the risk of nutritional consequences, thus selecting malabsorptive procedures. Among them, the one anastomosis gastric bypass has shown an exponential growth during the last decade, mainly based on excellent long-term results, technical ease, and low complication rate. At this moment, it represents the third most frequently performed bariatric procedure worldwide, after sleeve gastrectomy and RYGB (Ruiz-Tovar et al. 2019).

Metabolic Operations

They are performed on patients with class I obesity (BMI 30–35 kg/m^2), presenting associated metabolic comorbidities, mostly T2D. The American Society for Bariatric and Metabolic Surgery (ASMBS) has recently published their recommendations. Patients with class I obesity and obesity-related comorbidities, with previous failure of conservative therapies, are candidates. Surgical treatment is specially indicated in patients with T2D. Such interventions are as safe and efficient as on patients with greater BMI. However, they do not suggest an ideal procedure for metabolic surgery and defend that any bariatric procedure can be applied with these aims (Aminian et al. 2018).

Laparoscopic Adjustable Gastric Band (LAGB)

This was a technique of choice in the last decades of the twentieth century, due to low mortality rates and technical ease of performance. Absence of organ transection and alteration of the continuity of the intestinal transit are positive elements that promoted the reversibility of the technique.

LAGB involves placement of an adjustable silicone band around the upper portion of the stomach. This band is filled with sterile saline, which is injected through a special device in the adjacent abdominal wall. LAGB artificially creates a small gastric pouch above the band, and the size of the opening between the upper stomach and the lower chamber can be adjusted depending on the amount of injected fluid, in order to manipulate the restrictive level of ingested food. Adjustment of the band is performed in the outpatient clinic (Dixon et al. 2008).

Metabolic benefits are mainly restriction and satiety, which result in a decrease in calorie consumption. Accompanied by dietary control, LAGB can achieve acceptable rates of adiposity loss, modulating metabolic signals that improve the lipid and carbohydrate profile. However, long-term follow-up studies reveal late complications such as band erosion, slippage, gastro-esophageal reflux, esophageal dilatation, and port infection, along with high variability in weight loss, partly related to noncompliance with dietary measures. These were the most influential factors leading to the sharp decline in the use of LAGB in Europe, and at most US bariatric centers (Kindel et al. 2014; Angrisani et al. 2018).

Clinical Results

Excess BMI Loss (EBMIL) ranged from 35.5% to 43.3% at 1 year, with Excess Weight Loss (EWL) of 16–50% and Total Weight Loss (TWL) of 12–13.7%. Mean BMI reduction at 1 year ranged from 5 to 8.4 kg/m^2 (40.7 0 93.8 kg) (Panagiotou et al. 2018). Moon et al. (2016) evaluated weight-loss failure, defined as EWL less than 30%. At the time of last follow-up, 58.8% of the patients failed to achieve weight loss and 4.4% gained weight. According to Loy et al. (2014), EWL increased up to >40% at 2 years after surgery, then remained stable between years 2 and 3, and increased again in years 4 and 5. However, the number of patients followed 5 years after surgery declined up to less than 20%, so that these results are not fully representative.

Complete remission of T2D is one of the lowest, ranging between 24.4% and 29.4% at 1 year, but it increases up to 37.5% after 3 years (Panagiotou et al. 2018; Keogh et al. 2013). Remission of dyslipidemia was achieved in around 65–70% of the patients (Lee et al. 2016a).

Laparoscopic Sleeve Gastrectomy (LSG)

This modality was initially conceived as the first phase from a Duodenal Switch. However, at the first International Consensus Summit on this technique, its validity as a stand-alone technique was established (Deitel et al. 2008). Since 2014, LSG is the most performed procedure in the world, representing more than 50% of all primary bariatric interventions (Angrisani et al. 2018).

The basis of LSG performance includes the resection of the ~80% of the lateral side (major curvature) of the stomach in a vertical fashion, leaving a remaining long, tubular gastric tube (Hutter et al. 2011). Several metabolic benefits outside of restriction have been described in LSG, and various mechanisms of actions have been reported. The new stomach in sleeve shape signifies a substantially smaller volume of the stomach capacity compared to a normal one. This reduction helps to significantly reduce food and calories intake.

Another mechanism of action is the drop in ghrelin levels. Ghrelin is a hormone secreted mainly in the gastric fundus that is responsible for appetite stimulation, increases gastric motility and secretion, increases growth hormone secretion, and reduces fat utilization. If this hormone is reduced, effects fall drastically, mainly appetite stimulation (Madsbad et al. 2014). Another effect of this surgery is the acceleration of the passage of food into the intestine, which avoids the mechanic digestive processes that are present in the gastric camera. Absence of space for creation of chyme justifies some malabsorptive effects to LSG. If the nutrients are not properly transformed, they cannot be totally absorbed.

Resection of gastric tissue also predisposes to the deficit of certain molecules, such as Intrinsic Factor. Moreover, accelerated passage of the nutrients to the gut has been associated with increase of incretin secretion, bile acid signal changes, and modifications in microbiota density. However, these effects do not seem to be permanent, and adaptation to the initial changes can justify the loss of effectiveness in the long term (Cavin et al. 2016). In fact, the hormonal effects of LSG appear to be greater in the first 12–36 postoperative months, and then becomes more dependent on lifestyle changes. Thus, longer-term cohorts refer to weight regain, especially in those patients operated on for more than 5 years, if they have not been able to modify their eating habits (Felsenreich et al. 2019).

The main strengths of LSG are short operation time, absence of foreign material, elimination of intestinal anastomosis, and possibility of being converted into multiple other bariatric procedures, if necessary in the future. Conversely, the main disadvantage is irreversible removal of a substantial portion of the stomach, however with the long-term possibility of compensatory dilatation. In addition, LSG is associated with gastroesophageal reflux, which seems to have increased the incidence of esophageal metaplasia and Barrett's esophagus. In this sense, frequent endoscopic follow-up is recommended (Angrisani et al. 2018).

Clinical Results

Mean EBMIL is around 60% at 1 year (EWL 26%–74.3%, TWL 5.5%–26.5%) (Panagiotou et al. 2018). Lee et al. (2016b) reported mean BMI reduction at 1 year of 7.9 kg/m^2 (24.4 kg). According to Moon et al. (2016), 20.5% of the patients fail to achieve 30% weight loss (weight failure). Sherman et al. (2016) reported that weight loss was sustained up to 3 years after surgery; however, later on weight regain was observed. Lee et al. detected complete T2D remission of 30.2%, and resolution of dyslipidemia in 73.7% of the patients at 1 year follow-up. Our group has recently published a prospective randomized clinical trial comparing SG, RYGB, and OAGB. In our sample of SG patients, the complete remission rate of T2D at 1 year was 86.9% and at 5 years was 82%, while the remission of dyslipidemia was 41.4% and 28.6, respectively (Ruiz-Tovar et al. 2019).

Roux-en-Y Gastric Bypass (RYGB)

It is still considered the 'gold-standard' bariatric-metabolic surgery and was also the most frequently performed technique at the end of the twentieth century. The procedure consists of two phases: firstly, the performance of a small gastric pouch of ~30 cm^3 by separating the upper stomach, just below the gastro-esophageal junction, from the remaining stomach. The second phase includes the division of the small intestine: the first portion from 60 to 100 cm distal to the ligament of Treitz which represents the biliary limb (for the drainage of bile and pancreatic enzyme secretion) and a second segment 100–150 cm of the previous mark. This distal end of the divided bowel is brought up (alimentary limb or Roux limb) typically in an ante-colic manner (on top of the colon) and anastomosed to the constructed gastric pouch (Nguyen and Varela 2017).

RYGB is considered as a mixed technique, which works by several mechanisms that are independent of weight loss. The newly constructed gastric pouch is considerably smaller than the normal stomach, which facilitates the consumption of less food and fewer calories. Additionally, nutrient malabsorption occurs to some degree, which leads to less absorption. In the control of T2D, the exclusion of the duodenum and proximal jejunum has an undeniable benefit, as this intestinal portion is responsible for the rapid absorption of carbohydrates which determines many of the body's metabolic responses. On the other hand, early presence of undigested nutrients traveling to the distal small bowel can stimulate secretion of the incretin hormones, such as glucagon-like peptide 1, which leads to the improvement of different responses including gastric empting, satiety, and insulin resistance.

Other mechanisms of action unveiled during experimental analysis encompass modification of the entire hepatic cycle of bile acids, shifts in the composition and function of the microbiota, and above all, the increase in the intrinsic metabolic machinery of the intestine (Stefater et al. 2012).

The large number of patients operated with this technique display a solidity and reproducibility of the data and outcomes. However, more and more patients regain weight and relapse into their comorbidities. Certain degrees of vitamin and mineral deficiencies due to the malabsorptive component can occur. Likewise, patients with hyperinsulinemic hypoglycemia are occasionally reported, in some circumstances requiring reoperation and recovery of food transit through the excluded segments of the intestine (Malik et al. 2016).

Clinical Results

After RYGB, the mean EBMIL ranged from 73.1% to 80.6% at 1 year (EWL 41.4%–82.8%, TWL 25%–31.8%). The mean BMI reduction at 1 year ranged from 16.5 to 13.4 kg/m^2 (40.7–93.8 kg) (Panagiotou et al. 2018; Van Nieuwenhove et al. 2016). At last follow-up, 14% of the patients failed to achieve 30% weight loss (Moon et al. 2016). In our clinical trial, BMI was 28.7 kg/m^2 and EBMIL 81.2% 1 year after surgery, whereas 5 years postoperatively 29.9 kg/m^2 and 77.1% were registered (Ruiz-Tovar et al. 2019).

The complete remission rate of T2D was 65.9% in the manuscript of Lee et al., while dyslipidemia remission rate was 100% at 1 year. Our group obtained complete remission rate of T2D of 84.3% at 1 year and 73.5% at 5 years, while for dyslipidemia values were 79.7% and 71% at 1 and 5 years, respectively (Ruiz-Tovar et al. 2019).

One Anastomosis Gastric Bypass

The concept unifies a group of bariatric-metabolic techniques characterized by duodenal exclusion and the rapid transit of the nutrients to the distal intestine, however without a Roux (alimentary) limb. The techniques of one anastomosis has received different names, depending on where the digestive tract is interrupted and where the anastomosis is located: minigastric bypass, omega bypass and SASI (single anastomosis sleeve ileum bypass) (Mahawar et al. 2019). The International Federation for the Surgery of Obesity and Metabolic Disorders (IFSO) Scientific Committee in 2018 recommended that techniques derived from a single anastomosis procedure should be called OAGB (De Luca et al. 2018). However, the important anatomical differences between the diverse techniques within this group establish a great variability of the maneuvers to perform them, as well as generate difficulties to compare outcomes. Many of these differences are surgeon-dependent, and until now, none of the hypotheses proposed to defend one OAGB modality over another has been totally convincing. In spite of this, in the last IFSO survey, the OAGB techniques have increased in popularity, becoming the third most performed option in the world (Angrisani et al. 2018).

In Spain, the most widespread OAGB-type technique is the BAGUA ("bypass gastrico de una anastomosis") local style. This technique does not perform gastric resection during sleeve preparation (similarly to RYGB). The section for the gastric tube begins 3 cm before the pylorus, allowing a long tube of more than 18 cm. Then, the gastro-ileal anastomosis is located on the lateral side of the gastric tube. The length of the intestine to be excluded is determined by the total length of the intestine, maintaining a constant percentage: 55% biliary limb and 45% common limb (a long biliary limb is analogous to other malabsorptive procedures, such as biliopancreatic diversion). The same mechanisms of action that are attributed to RYGB occur in OAGB. The main difference is that OAGB involves large portions of excluded intestine, which′ generate significant malabsorption of nutrients. Therefore, intensive postoperative nutritional surveillance is mandatory (Carbajo et al. 2017).

Performing the OAGB techniques implies certain technical advantages, such as minimizing the risk of leakage, fistulas, stenosis, or anastomotic ulcers. With no mesenteric dissection, there is less risk of internal hernias and intestinal obstruction. The lower anastomosis decreases the tension between tissues, and the absence of an alimentary limb reduces operating times and complications of the loop foot (Carbajo et al. 2017).

However, the absence of the alimentary limb triggered the alarm about biliary reflux, which forced the abandonment of popular techniques, such as Billroth II subtotal gastrectomy, a long time ago. Different studies have confirmed the efficacy and safety of OAGB procedures, without higher incidence of gastric cancer, so there must be other elements in these new derivative techniques that are protective, unlike Billroth II gastrectomy. However, frequent endoscopic follow-up is recommended, as with LSG.

Clinical Results

In the series of Peraglie (2016), EWL at 1, 2, 3, 4, 5 and 6 years after surgery was 52%, 67%, 70%, 68%, 66%, 67%, and 72%, respectively. In our study, mean BMI was 25 kg/m^2 with an EBMIL of 100.4% at 1 year, 24.8 kg/m^2 and 104.3% at 2 years, and 25.1 kg/m^2 and 97.9% at 5 years, respectively (Ruiz-Tovar et al. 2019). Carbajo et al. (2017) reported a mean BMI of 29.9 kg/m^2 with a mean EWL of 70%, 12 years after surgery.

Our group documented complete remission of T2D of 94.2% at 1 year and 95.7% at 5 years. Resolution rate of dyslipidemia was 100% at both timepoints (Ruiz-Tovar et al. 2019). Carbajo et al. (2017) observed complete remission of T2D of 94% and of dyslipidemia of 96% at 12 years.

The Biliopancreatic Diversion (BPD)

BPD consists of a horizontal distal gastrectomy, closure of the duodenal stump, and a distal gastroileostomy creating a common 50 cm long channel. Some years later, the horizontal gastrectomy was modified by a LSG in an attempt to reduce some complications of BPD in relation to the tension produced by raising a short ileum to the supramesocolic area, especially in patients with mesentery thickened by fat infiltration. This new variant is known as Duodenal Switch (DS). The duodenum is divided just past the pylorus. A segment of the distal ileum is then divided at 250 cm proximal to the ileo-cecal valve and brought up and anastomosed to the duodenum in a Roux-en-Y configuration. Another anastomosis of the ileo-ileostomy is performed at 100 cm proximal to the ileo-cecal valve to complete the operation (Dapri et al. 2011).

In the 2016 IFSO survey, the percentage of both variants (separately or as BPD-DS) was less than 2% of the total number of surgeries performed in the world. The technical complexity in comparison with other bariatric techniques, the higher rates of complications and mortality, and especially, the refusal of patients to perform a second intervention when the surgery was scheduled in 2 times (first LSG and then intestinal time) made the technique lose followers. However, in all the surveys and in the different meta-analyses carried out so far, these procedures continue to have the best long-term results and are considered the best options for patients with refractory diabetes and in super-obese patients (Angrisani et al. 2018).

In both techniques, there could be a summing effect of performing a gastric resection (the same effects as LSG) to which are added the effects of deriving large portions of intestine that do not absorb. In addition to malabsorption, it must be remembered that this intestine is more demanding from a bioenergetics point of view. In fact, the main mechanism of action of the BPD has been recently described, comparing patients who underwent either BPD or RYGB and were matched for postoperative weight loss. The BPD patients resulted in an important reduction of the rate of appearance of ingested glucose into the circulation, and the postprandial increases in plasma glucose and insulin concentrations were markedly blunted after BPD compared to after RYGB. Insulin sensitivity, assessed as glucose disposal rate during insulin infusion, was 45% greater after BPD than RYGB, whereas ß-cell function was not different between groups (Harris et al. 2019).

However, nutritional deficits also increase, forcing a strict follow-up of these patients throughout their lives. The difference between the dimensions of the limbs excluded in the RYGB and in the BPD-DS has forced surgeons to reconsider which is the ideal limb length to exclude and the need to measure the entire bowel extension in each patient, since the variability of the total bowel length seems to be a determining factor when it comes to preventing complications due to malnutrition or preventing technical failures due to insufficient weight loss. Studies in this sense are being carried out in different parts of the world, coming to the conclusion that, more than specific lengths of intestine, percentages should be considered, bearing in mind that the common limb is the determining factor in the degree of aggressiveness of the derivative techniques (Murad et al. 2018).

Clinical Results

Michaud et al. (2016) described that EWL increased between surgery and the second postoperative year and then remained stable until 5 years. Over a mean follow-up of 7.1 years, the mean EWL was 72.2%. In addition, 82.9% of patients lost more than 50% of their initial excess weight (successful weight loss), and only 0.9% lost less than 25% percent of their initial excess weight. Baltasar et al. (2017), in a series of 1475 patients with 15-year follow-up, reported a mean EBMIL of 96% at 15 years. Bolckmans and Himpens described a TWL of 40.7% 10 years after surgery, confirming that this technique is the one obtaining greatest weight loss and maintained during long-term follow-up (Bolckmans and Himpens 2016).

Referring to remission of comorbidities, Bolckmans and Himpens reported a complete remission of T2D of 87.5% at 10 years after surgery. Dyslipidemia remission rates were 93.3% for total cholesterol, 89.7% for triglycerides, and 95% for low-density lipoprotein cholesterol (Bolckmans and Himpens 2016).

Comparative outcomes between the different bariatric techniques 1 year after surgery are summarized in Table 57.1.

Conclusion

There are diverse surgical bariatric techniques that can be offered to a patient. Given that there is no universally accepted gold-standard, the election of the technique must be based on the individual features of the patient, the experience of the surgeon, and the decision of the patient once he has received all the information about advantages and drawbacks of each technique.

Restrictive procedures are technically more simple and associated with lower complications rates and few nutritional deficiencies, but they obtained lower weight loss and remission of comorbidities and there is a risk of long-term weight regain. RYGB is a mixed procedure with restrictive and malabsorptive components. It obtains better weight loss outcomes and remission of comorbidities than pure restrictive techniques, but it is also associated with long-term weight regain. Malabsorptive approaches are the most effective techniques to achieve significant and maintained weight loss and resolution of comorbidities, but are associated with severe nutritional deficiencies.

Table 57.1 Comparative outcomes between the different bariatric techniques 1 year after surgery

	Weight loss	Remission of T2D	Remission of DL
AGB	EBMIL 35.5–43.3% EWL 16–50% TWL 12–13.7% Failure 58.8%	24.4–29.4%	65–70%
SG	EBMIL 60–64.6% EWL 26–74.3% TWL 5.5–26.5% Failure 20.5%	30.2–86.9%	41.4–73.7%
RYGB	EBMIL 73.1–80.6% EWL 41.4–82.8% TWL 25–31.8% Failure 14%	65.9–84.3%	73.5–100%
OAGB	EBMIL 52–100.4%	94.2%	100%
BPD-DS	EWL 72.2%	87.5%	93.3%

Acknowledgments To Maite Zubiaga, for the performance of graphical illustrations.

References

Aminian A, Chang J, Brethauer SA et al (2018) ASMBS updated position statement on bariatric surgery in class I obesity (BMI 30-35 kg/m^2). Surg Obes Relat Dis 14 (8):1071–1087

Angrisani L, Santonicola A, Iovino P et al (2018) IFSO worldwide survey 2016: primary, endoluminal, and revisional procedures. Obes Surg 28(12):3783–3794

Baltasar A, Bou R, Bengochea M et al (2017) Four decades of bariatric surgery in a community hospital of Spain. Nutr Hosp 34(4):980–988

Bolckmans R, Himpens J (2016) Long-term (>10 yrs) outcome of the laparoscopic biliopancreatic diversion with duodenal switch. Ann Surg 264(6):1029–1037

Carbajo MA, Luque-de-Leon E, Jimenez JM et al (2017) Laparoscopic one-anastomosis gastric bypass: technique, results, and long-term follow-up in 1200 patients. Obes Surg 27(5):1153–1167

Cavin JB, Couvelard A, Lebtahi R et al (2016) Differences in alimentary glucose absorption and intestinal disposal of blood glucose after Roux-en-Y gastric bypass vs sleeve gastrectomy. Gastroenterology 150 (2):454–464

Colquitt JL, Pickett K, Loveman E et al (2014) Surgery for weight loss in adults. Cochrane Database Syst Rev 8: CD003641

Daigle CR, Brethauer SA, Tu C et al (2018) Which postoperative complications matter most after bariatric surgery? Prioritizing quality improvement efforts to improve national outcomes. Surg Obes Relat Dis 14 (5):652–657

Dapri G, Cadiere GB, Himpens J (2011) Superobese and super-superobese patients: 2-step laparoscopic duodenal switch. Surg Obes Relat Dis 7(6):703–708

De Luca M, Tie T, Ooi G et al (2018) Mini gastric bypass-one anastomosis gastric bypass (MGB-OAGB)-IFSO position statement. Obes Surg 28(5):1188–1206

Deitel M, Crosby RD, Gagner M (2008) The first international consensus summit for sleeve gastrectomy (SG), New York City, October 25-27, 2007. Obes Surg 18 (5):487–496

Dixon JB, O'Brien PE, Playfair J et al (2008) Adjustable gastric banding and conventional therapy for type 2 diabetes: a randomized controlled trial. JAMA 299 (3):316–323

Felsenreich DM, Langer FB, Prager G (2019) Weight loss and resolution of comorbidities after sleeve gastrectomy: a review of long-term results. Scand J Surg 108(1):3–9

Harris LA, Kayser BD, Cefalo C et al (2019) Biliopancreatic diversion induces greater metabolic improvement than Roux-en-Y gastric bypass. Cell Metab. https://doi.org/10.1016/j.cmet.2019.09.002

Higa K, Ho T, Tercero F et al (2011) Laparoscopic Roux-en-Y gastric bypass: 10-year follow-up. Surg Obes Relat Dis 7(4):516–525

Hutter MM, Schirmer BD, Jones DB et al (2011) First report from the American College of Surgeons Bariatric Surgery Center Network: laparoscopic sleeve gastrectomy has morbidity and effectiveness positioned between the band and the bypass. Ann Surg 254 (3):410–420. discussion 20-2

Keogh JB, Turner KM, McDonald F et al (2013) Remission of diabetes in patients with long-standing type 2 diabetes following placement of adjustable gastric band: a retrospective case control study. Diabetes Obes Metab 15(4):383–385

Kim J, Brethauer S, Committee ACI et al (2015) Metabolic bone changes after bariatric surgery. Surg Obes Relat Dis 11(2):406–411

Kindel T, Martin E, Hungness E et al (2014) High failure rate of the laparoscopic-adjustable gastric band as a primary bariatric procedure. Surg Obes Relat Dis 10 (6):1070–1075. https://doi.org/10.1016/j.soard.2013.11.014

Lee SK, Heo Y, Park JM et al (2016a) Roux-en-Y gastric bypass vs. sleeve gastrectomy vs. gastric banding: the first multicenter retrospective comparative cohort study in obese Korean patients. Yonsei Med J 57 (4):956–962

Lee JH, Nguyen QN, Le QA (2016b) Comparative effectiveness of 3 bariatric surgery procedures: Roux-en-Y gastric bypass, laparoscopic adjustable gastric band, and sleeve gastrectomy. Surg Obes Relat Dis 12 (5):997–1002

Look ARG, Wing RR, Bolin P et al (2013) Cardiovascular effects of intensive lifestyle intervention in type 2 diabetes. N Engl J Med 369(2):145–154

Loy JJ, Youn HA, Schwack B et al (2014) Safety and efficacy of laparoscopic adjustable gastric banding in patients aged seventy and older. Surg Obes Relat Dis 10(2):284–289

Madsbad S, Dirksen C, Holst JJ (2014) Mechanisms of changes in glucose metabolism and bodyweight after bariatric surgery. Lancet Diabetes Endocrinol 2 (2):152–164

Magouliotis DE, Tasiopoulou VS, Svokos KA et al (2018) Banded vs. non-banded Roux-en-Y gastric bypass for morbid obesity: a systematic review and meta-analysis. Clin Obes 8(6):424–433

Mahawar KK, Parmar C, Graham Y (2019) One anastomosis gastric bypass: key technical features, and prevention and management of procedure-specific complications. Minerva Chir 74(2):126–136

Malik S, Mitchell JE, Steffen K et al (2016) Recognition and management of hyperinsulinemic hypoglycemia after bariatric surgery. Obes Res Clin Pract 10 (1):1–14

Michaud A, Marchand GB, Nadeau M et al (2016) Biliopancreatic diversion with duodenal switch in the elderly: long-term results of a Matched-Control Study. Obes Surg 26(2):350–360

Moon RC, Kreimer F, Teixeira AF et al (2016) Morbidity rates and weight loss after Roux-en-Y gastric bypass, sleeve gastrectomy, and adjustable gastric banding in patients older than 60 years old: which procedure to choose? Obes Surg 26(4):730–736

Murad AJ Jr, Cohen RV, de Godoy EP et al (2018) A prospective single-arm trial of modified long biliopancreatic and short alimentary limbs Roux-En-Y gastric bypass in type 2 diabetes patients with mild obesity. Obes Surg 28(3):599–605

Nguyen NT, Varela JE (2017) Bariatric surgery for obesity and metabolic disorders: state of the art. Nat Rev Gastroenterol Hepatol 14(3):160–169

Panagiotou OA, Markozannes G, Kowalski R et al (2018) Short- and long-term outcomes after bariatric surgery in the medicare population. Agency for Healthcare Research and Quality (US), Rockville (MD)

Peraglie C (2016) Laparoscopic mini-gastric bypass in patients age 60 and older. Surg Endosc 30(1):38–43

Rachlin E, Galvani C (2018) Metabolic surgery as a treatment option for type 2 diabetes mellitus: surgical view. Curr Diab Rep 18(11):113

Rodriguez-Carmona Y, Lopez-Alavez FJ, Gonzalez-Garay AG et al (2014) Bone mineral density after bariatric surgery. A systematic review. Int J Surg 12 (9):976–982

Ruiz-Tovar J, Carbajo MA, Jimenez JM et al (2019) Long-term follow-up after sleeve gastrectomy versus Roux-en-Y gastric bypass versus one-anastomosis gastric bypass: a prospective randomized comparative study of weight loss and remission of comorbidities. Surg Endosc 33(2):401–410

Salas-Salvado J, Rubio MA, Barbany M et al (2007) SEEDO 2007 Consensus for the evaluation of overweight and obesity and the establishment of therapeutic intervention criteria. Med Clin (Barc) 128 (5):184–196

Sherman RE, Anderson SA, Dal Pan GJ et al (2016) Real-world evidence - what is it and what can it tell us? N Engl J Med 375(23):2293–2297

Stefater MA, Wilson-Perez HE, Chambers AP et al (2012) All bariatric surgeries are not created equal: insights from mechanistic comparisons. Endocr Rev 33 (4):595–622

Tice JA, Karliner L, Walsh J et al (2008) Gastric banding or bypass? A systematic review comparing the two most popular bariatric procedures. Am J Med 121 (10):885–893

Van Nieuwenhove Y, Spriet E, Sablon T et al (2016) Metabolic surgery in patients over 60 years old: short- and long-term results. Acta Chir Belg 116 (6):362–366

One Anastomosis Gastric Bypass in the Treatment of Obesity: Effects on Body Weight and the Metabolome

58

Adriana Mika, Tomasz Sledzinski, Monika Proczko-Stepaniak, and Faidon Magkos

Contents

Introduction 778

Metabolic Abnormalities Associated with Obesity 778

Obesity Treatment and the Role of Bariatric Surgery 779

One Anastomosis Gastric Bypass 780
Surgical Technique, Weight Loss, and Complications 780
Surgical Complications 780
Standardized Interventions 781

Metabolic Effects 781
Fasting Lipid Profile and Postprandial Lipemia 781
Plasma Fatty Acid Profile 781
Excessive Free Fatty Acids 783
Plasma Amino Acid Profile 784
Bile Acids 786

Conclusions 787

References 787

A. Mika
Department of Environmental Analytics, Faculty of Chemistry, University of Gdansk, Gdansk, Poland

Department of Pharmaceutical Biochemistry, Faculty of Pharmacy, Medical University of Gdansk, Gdansk, Poland

T. Sledzinski
Department of Pharmaceutical Biochemistry, Faculty of Pharmacy, Medical University of Gdansk, Gdansk, Poland

M. Proczko-Stepaniak
Department of General, Endocrine and Transplant Surgery, Faculty of Medicine, Medical University of Gdansk, Gdansk, Poland

F. Magkos (✉)
Department of Nutrition, Exercise and Sports, Faculty of Science, University of Copenhagen, Frederiksberg C, Denmark
e-mail: fma@nexs.ku.dk

Abstract

Bariatric surgery is the most effective treatment for obesity, producing massive and durable weight loss. One anastomosis gastric bypass (OAGB) is a restrictive and malabsorptive procedure that is rapidly gaining in popularity; it is simpler technically than other common procedures and has a low rate of complications. OAGB has similar, if not better, weight loss results than Roux-en-Y gastric bypass and biliopancreatic diversion, likely because of the exclusion of a longer biliopancreatic limb. In addition to weight

J. Faintuch and S. Faintuch (eds.), *Obesity and Diabetes*, https://doi.org/10.1007/978-3-030-53370-0_58

loss, OAGB has been shown to improve many of the commonest metabolic abnormalities associated with obesity such as insulin resistance. Recent studies using traditional analytical techniques, however also more advanced metabolomics techniques, provide important insights into the potential mechanisms for the physiological effects of OAGB, including improved lipid profile (normalization of triacylglycerol and cholesterol levels), reduced concentrations of total free fatty acids however increased branched-chain fatty acids, decreased branched-chain amino acid levels, and increased circulating bile acids. All these changes likely contribute to improved glucose homeostasis and cardiometabolic health.

Keywords

Omega-loop gastric bypass · Mini gastric bypass · Fatty acid profile · Branched chain fatty acids · Branched chain amino acids · Bile acids

Introduction

Obesity has evolved into a global epidemic according to the World Health Organization (WHO), which reports that the worldwide prevalence of obesity nearly tripled between 1975 and 2016 (WHO 2018). Globally, a total of 1.9 billion adults were estimated to be overweight (609 million of whom were obese) in 2015, representing approximately 39% of the world's population (Chooi et al. 2019). Obesity is quite an expensive disease from both a "health currency" and a real currency perspective, for the patients themselves and the society as a whole. At the individual level, patients with obesity have increased risk to suffer from major health problems including hypertension, osteoarthritis, dyslipidemia, type 2 diabetes, coronary heart disease, stroke, gallbladder disease, sleep apnea, and respiratory problems, as well as some cancers (endometrial, breast, and colon) (Bhoyrul and Lashock 2018).

As a result, obesity increases mortality from many causes, and in fact, it is one of the most important reasons for reduced life expectancy in the modern world (Prospective Studies Collaboration et al. 2009). It is well-known that obesity results in suboptimal health and premature death (Tremmel et al. 2017). Obesity does not only impair the physical and mental health of people, but has many economic consequences as well. Obesity-related comorbidities require high medical costs for patient management and treatment, thereby directly increasing healthcare expenditures. On the other hand, obesity indirectly undermines economic wealth of the society, as it reduces effective labor power and timespan. Policy makers in health services and health professionals in medicine and other related fields, therefore, have important roles when it comes to the prevention and treatment of this contemporary epidemic (Çakmur 2016).

Metabolic Abnormalities Associated with Obesity

Dyslipidemia with increased triacylglycerol (TAG) and low-density lipoprotein (LDL) cholesterol concentrations, along with decreased high-density lipoprotein (HDL) cholesterol concentrations, dysregulation of glucose homeostasis with fasting and postprandial hyperglycemia and hyperinsulinemia, multi-organ insulin resistance, and liver steatosis, are the hallmark features of obesity-related metabolic dysfunction (Klein et al. 2002; Conte et al. 2012; Fabbrini and Magkos 2015). Currently, several other alterations in lipid and lipoprotein metabolism have been identified in obese individuals (Magkos et al. 2008). Quantitative and qualitative changes in lipid profile may be the result, however also the cause, of metabolic disorders or a side effect from various therapeutic activities (Bray 2004).

We have recently described the alterations of specific lipid classes in the serum of obese individuals (Mika and Sledzinski 2017). Obesity is associated with increased free fatty acid (FFA) concentrations and fatty acid release rates into the bloodstream (Mittendorfer et al. 2009) that may lead to insulin resistance, and eventually, type

2 diabetes (Carswell et al. 2016). A handful of studies have demonstrated obesity-related alterations in serum fatty acid profiles; however, also in adipose tissue fatty acid composition. Typically, increased content of monounsaturated fatty acids (MUFA) and decreased content of polyunsaturated fatty acids (PUFA) are reported (Rössner et al. 1989; Okada et al. 2005; Pakiet et al. 2020). Moreover, morbidly obese patients have a greater n-3/n-6 PUFA ratio that is highly correlated with systemic inflammation, e.g., increased circulating concentrations of C-reactive protein (CRP), which is very common in obesity (Kaska et al. 2014).

The analysis of complex groups of lipids revealed various alterations in the serum of obese subjects including changes in fatty acid composition of phospholipids and sphingolipids, increased levels of circulating ceramides, sphingosine-1-phosphate and, of course, TAG (Mika and Sledzinski 2017). However, it should be noted that some studies provided contradictory results, which may be the result of different populations studied or analytical techniques used. Another lipid class which may be of particular interest regarding obesity-related metabolic dysfunction are the bile acids (Kaska et al. 2016). Some data indicate that increased body weight is associated with decreased serum bile acid concentrations (Jahnel et al. 2015). There are also some metabolomic studies showing obesity-related alterations in serum amino acids concentrations (Pennetti et al. 1982; Zhou et al. 2013; Moran-Ramos et al. 2017). Interestingly, decreased concentrations of branched-chain fatty acids in obese subjects may be related to increased levels of branched-chain amino acids, since these amino acids may be converted into branched-chain fatty acids in adipose tissue (Pakiet et al. 2020).

Obesity Treatment and the Role of Bariatric Surgery

The cornerstone of obesity treatment consists of changes in lifestyle including adoption of an energy-prudent diet and increased physical activity, in order to promote a negative energy balance and induce weight loss (Klein et al. 2002; Bray et al. 2018). Despite the short-term success of many available diet and exercise regimens, their long-term efficacy remains limited due to the eventual weight regain and gradual return to baseline weight in most patients (Franz et al. 2007). Bariatric surgery is the most effective treatment for obesity, readily reducing weight and—despite some regain inevitably occurring—keeping much of it off for a prolonged period of time, often exceeding 10–20 years after surgery (Pories et al. 1995; Sjöström et al. 2012). Due to the global epidemic of obesity and the apparent long-term efficacy of bariatric surgery, demand for surgical treatment has increased significantly during the last decade (The MarketWatch News Department 2019).

There are many different surgical procedures, the commonest ones being Roux-en-Y gastric bypass (RYGB), adjustable gastric banding, sleeve gastrectomy (SG), and biliopancreatic diversion (Bradley et al. 2012), which differ in rates of weight loss and other outcomes. Overall, bariatric surgery appears to be a reasonably safe and cost-effective intervention for individuals with moderate to severe obesity compared with nonsurgical interventions (Gloy et al. 2013). Surgical treatments produce durable excess weight loss (EWL), reduce or at least improve comorbidities and significantly improve quality of life (QoL), and increase lifespan (Picot et al. 2009). In fact, bariatric surgery is considered a powerful "disease modifier" and not merely a strategy for weight loss, for example, given its widespread beneficial effects on metabolic homeostasis.

Many of the metabolic comorbidities of obesity and particularly type 2 diabetes improve after bariatric surgery, sometimes even before major weight loss occurs. However, why this happens remains unclear, and exploring the metabolic signatures after bariatric surgery is an area of active investigation in the attempt to better understand the mechanisms for the improvement in metabolic function after the surgical treatment of obesity (Palau-Rodriguez et al. 2018).

One Anastomosis Gastric Bypass

Surgical Technique, Weight Loss, and Complications

One anastomosis gastric bypass (OAGB) is a relatively new malabsorptive procedure that, according to IFSO Global Surgery Report, is the third commonest bariatric technique (Angrisani et al. 2017; Welbourn et al. 2019). OAGB was innovated by Robert Rutledge in 1997 as a modification of the Mason's loop gastric bypass (Rutledge 2001). Several other terms have been used in the past to describe the OAGB procedure such as omega-loop gastric bypass or mini gastric bypass. OAGB is a restrictive and malabsorptive surgical procedure with excellent weight loss results compared to biliopancreatic diversion (BPD) and even better results than RYGB. Improved weight loss efficacy is likely due to the exclusion of a longer biliopancreatic limb (Solouki et al. 2018).

Besides similar or better weight loss outcomes, OAGB is considered a simpler technique with shorter learning curve and operative time, similar rates of remission of comorbidities, and lower percentage of complications (Taha et al. 2017). This procedure involves the formation of a long narrow gastric pouch with a single gastrojejunal anastomosis (omega-loop). From the technical point-of-view, it is basically an omega-loop anastomosis between a long and vertical lesser curvature-based gastric pouch and the jejunum, at 150–200 cm from the ligament of Treitz.

During the initial stages of its clinical implementation, OAGB fell into disrepute due to high probability of bile reflux, leading to Barrett's esophagus and serious nutritional deficiencies (Fisher et al. 2001; Motamedi et al. 2017). Later studies have reported some advantages of OAGB over other bariatric procedures, such as one less anastomosis, shorter operative time and hospital stay, lower risk of anastomotic leakage, and easy reversibility. Despite severe criticisms, OAGB compares favorably among other bariatric procedures regarding effects on weight loss and obesity-related comorbidities (Chevallier et al. 2015) and should be considered for obese or super obese individuals who can be highly compliant to the postoperative nutritional supplementation.

A long-term observational study of 1054 morbidly obese patients undergoing OAGB reported mean excess weight loss (EWL) of 85% at the 6-year follow-up (Kular et al. 2014). In a total of 12,807 patients who underwent OAGB and were followed from 6 months to 12 years, mean EWL ranged from 60% to 78% (Parmar and Mahawar 2018). Alongside weight loss and minimal weight regain, OAGB leads to remission of type 2 diabetes in about 84% to 93% of patients, which is similar to other surgical procedures such as RYGB (remission of type 2 diabetes in 80–90% of patients) and somewhat better than SG (Kular et al. 2014; Parmar and Mahawar 2018). In parallel, OAGB results in significant improvements in the blood lipid profile and lowers risk of cardiovascular complications (Carbajo et al. 2017).

Surgical Complications

A comparative study of 1107 patients undergoing three different bariatric operations (339 had SG, 473 had OAGB, and 295 had RYGB) reported respective mortality rates of 2.1%, 0%, and 0.3% and respective leak rates of 1.5%, 0%, and 0.3% (Jammu and Sharma 2016). Among the OAGB group of patients, bile reflux was observed in fewer than 1% and no weight regain was observed (Bhandari et al. 2019). These results indicate that OAGB is an effective and safe surgical procedure and compared favorably to SG and RYGB. Among 1054 morbidly obese patients who underwent OAGB, 49 of them (4.6%) suffered early minor complications and only two exhibited major ones—anastomotic leak (0.2%) (Kular et al. 2014). For late complications, marginal ulceration was observed in 5 patients (0.6%) and anemia—the most frequent late complication—in 68 patients (7.6%) (Jammu and Sharma 2016). Likewise, an early (up to 30 days) mortality rate of 0.1%, leak rate of 1.0%, marginal ulcer rate of 2.7% (0–10%), and malnutrition rate of 0.7% were reported in a large series of OAGB patients (Parmar and Mahawar 2018).

The rate of nutritional deficiencies and complications depends on the length of the biliopancreatic limb. The operating procedure should be individualized based on the patient's ability to follow nutritional supplementation recommendations and attend long-term follow-up visits. Nevertheless, compliance is often challenging, and there are some instances when revision surgery is required, oftentimes involving conversion from OAGB to RYGB. Overall, both types of gastric bypass, OAGB and RYGB, are associated with low and comparable rates of postoperative complications. OAGB is associated with a lower incidence of leaks due to the longer and wider gastric tube, leading to low intramural pressure and reflux (Mahawar et al. 2018).

Standardized Interventions

Recent surgical guidelines have been published (Parmar and Mahawar 2018). A standard of anatomic configurations of surgical procedures was established in 2018 at the Bariatric Metabolic Surgery Standardization (BMSS) World Consensus Meeting. The BMSS process was undertaken as a first step in developing evidence-based standard bariatric metabolic operations, with the main aim of improving consistency in surgical techniques, data collection, comparison of procedures, and outcome reporting. Based on BMSS standards, the OAGB configuration should consist of a gastric pouch 12–19 cm long and 2.5–3.0 cm wide, with a volume of 50–75 ml, whereas the length of biliopancreatic limb should be based on the patient's body mass index (BMI): 200 cm for individuals with BMI >50 kg/m^2 and 150 cm for those with BMI <50 kg/m^2. A meta-analysis published in 2019 including 16 studies that comparatively evaluated RYGB and OAGB concluded that OAGB is associated with shorter operative time, better resolution of diabetes, and greater EWL at 1, 2, and 5 years, with similar perioperative and postoperative effectiveness, feasibility, and safety (Magouliotis et al. 2019).

Metabolic Effects

There is relatively little research on the metabolic effects of OAGB, whether using traditional or more advanced analytical chemistry methodologies.

Fasting Lipid Profile and Postprandial Lipemia

It is well-documented that all types of bariatric surgery lead to improvements in the fasting lipid profile, including a reduction in serum TAG and total and LDL cholesterol concentrations, and an increase in HDL cholesterol concentration. Studies evaluating the effects of OAGB on the standard lipid profile parameters confirm these beneficial changes (Carbajo et al. 2017; Nabil et al. 2019). Milone et al. (2015) compared SG and OAGB and found similar effects on serum lipid concentrations 1 year after surgery, and similar probability of normalization of the blood lipid profile. In our recent study, we also observed improvements in total and LDL cholesterol and TAG concentrations; however, we did not find any significant changes in HDL cholesterol levels 6–9 months after OAGB (Pakiet et al. 2020).

Plasma Fatty Acid Profile

Fatty acids comprise a largely heterogeneous group of molecules, varying in carbon chain length, the number and position of double bonds, and functional group(s), thereby making them particularly susceptible to the effects of various metabolic factors (Repetto et al. 2012). Not only elevated concentrations of FFA in plasma, but also alterations of fatty acid composition can occur in obesity and can increase risk for metabolic and cardiovascular diseases. Fatty acid composition in serum is the result of endogenous synthesis occurring in the body, dietary intake, and release of FFA from adipose tissue. The fatty acid profile of fasting blood reflects the composition of fat stored in adipose tissue, which in turn

Table 58.1 Short-term and long-term effects of one anastomosis gastric bypass (OAGB) on the serum fatty acid composition of obese patients; the corresponding effects of the presurgical low-calorie diet are also depicted for comparison (Mika et al. 2020; Pakiet et al. 2020)

	Before low-calorie diet vs. baseline	2 weeks after surgery vs. baseline	6–9 months after surgery vs. baseline
Myristic acid (14:0)	↓	↓	↑
Palmitic acid (16:0)	↔	↑	↔
Stearic acid (18:0)	↓	↓	↑
Arachidic acid (20:0)	↓	↔	↔
Behenic acid (22:0)	↓	↓	↓
Total even chain fatty acids	↔	↔	↔
Pentadecanoic acid (15:0)	↔	↔	↑
Heptadecanoic acid (17:0)	↔	↓	↔
Total odd-chain fatty acids	↔	↓	↔
12-methyl-14:0	↔	↔	↑
14-methyl-16:0	↔	↓	↑
Anteiso-branched-chain fatty acids	↔	↓	↑
13-methyl-14:0	↓	↔	↑
14-methyl-15:0	↔	↓	↑
15-methyl-16:0	↔	↔	↑
Iso-branched-chain fatty acids	↔	↔	↑
Total branched-chain fatty acids	↔	↔	↑
Total saturated fatty acids	↔	↔	↔
Myristoleic acid (14:1)	↔	↔	↑
Palmitoleic acid (16:1)	↔	↔	↔
Oleic acid (18:1)	↑	↔	↔
Eicosenoic acid (20:1)	↔	↓	↓
Erucic acid (22:1)	↔	↔	↔
Total monounsaturated fatty acids	↑	↔	↔
α-linolenic acid (ALA)	↓	↓	↔
Eicosapentaenoic acid (EPA)	↔	↓	↔
Docosahexaenoic acid (DHA)	↑	↔	↓
Total polyunsaturated fatty acids n-3	↔	↔	↓
Linoleic acid (LA)	↔	↓	↔
Arachidonic acid (ARA)	↔	↑	↓
Total polyunsaturated fatty acids n-6	↔	↔	↔
n-3/n-6 PUFA ratio	↔	↔	↔

"Baseline" refers to a sample collected on the day of surgery. Symbols: ↑ increase, ↓ decrease, ↔ no significant change

reflects long-term dietary fatty acid intake (Hodson et al. 2008). A diet rich in carbohydrate induces de novo synthesis of fatty acids in the liver (i.e., lipogenesis) and contributes to the formation and accumulation of TAG in plasma and adipose tissue and probably also to the subsequent expansion of adipose tissue when energy balance is positive (Mika and Sledzinski

2017). Despite the significance of fatty acid profile for the metabolic manifestations of obesity, only a handful of studies have been conducted to evaluate the effects of OAGB surgery. In a studies conducted in our laboratory (Pakiet et al. 2020) (Mika et al. 2020), we evaluated the short-term (after 2 weeks) and long-term (after 6–9 months) effects of OAGB surgery on fatty acid profile; Table 58.1 summarizes these findings, together with the effects of low-calorie diet that patients follow for 2–3 months before surgery.

Determination of the fatty acid profile in the serum of obese patients after OAGB revealed many statistically significant and potentially clinically relevant changes. Studies conducted 2 weeks after OAGB showed decreased levels of α-linolenic acid (ALA), eicosapentaenoic acid (EPA) from n-3 PUFA and linoleic acid (LA) from n-6 PUFA groups, total odd-chain fatty acids (OCFA), as well as stearic acid and some other acids present in smaller amounts in blood. By contrast, palmitic acid and arachidonic acid (ARA) were increased (Mika et al.2020). It should be noted that the fatty acid classes that are predominantly depleted from serum after OAGB surgery are essential PUFA (consumed from the diet) and OCFA (that originate from diet, i.e., dairy products) (Mika et al. 2016).

This suggests that the short-term effects of OAGB on serum fatty acid profile reflect the acute reduction in food intake. A decrease in PUFA levels has also been observed in the short term after SG, which may be the result of reduced dietary intake or malabsorption of fat (Lin et al. 2016). Reduced levels of some PUFA were also observed in the long-term, 6–9 months after OAGB (Table 58.1). The lowering of ARA after OAGB can be considered beneficial, given the pro-inflammatory properties of ARA derivatives (i.e., eicosanoids) and the link between inflammation and obesity (Shearer and Walker 2018). Accordingly, OAGB decreases serum CRP levels (Chiappetta et al. 2018; Pakiet et al. 2020), but not the n-3/n-6 ratio, which is also associated with systemic inflammation (Kaska et al. 2014). Decreases in ALA, LA, and EPA concentrations have also been observed 6 months after RYGB (Forbes et al. 2016), although some authors reported elevated levels of PUFA 12 months after RYGB (Walle et al. 2017). Regardless of the type of surgery and the time after surgery, changes in the fatty acid profile are quite variable depending on the characteristics of the patient as well as the postsurgical treatment and diet.

Increased levels of some fatty acids after OAGB may be the result of reverse transport of fatty acids from different body fat depots during weight loss (Lin et al. 2016). For instance, OAGB increases branched-chain fatty acids (BCFA) 6–-9 months after surgery (Table 58.1) (Pakiet et al. 2020), which may have health-promoting properties. BCFAs increase the fluidity of cell membranes, similarly to unsaturated fatty acids, but are less susceptible to oxidation (Vlaeminck et al. 2006). They are also endowed with antibacterial, anti-inflammatory, and anti-cancer properties (Mika et al. 2016). Su et al. (2015) also demonstrated an increase in adipose tissue BCFA levels 1 year after RYGB surgery and an inverse association between adipose tissue BCFA levels and obesity-related insulin resistance.

Excessive Free Fatty Acids

Esterification of fatty acids on a glycerol backbone and conversion to TAG for storage in adipose tissue depots is thought to be a protective metabolic response to excessive dietary calorie and/or fat intake. TAG are chemically inert molecules, whereas fatty acids are highly reactive and can cause oxidative stress and lipotoxicity. Excessive fat intake ultimately leads to excessive FFA release from increased adipose tissue lipolysis (Redinger 2007) and increased availability of FFA in the systemic circulation, which is known to cause insulin resistance (Boden 1997). Unpublished data from our laboratory demonstrate that OAGB significantly reduces FFA concentrations in serum by about 50% (Fig. 58.1). This effect may contribute to the increase in insulin sensitivity observed after OAGB (Kaska et al. 2015).

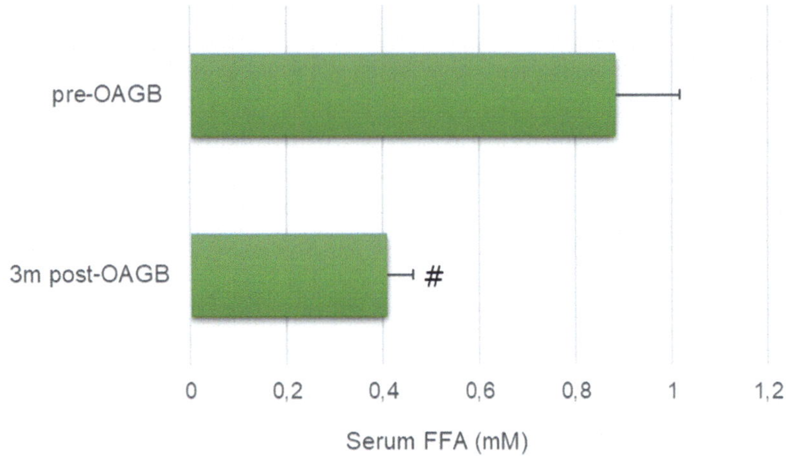

Fig. 58.1 Fasting plasma FFA concentrations before and 3 months after OAGB surgery (unpublished data). $^{\#}p < 0.01$ compared to before surgery

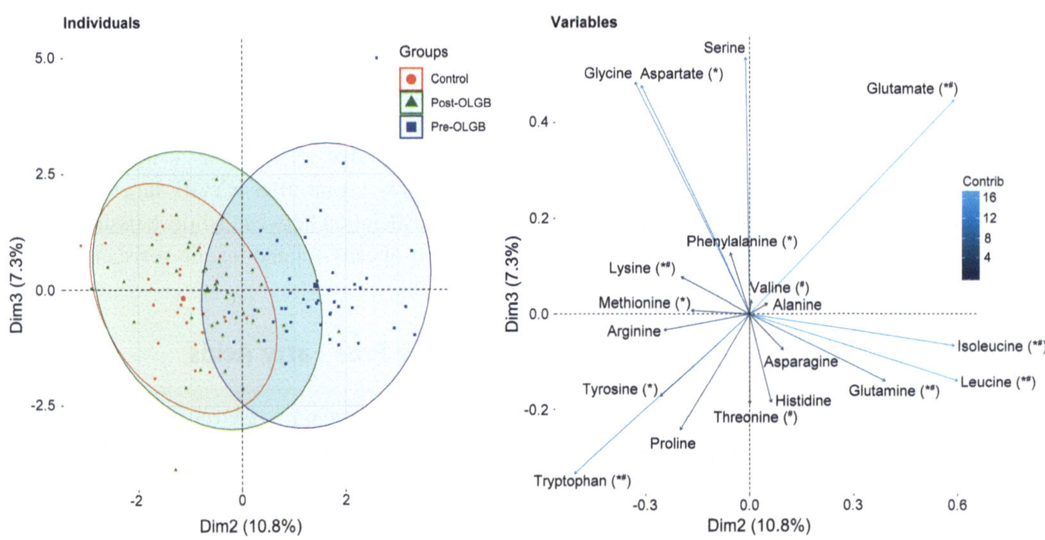

Fig. 58.2 Results of principal component analysis (PCA) based on the common amino acid profile: score plot of cases (left) and variables (right) for the second and third principal complement. Variables marked with (*) displayed statistically significant differences in their average concentrations between lean subjects (the control group) and OAGB patients before surgery, while variables marked with (#) displayed significant differences in patients before and after OAGB surgery (Halinski et al. 2020).

Interestingly, studies in animal models have shown that, among several bariatric surgical procedures applied to diet-induced obese Long-Evans rats—including OAGB, RYGB, SG, and single-anastomosis duodenal switch—the greatest decrease in circulating FFA was recorded at 15 weeks after OAGB (Arble et al. 2018).

Plasma Amino Acid Profile

Nonessential amino acids are synthesized de novo in the human body and their abundance does not depend on dietary intake. Essential amino acids, on the other hand, cannot be synthesized de novo and need to be obtained from the diet. However,

Table 58.2 Long-term effect of one anastomosis gastric bypass (OAGB) on plasma amino acid composition (Halinski et al. 2020)

Amino acid	6–9 months after surgery vs. lean controls	6–9 months after surgery vs. baseline
Alanine	↔	↔
Arginine	↔	↔
Asparagine	↓	↓
Aspartic acid	↔	↓
Glutamate	↓	↓
Glutamine	↔	↓
Glycine	↔	↑
Histidine	↓	↔
Isoleucine	↔	↓
Leucine	↓	↓
Lysine	↓	↓
Methionine	↓	↔
Phenylalanine	↓	↔
Proline	↔	↑
Serine	↔	↔
Threonine	↓	↓
Tryptophan	↓	↑
Tyrosine	↓	↔
Valine	↓	↓
Betaine	↔	↔
Carnosine	↔	↔
Citrulline	↔	↔
Creatinine	↓	↓
Ornithine	↓	↔
Taurine	↓	↔
DL-3-aminobutyric acid (BABA)	↔	↓
γ-aminobutyric acid (GABA)	↓	↓
L-2-aminobutyric acid (AABA)	↓	↓
β-alanine	↔	↔
Asymmetric dimethylarginine (ADMA)	↔	↓
Mono-L-methylarginine (NMMA)	↓	↔
Symmetric dimethylarginine (SDMA)	↓	↔
Branched-chain amino acids	↓	↓
Essential amino acids	↓	↓
Aromatic amino acids	↓	↔

"Baseline" refers to a sample collected on the day of surgery. Symbols: ↑ increase, ↓ decrease, ↔ no significant change

under conditions of stress or in catabolic states, some nonessential amino acids become essential—these are the "conditionally essential" amino acids (Morris et al. 2017). We have recently observed that OAGB surgery modified the serum concentrations of amino acids and normalized plasma amino acid profile in obese patients relative to lean controls (Fig. 58.2) (Halinski et al. 2020).

Essential amino acid concentrations decreased after OAGB (Table 58.2), which suggests that patients after this type of bariatric surgery should increase protein intake from food, at least in terms of quality if not quantity. Furthermore, we found

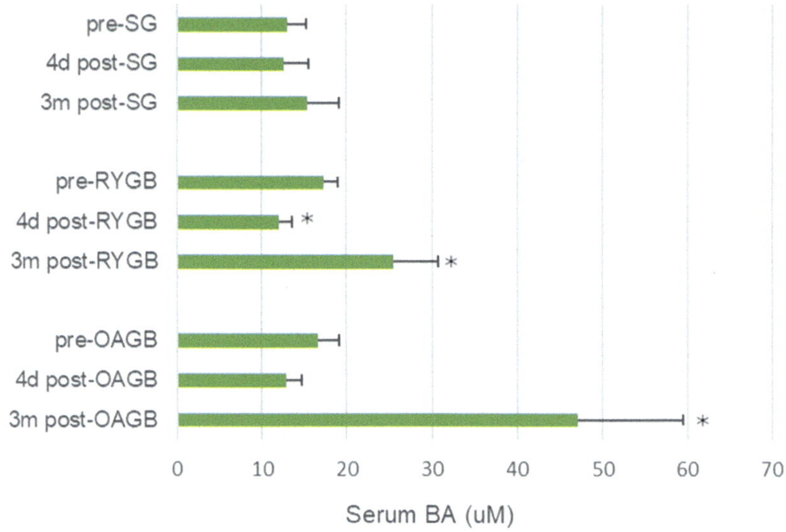

Fig. 58.3 Short-term and long-term effects of sleeve gastrectomy (SG), Roux en Y gastric bypass (RYGB), and one anastomosis gastric bypass (OAGB) on fasting serum concentration of total bile acids. $^{*}p < 0.05$ compared to before surgery (Mika et al. 2018).

decreased concentrations of branched-chain amino acids (BCAA) 6–9 months after OAGB, which may also contribute to a reduction in risk for developing type 2 diabetes. Reduction in BCAA levels was also observed after RYGB and SG, with patterns of change being similar to those after OAGB (Wijayatunga et al. 2018; Palau-Rodriguez et al. 2018).

Bile Acids

Resolution of insulin resistance and hyperglycemia, and remission of type 2 diabetes, can be observed even before the reduction of body weight, just within hours or days after the surgical intervention (Buchwald et al. 2009; Thaler and Cummings 2009; Arble et al. 2018). The exact mechanisms underlying the beneficial effects of bariatric surgery on glucose metabolism are not fully understood; however, some evidence suggests an important role for serum bile acids (Sinclair et al. 2018). Bile acids are a heterogeneous group of cholesterol derivatives with different biological functions and chemical properties. An increase in serum bile acid concentrations after bariatric surgery may be beneficial, as it has been associated with improvements in glucose metabolism (Kaska et al. 2016). Weight loss caused by calorie restriction also affects bile acid and glucose homeostasis (Jansen et al. 2011; Simonen et al. 2012).

Some studies indicated that RYGB may increase the concentrations of primary, secondary, and conjugated bile acids (Scholtz et al. 2014), whereas SG was reported to exert smaller effects or have no effect at all (Steinert et al. 2013). A recent study by our group demonstrated a greater improvement in insulin sensitivity after OAGB compared to RYGB (Kaska et al. 2015). What is more, OAGB provided better glycemic control than RYGB and SG (Lee et al. 2014; Deitel 2018; Mika et al. 2018). A comparative study of the short-term and long-term effects of OAGB, RYGB, and SG on circulating bile acids has demonstrated that differences in bile acid concentrations are greatest 3 months post-surgery; both RYGB and OAGB increased bile acid concentrations, but the increase after OAGB was almost twice as great as after RYGB, whereas SG had no effect (Fig. 58.3) (Mika et al. 2018).

Conclusions

We and others published studies on the metabolic effects of OAGB using more advanced metabolomics techniques, which provide important insights into the potential mechanisms for the physiological changes occurring after bariatric surgery. Serum fatty acid profile analysis revealed a decrease of essential PUFA that may be considered an adverse effect of OAGB. However, improved traditional lipid profile parameters (normalization of TAG and cholesterol levels), reduced circulating FFA as well as increased BCFA and bile acids, and decreased BCAA are all beneficial changes, which likely contribute to better cardiometabolic health. Results from a number of recent studies help increase our understanding of the cellular and molecular mechanisms underlying the beneficial effects of OAGB surgery; however, there are several unresolved issues. Future studies along these lines will with no doubt help improve the surgical treatment of obesity.

Funding Writing of this paper was partly supported by the Medical University of Gdansk (grant nos. ST40, and ST89) and the National Science Centre of Poland (grant no. NCN 2016/21/D/NZ5/00219).

References

Angrisani L, Santonicola A, Iovino P et al (2017) Bariatric surgery and endoluminal procedures: IFSO Worldwide Survey 2014. Obes Surg 27:2279–2289. https://doi.org/10.1007/s11695-017-2666-x

Arble DM, Evers SS, Bozadjieva N et al (2018) Metabolic comparison of one-anastomosis gastric bypass, single-anastomosis duodenal-switch, Roux-en-Y gastric bypass, and vertical sleeve gastrectomy in rat. Surg Obes Relat Dis 14:1857–1867. https://doi.org/10.1016/j.soard.2018.08.019

Bhandari M, Fobi MAL, Buchwald JN, Group: and the BMSS (BMSS) W (2019) Standardization of bariatric metabolic procedures: world consensus meeting statement. Obes Surg 29:309–345. https://doi.org/10.1007/s11695-019-04032-x

Bhoyrul S, Lashock J (2018) The physical and fiscal impact of the obesity epidemic: the impact of comorbid conditions on patients and payers. J Manag Care Med 11:10–17

Boden G (1997) Role of fatty acids in the pathogenesis of insulin resistance and NIDDM. Diabetes 46:3–10

Bradley D, Magkos F, Klein S (2012) Effects of bariatric surgery on glucose homeostasis and type 2 diabetes. Gastroenterology 143:897–912. https://doi.org/10.1053/j.gastro.2012.07.114

Bray GA (2004) Medical consequences of obesity. J Clin Endocrinol Metab 89:2583–2589. https://doi.org/10.1210/jc.2004-0535

Bray GA, Heisel WE, Afshin A et al (2018) The science of obesity management: an endocrine society scientific statement. Endocr Rev 39:79–132. https://doi.org/10.1210/er.2017-00253

Buchwald H, Estok R, Fahrbach K et al (2009) Weight and type 2 diabetes after bariatric surgery: systematic review and meta-analysis. Am J Med 122:248–256.e5. https://doi.org/10.1016/j.amjmed.2008.09.041

Çakmur H (2016) Obesity as a growing public health problem. INTECH i:13. https://doi.org/10.5772/65718

Carbajo MA, Fong-Hirales A, Luque-de-León E et al (2017) Weight loss and improvement of lipid profiles in morbidly obese patients after laparoscopic one-anastomosis gastric bypass: 2-year follow-up. Surg Endosc 31:416–421. https://doi.org/10.1007/s00464-016-4990-y

Carswell KA, Belgaumkar AP, Amiel SA, Patel AG (2016) A systematic review and meta-analysis of the effect of gastric bypass surgery on plasma lipid levels. Obes Surg 26:843–855. https://doi.org/10.1007/s11695-015-1829-x

Chevallier JM, Arman GA, Guenzi M et al (2015) One thousand single anastomosis (omega loop) gastric bypasses to treat morbid obesity in a 7-year period: outcomes show few complications and good efficacy. Obes Surg 25:951–958. https://doi.org/10.1007/s11695-014-1552-z

Chiappetta S, Schaack HM, Wölnerhannsen B et al (2018) The impact of obesity and metabolic surgery on chronic inflammation. Obes Surg 28:3028–3040. https://doi.org/10.1007/s11695-018-3320-y

Chooi YC, Ding C, Magkos F (2019) The epidemiology of obesity. Metabolism 92:6–10. https://doi.org/10.1016/j.metabol.2018.09.005

Conte C, Fabbrini E, Kars M et al (2012) Multiorgan insulin sensitivity in lean and obese subjects. Diabetes Care 35:1316–1321. https://doi.org/10.2337/dc11-1951

Deitel M (2018) A brief history of bariatric surgery to the present. In: Essentials of mini – one anastomosis gastric bypass. Springer, Cham, pp 1–15

Fabbrini E, Magkos F (2015) Hepatic steatosis as a marker of metabolic dysfunction. Nutrients 7:4995–5019. https://doi.org/10.3390/nu7064995

Fisher BL, Buchwald H, Clark W et al (2001) Mini-gastric bypass controversy. Obes Surg 11:773–777

Forbes R, Gasevic D, Watson EM et al (2016) Essential fatty acid plasma profiles following gastric bypass and adjusted gastric banding bariatric surgeries. Obes Surg

26:1237–1246. https://doi.org/10.1007/s11695-015-1876-3

Franz MJ, VanWormer JJ, Crain AL et al (2007) Weight-loss outcomes: a systematic review and meta-analysis of weight-loss clinical trials with a minimum 1-year follow-up. J Am Diet Assoc 107:1755–1767. https://doi.org/10.1016/j.jada.2007.07.017

Gloy VL, Briel M, Bhatt DL et al (2013) Bariatric surgery versus non-surgical treatment for obesity: a systematic review and meta-analysis of randomised controlled trials. BMJ 347:f5934–f5934. https://doi.org/10.1136/bmj.f5934

Halinski LP, Pakiet A, Jablonska P, Kaska L, Proczko-Stepaniak M, Slominska E, Sledzinski T, Mika A (2020) One anastomosis gastric bypass reconstitutes appropriate profile of serum amino acids in patients with morbid obesity. J Clin Med 9:100

Hodson L, Skeaff CM, Fielding BA (2008) Fatty acid composition of adipose tissue and blood in humans and its use as a biomarker of dietary intake. Prog Lipid Res 47:348–380. https://doi.org/10.1016/j.plipres.2008.03.003

Jahnel J, Zöhrer E, Alisi A et al (2015) Serum bile acid levels in children with nonalcoholic fatty liver disease. J Pediatr Gastroenterol Nutr 61:85–90. https://doi.org/10.1097/MPG.0000000000000774

Jammu GS, Sharma R (2016) A 7-year clinical audit of 1107 cases comparing sleeve gastrectomy, Roux-en-Y gastric bypass, and mini-gastric bypass, to determine an effective and safe bariatric and metabolic procedure. Obes Surg 26:926–932. https://doi.org/10.1007/s11695-015-1869-2

Jansen PLM, van Werven J, Aarts E et al (2011) Alterations of hormonally active fibroblast growth factors after Roux-en-Y gastric bypass surgery. Dig Dis 29:48–51. https://doi.org/10.1159/000324128

Kaska L, Mika A, Stepnowski P et al (2014) The relationship between specific fatty acids of serum lipids and serum high sensitivity C-reactive protein levels in morbidly obese women. Cell Physiol Biochem 34:1101–1108. https://doi.org/10.1159/000366324

Kaska Ł, Proczko M, Wiśniewski P et al (2015) A prospective evaluation of the influence of three bariatric procedures on insulin resistance improvement. Should the extent of undiluted bile transit be considered a key postoperative factor altering glucose metabolism? Videosurg Other Miniinvasive Tech 2:213–228. https://doi.org/10.5114/wiitm.2015.52062

Kaska L, Sledzinski T, Chomiczewska A et al (2016) Improved glucose metabolism following bariatric surgery is associated with increased circulating bile acid concentrations and remodeling of the gut microbiome. World J Gastroenterol 22:8698–8719. https://doi.org/10.3748/wjg.v22.i39.8698

Klein S, Wadden T, Sugerman HJ (2002) AGA technical review on obesity. Gastroenterology 123:882–932. https://doi.org/10.1053/gast.2002.35514

Kular KS, Manchanda N, Rutledge R (2014) A 6-year experience with 1,054 mini-gastric bypasses—first study from Indian subcontinent. Obes Surg 24:1430–1435. https://doi.org/10.1007/s11695-014-1220-3

Lee W-J, Chong K, Lin Y-H et al (2014) Laparoscopic sleeve gastrectomy versus single anastomosis (mini-) gastric bypass for the treatment of type 2 diabetes mellitus: 5-year results of a randomized trial and study of incretin effect. Obes Surg 24:1552–1562. https://doi.org/10.1007/s11695-014-1344-5

Lin C, Våge V, Mjøs SA, Kvalheim OM (2016) Changes in serum fatty acid levels during the first year after bariatric surgery. Obes Surg 26:1735–1742. https://doi.org/10.1007/s11695-015-1980-4

Magkos F, Mohammed BS, Mittendorfer B (2008) Effect of obesity on the plasma lipoprotein subclass profile in normoglycemic and normolipidemic men and women. Int J Obes 32:1655–1664. https://doi.org/10.1038/ijo.2008.164

Magouliotis DE, Tasiopoulou VS, Tzovaras G (2019) One anastomosis gastric bypass versus Roux-en-Y gastric bypass for morbid obesity: an updated meta-analysis. Obes Surg 29:2721–2730. https://doi.org/10.1007/s11695-019-04005-0

Mahawar KK, Himpens J, Shikora SA et al (2018) The first consensus statement on one anastomosis/mini gastric bypass (OAGB/MGB) using a modified Delphi approach. Obes Surg 28:303–312. https://doi.org/10.1007/s11695-017-3070-2

Mika A, Sledzinski T (2017) Alterations of specific lipid groups in serum of obese humans: a review. Obes Rev 18:247–272. https://doi.org/10.1111/obr.12475

Mika A, Stepnowski P, Kaska L et al (2016) A comprehensive study of serum odd- and branched-chain fatty acids in patients with excess weight. Obesity 24. https://doi.org/10.1002/oby.21560

Mika A, Kaska L, Proczko-Stepaniak M et al (2018) Evidence that the length of bile loop determines serum bile acid concentration and glycemic control after bariatric surgery. Obes Surg 28:3405–3414. https://doi.org/10.1007/s11695-018-3314-9

Mika A, Wilczynski M, Pakiet A, Kaska L, Proczko-Stepaniak M, Stankiewicz M, Stepnowski P, Sledzinski T (2020) Short-term of one anastomosis gastric bypass on essential fatty acids in the serum of obese patients. Nutrients 12(1):187

Milone M, Lupoli R, Maietta P et al (2015) Lipid profile changes in patients undergoing bariatric surgery: a comparative study between sleeve gastrectomy and mini-gastric bypass. Int J Surg 14:28–32. https://doi.org/10.1016/J.IJSU.2014.12.025

Mittendorfer B, Magkos F, Fabbrini E et al (2009) Relationship between body fat mass and free fatty acid kinetics in men and women. Obesity 17:1872–1877. https://doi.org/10.1038/oby.2009.224

Moran-Ramos S, Ocampo-Medina E, Gutierrez-Aguilar R et al (2017) An amino acid signature associated with obesity predicts 2-year risk of hypertriglyceridemia in school-age children. Sci Rep 7:5607. https://doi.org/10.1038/s41598-017-05765-4

Morris CR, Hamilton-Reeves J, Martindale RG et al (2017) Acquired amino acid deficiencies: a focus on

arginine and glutamine. Nutr Clin Pract 32:30S–47S. https://doi.org/10.1177/0884533617691250
Motamedi MAK, Barzin M, Ebrahimi M et al (2017) Severe fatal protein malnutrition and liver failure in a morbidly obese patient after mini-gastric bypass surgery: case report. Int J Surg Case Rep 33:71–74. https://doi.org/10.1016/j.ijscr.2017.02.033
Nabil TM, Khalil AH, Mikhail S et al (2019) Conventional versus distal laparoscopic one-anastomosis gastric bypass: a randomized controlled trial with 1-year follow-up. Obes Surg 29:3103–3110. https://doi.org/10.1007/s11695-019-03991-5
Okada T, Furuhashi N, Kuromori Y et al (2005) Plasma palmitoleic acid content and obesity in children. Am J Clin Nutr 82:747–750
Pakiet A, Wilczynski M, Rostkowska O et al (2020) The effect of one anastomosis gastric bypass on branched-chain fatty acid and branched-chain amino acid metabolism in subjects with morbid obesity. Obes Surg 30 (1):304–312. https://doi.org/10.1007/s11695-019-04157-z
Palau-Rodriguez M, Tulipani S, Marco-Ramell A et al (2018) Metabotypes of response to bariatric surgery independent of the magnitude of weight loss. PLoS One 13:e0198214. https://doi.org/10.1371/journal.pone.0198214
Parmar CD, Mahawar KK (2018) One anastomosis (mini) gastric bypass is now an established bariatric procedure: a systematic review of 12,807 patients. Obes Surg 28:2956–2967. https://doi.org/10.1007/s11695-018-3382-x
Pennetti V, Galante A, Zonta-Sgaramella L, Jayakar SD (1982) Relation between obesity, insulinemia, and serum amino acid concentrations in a sample of Italian adults. Clin Chem 28:2219–2224
Picot J, Jones J, Colquitt J et al (2009) The clinical effectiveness and cost-effectiveness of bariatric (weight loss) surgery for obesity: a systematic review and economic evaluation. Health Technol Assess (Rockv) 13:1–190, 215–357, iii–iv. https://doi.org/10.3310/hta13410
Pories WJ, Swanson MS, MacDonald KG et al (1995) Who would have thought it? An operation proves to be the most effective therapy for adult-onset diabetes mellitus. Ann Surg 222:339–352. https://doi.org/10.1097/00000658-199509000-00011
Prospective Studies Collaboration PS, Whitlock G, Lewington S et al (2009) Body-mass index and cause-specific mortality in 900 000 adults: collaborative analyses of 57 prospective studies. Lancet (London, England) 373:1083–1096. https://doi.org/10.1016/S0140-6736(09)60318-4
Redinger RN (2007) The pathophysiology of obesity and its clinical manifestations. Gastroenterol Hepatol 3:856–863. https://doi.org/10.1007/s00384-008-0605-y
Repetto M, Semprine J, Boveris A (2012) Lipid peroxidation: chemical mechanism, biological implications and analytical determination. J Free Radic Biol Med 1:3–30. https://doi.org/10.5772/45943
Rössner, Walldius G, Björvell H (1989) Fatty acid composition in serum lipids and adipose tissue in severe obesity before and after six weeks of weight loss. Int J Obes 13:603–612
Rutledge R (2001) The mini-gastric bypass: experience with the first 1,274 cases. Obes Surg 11:276–280. https://doi.org/10.1381/096089201321336584
Scholtz S, Miras AD, Chhina N et al (2014) Obese patients after gastric bypass surgery have lower brain-hedonic responses to food than after gastric banding. Gut 63:891–902. https://doi.org/10.1136/gutjnl-2013-305008
Shearer GC, Walker RE (2018) An overview of the biologic effects of omega-6 oxylipins in humans. Prostaglandins Leukot Essent Fat Acids 137:26–38. https://doi.org/10.1016/j.plefa.2018.06.005
Simonen M, Dali-Youcef N, Kaminska D et al (2012) Conjugated bile acids associate with altered rates of glucose and lipid oxidation after Roux-en-Y gastric bypass. Obes Surg 22:1473–1480. https://doi.org/10.1007/s11695-012-0673-5
Sinclair P, Docherty N, le Roux CW (2018) Metabolic effects of bariatric surgery. Clin Chem 64:72–81. https://doi.org/10.1373/clinchem.2017.272336
Sjöström L, Peltonen M, Jacobson P et al (2012) Bariatric surgery and long-term cardiovascular events. JAMA 307:56. https://doi.org/10.1001/jama.2011.1914
Solouki A, Kermansaravi M, Davarpanah Jazi AH et al (2018) One-anastomosis gastric bypass as an alternative procedure of choice in morbidly obese patients. J Res Med Sci 23:84. https://doi.org/10.4103/jrms.JRMS_386_18
Steinert RE, Peterli R, Keller S et al (2013) Bile acids and gut peptide secretion after bariatric surgery: a 1-year prospective randomized pilot trial. Obesity 21:E660–E668. https://doi.org/10.1002/oby.20522
Su X, Magkos F, Zhou D et al (2015) Adipose tissue monomethyl branched-chain fatty acids and insulin sensitivity: effects of obesity and weight loss. Obesity 23:329–334. https://doi.org/10.1002/oby.20923
Taha O, Abdelaal M, Abozeid M et al (2017) Outcomes of omega loop gastric bypass, 6-years experience of 1520 cases. Obes Surg 27:1952–1960. https://doi.org/10.1007/s11695-017-2623-8
Thaler JP, Cummings DE (2009) Minireview: hormonal and metabolic mechanisms of diabetes remission after gastrointestinal surgery. Endocrinology 150:2518–2525. https://doi.org/10.1210/en.2009-0367
Tremmel M, Gerdtham U-G, Nilsson P, Saha S (2017) Economic burden of obesity: a systematic literature review. Int J Environ Res Public Health 14:435. https://doi.org/10.3390/ijerph14040435
Vlaeminck B, Fievez V, Cabrita ARJ et al (2006) Factors affecting odd- and branched-chain fatty acids in milk: a review. Anim Feed Sci Technol 131:389–417. https://doi.org/10.1016/j.anifeedsci.2006.06.017

Walle P, Takkunen M, Männistö V et al (2017) Alterations in fatty acid metabolism in response to obesity surgery combined with dietary counseling. Nutr Diabetes:7. https://doi.org/10.1038/nutd.2017.33

Watch M Bariatric Surgery Market 2018 Exclusive analysis | Growing at a CAGR of 6.5% by 2023 | Rising prevalence of obesity and increasing demand for weight loss procedures – MarketWatch. https://www.marketwatch.com/press-release/bariatric-surgery-market-2018-exclusive-analysis-growing-at-a-cagr-of-65-by-2023-rising-prevalence-of-obesity-and-increasing-demand-for-weight-loss-procedures-2018-06-01. Accessed 5 Dec 2019

Welbourn R, Hollyman M, Kinsman R et al (2019) Bariatric surgery worldwide: baseline demographic description and one-year outcomes from the fourth IFSO Global Registry Report 2018. Obes Surg 29:782–795. https://doi.org/10.1007/s11695-018-3593-1

WHO int 16 February 2018 Key facts Obesity and overweight. https://www.who.int/news-room/fact-sheets/detail/obesity-and-overweight. Accessed 30 Oct 2019

Wijayatunga NN, Sams VG, Dawson JA et al (2018) Roux-en-Y gastric bypass surgery alters serum metabolites and fatty acids in patients with morbid obesity. Diabetes Metab Res Rev:1–9. https://doi.org/10.1002/dmrr.3045

Zhou Y, Qiu L, Xiao Q et al (2013) Obesity and diabetes related plasma amino acid alterations. Clin Biochem 46:1447–1452. https://doi.org/10.1016/j.clinbiochem.2013.05.045

Robotic Bariatric Surgery

59

Candice Silverman, Anna Isaacs, and Nan Guang Tan

Contents

Introduction ... 792

History and Evolution of Robotic Minimally Invasive Surgery ... 794

The da Vinci® Surgical System ... 795

Introduction of New Technology ... 797

Training Pathway for the da Vinci® Robot ... 797

The Advantages and Disadvantages of Robotic Surgery ... 799
Monetary Costs ... 799
Learning Curve/Operating Theater Time ... 800
Unique Hazards ... 800

Driving Forces Behind the Improvements in Bariatric Surgery ... 800
Enhanced Recovery After Surgery (ERAS) Programs ... 801
Databases ... 801

Robotic Bariatric Procedures ... 801
Robotic Sleeve Gastrectomy: The Evolution of Technique from Laparoscopic, Hybrid Procedures to Totally Robotic ... 802
Robotic Roux-en-Y Gastric Bypass: The Evolution of Technique: From Laparoscopic, to Hybrid Laparoscopic/Robotic, to Totally Robotic RYGB ... 804
Robotic Adjustable Gastric Band ... 805
Robotic Biliopancreatic Diversion with Duodenal Switch (BPD-DS) ... 805
Revisional Bariatric Procedures ... 806

Future Directions of Robotic Technology in Bariatric Procedures ... 806
Augmented Reality ... 806
Autonomous Surgery ... 806

C. Silverman (✉) · A. Isaacs
John Flynn Private Hospital, Griffith University, Bond University, Gold Coast, QLD, Australia
e-mail: drcsilverman@gmail.com

N. G. Tan
Singapore General Hospital, Duke-NUS Medical School, Singapore, Singapore

J. Faintuch and S. Faintuch (eds.), *Obesity and Diabetes*, https://doi.org/10.1007/978-3-030-53370-0_59

New Robotic Platforms and Manufacturers 807
Senhance™ (Transenterix Inc., Morrisville, NC, USA) 807
Versius (Cambridge Medical Research, Cambridge, UK) 807

References 810

Abstract

Obesity is a global problem and surgery remains the most effective tool to treat this chronic disease. Minimally invasive surgery is the standard of care for performing operations for metabolic control and weight loss. The robotic digital platform overcomes the limitations of traditional laparoscopy which are amplified in the morbidly obese population with improved visualization, precision, and surgeon ergonomics. Current robotic systems utilize a master-slave relationship where the surgeon is operating at a distance from the patient, known as telepresence. The da Vinci® robot from Intuitive Incorporated currently has the monopoly for robotic bariatric procedures performed worldwide, though this status is being challenged by emerging robotic platforms and manufacturers. New technology is costly and has its own distinctive learning curve. Despite the inherent benefits of the robotic platform, there is no conclusive evidence to suggest its superiority over conventional laparoscopy. Eagerly awaited is a cost reduction in robotic-assisted surgery driven by competition and the advent of artificial intelligence to provide safer surgery.

Keywords

Bariatric surgery · Obesity · Robotic surgery · Minimally invasive surgery · Learning curve · Training · Cost analysis · Technology · Safety

Introduction

Obesity is considered the most prevalent noninfectious epidemic of our generation. Currently, more than two thirds of the adult population are considered overweight or obese (Organisation WH 2018). Obesity is often a chronic, progressive disorder leading to poor health, unwarranted stigma, and increased mortality (Kleinert and Horton 2019). Surgery is the most effective tool for the long-term management of obesity and the reduction of obesity-related health conditions (Colquitt et al. 2014). Bariatric surgical procedures are almost entirely performed using a minimally invasive approach, with 99.1% of procedures performed laparoscopically according to current registry data (IFSO 2019a).

Operative technical difficulties for bariatric surgeons include dealing with thick abdominal walls, limited intra-abdominal space, enlarged fatty livers, and increased visceral adiposity. A robotic digital platform as an extension of the minimally invasive surgeon's armamentarium would seem to be ideally suited for this patient group (Wilson and Sudan 2013). The United States Food and Drug Administration (FDA) approved the da Vinci® Surgical System (Intuitive Surgical Inc., Sunnyvale California) for use in general laparoscopic surgery in 2000. This robotic platform and company retain the monopoly status on robotic bariatric procedures performed worldwide.

Technically, robotic surgery differs from standard laparoscopic surgery in that the console surgeon remotely controls camera and up to three instruments. Surgeon ergonomics are improved which results in less fatigue (Lee et al. 2014). Three-Dimensional (3D) vision is retained and the instruments are wristed with seven degrees of freedom (Fig. 59.1), allowing improved dexterity for surgical tasks such as intracorporeal suturing (Hernandez et al. 2004; Yohannes et al. 2002). There is, however, a loss of tactile feedback which is in part compensated by visual clues (Hagen et al. 2008).

While the clinical advantages of the robot are still being studied, the true advantage may be more evident for complex, lengthy procedures

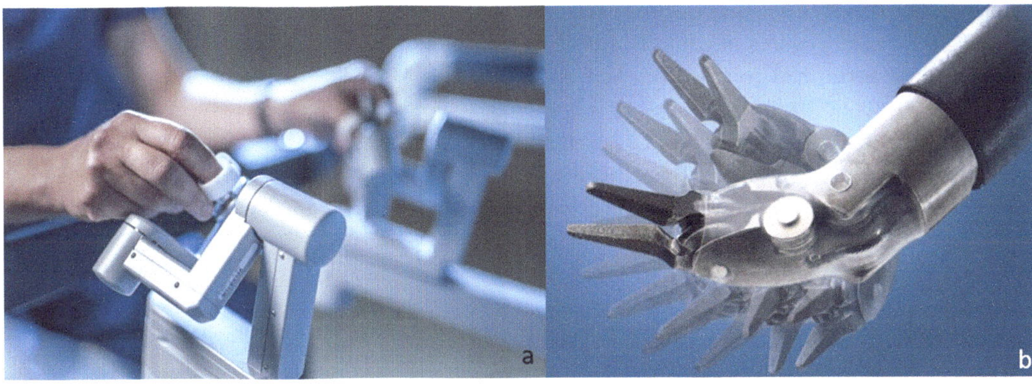

Fig. 59.1 da Vinci® instruments (**a**) ergonomically controlled at the surgeon console with (**b**) seven degrees of freedom (with permissions from Intuitive Surgical, Inc.)

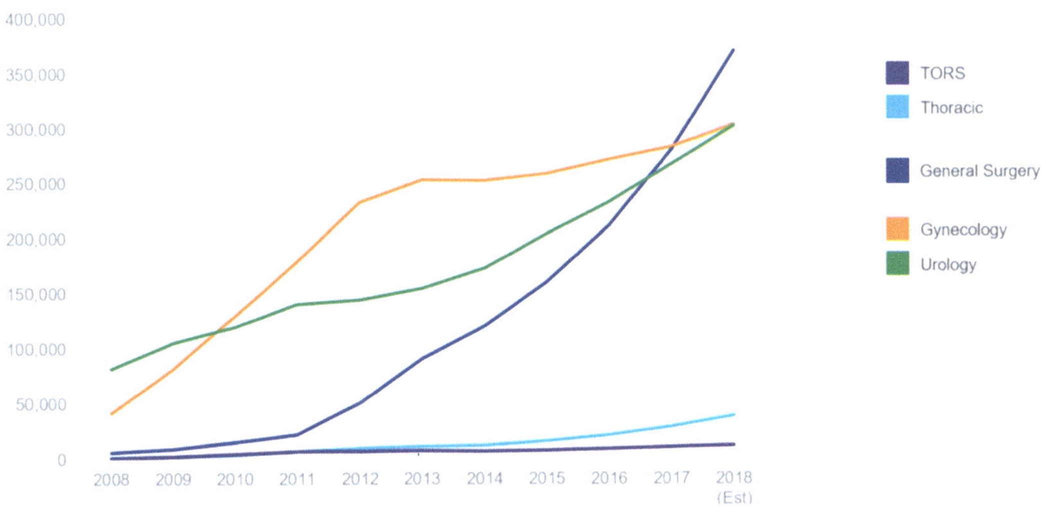

Fig. 59.2 Global robotic surgery load between 2008 and 2017 by surgical specialties (with permissions from Intuitive Surgical, Inc.)

that include a sutured anastomosis such as the Roux-en-Y gastric bypass (RYGB) (Tieu et al. 2013). Furthermore, the incidence of revisional surgery is on the rise, and therefore, more complex procedures may also benefit from a robotic-assisted approach (Tieu et al. 2013).

Unlike traditional surgery, robotic surgery requires a distinct learning process not limited to the operating surgeon, but also applies to the entire surgical team and institution. This new technology needs to be introduced safely to minimize patient harm, during the transition from traditional laparoscopic surgery to operating on a digital surgical platform. Currently, the manufacturer provides most of the training with formal robotic training, fellowships, and credentialing requirements still in its infancy.

Importantly, the widespread application of robotic bariatric surgery would have major cost implications unless it results in better long-term patient outcomes. The da Vinci® robotic system is now utilized in every continent with general surgical procedures (Fig. 59.2) having the highest rate of growth (Intuitive 2018).

In the United States of America, over 6% of primary bariatric procedures in centers of excellence are now performed robotically (Acevedo et al. 2019). Robotic bariatric surgery as a specialty is certainly well-established. Given this establishment, has the da Vinci® robot lived up to its hype as the dominant fully operational system? Are we being driven by industry, patient demand, or surgeon preference? Is robotic bariatric surgery enabling technology that advances human health care and surgical proficiency, or is it just a disruptive innovation?

History and Evolution of Robotic Minimally Invasive Surgery

Robotic surgery is a story of the amalgamation of the work of innovative surgeons, the military, NASA, entrepreneurs, and engineers to create useful and needed technology. The initial need for a robot was to provide remote surgical care in the battlefield for soldiers and in space for astronauts. Telepresence, which uses virtual technology to operate machinery by remote control, was deemed to be the solution. A further aim was to alleviate surgeon fatigue, while aiding procedures which require greater precision.

Historically, the first surgical procedure performed with a robot was for a neurosurgical biopsy using the Programmable Universal Manipulation Arm (PUMA), as shown in Fig. 59.3, in 1985.

The initial telepresence surgical system was developed with military funding by Stanford Research Institute (SRI). The SRI telepresence system had manipulators with force-sensing elements that transmitted pressure sensations, or haptic feedback, to the surgeon controllers (George et al. 2018). Dr. Fred Moll founded Intuitive Surgical Incorporated in 1995 having obtained SRI's intellectual property, with the focus shifting from telepresence to a platform that facilitated minimally invasive surgery. The first Intuitive surgical system to be used on humans was called MONA, upon which Dr. Jacques Himpens performed the first robotic cholecystectomy in Belgium in 1997. Prominent

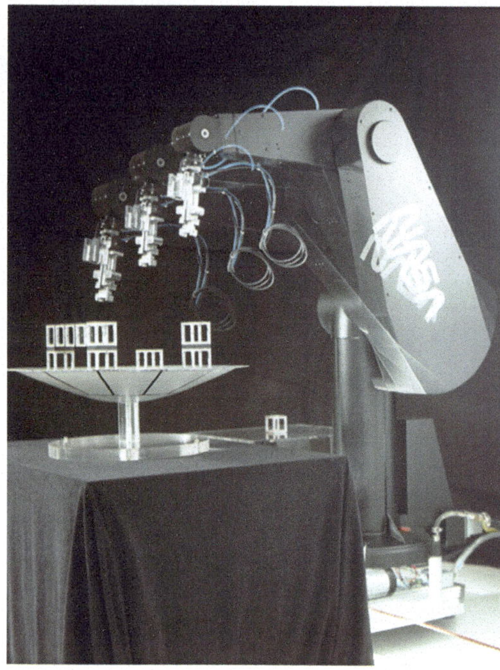

Fig. 59.3 PUMA robotic arm at NASA (no copyright restrictions) (NASA 1990)

journals refused to publish this landmark event due to disbelief in the robotic technology (George et al. 2018). Dr. Himpens was finally successful in publishing the inaugural robotic-assisted gastric band operation in the journal *Obesity Surgery* in 1998 (George et al. 2018). In 1999, the more recognisable da Vinci® robotic platform was introduced, differing from its predecessors in having a stand-alone cart for the patient (slave) and a surgeon console (master), connected by a fiber-optic cord (George et al. 2018).

In parallel to the work at Intuitive, Computer Motion Inc., founded by Dr. Yulin Wang, developed the Automated Endoscopic System for Optimal Positioning (AESOP) robotic device, which is a camera that could be moved by voice or foot pedal commands allowing for independent laparoscopic surgery. Dr. Wang received military funding to add additional arms to his robotic device which he later called ZEUS. ZEUS was used for the first ever trans-Atlantic robotic-assisted cholecystectomy telesurgery in 2001. Dr. Jacques Marescaux controlled the master console in New York, while the patient was operated

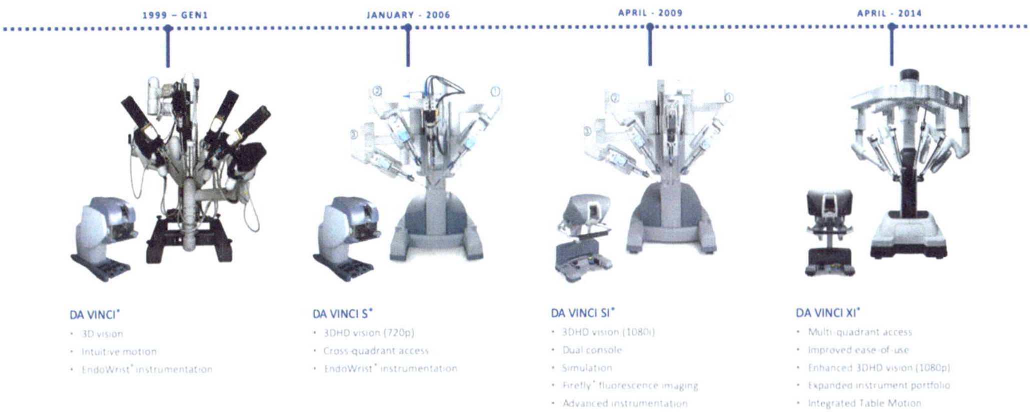

Fig. 59.4 Evolution of the da Vinci® Surgical System (with permissions from Intuitive Surgical, Inc)

on by the slave robot in Strasbourg, France (Kalan et al. 2010).

Shaping this robotic landscape was a seminal legal battle lasting three years between Computer Motion and Intuitive Surgical over the intellectual property rights of these parallel developments. Intuitive Surgical won the lawsuit, which resulted in a merger of the two companies in 2003. While ZEUS had been phased out of production, many of its elements have been integrated into later versions of the da Vinci® platform (George et al. 2018).

Figure 59.4 illustrates the da Vinci® robot evolution from its original prototype in 1999 to the Xi version with its ever-increasing abilities and features (Fig. 59.4). There is also a da Vinci® SP (single port) model which has a single boom for a single 2.5 cm cannula, which allows for control of 3 wristed instruments and a wristed endoscope. Currently, only the Si, Xi, and SP versions of the da Vinci® robot are in operation.

The da Vinci® Surgical System

The da Vinci® surgical system comprises of three integral parts; (1) the surgical console, (2) the patient-side cart, and (3) the electronic or vision cart as illustrated below (Fig. 59.5).

The Surgeon Console is the control center of the system that allows the surgeon to view the surgical field and control the movements of the camera and up to three instruments. The surgeon can adjust the console to suit personal ergonomics. This is a stable operating platform where the hand and the eyes of the surgeon are aligned in a single plane, with the surgeon's forearms being supported, therefore reducing visual and muscular fatigue. There is a magnified *3D* view with the *zoom* feature, *motion scaling* up to a 3:1 ratio and a *tremor reduction* feature. The surgeon can switch between white light modes to near infrared wavelength (i.e., *firefly mode*), to benefit from the fluorescence of a compound called Indocyanine Green (ICG).

As a safety feature, the instruments can only move when the surgeon's head is in the console as detected by infra-red sensors. The robotic instruments (see Fig. 59.6) are wristed (i.e., *Endowrist*®) and controlled with a pincer grip allowing for seven degrees of freedom.

The Patient-Side Cart has articulated mechanical arms which support the instruments and the camera arms. The robotic arms dock onto specific ports that feature a *Remote Center Control (RCC)*, a black marked zone on the port which should be placed in the patient's abdominal wall (Fig. 59.7). The RCC reduces the trauma to the abdominal wall caused by movements of the port during surgery and negates the effect of thick abdominal walls associated with a higher BMI.

The Electronic Cart or Vision Cart contains supporting hardware and software components, such as the electrical surgical unit, suction/

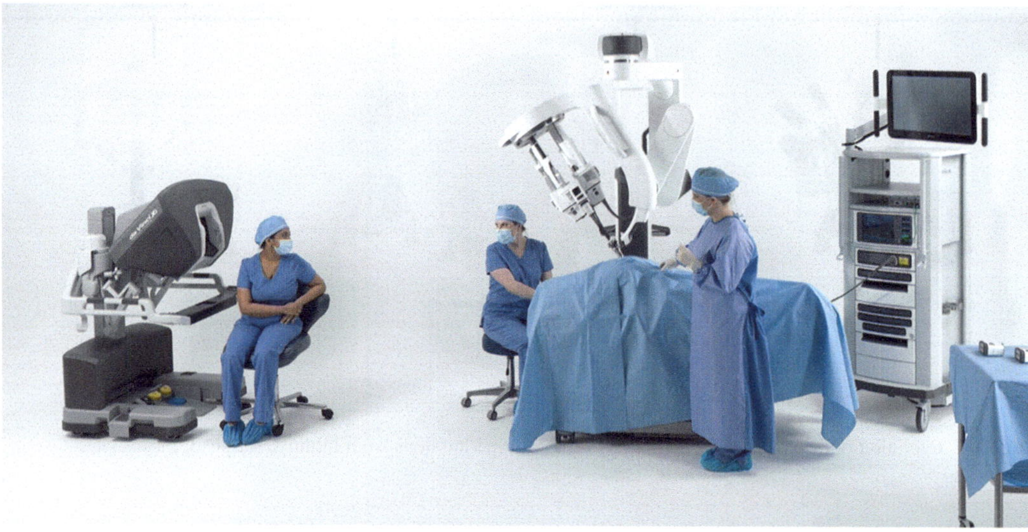

Fig. 59.5 The da Vinci® Surgical System has three main integrated subsystems: (L–R) the surgeon console, patient cart, and the vision cart (with permissions from Intuitive Surgical, Inc.)

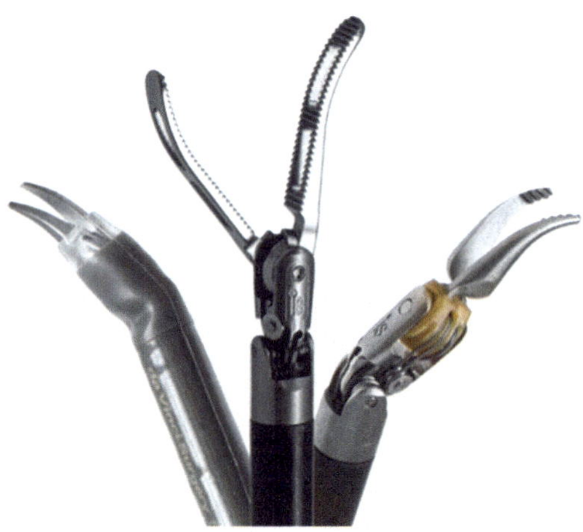

Fig. 59.6 da Vinci® instruments (with permissions from Intuitive Surgical, Inc.)

irrigation pumps, insufflator, and light source for the endoscope. It also has a touch screen that can receive auxiliary images or videos to use in multi-image mode called *Tilepro*. One benefit is that the supervisor can draw on the touch screen as a training tool.

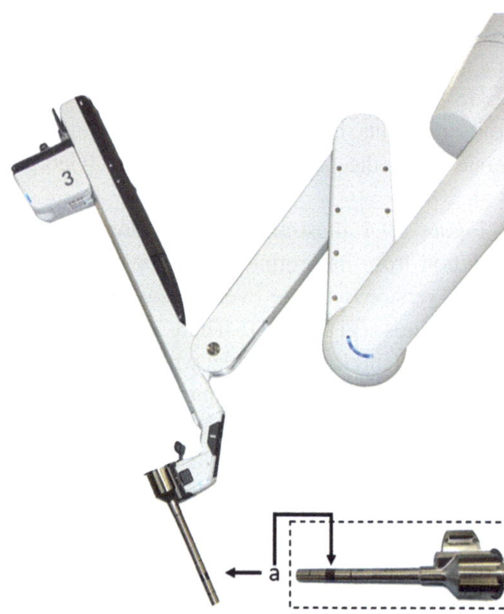

Fig. 59.7 Thick band denoted by arrows, a, that should be kept in the middle of the patient's abdominal wall when the robotic port (inset) is docked to robotic arm

The features of the Si version of the robot and then improvements of the Xi robot are listed in Table 59.1.

Table 59.1 Capability comparisons of the da Vinci® Surgical Si and Xi Systems

	Si (2009)	Xi (2014)
3DHD vision	1080i	1080p
Tremor Reduction	√	
Motion Scaling	√	√
Tilepro	√	√
Skills simulator	√	√
Dual surgeon console	√	√
Firefly	√	√
Single site ability	√	Sp
System Networking	√	√
Advanced instrumentation	√	√
Multi-quadrant access		√
Targeting		√
Chip-on-tip 3D HD camera		√
Wristed stapler with SmartFire technology		45 mm, 60 mm
Integrated Table Motion		√

Introduction of New Technology

The introduction of new technology in surgery is a careful balance between ensuring technology that can benefit patients is not delayed unnecessarily and preventing harm from its early adoption (Brown et al. 2019). The da Vinci® robot is an excellent illustration of outstanding technology introduced into an evidence-free environment (Maddern 2019). The da Vinci® robot was approved by the FDA in the USA in 2000, after it was determined that it did not represent a major deviation from current laparoscopic techniques. The Learning Curve as described for laparoscopic RYGB is that 30 to 70 cases are required for competency, 100 cases for proficiency, and up to 500 cases to become a master of that procedure (Wehrtmann et al. 2019). Similarly, hours of practice and training are required to overcome the learning curve for robotic procedures.

Safeguards to increase comfortability and remediate learning curve effects include having a constant surgical team for initial cases with an experienced bedside assistant. It is also important to start with less technical procedures such as the sleeve gastrectomy (SG) or to perform "hybrid" robotic/laparoscopic procedures, for example where the robot is only used for the gastro-jejunal anastomosis in a RYGB, with the remainder of the procedure performed laparoscopically.

Overall, it is important that industry has a responsibility to ensure safe integration of their device into medical institutions. Intuitive Surgical had termed this integration as the "da Vinci® Ecosystem" which is summarized in the diagram in Fig. 59.8.

Training Pathway for the da Vinci® Robot

There are several pathways for training on the da Vinci® robot. There is the *Video-based online training* where surgeons are required to log onto the web portal (i.e., www.davincisurgerycommunity.com/) and complete online modules to learn about the system components, its instruments, and advanced technologies. Online assessments need to be performed prior to progressing to the next training step to ensure knowledge competency. *Hands on training* is provided in-house by the da Vinci® representatives to practice docking, port placement, instrumentation, use of the surgeon console, and emergency procedures to power off the device. Ideally, this is performed as a team to gain

Fig. 59.8 The da Vinci® Ecosystem (with permissions from Intuitive Surgical, Inc)

proficiency and fluidity in docking prior to live cases. Additionally, *Simulation training* (Fig. 59.9) is becoming an increasingly popular training method, which may facilitate shortening of the learning curve without jeopardizing patient safety. Specific tasks can be practised which are relevant to bariatric surgery (Fantola et al. 2014).

Case observation involves watching an experienced robotic bariatric surgeon perform cases and provides the opportunity for real-time learning, while *Wet labs* (i.e., cadaver or animal) are supported by Intuitive personnel and are carried out at medical training facilities such as The Royal Prince Alfred Surgical and Robotic Training Institute in Sydney, Australia. *Proctored* Cases are essential and advantageous with the Proctor providing feedback to the administration on the safety and the quality of the care of the patient, in order to appropriately credential the surgeon for future robotic surgery (Heit 2014). Finally, *Continual Audit and Review* is a process of measuring outcomes such as operative time and patient complications, with an aim of improving in these parameters over time.

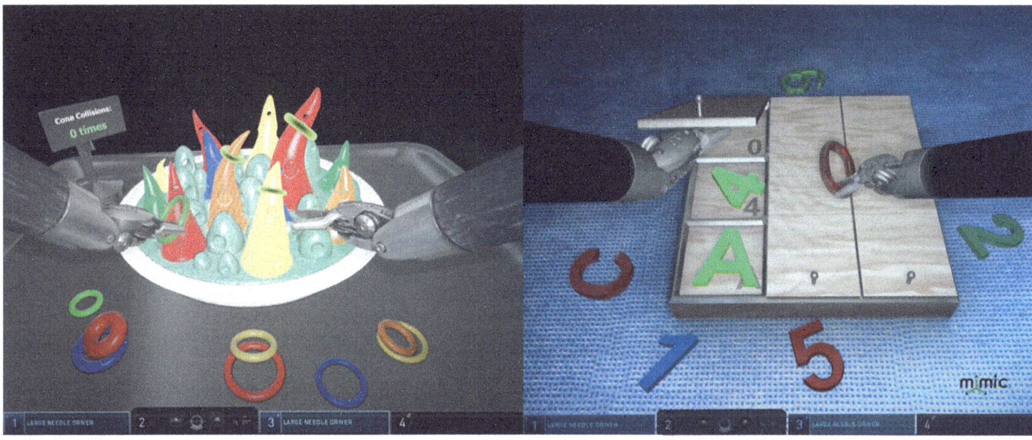

Fig. 59.9 Screenshots of simulation programs available when training with the da Vinci® surgeon console

The Advantages and Disadvantages of Robotic Surgery

The adoption of robotic devices has been reported as being somewhat cost-ineffective. This is due to several factors, which are not uncommon when adopting new technologies, including the initial capital outlay and the learning curves required for proficiency when doing bariatric cases. There are also some unique hazards that the robotic technology presents in this surgical field.

Monetary Costs

The initial capital outlay for the robotic system is 0.5 to 2.5 million US dollars (US$) dependant on the model, configuration, and geographical location of the da Vinci® System (Perez and Schwaitzberg 2019). Then there are additional costs for servicing and consumables. When examining data from the USA HealthCare Cost and Utilization Project, robotic procedures incurred an additional cost of US$1600. If one also counts the initial purchase price of the robot, this additional cost comes in at US$3200 compared to a standard laparoscopic procedure (Service USDoHaH 2019).

In addition to hardware expenses, the team learning curve also should be considered. Robotic procedures typically lead to increases in operating theater time. Applying estimates of operating room time costs of US$37 per minute is a further cost consideration, when evaluating the overall expenses of the robot (Childers and Maggard-Gibbons 2018).

Despite these cost disadvantages, they may be offset by various advantageous factors, including: the (i) patient's reduced length of stay, (ii) reduced surgical-related complications, (iii) avoiding the use of stapling devices, and (iv) operating theater time, which is also important to the health economist.

The cost factor analysis is both specific to the institution and health system. From the model of the Swiss Health Care system, Hagens et al. (2017) found a cost reduction for robotic RYGB when factoring the expense of complications such as leaks. Also, by utilizing hand-sewn anastomosis for the gastro-jejunostomy and jejuno-jejunostomy, there was cost avoidance for not utilizing the stapling device (Hagen et al. 2017).

One cost advantage, which is not easily measured, is the ergonomic improvement for surgeons. Laparoscopy is physically taxing, particularly for bariatric surgeons who have a 66% prevalence of musculoskeletal pain, which can hamper performance (AlSabah et al. 2019). Theoretically, surgeon attrition rates can be decreased by using a robotic platform.

A profitable robotic program would seek to use the robot to its full potential to offset its initial cost. This is analogous to airlines seeking to maximize revenue by keeping their aircrafts in the air as much as possible (Veilleux et al. 2019).

Learning Curve/Operating Theater Time

The learning curve effect has two realms in relation to robotic bariatric surgery. The first is the learning curve of introducing the robot to an established laparoscopic bariatric service. The second is how the learning curves of complex bariatric procedures such as the RYGB may be positively affected by surgeons using robotic technology.

Operating times for robotic RYGB have been found to be consistently longer than for laparoscopic cases (Acevedo et al. 2019; Celio et al. 2017; Li et al. 2016). Unfortunately, longer operations can cause more patient complications such as deep venous thrombosis and surgical site infections. Hence, shorter operative times are associated with better outcomes in bariatric surgery (Reames et al. 2015). Increases in operating time are due to the robot docking times and increased surgeon console times. Experienced teams can achieve the docking component in a matter of minutes. The learning curve for this component is typically achievable within 20 cases (Vilallonga et al. 2012). Robotic surgery may take longer if meticulous dissection is performed or if additional suturing of anastomosis or staple lines is applied.

The question remains whether the da Vinci® robot can aid a surgeon to transition through the learning curve for a complex procedure such as the RYGB. The literature is conflicted in this regard. The robotic learning curve for RYGB cases varies from 14 cases for an experienced laparoscopic surgeon new to RYGB (Buchs et al. 2012), to 84 cases for a surgeon experienced in laparoscopic RYGB (Renaud et al. 2013). In the latter series, the experience of the bedside assistant was an independent predictor of operative time.

Presently, there is only one randomized trial comparing robotic to laparoscopic RYGB. The new fellows' first 50 RYGBs were randomized to totally robotic or laparoscopic procedures, and both groups utilized a hand-sewn gastrojejunostomy. Overall, this study demonstrated a reduction in operating time for the robotic cases, which was amplified as the patient's BMI increased (Sanchez et al. 2005).

Unique Hazards

One hazard is that there is no haptic feedback in robotic surgery. Sutures can break and needles can be bent by the force of the robotic instruments. There has also been concerns raised that automated instruments can pose potential safety hazards. The literature reports robotic mechanical failure or malfunction rates between 0.4 to 4.6% (Agcaoglu et al. 2012). Most malfunctions can be avoided and detected early, through a comprehensive evaluation of the system before starting the procedure. Despite the complexity of the da Vinci robotic system, the fact to date that there have been no reported patient injuries or death due to robotic failure reflects positively on the safety of robotic surgery (Agcaoglu et al. 2012).

Driving Forces Behind the Improvements in Bariatric Surgery

The outcomes of bariatric surgery continue to improve (Wittgrove et al. 1994). The factors responsible for this include improvements in equipment, such as stapler and haemostatic devices, improvements in technique, post-fellowship training in bariatric surgery, surgery being performed in high volume centers by high volume surgeons, improvements in perioperative care including Enhanced Recovery After Surgery (ERAS) protocols, and learning curve effects for surgeons and their institutions. Most published articles comparing robotic to laparoscopic bariatric cases compare robotic cases to laparoscopic historical cohorts, so any advantage seen needs to be viewed with that perspective.

Two other important forces that are driving improved outcomes in bariatric surgery are ERAS programs and improved data collection systems with national and international registries.

Enhanced Recovery After Surgery (ERAS) Programs

ERAS describes multimodal perioperative pathways that are protocoled and provide benefits to the outcomes of surgery. ERAS protocols include a limit on preoperative fasting, deep neuromuscular blockade during general anaesthesia, avoidance of drains, avoidance of opiate analgesia, early post-surgical mobilization, and early return to oral nutrition. When delivered, they result in improved patient outcomes, particularly perioperative pain and nausea, with earlier patient discharge (Ruiz-Tovar et al. 2019). For robotic cases, anaesthetists are especially aware of the need for deep neuromuscular blockade, to avoid any patient movement while the robotic arms are docked. Deep neuromuscular blockade leads to improved surgical conditions and less postoperative pain. ERAS is considered the standard of care and should be implemented prior to deciding whether to replace standard laparoscopy with a robotic digital platform (Slim and Mattevi 2018).

Databases

The primary aim of databases and registries for bariatric surgical patients is to measure outcomes on bariatric surgery and provide quality and safety data. The International Federation for the Surgery of Obesity and Metabolic Disorders (IFSO) has just published its 2019 Global Registry Report, with the three most popular procedures being sleeve gastrectomy (SG) making up 47% of all procedures, RYGB 35%, and gastric band 8% (IFSO 2019b).

The Metabolic and Bariatric Surgery Accreditation and Quality Improvement Program (MBSAQIP) was rolled out in 2014, and it is the largest bariatric-specific prospective database in the world. It records 30-day outcomes from around 900 American Society of Metabolic and Bariatric Surgery (ASMBS) accredited centers in the USA and Canada, totalling 150,000 annual procedures. Its benefit relates to power and national representation of bariatric surgery. Its limitations are that the database lacks information on procedure-related costs, surgical techniques are not standardized, hybrid robotic/laparoscopic cases are not accounted for, and there is a lack of metrics on surgeon experience (Lundberg et al. 2019). Preceding the MBSAQIP database was the Bariatric Outcomes Longitudinal Database (*BOLD*) which ran from 2007 to 2013 (Celio et al. 2017).

Robotic Bariatric Procedures

The RYGB is historically the most established bariatric procedure, with the original open procedure performed and described by Mason and Ito in 1967 (Mason and Ito 1996), and the laparoscopic procedure first performed by Wittgrove in 1993 (Wittgrove et al. 1994). The RYGB, despite its popularity, is complex and difficult to master. The RYGB has only recently been overtaken by the SG as the most common obesity operation. The SG is popular due to its relative simplicity, low complication rates, and good long-term efficacy (Li et al. 2019). The LAGB has been declining in popularity secondary to the modest weight loss results of this procedure, high rates of revision, and long-term complications (Li et al. 2019). The Biliopancreatic diversion with Duodenal Switch (BPD-DS) is a rarely performed operation, as it is more malabsorptive than the other procedures and is considered the most technically difficult. These four procedures will be discussed below, illustrating the evolution of technique and also outcomes of robotic versus laparoscopic procedures. The SG, RYGB, LAGB, and BPD-DS procedures are illustrated in Fig. 59.10.

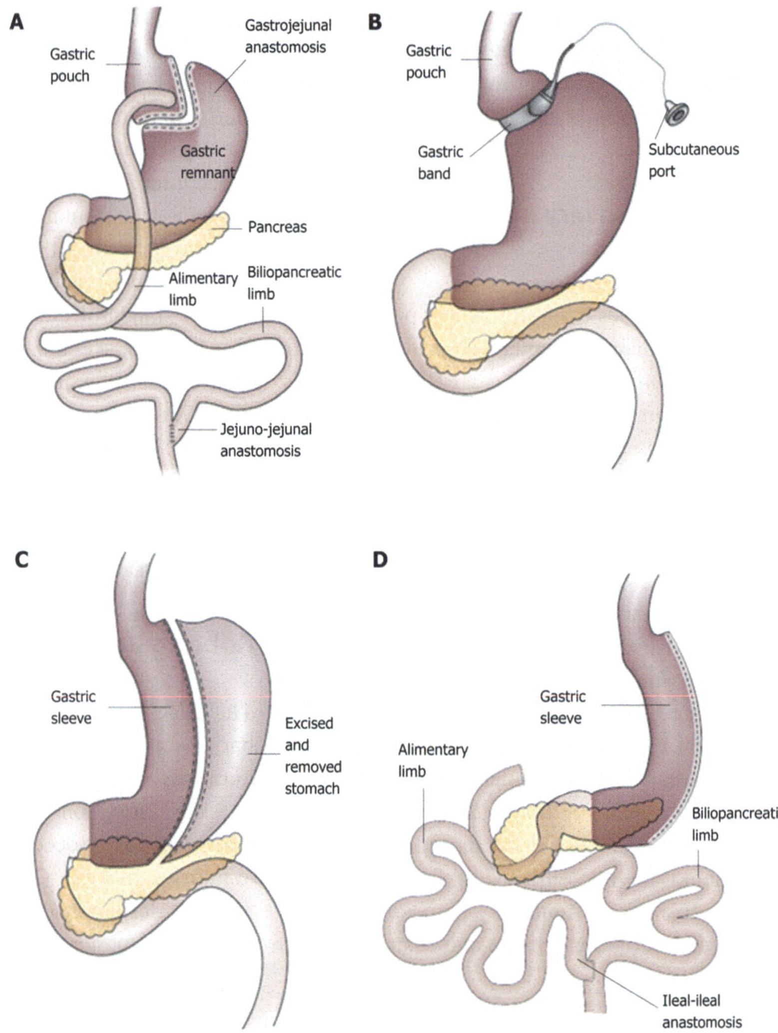

Fig. 59.10 Illustration of the four procedures: (**a**) RYGB, (**b**) LAGB, (**c**) SG, and (**d**) BPD-DS (with permission from Baishideng Publishing Group)

Robotic Sleeve Gastrectomy: The Evolution of Technique from Laparoscopic, Hybrid Procedures to Totally Robotic

The SG involves resecting most of the greater curve of the stomach after it has been devascularized. This leaves a narrow tube of stomach between the gastro-esophageal junction and the pylorus. The SG is considered an ideal training procedure for introducing the robot into a bariatric surgical practice (Silverman and Ghusn 2017) (Figs. 59.11 and 59.12). It is a relatively short operation, so increases in operating time during the learning curve are less clinically relevant. It has even been used as a training procedure for residents (Ecker et al. 2016).

Laparoscopic-assisted robotic SG can be performed with the robot docked for the entire procedure, although the stapling is performed by a skilled bedside assistant, with the console surgeon controlling the tension and alignment of the stomach. The robotic EndoWrist® Stapler was FDA-approved in 2012, initially with 45 mm length staplers. Recently, the more appropriately sized 60 mm staplers are available. The totally

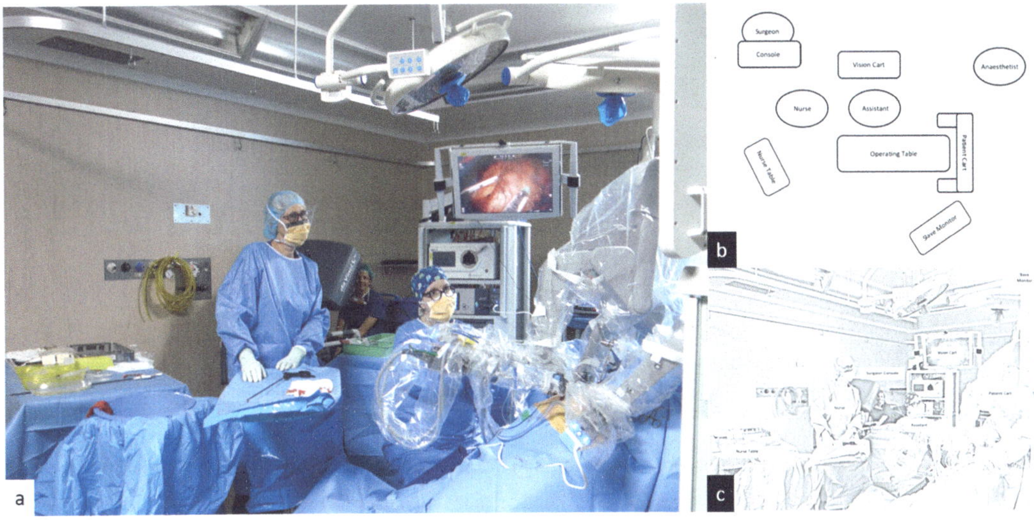

Fig. 59.11 (**a**) Actual OT setup, (**b**) schematic diagram, (**c**) labels on the actual setup

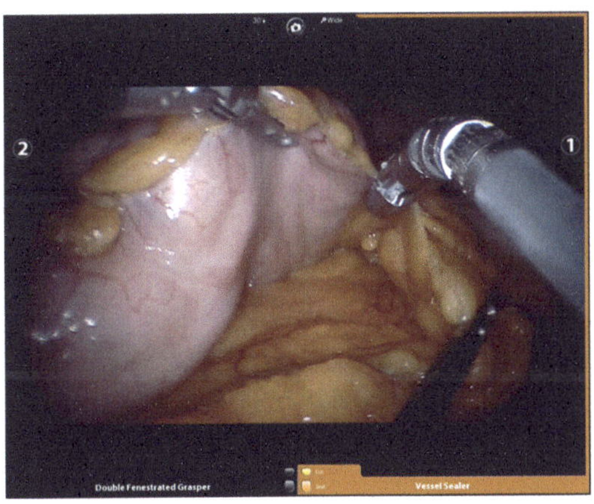

Fig. 59.12 The vessel sealer used to divide short gastric vessels in a sleeve gastrectomy

robotic SG is where the console surgeon can perform the stapling independent of the bed side assistant (Fig. 59.13).

Outcomes of Robotic Sleeve Gastrectomy

A recent systematic review and meta-analysis by Magouliotis et al. (2017) analyzed 16 studies and 29,787 patients and found a longer operating time and length of stay for patients who underwent robotic procedures. A higher cost of robotic SG compared to laparoscopy was demonstrated, without significant differences in patient outcomes including complications and excess weight loss (Moon et al. 2016). Typically, the Learning Curve for performing a SG on the *da Vinci*® system is around 20 cases for an experienced bariatric surgeon (Vilallonga et al. 2012).

Using the MBSAQIP (2016) database of 107,726 patients who underwent a SG, 7% were performed robotically. Overall, robotic-assisted procedures had a longer mean operating time of 89 min compared with 63 min for laparoscopic procedures. There was a doubling of the risk of organ space infection in the robotic group. Organ space infection includes staple line leaks which is a most feared complication. The limitations of this data are that variables of surgeon experience and operative technique such as stapler type (robotic *EndoWrist*® *stapler* or that applied by the bed side assistant) were not accounted for (Lundberg et al. 2019).

Overall, robotic SG can be considered an ideal training procedure to introduce the da Vinci® robot into a bariatric practice. However, increased cost and operative time impede generalized applicability. To date, an empirical benefit of robotic over laparoscopic SG has not been shown. Some caution is warranted with respect to increased infectious complications from recent MBSAQIP data (Lundberg et al. 2019).

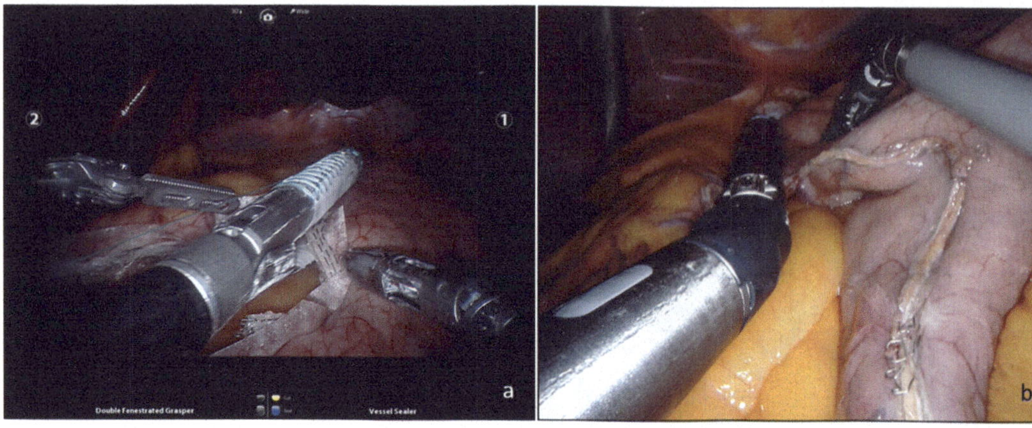

Fig. 59.13 A comparison of (**a**) laparoscopic stapler through the assistant port in the Si, and (**b**) wristed stapler through the robotic port in the Xi

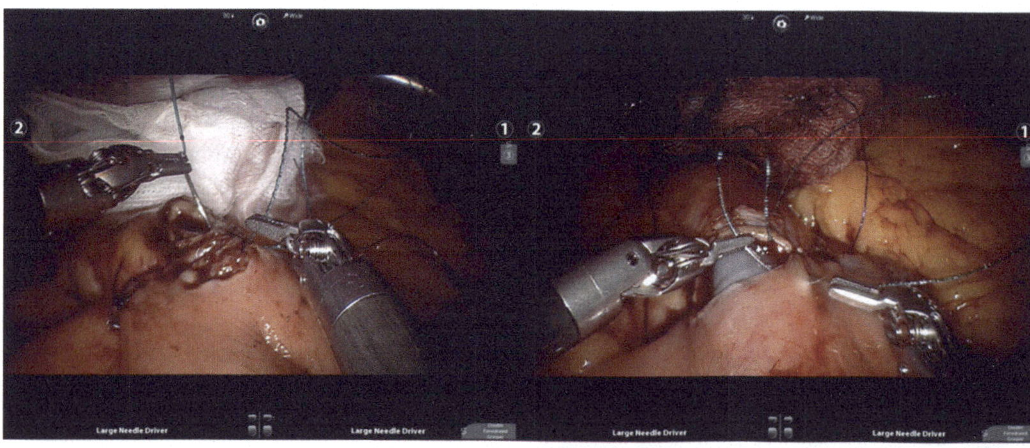

Fig. 59.14 Intraoperative picture of a robotic hand-sewn gastro-jejunostomy

Robotic Roux-en-Y Gastric Bypass: The Evolution of Technique: From Laparoscopic, to Hybrid Laparoscopic/Robotic, to Totally Robotic RYGB

The gastric bypass involves creating a small proximal gastric pouch and rerouting the small bowel (alimentary limb) to connect to this pouch via the gastro-jejunal anastomosis. Secretions of the gastric remnant mix with the bile and pancreatic juices via the biliopancreatic limb which joins the alimentary limb at the jejuno-jejunal anastomosis, forming the common channel. Today, there are many variations of this minimally invasive technique. In regard to the formation of the gastro-jejunal anastomosis, the circular stapler technique, although the most time-efficient, has been hampered by increased complications such as stricture, bleeding, and port site infections. The linear stapled technique and totally hand-sewn gastro-enterostomy have been demonstrated as a preferred technique (Rogula et al. 2018). A totally hand-sewn anastomosis (Fig. 59.14) is the most time-consuming and technically demanding laparoscopically, and this is where the robot excels.

A hybrid approach where the robot is used exclusively for the gastro-jejunal anastomosis

Table 59.2 Meta-analyses and systematic review results of robotic versus laparoscopic RYGB

Author	Patients	Findings
Marker et al. (2011)	1686	Reduction in stricture for robotic cases
Bailey et al. (2014)	2557	Equivalent outcomes through increased cost with robotic cases
Economopoulos et al. (2015)	5145 patients (3189 robotic, 1956 laparoscopic)	Robotic RYGB has lower rates of stricture, reoperations, and a reduced length of stay
Li et al. (2016)	27,997	Similar results through slightly improved stricture rates, reoperations, leak rate, and length of stay for robotic cases
Celio et al. (2017)	137,455 1.8% robotic	Increased morbidity in robotic group. Higher leak rate, stricture, reoperation, and 90-day reoperation rate with longer operating times for robotic procedures
Lundberg et al. (2018)	39,425 7.8% robotic	Equal safety through longer operating time for robotic procedures

has been a starting point for many robotic surgeons. These hybrid, robotic-assisted RYGB were described by Horgan and Vanuno in 2001 (Horgan and Vanuno 2001). Mohr et al. described the first "totally robotic" RYGB in 2005 (Mohr et al. 2005), though the first truly robotic approach would have only been feasible after the market launch of the Endowrist® stapler in 2014 (Jung et al. 2017).

Outcomes of Robotic RYGB

Many studies have compared robotic with laparoscopic RYGB, mostly from single institutions or meta-analysis of fewer than 3000 patients, with mixed results. *Comparative studies* show potential benefit of the robotic approach, especially when a hand-sewn gastro-jejunal anastomosis is performed (Jung et al. 2017). In the largest cohort series of two surgical sites, Tieu et al. (2013) reported 1100 patients operated on robotically with only one gastro-jejunal leak, which equated to a 0.09% leak rate (Tieu et al. 2013). Currently, there is only one single *randomized trial of RYGB outcomes* which was underpowered; however, did find a reduction in operative times during the learning curve on the robotic platform (Sanchez et al. 2005). Table 59.2 summarizes the current *meta-analyses* and *systematic reviews* on Robotic RYGB versus Laparoscopic RYGB.

In summary, RYGB robotic procedures now offer surgeons a safe modification to conventional laparoscopic RYGB, though at the expense of surgical time and money.

Robotic Adjustable Gastric Band

An adjustable gastric band is a restrictive bariatric procedure where a silicone band is placed one to two centimeters below the gastro-esophageal junction, with the device being secured with a suture on the gastric wall. Placement of an adjustable gastric band requires minimal dissection, tissue handling, and suturing. Due to its declining use in bariatric surgery, the number of robotic gastric band cases performed is limited, and robotic technology did not add any benefit to this procedure (Alibhai et al. 2015). However, a study by Edelson et al. did report an advantage when performing those procedures robotically in patients with a BMI over 50 kg/m^2 (Edelson et al. 2011).

Robotic Biliopancreatic Diversion with Duodenal Switch (BPD-DS)

Sudan and Podolsky reported the first BPD-DS as a hybrid robotic procedure in 2000, initially as a multiple docking procedure and finally with a single docking approach (Sudan and Podolsky 2015). The Xi version of the da Vinci® robot with its integrated table motion, robotic arm-controlled staplers, and improved mobility has enabled this single docking robotic procedure.

Revisional Bariatric Procedures

Revisional bariatric procedure numbers are increasing as a percentage of procedures performed, having doubled in number from 2011 to 2015 (El Chaar et al. 2018); MBSAQIP data show they represent 11.9% of bariatric procedures performed (Clapp et al. 2019). Revisional cases are challenging, take longer to perform, and are associated with higher patient morbidity (El Chaar et al. 2018). It is in this setting the da Vinci® robot would be expected to excel. Single site performance analyses have indicated excellent results, with complication rates from revisional RYGB being equivalent to outcomes for primary bariatric procedures (Snyder et al. 2013). MBSAQIP data from 2015, however, have not demonstrated an advantage for the robotic platform in revisional cases (Clapp et al. 2019).

Future Directions of Robotic Technology in Bariatric Procedures

We are on the precipice of new technologies in surgical robotics, with the promises of nanotechnology and the applications of big data. Big data is defined as a set of data so large that traditional data processing applications are ineffective (Taragona and Batista 2018). Big data is now driving Artificial Intelligence, Machine Learning, and Automaticity (Lazar 2019). Robotic surgery has an important role in big data acquisition. Every surgical movement on a robotic platform is a data point. Machine learning enables machines to make predictions by recognizing patterns (Hashimoto et al. 2018). This could effectively create a "*collective surgical consciousness*" as, while operating, you provide information which helps to decide the future (Rodriguez Luna and Vilallonga 2019).

Augmented Reality

Under this platform, the robot would anticipate the surgeon's next move with real-time prediction and avoidance of adverse events, with decision support akin to an intraoperative Global Positioning System. Advances in computer vision now enable augmented reality, created by fusing images and the use of fluorescence in real-time. Similarly, the worldwide introduction of 5G technology with wireless transmission and real-time data analytics may allow robotic surgery to be made available in remote areas to standardize surgical care. Data can aid in better value-based healthcare to determine what instruments and devices are actually required. The true potential of Artificial Intelligence is difficult to predict, and its ability is limited by the quality of the data analyzed (Dibbs and Hollier 2019).

Autonomous Surgery

Technical skill augmentation with task deconstruction and autonomous performance of tasks such as suturing is already here with robots that outperform humans in bowel anastomosis on experimental models (Hashimoto et al. 2018). Natural Orifice Transluminal Endoscopic Surgery (NOTES), also known as scar-less surgery, is where the abdominal cavity is entered via the stomach, vaginal vault, or rectum to perform procedures. Due to the technical limitations of current endoscopic systems, only transvaginal-assisted SG and LAGB have been performed with this approach (Erridge et al. 2016). Improved robotic platforms may allow NOTES anastomotic procedures to become feasible. In the future, "Nanobots" (nanosize robotic machines) could allow patients to swallow a surgical device and undergo genetic manipulation or molecular medicine diagnosis, with the help of novel miniaturized vehicles.

New Robotic Platforms and Manufacturers

With less than 5% of global procedures performed robotically, the space is open for the digitalization and mechanization of minimally invasive surgery, and the race is on for industry to fill this void. The global market for Robotic-Assisted Surgery is predicted to have a compound annual growth rate of 10.4%, from US$3.9 billion in 2018 to US$6.5 billion by 2023 (Review RB 2019). New robotic systems now display features of modularity, improved affordability, and data acquisition with consumer feedback using a decreased physical footprint (i.e., size reduction), when compared with the da Vinci® System. Two exciting new robots are the Senhance™ by Transenterix and the Versius™ by Cambridge Medical Research (CMR), both FDA-approved and available on the market (Perez and Schwaitzberg 2019).

Senhance™ (Transenterix Inc., Morrisville, NC, USA)

The Senhance™ has haptic feedback and an open footprint with the surgeon in a seated position, wearing 3D glasses. A stand-out feature of this system is that the endoscopic camera is controlled with the surgeon's eye movements. The instruments are reusable, though not wristed (Fig. 59.15). Currently, this robot has found its way into the Japanese market (Transenterix 2019). In the USA, it has limited (510 k) clearance by the FDA for some surgical modalities; however not including bariatric operations.

Versius (Cambridge Medical Research, Cambridge, UK)

The Versius™ is portable and modular, with the aim of working interchangeably with laparoscopic equipment. It has five arms which are self-contained, each weighing 25 kg, with intelligence and sensing capability for surgeon control and assistant staff. The surgeon console is open, slim, and portable, and the surgeon has the option of sitting or standing for optimal ergonomics. It is supplied as a service including the robot, instruments, and maintenance at a fixed annual cost, with the aim to keep costs similar to laparoscopic systems. Galaxy Care Hospital in Pune, India, was the first to use the Versius™ robot (Fig. 59.16). The company is committed to monitoring all outcomes on their system and has initiated the world's first clinical registry for surgical robotic systems (Network MD 2019).

Additionally, there are the eagerly anticipated offerings from the company VerbSurgical Inc. (Bayshore, CA, USA), which is a partnership

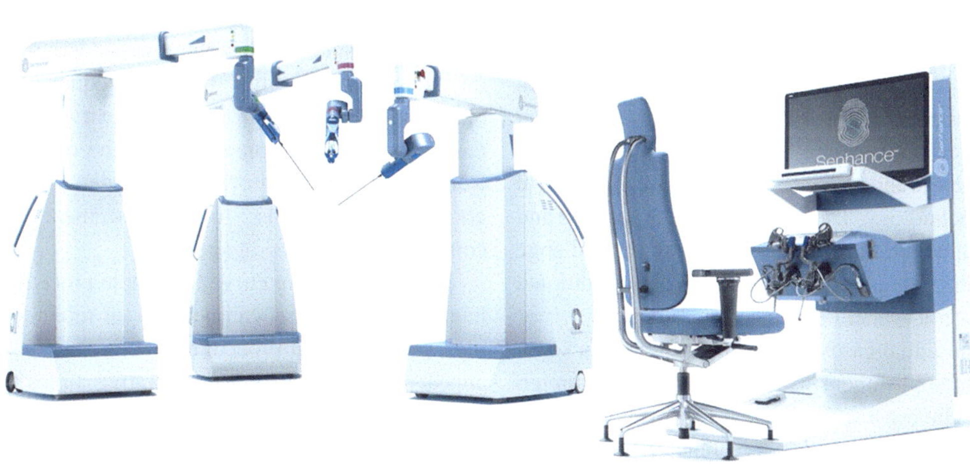

Fig. 59.15 Senhance™ robot operating room setup (with permission from Transenterix Inc.)

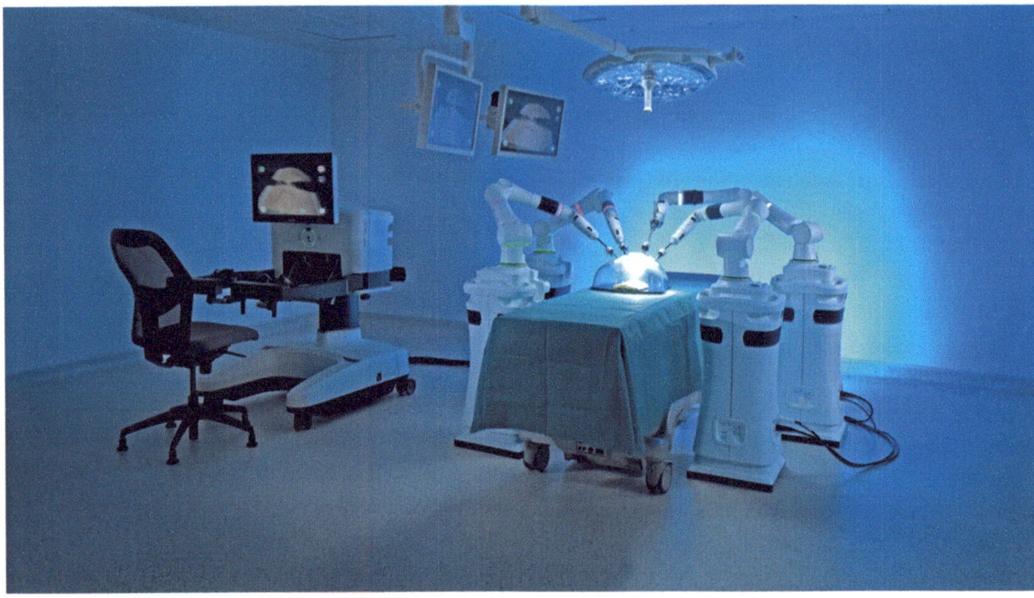

Fig. 59.16 Versius™ robotic system setup in the operating room (with permission from CMR)

between Johnson & Johnson and Google. This well-funded start-up company aims to create a new paradigm in digital surgery with a connected platform between the surgeon and patient, for all phases of patient care (VerbSurgical 2019). Despite these aspirations, they are yet to divulge information about their soft tissue robotic system, which is pegged for commercial release in 2020 (Biotech 2019).

Robotic devices beginning to emerge are summarized in Table 59.3.

Authors' Perspective

A surgeon's ability to transition from a pure laparoscopic surgeon to a robotic surgeon requires access to a da Vinci® robot at their institution and the intrinsic drive to immerse in this technology. The training phase takes time and commitment, though provides a unique challenge via the complete disruption of the traditional surgeon-to-patient interface.

The initial cases are exhilarating with a sense of accomplishment to be operating in a foreign way. Having a specialist proctor for the first few cases and performing less technically demanding procedures, such as the SG, can ease the stress during this transition. Every member of the team has their own robotic learning curve so that patience, leadership, and a focus on positive communication to support team dynamics are important.

Next comes a phase of disillusionment as realities set in. Administrative frustrations, such as a lack of access to the robot, additional "technology" fees, and insurance not covering costs for example, need to be overcome. Often business cases need to be drawn up and robot access issues resolved via consistent negotiations with hospital administration and other surgeons.

The time spent in this phase depends not just on volume alone, but on the frequency of operations performed, the complexity of the cases, and also the consistency of the operating team. Coaching, mentoring, and video analysis are important tools for advancement and competency.

After the learning curve for simple procedures has passed, the robot can be used for more complex and revisional procedures. These later cases appear more precise, and operative times are improving. Importantly, there is less physical fatigue after completing a robotic case.

The robotic platform can be used across various surgical specialty groups (e.g., general

Table 59.3 Summary table of emerging robotic-assisted surgery platforms

Product	Unique features	Prototype	Regulatory clearance
Single Port Orifice Robotic Technology SPORT™*, Titan Medical	– Single port system with flexible endoscopic 3D camera – Two wristed instruments can be used together with the camera	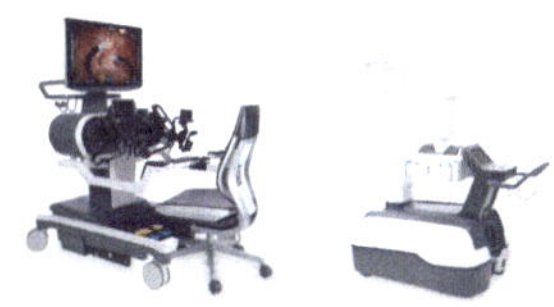	FDA in 2020
Hugo™ *, Medtronic	– Open surgeon console – Up to five minimally invasive robotic-assisted modular arms, incorporating elements of MiroSure – Tower and visualization system, generator, process, and endoscopy can be used for laparoscopy or open surgery	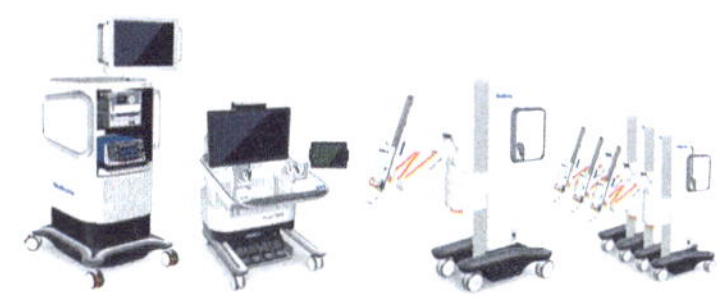	FDA in 2021
REVO-I™ *, Meere Company	– Similar in design and concept to da Vinci® Si – Haptic feedback to be introduced in new version	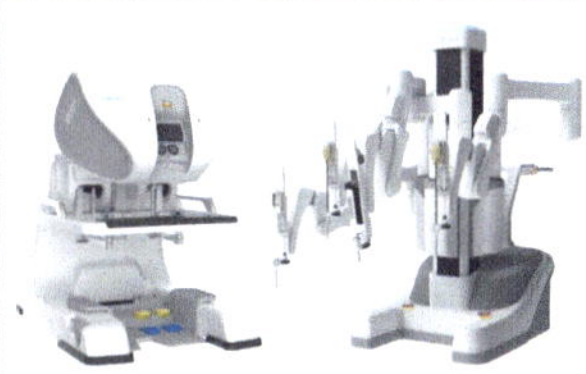	South Korea,2017
Verb Surgical™ *, Johnson & Johnson, Google			FDA in 2020
Vicarious™ *, Vicarious Surgical	– In-vivo robot fully inserted into patient through thumb-sized incision – Controlled wirelessly by surgeon wearing virtual reality goggles – Full suite of proprietary instruments including staplers, energy devices, and retractors		–

SPORT is a Trade Mark of Titan Medical; Hugo is a Trade Mark of Medtronic; REVO is a Trade Mark of Meere Company; Verb Surgical is a Trade Mark of Johnson & Johnson; Google Vicarious is a Trade Mark of Vicarious Surgical

surgery, colorectal surgery, hernia surgery, urology, gynaecology, and head and neck surgery), and hence, requires its own service line of administrative support. Efficiencies can be gained by working together, improving communication, collaboration, and having goals that are measurable and achievable (Fagin 2014).

Membership with the Clinical Robotics Surgery Association and the Facebook group Robotic Bariatric Collaboration ensures well-rounded learning with vicarious peer-to-peer interactions and video reviews (Myers et al. 2018).

Two-dimensional laparoscopic skills are required for port placement and division of adhesions and to allow the robotic ports to be inserted. There is also a need to train fellows and residents in laparoscopy. For most institutions, the da Vinci® robot is only available in usual working hours for elective cases. Emergency and weekend use of the robot is, however, possible with considered planning and availability of trained staff. This would expand the applications of the robot (Sudan and Desai 2012).

Presently, robotic surgery in bariatrics will not replace standard laparoscopy. Interestingly, it has been commented on that being a robotic surgeon can make you a better laparoscopic surgeon (Wilson and Sudan 2013), and this is certainly applicable to my robotic journey. Instead of viewing the robotic platform as one that needs to delineate traditional methods, the use of the robot can be viewed as a complementary adjunct to performing surgery on difficult cases, with the added benefit of less surgical fatigue and providing different and emerging stimuli for surgical skill acquisition (Wilson and Sudan 2013).

References

Acevedo E Jr, Mazzei M, Zhao H, Lu X, Soans R, Edwards MA (2019) Outcomes in conventional laparoscopic versus robotic-assisted primary bariatric surgery: a retrospective, case-controlled study of the MBSAQIP database. Surg Endosc 17:17

Agcaoglu O, Aliyev S, Taskin HE, Chalikonda S, Walsh M, Costedio MM et al (2012) Malfunction and failure of robotic systems during general surgical procedures. Surg Endosc 26(12):3580–3583

Alibhai MH, Shah SK, Walker PA, Wilson EB (2015) A review of the role of robotics in bariatric surgery. J Surg Oncol 112(3):279–283

AlSabah S, Al Haddad E, Khwaja H (2019) The prevalence of musculoskeletal injuries in bariatric surgeons. Surg Endosc 33(6):1818–1827

Bailey JG, Hayden JA, Davis PJ, Liu RY, Haardt D, Ellsmere J (2014) Robotic versus laparoscopic Roux-en-Y gastric bypass (RYGB) in obese adults ages 18 to 65 years: a systematic review and economic analysis. Surg Endosc 28(2):414–426

Biotech F. VerbSurgical taps new CEO as it inches its digital surgery robot toward market 2019. Available from https://www.fiercebiotech.com/medtech/verb-surgical-taps-new-ceo-as-it-inches-its-digital-surgery-robot-toward-market

Brown K, Solomon MJ, Young J, Seco M, Bannon PG (2019) Addressing the ethical grey zone in surgery: a framework for identification and safe introduction of novel surgical techniques and procedures. ANZ J Surg 89(6):634–638

Buchs NC, Pugin F, Bucher P, Hagen ME, Chassot G, Koutny-Fong P et al (2012) Learning curve for robot-assisted Roux-en-Y gastric bypass. Surg Endosc 26(4):1116–1121

Celio AC, Kasten KR, Schwoerer A, Pories WJ, Spaniolas K (2017) Perioperative safety of laparoscopic versus robotic gastric bypass: a propensity matched analysis of early experience. Surg Obes Relat Dis 13(11):1847–1852

Childers CP, Maggard-Gibbons M (2018) Understanding costs of care in the operating room

Clapp B, Liggett E, Jones R, Lodeiro C, Dodoo C, Tyroch A (2019) Comparison of robotic revisional weight loss surgery and laparoscopic revisional weight loss surgery using the MBSAQIP database. Surg Obes Relat Dis 15(6):909–919

Colquitt JL, Pickett K, Loveman E, Frampton GK (2014) Surgery for weight loss in adults. Cochrane Database Syst Rev 2014(8):Cd003641

Dibbs RPBA, Hollier LHJMDF (2019) Review of "artificial intelligence in surgery: promises and perils" by Hashimoto DA et al in Ann Surg 268;70–76, 2018. J Craniofac Surg 30(3):955

Ecker BL, Maduka R, Ramdon A, Dempsey DT, Dumon KR, Williams NN (2016) Resident education in robotic-assisted vertical sleeve gastrectomy: outcomes and cost-analysis of 411 consecutive cases. Surg Obes Relat Dis 12(2):313–320

Economopoulos KP, Theocharidis V, McKenzie TJ, Sergentanis TN, Psaltopoulou T (2015) Robotic vs. laparoscopic Roux-En-Y gastric bypass: a systematic review and meta-analysis. Obes Surg 25(11):2180–2189

Edelson PK, Dumon KR, Sonnad SS, Shafi BM, Williams NN (2011) Robotic vs. conventional laparoscopic gastric banding: a comparison of 407 cases. Surg Endosc 25(5):1402–1408

El Chaar M, Stoltzfus J, Melitics M, Claros L, Zeido A (2018) 30-Day outcomes of revisional bariatric stapling procedures: first report based on MBSAQIP data registry. Obes Surg 28(8):2233–2240

Erridge S, Sodergren MH, Darzi A, Purkayastha S (2016) Natural orifice translumenal endoscopic surgery: review of its applications in bariatric procedures. Obes Surg 26(2):422–428

Fagin R (2014) Challenges and Critical Elements of Setting Up a Robotics Program. 2014. In: Robotics in General Surgery [Internet]. [415–29]

Fantola G, Perrenot C, Frisoni R, Germain A, Ayav A, Bresler L et al (2014) Robotic Roux-en-Y gastric bypass surgical simulation curriculum. Obes Surg 24(10):1833–1834

George EI, Brand TC, LaPorta A, Marescaux J, Satava RM (2018) Origins of robotic surgery: from skepticism to standard of care. J Soc Laparoendosc Surg 22(4): Oct–Dec

Hagen ME, Meehan JJ, Inan I, Morel P (2008) Visual clues act as a substitute for haptic feedback in robotic surgery. Surg Endosc. 22(6):1505–1508

Hagen ME, Rohner P, Jung MK, Amirghasemi N, Buchs NC, Fakhro J et al (2017) Robotic gastric bypass surgery in the swiss health care system: analysis of hospital costs and reimbursement. Obes Surg 27(8):2099–2105

Hashimoto DAMDMS, Rosman GP, Rus DP, Meireles ORMDF (2018) Artificial intelligence in surgery: promises and perils. Ann Surg 268(1):70–76

Heit M (2014) Surgical proctoring for gynecologic surgery. Obstet Gynaecol 123:349–352

Hernandez JD, Bann SD, Munz Y, Moorthy K, Datta V, Martin S et al (2004) Qualitative and quantitative analysis of the learning curve of a simulated surgical task on the da Vinci system. Surg Endosc 18(3):372–378

Horgan S, Vanuno D (2001) Robots in laparoscopic surgery. J Laparoendosc Adv Surg Tech A 11(6):415–419

IFSO. IFSO Global Registry Report 2019a 2019. Available from https://www.ifso.com/pdf/5th-ifso-global-registry-report-september-2019.pdf

IFSO. International Federation for the Surgery of Obesity and Metabolic Disorders 2019b. Available from http://www.ifso.com/sleeve-gastrectomy/

Intuitive I. Annu Rep 2018. 2018

Jung MK, Hagen ME, Buchs NC, Buehler LH, Morel P (2017) Robotic bariatric surgery: a general review of the current status. Int J Med Robot + Comput Assist Surg: MRCAS. 13(4)

Kalan S, Chauhan S, Coelho RF, Orvieto MA, Camacho IR, Palmer KJ et al (2010) History of robotic surgery. J Robot Surg 4(3):141–147

Kleinert S, Horton R (2019) Obesity needs to be put into a much wider context. Lancet 393(10173):724–726

Lazar JF (2019) Cutting the optical leash: rethinking the "robot" in robotic surgery. J Robot Surg 13 (6):787–788

Lee GI, Lee MR, Clanton T, Sutton E, Park AE, Marohn MR (2014) Comparative assessment of physical and cognitive ergonomics associated with robotic and traditional laparoscopic surgeries. Surg Endosc 28 (2):456–465

Li K, Zou J, Tang J, Di J, Han X, Zhang P (2016) Robotic versus laparoscopic bariatric surgery: a systematic review and meta-analysis. Obes Surg 26 (12):3031–3044

Li L, Yu H, Liang J, Guo Y, Peng S, Luo Y et al (2019) Meta-analysis of the effectiveness of laparoscopic adjustable gastric banding versus laparoscopic sleeve gastrectomy for obesity. Medicine 98(9): e14735

Lundberg PW, Wolfe S, Seaone J, Stoltzfus J, Claros L, El Chaar M (2018) Robotic gastric bypass is getting better: first results from the metabolic and bariatric surgery accreditation and quality improvement program. Surg Obes Relat Dis 14(9):1240–1245

Lundberg PW, Stoltzfus J, El Chaar M (2019) 30-day outcomes of robot-assisted versus conventional laparoscopic sleeve gastrectomy: first analysis based on MBSAQIP. Surg Obes Relat Dis 15(1):1–7

Maddern GJ (2019) Introduction of new surgical techniques and technologies. ANZ J Surg 89 (6):625–626

Markar SR, Karthikesalingam AP, Venkat-Ramen V, Kinross J, Ziprin P (2011) Robotic vs. laparoscopic Roux-en-Y gastric bypass in morbidly obese patients: systematic review and pooled analysis. Int J Med Robot + Comput Assist Surg: MRCAS 7(4):393–400

Mason EE, Ito C (1996) Gastric bypass in obesity. Surg Clin N Am 47(6):1345–1351

Mohr CJ, Nadzam GS, Curet MJ (2005) Totally robotic Roux-en-Y gastric bypass. Arch Surg 140(8):779–786

Moon RC, Stephenson D, Royall NA, Teixeira AF, Jawad MA (2016) Robot-assisted versus laparoscopic sleeve gastrectomy: learning curve, perioperative, and short-term outcomes. Obes Surg 26 (10):2463–2468

Magouliotis, D. E., V. S. Tasiopoulou, E. Sioka & D. Zacharoulis (2017) Robotic versus Laparoscopic Sleeve Gastrectomy for Morbid Obesity: a systematic review and Meta-Analysis. Obesity Surgery 27, 245–253

Myers CG, Kudsi OY, Ghaferi AA (2018) Social media as a platform for surgical learning: use and engagement patterns among robotic surgeons. Ann Surg 267 (2):233–235

NASA (April 2, 1990) The Puma robotic sensor arm for use in virtual reality development and studies at the NASA Ames Research Center, Mountain View, California. Image # : AC90–0187-1. In: Commons NoT, editor.: Flickr; 1990

Network MD. Galaxy Care Hospital Acquires First Versius Surgical Robotic System 2019. Available from https://www.medicaldevice-network.com/news/galaxy-care-hospital-acquires-first-versius-surgical-robotic-system/

Organisation WH. Obesity and overweight 2018 [cited 2018 16 February 2018]. Available from https: //www.who.int/en/news-room/fact-sheets/detail/obesity-and-overweight

Perez RE, Schwaitzberg SD (2019) Robotic surgery: finding value in 2019 and beyond. Ann Laparosc Endosc Surg 4

Reames BN, Bacal D, Krell RW, Birkmeyer JD, Birkmeyer NJ, Finks JF (2015) Influence of median surgeon operative duration on adverse outcomes in bariatric surgery. Surg Obes Relat Dis 11 (1):207–213

Renaud M, Reibel N, Zarnegar R, Germain A, Quilliot D, Ayav A et al (2013) Multifactorial analysis of the learning curve for totally robotic Roux-en-Y gastric bypass for morbid obesity. Obes Surg 23 (11):1753–1760

Review RB. Surgical Robots 2019 2019. Available from https://www.roboticsbusinessreview.com/health-medical/5-surgical-robots-2019/

Rodriguez Luna MR, Vilallonga R (2019) RE: "a review of the role of robotics in bariatric surgery: finding our future?" by Veilleux et al. (J Laparoendosc Adv Surg Tech A [Epub ahead of print]; https://doi.org/10.1089/lap.2019.0419). J Laparoendosc Adv Surg Tech A 29 (11):1389–1390

Rogula T, Koprivanac M, Janik MR, Petrosky JA, Nowacki AS, Dombrowska A et al (2018) Does robotic Roux-en-Y gastric bypass provide outcome advantages over standard laparoscopic approaches? Obes Surg 28(9):2589–2596

Ruiz-Tovar J, Garcia A, Ferrigni C, Gonzalez J, Castellon C, Duran M (2019) Impact of implementation of an enhanced recovery after surgery (ERAS)

program in laparoscopic Roux-en-Y gastric bypass: a prospective randomized clinical trial. Surg Obes Relat Dis 15(2):228–235
Sanchez BR, Mohr CJ, Morton JM, Safadi BY, Alami RS, Curet MJ (2005) Comparison of totally robotic laparoscopic Roux-en-Y gastric bypass and traditional laparoscopic Roux-en-Y gastric bypass. Surg Obes Relat Dis 1(6):549–554
Service USDoHaH. USA HealthCare Cost and Utilization Project 2019
Silverman CD, Ghusn MA (2017) Early Australian experience in robotic sleeve gastrectomy: a single site series. ANZ J Surg 87(5):385–389
Slim K, Mattevi C (2018) Robotic surgery or enhanced recovery programs or both? And in which order? Surgery 164(5):937–938
Snyder B, Wilson T, Woodruff V, Wilson E (2013) Robotically assisted revision of bariatric surgeries is safe and effective to achieve further weight loss. World J Surg 37(11):2569–2573
Sudan R, Desai SS (2012) Emergency and weekend robotic surgery are feasible. J Robot Surg 6 (3):263–266
Sudan R, Podolsky E (2015) Totally robot-assisted biliary pancreatic diversion with duodenal switch: single dock technique and technical outcomes. Surg Endosc 29 (1):55–60
Taragona EMBA, Batista G (2018) Big data and surgery: the digital revolution continues. Cir Esp 96:247–249
Tieu K, Allison N, Snyder B, Wilson T, Toder M, Wilson E (2013) Robotic-assisted Roux-en-Y gastric bypass: update from 2 high-volume centers. Surg Obes Relat Dis 9(2):284–288
Transenterix. Transenterix 2019. Available from https://transenterix.com
Veilleux E, Ponce J, Lutfi R (2019) A review of the role of robotics in bariatric surgery: finding our future? J Laparoendosc Adv Surg Tech A 02:02
VerbSurgical. VerbSurgical 2019. Available from https://www.verbsurgical.com/physicians/
Vilallonga R, Fort JM, Gonzalez O, Caubet E, Boleko A, Neff KJ et al (2012) The initial learning curve for robot-assisted sleeve gastrectomy: a surgeon's experience while introducing the robotic technology in a bariatric surgery department. Minimal Invas Surg 2012:347131
Wehrtmann FS, de la Garza JR, Kowalewski KF, Schmidt MW, Muller K, Tapking C et al (2019) Learning curves of laparoscopic Roux-en-Y gastric bypass and sleeve gastrectomy in bariatric surgery: a systematic review and introduction of a standardization. Obes Surg 29:29
Wilson EB, Sudan R (2013) The evolution of robotic bariatric surgery. World J Surg 37(12):2756–2760
Wittgrove AC, Clark GW, Tremblay LJ (1994) Laparoscopic gastric bypass, Roux-en-Y: preliminary report of five cases. Obes Surg 4(4):353–357
Yohannes P, Rotariu P, Pinto P, Smith AD, Lee BR (2002) Comparison of robotic versus laparoscopic skills: is there a difference in the learning curve? Urology 60 (1):39–45; discussion

The Role of Obesity and Bariatric Surgery in the Management of Knee and Hip Osteoarthritis

60

Katelyn M. Mellion and Shanu N. Kothari

Contents

Introduction 814
Impact of Obesity on Lower Limb Osteoarthritis 814
Weight-Bearing Forces 815
Gait Changes 815
Association of Inflammation with Mechanical Stress 816
Leptin 816
Adiponectin 817
Resistin 818
Other Adipocytokines 818

Impact of Obesity on Arthroplasty 818
Late Postoperative Abnormalities 819
Functional Outcome 819
Impact of Weight Loss and Bariatric Surgery on Osteoarthritis 819

Impact of Bariatric Surgery on Arthroplasty 820
Bariatric Surgery and Total Arthroplasty: Which Comes First, the Chicken or the Egg? 821
Mobility and Access to Care 822
Financial Issues 822
Ongoing Trials and Future Perspectives 822
Future Directions 823

References 823

K. M. Mellion
Department of Medical Education, Gundersen Health System, La Crosse, WI, USA

S. N. Kothari (✉)
Department of Surgery, Prisma Health, Greenville, SC, USA
e-mail: shanu.kothari@prismahealth.org

J. Faintuch and S. Faintuch (eds.), *Obesity and Diabetes*, https://doi.org/10.1007/978-3-030-53370-0_60

Abstract

The prevalence of both obesity and osteoarthritis is increasing. Obesity is an independent risk factor for osteoarthritis. Both mechanical and inflammatory factors contribute to this risk. The gold-standard treatment for end-stage knee and hip osteoarthritis is total joint arthroplasty. Patients with obesity have worse outcomes than patients with normal weight following total joint arthroplasty. There is a dearth of information on the effects of bariatric surgery on the incidence of total joint arthroplasty, and the impact on outcomes is limited with mixed results. Further study on total joint arthroplasty in patients with obesity is needed to determine optimal risk stratification, bariatric procedure selection, and timing of bariatric surgery relative to total joint arthroplasty.

Keywords

Bariatric surgery · Lower extremity osteoarthritis · Arthroplasty · Outcomes

Introduction

In 2016, the prevalence of obesity in America was 39.8% among adults (Hales et al. 2017), with an estimated 7% reaching morbid obesity, defined as a body mass index (BMI) >40 kg/m^2 (Sturm and Hattori 2013). Osteoarthritis (OA) is also highly prevalent. It is estimated that 25.9% of the projected total adult population will have doctor-diagnosed arthritis by 2040 (Hootman et al. 2016). The lifetime risk of end-stage knee osteoarthritis alone is 13.3% for men and 18.8% for women (Ogden et al. 2014). Knee OA is the second most prevalent cause of disability in the world (GBD 2016 Disease and Injury Incidence and Prevalence Collaborators 2017). The risk of OA-associated disability is as elevated as of cardiac conditions (Wang et al. 2016) and is the most prevalent of all diseases in the elderly (Yokota et al. 2015). In terms of hospital charges, osteoarthritis is more expensive than pneumonia, stroke, or complications from diabetes (Teichtahl et al. 2009).

Obesity is an independent risk factor for osteoarthritis (Gu et al. 2019). Patients with grade II obesity have 4.7 times the likelihood of developing knee osteoarthritis compared to those with normal weights (Reyes et al. 2016). Research suggests this risk is due to both biomechanical and systemic inflammatory effects on joints (Koonce and Bravman 2013). These effects eventually lead to total joint arthroplasty (TJA) in an increasing number of patients with obesity (Belmont et al. 2014; Fehring et al. 2007; Kremers et al. 2014). In the United States, over one million total knee and total hip replacement procedures are performed yearly.

Patients with obesity are known to have a higher risk of complications for total joint arthroplasty (Salih and Sutton 2013; Belmont et al. 2014). The American Association of Hip and Knee Surgeons suggests weight loss prior to TJA for patients with BMI >40 kg/m^2 (Workgroup of the American Association of Hip and Knee Surgeons Evidence-Based Committee 2013). While diet and exercise are important for a healthy lifestyle, these methods will not achieve a significant sustained weight loss. Bariatric surgery has been shown to provide sustained weight loss, as well as resolution of many obesity-related comorbidities, including type II diabetes, gastroesophageal reflux, and hyperlipidemia (Sharples and Mahawar 2019). Data investigating the effect of prior bariatric surgery on postoperative TJA outcomes is mixed.

Impact of Obesity on Lower Limb Osteoarthritis

In the United States, 31% of those with obesity were diagnosed with arthritis, while only 16% of normal weight individuals carried the diagnosis (Center for Disease Control and Prevention 2018). Obesity has been shown to be an independent risk factor for osteoarthritis (Gu et al. 2019). In a Swedish study by Jarvholm et al.,

15–67 year-old male construction workers were found to be more likely to undergo knee or hip replacement as BMI increased, even when controlling for age and tobacco use (Jarvholm et al. 2008). Data from the Nurses' Health Study, which included over 121,000 female nurses, revealed those with a BMI >35 kg/m^2 were twice as likely to undergo total hip arthroplasty (THA) compared to those with BMI <22 kg/m^2. Interestingly, those with a higher BMI at age 18 years had the highest risk of undergoing THA (Karlson et al. 2003).

Weight-Bearing Forces

Joint "wear-and-tear" secondary to mechanical and structural factors has classically been suggested as the mechanism for osteoarthritis. While these factors surely contribute, recent studies suggest they cannot solely account for the relationship between obesity and osteoarthritis (Sowers and Karvonen-Gutierrez 2010). For example, in a systematic review and meta-analysis by Long et al., both lean and fat mass in subjects with knee osteoarthritis were higher than in those without osteoarthritis. While fat mass percentage was positively associated with knee osteoarthritis, lean mass percentage was negatively associated with knee osteoarthritis. Increased fat mass was also associated with hand joint osteoarthritis (Long et al. 2019). The increased incidence of non-weight-bearing joint osteoarthritis in individuals with obesity has led to investigation into inflammatory factors.

Mechanical Factors

Abnormal loading of the knee and hip joint can lead to changes in the composition, structure, and mechanical characteristics of articular cartilage (Maly et al. 2005; Mundermann et al. 2005).

Joint load during walking is directly related to body weight. Peak *in vivo* hip contact forces during normal walking have been noted to be 2–3 times body weight (Heywood et al. 2019). So, as body weight increases, so do joint contact forces. The knee joint is especially vulnerable in the population with obesity where there is a high fat:lean mass ratio. The quadriceps muscle may fail to adequately absorb forces on the knee joint, leading to an increased load on the articular cartilage and subsequent progressive degeneration (Maly et al. 2005). Knee joint loads during walking and stair climbing have been shown to be higher than loads across the hip joint (Taylor et al. 2004).

Gait Changes

Gait disturbances are also common in patients with obesity. In a systematic review, Runhaar et al. found that patients with obesity had altered lower extremity biomechanics for everyday movements including walking, standing, and rising from a sit-to-stand position. When walking, patients with obesity took shorter, wider steps and chose a slower pace. During standing, individuals with obesity had greater toe-out angles. When transitioning from sitting to standing, those with obesity had less hip flexion and greater foot displacement. These adjustments lead to alterations in the regions of the articular cartilage within the joint that bears the load and higher loads across the hip and knee joints (Runhaar et al. 2011). While gait alterations are likely adaptations adopted as a means to temporarily offload the hip and knee joints, studies show that joint loads in excess of 4 times body weight can occur regularly. Furthermore, stumbling can result in loads over 8 times body weight (Bergmann et al. 2001).

Inflammatory Factors

Obesity has been shown to be associated with a chronic inflammatory state. Specifically, where adipocytes were previously seen as solely storage vesicles, they are now known to be metabolically active cells that secrete a multitude of factors, termed adipocytokines. Many of these adipocytokines are thought to play a role in cartilage homeostasis. Because obese patients have a higher adipocyte mass, the secondary changes in the adipocytokine milieu are thought to contribute to the degradation of cartilage, and therefore, the development of osteoarthritis (Francisco et al. 2018b).

Research on adipocytes and adipocytokines aims to determine why obesity is associated with a range of comorbidities, including osteoarthritis. Adipocytokines of interest include leptin, adiponectin, resistin, visfatin, lipocalin-2, chemerin, and apelin (Fig. 60.1) (Sowers and Karvonen-Gutierrez 2010; Francisco et al. 2018a).

It is unknown why some patients with obesity develop osteoarthritis while others do not, despite all these patients experiencing increased joint loads and most having altered gait characteristics. However, the conglomerate of comorbidities associated with metabolic syndrome may have an interrelated etiology. Karvonen-Gutierrez et al. showed that cardiometabolic biomarkers are associated with knee osteoarthritis regardless of BMI. In this report, mid-aged women with both obesity and two or more cardiovascular risk factors were more than 6 times more likely to have knee osteoarthritis when compared to women without obesity or cardiometabolic risk factors (Karvonen-Gutierrez et al. 2012).

Association of Inflammation with Mechanical Stress

There is evidence that increased load on the knee joint experienced by individuals with obesity may lead to the release of cytokines, growth factors, and metalloproteinases triggered by mechanoreceptors on the surface of chondrocytes (Sowers and Karvonen-Gutierrez 2010), which may contribute to the inflammatory milieu in addition to the increased adipocytokine production by adipose tissue.

Leptin

One of the most studied adipokines that likely plays a role in the development of osteoarthritis is leptin. Leptin is classically known as a metabolic adipocytokine that acts to reduce food intake and increase energy expenditure. It is known to be present in both plasma and synovial fluid in higher concentrations in individuals with obesity (Dumond et al. 2003). Griffin et al. showed that leptin-deficient mice could develop an obese phenotype without increasing the incidence of osteoarthritis, suggesting a key role for leptin in the development of osteoarthritis (Griffin

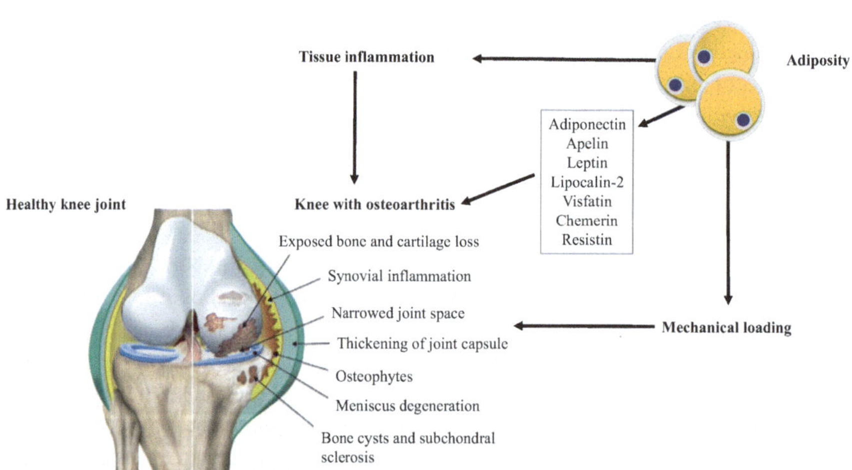

Fig. 60.1 Illustration depicting the effects of increased fat mass and dysregulation in cartilage degradation (Adapted from Uhalte E, Wilkinson JM, Southam L, Zeggini E. Pathways to understanding the genomic aetiology of osteoarthritis. Hum Mol Genet. 2017;26(R2):R193-R201)

et al. 2009). Leptin has a synergistic relationship with pro-inflammatory cytokines which leads to increased production of several inflammatory factors, growth factors, and matrix metalloproteinases by chondrocytes (Francisco et al. 2018a). Leptin also has been shown to affect chondrogenic progenitor cells by reducing their ability to migrate, changing their differentiation pathway, and modifying their cell cycle, ultimately altering their ability to maintain cartilage homeostasis and replace damaged tissue (Francisco et al. 2018b). Leptin likely has a dose-dependent relationship with manifestations of osteoarthritis. Leptin expression in cartilage, subchondral bone, synovial tissues, and osteophytes has been associated with the degree of cartilage degeneration and radiologic severity of osteoarthritis (Fig. 60.2) (Simopoulou et al. 2007; Francisco et al. 2018b).

Adiponectin

Adiponectin is a protein structurally similar to collagen that is synthesized by adipose tissue, although circulating levels are inversely related to weight. Using knockout mice, adiponectin has been associated with preventing insulin resistance and lipid accumulation in muscles when the mice are placed on a high fat/sucrose diet (Francisco et al. 2018a). Adiponectin receptors have been found in cartilage, bone, and synovial tissues (Sowers and Karvonen-Gutierrez 2010). Conflicting data exists regarding the role of adiponectin in the development of osteoarthritis with some studies suggesting a protective effect (such as inhibition of pro-inflammatory factors, stimulation of osteoclast proliferation, and mineralization), while others suggest a negative effect (such as increased production of

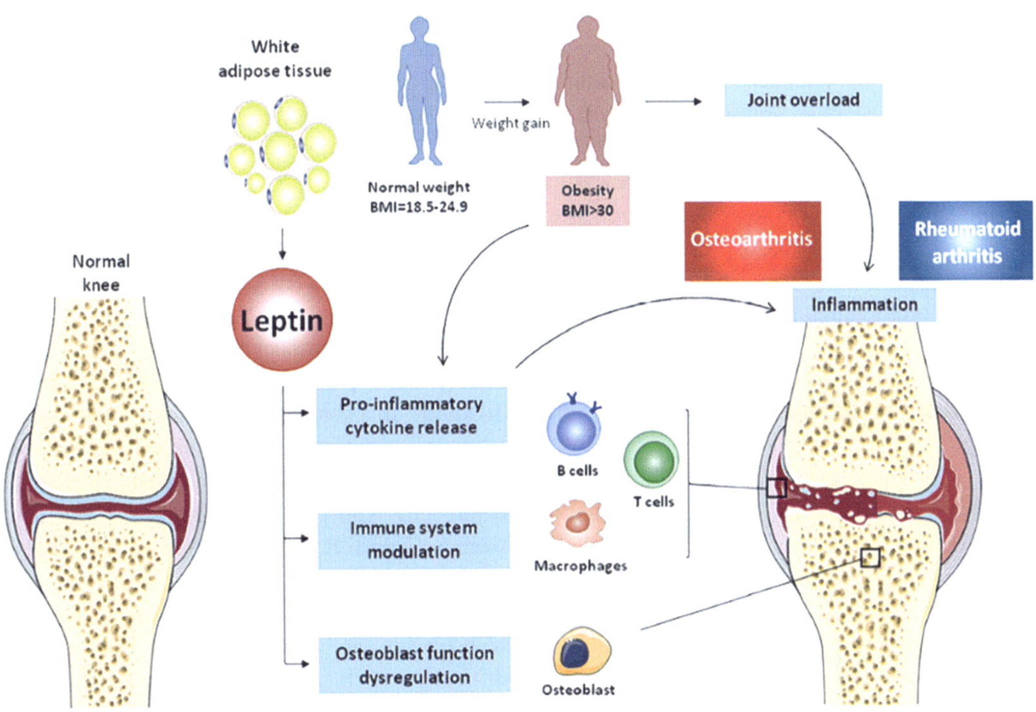

Fig. 60.2 Effects of adipose tissue-derived leptin on osteoarthritis and rheumatoid arthritis. Body weight gain, accompanied by white adipose tissue expansion, leads to obesity and subsequent increase of mechanical load, resulting in cartilage degradation and osteoarthritis onset. Adipose tissue-derived leptin causes osteoblast dysregulation in subchondral bone, thus promoting joint destruction. Additionally, leptin induces pro-inflammatory cytokine release from innate and adaptive immune cells, generating an inflammatory environment that prompts cartilage damage and rheumatoid arthritis (Reprinted from Francisco V, Pino J, Campos-Cabaleiro V, et al. Obesity, Fat Mass and Immune System: Role for Leptin. Front Physiol. 2018;9:640)

pro-inflammatory factors, inhibition of osteoclast differentiation, promotion of apoptosis) (Francisco et al. 2018a).

Studies have shown an increase in adiponectin in patients with osteoarthritis compared to heathy controls, as well as in patients with the radiologically most severe osteoarthritis disease. Both exercise and mechanical loading have been shown to increase adiponectin levels and its receptors in skeletal muscle, suggesting that adiponectin may play a protective role in preventing bone loss (Francisco et al. 2018a).

Resistin

Serum resistin levels have been shown to be increased in obesity and associated with adipose tissue inflammation. While in rodents the main source of resistin is adipocytes, in humans it is macrophages. Both serum and synovial fluid levels of resistin have been found in higher concentrations in patients with osteoarthritis compared to healthy controls; however, its correlation to radiographic damage is unclear, with conflicting evidence. Resistin is a pro-inflammatory cytokine that has been associated with increased levels of other cytokines and chemokines and with increased osteoblast proliferation (Francisco et al. 2018a).

Other Adipocytokines

Other adipokines that may contribute to the development of osteoarthritis in the population with obesity include visfatin, lipocalin-2, chemerin, and apelin. In patients with osteoarthritis, visfatin production was noted in several joint structures including the infrapatellar fat pad, synovium, and osteophytes. Visfatin is more prevalent in the serum and synovial fluid of osteoarthritis patients, where it increases both the production of pro-inflammatory factors and the degradation of joint connective tissue and extracellular matrix (Francisco et al. 2018a). Lipocalin-2 is a glycoprotein that circulates in a covalent complex with matrix metalloproteinase-9, whose main source is adipose tissue. Lipocalin-2 is expressed in joint tissues and found in elevated levels in patients with obesity as well as in patients with osteoarthritis. It has been shown to stimulate pro-osteoclastogenetic factors, inhibit anti-osteoclastogenic factors, and reduce chondrocyte proliferation. Levels have been correlated with fracture risk in the geriatric population (Francisco et al. 2018a).

Chemerin and its receptor are both expressed in adipose tissue, and levels are correlated with BMI in humans. The adipokine has been shown to be upregulated in adipose tissue of obese rats with type 2 diabetes mellitus. It is suspected to serve as a bridge between innate and adaptive immunity and possibly plays a role in osteoblast differentiation (Francisco et al. 2018a). Apelin is thought to be a pro-inflammatory adipokine that is found in higher levels in synovial fluid in individuals with osteoarthritis and has been positively correlated with severity of disease. It increases expression of catabolic factors in chondrocytes and decreases proteoglycan in articular cartilage (Francisco et al. 2018a).

Impact of Obesity on Arthroplasty

The risk of complications after total joint arthroplasty is particularly high once BMI reaches 40 kg/m^2. In a cohort followed for 10 years, mortality was higher in both genders, whereas risk of revision and dislocation increased in men only, contrasting with risk of reoperation, elevated in women (Tohidi et al. 2019). Short-term postoperative complications are likely related to the increased incidence of comorbidities in patients with obesity, as well as increased intraoperative difficulty, such as technical errors, surgeon reported difficulty, and problems that occur during surgery (Jarvenpaa et al. 2010; Nunez et al. 2011). Type II diabetes and obstructive sleep apnea, common comorbidities in the population with obesity, have both been associated with an increase in serious postoperative complications after TJA (Jamsen et al. 2012).

Severe obesity is noted to be associated with an increased risk of surgical site infections, respiratory complications, thromboembolic events, and hospital length of stay (Zusmanovich et al. 2018).

In the most comprehensive systematic review and meta-analysis on the effect of obesity on postoperative outcomes of TJA for osteoarthritis, 31 studies from 18 different countries were included. Patients without obesity were found to have fewer postoperative infections and fewer deep venous thromboses after either total hip arthroplasty (THA) or total knee arthroplasty (TKA) when compared to patients with morbid obesity (Pozzobon et al. 2018).

Late Postoperative Abnormalities

Long-term complications are likely associated with the mechanical and structural effects of obesity on artificial joints over time. The differences in joint load and daily movements between individuals with obesity versus normal weight, mentioned in the prior section, lead to increased risk for accelerated bare surface wear, early prosthesis failure, implant loosening, need for revisional joint surgery, and component malposition in the population with obesity after TJA. Likely due to the increased load in the knee compared to the hip joint, these effects are particularly evident after TKA (Abdel et al. 2015; Kerkhoffs et al. 2012).

Revisions after TJA have been shown to be higher in individuals with obesity versus those without obesity, with more discrepancies after TKA than THA (Chee et al. 2010; Foran et al. 2004). Patients with obesity are also at increased risk of prosthetic dislocation after THA (Watts et al. 2016). In a systematic review by Barrett et al., THA outcomes were compared between 66,238 patients who were morbidly obese and 705,619 patients with a BMI <30 kg/m^2. The overall revision rate was 7.99% in patients with morbid obesity versus 2.75% in patients without obesity (Barrett et al. 2018).

Functional Outcome

Poorer functional results are likely due to the conglomerate effect of short-term and long-term complications experienced. In the same comprehensive systematic review and meta-analysis mentioned above, participants without obesity reported less knee pain at both short-term (<6 months) and long-term (>6 months) follow-up after TKA, and less hip pain at short-term follow-up after THA. Patients without obesity also reported less disability at long-term follow-up for both TKA and THA (Pozzobon et al. 2018). The functional outcome, quantified using the Harris Hip Score, was comparable between patients with and without obesity. However, the individual studies that looked at quality of life did report lower Short-Form scores in patients with obesity postoperatively (Barrett et al. 2018).

Impact of Weight Loss and Bariatric Surgery on Osteoarthritis

A 5–10% weight loss significantly improves pain, self-reported disability, and quality of life in adults with both obesity (BMI 33–36 kg/m^2) and mild to moderate knee osteoarthritis (Chu et al. 2018).

In-hospital complications after hip and knee arthroplasty, and 90-day troubles after knee arthroplasty, diminish in patients who underwent prior bariatric surgery (McLawhorn et al. 2018a, 2018b). As many as 91% of patients undergoing gastric bypass or sleeve gastrectomy experienced a resolution of arthropathy by 21 months after surgery (Nelson et al. 2006). All patients undergoing sleeve gastrectomy had resolution of their joint pain 1 year after surgery, in another experience (Moon Han et al. 2005).

Studies assessing the specific effect of bariatric surgery on osteoarthritis progression suggest several benefits including: radiographic improvement of disease, decreased frequency and intensity of joint pain, improved physical function, and improved range of motion (Groen et al. 2015).

Improvements have also been noted in postural stability and sway during walking, which may decrease the magnitude and frequency of abnormal joint loads experienced by patients with obesity (Ponta et al. 2014).

Improved gait parameters including decreases in step width in the frontal plane, increases in step length, decreased torque around the hip and knee joint, and increased self-directed walking speed were noted after bariatric surgery. However, other studies suggest weight loss leads to increased torque across joints due to the increased stride length and gait velocity that accompany weight loss (Vartiainen et al. 2012; Vincent et al. 2012).

Impact of Bariatric Surgery on Arthroplasty

Because of the increased short-term and long-term complications, poorer functional outcomes, and increased cost of TJA in patients with morbid obesity compared to patients with normal BMI, the American Association of Hip and Knee Surgeons suggested to strongly recommend consideration for weight loss prior to TJA in patients with BMI > 40 kg/m^2 (Workgroup of the American Association of Hip and Knee Surgeons Evidence Based Committee 2013). Bariatric surgery has been shown to be the only reliable method of weight loss and comorbidity resolution for patients in this BMI range, though the effect of bariatric surgery on TJA outcomes shows mixed results.

The three most recent systematic reviews evaluating the effect of bariatric surgery prior to TJA include the same 13 studies in different combinations. Gu et al. reported conflicting evidence on the impact of bariatric surgery prior to TJA on in-hospital complications, 30-day complications, 90-day complications, revision rates, and hospital length of stay (Gu et al. 2019). Similarly, Stavrakis et al. reviewed 7 studies, 6 of which were included in the meta-analysis by Gu et al., to describe the conflicting evidence regarding the effect of bariatric surgery on TJA outcomes (Stavrakis et al. 2018). A third review and meta-analysis by Smith et al. which was included in the review by Stavrakis, included 5 of the studies used by Gu et al. No significant differences in the incidence of superficial wound infection, deep wound infection, deep vein thrombosis, pulmonary embolism, reoperation, joint revision, or mortality were found between the two groups in this meta-analysis. There were, however, more medical complications in patients who did not undergo bariatric surgery prior to TJA (Smith et al. 2016).

There are several individual studies reporting advantages to undergoing bariatric surgery prior to TJA. In-hospital complications after hip and knee arthroplasty, and 90-day troubles after knee arthroplasty, diminish in patients who underwent prior bariatric surgery (McLawhorn et al. 2018a, 2018b). A large study by Werner et al using a Medicare database with 78,036 patients compared the 90-day complication rate in three groups: (1) patients without obesity, (2) patients with morbid obesity who did not undergo bariatric surgery, and (3) patients with morbid obesity who underwent bariatric surgery prior to TKA. They found a decreased rate of both major and minor complications in patients with morbid obesity who underwent bariatric surgery when compared to patients with morbid obesity who did not undergo bariatric surgery prior to TKA (Werner et al. 2015). Watts et al compared the outcome of THA in patients with morbid obesity who did and did not previously undergo bariatric surgery in a matched cohort study. They found a decrease in reoperations and revisions for patients who underwent bariatric surgery prior to THA (Watts et al. 2016b). Kulkarni et al found that patients with obesity who underwent bariatric surgery prior to TJA had 3.5 times lower likelihood of wound infection and 7 times lower likelihood of hospital readmission compared to those who underwent TJA alone (Kulkarni et al. 2011). Nearing et al found operative time and length of hospital stay were decreased for patients who had arthroplasty performed after bariatric surgery. Early complications and late reinterventions were similar (Nearing et al. 2017).

While the decrease in joint loading and improvement in gait mechanics account for a portion of the improvement in joint pain and osteoarthritic progression, the changes in the hormonal milieu secondary to bariatric surgery are also likely to contribute. There have been studies investigating the changes in leptin secondary to bariatric surgery. Chen et al. found that levels of several adipocytokines and related factors decreased over the course of the first year after bariatric surgery. Even more interestingly, leptin reduction and stabilization corresponded closely with the initial reduction and stabilization in osteoarthritic knee pain (Chen et al. 2018). This suggests that a decrease in leptin, initiated by bariatric surgery, is directly associated with an improvement in osteoarthritis symptoms and potentially joint damage.

There are also a few studies that suggested worse postoperative TJA outcomes in patients who had prior bariatric surgery. A New York statewide cooperative study over a 10-year period revealed that 90-day postoperative complications after total joint arthroplasty were increased in obese patients who previously underwent bariatric operations. Immediate hospital costs were also more substantial (Liu et al. 2019). In a previous article with the same global database, they found that bariatric surgery was not a risk factor for nonelective readmissions at 30-days, 90-days, or 1 year and did not change overall cost. Nevertheless, it did predict an increase in elective admission for up to 1 year (Liu et al. 2018).

Within a similar perspective, a controlled investigation using the Medicare database unveiled about twice the risk of dislocation, as well as revision surgery, in the bariatric population compared to no weight-reducing operation in obese or lean controls. However, the bariatric patients in this study were significantly more comorbid than the non-bariatric population, calling to question the real reason for these differences in outcome (Nickel et al. 2018).

Lee et al. used Medicare 5% part B data from 1999–2012 to identify patients who underwent THA or TKA. A history of prior bariatric surgery as well as the presence of comorbid conditions such as diabetes, hypothyroidism, impaired renal function, osteomalacia, and Cushing's syndrome was identified for all patients. Kaplan-Meier risk of revision of THA or TKA within 0.5, 1, 2, and 5 years was evaluated for each condition, including the presence of prior bariatric surgery. Bariatric surgery prior to THA was not associated with an increased overall risk for revision, but was associated with an increased risk for revision for periprosthetic infection. Patients undergoing TKA following bariatric surgery were at increased overall risk for revision, but not at increased risk for revision for periprosthetic infection (Lee et al. 2018).

Bariatric Surgery and Total Arthroplasty: Which Comes First, the Chicken or the Egg?

While it is possible that immobility secondary to osteoarthritis contributes to the development of obesity, evidence does not suggest that TJA will lead to postoperative weight loss. Despite increased mobility after TJA, most patients maintain and many actually gain weight postoperatively (Springer et al. 2017). While in one study, postoperative weight loss was most strongly associated with preoperative BMI (Inacio et al. 2014), this may just indicate better results of TJA when concurrent metabolic syndrome is earlier in its course. TJA is not suggested as a treatment for significant weight loss.

The timing of bariatric surgery relative to TJA may have an effect on TJA outcomes in patients after bariatric surgery.

Schwarzkopf et al. used the Healthcare Cost and Utilization Project California State Inpatient Database to identify patients who underwent TJA following bariatric surgery from 2007–2011. There were 330 patients who underwent bariatric surgery followed by THA and 1017 patients who underwent bariatric surgery followed by TKA with 19% of patients having THA within 6 months of their bariatric procedure and 10% of patients having TKA within 6 months of their bariatric procedure. There was no association found between time of bariatric surgery and THA or TKA and 90-day complications in multivariate logistic regression analysis. However, patients undergoing THA more than 6 months after bariatric surgery were significantly less likely to have a 90-day readmission compared to patients undergoing THA within 6 months of bariatric surgery (Schwarzkopf et al. 2018).

Severson et al compared the outcomes of three groups of patients: (1) patients with TKA prior to bariatric surgery (n = 39) (2) patients who underwent TKA less than 2 years after bariatric surgery (n = 25) (3) patients who underwent TKA more than 2 years after bariatric surgery (n = 61). They found patients in group 3 had shorter anesthesia time, total operative time, and tourniquet time while 90-day complication rates and duration of hospital stay did not differ among groups (Severson et al. 2012).

More study is needed in this to determine the optimal timing between bariatric surgery and TJA.

Mobility and Access to Care

There are potential advantages to performing bariatric surgery prior to TJA other than postoperative outcomes, such as potential increased access to care and decreased costs. Many rural facilities will refer patients with obesity to tertiary medical centers to undergo TJA. This requires increased transportation costs and difficulty with follow-up. If a patient with osteoarthritis was able to lose weight with bariatric surgery, the patient may be able to undergo TJA at an institution closer to home, improving access to care and eliminating some of the complexities of follow-up (Springer et al. 2017).

Financial Issues

McLawhorn et al. compared the cost of performing bariatric surgery prior to TKA in patients with morbid obesity and osteoarthritis to TKA alone in BMI-matched patients. Costs included costs of treatment, complications, and 90-day follow-up. Quality-adjusted-life-years (QALY) were also calculated for both groups. While performing bariatric surgery prior to TKA was more expensive than TKA alone, the QALY was also higher. Given this, the incremental cost-effectiveness ratio between the two groups was found to be $13,910 per QALY. This value was significantly less than the $100,000 per QALY used as the threshold willingness to pay, leading to the conclusion that performing bariatric surgery prior to TKA is cost-effective (McLawhorn et al. 2016). Kremers et al. found that, for every 5-unit increase in BMI over 30 kg/m^2, hospitalization cost increased $300 for TKA and $650 for revisional TKA, even when adjusting for comorbidities and complications (Kremers et al. 2014).

Culler et al. used the Medicare Provider Analysis Review file to show that any adverse event for TJA increased mean hospital cost by $3429 and increased hospital length of stay by 1 day. Given these findings, in the setting of bundled payment for TJA, these authors suggested undertaking all possible complication-reducing measures to decrease the risk of complications and therefore reduce hospital cost (Culler et al. 2016; Kremers et al. 2014).

Conversely, Giori et al. demonstrated that for patients with BMI >40 kg/m^2 who undergo TJA, there is only one complication for every 14 complication-free operations. These authors suggested not using a strict BMI cutoff for patients who otherwise qualify for TJA. Similarly, Springer et al. followed 289 patients with BMI >40 kg/m^2 and end-stage osteoarthritis for 2 years. They found that requiring weight loss prior to TJA in this population most often leads to these patients remaining morbidly obese and never undergoing TJA. Only 20% of patients underwent bariatric surgery. These authors suggested increasing the resources and coordinated care to facilitate patients' weight loss (Giori et al. 2018).

Ongoing Trials and Future Perspectives

There is currently no consensus on the utility of bariatric surgery for weight loss prior to TJA, and no prospective or randomized controlled trials on the topic. In addition, no prospective studies have evaluated whether reduced biomechanical strain on lower extremity joints after bariatric surgery eliminates the need for TJA in certain patients, or if, conversely, the increased mobility and exercise capability of patients who lose significant weight from bariatric surgery lead to an increase in TJA in this population.

The Surgical Weight-Loss to Improve Functional Status Trajectories Following TKA (SWIFT) Trial (NCT02598531) is the first prospective, controlled, multicenter trial comparing postoperative outcomes in patients with severe obesity who undergo TKA with a well-matched cohort of patients who undergo weight loss surgery prior to elective TKA. This trial is currently recruiting patients (Surgical Weight-Loss to Improve Functional Status Trajectories Following Total Knee Arthroplasty/SWIFT Trial n.d.) (Benotti et al. 2018).

Future Directions

Important areas for study include: comparing the effects of surgical weight loss and medical weight loss on TJA necessity and outcome, comparing the effect of different surgical weight loss procedures on TJA necessity and outcome, investigating risk stratification protocols for patients with obesity and end-stage osteoarthritis, determining a safe and cost-effective BMI cutoff for TJA, and ultimately determining an appropriate weight management algorithm and multidisciplinary approach for obese patients with end-stage osteoarthritis. In addition to studies investigating the effect of bariatric surgery on osteoarthritis and TJA outcomes, further investigation into the mechanisms that regulate peripheral and central adipokine activity and their contribution to the development of osteoarthritis may lead to novel treatments for osteoarthritis in the morbidly obese population.

References

Abdel MP, Bonadurer GF 3rd, Jennings MT, Hanssen AD (2015) Increased aseptic tibial failures in patients with a BMI >/=35 and well-aligned total knee arthroplasties. J Arthroplast 30(12):2181–2184. https://doi.org/10.1016/j.arth.2015.06.057

Barrett M, Prasad A, Boyce L, Dawson-Bowling S, Achan P, Millington S, Hanna SA (2018) Total hip arthroplasty outcomes in morbidly obese patients: a systematic review. EFORT Open Rev 3(9):507–512. https://doi.org/10.1302/2058-5241.3.180011

Belmont PJ Jr, Goodman GP, Waterman BR, Bader JO, Schoenfeld AJ (2014) Thirty-day postoperative complications and mortality following total knee arthroplasty: incidence and risk factors among a national sample of 15,321 patients. J Bone Joint Surg Am 96(1):20–26. https://doi.org/10.2106/JBJS.M.00018

Benotti PN, Still CD, Craig Wood G, Seiler JL, Seiler CJ, Thomas SP, Petrick AT, Suk M, Irving BA, SWIFT Trial Investigators (2018) Surgical weight-loss to improve functional status trajectories following total knee arthroplasty: SWIFT trial: rationale, design, and methods. Contemp Clin Trials 69:1–9

Bergmann G, Deuretzbacher G, Heller M, Graichen F, Rohlmann A, Strauss J, Duda GN (2001) Hip contact forces and gait patterns from routine activities. J Biomech 34(7):859–871. https://doi.org/10.1016/s0021-9290(01)00040-9

Centers for Disease Control and Prevention (2018) Arthritis-related statistics. U.S. Department of Health & Human Services. https://www.cdc.gov/arthritis/data_statistics/arthritis-related-stats.htm. Accessed December 20 2019

Chee YH, Teoh KH, Sabnis BM, Ballantyne JA, Brenkel IJ (2010) Total hip replacement in morbidly obese patients with osteoarthritis: results of a prospectively matched study. J Bone Joint Surg Br 92(8):1066–1071. https://doi.org/10.1302/0301-620X.92B8.22764

Chen SX, Bomfim FA, Youn HA, Ren-Fielding C, Samuels J (2018) Predictors of the effect of bariatric surgery on knee osteoarthritis pain. Semin Arthritis Rheum 48(2):162–167. https://doi.org/10.1016/j.semarthrit.2018.02.001

Chu IJH, Lim AYT, Ng CLW (2018) Effects of meaningful weight loss beyond symptomatic relief in adults with knee osteoarthritis and obesity: a systematic review and meta-analysis. Obes Rev 19(11):1597–1607. https://doi.org/10.1111/obr.12726

Culler SD, Jevsevar DS, Shea KG, McGuire KJ, Wright KK, Simon AW (2016) The incremental hospital cost and length-of-stay associated with treating adverse events among medicare beneficiaries undergoing THA during fiscal year 2013. J Arthroplast 31(1):42–48. https://doi.org/10.1016/j.arth.2015.07.037

Dumond H, Presle N, Terlain B, Mainard D, Loeuille D, Netter P, Pottie P (2003) Evidence for a key role of leptin in osteoarthritis. Arthritis Rheum 48(11):3118–3129. https://doi.org/10.1002/art.11303

Fehring TK, Odum SM, Griffin WL, Mason JB, McCoy TH (2007) The obesity epidemic: its effect on total joint arthroplasty. J Arthroplast 22(6 Suppl 2):71–76. https://doi.org/10.1016/j.arth.2007.04.014

Foran JR, Mont MA, Rajadhyaksha AD, Jones LC, Etienne G, Hungerford DS (2004) Total knee arthroplasty in obese patients: a comparison with a matched control group. J Arthroplasty 19(7):817–824. https://doi.org/10.1016/j.arth.2004.03.017

Francisco V, Perez T, Pino J, Lopez V, Franco E, Alonso A, Gonzalez-Gay MA, Mera A, Lago F, Gomez R, Gualillo O (2018a) Biomechanics, obesity, and osteoarthritis. The role of adipokines: when the levee breaks. J Orthop Res 36(2):594–604. https://doi.org/10.1002/jor.23788

Francisco V, Pino J, Campos-Cabaleiro V, Ruiz-Fernandez C, Mera A, Gonzalez-Gay MA, Gomez R, Gualillo O (2018b) Obesity, fat mass and immune system: role for leptin. Front Physiol 9:640. https://doi.org/10.3389/fphys.2018.00640

Galea VP, Rojanasopondist P, Ingelsrud LH, Rubash HE, Bragdon C, Huddleston Iii JI, Malchau H, Troelsen A (2019) Longitudinal changes in patient-reported outcome measures following total hip arthroplasty and predictors of deterioration during follow-up: a seven-year prospective international multicentre study. Bone Joint J 101-B(7):768–778

GBD 2016 Disease and Injury Incidence and Prevalence Collaborators Global, regional, and national incidence, prevalence, and years lived with disability for 328 diseases and injuries for 195 countries, 1990–2016: a systematic analysis for the Global Burden of Disease Study 2016. Lancet. 2017;390 (10100):1211–1259

Giori NJ, Amanatullah DF, Gupta S, Bowe T, Harris AHS (2018) Risk reduction compared with access to care: quantifying the trade-off of enforcing a body mass index eligibility criterion for joint replacement. J Bone Joint Surg Am 100(7):539–545. https://doi.org/10.2106/JBJS.17.00120

Griffin TM, Huebner JL, Kraus VB, Guilak F (2009) Extreme obesity due to impaired leptin signaling in mice does not cause knee osteoarthritis. Arthritis Rheum 60(10):2935–2944. https://doi.org/10.1002/art.24854

Groen VA, van de Graaf VA, Scholtes VA, Sprague S, van Wagensveld BA, Poolman RW (2015) Effects of bariatric surgery for knee complaints in (morbidly) obese adult patients: a systematic review. Obes Rev 16 (2):161–170. https://doi.org/10.1111/obr.12236

Gu A, Cohen JS, Malahias MA, Lee D, Sculco PK, McLawhorn AS (2019) The effect of bariatric surgery prior to lower-extremity total joint arthroplasty: a systematic review. HSS J 15(2):190–200. https://doi.org/10.1007/s11420-019-09674-2

Hales CM, Carroll MD, Fryar CD, Ogden CL (2017) Prevalence of obesity among adults and youth: United States, 2015–2016. NCHS Data Brief, vol 288. National Center for Health Statistics, Hyattsville, MD

Heywood S, McClelland J, Geigle P, Rahmann A, Villalta E, Mentiplay B, Clark R (2019) Force during functional exercises on land and in water in older adults with and without knee osteoarthritis: Implications for rehabilitation. Knee 26(1):61–72

Hootman JM, Helmick CG, Barbour KE, Theis KA, Boring MA (2016) Updated projected prevalence of self-reported doctor-diagnosed arthritis and arthritis-attributable activity limitation among US adults, 2015–2040. Arthritis Rheumatol 68(7):1582–1587. https://doi.org/10.1002/art.39692

Inacio MC, Silverstein DK, Raman R, Macera CA, Nichols JF, Shaffer RA, Fithian D (2014) Weight patterns before and after total joint arthroplasty and characteristics associated with weight change. Perm J 18(1):25–31. https://doi.org/10.7812/TPP/13-082

Jamsen E, Nevalainen P, Eskelinen A, Huotari K, Kalliovalkama J, Moilanen T (2012) Obesity, diabetes, and preoperative hyperglycemia as predictors of periprosthetic joint infection: a single-center analysis of 7181 primary hip and knee replacements for osteoarthritis. J Bone Joint Surg Am 94(14):e101. https://doi.org/10.2106/JBJS.J.01935

Jarvenpaa J, Kettunen J, Kroger H, Miettinen H (2010) Obesity may impair the early outcome of total knee arthroplasty. Scand J Surg 99(1):45–49. https://doi.org/10.1177/145749691009900110

Jarvholm B, From C, Lewold S, Malchau H, Vingard E (2008) Incidence of surgically treated osteoarthritis in the hip and knee in male construction workers. Occup Environ Med 65(4):275–278. https://doi.org/10.1136/oem.2007.033365

Karlson EW, Mandl LA, Aweh GN, Sangha O, Liang MH, Grodstein F (2003) Total hip replacement due to osteoarthritis: the importance of age, obesity, and other modifiable risk factors. Am J Med 114(2):93–98. https://doi.org/10.1016/s0002-9343(02)01447-x

Karvonen-Gutierrez CA, Sowers MR, Heeringa SG (2012) Sex dimorphism in the association of cardiometabolic characteristics and osteophytes-defined radiographic knee osteoarthritis among obese and non-obese adults: NHANES III. Osteoarthr Cartil 20(7):614–621. https://doi.org/10.1016/j.joca.2012.02.644

Kerkhoffs GM, Servien E, Dunn W, Dahm D, Bramer JA, Haverkamp D (2012) The influence of obesity on the complication rate and outcome of total knee arthroplasty: a meta-analysis and systematic literature review. J Bone Joint Surg Am 94(20):1839–1844. https://doi.org/10.2106/JBJS.K.00820

Koonce RC, Bravman JT (2013) Obesity and osteoarthritis: more than just wear and tear. J Am Acad Orthop Surg 21(3):161–169. https://doi.org/10.5435/JAAOS-21-03-161

Kremers HM, Visscher SL, Kremers WK, Naessens JM, Lewallen DG (2014) The effect of obesity on direct medical costs in total knee arthroplasty. J Bone Joint Surg Am 96(9):718–724. https://doi.org/10.2106/JBJS.M.00819

Lee GC, Ong K, Baykal D, Lau E, Malkani AL (2018) Does prior bariatric surgery affect implant survivorship and complications following primary total hip arthroplasty/total knee arthroplasty? J Arthroplast 33 (7):2070–2074 e2071. https://doi.org/10.1016/j.arth.2018.01.064

Li S, Luo X, Sun H, Wang K, Zhang K, Sun X (2019) Does prior bariatric surgery improve outcomes following total joint arthroplasty in the morbidly obese? A Meta-Anal J Arthropl 34(3):577–585

Liu JX, Paoli AR, Mahure SA, Bosco J 3rd, Campbell KA (2018) Preoperative bariatric surgery and the risk of readmission following total joint replacement. Orthopedics 41(2):107–114. https://doi.org/10.3928/01477447-20180226-02

Liu JX, Paoli AR, Mahure SA, Bosco J 3rd, Campbell KA (2019). Preoperative bariatric surgery utilization is associated with increased 90-day postoperative complication rates after total joint arthroplasty. J Am Acad Orthop Surg 2019 Sep 16

Long H, Xie D, Zeng C, Wei J, Wang Y, Yang T, Xu B, Qian Y, Li J, Wu Z, Lei G (2019) Association between body composition and osteoarthritis: a systematic review and meta-analysis. Int J Rheum Dis 22(12):2108–2118. https://doi.org/10.1111/1756-185X.13719

Maly R, Costigan PA, Olney SJ (2005) Contribution of psychosocial and mechanical variables to physical performance measures in knee osteoarthritis. Phys Ther 85 (12):1318–1328

McLawhorn AS, Southren D, Wang YC, Marx RG, Dodwell ER (2016) Cost-effectiveness of bariatric surgery prior to total knee arthroplasty in the morbidly obese: a computer model-based evaluation. J Bone Joint Surg Am 98(2):e6. https://doi.org/10.2106/JBJS.N.00416

McLawhorn AS, Levack AE, Lee YY, Ge Y, Do H, Dodwell ER (2018a) Bariatric surgery improves outcomes after lower extremity arthroplasty in the morbidly obese: a propensity score-matched analysis of a new york statewide database. J Arthroplast. 33 (7):2062–2069.e

McLawhorn AS, Levack AE, Lee YY, Ge Y, Do H, Dodwell ER (2018b) Bariatric surgery improves outcomes after lower extremity arthroplasty in the morbidly obese: a propensity score-matched analysis of a New York Statewide Database. J Arthroplast 33 (7):2062–2069. e2064. https://doi.org/10.1016/j.arth.2017.11.056

Moon Han S, Kim WW, Oh JH (2005) Results of laparoscopic sleeve gastrectomy (LSG) at 1 year in morbidly obese Korean patients. Obes Surg 15 (10):1469–1475. https://doi.org/10.1381/096089205774859227

Mundermann A, Dyrby CO, Andriacchi TP (2005) Secondary gait changes in patients with medial compartment knee osteoarthritis: increased load at the ankle, knee, and hip during walking. Arthritis Rheum 52 (9):2835–2844. https://doi.org/10.1002/art.21262

Nelson LG, Lopez PP, Haines K, Stefan B, Martin T, Gonzalez R, Byers P, Murr MM (2006) Outcomes of bariatric surgery in patients > or =65 years. Surg Obes Relat Dis 2(3):384–388. https://doi.org/10.1016/j.soard.2006.02.012

Nickel BT, Klement MR, Penrose CT, Green CL, Seyler TM, Bolognesi MP (2016) Lingering risk: bariatric surgery before total knee arthroplasty. J Arthroplasty 31(9 Suppl):207–211. https://doi.org/10.1016/j.arth.2016.02.075

Nunez M, Lozano L, Nunez E, Sastre S, Del Val JL, Suso S (2011) Good quality of life in severely obese total knee replacement patients: a case-control study. Obes Surg 21(8):1203–1208. https://doi.org/10.1007/s11695-010-0197-9

Ogden CL, Carroll MD, Flegal KM (2014) Prevalence of obesity in the United States. JAMA 312(2):189–190. https://doi.org/10.1001/jama.2014.6228

Ponta ML, Gozza M, Giacinto J, Gradaschi R, Adami GF (2014) Effects of obesity on posture and walking: study prior to and following surgically induced weight loss. Obes Surg 24(11):1915–1920. https://doi.org/10.1007/s11695-014-1254-6

Pozzobon D, Ferreira PH, Blyth FM, Machado GC, Ferreira ML (2018) Can obesity and physical activity predict outcomes of elective knee or hip surgery due to osteoarthritis? A meta-analysis of cohort studies. BMJ Open 8(2):e017689. https://doi.org/10.1136/bmjopen-2017-017689

Reyes C, Leyland KM, Peat G, Cooper C, Arden NK, Prieto-Alhambra D (2016) Association between overweight and obesity and risk of clinically diagnosed knee, hip, and hand osteoarthritis: a population-based cohort study. Arthritis Rheumatol 68(8):1869–1875

Runhaar J, Koes BW, Clockaerts S, Bierma-Zeinstra SM (2011) A systematic review on changed biomechanics of lower extremities in obese individuals: a possible role in development of osteoarthritis. Obes Rev 12 (12):1071–1082. https://doi.org/10.1111/j.1467-789X.2011.00916.x

Salih S, Sutton P (2013) Obesity, knee osteoarthritis and knee arthroplasty: a review. BMC Sports Sci Med Rehabil 5(1):25. https://doi.org/10.1186/2052-1847-5-25

Schwarzkopf R, Lavery JA, Hooper J, Parikh M, Gold HT (2018) Bariatric surgery and time to total joint arthroplasty: does it affect readmission and complication rates? Obes Surg 28(5):1395–1401. https://doi.org/10.1007/s11695-017-3034-6

Sharples AJ, Mahawar K (2019) Systematic review and meta-analysis of randomised controlled trials comparing long-term outcomes of Roux-En-Y gastric bypass and sleeve gastrectomy. Obes Surg Nov 13

Simopoulou T, Malizos KN, Iliopoulos D, Stefanou N, Papatheodorou L, Ioannou M, Tsezou A (2007) Differential expression of leptin and leptin's receptor isoform (Ob-Rb) mRNA between advanced and minimally affected osteoarthritic cartilage; effect on cartilage metabolism. Osteoarthr Cartil 15 (8):872–883. https://doi.org/10.1016/j.joca.2007.01.018

Smith TO, Aboelmagd T, Hing CB, MacGregor A (2016) Does bariatric surgery prior to total hip or knee arthroplasty reduce post-operative complications and improve clinical outcomes for obese patients? Systematic review and meta-analysis. Bone Joint J 98-B (9):1160–1166. https://doi.org/10.1302/0301-620X.98B9.38024

Sowers MR, Karvonen-Gutierrez CA (2010) The evolving role of obesity in knee osteoarthritis. Curr Opin Rheumatol 22(5):533–537. https://doi.org/10.1097/BOR.0b013e32833b4682

Springer BD, Carter JT, McLawhorn AS, Scharf K, Roslin M, Kallies KJ, Morton JM, Kothari SN (2017) Obesity and the role of bariatric surgery in the surgical management of osteoarthritis of the hip and knee: a review of the literature. Surg Obes Relat Dis 13 (1):111–118. https://doi.org/10.1016/j.soard.2016.09.011

Stavrakis AI, Khoshbin A, McLawhorn AS, Parks ML (2018) Bariatric surgery prior to total joint arthroplasty, does it decrease the risk of obesity related perioperative complications? Curr Rheumatol Rep 20(2):7. https://doi.org/10.1007/s11926-018-0717-5

Sturm R, Hattori A (2013) Morbid obesity rates continue to rise rapidly in the United States. Int J Obes 37 (6):889–891. https://doi.org/10.1038/ijo.2012.159

Surgical Weight-Loss to Improve Functional Status Trajectories Following Total Knee Arthroplasty (SWIFT Trial) (n.d.). https://clinicaltrials.gov/ct2/show/NCT02598531. Accessed December 25 2019

Taylor WR, Heller MO, Bergmann G, Duda GN (2004) Tibio-femoral loading during human gait and stair climbing. J Orthop Res 22(3):625–632. https://doi.org/10.1016/j.orthres.2003.09.003

Teichtahl AJ, Wluka AE, Wang Y, Hanna F, English DR, Giles GG, Cicuttini FM (2009) Obesity and adiposity are associated with the rate of patella cartilage volume loss over 2 years in adults without knee osteoarthritis. Ann Rheum Dis 68(6):909–913. https://doi.org/10.1136/ard.2008.093310

Tohidi M, Brogly SB, Lajkosz K, Harrison MM, Campbell AR, VanDenKerkhof E, Mann SM (2019) Ten-year risk of complication and mortality after total hip arthroplasty in morbidly obese patients: a population study. Can J Surg 62(6):442–449

Vartiainen P, Bragge T, Lyytinen T, Hakkarainen M, Karjalainen PA, Arokoski JP (2012) Kinematic and kinetic changes in obese gait in bariatric surgery-induced weight loss. J Biomech 45(10):1769–1774. https://doi.org/10.1016/j.jbiomech.2012.05.002

Vincent HK, Ben-David K, Conrad BP, Lamb KM, Seay AN, Vincent KR (2012) Rapid changes in gait, musculoskeletal pain, and quality of life after bariatric surgery. Surg Obes Relat Dis 8(3):346–354. https://doi.org/10.1016/j.soard.2011.11.020

Wang H, Bai J, He B, Hu X, Liu D (2016) Osteoarthritis and the risk of cardiovascular disease: a meta-analysis of observational studies. Sci Rep 6:39672

Watts CD, Houdek MT, Wagner ER, Lewallen DG, Mabry TM (2016) Morbidly obese vs nonobese aseptic revision total hip arthroplasty: surprisingly similar outcomes. J Arthroplast 31(4):842–845. https://doi.org/10.1016/j.arth.2015.08.036

Workgroup of the American Association of Hip and Knee Surgeons Evidence Based Committee (2013) Obesity and total joint arthroplasty: a literature based review. J Arthroplast 28(5):714–721. https://doi.org/10.1016/j.arth.2013.02.011

Yokota RT, Van der Heyden J, Demarest S, Tafforeau J, Nusselder WJ, Deboosere P et al (2015) Contribution of chronic diseases to the mild and severe disability burden in Belgium. Arch Public Health 73(1):37

Zusmanovich M, Kester BS, Schwarzkopf R (2018) Postoperative complications of total joint arthroplasty in obese patients stratified by BMI. J Arthroplast 33 (3):856–864

Relapse of Diabetes After Metabolic/ Bariatric Surgery

61

Lívia Porto Cunha da Silveira, Tarissa Zanata Petry, and Ricardo Vitor Cohen

Contents

Introduction 827

Pathogenesis of T2D 828

How to Predict Metabolic/Bariatric Surgery Outcomes 828

Predictive Scores 829

Revisional Bariatric and Metabolic Surgery 830

References 832

Abstract

The use of nonclinical but also laboratory tools to predict type 2 diabetes (T2D) remission after metabolic/bariatric surgery has been extensively researched. Basal fasting C-peptide level is an accessible test, but it cannot predict the chances of remission alone and is biased in patients with more severe diabetes. So, other robust biological and clinical predictors of T2D remission after surgery have been proposed. After assessing the predictive factors and acknowledging the risk of non-remission and relapse, a pathway to adequate clinical management is necessary, particularly for cases with recurrent T2D. Revisional surgery has been widely discussed as an option for those cases, despite the lack of robust evidence. Interestingly, inadequate weight loss or weight regain that usually indicates the reoperation is not always linked with T2D recurrence. Pharmacological agents such as GLP1 analogs may be an add-on alternative to reach glycemic control, if revisional surgery is insufficient.

Keywords

Relapsing diabetes · Recurrent diabetes · Predictive scores · Sleeve gastrectomy · Diabetes remission · Incretins

Introduction

Type 2 Diabetes mellitus (T2D) is a chronic and progressive disease characterized by

L. P. C. da Silveira · T. Z. Petry
Endocrinologist, Center for Obesity and Diabetes, Oswaldo Cruz German Hospital, São Paulo, SP, Brazil

R. Vitor Cohen (✉)
Center for Obesity and Diabetes, Oswaldo Cruz German Hospital, São Paulo, SP, Brazil
e-mail: ricardo.cohen@haoc.com.br

J. Faintuch and S. Faintuch (eds.), *Obesity and Diabetes*, https://doi.org/10.1007/978-3-030-53370-0_61

hyperglycemia, insulin resistance, and relative impairment of insulin secretion. It is strongly associated with obesity in all ethnic groups (Mokdad et al. 2001).

The prevalence of T2D is increasing worldwide. The relationship between this disease and obesity is well-established as 80% of T2D patients are overweight or obese (Schauer et al. 2012).

As a multifactorial disease, understanding its pathogenesis and its treatment is challenging. Although insulin resistance can be traced back to its genetic background, the epidemic of diabetes is related to the epidemic of obesity and physical inactivity as both are insulin-resistant states (Sha and Laferrère 2017).

Pathogenesis of T2D

It is currently known that eight factors play a role in the development of glucose intolerance in T2D individuals (DeFronzo 2009). Genetically, individuals prone to develop T2D evolve to insulin resistance that presents itself in multiple organs. In the liver, it leads to an overproduction of glucose, despite fasting hyperinsulinemia and by reduced suppression of hepatic glucose production in response to insulin. In the muscle, it decreases glucose uptake after carbohydrate ingestion resulting in postprandial hyperglycemia. Thus, decreased insulin sensitivity stimulates insulin production by the pancreatic ß-cells. Over time, the ß-cells become fatigued, losing function progressively and reducing circulating insulin levels, consequently increasing plasma glucose concentration leading to the onset of diabetes. In addition to the muscle, liver, and ß-cell, several other cellular dysfunctions are part of T2D pathogenesis, with the involvemet of fat cells (accelerated lipolysis), gastrointestinal tract (incretin deficiency/resistance), alfa cells (hyperglucagonemia), kidney (increased glucose reabsorption), and brain (insulin resistance).

As complex as the pathogenesis, the treatment of this disease is also challenging. Randomized clinical trials have indicated that metabolic/bariatric surgery achieves rapid and better glycemic control than medical therapy alone in obese patients with T2D. Accordingly, metabolic/bariatric surgery is recommended in the treatment algorithm of T2D in patients with a body mass index of 30 kg/m^2 and over and endorsed by over 50 worldwide medical and scientific associations (Rubino et al. 2017).

How to Predict Metabolic/Bariatric Surgery Outcomes

As the goals of metabolic/bariatric surgery are the control of T2D and related metabolic conditions, high remission rates are expected. Due to the use of different remission criteria, rates are very variable in the literature. The rate of T2D remission varies from 24% to 95% at 2 years, depending on the type of surgery, the definition of remission, and the characteristics of patients enrolled in the studies (Yan et al. 2017; Maleckas et al. 2015).

The main goal of metabolic/bariatric surgery, as described above, is to keep the American Diabetes Association (ADA) triple endpoint under control (Table 61.1) (Standards of Medical Care in Diabetes 2020). Therefore, as T2D is a progressive disease, the ADA are targeted with the correct association of surgery and the best medical treatment available.

Several studies tried to identify the best protocols to manage these patients peri and postoperatively (Sjostrom 2013). The main objectives

Table 61.1 Summary of the American Diabetes Association for glycemic, blood pressure, and lipid control for adults with diabetes (Standards of Medical Care in Diabetes 2020)

A1c	<7%
Blood Pressure	<130/80 mmHg
Lipids	
LDL cholesterol	<100 mg/dl

were identifying responders and the development of preoperative scores that could shed a light on achieving the best outcomes.

Different trials reported some clinical issues that could predict T2D recurrence after surgery. One study showed that one third of severely obese diabetes adults who underwent gastric bypass surgery experience relapse after diabetes remission in five years (Schauer et al. 2014).

A retrospective cohort study that included 345 patients with obesity and type T2D, submitted to Roux-en-Y gastric bypass (RYGB), with a minimum 3-year follow-up since surgery, assessed the factors associated with T2D remission and relapse. The predictors of remission were younger age, preoperative glycemic control, and duration of diabetes at baseline. Of the patients in remission, 12% experienced T2D recurrence. Predictors of recurrence were preoperative use of insulin, or any anti-diabetes agents other than metformin (Mingrone et al. 2012; Arterburn et al. 2013a).

Predictive Scores

Despite some limitations, a model of score based on clinical variables for the preoperative prediction of T2D remission following RYGB surgery was proposed (DiaRem). In a retrospective cohort study of 2300 patients who underwent RYGB surgery, predictive factors of remission were evaluated to produce a score that predicts the probability of diabetes remission within 5 years. This score uses as prediction factors age, Hba1c, use of diabetes medications beyond metformin, and use of insulin (Oliveira et al. 2018; Dixon et al. 2013).

Young age and low BMI (25–35 kg/m^2) are also predictors of long-term T2D remission, while the use of insulin and high glycated hemoglobin A1c (HbA1c) levels are predictors of decreased rates of remission after surgery. Glycemic response to RYGB has also been correlated with BMI, duration of diabetes, fasting C-peptide, and weight loss (Brethauer et al. 2013; Arterburn et al. 2013b).

DiaRem is a well-validated tool with easy clinical use (Still et al. 2014). However, one of the drawbacks related to this score is the fact that it does not use the duration of diabetes to calculate the prediction of disease remission after surgery. The DiaRem 2, updated its first version. It added the time of history of T2D and changed the remission groups from 5 to 3 categories (low, intermediate, and high) which allowed the score to improve its accuracy (Still et al. 2019; Pucci et al. 2018).

A prospective single center, nonblinded, randomized controlled trial including 45 obese patients with T2D showed the potential use of baseline circulating succinate to predict T2D remission after metabolic/bariatric surgery. Succinate baseline concentrations were an independent predictor of T2D remission (Ceperuelo-Mallafre et al. 2019). Succinate is a metabolite that signals metabolic stress and inflammation. Microbiota-produced succinate has also been related to intestinal glucose metabolism and metabolic activity of brown adipose tissue. This study reported that patients who achieved remission after 1 year had lower levels of baseline succinate.

Nevertheless, more studies are needed to incorporate the use of succinate as a tool to help identifying remission chances. Moreover, the need to standardize this test in daily practice is also required (Ceperuelo-Mallafre et al. 2019).

A pilot study to access the genetic predisposition risk scores in T2D in order to predict the better response to metabolic/bariatric surgery in terms of either weight loss or diabetes remission showed highly sensitive and specific genetic predictive scores of responses to surgery in terms of weight loss and T2D remission and the long-term sustainability of these effects (Ciudin et al. 2019).

Further studies are needed to justify the applicability of genetic analysis as a predictor of metabolic response. However, the available current studies mentioned above have already given some clues towards precision medicine that can help clinicians to select even better candidates for metabolic/ bariatric surgery.

Revisional Bariatric and Metabolic Surgery

Metabolic surgery is the most effective treatment for medically uncontrolled type 2 diabetes (T2D), leading to a meaningful improvement in glycemic control as good as 70% of diabetes remission early after surgery (Panunzi et al. 2015). However, T2D recurrence rate five years after surgery is not low (Mingrone et al. 2015). The main apparent reason for those outcomes is that almost all metabolic surgery trials have been designed using T2D remission without antidiabetic medications as an endpoint (Gloy et al. 2013).

As T2D is a chronic disease, consequently it is not reasonable to have medication withdrawal as a treatment target. On the contrary, the main goal is the reduction of micro and macrovascular complications and mortality, regardless of medication use (Cohen et al. 2017).

Although there is some data supporting revisional surgery for insufficient weight loss or weight regain, the literature is scarce, with weaker levels of evidence about the benefits of revisional metabolic surgery. Because of potentially higher complication rates, and modest weight loss after bariatric surgical revisions (Mor et al. 2013; Sudan et al. 2014), a careful evaluation of its necessity is mandatory, and many times, the treatment with new antidiabetic medications, as glucagon peptide 1 (GLP-1) analogues, can be better options. Gravitas, a randomized controlled trial comparing liraglutide versus placebo in 80 patients with persistent or recurrent type 2 diabetes at least one year after surgery, demonstrated a decrease of 1.22% in HbA1c and 4.23 kg after 26 weeks of treatment (Miras et al. 2019).

Any given surgical technique has distinct mechanisms of action. It was demonstrated that rerouting the proximal bowel leads to a more powerful metabolic improvement after surgery. Therefore, depending on the index surgery and the choice of the second operative technique, it is expected that the metabolic outcomes potentially will change (Brethauer et al. 2013).

In 2017, a review (Yan et al. 2017) summarized all the evidences published to determine if there is metabolic improvement after revisional surgery. The authors demonstrated that converting restrictive procedures as the vertical banded gastroplasty (VBG) or adjustable gastric banding (AGB) to Roux-en-Y gastric bypass (RYGB) leads to T2D improvement or remission in 79% and 72%, as well as a decrease of 15–25% and 13–26% in BMI, respectively. Similarly, two studies reported outcomes converting AGB into sleeve gastrectomy (SG) with 65% of T2D improvement or remission.

Moreover, procedures as SG and RYGB are not purely restrictive because other mechanisms, besides the weight loss, underlie the improvement of glucose metabolism. Nevertheless, studies showed that, when converting SG into RYGB, T2D improves or remits in 62% among patients with residual disease. These results are even better when converting SG into duodenal switch (79% of diabetes improvement).

In patients submitted primarily to RYGB, reoperative techniques lack compelling positive data. Better results were reported after trimming/ redoing the gastric pouch and/or increasing the malabsorptive component. The first option shows heterogeneous results in different studies (50 to 100% of diabetes improvement), maybe because of the diversity of techniques employed. Only one study reported that all patients with T2D recurrence after RYGB ($n = 11$) presented improvement after distalization of the common channel.

The authors concluded some interesting points after that narrative review of the literature. Around 80% of patients are diabetes-free at the time of the surgical revision and inadequate weight loss or weight regain that usually indicates the reoperation is not always linked with T2D recurrence. In spite of the lack of quality data and the limitations of the included studies, revisional surgery seems to be effective in reducing weight and improving T2D among patients with inadequate weight loss or weight regain accompanied with T2D relapse.

Recently, Aleassa et al. (2019) published a retrospective study with the most significant series of 81 patients with T2D relapse after the primary bariatric operation submitted to revisional surgery in an academic medical center. T2D remission was defined as glycated hemoglobin (HbA1c) $< 6.5\%$ and fasting plasma glucose (FPG) < 126 mg/dL without diabetes medications. T2D improvement was considered when patients presented a reduction in HbA1c (by $>1\%$) and FPG (by >25 mg/dL), or reduction in HbA1c and FBG accompanied by a decrease in antidiabetic medications.

Patients were divided accordingly to the type of primary and revisional bariatric procedures. The median follow-up was 22 months. Among 20 patients submitted to a conversion of VBG to RYGB, a median additional total weight loss (TWL) was 20.5% and T2D remission and improvement rates were 35% and 55% with a 44.4% reduction of patients using insulin. Among AGB conversions, 14 were converted into RYGB with a median of 22.4% of additional TWL and 35.7% of T2D remission and 35.7% of improvement in 43.5 months of median follow-up. In 7 cases, the second procedure was a SG, with modest results and a median of 13.5% of additional TWL and HbA1c decrease from 7.2% at baseline to 6.7% after median of 38 months. Only 2 patients were converted to duodenal switch (DS) with an additional weight loss and T2D improvement, but in a short follow-up.

Sleeve gastrectomy was converted to RYGB in 13 patients with a median additional TWL of 11.3% and T2D remission and improvement rates of 23.1 and 30.8%, respectively, but the HbA1c change was not statistically significant. Conversion to DS was performed only in three patients leading to significant T2D improvement and weight loss.

In a median length of 12 months of follow-up, revising the pouch after RYGB resulted in a median of total weight loss of 14.1%, 22.7% of T2D remission, and 50% of improvement.

Analyzing 115 patients converted to RYGB (82 from AGB and 33 from SG) matched in a 1:1 ratio, including T2D as matching criteria, with control patients who underwent primary RYGB, Vallois et al. (2019) found similar early and late morbidity between groups and no mortality. Although the revisional group presented less weight loss compared with primary RYGB group (mean % excess weight loss 67.4 ± 20.7 vs. 72.7 ± 22.9, $p = 0.023$) after 1 year, improvement of hypertension (62.5 vs. 70.5%; $p > 0.05$), diabetes (73.7 vs. 79%; $p > 0.05$), and obstructive sleep apnea syndrome (100 vs. 97%; $p > 0.05$) was similar between groups. Differently from previous data, this study showed that the revisional procedure was as safe as the primary one and showed similar significant improvement in comorbidities, as T2D, even with less weight loss.

Finally, a systematic review and meta-analysis published in 2018 (Pędziwiatr et al. 2018) assessed 21 studies comparing outcomes after primary versus revisional RYGB and included 11,720 and 3043 patients, respectively. Seven studies reported results on T2D and hypertension remission, and there were no differences between groups ($p = 0.61$). It is important to notice that the majority of patients in the revisional group had a VBG or AGB as the primary procedure. Weight loss was significant in the primary RYGB group (20% more excess weight loss than revisions). The analysis showed higher rates of morbidity and mortality in revisional procedures, although the numbers are still relatively low (0.2 versus 0.6% of mortality). However, there were no differences in the case-matched subgroup.

Revisional surgery for Type 2 diabetes relapse still lacks quality data. Keeping in mind that T2D is a chronic and progressive disease, maintaining glycemic endpoints under the target is important to prevent micro and macrovascular complications. Metabolic surgery is recommended when the best medical treatment fails. However, the optimal strategy to mitigate recurrence, analog to cancer treatment, when oncological surgery is often associated to radio/chemotherapy, is to keep the best medical treatment after surgery. So far, there are no accurate scores to predict T2D outcomes after metabolic interventions. Those available are good tools to help the clinician to refer the best patients to surgery.

Operations that reroute the food through the gastrointestinal tract deliver long-term better outcomes than those that don't. Currently, based on the mechanisms behind the "bypass" operations, it seems that converting SG or AGB to RYGB or DS is efficacious to achieve glycemic control and added weight loss, but always keeping along the best medical treatment.

In most cases of postoperative T2D recurrence after RYGB or DS, if there is no anatomical complication to be fixed (as gastro-gastric fistulas), adding pharmacological agents as GLP1 analogs may be an alternative to reach glycemic control.

References

Aleassa EM, Hassan M, Hayes K, Brethauer SA, Schauer PR, Aminian A (2019) Effect of revisional bariatric surgery on type 2 diabetes mellitus. Surg Endosc 33 (8):2642–2648

Arterburn DE, Bogart A, Sherwood NE et al (2013a January) A multisite study of long-term remission and relapse of type 2 diabetes mellitus following gastric bypass. Obes Surg 23(1):93–102

Arterburn DE, Bogart A, Sherwood NE et al (2013b) A multisite study of long-term remission and relapse of type 2 diabetes mellitus following gastric bypass. Obes Surg 23:93–102

Brethauer SA, Aminian A, Romero-Talamás H et al (2013) Can diabetes be surgically cured? Long-term metabolic effects of bariatric surgery in obese patients with type 2 diabetes mellitus. Ann Surg 258:628–636

Ceperuelo-Mallafre V, Llaurado G, Keiran N et al (2019 Oct) Preoperative circulating succinate levels as a biomarker for diabetes remission after bariatric surgery. Diabetes Care 42(10):1956–1965

Ciudin A, Fidilio E, Ortiz A et al (2019 Jul) Genetic testing to predict weight loss and diabetes remission and long-term sustainability after bariatric surgery: a pilot study. J Clin Med 8(7):964

Cohen RV, Pereira TV, Aboud CM et al (2017) Microvascular Outcomes after Metabolic Surgery (MOMS) in patients with type 2 diabetes mellitus and class I obesity: rationale and design for a randomized controlled trial. BMJ Open 7:e013574

DeFronzo RA (2009 Apr) From the triumvirate to the ominous octet: a new paradigm for the treatment of type 2 diabetes mellitus. Diabetes 58(4):773–795

Dixon JB, Chuang LM, Chong K et al (2013 Jan) Predicting the glycemic response to gastric bypass surgery in patients with type 2 diabetes. Diabetes Care 36(1):20–26

Gloy VL, Briel M, Bhatt DL et al (2013) Bariatric Surgery versus non-surgery treatment for obesity: a systematic review and meta-analysis of randomized controlled trials. BMJ 347:f5934

Maleckas A, Venclauskas L, Wallenius V et al (2015 Mar) Surgery in the treatment of type 2 diabetes mellitus. Scand J Surg 104(1):40–47

Mingrone G, Panunzi S, De Gaetano A, Guidone C, Iaconelli A, Leccesi L, Nanni G, Pomp A, Castagneto M, Ghirlanda G, Rubino F (2012) Bariatric surgery versus conventional medical therapy for type 2 diabetes. N Engl J Med 366:1577–1585

Mingrone G, Panunzi S, De Gaetano A et al (2015) Bariatric–metabolic surgery versus conventional medical treatment in obese patients with type 2 diabetes: 5 year follow-up of an open-label, single-centre, randomized controlled trial. Lancet 386:964–973

Miras AD, Pérez-Pevida B, Aldhwayan M et al (2019) Adjunctive liraglutide treatment in patients with persistent or recurrent type 2 diabetes after metabolic surgery (GRAVITAS): a randomised, double-blind, placebo-controlled trial. Lancet Diabetes Endocrinol

Mokdad AH, Bowman BA, Ford ES et al (2001) The continuing epidemics of obesity and diabetes in the United States. JAMA 286:1195–1200

Mor A, Keenan E, Portenier D, Torquati A (2013) Case-matched analysis comparing outcomes of revisional versus primary laparoscopic Roux-en-Y gastric bypass. Surg Endosc 27:548–552

Oliveira VLP, Martins GP, Mottin CC et al (2018 Jan) Predictors of long-term remission and relapse of type 2 diabetes mellitus following gastric bypass in severely obese patients. Obes Surg 28(1):195–203

Panunzi S, De Gaetano A, Carnicelli A, Mingrone G (2015) Predictors of remission of diabetes mellitus in severely obese individuals undergoing bariatric surgery: do BMI or procedure choice matter? Ann Surg 261:459–467

Pędziwiatr M, Malczak P, Wierdak M et al (2018) Revisional Gastric bypass is inferior to primary gastric bypass in terms of short- and long-term outcomes—systematic review and meta-analysis. Obes Surg 28:2083–2091

Pucci A, Tymoszuk U, Cheung WH et al (2018 Mar) Type 2 diabetes remission 2 years post Roux-en-Y gastric bypass and sleeve gastrectomy: the role of the weight loss and comparison of DiaRem and DiaBetter scores. Diabet Med 35(3):360–367

Rubino F, Nathan DM, Eckel RH et al (2017) Metabolic surgery in the treatment algorithm for type 2 diabetes: a joint statement by international diabetes organizations. Obes Surg 27:2–21

Schauer PR, Kashyap SR, Wolski K et al (2012) Bariatric surgery versus intensive medical therapy in obese patients with diabetes. N Engl J Med 366:1567–1576

Schauer PR, Bhatt DL, Kirwan JP et al (2014) STAMPEDE investigators. Bariatric surgery versus intensive medical therapy for diabetes—3-year outcomes. N Engl J Med 370:2002–2013

Sha A, Laferrère B (2017 Aug) Diabetes after bariatric surgery. Can J Diabetes 41(4):401–406

Sjostrom L (2013) Review of the key results from the Swedish obese Subjects (SoS) trial—a prospective controlled intervention study of bariatric surgery. J Intern Med 273:219–234

Standards of Medical Care in Diabetes—2020—American Diabetes Association—Diabetes care Diabetes Care 2020 Jan; 43(Suppl. 1):S1–S2

Still CD, Wood GC, Benotti P et al (2014) Preoperative prediction of type 2 diabetes remission after Roux-en-Y gastric bypass surgery: a retrospective cohort study. Lancet Diabetes Endocrinol 2:38–45

Still CD, Benotti P, Mirshahi T et al (2019 May) DiaRem2: incorporating duration of diabetes to improve prediction of diabetes remission after metabolic surgery. Surg Obes Relat Dis 15(5):717–724

Sudan R, Nguyen NT, Hutter MM, Brethauer SA, Ponce J, Morton JM (2014) Morbidity, mortality, and weight loss outcomes after reoperative bariatric surgery in the USA. J Gastrointest Surg 19(1):171–178

Vallois A, Menahem B, Le Roux Y et al (2019) Revisional Roux-en-Y Gastric bypass: a safe surgical opportunity? Results of a case-matched study. Obes Surg 29:903–910

Yan J, Cohen R, Aminian A (2017) Reoperative bariatric surgery for treatment of type 2 diabetes mellitus. Surg Obes Relat Dis 00–00

Malnutrition in Obesity

62

Sonmoon Mohapatra and Capecomorin S. Pitchumoni

Contents

Introduction 836

Deficiencies Across All Classes 836
Definition of Obesity and Its Types 836

Metabolic Obesity 836

Obesity and Muscle Wasting 837

Types of Bariatric Surgery 837

Nutritional Deficiencies 837

Deficiencies Post-Bariatric Surgery 838

Nutritional Deficiencies Before and After Bariatric Management 838

Protein 838
Essential Fatty Acids 838
Vitamin D 841
Vitamin B1 (Thiamine) 842
Vitamin B12 and Folate 843
Vitamins A, C, E, K 843
Iron 843
Zinc 844
Copper 844

References 845

S. Mohapatra
Department of Gastroenterology and Hepatology, Saint Peter's University Hospital – Rutgers Robert Wood Johnson School of Medicine, New Brunswick, NJ, USA

C. S. Pitchumoni (✉)
Department of Gastroenterology, Hepatology and Clinical Nutrition, Saint Peter's University Hospital – Rutgers Robert Wood Johnson School of Medicine, New Brunswick, NJ, USA

New York Medical College, Valhalla, NY, USA

Robert Wood Johnson School of Medicine, Rutgers University, New Brunswick, NJ, USA
e-mail: pitchumoni@hotmail.com

J. Faintuch and S. Faintuch (eds.), *Obesity and Diabetes*, https://doi.org/10.1007/978-3-030-53370-0_62

Abstract

Obesity and malnutrition appear to be at the opposite ends of the spectrum. But in reality, obesity in itself is a form of malnutrition associated with nutritional deficiencies. Attempts to lose weight by various special diets complement the preexisting nutritional deficiencies. Bariatric surgical procedures are well-accepted for sustained remission of obesity. The physiological basis of these procedures is either gastric reduction (restrictive) or diminished absorption of nutrients (malabsorptive). In both types, the weight loss is associated with worsening of the preexisting nutrient deficiencies or creating new deficiencies of macro- and micronutrients. A well-defined protocol-based pre- and post-operative nutritional assessment is needed for minimal adverse effects. Clinical guidelines and recommendations are provided in the present chapter.

Keywords

Obesity · Bariatric surgery · Roux-en-Y gastric bypass · Sleeve gastrectomy · Biliopancreatic bypass with duodenal switch · Micronutrients · Vitamin B12 · Folate (vitamin B9) Vitamin B1 · Thiamine · Biotin · Wernicke's psychosis · Osteomalacia · Osteoporosis · Blind loop syndrome · SIBO · Anemia · Neuropathy · Magnesium · Copper · Selenium

Introduction

Obesity and malnutrition are presently considered as two sides of one crisis or "a double burden of malnutrition," the coexistence of nutritional deficiencies along with overweight and obesity.

A recent study has projected that nearly 1 in 2 US adults will be obese by 2030 (Ward et al. 2019). It also showed that severe obesity would affect nearly 1 in 4 adults by 2030, particularly women and certain disadvantaged populations. China and India also are facing the obesity crisis, especially in the middle and upper class, contrasting with a high prevalence of malnourished poor.

Deficiencies Across All Classes

A dogmatic view of the past is that malnutrition is the consequence of the chronic consumption of an insufficient quantity of inferior quality diet as a result of poverty or disease. However, it has become clear that malnutrition does not spare the rich. The excessive use of high calorie foods and drinks along with sedentarism is relevant. The available solutions to the problem of obesity are many, including various calorie-restricted diets and bariatric surgical procedures designed to induce malabsorption. All of the above cause malnutrition and malabsorption of nutrients with varying amounts as side effects.

Definition of Obesity and Its Types

In several Asiatic countries, BMI of 23.0–24.9 kg/m^2 *and* $\geq$ *25* kg/m^2 are adopted for *overweight and obesity, respectively (*World Health Organization 2014*;* Gronroos et al. 2010*;* Misra et al. 2009*). The traditional WHO classification is predominantly applicable to the people of European ancestry.*

Abdominal fat accumulation has a more ominous prognosis than gluteo-femoral fat, raising doubts about reliability of BMI alone (Adab et al. 2018; Kissebah et al. 1982; Després et al. 1990). Even visceral fat signals cardiovascular complications at different proportions among Asians and other Americans (Demerath et al. 2007; Lim et al. 2011).

Metabolic Obesity

New clinical modalities encompassing metabolically healthy but obese (MHO), metabolically obese but normal weight (MONW), and metabolic or central obesity have thus been devised (Karelis et al. 2004; Hamdy et al. 2006). A normal waist to hip ratio ($<$ 0.95 for men, $<$ 0.86 for women) rules out metabolic obesity. Based on waist-hip ratio, obesity is classified as pear-shaped when the waist-hip ratio is <0.8 in women and 0.9 in men. An apple-shaped obesity is characterized as a higher waist to hip ratio compared to the pear shape, implying that the accumulated fat on an apple-shaped body is mostly around the abdomen or waist (central obesity) and is more often related to cardiovascular and metabolic problems.

Obesity and Muscle Wasting

Sarcopenic obesity is not uncommon (Choi 2016), and aging, undernutrition, consumptive illnesses, or disuse atrophy are common precipitants. Functional and clinical aberrations are observed, including propensity towards increased adiposity (Dufour et al. 2013; Lee et al. 2016), whereas obesity itself as an inducer of sarcopenia has also been proposed, suggesting some common pathways (Gregor and Hotamisligil 2011). Glucose homeostasis and other metabolic conditions could be negatively affected, thus demanding a more structured approach.

Types of Bariatric Surgery

Bariatric interventions can be restrictive, malabsorptive, and mixed (restrictive and malabsorptive). Most operations are surgical (laparoscopic), along with emerging endoscopic maneuvers (Table 62.1). More aggressive modalities (malabsorptive) generally feature more nutritional deficits.

Nutritional Deficiencies

Among Western populations, industrialized meals are preferred, with an excessive amount of undesirable calories and sodium and insufficient fiber and micronutrients (Damms-Machado

Table 62.1 Types of bariatric surgical procedures

Laparoscopic or surgical procedures[89, 90] (Indication: BMI $\geq$ 40 or $\geq$ 35 kg/m^2 with clinically significant comorbidities)			
Type	Mechanism	Procedural details	% TWL (Mean)
Adjustable gastric banding	Restrictive	A balloon encircles the upper stomach	11.8
Vertical sleeve gastrectomy (VSG)	Restrictive	The entire greater curvature is resected	15.4
Roux-en-Y gastric bypass (RYGB)	(Restrictive and malabsorptive)	A proximal pouch is anastomosed to a gastrojejunostomy. A Roux-en-Y biliopancreatic conduit joins 75–100 cm downstream	20.1
Biliopancreatic bypass with duodenal switch (BPDS)	Restrictive and malabsorptive	After sleeve gastrectomy, the ileum is anastomosed to the duodenum	36.7
Endoscopic bariatric procedures[91] Indication: mild to moderate obesity (BMI 30–40 kg/m^2), or more severe cases who refuse surgery			
Space occupying devices	Restrictive	Promote gastric distension and reduced emptying	10.2 (Orbera balloon)
Aspiration assist device	Unique class	A gastrostomy for removing gastric contents after meals	14.2
Endoscopic sleeve gastroplasty	Restrictive	Endoscopic sutures to collapse the greater curvature and diminish gastric volume	18.6
Primary obesity surgery endoluminal	Restrictive	Endoscopic sutures that reduce accommodation of the gastric fundus.	4.9

%TWL (total weight loss) = % total weight loss which is computed as 100% × (pre-surgery BMI—post-surgery BMI at the time of measurement)/pre-surgery BMI et al. 2012). Obesity could by itself trigger certain deficits (Kaidar-Person et al. 2008), whereas crash diets and other radical attempts to lose weight could be equally deleterious.

Deficiencies Post-Bariatric Surgery

1. Restrictive interventions diminish nutrient intake and could lead to an imbalanced diet.
2. Vertical sleeve gastrectomy can be additionally followed by loss of appetite (reduced ghrelin), abnormal iron, and vitamin B12 absorption (less intrinsic factor, less acidification of the duodenum).
3. Up to one quarter of dietary protein and three quarters of fat intake can be lost with malabsorptive techniques, along with associated micronutrients. Frequent stools can further compromise nutritional status.
4. A combination of some of the features of restriction and malabsorption can be detected after RYGB (Roux-en-Y gastric bypass) (Kizy et al. 2017).

Nutritional Deficiencies Before and After Bariatric Management

Dietary shifts, clinical comorbidities and treatments, along with the profile of the operation itself modulate the advent and course of nutritional deficits. Current recommendations are listed in Tables 62.2 and 62.3 (Mechanick et al. 2013; Parrott et al. 2016; Thibault et al. 2016).

Protein

Protein is essential for the synthesis of tissues, enzymes, hormones, albumin, hemoglobin, and other biomolecules. Severe protein deficit can occur in 3–21% of patients after BPDS and up to 13% following RYGB, depending on the length of the Roux-limb (Faintuch et al. 2004). Some bariatric patients develop an intolerance to proteins due to altered gastrointestinal anatomy and function, with inadequate digestion and absorption. Early low-caloried liquid diets and eventual vomiting could further contribute to depletion.

Protein deficiency conducts to skin atrophy and decreased muscle mass, besides classic hypoalbuminemia and even generalized edema. Impairment of growth, wound healing, immune competence, and other essential functions can trigger multiple complications.

Current consensus guidelines suggest additional protein along with all modalities of exercise. Daily consumption of 60–120 g after RYGB, 60–80 g or 1.1 g/kg of ideal body weight after LSG, and 90 g following BPDS is advised. Enteral nutrition should be considered in case of significant undernutrition.

Essential Fatty Acids

Both omega 6 linoleic acid (LA) and omega 3 fatty acids, including alpha-linoleic acid (ALA), eicosapentaenoic acid (EPA), and docosahexaenoic acid (DHA), are considered essential for humans, even though some interconversions among the omega 3 family are recognized. Vegetable oils are the most important sources of omega 6 fatty acids (EFAs) and ALA, whereas EPA and DHA are usually ingested in the form of fish. EFAs are required for the formation of various eicosanoids (biologically active lipids). Patients after bariatric surgeries (particularly malabsorptive procedures) are susceptible to EFAs deficiency due to aberrant intestinal fat processing (Forbes et al. 2016).

Dermatologic disorders, slow wound healing, and increased susceptibility to infections are some of the features of EFAs deficiency. Associated zinc deficit (which also presents with dry scaly rash) is a possibility which should be promptly recognized (Mogensen 2017).

EFAs deficiency can be managed by supplementing unsaturated omega-6-rich

Table 62.2 Symptoms of micronutrient deficiencies and their laboratory indices

Micro nutrients	Recommended daily intake	Sources	Symptoms of deficiency	Lab range	Additional lab indices
Vitamin D	31–70 years: 15 mcg (600 IU) >70: 20 mcg	Sunlight, cod liver oil, egg yolk, fortified foods	Rickets in toddlers Osteomalacia in adults	25 (OH) D > 30 mg/mL (> 75 nmol/L)	↓ Serum phosphorus, urinary calcium ↑ Serum PTH, N-telopeptide, osteocalcin
Thiamine	M: 1.2 mg F: 1.1 mg	Enriched, fortified, or whole-grain products, cereals	Wet beriberi: heart failure; dry beriberi: increased reflexes, neuropathy, ataxia, convulsions Wernicke's encephalopathy	Plasma: 4–15 nmol/L; blood: 70–180 nmol/L Transketolase >150 nmol/L Erythrocyte trasketolase coefficient < 1.15	↑ pyruvate or ↓ lactate, urinary thiamin
Vitamin B12	2.4 µg	Fortified cereals, meat, fish, poultry	Pernicious anemia, paresthesia, sore tongue, demyelination and axonal degeneration tinnitus, dementia	Serum B12 (cobalamin) 200–1000 pg/mL	↑ Serum methyl malonic acid (MMA) ↑ Serum homocysteine (tHcy)
Folate	400 µg	Grains, whole grain products, fortified cereals, dark leaves	Macrocytic anemia, neural tube defects, skin and mucosa changes	Folate 340–1020 ng/mL for age ≥ 18 year	Normal MMA ↑ Serum tHcy
Vitamin B6	1.5 mg	Fortified cereals, chickpeas, meats	Dermatitis glossitis, cheilitis, conjunctivitisneurological symptoms	Pyridoxal-5′-phosphate: 5–24 ng/mL	RBC glutamic pyruvate oxaloacetic transaminase; urinary 4-pyridoxic acid
Vitamin A	M: 900 mcg (3000 IU) W: 700 mcg	Liver, dairy products, fish, dark fruits, leaves	Night blindness, endophthalmitis, other eye changes	Plasma retinol 20–80 µg/dL	Retinol-binding protein
Vitamin E	15 mg	Vegetable oils, cereals, nuts, fruits, meats	Myelopathy, polyneuropathy ophthalmoplegia, nystagmus	Plasma alpha-tocopherol	Plasma lipids
Vitamin K	M: 120 µg W: 90 µg	Green leaves, brussel sprouts, cabbage, plant oils, margarine	Easy bleeding	PT 10–13 s	↑ Des-gamma carboxyprothrombin; ↓ plasma phylloquinone
Iron	8 mg (Men and postmenopausal women) Females: 18 mg	Fruits, vegetables fortified foods; *Heme iron:* Meat and poultry	Microcytic anemia, immune depression, abnormal nails, glossitis, dysphagia	Serum iron: 60–170 µg/dL Transferrin 200–360 µg/dL saturation: 20–50% Ferritin: 12–300 ng/mL (male), 12–150 ng/mL (females)	↑ TIBC Soluble transferrin receptor

(continued)

Table 62.2 (continued)

Micro nutrients	Recommended daily intake	Sources	Symptoms of deficiency	Lab range	Additional lab indices
Calcium	31–50 M 1000 mg, F 1000 51–70 M 1000 mg, F: 1200 71+: M 1200 mg, F: 1200	Milk, cheese, calcium-set tofu, Chinese cabbage, kale, broccoli, Fortified foods and beverages	Low bone density, Osteoporosisincreased reflexes, muscle weakness, paresthesia	Serum PTH (PTH >) 65 pg/mL indicates ↓ calcium 25 (OH) D	Serum calcium, ionized calcium, bone densitometry
Copper	900 μg	Organ meats, seafood, nuts, wheat bran, whole grain cocoa	Anemia, leukopenia, dermatologic changes. Optic neuropathy, myelopathy myopathy,	Serum or plasma copper 11.8–22.8 mmol/L Ceruloplasmin 75–145 μg/dL	Decreased erythrocyte superoxide dismutase activity 24-hour urine copper
Zinc	M: 11 mg, F: 8 mg	Seafood, beans, nuts, whole-grain, red meat	Hypogeusia, skin lesions, diarrhea, poor wound healing, infections	Plasma zinc 60–130 μg/dL	↓ Serum zinc ↓ Erythrocyte zinc (RBC zinc) ↓ Urinary zinc
Magnesium	M: 420 mg F: 320 mg	Whole wheat, spinach, nuts, black Beans, avocado	Muscle cramps, arrythmias, paresthesia, tremors	Plasma magnesium 1.7–2.2 mg/dL	↓Potassium ↓Calcium
Selenium	55 μg	Brazil nuts, yellowfin tuna, meats	Cartilage damage, cardiomyopathy, fatigue	Plasma selenium: 53–109 mg/L blood 66.7–119 mg/L; glutathione peroxidase: 196–477 U/L	–

LAGB laparoscopic adjustable gastric banding, *RYGB* Roux-en-Y gastric bypass, *VSG* vertical sleeve gastrectomy, *BPDS* biliopancreatic bypass with duodenal switch, *RDA* recommended dietary allowance

Table 62.3 Micronutrient assessment before and after bariatric surgeries

Micronutrient	Deficiency in obesity	Deficiency after bariatric surgery	Screening post-bariatric surgery	Supplementation post-bariatric surgery	Comments
Thiamine	15.5–29%	<1% to 49%	High-risk patients	12–50 mg once or twice daily	Begin on day 1
Vitamin B12	2–18%	At 2–5 year of RYGB-(<20%) VSG- (4–20%)	Every 3 months for 1 year, and then yearly	IM or SQ: 1000 μg monthly Orally or sublingual: 350–500 μg daily	Begin 0–3 months after surgery
Folate	Up to 54%	up to 65%	Routine screening	400–800 μg daily 800–1000 μg for fertile women	>1000 mg could mask low vit. B12
Vitamin D/calcium	25–90%	20–100%	Routine screening	LAGB, VSG, RYGB: 1200–1500 mg/day BPDS: 1800–2400 mg/day; Vit. D3: 3000/day	Day 1–30 Calcium should be given in divided doses
Vitamins A, E, K	Vitamin A 14% Vitamin E 2.2%	Within 4 years of RYGB and BPDS—up to 70%	Vitamin A: first year RYGB, BPDS, undernutrition Vitamin E, K: symptomatic patients	LAGB: Vit A 5000 IU/d Vit K 90–120 μg/d; RYGB and VSG: Vit A 5000–10,000 IU/d, Vit K 90–120 μg/d; BPDS: Vit A 10,000 IU/d Vit K 300 μg/d Vit E 15 mg/d(all)	2–4 weeks; Higher doses with history of deficiencies
Iron	Up to 45%	AGB (14%); SG (Up to18%); RYGB (20–55%); BPD (13–62%); DS 8–50%	Every 3 months for 1 year, and then yearly	Males: 18 mg;menstruating females: 45–60 mg	Day 1; Avoid Simultaneous acid, calcium, Polyphenols, or phytates
Zinc	24–28% overall 74% BPDS	BPD (45 to 91%); RYGB (15 to 21%); LSG (11 to 14%).	Annually (RYGB or BPDS); anemia with normal iron	LABG/VSG: 8–11 mg/d; RYGB: 8–22 mg/d; BPDS: 16–22 mg	Day 1; Suggested 1 mg copper: 8–15 mg zinc (copper deficit risk)
Copper	Up to 70% BPD women	BPDS- (Up to 90%); RYGB- (10–20%)	Annually RYGB or BPDS	LAGB/VSG: 1 mg/d;RYGB or BPDS: 2 mg/d	Day 1 of discharge
Magnesium	35%	32%	BPDS (eventually others)	Magnesiumcitrate, 300 mg/day	–

25 (OH) D 25-hydroxyvitamin D, *PTH* parathyroid hormone, *MMA* serum methyl malonic acid, *tHcy* serum homocysteine, *PT* prothrombin time, *TIBC* total iron-binding capacity, *DXA* dual-energy X-ray absorptiometry

vegetable oils, as well as omega-3-rich fish oil, for patients tolerating oral diet. For intravenous nutrition-dependent patients, injectable fat emulsions containing both omega 6 and omega 3 fatty acids are available in the market (Mogensen 2017). The lipid injectable emulsion infusion can be considered periodically for these patients.

Vitamin D

This is a major fat-soluble nutrient, with relevance in a variety of benign and malignant conditions (Charoengam et al. 2019; Bouillon and Carmeliet 2018; Amrein et al. 2020). Deficiency is defined as levels of 25 (OH)D less than 20 ng/mL (50 nmol/L), whereas less than 12 ng/mL (30 nmol/L) characterize severe deficiency

(Bouillon and Carmeliet 2018; Amrein et al. 2020).

Diagnosis in populations with excessive adiposity is extremely common, with an estimated prevalence of 25–90% (Bouillon and Carmeliet 2018; Amrein et al. 2020; Xanthakos 2009; Lespessailles and Toumi 2017). As body fat accumulates, 25(OH)D proportionally moves down (Arunabh et al. 2003; Vimaleswaran et al. 2013). Volumetric dilution of the nutrient, and reduced exposure to sunlight on account of sedentarism and avoidance of beaches and swimming pools, could contribute to it. Importance of dietary intake, however, is unclear. Liver steatosis and certain cytokines could be detrimental for endogenous production as well (Mohapatra et al. 2019).

Following malabsorptive bariatric procedures, an estimated 17–52% (2 years) and 50–63% (4 years) exhibit deficiency (Hamoui et al. 2003; Newbury et al. 2003; Slater et al. 2004; Bloomberg et al. 2005). After bypass procedures, the altered anatomy and reduced biliary-pancreatic processing decrease the absorption of all fat-soluble vitamins. Rapid transit time after VSG could similarly be deleterious.

Severe deficiency is predominantly detected as rickets/osteomalacia (Charoengam et al. 2019). Asymptomatic or oligosymptomatic cases are the majority; however, secondary hyperparathyroidism can be damaging to bones, with the possibility of fractures. Such endocrine reaction is not uncommon with vitamin D levels $\leq$30 ng/mL. Dietary calcium processing is impaired when the upper small bowel is excluded in RYGB and BPDS (as little as 20% only) (Hamoui et al. 2003; De Prisco and Levine 2005), potentially resulting in calcium resorption and weakened bones.

Patients should be routinely monitored to inhibit secondary hyperparathyroidism. Yet, paracellular migration of calcium through the small bowel occurs even during vitamin D deficiency, raising questions about the real role of such prescriptions.

Baseline calcium and vitamin D are mandatory. Calcium citrate (1200 to 1500 mg) and 3000 IU of vitamin D, or more in case of deficiency, are necessary as well. As much as 50,000 units 1 to 3 times weekly, and eventually calcitriol (1,25—dihydroxy vitamin D), have been suggested for refractory cases. Associated hypophosphatemia is frequent and demands prevention (1.5 to 2.5 mg/dL). Besides vitamin D measurements, bone densitometry every two years is advocated after mixed or malabsorptive interventions.

Vitamin B1 (Thiamine)

The prevalence of thiamine deficiency is 16–29% in preoperative patients who undergo bariatric surgery (Carrodeguas et al. 2005; Flancbaum et al. 2006). Thiamine can get depleted rapidly within as little as 2–3 weeks (Lonsdale 2006). Large intakes of simple carbohydrates consume available thiamine, further precipitating its depletion.

Several case reports of Wernicke's encephalopathy are reported in post-bariatric patients (Mohapatra et al. 2019). Beriberi is not entirely unknown either. The prevalence of thiamine deficiency ranges from <1% to 49% according to population and intervention (Parrott et al. 2016). Thiamine deficiency can be attributed to duodenal malabsorption, metabolic demand, and decreased dietary intake, particularly with frequent vomiting. Altered intestinal bacterial composition could also enhance production of thiaminases (Lakhani et al. 2008).

Carbohydrate-rich prescriptions in the form of intravenous fluids as well as enteral or parenteral nutrition can also be risky in certain circumstances due to elevated thiamine consumption by such substrates. Routine monitoring and supplementation are therefore advised (Mechanick et al. 2013).

Acute cases may require thiamine, 500 mg IV/d, for 3 to 5 days, followed by 250 mg PO/d for 3 to 5 days or more. In these patients, long-term oral supplementation with 100 mg/d is advocated. For less stringent circumstances, 7–14 days of IV thiamine, 100 mg/d, is an alternative.

Individuals with irregular or insufficient diet and possibility of nutritional imbalances, digestive diseases, comorbidities including furosemide prescription, alcohol abuse, or neuropsychiatric conditions need follow-up for 6 months and for even longer if suspicions remain (every 3–6 months).

Vitamin B12 and Folate

About 2–18% of those with excessive adiposity suffer from these deficiencies, and it is debated whether there are links between the two conditions (Parrott et al. 2016; El-Qudah et al. 2013; Yagan et al. 2016). Vitamin B12 containing animal protein is not always appreciated by the obese, a possible contributing factor. Certain medications such as metformin and proton pump inhibitors affect B12 absorption and stores. Serum 5-methyltetrahydrofolate and unmetabolized folic acid are inversely associated with BMI; (Bradbury et al. 2014) however, not RBC folate (Bird et al. 2015). Obese pregnant women very often exhibit low serum folate (da Silva et al. 2013; Knight et al. 2015), which predisposes to obesity in the offspring (Wang et al. 2016).

Restrictive bariatric interventions are not associated with vitamin B12 or folate deficits (Alvarez-Leite 2004), contrasting with RYGB after which the opposite occurs with vitamin B12. Lack of intrinsic factor, achlorhydria, and insufficient intake of B12 are the putative mechanisms (Ponsky et al. 2005). Folate deficiency is registered after RYGB patients; however, it seems much less common (MacLean et al. 1983; Boylan et al. 1988). Available supplements usually prevent both abnormalities.

Pre- and postoperative vitamin B12 and folate screening are necessary every 3–6 months until 12 months or longer when required. Because serum B12 can be misleading, serum methylmalonic acid (MMA) with or without homocysteine should be checked.

Dosages of vitamin B12 for supplementation encompass 1000 mcg/day orally, 1000 mcg monthly (IM or SQ), or eventually nasal spray. Being water-soluble, it is promptly excreted with a low risk of overdose. About 400–800 μg of folic acid, or 800–1000 μg for women of reproductive age, should be added.

Vitamins A, C, E, K

It is still debated whether antioxidants such as beta-carotene, alpha-tocopherol, and vitamin C correlate or not with BMI (Stryker et al. 1988; Schectman et al. 1989; Drewnowski et al. 1997; Wallström et al. 2001). More evidence is available regarding retinol, which becomes depressed as BMI increases (Viroonudomphol et al. 2003; Switzer et al. 2005).

We recommend baseline screening of vitamins A, C, E, and K during the preoperative period. Regardless of the symptoms, those submitted to mixed or malabsorptive interventions deserve subsequent vitamin A monitoring, with more selective indication for the other, less frequently deranged vitamins.

Prescriptions after surgery are listed: LAGB: Vitamin A 5000 IU/d and vitamin K 90–120 μg/d. RYGB and SG: Vitamin A 5000–10,000 IU/d and vitamin K 90–120 μg/d. LAGB, SG, RYGB, BPDS: Vitamin E 15 mg/d DS: Vitamin A (10,000 IU/d) and vitamin K (300 μg/d). Personalized attention may be required for certain conditions.

Iron

Iron deficiency (ID) is the principal predisposing factor for anemia (Lopez et al. 2016), being detected in as much as one sixth of humankind (Lopez et al. 2016). A 35% prevalence is occasionally observed in the obese (Schweiger et al. 2010). Hepcidin, a pro-inflammatory adipokine promoted by obesity, could aggravate the problem by linking to iron transporter ferroportin, which is already less abundant in the small bowel of this population (Girelli et al. 2016). The opposite phenomenon, possibly triggered by hepcidin dysfunction as well, is "Dysmetabolic iron overload syndrome (DIOS)," featuring elevated ferritin with normal or mildly elevated transferrin saturation. This is not rare in metabolic syndrome (15%) and even more frequent (50%) with fatty liver disease (NAFLD) (Bozzini et al. 2005).

Also, after bariatric treatment, iron deficiency ranks among the most common troubles (30–60%) (Muñoz et al. 2009). Hypochlorhydria, duodenum bypass, dietary shifts, and bacterial interference (*H.pylori*, intestinal microbiota) are the most frequently mentioned drivers (Topart 2008; Hershko et al. 2005).

Baseline iron and transferrin saturation, total iron-binding capacity, and ferritin are advised. Daily iron supplementation should be given. Iron monitoring schedule should be every 3--6 months (first year) and annually thereafter, if there are no abnormalities. Complete blood count, iron panel, total iron-binding capacity, ferritin, and soluble transferrin receptor are useful as well.

Zinc

Zinc deficiency is not uncommon (up to 50%) in bariatric candidates (Mahawar et al. 2017). Insulin pathways require zinc, and there is experimental evidence that it could enhance leptin expression, as well as macrophage infiltration into adipose tissue (Liu et al. 2013). Weight loss is associated with restoration of zinc concentrations (Ishikawa et al. 2005).

All bariatric interventions have some negative repercussions on zinc measurements, often detected quite early due to lack of reserves, with emphasis on the malabsorptive modality (45–91%) (Gehrer et al. 2010; Sallé et al. 2010; Balsa et al. 2011; Saif et al. 2012). DMT-1 (divalent metal transporter-1) is the most relevant mediator of absorption, expressed in the duodenum and proximal jejunum; therefore circumstances that shorten or eliminate food passage at this level exert negative impact. In large doses, all major nutritional metals, namely zinc, copper, and iron, can compromise absorption of the others, therefore reciprocal monitoring is advised (Bowley et al. 2017).

The majority of patients with mild to moderate deficiency remain asymptomatic. Advanced cases could present with dermatologic changes, immune deficiency, neurohormonal disorders, diminished appetite, and slow wound healing.

Zinc should be tested in candidates for mixed or malabsorptive procedures, and every year thereafter. Low findings do not necessarily require intervention (Marreiro et al. 2002) unless there are symptoms. Anemia with normal iron demands investigation, along with persistent diarrhea. Routine prescription is advised after these operations.

Copper

This is a cofactor for redox enzymes and contributes to antioxidant defense, immune function, and neuropeptide synthesis (Uriu-Adams 2005). Ceruloplasmin, relevant for iron metabolism, is a Cu-dependent ferroxidase.

As occurs with iron, gastric acid and the duodenum are relevant for copper absorption. Consequently, the same malabsorptive operations previously alluded to represent risk factors for such deficiency (9.6%) (Gletsu-Miller et al. 2012). Unsupplemented intravenous nutrition could occasionally be a precipitant as well (Shike 2009). Anemia with neutropenia, myeloneuropathy, and unusually optic neuropathy have been documented in severe cases (Shah and Tamhankar 2014).

The same testing routine as alluded to for zinc should be adopted for copper and ceruloplasmin. During acute-phase conditions such as fever or inflammation, spurious elevation can occur. Erythrocyte superoxide dismutase changes earlier during copper deficiency and could represent an alternative. Non-iron deficiency anemia demands copper screening, along with other nutrients.

Oral (2 mg/d), or IV copper (2–4 mg/d), for advanced deficiency can be adopted until normalization. Around 1 mg of copper is suggested for each 8 to 15 mg of zinc prescription, as high zinc precipitates copper reduction.

Conflict of Interest
The authors declare no conflict of interest.

Funding
None.

References

Adab P, Pallan M, Whincup PH (2018) Is BMI the best measure of obesity? BMJ 360:k1274. https://doi.org/10.1136/BMJ.K1274

Alvarez-Leite JI (2004) Nutrient deficiencies secondary to bariatric surgery. Curr Opin Clin Nutr Metab Care 7:569–575

Amrein K, Scherkl M, Hoffmann M, Neuwersch-Sommeregger S, Köstenberger M, Tmava Berisha A, Martucci G, Pilz S, Malle O (2020) Vitamin D deficiency 2.0: an update on the current status worldwide. Eur J Clin Nutr. https://doi.org/10.1038/s41430-020-0558-y

Arunabh S, Pollack S, Yeh J, Aloia JF (2003) Body fat content and 25-hydroxyvitamin D levels in healthy women. J Clin Endocrinol Metab 88:157–161. https://doi.org/10.1210/jc.2002-020978

Balsa JA, Botella-Carretero JI, Gómez-Martín JM et al (2011) Copper and zinc serum levels after derivative bariatric surgery: differences between Roux-en-Y gastric bypass and biliopancreatic diversion. Obes Surg 21:744–750. https://doi.org/10.1007/s11695-011-0389-y

Bird JK, Ronnenberg AG, Choi S-W, Du F, Mason JB, Liu Z (2015) Obesity is associated with increased red blood cell folate despite lower dietary intakes and serum concentrations. J Nutr 145:79–86. https://doi.org/10.3945/jn.114.199117

Bloomberg RD, Fleishman A, Nalle JE, Herron DM, Kini S (2005) Nutritional deficiencies following bariatric surgery: what have we learned? Obes Surg 15:145–154. https://doi.org/10.1381/0960892053268264

Bouillon R, Carmeliet G (2018) Vitamin D insufficiency: definition, diagnosis and management. Best Pract Res Clin Endocrinol Metab 32(5):669–684. https://doi.org/10.1016/j.beem.2018.09.014

Bowley MP, David WS, Cho TA, Dighe AS (2017) Case 35-2017. A 57-year-old woman with hypoesthesia and weakness in the legs and arms. N Engl J Med 377:1977–1984. https://doi.org/10.1056/NEJMcpc1710564

Boylan LM, Sugerman HJ, Driskell JA (1988) Vitamin E, vitamin B-6, vitamin B-12, and folate status of gastric bypass surgery patients. J Am Diet Assoc 88:579–585

Bozzini C, Girelli D, Olivieri O et al (2005) Prevalence of body iron excess in the metabolic syndrome. Diabetes Care 28:2061–2063. https://doi.org/10.2337/DIACARE.28.8.2061

Bradbury KE, Williams SM, Mann JI, Brown RC, Parnell W, Skeaff CM (2014) Estimation of serum and erythrocyte folate concentrations in the New Zealand adult population within a background of voluntary folic acid fortification. J Nutr 144:68–74. https://doi.org/10.3945/jn.113.182105

Carrodeguas L, Kaidar-Person O, Szomstein S, Antozzi P, Rosenthal R (2005) Preoperative thiamine deficiency in obese population undergoing laparoscopic bariatric surgery. Surg Obes Relat Dis 1:517–522. https://doi.org/10.1016/j.soard.2005.08.003

Charoengam N, Shirvani A, Holick MF (2019) Vitamin D for skeletal and non-skeletal health: what we should know. J Clin Orthop Trauma 10(6):1082–1093. https://doi.org/10.1016/j.jcot.2019.07.004

Choi KM (2016) Sarcopenia and sarcopenic obesity. Korean J Intern Med 31:1054–1060. https://doi.org/10.3904/kjim.2016.193

da Silva VR, Hausman DB, Kauwell GPA et al (2013) Obesity affects short-term folate pharmacokinetics in women of childbearing age. Int J Obes 37:1608–1610. https://doi.org/10.1038/ijo.2013.41

Damms-Machado A, Weser G, Bischoff SC (2012) Micronutrient deficiency in obese subjects undergoing low calorie diet. Nutr J 11:34. https://doi.org/10.1186/1475-2891-11-34

De Prisco C, Levine SN (2005) Metabolic bone disease after gastric bypass surgery for obesity. Am J Med Sci 329:57–61

Demerath EW, Sun SS, Rogers N et al (2007) Anatomical patterning of visceral adipose tissue: race, sex, and age variation**. Obesity 15:2984–2993. https://doi.org/10.1038/oby.2007.356

Després JP, Moorjani S, Lupien PJ, Tremblay A, Nadeau A, Bouchard C (1990) Regional distribution of body fat, plasma lipoproteins, and cardiovascular disease. Arteriosclerosis 10:497–511. https://doi.org/10.1161/01.ATV.10.4.497

Drewnowski A, Rock CL, Henderson SA et al (1997) Serum β-carotene and vitamin C as biomarkers of vegetable and fruit intakes in a community-based sample of French adults. Am J Clin Nutr. https://doi.org/10.1093/ajcn/65.6.1796

Dufour AB, Hannan MT, Murabito JM, Kiel DP, McLean RR (2013) Sarcopenia definitions considering body size and fat mass are associated with mobility limitations: the Framingham Study. J Gerontol A Biol Sci Med Sci 68:168–174. https://doi.org/10.1093/gerona/gls109

El-Qudah JM, Dababneh BF, Al-Qudah MM, Haddad M (2013) Serum vitamin B12 levels related to weight status among healthy Jordanian students. Lab Med 44:34–39. https://doi.org/10.1309/LMG2E7PEDHET5QVI

Faintuch J, Matsuda M, Cruz MELF et al (2004) Severe protein-calorie malnutrition after bariatric procedures. Obes Surg 14:175–181. https://doi.org/10.1381/096089204322857528

Flancbaum L, Belsley S, Drake V, Colarusso T, Tayler E (2006) Preoperative nutritional status of patients undergoing Roux-en-Y gastric bypass for morbid obesity. J Gastrointest Surg 10:1033–1037. https://doi.org/10.1016/j.gassur.2006.03.004

Forbes R, Gasevic D, Watson EM et al (2016) Essential fatty acid plasma profiles following gastric bypass and adjusted gastric banding bariatric surgeries. Obes Surg 26:1237–1246. https://doi.org/10.1007/s11695-015-1876-3

Gehrer S, Kern B, Peters T, Christoffel-Courtin C, Peterli R (2010) Fewer nutrient deficiencies after laparoscopic sleeve gastrectomy (LSG) than after laparoscopic Roux-Y-Gastric bypass (LRYGB)—a prospective study. Obes Surg 20:447–453. https://doi.org/10.1007/s11695-009-0068-4

Girelli D, Nemeth E, Swinkels DW (2016) Hepcidin in the diagnosis of iron disorders. Blood 127:2809–2813. https://doi.org/10.1182/blood-2015-12-639112

Gletsu-Miller N, Broderius M, Frediani JK et al (2012) Incidence and prevalence of copper deficiency following roux-en-y gastric bypass surgery. Int J Obes (2005) 36:328–335. https://doi.org/10.1038/ijo.2011.159

Gregor MF, Hotamisligil GS (2011) Inflammatory mechanisms in obesity. Annu Rev Immunol 29:415–445. https://doi.org/10.1146/annurev-immunol-031210-101322

Gronroos JM, Salminen P, Laine S, Gullichsen R (2010) Feasibility of ERCP procedures in patients 90 years of age and older. J Clin Gastroenterol 44:227–228

Hamdy O, Porramatikul S, Al-Ozairi E (2006) Metabolic obesity: the paradox between visceral and subcutaneous fat. Curr Diabetes Rev 2:367–373

Hamoui N, Kim K, Anthone G, Crookes PF (2003) The significance of elevated levels of parathyroid hormone in patients with morbid obesity before and after bariatric surgery. Arch Surg 138:891. https://doi.org/10.1001/archsurg.138.8.891

Hershko C, Hoffbrand AV, Keret D et al (2005) Role of autoimmune gastritis, Helicobacter pylori and celiac disease in refractory or unexplained iron deficiency anemia. Haematologica 90:585–595

Ishikawa Y, Kudo H, Kagawa Y, Sakamoto S (2005) Increased plasma levels of zinc in obese adult females on a weight-loss program based on a hypocaloric balanced diet. In Vivo 19:1035–1037

Kaidar-Person O, Person B, Szomstein S, Rosenthal RJ (2008) Nutritional deficiencies in morbidly obese patients: a new form of malnutrition? Obes Surg 18:870–876. https://doi.org/10.1007/s11695-007-9349-y

Karelis AD, St-Pierre DH, Conus F, Rabasa-Lhoret R, Poehlman ET (2004) Metabolic and body composition factors in subgroups of obesity: what do we know? J Clin Endocrinol Metabol 89:2569–2575. https://doi.org/10.1210/jc.2004-0165

Kissebah AH, Vydelingum N, Murray R, Evans DJ, Kalkhoff RK, Adams PW (1982) Relation of body fat distribution to metabolic complications of obesity*. J Clin Endocrinol Metabol 54:254–260. https://doi.org/10.1210/jcem-54-2-254

Kizy S, Jahansouz C, Wirth K, Ikramuddin S, Leslie D (2017) Bariatric surgery: a perspective for primary care. Diab Spectrum 30:265–276. https://doi.org/10.2337/ds17-0034

Knight BA, Shields BM, Brook A et al (2015) Lower circulating B12 is associated with higher obesity and insulin resistance during pregnancy in a non-diabetic white British population. PLoS One 10:e0135268. https://doi.org/10.1371/journal.pone.0135268

Lakhani VS, Shah HN, Alexander K, Finelli FC, Kirkpatrick JR, Koch TR (2008) Small intestinal bacterial overgrowth and thiamine deficiency after Roux-en-Y gastric bypass surgery in obese patients. Nutr Res 28:293–298. https://doi.org/10.1016/j.nutres.2008.03.002

Lee J, Hong YP, Shin HJ, Lee W (2016) Associations of sarcopenia and sarcopenic obesity with metabolic syndrome considering both muscle mass and muscle strength. J Prev Med Public Health 49:35–44. https://doi.org/10.3961/jpmph.15.055

Lespessailles E, Toumi H (2017) Vitamin D alteration associated with obesity and bariatric surgery. Exp Biol Med (Maywood) 242:1086–1094. https://doi.org/10.1177/1535370216688567

Lim U, Ernst T, Buchthal SD et al (2011) Asian women have greater abdominal and visceral adiposity than Caucasian women with similar body mass index. Nutr Diabetes 1:e6. https://doi.org/10.1038/nutd.2011.2

Liu M-J, Bao S, Bolin ER et al (2013) Zinc deficiency augments leptin production and exacerbates macrophage infiltration into adipose tissue in mice fed a high-fat diet. J Nutr 143:1036–1045. https://doi.org/10.3945/jn.113.175158

Lonsdale D (2006) A review of the biochemistry, metabolism and clinical benefits of thiamin(e) and its derivatives. Evid Based Complement Alternat Med 3:49–59. https://doi.org/10.1093/ecam/nek009

Lopez A, Cacoub P, Macdougall IC, Peyrin-Biroulet L (2016) Iron deficiency anaemia. Lancet 387 (10021):907–916. https://doi.org/10.1016/S0140-6736(15)60865-0

MacLean LD, Rhode BM, Shizgal HM (1983) Nutrition following gastric operations for morbid obesity. Ann Surg 198:347–355

Mahawar KK, Bhasker AG, Bindal V et al (2017) Zinc deficiency after gastric bypass for morbid obesity: a systematic review. Obes Surg 27:522–529. https://doi.org/10.1007/s11695-016-2474-8

Marreiro DN, Fisberg M, Cozzolino SMF (2002) Zinc nutritional status in obese children and adolescents. Biol Trace Elem Res 86:107–122

Mechanick J, Youdim A, Jones D et al (2013) Clinical practice guidelines for the perioperative nutritional, metabolic, and nonsurgical support of the bariatric surgery patient—2013 update: cosponsored by American Association of Clinical Endocrinologists, The Obesity Society, and American Society for Metabolic & Bariatric Surgery. Endocr Pract 19:337–372. https://doi.org/10.4158/EP12437.GL

Misra A, Chowbey P, Makkar BM et al (2009) Consensus statement for diagnosis of obesity, abdominal obesity and the metabolic syndrome for Asian Indians and recommendations for physical activity, medical and surgical management. J Assoc Physicians India 57:163–170

Mogensen K (2017) Essential fatty acid deficiency. Pract Gastroenterol XLI:37–44

Mohapatra S, Gangadharan K, Pitchumoni CS (2019) Malnutrition in obesity before and after bariatric surgery. Dis Mon 100866. https://doi.org/10.1016/j.disamonth.2019.06.008

Muñoz M, Botella F, Gómez-Ramírez S, Campos A, García-Erce J (2009) Iron deficiency and anaemia in bariatric surgical patients: causes, diagnosis and proper management. Nutr Hosp 24:640–654. https://doi.org/10.3305/nh.2009.24.6.4547

Newbury L, Dolan K, Hatzifotis M, Low N, Fielding G (2003) Calcium and vitamin D depletion and elevated parathyroid hormone following biliopancreatic diversion. Obes Surg 13:893–895. https://doi.org/10.1381/096089203322618722

Parrott J, Frank L, Rabena R, Craggs-Dino L, Isom KA, Greiman L. American Society for Metabolic and Bariatric Surgery Integrated Health Nutritional Guidelines for the Surgical Weight Loss Patient 2016 Update: Micronutrients 2016. https://doi.org/10.1016/j.soard.2016.12.018

Ponsky TA, Brody F, Pucci E (2005) Alterations in gastrointestinal physiology after Roux-En-Y gastric bypass. J Am Coll Surg 201:125–131. https://doi.org/10.1016/j.jamcollsurg.2005.03.021

Saif T, Strain GW, Dakin G, Gagner M, Costa R, Pomp A (2012) Evaluation of nutrient status after laparoscopic sleeve gastrectomy 1, 3, and 5 years after surgery. Surg Obes Relat Dis 8:542–547. https://doi.org/10.1016/j.soard.2012.01.013

Sallé A, Demarsy D, Poirier AL et al (2010) Zinc deficiency: a frequent and underestimated complication after bariatric surgery. Obes Surg 20:1660–1670. https://doi.org/10.1007/s11695-010-0237-5

Schectman G, Byrd JC, Gruchow HW (1989) The influence of smoking on vitamin C status in adults. Am J Public Health. https://doi.org/10.2105/AJPH.79.2.158

Schweiger C, Weiss R, Berry E, Keidar A (2010) Nutritional deficiencies in bariatric surgery candidates. Obes Surg 20:193–197. https://doi.org/10.1007/s11695-009-0008-3

Shah AR, Tamhankar MA (2014) Optic neuropathy associated with copper deficiency after gastric bypass surgery. Retin Cases Brief Rep 8:73–76. https://doi.org/10.1097/ICB.0000000000000008

Shike M (2009) Copper in parenteral nutrition. Gastroenterology 137:S13–S17. https://doi.org/10.1053/j.gastro.2009.08.017

Slater GH, Ren CJ, Siegel N et al (2004) Serum fat-soluble vitamin deficiency and abnormal calcium metabolism after malabsorptive bariatric surgery. J Gastrointest Surg 8:48–55

Stryker WS, Kaplan LA, Stein EA, Stampfer MJ, Sober A, Willett WC (1988) The relation of diet, cigarette smoking, and alcohol consumption to plasma beta-carotene and alpha-tocopherol levels. Am J Epidemiol. https://doi.org/10.1093/oxfordjournals.aje.a114804

Switzer BR, Atwood JR, Stark AH et al (2005) Plasma carotenoid and vitamins A and E concentrations in older African American women after Wheat Bran supplementation: effects of age, body mass and smoking history. J Am Coll Nutr 24:217–226. https://doi.org/10.1080/07315724.2005.10719468

Thibault R, Huber O, Azagury DE, Pichard C (2016) Twelve key nutritional issues in bariatric surgery. Clin Nutr 35:12–17. https://doi.org/10.1016/j.clnu.2015.02.012

Topart P (2008) Iron deficiency and anemia after bariatric surgery. Surg Obes Relat Dis 4:719–720. https://doi.org/10.1016/j.soard.2008.05.006

Uriu-Adams JY (2005) Copper, oxidative stress, and human health. Mol Asp Med 26:268–298. https://doi.org/10.1016/J.MAM.2005.07.015

Vimaleswaran KS, Berry DJ, Lu C et al (2013) Causal relationship between obesity and vitamin D status: bi-directional mendelian randomization analysis of multiple cohorts. PLoS Med 10:e1001383. https://doi.org/10.1371/journal.pmed.1001383

Viroonudomphol D, Pongpaew PMS, Tungtrongchitr R et al (2003) The relationships between anthropometric measurements, serum vitamin A and E concentrations and lipid profiles in overweight and obese subjects. Asia Pacific J Clin Nutr

Wallström P, Wirfält E, Lahmann PH, Gullberg B, Janzon L, Berglund G (2001) Serum concentrations of β-carotene and α-tocopherol are associated with diet, smoking, and general and central adiposity. Am J Clin Nutr 73:777–785. https://doi.org/10.1093/ajcn/73.4.777

Wang G, Hu FB, Mistry KB et al (2016) Association between maternal prepregnancy body mass index and plasma folate concentrations with child metabolic health. JAMA Pediatr 170:e160845. https://doi.org/10.1001/jamapediatrics.2016.0845

Ward ZJ, Bleich SN, Cradock AL et al (2019) Projected U.S. State-level prevalence of adult obesity and severe obesity. N Engl J Med 381:2440–2450. https://doi.org/10.1056/NEJMsa1909301

World Health Organization. Obesity. WHO 2014. https://www.who.int/topics/obesity/en/. Accessed December 31, 2018

Xanthakos SA (2009) Nutritional deficiencies in obesity and after bariatric surgery. Pediatr Clin N Am 56:1105–1121. https://doi.org/10.1016/j.pcl.2009.07.002

Yagan FCC, Olcum GG, Kurtulus D, Tugtekin HB (2016) The relationship between obesity and serum Vitamin B12, folic acid, vitamin D concentrations in obes adults: a retrospective study. Endocr Abstr 41:EP183. https://doi.org/10.1530/endoabs.41.EP183

Part VIII

New Apps, Drugs, Artificial Intelligence and Public Health Initiatives

Sugar-Sweetened Beverage Taxes: Origins, Mechanisms, and Current Worldwide Status

63

Fabrizio Ferretti

Contents

Introduction 852
Drink Composition 852
Tax Strategies 853

The Demand for Sugar-Sweetened Beverages (SSBs) 853
Affordability and Risk Exposure 854
Practical Example 854
Consumption Scenarios 855
Tax and Soda Sales 855
Multi-pronged Strategy 856

Taxing Sugar-Sweetened Beverages (SSBs): Rationale 856
The Pitfalls of Wrong Perceptions 857

Taxing Sugar-Sweetened Beverages (SSBs): Mechanisms 857
A Comprehensive and Enlightened Approach 858
Sugar Tax in the Real World 858
Price Elasticity of Demand 858
Cross Price Elasticity of Demand 858
The General Economy 859
Behavior of the Industry and the Government 859

Current Worldwide Status 859
Sugared Drinks in Mexico 861
Other Consequences in Mexico 861
Unjust and Regressive Taxes? 861
The Broader Context 861

Conclusions 862

References 862

F. Ferretti (✉)
School of Social Sciences, University of Modena and Reggio Emilia, Modena, Italy

Department of Communication and Economics, Palazzo Dossetti, Reggio Emilia, Italy
e-mail: fabrizio.ferretti@unimore.it

J. Faintuch and S. Faintuch (eds.), *Obesity and Diabetes*, https://doi.org/10.1007/978-3-030-53370-0_63

Abstract

The emergence of modern food systems has been causing unhealthy changes in the dietary patterns of millions of consumers worldwide. Nowadays, ultra-processed foods and beverages represent an increasing fraction of the total daily energy intake in both advanced and emerging economies. Most of these products are typically low in nutrients but high in one or more added ingredients, such as in the case of sugar-sweetened beverages (SSBs) usually rich in free sugars. The regular consumption of SSBs has been associated with adverse health outcomes, including obesity, type 2 diabetes, and cardiovascular diseases. Taxes are an increasingly popular approach to attempt reducing the overconsumption of SSBs. This chapter provides a non-technical description of the rationale, mechanisms, and current worldwide status of taxes on SSBs.

Keywords

Tax strategies · Soda sales · Price elasticity · Industry sugar tax · Government sugar tax · Regressive taxes

Introduction

The production of ultra-processed foods and beverages is a common feature of modern food systems (Monteiro et al. 2018a). Nowadays, an increasing fraction of world population has access to a large variety of convenient (i.e., durable and ready-to-consume) products, in the form of hyper-palatable and relatively inexpensive calorie-dense foods and beverages (Baker and Friel 2016; Monteiro et al. 2018b; Pan American Health Organization 2019). Most of these products are typically low in nutrients but high in one or more added ingredients (such as sugar, salt, and industry-created fats). Various types of additives and preservatives, as well as neo-formed contaminants derived from the production processes and the packaging materials (e.g., acrylamide and bisphenol A), are also present (Monteiro et al. 2019).

A traditional balanced and diversified diet is based on the regular consumption of fresh and minimally processed foods. In contrast, a 'Western-style' dietary pattern relies heavily on ultra-processed foods and beverages and is regarded as one of the key drivers of the growing worldwide burden of the leading noncommunicable diseases, such as obesity, cardiovascular diseases, diabetes, and some types of cancer (WHO 2014; Chazelas et al. 2019; Lawrence and Baker 2019). In the current food environment, a popular source of 'empty calories' comes from soft (non-alcoholic) drinks and, more specifically, from sugar-sweetened beverages (SSBs), (Martinez Steele et al. 2016). The world's most consumed SSBs include regular (i.e., non-diet) sodas, fruit juices, fruit, and milk-flavored drinks, ready-to-drink tea and coffee, flavored water, and energy, functional, and sports drinks (Singh et al. 2015a).

SSBs are defined as all types of carbonated and non-carbonated beverages containing free sugars (British Medical Association 2015). The World Health Organization (WHO) defines free sugars as all monosaccharides (e.g., glucose and fructose) and disaccharides (e.g., sucrose) inherent in the food plus any added sugars (WHO 2015). A given amount of sugar is naturally present in certain types of SSBs (e.g., fruit juices). One key feature of SSBs is a large amount of free sugar added by the manufacturer (The Nutrition Source 2019). The beverage industry adds free sugars, usually in the form of refined beet and cane sugar, or by using high-fructose corn syrup, a liquid sweetener derived from the processing of corn starch (Public Health Law and Policy 2011). These added sugars are incorporated to improve the products' palatability (Goldfein and Slavin 2015). Hence, SSBs are also called 'sugary drinks'.

Drink Composition

There are notable differences in the sugar content of these products, both across and within beverage categories. In general, the total quantity of sugar may range from 1 g to more than 50 g per 8-oz serving. A typical 8-oz serving of regular soda contains about 29 g of free sugars. One teaspoon provides approximately 4.2 g of sugar. On this basis, drinking a standard 8-oz serving of soda is equivalent to eating more than 7 tsp. of table sugar (Harris and Schwartz 2014).

In the same 8-oz serving of other popular SSBs, for example, fruit and energy drinks, the average sugar content equates to about 22 g of free sugars (around 5 tsp). Similarly, in flavored water and ready-to-drink tea/coffee, there are typically 10 and 15 g of sugar, around 2.3 and 3.5 tsp. of table sugar, respectively. However, great diversity exists even within the same beverage category. Some regular sodas contain less than 10 g of sugar per 8-oz serving, and others, more than 45 g. Likewise, fruit drinks may contain from 1 g up to 57 g (about 13.5 tsp) of sugar per 8-oz serving (Harris and Schwartz 2014).

Despite the differences between products, the regular consumption of one or more SSBs in the everyday diet normally results in an abnormal high free-sugars intake (WHO 2015). Furthermore, studies on the health impacts of frequent SSB consumption address not only the total sugar content of these products, but also the proportion of the different types of sugars (e.g., the actual fructose content of high-fructose corn syrup) added to the beverages (Ventura et al. 2011). The most documented adverse health effects associated with SSB consumption include: (1) a greater likelihood of unhealthy weight gain (thus, an increased risk for becoming overweight or obese); (2) a greater incidence of type 2 diabetes; and (3) a higher risk of hypertension (Vartanian and Brownell 2006; Te Morenga et al. 2012).

During the last two decades, the daily consumption of SSBs is one of the modifiable risk factors responsible for the increase in global attributable deaths and disability-adjusted life-years (GBD 2017 Diet Collaborators 2019; Rico-Campà et al. 2019). In order to prevent the negative health effects of an excessive intake of free sugars, the WHO (2015) suggests reducing their consumption preferably to less than 5% of the daily energy intake. This limit corresponds to about 6 tsp. of table sugar for adults per day.

Tax Strategies

Substantial research supports the need for action to curb the unhealthy habit of consuming too much sugar from SSBs (Hu 2013; Tedstone et al. 2015). Indirect taxes are an increasingly popular approach to promoting healthy eating habits by changing the relative prices of the different foods and beverages available on the market (Mytton et al. 2012; Powell et al. 2013).

Excise taxes are taxes on the sale of a specific category of goods and services, such as tobacco, alcohol, and unhealthy foods and beverages. These taxes usually target products deemed harmful to human health, but that society does not want to completely ban (Samuelson and Nordhaus 2010). Specifically, taxes on SSBs (SSB taxes) are excise taxes imposed by national or local governments for three main reasons: One, to discourage the excessive consumption of free sugars from SSBs; Secondly, to encourage producers to develop healthier soft drinks; and thirdly, to raise revenue to finance comprehensive policy programs, devoted to promoting healthy eating habits (WHO 2017).

The main motivation behind implementing SSB taxes is that the tax will increase the shelf price. This price rise will result in a decline in regular consumption of SSBs, which, in turn, will reduce the incidence of the harmful effects of an excessive intake of free sugars. Regular sodas are usually the most common SSBs. Hence, these taxes are often called 'soda (or sugar) taxes'.

The Demand for Sugar-Sweetened Beverages (SSBs)

Due to the combination of rising disposable income and declining prices during the last 25 years, SSBs have become more affordable in more than 80 countries worldwide (Blecher et al. 2017). The affordability of SSBs is measured by the percentage of consumer's disposable income required to buy a given amount (e.g., 1 l) of product. As a result, it captures changes in SSB prices and consumers' income simultaneously (Ferretti and Mariani 2019). Considering the same share of income, a typical world consumer could buy about 70% more SSBs in 2016 than 1990. This percentage has risen above 85% in many emerging economies, in which the affordability of SSBs has nearly doubled in less than three decades (Blecher et al. 2017).

Affordability and Risk Exposure

According to the Global Burden of Disease (GBD) study, during around the same time frame (1990–2016), the 'summary exposure value' (SEV) of SSBs increased by more than 40% (GBD 2016 Risk Factors Collaborators 2017; Singh et al. 2015b). The SEV measures the population's exposure to a specific risk factor, by considering both the extent of exposure by risk level and the severity of that risk's contribution to disease burden (Institute for Health Metrics and Evaluation 2019). For a given content of free sugars, the SEV of products such as SSBs is mainly affected by the average level of SSBs' consumption among the general population. Research has consistently shown that the amount of SSBs consumed in each population primarily responds to SSB market prices, and all else being equal, prices and consumption are negatively related (Andreyeva et al. 2010; Goryakin et al. 2017). The curve labeled *D* in Fig. 63.1a illustrates this inverse relationship.

Practical Example

Let's assume that only one type of SSB is available in the market, for example, a carbonated soft drink containing 39 g of free sugars (about 9.3 tsp) per 12-oz (335-ml) serving. In Fig. 63.1a, the vertical axis of the graph shows the average retail price (*P*) of the soda, measured in U.S. dollars (USD). The horizontal axis shows the total quantity of soda demanded in the market (*Q*), measured in units per period (e.g., cans per day). Finally, *D* is the demand curve that shows how many cans of soda consumers are willing to buy per day as the price per unit changes, *ceteris paribus*.

The demand curve for SSBs slopes downward. A relatively lower price may encourage consumers who have already been buying the soda, to purchase more cans per day. Likewise, it may allow other consumers who were previously unable to afford the product to begin buying it. As shown in Fig. 63.1a, if the average price per decreases from P_1 to P_2, the quantity demanded increases from Q_1 to Q_2 and *vice versa*. Overall,

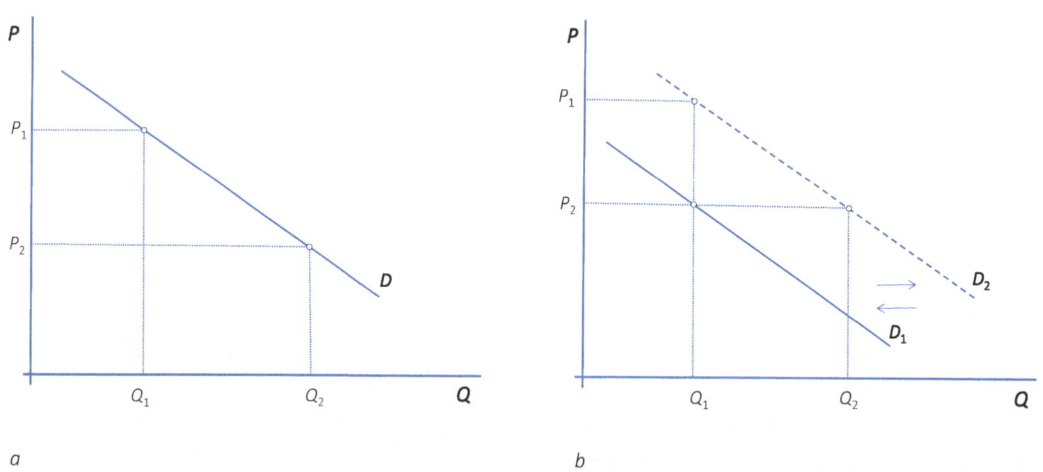

Fig. 63.1 (**a** and **b**) The demand curve for sugar-sweetened beverages (SSBs)

changes in price result in changes in the quantity demanded, that is, in movements along a given demand curve (such as *D*).

The quantity of soda that consumers are willing to buy in this market depends on many other factors besides its price, among which consumers' income is especially important. A good whose consumption increases when income rises is called a 'normal good'. Empirical evidence indicates that the consumption of SSBs varies substantially by income groups. Up to a threshold level of disposable income, low-income consumers tend to buy more SSBs when their incomes rise (Singh et al. 2015a; Allcott et al. 2019; Ferretti and Mariani 2019).

More generally, besides the market price, any other factor influencing purchasing behavior will result in a change in demand, and thus, in a shift of the entire demand curve. Understanding the difference between changes in the quantity demand and changes in demand is essential to improve a soda tax, and to design an effective and comprehensive policy program to reduce SSBs' consumption. Specifically, a change in the quantity demand is a movement along a given demand curve that occurs as a result of changes in the good's price. A change in demand is a shift of the entire demand curve, caused by a change in a determinant of demand other than the good's price.

Consumption Scenarios

The effects of changes in income are shown in Fig. 63.1b. If soda is a 'normal good', and there are no changes in the price of soda when consumers' income increases, then consumption also increases. This increased consumption occurs independently of the market price. Consequently, the demand for soda changes. That is, there is a shift to the right of the entire curve (in Fig. 63.1b, this is shown as a shift from the solid D_1 curve to the dashed D_2 curve). In other words, with higher incomes, consumers are willing to pay a higher price (P_1 instead of P_2) to buy any given quantity (Q_1) of sodas.

Along with disposable income, the market for soda is also shaped by consumers' tastes. Changes in taste will affect the number of sodas that consumers want to purchase, given their disposable income and the soda market price. If consumers become aware of the negative health effects of SSBs, then they will be willing to buy fewer sodas at any given price. As exemplified in Fig. 63.1b, this will shift the entire demand curve to the left (from D_2 to D_1). Conversely, a successful industry's advertising strategy shapes consumers' tastes and shifts the demand curve outward again to D_2.

Tax and Soda Sales

In Fig. 63.2, it is assumed that at price *P*, consumers buy *Q* cans of soda per day (point *A* on the solid demand curve D_2). To discourage SSBs' consumption, the government introduces an excise tax that raises the shelf price of each can of soda from *P* to P_{TAX}. Without changes occurring in their incomes and tastes, consumers respond to the higher price by reducing the quantity demanded from *Q* to Q_{TAX}. In Fig. 63.2, this corresponds to an upward movement along the D_2 demand curve until point *B*.

However, producers might be able to 'sterilize' the impact of the tax, for instance, by developing an aggressive price discount strategy to keep the retail price almost unchanged at the original level *P*. In this case, the quantity demand remains nearly *Q*. The tax has little or no impact on consumers' behavior, and so little or no impact on the population's intake of free sugars, because government policy only affects the quantity demand. Conversely, if the revenue collected with the tax is used to implement an effective program to enhance consumers' nutrition and health education, consumers' tastes are affected. As a result, the entire demand curve shifts to the left. This scenario is illustrated in Fig. 63.2. It

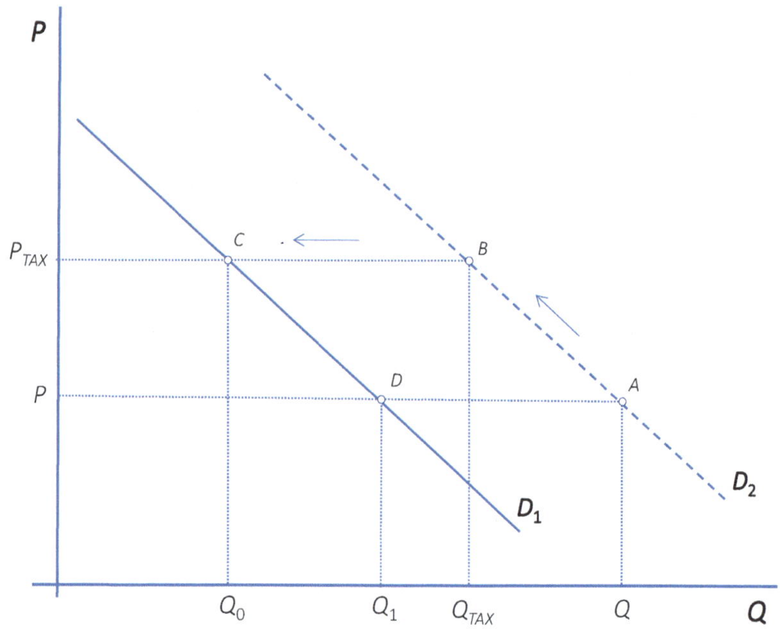

Fig. 63.2 The impact of an excise tax on sugar-sweetened beverages (SSBs)

shows that the tax decreases the quantity demand (from Q to Q_{TAX}), and the education program decreases the demand (from D_2 to D_1). Eventually, the daily consumption of soda becomes Q_0 (point C on the D_2 curve) as a result of changes in price and tastes, respectively.

Multi-pronged Strategy

If producers attempt to 'sterilize' the effect of the tax by reducing the price back to its pre-tax level P, then the strategy will have little effect on Q. The reason being that better-educated consumers are willing to buy fewer cans of soda per day at any given price. Thus, a decrease in price from P_{TAX} to its original level P increases the quantity demand. This effect is shown as a downward movement along the new D_1 curve from Q_0 only to Q_1 (point D), which is well below the initial level of consumption Q. In other words, the use of taxes as economic tools to affect health-related purchasing behavior should be part of a comprehensive policy approach.

Taxing Sugar-Sweetened Beverages (SSBs): Rationale

Sometimes consumers tend to ignore the adverse health effects of their consumption patterns (Bernheim 2016). There are two main reasons for the overconsumption of SSBs. Either the consumers are misinformed, or they deal with problems of self-control and time-inconsistency. In the former case, consumers do not know that free sugars can damage their health. In the latter case, consumers prioritize the immediate satisfaction of drinking SSBs, over future negative consequences.

In both cases, consuming high levels of free sugars from SSBs has two main effects. First, it harms consumers of SSBs by damaging their health. The direct negative effects of SSBs on consumers' health lead to various monetary and non-monetary costs, such as not-reimbursed out-of-pocket health care expenses, as well as wage loss and reduced income, due to poorer health conditions (for instance, obesity and diabetes). These are called 'internality costs' because they are borne directly by SSB consumers. Second, the unhealthy habit of consuming SSBs generates indirect costs that spread to the whole economy, *via* the public (or private) health system. These are called 'externality costs' because they are unintended and uncompensated costs that SSB consumers impose on the entire health system, primarily in the form of increased health care expenditure on treating SSB-attributable diseases (Allcott et al. 2019).

The Pitfalls of Wrong Perceptions

Internality and externality indicate that there is a divergence between the perceived and actual costs of consuming SSBs. This divergence leads individuals to consume more SSBs than they would if they were able to consider all the direct and indirect costs. As a result, society faces a market failure, namely a misleading market reality. A situation that occurs when the individual pursuit of self-interest leads to adverse results for the collectivity as a whole (Perloff 2015). The attempt to correct this market failure is the economic rationale underlying the government intervention in the market for SSBs. In this context, a 'sugar tax' is an economic tool used to close the gap between the price that the consumer pays for a can of sugary drink at the time of purchase, and the total (i.e., internal and external) cost associated with SSB consumption (American Heart Association 2016; Griffith et al. 2016).

These corrective taxes are also named 'Pigouvian taxes' after the Cambridge economist A.C. Pigou, who emphasized their usefulness to correct market failure, as early as the beginning of the twentieth century (Pigou 1920). SSB taxes are thus policy tools designed to discourage activities that yield internal and external costs. Essentially, by affecting the price paid by the consumers and received by the producer, these taxes are economic signals, a piece of information that helps people make better economic decisions (Perloff 2015).

Taxing Sugar-Sweetened Beverages (SSBs): Mechanisms

The mechanisms through which a 'sugar tax' is supposed to operate are illustrated in Fig. 63.3. The government decides to discourage SSB consumption by levying an excise tax on the beverage industry. Some fraction of the excise tax is passed on to distributors, and, in turn, on to the final consumers in the form of a higher retail price. The remaining fraction of the tax (along with its administrative costs) falls on producers. Consumers respond to the higher price by reducing the quantity demand (i.e., by moving upward along the demand curve). Given the SSBs' nutrition facts, a lower consumption means fewer calories from free sugars, resulting in better health outcomes (Institute of Economic Affairs 2016).

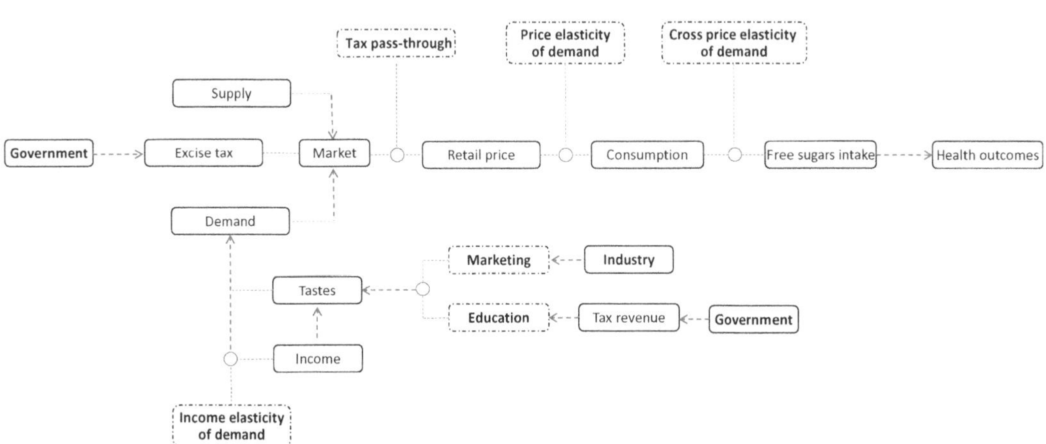

Fig. 63.3 Taxing sugar-sweetened beverages (SSBs): Mechanisms

A Comprehensive and Enlightened Approach

In addition to these basic static effects, the tax may help in triggering a virtuous cycle of a gradual reduction in SSB consumption. The revenue collected with the 'sugar tax' is used by the government to implement nutrition education programs. Increasing public awareness about the health implication of SSB consumption decreases the demand for SSBs (i.e., the entire demand curve shifts to the left). A declining demand encourages the beverage industry to develop healthier products, to avoid taxes and meet emerging trends in consumers' health-related lifestyles. As a result of all these demand and supply changes, the consumption of SSBs falls, reducing people's exposure to the harmful effects of free sugars and decreasing the incidence of SSB-related diseases (Heise et al. 2016).

However, as highlighted in Fig. 63.3, several links in this complex chain of events deserve careful consideration from policymakers (Allcott et al. 2019; Francis et al. 2016). The quantity demand decreases only if the tax results in a higher shelf price. In other words, the direct impact of the tax on SSB consumption depends primarily on the extent to which the tax is passed on to consumers. Besides the question of who will bear the administrative costs of the tax (i.e., the time and money spent to compute and pay the tax to the government), the tax incidence indicates how the economic burden (i.e., the amount of the tax) is divided between consumers and producers (Mankiw 2016).

Sugar Tax in the Real World

The burden of an excise tax is usually shared by both buyers and sellers. The tax drives a wedge between the price paid by consumers and the price received by producers. The consequence of this wedge is that consumers pay more, and producers receive less. As a general rule, the tax falls more heavily on the side of the market that is less responsive to price changes (Mankiw 2016). The beverage industry, however, is dominated by large multinational corporations that set the product prices based on their direct costs per unit of output, plus a mark-up to cover overhead costs and profits. In theory, these firms can absorb some or all of the tax, by offering discounts and reducing their profit margins (Powell and Maciejewski 2018).

Even when the tax is mainly (or fully) paid by the consumers, the quantity demand only decreases if consumers are sufficiently responsive to price changes. This responsiveness is usually measured by the price elasticity of demand, computed as the ratio of the percentage change in quantity demanded to the percentage change in price (Perloff 2015).

Price Elasticity of Demand

Specifically, the price elasticity of demand provides the percentage change in the quantity demanded that results from a 1% increase in price. Thus, it captures the willingness of consumers to leave the market when conditions become unfavorable. If consumers have particularly strong preferences for SSBs, the demand will be very inelastic (i.e., the price elasticity is well below one). In this instance, a tax-induced price change would be an ineffective strategy to nudge consumer behavior toward healthier dietary habits (Powell and Maciejewski 2018).

The impact of SSB taxes on health outcomes also depends on consumers' choice of alternative foods and beverages. Consumers might react to higher SSB prices by shifting their demand toward other unhealthy, but untaxed products. If consumers have developed strong preferences for sugary tastes, they could buy more untaxed sugar-sweetened foods and drinks (such as ultra-processed snacks, fruit juice or smoothies).

Cross Price Elasticity of Demand

The direction and strength of these 'cross effects' are captured by the cross-price elasticity of demand, which measures the percentage change in one good's quantity demanded (e.g., a fruit juice) that results from a 1% change in the price of another good (e.g., a regular soda). This coefficient can be a positive or negative number, depending on whether the two goods are substitutes or complements. For instance, if SSBs and bottled water are substitutes, a sugar tax that raises SSB price will increase the demand for bottled water. *Vice versa*, if two goods are complements, as the price of one rises (falls), the demand for the other falls (rises). Thus, if regular sodas and ultra-processed snacks are complements, an excise tax on soft drinks will decrease the demand for unhealthy snacks (Mankiw 2016).

The General Economy

The impact of a tax that increases the retail price may be offset by a rising disposable income. In emerging economies SSBs tend to be 'normal goods', and the income per capita increases rapidly. Hence, a small excise tax is likely to have only a limited and transitory effect on SSB consumption, and virtually no impact on health outcomes. The reason is that the increase in demand, due to the higher income, outweighs the decrease in the quantity demand due to the tax. In such situations, SSBs become more affordable, and the resulting net effect is an increase in SSB consumption despite the excise tax (Ferretti and Mariani 2019).

Behavior of the Industry and the Government

Producers may react with a collective advertising strategy designed to offset the impact of the tax. Producers may also choose to reformulate their products by switching from regular to diet (i.e., no calories) beverages, in order to avoid the tax. Although the diet versions are without added sugars, these products have no nutritional value, and their health effects are still uncertain. Finally, the government may promote healthy dietary patterns, by using the revenue collected with the tax to finance nutrition education programs. These programs tend to reinforce the impact of the tax, because better-educated consumers are normally less exposed to industry advertising strategies (Obesity Policy Coalition 2018).

Current Worldwide Status

In many countries, sugar has been heavily taxed, or even a state monopoly, for centuries. The origins of a tax on SSBs, however, date back to the beginning of the last century. A specific excise tax on soft drinks was first proposed in 1914 by U.S. President Woodrow Wilson, to replace declining revenue from import tariffs (Blakey 1917; Brownlee 1985). An excise tax on SSBs was then introduced in Denmark during the mid-1930s, and Finland in the 1940s (Global Food Research Program 2019).

At those times, the governments were not concerned about public health issues. These taxes were levied merely to collect revenue in order to balance the national budget. The modern history of SSB taxes began in the early 1980s in Norway, but most of the currently existing SSB taxes were established after 2010, notably during the last 5 years. Nowadays, SSB taxes have been implemented in more than 35 countries worldwide, and seven U.S. local municipalities (Global Food Research Program 2019; Wan et al. 2017).

Apart from minor differences, current SSB taxes share a common design. They are usually collected from producers or distributors (i.e., the legal burden to remit the tax normally falls on manufacturers or retailers), and consumers simply see the tax as somewhat higher retail price. Overall, an excise tax can be specific, or *ad valorem*—that is, a given amount per beverage unit (e.g., 10 cents per liter) or a given percentage of the beverage price (e.g., 10% of the price of 1 l). SSB taxes are mostly specific and typically computed on a volumetric basis (a constant rate per unit of product). In some U.S. cities, for example, the average SSB tax is 1.5 cents per ounce of liquid (Global Food Research Program 2019; Wan et al. 2017).

Less frequently, SSB taxes directly target the amount of sugar content (for example, 0.5 cents per gram of sugar) or adopt a tiered tax design, which applies different tax rates depending on the sugar content. This is the case, for example, in the U.K., which has two tax bands. One is for SSBs with more than 5 g of sugar per 100 ml and the other for more than 8 g of sugar per 100 ml, taxed at 18 and 24 pence per liter, respectively (Global Food Research Program 2019; Wan et al. 2017).

Over the past 5 years, seven U.S. cities have introduced an excise tax on SSBs. A tax was first established in 2014, in Berkeley, followed in November 2016, by three other Californian cities (Albany, Oakland, San Francisco), and then by Seattle (Washington) in 2017. Since 2016, Boulder (Colorado) and Philadelphia (Pennsylvania) have also implemented a tax on SSBs. All these cities use a tax rate that ranges from 1 to 2 cents per ounce, levied on all SSBs, except milk products and 100% fruit juice. Philadelphia is the only U.S. city that also taxes diet drinks (Allcott et al. 2019).

A picture of the worldwide status of SSB taxes is shown in Fig. 63.4. At the end of 2019, of the 39 countries that have implemented a tax on SSBs, ten are in Europe, seven in the Americas, eleven in the western Pacific region, and the remaining are spread throughout African, eastern Mediterranean, and Southeast Asian countries. The governments of several countries (such as Australia, Canada, Singapore, and Italy) have recently announced plans to imminently introduce a tax on SSBs. Besides the U.K., also in France, and Portugal taxes on SSBs are proportional to the sugar content, but other countries have typically set a constant rate per volume of product (Global Food Research Program 2019; Wan et al. 2017).

Empirical evidence indicates that producers usually do not attempt to 'sterilize' the tax. They normally pass along the supply chain down to the final consumers and raise the retail price by around the same amount of the tax. In countries such as Denmark, Mexico, and France, a 100% pass-through of the tax on the retail price was found. In some U.S. cities (Philadelphia, Boulder, and Berkeley), the percentage of the tax that falls on the consumer ranges from 75% to 90%, depending on the product variety and store type (Cawley et al. 2019; Powell et al. 2013).

Studies (Cawley et al. 2019) indicate that SSBs exhibit a price elasticity that varies widely with product types and consumer groups. Responsiveness of consumers is estimated to be, on average, around one (specifically, between 0.8

Fig. 63.4 The worldwide status of sugar-sweetened beverages taxes. Figure based on data from Global Food Research Program 2019 and Allcott et al. 2019

and 1.2). It means that an excise tax that raises the shelf price, for instance, by 20%, is expected to decrease the quantity demand for that good by 20% too, *ceteris paribus* (Muhammad et al. 2019; Powell et al. 2013).

Sugared Drinks in Mexico

In the Organization for Economic Cooperation and Development (OECD), Mexico is one with both high consumption of SSBs and high prevalence of obesity and diabetes (WHO 2011). In 2012, about 30% of the Mexican population was obese, and diabetes mellitus affected around 10% of the adult population. The prevalence of obesity and diabetes has been growing rapidly during the last decades, along with the consumption of SSBs, which exceeded 160 l per capita in 2013 (Colchero et al. 2017).

According to preliminary studies, an increase of 20% in the retail price of SSBs would have decreased consumption from 163 to 120 l per capita, which is a 26% reduction (Donaldson 2015). In January 2014, Mexican authorities passed a 1 peso (about 0.12 USD) per liter excise tax on all SSBs, which increased the price paid by the consumers by about 10%. During the following 2 years, the consumption of SSBs declined by 7.6% (5.5% in 2014 and 9.7% in 2015) among the general population, and by more than 11.5% among consumers with the fewest resources (Colchero et al. 2017).

Other Consequences in Mexico

These results imply a price elasticity of 0.76 and 1.5, respectively, and indicate greater price responsiveness in relatively poorer consumers (Pan American Health Organization 2015). Conversely, the consumption of untaxed beverages increased, on average, by more than 2% during the same period. Specifically, research has found a cross-price elasticity for water and unhealthy snacks of +0.10 and − 0.23, respectively. In other words, these 'cross effects' improved the positive effects of the tax on consumers' behavior. Moreover, in 2 years, the tax has generated revenue of more than 2.6 billion USD, part of which might be used to finance measures promoting healthy habits, such as the installation of drinking fountains at schools and public spaces (Colchero et al. 2017; Pan American Health Organization 2015).

Unjust and Regressive Taxes?

A regressive tax is one that takes a larger fraction of income from the underprivileged than from high-income groups. Overall, all excise taxes are inherently regressive because they do not vary with an individual's financial status. Furthermore, SSBs often behave as inferior goods, so that lower- and middle-income consumers spend a larger fraction of their income on SSBs than richer consumers. In the U.S., for instance, people with a household income above 75,000 USD per year, on average, consume about 117 cal per day from SSBs, roughly half of those consumed by people with a household income below 15,000 USD per year (Allcott et al. 2019).

The Broader Context

There are three main reasons why this regressive effect may be negligible. First, even in low-income groups, expenditure on SSBs is usually a very small part of household budgets (Duckett et al. 2016). Second, the responsiveness to SSB taxes tends to be greater in low-income consumers. In Berkeley (California), for example, the consumption of SSBs dropped by about 52% among low-income residents in the 3 years after the implementation of a tax of 1 cent per ounce on all SSBs, excluding diet drinks (Lee et al. 2019). Third, the health benefits of reducing SSBs tend to be greater for low-income individuals, who are usually the most disadvantaged in access to high-quality health care services. In other words, strong progressive health benefits are likely to overcome the 'mildly regressive' direct impact of the tax on low-income consumers' expenditures (Allcott et al. 2019; Grummon et al. 2019).

Conclusions

The first attempts to quantify tax effects indicate a persistent reduction in SSB consumption in favor of healthier alternatives. It thus suggests that fiscal policy may be an effective tool to promote healthy dietary patterns (Lee et al. 2019; Teng et al. 2019; Vecino-Ortiz and Arroyo-Ariza 2018; Veerman et al. 2016).

Some guiding principles should be followed (Allcott et al. 2019; Lloyd and MacLaren 2019; Jou and Techakehakij 2012). The excise tax should be levied on all kinds of soft drinks with added sugar and implemented statewide to avoid both substitution effects (increase in demand of untaxed SSBs), and cross-border shopping. As the harmful health effects of SSBs come from sugar, the tax should be based on the quantity of sugar contained in the beverages, rather than on the amount of liquid. This strategy will encourage consumers and producers to switch to low-sugar drinks (Grummon et al. 2019).

The tax should reduce as well the gap between the private and social costs of consuming SSBs. The socially optimal tax rate is the one needed to offset both the internality and the externality costs of consuming an additional unit of sugar from SSBs. This scenario usually implies an increase of at least 10% to 20% of the average retail price. The revenue collected by SSB taxes should be used to foster health and nutrition education among children and adults. Finally, SSB taxes should be a component of a more comprehensive policy . Such a plan should include effective controls to reduce people's exposure to SSB marketing, phasing out SSBs from schools, workplaces, and all public institutions, and social campaigns to promote healthy lifestyles (Allcott et al. 2019; Backholer et al. 2017; Lloyd and MacLaren 2019; Jou and Techakehakij 2012).

References

Allcott H, Lockwood BB, Taubinsky D (2019) Should we tax sugar-sweetened beverages? An overview of theory and evidence. J Econ Perspect 33(3):202–227. https://doi.org/10.1257/jep.33.3.202

American Heart Association (2016) Reducing sugar-sweetened beverage consumption. A focus on sugar-sweetened beverage taxes. Revised June 2016. American Heart Association, Dallas, TX

Andreyeva T, Long MW, Brownell KD (2010) The impact of food prices on consumption: a systematic review of research on the price elasticity of demand for food. Am J Public Health 100(2):216–222. https://doi.org/10.2105/AJPH.2008.151415

Backholer K, Blake M, Vandevijvere S (2017) Sugar-sweetened beverage taxation: an update on the year that was 2017. Public Health Nutr 20(18):3219–3224. https://doi.org/10.1017/S1368980017003329

Baker P, Friel S (2016) Food systems transformations, ultra-processed food markets and the nutrition transition in Asia. Glob Health 12(1). https://doi.org/10.1186/s12992-016-0223-3

Bernheim BD (2016) The good, the bad, and the ugly: a unified approach to behavioral welfare economics. J Benefit–Cost Anal 7(1):12–68. https://doi.org/10.1017/bca.2016.5

Blakey RG (1917) The War Revenue Act of 1917. Am Econ Rev 7(4):791–815

Blecher E, Liber AC, Drope JM et al (2017) Global trends in the affordability of sugar-sweetened beverages, 1990–2016. Prev Chronic Dis 14:E37. https://doi.org/10.5888/pcd14.160406

British Medical Association (2015) Food for thought: promoting healthy diets among children and young people. British Medical Association, London, UK

Brownlee WE (1985) Wilson and financing the modern state: the Revenue Act of 1916. Proc Am Philos Soc 129(2):173–210

Cawley J, Thow AM, Wen K et al (2019) The economics of taxes on sugar-sweetened beverages: a review of the effects on prices, sales, cross-border shopping, and consumption. Annu Rev Nutr 39:317–338. https://doi.org/10.1146/annurev-nutr-082018-124603

Chazelas E, Srour B, Desmetz E et al (2019) Sugary drink consumption and risk of cancer: results from NutriNet-Santé prospective cohort. BMJ 366(l):2408. https://doi.org/10.1136/bmj.l2408

Colchero MA, Rivera-Dommarco J, Popkin BM et al (2017) In Mexico, evidence of sustained consumer response two years after implementing a sugar-sweetened beverage tax. Health Aff 36(3):564–571. https://doi.org/10.1377/hlthaff.2016.1231

Donaldson E (2015) Advocating for sugar-sweetened beverage taxation—a case study of Mexico. Bloomberg School of Public Health, Johns Hopkins University, Baltimore, MD, USA

Duckett SH, Swerissen H, Wiltshire T (2016) A sugary drinks tax: recovering the community costs of obesity, Grattan Institute, Melbourne, Australia

Ferretti F, Mariani M (2019) Sugar-sweetened beverage affordability and the prevalence of overweight and obesity in a cross section of countries. Glob Health 15(1):30. https://doi.org/10.1186/s12992-019-0474-x

Francis N, Marron D, Rueben K (2016) The pros and cons of taxing sweetened beverages based on sugar content. Urban Institute, Washington, DC, USA

GBD 2016 Risk Factors Collaborators (2017) Global, regional, and national comparative risk assessment of

84 behavioral, environmental and occupational, and metabolic risks or clusters of risks, 1990–2016: a systematic analysis for the Global Burden of Disease Study 2016. Lancet 390(10100):1345–1422. doi: https://doi.org/10.1016/S0140-6736(17)32366-8

GBD 2017 Diet Collaborators (2019) Health effects of dietary risks in 195 countries, 1990–2017: a systematic analysis for the Global Burden of Disease Study 2017. Lancet 393(10184):1958–1972. https://doi.org/10.1016/S0140-6736(19)30041-8

Global Food Research Program (2019) Multi-country obesity prevention initiative. University of North Carolina, Chapel Hill, NC, viewed 5 Dec 2019. http://globalfoodresearchprogram.web.unc.edu/multi-country-initiative/

Goldfein KR, Slavin JL (2015) Why sugar is added to food. Compr Rev Food Sci Food Saf 14(5):644–656. https://doi.org/10.1111/1541-4337.12151

Goryakin Y, Monsivais P, Suhrcke M (2017) Soft drink prices, sales, body mass index and diabetes: evidence from a panel of low-, middle- and high-income countries. Food Policy 73:88–94. https://doi.org/10.1016/j.foodpol.2017.09.002

Griffith R, Lührmann M, O'Connell M et al (2016) Using taxation to reduce sugar consumption. IFS Briefing Note BN180. Institute for Fiscal Studies, London, UK

Grummon AH, Lockwood BB, Taubinsky D et al (2019) Designing better sugary drink taxes. Science 365 (6457):989–990. https://doi.org/10.1126/science.aav5199

Harris J, Schwartz M (2014) Sugary drink facts 2014. Rudd Center for Food Policy and Obesity, Hartford, CT

Heise TL, Katikireddi SV, Pega F et al (2016) Taxation of sugar-sweetened beverages for reducing their consumption and preventing obesity or other adverse health outcomes. Cochrane Database Syst Rev 8: CD012319. https://doi.org/10.1002/14651858.CD012319

Hu FB (2013) Resolved: there is sufficient scientific evidence that decreasing sugar-sweetened beverage consumption will reduce the prevalence of obesity and obesity-related diseases. Obes Rev 14(8):606–619. https://doi.org/10.1111/obr.12040

Institute for Health Metrics and Evaluation (2019). Summary exposure value (SEV). Terms defined. Seattle, WA, USA

Institute of Economic Affairs (2016) Sugar taxes: a briefing. Institute of Economic Affairs, London, UK

Jou J, Techakehakij W (2012) International application of sugar-sweetened beverage (SSB) taxation in obesity reduction: Factors that may influence policy effectiveness in country-specific contexts. Health Policy 107 (1):83–90. https://doi.org/10.1016/j.healthpol.2012.05.011

Lawrence MA, Baker PI (2019) Ultra-processed food and adverse health outcomes. BMJ 365:l2289. https://doi.org/10.1136/bmj.l2289

Lee MM, Falbe J, Schillinger D et al (2019) Sugar-sweetened beverage consumption 3 years after the Berkeley, California, sugar-sweetened beverage tax. Am J Public Health 109(4):637–639. https://doi.org/10.2105/AJPH.2019.304971

Lloyd P, MacLaren D (2019) Should we tax sugar and if so how? Aust Econ Rev 52(1):19–40. https://doi.org/10.1111/1467-8462.12299

Mankiw NG (2016) Microeconomics, 6th edn. Worth Publishers, New York, NY

Martinez Steele E, Baraldi LG, da Costa Louzada ML et al (2016) Ultra-processed foods and added sugars in the US diet: evidence from a nationally representative cross-sectional study. BMJ Open 6(3):e009892. https://doi.org/10.1136/bmjopen-2015-009892

Monteiro CA, Cannon G, Moubarac JC et al (2018a) The UN decade of nutrition, the NOVA food classification and the trouble with ultra-processing. Public Health Nutr 21(1):5–17. https://doi.org/10.1017/S1368980017000234

Monteiro CA, Moubarac JC, Levy RB et al (2018b) Household availability of ultra-processed foods and obesity in nineteen European countries. Public Health Nutr 21(1):18–26. https://doi.org/10.1017/S1368980017001379

Monteiro CA, Cannon G, Lawrence M et al (2019) Ultra-processed foods, diet quality, and health using the NOVA classification system. FAO, Food and Agriculture Organization of the United Nations, Rome, Italy

Muhammad A, Meade B, Marquardt DR et al (2019) Global patterns in price elasticities of sugar-sweetened beverage intake and potential effectiveness of tax policy: a cross-sectional study of 164 countries by sex, age and global-income decile. BMJ Open 9(8):e026390. https://doi.org/10.1136/bmjopen-2018-026390

Mytton OT, Clarke D, Rayner M (2012) Taxing unhealthy food and drinks to improve health. BMJ 344:e2931–e2931. https://doi.org/10.1136/bmj.e2931

Obesity Policy Coalition (2018) Policy Brief. A comprehensive policy program to reduce consumption of sugar sweetened beverages in Australia. Obesity Policy Coalition, Melbourne, Australia

Pan American Health Organization (2015) Taxes on sugar-sweetened beverages as a public health strategy: the experience of Mexico. Pan American Health Organization, Mexico D.F., Mexico

Pan American Health Organization (2019) Ultra-processed food and drink products in Latin America: Sales, sources, nutrient profiles, and policy implications. PAHO, Washington, DC, USA

Perloff JM (2015) Microeconomics, 7th edn. Pearson Education, Harlow

Pigou AC (1920) The economics of welfare, 1st edn. Macmillan, London

Powell LM, Maciejewski ML (2018) Taxes and sugar-sweetened beverages. JAMA 319(3):229–230. https://doi.org/10.1001/jama.2017.19522

Powell LM, Chriqui JF, Khan T et al (2013) Assessing the potential effectiveness of food and beverage taxes and subsidies for improving public health: a systematic review of prices, demand and body weight outcomes.

Obes Rev 14(2):110–128. https://doi.org/10.1111/obr.12002
Public Health Law & Policy (2011) Breaking down the chain: a guide to the soft drink industry. NPLAN National Policy and Legal Analysis Network to Prevent Childhood Obesity, Oakland, CA
Rico-Campà A, Martínez-González MA, Alvarez-Alvarez I et al (2019) Association between consumption of ultra-processed foods and all cause mortality: SUN prospective cohort study. BMJ 365:l1949. https://doi.org/10.1136/bmj.l1949
Samuelson PA, Nordhaus WD (2010) Economics, 19th edn. McGraw-Hill, Boston, NY
Singh GM, Micha R, Khatibzadeh S et al (2015a) Global, regional, and national consumption of sugar-sweetened beverages, fruit juices, and milk: a systematic assessment of beverage intake in 187 countries. PLoS One 10 (8):e0124845. https://doi.org/10.1371/journal.pone.0124845
Singh GM, Micha R, Khatibzadeh S et al (2015b) Estimated global, regional, and national disease burdens related to sugar-sweetened beverage consumption in 2010. Circulation 132(8):639–666. https://doi.org/10.1161/CIRCULATIONAHA.114.010636
Te Morenga L, Mallard S, Mann J (2012). Dietary sugars and body weight: systematic review and meta-analyses of randomized controlled trials and cohort studies. BMJ, 346(jan15 3), e7492–e7492. https://doi.org/10.1136/bmj.e7492
Tedstone A, Targett V, Allen R, et al (2015) Sugar reduction: the evidence for action. Public Health England, London
Teng AM, Jones AC, Mizdrak A et al (2019) Impact of sugar-sweetened beverage taxes on purchases and dietary intake: systematic review and meta-analysis. Obes Rev 20(9):1187–1204. https://doi.org/10.1111/obr.12868
The Nutrition Source (2019) Sugary Drinks. Harvard School of Public Health, Boston, MA, USA
Vartanian LR, Brownell KD (2006) Effects of soft drinks consumption on nutrition and health: a systematic review and meta-analysis. Am J Pub Health 97(4):667–675. https://doi.org/10.2105/AJPH.2005.083782
Vecino-Ortiz AI, Arroyo-Ariza D (2018) A tax on sugar sweetened beverages in Colombia: estimating the impact on overweight and obesity prevalence across socio economic levels. Soc Sci Med 209:111–116. https://doi.org/10.1016/j.socscimed.2018.05.043
Veerman JL, Sacks G, Antonopoulos N et al (2016) The impact of a tax on sugar-sweetened beverages on health and health care costs: a modelling study. PLoS One 11(4):e0151460. https://doi.org/10.1371/journal.pone.0151460
Ventura EE, Davis JN, Goran MI (2011) Sugar content of popular sweetened beverages based on objective laboratory analysis: focus on fructose content. Obesity 19 (4):868–874. https://doi.org/10.1038/oby.2010.255
Wan L, Watson E, Arthur R (2017) Sugar taxes: the global picture in 2017. The Beverage Daily, viewed 20 Oct 2019. https://www.beveragedaily.com/Article/2017/12/20/Sugar-taxes-The-global-picture-in-2017
WHO (2011) Noncommunicable diseases country profiles 2011. World Health Organization, Geneva, Switzerland
WHO (2014) Global status report on noncommunicable diseases 2014. World Health Organization, Geneva, Switzerland
WHO (2015) Guideline: sugars intake for adults and children. World Health Organization, Geneva, Switzerland
WHO (2017) Taxes on sugary drinks: why do it? World Health Organization, Geneva, Switzerland

Mobile Health Interventions for Weight Management in Overweight and Obese Populations

64

Lynnette Lyzwinski

Contents

Introduction 866
mHealth Overview 866
Usage and Trends 867
mHealth for Weight Loss in General Populations 867

Special Populations 867
Pregnancy 867
Adolescents 869

Quality Assessment 869

Consideration of Publication Bias 870

Integration of Behaviour Change Techniques 870

Use of Theory 871

Emerging Research in Other Populations 871

Commercial Weight Loss Apps 871
App Quality 872
Mindfulness and Weight Loss 874
Top-Ranked Apps 875

Evidence for Effectiveness 875

Privacy Issues 875

Consumer Preferences for mHealth 875
User-Friendly Features 875
Barriers 876

Challenges with mHealth 876

New Areas of mHealth Research 877
Preliminary Student Analysis 877

L. Lyzwinski (✉)
Centre for Online Health (PA Hospital) and Department of Psychiatry (RBWH Hospital), School of Medicine, University of Queensland, Woolloongabba, Australia
e-mail: Lynnette.Lyzwinski@uqconnect.edu.au

J. Faintuch and S. Faintuch (eds.), *Obesity and Diabetes*, https://doi.org/10.1007/978-3-030-53370-0_64

Randomized Trial of a Mindfulness App 877
Protocol Inclusion Criteria and Methods 878
Perspectives for Future Improvement 878
Relevance of the e-Diary 878
Conclusion 879
References 879

Abstract

Overweight and obesity are significant global public health problems. Developing evidence-based and novel health promotion interventions is greatly needed. One emerging area of research that holds promise for managing weight in overweight and obese populations involves mHealth. This chapter reviews the current evidence-base for mHealth for weight loss including the latest studies in different populations, commercial weight loss apps, use of theory and behaviour change techniques, consumer preferences, and emerging areas of research. mHealth holds promise for weight loss and weight-related behaviours in diverse populations. Current commercial weight loss apps are suboptimal in quality and lack empirical research on evidence of effectiveness. To ensure consumer satisfaction and sustainable use, considering consumer mHealth preferences and barriers is important. New and exciting areas of research merge the mind with the body, through mindfulness delivered via mHealth for weight loss. Current evidence indicates that this new area of research holds promise for targeting stress-related weight behaviours, which may in turn lead to weight loss.

Keywords

mHealth · Weight loss · Diet · Exercise · Behaviour change · Eating behaviour · Obesity · Mindfulness

Introduction

mHealth Overview

Overweight and obesity are highly prevalent global public health problems, that are significant risk factors for premature mortality and morbidity from diseases such as cancer, cardiovascular disease, and diabetes (CDC 2015; WHO—Obesity and Overweight Fact Sheet 2015). They are also risk factors for osteoarthritis and mental illness (CDC 2015; WHO—Obesity and Overweight Fact Sheet 2015). The global mortality attributable to obesity and overweight was 3.4 million in 2010 (The G. B. D. Obesity Collaboration et al. 2014). Developing novel and effective health promotion interventions to tackle overweight and obesity through behaviour change interventions, that target their determinants such as diet and exercise (WHO—Healthy Diet Fact Sheet 2015; WHO—Physical Activity Fact Sheet 2015) are greatly needed.

One novel method for delivering a behavioural change intervention for weight loss is through mobile health. mHealth is mobile healthcare on the go, which has the convenience of portability and increased accessibility to valuable health information, and monitoring of health behaviours at any time and place for consumers (Lefebvre 2009; World Health Organization 2011). Users can gain instant access to health promotion educational content and interventions even with a busy schedule. The devices that fall under the umbrella of mHealth include smartphones, personal digital assistants (PDAs), iPads, and iPods or MP3 players. Smartphone interventions can include apps, text messages or short message service, and phone coaching calls (Lefebvre 2009; World Health Organization 2011).

Usage and Trends

Delivering health promotion interventions through mHealth is particularly scalable due to their ubiquitous nature (Lefebvre 2009; World Health Organization 2011). The majority of people across the globe use a mobile phone and send daily text messages. There were seven billion people who subscribed to a mobile phone in 2017 (ITU 2015). Research also indicates that the number of text messages that users sent per month totalled 350 billion around the world in 2014 alone (Open University 2014).

Recent reports indicate that half of all smartphone users regularly use a health app, mostly in the wellness lifestyle category (Krebs and Duncan 2015; State L 2017). According to the mHealth economics report, there were 78,000 new health apps that were made in 2017 alone ((R2G) RG 2017). There is a massive market of health apps worth 23 billion US dollars (Reuters 2017) due to the demand and drive by consumers, with a 16% increase in health app downloads totalling 3.7 billion downloads in 2017 ((R2G) RG 2017). Findings in the mHealth economics report indicate that the leading health apps that are being made by developers based on demand are for diabetes, depression, and obesity ((R2G) RG 2017).

Research indicates that mHealth can save international health systems 370 billion US dollars a year by managing chronic health conditions on a global scale (Reuters 2017). Although there was mostly interest from non-health industry personnel when mHealth first emerged in 2008, there has been a recent increase in the enthusiasm and interest of healthcare practitioners in adopting and recommending mHealth for their patients ((R2G) RG 2017). The leading countries where health practitioners are interested in using mHealth for their patients are the UK (European share 47%), USA (North American share 36%), and the Asia Pacific region (11%) ((R2G) RG 2017).

Thus, given the massive use of smartphones, willingness of users to download health apps, and the rising willingness of health practitioners to become involved, there is much potential to deliver weight loss and weight-related behaviour change interventions via the mobile phone medium.

mHealth for Weight Loss in General Populations

Recent meta-analyses have found that mobile devices are effective for assisting with weight loss or weight-related behaviours, such as dietary intake and physical activity in various populations (Lyzwinski 2014). Table 64.1 summarizes the major meta-analyses for various populations. My previous meta-analysis found a medium effect size (Cohen's $D = 0.43$; p-value<0.01) (95% CI = 95% CI = 0.252–0.609), which favoured the mHealth interventions for weight loss over the control groups (Lyzwinski 2014). My meta-analysis included all interventions that targeted diet and exercise for weight in different populations.

Since my work on mHealth, some newer studies have also confirmed that mHealth is effective for assisting with weight loss and maintaining weight (preventing weight gain) after behaviour change (Chin et al. 2016). A recent review found a modest significant effect of mHealth for weight loss in obese and overweight populations, with a mean weighted standardized weight loss of 2.66 kg in the intervention, compared with the control at six months (Park et al. 2019).

Special Populations

Pregnancy

The study by Chan and Chen (2019) similarly found a medium effect size of mobile devices

Table 64.1 Summary of major meta-analyses of mHealth for weight loss in populations

Population, design, reference	Weight and Weight-related Behavioural Outcomes	Quality Appraisal, Publication Bias	Studies used Theory and BCTs with a Clear description of the content?
All General Adults >18, RCTs (Lyzwinski 2014)	The results favoured mHeath for weight loss Improvements in dietary and physical activity indices	Cochrane Risk of Bias Grading Tool Overall Quality = Moderate Funnel Plot: Studies with small sample sizes and + findings more likely to be published Effect would be smaller but still significant	Yes Most common BCTs: Goal setting, self-monitoring, feedback, tips, prompts, encouragement, social support, health education Most common theories: Social Cognitive Theory Others: Self-monitoring theory, Implementation Intentions, Self-Regulation Model, Action Belief Model
Pregnant and Post-partum Women, RCTs (Chan and Chen 2019)	The results favoured mHealth for weight loss	Yes 8-item checklist: the study design, sample size calculation, form of randomization, details of the intervention, eligibility (participant details), outcome measures, and statistical analysis Quality = Moderate Low Risk of Bias Funnel Plot: Studies with small sample sizes and + findings more likely to be published	Yes Most Common Theories: social cognitive theory social learning theory, combined models of behaviour prediction and behaviour persuasive design, and stage theories of decision-making Most Common BCTs: were encouragement, reminders, goal setting self-monitoring feedback health education, and social support
Adolescents ages 10–19, RCTs, and non-RCTs (Shin et al. 2019)	No significant weight loss compared with the controls Significant reduction in sugar-sweetened beverages Significant reduction in screen time Significant increase in physical activity	Cochrane Risk of Bias Grading Tool of RCTs And Risk of Bias for Non-Randomized Controlled Trials Quality = Weak High risk of Bias No publication bias assessment	No A few studies reported details of the intervention content Theory used and BCTs only in a few studies Theories: BCTs:

for assisting with weight loss in pregnant and in post-partum women (Cohen's $D = 0.45$) (p-value<0.05) (95% 0.16–0.074). Interestingly, they found that both mHealth (9/15 studies) and social media interventions (6/15) were equally effective for weight loss. Additional professional consultations or support did not modify the effect, and it was postulated that this was likely due to the fact that the mHealth interventions already had a strong health educational component, which did not require further professional guidance. However, they found that studies which utilized a self-monitoring wearable band in addition to the use of mHealth, were more effective for weight loss (Cohen's $D = 0.97$). Thus, it could be that additional behavioural self-monitoring increases motivation and reminds users to continue to engage in behaviour change.

Adolescents

There was a recent meta-analysis of mHealth for weight and weight-related behaviours in adolescents ages 10–19 years by Shin et al. (2019). While they did not find a significant effect of mHealth on weight loss, there were only three studies that assessed this outcome in the teen population. Thus, more studies are needed that may better ascertain whether mHealth is feasible and acceptable in adolescents. Their meta-analysis of nine studies that examined the effects of mHealth on BMI also did not have significant effect sizes. Given the large age group gap, it could be that the effectiveness of mHealth is contingent upon the involvement of parents for younger children, compared with those in their mid to late teens. More research is needed to better understand this.

However, they found support for the effectiveness of mHealth for physical activity, screen time, and sugar-sweetened beverage intake, which are weight-related behaviours. They found a moderate effect size for physical activity in six studies (standard differences in means = 0.341) (95% CI = 0.019–0.664). They found larger effect sizes for SMS interventions over app interventions, in tandem with single over multiple complex interventions involving different elements besides simple SMS (Shin et al. 2019).

They further found that mHealth is effective for reducing screen time in their pooled analysis of five studies (Cohen's *D* = −0.198) (95% CI = −0.317 to −0.078). Effects were larger for app interventions over SMS, and interventions involving multiple components rather than singular ones (Shin et al. 2019).

They additionally found support for mHealth for reducing sugar-sweetened beverages, in their pooled analysis of three studies. The standardized difference in means (SMD) was −0.40 (95% CI = −0.759 to −0.040), and shorter interventions were more effective than lengthier ones (over six months) (Shin et al. 2019). Given the positive findings for weight-related behaviours, studies of longer duration could be needed in order for the weight-related behaviours to influence weight (Shin et al. 2019).

An earlier review found equally little studies, with mixed results for weight loss in adolescents and children. They did find some support for goal setting and motivational behaviour (Quelly et al. 2016). As attitudes precede behaviours, which in turn translate into weight, it is plausible that more time is needed before the benefits in weight may be observed (Barber 2011).

Quality Assessment

The quality of the studies across the major meta-analyses was variable. My previous meta-analysis in the general adult population only included randomized controlled trials (Lyzwinski 2014). They were appraised using the Cochrane risk of bias grading tool, which includes a consideration of blinding (participants and outcome assessors), quality of randomization, allocation concealment, and attrition (and intention to treat analysis) (Higgins and Green 2011; Higgins et al. 2011; Hannes 2011). Overall, most of the studies in my meta-analysis were of moderate quality, meeting 3/5 of the main domains (Lyzwinski 2014). The meta-analysis in adolescents (Shin et al. 2019) also used the Cochrane risk of bias grading tool to appraise the studies (Shin et al. 2019). They also assessed non-randomized trials, using the risk of bias grading for non-randomized controlled trials (Shin et al. 2019). They found that their studies were weak in quality, as many were lacking in the following domains: allocation concealment, blinding, attrition, and not controlling for cofounding (in non-RCTs). This may possibly explain some of the discrepancies in the negative findings relative to the positive findings in other reviews.

The meta-analysis in pregnant and postpartum women used a checklist of eight items to appraise the studies, which included a consideration of the study design, sample size calculation, form of randomization, details of the intervention, eligibility including participant details, outcome measures, and statistical analyses. They found that the quality of the studies was moderate and that there was a low risk of bias (Chan and Chen 2019).

Consideration of Publication Bias

My previous systematic review found some evidence of publication bias of studies, with small sample sizes with positive results (Lyzwinski 2014). However, the funnel plot also indicated that without the bias, the effect would have still been significant, though smaller (Lyzwinski 2014). The meta-analysis by Chan et al. also found that there was some publication bias to keep in mind when interpreting the results, as studies with smaller sample sizes and positive findings were also more likely to be published (Chan and Chen 2019).

Integration of Behaviour Change Techniques

Abraham and Michie developed a list of 26 behaviour change techniques in interventions that shape behaviour (Abraham and Michie 2008). Many successful interventions in my meta-analysis had integrated a wide range of behaviour change techniques, highlighting their importance. The most common ones were health education (Brindal et al. 2013; Haapala et al. 2009; Hurling et al. 2007; Patrick et al. 2009; Prestwich et al. 2010; Shapiro et al. 2012; Turner-McGrievy and Tate 2011; Turner-McGrievy et al. 2009), goal setting (Chan and Chen 2019; Brindal et al. 2013; Haapala et al. 2009; Hurling et al. 2007; Prestwich et al. 2010; Shapiro et al. 2012; Burke et al. 2012; Carter et al. 2013; Napolitano et al. 2013; Spring et al. 2013), self-monitoring (Brindal et al. 2013; Haapala et al. 2009; Hurling et al. 2007; Prestwich et al. 2010; Shapiro et al. 2012; Turner-McGrievy and Tate 2011; Turner-McGrievy et al. 2009; Burke et al. 2012; Carter et al. 2013; Napolitano et al. 2013; Spring et al. 2013), and feedback (Brindal et al. 2013; Haapala et al. 2009; Hurling et al. 2007; Prestwich et al. 2010; Shapiro et al. 2012; Burke et al. 2012; Carter et al. 2013; Napolitano et al. 2013; Spring et al. 2013). Other important techniques included intention formation (Brindal et al. 2013; Hurling et al. 2007; Prestwich et al. 2010; Napolitano et al. 2013), social support (Hurling et al. 2007; Turner-McGrievy and Tate 2011; Turner-McGrievy et al. 2009; Carter et al. 2013; Napolitano et al. 2013), encouragement (Brindal et al. 2013; Hurling et al. 2007; Patrick et al. 2009; Shapiro et al. 2012; Carter et al. 2013; Napolitano et al. 2013), barrier identification (Brindal et al. 2013; Patrick et al. 2009; Shapiro et al. 2012; Napolitano et al. 2013), prompts (Brindal et al. 2013; Haapala et al. 2009; Hurling et al. 2007; Patrick et al. 2009; Prestwich et al. 2010; Shapiro et al. 2012; Carter et al. 2013), and tips (Brindal et al. 2013; Patrick et al. 2009; Turner-McGrievy and Tate 2011; Turner-McGrievy et al. 2009; Burke et al. 2012; Spring et al. 2013), One used rewards (Brindal et al. 2013) and one used stress management strategies (Napolitano et al. 2013).

The meta-analysis in pregnant women (Chan and Chen 2019) also used behaviour change techniques; however, they were not as extensive. The most common ones across the included studies were encouragement (Choi et al. 2016; Dodd et al. 2018; Fiks et al. 2017; Kennelly et al. 2018; Mackillop et al. 2018; Olson et al. 2018), reminders (Dodd et al. 2018; Mackillop et al. 2018; Olson et al. 2018; Herring et al. 2014), goal setting (Dodd et al. 2018; Mackillop et al. 2018; Olson et al. 2018; Herring et al. 2014), self-monitoring (Choi et al. 2016; Dodd et al. 2018; Fiks et al. 2017; Kennelly et al. 2018; Mackillop et al. 2018; Olson et al. 2018; Herring et al. 2014, 2016; Gilmore et al. 2017; Redman et al. 2017; Yang et al. 2018), feedback (Choi et al. 2016; Mackillop et al. 2018; Olson et al. 2018; Gilmore et al. 2017; Herring et al. 2016; Redman et al. 2017), health education (Choi et al. 2016; Dodd et al. 2018; Fiks et al. 2017; Kennelly et al. 2018; Olson et al. 2018; Herring et al. 2014, 2016; Gilmore et al. 2017; Redman et al. 2017; Yang et al. 2018), and social support (Choi et al. 2016; Fiks et al. 2017; Herring et al. 2016).

The authors of the meta-analysis in adolescents did not identify many intervention details nor behaviour change techniques (Shin et al. 2019). Upon reviewing the independent studies, a few studies used some techniques, though the details of the interventions were scarce. A few used the following techniques: goal setting (Abraham et al. 2015; Babic et al. 2016; Mameli et al. 2018; Thompson et al. 2016), self-monitoring (Abraham et al. 2015; Babic et al. 2016; Mameli et al. 2018; Thompson et al. 2016; Chen et al. 2017a, b; Vidmar et al. 2019), feedback (Abraham et al. 2015; Mameli et al. 2018; Thompson et al. 2016; Chen et al. 2017b), and education (Abraham et al. 2015; Babic et al. 2016; Mameli et al. 2018; Chen et al. 2017b; Vidmar et al. 2019; Direito et al. 2015). One used semi-personal tailoring (Abraham et al. 2015). It is important to understand what features and techniques were used in an intervention, in order to better understand if it was based on evidence-based features, and what features worked or need to be improved.

Use of Theory

The use of theory to inform an intervention is important, as psychological theories are the foundations for testing whether certain targets such as attitudes, motivation, and other relevant constructs will impact behaviour change (Lyzwinski 2014). Many studies in my meta-analysis were informed by an explicit theory, while the rest used implicit theories. The most common explicit theory was social cognitive theory (Shapiro et al. 2012; Turner-McGrievy and Tate 2011; Turner-McGrievy et al. 2009). Other theories included self-monitoring theory (Patrick et al. 2009), implementation intentions (Prestwich et al. 2010), self-regulation model (Burke et al. 2012), and the health action model (Brindal et al. 2013). A few combined several theories (Haapala et al. 2009; Hurling et al. 2007).

Theories were also used in the studies in the meta-analysis of mHealth in pregnant and post-partum women (Chan and Chen 2019). These included social cognitive theory (Choi et al. 2016; Kennelly et al. 2018; Herring et al. 2016), social learning theory (Fiks et al. 2017), combined models of behaviour prediction and behaviour persuasive design (Olson et al. 2018), and stage theories of decision-making (Dodd et al. 2018).

The use of explicit theory was not often mentioned in the studies in the meta-analysis of mHealth in adolescents (Shin et al. 2019). Four studies mentioned the use of an explicit theory which included social cognitive theory (Chen et al. 2017b; Lubans et al. 2016) and self-determination theory (Babic et al. 2016; Thompson et al. 2016).

Emerging Research in Other Populations

There have been emerging studies in other populations including paediatric obesity prevention. One SMS study found that mothers significantly engaged in greater breastfeeding, had increasing awareness of the WHO's breastfeeding guidelines, and were less prone to introduce solid foods to their infants before four months, compared with the control (Jiang et al. 2014). However, a follow-up study did not find any effect of mHealth on an infant's BMI or weight relative to the control (Jiang et al. 2019).

Studies in older adults have been scarce. There is a current protocol for a study in a senior population of adults that will examine the effectiveness of mHealth for lifestyle, weight-related behaviours, and weight in adults over the age of 60 (Fuchs et al. 2018).

In postmenopausal women, two studies observed that mHealth assists with weight loss in this population. The 12-week study by Buckworth et al. (David et al. 2012) which combined daily phone calls with a pedometer found significant weight loss, with a mean reduction of 0.93 kg and a mean BMI reduction of 0.28 kg/m^2. Another twelve-week SMS study by Park and Kim (2012) also found weight loss of 2 kg and reduction in waist circumference by 3 cm, while the controls gained weight and experienced an increase in their waist circumferences.

Commercial Weight Loss Apps

Reviews of commercially available weight loss apps have been undertaken in the adult and adolescent populations, available in the iTunes and Google Play stores in different countries. The results are summarized in Table 64.2.

App Quality

Chen et al. (2015) preliminarily screened the top 200 rated apps in the health and fitness app category in both stores in Australia. They appraised 28 apps against a set of criteria which included accountability, use of technology-enhanced features, and scientific validity of information. Most apps were weak overall in quality, with weak scores for accountability as there was no information about whether the developers had medical research backgrounds, or their affiliations. There was also little scientific information as well as references that backed up the educational content.

Most apps utilized some technology-enhanced features, such as use of energy intake and expenditure input methods with graphic feedback (79%). Social media and support were integrated in 43% of apps. They gave the apps a score of 5 out of 14 for usage of technology-enhanced features. Usability wise, they received suboptimal quality scores of 13.5/20 (Chen et al. 2015).

Additionally, few incorporated behaviour change techniques (BCTs), with a mean score of 6.3/26 in total. The main BCTs used were self-monitoring with feedback. They lacked stress management, prompts, and personal tailoring among other important techniques (Chen et al. 2015).

Another review by Breton et al. appraised the top 204 apps in the app store (Breton and Abroms 2011). Criteria included a comprehensive set of evidence-informed practice guidelines, informed by the Centre for Disease Control, The National Institutes of Health, The Food and Drug Administration, and the US Department of Agriculture. They assessed whether the apps had a food diary, assessed weight, helped maintain calorie balance, encouraged journaling, included portion control, included nutrition facts and encouraged reading labels, encouraged regular physical activity, and whether they educated people about a diet rich in fruits and vegetables. They further assessed whether the apps had meal planning, promoted drinking water over soda, assisted with loss of 1–2 pounds per week, and integrated social support.

Most apps had a food diary (43%), assessed weight (36%), and had info on maintaining caloric balance (34%). Around a quarter of the apps had info on physical activity, including encouraging the use of a journal and regular exercise, as well as information on portion control and nutrition labels. Only 12% of apps had information on eating a diet rich in fruit and vegetables, and less than 10% of apps had information on meal planning. Only seven per cent encouraged water drinking over soda. Only 6% encouraged weight loss of 1–2 pounds per week, and just 3% had any kind of social support in them (Breton and Abroms 2011).

The appraisal by Schoffman et al. reviewed 57 commercially available weight loss apps for teens and children in iTunes (Schoffman et al. 2013). A set of expert-recommended strategies for paediatric obesity prevention were used. These included calculating and plotting BMI over time, incorporating strategies for obesity treatment, incorporating motivation to change behaviours, personally tailoring the intervention, setting goals, discussing the influence of the built environment, incorporating the whole family in the intervention, and setting behavioural targets for sugar-sweetened beverage reduction and consumption of fruit and vegetables. They further assessed whether the apps targeted both diet and exercise, encouraged a reduction of screen time, home-made meals over fast food, family meal eating and daily breakfast, and physical activity for over an hour a day (Schoffman et al. 2013).

The overall quality of the apps was very poor, as 61.4% did not utilize any expert-recommended strategies. Over a quarter of the apps set goals like screen time limits. Only 5% plotted BMI and calculated it. None of the apps discussed the role of environmental determinants. Less than 10% of apps encouraged physical activity for more than an hour a day in children. Only 12% of the apps involved the family, and less than 2% encouraged eating with the family at home and preparing

Table 64.2 Summary of critical appraisals of commercial weight loss apps in different populations

Population, app type, platform, reference	Appraisal criteria	Appraisal findings	Top Rated Apps
General adults Australia, weight loss, iTunes and Google Play (Chen et al. 2015)	Accountability, use of technology-enhanced features, and scientific validity of information Incorporation of behaviour change techniques (BCTs)	Weak Overall Weak areas include: lack of professional references and referenced information Poor use of BCTs (score = 6.3/26) Usability Weak (Score = 13.5/20) Technology Enhanced Features Weak (Score = 5/14)	Noom Weight Loss Coach, Calorie Counter Pro and Control my Weight by Calorie King
General adults, weight loss, iTunes (Breton and Abroms 2011)	Evidence-informed practice guidelines Criteria: food diary, assessed weight, calorie balance, journaling, portion control, nutrition facts, reading labels, regular physical activity, info about a diet rich in fruits and vegetables, meal planning, drinking water over soda, weekly weight loss goals, and social support.	Weak Strong areas: assessed weight (43%) Info on maintaining caloric balance (36%) ~25% info on physical activity including, journal regular exercise, portion control and nutrition labels. Diet rich in fruit and vegetables (12%) <10% of apps information on meal planning. Water drinking over soda (7%) Social Support (3%)	Calorie Tracker Livestrong (9/13) Tap and Track Weight Calorie Exercise Tracker (8/13)
Adolescents, weight loss and weight-related behaviours, iTunes (Schoffman et al. 2013)	Expert Recommended Strategies Plotting BMI over time, Strategies obesity treatment Motivation to change behaviours, personal tailoring Setting goals The built environment The whole family Involved Sugar-sweetened beverage, fruit and vegetables Target both diet and exercise, reduction of screen time, home-made meals, family meal eating and daily breakfast, physical activity >1 hour/ day	Weak 61.4% did not adopt expert strategies >25% of the apps set goals (screen time limits) 5% plotted BMI and calculated it. Zero discussed the role of environmental determinants. <10% physical activity > hour a day 12% involved the family Family meal eating and planning = 2% Reduce sugar-sweetened beverages and increasing fruit and vegetable intake (15%)	HyperAnt, Ideserve2, and Smash Four Food
Mindful eating apps for all populations, iTunes (Lyzwinski et al. 2019a)	Mobile App Rating Scale (MARS): aesthetics, functionality, engagement, subjective assessment, information Therapeutic effectiveness Royal College of Physicians Unit Informatics Guidelines Privacy from the Enlight Criteria	Moderate for MARS (72% of apps) Mean global score range: 1.96–3.75/5 Weak MARS areas: information, engagement, privacy policy, and evidence 85% Weak for Mindful Eating Specific Assessment Weak areas: guided mindful	Am I Hungry In the Moment Eat C

(continued)

Table 64.2 (continued)

Population, app type, platform, reference	Appraisal criteria	Appraisal findings	Top Rated Apps
	Guided eating examples, increased self-awareness of hunger and fullness cues (over extra stressor triggers), mindfulness through a diversity of media. Real-life mindful eating examples and tips, dietary guidelines	eating examples, real-life eating tips and examples, tuning in with one's body and hunger, and little integration of BCTs	

meals at home. Only 15% focused on reducing sugar-sweetened beverages and increasing fruit and vegetable intake (Schoffman et al. 2013).

Mindfulness and Weight Loss

Lyzwinski et al. reviewed mindful eating apps in the iTunes store (Lyzwinski et al. 2019a). Mindfulness is a form of present moment meditation which involves both formal meditation practice and informal practice, where one is actively present in one's daily interactions (Chiesa and Malinowski 2011). Mindfulness and mindful eating have been linked with healthier weight-related behaviours such as lower binge eating, emotional eating, and weight loss (O'Reilly et al. 2014; Olson and Emery 2015). Given the high number of commercial apps, my colleagues and I decided to appraise their quality. The Mobile App Rating Scale (MARS) (Stoyanov et al. 2015) was used to assess the apps across the following domains: functionality, aesthetics, information, engagement, and a subjective score of whether one would actually get the app (not included in the total score). In addition to this, privacy policy was added as an extra domain, adapted from the Enlight criteria (Baumel et al. 2017). Therapeutic effectiveness was also adapted from the Royal College of Physicians and surgeons health informatics guidelines (Physicians RCo 2019).

Additionally, my colleagues and I developed a set of appraisal criteria specific to mindful eating, including whether the apps offered guided eating examples, whether they increased self-awareness of hunger and fullness cues (over extra stressor triggers), and whether they taught mindfulness through a diversity of media. Other domains that were assessed included whether the apps offered real-life mindful eating examples and tips, whether they utilized mindful eating self-monitoring techniques, whether they had information about balanced dietary guidelines, and whether they integrated a range of behaviour change techniques (BCTs) (Lyzwinski et al. 2019a).

The majority of the mindful eating apps received moderate scores in the MARS scale (72% moderate). The scores ranged from 1.96 to 3.75 out of 5. The apps had average scores for aesthetics and functionality. They had weak scores in the domain of quality of information about mindful eating, which was often limited and not backed up by scientific research.

The majority of the mindful eating apps (86%) were rated as weak with regard to the mindful eating specific appraisal. Many did not teach mindful eating using guided audios and a range of media. They also did not tune users in with their hunger and fullness cues. Many did not incorporate behaviour change techniques. Most did not offer any real-life tips or advice about how to eat mindfully (Lyzwinski et al. 2019a). Improvements need to be made in particular in the areas of evidence-based information, expert recommendations, and behaviour change techniques (BCTs).

Top-Ranked Apps

The top-ranked commercial apps according to the appraisal by Chen et al. were Noom Weight Loss Coach, Calorie Counter Pro, and Control my Weight by Calorie King (Chen et al. 2015). Calorie Tracker by Livestrong (9/13), and Tap and Track Weight, Calorie, and Exercise Tracker (8/13) were the first in the review by Breton and Abroms (2011). The top-ranked apps for weight in teens and children in the appraisal by Schoffman et al. (2013) were HyperAnt, Ideserve2, and Smash Four Food. These apps had information on healthy dietary intake and servings for children; however, the reviewers argued that the apps could have had more healthy tips and information in them. The top-ranked mindful eating apps in the review by Lyzwinski et al. (2019a) were Am I Hungry by Dr. May, In the Moment, and Eat C. None of the apps in any of the reviews received a perfect score in any of the appraisals.

Evidence for Effectiveness

The scientific evidence on the effectiveness of these commercial apps is scarce. The researchers argue that while apps like My FitnessPal or Lose It! are popular, they were no more effective than the control in randomized controlled trials, highlighting the discrepancy between evidence and popularity (Chen et al. 2015). Likewise, there was no evidence of any effectiveness studies of the apps that were reviewed in the appraisal by Breton and Abroms (2011). They also critiqued the apps for not including any SMS or notifications, given that the prevailing literature had found that SMS is effective for weight loss (Breton and Abroms 2011). Likewise, the review by Schoffman found that none of the apps for children were tested in trials for effectiveness, and they felt that the lack of sufficient information and behaviour change in the apps would make them unlikely to be effective when tested in trials (Schoffman et al. 2013).

Similarly, none of the mindful eating apps in our appraisal were empirically tested in randomized controlled trials (Lyzwinski et al. 2019a). The lack of scientific validity coupled with low quality of information and aesthetics was also found in a review of general mindfulness apps (Mani et al. 2015). Thus, there is insufficient evidence to recommend the commercially available apps until they are proven to be effective in quality randomized controlled trials.

Privacy Issues

My review of mindful eating apps found that privacy is a major issue with existing apps (Lyzwinski et al. 2019a). There is a need for clear and explicit privacy information in apps for weight loss and weight-related behaviours, in order to protect consumers (Lyzwinski et al. 2019a).

Consumer Preferences for mHealth

Gaining insight into consumer preferences for mHealth for weight loss is valuable for promoting continued interest and use. Thus, my colleagues and I undertook the first meta-synthesis of qualitative studies that examined user perceptions, barriers, and facilitators of use (Lyzwinski et al. 2018a). Figure 64.1 Illustrates the benefits that should be maximized, and the barriers that should be minimized in consumers.

User-Friendly Features

Three themes were identified for app preferences which included simplicity in design and ease of use, personalization, and engagement or entertainment. Users wanted apps that would be user friendly to navigate without any major hindrances or difficulties. They also wanted personal tailoring in the apps. They further wanted the apps to be fun or engaging for them.

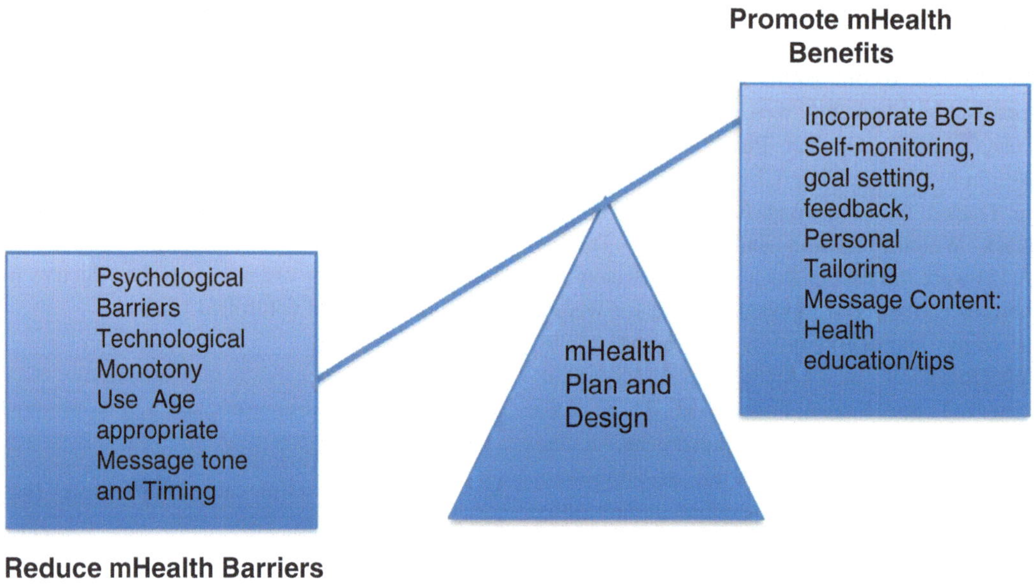

Fig. 64.1 mHealth for weight loss consumer perspectives model

Five themes were identified for consumer preferences for text messages in mHealth weight-loss interventions (Lyzwinski et al. 2018a). Consumers wanted personally tailored messages. They further wanted messages to be considerate of tone. An overtly judgemental tone or one that pressured them could instil psychological roadblocks. These included guilt or fear of failing. The importance of message phrasing and structure was also brought up as a theme for users. They wanted the messages to be short and professional. Teenagers had a dislike of imperative wording, colloquial abbreviations, and phrasing messages in a non-professional way, that sounded like the way their peers would message one another. Message content was also raised as an important theme. Consumers wanted the text messages to be filled with educational material and practical health tips.

Barriers

Several barriers and benefits of mHealth for weight loss were identified as themes by the consumers (Lyzwinski et al. 2018a). Consumers liked the integration of a wide variety of behaviour change techniques, which were viewed positively by them. They found that goal setting, self-monitoring, encouragement, social support, reminders/prompts, and health education were valuable. A few barriers were also identified which were psychological, technological, and engagement related. Consumers wanted encouraging messages over ones that could create excessive pressure or guilt. They also expressed technological concerns in remote areas. Older adults felt that modern technology could be a barrier. Some consumers expressed privacy concerns. Finally, excessive engagement through sending too many messages, or monotonous apps that lacked engagement, was viewed as barriers to continued use.

Challenges with mHealth

Research indicates that most users cease using health apps after a period of time (Krebs and Duncan 2015).

The main reasons are usually the cost of the app if it requires a subscription, and lack of sufficient engagement in order to keep them interested in sustained use (Krebs and Duncan 2015). This is in congruence with research which found that brief SMS messages for sugar-sweetened beverages over six months, were more effective than longer ones (Shin et al. 2019).

Another factor to consider involves the psychological and mental health component. Only one study in my previous meta-analysis had anything for stress management (Napolitano et al. 2013). Likewise, one review also found that there were not any studies that employed stress management techniques (Chen et al. 2015). My previous research has linked stress with maladaptive coping behaviours that are associated with weight which include binge eating, emotional eating, craving junk food, eating fast food, and low levels of exercise (Lyzwinski et al. 2018b).

New Areas of mHealth Research

Research has found that mindfulness is beneficial for mental health (Hofmann et al. 2010) as well as stress, weight, and weight-related behaviours (O'Reilly et al. 2014; Olson and Emery 2015; Chiesa and Serretti 2009; Lyzwinski et al. 2018c). Given that mindfulness targets stress, it is possible that behaviour change and app usage may be sustained over longer periods of time during fluctuating periods of stress.

Despite the large number of studies involving face-to-face interventions and the large number of mindfulness apps in the app stores, my colleagues and I found that there were no RCTs that examined mindfulness for weight and weight-related behaviours using mHealth (apps or SMS), in our previous systematic review (Lyzwinski et al. 2018d).

Preliminary Student Analysis

Prior to undertaking a trial, exploratory focus groups were undertaken to better understand consumer preferences, and any potential barriers and facilitators of use (Lyzwinski et al. 2018e). Most students wanted a mindfulness weight loss app that would be simple and easy to use. Their preferred colour was blue like the sky or ocean for a splash screen. They also wanted the app to have plenty of educational material and health tips. Certain barriers were identified which were technological, psychological, and time related. Students were concerned about technological distractions such as the phone ringing or having pop up ads. They also did not want the app to pressure them into changing dramatically. Finally, they were concerned that some mindfulness exercises might be too long, and they would not do them given their busy schedules. Thus, the app was made with these suggestions in mind, along with consulting key evidence-based mindfulness-based stress reduction (MBSR) and mindful eating books (Albers 2006; Kabatt-Zinn 2016; Somov 2008; Stahl and Elisha 2010, 2015).

Randomized Trial of a Mindfulness App

Following this, my colleagues and I undertook the first randomized controlled trial of a mindfulness-based mHealth app for weight, weight-related behaviours, and stress in university students at the University of Queensland in Australia (Lyzwinski et al. 2019b). The RCT was 11 weeks in duration.

The app had mindfulness-based stress reduction techniques, derived from MBSR and mindful eating as well as mindful exercise (Albers 2006; Kabatt-Zinn 2016; Somov 2008; Stahl and Elisha 2010, 2015). It had formal mindfulness meditation including the body scan, breathing exercises, sitting meditation (loving kindness and concentrative), walking meditation, and choiceless awareness mindfulness meditation, in addition to informal exercises. There were also tailored eating meditation audios. The app used a range of behaviour change techniques including sending push notifications that reminded, educated, and prompted students to practice mindfulness in addition to health education tips, self-monitoring, and stress management. The app taught mindfulness through a diversity of media including educational articles, audios, videos, and games.

Protocol Inclusion Criteria and Methods

All healthy students who wished to lose weight were eligible to participate. A priori sample sizes were calculated and 90 students enrolled in the study. Weight was measured objectively with a digital scale at baseline and follow-up. Weight-related behavioural patterns were assessed using valid and reliable surveys. Students were randomized to either the mindfulness app or to a control group, which received an electronic diary for self-monitoring of diet and exercise. Allocation was concealed as an independent statistician randomized the participants using a 1:1 simple parallel randomization sequence.

There was no weight loss in either group. However, neither the mindfulness app group nor the e-diary group gained weight either. This finding is nonetheless relevant, as university students are at risk of weight gain during their studies (Crombie et al. 2009). Behavioural self-monitoring using a diary has long been established as a method of weight control, which explains why the control group also did not gain weight (Burke et al. 2011).

When compared with the control, the mindfulness app intervention had lower levels of emotional eating by 1.088 points ($F = 5.893$; p-value < 0.05) in the analysis of covariance. The app intervention also had significantly lower levels of uncontrolled eating compared with the control, by 2.047 points ($F = 5.974$; p-value < 0.05) (Lyzwinski et al. 2019b).

Furthermore, mindfulness app intervention also had lower levels of stress in the adherer analysis by 3.21 points ($F = 5.943$; p-value < 0.05). In addition to this, they also had higher levels of mindful eating by 0.295 points ($F = 21.035$; p-value < 0.01) and mindfulness by 3.104 points ($F = 14.580$; p-value < 0.01) than the control.

It could be that changes in weight will be observed in studies of longer duration, as it takes times for weight-related behaviours and changes in attitudes to lead to weight changes. Thus, there is a need for studies of longer duration.

Most students enjoyed the app (94%), and there was a high retention of 80%. Adherence was a challenge as regular usage of the app was low, and most students used it periodically (61%).

Perspectives for Future Improvement

Several themes were identified from the post-trial focus groups that could be used to ameliorate the app (Lyzwinski et al. 2018f). The first few themes were about the control group e-diary. Many found the diary to be helpful and that it increased accountability and knowledge about guidelines; however, time was a barrier. Most students enjoyed using the app and found that it made mindfulness much more accessible to them. Previous myths about meditation being difficult were debunked. Students also felt that their attitudes towards a healthy lifestyle (diet and exercise) had improved as a result. The short breathing audios were preferred by students, due to their time convenience. Many felt that the app reduced their stress. They also felt that they were more mindful as a result of the app. Barriers included finding the time to meditate and getting into the habit of meditating. Students wanted more push notifications in the app and more healthy tips.

Relevance of the e-Diary

Most students wanted a mindfulness weight loss app that would have an e-diary for diet and exercise like the control group had. The control diary group had higher levels of moderate to vigorous MET physical activity ($F = 10.016$; p-value < 0.01 than the app group, highlighting the potential benefit of incorporating exercise self-monitoring into the mindfulness app (Lyzwinski et al. 2018f). This would likely increase the benefits.

Since this trial, a few other trials have been undertaken. There was an Acceptance and Commitment Therapy (ACT) app study, which found that using food as a reward decreased and intuitive eating increased. They did not find that stress modified this relationship, though the app was not

MBSR based which targets stress (Jarvela-Reijonen et al. 2018). Another study in teens found that the mindfulness app increased awareness of weight-related behaviours (Turner and Hingle 2017). There was also a mobile phone mindfulness coaching study, which found significant decreases in binge eating and experiential avoidance (indicator of emotional eating). Both the mindfulness intervention and the active control group (which also received a weight loss intervention) lost weight, with no differences between groups. Adherers of the mindfulness intervention lost the most weight (Carpenter et al. 2019). Indeed, adherence is an issue for mindfulness weight-loss interventions which needs to be tackled in the future.

Conclusion

mHealth holds great promise for weight loss across different populations. More studies are needed in teens and children in order to ascertain their effectiveness in this population, and whether other strategies such as parental involvement need to be implemented. Current commercial mHealth weight loss apps are weak in quality and lack an evidence base. More work needs to be done to create apps that are of higher quality and that are empirically tested for effectiveness in trials. It is important to consider consumer preferences and minimize barriers, in order to ensure long-term sustainable use and behaviour change as well as consumer satisfaction. Mobile mindfulness weight-loss interventions are a new area of mHealth which is very exciting and filled with potential, as they target the underlying psychological factors involved with weight-related behaviours. More studies of longer duration are needed to confirm their effectiveness.

References

CDC. Overweight and obesity. Adult obesity causes and consequences 2015. Retrieved from https://www.cdc.gov/obesity/adult/causes.html. Accessed June 2016

WHO. Obesity and Overweight Fact Sheet (2015). Retrieved from http://www.who.int/mediacentre/factsheets/fs311/en/. Accessed June 2016

The G. B. D. Obesity Collaboration, Ng M, Fleming T et al (2014) Global, regional and national prevalence of overweight and obesity in children and adults 1980–2013: a systematic analysis. Lancet (London, England) 384(9945):766–781

WHO. Healthy Diet Fact Sheet (2015). Retrieved from http://www.who.int/mediacentre/factsheets/fs394/en/ (accessed June 2016)

WHO. Physical Activity Fact Sheet (2015). Retrieved from http://www.who.int/mediacentre/factsheets/fs385/en/. Accessed June 2016

Lefebvre C (2009) Integrating cell phones and mobile technologies into public health practice: a social marketing perspective. Health Promot Pract 10 (4):490–494

World Health Organization (2011) mHealth new horizons for health through mobile technologies. Global Observatory for E Health Series 3

ITU (2015) World Telecommunications/ICT Database. Key ICT indicators for developed and developing countries and the world (totals and penetration rates)

Open University (2014) Text messaging satistics. http://www.openuniversity.edu/news/news/2014-text-messaging-usage-statistics

Krebs P, Duncan DT (2015) Health app use among us mobile phone owners: a national survey. JMIR Mhealth Uhealth 3(4):e101

State L. Why Healthcare Needs Digital Health Apps (2017) Retrieved on March 27th 2018 from https://liquid-state.com/healthcare-needs-digital-health-apps/

(R2G) RG (2017) mHealth App Economics. Current Trends and Future and Trends in Mobile Health. Retrieved on March 27th 2018 from http://www.uzelf.org/wp-content/uploads/2017/12/R2G-mHealth-Developer-Economics-2017-Status-And-Trends.pdf

Reuters (2017) mHealth market worth 23 billion in 2017 and estimated to grow at a CAGR of more than 35% over the next three years. Retrieved on March 27th 2018 from https://www.reuters.com/brandfeatures/venture-capital/article?id=4640

Lyzwinski LN (2014) A systematic review and meta-analysis of mobile devices and weight loss with an intervention content analysis. J Personal Med 4 (3):311–385

Chin SO, Keum C, Woo J et al (2016) Successful weight reduction and maintenance by using a smartphone application in those with overweight and obesity. Sci Rep 6:34563

Park SH, Hwang J, Choi YK (2019) Effect of mobile health on obese adults: a systematic review and meta-analysis. Healthcare Inform Res 25(1):12–26

Chan KL, Chen M (2019) Effects of social media and mobile health apps on pregnancy care: meta-analysis. JMIR Mhealth Uhealth 7(1):e11836

Shin Y, Kim SK, Lee M (2019) Mobile phone interventions to improve adolescents' physical health:

a systematic review and meta-analysis. Public Health Nursing (Boston, MA) 36(6):787–799
Quelly SB, Norris AE, DiPietro JL (2016) Impact of mobile apps to combat obesity in children and adolescents: a systematic literature review. J Special Pediatric Nursing JSPN 21(1):5–17
Barber JS (2011) The theory of planned behaviour: considering drives, proximity and dynamics. Popul Res 9:31–35. https://doi.org/10.1553/populationyearbook2011s31
Higgins JP, Green S (2011) Cochrane Handbook for Systematic Reviews of Interventions Version 5.1.0 [updated March 2011]. Cochrane Collab
Higgins JPT, Altman DG, Gøtzsche PC et al (2011) The Cochrane collaboration's tool for assessing risk of bias in randomised trials. BMJ 343
Hannes K (2011) Critical appraisal of qualitative research. In: Noyes J, Booth A, Hannes, K, Harden A. Cochrane handbook for systematic reviews of interventions. The Cochrane Collaboration 1(Supplementary Guidance for Inclusion of Qualitative Research in Cochrane Systematic Reviews of Interventions. Cochrane Collaboration Qualitative Methods Group)
Abraham C, Michie S (2008) A taxonomy of behavior change techniques used in interventions. Health Psychol 27(3):379–387
Brindal E, Hendrie G, Freyne J, Coombe M, Berkovsky S, Noakes M (2013) Design and pilot results of a mobile phone weight-loss application for women starting a meal replacement programme. J Telemed Telecare 19 (3):166–174
Haapala I, Barengo NC, Biggs S, Surakka L, Manninen P (2009) Weight loss by mobile phone: a 1-year effectiveness study. Public Health Nutr 12(12):2382–2391
Hurling R, Catt M, Boni MD et al (2007) Using internet and mobile phone technology to deliver an automated physical activity program: randomized controlled trial. J Med Internet Res 9(2):e7
Patrick K, Raab F, Adams MA et al (2009) A text message-based intervention for weight loss: randomized controlled trial. J Med Internet Res 11(1):e1
Prestwich A, Perugini M, Hurling R (2010) Can implementation intentions and text messages promote brisk walking? A randomized trial. Health Psychol 29 (1):40–49
Shapiro JR, Koro T, Doran N et al (2012) Text4Diet: a randomized controlled study using text messaging for weight loss behaviors. Prev Med 55(5):412–417
Turner-McGrievy G, Tate D (2011) Tweets, apps, and pods: results of the 6-month mobile pounds off digitally (mobile POD) randomized weight-loss intervention among adults. J Med Internet Res 13(4):e120
Turner-McGrievy GM, Campbell MK, Tate DF, Truesdale KP, Bowling JM, Crosby L (2009) Pounds off digitally study: a randomized podcasting weight-loss intervention. Am J Prev Med 37(4):263–269
Burke LE, Styn MA, Sereika SM et al (2012) Using mHealth technology to enhance self-monitoring for weight loss: a randomized trial. Am J Prev Med 43 (1):20–26
Carter MC, Burley VJ, Nykjaer C, Cade JE (2013) Adherence to a smartphone application for weight loss compared to website and paper diary: pilot randomized controlled trial. J Med Internet Res 15(4):e32
Napolitano MA, Hayes S, Bennett GG, Ives AK, Foster GD (2013) Using Facebook and text messaging to deliver a weight loss program to college students. Obesity (Silver Spring, MD) 21(1):25–31
Spring B, Duncan JM, Janke EA et al (2013) Integrating technology into standard weight loss treatment: a randomized controlled trial. JAMA Intern Med 173 (2):105–111
Choi J, Lee JH, Vittinghoff E, Fukuoka Y (2016) mHealth physical activity intervention: a randomized pilot study in physically inactive pregnant women. Matern Child Health J 20(5):1091–1101
Dodd JM, Louise J, Cramp C, Grivell RM, Moran LJ, Deussen AR (2018) Evaluation of a smartphone nutrition and physical activity application to provide lifestyle advice to pregnant women: The SNAPP randomised trial. Matern Child Nutr 14(1)
Fiks AG, Gruver RS, Bishop-Gilyard CT et al (2017) A social media peer group for mothers to prevent obesity from infancy: the grow2gether randomized trial. Childhood Obes (Print) 13(5):356–368
Kennelly MA, Ainscough K, Lindsay KL et al (2018) Pregnancy exercise and nutrition with smartphone application support: a randomized controlled trial. Obstet Gynecol 131(5):818–826
Mackillop L, Hirst JE, Bartlett KJ et al (2018) Comparing the efficacy of a mobile phone-based blood glucose management system with standard clinic care in women with gestational diabetes: randomized controlled trial. JMIR Mhealth Uhealth 6(3):e71
Olson CM, Groth SW, Graham ML, Reschke JE, Strawderman MS, Fernandez ID (2018) The effectiveness of an online intervention in preventing excessive gestational weight gain: the e-moms roc randomized controlled trial. BMC Pregnancy Childbirth 18(1):148
Herring SJ, Cruice JF, Bennett GG, Davey A, Foster GD (2014) Using technology to promote postpartum weight loss in urban, low-income mothers: a pilot randomized controlled trial. J Nutr Educ Behav 46 (6):610–615
Gilmore LA, Klempel MC, Martin CK et al (2017) Personalized mobile health intervention for health and weight loss in postpartum women receiving women, infants, and children benefit: a randomized controlled pilot study. J Womens Health (Larchmt) 26 (7):719–727
Herring SJ, Cruice JF, Bennett GG, Rose MZ, Davey A, Foster GD (2016) Preventing excessive gestational weight gain among African American women: a randomized clinical trial. Obesity (Silver Spring, MD) 24(1):30–36
Redman LM, Gilmore LA, Breaux J et al (2017) Effectiveness of SmartMoms, a Novel eHealth intervention

for management of gestational weight gain: randomized controlled pilot trial. JMIR Mhealth Uhealth 5(9):e133

Yang P, Lo W, He ZL, Xiao XM (2018) Medical nutrition treatment of women with gestational diabetes mellitus by a telemedicine system based on smartphones. J Obstet Gynaecol Res 44(7):1228–1234

Abraham AA, Chow WC, So HK et al (2015) Lifestyle intervention using an internet-based curriculum with cell phone reminders for obese Chinese teens: a randomized controlled study. PLoS One 10(5): e0125673

Babic MJ, Smith JJ, Morgan PJ et al (2016) Intervention to reduce recreational screen-time in adolescents: outcomes and mediators from the 'Switch-Off 4 Healthy Minds' (S4HM) cluster randomized controlled trial. Prev Med 91:50–57

Mameli C, Brunetti D, Colombo V et al (2018) Combined use of a wristband and a smartphone to reduce body weight in obese children: randomized controlled trial. Pediatr Obes 13(2):81–87

Thompson D, Cantu D, Ramirez B et al (2016) Texting to increase adolescent physical activity: feasibility assessment. Am J Health Behav 40(4):472–483

Chen J, Lieffers J, Bauman A, Hanning R, Allman-Farinelli M (2017a) The use of smartphone health apps and other mobile health (mHealth) technologies in dietetic practice: a three country study. J Hum Nutr Diet 30(4):439–452

Chen JL, Guedes CM, Cooper BA, Lung AE (2017b) Short-term efficacy of an innovative mobile phone technology-based intervention for weight management for overweight and obese adolescents: pilot study. Interact J Med Res 6(2):e12

Vidmar AP, Pretlow R, Borzutzky C et al (2019) An addiction model-based mobile health weight loss intervention in adolescents with obesity. Pediatr Obes 14 (2):e12464

Direito A, Jiang Y, Whittaker R, Maddison R (2015) Apps for IMproving FITness and increasing physical activity among young people: the AIMFIT pragmatic randomized controlled trial. J Med Internet Res 17(8): e210

Lubans DR, Smith JJ, Plotnikoff RC et al (2016) Assessing the sustained impact of a school-based obesity prevention program for adolescent boys: the ATLAS cluster randomized controlled trial. Int J Behav Nutr Phys Act 13:92

Jiang H, Li M, Wen LM et al (2014) Effect of short message service on infant feeding practice: findings from a community-based study in Shanghai, China. JAMA Pediat 168(5):471–478

Jiang H, Li M, Wen LM et al (2019) A community-based short message service intervention to improve mothers' feeding practices for obesity prevention: quasi-experimental study. JMIR Mhealth Uhealth 7(6):e13828

Fuchs SC, Harzheim E, Iochpe C et al (2018) Technologies for innovative monitoring to reduce blood pressure and change lifestyle using mobile phones in adult and elderly populations (TIM Study): protocol for a randomized controlled trial. JMIR Res Protocols 7(8):e169

David P, Buckworth J, Pennell ML, Katz ML, DeGraffinreid CR, Paskett ED (2012) A walking intervention for postmenopausal women using mobile phones and interactive voice response. J Telemed Telecare 18(1):20–25

Park MJ, Kim HS (2012) Evaluation of mobile phone and Internet intervention on waist circumference and blood pressure in post-menopausal women with abdominal obesity. Int J Med Inform 81(6):388–394

Chen J, Cade JE, Allman-Farinelli M (2015) The most popular smartphone apps for weight loss: a quality assessment. JMIR Mhealth Uhealth 3(4):e104

Breton ERFB, Abroms LC (2011) Weight loss-there is an app for that! But does it adhere to evidence-informed practices? Transl Behav Med 4:523–529. https://doi.org/10.1007/s13142-011-0076-5

Schoffman DE, Turner-McGrievy G, Jones SJ, Wilcox S (2013) Mobile apps for pediatric obesity prevention and treatment, healthy eating, and physical activity promotion: just fun and games? Transl Behav Med 3 (3):320–325

Lyzwinski LN, Caffery LC, Bambling M, Edirippulige S (2019a) Mindful eating mobile health apps: review and appraisal. JMIR Mental Health 6(8):e12820

Chiesa A, Malinowski P (2011) Mindfulness-based approaches: are they all the same? J Clin Psychol 67 (4):404–424

O'Reilly GA, Cook L, Spruijt-Metz D, Black DS (2014) Mindfulness-based interventions for obesity-related eating behaviours: a literature review. Obes Rev 15 (6):453–461

Olson KL, Emery CF (2015) Mindfulness and weight loss: a systematic review. Psychosom Med 77(1):59–67

Stoyanov SR, Hides L, Kavanagh DJ, Zelenko O, Tjondronegoro D, Mani M (2015) Mobile App rating scale: a new tool for assessing the quality of health mobile apps. JMIR Mhealth Uhealth 3(1):e27

Baumel A, Faber K, Mathur N, Kane JM, Muench F (2017) Enlight: a comprehensive quality and therapeutic potential evaluation tool for mobile and web-based eHealth interventions. J Med Internet Res 19(3):e82

Physicians RCo (2019) Privacy statement https://www.rcplondon.ac.uk/privacy-statement

Mani M, Kavanagh DJ, Hides L, Stoyanov SR (2015) Review and evaluation of mindfulness-based iPhone apps. JMIR Mhealth Uhealth 3(3):e82

Lyzwinski LN, Caffery LJ, Bambling M, Edirippulige S (2018a) Consumer perspectives on mHealth for weight loss: a review of qualitative studies. J Telemed Telecare 24(4):290–302

Lyzwinski LN, Caffery L, Bambling M, Edirippulige S (2018b) The relationship between stress and maladaptive weight-related behaviors in college students: a review of the literature. Am J Health Educ 49 (3):166–178

Hofmann SG, Sawyer AT, Witt AA, Oh D (2010) The effect of mindfulness-based therapy on anxiety and depression: a meta-analytic review. J Consult Clin Psychol 78(2):169–183

Chiesa A, Serretti A (2009) Mindfulness-based stress reduction for stress management in healthy people: a review and meta-analysis. J Alternat Complement Med (New York, NY) 15(5):593–600

Lyzwinski LN, Caffery LC, Bambling M, Edirripulige S (2018c) The relationship betwen mindfulness and weight related behaviours in college students: a systematic review. Alternat Complement Therap

Lyzwinski LN, Caffery L, Bambling M, Edirippulige S (2018d) A systematic review of electronic mindfulness-based therapeutic interventions for weight, weight-related behaviors, and psychological stress. Telemed J e-Health 24(3):173–184

Lyzwinski LN, Caffery L, Bambling M, Edirippulige S (2018e) University students' perspectives on a mindfulness app for weight management, weight related behaviours, and stress: a qualitative focus group study. Am J Health Educ 49(6):341–353

Albers S (2006) MIndful eating 101: a guide to healthy eating in COllege and beyond. Routledge

Kabatt-Zinn J (2016) Mindfulness for beginners: reclaiming the present moment and your life. Sounds True

Somov P (2008) Mindful practices to overcome overeating one meal at a time. New Harbinger Publications

Stahl BG, Elisha (2010) A mindfulness-based stress reduction workbook. New Harbinger Publications

Stahl BG, Elisha (2015) MBSR everyday: daily practices from the heart of MBSR. New Harbinger Publications

Lyzwinski LN, Caffery L, Bambling M, Edirippulige S (2019b) The mindfulness app trial for weight, weight-related behaviors, and stress in University students: randomized controlled trial. JMIR Mhealth Uhealth 7 (4):e12210

Crombie AP, Ilich JZ, Dutton GR, Panton LB, Abood DA (2009) The freshman weight gain phenomenon revisited. Nutr Rev 67(2):83–94

Burke LE, Wang J, Sevick MA (2011) Self-monitoring in weight loss: a systematic review of the literature. J Am Diet Assoc 111(1):92–102

Lyzwinski LN, Caffery LC, Bambling M, Edirippulige S (2018f) Student attitudes and experiences using a mindfulness app versus an e-diary for lifestyle related weight behaviours and wellbeing in the mindfulness app trial: a qualitative comparative analysis. JMIR. Under Review

Jarvela-Reijonen E, Karhunen L, Sairanen E et al (2018) The effects of acceptance and commitment therapy on eating behavior and diet delivered through face-to-face contact and a mobile app: a randomized controlled trial. Int J Behav Nutr Phys Act 15(1):22

Turner T, Hingle M (2017) Evaluation of a mindfulness-based mobile app aimed at promoting awareness of weight-related behaviors in adolescents: a pilot study. JMIR Res Protoc 6(4):e67

Carpenter KM, Vickerman KA, Salmon EE, Javitz HS, Epel ES, Lovejoy JC (2019) A randomized pilot study of a phone-based mindfulness and weight loss program. Behav Med (Washington, DC) 45(4): 271–281

Biomarkers and Machine Learning Applications in Obesity

65

Olga Tsave and Ioannis Kavakiotis

Contents

Introduction 884

Background Knowledge 884
Knowledge Discovery in Databases (KDD) and Data Mining 884
Machine Learning 885

Biomarkers 886
Anthropometric Predictive Indices 886
Genetics 886
Epigenetics 887
Biochemical Markers and Proteins 887
The Gut Microbiome 887
Metabolomics 888

Machine Learning Tools: Obesity 888
Predictive Tools in Obesity 888
Feature Extraction and Closely Related Factor Evaluation 889
Diet and Food-Driven Applications 889
Bariatric Surgery 889
Childhood Obesity 890

References 890

Abstract

Next-generation sequencing, multi-omics, high-resolution medical imaging technologies, and a wide range of (smart) sensors have led the biosciences into the big data era. To this end, machine learning approaches are more than ever indispensable to transform all this data into valuable knowledge. This chapter aims to address the current state, progress, and challenges in biomarkers discovery in obesity with respect to anthropometric measurements, biochemical and omics approaches, and to discuss the most representative studies relative to the application of machine learning emphasizing prediction tools, feature extraction and closely related factor evaluation, nutrition-driven applications, bariatric surgery, and

O. Tsave
Department of Chemical Engineering, Aristotle University of Thessaloniki, Thessaloniki, Greece

I. Kavakiotis (✉)
Department of Computer Science and Biomedical Informatics, University of Thessaly, Lamia, Greece
e-mail: ikavakiotis@gmail.com

J. Faintuch and S. Faintuch (eds.), *Obesity and Diabetes*, https://doi.org/10.1007/978-3-030-53370-0_65

childhood obesity. The mapping, and hence understanding, of invasive and noninvasive biomarkers extraction and utility along with its limitations, could offer valuable insights into pathophysiology, prevention, and treatment.

Keywords

Machine learning · Data mining · Obesity · Childhood obesity · Disease prediction models · Biomarker(s) identification

Introduction

Recent advances in both analytical techniques and artificial intelligence have given rise to time- and cost-efficient tools. Numerous sensor-based devices such as smartphones (utilizing gyroscope, cameras, accelerometers), smartwatches, and several wireless (nano)networks enable continuous health monitoring (Majumder and Deen 2019; Tsave et al. 2019). High-throughput data derived from genomics, transcriptomics, proteomics, metabolomics, lipidomics, and from medical imaging techniques such as positron emission tomography (PET), single-photon emission computed tomography (SPECT), and magnetic resonance imaging (MRI) have led to the development of integrated strategies that give systematic and mechanistic insight in numerous pathophysiological conditions (Kavakiotis et al. 2017; Severin et al. 2019).

The management of obesity and/or its related complications along with the delineation of its causal mechanisms stand as a challenge in the twenty-first century (World Health Organization 2016; Nguyen et al. 2012). Machine learning, an artificial intelligence field utilizing methodologies for extracting useful knowledge from databases, could be helpful to this aim.

Background Knowledge

Knowledge Discovery in Databases (KDD) and Data Mining

KDD is a multistep process (Fayyad and Piatetsky-Shapiro 1996), depicted in Fig. 65.1, focusing on data management and extraction of useful information. A definition of KDD is given by Fayyad et al. (Fayyad and Piatetsky-Shapiro 1996): *KDD is the nontrivial process identifying valid, novel, potentially useful, and ultimately understandable patterns in data.* The most important step is data mining, in which algorithms from machine learning and statistics are employed in order to extract patterns from data. Alpaydin (Alpaydin 2004) states that application of machine learning methods to large databases is called data mining.

At the very beginning of the KDD process, stands the definition of the problem and the understanding of the field, which is performed in close collaboration between the application's field and KDD experts. Moreover, various aspects of the analysis are sought such as restrictions that may apply in the application domain.

Data Selection Nowadays, in the big data era, it is common that heterogeneous data are included in a study, such as electronic health records

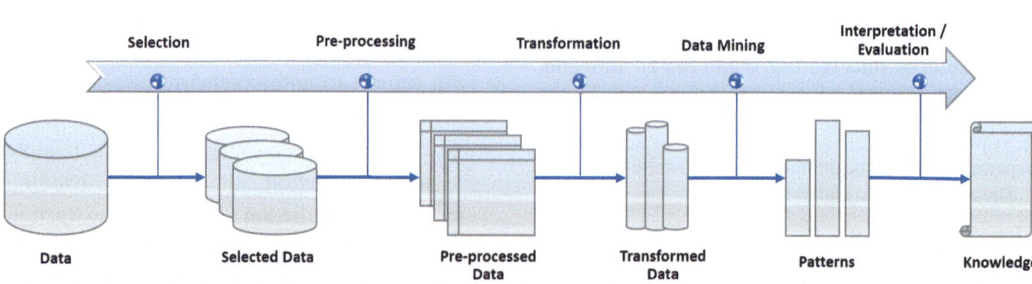

Fig. 65.1 The basic steps of the knowledge discovery in databases (KDD) process (Kavakiotis et al. 2017)

(EHR), next-generation sequencing of DNA/RNA, and metabolomic data. Those data are not organized in a way that expedites the KDD process. The first step is to select and export the appropriate data from the originating resources and to organize them into simpler, appropriate structure.

Pre-processing This is also called the data cleaning step. Collecting data from different resources may result in inconsistencies concerning the features (i.e., the descriptors) of the data. Such inconsistencies include different measurement units for the same quality of information (e.g., blood glucose levels), or even missing essential data. The preprocessing step aims at the correction or removal of erroneous data and acquisition or prediction of missing data.

Data Transformation Data are transformed in an effort to expedite the extraction process either by reducing the number of features (dimensionality reduction) or the number of values that a feature can have (variability reduction), thus providing more comprehensible results. Feature selection is the reduction of the number of features (dimensionality reduction). This technique embodies many advantages, such as a) better visualization and understanding of the data, b) reduction of the analysis duration, and c) better prediction accuracy (Guyon and Elisseeff 2003; Witten et al. 2011). It is the process of selecting a subset from the available features of the dataset, which are more relevant and informative for the construction of a model. Other data transformation techniques include discretization of the data, which is the conversion of integer or real values to categorical values, and the integration of variables, to reduce the complexity of the data.

Data Mining The appropriate machine learning algorithm is applied to the transformed data in order to produce the desired results. There are two general categories of outcomes from a data mining process: informative patterns and predictive models.

Interpretation—Evaluation In the last step of the KDD process, various visualization techniques are employed for the presentation of results (e.g., graphs, plots). These techniques offer to the field expert, e.g., medical doctor or biologist, the flexibility to summarize and draw more solid conclusions.

Machine Learning

For the data mining step, information is extracted through predictive models, utilizing supervised learning, and informative patterns through unsupervised learning. A formal definition of machine learning is given by Mitchell (1997): A computer program is set to "learn" from experience (E) based on some class of tasks (T) and performance measure (P), if its performance at tasks in (T), as measured by (P), improves with experience (E). Different protocols can be adopted (Severin et al. 2019; Witten and Frank 1999; Russell and Norvig 2003): (a) supervised learning, inferring a predictive model from labeled data; (b) unsupervised learning, during which unlabeled data are employed; (c) reinforcement learning, based on a dynamic environment.

Supervised learning Input-output pairs, analyzed by an algorithm, are used for the development of a predictive model (function), with the help of labeled examples (training data). The resulting model is subsequently applied to unknown examples (input features only). In supervised learning, learning tasks are divided into classification (predicting classes) and regression (predicting numbers) models. Decision trees (DT), rule learning, and instance-based learning (IBL) such as *k*-nearest neighbors (*k*-NN), genetic algorithms (GA), artificial neural networks (ANN), support vector machines (SVM), all belong to the supervised category.

Unsupervised Learning Clusters are informative patterns occurring through clustering, i.e. the separation of a whole dataset into groups of data, so that instances belonging to the same group are as similar as possible, and instances belonging to different groups differ as much as possible (Han et al. 2011).

Biomarkers

Nutritional strategies against obesity encompass a) general dietary interventions, b) control of macronutrient composition (e.g., low-fat diets), c) calorie restriction (low and very low-calorie diets), and d) lifestyle interventions (weight loss-physical exercise along with beneficial dietary styles) (Parsanathan and Jain 2019). Pharmacological options are also available, as well as bariatric surgery and endoscopically introduced intragastric balloons (Lukas et al. 2019). Current available therapies remain important in treating obesity but are still limited.

Since obesity has a fundamental, causal effect in developing numerous complications and comorbidities, robust and reliable clinical biomarkers would be welcome to improve (a) diagnosis and classification of the disease, (b) early diagnosis or prediction of the risk to develop secondary events, and (c) more personalized and precise medical treatment.

Biomarkers are defined as biological measures of a certain clinical state. The ideal biomarker should be easy to measure, safe (non-invasive), cost-efficient, specific, sensitive, and predictive. Biomarkers are used as monitoring, screening, diagnostic, and prognostic tools, whereas they can also be considered as potential targets of drugs or other therapies (Lukas et al. 2019; Kosaka et al. 2010). Currently available biomarkers suffer from low specificity, sensitivity, and false-positive outcomes. New molecules (such as miRNAs, metabolites etc.) gain significance in this context (Shih and Lin 2018; Musaad and Haynes 2007). Several circulating molecules have been proposed as mediators of obesity, namely adipose-derived hormones and adipokines such as adiponectin and leptin (Pischon 2009; Pan and Yeh 2008; Nimptsch et al. 2019).

Anthropometric Predictive Indices

Indicative anthropometric measurements or calculations that have been used as markers or features at the clinical level include Visceral Adiposity Index (VAI), epicardial fat thickness (EFT), the density of both visceral adipose tissue (VAT) and subcutaneous adipose tissue (SAT), body fat distribution (BFD), waist circumference, waist-to-hip ratio, and waist-to-height (Okorodudu et al. 2010; Tsave et al. 2018a, b). Most of the aforementioned indicators are used in combination with specific phenotypes such as the presence of hypertension, hypertriglyceridemic waist (HTGW) phenotype, or blood glucose levels, in an effort to improve diagnostic and prognostic value (Okorodudu et al. 2010; Tsave et al. 2018a, b).

Genetics

Genes involved with obesity include human leptin receptor gene, proprotein convertase subtilisin/kexin type 1 (PCSK1; also known as PC1) gene, brain-derived neurotrophic factor (BDNF) and its receptor, tyrosine receptor kinase B (encoded by NTRK2), genes encoding the glutamic acid decarboxylase enzyme (GAD2), peroxisome proliferator-activated receptor gamma(PPAR-γ) gene, cannabinoid receptor 1 (CNR1), dopamine receptor 2 (DRD2), serotonin receptor 2C (HTR2C), and SLC6A4 genes among others (Fall and Ingelsson 2014; Jackson et al. 1997; O'Rahilly et al. 1995; Deeb et al. 1998). As a chronic inflammatory condition, inflammation-related genes are generally upregulated in obesity. Adipogenic genes are often downregulated since the physiology of adipocytes is altered (lipotoxicity, adipose tissue excess) leading to cardiometabolic disorders and the onset of type 2 diabetes (T2D) (Wells 2012).

Epigenetics

Epigenetic patterns, reversible impacts of chromatin, encompass DNA methylation, acetylation, phosphorylation, histone modification, and noncoding RNAs (ncRNAs), and can be passed on mitotically (cell division) or meiotically (transgenerational inheritance) (Lopomo et al. 2016). Among individuals with excessive adiposity, they depend on alimentation patterns and other environmental influences. Multiple elevations of cellular micro-RNAs such as miR-148a, miR-26b, miR-30, and miR-199a were detected in specialized mesenchymal or adipose cells in patients with excessive body weight (Van Dijk et al. 2015; Zhang et al. 2012; Graham et al. 2015; Ouyang et al. 2017). Also in pediatric populations with adipose tissue accumulation, elevated miRNAS (miR-199a-3p/miR-199b-3p and miR-4454) regulating hundreds of genes were unearthed, which could feed a machine learning protocol.

Biochemical Markers and Proteins

Alterations in protein and metabolite profile (e.g., insulin, blood glucose, triglycerides, cholesterol, and related pathways) also provide mechanistic insight in the underlying mechanisms that lead to obese phenotype. Proteomics is defined as the mapping of the expression patterns of proteins at a given time in response to a specific stimulus or disease state and is used for the determination of the functional protein networks that exist at the cellular, tissue, or organism level (Masood et al. 2018). The construction of the proteome is a data-driven approach applied in both untargeted and targeted proteomics workflows, generating qualitative and quantitative datasets of protein expression levels, composition, and post-translational alterations.

In a cohort of 36 male patients, with or without abdominal obesity and diabetes, seven matched protein spots (alpha-1-antichymotrypsin, alpha-1-antitrypsin, apolipoprotein A-I, haptoglobin, retinol-binding protein 4, transthyretin, and zinc-alpha2-glycoprotein) were different between normal and prediabetes/diabetes patients. Network analysis followed by functional annotation revealed that most proteins were involved in lipid transport, lipid localization, and the regulation of serum lipoprotein particles (Kim et al. 2019).

Two large series of roughly 500 subjects each were proteomically analyzed. Plasma protein pattern for age, gender, and general variability was defined, and associations with insulin, glucose, and other markers were unearthed or confirmed (Cominetti et al. 2018). Matrix correlation, one of the available preliminary tools for machine learning development, was applied to 200,000 relationships of 1700 proteins, detected in a cohort of 47 obese patients submitted to bariatric surgery (Roux en Y gastric bypass). Certain proteins changed as early as one week after surgery, whereas in others that occurred as late as after more than 2 years, indicating quite specific patterns for each of them. Elevations, reductions and other behaviors were demonstrated, depending on the analyzed molecule. Altered proteins were mostly markers of systemic inflammation and lipid metabolism (Wewer Albrechtsen et al. 2018).

The Gut Microbiome

In the review of Castaner et al. (Castaner et al. 2018), it was shown that species from the Firmicutes phylum namely *Blautia hydrogenotrophica, Coprococcus catus, Eubacterium ventriosum, Ruminococcus bromii, and Ruminococcus obeum* were associated with obesity. In a systematic review regarding bariatric surgery, some representatives from the phyla Bacteroidetes, Fusobacteria, and Verrucomicrobia increased after surgery, whereas the Proteobacteria phylum encompassed both elevated and diminished components. Reduction of

Firmicutes was also evidenced. *Faecalibacterium prausnitzii, Lactobacillus,* and *Coprococcus* are highlighted as often beneficial (Guo et al. 2018). Although machine learning was not employed in these protocols, the massive numbers of the gut microbiome are good candidates for such technique. Incidentally, *F. prausnitzii* along with *Akkermansia muciniphila* are being investigated as novel probiotics, with potential indications in obesity.

One protocol with 75 men suggests that circulating gut permeability biomarkers, such as lipopolysaccharide-binding protein (LBP) and soluble CD14, together with bacterial metabolites (short-chain fatty acids), more robustly correlate with obesity than fecal microbial community structure, as evidenced by sequencing of 16S rRNA originated from the gut microbiome. LBP, LBP/CD14 ratio, propionic, and butyric acids were independent determinants of BMI. These and other variables explained 39–47% of BMI variability, consistent with the hypothesis that both microbiome-related and independent biomarkers are involved in obesity risk (Barengolts et al. 2019).

Metabolomics

Metabolomic elucidation of the underlying pathways in obesity and obesity-associated comorbidities is currently one of the most actively pursued roads. The metabolome, analogously to several omics techniques, can be constructed through the employment of high-resolution analytical chemistry, for instance, nuclear magnetic resonance (NMR) or liquid chromatography in tandem with mass spectrometry (LC-MS), combined with multivariate biostatistics such as pattern recognition techniques, including principal component analysis (PCA).

Serum metabolomics in obesity usually reveals correlations with blood glucose and cholesterol profile, apart from body weight and dietary regimen, consistent with metabolic pathway changes in lipogenesis, lipid oxidation, energy utilization, protein and amino acid metabolism (Palau-Rodriguez et al. 2020). Given the growing interest in multi-omics data, integrating inputs from metabolomics, genomics, microbiomics, and other fields, automated pipelines/approaches are already emerging. In a large series, the system was able to demonstrate two additional metabolic pathways, namely benzoate degradation and phosphotransferase system, as linked to obesity (Ni et al. 2020). The M2IA platform streamlines the integrative data analysis between metabolome and microbiome, including statistical analysis and biological interpretation (M2ia 2020). Within an experimental setting, machine learning analysis applied to serum metabolomics pointed to deoxyhexose sugars as predictors of gradual loss of β-cells, in prediabetic mice that genetically progress to diabetes. The deoxyhexose 1,5-anhydroglucitol displayed the most substantial changes. Robust markers of β-cell mass in obesity and diabetes are not currently available and would have important clinical applications. Further studies in humans are eagerly anticipated (Li et al. 2019).

Machine Learning Tools: Obesity

Predictive Tools in Obesity

The most diverse and common task found in the applications of machine learning in obesity research is prediction. Gerl et al. (2019) aimed at predicting different measures of obesity, i.e., body mass index (BMI), waist circumference (WC), waist-hip ratio (WHR), and body fat percentage (BFP) based on the plasma lipidome employing several models (Cubist, Lasso, partial least squares, stochastic gradient boosting, and random forest). Lasso was finally employed for the analysis as it exhibited the best performance. Childhood clinical, environmental, and genetic risk factors served as background for early prediction of adulthood obesity. The genetic risk factors were 97 single-nucleotide polymorphisms (SNPs) associated with BMI in over 100,000 genotyped individuals or a 19 SNPs subset of the initial superset. The analysis was performed using gradient boosting machines which are a representative example of ensemble learning (Seyednasrollah et al. 2017).

In a protocol addressing serum FFA levels, gut microbiota, diet, and obesity through a machine learning model based on a regression tree, a gender-independent non-obese profile was generated, with serum EPA > 0.235 μg/mL and *Bacteroides* > 9.1 log cells per g of feces. Regarding obesity, *Bacteroides* group was much less involved (20% importance). Serum EPA and gender (100% and 80%, respectively), along with palmitic acid, *Bifidobacterium*, and *Faecalibacterium* explained >30% (Fernandez-Navarro et al. 2019). In another related work that utilizes both genetic sequencing and support vector machines (SVM), the model exhibited 70.8% accuracy, 80.1% sensitivity, and 63.0% specificity. The specific work showed that selected SNPs were effective in the detection of obesity risk. Additionally, the ML-based method provides a feasible mean for conducting preliminary analyses of genetic characteristics of obesity (Wang et al. 2018).

Feature Extraction and Closely Related Factor Evaluation

Easily available scanned facial data, submitted to feature extraction, have shown robust correlation with anthropometrically relevant markers of excessive body weight. With a different approach (deep learning/convolutional neural networks), satellite geographical images (Google Static Maps API) were analogously used to trace populations with increased adiposity (Pascali et al. 2016; Maharana and Nsoesie 2018). Demographic, social-economical, and lifestyle variables also conducted, by means of machine-learning tools, to findings concerning regional distribution of elevated body weight populations (Scheinker et al. 2019).

Diet and Food-Driven Applications

Personal diet management, automated monitoring, and analysis of eating behavior patterns can be useful in fighting the obesity epidemic and its comorbidities. The advances in smartphones and wearable sensor technologies, including machine learning systems, can be used for meal micro-structure analysis (Papadopoulos et al. 2018), for tracking the user's dietary and physical activities (Silva et al. 2018), to monitor dietary intake using automate image classification (McAllister et al. 2018), or to provide just-in-time adaptive interventions to prevent dietary lapses (Forman et al. 2019).

Country-level food sales information can be obtained from suppliers inexpensively, rapidly, and on a regular basis. Only a small subset of food and beverage categories can be used, utilizing support vector machines, random forests, and extreme gradient boosting to predict country-level obesity prevalence. The top three categories in predicting country-level obesity were baked goods/flours, cheese, and carbonated drinks (Dunstan et al. 2019).

Bariatric Surgery

Algorithms (Sheikhtaheri et al. 2019; Cao et al. 2019) including logistic regression, linear discriminant analysis (LDA), quadratic discriminant analysis (QDA), decision trees, k-nearest neighbor (KNN), SVM, multilayer perceptron (MLP), and deep neural networks can be selected for diagnosis, prognosis, or complication prediction have extensively been used in bariatric cohorts. Ensemble methods should be mentioned as well, namely adaptive boosting (AdaBoost) logistic regression, extra trees, gradient trees, or support vector machine (SVM), bagging LDA, bagging QDA, random forest, extremely randomized (Extra) trees, gradient regression tree, bagging KNN, and bagging MLP. Deep learning showed superior performance in comparison to other machine learning algorithms.

Childhood Obesity

Electronic health records (EHR) are a valid substrate for tracking information for the specific population (obese chirdren) (Lingren et al. 2016; Hammond et al. 2019), as well as neuroimaging features extracted from resting-state functional magnetic resonance imaging (rs-fMRI) datasets (Park et al. 2019). Apparently, nonspecific demographic, psychological, behavioral, and cognitive variables could also allow for excessive weight surveillance (Gray et al. 2019). Social and physical attributes of the school environment similarly impact such disease and could be endowed with predictive value (Hinojosa et al. 2018). The importance of potentially modifiable psychosocial, dietary, and environmental mechanisms and their influence on weight gain in adolescent girls, addressed in a large-scale regression model, is the fact that they are potentially amenable to change, thus pointing towards prophylactic interventions (Narla and Rehkopf 2019).

Acknowledgments Ioannis Kavakiotis's postdoctoral research was carried out under the call "call for interest for postdoctoral researchers, scholarship for postdoctoral research" of University of Thessaly, that is implemented by University of Thessaly and funded by the "Stavros Niarchos Foundation".

References

Alpaydin E (2004) Introduction to machine learning. The MIT Press, Cambridge

Barengolts E, Green SJ, Chlipala GE, Layden BT, Eisenberg Y, Priyadarshini M, Dugas LR (2019) Predictors of obesity among gut microbiota biomarkers in African American Men with and without diabetes. Microorganisms 2019 Sep 5;7(9)

Cao Y, Fang X, Ottosson J, Näslund E, Stenberg E (2019) A comparative study of machine learning algorithms in predicting severe complications after bariatric surgery. J Clin Med 8(5):668

Castaner O, Goday A, Park YM, Lee SH, Magkos F, Shiow SATE, Schröder H (2018) The gut microbiome profile in obesity: a systematic review. Int J Endocrinol

Cominetti O, Nunez Galindo A, Corthesy J, Valsenia A, Irincheeva I, Kussmann M, Saris WHM, Astrup A, McPherson R, Harper ME, Dent R (2018) Obesity shows preserved plasma proteome in large independent clinical cohorts. Sci Rep 8(1):16981

Deeb SS, Fajas L, Nemoto M, Pihlajamäki J, Mykkänen L, Kuusisto J et al (1998) A Pro12Ala substitution in PPARγ2 associated with decreased receptor activity, lower body mass index and improved insulin sensitivity. Nat Genet 20(3):284–287

Dunstan J, Aguirre M, Bastías M, Nau C, Glass TA, Tobar F (2019) Predicting nationwide obesity from food sales using machine learning. Health Informatics J 1460458219845959

Fall T, Ingelsson E (2014) Genome-wide association studies of obesity and metabolic syndrome. Mol Cell Endocrinol 382(1):740–757

Fayyad U, Piatetsky-Shapiro G (1996) P. Smyth From data mining to knowledge discovery in databases. AI Mag 17:37–54

Fernandez-Navarro T, Diaz I, Gutierrez-Diaz I, Rodriguez-Carrio J, Suarez A, de los Reyes-Gavilan CG, Gueimonde M, Salazar N, Gonzalez S (2019) Exploring the interactions between serum free fatty acids and fecal microbiota in obesity through a machine learning algorithm. Food Res Int 121:533–541

Forman EM, Goldstein SP, Crochiere RJ, Butryn ML, Juarascio AS, Zhang F, Foster GD (2019) Randomized controlled trial of OnTrack, a just-in-time adaptive intervention designed to enhance weight loss. Transl Behav Med 9(6):989–1001

Gerl MJ, Klose C, Surma MA, Fernandez C, Melander O, Männistö S, Borodulin K, Havulinna AS, Salomaa V, Ikonen E, Cannistraci CV (2019) Machine learning of human plasma lipidomes for obesity estimation in a large population cohort. PLoS Biol 2019 Oct;17(10)

Graham C, Mullen A, Whelan K (2015) Obesity and the gastrointestinal microbiota: a review of associations and mechanisms. Nutr Rev 73(6):376–385

Gray JC, Schvey NA, Tanofsky-Kraff M (2019 Jul) Demographic, psychological, behavioral, and cognitive correlates of BMI in youth: findings from the Adolescent Brain Cognitive Development (ABCD) study. Psychol Med 10:1–9

Guo Y, Huang ZP, Liu CQ, Qi L, Sheng Y, Zou DJ (2018) Modulation of the gut microbiome: a systematic review of the effect of bariatric surgery. Eur J Endocrinol 178 (1):43–56

Guyon I, Elisseeff A (2003) An introduction to variable and feature selection. J Mach Learn Res 3:1157–1182

Hammond R, Athanasiadou R, Curado S, Aphinyanaphongs Y, Abrams C, Messito MJ, ... Elbel B (2019) Predicting childhood obesity using electronic health records and publicly available data. PLoS One 14(4)

Jiawei Han, Micheline Kamber, Jian Pei (2011) Data mining: concepts and techniques, 3rd edn (The Morgan Kaufmann Series in Data Management Systems)

Hinojosa AMO, MacLeod KE, Balmes J, Jerrett M (2018) Influence of school environments on childhood obesity in California. Environ Res 166:100–107

Jackson RS, Creemers JW, Ohagi S, Raffin-Sanson ML, Sanders L, Montague CT et al (1997) Obesity and

impaired prohormone processing associated with mutations in the human prohormone convertase 1 gene. Nat Genet 16(3):303–306
Kavakiotis I, Tsave O, Salifoglou A, Maglaveras N, Vlahavas I, Chouvarda I (2017) Machine learning and data mining methods in diabetes research. Comput Struct Biotechnol J 15:104–116
Kim SW, Choi JW, Yun JW, Chung IS, Cho HC, Song SE, Im SS, Song DK (2019) Proteomics approach to identify serum biomarkers associated with the progression of diabetes in Korean patients with abdominal obesity. PLoS One 14(9):e0222032
Kosaka N, Iguchi H, Ochiya T (2010) Circulating microRNA in body fluid: a new potential biomarker for cancer diagnosis and prognosis. Cancer Sci 101 (10):2087–2092
Li L, Krznar P, Erban A, Agazzi A, Martin-Levilain J, Supale S, Kopka J, Zamboni N, Maechler P (2019) Metabolomics identifies a biomarker revealing in vivo loss of functional β-cell mass before diabetes onset. Diabetes 68(12):2272–2286
Lingren T, Thaker V, Brady C, Namjou B, Kennebeck S, Bickel J et al (2016) Developing an algorithm to detect early childhood obesity in two tertiary pediatric medical centers. Appl Clin Inform 7(03):693–706
Lopomo A, Burgio E, Migliore L (2016) Epigenetics of obesity. In: Progress in molecular biology and translational science, vol 140. Academic Press, New York, pp 151–184
Lukas A, Heinzel A, Mayer B (2019) Biomarkers for capturing disease pathology as molecular process hyperstructure. bioRxiv 573402
M2ia. http://m2ia.met-bioinformatics.cn. Accessed March 19th, 2020
Maharana A, Nsoesie EO (2018) Use of deep learning to examine the association of the built environment with prevalence of neighborhood adult obesity. JAMA Netw Open 1(4):e181535–e181535
Majumder S, Deen MJ (2019) Smartphone sensors for health monitoring and diagnosis. Sensors 19(9):2164
Masood A, Benabdelkamel H, Alfadda AA (2018) Obesity proteomics: an update on the strategies and tools employed in the study of human obesity. High-throughput 7(3):27
McAllister P, Zheng H, Bond R, Moorhead A (2018) Combining deep residual neural network features with supervised machine learning algorithms to classify diverse food image datasets. Comput Biol Med 95:217–233
Mitchell TM (1997) Machine learning
Musaad S, Haynes EN (2007) Biomarkers of obesity and subsequent cardiovascular events. Epidemiol Rev 29 (1):98–114
Narla A, Rehkopf DH (2019) Novel ranking of protective and risk factors for adolescent adiposity in US females. Obes Sci Pract 5(2):177–186
Nguyen N, Champion JK, Ponce J, Quebbemann B, Patterson E, Pham B et al (2012) A review of unmet needs in obesity management. Obes Surg 22 (6):956–966
Ni Y, Yu G, Chen H, Deng Y, Wells PM, Steves CJ, Ju F, Fu J (2020) M2IA: a web server for microbiome and metabolome integrative analysis. Bioinformatics 2020 Mar 16
Nimptsch K, Konigorski S, Pischon T (2019) Diagnosis of obesity and use of obesity biomarkers in science and clinical medicine. Metabolism 92:61–70
O'Rahilly S, Gray H, Humphreys PJ, Krook A, Polonsky KS, White A et al (1995) Impaired processing of prohormones associated with abnormalities of glucose homeostasis and adrenal function. N Engl J Med 333 (21):1386–1391
Okorodudu DO, Jumean MF, Montori VM, Romero-Corral A, Somers VK, Erwin PJ, Lopez-Jimenez F (2010) Diagnostic performance of body mass index to identify obesity as defined by body adiposity: a systematic review and meta-analysis. Int J Obes 34 (5):791–799
Ouyang S, Tang R, Liu Z, Ma F, Li Y, Wu J (2017) Characterization and predicted role of microRNA expression profiles associated with early childhood obesity. Mol Med Rep 16(4):3799–3806
Palau-Rodriguez M, Garcia-Aloy M, Miñarro A, Bernal-Lopez MR, Brunius C, Gómez-Huelgas R, Landberg R, Tinahones FJ, Andres-Lacueva C (2020) Effects of a long-term lifestyle intervention on metabolically healthy women with obesity: Metabolite profiles according to weight loss response. Clin Nutr 39(1):215–224
Pan WH, Yeh WT (2008) How to define obesity? Evidence-based multiple action points for public awareness, screening, and treatment: an extension of Asian-Pacific recommendations. Asia Pac J Clin Nutr 17(3):370
Papadopoulos A, Kyritsis K, Sarafis I, Delopoulos A (2018) Personalised meal eating behaviour analysis via semi-supervised learning. In 2018 40th Annual International Conference of the IEEE Engineering in Medicine and Biology Society (EMBC) (pp. 4768–4771). IEEE
Park BY, Chung CS, Lee MJ, Park H (2019) Accurate neuroimaging biomarkers to predict body mass index in adolescents: a longitudinal study. Brain Imaging Behav 1–14
Parsanathan R, Jain SK (2019) Novel invasive and noninvasive cardiac-specific biomarkers in obesity and cardiovascular diseases. Metab Syndr Relat Disord
Pascali MA, Giorgi D, Bastiani L, Buzzigoli E, Henríquez P, Matuszewski BJ, Morales MA, Colantonio S (2016 Sep 1) Face morphology: can it tell us something about body weight and fat? Comput Biol Med 76:238–249
Pischon T (2009) Use of obesity biomarkers in cardiovascular epidemiology. Dis Markers 26(5–6):247–263
Russell S, Norvig P (2003) [1995] Artificial intelligence: a modern approach (2nd ed.). Prentice Hall. ISBN 978-0137903955
Scheinker D, Valencia A, Rodriguez F (2019) Identification of factors associated with variation in US County-level obesity prevalence rates using epidemiologic vs

machine learning models. JAMA Netw Open 2(4): e192884–e192884

Severin R, Sabbahi A, Mahmoud AM, Arena R, Phillips SA (2019) Precision medicine in weight loss and healthy living. Prog Cardiovasc Dis 62(1):15–20

Seyednasrollah F, Mäkelä J, Pitkänen N, Juonala M, Hutri-Kähönen N, Lehtimäki T, Viikari J, Kelly T, Li C, Bazzano L, Elo LL (2017 Jun) Prediction of adulthood obesity using genetic and childhood clinical risk factors in the cardiovascular risk in young Finns study. Circ Cardiovasc Genet 10(3):e001554

Sheikhtaheri A, Orooji A, Pazouki A, Beitollahi M (2019) A clinical decision support system for predicting the early complications of one-anastomosis gastric bypass surgery. Obes Surg 29(7):2276–2286

Shih WJ, Lin Y (2018) Relative efficiency of precision medicine designs for clinical trials with predictive biomarkers. Stat Med 37(5):687–709

Silva BVR, Rad MG, Cui J, McCabe M, Pan K (2018) A mobile-based diet monitoring system for obesity management. J Health Med Inform 9(2)

Tsave O, Yavropoulou MP, Kafantari M, Gabriel C, Yovos JG, Salifoglou A (2018a) Comparative assessment of metal-specific adipogenic activity in zinc and vanadium-citrates through associated gene expression. J Inorg Biochem 186:217–227

Tsave O, Kavakiotis I, Vlahavas I, Salifoglou A (2018b) Adipose tissue as a biomarker in data mining predictive models of metabolic pathophysiologies. In: Precision medicine powered by pHealth and connected health. Springer, Singapore, pp 105–108

Tsave O, Kavakiotis I, Kantelis K, Mavridopoulos S, Nicopolitidis P, Papadimitriou G, … Salifoglou A (2019) The anatomy of bacteria-inspired nanonetworks: molecular nanomachines in message dissemination. Nano Commun Netw

Van Dijk SJ, Molloy PL, Varinli H, Morrison JL, Muhlhausler BS (2015) Members of EpiSCOPE. Epigenetics and human obesity. Int J Obes 39(1):85–97

Wang HY, Chang SC, Lin WY, Chen CH, Chiang SH, Huang KY et al (2018) Machine learning-based method for obesity risk evaluation using single-nucleotide polymorphisms derived from next-generation sequencing. J Comput Biol 25(12):1347–1360

Wells JC (2012) The evolution of human adiposity and obesity: where did it all go wrong? Dis Model Mech 5 (5):595–607

Wewer Albrechtsen NJ, Geyer PE, Doll S, Treit PV, Bojsen-Moller KN, Martinussen C, Jorgensen NB, Torekov SS, Meier F, Niu L, Santos A, Keilhauer EC, Holst JJ, Madsbad S, Mann M (2018) Plasma proteome profiling reveals dynamics of inflammatory and lipid homeostasis markers after Roux-En-Y gastric bypass surgery. Cell Syst 7(6):601–612.e3

Witten IH, Frank E (1999) Data mining: practical machine learning tools and techniques with java implementation. Morgan Kaufmann

Witten IH, Frank E, Hall MA (2011) Data mining: practical machine learning tools and techniques, 3rd edn. Morgan Kaufmann, Burlington, MA

World Health Organization (2016) World Health Organization obesity and overweight fact sheet

Zhang Q, Ramlee MK, Brunmeir R, Villanueva CJ, Halperin D, Xu F (2012) Dynamic and distinct histone modifications modulate the expression of key adipogenesis regulatory genes. Cell Cycle 11(23): 4310–4322

Current Approaches in Diabetes Mellitus Prediction: Applications of Machine Learning and Emerging Biomarkers

66

Sergey A. Solodskikh, Vladimir M. Dudenkov, Viktor Yu. Glanz, Anna V. Panevina, Vasily N. Popov, and Alexey S. Velikorodny

Contents

Introduction 894
Diabetes Etiology and Biochemistry 894
Biomarkers of Diabetes 895
Genetics of Diabetes 895
Epigenetic Biomarkers of Diabetes 895
Statistical Methods (Logistic Regression) 896
Machine Learning 897
Deep Learning 897
Personal Experience 898
Biomarkers and Their Derivatives 898
Dataset Preparation and Experimental Design 898
Computational Methods and Software 898
Study Participants and Experimental Groups 899
Training and Testing Subset Size Selection 899
Classification Performance and Model Comparison 899

S. A. Solodskikh
Voronezh State University, Voronezh, Russian Federation

Voronezh State University of Engineering Technologies, Voronezh, Russian Federation

BioME, LLC, Voronezh, Russian Federation

V. M. Dudenkov · V. Y. Glanz
BioME, LLC, Voronezh, Russian Federation

A. V. Panevina
Voronezh State University, Voronezh, Russian Federation

BioME, LLC, Voronezh, Russian Federation

V. N. Popov
Voronezh State University of Engineering Technologies, Voronezh, Russian Federation

A. S. Velikorodny (✉)
Voronezh State University of Engineering Technologies, Voronezh, Russian Federation

BioME, LLC, Voronezh, Russian Federation
e-mail: asvelikorodny@gmail.com

J. Faintuch and S. Faintuch (eds.), *Obesity and Diabetes*, https://doi.org/10.1007/978-3-030-53370-0_66

Calibration Plots and Cut-Off Threshold Selection 901
Probability of T2DM Development 901
Comments 901
Conclusions 903
References 904

Abstract

The purpose of the present chapter was to compare performance and accuracy of three different approaches to diabetes mellitus risk prediction. It was shown, in particular, that a multilayer perceptron, logistic regression, and random forest classifier can be successfully employed for prediction of the T2DM risk using the electronic medical record (EMR) system data with a relatively large number of individuals, albeit a limited dimension of predictor parameters space. Further improvement of the models should be addressed through the following avenues: application of machine and deep learning models for analysis of greater set of factors and derived parameters from the EMR system, including biometric, socioeconomic, geographical, ethnical, and genetic parameters.

Keywords

Big data · Machine learning · Risk prediction · Diabetes mellitus

Introduction

To date, a vast number of predictive models of type 2 diabetes mellitus (T2DM) were reported. Most of them use basic logistic regression (LR) as the underlying mathematical method for obtaining the final model, and include several biochemical, anthropometric, and environmental parameters of the individual as predictor variables. During past decades, a number of machine learning methods were implemented in diabetes risk prediction. However, most of the models widely used in clinical practice were developed based on classical LR and employ a limited number of predictor parameters.

A wider spectrum of parameters and further optimizing computational methods are currently needed. Increasing the reliability and reproducibility of medical data employed for model training is another priority. This can be achieved by extensive application of data stored in electronic medical record (EMR) systems, which become increasingly popular both in medical institutions and in general practices. Measurements being uploaded directly from a certain instrument/device to a data cloud are less prone to errors due to human factors, thus making it possible to develop a more accurate and sensitive model for risk predictions.

Diabetes Etiology and Biochemistry

According to the American Diabetes Association, there are four general types of diabetes (Standards of Medical Care in Diabetes 2018):

- Type 1 diabetes (T1DM), that is a manifestation of autoimmune pancreatic β-cell destruction, ultimately leading to complete loss of insulin secretion. A key biochemical feature of T1DM is presence of autoantibodies against pancreatic antigens, such as insulin, glutamic acid decarboxylase (GADA), insulinoma-associated autoantigen 2 (IA2A), and zinc transporter 8 (ZnT8A) (Atkinson et al. 2014).
- Type 2 diabetes, a condition associated with a progressive impairment of β-cell insulin secretion ability and general insulin resistance. Type 2 diabetes dominant biochemical aspects include lipid metabolism dysfunction,

inflammation, oxidative stress, and impaired mitochondrial metabolism (Ma et al. 2011).

- Gestational diabetes mellitus (GDM), a type of diabetes, diagnosed during pregnancy (the second or third trimester). This condition includes undiagnosed preexisting diabetes and so-called true GDM (Mirghani Dirar and Doupis 2017).
- Other types of diabetes that develop as a result of action of various specific causes (monogenic diabetic conditions, malfunction of exocrine pancreas, chemically-induced diabetes) (Schwitzgebel 2014).

All types of diabetes are traditionally diagnosed by fasting plasma glucose (FPG) or the 2-h plasma glucose (2-h PG) combined with oral glucose tolerance test (OGTT), or A1C glycated hemoglobin levels. If necessary to specify the diagnosis, additional tests (for example, genetic testing) may be performed.

Biomarkers of Diabetes

Despite the fact that T2DM risk factors are well-established, risk assessment and stratification are still a challenge. Traditional methods fail to identify individuals that display low risk according to conventional scales, but nevertheless are prone to development of T2DM. Use of novel biomarkers (genomic, genetic, epigenetic, metabolomic) and computational methods is a promising direction of future research. Currently used clinical biomarkers include:

- anthropometric measurements (weight, BMI, waist circumference, waist-to-hip ratio),
- biochemical, metabolic, and clinical markers (triglycerides, total cholesterol, HDL (high-density lipoprotein) cholesterol, LDL (low-density lipoprotein) cholesterol, use of lipid-lowering drugs, fasting glucose, 2-h glucose, fasting insulin, estimates of insulin resistance, estimates of β-cell function, HbA1c, liver enzymes, uric acid, leukocyte count, C-reactive protein, adiponectin, systolic blood pressure, diastolic blood pressure, use of antihypertensive drugs.
- lifestyle variables: physical inactivity, dietary components (red meat, fats, fruits, vegetables, fiber, coffee), smoking, alcohol consumption.
- socioeconomic factors (Herder et al. 2011).

Genetics of Diabetes

Both type 1 and type 2 diabetes are polygenic disorders. At least 40 genetic loci are described as responsible for T1DM development (Atkinson et al. 2014), and the contribution of genetic component to T2DM pathogenesis is estimated to be 40–80% (Dorajoo et al. 2015; Florez 2016; Talmud et al. 2010). Genetic mechanisms of T2DM pathogenesis include differentially expressed genes, genetic variants (SNPs or single-nucleotide polymorphisms), and the combined effect of SNPs and differentially expressed genes (Das and Rao 2007). The knowledge of SNPs in T2DM has been expanded by genome-wide association studies (GWAS) (Mahajan et al. 2014).

Epigenetic Biomarkers of Diabetes

All known epigenetic hallmarks, such as chromatin modification, DNA methylation, and non-coding RNA involvement (Allis and Jenuwein 2016), are shown to have an effect on T2DM development. It has been shown that chromatin activity can be modulated by certain metabolic signals in diabetes, determining the reaction of genome to stimuli (Rosen et al. 2018). Expression of genes involved in glucose metabolism pathways is greatly influenced by their methylation state that affects overall pancreatic activity; moreover, epigenetic patterns may influence the inheritance of T2DM (Rosen et al. 2018; Sommese et al. 2017; Jerram et al. 2017). MicroRNAs, small non-coding RNAs that regulate gene expression, are essential players in diabetes pathogenesis. Of all several thousands known human microRNAs (miRs), some were shown to be critical for maintaining pancreatic β-cell function, proliferation, and survival (Backe et al. 2014; Filios and Shalev 2015; Avnit-Sagi et al. 2012). Due to relatively simple mechanism of action (complementary inhibition of target mRNA), they possess qualities of a promising biomarker. For example, by manipulating expression of certain miRs, it has been shown possible to enhance insulin secretion in response to glucose (Soni et al. 2014).

Statistical Methods (Logistic Regression)

Logistic regression (Bagley et al. 2001) is a statistical method for predicting binary outcome (dependent or response variable), based on independent variables (predictors). Logistic regression is applicable in cases when dependent variable has only two possible values. The goal of regression analysis is to describe the relationship between a set of parameters and an outcome, and to assess the fitness of the resulting model. This approach assumes that all predictors are independent from each other, and that the outcome is a linear combination of all predictors taken with some coefficient. The latter is called regression coefficient or weight of the variable. One can describe logistic regression by the following equation

$$Logit(p) = \propto + \sum_{i=1} \beta_i x_i$$

where α is the intercept of the logistic regression model, x_i is the predictor variable, and β_i is the regression coefficient for each predictor variable.

Regression coefficients are usually initialized with zero values and then are repeatedly optimized using cost function. Several methods of optimization of variable weight using different cost functions were described. Log-loss (or cross-entropy), Huber, and Kullback-Leibler loss functions along with the gradient descent are commonly used for weights optimization.

In order to obtain predictions, which are probabilities that the entity described by the given set of variable values belongs to one of two classes, one should use sigmoid function that maps output of the logistic regression to the probability. Sigmoid function is described by the following equation:

$$S(p) = \frac{1}{1 + e^{-p}}$$

where p is the output of logistic regression, e.g., $\propto + \beta_1 x_1 + \beta_2 x_2 + \beta_3 x_3$, $S(p)$ is the probability estimate, and e is the base of natural logarithm.

The process of classification itself relies on the use of a threshold parameter. The latter allows identifying a class to which a given object is supposed to belong. This parameter is 0.5 by default, which means that if S(p) is equal or greater than 0.5, the object belongs to one class, and to another if otherwise. However, this parameter for the particular model could be adjusted in order to obtain a desired specificity and sensitivity. This can be done by employing a calibration plot, which allows visualizing dependency between specificity, sensitivity, and a threshold value. Adjusting threshold is important for development of predictive models in a healthcare sector because, depending on the application, one would like to maximize sensitivity by sacrificing specificity and vice versa.

Performance of the model is usually reported in literature using C-statistic and area under receiver operational curve (AUC). The C-statistic provides a general idea of the model, while AUC combines information about specificity and sensitivity of the model.

Logistic regression has been widely used in medical studies for developing risk models of important groups of conditions. Abassi et al. showed that logistic regression is one of the most frequently used techniques for assessing risk of diabetes mellitus (Abbasi et al. 2016). However, choice of computational method and the set of predictors may dictate different approaches, which can perform better.

More than 170 diabetes mellitus risk models have been developed and published from 1993 to

2019. Most of the papers describing these models report performance in terms of C-statistic or D-statistic, as well AUC, with the latter being in the range from 0.61 to 0.91. Most of them are based on logistic regression and use both invasive and noninvasive biomarkers along with reports of questionnaires as independent variables. One of the most comprehensive meta-reviews of diabetes mellitus risk prediction models was performed by Lucaroni et al. (Lucaroni et al. 2019). It is worth noting that age and fasting glucose levels are the main predictors in most regression-based models, displaying the best comparable performance (measured by AUC or C-statistic).

Machine Learning

Artificial neural networks (ANN) (Haykin 1999) are mathematical models inspired by organization of neurons in brain. ANNs form layers and nodes. These nodes take some value as an input, pass it through a hidden activation function provided with weighted links, and use its result as an output.

Weights of the connections are generally calibrated to the input patterns, and ANNs learn with growing experience. Among multiple designs of ANNs, we adopted a multilayer perceptron-based ANN (MLP-ANN) (Ncibi et al. 2017). Within this system, there are only connections between adjacent layers (Heidari et al. 2016). Single-layer MLP-ANN consists of one layer of adaptive weights, with full connectivity between inputs and hidden units, and between hidden units and output.

Let $feature_i$ be the variables given to ANN. The first layer forms M linear combinations, which result in the activation a_j

$$a_j = \sum_{i=1} w_{ji} feature_i + w_{j0}$$

where $j = 1, \ldots, M$. w_{ji} represents the first layer weight matrix, and w_{j0} is the bias figures of the hidden units j. The variables a_j are converted by the activation. Employing ReLU activation function, the outputs correspond to $z_j = \max(0, a_j)$ where $j = 1, \ldots, M$.

The model described above is relevant for the hidden and the output layers. Each unit in one layer connects with a certain weight matrix, to every unit in the following layer. There is one output in the last layer of ANN. After passing through the activation function, it returns the final value to be between 0 and 1. In medical applications, this is the disease risk value. Similar to logistic regression, output of the ANN is the probability that an entity described by a set of variable values, belongs to one of two classes (e.g., "high risk"-"low risk"). Therefore, similar adjustment of a threshold can be performed, in order to achieve a desired balance of sensitivity and specificity.

Deep Learning

A class of deep learning methods can also be used for diagnosis and treatment. To date, several deep learning strategies were implemented for type 2 diabetes mellitus (T2DM) (Talaei-Khoei and Wilson 2018; Miotto et al. 2016; Bernardini et al. 2019; Nguyen et al. 2019; Spänig et al. 2019), with limited success. Methods such as Adaboost, decision trees, and bagging were employed for T2DM risk prediction, as reported by Perveen et al. (2016).

Various deep learning approaches were employed for detection, prediction, and classification of diabetic retinopathy (DR), mainly via analyzing retinal images (Ting et al. 2017; Gargeya and Leng 2017; Abràmoff et al. 2016). The utility of deep learning in screening and classification of diabetic retinopathy is reviewed by Grzybowski et al. (2019; Cheung et al. 2010). However, to our best knowledge, currently there are no published studies describing use of machine learning approaches for prediction of DR based on noninvasive or biochemical biomarkers, or on data extracted from EMR.

Deep learning-based natural language processing system HYPE was developed for prediction of hypoglycemic events, in patients treated for diabetes (Jin et al. 2019). This system uses unstructured data stored in electronic medical records. Several studies describe use of convolutional neural networks, support vector machines for forecasting blood glucose levels in diabetics for short-term (30–60 min) (Li et al. 2019a; Li et al. 2019b) and mid-term (up to one day) periods (Faruqui et al. 2019). Current advances are reviewed by Woldaregay et al. (2019)

Personal Experience

Study participants were recruited among patients of MedExpert medical center (Voronezh, Russian Federation), examined within a 7-year period from 2010 to 2017. Informed consent was obtained from all the participants, and the research was approved by Voronezh State University Ethics Committee. All measurement results, biometric data, and diagnoses were extracted from the Numedy medical information system database, which is used in MedExpert medical centers.

Biomarkers and Their Derivatives

We computed the rate of change over time (if applicable) for all biomarkers controlled in this study. Thus, time-dependent change in biomarker values, expressed in days, went through a simple linear regression (i.e., $f(x) = kx + b$), with a slope coefficient being representative of this biomarker rate of change, to be used in a subsequent analysis.

Dataset Preparation and Experimental Design

The aim of the present study was to compare performance and accuracy of three different approaches to diabetes risk prediction. In order to do this, we compared artificial neural network, namely multilayer perceptron (MLP), logistic regression (LR), and random forest (RF) (Breiman 2001) classifiers. Same training and testing subsets of the dataset were used for all three algorithms. The data were shuffled and randomly assigned into two subgroups based on the selected size of training subset (SEP parameter).

In order to ensure best performance of the classifiers, data were preprocessed in the following manner: the missing values were replaced with the median value of the variable across the subset, and all variables were subsequently normalized using the formula

$$z = \frac{x - u}{s}$$

where z is the normalized value, x is the current value, u is the mean value of the variable, and s is its standard deviation.

Computational Methods and Software

All experiments regarding machine learning were performed using Python platform v. 3.7.4. We used scikit-learn v.0.21.3 library in Python, for training logistic regression, MLP, and RF models. The matplotlib library for Python was employed to visualize results of the model performance. Calibration plots were created using Orange open-source software v. 3.24.

Logistic regression with iteratively reweighted least-squares optimization and 100 decision trees in the random forest classifier were employed.

Hyperparameters of multilayer perceptron were chosen automatically. All combinations of hyperparameters from predefined set of possible options were tested, and trained neural networks were compared in terms of their classification accuracy. It was shown that MLP provides an optimal performance with the following parameters: ReLU activation function $f(x) = max(0,x)$ (Hahnioser et al. 2000), L-BFGS optimization (Nocedal and Wright 1999), and 1 hidden layer with 8 neurons in it. Training process was limited to 400 iterations.

Measurements of classification accuracy, mean absolute error (MAE), area under receiver operation curve (ROC), sensitivity, and specificity of all classifiers were used to estimate and compare their relative performance. As long as the process of the dataset division into training and testing subsets is random, the resulting parameters of classifiers slightly change in different experiments. Therefore, it was decided to consider performance metrics of classifiers as random and report the averaged values as a final result of analysis.

Let X_{ac}, X_{auc}, X_{mae}, X_{sen}, and X_{spec} be random values of accuracy, AUC, MAE, sensitivity, and specificity, respectively, for a given random training subset. Then, $\overline{X_{ac}}$, $\overline{X_{auc}}$, $\overline{X_{mae}}$, $\overline{X_{sen}}$, and $\overline{X_{spec}}$ are sample means for these values, and $S_{X_{ac}}$, $S_{X_{auc}}$, $S_{X_{mae}}$, $S_{X_{sen}}$, and $S_{X_{spec}}$ are sample standard deviations. In order to obtain means and standard deviations for classifiers parameters, we repeated the process of dataset subdivision and model training for 100 times.

Desired specificity and sensitivity of the models were obtained in the following manner: calibration curves were calculated, and the threshold of classification was adjusted for the highest possible sensitivity with reasonable decrease in specificity.

Study Participants and Experimental Groups

Participants for the further analysis were chosen based on the following criteria:

1. Fasting glucose should be measured at least twice during the period of examination.
2. At least 3 of 4 biomarkers controlled in this study, namely bilirubin, total cholesterol, alanine transaminase (ALT), and aspartate transaminase (AST), should be measured at least once.
3. Selected individual should have follow-up, namely two or more appointments during past 6 years.

After filtering out all individuals who did not meet these criteria, we included 2963 patients in our dataset for further analysis. 885 of them were diagnosed with type 2 diabetes mellitus.

We did not use anthropometric and clinical parameters such as BMI, waist circumference, and blood pressure, as these are measured manually and therefore might be unreliable.

Training and Testing Subset Size Selection

First, we determined the optimal size of training subset of our dataset. It had to be done in this particular case because of a relatively small size of the dataset, in order to achieve the best performance of models. These results are shown in Table 66.1.

It was determined that with this particular dataset, the best performance for all three classifiers was achieved with SEP parameter equal to 0.9.

Classification Performance and Model Comparison

Performance parameters of all three models are summarized in Fig. 66.1. One could observe that both random forest and MLP classifiers demonstrate better results compared to logistic regression. With the default value of classification threshold $p = 0.5$, classification accuracy of these models is in average 4% higher compared to LR, while sensitivity is almost 20% higher. However, specificity of both MLP and random forest was 3% lower than that of LR. Mean values of performance metrics of all three models are shown in Table 66.2 and in Fig. 66.2.

Overall, ANN displays better AUC (0.812 vs 0.76 and 0.802) and accuracy (0.764 vs 0.716 and 0.761) compared to other models.

Table 66.1 Performance of classifiers with different training subset size

	Artificial neural network			Logistic Regression			Random Forest		
Parameter	SEP = 0.75	SEP = 0.8	SEP = 0.9	SEP = 0.75	SEP = 0.8	SEP = 0.9	SEP = 0.75	SEP = 0.8	SEP = 0.9
MAE	0.31	0.32	0.30	0.36	0.36	0.36	0.32	0.32	0.31
AUC	0.81	0.80	0.81	0.76	0.76	0.73	0.80	0.79	0.81
Accuracy	0.77	0.78	0.77	0.71	0.73	0.72	0.77	0.76	0.78
Sensitivity	0.50	0.40	0.51	0.30	0.35	0.37	0.49	0.51	0.54
Specificity	0.88	0.93	0.87	0.90	0.90	0.86	0.89	0.86	0.88

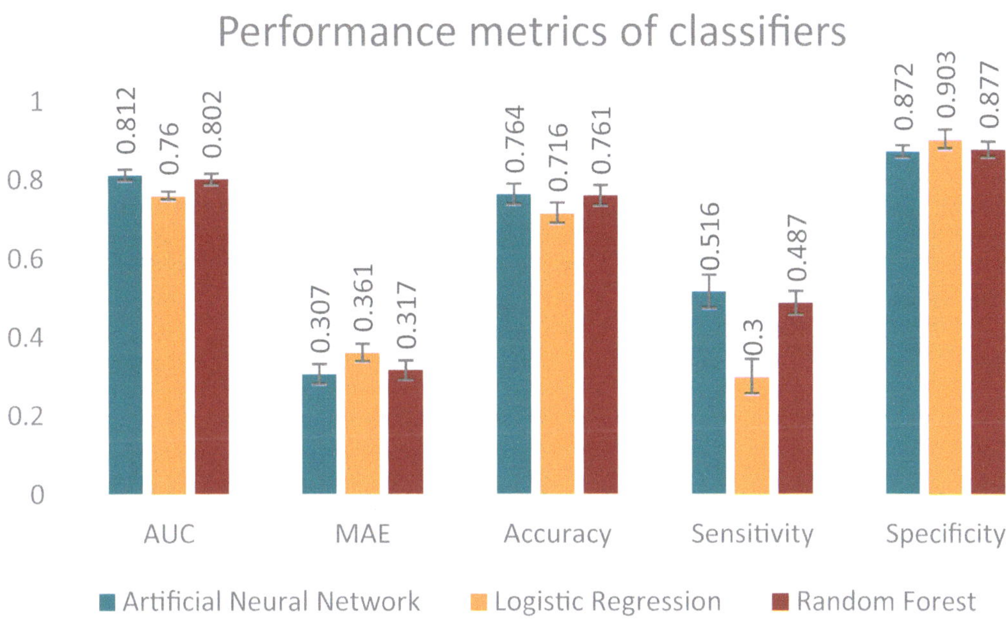

Fig. 66.1 Performance metrics of ANN, logistic regression, and random forest models. A—accuracy distributions, B—AUC distributions; C—sensitivity distributions; D—specificity distributions

Calibration Plots and Cut-Off Threshold Selection

In order to obtain the final model with the desired specificity and sensitivity, a calibration plot was built (see Fig. 66.3). As long as the main goal of the model is the prediction of the T2DM risk, it should detect as many cases of individuals at high risk of the disease as possible, therefore making sensitivity the highest priority.

Optimal probability threshold for clinical applications was found to be close to the value of 0.2, which allows one to achieve high sensitivity with reasonable decrease in specificity for all three models (see Table 66.3).

Probability of T2DM Development

In most regression-based models, age of a patient makes the greatest contribution to the probability of developing the disease. Figure 66.4 shows dependence of the probability to develop diabetes on the age of an individual. According to the published data and results observed in Fig. 66.4, males are more likely to develop T2DM compared to females, especially in the range from 25 to 50 years (Kautzky-Willer et al. 2016).

Comments

A large number of tools for assessing the risk of diabetes have already been developed. Some of them are designed as risk calculators that predict a risk of diabetes for various time periods. Most risk models rely on combination of both invasive biomarkers (levels of various metabolites in blood) and noninvasive anthropometric variables, such as height, weight, and wrist circumference, as well as personal data on the patient's lifestyle obtained via the questionnaire.

One can easily observe that in the vast majority of cases, the best predictors are age, body mass index, blood glucose and glycated hemoglobin levels. As a result of applying various methods of data analysis (often representing a classic logistic regression), models from reasonable to

Table 66.2 Performance metrics of classifiers

Parameter (mean ± SD)	Artificial neural network	Logistic regression	Random forest
AUC	0.812 ± 0.014	0.760 ± 0.010	0.802 ± 0.013
MAE	0.307 ± 0.025	0.361 ± 0.022	0.317 ± 0.024
Accuracy	0.764 ± 0.026	0.716 ± 0.026	0.761 ± 0.025
Sensitivity	0.516 ± 0.042	0.300 ± 0.044	0.487 ± 0.029
Specificity	0.872 ± 0.015	0.903 ± 0.024	0.877 ± 0.019

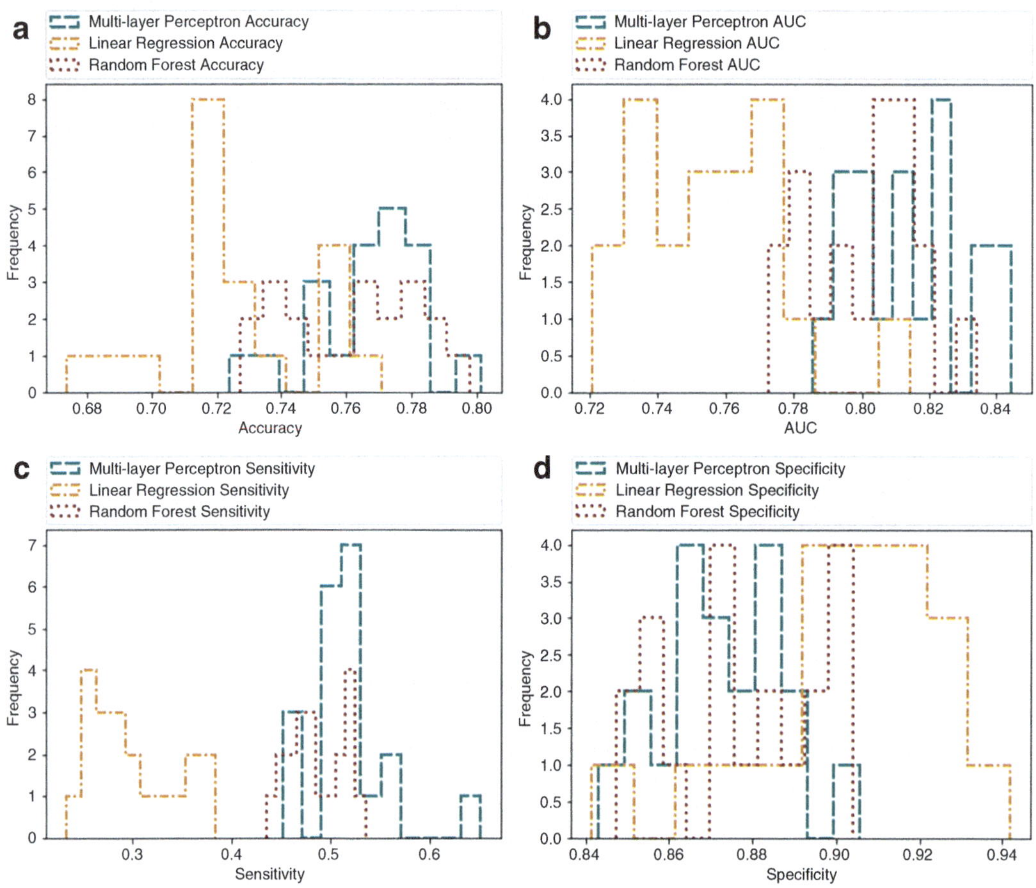

Fig. 66.2 Performance metrics of classifiers. Error bars indicate standard deviations

high values of accuracy, specificity, and sensitivity, with AUC of 0.8 and more were proposed.

Despite the fact that these results indicate a good predictive power of the model, from the point of view of statistics, in most cases, this approach is not entirely correct, since the diagnosis is made by the physician on the basis of the same biomarkers that are predictor variables in the model. In other words, a model uses these biomarkers to predict a response variable with a value, which had already been selected based on these predictors. This approach is similar to predicting a variable by the variable itself and therefore should be avoided if possible.

One way to resolve this problem is to search for new biomarkers, for example, using high-throughput platforms for analyzing "omics" data, or by using data mining in databases, which store EMRs. Several studies have shown the utility of deep learning for risk prediction of

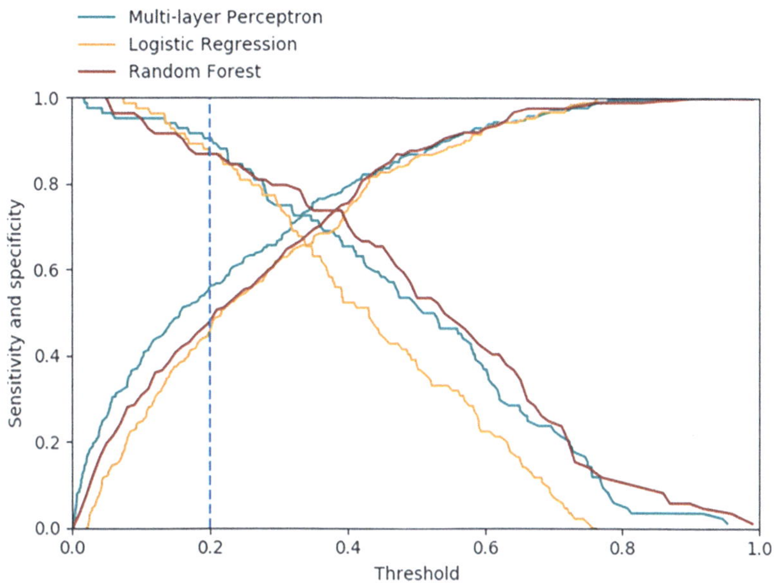

Fig. 66.3 Calibration plot for three models with sensitivity corresponding to curves falling from left to right, and specificity—to the rising curves

Table 66.3 Sensitivity/specificity of the models after threshold adjustment

Model	Sensitivity	Specificity
Logistic Regression	0.898	0.476
Artificial Neural Network	0.884	0.570
Random Forest	0.861	0.560

diabetes mellitus, based on both supervised and unsupervised processing of electronic medical records (Miotto et al. 2016; Bernardini et al. 2019; Nguyen et al. 2019; Spänig et al. 2019).

Another promising approach is to derive new types of biomarkers based on existing ones, for example, by calculating dynamic biomarkers, such as the rate of change of the biomarker over time. It was shown earlier that biomarker dynamics can significantly increase performance of a risk prediction model in terms of T2DM (Solodskikh et al. 2019), thus making them a promising tool in clinical practice.

One of the major drawbacks of existing approaches to diabetes risk assessment is the fact that the vast majority of measurements, which are stored in the electronic databases (employed for model training), are entered there manually. This makes such data less reliable compared to the data collected directly via program interfaces connected to a measurement instrument/device.

Conclusions

Introduction of EMR and EMR storage systems, e.g., Egton Medical Information Systems (EMIS Web) in the United Kingdom, or Veterans Health Information Systems and Technology Architecture (VistA), developed and implemented in Department of Veterans Affairs in the United States, resulted in accumulation of medical databases, which are accessible for both physicians and patients. More than 80 percent of hospitals in the U.S. claim that they have adopted some type of EMR management system. However, even within a hospital, the type of EMR data varies significantly. Types of EMR data used in hospitals include structured data (e.g., medication

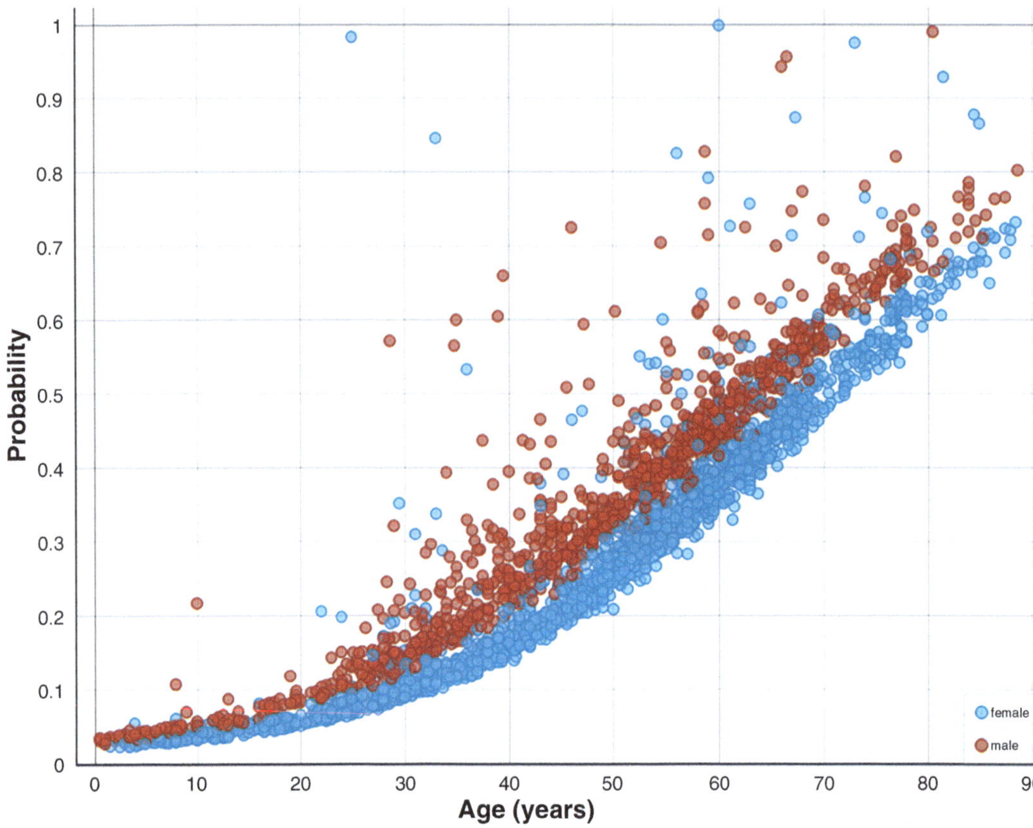

Fig. 66.4 Dependence of T2DM probability development on age, gender

information or results of measurements performed by various devices) and unstructured data (e.g., clinical notes).

Data stored in these databases can be potentially analyzed in order to obtain useful insights about certain relationships between measurement results and health conditions, drug intervention results and overall treatment outcomes, and prognosis. These analytical procedures can be performed in a background manner, using various unsupervised machine learning approaches.

Acknowledgments BioME LLC funded the study design, collection and analysis, of MedExpert data, preparation and submission of the manuscript. BioME, LLC. non-research staff had no role in the study design, data collection, analysis, or preparation of the manuscript. Dataset for analysis was provided by MedExpert, LLC.

Conflicts of interests Alexey S. Velikorodny, Sergei A. Solodskikh, Anna V. Panevina, Vladimir M. Dudenkov, and Viktor Yu. Glanz are currently employees of BioME, LLC. Other authors declare no conflict of interests that are directly relevant to this study.

References

Abbasi A et al (2016) A systematic review of biomarkers and risk of incident type 2 diabetes: an overview of epidemiological, prediction and aetiological research literature. PLoS One 11(10):e0163721

Abràmoff MD et al (2016) Improved automated detection of diabetic retinopathy on a publicly available dataset through integration of deep learning. Invest Ophthalmol Vis Sci

Allis CD, Jenuwein T (Aug 2016) The molecular hallmarks of epigenetic control. Nat Rev Genet 17 (8):487–500

Atkinson MA, Eisenbarth GS, Michels AW (2014) Type 1 diabetes. Lancet 383:69–82

Avnit-Sagi T, Vana T, Walker MD (Jan 2012) Transcriptional mechanisms controlling miR-375 gene expression in the pancreas. Exp Diabetes Res 2012:891216

Backe MB, Novotny GW, Christensen DP, Grunnet LG, Mandrup-Poulsen T (Jan 2014) Altering β-cell number through stable alteration of miR-21 and miR-34a expression. Islets 6(1):e27754

Bagley SC, White H, Golomb BA (2001) Logistic regression in the medical literature: standards for use and reporting, with particular attention to one medical domain. J Clin Epidemiol 54(10):979–985

Bernardini M, Romeo L, Misericordia P, Frontoni E (2019) Discovering the type 2 diabetes in electronic health records using the sparse balanced support vector machine. IEEE J Biomed Heal Inform

Breiman L (2001) Random forests. Mach Learn

Cheung N, Mitchell P, Wong TY (2010) Diabetic retinopathy. The Lancet

Das UN, Rao A (Jan 2007) Gene expression profile in obesity and type 2 diabetes mellitus. Lipids Health Dis 6:35

Dorajoo R, Liu J, Boehm BO (2015) Genetics of type 2 diabetes and clinical utility. Genes (Basel) 6 (2):372–384

Faruqui SHA et al (2019) Development of a deep learning model for dynamic forecasting of blood glucose level for type 2 diabetes mellitus: Secondary analysis of a randomized controlled trial. JMIR Health

Filios SR, Shalev A (Nov 2015) β-Cell MicroRNAs: Small but Powerful. Diabetes 64(11):3631–3644

Florez JC (Mar 2016) Found in translation: a type 1 diabetes genetic risk score applied to clinical diagnosis. Diabetes Care 39(3):330–332

Gargeya R, Leng T (2017) Automated identification of diabetic retinopathy using deep learning. Ophthalmology

Grzybowski A et al (2019) Artificial intelligence for diabetic retinopathy screening: a review. Eye (Basingstoke)

Hahnioser RHR, Sarpeshkar R, Mahowald MA, Douglas RJ, Seung HS (2000) Digital selection and analogue amplification coexist in a cortex- inspired silicon circuit. Nature

Haykin S (1999) Neural networks: a comprehensive foundation (3rd edn)

Heidari E, Sobati MA, Movahedirad S (2016) Accurate prediction of nanofluid viscosity using a multilayer perceptron artificial neural network (MLP-ANN). Chem Intel Lab Syst 15:73–85

Herder C, Karakas M, Koenig W (Jul 2011) Biomarkers for the prediction of type 2 diabetes and cardiovascular disease. Clin Pharmacol Ther 90(1):52–66

Jerram ST, Dang MN, Leslie RD (Oct 2017) The role of epigenetics in type 1 diabetes. Curr Diab Rep 17(10)

Jin Y, Li F, Vimalananda VG, Yu H (2019) Automatic detection of hypoglycemic events from the electronic health record notes of diabetes patients: empirical study. JMIR Med Inf

Kautzky-Willer A, Harreiter J, Pacini G (2016) Sex and gender differences in risk, pathophysiology and complications of type 2 diabetes mellitus. Endocr Rev

Li K, Liu C, Zhu T, Herrero P, Georgiou P (2019a) GluNet: a deep learning framework for accurate glucose forecasting. IEEE J Biomed Heal Inform

Li K, Daniels J, Liu C, Herrero-Vinas P, Georgiou P (2019b) Convolutional recurrent neural networks for glucose prediction. IEEE J Biomed Heal Inform

Lucaroni F et al (Dec 2019) Can risk be predicted? An umbrella systematic review of current risk prediction models for cardiovascular diseases, diabetes and hypertension. BMJ Open 9(12):e030234

Ma Z, Zhao Z, Turk J (2011) Mitochondrial dysfunction and β-cell failure in type 2 diabetes mellitus. Exp Diabetes Res 2012

Mahajan A et al (2014) Genome-wide trans-ancestry meta-analysis provides insight into the genetic architecture of type 2 diabetes susceptibility. Nat Genet 46 (3):234–244

Miotto R, Li L, Kidd BA, Dudley JT (2016) Deep patient: an unsupervised representation to predict the future of patients from the electronic health records. Sci Rep

Mirghani Dirar A, Doupis J (Dec 2017) Gestational diabetes from A to Z. World J Diabetes 8(12):489–511

Ncibi K, Sadraoui T, Faycel M, Djenina A (2017) A Multilayer perceptron artificial neural networks based a preprocessing and hybrid optimization task for data mining and classification. Int J Econ Financ Manag 5:12–21

Nguyen BP et al (2019) Predicting the onset of type 2 diabetes using wide and deep learning with electronic health records. Comput. Methods Progr Biomed

Nocedal J, Wright SJ (1999) Numerical optimization. Springer, New York

Perveen S, Shahbaz M, Guergachi A, Keshavjee K (2016) Performance analysis of data mining classification techniques to predict diabetes. In Procedia computer science

Rosen ED et al (Oct 2018) Epigenetics and epigenomics: Implications for diabetes and obesity. Diabetes 67 (10):1923–1931

Schwitzgebel VM (Mar 2014) Many faces of monogenic diabetes. J Diab Invest 5(2):121–133

Solodskikh SA, Velikorondy AS, Popov VN (Oct 2019) Predictive estimates of risks associated with type 2 diabetes mellitus on the basis of biochemical biomarkers and derived time-dependent parameters. J Comput Biol 26(10):1041–1049

Sommese L, Zullo A, Mancini FP, Fabbricini R, Soricelli A, Napoli C (Jun 2017) Clinical relevance of epigenetics in the onset and management of type 2 diabetes mellitus. Epigenetics 12(6):401–415

Soni MS et al (Nov 2014) Downregulation of carnitine acyl-carnitine translocase by miRNAs 132 and 212 amplifies glucose-stimulated insulin secretion. Diabetes 63(11):3805–3814

Spänig S, Emberger-Klein A, Sowa JP, Canbay A, Menrad K, Heider D (2019) The virtual doctor: an interactive clinical-decision-support system based on deep learning for non-invasive prediction of diabetes. Artif Intel Med 100:101706

Standards of Medical Care in Diabetes—2018 Abridged for Primary Care Providers. Clin Diab 36(1):14–37 (Jan 2018)

Talaei-Khoei A, Wilson JM (2018) Identifying people at risk of developing type 2 diabetes: A comparison of predictive analytics techniques and predictor variables. Int J Med Inform 119, 22:–38

Talmud P, Hingorani A, Cooper J (2010) Utility of genetic and non-genetic risk factors in prediction of type 2 diabetes: Whitehall II prospective cohort study. BMJ

Ting DSW et al (2017) Development and validation of a deep learning system for diabetic retinopathy and related eye diseases using retinal images from multiethnic populations with diabetes. JAMA J Am Med Assoc

Woldaregay AZ et al (2019) Data-driven modeling and prediction of blood glucose dynamics: machine learning applications in type 1 diabetes. Artif Intel Med

67 Lifestyle Interventions for Sarcopenic Obesity in Polycystic Ovary Syndrome

Philip D. Chilibeck, Maryam Kazemi, Laura E. McBreairty, and Gordon A. Zello

Contents

General Features of Polycystic Ovary Syndrome 908
Sarcopenic Obesity 909
Mechanisms for Sarcopenic Obesity 909
Lipid Metabolism 909
Myokines 910
Dietary Interventions: Vitamin D 910
Omega-3 Fatty Acids 911
Antioxidant Supplementation 911
Selenium Supplementation 912
Chromium Picolinate 912
Carnitine 912
Combinations of Nutritional Supplements 912
Dietary Approaches 912
DASH Diet 912
Low-Glycemic Index Diets 913
Low Carbohydrate Diets 913
The Mediterranean Diet and the Portfolio Diet 913
Exercise 914

P. D. Chilibeck (✉) · M. Kazemi · L. E. McBreairty · G. A. Zello
Colleges of Kinesiology and Pharmacy and Nutrition, University of Saskatchewan, Saskatoon, SK, Canada
e-mail: phil.chilibeck@usask.ca

J. Faintuch and S. Faintuch (eds.), *Obesity and Diabetes*, https://doi.org/10.1007/978-3-030-53370-0_67

Aerobic Training 914
High-Intensity Interval Training (HIIT) 914
Resistance Training 915
Summary/Overall Recommendations 915
References 915

Abstract

Polycystic ovary syndrome (PCOS) is a common endocrine disorder in women of reproductive age and is characterized by a combination of hyperandrogenism, ovulation dysfunction, and/or polycystic ovarian morphology. Low muscle mass relative to fat mass, described as "sarcopenic obesity," is a condition which leads to metabolic disorders to a greater degree than either sarcopenia or obesity alone. About 58% of women with PCOS have sarcopenic obesity; this may contribute the high degree of metabolic disorders (i.e., hyperglycemia, hyperinsulinemia, dyslipidemia). Diets designed to improve risk factors for cardio-metabolic disease (DASH, Mediterranean, Portfolio, low-glycemic index, low-carbohydrate diets), as well as supplements (vitamin D, omega-3 fatty acids, antioxidants, carnitine) have potential to be effective for women with PCOS. Low-moderate intensity aerobic training or high-intensity interval training combined with resistance training improves metabolic aberrations as well. Women with PCOS should be offered a wide variety of effective interventions to choose from since adherence to any one intervention tends to be low.

Keywords

Muscle · Fat · Lipids · Glucose · Insulin · Androgens · Testosterone

General Features of Polycystic Ovary Syndrome

Approximately 18% of women worldwide suffer from polycystic ovary syndrome (PCOS), (Carmina and Lobo 1999); however, estimates vary with the definition and proper diagnosis of PCOS. Hyperandrogenism, ovulation dysfunction, and/or aberrant polycystic ovarian morphology are the cardinal features (Teede et al. 2018). Hyperandrogenism appears as hirsutism (unwanted or excessive hair growth) accompanied by elevation in total, free, and bioavailable testosterone, or free androgens (i.e., free androgen index) (Teede et al. 2018). Hirsutism is usually determined by the modified Ferriman-Gallwey Index. A score of ≥4–6 is considered diagnostic (Teede et al. 2018). The ovulation/ovulatory dysfunctions typically feature oligomenorrhea, whereas ≥20 follicles per ovary or mean ovarian volume ≥10 mL in the absence of corpora lutea, cysts, or dominant follicles characterize the polycystic ovaries (Teede et al. 2018).

8–46% in women with PCOS are affected by metabolic syndrome (Kazemi et al. 2019a). In our experience, compared to age-matched controls, patients had the greatest elevation of triglycerides (by 63%), followed by waist circumference (21%), reduced high-density lipoproteins (19%), elevated systolic blood pressure (5%) and glucose (4%) (Kazemi et al. 2019a). Other risk factors for diabetes or cardiovascular disease were also elevated. Elevated low-density lipoproteins (by 16%), total cholesterol to high-density lipoprotein ratio (by 31%), fasting insulin (by 122%), 2-hour oral glucose tolerance test glucose level (29%), and homeostatic model of insulin resistance (HOMA-IR) (elevated by 200%) were also documented, along with increased high-sensitivity C-reactive protein (by 413%), total testosterone (by 25%), whereas sex-hormone-binding globulin (SHBG) was reduced by 37% (Kazemi et al. 2019a). Pro-inflammatory status and hyperandrogenism in women with PCOS promote atherosclerosis and endothelial dysfunction (Chiu et al. 2017).

Sarcopenic Obesity

Sarcopenic obesity signals low muscle mass (sarcopenia) combined with high-fat mass (obesity). Besides expected dyslipidemia, insulin resistance, diabetes, osteoarthritis, and depression, also mortality, reduced physical function, falls, and quality of life can be further impaired (Koliaki et al. 2019).

Defining features are low appendicular muscle mass relative to total body weight, two standard deviations below the age- and sex-matched mean of the population (McBreairty et al. 2019). In our experience, 53% of the women with PCOS had sarcopenic obesity (McBreairty et al. 2019). The percent appendicular skeletal muscle mass was negatively correlated with C-reactive protein ($r = -0.61$) and positively correlated to serum vitamin D ($r = 0.40$), suggesting that inflammation and low vitamin D could be relevant. Negative correlation with HOMA-IR ($r = -0.41$) and glycosylated hemoglobin ($r = -0.43$) further pointed toward the adverse effects of insulin resistance and poor glucoregulatory status. In a larger study, our observed incidence of sarcopenic obesity was slightly higher (i.e., 46/80 women or 58%) (Kazemi et al. 2019b), and appendicular skeletal muscle mass was positively associated with protein, monounsaturated fat, and omega-3 polyunsaturated fat intake, with potential implications for unique dietary recommendations in this setting.

Mechanisms for Sarcopenic Obesity

Women with PCOS have increased visceral adiposity (Kazemi et al. 2019b), a source of "adipokines" (i.e., leptin, adiponectin) and cytokines (Chilibeck et al. 2014). Elevated cytokines such as interleukin-6 (IL-6) and tumor necrosis factor-alpha (TNF-alpha) and leptin, along with reduced adiponectin (which tends to be anti-inflammatory) impact on muscle tissue. For example, leptin, IL-6, and TNF-alpha decrease the anabolic activity of insulin-like growth factor-1 (IGF-1), which is important for muscle growth and development (Koliaki et al. 2019). Inflammatory cytokines can directly increase protein degradation, leading to the loss of muscle mass (Reid and Li 2001). They also increase insulin resistance at muscle (Chilibeck et al. 2014), an anabolic and anti-catabolic hormone (Lee et al. 2015). Adiponectin improves insulin sensitivity (Chilibeck et al. 2014) potentially magnifying the negative muscular consequences. In women with PCOS, blood adiponectin levels are reduced by 40%, and this coincides with a 25% reduction in whole-body insulin sensitivity (Hansen et al. 2019).

Lipid Metabolism

Increased intra-muscular lipids are typical of obesity, associated with derivatives such as diacylglycerols and ceramides, which are considered to possess "lipotoxicity," increasing inflammation, and inducing insulin resistance (Hansen et al. 2019). Ceramides may cause mitochondrial dysfunction, leading to defects in the electron transport chain and increased oxidative stress. The oxidative stress leads to the degradation of muscular proteins (Kohara 2014). An increase in diacylglycerols and ceramides is associated with increased levels of myostatin (Consitt and Clark 2018) which diminishes satellite cell numbers and activation (Thornell 2011).

Satellite cells are stem cells associated with muscle that is important for repair of muscle damage and are incorporated into muscle fibers as new myonuclei when activated, leading to enhanced capacity for protein synthesis. Muscle tissue assessed from women with PCOS indicates an elevation of intramuscular triglycerides, diacylglycerol, and ceramide by 40%, 50%, and 300%, respectively. In addition, the regulation of pyruvate dehydrogenase, an enzyme important in mitochondrial metabolism of carbohydrates, was negatively affected, implying mitochondrial dysfunction (Hansen et al. 2019).

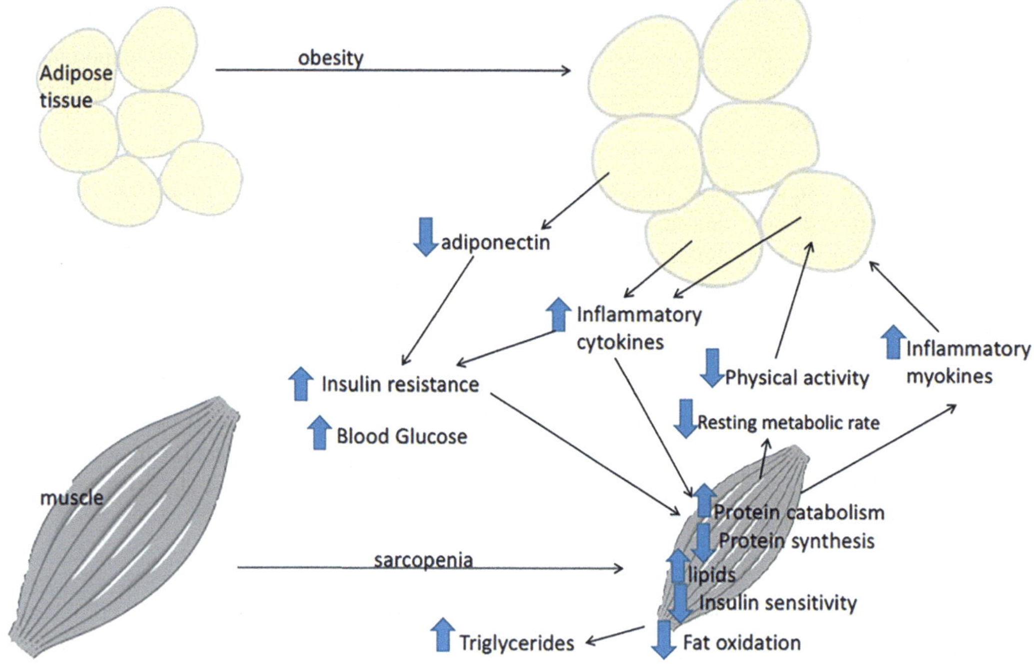

Fig. 67.1 Mechanisms and metabolic consequences of sarcopenic obesity

Myokines

The increased inflammation may result in the release of pro-inflammatory "myokines" from the muscle itself (Koliaki et al. 2019). These myokines may then circulate to adipose tissue, causing further inflammation in adipocytes, leading to a vicious cycle (Fig. 67.1).

Dietary Interventions: Vitamin D

In a previous review (Candow et al. 2012), we highlighted the anti-sarcopenic interest of vitamin D and omega-3 fatty acids, both of which reduce inflammation. In older adults, low vitamin D levels are associated with lower muscle mass, impaired strength, and reduced functional abilities (Visser et al. 2003; Snijder et al. 2006). There are receptors for vitamin D on muscle (Bischoff-Ferrari et al. 2004). Vitamin D has anti-inflammatory properties and decreases C-reactive protein, and several inflammatory cytokines (Schleithoff et al. 2006). In our previously alluded to study, vitamin D deficiency was frequent (41% in PCOS) and correlated with several indexes of obesity (McBreairty et al. 2019).

High-dose vitamin D supplementation (4000–8500 IU per day for 2–3 months) was followed by decreased testosterone levels, free androgen index, and androstenedione, and increased SHBG in women with PCOS, most of who presented with vitamin D deficiency (Jamilian et al. 2017a; Pal et al. 2012); increased menstrual cyclicity (Jafari-Sfidvajani et al. 2018) and reduced hirsutism scores (Jamilian et al. 2017a) were also observed. Improvements in metabolic profiles included increased antioxidant capacity and adiponectin levels, and reduced inflammation (i.e., C-reactive protein), fasting insulin, fasting glucose, HOMA-IR, glucose response to an oral glucose tolerance test, triglycerides, very low-density lipoproteins, low-density lipoproteins, total cholesterol, blood pressure, and alanine aminotransferase levels (Asemi et al. 2015; Dastorani et al. 2018; Jamilian et al. 2017a; Javed et al. 2019; Pal et al. 2012;

Seyyed Abootorabi et al. 2018; Trummer et al. 2019).

The latter improved liver enzyme could indicate reduced liver damage and possibly the alleviation of nonalcoholic fatty liver disease, which is common with obesity and in conditions such as PCOS (Javed et al. 2019).

We recommend that women with PCOS who have deficient vitamin D levels should be considered for high-dose vitamin D supplementation (i.e., 4000 to 8500 IU per day). Meta-analyses show that even with doses less than 4000 IU per day, it reduces inflammation, oxidative stress (Akbari et al. 2018), HOMA-IR (Łagowska et al. 2018), harmful lipids, testosterone levels, and hirsutism and improves menstrual regularity and follicular maturation (Shojaeian et al. 2019).

Omega-3 Fatty Acids

Omega-3 fatty acids (mostly from fish oils) compete with some of the same enzymes that are involved in the metabolism of omega-6 fatty acids, to produce eicosanoids which are less inflammatory (Calder and Grimble 2002). The end effect is that improving omega-3 to omega-6 ratio in the diet is likely to reduce inflammation. Omega-3 fatty acid intake is associated with enhanced strength and functional ability in older adults (Rodacki et al. 2012) and, therefore, has been promoted for preventing sarcopenia (Candow et al. 2012). Supplementation with 1.2 to 2.4 g/d omega-3 fatty acids derived from fish oil for 6–12 weeks in women with PCOS decreased free testosterone, luteinizing hormone to follicle-stimulating hormone ratio, hirsutism score, oxidative stress, C-reactive protein, fasting glucose, fasting insulin, and HOMA-IR, and increased adiponectin levels (Amini et al. 2018; Nadjarzadeh et al. 2015; Phelan et al. 2011; Rafraf et al. 2012). A longer supplementation intervention (i.e., 2 g/d for 6 months) was effective for reducing waist circumference, low-density lipoproteins, total cholesterol, and triglycerides, and increasing high-density lipoproteins and menstrual regularity (Khani et al. 2017).

Higher-dose supplementation (i.e., 4 g/d for 8 weeks) reduced liver fat, triglycerides, and systolic and diastolic blood pressure (Cussons et al. 2009). A less bioactive form of omega-3 fatty acid from flaxseed is also effective: Supplementation with 2 g/d for 12 weeks reduced fasting insulin, HOMA-IR, triglycerides, very low-density lipoprotein, and C-reactive protein, and increased peroxisome proliferator-activated receptor-gamma (PPAR-gamma) gene expression (i.e., mRNA) from blood mononuclear cells in women with PCOS (Mirmasoumi et al. 2018; Nasri et al. 2017). Peroxisome proliferator-activated receptor-gamma activation is important for improving glucose metabolism and reducing muscle insulin resistance (Janani and Ranjitha Kumari 2015). The most recent meta-analysis indicated that omega-3 supplementation in women with PCOS reduces HOMA-IR and harmful lipids and increases adiponectin levels (Yang et al. 2018). The only meta-analysis to assess hormone levels indicated a small effect for reducing testosterone levels, and that only longer-term studies (i.e., longer than 6 weeks) were effective for reducing DHEAS levels (Hajishafiee et al. 2016).

Antioxidant Supplementation

Sarcopenia may be linked to defects in the electron transport chain of mitochondria, which produces free radicals and oxidative damage to tissues (Johnston et al. 2008). Supplementation with antioxidants could alleviate inflammation associated with oxidative damage and its metabolic consequences. In a rodent model of PCOS, co-enzyme Q-10 supplementation reversed impaired mitochondrial function, reduced insulin resistance, decreased proteins related to apoptosis, and increased those with anti-apoptosis properties (Ding et al. 2019). In a clinical randomized series (Izadi et al. 2019), the combination of Coenzyme Q-10 and vitamin E reduced visceral adiposity, total cholesterol, low-density lipoproteins, and diastolic blood pressure, and increased high-density lipoproteins and SHBG. Fasting glucose and HOMA-IR were ameliorated in patients managed with Coenzyme Q-10, whereas triglycerides diminished in all treatment groups.

Increased gene expression of PPAR-gamma, ovulation frequency, and pregnancy, along with attenuated gene expression for inflammatory cytokines, was documented in Coenzyme Q-10-treated cases (El Refaeey et al. 2014; Rahmani et al. 2018).

Selenium Supplementation

Selenium given in small amounts (200 μg/d for eight weeks) decreased dehydroepiandrosterone, and high-sensitivity C-reactive protein, and was beneficial for hirsutism, acne, and fertility rate (19% pregnancy vs. 3% in controls) (Razavi et al. 2016).

Chromium Picolinate

Chromium has long been associated with benefits in diabetes, and supplements in the form of chromium picolinate. Nevertheless, meta-analyses failed to confirm significant impact on insulin and other metabolic variables. Paradoxically, total and free testosterone became more elevated (Heshmati et al. 2018; Tang et al. 2018). Within such circumstances, there is doubt whether it could be recommended for PCOS.

Carnitine

Carnitine is an amino acid derivative that helps to shuttle fatty acids across the mitochondrial membrane so they can be oxidized. Supplementation of 200 mg/d carnitine for 12 weeks in women with PCOS was effective for reducing body weight, waist circumference, fasting glucose, fasting insulin, HOMA-IR, DHEAS, oxidative stress, and carotid intima-media thickness (Jamilian et al. 2017b; Samimi et al. 2016; Talari et al. 2019). Meta-analyses are not available yet.

Combinations of Nutritional Supplements

A combination of chromium picolinate (200 μg/d) and carnitine (1000 mg/d) for 12 weeks in women with PCOS reduced body weight, testosterone, C-reactive protein, markers of oxidative stress, hirsutism, fasting glucose, fasting insulin, triglycerides, low-density lipoproteins, and increased PPAR-gamma gene expression (Jamilian et al. 2019, 2020). Combining chromium picolinate, antioxidants (i.e., lycopene, n-acetylcysteine), and vitamins (i.e., vitamin D) for 12 weeks decreased body weight, improved menstrual cyclicity, and reduced acne and hirsutism in obese women with PCOS (Advani et al. 2020). Vitamin D in conjunction with omega-3 fatty acids for 12 weeks reduced total cholesterol, C-reactive protein, and enhanced antioxidant capacity (Jamilian et al. 2018). Also omega-3 fatty acids with vitamin E for 8–12 weeks improved antioxidant capacity, reduced C-reactive and carotid intima-media thickness (Sadeghi et al. 2019; Talari et al. 2018). Further studies on the impact on metabolic and reproductive phenotypes of PCOS would be advisable.

Dietary Approaches

Several diets such as the "dietary approaches to stop hypertension" (DASH), low glycemic index diets, low carbohydrate diets, Mediterranean Diet, and Portfolio Diet look effective for reducing risk factors for cardiovascular disease, occasionally encompassing women with PCOS.

DASH Diet

It emphasizes reduced sodium, saturated fat, refined grains and sweets, with more generous ingestion of potassium, calcium, and magnesium. Fruits, vegetables, low-fat dairy products, and moderate amounts of fish, poultry, whole grains, and nuts are recommended. Two to three months of the DASH diet for women with PCOS reduced body weight, fat mass, waist circumference, and androstenedione levels while increasing SHBG (Asemi et al. 2014; Asemi and Esmaillzadeh 2015; Azadi-Yazdi et al. 2017; Foroozanfard et al. 2017). This alleviated oxidative stress and inflammation (i.e., C-reactive protein) and improved metabolic profile, as indicated by decreased fasting insulin, HOMA-IR, triglycerides, and very low lipoproteins (Asemi et al. 2014; Asemi and Esmaillzadeh 2015; Azadi-Yazdi et al. 2017; Foroozanfard et al. 2017).

Low-Glycemic Index Diets

The glycemic index (GI) of foods is determined by the 2-hour blood glucose response to 50 g available carbohydrate from a test food (Jenkins et al. 1981). Twelve weeks of a low GI diet in women with PCOS was effective for increasing insulin sensitivity and reducing HOMA-IR and C-reactive protein (Barr et al. 2013; Mehrabani et al. 2012), and after 12 months, menstrual cycles turned more regular (Marsh et al. 2010). We showed that 16 weeks of a pulse-based diet (composed of high-protein, high-fiber, low glycemic index legumes such as lentils, chickpeas, beans, and peas) reduced insulin response to an oral glucose tolerance test, low-density lipoproteins, total cholesterol/high-density lipoprotein ratio, and diastolic blood pressure, and increased high-density lipoprotein levels compared to a healthy diet designed for reducing lipids (the Therapeutic Lifestyle Changes diet) (Kazemi et al. 2018). A problem with many diets is long-term adherence. Indeed, in a 12-month follow-up after completion of our intervention, only HDL and total cholesterol/HDL improvements remained evident.

Low Carbohydrate Diets

They have reduced total fat, intra-abdominal fat, body weight, BMI, waist circumference, fasting glucose and insulin, HOMA-IR, total and free testosterone, and hirsutism, and preserved lean tissue mass loss in women with PCOS (Goss et al. 2014; Phy et al. 2015; Pohlmeier et al. 2014). The most recent meta-analysis on the topic indicated that low carbohydrate diets in women with PCOS are effective for reducing BMI, HOMA-IR, total cholesterol, and low-density lipoproteins (Zhang et al. 2019). More than four months were necessary for reducing testosterone and increasing SHBG. Diets lower in fat were best for improving hormone profiles and should be preferred.

The Mediterranean Diet and the Portfolio Diet

The lipid-lowering effect of these diets is comparable to that of certain pharmaceuticals (i.e., first-generation statins), without the side effects (Jenkins et al. 2003). A recent cross-sectional study indicated that women with PCOS consume smaller amounts of olive oil, legumes, fish, and nuts (Barrea et al. 2019). Fiber and magnesium are similarly scarce (Cutler et al. 2019). A randomized controlled study emphasizing increased intake of legumes (i.e., beans, lentils, chickpeas, peas) showed this diet was effective for reducing insulin levels during an oral glucose tolerance test, triglycerides, low-density lipoproteins, total cholesterol to high-density lipoprotein levels, and diastolic blood pressure, and increased high-density lipoproteins (Kazemi et al. 2018). Also walnuts decreased low-density lipoproteins, and glycosylated hemoglobin, and increased SHBG levels, whereas both walnuts and almonds increased adiponectin levels, and almonds reduced androgens in women with PCOS (Kalgaonkar et al. 2011).

The Portfolio diet includes viscous fibers, plant sterols, soy protein, and nuts, which could be useful for PCOS (Cutler et al. 2019; Kalgaonkar et al. 2011; Kazemi et al. 2018). An eight-week randomized controlled trial of a diet high in soy (i.e., 35% animal protein, 35% soy, 30% vegetable protein versus a diet of 70% animal protein and 30% vegetable protein) was effective for reducing BMI, glucose, total testosterone, insulin, insulin resistance, and triglycerides in women with PCOS (Karamali et al. 2018). Yet at present, there are no interventions that have actually assessed the effectiveness of the Mediterranean or the Portfolio diet in women with PCOS.

Exercise

Exercise can target muscle and/or adipose tissue. Resistance (strength) training is best for increasing muscle mass, whereas aerobic training could reduce adipose tissue and change the adipokines released by adipose tissue (Chilibeck et al. 2014). Regarding aerobic exercise, there are traditional aerobic prescriptions (i.e., low- to moderate-intensity, long duration) and high-intensity interval training (high-intensity, shorter duration). There is interest in the latter form of exercise because it seems to be highly effective for inducing mitochondrial adaptations (Chilibeck et al. 1998).

Aerobic Training

This is the most common intervention for women with PCOS because of alleviation of metabolic and cardiovascular consequences of obesity. Simply encouraging women with PCOS to increase their step count by 1000 steps/day, or about 0.8 km/day walking over six months, significantly reduced inflammation by 13% as indicated by reduced interleukin-6 (IL-6) and C-reactive protein (Webb et al. 2018). Moderate-intensity aerobic training (i.e., 40–75% of peak heart rate, 60–70% of peak aerobic capacity) at three times per week for 30–60 min per session that involved walking, jogging, stationary cycling, or exercise on an elliptical trainer for 8–26 weeks reduced waist circumference, BMI, fat mass, ectopic fat in the lower leg, harmful lipids (i.e., triglycerides, cholesterol, low-density lipoproteins, very low-density lipoproteins), insulin resistance, fasting insulin, area under the insulin curve during oral glucose tolerance tests, systolic and diastolic blood pressure, intima-media thickness, testosterone, and number of ovarian follicles, and increased high-density lipoproteins, insulin sensitivity, flow-mediated dilation (indicating improved endothelial function), SHBG, and menstrual regularity (Costa et al. 2018; Orio et al. 2016; Sprung et al. 2013). Meta-analyses conclude that moderate-intensity aerobic exercise with a frequency of three times per week for a duration of at least 30 min/session and greater than or equal to three-month duration had favorable effects on waist circumference, body fat percentage, testosterone, SHBG, fasting glucose and insulin, HOMA-IR, blood lipids, systolic blood pressure, and C-reactive protein (Kite et al. 2019; Woodward et al. 2019), even though changes could be modest (Kite et al. 2019). Aerobic training interventions may, therefore, be best combined with other lifestyle interventions such as changes in diet (Kazemi et al. 2018).

High-Intensity Interval Training (HIIT)

HIIT is more effective than standard continuous moderate-intensity aerobic training for improving mitochondrial function, despite the fact that it is done for shorter durations and lower volume (Chilibeck et al. 1998). This is important considering that time commitment is often a major barrier to performing physical activity. Samadi et al. (2019) used a protocol where twelve weeks of aquatic HIIT involved 20 min of exercise three times per week. High-intensity interval training sessions involved 4×4 min bouts with 1-min rest between bouts; each bout involved 8×20 s "all out" intervals with 10 s rest between intervals. This form of HIIT reduced BMI, fat mass, HOMA-IR, hirsutism, and increased SHBG and menstrual cycle regularity.

Aktaş et al. (2019) compared HIIT to standard moderate-intensity aerobic training. Both programs involved 30 min of exercise done three times per week for 12 weeks. The HIIT sessions simply involved 2-min intervals of running alternated with 2 min of walking, whereas the standard aerobic program involved 30 min of moderate tempo running for 30 min. Body mass index decreased in both groups; however, only the HIIT increased adiponectin and high-density lipoproteins and reduced insulin, triglycerides, total cholesterol, and low-density lipoproteins. Almenning et al. (2015) compared thrice-weekly HIIT to resistance training over 10 weeks. The HIIT involved twice a week training with 4 × 4 min exercise bouts at 90–95% maximal heart rate with each bout separated by 3 min moderate-intensity exercise at 70% maximal heart rate, and one session per week with 10 intervals of maximal exercise, with each interval separated by 1-min rest or low-intensity exercise. The mode of HIIT was on a treadmill, outdoor walking/running, or on a stationary bike depending on participants' preferences. The resistance training involved eight exercises done for 3 sets × 10 repetitions with 1-min rest between sets and resistance at 75% of 1RM, where 1RM is defined as the maximal strength or the maximal amount of weight that can be lifted once.

The resistance training was superior for increasing muscle mass and SHBG, and decreasing free androgen index, whereas the HIIT was superior for decreasing insulin, HOMA-IR, and DHEAS, and increasing high-density lipoproteins and flow-mediated dilation and therefore endothelial function. Both programs were effective for reducing percent body fat. Ideal results may be derived where HIIT training is combined with a resistance training program.

Resistance Training

It has the greatest potential for increasing muscle mass and, therefore, may be most effective for alleviating sarcopenia. One study indicated that 4 months of resistance training, done 3 times per week with 10 exercises induced favorable changes in body composition and strength, and reduced testosterone and fasting glucose levels, but had no effect on fasting insulin or HOMA-IR (Kogure et al. 2016, 2018). A small study randomized protocol of resistance training ($n = 7$) and controls ($n = 6$) revealed similar results (Vizza et al. 2016). While resistance training is effective for increasing muscle mass and physical function in women with PCOS, aerobic training may need to be added to achieve more marked improvements in metabolic measures. Turan et al. (2015) submitted women with PCOS to combined aerobic and resistance training. The duration of training was 8 weeks, 3 times per week. Compared to controls, the exercise group reduced waist circumference, diastolic blood pressure, low-density lipoproteins, total cholesterol, fasting insulin and glucose, HOMA-IR, and increased high-density lipoproteins and menstrual cycle regularity. A limitation of this study is that resistance training only or aerobic training only groups were not included; therefore, it is difficult to discern whether metabolic benefits were due to resistance training or aerobic programs.

A restricted diet, aerobic training, or the association of both were investigated in a cohort of PCOS. Exercise was superior for inducing changes in body composition. All groups showed metabolic and hormonal benefits, decreased waist circumference, and abdominal fat mass. The aerobic training group increased the number of ovulatory cycles compared to the diet-only group (Thomson et al. 2008, 2012). It appears diet and/or exercise are effective for inducing metabolic benefits; however, exercise interventions are best for improving body composition in women with PCOS.

Summary/Overall Recommendations

Dietary and exercise intervention studies in women with PCOS are characterized by low adherence rates; therefore, it would be practical to offer a variety of diet and exercise options so women can choose options they consider most enjoyable and feasible.

References

Advani K, Batra M, Tajpuriya S et al (2020) Efficacy of combination therapy of inositols, antioxidants and vitamins in obese and non-obese women with polycystic ovary syndrome: an observational study. J Obstet Gynaecol 40(1):96–101. https://doi.org/10.1080/01443615.2019.1604644

Akbari M, Ostadmohammadi V, Lankarani KB et al (2018) The effects of vitamin D supplementation on biomarkers of inflammation and oxidative stress among women with polycystic ovary syndrome: a systematic review and meta-analysis of randomized controlled trials. Horm Metab Res 50(4):271–279. https://doi.org/10.1055/s-0044-101355

Aktaş HŞ, Uzun YE, Kutlu O et al (2019) The effects of high intensity-interval training on vaspin, adiponectin and leptin levels in women with polycystic ovary syndrome. Arch Physiol Biochem https://doi.org/10.1080/13813455.2019.1662450

Almenning I, Rieber-Mohn A, Lundgren KM et al (2015) Effects of high intensity interval training and strength training on metabolic, cardiovascular and hormonal outcomes in women with polycystic ovary syndrome: a pilot study. PLoS One 10(9):e0138793. https://doi.org/10.1371/journal.pone.0138793

Amini M, Bahmani F, Foroozanfard F et al (2018) The effects of fish oil omega-3 fatty acid supplementation on mental health parameters and metabolic status of patients with polycystic ovary syndrome: a randomized, double-blind, placebo-controlled trial. J Psychosom Obstet Gynaecol. https://doi.org/10.1080/0167482X.2018.1508282

Asemi Z, Esmaillzadeh A (2015) DASH diet, insulin resistance, and serum hs-CRP in polycystic ovary syndrome: a randomized controlled clinical trial. Horm Metab Res 47(3):232–238. https://doi.org/10.1055/s-0034-1376990

Asemi Z, Samimi M, Tabassi Z et al (2014) Effects of DASH diet on lipid profiles and biomarkers of oxidative stress in overweight and obese women with polycystic ovary syndrome: a randomized clinical trial. Nutrition 30(11–12):1287–1293. https://doi.org/10.1016/j.nut.2014.03.008

Asemi Z, Foroozanfard F, Hashemi T et al (2015) Calcium plus vitamin D supplementation affects glucose metabolism and lipid concentrations in overweight and obese vitamin D deficient women with polycystic ovary syndrome. Clin Nutr 34(4):586–592. https://doi.org/10.1016/j.clnu.2014.09.015

Azadi-Yazdi M, Karimi-Zarchi M, Salehi-Abargouei A et al (2017) Effects of Dietary Approach to Stop Hypertension diet on androgens, antioxidant status and body composition in overweight and obese women with polycystic ovary syndrome: a randomised controlled trial. J Hum Nutr Diet 30(3):275–283. https://doi.org/10.1111/jhn.12433

Barr S, Reeves S, Sharp K et al (2013) An isocaloric low glycemic index diet improves insulin sensitivity in women with polycystic ovary syndrome. J Acad Nutr Diet 113(11):1523–1531. https://doi.org/10.1016/j.jand.2013.06.347

Barrea L, Arnone A, Annunziata G et al (2019) Adherence to the mediterranean diet, dietary patterns and body composition in women with Polycystic Ovary Syndrome (PCOS). Nutrients 11(10). pii: E2278. https://doi.org/10.3390/nu11102278

Bischoff-Ferrari HA, Borchers M, Gudat F et al (2004) Vitamin D receptor expression in human muscle tissue decreases with age. J Bone Miner Res 19:265–269

Calder PC, Grimble RF (2002) Polyunsaturated fatty acids, inflammation and immunity. Eur J Clin Nutr 56:S14–S19

Candow DG, Forbes SC, Little JP et al (2012) Effect of nutritional interventions and resistance exercise on aging muscle mass and strength. Biogerontology 13 (4):345–358. https://doi.org/10.1007/s10522-012-9385-4

Carmina E, Lobo RA (1999) Polycystic ovary syndrome (PCOS): arguably the most common endocrinopathy is associated with significant morbidity in women. J Clin Endocrinol Metab 84:1897–1899

Chilibeck PD, Bell GJ, Farrar RP et al (1998) Higher mitochondrial fatty acid oxidation following intermittent versus continuous endurance exercise training. Can J Physiol Pharmacol 76(9):891–894

Chilibeck PD, Pérez-López FR, Bodary P et al (2014) Adipocytokines, metabolic syndrome, and exercise. Int J Endocrinol 2014:597162. https://doi.org/10.1155/2014/597162

Chiu W-L, Boyle J, Vincent A, et a (2017) Cardiometabolic risks in polycystic ovary syndrome: non-traditional risk factors and the impact of obesity. Neuroendocrinology 104:412–424

Consitt LA, Clark BC (2018) The vicious cycle of myostatin signaling in sarcopenic obesity: myostatin role in skeletal muscle growth, insulin signaling and implications for clinical trials. J Frailty Aging 7:21–27

Costa EC, DE Sá JCF, Stepto NK et al (2018) Aerobic training improves quality of life in women with polycystic ovary syndrome. Med Sci Sports Exerc 50 (7):1357–1366. https://doi.org/10.1249/MSS.0000000000001579

Cussons AJ, Watts GF, Mori TA et al (2009) Omega-3 fatty acid supplementation decreases liver fat content in polycystic ovary syndrome: a randomized controlled trial employing proton magnetic resonance spectroscopy. J Clin Endocrinol Metab 94(10):3842–3848. https://doi.org/10.1210/jc.2009-0870

Cutler DA, Pride SM, Cheung AP (2019) Low intakes of dietary fiber and magnesium are associated with insulin resistance and hyperandrogenism in polycystic ovary syndrome: a cohort study. Food Sci Nutr 7 (4):1426–1437. https://doi.org/10.1002/fsn3.977

Dastorani M, Aghadavod E, Mirhosseini N et al (2018) The effects of vitamin D supplementation on metabolic profiles and gene expression of insulin and lipid metabolism in infertile polycystic ovary syndrome candidates for in vitro fertilization. Reprod Biol Endocrinol 16 (1):94. https://doi.org/10.1186/s12958-018-0413-3

Ding Y, Jiang Z, Xia B et al (2019) Mitochondria-targeted antioxidant therapy for an animal model of PCOS-IR. Int J Mol Med 43(1):316–324. https://doi.org/10.3892/ijmm.2018.3977

El Refaeey A, Selem A, Badawy A (2014) Combined coenzyme Q10 and clomiphene citrate for ovulation induction in clomiphene-citrate-resistant polycystic ovary syndrome. Reprod Biomed Online 29 (1):119–124. https://doi.org/10.1016/j.rbmo.2014.03.011

Foroozanfard F, Rafiei H, Samimi M et al (2017) The effects of dietary approaches to stop hypertension diet on weight loss, anti-Müllerian hormone and metabolic profiles in women with polycystic ovary syndrome: A randomized clinical trial. Clin Endocrinol 87(1):51–58. https://doi.org/10.1111/cen.13333

Goss AM, Chandler-Laney PC, Ovalle F et al (2014) Effects of a eucaloric reduced-carbohydrate diet on body composition and fat distribution in women with PCOS. Metabolism 63(10):1257–1264. https://doi.org/10.1016/j.metabol.2014.07.007

Hajishafiee M, Askari G, Iranj B et al (2016) The effect of n-3 polyunsaturated fatty acid supplementation on androgen status in patients with polycystic ovary syndrome: a systematic review and meta-analysis of clinical trials. Horm Metab Res 48(5):281–289. https://doi.org/10.1055/s-0042-105288

Hansen SL, Svendsen PF, Jeppesen JF et al (2019) Molecular mechanisms in skeletal muscle underlying insulin resistance in women who are lean with polycystic ovary syndrome. J Clin Endocrinol Metab 104 (5):1841–1854. https://doi.org/10.1210/jc.2018-01771

Heshmati J, Omani-Samani R, Vesali S et al (2018) The effects of supplementation with chromium on insulin resistance indices in women with polycystic ovarian syndrome: a systematic review and meta-analysis of randomized clinical trials. Horm Metab Res 50 (3):193–200. https://doi.org/10.1055/s-0044-101835

Izadi A, Ebrahimi S, Shirazi S, Taghizadeh et al (2019) Hormonal and metabolic effects of coenzyme Q10 and/or vitamin E in patients with polycystic ovary syndrome. J Clin Endocrinol Metab 104(2):319–327. https://doi.org/10.1210/jc.2018-01221

Jafari-Sfidvajani S, Ahangari R, Hozoori M et al (2018) The effect of vitamin D supplementation in combination with low-calorie diet on anthropometric indices and androgen hormones in women with polycystic ovary syndrome: a double-blind, randomized, placebo-controlled trial. J Endocrinol Investig 41 (5):597–607. https://doi.org/10.1007/s40618-017-0785-9

Jamilian M, Foroozanfard F, Rahmani E et al (2017a) Effect of two different doses of vitamin D supplementation on metabolic profiles of insulin-resistant patients with polycystic ovary syndrome. Nutrients 9(12). pii: E1280. https://doi.org/10.3390/nu9121280

Jamilian H, Jamilian M, Samimi M et al (2017b) Oral carnitine supplementation influences mental health parameters and biomarkers of oxidative stress in women with polycystic ovary syndrome: a randomized, double-blind, placebo-controlled trial. Gynecol Endocrinol 33(6):442–447. https://doi.org/10.1080/09513590.2017.1290071

Jamilian M, Samimi M, Mirhosseini N et al (2018) The influences of vitamin D and omega-3 co-supplementation on clinical, metabolic and genetic parameters in women with polycystic ovary syndrome. J Affect Disord 238:32–38. https://doi.org/10.1016/j.jad.2018.05.027

Jamilian M, Foroozanfard F, Kavossian E et al (2019) Carnitine and chromium co-supplementation affects mental health, hormonal, inflammatory, genetic, and oxidative stress parameters in women with polycystic ovary syndrome. J Psychosom Obstet Gynaecol. https://doi.org/10.1080/0167482X.2018.1557144

Jamilian M, Foroozanfard F, Kavossian E et al (2020) Effects of chromium and carnitine co-supplementation on body weight and metabolic profiles in overweight and obese women with polycystic ovary syndrome: a randomized, double-blind, placebo-controlled trial. Biol Trace Elem Res 193 (2):334–341. https://doi.org/10.1007/s12011-019-01720-8

Janani C, Ranjitha Kumari BD (2015) PPAR gamma gene – a review. Diabetes Metab Syndr 9(1):46–50. https://doi.org/10.1016/j.dsx.2014.09.015

Javed Z, Papageorgiou M, Deshmukh H et al (2019) A randomized, controlled trial of vitamin D supplementation on cardiovascular risk factors, hormones, and liver markers in women with polycystic ovary syndrome. Nutrients 11(1). pii: E188. https://doi.org/10.3390/nu11010188

Jenkins DJ, Wolever TM, Taylor RH et al (1981) Glycemic index of foods: a physiological basis for carbohydrate exchange. Am J Clin Nutr 34(3):362–366

Jenkins DJ, Kendall CW, Marchie A et al (2003) Effects of a dietary portfolio of cholesterol-lowering foods vs lovastatin on serum lipids and C-reactive protein. JAMA. 290(4):502–510

Johnston AP, De Lisio M, Parise G (2008) Resistance training, sarcopenia, and the mitochondrial theory of aging. Appl Physiol Nutr Metab 33(1):191–199. https://doi.org/10.1139/H07-141

Kalgaonkar S, Almario RU, Gurusinghe D et al (2011) Differential effects of walnuts vs almonds on improving metabolic and endocrine parameters in PCOS. Eur J Clin Nutr 65(3):386–393. https://doi.org/10.1038/ejcn.2010.266

Karamali M, Kashanian M, Alaeinasab S et al (2018) The effect of dietary soy intake on weight loss, glycaemic control, lipid profiles and biomarkers of inflammation and oxidative stress in women with polycystic ovary syndrome: a randomised clinical trial. J Hum Nutr Diet 31(4):533–543. https://doi.org/10.1111/jhn.12545

Kazemi M, McBreairty LE, Chizen DR et al (2018) A comparison of a pulse-based diet and the therapeutic lifestyle changes diet in combination with exercise and health counselling on the cardio-metabolic risk profile in women with polycystic ovary syndrome: a

randomized controlled trial. Nutrients 10(10). pii: E1387. https://doi.org/10.3390/nu10101387
Kazemi M, Pierson RA, Lujan ME et al (2019a) Comprehensive evaluation of type 2 diabetes and cardiovascular disease risk profiles in reproductive-age women with polycystic ovary syndrome: a large canadian cohort. J Obstet Gynaecol Can 41(10):1453–1460. https://doi.org/10.1016/j.jogc.2018.11.026
Kazemi M, Zello G, McBreairty L et al (2019b) Sarcopenia is positively associated with poor dietary intakes of women with polycystic ovary syndrome. Appl Physiol Nutr Metab 44:S26
Khani B, Mardanian F, Fesharaki SJ (2017) Omega-3 supplementation effects on polycystic ovary syndrome symptoms and metabolic syndrome. J Res Med Sci 22:64. https://doi.org/10.4103/jrms.JRMS_644_16
Kite C, Lahart IM, Afzal I (2019) Exercise, or exercise and diet for the management of polycystic ovary syndrome: a systematic review and meta-analysis. Syst Rev 8 (1):51. https://doi.org/10.1186/s13643-019-0962-3
Kogure GS, Miranda-Furtado CL, Silva RC et al (2016) Resistance exercise impacts lean muscle mass in women with polycystic ovary syndrome. Med Sci Sports Exerc 48(4):589–598. https://doi.org/10.1249/MSS.0000000000000822
Kogure GS, Silva RC, Miranda-Furtado CL et al (2018) Hyperandrogenism enhances muscle strength after progressive resistance training, independent of body composition, in women with polycystic ovary syndrome. J Strength Cond Res 32(9):2642–2651. https://doi.org/10.1519/JSC.0000000000002714
Kohara K (2014) Sarcopenic obesity in aging population: current status and future directions for research. Endocrine 45:15–25
Koliaki C, Liatis S, Dalamaga M et al (2019) Sarcopenic obesity: epidemiologic evidence, pathophysiology, and therapeutic perspectives. Curr Obes Rep 8 (4):458–471. https://doi.org/10.1007/s13679-019-00359-9
Łagowska K, Bajerska J, Jamka M (2018) The role of vitamin D oral supplementation in insulin resistance in women with polycystic ovary syndrome: a systematic review and meta-analysis of randomized controlled trials. Nutrients 10(11). pii: E1637. https://doi.org/10.3390/nu10111637
Lee SW, Youm Y, Lee WJ et al (2015) Appendicular skeletal muscle mass and insulin resistance in an elderly korean population: the Korean social life, health and aging project-health examination cohort. Diabetes Metab J 39:37–45
Marsh KA, Steinbeck KS, Atkinson FS et al (2010) Effect of a low glycemic index compared with a conventional healthy diet on polycystic ovary syndrome. Am J Clin Nutr 92(1):83–92. https://doi.org/10.3945/ajcn.2010.29261
McBreairty LE, Chilibeck PD, Gordon JJ et al (2019) Polycystic ovary syndrome is a risk factor for sarcopenic obesity: a case control study. BMC Endocr Disord 19(1):70. https://doi.org/10.1186/s12902-019-0381-4
Mehrabani HH, Salehpour S, Amiri Z et al (2012) Beneficial effects of a high-protein, low-glycemic-load hypocaloric diet in overweight and obese women with polycystic ovary syndrome: a randomized controlled intervention study. J Am Coll Nutr 31 (2):117–125
Mirmasoumi G, Fazilati M, Foroozanfard F et al (2018) The effects of flaxseed oil omega-3 fatty acids supplementation on metabolic status of patients with polycystic ovary syndrome: a randomized, double-blind, placebo-controlled trial. Exp Clin Endocrinol Diabetes 126(4):222–228. https://doi.org/10.1055/s-0043-119751
Nadjarzadeh A, Dehghani-Firouzabadi R, Daneshbodi H et al (2015) Effect of omega-3 supplementation on visfatin, adiponectin, and anthropometric indices in women with polycystic ovarian syndrome. J Reprod Infertil 16(4):212–220
Nasri K, Hantoushzadeh S, Aghadavod E et al (2017) The effects of omega-3 fatty acids supplementation on gene expression involved in the insulin and lipid signaling pathway in patients with polycystic ovary syndrome. Horm Metab Res 49(6):446–451. https://doi.org/10.1055/s-0042-122782
Orio F, Muscogiuri G, Giallauria F et al (2016) Oral contraceptives versus physical exercise on cardiovascular and metabolic risk factors in women with polycystic ovary syndrome: a randomized controlled trial. Clin Endocrinol 85(5):764–771. https://doi.org/10.1111/cen.13112
Pal L, Berry A, Coraluzzi L et al (2012) Therapeutic implications of vitamin D and calcium in overweight women with polycystic ovary syndrome. Gynecol Endocrinol 28(12):965–968. https://doi.org/10.3109/09513590.2012.696753
Phelan N, O'Connor A, Kyaw Tun T et al (2011) Hormonal and metabolic effects of polyunsaturated fatty acids in young women with polycystic ovary syndrome: results from a cross-sectional analysis and a randomized, placebo-controlled, crossover trial. Am J Clin Nutr 93(3):652–662. https://doi.org/10.3945/ajcn.110.005538
Phy JL, Pohlmeier AM, Cooper JA et al (2015) Low starch/low dairy diet results in successful treatment of obesity and co-morbidities linked to Polycystic Ovary Syndrome (PCOS). J Obes Weight Loss Ther 5(2). pii: 259
Pohlmeier AM, Phy JL, Watkins P et al (2014) Effect of a low-starch/low-dairy diet on fat oxidation in overweight and obese women with polycystic ovary syndrome. Appl Physiol Nutr Metab 39(11):1237–1244. https://doi.org/10.1139/apnm-2014-0073
Rafraf M, Mohammadi E, Asghari-Jafarabadi M et al (2012) Omega-3 fatty acids improve glucose metabolism without effects on obesity values and serum visfatin levels in women with polycystic ovary syndrome. J Am Coll Nutr 31(5):361–368

Rahmani E, Jamilian M, Samimi M et al (2018) The effects of coenzyme Q10 supplementation on gene expression related to insulin, lipid and inflammation in patients with polycystic ovary syndrome. Gynecol Endocrinol 34(3):217–222. https://doi.org/10.1080/09513590.2017.1381680

Razavi M, Jamilian M, Kashan ZF et al (2016) Selenium supplementation and the effects on reproductive outcomes, biomarkers of inflammation, and oxidative stress in women with polycystic ovary syndrome. Horm Metab Res 48(3):185–190. https://doi.org/10.1055/s-0035-1559604

Reid MB, Li YP (2001) Tumor necrosis factor-alpha and muscle wasting: a cellular perspective. Respir Res 2:269–272

Rodacki CLN, Rodachi ALF, Pereira G et al (2012) Fish-oil supplementation enhances the effects of strength training in elderly women. Am J Clin Nutr 95:428–436

Sadeghi F, Alavi-Naeini A, Mardanian F et al (2019) Omega-3 and vitamin E co-supplementation can improve antioxidant markers in obese/overweight women with polycystic ovary syndrome. Int J Vitam Nutr Res. https://doi.org/10.1024/0300-9831/a000588

Samadi Z, Bambaeichi E, Valiani M et al (2019) Evaluation of changes in levels of hyperandrogenism, hirsutism and menstrual regulation after a period of aquatic high intensity interval training in women with polycystic ovary syndrome. Int J Prev Med 10:187. https://doi.org/10.4103/ijpvm.IJPVM_360_18

Samimi M, Jamilian M, Ebrahimi FA et al (2016) Oral carnitine supplementation reduces body weight and insulin resistance in women with polycystic ovary syndrome: a randomized, double-blind, placebo-controlled trial. Clin Endocrinol 84(6):851–857. https://doi.org/10.1111/cen.13003

Schleithoff SS, Zittermann A, Tenderich G et al (2006) Vitamin D supplementation improves cytokine profiles in patients with congestive heart failure: a double-blind, randomized, placebo-controlled trial. Am J Clin Nutr 83:754–759

Seyyed Abootorabi M, Ayremlou P, Behroozi-Lak T et al (2018) The effect of vitamin D supplementation on insulin resistance, visceral fat and adiponectin in vitamin D deficient women with polycystic ovary syndrome: a randomized placebo-controlled trial. Gynecol Endocrinol 34(6):489–494. https://doi.org/10.1080/09513590.2017.1418311

Shojaeian Z, Sadeghi R, Latifnejad Roudsari R (2019) Calcium and vitamin D supplementation effects on metabolic factors, menstrual cycles and follicular responses in women with polycystic ovary syndrome: a systematic review and meta-analysis. Caspian J Intern Med 10(4):359–369. https://doi.org/10.22088/cjim.10.4.359

Snijder MB, van Schoor NM, Pluijm SM et al (2006) Vitamin D status in relation to one-year risk of recurrent falling in older men and women. J Clin Endocrinol Metab 91:2980–2985

Sprung VS, Cuthbertson DJ, Pugh CJ et al (2013) Exercise training in polycystic ovarian syndrome enhances flow-mediated dilation in the absence of changes in fatness. Med Sci Sports Exerc 45(12):2234–2242. https://doi.org/10.1249/MSS.0b013e31829ba9a1

Talari HR, Poladchang S, Hamidian Y et al (2018) The effects of omega-3 and vitamin E co-supplementation on carotid intima-media thickness and inflammatory factors in patients with polycystic ovary syndrome. Oman Med J 33(6):473–479. https://doi.org/10.5001/omj.2018.88

Talari HR, Azad ZJ, Hamidian Y et al (2019) Effects of carnitine administration on carotid intima-media thickness and inflammatory factors in patients with polycystic ovary syndrome: a randomized, double-blind, placebo-controlled trial. Int J Prev Med 10:89. https://doi.org/10.4103/ijpvm.IJPVM_2_18

Tang XL, Sun Z, Gong L (2018) Chromium supplementation in women with polycystic ovary syndrome: Systematic review and meta-analysis. J Obstet Gynaecol Res 44(1):134–143. https://doi.org/10.1111/jog.13462

Teede HJ, Misso ML, Costello MF et al (2018) Recommendations from the international evidence-based guideline for the assessment and management of polycystic ovary syndrome. Hum Reprod 33 (9):1602–1618. https://doi.org/10.1093/humrep/dey256

Thomson RL, Buckley JD, Noakes M et al (2008) The effect of a hypocaloric diet with and without exercise training on body composition, cardiometabolic risk profile, and reproductive function in overweight and obese women with polycystic ovary syndrome. J Clin Endocrinol Metab 93(9):3373–3380. https://doi.org/10.1210/jc.2008-0751

Thomson RL, Brinkworth GD, Noakes M et al (2012) The effect of diet and exercise on markers of endothelial function in overweight and obese women with polycystic ovary syndrome. Hum Reprod 27 (7):2169–2176. https://doi.org/10.1093/humrep/des138

Thornell LE (2011) Sarcopenic obesity: satellite cells in the aging muscle. Curr Opin Clin Nutr Metab Care 14:22–27

Trummer C, Schwetz V, Kollmann M et al (2019) Effects of vitamin D supplementation on metabolic and endocrine parameters in PCOS: a randomized-controlled trial. Eur J Nutr 58(5):2019–2028. https://doi.org/10.1007/s00394-018-1760-8

Turan V, Mutlu EK, Solmaz U et al (2015) Benefits of short-term structured exercise in non-overweight women with polycystic ovary syndrome: a prospective randomized controlled study. J Phys Ther Sci 27 (7):2293–2297. https://doi.org/10.1589/jpts.27.2293

Visser M, Deeg DJ, Lips P et al (2003) Low vitamin D and high parathyroid hormone levels as determinants of loss of muscle strength and muscle mass (sarcopenia): the Longitudinal Aging Study Amsterdam. J Clin Endocrinol Metab 88:5766–5772

Vizza L, Smith CA, Swaraj S et al (2016) The feasibility of progressive resistance training in women with polycystic ovary syndrome: a pilot randomized controlled trial. BMC Sports Sci Med Rehabil 8:14. https://doi.org/10.1186/s13102-016-0039-8

Webb MA, Mani H, Robertson SJ et al (2018) Moderate increases in daily step count are associated with reduced IL6 and CRP in women with PCOS. Endocr Connect 7(12):1442–1447. https://doi.org/10.1530/EC-18-0438

Woodward A, Broom D, Harrop D et al (2019) The effects of physical exercise on cardiometabolic outcomes in women with polycystic ovary syndrome not taking the oral contraceptive pill: a systematic review and meta-analysis. J Diabetes Metab Disord 18(2):597–612. https://doi.org/10.1007/s40200-019-00425-y

Yang K, Zeng L, Bao T et al (2018) Effectiveness of Omega-3 fatty acid for polycystic ovary syndrome: a systematic review and meta-analysis. Reprod Biol Endocrinol 16(1):27. https://doi.org/10.1186/s12958-018-0346-x

Zhang X, Zheng Y, Guo Y et al (2019) The effect of low carbohydrate diet on polycystic ovary syndrome: a meta-analysis of randomized controlled trials. Int J Endocrinol 2019:4386401. https://doi.org/10.1155/2019/4386401

Drug Pipeline for Obesity Therapy 68

Azania T. Panicker, Priyanshi Desai, Sanika Karnik, and Maushmi S. Kumar

Contents

Introduction 922

Current Drugs 923

Drugs in Pipeline 924
Centrally Acting Drugs 924

Gut Hormones and Incretin Targets 927
Glucagon-Like Peptide-1 (GLP-1) Analogs: Semaglutide (Ozempic) 927
Dual Action GLP-1/Glucagon Receptor Agonists 928

Novel Targets 929
Methionine Aminopeptidase 2 (MetAP2) Inhibitors: ZGN-1061 929
Lipase Inhibitor: Cetilistat 930
Triple Monoamine Reuptake Inhibitor: Tesofensine 930
Fibroblast Growth: Factor 21 930

Anti-obesity Vaccines 930
Ghrelin 930
Somatostatin 931
Adenovirus 36 931

Conclusion 931

References 931

Abstract

Obesity contributes more to the mortality rates than malnourishment in countries with the majority of the world's population. First-line methods with lifestyle modifications may not be enough for the complete treatment of obesity, especially in genetic disorders. In the last two decades, several new therapies have emerged and many are still in development, to provide potential alternatives to currently available anti-obesity drugs. They belong to the class of centrally acting drugs, gut hormones, and incretin targets. Additionally, we have listed and described the status of novel anti-obesity agents targeting methionine

A. T. Panicker · P. Desai · S. Karnik · M. S. Kumar (✉)
Shobhaben Pratapbhai Patel School of Pharmacy and Technological Management, SVKM's NMIMS, Mumbai, India
e-mail: maushmiskumar@gmail.com

J. Faintuch and S. Faintuch (eds.), *Obesity and Diabetes*, https://doi.org/10.1007/978-3-030-53370-0_68

aminopeptidase 2 (MetAP2), lipase, triple monoamine reuptake, fibroblast growth factor 21, and anti-obesity vaccines. With an advancement in the development of these new drugs, anti-obesity drugs have the potential to provide better treatment opportunities and personalized care.

Keywords

Anti-obesity · Pharmacotherapy · Ghrelin · Vaccines · Neuropeptide Y · Mimetics · Leptin

Introduction

In 2010, obesity caused more than 3.4 million deaths, along with heavy loss of years of life and disability-adjusted life years/DALYs worldwide (Ng et al. 2014).

It is characterized by an increase in body fat content, when energy intake exceeds energy expenditure. Obesity has pernicious effects on nearly all the organ systems, which exacerbates the disease state and also increases the cost of care (Srivastava and Apovian 2018). Body mass index is widely used for diagnosis; however, this measurement does not differentiate between lean mass and fat mass and can thus be misleading (Table 68.1).

When WHO estimated a threefold increase in the worldwide prevalence of obesity between 1975 and 2016, it drew attention to lifestyle. Most of the world's population lives in countries where overweight and obesity kill more than underweight. Monitoring of diet and exercise are considered to be the first-line methods over drugs (WHO). However, these methods may not be enough. Prader-Willi syndrome (PWS) is an example of genetic obesity, with upper body fat accumulation, hypogonadism, and mental retardation. It is due to the loss of expression of genes from paternal inheritance on chromosome 15, deletion of chromosome 15q11-q13. Type II diabetes and thromboembolism are related concerns. Drugs that act on the actual body mechanisms (ghrelin) or on genes (melanocortin-4 receptor) are recommended (Duis et al. 2019).

Even alterations in the circadian cycle and nocturnal behavior (late awakening and late breakfast) can have an effect on the plasma glucose levels. Moreover, cardiovascular diseases, type II diabetes, and obesity are partially interdependent. Insulin is involved in promotion of lipolysis; therefore, decrease in insulin levels causes reduction in lipolysis. Thus, some of the target molecules in the pharmacological interventions are hormones and gut peptides, involved in long-term energy balance for, e.g., leptin and insulin, along with acute modulators like ghrelin, peptide YY (PYY), glucagon-like peptide 1 (GLP-1), and cholecystokinin (CCK) (Zhang et al. 2014).

Given that different types of obesity based on BMIs do not subject to the same risk of comorbidities (Table 68.1), there is a need for improved strategies for the identification of predictive biomarkers for these comorbidities (e.g., diabetes, hypertension, dyslipidemia, and cardiovascular disease). Likewise, effective preventive interventions with improved outcomes, by identifying what intervention is best suited for each individual (Zhang et al. 2014).

In this chapter, we will mainly focus on drugs in different stages of the research pipeline, which can be potential alternatives to currently available

Table 68.1 Classification of body weight in adults according to BMI (WHO)

Classification	BMI (kg/m^3)	Risk of obesity comorbidities
Underweight	<18.50	Low (however risk of other clinical problems is increased)
Normal range	18.50–24.99	Average
Overweight	25.00–29.99	Mild
Obese class I	30.00–34.99	Moderate
Obese class II	35.00–39.99	Severe
Obese class III	≥40.00	Very severe

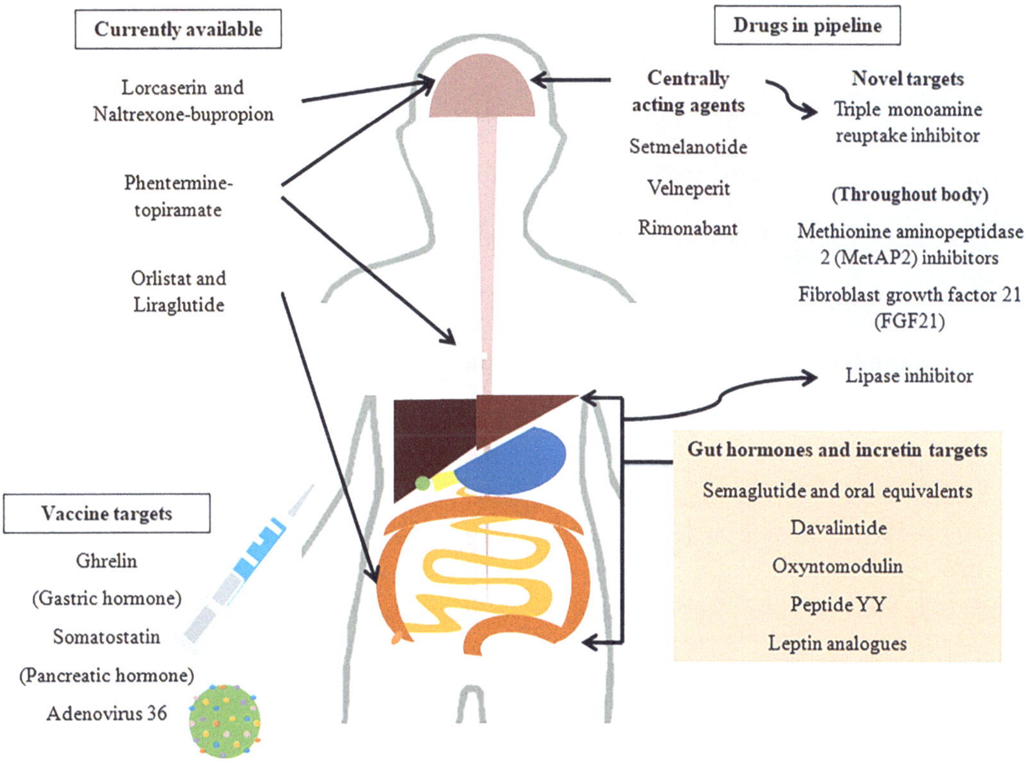

Fig. 68.1 Anti-obesity drugs, their targets, and locations

anti-obesity drugs (Fig. 68.1). The centrally acting drugs can be of potential use in the treatment of genetically acquired obesity. These drugs are known to mainly target receptors which influence the food intake and energy expenditure centrally. Novel anti-obesity targets mainly acting as inhibitors for methionine aminopeptidase 2 (MetAP2), lipase, and triple monoamine reuptake along with fibroblast growth factor 21 and anti-obesity vaccines which include ghrelin, somatostatin, and adenovirus 36 will also be addressed.

Current Drugs

There are currently five FDA-approved drugs used as prescription medications for weight loss namely Orlistat (Alli and Xenical), Lorcaserin (Belviq), Phentermine-topiramate (Qsymia), Naltrexone-bupropion (Contrave), and Liraglutide (Victoza) (Fig. 68.2). Given the magnitude of the obesity epidemic, they are considered as underutilized, in the USA and other countries. Table 68.2 compiles their information target sites, mechanism of action, dosage forms, and adverse effects.

Orlistat and liraglutide theoretically have high safety margin, as they have no central activity. Yet, there are indications that liver and kidney damage due to Orlistat along with faster multiplication of cancer cells and liraglutide 3.0 mg has been associated with gallstone disease and pancreatitis (Patel and Stanford 2018). Lorcaserin can induce depression and suicidal tendencies and can worsen preexisting depression. The rationale for a combination medication is that appetite regulation involves multiple pathways, so targeting different mechanisms simultaneously can have an additive effect on body weight. Another benefit is that the smaller dose of each medication reduces the risk of adverse effects.

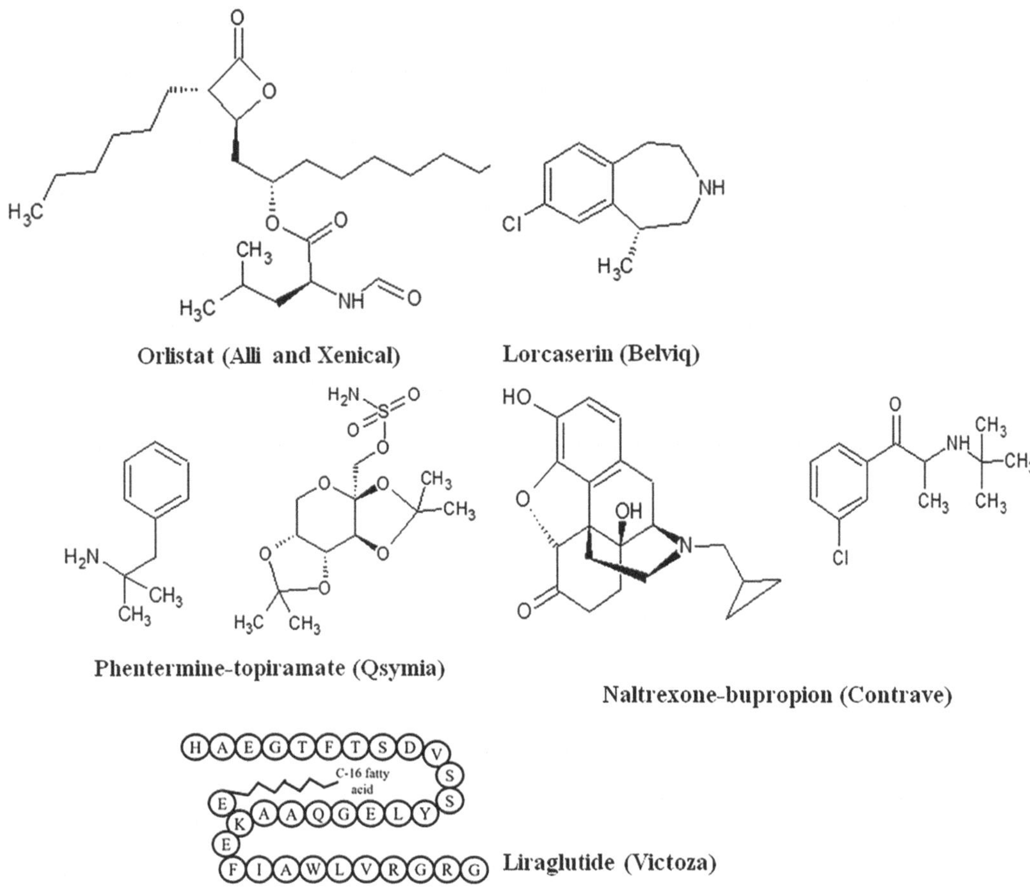

Fig. 68.2 Currently available FDA-approved prescription medications

Still naltrexone, an opioid receptor antagonist, can lead to drug abuse and dependency. Phentermine-topiramate can be used in combination therapy with other drugs as it is not likely to cause any clinically significant drug interactions.

Drugs in Pipeline

Centrally Acting Drugs

It has been found that food intake is controlled by receptors like melanocortin-4-receptor (MC4R), neuropeptide Y (NPY), and cannabinoid type-1 receptor (CB1R) which are present centrally, mainly in the hypothalamus. Drugs targeting these receptors are being researched. Some of these drugs have also shown positive results in clinical trials for the treatment of obesity that results from genetic disorders. However, since these drugs act centrally, their side effects can be more adverse compared to other anorectic drugs.

Melanocortin Receptor Agonist

The melanocortin-4-receptor (MC4R), which serves as a key regulator in the appetite controlling pathway within the hypothalamus, is a 332 amino acid GPCR (Garfield et al. 2015). When hunger signals are produced, they bind to their respective receptors and act on orexigenic neurons. This process leads to the release of melanocortin inverse agonist, agouti-related protein (AgRP). AgRP lowers the MC4R activation, thus inhibiting satiety signals and inducing the perception of hunger. The endogenous agonist

Table 68.2 Currently available FDA-approved anti-obesity medications

Drug, Class, Reference	Mechanism of action	Dosage form	Pharmacokinetics	Interactions	Most common adverse effects
Orlistat (Alli and Xenical) Lipase inhibitor Heck et al. (2000)	Inactivation of gastric and pancreatic lipase; different from other drugs as it does not target the feeding center but reduces fat absorption from food	Capsules (Alli—60 mg and Xenical—120 mg)	Half-life is 14–19 h. Two metabolites mainly M1 and M3 are excreted in bile having half-life of 2 h and 3 h, respectively; the drug undergoes extensive first pass metabolism.	Cyclosporine, warfarin, amiodarone, antiepileptic agents, and antiretroviral drugs. Malabsorption of fat-soluble vitamins (A, D, E, and K)	Abdominal discomfort, fecal incontinence, liquid stools, soft stools, and fecal urgency.
Lorcaserin (Belviq) Selective serotonin (5-HT_{2C}) agonist Brashier et al. (2014)	Stimulates the release of POMC from the arcuate nucleus which further causes the release of α-MSH which suppress the appetite	Tablet (10 mg or 20 mg)	Rapidly absorbed, delayed with fatty meals, peak concentration: 1.5–2 h; half-life: 11 h, distributed in the CNS, 70% plasma protein bound; metabolized in liver, major metabolite lorcaserin sulfamate, excretion mainly through urine (92%)	Risk of serotonin syndrome or neuroleptic malignant syndrome-like reactions with SSRIs, SNRIs, and MAOIs along with CYP450 2D6 inhibition; increase blood levels of drugs metabolized by this enzyme	Nausea, vomiting, constipation, diarrhea, fatigue, upper respiratory tract infection, urinary tract infections, back pain, headache, dizziness, and rash. Attention and memory deficit rare
Phentermine-topiramate (Qsymia) Adrenergic agonist/ neurostabilizer Shyh and Cheng-Lai (2014)	Phentermine is an agonist at TAAR1 receptor site that stimulates the release of norepinephrine and epinephrine. Topiramate modulates through GABA receptors, inhibiting carbonic anhydrase, antagonism of glutamate.	Capsule (3.75/23 mg, 7.5/46 mg, 11.25/69 mg, and 15.0/92 mg)	Phentermine undergoes p-hydroxylation on the aromatic ring and N-oxidation on the aliphatic side chain. 70–80% excreted in urine, half-life 20 hours. Topiramate does not undergo extensive metabolism, 70% excreted in urine; half-life: about 65 hrs	Topiramate may induce metabolism of ethinyl estradiol, thus decreasing contraceptive efficacy. Contraindications include pregnancy, glaucoma, hyperthyroidism, and MAOIs	Paresthesia, dizziness, dysgeusia, insomnia, constipation, and dry mouth.
Naltrexone-bupropion (Contrave) Opioid receptor antagonist/ NDRI Saunders et al. (2018)	Acts on the arcuate nucleus of the hypothalamus where the firing rate of POMC neurons is increased and the mesolimbic dopamine reward	Tablet (8 mg or 90 mg)	It is metabolized in the liver and has a bioavailability of 5–40%, half-life: 4 hrs	It shows interactions with MAOIs opioids, disulfiram, and thioridazine	Nausea, constipation, headache, dizziness, insomnia, dry mouth, hallucinations, blurred vision, angioedema

(continued)

Table 68.2 (continued)

Drug, Class, Reference	Mechanism of action	Dosage form	Pharmacokinetics	Interactions	Most common adverse effects
	circuit is modulated.				
Liraglutide (Victoza) GLP-1 analog Bode (2012), Saunders et al. (2018)	It increases secretion of insulin from beta-cells, reduces postprandial glucagon secretion, delays gastric emptying, and improves the function of beta-cells.	Subcutaneous (dose increased 0.6 mg daily until 3 mg	Slowed absorption from the subcutis and decreased rate of elimination due to chemical difference from the endogenous peptide GLP-1 and protein binding leads to a longer half-life of approx. 13 h	Contraindicated in pregnancy or those with personal or family history of medullary thyroid carcinoma or multiple endocrine neoplasia syndrome type 2, with insulin or insulin secretagogues; risk of hypoglycemia.	Diarrhea, dyspepsia, abdominal pain, nausea, vomiting, constipation

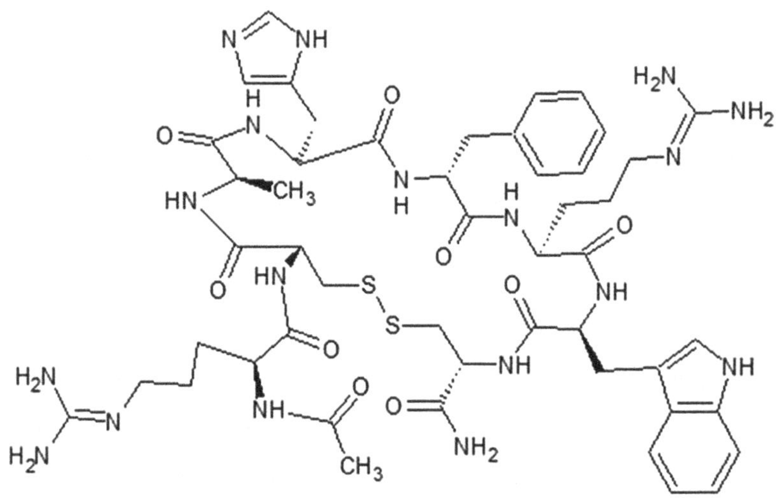

Fig. 68.3 Setmelanotide, an MC4R agonist

of MC4R is α-melanocortin-stimulating hormone (α-MSH) and is released when satiety signals bind to their respective receptors on anorexigenic neurons (Kim et al. 2011).

About 6–8% cases of obesity have been found to be a result of dysregulation or mutation of MC4R. Mutations can result in decreased affinity of endogenous agonist toward this receptor and decreased surface expression of the receptor or impaired signal transduction (Santini et al. 2009). Ser127Leu, Ser58Leu, Ile102Thr, and Gly252Ser are the four-point mutations that occur in MC4R and are known to affect the potency of this receptor (Roubert et al. 2010).

Setmelanotide

Setmelanotide, also known as RM-493 or BIM-22493, is a melanocortin-4-receptor (MC4R) agonist. It is a synthetic 8-amino acid cyclic peptide and has a molecular formula of $C_{49}H_{68}N_{18}O_9S_2$, with a molecular weight of 1117.3 Da (Fig. 68.3). Setmelanotide has a high affinity toward human MC4R receptors (inhibitory constant $K_i = 2.1$ nmol/l) (Kievit et al. 2013).

Setmelanotide was studied for its action on normal MC4R (wild-type model) as well as on mutated MC4R. In the binding assays, setmelanotide showed 37-fold higher affinity than α-MSH (endogenous agonist). In the activation assays, it showed 20-fold more potency (Falls and Zhang 2018). In vivo studies were conducted on normal and diet-induced obese (DIO) mice. In normal mice, it reduced food intake and weight gain. In DIO mice, it reduced fatty liver disease and hyperinsulinemia (Roubert et al. 2010; Kumar et al. 2009). When setmelanotide was introduced in the wild-type model, it exhibited interactions with several residues that were known to be critical for α-MSH.

According to NMR and modeling studies, the endogenous agonist α-MSH binds to the receptor in a hairpin loop. The tetra peptide messaging sequence binds to the active site, and the rest of the peptide points out in the extracellular space. In contrast to α-MSH, setmelanotide is fully accommodated in the binding site, which is attributed to its better affinity and potency in the wild-type protein. The additional interactions of the amino acids present in setmelanotide are responsible for its accommodation in the binding site (Falls and Zhang 2018). When setmelanotide was introduced in the mutated model, it was observed that the mutated residues were not directly involved in the binding process. The mutated residues were found to be located in the proximity to those residues which are important in the MC4R activation by setmelanotide. However, in case of α-MSH, Ile102Thr and Ser127Leu are located in such a way that they can cause a shift of Glu100, Asp122, and Asp126, which are considered to be necessary for receptor activation.

Ser58Cys and Gly252Ser are located away from the binding site but are known to induce conformational changes, important for MC4R activation (Falls and Zhang 2018). Hence, setmelanotide due to its cyclic structure and extensive interactions has the ability to stabilize and activate the receptor even in the presence of polymorphisms. It can act as a positive allosteric modulator. Such properties of setmelanotide make it a non-traditional bitopic ligand. It has been termed as non-traditional because it can interact with two sites without requiring traditional linker and allosteric building blocks (Falls and Zhang 2018).

Setmelanotide has entered two phase III clinical trials in 2017. Indications were pro-opiomelanocortin (POMC) deficiency obesity and leptin receptor (LEPR) deficiency obesity. According to the manufacturer (Rhythm), both one-year-long trials met primary and secondary end points, with effects on weight loss, insatiable hunger, or hyperphagia (Globenewswire.com 2019).

Gut Hormones and Incretin Targets

Glucagon-Like Peptide-1 (GLP-1) Analogs: Semaglutide (Ozempic)

Glucagon-like peptide-1 (GLP-1), an incretin hormone, is secreted from L cells in the small intestine. Semaglutide is a GLP-1 receptor agonist, and it has more than two amino acids substitutions at position 8 (alanine to α amino butyric acid) and 34 (lysine to arginine) (Fig. 68.4). The substitution at eighth position provides resistance to DPP-4 (dipeptidyl-peptidase) degradation. The modification in the structure prolongs its half-life which in turn increases the efficacy of the drug (Christou et al. 2019).

It acts by releasing insulin from β pancreatic cells and reducing the production of glucagon from the α-pancreatic cells. It decreases both fasting and postprandial plasma glucose levels; therefore, it was approved in Dec 2017 for the treatment of Type-2 Diabetes Mellitus (T2DM). It also acts by promoting weight loss. Upon administration, it decreases the total energy intake followed by suppression of appetite and less food cravings for high-fat foods, along with better control of eating and meal portion size (Christou et al. 2019).

In the presence of GLP-1 agonist, the glucose activity is regulated by inhibiting the K^+ ATP channels. Further opening of voltage-dependant

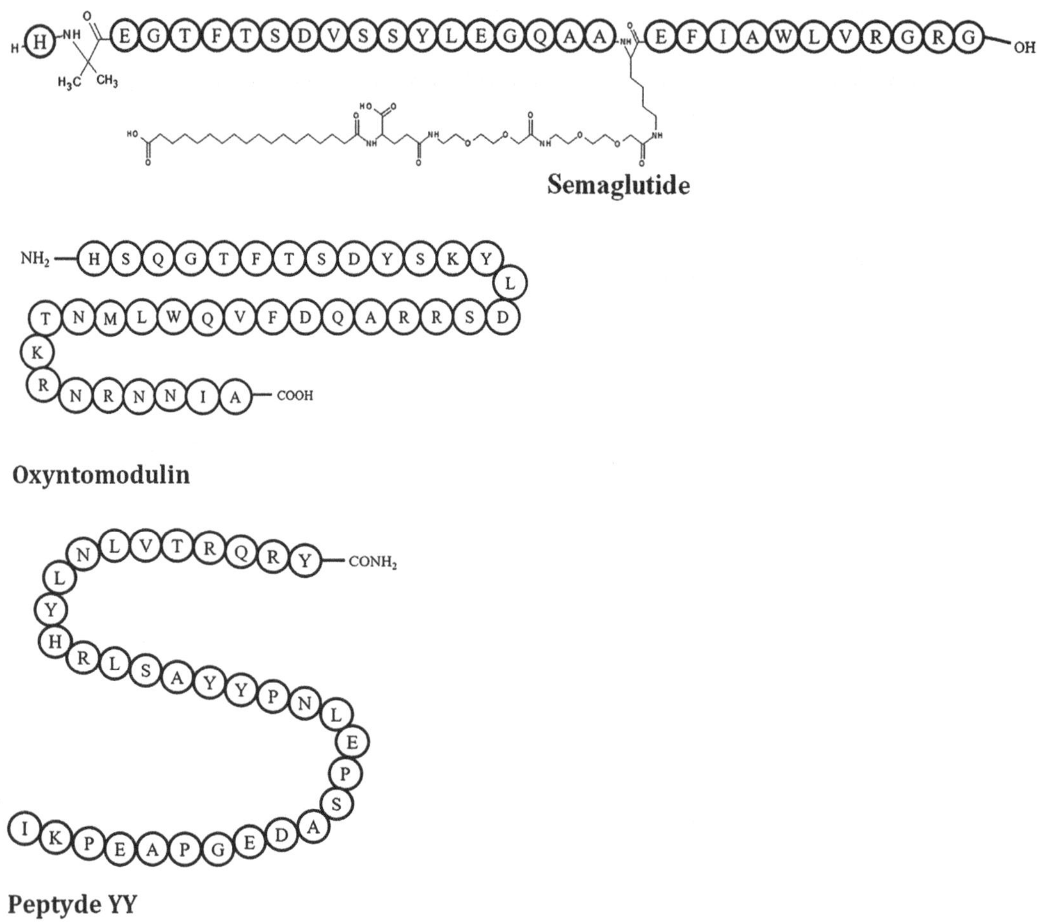

Fig. 68.4 Structure of Semaglutide and oxyntomodulin

calcium channels (VDCC) inhibits the K^+ channel repolarization. GLP-1 stimulates IP3 and ryanodine receptors (RYR) facilitating the influx of Ca^{++} ions, which further promote insulin secretion (Meloni et al. 2013). Semaglutide-induced weight loss was associated with a threefold greater loss of mean fat over lean body mass (Blundell et al. 2017).

The half-life of the drug is 7 days (183 h), in case of normal kidney function, 201 h in moderate kidney dysfunction and 221 h in severe kidney failure. Catabolism mainly occurred through neutral endopeptidase (NEP), a membrane-bound enzyme located in the kidneys (Christou et al. 2019). Cholelithiasis is a side effect of semaglutide (Chamberlin and Dabbs 2019) (Fig. 68.4).

GLP-1 agonists are not advised as first-line treatment due to potential risk of thyroid carcinoma or multiple endocrine neoplasia nor in patients with a history of pancreatitis. Gastrointestinal adverse effects limit use in some patients. Patients having a history of diabetic retinopathy should be constantly monitored for disease progression.

Dual Action GLP-1/Glucagon Receptor Agonists

Oxyntomodulin

Oxyntomodulin (OXM) is a 37-amino acids peptide (Fig. 68.4) and influences energy intake. On central administration of oxyntomodulin, its effects are canceled by exendin which is an antagonist of GLP-1 receptors (Baggio et al. 2004). The effects of oxyntomodulin also include decreased heart rate and energy expenditure. Similar effects were observed on activation of glucagon receptor (Sowden et al. 2007). These observations hence indicated that oxyntomodulin shows a dual action by acting as an agonist for both GLP-1 as well as glucagon receptor.

A phase 2b trial with an analog (OPK 88003) was finished in 2019 with obese diabetic patients. Weight loss, lipid profile, and HbA1c were more improved than with metformin. Further studies are expected (www.opko.com, 2019).

Peptide YY

Peptide YY (PYY) is a 36 amino acid peptide and is synthesized and discharged from specific endocrine cells called L cells, which are found essentially in the ileum. GLP-1, GLP-2, and oxyntomodulin are the co-secretions with PYY. PYY_{1-36} and PYY_{3-36} are the two forms of PYY, and PYY_{3-36} is the most common form, which binds to Y2 receptor (Murphy and Bloom 2006). It acts through NPY receptors and reduces the secretion of gastric acid and GIT motility when the level increases during fasting and after feeding. One of the pathogenesis of obesity is known to be a decrease in PYY levels (Batterham et al. 2003). PYY_{3-36} does not show any effect in Y2-deficient mouse, which is indicative of the fact that it shows its action through Y2 receptors (Batterham et al. 2003). A silent single nucleotide polymorphism (SNP), 585 T $>$ C, is present on the Y2 receptors and was found to be associated with obesity. The frequency of 585 T $>$ C was lower in obese men compared to the lean group (Lavebratt et al. 2006).

PYY Analogs

Native PYY has a short life; thus, several long-acting analogs are being developed. One of these, Y242, has demonstrated encouraging results in animal studies, with at least 72 hours action in rodents, after subcutaneous injection. Given the inhibition in food intake and significant weight loss associated with the product, toxicologic studies are going on, in anticipation of a human phase I trial (www.ukri.org 2019).

Novel Targets

Methionine Aminopeptidase 2 (MetAP2) Inhibitors: ZGN-1061

Methionine aminopeptidase 2 (MetAP2) is an enzyme responsible for the co-translational cleavage of N-terminal methionine, from a subset of newly made proteins important for a broad range of cellular functions, including metabolism, growth, and proliferation (Warder et al. 2008). It also binds to other important regulatory proteins such as extracellular signal-regulated kinase 1/2 (Erk 1/2) and elongation initiation factor 2 α (EIF2α) (Datta et al. 2004). Originally intended to be used in the treatment of cancer due to their anti-angiogenic property (Yeh et al. 2006), MetAP2 inhibitors were found to have marked anti-obesity effects in animal models and humans (Siddik et al. 2019).

This is a potent, selective, and covalent inhibitor of the MetAP2 enzyme that was developed to minimize endothelial cell exposure and exhibit an improved safety profile regarding platelet aggregation, neutrophil adhesion or NETs formation, clotting factor levels or function or blood clotting or clot lysis. After a 12-week treatment (Fig. 68.5) with 1.8 mg of ZGN-1061, subcutaneously injected every three days, part of a phase II clinical trial, body weight, and HbA1c favorably

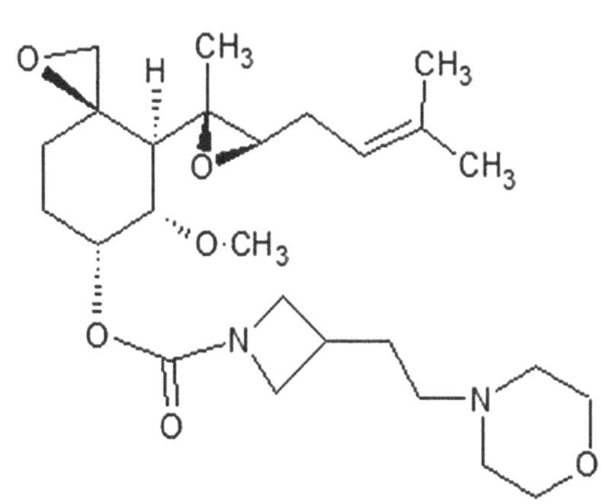

ZGN- 1061

Fig. 68.5 Structures of the natural (Fumagillin) and synthetic (Beloranib and ZGN-1061) methionine aminopeptidase 2 (MetAP2) inhibitors

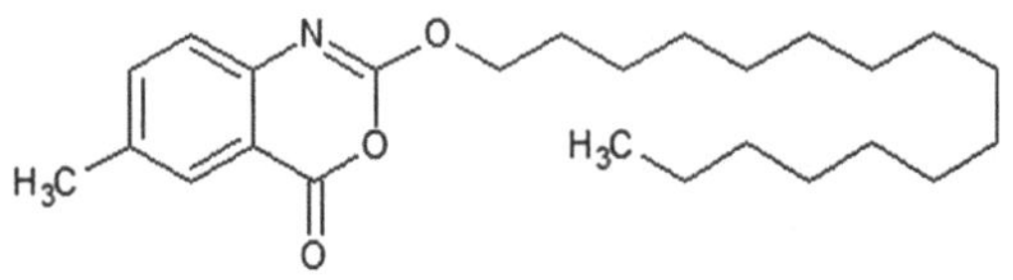

Fig. 68.6 Cetilistat, a lipase inhibitor

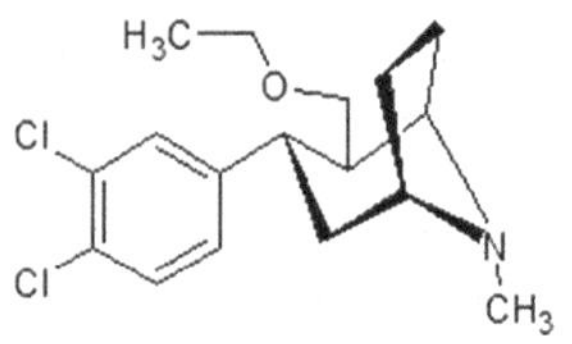

Fig. 68.7 Tesofensine, a triple monoamine reuptake inhibitor

responded in obese diabetic patients. Safety and tolerability were also encouraging, with no treatment-related serious adverse events and no cardiovascular (CV) safety signals observed. Additional studies are expected in the near future (Globenewswire 2019).

Lipase Inhibitor: Cetilistat

Cetilistat is a lipase inhibitor having the same pharmacological activity as orlistat and it inhibits the absorption of fat (Fig. 68.6). In phase II trial, there was an increase in fecal fat comparable to orlistat, and also, 11% of the subjects had oilier stools. It has less gastrointestinal complications than orlistat and is therefore more reliable (Hossain et al. 2019).

Triple Monoamine Reuptake Inhibitor: Tesofensine

Tesofensine (TE) inhibits appetite and stimulates thermogenesis. Devised for neurological diseases, weight loss was incidentally observed (Astrup et al. 2008b). Experimentally, it is associated with dopaminergic pathways in the brain (Hansen et al. 2013). A clinical trial confirmed weight loss (Astrup et al. 2008a); however, depending on the dose, major adverse effects were documented (Doggrell 2009; Astrup et al. 2008a). After a publication retraction (Retractionwatch 2013), further trials followed. Yet, long-term effects were not entirely satisfactory (Gilbert et al. 2012).

A new more successful trial was recently conducted, and the manufacturer intends to market the product in Mexico by 2020 (https://saniona.com 2019). Tesomet, which includes tesofensine and metoprolol, looks promising for Prader-Willi syndrome and other forms of obesity. Trials by the same manufacturer are going on (https://saniona.com 2019) (Fig. 68.7).

Fibroblast Growth: Factor 21

Fibroblast growth factor 21 (FGF21), analogously to other FGFs, stimulates cell growth and other functions in liver, adipose tissue, skeletal muscle, and pancreas (Giralt et al. 2015). Among other effects, glucose uptake, increased thermogenesis, and modulation of lipid metabolism experimentally occur (Cuevas-Ramos and Aguilar-Salinas 2016; Fisher and Maratos-Flier 2016; Domouzoglou and Maratos-Flier 2011; Badman et al. 2009).

Peroxisome-proliferator-activated receptor-α (PPARα) is involved in hepatic expression of FGF21 p (Inagaki et al. 2007; Lundåsen et al. 2007). Anti-inflammatory and anti-oxidative stress properties of FGF21 have been reported (Gómez-Sámano et al. 2017).

Preliminary observations in obese diabetics did not confirm clear benefits for either disease; however, lipid metabolism improved, along with markers of liver fibrosis. Steatohepatitis could therefore be a future target (Charles et al. 2019).

Anti-obesity Vaccines

Ghrelin

Under this section we will discuss the normal function of the target molecule and the results of various studies on vaccines against these targets that have been carried out. Ghrelin is converted from the precursor proghrelin. As a major regulator of appetite, it is relevant for obesity and metabolic diseases. The x/A cells of the oxyntic glands in the gastric fundus are the main source, with minor contributions from pancreas, hypothalamus, and kidney glomeruli (Sakata et al. 2009; Kojima et al. 1999).

Ghrelin undergoes subsequent acylation to form acyl ghrelin (AG), with the help of ghrelin O-acyltransferase (GOAT), which occurs at the same gastrointestinal sites as ghrelin (stomach, pancreas, and gut) (Sakata et al. 2009; Yang et al. 2008; Gutierrez et al. 2008). In rodents, after ghrelin vaccines were applied, weight gain was inhibited along with less feed efficiency, whereas lean body mass was mostly conserved, a response consistent with diminished leptin. The ratio of brain/serum ghrelin signaled the anti-ghrelin effect (Zorrilla et al. 2006). Also in pigs, conjugated ghrelin combined with an adjuvant provided a successful vaccine, which induced mild anorexia and weight loss, with elevated growth hormone levels (Vizcarra et al. 2007).

Virus-like proteins (VLPs) are highly immunogenic viral proteins devoid of genetic material therefore safe and effective for vaccination studies (Lechner et al. 2002). Experimental combination with ghrelin confirmed weight loss; however, human trials were not successful (Vizcarra et al. 2007).

Somatostatin

Growth hormone (GH) and its effector insulin-like growth factor-I (IGF-I) stimulate positive nitrogen balance and tissue growth in most organs, whereas lipids tend to undergo catabolism. Deficiency has been deemed a cause of fat accumulation (Berryman and List 2017; Lewitt 2017). The aim of vaccination is to enhance GH and IGF-1. In rodents, weight loss was achieved; however, no human trials are available (Haffer 2012).

Adenovirus 36

Human adenovirus 36 (Ad36) is not uncommon, differentially affecting the obese (30%) and non-obese population (11%) in the USA (Atkinson et al. 2005), however, not in all countries (Abdel-Moniem et al. 2018). Experimentally, it induces fat accumulation; however, insulin sensitivity increases. In mice, vaccination followed by virus inoculation prevented weight gain and elevation of pro-inflammatory cytokines (Na and Nam 2014). Human experience is not available.

Conclusion

Currently, unaffordable drugs continue to be a challenge, and available pharmacotherapies are not efficient enough to overcome obesity. People with obesity and failing lifestyle therapies need to be escalated to the required pharmacological treatment, as bariatric surgery has strict criteria and is not appropriate for mass therapy.

The most promising of these are co-agonists for multiple gut hormones including glucagon, GLP-1, and GIP. They mainly affect food intake; hence, their analogs could prove to be a better line of therapy for obesity. Unlike other centrally acting agents, setmelanotide has shown comparatively less severe side effects, and therefore, it could have a future in the therapy of genetically acquired obesity.

References

Abdel-Moniem MM, Ibrahim OM, Gabr AA, EL-Sonousy WM, Mohsen MA, Hassaballah FA, Aziz AA, El Malt H (2018) Lack of evidence for the role of human adenovirus 36 in obesity of Egyptian children. Arch Med Sci Civil Dis 3(1):210–216

Astrup A, Madsbad S, Breum L, Jensen TJ, Kroustrup JP, Larsen TM (2008a) Effect of tesofensine on bodyweight loss, body composition, and quality of life in obese patients: a randomised, double-blind, placebo-controlled trial. Lancet 372(9653):1906–1913

Astrup A, Meier DH, Mikkelsen BO, Villumsen JS, Larsen TM (2008b) Weight loss produced by tesofensine in patients with Parkinson's or Alzheimer's disease. Obesity 16(6):1363–1369
Atkinson RL, Dhurandhar NV, Allison DB, Bowen RL, Israel BA, Albu JB, Augustus AS (2005) Human adenovirus-36 is associated with increased body weight and paradoxical reduction of serum lipids. Int J Obes 29(3):281
Badman MK, Koester A, Flier JS, Kharitonenkov A, Maratos-Flier E (2009) Fibroblast growth factor 21-deficient mice demonstrate impaired adaptation to ketosis. Endocrinology 150(11):4931–4940
Baggio LL, Huang Q, Brown TJ, Drucker DJ (2004) Oxyntomodulin and glucagon-like peptide-1 differentially regulate murine food intake and energy expenditure. Gastroenterology 127(2):546–558
Batterham RL, Cohen MA, Ellis SM, Le Roux CW, Withers DJ, Frost GS, Ghatei MA, Bloom SR (2003) Inhibition of food intake in obese subjects by peptide YY3–36. N Engl J Med 349(10):941–948
Berryman DE, List EO (2017) Growth hormone's effect on adipose tissue: quality versus quantity. Int J Mol Sci 18(8):1621
Blundell J, Finlayson G, Axelsen M, Flint A, Gibbons C, Kvist T, Hjerpsted JB (2017) Effects of once-weekly semaglutide on appetite, energy intake, control of eating, food preference and body weight in subjects with obesity. Diabetes Obes Metab 19(9):1242–1251
Bode B (2012) An overview of the pharmacokinetics, efficacy and safety of liraglutide. Diabetes Res Clin Pract 97(1):27–42
Brashier DB, Sharma AK, Dahiya N, Singh SK, Khadka A (2014) Lorcaserin: a novel antiobesity drug. J Pharmacol Pharmacother 5(2):175
Chamberlin S, Dabbs W (2019) Semaglutide (ozempic) for type 2 diabetes mellitus. Am Fam Physician 100 (2):116–117
Charles ED, Neuschwander-Tetri BA, Frias JP, Kundu S, Luo Y, Tirucherai GS, Christian R (2019) Pegbelfermin (BMS 986036), PEGylated FGF21, in patients with obesity and type 2 diabetes: results from a randomized phase 2 study. Obesity 27(1):41–49
Christou GA, Katsiki N, Blundell J, Fruhbeck G, Kiortsis DN (2019) Semaglutide as a promising anti obesity drug. Obes Rev 20(6):805–815
Cuevas-Ramos D, Aguilar-Salinas CA (2016) Modulation of energy balance by fibroblast growth factor 21. Horm Mol Biol Clin Invest 30(1)
Datta B, Majumdar A, Datta R, Balusu R (2004) Treatment of cells with the angiogenic inhibitor fumagillin results in increased stability of eukaryotic initiation factor 2-associated glycoprotein, p67, and reduced phosphorylation of extracellular signal-regulated kinases. Biochemistry 43(46):14821–14831
Doggrell SA (2009) Tesofensine–a novel potent weight loss medicine: evaluation of: Astrup A, Breum L, Jensen TJ, Kroustrup JP, Larsen TM. Effect of tesofensine on bodyweight loss, body composition, and quality of life in obese patients: a randomised, double-blind, placebo-controlled trial. Lancet 2008;372:1906–13. Expert opinion on investigational drugs 18(7):1043–1046
Domouzoglou EM, Maratos-Flier E (2011) Fibroblast growth factor 21 is a metabolic regulator that plays a role in the adaptation to ketosis. Am J Clin Nutr 93 (4):901S–905S
Duis J, van Wattum PJ, Scheimann A, Salehi P, Brokamp E, Fairbrother L, Childers A, Shelton AR, Bingham NC, Shoemaker AH, Miller JL (2019) A multidisciplinary approach to the clinical management of Prader-Willi syndrome. Mol Genet Genomic Med 7 (3):e514. https://doi.org/10.1002/mgg3.514
Falls BA, Zhang Y (2018) Insights into the allosteric mechanism of setmelanotide (RM-493) as a potent and first-in-class melanocortin-4 receptor (MC4R) agonist to treat rare genetic disorders of obesity through an in silico approach. ACS Chem Neurosci 10(3):1055–1065
Fisher FM, Maratos-Flier E (2016) Understanding the physiology of FGF21. Annu Rev Physiol 78:223–241
Garfield AS, Li C, Madara JC, Shah BP, Webber E, Steger JS, Campbell JN, Gavrilova O, Lee CE, Olson DP, Elmquist JK (2015) A neural basis for melanocortin-4 receptor-regulated appetite. Nat Neurosci 18(6):863
Gilbert JA, Gasteyger C, Raben A, Meier DH, Astrup A, Sjödin A (2012) The effect of tesofensine on appetite sensations. Obesity 20(3):553–561
Giralt M, Gavaldà-Navarro A, Villarroya F (2015) Fibroblast growth factor-21, energy balance and obesity. Mol Cell Endocrinol 418:66–73
Globenewswire (2019) www.globenewswire.com/news-release/2019/06/10/1866194/0/en/Zafgen-Presented-Full-Results-of-Phase-2-Clinical-Trial-for-ZGN-1061-at-the-American-Diabetes-Association-s-79th-Scientific-Sessions.html
Gómez-Sámano MÁ, Grajales-Gómez M, Zuarth-Vázquez JM, Navarro-Flores MF, Martínez-Saavedra M, Juárez-León ÓA, Morales-García MG, Enríquez-Estrada VM, Gómez-Pérez FJ, Cuevas-Ramos D (2017) Fibroblast growth factor 21 and its novel association with oxidative stress. Redox Biol 11:335–341
Gutierrez JA, Solenberg PJ, Perkins DR, Willency JA, Knierman MD, Jin Z, Witcher DR, Luo S, Onyia JE, Hale JE (2008) Ghrelin octanoylation mediated by an orphan lipid transferase. Proc Natl Acad Sci 105 (17):6320–6325
Haffer KN (2012) Effects of novel vaccines on weight loss in diet-induced-obese (DIO) mice. J Anim Sci Biotechnol 3(1):21
Hansen HH, Jensen MM, Overgaard A, Weikop P, Mikkelsen JD (2013) Tesofensine induces appetite suppression and weight loss with reversal of low forebrain dopamine levels in the diet-induced obese rat. Pharmacol Biochem Behav 110:265–271
Heck AM, Yanovski JA, Calis KA (2000) Orlistat, a new lipase inhibitor for the management of obesity.

Pharmacother: J Hum Pharmacol Drug Ther 20 (3):270–279

Hossain MA, Pervin R, Debnath D, Bhuiyan MA (2019) Therapeutic treatment for controlling childhood obesity. In: Global perspectives on childhood obesity. Academic Press, New York, pp 377–385

Inagaki T, Dutchak P, Zhao G, Ding X, Gautron L, Parameswara V, Li Y, Goetz R, Mohammadi M, Esser V, Elmquist JK (2007) Endocrine regulation of the fasting response by PPARα-mediated induction of fibroblast growth factor 21. Cell Metab 5(6):415–425

Kievit P, Halem H, Marks DL, Dong JZ, Glavas MM, Sinnayah P, Pranger L, Cowley MA, Grove KL, Culler MD (2013) Chronic treatment with a melanocortin-4 receptor agonist causes weight loss, reduces insulin resistance, and improves cardiovascular function in diet-induced obese rhesus macaques. Diabetes 62 (2):490–497

Kim GW, Lin JE, Valentino MA, Colon-Gonzalez F, Waldman SA (2011) Regulation of appetite to treat obesity. Expert Rev Clin Pharmacol 4(2):243–259

Kojima M, Hosoda H, Date Y, Nakazato M, Matsuo H, Kangawa K (1999) Ghrelin is a growth-hormone-releasing acylated peptide from stomach. Nature 402 (6762):656

Kumar KG, Sutton GM, Dong JZ, Roubert P, Plas P, Halem HA, Culler MD, Yang H, Dixit VD, Butler AA (2009) Analysis of the therapeutic functions of novel melanocortin receptor agonists in MC3R- and MC4R-deficient C57BL/6J mice. Peptides 30:1892–1900

Lavebratt C, Alpman A, Persson B, Arner P, Hoffstedt J (2006) Common neuropeptide Y2 receptor gene variant is protective against obesity among Swedish men. Int J Obes 30(3):453

Lechner F, Jegerlehner A, Tissot AC, Maurer P, Sebbel P, Renner WA, Jennings GT, Bachmann MF (2002) Virus-like particles as a modular system for novel vaccines. Intervirology 45(4–6):212–217

Lewitt MS (2017) The role of the growth hormone/insulin-like growth factor system in visceral adiposity. Biochem Insights 10, 1178626417703995

Lundåsen T, Hunt MC, Nilsson LM, Sanyal S, Angelin B, Alexson SE, Rudling M (2007) PPARα is a key regulator of hepatic FGF21. Biochem Biophys Res Commun 360(2):437–440

Meloni AR, DeYoung MB, Lowe C, Parkes DG (2013) GLP-1 receptor activated insulin secretion from pancreatic β-cells: mechanism and glucose dependence. Diabetes Obes Metab 15(1):15–27

Murphy KG, Bloom SR (2006) Gut hormones and the regulation of energy homeostasis. Nature 444 (7121):854

Na HN, Nam JH (2014) Proof-of-concept for a virus-induced obesity vaccine; vaccination against the obesity agent adenovirus 36. Int J Obes 38 (11):1470

Ng M, Fleming T, Robinson M, Thomson B, Graetz N, Margono C et al (2014) Global, regional, and national prevalence of overweight and obesity in children and adults during 1980–2013: a systematic analysis for the global burden of disease study 2013. Lancet 384 (9945):766–781

Patel DK, Stanford FC (2018) Safety and tolerability of new-generation anti-obesity medications: a narrative review. Postgrad Med 130(2):173–182

Retractionwatch. http://retractionwatch.wordpress.com/2013/04/09/trial-irregularities-earn-lancet-study-of-potential-weight-loss-drug-tesofensine-expression-of-concern/. Accessed 19 October 2019

Roubert P, Dubern B, Plas P, Lubrano-Berthelier C, Alihi R, Auger F, Deoliveira DB, Dong JZ, Basdevant A, Thurieau C, Clément K (2010) Novel pharmacological MC4R agonists can efficiently activate mutated MC4R from obese patient with impaired endogenous agonist response. J Endocrinol 207(2):177–183

Sakata I, Yang J, Lee CE, Osborne-Lawrence S, Rovinsky SA, Elmquist JK, Zigman JM (2009) Colocalization of ghrelin O-acyltransferase and ghrelin in gastric mucosal cells. Am J Physiol Endocrinol Metab 297(1): E134–E141

Santini F, Maffei M, Pelosini C, Salvetti G, Scartabelli G, Pinchera A (2009) Melanocortin-4 receptor mutations in obesity. Adv Clin Chem 48:95–109

Saunders KH, Umashanker D, Igel LI, Kumar RB, Aronne LJ (2018) Obesity pharmacotherapy. Med Clin 102 (1):135–148

Shyh G, Cheng-Lai A (2014) New antiobesity agents: lorcaserin (Belviq) and phentermine/topiramate ER (Qsymia). Cardiol Rev 22(1):43–50

Siddik AB, Das BC, Weiss L, Dhurandar NV, Hegde V (2019) A MetAP2 inhibitor blocks adipogenesis, yet improves glucose uptake in cells. Adipocytes 8 (1):240–253

Sowden GL, Drucker DJ, Weinshenker D, Swoap SJ (2007) Oxyntomodulin increases intrinsic heart rate in mice independent of the glucagon-like peptide-1 receptor. Am J Phys Regul Integr Comp Phys 292(2):R962–R970

Srivastava G, Apovian C (2018) Future pharmacotherapy for obesity: new anti-obesity drugs on the horizon. Curr Obes Rep 7(2):147–161

Vizcarra JA, Kirby JD, Kim SK, Galyean ML (2007) Active immunization against ghrelin decreases weight gain and alters plasma concentrations of growth hormone in growing pigs. Domest Anim Endocrinol 33 (2):176–189

Warder SE, Tucker LA, McLoughlin SM, Strelitzer TJ, Meuth JL, Zhang Q, Sheppard GS, Richardson PL, Lesniewski R, Davidsen SK, Bell RL (2008) Discovery, identification, and characterization of candidate pharmacodynamic markers of methionine aminopeptidase-2 inhibition. J Proteome Res 7(11):4807–4820

Yang J, Brown MS, Liang G, Grishin NV, Goldstein JL (2008) Identification of the acyltransferase that octanoylates ghrelin, an appetite-stimulating peptide hormone. Cell 132(3):387–396

Yeh JRJ, Ju R, Brdlik CM, Zhang W, Zhang Y, Matyskiela ME, Shotwell JD, Crews CM (2006) Targeted gene disruption of methionine aminopeptidase 2 results in an embryonic gastrulation defect and endothelial cell growth arrest. Proc Natl Acad Sci 103 (27):10379–10384

Zhang Y, Liu J, Yao J, Ji G, Qian L, Wang J, Zhang G, Tian J, Nie Y, Zhang YE, Gold MS, Liu Y (2014) Obesity: pathophysiology and intervention. Nutrients 6(11):5153–5183. https://doi.org/10.3390/nu6115153

Zorrilla EP, Iwasaki S, Moss JA, Chang J, Otsuji J, Inoue K, Meijler MM, Janda KD (2006) Vaccination against weight gain. Proc Natl Acad Sci 103 (35):13226–13231

MC4R as a Target for Pharmacotherapeutic Treatment of Obesity and Type 2 Diabetes

69

Juliana Pereira Lopes Gonçalves, Daniel Palmer, and Morten Meldal

Contents

Incidence and Societal Burden of Obesity 936

Challenges in Pharmacological Management of Obesity 936

Melanocortin Receptors (MCRs) 937

Melanocortins and Feeding Behavior 938

Hormones and Biomolecules 939

Signaling Pathways 940

Agonists with Potential Therapeutic Interest and Clinical Trials 941

Diabetes as a Consequence of Obesity 942

Perspectives 943

References 943

J. P. L. Gonçalves (✉)
Center for Evolutionary Chemical Biology, Department of Chemistry, University of Copenhagen, Copenhagen, Denmark

Pathology Institute, School of Medicine, Technical University of Munich, Munich, Germany
e-mail: juliana.goncalves@tum.de

D. Palmer
Center for Evolutionary Chemical Biology, Department of Chemistry, University of Copenhagen, Copenhagen, Denmark

Novo Nordisk, Copenhagen, Denmark

M. Meldal
Center for Evolutionary Chemical Biology, Department of Chemistry, University of Copenhagen, Copenhagen, Denmark

J. Faintuch and S. Faintuch (eds.), *Obesity and Diabetes*, https://doi.org/10.1007/978-3-030-53370-0_69

Abstract

The melanocortin family of G-protein-coupled receptors is the main group of receptors to control appetite, energy homeostasis, and adipose tissue accumulation. In particular, the melanocortin-4 receptor (MC4R) is believed to have a central role in the management of food intake and energy expenditure. Mutations in this gene often lead to an increase in weight, therefore being considered the monogenic target for the treatment of obesity. Obesity is also often the cause of other onerous conditions, such as type 2 diabetes. Efforts to develop a novel pharmacotherapy for this condition have disclosed the complexity of physiological roles and activation mechanisms of this receptor. A promising drug is currently on clinical trials, but efforts to develop a more selective and potent compound have not ceased.

Keywords

Melanocortin receptor · Appetite hormones · Melanocortin pathways · Metanocortin agonists · Melanocortin diabetes

Incidence and Societal Burden of Obesity

The World Health Organization (WHO) data report that the state of being overweight and obese constitutes the fifth leading risk for global deaths. At least 2.8 million adults die each year as a result of being overweight or obese. In addition, 44% of the diabetes mellitus (DM) burden is attributed to excess weight. This increase in obesity rates is a great concern for governments worldwide, and it is also a huge economic burden for the society (De Carvalho et al. 2017; Leitner et al. 2017).

Obesity is defined by the WHO as abnormal or excessive fat accumulation that presents a risk to health. A person with a body mass index (BMI) of 30 or more is generally considered obese. Overweight and obese individuals are at greater risk for a number of chronic diseases, including cardiovascular diseases, type 2 diabetes mellitus, osteoarthritis, nonalcoholic fatty liver (NAFL) disease, respiratory complications, and certain forms of cancer. Cytokines, such as interleukin-6, resistin, and plasminogen activation inhibitor 1, derived from adipose tissue, have been implicated in the pathogenesis of several of the diseases mentioned (Ellacott et al. 2007; Makki et al. 2013). Therefore, when researching pharmacotherapies for this epidemic, the focus needs to expand beyond just weight loss and keep in mind all the other associated diseases (hyperglycemia, hyperlipidemia, and cardiovascular diseases). The healthcare system, the physicians, and the patients need to envisage obesity as being a chronic disease that will require a long-term treatment and follow-up. A reduction in bodyweight as low as 5–10% has a major impact in the levels of inflammation and pro-thrombotic markers, as well as chronic disease incidence (Guh et al. 2009).

Challenges in Pharmacological Management of Obesity

Currently, there are three main treatments for the management of obesity: behavioral modification, pharmacological therapy, and bariatric surgery. Despite being the most effective option, bariatric surgery is the least recommended due to the invasive nature of the procedure and the relative high costs associated with the surgery (Velazquez and Apovian 2018). Rather, behavioral modification is preferred as the first choice for therapeutic intervention. Even though different approaches are available, the success of the commonly applied medicinal or behavioral modifications methodologies is often transient, with most patients having regained the lost weight only two years after the treatment. To reduce the increasing incidence of obesity worldwide, new and more effective ways of controlling appetite and adipose tissue accumulation are, therefore, a pressing need.

Mammalian metabolism provides several protective mechanisms against weight loss (e.g., regulation of neuroendocrine hormones, leptin and ghrelin) (Wren et al. 2001; Batterham et al. 2002; Cowley et al. 2001). In addition, energy expenditure is associated with increased sympathetic nervous system activity and increase of body temperature. Both responses present major challenges for the success and safety of pharmacological treatments, along with drug side effects.

Causes for obesity encompass primary causes derived from monogenic diseases and syndromes, and secondary ones that result from psychological, neurological, and drug-induced stress (Butler et al. 2000). Lifestyle adjustments are usually insufficient for the first category, whereas unhealthy habits and other environmental mechanisms are in principle amenable to behavior change (see Table 69.1).

Different pharmacological approaches for weight loss are available (Van Gaal and Dirinck 2016); however, undesired side effects have to be accounted for, besides the need for long-term safety and efficacy monitoring.

Melanocortin Receptors (MCRs)

These are a part of the G-protein-coupled membrane receptors (GPCRs), the largest protein superfamily in mammalian genomes. The MCRs cognate, physiological ligands are variants of melanocyte-stimulating hormone (MSH) and processed products of the proopiomelanocortin (POMC) gene. Following the discovery of these peptide hormones, denominated melanocortins, and identification of a cognate receptor (later determined to be the melanocortin-1 receptor, MC1R), Hruby and co-workers conducted comprehensive work in determining the physiological roles of this receptor (Sawyer et al. 1982; Castrucci et al. 1984; Haskell-Luevano et al. 1997). Subsequently, functional and structural diversity of the MCR family of receptors was revealed, including the four additional melanocortin receptors MC2R, MC3R, MC4R, and MC5R, as well as their distinct tissue

Table 69.1 Causes of obesity

Causes of obesity	
Primary Causes	Secondary Causes
Genetic causes	Neurological
Monogenic disorders	• Endocrine
• Melanocortin-4 receptor mutation	• Hypothyroidism[a]
• Leptin deficiency	• Cushing syndrome
• POMC deficiency	• GH deficiency
Syndromes	• Pseudohypoparathyroidism
• Prader-Willi	Psychological
• Bardet-Biedl	• Depression[b]
• Cohen	• Eating disorders
• Alström	Drug-Induced
• Froehlich	• Tricyclic antidepressants
	• Oral contraceptives
	• Antipsychotics
	• Anticonvulsants
	• Glucocorticoids
	• Sulfonylureas
	• Glitazones
	• β-blockers

[a]Controversial whether hypothyroidism causes or exacerbates obesity
[b]Depression associated with overeating or binging

distributions and physiologic roles (Halem et al. 2008; Pantel et al. 2011). With respect to MC4R, this receptor was identified by northern blot analysis and its tissue distribution was characterized via *in situ* hybridization by Gantz et al. (1993). While several melanocortin agonists were developed initially, with some of them even completing clinical trials (as tanning agents), it was not until after the identification of the different melanocortin receptors that further testing for receptor selectivity took place (Sawyer et al. 1982; Castrucci et al. 1984; Hruby et al. 1995). This stimulated the development of specific agonists and investigations into possible applications for the previously developed MCR agonists.

Melanocortins and Feeding Behavior

In 1986, it was already suggested that melanocortin peptides inhibit food intake (Poggioli et al. 1986). Obesity was subsequently linked to disruption of the MC4R (Huszar et al. 1997). Mice with MC4R deficiency display elevated body fat tissue, hyperphagia, and hyperinsulinemia, and ectopic natural antagonists of the MC4R ligand are relevant in this context (Yeo et al. 2003). These findings were simultaneously corroborated in studies where MC4R mutations in mice resulted in a non-syndromic form of obesity (Vaisse et al. 1998). When increased the content of fat in a diet, wild-type mice tend to increase diet-induced thermogenesis and their physical activity and at the same time decrease the food intake in a few days, opposed to knockout mice that increase the food intake (Butler et al. 2000; Litt et al. 2017; Panaro and Cone 2013).

Heritability of fat mass ranges between 70–90% in monozygotic and 35–45% in dizygotic twins (Berrettini 2004; Maes et al. 1997). Not to exclude environmental factors, but when analyzed different humans with different ethnic backgrounds, the most prevalent single gene mutation that induces obesity is in MC4R (Yeo et al. 2003; Nogueiras et al. 2007; O'Rahilly et al. 2003). Over 1000 mutations are linked with adipose tissue and energy balance, encompassing different phenotypes (e.g. complete loss of receptor function to polymorphism, obesity, and even difficulty in gaining weight) (Santini et al. 2009); an MC4R website with this information has been compiled (Melanocortin 4 Receptor 2018).

Research on the other members of the melanocortin family describes the involvement of MC2 and MC3 receptors in the management of caloric intake and food consumption. MC2R, which is mainly located in the adrenal cortex (Table 69.2), is believed to be involved in the control of the glucocorticoids—so-called "stress hormones" that are responsible for helping the immune system, regulating the blood sugar level and help triggering nerve cell signaling in the brain (Webb and Clark 2010). Of the melanocortin receptors involved in weight regulation, MC3R remains the least characterized in terms of its physiological role(s).

Reports on knockout MC3R mice showed an increase in adiposity, which is believed to be

Table 69.2 Melanocortin Receptor Expression: The affinity of natural melanocortin ligands for their receptors, measured by radioligand binding, plus the distribution of melanocortin receptors in rodent tissues (inspired by (Millington 2006; Abdel-Malek 2001))

Ligand	Melanocortin receptor affinities	Melanocortin receptor expression
MC1R	αMSH >> ACTH, βMSH, γMSH	Melanocytes, macrophages, adipocytes, keratinocytes, fibroblasts
MC2R	ACTH	Adrenal gland, adipocytes
MC3R	γMSH > αMSH and βMSH	Brain, placenta, duodenum, pancreas, stomach
MC4R	βMSH > αMSH >>> γMSH	Brain, spinal cord, muscle
MC5R	as for MC1R, with lower affinity	Brain, sebaceous glands, adrenal, spleen, thymus, testis, ovary, muscle, adipocytes, mast cells

unrelated to the obesity phenotype of MC4R since investigations on the deletions of both MC3R and MC4R revealed an accumulative effect of adipose tissue. Contrary to what was observed for MC4R, MC3R knockout mice showed a susceptibility to weight loss, indicated by a decrease in lean mass (Huszar et al. 1997; Wang et al. 2000; Marks et al. 2006). Separate studies on MC3R suggest that this receptor has a meaningful role in energy homeostasis, immune response, and circadian rhythm (Leoni et al. 2008; Getting et al. 2008). Loss of MC3R has also been suggested to decrease the non-resting energy expenditure. Knockout MC3R mice exhibited a higher respiratory exchange ratio. On another report, it was shown that MC3R−/− mice become glucose intolerant, hyperglycemic, and hyper-insulinemic, despite normal response of the liver to an elevated insulin, indicating a unique diabetic phenotype (further reports on MC3R (Renquist et al. 2011; Begriche et al. 2009)). No obvious role for MC3R haploinsufficiency in human obesity has been elucidated so far (Girardet and Butler 2014).

Hormones and Biomolecules

The melanocortin pathway is highly relevant for CNS management of food intake. Two neuronal populations in the arcuate nucleus of the hypothalamus (ARH) are involved: those producing neuropeptides and those expressing the melanocortin-4 receptor. The anorexigenic POMC neurons release α-melanocyte-stimulating hormone (α-MSH) to stimulate MC4Rs, while the orexigenic agouti-related (AgRP) neurons release AgRP, with opposite properties. AgRP co-localizes with neuropeptide Y (NPY) (NPY/AgRP neurons) (Fig. 69.1) (Xiang et al. 2010;

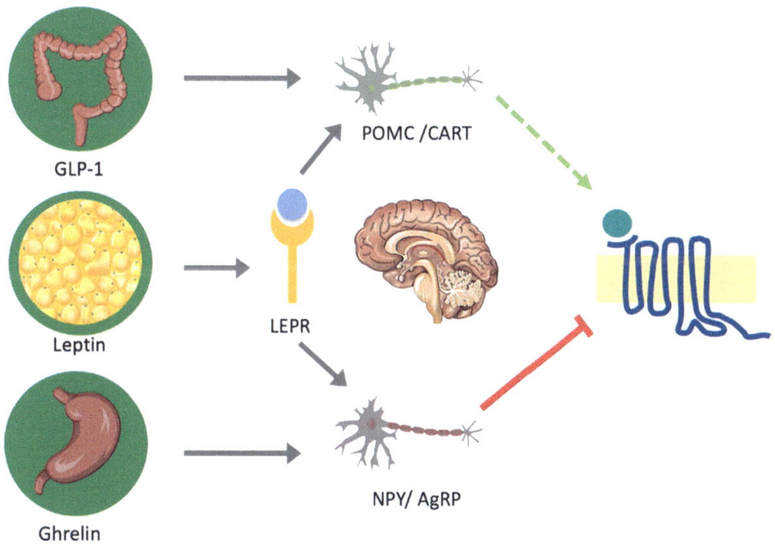

Fig. 69.1 Central regulation of appetite control and energy expenditure. Numerous peripheral signals, such as leptin, ghrelin, NPY, and glucagon-like peptide 1 (GLP-1), activate neurons in the ARH. Leptin is secreted by adipose tissue and stimulates leptin receptors (LEPR) in the arcuate nucleus. Thereby, POMC-expressing neurons are stimulated, promoting secretion of MSH. MSH triggers MC4R expressed in the paraventricular nucleus, regulating satiety, energy expenditure, blood pressure, and growth. In parallel, LEPR inhibits AgRP-expressing neurons, which are also localized in the ARH. AgRP has been suggested as an inverse agonist or antagonist at the MC4R (Kühnen et al. 2019)

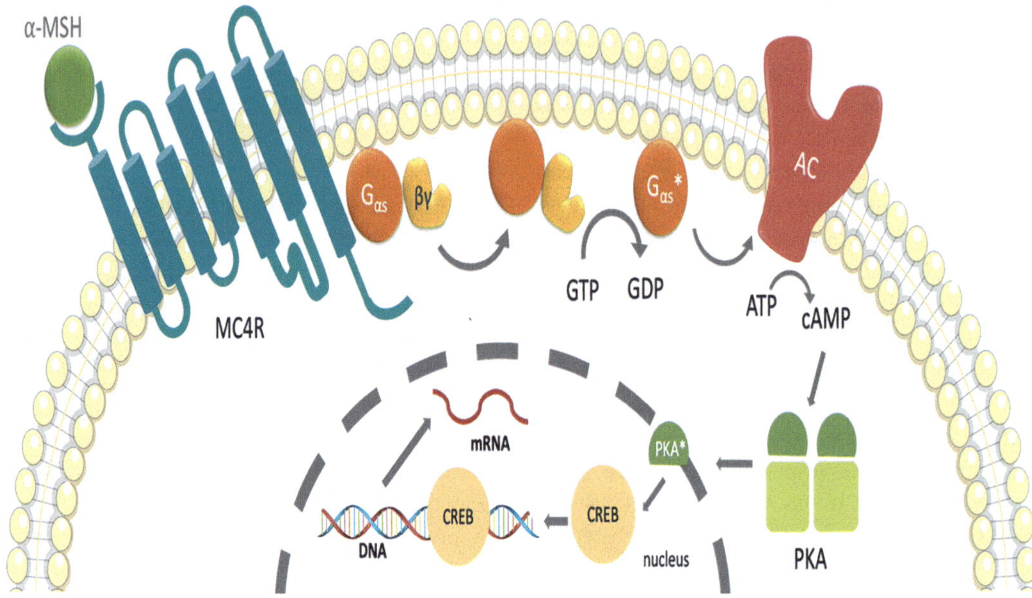

Fig. 69.2 Schematic representation of the GPCR G_s-mediated signaling cascade: Schematics for canonical cAMP signaling cascade as a mechanism for melanocortin-mediated regulation of gene expression. α-MSH stimulation of melanocortin receptors leads to activation of the α-subunit of the stimulatory heterotrimeric G-protein, $G_{\alpha s}$, and the subsequent activation of adenylyl cyclase (AC) enzymes, which convert ATP in cAMP. cAMP-mediated activation of protein kinase A (PKA) leads to the dissociation of the catalytic subunit of the PKA complex. Upon translocation of the active PKA catalytic subunit into the nucleus, PKA can phosphorylate the transcription factor CREB (cAMP Response Element-Binding Protein; CREB-P denoting the phosphorylated variant). When bound to an upstream consensus, DNA sequence called cAMP Response Elements (CREs) CREB-P can facilitate transcription of a gene by interaction with the core transcriptional machinery (adapted with permission from Gonçalves JPL, Palmer D, Meldal M, MC4R Agonists: Structural Overview on Antiobesity Therapeutics, https://doi.org/10.1016/j.tips.2018.01.004)

Millington 2007; Anderson et al. 2016; Williams and Elmquist 2012).

The most frequent genetic causes of severe human obesity are heterozygous mutations in the coding sequence of the serpentine melanocortin-4 receptor, which often causes total or partial loss of receptor function and ambiguous receptor activity (Farooqi et al. 2003); therefore, the MC4R system is often targeted for pharmacotherapy (Panaro and Cone 2013; Gonçalves et al. 2018). The natural regulating hormones (Table 69.2), such as the melanocyte-stimulating hormones, were the first studied. However, imbalance in MSH can result in hyperpigmentation and Addison's disease (Gonçalves et al. 2018; Tao 2010).

Signaling Pathways

Among the signaling pathways of melanocortins, the G-protein cascade deserves emphasis (Fig. 69.2) (Eves and Haycock 2010; Tao 2014). The bound receptor activates the alpha subunit of the G-protein complex, whereby effector enzymes release second messengers sequentially mobilizing effector proteins, including kinases, ion channels, and additional enzymes (Calebiro et al. 2009).

Despite the varied nature and function of GPCRs, only four G-protein paths have been assigned a nomenclature based on the subset of $G_{\beta\gamma}$ subunits activating the pathway: with the subsets being G_s (stimulatory), $G_{i/o}$ (inhibitory), $G_{q/11}$, and $G_{12/13}$. Of particular relevance to

MC4R is the G_s family, which is based on cAMP synthesis, conducts to both short- and long-term effectors (Hepler and Gilman 1992; Lalli and Sassone-Corsi 1994).

Protein kinases (PKA) are effectors relevant for ion channels, kinases, enzymes, and scaffolding/structural proteins, influencing transcription and other cell functions. cAMP response element-binding protein (CREB) and related proteins are involved in the transcription pathway (Imamura et al. 2017).

In addition to PKA isoforms, cyclic nucleotide-gated ion channels, such as Kv11.1, can modulate membrane polarity. Within such circumstances, cAMP electrically modulates sensitive cells. cAMP also binds strongly to guanine nucleotide exchange factors that facilitate the conversion of other small G-proteins, a class distinct from the receptor-coupled heterotrimeric G-proteins, to an activated GTP-bound state. Mesuring receptor activation through cAMP or some of its downstream signaling events is the most usual monotoring approach (Gonçalves et al. 2018).

Over the last 5 years, there have been an increasing number of reports confirming existence of different signaling cascades for MC4R activation (Tao 2014; Lotta et al. 2019; Yang and Tao 2017).

Interpretation of pharmacological studies can be biased when cell-specific effects on differential propagation of signaling responses are not taken into consideration. Some mutations causing gain of function are associated with protection from obesity (Lotta et al. 2019). Enhanced β-arrestin-mediated signaling rather than $G_s\alpha$ signaling was detected with some of such mutations. Natural variants of MC4R affecting signaling and the degree of β-arrestin recruitment to MC4R account for a large proportion of the variation in genetic association of these variants with obesity and diabetes risk.

In another study, seven human obesity phenotypes were associated as class V variants of the MC4R (Gillyard et al. 2019) providing evidence of other G-protein-dependent and G-protein-independent signaling cascades. The authors also underlined the impact of loss of β-MSH for the normal regulation of energy homeostasis, and advance a model where the recruitment of β-arrestin is a marker of normal MC4R function, in contrast to the results of Lotta *et al.* These results suggest a broader range of signaling responses than what it is known or assumed today.

MC4R exhibits a significant degree of constitutive activation of the $G_s\alpha$ signaling pathway, by the receptor's amino terminal domain, potentially inhibited by the melanocortin-2 receptor accessory protein (MRAP2), which also enhances response to α-MSH action (Anderson et al. 2016; Tao 2014; Srinivasan et al. 2004).

Like many other GPCRs, MCRs likely exist as a higher ordered multimers and particularly as dimers (Mondal et al. 2013; Piechowski et al. 2013). This has inspired the development of bivalent ligands that presumably can activate dimeric receptors and may yield improved agonists and antagonists (Fernandes et al. 2014; Lensing et al. 2016).

It is speculated that MC3R, MC4R, and MRAP2, all highly expressed in the hypothalamus, coordinate to regulate the energy homeostasis, yet further studies are necessary to develop successful pharmacologic interventions (Ghamari-Langroudi and Cone 2011).

Agonists with Potential Therapeutic Interest and Clinical Trials

As previously mentioned, MC4R is activated by POMC-derived peptides (Anderson et al. 2016). These peptide ligands derive from the proteolytic processing of POMC, including adenotropic hormone (ACTH) and the melanocortin-stimulating hormones beta (β-MSH) and gamma (γ-MSH). ACTH can be further processed to α-MSH which has a prominent role in the regulation of MC4R.

α-MSH is predicted to bind in a β-turn conformation, which presents the His-Phe-Arg-Trp motif for optimal binding with the receptor binding pocket, specifically with the transmembrane domains between TM2 and TM3 (Haskell-Luevano et al. 2001; Igel et al. 2017). This motif appears in several synthetic agonists. Cyclic peptidomimetic structures have also been developed yielding a higher degree of potency, resembling the β-turn conformation of the natural agonist (α-MSH), providing a good fit in the receptor pocket. Despite having less potency, non-peptide agonists showed better selectivity, paving the way for additional molecular structures (Gonçalves et al. 2018).

Prior to 2012, few drugs had been approved by the Federal Drug Administration for obesity management. Often obesity was associated with an unhealthy lifestyle choice and not fully accepted as a disease. As the endogenous pathways governing these processes have been elucidated, obesity has been better established as a serious disorder and as such, pharmacotherapeutics have been accepted as a treatment approach with several more potential pharmacophores under evaluation in clinical trials. Currently, clinical trials evaluating Setmelanotide (RM-493 developed by Rhythm Pharma) are underway for the treatment of Leptin Receptor Deficiency Obesity, Pro-opiomelanocortin (POMC) Deficiency Obesity, Bardet-Biedl Syndrome, Alström Syndrome, and Smith-Magenis Syndrome. Unlike other drugs based on Melanotan-II, Setmelanotide does not increase blood pressure.

Some of the most promising structures for activation of MC4R developed over the last decade are summarized in Table 69.3.

Upon closer look at the most potent and selective candidates, it is possible to see the repetition of the motif, -His-D-phe-Arg-Trp-. Generally, the use of positively charged groups and three aromatic moieties, including the -His-D-phe-Arg-Trp- motif, are typical of synthetic MC4R agonists, in order to retain the core from the natural pharmacophore. In recent structures, however, more charged residues have been employed and have yielded enhanced selectivity, potency, and conformational preference.

Taking into account that MC4R also governs other body activities, namely sexual desire and neuropeptide regulation, any therapeutic targeted toward MC4R will need to find a way to minimize the activation of these additional pathways while maintaining robust activation of pathways shaping metabolism (Tao 2010).

Despite several efforts, there is still not a drug in the market targeting MC4R for the treatment of obesity. Nonetheless, advances in identifying optimal structural properties for ligands, paired with a deeper understanding of the signaling mechanisms that best shape desired physiological responses, promise to offer better MC4R-targeted therapeutics.

Diabetes as a Consequence of Obesity

Roughly 30% of overweight people have diabetes, whereas 85% of diabetics are overweight. In the USA, diabetes affects nearly 1 in 10 adults, with a majority (90%–95%) of the cases being type 2 diabetes (T2D). Diabetes induced by excess weight often requires lifelong chemical treatment, besides the possibility of cardiovascular complications, microvascular vision

Table 69.3 Potent (EC_{50} < 5 nM) and selective (> 50-fold EC_{50}) MC4R agonists

Structure/Reference	EC_{50} (nM)
(*S*)-2-Acetamido-*N*-((*R*)-1-((2*R*,4*S*)-2-(3-guanidinopropyl)-4-(naphthalen-2-ylmethoxy)pyrrolidin-1-yl)-1-oxo-3-phenylpropan-2-yl)-3-(1H-imidazol-4-yl)propanamide (Hong et al. 2011)	4.9
c[Hydantoin(C(O)-(Asp-D-Ala))-His-D-Phe-Arg-Trp-Dap]-NH_2 (Halem et al. 2008)	0.177
Hydantoin(C(O)-(D-Ala-Gly))-c(Cys-D-Ala-His-D-Phe-Arg-Trp-Cys)-NH_2 (Halem et al. 2008)	0.16
Hydantoin(C(O)-(D-Arg-Gly))-c(Cys-Glu-His-D-Phe-Arg-Trp-Cys)-NH_2 (Halem et al. 2008)	0.125
Hydantoin(C(O)-(Gly-D-Arg))-c(Cys-Glu-His-D-Phe-Arg-Trp-Cys)-NH_2 (Halem et al. 2008)	0.0979
c-Lys-DapN_3_His-D-Phe-Arg-Trp-Pra-Met-NH_2 (Palmer et al. 2017)	1.8
c-Lys-DapN_3_His-D-Phe-Arg-Trp-Pra-Ala-NH_2 (Palmer et al. 2017)	2.3
c-Lys-DapN_3_Ala-D-Phe-Arg-Trp-Pra-Met-NH_2 (Palmer et al. 2017)	2.8

impairments (glaucoma, retinopathia, cataracts), and kidney dysfunction, among others, justifying the investigation of MC4R and insulin resistance (De Carvalho et al. 2017; Chen et al. 2019).

Nucleotide polymorphisms near the MC4R gene have been strongly associated with T2D. There is robust epidemiological evidence that rs17782313 polymorphism is significantly associated with increasing risk of T2D (Xi et al. 2012). In mice with a homozygous $G_s\alpha$ deficiency in MC4R-expressing cells, impaired glucose tolerance and elevated insulin response were unearthed, suggesting that $G_s\alpha$ signaling in MC4R-expressing cells directly disturbs glucose tolerance and insulin sensitivity (Podyma et al. 2018; Al-Goblan et al. 2014).

Mice with POMC and MC4R deficiency released epinephrine and glucagon as response to hypoglycemia and glucopenia. In addition, diabetic mice showed decreased POMC and MC4R expression and a deficient counter regulatory response (Tooke et al. 2019) and this phenotype was not normalized with insulin treatment. On the other hand, when injected in the paraventricular hypothalamic nucleus with a MC4R agonist, the counter regulatory response was restored.

Perspectives

The melanocortin-4 receptor has a pivotal role in energy homeostasis and adipose tissue regulation. Nevertheless, this is a complex receptor class with diverse signaling cascades. There is not enough consensus about the role of certain proteins, such as β-arrestin, in the signaling mechanism. Understanding about GPCRs signaling pathways and specifically, how melanocortin receptors can be controlled are clear priorities, along with elucidation of the effects of mutations on MC2R and MC3R in the feeding behavior.

Melanotan-II and Bremelanotide did not survive clinical trials. Melanotan-II promoted intracellular synthesis of eumelanin in favor of pheomelanin which gave a darker skin pigmentation and also caused penile erection, while Bremelanotide is now being tested for the treatment of female sexual dysfunction. Currently, Setmelanotide is undergoing phase-III clinical trials.

New agonists with charged residues and aromatic moieties provided the most promising agonists, while non-peptide-derived candidates allowed for more selective ligands. Additionally, cyclic structures should be further considered because of the conformational preference for the receptor's binding site. Activation of the most appropriate signaling pathway for the physiologic effect should be highlighted, as biased agonists may offer the means for avoiding unwanted effects.

Supplementary evidence on the existence of multiple binding sites or receptor multimers is also needed, alongside with the receptor crystal structures.

References

Abdel-Malek ZA (2001) Melanocortin receptors: their functions and regulation by physiological agonists and antagonists. Cell Mol Life Sci 58(3):434–441. https://doi.org/10.1007/PL00000868

Al-Goblan AS, Al-Alfi MA, Khan MZ (2014) Mechanism linking diabetes mellitus and obesity. Diabetes Metab Syndr Obes Targets Ther 7:587–591. https://doi.org/10.2147/DMSO.S67400

Anderson EJP, Çakir I, Carrington SJ et al (2016) 60 years of POMC: regulation of feeding and energy homeostasis by α-MSH. J Mol Endocrinol 56(4):T157–T174. https://doi.org/10.1530/JME-16-0014

Batterham RL, Cowley MA, Small CJ et al (2002) Gut hormone PYY3-36 physiologically inhibits food intake. Nature 418(6898):650–654. https://doi.org/10.1038/nature00887

Begriche K, Sutton GM, Fang J, Butler AA (2009) The role of melanocortin neuronal pathways in circadian biology: a new homeostatic output involving melanocortin-3 receptors? Obes Rev 10(s2):14–24. https://doi.org/10.1111/J.1467-789X.2009.00662.X

Berrettini W (2004) The genetics of eating disorders. Psychiatry (Edgmont) 1(3):18–25. http://www.ncbi.nlm.nih.gov/pubmed/21191522. Accessed September 18, 2017

Butler AA, Kesterson RA, Khong K et al (2000) A unique metabolic syndrome causes obesity in the melanocortin-3 receptor-deficient mouse. Endocrinology 141(9):3518–3521. https://doi.org/10.1210/endo.141.9.7791

Calebiro D, Nikolaev VO, Gagliani MC et al (2009) Persistent cAMP-signals triggered by internalized G-protein-coupled receptors. PLoS Biol 7(8):e1000172. https://doi.org/10.1371/journal.pbio.1000172

Castrucci AM, Hadley ME, Sawyer TK, Hruby VJ (1984) Enzymological studies of melanotropins. Comp Biochem Physiol B 78(3):519–524. https://doi.org/10.1016/0305-0491(84)90090-7

Chen Y, Liu H, Wang L et al (2019) Lifestyle intervention modifies the effect of the MC4R genotype on changes in insulin resistance among women with prior gestational diabetes: Tianjin Gestational Diabetes Mellitus Prevention Program. Am J Clin Nutr 110(3):750–758. https://doi.org/10.1093/ajcn/nqz121

Cowley MA, Smart JL, Rubinstein M et al (2001) Leptin activates anorexigenic POMC neurons through a neural network in the arcuate nucleus. Nature 411:480–484. https://doi.org/10.1038/35078085

De Carvalho AM, Shao P, Liu H et al (2017) The MC4R genotype is associated with postpartum weight reduction and glycemic changes among women with prior gestational diabetes: longitudinal analysis. Sci Rep 7 (1):1–6. https://doi.org/10.1038/s41598-017-10101-x

Ellacott KLJ, Murphy JG, Marks DL, Cone RD (2007) Obesity-induced inflammation in white adipose tissue is attenuated by loss of melanocortin-3 receptor signaling. Endocrinology 148(12):6186–6194. https://doi.org/10.1210/en.2007-0699

Eves PC, Haycock JW (2010) Melanocortin signalling mechanisms. In: Catania A (ed) Melanocortins: multiple actions and therapeutic potential, vol 681. Springer Science, New York, NY, pp 19–28. https://doi.org/10.1007/978-1-4419-6354-3_2

Farooqi IS, Keogh JM, Yeo GSH, Lank EJ, Cheetham T, O'Rahilly S (2003) Clinical spectrum of obesity and mutations in the Melanocortin 4 Receptor gene. N Engl J Med 348(12):1085–1095. https://doi.org/10.1056/NEJMoa022050

Fernandes SM, Lee YS, Gillies RJ, Hruby VJ (2014) Synthesis and evaluation of bivalent ligands for binding to the human melanocortin-4 receptor. Bioorg Med Chem 22(22):6360–6365. https://doi.org/10.1016/j.bmc.2014.09.055

Gantz I, Miwa H, Konda Y et al (1993) Molecular cloning, expression, and gene localization of a fourth melanocortin receptor. J Biol Chem 268 (20):15174–15179. http://www.ncbi.nlm.nih.gov/pubmed/8392067. Accessed March 15, 2017

Getting SJ, Riffo-Vasquez Y, Pitchford S et al (2008) A role for MC3R in modulating lung inflammation. Pulm Pharmacol Ther 21(6):866–873. https://doi.org/10.1016/j.pupt.2008.09.004

Ghamari-Langroudi M, Cone RD (2011) Shining a light on energy homeostasis. Cell Metab 13(3):235–236. https://doi.org/10.1016/j.cmet.2011.02.007

Gillyard T, Fowler K, Williams SY, Cone RD (2019) Obesity-associated mutant melanocortin-4 receptors with normal Gαs coupling frequently exhibit other discoverable pharmacological and biochemical defects. J Neuroendocrinol 31(10):e12795. https://doi.org/10.1111/jne.12795

Girardet C, Butler AA (2014) Neural melanocortin receptors in obesity and related metabolic disorders. Biochim Biophys Acta Mol basis Dis 1842 (3):482–494. https://doi.org/10.1016/j.bbadis.2013.05.004

Gonçalves JPL, Palmer D, Meldal M (2018) MC4R agonists: structural overview on antiobesity therapeutics. Trends Pharmacol Sci 39(4):402–423. https://doi.org/10.1016/j.tips.2018.01.004

Guh DP, Zhang W, Bansback N, Amarsi Z, Birmingham CL, Anis AH (2009) The incidence of co-morbidities related to obesity and overweight: a systematic review and meta-analysis. BMC Public Health 9(1):88. https://doi.org/10.1186/1471-2458-9-88

Halem HA, Culler MD, Butler AA (2008) Use melanocortins to treat insulin sensitivity. http://www.google.tl/patents/WO2009061411A2?cl=en&hl=pt-PT. Accessed October 16, 2017

Haskell-Luevano C, Hendrata S, North C et al (1997) Discovery of prototype peptidomimetic agonists at the human melanocortin receptors MC1R and MC4R. J Med Chem 40(14):2133–2139. https://doi.org/10.1021/jm960840h

Haskell-Luevano C, Cone RD, Monck EK, Wan Y-P (2001) Structure activity studies of the melanocortin-4 receptor by in vitro mutagenesis: identification of agouti-related protein (AGRP), Melanocortin Agonist and Synthetic Peptide Antagonist Interaction Determinants †. Biochemistry 40(20):6164–6179. https://doi.org/10.1021/bi010025q

Hepler JR, Gilman AG (1992) G proteins. Trends Biochem Sci 17(10):383–387. http://www.ncbi.nlm.nih.gov/pubmed/1455506. Accessed September 26, 2017

Hong Q, Bakshi RK, Palucki BL et al (2011) Discovery of a piperazine urea based compound as a potent, selective, orally bioavailable melanocortin subtype-4 receptor partial agonist. Bioorg Med Chem Lett 21 (8):2330–2334. https://doi.org/10.1016/j.bmcl.2011.02.090

Hruby VJ, Lu D, Sharma SD et al (1995) Cyclic lactam alpha-melanotropin analogues of Ac-Nle4-cyclo [Asp5, D-Phe7,Lys10] alpha-melanocyte-stimulating hormone-(4-10)-NH2 with bulky aromatic amino acids at position 7 show high antagonist potency and selectivity at specific melanocortin receptors. J Med Chem 38(18):3454–3461. http://www.ncbi.nlm.nih.gov/pubmed/7658432. Accessed March 14, 2017

Huszar D, Lynch CA, Fairchild-Huntress V et al (1997) Targeted disruption of the melanocortin-4 receptor results in obesity in mice. Cell 88(1):131–141. https://doi.org/10.1016/S0092-8674(00)81865-6

Igel LI, Kumar RB, Saunders KH, Aronne LJ (2017) Practical use of pharmacotherapy for obesity. Gastroenterology 152(7):1765–1779. https://doi.org/10.1053/j.gastro.2016.12.049

Imamura H, Wagih O, Niinae T, Sugiyama N, Beltrao P, Ishihama Y (2017) Identifications of putative PKA substrates with quantitative phosphoproteomics and primary-sequence-based scoring. J Proteome Res 16

(4):1825–1830. https://doi.org/10.1021/acs.jproteome.7b00087

Kühnen P, Krude H, Biebermann H (2019) Melanocortin-4 Receptor signalling: importance for weight regulation and obesity treatment. Trends Mol Med 25 (2):136–148. https://doi.org/10.1016/j.molmed.2018.12.002

Lalli E, Sassone-Corsi P (1994) Signal transduction and gene regulation: the nuclear response to cAMP. J Biol Chem 269(26):17359–17362. http://www.ncbi.nlm.nih.gov/pubmed/8021233. Accessed September 26, 2017

Leitner DR, Frühbeck G, Yumuk V et al (2017) Obesity and type 2 diabetes: two diseases with a need for combined treatment strategies – EASO can lead the way. Obes Facts 10(5):483–492. https://doi.org/10.1159/000480525

Lensing CJ, Freeman KT, Schnell SM, Adank DN, Speth RC, Haskell-Luevano C (2016) An in vitro and in vivo investigation of bivalent ligands that display preferential binding and functional activity for different melanocortin receptor homodimers. J Med Chem 59 (7):3112–3128. https://doi.org/10.1021/acs.jmedchem.5b01894

Leoni G, Patel HB, Sampaio ALF et al (2008) Inflamed phenotype of the mesenteric microcirculation of melanocortin type 3 receptor-null mice after ischemia-reperfusion. FASEB J 22(12):4228–4238. https://doi.org/10.1096/fj.08-113886

Litt MJ, Okoye GD, Lark D et al (2017) Loss of the melanocortin-4 receptor in mice causes dilated cardiomyopathy. elife 6. https://doi.org/10.7554/eLife.28118

Lotta LA, Mokrosiński J, Mendes de Oliveira E, et al. Human gain-of-function MC4R variants show signaling bias and protect against obesity. Cell. 2019;177 (3):597-607.e9. doi:https://doi.org/10.1016/j.cell.2019.03.044

Maes HH, Neale MC, Eaves LJ (1997) Genetic and environmental factors in relative body weight and human adiposity. Behav Genet 27(4):325–351. http://www.ncbi.nlm.nih.gov/pubmed/9519560. Accessed September 18, 2017

Makki K, Froguel P, Wolowczuk I (2013) Adipose tissue in obesity-related inflammation and insulin resistance: cells, cytokines, and chemokines. ISRN Inflamm 2013:1–12. https://doi.org/10.1155/2013/139239

Marks DL, Hruby V, Brookhart G, Cone RD (2006) The regulation of food intake by selective stimulation of the type 3 melanocortin receptor (MC3R). Peptides 27 (2):259–264. https://doi.org/10.1016/j.peptides.2005.01.025

Melanocortin 4 Receptor. https://www.mc4r.org.uk/. Published 2018. Accessed March 13, 2020

Millington GWM (2006) Pro-opiomelanocortin (POMC): the cutaneous roles of its melanocortin products and receptors. Clin Exp Dermatol 31(3):407–412. https://doi.org/10.1111/j.1365-2230.2006.02128.x

Millington GW (2007) The role of proopiomelanocortin (POMC) neurones in feeding behaviour. Nutr Metab (Lond) 4(1):18–34. https://doi.org/10.1186/1743-7075-4-18

Mondal S, Johnston JM, Wang H, Khelashvili G, Filizola M, Weinstein H (2013) Membrane driven spatial organization of GPCRs. Sci Rep 3:2909. https://doi.org/10.1038/srep02909

Nogueiras R, Wiedmer P, Perez-Tilve D et al (2007) The central melanocortin system directly controls peripheral lipid metabolism. J Clin Invest 117 (11):3475–3488. https://doi.org/10.1172/JCI31743

O'Rahilly S, Farooqi IS, Yeo GSH, Challis BG (2003) Minireview: human obesity-lessons from monogenic disorders. Endocrinology 144(9):3757–3764. https://doi.org/10.1210/en.2003-0373

Palmer D, Gonçalves JPL, Hansen LV et al (2017) Click-chemistry-mediated synthesis of selective Melanocortin Receptor 4 agonists. J Med Chem 60 (21):8716–8730. https://doi.org/10.1021/acs.jmedchem.7b00353

Panaro BL, Cone RD (2013) Melanocortin-4 receptor mutations paradoxically reduce preference for palatable foods. Proc Natl Acad Sci USA 110 (17):7050–7055. https://doi.org/10.1073/pnas.1304707110

Pantel J, Williams SY, Mi D et al (2011) Development of a high throughput screen for allosteric modulators of melanocortin-4 receptor signaling using a real time cAMP assay. Eur J Pharmacol 660(1):139–147. https://doi.org/10.1016/j.ejphar.2011.01.031

Piechowski CL, Rediger A, Lagemann C et al (2013) Inhibition of melanocortin-4 receptor dimerization by substitutions in intracellular loop 2. J Mol Endocrinol 51(1):109–118. https://doi.org/10.1530/JME-13-0061

Podyma B, Sun H, Wilson EA et al (2018) The stimulatory G protein Gsα is required in melanocortin 4 receptor-expressing cells for normal energy balance, thermogenesis, and glucose metabolism. J Biol Chem 293 (28):10993–11005. https://doi.org/10.1074/jbc.RA118.003450

Poggioli R, Vergoni AV, Bertolini A (1986) ACTH-(1–24) and α-MSH antagonize feeding behavior stimulated by kappa opiate agonists. Peptides 7 (5):843–848. https://doi.org/10.1016/0196-9781(86)90104-X

Renquist BJ, Lippert RN, Sebag JA, Ellacott KLJ (2011) Physiological roles of the melanocortin MC3 receptor. Eur J Pharmacol 660(1):13–20. https://doi.org/10.1016/j.ejphar.2010.12.025

Santini F, Maffei M, Pelosini C, Salvetti G, Scartabelli G, Pinchera A (2009) Melanocortin-4 receptor mutations in obesity. Adv Clin Chem 48:95–109. http://www.ncbi.nlm.nih.gov/pubmed/19803416. Accessed January 19, 2018

Sawyer TK, Hruby VJ, Wilkes BC, Draelos MT, Hadley ME, Bergsneider M (1982) Comparative biological

activities of highly potent active-site analogs of alpha-melanotropin. J Med Chem 25(9):1022–1027. https://doi.org/10.1021/jm00351a004
Srinivasan S, Lubrano-Berthelier C, Govaerts C et al (2004) Constitutive activity of the melanocortin-4 receptor is maintained by its N-terminal domain and plays a role in energy homeostasis in humans. J Clin Invest 114(8):1158–1164. https://doi.org/10.1172/JCI21927
Tao Y-X (2010) The Melanocortin-4 Receptor: physiology, pharmacology, and pathophysiology. Endocr Rev 31(4):506–543. https://doi.org/10.1210/er.2009-0037
Tao YX (2014) Constitutive activity in melanocortin-4 Receptor: biased signaling of inverse agonists. In: Advances in pharmacology, vol 70. Elsevier, San Diego, CA, pp 135–154. https://doi.org/10.1016/B978-0-12-417197-8.00005-5
Tooke BP, Yu H, Adams JM et al (2019) Hypothalamic POMC or MC4R deficiency impairs counterregulatory responses to hypoglycemia in mice. Mol Metab 20:194–204. https://doi.org/10.1016/j.molmet.2018.11.004
Vaisse C, Clement K, Guy-Grand B, Froguel P (1998) A frameshift mutation in human MC4R is associated with a dominant form of obesity. Nat Genet 20(2):113–114. https://doi.org/10.1038/2407
Van Gaal L, Dirinck E. Pharmacological approaches in the treatment and maintenance of weight loss. Diabetes Care. 2016;39(Supplement 2):S260-S267. doi:https://doi.org/10.2337/dcS15-3016
Velazquez A, Apovian CM (2018) Pharmacological management of obesity. Minerva Endocrinol 43(3):356–366. https://doi.org/10.23736/S0391-1977.17.02654-2
Wang GT, Bent SF, Russell JN, Butler JE, D'Evelyn MP (2000) Functionalization of Diamond(100) by Diels–Alder chemistry. J Am Chem Soc 122(4):744–745. https://doi.org/10.1021/ja993024i
Webb TR, Clark AJL (2010) Minireview: the Melanocortin 2 Receptor accessory proteins. Mol Endocrinol 24(3):475–484. https://doi.org/10.1210/me.2009-0283
Williams KW, Elmquist JK (2012) From neuroanatomy to behavior: central integration of peripheral signals regulating feeding behavior. Nat Neurosci 15(10):1350–1355. https://doi.org/10.1038/nn.3217
Wren AM, Seal LJ, Cohen MA et al (2001) Ghrelin enhances appetite and increases food intake in humans. J Clin Endocrinol Metab 86(12):5992–5992. https://doi.org/10.1210/jcem.86.12.8111
Xi B, Takeuchi F, Chandak GR et al (2012) Common polymorphism near the MC4R gene is associated with type 2 diabetes: data from a meta-analysis of 123,373 individuals. Diabetologia 55(10):2660–2666. https://doi.org/10.1007/s00125-012-2655-5
Xiang Z, Proneth B, Dirain ML, Litherland SA, Haskell-Luevano C (2010) Pharmacological characterization of 30 human Melanocortin-4 Receptor polymorphisms with the endogenous proopiomelanocortin-derived agonists, synthetic agonists, and the endogenous agouti-related protein antagonist. Biochemistry 49(22):4583–4600. https://doi.org/10.1021/bi100068u
Yang LK, Tao YX (2017) Biased signaling at neural melanocortin receptors in regulation of energy homeostasis. Biochim Biophys Acta Mol basis Dis 1863(10):2486–2495. https://doi.org/10.1016/j.bbadis.2017.04.010
Yeo GSHH, Lank EJ, Farooqi IS et al (2003) Mutations in the human melanocortin-4 receptor gene associated with severe familial obesity disrupts receptor function through multiple molecular mechanisms. Hum Mol Genet 12(5):561–574. https://doi.org/10.1093/hmg/ddg057

Nanotechnology: Can It Be a Crusader in Diabesity?

70

Alexis Marie Speer and Mahua Choudhury

Contents

Introduction 948

Limitations of Current Pharmacological Therapies 948
Diabetes Overview 948

Nanotechnology Promise 949
Pharmacological Applications 949

Defining 'Nanoscale' 949

Origins of Nanotechnology 950

Nanotechnology Advances for the Treatment of Diabetes 951
Artificial Pancreas 951

Biological Nanosensors and Diabetes Monitoring 951
Smart Tattoos 952
Pancreatic Beta Cell Mass 953

Nanotechnology Targeted Treatment for Obesity 953
Lipid Nanoparticles 953

Conclusion 953

References 954

Abstract

Research in nanotechnology has been slowly expanding. It allows for expansion of work on a molecular scale. There has been success in targeting adipocytes, macrophages, fibroblasts, and vascular cells, which are related to diabetes and obesity pathophysiology. Development in nanosensors for usage in diabetic monitoring could provide better real-time monitoring and early detection of progression for patients, and this technology could help in earlier, life-saving intervention. In conclusion, nanotechnology could lead to the development of new treatments or novel prevention of diabesity in the near future.

A. M. Speer · M. Choudhury (✉)
Department of Pharmaceutical Sciences, School of Pharmacy, Texas A & M University, College Station, TX, USA
e-mail: mchoudhury@tamu.edu

J. Faintuch and S. Faintuch (eds.), *Obesity and Diabetes*, https://doi.org/10.1007/978-3-030-53370-0_70

Keywords

Diabetes · Obesity · Nanoscale · Nanosensors · Smart tattoos · Artificial pancreas · Lipid nanoparticles

Introduction

Current treatments for obesity and diabetes (diabesity) rely on lifestyle changes, surgery, and pharmacotherapy. There is an emphasis on diet modifications and increased exercise; yet, few American citizens have healthy nutrition or activity levels, with only 28% women and 20% men eating at least five servings of fruit and vegetables per day. Additionally, approximately 60% of adults are believed to not be engaged in sufficient levels of physical activity to provide health benefits. Due to low adherence to healthy living, there is a need to search for alternative therapy approaches to treat or prevent obesity, diabetes, and metabolic syndrome (Carbone et al. 2019).

Limitations of Current Pharmacological Therapies

Pharmaceuticals on the market for obesity treatments target weight management and progression prevention of its related diseases. These drugs work on the central nervous system (CNS) and gastrointestinal system (GI) to either suppress appetite, inhibit fat absorption, or increase energy expenditure and consist of both short-term and long-term therapies (Sibuyi et al. 2019). However, the effectiveness of obesity pharmaceutical treatments is questionable due to the side effects seen in these therapies. This has resulted in withdrawing patients from certain drugs and removal of others from the market due to their poor efficacy and nonspecificity. An example of this was the use of amphetamine analogs, sympathomimetics, and cannabinoid agents as anti-obesity therapies, where their side effects of cardiac valve defects, pulmonary hypertension, stroke, addiction liability, abuse potential, and psychotic behaviors resulted in withdrawal from the market (Sibuyi et al. 2019; Smith et al. 2020). Some agents used for anti-obesity purposes that remain on the market are Orlistat, a lipase inhibitor, and Liraglutide, a GLP-1 receptor agonist, as long-term treatment options [40]. Phentermine, diethylpropion, and benzphetamine are some examples of obesity short-term treatments [40]. Undesirable side effects of these medications that can lead to reduced efficacy and adherence include oily spotting, fecal urgency, pancreatitis, increased heart rate, fecal incontinence, insomnia, and restlessness. The side effects of Orlistat have caused a recommended limited use of not more than 2 years. Another shortcoming of current anti-obesity drugs is the failure to maintain a weight loss of over 10% in a year, with most FDA-approved drugs being discontinued due to the failure for patients to lose at least 5% of their body weight within 12 weeks of initiation of therapy (Smith et al. 2020; Dehdari and Dehdari 2019).

Diabetes Overview

Type 2 diabetes (T2D) has a variety of oral antidiabetic medication treatment options such as Metformin, Glinides, α-Glucosidase inhibitors, Thiazolidinediones, and Sulfonylureas, which carry a risk of side effects including hypoglycemia, β-cell death, pancreatitis, GI reactions, cardiovascular effects, and central nervous system effects (Smith et al. 2020; Vasan et al. 2018; Shi et al. 2010). T2D treatments usually begin as a monotherapy if nonpharmacological measures alone fail, followed by combination therapy if monotherapy proved to be insufficient (Veiseh et al. 2015).

Diabetes management can be complex and challenging. The goal of therapy is to prevent or delay further complications because of uncontrolled blood sugar levels, but the medication burden, monitoring schedules, and lifestyle change requirements make treatment plans strenuous. Adherence to oral antidiabetic medications ranges from 36% to 93% and to insulin approximately 60% (Dehdari and Dehdari 2019;

American Diabetes Association (ADA) 2019). Poor adherence could be contributed to many factors such as medication cost, medication regimen complexity, and side effects, as seen in studies linking low treatment adherence to oral antidiabetic medications and to the fear of side effects and to insulin, due to the injection itself and its storage being very difficult (Ash et al. 2019; Capoccia et al. 2016). Therefore, alternative therapies are warranted to prevent the diabesity epidemic.

Nanotechnology Promise

Scientists and engineers of varying backgrounds have discovered ways to manipulate materials down to the atomic level. Since these newer methods and technology involve working on such a minute scale, the terms 'nanotechnology,' 'nanoscale,' and 'nanoscience' have come into creation. Emerging fields involving nanotechnology hold promise of a wide range of biomedical applications including enhancing current treatment methods or creating new, more efficient therapies for many chronic disease states including diabetes and obesity (Sibuyi et al. 2019; Shi et al. 2010; Veiseh et al. 2015; Ash et al. 2019; Devadasu et al. 2013, 2017).

Pharmacological Applications

Potential nanotechnological impacts in treatments could be to improve drug encapsulation, protect the payload from degradation, improve solubility, and target specific disease areas (Ash et al. 2019; Devadasu et al. 2013, 2017). Nanodrugs and nanotechnology developments have been used in successful treatment and detection of chronic diseases and conditions (Sibuyi et al. 2019; Shi et al. 2010; Veiseh et al. 2015; Ash et al. 2019; Devadasu et al. 2013, 2017). Fields of interest encompass emesis, infectious diseases, autoimmune diseases, lipid regulation, immunosuppression, and especially cancer (Marta et al. 2016). Nanostructured contrast agents have been used in cases of cancer for targeted specificity and optimal sensitivity for diagnosis, using imaging techniques like optical imaging, radionuclide-based imaging, and computer tomography or magnetic resonance imaging (Shi et al. 2010; Devadasu et al. 2013; Marta et al. 2016).

Cancer, diabetes, and obesity can share a few characteristics such as hyperinsulinemia, hyperglycemia, inflammation, and compromised vascular system angiogenesis, suggesting the potential for nanomedicine to also work in cases of obesity and diabetes (Cui et al. 2015). Obesity and diabetes treatments have benefited thus far from strategies used in cancer treatment, and so there is promise that obesity treatments could benefit from nano-based antiangiogenic, antineoplastic, and transdermal therapies that have been originally used in cancer treatments (Sibuyi et al. 2019; Shi et al. 2010; Veiseh et al. 2015; Devadasu et al. 2013, 2017; Cui et al. 2015; Giovannucci et al. 2010; Lee et al. 2017; Sheng et al. 2014).

Defining 'Nanoscale'

The prefix 'micro-' originates from the Greek word *mikros* meaning "small" and is valued at 10^{-6} m. Here, on this scale, things begin to fall out of focus of the human eye. The diameter of a typical bacterium can range from 10^{-5} m to 10^{-6} m, the mean width of human hair is 8×10^{-6} m, and a nucleus of a typical eukaryotic cell is 7×10^{-6} m (Deza and Deza 2006).

The prefix 'nano-', from the Greek word *nannos*, means "dwarf". A nanometer is 10^{-9} m and can be used to refer to the size of molecules. The diameter of the DNA helix is 2×10^{-9} m, the diameter of known viruses range from 2×10^{-8} m and 8×10^{-7} m, and the limit of resolution of the light microscope is 2×10^{-7} m (Deza and Deza 2006). 'Nano' is used to indicate a scale at one-billionth of the *SI base units*, or the size in which nanotechnology functions, and operates (Allhoff et al. 2010).

Origins of Nanotechnology

One of the first traces of the concepts of nanotechnology is found in 1959 in a speech given by American physicist Richard Feynman. He discusses *"a field, in which little has been done, but in which an enormous amount can be done in principle."* His talk, called "There's Plenty of Room at the Bottom," introduces how it could be possible to manipulate and control things on a small scale or more specifically on a *nanoscale*. Though not all concepts and ideas Feynman introduced in his speech were how the nanotechnological field expanded, he is still considered one of the many founders of nanotechnology (Allhoff et al. 2010).

The actual term 'nanotechnology' was not first coined, though, until 1974 by a professor of the Tokyo University of Science named Norio Taniguchi. Taniguchi' s paper "On the Basic Concept of 'Nano-Technology'" discusses the use of nanotechnology in semiconductor processes such as thin film deposition and ion beam milling (Taniguchi 1974).

Then, in 1987, American engineer K. Eric Drexler released his book "The Engines of Creation: The Coming Era of Nanotechnology," where he offers an introduction to the field of nanotechnology and where he sees the opportunities for the field to expand (Drexler 2000) (Fig. 70.1).

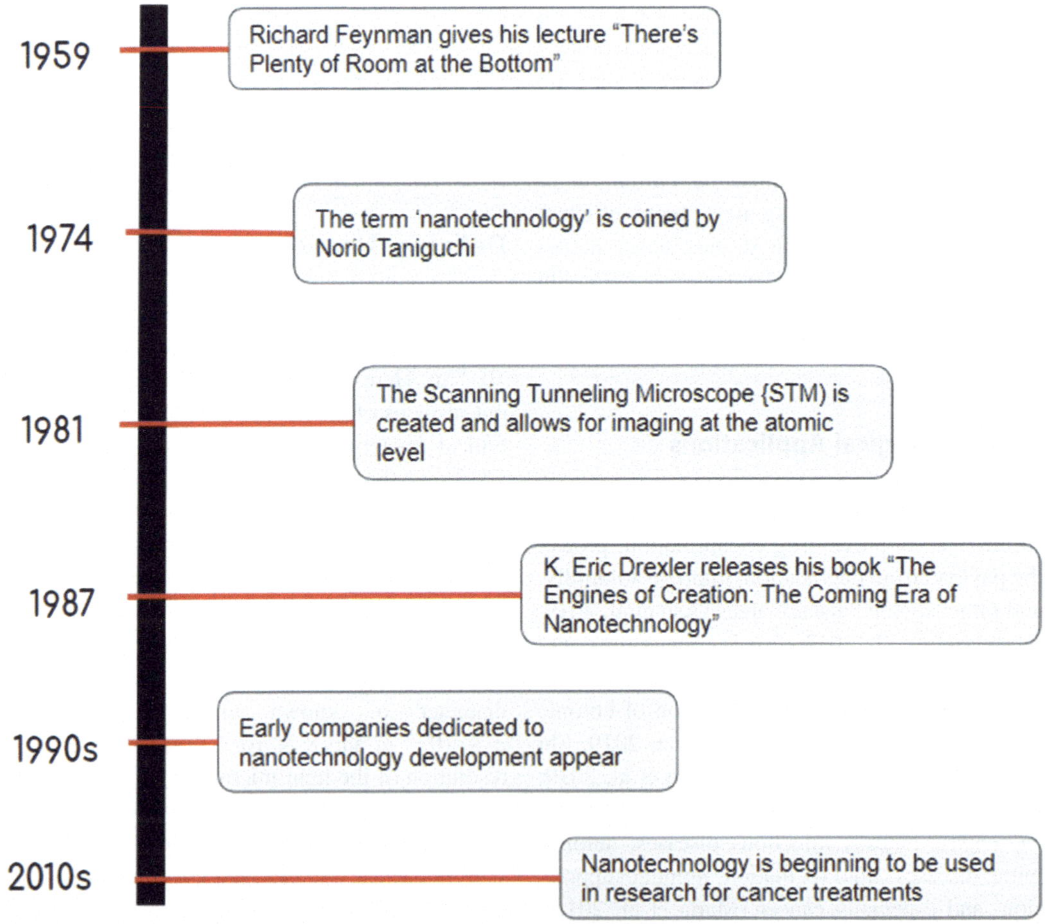

Fig. 70.1 A short preview timeline of the usage and development of nanotechnology

Nanotechnology Advances for the Treatment of Diabetes

Polymer therapeutics, micelles, liposomes, solid lipid nanoparticles, and nanoparticles of biodegradable polymers are examples of current nanotechnology-based insulin delivery methods that have caught many researcher's attention as a potential way to deliver noninvasively. There is also promise of prolonging insulin's duration of effect through nano-technological methods. Liposomal insulin has been tested in animals, and it has been tested orally. Compared to its free form, liposomal insulin has shown to be more bioavailable due to protecting insulin molecules from gastrointestinal degradation (Devadasu et al. 2013, 2017; Cui et al. 2015; Disanto et al. 2015; Liu et al. 2007).

Artificial Pancreas

The artificial pancreas is a closed loop delivery of insulin intended to provide constant measures of insulin throughout the day, triggering the release of insulin when sensors detect high glucose levels. This could offer a permanent solution for T1D patients or some cases of T2D. Setbacks of this development include the reliance on a sensor electrode that is capable of repeatedly measuring blood glucose along with the need to feed the information to a computer database that can activate the insulin release at a calculated amount based on the body's need (Cash and Clark 2010). An alternative to an artificial pancreas would involve prolonging the insulin delivery to the body's changing levels of glucose through the development of nanodrug delivery systems.

One such system is theorized to be composed of glucose-sensitive nanoparticles, which holds the possibility of releasing insulin in response to undesired high glucose levels (Rodbard 2016). Through this controlled release provided by nanoparticles, insulin's action could be extended, replacing the need for multiple injections (Cash and Clark 2010; Rodbard 2016; Wang and Lee 2015; Chen et al. 2017; Scognamiglio 2013; Rahiman 2012).

Biological Nanosensors and Diabetes Monitoring

Glucose monitoring devices have been developed through the usage of nanotechnology, which consist of a detector, a transducer, and a reporter that operate through the use of canulae or sensors placed under the skin, which monitor glucose multiple times throughout the day, with only needing a single insertion (Cash and Clark 2010; Rodbard 2016; Wang and Lee 2015; Chen et al. 2017). These devices hold the potential to provide real-time measurements of glucose levels in the blood or interstitial fluid, which can be more beneficial and accurate for diabetic patients, but there is room for improvement in the accuracy and specificity to properly achieve real-time detection, which can be offered through developments with nanotechnology (Cash and Clark 2010; Rodbard 2016; Wang and Lee 2015; Chen et al. 2017; Scognamiglio 2013; Rahiman 2012) (Fig. 70.2).

Current sensor devices used for diabetic monitoring also have limitations of surface area, catalytic activity, and size that can be addressed through the usage of nanotechnology, which could be a means to increase the surface area for the sensors and increase the catalytic activity of

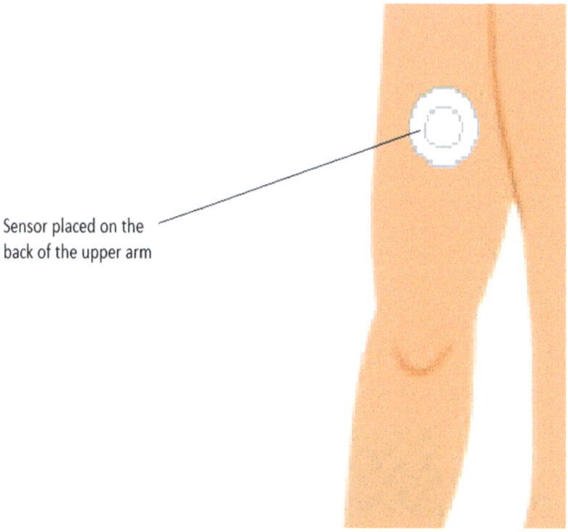

Fig. 70.2 Sensors are placed on areas of the back of the upper arms and the abdomen. These sensors measure interstitial glucose levels every few minutes throughout the day, providing continuous readings

the electrodes. Minimally invasive real-time glucose sensing electrochemical devices, intended to be a nanoscale implant under the skin, are being investigated [36]. Nanoparticle-based glucose sensors use enzymes such as glucose oxidase, or glucose binding proteins, and glucose binding small molecules as transducers. This enables a more convenient, precise, and rapid sensor method desired for the real-time measurements of glucose monitoring, by achieving larger signal enhancements and lower detection limits (Devadasu et al. 2017; Cao et al. 2016).

Smart Tattoos

There has also been development in using nanofibers as a means to provide noninvasive glucose readings in the form of smart tattoos, which are designed as skin implants that can provide continuous and reliable glucose monitoring for diabetic patients (Meetoo et al. 2019). Smart tattoos consist of biosensors made of nanofibers created primarily from electrospinning technology and alternating layer-by-layer deposition technique (Sibuyi et al. 2019; Shi et al. 2010; Veiseh et al. 2015; Devadasu et al. 2013; Cui et al. 2015; Giovannucci et al. 2010; Lee et al. 2017; Sheng et al. 2014). Usage of the layer-by-layer technique requires deposition of alternating layers of positively and negatively charged polymers, which results in a thin film with high activity. Six biolayers often achieve a thickness of 10 nm, implanted into the subcutaneous layer as seen in Fig. 70.3 (Meetoo et al. 2019).

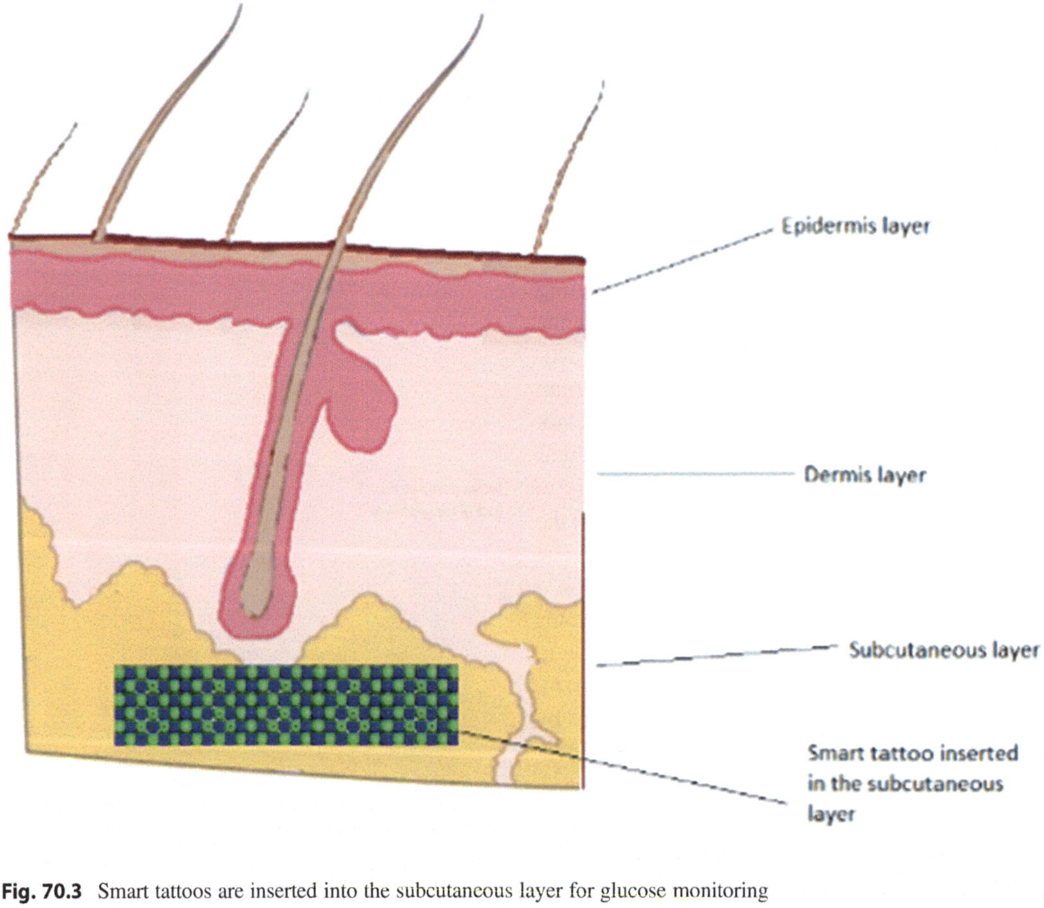

Fig. 70.3 Smart tattoos are inserted into the subcutaneous layer for glucose monitoring

Pancreatic Beta Cell Mass

Currently, the only way to measure β cell mass is by excising tissue samples from patients. A new development proposes a noninvasive method of quantifying β cell mass loss. This could be done by targeting β cells with superparamagnetic iron oxide nanoparticles, developed by the attachment of targeting moieties that can detect β cell mass noninvasively through magnetic resonance (Laurent et al. 2016). Using this method would allow for better monitoring of disease progression and can better prepare physicians for future treatment plans. For example, some medications have been known to exacerbate the death of β cells. Having the knowledge of the extent of death as a patient's disease state progresses could help signal to physicians when to change or avoid certain therapies.

Nanotechnology Targeted Treatment for Obesity

Interventions would target the various disordered molecules secreted or expressed by the cells that make up white adipose tissue (WAT), namely, adipocytes, macrophages, fibroblasts, and vascular cells. This could also potentially lead to further insight into the pathophysiology, prevention, and management of obesity and its comorbidities and provide improvement of the efficacy of existing therapies on and off the market (Sibuyi et al. 2019; Shi et al. 2010; Veiseh et al. 2015; Ash et al. 2019; Devadasu et al. 2013, 2017).

Inorganic nanoparticles could serve as a vector for delivery or as a treatment for obesity due to their physicochemical properties. Examples of inorganic nanoparticles include gold GNPs, iron oxide, mesoporous silica (MSNs), and calcium phosphate (Lee et al. 2017). Cerium oxide nanoparticles have been investigated, as they are linked to inhibiting adipogenesis through the reduction of mRNA transcription of obesogen genes and triglyceride accumulation. This process is expanding on the idea that oxidative stress contributes to the development of obesity, and so through the approach of inhibition, weight loss and a decrease in levels of insulin, leptin, glucose, and triglycerides in plasma have been demonstrated (Rocca et al. 2015).

Lipid Nanoparticles

Therapeutic characteristics are also present in lipid nanoparticles, which are biocompatible, form bilayers for encapsulation, and offer a method of fusion with the cell membrane for delivery (Zu et al. 2018). Lipid nanoparticles also allow for improved solubility due to their amphiphilic components. For instance, there have been investigations in using trans-resveratrol (R), an anti-obesity drug in development that induces browning in WAT. Figure 70.4 shows the theoretical process of how these nanocarriers assist in WAT browning.

Prohibitin (PHB)-targeted therapy (Veiseh et al. 2015) employing nanoparticles is also investigated. It is a 30-kDa protein that plays roles in cell proliferation, endothelial reticulum (ER) stress, transcription, apoptosis, and tumor suppression and is expressed in various compartments of the cell such as the mitochondria, nucleus, ER, the Golgi complex, and plasma membrane. PHB has been found in several reports to be overexpressed in obesity, cancer, and even in diabetes, suggesting a pleiotropic effector role in immunometabolism for these conditions (Cao et al. 2016).

Conclusion

In conclusion, nanotechnology provides healthcare professionals with the opportunity to provide a higher level of patient care by providing a wider range of treatment options that would resolve issues such as compliance. There is also a significant chance to halt progression of comorbidities seen with obesity and diabetes, as

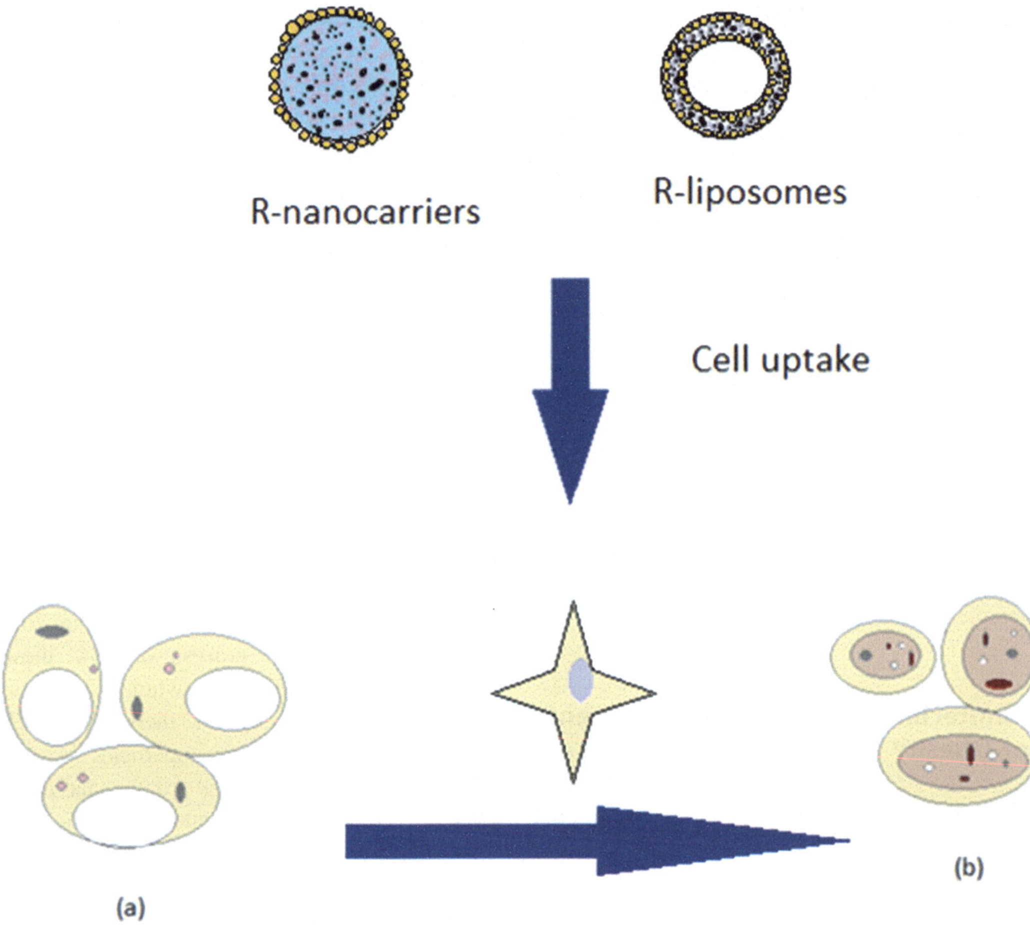

Fig. 70.4 Larger, white adipose tissue with noticeable single lipid droplet (**a**) is converted to smaller, brown adipose tissue with small lipid droplets and more mitochondria (**b**)

many researchers are looking to develop early detection techniques for these double epidemics .

Acknowledgement Morris L. Lichtenstein Jr. Medical Research Foundation supports Mahua Choudhury.

References

Carbone S, Del Buono MG, Ozemek C, Lavie CJ (2019) Obesity, risk of diabetes and role of physical activity, exercise training and cardiorespiratory fitness. Prog Cardiovasc Dis 62(4):327–333. https://doi.org/10.1016/j.pcad.2019.08.004

Sibuyi NRS et al (2019) Nanotechnology advances towards development of targeted-treatment for obesity. J Nanobiotechnol 17(1). https://doi.org/10.1186/s12951-019-0554-3. PMID: 31842876

Smith CJ, Perfetti TA, Hayes AW, Berry SC (2020 Mar 24) Obesity as a source of endogenous compounds associated with chronic disease: a review. Toxicol Sci. pii: kfaa042. https://doi.org/10.1093/toxsci/kfaa042

Dehdari L, Dehdari T (2019) The determinants of anti-diabetic medication adherence based on the experiences of patients with type 2 diabetes. Arch Publ Health 77(1). https://doi.org/10.1186/s13690-019-0347-z. PMID: 31123588

Vasan SK, Osmond C, Canoy D, Christodoulides C, Neville MJ, Di Gravio C, Fall CHD, Karpe F (2018) Comparison of regional fat measurements by dual-energy X-ray absorptiometry and conventional anthropometry and their association with markers of diabetes and cardiovascular disease risk. Int J Obes 42(4):850–857. https://doi.org/10.1038/ijo.2017.289

Shi J et al (2010) Nanotechnology in drug delivery and tissue engineering: from discovery to applications. Nano Lett 10(9):3223–30. https://doi.org/10.1021/nl102184c. PMID: 20726522

Veiseh O, Tang BC, Whitehead KA, Anderson DG, Langer R (2015) Managing diabetes with nanomedicine: challenges and opportunities. Nat Rev Drug Discov 14:45–57. PMID: 25430866

American Diabetes Association (ADA) (2019) Diabetes care. 42(suppl 1): S1–159. http://care.diabetesjournals.org/content/41/Supplement 1

Ash GI et al (2019) Promises of nanotherapeutics in obesity. Trends Endocrinol Metab 30(6):369–383. https://doi.org/10.1016/j.tem.2019.04.004

Capoccia K, Odegard PS, Letassy N (2016 Feb) Medication: a systematic review of the literature. Diabetes Educ 42(1):34–71. https://doi.org/10.1177/0145721715619038

Devadasu VR, Bhardwa V, Kumar MNVR (2013) Can controversial nanotechnology promise drug delivery? Chem Rev 1686–735. PMID: 23276295

Devadasu VR et al (2017) Current advances in the utilization of nanotechnology for the diagnosis and treatment of diabetes. Int J Diab Dev Countries 38(1):11–19. https://doi.org/10.1007/s13410-017-0558-1

Marta T, Luca S, Serena M, Luisa F, Fabio C (2016) What is the role of nanotechnology in diagnosis and treatment of metastatic breast cancer? Promising Scenarios for the Near Future. J Nanomater Hindawi Publishing Corporation 2016

Cui, M., Wu, W., Hovgaard L, Lu Y, Chen D, Qi J (2015) Liposomes containing cholesterol analogues of botanical origin as drug delivery systems to enhance the oral absorption of insulin. Int J Pharm 489:277–284. PMID: 25957702

Giovannucci E et al (2010) Consensus report explores connection between diabetes and cancer. Oncol Times 32(14):40–41

Lee JH, Jeong HS, Lee DH, Beac S, Kim T, Lee GH et al (2017) Targeted hyaluronate–hollow gold nanosphere conjugate for anti-obesity photothermal lipolysis. ACS Biomater Sci Eng 3(12):3646–3653

Sheng W, Alhasan AH, DiBernardo G, Almutairi KM, Rubin JP, DiBernardo BE et al (2014) Gold nanoparticle-assisted selective photothermolysis of adipose tissue (NanoLipo). Plast Reconstr Surg Glob Open 2(12):e283. PMID: 25587517

Deza E, Deza M (2006) Dictionary of distances. Elsevier

Allhoff F et al (2010) What is nanotechnology and why does it matter?: from science to ethics. Wiley-Blackwell

Taniguchi N (1974) On the basic concept of nanotechnology. Proceedings of the international conference of production engineering, London, Part II. British Society of Precision Engineering

Drexler KE (2000) Engines of creation: the coming era of nanotechnology. Eric Drexler

Disanto RM et al (2015) Recent advances in nanotechnology for diabetes treatment. Wiley Interdiscip Rev Nanomed Nanobiotechnol 7(4), 548–564., https://doi.org/10.1002/wnan.1329. PMID: 25641955

Liu J, Zhang SM, Chen P, Cheng L, Zhou W, Tang WX et al (2007) Controlled release of insulin from PLGA nanoparticles embedded within PVA hydrogels. J Mater Sci Mater Med 18:2205–2210. PMID: 17668296

Cash KJ, Clark HA (2010) Nanosensors and nanomaterials for monitoring glucose in diabetes. Trends Mol Med 16:584–93c. https://doi.org/10.1016/j.molmed.2010.08.002. PMID: 20869318

Rodbard D (2016) Continuous glucose monitoring: a review of successes, challenges, and opportunities. Diabetes Technol Ther 18:S2–3. PMID: 26784127

Wang H-C, Lee A-R (2015) Recent developments in blood glucose sensors. J Food Drug Anal 23(2):191–200. https://doi.org/10.1016/j.jfda.2014.12.001

Chen C, Zhao XL, Li ZH, Zhu ZG, Qian SH, Flewitt AJ (2017) Current and emerging technology for continuous glucose monitoring. Sensors (Basel) 17(1). pii: E182. https://doi.org/10.3390/s17010182

Scognamiglio V (2013) Nanotechnology in glucose monitoring: advances and challenges in the last 10 years. Biosens Bioelectron 47:12–25. PMID: 23542065

Rahiman S (2012) Nanomedicine current trends in diabetes management. J Nanomed Nanotechnol 3:3–8

Meetoo D, Wong L, Ochieng B (2019) Smart Tattoo: technology for monitoring blood glucose in the future. Br J Nurs 28(2):110–115

Laurent D, Vinet L, Lamprianou S, Daval M, Filhoulaud G, Ktorza A et al (2016) Pancreatic β-cell imaging in humans: fiction or option? Blackwell 18:6–15. https://doi.org/10.1111/dom.12544. PMID: 26228188

Rocca A et al (2015) Pilot in vivo investigation of cerium oxide nanoparticles as a novel anti-obesity pharmaceutical formulation. Nanomedicine 11(7):1725–1734. https://doi.org/10.1016/j.nano.2015.05.001

Zu Y et al (2018) Resveratrol liposomes and lipid nanocarriers: comparison of characteristics and inducing browning of white adipocytes. Colloids Surf B: Biointerfaces 164:414–423. https://doi.org/10.1016/j.colsurfb.2017.12.044. PMID: 29433059

Cao Y, Liang H, Zhang F et al (2016) Prohibitin overexpression predicts poor prognosis and promotes cell proliferation and invasion through ERK pathway activation in gallbladder cancer. J Exp Clin Cancer Res 35(35):68. PMID: 27084680

Part IX

Obesity Devices and Supplemental Material

Old and New Anti-obesity Devices for Medical, Surgical, and Endoscopical Use

71

Joel Faintuch and Salomao Faintuch

Contents

Introduction 960
Dental Wiring and Oral Intake Restriction 960
SMARTByte Palate Prosthesis Device 960
Intragastric Balloon 961
Gastric Aspiration 961
Nonsurgical Bypass of the Upper Gut 963
Plenity Capsule 963
Final Considerations 964
References 964

Abstract

The advent of surgical interventions in the 1950s to mechanically (and nonmechanically) prevent and treat obesity was received by certain health care professionals with shock, looking risky and disproportionate for an imbalance between calorie intake and expenditure. Newer anti-obesity devices run the gamut from merely curious to relative aggressive, sometimes reenacting a similar degree of surprise and disbelief. Nevertheless, they are not mere curiosities or experimental musings, but actual therapies are available for selected candidates. This chapter reviews a current sample of oral, intragastric, intrajejunal, and swallowable products, with the corresponding usage and contraindications.

Keywords

Dental wiring · Palate prosthesis · Intragastric devices · Temporary duodenojejunal bypass · Space-occupying swallowable capsule

J. Faintuch (✉)
Hospital das Clinicas, Sao Paulo University Medical School, Sao Paulo, Brazil
e-mail: j.faintuch@hc.fm.usp.br

S. Faintuch
Harvard Medical School, Beth Israel Deaconess Medical Center, Boston, MA, USA

J. Faintuch and S. Faintuch (eds.), *Obesity and Diabetes*, https://doi.org/10.1007/978-3-030-53370-0_71

Introduction

Anti-obesity devices seem to be as novel as the obesity epidemic itself, with a history of commercialized products spanning a couple of decades at most. Indeed, those in current use are mostly recent, given the shortcomings of historical models. Nevertheless, mechanical attempts to reverse overweight and obesity, or more precisely the ideas leading to the mental framework about how to accomplish it, have a long past.

Dental Wiring and Oral Intake Restriction

Few treatments seem as outlandish, or even cruel, as mechanical closing the mouth, in order to diminish food intake. No appliance ever became popular. In turn, the procedure was never entirely abominated. As officially approved devices or techniques are lacking, application is customized on a case-by-case basis. Yet, since the 1970s, reports keep appearing in the literature, even if in small numbers (Rodgers et al. 1977; Al-Dhubhani and Al-Tarawneh 2015).

Over 2400 years ago, Hippocrates already recommended jaw immobilization with gold wire (Ebooks 2019), although the indication was trauma, not adiposity. Only with the advent of modern orthodontics in the 1970s, based on brackets glued to the dental surface, easily and painlessly manipulated with elastic bands and other materials, were dentists attracted to the new possibility.

Candidates are morbidly obese subjects refractory to medical therapy, with concomitant surgical contraindications, or who abhor invasive operative or endoscopic interventions, yet are still willing to undergo much annoyance in order to improve body composition. Inconvenient the treatment unquestionably is, as it interferes not only with mastication but also with speech, jaw mobility, dental cleaning, and social life. Moreover eventual vomiting becomes risky and could lead to potentially fatal lung aspiration, even though no such cases have been registered so far (Al-Dhubhani and Al-Tarawneh 2015).

Regular visits to the dentist are mandatory, not only to prevent tooth and gum complications but also to unwire the brackets every 30 days and exercise the temporomandibular joints for a couple of days. And yes, the treatment cannot be indefinitely extended, with typical duration being around 6 months. Still, weight loss has been described as comparable to standard bariatric procedures, something potentially able to jumpstart more durable dieting and lifestyle changes (Rodgers et al. 1977; Al-Dhubhani and Al-Tarawneh 2015).

SMARTByte Palate Prosthesis Device

SMARTByte system (Sensor Monitored Alimentary Restriction Therapy) does not lock the teeth or close the mouth; however, its flattened palate cushion diminishes the volume of the oral cavity, so that smaller bites occur. The trick is accomplished by means of an intraoral tool, custom made to precisely fit the palate anatomy of the patient, so that it snuggly stays in place. It does not move nor is it swallowed, something that could otherwise precipitate choking. It is introduced and positioned by the patient himself or herself during mealtime, three times a day, being removed afterward. An embedded sensor documents compliance.

A pilot study with 76 adults, suffering from overweight or Class I obesity, was conducted during 16 weeks. Video lifestyle instruction was simultaneously provided. The protocol required at least 7 times/week use of the SmartByte, which would represent an average of one meal/day or more. At the end of the study, those who followed the protocol achieved 2.9% weight loss, contrasting with 1.5% reduction in noncompliers. One palate abrasion and two tongue lacerations were registered; however, they spontaneously resolved. The authors highlight that more frequent use than the minimum recommendation could be associated with 4% and even 5% weight reduction within a relatively short period, involving little risk (Ryan et al. 2017).

Intragastric Balloon

A well-established option for kick-starting weight loss, combined with lifestyle reeducation, or as preparation for future bariatric procedures in the high-risk super-super-obese, and its rationale are not as recent as usually believed. In the 1930s, gastrointestinal surgery was still infrequent and dangerous, as anesthesia, fluid and blood replacement as well as antimicrobials was rather primitive. Yet, emergencies such as gastrointestinal obstructions were already dealt with. Some surgeons were particularly impressed by gastric bezoars, rare and unexpected concretions formed by large amounts of swallowed hair or other nondigestible materials, often in compulsive patients with behavior disorders.

Such partially floating balls could become big enough, to intermittently impair food passage through the antrum and pylorus. Even if gastric obstruction was incomplete, vomiting, appetite reduction, and weight loss were prominent, as emphasized in surgical reports (DeBakey and Ochsner 1938). It was no coincidence that in the early 1980s, an "artificial bezoar" was nominally proposed for the handling of obesity (Nieben and Harboe 1982).

Of course, more than three decades later, materials, models, indications, and endoscopic introduction techniques are conspicuously more advanced. Currently available products are usually represented by a silicone 500 ml bag, which is endoscopically filled with saline after positioning, by means of a valve. A swallowable type also exists (Mion et al. 2013).

The swallowable device is smaller (250 ml), and covered with gelatin, for easier ingestion. It includes a thin catheter connecting the gelatin capsule to the exterior, for subsequent injection of gas. After gastric positioning is fluoroscopically confirmed, the balloon is insufflated with a special nitrogen gas mixture and the catheter is detached and removed, leaving the free-floating device in the stomach. Additional balloons (up to a total of three, amounting to 750 ml) can be swallowed in subsequent weeks, if weight loss is insufficient.

Another modality is the transpyloric shuttle device (Baronova 2019). Also much smaller than conventional gastric balloons, its main difference is a transpyloric extension till the first part of the duodenum. The purpose of this "tail" is to keep the balloon anchored in the prepyloric region, where it works as an intermittent embolus, blocking chyme passage into the intestine.

A controlled trial with 302 participants indicated substantially higher weight loss than in the sham control group, with good tolerance of the device. Of those who carried the shuttle, 66% achieved loss of 5% body weight or more (total reduction 9.3%), contrasting with just 30% exhibiting over 5% reduction among controls (total weight shift 2.8%). The product has an FDA initial approval.

Gastric balloons are temporary aids, as many other devices. After 6 or 12 months, depending on the model, they are removed. Their purpose is not to be a standalone therapy. Lifestyle interventions are mandatory, so that after loss of weight is jumpstarted, the patient will presumably comply with recommendations, adhering to a healthy diet and exercise.

Gastric Aspiration

Gastrointestinal fluid losses, as nowadays induced by gastric aspiration devices for weight-loss purposes, have a remote indirect early history. Sushruta, the author of a treatise on general and plastic surgery, lived in India around 600 BCE or 150 years before Hippocrates. He already reported some features of external gastrointestinal fistulas, indicating how important it was to repair the wound, with the help of suture materials such as tendon, hair, and silk (Saraf and Parihar 2006).

Of course, unplanned, spontaneous fistulas carry heavy morbidity and mortality, on account of unbridled dehydration, electrolyte imbalance, sepsis, tissue destruction, and malnutrition. Specifically, gastric fistulas are burdened by hypokalemic metabolic alkalosis, a condition that can lead to reduced muscle strength and nonperiodic paralysis, not unlike severe bulimia (Sung et al.

2015). The modern anti-obesity device bypasses those pitfalls by means of clean, aseptic, and titrated drainage of gastric juice, through an endoscopically introduced tube, connected to the abdominal wall.

Tube positioning is reminiscent of percutaneous endoscopic gastrostomy (PEG) placement. Indeed, a PEG tube could be used as well, although the purpose-built device is more efficient. Comparison of tube opening with bulimic episodes, during which certain patients forcefully eject their recent dietary intake by vomiting, is inevitable. However, these maneuvers are not associated with shame and guilt, analogous to other disordered eating behaviors, as they are medically directed and supervised.

The tube is conventionally opened three times a day, 20–30 min after meals, allowing food and fluid loss for about 10–15 min. The target is to remove 30% of ingested calories, thus inducing up to 20% BMI reduction in 12 months (Figs. 71.1, 71.2).

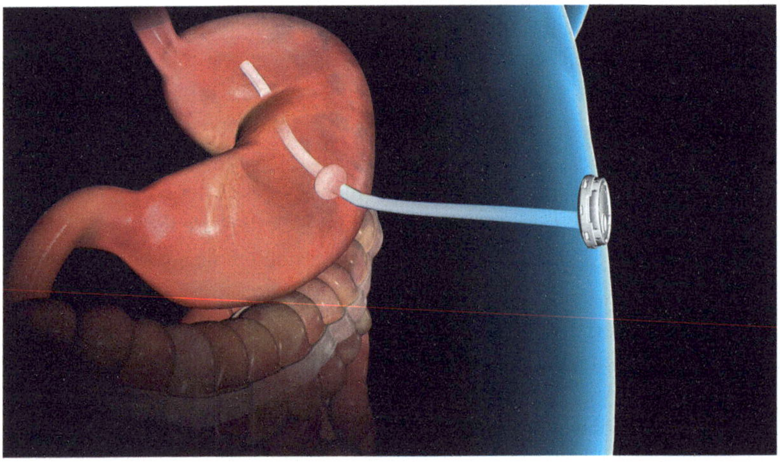

Fig. 71.1 Anatomic outline of the gastric drainage device. The tube can be accessed through an abdominal button, not unlike a percutaneous endoscopic gastrostomy (PEG). Reproduced by permission from Aspire Bariatrics.com

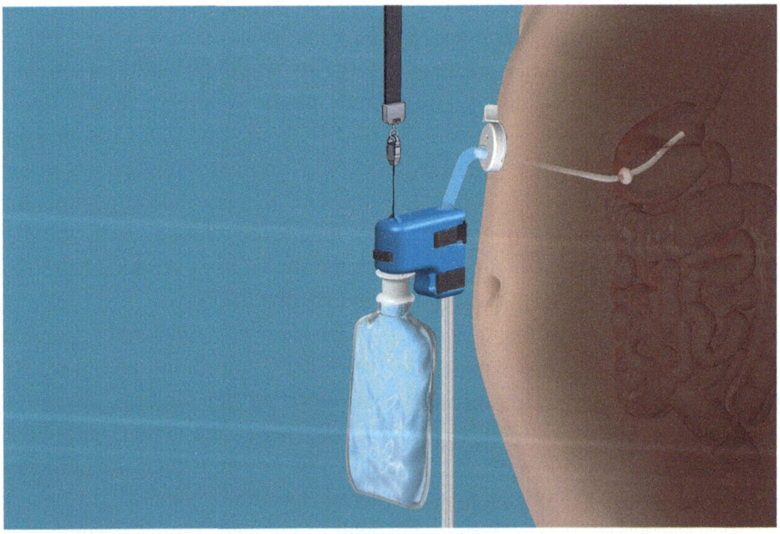

Fig. 71.2 The drainage reservoir is periodically connected to the intragastric tube, for postcibal partial evacuation of ingested foodstuffs. Reproduced by permission from Aspire Bariatrics.com

A 4-year trial has indicated safe and sustained weight loss for class II and III obesity patients (above 35.0 and 40.0 kg/m^2, respectively). There was no mortality, serious metabolic and electrolyte disorders did not occur, and tube removals as well as other complications were similarly uncommon (Thompson et al., 2019).

Nonsurgical Bypass of the Upper Gut

Operative jejunoileal bypass for obesity management was introduced by Payne and DeWind, in 1969. The roots of the malabsorptive procedure are more ancient, partly stemming from inadvertent gastroileal anastomosis in the management of peptic ulcers. As originally reported by Martin and Carroll in 1915 (Anonymous 1951), in the trying times of abdominal surgery in the early twentieth century, occasional surgeons catastrophically reconstructed the gastrointestinal continuity after Billroth II gastrectomy, using the terminal ileum instead of the first loop of jejunum. As both extremities of the small bowel are fixed, one to the colon, and the other by the ligament of Treitz, in a haste, an inexperienced surgeon can pull one instead of the other.

Such accident was followed by severe fecal loss of undigested food directly from the stomach, resulting in extreme undernutrition, requiring surgical correction. The Payne procedure was less radical, leaving 12 inches of functional ileum for a more reasonable degree of absorption. Still, the operation was virtually abandoned in the 1980s, in favor of similarly effective and better tolerated stomach-reducing alternatives.

An even more attenuated and safer alternative has now been devised, namely, the endoscopic temporary liner. The duodenojejunal bypass liner, as it is usually designated, is an impermeable, light and flexible plastic sleeve, endoscopically introduced via the stomach, so as to isolate food content from the surrounding mucosa. Digestion and absorption are consequently halted, until resumption some time later. Anchored at the pylorus by tiny hooks, and with a total length of 60 cm, it excludes the entire duodenum (roughly 30 cm), plus a couple of loops of upper jejunum. This provides a noninvasive gastrojejunal bypass, less aggressive than the alluded to inadvertent gastroileostomy. Standard implant duration is 12 months. Multicenter trials have pointed out not uncommon adverse effects, including hepatic abscess, eventually leading to early removal. Yet, results are encouraging, with significant weight loss and reduction of insulin consumption (Betzel et al. 2019). Weight loss is partially maintained 2 years after barrier explanation, however, not after 4 years (Betzel et al. 2019; van Rijn et al. 2019).

Plenity Capsule

A magic pill, which instantly fills the stomach with gases or liquids at mealtime and inhibits food ingestion, was probably one of the wildest dreams of early anti-obesity pioneers. Indeed, when one drinks highly carbonated sodas, and these have been in the market for a long time, the abdomen sometimes fills up to the point of discomfort. It is attributed to Joseph Priestley (1733–1804), the discoverer of oxygen, also the infusion of carbon dioxide from a beer vat into a bowl of water, thus generating a bubbly water drink. However, even in modern, highly carbonated soda, the impact on consumers is typically short lived, as it all vanishes with eructation and elimination of the fizz.

Plenity (Gelesis 2019) survives longer than an instant and represents more than hot air. This hydrophilic matrix of cross-linked cellulose and citric acid absorbs water in the stomach, converting into multiple tiny fragments of space-occupying gel, which generates satiety and inhibits food ingestion. After intestinal migration, it is partially digested and the remainder is eliminated, with virtually no systemic impact on the host, except for the intended one of appetite restriction and weight loss. In a randomized 24-week trial (GLOW protocol), 59% of the patients achieved weight loss of $\geq$5%, with tolerance comparable to placebo (Greenway et al. 2019). Prediabetics and drug-naive type 2 diabetics were even more benefited, with a six times higher chance of responding with $\geq$10% weight loss.

Within such circumstances, Plenity is considered a device, not a drug, being theoretically exempt from the safety constrains of a pharmacologic agent. The FDA (USA) has approved it for overweight and obese adults with a BMI of 25–40 kg/m^2, in conjunction with diet and exercise. This is a first for the overweight population, as most devices are recommended for the higher strata of adiposity.

Final Considerations

Ingenuity has always been considered a virtue in medicine and even more so in surgical specialties. Especially at times of epidemics, wars, and other calamities, inventiveness has been a major route to make do in face of shortages and lack of resources. Nevertheless, it comes handy on a daily basis to every professional, even with the risk of being considered insane on account of his or her bold steps. Endoscopy provides stark examples. An important advance toward modern techniques occurred in 1868, when Adolph Kussmaul (1822–1902) used the abilities of a sword swallower to test and improve the apparatus he had developed, for in vivo viewing of the esophagus and stomach.

His highly successful career was enhanced, different from the earlier pioneer Philipp Bozzini (1773–1809), who created the *Lichtleiter* (light conductor), a smaller device that could be used for not so deep, however, still useful and effective endoscopic exploration, of such diverse sites as the upper and lower digestive system, nose, and ears, as well as urinary and genital tracts. He dedicated his life to the improvement of the instrument, aiming at its recognition and adoption by Austrian military hospitals. Due to intrigues, this never occurred. He died in poverty, and his three children were given over to friends.

Obesity and diabetes are not the only challenges in a world dilacerated by medical and social conflicts and inequalities. Yet, a dose of therapeutic ingenuity would be welcome, given the estimate that as little as 2% of obese patients are currently being managed by officially mandated lifestyle, pharmacologic or surgical interventions, and not more than 0.1% by available endoscopic alternatives (Kahan 2018).

References

Al-Dhubhani MK, Al-Tarawneh AM (2015) The role of dentistry in treatment of obesity – review. Saudi J Dental Res 6(2):152–156

Anonymous (1951) Surgical error of gastroileostomy. US Navy Medical News Lett 18(4):8–10

Baronova. https://baronova.com. Accessed on November 3, 2019

Betzel B, Cooiman MI, Aarts EO, Janssen IMC, Wahab PJ, Groenen MJM, Drenth JPH, Berends FJ. Clinical follow-up on weight loss, glycemic control, and safety aspects of 24 months of duodenal-jejunal bypass liner implantation. Surg Endosc 2019 Mar 14

DeBakey M, Ochsner A (1938) Bezoars and concretions. A comprehensive review of the literature with an analysis of 303 collected cases and a presentation of 8 additional cases. Surgery 4:934–963

Ebooks. http://ebooks.adelaide.edu.au/h/hippocrates/articulation/complete.html. Accessed on November 3, 2019

Gelesis. www.gelesis.com. Accessed on November 3, 2019

Greenway FL, Aronne LJ, Raben A, Astrup A, Apovian CM, Hill JO, Kaplan LM, Fujioka K, Matejkova E, Svacina S, Luzi L, Gnessi L, Navas-Carretero S, Alfredo Martinez J, Still CD, Sannino A, Saponaro C, Demitri C, Urban LE, Leider H, Chiquette E, Ron ES, Zohar Y, Heshmati HM (2019) A randomized, double-blind, placebo-controlled study of gelesis100: a novel nonsystemic oral hydrogel for weight loss. Obesity (Silver Spring) 27(2):205–216

Kahan S (2018) Practical strategies for engaging individuals with obesity in primary care. Mayo Clin Proc 93:351–359

Mion F, Ibrahim M, Marjoux S, Ponchon T, Dugardeyn S, Roman S, Deviere J (2013) Swallowable Obalon® gastric balloons as an aid for weight loss: a pilot feasibility study. Obes Surg 23(5):730–733

Nieben OG, Harboe H (1982) Intragastric balloon as an artificial bezoar for treatment of obesity. Lancet 319 (8265):198–199

Rodgers S, Burnet R, Goss A, Phillips P, Goldney R, Kimber C, Thomas D, Harding P, Wise P (1977) Jaw wiring in treatment of obesity. Lancet 1 (8024):1221–1222

Ryan DH, Parkin CG, Longley W, Dixon J, Apovian C, Bode B (2017) Efficacy and safety of an oral device to reduce food intake and promote weight loss. Obes Sci Pract 4(1):52–61

Saraf S, Parihar R (2006) Sushruta: The first Plastic Surgeon in 600 B.C. Internet J Plastic Surg 4(2):8232

Sung CC, Cheng CJ, Chiang WF, Chau T, Hsu YJ, Yang SS, Lin SH (2015) Etiologic and therapeutic analysis in patients with hypokalemic nonperiodic paralysis. Am J Med 128(3):289–296

Thompson CC, Abu Dayyeh BK, Kushnir V, Kushner RF, Jirapinyo P, Schorr AB, Aronne LJ, Amaro A, Jaffe DL, Schulman AR, Early D, Stein AC, Sharaiha R, Edmundowicz SA, Bohning JM, Jensen MD, Shukla AP, Apovian C, Kim DW, Tran D, Zarrinpar A, Ryan MB, Young M, Lowe A, Haas M, Goldsmith H, McCrea J, Sullivan S (2019 Aug) Aspiration therapy for the treatment of obesity: 4-year results of a multicenter randomized controlled trial. Surg Obes Relat Dis 15(8):1348–1354

van Rijn S, Roebroek YGM, de Jonge C, Greve JWM, Bouvy ND (2019) Effect of the EndoBarrier device: a 4-year follow-up of a multicenter randomized clinical trial. Obes Surg 29(4):1117–1121

Electroceutical Approaches for Gastroparesis

72

Niranchan Paskaranandavadivel, Recep Avci,
Nipuni Nagahawatte, Aydin Farajidavar, and Leo K. Cheng

Contents

Introduction 968

Electrophysiology 968
ICC Regulation of Slow Waves 968
Measurement of Slow Waves In Vivo 969

Gastroparesis 970
Diagnostic Scintigraphy 970
Pathophysiology 972
Differential Diagnosis 972

Low-frequency Pacing and High-Frequency Stimulation 973
Pacing 973
Gastric Pacing for Obesity Treatment 975
High Frequency Stimulation 976
Human Trials 976
Clinical Protocols 977

Future Outlook 978

References 978

N. Paskaranandavadivel (✉) · R. Avci · N. Nagahawatte
Auckland Bioengineering Institute, University of Auckland, Auckland, New Zealand
e-mail: nira.pask@auckland.ac.nz

A. Farajidavar
Department of Electrical and Computer Engineering, New York Institute of Technology, New York, USA

L. K. Cheng
Auckland Bioengineering Institute, University of Auckland, Auckland, New Zealand

Department of Surgery, Vanderbilt University, Nashville, TN, USA

J. Faintuch and S. Faintuch (eds.), *Obesity and Diabetes*, https://doi.org/10.1007/978-3-030-53370-0_72

Abstract

Gastroparesis is a symptomatic chronic clinical syndrome where there is delayed emptying of the stomach in the absence of mechanical obstruction. Typical symptoms of gastroparesis include early satiety, nausea and vomiting and it severely affects nutritional intake. For patients with diabetes, glycaemic control will be negatively impacted, with possible secondary effects on other organs that affect mortality rate. Conventional drug treatment and nutritional support fail some patients and new therapy options are required. High-frequency electrical stimulation is a treatment strategy that is employed in severe cases of gastroparesis. In this chapter, an overview of the electrophysiological slow waves is given, followed by emerging trends in novel electrical stimulation therapies in the stomach, with an emphasis on gastroparesis.

Keywords

Diabetic gastroparesis · Obesity control · Gastric pacing · Gastric slow waves · Low frequency stimulation · High frequency stimulation

Introduction

Electroceuticals are an all-encompassing term coined for bio-electrical medicine to influence and modify biological and physiological function. The use of exogenous electrical impulses is a type of electroceutical and has been extensively used in clinical electrocardiology, neurology and pain management to offer significant therapeutic benefit to patients with disordered electrical rhythms, such as atrial fibrillation and epilepsy (Cingolani et al. 2018; Frequin et al. 2020; Merin et al. 2009; Moreno-Duarte et al. 2014). Cardiac pacing has been used to treat arrhythmias in the heart, where the rhythm of the heart is fast, slow or irregular (Abrich et al. 2020; Dallaglio et al. 2020). Also, cardiac defibrillators are common in public places, which use high-energy impulses to restore the natural rhythm of the heart. More recently, electroceuticals have been used to alleviate or minimise chronic ailments (such as failed back surgery syndrome (Reverberi et al. 2013)), where previously pharmaceuticals were the primary treatment strategy. Furthermore, technological advances in real-time monitoring and miniaturisation of electrodes for brain surgery have advanced deep brain stimulation to be a viable treatment option for Parkinson's disease, epilepsy and tremors (Frequin et al. 2020). Studies have shown that implantable electroceutical devices both extend people lives and improve their quality of life (Marini et al. 2019; Merin et al. 2009).

The concept of applying electroceuticals to the stomach has been investigated by many researchers (Familoni et al. 1997b; Kelly and La Force 1972; O'Grady et al. 2010b). However, to date, its clinical use has been limited to a number of specialised centres (Abell et al. 2003, 2011; McCallum et al. 2011). For other regions of the gut, there is increasing use, such as stimulation of sacral nerves to control urinary and faecal incontinence (Rosen et al. 2001; Yamanishi et al. 2008), and stimulation of the lower oesophageal sphincter to increase resting lower oesophageal sphincter pressure to treat reflux disease (Soffer and Kim 2016). In this chapter, we will focus on the use of electrical stimulation targeted for functional gastric motility disorders such as gastroparesis, where there is an increasing level of research and clinical evidence to suggest that electroceuticals could provide therapeutic benefit. We will first introduce the gastric slow wave electrophysiological mechanisms and two types of gastric electroceutical approaches that have been used to alleviate symptoms and/or treat a gastric motility and emptying disorder known as gastroparesis.

Electrophysiology

ICC Regulation of Slow Waves

Gastric contractions are coordinated in part by bio-electrical slow waves (Huizinga and

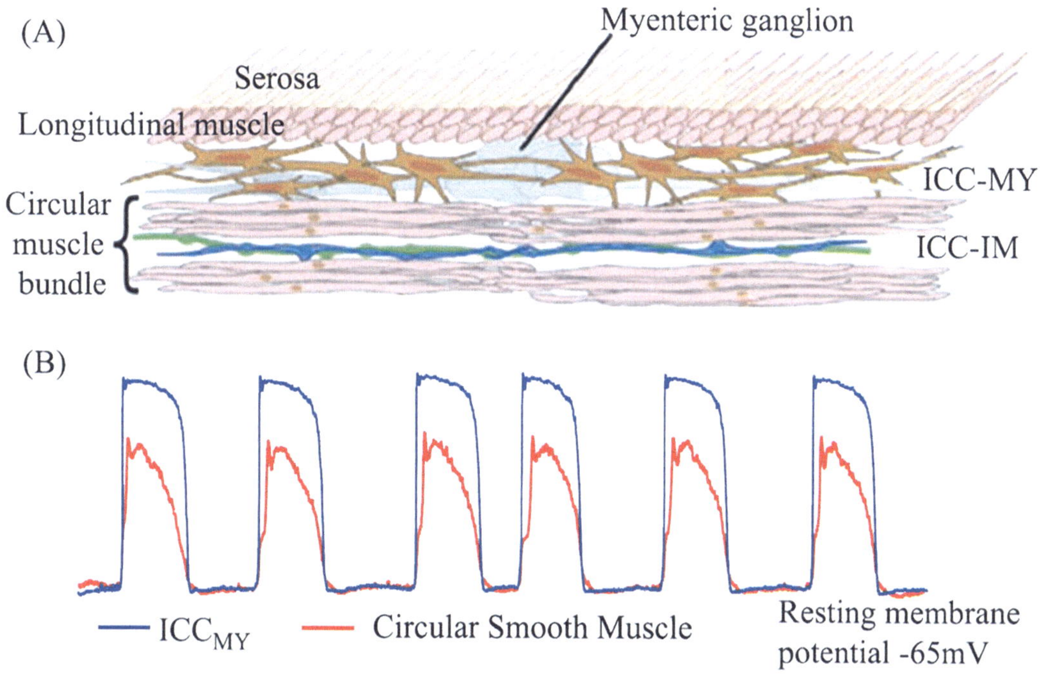

Fig. 72.1 Slow wave initiation in ICC and conduction pathway to smooth muscles. (**a**) Schematic of ICC and gastrointestinal smooth muscle architecture. ICC generates and propagates slow wave activity, which depolarises to the smooth muscles and initiates muscular contraction through Ca^{2+} channels in smooth muscles. (**b**) Shows a representative intracellular trace from ICC-MY and circular smooth muscle illustrating the action potential dynamics. Adapted from Hirst (2001) and Hirst and Edwards (2006)

Lammers 2009). Slow waves are generated and propagated through a network of interstitial cells of Cajal (ICC) present in the gastric musculature (Huizinga et al. 1995). A schematic architecture of the ICC networks and smooth muscles is shown in Fig. 72.1a. The ICC in the myenteric plexus (ICC-MY) is responsible for generating the slow wave action potential, which then passively conducts to the smooth muscle to induce slow wave tissue depolarisation (Farrugia 1999; Hirst and Edwards 2006). Intra-muscular ICC (ICC-IM), another class of ICC, lies within the circular muscle and is responsible for augmenting the action potential generated by ICC-MY to cause the L-type calcium channels in smooth muscles to open for induction of muscular contraction (Hirst 2001; Hirst and Edwards 2006). Figure 72.1b shows the time course of representative slow wave action potentials generated in the ICC-MY, which are then transduced to the neighbouring circular smooth muscles. However, as there is a decreasing intrinsic frequency across the stomach, the ICC with the highest frequency entrains the whole network (Hinder and Kelly 1977; Hirst and Edwards 2006).

Measurement of Slow Waves In Vivo

In vitro studies are attractive to investigate the cellular and ionic mechanisms and dynamics of slow wave action potential. However, findings from studies using excised tissue blocks are inherently different from in vivo whole organ studies due to the loss of communication in the enteric neural system and loss of intrinsic entrainment mechanism from neighbouring ICC, which are critical for normal function (Diamant and Bortoff 1969; Hinder and Kelly 1977; O'Grady et al. 2012b; Xue et al. 1995). Therefore, to get a physiologically realistic understanding of bio-electrical slow wave activity during normal and abnormal function of the gut, organ level in vivo studies are required.

Traditionally, sparse recording sites were utilised in vivo to understand electrophysiological gastric slow wave activity, but they lacked spatio-temporal detail (O'Grady et al. 2018). Lammers et al. translated high-resolution (HR) mapping from the cardiac field to the gastrointestinal field (Lammers et al. 1993, 1996) and similar techniques were later adapted for use in human studies (Berry et al. 2016; O'Grady et al. 2010a). The application of HR mapping techniques in diseased cases, such as gastroparesis and chronic unexplained nausea and vomiting, has offered new insight into the dynamics of dysrhythmic slow wave propagation (See 'Gastroparesis' section for more details) (Angeli et al. 2015; O'Grady et al. 2012a).

With sparse recordings, typically 4 to 8 electrode sites are placed 1–3 cm apart either longitudinally or circumferentially (Alvarez and Mahoney 1922; Kelly et al. 1969), while with HR mapping, electrodes are densely spaced in a 2D configuration with inter-electrode distances of 1 to 7 mm. HR mapping allows for detailed tracking of spatial dynamics and these studies have provided a detailed understanding of the critical role that spatio-temporal slow wave dynamics play in health and disease (Lammers et al. 1993; O'Grady et al. 2018).

In the normal human stomach, the normal slow wave frequency is around 3 cycles per minute (cpm) and there are 3 or more concurrent wavefronts present at any one time (O'Grady et al. 2010a). The slow waves initiate in the mid- to upper-corpus region of the stomach on the greater curvature and propagates rapidly towards the lesser curvature and distally towards the antrum, in a ring formation (Kelly and Code 1971; Lammers et al. 2009; O'Grady et al. 2010a). Figure 72.2a shows an experimental recording of a slow wave propagation profile from a human stomach. The slow waves originate in the mid-upper corpus region of the stomach, where the signals are of a higher amplitude with faster propagation primarily in the circumferential direction (~0.6 mV, 8 mm/s). When the slow wave activity traverses the corpus-antrum border, there is an increase in slow wave amplitude and velocity as the propagation terminates at the pylorus (Berry et al. 2016) (Fig. 72.2b). This increase in slow wave activity is hypothesised to correlate with terminal antral contractions that aid in retropulsion of the food and allow recirculation and trituration (Berry et al. 2016).

Gastroparesis

Gastroparesis is defined as delayed gastric emptying in the absence of mechanical obstruction (Camilleri et al. 2013). Gastroparesis has a prevalence of 24.2 per 100,000 inhabitants and an incidence of 6.3 per 100,000 persons per year (Bharucha 2015; Rey et al. 2012). However, not everyone in the population undergoes a gastric emptying test, and regression models estimate the prevalence of gastroparesis at 1.8% of the population (Rey et al. 2012). The aetiology of gastroparesis is varied with around 36% of patients having an unidentified cause or idiopathic gastroparesis (Camilleri et al. 2013). Diabetes is the main cause and accounts for 29% of patients, followed by post-surgical causes, which account for 13% of patients (Camilleri et al. 2013). Other disorders such as Parkinson's, connective tissue disease and enteroviruses have also been identified as causes for gastroparesis (Camilleri 2007).

Diagnostic Scintigraphy

The conventional test for gastroparesis diagnosis is gastric emptying scintigraphy of a solid-phase meal (Camilleri et al. 2013). Diagnosis is confirmed if more than 10% of the meal remains in the stomach after 4 h, along with presentation of typical gastroparesis symptoms (Abell et al. 2008; Guo et al. 2001). The symptoms are multifactorial and include early satiety, post-prandial fullness, vomiting, nausea and epigastric and/or abdominal pain (Rey et al. 2012). It has a deleterious effect on acquiring adequate nutrition and impacts glycaemic control. It can induce secondary effects on other organs that reduce the quality of life and

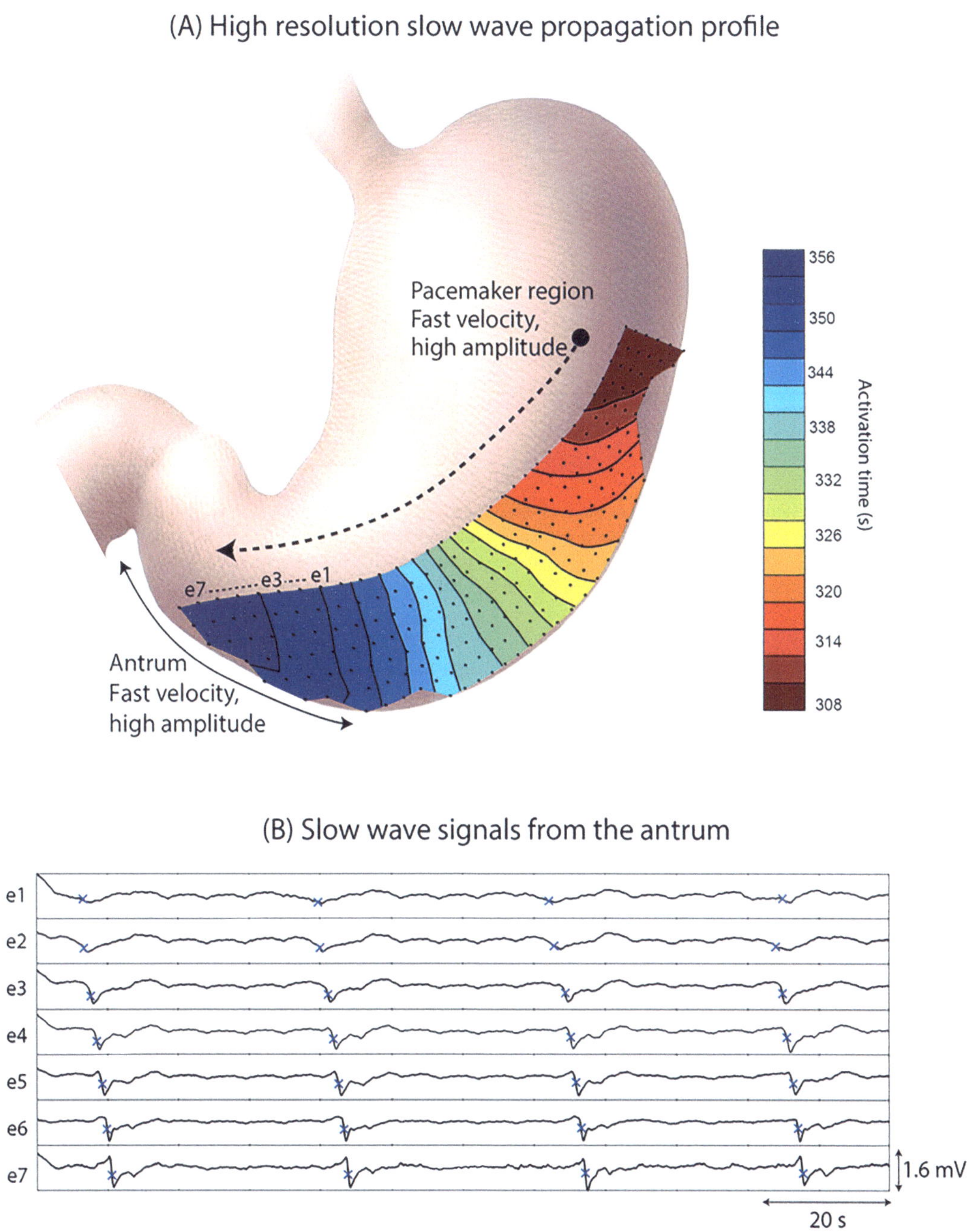

Fig. 72.2 Normal human gastric slow wave activity. (**a**) Shows an experimental recording of the slow wave conduction from the corpus to the antrum. The activation map is illustrated as an isochronal map, where the pacemaker initiates in the greater curvature of the mid-upper corpus and propagates towards the antrum. As the slow waves travel through the corpus and antrum border, the slow wave travels rapidly with increased amplitude. (**b**) Shows the recorded slow wave signals from the antrum, where the amplitude of slow waves is increasing from e1 to e7. The blue cross denotes the activation of the slow wave event

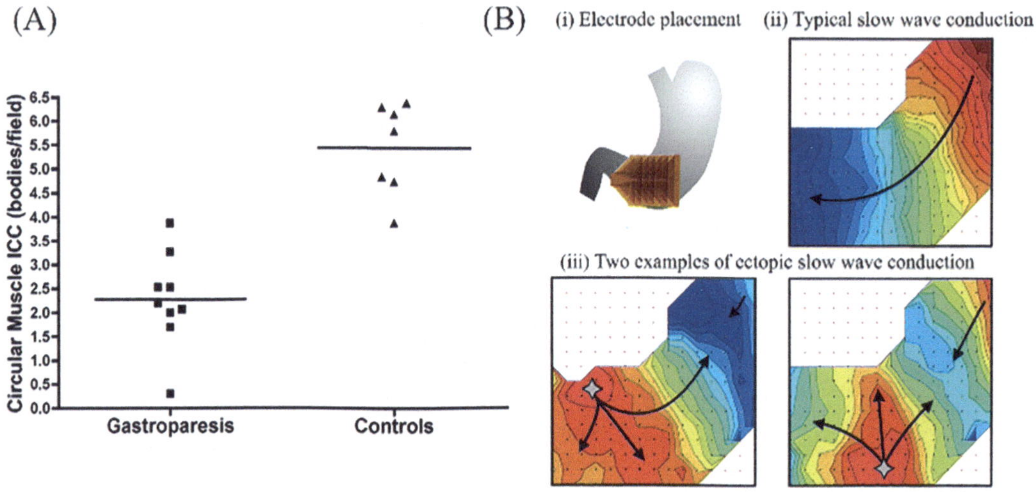

Fig. 72.3 (**a**) ICC count for gastroparetic patients and control subjects. (**b**) HR slow wave mapping in a gastroparetic patient. The placement of the HR electrode array on the distal stomach is shown in (**b**—i). A typical pattern of normal propagation is shown in (**b**—ii), and two examples of ectopic slow wave conduction were also recorded (**b**—iii). Reproduced from O'Grady et al. (2012a)

increase mortality in severe cases (Camilleri et al. 2013). The symptoms of gastroparesis overlap with other disorders such as functional dyspepsia and gastro-oesophageal reflux disease, making treatment stratification difficult for these patient groups.

Pathophysiology

The common denominator for gastroparesis is the loss and degradation of ICC (Farrugia 2008; Grover et al. 2011). Compared to control subjects, a study found significantly reduced levels of ICC in gastroparetic patients (Fig. 72.3a). Furthermore, HR mapping studies performed in patients with gastroparesis have revealed that these patients have abnormal patterns of slow wave propagation (O'Grady et al. 2012a). An example is shown in Fig. 72.3b, where in a gastroparetic patient, the slow wave ectopically activated in the distal corpus, propagating towards the proximal corpus, thus generating competing pacemakers. An important observation, however, was that despite the presence of abnormal patterns of activation, the frequency of slow wave was in the normal range of 3 cpm. Therefore, the use of spatio-temporal mapping and development of efficient analytical frameworks are pertinent to developing a better understanding of functional gastric disorders (Lammers et al. 1993; Paskaranandavadivel et al. 2017).

Differential Diagnosis

Similar patterns of dysrhythmic slow wave activity were also present in patients diagnosed with chronic unexplained nausea and vomiting (CUNV). CUNV patients suffer from the same symptoms as gastroparesis patients, but they have adequate gastric emptying, with less than 10% of food retained after 4 h. In these patients, a moderate ICC loss was observed (i.e. they had less ICC numbers than control subjects, but more than gastroparesis patients) (Angeli et al. 2015).

The ability to accurately quantify the structure of ICC networks has the potential to provide improved pathophysiological understanding (Krohn et al. 2017). Currently, histopathological classification and analysis of gastrointestinal tissue biopsies involve manual visual assessment of small tissues samples. This process is subjective and prone to bias (Grover et al. 2012; Knowles et al. 2011). For example, as shown in Fig. 72.3a, in some control and gastroparesis subjects, 3–4 ICC bodies/field were observed, leading to uncertainties in classification when an intermediate level of ICC numbers was observed. More sophisticated techniques have been proposed where the spatial properties of the ICC networks are quantified by analysing the structure of the ICC network (Gao et al. 2013). Numerical metrics (e.g. ICC density, process thickness, whole size, etc.) were able to effectively differentiate between wild-type and 5-HT2B serotonin receptor knockout jejunal tissue samples that had compromised ICC networks. In addition, the functional properties of the ICC networks can also be quantified using biophysical simulations, to help distinguish between normal and ICC depleted tissue samples (Krohn et al. 2017; Sathar et al. 2015).

Management of gastroparesis begins by restoring hydration and nutrition, followed by prokinetics pain relief (Camilleri et al. 2013). It has been reported that around 30% of patients with gastroparesis will fail conservative pharmaceutical therapy. High-frequency gastric electrical stimulation, marketed as Enterra Therapy by Medtronic (Minneapolis, MN, USA), is another option which could be undertaken if patients have failed conservative treatment (Abell et al. 2002). Enterra Therapy was approved by the Food and Drug Administration (FDA) in 2000 as a humanitarian device exemption (HDE).

Low-frequency Pacing and High-Frequency Stimulation

There are two main classifications of electroceuticals in the stomach: (i) low-frequency (LF) pacing and (ii) high-frequency (HF) stimulation. With LF pacing, the frequency of the delivered impulses is typically within the physiological ranges of slow wave activity (generally on the order of 2–12 cpm), while with HF stimulation, the delivered impulses are significantly higher than the native slow wave frequency range. It is thought that HF stimulation affects the neural afferents in the vagus nerve, while LF pacing modulates the slow wave activity.

The impulse properties that could be applied are generally in three types of waveforms (i) monophasic, (ii) biphasic and (iii) modulated (Fig. 72.4). Monophasic and biphasic waveforms are typically used for pacing, while modulated monophasic impulses are used for HF stimulation. LF pacing can be classified as high-energy stimulation and HF stimulation as low-energy stimulation based primarily on the amplitude and width of the applied impulse.

Pacing

An overview of gastric pacing is given followed by recent advancements in HR mapping and closed loop systems, which are allowing renewed interest in the field.

In the early 1970s, gastric electrical stimulation studies were conducted in animals, where Kelly et al. were able to pace slow waves from 4.2 to 8 cpm and reverse spontaneous dysrhythmias (Kelly and La Force 1972; Kuramoto et al. 1979; Yamashita 1980). Canine trials may have shown increased emptying rates with pacing (Bellahsène et al. 1992), but when pacing was applied in human trials, it entrained slow waves but did not affect gastric emptying rates (Hocking et al. 1992). Early animal studies also showed that the antrum could be paced at higher frequencies than proximal regions of the stomach (Kelly 1974) and is likely due to a consequence of the intrinsic slow wave frequency gradient across the stomach. Gastric pacing is also likely to affect other gastrointestinal control mechanisms, rather than just myogenic. A canine study showed that long pulses reduced the fundic tone and increased the gastric volume with pacing protocols (Xing and Chen 2006).

A variety of pacing parameters have been applied to control gastric slow wave in animals and humans, as shown in Table 72.1. The seminal study by Kelly et al. applied a range of pulse widths and impulse frequencies from 2 to 12 cpm. They suggested that a pulse width of 500 ms at 4 mA was able to entrain slow wave

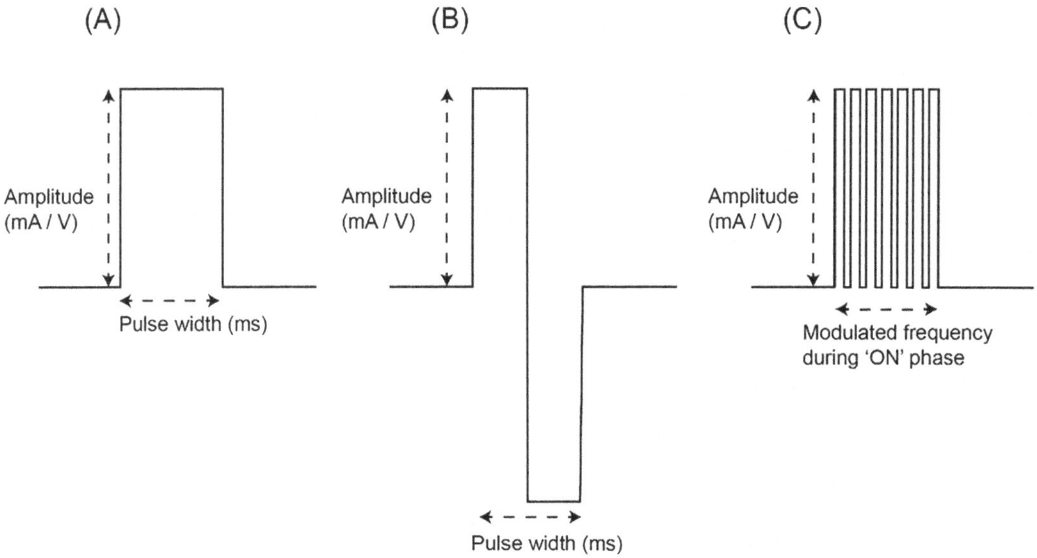

Fig. 72.4 Types of impulses for pacing and stimulation. (**a**) Shows a monophasic impulse, while (**b**) shows a biphasic impulse. Both (**a**) and (**b**) are typical waveforms used for slow wave pacing. (**c**) Shows a modulated impulse during the 'ON' phase, which is typically used for HF stimulation

Table 72.1 Impulse properties applied to control gastric slow waves

Article	Subject	Amplitude	Pulse width	Entrained frequency
Kelly and La Force (1972)	Dog	1–8 mA	100–200 ms	4–8 cpm
Lin et al. (1998) McCallum et al. (1998)	Human	1–4 mA	3–300 ms	4–12 cpm
O'Grady et al. (2010b)	Pig	2–4 mA	400 ms	3.2–3.52 cpm
Wang et al. (2018)	Pig	5–8 mA	500–900 ms	2.72–6 cpm
Alighaleh et al. (2019)	Pig	4 mA	400 ms	2–4.3 cpm

from a range of 4–8 cpm (Kelly and La Force 1972). Furthermore, they reported that a period of pacing interval shorter than 7.5 s (8 cpm) was unstable as it would initiate during the refractory period of slow waves. Consequent studies also varied the parameters but have found that slow waves cannot entrain beyond its physiological slow wave range (Alighaleh et al. 2019; Wang et al. 2018).

A significant long-term human pacing study was performed where pacing and the effects of slow waves were monitored over a 1-month period in gastroparetic subjects (McCallum et al. 1998). Slow wave entrainment was achieved in all subjects, where gastric retention reduced from 77% to 57% and induced symptom relief by 50%. The optimal pacing parameters were also investigated in human trials to apply appropriate impulses to modulate slow waves (Lin et al. 1998). The study showed that optimal pacing can be performed at 3.3 cpm at a current of 4 mA and pulse width of 300 ms, while decreasing the current to 1 mA reduces efficacy of successful entrainment by 50 $\pm$ 11% and decreasing the pulse width to 30 ms results in loss of slow wave entrainment (Lin et al. 1998).

Combining HR slow wave mapping with pacing allows a better understanding of spatio-temporal characteristics of the slow wave responses. The first HR pacing study was performed by Lammers et al. in an in vitro rabbit duodenum, illustrating that the site of slow wave initiation was likely 2 mm away from the electrodes (Lammers et al. 1993). These methods

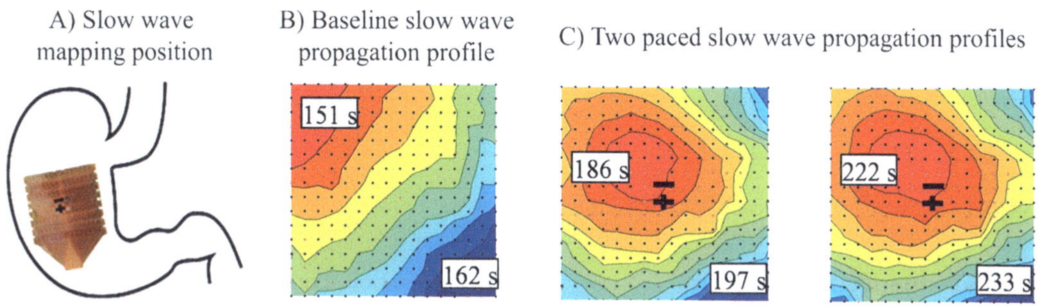

Fig. 72.5 Paced slow wave propagation. (**a**) Shows the area of the serosal region over which slow wave activity was mapped. (**b**) Shows the normal conduction profile, while (**c**) shows two slow wave propagation patterns due to the application of an impulse. The black plus and minus represent the positive and negative leads of the pacing leads

were also applied in an in vivo gastric pig model to reveal that pacing induces a new site of slow wave initiation around 8 mm away from the stimulus site (Alighaleh et al. 2019; O'Grady et al. 2010b). Figure 72.5 shows that the baseline slow wave propagation profile and the induction of a slow wave pacemaker activity near the pacing leads with the application of impulses. Future detailed analyses are required to characterise the regional variation of initiated slow wave activity to improve physiological understanding and clinical trials.

Despite successful trials, pacing has not yet been translated into clinical trials, in part due to high energy requirements, energy consumption in implants and electrode stability over time. Nonetheless, recent advances have shown that devices can be miniaturised and pacing long term is becoming a viable option. In particular, the use of application specific integrated circuits (ASICs) and wireless technologies for data communication and power transfer are being adopted for pacing applications and are improving the form factor and power utilisation (Ibrahim et al. 2015; Javan-Khoshkholgh et al. 2019a, b; Javan-Khoshkholgh and Farajidavar 2019; Paskaranandavadivel et al. 2015; Wang et al. 2018). The dimensions of one of the first reported portable gastric pacemakers were around 13 × 9 × 3 cm (Chen et al. 1995). This form factor has significantly shrunk in size to 3 × 2 × 0.5 cm (Javan-Khoshkholgh et al. 2019a, b). The combination of power efficient electronics and wireless power transfer has resulted in electroceutical devices that do not rely on batteries, circumventing a bottleneck in medical implants.

Gastric Pacing for Obesity Treatment

The concept of closed-loop pacing could further improve power consumption and enable further clinical interest in gastric slow wave pacing. The first reported use of closed-loop stimulation in the gut was in 1973, where stimulation was applied only if the upstroke of recorded bio-signals did not surpass a defined threshold (Sobakin and Shepelev 1973). In recent years, this concept has also been used for gastric stimulation to treat obesity. For obesity treatment, higher energy implantable stimulation devices are required where leads are placed in the fundus to detect food intake, after which stimulation is applied distally (Bohdjalian et al. 2006; Horbach et al. 2015). Meal intake is detected via a transient reduction in slow wave activity and rise in fundus impedance (Bohdjalian et al. 2006). The intrinsic slow wave frequency can also be directly used as an input to control the pacing stimulus. Recent hardware developments have allowed for recording and stimulation on a single module (Wang et al. 2018), while software developments have utilised hybrid automata methods to enable verifiable systems for real-time applications (Wang et al. 2019, 2020).

Table 72.2 High-frequency stimulation settings

Article	Subject	Amplitude	Pulse width	Modulated ON phase	Applied frequency
Bilgutay et al. (1963)	Human	1–10 mA	5 s	50 Hz	0.2–1 cpm
Familoni et al. (1997a)	Human	2 mA	300 μs	Nil	12 cpm
Song et al. (2006)	Dogs	5 mA	0.1, 0.3, 0.6 ms	14, 40 Hz	12 cpm
Angeli et al. (2016)	Human	5–19.2 mA	330, 450 μs	14–130 Hz	3.5–60 cpm

High Frequency Stimulation

The excessive energy requirements for long-term pacing motivated the use of HF stimulation, where lower energy impulses were applied at higher frequencies. The evolution of the HF stimulation is discussed along with emerging practice. Electrical stimulation in the gastrointestinal tract was initially trialled by Bilguaty et al. in 1963 to stimulate the stomach and intestine for the treatment of ileus (Bilgutay et al. 1963) (see Table 72.2 for stimulation setting used). However, following studies reported conflicting results and thus were not translated for further clinical use (Berger et al. 1966; Quast et al. 1965). About three decades later, Familoni et al. utilised HF stimulation, with shorter pulse widths of 0.3 ms applied at 0.6–6 times the intrinsic slow wave frequency in animals (Familoni et al. 1997a). They showed that a stimulation at 30 cpm, as compared to 3 cpm, elicited more mechanical activity, which was measured using a strain gauge on the serosal surface (Familoni et al. 1997a). A human case report was published by the same group where they performed HF stimulation on a gastroparetic patient at 12 cpm (pulse width: 300 μs; amplitude: 2 mA), and illustrated symptom relief and improvements in gastric emptying rates (Familoni et al. 1997b). For this subject, it was observed that the distal corpus and proximal antrum achieved the most effective stimulation response.

Consequent studies focussed closely in this range and utilised pulse width parameters from 0.1–0.6 μs at a modulated frequency of 14 Hz (Chen et al. 2003; Song et al. 2006). The mechanism of action for HF stimulation is not well understood but is likely to involve indirect neuromodulation of the parasympathetic nerves and/or ganglion, which is known to regulate gastric function. Early studies demonstrated that the HF stimulation did not affect gastric slow waves but caused a potent anti-emetic to the effects of vasopressin, which indicated that vagal pathways were being modulated (Chen et al. 2003; Song et al. 2006). HF stimulation has also been shown to modulate the thoracic spinal neurons, which are responsible for gastric distention (Qin et al. 2007). Results from position emission tomography of gastroparetic patients with HF stimulation indicated that there was an increase in the thalamus region of the brain (McCallum et al. 2010). Further studies in rats pinpointed that the increased neuronal activity from HF stimulation was from the paraventricular nucleus in the hypothalamus, which regulates food intake (Tang et al. 2006).

Human Trials

Two multi-centre human trials using HF stimulation (12 cpm, 5 mA and pulse width 0.3 ms, 14 Hz modulated impulse) were performed on refractory-diabetic or idiopathic gastroparetic subjects; GEMS (Gastric Electrical Mechanical Study) (Abell et al. 2002) and WAVESS (Worldwide Anti-Vomiting Electrical Stimulation Study) (Abell et al. 2003). Over 70% of subjects reported improvements in symptoms with reduced nausea and vomiting scores at 12 month follow-up and 73% of subjects experienced an improvement in the quality of life (Abell et al. 2003). Based on the WAVESS study, Enterra Therapy system was approved for use under the HDE by the FDA.

The Enterra Therapy system consists of an implantable pulse generator, a pair of leads and a programming system. A radio-telemetry link from the programming system to the pulse

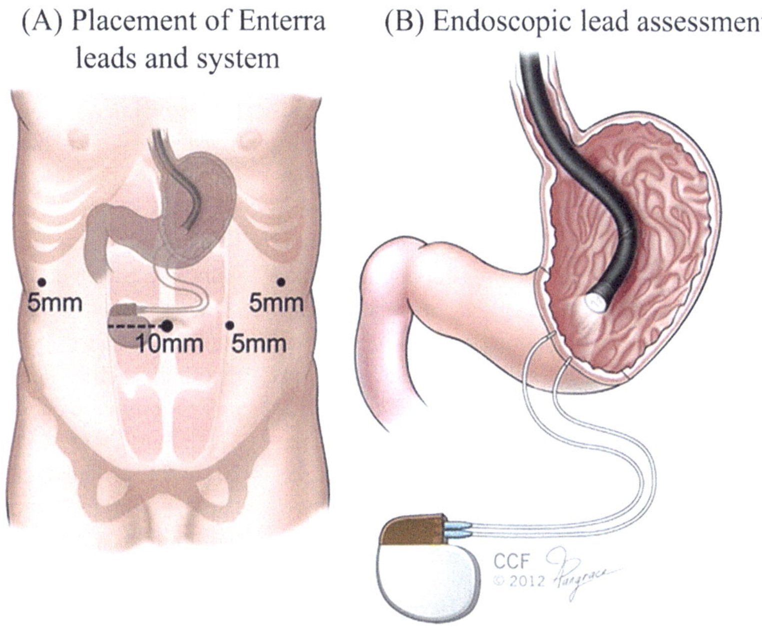

Fig. 72.6 Placement of Enterra system and verification of stimulation leads via the lumen wall. (**a**) Shows the position of where the leads are placed in the stomach guided via four laparoscopic ports. (**b**) Shows the use of endoscopy to assess that the leads are at the right position in the stomach as visualised from the lumen wall. Reproduced with permission from Timratana et al. (2013)

generator controls the applied stimulation protocols. The pair of leads is surgically placed in the serosa-muscular layer of the gastric wall, around 10 cm proximal from the pylorus (Fig. 72.6a), and connected to the implantable pulse generator. The placement of leads is visually assessed endoscopically to ensure that it has not penetrated the lumen wall and is at the right position in the stomach (Fig. 72.6b). A subcutaneous pocket is made for the pulse generator to be implanted in the body. The pulse generator has been estimated to have a battery life of 5 to 10 years, dependent on the stimulation parameters (Medtronic 2020).

There have been a number of studies with varying study design and quality to assess the efficacy of HF stimulation for gastroparesis. One meta-analysis, which included open-labelled studies, reported that the total symptom severity score (TSS) significantly decreased by a mean difference of 6.52 (O'Grady et al. 2009). However, a recent meta-analysis revealed that the TSS score did not differ in randomly allocated design groups (TSS difference of 0.17, $P = 0.15$), but there was a significant decrease in TSS in open-label studies (TSS difference of 2.68, $P < 0.001$) (Levinthal and Bielefeldt 2017). Therefore, there is a significant need for randomised double blind, cross-over, placebo-controlled studies to ascertain efficacy of HF stimulation. According to the American Gastroenterological Association, HF stimulation has moderate level of evidence for use (Camilleri et al. 2013; Levinthal and Bielefeldt 2017). A recent randomised cross-over trial reported that although HF stimulation did not increase the quality of life or rate of gastric emptying, it reduced the frequency of refractory vomiting in both patients with and without diabetes (Ducrotte et al. 2020).

Clinical Protocols

Typically for Enterra system, HF stimulators are implanted permanently, after which the clinician might optimise stimulation parameters. Another approach that is adopted by some centres is to perform temporary stimulation, which can allow for better patient selection and optimisation of stimulation location (mid-corpus vs antrum) and parameters (Abell et al. 2011; Ayinala et al. 2005; Corvinus et al. 2018). This is a similar approach to what is undertaken for sacral nerve stimulator implantation for faecal incontinence where a testing phase is standard (Dudding et al. 2011). Corvinus et al. used a standard gastroparesis symptom index, before and after temporary stimulation to assess proceeding to permanent implantation (Corvinus et al. 2018). A 50% decrease in symptom index was considered as the threshold to continue to permanent implantation and illustrated economic benefit from a case series. The temporary stimulation leads were inserted laparoscopically and subsequent surgery was required for permanent lead and device implantation (Corvinus et al. 2018). This approach may lead to extended hospital stays and increases likelihood of infection.

Another approach for temporary stimulation is to place the stimulation leads mucosally via endoscopy (Abell et al. 2011). Here, the stimulation leads, which are extending from the nasal cavity, are connected to the pulse generator secured in an external pouch. Abell et al. illustrated that temporary stimulation can predict the outcomes of permanent HF stimulation, and reported that baseline symptom status and mucosal slow waves should be taken into account prior to permanent implantation (Abell et al. 2011). Recent advances in recording methods have allowed mucosal slow wave recordings to be obtained simultaneously along-side the temporary stimulation process (typically 5 days) and could be integrated alongside patient symptoms for future optimisation of stimulation settings (Paskaranandavadivel et al. 2019).

Future Outlook

Detailed comparative HF stimulation and LF pacing studies are required to illustrate efficacies of each method. To date, the only trial was conducted by Sobocki et al. where they showed that both LF and HF stimulation provided improvements to subjects with symptom score and gastric retention rates (Sobocki et al. 2003). Applying pacing and HF stimulation in tandem has been hypothesised to yield superior performance (Song et al. 2008). It is thought that using this approach, HF stimulation could provide relief from symptoms, while pacing could improve emptying rates; but to date, this has only been performed in an animal study (Song et al. 2008). The use of multi-channel stimulation should also be further explored, as it has been indicated that about 1% of energy of the single channel protocol is required to modulate slow waves for gastric emptying (Chen et al. 2005).

There are possible neuromodulation applications to obesity and gastrointestinal disorders such as ileus, irritable bowel syndrome and abdominal pain (Kovacic et al. 2017; Lin and McCallum 2015; Liu et al. 2006; Zhang et al. 2014). In particular, efforts are underway to translate HF stimulation to treat obesity by using a modified set of stimulation parameters (Cigaina 2002), but a recent clinical trial did not show significant improvements (Shikora et al. 2009). The use of stimulation has been highly successful in the treatment of gastroesophageal reflux disease (GERD) (Soffer and Kim 2016). The use of percutaneous electrical nerve field stimulation (PENFS) in the external ear can modulate pain pathways and a recent randomised double-blind, sham-controlled study has shown sustained efficacy in pain reduction (Kovacic et al. 2017).

Acknowledgements The authors acknowledge funding support from the Health Research Council of New Zealand, Marsden Fund Council, Medical Technologies Centre of Research Excellence (MedTechCoRE) and the National Institutes of Health (R01 HD 088662 and U18EB021789).

References

Abell TL, van Cutsem E, Abrahamsson H, Huizinga JD, Konturek JW, Galmiche JP, Voeller G, Filez L, Everts B, Waterfall WE, Domschke W, Bruley des Varannes S, Familoni BO, Bourgeois IM, Janssens J, Tougas G (2002) Gastric electrical stimulation in intractable symptomatic gastroparesis. Digestion 66:204–212

Abell T, McCallum R, Hocking M, Koch K, Abrahamsson H, Leblanc I, Lindberg G, Konturek J, Nowak T, Quigley EMM, Tougas G, Starkebaum W (2003) Gastric electrical stimulation for medically refractory gastroparesis. Gastroenterology 125:421–428

Abell TL, Camilleri M, Donohoe K, Hasler WL, Lin HC, Maurer AH, McCallum RW, Nowak T, Nusynowitz ML, Parkman HP, Shreve P, Szarka LA, Snape WJ, Ziessman HA (2008) Consensus recommendations for gastric emptying scintigraphy: a joint report of the American neurogastroenterology and motility society and the society of nuclear medicine. J Nucl Med Technol 36:44–54

Abell TL, Johnson WD, Kedar A, Runnels JM, Thompson J, Weeks ES, Minocha A, Griswold ME (2011) A double-masked, randomized, placebo-controlled trial of temporary endoscopic mucosal gastric electrical stimulation for gastroparesis. Gastrointest Endosc 74:496–503.e3

Abrich VA, Le RJ, Mulpuru SK, Friedman PA, Barsness GW, Cha Y-M, Lennon RJ, Lewis BR, Yang EH (2020) Clinical outcomes of various management strategies for symptomatic bradycardia. Clin Med Res 18 (2–3):75–81. https://doi.org/10.3121/cmr.2019.1507

Alighaleh S, Cheng LK, Angeli TR, Amiri M, Sathar S, O'Grady G, Paskaranandavadivel N (2019) A novel gastric pacing device to modulate slow waves and assessment by high-resolution mapping. IEEE Trans Biomed Eng 66:2823–2830

Alvarez WC, Mahoney LJ (1922) Action currents in stomach and intestine. Am J Physiol 58:476–493

Angeli TR, Cheng LK, Du P, Wang THH, Bernard CE, Vannucchi MG, Faussone-Pellegrini MS, Lahr C, Vather R, Windsor JA, Farrugia G, Abell TL, O'Grady G (2015) Loss of interstitial cells of cajal and patterns of gastric dysrhythmia in patients with chronic unexplained nausea and vomiting. Gastroenterology 149:56–66

Angeli TR, Du P, Midgley D, Paskaranandavadivel N, Sathar S, Lahr C, Abell TL, Cheng LK, O'Grady G (2016) Acute slow wave responses to high-frequency gastric electrical stimulation in patients with gastroparesis defined by high-resolution mapping. Neuromodulation 19:864–871

Ayinala S, Batista O, Goyal A, Al-Juburi A, Abidi N, Familoni B, Abell T (2005) Temporary gastric electrical stimulation with orally or PEG-placed electrodes in patients with drug refractory gastroparesis. Gastrointest Endosc 61:455–461

Bellahsène BE, Lind CD, Schirmer BD, Updike OL, McCallum RW (1992) Acceleration of gastric emptying with electrical stimulation in a canine model of gastroparesis. Am J Phys 262:G826–G834

Berger T, Kewenter J, Kock NG (1966) Response to gastrointestinal pacing: antral, duodenal and jejunal motility in control and postoperative patients. Ann Surg 164:139–144

Berry R, Miyagawa T, Paskaranandavadivel N, Du P, Angeli TR, Trew ML, Windsor JA, Imai Y, O'Grady G, Cheng LK (2016) Functional physiology of the human terminal antrum defined by high-resolution electrical mapping and computational modeling. Am J Physiol Gastrointest Liver Physiol 311:G895–G902

Bharucha AE (2015) Epidemiology and natural history of gastroparesis. Gastroenterol Clin N Am 44:9–19

Bilgutay AM, Wingrove R, Griffen WO, Bonnabeau RC, Lillehei WC (1963) Gastro-intestinal pacing: a new concept in the treatment of ileus. Ann Surg 158:338–348

Bohdjalian A, Prager G, Aviv R, Policker S, Schindler K, Kretschmer S, Riener R, Zacherl J, Ludvik B (2006) One-year experience with Tantalus™: a new surgical approach to treat morbid obesity. Obes Surg 16:627–634

Camilleri M (2007) Diabetic gastroparesis. N Engl J Med 356:820–829

Camilleri M, Parkman HP, Shafi MA, Abell TL, Gerson L, and American College of Gastroenterology (2013) Clinical guideline: management of gastroparesis. Am J Gastroenterol 108:18–37; quiz 38

Chen J, Lin Z, Schirmer BD, William RD, Ross R, McCallum RW (1995) Long-term gastric pacing with a portable gastric pacemaker to aid gastric emptying in humans

Chen JDZ, Qian L, Ouyang H, Yin J (2003) Gastric electrical stimulation with short pulses reduces vomiting but not dysrhythmias in dogs. Gastroenterology 124:401–409

Chen JDZ, Xu X, Zhang J, Abo M, Lin X, Mccallum RW, Ross B (2005) Efficiency and efficacy of multi-channel gastric electrical stimulation. Neurogastroenterol Motil 17:878–882

Cigaina V (2002) Gastric pacing as therapy for morbid obesity: preliminary results. Obes Surg 12:12S–16S

Cingolani E, Goldhaber JI, Marbán E (2018) Next-generation pacemakers: from small devices to biological pacemakers. Nat Rev Cardiol 15 (3):139–150

Corvinus FM, Heinrich S, Neumann H, Hadzijusufovic E, Babic B, Lang H, Grimminger PP (2018) Minimally-invasive temporary gastric stimulation: a pilot study to predict the outcome of electronic gastric stimulation with the Enterra™ system. Dig Liver Dis 50:1030–1034

Dallaglio PD, di Marco A, Weidmann ZM, Perez L, Alzueta J, García-Alberola A, Fernandez-Lozano I, Díaz-Infante E, Rodriguez A, Basterra N, Calvo D, Garcia MR, Aceña M, Anguera I (2020) Antitachycardia pacing for shock prevention in patients with hypertrophic cardiomyopathy and ventricular tachycardia. Heart Rhythm 17(7):1084–1091. https://doi.org/10.1016/j.hrthm.2020.02.024

Diamant N, Bortoff A (1969) Effects of transection on the intestinal slow-wave frequency gradient. Am J Physiol 216:734–743

Ducrotte P et al (2020) Gastric electrical stimulation reduces refractory vomiting in a randomized crossover trial. Gastroenterology 158:506–514.e2

Dudding TC, Hollingshead JR, Nicholls RJ, Vaizey CJ (2011) Sacral nerve stimulation for faecal incontinence: patient selection, service provision and operative technique. Color Dis 13:e187–e195

Familoni BO, Abell TL, Nemoto D, Voeller G, Johnson B (1997a) Efficacy of electrical stimulation at frequencies higher than basal rate in canine stomach. Dig Dis Sci 42:892–897

Familoni BO, Abell TL, Voeller G, Salem A, Gaber O (1997b) Electrical stimulation at a frequency higher than basal rate in human stomach. Dig Dis Sci 42:885–891

Farrugia G (1999) Ionic conductances in gastrointestinal smooth muscles and interstitial cells of Cajal. Annu Rev Physiol 61:45–84

Farrugia G (2008) Interstitial cells of Cajal in health and disease. Neurogastroenterol Motil 20:54–63

Frequin HL, Bot M, Dilai J, Scholten MN, Postma M, Bour LJ, Contarino MF, de Bie RMA, Schuurman PR, van den Munckhof P (2020) Relative contribution of magnetic resonance imaging, microelectrode recordings, and awake test stimulation in final lead placement during deep brain stimulation surgery of the subthalamic nucleus in Parkinson's disease. Stereotact Funct Neurosurg 98:118–128. https://doi.org/10.1159/000505710

Gao J, Du P, O'Grady G, Archer R, Farrugia G, Gibbons SJ, Cheng LK (2013) Numerical metrics for automated quantification of interstitial cell of Cajal network structural properties. J R Soc Interface 10:20130421

Grover M, Farrugia G, Lurken MS, Bernard CE, Faussone-Pellegrini MS, Smyrk TC, Parkman HP, Abell TL, Snape WJ, Hasler WL, Ünalp-Arida A, Nguyen L, Koch KL, Calles J, Lee L, Tonascia J, Hamilton FA, Pasricha PJ (2011) Cellular changes in diabetic and idiopathic gastroparesis. Gastroenterology 140:1575–1585.e8

Grover M, Bernard CE, Pasricha PJ, Lurken MS, Faussone-Pellegrini MS, Smyrk TC, Parkman HP, Abell TL, Snape WJ, Hasler WL, McCallum RW, Nguyen L, Koch KL, Calles J, Lee L, Tonascia J, Ünalp-Arida A, Hamilton FA, Farrugia G (2012) Clinical-histological associations in gastroparesis: results from the Gastroparesis Clinical Research Consortium. Neurogastroenterol Motil 24:531–e249

Guo JP, Maurer AH, Fisher RS, Parkman HP (2001) Extending gastric emptying scintigraphy from two to four hours detects more patients with gastroparesis. Dig Dis Sci 46:24–29

Hinder RA, Kelly KA (1977) Human gastric pacesetter potential. Site of origin, spread, and response to gastric transection and proximal gastric vagotomy. Am J Surg 133:29–33

Hirst GDS (2001) An additional role for ICC in the control of gastrointestinal motility? J Physiol 537:1–1

Hirst GDS, Edwards FR (2006) Electrical events underlying organized myogenic contractions of the guinea pig stomach. J Physiol 576:659–665

Hocking MP, Vogel SB, Sninsky CA (1992) Human gastric myoelectric activity and gastric emptying following gastric surgery and with pacing. Gastroenterology 103:1811–1816

Horbach T, Thalheimer A, Seyfried F, Eschenbacher F, Schuhmann P, Meyer G (2015) Abiliti® closed-loop gastric electrical stimulation system for treatment of obesity: clinical results with a 27-month follow-up. Obes Surg 25:1779–1787

Huizinga JD, Lammers WJEP (2009) Gut peristalsis is governed by a multitude of cooperating mechanisms. Am J Physiol Gastrointest Liver Physiol 296:G1–G8

Huizinga JD, Thuneberg L, Klüppel M, Malysz J, Mikkelsen HB, Bernstein A, Hulzinga JD, Thuneberg L, Klüppel M, Malysz J, Mikkelsen HB, Bernstein A, Huzinga JD, Thuneberg L, Klüppel M, Malysz J, Mikkelsen HB, Bernstein A (1995) W/kit gene required for interstitial cells of Cajal and for intestinal pacemaker activity. Nature 373:347–349

Ibrahim A, Farajidavar A, Kiani M (2015) Towards a highly-scalable wireless implantable system-on-a-chip for gastric electrophysiology. Annual International Conference in IEEE Eng Med Biol Soc. 2689–2692. https://doi.org/10.1109/EMBC.2015.7318946

Javan-Khoshkholgh A, Farajidavar A (2019) An implantable inductive near-field communication system with 64 channels for acquisition of gastrointestinal bioelectrical activity. Sensors 19:2810

Javan-Khoshkholgh A, Alrofati W, Naidu-Naidugari S, Sassoon J, Gharibani P, Chen J, Farajidavar A (2019a) System and methodology to study and stimulate gastric electrical activity in small freely behaving animals. Bioelectr Med. https://doi.org/10.2217/bem-2019-0019

Javan-Khoshkholgh A, Sassoon JC, Farajidavar A (2019b) A wireless rechargeable implantable system for monitoring and pacing the gut in small animals. IEEE Biomed Circ Syst Conf (BioCAS) 2019:1–4. https://doi.org/10.1109/BIOCAS.2019.8919125

Kelly K (1974) Differential responses of the canine gastric corpus and antrum to electric stimulation. Am J Physiol 226:230–234

Kelly K, Code C (1971) Canine gastric pacemaker. Am J Physiol 220:112–118

Kelly KA, La Force RC (1972) Pacing the canine stomach with electric stimulation. Am J Phys 222:588–594

Kelly KA, Code CF, Elveback LR (1969) Patterns of canine gastric electrical activity. Am J Physiol 217:461–470

Knowles CH, Veress B, Kapur RP, Wedel T, Farrugia G, Vanderwinden J-M, Geboes K, Smith VV, Martin JE, Lindberg G, Milla PJ, De Giorgio R (2011) Quantitation of cellular components of the enteric nervous system in the normal human gastrointestinal tract – report on behalf of the Gastro 2009 International Working Group. Neurogastroenterol Motil 23:115–124

Kovacic K, Hainsworth K, Sood M, Chelimsky G, Unteutsch R, Nugent M, Simpson P, Miranda A (2017) Neurostimulation for abdominal pain-related functional gastrointestinal disorders in adolescents: a

randomised, double-blind, sham-controlled trial. Lancet Gastroenterol Hepatol 2:727–737
Krohn B, Sathar S, Rohrle O, Vanderwinden J-M, O'Grady G, Cheng LK (2017) A framework for simulating gastric electrical propagation in confocal microscopy derived geometries. Annual International Conference in IEEE Eng Med Biol Soc. 4215-4218. https://doi.org/10.1109/EMBC.2017.8037786
Kuramoto M, Kuwashima T, Morishita K, Fukuda T, Kawauchi M, Kita T, Komi N (1979) Electrical pacing in the gastroduodenal junction area of conscious dogs (author's transl). *Nihon seirigaku zasshi*. J Physiol Soc Jpn 41:1–5
Lammers WJEP, Al-Kais A, Singh S, Arafat K, El-Sharkawy TY (1993) Multielectrode mapping of slow-wave activity in the isolated rabbit duodenum. J Appl Physiol 74:1454–1461
Lammers WJEP, Stephen B, Arafat K, Manefield GW (1996) High resolution electrical mapping in the gastrointestinal system: initial results. Neurogastroenterol Motil 8:207–216
Lammers WJEP, Ver Donck L, Stephen B, Smets D, Schuurkes JAJ (2009) Origin and propagation of the slow wave in the canine stomach: the outlines of a gastric conduction system. Am J Physiol-Gastrointest Liver Physiol 296:G1200–G1210
Levinthal DJ, Bielefeldt K (2017) Systematic review and meta-analysis: gastric electrical stimulation for gastroparesis. Auton Neurosci 202:45–55
Lin Z, McCallum RW (2015) Implantable neurostimulator for gut function. Elsevier, 251–266. https://doi.org/10.1016/B978-1-78242-101-6.00011-2
Lin ZY, McCallum RW, Schirmer BD, Chen JDZ (1998) Effects of pacing parameters on entrainment of gastric slow waves in patients with gastroparesis. Am J Phys 274:G186–G191
Liu S, Liu J, Chen JDZ (2006) Neural mechanisms involved in the inhibition of intestinal motility induced by intestinal electrical stimulation in conscious dogs. Neurogastroenterol Motil 18:62–68. https://doi.org/10.1111/j.1365-2982.2005.00739.x
Marini M, Martin M, Saltori M, Quintarelli S, Zilio F, Guarracini F, Coser A, Valsecchi S, Bonmassari R (2019) Pacemaker therapy in very elderly patients: survival and prognostic parameters of single center experience. J Geriatr Cardiol 16:880–884
McCallum RW, Chen JDZ, Lin Z, Schirmer BD, Williams RD, Ross RA (1998) Gastric pacing improves emptying and symptoms in patients with gastroparesis. Gastroenterology 114:456–461
McCallum RW, Dusing RW, Sarosiek I, Cocjin J, Forster J, Lin Z (2010) Mechanisms of symptomatic improvement after gastric electrical stimulation in gastroparetic patients. Neurogastroenterol Motil 22:167–167
McCallum RW, Lin Z, Forster J, Roeser K, Hou Q, Sarosiek I (2011) Gastric electrical stimulation improves outcomes of patients with gastroparesis for up to 10 years. Clin Gastroenterol Hepatol 9:314–319
Medtronic (2020) Enterra II neurostimulator: gastric electrical stimulation. https://www.medtronic.com/us-en/healthcare-professionals/products/digestive-gastrointestinal/gastric-electrical-stimulation-systems/enterra-ii-neurostimulator.html
Merin O, Ilan M, Oren A, Fink D, Deeb M, Bitran D, Silberman S (2009) Permanent pacemaker implantation following cardiac surgery: indications and long-term follow-up. Pacing Clin Electrophysiol 32:7–12
Moreno-Duarte I, Morse LR, Alam M, Bikson M, Zafonte R, Fregni F (2014) Targeted therapies using electrical and magnetic neural stimulation for the treatment of chronic pain in spinal cord injury. NeuroImage 85:1003–1013
O'Grady G, Egbuji JU, Du P, Cheng LK, Pullan AJ, Windsor JA (2009) High-frequency gastric electrical stimulation for the treatment of gastroparesis: a meta-analysis. World J Surg 33:1693–1701
O'Grady G, Du P, Cheng LK, Egbuji JU, Lammers WJEP, Windsor JA, Pullan AJ (2010a) Origin and propagation of human gastric slow-wave activity defined by high-resolution mapping. Am J Physiol Gastrointest Liver Physiol 299:G585–G592
O'Grady G, Du P, Lammers WJEP, Egbuji JU, Mithraratne P, Chen JDZ, Cheng LK, Windsor JA, Pullan AJ (2010b) High-resolution entrainment mapping of gastric pacing: a new analytical tool. Am J Physiol Gastrointest Liver Physiol 298:G314–G321
O'Grady G, Angeli TR, Du P, Lahr C, Lammers WJEP, Windsor JA, Abell TL, Farrugia G, Pullan AJ, Cheng LK (2012a) Abnormal initiation and conduction of slow-wave activity in gastroparesis, defined by high-resolution electrical mapping. Gastroenterology 143:589–598
O'Grady G, Pullan AJ, Cheng LK (2012b) The analysis of human gastric pacemaker activity. J Physiol 590:1299–1300
O'Grady G, Angeli TR, Paskaranandavadivel N, Erickson JC, Wells CI, Gharibans AA, Cheng LK, Du P (2018) Methods for high-resolution electrical mapping in the gastrointestinal tract. IEEE Rev Biomed Eng 12:287–302
Paskaranandavadivel N, Wang R, Sathar S, O'Grady G, Cheng LK, Farajidavar A (2015) Multi-channel wireless mapping of gastrointestinal serosal slow wave propagation. Neurogastroenterol Motil 27:580–585
Paskaranandavadivel N, O'Grady G, Cheng LK (2017) Time-delay mapping of high-resolution gastric slow-wave activity. IEEE Trans Biomed Eng 64:166–172
Paskaranandavadivel N, Angeli TR, Manson T, Stocker A, McElmurray L, O'Grady G, Abell T, Cheng LK, O'Grady G, Abell T, Cheng LK (2019) Multi-day, multi-sensor ambulatory monitoring of gastric electrical activity. Physiol Meas 40:25011
Qin C, Chen JDZ, Zhang J, Foreman RD (2007) Modulatory effects and afferent pathways of gastric electrical stimulation on rat thoracic spinal neurons receiving input from the stomach. Neurosci Res 57:29–39

Quast DC, Beall AC, Debakey ME (1965) Clinical evaluation of the gastrointestinal pacer. Surg Gynecol Obstet 120:35–37

Reverberi C, Dario A, Barolat G (2013) Spinal cord stimulation (SCS) in conjunction with peripheral nerve field stimulation (PNfS) for the treatment of complex pain in failed back surgery syndrome (FBSS). Neuromodulation 16:78–82; discussion 83

Rey E, Choung RS, Schleck CD, Zinsmeister AR, Talley NJ, Locke GR (2012) Prevalence of hidden gastroparesis in the community: the gastroparesis "iceberg". J Neurogastroenterol Motil 18:34–42

Rosen HR, Urbarz C, Holzer B, Novi G, Schiessel R (2001) Sacral nerve stimulation as a treatment for fecal incontinence. Gastroenterology 121:536–541

Sathar S, Trew ML, Cheng LK (2015) Tissue specific simulations of interstitial cells of cajal networks using unstructured meshes. Annual International Conference in IEEE Eng Med Biol Soc. 8062–8065. https://doi.org/10.1109/EMBC.2015.7320264

Shikora SA, Bergenstal R, Bessler M, Brody F, Foster G, Frank A, Gold M, Klein S, Kushner R, Sarwer DB (2009) Implantable gastric stimulation for the treatment of clinically severe obesity: results of the SHAPE trial. Surg Obes Relat Dis 5:31–37

Sobakin MA, Shepelev VA (1973) Apparatus for electrical stimulation of weakened peristaltic activity of the stomach (experimental investigation). Biomed Eng 7:90–93

Sobocki J, Thor PJ, Krolczyk G (2003) High frequency electrical stimulation of the stomach is more effective than low frequency pacing for the treatment of postoperative functional gastric stasis in humans. Neuromodulation 6:254–257

Soffer E, Kim S (2016) Electrical stimulation for gastroesophageal reflux disease: current state of the art. Clin Exp Gastroenterol 9:11

Song G, Hou X, Yang B, Sun Y, Liu J, Qian W, Chen JDZ (2006) Efficacy and efficiency of gastric electrical stimulation with short pulses in the treatment of vasopressin-induced emetic responses in dogs. Neurogastroenterol Motil 18:385–391

Song G-Q, Hou X, Yang B, Sun Y, Qian W, Chen JDZ (2008) A novel method of 2-channel dual-pulse gastric electrical stimulation improves solid gastric emptying in dogs. Surgery 143:72–78

Tang M, Zhang J, Chen JDZ (2006) Central mechanisms of gastric electrical stimulation involving neurons in the paraventricular nucleus of the hypothalamus in rats. Obes Surg 16:344–352

Timratana P, El-Hayek K, Shimizu H, Kroh M, Chand B (2013) Laparoscopic gastric electrical stimulation for medically refractory diabetic and idiopathic gastroparesis. J Gastrointest Surg 17:461–470

Wang R, Abukhalaf Z, Javan-Khoshkholgh A, Wang TH, Sathar S, Du P, Angeli TR, Cheng LK, O'Grady G, Paskaranandavadivel N, Farajidavar A (2018) A miniature configurable wireless system for recording gastric electrophysiological activity and delivering high-energy electrical stimulation. IEEE J Emerg Select Topics Circuits Syst 8:221–229

Wang L, Malik A, Roop PS, Cheng LK, Paskaranandavadivel N (2019) A formal approach for scalable simulation of gastric ICC electrophysiology. IEEE Trans Biomed Eng 66:3320–3329

Wang L, Malik A, Roop PS, Cheng LK, Paskaranandavadivel N, Ai W (2020) A novel approach for model-based design of gastric pacemakers. Comput Biol Med 116:103576

Xing JH, Chen JDZ (2006) Effects and mechanisms of long-pulse gastric electrical stimulation on canine gastric tone and accommodation. Neurogastroenterol Motil 18:136–143

Xue S, Valdez DT, Tremblay L, Collman PI, Diamant NE (1995) Electrical slow wave activity of the cat stomach: its frequency gradient and the effect of indomethacin. Neurogastroenterol Motil 7:157–167

Yamanishi T, Kamai T, Yoshida K (2008) Neuromodulation for the treatment of urinary incontinence. Int J Urol 15:665–672

Yamashita M (1980) Studies of pacing the canine stomach with electric stimulation for its possible clinical application (author's transl). Nihon Heikatsukin Gakkai zasshi 16:315–329

Zhang Y, Du S, Fang L, Yao S, Chen JDZ (2014) Retrograde gastric electrical stimulation suppresses calorie intake in obese subjects. Obesity 22:1447–1451

Useful Online Resources Hosting Information on Obesity and Diabetes

73

Joel Faintuch and Jacob J. Faintuch

Abstract

The internet has become such an omnipresent service, which, in the view of many, has already upgraded from slave to master and quite a tyrannical one. Whatever the feelings, scientific information would not flow worldwide without the tool, in such unlimited amounts. Just as an example, gene sequencing and metabolomics as well as the related precision medicine approach as now envisaged would be inconceivable. Each omics analysis and clinical correlation would be labor intensive and take weeks to be conducted, instead of minutes or seconds and rather negligible expenditures, without internet-available search engines and reference libraries. The same applies for large biobank screening and biostatistical interventions, for both clinical and research purposes. This chapter has listed a large choice of useful obesity, bariatric, type 1 and 2 diabetes sites, as well as general internet sources, most of them costless, geared at the needs of office or bedside assistance, laboratory investigation, and educational or public health initiatives.

Keywords

Diabetes sites · Obesity sites · Bariatric sites · Trial database sites · Endoscopic intervention sites

J. Faintuch (✉) · J. J. Faintuch
Hospital das Clinicas, Sao Paulo University Medical School, Sao Paulo, Brazil
e-mail: blfaintuch@hotmail.com

J. Faintuch and S. Faintuch (eds.), *Obesity and Diabetes*, https://doi.org/10.1007/978-3-030-53370-0_73

Table 73.1 General diabetes sites

cdc.gov/diabetes/	A large and authoritative USA government site
who.int/health-topics/diabetes	The official diabetes homepage of the World Health Organization
joslin.org	Up to date information for healthcare providers and patients
diabetesresearch.org	Ongoing research for diabetes therapy
niddk.nih.gov/health-information/diabetes	Official website of the USA National Institutes of Health for diabetes
care.diabetesjournals.org/content/42/Supplement_1/S81	Obesity management for the treatment of type 2 diabetes
annals.org/aim/fullarticle/2653838/treatment-type-1-diabetes-synopsis-2017-american-diabetes-association-standards	American Diabetes Association standards for type 1 diabetes care
aace.com/pdfs/diabetes/AACE_2019_Diabetes_Algorithm_FINAL_ES.pdf	A practical algorithm for type 2 diabetes management
aace.com	General site of the American Association of Clinical Endocrinologists
ndep.nih.gov	An educational site by the USA Department of Health
Diabetesincontrol.com	Information for professionals
diabetes.org	Official site of the American Diabetes Organization
Idf.org	Official site of the International Diabetes federation
oecd-ilibrary.org/social-issues-migration-health/health-at-a-glance-2017/diabetes-prevalence_health_glance-2017-15-en	Detailed information about diabetes prevalence in OECD countries
IDF-T2D-CPR-2017-print.pdf	Type 2 diabetes guidelines by the International Diabetes Federation
pathways.nice.org.uk/pathways/preventing-type-2-diabetes	NICE guidelines- Preventing type 2 diabetes
pathways.nice.org.uk/pathways/type-2-diabetes-in-adults	NICE guidelines for type 2 diabetes in adults
pathways.nice.org.uk/pathways/type-1-diabetes-in-adults	NICE guidelines for type 1 diabetes in adults
eatright.org	Tips from the American Academy of Nutrition/Dietetics
diabetes educator.org	Self-care recommendations for diabetics

OECD Organization for Economic Cooperation and Development

Table 73.2 Diabetes information sites in various languages

Language/Url	Sponsor/Features
English and Spanish learningaboutdiabetes.org	General information
diabetes.org	American Diabetes Association
Spanish feaed.org	Spanish Educators in Diabetes
alad.org	Latin American Association of Diabetes
fdc.org.co	Colombian Federation of Diabetes
French diabetenet.com	French Diabetes Association
sante.ujf-grenoble.fr/SANTE/alfediam/mellitis.html	Endorsed by the French Diabetes Association
diabete.qc.ca	Diabetes Association of Quebec (CA)
German uni-duesseldorf.de/diabetes/index.htm	German Ministry of Health
diabetes-austria.com	Austrian Diabetes Association
Norwegian dianet.no	Norwegian Diabetes Association
Dutch dvn.nl	Dutch Diabetes Association
Hebrew sukeret.co.il	Israel Diabetes Association
Arabic aasdonline.com	Arabic Assn. for the Study of Diabetes and Metabolism
Multilingual—Type 1 diabetes jdrf.org/international	Juvenile Diabetes Research Foundation (International sites)
Multilingual—General diabetes guidelines diabetes-forum.com	General information
diabetesaustralia.com.au/multilingualdiabetes	National Diabetes Service Scheme
diabtrends.com	Diabetes FIT Foundation

Table 73.3 Diabetes trial databases

clinicaltrials.gov/ct2/results?cond=Type+2+Diabetes&recrs=a&age_v=&gndr=&type=Intr&rslt=&fund=0&fund=1&Search=Apply	Registered clinical trials on type 2 diabetes
clinicaltrials.gov/ct2/results?cond=Gestational+diabetes&term=&cntry=US&state=&city=&dist=&Search=Search&recrs=a&type=Intr&fund=0&fund=1&fund=3	Registered clinical trials on gestational diabetes
clinicaltrials.ucsf.edu/diabetes	Diabetes trials of the University of California San Francisco
centerwatch.com/clinical-trials/listings/condition/947/diabetes-mellitus-type-2/	A vast list of clinical trials on type 2 diabetes, in the USA
clinicaltrialsregister.eu/ctr-search/search?query=Diabetes	Diabetes trials registered in Europe
clinicaltrials.gov/ct2/results?cond=Type+1+Diabetes&term=&cntry=&state=&city=&dist=&Search=Search&recrs=a&type=Intr&fund=0&fund=1	Registered clinical trials on type 1 diabetes

Table 73.4 General obesity sites

surgeongeneral.gov/library/calls/obesity/fact_consequences.html	Surgeon General (USA) recommendations for obesity and weight loss
ec.europa.eu/health/nutrition_physical_activity/platform_en	European Union platform for diet, physical activity, and health
Obesityaction.org	A nonprofit organization for obesity
heart.org/en/healthy-living/healthy-eating/losing-weight?uid=1956	American Heart Association advice on obesity and weight loss
easo.org/education/guidelines	European guidelines for obesity management in adults
karger.com/Article/FullText/496183	European guidelines for adult obesity management in primary care
pathways.nice.org.uk/pathways/obesity	Obesity management adults- NICE pathways
care.diabetesjournals.org/content/42/Supplement_1/S81	Obesity management for the treatment of type 2 diabetes
healthdirect.gov.au/obesity	A well-organized official Australian site
Obesitycanada.ca	A well-organized official Canadian site
nutrition.gov/diet-an-health-conditions	A US Department of Agriculture guide for obesity, diabetes, and other conditions
gov.uk/government/publications/the-eatwell-guide	An official British guide for healthy eating
nationaleatingdisorders.org	A large nonprofit for eating disorders
who.int/news-room/fact-sheets/detail/obesity-and-overweight	Information about obesity from the World Health Organization
who.int/nutrition/topics/new-release-guideline-obesity-children/en/	Guidelines for overweight and obesity management in children
oecd.org/els/health-systems/Obesity-Update-2017.pdf	A detailed profile of the obesity burden in OECD countries
eatright.org	Tips from the American Academy of Nutrition and Dietetics

OECD Organization for Economic Cooperation and Development, *BMI* Body mass index

Table 73.5 Useful obesity pages from the National Institutes of Health (NIH)

win.niddk.nih.gov/index.htm	Weight control information network
nhlbi.nih.gov/health/health-topics/topics/obe/risks.html	Health risks of overweight
win.niddk.nih.gov/publications/understanding.htm	Understanding adult obesity
win.niddk.nih.gov/publications/tools.htm	Weight and waist measurement—adults
win.niddk.nih.gov/statistics/	Statistics about obesity
nhlbi.nih.gov/health/educational/lose_wt/BMI/bmicalc.htm	BMI calculator for adults

Table 73.6 Obesity pages from the Communicable Diseases Center—CDC

cdc.gov/obesity	General site
cdc.gov/obesity/adult/causes/index.html	Causes and consequences of obesity
cdc.gov/obesity/data/trends.html	Obesity trends in USA
cdc.gov/healthyweight/bmi/calculator.html	BMI percentile calculator for children

Table 73.7 Obesity material from the Obesity Society

worldobesity.org	Official site of the World Obesity Federation
obesity.org/resources-for/your-weight-and-diabetes.htm	Obesity and diabetes
obesity.org/resources-for/cancer-and-obesity.htm	Obesity and cancer
obesity.org/resources-for/childhood-overweight.htm	Childhood overweight
obesity.org/resources-for/obesity-bias-and-stigmatization.htm	Bias and stigmatization

Table 73.8 Obesity trial databases

clinicaltrials.gov/ct2/results?recrs=ab&cond=Obesity&term=&cntry=US&state=&city=&dist=	Registered obesity trials in the USA
clinicaltrials.gov/ct2/results?cond=Obesity%2C+Childhood&term=&cntry=US&state=&city=&dist=	Registered childhood obesity trials – USA
centerwatch.com/clinical-trials/listings/condition/107/obesity/	A vast list of clinical trials on all modalities of obesity, in the USA
pmg-research.com/indications/weight-loss-obesity-clinical-trials.aspx	Overweight and obesity trials in the USA
clinicaltrials.ucsf.edu/obesity	Obesity trials conducted by the University of California – San Francisco
clinicaltrialsregister.eu/ctr-search/search?query=Overweight+and+Obese+by+BMI	Obesity trials registered in Europe

Table 73.9 Bariatric, metabolic, and endoscopic intervention sites

soard.org	Site of the American Society for Metabolic and Bariatric Surgery/ASMBS
facs.org/search/bariatric-surgery-centers?allresults=	Bariatric centers accredited by the American College of Surgeons
.facs.org/quality-programs/mbsaqip	Site of the accreditation program of the American Society for Metabolic and Bariatric Surgery
ohsu.edu/bariatric-services	A well-organized site by the Oregon Health and Science University
muhc.ca/chirurgie-gen-bariatrique/profile	A useful site by the Bariatric Surgery Department, McGill University, Montreal, Canada
asmbs.org/resources/bariatric-surgery-guidelines-and-recommendations	Bariatric surgery guidelines by the American Society for Metabolic and Bariatric Surgery
ifso.com/pdf/easo-guidelines-practical-recommendations-of-the-obesity.pdf	Guidelines for post bariatric surgery medical management
asge.org/home/about-asge/newsroom/media-backgrounders-detail/endoscopic-bariatric-therapies	Recommendations about endoscopic bariatric therapies by the American Society of Gastrointestinal Endoscopy
researchgate.net/publication/327975893_Training_in_Bariatric_and_Metabolic_Endoscopic_Therapies	Training in bariatric and metabolic endoscopic therapies

Correction to: Obesity and Diabetes

Joel Faintuch and Salomão Faintuch

Correction to:
J. Faintuch and S. Faintuch (eds.), *Obesity and Diabetes*,
https://doi.org/10.1007/978-3-030-53370-0

The book was inadvertently published without the co-author names Yue Liao and Michelle R. Jospe. This has been corrected now.

In addition to this, the new co-author "Sandra Vazquez-Diaz" was included in Chapter 27.

The updated versions of the chapters can be found at
https://doi.org/10.1007/978-3-030-53370-0_27
https://doi.org/10.1007/978-3-030-53370-0_43

J. Faintuch and S. Faintuch (eds.), *Obesity and Diabetes*, https://doi.org/10.1007/978-3-030-53370-0_74

Glossary[1]

Diabetes Glossary

A1C (HbA1C) Glycosylated hemoglobin concentration reflects blood glucose over the past 2 to 3 months. There are cut-off points for prediabetes and diabetes.

Acanthosis nigricans Dark skin patches in armpits, groin and neck. Not uncommon in diabetics, can be associated with insulin resistance.

Acarbose See Alpha glucosidase inhibitors.

Alpha glucosidade inhibitors Oral antidiabetic drugs which inhibit an intestinal enzyme responsible for carbohydrate digestion. Less glucose is produced and absorbed, benefiting the disease. Clinical efficacy is nor remarkable, and gastrointestinal side effects can occur. Acarbose and miglitol are best known examples.

Artificial pancreas Any system or device that simultaneously monitors blood glucose, and pumps into the organism appropriate amounts of insulin. The aim is to mimick pancreatic function and improve quality of life, mostly in type 1 diabetes, however also in severe forms of type 2 diabetes.

Beta cells One type of cells present in pancreatic islets of Langerhans. They are responsible for the synthesis of insulin.

Biguanides Oral antidiabetic drugs with actions on liver and muscle glucose metabolism, as well as possible indirect incretin properties, via modulation of bile acid resorption. The principal representative is Metformin.

Body mass index (BMI) The ratio between body weight (in kilograms), and the square of the height (in meters). Used for both adults and children to diagnose obesity, Normal range is 18.5–24.9 kg/m^2, whereas values up to 29.9 signal overweight, and above that, obesity.

Canaglifozin See Glifozins

DPP4 Inhibitors These antidiabetic drugs inhibit the enzyme dipeptidyl peptidase 4 (DPP 4), responsible for degradation of GLP-1 (Glucagon like peptide 1), a major endogenous incretin. As a consequence, endogenous insulin production is enhanced. See also Gliptins.

Dapaglifozin See Glifozins.

Diabetic ketoacidosis See Ketoacidosis.

Diabetic Retinopathy See Retinopathy.

Dulaglutide See Glucagon-like peptide 1 receptor agonists.

Empaglifozin See Glifozins.

Exenatide See Glucagon-like peptide 1 receptor agonists.

Fasting blood glucose The most widely used test for diabetes screening and monitoring. Can be performed in the laboratory, or at home, with paper strips or electronic devices.

Gastroparesis A gastric neuromuscular complication mostly seen after many years of type 2 diabetes. The stomach and small bowel nerves become dysfunctional, resulting in disordered peristalsis, poor gastric and/or intestinal emptying, nausea, vomiting, and weight loss.

Gestational diabetes Diabetes that is precipitated by pregnancy. Both early and late modalities are recognized. Spontaneous

[1] This Glossary was compiled by Joel Faintuch and Jacob J. Faintuch.

J. Faintuch and S. Faintuch (eds.), *Obesity and Diabetes*, https://doi.org/10.1007/978-3-030-53370-0

cure after delivery is common, however type 2 diabetes may subsequently ensue.

Ghrelin A major appetite hormone, produced mostly at the gastric fundus. Diminishes after bariatric operations which remove the fundus, such as gastric bypass and vertical gastroplasty, thus contributing to the favorable weight loss curve. Long term recovery after bariatric intervention is reported.

GLP1-RA See Glucagon-like peptide 1 receptor agonists.

Glifozins Oral antidiabetic drugs, inhibiting glucose reabsorption in the kidney tubuli. Canagliflozin, dapagliflozin, and empagliflozin belong to this family.

Glimepiride See Sulfonylureas.

Glipizide See Sulfonylureas.

Gliptins Oral antidiabetic drugs (See DPP 4 inhibitors). Alogliptin, linagliptin, saxagliptin, and sitagliptin are available in the market.

Glitazones See Tiazolidinediones.

Glucagon A glucose-elevating hormone secreted by alpha cells of the islets of Langerhans. Together with insulin, it is a key regulator of glucose homeostasis. Intranasal glucagon can be used for emergency therapy of hypoglycemia.

Glucagon-like peptide 1 receptor agonists A recent family of antidiabetic, and antiobesity injectable drugs, with strong incretin properties (incretin mimetics). Principal agents are Exenatide, Semaglutide, Dulaglutide and Liraglutide.

Glucose homeostasis The association of biochemical reactions, metabolic pathways, regulatory molecules and feed-back mechanisms, responsible for assuring stable glucose concentrations along the 24 h. It is conjectured that the aim is to provide a steady flux of glucose to the brain, despite fluctuations resulting from such potentially disruptive phenomenons as meal ingestion, physical exercise, fasting periods, and night sleep. Diabetes does impact glucose homeostasis.

Glyburide See Sulfonylureas.

Glycemic index (GI) A measure of the ability of ingested foods to raise blood glucose levels. Foods with high GI, such as white bread or ordinary rice, tend to be associated with obesity and diabetes. The opposite should occur with whole grain, fiber rich products, considered more desirable in these contexts.

Hypoglycemia Blood glucose concentration below 80 mg/dl. It is typically associated with excessive insulin or insulin secretagogue administration, or insufficient food intake. It becomes a life-threatening condition below 50 mg/dl, requiring immediate glucose replenishment, via sugar rich food (if the patient is conscious), intravenous glucose, or nasal glucagon powder.

Incretins Hormones (endogenous or exogenous) as well as synthetic pharmacological agents, with the ability of enhancing beta-cell mass and function, or inhibiting their apoptosis, thus improving glucose homeostasis. Principal physiological incretins are GLP-1 (glucagons like peptide 1), and GIP (glucose dependent insulinotropic polypeptide).

Induced pluripotent stem cells Adult cells from skin and other sources. After laboratory processing, they can become pluripotent, mimicking embryonic stem cells. See also Stem cells.

Insulin The most important hormone produced by the endocrine pancreas, and also the crucial molecule of glucose homeostasis, along with glucagon. It is lacking or inefficient in all types of diabetes. See also Beta cells, and Glucose homeostasis.

Insulin resistance Any condition that hampers response to insulin, resulting in impaired cell entrance and metabolization of glucose. One of the first derangements in severe obesity and type 2 diabetes. During many years it was a synonym for prediabetes. See also Prediabetes.

Ketones/Ketone bodies See ketoacidosis.

Ketoacidosis A condition that signals serious decompensation of diabetes, either type 1 or 2. Lack of sufficient insulin, be it endogenous (synthesized by the pancreas) or exogenous (pharmacologically administered), leads to high glucose concentration in the blood, along with accumulation of endogenous acids, and small molecules called ketone bodies. It can result in coma and death, and requires emergency treatment.

LADA See Latent Autoimmune Diabetes in Adults.

Latent Autoimmune Diabetes in Adults A modality of autoimmune diabetes, exhibiting islet autoantibodies. Beta-cell destruction tends to be slower than in typical autoimmune disease. Despite the name, it is not uncommon in children and adolescents, as well as in adults with type 2 diabetes.

Liraglutide See Glucagon-like peptide 1 receptor agonists.

Maturity Onset Diabetes of the Young An uncommon familial form of diabetes, with features of both type 1 and 2 of the disease. The most common involved gene is HNF1-alpha, and onset is usually in adolescence. It may be insulin-dependent or not, and association with obesity is not obligatory. See also Type 2 diabetes of children and adolescents.

Mediterranean-type diet A popular diet for metabolic diseases, and prevention of cardiovascular troubles. It is rich in fruits, vegetables, nuts, grains and omega-9 fatty acid (olive oil).

Meglitinides Short-action oral insulin secretagogues, used in type 2 diabetes therapy. Examples are Nateglinide and Repaglinide.

Mesenchymal stem cells Mostly collected from adipose tissue. Analogously to other stem cells, they can differentiate into several types of cells. See also Stem cells.

Metabolic syndrome The association of abdominal obesity, arterial hypertension, dyslipidemia, and glucose intolerance. Other derangements are sometimes included in the cluster, and diagnostic criteria vary. Typically linked to fatty liver disease, diabetes, and cardiovascular complications.

Metformin See Biguanides.

Miglitol See Alpha glucosidase inhibitors.

MODY See Maturity Onset Diabetes of the Young.

Nateglinide See Meglitinides.

Pioglitazone See Thiazolidinediones.

Prediabetes Impaired fasting blood glucose, or glucose tolerance test. Increases risk for type 2 diabetes, as well as for cardiovascular complications. Formerly known as insulin resistance.

Repaglinide See Meglitinides.

Retinopathy A serious ophtalmological complication of diabetes, especially in the elderly, however also affecting children and young adults. It can progress from blorred vision to total blindness. Requires periodic ophtalmological evaluations for early diagnosis and treatment.

Rosiglitazone See Thiazolidinediones.

Semaglutide See Glucagon-like peptide 1 receptor agonists.

SGLT2 inhibitors See Glifozins.

Sodium glucose co-transporter 2 inhibitors See Glifozins.

Stem cell A pluripotent cell able to differentiate into different tissues, given the right stimulation. This includes cells of pancreatic islets of Langerhans, or more specifically, insulin-secreting beta cells, rendering them attractive for cell therapy of type 1 diabetes. Embryonic stem cells can be collected from cord blood, whereas adult stem cells are often harvested in blood, bone marrow, or muscle tissue.

Sulfonylureas Oral antidiabetic drugs which stimulate insulin secretion. Hypoglycemia can be a side effect. Glimepiride, glipizide, and glyburide are better tolerated, second generation products.

Thiazolidinediones Oral antidiabetic drugs which are not insulin secretagogues, however improve insulin sensitivity. Examples are troglitazone, pioglitazone and rosiglitazone.

Trans fats An ubiquitous type of fatty acids, found in nature only in small concentrations, in meats and dairy products. Significant amounts in dozens of industrialized or processed foods, such as cakes, cookies, margarines, ice cream, fried potatoes, snacks, and fast food. Formed by hydrogenation of preexistent unsaturated fatty acids. They raise

LDL cholesterol and diminish HDL cholesterol, a double jeopardy for lipid blood profile, and should be strictly avoided.

Troglitazone See Thiazolidinediones.

Type 1 Diabetes Previously known as juvenile diabetes, onset is more common at early age, however it affects adults and elderly as well. Beta cell destruction is the rule, and insulin therapy is almost always required.

Type 2 diabetes By far the most common form of the disease. Previously known as maturity onset diabetes, it progressively affects young people as well, on account of obesity and metabolic syndrome prevalence at young age. Confusion should be avoided with Maturity onset diabetes of the young, a hereditary modality.

TZD See Thiazolidinediones.

Obesity and Bariatric Surgery Glossary (See also Diabetes Glossary)

Anthropometric evaluation Measurement of external variables such as weight, height, and waist circumference, in order to gain insight about nutritional and metabolic status. Routinely conducted in both adult and pediatric patients, even though variables are not always the same.

Body mass index (BMI) The ratio between body weight (in kilograms), and the square of the height (in meters). Used for both adults and children to diagnose obesity, Normal range is 18.5–24.9 kg/m^2, whereas values up to 29.9 signal overweight, and above that, obesity.

Calorie dense foods Foods, snacks and drinks rich in simple carbohydrates and fats. They can be highly processed, with elimination of useful nutrients, or just overcaloric. They tend to precipitate obesity.

DALY See Disability adjusted life year.

Disability adjusted life years A modern concept to estimate the burden of chronic diseases, like obesity and diabetes. It basically multiplies the number of years the individual is expected to suffer from disability, by a coefficient related to its severity.

Duodenal switch See Biliopancreatic diversion.

Duodeno-jejunal bypass barrier An endoscopic modality of bariatric/metabolic treatment for obesity and diabetes. A long flexible implant (sleeve or liner) is anchored at the pylorus, and positioned inside the duodenum and part of the jejunum. The aim is to produce limited malabsorption, preventing contact of the food bolus with the lined mucosa. Usually limited for 6 months utilization only.

Duodeno-jejunal bypass liner See Duodeno-jejunal bypass barrier.

Endoluminal bariatric techniques See Duodeno-jejunal bypass barrier, see Endoscopic sleeve gastrectomy.

Endoscopic sleeve gastroplasty An endoscopic procedure that mimicks classic sleeve gastrectomy. A suturing device applied to the gastric mucosa, without puncturing the gastric wall, creates a series of internal folds or plications, which substantially reduce stomach capacity.

Epigenome Assembly or set of chemical changes of DNA which regulate gene expression and consequently phenotype (disease). Even though they are environmental, not primarily inherited, they can eventually be passed down to future generations by epigenetic inheritance.

Fish oil A popular designation for omega-three-fatty-acids rich seafood. Best sources are salmon, menhaden, mackerel, herring, sardines, cod liver oil, and krill. Consumption is not associated with elevated cardiovascular risk, however purported protective actions in adults are still debated. Supplementation is recommended for infants (central nervous system development).

Gastric balloon A soft, fluid-filled balloon that is positioned in the stomach. By occupying space, it prevents overeating and promotes early satiety. It is usually introduced via endoscope, however a model than can be swallowed is aleady available. Recommendations are for use up to 6 months.

Gastric Banding A bariatric operation which reduces food intake (restrictive procedure), by means of a silicone band placed around the stomach, a few centimeters below the esophagogastric junction. Very popular during decades in some parts of the world, it is being progressively abandoned, due to unreliable long term results.

Gastric Bypass One of the most popular bariatric interventions. It markedly reduces gastric volume to just a small pouch (restrictive feature), at the same time shunting food down to the small bowel (malabsorptive characteristic). Also known as Roux-en-Y gastric bypass.

Genome The entire complement of genetic material, in a human or animal sample.

Genotype The entire set of genes and their characteristics, which determine the structure and function of an organism.

High-density lipoprotein/High density cholesterol (HDL) A fraction of plasma cholesterol that is associated with improved cardiovascular prognosis. Also known as "good cholesterol".

Highly processed foods Those that undergo multiple industrialization steps, typically removing useful nutrients such as fibre, vitamins and minerals. Sugar, fat and salt are often added as well. They tend to precipitate obesity. See also Calorie dense foods.

Intragastric balloon See Gastric balloon.

Leptin A central hormone for appetite and obesity. Manufactures by adipose cells as well as other sources, it inhibits appetite at high concentrations, thus preventive excessive fat accumulation. In obese humans, leptin administration doesn't lead to weight loss, suggesting decreased sensitivity.

Low-density lipoprotein/Low density cholesterol (LDL) The main fraction of plasma cholesterol that is associated with elevated cardiovascular risk. Also known as "bad cholesterol".

Metabolic syndrome The association of abdominal obesity, arterial hypertension, dyslipidemia, and glucose intolerance. Other derangements are sometimes included in the cluster, and diagnostic criteria vary. Typically linked to fatty liver disease, diabetes, and cardiovascular complications.

Monogenetic A disease or trait that is triggered by a single, well-known gene. A few uncommon modalities of obesity and diabetes are monogenetic.

Monounsaturated fatty acid Mainly consumed as olive oil, it is one of the pillars of the Mediterranean diet. Also present in canola oil, avocados and other sources.

Morbid Obesity A Body Mass Index of 40 kg/m^2 or greater. Also known as Class III obesity. The concept is intimately associated with indication for bariatric surgery, even though a BMI of 35 kg/m^2 with comorbidities, or in certain countries <35 kg/m^2, can be a legitimate indication as well.

Nutrient dense foods and drinks Usually designates natural foods, with a useful composition. Contrasts with Highly processed foods or Calorie dense foods.

Obstructive sleep apnea See Sleep apnea.

Omega-nine fatty acids See Monounsaturated fatty acids.

Omega-six fatty acids See Vegetable oils, See Polyunsaturated fatty acids.

Omega-three fatty acids See Fish oil, See Polyunsaturated fatty acids.

Phenotype Physical and metabolic makeup of an organism, which is controlled by genes. Can be divided in conventional (external appearance) and deep phenotype (proteins, enzymes, metabolites), as identified by biochemical and metabolomics studies. See also Genotype.

Polygenetic Disease or trait related to the interaction of many genes. Most types of obesity and diabetes undergo polygenetic influences.

Polyunsaturated fatty acids Type of fat which is chemically unsaturated, with multiple double or triple bonds. Typically liquid at room temperature, they comprise omega-6 fatty acids (soybean oil, corn oil, safflower oil and some other vegetable oils) as well as omega3 fatty acids (basically fish oil, however also flaxseed oil, canola oil and soybean oil).

Roux-en-Y gastric bypass See Gastric bypass.

Satiety The sensation of a fed state. It is mediated by hormones like ghrelin, leptin, PYY and other appetite regulating biomolecules. Derangements in obese subjects have been reported.

Saturated fatty acids Type of fat that is chemically saturated, or that has been saturated by industrial hydrogenation. Typically solid at room temperature. Representative examples are milk fats, other animal fats (lard, chicken fat), coconut and palm oil, any hydrogenated vegetable fat (margarine, mayonnaise, cookies).

Sedentarism A lifestyle modality featuring little or no physical activity or exercise. It is associated with watching television and other screens. A serious predisposing factor for obesity, in both children and adults.

Scopinaro's operation See Biliopancreatic diversion.

Sleep apnea A respiratory complication of severe obesity. It can be associated with cardiovascular complications, including hypertension, coronary heart disease, arrhythmia, and heart failure. Response to bariatric weight loss is usually favorable.

Sleeve gastrectomy One of the most popular bariatric interventions. Originally part of the biliopancreatic diversion, it is performed as a standalone procedure. The stomach is divided along the smaller curvature, creating a long and narrow channel, which inhibits large meals (restrictive modality).

Subcutaneous fat The most important fat depot, with over 95% of total body fat. Has much less implications regarding cardiovascular risk than visceral fat. Increased thigh fat in postmenopausal women is indeed associated with less myocardial infarction and stroke.

Vertical gastroplasty The vertical banded gastroplasty was a purely restrictive operation, which partially divided the stomach, creating a small vertical pouch. This intervention has been nearly completely abandoned. See also Sleeve gastrectomy.

Visceral fat Body fat that accumulates inside the abdominal cavity, mostly on the greater omentum, and around viscera. A metabolically active fat compartment associated with inflammation, hormone and coagulation factor production, and cardiovascular complications. See also Waist circumference.

Waist circumference A surrogate (anthropometric) measurement of visceral fat. Associated with metabolic syndrome and cardiovascular complications.

Wild type The genotype or phenotype of an experimental animal that is found in nature. Non genetically manipulated.

The manufacturer's authorised representative in the EU is Springer Nature Customer Service Centre GmbH, Europaplatz 3, 69115 Heidelberg, Germany. If you have any concerns regarding our products, please contact ProductSafety@springernature.com

Printed and bound by CPI Group (UK) Ltd, Croydon, CR0 4YY

30/09/2026

02225511-0003